D0151720

DICTIONARY OF
BRITISH AND IRISH BOTANISTS
AND HORTICULTURISTS

DICTIONARY OF BRITISH AND IRISH BOTANISTS AND HORTICULTURISTS

including Plant Collectors and Botanical Artists

by

RAY DESMOND

with historical introduction by William T. Stearn

TAYLOR & FRANCIS LTD
London
1977

First published 1893 under the title *A Biographical Index of British and Irish Botanists* by West, Newman & Company, 54 Hatton Garden, London EC1.

Second edition published 1931 by Taylor & Francis Ltd, Red Lion Court, Fleet Street, London EC4.

This edition published 1977 by Taylor & Francis Ltd, 10–14 Macklin Street, London WC2B 5NF.

ISBN 0 85066 089 0

Printed and Bound in Great Britain by Taylor & Francis (Printers) Ltd, Rankine Road, Basingstoke, Hampshire.

Distributed in the United States of America and its territories by Roman & Littlefield, 81 Adams Drive, Totowa, New Jersey 07512.

Contents

Preface

The Swedish traveller, Peter Kalm, writing in 1751 to the American botanist Cadwallader Colden observed that "there is nothing we want so much as a Biographica Botanicorum; the old were very negligent in that: there are many of which we hardly know any other thing but the name; nay, if we seek yet to the history of their life, we are obliged to pick up here and there a word in the writings of their contemporaries". Many men and women survive only in the names of the plants commemorating them; apart from this brief reminder of their existence they have disappeared without trace. We may never know why they were singled out for this honour. The pages of Curtis's *Botanical Magazine* frequently record the names of soldiers, sailors and merchants who during their travels collected new and unusual plants, which they brought back to eager gardeners and nurserymen in England.

It was in a commendable attempt to provide a biographical guide not only for such plant collectors, but also for the more important figures in botany, that James Britten and George E. S. Boulger collaborated in a series of regular instalments in the *Journal of Botany* from 1888 to 1891, which were revised and published in 1893 as *A Biographical Index of British and Irish Botanists.* Its compilers intended it "mainly as a guide to further information, and not as a bibliography or biography". Its proven value as a work of reference encouraged A. B. Rendle of the Natural History Museum to revise it in 1931. He preserved the conciseness of the original work which he enlarged from 1825 entries to over 2700. In this revision of the 1931 edition the number of entries has been almost quadrupled. This has been achieved not only by scanning books and periodicals published up to 1975, but also by substantially enlarging the scope of the work. Rendle admitted he found it "hard to draw a line of exclusion especially on the side of horticulture and in dealing with plant collectors". He solved his dilemma by rigorously excluding horticulturists. I have not sought to make such distinctions, and so for the first time nurserymen, gardeners and horticultural writers join botanists, plant collectors and botanical artists. All recipients of the Victoria Medal of Honour awarded by the Royal Horticultural Society have also been included. The entries relate mainly to people born in the British Isles but occasionally other nationals who made significant contributions to British plant science have been added. Some of the entries are brief due to the lack of any further information, but as I know from personal experience even the barest details (merely approximate dates when flourishing and place of residence) can be a useful starting point in research.

The sequence of the data given in the entries follows the pattern established by the previous editions:

1. Name (maiden name and title where applicable).
2. Dates and places of birth and death.
3. Education. Qualifications. Honours. Officership in relevant societies. Brief details of career. Selected publications. Editorship of periodicals.
4. Biographical references in books and periodicals.
5. Location of plant collections, herbaria, manuscripts, drawings, portraits. Sale of collections.
6. Any plant commemorating the individual followed by the abbreviated name of the person responsible for the plant name.

For reasons of economy abbreviations have been used but they are largely confined to the biographical references in books and periodicals. The author's own works are cited in full except for the use of the ubiquitous 'Fl.' for 'Flora'. The Hunt Botanical Library in Pittsburgh is performing a valuable service in assembling a collection of photographs of portraits of botanists. Since copies of all these photographs are readily available on request I thought it would be useful to identify them by the phrase "Portr. at the Hunt Library". The brief statement of the location of plant collections (Cambridge, Kew, Oxford, etc.) usually refers to the botanical institution in the places mentioned. It has not been possible to list all known locations or to confirm that plant collections are still in existence. Some were destroyed during the last war or have been lost through neglect or wilful destruction.

I have been surprised and sometimes confused by the errors that creep into obituaries and biographies. For instance, James Ramsay Drummond was reported by the *Journal of Botany* and the *Kew Bulletin* as having been born in Scotland on 13 May 1851; the *Proceedings of the Linnean Society* however preferred the more romantic setting of a maritime birth off the coast of Madras. There being far too many of these contradictions for me to investigate I must leave it to other research workers to establish the true facts. What is particularly disconcerting is to encounter obituaries of people still very much alive. In September 1912 the *Journal of Botany* lamented the demise of the eminent mycologist, Dr. Mordecai Cooke on 3 July. The *Kew Bulletin* which had reported that he had peacefully passed away at his residence on 19 August 1912 announced sheepishly in a later issue that it gave them "much pleasure to learn from Dr. Cooke himself that he is in excellent health". Dr. Cooke, who had this dubious pleasure of reading his own obituary, defiantly survived the shock for another two years. I have always given the facts as recorded, correcting them when I knew them to be wrong. Errors there must be but even the august *Dictionary of National Biography* is not without blemishes.

The Subject Index is a new feature classifying many of the entries under profession, plants or the country where the flora has been studied and collected. The chronological arrangement under each heading could serve as a useful introduction to the study of the historical development of a discipline, profession or trade. I also thought it would be of some interest to list English botanists and nurserymen under counties (the new administrative counties have not been adopted in this book).

I have spent the best part of eight years scanning books and long runs of periodicals to collect the raw material for this book. I believe I have searched most of the important published sources of information, but a number of marginal books and periodicals have not been consulted simply because there was not sufficient time to do so. Also unfortunately two major works, J. Harvey's *Early Nurserymen* (Phillimore, 1974 [1975]) and Miss B. Henrey's *British Botanical and Horticultural Literature before 1800* (Oxford University Press, 1975, 3 vols), appeared too late for consultation. Nevertheless I hope this new compilation will eventually prove to be indispensable for historians working in the fields of British botany, horticulture, botanical art and local history.

I wish to acknowledge my indebtedness to Mr. J. Harvey with whom I have exchanged information on nurserymen over a number of years. He has generously supplied me with details of many early nurserymen which he has obtained from scanning contemporary trade directories. Professor J. Ewan, Dr. G. Ainsworth, Mrs. B. Bolt, Mr. G. Claxton, Mr. R. Davidge, Mr. D. Kent, Miss S. Raphael, Dr. W. T. Stearn, Dr. T. D. Whittet, Mrs. S. Wickens and many others have kindly provided me with references. Above all I must thank my wife without whose patience and encouragement the work would never have been completed.

Historical Introduction

BY WILLIAM T. STEARN

In the Linnaean dissertation *Flora Anglica* publicly defended at Uppsala on 3 April 1754 by a student I. A. Grufberg, who may have done little more towards its production than pay for the printing, Linnaeus stated that, whereas at the beginning of the previous century the English nation seemed little suited, indeed alien, to the study of botany, it had produced by the end of that century as many botanists as the whole of the rest of Europe. Moreover they had studied the plants not only of Britain but of the Indies. He listed 25 who had contributed to the study of the British flora as enumerated in John Ray's *Synopsis methodica Stirpium Britannicarum* (3rd ed., revised and enlarged by Dillenius; 1724) which was long the best guide to British plants. In fact Ray and Dillenius named at least 42 contributors, without including such notable predecessors as William Turner, John Gerard and Thomas Johnson. Whether judged by their callings or their social standing, these co-operative enthusiasts for plants were a motley crowd; among them were apothecaries such as Samuel Dale, Samuel Doody, James Petiver and Isaac Rand, physicians such as John Martyn, Charles Preston, Richard Richardson, Robert Sibbald and Hans Sloane, antiquaries such as Edward Lhwyd and Robert Plot, clergymen such as Adam Buddle, Thomas Lawson and John Ray, gardeners such as Jacob Bobart and Thomas Knowlton, merchants such as Joseph Dandridge and Charles Dubois, soldiers such as Gedeon Bonnivert and Thomas Willisel and a writer with such miscellaneous interests as John Aubrey. Some like Dubois, the Sherards and Sloane became very rich, some such as Willisel and Lhwyd remained always poor. Ray, Willisel and Lhwyd made long and extensive herborising journeys, others did not go far afield but recorded the plants of their immediate neighbourhood. Indeed all they seem to have had in common was their enthusiastic interest in the plants of Britain.

This interest had grown slowly since the 16th century when William Turner published his *Libellus de Re Herbaria* (1538), his *Names of Herbes* (1548) and his *New Herball* (1551–68). It had its origin in the study of plants for medicinal purposes, a study which can be traced back to the herbalists of Ancient Greece. This accounts for the large number of medical men recorded in Britten and Boulger's *Biographical Index of British and Irish Botanists* (1891) and the present work. Turner's remark that he was "unworthy to act as a bottle-washer to the most learned Doctor Clement" indicates the existence of others in the 16th century who were equally interested although they published nothing. Without a continuously growing body of people united by such an interest, Turner, Ray and others would have published in vain. Physicians and apothecaries and indeed clergymen, who then and much later "grudged no personal fatigue to attend the sick-bed of the peasant, in the double capacity of physician and priest", as was said of George Crabbe, had a professional interest in plants for their uses, but there were undoubtedly many others whose interest sprang purely from delight in the beauty and diversity of flower and leaf. Ray himself recorded that he began his study of plants at Cambridge in 1650 when convalescence after illness gave him leisure to observe in detail "the cunning craftmanship of nature" manifested in the wild plants of Cambridgeshire. Gardeners too had a professional interest in distinguishing the kinds of plants. What they all needed were practical classifications and convenient names, but to achieve these demanded not only much ingenuity but also a much more intimate knowledge of plants than anyone could then acquire; not until the 19th century did the microscope in vastly improved forms come to importance for such studies. The procedure in making a classification is to group plants according to their major resemblances and then to name them as members of the groups thus formed; defining such groups has always been the chief difficulty in creating acceptable lasting systems. The change from Turner's essentially alphabetical arrangement of 1538–68, by way of John Gerard's essentially utilitarian one of 1597 with "all sorts of herbes for meate, medicine or sweet smelling use" grouped together, to Ray's scientific one of 1698 with dicotyledons and monocotyledons separated as major groups, was a lengthy and gradual process with its further development interrupted in the 18th century by the success of Linnaeus's convenient but admittedly artificial "sexual system". Every new system called for further observation of characters to which little attention or emphasis had been given before; thus the foundations for yet other systems, of which Bentham and Hooker's in the 19th century was the most influential for the English-speaking world, became broader and more solid and this affected botany as a whole. Botany is co-ordinated systematic knowledge of the plant kingdom; thus it covers not only what plants are but what they do, in other words, their physiology, ecology and distribution as well as their morphology, anatomy, biochemistry, cytology and genetics; all provide some information which a taxonomist can use. To such knowledge the amateur and the professional, the gardener and the botanist can contribute and indeed in Britain have often jointly contributed.

The distinctions now made in our age of specialisation between different kinds of botanists and horticulturists, sometimes with very invidious results, become meaningless when applied to their predecessors. Thus Gerard (1545–1612) was a barber-surgeon, a herbalist and a gardener. The present *Dictionary* unlike its predecessors, Britten & Boulger's *Biographical Index*, of 1893 and 1931, has not attempted to distinguish between botanists and horticulturists by excluding the latter. With a liberality of approach, which has immensely increased the labour of its compiler and has equally increased the reference value of the book, it includes them all, nurserymen and seedsmen along with botanical professors. An intermediate group, the gardener-botanists, including A. H. Haworth, William Herbert, George Maw, H. J. Elwes, W. R. Dykes and E. A. Bowles, has made outstanding contributions to knowledge of such groups as succulent plants and petaloid monocotyledons, e.g. *Crocus*, *Iris* and *Lilium*, which cannot be understood from herbarium material alone. Moreover a number of distinguished botanists, such as W. B. Hemsley, R. A. Rolfe, E. H. Wilson and John Hutchinson, began their careers as working gardeners. Philip Miller was professionally a gardener all his life but the lasting importance of his great *Gardeners Dictionary* rests upon its botanical content. Botany and horticulture played equal parts in the astonishing career of John Lindley. Because of their botanical interests such men have gained places in reference works, but contemporary nurserymen and seedsmen, despite their importance for the spread and maintenance of horticulture, have until now lacked any comprehensive listing. One of the especial merits of the present *Dictionary* is that it brings them together for the first time. By coming down to 1975 and thus covering the very fruitful 19th century, it supplements Blanche Henrey's elaborate three-volume *British Botanical and Horticultural Literature before 1800* (1975).

A biographical dictionary with its entries alphabetically arranged for convenient reference cannot, of course, by itself give any impression of the continuous growth and achievements of botany and gardening in the British Isles but the subject index added to the present work, with entries in chronological order, will provide valuable guidance to anyone investigating its subjects historically. The entries in their turn provide concise guides to widely scattered biographical and bibliographical information never before so thoroughly correlated and indexed. The general history of a subject can thus be built out of these records of the labours of the individuals who forwarded it. For taxonomic botanists, whose labours are of necessity historically based and dependent upon those of their predecessors, since the number of plants in the world is far too great for any one generation to be able to describe and classify them anew, the *Dictionary* has practical utility by indicating not only the major publications of authors but also the location of their herbaria and manuscripts. Such information is indeed internationally important. Linnaeus, as mentioned above, referred to the interest of British botanists not only in the plants of the British Isles but also in the Indies. There is now hardly a region of the world to which they have not paid attention in floristic or monographic studies or else have visited and made collections.

British botany has indeed had an international aspect from its beginning and has never been isolated from the mainstream of European botanical learning. The "father of British botany", William Turner (*d.* 1568), had twice to leave England for his personal safety in periods of lethal religious persecution; during his exile, however, he profited by travel in Germany, Italy, Switzerland and the Low Countries, by study under Italian teachers of such importance as Brasavola at Ferrara and Ghini at Bologna and by friendship with Fuchs, Gessner and Valerius Cordus. The effect of this Continental experience is evident in his publications from 1548 onwards: thus of one plant he remarked "I have sene it in east Freslande [Friesland] by the sea syde", of another "it groweth very little in Englande that I have sene, but about Bon [Bonn] it groweth in many hedges by the vineyards", of another "it groweth in gardines in Bonony [Bologna]", of yet another "it is very much in highe Germany besyde Embis [Ems] bath, and besyde S. Goweris [St. Goar]". Turner's *Names* and *Herball* abound with such references to the Continental occurrence of plants based upon his own observation. "I wente", he wrote, "into Italye and into diverse partes of Germany to knowe and se the herbes my selfe and to knowe by practise their powers and workinge not trusting onlye to the olde herbe wives and apothecaryes (as many Physicianes have done of late yeres) but in the mater of simples myne owne eyes and knowledge". Ray's Continental travels in the 18th century took him even further and gave him a much wider and more detailed knowledge of European plants. Continuance of such a British interest in the European flora, often resulting from holiday travel, led to the inception at Leicester in 1956 of the *Flora Europaea* project. This has proved a very successful international co-operative undertaking to which, for volumes 1–4 (1964–76) including all the vascular cryptogams, gymnosperms and dicotyledons, 149 botanists representing 27 countries have contributed; of these Great Britain has provided 62 contributors, followed by Germany with 10, Austria with 9, Czechoslovakia with 7, the others with 1 to 5.

The overseas expansion of the British people by trading, settlement, conquest and colonisation from the early 17th century onwards led to the introduction of many exotic plants into British and thence into Continental gardens and was reflected in an increase of botanical and horticultural literature. This took many forms, from concise octavo lists of plants in botanic gardens, such as Bobart's Oxford catalogue of 1648 and Sutherland's Edinburgh catalogue of 1683, to the great comprehensive folio works of Morison and Ray late in the 17th century, followed by the many editions of Miller's *Gardeners Dictionary* in the 18th century. Miller's works are valuable not only for recording the plants in cultivation in Britain, which more than doubled during the 18th century but also for accounts of the techniques of their cultivation. Botanical exploration in the British Isles and overseas also provided the material and the stimulus for Floras, i.e. works devoted to listing the plants of a given area and often indeed describing and illustrating them. Among the most impressive examples of these are Sloane's *Voyage to the Islands Madera* etc. (1707–25), which includes the natural history of Jamaica, in the 18th century and Sibthorp and Smith's *Flora Graeca* (1806–40) in the 19th century. During the 19th century the efforts of local almost entirely amateur botanists resulted in a succession of county Floras, which have made the flora of the British Isles the best-documented one in Europe. Thus by the time of his death in 1974 N. D. Simpson had collected a library of over 6,000 books relating to British and Irish plants. His *Bibliographical Index of the British Flora* (1960) has some

65,000 entries. Outside the British Isles, the economic and scientific needs of the expanding 19th-century British Empire resulted in the production of colonial Floras, the most extensive being Bentham's *Flora Australiensis* (1863–78) and J. D. Hooker's *Flora of British India* (1872–97) based for the most part on herbarium material accumulated at the Royal Botanic Gardens, Kew. During the first half of the 19th century, many botanists, either for research purposes or the satisfaction given by collecting, assembled large private herbaria. During the second half of the century, such collections passed more and more into the keeping of institutions. This transfer of herbaria from private to public ownership and responsibility is discussed in *Natural History Auctions 1700–1972; a Register of Sales in the British Isles* (1976) edited by J. M. Chalmers-Hunt. Wherever known the present location of these collections is indicated in the *Dictionary*.

Since modern botany grew out of the study of medicinal plants, for which correct identification was essential, matters of classification and nomenclature accordingly formed the major part of botanical activity throughout Europe well into the 19th century until progress in other disciplines, particularly chemistry, made possible physiological investigations; at the same time the development of microscopes and staining techniques provided adequate tools for anatomical and cytological work. Towards the end of the 19th century in Britain and elsewhere the supremacy of taxonomic work gave place in academic esteem, though hardly in achievement, to these more stimulating new fields which had grown so lustily in Germany with its numerous mostly small universities providing scope for academic diversity and innovation. Their effect on botany in British universities was profound. As F. O. Bower (1855–1948) wrote in his autobiographical *Sixty Years of Botany in Britain (1875–1935)*, "at the middle of the nineteenth century British botany was marked by strange contrasts. It witnessed the production by its seniors of descriptive works of fundamental importance, such as the *Species filicum* and the *Genera plantarum*; overshadowing all was the *Origin of species*. But up to 1875 the universities almost entirely failed to train students who as juniors should be preparing to take the place of the veteran authors. Moreover it was collecting, classifying and recording that were the order of the day; while anatomy, physiology and the study of the complete life-cycle, especially in the lower forms, were given only a minor place in the university curricula, if indeed they were not wholly neglected. It seems strange now to look back upon the deadness of botanical teaching in the universities in the years prior to 1875, though this was a period of extreme brilliance of individual production; read the *Life and Letters* of Darwin, Hooker and Huxley, books which reveal the intense personal activity of these giants." To the younger British academic botanists in the 1870s, publications coming from Germany revealed fascinating new fields of enquiry; they quickly learned, as Robert Brown had been well aware so much earlier when in 1800 he was industriously studying the genders of German nouns at breakfast time, "knowledge of German was essential for the pursuit of botany". Two of the most brilliant teachers and investigators were Anton de Bary (1831–1888), professor of botany from 1872 to 1888 at Strassburg (Strasbourg), and Julius Sachs (1832–1897), professor of botany from 1866 to 1897 at Würzburg. To their laboratories in the 1870s and '80s went from Oxford and Cambridge a succession of promising young men, notably S. H. Vines, Francis Darwin, F. O. Bower, Walter Gardiner, D. H. Scott and Marshall Ward, who later introduced new attitudes, some of them long-lastingly detrimental to taxonomic teaching in England, new fields of enquiry and new research methods and techniques to Oxford, Cambridge, Glasgow and Kew. They had little time or interest for the "collecting, classifying and recording" of their predecessors in which so many amateur botanists had taken part. The new botany of their new laboratories became less and less open to contributions by amateurs on account of the facilities and specialised knowledge required and the great British tradition of taxonomic botany might almost have perished but for the Royal Botanic Gardens of Kew and Edinburgh and the British Museum (Natural History). However, the separation of laboratory and field work which might have made a great gulf between professional and amateur has fortunately never in Britain become very rigid and, through the development in particular of ecology, has often diminished. Moreover its foremost biological society, the Linnean Society of London founded in 1788, has always welcomed both in its promotion of natural history. It was possible for J. Reynolds Green to provide in 1914 *A History of Botany in the United Kingdom from the earliest Times to the End of the* 19*th Century* which gave a reasonably detailed view of the progress in all botanical disciplines in the British Isles. Now each discipline requires its own history. Through the references in the *Dictionary* specialist historians of both botany and horticulture can trace their way backwards to earlier and firsthand sources of information.

In 1863 Berthold Carl Seemann, an extraordinarily enterprising German botanist over many years in the service of the British Government as naturalist on exploring voyages, founded *The Journal of Botany British and Foreign*. The inclusion of the words "British and Foreign" in its title is significant. The word "British" was a tribute to the need of amateur botanists concerned with British plants to have a medium for publication. The word "Foreign" indicated a breadth of coverage more suited to the professional botanist. Until its much regretted and certainly avoidable death in 1945 the *Journal of Botany* provided convenient and esteemed means of publishing biographical and bibliographical as well as botanical information of great diversity, although primarily taxonomic; for amateur and professional alike, it indeed "filled a position which, even now, is covered by no other periodical". The list of contributors to volume 1 (1863) includes many distinguished names, among those who were then amateurs being J. G. Baker, then a draper, later Keeper of the Herbarium of the Royal Botanic Gardens, Kew, and James Britten (1846–1924), then apparently a medical student, later Senior Assistant in the Department of Botany, British Museum (Natural History) and from 1880 until 1924 editor of the *Journal of Botany* itself. Although a competent botanist "with a good eye for plants and a wonderful memory", Britten was above all interested in the bibliographical and biographical side of botany. His interests coincided in this with those of the many-sided George Simonds Boulger (1853–1922), at first professor of natural history at the Royal Agricultural College, Cirencester, later professor of botany at the City of London College, and the author of well-prepared "solid and good" books on trees, wood, elementary geology and British wild plants. Both felt the need in the 1880s of "some reference-list of byegone workers in Botany". After several years of preparation they began the publication of their biographical list in the *Journal of Botany* in 1888 and completed this in 1891. It included 1,619 names. In 1893 they issued it in book form as *A Biographical Index of British and Irish Botanists* with 1,825 names, their aim being to include "all who have in any way contributed to the literature of Botany, who have made scientific collections of plants, or have otherwise assisted directly in the progress of Botany, exclusive of pure Horticulture". They issued supplements in 1899, 1905 and 1908, but never lived to produce the contemplated new edition, for which however Britten bequeathed money in his will towards the cost of printing. Edited by A. B. Rendle, after his retirement as Keeper of Botany in the British Museum (Natural History), the second edition appeared in 1931 with about 2,700 names. The present work has more than 8,000 entries, many of them more detailed than their predecessors.

Much side information can be gleaned from these entries even as they stand, e.g. the longevity of botanists in different fields. To take successive Keepers of the Kew Herbarium, Daniel Oliver (1830–1916) died at the age of 85, J. G. Baker (1834–1920) at 86, W. B. Hemsley (1843–1924) at 80, O. Stapf (1859–1933) at 76, A. D. Cotton (1879–1962) at 82 and W. B. Turrill (1890–1961) at 71. A more healthy appointment would seem to be that of Director of the Singapore Botanic Garden, H. W. Ridley (1855–1956) dying at the age of 100, I. H. Burkill (1870–1965) at 94, and their successor Eric Holttum (b.1895) celebrating his 81st birthday in 1976 and publishing six pteridological papers this year.

The achievements of the present rest upon the achievements of the past to a much greater extent than is commonly appreciated and can often only be understood in relation to them. For those who would learn about the British and Irish botanists and gardeners, men and women, amateurs and professionals, who by their collective endeavours over four centuries have gathered the specimens preserved in the herbaria of the British Isles, have introduced, maintained and propagated the many plants grown in these islands, have written and illustrated innumerable books and articles about them and the plants of other lands, have made notable contributions to paleobotany and the manifold other disciplines of modern botany, for all such enquirers the brief records of their lives and achievements in this *Dictionary* will be found an indispensable time-saving guide often leading to obscure sources of information which otherwise could be traced with difficulty or not at all. Its scope is, however, far from being insular. Names familiar to workers on the plants of Canada, the United States, the West Indies, southern Africa, India, Malaya, Australia, New Zealand and China as the names of collectors or investigators occur throughout its pages. It should accordingly be destined for as much or more wear and tear by grateful users as its predecessors of 1893 and 1931.

Bibliography of Works Consulted

Books

——— *Australian Dictionary of Biography*, vols 1 and 2: 1788–1850; vol 3: A–C, 1851–1890. Melbourne, 1966–69.

——— *Comptes Rendus de la IVe Reunion Plénière de l'Association pour l'Étude Taxonomique de la Flore d'Afrique Tropicale, 16–23 Sept., 1960*. Lisbon, 1962.

——— *Dictionary of National Biography*. London, 1885–1904. 64 vols; *Supplement*. 1901. 3 vols; *Second Supplement, 1901–1911* 1920; *1912–1921* 1927; *1922–1930* 1937; *1931–1940* 1949; *1941–1950* 1959; *1951–1960* 1971.

Allan, M. *Hookers of Kew, 1785–1911*. London, 1967.

Amherst, A. *afterwards Lady* E. Cecil *History of Gardening in England*. London, 1895; ed. 3 1910.

Anderson, A. W. *Coming of the Flowers*. London, 1950.

Andrews, H. C. *Botanists' Repository*. London, 1797–1814. 10 vols; ed. 2 1816. 10 vols.

Arber, A. *Herbals: their Origin and Evolution*. Ed. 2. Cambridge, 1938.

Archer, M. *Natural History Drawings in the India Office Library*. London, 1962.

Babington, C. C. *Manual of British Botany*. London, 1843; ed. 8 1881.

Babington, C. C. *Memorials, Journals and Botanical Correspondence*. Cambridge, 1897.

Bagnall, J. E. *Flora of Warwickshire*. Birmingham, 1891.

Bailey, K. *Irish Flora*. Dublin, 1833.

Baillon, H. E. *Dictionnaire de Botanique*. Paris, 1876–92. 4 vols.

Baines, H. *Flora of Yorkshire*. London, 1840.

Baker, J. G. *Flora of the English Lake District*. London, 1885.

Baker, J. G. *Flora of Mauritius and the Seychelles*. London, 1877.

Balfour, E. G. *Cyclopaedia of India*. Ed. 2. Madras, 1871–73. 5 vols.

Barnhart, J. H. *Biographical Notes upon Botanists*. Boston, Mass., 1965. 3 vols.

Bentham, G. *Flora Australiensis*. London, 1863–78. 7 vols.

Bentham, G. *Flora Hongkongensis*. London, 1861.

Berkeley, E. *and* D. S. *John Clayton; Pioneer of American Botany*. N. Carolina, 1963.

Berkenhout, J. *Synopsis of the Natural History of Great Britain and Ireland...* Ed. 3. London, 1795.

Bingley, W. *Practical Introduction to Botany*. London, 1817.

Bladen, F. M., *et al., ed. Historical Records of New South Wales*. Sydney, 1893–98. 6 vols.

Blunt, W. *Art of Botanical Illustration*. London, 1950; ed. 3 1955.

Boase, F. *Modern English Biography*. Truro, 1892–1921. 6 vols.

Boase, G. C. *and* Courtney, W. P. *Bibliotheca Cornubiensis*. London, 1874–82. 3 vols.

Bolton, J. *History of Fungusses growing about Halifax*. Halifax, 1788–89. 3 vols. *Appendix*. Huddersfield, 1791.

Bower, F. O. *Sixty Years of Botany in Britain (1875–1935): Impressions of an Eye-witness*. London, 1938.

Bretschneider, E. *History of European Botanical Discoveries in China*. London, 1898. 2 vols.

Brett-James, N. G. *Life of Peter Collinson, F.R.S., F.S.A.* London, 1926.

Brewer, J. A. *Flora of Surrey*. London, 1863.

Briggs, T. R. A. *Flora of Plymouth*. London, 1880.

British Museum (Natural History) *History of the Collections contained in the Natural History Departments of the British Museum*. London, 1904–12. 2 vols.

Brown, R. *Prodromus Florae Novae Hollandiae et Insulae Van-Diemen*. London, 1810.

Browne, P. *Civil and Natural History of Jamaica in Three Parts*. London, 1756; ed. 2 1789.

Buckland, C. E. *Dictionary of Indian Biography*. London, 1906.

Burkill, I. H. *Chapters on the History of Botany in India*. Delhi, 1965.

Burns, T. E. *and* Skemp, J. R. *Van Diemen's Land Correspondents...1827–1849*. Launceston, 1961.

Buxton, R. *Botanical Guide to the Flowering Plants, Ferns, Mosses and Algae, found Indigenous within Sixteen Miles of Manchester*. London, 1849.

Cadbury, D. A., Hawkes, J. G. *and* Readett, R. C. *A Computer-mapped Flora: a Study of the County of Warwickshire*. London, 1971.

Cash, J. *Where there's a Will there's a Way! or, Science in the Cottage, an Account of the Labours of Naturalists in Humble Life*. London, 1873.

Chambers, R. *Book of Days*. London, 1864. 2 vols.

Cheeseman, T. F. *Manual of the New Zealand Flora*. Wellington, 1906; ed. 2 1925.

Christy, R. M. *Birds of Essex*. Chelmsford, 1890.

Clark-Kennedy, A. E. *Stephen Hales, D.D., F.R.S.* Cambridge, 1929.

Clokie, H. N. *Account of the Herbaria of the Department of Botany in the University of Oxford*. Oxford, 1964.

Coats, A. M. *Flowers and their Histories*. London, 1968.
Coats, A. M. *Garden Shrubs and their Histories*. London, 1963.
Coats, A. M. *Quest for Plants*. London, 1969.
Colgan, N. *Flora of the County Dublin*. Dublin, 1904.
Colgan, N. *and* Scully, R. W. *Contributions towards a Cybele Hibernica*. Ed. 2. Dublin, 1898.
Colmeiro, M. *La Botánica y los Botánicos de la Peninsula Hispano-Lusitana. Estudios Bibliográficos y Biográficos*. Madrid, 1858.
Copeman, W. S. C. *Worshipful Society of Apothecaries of London: a History, 1617–1967*. Oxford, 1967.
Cosson, E. St.-C. *Compendium Florae Atlanticae*. Paris, 1881–87. 2 vols.
Cox, E. H. M. *Plant Hunting in China*. London, 1945.
Coxe, W. *Literary Life and Select Works of Benjamin Stillingfleet*. London, 1811. 2 vols.
Crawford, D. G. *History of the Indian Medical Service, 1600–1913*. London, 1914. 2 vols.
Crossland, C. *An Eighteenth Century Naturalist: James Bolton, Halifax*. Halifax, 1910.
Crump, W. B. *and* Crossland, C. *Flora of the Parish of Halifax*. Halifax, 1904.
Cundall, F. *Historic Jamaica*. Kingston, 1915.
Currey, J. E. B., *ed. Reflections on the Colony of New South Wales*. London, 1967.
Dandy, J. E., *ed. Sloane Herbarium*. London, 1958.
Darlington, W. *Memorials of John Bartram and Humphry Marshall*. Philadelphia, 1849. (Reprint 1967).
Darlington, W. *Reliquiae Baldwinianae*. Philadelphia, 1843. (Reprint 1969).
Darwin, F. *and* Seward, A. C. *More Letters of Charles Darwin*. London, 1903. 2 vols.
Davey, F. H. *Flora of Cornwall*. Penryn, 1909.
Davies, H. *Welsh Botanology*. London, 1813.
Dickie, G. *Flora of Ulster and Botanist's Guide to the North of Ireland*. Belfast, 1864.
Dillenius, J. J. *Historia Muscorum*. Oxford, 1741.
Dillenius, J. J. *Hortus Elthamensis*. London, 1732. 2 vols.
Dillwyn, L. W. *British Confervae*. London, 1802–9.
Dillwyn, L. W. *Hortus Collinsonianus. An Account of the Plants cultivated by the Late Peter Collinson Esq*. Swansea, 1843.
Dillwyn, L. W. *Materials for a Fauna and Flora of Swansea and the Neighbourhood*. Swansea, 1848.
Donaldson, J. *Agricultural Biography: Containing a Notice of the Life and Writings of the British Authors on Agriculture, from the Earliest Date in 1480 to the Present Time*. London, 1854.
Dony, J. G. *Flora of Bedfordshire*. Luton, 1953.
Dony, J. G. *Flora of Hertfordshire*. Hitchin, 1967.
Druce, G. C. *Account of the Herbarium of the University of Oxford*. Oxford, 1897.
Druce, G. C. *Comital Flora of the British Isles*. Arbroath, 1932.
Druce, G. C. *Dillenian Herbaria*. Edited by S. H. Vines. Oxford, 1907.
Druce, G. C. *Flora of Berkshire*. Oxford, 1897.
Druce, G. C. *Flora of Buckinghamshire*. Arbroath, 1926.
Druce, G. C. *Flora of Northamptonshire*. Arbroath, 1930.
Druce, G. C. *Flora of Oxfordshire*. Oxford, 1886; ed. 2 1927.
Dunthorne, G. *Flower and Fruit Prints of the 18th and Early 19th Centuries*. London, 1938.
Edees, E. S. *Flora of Staffordshire*. Newton Abbot, 1972.
Erickson, R. *Drummonds of Hawthornden*. Osborne Park, W. Australia, 1969.
Evans, A. H. *Flora of Cambridgeshire*. London, 1939.
Ewan, J. *Rocky Mountain Naturalists*. Denver, 1950.
Faulkner, T. *Historical and Topographical Description of Chelsea and its Environs*. London, 1829. 2 vols.
Faulkner, T. *History and Antiquities of Kensington*. London, 1820.
Fawcett, W. *and* Rendle, A. B. *Flora of Jamaica*. London, 1910–36. 5 vols.
Fée, A. L. A. *Vie de Linné*. Paris, 1832.
Felton, S. *On the Portraits of English Authors on Gardening*. Ed. 2. London, 1830.
Field, H. *Memoirs Historical and Illustrative of the Botanick Garden at Chelsea belonging to the Society of Apothecaries of London*. London, 1820.
Fletcher, H. R. *Story of the Royal Horticultural Society, 1804–1968*. Oxford, 1969.
Fletcher, H. R. *and* Brown, W. H. *Royal Botanic Garden Edinburgh, 1670–1970*. Edinburgh, 1970.
Forster, J. R. *Characteres Generum Plantarum quas in Itinere ad Insulas Maris Australis Collegerunt, Descripserunt, Delinearunt Annis 1772–75*. London, 1776.
Foster, J. *Alumni Oxonienses*. Oxford, 1887–92. 8 vols.
Fox, R. H. *Dr. John Fothergill and his Friends*. London, 1919.
Gage, A. T. *History of the Linnean Society of London*. London, 1938.
George, W. *Biologist Philosopher* [*A. R. Wallace*]. New York, 1964.
Gerard, J. *Herball or Generall Historie of Plantes*. London, 1597; ed. 2 1633.
Gibbons, E. J. *Flora of Lincolnshire*. Lincoln, 1975.
Gibson, G. S. *Flora of Essex*. London, 1862.
Gillispie, C. C., *ed. Dictionary of Scientific Biography*. New York, 1970–76. 14 vols.
Gillow, J. *Literary and Biographical History, or Bibliographical Dictionary of English Catholics*. London, 1885–1902.

Glenn, R. *Botanical Explorers of New Zealand*. Wellington, 1950.
Godman, F. D. *and* Salvin, O. *Biologia Centrali-Americana. Botany by W. B. Hemsley*. London, 1879–88. 4 vols.
Gorham, G. C. *Memoirs of John Martyn, F.R.S. and of Thomas Martyn*. London, 1830.
Graustein, J. E. *Thomas Nuttall, Naturalist: Explorations in America, 1808–1841*. Cambridge, Mass., 1967.
Green, J. R. *History of Botany in the United Kingdom from the Earliest Times to the End of the 19th Century*. London, 1914.
Greville, R. K. *Algae Britannicae*. Edinburgh, 1830.
Griffith, W. *Posthumous Papers: Journals of Travels; Itinerary Notes*. Calcutta, 1847–48. 2 vols.
Grigor, J. *Eastern Arboretum; or, Register of Remarkable Trees, Seats, Gardens, etc. in the County of Norfolk*. London, 1841.
Grose, D. *Flora of Wiltshire*. Devizes, 1957.
Guilding, L. *Account of the Botanic Garden in the Island of St. Vincent*. Glasgow, 1825.
Gunther, R. W. T. *Early British Botanists and their Gardens*... Oxford, 1922.
Hadfield, M. *Gardening in Britain*. London, 1960.
Hadfield, M. *Pioneers in Gardening*. London, 1955.
Haller, A. von *Bibliotheca Botanica*. London, 1771–72. 2 vols.
Hanbury, F. J. *and* Marshall, E. S. *Flora of Kent*. London, 1899.
Harshberger, J. W. *Botanists of Philadelphia and their Work*. Philadelphia, 1899.
Harvey, J. *Early Gardening Catalogues*. Chichester, 1972.
Harvey, W. H. *Manual of the British Algae*. London, 1841.
Harvey, W. H. *Phycologia Australica; or, a History of the Australian Seaweeds*. London, 1858–63. 5 vols.
Harvey, W. H. *Phycologia Britannica; or, a History of the British Seaweeds*. London, 1846–51. 4 vols.
Harvey, W. H. *Thesaurus Capensis*. Dublin, 1859–63. 2 vols.
Harvey, W. H., Sonder, O. W., *et al. Flora Capensis*. London, 1859–1925. 7 vols.
Hasluck, A. *Portrait with Background: a Life of Georgiana Molloy*. Melbourne, 1955.
Hawks, E. *Pioneers of Plant Study*. London, 1928.
Haworth, A. H. *Miscellanea Naturalia sive Dissertationes Variae ad Historiam Naturalem Spectantes*. London, 1803.
Hedrick, U. P. *History of Horticulture in America to 1860*. New York, 1950.
Hemsley, W. B. *Report on the Scientific Results of the Voyage of H.M.S. Challenger during the Years 1873–76...Botany*. London, 1885–86. 2 vols.
Hepper, F. N. *and* Neate, F. *Plant Collectors in West Africa*. London, 1971.
Herklots, G. A. C. *Hong Kong Countryside*. Hong Kong, 1951.
Hill, A. W. *Henry Nicholson Ellacombe...a Memoir*. London, 1919.
Hill, R. *Biographical Dictionary of the Sudan*. Ed. 2. London, 1967.
Hind, W. M. *Flora of Suffolk*. London, 1889.
Hodgson, W. *Flora of Cumberland*. Carlisle, 1898.
Hoeniger, F. D. *and* J. F. M. *Growth of Natural History in Stuart England from Gerard to the Royal Society*. Virginia, 1969.
Hooker, J. D. *Flora of British India*. London, 1872–97. 7 vols.
Hooker, J. D. *Flora Tasmaniae*. London, 1855–59.
Hooker, J. D. *and* Thomson, T. *Flora Indica*. London, 1855.
Hooker, W. J. *Exotic Flora*. Edinburgh, 1823–27. 3 vols.
Hooker, W. J. *Flora Boreali-americana, or the Botany of the Northern Parts of British America*. London, 1829–40. 2 vols.
Horwood, A. R. *and* Noel, C. W. F. *Flora of Leicestershire and Rutland*. Oxford, 1933.
Howitt, R. C. L. *and* B. M. *Flora of Nottinghamshire*. Nottingham, 1963.
Humphrey, H. B. *Makers of North American Botany*. New York, 1961.
Hunt, P., *ed. Shell Gardens Book*. London, 1964.
Hunt, T. *Medical Society of London, 1773–1973*. London, 1972.
Hunt Botanical Library *Biographical Dictionary of Botanists represented in the Hunt Institute Portrait Collection*. Boston, Mass., 1972.
Hutchinson, J. *A Botanist in Southern Africa*. London, 1946.
Huxley, L. *Life and Letters of Sir Joseph Dalton Hooker*. London, 1918. 2 vols.
Jackson, B. D. *Guide to the Literature of Botany*. London, 1881.
Jeffers, R. H. *Friends of John Gerard (1545–1612): Surgeon and Botanist*. Connecticut, 1967. *Biographical Appendix*. 1969.
Jermyn, S. T. *Flora of Essex*. Colchester, 1974.
Johnson, G. W. *History of English Gardening*. London, 1829.
Johnston, G. *Flora of Berwick-upon-Tweed*. Edinburgh, 1829–31. 2 vols.
Johnston, G. *Terra Lindisfarnensis. The Natural History of the Eastern Borders*. London, 1853.
Jolly, W. *Life of John Duncan, Scotch Weaver and Botanist*... London, 1883.
Jones, J. P. *Botanical Tour through...Devon and Cornwall*. Exeter, 1820.
Jones, J. P. *and* Kingston, J. F. *Flora Devoniensis*. London, 1829.

Kent, D. H. *British Herbaria*. London, 1957.
Kent, D. H. *Historical Flora of Middlesex: an Account of the Wild Plants found in the Watsonian Vice-county 21 from 1548 to the Present Time*. London, 1975.
Kew, H. W. *and* Powell, H. E. *Thomas Johnson: Botanist and Royalist*. London, 1932.
Kirby, M. *Flora of Leicestershire*. Leicester, 1848; ed. 2 1850.
Lankester, E., *ed. The Correspondence of John Ray*. London, 1848.
Lankester, E., *ed. Memorials of John Ray*. London, 1846.
Lasègue, A. *Musée Botanique de B. Delessert*. Paris, 1845.
Lees, E. *Botany of Worcestershire*. Worcester, 1867.
Lees, F. A. *Flora of West Yorkshire*. London, 1888.
Lemmon, K. *Golden Age of Plant Hunters*. London, 1968.
Lettsom, J. C. *Memoirs of John Fothergill, M.D., etc.* Ed. 4. London, 1786.
Lindau, G. *and* Sydow, P. *Thesaurus Litteraturae Mycologicae et Lichenologicae*... Lipsiis, 1908–17. 5 vols.
Lindley, J. *Collectanea Botanica*. London, 1821.
Lindley, J. *Genera and Species of Orchidaceous Plants*. London, 1830–40.
Lindley, J. *and* Moore, T. *Treasury of Botany*. London, 1866. 2 vols.
Liverpool Museums *Handbook and Guide to the Herbarium Collections in the Public Museums*. London, 1935.
Lobel, M. De *Stirpium Illustrationes*. London, 1655.
Loudon, J. C. *Encyclopaedia of Gardening*. London, 1822.
Loudon, J. C. *Arboretum et Fruticetum Britannicum*. London, 1838. 8 vols.
Lousley, J. E. *Flora of the Isles of Scilly*. Newton Abbot, 1972.
Lowe, E. J. *Fern Growing*. London, 1895.
Lyell, K. M. *Life of Sir Charles J. F. Bunbury, Bart*. London, 1906. 2 vols.
McClintock, D. *Companion to Flowers*. London, 1966.
McDonald, D. *Agricultural Writers*. London, 1908.
Macfadyen, J. *Flora of Jamaica*. London, 1837–50. 2 vols.
McKelvey, S. D. *Botanical Exploration of the Trans-Mississippi West, 1790–1850*. Cambridge, Mass., 1955.
Maiden, J. H. *Sir Joseph Banks, the Father of Australia*. Sydney, 1909.
Marloth, R. *Flora of South Africa*. Capetown, 1913–32. 4 vols.
Mansel-Pleydell, J. C. *Flora of Dorsetshire*. London, 1874.
Martin, W. Keble *and* Fraser, G. T. *Flora of Devon*. Arbroath, 1939.
Martius, C. F. P. de *Flora Brasiliensis*. Leipzig, 1840–1906. 15 vols.
Matthew, L. G. *Royal Apothecaries*. London, 1967.
Menezes, C. A. de *Flora do Archipelago da Madeira*. Funchal, 1914.
Mennell, P. *Dictionary of Australasian Biography*. London, 1892.
Merrett, C. *Pinax Rerum Naturalium Britannicarum*... London, 1666.
Milner, J. D. *Catalogue of Portraits of Botanists exhibited in the Museums of the Royal Botanic Gardens* (*Kew*). London, 1906.
Mueller, F. von *Fragmenta Phytographiae Australiae*. Melbourne, 1858–82. 12 vols.
Munk, W. *Roll of Royal College of Physicians of London*. Ed. 2. London, 1878. 3 vols. *Lives of Fellows of Royal College of Physicians of London, 1826–1925*, by G. H. Brown. 1955. ...*1926–1965*, by R. R. Trail. 1968.
Murray, A. *Northern Flora; or a Description of the Wild Plants belonging to the North and East of Scotland*. Edinburgh, 1836.
Nichols, J. *Illustrations of the Literary History of the Eighteenth Century*. London, 1817–58. 8 vols.
Nichols, J. *Literary Anecdotes of the Eighteenth Century*. London, 1812–15. 9 vols.
Nuttall, T. *Genera of North American Plants*. Philadelphia, 1818. 2 vols.
Oliver, D. *et al. Flora of Tropical Africa*. London, 1868–1937. 10 vols.
Oliver, F. W. *Makers of British Botany*. Cambridge, 1913.
Parkinson, J. *Paradisi in Sole Paradisus Terrestris, or a Garden of all Sorts of Pleasant Flowers*. London, 1629.
Parkinson, J. *Theatrum Botanicum*. London, 1640.
Petch, C. P. *and* Swann, E. L. *Flora of Norfolk*. Norwich, 1968.
Petiver, J. *Musei Petiveriani*. London, 1695–1703.
Plukenet, L. *Almagesti Botanici Mantissa*. London, 1700.
Plukenet, L. *Almagestum Botanicum*. London, 1696.
Plukenet, L. *Amaltheum Botanicum*. London, 1705.
Praeger, R. L. *Some Irish Naturalists: a Biographical Notebook*. Dundalk, 1949.
Pritzel, G. A. *Thesaurus Literaturae Botanicae*. Ed. 2. Lipsiae, 1872.
Pryor, A. R. *Flora of Hertfordshire*. Hertford, 1887.
Pulteney, R. *Historical and Biographical Sketches of the Progress of Botany in England*. London, 1790. 2 vols.
Purchas, W. H. *and* Ley, A. *Flora of Herefordshire*. Hereford, 1889.
Purton, T. *Botanical Description of British Plants in the Midland Counties*. Stratford-upon-Avon, 1817–21. 3 vols.

Raven, C. E. *English Naturalists from Neckam to Ray*. Cambridge, 1947.
Raven, C. E. *John Ray, Naturalist: his Life and Works*. Cambridge, 1942; ed. 2 1950.
Ray, J. *Historia Plantarum*. London, 1686–1704. 3 vols.
Ray, J. *Synopsis Methodica Stirpium Britannicarum*. London, 1690; ed. 2 1696; ed. 3 1724.
Rees, A. *Cyclopaedia; or, Universal Dictionary of Arts, Sciences and Literature*. London, 1802–20. 45 vols. (J. E. Smith contributed most of the botanical articles and biographies).
Reeve, L. A. *and* Walford, E. *Portraits of Men of Eminence in Literature, Science and Art*. London, 1863–67. 6 vols.
Rehder, A. *Bradley Bibliography. A Guide to the Literature of the Woody Plants of the World...* Cambridge, Mass., 1911–18. 5 vols.
Reinikka, M. A. *History of the Orchid*. Miami, 1972.
Retzius, A. J. *Observationes Botanicae Sex Fasciculis Comprehensae*. Lipsiae, 1779–91.
Riddelsdell, H. J., Hedley, G. W. *and* Price, N. R. *Flora of Gloucestershire*. Cheltenham, 1948.
Robbins, C. C. *David Hosack: Citizen of New York*. Philadelphia, 1964.
Rohde, E. S. *Old English Herbals*. London, 1922.
Roxburgh, W. *Hortus Bengalensis*. Serampore, 1814.
Roxburgh, W. *Plants of the Coast of Coromandel; selected from Drawings and Descriptions presented to the Hon. Court of Directors of the East India Company*. London, 1795–1819. 3 vols.
Royal Horticultural Society *The Lindley Library. Catalogue...* London, 1927.
Royal Society *Catalogue of Scientific Papers, 1800–1900*. London, 1867–1925. 19 vols.
Sachs, F. G. J. von *History of Botany (1530–1860)...Translation by H. E. F. Garnsey, Revised by I. B. Balfour*. Oxford, 1890.
Salmon, C. E. *Flora of Surrey*. London, 1931.
Scott, E. J. L. *Index to the Sloane Manuscripts in the British Museum*. London, 1904.
Seemann, B. C. *Botany of the Voyage of H.M.S. "Herald" under...Capt. Henry Kellett during the Years 1845–51*. London, 1852–57.
Seemann, B. C. *Flora Vitiensis*. London, 1865–73.
Semple, R. H. *Memoirs of the Botanic Garden at Chelsea...by the Late Henry Field*. London, 1878.
Simmonds, A. *Horticultural Who was Who*. London, 1948.
Smith, E. *Life of Sir Joseph Banks...with some Notices of his Friends and Contemporaries*. London, 1911.
Smith, J. *Records of the Royal Botanic Gardens, Kew*. London, 1880.
Smith, J. E. *Exotic Botany*. London, 1804–5. 2 vols.
Smith, J. E., *ed*. *Selection from the Correspondence of Linnaeus and other Naturalists*. London, 1821. 2 vols.
Smith, P. *Memoir and Correspondence of...Sir James Edward Smith*. London, 1832. 2 vols.
Society of Friends *Biographical Catalogue*. London, 1888.
Sowerby, J. *and* Smith, J. E. *English Botany, or Coloured Figures of British Plants, with their Essential Characters, Synonyms and Places of Growth*. London, 1790–1814. 36 vols.
Sowerby, A. de C., Sowerby, A. M. *and* Stone, J. E. *Sowerby Saga*. Washington, 1952.
Spence, M. *Flora Orcadensis*. Kirkwall, 1914.
Stafleu, F. A. *Taxonomic Literature: a Selective Guide to Botanical Publications with Dates, Commentaries and Types*. Utrecht, 1967.
Stewart, S. A. *and* Corry, T. H. *Flora of the North-East of Ireland*. Belfast, 1888.
Stokes, J. *Botanical Commentaries*. London, 1830.
Sweet, R. *British Flower Garden*. London, 1823–38. 7 vols.
Swem, E. G. *Brothers of the Spade: Correspondence of Peter Collinson of London, and of John Custis, of Williamsburg, Virginia, 1734–1746*. (*Proc. Amer. Antiq. Soc.*, 1948; reprinted 1949).
Symons, J. *Synopsis Plantarum Insulis Britannicis Indigenarum*. London, 1798.
Taylor, G. *Some Nineteenth Century Gardeners*. London, 1951.
Taylor, R. V. *Biographia Leodiensis*. London, 1865–67.
Townsend, F. *Flora of Hampshire including the Isle of Wight*. London, 1883; ed. 2 1904.
Trimen, H. *Hand-book to the Flora of Ceylon*. London, 1893–1931. 6 vols.
Trimen, H. *and* Thiselton-Dyer, W. T. *Flora of Middlesex*. London, 1869.
Turner, D. *Extracts from the Literary and Scientific Correspondence of Richard Richardson M.D., F.R.S. of Bierley, Yorkshire*. Yarmouth, 1835.
Turner, D. *Fuci*. London, 1808–19. 4 vols.
Turner, D. *Muscologiae Hibernicae Spicilegium*. Yarmouth, 1804.
Turner, D. *and* Dillwyn, L. W. *Botanist's Guide through England and Wales*. London, 1805. 2 vols.
Urban, I. *Symbolae Antillanae seu Fundamenta Florae Indiae Occidentalis*. Berlin, 1898–1913. 7 vols.
Van Steenis, C. G. G. J., *ed*. *Flora Malesiana*. Vol. 1. [*Cyclopaedia of Collectors*.] Leiden, 1950; Vol. 5. 1958.
Veitch, J. H. *Hortus Veitchii*. London, 1906.
Venn, J. *and* J. A. *Alumni Cantabrigienses*. Cambridge, 1922.
Vines, S. H. *and* Druce, G. C. *Account of the Morisonian Herbarium in the Possession of the University of Oxford*. Oxford, 1914.
Wade, A. E. *Flora of Monmouthshire*. Cardiff, 1970.
Wallich, N. *Plantae Asiaticae Rariores*. London, 1830–32. 3 vols.
Wallich, N. *Tentamen Florae Napalensis Illustratae*. Calcutta, 1824–26.

Warren, J. B. L., *3rd Baron de Tabley*. *Flora of Cheshire*. London, 1899.
Watson, H. C. *Cybele Britannica, or British Plants and their Geographical Relations*. London, 1847–72. 8 vols.
Watson, H. C. *London Catalogue of British Plants*. London, 1844.
Watson, H. C. *New Botanist's Guide to the Localities of the Rarer Plants of Britain*. London, 1835–37. 2 vols.
Watson, H. C. *Topographical Botany*. Ed. 2. London, 1883.
Webber, R. *Early Horticulturists*. Newton Abbot, 1968.
Weston, R. *Tracts on Practical Agriculture and Gardening; Added, a Chronological Catalogue of English Authors on Agriculture, Gardening...* London, 1769; 1773.
White, A., Dyer, R. A. *and* Sloane, B. L. *Succulent Euphorbieae* (*Southern Africa*). Pasadena, 1941. 2 vols.
White, A. *and* Sloane, B. L. *Stapelieae*. Pasadena, 1937. 3 vols.
White, F. B. W. *Flora of Perthshire*. Edinburgh, 1898.
White, J. W. *Flora of Bristol*. Bristol, 1912.
White, J. W. *Flora of the Bristol Coal-field*. Bristol, 1887.
Wight, R. *Icones Plantarum Indiae Orientalis*. Madras, 1840–53. 6 vols.
Wight, R. *and* Arnott, G. A. W. *Prodromus Florae Peninsulae Indiae Orientalis*. London, 1834.
Withering, W. *Botanical Arrangement of all the Vegetables naturally growing in Great Britain*. London, 1776. 2 vols.
Withering, W. *junior*. *Miscellaneous Tracts of William Withering*. London, 1822.
Wolley-Dod, A. H. *ed*. *Flora of Sussex*. Hastings, 1937.
Wood, A. à *Athenae Oxonienses*. Edited by P. Bliss. London, 1813–20. 4 vols.
Woodward, H. B. *ed*. *History of Geological Society of London*. London, 1907.
Woolls, W. *Progress of Botanical Discovery in Australia: a Lecture*. Sydney, 1869.
Wyatt, W. *and* Thornton, C. G. *Flora Repandunensis. The Wild Flowers of Repton*. London, 1866; ed. 2 1881.

Bibliography of Works Consulted

Periodicals

Advancement of Science see *British Association*
Agricultural Bulletin of Straits and Federated Malay States. Kuala Lumpur. 1901–12; C. as: *Gardens' Bulletin, Straits Settlements*. 1913–41; C. as: *Gardens' Bulletin, Singapore*. 1947– .
American Naturalist. Salem, Mass. 1867–1911.
American Orchid Society Bulletin. New York. 1932– .
Annals of Andersonian Naturalists' Society. Glasgow. 1893–1914.
Annals of Applied Biology. London. 1914– .
Annals of Bolus Herbarium of South African College. Cape Town. 1914–28.
Annals of Botany. London. 1804–6.
Annals of Botany. London. 1887– .
Annals of Natural History. London. 1838–40; C. as: *Annals and Magazine of Natural History*. 1841–85.
Annals of Royal Botanic Gardens of Peradeniya. Colombo. 1901–32.
Annals of Scottish Natural History. Edinburgh. 1892–1911.
Asiatick Researches. Calcutta. 1788–1839.
Bartonia. Philadelphia. 1908– .
Bedfordshire Naturalist. Bedford. 1946–69.
Botanic Garden consisting of Highly Finished Representations of Hardy Ornamental Flowering Plants... by B. Maund. London. 1825–50.
Botanical Cabinet consisting of Coloured Delineations of Plants...by Conrad Loddiges and Sons. London. 1817–33.
Botanical Gazette. London. 1849–51.
Botanical Gazette. Chicago. 1875– .
Botanical Miscellany. London. 1830–33; C. as: *Journal of Botany* (Hooker). 1834–42; C. as: *London Journal of Botany*. 1842–48; C. as: *Hooker's Journal of Botany and Kew Garden Miscellany*. 1849–57.
Botanical Register consisting of Coloured Figures of Exotic Plants. London. 1815–47.
Botanical Society and Exchange Club of British Isles Report. London. 1879–1947; C. as: *Watsonia*. 1949– .
Botanist's Chronicle. London. 1863–65.
British Association for Advancement of Science. Report. London. 1831–1938; C. as: *Advancement of Science*. 1939– .
British Bryological Society. Report see *Moss Exchange Club. Report*
British Fern Gazette. Kendal. 1909– .
British Journal for the History of Science. Oxford. 1962– .
British Phycological Bulletin. Glasgow. 1952–66.
Bulletin of Alpine Garden Society. London. 1930–33; C. as: *Quarterly Bulletin of Alpine Garden Society*. 1933– .
Bulletin of Torrey Botanical Club. New York. 1870– .
Calcutta Journal of Natural Science. 1840–47.
Canadian Record of Science. Montreal. 1884–1905.
Chronica Botanica. New York. 1935–59.
Commonwealth Forestry Review see *Empire Forestry Journal*
Companion to the Botanical Magazine. London. 1835–37.
Cottage Gardener. London. 1848–61; C. as: *Journal of Horticulture, Cottage Gardener,...* 1861–1915.
Curtis's Botanical Magazine. London. 1787– .
Daffodil (and Tulip) Year Book. London. 1933– .
Edinburgh Journal of Science. 1824–32.
Edinburgh Philosophical Journal. 1819–26; C. as: *Edinburgh New Philosophical Journal*. 1826–64.
Empire Forestry Journal. London. 1922–45; C. as: *Empire Forestry Review*. 1946–62; C. as: *Commonwealth Forestry Review*. 1962– .
Essex Naturalist see *Transactions of Epping Forest and County of Essex Naturalists' Field Club*
Flora Malesiana Bulletin. Leyden. 1947– .
Floricultural Cabinet and Florist's Magazine. London. 1833–59.
Florist. London. 1848; C. as: *Florist and Garden Miscellany*. 1849–50; C. as: *Florist, Fruitist and Garden Miscellany*. 1851–61; C. as: *Florist and Pomologist*. 1862–84.
Forestry. London. 1927– .
Garden. London. 1871–1927.
Garden see *Journal of Royal Horticultural Society*
Garden Journal of New York Botanical Garden. 1951– .

Gardeners' Chronicle. London. 1841– .
Gardener's Magazine. London. 1826–44.
Gardeners' Magazine. London. 1895–1916.
Gardens' Bulletin, Straits Settlements see *Agricultural Bulletin*
Gentleman's Magazine. London. 1731–1907.
Geographical Journal see *Journal of Royal Geographical Society*
Glasgow Naturalist. 1916–64.
Hardwicke's Science-Gossip. London. 1865–92; C. as: *Science Gossip*. 1894–1902.
Hastings and East Sussex Naturalist. 1906– .
History of Berwickshire Naturalists' Club. Edinburgh. 1831– .
Hooker's Journal of Botany see *Botanical Miscellany*
Horticultural Register and General Magazine. London. 1831–36.
Huntia. Pittsburgh. 1964–65.
Icones Plantarum; or Figures with Brief Descriptive Characters and Remarks of New or Rare Plants. London. 1836– .
Indian Forester. Dehra Dun. 1875– .
Iris Year Book. London. 1930– .
Irish Naturalist. Dublin. 1892–1924.
Irish Naturalists' Journal. Belfast. 1925– .
Isis. Brussels. 1913– .
Journal of Botany, British and Foreign. London. 1863–1942.
Journal of Botany (Hooker) see *Botanical Miscellany*
Journal of Bryology see *Transactions of British Bryological Society*
Journal of Ecology. London. 1913– .
Journal of Horticulture see *Cottage Gardener*
Journal of Kew Guild. Kew. 1893– .
Journal of Linnean Society see *Journal of Proceedings of Linnean Society. Botany*
Journal of Natural History and Science Society of Western Australia see *Journal of Proceedings of Müller Botanic Society of Western Australia*
Journal of Natural Philosophy, Chemistry and the Arts. London. 1797–1818.
Journal of Proceedings of Linnean Society. Botany. London. 1856–64; C. as: *Journal of Linnean Society. Botany*. 1865–1968; C. as: *Botanical Journal Linnean Society*. 1969– .
Journal of Proceedings of Müller Botanic Society of Western Australia. Perth. 1899–1903; C. as: *Journal of Western Australian Natural History Society*. 1904–9; C. as: *Journal of Natural History and Science Society of Western Australia*. 1910–14; C. as: *Journal and Proceedings of Royal Society of Western Australia*. 1914– .
Journal of Quekett Microscopical Club. 1868–1966.
Journal of Royal Geographical Society of London. 1831–80; C. as: *Geographical Journal*. 1893– .
Journal of Royal Horticultural Society. London. 1846–1975; C. as: *Garden*. 1975– .
Journal of Royal Institution of Cornwall. Truro. 1864– .
Journal of Royal Microscopical Society. London. 1878–1968.
Journal of Society for Bibliography of Natural History. London. 1936– .
Journal of South African Botany. Kirstenbosch. 1935– .
Journal of Western Australian Natural History Society see *Journal of Proceedings of Müller Botanic Society of Western Australia*
Kew Bulletin. Kew. 1898– .
Kirkia. Salisbury. 1960– .
Lancashire Naturalist. Darwen. 1907–14; C. as: *Lancashire and Cheshire Naturalist*. 1914–25.
Lichenologist. London. 1958– .
Lily Year-Book. London. 1932– .
London Journal of Botany see *Botanical Miscellany*
London Naturalist. 1934– .
Memoirs of Literary and Philosophical Society of Manchester. 1789–1879; C. as: *Memoirs and Proceedings of Manchester Literary and Philosophical Society*. 1882– .
Memoirs of Wernerian Natural History Society. Edinburgh. 1808–39.
Midland Naturalist. London. 1878–93.
Moss Exchange Club. Report. Stroud. 1896–1922; C. as: *British Bryological Society. Report*. 1923–45.
National Cactus and Succulent Journal. Leeds. 1949– .
Naturalist. Leeds. 1886– .
Nature. London. 1869– .
North Western Naturalist. Arbroath. 1926–55.
Notes and Records [Royal Society of London]. 1938– .
Notes from Royal Botanic Garden, Edinburgh. 1900– .
Orchid Review. London. 1893– .
Pharmaceutical Journal. London. 1841– .
Philosophical Transactions of Royal Society. London. 1665– .
Proceedings of Belfast Natural History and Philosophical Society. 1852–61, 1871–82; C. as: *Report and Proceedings of Belfast Natural History and Philosophical Society*. 1882–1920; C. as: *Proceedings and Report of Belfast Natural History and Philosophical Society*. 1920– .

Proceedings Belfast Naturalist's Field Club. 1866–1941.
Proceedings Berwickshire Naturalists' Club see *History of Berwickshire Naturalists' Club*
Proceedings Birmingham Natural History and Microscopical Society. 1869–70; C. as: *Report and Transactions Birmingham Natural History and Microscopical Society*. 1872–86.
Proceedings of Botanical Society of British Isles. London. 1954–69; C. in: *Watsonia*. 1970– .
Proceedings of Bournemouth Natural Science Society. 1908– .
Proceedings of Bristol Naturalists' Society. 1863– .
Proceedings of Cotteswold Naturalists' Field Club. Gloucester. 1853– .
Proceedings Coventry Natural History and Scientific Society. 1909–17; 1930– .
Proceedings of Dorset Natural History and Antiquarian Field Club. Dorchester. 1877–1928; C. as: *Proceedings Dorset Natural History and Archaeological Society*. 1928– .
Proceedings of Linnean Society of London. 1838–1968.
Proceedings of Royal Society of Edinburgh. 1832–1940; obituaries continued in *Year Book of Royal Society of Edinburgh*. 1940– .
Proceedings of Royal Society of London. 1854– .
Quarterly Bulletin of Alpine Garden Society see *Bulletin of Alpine Garden Society*
Quarterly Journal of Forestry. London. 1907– .
Report of Australasian Association for Advancement of Science. Sydney. 1888–1928; C. as: *Report of Australian and New Zealand Association for Advancement of Science*. 1930– .
Report North Staffordshire Field Club and Archaeological Society. 1883–86; C. as: *Report and Transactions North Staffordshire Field Club*. 1887–1915; C. as. *Transactions and Annual Report. North Staffordshire Field Club*. 1915–60.
Report and Transactions Cardiff Naturalists' Society. 1867–1902; C. as: *Transactions Cardiff Naturalists' Society*. 1902– .
Rhododendron Year Book. London. 1946–53; C. as: *Rhododendron and Camellia Year Book*. 1954–71.
Rhodora. Boston, Mass. 1899– .
Scientific Proceedings of Royal Dublin Society. 1877– .
Scientific Transactions of Royal Dublin Society. 1877–1909.
Scottish Geographical Magazine. Edinburgh. 1885– .
Scottish Naturalist. Perth. 1871–91, 1912–64.
Sorby Record. Sheffield. 1958–67.
South Australian Naturalist. Adelaide. 1919– .
South-Eastern Naturalist. London. 1900–27; C. as: *South-Eastern Naturalist and Antiquary*. 1928– .
Taxon. Utrecht. 1951– .
Timehri. Demerara. 1882–1912.
Transactions of Botanical Society. Edinburgh. 1839–90; C. as: *Transactions and Proceedings of Botanical Society of Edinburgh*. 1890– .
Transactions of British Bryological Society. Cambridge. 1947–71; C. as: *Journal of Bryology*. 1972– .
Transactions of British Mycological Society. London. 1896– .
Transactions Cardiff Naturalists' Society see *Report and Transactions Cardiff Naturalists' Society*
Transactions of Cumberland Association for Advancement of Literature and Science. Keswick. 1875–93.
Transactions of Epping Forest and County of Essex Naturalists' Field Club. Buckhurst Hill. 1880–82; C. as: *Transactions of Essex Field Club*. 1883–87; C. as: *Essex Naturalist*. 1887– .
Transactions of Essex Field Club see *Transactions of Epping Forest and County of Essex Naturalists' Field Club*
Transactions of Guernsey Society of Natural Science and Local Research. 1882–1921; C. as: *Report and Transactions of Société Guernésiaise*. 1922–72.
Transactions of Hertfordshire Natural History Society and Field Club. 1879– .
Transactions of Horticultural Society of London. 1807–48.
Transactions Inverness Scientific Society and Field Club. 1875–1925.
Transactions and Journal of Proceedings of Dumfriesshire and Galloway Natural History and Antiquarian Society. Edinburgh. 1862–1966.
Transactions. Lincolnshire Naturalists' Union. 1893– .
Transactions of Linnean Society of London. 1791–1875. *Botany*. 1875–1922.
Transactions Natural History Society of Northumberland, Durham and Newcastle-upon-Tyne. 1831– .
Transactions of Norfolk and Norwich Naturalists' Society. Norwich. 1869– .
Transactions and Proceedings of the Botanical Society, Edinburgh. 1844– .
Transactions and Proceedings. Perthshire Society of Natural Science. Perth. 1886– .
Transactions of Woolhope Naturalists' Field Club. Hereford. 1852– .
Transactions Worcestershire Naturalists' Club. Worcester. 1847– .
Tropical Agriculturist. Colombo. 1881– .
Vasculum. Newcastle. 1929– .
Victorian Naturalist. Melbourne. 1884– .
Watsonia see *Botanical Society and Exchange Club of British Isles*
Who was Who, 1897–1915; 1916–1928; 1929–1940; 1941–1950; 1951–1960.
Year Book Royal Society of Edinburgh. 1940– .

Abbreviations

Acad.	Academy
Agric.	Agricultural
ALS	Associate of Linnean Society
Amer.	American
Ann.	Annals
Antiq.	Antiquarian
Arb.	Arboretum
Archaeol.	Archaeological
Assoc.	Association
Aug.	August
Austral.	Australian
b.	born
Beds	Bedfordshire
Berks	Berkshire
Bibl.	Bibliography
Biogr.	Biography
Biol.	Biological, Biology
BM(NH)	British Museum (Natural History)
Bot.	Botanical, Botany
Br.	British
Bryol.	Bryology
Bucks	Buckinghamshire
Bull.	Bulletin
C. as	Continued as
Cambs	Cambridgeshire
Cantab	Cambridge
Carms	Carmarthenshire
Cat.	Catalogue
Ches	Cheshire
Chron.	Chronicle
c.	circa
Circ.	Circular
Contrib.	Contributed
d.	died
Dec.	December
Dept.	Department
Dict.	Dictionary
DNB	Dictionary of National Biography
E.	East
Ecol.	Ecological
ed.	edition
Educ.	Educated at
Encyclop.	Encyclopaedia
Entomol.	Entomological
Exch.	Exchange
Feb.	February
FGS	Fellow of Geological Society
Fl.	Flora
fl.	floruit
FLS	Fellow of Linnean Society
For.	Forestry
front.	frontispiece
FRS	Fellow of Royal Society
FRSE	Fellow of Royal Society of Edinburgh
FZS	Fellow of Zoological Society
Gaz.	Gazette
Gdning	Gardening
Gdnrs	Gardeners
Geneal.	Genealogical
Geogr.	Geographical
Geol.	Geological
Glam	Glamorganshire
Glos	Gloucestershire
Hants	Hampshire
Herb.	Herbarium
Herts	Hertfordshire
Hist.	Historical
Hort.	Horticultural, Horticulture
Hunts	Huntingdonshire
Illus.	Illustrated, Illustration
Inst.	Institute, Institution
Int.	International
Isl.	Island, Isles
J.	Journal
Jan.	January
Kew	Royal Botanic Gardens, Kew
Lancs	Lancashire
Leics	Leicestershire
Lincs	Lincolnshire
Linn.	Linnean
Lit.	Literary
Mag.	Magazine
Med.	Medical
Mem.	Memoir(s)
Microsc.	Microscopical
Middx	Middlesex
Misc.	Miscellaneous, Miscellany
Mon.	Monthly
MP	Member of Parliament
MS(S).	Manuscript(s)
Mus.	Museum
Mycol.	Mycological, Mycology
N.	North
Nat.	Natural, Naturalist
no.	number
Northants	Northamptonshire
Notts	Nottinghamshire
Nov.	November
N.S.W.	New South Wales
N.Y.	New York
N.Z.	New Zealand
Obit.	Obituary
Occas.	Occasional
Oct.	October
Oxon	Oxford

ABBREVIATIONS

Pharm.	Pharmaceutical
Philos.	Philosophical
Photogr.	Photographical, Photography
Physiol.	Physiology
Phytol.	Phytologist
Portr.	Portrait
Proc.	Proceedings
Publ.	Publication
Quart.	Quarterly
R.	Royal
Rec.	Records
Rep.	Report
Res.	Research
Rev.	Review
R.S.C.	Royal Society Catalogue
S.	South
Sci.	Science, Scientific
Scott.	Scottish
Sept.	September
Ser.	Series
Soc.	Society
Staffs	Staffordshire
Supplt.	Supplement
t.	tabula
Trans.	Transactions
U.K.	United Kingdom
VMH	Victoria Medal of Honour
VMM	Veitch Memorial Medal
v.	volume
vol(s)	volume(s)
W.	West
Wilts	Wiltshire
Worcs	Worcestershire
Yb.	Yearbook
Yorks	Yorkshire
Zool.	Zoological, Zoology

ABBEY, George (1835–)
b. Stillingfleet, Yorks 2 Sept. 1835
Gardener and horticultural journalist. Contrib. to *J. Hort.*
J. Hort. Home Farmer v.63, 1911, 54–56 portr., 100.

ABBISS, Harry Walter (1891–1965)
b. 26 Aug. 1891 *d.* Truro, Cornwall 17 Nov. 1965
FLS 1948. Horticultural superintendent, Cornwall, 1923–46. Senior horticultural officer, S. West Province, Cornwall, 1946–52.
Gdnrs Chron. 1955 i 78 portr. *Daffodil Tulip Yb.* 1967, 180–82. *Proc. Linn. Soc.* 1967, 89.

ABBOT, Rev. Charles (*c.* 1761–1817)
b. Blandford, Dorset 24 March 1761 *d.* Bedford 8 Sept. 1817
MA Oxon 1788. DD Oxon 1802. FLS 1793. Vicar of Oakley Reynes, 1798 and Goldington, Beds., 1803. Botanist and lepidopterist. Discovered *Epipactis purpurata* 1807. *Fl. Bedfordiensis* 1798.
P. Smith *Mem. and Correspondence of...Sir James Edward Smith* v.2, 1832, 84. *D.N.B.* v.1, 3. *J. Bot.* 1881, 40–46, 67–75. *Bot. Soc. Exch Club Br. Isl. Rep.* 1935, 57–59. *Bedfordshire Nat.* 1948, 38–42; 1968, 27–29. J. G. Dony *Fl. Bedfordshire* 1953, 18–19. *Bedfordshire Mag.* 1967, 69–72.
Herb. at Luton Museum. MS. cat. of Bedford plants at Linnean Society. Letters at BM(NH).

ABBOT, John (1751–*c.* 1840)
b. London 1 June 1751 *d.* London or in Georgia?
Went to Virginia in 1773 for Royal Society and Chetham's Library, Manchester. Executed natural history drawings with plant background. Collected plants in Savannah and Bulloch County, Georgia.
J. E. Smith *Nat. Hist. Rarer Lepidopterous Insects of Georgia collected from Observations by John Abbot* 1787. *Auk* v.13, 1896, 204–15; v.35, 1918, 271–86; v.55, 1938, 244–54; v.59, 1942, 563–71. *Georgia Hist. Quart.* v.41, 1957, 141–57.
Plants in Elliott Herb., Charleston Museum. Drawings at BM(NH), Harvard and library of Earl of Derby.

ABBOT, Rev. Robert (*c.* 1560–1618)
b. Guildford, Surrey *c.* 1560 *d.* 2 March 1618
BA Oxon 1579. MA 1583. DD 1596. Rector of Bishop's Hatfield, Herts, 1584. Master of Balliol, 1609. Bishop of Salisbury, 1615. "Excellent and diligent herbarist."
J. Gerard *Herball* 1597, 166, 175. A. Wood *Athenae Oxonienses* v.2, 1813–20, 224, 859. R. Pulteney *Hist. Biogr. Sketches of...Bot. in England* v.1, 1790, 137. *D.N.B.* v.1, 24–25. C. E. Raven *English Nat.* 1947, 292.

ABBOT-ANDERSON, Sir Maurice (1861–1938)
d. 3 May 1938
Educ. London and Durham Univ. Physician. Founded Flora's League for conservation of native flora in 1925.
Countryside 1938, 308. *Bot. Soc. Exch. Club Br. Isl. Rep.* 1938, 17–18. *Times* 4 May 1938. *Who was Who, 1929–1940* 25.

ABBOTT, Charles, 1st Baron Tenterden (1762–1832)
b. Canterbury, Kent 7 Oct. 1762 *d.* London 4 Nov. 1832
BA Oxon 1785. Lord Chief Justice, 1818. Baron Tenterden, 1827. Studied botany and wrote Latin odes to flowers. Correspondent of Sir J. E. Smith.
D.N.B. v.1, 26.
Portr. at National Portrait Gallery, London.

ABBOTT, Clementina (*afterwards* **Elphinstone**) (*fl.* 1820s)
Drew plants. Volume of drawings of plants in Calcutta Botanic Garden for sale at Dulau, 1936.

ABBOTT, Francis (1834–1903)
b. Derby 18 June 1834 *d.* Hobart, Tasmania 22 Nov. 1903
Gardener, Botanical Gardens, Hobart, 1851; Superintendent, 1859–1903.
Proc. R. Soc. Tasmania 1909, 10–11. *J. Arnold Arb.* v.3, 1921, 52. *R.S.C.* v.9, 3.
Abbottia F. Muell.

ABBOTT, George (*fl.* 1830s)
Nurseryman, High Street, Knaresborough, Yorks.

ABBOTT, Thomas (*fl.* 1820s)
Nurseryman, High Street, Knaresborough, Yorks.

ABEL, Clarke (1780–1826)
b. 1780 *d.* Cawnpore, India 14 Nov. 1826
MD. FRS 1819. FLS 1818. Surgeon, Norwich. In China with Lord Amherst, 1816–17. *Narrative of Journey in Interior of China...in... 1816–17*, 1818 (botanical appendix by R. Brown, based on Abel's plant collections that survived the wreck of the Alceste).
E. Bretschneider *Hist. European Bot. Discoveries in China* 1898, 225–37. *D.N.B.* v.1, 32–33. D. G. Crawford *Hist. Indian Med. Service, 1600–1913* v.2, 1914, 146. E. H. M. Cox *Plant Hunting in China* 1945, 50–51. *Gdnrs Chron.* 1964 ii 629 portr. A. M. Coats *Quest for Plants* 1969, 94–95 portr.
Chinese plants at BM(NH). Portr. at Hunt Library.
Abelia R. Br.

ABEL, John (1740–1810)
d. York 16 Nov. 1810
Gardener, Castle Howard, Yorks. Seedsman, York.

ABELL, Miss E. (*fl.* 1850s)
Of Gloucester. Plants in Druce Herb., Oxford.
H. N. Clokie *Account of Herb. Dept. Bot., Oxford* 1964, 120.

ABERCONWAY *see* McLaren, H. D., *2nd Baron* Aberconway

ABERCROMBIE, John (1726–1806)
b. Prestonpans, E. Lothian 1726 *d.* London 2 May 1806
Kew gardener. Market gardener, London. Prolific writer of horticultural books. *Every Man his own Gardener* 1767 (Thomas Mawe was not the author). *Universal Gardener and Botanist* 1778. *British Fruit-Gardener* 1779. *Complete Kitchen Gardener* 1789. *Hot-house Gardener* 1789, etc.
J. C. Loudon *Encyclop. Gdning* 1822, 1273–74. G. W. Johnson *Hist. English Gdning* 1829, 219–24. *Cottage Gdnr* v.4, 1850, 65. *J. Hort.* v.55, 1876, 469–71 portr.; v.48, 1904, 258–59 portr. *Gdnrs Chron.* 1904 i 233; 1906 ii 437–38; 1926 ii 173, 216; 1932 ii 453; 1948 ii 108–9. *J. R. Hort. Soc.* 1947, 245–48. *D.N.B.* v.1, 36–37.
Portr. at Hunt Library.

ABRAHAM, John (*fl.* 1800s–1830s)
Nurseryman, Windmill Lane, Chester.

ABRAHAM, Samuel (*fl.* 1840s–1850s)
Nurseryman, Dee Lane, Chester.

ABRAMS, J. (*fl.* 1880s–1910s)
Sergeant, Forest Guards; later forest ranger, Penang, 1888–1910. Collected for C. Curtis.
Gdns Bull. Straits Settlements, 1927, 116.

ABSOLON, William (*fl.* 1780s)
Florist, Great Yarmouth, Norfolk.

ACKERMAN, G.
"On his return from his travels in South America, presented a fine collection of new and undescribed plants" to Tate's nursery, Sloane Street, Chelsea.
T. Faulkner *Hist. and Topographical Description of Chelsea and Environs* v.2, 1829, 347.

ACLAND, Sir Thomas Dyke (1787–1871)
b. London 29 March 1787 *d.* Killerton, Devon 22 July 1871
MA Oxon 1814. MP for Devon, 1812–18, 1820–31; N. Devon, 1837–57. Began the planting of his estate at Killerton in which his wife Lady Lydia Elizabeth Acland took a keen interest.
Bot. Register 1840, t.48. *Bot. Mag.* 1858, t.5039. *D.N.B.* v.1, 62.

ACTON, Edward Hamilton (1862–1895)
b. Wrexham, Denbighshire 16 Nov. 1862 *d.* Cambridge 15 Feb. 1895
BA Cantab 1885. Chemist and plant physiologist. *Practical Physiology of Plants* (with Francis Darwin) 1894.
J. Bot. 1895, 127–28. *R.S.C.* v.9, 10.

ACTON, Frances (*née* Knight)
(*c.* 1793–1881)
b. Elton *c.* 1793 *d.* Acton Scolt, Shropshire 24 Jan. 1881
Eldest daughter of T. A. Knight. Married T. P. Stackhouse Acton, 1812. Helped her father in his experiments and drew 3 plates for his *Pomona Herefordiensis* 1811 and a plate for Hogg and Bull's *Herefordshire Pomona* 1876–85.
R. Hogg and H. G. Bull *Herefordshire Pomona* v.1, 1876, iii. *Gdnrs Chron.* 1888 i 182–83. *J. Bot.* 1931, 77.

ADAIR, Patrick (*fl.* 1670s–1690s)
MD. FRS 1688. Surgeon at Naval Hospital, Chatham. Sent British seaweeds from Gosport to S. Doody (J. Ray *Synopsis Methodica* ed. 2 1696, 327, 328, 330), Indian drugs to Sir H. Sloane and Cape plants to L. Plukenet. "Medicinae et chirurgiae facultatibus exercitatissimus."
L. Plukenet *Almagestum Botanicum* 1696, 45, 115; *Almagesti Botanici Mantissa Plantarum* 1700, 167. *Index to Sloane Manuscripts* 1904, 3. J. E. Dandy *Sloane Herb.* 1958, 81.
Johanna Island plants in Sloane herb. at BM(NH).

ADAM, Sir Frederick (*c.* 1781–1853)
d. Greenwich, London 17 Aug. 1853
Ensign, 1795. Governor of Madras, 1832–37. Sent plants collected in Nilgiri Hills to Sir W. Hooker in 1837.
D.N.B. v.1, 85–86. I. H. Burkill *Chapters on Hist. Bot. in India* 1965, 55.
Plants at Kew.

ADAM, Robert Moyes (1885–1967)
b. Carluke, Lanarkshire 1885 *d.* Edinburgh Nov. 1967
Photographer, Royal Botanic Garden, Edinburgh, 1914–49. Botanical artist to Botanical Society of Edinburgh.
Trans. Bot. Soc. Edinburgh v.40, 1968, 493.

ADAMS, Miss (*fl.* 1830s)
Of Witham, Essex. Drew plates 3475 and 3490 for *Bot. Mag.* 1836.

ADAMS, Alfred (1866–1919)
b. Stockcross, Berks 12 Sept. 1866 *d.* Blyth, Northumberland 22 Oct. 1919

Of Looe, Cornwall. Studied and cultivated Mycetozoa, 1911–19. Found *Physarum nucleatum*.
J. Bot. 1920, 127–30.

ADAMS, Arthur (*c.* 1820–1878)
d. Honor Oak, Kent 16 Oct. 1878
FLS. Assistant Surgeon, HMS 'Samarang', 1843–46. Possibly collected plants in Eastern Archipelago. *Travels of Naturalist in Japan and Manchuria* 1870 includes floristic notes.
Fl. Malesiana v.1, 1950, 6.

ADAMS, C. (*fl.* 1830s)
Nurseryman, Nottingham.

ADAMS, Rev. Daniel Charles Octavius (1822–1914)
b. Anstey, Warwickshire 1822 *d.* 1914
BA Oxon 1845. Ordained 1846. Studied *Rubus*. Collected plants with Moyle Rogers.
Bot. Soc. Exch. Club Br. Isl. Rep. 1915, 246–47. H. N. Clokie *Account of Herb. Dept. Bot., Oxford* 1964, 120.
Rubi at Oxford.

ADAMS, Miss E. G. (*fl.* 1890s)
Collected plants at Cowhill, Dumfriesshire.
Proc. and Trans. Dumfriesshire and Galloway Nat. Hist Antiq. Soc., 1891–92, 12–13.

ADAMS, Francis (1796–1861)
b. Lumphanan, Aberdeen 13 March 1796 *d.* Banchory Ternan, Kincardineshire 26 Feb. 1861
MA Aberdeen. LLD Glasgow 1846. Physician and classical scholar. Translated Hippocrates 1849. 'Notes from the Ancients on certain Indigenous Species' (A. Murray *Northern Fl.* 1836, appendix i–iv).
D.N.B. v.1, 95–96.
Bust at Aberdeen University.

ADAMS, Frederick (1867–1938)
b. 30 Jan. 1867 *d.* 14 Nov. 1938
Worked as engineer in West Indies, Mexico, etc. Settled in Jersey in 1931. Studied diatoms.
Kew Bull. 1932, 250–51. *J. Bot.* 1939, 154–56. *J. Quekett Microsc. Club* v.1, 1939, 128–29.
Diatoms at BM(NH).

ADAMS, George (1720–1786)
b. London 1720 *d.* London 5 March 1786
Mathematical instrument maker. *Micrographia Illustrata* 1746 (anatomy of plants on p. 165–235); ed. 4 1771.
D.N.B. v.1, 97.
Manuscripts at Royal Society.

ADAMS, George (1750–1795)
b. London 1750 *d.* Southampton 14 Aug. 1795
FLS 1788. Son of G. Adams (1720–1786). *Essays on the Microscope...with a View of the Organization of Timber* 1787; ed. 2 1798.
D.N.B. v.1, 97.

ADAMS, Henry Gardiner (*c.* 1811–1881)
d. Gillingham, Kent 1 May 1881
Druggist at Canterbury. *Flowers; their Moral Language and Poetry* 1844. *Wild Flowers...of the Months* 1862; new ed. 1868.
F. Boase *Modern English Biogr., Supplt.* v.1, 1908, 32.

ADAMS, James (1839–1906)
b. near Killarney, County Kerry 12 May 1839
d. Thames, New Zealand 1906
BA London. Schoolmaster, Isle of Man, 1865. Emigrated to New Zealand, 1870. Schoolmaster, Thames. Botanised on Coromandel Peninsula where he discovered *Celmisia adamsii* and *Elytranthe adamsii*. Also made botanical trips to other parts of New Zealand.
Tuatara 1972, 53–56 portr.
Herb. at Auckland Institute.

ADAMS, John (*fl.* 1690s)
Gardener to Duke of Beaufort.
Index to Sloane Manuscripts 1904, 4. J. E. Dandy *Sloane Herb.* 1958, 81.

ADAMS, John (*fl.* 1790s)
Ironmonger and seedsman, Northampton.

ADAMS, John (–1798)
d. drowned off Pembrokeshire, 21 Feb. 1798
FLS 1795. Conchologist. Correspondent of J. E. Smith who described some of his plant discoveries in *English Bot.*, t.111, 248 and 462.
E. Lees *Bot. Looker-out* 1842, 216–17. *R.S.C.* v.1, 17.

ADAMS, John (1872–1950)
b. Ballymena, County Antrim 20 Jan. 1872
d. Ottawa 18 May 1950
MA Cantab. Lecturer in Botany, Dublin Municipal Technical Schools. Assistant Professor in Botany, Royal College of Science, Dublin. Professor of Botany, Royal Veterinary College, Dublin. Dominion Botanist, Ottawa, 1914–38. Studied photoperiodism and germination. Collected plants in Anticosti, 1933–36; Gaspe, 1934–35; Prince Edward Island, 1936. *Guide to Principal Families of Flowering Plants after Engler's System* 1906. *Bibliography of Canadian Plant Geography* (*Trans. R. Canadian Inst.*, 1928–36). *Student's Illustrated Irish Fl.* 1931. Contrib. lists of seed plants, algae and fungi to *Canadian Field Nat.* 1935–40.
R. L. Praeger *Some Irish Nat.* 1949, 38–39. *Who was Who, 1941–1950* 6.
Plants at Ottawa.

ADAMS, Patrick (*fl.* 1760s–1790s)
Nurseryman, Gormanstown, County Meath.
Irish For. 1967, 56.

ADAMS, Sarah Coker *see* Beck, S. C.

ADAMS, T. (*fl.* 1830s)
Nurseryman, Newport, Shropshire.

ADAMS, Thomas William (1841–1919)
b. Gravely, Hunts 1841 *d.* New Zealand 1 June 1919
Emigrated in 1862 to Canterbury Plain, New Zealand, where he formed large collection of exotic trees and shrubs on his estate at Greendale. Contrib. papers on tree culture to *Trans. N.Z. Inst.*
Trans. N. Z. Inst. 1919 xi–xii portr. G. H. Scholefield *Dict. N. Z. Biogr.* 1940, 5–6.

ADAMSON, Mr. (*fl.* 1840s)
Collected on Clyde Islands, Scotland in 1840. First found *Pleuroclada albescens* var. *islandica* in Scotland.
Trans. Bot. Soc. Edinburgh v.25, 1910, 4.

ADAMSON, Frederick M. (*fl.* 1830s–1850s)
Settler in Victoria.
J. D. Hooker *Fl. Tasmaniae* v.1, 1859, cvi, cxxxvii. *Victorian Nat.* v.25, 1908, 102.
Letters in Sir W. J. Hooker's correspondence at Kew. Plants at Kew and Melbourne.

ADAMSON, George (*fl.* 1890s)
Collected plants in Nyasaland and Zambesiland, now at Kew and Edinburgh.

ADAMSON, Robert Stephen (1885–1965)
b. Manchester 2 March 1885 *d.* Jedburgh, Roxburghshire 6 Nov. 1965
MA. DSc Edinburgh. FLS 1956. Lecturer in Botany, Manchester University, 1912–23. Professor of Botany, Capetown University, 1923–50. Returned to U.K. 1955. *Vegetation of South Africa* 1938. *Fl. of Cape Peninsula* 1950.
S. African J. Sci. 1965, 443. *S. African Forum Botanicum* v.3 (11), 1965, 1–3. *Proc. Linn. Soc.* 1967, 89. *Who was Who, 1961–1970* 7.
Portr. at Hunt Library.

ADAMSON, William (*fl.* 1840s)
Nurseryman, Stoke Newington, London.

ADCOCK, William (*fl.* 1820s)
Gardener and nurseryman, Allhallowgate, Ripon, Yorks.

ADDISCOTT, William (*fl.* 1820s–1850s)
Nurseryman, Alphington Road, Exeter, Devon.

AFZELIUS, Adam (1750–1837)
b. Larf, W. Gothland 8 Oct. 1750 *d.* Upsala, Sweden 20 Jan. 1837
FRS 1798. FLS 1790. Studied under Linnaeus. Demonstrator in Botany, Upsala, 1785. Botanist, Sierra Leone Co., 1792. Secretary, Swedish Embassy, London. Professor of Mathematics and Medicine, Upsala, 1812. *Genera Plantarum Guineensium* 1804.
P. Smith *Mem. and Correspondence* of *Sir J. E. Smith* 1832 2 vols *passim. R.S.C.* v.1, 22.
Herb. at Upsala University. Plants at BM (NH). Portr. at Hunt Library.
Afzelia Smith.

AGNIS, John (*fl.* 1770s–1808)
d. 13 Sept. 1808
Gardener and nurseryman, East Hill, Colchester, Essex.

AHERN, Garret (*fl.* 1760s–1780s)
Nurseryman, Cork City; joined by John Ahern.
Irish For. 1967, 50.

AIKEN, Rev. James John Marshall Lang (1857–1933)
b. Aberdeen 1857 *d.* 27 Nov. 1933
MA. BD Aberdeen. Pastor, Ayton, Berwickshire, 1882–1933. Studied local flora.
Proc. Berwickshire Nat. Club 1934, 261–65 portr.

AIKIN, Arthur (1773–1854)
b. Warrington, Lancs 19 May 1773 *d.* London 15 April 1854
FLS 1818. Son of John Aikin (1747–1822). Chemist. Secretary, Geological Society, 1811. Secretary, Society of Arts, 1817–40. Discovered *Astrantia major* in Shropshire, 1824. Contrib. Shropshire plants to D. Turner and L. W. Dillwyn's *Botanist's Guide*, 1805.
Trans. Linn. Soc. v.15, 1827, 507. *Proc. Linn. Soc.* 1854, 304–6. *Proc. Geol. Soc.* 1855, xli. J. B. L. Warren *Fl. Cheshire* 1899, lxxvii. *Trans. Liverpool Bot. Soc.* 1909, 56. *D.N.B.* v.1, 184. *R.S.C.* v.1, 28.
Aikinia Wall.

AIKIN, John (1747–1822)
b. Kibworth Harcourt, Leics 15 Jan. 1747
d. Stoke Newington, London 7 Dec. 1827
MD Leyden 1784. FLS 1795. Physician at Great Yarmouth, 1784–92. Taught R. A. Salisbury (*J. Bot.* 1904, 294). Friend of E. Forster, T. Pennant and W. Roscoe. *Woodland Companion*, 1802.
L. Aikin *Mem. of John Aikin* 1823. *Gent. Mag.* 1823 i 160. *Trans. Liverpool Bot. Soc.* 1909, 56. *Trans. Norfolk Norwich Nat. Soc.* v.9, 1914, 685. *D.N.B.* v.1, 185–86. *R.S.C.* v.1, 28.
Silhouettes of John and Arthur Aikin in Kendrick's *Profiles of Warrington Worthies.*

AIKMAN, John (1867–1938)
b. Grantshouse, Berwickshire 10 Oct. 1867 *d.* Kew, Surrey 22 Dec. 1938
Gardener, Kew, 1888. Assistant in Herb. and Director's Office, 1891–1932.
Times 24 and 31 Dec. 1938. *Gdnrs Chron.* 1939 i 47. *J. Kew Guild* 1939–40, 915. *Kew Bull.* 1939, 100–1.

AINSLIE, Sir Whitelaw (1767–1837)
b. Duns, Berwickshire 17 Feb. 1767 *d.* London 29 April 1837
MD Leyden 1786. FRSE. Knighted 1835. Surgeon, East India Company, 1788–1815. *Materia Medica of Hindostaan* 1813; ed. 2 as *Materia Indica* 1826 2 vols.
D.N.B. v.1, 190. D. G. Crawford *Hist. Indian Med. Service 1600–1913* v.2, 1914, 22. *R.S.C.* v.1, 30.
Ainsliaea DC.

AINSWORTH, Ralph Fawsett (1811–1890)
b. Manchester 1811 *d.* Lower Broughton, Lancs 6 March 1890
FLS 1869. Surgeon. Cultivated orchids. On Management Committee of Manchester Botanic Garden
Garden v.37, 1890, 285. *Proc. Linn. Soc.* 1889–90, 90.

AINSWORTH, Samuel (*c.* 1834–1904)
d. 12 March 1904
Seedsman, High Holborn, London.
Garden v.65, 1904, 228–29.

AINSWORTH-DAVIS, James Richard (1861–1934)
b. Bristol 8 April 1861 *d.* 7 April 1934
Zoologist. *Crops and Fruits* 1924. Translated P. Knuth's *Handbook of Flower Pollination* 1906–9 3 vols and W. F. Bruck's *Plant Diseases* 1912.

AIREY, James *see* Ayrey, J.

AITCHISON, James Edward Tierney (1836–1898)
b. Nimach, India 28 Oct. 1836 *d.* Mortlake, Surrey 30 Sept. 1898
MD Edinburgh 1858. FRS 1883. FLS 1863. Bengal Medical Service, 1858. Collected plants in India, 1861–72, Afghanistan, 1879–85; also in Ireland, 1867–69. *Catalogue of plants of Punjab and Sindh* 1869. 'On the flora of Kuram Valley, etc...Afghanistan' (*J. Linn. Soc.* 1880–81, 1–113, 139–200). Contrib. to *J.* and *Trans. Linn. Soc.*
Proc. Linn. Soc. 1898–99, 40–41. *J. Bot.* 1898, 463. *Nature* v.58, 1898, 578. *Kew Bull.* 1898, 310–11. *Trans. Bot. Soc. Edinburgh* v.21, 1899, 224–29. *Proc. R. Soc.* v.64, 1899, xi-xiii. D. G. Crawford *Hist. Indian Med. Service, 1600–1913* v.2, 1914, 145. *R.C.S.* v.7, 16; v.9, 21.
Plants at Kew, Calcutta, Dehra Dun. MSS. at Kew.
Aitchisonia Hemsley.

AITKEN, Andrew Peebles (–1904)
b. Edinburgh *d.* Edinburgh 17 April 1904
MA Edinburgh 1867. DSc 1873. FRSE. Professor of Chemistry and Toxicology, Royal Veterinary College, Edinburgh, 1875. President, Botanical Society of Edinburgh, 1895–97. Contrib. to *Trans Bot. Soc. Edinburgh.*
Trans. Bot Soc. Edinburgh v.23, 1904, 47–53.

AITKEN, Robert Turnbull (1843–1915)
b. Jedburgh, Roxburghshire 7 Sept. 1843 *d.* 4 March 1915
Shoe-maker. Botanist.
Proc. Berwickshire Nat. Club 1917, 377–78.

AITON, John Townsend (1777–1851)
d. Kensington, London 4 July 1851
Younger son of W. Aiton (1731–1793). Royal gardener at Windsor, Frogmore and Kensington Palace.
Gent. Mag. 1851 ii 223.

AITON, William (1731–1793)
b. Boghall, Carnwath, Lanarkshire 1731 *d.* Kew, Surrey 2 Feb. 1793
Came south to England in 1754. Trained under Philip Miller at Chelsea Physic Garden. Gardener at Kew to Princess Augusta and George III, 1759–93. *Hortus Kewensis*, 1789 3 vols (*J. Bot.* 1897, 481–85; 1912, Supplt. 16p.)
Gent. Mag. 1793, 391. G. W. Johnson *Hist. English Gdning* 1829, 298–99. W. Aiton *Enquiry into Origins of Aiton family* 1830. *Cottage Gdnr* v.5, 1850, 263–64. *Kew Bull.* 1891, 298–99. *J. Kew Guild* 1902, 87–90. *D.N.B.* v.1, 207. *Huntia* 1965, 185–87 portr. F. A. Stafleu *Taxonomic Literature* 1967, 4. C. C. Gillispie *Dict. Sci. Biogr.* v.1, 1970, 88–89. *Taxon* 1972, 26 portr.
Portr. attributed to Zoffany at Kew. Portr. at Hunt Library.
Aitonia Thunb.

AITON, William Townsend (1766–1849)
b. Kew, Surrey 2 Feb. 1766 *d.* Kensington, London 9 Oct. 1849
FLS 1797. Son of W. Aiton (1731–1793) whom he succeeded as superintendent at Kew in 1793. Relinquished control of botanic garden at Kew when Sir W. J. Hooker became Director in 1841; retired in 1845. Founder member of Horticultural Society of London. *Hortus Kewensis* ed. 2 1810–13 5 vols; epitome vol. 1814 (*J. Bot.* 1912, Supplt. 16p.)
Gdnrs Mag. 1830, 731; 1838, 194; 1840, 366–68. *Trans. Hort. Soc. London* v.1, 1812, 262. *Proc. Linn. Soc.* v.2, 1850, 82–83. *Kew Bull.* 1891, 304–5; 1910, 306–8. *J. Kew Guild* 1894, 36; 1966, 688–93 portr. *D.N.B.* v.1, 208. *Curtis's Bot. Mag. Dedications, 1827–1927*, 7–8 portr. C. C. Gillispie *Dict. Sci. Biogr.* v.1, 1970, 89–90.
Letters in Sir J. Banks and R. Brown correspondence at BM(NH). Lithograph portr. by L. Poyot at Kew. Portr. at Hunt Library. Library and Herb. sold by Foster and Son, London, 4 Sept. 1851.

AKEROYD, James Walter (1906–1944)
d. 6 Nov. 1944
LLB Leeds. Solicitor in Reading, Richmond (Yorks), Norwich, Leeds. Botanist.
Naturalist 1945, 116.

ALABASTER, Sir Chaloner (1838–1898)
d. 28 June 1898
To China as student interpreter, 1855. Consul in Chinese ports. Collected some plants for H. F. Hance. Returned to England, 1893.
E. Bretschneider *Hist. European Bot. Discoveries in China* 1898, 698.

ALABASTER, Henry (–1884)
d. Bangkok 8 Aug. 1884
Student interpreter, British Consulate, Bangkok, 1858. Director, Royal Museum and Garden, Siam. Studied flora of Siam; sent orchids, etc. to Kew.
J. Thailand Res. Soc. Nat. Hist. Supplt. v.12, 1939, 19–20.

ALCHORNE, Stanesby (1727–1800)
b. 1727 *d.* 5 Nov. 1800
Assay-master in the Mint, 1789. Demonstrator at Chelsea Physic Garden, 1771–73. "Plantarum britannicarum collector et explorator solertissimus, cui multum debent rei herbariae amatores" (Solander).
Philos. Trans. R. Soc. v.61, 1771, 390–96; v.63, 1773, 30–37. *Gent. Mag.* 1789 i 181; 1800 ii 215. J. E. Smith *Correspondence of Linnaeus* v.2, 1821, 4–7. *Phytologist* 1848, 166–70, 189–90. H. Trimen and W. T. T. Dyer *Fl. Middlesex* 1869, 392. R. H. Semple *Mem. of Bot. Gdn at Chelsea* 1878, 95–96.
Plants at BM(NH).
Alchornea Sweet.

ALCOCK, Alfred William (1859–)
Surgeon Captain. Collected plants in the Pamirs. *Report of Scientific Results of Pamir Boundary Commission of 1898.*

ALCOCK, J. (*fl.* 1670s)
Nurseryman. Supplied fruit trees for Sir Roger Pratt's gardens at Ryston Hall, Norfolk.
R. T. Gunther *Architecture of Sir Roger Pratt* 1928, 308.

ALCOCK, Mrs. Nora Lilian Leopard (*c.* 1875–1972)
d. 1 April 1972
Mycologist at Plant Pathology Laboratory, Kew and Harpenden, 1917–24. Plant pathologist, Dept. Agriculture for Scotland, 1924–37. Did pioneer research work on red core disease of strawberries. Contrib. to *Trans. Bot. Soc. Edinburgh.*
Bull. Br. Mycol. Soc. 1972, 81. *B.S.E. News* no. 10, 1973, 10–11. *J. Kew Guild* 1974, 342.

ALCOCK, Randal Hibbert (1833–1885)
b. Gatley, Cheshire 21 July 1833 *d.* Didsbury, Lancs. 9 Nov. 1885
Of Bury. FLS 1876. Cottonspinner. Founder and President, Bury Natural History Society. *Botanical Names for English Readers* 1876. List of plants of Bury district in *Bury Nat. Hist. Soc. Rep.* 1872, 14–23. *Fl. of Virgil* (unpublished).
Proc. Linn. Soc. 1883–86, 137. *J. Bot.* 1886, 160. *Trans. Liverpool Bot. Soc.* 1909, 57.

ALCOCK, Sir Rutherford (1809–1897)
b. Ealing, Middx 1809 *d.* London 2 Nov. 1897
KCB 1862. DCL Oxon 1863. Army surgeon in Portugal, 1832–36 and Spain, 1836. Consul at Fuchow, China, 1844 and at Shanghai, 1846. Minister-plenipotentiary at Peking, 1865–71. President, Royal Geographical Society, 1876–78.
Geogr. J. v.10, 1897, 642–45 portr. *Scott. Geogr. Mag.* 1898, 90–91. *D.N.B. Supplt. 1* v.1, 29. A. Michie *Englishman in China* 1900, portrs. 2 vols. *R.S.C.* v.7, 18; v.9, 26; v.13, 55.
Peking plants at Kew.
Picea alcockiana Carr.

ALDERSEY, John (*fl.* 1790s)
Seedsman, Milton next Sittingbourne, Kent.

ALDERSON, John (1757–1829)
b. Lowestoft 4 June 1757 *d.* Hull 16 Sept. 1829
MD. Physician to Hull infirmary. A founder of Hull Botanic Garden. First President, Hull Literary and Philosophical Society. *Rhus toxicodendron* 1794; ed. 4 1811.
C. F. Corlass and W. Andrews. *Sketches of Hull Authors* 1879. *D.N.B.* v.1, 243. *Notes and Queries* v.153, 1927, 485; v.154, 1928, 15.

ALDRIDGE, George (*fl.* 1830s)
Kew gardener. Went to Trinidad in 1831; returned in 1833 due to ill-health bringing a collection of living orchids for Kew.
Gdnrs Mag. 1835, 5. *Kew Bull.* 1891, 320. *Taxon* 1970, 511.

ALDUS, John (–1767)
d. Jan. 1767
Gardener and nurseryman, East Hill, Colchester, Essex. Business carried on by his widow and his son, John.

ALEXANDER, Edwin (1870–)
b. Edinburgh 1870
Botanical artist. Illustrated J. H. Crawford's *Wild Flowers* 1909.

ALEXANDER, H. G. (*c.* 1875–1972)
b. Bath, Somerset *c.* 1875 *d.* 17 Feb. 1972
VMH 1926. Orchid foreman, Blenheim Palace, 1897. In charge of orchid collection

belonging to Sir George Holford at Westonbirt from 1899. Hybridised cymbidiums. On Sir George Holford's death Alexander transformed the collection into a commercial enterprise.

Gdnrs Mag. 1908, 331–32 portr. *Amer. Orchid Soc. Bull.* 1965, 301 portr.; 1972, 415. *Gdnrs Chron.* 1965 i 399. *Orchid Rev.* 1972, 66–68 portr. *J. R. Hort. Soc.* 1973, 277.

Cymbidium X alexanderi.

ALEXANDER, H. T. *see* Alexander, W. T.

ALEXANDER, James (*c.* 1805–1881)
b. Banffshire *d.* Edinburgh 12 March 1881

Senior partner, Messrs Dickson and Co., nurserymen, Edinburgh. Secretary and Treasurer, Scottish Arboricultural Society.

J. Hort. Cottage Gdnr. v.2, 1881, 233.

ALEXANDER, James (*c.* 1845–1890)
d. Hastings, Sussex 21 April 1890

Nurseryman, Edinburgh. Nephew of A. James (*c.* 1805–1881).

Gdnrs Chron. 1890 ii 529.

ALEXANDER, Sir James Edward (1803–1885)
b. Powis, Clackmannanshire 16 Oct. 1803 *d.* Ryde, Isle of Wight 2 April 1885

General. To Madras as army cadet, 1820. Left East India Company's army in 1825; saw active service in the Balkans, Portugal, S. Africa, Canada and Crimea. Collected plants in S. Africa. *Passages in Life of a Soldier* 1857 *Expedition...into Interior of Africa*, 1838 2 vols.

Trans. Linn. Soc. v.18, 1841, 305. *D.N.B. Supplt. 1* v.1, 31–32. C. E. Buckland *Dict. Indian Biogr.* 1906, 10. *Comptes Rendus AETFAT, 1960* 1962, 124.

Plants at Cambridge.

Catophractes alexandri D. Don.

ALEXANDER, John Abercromby (1854–)
b. Edinburgh 9 Oct. 1854

Educ. Edinburgh Univ. Forester in India and Mozambique. 'Notes on Fl. of Coast and Islands of Portuguese East Africa' (*Trans. Proc. Bot. Soc. Edinburgh* 1906, 167–85, 277–78).

Moçambique 1941, 23–24.

ALEXANDER, Richard Chandler *see* Prior, R. C. A.

ALEXANDER, Stanley George (1910–1944)
b. Westonbirt, Glos 19 March 1910 *d.* 26 Sept. 1944

Director of H. G. Alexander, Westonbirt, orchid growers.

Orchid Rev. 1944, 149–50 portr.

ALEXANDER, W. (*fl.* 1830s)

Nurseryman, Kingsland, Islington, London.

ALEXANDER, William Thomas (1818–1872)
b. Cork, Ireland 23 June 1818 *d.* 31 May 1872

Surgeon, Royal Navy on HMS 'Plover' in E. Indies and China, 1845–46. Collected ferns, mosses, etc. on Chinese coast and Ryukyu Islands. Studied mosses of Co. Cork; some of his records appear in T. Power's *Botanist's Guide for County of Cork*, 1845 (appears here as H. T. Alexander). 'Fungi of Cloyne' (*Phytologist* v.4, 1852, 727–34).

Hooker's J. Bot. 1848, 273–78. *J. Bot.* 1894, 294–95, 299. G. Bentham *Fl. Hongkongensis* 1861, pref. 11*. E. Bretschneider *Hist. European Bot. Discoveries in China* 1898, 360–62. *Proc. R. Irish Acad. B.* 1915, no. 7, 71. R L.. Praeger. *Some Irish Nat.*, 1949, 39. *R.S.C.* v.1, 44.

Chinese plants at Kew. Mosses at BM(NH). Letters in Sir W. J. Hooker's correspondence at Kew

ALFREY, George (*fl.* 1690s)

Surgeon on Halley's expedition of 1699. Collected for J. Petiver on coast of Brazil, etc.

J. Petiver *Museii Petiveriani* 1695–1703, 43, n. 347. J. E. Dandy *Sloane Herb.* 1958, 81–82.

Plants at BM(NH).

ALLAN, James (1825–1866)

Professor of Chemistry. Pilloried in *Gdnrs Chron.* 1853, 791. *Botanists' Word-book* (with G. Macdonald) 1853

Trans. Liverpool Bot. Soc. 1909, 57.

ALLAN, John (1845–1888)
b. Carluke, Lanarkshire 12 July 1845 *d.* Easter Middleton 2 Dec. 1888

Solicitor's clerk, Edinburgh. 'Report on Visit to Applecross' (*Trans. Bot. Soc. Edinburgh* v.17, 1886–87, 117–21).

Trans. Bot. Soc. Edinburgh v.17, 1886–87, 524. *Trans. Edinburgh Field Nat. Microsc. Soc.* 1888–89, 275–79.

ALLARD, Edgar John (*c.* 1877–1918)
d. Merton, Surrey 23 Oct. 1918

Kew gardener, 1898–99. Foreman, Cambridge Botanic Garden, 1899. Superintendent, John Innes Horticultural Institution where he did important research on hybridisation.

Gdnrs Mag. 1910, 237–38 portr. *Kew Bull.* 1918, 342. *Gdnrs Chron.* 1918 ii 182 portr. *Garden* 1918, 414 portr. *J. Kew Guild* 1919, 464–65 portr.

ALLCARD, John (*c.* 1777–1855)
d. London 9 April 1855

FLS 1844. Had outstanding collection of plants including epiphytes and exotic ferns at Stratford Green, London.

Bot. Mag. 1839, t.3767. *Proc. Linn. Soc.* 1856, xxxv.

ALLCHIN, Sir William Henry (*c.* 1828–1891)
d. London 24 Nov. 1891

MB London. Practised medicine in Bayswater, London. Grew varieties of British ferns. 'Classification of fern varieties' (unpublished).

Br. Fern Gaz. 1909, 43.

ALLEN, C. Ernest F. (*fl.* 1900s–1920s)

Kew gardener until 1904. Curator, Botanic Garden, Darwin, N. Australia, 1913. Collected plants at Victoria Falls, Rhodesia, 1904, Mozambique, 1912, N. Australia, 1924–28.

Kew Bull. 1913, 417. J. D. Clark *Victoria Falls* 1952, 121–60.

Plants and a few letters at Kew.

ALLEN, Charles Grant Blairfindie (1848–1899)

b. Alwington, Kingston, Canada 24 Feb. 1848 *d.* Hindhead, Surrey 25 Oct. 1899

BA Oxon 1871. Novelist and scientist. *Colours of Flowers* 1882. *Flowers and their Pedigrees* 1883.

J. Bot. 1900, 62–63. *D.N.B. Supplt. 1* v.1, 36.

ALLEN, David (*fl.* 1800s)

Florist and seedsman, Hammersmith, London. In partnership with Newman and Gyet.

ALLEN, Edgar Johnson (1866–1942)

b. Preston, Lancs 1866 *d.* 7 Dec. 1942

BSc London 1885. DSc 1900. FRS 1914. FLS 1929. Zoologist. Studied diatoms. Secretary, Marine Biological Association of U.K. and Director, Marine Biological Laboratory, Plymouth, 1895–1936.

Proc. Linn. Soc. 1942–43, 290–92. *Who was Who, 1941–1950* 16.

ALLEN, Eliza (*née* **Stevens**) (*c.* 1842–)

Botanical artist. Illustrated C. A. Johns's *Flowers of the Field* 1853.

ALLEN, Guy Oldfield (1883–1963)

b. London 1883 *d.* Godalming, Surrey 11 April 1963

Read law at Oxford. FLS 1949. Indian Civil Service, 1908–33. Studied Charophyta. Inherited J. Groves's library and charophyta collection in 1933. *British Stoneworts* 1950.

Proc. Linn. Soc. 1964, 187–88. *Proc. Bot. Soc. Br. Isl.* 1965, 97–99. *Taxon* 1965, 206–7.

Library and Herb. at BM(NH).

ALLEN, Henry George (1890–1965)

b. Northampton Oct. 1890 *d.* 15 Jan. 1965

BSc London. Teacher at Northampton. Secretary, Botanical Section, Northampton Natural History Society, 1925; President, 1945; General Secretary, 1960. Had knowledge of local flora.

J. Northampton Nat. Hist. Soc. Field Club 1965, 359–60 portr.

ALLEN, James (–1906)

d. Shepton Mallet, Somerset 8 March 1906

Hybridised snowdrops, anemones, etc. 'Snowdrops' (*J. R. Hort. Soc.* 1891, 172–219).

Gdnrs Chron. 1906 i 175. *J. Hort. Cottage Gdnr* v.52, 1906, 230.

ALLEN, John (*fl.* 1820s)

Nurseryman, Union Nursery, King's Road, Sloane Square, London.

ALLEN, Joseph (*fl.* 1760s–1770s)

Apothecary and seedsman, Limerick, Ireland.

ALLEN, Paul Henry (1890–1914)

b. New Zealand 1890 *d.* Oxford 6 Aug. 1914

BA Cantab 1913. Studied under Professor Czapek at Prague, 1913–14. Demonstrator at Cambridge, 1914. Made collection of British plants.

J. Ecol. 1915, 241–42.

ALLEN, Thomas (fl. 1820s–1840s)

Nurseryman and seedsman, Oswestry, Shropshire.

ALLEN, William (*fl.* 1830s–1840s)

Gardener, seedsman and nurseryman, London End, Wycombe, Bucks.

ALLEN, William Beriah (1875–1922)

b. Benthall, Shropshire 1875 *d.* Benthall 20 Nov. 1922

Potter. Mycologist. Contrib. to *Trans. Br. Mycol. Soc.*

Trans. Br. Mycol. Soc. v.8, 1923, 191–92.

Cudionella allenii A. L. Smith.

ALLERTON, Joseph (*fl.* 1730s)

Nurseryman, Old Spring Gardens, Knightsbridge, London.

ALLGROVE, J. C. (–1930)

d. 4 Sept. 1930

Nurseryman, Langley, Slough, Bucks. Specialised in fruit-growing.

J. Hort. Home Farmer v.61, 1910, 369, 371 portr. *Gdnrs Chron.* 1925 ii 162 portr.; 1930 ii 223.

ALLIN, Rev. Thomas (–*c.*1909)

BD Dublin 1859. Curate, Lickmolassy, Co. Galway, 1864–65; Fenagh, Co. Carlow, 1865–66; Middleton, Co. Cork, 1870–74; Myross, Co. Cork, 1874–77. Retired to Weston-super-Mare. *Flowering Plants and Ferns of County Cork* 1883. Contrib. to *J. Bot.*

J. Bot. 1884, 57–58. R. L. Praeger *Some Irish Nat.* 1949, 40.

ALLITT, William (1828–1893)

b. County Kerry *d.* Victoria, Australia 1893

Curator, Portland Botanic Gardens, *c.* 1863–85. Nurseryman, Tyrendarra. Collected plants in Victoria and S. Australia border. Sent plants to F. von Mueller and G. Bentham.

Gdnrs Chron. 1893 ii 538. *Victorian Nat.* 1908, 102; 1949, 123.

Plants at Melbourne.

Leucopogon allittii F. Muell.

ALLMAN, George James (1812–1898)
b. Cork 1812 *d.* Parkstone, Dorset 24 Nov. 1898

MD Dublin 1844. FRS 1854. FLS 1872. Professor of Botany, Dublin, 1844–55. Professor of Natural History, Edinburgh, 1855–70. President, Linnean Society, 1874–81. 'Vegetation of Riviera' (Barėty *Nice and its Climate* 1882).

Biogr. Mag. 1889, 109–13 portr. *Nature* v.59, 1898, 202–4, 269–70. *Proc. Linn. Soc.* 1898–99, 41–43. *Irish Nat.* 1899, 104. *Notes Bot. School Dublin* v.1, 1901, 157–59. *Proc. R. Soc.* v.75, 1904, 25–27. *D.N.B. Supplt 1* v.1, 40. R. L. Praeger *Some Irish Nat.* 1949, 40 portr. *Who was Who, 1897–1916* 13. *R.S.C.* v.1, 48; v.7, 24; v.9, 33; v.12, 10; v.13, 71.

Plants at Dublin. Sale at Sotheby 11 July 1899. Portr. at Linnean Society and National Gallery, Dublin.

ALLMAN, William (1776–1846)
b. Kingston, Jamaica 7 Feb. 1776 *d.* Dublin 8 Dec. 1846

MD Dublin 1804. Professor of Botany, Dublin, 1809–44. *Analysis ... Generum Plantarum Phanerostemonum* 1828. *Familiae Plantarum* 1836. MS. on 'Mathematical Connection between Parts of Vegetables', 1811 at BM(NH).

Notes Bot. School Dublin v.1, 1896, 3. *D.N.B.* v.1, 335. R. L. Praeger *Some Irish Nat.* 1949, 40–41. *R.S.C.* v.1, 51; v.13, 71.

Allmania R. Br.

ALLNUT, Mr. (*fl.* 1810s–1820s)

Nurseryman, Clapham Common, London.

E. Bretschneider *Hist. European Bot. Discoveries in China* 1898, 282.

ALLOM, Mrs. Elizabeth Anne (*fl.* 1840s–1870s)

Of Margate Kent. *Seaweed Collector* (with specimens) 1841; reissued 1845. List of algae in T. B. Flower's *Fl. Thanetensis* 1847.

ALLPORT, George (*fl.* 1810s–1820s)

Partner (probably son) of John Allport, nurseryman, Hackney Road, London.

ALLPORT, John (*fl.* 1790s)

Seedsman and nurseryman, The Pine Apple, Hackney Road, Shoreditch, London. *Trans. London Middlesex Archaeol. Soc.* v.24, 1973, 192.

ALLPORT, John (*fl.* 1800s–1810s)

Seedsman, 39 Holborn Hill, London.

ALLPORT, Joseph (1800–1877)
b. Aldridge, Staffs 1800

Emigrated to Tasmania, 1831. Interested in natural history and horticulture. Founder member of Tasmanian Society of Natural History.

Austral. Dict. Biogr. v.1, 1966, 9–10.

ALLPORT, Mrs. Julia (1796–*c.* 1846/7)
b. Mahé, India 23 Nov. 1796 *d.* Malvern, Worcs *c.* 1846/7

Wife of English businessman in India and China, *c.* 1813–44. Her paintings of flora of these countries are now at Kew.

Kew Bull. 1916, 164.

ALLPORT, Morton (1830–1878)
b. West Bromwich, Staffs 4 Dec. 1830 *d.* Hobart, Tasmania 10 Sept. 1878

FLS 1867. Son of J. Allport (1800–1877). To Tasmania with his parents, 1831. Ichthyologist. Also interested in botany. Vice-President, Royal Society of Tasmania.

Papers Proc. R. Soc. Tasmania 1878, 12–13. P. Mennell *Dict. Austral. Biogr.* 1892. A. Wilberforce *Austral. Encyclop.* v.1, 1927, 52.

ALLPORT, William (*fl.* 1800s)

Nurseryman, Hackney Road, Shoreditch, London.

ALLSOPP, Allan (1914–1971)
b. Darwen, Lancs 9 April 1914 *d.* Isle of Man 19 April 1971

BSc Manchester 1934. PhD 1938. Assistant lecturer, Cryptogamic Botany, Manchester University, 1946; Lecturer, 1948; Reader, 1966. Studied plant morphogenesis.

Phytomorphology 1971, 251–52 portr.

ALLWOOD, Montagu Charles W. (*c.* 1879–1958)
d. Wivelsfield Green, Sussex 24 July 1958

FLS 1922. VMH 1949. Nurseryman, Haywards Heath, Sussex, specialising in dianthus. *Perpetual Flowering Carnation* 1907. *Carnations, Pinks and all Dianthus* 1926. *Carnations for every Garden and Greenhouse* 1926. *Carnations for Everyone* 1931. *Third and Fourth Generation* 1940 2 vols. (autobiography).

Gdnrs Mag. 1911, 951–52 portr. *J. Hort. Home Farmer* v.64, 1912, 259–60 portr. *Gdnrs Chron.* 1950 i 22 portr.; 1958 ii 91; 1960 i 524–25 portr. *Sunday Times* 17 June 1951 portr. *Proc. Linn. Soc.* 1957–58, 127.

ALSTON, Arthur Hugh Garfit (1902–1958)
b. West Ashby, Lincs 4 Sept. 1902 *d.* Barcelona, Spain 17 March 1958
BA Oxon 1924. FLS 1927. Assistant, Kew Herb., 1924. In charge of Herb., Botanic Gardens, Peradeniya, Ceylon, 1925–30. Assistant Keeper, BM(NH), 1930 where he specialised in Pteridophyta, particularly *Selaginella*. Collected plants in Ceylon, Central America, 1938–39, Indonesia, 1953–54. President, British Pteridological Society, 1947–58. Supplement to H. Trimen's *Fl. of Ceylon* 1931. *Kandy Fl.* 1938. *Ferns and Fern-allies of West Tropical Africa* 1959. Editor, *British Fern Gazette*, 1937–49.
J. Bot. 1930, 96. *Bol. Soc. Argent. Bot.* 1958, 136–37. *Br. Fern Gaz.* 1958, 220–21. *Fl. Malesiana* v.5, 1958, 19–20 portr. *J. Kew Guild* 1958, 587 portr. *Times* 28 March 1958. *Proc. Bot. Soc. Br. Isl.* 1959, 361–65. *Amer. Fern J.* 1959, 1–2, *Taxon* 1959, 83–86 portr. *J. Soc. Bibl. Nat. Hist.* 1960, 383–404; 1969, 240–48.
Plants at BM(NH) and Kew.

ALSTON, Charles (1683–1760)
b. Eddlewood, Lanarkshire 24 Oct. 1683 *d.* Edinburgh 22 Nov. 1760
MD Leyden 1719. Professor of Botany, Edinburgh, 1738–60. *Tirocinium Botanicum Edinburgense* 1753. *Dissertation on Botany* 1754.
R. Pulteney *Hist. Biogr. Sketches of Progress of Bot. in England* v.2, 1790, 9. J. E. Smith *Correspondence of Linnaeus* v.1, 1821, 510. D. Turner *Extracts from Lit. and Sci. Correspondence of Richard Richardson* 1835, 275. *Index to Sloane Manuscripts* 1904, 12. *D.N.B.* v.1, 346–47. F. W. Oliver *Makers of British Bot.* 1913, 283–84. H. R. Fletcher and W. H. Brown *R. Bot. Gdn, Edinburgh, 1670–1970* 1970, 37–45.
MSS. at Royal Society.
Alstonia R. Br.

ALSTON, Emma M. and **Caroline M.** (*fl.* 1830s)
Drawings of plants from Odell Wood, Beds at Bedford Record Office.

ALSTON, John (*fl.* 1730s)
Nurseryman, near Chelsea College, London. Member of London Society of Gardeners.

ALUN ROBERTS, Robert *see* Roberts, R. A.

ALVINS, M. V. (*fl.* 1880s)
Plant collector for Forest Dept., Straits Settlements in Malacca, 1884–88.
Gdns Bull. Straits Settlements 1927, 116. *Fl. Malesiana* v.1, 1950, 13.
Plants at Singapore.

AMANN, J. *pseudonym see* Kurz, W. S.

AMHERST *see* Rockley, *Lady* A. M.

AMHERST, Countess Sarah (*née* **Archer**) (1762–1838)
b. London 19 July 1762 *d.* London 27 May 1838
Wife of Lord Amherst, Governor-General of India, 1823–28. Collected plants on her travels. Introduced *Clematis montana* and other plants to Europe. "Right Honourable Countess Amherst and her daughter Lady Sarah Amherst, the zealous friends and constant promoters of all branches of Natural History, especially botany" (Wallich).
R. T. Ritchie *Lord Amherst* 1894. *Taxon,* 1970, 511.
Indian plants at Kew.

AMORY, Andrew (1841–1921)
b. Alnwick, Northumberland 23 Nov. 1841 *d.* 6 April 1921
Keeper of objets d'art at Alnwick Castle. Studied marine algae.
Proc. Berwickshire Nat. Club 1921, 316–17.

AMORY, Sir John Heathcoat (1894–1972)
b. 2 May 1894 *d.* Nov. 1972
VMH 1966. Developed garden at Knightshayes, Tiverton, Devon.
Yb. Int. Dendrology Soc. 1972, 98–99. *J. R. Hort. Soc.* 1973, 188–89.

AMOS, William (*fl.* 1800s)
Of Brotherstoft, Boston, Lincs. Gardener and bailiff; later a farmer. *Minutes of Agriculture and Planting, with Specimens of Grasses* 1804.

AMPHLETT, John (1845–1918)
b. Clent House, Worcs 22 March 1845 *d.* Whitehall, Hayes, Worcs 23 June 1918
MA. SCL Oxon. Barrister. *Botany of Worcestershire* (with C. Rea) 1909.
Bot. Soc. Exch. Club Br. Isl. Rep. 1918, 349. *Trans. Worcs. Nat. Club* v.7, 1918, 46–56.

AMSLER, Maurice (*c.* 1876–1952)
d. 11 June 1952
VMH 1951. Medical officer, Eton College. Specialist on lilies.
Gdnrs Chron. 1952 ii 9 portr. *Times* 31 June 1952. *Lily Yb.* 1953, 153.

AMYOT, Thomas Edward (*c.* 1817–1895)
b. London *c.* 1817 *d.* 15 Dec. 1895
FRCS. Of Diss, Norfolk. Microscopist. Botanist.
Trans. Norfolk Norwich Nat. Soc. v.6, 1896, 124.

ANDERSON, Capt. (*fl.* 1810s)
Brought seeds from Ambon to W. Roxburgh before 1814.
Bull. Jard. Bot. Buitenzorg v.14, 1936, 33. *Fl. Malesiana* v.1, 1950, 15.

ANDERSON, Adam (*c.* 1850–1932)
b. Bunkle, Berwickshire *c.* 1850 *d.* Galashiels, Selkirk 27 March 1932
Had good knowledge of flowering plants and mosses of Berwickshire.
Proc. Berwickshire Nat. Club 1932, 112–14 portr.

ANDERSON, Alexander (–1811)
d. St. Vincent 8 Sept. 1811
MD. FLS 1808. FRSE. Curator, St. Vincent Botanic Garden, 1783. In Guiana, 1791.
L. Guilding *Account of St. Vincent Garden* 1825, 8. *Gent. Mag.* 1811 ii 657. A. B. Lambert *Description of Genus Pinus* v.2, 1828–37, 14. *Gdnrs Mag.* 1826, 194. Rees *Cyclop.* v.39, addenda. *Cottage Gdnr* v.8, 1852 (letters *passim*). *D.N.B.* v.1, 372. *Kew Bull.* 1892, 94–97. *Taxon* 1970, 511–12.
Plants at BM(NH) and Cambridge. Drawings and MSS. at Linnean Society. Letters at BM(NH), Kew and Linnean Society.
Cyrtopodium andersonii R. Br.

ANDERSON, Alexander (1820–)
b. Redgorton, Perthshire 24 July 1820
Gardener to Earl of Stair at Oxenford Castle, Edinburgh.
Gdnrs Chron. 1877 i 529 portr.

ANDERSON, Edward Bertram (*c.* 1895–1971)
d. 29 July 1971
VMM 1952. VMH 1960. Hybridised alpines and irises in his garden at Lower Slaughter, Glos. President, Alpine Garden Society. *Rock Gardens* 1959. *Hardy Bulb Species* 1964. *Dwarf Bulbs for Rock Garden* 1959. Contrib. to *Quart. Bull. Alpine Gdn Soc.* and *J. R. Hort. Soc.*
Iris Yb. 1971, 17. *J. R. Hort. Soc.* 1971, 553–54. *Quart. Bull. Alpine Gdn Soc.* 1971, 267–68 portr.; 1973, 138–48. *Times* 6 Aug. 1971. *Lilies 1972 and Allied Plants*, 50. *Rea* 1972, 44 portr.

ANDERSON, Frederick William (1866–1891)
b. Wisbech, Cambridge 22 June 1866 *d.* New York 22 Dec. 1891
DSc Montana 1890. To America, 1881. Botanised chiefly in Montana. Contrib. to *Bull. Torrey Bot. Club.*
Bot. Gaz. 1892, 78–81 portr. *Bull Torrey Bot. Club* v.19, 1892, 73, 74. J. W. Harshberger *Botanists of Philadelphia and their Work* 1899, 266. J. Ewan *Rocky Mountain Nat.* 1950, 150.
Plants at College of Montana and New York.
Portr. at Hunt Library.

ANDERSON, G. C. (*fl.* 1880s)
Marine surveyor, Hong Kong. Collected ferns near Fuchow.
E. Bretschneider *Hist. European Bot. Discoveries in China* 1898, 755–56.

ANDERSON, George (*fl.* 1800s–1817)
d. 10 Jan. 1817
Of West Ham, London. FLS 1800. Had a salicetum. Grew narcissi. Collected plants in Brazil and Barbados, 1815. Collected with J. P. Jones in Devon and Cornwall (Jones *Botanical Tour* 1820). 'British Plants' (*Trans. Linn. Soc.* v.11, 1815, 216–26). 'Monograph of Genus Paeonia' (*Trans. Linn. Soc.* v.12, 1817, 248–90). Contrib. willows to J. Sowerby and J. E. Smith's *English Bot.*
Gent. Mag. 1817 i 92. Rees *Cyclop.* v.39, addenda. *Trans. Linn. Soc.* v.12, 1817, 283–84. *R.S.C.* v.1, 63.
Salix Andersoniana Smith.

ANDERSON, James (–1809)
d. Madras, India 5 Aug. 1809
MD. FRSE. Surgeon, East India Company, 1762. Physician-General to Forces, 1786. *Varnish and Tallow-trees* 1791. *Culture of Bastard Cedar Trees on Coast of Coromandel* 1794.
Gent. Mag. 1810 ii 180. M. Graham *J. of Residence in India* 1812, 125. *J. Bot.* 1884, 359. *D.N.B.* v.1, 382. D. G. Crawford. *Hist. Indian Med. Service 1600–1913* v.2, 1914, 14, 148 portr.
Letters in Banks correspondence at BM (NH).
Andersonia Roxb.

ANDERSON, James (1739–1808)
b. Hermiston, Edinburgh 1739 *d.* Isleworth, Middx 15 Oct. 1808
LLD Aberdeen 1780. Economist. Farmed for 20 years in Aberdeenshire. *Miscellaneous Observations on Planting and Training Timber-trees* 1777. *Description of Patent Hot-House* 1804.
Gent. Mag. 1808 ii 1051–54. J. C. Loudon *Encyclop. Gdning* 1822, 1277. G. W. Johnson *Hist. English Gdning* 1829, 236–37. *Cottage Gdnr* v.5, 1851, 1; v.7, 1852, 364–65, 377–79. J. Donaldson *Agric. Biogr.* 1854, 59–60. *D.N.B.* v.1, 381.
Medallion by Tassie. Portr. at Hunt Library.

ANDERSON, James (1797–1842)
b. Boquhan, Stirling 1797 *d.* Sydney 22 April 1842

Collected plants on west coast of Africa and Ascension Island, 1821; Rio de Janeiro and Valparaiso, 1823–24. Botanical collector on HMS 'Adventure' during voyage to S. America and survey of Straits of Magellen, 1825–30. At Port Jackson, 1832, whence he sent plants to W. J. Hooker at Glasgow. Collected for Low and Mackay of Clapton. Superintendent, Botanic Garden, Sydney, 1838–42.

A. B. Lambert *Description of Genus Pinus* v.2, 1828–37 appendix, 24. *Gdnrs Mag.* 1833, 469–70; 1840, 116. J. Fitzroy *Voyage of 'Adventure'* v.1, 1839 preface, xii. J. D. Hooker *Fl. Tasmaniae* 1859, cxxiii. W. Woolls *Lectures on Vegetable Kingdom* 1879, 58. J. H. Maiden *James Anderson* 1903. *Kew Bull.* 1906, 217. *J. R. Soc. N.S.W.* 1908, 82–83. *Taxon* 1970, 512. *J. Proc. R. Austral. Hist. Soc.* v.18, 119–22.

MS. lists and S. American plants at BM (NH).

Carex andersoni Boott.

ANDERSON, James (*c.* 1832–1899)
d. Glasgow 16 June 1899

Gardener to T. Dawson of Uddingston. Landscape gardener at Glasgow and Manchester. *New Practical Gardener and Modern Horticulturist* 1873. Edited *Northern Gdnr.* Contrib. to *Gdnrs Chron.*

Gdnrs Chron. 1899 i 414. *Orchid Rev.* 1899, 198.

Odontoglossum X Andersonianum.

ANDERSON, James (*c.* 1802–1871)
d. York 22 Dec. 1871

Nurseryman and seedsman, York.

ANDERSON, James (*fl.* 1900s)

Kew gardener. Curator, Botanical Station, Gold Coast, 1905.

Kew Bull. 1905, 61.

Plants at Kew.

ANDERSON, James Webster
(*fl.* 1910s–1920s)

Assistant Curator, Gardens Dept. Straits Settlements, 1910–17; later a planter. Became private gardener on return to U.K. *Index of Plants, Botanic Gardens, Singapore* 1912.

Gdns Bull. Straits Settlements 1927, 116. *Fl. Malesiana* v.1, 1950, 15.

Plants at Singapore and Kew.

ANDERSON, John (*fl.* 1780s)

Nurseryman, Co. Westmeath, Ireland.

ANDERSON, John (–1847)
d. Sydney 1847

Gardener to Earl of Essex at Cassiobury. Collected plants at Tierra del Fuego, 1829–30, on HMS 'Adventure' under Capt. P. P. King. Superintendent, Botanic Garden, Sydney, 1840–42. 'Account of New Esculent Vegetable called Tetragonia' (*Trans. Hort. Soc.* 1822, 488–94).

J. D. Hooker *Fl. Tasmaniae* v.1, 1859, cxxiii. *Proc. Linn. Soc. N.S.W.* v.25, 1900, 789.

Chile plants and MS. list at BM(NH).

ANDERSON, John (1833–1900)
b. Edinburgh 4 Oct. 1833 *d.* Matlock, Derbyshire 15 Aug. 1900

MD Edinburgh 1862. FRS 1879. FLS 1862. Professor of Natural History, Free Church College, Edinburgh before going to India in 1865. Superintendent, Calcutta Indian Museum, 1865–66. Medical officer and naturalist on expeditions to Upper Burma and Yunnan, 1867–68 and 1875–76; also to Mergui Archipelago, 1881–82. *Report on Expedition to Western Yunnan via Bhamo* 1871.

J. Bot. 1873, 193–95; 1875, 160. E. Bretschneider *Hist. of European Bot. Discoveries in China* 1898, 692–93. *Proc. Linn. Soc.* 1900–1, 38–40. *D.N.B. Supplt 1* v.1, 46–47. C. E. Buckland *Dict. Indian Biogr.* 1906, 14. D. G. Crawford *Hist. Indian Med. Service, 1600–1913* v.2, 1914, 371. *R.S.C.* v.1, 63; v.7, 30; v.9, 42.

Plants at Calcutta and Kew.

ANDERSON, John Anderson
(*c.* 1811–1891)
d. Perth 6 Dec. 1891

Nurseryman at Perth, trading under name of Dickson and Turnbull. Sold the business in 1883 to J. McLeod, nurseryman of Crieff.

Garden v.40, 1891, 599.

ANDERSON, Mark Louden (1895–1961)
b. Kinneff, Kincardine 16 April 1895 *d.* Bristol 6 Sept. 1961

BSc Edinburgh 1919. DSc 1924. Director, Forest Service, Eire, 1931–46. Lecturer, Imperial Forestry Institute, Oxford, 1946–51. Professor of Forestry, Edinburgh University, 1951–61. *Natural Woodlands of Britain and Ireland* 1932. *Selection of Tree Species* 1950.

Yb. R. Soc. Edinburgh 1962, 21–22. *Who was Who 1961-1970* 23.

ANDERSON, Sir Maurice Abbot– *see* Abbot-Anderson, *Sir* M.

ANDERSON, Robert

According to J. H. Burkill this name is a confusion between Robert Scott and Thomas Anderson, both of whom served in the Botanic Gardens, Calcutta.

I. H. Burkill *Chapters on Hist. Bot. in India* 1965, 131.

ANDERSON, Robert (1760s–1800)
Nurseryman and seedsman, Edinburgh.
J. Harvey *Early Hort. Cat.* 1973, 1.

ANDERSON, Robert (1818–1856)
b. Fettercairn, Aberdeenshire *d.* June 1856
Assistant Surgeon, Royal Navy. Surgeon on HMS 'Investigator' and 'Enterprise'. Collected plants in Straits of Magellen region. 1850–53.
F. Boase *Modern English Biogr.* v.1, 1892, 65.
Plants at Kew.

ANDERSON, Robert (1859–1939)
d. Dublin 17 March 1939
Superintendent, Phoenix Park, Dublin.
Gdnrs Chron. 1923 ii 154 portr., 223.

ANDERSON, Samuel (–1878)
FLS 1854. Of Whitby, Yorks. Bryologist and hepaticologist. Discovered *Sphagnum molle*. Contrib. to 'Sphagnaceae Exsiccatae' (*J. Bot.* 1919, 142).
J. Bot. 1878, 64.

ANDERSON, Thomas (1832–1870)
b. Edinburgh 26 Feb. 1832 *d.* Edinburgh 26 Oct. 1870
MD Edinburgh 1853. FLS 1859. Bengal Medical Service, 1854. Superintendent, Calcutta Botanic Gardens, 1861–68. 'Florula Adenensis' (*J. Linn. Soc. Supplt.* 1860, 1–47). 'Acanthaceae' (*J. Linn. Soc.* 1864, 13–54, 111–118; 1867, 425–526). *Catalogue of Plants... Calcutta* 1865.
J. Bot. 1870, 368. *Gdnrs Chron.* 1870, 1478. *Proc. Linn. Soc.* 1870–1, lxxx–lxxxii. *Trans. Bot. Soc. Edinburgh* v.11, 1870, 41–45. *D.N.B.* v.1, 392. C. E. Buckland *Dict. Indian Biogr.* 1906, 14. D. G. Crawford *Hist. Indian Med. Service, 1600–1913* v.2, 1914, 145. *Rec. Bot. Survey India* v.7, 1914, 11–14. *Gdns Bull. Straits Settlements* 1927, 116. *Curtis's Bot. Mag. Dedications, 1827–1927*, 151–52 portr. *Fl. Malesiana* v.1, 1950, 15. I. H. Burkill *Chapters on Hist. Bot. in India* 1965, *passim*. *R.S.C.* v.1, 65; v.7, 33; v.12, 16.
Plants at Kew, BM(NH), Calcutta, Leiden.
Letters in Sir W. J. Hooker's correspondence at Kew.
Strobilanthes andersonii Bedd.

ANDERSON, William (1750–1778)
b. North Berwick 28 Dec. 1750 *d.* at sea off Anderson's Island 3 Aug. 1778
Studied medicine at Edinburgh University, 1766–69. On Capt. James Cook's 2nd (1772–75) and 3rd (1776–79) voyages.
Rees *Cyclop.* v.39, addenda. J. D. Hooker *Fl. Tasmaniae* v.1, 1859, cxiii. *Proc. R. Soc. Tasmania* 1909, 11. *D.N.B.* v.1, 393. *J. Bot.* 1916, 345; 1917, 54. *J. R. Soc. N.S.W.* v.55, 1921, 150. *Ann. Med. Hist.* 1933, 511–24. *Emu* 1938, 60–62. *Fl. Malesiana* v.1, 1950, 15.
MS. and plants from Australia, N. Caledonia, Pacific Islands at BM(NH). Letters in Sir J. Banks's correspondence at Kew.
Andersonia R. Br.

ANDERSON, William (1766–1846)
b. Easter Warriston, Edinburgh 1766 *d.* Chelsea, London 6 Oct. 1846
ALS 1798. FLS 1815. Botanic gardener, 1793–1814, to James Vere at Kensington Gore "whose collection he manages with superior skill and more scientific knowledge than is often met with." (*Bot. Mag.* 1808, t.1136). Curator, Chelsea Physic Garden, 1815. Studied *Stapelia* (A. H. Haworth *Synopsis Plantarum Succulentarum* 1812, 25).
Rees *Cyclop.* v.39, addenda, H. C. Andrews *Botanist's Repository* 1801–2, t.184, t.217. *Bot. Misc.* 1829–30, 67. *Proc. Linn. Soc.* 1847, 331–32. E. Bretschneider *Hist. of European Bot. Discoveries in China* 1898, 210. R. H. Semple *Mem. of Bot. Gdn at Chelsea* 1878, 119–20, 203–5. *D.N.B.* v.1, 393.
British plants at Kew. Letters in M. T. Ellacombe correspondence at Kew.

ANDERSON-HENRY, Isaac
(*olim* **Anderson**) (1800–1884)
b. Caputh, Perthshire 14 July 1800 *d.* 21 Sept. 1884
FLS 1865. Bred new varieties of plants in his garden at Hay Lodge. President, Botanical Society of Edinburgh, 1867–68.
Gdnrs Chron. 1873, 399 portr.; 1884 ii 400. *Garden* v.26, 1884, 288. *Proc. Linn. Soc.* 1886–87, 42–44. *Trans. Bot. Soc. Edinburgh* 1886, 189–90. *Notizblatt Bot. Gart. Berlin* v.4, 1906, 243–45. *Curtis's Bot. Mag. Dedications, 1827–1927*, 175–76 portr. *R.S.C.* v.7, 951; v.12, 16, 325.
Cultivated plants at Kew.

ANDERTON SMITH, Mrs. W. (*fl.* 1850s)
Of Tedstone Delamere, Herefordshire. Discovered *Epipogium gmelini* in Herefordshire in 1854.
Hooker's J. Bot. 1854, 318–19. *Bot. Mag.* 1854, t.4821.

ANDREW, Thomas (*c.* 1773–1827)
d. June 1827
Florist, Coggeshall, Essex.
Gdnrs Mag. 1827, 255.

ANDREWS, Cecil Rollo Payton (1870–1951)
b. London 2 Feb. 1870 *d.* London 14 June 1951

BA Oxon. Schoolteacher, London. Principal, Claremont Training College, W. Australia, 1901–3. Director of Education, Claremont, 1903–29. Collected plants in W. Australia, 1901–6. President, Western Australia Natural History Society, 1904. Contrib. to *J. Proc. Mueller Bot. Soc.* and *J. W. Austral. Nat. Hist. Soc.*

Rep. Bot. Gdn. Sydney 1915, 16. *Watsonia* v.2, 1951–53, 359. *Who was Who, 1951–1960* 26.

Plants at Perth and Kew.

Hibbertia andrewsiana Diels.

ANDREWS, Charles William (1866–1924)
b. Hampstead, London 1866 *d.* 25 May 1924

Educ. London University. FRS 1906. Assistant, Geology Dept. BM(NH), 1892; Assistant Keeper. On expedition to Christmas Island (Indian Ocean), 1897–98, 1909. *Monograph of Christmas Island* 1900.

Nature v.113, 1924, 827–28. *Who was Who, 1916–1928* 23. *Fl. Malesiana* v.1, 1950, 16–17.

Christmas Island plants at BM(NH) and Kew.

ANDREWS, F. W. (–1961)
d. March 1961

Chief Economic Botanist in Sudan where he collected plants. East Malling Research Station, 1953–58. *Flowering Plants of Sudan* 1950–56 3 vols.

ANDREWS, Henry Charles (*fl.*1790s–1830s)

Of Knightsbridge, London. Son-in-law of J. Kennedy (1759–1842). Botanical artist and engraver. *Botanists' Repository* 1797–1811 10 vols.; ed. 2 1816 10 vols. *Coloured Engravings of Heaths* 1794–1830 4 vols. *Heathery* 1804–12 4 vols.; ed. 2. 1845 6 vols. *Geraniums* 1805. *Roses* 1805–28 2 vols.

Gdnrs Chron. 1855, 237. *D.N.B.* v.1, 406–7. *J. Bot.* 1916, 236–46. W. Blunt *Art of Bot. Illus.* 1950, 209–11. F. A. Stafleu *Taxonomic Literature* 1967, 6–8.

Andrewsia Vent.

ANDREWS, Isaac (*fl.* 1820s)

Fruit-grower, Lambeth, London; "famous for his pine-apples."

ANDREWS, James (*c.* 1801–1876)
b. c. 1801 *d.* Walworth, London 17 Dec. 1876

Prolific botanical artist. Illustrated a number of books by R. Tyas. Contrib. many plates to *Floral Mag.* and *Florist*. *Lessons in Flower Painting* 1836.

Gdnrs Chron. 1877 i 24. E. Mannering *Flower Portraits* 1961, 11.

ANDREWS, Joseph (*fl.* 1710s–1760s)

Of Sudbury, Suffolk. FRS 1727. Apothecary. Friend of S. Dale. Account of his herb. in *J. Bot.* 1918, 257–61, 294–98, 346–54; 1919, 337–40.

J. E. Dandy *Sloane Herb.* 1958, 83.

Herb. at BM(NH).

ANDREWS, R. Thornton (1839–1928)

Founder member of Hertford Museum. Collected Hertfordshire plants.

J. G. Dony *Fl. Hertfordshire* 1967, 18.

ANDREWS, William (1802–1880)
b. Chichester, Sussex 1802 *d.* Dublin 11 March 1880

President and Secretary, Dublin Nat. Hist. Soc. Discovered *Trichomanes andrewsii*. 'Botany of Great Arran Island' (Hooker's *London J. Bot.* 1845, 569–70). 'Hymenophylla' (*J. Bot.* 1871, 188–89). 'Irish Saxifrages' (*J. Bot.* 1871, 253–55).

Ann. Mag. Nat. Hist. v.6, 1841, 382–85. *J. Bot.* 1880, 256, 286; 1883, 181; 1926, 16. *Proc. R. Irish Acad.* v.3, 1880, 131. *D.N.B.* v.1, 409. R. L. Praeger. *Some Irish Nat.* 1949, 41. *R.S.C.* v.1, 70; v.7, 36; v.9, 49.

Plants at Kew, Dublin, Manchester. Letters in Sir W. J. Hooker's correspondence at Kew.

ANGAS, George French (1822–1886)
b. Newcastle, Northumberland 25 April 1822 *d.* London 4 October 1886

FLS 1866. Studied drawing and lithography. Arrived in Adelaide, S. Australia, 1844. Made drawings during his travels to mouth of River Murray and on south-east coast of S. Australia. Returned to England, 1846. In S. Africa, returning to Australia in 1850. Secretary, Australian Museum, Sydney, 1853–60. Returned to England, 1863. *South Australia Illustrated* 1847. *New Zealanders Illustrated* 1847. *Savage Life and Scenes in Australia and New Zealand* 1847 includes some flower drawings. "In his Australian landscapes he took care to depict native vegetation accurately."

Rep. ... Austral. Assoc. Advancement Sci. 1907, 172. *Austral. Zool.* v.12, 1952–59, 362–63. *Austral. Dict. Biogr.* v.1, 1966, 18–19.

Drawings at National Gallery of S. Australia, S. Australian Museum and Royal Geographical Society, London.

ANLEY, Mrs. Gwendolyn (*c.* 1876–1968)
d. Jan. 1968

VMH 1964. Amateur gardener with interest in irises. *Alpine House Culture for Amateurs* 1938. *Irises, their Cultivation and Selection* 1946. Edited *Iris Yb.*, 1940–44.

Iris Yb. 1968, 133–34. *Quart. Bull. Alpine Gdn Soc.* 1968, 203–4.

ANNANDALE, Thomas Nelson (1876–1924)
b. Edinburgh 1876 *d.* Calcutta 10 April 1924
BA Oxon. FLS 1907. Zoologist and anthropologist. Malay Peninsula, 1899; attached to Skeat expedition. Deputy Superintendent, Indian Museum, Calcutta, 1904; Superintendent, 1906. Director, Zoological Survey of India. Collected plants. 'Vegetation of an Island in Chilka Lake on East Coast of India' (*Proc. Linn. Soc.* 1920–21, 63).
Proc. Linn. Soc. 1923–24, 45–46. *Gdns Bull. Straits Settlements* 1927, 116. *Rec. Indian Mus. Calcutta* v.27, 1925, 1–28. *J. Bombay Nat. Hist. Soc.* v.30, 1925, 213–14. *Fl. Malesiana* v.1, 1950, 17.
Plants at Singapore.

ANNESLEY, George, 2nd Earl of Mountnorris, Viscount Valentia (1769–1844)
b. 2 Nov. 1769 *d.* 23 July 1844
FRS 1796. FLS 1796. Succeeded to earldom, 1816. Had "famed collection" of plants at Arley, near Bewdley. *Voyages and Travels to India, Ceylon, the Red Sea, Abyssinia and Egypt* 1809 3 vols.
H. C. Andrews *Botanist's Repository* 1801, t.145.
Letters in Sir J. Banks's correspondence at BM(NH) and Sir J. E. Smith's correspondence at Linnean Society.
Anneslea Wall.

ANNESLEY, Hugh, 5th Earl Annesley (1831–1908)
b. 26 Jan. 1831 *d.* Castlewellan, County Down 15 Dec. 1908
Created notable garden at Castlewellan. *Beautiful and Rare Trees and Shrubs* 1903. Contrib. to *Gdnrs Chron.*
Gdnrs Chron. 1908 ii 440. *Kew Bull.* 1909, 72–73. *Who was Who, 1897–1916* 18.

ANNESLEY, Oliver Francis Theodore (*fl.* 1870s)
Captain. Collected some plants at Aden, 1875, now at Kew.
Rec. Bot. Survey India v.7, 1914, 15.

ANSELL, John (–1847)
b. Hertford *d.* Hertford 1847
Gardener. Collected plants in Derbyshire and Herts *c.* 1820. Accompanied T. Vogel on expedition up Niger, 1841–42. At R. Hort. Soc. garden, Chiswick, 1842. Afterwards of Chislehurst. 'Juncus diffusus' (*Phytologist* v.2, 1846, 662–63).
British and Ivory Coast plants at Kew; plants at Oxford. Letters in Sir W. J. Hooker's correspondence at Kew.
Ansellia Lindl.

ANSELL, Thomas (*c.* 1793–1867)
b. Bedfordshire *d.* Camden Town, London 26 Dec. 1867
Nurseryman and florist at Camden Town for 50 years. Raised dahlia 'Unique'.
Gdnrs Chron. 1868, 10.

ANSON, Thomas (*fl.* 1810s–1830s)
Nurseryman, Shaddongate, Carlisle, Cumberland.

ANSTEAD, Rudolph David (1876–1962)
b. Wisbech, Cambridgeshire 2 June 1876 *d.* 6 Jan. 1962
Educ. Cambridge. Sugar chemist, British West Indies, 1902. Superintendent of Agriculture, Grenada, 1906. Deputy Director of Agriculture, India, 1909. Director, Madras Agricultural Department, 1922–31.
Who was Who, 1961–1970 27.
Plants at Arnold Arboretum; lichens at BM(NH).

ANSTED, David Thomas (1814–1880)
b. London 5 Feb. 1814 *d.* Melton, Suffolk 13 May 1880
MA Cantab. FRS 1844. Professor of Geology, King's College, London, 1840–53. Consulting geologist and mining engineer. In West Indies, 1855.
Nature v.22, 1880, 86. *Proc. R. Soc.* 1880–81, i–ii.
West Indian plants at Kew.

ANTHONY, John (1891–1972)
d. Edinburgh 18 June 1972
MA Edinburgh 1924. FRSE. Scientific post on rubber estate in Malaya. Assistant lecturer in Botany, Dundee University College, 1932. Lecturer in Forest Botany, Edinburgh University, 1934–58. Contrib. to *Notes R. Bot. Gdn Edinburgh.* Was preparing a flora of Sutherland.
B.S.E. News Oct. 1972, 2–3.

ANTISELL, Thomas (1817–1893)
b. Dublin 16 Jan. 1817 *d.* Washington, U.S.A. 14 June 1893
Geologist and chemist. Lecturer in Botany, Peter St. School of Medicine, Dublin. To New York, 1848. Professor of Chemistry and Toxicology, Georgetown University. Chief chemist for U.S.D.A. Geologist on railroad survey in California and Arizona. Botanised in Burro Mountains, 1854. 'Synoptical tables of Botanical Localities' (J. Torrey. *Pacific Railroad Survey* v.7, pt. 3, 1856, 23–26).
Philos. Soc. Washington Bull. 1896, 367–70. *Rep. U.S. National Mus.* 1904, 953–55 portr. *Irish Bk-Lover* v.6, 1915, 118. H. A. Kelly and W. L. Burrage. *Dict. Amer. Med. Biogr.* 1928, 32–33. R. L. Praeger *Some Irish*

Nat. 1949, 41. J. Ewan *Rocky Mountains Nat.* 1950, 151.
Astragalus antisellii A. Gray.

ANTROBUS, Robert
Eton. Peterhouse College, Cambridge.
J. E. Dandy *Sloane Herb.* 1958, 83.

APJOHN, Mrs. (*fl.* 1850s)
Wife of J. Apjohn, Professor of Chemistry, Dublin. "A zealous collector and observer of British algae."
Ann. Mag. Nat. Hist. v.15, 1855, 335–36.
Apjohnia Harvey.

APLIN, Oliver Vernon (*fl.* 1880s?)
Collected plants in Uruguay; acquired by BM(NH) in 1883.

APPLEBY, Samuel (1806–1870)
b. Egmanton, Notts May 1806
FLS 1831. Nurseryman, St. James's Gardens, Doncaster until 1855 when he opened new nursery at Balby. Studied flora of S. Yorkshire. Had herb. 'Rare Plants of Doncaster District' (*Ann. Mag. Nat. Hist.* 1832, 556–58). 'Old Hortulan Grounds of Doncaster' (*Doncaster Gaz.* 18 May 1866).
Naturalist 1972, 55–57.

APPLEBY, Thomas (*c.* 1795–1875)
d. Manchester 20 Oct./Nov. 1875
Gardener to T. Brocklehurst, Macclesfield. Manager, Pine Apple Place nursery, Edgeware Road, London. Cultivated orchids. *Orchid Manual* 1861. Contrib. to *Cottage Gdnr.*
Florist and Pomologist 1875, 264. *Gdnrs Yb. and Almanack* 1876, 172.

APPLETON, A. F. (*fl.* 1900s)
Lieut.-Colonel, Army Veterinary Dept. Collected grasses in Somaliland, 1902–4; S. Africa, Rhodesia and Mozambique, 1911.
Kew Bull. 1907, 204–5.
Plants at Kew.

APPLETON, Henry (*c.* 1855–1930)
b. Manchester *c.* 1855 *d.* March 1930
Lieut.-Colonel, Royal Engineers. Collected plants in Baluchistan, *c.* 1884; Chinese Turkestan, 1906. Retired to Victoria, British Columbia.
Times 7 March 1930. *Pakistan J. For.* 1967 339.
Plants at Kew and Poona.

ARAM, George (*fl.* 1790s)
Gardener and nurseryman, Mansfield, Notts.

ARAM, John (*fl.* 1790s–1800s)
Gardener, Audley End House, Essex. Nurseryman, Mansfield, Notts.

ARAM, Peter (–1735)
d. Ripley, Yorks June 1735
Of Nottinghamshire. Trained as gardener under George London at Fulham Palace. Gardener to Henry Compton, Bishop of London, to Sir Edward Blackett, Newby Hall, Yorks 1694–1716 and to Sir John Ingilby, Ripley Castle, Yorks, *c.* 1729. Went to Holland to collect plants in 1695. MS. *Treatise of Flowers.*
J. Harvey *Early Gdning. Cat.* 1972, 28.

ARAM, Robert (*fl.* 1830s)
Nurseryman and seedsman, 8 Manchester Street, Huddersfield, Yorks.

ARAM, Thomas (*fl.* 1820s–1830s)
Nurseryman, Manchester Street, Huddersfield, Yorks.

ARAM, William (*fl.* 1770s)
Member of Botanic Society of Norwich. List of Norfolk plants in *Description of England and Wales* v.6, 1769–70, 239–48.
B. D. Jackson *Guide to Lit. Bot.* 1881, 503.

ARAM, William (*fl.* 1860s)
Nurseryman, 13 Manchester Street, Huddersfield, Yorks.

ARBER, Agnes (*née* **Robertson**) (1879–1960)
b. London 23 Feb. 1879 *d.* Cambridge 22 March 1960
Educ. Cambridge. DSc London 1905. FRS 1946. FLS 1908. Research assistant to Ethel Sargant, 1902. Lecturer in Botany, University College, London, 1908–9. Married E. A. N. Arber in 1909. Converted part of her house into private laboratory, 1926. Plant morphologist. *Herbals* 1912; ed. 2 1938. *Water Plants* 1920 (biography in 1968 reprint, i–xii). *Monocotyledons* 1925. *Gramineae* 1934. *Natural Philosophy of Plant Form* 1950. *Mind and the Eye* 1954.
Biogr. Mem. Fellows R. Soc. 1960, 1–11 portr. *Proc. Linn. Soc.* 1959–60, 128. *Nature* v.186, 1960, 847–48. *Taxon* 1960, 261–63 portr. *Times* 24 March 1960. *Archs. Int. Hist. Sci.* 1960, 118–19. *Publ. English Goethe Soc.* 1960, 94–95. *Phytomorphology* 1961, 197–98 portr. *Huntia* 1964, 169–71. *J. Soc. Bibl. Nat. Hist.* 1968, 571–79 portr. C. C. Gillispie *Dict. Sci. Biogr.* v.1, 1970, 205–6. *D.N.B. 1951–1960* 1971, 28–30. *Who was Who, 1951–1960* 31.
MSS. and portr. at Hunt Library.

ARBER, Edward Alexander Newell (1870–1918)
b. London 5 Aug. 1870 *d.* Cambridge 14 June 1918
BA Cantab 1898. DSc 1912. FLS 1903. FGS. Demonstrator in Palaeobotany, Cambridge 1899. *Catalogue of Glossopteris Fl. in British Museum* 1905. *Fossil Plants* 1909.

Plant Life in Alpine Switzerland 1910. *Natural History of Coal* 1911.

Ann. Bot. 1918, vii–ix. *Cambridge Rev.* 1918, 59. *Geol. Mag.* 1918, 426–31 portr. *J. Bot.* 1918, 305–8 portr. *Nature* v.101, 1918, 328–29. *Proc. Linn. Soc.* 1918–19, 39–44. *Times* 18 June 1918. *Who was Who 1916–1928* 27. *J. Soc. Bibl. Nat. Hist.* 1968, 371–72, 379–83 portr.

Portr. at Hunt Library.

ARCHER, John (*fl.* 1660s–1680s)

Physician; practised in Dublin, 1660. Court physician to Charles II, 1671. *Everyman his own Doctor; compleated with a Herbal* 1673.

D.N.B. v.2, 71–72. *Gdnrs Chron.* 1922 ii 352. C. E. Raven *John Ray* 1942, 261.

ARCHER, Sarah *see* Amherst, *Countess* S.

ARCHER, Thomas Croxen (1817–1885)

b. Northampton 1817 *d.* Edinburgh 19 Feb. 1885

Surgeon. FRSE. Liverpool Custom House, 1841. Professor of Botany, Queen's College, Liverpool, 1857. Director, Edinburgh Museum of Science and Art, 1860–85. President, Botanical Society of Edinburgh. 1862. *Popular Economic Botany* 1853 (reissued as *Profitable Plants* 1865). *On the Study of Botany* 1857. Contrib. to *Liverpool Lit. Philos. Soc. Proc.*

Pharm. J. v.15, 1884, 709. *Gdnrs Chron.* 1885 i 321. *Proc. R. Soc. Edinburgh* v.14, 1886–87, 110–14. *Trans. Bot. Soc. Edinburgh* v.16, 1886, 272–76. F. Boase *Modern English Biogr.* v.1, 1892, 82. *Trans. Liverpool Bot. Soc.* 1909, 57–58. *R.S.C.* v.1, 85; v.6, 567; v.7, 42; v.13, 142.

Letters at Kew.

ARCHER, William (1820–1874)

b. Launceston, Tasmania 16 May 1820 *d.* Longford, Tasmania 14 Oct. 1874

FLS. Qualified in England as architect. Studied Tasmanian plants in Kew Herb., 1856–58. Drew and analysed orchids for *Fl. Tasmaniae*. J. D. Hooker acknowledged the help of "his observations and collections." Sent algae to W. H. Harvey. Contrib. appendix on 'Vegetable Products' to G. Whiting's *Products and Resources of Tasmania* 1862. Contrib. to *Papers Proc. R. Soc. Tasmania*.

J. D. Hooker *Fl. Tasmaniae* v.1, 1859, cxxvii. *Proc. Linn. Soc. N.S.W.* v.25, 1900, 778. *Papers Proc. R. Soc. Tasmania* 1909, 11–12; 1913 t.xx portr. *Austral. Dict. Biogr.* v.3, 1969, 40–41.

Herb. and letters at Kew. Drawings, 1848–56, at Linnean Society. MSS. at Tasmania University. Portr. at Hunt Library.

Archeria Hook. f.

ARCHER, William (1830–1897)

b. Maghera, County Down 6 May 1830 *d.* Dublin 14 Aug. 1897

FRS 1875. Librarian, Royal Dublin Society, 1876–95. Papers on Desmids in *Proc. Dublin Nat. Hist. Soc.* v.3–v.5, 1859–65 and *J. Bot.* 1874. Contrib. to A. Pritchard *History of Infusoria* ed. 4 1861.

Ann. Scott. Nat. Hist. 1898, 8. *Irish Nat.* 1897, 253–57 portr. *J. Bot.* 1897, 501–2. *Notes Bot. School, Dublin* 1898, 123–26. *Proc. R. Soc.* v.62, 1898, xl–xlii. *D.N.B. Supplt. 1* v.1, 57–58. R. L. Praeger. *Some Irish Nat.* 1949, 41–42 portr. *R.S.C.* v.1, 86; v.7, 42; v.9, 62; v.13, 142.

ARCHER-HIND, Thomas Hodgson (1814–1911)

FLS 1834. Of Combe Fishacre House, S. Devon. Amateur gardener; hybridised hellebores.

Garden v.57, 1900, 264–65 portr.

ARCHIBALD, Joseph (1784–1874)

b. Edinburgh 5 Nov. 1784 *d.* Deescart, County Monaghan, 16 Dec. 1874

Son of Edinburgh nurseryman. Gardener to Lord Dalhousie at Dalhousie Castle, Lord Farnham at Farnham Castle, Ireland and Col. Pratt at Kingscourt, County Monaghan.

Gdnrs Yb. and Almanack 1876, 170.

ARDAGH, John (1885–1949)

b. Fulham, London 15 June 1885 *d.* Maidstone, Kent 8 March 1949

Appointed Herb. BM(NH), 1919; Clerk-in-charge of Botanical Library, 1922–47. Contrib. to *J. Bot.*

Nature v.163, 1949, 557. *Times* 18 March 1949.

ARDEN, Lady Margaret Elizabeth (*née* Spencer-Wilson) (–1851)

d. 20 May 1851

Of Nork, Epsom, Surrey. Mycologist. Contrib. to J. Sowerby and J. E. Smith's *English Bot.*, 461, 2659.

ARDEN, Stanley (1874–1942)

b. Heaton Norris, Cheshire Sept. 1874 *d.* Brighton, Sussex 5 May 1942

Kew gardener, 1898. Rubber plantations, Malaya, 1900–12. Adelaide, S. Australia, 1925–35. Sent plants to Singapore.

Kew Bull. 1906, 383. *Gdns Bull. Straits Settlements* 1927, 116. *J. Kew Guild* 1942, 191. *Fl. Malesiana* v.1, 1950, 598.

ARDERNE, John (*fl.* 1307–1380)

d. London

First great English surgeon. At Newark,

1349–70; London after 1370. *De re Herbaria, Physica et Chirurgica* (MS. in Sloane collection).
Philos. Trans. R. Soc. v.63, 1773, 81. R. Pulteney *Hist. Biogr. Sketches of Progress of Bot. in England* v.1, 1790, 23. Rees *Cyclop.* v.2, 1819. *D.N.B.* v.2, 76–77. E. J. L. Scott *Index to Sloane Manuscripts* 1904, 19.
Ardernia Salisb.

ARGENT, John (*fl.* 1597–1643)
d. May 1643
MD. FRCP 1597; President, 1625–27, 1629–33. During his Presidency he gave T. Johnson the first bunch of bananas ever exhibited in a London shop.

ARGYLL, Archibald, 3rd Duke *see* Campbell, A., *3rd Duke of Argyll*

ARKWRIGHT, Sir John Stanhope (1872–1954)
d. Sept. 1954
FLS 1925. MP for Hereford, 1900–12. Grew and bred daffodils.
Proc. Linn. Soc. 1953–54, 45–46. *Who was Who, 1951–1960* 34.

ARMISTEAD, Wilson (1819–1868)
d. Leeds, Yorks 18 Feb. 1868
Meteorologist and entomologist. Had herb.
T. B. Hall *Fl. Liverpool* 1839, vii. J. Smith *Cat. of Friends' Books* v.1, 1867, 124–31. *Trans. Liverpool Bot. Soc.* 1909, 58. F. Boase *Modern English Biogr.* v.1, 1892, 83. *R.S.C.* v.7, 46.

ARMITAGE, Edward (1822–1906)
b. Maidstone, Kent 17 May 1822 *d.* Cheltenham, Glos 22 Feb., 1906
'Observations on Botany of Natal' (J. Chapman *Travels in Interior of S. Africa* 1868, appendix). *Catalogo Nominale della Piante Vascolari* (with E. Weiss) 1891.
P. A. Saccardo *La Botanica in Italia* Pt. 1, 1895, 17.

ARMITAGE, Miss Eleanora (1865–1961)
b. Dadnor 11 Dec. 1865 *d.* 24 Oct. 1961
Of Bridstow, Ross. Studied Herefordshire flora. Specialised in *Iris*. Contrib. to *J. R. Hort. Soc.*
Trans. Woolhope Nat. Field Club v.34, 1954, 257. *Trans. Br. Bryol. Soc.* 1962, 338–40 portr. *Rev. Bryol. Lichenol.* 1963, 296–97 portr.
Plants at Bristol and Manchester Universities.

ARMITAGE, Elijah (*fl.* 1820s)
Gardener, nurseryman and seedsman, Kirkburton, Yorks.

ARMITAGE, James (–1834/5)
Of Birmingham. Treasurer and one of founders of Birmingham Botanical and Horticultural Society, 1830.
Gdnrs Mag. 1835, 525. *Floral Cabinet* v.3, 1840, 81–82.
Lathyrus armitageanus Westc.

ARMITAGE, Joseph (1853–1897)
d. at sea 23 Oct 1897
Oxford, 1876. MRCS. FLS 1881. Physician, St. Bartholomew's Hospital. Lived at Emu Bay, Tasmania, 1881–97. Interested in botany and geology.
Proc. Linn. Soc. 1897–98, 33–34.

ARMSTRONG, Miss (*fl.* 1860s)
Collected lichens in Natal, 1864. Also collected plants in New Zealand, 1867.
Bothalia 1950, 29.
Plants at Kew.

ARMSTRONG, Sir Alexander (1818–1899)
d. 5 July 1899
Educ. Dublin and Edinburgh Universities. FRS. Medical Dept., Royal Navy; Director, 1869–80. Collected plants in Banks Island, New Hebrides; Port Essington, N. Australia, 1842.
Nature v.60, 1899, 257. *Who was Who, 1897–1916* 23–24.
Plants at Kew.

ARMSTRONG, Edward Frankland (1878–1945)
d. 14 Dec. 1945
DSc London 1905. FRS 1920. Son of H. E. Armstrong (1848–1937). Chemist. Contrib. to *Ann. Bot.*, 1911.
Who was Who, 1941–1950 32.

ARMSTRONG, Henry Edward (1848–1937)
b. Lewisham, London 6 May 1848 *d.* 13 July 1937
FRS 1876. Lecturer in Chemistry, London Institution, City and Guilds College, etc. Researched in agricultural chemistry. Contrib. to *Ann. Bot.*
Nature v.140, 1937, 140–42. *Chronica Botanica* 1938, 262 portr. *Who was Who, 1929–1940* 35.

ARMSTRONG, John (*fl.* 1780s–1790s)
Nurseryman, North Warnborough, Hants.

ARMSTRONG, John (–1847)
d. Coepang, Timor 21 Jan. 1847
Of Belize, Honduras. Established Government Garden at Port Essington, N. Australia, 1838. Kew collector in Timor, 1840–45.
J. D. Hooker *Fl. Tasmaniae* v.1, 1859, cxvii. *Rep. Austral. Assoc. Advancement Sci.* 1907, 199.
Plants and letters at Kew and BM(NH).
Eugenia armstrongii Benth.

ARMSTRONG, Thomas (-1944)
d. 8 July 1944
Founded Armstrong and Brown, Tunbridge Wells, Kent, orchid nursery.
Orchid Rev. 1944, 124. *Amer. Orchid Soc. Bull.* 1958, 736–37.

ARMYTAGE-MOORE, Hugh (*c.* 1873–1954)
d. 4 Dec. 1954
VMH 1942. Created garden at Rowallane, County Down.
Gdnrs Chron. 1954 ii 248; 1955 i 18.

ARNELL, Jonah (*fl.* 1790s)
Seedsman, Winchester, Hants.

ARNOLD, Charles (1818–1883)
b. Bedfordshire 1818 *d.* Paris, Ontario 1883
Plant breeder. Went to Paris, Canada in 1833 where he established a nursery.
L. H. Bailey *Standard Cyclop. Hort.* v.2, 1939, 1564.
Portr. at Hunt Library.

ARNOLD, Rev. Frederick Henry (1831–1906)
b. Petworth, Sussex 18 Feb. 1831 *d.* Emsworth, Sussex 4 May 1906
BA Dublin 1859. LLD 1892. *Fl. Sussex* 1887; ed. 2 1907. Contrib. Sussex botany to *Victoria County Hist. Sussex* v.1, 1905, 41–67.
J. Bot. 1906, 135–36, 287–88; 1907, 287–88. A. H. Wolley-Dod *Fl. Sussex* 1937, xlv–xlvii. *R.S.C.* v.13, 159.
Plants at Chichester Museum and Manchester University. MSS. at BM(NH).

ARNOLD, Joseph (1782–1818)
b. Beccles, Suffolk 28 Dec. 1782 *d.* Padang, Sumatra 26 July 1818
MD Edinburgh 1807. FLS 1815. Surgeon, Royal Navy, 1808–16. To Botany Bay, 1815, in charge of convict ship. Naturalist under Raffles, 1818, upon recommendation by Sir J. Banks. Discovered *Rafflesia arnoldi* in Sumatra. Java journal, 1815 published in *J. Malayan Branch R. Asiatic Soc.* v.46 (1), 1973, 1–92 portr.
Flora 1821, 637–41. *Trans. Linn. Soc.* v.13, 1822, 201–5. D. Turner *Mem. of Joseph Arnold* 1849. *Beccles and Bungay Weekly News* 10 and 17 Dec. 1861. D. C. Boulger *Life of Sir Stamford Raffles* 1899, 282–83 portr. *Br. Med. J.* 2 Jan. 1915, 27–28. *D.N.B.* v.2, 110. G. P. Whitley *Some Early Nat. and Collectors in Australia* 1933, 15–18. *J. Proc. R. Austral. Hist. Soc.* v.19, 1934, 303–5. *Fl. Malesiana* v.1, 1950, 23–24. *Austral. Dict. Biogr.* v.1, 1966, 29–30. *J. Soc. Bibl. Nat. Hist.* 1973, 309–72 portr.
Plants at BM(NH). MS. journals, 1809–15, and letters at Mitchell Library, Sydney. Self portr. miniature at Kew. Portr. at Hunt Library.
Arnoldia Blume.

ARNOLD, Ralph Edward (*c.* 1891–1962)
d. Cirencester, Glos 23 Oct. 1962
Gardener at Bathurst Estate, Kiftsgate Court and Moonwood Estate. Employed by Cirencester U.D.C., 1930. Contrib. to *Wilts and Glos Standard* and *Orchid Rev.*
Orchid Rev. 1962, 394.

ARNOLD, Thomas (1742–1816)
b. Leicester 1742
Doctor in Leicester. Owner of private lunatic asylum. Botanist.
A. R. Horwood and C. W. F. Noel *Fl. Leicestershire.* 1933, cxcvi.

ARNOT, D. B. (-1942)
d. Java 1942
Forest Dept., Malay Peninsula, 1925. Collected plants.
Fl. Malesiana v.1, 1950, 24.
Plants at Kuala Lumpur.

ARNOT, Robert (*fl.* 1790s)
Gardener and seedsman, Cowbridge, Glam.

ARNOTT, George Arnott Walker (1799–1868)
b. Edinburgh 6 Feb. 1799 *d.* Glasgow 17 May or June 1868
MA Edinburgh 1818. LLD Aberdeen 1837. FLS 1825. Regius Professor of Botany, Glasgow, 1845. *Analytical Botanical Tables* 1842? *Prodromus Florae Peninsulae Indiae Orientalis* v.1, 1834 (with R. Wight). *Botany of Captain Beechey's Voyage* 1830–41 (with W. J. Hooker). *British Fl.* 1850, 1855, 1860 (with W. J. Hooker).
Gdnrs Chron. 1868, 662, 683. *J. Bot.* 1868, 223–24. *Proc. Linn. Soc.* 1868–69, ci–cii. *Trans. Bot. Soc. Edinburgh* v.9, 1868, 414–26. *D.N.B.* v.2, 120. *Meded. Bot. Mus. Herb. Rijksunivers. Utrecht* no. 283, 1968, 36–49. *R.S.C.* v.1, 98; v.6, 568; v.12, 23.
Herb. at Glasgow. Mosses at BM(NH). Letters at BM(NH), Kew and Linnean Society. Crayon portr. by Sir D. Macnee at Kew. Portr. at Hunt Library.
Arnottia A. Rich.

ARNOTT, Neill (*c.* 1788–1874)
b. Upper Dysart, near Montrose *d.* Regent's Park, London 2 March 1874
Educ. Aberdeen University. Physician in East India Company. Practised in London from 1811. Invented 'Arnott's stove'.
J. Hort. Cottage Gdnr v.51, 1874, 224.

ARNOTT, Robert (*fl.* 1840s)
Nurseryman, Cambrian Nursery, Charlton Kings, near Cheltenham, Glos.

ARNOTT, Samuel (1852–1930)
b. Dumfries 1852 *d.* Maxwelltown, Dumfries 17 Feb. 1930
Horticultural writer. *Book of Bulbs* 1901. *Book of Climbing Plants and Wall Shrubs* 1903. *Gardening in the North* 1909 (with R. P. Brotherton). Contrib. to *Gdnrs Chron.*, *Cassell's Dict. of Popular Gdning* 1902 2 vols., *Black's Gdning Dict.* 1921.
J. Hort. Cottage Gdnr v.43, 1901, 558–59 portr. *Gdnrs Mag.* 1907, 561 portr., 562. *Gdnrs Chron.* 1926 ii 142 portr.; 1930 i 177 portr.

ARSENE, Brother Louis (1875–1959)
b. 5 Aug. 1875 *d.* Josselin, France 25 Jan. 1959
Collected plants in Channel Islands where he spent 30 years in the service of Institut des Frères de l'Instruction Chrétienne.
Proc. Bot. Soc. Br. Isl. v.3, 1959, 360–61.
Plants at Ploermel, France.

ARTHUR, John (1804–1849)
b. Dunkeld, Perthshire 1804 *d.* Melbourne Jan. 1849
Gardener. Emigrated to Australia, 1839. First Superintendent, Botanic Garden, Melbourne, 1846–49.
Victorian Nat. v.25, 1908, 102.

ARTHUR, William (1680–1716)
b. Elie, Fife Sept. 1680 *d.* Rome 1716
MD Utrecht 1701. FRCPE 1714. King's Botanist in Scotland. Professor of Botany and Materia Medica and Keeper of Royal Physick Garden, Edinburgh, 1715.
Trans. Proc. Bot. Soc. Edinburgh 1914–15, 375–404; H. R. Fletcher and W. H. Brown *R. Bot. Gdn Edinburgh* 1970, 20–25.

ARTIS, Edmund Tyrrell (1789–1847)
b. Sweflin, Suffolk 1789 *d.* Doncaster, Yorks 24 Dec. 1847
Antiquarian. Collected fossil plants in Yorkshire and Derbyshire coalfields. *Antediluvian Phytology* 1825.
Quart. J. Geol. Soc. 1849, xxii–xxiii.
Drawings at BM(NH).

ARVIEL, Henry (*fl.* 1280s)
Resided at Bologna. *De Botanica, sive Stirpium Varia Historia* (MS.) 'Varia itinera susceperat' (A. Haller *Bibliotheca Botanica* v.1, 1771, 219).
R. Pulteney *Hist. Biogr. Sketches of Progress of Bot. in England* v.1, 1790, 22.
Arviela Salisb.

ASCHAM, Rev. Anthony (*fl.* 1550s)
MB Cantab 1540. Vicar, Burniston near Bedale, Yorks. *A Little Herbal* 1550.
R. Pulteney *Hist. Biogr. Sketches of Progress of Bot. in England* v.1, 1790, 50–51. *D.N.B.* v.2, 149–50.
Aschamia Salisb.

ASH, Gerald Mortimer (1900–1959)
b. London 1900 *d.* Guildford, Surrey 5 Sept. 1959
Educ. St. Paul's School and Wye Agricultural College. Botanist with interest in *Epilobium*.
Proc. Bot. Soc. Br. Isl. 1960, 106–7.
Herb. at Haslemere Museum; *Epilobium* at BM(NH).

ASHBY, Edwin (1861–1941)
b. Capel, Surrey 1861 *d.* Blackwood, S. Australia 8 Jan. 1941
FLS 1919. Emigrated to S. Australia, 1888. Land and estate agent. Keen naturalist. Published a few papers on S. Australian flora. Vice-President, Royal Society of S. Australia, 1919–21.
Proc. Linn. Soc. 1940–41, 287–88. *Emu* v.40, 1941, 409. *Proc. R. Zool. Soc. N.S.W.* 1941, 45. *Trans. R. Soc. S. Austral.* v.65, 1941, 1–2 portr.

ASHBY, John (1754–1828)
d. Bungay, Suffolk 24 Nov. 1828
Grocer and draper in Bungay. Had herb. of British plants. Contrib. to J. E. Smith's *Fl. Britannica* 1800–4 and D. Turner and L. W. Dillwyn's *Botanist's Guide* v.2, 1805, 547.
Mag. Nat. Hist. v.2, 1829, 120.

ASHBY, Sydney Francis (1874–1954)
b. Rockferry, Cheshire 31 Dec. 1874 *d.* 6 March 1954
BSc Glasgow. Research Fellow, Rothamsted Experimental Station. Microbiologist, Dept. of Agriculture, Jamaica, 1906–10, 1912–21. Professor of Mycology, Imperial College of Tropical Agriculture, Trinidad, 1922–26. Mycologist, Imperial Bureau of Mycology, 1926; Director, 1935–39. Studied *Phytophthora.*
Chronica Botanica 1936, 184 portr. *Nature* v.173, 1954, 802–3.
Fungi at Commonwealth Mycological Institute.

ASHE, John (*fl.* 1780s–1790s)
Nurseryman, Twickenham, Middx.

ASHE, Thomas (*fl.* 1740s–1750s)
Nurseryman, Strawberry Hill, Twickenham, Middx. Supplied plants to A. Pope and Horace Walpole who lived nearby. "Mr. Walpole telling [Ashe] he would have his trees planted irregularly, [Ashe] said 'Yes Sir, I understand; you would have them hang down somewhat poetical.' "
Trans. London Middlesex Archaeol. Soc. v.24, 1973, 194.

ASHE, William (*fl.* 1780s)
Nurseryman, Twickenham, Middx. Successor to T. Ashe and succeeded by J. Ashe.

ASHFIELD, Charles Joseph (*c.* 1817–1877)
b. Norfolk *c.* 1817 *d.* Preston, Lancs 9 Aug. 1877
Schoolmaster at Preston. 'Flora of Preston' (*Trans. Hist. Soc. Lancs and Cheshire* 1858, 143–64). Contrib. to *Botanists' Chron.* and *Phytologist* v.5–v.6, 1861–63. Discovered *Pulmonaria officinalis* in Suffolk.
Trans. Liverpool Bot. Soc. 1909, 58. G. C. Druce *Fl. Buckinghamshire* 1926, ci.
Herb. at Preston; plants and MS 'Fl. of Preston' at Liverpool Museum.

ASHFORD, Frederic F. (*fl.* 1830s)
Gardener to P. L. Brooke at Mere Hall, Knutsford, Cheshire; at Colston Hall, Birmingham, 1834. Contrib. to *Hort. Reg.* 1831–36 and *Floricultural Cabinet* 1833–35.

ASHLEY, R. (1682–1782)
d. Streatham, London 7 June 1782
Gardener to Mr. Stallard.
Gent. Mag. 1782, 310.

ASHLEY, W. H. (*fl.* 1850s)
MD. Of London. Botanised in Devonshire before 1851.
H.R.P.F. Halle. *Letters, Hist. and Bot.* 1851, preface.

ASHMOLE, Elias (1617–1692)
b. Lichfield, Staffs 23 May 1617 *d.* Lambeth, London 26 May 1692
Hon. MD Oxon 1669. FRS 1663. Antiquary and astrologer. Windsor Herald 1660.
A. à Wood *Athenae Oxoniensis* v.3, 1813–20, 354. *Mag. Nat. Hist.* 1837, 272. *Cottage Gdnr* v.4, 1849, 269. *Notes and Queries* 1852, 367, 385. G. C. Druce *Fl. Berkshire* 1897, cix–cxi. E. L. J. Scott *Index to Sloane Manuscripts* 1904, 25. *D.N.B.* v.2, 172–74. *J. Bot.* 1918, 197–202. *Nature* v.99, 1917, 234–35. G. C. Druce *Fl. Oxfordshire* 1927, lxviii–lxix. *Notes R. Soc. London* v.15, 1960, 221–30. M. Allan. *The Tradescants* 1964 *passim* portr. C. H. Josten *Elias Ashmole* 1967 5 vols.
Portr at Ashmolean Museum, Oxford. Portr. at Hunt Library.

ASHTON, Ernest Russell (*c.* 1860–1951)
d. Tunbridge Wells, Kent 11 Nov. 1951
Grew and exhibited orchids. Contrib. floral photographs to *Orchid Rev.*
Orchid Rev. 1952, 15–16.

ASHTON, Squire (1827–1897)
Plants at Werneth Museum, Oldham, Lancs.

ASHWORTH, Dorothy (1908–1944)
b. Prestwich, Lancs 13 Jan. 1908 *d.* 4 Oct. 1944
BSc London 1929. Assistant mycologist, R. Hort. Soc. Gardens, Wisley, 1935–44.
Nature v.154, 1944, 571–72.

ASHWORTH, Elijah (*c.* 1840–1917)
d. 18 Oct. 1917
Of Harefield Hall, Wilmslow, Cheshire. Cultivated orchids.
Orchid Rev. 1917, 251.
Cypripedium ashworthiae X Hort.

ASHWORTH, Richard (*c.* 1848–1928)
d. Manchester 30 June 1928
Cultivated and exhibited orchids. His collection described in *Orchid World* 1913, 175–79 portr.
Orchid Rev. 1928, 247–48.

ASKEW, John (*fl.* 1790s)
Corn dealer, gardener and seedsman, Malden, Essex.

ASKEW, W. F. (*c.* 1858–1949)
d. Grange, Borrowdale, Cumberland 1949
Nurseryman. Fern-grower.
Br. Fern Gaz. 1949, 262.

ASLET, James? (*fl.* 1760s–1770s)
Nurseryman, Isleworth, Middx.

ASTLEY, Francis Duckenfield (*fl.* 1790s)
Of Duckinfield Hall near Aston, Cheshire. Member of Manchester Agricultural Society. *A Few Minutes Advice to Gentlemen of Landed Property* 1797. *Hints to Planters* 1797.
J. B. L. Warren *Fl. Cheshire* 1899, lxxvii.

ASTLEY-BELL, H. (–1937)
d. Preston, Lancs 14 Dec. 1937
Amateur orchid-grower. President, Manchester and North of England Orchid Society.
Orchid Rev. 1937, 31–32.

ASTON, Bernard Cracroft (1871–1951)
b. Beckenham, Kent 1871 *d.* New Zealand 31 May 1951
Chief Agricultural Chemist, New Zealand. President, New Zealand Forest and Bird Protection Society. Contrib. to *Trans. N. Z. Inst.*
T. F. Cheeseman *Manual of New Zealand Fl.* 1906, *passim*. *Pharm. J.* 1951, 487. *Who was Who, 1951–1960* 42.
Plants at Dominion Museum, New Zealand.

ATCHLEY, Shirley Clifford (*c.* 1871–1936)
d. Mt. Kyllene, Greece 20 June 1936
First Secretary, British Legation, Athens. Collected plants in Balkans, 1930–36 and in Portugal, 1933–35. *Wild Flowers of Attica* 1938 (ed. by W. B. Turrill).
Balkan Herald July 1936. *Gdnrs Chron.*

1936 i 425. *Kew Bull.* 1936, 336–37; 1939, 45–46. *Times* 22 June 1936. *Chronica Botanica* 1938, 152. *J. R. Hort. Soc.* 1939, 149.
Plants at Kew and BM(NH).

ATHERSTONE, William Guybon (1814–1898)
b. Nottingham 27 May 1814 *d.* Grahamstown, S. Africa 26 June 1898
MD Heidelburg 1839. FRCS 1863. FGS 1864. Madras Civil Service, 1838. Surgeon, Grahamstown, 1839–98.
Kew Bull. 1904, 4. F. Boase *Modern English Biogr., Supplt.* v.1, 1908, 190–91. *S. African Med. J.* v.4, 256–58. *Dict. S. African Biogr.* v.1, 1968. *R.S.C.* v.1, 109.
African plants and letters at Kew.
Atherstonea Pappe.

ATKIN, George (–*c.* 1862)
MD. Of Hull, Yorks. Had herb. which was sold 1 July 1862.

ATKIN, William (*fl.* 1790s)
Gardener and seedsman, Alford, Lincs.

ATKINS, Anna (*née* **Children**) (1797–1871)
b. Tonbridge, Kent 1797 *d.* Halstead, Kent 9 June 1871
Daughter of J. G. Children of BM. *Photographs of British Algae* 1843–59; these are reproduced by the cyanotype or blueprint process.
J. Photography 1889, 702–3, 787. *Proc. Philos. Soc. Glasgow* v.21, 1889–90, 155–57. F. Boase *Modern English Biogr.* v.1, 1908, 192.
British plants at BM(NH).

ATKINS, James (*c.* 1802–1884)
d. Painswick, Glos 2 April 1884
Nurseryman, Kingsthorpe Road, Northampton. In partnership with Jeyes from *c.* 1841. Specialised in alpines. Contrib. note on fungus on pear tree roots to *Gdnrs Chron.* 1872, 40.
J. Hort. Cottage Gdnr v.8, 1884, 304.

ATKINS, Sarah *see* Wilson, L. S.

ATKINS, William Ringrose Golston (1884–1959)
b. 4 Sept. 1884 *d.* 4 April 1959
Educ. Trinity College, Dublin. FRS 1925. Assistant to Professor of Botany, Trinity College, Dublin, 1911–20. Indigo research botanist, Dept. of Agriculture, India, 1920. Head of General Physiology, Marine Biological Association Laboratory, 1921–55. *Some Recent Researches in Plant Physiology* 1916.
Who was Who, 1951–1960 45.

ATKINSON, Caleb (*fl.* 1790s)
Nurseryman, Keswick, Cumberland.

ATKINSON, Caroline Louisa Waring *see* Calvert, C. L. W.

ATKINSON, Edwin Felix Thomas (1840–1890)
b. Tipperary 6 Sept. 1840 *d.* Calcutta 18 Sept. 1890
BA. Entomologist. Indian Civil Service, 1862–90. Had herb. *Himalayan Districts of the N.W. Provinces* v.1, 1882 (*N.W. Provinces Gazetteer* v.10, 1882. botany on pp. 299–946 with list by G. King and W. Watson). *Economic Products of the N.W. Provinces* 1876–81.
C. E. Buckland *Dict. Indian Biogr.* 1906, 18–19. *Insect Life* v.3, 303–4. *Entomologists Mon. Mag.* v.26, 329.

ATKINSON, Gerald (1893–1971)
b. Kingston-upon-Hull, Yorks 26 Aug. 1893 *d.* Cornwall 1 July 1971
Studied at Hull College of Art. ALS 1955. Botanical artist and photographer, Kew Gardens, 1922–59. Contrib. line-drawings to *Hooker's Icones Plantarum* and *Kew Bull.* and a few colour plates to *Bot. Mag.*
Kew Bull. 1922, 349. *Hull Daily Mail* July 1971. *J. Kew Guild* 1972, 59–60 portr.
Portr. at Hunt Library.

ATKINSON, Henry (*c.* 1872–1954)
d. 24 Dec. 1954
FLS 1907. Engineer. *Life Magnificent* 1941
Proc. Linn. Soc. v.166, 1956, 45.

ATKINSON, Rev. Henry Dresser (*fl.* 1860s–1890s)
Educ. Cambridge and Trinity College, Dublin. BA 1864. Ordained priest, 1866. To Tasmania, *c.* 1868. Vicar, Stanley, Tasmania, 1877–90. Rural Dean, Evandale until 1895. Studied local flora.
Rec. Queen Victoria Mus., Launceston no. 21, 10.

ATKINSON, John (1787–1828)
b. Leeds, Yorks 29 May 1787 *d.* Leeds 3 Oct. 1828
FLS 1812. Surgeon. Leeds plants listed in T. D. Whitaker's *Loidis and Elmete* 1816. 'Geographical Distribution of Plants in Yorkshire' (*Mem. Wernerian Nat. Hist. Soc.* v.5, 1824, 277–86).
Leeds Worthies, 311–13. R. Thoresby *Ducatus Leodiensis*; ed. 2, 1816 by T. D. Whitaker, 76.
Oil painting portr. by G. Richmond.

ATKINSON, Richard (*c.* 1656–1746)
Gardener of Southwark, London; "a batchelor worth £30000."
Gent. Mag. 1746, 328.

ATKINSON, Thomas Witlam (1799–1861)
b. Cawthorne, Yorks 6 March 1799 *d.* Lower Walmer, Kent 13 Aug. 1861
Architect, artist and traveller. *Travels in Regions of Upper and Lower Amoor* 1860 (includes 3 plant lists).
D.N.B. Supplt. 1 v.1, 84–85.

ATKINSON, William (1765–1821)
b. Dalton-in-Furness, Lancs 3 May 1765 *d.* Dalton 8 Dec. 1821
Solicitor of Dalton. Contrib. to W. Withering's *Botanical Arrangement* ed. 3, 1796. List of plants in T. West's *Antiquities of Furness* 1805.
Trans. Liverpool Bot. Soc. 1909, 58.

ATKINSON, William (–1933)
d. 7 June 1933
VMH 1928. Managing Director, Messrs. Fisher, Son and Sibrary, Sheffield, Yorks.
Gdnrs Chron. 1929 i 174 portr.; 1933 i 426 portr.

ATKINSON, William Sackston
(–1878/9)
Entomologist. Collected plants in Kashmir. Many of his specimens reached Kew through C. B. Clarke.
Pakistan J. For. 1967, 339.

ATTHEY, Thomas (1814–1880)
b. Kenton, Northumberland 1814 *d.* Gosforth, Northumberland April 1880
ALS 1875. Collected *Diatomaceae* and plants of coal measures. Papers on diatoms in *Ann. Mag. Nat. Hist.*
J. Bot. 1880, 224. *Trans. Northumberland Durham Nat. Hist. Soc.* v.8, 1884, 88–90. *R.S.C.* v.1, 111; v.7, 55; v.9, 80.
Carboniferous fossils at Newcastle Museum.
Attheya T. West.

ATWOOD, Martha Maria
(*fl.* 1850s–1860s)
d. Worcester?
Of Clifton, Bristol and Bath. Bryologist and lichenologist. Contrib. to E. H. Swete's *Fl. Bristoliensis* 1854, W. A. Leighton's *Lichen-flora* 1879 and *Phytologist*.
J. W. White *Fl. Bristol* 1912, 96–97. H. J. Riddelsdell *Fl. Gloucestershire* 1948, cxxix.
Letters in W. Wilson's correspondence at BM(NH).

AUBREY, John (1626–1697)
b. Easton Piercy, Kington St. Michael, Wilts 12 March 1626 *d.* Oxford June 1697
Educ. Oxford. FRS 1663. Antiquarian. Nephew to Henry Lyte. Suggestion made that the John Aubrey referred to by Dillenius in J. Ray's *Synopsis* ed. 3 1724, 131, 439, 445 is another person. (H. N. Clokie *Account of Herb. Dept. Bot. Oxford* 1964, 122–23).
J. Britton *Mem. of John Aubrey* 1845. A. Powell *John Aubrey and his Friends* 1948. D. Grose *Fl. Wiltshire* 1957, 30–33.

AUDLEY, James Aloysius (1859–1938)
b. Liverpool 1 Oct. 1859 *d.* Stoke, Staffs 15 July 1938
BSc London. Ceramics chemist, Hanley. Botanist.
Trans. N. Staffordshire Field Club 1938–39, 69–71 portr.

AUGHTIE, Robert (1823–1901)
Gardener to Duke of Devonshire at Chiswick and Chatsworth. Head gardener, Smethwick, Staffs.
Gdnrs Chron. 1963 i 430.

AUSTEN, George (–1876)
d. Tresco, Isles of Scilly 1876
Gardener to Augustus Smith, Tresco. Raised melon, 'Austen's Incomparable'.
Gdnrs Chron. 1876 ii 597.

AUSTEN, Henry Haversham Godwin-
see Godwin-Austen, H. H.

AUSTEN, Ralph (–1676)
b. Staffordshire *d.* Oxford Oct. 1676
Proctor of Oxford University, 1630. "Practised in the Art of Planting." Writer on gardening. *Treatise of Fruit-trees* 1653; ed. 2 1657. *Observations on Some Parts of Sir Francis Bacon's Naturall History as it Concernes Fruit-trees, Fruits and Flowers* 1658. *A Dialogue...betweene the Husbandman and Fruit-trees in his Nurseries* 1676.
Cottage Gdnr. v.7, 1852, 363–64, 391. *D.N.B.* v.2, 260–61. *Gdnrs Chron.* 1922 i 90.

AUSTIN, Benjamin James (1829–1912)
b. Horsleydown, London 5 April 1829 *d.* Reading, Berks 2 June 1912
FLS 1892. Science teacher, Reading, 1871. Lecturer, Reading University College, 1892; Emeritus Professor of Botany, 1907.
Nature v.89, 1912, 352. *Proc. Linn. Soc.* 1912–13, 51–52.
Portr. at Reading University.

AUSTIN, Hugh (*c.* 1849–1894)
d. Cathcart, Glasgow 1894
Partner in nursery firm of Austin and McAslan, Glasgow.
J. Hort. Cottage Gdnr. v.29, 1894, 9.

AUSTIN, James (1776–1849)
b. Craigton, Milngavie, Dunbartonshire 1776
Employed by Ronalds's nursery at Brentford and Royal Gardens, Kew. Returned to Austin and McAslan nursery, Glasgow about 1800; admitted as partner, 1812.
Gdnrs Chron. 1917 ii 147–48 portr.

AUSTIN, James (*fl.* 1840s)
Nurseryman and seedsman, Burnham, Bucks.

AUSTIN, Richard (*fl.* 1830s)
Nurseryman and seedsman, Burnham, Bucks.

AUSTIN, Robert (*c.* 1754–1830)
b. Craigton, Milngavie, Dunbartonshire 1754 *d.* Glasgow 14 March 1830
Gardener under W. Aiton at Royal Gardens, Kew. Foreman in J. McAslan's nursery, Glasgow; made partner, 1782. Nursery moved from Tron Gate to Little Govan in 1798 and to Coplawhill in 1828.
Gdnrs Mag. 1830, 384. *Gdnrs Chron.* 1917 ii 147–48 portr.

AUSTIN, Robert (1826/7–1905)
b. Essex 1826/7 *d.* Thornborough, N. Queensland Feb. 1905
Emigrated to W. Australia, *c.* 1840. Surveyor, Lands Dept., W. Australia, 1846. Made several explorations and collected plants. Surveyor, Queensland, 1860. *Journal of Assistant Surveyor R. Austin* 1855. 'Report of Assistant Surveyor R. Austin of an Expedition to explore Interior of W. Australia' (*J. R. Geogr. Soc.* 1856, 235–74).
J. E. T. Woods *Hist. of Discovery and Exploration of Australia* v.2, 1865, 212–27. *J. Proc. R. Soc. N.S.W.* 1921, 151–52.
MS. field books at Dept. of Lands, Brisbane.

AUSTIN, William (*fl.* 1850s–1870s)
Nurseryman at Partick, Glasgow in firm of Brown and Austin. Partner in nursery of Austin and McAslan, Glasgow, 1859.
Gdnrs Chron. 1917 ii 147–48 portr.

AUSTON, Edward (*c.* 1738–1806)
d. 25 Jan. 1806
Nurseryman, Land Lane, East Hill, Colchester, Essex.

AUSTON, Edward (*c.* 1765–1820)
d. 16 Dec. 1820
Son of E. Auston (*c.* 1738–1806). Nurseryman, Land Lane, East Hill, Colchester, Essex.

AUSTON, Edward (1796–1877)
d. 9 Feb. 1877
Nurseryman, Land Lane, East Hill, Colchester, Essex.

AUSTON, George Edward (–1914)
Son of E. Auston (1796–1877). Nurseryman, Colchester, Essex. Property sold by Edward and George Auston in 1922.

AUTON, William James (1875–1931)
b. Compton Chamberlayne, Wilts 1875 *d.* Botley, Hants 1 April 1931
Kew gardener, 1895–97. Gardener, Pyrford Court near Woking, Surrey. Nurseryman, Botley, Hants, 1929. Contrib. to *Gdnrs Chron.*
Gdnrs Chron. 1925 i 4 portr.; 1926 i 5 portr.; 1931 i 323. *J. Kew Guild* 1931, 74.

AVEBURY, John 1st Baron *see* Lubbock, J.

AVELING, Edward Bibbins (1851–1898)
b. Stoke Newington, London 1851 *d.* Sydenham, London 2 Aug. 1898
DSc London 1876. *Botanical Tables* 1874. *Introduction to Study of Botany* 1891; ed. 2 1897.

AVERY, Charles (1880–1960)
d. Aug. 1960
Of Fulham, London. Botanist with interest in *Rubi.*
Proc. Bot. Soc. Br. Isl. 1961, 351.
Herb. at South London Botanical Institute.
Rubus averyanus W. C. R. Watson.

AXFORD, Walter Godfrey (*c.* 1861–1942)
d. Windsor, Berks 23 Feb. 1942
FLS 1893. Rear-Admiral. Botanist. Active member of Haslemere Natural History Society.
Proc. Linn. Soc. 1942–43, 293.

AXTON, A. (*fl.* 1820s)
Seedsman and florist, 19 Chapel Street, Lisson Grove, London.

AYLESFORD, Countess of *see* Finch, L. *Countess of Aylesford*

AYLING, Edmund (1841–1931)
b. Privett, Hants 1841 *d.* Nov. 1931
Gardener to A. J. Hollington at Forty Hill, Enfield. Hybridised orchids.
Gdnrs Chron. 1931 ii 480 portr. *Orchid Rev.* 1932, 23.
Cypripedium aylingii (*ciliolare X niveum*).

AYLMER, Gerald Percy Vivian (1856–before 1954)
Army Major. Collected plants in Sierra Leone, 1914–19. Forestry Section, Sudan, 1920; retired 1937. Collected plants in Sudan, 1922–34.
Ann. Rep. Dept. Agric. and Forests, Sudan 1937, 26.
Plants at Kew.

AYRES, Philip Burnard (1813–1863)
b. Thame, Oxfordshire 12 Dec. 1813 *d.* Port Louis, Mauritius 23 April 1863
MD London 1841. Pupil of J. Lindley. Physician, Islington Dispensary 1851. Superintendent of Quarantine, Mauritius, 1856–63. Intended writing a flora of Mauritius (MS. at Kew). Cryptogamist. *Mycologia*

Britannica 1845 (exsiccatae). Contrib. to *Phytologist*.

J. Bot. 1863, 224; 1865, 191. *Lancet* 1863 i 707. J. G. Baker *Fl. Mauritius and Seychelles* 1877, 10. F. Boase *Modern English Biogr.* v.1, 1892, 114. G. C. Druce *Fl. Buckinghamshire* 1926, c. G. C. Druce *Fl. Oxfordshire* 1927, cxiii. *R.S.C.* v.1. 129.

Herb., MSS. and letters at Kew. Plants at Oxford.

AYRES, William Port (1815–1875)
d. Forest Hill, London 14 Jan. 1875

Nurseryman, Blackheath, specialising in pelargoniums. Moved to Nottingham. Gardening editor, *Nottingham Guardian*. Patented hothouses. *Cultivation of the Cucumber in Pots* 1841. Co-editor with T. Moore of *Gdnrs Mag. Bot., Hort., Floriculture and Nat. Sci.*, 1850–51.

Florist and Pomologist 1875, 48. *Gdnrs Chron.* 1875 i 119.

AYREY, James (*fl.* 1690s–1700s)

London merchant. Friend of A. Buddle, J. Petiver, L. Plukenet, etc. Sent plants to Petiver.

E. J. L. Scott *Index to Sloane Manuscripts* 1904, 32. J. E. Dandy *Sloane Herb.* 1958, 83–84.

AYRTON, Matilda (*née* **Chaplin**)
(1846–1883)
b. Honfleur, France 1846 *d.* London 19 July 1883

MD Paris 1879. Lived for some years in Japan. 'Plants used in New Year Celebrations by the Japanese' (*Trans. Bot. Soc. Edinburgh* v.13, 1879, xiv–xvi).

Nature v.28, 1883, 306. *D.N.B.* v.2, 292–93.

BABBAGE, Benjamin Herschel (1815–1878)

Engineer. Emigrated to S. Australia, 1851. Chief Engineer of City and Port Adelaide railway, 1853. Led two expeditions in S. Australia, 1856 and 1858, during which he collected plants which he gave to F. von Mueller.

J. E. T. Wood *Hist. of Discovery and Exploration of Australia* v.2, 1865, 260–79. *South Austral. Land Exploration* v.1, 1922, 4–5, 11–27. *Austral. Encyclop.* v.1, 1962, 384–85.

Plants at Kew and Melbourne. Portr. at Hunt Library.

Babbagia F. Muell.

BABINGTON, Benjamin Guy (1794–1866)
b. Guy's Hospital, London 1794 *d.* London 8 April 1866

MD Cambridge 1830. FRCP 1831. FRS. Midshipman, Royal Navy. Indian Civil Service. Collected plants in Mauritius, 1811–12 (*Trans. Medico-Botanical Soc. London* 1829, appendix 11). Physician, Guy's Hospital, 1840–58.

Proc. R. Soc. v.16, 1867, i–ii. *D.N.B.* v.2, 311. *R.S.C.* v.1, 136.

BABINGTON, Charles Cardale
(1808–1895)
b. Ludlow, Shropshire 23 Nov. 1808 *d.* Cambridge 22 July 1895

BA Cantab 1830. FRS 1851. FLS 1830. Professor of Botany, Cambridge, 1861. *Fl. Bathoniensis* 1834; Supplement 1839. *Primitiae Fl. Sarnicae* 1839. *Manual of British Botany* 1843; ed. 10 by A. J. Willmott 1922. *Fl. Cambridgeshire* 1860. *British Rubi* 1869.

Portr. of Men of Eminence in Lit., Sci. and Art v.3, 1865, 51–52 portr. *Cambridge Chronicle* 30 Aug. 1895. *J. Bot.* 1895, 257–66. *Nature* v.52, 1895, 371–72. *Proc. Linn. Soc.* 1895–96, 30–32. *Proc. R. Soc.* v.59, 1896, viii. *Memorials, Journal and Bot. Correspondence of C. C. Babington* 1897 portr. *D.N.B. Supplt. 1.* v.1, 90–92. J. W. White *Fl. Bristol* 1912, 91–93. C. C. Gillispie *Dict. Sci. Biogr.* v.1, 1970, 358–59. *R.S.C.* v.1, 136; v.7, 62; v.9, 91; v.12, 33; v.13, 220.

Herb. at Cambridge. Plants at BM(NH) and Kew. Letters at Kew. Portr. at St. John's College, Cambridge and Hunt Library.

Babingtonia Lindl.

BABINGTON, Rev. Churchill (1821–1889)
b. Roecliffe, Leics 11 March 1821 *d.* Cockfield, Suffolk 12 Jan. 1889

MA Cantab 1846. DD 1879. FLS 1853. Professor of Archaeology, Cambridge, 1865. Rector, Cockfield, 1866. Cousin of C. C. Babington (1808–1895). Lichenologist. Contrib. lichens to J. D. Hooker *Fl. Tasmaniae* 1844–60. Contrib. to W. M. Hind *Fl. Suffolk* 1889. Corresponded with H. C. Watson.

Ann. Bot. v.3, 1889–90, 449. *Gdnrs Chron.* 1889 i 89. *J. Bot.* 1889, 110–11. *Proc. Berwickshire Nat. Club* 1898, 313–21. *Trans. Bot. Soc. Edinburgh* v.17, 1889, 519–21. *D.N.B. Supplt. 1* v.1, 92. *R.S.C.* v.1, 139; v.13, 221.

Lichen herb. at Cambridge. Letters at Kew. Portr. at Hunt Library.

Strigula babingtonii Berkeley.

BABINGTON, Rev. Joseph (1768–1826)
b. Rothley Temple, Leics 2 Jan. 1768 *d.* Bath, Somerset 15 Dec. 1826

BA Cantab 1791. MB Oxon 1795. Father of C. C. Babington (1808–1895). Physician at Ludlow. Rector, Broughton Gifford. Contrib. Shropshire plants to J. Plymley's *General View of Agriculture of Shropshire* 1803, 180–211. Contrib. to J. Sowerby and

J. E. Smith's *English Bot.* 450, 740, 887. Plants at Cambridge.

BACK, Sir George (1796–1878)
b. Stockport, Cheshire 6 Nov. 1796 *d.* London 23 June 1878
Admiral, Royal Navy 1857. FRS 1847. Arctic explorer. Led exploring expedition to Great Fish River, 1833–35; collected plants in Hudson Bay, now at BM(NH). *Narrative of Arctic Land Expedition...1833–35* 1836.
D.N.B. v.2, 318–20.

BACKHOUSE, Edward (*fl.* 1890s)
Local plants at Sunderland Museum.

BACKHOUSE, James (1794–1869)
b. Darlington, Durham 8 July 1794 *d.* York 20 Jan. 1869
With his brother, Thomas, bought Telford's nursery, York, 1816. Sailed to Australia as Quaker missionary in 1831. Returned to England and nursery at York, 1841. *Extracts from Letters...* 1838–41. *Narrative of Visit to Australian Colonies* 1843. *Narrative of Visit to Mauritius and South Africa* 1844. 'Indigenous Plants of Van Diemen's Land' (*Ross's Hobart Town Almanack and Van Diemen's Land Annual 1835*, 61–114).
J. D. Hooker *Fl. Tasmaniae* v.1, 1859, cxxv-cxxvi. *Gdnrs Chron.* 1869, 136. *J. Bot.* 1869, 51–58 portr. *J. Hort. Cottage Gdnr* v.41, 1869, 82. S. Backhouse *Mem. of James Backhouse* 1870. *Ann. Monitor* 1870, 6–14. Friends' Institute *Biogr. Cat.* 1888, 29–34. P. Mennell *Dict. Austral. Biogr.* 1892. J. Foster *Descendants of John Backhouse* 1894. *Rep. Austral. Assoc. Advancement Sci.* 1907, 172–73. *J. Proc. R. Soc. N.S.W.* 1908, 62–63. *Austral. Dict. Biogr.* v.1, 1966, 45–46. *R.S.C.* v.1, 147; v.6, 573; v.7, 65; v.13, 227.
Plants at Edinburgh. MS. Flora of N.S.W and letters at Kew. MS. Journals and MS. list of Norfolk Island plants 1835 at Mitchell Library, Sydney.
Backhousia Hook. & Harv.

BACKHOUSE, James (1825–1890)
b. York 22 Oct. 1825 *d.* York 31 Aug. 1890
FLS 1885. Son of James Backhouse (1794–1869) with whom he botanised in Teesdale, Norway 1851; Ireland, 1854; and Scotland, 1859. Discovered *Viola arenaria* in Teesdale, 1862. *Monograph of British Hieracia* 1856. Contrib. to *Phytologist.*
Garden v.38, 1890, xii, 239 portr. *Gdnrs Chron.* 1890 ii 310, 332, 661. *J. Bot.* 1890, 353–56 portr. *Proc. Linn. Soc.* 1890–91, 20–21. *R.S.C.* v.1, 147; v.13, 227.
Herb. at York. Plants at Cambridge and Kew. Letters at Kew. Portr. at Hunt Library.

BACKHOUSE, James (1861–1945)
b. 14 April 1861 *d.* 1 Jan. 1945
Nurseryman, York. Son of James Backhouse (1825–1890).

BACKHOUSE, Nathan (*fl.* 1780s–1800s)
Herb. at BM(NH).

BACKHOUSE, Robert Ormston (*c.* 1854–1940)
d. Sutton Court, Hereford 10 April 1940
With his wife, Mrs. R. O. Backhouse (1857–1921), hybridised narcissi at Sutton Court.
Daffodil Yb. 1933, 30. *Daffodil Tulip Yb.* 1946, 1–2 portr. *Gdnrs Chron.* 1940 i 194.

BACKHOUSE, Thomas (1792–1845)
b. 15 June 1792 *d.* 21 March 1845
Nurseryman, York. Brother of James Backhouse (1794–1869).

BACKHOUSE, William (1779–1844)
b. Darlington, Durham 17 Nov. 1779 *d.* Darlington 9 June 1844
Banker. Cousin of James Backhouse (1794–1869). List of plants in W. H. D. Longstaffe's *History of Darlington* 1854, xciii–xciv. Plant records in J. Sowerby and J. E. Smith's *English Bot.* 1984, 2529, 2922, etc.
Naturalist 1864, 42. *Trans. Northumberland Durham Nat. Hist. Soc.* 1903, 80.
Plants at Manchester University (herb. destroyed by fire in 1864). Letters in Winch correspondence at Linnean Society.

BACKHOUSE, William (1807–1869)
b. Darlington, Durham 1807 *d.* Leeds 1869
Banker of St. John's, Walsingham, County Durham. Cultivated narcissi.
Daffodil Yb. 1933, 25–26 portr.

BACKHOUSE, William Ormston (1885–1962)
b. 20 Feb. 1885 *d.* Sutton Court, Hereford 7 Aug. 1962
BA Cantab. Assistant, Plant Breeding Institute, Cambridge and John Innes Institute. Farmed in Argentina. Raised narcissi.
Gdnrs Chron. 1962 ii 305 portr. *Daffodil Tulip Yb.* 1964, 182–83. M. J. Jefferson-Brown *Daffodils and Narcissi* 1969, 33–34.
Portr. at Hunt Library.

BACON, Alice Sophia *see* Cooke, A. S.

BACON, Francis, 1st Baron Verulam and Viscount St. Albans (1561–1626)
b. Strand, London 22 Jan. 1561 *d.* Highgate, London 9 April 1626
Lord Chancellor, 1619–21. *Sylva Sylvarum* 1627. *Of Gardens* 1625.
D. Clos *La Botanique dans Francis Bacon*

1875. W. C. Hazlitt *Gleanings in Old Garden Literature* 1887, 90–111. *D.N.B.* v.2, 328–60. *Chronica Botanica* 1944, 34–40. *J. R. Hort. Soc.* 1971, 462–66.

BACON, Gertrude *see* Foggitt, G.

BACON, Stephen (1709–1734)
b. Aldbourne? Wilts 1709
Nurseryman, Hoxton, Shoreditch, London. Member of Society of Gardeners. Grew Mark Catesby's American seeds.
J. R. Hort. Soc. 1938, 424.

BACON, Vincent (–1739)
d. 6 April 1739
FRS 1732. Surgeon-apothecary. Of London, afterwards of Grantham. Member of J. Martyn's Botanical Society of London, 1721.
Philos. Trans. R. Soc. v.38, 1735, 287–91. *Gent. Mag.* 1739, 216. G. C. Gorham *Mem. of John and Thomas Martyn* 1830, 19. *Naturalist* 1898, 177–79 *Proc. Bot. Soc. Br. Isl.* 1967, 310.

BACSTROM, Sigismund (*fl.* 1770s–1790s)
MD. Ships' Surgeon. In England from 1770. Accompanied Banks to Iceland, 1772, and engaged for his contemplated second voyage; also employed in Banks's herb., 1773–75.
Notes and Queries v.9, 1866, 238. *J. Bot.* 1911, 92–97.
Letters in Banks's correspondence at Kew.

BADCOCK, Richard (*fl.* 1790s)
Gardener and seedsman, Abingdon, Berks.

BADDELEY, John (1846–1868)
b. at sea, Bay of Bengal 22 Jan. 1846 *d.* Edinburgh 29 Feb. 1868
MB Edinburgh 1867. Member of Botanical Society of Edinburgh.
Trans. Proc. Bot. Soc. Edinburgh v.9, 1868 304–12.

BADHAM, Rev. Charles David (1805–1857)
b. London 27 Aug. 1805 *d.* East Bergholt, Suffolk 14 July 1857
BA Cantab 1826. MD Oxon 1833. Practised in Rome and Paris. Curate, East Bergholt, 1849–55. *Treatise on Esculent Funguses of England* 1847; ed. 2 1863.
D.N.B. v.2, 387–88. G.H. Brown *Lives of Fellows of R. College of Physicians of London 1826–1925* 1955, 9.
Badhamia Berkeley.

BAGNALL, James Eustace (1830–1918)
b. Birmingham 7 Nov. 1830 *d.* Birmingham 3 Sept. 1918
ALS 1885. Specialized in mosses. Contrib. botany to *Victoria County History of Warwickshire* 1904 and *Staffordshire* 1910. *Notes on Sutton Park* 1877. *Handbook of Mosses* 1886. *Fl. Warwickshire* 1891. *Fl. Staffordshire* (*J. Bot.* 1901, Supplt.). Contrib. to *J. Bot* from 1874.
J. Bot. 1918, 354–56. *Rep. Bot. Soc. Exchange Club Br. Isl.* 1918, 349–52. *Kew Bull.* 1901, 5. A. R. Horwood and C. W. F. Noel *Fl. Leicestershire* 1933, ccxix–ccxx. E. S. Edees *Fl. Staffordshire* 1972, 21. D. A. Cadbury et al. *Computer-mapped Fl.... Warwickshire* 1971, 57–58. *R.S.C.* v.7, 68; v.9, 97; v.12, 35.
Herb. at Birmingham.

BAGOT, William, 2nd Baron Bagot (1773–1856)
b. London 11 Sept. 1773 *d.* Blithfield, Staffs 12 Feb. 1856
Educ. Oxford. FLS 1798. DCL Oxon 1834. Contrib. to W. Withering's *Bot. Arrangement* ed. 3 1796.
J. Bot. 1901, Supplt. 71. *D.N.B.* v.2, 400.

BAGSHAWE, Sir Arthur William Garrard (1871–1950)
b. 20 July 1871 *d.* 24 March 1950
Educ. Cambridge. Medical Officer, Uganda 1900–12; Director, Bureau of Hygiene and Tropical Medicine, 1912–35. Collected plants in Uganda.
Kew Bull. 1907, 234. *Who was Who, 1941–1950* 47.
Plants at BM(NH) and Kew.

BAIKIE, William Balfour (1825–1864)
b. Kirkwall, Orkney 27 Aug. 1825 *d.* Sierra Leone 12 Dec. 1864
MD Edinburgh. Surgeon, Royal Navy. Surgeon and naturalist to Niger Expedition, 1854, 1857–59. *Narrative of Exploring Voyage up...the Niger and Tsadda in 1854* 1856.
Hooker's J. Bot. Kew Gdn Misc. 1857, 357–58. S. Crowther and J. C. Taylor *Niger Expedition of 1857–1859* 1859. *J. Bot.* 1865, 71. *Trans. Bot. Soc. Edinburgh* v.8, 1866, 336. D. Oliver et al. *Fl. Tropical Africa* v.1, 1868, 8*. *D.N.B.* v.2, 406–7. *R.S.C.* v.1, 154; v.7, 68; v.12, 36.
Niger plants at Kew. Letters at Kew. Monument in Kirkwall Cathedral. Property sold at Sotheby, 7 July 1866.
Baikiaea Benth.

BAILEY, Rev. Benjamin (1791–1871)
Missionary in Travancore for 40 years. Linguist and botanist.
C. E. Buckland *Dict. Indian Biogr.* 1906, 22.

BAILEY, Charles (1838–1924)
b. Atherstone, Warwickshire 14 June 1838 *d.* St Marychurch, Torquay, Devon 14 Sept. 1924
FLS 1878. Employed by Ralli Brothers,

East India merchants, Manchester. President, Manchester Literary and Philosophical Society, 1901–3. 'Structure... of *Naias graminea* var. *Delilei* Magn' (*J. Bot.* 1884, 305–33). 'Notes on Adventitious Vegetation of the Sand-hills at St. Anne's-on-Sea' (*Mem. Proc. Manchester Lit. Philos. Soc.* v.47, 1902–3, 1–8; v.51, 1906–7, 1–16; v.54, 1909–10, 1–11).

Bot. Soc. Exch. Club Br. Isl. Rep. 1924, 526–34 portr. *Rep. Watson Bot. Exch. Club* 1924–25, 281–82. *Proc. Linn. Soc.* 1924–25, 62–64. *J. Bot.* 1925, 23–25. *N. Western Nat.* 1930, 81–86 portr. *Proc. Manchester Lit. Philos. Soc.* v.69, 1924–25, i–ii. *Notes Manchester Mus.* v.33, 1930, 81–86 portr. Liverpool Museums *Handbook and Guide to Herb. Collections* 1935, 22–23 portr.

Herb. and library at Manchester University. Plants at Belfast, Liverpool, Oxford. Portr. at Hunt Library.

BAILEY, Frederick (1840–1912)
d. Edinburgh 21 Dec. 1912

Royal Engineers, 1859. Indian Forest Service, 1871. Director, Central Forestry School, Dehra Dun until 1884. Inspector-General of Forests. Lecturer in Forestry, Edinburgh University until 1907. President, Royal Scottish Arboricultural Society, 1898.

Indian For. 1913, 206–9 (reprint of *Scotsman*, 23 Dec. 1912). *Garden* 1913, xvi. *Nature* v.90, 1913, 577.

MSS. at India Office Library.

BAILEY, Frederick Manson (1827–1915)
b. Hackney, London 8 March 1827 *d.* Brisbane, Queensland 25 June 1915

FLS 1878. CMG 1911. To S. Australia in 1839 when his father John Bailey (1800–1864) was appointed Govt. Botanist. Partner with his father in nursery in Adelaide. Opened seed store in Brisbane, 1861. Appointed botanist by Queensland Govt. to board investigating plant and animal diseases. Colonial Botanist, Queensland, 1881–1915. Made many plant collecting expeditions, most notable being Bellenden-Ker expedition to N. Queensland, 1889. *Handbook to Ferns of Queensland* 1874. *Synopsis of Queensland Fl.* 1888–90. *Queensland Fl.* 1899–1905. *Comprehensive Catalogue of Queensland Plants* 1900–13.

P. Mennell *Dict. Austral. Biogr.* 1892, 22. *Gdnrs Chron.* 1915 ii 136 portr. *J. Bot.* 1915, 275–77. *Kew Bull.* 1915, 356–57. *Nature* v.96, 1915, 10–11. *Proc. Linn. Soc.* 1915, 55–56. *Proc. R. Soc. Queensland* 1916, 3–10; 1949, 105–14. *Proc. Linn. Soc. N.S.W.* 1916, 7–9; 1921, 152–53. *J. Hist. Soc. Queensland* 1945, 362–83. *Austral. Dict. Biogr.* v.3, 1969, 73–74. F. A. Stafleu *Taxonomic Literature* 1967, 11–12. *R.S.C.* v.9, 98; v.13, 238.

Plants at Brisbane, BM(NH), Kew. MS. diaries at Brisbane. Letters at Mitchell Library, Sydney and BM(NH). Portr. at Hunt Library.

Dendrobium baileyi F. Muell.

BAILEY, Frederick Marshman (1882–1967)

Indian Army. Lieut.-Colonel. Explorer of western China, western and south-eastern Tibet. Discovered *Meconopsis betonicifolia*. Collected plants in Nepal before 1936, Sikkim, 1924, etc. *No Passport to Tibet* 1957.

Plants at BM(NH) and Edinburgh.

BAILEY, Isaac (–*c.* 1793)

Seedsman, at the sign of the Auriculas, Bishopsgate Street, London; succeeded in 1794 by Elizabeth Bailey. *A Catalogue of Seeds, Plants, etc. c.* 1790.

J. Harvey *Early Hort. Cat.* 1973, 1.

BAILEY, John (1800–1864)
b. Hackney, London 1800 *d.* Adelaide 1864

Gardener with Loddiges at Hackney. Colonial Botanist and Curator of Botanic Garden, S. Australia, 1839–41. Nurseryman, Adelaide.

Proc. R. Soc. Queensland v.8, 1890, 25–26. *Proc. Linn. Soc. N.S.W.* 1900, 774, 793. *Proc. Geogr. Soc. S. Australia* v.35, 1933–34, 38a, 41–44. *J. Hist. Soc. Queensland* 1945, 362–68. *Austral. Encyclop.* v.1, 1962, 389.

BAILEY, Katherine Sophia *see* Kane, *Lady* K. S.

BAILEY, Maurice Armand (*c.* 1890–1939)
d. Cambridge 16 Oct. 1939

MA Cantab. Research student, John Innes Horticultural Institution, 1911–15. Senior botanist, Ministry of Agriculture, Egypt, 1919. Director of Agriculture, Sudan, 1931–38. Director, National Institute of Agricultural Botany, Cambridge, 1938–39.

Kew. Bull. 1939, 669–70. *Nature* v.144, 1939, 890–91.

BAILEY, Thomas (*fl.* 1800s)

Seedsman and herbalist, Covent Garden, London.

BAILEY, Thomas (1806–1887)
b. Croydon, Surrey 13 Oct. 1806 *d.* Amersham, Bucks 17 April 1887

Gardener to E. Bouverie at Delapré Abbey, to Duke of Devonshire at Chiswick, 1835 and to T. T. Drake at Shardloes, Amersham, 1838. Raised new varieties of melons, e.g. 'Bailey's Greenflesh', and also vegetables, e.g. 'Bailey's Superb Cabbage'.

Gdnrs Chron. 1874, 527; 1887 i 552 portr.

BAILLIE, Charles Wallace Alexander Cochrane– *see* Cochrane–Baillie, C. W. A.

BAILLIE, Edmund John (1851–1897)
b. Hawarden, Cheshire 4 May 1851 *d.* Chester 18 Oct. 1897
FLS 1883. Seedsman of Chester. 'The City Flora' (*Proc. Chester Soc. Nat. Sci* no. 3, 1878, 67–68).
Gdnrs Mag. 1897, 676. *J. Bot.* 1897, 464. *J. Hort. Cottage Gdnr* v.22, 1897, 7 portr. *Proc. Linn. Soc.* 1897–98, 34.

BAILY, William Hellier (1819–1888)
b. Bristol 7 July 1819 *d.* Rathmines, Dublin 6 Aug. 1888
FLS 1863. FGS. Assistant Curator, Bristol Museum. Paleontologist, Geological Survey of Ireland, 1857–88. Wrote papers on palaeobotany.
Geol. Mag. 1888, 431–32. *Nature* v.38, 1888, 396. *Proc. Linn. Soc.* 1888–89, 47–48. *Quart. J. Geol. Soc.* 1889, 39–41. R. L. Praeger *Some Irish Nat.* 1949, 43–44. *R.S.C.* v.1, 160; v.7, 72; v.9, 101.

BAIN, John (1815–1903)
b. Ireland 9 May 1815 *d.* Holyhead, Anglesey 28 April 1903
ALS 1863. Curator, Trinity College Garden, Dublin, 1862–78. Discovered *Hordeum sylvaticum* in Ireland (*Proc. Dublin Nat. Soc.* v.1, 1853, 45).
Garden v.35, 1889 portr. *Gdnrs Chron.* 1903 i 299–300. *Irish Nat.* 1903, 192–93. *Proc. Linn. Soc.* 1902–3, 26–27.

BAIN, Thomas C. (*fl.* 1870s)
Son of A. G. Bain (1797–1864), "father of S. African geology." Collected *Hoodia bainii* in Cape in 1876. Assisted Sir Henry Barkly in his study of stapeliads.
Bot. Mag. 1878, t.6348. A. White and B. L. Sloane *Stapelieae* v.1, 1937, 110.

BAINBRIDGE, Richard (*fl.* 1840s)
Gardener to Lord Wenlock at Escrick, Yorks. *Guide to the Conservatory* 1842. Prepared Bainbridge's Alkaline Extract for destroying red spider and mildew.

BAINES, Henry (1793–1878)
b. York 15 May 1793 *d.* York 3 April 1878
To Halifax as gardener. Returned to York to work for Messrs Thomas and James Backhouse. Sub-Curator, Yorkshire Museum Gardens, 1829–70. *Fl. of Yorkshire* 1840.
Annual Rep. Yorkshire Philos. Soc. 1878, 17; 1906, 68–69; 1971, 64–65. *Garden* v.13, 1878, 374.
Plants at York Museum. Portr. by T. Banks at York City Art Gallery. Portr. at Hunt Library.

BAINES, John (1787–1838)
b. Horbury, Yorks 1787 *d.* Thornhill, Yorks 1 May 1838
Mathematician. 'In herbis decernendis peritus' (inscription on his tombstone).
D.N.B. v.2, 439.

BAINES, Thomas (1820–1875)
b. King's Lynn, Norfolk 1820 *d.* Durban, Natal 8 May 1875
Artist and explorer. To Cape Colony, 1842. Artist with British army in Kafir war, 1848–51. In N.W. Australia in 1855. On Zambesi Expedition, 1858. *Explorations in South-West Africa* 1864. *Victoria Falls, Zambesi River* 1865. *Gold Regions of South Eastern-Africa* 1877.
Gdnrs Chron. 1874 i 568. *Proc. R. Geogr. Soc.* v.20, 1876, 289–392. *D.N.B.* v.2, 441. J. P. R. Wallis *Thomas Baines of King's Lynn* 1941 portr. J. P. R. Wallis *Northern Goldfield Diaries of Thomas Baines, 1869–1872*, 1946. *Comptes Rendus AETFAT 1960* 1962, 161–63. *Kirkia* 1967, 104 portr. *Dict. S. African Biogr.* v.1, 1968. *Quart. Bull. S. African Library* 1969, 181–90; 1971, 87–99. *See also* Addendum.
Plants and letters at Kew. MSS. at Government Archives, Salisbury, Rhodesia and Royal Geographical Society. Portr. at Kew and Hunt Library.
Aloë bainesii Dyer.

BAINES, Thomas (1823–1895)
b. Claughton Hall, Lancs 9 July 1823 *d.* Palmer's Green, London 2 March 1895
Gardener to S. L. Behrens at Catteral House near Garstang, Lancs, and to H. L. Nicholls at Southgate House, London. *Greenhouse and Stove Plants* 1885.
Gdnrs Chron. 1875 i 180 portr.; 1895 i 307. *Garden* v.47, 1895, 174; v.49, 1896 portr. *Gdnrs Mag.* 1895, 138. *J. Hort. Cottage Gdnr* v.30, 1895, 295 portr.

BAIRD, Rev. Andrew (1800–1845)
b. Eccles, Berwickshire 16 Nov. 1800 *d.* Old Hamstocks, E. Lothian 22 June 1845
Minister of Cockburnspath, Berwickshire.
G. Johnston *Terra Lindisfarnensis* 1853, 110–11. G. Johnston *Fl. Berwick-upon-Tweed* 1829 preface xxiv.
Letters in N. J. Winch correspondence at Linnean Society.

BAIRD, James (*fl.* 1830s)
Collected plants around Buenos Aires and in Uruguay, 1829–30.
Bot. Mag. 1831, t.3113. *Kew Bull.* 1920, 60.
Plants at Kew.

BAIRSTOW, Uriah (1847–1914)
d. Sept. 1914
Studied flora of Halifax, Yorks area.
Naturalist 1914, 325–26.

BAKER, Anne Elizabeth (1786–1861)
b. Northampton ? 16 June 1786 *d.* Northampton 22 April 1861
Supplied botanical notes and many of the drawings for her brother George Baker's *Hist. and Antiq. of County of Northampton* 1822–41.
D.N.B. v.3, 1. G. C. Druce *Fl. Northamptonshire* 1930, lxxxix–xc.

BAKER, Charles (*fl.* 1830s)
Nurseryman, Bedminster, Bristol.

BAKER, Edmund Gilbert (1864–1949)
b. Thirsk, Yorks 9 Feb. 1864 *d.* Kew, Surrey 17 Dec. 1949
Educ. Pharmaceutical College, London. FLS 1887. Son of J. G. Baker (1834–1920). Assistant, later Assistant Keeper, Department of Botany, BM(NH), 1887–1924. Collected plants in N. Africa and on continent of Europe. *Leguminosae of Tropical Africa* 1926–30.
Pharm. J. 1949, 493. *Times* 20 Dec. 1949. *Nature* v.165, 1950, 98. *Gdnrs Chron.* 1887 i 581; 1950 i 10. *Watsonia* 1949–50, 343–44.
Plants at BM(NH), Kew, Reigate Museum.

BAKER, George (1781–1851)
b. Northampton 1781 *d.* Northampton 12 Oct. 1851
His *History and Antiquities of County of Northampton* 1815–30 contains plant lists compiled by his sister.
Northampton Mercury 13 Oct. 1851. G. C. Druce *Fl. Northamptonshire* 1930, lxxxviii–lxxxix.

BAKER, George (–1885)
d. Reigate, Surrey May 1885
Rosarian. Contrib. to *Rosarian's Yb.*
J. Hort. Cottage Gdnr v.10, 1885, 444.

BAKER, George Percival (1856–1951)
b. Constantinople 1856 *d.* 29 Dec. 1951
VMH 1934. Land agent, Constantinople and head of textile printing firm at Crayford. Mountaineer. Collected plants in Pyrenees, High Atlas, Morocco, Mt. Olympus. President, Iris Society.
Gdnrs Chron. 1934 i 88 portr. *Quart. Bull. Alpine Gdn Soc.* v.20, 1952, 51–54.

BAKER, Mrs. H. Wright *see* Drew, K. M.

BAKER, Harold Trevor (1877–1960)
b. 22 Jan. 1877 *d.* Winchester, Hants 12 July 1960
Educ. Oxford. MP Accrington, 1910. Warden and Bursar, Winchester College. Chairman, Advisory Committee, Botanical Society of British Isles, 1932–47.
Proc. Bot. Soc. Br. Isl. 1961, 351–53.

BAKER, Henry (1698–1774)
b. London 8 May 1698 *d.* London 25 Nov. 1774
FRS 1740. Naturalist and poet. Described embryo of *Briza* (*Philos. Trans. R. Soc.* v.41, 1744, 448–55). Discovered cilia in *Volvox*, 1753. Introduced *Rheum palmatum*. *Microscope Made Easy* 1742.
J. Nichols *Lit. Anecdotes of Eighteenth Century* v.5, 272–78 portr. *D.N.B.* v.3, 10. J. E. Dandy *Sloane Herb.* 1958, 84. *Notes Rec. R. Soc. London* 1974, 53–77 portr.
MS. at Royal Society. Letters in Forster collection, Victoria and Albert Museum. Library and letters in Dawson Turner's sale at Puttock and Simpson, 6 June 1859.

BAKER, Henry (–1958)
MA 1918. Demonstrator and acting librarian, Oxford. Technical Assistant, Dept. of Botany, Oxford.
H. N. Clokie *Account of Herb. Dept. Bot., Oxford* 1964, 124.
Local plants at Oxford.

BAKER, Henry William Clinton- *see* Clinton-Baker, H. W.

BAKER, John Gilbert (1834–1920)
b. Guisborough, Yorks 13 Jan. 1834 *d.* Kew, Surrey 16 Aug. 1920
FRS 1878. FLS 1866. VMH 1897. Assistant, Kew Herb. 1866; Keeper, 1890–99. *North Yorkshire* 1863; ed. 2 1906. *Fl. of Mauritius and Seychelles* 1877. *Fl. of English Lake District* 1885. *Handbook of Fern-allies* 1887. *Handbook of Amaryllideae* 1888. *Handbook of Bromeliaceae* 1889. *Handbook of Irideae* 1892. Contrib. to *J. Bot.* from 1863.
Gdnrs Chron. 1864, 631; 1893 i 746–47 portr.; 1920 ii 102 portr. *Garden* v.52, 1897, xii portr.; v.60, 1901, 315 portr. *J. Kew Guild* 1897, 1 portr.; 1917, 365–66 portr. *Gdning World* 1899, 134–35, 184 portr. *Kew Bull.* 1899, 17–18; 1920, 319–20. *Naturalist* 1893, 27; 1907, 5–8; 1920, 367–68. *J. Bot.* 1893, 243–44 portr.; 1920, 233–38. *Proc. Linn. Soc.* 1920–21, 41–44. *Bot. Soc. Exch. Club Br. Isl. Rep.* 1920, 93–100; 1933, 289–97 portr. *Proc. R. Soc. B.* 1921, xxiv–xxx. *Curtis's Bot. Mag. Dedications, 1827–1927*, 211–12 portr. *Lily Yb.* 1937, 3–4 portr. *Watsonia* 1950, 343–44 portr. C. C. Gillispie *Dict. Sci. Biogr.* v.1, 1970, 412–13. *R.S.C.* v.1, 164; v.7, 74; v.9, 102; v.12, 41.
Plants at BM(NH), Kew, etc. Fern MSS. at Kew. Portr. at Kew and Hunt Library.
Bakeria André; *Bakerella* van Tieghem.

BAKER, Richard Eric Defoe (1908–1954)
d. Trinidad 19 Nov. 1954
Educ. Cambridge. Lecturer in Mycology, Imperial College of Tropical Agriculture, Trinidad, 1933; Professor, 1945; Professor

of Botany and Mycology, 1949. Collected plants in East Africa, 1948 and in Colombia, 1952. Edited *Fl. of Trinidad and Tobago* from 1947.

Nature v.174, 1954, 1128–29. *Amer. Orchid Soc. Bull.* 1958, 83 portr.

Plants at Commonwealth Mycological Institute.

BAKER, Richard Thomas (1854–1941)
b. Woolwich, London 1 Dec. 1854 *d.* Sydney 14 July 1941

To Australia, 1879. Science and Art Master, Newington College, Sydney, 1880. Sydney Technological Museum, 1888; Curator and Economic Botanist, 1901–21. *Australian Fl. in Applied Art. Hard Woods of Australia* 1919. *Research on Eucalypts* (with H. G. Smith) 1902. *Research on Pines of Australia* (with H. G. Smith) 1910. *Research on Eucalypts of Tasmania* (with H. G. Smith) 1912.

Sydney Morning Herald 20 Aug. 1941. *J. Proc. R. Soc. N.S.W.* v.76, 1942, 5–6. *Proc. R. Austral. Chem. Inst.* 1960, 309–16. *Who was Who, 1941–1950* 51. P. Serle *Dict. Austral. Biogr.* v.1, 1949.

Plants at BM(NH), Kew, Melbourne, Sydney.

BAKER, Robert (*c.* 1824–1885)
d. Leamington, Warwickshire 1885

MD. Of Birmingham and afterwards of Leamington. 'Catalogue of Warwickshire Plants' (with J. R. Young) (*Proc. Warwickshire Nat. Hist. Soc.* 1874, 56–57).

J. E. Bagnall *Fl. Warwickshire* 1891, 507.

Plants at Birmingham University.

BAKER, Sarah Martha (1887–1917)
b. 4 June 1887 *d.* 29 May 1917

DSc London 1913. FLS 1914. 'Fucaceae of the Salt Marsh' (*J. Linn. Soc.* 1911–12, 275–91; 1916, 325–80). 'Vegetable Dyes' (F. W. Oliver *Exploitation of Plants* 1917, 99–119).

Nature v.99, 1917, 329. *J. Ecol.* 1917, 222–23. *Proc. Linn. Soc.* 1916–17, 41–42.

BAKER, Rev. Thomas (*fl.* 1830s)

Rector, Whitburn, Durham. Collected plants in France, Switzerland and Italy.

Letters in N. J. Winch correspondence at Linnean Society.

J. Bot. 1914, 317.

BAKER, William Bennett (*fl.* 1850s–1860s)
b. Gloucestershire

To Stockton-on-Tees *c.* 1850. Journalist. *Field and Flowers* 1867 (reprinted from *Stockton Herald* of which he was proprietor).

BAKER, William G. (1861–1945)
b. Kent 1861 *d.* 11 Jan. 1945

Kew gardener. Curator, Botanic Garden, Oxford, 1887–1943.

Gdnrs Chron. 1942 i 227; 1945 i 42. *J. Kew Guild* 1945, 467.

BAKER, Rev. William Lloyd (1752–1830)
b. Bibury, Glos 1752

FLS 1793. Of Stoutshill, Uley, Glos. Discovered *Cephalanthera rubra*. Had herb. Contrib. to J. Sowerby and J. E. Smith's *English Bot.* (68, 483, 550, etc.) and W. Withering's *Bot. Arrangement*.

J. Bot. 1864, 346. *Proc. Cotteswold Nat. Field Club* 1940, 85–86. H. J. Riddelsdell *Fl. Gloucestershire* 1948, cxiv–cxv.

Herb. at Gloucester Museum. Letters at BM(NH) and Linnean Society. Portr. at Hardwicke House, Glos.

BALAM, Alexander (*fl.* 1650s–1690s)

Merchant. Collected plants and seeds for Gaston, Duke of Orleans, Tuscany in 1656, and in Tangier, and sent seeds to C. Merrett, L. Plukenet and R. Morison.

G. Zanoni *Istoria Botanica* 1675, 73. E. S.-C. Cosson, *Compendium Florae Atlanticae* v.1, 1881, 7. J. E. Dandy *Sloane Herb.* 1958, 84.

BALCHIN, William (*c.* 1824–1901)
d. Hassocks, Sussex 16 Nov. 1901

Nurseryman, Brighton and Hassocks.

Gdnrs Chron. 1901 ii 383 portr. *Gdnrs Mag.* 1901, 768. *J. Hort. Cottage Gdnr* v.43, 1901, 461.

BALDING, A. (*fl.* 1880s)

Plant specimens from East Anglia and Hebrides at Oxford.

BALDOCK, Robert (–*c.* 1860)
d. High Holborn, London *c.* 1860

Bookseller, 85 High Holborn. Interested in London flora. MS. biographical account by W. Pamplin in copy of H. Trimen and W. T. T. Dyer's *Fl. Middlesex* 1869 at Kew Library.

BALDWIN, Thomas (*fl.* 1810s)

Gardener to Marquis of Hertford, Ragley, Warwickshire. *Short Practical Directions for Culture of Ananas, or Pine Apple* 1818. Contrib. to *Trans. Hort. Soc. London*.

J. C. Loudon *Encyclop. Gdning* 1822, 1290.

BALE, Rev. Sackville (*fl.* 1800s)

Had garden at Withyham near Tunbridge, Kent.

Bot. Mag. 1802, t.561.

BALFOUR, Sir Andrew (1630–1694)
b. Denmiln, Fife 18 Jan. 1630 *d.* London 9 Jan. 1694

MD Caen 1661. Practised as physician at London, St. Andrews and Edinburgh. Founded Edinburgh Botanic Garden with Sir R. Sibbald. Discovered *Ligusticum scoticum*.

R. Sibbald *Memoria Balfouriana* 1699. R. Pulteney *Hist. Biogr. Sketches of Progress of Bot. in England* v.2, 1790, 3–4. *Scots Mag.* 1803, 747–60. J. Walker *Essays on Nat. Hist.* 1808, 347–69. *D.N.B.* v.3, 48–49. H. R. Fletcher and W. H. Brown *R. Bot. Gdn Edinburgh, 1670–1970* 1970, 4–9.

Balfouria R. Br.

BALFOUR, Andrew Francis (*fl.* 1870s)

Lieutenant, Royal Navy. Collected a few plants on Mt. Ternati, Philippines, now at Edinburgh.

Trans. Bot. Soc. Edinburgh v.12, 1876, lxi. *Fl. Malesiana* v.1, 1950, 598.

BALFOUR, Archibald Park (*c.* 1887–1973)
b. Edinburgh *c.* 1887 *d.* Jan. 1973

VMH 1953. Horticulturalist with Sutton and Sons Ltd., seedsmen, Reading, 1907. Raised many new plant varieties including Charm chrysanthemums and bigeneric hybrid *Venidioarctotis*.

J. R. Hort. Soc. 1973, 413. *Times* 15 Jan. 1973.

BALFOUR, Edward Green (1813–1889)
b. Montrose, Angus 6 Sept. 1813 *d.* London 8 Dec. 1889

LRCS Edinburgh 1833. Surgeon, Madras Army, 1836–71. Formed Government Central Museum, 1850. *Timber Trees...also the Forests of India* 1858; ed. 3 1870. *Agricultural Pests of India* 1887. *Cyclopaedia of India* 1857; ed. 2 1871–73 5 vols.

F. Boase *Modern English Biogr.* v.1, 1908, 246–47. *D.N.B. Supplt. 1* v.1, 113–15. *R.S.C.* v.1, 170.

BALFOUR, Frederick Robert Stephen (1873–1945)
b. Mount Alyn, Denbighshire 11 March 1873 *d.* London 2 Feb. 1945

BA Oxon 1896. FLS 1924. VMH 1927. During youth spent 4 years on Pacific coast of N. America where he acquired an intimate knowledge of forest trees. Had arboretum at Dawyck near Peebles. Contrib. chapter on botany to J. W. Buchan and H. Paton's *History of Peebleshire* 1925. Contrib. to *Quart J. For.*

Gdnrs Chron. 1931 ii 342 portr.; 1945 i 86. *Nature* v.155, 1945, 357–58. *Proc. Linn. Soc.* 1944–45, 63–65. *Times* 8 Feb. 1945. *J. R. Hort. Soc.* 1947, 5–12. *Who was Who, 1941–1950* 52–53.

BALFOUR, Mrs. Frederick (*née* Norman) (1877–1970)
d. 4 March 1970

Collected conifers in N.W. America with her husband, F. R. S. Balfour. Gardened at Dawyck and Kemsing.

Times 11 March 1970.

BALFOUR, Sir Isaac Bayley (1853–1922)
b. Edinburgh 31 March 1853 *d.* Haslemere, Surrey 30 Nov. 1922

BSc Edinburgh 1873. DSc 1875. MD 1877. FRS 1884. FLS 1875. KBE 1920. VMH 1897. Son of J. H. Balfour (1808–1884). Botanist on Transit of Venus Expedition to Rodriquez Island, 1874. Collected plants in Socotra, 1879–80. Professor of Botany, Glasgow, 1879. Professor of Botany, Oxford 1884. Professor of Botany and Keeper, Royal Botanic Garden, Edinburgh, 1888. Authority on Asiatic plants, particularly *Primula* and *Rhododendron*. *Botany of Socotra* (*Trans. R. Soc. Edinburgh* v.31, 1888, 1–446). Editor, *Ann. Bot.*, 1887–1912.

Gdnrs Chron. 1891 ii 275 portr.; 1922 i 161 portr.; 1922 ii 346, 356–57; 1923 ii 336–37; 1926 ii 491. *Garden* v.60, 1901, 48–49 portr.; v.70, 1906, iv portr.; 1922, 638. *Gdnrs Mag.* 1908, 1–2 portr. *Bot. Soc. Exch. Club Br. Isl. Rep.* 1922, 690–92. *Nature* v.110, 1922, 816–17. *Times* 1 Dec. 1922. *Proc. Linn. Soc.* 1922–23, 36–37. *Proc. R. Soc. Edinburgh* 1922–23, 230–36. *Trans. Bot. Soc. Edinburgh* v.28, 1922–23, 192–96 portr. *J. Bot.* 1923, 23–26 (reprinted from *Glasgow Herald* 5 Dec. 1922). *Kew Bull.* 1923, 30–35. *Ann. Bot.* 1923, 335–39 portr. *Empire For. J.* 1924, 22–23. *Proc. R. Soc. B.* v.96, 1924, i–xvii portr. *D.N.B. 1922–1930* 56–57. *Curtis's Bot. Mag. Dedications, 1827–1927* 247–48 portr. F. O. Bower *Sixty Years of Bot. in Britain* (*1875–1935*) 1938, 57–59 portr. H. R. Fletcher and W. H. Brown *R. Bot. Gdn Edinburgh, 1670–1970* 1970, 195–232 portr. *R.S.C.* v.7, 77; v.9, 108; v. 12, 45.

Plants at BM(NH), Kew. Portr. at Hunt Library.

BALFOUR, John Hutton (1808–1884)
b. Edinburgh 15 Sept. 1808 *d.* Edinburgh 11 Feb. 1884

MA and MD Edinburgh 1832. FRS 1856. FLS 1844. Professor of Botany, Glasgow, 1841. King's Botanist, Edinburgh, 1845–79. Known to generations of students as 'Woody Fibre'. Founded Botanical Society of Edinburgh, 1836. *Manual of Botany* 1849. *Class Book of Botany* 1852–54. *Outlines of Botany* 1854. *Fl. of Edinburgh* 1863.

Proc. Linn. Soc. 1883–84, 30–31. *Gdnrs Chron.* 1884 i 220. *J. Bot.* 1884, 128. *Nature* v.29, 1884, 385–87. *Scott. Nat.* 1884, 160–62. *Hist. Berwickshire Nat. Club* v.11, 1885,

218–26. *Trans. Bot. Soc. Edinburgh* v.16, 1886, 187–89. *D.N.B.* v.3, 56. *Notes Bot. Gdn Edinburgh* 1902, 21–23. F. W. Oliver *Makers of British Bot.* 1913, 293–300 portr. *Curtis's Bot. Mag. Dedications, 1827–1927* 179–80 portr. H. R. Fletcher and W. H. Brown *R. Bot. Gdn Edinburgh, 1670–1970* 1970, 125–37 portr. C. C. Gillispie *Dict. Sci. Biogr.* v.1, 1970, 423. *R.S.C.* v.1, 170; v.7, 77; v.9, 108; v.13, 262.

Plants at Edinburgh, Glasgow, Kew. Letters at Kew. Portr. at Hunt Library.

Balfourodendron Mello.

BALFOUR, Marian *see* Busk, *Lady* M.

BALFOUR, Thomas Alexander Goldie (1825–1895)

b. Edinburgh 1825 *d.* Edinburgh 10 March 1895

MD Edinburgh 1851. FRSE 1870. Brother of J. H. Balfour (1808–1884). President, Botanical Society of Edinburgh, 1877–79. 'Dionaea' (*Trans. Bot. Soc. Edinburgh* v.12, 1876, 334–69; v.13, 1879, 353–77).

Br. Med. J. 1895 i 679. *Trans. Bot. Soc. Edinburgh* v 20, 1896, 449–51. *R.S.C.* v.9, 109.

BALFOUR-BROWNE, William Alexander Francis (1874–1967)

b. London 27 Dec. 1874 *d.* 28 Sept. 1967

FLS 1928. FES. FZS. Held posts in botany, entomology, zoology at Belfast, Cambridge and London.

R. L. Praeger *Some Irish Nat.* 1949, 44. *Who was Who, 1951–1960* 53.

BALFOUR-GOURLAY, William (*c.* 1879–1966)

Collected plants with E. K. Balls in Turkey, 1933–35, and with C. Elliott in Chile, 1927–30.

Nature in Cambridgeshire no. 10, 1967, 3.

Plants at Kew.

BALL, Anne E. (–1872)

Of Youghal, County Cork. Algologist. Assisted W. H. Harvey in *Phycologia Britannica* 1846–51, t.356.

J. Bot. (Hooker) 1840, 191.

Algae at National Museum, Dublin. Letters and Irish plants at Kew.

Ballia Harvey.

BALL, Charles Frederick (1879–1915)

b. Loughborough, Leics 13 Oct. 1879 *d.* Suvla Bay, Gallipoli Sept. 1915

Kew gardener, 1900–6. Assistant to Keeper, Botanic Garden, Glasnevin, Dublin. Collected plants in Switzerland and Bulgaria. Edited *Irish Gdning.*

Gdnrs Chron. 1912 i 252, 274–75; 1915 ii 239. *Garden* 1915, 514 portr. *Irish Gdning* 1915, 161–62 portr. *Nature* v.96, 1915, 207. *Orchid Rev.* 1915, 348. *J. Kew Guild* 1916, 307 portr.

BALL, Francis (*fl.* 1670s)

Gardener and nurseryman, Brentford, Middx.

BALL, Henry (1857–1925)

b. Southport, Lancs 11 Jan. 1857 *d.* Plymouth, Devon 10 May 1925

FLS 1921. Pharmaceutical chemist, Southport. Local botanist.

Proc. Linn. Soc. 1924–25, 64. *Bot. Soc. Exch. Club Br. Isl. Rep.* 1925, 843–44.

BALL, John (*fl.* 1850s)

Employee at Charles Turner's nursery, Slough, Bucks. Mr. McClean raised laced pink, 'John Ball', in mid 19th century.

C. O. Moreton *Old Carnations and Pinks* 1955, 34.

BALL, John (1818–1889)

b. Dublin 20 Aug. 1818 *d.* S. Kensington, London 21 Oct. 1889

MA Dublin. FRS 1868. FLS 1856. Collected plants with J. D. Hooker in Morocco, 1871. Collected in Rocky Mountains, U.S.A., 1884. First President, Alpine Club, 1858–60. *Journal of Tour in Marocco and Great Atlas* (with J. D. Hooker) 1878. 'Spicilegium Florae Maroccanae' (*J. Linn. Soc.* v.16, 1877, 281–742). *Notes of a Naturalist in South America* 1887.

Ann. Bot. 1889–90, 450–51. *J. Bot.* 1889, 365–70; 1904, 295. *Proc. R. Soc.* v.47, 1890, v–ix. *Proc. Linn. Soc.* 1888–89, 90–92. *Trans. Bot. Soc. Edinburgh* v.19, 1893, 3. *Kew Bull.* 1896, 151–52. N. Colgan *Fl. Dublin* 1904, xxvii–xxviii. *D.N.B. Supplt. 1* v.1, 115–18. *Curtis's Bot. Mag. Dedications, 1827–1927* 227–28 portr. *Science* v.101, 1945, 678. R. L. Praeger *Some Irish Nat.* 1949, 44–45. *R.S.C.* v.1, 171; v.7, 78; v.9 109; v.13, 263.

Plants at Edinburgh, Kew. Letters at Kew. Portr. at Hunt Library.

BALL, Robert (1802–1857)

b. Cobh, County Cork 1 April 1802 *d.* Dublin 30 March 1857

Zoologist. Collected algae on Arran. Discovered with W. Thompson *Astragalus danicus* and *Allium babingtonii* in Ireland.

Nat. Hist. Rev. 1858, 134. *Proc. Dublin Univ. Zool. Bot. Assoc.* v.1, 7–48. *D.N.B.* v.2, 77–78.

BALL, Samuel (1780–1874)

d. Wolverley, Worcs 5 March 1874

Inspector of teas, United East India Company in China, 1804–26. *Account of Cultivation and Manufacture of Tea in China* 1848.

F. Boase *Modern English Biogr.* v.1, 1892, 146.

BALL, Valentine (1843–1895)
b. Dublin 14 July 1843 *d.* Dublin 15 June 1895

LLD Dublin. FRS 1882. On staff of Geological Survey of India, 1864–81. Collected plants in India. Director, Science and Art Museum, Dublin, 1883. 'Fl. of Manbhum' (*Proc. Bengal Asiatic Soc.* 1868, 254–55).

Irish Nat. 1895, 169–71 portr. *Proc. R. Soc.* v.58, 1895, xlvii–xlix. *Science-Gossip* 1895, 132. R. L. Praeger *Some Irish Nat.* 1949, 45–46 portr. *R.S.C.* v.7, 79; v.9, 111; v.12, 45; v.13, 265.

Plants and bust at National Museum, Dublin. Letters at Kew.

BALLANTINE, Henry (1833–1929)
b. Monzie, Perthshire 1833 *d.* Egham, Surrey 27 June 1929

VMH 1907. Gardener to Baron Schröder at Englefield Green, Surrey. Raised orchids.

Gdnrs Chron. 1907 ii 200–1 portr.; 1917 i 141 portr.; 1929 ii 19. *Gdnrs Mag.* 1908, 78–79 portr. *Orchid Rev.* 1929, 250.

Cattleya ballantineana (*Trianae X Warscewiczii*).

BALLARD, Edward (1820–1897)
b. Islington, London 1820 *d.* Islington 19 Jan. 1897

MB London 1843. MD 1844. LRCP 1853. FRCP 1872. FRS 1889. Medical Officer of Health, Islington, 1856. Vice-President, Medico-Chirurgical Society. MS. list of Islington and vicinity plants at Stoke Newington Public Library (*J. Bot.* 1928, 185–94).

Med. Circ. 1852 i 151. *Br. Med. J.* 1897 i 281–82. *Proc. R. Soc.* 1897–98, iii–v. *D.N.B.* v.3, 83–84. D. H. Kent *Hist. Fl. Middlesex* 1975, 21–22.

British herb. at BM(NH).

BALLARD, Ernest (*c.* 1871–1952)
b. Herefordshire *c.* 1871 *d.* 30 March 1952

VMH 1949. Educ. Birmingham University. Raised new varieties of Michaelmas daisies. *Days in my Garden* 1919.

Gdnrs Mag. 1912, 951–52 portr. *Gdnrs Chron.* 1950 i 82 portr.; 1952 i 126.

BALLARD, Robert (*fl.* 1780s–1790s)

ALS 1793. Surgeon of Hanley and Malvern. Had herb. Contrib. to W. Withering's *Bot. Arrangement* (ed. 2 1787, preface xi).

E. Lees *Bot. Worcestershire* 1867, lxxxviii. *J. Bot.* 1914, 321.

BALLS, Matthew (*fl.* 1860s)

Of Hitchin and Hertford. Contrib. botanical notes to *Phytologist* 1869.

A. R. Pryor *Fl. Hertfordshire* 1887, xlvii.

BALLS, William Lawrence (1882–1960)
b. Garboldisham, Norfolk 3 Sept. 1882 *d.* 18 July 1960

Educ. Cambridge. FRS 1928. Botanist, Khedivial Agricultural Society, 1904–10. Egyptian Ministry of Agriculture, 1911–13, 1927–33; cotton technologist, 1934–47. *Cotton Plant in Egypt* 1912. *Development and Properties of Raw Cotton* 1915. *Yield of a Crop* 1953.

Nature v.187, 1960, 989–90. *Who was Who, 1951–1960* 59. *Biogr. Mem. Fellows R. Soc.* 1961, 1–16 portr.

BALOE, William *see* Boylan, W.

BALSTON, R. J. (*fl.* 1880s–1900s)

Plants at Maidstone Museum, Kent.

BAMBER, Charles James (1855–1941)
b. 14 July 1855 *d.* 9 Jan. 1941

FLS. Indian Medical Service, 1878. Inspector-General of Civil Hospitals, Punjab, 1910–15. 'Plants of the Punjab' (*J. Bombay Nat. Hist. Soc.* 1908–13).

Who was Who, 1941–1950 55.

BANBURY, Arnold (1598–1664/5)
d. Feb. 1664/5

Nurseryman, Tothill Street, Westminster, London. Son of H. Banbury (1540–1609/10).

BANBURY, Henry (1540–1609/10)
d. Feb. 1609/10

Nurseryman, Tothill Street, Westminster, London; "excellent graffer and painfull planter" (Gerard).

BANCKES, Richard (*fl.* 1520s)

BA Oxon 1515? A London printer who printed in 1525 an anonymous work now known as 'Banckes's Herball'.

E. S. Rohde *Old English Herbals* 1922, 55–65, 204. A. Arber *Herbals* ed. 2, 1938, 41–44. *Bull. Hist. Med.* v.15, 1944, 246–60.

BANCROFT, Claude Keith (1885–1919)
b. Barbados 30 Oct. 1885 *d.* Toronto 11 Jan. 1919

BA Cantab 1908. FLS 1911. Mycologist, Federated Malay States, 1910–13. Government Botanist, British Guiana, 1913–19.

Kew Bull. 1910, 253; 1913, 91; 1919, 86. *Nature* v.103, 1919, 191. *Who was Who, 1916–1928* 48.

BANCROFT, Edward Nathaniel (1772–1842)
b. London 1772 *d.* Kingston, Jamaica 18 Sept. 1842

MB Cantab 1794. MD 1804. Physician to Forces in Jamaica from 1811. 'Cuichunchulli' (*Companion Bot. Mag.* v.1, 1836, 277–82). Contrib. to J. Lunan *Hortus Jamaicensis* 1814.

Bot. Mag. 1831, t.3059, 3092. J. Macfadyen *Fl. of Jamaica* v.1, 1837, 112–13. W. Munk *Roll of Royal College of Physicians* v.3, 1878, 31–32. I. Urban *Symbolae Antillanae* v.3, 1902, 19. *D.N.B.* v.3, 106–7. F. Cundall *Historic Jamaica* 1915, 176–77. *R.S.C.* v.1, 75.

Letters and Jamaican plants at Kew.

Bancroftia Macfad.

BANCROFT, Joseph (1836–1894)
b. Stretford, Manchester 21 Feb. 1836 *d.* Brisbane 16 June 1894

MD St. Andrews 1859. Pharmacologist. To Brisbane 1864. Surgeon, Brisbane Hospital, 1867. *Pituri and Duboisia* 1877. *Further Remarks on Pituri Group of Plants* 1878. *Contributions to Pharmacy from Queensland* 1886.

Proc. R. Soc. Queensland v.8, 1890, 36; v.10, 1892–94, 102. P. Mennell *Dict. Austral. Biogr.* 1892, 27. *Br. Med. J.* 1894 ii 395. *Gdnrs Chron.* 1894 ii 255. *J. Bot.* 1894, 288. *Trans. Liverpool Bot. Soc.* 1909, 58. *Proc. Linn. Soc. N.S.W.* 1900, 785. *Rep. Austral. Assoc. Advancement Sci.* 1909, 374–75 portr. *Med. J. Austral.* 1948, 621–27; 1961, 153–70. P. Serle *Dict. Austral. Biogr.* v.1, 1949, 39–40. *R.S.C.* v.13, 278.

Letters and plants at Kew.

Strychnos bancroftiana F. M. Bailey.

BANCROFT, Thomas Lane (1860–1933)
b. Nottingham 2 Jan. 1860 *d.* Wallaville, Queensland 12 Nov. 1933

To Brisbane, 1864. Graduated in medicine, Edinburgh University, 1883. Undertook medical research on return to Australia. Hybridised plants. Contrib. botanical papers to *Proc. R. Soc. Queensland.*

J. Bot. 1934, 141. *Queensland Nat.* v.9, 1934, 25. *Austral. Encyclop.* v.1, 1962, 408.

BANFIELD, Frederick Sydney (–1967)
b. Sussex *d.* 28 June 1967

Kew gardener. Horticultural Assistant, Malaya, 1927–34. Horticultural Officer, London County Council. Malaya, 1945–54; in charge of Botanic Garden, Penang, 1946.

J. Kew Guild 1967, 823–24 portr.

BANISTER, Rev. John (1650–1692)
b. Twigworth, Glos 1650 *d.* Roanoke River, Virginia May 1692

BA Oxon 1671. MA 1674. Sent by Bishop Compton as Anglican minister to Virginia, also to collect plants for R. Morison. 'Catalogue of Plants Observed in Virginia' (J. Ray *Historia Plantarum* v.2, 1688, 1928). Correspondent of J. Ray, R. Morison, S. Doody, L. Plukenet, M. Lister and Bishop Compton. Excerpts of letters to Lister in *Philos. Trans. R. Soc.* v.17, 1693, 667–72; v.22, 1701, 814. Helped R. Plot (*Nat. Hist. Oxfordshire* 1677, 153).

J. Ray *Historia Plantarum* v.3, 1704, preface iv. *Philos. Trans. R. Soc.* v.17, 1693, 667–72; v.28, 1714, 188. R. Pulteney *Hist. Biogr. Sketches of Progress of Bot. in England* v.2, 1790, 55–57. S. Ayscough *Cat. of Manuscripts...in British Museum* 1782, 725. *D.N.B.* v.3, 119–20. E. J. L. Scott *Index to Sloane Manuscripts* 1904, 38. *Amer. Midland Nat.* 1910, 195–96. J. E. Dandy *Sloane Herb.* 1958, 84–87. *Amer. Fern J.* 1963, 138–44. H. N. Clokie *Account of Herb. Dept. Bot., Oxford* 1964, 125–26. J. and N. Ewan *John Banister and his Nat. Hist. of Virginia* 1970. C. C. Gillispie *Dict. Sci. Biogr.* v.1, 1970, 431–32.

Plants at BM(NH) and Oxford. MSS. at BM(NH).

Banisteria L.

BANKER, J. (–*c.* 1866)

Of Plymouth, Devon. Had herb.

I. W. N. Keys. *Fl. Devon Cornwall* 1866, preface.

BANKS, George (*fl.* 1820s–1830s)
d. Devonport

Silversmith and engraver, Devonport. Lectured on botany at Devonport. *Introduction to Study of English Botany* 1823 (plates by author). *Plymouth and Devonport Fl.* 1830–32. 'Indigenous Fl. of London and Plymouth' (*Mag. Nat. Hist.* 1829, 265–66).

J. Bot. 1872, 153. F. H. Davey *Fl. Cornwall* 1909, xxxix–xl. W. Keble Martin and G. T. Fraser *Fl. Devon* 1939, 772.

Plants and letters at Kew.

BANKS, George Henry (*c.* 1882–1948)
d. Histon, Cambridge 11 Nov. 1948

Kew gardener, 1905. Foreman, Botanic Garden, Cambridge, 1906. Curator, Botanic Garden, Glasgow, 1915–47.

Gdnrs Chron. 1948 ii 186. *J. Kew Guild* 1948, 694.

Plants at Glasgow.

BANKS, Sir Joseph (1743–1820)
b. London 13 Feb. 1743 *d.* Spring Grove, Isleworth, Middx 19 June 1820

MA Oxon 1763. DCL 1771. FRS 1766. FLS 1788. KCB 1795. To Newfoundland with Lt. Phipps, 1766–67 (*J. Bot.* 1904, 84–86, 352; 1905, 248). Eastbury and Bristol, May–June 1767 (Journal published in *Proc. Bristol Nat. Soc.* v.9, 1899, 6–37). Around the world with Capt. Cook on 'Endeavour', 1768–71. (J. C. Beaglehole *Endeavour Journal of Joseph Banks 1768–1771* 1962 2 vols.) To Iceland with D. C. Solander, 1772. Purchased Clifford's herb., 1791. Established international role of Kew Gardens. President, Royal Society, 1778–1820.

J. Nichols *Lit. Anecdotes of Eighteenth Century* v.7, 20–21, 509. *The Sun* 21 Sept. 1820 (contains abstract of Banks's Will). *Gent. Mag.* 1820 i 637–38; 1820 ii 86–88. J. E. Smith *Selections of Correspondence of Linnaeus* v.2, 1821, 574–80. A. Duncan *Short Account of Life of Sir Joseph Banks* 1821. *Asiatic Observer* 1823, 125–52. *Mém. Acad. Sci. Inst. Fr. (Hist.)* 1826, 204–30. C. Tomlinson *Sir Joseph Banks and the Royal Society* 1844. G. Suttor *Mem., Hist. and Sci. of...Sir Joseph Banks* 1855 portr. *J. Hort. Cottage Gdnr* v.56, 1876, 308 portr. *Proc. Linn. Soc.* 1889–90, 31; 1919–20, Supplt. 1–21; 1950–51, 21–26. *Kew Bull.* 1891, 305–9; 1960, 416–17. *Gdnrs Chron.* 1897 i 15, 36; 1909 ii 248–49 portr.; 1945 ii 96–97. *D.N.B.* v.3, 129–33. *J. Bot.* 1902, 388–90; 1905, 122–24, 284–90. *Proc. Linn. Soc. N.S.W.* 1905, 34–39. J. H. Maiden *Sir Joseph Banks* 1909 portr. E. Smith *Life of Sir Joseph Banks* 1911 portr. J. W. White *Fl. Bristol* 1912, 67–71. *Nature* v.122, 1925, 815–16; v.151, 1943, 181–83. H. Hermannsson *Sir Joseph Banks and Iceland* 1928. G. MacKaness *Sir Joseph Banks and his Relations with Australia* 1936. *Austral. National Rev.* 1938, 13–23. *Notes and Rec. R. Soc.* 1940, 85–87; 1946, 49–57; 1954, 91–99. *Chronica Botanica* v.9, 1945, 94–106. *J. R. Austral. Hist. Soc.* 1949, 116–30; 1971, 10–16. *Fl. Malesiana* v.1, 1950, 34–35 portr. *Geogr. J.* 1950, 49–54. H. C. Cameron *Sir Joseph Banks* 1952 portr. W. R. Dawson *Banks Letters* 1958; Supplts. in *Bull. BM(NH) Hist. Ser.* v.3 (2) and v.3 (3), 1962–69. W. P. Morrell *Sir Joseph Banks in New Zealand from his Journal* 1958 portr. *J. Soc. Bibl. Nat. Hist.* v.4, 1962, 57–62. *Lychnos* 1962, 200–11. S. Ryden *Banks Collection* 1963 portr. *Isis* 1964, 62–67. *Austral. Dict. Biogr.* v.1, 1966, 52–55. *Huntia* 1965, 187–89 portr. F. A. Stafleu *Taxonomic Literature* 1967, 14–15. *Endeavour* 1968, 3–10. *Trans. Amer. Philos. Soc.* v.58, 1968, 26–58. K. Lemmon *Golden Age of Plant Hunters* 1968, 17–42 portr. *Austral. Nat. Hist.* 1969, 251–54 portr. H. R. Fletcher *Story of R. Hort. Soc.* 1969 *passim* portr. *Country Life* 1970, 930–32 portr. C. C. Gillispie *Dict. Sci. Biogr.* v.1, 1970, 433–37. A. M. Lysaght *Joseph Banks in Newfoundland and Labrador* 1971 portr. *Proc. R. Soc. Queensland* 1971, 1–19 portr. M. A. Reinikka *Hist. of the Orchid* 1972, 116–21 portr. *BSBI News* v.2 (1), 1973, 15–16. *Proc. Amer. Philos. Soc.* v.117, 1973, 186–226 portr. *Notes Rec. R. Soc.* v.28 (2) 1974, 221–34. *Bull. BM(NH) Hist. Ser.* v.4 (5) 1974, 283–385. *J. R. Hort. Soc.* 1974, 339–47 portr. *See also* Addendum.

Herb. at BM(NH). Library at BM. Letters at BM(NH), Kew, Sutro Library, Mitchell Library, Sydney; photocopies at American Philosophical Library. Portr. at National Portrait Gallery (Phillips and Lawrence), Royal Society, Royal Horticultural Society, Linnean Society (marble bust by Chantrey, copy at Kew and BM(NH)), Parham Park, Sussex (Reynolds).

Banksia L. f.; *Josephia* Salisb.

BANKS, Peter (–1838)
d. drowned on Columbia River, British Columbia 22 Oct. 1838

Gardener at Chatsworth, Derbyshire. Sent by J. Paxton with Robert Wallace to collect plants in north-west corner of N. America, 1838. Both drowned in boating accident.

Gdnrs Mag. 1839, 479–80. V. R. Markham *Paxton and the Bachelor Duke* 1935, 63–72. J. Ewan *Rocky Mountain Nat.* 1950, 157. S. D. McKelvey *Bot. Exploration of the Trans-Mississippi West, 1790–1850* 1955, 797.

BANKS, Thomas (*fl.* 1830s–1840s)

Plants of London area at Melbourne, Australia.

BANNERMAN, Thomas (1840–1920)
b. Golspie Tower, Sutherland 1840 *d.* Nov. 1920

Gardener to Lord Bagot at Blithfield, Staffs for over 60 years.

Gdnrs Chron. 1877 ii 52 portr.; 1920 ii 282 portr.

BANNING, William (*fl.* 1820s)

Nurseryman and seedsman, Pickhill, Yorks.

BANTON, E. (*fl.* 1830s)

Nurseryman, Oakham, Rutland.

BARBER, Charles Alfred (1860–1933)
b. Wynberg, S. Africa 10 Nov. 1860 *d.* Cambridge 23 Feb. 1933

BA Cantab. FLS 1908. Demonstrator in Botany, Cambridge, 1889. Superintendent, Botanical and Agricultural Dept., Leeward Islands, 1891–95. Professor of Botany, Cooper's Hill, 1895–98. Government Botanist, Madras, 1898–1912. Sugar-cane consultant to Government of India, 1912–19. *Studies in Root-parasitism* 1906–8. *Studies in Indian Sugar-cane* 1915–19. *Tropical Agricultural Research in the Empire* 1927.

Kew Bull. 1898, 277; 1913, 48; 1933, 157–58. *Kew Bull. Additional Ser. I* 1898, 119–22. I. Urban *Symbolae Antillanae* v.3, 1902, 19. *J. Bot.* 1933, 102–3. *Nature* v.131, 1933, 389. *Times* 24 Feb. 1933. *Who was Who, 1929–1940*, 62. I. H. Burkill *Chapters on Hist. Bot. in India* 1965, 139.

Plants at Calcutta, Kew. Indian notes and drawings at Kew.

BARBER, Edmund Scott (*fl.* 1850s)
Administrator of Labuan, Borneo, 1853–54. Purchased J. Motley's herb., now at Kew.
Fl. Malesiana v.1, 1950, 36.

BARBER, Horace Newton (1914–1971)
b. Warburton, Cheshire 26 May 1914 *d.* Sydney 16 April 1971
Educ. Cambridge. FRS 1963. Researched on chromosomes at John Innes Institute, 1936. Lecturer in Botany, Sydney University, 1946. Professor of Botany, Tasmania University, 1947. Professor of Botany, Sydney, 1964. Cytologist.
Biogr. Mem. Fellows R. Soc. 1972, 21–33 portr.
Portr. at Hunt Library.

BARBER, Leal Mitford (*fl.* 1890s)
Plants collected in S. Africa, now at Kew and Lyon.

BARBER, Mary Elizabeth (*née* **Bowker**) (1818–1899)
d. Pietermaritzburg, S. Africa Aug. 1899
Came to S. Africa with her parents in 1820. Correspondent of C. Darwin, W. H. Harvey and the Hookers. Collected plants in S. Africa. 'Stapelias' (*Kew Bull.* 1903, 17–19) Contrib. to *J. Linn. Soc.* 1869–71. Discovered *Stapelia glabricaulis* and *S. jucunda.*
W. H. Harvey *Thesaurus Capensis* v.1, 1859, 24–25. *Rec. Albany Mus.* 1903, 95–108. I. Mitford-Barberton *Barbers of the Peak* 1934. A. White and B. L. Sloane *Stapelieae* v.1, 1937, 100 portr. E. Rosenthal *Southern African Dict. National Biogr.* 1966, 17. *R.S.C.* v.5, 89; v.9, 118; v.12, 48.
Herb. at Albany Museum, Grahamstown. Drawings of stapelias at Kew and Albany Museum.
Bowkeria Harvey.

BARBOUR, H. (*fl.* 1890s–1900s)
Contrib. to *Fl. Preston and Neighbourhood* 1903. Member of Botanical Section of Preston Scientific Society.

BARCLAY, Arthur (1852–1891)
b. Edinburgh 3 Aug. 1852 *d.* Simla, India 2 Aug. 1891
MB Glasgow 1874. FLS 1890. Indian Medical Service. Professor of Physiology, Medical College, Calcutta, 1874. Plant pathologist, Indian Forest Service. Worked on *Uredineae*. Contrib. to *Trans. Linn. Soc., Ann. Bot., J. Bot.*, etc.
Indian For. 1891, 303–4. *J. Bot.* 1891, 384. *Proc. Linn. Soc.* 1891–92, 60–61. D. G. Crawford *Hist. Indian Med. Service, 1600–1913* v.2, 1914, 145–46. *R.S.C.* v.9, 121; v.13, 293.
Fungi at Kew. MSS. and drawings at BM(NH).

BARCLAY, Francis Herbert (1870–1935)
Plants at Norwich Museum, Norfolk.

BARCLAY, George (*fl.* 1830s–1840s)
b. Huntley, Aberdeenshire *d.* Buenos Aires, Argentine
Kew gardener and collector. On HMS 'Sulphur', 1836–41, to Chile, Peru, Panama, Sandwich Isles, etc. Plants from Fiji in Hooker's *London J. Bot.* 1843, 211–40.
Gdnrs Chron. 1882 i 305–6. *Kew Bull.* 1891, 321. *Fl. Malesiana* v.1, 1950, 36. *Aliso* v.5, 1964, 469–77. *Contrib. Univ. Michigan Herb.* 1972, 216–21.
Plants and MS. 'Journal of Voyage round the World' at BM(NH).
Sida barclayi Baker f.

BARCLAY, Robert (1751–1830)
b. Philadelphia 15 May 1751 *d.* Bury Hill, Surrey 22 Oct. 1830
FLS 1788. Of Clapham, 1781 and Bury Hill, 1805. Had large garden. Advised Curtis to publish *Bot. Mag.* (see dedication to v.54).
W. J. Hooker *Exotic Fl.* v.3, 1827, 166. *Gdnrs Mag.* v.2, 1827, 297–303; 1828, 1–6. *Bot. Misc.* v.2, 1830–31, 122–25. *Bot. Gdn* (Maund) v.4, 1834–35, 337. W. Darlington *Memorials of John Bartram* 1849, 531–32. E. Bretschneider *Hist. of European Bot. Discoveries in China* 1898, 285–86. *Curtis's Bot. Mag. Dedications, 1827–1927* 3–4 portr.
Lithograph after portr. by Raeburn at Kew. Portr. at Hunt Library. Library sold by R. H. Evans, 1831.

BARCLAY, William (1846–1923)
b. Tulloch, Perth 19 March 1846 *d.* Perth 10 May 1923
ALS 1923. Schoolmaster, Perth, 1871. President, Perthshire Society of Natural Science, 1907–18. Studied British roses from 1894. Discovered *Potamogeton venustus* 1915. Assisted F. B. White in *Fl. Perthshire* 1898, edited by J. W. H. Trail. Contrib. to *Ann. Scott Nat. Hist., Trans. Proc. Perthshire Soc. Nat. Sci., J. Bot.* Edited *Trans. Proc. Perthshire Soc. Nat. Sci.* 1895–1907.
J. Bot. 1923, 234–37. *Proc. Linn. Soc.* 1923–24, 46. *Bot. Soc. Exch. Club Br. Isl. Rep.* 1923, 145–48. *Watson Bot. Exch. Club Rep.* 1923–24, 241–43. *Trans. Proc. Perthshire Soc. Nat. Sci.* 1924, 1–3.
Plants at Nottingham Museum.

BARD, Samuel (*fl.* 1760s)
MD Edinburgh 1765. Sent plants to J. Hope (1725–1786).
Notes R. Bot. Gdn Edinburgh v.4, 1907, 125.

BARFOOT, Joseph Lindsey (1816–1882)
b. Warwick 29 March 1816 *d.* Salt Lake City, U.S.A. 25 April 1882

Royal Marines, 1834. To Salt Lake City, 1865 where he became Curator of Desert Museum.
Desert News Weekly 26 April 1882. J. Ewan *Rocky Mountain Nat.* 1950, 157.

BARHAM, Henry (1670–1726)
d. Spanish Town, Jamaica May 1726
FRS 1717. Surgeon, Royal Navy. *Hortus Americanus* 1794 (written 1711).
H. Sloane *Voyage to...Madeira...and Jamaica...* v.2, 1725, viii–x. J. Lunan *Hortus Jamaicensis* v.1, 1814, ii–iii. J. H. L. Archer *Monumental Inscriptions of British West Indies* 1875, 29, 51. W. A. Feurtado *Official and other Personages of Jamaica* 1896, 6. *D.N.B.* v.3, 186–87. I. Urban *Symbolae Antillanae* v.3, 1902, 19. E. J. L. Scott *Index to Sloane Manuscripts* 1904, 39. J. E. Dandy *Sloane Herb.* 1958, 87.
Plants at BM(NH).
Barhamia Klotz.

BARING, Olive Alethea (*née* Smith) (–1964)
d. Liss, Hants 5 March 1964
Close friend of G. C. Druce.
Proc. Bot. Soc. Br. Isl. 1964, 417.

BARKER, Mrs. (–1876)
d. 9 March 1876
Of Cape Schank, Victoria, Australia. Gave collection of algae to W. H. Harvey who included them in his *Phycologia Australica* 1858–63.
Victorian Nat. v.25, 1909, 102.
Rhodophyllis barkeriae Harvey.

BARKER, Bertie Thomas Percival (1877–1961)
b. Cambridge 9 Aug. 1877 *d.* 19 Dec. 1961
Educ. Cambridge. VMM 1952. Demonstrator in Botany, Cambridge. Assistant Director, National Fruit and Cider Institute, Long Ashton, 1904; Director, 1905–43. Professor of Agricultural Biology, Bristol, 1912–43. Joint editor, *J. Pomology and Hort. Sci.*, 1924–44. *Cider Apple Production* 1954.
Ann. Appl. Biol. 1962, 371. *Nature* v.193, 1962, 526. *Who was Who, 1961–1970* 57–58.

BARKER, Frank (1895–1954)
b. Calverley, Yorks 1895 *d.* Jan. 1954
VMM 1954. Nurseryman, Six Hills Nurseries, Leeds, Yorks.
Quart. Bull. Alpine Gdn Soc. v.22, 1954, 189–90.

BARKER, George (1776–1845)
d. 6 Dec. 1845
FRS 1839. Solicitor, Birmingham. Founded Philosophical Society in Birmingham. Had large collection of orchids, etc.
Floricultural Cabinet v.2, 1838, 7–8. *Gent. Mag.* 1846, 324–25. *D.N.B.* v.3, 200.
Cultivated plants at Kew.
Barkeria Knowles & Westc.

BARKER, John (1901–1970)
b. Finchley, London 24 April 1901 *d.* 30 Dec. 1970
Educ. Cambridge. FRS 1953. Researched on plant physiology under F. F. Blackman at Cambridge. On staff of Low Temperature Research Station, 1922. Reader in Botany and Head of Plant Physiology, Cambridge, 1952. Contrib. to *Proc. R. Soc.*, *New Phytol.*
Times 2 Jan. 1971. *Biogr. Mem. Fellows R. Soc.* 1972, 35–42 portr.

BARKER, Rev. John Theodore (–1883)
d. Leeds, Yorks 23 Sept. 1883
Of Louth and Bath. *Beauty of Flowers in Field and Wood* 1852; ed. 2 1857. Taught botany at Louth, Lincolnshire during 1850s.
Bot. Soc. Exch. Club Br. Isl. Rep. 1929, 91.

BARKER, Robert (*fl.* 1690s)
Of Beccles, Suffolk. "An industrious botanist who without banter knows to a yard square of ground where every rare plant of ye Island grows, having search'd it for these several years past" (A. Buddle in H. Sloane's Herb. at BM(NH) 150, f.46).
J. E. Dandy *Sloane Herb.* 1958, 87.

BARKER, Thomas (1838–1907)
b. Balgonie near Aberdeen 9 Sept. 1838 *d.* Buxton, Derbyshire 20 Nov. 1907
MA Aberdeen 1857. MA Cantab 1862. Professor of Mathematics, Owen's College, Manchester, 1865–85. Bequeathed £40,000 to found Professorship of Cryptogamic Botany at Manchester. Had alpine garden. Bryologist.
J. Bot. 1897, 91. *Trans. Liverpool Bot. Soc.* 1909, 59. *Ann. Scott. Nat. Hist.* 1908, 121–22. *D.N.B. Supplt. 2* v.1, 96–97.
Bryophytes at Manchester.

BARKLAY (Barckley) (*fl.* 1690s)
Surgeon. Correspondent of J. Petiver.
J. E. Dandy *Sloane Herb.* 1958, 88.

BARKLY, Lady Anne Maria (*née* Pratt) (1838–1932)
d. Sept. 1932
Second wife of Sir H. Barkly (1815–1898). Collected plants in Mauritius, Bourbon, Cape, 1867–78. 'Revised List of Ferns of South Africa' (*Cape Mon. Mag.* 1875).
Times 10 Sept. 1932.
Ferns at BM(NH) and Kew.

BARKLY, Lady Elizabeth Helen (*née* **Timins**) (–1857)
d. Melbourne 17 April 1857

First wife of Sir H. Barkly (1815–1898). Drawings of plants of British Guiana and Jamaica at Kew. Her daughter, E. B. Barkly, also drew plants. Collected ferns in Jamaica.

Proc. R. Soc. v.75, 1904, 23–25. *Kew Bull.* 1918, 342.

BARKLY, Sir Henry (1815–1898)
b. Monteagle, Ross-shire 24 Feb. 1815 *d.* S. Kensington, London 21 Oct. 1898

FRS 1864. KCB 1853. GCMG 1874. Governor, British Guiana, 1848; Jamaica, 1853; Victoria, 1856; Mauritius, 1863–70; Cape, 1870–77. *Notes on Fl. and Fauna of Round Island* 1870. "Patronised and encouraged horticulture and botany in our colonies" (*Bot. Mag.* v.79, 1853, Dedication). Collected plants which he sent to Kew. Collected and cultivated stapelias of which Lady Barkly and her daughter made drawings. '*Stapeliae Barklyanae*' (*Hooker's Icones Plantarum* v.20, 1890, t.1909).

J. Bot. 1873, 353. J. G. Baker *Fl. Mauritius* 1877, 9*, 10*–11*. *Kew Bull.* 1898, 335–36. *Britannia* Jan. 1899, 2–5 portr. *Yb. R. Soc.* 1900, i–ii. *D.N.B. Supplt 1* v.1, 124–26. I. Urban *Symbolae Antillanae* v.3, 1902, 19. *Who was Who, 1897–1916* 40. *Curtis's Bot. Mag. Dedications, 1827–1927* 103–4 portr. A. White and B. L. Sloane *Stapelieae* v.1, 1937, 108–10 portr. M. Macmillan *Sir Henry Barkly* 1970 portr. *R.S.C.* v.9, 124; v.12, 49.

Plants and letters at Kew. Portr. at Hunt Library.

Barklya F. Muell.

BARLOW, Samuel (1795–1855)
b. Blackley, Lancs 1795 *d.* Castleton near Manchester 4 Dec. 1855

Son of Joseph Barlow, a silk weaver and gardener. Bleach works manager at Stakehill, Castleton. "An ardent botanist, entomologist and gardener."

Trans. Liverpool Bot. Soc. 1909, 59.

BARLOW, Samuel (1825–1893)
b. Medlock Vale, Lancs 1825 *d.* Manchester 28 May 1893

Son of S. Barlow (1795–1855). Proprietor, Stakehill bleach works, Castleton. Cultivated florists' flowers in his garden at Stakehill. Vice-president, Manchester Botanical and Horticultural Society.

Florist and Pomologist 1883, 100–2 portr. *Gdnrs Chron.* 1883 i 531–32, 537 portr.; 1893 i 668–69. *Garden* 1893, 472. *J. Hort. Cottage Gdnr* 1893, 433, 502. R. Genders *Collecting Antique Plants* 1971, 18, 34.

BARLOW, Thomas Worthington (1823–1856)
b. Cranage, Cheshire 1823 *d.* Freetown, Sierra Leone 10 Aug. 1856

FLS 1848. Lawyer in Manchester. Antiquarian and naturalist. Went to Sierra Leone as Queen's Advocate, 1856. *Field Naturalist's Note Book* 1848.

Proc. Linn. Soc. 1856, xxi–xxii. *D.N.B.* v.3, 229. *Trans. Liverpool Bot. Soc.* 1909, 59.

BARNARD, Alicia Mildred (1825–1911)
b. 22 March 1825 *d.* Norwich, Norfolk 1 May 1911

Grand-niece of Sir J. E. Smith (1759–1828). Drew plants. List of bryophytes in R. H. Mason's *Hist. of Norfolk* 1884. '*Bromus pseudo-velutinus*' (*Phytologist* v.3, 1850, 807–8).

Herb. at Norwich Museum. Letters at Kew.

BARNARD, Anne (*née* **Henslow**) (–1899)
d. 19 Jan. 1899

Daughter of J. S. Henslow (1796–1861). Contrib. plates to *Bot. Mag.*, 1879–1886; also illustrated D. Oliver's *Lessons in Elementary Botany* 1864.

Kew Bull. 1899, 19–20.

BARNARD, Lady Anne (*née* **Lindsay**) (*fl.* 1790s–1800s)

Eldest daughter of James, Earl of Balcarres. Married Andrew Barnard who was appointed Colonial Secretary to the Cape, 1797–1802. Botanised on Table Mountain and drew Cape flora.

A. White and B. L. Sloane *Stapelieae* v.1, 1937, 88. D. Fairbridge *Lady Anne Barnard at the Cape of Good Hope, 1797–1802* 1924. *Country Life* 1963, 360–61 portr.

BARNARD, Basil Henry Francis (1874–1953)
b. Yatton, Somerset 1874 *d.* Clevedon, Somerset 20 Aug. 1953

Educ. Oxford. Forest Department, Malaya, 1898–1929. Chiefly in Perak where he collected forest trees.

J. Fed. Malay States Mus. 1915, 43–62. *Gdns Bull. Straits Settlements* 1927, 117. *Indian For.* 1930, 191–96. *Fl. Malesiana* v.1, 1950, 37; v.5, 1958, 23. *Malayan For.* 1954, 3–4.

Plants at Kuala Lumpur, Penang, Singapore.

BARNARD, Edward (1786–1861)
b. 14 March 1786 *d.* 13 Dec. 1861

FRS 1828. FLS 1818. Agent-General for Crown Colonies. Vice-secretary, Horticultural Society of London. 'Notes made in Garden of Horticultural Society upon Rate of Growth by Plants at Different Periods of the Day' (*Trans. Hort. Soc.* v.3, 1843, 103–13).

Proc. Linn. Soc. 1861–62, lxxxv. *Bot. Register* v.12, 1826 t.1029. *R.S.C.* v.1, 184.
Barnardia Lindl.

BARNARD, Francis (1823–1912)
d. Melbourne 21 Sept. 1912
Microscopist. Studied microscopic fungi.
Victorian Nat. 1912, 101.
Phragmidium barnardi Plowright.

BARNARD, Henry (*fl.* 1820s)
Sent *Peristeria* sp. collected in Panama to R. Harrison at Liverpool in 1826.
Bot. Mag. 1831, t.3116.

BARNARD, Robert Cary (1827–1906)
b. Cheltenham, Glos 13 Dec. 1827 *d.* Cheltenham 22 Dec. 1906
Educ. Cambridge. FLS 1861. British army officer. Botany master, Cheltenham College, 1876.
Proc. Linn. Soc. 1906–7, 37–38. H. J. Riddelsdell *Fl. Gloucestershire* 1948, cxxviii–cxxix.
Herb., formerly at Cheltenham School, now missing.

BARNES, Bertie Frank (1888–1965)
b. 19 July 1888 *d.* Midhurst, Sussex 19 March 1965
PhD 1928. FLS 1925. Demonstrator, then Lecturer in Botany, Birkbeck College, 1922–33. Head, Department of Biology, Chelsea Polytechnic, 1934–52. President, British Mycological Society, 1934. Botanical Secretary, Linnean Society, 1944–51. *Structure and Development of Fungi* (with H. Gwynne-Vaughan) 1930; ed. 2 1937. Contrib. to *J. Bot., New Phytol., Trans. Br. Mycol. Soc.*
Proc. Linn. Soc. 1966, 119. *Who was Who, 1961–1970* 59.

BARNES, Edward (*c.* 1892–1941)
d. Madras, India 31 May 1941
Professor of Chemistry, Madras Christian College. Collected plants in Madras Presidency for Kew. Particularly interested in *Arisaema, Impatiens* and *Sonerila.* Contrib. to *J. Nat. Hist. Soc. Bombay.*
Kew Bull. 1941, 240–41. *Nature* v.148, 1941, 221–22.
Herb. at Kew.

BARNES, Eliza Standerwick *see* Gregory, E. S.

BARNES, Isaac (*fl.* 1790s)
Seedsman, Northwick, Cheshire.

BARNES, James (1806–1877)
b. Farnham, Surrey 1806 *d.* Exmouth, Devon 23 May 1877
Gardener to Lord Rolle at Bicton, Devon, 1839–69. Contrib. to *Florist, J. Hort. Soc. Gdnrs Chron.* 1874 i 655–56 portr.; 1877 i 700. *Garden* v.11, 1877, 437. *J. Hort. Cottage Gdnr* v.38, 1899, 180. *R.S.C.* v.1, 185.

BARNES, James Martindale (1814–1890)
b. Selside, Westmorland 10 Feb. 1814 *d.* Levens, Westmorland 9 May 1890
Edited W. J. Linton's *Ferns of English Lake Country* 1865.
Kendal Times 16 May 1890. *J. Bot.* 1910, 161–62.
Mosses at Kendal Museum. Letters at BM (NH).
Bryum barnesii Schimper.

BARNES, John (*fl.* 1820s)
Gardener and seedsman, Potternewton, Leeds, Yorks.

BARNES, Nicholas F. (*c.* 1865–1950)
d. Chester 16 Jan. 1950
VMH 1924. Gardener to Duke of Westminster at Eaton Hall, Chester. Raised 'Arthur W. Barnes' apple in 1902.
Gdnrs Mag. 1907, 877–78 portr. *Gdnrs Chron.* 1950 i 38.

BARNES, Philip (1791–1874)
d. 24 Feb. 1874
FLS 1824. Founder member of Royal Botanical Society of London, 1839.
Gdnrs Chron. 1909 ii 234.

BARNES, Richard (1851–1918)
b. Thirsk, Yorks 6 Aug. 1851 *d.* Harrogate, Yorks 7 Nov. 1918
Kew gardener. Superintendent, Public Parks, Saltburn. In business in Harrogate. Bryologist. Contrib. to R. Braithwaite's *British Moss-flora* 1879–1905, J. G. Baker's *North Yorkshire* ed. 2 1906, and *Naturalist.*
Naturalist 1919, 44–45 portr.; 1973, 1–2.

BARNES, Thomas (1750s–1790s)
Gardener to William Thompson at Elsham, Lincs. Nurseryman at Briggate, Leeds. In partnership with Ebenezer Romain Calender in 1795. *New Method of Propagating Fruit-trees and Flowering Shrubs* 1758-59.

BARNES, Warren Delabere (1865–1911)
d. Hong Kong 1911
Educ. Cambridge. Malayan Civil Service, 1888–1910; Colonial Secretary, Hong Kong. Conducted expedition to Pahang, 1900 where he collected plants. 'Notes on Trip to Gunong Benom in Pahang' (*J. Straits Branch R. Asiatic Soc.* 1903, 1–10).
J. Straits Branch R. Asiatic Soc. 1911, 4. *Gdns Bull. Straits Settlements* 1927, 117. *Fl. Malesiana* v.1, 1950, 37.
Elytranthe barnesii Gamble.

BARNES, William (1809–1869)
b. Surrey 23 Oct. 1809 *d.* 22 Dec. 1869
Gardener to G. W. Norman at Bromley Common, Kent. Excelled in growing greenhouse plants. Nurseryman, Camden Nursery, Camberwell, London.
Florist and Pomologist 1870, 24. *Gdnrs Chron.* 1870, 14–15, 43.

BARNETT, Euphemia Cowan (–1970)
d. 12 March 1970
FLS 1938. Lecturer in Botany, Robert Gordon Technical College, Aberdeen. Assistant to W. G. Craib and worked on A. F. G. Kerr's Siamese collections. Contrib. to *Fl. Siam* (*Kew Bull.*)

BARON, Rev. Richard (1847–1907)
b. Kendal, Westmorland 8 Sept. 1847 *d.* Morecambe, Lancs 12 Oct. 1907
FLS 1882. Missionary in Madagascar, 1872–1907. Plants collected by him described by J. G. Baker in *J. Bot.* 1882, *J. Linn. Soc. Bot.* 1883 and 1889.
Kendal Mercury 18 Oct. 1907. *Kew Bull.* 1908, 45–46. *Proc. Linn. Soc.* 1907–8, 44–45. *Quart. J. Geol. Soc.* 1908, lxiv. *Dansk Bot. Archiv* v.7, 1932, x. *Comptes Rendus AETFAT 1960* 1962, 133.
Madagascar plants at Kew, BM(NH), Paris. Portr. at Hunt Library.
Baronia Baker. *Neobaronia* Baker.

BARR, Peter (1826–1909)
b. near Govan, Lanarkshire 20 April 1826 *d.* London 17 Sept. 1909
VMH 1897. Florist with Sugden at Covent Garden, 1861. Specialised in narcissi. *Ye Narcissus, or Daffodyl Flower* 1884. *Chat on Ancient and Modern Daffodils* 1900.
Gdnrs Chron. 1901 ii 46; 1909 ii 216–17. *J. Hort. Cottage Gdnr.* v.47, 1903, 241 portr.; v.57, 1908, 448–50 portr.; v.59, 1909, 307 portr. *Garden* v.42, 1892, xii portr.; 1909, 475 portr. *Gdnrs Mag.* 1909, 744 portr. *Nature* v.81, 1909, 400. *J. Bot.* 1915, Supplt 2, 10. *Daffodil Yb.* 1933, 24–34 portr. *Daffodil Tulip Yb.* 1950, 9–10 portr.; 1971, 11–17.
Portr. at Hunt Library.

BARR, Peter Rudolph (*c.* 1862–1944)
d. Hendon, London 25 Nov. 1944
VMH 1931. Nurseryman, Covent Garden and Taplow, Bucks. Authority on daffodils and tulips.
Garden 1909, 587 portr. *Gdnrs Mag.* 1911, 257–58 portr. *Gdnrs Chron.* 1932 i 18 portr.; 1944 ii 222. *Daffodil Tulip Yb.* 1946, 89.

BARR, Thomas (*fl.* 1790s)
Nurseryman, Northampton Nursery, Ball's Pond, Islington, 1791. Samuel Brookes became a partner in 1819.
Trans. London Middlesex Archaeol. Soc. v.24, 1973, 185.

BARRATT, John (*c.* 1770–*c.* 1829)
Gardener and nurseryman, Market Place, Wakefield, Yorks. Went bankrupt in 1803. Soon re-established business with aid of his son, William.

BARRATT, Joseph (1796–1882)
b. Little Hallam, Derbyshire 7 Jan. 1796 *d.* Middletown, Conn. U.S.A. 25 Jan. 1882
Studied medicine in London. Practised at Leicester, 1816–19. Went to Philipstown, N.Y., 1819. Collected in vicinity cited by J. Torrey in *Fl. State of New York*. Collected at Mt. Washington, Conn., 1824. Professor of Botany, Chemistry and Mineralogy, Military Academy, Middletown, Conn., 1826. *Salices Americanae* 1840. Willows in Sir W. J. Hooker's *Fl. Boreali-americana* v.2, 1840, 144–60.
Rhodora 1921, 121–25, 171–77, 300–1. C. S. Sargent *Silva of North America* v.14, 1902, 64.
Herb. at Wesleyan University. Correspondence with J. Torrey, 1827–40, at New York Botanic Garden.
Barrattia A Gray.

BARRATT, Joseph (*fl.* 1820s)
Nurseryman and seedsman, 44 Mornington Place, London.

BARRATT, Thomas J. (1841–1914)
Managing director, A. and F. Pears. Studied diatoms.
J. R. Microsc. Soc. 1914, 237–38.

BARRATT, William (*fl.* 1800s–1860s)
Son of J. Barratt (*c.* 1770–*c.* 1829). Nurseryman. Opened Subscription Botanic Gardens in St. John's, Wakefield, Yorks., 1833.

BARRETT, Richard (*fl.* 1800s)
Nurseryman and seedsman, Hornsey, Middx.

BARRETT, William Bowles (1833–1915)
b. 17 March 1833 *d.* Weymouth, Dorset 17 April 1915
FLS 1883–93. Solicitor. 'Contribution towards Fl. Breconshire' (*J. Bot.* 1885 *passim*). 'Notes on Fl. Chesil Bank and the Fleet' (*Proc. Dorset Field Club* 1905, 251–65).
Proc. Linn. Soc. 1883, 6.

BARRETT-HAMILTON, Gerald Edwin Hamilton (1871–1914)
b. India 1871 *d.* South Georgia 17 Jan. 1914
Educ. Cambridge. Zoologist. Contrib. papers on Wexford plants to *J. Bot.* and *Irish Nat.*, 1887–1908.
Irish Nat. 1914, 81–93 portr.

BARRINGTON, Arthur Harry Manliffe (1886-1932)
d. 11 June 1932

Indian Forest Service in Burma, 1907. *Note on Vegetation on Forest Soils* 1929. *Forest Soils and Vegetation in Hlaing Circle, Burma* 1931.

Indian For. 1932, 464–65.

BARRINGTON, Hon. Daines (1727–1800)
d. London 14 March 1800

FRS 1767. Lawyer, antiquarian and naturalist. Correspondent of Gilbert White. Close friend of Sir Joseph Banks. *Naturalists' Calendar* 1817. Contrib. papers on indigenous trees to *Philos. Trans. R. Soc.* 1770 and 1772.

Gent. Mag. 1800, 291. J. Nichols *Lit. Anecdotes of Eighteenth Century* v.3, 3; v.6, 385–86. J. Nichols *Illus. Lit. Hist. of Eighteenth Century* v.5, 582–607. S. Felton *Portr. of English Authors on Gdning* 1830, 177–79. *D.N.B.* v.3, 286–88. *Dict. Welsh Biogr. down to 1940* 1959, 27.

MS. at Royal Society. Letters at National Museum of Wales.

Barringtonia Forster.

BARRINGTON, Richard Manliffe (1849–1915)
b. Fassaroe, Bray, County Wicklow 22 May 1849 *d.* Fassaroe 15 Sept. 1915

LLB Dublin. MA Cantab. FLS 1883. Land valuer. Explored botanically the Irish lakes; also visited Iceland, 1881; St. Kilda, 1883; Rocky Mountains, 1884. Discovered *Caltha radicans* in Ireland. Contrib. to *J. Bot.*, 1872–92 and *Proc. R. Irish Acad.*

N. Colgan and R. W. Scully *Cybele Hibernica* 1898, xxi. *Irish Nat.* 1915, 193–206 portr. *J. Bot.* 1915, 364–67 portr. *Proc. Linn. Soc.* 1915–16, 56–57. *Bot. Soc. Exch. Club Br. Isl. Rep.* 1915, 247–49. *Nature* v.96, 1915, 119. R. W. Scully *Fl. Kerry* 1916, xvi. R. L. Praeger *Some Irish Nat.* 1949, 47–48 portr. *R.S.C.* v.7, 91; v.9, 128

Herb. at Dublin National Museum.

BARRINGTON, Rev. Shute (1734–1826)
b. Beckett, Berks 26 May 1734 *d.* London 25 March 1826

MA Oxon 1757. DCL 1762. FLS 1812. Bishop of Llandaff, 1769; Salisbury, 1782; Durham, 1791. "Much devoted to the study of Botany." Lord Bute presented his wife with one of his twelve copies of *Botanical Tables.*

Proc. Linn. Soc. 1888–89, 31–32. *D.N.B.* v.3, 294.

Bust by A. Behnes at Linnean Society.

BARRITT, Peter (*c.* 1748–1798)

Nurseryman, Old Brompton, Kensington, London. Partner of D. Grimwood (–1796)

BARRON, Archibald F. (1835–1903)
b. Banchory near Aberdeen 1835 *d.* Chiswick, Middx 15 April 1903

VMH 1897. Gardener on numerous estates. Superintendent, Royal Horticultural Society gardens at South Kensington and Chiswick. *Vines and Vine Culture* 1883; ed. 4 1900.

Gdnrs Chron. 1872, 74 portr.; 1893 i 635 portr.; 1903 i 265–66 portr. *Gdnrs Mag.* 1894, 601 portr.; 1903, 280 portr. *Garden* v.63, 1903, 286–87 portr., 305–6. *J. Hort. Cottage Gdnr* v.13, 1886, 7 portr.; v.32, 1896, 363–65, 375 portr.; v.46, 1903, 367–69 portr. *J. R. Hort. Soc.* 1903–4, 181–82 portr. H. R. Fletcher *Story of R. Hort. Soc.* 1969, *passim* portr.

BARRON, James (*fl.* 1830s)

Nurseryman, 10 Haymarket, Sheffield, Yorks.

BARRON, John (1844–1906)
b. Elvaston, Derbyshire 8 June 1844 *d.* 7 May 1906

Nurseryman. Joined his father's nursery which had been established at Elvaston in 1851. Like his father, W. Barron, he specialised in transplanting trees, their most successful joint projects being the moving of the Buckland Yew near Dover and the John Knox Yew at Longbank, Scotland.

Gdnrs Chron. 1906 i 303, 319.

BARRON, Leonard (1868–1938)
b. Chiswick, Middx 29 Sept. 1868 *d.* Rockville, New York, U.S.A. April 1938

Horticultural writer. *Pocket Garden Library* 1917 4 vols. *American Home Book of Gardening* 1931. *Gardening for the Small Place* 1935.

American Men of Sci. ed. 2, 3–4.

Portr. at Hunt Library.

BARRON, William (1800–1891)
b. Eccles, Berwickshire 7 Sept. 1800 *d.* 8 April 1891

Gardener to Lord Harrington at Elvaston Castle, Derbyshire. Later nurseryman at Borrowash and Nottingham. Skilled in moving large trees. *British Winter Garden* 1852.

Garden v.39, 1891, 401. *Gdnrs Chron.* 1891 i 505, 522–24, 531 portr.

BARROW, Sir John (1764–1848)
b. Ulverston, Lancs 19 June 1764 *d.* London 23 Nov. 1848

LLD Edinburgh. FRS 1805. FLS 1810. Secretary to Admiralty, 1804. *Autobiographical Memoir* 1847 portr.

Proc. Linn. Soc. 1848, 38–39, *D.N.B.* v.3, 305–7. *Trans. Liverpool Bot. Soc.* 1909, 59.

British plants at Manchester. Portr. at Admiralty.

Barrowia Decaisne.

BARROW, John (1822–1890)
d. Rhos, Colwyn, Denbigh 19 Oct. 1890
Microscopist. Original member, Leeuwenhoek Microscopical Club, Manchester, 1867; President, 1883–84. '*Tricophyta tonsurans*' (*Proc. Lit. Philos. Soc., Manchester* v.11, 1871–72, 29–32, 61–62).
Herb. at Manchester.

BARROW, Robert (*fl.* 1780s)
Gardener and nurseryman, Windsor, Berks.

BARRY, Rev. George *see* Low, G.

BARRY, Martin (1802–1855)
b. Fratton, Hants 28 March 1802 *d.* Beccles, Suffolk 27 April 1855
MD Edinburgh 1833. FRS 1840. FRSE. Physician, Edinburgh. Wrote several papers on plant cells and fibres.
D.N.B. v.3, 326–27.
Plants at Kew.

BARRY, Patrick (1816–1890)
b. near Belfast May 1816 *d.* Rochester, New York, U.S.A. 23 June 1890
Schoolteacher. Emigrated to U.S.A., 1836. In 1840 founded with G. Ellwanger Mount Hope Nurseries at Rochester. *Treatise on Fruit Garden* 1851; new ed. as *Barry's Fruit Garden* 1872. Edited *Horticulturist*.
Appleton's Cyclop. Amer. Biogr. v.1, 1887, 181. *Gdn and For.* v.3, 1890, 328. *Ann. Hort.* 1890, 287–90. *J. Hort. Cottage Gdnr* v.21, 1890, 115. L. H. Bailey *Standard Cyclop. Hort.* v.2, 1939, 1564 portr.
Portr. at Hunt Library.

BARTAR, Edward (*fl.* 1690s)
Employee of Royal African Company in Guinea whence he sent plants to J. Petiver (1663–1718).
J. E. Dandy *Sloane Herb.* 1958, 88.

BARTER, Charles (–1859)
d. Rabba, W. Africa 15 July 1859
ALS 1858. Kew gardener, 1849–51. Foreman, Royal Botanic Society, Regent's Park, London, 1851–57. On Niger Expedition, 1857–59. Letters published in *J. Linn. Soc.* v.4, 1879, 17–26 and *Gdnrs Chron.* 1858–59.
Proc. Linn. Soc. 1859, xx–xxi. S. Crowther and J. C. Taylor *Niger Expedition of 1857–1859* 1859. *J. Linn. Soc.* v.5, 1860, 14–16. *Garden* v.5, 1874, 216. *Bull. Soc. Bot. France* 1882, 172–73. A. C. C. Hastings *Voyage of the Daysprig* 1926. *Comptes Rendus AETFAT, 1960* 1962, 98. *R.S.C.* v.1, 196; v.12, 53.
Letters and Nigerian plants at Kew.
Barteria Hook.f.

BARTHOLOMAEUS Anglicus *see* Glanville, Bartholomaeus de

BARTHOLOMEW, Arthur Churchill (*c.* 1846–1940)
d. Reading, Berks 29 March 1940
VMH 1923. Noted dahlia grower. Received the Blue Riband of British Dahlia Growers' Association 3 times.
Gdnrs Chron. 1940 i 170.

BARTHOLOMEW, Valentine (1799–1879)
b. Clerkenwell, London 18 Jan. 1799 *d.* 21 March 1879
Artist. Member of Society of Painters in Water-colours. Flower-painter-in-ordinary to Queen Victoria. *Selection of Flowers* 1822. *Group of Flowers.* Contrib. t.3479 to *Bot. Mag.*
Athenaeum 29 March 1879. *Portr. of Men of Eminence in Lit., Sci. and Art* v.4, 1866, 89–92 portr. *Bryan's Dict. of Painters and Engravers* v.1, 1903–4, 88. *Notes and Queries* v.158, 1930, 223, 261–62.
Paintings at Victoria and Albert Museum.

BARTLETT, Albert William (1875–1943)
b. 2 June 1875 *d.* 4 March 1943
BSc London 1898. BA Cantab 1903. FLS 1903. Science master, Henley-on-Thames, 1898–1901. Superintendent, Botanic Garden, Georgetown and Government Botanist, British Guiana, 1903–8. Founded and edited *J. Board Agric. Br. Guiana.* Assistant lecturer in Botany, Sheffield University, 1909–20. Lecturer in Botany, Armstrong College, Newcastle-on-Tyne, 1920–39. Mycologist. President, Northern Naturalists Union, 1938. Contrib. to *Trans. Br. Mycol. Soc.* and *Vasculum.*
Proc. Linn. Soc. 1942–43, 293–94. *Vasculum* v.28, 1943, 11–12.
British Guiana plants at Kew.

BARTLETT, Aubrey Cecil (–1950)
b. Devon
Kew gardener. Contrib. reports on Royal Horticultural Society shows to *Gdnrs Chron.* for 40 years. Secretary, National Sweet Pea Society, National Dahlia Society. For short while had small market garden at Hampton, Middx.
Gdnrs Chron. 1950 i 69.

BARTLETT, Edward (*fl.* 1890s)
Curator, Sarawak Museum, Kuching, 1895. Collected plants.
Fl. Malesiana v.1, 1950, 38.
Plants at BM(NH).

BARTON, Benjamin Herbert (*fl.* 1830s–1840s)
FLS 1835. Of Great Missenden, Bucks. *British Fl. Medica* (with T. Castle) 1836–38; new ed. 1877.
Gdnrs Mag. 1836, 263.

BARTON, Ethel Sarel *see* Gepp, E. S.

BARTON, Francis Rickman (1865–1947)
b. 4 Jan. 1865 *d*. 4 Oct. 1947
Army Captain; served in Sierra Leone and Barbados. Lieut.-Governor, New Guinea, 1899; Magistrate, 1903; Administrator, Papua, 1904–7; First Minister, Zanzibar, 1908. Collected plants.
Geogr. J. 1900–1, 63–68. *Fl. Malesiana* v.1, 1950, 39–40. *Who was Who, 1941–1950* 69.
Plants at BM(NH), Brisbane.
Platyclinis bartoni Ridley.

BARTON, John (*fl*. 1810s–1830s)
Of Stoughton, Chichester, Sussex. *Lecture on Geography of Plants* 1827.
Gdnrs Mag. 1828, 200. *R.S.C.* v.1, 200.

BARTON, Rev. William (*c*. 1754–1829)
b. Preston, Lancs *c*. 1754 *d*. Lytham, Lancs 21 March 1829
MD Edinburgh. BA Cantab 1795. Incumbent of Samlesbury, Lango and Great Harwood. "An accurate naturalist, botanist, poet."
P. Whittle *Topographical...and Hist. Account of...Preston* v.2, 272. *Trans. Liverpool Bot. Soc.* 1909, 59.

BARTON, William Charles (–1955)
H. N. Clokie *Account of Herb. Dept. Bot., Oxford* 1964, 126.
Herb. at BM(NH). Plants at Kew and Oxford.

BARTRAM, John (1699–1777)
b. Darby, U.S.A. 23 March 1699 *d*. Philadelphia, U.S.A. 22 Sept. 1777
King's botanist in America, 1765. "The greatest natural botanist in the world" (Linnaeus). "The earliest native-born American botanist" (Asa Gray). Correspondent of Sir H. Sloane, J. Hill, J. Ellis and P. Collinson. 'Diary of Journey through the Carolinas, Georgias and Florida, 1765–1766' (*Trans. Amer. Philos. Soc.* v.33 (1), 1942 port.).
Med. Physical J. v.1, 1804, 115–24. J. Nichols *Lit. Anecdotes of Eighteenth Century* v.5, 485. J. E. Smith *Selections from Correspondence of Linnaeus* v.1, 1821, 536–38. *Gdnrs Mag.* 1831, 665–66. W. Darlington *Memorials of John Bartram and Humphry Marshall* 1849. *Cottage Gdnr* v.5, 1851, 327. J. Smith *Cat. of Friends' Books* v.1, 1867, 201–2. *Harper's Mag.* v.60, 1880, 321–30. *Gdn and For.* 1896, 121–24. *Meehan's Mon.* 1896, 17; 1899, 96. J. W. Harshberger *Botanists of Philadelphia and their Work* 1899, 46–76. E. O. Abbot *Bartram's Gdn* 1904 portr. E. J. L. Scott *Index to Sloane Manscripts* 1904, 42. *J. Bot.* 1906, 213–14; 1915, 255. *Friends' Quart. Examiner* 1915, 146–53. H. A. Kelly *Some American Med. Botanists* 1915, 49–59. *Torreya* v.16, 1916, 116–19. R. H. Fox *Dr. John Fothergill and his Friends* 1919, 169–71. *Sci. Mon.* v.21, 1925, 191–216. *Nat. Hist. Mag.* v.2, 1929, 50–58. *Bartonia* no. 12, 1931 (special issue, 68p.) *Notes and Queries* v.161, 1931, 75–76. *Bull. Soc. Dendrol. France* no. 80, 1931, 89–91. *Gdnrs Chron.* 1932 i 182–83. *J. Heredity* 1932, 443–45. *Bull. Gdn Club Amer.* 1939, 56–64. E. Earnest *John and William Bartram, Botanists and Explorers, 1699–1823* 1940 portr. *Herbarist* no. 15, 1949, 24–30. J. Herbst *New Green World. John Bartram and Early Nat.* 1954. *Classical J.* v.50, 1955, 167–70. H. G. Cruickshank *John and William Bartram's America* 1957. *J. Soc. Bibl. Nat. Hist.* v.3, 1957, 263–72. *Penn. Genealogical* Mag. v.20, 1957, 253–55. J. E. Dandy *Sloane Herb.* 1958, 88–89. *Country Life* 1956, 549–50; 1966, 556–59 portr. *Gdn J.* 1967, 11–14 portr. *Morris Arboretum Bull.* 1967, 75–81. A. M. Coats *Quest for Plants* 1969, 273–76. C. C. Gillispie *Dict. Sci. Biogr.* v.1, 1970, 486–88. H. Savage *Lost Heritage* 1970, 93–132.
Plants at BM(NH). Letters to J. Fothergill at BM(NH). MS. at Royal Society. Photocopies of MSS. at American Philosophical Society. Portr. at Hunt Library.
Bartramia Salisb.

BARTRAM, William (1739–1823)
b. Botanic Garden, Kingsessing, Philadelphia, U.S.A. 9 Feb. 1739 *d*. Philadelphia 22 July 1823
Son of J. Bartram (1699–1777). Drew plates for B. S. Barton's *Elements of Botany* 1803. Professor of Botany, Pennyslvania University, 1782. *Travels through North and South Carolina, Georgia, East and West Florida* 1791. 'Travels in Georgia and Florida, 1773–74' (*Trans. Amer. Philos. Soc.* v.33 (2), 1943 portr.).
Cabinet Nat. Hist. and Amer. Rural Sports v.2, 1832, i–vii. W. Darlington *Reliquiae Baldwinianae* 1843 *passim*. W. Darlington *Memorials of John Bartram and Humphry Marshall* 1849, 288–90, etc. J. W. Harshberger *Botanists of Philadelphia and their Work* 1899, 86–88. E. O. Abbot *Bartram's Gdn* 1904 portr. *A.L.A. Portr. Index* 1906, 98. *Cassinia* v.10, 1906, 2–9. R. H. Fox *Dr. John Fothergill and his Friends* 1919, 185–92. *Bartonia* no. 12, 1931 (special issue, 68p.); no. 21, 1940–41, 6–8; no. 23, 1944–45, 10–35. N. B. Fagin *William Bartram* 1933, portr. *Bull. Gdn Club Amer.* 1939, 54–64. *Sci. Mon.* v.48, 1939, 380–84 portr. *Amer. Midland Nat.* v.23, 1940, 692–723. E. Earnest *John and William Bartram, Botanists and Explorers, 1699–1823* 1940, portr. *Notes and Queries* v.184, 1943, 154. *Rhodora* 1944, 389–91. *Amer. Fern J.* 1945, 23–25. *Proc. Amer. Philos. Soc.* 1945, 27–38; 1953, 571–77. *Chronica Botanica* 1946, 377–

85. *Country Life* 1956, 549–50 portr. H. G. Cruickshank *John and William Bartram's America* 1957. F. Harper *Travels of William Bartram* 1958. F. A. Stafleu *Taxonomic Literature* 1967, 16. J. Ewan *William Bartram: Bot. and Zool. Drawings, 1756–1788* 1969. C. C. Gillispie *Dict. Sci. Biogr.* v.1, 1970, 488–90. H. Savage *Lost Heritage* 1970, 133–78 portr. *Taxon* 1974, 462 portr.

Plants, MSS. and drawings at BM(NH). Portr. at Hunt Library.

Lantana bartramii Baldw.

BARTY, Rev. James Strachan (1805–1875)
b. Bendochy, Perthshire 1805 *d.* Bendochy 1875

DD St. Andrews 1852. Minister, Bendochy, 1832. List of plants for Bendochy in *New Statistical Account of Scotland.*

Perthshire Soc. Nat. Sci. v.4, 1907–8, cxciii.

BARWISE, Joseph F. (*c.* 1874–1965)
d. 12 Feb. 1965

Nurseryman, Barnsley, Yorks. Hybridised dahlias. Received the Blue Riband of British Dahia Growers' Association 3 times.

Gdnrs Chron. 1965 i 295 portr.

BASSINGTON, George Henry
(*fl.* 1810s–1820s)

Nurseryman, Kingsland, Islington, London. Nursery comprised the grounds formerly belonging to Thomas Fairchild and John Cowell. In partnership with Bunney by 1825.

Bot. Mag. 1813, t.1583.

BASSINGTON, James (*fl.* 1800s)

Seedsman, 68 West Smithfield, London.

BASSINGTON, Thomas (*fl.* 1800s–1840s)

Nurseryman, Kingsland Nursery, Balls Pond Road, Islington. Successor to Mackie's nursery, 1800. Bunney became a partner in 1825.

BASTER, Job (1711–1775)
b. Zierikzee, Netherlands 2 April 1711 *d.* Leyden, Netherlands 6 March 1775

MD. FRS 1738. Algologist. Friend of Philip Miller. *Opulusca Subseciva* 1762–65. *Verhandeling over de voorttelling der dieren en planten* 1768. Contrib. papers on Corallines to *Philos. Trans. R. Soc.* v.50, 1758, 258–87; v.52, 1762, 108–18.

P. Miller *Gdnrs Dict.* 1759 (under *Basteria*). E. J. L. Scott *Index to Sloane Manuscripts* 1904, 42.

Basteria Miller.

BASTIAN, Henry Charlton (1837–1915)
b. Truro, Cornwall 26 April 1837 *d.* Chesham Bois, Bucks 17 Nov. 1915

MA London 1861. MD 1866. FRS 1868. FLS 1863. Professor, University College, London, 1887. *Beginnings of Life* 1872. 'Fl. Falmouth and Surrounding Parishes' (*Rep. Cornwall Polytechnic Soc.* 1856, 83–112).

F. H. Davey *Fl. Cornwall* 1909, l–li. *Nature* v.96, 1915, 347–48. *Proc. Linn. Soc.* 1915–16, 57–58. *Proc. R. Soc.* v.89, 1916, xxi–xxiv. *R.S.C.* v.1, 204; v.7, 97; v.9, 137; v.12, 54.

BASTOW, Richard Austin (1839–1920)
b. Edinburgh 14 May 1839 *d.* St. Kilda, Australia 14 May 1920

FLS 1885–89. Architect. To Tasmania, 1884. Paper on mosses and algae in *Proc. R. Soc. N.S.W.* Contrib. to *Victorian Nat.*

Lancashire and Cheshire Nat. v.15, 1922–23, 88–89. *Victorian Nat.* v.37, 1920, 27; v.66, 1949, 106 portr.

Herb. and notebooks on bryophyta at Melbourne.

Jungermannia bastovii Carr. & Pears.

BATCHELDER, Stephen John (1870–1949)
b. Great Yarmouth, Norfolk 31 May 1870 *d.* Ipswich, Suffolk 9 Nov. 1949

FLS 1928. Science master, Ipswich. Studied flora of Suffolk. 'Wild Plants growing on Waste Ground at Ipswich' (*Trans. Suffolk Nat. Soc.* 1949).

Proc. Linn. Soc. 1949–50, 99.

BATCHELOR, Rev. John (*fl.* 1880s–1910s)

DD. Missionary to the Ainu, 1887–1911. Archdeacon of Hokkaido, 1911. 'China Economic Plants' (with Kingo Miyabe) (*Trans. Asiatic Soc. Japan* 1893).

BATEMAN, Edward La Trobe (*c.* 1815–1897)
b. Derbyshire ? *d.* Rothesay, Bute 30 Dec. 1897

Book illuminator, draughtsman, architectural decorator and garden designer. Contrib. flower illustrations to M. A. Bacon's *Fruits from Garden and Field* 1837 and *Winged Thoughts* 1851. Joined gold rush in Victoria, 1852. It has been said that he had a part in planning Botanic Garden, Melbourne. Left Australia, 1867. Landscape gardener to Marquess of Bute.

Austral. Dict. Biogr. v.3, 1969, 117–18.

Drawings at National Gallery of Victoria.

BATEMAN, James (1811–1897)
b. Redivals, Bury, Lancs 18 July 1811 *d.* Worthing, Sussex 27 Nov. 1897

BA Oxon 1834. MA 1845. FRS 1838. FLS 1833. Cultivated tropical plants, especially orchids. Employed T. Colley to collect orchids in British Guiana. President, North Staffordshire Field Club, 1865–70. *Orchidaceae of Mexico and Guatemala* 1837–43. *Guide to Cool Orchid Growing* 1864. *Monograph of Odontoglossum* 1864–74. *Second Century of Orchidaceous Plants* 1867.

Gdnrs Chron. 1871, 1514–15 portr.; 1897 ii 400–3, portr. 410, 436, 446. *Orchid Rev.* 1898, 10–14, 56–57. *Garden* v.52, 1897, 454. *Times* 2 Dec. 1897. *Proc. Linn. Soc.* 1897–98, 34–35. *D.N.B. Supplt. 1* v.1, 137. *Trans. Liverpool Bot Soc.* 1909, 59–60. *Curtis's Bot. Mag. Dedications, 1827–1927* 43–44 portr. *Trans. N. Staffordshire Field Club* v.67, 1932–33, 38–40 portr. *Amer. Orchid Soc. Bull.* 1964, 297–98 portr. F. A. Stafleu *Taxonomic Literature* 1967, 17. M. A. Reinikka *Hist. of the Orchid* 1972, 184–85 portr.

Plants, drawings and letters at Kew. Portr. at Hunt Library.

Batemannia Lindl.

BATEMAN, Rev. John (*fl.* 1665–1720s)
b. Sittingbourne, Kent 1665

BA Oxon 1687. MA 1690. List of his Faversham plants in Sloane Herb. (317); basis of E. Jacob's *Plantae Favershamienses* 1777 (preface).

R. Pulteney *Hist. Biogr. Sketches of Progress of Bot. in England* v.2, 1790, 272. D. Turner *Extracts from Lit. and Sci. Correspondence of Richard Richardson* 1835, 140. F. J. Fawcett and A. B. Rendle *Fl. Kent* 1899, lxvi. *Proc. Bot. Soc. Br. Isl.* 1963, 20–22; 1966, 226–28.

Herb. at Pharmaceutical Society, London.

BATES, George (*c.* 1880–1971)
b. Ware, Herts *c.* 1880 *d.* 16 May 1971

Kew gardener, 1902–4. Established carnation nursery in 1913 in partnership with Mr. Willis at Sawbridgeworth, Herts; dissolved in 1919. Subsequently manager of nurseries at Winchmore Hill, Walthamstow and Westerham. Returned to private estate gardening.

J. Kew Guild 1972, 61 portr.

BATES, George F. (*c.* 1868–1933)
b. Kendal, Westmorland *c.* 1868 *d.* 10 Feb. 1933

BA. BSc. FRMS 1920. Schoolteacher at various schools in Perthshire. President, Perthshire Society of Natural Science, 1918–26. Collected and studied diatoms. Contrib. to *Trans. Proc. Perthshire Soc. Nat. Sci.*

Trans. Proc. Perthshire Soc. Nat. Sci. v.9, 1937–38, xliv–xlv.

BATES, Henry Walter (1825–1892)
b. Leicester 8 Feb. 1825 *d.* London 16 Feb. 1892

FRS 1881. FLS 1871. Traveller, naturalist and entomologist. Explored Amazon Valley, 1848–59. Assistant Secretary, Royal Geographical Society, 1864–92. *Naturalist on the Amazons* 1863.

Nature v.45 1891–92, 398–99. *Proc. Linn. Soc.* 1891–92, 61–62. *Trans. R. Geogr. Soc.* 1892, 177, 190–93, 245–57. *D.N.B. Supplt 1.* v.1, 141–44. *Adelphi* 1945, 115–21. *Geogr. J.* 1949, 1–3 portr. G. Woodcock *Henry Walter Bates* 1969 portr. C. C. Gillispie *Dict. Sci. Biogr.* v.1, 1970, 500–4.

Portr. at Hunt Library.

Batesia Spruce.

BATES, J. (*fl.* 1830s)

Nurseryman, Oxford.

BATES, John Thomas (*c.* 1884–1966)
b. Hounslow, Middx *c.* 1884 *d.* Hounslow 23 July 1966

Had large collection of succulents.

Cactus and Succulent J. Great Britain 1966, 63 portr.

BATES, Richard (*fl.* 1680s)

Gardener and nurseryman, Ashley, Staffs. Staffs.

BATESON, Mrs. Beatrice (–1941)

Widow of W. Bateson (1861–1926). *William Bateson, F.R.S., Naturalist* 1928.

Nature v.147, 1941, 703.

BATESON, William (1861–1926)
b. Whitby, Yorks 8 Aug. 1861 *d.* Merton, Surrey 8 Feb. 1926

BA Cantab 1883. FRS 1894. FLS 1909. VMH 1901. Geneticist. Professor of Biology, Cambridge, 1908–9. Director, John Innes Horticultural Institution, 1910–26. Founded *J. Genetics*. *Materials for Study of Variation*, 1894. *Variation and Differentiation in Parts and Brethren* 1903. *Mendel's Principles of Heredity* 1913. *Problem of Genetics* 1913. *Scientific Papers of William Bateson* (edited by R. C. Punnett) 1928 2 vols. *Letters from the Steppe...1886–1887* 1928.

Nature v.50, 1894, 55; v.71, 1904–5, 110; v.117, 1926, 239, 312–13. *J. Hort. Cottage Gdnr* v.45, 1902, 310–11. *Gdnrs Chron.* 1909 ii 360 portr.; 1926 i 125, 127, 128 portr., 141; 1929 i 62–63. *Ann. Rep. Smithsonian Inst.* 1926, 521–32. *J. Bot.* 1926, 78–80; 1929, 213–15. *J. Heredity* 1926, 433–49 portr. *Priroda* no. 11–12, 1926, 69–74. *Proc. Linn. Soc.* 1925–26, 66–74. *Bot. Soc. Exch. Club Br. Isl. Rep.* 1926, 87–88. *Science* v.63, 1926, 531–35. *Times* 9 Feb. 1926. *Proc. R. Soc. B.* v.101, 1927, i–v portr. *J. R. Hort. Soc.* 1928, 399–402. B. Bateson *William Bateson* 1928 portr. *D.N.B. 1922–1930* 67–68. *John Innes Inst. 1910–1935* 2, 3, 22, 37–39. *Notes and Records R. Soc.* 1952, 336–47. C. C. Gillispie *Dict. Sci. Biogr.* v.1, 1970, 505–7.

Portr. by W. A. Foster at National Portrait Gallery, London. Portr. at Hunt Library.

BATHURST, Rev. Ralph (1620–1704)
b. Hothorpe, Theddingworth, Northants 1620 *d.* Oxford 8 June 1704

BA Oxon 1638. MD 1654. FRS 1663. Dean of Wells, 1670. *Praelectiones tres de Respiratione* 1654.

T. Warton *Life and Lit. Remains of Ralph Bathurst* 1761 portr. K. Sprengel *Historia Rei Herbariae* v.2, 1808, 9. J. Foster *Alumni Oxonienses* v.1, 1891–92, 2. *D.N.B.* v.3, 409–11. E. J. L. Scott *Index to Sloane Manuscripts* 1904, 43. A. R. Horwood and C. W. F. Noel *Fl. Leicestershire* 1933, clxxxvi.

BATLEY, James (*c.* 1820–1902)
d. Wentworth, Yorks 11 May 1902

Gardener to Vernon-Wentworth family at Wentworth Castle.

Gdnrs Chron. 1902 i 360.

BATT, George (*fl.* 1820s–1840s)

Seedsman and florist, 412 Strand, London.

BATT, James (*fl.* 1800s)

Gardener and florist, Dog Row, Bethnal Green, London.

BATTERS, Edward Arthur Lionel
b. Enfield, Middx 26 Dec. 1860 *d.* Gerrard's Cross, Bucks 11 Aug. 1907

BA Cantab 1882. FLS 1883. Algologist. *List of Marine Algae of Berwick-on-Tweed* 1889. 'Catalogue of British Marine Algae' (*J. Bot.* 1902, Supplt.). 'Algae of Clyde Sea-area' (*J. Bot.* 1891–92).

J. Bot. 1907, 385–88 portr.; 1937, 327–28. *Bot. Soc. Exch. Club Br. Isl. Rep.* 1907, 263. *Proc. Linn. Soc.* 1907–8, 45–46. *Proc. Berwickshire Nat. Club* 1907, 215–16. *Irish Nat.* 1908, 43. Devonshire Association *Fl. Devon* v.2(1), 1952, 69. *R.S.C.* v.9, 141; v.13, 346.

Herb., microscope slides and letters at BM (NH). Portr. at Hunt Library.

Battersia Reinke.

BATTERSBY, Charles Henry (1836–)
b. Kells, County Meath 16 Jan. 1836

Physician at Cannes, *c.* 1883–1909. Made herb. of plants collected in neighbourhood of Cannes.

Bull. Soc. Bot. France v.30, 1883, cxi. *Boissiera* v.5, 1941, 16.

BATTISCOMBE, Edward (*c.* 1875–1971)
d. Glasbury-on-Wye, Radnorshire 1971

Conservator of Forests, Kenya, 1906–25. *Descriptive Catalogue of some of the Common Trees and Woody Plants of Kenya Colony* 1926.

Commonwealth For. Rev. 1972, 186–87.

Kenya plants at Kew.

BATTS, Charles C. V. (–1960)
d. Dec. 1960

Mycologist, National Institute of Agricultural Botany, Cambridge, 1951–56. Lecturer in Plant Pathology, Imperial College of Science, London, 1956–60.

BATTY, John (*fl.* 1780s)

Seedsman and nurseryman, Worcester.

BAUER, Ferdinand Lucas (1760–1826)
b. Feldsberg, Austria 20 Jan. 1760 *d.* Hietzing, Vienna 17 March 1826

Botanical artist. With his brother, Franz, employed by N. von Jacquin to illustrate his *Icones Plantarum Rariorum* 1781–93. Accompanied J. Sibthorp to Greece, 1784; illustrated Sibthorp's *Fl. Graeca* 1806–40. Botanical draughtsman on M. Flinders's expedition to Australia, 1801–5. Cape Bauer named after him by Flinders. *Illustrationes Florae Novae Hollandiae* 1813. Contrib. plates to A. B. Lambert's *Genus Pinus* and J. Lindley's *Digitalium Monographia* 1821.

J. Nichols *Illus. of Lit. Hist. of Eighteenth Century* v.6, 838. *Bot. Mag.* 1834, t.3313. *Bot. Register* 1839, Misc. 41. *Proc. Linn. Soc.* 1839, 39–40; 1931–32, 42. *Ann. Nat. Hist.* v.4, 1840, 67–68. *London J. Bot.* 1843, 106–13. F. M. Bladen *Hist. Rec. N.S.W.* v.6, 1898, 11, 16–19. *J. Bot.* 1909, 140–46. *Bot. Soc. Exch. Club Br. Isl. Rep.* 1917, 143–44. *Beiträge zum landwirtsch Pflanzenbau insbesondere Getreidebau* 1924, 40–42. *Kew Bull.* 1934, 455–56. *J. W. Austral. Hist. Soc.* 1949–54, 59–73. *Fl. Malesiana* v.1, 1950, 41–42. W. Blunt *Art of Bot. Illus.* 1955, 195–202. *Endeavour* 1960, 27–35. *Austral. Dict. Biogr.* v.1, 1966, 73. F. A. Stafleu *Taxonomic Literature* 1967, 19–20. C. C. Gillispie *Dict. Sci. Biogr.* v.1, 1970, 520. *Australian Flower Paintings of F.L. Bauer* 1976.

Norfolk Island plants at Vienna. Drawings of Greek plants at Oxford. Australian drawings at BM(NH) and Vienna (*J. Bot.* 1909, 140–46). Letters at BM(NH).

Murucuja baueri Lindl.

BAUER, Franz Andreas (1758–1840)
b. Feldsberg, Austria 4 Oct. 1758 *d.* Kew, Surrey 11 Dec. 1840

FRS 1821. FLS 1804. Brother of Ferdinand Bauer (1760–1826). Came to England in 1788 and in 1790 was engaged by Sir J. Banks as botanical artist at Kew Gardens where he remained until his death. Illustrated *Delineations of Exotick Plants Cultivated in Royal Garden at Kew* 1796 (*J. Bot.* 1899, 181–83; 1901, 107–8); *Strelitzia Depicta* 1818; J. Lindley's *Illustrations of Orchidaceous Plants* 1830–38.

H. C. Andrews *Botanist's Repository* 1801, t.198. *Ann. Nat. Hist.* v.5, 1840, 47; v.7, 1841, 77–78, 439–41. *Bot. Register* 1841, Misc. Notes 85. *Gdnrs Chron.* 1841, 22. *Gdnrs Mag.* 1841, 186–88. *Proc. Linn. Soc.* 1841, 101–4. *Proc. R. Soc.* v.4, 1843, 342–44. *Trans. Linn. Soc.* v.19, 1845, 222. W. Griffith *Journals of Travels* 1847, vi–viii. *Kew Bull.* 1891, 302–3; 1934, 455–56. J. H. Maiden *Sir Joseph Banks* 1909, 70–72. *Nature* v.167, 1951, 457–60. W.

Blunt *Art of Bot. Illus.* 1955, 195–202. *Endeavour* 1960, 27–35 portr. F. A. Stafleu *Taxonomic Literature* 1967, 20–21. C. C. Gillispie *Dict. Sci. Biogr.* v.1, 1970, 520–21. M. A. Reinikka *Hist. of the Orchid* 1972, 122–26 portr.

Drawings at BM(NH) and a few at Kew. Letters at BM(NH) and Kew; MS. of 'Diseases of Cereals' at Kew. Portr. in oils at Kew. Portr. at Hunt Library. Sale of drawings at Christie's, 1841. (*Bot. Register* 1841, 85; *Gdnrs Chron.* 1841, 783).

Bauera Banks (named in honour of both brothers).

BAUSE, Christian Frederick (*c.* 1839–1895)
b. Rödichen, Saxe Coburg Gotha *d.* Norwood, Middx 23 Oct. 1895

Came to England and employed in Royal Horticultural Society's gardens at Chiswick; later with J. Veitch and Son at Anerley. Nurseryman, Norwood.

J. Hort. Cottage Gdnr v.31, 1895, 413 portr.

BAWDEN, Sir Frederick Charles (1908–1972)
b. North Tawton, Devon 1908 *d.* 8 Feb. 1972

MA Cantab. FRS 1949. Knighted 1967. Research Assistant to R. N. Salaman, Potato Virus Research Station, Cambridge, 1930–36. Virus physiologist. Rothamsted Experimental Station, 1936–40; Head, Plant Pathology Dept., 1940; Director, 1958. President, Association of Applied Biologists, 1965. President, Society for General Microbiology, 1959–61. *Plant Viruses and Virus Diseases* 1939; ed. 3 1950. *Plant Diseases* 1948; ed. 2 1950.

Nature v.236, 1972, 128–29 portr. *Span* v.15 (1), 1972, 14. *Times* 11 and 17 Feb. 1972. *Biogr. Mem. Fellows R. Soc.* 1973, 19–63 portr.

BAXTER, John (1836–1902)
b. Dunblane, Perthshire 1836 *d.* 23 Nov. 1902

Gardener, Royal Botanic Garden, Edinburgh. Gardener to Col. McCall at Daldowie. Raised violas.

Gdnrs Chron. 1900 ii 362 portr.; 1902 ii 331 portr. *Garden* v.58, 1900, 414–15 portr.; v.62, 1902, 318 portr.

BAXTER, William (1787–1871)
b. Rugby, Warwickshire 15 Jan. 1787 *d.* Oxford 1 Nov. 1871

ALS 1817. Head gardener, Botanic Garden, Oxford, 1813–51. *Stirpes Cryptogamicae Oxonienses* 1825 (exsiccatae). *British Phaenogamous Botany* 1834–43 6 vols. (*J. Bot.* 1919, 58–63).

Bot. Misc. v.1, 1830, 58. *Gdnrs Mag.* 1828, 490–92; 1834, 110–13. *Gdnrs Chron.* 1871, 1426–27; 1924 i 106, 120-21. *J. Bot.* 1871, 380–81. *J. Hort. Cottage Gdnr* v.46, 1871, 362. *Proc. Linn. Soc.* 1871–72, lx–lxi. G. C. Druce *Fl. Oxfordshire* 1886, 392–94. J. E. Bagnall *Fl. Warwickshire* 1891, 498–99. G. C. Druce *Fl. Berkshire* 1897, clxii–clxiv. *D.N.B.* v.3, 438–39. *Rep. Ashmolean Nat. Hist. Soc.* 1903, 22–25. B. Lynge *Index...Lichenum Exsiccatorum* 1915–19, 97. H. N. Clokie *Account of Herb. Dept. Bot., Oxford* 1964, 127–28. F. A. Stafleu *Taxonomic Literature* 1967, 23. *Meded. Bot. Mus. Herb. Utrecht* no. 283, 1968, 53–54. D. A. Cadbury *Computer-mapped Fl. Warwickshire* 1971, 54–55.

Plants at Oxford. Original drawings for *British Phaenogamous Bot.* at Royal Horticultural Society. Engraving after portr. by A. Burt at Kew. Portr. at Hunt Library.

BAXTER, William (*fl.* 1820s–1830s)
d. before 1836

Collected plants in W. Australia 1823, 1825, 1828–29 for John MacKay.

Bot. Mag. 1825, t.2600. C. McIntosh *Flora and Pomona* 1829, t.23. R. Brown *Prodromus Florae Novae Hollandiae, Supplementum* 1830. *Gdnrs Mag.* 1830, 507. A. B. Lambert *Description of Genus Pinus* v.2, 1832, 23. *London J. Bot.* 1843, 492–95. J. D. Hooker *Fl. Tasmaniae* v.1, 1859, cxxiv. J. Smith *Rec. R. Bot. Gdns Kew* 1880, 10. *Proc. Linn. Soc. N.S.W.* 1900, 772. *Rep. Austral. Assoc. Advancement Sci.* 1907, 173. *J. W. Austral. Nat. Hist. Soc.* no. 6, 1909, 6–10. *W. Austral. Nat.* 1947–49, 115–16.

Plants at BM(NH), Kew. Letters at Kew.

Baxteria R.Br.

BAXTER, William Hart (*c.* 1816–1890)
d. Oxford 19 June 1890

Son of W. Baxter (1787–1871). Curator, Bath Botanic Garden, 1839; Oxford Botanic Garden, 1854. Contrib. to J. C. Loudon's *Hortus Britannicus* and *Encyclop. Plants.*

Gdnrs Mag. 1841, 283–84. *Gdnrs Chron.* i 1890, 49, 797–98. *Garden* v.47, 1895, 124.

BAXTER, Wynne Edwin (1844–1920)
b. Lewes, Sussex 1844 *d.* London 1 Oct. 1920

Solicitor. High Constable of Lewes, 1877–81. Diatomist. Translated H. van Heurck's *Treatise on Diatomaceae* 1896.

J. R. Microsc. Soc. 1921, 109. *Who was Who, 1916–1928* 68.

Diatom collection at BM(NH).

BAYLEY, Walter (1529–1592)

Professor of Medicine, Oxford, 1561–82. Physician to Queen Elizabeth, 1581. Suggested by Gunther as the possible author of some Hampshire plant records, 1570–72.

R. T. Gunther *Early English Botanists* 1922, 235–36.

BAYLIS, Charles (*c.* 1838–1904)
d. 21 Oct. 1904
Nurseryman, Bourton-on-the-Water, Glos.

BAYLIS, Edward (*fl.* 1790s)
MD? of Bristol. Professor of Botany, Physic Gardens, Clifton. *New and Compleat Body of Practical Botanic Physic* 1791–93.
J. Bot. 1918, 52–54.

BAYLISS-ELLIOTT, Jessie (1869–*c.* 1957)
b. 25 Aug. 1869
Mycologist. Lecturer, Birmingham University, 1908–30.

BEACH, George (*fl.* 1790s)
Seedsman, The Wheatsheaf, 32 Blackman Street, Southwark, London.

BEACH, H. (*fl.* 1860s)
Surgeon of Cheltenham, Glos. Studied mosses.
H. J. Riddelsdell *Fl. Gloucestershire* 1948, cxxviii.
MS., 'Mosses of Cheltenham District' 1861 at Cheltenham Public Library.

BEADLE, Clayton (–1917)
d. 16 Aug. 1917
'Hedychium coronarium in Brazil' (*Kew Bull.* 1917, 104–5).
Kew Bull. 1917, 211.

BEALE, Edward John (1835–1902)
d. London 8 Jan. 1902
FLS 1871. VMH 1897. Senior partner, Carter and Co., seedsmen, High Holborn. *English Tobacco Culture* 1887. Edited *Carter's Practical Gardener* ed. 16 1896.
Garden v.61, 1902, 48. *Gdnrs Chron.* 1902 i 49 portr. *Gdnrs Mag.* 1902, 20 portr. *Proc. Linn. Soc.* 1901–2, 25–26.

BEALE, Rev. John (1603–1682)
b. Herefordshire 1603 *d.* Yeovil, Somerset 1682
BA Cantab 1632. DD. FRS 1663. Rector, Yeovil, 1660–82. Chaplain to Charles II, 1665, Correspondent of Boyle and Hartlib. *Treatise on Fruit Trees* 1653. *Herefordshire Orchards* 1657. *Nurseries, Orchards, Profitable Gardens and Vineyards Encouraged* (with A. Lawrence) 1677. Contrib. to *Philos. Trans. R. Soc.* 1669–71.
J. Nichols *Lit. Anecdotes of Eighteenth Century* v.1, 447; v.4, 256. J. C. Loudon *Encyclop. Gdning* 1822, 1265–66. G. W. Johnson *Hist. of English Gdning* 1829, 101–2. S. Felton *Portr. of English Authors on Gdning* 1830, 21–22. *Cottage Gdnr* v.8, 1852, 44. J. von Sachs *Hist. Bot.* 1890, 472. *D.N.B.* v.4, 1–2. *J. Hort. Cottage Gdnr* v.43, 1901, 417–18.

BEALE, Lionel Smith (1828–1906)
b. London 5 Feb. 1828 *d.* Weybridge, Surrey 28 March 1906
MB London 1851. FRS 1857. Professor of Physiology and Anatomy, King's College, London, 1853–69; Pathological Anatomy, 1869–76; Medicine, 1876–96. *The Microscope* 1854. *Protoplasm* 1870. *Bioplasm* 1872.
Proc. R. Soc. B. v.79, 1907, lvii–lxiii. *D.N.B. Supplt. 2* v.1, 118–20. *Who was Who, 1897–1916* 49. *R.S.C.* v.7, 111; v.9, 152.

BEALE, Reginald Evelyn Child (*c.* 1877–1952)
d. 14 Nov. 1952
FLS 1907. Director, Carters Seeds, Ltd. *Practical Greenkeeper. Lawns for Sports* 1924. *Book of the Lawn 1931.*
Gdnrs Chron. 1952 ii 228.

BEALE, Thomas (*c.* 1775–1842)
d. Macao, China Dec. 1842
To China, 1792. He and his brother, Daniel, were eastern merchants. Lived in Macao for 50 years and had a very fine garden. Friend of J. Reeves (1774–1856) and sent plants to England. Just before his death his collections were taken to form nucleus of Dent's famous garden at Green Bank.
E. Bretschneider *Hist. European Bot. Discoveries in China* 1898, 306–7. E. H. M. Cox *Plant Hunting in China* 1945, 51–52.

BEALE, William (*c.* 1865–1903)
d. 25 Oct. 1903
Gardener to E. A. Hambro at Hayes Place, Kent.
Garden v.64, 1903, 308 portr.

BEAMISH, Richard Henrik (1862–1938)
d. Weybridge, Surrey 23 Feb. 1938
FLS 1916. Businessman, Cork. Created garden at Ashbourne, County Cork.
Proc. Linn. Soc. v.150, 1937, 309.

BEAN, William (*fl.* 1780s–1800s)
Seedsman, fruiterer and florist of Scarborough, Yorks.
J. Soc. Bibl. Nat. Hist. 1972, 154.

BEAN, William (1787–1866)
d. Scarborough, Yorks 22 Dec. 1866
Of Scarborough. Son of W. Bean (*fl.* 1780s–1800s). Interested in natural history, especially conchology and geology. One of the promoters of Scarborough Museum. His botanical observations acknowledged in J. Lindley and T. Moore's *Nature Printed Ferns* 1859 and J. G. Baker's *North Yorkshire* 1863.
Scarborough Gaz. 27 Dec. 1866. *J. Soc. Bibl. Nat. Hist.* 1972, 154–57.
Algae and ferns at Yorkshire Museum, York.

BEAN, William (1817–1864)
d. Scarborough, Yorks 27 March 1864
Son of W. Bean (1787–1866). Clerk, H.M. Customs, Kirkaldy; then in Liverpool, 1841. Friend of F. Brent (1816–1903) and T. Sansom (1816–1872) with whom he formed herb. and presented to Historic Society of Lancashire and Cheshire, 1855, now at Liverpool Museums. Contrib. to G. J. Dickinson's *Fl. Liverpool* 1851.
Proc. Bot. Soc. Br. Isl. 1968, 170. *J. Soc. Bibl. Nat. Hist.* 1972, 157–60.

BEAN, William Jackson (1863–1947)
b. Malton, Yorks 26 May 1863 *d.* Kew, Surrey 19 April 1947
VMH 1917. VMM 1922. Kew gardener, 1883; Assistant Curator, 1900; Curator, 1922–29. *Royal Botanic Gardens, Kew* 1908. *Trees and Shrubs Hardy in British Isles* 1914 2 vols; ed. 8 1970. *Shrubs for Amateurs* 1924. *Ornamental Trees for Amateurs* 1925. *Wall Shrubs and Hardy Climbers* 1939. *Orchids: their Culture and Management* (with W. Watson) 1890. Contrib. to *Gdnrs Chron.*
Gdnrs Chron. 1927 ii 2 portr.; 1947 i 162. *J. Kew Guild* 1914, 133 portr.; 1928, 575–76; 1930, 771–72; 1946–47, 598–600 portr. *Kew Bull.* 1929, 139–40; 1947, 95–96. *Bot. Mag.* 1938 portr. *Nature* v.159, 1947, 667–68. *Quart. J. For.* 1947, 65–66. *Times* 22 April 1947. *Who was Who, 1941–1950* 76.
Portr. by E. Moore at Kew. Portr. at Hunt Library.

BEANLAND, Joseph (1857–1932)
b. Bradford, Yorks 1857 *d.* 18 Sept. 1932
President and Secretary, Bradford Naturalists. Botanist.
Naturalist 1933, 125 portr.
Herb. at Bradford University.

BEARD, John (1858–1934)
b. Heaton Norris, Cheshire 11 Nov. 1858 *d.* Edinburgh 2 Dec. 1934
BSc Manchester 1885. DSc 1894. PhD Freiburg. Lecturer in Zoology, Edinburgh University. 'On the Phenomenon of Reproduction in Animals and Plants' (*Ann. Bot.* 1895, 441–68).
Nature v.114, 1924, 904.

BEARDSLEY, A. F. (*fl.* 1850s–1860s)
b. Scotland?
Collected seeds in N. America for Lawson and Co., Edinburgh, 1852–54. Orchard supervisor in Napa Valley, California.
W. H. Brewer, S. Watson, A. Gray *Bot. of California* v.2, 1880, 557. F. P. Farquhar *Up and Down California* 1930, 218–19. *California Hist. Soc. Quart.* 1942, 366.
Pinus beardsleyi A. Murray.

BEARPARK, Benjamin (1760–1826)
Gardener and nurseryman, Bootham, York. Son of C. Bearpark (*c.* 1724–1798).

BEARPARK, Christopher (*c.* 1724–1798)
Gardener and seedsman, Bootham, York.

BEARPARK, Edward (*fl.* 1810s–1820s)
Gardener to G. J. Legh, High Legh, Cheshire. Hybridised amaryllis.
Bot. Mag. 1822, t.2315.

BEARPARK, Henry Edmund (*c.* 1829–1861)
Nurseryman, Bootham, York.

BEARPARK, James (1773–1836)
Gardener and nurseryman, Bootham, York.

BEARPARK, Robert (1797–1869)
Nurseryman and florist, 1 Lord Mayor's Walk, York. Son of R. Y. Bearpark (1757–1827).

BEARPARK, Robert Young (1757–1827)
d. York Sept. 1827
Gardener and nurseryman, York.

BEATON, Donald (1802–1863)
b. Urray, Ross-shire 8 March 1802 *d.* Surbiton, Surrey Oct. 1863
Gardener to W. Gordon at Hatfield, Ledbury, 1829–37. Moved to Kilburn, 1837. Interested in hybridisation. "A clever fellow and a damned cocksure man" (C. Darwin). 'Cereus senilis' (*Gdnrs Mag.* 1839, 549–52). Contrib. to *Cottage Gdnr* and *Gdnrs Mag.*
Cottage Gdnr v.13, 1854, 153–58 portr. *J. Hort. Cottage Gdnr* v.30, 1863, 349, 415–16 portr.; v.38, 1899, 181 portr. F. Darwin and A. C. Seward *More Letters of Charles Darwin* v.2, 1903, 268–69.
Beatonia Herbert.

BEATSON, Alexander (1759–1833)
b. Kilrie, Fife 1759 *d.* 14 July 1833
Ensign, Madras Infantry, 1776. Colonel, 1801. Lieut.-General, 1814. Governor, St. Helena, 1808–13. *Tracts Relative to Island of St. Helena* 1816. *Fl. Sta. Helenica* 1825.
D.N.B. v.4, 20.
Beatsonia Roxb.

BEATTIE, James (1735–1810)
b. Laurencekirk, Kincardine 1735 *d.* Aberdeen 5 Oct. 1810
ALS 1807. Professor of Natural History, Aberdeen, 1788. Discovered *Linnaea* in Britain (J. Sowerby and J. E. Smith *English Bot.* 433).
P. Smith *Mem. and Correspondence of... Sir James Edward Smith* v.1, 1832, 441–43. *Ann. Scott. Nat. Hist.* 1902, 167–69; 1911, 178–79. W. Dawson *Smith Papers in Linnean Soc.* 1934, 13.
Herb. formerly at Linnean Society, London.

BEATTIE, William (1758–1839)
b. Wasthill, Aberdeenshire 1758 *d.* 2 April 1839
Gardener to Earl of Mansfield at Scone, Perth. Contrib. to *Trans. Hort. Soc. London.*
Gdnrs Mag. 1841, 237–38.

BEAUFORT, Mary, Duchess of *see* Somerset, Mary, *Duchess of Beaufort*

BEAUMONT, Diana (–1831)
Of Bretton Hall near Sheffield, Yorks. Sent plants to N. Wallich at Calcutta Botanic Garden.
Beaumontia Wall.

BEAUMONT, Guillaume (*fl.* 1680s–*c.* 1727)
Gardener to James Graham at Levens Hall, Westmorland. Believed to have trained under Le Nôtre. Designed garden and park at Levens Hall. Advised on the landscaping of other gardens.
Portr. at Levens Hall.

BEAUMONT, John (–1731)
d. Stone-Easton, Somerset March 1731
FRS 1685. Surgeon at Stone-Easton, Somerset. Geologist and writer on spiritualism. 'Rock-plants in Lead Mines of Mendip Hills' (Sloane MSS. 4037, 128–32).
Gent. Mag. 1731, 34. E. Lankester *Correspondence of John Ray* 1848, 262. *D.N.B.* v.4, 60.

BEAUMONT, William (*fl.* 1860s–1890s)
b. Newton near Hyde, Cheshire
Member of Ashton-under-Lyne Linnean Society. Founder member of Ashton Field Naturalist Society, 1861. Workingman naturalist.

BEAZELEY, M. (*fl.* 1880s)
Light Inspector, Amoy, China. Sent Chinese plants to Kew, 1884.
E. Bretschneider *Hist. European Bot. Discoveries in China* 1898, 707–8. *Kew Bull.* 1901, 7.

BECHER, H. M. (–1893)
d. near Gunong Tahan, Malaya 1893
Collected plants in neighbourhood of Kuala Tembeling, Pahang.
Gdns Bull. Straits Settlements 1927, 117.

BECHER, Rev. John Thomas (1770–1848)
d. Hill House, Southwell, Notts 3 Jan. 1848
MA Oxon 1795. Vicar, Rumpton, Midsomer Norton. Vicar-General, Southwell. Rector, Barnborough, 1830. Discovered *Crocus nudiflorus* (J. Sowerby and J. E. Smith *English Bot.* 491).
T. Ordoyno *Fl. Nottinghamiensis* 1807, preface, v. *D.N.B.* v.4, 75–76.

BECK, Christabel (1884–1960)
b. Stamford Hill, London 1884 *d.* June 1960
Collected plants in America, Palestine, Sweden, etc. *Fritillaries* 1953.
Lily Yb. 1959, 10–11 portr.; 1961, 67–69.
Plants at Kew.

BECK, Edward (–1861)
d. 15 Jan. 1861
Nurseryman, Isleworth, Middx. Proprietor of *Florist* 1848–50. *A Packet of Seeds Saved by Old Gardener. Treatise on Cultivation of Pelargonium* 1847.
Cottage Gdnr v.25, 1861, 255–56. *Florist* 1861, 37–38. *Garden* v.30, 1886, 321–22 portr.

BECK, Sarah Coker (*née* **Adams**) (1821–1915)
b. Anstey, Warwickshire 20 Jan. 1821 *d.* Monk's Risborough, Bucks 8 Nov. 1915
Sister of Rev. D. C. O. Adams. Contrib. records of fungi to G. C. Druce's *Fl. Oxfordshire* 1886.
Bot. Soc. Exch. Club Br. Isl. Rep. 1915, 249–50.

BECKER, Hermann Franz (1838–1917)
d. Grahamstown, S. Africa 3 April 1917
MD Jena. FLS 1866. Surgeon, Capetown, 1870; moved to Grahamstown, 1874. Algologist.
Proc. Linn. Soc. 1917–18, 34–35.

BECKER, Ludwig (1808–1861)
b. Darmstadt, Germany 1808 *d.* Cooper's Creek, N. Australia 28 April 1861
Left Germany in 1840s, going first to Brazil, then about 1850 to Australia. In Tasmania, 1851–52. Did drawings for F. von Mueller in Melbourne. Collector and artist to Burke and Wills Expedition, 1860. Contrib. to *Trans. Philos. Inst. Victoria* 1855–56, 14–18.
F. J. H. von Mueller *Fragmenta Phytographiae Australiae* v.1, 1858, 156. *Victorian Nat.* v.25, 1908, 103; v.48, 1932, 244–45. *J. Proc. R. Austral. Hist. Soc.* v.9, 166–67. *Austral. Encyclop.* v.1, 1962, 471.
Drawings at Public Library, Victoria; Gallery, Bendigo, Victoria. Diary at Public Library, Victoria.
Hovea beckeri F. Muell.

BECKER, Lydia Ernestine (1827–1890)
b. Manchester 24 Feb. 1827 *d.* Geneva 18 July 1890
Advocate of women's suffrage. Lectured on botany. *Botany for Novices* 1864. Horticultural Society Gold Medal, 1865.
J. Bot. 1865, 264; 1890, 320. *Trans. Liverpool Bot. Soc.* 1909, 60. *D.N.B. Supplt 1* v.1, 159–60. *R.S.C.* v.7, 118.

BECKETT, Edwin (*c.* 1853–1935)
b. Remenham near Henley-on-Thames, Berks *c.* 1853 *d.* Radlett, Herts 6 Feb. 1935

VMH 1906. VMM 1930. Gardener to Vicary Gibbs at Aldenham House, Herts. *Vegetables for Exhibition and Home Consumption* 1899. *Book of the Strawberry* 1902. *Pruning Hardy Shrubs* 1916. *Roses and how to grow them* 1918. *Potatoes* 1922.

Gdnrs Chron. 1921 ii 132 portr.; 1935 i 117. *Gdnrs Mag.* 1908, 295–96 portr.

BECKETT, John Edgar (1878–1934)
b. British Guiana 1 July 1878 *d.* Georgetown, British Guiana 24 Feb. 1934

Agricultural adviser in British Guiana where he also collected plants.

Proc. Linn. Soc. 1933–34, 142–43.

Plants at Georgetown, Kew and New York.

BECKETT, Thomas W. Naylor (1839–1906)
d. Christchurch, New Zealand 5 Dec. 1906

Coffee planter in Ceylon. Collected plants in Ceylon, 1882–83; N. W. Himalayas *c.* 1900; New Zealand, 1903–7. Fern herb. which included specimens from eminent pteridologists such as G. H. K. Thwaites, R. H. Beddome, H. Cuming and J. Buchanan, now at Liverpool Museums.

I. H. Burkill *Chapters on Hist. Bot. in India* 1967, 56, 181, 182.

Plants and letters at Liverpool. Plants at Berlin and Copenhagen. Portr. at Hunt Library.

BECKLER, Herman (*fl.* 1860s)
b. Germany

Medical officer and botanist to Burke and Wills Expedition, 1860. Collected plants around Moreton Bay, Queensland and in New South Wales.

J. H. Maiden *For. Fl. N.S.W.* v.3, 1907, 99–102. *J. R. Soc. N.S.W.* 1908, 84. *Med. J. Austral.* 1938, 879.

Plants at Melbourne and Munich.

Hibiscus beckleri F. Muell.

BECKTON, Bertram J. (–1943)
d. 29 Jan. 1943

Cultivated orchids. Contrib. photographs of orchids to *Orchid Rev.*

Orchid Rev. 1943, 65–66.

BECKWITH, William Edmund (1844–1892)
b. Eaton Constantine, Shropshire 1844 *d.* Shrewsbury, Shropshire 22 July 1892

Ornithologist. Contrib. papers on Shropshire plants to *J. Bot.* 1881–82.

R.S.C. v.9, 164; v.12, 62; v.13, 394.

Plants at BM(NH).

BEDDOME, Richard Henry (1830–1911)
b. 11 May 1830 *d.* Wandsworth, London 23 Feb. 1911

Educ. Charterhouse. FLS 1882. Entered Indian Army, 1848. Forestry Dept., India, 1857. Head of Forestry Dept., S. India, 1860–82. Collected plants in India and Ceylon. *Ferns of Southern India* 1863, 1876. *Trees of Madras Presidency* 1863. *Ferns of British India* 1865–70. *Fl. Sylvatica of Southern India* 1869–74 2 vols. *Icones Plantarum Indiae Orientalis* 1869–74. *Handbook to Ferns of British India* 1883, 1892.

Kew Bull. 1911, 164–65. *Proc. Linn. Soc.* 1910–11, 32–34. *Gdnrs Chron.* 1911 i 143. C. E. Buckland *Dict. Indian Biogr.* 1906, 33. *Who was Who, 1897–1916* 51. F. A. Stafleu *Taxonomic Literature* 1967, 24–25. *R.S.C.* v.1, 242; v.7, 122; v.9, 168; v.13, 400.

Plants at BM(NH), Kew.

Beddomea Hook f. *Beddomiella* Dixon.

BEDFORD, Arthur (–1934)
d. June 1934

Gardener to Lionel de Rothschild at Gunnersbury Park, Middx.

Gdnrs Chron. 1934 ii 398.

BEDFORD, Edward John (1866–1953)

Headmaster, Art School, Eastbourne. Curator, Lewes Natural History Museum. Contrib. records to A. H. Wolley-Dod's *Fl. Sussex* 1937.

J. Trans. Eastbourne Nat. Hist. Archaeol. Soc. 1953, 22.

Drawings of British orchids at BM(NH); some have been reproduced as postcards.

BEDFORD, John, Duke of *see* **Russell, J.** *6th Duke of Bedford*

BEDFORD, T. H. B. (–1961)
d. Dec. 1961

Investigated distribution of *Seligeria* in N. England.

Naturalist 1963, 62, 106.

BEDFORD, William (1700–)

MD Cantab. FRS. Possibly member of Botanical Society of London.

Proc. Bot. Soc. Br. Isl. 1967, 319.

BEEBY, William Hadden (1849–1910)
b. 9 June 1849 *d.* Long Ditton, Surrey 4 Jan. 1910

ALS 1887. FLS 1890. Discovered *Sparganium neglectum* (*J. Bot.* 1885, 193). Contrib. papers on Shetland plants to *Scott. Nat.* from 1887. Contrib. botany to *Victoria County Hist. Surrey* v.1, 1902, 56–57. Contrib. to *J. Bot.*

Ann. Scott. Nat. Hist. 1910, 129–32. *Bot. Soc. Exch. Club Br. Isl. Rep.* 1910, 531–32. *J. Bot.* 1910, 121–23; 1932, 53. *Proc. Linn. Soc.* 1909–10, 86. C. E. Salmon *Fl. Surrey* 1931, 55. *R.S.C.* v.7, 123; v.12, 62; v.13, 403.

Herb. at S. London Botanical Institute. Portr. at Hunt Library.

BEECH, Dr. (*fl.* 1850s)

Plants at Westonbirt, Glos.

BEECHEY, Frederick William (1796–1856)
b. London 17 Feb. 1796 *d.* London 29 Nov. 1856
Rear-Admiral and geographer. Capt., HMS 'Blossom', which explored Pacific, 1825–28; A. Collie and G. T. Lay were the 2 naturalists on board. *Narrative of Voyage to Pacific and Bering's Strait, 1825–1828* 1831 2 vols.
E. Bretschneider *Hist. European Bot. Discoveries in China* 1898, 288–89. *D.N.B.* v.4, 121–22. G. A. C. Herklots *Hongkong Countryside* 1951, 162–63. *Contrib. Univ. Michigan Herb.* 1972, 222–23.
Plants at BM(NH), Geneva, Kew.
Ficus beecheyana Hook. & Arn.

BEEKE, Rev. Henry (1751–1837)
b. Kingsteignton, Devon 6 Jan. 1751 *d.* Torquay, Devon 9 March 1837
BA Oxon 1773. DD 1880. FLS 1800. Professor of Modern History, 1801. Vicar, Oxford, 1782. Dean of Bristol, 1814.
J. Sowerby and J. E. Smith *English Bot.* 2468. D. Turner and L. W. Dillwyn *Botanist's Guide through England and Wales* 1805, 527–28. G. C. Druce *Fl. Berkshire* 1897, cxlvii–cxlix. *D.N.B.* v.4, 124–25.
Letters at Linnean Society.

BEER, Rudolf (1873–1940)
b. Bickley, Kent 29 Nov. 1873 *d.* Crowborough, Sussex 23 Sept. 1940
BSc London 1903. FLS 1895. On staff of Ministry of Agriculture Plant Pathology Laboratory, 1914–18. *Natural Science* 1895. Contrib. to *Ann. Bot.*
John Innes Inst. 1910–1935 22, 39. *Proc. Linn. Soc.* 1940–41, 288–89.

BEESLEY, H. (*fl.* 1900s)
Railway agent. Member of Botanical Section of Preston Scientific Society. Studied Isle of Man bryophytes, 1902–8. Contrib. to *Fl. Preston and Neighbourhood* 1903 and *Fl. W. Lancashire* 1907. Contrib. mosses of Isle of Man to *Proc. Isle of Man Nat. Hist. Antiq. Soc.* v.1, 1908, 164–65.
Plants at Liverpool, Oxford.

BEESLEY, Thomas (1818–1896)
b. Banbury, Oxfordshire 28 March 1818 *d.* Banbury 15 May 1896
Chemist and druggist, Chipping Norton, then Banbury, Oxfordshire. Contrib. 'Botany and Geology of Neighbourhood of Banbury' to *Hist. Banbury* 1841, 571–99, 626–27 by his uncle, A. Beesley. President, Banbury Natural History Society.
G. C. Druce *Fl. Oxfordshire* 1886, 395–96. *J. Bot.* 1896, 440. *Pharm. J.* 1896, 201–2. H. B. Woodward *Thomas Beesley* 1897. G. C. Druce *Fl. Northamptonshire* 1930, cxvi–cxvii.
Plants at Oxford.

BEESTON, William (*c.* 1672–1732)
d. Bentley, Suffolk 4 Dec. 1732
MB Cantab 1692. MD 1702. Of Ipswich. "He is very curious and knowing in plants, and has a fine collection of exotics" (W. Sherard).
J. Nichols *Illus. Lit. Hist. of Eighteenth Century* v.6, 879. D. Turner *Extracts from Lit. and Sci. Correspondence of Richard Richardson* 1835, 184.

BEEVER, Mary (*c.* 1800–1883)
b. Ardwick, Manchester *c.* 1800 *d.* Coniston, Lancs 31 Dec. 1883
Sent *Gentiana pneumonanthe* to W. Baxter in 1836; also ferns to E. Newman and E. J. Lowe. Correspondent of J. G. Baker (*Fl. English Lake District* 1885) and John Ruskin (*Hortus Inclusus* 1887).
W. Baxter *British Phaenogamous Bot.* 1834–43, t.185–87. W. G. Collingwood *John Beever's Practical Fly-fishing* 1893, preface. *Trans. Liverpool Bot. Soc.* 1909, 60.
Lastraea felix-mas var. *Beevorii* Lowe.

BEEVER, Susan (1805–1893)
b. Manchester? 27 Nov. 1805 *d.* Coniston, Lancs 29 Nov. 1893
Sent *Radiola linoides* to W. Baxter. Letters to John Ruskin in his *Hortus Inclusus* 1887, 155–72.
W. Baxter *British Phaenogamous Bot.* 1834–43, t.188. *Naturalist* 1894, 290. *Trans. Liverpool Bot. Soc.* 1909, 60.

BEEVOR, Sir Hugh Reese (1858–)
b. 31 Oct. 1858
Collected plants in Aden, 1884, now at Kew.
Kew Bull. 1901, 7. *Rec. Bot. Survey India* v.7, 1914, 17.

BEGG, Mrs. Catherine Clarke (–1928)
b. Peebles *d.* Dunedin, New Zealand 1 Oct. 1928
Collected plants for National Museum of Wales.
Proc. Swansea Sci. Field Nat. Soc. v.1, 1929, 59.

BEGG, William Robertson (1833–1923)
d. Aberdeen 31 Oct. 1923
Nurseryman, Burnside Nurseries, Aberdeen.
Gdnrs Chron. 1923 ii 286.

BEHR, Hermann Hans (1818–1904)
b. Coethen 18 Aug. 1818 *d.* San Francisco, U.S.A. 6 March 1904

Doctor. Arrived in Australia in 1844. Lived with natives of S. Australia studying their language and customs. Collected plants which he gave to F. von Mueller. Returned to Germany in 1847, later went to California. 'On Character of South Australian Flora' (*Hooker's J. Bot.* 1851, 129–34; translated from *Linnaea* 1847, 545–58). *Synposis of Genera of Vascular Plants...of San Francisco* 1884.

Rep. Austral. Assoc. Advancement Sci. 1907, 173–74. *Zoe* 1891, 2–6. *R.S.C.* v.9, 171.

Aristida behriana F. Muell.

BEISLY, Sidney (*fl.* 1850s–1860s)

Of Sydenham. *Shakespere's Garden* 1864. Contrib. to *Phytologist* 1857–63 as 'S.B.'; his mother Harriet, also contributed, sometimes as 'H.B.'.

F. J. Hanbury and E. S. Marshall *Fl. Kent* 1899, lxxiii, G. C. Druce *Fl. Buckinghamshire* 1926, ci.

BELCHER, Sir Edward (1799–1877)

d. 18 March 1877

Entered Royal Navy, 1812; Admiral, 1872. On W. African survey, 1830–33; collected plants in French Guinea, now at BM(NH). Plants collected in Arctic N. America now at Kew. *Narrative of Voyage...in HMS 'Sulphur' ...1836–42* 1843. *Narrative of Voyage of HMS 'Samarang'...1843–46* 1848.

J. R. Geogr. Soc. 1832, 278–304; 1877, cxxxvi–cxlii. *D.N.B.* v.4, 142–43.

BELL, Miss A. (*fl.* 1830s)

Of Stow Bardolph Vicarage, Norfolk. Correspondent of J. S. Henslow and H. C. Watson.

H. C. Watson *New Botanists' Guide* 1835–37, 125. H. C. Watson *Topographical Bot.* 1883, 537. C. C. Babington *Memorials* 1897, 284.

BELL, Alexander James Montgomerie (1845–1920)

b. Edinburgh 1845 *d.* Oxford 3 July 1920

BA Oxon 1869. MA 1871. Master, Marlborough College, 1870. Brother of J. M. Bell (1837–1910). Made many drawings of British plants.

Bot. Soc. Exch. Club Br. Isl. Rep. 1920, 100–101. *Nature* v.105, 1920, 721.

BELL, Alfred Redmayne (1892–1925)

b. Lake District 4 April 1892 *d.* Onitsha, Nigeria 1 Feb. 1925

Kew gardener, 1913. Curator, Dept. of Agriculture, S. Provinces, Nigeria, 1914; later Superintendent.

Kew Bull. 1914, 227. *J. Kew Guild* 1926, 425–26 portr.

BELL, Rev. Edward (1829–1904)

b. Uppingham, Rutland 26 Jan. 1829 *d.* Poole, Dorset 5 March 1904

BA Cantab 1858. Vicar, Wakefield, 1868–90. *Primrose and Darwinism* 1902.

J. Bot. 1904, 159–60. A. R. Horwood and C. W. F. Noel *Fl. Leicestershire* 1933, cclxxvi.

BELL, Elizabeth (–1876)

Of Coldstream, Berwick. Correspondent of G. Johnston.

Ann. Mag. Nat. Hist. 1841, 356–57. *Hist. Berwick Field Club* v.8, 1876, 31.

Agaricus belliae Johnston ex Berkeley.

BELL, Frank (*fl.* 1880s)

Major. Collected plants in Persia, 1884, now at Kew.

BELL, George (*fl.* 1730s)

MD. Collected plants at Cape of Good Hope.

J. E. Dandy *Sloane Herb.* 1958, 90.

BELL, George (1755–1784)

b. Greenhill, Dumfries 1755 *d.* Manchester 3 Feb. 1784

MD Edinburgh 1777. *De Physiologia Plantarum* 1777 (translated in *Mem. Manchester Lit. Philos. Soc.* v.2, 1789, 410–35; memoir of Dr. Bell, pp. 397–409).

BELL, H. Astley– *see* Astley-Bell, H.

BELL, Isaac (*fl.* 1790s)

Gardener and seedsman, Downham, Norfolk.

BELL, J. (*fl.* 1790s)

Gardener and seedsman, Boughton, Chester.

BELL, John (*fl.* 1780s–1790s)

Probably nurseryman at Isleworth, Middx. Introduced *Campanula grandiflora* 1782 and *Paeonia daurica* 1790.

Bot. Mag. 1794, t.252; 1812, t.1441.

BELL, John (*fl.* 1840s–1850s)

Nurseryman, 3 Exchange Street, Norwich.

BELL, John Montgomerie (1837–1910)

b. Edinburgh Feb. 1837 *d.* Edinburgh June 1910

Pupil of J. H. Balfour (1808–1884). 'Notes on Visit to the Dovrefjeld, Norway' (*Trans. Bot. Soc. Edinburgh* v.21, 1900, 281–90).

Trans. Bot. Soc. Edinburgh v.24, 1912, 99–100. *Bot. Soc. Exch. Club Br. Isl. Rep.* 1920, 177. *R.S.C.* v.13, 422.

Plants at Oxford.

BELL, Mary Louisa *see* Wedgwood, *Mrs.* M. L.

BELL, Thomas (1792–1880)
b. Poole, Dorset 11 Oct. 1792 *d.* Selborne, Hants 13 March 1880
FRS 1828. FLS 1815. Dental surgeon. Professor of Zoology, King's College, London, 1836. President, Linnean Society, 1853–61. Studied birds and plants of Selborne.
J. Bot. 1880, 128. *Nature* v.21, 1880, 473, 499–500. *J. R. Microsc. Soc.* 1895, 9. *Curtis's Bot. Mag. Dedications, 1827–1927* 139–40 portr.

BELL, Thomas (1836–1914)
Physician at Uppingham, Rutland. Collected Rutland plants. Had herb.
A. R. Horwood and C. W. F. Noel *Fl. Leicestershire* 1933, cclxxvii.

BELL, Thomas Bush (*fl.* 1810s)
Nurseryman, Brentford End, Isleworth, Middx.
Trans. London Middlesex Archaeol. Soc. v.24, 1973, 184.

BELL, Thomas Reid (1863–1948)
b. 2 May 1863 *d.* 24 June 1948
Indian Forest Service. Entomologist. Collected plants in Kashmir.

BELL, William (*fl.* 1780s)
Nurseryman, County Down, Ireland.

BELL, William (*fl.* 1810s)
Nurseryman and seedsman, Westgate, Bradford, Yorks.

BELL, William (*c.* 1833–1916)
b. Dumfriesshire *c.* 1833 *d.* Christchurch, New Zealand 17 March 1916
Gardener, Royal Botanic Garden, Edinburgh. Head gardener, Saharanpur Botanic Garden, India, 1862. Sent plants to Royal Botanic Garden, Edinburgh to which he returned in 1869. In New Zealand in 1872 with visits to India.
Tuatara v.19 (1), 1971, 1–5.
Indian plants at Otago Museum, New Zealand. New Zealand mosses at Helsinki University.
Bellia Broth. *Eulophia campanulata* Duthie.

BELL, William (*c.* 1840–1920)
d. Barrow in Furness, Lancs 28 May 1920
Collected British ferns.
Br. Fern Gaz. 1920, 81.

BELL, William (1862–1925)
Of Leicester. Studied local flora. 'Phanerogams of Leicester and District' (British Association Guide to Leicester 1907, 334–45).
Bot. Soc. Exch. Club Br. Isl. Rep. 1925, 844. A. R. Horwood and C. W. F. Noel *Fl. Leicestershire* 1933, ccxix–ccxx.
Plants at Oxford and Leicester Museums.

BELL, William Abraham (1841–1921)
b. Clonmel, County Tipperary 26 April 1841 *d.* Bletchingly, Surrey 6 June 1921
MD. Collected plants in Colorado in 1867. Vice-President, Denver and Rio Grande Western R. R. *New Tracks in N. America* 1869 (ed. 2 contains list of plants, 521–33).
J. Ewan *Rocky Mountain Nat.* 1950, 160. *National Cyclop. Amer. Biogr.* v.24, 404 portr.
Plants at BM(NH), Philadelphia.

BELLAIRS, Nona Maria Stephenson (*fl.* 1850s–1890s)
d. Bournemouth, Hants 14 May 1897
Hardy Ferns 1865. *Wayside Flora* 1866.

BELLAMY, Frank Arthur (*fl.* 1870s–1920s)
Pharmacist in Oxford. Assistant, University Observatory, Oxford. Friend of G. C. Druce. 'Phenological Observations made in Oxford' (*Bot. Soc. Exch. Club Br. Isl. Rep.* 1927, 534). 'Notes on Flowers in Oxon and Glos' 1901–3; MS. at Oxford.
H. N. Clokie *Account of Herb. Dept. Bot., Oxford* 1964, 130.
Canary Island and Egyptian plants at Oxford.

BELLERBY, William (1852–1936)
b. York 11 April 1852 *d.* York 27 March 1936
Director of family firm of timber merchants Bryologist and mycologist.
N. Western Nat. 1936, 166–69 portr. *Br. Bryol. Soc. Rep.* 1936, 394–95.
Drawings of mosses and moss collection at BM(NH).

BELLERS, Fettiplace (*fl.* 1720s)
Of Gloucester. Gloucestershire and Oxfordshire plants at Oxford.
Bot. Soc. Exch. Club Br. Isl. Rep. 1926, 463. H. N. Clokie *Account of Herb. Dept. Bot., Oxford* 1964, 130.

BELLEW, Henry Walter (1834–1892)
MD. Surgeon-major, Bengal Staff Corps. Surgeon on 2nd D. Forsyth mission to Turkestan, 1874. Collected plants in Afghanistan, 1858, Kashmir, Kashgar, etc., 1875. *Kashmir and Kashgar, 1873–1874* 1875.
E. Bretschneider *Hist. of European Bot. Discoveries in China* 1898, 798–99. *Kew Bull.* 1901, 7. *Pakistan J. For.* 1967, 340.
Plants at Calcutta, Kew.
Nepeta bellevii Prain.

BELLING, John (1866–1933)
b. Aldershot, Hants 7 Oct. 1866 *d.* San Francisco, U.S.A. 28 Feb. 1933
BSc London. Lecturer, Horticultural College, Swanley, then Llanidloes, 1900–1. Investigator, Dept. of Agriculture, British West Indies. Assistant botanist, Florida Agricultural Experiment Station, 1907. Cytologist,

Carnegie Institution of Washington. Studied chromosomes of *Datura* with A. F. Blakeslee.
Nature v.131, 1933, 575–76, 390. *Science* v.77, 1933, 250–51.

BELLINGHAM, O'Bryen (*fl.* 1840s)
Professor of Botany, Royal College of Surgeons of Ireland.
Proc. R. Irish Acad. B. v.39, 1929, 55.

BELSON, Nicholas (*fl.* 1560s–1600s)
Fellow, King's College, Cambridge, 1577–79. Schoolmaster, Suffolk. Friend of J. Gerard who called him "a diligent searcher of nature."
J. Gerard *Herball* 1597, 116, 915, 977. C. E. Raven *British Nat.* 1947, 212–13.

BELT, Thomas (1832–1878)
b. Newcastle, Northumberland 1832 *d.* Denver, Colorado, U.S.A. 21 Sept. 1878
Son of George Belt, nurseryman at Newcastle-upon-Tyne. Joined Australian gold rush, 1852. Prospected in Central America and Russia. *Naturalist in Nicaragua* 1874.
Gdnrs Chron. 1878 ii 478–79. *Proc. Geol. Soc.* 1878–79, 48–50. *Trans. Northumberland Durham Nat. Hist. Soc.* v.7, 1880, 235–40. *D.N.B.* v.4, 204. *R.S.C.* v.1, 264; v.7, 132; v.9, 182; v.12, 66.
Bargellinia belti Speg.

BENBOW, John (1821–1908)
b. Maidenhead, Berks 6 March 1821 *d.* Uxbridge, Middx 10 Feb. 1908
FLS 1887. Collected ferns in Devon and Cornwall, particularly *Polystichum*. Contrib. papers on Middx and Herts plants to *J. Bot.* from 1884. 'Middlesex Mosses' (*J. Bot.* 1894, 106–7, 369–70.)
Bot. Soc. Exch. Club Br. Isl. Rep. 1908, 345–46. *Proc. Linn. Soc.* 1907–8, 46. *Gdnrs Chron.* 1911 i 346. *J. Bot.* 1909, 154. *Br. Fern Gaz.* 1911, 189–92, 201–2. G. C. Druce *Fl. Buckinghamshire* 1926, ciii–civ. D. H. Kent *Hist. Fl. Middlesex* 1975, 24.
Herb. and MS. flora of Uxbridge at BM (NH). Portr. at Hunt Library.

BENCHLEY, Winifred Elsie (1883–1953)
d. Harpenden, Herts 27 Oct. 1953
DSc. FLS 1910. Rothamsted Experimental Station, 1906–48; Head of Botany Department.
Pharm. J. 1953, 347.

BENDEL, John (*fl.* 1780s)
Nurseryman, Montpelier Gardens, Walworth, London.

BENNET, Mr. *see* Bennett, W.

BENNET, James Henry (1816–1891)
b. Manchester 1816 *d.* La Bollène, Alpes Maritimes, France 26 July 1891
Physician at Mentone where he created a hillside garden. *Winter and Spring on Shores of Mediterranean* ed. 5, 1875. Contrib. to *Gdnrs Chron.*
Gdnrs Chron. 1874 ii 424–25; 1891 ii 170.

BENNET, Rev. William (–1805)
d. drowned in Duddingston Loch near Edinburgh 1805
Minister of Duddingston, Edinburgh, 1786–1805. 'Duddingston' (J. Sinclair *Statistical Account of Scotland* v.18, 1796, 372–76 containing an account of the flora).

BENNETT, Alfred William (1833–1902)
b. Clapham, London 24 June 1833 *d.* London 23 Jan. 1902
BA London 1853. BSc 1868. FLS 1868. Bookseller and publisher, 1858–68. Lecturer on Botany, St. Thomas's Hospital and Bedford College. *Fl. of Alps* 1892 2 vols. Translated and edited J. Sach's *Lehrbuch der Botanik* (with W. T. Thiselton-Dyer) 1875. *Handbook of Cryptogamic Botany* (with G. Murray) 1889. Contrib. to *J. Bot.*, etc.
Garden v.61, 1902, 81–82. *Gdnrs Chron.* 1902 i 85. *J. Bot.* 1902, 113–15. *J. R. Microsc. Soc.* 1902, 155–57 portr. *Proc. Linn. Soc.* 1901–2, 26–27. *D.N.B. Supplt. 2* v.1, 143–44. *Who was Who, 1897–1916* 57. *Tribuna Farmaceutica* v.10, 1942, 231. *R.S.C.* v.7, 137; v.9, 189; v.13, 444.
Plants at BM(NH), Kew, Oxford. Letters at Kew. Portr. at Hunt Library.

BENNETT, Arthur (1843–1929)
b. Croydon, Surrey 19 June 1843 *d.* Croydon 2 May 1929
FLS 1881–1905. ALS 1910. Builder and house decorator, Croydon. Authority on *Potamogeton* and *Cyperaceae*. *Potamogetons of British Isles* (with A. Fryer) 1915. Contrib. supplements to H. C. Watson's *Topographical Botany* (*J. Bot.* 1905; 1929, 1–56). *Naiadaceae* in *Fl. Capensis* 1897, *Fl. Tropical Africa* 1901. Prolific contributor to many periodicals including *J. Bot.*, *Ann. Scott. Nat. Hist.*, *Trans. Bot. Soc. Edinburgh*, etc.
Bot. Soc. Exch. Club Br. Isl. Rep. 1930, 81–89; 1936, 215–16. *Essex Nat.* v.29, 1954, 192. *J. Bot.* 1929, 184, 217–21 portr. *Kew Bull.* 1929, 302–3. *Proc. Linn. Soc.* 1929–30, 187–89. *Trans. Norfolk Norwich Nat. Soc.* 1929–30, 84–86. *Trans. Proc. Bot. Soc. Edinburgh* v.30, 1929, 180–82. *Watson Bot. Exch. Club Rep.* 1929–30, 5–7 portr. C. P. Petch and E. L.

Swann *Fl. Norfolk* 1968, 17. C. E. Salmon *Fl. Surrey* 1931.

British Potamogetons at BM(NH), foreign ones at Kew. British *Characeae* at Essex Field Club. Notebooks of Norfolk plants at BM (NH). Portr. at Hunt Library.

BENNETT, Charles *pseudonym see* Davis, Walter

BENNETT, Edward (*c.* 1826–1904)
d. Ash Vale near Aldershot, Hants 8 Oct. 1904

Gardener at Osberton, Notts and Enville Hall, Staffs and to Lord Salisbury at Hatfield, Herts. Contrib. to *Gdnrs Chron.*

Gdnrs Chron. 1904 ii 279.

BENNETT, Edward Trusted (1831–1908)
b. London 1 July 1831 *d.* 16 Nov. 1908

Collected plants in Cornwall and New Forest. 'Restoration of Euphorbia peplus to One of its Old Localities' (*Phytologist* v.4, 1851, 1–5). 'Three Days' Walk in New Forest' (*Phytologist* v.4, 1852, 753–56).

J. Bot. 1909, 39. F. H. Davey *Fl. Cornwall* 1909, xlix. *R.S.C.* v.1, 273.

BENNETT, Edward Turner (1797–1836)
b. Hackney, London 6 Jan. 1797 *d.* 21 Aug. 1836

Brother of J. J. Bennett (1801–1876). Surgeon in London. Secretary, Zoological Society of London, 1833–36. Zoologist. Assisted S. F. Gray in *Hepaticae* in *Natural Arrangement of British Plants* v.1, 1821, xiii.

J. Bot. 1872, 223.

BENNETT, Frederick Debell (*fl.* 1830s-1840s)

Brother of G. Bennett (1804–1893). *Narrative of Whaling Voyage round the Globe... 1833–1836* v.2, 1840, 327–95 contains a list of Sandwich Islands plants. Sent plants to A. B. Lambert.

Fl. Malesiana v.1, 1950, 49–50. *Taxon* 1970, 514.

Plants at Berlin-Dahlem.

BENNETT, George (1804–1893)
b. Plymouth, Devon 31 Jan. 1804 *d.* Sydney 29 Sept. 1893

MRCS 1828. MD Glasgow 1859. FLS 1831. In Ceylon, 1819. Visited Australia in 1829 and 1832 and also in the Pacific making a large collection of plants. Settled in Sydney, 1836, as physician. "The greatest of the physician-naturalists of Australia." *Wanderings in New South Wales...during 1832–1834* 1834 2 vols. *Gatherings of a Naturalist in Australasia* 1860. *On the Introduction...and...Uses of the Orange and Others of the Citron Tribe in New South Wales* 1871. Contrib. to Loudon's *Mag. Nat. Hist.*, 1832; *J. Bot.*, 1866–72.

Proc. Linn. Soc. N.S.W. 1893, 542–43. *Proc. Linn. Soc.* 1893-94, 27–28. *J. Bot.* 1894, 191. P. Mennell *Dict. Austral. Biogr.* 1892. *Sydney Morning Herald* 30 Sept. 1893. *Rep. Austral. Assoc. Advancement Sci.* 1911, 225. Bladen *Notes on Library N.S.W.* ed. 2 1911, 21 portr. *Victorian Nat.* v.45, 1928, 207–8 portr. *J. Proc. R. Austral. Hist. Soc.* 1933, 321–22. *Fl. Malesiana* v.1, 1950, 50. *Med. J. Austral.* 1950, 554–55 *Bull. Post-Graduate Comm. Med., Univ. Sydney* 1955, 207–64 portr. *Austral. Dict. Biogr.* v.1, 1966, 85–86. *R.S.C.* v.1, 273; v.7, 138; v.9, 190; v.12, 69.

Australian plants at BM(NH) and Kew. New Zealand and Tahiti plants at Linnean Society. Letters at Kew. Journal and letters at Mitchell Library, Sydney.

Antiaris bennettii Seem.

BENNETT, Henry (–1890)

Farmer near Salisbury, Wilts then rose grower at Shepperton, Middx. Raised many new rose varieties including 'Mrs John Laing'.

Garden v.38, 1890, 191. *J. Hort. Cottage Gdnr* v.21, 1890, 159.

BENNETT, John Joseph (1801–1876)
b. Tottenham, Middx 8 Jan. 1801 *d.* Maresfield, Sussex 29 Feb. 1876

MRCS 1825. FRS 1841. FLS 1828. Assistant Keeper, Dept. of Botany, BM(NH) 1827; Keeper, 1859–70. Secretary, Linnean Society, 1838–52. *Plantae Javanicae Rariores* 1838–52.

J. Bot. 1876, 95–105. *Proc. Linn. Soc.* 1888–89, 32 portr. *Trans. Bot. Soc. Edinburgh* v.13, 1879, 3–5. *D.N.B.* v.4, 1876, 97–105. A. T. Gage *Hist. of Linn. Soc. of London* 1938 portr. *Proc. Bot. Soc. Br. Isl.* 1955, 490–91. *R.S.C.* v.1, 275.

Herb. at Croydon Natural History Society. Portr. by Eddis at Linnean Society. Bust at Linnean Society, BM(NH) and Kew. Portr. at Hunt Library. Letters at Kew. Part of library sold at Sotheby, 4 Dec. 1871.

Bennettia Miquel. *Bennettites* Carr.

BENNETT, John Whitworth (*fl.* 1800s–1840s)

FLS 1828. Of Ceylon civil establishment. *Coco-nut Palm* 1831. *Selection of Rare and Curious Fruits Indigenous to Ceylon* 1850? for which he also drew the plates.

BENNETT, Kenrick Harold (*c.* 1840–1891)
d. Ivanhoe, N.S.W. 30 June 1891

Bush naturalist, New South Wales. Collected plants for J. H. Maiden, 1886–87. Contrib. to *Proc. Linn. Soc. N.S.W.*

J. R. Soc. N.S.W. 1908, 84–85.

BENNETT, Stephen Allen (1868–1934)
b. Burslem, Staffs 13 Dec. 1868 *d.* Burslem 8 Feb. 1934

MA Cantab. BSc London. Science master, Campbell College, Belfast, 1898–1926. President, Belfast Naturalists' Field Club, 1920–22. Stewart and Corry's *Fl. N.E. Ireland Second Supplt.* 1923 contains many of his plant records.

Irish Nat. J. v.3, 1931, 250–51 portr.; v.5, 1934, 37–38. R. L. Praeger *Some Irish Nat.* 1949, 50–51.

BENNETT, T. E. (*fl.* 1840s)
Of Bletchworth, Surrey. Found *Polypodium* var. *omnilacerum* in 1848.
Br. Fern Gaz. 1909, 43.

BENNETT, William (–*c.* 1765)
Corn merchant and biscuit maker of Whitechapel, London. Had gardens of exotic plants at Shadwell and Whitechapel. His plants sold 27 March 1766 by Langford and Son.
Trans. Linn. Soc. v.10, 1811, 279–80. *Gdnrs Chron.* 1918 ii 245–46.

BENNETT, William (1804–1873)
b. London 29 Feb. 1804 *d.* London 1873
Father of E. T. Bennett (1831–1908) and A. W. Bennett (1833–1902). 'Ferns of North Wales' (*Phytologist* v.3, 1849, 709–15). Discovered *Teucrium botrys* (*Phytologist* v.3, 1849, 737–38).
W. H. Purchas and A. Ley *Fl. Herefordshire* 1899, preface, v. *R.S.C.* v.1, 275; v.7, 138.

BENNETT-POË, John T. (1846–1926)
b. County Tipperary 1846 *d.* Westminster, London 14 May 1926
MA Dublin. VMH 1902. Cultivated florists' flowers and orchids.
Garden v.61, 1902, iv portr. *J. Hort. Cottage Gdnr* v.45, 1902, 310–11 portr. *Gdnrs Mag.* 1907, 243–44 portr. *Gdnrs Chron.* 1926 i 394. *Orchid Rev.* 1926, 164.

BENSON, Margaret Jane (1859–1936)
b. London 20 June 1859 *d.* Hertford 20 June 1936
BSc London 1891. DSc 1894. FLS 1904. Head, Dept. of Botany, Royal Holloway College, 1893–1912; Professor, 1912–22. Collaborated with D. H. Scott and F. W. Oliver. Palaeobotanist. Studied embryology of *Amentiferae*. Contrib. to *Ann. Bot.*, *New Phytol.*
Nature v.138, 1936, 17. *Chronica Botanica* 1937, 159–60 portr. *Proc. Linn. Soc.* 1937, 186–89. *Who was Who, 1929–1940* 97.
Herb., fossil slides and portr. at Royal Holloway College.

BENSON, Richard de Gylpyn (1856–1904)
b. Church Pulverbatch, Shropshire 25 June 1856 *d.* Church Pulverbatch 24 Feb. 1904
Solicitor. Bryologist. 'Shropshire Mosses' (*J. Bot.* 1893, 257–65).
J. Bot. 1904, 128.
Herb. at Shrewsbury Museum. British mosses at BM(NH).

BENSON, Robson (1822–1894)
b. 5 Jan. 1822 *d.* Bath, Somerset 22 Oct. 1894
FLS 1870. General. Superintendent of gardens of Agri-Horticultural Society, Rangoon, 1865–69 and Madras, 1872–76. Collected orchids in Burma.
Proc. Linn. Soc. 1894–95, 30. *Times* 24 Oct. 1894. F. Boase *Modern English Biogr. Supplt.* v.1, 1908, 369. *R.S.C.* v.12, 69.
Letters at Kew.
Vanda bensoni Bateman.

BENSON, Rev. Thomas (1802–1887)
b. Cockermouth, Cumberland Oct. 1802 *d.* Great Fambridge, Essex 9 June 1887
BA Cantab 1824. Rector, North Fambridge, 1832–87. Contrib. to G. S. Gibson's *Fl. Essex* 1862.
Essex Nat. v.1, 1887, 138–39; v.29, 1954, 189. S. T. Jermyn *Fl. Essex* 1974, 17–18.
Herb. at Essex Field Club.

BENT, James Theodore (1852–1897)
b. Baildon, Leeds 30 March 1852 *d.* London 5 May 1897
BA Oxon 1875. Traveller and archaeologist. Travelled widely in Middle East. Collected plants in Arabia, Sudan and Socotra, 1893–97.
Kew Bull. 1894, 328; 1895, 158–59, 180; 1897, 206. *D.N.B. Supplt. 1* v.1, 179. *Rec. Bot. Survey India* v.8, 1933, 487–88. *R.S.C.* v.13, 450.
Plants and letters at Kew.
Bentia Rolfe.

BENTALL, Thomas (*fl.* 1840s–1860s)
Of Halstead, Essex. Manufactured a drying paper. Contrib. to *Phytologist* v.2, 1845–47 and v.3, 1848; also to G. S. Gibson's *Fl. Essex* 1862.
R.S.C. v.1, 280.

BENTHALL, Rev. Charles Francis (1861–1936)
d. Bishopsteignton, Devon 1936
Vicar, Cofton, Devon, 1891–1918; Rural Dean, Kenn, 1909–12. Contrib. records of local plants.
W. K. Martin and G. T. Fraser *Fl. Devon* 1939, 778.

BENTHAM, George (1800–1884)
b. Stoke, Plymouth, Devon 22 Sept. 1800 *d.* London 10 Sept. 1884
LLD Cantab 1874. FRS 1862. FLS 1826. Nephew of Jeremy Bentham. Secretary, Horticultural Society, 1829–40. President, Linnean Society, 1861–74. Leading taxonomist. *Catalogue des Plantes Indigènes des Pyrénées* 1826. *Labiatarum Genera et Species* 1832–36.

Scrophularineae Indicae 1835. *Plantae Hartwegianae* 1839–57. *Handbook of British Fl.* 1858; ed. 4 1878. *Fl. Hongkongensis* 1861. *Fl. Australiensis* 1863–78 7 vols. *Genera Plantarum* (with J. D. Hooker) 1862–83 3 vols.

Gdnrs Chron. 1884 ii 336, 368, 370, 432, 464. *J. Bot.* 1884, 353–56 portr.; 1906, 397–401; 1918, 241. *Magyar Növenytani Lapok* 1884, 97–108. *Nature* v.30, 1884, 539–43. *Proc. Linn. Soc.* 1884–85, 90–104; 1887–88, 71–79; 1888–89, 32–33. *Amer. J. Sci.* 1885, 103-18. *Proc. R. Soc.* v.38, 1885, i–v. *Trans. Bot. Soc. Edinburgh* v.16, 1886, 190–92. *Life and Letters of Sir C. J. F. Bunbury* 1895 *passim. Ann. Bot.* 1898, ix–xxx portr. E. Bretschneider *Hist. European Bot. Discoveries in China* 1898, 401–3. *D.N.B.* v.4, 263–67. B. D. Jackson *George Bentham* 1906 portr. *Kew Bull.* 1906, 187–88. *J. R. Soc. N.S.W.* 1908, 64–65. J. R. Green *History of Botany* 1914, 492–99. *Proc. R. Soc. Queensland* 1954, 8–10. *J. Soc. Bibl. Nat. Hist.* 1956, 127–32. *Taxon* 1966, 37–39. F. A. Stafleu *Taxonomic Literature.* 1967, 26–28. *Austral. Dict. Biogr.* v.3, 1969, 146–47. C. C. Gillispie *Dict. Sci. Biogr.* v.1, 1970, 614–15. M. A. Reinikka *Hist. of the Orchid* 1972, 160–63 portr. *R.S.C.* v.1, 280; v.7, 240; v.9, 192; v.12, 10; v.13, 451.

Herb., letters, diaries and library at Kew. Portr. by D. Macnee at Kew; portr. by L. Dickinson at Linnean Society. Portr. at Hunt Library.

Benthamia Lindl. *Neobenthamia* Rolfe. *Benthamantha* Alef.

BENTHAM, Lady Mary Sophia (*née* **Fordyce**) (*c.* 1765–1858)
d. 18 May 1858

Daughter of G. Fordyce (1736–1802). Married General Sir General Samuel Bentham, 1796. Mother of G. Bentham (1800–1884). "A very good botanist" (A. Gray *Letters* v.1, 1893, 188). Had herb.

J. Bot. 1894, 315. *Ann. Bot.* v.12, 1898, xiii.

BENTICK, Mr. (*fl.* 1690s)

Introduced *Ixora coccinea* and *Cotyledon orbiculata* in 1690.

Bot. Mag. 1791, t.169; 1795, t.321.

BENTINCK, Margaret Cavendish, Duchess of Portland (*née* **Harley**) (1715–1785)
b. London 11 March 1715 *d.* Bulstrode, Bucks 7 July 1785

Married William, 2nd Duke of Portland, 1734. "Well acquainted with English plants" (P. Browne *Civil and Natural History of Jamaica* 1756, 165). J. Lightfoot, who produced a catalogue of her museum, dedicated his *Fl. Scotica* 1777 to her. She had a botanic garden as well as a museum at Bulstrode. Her herb. was purchased by A. B. Lambert (J. E. Smith *Selection from Correspondence of Linnaeus* v.2, 1821, 44–45; P. Smith *Mem. and Correspondence of...Sir J. E. Smith* v.1, 1832, 135–36).

Gent. Mag. 1785, 575. *Bot. Mag.* 1795, t.286. J. Lightfoot *Cat. of Portland Mus.* 1786. E. J. L. Scott *Index to Sloane Manuscripts* 1904, 50. G. C. Druce *Fl. Buckinghamshire* 1926, xci. *J. Soc. Bibl. Nat. Hist.* 1962, 30–34. S. P. Dance *Shell Collecting* 1966, 103–7 portr.

Sale of collections (*Gent. Mag.* 1786, 526–27).

Portlandia R. Br.

BENTLEY, Bertram Henry (1873–1946)
b. Manchester 2 March 1873 *d.* Cheltenham, Glos 24 June 1946

BA Oxon 1896. FLS 1899. Lecturer in Botany, Sheffield University, 1896–1931; Professor of Botany, 1931–39.

Nature v.158, 1946, 263–64. *Proc. Linn. Soc.* 1945–46, 131. *Times* 26 June 1946. *Who was Who, 1941–1950* 91.

BENTLEY, James W. (1859–1924)
b. Royton near Manchester 1859

Of Stakehill, Manchester. Chemist and colourist. Had keen interest in florists' flowers. President, Wakefield and North of England Tulip Society. President, Northern Carnation Society, 1893–1910.

Gdnrs Mag. 1911, 441–42 portr. *Garden* 1924, 272 (ix).

BENTLEY, Robert (1821–1893)
b. Hitchin, Herts 25 March 1821 *d.* London 24 Dec. 1893

MRCS 1847. FLS 1849. Lecturer in Botany, London Hospital. Professor of Botany at London Institution and King's College and of Botany and Materia Medica to Pharmaceutical Society. *On Advantages of Study of Botany to Student of Medicine* 1860. *Manual of Botany* 1861; ed. 5 1887. *On the Characters, Properties, and Uses of Eucalyptus globulus* 1874. *Medicinal Plants* (with H. Trimen) 1875–80 4 vols. Editor, *Pharm. J.*

Pharm. J. 1893–94, 559–60. *Proc. Linn. Soc.* 1893–94, 28. *D.N.B. Supplt. 1* v.1, 181–82. *Notes R. Bot. Gdn Edinburgh* 1953, 157–62. F. A. Stafleu *Taxonomic Literature* 1967, 29–30. *R.S.C.* v.1, 282; v.9, 192.

Portr. at Hunt Library.

BENTLEY, Walter (*c.* 1875–1953)
d. London 14 April 1953

VMH 1952. Had garden at Quarry Wood, Newbury where he specialised in lilies.

Gdnrs Chron. 1953 i 146 portr. *Lily Yb.* 1954, 9–10 portr.

BENTLEY, William (1815–1881)
b. Royton, Lancs 1815 *d.* Castleton near Manchester 21 Dec. 1881

Blacksmith at Royton and later rate collector at Oldham. Friend of Mellor, Binney, Buxton, Jethro Tinker, Stansfield, Nowell, Percival and the Horsefields. President, Royton Botanical Society, 1854.

Trans. Liverpool Bot. Soc. 1909, 60.

BENWELL, James (*c.* 1735–1819)
d. Oxford 7 Oct. 1819

Employed in Oxford Botanic Garden for more than 40 years. Accompanied J. Sibthorp on botanical excursions.

W. Baxter *Br. Phaenogamous Bot.* v.6, 1843, 415. G. C. Druce *Fl. Berkshire* 1897, clvii. G. C. Druce *Fl. Oxfordshire* 1927, xciii–xciv.

Portr. at Hunt Library.

BERESFORD, Sir John (*fl.* 1810s)

Sent *Bignonia* sp. from S. America in 1815 to Lord Liverpool.

Bot. Mag. 1819, t.2050.

BERESFORD, Marshal (*fl.* 1820s)

Sent *Franciscea* sp. from Brazil to his sister, Mrs. T. Hope, at Deepdene, Surrey.

Bot. Mag. 1828, t.2829.

BERKELEY, Cecilia E.

Rendered services to "mycology by many excellent illustrations and in other ways" (M. J. Berkeley *Ann. Mag. Nat. Hist.* v.13, 1854, 396–97).

Agaricus ceciliae Berkeley.

BERKELEY, Emeric Streatfield (*c.* 1823–1898)
d. Bitterne, Southampton, Hants Dec. 1898

Son of Rev. M. J. Berkeley (1803–1889). Major-General. Collected orchids, etc. in India, Andaman and Nicobar Islands. Discovered the fungi, *Emericella*, in India.

Gdnrs Chron. 1898 ii 427–28. *Orchid Rev.* 1899, 9.

Plants at Vienna. Kew. Letters at Kew.

Emericella Berkeley.

BERKELEY, Rev. Miles Joseph (1803–1889)
b. Biggin, Oundle, Northampton 1 April 1803 *d.* Sibbertoft, Market Harborough, Leics 30 July 1889

BA Cantab 1825. FRS 1879. FLS 1836. Curate, Margate, 1828. Rector, Sibbertoft, 1868–89. Mycologist. *Gleanings of British Algae* 1833. *British Fungi* 1836–43 (exsiccatae). *British Fungi* 1836 (in W. J. Hooker's *English Fl.*) *Introduction to Cryptogamic Botany* 1857. *Outlines of British Fungology* 1860. *Handbook of British Mosses* 1863. Regular contributor to *Gdnrs Chron.* almost from its beginning in 1841. Editor, *J. R. Hort. Soc.* 1866–77.

L. Reeve *Portr. of Men of Eminence in Literature, Sci. and Art* v.2, 1864, 5–8 portr. *Gdnrs Chron.* 1871, 271 portr.; 1879 i 789 portr.; 1889 ii 141–42. *J. Bot.* 1879, 345–46; 1889, 305–8. *Trans. Crypt. Soc. Scotland* 1888, 92–95. *Northampton Nat. Hist. Soc. J.* v.5, 1888–89, 345–49. *Proc. Linn. Soc.* 1888–89, 33, 93–94. *Gdn For.* v.2, 1889, 410–11. *J. Hort. Cottage Gdnr* v.19, 1889, 84, 113–14 portr. *Nature* v.40, 1889, 371–72. *Scott. Nat.* 1889, 145–48. *Ann. Bot.* 1889–90, 451–56; 1897, ix–xi. *Grevillea* v.18, 1890, 17–19. *Proc. R. Soc.* v.47, 1890, ix–xii. *Northampton Notes and Queries* v.4, 1892, 25–37, 221–24. *D.N.B. Supplt* v.1, 183. G. Lindau and P. Sydow *Thesaurus Litteraturae Mycologicae* v.1, 1907, 121–35. F. W. Oliver *Makers of Br. Bot.* 1913, 225–32 portr. *Mycologia* 1919, 194. *Curtis's Bot. Mag. Dedications, 1827–1927* 107–8 portr. G. C. Druce *Fl. Northamptonshire* 1930, cxvii–cxx. *Phytopathol. Classics* no. 8, 1948, 5–12. C. C. Gillispie *Dict. Sci. Biogr.* v.2, 1970, 18–19. *Country Life* 1974, 639–40 portr. *Taxon* 1974, 324 portr. *R.S.C.* v.1, 295; v.7, 144; v.9, 200; v.12, 73; v.13, 475.

Fungi, MSS. and letters at Kew. Drawings and letters at BM(NH). Algae at Cambridge.

Portr. by J. Peel at Linnean Society. Portr. at Kew, Hunt Library.

Berkeleya Grev.

BERKENHOUT, John (1726–1791)
b. Leeds, Yorks 8 July 1726 *d.* Besselsleigh, Berks 3 April 1791

MD Leyden 1765. Accompanied government commissioners to America, 1778–80. *Clavis Anglica Linguae Botanicae; or, Botanical Lexicon* 1764; ed. 2 1789. *Outlines of Natural History of Great Britain and Ireland* 1770.

Gent. Mag. 1791, 388, 485. R. V. Taylor *Biographia Leodiensis* 1865–67, 187–91. *D.N.B.* v.4, 369–70.

BERNAYS, Lewis Adolphus (1831–1908)
b. London 3 May 1831 *d.* Brisbane, Queensland 22 Aug. 1908

FLS 1871. CMG 1892. To New Zealand 1850. Clerk to Legislative Assembly, Queensland, 1859. Interested in plants of economic importance. *Olive and its Products* 1872. *Cultural Industries for Queensland...Useful Plants* ...1883. Contrib. to *Proc. R. Soc. Queensland.*

P. Mennell *Dict. Austral. Biogr.* 1892, 37. *Brisbane Courier* 24 Aug. 1908. *Proc. Linn. Soc.* 1908–9, 34–35. *Rep. Austral. Assoc. Advancement Sci* 1909, 375. *Who was Who, 1897–1916* 59. P. Serle *Dict. Austral. Biogr.* v.1, 1949, 73–74. *R.S.C.* v.9, 204; v.13, 482.

Nepenthes bernaysii Bailey.

BERNHARD-SMITH, William Arthur Hans (–1927)
d. 24 Jan. 1927

Surgeon, King's College Hospital. *Poisonous Plants of All Countries* 1905; ed. 2 1923.
Pharm. J. 1927 i 155. *Times* 31 Jan. 1927.

BERRIDGE, Emily Mary (1872–1947)
b. Bromley, Kent 20 Feb. 1872 *d.* 8 Oct. 1947
BSc London 1898. DSc 1914. FLS 1905. Began her researches on fossil plants, Gnetales and *Amentiferae* under Professor Margaret Benson at Royal Holloway College. Contrib. to *Ann. Bot.*, *New Phytol.* and *Ann. Applied Biol.*
Nature v.161, 1948, 87. *Proc. Linn. Soc.* 1947–48, 68–70.

BERRISFORD, Samuel (1859–1938)
Blacksmith of Oakamoor, Staffs. Made a collection of Staffordshire plants now in the possession of E. S. Edees.
Gdnrs Chron. 1938 ii 380. E. S. Edees *Fl. Staffordshire* 1972, 20.

BERRY, Andrew (*fl.* 1780s–1810s)
MD. FRSE 1796. Nephew of J. Anderson (–1809). Madras Medical Service, 1784–1814. Laid out garden in Madras. Friend of W. Roxburgh and sent plants to Calcutta Botanic Garden. 'Account of...Colomba Root' (*Asiatick Researches* v.10, 1808, 385–88).
W. Roxburgh *Plants of Coast of Coromandel* v.3, 1819, 60–61. D. G. Crawford *Hist. Indian Med. Service, 1600–1913* v.2, 1914, 20–23. *Fl. Malesiana* v.5, 1958, 24. *R.S.C.* v.1, 70.
Berria Roxb.

BERRY, Edward (*c.* 1842–1903)
b. Rathfarnham Castle, Dublin *c.* 1842 *d.* 25 May 1903
Gardener to Lord Leven at Roehampton House, Wimbledon, 1864.
Gdnrs Chron. 1903 i 350.

BERRY, Elihu (1812–1869)
b. Darton near Barnsley, Yorks 3 Sept. 1812 *d.* Dublin 22 Oct. 1869
Gardener at Park House, Ardsley near Barnsley. Studied local flora.
Naturalist 1946, 113–14.
Plants at Barnsley Museum.

BERRY, Grimshaw Heyes (1880–1956)
b. London 30 May 1880 *d.* 20 Aug. 1956
Pianoforte maker. Amateur gardener with interest in plant physiology. Contrib. to *Quart. Bull. Alpine Gdn Soc.*
Quart. Bull. Alpine Gdn Soc. v.24, 1956, 360–62.

BERRY, John (–1727)
Shopkeeper, nurseryman and gardener, Tytherington, Glos.
Agric. Hist. Rev. 1974, 21, 24.

BERRY, W. G. (–1964)
d. June 1964
Amateur orchid grower. President, Manchester and North of England Orchid Society.
Orchid Rev. 1964, 307.

BESANT, John William (1878–1944)
b. Longforgan, Perthshire Aug. 1878 *d.* 18 Sept. 1944
Kew gardener, 1901–5. Foreman, Botanic Garden, Glasnevin, 1907; Assistant Keeper, 1920; Keeper, 1922. Editor, *Irish Gdning*, 1914–22. Contrib. to *Gdnrs Chron.*
Gdnrs Chron. 1944 i 124. *J. Kew Guild* 1943, 217–18 portr. *Richmond & Twickenham Times* 30 Sept. 1944.

BESANT, W. D. (*c.* 1886–1946)
VMH 1944. Director, Parks Dept., Glasgow.

BEST, George Arnold (–1937)
d. Birmingham 20 Aug. 1937
Assistant Curator, Gardens Dept., Straits Settlements, 1921. Collected plants in Malaya, 1922–28; W. Java, 1928.
Gdns Bull. Straits Settlements 1927, 117. *Fl. Malesiana* v.1, 1950, 53.
Plants at Singapore.

BESTER, John (1847–)
Nurseryman, Granchester, 1868; moved to Chesterton, Cambridge, 1875. President, Chesterton Horticultural Society.
Gdnrs Mag. 1913, 593–94 portr.

BETCHE, Ernst (1851–1913)
b. Potsdam, Germany 31 Dec. 1851 *d.* Sydney 28 June 1913
In Municipal Gardens, Berlin. Gardener, Van Houtte nursery, Ghent, 1874. Collected plants in Samoa, 1880–81 and for Sydney Botanic Garden, 1881–97. Botanical Assistant, Sydney, 1897. *Handbook Fl. N.S.W.* (with C. Moore) 1893. *Census of N.S.W. Plants* (with J. H. Maiden) 1916.
Rep. Bot. Gdn N.S.W. 1913, 14–15. *Proc. Linn. Soc. N.S.W.* 1914, 3–4. *J. R. Soc. N.S.W.* 1921, 153 portr. *J. Proc. R. Austral. Hist. Soc.* 1933, 130–31. *Austral. Encyclop.* v.1, 1962, 498. *Austral. Dict. Biogr.* v.3, 1969, 158.
Plants at Melbourne, Sydney.

BETTANY, George Thomas (1850–1891)
b. Penzance, Cornwall 30 March 1850 *d.* Dulwich, London 2 Dec. 1891
BSc 1871. BA Cantab 1873. FLS 1880. Lecturer in Botany, Guy's Hospital, 1877–86. *Simple Lessons in Botany* 1878. *First Lessons in Practical Botany* 1881. *Botany* 1882. *R.S.C.* v.7, 165; v.9, 231.
Illus. London News 12 Dec. 1891, 758 portr. F. Boase *Modern English Biogr. Supplt.* v.1, 1908, 389–90. *Guy's Hospital Rep.* 1971, 70–71.

BETTRIDGE, William (*fl.* 1790s)
Nurseryman and seedsman, Warwick.

BEVAN, David William (1860–1944)
d. Scarborough, Yorks 14 Nov. 1944
Schoolteacher, Scarborough. Taught botany.
Naturalist 1945, 71.

BEVAN, Theodore F. (1860–)
b. London 1860
Explored and collected plants in New Guinea, 1884–87. *Toil, Travel and Trouble in British New Guinea* 1890.
Proc. Linn. Soc. N.S.W. 1887, 419–22. *Fl. Malesiana* v.1, 1950, 55.
New Guinea plants at Melbourne.
Mussaenda bevani F. Muell.

BEWICK, Mr. (*fl.* 1780s)
Of Clapham, London. "Celebrated for collections of American plants." Plants sold in 1782.
Bot. Mag. 1792, t.180.

BEWS, John William (1884–1938)
b. Kirkwall, Orkney 16 Dec. 1884 *d.* Maritzburg, Natal 10 Nov. 1938
MA Edinburgh 1906. BSc 1907. DSc 1912. FLS 1926. Lecturer in Economic Botany, Manchester, 1907–8. Lecturer in Plant Physiology, Edinburgh, 1908–10. Professor of Botany, Natal, 1910–38 excluding 1925–27 when he was Professor of Botany, Durham. Travelled in Portuguese East Africa, Rhodesia, U.S.A. and Canada. Pioneer of plant ecology in S. Africa. *Grasses and Grasslands of South Africa* 1918. *Fl. Natal and Zululand* 1921. *Plant Forms and their Evolution in South Africa* 1925. *Ecological Evolution of the Angiosperms* 1927. *World's Grasses* 1929.
Nature v.142, 1938, 1066–67. *Proc. Linn. Soc.* 1938–39, 230–31. *Chronica Botanica* 1939, 282 portr. *Kew Bull.* 1939, 32–33. *Who was Who, 1929–1940* 107–8. G. W. Gale *John William Bews* 1954. *S. African J. Sci.* 1971, 409–10.
Herb. at Natal University.

BICHENO, James Ebenezer (1785–1851)
b. Newbury, Berks 25 Jan. 1785 *d.* Hobart, Tasmania 25 Feb. 1851
FRS 1827. FLS 1812. Called to the Bar, 1822. Secretary, Linnean Society, 1824–32. Colonial Secretary, Van Dieman's Land, 1842. Contrib. to J. Sowerby and J. E. Smith's *English Bot.* 2631, 2675, 2680. 'British Species of Juncus' (*Trans. Linn. Soc.* v.12, 1818, 291–339).
Hooker's J. Bot. 1851, 250–51. *Proc. Linn. Soc.* 1852, 180–82; 1888–89, 33. *J. Bot.* 1898, 275. G. C. Druce *Fl. Berkshire* 1897, clii–clv. *D.N.B.* v.5, 1–2. *Naturalist* 1902, 337–42. *Austral. Encyclop.* v.1, 1962, 499. *Austral. Dict. Biogr.* v.1, 1966, 97–98. *R.S.C.* v.1, 358.
Herb. at Swansea Museum. Plants at BM (NH). Letters at BM(NH), Kew and Linnean Society. Portr. by Eddis at Linnean Society. Library at Tasmanian Public Library.
Bichenia D. Don.

BICKHAM, Spencer Henry (1841–1933)
b. near Manchester 1841 *d.* Ledbury, Herefordshire 7 April 1933
FLS 1896. Treasurer, Watson Botanical Exchange Club, 1907–20. Had garden at 'Underdown', Ledbury. Had herb. of British plants. Contrib. to *J. Bot.*
J. Bot. 1933, 202, 265; 1934, 177. *Br. Fern Gaz.* 1933, 195. *Mem. Proc. Manchester Lit. and Philos Soc.* v.77, 1933, iv. *Proc. Linn. Soc.* 1932–33, 183–84. *Watson Bot. Exch. Club Rep.* 1932–33, 163–64 portr. *Trans. Woolhope Nat. Field Club* v.34, 1954, 254.
Herb. at Cambridge.

BICKNELL, Algernon Sidney (1832–1911)
b. Herne Hill, Surrey 9 Oct. 1832 *d.* Brighton, Sussex 26 Oct. 1911
FLS 1877. Interested in botany, astronomy and alpine exploration. Collected fungi.
Proc. Linn. Soc. 1911–12, 42.

BICKNELL, Rev. Clarence (1842–1918)
b. Herne Hill, Surrey 27 Oct. 1842 *d.* Val Casterino di Tenda, Italy 17 July 1918
BA Cantab 1865. To Bordighera, 1879. *Flowering Plants and Ferns of Riviera* 1885. *Fl. Bordighera and San Remo* 1896. 'Ranunculus lacerus' (*J. Bot.* 1891, 21–22).
J. Bot. 1918, 303. *Bot. Soc. Exch. Club Br. Isl. Rep.* 1918, 352. *Boissiera.* fasc. 5, 1941, 21–22.
Plants at Genoa and Turin. Letters at BM (NH). Portr. at Hunt Library.
Bicknellii Knoche.

BIDDER, Rev. Henry Jardine (1847–1923)
b. Mitcham, Surrey 1847 *d.* Oxford 19 Oct. 1923
Educ. Oxford. Vicar, St. Giles, Oxford, 1887–1903. Fellow and Burser, St. John's College whose garden he created.
Gdnrs Chron. 1923 ii 270. *Nature* v.112, 1923, 663. *Bot. Soc. Exch. Club Br. Isl. Rep.* 1923, 148–49.

BIDDULPH, Harriet Sophia (*née* **Foot**) (1839–1940)
b. Singleton, N.S.W. 22 Sept. 1839 *d.* 11 Sept. 1940
Collected plants in Mt. Playfair, Central Queensland. Sent plants to F. von Mueller at Melbourne Botanic Garden.
Victorian Nat. 1970, 146.
Hemigenia biddulphiana F. Muell.

BIDDULPH, John (*fl.* 1840s–1870s)
b. 25 July 1840
Colonel, Indian Army. Collected plants on Gilgit expedition, 1879–81.
Pakistan J. For. 1967, 340.
Plants at Calcutta, Dehra Dun.

BIDDULPH, Susanna (*fl.* 1790s–1800s)
Of Southampton. Contrib. algae records to J. Sowerby and J. E. Smith's *English Bot.* 1762, etc.
Biddulphia Gray.

BIDGOOD, John (1853–1905)
b. Gateshead-upon-Tyne, Durham 1853 *d.* Bournemouth, Hants 6 Oct. 1905.
BSc London 1887. FLS 1889. Schoolmaster at Gateshead. *Textbook of Biology* 1893. 'Floral Colours and Pigments' (*J. R. Hort. Soc.* v.29, 1905, 463–80).
Gdnrs Chron. 1905 ii 287. *Proc. Linn. Soc.* 1905–6, 32. *J. R. Hort. Soc.* v.31, 1906, 189–90 portr.
Gloesporium bidgoodii Cooke.

BIDIE, George (1830–1913)
b. Backies, Banffshire 3 April 1830 *d.* 19 Feb. 1913
MB Aberdeen 1853. LRCS. Madras Medical Service 1856. Professor of Botany and Materia Medica, Madras Medical College. Surgeon-General, Madras, 1886–90. Collected plants in the Nilgiris, 1873. *Report on Neilgherry Loranthaceous Parasitical Plants Destructive to Exotic Forest and Fruit Trees* 1874. *Cinchona Culture in British India* 1879. *Sand-binding Plants of Southern India* 1883.
Who was Who, 1897–1916 62. C. E. Buckland *Dict. Indian Biogr.* 1906, 41.

BIDWELL, Henry (1816–1868)
b. Albrighton, Shropshire 8 July 1816 *d.* Albrighton 13 March 1868
MD. Local Secretary, Botanical Society of London. Fellow, Botanical Society of Edinburgh. Had herb.
H. N. Clokie *Account of Herb. Dept. Bot., Oxford* 1964, 132.
Herb. at Oxford.

BIDWELL, W. H. (–1909)
b. Norwich, Norfolk *d.* 25 April 1909
President, Norfolk and Norwich Naturalists Society 1901–2. Contrib. some botanical papers to their *Transactions.*
Trans. Norfolk Norwich Nat. Soc. v.9, 1909–10, 138.

BIDWILL, John Carne (1815–1853)
b. Exeter, Devon 1815 *d.* Tinana, Wide Bay, N.S.W. 16 March 1853
Arrived Sydney 1838. Travelled in North Island, New Zealand and returned to England with plants for Kew Gardens. Back in Sydney, 1844. First Director, Sydney Botanic Garden 1847. *Rambles in New Zealand* 1841. Contrib. to *Gdnrs Chron.*
Ann. Mag. Nat. Hist. 1842, 438–39. *Gdnrs Chron.* 1953, 438; 1856, 20–21; 1958 ii 28–29, 31. *Hooker's J. Bot.* 1853, 252. *Fl. Tasmaniae* v.1, 1859, cxxvi. J. Smith *Rec. R. Bot. Gdns, Kew* 1880, 67. *D.N.B.* v.5, 18–19. *Proc. Linn. Soc. N.S.W.* v.25, 1900, 778, 790. J. H. Maiden *Sydney Bot. Gdns Biogr. Notes* no. 8. *Kew Bull.* 1906, 217–18. *J. R. Soc. N.S.W.* 1908, 85–93. *J. Proc. R. Austral. Hist. Soc.* 1932, 124–26. *J. Hist. Soc. Queensland* 1939–41, 97–99. R. Glenn *Bot. Explorers of N.Z.* 1950, 63–68.
Plants at Kew. Letters at Kew, Sydney.
Bidwillia Herbert.

BIFFEN, Sir Rowland Harry (1874–1949)
b. Cheltenham, Glos 28 May 1874 *d.* Cambridge 12 July 1949
BSc Cantab 1896. FRS 1914. Knighted 1925. In Brazil and West Indies, 1897–98, to study sources of rubber. Demonstrator in Botany, Cambridge, 1898. Lecturer in Botany, Dept. of Agriculture, Cambridge, 1899; Professor of Agricultural Botany, 1908–31. Director, Plant Breeding Institute, Cambridge, 1912–36. Contrib. to *Ann. Bot.*, *Trans. Br. Mycol. Soc.*, etc.
Times 15 July 1949. *Nature* v.164, 1949, 305. *Obit. Notices Fellows R. Soc.* 1950, 9–24 portr. *Trans. Br. Mycol. Soc.* 1950, 167 portr. *D.N.B. 1941–1950* 76–77.
British and S. American fungi at Cambridge.

BIGG, Charles (*fl.* 1820s–1830s)
Nurseryman and seedsman, Shrewsbury, Shropshire.

BIGGE, Rev. John Frederick (1814–1885)
b. Linden, Northumberland 12 July 1814 *d.* Newcastle, Northumberland 28 Feb. 1885
BA Durham 1839. Vicar of Stamfordham. A founder of Tyneside Naturalists Club; President, 1847.
Hist. Berwick Nat. Club v.11, 1885, 207–17.
Herb. at Natural History Society Museum, Newcastle.

BIGGS, Arthur (1765–1848)
d. Cambridge 27 Jan. 1848
ALS 1815. FLS 1815. Gardener to Isaac Swainson at Twickenham. Raised 'Bigg's Nonsuch' apple. Curator, Cambridge Botanic Garden. 'Account of some New Apples' (*Trans. Hort. Soc.* 1807, 63–70).
Bot. Mag. 1816, t.1863, t.1875. *Proc. Linn. Soc.* 1848, 371–72.

BIGGS, James (*fl.* 1790s)
Seedsman, nurseryman and florist, 27 Mealcheapen Street, Worcester.

BIGGS, Thomas (*fl.* 1780s)
Florist at 'Fishburton' near Salisbury. Issued catalogue of 251 named varieties of auricula in 1782.
C. O. Moreton *The Auricula* 1964, 36–37.

BIGSBY, John Jeremiah (1792–1881)
b. Nottingham 14 Aug. 1792 *d.* London 10 Feb. 1881
MD Edinburgh 1814. FRS 1869. FLS 1823. Army medical officer at Cape, 1817. British Secretary and Medical Officer, Canadian Boundary Commission, 1822. Practised medicine at Newark, 1827–46, London, 1846–81. *Thesaurus Siluricus* 1868. *Thesaurus Devonico-Carboniferus* 1878.
J. Bot. 1881, 96. *Proc. Geol. Soc.* 1880–81, 39–41. *Proc. R. Soc.* v.33, 1882, xvi–xvii. *D.N.B.* v.5, 27. H. B. Woodward *Hist. Geol. Soc. London* 1907, 252 portr.

BILLIALD, R. A.
Surgeon of Kington. Collected Herefordshire plants.
W. H. Purchas and A. Ley. *Fl. Herefordshire* 1899, vi.

BILLINGS, John (*fl.*1790s)
Seedsman, Hinckley, Leics.

BILLINGSLEY, Lavinia (–1819)
Painted botanical decoration on Nantgarw china, Swansea, Glam.
Country Life 1964, 89.

BILLINGTON, Horace Walter Leighton (1867–1897)
b. Chalbury, Dorset 27 July 1867 *d.* Old Calabar 2 Nov. 1897
First Curator, Botanic Station, Old Calabar, Niger Coast, 1893–97. Reports on botany, etc. in *Parliamentary Papers, Africa* no. 1, 1895.
Gdnrs Chron. 1897 ii 406, 421. *Kew Bull.* 1897, 424. *Daily Telegraph* 20 Nov. 1897.
Letters and MSS. at Kew.

BILLINGTON, William (1776–1861)
b. Hadnall, Shropshire 27 July 1776 *d.* Belmullet, County Mayo 31 Oct. 1861
Gardener and arboriculturist. *Different Modes of Raising Young Plantations of Oaks* 1825. *Exposure of Misrepresentations of Author's Treatise on Planting* 1830.
Gdnrs Chron. 1861, 1047.

BILNEY, W. A. (*c.* 1853–1939)
d. Newbury, Berks 13 July 1939
VMH 1922. Solicitor. Keen gardener and active supporter of Royal Horticultural Society.
Gdnrs Chron. 1922 i 122 portr.; 1939 ii 107.

BINCKS, William (*fl.* 1730s)
Seedsman, Thames Street, London.

BINFIELD, Rev. Edward (*d.* before 1813)
ALS 1802. Of Spettisbury, Dorset. List of Dorset plants in R. Pulteney's *Cat. of Birds, Shells and some of the more Rare Plants of Dorsetshire* 1813, 62.

BINGHAM, John (*fl.* 1790s)
Gardener and seedsman, Eastbourne, Sussex.

BINGLEY, Rev. William (1774–1823)
b. Doncaster, Yorks Jan. 1774 *d.* London 11 March 1823
BA Cantab 1799. FLS 1800. Curate, Christchurch Priory, Hants, 1802–16. Minister, Fitzroy Chapel, Charlotte Street, London, 1816–23. Catalogue of Welsh plants in *Tours round North Wales during Summer of 1798* 1800 2 vols. *Practical Introduction to Botany* 1817. Contrib. to *Mon. Mag.*
J. Sowerby and J. E. Smith *English Bot.* 663, 673, 675. D. Turner and L. W. Dillwyn *Botanist's Guide* 1805, 166. *Mag. Zool. Bot.* 1838, 170. *J. Bot.* 1898, 14. *D.N.B.* v.5, 55. *Hist. Account of Herb. Yorkshire Philos. Soc.* 1907, 12–13. *Ann. Rep. Yorkshire Philos. Soc.* 1966, 56–57. *Country Life* 1974, 578–79.
Plants at BM(NH), York Museum.

BINNEY, Edward William (1812–1881)
b. Morton, Notts 1812 *d.* Manchester 19 Dec. 1881
FRS 1856. FGS 1853. Solicitor in Manchester. Palaeobotanist. President, Manchester Geological Society, 1857–59, 1865–67. *Observations on Structure of Fossil found in Carboniferous Structure* 1868–75.
Nature v.25, 1881–82, 293–94. *Quart. J. Geol. Soc. London* 1882, 58–59. *Mem. Manchester Lit. Philos. Soc.* 1883, 447–64. W. C. Williamson *Reminiscences of a Yorkshire Nat.* 1896, 62, 78. *D.N.B.* v.5, 56–57. *Trans. Liverpool Bot. Soc.* 1909, 60. J. Binney *Centenary of Nineteenth Century Geologist, Edward William Binney* 1912. F. W. Oliver *Makers of Br. Bot.* 1913, 245–46. *R.S.C.* v.1, 372; v.6, 589; v.7, 174; v.9, 243; v.13, 562.
Portr. at Manchester Literary and Philosophical Society.
Volkmannia binneyi Carr.

BINSTEAD, Rev. Charles Herbert (*c.* 1862–1941)
d. Hereford 10 Jan. 1941
MA. FLS 1925. Vicar, Breinton, 1897. Rector, Whitbourne, 1906; Mordiford, 1915–23. Had herb. of Hereford mosses. Also collectedmosses in Penang, Singapore and Borneo, 1913 (*J. Linn. Soc.* v.43, 1916, 291–322). Contrib. plant records to W. H. Purchas and

A. Ley's *Fl. Herefordshire* 1889 and *Additions to Fl. Herefordshire* 1894. Contrib. mosses to *Victoria County Hist. Herefordshire* 1908. *Mosses of English Lake District* 1922. *Mosses of Herefordshire* 1940.

Rep. Br. Bryol. Soc. 1940–43, 230. *Trans. Woolhope Nat. Field Club* 1939–41, 201; 1954, 256–57. *Fl. Malesiana* v.1, 1950, 58. H. N. Clokie *Account of Herb. Dept. Bot., Oxford* 1964, 132.

Plants at BM(NH), Oxford, Woolhope Naturalists' Field Club.

Lophozia binsteadii.

BIOLETTI, Frederic Theodore (1865–1939)
b. Liverpool 21 July 1865 *d.* Redwood City, California, U.S.A. 12 Sept. 1939

BS Berkeley 1894. Thesis on *Notes on Genus Nemophila* 1895. Collected plants in San Francisco area. Viticulturist. Contrib. to L. H. Bailey's *Standard Cyclop. Hort.* 1914–17.

Leaflets of Western Bot. v.8, 1957, 85.

Mimulus biolettii Eastwood.

BIRD, Frederick John (1818–1874)
d. 28 April 1874

MD. ALS 1840. FLS 1862. Brother of Golding Bird. 'Artificial Arrangement of British Plants' (*Mag. Nat. Hist.* 1838, 604–9; 1839, 181–84).

Proc. Linn. Soc. 1873–74, xlvi.

BIRD, George William (1855–1928)
d. Kent 5 Dec. 1928

Hybridised orchids. Raised *Odontioda* 'Stone Age' in 1925.

Orchid Rev. 1928, 16.

BIRD, Golding (1814–1854)
b. Downham, Norfolk 9 Dec. 1814 *d.* Tunbridge Wells, Kent 27 Oct. 1854

MD St. Andrews 1838. MA 1840. FRS 1846. FLS 1836. Lecturer in Natural Philosophy, Guy's Hospital, 1836–53. Lecturer in Materia Medica, College of Physicians, 1847. Contrib. to *London Fl.* 'Equisetum hyemale' (*Proc. Linn. Soc.* 1846, 290–92).

Proc. Linn. Soc. 1855, 404–5. *D.N.B.* v.5, 74–75. *Guy's Hospital Rep.* 1971, 68–69. *R.S.C.* v.1, 386.

Sale at Puttock and Simpson, 16 Jan. 1856.

BIRD, Isabella Lucy *see* Bishop, I. L.

BIRD, Rev. Maurice Charles Hilton (1857–1924)
b. 28 March 1857 *d.* Brunstead, Norfolk 18 Oct. 1924

BA Cantab 1879. MA 1884. Rector, Brunstead, 1887–1924. Naturalist. Contrib. papers to *Trans. Norfolk Norwich Nat. Soc.*, 1894–1922.

Trans. Norfolk Norwich Nat. Soc. v.11, 1923–24, 608–11 portr.

BIRDWOOD, Sir George Christopher Molesworth (1832–1917)
b. Belgaum, Bombay 8 Dec. 1832 *d.* Ealing, Middx 28 June 1917

MD Edinburgh 1854. Bombay Medical Service, 1854. Professor of Anatomy and Botany, Grant Medical College, Bombay. *Catalogue of Economic Products of Presidency of Bombay* 1862. 'Boswellia' (*Trans. Linn. Soc.* v.27, 1871, 111–48). Supplied botanical footnotes for Markham's translation of Garcia da Orta, 1914.

C. E. Buckland *Dict. Indian Biogr.*, 1906, 43. Nature v.99, 1917, 370–71. *J. Bot.* 1918, 87–90, *Who was Who, 1916–1928* 94–95. *R.S.C.* v.7, 178.

Plants at Kew.

BIRDWOOD, Herbert Mills (1837–1907)
b. Belgaum, Bombay 20 May 1837 *d.* Twickenham, Middx 21 Feb. 1907

BA Cantab 1858. LLD Cantab 1889. Brother of Sir G. C. M. Birdwood. Pupil of J. H. Balfour at Edinburgh. Indian Civil Service, 1858. Judge of High Court, 1881–85. Governor of Bombay, 1895. 'Catalogue of Fl. Matheran and Mahableshwar' (*J. Bombay Nat. Hist. Soc.* 1887, 107–72; 1897, 394–448). 'Indian Timbers,' edited by W. Griggs with biography (*J. Indian Art Industry* 1910, 24-32).

Who was Who, 1897–1916 65.

BIRDWOOD, William Spiller (*fl.* 1890s)

Lieut.-Colonel. He and his wife collected plants in Aden, 1897–98. From this collection Blatter compiled his list of flora of Aden.

Kew Bull. 1901, 8. *Rec. Bot. Survey India* v.7, 1914, 21.

Plants at Kew, Bombay Natural History Society.

BIRKETT, Rev. Robert (*fl.* 1810s–1820s)

Plants at Ipswich Museum.

BIRNIE, Mr. (*fl.* 1860s)

English missionary. Made small collection of plants at Ta lien wan Bay, China *c.* 1860.

E. Bretschneider *Hist. European Bot. Discoveries in China* 1898, 679.

BIRSCHEL, F. W. (*fl.* 1850s)
b. Hanover, Germany *d.* Liverpool

Kew gardener, 1852–53. Collected plants in Caracas, 1854. Gardener at Chatsworth under Paxton for 2 years. Curator, Liverpool Botanic Garden, 1858.

Bonplandia v.2, 1854, 239–42; v.3, 1855, 94, 229, 246, 257–58, 327. *J. Kew Guild* 1901, 38 *Kew Bull.* 1923, 351.

Plants at Kew.

BISBY, Guy Richard (1889–1958)
b. Brookings, S. Dakota 17 Aug. 1889 *d.* Staines, Middx 3 Sept. 1958
BSc S. Dakota State College 1912. PhD Minnesota University. Assistant Professor, then Professor of Plant Pathology, Manitoba Agricultural College, 1920–36. Mycologist, Commonwealth Mycological Institute, Kew, 1937–54. Studied Hyphomycetes. *Fungi of Manitoba and Saskatchewan* 1938. *Introduction to Taxonomy and Nomenclature of Fungi* 1945; ed. 2 1953. *Fungi of Manitoba* (with A. H. R. Buller and J. Dearness) 1929. *Fungi of India* (with E. J. Butler) 1931. *Fungi of Ceylon* (with T. Petch) 1950. *Dictionary of Fungi* (with G. C. Ainsworth) 1943; ed. 4 1954.
Nature v.182, 1958, 987. *Phytopathology* 1959, 323 portr. *Trans. Br. Mycol. Soc.* 1959, 129 portr.
Fungi at Commonwealth Mycological Institute. Portr. at Hunt Library.

BISHOP, David (*c.* 1788–1849)
b. Scone, Perthshire *c.* 1788 *d.* Malone, Belfast 4 Aug. 1849
Gardener. Curator, Belfast Botanic Garden, 1830. Discovered *Juncus effusus* var. *spiralis*. *Causal Botany* 1829. Account of plants found in Perthshire (*Edinburgh Philos. J.* v.14, 1826, 180–81).
Cottage Gdnr, v.2, 1849, 306–7. *Trans. Bot. Soc. Edinburgh* v.11, 1873, 502–4.

BISHOP, Edmund Browne (1864–1947)
b. Bradpole, Dorset 23 Sept. 1864 *d.* 2 Nov. 1947
Civil servant. President, London Natural History Society, 1921–24. Studied *Rosa*. 'Notes on Roses of Bedfordshire' (*Bot. Soc. Exch. Club Br. Isl. Rep.* 1938, 84–92).
London Nat. 1947, 112–13. *Watsonia* 1949–50, 185–86.
Plants at BM(NH), London Natural History Society.

BISHOP, George A. (*fl.* 1890s)
Head gardener, Wightwick Manor near Wolverhampton. Superintendent, Public Garden, Bermuda, 1898.
Kew Bull. 1898, 96.

BISHOP, Isabella Lucy (*née* **Bird**) (*c.* 1832–1904)
b. Boroughbridge Hall, Yorks. 15 Oct. *c.* 1832 *d.* Edinburgh 7 Oct. 1904
Travelled extensively in N. America and Asia. First lady fellow of R.G.S., 1892. Collected plants in Luristan, Persia, *c.* 1891. *A Lady's Life in Rocky Mountains* 1879 (plants of Estes Park, p. 125).
A. M. Stoddart *Life of Isabella Bird* 1906. *Who was Who, 1897–1916* 65–66. J. Ewan *Rocky Mountains Nat.* 1950, 164–65.
Plants at Kew.

BISSET, James (1843–1911)
b. 4 June 1843 *d.* Edinburgh 3 April 1911
MA Oxon 1899. FLS 1881. FRSE 1900. Collected plants in Japan, 1866–86 (*J. Bot.* 1877–78). Correspondent of C. J. Maximowicz. 'Desmidiaceae in Windermere' (*J. R. Microsc. Soc.* 1884, 192–97).
Proc. Linn. Soc. 1910–11, 34–35.
Plants at BM(NH), Edinburgh, Kew.
Viola bisseti Maxim.

BISSET, John Petrie (1839–1906)
b. Inverurie, Aberdeenshire 1839 *d.* Banchory-Ternan, Kincardineshire 17 April 1906
Brother of J. Bisset (1843–1911). Desmidiologist. 'Notes on Japanese Desmids' (*J. Bot.* 1886, 193–96, 237–42). Accounts of Scottish Desmids in *Ann. Scott. Nat. Hist.* 1893–94; 1906, 187–88.
Gdnrs Chron. 1906 i 272. *R.S.C.* v.13, 573.
Cosmarium bissetii W. B. Turn.

BLACK, Alexander Osmond (–*c.* 1864)
Medical student. Found *Alchemilla conjuncta* on Clova in 1853.
Gdnrs Chron. 1853, 724. H. C. Watson *Cybele Britannica. Compendium* 1870, 34–35. C. C. Babington *Memorials* 1897, 323.
Mosses at BM(NH). Letters at Kew.

BLACK, Allan A. (1832–1865)
b. Forres, Morayshire 16 Sept. 1832 *d.* Bay of Bengal 4 Dec. 1865
ALS 1858. First Curator, Kew Herb., 1853–64. Superintendent, Botanic Garden, Bangalore, 1864. 'Catalogue of Japan Plants Systematically Arranged' (C. P. Hodgson *A Residence at Nagasaki and Hakodate in 1859–60* 1861, 327–50). List of Japanese plants in *Bonplandia* 1862, 88–100. Contrib. to J. Lindley and T. Moore's *Treasury of Bot.* 1866.
Gdnrs Chron. 1863, 1203; 1866, 102. *J. Bot.* 1866, 64. *Proc. Linn. Soc.* 1865–66, lvi–lvii. *J. Kew Guild* 1895, 29–30.
Plants and letters at Kew.
Allanblackia Oliver.

BLACK, Charles (1813–)
b. Mains of Pitcaple, Aberdeenshire 1 July 1813
Gardener and botanist. Friend of J. Duncan. Had herb.
W. Jolly *Life of John Duncan* 1883, 135–69, 182–96.

BLACK, J. Mackenzie (1870–1946)
b. Glen of Ogilvie, Forfarshire 1870 *d.* Langley, Slough, Bucks 30 Aug. 1946
In charge of Baron de Rothschild's orchid collection at Vienna and Paris. Orchid grower to R. G. Thwaites, Streatham, 1898–1913. Entered into partnership with S. W. Flory. Acquired orchid establishment of J. Veitch and Son, Langley, 1913. Raised many hybrid orchids.
Orchid Rev. 1946, 153–54. *Amer. Orchid Soc. Bull.* 1958, 735–36.

BLACK, John McConnell (1855–1951)
b. Wigtown 28 April 1855 *d.* Adelaide 2 Dec. 1951
ALS 1930. Emigrated to S. Australia, 1877, spending 5 years farming. Newspaper reporter, Adelaide, until 1902. President, Royal Society, S. Australia, 1933–34. *Naturalised Fl. S. Australia* 1909. *Fl. S. Australia* 1922–29 4 vols; ed. 2 1943–57. Contrib. to *Trans. R. Soc. S. Austral.*
Austral. Herb. News 1952 (10), 1–5. *Fl. Malesiana Bull.* 1952, 283; 1957, 547–49. *Nature* v.169, 1952, 94. *Proc. Linn. Soc.* v.163, 1952, 260–61. *Taxon* 1952, 62–63. *Victorian Nat.* v.68, 1952, 187–88.
Plants at Adelaide, Melbourne, Kew.

BLACK, Joseph (1728–1799)
b. Bordeaux, France 1728 *d.* 6 Dec. 1799
Educ. Glasgow and Edinburgh Universities. Professor of Medicine, Glasgow, 1756–66. Professor of Medicine and Chemistry, Edinburgh, 1766–97. Investigated lime and various calcareous manures. His work contains one of the earliest references to toxicity of magnesium compounds to vegetation. *Lectures on Elements of Chemistry* 1803.
W. Ramsay *Essays Biogr. and Chemical* 1909, 67–87. *D.N.B.* v.5, 109–12. *Chronica Botanica* 1944, 135–38.

BLACKALL, William Edward (1876–1941)
MB Oxon. Arrived in W. Australia, 1904. Botanist. Friend of C. A. Gardner, Government Botanist, W. Australia. *How to know Western Australian Wildflowers* (with B. J. Grieve) 1954–65 3 vols.
Herb. and MSS. at Perth.

BLACKBURN, Edward Bevens (–1839)
b. Bush Hill, Middx *d.* Alnwick, Northumberland 7 Aug. 1839
BA Cantab 1809. Chief Justice, Mauritius, 1824–35.
Mauritius and Madagascar plants at Kew.
Trochetia blackburniana Bojer.

BLACKBURN, Kathleen Bever (*c.* 1892–1968)
d. 20 Aug. 1968
DSc London 1924. FLS 1922. Lecturer in Botany, Southlands Training College, Battersea, 1914–18. Lecturer, King's College, Newcastle-upon-Tyne, 1918–57. Cytologist.
Watsonia 1970, 69–70.
Portr. at Hunt Library.

BLACKBURNE, Anna (1740–1793)
b. Orford Hall, Warrington, Lancs 1740 *d.* Fairfield, Warrington 30 Dec. 1793
Daughter of J. Blackburne (1693–1786). Correspondent of Linnaeus.
Gent. Mag. v.64, 1794, 180. J. Kendrick *Profiles of Warrington Worthies* 1853 portr. *Trans. Liverpool Bot. Soc.* 1909, 61. *D.N.B.* v.5, 121. *J. Bot.* 1914, 321.
Blackburnia Forster.

BLACKBURNE, John (1693–1786)
d. 20 Dec. 1786
Built first hot-house in N. England and ripened pine-apples. Catalogue of his garden at Orford Hall, Warrington by his gardener, Adam Neal, published 1779. Friend of T. Pennant, J. Aikin, etc.
J. Nichols *Illus. of Lit. Hist. of Eighteenth century* v.1, 238. D. Turner *Extracts from Lit. and Sci. Correspondence of Richard Richardson* 1835, 324–25. *Gdnrs Mag.* 1829, 53. *D.N.B.* v.5, 123. *Trans. Liverpool Bot. Soc.* 1909, 61. *J. Bot.* 1914, 321.
Letters at S.P.C.K.
Blackburnia Forster.

BLACKIE, George Stodart (1834–1881)
b. Aberdeen 10 April 1834 *d.* Nashville, Tenn., U.S.A. 19 June 1881
MD Edinburgh 1855. Curator, Botanical Museum, Botanical Society of Edinburgh. Professor of Botany, Nashville, Tennessee, 1856.
Trans. Bot. Soc. Edinburgh v.14, 1883, 282–84. H. A. Kelly and W. L. Burrage *Dict. Amer. Med. Biogr.* 1928, 646.

BLACKMAN, Frederick Frost (1866–1947)
b. Lambeth, London 25 July 1866 *d.* Cambridge 30 Jan. 1947
BSc London 1885. FRS 1906. Demonstrator in Botany, Cambridge, 1891; Lecturer, 1897; Reader, 1904–36. Plant physiologist.
Nature v.159, 1947, 394–95. *Plant Physiol.* 1947, i–viii. *Times* 1 Feb. 1947. *Obit. Notices Fellows R. Soc.* 1948, 651–57. *D.N.B. 1941–1950* 82–83. C. C. Gillispie *Dict. Sci. Biogr.* v.2, 1970, 183–85.
Portr. by H. Lamb at Cambridge.

BLACKMAN, Vernon Herbert (1872–1967)
b. Lambeth, London 8 Jan. 1872 *d.* Wimbledon, Surrey 1 Oct. 1967
FRS 1913. FLS 1896–1913, 1919. Assistant,

Dept. of Botany, BM(NH), 1896–1906. Professor of Botany, Leeds, 1907–11. Professor of Plant Physiology and Pathology, Imperial College, London, 1911–37. Director, Research Institute of Plant Physiology, Imperial College, 1913–43. Edited *Ann. Bot.*

Gdnrs Chron. 1937 i 234 portr. *Times* 3 Oct. 1967. *Ann. Bot.* v.32, 1968, 233–35 portr. *Biogr. Mem. Fellows R. Soc.* 1968, 37–60 portr. *Trans. Br. Mycol. Soc.* 1968, 351–52. *Who was Who, 1961–1970* 102.

Portr at Hunt Library.

BLACKMORE, J. B. (*fl.* 1870s–1900s)
b. Newton St. Loe, Somerset

Engineer who changed to horticulture in 1900. Started nursery at Twerton-on-Avon in 1901 in partnership with C. F. Langdon. Specialised in begonias and carnations.

Gdnrs Mag. 1909, 817–18 portr.

BLACKMORE, Richard Doddridge (1825–1900)
b. Longworth, Berks 7 June 1825 *d.* Teddington, Middx 20 Jan. 1900

MA Oxon 1852. Classical master, Wellesley House School, Twickenham, 1853. Fruit grower at Teddington, Middx. *The Farm and Fruit of Old* 1862 (translation of part of Virgil's *Georgics*). Novelist. *Lorna Doone* 1869.

D.N.B. Supplt. v.1, 207–10. D. Macleod *Gdnr's London* 1972, 205–12.

BLACKSTONE, John (1712–1753)
b. London 1712 *d.* Harefield, Middx 11 March 1753

Apothecary, Fleet Street, London. Botanist. *Fasciculus Plantarum circa Harefield sponte Nascentium* 1737. *Specimen Botanicum* 1746.

R. Pulteney *Hist. Biogr. Sketches of Progress of Bot. in England* v.2, 1790, 270–74. D. Turner *Extracts from Lit. and Sci. Correspondence of Richard Richardson* 1835, 351–55. H. Trimen and W. T. T. Dyer *Fl. Middlesex* 1869, 389–91. G. C. Druce *Fl. Berkshire* 1897, cxxxvi–cxxxvii. G. C. Druce *Fl. Oxfordshire* 1927, lxxxviii–lxxxix. *D.N.B.* v.5, 132. E. J. L. Scott *Index to Sloane Manuscripts* 1904, 59. *Watsonia* 1949–50, 141–48. J. E. Dandy *Sloane Herb.* 1958, 90–91. H. N. Clokie *Account of Herb. Dept. Bot., Oxford* 1964, 133. *See also* Addendum.

Plants and MS. list of Oxford flora at BM (NH). Plants at Ripon Museum.

Blackstonia Huds.

BLACKWELL, Alexander (1709–1747)
b. Aberdeen 1709 *d.* Stockholm, Sweden 29 July 1747

MD Leyden. Printer in London, 1730. Physician in ordinary to King of Sweden. Involved in Swedish political intrigue and executed without public trial. Wrote books on agriculture. Abridged P. Miller's *Catalogus Plantarum Officinalium* 1730 for his wife's *A Curious Herbal* 1737–39.

Bath J. 14 Sept. 1747. *Gent. Mag.* 1747, 424–27. R. Pulteney *Hist. Biogr. Sketches of Progress of Bot. in England* v.2, 1790, 251–56. J. Nichols *Lit. Anecdotes of Eighteenth Century* v.2, 93–95. T. Faulkner *Hist. Topographical Description of Chelsea* v.2, 1829, 207–9. *D.N.B.* v.5, 142. *J. Bot.* 1910, 193–95.

BLACKWELL, Charles (*fl.* 1680s)

Seedsman, King's Head near Fetter Lane, Holborn. Catalogue in J. Worlidge's *Systema Horti-culturae* ed. 3 1688, 271–78.

BLACKWELL, Elizabeth (*c.* 1700–1758)
b. Aberdeen *c.* 1700 *d.* Chelsea, London Oct. 1758

Wife of A. Blackwell (1709–1747) whose debts she cleared by publishing *A Curious Herbal* 1737–39, drawn, engraved and coloured by herself. A German edition, *Herbarium Blackwellianum*, 6 vols. appeared 1750–73.

Gent. Mag. 1747, 425. R. Pulteney *Hist. Biogr. Sketches of Progress of Bot. in England* v.2, 1790, 251-57. Jourdan *Dict. Sci. Med. Biogr.* v.2, 1820, 275–76. T. Faulkner *Hist. Topographical Description of Chelsea* v.2, 1829, 208–9. *D.N.B.* v.5, 144. *Herbarist* 1942, 24–26. W. Blunt *Art of Bot. Illus.* 1950, 136–37. F. A. Stafleu *Taxonomic Literature* 1967, 35. *Taxon* 1972, 147–52.

Original drawings for *A Curious Herbal* sold by Leigh and Sotheby, 27 Oct. 1796.

Blackwellia Comm.

BLACKWELL, Elizabeth Marianne (1889–1973)
b. 8 Jan. 1889 *d.* 25 May 1973

BSc Liverpool. MSc 1912. Demonstrator in Botany, Liverpool, 1912. Head of Botany Dept., Royal Holloway College, 1922–49. Studied *Phytophthora*. President, British Mycological Society, 1942. Contrib. to *Trans. Br. Mycol. Soc.*

Naturalist 1973, 113–14. *Trans. Br. Mycol. Soc.* 1973, 611–14 portr.

BLACKWELL, John (*fl.* 1790s)

Seedsman, Covent Garden, London.

BLACKWELL, William (*fl.* 1770s)

Herbalist and seedsman, Buckthorn Tree, Covent Garden and his garden, South Lambeth. Trade card at BM.

BLADON, James (*fl.* 1840s–1850s)

Of Pontypool, Monmouthshire. Contrib. papers on *Hieracia*, etc. to *Phytologist* 1845–50. Member of Botanical Society of London.

A. E. Wade *Fl. Monmouthshire* 1970, 13. *R.S.C.* v.1, 405.

Plants at Manchester University.

BLAIKIE, Francis (*fl.* 1810s–1820s)
b. Tweedside
Kew gardener. Farm and horticultural manager to Earl of Chesterfield at Bretby Park, Derbyshire. Land steward to J. Coke of Holkham, Norfolk until 1815. *Treatise on Management of Hedgerows and Hedgerow Timber* 1820.*Treatise on Mildew and Cultivation of Wheat* 1821. *On Smut in Wheat* 1822.
J. Donaldson *Agric. Biogr.* 1854, 127.

BLAIKIE, John (1837–1912)
d. 29 May 1912
FLS 1882. Proprietor, White Barn Colliery, Knutton. Involved in botanical activities of North Staffordshire Field Club.
Proc. Linn. Soc. 1912–13, 52.

BLAIKIE, Thomas (1750–1838)
b. near Edinburgh 1750 *d.* Paris 19 July 1838
Gardener to J. Fothergill (1712–1780) at Upton House, East Ham. Collected plants in Swiss Alps in 1775 for Fothergill and W. Pitcairn (1711–1791). Entered the service of Comte de Lauragais and spent the rest of his life in France. *Diary of a Scotch Gardener at the French Court* 1931. His method of grafting was known in France as 'graffe Blaikie' (*Floricultural Cabinet* 1836, 196–98).
Gdnrs Mag. 1838, 448. *Berichte Schweizerischen Botanischen Gesellschaft* v.50A, 1940, 43–44.

BLAIR, Mr. (*fl.* 1830s)
Sent seeds from Mexico to Glasgow Botanic Garden.
Bot. Mag. 1842, t.3927.

BLAIR, Patrick (*c.* 1666–1728)
b. Dundee, Angus *c.* 1666 *d.* Boston, Lincs Feb. 1728
MD Aberdeen 1712. FRS 1712. Practised medicine at Dundee, London and from 1720 at Boston. *Miscellaneous Observations in the Practise of Physic...with Remarks in Botany* 1718. *Botanik Essays* 1720. *Pharmaco-botanologia* 1723–28.
R. Pulteney *Hist. Biogr. Sketches of Progress of Bot. in England* v.2, 1790, 134–49. G. C. Gorham *Mem. of John and Thomas Martyn* 1830, 7–9. *D.N.B.* v.5, 163. *Trans. Bot. Soc. Edinburgh* v.23, 1908, 259–76. Br. Assoc. Advancement of Sci. *Dundee Handbook* 1912, 441–46. E. J. L. Scott *Index to Sloane Manuscripts* 1904, 59. J. E. Dandy *Sloane Herb.* 1958, 91.
Letters at Royal Society. MSS. at BM(NH) and Royal Society. Plants at Cambridge.
Bloeria L.

BLAIR, Peter (*c.* 1854–1936)
d. Oct. 1936
Gardener to Duke of Sutherland at Trentham Hall, Staffs.
Gdnrs Mag. 1910, 865–66 portr. *Gdnrs Chron.* 1936 ii 327.

BLAIR, Thomas (*fl.* 1820s–1830s)
Gardener to Mr. Clay of Stamford Hill, London. Collected plants in Canada, 1823; White Mountains, N. H., 1826. Contrib. to *Gdnrs Mag.* 1830–33.

BLAIR, Thomas (1819–)
b. Aberdour, Fife 1819
Gardener to Sir G. N. B. Middleton at Shrubland Park, Suffolk, 1862.
Gdnrs Chron. 1875 ii 393 portr.

BLAKE, Mr. (*fl.* 1830s)
Nurseryman, Potterne Road and Hartmoor, Devizes, Wilts.

BLAKE, Miss Caroline (*fl.* 1830s–1850s)
Collected British and European plants, now at Oxford.
H. N. Clokie *Account of Herb. Dept. Bot., Oxford* 1964, 133.

BLAKE, Lady Edith (*née* **Osbourne**) (*fl.* 1870s–1890s)
Married Sir Henry Arthur Blake, 1874. Collected plants in Jamaica while her husband was Governor, 1889–97.
W. Fawcett and A. B. Rendle *Fl. Jamaica* v.1, 1910, xix.

BLAKE, Sir Henry Arthur (1840–)
b. Limerick 18 Jan. 1840
Collected plants in Jamaica while Governor, 1889–97.
W. Fawcett and A. B. Rendle *Fl. Jamaica* v.1, 1910, xix.

BLAKE, John Bradby (1745–1773)
b. London 4 Nov. 1745 *d.* Canton, China 16 Nov. 1773
Supercargo to East India Company. Sent many plants and seeds to Kew Gardens and Chelsea Physic Garden. Introduced Cochin-China rice and tallow tree to Jamaica and Carolina.
Gent. Mag. 1776, 348–51. *Trans. Linn. Soc.* v.1, 1791, 172. E. Bretschneider *Hist. European Bot. Discoveries in China* 1898, 152. *J. Bot.* 1899, 87. *D.N.B.* v.5, 170. *Notes and Queries* v.170, 1936, 2. E. H. M. Cox *Plant Hunting in China* 1945, 43.
Chinese plant drawings and plants at BM (NH).

BLAKE, Martin
Of Antigua. "A gentleman to whose friendship the Natural History of Jamaica owes its early appearance."
P. Brown *Hist. Jamaica* 1756, 323. *Bot. Mag.* 1799, t.451.
Blakea R. Br.

BLAKE, Stephen (*fl.* 1660s)
Gardener to W. Ouglander, MP. *Compleat Gardener's Practice* 1664.
Garden 1920, 6–7, 20–21.

BLAKE, William John (1805–1875)
Of Danesbury, Welwyn, Herts. MP. Had herb., chiefly of Hertfordshire plants, which was given to G. C. Druce. Friend and correspondent of Sir W. J. Hooker, E. Forster and G. Don.
G. C. Druce *Fl. Buckinghamshire* 1926, xcviii. G. C. Druce *Comital Fl. Br. Isles* 1932, xi. H. N. Clokie *Account of Herb. Dept. Bot., Oxford* 1964, 133. J. G. Dony *Fl. Hertfordshire* 1967, 13.
Plants at Oxford.

BLAKELOCK, Ralph Antony (1915–1963)
d. London 31 May 1963
BSc. FLS 1948. Assistant, Kew Herb., 1937–63. Contrib. papers on flora of Iraq to *Kew Bull.*, 1948–58. Revised C. A. Johns's *Flowers of the Field* 1949 and E. Step's *Wayside and Woodland Blossoms* 1963.
J. Kew Guild 1963, 323–24 portr. *Taxon* 1963, 297–98.
Portr. at Hunt Library.

BLAKESLEY, John (*fl.* 1790s)
Nurseryman, Birmingham.

BLANCHARD, Douglas (1887–1969)
b. Poole, Dorset 1887
Solicitor. Had garden at Blandford, Dorset; specialised in daffodils.
Daffodil Tulip Yb. 1970, 188–89.

BLAND, J. Edmund (*fl.* 1880s)
Gardener on several estates in Richmond, Surrey area. Gardener to S. K. Mainwaring at Oteley Park, Shropshire, 1881. Raised many fuchsia varieties.
Gdnrs Chron. 1894 i 146 portr.

BLAND, Mrs. Laura Shelford (*fl.* 1900s)
Wife of R. N. Bland, Resident Councillor, Penang. Collected plants on Taiping Hills, 1905.
Gdns Bull. Straits Settlements 1927, 118.
Plants at Singapore.

BLAND, Michael
Raised 'Bland's jubilee' apple in Norwich in early 19th century.

BLANDFORD, George, Marquis of *see* Spencer-Churchill, G. *5th Duke of Marlborough*

BLANE, Sir Gilbert (1749–1834)
b. Blanefield, Ayrshire 29 Aug. 1749 *d.* London 26 June 1834
MD Glasgow 1778. FRS. Baronet 1812. Physician to the Fleet, 1779–83. Physician in ordinary to George IV. 'Nardus indica or Spikenard' (*Philos. Trans. R. Soc.* 1790, 284–92).
W. Munk *Roll of R. College of Physicians* v.2, 1878, 325. *D.N.B.* v.5, 202–4.
Portr. by Shee at Royal College of Physicians.

BLATTER, Ethelbert (1877–1934)
b Rebstein, Switzerland 15 Dec. 1877 *d.* Poona, India 26 May 1934
FLS 1913. Jesuit priest. Lecturer in Botany and later Principal, St. Xavier's College, Bombay, 1903-8, 1915-26. Chaplain at Panchgani. Collected plants in Sind and Waziristan. *Palms of British India and Ceylon* 1926. *Beautiful Flowers of Kashmir* 1927-28 2 vols. *Ferns of Bombay* (with J.F. D'Almeida) 1932. Contrib. to *J. Bombay Nat. Hist. Soc.*
Examiner 9 June 1934. *J. Bot.* 1934, 318–19. *J. Bombay Nat. Hist. Soc.* 1934, 466–73 portr.
Plants at St. Xavier College, Bombay and Kew.

BLEASDALE, Rev. John Ignatius (1822–1884)
b. Lancs 1822
English College, Lisbon. Ordained R.C. priest, 1844. Teacher in Experimental Physics, St. Patrick's College, Victoria, Australia. President, Royal Society of Victoria, 1865. Interested in viticulture.
Trans. Proc. R. Soc. Victoria 1885, 148.

BLICK, Charles (*c.* 1856–1919)
d. 30 April 1919
Gardener to M. R. Smith at Warren House, Hayes, Kent for over 20 years. Nurseryman, Warren Nurseries, Hayes, 1908.
Gdnrs Chron. 1919 i 232.

BLIGH, William (1754–1817)
b. St. Tudy, Cornwall 9 Sept. 1754 *d.* London 7 Dec. 1817
Vice-Admiral, 1814. Sailing master to J. Cook on his 2nd voyage round world, 1772-75. Captain of HMS 'Bounty,' 1787 sailing to Tahiti to obtain bread-fruit plants; cast adrift in open boat by mutinous crew. Commanded HMS 'Providence', 1791–93, on successful voyage to transport bread-fruit from Pacific to West Indies; brought back plants to Kew. *Voyage to South Sea* 1792.
Ann. Bot. v.2, 1806, 571. *D.N.B.* v.5, 219–20. F. M. Bladen *Hist. Rec. N.S.W.* v.6, 1898 passim portr. I. Lee *Captain Bligh's Second Voyage to the South Sea*, 1920. *J. Bot.* 1922, 22–25. *Papers R. Soc. Tasmania* 1922, 1–22. G. Mackaness *Life of Vice-Admiral William Bligh* 1931 2 vols. *Fl. Malesiana* v.1, 1950, 61–62.

Austral. Dict. Biogr. v.1, 1966, 118–22. K. Lemmon *Golden Age of Plant Hunters* 1968, 93–106 portr.

Letters at BM(NH). Logbooks of 'Providence' at Mitchell Library, Sydney.

Blighia Koenig.

BLIGHT, Rev. Robert B. (–1907)
d. Philadelphia, U.S.A. 28 March 1907

BA London 1870. Of Lewes until 1890. Curate, Bredwardine, Herefordshire. 'Reproduction and Growth of Mistletoe' (*Trans. Woolhope Club* 1870, 16–24).

R.S.C. v.9, 268.

BLINKWORTH, Robert (*fl.* 1830s)

Collected plants at Kumaon and Rangoon. Correspondent of N. Wallich.

Plants at BM(NH), Kew.

Blinkworthia Choisy.

BLISS, Arthur John (1859–1931)
d. Tavistock, Devon 17 Feb. 1931

Mining engineer. Hybridised narcissi, gladioli and irises.

Gdnrs Chron. 1931 i 174. *Iris Yb.* 1931, 12–13 portr.

BLISS, Daniel (*c.* 1871–1939)
b. Forres, Morayshire *c.* 1871 *d.* 15 Aug. 1939

VMH 1932. Kew gardener. Superintendent, Swansea Public Parks, Glam.

Gdnrs Chron. 1921 ii 144 portr.; 1939 ii 169.

BLOMEFIELD, Rev. Leonard (*né* **Jenyns**) (1800–1893)
b. London 25 May 1800 *d.* Bath, Somerset 1 Sept. 1893

BA Cantab 1822. FLS 1822. Curate, 1823, and Vicar, 1828–49 of Swaffham Bulbeck, Cambridge. Moved to South Stoke, 1850, Swainswick, 1852, and Bath, 1860. Founder and first President of Bath Natural History and Antiquarian Field Club. Wrote on Bath flora in *Proc. Bath Nat. Hist. Ant. Field Club* 1865–66, 23–60. *Chapters in My Life* 1887. *Naturalist's Calendar* 1903.

F. Darwin *Life and Letters of Charles Darwin* v.1, 1887, 54. *Nature* v.47, 1892–93, 85. *Gdnrs Chron.* 1893 ii 307. *J. Bot.* 1893, 320. *Proc. Linn. Soc.* 1893–94, 29. *Hist. Berwick Nat. Club* v.14, 1895, 347–58. C. C. Babington *Memorials* 1897, 267–68. *D.N.B. Supplt. 1* v.1, 221–22. H. J. Riddelsdell *Fl. Gloucestershire* 1948, cxxii.

Letters at Bath Public Library, BM(NH) and Kew. Herb. was at Bath Royal Literary and Scientific Institution.

BLOMFIELD, Sir Richard Massie (1835–1921)
b. 3 March 1835 *d.* London 26 June 1921

Rear-Admiral. KCMG. Collected plants in Japan, 1873, and in Egypt from 1879. 'Wild Flowers around Alexandria' (*Bull. Alexandria Hort. Soc.* no. 4, 1909, 4–16 portr.).

Kew Bull. 1921, 221.

Plants at Kew.

BLOOMER, Harry Howard (1866–1960)
b. 28 Oct. 1866 *d.* 15 June 1960

FLS 1907. Accountant. President, Birmingham Natural History and Philosophical Society, 1927. Collected shells and botanised in Queensland with J. Shirley, 1911. Co-editor with W. B. Grove of *Fauna and Fl. of Sutton Park*.

Proc. Linn. Soc. v.172, 1961, 129–30.

BLOOMFIELD, Rev. Edwin Newson (1827–1914)
b. Great Glenham, Suffolk 25 Sept. 1827 *d.* Guestling, Sussex 29 April 1914

BA Cantab 1850. Rector, Guestling, 1862. Collected local plants and contrib. botanical notes to *Proc. Hastings Nat. Hist. Soc.* Hastings plants listed in his *Natural History of Hastings and St. Leonards* 1888. Suffolk plants in *Victoria County Hist. Suffolk* v.1, 1911, 47–84 and *Trans. Norfolk Norwich Nat. Soc.* 1904–9. Contrib. to *J. Bot.* from 1884.

Proc. Hastings Nat. Hist. Soc. v.2, 1903, 91–93 portr. A. H. Wolley-Dod *Fl. Sussex* 1937, xliii. *R.S.C.* v.7, 200; v.9, 270; v.13, 615.

Plants at Kew, Oxford.

BLOW, Thomas Bates (1854–1941)
b. Welwyn, Herts 1854 *d.* Welwyn 16 Jan. 1941

FLS 1884. At one time made and exported bee-keeping equipment. Collected charophytes in British Guiana, Ceylon, Japan, Madagascar, etc. Also collected ferns.

Gdns Bull. Straits Settlements 1927, 118. *J. Bot.* 1938, 295–98. *Br. Fern Gaz.* 1939, 180–82; 1948, 205–6. *Hertfordshire Express* 25 Jan. 1941. *Bot. Soc. Exch. Club Br. Isl. Rep.* 1941–42, 450–51. *Proc. Linn. Soc.* 1940–41, 289–90. R. L. Hine *Thomas Bates Blow* 1941 portr. J. G. Dony *Fl. Bedfordshire* 1953, 25. H. N. Clokie *Account of Herb. Dept. Bot., Oxford* 1964, 133.

Plants at BM(NH), Kew, Oxford.

Nitella blowiana Groves.

BLOXAM, Rev. Andrew (1801–1878)
b. Rugby, Warwickshire 22 Sept. 1801 *d.* Harborough Magna, Warwickshire 2 Feb. 1878

Naturalist on 'Blonde' on voyage to Sandwich Islands, 1824. Diary of voyage in *B. P. Bishop Museum Special Publ.* 10, 1925. Rector, Twycross, Leics., then Harborough. Studied *Rubi* and fungi. 'Botany of Twycross' (*Phytologist* v.2, 1846, 640–42). 'Botany of Charnwood Forest' (with C. Babington) (T. R. Potter *Hist. and Antiquities of Charnwood Forest* 1842).

Gdnrs Chron. 1878 i 311; 1924 i 120–21. *J. Bot.* 1878, 96. *Midland Nat.* 1878, 88–90. *Rugby Advertiser* 16 Feb. 1878. J. E. Bagnall *Fl. Warwickshire* 1891, 501–2. *D.N.B.* v.5, 264. *Acta Horti Bergiani* v.3(2), 1905, t.145 portr. *Bot. Soc. Exch. Club Br. Isl. Rep.* 1932, 283. A. R. Horwood and C. W. F. Noel *Fl. Leicestershire* 1933, ccviii portr. D. A. Cadbury *Computer-mapped Fl.... Warwickshire* 1971, 55–56.

Herb. formerly at Calke Abbey, Derbyshire (*J. Bot.* 1887, 40–41, 145). Fungi and letters at Kew. MS. and letters at BM(NH). Portr. by Sir T. Laurence at National Gallery.

Rubus bloxamianus Coleman.

BLUNDELL, James (–*c.* 1798)

Nurseryman, Ormskirk, Lancs. In partnership with Hankin, 1795.

BLUNT, Lady Anne Isabella (1837–1917)

b. London 22 Sept. 1837 *d.* Cairo 15 Dec. 1917

Wife of Wilfrid Scawen Blunt, traveller and writer. Collected plants at Nejd in Arabia, 1880. *Pilgrimage to Nejd* 1881 2 vols.

D.N.B. 1922–1930 86.

Plants at Kew.

BLUNT, Henry (*fl.* 1860s)

Orchid collector in Colombia, 1863, and in Brazil for Messrs H. Low and Son and R. Bullen.

Gdnrs Chron. 1924 ii 432–33 portr.

Miltonia bluntii Reichenb. f.

BLUNT, Robert *see* Dalby, R.

BLUNT, Thomas Porter (*c.* 1842–1929)

MA Oxon. Chemist, Wyle Cop. Curator, Botanical Section, Shrewsbury Museum. Studied Shropshire flora.

N. Western Nat. 1929, 74–76.

BOARDMAN, Richard (*fl.* 1810s–1840s)

Gardener and seedsman, Hallgate, Wigan, Lancs.

BOBART, Jacob (1599–1680)

b. Brunswick, Germany 1599 *d.* Oxford 4 Feb. 1680

Father of Jacob Bobart (1641–1719). First gardener and Horti Praefectus of Botanic Garden, Oxford. *Catalogus Plantarum Horti Medici Oxoniensis* 1648; ed. 2 (with P. Stephens and W. Brown) 1658.

R. Pulteney *Hist. Biogr. Sketches of Progress of Bot. in England* v.1, 1790, 312–13. S. Felton *Portr. of English Authors on Gdng* 1830, 108–9. *Notes and Queries* v.7, 1853, 428, 578; v.8, 1853, 37, 159, 344; v.3, 1863, 150, 180. H. T. Bobart *Biogr. Sketch of Jacob Bobart...* 1884. *Gdnrs Chron.* 1885 ii 208–9 portr. *Proc. Linn. Soc.* 1888–89, 33. G. C. Druce *Fl. Berkshire* 1897, c–ci. *D.N.B.* v.5, 285–86. F. W. Oliver *Makers of Br. Bot.* 1913, 17–18 portr. S. H. Vines and G. C. Druce *Account of Morisonian Herb. in...Oxford* 1914 xv–xxii portr. *Bot. Soc. Exch. Club Br. Isl. Rep.* 1923, 344–45 portr.

Plants at Oxford. Engraved portr by D. Loggan at Kew. Portr. at Hunt Library.

Bobartia L.

BOBART, Jacob (1641–1719)

b. Oxford 2 Aug. 1641 *d.* Oxford 28 Dec. 1719

Succeeded his father, J. Bobart (1599–1680), as Horti Praefectus, Botanic Garden, Oxford, 1680. Professor of Botany, Oxford, 1684. Edited part 3 of R. Morison's *Plantarum Historiae Universalis Oxoniensis* 1699.

P. Blair *Botanick Essays* 1720, 243. R. Pulteney *Hist. Biogr. Sketches of Progress of Bot. in England* v.1, 1790, 312–13. D. Turner *Extracts from Lit. and Sci. Correspondence of Richard Richardson* 1835, 10–11, 152. H. T. Bobart *Biogr. Sketch of Jacob Bobart...* 1884. *Gdnrs Chron.* 1885 ii 209 portr. *J. Hort Cottage Gdnr* v.54, 1875, 363–64 portr. *J. Bot.* 1874, 37–38. G. C. Druce *Fl. Berks.* 1897, cxxv. *D.N.B.* v.5, 286, E. J. L. Scott *Index to Sloane Manuscripts* 1904, 61. F. W. Oliver *Makers of Br. Bot.* 1913, 18–19. S. H. Vines and G. C. Druce *Account of Morisonian Herb. in... Oxford* 1914, lii–lxv, portr. G. C. Druce *Fl. Oxfordshire* 1927, lxxx–lxxxii. J. E. Dandy *Sloane Herb.* 1958, 91–92.

Herb., MSS. and letters at Oxford. Portr. at Hunt Library.

Bobartia L.

BOBART, Tilleman (*fl.* 1650s–1720s)

Son of Jacob Bobart (1599–1680). Employed in Oxford Botanic Garden, Hampton Court, Blenheim and Cannons. Friend of W. Sherard. Found *Mentha sativa* (J. Ray *Synopsis Methodica* 1696, 124).

R. Pulteney *Hist. Biogr. Sketches of Progress of Bot. in England* v.1, 1790, 313. J. Nichols *Illus. of Lit. Hist. of Eighteenth Century* v.1, 357, 366, 369. H. T. Bobart *Biogr. sketch of Jacob Bobart...* 1884. G. C. Druce *Fl. Berkshire* 1897, cxxv. *D.N.B.* v.5, 285. E. J. L. Scott *Index to Sloane Manuscripts* 1904, 62. G. C. Druce *Fl. Oxfordshire* 1927, lxxxii.

Letters at Oxford.

BODDY, J. H. (–1894)

Nurseryman, Land's End Vineries, Cornwall.

Gdnrs Chron. 1894 ii 350.

BODDY, James (*fl.* 1830s–1840s)

Nurseryman and seedsman, London Street, Beaconsfield, Bucks.

BODGER, John (1846–1924)
b. Somerset 1846 *d.* California, U.S.A. 1924
To California, 1891, and became flower-seed merchant, Santa Paula, Ventura Co. At El Monte, 1916 and Lompoc.
V. Padilla *Southern California Gdns* 1961, 181–86 portr.

BODGER, John William (1865–1939)
b. Peterborough, Northants 14 March 1865 *d.* Peterborough 12 Feb. 1939
FLS 1920. Pharmaceutical chemist, Peterborough. Secretary and Treasurer, Peterborough Natural History, Scientific and Archaeological Society. Founder of Peterborough Museum. Good knowledge of local flora.
Proc. Linn. Soc. 1938–39, 151, 231.

BOFF, John (*fl.* 1840s)
Nurseryman, Balls Pond, Islington, London.

BOGG, Edward B. T. (*fl.* 1850s)
Herb. at Lincoln Museum.

BOGG, John (1799–1866)
b. Louth, Lincs 1799 *d.* Louth 15 April 1866
Surgeon at Louth, 1821–1866. "The first native of Louth who made a collection of the wild-flowers of the district." (R. W. Goulding).
Bot. Soc. Exch. Club Br. Isl. Rep. 1929, 91.

BOGG, Thomas Wemyss (*fl.* 1850s–1870s)
Supplied H. C. Watson with Louth plants for his *London Catalogue of British Plants*.
Bot. Soc. Exch. Club Br. Isl. Rep. 1929, 91.
Herb. at Lincoln Museum.

BOGUE, George (*c.* 1807–1893)
b. Lanarkshire *c.* 1807 *d.* St. Albans, Herts 19 Sept. 1893
Gardener to Earl of Verulam at Gorhambury. Contrib. to *Cottage Gdnr.*
Gdnrs. Chron. 1893 ii 408. *J. Hort. Cottage Gdnr* v.27, 1893, 317.

BOHLER, John (1797–1872)
b. S. Wingfield, Derbyshire 31 Dec. 1797 *d.* Sheffield, Yorks 24 Sept. 1872
Stocking weaver. Collector of medicinal plants for physicians. *Lichenes Britannici* 1835–37 (exsiccatae). 'Flora of Sherwood Forest' (R. White *Nottinghamshire, Worksop, 'the Dukery', and Sherwood Forest* 1875).
Gdnrs Mag. 1835, 593. *Naturalist* 1838–39, 206; 1946, 48; 1961, 54. *Reliquary* 1870, 212. *J. Bot.* 1872, 384. *Pharm. J.* v.3, 1872, 394–95. *D.N.B.* v.5, 304. B. Lyne *Index...Lichenum Exsiccatorum* 1915–19, 98–100. *Sorby Rec.* v.2(2), 1966, 32.

BOHN, Henry George (1796–1884)
b. London 4 Jan. 1796 *d.* Twickenham, Middx 22 Aug. 1884
Bookseller and publisher. Interested in growing conifers. Edited G. Gordon's *Pinetum* ed. 2 1875. With E. C. Otte translated F. H. A. von Humboldt's *Views of Nature* 1850. Contrib. to public awareness of Humboldt by sponsoring inexpensive editions of his works.
Athenaeum 30 Aug. 1884. *Garden* v.26, 1884, 194. *Gdnrs Chron.* 1884 ii 283. *Times* 25 Aug. 1884. *D.N.B.* v.5, 304–6. F. A. Munby *Romance of Book Selling* 1910, 397–40 portr.

BOHUN, Edmund (1672–1734)
b. Westhall, Suffolk 25 March 1672 *d.* Westhall 13 Oct. 1734
To Charleston, S. Carolina *c.* 1696, returning to England in 1701. Collected plants for J. Petiver and Sir H. Sloane.
J. Petiver *Museii Petiveriani* 1695, 79–80, 94. S. W. Rix *Diary and Autobiography of Edmund Bohun* 1853. E. J. L. Scott *Index to Sloane Manuscripts* 1904, 66. J. E. Dandy *Sloane Herb.* 1958, 93. R. P. Stearn *Sci. in British Colonies of America* 1970, 295–97, 305, 306.

BOIS, Charles du *see* Du Bois, C.

BOJER, Wenceslaus (1795–1856)
b. Resanice, Bohemia 23 Sept. 1795 *d.* Port Louis, Mauritius 4 June 1856
To Mauritius with Hilsenberg, 1820. Founder member of Royal Society of Mauritius, 1830. Curator, Mauritius Museum. Professor of Botany, Royal College, Port Louis, 1828. Director, Royal Botanic Garden, Pamplemousses. Collected plants in Mauritius, 1820–56; Madagascar, 1822–23, 1835; Zanzibar. Sent drawings to W. J. Hooker at Kew. *Hortus Mauritianus* 1837. 'Description des Plantes Recueillies en Madagascar' (*Rapports Ann. Ile Maurice* nos. 12–13, 1843, 13–21, 43–54). 'Sketch of Emerina' (*Bot. Misc.* 1832–33, 246–75).
Bot. Mag. 1828, t.2817; 1829, t.2884. *Hooker's J. Bot.* 1856, 312–17. J. G. Baker *Fl. Mauritius and Seychelles* 1877, 9–10*. V. Maiwald *Geschichte Bot. Böhm* 1904, 116–17. *Proc. R. Soc. Arts Sci. Mauritius* v.2, 1958, 73–98. *R.S.C.* v.1, 463.
Plants at Kew, BM(NH), Vienna. Letters at Kew.
Bojeria DC.

BOLEY, Mrs. Gertrude Maud (*c.* 1877–1965)
d. 15 June 1965
Active in botanical affairs of Bristol Naturalists' Society. Contrib. on plant ecology of N. Somerset to *Proc. Bristol Nat. Soc.*
Proc. Bristol Nat. Soc. v.31, 1965, 125.

BOLTON, Daniel(*fl.* 1850s)
Colonel, Royal Engineers. Collected plants in Auckland Islands, New Zealand, 1850. Correspondent of Sir W. J. Hooker.
Tuatara 1970, 49–50.

BOLTON, Henry (*c.* 1847–1939)
Fern collector of Warton, Lancs.
Br. Fern Gaz. 1932, 145 portr.

BOLTON, J. J. (1844–1928)
d. Claygate, Surrey 2 Nov. 1928
Cultivated orchids at Claygate Lodge.
Orchid Rev. 1922, 302–6; 1928, 374.

BOLTON, James (*fl.* 1750s–1799)
b. Halifax, Yorks *d.* Luddenden, Halifax 7 Jan. 1799
Self-taught botanist, artist and engraver of humble birth. *Filices Britannicae* 1785–90. *History of Funguses growing about Halifax* 1788–91 (German translation by Willdenow, 1795–1820). 'Catalogue of Plants growing in Parish of Halifax' (J. Watson *History of Parish of Halifax* 1775, 729–64). Drew plates for R. Relhan's *Fl. Cantabrigiensis* 1785.
W. Withering *Miscellaneous Tracts* v.1, 1822, 9. *Notes and Queries* v.5, 1864, 345; v.12, 1897, 468. *Naturalist* 1900, 165–72, 225–26; 1940, 105–6. W. B. Crump and C. Crossland *Fl. of Parish of Halifax* 1904, xlviii–liv. *Lanc. Nat.* v.2, 1909, 120–21, 169–70, 205–6, 255–56. C. Crossland *Eighteenth Century Nat.* 1910. *D.N.B.* v.5, 327–28. *Trans. Br. Mycol. Soc.* 1933, 302–7. *J. Bot.* 1941, 156–58.
Drawings at BM(NH). MSS. and drawings at U.S.D.A. Library. Fungi drawings sold at Sotheby 16 May 1939 and Christie 17 Dec. 1969.
Boltonia L'Hérit.

BOLTON, Robert (*c.* 1869–1949)
d. Birdbrook, Essex 26 March 1949
VMH 1949. Hybridised sweet peas. Cultivated ferns. President, British Pteridological Society, 1948–49.
Br. Fern Gaz. 1949, 222. *Gdnrs Chron.* 1949 i 128.

BOLTON, Thomas (–1778)
Brother of J. Bolton (*fl.* 1750s–1799) with whom he collaborated in 'Catalogue of Plants growing in Parish of Halifax' (Watson's *History of...Halifax* 1775, 729–64).
Naturalist 1940, 105–6. *J. Bot.* 1941, 156–58.

BOLTON, Thomas (–1923)
d. Warton, Lancs 1923
Cultivated British ferns.
Br. Fern Gaz. 1923, 9.

BOLTON, William (1852–1921)
b. Warrington, Lancs 1852 *d.* Warrington 18 Nov. 1921
Brewer, Warrington. Collected and cultivated orchids. Sent Kromer to Brazil and Jensen to Colombia for orchids.
Gdnrs Chron. 1921 ii 282. *Orchid Rev.* 1921, 182–84 portr.

BOLUS, Harry (1834–1911)
b. Nottingham 28 April 1834 *d.* Oxted, Surrey 25 May 1911
DSc 1902. FLS 1873. To Cape 1850. Collected plants extensively in S. Africa. Founded Chair of Botany at S. African College, 1902. *Orchids of Cape Peninsula* (*Trans. S. African Philos. Soc.* v.5, 1888). *Icones Orchidearum Austro-africanarum Extra-tropicarum* 1893–1913 4 vols. *List of Flowering Plants and Ferns of Cape Peninsula* (with A. H. Wolley-Dod) (*Trans. S. African Philos. Soc.* v.14, 1903).
Gdnrs Chron. 1911 i 358–59. *J. Bot.* 1911, 241–43; 1912, 32. *Kew Bull.* 275–77, 319–22. *Proc. Linn. Soc.* 1911–12, 42–44. *S. African J. Sci.* 1911, 69–79 portr. R. Marloth *Fl. S. Africa* v.1, 1913, x portr. *Curtis's Bot. Mag. Dedications, 1827–1927* 271–72 portr. A. White and B. L. Sloane *Stapelieae* v.1, 1937, 105–6 portr. *Moçambique* 1940, 123–26 portr. C. Lighton *Cape Floral Kingdom* 1960, 35–37. H. A. Baker and E. G. H. Oliver *Ericas in Southern Africa* 1967, xii–xiv portr. *Dict. S. African Biogr.* v.1, 1968. M. R. Levyns *A Botanist's Mem.* 1968, 28–30. M. A. Reinikka *Hist. of the Orchid* 1972, 241–43 portr. *R.S.C.* v.9, 288; v.13, 665.
Plants at BM(NH), Cape Town, Durban, Kew, etc. Letters at Kew.
Bolusia Benth. *Bolusanthus* Harms. *Neobolusia* Schlechter.

BONAVIA, Emanuel (1826–1908)
d. Worthing, Sussex 1908
MD. Bengal Medical Service, 1857. Superintendent, Lucknow Gardens, 1876. At Calcutta, 1885. *Future of Date Palm in India* 1885. *Cultivated Oranges and Lemons, etc. of India and Ceylon* 1890. *Fl. of Assyrian Monuments* 1894. *Philosophical Notes on Botanical Subjects* 1892. Contrib. botanical articles to *Babylonian and Oriental Rec.*
Gdnrs Chron. 1908 ii 385. *R.S.C.* v.9, 289; v.12, 99; v.13, 669.

BOND, George (*c.* 1806–1892)
d. Lydbury, Shropshire 6 Sept. 1892
Kew gardener. During 1826–35 drew 1700 plants (MS. list of drawings at Kew). Gardener to Earl of Powis at Walcot, Shropshire, 1835–80.

Garden v.42, 1892, 309. *Gdnrs Chron.* 1892 ii 381. *J. Bot.* Supplt. 3, 15. *Kew Bull.* 1916, 164.

BONHAM, Thomas (–*c.* 1627)
BA Cantab 1584. MD Oxon 1611. Physician in London. Notes incorporated in J. Parkinson's *Theatrum Botanicum* 1640 (title-page, 745, etc.).
D.N.B. v.5, 345. *J. Bot.* 1915, 67–68, 179–80.

BONING, Miss
Her collection of rare marine algae frequently cited by E. A. L. Batters and preserved in his herb., now at BM(NH).
Plants at Torquay Museum, Devon.

BONNIVERT, Gedeon (*fl.* 1670s–1700s)
b. Sedan, France
Lieut. in English army, 1690. Dutch and English plants in Sloane herb., BM(NH). Sent Irish plants to L. Plukenet. Correspondent of Sir Hans Sloane.
L. Plukenet *Almagestum Botanicum* 1696, 284, 312, 344. *Ancestor* 1903, 26–32. E. J. L. Scott *Index to Sloane Manuscripts* 1904, 66. *J. Bot.* 1915, 107–12. J. E. Dandy *Sloane Herb.* 1958, 93–94.

BOODLE, Leonard Alfred (1865–1941)
b. St. John's Wood, London 5 May 1865 *d.* 22 Aug. 1941
ARCS. FLS 1888. Demonstrator at Royal College of Science for 7 years under D. H. Scott whose assistant he became when Scott was Hon. Keeper, Jodrell Laboratory, Kew Gardens. Assistant Keeper, Jodrell Laboratory, 1909–30. Published anatomical papers, especially on vascular system of Pteridophytes. Translated H. Solereder's *Systematic Anatomy of Dicotyledons* (with F. E. Fritsch) 1908.
J. Kew Guild 1941, 82–83 portr. *Kew Bull.* 1930, 328; 1941, 236–39. *Nature* v.148, 1941, 402.
Cape algae at BM(NH).
Boodlea Murray.

BOOKER, James (*fl.* 1840s)
Florist and seedsman, Salterhebble, Halifax, Yorks.

BOORMAN, John Luke (1864–1938)
d. Liverpool, N.S.W. 18 Nov. 1938
Kew gardener. Gardener, Sydney Botanic Garden, *c.* 1887; appointed plant collector, 1901–30.
J. Kew Guild 1967, 817–19.

BOORN, John (*fl.* 1830s)
Nurseryman and seedsman, Blue Anchor Road, Bermondsey, London.

BOOTH, James (*fl.* 1830s)
Nurseryman, Flottbeck Nurseries, Hamburg, Germany. *List of Trees and Shrubs...* 1838 (reprinted in J. C. Loudon *Arboretum et Fruticetum Britannicum* v.4, 1838, 2646–54).

BOOTH, John (*fl.* 1720s)
Seedsman and netmaker, The Golden Lion, Fleet Street, London.

BOOTH, Thomas Jonas (1829–)
Nephew of T. Nuttall (1786–1859). Collected plants in Assam and Bhutan, 1850–51 (*Hooker's J. Bot.* 1853, 353–67).
Bot. Mag. 1854, t.4798; 1855, t.4875; 1858, t.5061; 1859, t.5103. J. E. Graustein *Thomas Nuttall* 1967, 379–85. *Trans. Bot. Soc. Edinburgh* 1972, 351–63.
Rhododendron boothii Nutt.

BOOTH, William Beattie (*c.* 1804-1874)
b. Scone, Perthshire *c.* 1804 *d.* London? 18 June 1874
ALS 1825. Gardener, Horticultural Society of London, Chiswick, 1824–30. Gardener to Sir C. Lemon at Carclew, Cornwall. Assistant Secretary, Horticultural Society, 1858–59. *Illustrations and Descriptions of...Camellieae* 1831. Contrib. to *Bot. Mag.*, *Bot. Reg.* and *Gdnrs Mag.*
Bot. Register v.24, 1838, Misc. Notes 5–7. *Gdnrs Chron.* 1874 i 838. *J. Hort. Cottage Gdnr* v.51, 1874, 511. *Proc. Linn. Soc.* 1874–75, xxxvii–xxxviii, *R.S.C.* v.12, 101.
Epidendrum boothianum Lindl.

BOOTHBY, Sir Brooke (1743–1824)
b. Ashbourne Hall, Derbyshire 1743 *d.* Boulogne, France 23 Jan. 1824
Formed with Erasmus Darwin and John Jackson the Botanical Society of Lichfield which published translations of some of Linnaeus's works such as *System of Vegetables* 1783. Boothby contributed little to the activities of this Society.
Trans. Hort. Soc. London v.1, 1812, 269. J. Stokes *Bot. Commentaries* 1830, cxxvi. P. Smith *Correspondence of Sir J. E. Smith* v.1, 1832, 401. R. Garner *Nat. Hist. County of Staffordshire* 1844, 7. *D.N.B.* v.5, 391.

BOOTT, Francis (1792–1863)
b. Boston, Mass., U.S.A. 26 Sept. 1792 *d.* London 25 Dec. 1863
BA Harvard 1810. Went to Liverpool, 1811. MD Edinburgh, 1824. FLS 1819. Lecturer in Botany, Webb Street School of Medicine, London, 1825. Practised medicine in London, 1825. Secretary, Linnean Society, 1832–39. *Two Introductory Lectures on Materia Medica* 1827. *Illustrations of Genus Carex* 1858–67 (vol. 4 ed. by J. D. Hooker).

Amer. J. Sci. 1864, 288–92. *Gdnrs Chron.* 1864, 51–52, *J. Bot.* 1864, 61–62; 1865, 191. *Proc. Linn. Soc.* 1863–64, xxiii–xxvii; 1888–89, 33. C. S. Sargent *Sci. Papers of Asa Gray* v.2, 1889, 315–20. A. T. Gage *Hist. of Linnean Society of London* 1938 portr. *Harvard Med. Alumni Bull.* v.42, 1968, 23–25 portr. *Taxon* 1970, 515.

Carex herb. and Massachusetts plants at Kew. Letters at BM(NH), Kew and Linnean Society. MS. of Carex at Kew. Portr. by Gambardella at Linnean Society; portr. at Gray Herb., Kew and Hunt Library.

Boottia Wallich.

BOR, Norman Loftus (1893–1972)
b. Ireland 4 May 1893 *d.* London 22 Dec. 1972

MA Dublin. DSc Edinburgh. FRSE. FLS 1931. Indian Forest Service, 1921; Deputy Conservator, 1924; Botanical Officer, Assam, 1936; Forest Botanist, Forest Research Institute, Dehra Dun, 1937–42. Assistant Director, Kew, 1948–59. President, Indian Botanical Society, 1945. Collected plants in Sikkim, Assam, Manipur, Tibet and S. India. Authority on taxonomy of grasses. *Manual of Indian Forest Botany* 1953. *Beautiful Indian Shrubs and Climbers* (with M. B. Raizada) 1954. *Grasses of Burma, Ceylon, India and Pakistan* 1960. *Fl. Iraq-Gramineae* 1968. *Fl. Iranica-Gramineae* 1970. Contrib. to *Indian Forester*.

J. Kew Guild 1948, 650 portr. *Indian Forester* 1948, 342–43 portr.; 1973, 184 portr. *Kew Bull.* 1948, 10. *Nature* v.161, 1948, 840. E. Bor. *Adventures of a Botanist's Wife* 1952, portr. *Commonwealth For. Rev.* 1973, 116–17. *Times* 1 Jan. 1973. *See also* Addendum.

Plants and MSS. at Kew. Portr. at Kew and Hunt Library.

BORLASE, Rev. William (1696–1772)
b. Pendeen, Cornwall 2 Feb. 1696 *d.* Ludgvan, Cornwall 31 Aug. 1772

MA Oxon. FRS 1750. Vicar, Ludgvan, 1722. Antiquary. *Natural History of Cornwall* 1758 (includes brief list of plants).

R. Pulteney *Hist. Biogr. Sketches of Progress of Bot. in England* v.1, 1790, 355. *D.N.B.* v.5, 398–99. F. H. Davey *Fl. Cornwall* 1909, xxxii–xxxiv.

BORLASE, William (1860–1948)
b. Liskeard, Cornwall 23 March 1860 *d.* Truro, Cornwall 14 Dec. 1948

FLS 1932. Studied Cornish flora.

Proc. Linn. Soc. 1948–49, 84–85.

Plants at Royal Institution of Cornwall.

BORRER, William (1781–1862)
b. Henfield, Sussex 13 June 1781 *d.* Henfield 10 Jan. 1862

FRS 1835. FLS 1805. Extensive knowledge of British botany, in particular *Salix*, *Rubus* and *Rosa*. *Specimen of Lichenographia Britannica* (with D. Turner) 1839. Contrib. to J. Sowerby and J. E. Smith's *English Bot.*

D. Turner and L. W. Dillwyn *Botanist's Guide through England and Wales* 1805, 596–632. *Gdnrs Mag.* 1838, 501–3. *Proc. Linn. Soc.* 1861–62, lxxxv–xc. *J. Bot.* 1863, 31–32. *Phytologist* v.6, 1862–63, 70–83. *Proc. R. Soc.* v.12, 1863, xlii. M. A. Lower *Worthies of Sussex* 1865, 71–73. *Trans. Penzance Nat. Hist. Soc.* 1890–91, 236. *D.N.B.* v.5, 406. *Kew Bull.* 1924, 78. A. H. Wolley-Dod *Fl. Sussex* 1937, xl. *Shooting Times and Country Mag.* 16 Jan. 1959, 59 portr.; 9 June 1961, 648 portr. *Watsonia* 1974, 55–60 portr. *R.S.C.* v.1, 499.

Herb. at Kew. Letters and MS. biography by L. Fleming (1966) at Kew. Several portr. at Kew, also at Hunt Library. Library sold at Sotheby 25 July 1899, 22 Feb. 1921.

Borreria G. F. W. Meyer. *Borreria* Achar.

BORTHWICK, Albert William (1872–1937)
b. Midlothian, 16 Oct. 1872 *d.* Aberdeen 19 April 1937

BSc St. Andrews 1895. DSc 1904. FRSE 1925. Assistant to J. Bayley Balfour, 1899. Lecturer in Forestry, Edinburgh and East of Scotland College of Agriculture, 1905. Lecturer in Forest Botany, Edinburgh University, 1908–14. Chief Research Officer, Forestry Commission, 1919–26. Professor of Forestry, Aberdeen, 1926–37. Editor, *J. R. Scott. Arbor. Soc.* 1913–29. President Royal Scottish Arboricultural Society, 1930, 1931.

J. Bot. 1937, 176–77. *Kew Bull.* 1937, 273–74. *Nature* v.139, 1937, 914. *Proc. R. Soc. Edinburgh* v.47, 1936–37, 404. *Scott. For. J.* 1937, 218–19 portr. *Times* 21 April 1937. *Chronica Botanica* 1938, 74 portr.

Herb. at Forestry Dept., Aberdeen.

BORTHWICK, William (*fl.* 1750s)

Merchant and seedsman, Lawnmarket and later Bridge Street, Edinburgh. His seed catalogues appear in James Justice's *Br. Gdnr's Calendar* 1759.

BOSANQUET, Rev. Edwin (*c.* 1800–1872)
b. London *c.* 1800 *d.* 30 Aug. 1872

MA Oxon 1826. Rector, Forscote, Somerset, 1848–70. *Plain Account of British Ferns* 1854; ed. 2 1855.

BOSANQUET, Louis Percival (1865–1930)
b. Southgate, Middx 20 July 1865 *d.* Lake County, Florida, U.S.A. 19 April 1930

To USA, 1888. Horticulturist. Hybridised *Crinum*.

Yb. Amer. Amaryllis Soc. 1935, 100–1.

BOSCAWEN, Rev. Arthur Townshend (*c*. 1862–1939)
d. Ludgvan, Cornwall 17 July 1939
VMH 1922. Rector, Ludgvan, 1893–1939. Had notable garden. One of the pioneers of Cornish fruit, flower and vegetable-growing industry.
Gdnrs Chron. 1939 ii 71. *Kew Bull*. 1939, 339. *J. R. Inst. Cornwall* 1942, appendix.

BOSCAWEN, John R. de C. (–1915)
d. 12 Dec. 1915
Of Tregge, Perranwell, Cornwall. Keen gardener. Secretary, Cornwall Daffodil Society.
Garden 1915, 632 portr. *Gdnrs Mag*. 1915, 582 (v).

BOSCAWEN, Rev. John Townshend (1820–1889)
d. 6 July 1889
BA Cantab 1845. Rector, Lamorran, Cornwall, 1849. Had garden with many Australian plants.
Gdnrs Chron. 1881 i 751–52; 1889 ii 46. *J. Hort. Cottage Gdnr* v.19, 1889, 17. *Proc. Linn. Soc*. 1889–90, 94. *J. R. Hort. Soc*. 1943, 265.

BOSISTO, Joseph (1827–1898)
b. 21 March 1827 *d*. Melbourne 8 Nov. 1898
Educ. Leeds School of Medicine. To Australia, 1848 where he opened a pharmacy in Richmond, Victoria. Did research on chemistry of Australian plants, especially *Eucalyptus*.
Jubilee Hist. Victoria and Melbourne v.2(3), 1888, 47. P. Mennell. *Dict. Austral Biogr*. 1892, 48–49. *Pharm. J*. v.8, 1899, 71. *Victorian Nat*. 1908, 103. *Austral. Encyclop*. v.2, 1962, 54. *R.S.C*. v.1, 503; v.9, 300; v.13, 704.
Bosistoa F. Muell.

BOSSEY, Francis (1809–1904)
b. Sutton-at-Hone, Kent 21 Oct. 1809 *d*. Redhill, Surrey 27 Sept. 1904
MD Glasgow. Practised at Woolwich until 1867. 'Fungi which attack Cereals' (*Proc. Bot. Soc. London* 1839, 50). 'Kent Plants' (*Ann. Mag. Nat. Hist*. v.3, 1839, 272–73). Contrib. to G. S. Gibson's *Fl. Essex* 1862.
J. Bot. 1904, 358.
Herb. with C. E. Salmon; plants at Holmesdale Club Museum, Reigate. Letters at Kew.

BOSTOCK, E. D. (–1953)
Herb. at Leicester University.

BOSTOCK, Frederick (–1940)
d. Pitsford House, Northants 18 May 1940
Boot and shoe manufacturer, Northampton. President, Northamptonshire Natural History Society, 1921, 1938. Botanist.
J. Northamptonshire Nat. Hist. Soc. Field Club v.29, 1940, 159–60.

BOSTOCK, John (1774–1846)
b. Liverpool 1774 *d*. London 6 Aug. 1846
MD Edinburgh 1798. FRS 1818. FLS 1819. Practised medicine in Liverpool; moved to London in 1817 when he abandoned medicine for general science. Lectured on chemistry at Guy's Hospital. Friend of W. Roscoe. One of founders of Liverpool Literary and Philosophical Society. Found *Erythraea latifolia*. With J. Shepherd edited Pliny's *Nat. Hist*.
J. E. Smith *Fl. Britannica* v.3, 1804, 1393. P. Smith *Correspondence of Sir J. E. Smith* v.2, 1832, 328. *Gent. Mag*. 1846 ii 653. *Proc. R. Soc*. 1846, 636–38. *Proc. Linn. Soc*. 1847, 333–34. *D.N.B*. v.5, 422–23. *Trans. Liverpool Bot. Soc*. 1909, 61. *R.S.C*. v.1, 505.

BOSTON, William (*c*. 1830–)
Nurseryman, Bedale, Yorks.
J. Hort. Home Farmer v.66, 1913, 28–29 portr.

BOSWARVA, John (*c*. 1791–1869)
d. Plymouth, Devon 30 Nov. 1869
Of Plymouth. Algologist. 'Catalogue of Marine Algae of Plymouth' (*Plymouth Inst. Rep*. 1861–62, 65–86; *U.K. Marine Biol. Ass. J*. 1887–88, 153–63).
Trans. Plymouth Inst. 1873, 28.

BOSWELL, Eliza and **Marianne** (*fl*. 1820s)
Of Balmuto, Fife. Aunts of J. T. I. Boswell (1822–1888). Assisted J. R. Scott and W. Jameson in *Herbarium Edinense* 1820 (exsiccatae) which is dedicated to them. Presented specimens and MS. catalogue of Balmuto plants to Linnean Society, 1822.

BOSWELL, Henry (1837–1897)
b. Oxford 27 Jan. 1837 *d*. Headington, Oxford 4 Feb. 1897
Hon. MA Oxon 1881. Portmanteau-maker, Oxford. Bryologist. *London Catalogue of British Mosses* (with C. C. P. Hobkirk) 1877; ed. 2 with F. A. Lees, 1881. Moss records in G. C. Druce's *Fl. Oxfordshire* 1886. Contrib. to *Phytologist* 1860; to *J. Bot*. from 1872.
G. C. Druce *Fl. Berkshire* 1897, clxxxi–clxxxiii. *J. Bot*. 1897, 132–37 portr. *Naturalist* 1897, 173–76. G. C. Druce *Fl. Oxfordshire* 1927, cxvi–cxvii. *R.S.C*. v.1, 50; v.7, 222; v.9, 301; v.13, 708.
Herb. at Oxford. Letters at BM(NH).

BOSWELL, John (1710–1780)
b. Auchinleck, Ayrshire 12 July 1710 *d*. Edinburgh 15 May 1780
MD Leiden 1733. FRCPE 1748. Uncle of James Boswell, the diarist and biographer of Samuel Johnson. *Dissertatio Inauguralis de Ambra* 1736.
Proc. R. Soc. Med. 1929, 283. *Gdnrs Chron* v. 160 (20), 1966, 14.
Boswellia Roxb.

BOSWELL, John Thomas Irvine (*né* **Syme** *afterwards* **Boswell-Syme**) (1822–1888)
b. Edinburgh 1 Dec. 1822 *d.* Balmuto, Fife 31 Jan. 1888

LLD St. Andrews 1875. FLS 1854. Son of Patrick Syme, flower-painter. Curator, Botanical Society, Edinburgh, 1850. Botanical Society London, 1851. Lecturer at Charing Cross and Westminster Hospitals. Edited J. Sowerby and J. E. Smith's *English Bot.* 1863–72. 'Pyrus aria' (*J. Bot.* 1875, 281–88).

Proc. Linn. Soc. 1887–88, 84–85. *Illus. London News* 11 Feb. 1888, 142 portr. *J. Bot.* 1872, 65–66; 1888, 82–84; 1889, 192; 1938, 212. *Scott. Nat.* 1888, 243–45. *Ann. Bot.* v.2, 1888–89, 430. *Trans. Bot. Soc. Edinburgh* v.17, 1889, 516–19. F. Boase *Modern English Biogr. Supplt* v.1, 1908, 458. M. Spence *Fl. Orcadensis* 1914, xliii–xlv. G. C. Druce *Fl. Oxfordshire* 1927, cxiv. *R.S.C.* v.5, 904; v.8, 1051; v.9, 301.

Herb. at BM(NH). Letters and MS. list of Orkney plants at Kew. Portr. and Hunt Library.

Symea Baker.

BOTHAM, Mary *see* Howitt, M.

BOTTOMLEY, William Beechcroft (1863–1922)
b. Apperley Bridge, Yorks 26 Dec. 1863 *d.* Huddersfield, Yorks 24 March 1922

BA Cantab 1891. FLS 1892. Lecturer in Biology, St. Mary's Hospital, 1886–91. Professor of Biology, Royal Veterinary College. Professor of Botany, King's College, 1893–1921. Discovered auximones in plants.

Bot. Soc. Exch. Club Br. Isl. Rep. 1922, 692. *Gdnrs Chron.* 1922 i 160. *J. Bot.* 1922, 157–58. *Nature* v.109, 1922, 524–25. *Proc. Linn. Soc.* 1921–22, 40. *Who was Who, 1916–1928* 109.

BOUCHER, George (*fl.* 1680s–1710s)

Surgeon at Port Mahon, Minorca. Sent plants to J. Petiver.

J. E. Dandy *Sloane Herb.* 1958, 94.

BOULGER, George Albert (1858–1937)
b. Brussels 19 Oct. 1858 *d.* St. Malo, France 23 Nov. 1937

FRS 1894. Zoologist. Head of Dept. of Reptiles, BM(NH). Studied roses at Crépin Herb., Brussels. 'Les Roses d'Europe de l'Herbier Crépin' (*Bull. Jardin Bot. de l'Etat* 1924–31).

Times 26 Nov. 1937. *Bull. Jardin Bot. de l'Etat* v.15, 1938, 1–24 portr. *Bull. Soc. Bot. Belg.* v.71, 1938, 9–16. *Kew Bull.* 1938, 84–85. *Proc. Amer. Acad. Sci.* v.73, 1939, 132–33. *Who was Who, 1929–40* 139.

Plants at BM(NH).

BOULGER, George Edward Simonds (1853–1922)
b. Bletchingley, Surrey 5 March 1853 *d.* Richmond, Surrey 4 May 1922

FLS 1877. Professor of Natural History, Cirencester, 1876. *Physiological Unity of Plants and Animals* 1881. *Uses of Plants* 1889. *Familiar Trees* 3 vols. *Plant Geography* 1912. *British Flowering Plants* (with I. S. Perrin) 1914 4 vols. *Biographical Index of British and Irish Botanists* (with J. Britten) 1893. *Name this Flower* 1917.

Bot. Soc. Exch. Club Br. Isl. Rep. 1922, 692–93. *J. Bot.* 1922, 232–36. *Nature* v.109, 1922, 653–54. *Proc. Linn. Soc.* 1921–22, 40–41. *S. E. Nat.* 1922, xxxiv–xxxv. *Times* 5 May 1922. *Who was Who, 1916–1928* 110–11. H. J. Riddelsdell *Fl. Gloucestershire* 1948, cxxxii–cxxxiv. S. T. Jermyn *Fl. Essex* 1974, 13. *R.S.C.* v.9, 312; v.12, 103; v.13, 726.

BOULT, Mrs. Ellen *see* Grundy, E.

BOULT, Mrs. Maria Ann *see* Grundy, M. A.

BOULT, Walter (*fl.* 1860s–1890s)

Of Birkenhead, Cheshire. Collected plants of Merseyside and Wirral Peninsula.

Liverpool Bull. 1967, 32–35.

Herb. at Liverpool Museums.

BOULTBEE, Mr. (*fl.* 1830s)

Sent *Caladium* sp. from Fernando Po to his father at Birmingham.

Bot. Mag. 1839, t.3728.

BOULTON, Mary Anne *see* Robb, M. A.

BOUND, William Pascoe (1870–1931)
d. Redhill, Surrey 27 Nov. 1931

Gardener to Sir J. Colman at Gatton Park, Reigate, 1900–8. Set up own business at Redhill. Contrib. to *Orchid Rev.*

Gdnrs Chron. 1931 ii 436. *Orchid Rev.* 1932, 23.

BOURDILLON, Thomas Fulton (1849–1930)
b. Palmaner, Madras 1 May 1849 *d.* Bexhill-on-Sea, Sussex 19 Dec. 1930

FLS 1891. Coffee planter, Travancore, *c.* 1871. Forest officer, Travancore, 1886; Conservator of Forests, 1891–1908. Collected plants in Travancore. Established Forest Herb. at Trivandrum. *Report on Forests of Travancore* 1893. *Forest Trees of Travancore* 1908.

Kew Bull. 1901, 10. *Proc. Linn. Soc.* 1930–31, 169–70. *Indian For.* 1931, 254–55.

Plants at Kew.

BOURKE, Patrick (*fl.* 1770s)

Apothecary and seedsman, Nenagh, County Tipperary, Ireland.

BOURKE, Thomas (*fl.* 1770s–1790s)

Nurseryman, County Mayo, Ireland.

BOURNE, Sir Albert Gibbs (1859–1940)
b. Lowestoft, Suffolk 8 Aug. 1859 *d.* 14 July 1940

DSc London. Assistant to E. Ray Lankester, 1879–85. To Madras, 1885. Registrar, Madras University, 1891–99. Botanist to Govt. of Madras, 1897–98. Professor of Biology, Presidency College, Madras, 1886–1903. Director of Institute of Science, Bangalore, 1915–21. Collected plants in India, Burma, 1896–1914 and Siam, 1902.

Who was Who, 1929–1940 141.

Plants at Kew.

BOURNE, Christopher (*fl.* 1910s)
Son of Rev. S. E. Bourne (*c.* 1845–1907). Nurseryman, Simpson near Bletchley, Bucks; specialised in daffodils.

Gdnrs Mag. 1911, 673–74 portr.

BOURNE, Edward (*fl.* 1790s)
MD. Of Cheadle, later Atherstone, Warwickshire.... *De Plantarum Irritabilitate* 1794.

J. Bot. 1901, Supplt., 71.

BOURNE, Frederick Samuel Augustus (1854–)
b. 3 Oct. 1854

Interpreter, British Consular Service, China, 1876. Vice-Consul, Canton. In Chungking, 1884–87. Collected plants, especially those of economic importance, in China and Formosa. *Lo-Fou Mountains* 1895.

Hooker's Icones Plantarum 1893, t.2254. E. Bretschneider *Hist. European Bot. Discoveries in China* 1898, 771–74. E. H. M. Cox *Plant Hunting in China* 1945, 106–7.

Plants at Kew.

Bournea Oliver.

BOURNE, Gilbert Charles (1861–1933)
d. 9 March 1933

FRS 1895. FLS 1886. Director, Marine Biological Association, Plymouth, 1887. Assistant, Zoology Dept., Oxford, 1892. Professor of Zoology, Oxford, 1906–21. Collected plants at Diego Garcia, 1885–86.

Kew Bull. 1901, 10. *Proc. Linn. Soc.* 1932–33, 184–85.

Plants at Kew.

BOURNE, Rev. S. Eugene (*c.* 1845–1907)
d. 11 May 1907

Vicar, Dunstan, Lincs. Grew narcissi. *Book of the Daffodil* 1903.

Gdnrs Chron. 1907 i 323. *Gdnrs Mag.* 1907, 360. *Daffodil Yb.* 1933, 35.

BOUTCHER, William (*fl.* 1710s–1730s)
Nurseryman and seedsman, Comely Garden, Edinburgh.

BOUTCHER, William (*fl.* 1730s–1780s)
Son of W. Boutcher (*fl.* 1710s–1730s). Nurseryman, Comely Garden, Edinburgh. *Treatise on Forest-trees* 1775; ed. 3 1774.

Gdnrs Chron. 1898 ii 189–90.

BOUTON, Louis (–1879)
Succeeded Bojer as Curator, Museum, Port Louis, Mauritius, 1858–65. *Medicinal Plants growing...in...Mauritius* 1857; ed. 2 1864.

Bot. Misc. 1832–33, 212–15. J. G. Baker *Fl. Mauritius and Seychelles* 1877, 10*.

Plants and letters at Kew.

Boutonia Bojer. *Neoboutonia* Muell. Arg.

BOVELL, John Redman (1855–1928)
b. Barbados? 31 Dec. 1855 *d.* Barbados 23 Nov. 1928

FLS 1895. Sugar planter. Superintendent, Botanical Station, Barbados, 1886. Superintendent of Agriculture, Barbados, 1898–1925.

Kew Bull. 1929, 63. *Who was Who, 1916–1928* xvi. *Who was Who, 1929–1940* 143.

BOWDEN, Athelstan (*fl.* 1890s)
Sent nerine bulbs from Cape to his mother at Newton Abbot, Devon at end of 19th century.

Gdnrs Chron. 1904 ii 365. *Bot. Mag.* 1907, t.8117.

Nerine bowdenii W. Watson.

BOWDICH, Sarah *see* Lee, S.

BOWDICH, Thomas Edward (1791–1824)
b. Bristol 20 June 1791 *d.* Bathurst, Gambia 10 Jan. 1824

African traveller and zoologist. In service of African Company and went to Cape Coast Castle, 1814. *Excursions in Madeira and Porto Santo...in 1823* 1825. *Mission from Cape Coast Castle to Ashantee* 1849.

A. Bonpland, F. Humboldt and S. Kunth *Nova Genera et Species Plantarum* v.6, 1823, 376–77. *D.N.B.* v.6, 41–43. *Bull. Herb. Boissier* 1907, 84–85. *Notes and Queries* v.125, 1917, 176–77. *R.S.C.* v.1, 550.

Ashantee plants at BM(NH).

Bowdichia H. B. & K.

BOWEN, Sir George Ferguson (1821–1899)
b. Ireland 2 Nov. 1821 *d.* Brighton, Sussex 21 Feb. 1899

BA Oxon 1844. Governor of Queensland, 1859; of New Zealand, 1867; of Victoria, 1872–87. *Bowenia* was named "in recognition of...his liberal encouragement of botany" (*Bot. Mag.* 1863 t.5398).

D.N.B. Supplt. v.1, 240–42.

BOWER, Frederick Orpen (1855–1948)
b. Ripon, Yorks 4 Nov. 1855 *d.* Ripon 11 April 1948

DSc. FRS 1891. FLS 1882. Studied under de Bary at Strasburg, 1879. Lecturer in Botany, S. Kensington, 1882. Professor of Botany, Glasgow, 1885–1925. Studied alternation of generations and evolutionary morphology of archegoniate plants. President, Royal Society of Edinburgh, 1919–24. *Origin of Land Flora* 1908. *Ferns* 1923–28 3 vols. *Size and Form in Plants* 1930. *Primitive Land Plants* 1935. *Sixty Years of Botany in Britain, 1875–1935* 1938.

Nature v.44, 1891, 15; v.161, 1948, 753–55. *Proc. Linn. Soc.* 1908–9, 32–34. *Gdnrs Chron.* 1910 ii 374 portr. *Obit. Notices Fellows R. Soc.* 1949, 347–74 portr. *Times* 12 April 1948. *Yb. R. Soc. Edinburgh* 1948–49, 8–9. *Naturalist* 1948, 111–14 portr. *D.N.B. 1941–1950* 94–95. *Who was Who, 1941–1950* 126. C. C. Gillispie *Dict. Sci. Biogr.* v.2, 1970, 370–72.

Plants at Glasgow. Portr. by W. Orpen at Glasgow. Portr. at Hunt Library.

BOWER, Hamilton (*fl.* 1890s)

Capt. 17th Bengal Cavalry. Journeyed across Tibet in 1891–92 with W. G. Thorold who collected plants.

E. Bretschneider *Hist. European Bot. Discoveries in China* 1898, 806–7. *J. Linn. Soc.* v.35, 1902, 142–48.

Corydalis boweri Hemsley.

BOWERBANK, James Scott (1797–1877)
b. London 14 July 1797 *d.* 8 March 1877

LLD. FRS 1842. FLS 1845. Geologist. Partner in London distillery. Lectured on botany, 1822–24. *History of Fossil Fruits and Seeds of London Clay* 1840.

Photographic Portr. of Men of Eminence in Literature, Sci. and Art no. 16, 1864, 133–36 portr. *Geol. Mag.* 1877, 191–92. *Proc. Geol. Soc.* 1877–78, 36–37. *Trans. Bot. Soc. Edinburgh* v.13, 1879, 123. *J. R. Microsc. Soc.* v.1, 1895, 28–30. *D.N.B.* v.6, 53–54.

Bust at BM(NH). Sale of library by Puttick and Simpson, 5 June 1877.

BOWERS, William (*fl.* 1770s)

Nurseryman, County Armagh, Ireland.

BOWES-LYON, Sir David (1902–1961)
b. 2 May 1902 *d.* Ballater, Aberdeenshire 13 Sept. 1961

VMH 1953. FLS 1948. Had garden at St. Paul's Walden Bury near Hitchin. President, Royal Horticultural Society, 1953–61.

Gdnrs Chron. 1961 ii 225 portr. *J. R. Hort. Soc.* 1961, 421–22 portr. *Proc. Linn. Soc.* 1960–61, 137–38. *Orchid Rev.* 1961, 341. *Times* 14 Sept. 1961 portr. *Bot. Mag.* 1964–65 portr. *Who was Who, 1961–1970* 121–22.

Portr. at Hunt Library.

BOWIE, Alexander (*fl.* 1790s–1820s)

Seedsman, 8 Edward Street, Portman Square and 241 Oxford Street, London.

BOWIE, James (*c.* 1789–1869)
b. London *c.* 1789 *d.* Claremount, Cape Town 2 July 1869

Kew gardener, 1810. Sent to Brazil to collect plants with A. Cunningham, 1814–17, and to Cape, 1818–23; at Cape again, 1827. Superintendent of Ludwig's garden at Cape.

W. J. Hooker *Exotic Fl.* v.3, 1827, 227. *Philos. Mag.* v.64, 1824, 185, 191. *Gdnrs Mag.* 1826, 473; 1830, 322–24; 1831, 490–96; 1832, 5–9, 718. *S. African Quart. J.* 1829–30. *Bot. Mag.* 1828, t.2856. *London J. Bot.* 1842, 20, 232–38. W. H. Harvey *Genera S. African Plants* 1838, xii. *Gdnrs Chron.* 1881 ii 568–69. *Trans. S. African Philos. Soc.* v.2, 1880–81, 123; v.4, 1884–86, xlii–xliii. *J. Bot.* 1889, 93–94. *J. Kew Guild* 1897, 27. *D.N.B.* v.6, 65. J. Hutchinson *Botanist in Southern Africa* 1946, 623–24. G. W. Reynold *Aloes of S. Africa* 1969, 57–58. A. M. Coats *Quest for Plants* 1969, 263–65.

Plants at BM(NH), Kew. Letters and journals at Kew.

Bowiea Harvey. *Bowiesia* Grev.

BOWIE, Robert (1817–1901)
b. Hepburn Hall, Northumberland 29 Aug. 1817 *d.* Wooler, Northumberland 1 Sept. 1901

Gardener to Earl of Tankerville at Chillingham, Northumberland.

Gdnrs Chron. 1876 i 300–2 portr.; 1901 ii 267.

BOWIE, William (1853–1931)
d. Glasgow 21 Sept. 1931

Chemist. Head, Pharmaceutical Laboratory, Glasgow Apothecaries Co. Had knowledge of Clydeside flora.

Pharm J. 1931, 272.

BOWKER, James Henry (–1900)
b. Tharfield, Lower Albany, Cape Colony *d.* Malvern, Natal 27 Oct. 1900

FLS 1889. Colonel, S. African Police. Collected plants in Cape and Natal, 1853–83. Correspondent of W. H. Harvey.

W. H. Harvey and O. W. Sonder *Fl. Capensis* v.1, 1859, 9*. W. H. Harvey *Thesaurus Capensis* v.1, 1859, 24–25. *Trans. S. African Philos. Soc.* v.3, 1885, 68–73. *Kew Bull.* 1901, 10. *Proc. Linn. Soc.* 1900–1, 40–41. *R.S.C.* v.13, 751.

Plants at BM(NH), Kew, Dublin.

Bowkeria Harvey.

BOWKER, Mary Elizabeth *see* Barber, M. E.

BOWLES, Edward Augustus (1865–1954)
b. Enfield, Middx 14 May 1865 *d.* Enfield 7 May 1954

BA Cantab 1887. FLS 1902. VMH 1916. VMM 1923. Had notable garden at Myddelton House, Enfield where he grew many bulbous plants. *My Garden in Spring* 1914. *My Garden in Summer* 1914. *My Garden in Autumn and Winter* 1915. Provided his own illustrations for *Handbook of Crocus and Colchicum for Gardeners* 1924; ed. 2 1952 and *Handbook of Narcissus* 1934.

Gdnrs Mag. 1907, 689–90 portr. *Garden* 1918, iii–iv portr. *Gdnrs Chron.* 1924 i 86 portr.; 1954 i 194 portr. *Curtis's Bot. Mag. Dedications, 1827–1927* 389–90 portr. *J. R. Hort. Soc.* 1954, 290–93 portr.; 512–19 portr.; 1955, 317–26, 367–76. *Daffodil Tulip Yb.* 1955, 10–12 portr. H. R. Fletcher *Story of R. Hort. Soc.* 1969, 373–78 portr. D. Macleod *Gdnr's London* 1972, 165–75 portr. M. Allan *E. A. Bowles and his Gdn at Myddelton House* 1973 portr.

Drawings of *Crocus* and *Iris* at BM(NH), of *Narcissus* and *Galanthus* at Royal Horticultural Society. Portr. at Hunt Library.

BOWLES (Bowle), George (*c.* 1604–1672)
b. Chislehurst, Kent *c.* 1604 *d.* 4 April 1672

MD Leyden 1640. FRCP 1664. Discovered *Impatiens noli-tangere*, etc.

J. Gerard *Herball* 1633, *passim.* J. Parkinson *Theatrum Botanicum* 1640, 297, 954. W. How *Phytologia Britannica* 1650, 5, 54. C. Merrett *Pinax* 1667, preface, 27. R. Pulteney *Hist. Biogr. Sketches of Progress of Bot. in England* v.1, 1790, 136. W. Munk *Roll of R. College of Physicians* v.1, 1878, 332. *J. Bot.* 1898, 18. C. E. Raven *English Nat.* 1947, 263, 294–97.

BOWLES, William (1705–1780)
b. near Cork 1705 *d.* Madrid 25 Aug. 1780

Travelled in France and Spain from 1740. Superintendent of state mines in Spanish service. *Introduccion a la Historia Natural...de España* 1775 (botany, 215–38); ed. 2 1782.

L. Ruiz and J. Pavon *Fl. Peruvianae et Chilensis Prodromus* 1794, 44. L. W. Dillwyn *Hortus Collinsonianus* 1843, 27, 39. *D.N.B.* v.6, 69. *Irish Nat.* 1911, 1–5. *Garden* 1975, 318–19.

Bowlesia Ruiz & Pavon.

BOWMAN, David (1838–1868)
b. Arniston, Edinburgh 3 Sept. 1838 *d.* Bogota, Colombia 25 June 1868

Foreman, Horticultural Society Gardens, Chiswick. Went to S. America, 1865. Collected plants in Brazil and Colombia.

Gdnrs Chron. 1868, 924, 942–43. J. H. Veitch *Hortus Veitchii* 1906, 53.

Plants at Kew, Vienna.

BOWMAN, Edward Macathur (1826–1872)
b. Sydney 1826 *d.* Peak Downs, Queensland 30 June 1872

Nephew of Sir W. Macarthur (1800–1882). Collected plants in Queensland for F. von Mueller and other botanists.

Gdnrs Chron. 1873, 177. *Rep. Austral. Assoc. Advancement Sci.* 1909, 375–76. *Proc. R. Soc. Queensland* 1954, 14–15.

Eucalyptus bowmanii F. Muell.

BOWMAN, John Eddowes (1785–1841)
b. Nantwich, Cheshire 30 Oct. 1785 *d.* Manchester 4 Dec. 1841

FLS 1828. Banker of Wrexham. To Manchester, 1837. Discovered *Cuscuta epilinum* (J. Sowerby and J. E. Smith *English Bot.* 2850). Drew and described *Elatine hydropiper* for *English Bot. Supplt.* 2670. 'Fossil Trees near Manchester' (*Trans. Manchester Geol. Soc.* v.1, 1841, 112–40). Contrib. to *Mag. Nat. Hist.* v.2, 1829.

Proc. Linn. Soc. 1842, 135–37. H. C. Watson *New Botanists' Guide* 1835–37, 209, 244. *Mem. Manchester Lit. Philos. Soc.* 1846, 45–85. H. C. Watson *Topographical Bot.* 1883, 539. *D.N.B.* v.6, 73. *Trans. Liverpool Bot. Soc.* 1909, 61–62. *Med. Hist.* 1966, 245–56.

Moss herb. at Oxford. Plants and letters at Kew.

Bowmania Gardner.

BOWMAN, John Herbert (1847–1932)
d. Newbury, Berks 5 Sept. 1932

Of Greenham Common near Newbury. Deputy Secretary, Bank of England. Had garden of rare and interesting plants.

Br. Fern Gaz. 1932, 165–66. *N. Western Nat.* 1933, 332–34.

BOWMAN, Robert Benson (*c* 1807–1882)
d. Newcastle, Northumberland 24 Nov. 1882

Of Newcastle and Richmond, Yorks. Druggist. Correspondent of H. C. Watson.

J. Sowerby and J. E. Smith *English Bot.* 2678, 2737, 2887. *Mag. Zool. Bot.* 1837, 205. *Trans. Northumberland Durham Nat. Hist. Soc.* v.8, 1886, 209. H. C .Watson *Topographical Bot.* 1883, 539. G. Johnston *Selections from Correspondence* 1892, 17.

Herb. at Newcastle-upon-Tyne Museum. Plants and letters at Kew.

BOWMAN, Sir William F. (1816–1892)
b. 20 July 1816 *d.* Dorking, Surrey 29 March 1892

Ophthalmic surgeon. Microscopist. Had garden at Joldwynds, Surrey.

Gdnrs Chron. 1892 i 440. *Nature* v.45, 1892, 517, 518, 564–66.

BOWRA, Edward Charles Mackintosh (1841–1874)
b. London 18 Oct. 1841 *d.* Halstead, Kent 15 Oct. 1874
Maritime Customs, Canton, 1863. Sent plants to H. F. Hance. Compiled 'Index Plantarum Sinice et Latine' for Doolittle's *Vocabulary and Handbook of Chinese Language* 1872.
E. Bretschneider *Hist. European Bot. Discoveries in China* 1898, 695.

BOWRING, Sir John (1792–1872)
b. Exeter, Devon 17 Oct. 1792 *d.* Exeter 23 Nov. 1872
LLD Groningen 1829. FRS 1856. FLS 1820. Knighted 1854. Linguist and traveller. British Consul, Canton, 1847. Governor, Hong Kong, 1854. Collected plants in China, 1850s–90s. *Autobiographical Recollections* 1877.
Hooker's J. Bot. 1853, 237–38. E. Bretschneider *Hist. European Bot. Discoveries in China* 1898, 381–83. *D.N.B.* v.6, 76–80.
Plants and letters at Kew. Library sold by Puttick and Simpson, 6 July 1874.
Bowringia Hook.

BOWRING, John Charles (1821–1893)
b. 24 March 1821 *d.* Windsor, Berks 20 Aug. 1893
FLS 1873. Son of Sir J. Bowring (1792–1872). Partner in Jardine, Matheson and Co., China merchants. Collected plants in Hong Kong and Hainan, 1852–91. Cultivated and hybridised orchids. 'Rice-paper' (*Trans. China Branch R. Asiatic Soc.* 1852, 37).
Hooker's J. Bot. 1852, 75; 1853, 237. *Orchid Rev.* 1893, 213. *Proc. Linn. Soc.* 1893–94, 29–30. E. Bretschneider *Hist. European Bot. Discoveries in China* 1898, 381–83. G. A. C. Herklots *Hong Kong Countryside* 1951, 165.
Plants and letters at Kew.
Bowringia Champ.

BOWSTREAD, William (*fl.* 1760s–1800s)
Nurseryman, Angel Row, Highgate, London and succeeded by W. Cutbrush.

BOX, James (*c.* 1839–1912)
d. 25 April 1912
Nurseryman and cattle dealer, Haywards Heath, Sussex.
J. Hort. Home Farmer v.64, 1912, 401.

BOXALL, William (1844–1910)
d. Clapton, London 28 Aug. 1910
VMH 1897. In 1860s employed by Veitch firm at Chelsea; at Earl Radnor's garden at Highwood; later foreman with Hugh Lowe & Co., Clapton. Collected orchids in Burma, Philippines, Borneo, Java, Brazil and Central America.
Gdnrs Mag. 1901, 795, 796 portr. *Gdnrs Chron.* 1910 ii 192 portr. *J. Hort. Home Farmer* v.61, 1910, 213. *Orchid Rev.* 1910, 346–47. *Orchid World* 1910, 70–71. *Fl. Malesiana* v.1, 1950, 74. R. S. Davis and M. L. Steiner *Philippine Orchids* 1952, 6, 8. M. A. Reinikka *Hist. of the Orchid* 1972, 253–54.
Dendrobium boxallii Reichenb. f.

BOYD, Daniel Alexander (1855–1928)
b. West Kilbride, Ayr Jan. 1855 *d.* Saltcoats, Ayr 8 Oct. 1928
Specialised in microfungi of Clyde area. President, Cryptogamic Society of Scotland. Contrib. to *Glasgow Nat.*
Glasgow Nat. 1930, 60–62. *J. Bot.* 1929, 52–53.
Fungi at Kew.
Coccomyces boydii A. L. Sm.

BOYD, Thomas (*c.* 1847–1900)
b. Hopetown near Edinburgh *c.* 1847 *d.* 19 March 1900
VMM. Gardener to W. Forbes at Callendar Park, Stirlingshire.
Gdnrs Chron. 1900 i 189. *J. Hort. Cottage Gdnr* v.40, 1900, 252 portr.

BOYD, William Brack (1831–1918)
b. Morebattle, Roxburghshire 22 Feb. 1831 *d.* Upper Faldonside, Roxburghshire 16 March 1918
Farmer, Eckford and Faldonside, Roxburghshire. President, Berwickshire Naturalists Club, 1905. Botanised in Grampians and studied British ferns. Discovered *Sagina boydii.*
Trans. Bot. Soc. Edinburgh v.17, 1889, 33–35; v.37, 1919, 344–45. *Bot. Soc. Exch. Club Br. Isl. Rep.* 1918, 352–54. *Br. Fern Gaz.* 1918, 276–78. *Garden* 1918, 143. *J. Bot.* 1918, 221–22. *Proc. Berwickshire Nat. Club* 1918, 423–25. *Quart. Bull. Alpine Gdn Soc.* 1940, 28–43 portr.
Salix boydii F. B. White.

BOYD WATT, Winifred (*née* **De Lisle**) (1878–1968)
d. 8 Jan. 1968
President, Bournemouth Science Society. Interested in botany.
Proc. Bot. Soc. Br. Isl. 1969, 624–25.

BOYDEN, Rev. Henry (*fl.* 1880s–1900s)
Collected plants in Scilly Isles, 1889. 'Fl. Isles of Scilly' (J. C. Tomkin and P. Row *Guide to Isles of Scilly* 1893, 1906).
Trans. Penzance Nat. Hist. Antiq. Soc. v.3, 1890, 186.
Plants at Exeter Museum.

BOYLAN (or Baloe), William (*fl.* 1800s)
Nurseryman, County Cork.
Irish For. 1967, 50.

BOYNTON, John (*fl.* 1820s)
Nurseryman, Walkington near Beverley, Yorks.

BOYS, William (1735–1803)
b. Deal, Kent 7 Sept. 1735 *d.* 15 March 1803
Surgeon, Sandwich, Kent. FSA 1776. Antiquary. Mayor of Sandwich, 1767 and 1782. Correspondent of W. Curtis.
D.N.B. v.6, 132–33.

BOYS-SMITH, Winifred Lily (–1939)
d. 1 Jan. 1939
Worked at scientific drawing, Cambridge, 1895. Science lecturer, Cheltenham Ladies College, 1896–1910. Professor of Home Science, Otago University, 1910–20. *Nature Note-book* 1902. Illustrated C. L. Laurie's *Elementary Bot.* 1901, *Flowering Plants* 1903, *Introduction to Elementary Bot.* 1907 and A. Johnson's *Bot. for Students* 1909.
Who was Who, 1929–1940 149.

BRACE, Josh (1856–1950)
b. 15 Oct. 1856 *d.* 30 Jan. 1950
Gardener at T. Rivers and Son, Sawbridgeworth, Herts. *Culture of Fruit Trees in Pots* 1904.

BRACE, Lewis Jones Knight (1852–1938)
b. Nassau, Bahamas 11 Nov. 1852 *d.* Nassau 15 Aug. 1938
FLS 1883. Collected plants in Bahamas, 1875–80, 1904–19. Curator, Herb., Royal Botanic Garden, Calcutta, 1882–86.
I. Urban *Symbolae Antillanae* v.3, 1902, 27. *Field Mus. Publ. Bot.* v.2, 1906, 138–39, 289. N. L. Britton and C. F. Millspaugh *Bahama Fl.* 1920, 648–49, 652, 654. *Proc. Linn. Soc.* 1938–39, 233–34.
Plants at Chicago, Kew, New York Botanic Garden.
Lonicera braceana Hemsley.

BRACEY, B. O. (–1956)
Raised orchids at Messrs Armcost and Royston, U.S.A. Started his own orchid business.
Orchid. Rev. 1956, 218.

BRACHER, Rose (1894–1941)
b. Salisbury, Wilts 1894 *d.* Bristol 15 July 1941
BSc Bristol 1917. MSc 1918. PhD 1927. FLS 1938. Demonstrator, London School of Medicine for Women, 1918–20. Lecturer, East London College, 1921–24. Lecturer, Bristol University, 1924–41. Mycologist. Studied ecology of mud-banks of Avon and species of *Euglena*. *Field Studies in Ecology* 1934. *Ecology in Town and Classroom* 1937. *Book of Common Flowers* 1941. Contrib. to *Proc. Bristol Nat. Soc.*, *J. Linn. Soc.*, *Ann. Bot.*
Nature v.148, 1941, 134. *Proc. Linn. Soc.* 1941–42, 270.

BRACKENRIDGE, William Dunlop (1810–1893)
b. Ayr 10 June 1810 *d.* Baltimore, Maryland 3 Feb. 1893
Gardener to Patrick Neill, Edinburgh. To Philadelphia, 1837. Assistant naturalist on Wilkes Expedition, 1838–42. Superintendent, National Botanic Garden, Washington. Nurseryman, Govanstown, Maryland. Contrib. *Filices* to *United States Exploring Expedition, 1838–42* 1855.
Gdnrs Mon. 1884, 375–76 portr. *J. Hort. Cottage Gdnr* v.53, 1906, 574. *J. New York Bot. Gdn* 1919, 117–24. *J. Bot.* 1919, 263. *Washington Hist. Quart.* 1930, 218–19, 298–305; 1931, 42–58, 129–45, 216–27. *Dict. Amer. Biogr.* v.2, 1929, 545–46. *Proc. Amer. Philos. Soc.* 1940, 673–79. *California Hist. Soc. Quart.* 1945, 321–25 portr., 337–42. *Fl. Malesiana* v.1, 1950, 74–75. S. D. McKelvey *Bot. Exploration of Trans. Mississippi West, 1790–1850* 1955, 685–730.
Plants at Washington. MS. Journal at Maryland Historical Society. Portr. at Hunt Library.
Brackenridgea A. Gray.

BRADBURY, George A. (–1962)
Pig farmer in Derbyshire. Cultivated orchids.
Orchid Rev. 1962, 201.

BRADBURY, Henry (1831–1860)
d. 2 Sept. 1860
Eldest son of William Bradbury, printer and publisher. Introduced to England nature-printing as patented in Vienna. *Nature-printing and its Origins and Objects* 1856. Nature-printed plates in T. Moore and J. Lindley's *Ferns of Great Britain and Ireland* 1855; W. G. Johnstone and A. Croall's *Nature-printed British Sea-weeds* 1859 4 vols.
D.N.B. v.6, 150. W. Blunt *Art of Bot. Illus.* 1955, 138–41. R. Cave and G. Wakeman *Typographia naturalis* 1967, 22–28.

BRADBURY, John (1768–1823)
b. Stalybridge, Cheshire 20 Aug. 1768 *d.* Middletown, Kentucky, U.S.A. 16 March 1823
FLS 1793. Plant collector in N. America for Liverpool Botanic Garden, 1809–11. Contrib. Cheshire plant localities to D. Turner and L. W. Dillwyn's *Botanist's Guide* 1805. *Travels in Interior of America* 1817 (plants, pp. 335–38).
W. Darlington *Reliquiae Baldwinianae* 1843, 316, 319. R. E. Call *Life of Rafinesque* 1895, 190. J. B. L. Warren *Fl. Cheshire* 1899, lxxix. *Lancashire Nat.* v.1, 1908, 191; v.2, 1909, 16–17. *Popular Sci. Mon.* 1908, 493–95. *Proc. Amer. Philos. Soc.* 1929, 133–50; 1950, 59–89. *Kew Bull.* 1934, 49–61. *Bartonia* 1936, 1–51. J. Ewan *Rocky Mountain Nat.* 1950, 169. *Herbarist* no. 18, 1952, 12–15. S. D. McKelvey *Bot. Exploration of Trans-Mississippi West, 1790–1850* 107–37. A. M. Coats *Quest for Plants* 1969, 296–99.
Plants and letters at Liverpool Public Libraries.
Bradburia Torrey & Gray.

BRADDICK, John (*c.* 1765–1828)
d. Maidstone, Kent 14 April 1828
Raised 'Braddick Nonpareil' apple at Thames Ditton in early 19th century. Contrib. to *Trans. Hort. Soc.* and *Gdnrs Mag.* on fruit culture.
Gdnrs Mag. 1828, 192.

BRADFORD, Edward (*fl.* 1840s–1890s)
MD. Army Surgeon. Collected plants, especially orchids, in Trinidad, 1845–46 and in China, 1879–99.
I. Urban *Symbolae Antillanae* v.3, 1902, 27. *Amer. Orchid Soc. Bull.* 1958, 78.
Chinese ferns at Kew. Orchids at Kew, Trinidad Herb.
Epidendrum bradfordii Griseb.

BRADFORD, Samuel Clement (1878–1948)
b. London 10 Jan. 1878 *d.* 13 Nov. 1948
Chief librarian, Science Museum, London, 1925–38. *Romance of Roses* 1946.
Who was Who, 1941–1950 133.

BRADLEY, Arnold Eastwood (1873–1944)
b. Sheffield, Yorks 18 Feb. 1873 *d.* 13 Nov. 1944
Bank of England employee. Interested in brambles and willows and their distribution in Yorkshire. 'West Yorkshire Brambles' (*Naturalist* 1937, 290–96).
Naturalist 1945, 115–16 portr. *Sorby Rec.* v.2 (2), 1966, 33.
Herb. at BM(NH).

BRADLEY, Henry (*fl.* 1770s)
Collected plants in China, 1779.
J. Lindley *Collectanea Botanica* 1821–24, t.3. E. Bretschneider *Hist. European Bot. Discoveries in China* 1898, 153–54.
Plants at BM(NH).

BRADLEY, John (*fl.* 1790s)
Seedsman, Leek, Staffs.

BRADLEY, John (*fl.* 1790s)
Gardener and seedsman, Dorking, Surrey.

BRADLEY, Richard (*c.* 1688–1732)
d. Cambridge 5 Nov. 1732
FRS 1712. Professor of Botany, Cambridge, 1724–32. Popular writer on horticulture. *History of Succulent Plants* 1716–27. *New Improvements of Planting and Gardening* 1717. *General Treatise of Husbandry and Gardening* 1724. *New Experiments...relating to the Generation of Plants* 1724. *Dictionarum Botanicum* 1728 2 vols. *Riches of a Hop-Garden Explain'd* 1729.
Philos. Trans. R. Soc. v.29, 1717, 486–92. *Gent. Mag.* 1732, 1082. R. Pulteney *Hist. Biogr. Sketches of Progress of Bot. in England* v.2, 1790, 129–33. J. Nichols *Lit. Anecdotes of Eighteenth Century* v.1, 446–51, 709. G. W. Johnson *Hist. English Gdning* 1829, 182–89. G. C. Gorham *Mem. of John and Thomas Martyn* 1830, 32–33. S. Felton *Portr. of English Authors on Gdning* 1830, 54–57. P. Smith *Correspondence of Sir J. E. Smith* v.2, 1832, 387. D. Turner *Extracts from Lit. and Sci. Correspondence of Richard Richardson* 1835, 283. *Cottage Gdnr* v.5, 1849, 93. *D.N.B.* v.6, 172. E. J. L. Scott *Index to Sloane Manuscripts* 1904, 73. D. McDonald *Agric. Writers* 1908, 170–76. *Gdnrs Chron.* 1910 i 33–34; 1912 ii 41–42, 391–92. *J. R. Hort. Soc.* 1939, 164–74. *Agric. Hist.* 1947, 1–2. *Cactus Succulent J. Great Britain* 1954, 30–31, 54–55, 78–81. J. E. Dandy *Sloane Herb.* 1958, 95. *Bull. Br. Soc. Hist. Sci.* v.1 (7), 1952, 176–78. *Med. Hist.* 1970, 53–62. *Notes Rec. R. Soc.* 1970, 59–77. H. Herre *Genera of Mesembryanthemaceae* 1971, 40. *Bull. African Succulent Plant Soc.* v.8, 1973, 132–35, 224–27; v.9, 1974, 5–12, 57–60, 85–86, 143–47; v.10, 1975, 21–22, 51–53, 76–77, 105–6.
MS. lectures at Cambridge Herb.
Bradleia Banks.

BRADLEY, Samuel (*fl.* 1850s)
Raised 'Baron Ward' apple at Elton Manor, Nottingham, 1850.

BRADSHAUGH, Roger (*fl.* 1610s–1640s)
Of Haigh, Lancs. Sent *Rubus chamoemorus* to T. Johnson and J. Parkinson.
J. Gerard *Herball* 1633, 1629. J. Parkinson *Theatrum Botanicum* 1640, 1015.

BRADSHAW, David Bigham (1869–1944)
b. Ballyshannon, County Donegal 13 April 1869 *d.* Dublin 5 Jan. 1944
Employee, Provincial Bank of Ireland until 1936. Studied Irish bryophyta.
Irish Nat. J. 1944, 166 portr. R. L. Praeger *Some Irish Nat.* 1949, 53–54.
Herb. at Ulster Museum.

BRADY, Henry Bowman (1835–1891)
b. Gateshead, Durham 23 Feb. 1835 *d.* Bournemouth, Hants 3 Jan. 1891
FRS 1874. FLS 1859. Lecturer in Botany, Durham College of Medicine. Pharmacist, Newcastle-upon-Tyne, 1855–76. *Report on Foraminefera collected by HMS Challenger* 1884 2 vols.
Pharm. J. v.21, 1891, 634–35, 699. *D.N.B. Supplt I* v.1, 254.
Bradyina von Möller.

BRAGG, Thomas (*fl.* 1790s)
Nurseryman and gardener, Honiton, Devon.

BRAGG, W. R. (*c.* 1805–1875)
b. Windsor, Berks *c.* 1805 *d.* Slough, Bucks 4 Feb. 1875
Publican, North Star Tavern, Slough. Florist.
Gdnrs Chron. 1875 i 213–14.

BRAIN, Lawrence Lewton- *see* Lewton-Brain, L.

BRAINE, Arthur Belgrave (1854–1945)
d. Melbourne 6 Oct. 1945
Coffee planter in Ceylon. In Australia, 1914. Teacher in Victoria where he collected orchids. Discovered *Chiloglottis pescottiana.*
Victorian Nat. 1945, 124; 1946, 166; 1949, 103.
Prasophyllum brainei R. D. Rogers.

BRAINE, C. J. (*fl.* 1840s–1850s)
Merchant at Hong Kong. Collected plants in Chusan. Sent ferns to Sir W. J. Hooker.
Hooker's J. Bot. 1850, 251. *J. Bot.* 1866, 15–17. E. Bretschneider *Hist. European Bot. Discoveries in China* 1898, 383–84. G. A. C. Herklots *Hong Kong Countryside* 1951, 165.
Plants and letters at Kew.
Brainea J. Smith.

BRAITHWAITE, John Oldham (–1937)
d. 20 Nov. 1937
Industrial chemist. Botanist and entomologist.
London Nat. 1937, 23.

BRAITHWAITE, Robert (1824–1917)
b. Ruswarp, Yorks 10 May 1824 *d.* Brixton, London 20 Oct. 1917
MRCS 1858. MD St. Andrews 1865. FLS 1863. Physician at Clapham. President, Queckett Microscopical Club, 1872–73. President, Royal Microscopical Society, 1892–93. President, Yorkshire Naturalists Union, 1895. *British Moss-flora* 1879–1905. *Sphagnaceae or Peat-mosses of Europe and North America* 1880. 'Sphagnaceae Exsiccatae' (*J. Bot.* 1919, 142–47).
Nature v.100, 1917, 150. *Naturalist* 1917, 361–63 portr. *J. R. Microsc. Soc.* 1917, 560. *Bot. Soc. Exch. Club Br. Isl. Rep.* 1917, 86–87. *J. Bot.* 1918, 23–25; 1926, 200. *Lancashire Nat.* v.10, 1918, 311–12. *Proc. Linn. Soc.* 1917–18, 35–37. *Bryologist* 1926, 77. *R.S.C.* v.7, 241; v.9, 328.
Herb. and MS. 'Flora Whitbyensis' at BM(NH). Portr. at Hunt Library.
Braithwaitea Lindl.

BRAMALL, John (*fl.* 1790s)
Seedsman, Lichfield, Staffs.

BRAMES, Peter (–1834)
Nurseryman, Old Brompton, London. Partner with R. Whitley (*c.* 1754–1835) and T. Milne (*c.* 1767–1838) at Fulham Nursery, 1810.

BRANCH, Christopher (–1958)
b. Warnham, Sussex *d.* Warnham 21 Oct. 1958
Gardener. Employed by Messrs. Charlesworth and private individuals. Started nursery at Warnham.
Orchid Rev. 1958, 285.

BRANCH, Harold (*c.* 1914–1947)
Grew alpine plants. Contrib. to *Quart. Bull. Alpine Gdn Soc.*
Quart. Bull. Alpine Gdn Soc. 1947, 198.

BRAND, William (1807–1869)
b. Blackhouse, Peterhead, Aberdeenshire 1807 *d.* 15 Oct. 1869
Pupil of R. Graham at Edinburgh. Discovered *Astragalus alpinus* 1831. 'On Systematic Arrangement in Formation of Natural History Collections' (*Trans. Bot. Soc. Edinburgh* v.1, 1844, 39–44).
Proc. Bot. Soc. Edinburgh 1838–39, 58–66, 108–12. *Trans. Bot. Soc. Edinburgh* v.10, 1870, 284–88. *R.S.C.* v.1, 564.
Plants at BM(NH), Edinburgh, Kew. Letters at Linnean Society.

BRANDIS, Sir Dietrich (1824–1907)
b. Bonn, Germany 31 March 1824 *d.* Bonn 28 May 1907
PhD Bonn 1848. LLD Edinburgh 1889. FRS 1875. FLS 1860. KCIE 1887. Superintendent of Forests, Pegu, 1856. Inspector-General of Indian Forests, 1864–83. Founded forestry school at Dehra Dun, 1878. *Forest Fl. N.W. and Central India* 1874. *Vegetation and Country from Narkanda to Pangi* 1879. *Indian Trees* 1906.
Gdnrs Chron. 1875 i 13 portr.; 1907 i 376. *Indian For.* 1907 *passim*; 1909 *passim*; 1912, 468–69. *J. Bot.* 1907, 288. *Nature* v.76, 1907, 131–32. *Proc. Linn. Soc.* 1907–8, 46–48. *Proc. R. Soc.* 1908, iii–vi. *Times* 31 May 1907. *Trans. Bot. Soc. Edinburgh* v.23, 1908, 363–65. *D.N.B. Supplt. 2* v.1, 217–18. *Rec. Bot. Survey India* v.7, 1914, 10–11. *Who was Who, 1897–1916* 84. I. H. Burkill *Chapters on Hist. Bot. in India* 1965, 158–67. *R.S.C.* v.7, 242; v.13, 765.
Plants, letters and portr. at Kew.
Brandisia Hook. f. & Thomson.

BRANSON, Ferguson (1809–1895)
MB 1834. Physician at Sheffield Infirmary. President, Literary and Philosophical Society, Sheffield, 1851. Experimented with nature-printing on which he contrib. papers to *Gdnrs Chron.* 1850, 824, *Athenaeum* 1850, 1350, *Trans. Soc. Arts* 1851, 96–98.
Sorby Rec. v.2 (2), 1966, 39–40.

BRANSON, Frederick Woodward (1851–1933)
d. Leeds, Yorks 30 Nov. 1933
MPS. FICh 1888. Pharmaceutical chemist. Active member of Leeds Naturalists' Club. Collected plants with G. C. Druce.
Pharm. J. 1933, 693 portr.

BRASS, William (–1783)
d. at sea 1783
Gardener to 1st Duke of Northumberland. To Cape 1780 where he collected plants for

Sir J. Banks, J. Fothergill and W. Pitcairn. Drew plants for Banks.

A. Rees *Cyclop*. v.39, Addenda. *Comptes Rendus AETFAT 1960* 1962, 57–58. *Gdnrs Chron*. v. 163 (17), 1968, 15. A. M. Coats *Quest for Plants* 1969, 245–47.

Plants at BM(NH). Letters at Kew.

Brassia R. Br.

BRAUNTON, Ernest (1867–1945)
b. London 21 Aug. 1867 *d*. Los Angeles, U.S.A. 22 March 1945

Horticulturist. To S. California from Iowa, 1887. Lecturer, University of California, 1901–13. Professor of Landscape Gardening, University of S. California, 1914–19. *Garden Beautiful in California* 1915. Contrib. to L. H. Bailey's *Standard Cyclop. Hort*. 1914.

Pacific Outlook 16 March 1907, 28.

Portr. at Hunt Library.

Astragalus brauntoni Parish.

BRAY, Miss E. A. (*c*. 1844–1938)
d. 16 April 1938

Conducted private school at Hailsham, Sussex. Contrib. to A. H. Wolley-Dod's *Fl. Sussex* 1937.

Trans. J. Eastbourne Nat. Hist. Photogr. Archaeol. Soc. 1938, 26.

BRAY, Edward (*fl*. 1770s–1800s)

Nurseryman, Merchant's Quay and Island Bridge, Dublin. Supplied plant material for Botanic Garden, Glasnevin.

Irish For. 1967, 45.

BRAY, John (*fl*. 1370s–1381)
d. 1381

Physician to W. de Montacute, Earl of Salisbury; to Edward III. Pensioned by Richard II. Author of MS. list of herbs in Latin, French and English: *Synonyma de Nominibus Herbarum* (Sloane MS. 282, ff. 167v–173v).

R. Pulteney. *Hist. Biogr. Sketches of Progress of Bot. in England* v.1, 1790, 22. *D.N.B*. v.6, 237. E. J. L. Scott *Index to Sloane Manuscripts* 1904, 75.

BRAYLSFORD, James (*fl*. 1700s)

"A Turkey merchant." Gave four books of Palestine plants collected by himself to J. Petiver, now at BM(NH).

J. Petiver *Museii Petiveriani* 1695, 79.

BRAZIL, Walter Henry (1861–1947)
b. Chorley, Lancs 25 Jan. 1861 *d*. 3 May 1947

MD. Practised medicine at Coventry. 'True Oxlip and Pseudo-oxlip' (*Proc. Coventry Nat. Hist. Sci. Soc*. v.2, 1948, 26).

Proc. Coventry Nat. Hist. Sci. Soc. 1934, 78–81.

BREACH, J. (*fl*. 1830s)

Herb. at Croydon Museum, Surrey.

BREAREY, Mrs. (*fl*. 1860s)

Herb. at Manx Museum, Douglas, Isle of Man.

BREBNER, George (*c*. 1855–1904)
b. Aberdeen *c*. 1855 *d*. Bristol 23 Dec. 1904

Lecturer in Botany, University College, Bristol. Worked at Plymouth Laboratory of Marine Biological Association. Cytologist. Algologist. Draughtsman. Pupil and collaborator of D. H. Scott. Illustrated 'Plants of Coal-measures' (*Philos. Trans. R. Soc*. v.185–v.186, 1894–95).

J. Bot. 1905, 60–61. *R.S.C*. v.42, 783.

Algae collection at Marine Biological Association.

Rhodochorton brebneri Batters.

BREDWELL, Mr. (*fl*. 1600s)

"Practitioner in physic, a learned and diligent searcher of simples" (R. Pulteney *Hist. Biogr. Sketches of Progress of Bot. in England* v.1, 1790, 125).

BREE, Rev. Robert Francis (*c*. 1746–1842)
d. London 28 Jan. 1842

FLS 1815. ALS 1827. Of Camberwell, Surrey and Chichester, Sussex. Collected plants (J. Sowerby and J. E. Smith's *English Bot*. 2809, 2934). Presented his herb. to a Boulogne Society (*Phytologist* v.2, 1845, 3).

Trans. Linn. Soc. v.12, 1818, 123. C. F. Lessing *Synopsis Generum Compositarum* 1832, 9–10. *Proc. Linn. Soc*. 1842, 145.

BREE, Rev. William (*fl*. 1770s–1820s)

Father of Rev. W. T. Bree. Contrib. to T. Purton's *Midland Fl*. and drew plates for it.

BREE, Rev. William Thomas (1787–1863)
b. Coleshill, Warwickshire 1787 *d*. Allesley, Warwickshire 25 Feb. 1863

BA Oxon 1808. Rector, Allesley. Discovered *Lastraea rigida* (J. Sowerby and J. E. Smith's *English Bot*. 2724). Collected and grew British saxifrages. Contrib. to *Gdnrs Mag*., *Mag. Nat. Hist*., *Phytologist*, T. Purton's *Bot. Description of British Plants in Midland Counties* 1817.

A. H. Haworth *Saxifragearum Enumeratio* 1821, xiii–xv. B. Maund *Bot. Gdn* v.12, 1847–48, 269. *Ann. Mag. Nat. Hist*. v.6, 1841, 401–2. *J. Bot*. 1863, 160. J. E. Bagnall *Fl. Warwickshire* 1891, 497–98. G. C. Druce *Fl. Berkshire* 1897, clvii–clviii. F. H. Davey *Fl. Cornwall* 1909, xl. D. A. Cadbury *Computer-mapped Fl.... Warwickshire* 1971, 52–54. *R.S.C*. v.1, 593.

Plants at Kew, Oxford.

BREEDON, Rev. J. Symonds

Raised 'Breedon' pippin and 'Bere Court' pippin at Bere Court, Pangbourne, Berks in early 19th century.

BRÉHAUT, Rev. Thomas Collings (1820–1880)
d. Guernsey, Channel Islands 4 Nov. 1880
Prison chaplain, Guernsey. Horticulturist. *Cordon Training of Fruit Trees* 1860. *Modern Peach Pruner* 1866.
Gdnrs Chron. 1880 ii 636.

BRENAN, Rev. Samuel Arthur (1837–1908)
d. Cushendun, County Antrim Jan. 1908
BA Dublin. Contrib. papers on *Rubi*, *Hieracia*, etc. to *J. Bot.* and *Irish Nat.*, 1884–98.
Irish Nat. 1908, 43; 1913, 30. *Proc. Belfast Nat. Field Club* v.6, 1913, 625. R. L. Praeger *Some Irish Nat.* 1949, 54. *R.S.C.* v.13, 789.
Herb. at Belfast Museum.

BRENCHLEY, Winifred Elsie (1883–1953)
b. 10 Aug. 1883 *d.* Harpenden, Herts 27 Oct. 1953
BSc London 1905. FLS 1910. Rothamsted Experimental Station, 1907; Head of Botanical Dept. until 1948. *Inorganic Plant Poisons and Stimulants* 1914; ed. 2 1927. *Weeds of Farm Land* 1920. Contrib. to *Ann. Bot.*, *New Phytol.*, *J. Linn. Soc.*
Nature v.172, 1953, 936. *Times* 29 and 31 Oct. 1953. *Who was Who, 1951–1960* 133.

BRENT, Francis (1816–1903)
b. 4 Dec. 1816 *d.* Aug. 1903
Clerk in H.M. Customs, Liverpool and Folkestone. Friend of T. Sansom (1816–1872) and W. Bean (1817–1864) with whom he formed a herb. 'Mosses of Devon and Cornwall' (*Trans. Plymouth Inst.* v.3, 1868–69, 305–45). Contrib. to J. Dickinson's *Fl. Liverpool* 1851 and J. B. Rowe's *Perambulation of Forest of...Dartmoor* ed. 3 1896.
W. K. Martin and G. T. Fraser *Fl. Devon* 1939, 775. *J. Soc. Bibl. Nat. Hist.* 1972, 157. *R.S.C.* v.1, 605; v.7, 254; v.9, 342.
Plants at Exeter, Liverpool Museums.

BRENTON, Mary E. (*fl.* 1830s)
Collected plants in Newfoundland and Labrador, 1830–31.
W. J. Hooker *Fl. Boreali-americana* v.1, 1831, 265 *et seq.*
Plants at Kew.

BREWER, James Alexander (1818–1886)
b. Reigate, Surrey 25 Feb. 1818 *d.* Tonbridge, Kent Jan. 1886
Collected plants in Australia. *New Fl. of Neighbourhood of Reigate, Surrey* 1856. *Fl. Surrey* 1863.
C. E. Salmon *Fl. Surrey* 1931, 52. *R.S.C.* v.7, 255.
Plants at BM(NH), Kew; herb. at Holmesdale Natural History Club, Reigate.

BREWER, Michael (*fl.* 1830s)
Nurseryman, Cambridge.

BREWER, Samuel (1670–1743)
b. Trowbridge, Wilts 1670 *d.* Bierley, Yorks 1743
In woollen manufacture in Trowbridge. In 1726 botanised with Dillenius in Mendips, N. Wales and Anglesey. MS. 'Botanical Journey, through Wales' at BM(NH). Went to live at Bierley, 1727. Gardener to Duke of Beaufort at Badminton. Discovered *Helianthemum breweri.*
R. Pulteney *Hist. Biogr. Sketches of Progress of Bot. in England* v.2, 1790, 188–90. J. Nichols *Illus. Lit. Hist. of Eighteenth Century* v.1, 261, 288. J. E. Smith *Selection of Correspondence of Linnaeus* v.2, 1821, 144, 146–47. D. Turner *Extracts from Lit. and Sci. Correspondence of Richard Richardson* 1835, 276–77. J. Cash *Where there's a Will there's a Way* 1873, 5–6. *Wiltshire Archaeol. Nat. Hist. Mag.* v.18, 1879, 71–80. *J. Bot.* 1898, 12; 1915, 68–69. *D.N.B.* v.6, 295–96. E. J. L. Scott *Index to Sloane Manuscripts* 1904, 76. G. C. Druce and S. H. Vines *Dillenian Herb.* 1907, lviii–lxix. *Bot. Soc. Exch. Club Br. Isl. Rep.* 1931 Supplt., 1–30. J. E. Dandy *Sloane Herb.* 1958, 95. H. N. Clokie *Account of Herb. Dept. Bot., Oxford* 1964, 138. C. O. Moreton *Auricula* 1964, 27.
Letters from Dillenius and plants at BM(NH). MS. account of journey from Yorkshire to London, 1691 at Kew.
Breweria R. Br.

BREWER, Thomas (*fl.* 1790s)
Gardener and seedsman, Chippenham, Wilts.

BREWER, William A. (*fl.* 1850s–1860s)
Plants at Holmesdale Natural History Club, Reigate, Surrey.

BRICKELL, John (*fl.* 1730s–1740s)
b. Ireland
MD. Of Edenton, N. Carolina. *Natural History N. Carolina* 1739 (plants, 57–106). *Catalogue of American Trees and Shrubs* 1739.
Rhodora 1916, 225–30. *J. Bot.* 1917, 118.

BRICKELL, John (*c.* 1749–1809)
b. County Louth, Ireland *c.* 1749 *d.* Savannah, Georgia 22 Dec. 1809
MD. At Savannah from about 1779. Acute observer of local flora. Sent plants to H. Muhlenberg. Contrib. to *New York Med. Repository* 1798–1809.
S. Elliott *Sketch of Bot. of S. Carolina and Georgia* v.2, 1824, 290. *Rhodora* 1916, 225–30.
Brickellia Elliott.

BRIDGE, Cyprian A. G. (*fl.* 1880s)
Capt., Royal Navy. Sailed in Pacific, 1882–85. Collected plants, partly with J. B. Chalmers, in Bismark Archipelago and New Guinea.
Victorian Nat. 1885, 168. *Fl. Malesiana* v.1, 1950, 79.
Plants at Melbourne.

BRIDGE, Thomas (*fl.* 1790s–1820s)
Seedsman and seed factor, 16 Trinity Square, London.

BRIDGE, William (*fl.* 1780s–1790s)
Gardener and seedsman, Tytheban Street, Liverpool.

BRIDGEFORD, John (*fl.* 1810s)
Nurseryman and seedsman, 2 Smithy Door, Manchester.

BRIDGEMAN, Charles (*fl.* 1790s)
Nurseryman, Hertford, Herts.

BRIDGEMAN, Thomas (–1850)
b. Berkshire
To U.S.A., 1824. Seedsman, New York. *Florist's Guide* 1835. *Fruit Cultivator's Manual* 1857. *Kitchen Gardener's Instructor* 1858. *Young Gardener's Assistant* 1829.
L. H. Bailey *Standard Cyclop. Hort.* v.2, 1939, 1564 portr.

BRIDGER, Frederick (*fl.* 1840s–1910s)
Gardener to Lord de l'Isle at Penshurst Place, Kent, 1849.
Gdnrs Mag. 1911, 489–90 portr.

BRIDGER, James (*c.* 1806–1885)
d. 4 May 1885
"Physic Gardener and Farmer", Mitcham, Surrey. Grew medicinal herbs. "The Mitcham lavender and peppermint that have passed through his stills have a world-wide reputation" (*Croydon Advertiser* 9 May 1885).

BRIDGES, Thomas (1807–1865)
b. Lilly, Herts, 22 May 1807 *d.* in Pacific 9 Nov. 1865
FLS 1844. Son-in-law of H. Cuming (1791–1865). Collected plants in Chile, Peru, Bolivia and California, 1827–65. Introduced *Victoria amazonica.*
Gdnrs Mag. 1831, 95–96; 1840, 116. *J. Bot.* (Hooker) 1834, 177–78. *London J. Bot.* 1845, 571–77. *J. Bot.* 1866, 64. *Gdnrs Chron.* 1865, 1226. *Proc. California Acad. Nat. Sci.* v.3, 1866, 236–37. *Trans. Bot. Soc. Edinburgh* v.8, 1866, 434–35. *West. Amer. Sci.* v.3, 1887, 223–27. *Kew Bull.* 1920, 59. *Contrib. Gray Herb.* no. 81, 1928, 98–106. *Madrono* 1933, 84–88. A. M. Coats *Quest for Plants* 1969, 371–73. *Taxon* 1970, 516. *R.S.C.* v.7, 259.
Plants at Kew, BM(NH), Oxford. Letters at Kew. Portr. at Hunt Library.
Bridgesia Bert. ex Cambess.

BRIERLEY, William Broadhurst (1889–1963)
b. Manchester 19 Feb. 1889 *d.* Keswick, Cumberland 20 Feb. 1963
DSc. FLS 1919. School-teacher, Manchester, 1907–9. Lecturer in Economic Botany, Manchester, 1911–14. Head, Dept. of Mycology, Rothamsted Experimental Station, 1918–32. Professor of Agricultural Botany, 1932–54. President, Association of Applied Biologists, 1932–33. Assisted in translation of E. Gäumann's *Principles of Plant Pathology* 1950. Edited *Ann. Applied Biol.* 1921–45.
Gdnrs Chron. 1925 i 104 portr. *Rothamsted Rep.* 1937, 60. *Ann. Applied Biol.* 1963, 509–10. *Nature* v.198, 1963, 133. *Times* 1 March 1963. *Who was Who, 1961–1970* 134.
Portr. at Hunt Library.

BRIGGS, Guy W. G. (*c.* 1904–1961)
d. Cambridge 7 March 1961
Colonial Agricultural Service, Nigeria, 1926–45. National Institution of Agricultural Botany, 1945; Head of Seed Production Branch, 1948; Assistant Director, 1960.
Nature v.190, 1961, 861–62.

BRIGGS, Thomas Richard Archer (1836–1891)
b. Fursdon, Plymouth, Devon 7 May 1836 *d.* Fursdon 23 Jan. 1891
FLS 1872. Botanised in Teign and Upper Tamar valleys with Rev. W. Moyle Rogers. Authority on *Rosa* and *Rubi. Fl. Plymouth* 1880. Contrib. to *Phytologist* and *J. Bot.*
H. C. Watson *Topographical Bot.* 1883, 539–40. *J. Bot.* 1891, 97–106 portr. *Proc. Linn. Soc.* 1890–91, 21. F. H. Davey *Fl. Cornwall* 1909, li–lii portr. W. K. Martin and G. T. Fraser *Fl. Devon* 1939, 774. *R.S.C.* v.7, 259; v.9, 349.
Herb. at Kew. Plants at BM(NH). Portr. at Hunt Library.
Rubus briggsii H. C. Watson.

BRIGHT, Frederick (1853–1934)
b. Weston near Bath, Somerset 1853 *d.* Shinfield, Berks 27 Jan. 1934
Gardener to P. Carslake and J. Friedlander at Whiteknights Park, Reading, 1883–1920. Noted for his fuchsias.
Gdnrs Chron. 1929 ii 218 portr.; 1934 i 103.

BRIGHT, Henry Arthur (1830–1884)
b. Liverpool 9 Feb. 1830 *d.* Knotty Ash, Liverpool 5 May 1884
BA Cantab 1857. Partner in shipping firm. *A Year in a Lancashire Garden* 1879. *English Flower Garden* 1881.
Gdnrs Chron. 1884 i 620. *Times* 10 May 1884. *D.N.B.* v.6, 331–33. *Trans. Liverpool Bot. Soc.* 1909, 62.

BRIGHTWELL, Thomas (1787–1868)
b. Ipswich, Suffolk 18 March 1787 *d.* Norwich, Norfolk 17 Nov. 1868
FLS 1821. Diatomist.
C. L. Brightwell *Memorials* 1869 portr. *Proc. Linn. Soc.* 1868–69, cv–cix. *D.N.B.* v.6, 340. *Trans. Norfolk Norwich Nat. Soc.* v.9, 1914, 677. *R.S.C.* v.1, 627.
Letters at BM(NH).
Brightwellia Ralfs.

BRIGHTWEN, Eliza (*née* **Elder)** (1830–1906)
b. Banff 30 Oct. 1830 *d.* Stanmore, Middx 5 May 1906

Studied natural history at her home at Stanmore. *Glimpses into Plant Life* 1898. *Life and Thoughts of a Naturalist* 1909 portr.

Gdnrs Chron. 1906 i 303. *J. Bot.* 1906, 216; 1909; 275–77. *Nature Notes* 1906, 113. *D.N.B. Supplt. 2* v.1, 225–26, *Country Life* 1969, 1194–95 portr.

BRIMBLE, Lionel John Farnham (1904–1965)
b. Radstock, Somerset 16 Jan. 1904 *d.* London 15 Nov. 1965

BSc Reading. FLS 1938. FRSE 1953. Lecturer in Plant Physiology, Glasgow, 1926–27. Lecturer, Manchester, 1927–30. Assistant Editor, *Nature* 1930; Editor, 1961. *Everyday Botany* 1934. *Intermediate Botany* 1936; ed. 3 1946. *Flowers in Britain* 1944. *Trees in Britain* 1946. *Floral Year* 1949.

Nature v.208, 1965, 819–21. *Yb. R. Soc. Edinburgh* 1967, 10–13. *Who was Who, 1961–1970* 134.

BRINKLEY, John (1763–1835)
b. Woodbridge, Suffolk 1763 *d.* Dublin 14 Sept. 1835

MA Cantab 1791. DD 1806. FRS 1803. Professor of Astronomy, Dublin, 1792. Bishop of Cloyne, 1826. Reputed to have discovered *Hypericum hirsutum* and *Oenanthe fistulosa* in Ireland.

D.N.B. v.6, 347–48. N. Colgan *Fl. County Dublin* 1904, xxv.

BRINKMAN, Alfred Henry (1873–)
b. Poplar, London 28 Feb. 1873

Lighthouse service, 1895–1908. Surveyor and farmer in Canada, 1908. Collected mosses, hepatics and flowering plants. Contrib. to *Bryologist*, *Canadian Field Nat.*, etc.

Canadian plants at BM(NH). British plants at Cambridge.

BRINKWORTH, Mr. (*fl.* 1780s)

Nurseryman, Ducking Pond Lane, near Whitechapel, London.

BRINSDEN, John (*fl.* 1790s)

Seedsman, Marlborough, Wilts.

BRISBANE, Thomas (–1742)
d. 3 April 1742

Professor of Botany and Anatomy, Glasgow.

Gent. Mag. 1742, 218. J. R. Green *Hist. Bot.* 1914, 189. *Gdnrs Chron.* 1919 i 147.

BRISTOW, Ernest (*c.* 1863–1934)
d. 12 March 1934

Cultivated orchids for Lee of Leatherhead, Surrey and J. W. Temple of Groombridge, Sussex.

Orchid. Rev. 1934, 127–28.

BRISTOW, William (*fl.* 1810s–1830s)

Nurseryman, Old Brompton, Kensington, London. Entered into partnership with Samuel Harrison at Old Brompton and St. Luke, Chelsea. Nursery went bankrupt in 1833.

BRITON-JONES, Harry Richard (1893-1936)
d. Trinidad 3 Nov. 1936

DSc London 1926. Course in tropical mycology at Kew, 1919. Mycologist, Ministry of Agriculture, 1920–23. Mycologist, Long Ashton Research Station, 1923–26. Professor of Mycology, Imperial College of Tropical Agriculture, 1926–36. *Diseases and Curing of Cacoa* 1934.

Kew Bull. 1926, 142; 1936, 530–31. *Nature* v.138, 1936, 913. *Times* 4 Nov. 1936. *Chronica Botanica* 1937, 245 portr.

Fungi at Commonwealth Mycological Institute, Kew.

BRITTAIN, Thomas (1806–1884)
b. Sheffield, Yorks 2 Jan. 1806. *d.* Urmston, Lancs 23 Jan. 1884

Accountant. President, Manchester Microscopical Society, 1882. *Micro-fungi, When and Where to Find Them* 1882.

Gdnrs Chron. 1884 i 155. *Manchester Microsc. Soc.* 1891 front. portr. *D.N.B.* v.6, 359–60. *Trans. Liverpool Bot. Soc.* 1909, 62.

Plants at Manchester Museum.

BRITTEN, Harold (1870–1954)
b. Wilts 1870 *d.* 31 Jan. 1954

ALS 1950. Game-keeper, Great Salkeld. Assistant to Professor of Zoology, Oxford. Employed by Flatters and Garnett, Manchester.

N. Western Nat. v.2, 1954, 324–29 portr.

Herb. at Carlisle Museum. Yorks, Westmorland and Cumberland plants at Oxford.

BRITTEN, James (1846–1924)
b. Chelsea, London 3 May 1846 *d.* London 8 Oct. 1924

FLS 1870. Assistant, Kew Herb., 1869–71. Assistant, Botany Dept., BM(NH), 1871–1909. *Dictionary of English Plant-names* (with R. Holland) 1878–86. *European Ferns* 1879–81. *Old Country and Farming Words* 1880. Edited W. Turner's *Names of Herbes* 1881. *Biographical Index of British and Irish Botanists* (with G. S. Boulger) 1893–1908. *Illustrations of Australian Plants collected in 1770...* 1900–5. *Popular British Fungi.* Edited *J. Bot.* 1880–1924; *Nature Notes* 1890–97.

J. Bot. 1912 portr.; 1924, 337–43 portr., 355, 358; 1925, 27. *Bot. Soc. Exch. Club Br. Isl. Rep.* 1924, 536–37. *Gdnrs Chron.* 1909 ii 398; 1924 ii 275, 400. *Kew Bull.* 1924, 392–93. *Nature* v.114, 1924, 583. *Proc. Linn. Soc.* 1924–25, 50, 64–66. *Times* 10 Oct. 1924. G. C. Druce *Fl. Buckinghamshire* 1926, cii–ciii. *Who was who, 1916–1928* 126–27. C. G. Gillispie *Dict. Sci. Biogr.* v.2, 1970, 475–76. *R.S.C.* v.7, 267; v.9, 356; v.12, 122; v.13, 816.

Plants at BM(NH). MS. list of Merionethshire plants 1877, at Kew. Portr. at Hunt Library.

Jamesbrittenia Kuntze. *Brittenia* Cogn. ex Boerlage.

BRITTON, Charles Edward (1872–1944)
b. Lambeth, London 2 Dec. 1872 *d.* Redhill, Surrey 23 March 1944

ALS 1938. G.P.O. employee. Studied flora of London area. Contrib. to *J. Bot.*

Bot. Soc. Exch. Club Br. Isl. Rep. 1943–44, 639–41. H. N. Clokie *Account of Herb. Dept. Bot., Oxford* 1964, 139.

Herb. at Kew. Plants at South London Botanical Institute, Edinburgh, Oxford.

Rubus brittonii Barton & Riddelsdell.

BROAD, William (*fl.* 1620s–1630s)
Apothecary. Accompanied T. Johnson on his visits to Kent in 1629 and 1632, and on his tour to West of England in 1634. Contrib. verses to J. Parkinson's *Paradisus* 1629.

J. Gerard *Herball* 1633, 825.

BROADBENT, Albert (1891–1962)
d. 26 Dec. 1962

Tailor at Huddersfield, Yorks. Studied ecology of Dean Nick near Huddersfield. Interested in Myxomycetes.

Naturalist 1963, 70.

BROADHEAD, Charles Herman (*c.* 1860–1920)
d. 27 July 1920

Nurseryman, Wooldale Nurseries, Thongsbridge, Yorks.

Naturalist 1920, 323 portr.

BROADWAY, Walter Elias (1863–1935)
b. Exbury, Hants 3 May 1863 *d.* Trinidad 1 Jan. 1935

Kew gardener. Assistant Superintendent, Botanic Gardens, Trinidad, 1888. Curator, Botanic Gardens, Grenada, 1894–1905. Curator, Botanic Garden, Tobago, 1910–15. Assistant Botanist, Trinidad, 1915–22. Collected in West Indies, Venezuela and French Guiana. Contrib. to *Orchid Rev.*

Kew Bull. 1894, 192. I. Urban *Symbolae Antillanae* v.3, 1902, 28. *Chronica Botanica* 1930, 280 portr. *J. Kew Guild* 1935, 460 portr. *Orchid Rev.* 1935, 354–55 portr. *Trinidad and Tobago Yb.* v.53, 161–62. *Amer. Orchid Soc. Bull.* 1958, 80.

Plants at BM(NH), Berlin, Kew. Portr. at Hunt Library.

Citharexylum broadwayi O. E. Schulz.

BROCAS, Frederick Yorke (*fl.* 1840s–1850s)
Supplier of botanical specimens at 85 St. Martin's Lane, London. *British Mosses* (catalogue) 1852.

Gdnrs Chron. 1858, 635. D. H. Kent *Br. Herb.* 1957, 45.

Plants at Charterhouse School Museum, Godalming.

BROCKBANK, Thomas (*fl.* 1790s)
Gardener and nurseryman, Hawkshead, Lancs.

BROCKBANK, William (1830–1896)
d. Didsbury, Manchester 25 Sept. 1896.

FLS 1884. Surveyor. Grew primulas, saxifrages, narcissi, etc. Experimented on doubling and colouring of flowers. *Notes on Seedling Saxifrages grown at Brockhurst c.* 1889.

Garden v.50, 1896, 282. *Gdnrs Chron.* 1896 ii. 409. *Gdnrs Mag.* 1896, 684. *Proc. Linn. Soc.* 1896–97, 57. *Trans. Liverpool Bot. Soc.* 1909, 62. *R.S.C.* v.7, 269; v.9, 361.

BROCKLEHURST, Mr. (*fl.* 1830s)
Of Macclesfield, Cheshire. First flowered *Houlletia brocklehurstiana.*

Bot. Mag. 1844, t.4072.

BROCKMAN, Ralph Evelyn Drake- *see* Drake-Brockman, R. E.

BROCQ, Rev. Philip le (*fl.* 1780s–1790s)
MA Cantab. Curate, Ealing, Middx. Chaplain to Duke of Gloucester. *Description...of Certain Methods of Planting...Fruit Trees...* 1786. *Plan for...the New Forest* 1793.

J. C. Loudon *Encyclop. Gdning* 1822, 1278. G. W. Johnson *Hist. English Gdning* 1829, 247–48.

BRODIE, Ian, of Brodie (1868–1943)
d. Feb. 1943

VMH 1942. Of Brodie Castle, Elgin. Hybridised narcissi.

Daffodil Tulip Yb. 1946, 84–87; 1948, 7–8 portr.; 1950, 183–99; 1951, 101–4.

BRODIE, James (*fl.* 1690s)
Brought plantain from Virginia to J. Petiver. Travelled in West Indies.

Philos. Trans. R. Soc. no. 229, 1697, 580–82. J. E. Dandy *Sloane Herb.* 1958, 98.

BRODIE, James (1744–1824)
b. 31 Aug. 1744 *d.* Brodie Castle, Elgin 17 Jan. 1824

FRS 1797. FLS 1795. MP for Elgin, 1796 Algologist. Discovered *Moneses.*

L. W. Dillwyn *British Confervae* 1802–9, 35. J. Sowerby and J. E. Smith *English Bot.* 146, 1966, 2589, etc. D. Turner *Fuci* v.2, 1808–19, 2. *Trans. Linn. Soc.* v.10, 1810, 1–5. W. J. Hooker *Musci Exotici* 1818 dedication. R. K. Greville *Algae Britannicae* 1830, vi. W. H.

Harvey *Phycologia Britannica* 1846–51, t.cxxix.
Herb. at Edinburgh.
Brodiaea Smith.

BRODIGAN, Thomas (–1849)
b. Drogheda, Louth *d.* Drogheda 4 Dec. 1849
To America in 1817 to study tobacco-growing. *Botanical, Historical, and Practical Treatise on Tobacco Plant* 1830 (reviewed in *Gdnrs Mag.* 1830, 270–74).

BRODRICK, Thomas (*fl.* 1670s)
Formed herb., dated 1672, formerly at Peper Harow, Surrey.
J. Bot. 1904, 295–96; 1919, 197.

BROGDEN, James (*fl.* 1850s?)
84 Australian plants collected by Brogden for sale at Samuel Stevens, London, 1857.
Hooker's J. Bot. 1857, 190.

BROKENSHIRE, Frederick Adolphus (1866–1957)
b. 3 Feb. 1866 *d.* Barnstaple, Devon 10 June 1957
Teacher, Barnstaple. Secretary, Botanical Section, Devonshire Association, 1931–47. Recorder of Cryptogams for Association's *Trans.*, 1942–53.
Rep. Trans. Devonshire Assoc. 1957, 248. *Proc. Bot. Soc. Br. Isl.* 1958, 129–30.

BROMEHEAD, Miss
Had garden between Chesterfield and Dronfield, Derbyshire. "An investigator and collector of plants" (Jonathon Stokes).

BROMFIELD, William Arnold (1801–1851)
b. Boldre, Hants 4 July 1801 *d.* Damascus, Syria 9 Oct. 1851
MD Glasgow 1823. FLS 1836. Settled in 1836 at Ryde, Isle of Wight where he spent the rest of his life preparing his *Fl. Vectensis* 1856. Travelled on Continent, 1826–30, Ireland, West Indies, 1844, N. America, 1846, and Egypt 1850. Distinguished *Calamintha sylvatica*. *List of Plants...in the Isle of Wight* 1840. *Catalogue of...Plants...in the Isle of Wight...as an Index to Herbarium of Dr. Bromfield* (by A. G. More) 1859. Contrib. to J. Sowerby and J. E. Smith's *English Bot.* 2812, 2863, 2897 and *Phytologist*.
Hooker's J. Bot. 1851, 373–82. *Phytologist* v.4, 1851, v–xiii. *Proc. Linn. Soc.* 1852, 182–83. *J. Bot.* 1870, 88, 191. *D.N.B.* v.6, 398–99. I. Urban *Symbolae Antillanae* v.3, 1902, 28–29. A. H. Wolley-Dod *Fl. Sussex* 1937, xli. *Bot. Soc. Exch. Club Br. Isl. Rep.* 1943–44, 655. *R.S.C.* v.1, 164; v.12, 124.
Herb. at Isle of Wight Philosophical Society. Plants, MSS., letters, portr. and library at Kew. Portr. at Hunt Library.

BROMHEAD, Sir Edward Thomas Ffrench (1789–1855)
b. Dublin 26 March 1789 *d.* Thurlby Hall, Newark, Notts 14 March 1855
MA Cantab 1815. FRS 1817. FLS 1844. High Steward of Lincoln. 'Botanical Alliances' (*Edinburgh New Philos. J.* v.24, 1838, 408–18).
Proc. Linn. Soc. v.2, 1855, 405–6. F. Boase *Modern English Biogr.* v.1, 1892, 412. *R.S.C.* v.1, 644.
Bromheadia Lindl.

BROMILOW, Henry J. (1850–1930)
d. Rainhill, Lancs 11 Aug. 1930
Cultivated orchids
Orchid Rev. 1930, 280.
Cypripedium charlesworthii var. *Bromilowianum*.

BROMWICH, Henry (1828–1907)
b. Warwick 10 March 1828 *d.* Milverton, Warwick 28 May 1907
Gardener at Wroxall Abbey, Warwickshire. Hon. Curator, Botanical Dept., Warwick Museum. Contrib. to J. E. Bagnall's *Fl. Warwickshire* 1891.
Bot. Soc. Exch. Club Br. Isl. Rep. 1907, 262. *J. Bot.* 1908, 304. A. R. Horwood and C. W. F. Noel *Fl. Leicestershire* 1933, ccxix. D. A. Cadbury *Computer-mapped Fl...Warwickshire* 1971, 57.
Herb. at Warwick Museum.

BROOK, George (1857–1893)
b. Huddersfield, Yorks 17 March 1857 *d.* near Newcastle, Northumberland 12 Aug. 1893
FLS 1879. Entomologist. Marine zoologist. Worked at *Saprolegnia*. Assistant to Scottish Fisheries Board, 1884–87. Secretary, Huddersfield Naturalists Society and Scottish Microscopical Society.
Proc. Linn. Soc. 1893–94, 30–31. *R.S.C.* v.9, 365; v.12, 125; v.13, 834.

BROOKE, Sir James (1803–1868)
b. Benares, India April? 1803 *d.* Burrator, Devon 1868
Cadet, Bengal Army, *c.* 1819. To Borneo, 1838. Rajah of Sarawak, 1842. Governor of Labuan and Consul General of Borneo 1847. Collected plants in Sarawak, now at Kew.
G. L. Jacob *Raja of Sarawak* 1876 2 vols. S. St. John *Life of Sir J. Brooke* 1879. *D.N.B.* v.6, 428–30. *Fl. Malesiana* v.1, 1950, 81–82.

BROOKE, Jocelyn (1908–1966)
d. 29 Oct. 1966
Wild Orchids of Britain 1950.
Daily Telegraph 1 Nov. 1966.

BROOKE, T. (*fl.* 1790s)
Herb. at Ipswich Museum, Suffolk.

BROOKES, Miss E. Muriel (–1950)
b. Lucknow, India *d.* Uffculme, Devon Nov. 1950

Of Cheltenham, Glos. Schoolteacher in Canada,1913–27. Contrib. to H. J. Riddelsdell's *Fl. Gloucestershire* 1948.

Proc. Cotteswold Nat. Field Club 1950, 239–40.

BROOKES, Richard (*fl.* 1720s–1760s)

MD. Physician in Surrey. Travelled in Africa and America. *Natural History of Vegetables* 1763.

D.N.B. v.6, 436.

BROOKES, Samuel (*fl.* 1820s–1830s)

Nurseryman, Turnpike, Balls Pond, Islington, London. In partnership with Thomas Barr, 1819. Imported Chinese plants. Emigrated to Chicago, U.S.A., 1832.

E. Bretschneider *Hist. European Bot. Discoveries in China* 1898, 221–22.

BROOKES, William Penny (1809–1895)
b. Much Wenlock, Shropshire 1809 *d.* Much Wenlock 11 Dec. 1895

MRCS 1870. Contrib. to W. A. Leighton's *Fl. Shropshire* 1841.

Herb. at Much Wenlock Museum.

BROOKHOUSE, Joseph (*c.* 1757–1831)
d. Warwick 15 May 1831

Of Warwick. 'Method of Cultivating Cucumbers in a Peach-house' (*Trans. Hort. Soc.* v.5, 1824, 487–88).

Gdnrs Mag. 1831, 512.

BROOKS, Archibald Joseph (1881–)
b. Bromham, Wilts 18 Jan. 1881

FLS 1912. Kew gardener, 1902. Officer in charge, Agricultural School, Dominica, 1903. Assistant Agricultural Superintendent, 1911; Superintendent and Curator, Botanic Gardens, 1913. Director of Agriculture, Gambia, 1923–33.

J. Kew Guild 1942, 126–28 portr.

Herb. and MSS. at Kew.

BROOKS, Cecil Joslin (1875–*c.* 1953)
b. Cambridge 1875 *d.* Hampstead, London *c.* 1953

Metallurgical chemist with Borneo Co. Ltd. in Sarawak, 1909–10. Sumatra, 1912–23. Collected natural history specimens including plants in Amboina, Borneo, Canary Islands, Sumatra, New Zealand.

Fl. Malesiana v.1, 1950, 82; v.5, 1958, 31. *Entomologist* 1955, 88.

Plants at BM(NH), Kew, Paris, Singapore, etc.

BROOKS, Frederick Tom (1882–1952)
b. Wells, Somerset 17 Dec. 1882 *d.* Cambridge 11 March 1952

MA Cantab 1909. FRS 1930. FLS 1922. Demonstrator in Botany, Cambridge, 1905. Government Mycologist, Federated Malay States, 1914 where he collected plants around Kuala Lumpur, Province Wellesley, etc. Lecturer in Botany, Cambridge, 1919; Reader in Mycology, 1931. Professor of Botany, Cambridge, 1936–48. President, British Mycological Society, 1922. *Plant Diseases* 1928. *Insect and Fungus Pests of the Farm* (with J. C. F. Fryer) 1928. Re-edited D. H. Scott's *Flowerless Plants* 1932.

Gdns Bull. Straits Settlements 1927, 118. *Chronica Botanica* 1937, 156 portr. *Ann. Applied Biol.* 1952, 617–19. *Fl. Malesiana* v.1, 1950, 82. *Forestry* 1952, 142. *Gdnrs Chron.* 1952 i 100. *Nature* v.169, 1952, 606–7. *Proc. Linn. Soc.* 1952, 254–56. *Times* 13 March 1952. *Obit. Notices Fellows R. Soc.* 1953, 341–54 portr. *Trans. Br. Mycol. Soc.* 1953, 177–79 portr. *Yb. R. Soc. Edinburgh* 1953, 11–12. *Who was Who, 1951–1960* 141–42.

Fungi at Cambridge, Kew.

BROOKS, Miss Sarah (1870–)

Collected plants in Mt. Baggett area, W. Australia. Correspondent of F. von Mueller.

Victoria Nat. 1970, 145.

Scaevola brooksiana.

BROOKS, Thomas (*fl.* 1800s)

Seedsman and florist, 92 New Bond Street, London.

BROOKSHAW, George (*fl.* 1810s–1820s)

Artist and teacher of flower-painting. Exhibited at Royal Academy, 1819. *Pomona Britannica* 1812; 1817 2 vols. *Horticultural Repository* 1823 2 vols. *New Treatise on Flower Painting* 1816. *Groups of Flowers*, 1819. *Groups of Fruit* 1819.

BROOME, Christopher Edmund (1812–1886)
b. Berkhampstead, Herts 24 July 1812 *d.* London 15 Nov. 1886

BA Cantab 1836. FLS 1866. Mycologist. 'Notes of British Fungi' (with M. J. Berkeley) (*Ann. Mag. Nat. Hist.* 1848–85).

Grevillea v.15, 1887, 63. *J. Bot.* 1887, 148–50; 1899, 398–99. *Proc. Nat. Hist. Club Bath* 1886–87, 202–3; 1889, 144–53. *Proc. Linn. Soc.* 1886–87, 34–35. C. C. Babington *Memorials* 1897, 269. J. W. White *Fl. Bristol* 1912, 83. H. J. Riddelsdell *Fl. Gloucestershire* 1948, cxx. *R.S.C.* v.1, 655; v.7, 274; v.9, 366; v.12, 125; v.13, 838.

Library and fungi at Bath Literary and Scientific Institute. Fungi at Kew. Letters and drawings at BM(NH).

BROOME, Joseph (1825–1907)
b. Preston Brook near Frodsham, Cheshire May 1825 *d.* 25 Jan. 1907
Had orchid collection.
Orchid Rev. 1907, 65–66.

BROOME, Samuel (1806–1870)
b. Weston, Staffs 29 June 1806 *d.* 22 Jan. 1870
Gardener, Inner Temple, London.
Gdnrs Mag. 1870, 44. *Garden* v.1, 1872, 198.

BROPHY, James (*fl.* 1790s)
Nurseryman, County Kilkenny.

BROTHERSTON, Andrew (1834–1891)
b. Eccles, Berwickshire 28 March 1834 *d.* Kelso, Roxburgh 16 March 1891
Gardener. 'On Poa sudetica as a British Plant' (*Hist. Berwick. Nat. Club* v.7, 1873, 129–30). 'List of Tweedside Plants mostly of Recent Introduction' (*Hist. Berwick. Nat. Club* v.7, 1873, 132–35). Had herb. purchased by Rev. G. Gunn.
Hist. Berwick. Nat. Club v.13, 1891, 399–402. *Trans. Bot. Soc. Edinburgh* v.21, 1900, 279. *R.S.C.* v.7, 275; v.9, 367.
Plants at Kew, Oxford.

BROTHERSTON, Robert Pace (1848–1923)
b. Ednam, Berwickshire 7 Feb. 1848 *d.* Old Scone, Perth 21 Dec. 1923
Brother of A. Brotherston (1834–1891). Gardener to Earl of Haddington at Tyninghame, E. Lothian, 1874–1923. *Book of Carnation* 1904. *Book of Cut Flowers* 1906. *Stocks. Gardening in the North* (with S. Arnott) 1909. Contrib. to *Gdnrs Chron.*, *Garden.*
Gdnrs Mag. 1911, 637–38 portr. *Gdnrs Chron.* 1923 ii 126; 1924 i 13 portr. *Garden* 1924, 27. *J. Bot.* 1924, 149–50. *Bot. Soc. Exch. Club Br. Isl. Rep.* 1924, 537.

BROTHERTON, Thomas (*fl.* 1690s)
Experimented on descent of sap (*Philos. Trans. R. Soc.* v.16, 1688, 307–13).

BROUGHTON, Arthur (–1796 or 1803)
b. Bristol *d.* Kingston, Jamaica 30 July 1796/1803
MD Edinburgh 1779. Physician, Royal Infirmary, Bristol, 1780–86. To Jamaica, 1779. *Enchiridion Botanicum* 1782. *Hortus Eastensis* 1792; ed. 2 1794. Linnaean Index to H. Barham's *Hortus Americanus* 1794.
I. Urban *Symbolae Antillanae* v.1, 1898, 17–18. *D.N.B.* v.6, 459. *Bull. Agric. Dept. Jamaica* 1903, 155. J. W. White *Fl. Bristol* 1912, 66–67. *J. Bot.* 1915, 104–5. W. Fawcett and A. B. Rendle *Fl. Jamaica* v.3, 1914, v. *Proc. Bristol Nat. Soc.* v.32 (1), 53–57. H. J. Riddelsdell *Fl. Gloucestershire* 1948, cxiv.
Jamaica herb. at Bristol University.
Broughtonia R. Br.

BROUGHTON, Henry Rogers, 2nd Baron Fairhaven (1900–1973)
b. 1 Jan. 1900 *d.* April 1973
Major, Royal Horse Guards, 1919–33, 1939–45. Created garden at South Walsham Hall near Norwich. Assembled fine collection of botanical art which he bequeathed to the Fitzwilliam Museum, Cambridge. Collection catalogued by M. H. Grant in *Flower Paintings through Four Centuries* 1952.

BROUN, Alfred Forbes (1858–)
Conservator of Forests, Ceylon. Director of Woods and Forests, Sudan until 1910. Collected plants in Sudan, now at Kew. *Catalogue of Sudan Flowering Plants* 1906. *Silviculture in Tropics* 1912. *Fl. Sudan* (with R. E. Massey) 1929.
Kew Bull. 1907, 234.

BROWN, Adrian John (1852–1919)
b. Burton-on-Trent, Staffs 27 April 1852 *d.* 2 July 1919
MSc Birmingham 1901. FRS 1911. Professor of Biology and Chemistry of Fermentation, Director, School of Brewing, University of Birmingham, 1899–1919. Contrib. to *Ann. Bot.* 1907, 21.
Nature v.103, 1919, 369. *Proc. R. Soc.* v.93, 1922, iii–ix. *Who was Who, 1916–1928* 132.

BROWN, Alexander (*fl.* 1690s)
Ship's surgeon. Collected plants for L. Plukenet, J. Petiver and J. Bobart in Cape and other places.
R. Morison *Plantarum Historiae Universalis Oxoniensis* v.3, 1699, 561, 614. L. Plukenet *Almagesti Botanici Mantissa Plantarum* 1700, 69. R. Pulteney *Hist. Biogr. Sketches of Progress of Bot. in England* v.2, 1790, 62–63. *J. Bot.* 1905, 256. J. E. Dandy *Sloane Herb.* 1958, 98. H. N. Clokie *Account of Herb. Dept. Bot., Oxford* 1964, 139.
Plants at BM(NH), Oxford.
Eriocephalos bruniades Plukenet.

BROWN, Charles (–1836)
Acquired nursery at Slough started by Thomas Brown (–1814). Edward and Thomas Brown took control of nursery after him.
Gdnrs Chron. 1883 ii 134.

BROWN, Edward (*fl.* 1620s)
Apothecary. Accompanied T. Johnson on his visit to Kent in 1629.

BROWN, Edwin (1818–1876)
d. Tenby, Pembrokeshire 1 Sept. 1876
Of Burton-on-Trent. Entomologist. 'Flora of...Tutbury and Burton' (O. Mosley *Nat. Hist. Tutbury* 1863, 233–364).
Entomologist 1876, 240. *Entomol. Monthly Mag.* v.13, 1876–77, 116–17, 257–58. *J. Bot.* 1901 Supplt., 73. A. R. Horwood and C. W. F. Noel *Fl. Leicestershire* 1933, ccxvii–ccxviii.

BROWN, Ernest George Salt (1910–1939)
b. Edinburgh 31 Jan. 1910 *d.* 5 April 1939
BSc Edinburgh 1934. PhD 1936. Botanist, Kew Herb., 1937.
Kew Bull. 1939, 252.

BROWN, George (1797–1874)
b. Dalhousie, Midlothian 8 July 1797 *d.* Powis Castle, Montgomeryshire 24 Oct. 1874
Gardener to Earl of Powis at Powis Castle, 1829–74.
Gdnrs Chron. 1874 ii 626.

BROWN, George (1835–)
b. Barnard Castle, Durham 1835
Australian Wesleyan Methodist Mission, Samoa, 1860–75. Secretary, Foreign Missionary Society, Sydney, 1887–1908. Collected zoological specimens and a few plants in Bismark Archipelago. *Pioneer Missionary and Explorer; an Autobiography* 1908 portr.
Fl. Malesiana v.1, 1950, 82–83.
Plants at Melbourne.

BROWN, George Charles (1889–1969)
b. Colchester, Essex 1889 *d.* Watford, Herts Aug. 1969
Studied flora of Colchester area. Botanised in Essex with his friend G. C. Druce. Contrib. to *Essex Nat.*
S. T. Jermyn *Fl. Essex* 1974, 20.
Herb. at BM(NH).

BROWN, George Dransfield (*c.* 1828–1885)
d. Ealing, Middx 17 July 1885
MRCS. FLS 1876. Physician at Ealing. Interested in cryptogamic botany.
Proc. Linn. Soc. 1883–86, 138.

BROWN, H. (*fl.* 1900s)
Collected plants in Sudan, 1905, now at Kew.
Kew Bull. 1907, 234.

BROWN, Harold (*c.* 1863–1936)
d. 3 March 1936
On staff of Chemistry Dept., Imperial Institute. Did research on alkaloids of *Hyoscyamus muticus* and *Datura stramonium*. *Rubber; its Sources, Cultivation and Preparation* 1914.
Pharm. J. v.136, 1936, 300.

BROWN, Henry (1824–1892)
Timber merchant, Luton, Beds. Had herb. of plants collected in Luton area.
Mus. J. v.31, 1932, 554. J. G. Dony *Fl. Bedfordshire* 1953, 21.
Herb. at BM(NH).

BROWN, Horace Tabberer (1848–1925)
b. Burton-on-Trent, Staffs 20 July 1848 *d.* London 6 Feb. 1925
FCS. FRS 1889. FLS 1898. Brewer and bio-chemist; studied especially bio-chemistry of brewing and fermentation processes and chemistry of carbohydrates.
J. Bot. 1925, 85. *Kew Bull.* 1925, 96. *Nature* v.115, 1925, 307–8. *R.S.C.* v.7, 278.

BROWN, Isaac (1803–1895)
b. Amwellbury, Herts 1803 *d.* Kendal, Westmorland 3 Nov. 1895
School-master, Hitchin, Herts. Contrib. to R. H. Webb and W. H. Coleman's *Fl. Hertfordiensis* 1849.
Trans. Herts. Nat. Hist. Soc. v.9, 1898, liv–lv. F. Boase *Modern English Biogr.* v.1, 1908, 512. *Herts Express* 14 March 1925. J. G. Dony. *Fl. Bedfordshire* 1953, 21. J. G. Dony *Fl. Hertfordshire* 1967, 14.
Herb. formerly at Herts Natural History Society. Letters at BM(NH).

BROWN, James (*fl.* 1810s)
Nurseryman, Ranelagh Gardens, Leamington Spa, Warwickshire. In 1813 became partner to Mackie until John Cullis succeeded to the business in 1816.

BROWN, James Campbell (1843–1910)
b. Aberdeenshire 1843 *d.* Liverpool 14 March 1910
DSc London 1870. Professor of Chemistry, Liverpool, 1883–1910. Interested in botany.
Nature v.83, 1910, 102. *Proc. Liverpool Bot. Soc.* 1910–11, 5–6. *Who was Who, 1897–1916* 93.

BROWN, James Meikle (1875–1951)
b. London 1875 *d.* 24 March 1951
BSc London. FLS 1910. Teacher, High Storrs School, Sheffield. All-round naturalist with an interest in botany. President, Yorkshire Naturalists' Union, 1933.
Naturalist 1951, 147–48 portr. *Sorby Rec.* v.2 (2), 1966, 33–34.

BROWN, John (–1851)
d. Boston, Lincs 30 Jan. 1851
MD. FLS 1826. Physician, Boston. Keen botanist and gardener.
Proc. Linn. Soc. v.2, 1851, 132.

BROWN, John (–1873)
d. Edinburgh July 1873
In charge of West Princes Street Gardens, Edinburgh. Secretary, Edinburgh Naturalists' Field Club. Studied local flora. List of Edinburgh plants in *Trans. Bot. Soc. Edinburgh* v.11, 1873, 472–74; v.12, 1876, 19.
Trans. Bot. Soc. Edinburgh v.12, 1876, 19. *R.S.C.* v.7, 278.

BROWN, Rev. John Croumbie (1808–1895)
b. 16 May 1808 *d.* Haddington, E. Lothian 17 Sept. 1895

LLD Aberdeen 1858. FLS 1867. Congregational minister in Cape Town, 1844. Presbyterian minister, Aberdeen, 1849. Lecturer in Botany, Aberdeen, 1853. Colonial botanist at Cape, 1863. Professor of Botany, S. African College. *Pine Plantations on Sandwastes of France* 1878. *Introduction to Study of Modern Forest Economy* 1884. *Forests of England* 1883. *Forests and Moisture* 1877. *Management of Crown Forests of Cape of Good Hope*.

Gdn J. New York Bot. Gdn 1958, 170–71 portr.

Letters at Kew.

BROWN, John Ednie (1848–1899)
b. Scotland 1848 *d.* Perth, W. Australia 1899

FLS 1879. Visited Canada and U.S.A., 1871–72. Conservator of Forests, S. Australia, 1878. Director-General of Forestry, New South Wales, 1890–92. Conservator of Forests, W. Australia, 1895–99. *Timber Trees of S. Australia* 1880. *Practical Treatise on Tree Culture in S. Australia* 1881. *Forest Fl. of S. Australia* 1882–90. *Report on Forests of W. Australia* 1895.

Rep. Austral. Assoc. Advancement Sci. 1907, 174. *Austral. Nat.* 1962, 39–42. *R.S.C.* v.7, 279; v.9, 371.

Plants at Adelaide. Drawings at Kew.

BROWN, John Peter (1785–1842)
b. Oxfordshire 1785 *d.* Thun, Switzerland 17 July 1842

Colonel. Friend of R. J. Shuttleworth. Lived near Thun from 1824. *Catalogue des Plantes...de Thoune* 1843. 'Preservation of Botanical Specimens' (*Mag. Nat. Hist.* 1837, 311–15).

Verh. Schweiz. Nat. Ges. v.27, 257–59.

Plants at BM(NH).

BROWN, John William (1834–1859)
b. Otley, Yorks 1834 *d.* Otley 19 Aug. 1859

Had herb.

T. Parkinson *Lays and Leaves of Forest* 1882, 252. *Leeds Mercury* 18 May 1889.

BROWN, John Wright (1836–1863)
b. Edinburgh 19 Jan. 1836 *d.* Edinburgh 23 March 1863

Assistant, Herb. Royal Botanic Garden, Edinburgh. 'Plants...of Elie, Fife' (*Trans. Bot. Soc. Edinburgh* v.7, 1863, 430).

Trans. Bot. Soc. Edinburgh v.7, 1863, 519–20. *D.N.B.* v.7, 21.

BROWN, Rev. Littleton (1699–1749)
b. Bishop's Castle, Shropshire 1699

MA Oxon 1722. FRS 1729. Of Shropshire. Helped J. J. Dillenius in *Historia Muscorum* 1741, viii. Corresponded with Dillenius (G. C. Druce and S. H. Vines *Dillenian Herbaria* 1907, lxix–lxxvii).

J. E. Smith *Selection of Correspondence of Linnaeus* v.2, 1821, 145. D. Turner *Extracts from Lit. and Sci. Correspondence of Richard Richardson* 1835, 233. G. C. Druce *Fl. Northamptonshire* 1930, lxxiv. *Trans. Woolhope Nat. Field Club* v.34, 1954, 235–36.

BROWN, Maitland (1843–1905)
b. July 1843 *d.* Perth, Australia 1905

Resident Magistrate, Geralton, W. Australia. Collected plants on F. T. Gregory's expedition to N.W. Australia, 1861.

G. Bentham *Fl. Australiensis* v.1, 1863, 13*. *J. W. Austral. Nat. Hist. Soc.* no. 6, 1909, 10. *Handbook and Review Austral. Assoc. Advancement Sci.* 1926, 44–45.

Acacia maitlandi F. Muell.

BROWN, Mrs. Margaret Read (*fl.* 1860s)

Wife of Major-General J. Read Brown, Madras Cavalry. Drew illustrations for her book, *Wild Flowers of Southern and Western India* 1868.

Art J. 1866, 159.

Drawings at Kew.

BROWN, Nicholas Edward (1849–1934)
b. Redhill, Surrey 11 July 1849 *d.* Kew, Surrey 25 Nov. 1934

ALS 1879. Curator, W. Wilson Saunders's Museum, Reigate. Assistant, Kew Herb., 1873; Assistant Keeper, 1909–14. Authority on succulents, asclepiads and Cape plants. *Mesembryanthema* (with A. Tischer and M. C. Karsten) 1931. *Arachroidiscus* 1933. Completed vol. 12 of Boswell-Syme's edition of *English Botany* 1886. Contrib. to *Fl. Capensis*, *Fl. Tropical Africa*, *Kew Bull.*, *Gdnrs Chron.*, *J. Linn. Soc.*

J. Kew Guild 1904, 167 portr.; 1935, 460–61. *Gdnrs Chron.* 1922 i 190 portr.; 1932 i 268 portr.; 1934 ii 417. *Nature* v.134, 1934, 961. *Chronica Botanica* 1935, 168 portr. *J. Bot.* 1935, 19–21. *Kew Bull.* 1935, 58–61. *Proc. Linn. Soc.* 1934–35, 165–67. *Kakteenkunde* Heft 8, 1935, 159–60 portr. A. White *et al.* *Succulent Euphorbieae* v.1, 1941, 47–49 portr. A. White and B. L. Sloane *Stapelieae* v.1, 1937, 115–17 portr. H. Herre *Genera of Mesembryanthemaceae* 1971, 46 portr. *Cactus* (Brussels) 1972, 119–21.

Herb. and MSS. at Kew. Portr. at Hunt Library.

BROWN, Peter (*fl.* 1770s–90s)

Flower painter. May have studied flower painting under G. D. Ehret. Court painter to Prince of Wales. Exhibited at Royal Academy, 1770–91.

Kew Bull. 1916, 164.

Four flower paintings at Kew.

BROWN, Philip (–1779)
d. Manchester 1779

MD. Practised medicine in Manchester. Grew exotic plants. *Catalogue of very Curious Plants collected by late Philip Brown MD, lately Deceased, to be sold...at his Garden near Manchester* 1779.

D.N.B. v.7, 23–24. *Trans. Liverpool Bot. Soc.* 1909, 62.

BROWN, Robert (–*c.* 1837)

Nurseryman, Brompton Park, Kensington, London. A partner with James Gray and Son 1827–*c.* 1837.

BROWN, Robert (*c.* 1767–1845)
b. Perth? *c.* 1767 *d.* near Philadelphia, U.S.A. 20 Sept. 1845

Nurseryman, Perth. Found *Bryanthus taxifolius* 1812. Visited N. America with James McNab, 1834.

J. E. Smith *English Fl.* v.2, 1824, 222. *Trans. Hort. Soc.* v.4, 1822, 285–86. J. C. Loudon *Arboretum et Fruticetum Britannicum* v.1, 1838, 182. *Gdnrs Chron.* 1845, 755–56.

BROWN, Robert (1773–1858)
b. Montrose, Angus 21 Dec. 1773 *d.* London 10 June 1858

Studied medicine at Edinburgh. DCL Oxon 1832. FRS 1811. ALS 1798. FLS 1822. Surgeon's mate, Fifeshire. Naturalist on Flinders's Australasian Expedition, 1801–5. Clerk and Librarian to Linnean Society, 1805–22. Librarian to Sir J. Banks, 1810–20. Banks bequeathed him a life-interest in his library and collections which Brown handed over to BM in 1827 when he was made first Keeper of Botany Dept., BM. President, Linnean Society, 1849–53. Made several discoveries including "Brownian movement". "Botanicorum facile princeps" (Humboldt). *Prodromus Florae Novae Hollandiae* 1810 (reprinted 1960 with biogr. introduction); *Supplementum* 1830. *Miscellaneous Botanical Works* 1866–67 2 vols.

Cottage Gdnr v.20, 1858, 176–77. *Belgique Hort.* v.7, 1858, i–xii portr. *Gdnrs Chron.* 1858, 493–94, 701, 732–33; 1901 i 176–78; 1945 i 136–37. *Times* 17 June 1858. *Ann. Mag. Nat. Hist.* 1859, 321–31. *Flora* 1859, 10–15, 25–31. *Proc. Linn. Soc.* 1859, xxv–xxx; 1887–88, 54–67; 1888–89, 34; 1931–32, 17–54 portr. *Proc. R. Soc.* v.9, 1859, 527–32. *Trans. Bot. Soc. Edinburgh* v.6, 1860, 119–28. *J. Bot.* 1896, 26–28; 1922, 177–84; 1932, 13–16. *D.N.B.* v.8, 25–27. C. J. F. Bunbury *Life and Letters and Journals* 1906, 145–49. *Rep. Austral. Assoc. Advancement Sci.* 1907, 161–67. J. H. Maiden *Sir Joseph Banks* 1909, 100–23 portr. R. S. Rogers *Introduction to Study of S. Austral. Orchids* 1911, 40–47 portr. F. W. Oliver *Makers of Br. Bot.* 1913, 108–25. J. R. Green *Hist. Bot.* 1914, 309–35. *Nat. Hist. Mag.* v.1, 1928, 158–62. *Papers Proc. R. Soc. Tasmania* 1929, 25–32. *Amer. Fern J.* v.40, 1950, 83–88. *Fl. Malesiana* v.1, 1950, 83–84 portr. *Proc. R. Soc. Queensland* 1954, 6–8. *Muelleria* 1955, 46–54. *Austral. J. Sci.* 1958, 127–30. *Austral. Orchid. Rev.* v.29, 1964, 16–19, 124–27. *Austral. Dict. Biogr.* v.1, 1966, 166–67. F. A. Stafleu *Taxonomic Literature* 1967, 54–55. C. C. Gillispie *Dict. Sci. Biogr.* v.2, 1970, 516–22. M. A. Reinikka *Hist. of the Orchid* 1972, 130–35 portr. *Taxon* 1972, 104 portr. *J. S. African Bot.* 1974, 47–60 portr.

Herb., diaries and letters at BM(NH). Plants and letters at Kew. Plants at Edinburgh. Portrs. at BM(NH), Kew (S. Pearce), Linnean Society (H. W. Pickergill). Bust by P. Slater at Linnean Society. Portr. at Hunt Library.

Brunonia Smith.

BROWN, Robert (*c.* 1824–1906)
d. St. Albans, Christchurch, New Zealand 13 Dec. 1906

Shoemaker. Bryologist. Described and figured mosses in *Trans. New Zealand Inst.* from 1892.

Trans. New Zealand Inst. v.36, 1903, 345–46. *J. Bot.* 1907, 126. *R.S.C.* v.13, 849.

BROWN, Robert (1839–1901)
b. Liverpool 27 Sept. 1839 *d.* Liverpool 6 April 1901

Book-keeper. President, Liverpool Field Club, 1896–98. 'Flintshire Plants' (*J. Bot.* 1875, 178; 1885, 357–60). Edited appendices to *Fl. Liverpool* 1875 and 1887. Contrib. to J. B. L. Warren's *Fl. Cheshire* 1899. 'Botany of Liverpool District' (*Handbook Br. Assoc., Liverpool* 1896, 73–87).

J. Bot. 1902, 236. *Trans. Liverpool Bot. Soc.* 1909, 62. *Handbook Guide Herb. Collections... Liverpool* 1935, 19–21 portr. *R.S.C.* v.13, 849.

Herb. at Liverpool Museums.

BROWN, Robert (1842–1895)
b. Campster, Caithness 23 March 1842 *d.* Streatham, Surrey 26 Oct. 1895

BA Edinburgh 1860. PhD Rostock. FLS 1873. Botanist to British Columbia Expedition, 1863–66. Collected plants in Greenland, 1867. Lecturer in Natural History, High School and Heriot-Watt College, Edinburgh, 1869–76. To London, 1876. 'Florula Discoana' (*Trans. Bot. Soc. Edinburgh* v.9, 1868, 430–64). *Manual of Botany* 1874.

Gdnrs Chron. 1863, 989; 1872, 464–65, 573–74, 636–37; 1874, 52–54. *Trans. Bot. Soc. Edinburgh* v.11, 1873, 322–38. *J. Bot.* 1895, 384. *Proc. Linn. Soc.* 1895–96, 34–35, C. S. Sargent *Silva of North America* v.8, 1895, 62. *D.N.B. Supplt. 1* v.1, 302–3. *R.S.C.* v.1, 661; v.7, 279; v.9, 371; v.13, 848.

Greenland plants at Kew.

Lecidea campsteriana Lindsay.

BROWN, Robert (–1949)
Herb. at Glasgow.

BROWN, Robert Drummond (1875–1936)
b. Liverpool 17 May 1875 *d.* West Kirby, Cheshire 2 Aug. 1936
Son of Robert Brown (1839–1901). Employee of Liverpool Insurance Company. President, Liverpool Naturalists' Field Club, 1936. Botanist.
N. Western Nat. 1937, 410.

BROWN, Robert N. (–*c.* 1862)
b. Scotland *d.* India *c.* 1862
Pupil of J. H. Balfour. Superintendent, Agri-horticultural Society Garden, Madras, 1857–62. *Hand Book of Trees, Shrubs and Herbaceous Plants...in Madras Agri-horticultural Society's Garden* 1862; ed. 2 1866.

BROWN, Robert Neal Rudmose (1879–1957)
b. London 13 Sept. 1879 *d.* 27 Jan. 1957
BSc Aberdeen 1900. Botanist to Scottish National Antarctic Expedition, 1902. Head, Geography Dept., Sheffield University, 1908; Professor of Geography, 1931. Expeditions in 1909, 1912, 1914, 1919, 1920. *Naturalist at the Poles* 1923. *Polar Regions* 1927.
Scott. Geogr. Mag. 1957, 123–24.
Antarctic plants at Edinburgh and Kew.

BROWN, T. (*fl.* 1820s)
Nurseryman, Measham near Ashby-de-la-Zouch, Leics. Raised 'Queen Caroline' apple, *c.* 1820.

BROWN, Thomas (1785–1862)
b. Perth 1785 *d.* Manchester 8 Oct. 1862
FLS 1816. Captain, Forfarshire Militia. Curator, Manchester Museum, 1838–62. Added notes, etc. to G. White's *Selborne* 1833. Contrib. to G. Johnston's *Fl. Berwick-upon-Tweed* 1829–30 2 vols.
Hist. Berwickshire Field Club v.1, 1834, 9–10. *Mem. Manchester Lit. Philos. Soc.* v.85, 1944.

BROWN, Thomas (*fl.* 1790s–1800s)
MD Edinburgh 1798. Taught botany at Glasgow University, 1799–*c.* 1808.
J. Sowerby and J. E. Smith *English Bot.* 1591. *Notes R. Bot. Gdn Edinburgh* v.3, 1904, 57.

BROWN, Thomas (1804–1886)
b. Slough, Bucks 27 July 1804 *d.* Honolulu, Hawaii 22 Oct. 1886
Nurseryman at Slough in partnership with his brother Edward who retired in 1837. Thomas Brown sold business in 1840 to W. Cutter and emigrated to Honolulu *c.* 1846. About 1844 the nursery was bought by Charles Turner.
Garden v.30, 1886, 598. *Gdnrs Chron.* 1886 ii 817.

BROWN, Thomas (*fl.* 1820s)
Nurseryman, Bedford Nursery, New Road, St. Pancras, London.

BROWN, Rev. Thomas (1811–1893)
b. Langton, Berwickshire 23 April 1811 *d.* Edinburgh 4 April 1893
DD Edinburgh 1888. FRSE 1861. Moderator of Free Church Assembly 1890. Botany of Langton in *New Statistical Account of Scotland* 1834.
Hist. Berwickshire Nat. Club v.14, 1893, 339–46. *Proc. R. Soc. Edinburgh* 1894–95, xxix–xxxv. *R.S.C.* v.1, 662; v.7, 280; v.9, 372.

BROWN, Thomas William (–*c.* 1951)
Kew gardener, 1899. Assistant Curator, Botanic Station, Aburi, Gold Coast, 1899–1902. Gardener to Sultan of Morocco at Fez. Director of Horticulture, Giza, Egypt.
Kew Bull. 1899, 221; 1902, 23. C. S. Jarvis *Desert and Delta* 1938, 90–93.
Egyptian plants at Kew.

BROWN, William (*fl.* 1700s)
Surgeon. Sent plants to J. Petiver from Virginia, Gibraltar and the Cape.
J. E. Dandy *Sloane Herb.* 1958, 99.

BROWN, William (*c.*1761–*c.* 1793)
d. Pitcairn Island
Gardener. With David Nelson on HMS 'Bounty', 1787 to take care of bread fruit trees which were collected in Pacific. Joined mutineers at time of celebrated mutiny and subsequently died during uprising of natives on Pitcairn Island.
W. Bligh *Narrative of Mutiny on Board His Majesty's Ship Bounty...* 1790, 7.

BROWN, William (*c.* 1838–1893)
d. 13 Jan. 1893
Nurseryman, Richmond, Surrey.
Gdnrs Mag. v.36, 1893, 42.

BROWN, William (1888–1975)
b. Middlebie, Dumfriesshire 1888 *d.* Stalybridge, Cheshire 19 Jan. 1975.
MA Edinburgh 1908. DSc London 1916. FRS 1938. Lecturer in Plant Physiology, Edinburgh, 1910. Associate Professor of Plant Pathology, Imperial College, London, 1923; Professor, 1928–53; Head of Dept. of Botany, 1938. President, British Mycological Society, 1934. President, Association of Applied Biology, 1955.
Bull. Br. Mycol. Soc. v.9, 1975, 58. *Trans. Br. Mycol. Soc.* 1975, 343–45 portr.

BROWN, William Henry (*fl.* 1860s)
North Shields plants at Botanic Gardens, Brussels.

BROWN, William Lindsay (1842–1900)
b. Kirkcudbright 14 Oct. 1842 *d.* 26 July 1900
FLS 1891. Travelled extensively and collected plants in N. America, Europe and Middle East. "A good botanist...left a very fine collection of plants...never had time to publish anything."
Proc. Linn. Soc. 1900–1, 42.

BROWNE, Charles (1821–1895)
b. London 20 April 1821 *d.* 1 Nov. 1895
MA Oxon 1845. 'Abnormal Forms of Vegetation' (*Essex Nat.* v.3, 1889, 168).
J. Foster *Men-at-the-Bar* 1885, 60.
Teratological herb. at Museum, West Ham.

BROWNE, Edward (1644–1708)
b. Norwich, Norfolk 1644 *d.* Northfleet, Kent 28 Aug. 1708
MD Oxon 1667. MD Cantab 1670. FRS 1667. Eldest son of Sir Thomas Browne (1605–1682). Travelled in Europe, 1664–73. *Brief Account of Some Travels in...Europe* 1685.
W. Munk *Roll of R. College of Physicians* v.1, 1878, 372–77. *D.N.B.* v.7, 42–43. E. J. L. Scott *Index to Sloane Manuscripts* 1904, 80. J. E. Dandy *Sloane Herb.* 1958, 99.
Plants at BM(NH).

BROWNE, Lady Isabel Mary Peyronnet (1881–1947)
b. London 6 Nov. 1881 *d.* 8 June 1947
Educ. London University. FLS 1909. Contrib. papers on *Equisetum* to *Ann. Bot.* and *Bot. Gazette.*
Proc. Linn. Soc. 1946–47, 154–58.

BROWNE, Patrick (*c.* 1720–1790)
b. Woodstock, County Mayo *c.* 1720 *d.* Rushbrook, County Mayo 29 Aug. 1790
MD Leyden 1743. In Antigua, 1737; Jamaica, 1746–55. *Civil and Natural History of Jamaica* 1756; ed. 2 1789.
R. Pulteney *Hist. Biogr. Sketches of Progress of Bot. in England* v.2, 1790, 349. *Trans. Linn. Soc.* v.4, 1818, 31–34. J. E. Smith *Selection of Correspondence of Linnaeus* v.1, 1821, 42–44. P. Smith *Correspondence of Sir J. E. Smith* v.2, 1832, 126. *Bot. Mag.* 1842, t.3964. *Notes and Queries* 1852, 518; 1858, 310; 1865, 316. *Proc. Linn. Soc.* 1894–95, 55. *D.N.B.* v.7, 53. I. Urban *Symbolae Antillanae* v.1, 1898, 18–28; v.3, 1902, 29. *J. Bot.* 1902, 139; 1912, 129; 1924, 351. J. M. Hulth *Bref. och Skriv Linné* v.2 (1), 1916, 339–62. R. L. Praeger *Some Irish Nat.* 1949, 55. F. A. Stafleu *Taxonomic Literature* 1967, 56. *Taxon* 1970, 517.
Plants and MSS. ('Fasciculus Pl. Hiberniae', 'Fl. Indiae Occidentalis', 'Cat. Pl. Jamaica') at Linnean Society. MS. 'Catalogue of Plants of English Sugar Colonies' and drawings of British plants at BM(NH).
Brownea Jacq.

BROWNE, Robert (*fl.* 1780s)
Gardener to Sir H. Harbord at Gunton, Norfolk. *Method to Preserve Peach and Nectarine Trees from the Effects of Mildew* 1786.
J. C. Loudon *Encyclop. Gdning* 1822, 1278.

BROWNE, Robert Clayton– *see* Clayton-Browne, R.

BROWNE, Samuel (–1698)
d. Madras, India 22 Sept. 1698
MD. Surgeon, East India Company, Madras, 1688. Sent plants to J. Petiver. Contrib. accounts of Indian plants to *Philos. Trans. R. Soc.* v.20–v.23, 1698–1702.
R. Pulteney *Hist. Biogr. Sketches of Progress of Bot. in England* v.2, 1790, 38, 39, 62–63. J. E. Smith *Selection of Correspondence of Linnaeus* v.2, 1821, 165. D. Turner *Extracts from Lit. and Sci. Correspondence of Richard Richardson* 1835, 76. E. J. L. Scott *Index to Sloane Manuscripts* 1904, 81. D. G. Crawford *Hist. Indian Med. Service, 1600–1913* v.1, 1914, 88–93 *passim.* J. E. Dandy *Sloane Herb.* 1958, 99–102. I. H. Burkill *Chapters on Hist. Bot. in India* 1965, 8–9.
Plants at BM(NH).

BROWNE, Sir Thomas (1605–1682)
b. London 19 Oct. 1605 *d.* Norwich, Norfolk 19 Oct. 1682
BA Oxon 1627. MD Oxon 1637. Knighted, 1671. Physician and author. *Garden of Cyrus* 1658. 'Observations upon Several Plants Mention'd in Scripture' (*Miscellany Tracts* 1683).
Trans. Linn. Soc. v.7, 1804, 296. S. Felton *Portraits of English Authors on Gdning* 1830, 94–97. *Cottage Gdnr* v.5, 1850, 15. W. Munk *Roll of R. College of Physicians* v.1, 1878, 321–27. *D.N.B.* v.7, 64–72. *Trans. Norfolk Norwich Nat. Soc.* v.7, 1899–1900, 72–89. E. J. L. Scott *Index to Sloane Manuscripts* 1904, 81. G. Keynes *Bibl. of Sir Thomas Browne* 1924 portr. *Osiris* 1936, 28–79. J. S. Finch *Sir Thomas Browne* 1950. *Isis* 1956, 161–71. J. E. Dandy *Sloane Herb.* 1958, 102. J. Bennett *Sir Thomas Browne* 1962. F. L. Huntley *Sir Thomas Browne* 1962. R. Cawley and R. G. Yost *Studies in Sir Thomas Browne* 1965.
Plants at BM(NH). Portr. at St. Peter Mancroft, Norwich, Royal College of Physicians.

BROWNE, Rev. William (*c.* 1628–1678)
b. Oxford *c.* 1628 *d.* Oxford 25 March 1678
BA Oxon 1647. BD 1665. Assisted R. Plot (*Natural History of Oxfordshire* 1677, 150). "Peritissimus botanicus" (J. Ray). *Catalogus Horti Botanici Oxoniensis* (with P. Stephens) 1658.
J. Ray *Synopsis* 1724, 265, 373, 437. R. Pulteney *Hist. Biogr. Sketches of Progress of Bot. in England* v.1, 1790, 167. A. Wood *Athenae Oxonienses* v.2, 1813–20, 282. *D.N.B.* v.7, 75. G. C. Druce *Fl. Berkshire* 1897, cvi–cvii. R. T. Gunther *Early Br. Botanists* 1922, 298–302. G. C. Druce *Fl. Oxfordshire* 1927, lxix–lxxiii.

BROWNE, William Alexander Francis Balfour- *see* Balfour-Browne, W. A. F.

BROWNE, Rev. William Bevil (1845–1928)
d. Exeter, Devon 1928
MA. Curate, South Huish, then Salcombe, Devon. Vicar, Churchstow, 1889–95. Retired to Sheldon. Contrib. plant lists to J. Fairweather's *Salcombe...and Neighbourhood c.* 1897.
W. K. Martin and G. T. Fraser *Fl. Devon* 1939, 776.

BROWNJOHN, John (*fl.* 1790s)
Seedsman, Salisbury, Wilts.

BROWNLEE, Rev. James (*fl.* 1840s)
Missionary, King William's Town, S. Africa. Sent plants to W. H. Harvey.
London J. Bot. 1842, 16. *Dict. S. African Biogr.* v.1, 1968.
Brownleea Harvey ex Lindl.

BROWNLOW, Lady (*fl.* 1810s)
Daughter of Lady Amelia Hume: "by whose premature decease botany has lost one of her best and most powerful patronesses" (W. Roxburgh *Plants of Coast of Coromandel* v.3, 1819, 61).
Brownlowia Roxb.

BROWNRIGG, William (1711–1800)
b. High Close Hall, Cumberland 24 March 1711 *d.* Ormathwaite, Keswick 6 Jan. 1800
MD Leyden 1735. FRS 1742. Practised medicine at Whitehaven, Cumberland. Sent mosses to J. J. Dillenius (*Historia Muscorum* 1741, viii). Friend of Sir H. Sloane and S. Hales.
J. Dixon *Lit. Life of W. Brownrigg* 1801. *D.N.B.* v.7, 85–86.
MS. at Royal Society.

BRUCE, Alexander James Adams (*c.* 1843–)
Nurseryman, Edge Lane Nursery, Chorlton-cum-Hardy, Lancs, 1881.
J. Hort. Home Farmer v.64, 1912, 73–74 portr.

BRUCE, Arthur (*c.* 1725–1805)
Land surveyor. Secretary, Natural History Society, Edinburgh. Discovered *Eriophorum pubescens*. Contrib. to James Sowerby and J. E. Smith's *English Bot.* 128, 1908, etc.
P. Smith *Correspondence of Sir J. E. Smith* v.1, 1832, 432–33. G. Johnston *Terra Lindisfarnensis* 1853, 196–97. *J. Bot.* 1863, 359–60.
Plants at BM(NH).

BRUCE, Eileen Adelaide (1905–1954)
b. 15 Feb. 1905 *d.* Kingston-upon-Thames, Surrey 6 Oct. 1954
BSc London. Assistant botanist, Herb., Kew, 1930. National Herb., Pretoria, 1945–52. Herb., Kew, 1952–54. Contrib. to *Kew Bull.*, *Flowering Plants of Africa*, *Fl. Tropical E. Africa*.
Kew Bull. 1956, 39–40. *Bothalia* 1968, 452 portr.
Portr. at Hunt Library.

BRUCE, George (*fl.* 1740s–1760s)
Nurseryman, Dublin. Succeeded by his children, George and Elizabeth.
Irish For. 1967, 46.

BRUCE, Henry (*fl.* 1830s)
Plant collector in India for N. Wallich.
N. Wallich *List of Specimens of Indian Plants* 1828, 24.

BRUCE, James (1730–1794)
b. Kinnaird, Stirling 14 Dec. 1730 *d.* Kinnaird 27 April 1794
FRS 1776. African explorer. Travelled in Abyssinia, 1769–71. Plants collected by him went mainly to Sir J. Banks and J. Lamarck. *Travels to Discover Source of Nile in...1768–1773* 1790 5 vols.; ed. 2 1805 (with biogr. and portr.)
P. Smith *Correspondence of Sir J. E. Smith* v.2, 1832, 293. J. Nichols *Illus. Lit. Hist. of Eighteenth Century* v.7, 4; v.8, 273–75. *D.N.B.* v.7, 98–102. *J. Bot.* 1913, 256. *Atti Accad. Ital. Rendic. Cl. Sci. Fis. Mat. Natur* v.2, 1941, 439–96. F. A. Stafleu *Taxonomic Literature* 1967, 56. J. M. Reid *Traveller Extraordinary* 1968, portr.
Plants, MSS. and drawings at Royal Society. Letters at BM(NH).
Brucea J. F. Mill.

BRUCE, Joseph (*fl.* 1790s)
Gardener and seedsman, Dedham, Essex.

BRUCE, William Spiers (1867–1921)
b. 1 Aug. 1867 *d.* Edinburgh 28 Oct. 1921
FRSE. Polar explorer. Organised Scottish National Antarctic Expedition in 'Scotia' to Weddell Sea where he collected plants now at Botanic Garden, Edinburgh.
Proc. R. Soc. Edinburgh v.42, 1921–22, 362–68.

BRÜHL, Paul (1855-)
b. Weifa, Saxony 25 Feb. 1855
Travelled in Turkey, Asia Minor and Armenia, 1878–81. Lecturer and Professor of Natural Sciences, Rajshahi College, Sibpur, 1882–1912. Professor of Botany, Calcutta, 1918–28. *Guide to Orchids of Sikkim* 1926. Contrib. to *Ann. Bot. Gdn Calcutta.*
Who was Who, 1929–1940 179.

BRUNKER, James Ponsonby (1885–1970)
Of Dublin. Botanist and ornithologist. Studied flora of County Wicklow.
R. L. Praeger *Some Irish Nat.* 1949, 55.
Herb. in possession of H. J. Hudson, Dublin.

BRUNNING, George (1830–1893)
b. Lowestoft, Suffolk 1830 *d.* St. Kilda, Australia 5 July 1893
Gardener. To Australia, 1853. Nurseryman, St. Kilda, Victoria, 1860.
Austral. Dict. Biogr. v.3, 1969, 279–80.

BRUNTON, John (*c.* 1721–1803)
d. Halesowen, Worcs April 1803
Nurseryman, 83 High Street, Birmingham with nursery at Perry Hill, Halesowen. *Catalogue of Plants Botanically Arranged According to System of Linnaeus...sold by John Brunton* 1777. Contrib. to W. Withering's *Bot. Arrangement of Br. Plants* ed. 2, v.1, 1787 (preface xi, 228). Had herb. which passed to Sir J. E. Smith. J. A. Forbes and Hunter later became partners in the nursery.
Gent. Mag. 1803 i 382.

BRUNTON, William (1775–1806)
b. 21 Oct. 1775 *d.* Ripon, Yorks 23 June 1806
FLS 1806. "Devoted much of his time to botanical and chemical pursuits". Found *Hypnum squarrulosum.* Contrib. to J. Sowerby and J. E. Smith's *English Bot.*
D. Turner and L. W. Dillwyn *Botanist's Guide through England and Wales* 1805, 663. *Gent. Mag.* 1807 i 376. D. Turner *Extracts from Lit. and Sci. Correspondence of Richard Richardson* 1835, 279.
Herb. at York Museum. Letters at Linnean Society.
Dicranum bruntoni Smith.

BRYAN, George Hartley (1864–1928)
b. Cambridge 1 March 1864 *d.* Bordighera, Italy 13 Oct. 1928
Professor of Pure and Applied Mathematics, University College, Bangor, 1896–1926. Microscopist, interested in diatoms and desmids.
Nature v.122, 1928, 849–50.

BRYANT, Alfred Thomas (*fl.* 1880s–1910s)
Malayan Civil Service, 1883–1918. Collected some plants in the Dindings, Malaya, 1890.
Gdns Bull. Straits Settlements 1927, 118. *Fl. Malesiana* v.1, 1950, 86.
Plants at Singapore.

BRYANT, Charles (–1799)
d. Norwich, Norfolk 11 Aug. 1799
Brother of Rev. H. Bryant (1721–1799). *Historical Account of Two Species of Lycoperdon* 1782. *Fl. Diaetetica* 1783. *Dictionary of Ornamental Trees, Shrubs and Plants... Cultivated in ...Great Britain* 1790.
Trans. Linn. Soc. v.7, 1804, 299–300. *Trans. Norfolk Norwich Nat. Soc.* 1914, 649; 1937, 196–98.
Sale at Sotheby, 8 March 1805.

BRYANT, Frederick Beadon (1859–1922)
d. 28 Nov. 1922
Indian Forest Service, 1881; Inspector-general of Forests, 1908–13. Collected plants in N.W. Provinces, 1890.
Nature v.110, 1922, 882–83.
Plants at Oxford.

BRYANT, Rev. Henry (1721–1799)
d. Colby, Norfolk 4 June 1799
BA Cantab 1749. ALS 1795. Rector, Colby. Discovered *Tillaea muscosa. Particular Enquiry into...Disease in Wheat...called Brand* 1784. Contrib. to J. Sowerby and J. E. Smith's *English Bot.* 221, etc.
Trans. Linn. Soc. v.7, 1804, 297–98. P. Smith *Correspondence of Sir J. E. Smith* v.1, 1832, 33–34; v.2, 475. *Trans. Norfolk Norwich Nat. Soc.* 1874–75, 27–29; 1914, 648. *D.N.B.* v.7, 155.

BRYCE, George (1885-)
b. Glasgow 1885
DSc Edinburgh 1922. Assistant Mycologist and Botanist, Dept. of Agriculture, Ceylon, 1913. Director of Agriculture, New Guinea, 1923–26. Dept. of Agriculture, Malaya, 1926–29, 1946- . Collected plants in New Guinea.
Kew Bull. 1913, 358. *Fl. Malesiana* v.1, 1950, 86.
New Guinea plants at Kew.

BRYCE, James, 1st Viscount Bryce (1838–1922)
b. Belfast 10 May 1838 *d.* 22 Jan. 1922
DCL. LLD. FRS 1893. Scholar, statesman and traveller. Regius Professor of Civil Law, Oxford, 1870. Collected plants in Western N. America, 1884 and in S. Africa, 1896. Contrib. account of flora of Arran to his father's *Geology of Clydesdale and Arran* 1859.
Bot. Soc. Exch. Club Br. Isl. Rep. 1922, 693–97. *Nature* v.109, 1922, 113–14. *Belfast Nat. Hist. Soc. Centenary Vol.* 1924 65–66. *D.N.B. 1922–1930* 127–35. *Who was Who, 1916–1928* 141–42. *Lesotho: Basutoland Notes and Records* v.5, 1965–66, 24, 26.
American and African plants at Kew.

BUCHAN, Alexander (1829–1907)
b. Kinnesswood, Kinross 11 April 1829 *d.* Edinburgh 13 May 1907

MA Edinburgh. LLD Glasgow. FRS 1898. Meteorologist. 'Plants...of Blackford' (*Trans. Bot. Soc. Edinburgh* v.5, 1858, 162). Presidential Address (*Trans. Bot. Soc. Edinburgh* v.11, 1871, 261–74).

Trans. Bot. Soc. Edinburgh v.23, 1908, 358–61. *D.N.B. Supplt. 2* v.1, 243–44. *R.S.C.* v.7, 289; v.9, 385; v.13, 884.

BUCHAN-HEPBURN, Sir Archibald (1852–1929)
d. 17 May 1929

Judge. Hybridised *X Orchis hepburnii*.

Bot. Soc. Exch. Club Br. Isl. Rep. 1929, 93. *Br. Fern Gaz.* 1929, 268.

BUCHANAN, Angus (1886–1954)
b. Kirkwall, Orkney 5 May 1886 *d.* 5 Feb. 1954

Collected plants in French Sudan, 1920, now at BM(NH).

Who was Who, 1951–1960 152.

BUCHANAN, Francis *afterwards* **Hamilton** *see* Hamilton, F.

BUCHANAN, Isaac (1808–1893)
b. Cardross, Perthshire

Gardener, Botanic Garden, Edinburgh and at Chiswick. To Philadelphia, 1836. Nurseryman, New York.

J. Hort. Cottage Gdnr v.53, 1906, 547. *Gdnrs Mag.* 1907, 48.

BUCHANAN, James (*fl.* 1790s–1820s)

Gardener and florist, Kew. Nurseryman, 7 Bowyer Place, Camberwell Road, 1795. In partnership with Henry Oldroyd 1812. "Very industrious in collecting rare plants, and obligingly communicative to scientific inquirers" (*Bot. Mag.* 1806, t.913, 966).

Trans. London Middlesex Archaeol. Soc. v.24, 1973, 181.

BUCHANAN, John (1819–1898)
d. New Zealand 18 Oct. 1898

FLS 1880. In New Zealand from 1859. Botanist to Geological Survey, Otago, 1862; to Geological Survey, New Zealand 1866. *Indigenous Grasses of New Zealand* 1880. 'Botany of Otago' (*Trans. N.Z. Inst.* 1868, 22–53).

J. Bot. 1869, 331–32. *Trans. Proc. N.Z. Inst.* 1898, 718–19, 744. *Proc. Ann. Rep. Wellington Philos. Soc.* 1898–99, 9–10. T. F. Cheeseman *Manual N.Z. Fl.* 1906, xxvii–xxviii. R. Glenn *Bot. Explorers of New Zealand* 1950, 135–38. *Tuatara* 1970, 66–67. *R.S.C.* v.7, 291; v.9, 385; v.12, 129; v.13, 886.

Plants, MSS. and drawings at Otago University. Plants at Kew.

Ranunculus buchanani Hook. f.

BUCHANAN, John (1855–1896)
b. Muthill, Perth 1855 *d.* Chinde, Zambesi Valley 9 March 1896

Founded Mission Station, Zomba, Nyasaland, 1876. Vice-Consul, Zomba, 1877. Started coffee plantation, 1881. Pioneered tobacco industry in 1890s. Collected plants in Ascension Island, Nyasaland and Rhodesia.

Kew Bull. 1891, 183–90; 1892, 249; 1896, 148. *Central African Planter* 1896, 99. *J. Bot.* 1896, 192. *Comptes Rendus AETFAT 1960* 1962, 163–64.

Plants and letters at Kew. Plants at BM(NH).

Albuca buchanani Baker.

BUCHANAN, Rev. John (*fl.* 1860s–1880s)

Minister, Free Church of Scotland, Bothwell. Minister at Durban, 1861–74. Collected plants in Natal. *Revised List of Ferns of Natal* 1875.

Kew Bull. 1901, 12.

Plants at Kew, Liverpool Museums and Oxford.

BUCHANAN, Robert MacNeill (1861–1931)

Herb. at Glasgow.

BUCHANAN, William Cullen (1887–1964)
b. Bearsden, Dunbartonshire 27 Jan. 1887 *d.* Bearsden 13 Sept. 1964

Interested in ferns and alpine plants. 'William Buchanan lecture' inaugurated by Scottish Rock Garden Club. Contrib. to *J. Scott. Rock Gdn Club.*

Br. Fern Gaz. 1964, 194–95.

BUCHANAN, William J. (*fl.* 1830s–1840s)

Succeeded to nursery of J. Buchanan and H. Oldroyd at 7 Bowyer Place, Camberwell Road, London.

BUCKELL, Francis (1818–1897)
b. Chichester, Sussex 28 May 1818 *d.* Romsey, Hants 3 April 1897

Medical Officer, Romsey. Contrib. to F. Townsend's *Fl. Hampshire* 1883, xvi.

Drawings and notes on pollen at BM(NH).

BUCKELL, William Robert (1856–1956)
b. Romsey, Hants 19 May 1856 *d.* Coyichan, Canada 2 Jan. 1956

FRCS. Cheltenham surgeon. Discovered *Ranunculus ophioglossifolius* near Cheltenham in 1890.

H. J. Riddelsdell *Fl. Gloucestershire* 1948, cxl.

BUCKINGHAM, E. B. (*fl.* 1810s–1830s)

Nurseryman in partnership with Chandler at Vauxhall Nursery, Lambeth, London.

BUCKLAND, Rev. William (1784–1856)
b. Axminster, Devon 12 March 1784 *d.* 15 Aug. 1856.

BA Oxon 1805. DD 1825. FRS 1818. FLS 1818. Professor of Mineralogy, Oxford, 1813. Dean of Westminster, 1845–56. President, Geological Society, 1824 and 1840. 'On Cycadeoideae' (*Geol. Trans.* v.2, 1829, 395–401).

Proc. R. Soc. 1856–57, 264–69. *Quart. J. Geol. Soc.* 1857, xxvi–xlv. E. O. Gordon *Life and Correspondence of William Buckland* 1894. *D.N.B.* v.7, 206–8. *Nature* v.178, 1956, 290–91. C. C. Gillispie *Dict. Sci. Biogr.* v.2, 1970, 566–72.

Letters at BM(NH), Kew. Portr. at Geological Society; bust at National Portrait Gallery. Portr. in Ipswich Museum Series, 1852. *Bucklandia* Brongn.

BUCKMAN, James (1814–1884)
b. 20 Nov. 1814 *d.* Bradford Abbas, Dorset 23 Nov. 1884

FLS 1850. Professor, Cirencester Agricultural College, 1848–63. Farmed in Dorset, 1863–84. *Botanical Guide to Environs of Cheltenham* 1844. *Natural History of British Meadow and Pasture Grasses* 1858. *Agricultural Dodders* 1876.

Proc. Linn. Soc. 1884–85, 104. *Quart. J. Geol. Soc.* 1884–85, 43–44. *Dorset Nat. Hist. Antiq. Field Club Proc.* 1886, 1–4. *D.N.B.* v.7, 216. H. J. Riddelsdell *Fl. Gloucestershire* 1948, cxxiii–cxxiv. *R.S.C.* v.1, 705; v.6, 611; v.7, 298; v.9, 393; v.12, 130.

Portr. at Hunt Library.

BUCKNALL, Cedric (1849–1921)
b. Bath, Somerset 2 May 1849 *d.* Clifton, Bristol 12 Dec. 1921

Mus B Oxon 1873. Organist at Clifton, 1876–1921. Collected plants in Spain and Portugal. Discovered *Stachys alpina* in Britain. 'Fungi of Bristol' (*Proc. Bristol Nat. Soc.* v.3–v.6, 1878–91. 'Revision of *Symphytum*' (*J. Linn. Soc.* v.41, 1913, 491–555). 'British *Euphrasiae*' (*J. Bot.* 1917 Supplt. 1–28). 'Balearic Plants' (*J. Bot.* 1907, 53–59).

Bot. Soc. Exch. Club Br. Isl. Rep. 1921, 355–56. *J. Bot.* 1922, 65–67 portr. *Proc. Bristol Nat. Soc.* v.5 (5), 1923, 243–44. H. J. Riddelsdell *Fl. Gloucestershire* 1948, cxxxiv–cxxxv.

Plants at Bristol, Kew. Portr. at Hunt Library.

Agaricus bucknalli B. & Br.

BUCKNALL, Thomas Skip Dyot *see* Addendum

BUCKNER, Leonard (*fl.* 1620s–1650s)

London apothecary. Collected plants with T. Johnson in Kent, 1629 and 1632.

J. Gerard *Herball* 1633, 697, 1115.

BUDDIN, Walter (1890–1962)
b. London 5 Nov. 1890 *d.* Reading, Berks 14 Aug. 1962

MA Cantab. Mycologist. Cheshunt Experiment Station, 1919. Reading University, 1921–46. Plant pathologist, Ministry of Agriculture, 1946–56.

Nature v.196, 1962, 811–12. *Ann. Applied Biol.* 1963, 525. *Trans. Br. Mycol. Soc.* 1963, 295 portr.

BUDDLE, Rev. Adam (*c.* 1660–1715)
b. Deeping St. James, Lincoln *c.* 1660 *d.* Gray's Inn, London 15 April 1715

BA Cantab 1681. MA 1685. Lived at Henley, Suffolk in 1698. Rector, Great Fambridge, Essex, 1703. Reader at Gray's Inn. Had collection of grasses and mosses (*J. Bot.* 1874, 36–47). 'Methodus Nova Stirpium Britannicarum' (Sloane MSS. 2970–2979). "The top of all the moss-croppers" (W. Vernon).

J. J. Dillenius *Historia Muscorum* 1741, ix. D. Turner *Extracts from Lit. and Sci. Correspondence of Richard Richardson* 1835, 87–89, 95, 102–4, 151. H. Trimen *Fl. Middlesex* 1869, 386–88. W. M. Hind *Fl. Suffolk* 1889, 475. G. C. Druce *Fl. Berkshire* 1897, cxxxi. *D.N.B.* v.7, 222. E. J. L. Scott *Index to Sloane Manuscripts* 1904, 85. C. E. Raven *John Ray* 1942, 393. J. E. Dandy *Sloane Herb.* 1958, 102–8. H. N. Clokie *Account of Herb. Dept. Bot., Oxford* 1964, 140–41. D. H. Kent *Hist. Fl. Middx* 1975, 14–15.

Herb. at BM(NH). Plants at Oxford.

Buddleia L.

BUFFHAM, Thomas Hughes (1840–1896)
b. Long Sutton, Lincolnshire 24 Dec. 1840 *d.* Walthamstow, London 9 Feb. 1896

ALS 1891. Clerk in London. Algologist. Contrib. to *J. Quekett Club.*

J. Bot. 1896, 170–71; 1902 Supplt. 75. *Proc. Linn. Soc.* 1895–96, 35. *R.S.C.* v.9, 395; v.13, 904.

Algae at BM(NH) and National Museum, Dublin.

Gonimophyllum buffhami Batters.

BUGGS, John (*fl.* 1620s–1630s)

Apothecary. Accompanied T. Johnson on his visit to Kent in 1629 and to West of England in 1634.

BUIST, Robert (1805–1880)
b. Cupar Fyfe near Edinburgh 14 Nov. 1805 *d.* Philadelphia, U.S.A. 13 July 1880

Gardener, Royal Botanic Garden, Edinburgh. Emigrated to U.S.A. in 1828. In 1830 became partner with T. Hibbert, seedsman and florist, Philadelphia. *American Flower-garden Directory* 1832. *Rose Manual* 1844. *Family Kitchen-gardener* 1847.

Garden 1880, 144. *Gdnrs Mon.* 1888, 372 portr. J. W. Harshberger *Botanists of Philadelphia and their Work* 1899, 193–95. L. H. Bailey *Standard Cyclop. Hort.* v.2, 1939, 1567. U. P. Hedrick *Hist. Hort. in America to 1860* 1950, 248.

BULGER, George Ernest (–1885)
FLS 1864. Lieut.-Colonel. Collected plants in India, 1867–69. In Burma, 1873. 'Fl. Windvogelberg' (*Student Intell. Observer* v.4, 1870, 275–78).
R.S.C. v.7, 301; v.9, 396.
Letters and Burmese plants at Kew.

BULKLEY, Edward (*c.* 1651–1714)
d. Madras, India 8 Aug. 1714
MD. In India, 1682. Succeeded S. Browne as Surgeon to East India Company, Madras, 1692. Collected plants in Bengal and Burma, 1702–8. Sent plants to J. Petiver, L. Plukenet and C. Du Bois. Correspondent of Camel.
J. Petiver *Musei Petiveriani* 1695, 43–44, 94. E. J. L. Scott *Index to Sloane Manuscripts* 1904, 85. D. G. Crawford *Hist. Indian Med. Service, 1600–1913*, v.1, 1914, 88–90. C. E. Raven *John Ray* 1942, 302, 333. J. E. Dandy *Sloane Herb.* 1958, 108. H. N. Clokie *Account of Herb. Dept. Bot., Oxford* 1964, 141.
Plants at BM(NH), Oxford.

BULL, Daniel (*fl.* 1760s–1770s)
Nurseryman, Dublin.
Irish For. 1967, 46.

BULL, Miss Edith E. (*fl.* 1840s–1880s)
Daughter of H. G. Bull (*c.* 1818–1885). Drew 19 plates for R. Hogg and H. G. Bull's *Herefordshire Pomona* 1876–85.
Portr. at Hunt Library.

BULL, Edward (*fl.* 1900s)
Nurseryman, King's Road, Chelsea, London. Son of W. Bull (1828–1902).
J. Hort. Home Farmer v.63, 1911, 538–39 portr. *Gdnrs Mag.* 1912, 331 portr. 342. *Garden* 1916, 163.

BULL, Henry Graves (*c.* 1818–1885)
b. Northamptonshire *c.* 1818 *d.* Hereford 31 Oct. 1885
MD Edinburgh. Physician, Hereford Infirmary. Mycologist. Established fungus forays, 1867. *Herefordshire Pomona* (with R. Hogg) 1876–85 2 vols.
J. Bot. 1871, 308; 1886, 62–64. *Trans. Woolhope Nat. Field Club* 1883–85, 373–76 portr.; 1954, 243–46. *Gdnrs Chron.* 1885 ii 583–84. *R.S.C.* v.1, 715; v.7, 301; v.9, 396; v.13, 911.
Letters at BM(NH). Drawings of Hereford fungi at Kew.

BULL, Martin M. (–1879)
d. Jersey 17 Aug. 1879
MD. Of Jersey. Collected plants on Sark (*J. Bot.* 1872, 199–203; 1874, 83).
J. Bot. 1879, 288. *R.S.C.* v.7, 302.

BULL, W. W.
Farmer, Billericay, Essex. Raised 'Aurora' and 'Queen' apples in late 19th century.

BULL, William (1828–1902)
b. Winchester, Hants 1828 *d.* Chelsea, London 1 June 1902
FLS 1866. VMH 1897. Acquired nursery and stock of John Weeks and Co., King's Road, Chelsea, 1861. Sent Shuttleworth and Corder to Colombia for orchids. Hybridised orchids. Introduced *Coffea liberica*.
Garden v.61, 1902, 386–87 portr. *Gdnrs Chron.* 1902 i 381 portr. *Gdnrs Mag.* 1902, 356 portr. *J. Hort. Cottage Gdnr* v.44, 1902, 491, 521 portr. *Orchid Rev.* 1902, 207–8. *Proc. Linn. Soc.* 1902–3, 27–28.

BULL, William (–1913)
d. Brighton, Sussex 8 Sept. 1913
Son and successor of W. Bull (1828–1902), founder of nursery at Chelsea.
Orchid Rev. 1913, 303.

BULLEN, Robert (–1892)
d. Glasgow 5 Oct. 1892
Curator, Glasgow Botanic Garden, 1868–92.
Gdnrs Chron. 1892 ii 473, 500.

BULLER, Arthur Henry Reginald (1874–1944)
b. Birmingham 19 Aug. 1874 *d.* Winnipeg, Canada 3 July 1944
Educ. Mason College, Birmingham. BSc London 1896. FRS 1929. Lecturer in Botany, Birmingham, 1901–4. Professor of Botany, University of Manitoba, 1904–36. Mycologist. President, British Mycological Society, 1913. President, Botanical Society of America, 1928. *Researches on Fungi* 1909–34 6 vols. *Essays on Wheat* 1919. *Practical Botany* 1929. *Fungi of Manitoba* (with G. R. Bisby and J. Dearness) 1929.
Nature v.154, 1944, 173. *Science* 1944, 305–7. *Times* 6 July 1944. *Mycologia* v.37, 1945, 275–77. *Phytopathology* 1945, 273, 577–84 portr. *Obit. Notices Fellows R. Soc.* v.5, 1945, 51–59 portr. *Friesia* 1946, 221–22 portr. *Trans. R. Soc. Canada* 1945, 79–81. *D.N.B. 1941–1950* 116–17. *Who was Who, 1941–1950* 159–60. H. B. Humphrey *Makers of North American Bot.* 1961, 42–45. C. C. Gillespie *Dict. Sci. Biogr.* v.2, 1970, 582–83.
MS. History of fungi, etc. at Kew. Portr. at Hunt Library.

BULLER, R. Hughes- *see* Hughes-Buller, R.

BULLER, William (-1757)
Nurseryman, Tinnahinch, County Wicklow, Ireland.

BULLEY, Arthur Kilpin (1861–1942)
b. New Brighton, Cheshire 10 Jan. 1861 *d.* Neston, Cheshire 3 May 1942
Cotton-broker, Liverpool. Founder of Bees Ltd., seeds and nursery firm which originally occupied part of Bulley's garden at Neston before moving to Sealand in 1911. Engaged G. Forrest, F. Kingdon Ward and R. E. Cooper to collect plants for his garden which is now the Botanic Garden of Liverpool University.
Gdnrs Mag. 1909, 429–30 portr. *Quart. Bull. Alpine Gdn Soc.* v.10, 1942, 248–49 portr. *N. Western Nat.* 1942, 396–98. M. Hadfield *Gdning in Britain* 1960, 407–9. *House Gdn* Jan. 1973, 110–11 portr. *J. Scott. Rock Gdn Club* 1973, 216–19.
Bulleya Schlechter.

BULLEYN, Rev. William (1500–1576)
b. Isle of Ely, Cambridgeshire 1500 *d.* London 7 Jan. 1576
Physician, Durham. Rector, Blaxhall, Suffolk, 1550–54. Travelled in Scotland and Germany. *Booke of Simples* (part of *Bulwarke of Defence against all Sicknes* 1562).
R. Pulteney *Hist. Biogr. Sketches of Progress of Bot. in England* v.1, 1790, 77–83. G. W. Johnson *Hist. English Gdning* 1829, 51–53. S. Felton *Portr. of English Authors on Gdning* 1830, 84–85. *Cottage Gdnr* v.5, 1850, 207. *J. Hort. Cottage Gdnr* v.55, 1876, 373–74 portr. *D.N.B.* v.7, 244–46.

BULLMORE, Ernest (*fl.* 1880s)
Solicitor, Falmouth, Cornwall. Studied flora of Falmouth.
F. H. Davey *Fl. Cornwall* 1909, lix.

BULLOCK, Thomas Lowndes (1845–1915)
b. Radwinter, Essex 27 Sept. 1845 *d.* Oxford 20 March 1915
BA Oxon. British Consular Service in China, 1869. Professor of Chinese, Oxford, 1899. Collected Chinese plants which he sent to H. F. Hance for identification.
E. Bretschneider *Hist. of European Bot. Discoveries in China* 1898, 736–38. *Kew Bull.* 1893, 369. *Bot. Soc. Exch. Club Br. Isl. Rep.* 1915, 250. *Oxford Times* 27 March 1915. *Who was Who, 1897–1916* 101. *R.S.C.* v.9, 398.
Plants at BM(NH), Kew, Oxford.
Eugenia bullockii Hance.

BULLOCK, William (*fl.* 1790s–1840s)
FLS 1810. Goldsmith, Liverpool, 1795. Proprietor of Museum of Natural History at Liverpool, 1799–1811; opened museum in London. Collected plants and birds in Mexico, 1822–23, 1827. Had museum in Cincinnati, Ohio. *A Companion to Liverpool Museum* 17 eds. 1801–16. *Six months Residence and Travel in Mexico* ed. 2 1825. *Sketch of Journey through Western States of North America* 1827.
W. Jordan *Men I have known* 1866, 67–82. *Mus. J.* 1917, 51–56; 1918, 132–37, 180–87. *Occas. Papers California Acad. Sci.* 1943, 29–30. *Bull. Hist. Philos. Soc. Ohio* v.19, 1961, 144–52.
Letters at Linnean Society.

BULLOCK-WEBSTER, Rev. George Russell (1858–1934)
b. Hereford 7 July 1858 *d.* Parkstone, Dorset 16 Feb. 1934
BA Cantab 1879. MA 1887. FLS 1918. Rector, St. Michael Paternoster Royal, London, 1910–32. Canon of Ely Cathedral. Studied Charophytes. *British Charophyta* (with J. Groves) 1920–24 2 vols. Discovered *Nitella mucronata* in Ireland, 1901. Contrib. to *J. Bot.* and *Irish Nat.*
J. Bot. 1934, 111–12. *Proc. Linn. Soc.* 1933–34, 143–44. *Times* 20 Feb. 1934. *Who was Who, 1929–1940* 188. R. L. Praeger *Some Irish Nat.* 1949, 56.
Characeae at BM(NH), Oxford, South London Botanical Institute.

BULMER, Rev. Charles Henry (*c.* 1833–1918)
d. 13 Feb. 1918
MA. Rector, Credenhill, Herefordshire. Pomologist and rose-grower. Contrib. to *J. Hort.*
J. Hort. Cottage Gdnr v.47, 1903, 397, 399 portr. *Gdnrs Chron.* 1918 i 83.

BUNBURY, Sir Charles James Fox (1809–1886)
b. Messina, Italy 4 Feb. 1809 *d.* Bury St. Edmunds, Suffolk 18 June 1886
FRS 1851. FLS 1833. Collected plants in S. Africa, 1838–39, which he sent to W. H. Harvey. Also collected in S. America, 1833–34; Madeira, Tenerife, 1853. *Journal of Residence at Cape of Good Hope* 1848. *Botanical Fragments* 1883. *Botanical Notes at Barton and Mildenhall, Suffolk* 1889. 'Plants of Brazil' (*Proc. Linn. Soc.* 1841, 108–10). 'Botanical Excursions in South Africa' (*London J. Bot.* 1842, 549–66; 1843, 15–41; 1844, 242–63).
Ann. Mag. Nat. Hist. v.7, 1841, 439. *Proc. Linn. Soc.* 1886–87, 35–36. *J. Bot.* 1888, 69. *Proc. R. Soc.* v.46, 1889, xiii–xiv. K. M. Lyell *Life of Sir Charles J. F. Bunbury* 1906 portr. 2 vols. C. F. P. von Martius *Fl. Brasiliensis* v.1 (1), 1906, 8–9. *Suffolk Nat. Hist.* v.15, 1971, 275–77.
Herb. and fossil plants at Cambridge. Letters and Brazilian plants at Kew. Portr. at Hunt Library.
Bunburia Harvey.

BUNBURY, Louisa Emily (*née* Fox) (–1828)
d. Nervi, Italy 15 Sept. 1828
Married Col. (afterwards Sir) Henry Bunbury, 1807. Mother of Sir C. J. F. Bunbury. Knew British and other plants and had a herb.
K. M. Lyell *Life of Sir Charles J. F. Bunbury* 1906, 45.

BUNCE, Daniel (1813–1872)
b. near Rickmansworth, Herts 18 March 1813 *d.* Geelong, Australia 2 June 1872
Gardener. Emigrated in 1835 to Hobart where he worked at Lightfoot's nursery which he eventually bought. Opened nursery at Launceston, 1839 and another at St. Kilda. Unsuccessfully applied for directorship of Melbourne and Adelaide Botanic Gardens. Collected plants on Leichhardt's second expedition, 1846. Collected near Murray River, 1848. First Curator, Geelong Gardens, 1858. *Manual of Practical Gardening for Van Dieman's Land* 1837–38; ed. 3 1851. *Hortus Tasmanensis* 1851 (all copies destroyed but one which is in National Library, Canberra). *Hortus Victoriensis* 1851. *Twenty-three Years Wanderings in the Australias and Tasmania* 1857. *Travels with Leichhardt* 1859.
Victorian Nat. 1908, 103–4. *Rep. Austral. Assoc. Advancement Sci.* 1911, 225. *Victorian Hist. Mag.* v.18 (1), 1940; v.23 (3), 1950–51. *Austral. Dict. Biogr.* v.1, 1966, 176–77. *Austral. Encyclop.* v.2, 182–83.
Plants at Melbourne Botanic Gardens. Portr. at Historical Society of Victoria.
Panicum buncei F. Muell.

BUNCH, Rev. Robert James (*fl.* 1790s–1830s)
Herb. at Leicester University.

BUNCLE, Margaret *see* Corstorphine, M.

BUNKER, Herbert Edwin (1899–1969)
b. Catford, London 11 Aug. 1899. *d.* Preston, Lancs 16 Jan. 1969
Manager, Leyland rubber company. Contrib. to *Travis's Fl. S. Lancashire* 1963.
Lancashire Evening Post 17 Jan. 1969.

BUNNETT, E. J. (–*c.* 1944)
Herb. at Juniper Hall Field Centre, Dorking, Surrey.

BUNNEY, George Hockley (*fl.* 1820s–1840s)
Succeeded George Bassington at nursery, Kingsland Road, Islington, London.
W. Robinson *Hist. Hackney* v.1, 1842, 6–7.

BUNNEY, John (*fl.* 1720s)
Seedsman, White Horse, Uxbridge, Middx.

BUNNY, Joseph (1798–1885)
b. Newbury, Berks 1798 *d.* Newbury 2 June 1885
MD Edinburgh 1823. Physician, Newbury, 1823. Contrib. to *Catalogue of Newbury Plants* 1839.
Trans. Newbury District Field Club 1875–86, xvi–xvii. G. C. Druce *Fl. Berkshire* 1897, clxviii.

BUNTING, Robert Hugh (1879–1966)
b. Swaffham, Norfolk 26 Nov. 1879 *d.* Jan. 1966
FLS 1916. At BM(NH), 1894–1910. Botanical collector to Liberian Rubber Co., 1910–13. Assistant Director, Agricultural Research, Sierra Leone 1913; Gold Coast, 1914–29. Imperial College Biological Field Station, 1929–34. Curator, Dorset County Museum, 1940–44. Collected plants in Sierra Leone and Gold Coast. *Gold Coast Plant Diseases* (with H. A. Dade) 1924.
Proc. Linn. Soc. 1967, 89.
Plants and MSS. at BM(NH). Fungi at Commonwealth Mycological Institute, Kew.

BUNYARD, E. (*fl.* 1830s)
Nurseryman and seedsman, 9 Commercial Place, City Road, London.

BUNYARD, Edward Ashdown (1878–1939)
b. Maidstone, Kent 1878 *d.* 19 Oct. 1939
FLS 1914. Son of George Bunyard (1841–1919). Nurseryman, Maidstone. *Handbook of Hardy Fruits* 1920. *Stone and Bush Fruits* 1925. *Old Garden Roses* 1936. Founder and Editor of *J. Pomology*, 1919.
J. Hort. Home Farmer v.61, 1910, 369–70 portr. *Gdnrs Mag.* 1912, 1–2 portr. *Gdnrs Chron.* 1939 ii 274. *Proc. Linn. Soc.* 1939–40, 362. *Chronica Botanica* 1940–41, 43.

BUNYARD, George (1841–1919)
b. Maidstone, Kent 1841 *d.* 22 Jan. 1919
VMH 1897. Fruit-grower at Royal Nurseries, The Triangle, Maidstone and at Allington. *Fruit Farming for Profit* 1881; ed. 5 1907. *Handbook of Hardy Trees and Shrubs* 1907. *Apples and Pears* 1911. *Fruit Garden* (with O. Thomas) 1904. Contrib. to *Century Book of Gardening* and *Gdnrs Chron.*
Gdnrs Mag. 1894, 581–82 portr.; 1908, 655–56 portr. *J. Hort. Cottage Gdnr* v.60, 1910, 197–98 portr. *Garden* 1901, 95 portr.; 1919, 48, 63 portr. *Gdnrs Chron.* 1919 i 59–60 portr.

BUNYARD, George Norman (1886–1969)
Nurseryman. Secretary, British Iris Society, 1922–27. Hybridised plants, especially irises.
Iris Yb. 1969, 103.

BUNYARD, James (*fl.* 1790s–1810s)
Founded nursery at Maidstone, Kent in 1796.
J. Hort. v.70, 1915, 141–42 portr.

BUNYARD, Thomas (*c.* 1804–1880)
d. Oct. 17 1880
Nurseryman, Maidstone, Kent.

BUONAIUTI, S. (*fl.* 1800s–1820s)
Of Kensington, London. Librarian to Lord Holland, 1807. 'On Dahlia' (R. W. Dickson *Complete Dictionary of Practical Gardening* v.2, 1807, Appendix).
J. Bot. 1918, 34.

BURBIDGE, Frederick William Thomas (1847–1905)
b. Wymeswold, Leics 21 March 1847 *d.* Dublin 24 Dec. 1905
Hon. MA Dublin 1889. VMH 1897. Gardener, Kew, 1868–70. On staff of *Garden*, 1873–77. Collected plants in Borneo for James Veitch and Sons, 1877–78. Introduced *Nepenthes rajah.* Curator, Trinity College Garden, Dublin, 1879–1905. Collected in Sierra Leone, 1913–19. *Cool Orchids and how to Grow Them* 1874. *Art of Botanical Drawing* 1873. *Domestic Floriculture* 1875. *Narcissus* 1875. *Cultivated Plants* 1877. *Horticulture* 1877. *Gardens of the Sun* 1880. *Chrysanthemum* 1884. *Wild Flowers in Art and Nature* (with J. G. L. Sparkes) 1894. *Book of Scented Garden* 1905.
Bot. Mag. 1879, t.6403. *Gdnrs Chron.* 1889 ii 212–13 portr.; 1896 i 736 portr.; 1905 ii 460 portr.; 1906 i 10. *Garden* v.66, 1904, iv portr.; v.69, 1906, 16 portr. *J. Hort. Cottage Gdnr.* v.51, 1905, 588. *J. Kew Guild* 1905, 269; 1906, 326–27 portr. *D.N.B. Supplt. 2* v.1, 257–58. *J. Bot.* 1906, 80. *Orchid Rev.* 1906, 8. J. H. Veitch *Hortus Veitchii* 1906, 75–78. *Notes Bot. School Dublin* v.2, 1909, 44–46. *Who was Who, 1897–1916* 102–3. *Daffodil Yb.* 1933, 26 portr. R. L. Praeger *Some Irish Nat.* 1949, 57. *Fl. Malesiana* v.1, 1950, 88–89 portr. A. M. Coates *Quest for Plants* 1969, 207–9. M. A. Reinikka *Hist. of the Orchid* 1972, 258–59. *R.S.C.* v.9, 399; v.12, 133; v.13, 917.
Plants and drawings at Kew, BM(NH). Portr. at Hunt Library. Library sold by Hazley and Co., Dublin, 23–24 May 1906.
Burbidgea Hook. f.

BURCH, George (*fl.* 1890s)
Nurseryman, Padholm Road Nurseries, Peterborough, Northants, 1886. In partnership with his brother, W. H. Burch. Specialised in roses.
Gdnrs Mag. 1897, 412 portr.

BURCHELL, Matthew (*c.* 1753–1828)
d. Fulham, Middx 12 July 1828
Nurseryman, King's Road, Fulham, London. Acquired business from his uncle, William Burchell (*c.* 1725–1800). In 1810 nursery was taken over by R. Whitley, P. Brames and T. Milne.

BURCHELL, William (*c.* 1725–1800)
d. 25 Feb. 1800
Acquired Christopher Gray's nursery at Fulham, London.
Trans. London Middlesex Archaeol. Soc. v.24, 1973, 183.

BURCHELL, William John (1781–1863)
b. Fulham, London 23 July 1781 *d.* Fulham 23 March 1863
FLS 1808. DCL Oxon 1834. Son of Matthew Burchell (*c.* 1753–1828). Schoolmaster and acting botanist to East India Company at St. Helena, 1805–10 (MS. sketches of St. Helena at Kew). Travelled in S. Africa, 1811–15. Explored Brazil, 1825–29 (itinerary in *Phytologia* v.14, 1967, 492–505). Heir to R. A. Salisbury (1761–1829). Collected plants on his travels. *Travels in Interior of S. Africa* 1822–24 2 vols.
Bot. Register 1816, t.139; 1820, t.466. *Bot. Misc.* 1832–33, 128–33. W. H. Harvey *Genera of S. African Plants* 1838, xii. *Cape Mon.* v.5, 1859, 356–63; v.6, 213–17. *Proc. Linn. Soc.* 1862–63, xxxiv–v; 1906–7, 64–65; 1946–47, 141–46. *J. R. Geogr. Soc.* v.33, 1863, cxxiv–cxxv. *Times* 27 March 1863. *Kew Gdns Rep.* 1865, 6–7. *D.N.B.* v.7, 290–91. *Notes and Queries* v.2, 1904, 486; v.3, 1905, 77. *Addresses Br. S. Africa Assoc. 1905* v.3, 57–110 portr. E. B. Poulton *William John Burchell* 1905. H. C. Notcutt *Pioneers* 1924, 93–159. *S. African J. Sci.* 1934, 481–89; 1935, 680–83, 689–95; 1937, 346–50. *J. S. African Bot.* 1941, 1–18, 61–76, 115–30; 1943, 27–78; 1944, 145–61. J. Hutchinson *Botanist in S. Africa* 1946, 625–41. *Bothalia* 1950, 23–27. H. M. McKay *S. African Drawings of William John Burchell* 1938–52 2 vols. A. M. Coats *Quest for Plants* 1969, 261–63, 354. G. W. Reynolds *Aloes of S. Africa* 1969, 53–56. *Taxon* 1970, 517.
Herb, letters and MSS. at Kew. Portr. at Botanic Garden, Kirstenbosch, Kew, Hunt Library. Library sold by Messrs. Foster, 54 Pall Mall, 5 Dec. 1865.
Burchellia R. Br.

BURDEN, Elizabeth Raymond *see* Ewing, E. R.

BURDON, Mrs. (–1885)
d. Castle Eden, Durham 10 March 1885
"An excellent botanist...Her collection of living alpine plants was one of the finest in the North of England" (*Naturalist* 1884–85, 235).

BURDON, Rev. Rowland John (*fl.* 1870s–1930s)
BA Oxon 1879. Curate, Midhurst, 1881–85; Brighton, 1885–95. Vicar, Oving, 1895–1900; Arundel, 1900–5; Chichester, 1905–23.
Plants at Oxford.

BURDON-SANDERSON, Sir John Scott (1828–1905)
b. Newcastle, Northumberland 21 Dec. 1828 *d.* Oxford 24 Nov. 1905
MD Edinburgh 1851. MA Oxon 1883. FRS 1867. Pupil of J. H. Balfour. Lecturer in Botany, St. Mary's Hospital, 1855. Professor of Physiology, Oxford, 1883–95. 'Vegetable Physiology' (*Todd's Cyclop.* v.5, 1859, 211–56). 'On Electromotive Properties of Leaf of Dionaea in Excited and Unexcited States' (*Philos. Trans. R. Soc.* v.173, 1882, 1–55; v.179, 1888, 417–49).
Proc. R. Soc. v.79, 1907, iii–xviii portr. *D.N.B. Supplt. 2* v.1, 267–69. *Carnivorous Plant Newsletter* v.2 (3), 1973, 41–42. *R.S.C.* v.5, 392; v.8, 827; v.11, 277.

BURGES, Richard Charles L'Estrange (1900–1959)
d. Birmingham July 1959
Educ. Cambridge. FLS 1939. Physician. Good knowledge of flora of Midlands. President, Birmingham Natural History and Philosophical Society, 1955–56.
Proc. Birmingham Nat. Hist. Philos. Soc. v.19, 1957–58, 55–56. *Proc. Linn. Soc.* 1958–59, 137–38. *Proc. Bot. Soc. Br. Isl.* 1960, 102–3.
Herb. and Library at Birmingham Natural History Society.

BURGESS, Henry W. (*fl.* 1820s–1830s)
Landscape painter to William IV. *Eidodendron: Views of...Character...of Trees* 1827–31.
J. Bot. 1919, 223–24.

BURGESS, James John (1863–1934)
b. Inveraven, Banff 19 Feb. 1863 *d.* Rosyth, Dunfermline, Fife 28 Feb. 1934
MA. Headmaster, Dyke School, Morayshire. Edited *Fl. Moray* 1935.
Weekly Scotsman 10 March 1934.

BURGESS, Rev. John (1725–1795)
b. Holywood, Dumfriesshire 1725 *d.* Kirkmichael, Dumfriesshire 2 Sept 1795
MA Glasgow 1747. DD Edinburgh 1769. Minister, Kirkmichael, 1759–95. He and his son, James Burgess MacGarroch, were lichenologists. Contrib. plant records to J. Sowerby and J. E. Smith's *English Bot.* 1790–1814.
J. Lightfoot *Fl. Scotica* v.1, 1777, xiii.
Leptogium burgessii Mont.

BURGESS, Joseph Tom (1828–1886)
b. Cheshunt, Herts 17 Feb. 1828 *d.* Leamington, Warwickshire 4 Oct. 1886
Wood-engraver at Northampton, *c.* 1844. Editor, *Clare Journal* and local English newspapers. Antiquary. *Old English Wild Flowers* 1868.
D.N.B. Supplt. 1 v. 1, 335.

BURGIS, Thomas (*fl.* 1760s)
Employed by Sir J. Banks to make finished drawings of plants from sketches by S. Parkinson, J. F. Miller and others.
Drawings at BM(NH).

BURKE, David (1854–1897)
b. Kent 1854 *d.* Amboina, Moluccas 11 April 1897
Entered Messrs Veitch's Chelsea nursery as young gardener. Collected plants with Charles Curtis in Borneo, 1880. Also collected in British Guiana, Philippines, New Guinea, Burma, Colombia, Moluccas.
J. H. Veitch *Hortus Veitchii* 1906, 87–88. *J. R. Hort. Soc.* 1948, 289. *Fl. Malesiana* v.1, 1950, 91–92. M. A. Reinikka *Hist. of the Orchid* 1972, 269–70.
Plants at Kew.
Didymocarpus burkei W. W. Smith.

BURKE, Joseph (*fl.* 1830s–1840s)
Collector of live animals for Earl of Derby. In S. Africa, 1839–42 where he joined Carl L. P. Zeyher in Transvaal, 1840–41, collecting plants. Discovered new species of stapelias. In N. America, 1844–46, where he collected plants in Snake River region of Idaho and upper valleys of Platte in the Rocky Mountains.
London J. Bot. 1843, 163–65; 1845, 643–53; 1846, 14–22, 430–35. *Kew Bull.* 1901, 84. A. White and B. L. Sloane *Stapelieae* v.1, 1937, 99. S. D. McKelvey *Bot. Exploration of Trans-Mississippi West, 1790–1850* 1955, 792–817. *Madrono* v.13, 1956, 260.
Plants at BM(NH), Kew, Oxford, Gray Herb., Boston. MS. Journal and letters at Kew.
Burkea Benth.

BURKILL, Ethel Maud (*née* **Morrison**) (1874–1970)
b. Wakefield, Yorks 28 Feb. 1874 *d.* Dorking, Surrey 20 July 1970
Wife of I. H. Burkill (1870–1965). Collected plants in Calcutta, 1911; Malay Peninsula, 1912–21; Sumatra, 1921. Collected and drew fungi for Botanic Gardens, Singapore.
Gdns Bull. Straits Settlements 1927, 118. *Fl. Malesiana* v.1, 1950, 92. *Gdns Bull. Singapore* 1967, 105.
Annularia burkillae Massee.

BURKILL, Harold John (1871–1956)
b. Chapel Allerton, Leeds, Yorks 24 Dec. 1871 *d.* 17 March 1956

BA Cantab 1895. MA 1899. Brother of I. H. Burkill (1870–1965). Stockbroker. Studied plant galls.

H. J. Riddelsdell *Fl. Gloucestershire* 1948, cxxxix. *London Nat.* 1956, 104–6.

BURKILL, Isaac Henry (1870–1965)
b. Chapel Allerton, Leeds, Yorks 18 May 1870 *d.* Leatherhead, Surrey 8 March 1965

BA Cantab 1891. MA 1895. FLS 1894. Assistant Curator, Herb. Cambridge, 1891–97. Assistant, Herb. Kew, 1897–1901. Indian Museum, Calcutta, 1901. Director, Botanic Gardens, Singapore, 1912–25. Collected plants in India, 1901–12, Malaya and E. Asia. 'Notes from Journey to Nepal' (*Rec. Bot. Survey India* 1910, 59–140). *Dictionary of Economic Plants of Malay Peninsula* 1935 2 vols.; ed. 2 1966. *Account of Genus Dioscorea* (with D. Prain) (*Ann. R. Bot. Gdn Calcutta* v.14, 1936–38). *Chapters on History of Botany in India* 1965.

Fl. Malesiana v.1, 1950, 92–93 portr. *Gdns Bull. Singapore* 1960, 341–56 portr.; 1967, 67–105. *J. Agric. Tropical Bot. Appliquée* v.12, 1965, 221–22. *Malayan Agric. J.* 1965, 216. *Nature* v.206, 1965, 871. *Times* 9 March 1965. *Who was Who, 1961–1970* 158.

Plants at Singapore and Kew. MSS. at Kew. Portr at Hunt Library.

BURKINSHAW, William Parker (*c.* 1835–1918)
d. Hessle, Yorks 30 Sept. 1918

Of West Hill, Hessle. Cultivated orchids.

Orchid Rev. 1918, 227.

BURLEY, James (*fl.* 1830s–1840s)

Florist, Limpsfield near Godstone, Surrey.

BURLEY, John (*fl.* 1860s)

Nurseryman, Albert Nursery, Pembridge Place, Bayswater, London.

BURLINGHAM, Daniel Catlin (1823–1901)
b. King's Lynn, Norfolk 1823 *d.* King's Lynn 1 April 1901

Watchmaker. Collected plants in Norfolk. Had herb.

Trans. Norfolk Norwich Nat. Soc. v.7, 1901–2, 414–21 portr.

BURN, James (*fl.* 1780s–1790s)

Gardener and nurseryman, Windsor, Berks.

BURN-MURDOCH, Alfred M. (1868–1914)
b. Scotland 1868 *d.* Klang, Selangor, Malaya 6 March 1914

Educ. Loretto School, Edinburgh. Indian Forest Dept., Burma, 1891. Conservator of Forests, Federated Malay States, 1901–14. *Trees and Timbers of Malay Peninsula* 1912. Collected plants in Pahang (*Gdns Bull. Straits Settlements* v.1, 1915, 310–18).

Indian For. 1914, 155–56. *Fl. Malesiana* v.1, 1950, 93–94.

Plants at Kuala Lumpur, Singapore and Kew.

BURNET, John (*fl.* 1710s–1730s)

Surgeon to South Sea Company in W. Indies. Afterwards physician to Philip V of Spain. Sent plants from Porto Bello to J. Petiver.

E. J. L. Scott *Index to Sloane Manuscripts* 1904, 87. J. E. Dandy *Sloane Herb.* 1958, 109–10.

Plants at BM(NH).

BURNET, Robert (1823–1889)
b. Lady Kirk, Berwickshire 1823 *d.* Hamilton, Ontario 1889

Minister, St. Andrew's Church, Hamilton, Ontario. Keen horticulturist.

L. H. Bailey *Standard Cyclop. Hort.* v.2, 1939, 1567.

BURNETT, Gilbert Thomas (1800–1835)
b. 15 April 1800 *d.* 27 July 1835

FLS 1832. Professor of Botany, King's College, London, 1831 and Chelsea, 1835. *Outlines of Botany* 1835. *Plantae Utiliores* 1839–50 4 vols (with plates by his sister, M. A. Burnett). Edited J. Stephenson and J. M. Churchill's *Medical Botany* 1834–36 3 vols.

Gdnrs Mag. 1840, 297. R. H. Semple *Mem. of Bot. Gdn at Chelsea* 1878, 186–88. *D.N.B.* v.7, 412. E. Bretschneider *Hist. European Bot. Discoveries in China* 1898, 280. *R.S.C.* v.1, 735.

Burnettia Lindl.

BURNETT, Richard (*fl.* 1790s)

Nurseryman, Richmond near Dublin, Ireland.

BURNETT, Stuart Mowbray (*c.* 1826–1893)
b. Kemnay, Aberdeenshire *c.* 1826 *d.* Aberdeen 23 Jan. 1893

Of Balbithan, Keithhall, Aberdeen. Had a herb. Read papers to Aberdeen Natural History Society.

Gdnrs Chron. 1893 i 112. *Gdnrs Mag.* 1893, 56.

BURNETT, William (*c.* 1800–1877)
d. 19 May 1877

Florist of Monkgate, York.

BURNS, John Sanderson (*fl.* 1860s)

History of Henley 1861 contains a few Buckinghamshire plants.

G. C. Druce *Fl. Buckinghamshire* 1926, ci.

BURNS, William (1884–1970)
b. 6 July 1884 *d.* 8 April 1970
Educ. Edinburgh. Assistant Lecturer in Botany, Reading, 1907–8. Economic Botanist, Bombay, 1908. Principal, Poona College of Agriculture, 1922–32. Director of Agriculture, Bombay, 1932–36. Agricultural Commissioner, Government of India, 1939–43.
Who was Who, 1961–1970 160.

BURNSIDE, Rev. F. R. (*fl.* 1880s–1890s)
Rector, Great Stambridge, Essex. Grew roses.
Garden v.62, 1902, 133–34 portr.

BURRELL, William Holmes (1865–1945)
b. London 1865 *d.* 30 March 1945
FLS 1907. Pharmacist. Bryologist. Lived in Yorkshire, 1914–45. Curator, Ingham Herb., Leeds University. Contrib. to *Naturalist*.
Naturalist 1945, 113–14 portr.; 1961, 159–60; 1973, 8–9. *Proc. Linn. Soc.* 1944–45, 111–12. *Rep. Br. Bryol. Soc.* 1945, 316.

BURTON, David (–1792)
d. Parramatta, Australia 13 April 1792
Gardener. Sent to Port Jackson, Australia, by Sir J. Banks who commissioned him to collect plants and seeds. Superintendent of convicts at Parramatta, *c.* 1791.
D. Collins *Account of English Colony in N.S.W.* 1798–1802 2 vols. W. T. Aiton *Hortus Kewensis* v.3, 1811, 12. *Hist. Rec. N.S.W.* v.1 (2), 1892, 599–600. *J. R. Soc. N.S.W.* 1908, 93–94. J. H. Maiden *Sir Joseph Banks* 1909, 156. *Hist. Rec. Austral.* Series I v.1, 1914, 250, 295, 297, 371, 435, 538. *J. Proc. R. Austral. Hist. Soc.* v.25 (6), 1939, 455. *Austral. Dict. Biogr.* v.1, 1966, 183–84.
Plants at BM(NH).
Burtonia R. Br.

BURTON, Esther *see* Hopkins, E.

BURTON, George (*c.* 1882–1960)
b. Wymondham, Norfolk 15 Dec. 1882 *d.* 3 May 1960
Archdeacon, Church Missionary Society. Collected plants in S. Nigeria, *c.* 1907–57, now at Kew.

BURTON, Richard Francis (1864–1922)
d. Shrewsbury, Shropshire 8 Jan. 1922
Entomologist. Cultivated British orchids.
Orchid Rev. 1922, 103.

BURTON, Sir Richard Francis (1821–1890)
b. Elstree, Herts 19 March 1821 *d.* Trieste, Italy 20 Oct. 1890
Explorer and scholar. Indian Army, 1842–49. Expedition to discover source of Nile, 1856–59. British Consul, Fernando Po, 1861–65; Santos, 1865–69. Collected plants in Dahomey, Arabia, etc., 1864–78; Gold Coast (with V. L. Cameron) 1882. *Lake Regions of Central Africa* 1860. *Abeokuta and Camaroons Mountains* 1863 2 vols. *Gold Mines of Midian* 1878 (contains plant list).
I. Burton *Life of Captain Sir R. F. Burton* 1893 portr. *D.N.B. Supplt 1* v.1, 349–56. H. J. Schonfield *Richard Burton* 1936 portr. *Tanganyika Notes Rec.* 1957, 257–97. B. Farwell *Burton* 1963 portr.
Plants and letters at Kew. Portr. at National Portrait Gallery.
Vernonia burtoni O. & H.

BURTON, W. (*fl.* 1870s)
Of Liverpool. Friend of W. Curnow.
Plants at Oxford.

BURTT, Bernard Dearman (1902–1938)
b. York 14 June 1902 *d.* Singida, Tanganyika 9 June 1938
Educ. Aberystwyth and Reading. FLS 1933. Cousin of J. Burtt Davy. Assistant, Kew Herb., 1922–25. District Reclamation Officer, Tanganyika, 1925. Survey Botanist, Tsetse Research Dept. Collected plants, mainly in Tanganyika but also N. Rhodesia, Nyasaland, Uganda, Belgian Congo. Contrib. to J. Burtt Davy's *Check Lists of Forest Trees and Shrubs*...: *Tanganyika Territory* 1940–49. Contrib. to *Kew Bull.*
Gdnrs Chron. 1938 ii 108. *J. Bot.* 1938, 213–14. *Kew Bull.* 1938, 301–3. *Empire For. J.* 1938, 7–8. *Nature* v.142, 1938, 199–200. *Proc. Linn. Soc.* 1938–39, 234–36. *Times* 15 and 18 June 1938. *Tanganyika Notes Rec.* no. 6, 1938, 3–4.
Plants at Kew, Forestry Institute Oxford, BM(NH).

BURTT DAVY, Joseph (1870–1940)
b. Findern, Derbyshire 7 March 1870 *d.* Birmingham 20 Aug. 1940
PhD Cantab 1924. FLS 1903. Assistant, Kew, 1891–92. Lecturer in Botany, California, 1893–96. Curator, U.S. Dept. Agriculture, 1902–3. Botanist and Agrostologist, Dept. of Agriculture, Pretoria, 1903–13. Farmed in Transvaal. Collected plants which formed basis of National Herb., Pretoria. *Manual of Flowering Plants and Ferns of Transvaal with Swaziland* 1926. Lecturer in Tropical Forestry Botany, Oxford, 1925–39. *Maize* 1914. Assistant editor of *Check Lists of Forest Trees*...*of British Empire* 1935.
Gdnrs Chron. 1924 i 266 portr. *Kew Bull.* 1925, 318; 1935, 588. *Chronica Botanica* 1940–41, 235. *J. S. African For. Assoc.* no. 5, 1940, 10–11. *Empire For. J.* 1940, 173–74. *Nature* v.146, 1940, 424. *Proc. Linn. Soc.* 1940–41, 291–93. *Who was Who, 1929–1940* 198–99. *Comptes Rendus AETFAT 1960* 1962, 187–88.
Plants at Kew. Portr. at Hunt Library.
Clarkia davyi (Jepson) Lewis & Lewis.

BURY, Mrs. Edward (*née* Falkner, Priscilla Susan) (*fl.* 1820s–1860s)

Of Liverpool. Drew plates for *Selection of Hexandrian Plants* 1831–34 (original drawings in Dumbarton Oaks Library, Washington). Contrib. 8 plates to Maund's *Botanist* 1836–41.

W. Blunt *Art of Bot. Illus.* 1955, 213. *J. Soc. Bibl. Nat. Hist.* 1968, 71–75.

BURY, Lindsay (–1935)
d. Bradfield, Berks 30 Oct. 1935

Grew ferns in garden at Stamford Wood, Bradfield.

Br. Fern Gaz. 1935, 12–13.

BUSBY, James (1801–1871)
b. Edinburgh 7 Feb. 1801 *d.* 15 July 1871

Emigrated with parents to New South Wales, 1824. *Treatise on Culture of Vine* 1825. Returned to England, 1831 and toured French and Spanish vineyards. *Manual of Plain Directions for Planting and Cultivating Vineyards and for Making Wine in New South Wales* 1830. Appointed British Resident in New Zealand, 1832.

J. Proc. R. Austral. Hist. Soc. v.26 (5), 1940. *Austral. Dict. Biogr.* v.1, 1966, 186–88.

BUSCH, John (*fl.* 1730s–1790s)
b. Lüneburg, Hanover *c.* 1730. *d.* Isleworth, Middx

Came to England, 1744. Nurseryman, Hackney. Nursery acquired by Conrad Loddiges in 1771. Gardener to Catherine II, Empress of Russia; laid out gardens of Tsarskoe Selo, 1772. Returned to England, 1789.

Gdnrs Mag. 1827, 386. *J. Hort. Cottage Gdnr* v.55, 1876, 371.

BUSCH, Joseph Charles (*c.* 1759–1838)
b. Hackney, London *d.* St. Petersburg, Russia 1838

Son of John Busch. Engaged by Czar Alexander I to lay out the Imperial Park on Yelagin Island in the Neva. Married twice in St. Petersburg where he died of cholera.

W. Dawson *Banks Letters* 1958, 645.

BUSH, Joseph (*fl.* 1800s)

Introduced *Serratula* sp. from N. Persia in 1804.

Bot. Mag. 1816, t.1871.

BUSH, William Maddocks (*c.* 1813–1857)
d. Weston-super-Mare, Somerset 17 Dec. 1857

MD. FLS 1843. Surgeon, Marylebone Infirmary. "Acquired considerable knowledge of botany."

Proc. Linn. Soc. 1857–58, xxv–xxvi.

BUSHELL, Stephen Woolton (*fl.* 1860s–1890s)

MD London. Physician, British Legation, Peking, 1868. Sent Chinese plants to Kew, 1874–82. 'Notes on Production of Insect White Wax in China' (*Kew Bull.* 1893, 101–7).

E. Bretschneider *Hist. European Bot. Discoveries in China* 1898, 703. *Kew Bull.* 1901, 13.

BUSHELL, Thomas (1594–1674)
d. April 1674

Page to Sir Francis Bacon. Speculator and mining engineer. Created garden at Neat Enstone, Oxfordshire where he installed some ingenious waterworks.

D.N.B. v.8, 35–37.

BUSK, Lady Marian (*née* Balfour) (1861–1941)

BSc London 1883. FLS 1905. Assisted F. W. Oliver in translation of *Natural History of Plants* 1894, by A. Kerner von Marilaun.

Proc. Linn. Soc. 1941–42, 271–72.

BUSSEY, Winifred (*née* Simmons) (1884–1969)

Studied art at Goldsmith College, London. Did not begin drawings of British flora until she was 71.

Drawings at BM(NH).

BUTCHER, Roger William (1897–1971)
b. Ashbourne, Derbyshire 25 July 1897 *d.* Weymouth, Dorset 13 Oct. 1971

PhD London 1931. FLS 1927. Assistant, Kew Herb. Assistant Naturalist, Ministry of Agriculture, 1925; later worked on marine algae at Fisheries Laboratory, Burnham-on-Crouch; retired in 1963. President, British Mycological Society, 1963–64. *Further Illustrations of British Plants* (with F. Strudwick) 1930. *A New Illustrated British Fl.* 1961 2 vols.

Proc. Dorset Nat. Hist. Archaeol. Soc. 1971, 29–30 portr. *Watsonia* 1972, 175–77. S. T. Jermyn *Fl. Essex* 1974, 20–21.

Plants at Kew.

BUTCHER, William *see* Boutcher, W.

BUTE, John, 3rd Earl of *see* Stuart, J.

BUTLER, Sir Edwin John (1874–1943)
b. Kilkee, County Clare 13 Aug. 1874 *d.* Esher, Surrey 4 April 1943

MB Cork 1892. FRS 1926. FLS 1902. Knighthood 1939. Cryptogamic botanist, India, 1901. Mycologist, Pusa, 1905. Agricultural Adviser to Government of India, 1920. Director, Imperial Bureau of Mycology, Kew, 1920–35. Secretary, Agricultural Research Council, 1935–41. President, Association of Applied Biologists, 1928–29. President, British Mycological Society, 1927. *Monograph on Pythium* 1907. *Fungi and Disease in Plants* 1918. *Fungi of India* (with G. R. Bisby) 1931. *Plant Pathology* (with S. G. Jones) 1949. Edited *Rev. Applied Mycol.*

Gdnrs Chron. 1921 i 26 portr.; 1943 i 180. *Nature* v.151, 1943, 552–53. *Obit. Notices Fellows R. Soc.* v.4, 455–74 portr. *Proc. Linn. Soc.* 1942–43, 294–96. *Times* 6 April 1943. *Palestine J. Bot.* 1944, 209–10. *Phytopathology* 1944, 149–50 portr. *Ann. Applied Biol.* 1944, 168.

Portr. at Hunt Library.

Sclerospora butleri Weston.

BUTLER, Frank (–1936)
Of Reading, Berks. Grew ferns.
Br. Fern Gaz. 1936, 50–51.

BUTLER, Frederick Berry-Lewis (–1941)
d. Cyprus 9 Aug. 1941
FLS 1922. Kew gardener, 1913. Botanical field assistant, Dept. of Agriculture, Kenya, *c.* 1919–22. Chief Grader of Produce, Cyprus, 1935.
Proc. Linn. Soc. 1941–42, 271.

BUTLER, Isaac (1689–1755)
d. Dublin 1755
"Judicial Astrologer." Collector for Physico-Historical Society of Dublin from 1744 in Leinster. Catalogue of plants in County Down in Harris's *Antient...State of Co. Down* 1744.
N. Colgan and R. W. Scully *Cybele Hibernica* 1898, xxv. N. Colgan *Fl. County Dublin* 1904, xxii–xxiii. R. L. Praeger *Some Irish Nat.* 1949, 58.

BUTLER, James (1810s)
Seedsman, Thistle and Crown, Covent Garden, London.

BUTT, Rev. John Martin (1774–1827)
b. Stanford, Worcs 1774
BA Oxon 1796. FLS 1797. Vicar, E. Garston, Berks, 1822. *Botanical Primer* 1825.
P. Smith *Mem. and Correspondence of Sir J. E. Smith* v.1, 1832, 440.

BUTT, Richard (*fl.* 1730s–1750s)
Had nursery at Kew Green, formerly the property of William Cox the elder. Butt supplied plants to Frederick, Prince of Wales at Kew House, the forerunner of Kew Gardens.
Trans. London Middlesex Archaeol. Soc. v.24, 1973, 192.

BUTT, Rev. Thomas (1776–1841)
b. Lichfield, Staffs 30 Oct. 1776 *d.* Trentham, Staffs 14 June 1841
BA Oxon 1799. FLS 1799. Brother of Rev. J. M. Butt (1774–1827). Of Arelay, Shropshire. Rector, Trentham, Staffs and Kinnersley, Shropshire, 1820. Had extensive collection of hardy plants at Trentham. Botanised in Ireland. Contrib. to J. Sowerby and J. E. Smith's *English Bot.* 662, 2018, etc.
P. Smith *Mem. and Correspondence of Sir J. E. Smith* v.1, 1832, 435–41. *J. Bot.* 1914, 323.

BUTT, Rev. Walter (*c.* 1850–1917)
d. Chepstow, Mon 14 July 1917
Vicar of Kempsford, Glos 1904–9. Had herb. of Gloucestershire plants. Worked on flora of Gloucestershire, later edited by his son-in-law, Rev. H. J. Riddelsdell. President, Cotteswold Naturalists' Field Club, 1906, 1908 and 1912.
Bot. Soc. Exch. Club Br. Isl. Rep. 1917, 87. *Proc. Cotteswold Nat. Field Club* 1917, 236, 239. H. J. Riddelsdell *Fl. Gloucestershire* 1948, cxlv–cxlvii.
Herb. at Gloucester Museum.

BUTTENSHAW, William Robert (*c.* 1877–1907)
d. Calcutta, India 9 Sept. 1907
MA. BSc. Lecturer in Agriculture, Jamaica, 1899. Assistant, Dept. of Agriculture, West Indies, 1903. Economic Botanist, Bengal, 1907.
Kew Bull. 1907, 404.

BUTTERFIELD, William (*fl.* 1820s)
Nurseryman and seedsman, Leyburn, Wensley, Yorks.

BUTTERWORTH, Alan (1864–1937)
b. 25 July 1864 *d.* 25 May 1937
Indian Civil Service, 1883 in Madras and Bombay Presidencies and Central Provinces. *Some Madras Trees* 1911.
Who was Who, 1929–1940 203.

BUTTLE, John (*fl.* 1850s–1860s)
Member of British Columbia Boundary Commission with David Lyall, 1858–61. Collected plants.
Contrib. U.S. National Herb. v.11, 1906, 11.

BUXTON, Bertram Henry (1852–1934)
d. Devon 5 Dec. 1934
Educ. at Cambridge. Bacteriologist in U.S.A., 1892–1912. On return to England worked with F. V. Darbishire on effect of varying hydrogen ion concentrations on colour pigments of plants.
Times 14 Dec. 1934. *Nature* v.135, 1935, 14–15.

BUXTON, E. Charles (*c.* 1838–1925)
d. Bettws-y-Coed, Caernarvonshire 11 July 1925
Amateur gardener who bred many new varieties of flowers.
Garden 1925, 425.

BUXTON, Eric William (1926–1964)
b. Ripley, Derbyshire 9 June 1926 *d.* Harpenden, Herts 4 Aug. 1964
BSc Cantab 1951. PhD 1954. Plant Pathologist, Rothamsted Experimental Station, 1954. Studied genetic variability.
Nature v.204, 1964, 426–27. *B.M.S. News Bull.* no. 23, 1964–65, 29.

BUXTON, Richard (1786–1865)
b. Prestwich, Manchester 15 Jan. 1786 *d.* Manchester 2 Jan. 1865
Shoe repairer and newsman. *Botanical Guide to...Manchester* 1849; ed. 2 1859. Contrib. to J. B. Wood's *Fl. Mancuniensis* 1840.

Hardwicke's Sci. Gossip 1865, 66. *J. Bot.* 1865, 71–72. *D.N.B.* v.8, 106–7. J. Cash *Where There's a Will There's a Way* 1873, 94–107. J. B. L. Warren *Fl. Cheshire* 1899, lxxix. *Trans Liverpool Bot. Soc.* 1909, 63. *Heywood Advertiser* 26 April 1918. *N. Western Nat.* 1931, 18–21 portr. *Manchester Guardian* 6 June 1944. *Sorby Rec.* v.2 (2), 1966, 34.
Plants at Oxford.

BYERLEY, Isaac (*c.* 1814–1897)
b. Isle of Wight *c.* 1814 *d.* 20 June 1897
FRCS 1857. FLS 1854. Physician at Upton until 1854, then Seacombe. Edited flora of Upton and neighbourhood, Cheshire.
Proc. Linn. Soc. 1897–98, 36.

BYNOE, Benjamin (*c.* 1803–1865)
d. 13 Nov. 1865
Surgeon, Royal Navy, 1825–63; on 'Beagle', 1837–43. Collected plants in New South Wales and Victoria.
J. D. Hooker *Fl. Tasmaniae* v.1, 1859, cxvii. *J. W. Austral. Nat. Hist. Soc.* no. 6, 1909, 10–11. *J. Linn. Soc.* v.45, 1920, 194. *Emu* v.38 (2), 1938, 154.
Plants at BM(NH), Kew.
Acacia bynoeana Benth.

BYRNE, Peter (*fl.* 1780s)
Nurseryman, County Kilkenny, Ireland.

CADELL, Henry Moubray (1860–1934)
b. 30 May 1860 *d.* Edinburgh 10 April 1934
Educ. Edinburgh. Geologist, H.M. Geological Survey. Palaeobotanist.
Nature, v.133, 1934, 822–23.

CADMAN, Colin Houghton (1916–1971)
b. Glasgow *d.* Dundee, Angus 27 Sept. 1971
BSc Liverpool. PhD Edinburgh 1942. FRSE 1950. Plant Pathologist, Scottish Raspberry Investigation, 1943. Head, Plant Pathology Section, Scottish Horticultural Research Institute, 1951; Director, 1965–71. President, Association of Applied Mycologists, 1971.
Ann. Applied Biol. 1971, 277–78 portr. *Hort. Res.* 1971, 125–26. *Annual Rep. Scott. Hort. Res. Inst.* 1971, 9–10 portr. *Yb. R. Soc. Edinburgh* 1971–72, 34–35. *Chronica Horticulturae* 1972, 23–24 portr.

CAFFREY, Michael (*c.* 1889–1959)
b. Lughill, County Kildare *d.* 17 Sept. 1959
Educ. Royal College of Science, Dublin. Assistant, Plant Breeding Station, Dept. of Agriculture, Dublin; Head Lecturer in Plant Breeding, National University of Ireland, 1927; Professor, 1938–59.
Nature v.184, 1959, 1359–60.

CAIE, John (1811–1879)
b. Renton, Dunbartonshire 15 April 1811 *d.* Dunoon, Argyll 22 Sept. 1879
Gardener, Glasgow Botanic Gardens where he attended botanical lectures and excursions by W. J. Hooker. Gardener to Duke of Bedford then to Duke of Argyll at Inverary Castle, 1856–79. Pioneer in "bedding sytem" of flower gardening.
Gdnrs Chron. 1875 ii 453 portr.; 1879 ii 442, 489. *Gdnrs Yb. Almanack* 1880, 192.

CAIRNS, Hugh (1904–1941)
d. Ballyrickard, Comber, County Down 1941
BSc Belfast 1927. B.Agr 1928. Assistant to Head of Plant Diseases Division, Ministry of Agriculture. Assistant, Dept. of Agricultural Botany, Belfast University; Lecturer, Botany Dept. Plant pathologist.
Irish Nat. 1942, 15.

CAIRNS, John (*c.* 1839–1906)
b. near Colquhoun, Loch Lomond *c.* 1839 *d.* 15 March 1906
Gardener to Lord Home at Hirsel, Berwickshire.
Gdnrs Chron. 1906 i 206 portr.

CAIUS, John *alias* Key (1510–1573)
b. Norwich, Norfolk 6 Oct. 1510 *d.* London 29 July 1573
MA Cantab 1535. MD Padua 1541. Lectured on anatomy in London, 1544–64. Physician to Edward VI, Mary and Elizabeth. Communicated histories of rare plants to C. Gesner. *De Rariorum Animalium et Stirpium Historia* 1570.
W. Munk *Roll of R. College of Physicians* v.1, 1878, 37–49. *D.N.B.* v.8, 221. E. J. L. Scott *Index to Sloane Manuscripts* 1904, 91.
Three portr. at Caius College, Cambridge.

CALCOENSIS, Henricus (*fl.* 1490s)
Scotch? Benedictine prior. *Synopsis Herbaria* MS. Translated Palladius *De Re Rustica* into Gaelic, *c.* 1493.
R. Pulteney *Hist. Biogr. Sketches of Progress of Bot. in England* v.1, 1790, 24. G. W. Johnson *Hist. English Gdning* 1829, 46.

CALDCLEUGH, Alexander (*fl.* 1820s–1858)
d. Valparaiso, Chile 11 Jan. 1858
FRS 1831. FLS 1823. At Croydon, 1823–35; Coquimbo, 1836–51. Attached to British Embassy at Rio. Visited Buenos Aires, Chile and Peru, 1819–21. Collected plants at Santiago and Coquimbo. Sent plants to A. B. Lambert. *Travels in South America...1819–1821* 1825 2 vols.
Edinburgh New Philos. J. v.9, 1830, 92. *Bot. Misc.* 1832–33, 303–4. A. Lasègue *Musée Botanique de B. Delessert* 1845, 259. *Taxon* 1970, 517.
Letters and Chile plants at Kew.
Caldcluvia D. Don.

CALDER, Charles Cumming (1884–1962)
b. Edinkillie, Morayshire Dec. 1884 *d.* Aberdeen April 1962

BSc Aberdeen 1908. FLS 1912. Assistant to J. W. H. Trail, Aberdeen University. Curator, Herb. Calcutta Botanic Gardens, 1912–23. Director, Botanical Survey of India, 1923–37. *Handbook of Common Water and Marsh Plants of India and Burma* (with K. P. Biswas) 1936.

Kew Bull. 1912, 109; 1925, 433. *Nature* v.196, 1962, 518–19.

Iraq plants at BM(NH).

CALDER, John (*fl.* 1880s)
Captain in Imperial Chinese Navy. Accompanied Rev. B. C. Henry to Lo Fou Shan Mountains, 1883.

E. Bretschneider *Hist. European Bot. Discoveries in China* 1898, 763–64.

CALDER, Marcus (*fl.* 1880s–1910s)
MD. Secretary, Greenock Natural History Society. Botanised in Greenock, Argyllshire, Bute and Cumbrae. Specialised in lichens.

Glasgow Nat. 1919–30, 52.

Herb. at Greenock Museum.

CALDWELL, Alfred (1852–1934)
Nurseryman, Knutsford, Cheshire. Son of William George Caldwell (1824–1873); succeeded to his father's business in 1873; later retired in favour of William Caldwell (1855–1918).

CALDWELL, Andrew (1733–1808)
b. Dublin 19 Dec. 1733 *d.* Bray, County Wicklow 2 July 1808

FLS 1796. Irish barrister, 1760. Corresponded with J. E. Smith on Irish plants.

P. Smith *Correspondence of Sir J. E. Smith* v.2, 1832, 123–66. *D.N.B.* v.8, 247–48. *J. Bot.* 1916, 173–80.

CALDWELL, Arthur (1865–1939)
Nurseryman, Knutsford, Cheshire. Son of William Caldwell (1824–1873). Joined the family nursery in 1881.

CALDWELL, James (–1795)
Nurseryman, Wavertree, Liverpool.

CALDWELL, John (1797–1840)
Nurseryman, Knutsford, Cheshire. Son of William Caldwell (1766–1844). Partner with his brother, William Caldwell (1789–1852) in 1825 in William Caldwell and Sons, High Street, Knutsford.

CALDWELL, John (1903–1974)
b. 8 May 1903 *d.* 26 Aug. 1974

PhD Cantab 1929. Virus physiologist, Rothamsted Experimental Station, Harpenden, 1929–35. Professor of Botany, Exeter, 1935–69. Published papers on plant physiology.

Times 30 Aug. 1974.

CALDWELL, Thomas (*fl.* 1790s)
Nurserymen, Wavertree, Liverpool. Succeeded James Caldwell (–1795). Bought Radshaw Nook Nursery, Knowsley, Lancs from his brother William Caldwell, 1796.

CALDWELL, William (–*c.* 1813)
Corn-dealer and nurseryman, Knowsley, Lancs. In partnership with John Nickson and John Carr.

CALDWELL, William (1766–1844)
b. 29 Sept. 1766

Nurseryman, High Street, Knutsford, Cheshire. In partnership with John Carr, later with Joseph Picken.

CALDWELL, William (1789–1852)
Nurseryman, Knutsford, Cheshire. Son of William Caldwell (1766–1844).

CALDWELL, William (1855–1918)
Nurseryman, Knutsford, Cheshire. Son of W. G. Caldwell (1824–1873).

CALDWELL, William (1887–1953)
Nurseryman, Knutsford, Cheshire. Son of William Caldwell (1855–1918).

CALDWELL, William George (1824–1873)
Nurseryman, Knutsford, Cheshire. Son of William Caldwell (1789–1852).

CALEY, George (1770–1829)
b. Craven, Yorks 10 June 1770 *d.* Bayswater, London 23 May 1829

Began life as stable-boy. Was introduced to Manchester School of Botanists through W. Withering. Found *Discelium nudum* in 1795 near Manchester. Appointed by Sir J. Banks as a botanical collector in 1799. In New South Wales, 1800–10. Superintendent, Gardens St. Vincent, W. Indies, 1816–22. *Reflections on Colony of New South Wales* (edited by J. E. B. Currey) 1967.

R. Brown *Prodromus Florae Novae Hollandiae* 1810, 329. *Trans. Linn. Soc.* v.15, 1827, 176. *Mag. Nat. Hist.* 1829, 310–12; 1830, 226–29. *Mem. Lit. Philos. Soc. Manchester* 1842, 313–16. A. Lasègue *Musée Botanique de B. Delessert* 1845, 278–79. J. D. Hooker *Fl. Tasmaniae* v.1, 1859, cxxiv. J. Cash *Where There's a Will There's a Way* 1873, 21–40. *Gdns Chron.* 1885 ii 263–64; 1946 i 6. *Kew Bull.* 1891, 303; 1892, 97–98. F. M. Bladen *Hist. Rec. N.S.W.* v.3, 1895 *passim*; v.5, 1897, 718–27. I. Urban *Symbolae Antillanae* v.3, 1902, 30–31. *Agric. Gaz. N.S.W.* 1903, 988–96. *J. R. Soc. N.S.W.* 1908, 94–95; 1921, 154–55. J. H. Maiden *Sir Joseph Banks* 1909, 57, 127–41, 210. *Trans. Liverpool Bot. Soc.* 1909, 62–63. T. Whitley *Blue Mountains Explorations* 1909, 15–16. I. Lee *George Caley* 1910. E. Smith *Life of Sir Joseph Banks* 1911, 224–29. I. Lee *Early Explorers in Australia* 1925, 131–53. *J. Proc. Parrametta District Hist. Soc.* v.3,

1926, 39–48. *J. Proc. R. Austral. Hist. Soc.* 1934, 301–2; 1939, 437–542; v.20, 164–96. G. Mackaness *Sir Joseph Banks* 1936, 115–24. *Austral. Dict. Biogr.* v.1, 1966, 194–95. A. M. Coats *Quest for Plants* 1969, 212–15.

MS. Journal, letters, Australian and W. Indian plants at BM(NH). MS. Journal at Mitchell library, Sydney. Library and plants sold by Christie 19 June 1829.

Caleana R. Br.

CALLCOTT, Lady Maria (*formerly* **Graham** *née* **Dundas**) (1785–1842)
b. Cockermouth, Cumberland 19 July 1785 *d.* Kensington, London 28 Nov. 1842

Married Thomas Graham, 1809; married A. W. Callcott, 1827. Travelled in India. Collected plants in Brazil, now at Kew. *Journal of Residence in India* 1812. *Letters on India* 1814. *Journal of Voyage to Brazil* 1824. *Journal of Residence in Chile* 1824. *Scripture Herbal* 1842.

Bot. Mag. 1826, t.2644. K. M. Lyell *Life and Letters and Journals of Sir C. J. F. Bunbury* v.1, 1906, 79. *D.N.B.* v.8, 258. R. B. Gotch *Maria, Lady Callcott* 1937. B. Howe *A Galaxy of Governesses* 1954.

Letters and drawings at Kew. Portr. at National Portrait Gallery.

Graemia Hook. *Escallonia callcottiae* H. & A.

CALLEN, Eric Ottleben (1912–1970)
b. Erfurt, Germany 21 Oct. 1912 *d.* Peru 22 Aug. 1970

BSc Edinburgh 1936. PhD 1939. FLS 1941. Lecturer, Edinburgh University, 1938. Dept. of Plant Pathology, McDonald College, Quebec, 1947. Studied taxonomy of *Lotus*. Also interested in ethnobotany.

Watsonia 1972, 53.

CALLENDER, Ebenezer Romain (*fl.* 1790s)
Nurseryman and seedsman, Brig-gate, Leeds, Yorks.

CALLENDER, Michael, John and **William R.** (*fl.* 1790s)
Nurserymen, The Orange Tree, Middle Street, Newcastle, Northumberland.

CALVERT, Caroline Louisa Waring (*née* **Atkinson**) (1834–1872)
b. Oldbury, Argyle County, N.S.W. 25 Feb. 1834 *d.* Sutton Forest, N.S.W. 28 April 1872

Married J. S. Calvert, 1870, fellow-traveller of Leichhardt. Drew Australian plants. Sent plants to William Woolls and F. von Mueller.

F. von Mueller *Fragmenta Phytographiae Australiae* v.5, 1865, 34; v.8, 1872, 52–53. F. Boase *Modern English Biogr.* v.1, 1892, 519. *Proc. R. Soc. N.S.W.* 1908, 83 portr. P. Mennell *Dict. Austral. Biogr.* 1892, 76–77. *J. Proc. R. Austral. Hist. Soc.* 1929, 1–29. *Dict. Austral. Biogr.* v.1, 1949, 140–41. *Austral. Encyclop.* v.1, 1962, 294–95. *Austral. Dict. Biogr.* v.3, 1969, 59–60.

Atkinsonia F. Muell. *Epacris calvertiana* F. Muell.

CALVERT, E. (*fl.* 1860s)
Collected plants in Rumania, Bulgaria and Asia Minor, now at Oxford.

CALVERT, Henry Hunter (*c.* 1816–1882)
d. Dardanelles, Turkey 29 July 1882

Vice-Consul, Alexandria, 1857. Collected plants with J. Zohrab in Erzurum and Jeddah in 1848–57. Sent plants to J. Lindley.

Bull. Inst. Egypt v.3, 1882, 77–80.

Plants at Cambridge, Kew, Oxford. Letters at Kew.

Senecio calverti Boiss.

CALVERT, James Snowden (1825–1884)
b. Otley, Yorks? 13 July 1825 *d.* Sydney 22 July 1884

Emigrated to Australia, 1841. Joined F. W. L. Leichhardt's expedition to Port Essington, 1844, during which he collected plants.

J. Proc. R. Austral. Hist. Soc. 1929, 20. A. H. Chisholm *Strange New World* 1941, 75 *et seq. Austral. Encyclop.* v.1, 1962, 241. *Austral. Dict. Biogr.* v.3, 1969, 333.

CALVERT, John (*fl.* 1780s)
Seedsman, Bedale, Yorks.

CAMBRIDGE, O. P. (*fl.* 1870s)
Bryophytes from Dorset at Oxford.

CAMERON, David (*c.* 1787–1848)
d. Shrawley, Worcs 25 June 1848

ALS 1827. Gardener to R. Barclay at Buryhill, Surrey until 1829. First Curator, Birmingham Botanic Garden, 1831–47. Contrib. to *Phytologist*, *Gdnrs Mag.* and *Floral Cabinet*.

Proc. Linn. Soc. 1849, 50–52. *Gdnrs Chron.* 1848, 435.

Letters and cultivated plants at Kew.

Hibiscus cameroni Knowles & Westc.

CAMERON, Donald (*fl.* 1830s–1850s)
MA. LLD. Of Rodney Street, Liverpool. Collected plants in Pyrenees and S. France, 1835–55, now at Liverpool Museums.

Liverpool Museums *Handbook and Guide to Herb. Collections* 1935, 70.

CAMERON, James (*fl.* 1830s)
Nurseryman, Uckfield, Sussex.

CAMERON, Kenneth J. (–1918)
d. Zomba, Nyasaland 26 Jan. 1918

African Lakes Corporation. Had estate at Ntondure. Collected plants in Nyasaland, 1896–99, 1905, and the Cape, 1913.

Kew Bull. 1907, 234. *Comptes Rendus AETFAT 1960* 1962, 164.

Plants at Kew, BM(NH).

CAMERON, Robert (1860–1957)
b. Inverness-shire 27 Feb. 1860 *d.* Winchester, Mass. 29 Aug. 1957
Kew gardener. To U.S.A., 1887. Curator, Botanic Garden, Harvard University, 1887–1912. Superintendent, Richard T. Crane Estate, Ipswich, Mass., 1919–34.
J. Kew Guild 1894, 43–45; 1957, 489–90.

CAMERON, Verney Lovett (1844–1894)
b. Radipole, Dorset 1 July 1844 *d.* Leighton Buzzard, Beds 27 March 1894
Midshipman, Royal Navy, 1860. Employed in suppression of slave trade in E. Africa. Leader of expedition to aid Livingstone, 1873. Collected plants near Lake Tanganyika, 1875. 'Enumeration of Plants collected by V. Lovett Cameron...about Lake Tanganyika' (*J. Linn. Soc.* v.15, 1876, 90–97). *Across Africa* 1877 2 vols.
Geogr. J. 1894, 429–31. *Scott. Geogr. Mag.* 1895, 22–23. *D.N.B. Supplt. 1* v.1, 379–81. *Kew Bull.* 1960, 319. *Comptes Rendus AETFAT 1960* 1962, 176–77. *R.S.C.* v.14, 29.
Plants at Kew.
Indigofera cameroni Baker.

CAMERON, William (*fl.* 1850s)
Kew gardener, March–Aug. 1857. To Ceylon in 1857 to be Superintendent of Peradeniya Garden under G. H. K. Thwaites. Left government service in 1860 for coffee planting.

CAMFIELD, Julius Henry (1852–1916)
b. Islington, London 30 March 1852 *d.* Sydney 26 Nov. 1916
Gardener. To Sydney, 1882. Overseer, Sydney Botanic Garden. Collected plants in New South Wales, 1895, 1906–*c.* 1910.
Rep. Sydney Bot. Gdn 1916, 12. *J. Proc. R. Austral. Hist. Soc.* 1932, 132.
Plants at Sydney Botanic Garden.

CAMPBELL, Alexander (*c.* 1795–1877)
d. Manchester 1877
Curator, Manchester Botanic Garden, 1831.
Gdnrs Yb. Almanack 1878, 191.

CAMPBELL, Alexander (*c.* 1804–1871)
d. Glasnevin, Dublin 30 Oct. 1871
Seedsman and florist, Dublin.

CAMPBELL, Rev. Alfred John (–*c.* 1932)
Minister, St. Annes Primitive Methodist Church.
Minutes Methodist Conference 1932, 302–4.
Herb. at Lytham St. Annes Library.

CAMPBELL, Andrew (1845–1947)
b. Ardross, Fife March 1845
Nurseryman, Harrogate, Yorks. Specialised in dwarf conifers.
Quart. Bull. Alpine Gdn Soc. 1948, 6.

CAMPBELL, Archibald, 3rd Duke of Argyll (1682–1761)
b. Petersham, Surrey June 1682 *d.* London 15 April 1761
Planted his estate at Whitton in Twickenham with many unusual trees. Called "tree-monger" by Horace Walpole. His nephew, Lord Bute, removed some of the trees to Princess Augusta's garden at Kew in 1762.
Trans. Linn. Soc. v.10, 1811, 275. J. C. Loudon *Arboretum et Fruticetum Britannicum* v.1, 1838, 57–58. L. W. Dillwyn *Hortus Collinsonianus* 1843, 32. *Kew Bull.* 1891, 292. J. Lucas *Kalm's Account of his Visit to England on his Way to America in 1748* 1892, 57–59. *D.N.B.* v.8, 341–42. *Country Life* 1972, 142–45 portr. E. and D. S. Berkeley *Dr. John Mitchell* 1974, 125–43.

CAMPBELL, Archibald (1805–1874)
b. Kildalton, Argyll 20 April 1805 *d.* 5 Nov. 1874
Assistant Surgeon, East India Company, 1827. Superintendent of Darjeeling in charge of relations with Sikkim, 1840–62. Collected with J. D. Hooker in Sikkim in 1849.
L. Huxley *Life and Letters of Sir Joseph Dalton Hooker* 1918 2 vols. *passim.*
Plants at Calculta, Kew.

CAMPBELL, Charles (*c.* 1766–1808)
b. Glasgow *c.* 1765 *d.* Calcutta, India 20 Jan 1808
Assistant Surgeon, Bengal, 1791. Superintendent, East India Company's spice plantations, Fort Marlborough, Sumatra, 1793. Supplied William Roxburgh at Calcutta Botanic Garden with Sumatran plants and seeds.
Fl. Malesiana v.5, 1958, 33. *J. Soc. Bibl. Nat. Hist.* 1973, 364–65.
Plants at Calcutta Botanic Garden.

CAMPBELL, Eugène J. F. (*fl.* 1880s–1920s)
Superintendent, King's House Garden, Jamaica, 1894–96. Curator, Botanic Station, British Honduras, 1896–1921. Collected plants.
I. Urban *Symbolae Antillanae* v.3, 1902, 31. *Kew Bull.* 1921, 222.
Plants at Kew, New York Botanic Garden.

CAMPBELL, Francis Maule (1843–1920)
b. Edmonton, Middx Aug. 1843 *d.* Nutfield, Surrey 31 Dec. 1920
FLS 1878. Entomologist. 'Means of Protection Possessed by Plants' (*Trans. Herts Nat. Hist. Soc.* v.5). President, Herts Natural History Society.
Proc. Linn. Soc. 1920–21, 44–45.

CAMPBELL, George (–1780)
d. Madras, India 18 Sept. 1780
Surgeon on Madras establishment. To Pullicate Hills with J. G. Koenig, 1766. MS. descriptions in Koenig's MS. in BM(NH).
W. Roxburgh *Plants of Coast of Coromandel* v.1, 1795 preface, ii.

CAMPBELL, James (*fl.* 1710s–1720s)
Surgeon at Port Mahon, Minorca. Sent plants to J. Petiver.
J. E. Dandy *Sloane Herb.* 1958, 110.
Plants at BM(NH).

CAMPBELL, James John (1884–1949)
b. Perthshire *d.* 27 Jan. 1949
Gardener, Botanic Garden, Edinburgh, 1905; foreman, 1911; Assistant Curator.
Gdnrs Chron. 1949 i 76.

CAMPBELL, John (–1804)
Nurseryman, High Street, Hampstead, London.

CAMPBELL, John (*fl.* 1830s)
Brother of W. H. Campbell (1814–1883). Army officer in India. Collected plants in Hyderabad and Circars, 1835–37. Sent plants to R. Wight.
I. H. Burkill *Chapters on Hist. Bot. in India* 1967, 56.
Campbellia Wight.

CAMPBELL, John William (1878–1929)
b. Killarney, County Kerry 17 Dec. 1878 *d.* Sidcup, Kent 22 May 1929
Kew gardener, 1902. Superintendent, Government Gardens and Experimental Plantations, Perak, 1906–1910. Estate Manager, Malacca.
Kew Bull. 1904, 13; 1906, 383; 1909, 342. *J. Kew Guild* 1931, 75–76 portr. *Times* 9 April 1945.

CAMPBELL, Matthew (*c.* 1835–1915)
d. 20 Aug. 1915
Nurseryman, Blantyre, Lanarkshire.
Garden 1915, 430.

CAMPBELL, Robert (*fl.* 1670s)
Collected plants in Oxford and London, now at BM(NH).
J. E. Dandy *Sloane Herb.* 1958, 110.

CAMPBELL, Rev. W. (*fl.* 1870s)
Joined English Presbyterian Mission in China, 1871. Collected plants in Formosa, now at BM(NH).
Kew Bull. 1896, 66. E. Bretschneider *Hist. European Bot. Discoveries in China* 1898, 702–3.

CAMPBELL, William Hunter (1814–1883)
b. Edinburgh 1814 *d.* London 3 Nov. 1883
LLD Edinburgh. First Secretary, Botanical Society of Edinburgh, 1836. Collected Scottish plants. *Catalogue of British Plants* (with J. H. Balfour and C. C. Babington) 1841. To Demerara, 1836; attorney in Georgetown. 'Vegetation of Georgetown' (*Ann. Mag. Nat. Hist.* v.10, 1842, 349–52).
Trans. Bot. Soc. Edinburgh v.5, 1858, 25. *Timehri* 1883, 366–70; 1886, 128.
Plants at BM(NH), Edinburgh, Kew. Letters at Kew. Bust at Botanic Garden, Georgetown.

CAMPBELL, William Macdonald (1900–1964)
b. May 1900 *d.* 3 Oct. 1964
VMH 1957. Kew gardener, 1922–24. Superintendent, Parks Dept., Southend-on-Sea, 1932. Curator, Kew Gardens, 1937–60.
Gdnrs Chron. 1932 i 214 portr.; 1937 ii 264 portr. *Daily Telegraph* 6 Oct. 1964. *J. Kew Guild* 1964, 456–57 portr.

CANDI, Thomas (*also* **Candish, Cavendish**) (*c.* 1555–1592)
b. Trimley, Suffolk *c.* 1555 *d.* on voyage home June 1592
Circumnavigator. Sailed in 'Tiger', 1586–88 to S. America, East Indies, Cape of Good Hope. Collected plants in Manila?
D.N.B. v.9, 358. *Fl. Malesiana* v.1, 1950, 99–100.

CANDISH, Thomas *see* Candi, T.

CANDLER, Edmund (*fl.* 1890s)
Traveller. Collected plants in Siam and Burma, now at Kew. *Vagabond in Asia* 1900.
J. Thailand Res. Soc. Nat. Hist. Supplt v.12, 1939, 23–24.

CANDLER, Howard (1838–1916)
b. 3 April 1838
Mathematics master, Uppingham School, 1860–1900. Made botanical contributions to *Victoria County History of Rutland* 1908.
A. R. Horwood and C. W. F. Noel *Fl. Leicestershire* 1933, cclxxviii–cclxxix.

CANNELL, Henry (1833–1914)
b. Swardeston, Norfolk March 1833 *d.* Swanley, Kent 25 Oct. 1914
VMH 1902. Nurseryman, first at Woolwich then at Swanley. Specialised in chrysanthemums and pelargoniums.
J. Hort. Cottage Gdnr v.21, 1890, 383 portr.; v.45, 1902, 219, 221 portr.; v.59, 1909, 104–6 portr.; v.69, 1914, 280. *Garden* v.63, 1903, 59 portr. *Gdnrs Chron.* 1914 ii 300–1. *Gdnrs Mag.* 1914, 776.

CANNELL, Robert (*fl.* 1910s)
Son of H. Cannell (1833–1914). Nurseryman, Swanley, Kent.
J. Hort. Home Farmer v.61, 1910, 370–71 portr.

CANNING, Edward J. (1863–1921)
b. Stratford-on-Avon, Warwickshire 19 Jan. 1863 *d.* Northampton, Mass. 1 Nov. 1921
Kew gardener, 1887. Gardener, Messrs T. Meehan and Sons, Philadelphia. Curator, Botanic Garden, Smith College, Northampton, Mass., *c.* 1889. Nurseryman.
J. Kew Guild 1922, 114.

CANNON, Martin (*fl.* 1790s)
Seedsman, Louth, Lincs.

CANT, Benjamin Revett (1827–1900)
d. Colchester, Essex 17 July 1900
Nurseryman, Colchester, Essex. Inherited nursery started by his grandfather in 1766. Specialised in roses.
Garden v.58, 1900, 73 portr., 87. *Gdnrs Chron.* 1900 ii 79. *Gdnrs Mag.* 1900, 447 portr., 482. *J. Hort. Cottage Gdnr* v.41, 1900, 85 portr.

CANT, Cecil E. (*fl.* 1890s–1910s)
Son of Benjamin Cant. Rose nurseryman, Colchester, Essex.
Gdnrs Mag. 1894, 397 portr. *J. Hort. Home Farmer* v.63, 1911, 417–18 portr.

CANT, Frank (*c.* 1857–1928)
d. Colchester, Essex 22 Aug. 1928
Founded Braiswick Rose Nurseries, Colchester.
J. Hort. Cottage Gdnr v.45, 1902, 474, 475 portr. *Gdnrs Mag.* 1909, 463–64 portr. *Gdnrs Chron.* 1922 ii 104 portr.; 1928 ii 180 portr.

CANT, George (–1805)
d. 19 June 1805?
Gardener and seedsman, Colchester, Essex.

CANT, William (1742–1805)
d. 27 Oct. 1805
Gardener and seedsman, Gutter Street, Colchester, Essex.

CANT, William (1779–1831)
d. 15 Jan. 1831
Nurseryman, Colchester, Essex.

CANTLEY, Nathaniel (–1888)
b. Thurso, Caithness *d.* Tasmania 29 Feb. 1888
Kew gardener, 1869–72. Assistant director, Mauritius Botanic Garden, 1873. Superintendent, Singapore Botanic Gardens, 1880–88.
J. Kew Guild 1898, 37 portr. *Gdns Bull. Straits Settlements* v.2, 1918, 101–3. H. N. Ridley *Fl. Malay Peninsula* v.1, 1922, xvii, 436. *Fl. Malesiana* v.1, 1950, 100. *R.S.C.* v.14, 48.
Herb. at Singapore. Letters and plants at Kew.
Cantleya Ridley.

CANTOR, Theodore Edward (1809–1854)
b. Copenhagen 1809
MD Halle 1833. Bengal Medical Service, 1835–39. Zoologist. Collected plants in Malaya and China, 1840–41. 'General Features of Chusan with Remarks on Flora and Fauna of that Island' (*Ann. Nat. Hist.* 1842, 265–78, 361–71, 481–94).
A Lasègue *Musée Botanique de B. Delessert* 1845, 436. *Botanisk Tidsskrift* (Copenhagen) v.12, 1880–81, 184. E. Bretschneider *Hist. European Bot. Discoveries in China* 1898, 359. D. G. Crawford *Hist. Indian Med. Service, 1600–1913* v.1, 1914, 504. M. Archer *Nat. Hist. Drawings in Inida Office Library* 1962, 51–53, 76–77. *R.S.C.* v.1, 779.
Chusan plants at Kew. Chinese flower drawings at India Office Library.
Bambusa cantori Munro.

CAPARNE, W. J. (1855–1940)
b. Newark, Notts 1855 *d.* Guernsey, Channel Islands 31 Jan. 1940
Botanical artist and plant hybridist. Illustrated Sir M. Foster's *Bulbous Irises* 1893.
Gdnrs Chron. 1940 i 121–22.

CAPEL, Sir Henry, Baron Capel of Tewkesbury (–1696)
Lord Deputy of Ireland, 1695. Had celebrated garden at Kew with "the choicest fruit of any plantation in England" (Evelyn). His Kew property was leased to Frederick, Prince of Wales in 1730, and later transformed into part of Royal Botanic Gardens, Kew.
Archaeologia v.12, 1794, 185. *Kew Bull.* 1891, 288–89. *D.N.B.* v.9, 17. M. Hadfield *Gdning in Britain* 1960, 129–30.

CAPEL, Mary *see* Somerset, Mary, *Duchess of Beaufort*

CAPPER, Walter William (1772–1834)
d. Hanley Castle, Malvern, Worcs 15 Oct. 1834
Of Bath and Hanley Castle. 'On the Anatomy of the Vine' (*Gdnrs Mag.* 1830, 12–25).
Gdnrs Mag. 1835, 56.

CAPRON, Edward (–1907)
d. Winterbourne, Glos 1907
MD. Of Shere, Surrey. Had herb. Drew fungi. Contrib. to J. A. Brewer's *Fl. Surrey* 1863. 'Mosses of Dorking' (*Sci. Gossip* 1872, 35).
Mosses at BM(NH).
Sphoerella capronii Sacc.

CARBONELL, William Charles (1887)
Of Usk, Monmouthshire. Fern pupil of A. M. Jones (1826–1889). Raised angulares.
Br. Fern Gaz. 1909, 43. *Garden* 1883, 359–60.
Fern collection bequeathed to Kew.

CARDER, John (–1908)
d. 7 Dec. 1908
Collected orchids for William Bull of Chelsea. In partnership with Shuttleworth, orchid nurseryman.
Gdnrs Mag. 1908, 955. *Orchid Rev.* 1909, 18.
Masdevallia carderi Reichenb. f.

CAREW, Sir Francis (*c.* 1530–1611)
d. 16 May 1611
Had garden at Beddington near Croydon, Surrey where he cultivated some of the first oranges in England.
M. Hadfield *Gdning in Britain* 1960, 52–53.

CAREW, Richard (1555–1620)
b. East Antony, Cornwall 1555 *d.* East Antony 6 Nov. 1620

His *Survey of Cornwall*, 1602 contained records of Cornish plants.
F. H. Davey *Fl. Cornwall* 1909, xxx.

CAREY, Rev. Felix (–1822)
d. 1822
Missionary. Son of Rev. W. Carey (1761–1834). Sent Burmese plants to W. Roxburgh at Calcutta in 1809.
S. P. Carey *William Carey* 1824 *passim.*

CAREY, John (1797–1880)
b. Camberwell, London 21 June 1797 *d.* Blackheath, Kent 26 March 1880
FLS 1828. To U.S.A., 1830. With Asa Gray, American botanist, in N. Carolina, 1841. Returned to England, 1852. Contrib. *Salix* and *Carex* to A. Gray's *Manual of Botany of Northern United States* 1848. Contrib. to *Amer. J. Sci.*, 1847–53.
J. Torrey *Fl. State of New York* v.1, 1843, viii. *Bot. Gaz.* 1880, 61, *Amer. J. Sci.* v.19, 1880, 421–23. *Rhodora* 1901, 207. C. S. Sargent *Silva of N. America* v.1, 1891, 115–16. *R.S.C.* v.1, 785.
U.S. herb. at Kew. Plants at Winchester College. Portr. at Hunt Library.
Saxifraga careyana A. Gray.

CAREY, Rev. William (1761–1834)
b. Paulerspury, Northants 17 Aug. 1761 *d.* Serampore, India 9 June 1834
FLS 1823. Baptist missionary in India, 1794. Moved to Serampore, 1799. Professor of Sanskrit at Fort William College, 1801. Founded Botanic Garden, Serampore. Joint editor of W. Roxburgh's *Fl. Indica* 1820–24 2 vols.
Gent. Mag. 1835 i 547–53. E. Carey *Mem. of William Carey* 1837. G. Smith *Life of William Carey* 1885 *D.N.B.* v.9, 77. *J. Bot.* 1904, 296. C. E. Buckland *Dict. Indian Biogr.* 1906, 73. *Notes and Queries* 1914, 103–4, 177–78. S. P. Carey *William Carey* 1923 portr. F. D. Walker *William Carey* 1926 portr. *Sci. Culture* v.13, 1947, 218–25. *Bull. Bot. Survey India* v.3, 1961, 1–10. *Fl. Malesiana* v.5, 1958, 33. I. H. Burkill *Chapters on Hist. Bot. in India* 1965, 44–46.
Portr. at Kew, Hunt Library.
Careya Roxb.

CARGILL, James (*c.* 1565–1616)
Studied botany at Basle under Caspar Bauhin. Practised medicine in Aberdeen. Correspondent of C. Gesner, M. Lobel and C. Bauhin. Discovered *Trientalis*. Described *Fuci*.
M. de Lobel *M.G. Rondelletii...Animadversiones* 1605, 485, 507. C. Bauhin *Prodromus Theatri Botanici* 1620, 100, 155. R. Pulteney *Hist. Biogr. Sketches of Progress of Bot. in England* v.2, 1790, 507. *D.N.B.* v.9, 80.
Cargillia R. Br.

CARLES, William Richard (1848–1929)
b. 1 June 1848
FLS 1898. CMG 1901. British Consular Service, China, 1867. Consul-General in Tiensin and Peking in 1900. Collected plants in China, 1867–83; Korea, 1883–85. Botanised with F. B. Forbes and sent plants to H. F. Hance. *Life in Corea* 1888.
E. Bretschneider *Hist. European Bot. Discoveries in China* 1898, 755. *Gdnrs Chron.* 1929 i 338. *Lingnan Sci. J.* v.7, 1929, 122–23. *Proc. Linn. Soc.* 1929–30, 190–91.
Plants at Kew.
Vaccinium carlesii Dunn.

CARLISLE, Sir Anthony (1769–1840)
b. Stillington, Durham 8 Feb. 1769 *d.* London 2 Nov. 1840
FRS 1800. Surgeon, Westminster Hospital, 1793–1840. 'On Connection between Leaves and Fruit of Vegetables' (*Trans. Hort. Soc.* v.2, 1817, 184–88).
Gent. Mag. 1840 ii 660. *D.N.B.* v.9, 103–4.

CARLTON, Henry (*né* Schneider)
(–1917)
d. Queensland 9 Feb. 1917
BA Oxon. To Queensland in 1866 where he farmed sugar-cane. Studied local flora and fauna.
Asplenium attenuatum var. *Schneideri* Bailey.

CARLYON, Miss Harriet (*c.* 1853–1924)
Of Yarmouth, Isle of Wight. Drew British orchids, 1900–14. Album of 39 water-colours for sale at Christie, 9 May 1973.

CARMICHAEL, Dugald (1772–1827)
b. Lismore, Hebrides 1772 *d.* Appin, Argyll Sept. 1827
FLS 1818. Capt., 73rd Regiment. At the Cape, 1806–10, 1814–15. Collected plants at Mauritius and Bourbon, 1810–14; Tristan da Cunha, 1817. India, 1815–17. 'Tristan da Cunha' (*Trans. Linn. Soc.* v.12, 1818, 483–99, 502–13).
Bot. Register 1825, t.912. *Bot. Misc.* 1830–31, 1–59, 258–343; 1832–33, 23–76. W. H. Harvey *Manual of British Algae* 1841, 49. W. J. Hooker *English Fl. of Sir J. E. Smith* v.5 (1), 1833, 284. J. G. Baker *Fl. Mauritius and Seychelles* 1877, 8*–9*. *R.S.C.* v.1, 791.
MS. Catalogue and plants, including algae, at BM(NH). MS. and plants at Kew.
Carmichaelia R. Br.

CARMICHAEL, J. R. (1838–1870s)
b. Blyth, Northumberland 1838
In charge of London Missionary Society Hospital at Canton, 1862. In private medical practice at Che foo, China.
E. Bretschneider *Hist. European Bot. Discoveries in China* 1898, 723.
Aconitum carmichaeli Debeaux.

CARMICHAEL, William (*c.* 1816–1904)
b. Comrie, Perthshire *d.* Edinburgh 6 April 1904
Gardener to Prince of Wales at Sandringham, Norfolk.
Garden v.65, 1904, 278.

CARNEGIE, David Wynford (1871–1900)
b. London 23 March 1871 *d.* Nigeria 27 Nov. 1900
Tea planter, Ceylon. To W. Australia, 1892, prospecting for gold. Assistant Resident, Nigeria, 1899. *Spinifex and Sand* 1898.
Icones Plantarum 1899, t.2582. *Kew Bull.* 1901, 169–70. *J. W. Austral. Hist. Soc.* 1948, 17–20. *Austral. Encyclop.* v.2, 1962, 267–68; v.3, 453.
Australian plants at Kew. Portr. at Hunt Library.
Dicrastylis carnegiei Hemsley.

CAROL, Patrick *see* Carroll, P.

CARPENTER, William Benjamin (1813–1885)
b. Exeter, Devon 29 Oct. 1813 *d.* London 10 Nov. 1885
MD Edinburgh 1839. LLD 1871. FRS 1844. FLS 1856. Lecturer, Bristol Medical School. Registrar, London University, 1856–79. *Botany* 1844. *Vegetable Physiology and Systematic Botany* 1844; 1858. *Microscope and its Revelations* 1856; ed. 6 1881.
Ipswich Mus. Portr. Ser. 1852. *Nature* v.33, 1885, 83–85. *Proc. Linn. Soc.* 1885–86, 138–40. *Proc. R. Soc.* v.41, 1886, ii–ix. *Quart. J. Geol. Soc. London* 1886, 40–43. *Trans. Bot. Soc. Edinburgh* v.16, 1886, 303–8. *D.N.B.* v.9, 166–68. C. C. Gillispie *Dict. Sci. Biogr.* v.3, 1971, 87–89.

CARR, Amos (*c.* 1829–1884)
b. Frant, Sussex *c.* 1829 *d.* Sheffield, Yorks 29 April 1884
Postman and shoe-maker, Sheffield. Knew local plants well and botanised in W. Lancashire. Herb. acquired by Jonathan Salt.
F. A. Lees *Fl. West Yorkshire* 1888, 359. *Naturalist* 1884, 71. *Trans. Liverpool Bot. Soc.* 1909, 63. *Sorby Rec.* v.2 (2), 1966, 35.

CARR, Cedric Erroll (1892–1936)
b. Napier, New Zealand 16 Nov. 1892 *d.* Port Moresby, New Guinea 3 June 1936
FLS 1930. Came to England *c.* 1899. Employed on rubber plantations in Malaya, 1913–31. At Kew Herb., 1933–34. Collected plants, especially orchids, in Malaya, 1930–32, N. Borneo, 1933, New Guinea, 1935–36. Contrib. to *Gdns Bull. Straits Settlements.*
Kew Bull. 1936, 531–32; 1974, 18. *Proc. Linn. Soc.* 1936–37, 189–91. *Chronica Botanica* 1937, 235 portr. *J. Bot.* 1937, 143–44. *Fl. Malesiana* v.1, 1950, 100–101 portr.
Plants at BM(NH), Kew, Singapore.

CARR, John (–*c.* 1803)
Nurseryman, Knutsford, Cheshire. In partnership with John Nickson until 1797; then William Caldwell.

CARR, John (*fl* 1820s)
Educ. Cambridge. Headmaster, Durham grammar school. Lived in Stackhouse near Settle, Yorks. Botanist.
J. Windsor *Fl. Cravoniensis* 1873, iii, viii.

CARR, John Wesley (1862–1939)
b. Wetherby, Yorks 26 Nov. 1862 *d.* Hastings, Sussex Jan. 1939
BA Cantab 1886. MA 1890. FLS 1895. Lecturer in Biology, University College, Nottingham, 1886; Professor of Biology, 1893–1927. Contrib. botanical section of *Victoria County Hist. Nottingham* 1906. MS. flora of Nottingham.
N. Western Nat. 1939, 137–38 portr. *Proc. Linn. Soc.* 1939, 237–38. *Who was Who, 1929–1940* 224. R. C. L. and B. M. Howitt *Fl. Nottinghamshire* 1963, 18–19. *Sorby Rec.* v.2 (2), 1966, 35.
Plants at Derby and Nottingham Museums.

CARR, Richard (*c.* 1830–1887)
d. 13 April 1887
Gardener to Duke of Portland, Welbeck Abbey, Notts.
Gdnrs Chron. 1887 i 556. *J. Hort. Cottage Gdnr* v.14, 1887, 312.

CARRINGTON, Benjamin (1827–1893)
b. Lincoln 18 Jan. 1827 *d.* Brighton, Sussex 18 Jan. 1893
MD Edinburgh 1851. FLS 1861. FRSE. Practised medicine at Radcliffe, Lincoln, Yeadon and Southport. Medical Officer of Health at Eccles for 18 years. Authority on hepatics. President, Manchester Cryptogamic Society. *Fl. W. Riding...of Yorkshire* (with L. C. Miall) 1862. *British Hepaticae* 1874–75. *Hepaticae Britannicae* 1878–90 (exsiccatae).
J. Bot. 1893, 120–22. *Revue Bryologique* 1893, 62–64. *Proc. R. Soc. Edinburgh* v.21, 1895–97, i–v. *Trans. Liverpool Bot. Soc.* 1909, 63–64. *Trans. Bot. Soc. Edinburgh* v.25, 1910, 4–5. *Proc. Irish Acad. B.* v.32, 1915, 75–76. *Naturalist* 1973, 6–7.
Herb. and portr. at Manchester University. Letters at Kew, BM(NH).
Radula carringtoni Jack.

CARRINGTON, Samuel (*fl* 1830s–1870s)
Schoolmaster, Wetton, Staffs. Geologist and botanist. "He was accustomed to make drawings of almost every wild plant he met with. They are extremely accurate."
Trans. N. Staffs. Field Club 1946–47, 123;

1949–50, 93–94. E. S. Edees *Fl. Staffordshire* 1972, 19.
Sketch book at Stafford Public Library.

CARROLL, Isaac (1828–1880)
d. Aghada, County Cork 17 Sept. 1880
Lichenologist. Visited Lapland, 1864, and Iceland. Contrib. to *Cybele Hibernica* 1866, *J. Bot.* 1865–68, *Phytologist*, *Nat. Hist. Rev. Lichenes hibernici* 1859 (exsiccatae).
J. Bot. 1881, 128. B. Lynge *Index... Lichenum Exsiccatorum* 1915, 121–22. *Proc. Irish Acad. B.* 1929, 184–85. *Irish Nat. J.* v.3, 1931, 237. R. L. Praeger *Some Irish Nat.* 1949, 60. *R.S.C.* v.1, 801; v.7, 339.
Herb. and MS. flora of Cork at University College, Cork. Irish and Iceland plants at BM(NH). Portr. at National Museum, Dublin.

CARROLL, Patrick (*fl.* 1780s–1800s)
Nurseryman, Dublin.
Irish For. 1967, 47.

CARRON, William (1823–1876)
b. Norfolk 18 Dec. 1823 *d.* Grafton, N.S.W. 25 Feb. 1876
Gardener. In Sydney, 1843. Botanist to E. B. Kennedy's exploration of Cape York Peninsula, 1848. Plant collector for Sydney Botanic Gardens, 1866–75. Forester on Clarence River, 1876.
E. B. Kennedy's Expedition in J. Macgillivray's *Narrative of Voyage of HMS Rattlesnake* v.2, 1852, 119–227. *J. R. Soc. N.S.W.* 1908, 95–97 portr. *Proc. Linn. Soc.* v.25, 1900, 777–78. *Victorian Nat.* v.48, 1932, 235–36. *J. Proc. R. Austral. Hist. Soc.* 1961, 292–311. *Austral. Encyclop.* v.1, 1962, 270. *Austral. Dict. Biogr.* v.3, 1969, 360–61.
Portr. at Hunt Library.
Carronia F. Muell.

CARROTHERS, Nathaniel (1852–1930)
b. Farmagh Lisbellaw, County Fermanagh 9 Jan. 1852 *d.* Belfast 24 April 1930
Studied flora of Antrim and Down. Herb. acquired by W. J. C. Tomlinson (1863–1921). Contrib. to *Irish Nat.* 1903–13.
Irish Nat. J. 1930, 57 portr. *Bot. Soc. Exch. Club Br. Isl. Rep.* 1930, 322. R. L. Praeger *Some Irish Nat.* 1949, 61.

CARRUTHERS, John (*fl.* 1820s–1830s)
Nurseryman, Botchergate, Carlisle, Cumberland.

CARRUTHERS, John Bennett (1869–1910)
b. Islington, London 19 Jan. 1869 *d.* Trinidad 17 July 1910
FLS 1890. FRSE 1906. Son of W. Carruthers (1830–1922). Demonstrator in Botany, Royal Veterinary College, London, 1893. Professor of Botany, Downton College of Agriculture, 1895. Cryptogamist to Ceylon Planters' Association, 1897–99. Published a series of reports on cacao canker. Government Mycologist and Assistant Director, Botanic Gardens, Peradeniya, Ceylon, 1900–5. Director of Agriculture, Malay States, 1905–9. Assistant Director, Department of Agriculture, Trinidad, 1909. Contrib. to *J. Linn. Soc.*, *J. R. Hort. Soc.*, *Tropical Agriculturist*.
Acta Horti Bergiani v.3 (2), 1905, t.146 portr. *Agric. Bull. Straits Fed. Malay States* 1910, 329. *J. Bot.* 1910, 217–19 portr. *Kew Bull.* 1910, 254. *Nature* v.84, 1910, 114–15. *India-Rubber J.* 1910, 118 portr. *Proc. Linn. Soc.* 1910–11, 35–36. *Who was Who, 1897–1916* 122. *R.S.C.* v.14, 81.
Indian plants at Manchester. Portr. at Hunt Library.

CARRUTHERS, Samuel William (1866–1962)
d. 6 April 1962
Son of W. Carruthers (1830–1922). Studied flora of Surrey.
Proc. Bot. Soc. Br. Isl. 1962, 505.

CARRUTHERS, William (1830–1922)
b. Moffat, Dumfriesshire 29 May 1830 *d.* Norwood, Surrey 2 June 1922
FRS 1871. FLS 1861. Assistant, Botany Dept., BM(NH), 1859; Keeper, 1871–95. Botanist, Royal Agricultural Society, 1871–1909. Contrib. *Diatomaceae* to J. E. Gray's *Handbook of British Water-weeds* 1864. Contrib. to *J. Bot.* 1863–1900. Edited *Ann. Mag. Nat. Hist.* President, Linnean Society, 1886–90.
J. Bot. 1895, 182–85; 1922, 249–56 portr. *Gdnrs Chron.* 1895 i 688; 1922 i 312. *Geol. Mag.* 1912, 193–99. *Trans. Proc. Bot. Soc. Edinburgh* v.28, 1921–22, 118–21. *Bot. Soc. Exch. Club Br. Isl. Rep.* 1922, 697–98. *Nature* v.109, 1922, 787–88. *Proc. Linn. Soc.* 1922–23, 38–40. *Who was Who, 1916–1928* 178.
Portr. at Hunt Library.
Carruthersia Seem.

CARSE, Harry (1857–1930)
b. Leek, Staffs 1857 *d.* 25 Nov. 1930
To New Zealand, 1885. Teacher at Helensville, Chelsea and elsewhere in Auckland. Collected plants mainly in Auckland. Interested in *Filices* and *Cyperaceae*. Had herb. Contrib. to *Trans. N.Z. Inst.*
T. F. Cheeseman *Manual of New Zealand Fl.* 1925, xxxv. *Trans. N.Z. Inst.* 1931, 176–78 portr.
Plants at Auckland University.

CARSON, Alexander (1850–1896)
b. Stirling 1850 *d.* Fwambo, Rhodesia 28 Feb. 1896
BSc Glasgow 1883. Engineer. To Tanganyika, 1888. Sent plants to Kew.
Kew Bull. 1893, 343–44; 1895, 46, 63–75, 288–93; 1896, 148–49. *Comptes Rendus AETFAT, 1960* 1962, 164, 180. *R.S.C.* v.14, 82.
Gloriosa carsoni Baker.

CARSON, S. M. (*c.* 1814–1881)
b. Gatehouse-of-Fleet, Kircudbrightshire *d.* Gatehouse 22 May 1881
Gardener to T. Gamul Farmer at Nonsuch Park, Cheam, Surrey, 1842–45.
Gdnrs Chron. 1881 i 825.

CARTER, Arthur H. (*c.* 1867–1939)
Sanitary engineer. Collected British plants.
Essex Nat. v.29, 1954, 192.
Plants at Essex Field Club.

CARTER, Charles (*fl.* 1720s)
His Majesty's gardener at Chapelizod, County Dublin. "Having since the decease of Robert Moody his late partner carried on the seed trade by himself."
Irish For. 1967, 51.

CARTER, Daniel (*fl.* 1760s)
Gardener and market florist, Battersea. *Modern Eden; or Gardener's Universal Guide* (with J. Rutter) 1767.
J. C. Loudon *Encyclop. Gdning* 1822, 1275.

CARTER, George (1838–1903)
b. Kendal? 1838 *d.* Ashton-upon-Ribble, Lancs 7 Sept. 1903
Draper, Preston, Lancs. Member of Botanical Section of Preston Scientific Society. Assisted in compilation of *Fl. Preston and Neighbourhood* 1903.
Trans. Liverpool Bot. Soc. 1909, 64.

CARTER, Sir Gilbert (*fl.* 1890s)
Governor, Gambia, 1890; Lagos, Nigeria, 1892. Collected plants, now at Kew with his MSS.

CARTER, Henry John (1813–1895)
b. Budleigh Salterton, Devon 18 Aug. 1813 *d.* Budleigh Salterton 4 May 1895
FRS 1859. Geologist. Diatomist. Bombay Medical Service, 1842–62. 'Frankincense Tree' (*J. Bombay Branch R. Asiatic Soc.* v.2, 1847, 380–90). '*Hildebrandtia fluviatilis*' (*J. Bot.* 1864, 225–28).
Trans. Linn. Soc. 1871, 143–45. *Intellectual Observer* v.2, 1863, 251. *Proc. R. Soc.* v.58, 1895, liv–lvii. D. G. Crawford *Hist. Indian Med. Service, 1600–1913* v.2, 1914, 148. *R.S.C.* v.1, 802; v.7, 341; v.9, 454; v.12, 144; v.14, 83.
Boswellia carteri Birdwood.

CARTER, Henry Vandyke (1831–1897)
d. Scarborough, Yorks 4 May 1897
MD London. Bombay Medical Service, 1858–88. Professor of Physiology, Bombay. 'On Mycetoma or Fungus-Disease of India' (*Trans. Bombay Med. Phys. Soc.* v.7, 1861, 206–21).
Intellectual Observer 1863, 248, 251. D. G. Crawford *Hist. Indian Med. Service, 1600–1913* v.2, 1914, 369.

CARTER, Humphrey Gilbert—*see* Gilbert-Carter H.

CARTER, James (*fl.* 1830s–*c.* 1856)
Seedsman, 238 High Holborn, London, *c.* 1836. Later acquired nursery at Raynes Park. From 1837 appeared a regular annual list of seeds.
Gdnrs Mag. 1911, 37. J. Harvey *Early Hort. Cat.* 1973, 2–4.

CARTER, John (*c.* 1825–1894)
d. 24 Dec. 1894
Nurseryman, Keighley, Yorks. Introduced raspberry, 'Carter's Prolific'.
J. Hort. Cottage Gdnr. v. 30, 1895, 7.

CARTER, John William (1852–1920)
d. Bradford, Yorks 15 Dec. 1920
President, Bradford Natural History Society. Entomologist.
Naturalist 1921, 103–6 portr.
Herb. at Keighley Museum.

CARTER, Thomas (*c.* 1840–1910)
Headmaster, Mill Hill School, Leicester, 1872–95. Botanist.
A. R. Horwood and C. W. F. Noel *Fl. Leicestershire* 1933, ccxxx.

CARTWRIGHT, Kenneth St. George (1891–1964)
d. Towersey, Oxford 17 Oct. 1964
MA Oxon. FLS 1929. Mycologist, Forest Products Research Laboratory, 1927–48. President, British Mycological Society, 1937. *Decay of Timber and its Prevention* (with W. P. K. Findlay) 1946.
Trans. Br. Mycol. Soc. v.48, 1965, 151.

CARTWRIGHT, Thomas (–1951)
d. Mouldsworth, Cheshire Jan. 1951
Kew gardener, 1905. Superintendent, rubber plantation, Jebelein, Sudan, 1908. Collected plants for Kew in Yei Valley, 1919.
Kew Bull. 1908, 421. *J. Kew Guild* 1951, 42.

CARTWRIGHT, Thomas Barclay (1856–1896)
d. Montreal, Canada 1896
Of Aynho near Banbury, Oxford. Travelled extensively and collected plants in Europe, 1872–90; N. America, 1886–88; Friendly Isles, 1888; New Zealand, 1889; Trinidad, 1890.
H. N. Clokie *Account of Herb. Dept. Bot., Oxford* 1964, 144.
Herb. at Oxford.

CARVER, William Robert (1860–1918)
b. Marylebone, London 25 Feb. 1860 *d.* Fulham, London 6 Sept. 1918
Attendant, BM(NH), 1880; later employed in Cryptogamic Herb. and Library.
J. Bot. 1918, 334–35.

CASBORNE, Mrs. (*née* **Lofft**) (-1884)
d. Pakenham, Suffolk 1884
Of Pakenham. Herb., 1819–50, in 25 vols. at Cambridge University.
W. M. Hind *Fl. Suffolk* 1889, 489.

CASEMENT, Roger David (1864–1916)
b. Kingstown, County Dublin 1 Sept. 1864 *d.* Pentonville, London 3 Aug. 1916
British Consul and Irish patriot. Commissioner, Niger Coast Protectorate, 1892. British Consul, Lourenço Marques, 1895. Consul-General, Rio de Janeiro, 1909. Collected ethnobotanical and economic botanical materials in Nigeria, 1892 (now at National Museum, Dublin); also in Putumayo, upper Amazon, 1910.
D.N.B. 1912–1921 95–97. *Eire–Ireland* no. 3, 1968, 46–54.

CASEY, George Edward Comerford (1846–1912)
b. Everton, Liverpool 19 March 1846 *d.* Parkstone, Dorset 4 Feb. 1912
MA Oxon 1873. FLS 1879. Schoolmaster, Nottingham, 1874–79. At Nice, 1882–94. *Riviera Nature Notes* 1898; ed. 2 1903.

CASEY, James (*fl.* 1800s–1830s)
Nurseryman, 2 South Mall, Cork, Ireland.

CASH, James (1839–1909)
b. Great Sankey, Warrington, Lancs 14 Feb. 1839 *d.* 20 Feb. 1909
Journalist on *Manchester Guardian.* Keen naturalist; studied freshwater Rhizopoda. President, Manchester Cryptogamic Society. *Notes on Some Rare British Mosses. Where There's a Will There's a Way* 1873. *The Late William Wilson* 1886.
N. Western Nat. 1941, 29–33 portr.
Mosses and MSS. at Manchester Museum.
Amblystegium cashii Buyss.

CASH, William (1843–1914)
b. Halifax, Yorks 1843 *d.* Halifax 16 Dec. 1914
Palaeobotanist. Collaborated with W. C. Williamson and Thomas Hick.
Naturalist 1915, 28–30 portr. *R.S.C.* v.9, 460; v.14, 91.
Micro-preparations at Manchester.

CASTELNAU, François L. de Laporte de, Comte (1810–1880)
b. London 15 Dec. 1810 *d.* Melbourne 4 Feb. 1880
Collected plants in Florida, *c.* 1838. Leader of French expedition to S. America, 1843–47; collected plants in Bolivia and Peru. Also collected at Cape, S. Africa. French Consul, Melbourne, 1862.
Bull. Soc. Bot. France v.27, 1880, 44–45. *Nature* v.21, 1880, 500. *Les Botanistes Français en Amerique du Nord avant 1850* 1957, 31.

CASTLE, R. Lewis (1854–1922)
b. Chelsea, London 3 May 1854 *d.* Kingston, Surrey Oct. 1922
Kew gardener, 1874–77. On editorial staff of *J. Hort.* Manager, Woburn Experimental Fruit Farm. *Cactaceous Plants* 1884. *Orchids* 1885; ed. 2 1887. *Flower Gardening for Amateurs* 1888. *Book of Market Gardening* 1906. Edited *Chrysanthemum Annual* 1887–90.
Gdnrs Mag. 1903, 37 portr. *Gdnrs Chron.* 1922 ii 276. *J. Kew Guild* 1923, 175 portr. *Orchid Rev.* 1928, 2.

CASTLE, Thomas (*c.* 1804–*c.* 1840)
b. Kent *c.* 1804 *d.* Brighton? *c.* 1840
MD Cantab. FLS 1827. Practised medicine in Bermondsey. *Introduction to Systematical and Physiological Botany* 1829. *Synopsis of Systematic Botany* 1833. *Introduction to Medical Botany* 1829; ed. 3 1837. *British Fl. Medica* (with B. H. Barton) 1837. *Linnean Artificial System of Botany* 1837.
D.N.B. v.9, 275.

CASTLES, Robert (*fl.* 1840s)
Manager of botanic garden, Twickenham, Middx formerly belonging to Isaac Swainson (1746–1812). 'Description of Species of Rose (*R. erecta*) New to British Fl.' (*Proc. Sci. Soc. London* 1840 ii 36). Herb. acquired by H. C. Watson (1804–1881).
Gdnrs Mag. 1832, 521–23.

CATESBY, Mark (1682–1749)
b. Castle Hedingham? Essex 24 March 1682 *d.* London 23 Dec. 1749
FRS 1733. Of Hoxton and Fulham. Naturalist and artist. Collected plants in Virginia, 1712–19; Jamaica, 1715. Sponsored by Sherard and others to collect in Carolina, Florida and Bahamas, 1722–26. Drew and engraved plates in his *Natural History of Carolina, Florida and Bahama Islands* 1730–47 2 vols.; ed. 2 1754; ed. 3 1771. *Hortus Britanno-americanus* 1763 (*Hortus Europae Americanus* 1767).
Gent. Mag. 1749, 446–47, 573; 1750, 30–31. R. Pulteney *Hist. Biogr. Sketches of Progress of Bot. in England* v.2, 1790, 219–30. J. Nichols *Illus. Lit. Hist. of Eighteenth Century* 371–92. F. Pursh *Fl. Americae Septentrionalis* v.1, 1814, xviii. *Rees' Cyclopaedia* v.7. J. E. Smith *Selections of Correspondence of Linnaeus* v.2. 1821, 440–41. D. Turner *Extracts from Lit. and Sci. Correspondence of Richard Richardson* 1835, 206, 401–2. J. C. Loudon *Arboretum et Fruticetum Britannicum* v.1, 1838, 68–70. W. Darlington *Reliquiae Baldwinianae* 1843, 64, 67, 96, 159. I. Urban *Symbolae Antillanae* v.1, 1898, 29–30; v.3, 1902, 31. *D.N.B.* v.9, 281. E. J. L. Scott *Index to Sloane Manuscripts* 1904, 101. *Auk* v.54, 1937, 349–63. *Connoisseur* 1948,

47–52. *Tyler's Quart. Hist. Geneal. Mag.* 1948, 167–80. *J. New York Bot. Gdn* v.50, 1949, 1–24. *Trans. Amer. Philos. Soc.* 1951, 63–78. *Gdnrs Chron.* 1952 ii 24. *J. Soc. Bibl. Nat. Hist.* v.3, 1957, 177–94. J. E. Dandy *Sloane Herb.* 1958, 110–13. *Papers Bibl. Soc. America* v.54, 1960, 163–75. G. F. Frick and R. P. Stearns *Mark Catesby* 1961. H. N. Clokie *Account of Herb. Dept. Bot., Oxford* 1964, 144. F. A. Stafleu *Taxonomic Literature* 1967, 78–79. A. M. Coats *Quest for Plants* 1969, 268–71. H. Savage *Lost Heritage* 1970, 57–92. C. C. Gillispie *Dict. Sci. Biogr.* v.3, 1971, 129–30. *Gdn J.* (New York) 1973, 71–81.

Plants at BM(NH), Oxford. MSS. and letters at Royal Society.

Catesbaea L.

CATHCART, John Ferguson (1802–1851)
b. Edinburgh 19 Feb. 1802 *d.* Lausanne, Switzerland 8 July 1851

Bengal Civil Service. To Calcutta, 1822. Collected plants at Darjeeling; also at Cape, 1839. Engaged Indian artists to draw flowers, a selection of which was reproduced by W. H. Fitch in J. D. Hooker's *Illustrations of Himalayan Plants* 1855.

Bot. Mag. 1851, t.4596.

Indian plants at Botanic Garden, Edinburgh. Indian flower drawings at Kew.

Cathcartia Hook. f.

CATLEUGH, W. (*fl.* 1830s–1840s)

Nurseryman, 27 New Road, Chelsea; nursery at Elizabeth Street, London.

CATLOW, Agnes (*c.* 1807–1889)
d. Addlestone, Surrey 10 May 1889

Popular Field Botany 1847; ed. 3 1849. *Popular Garden Botany* 1855. *Popular Greenhouse Botany* 1857.

CATTELL, J. (*fl.* 1840s)

Nurseryman, Westerham, Kent.

CATTELL, William (*fl.* 1870s–1880s)

FLS 1878. Surgeon, 10th Hussars. In India, *c.* 1871–80. Collected plants in Mauritius, Simla and Afghanistan.

Pakistan J. For. 1967, 341.

Plants at Kew.

CATTLEY, William (–1832)

FLS 1821. Of Barnet. Merchant. Benefactor of horticulture and J. Lindley's first patron, paying him a regular salary for drawing and describing new plants. "One of the most ardent collectors of rare plants of his day." Had large collection of drawings of plants. 'A New *Psidium*' (*Trans. Hort. Soc.* v.4, 1822, 315–17).

J. Lindley *Digitalium Monographia* 1821, preface. *J. Bot.* 1865, 385; 1893, 281. E. Bretschneider *Hist. European Bot. Discoveries in China* 1898, 187, 255–56. *Gdnrs Chron.* 1898 i 93.

Letters at Kew.

Cattleya Lindl.

CAVANAGH, Bernard (*fl.* 1890s)

Kew gardener. Superintendent, Gardens of Agri-Horticultural Society, Madras, 1899.

Kew Bull. 1899, 221.

CAVE, Edward (*c.* 1822–1905)
b. Carisbrooke, Isle of Wight *c.* 1822 *d.* 2 Aug. 1905

Acquired Ford Nurseries, Newport, Isle of Wight, formerly belonging to W. Wilkins.

J. Hort. Cottage Gdnr. v.51, 1905, 140.

CAVE, George H. (*c.* 1870–1965)

Kew gardener, 1894. Assistant, Botanic Gardens, Calcutta, 1896. Government Cinchona Plantations, Mungpoo, 1900. Curator, Lloyd Botanic Garden, Darjeeling, 1904. Collected plants in Tibet, Nepal and Sikkim. Contrib. to *Rec. Bot. Survey India.*

I. H. Burkill *Chapters on Hist. Bot. in India* 1965, 141. *J. Kew Guild* 1966, 708–9 portr.

CAVE, Norman Leslie *see* Addendum

CAVEN, John (*fl.* 1820s–1840s)

Nurseryman, Aiskew in Bedale, Yorks.

CAVENDISH, Hon. Henry (1731–1810)
b. Nice, France 10 Oct. 1731 *d.* 10 March 1810

Educ. Cambridge. FRS. Chemist and physicist. Made the first determinations of comparative weight and density of different gases produced by decomposition of plant substances. Contrib. to *Philos. Trans. R. Soc.*

G. Wilson *Life of Honourable Henry Cavendish* 1851. T. E. Thorpe *Essays in Hist. Chemistry* 1894, 70–86. *D.N.B.* v.9, 348–53. W. Ramsay *Essays Biogr. and Chemical* 1908, 19–41. *Chronica Botanica* 1944, 152–57.

CAVENDISH, Thomas *see* Candi, T.

CAVENDISH, William George Spencer, 6th Duke of Devonshire (1790–1858)
b. Paris 21 May 1790 *d.* Hardwick Hall, Derbyshire 17 Jan. 1858

BA Cantab 1811. LLB 1812. President, Horticultural Society of London, 1838–58. Employed Joseph Paxton to manage his celebrated gardens at Chatsworth, Derbyshire. Played a major role in the establishment of Kew as a national botanic garden. "The greatest encourager of gardening in England at the present time."

Curtis's Bot. Mag. Dedications, 1827–1927 31–32 portr. M. A. Reinikka *Hist. of the Orchid* 1972, 136–40 portr.

CAVERS, Francis (1876–1936)
b. Hawick, Roxburgh 5 Sept. 1876 *d.* 26 May 1936
BSc London 1901. MRCS 1918. FLS 1903–17. Demonstrator, Owens College, Manchester, 1900–1. Lecturer in Biology, Leeds, 1901–3. Lecturer in Biology, Plymouth Technical College, 1903–4. Professor of Biology, Southampton, 1904–11. Reader in Botany, Goldsmith College, London, 1911–15. In medical practice, Highbury, 1918–25. Studied Bryophyta, especially Hepaticae. *Life Histories of Common Plants* 1908. *Botany for Matriculation* 1909. *Senior Botany* 1910. *Practical Botany* 1911. Assistant Editor, *New Phytol.* First Editor, *J. Ecol.*
Ann. Bryologici 1936, 154 portr. *Nature* v.137, 1936, 1022. *Chronica Botanica* 1937, 154 portr. *Naturalist* 1969, 59–62.

CAWSWAY, Mr (*fl.* 1500s)
Market gardener, Houndsditch, London.
R. Webber *Early Horticulturists* 1968, 43–49.

CAYLEY, Dorothy Mary (1874–1955)
Mycologist, John Innes Horticultural Institution, 1910–38.
John Innes Hort. Inst. Rep. 1938.

CAYLEY, George *see* Caley, G.

CECIL, Hon. Mrs. Alicia Margaret *see* Rockley, A. M.

CHADBURN, George Haworthe (1870–1950)
d. 29 Jan. 1950
Artist. Hybridised irises.
Iris Yb. 1950, 12–14 portr.

CHALLIS, Thomas (1835–1923)
b. Barton Seagrave, Northants 19 June 1835 *d.* Wilton, Wilts 8 June 1923
VMH 1904. VMM. Gardener to Lord Pembroke at Wilton House, 1860–1915.
Gdnrs Chron. 1923 i 358; 1923 ii 13–14.

CHALMERS, Albert John (1870–1920)
b. London 1870 *d.* Calcutta, India 5 April 1920
MD Liverpool. Authority on tropical diseases; especially interested in disease-causing fungi. Director, Wellcome Research Laboratories, Khartoum, 1913–20. *Fungi Imperfecti in Tropical Medicine* 1916. *Manual of Tropical Medicine* (with A. Castellani) ed. 3 1919.
J. Bot. 1920, 158.

CHALMERS, Alexander (*fl.* 1790s)
Gardener and seedsman, Newtown, Montgomeryshire.

CHALMERS, James (–before 1834)
b. Dundee, Angus
"Manipulator" in W. J. Hooker's herb. at Glasgow in 1827. Algologist. Published fasciculi of *Algae Scoticae* 1826.
Ann. Bot. 1902, xxxiii, cxx.

CHALMERS, Rev. James B. (1841–1901)
b. Ardrishaig, Argyll 1841 *d.* Island Goaribavi, New Guinea 8 April 1901
Missionary, Raratongo, 1866. Collected plants in New Guinea and neighbouring islands, 1877–1901. *Adventures in New Guinea* 1886. *Pioneering in New Guinea* 1887.
W. Robson *James Chalmers* ed. 2 1887 portr. C. Lennox *James Chalmers of New Guinea* 1902. E. H. Hayes *Chalmers of Papua* 1930. *Fl. Malesiana* v.1, 1950, 103–4.
Plants at Melbourne.

CHAMBERLAIN, Houston Stewart (1855–1927)
b. Portsmouth, Hants 9 Sept. 1855 *d.* Bayreuth, Germany 9 Jan. 1927
BSc Geneva 1881. Left Switzerland in 1889 to live in Vienna where the botanist, J. Wiesner, encouraged him to publish his researches on ascent of sap. *Recherche sur la Sève Ascendante* 1897.
Berichte Schweizerischen Bot. Gesellschaft v.50A, 1940, 171–73.

CHAMBERLAIN, James Slade Hester (*fl.* 1840s–1920s)
Herb. at University College, Bangor, Caernarvonshire.

CHAMBERLAIN, Right Hon. Joseph (1836–1914)
b. London 8 July 1836 *d.* London 2 July 1914
Statesman. Had collection of orchids. Contrib. to *Orchid Rev.*
Orchid Rev. 1914, 238–39. *D.N.B. 1912–1921* 102–18.
Cattleya chamberlainiana Reichenb. f.

CHAMBERLAYNE, Charles (*fl.* 1820s)
Consul-General, Brazil. "A gentleman who has always been a zealous promoter of the comforts of such naturalists as have gone thither from this country." Sent *Bignonia* to the nursery firm of Lee and Kennedy at Hammersmith, London.
Bot. Mag. 1820, t.2148; 1829, t.2920.

CHAMBERS, B. E. C. (–1911)
d. Hèyères, France 23 March 1911
Cultivated a large collection of trees and shrubs at Grayswood Hill, Haslemere, Surrey.
Kew Bull. 1911, 166

CHAMBERS, John (*fl.* 1800s–1810s)
MD. Of East Dereham, Norfolk. "Very curious in botany." *Pocket Herbal* 1800.
J. Nichols *Illus. Lit. Hist. of Eighteenth Century* v.1, 1817, 263–64, 387, 392, 394.

CHAMBERS, Richard (1784–1858)
b. London 1784 *d.* Balderton, Notts 20 Dec. 1858
FLS 1822. Schoolmaster. *Introduction to Study of Botany* 1847. 'Catalogue of Plants of Tring' (*Mag. Nat. Hist.* 1838, 38–40).
Proc. Linn. Soc. 1859, xxx–xxxi. *R.S.C.* v.1, 868.

CHAMBERS, S. A. (*fl.* 1870s–1930s)
Herb. at South London Botanical Institute.

CHAMPION, John George (1815–1854)
b. Edinburgh 5 May 1815 *d.* Scutari, Turkey 30 Nov. 1854
Lieut-Colonel, 95th Regiment. In Ceylon, 1838–47; Hong Kong, 1847–50. 'Ternstroemiaceae of Hong Kong' (*Trans. Linn. Soc.* v.21, 1855, 111–16). Sent plants to G. Gardner (1812–1849) and to Kew. Hong Kong plants described in *Hooker's J. Bot.* 1849, 240–46, 308–20, 321–28.
Bot. Mag. 1851, t.4609. *Gdnrs Chron.* 1854, 819–20. G. Bentham *Fl. Hongkongensis* 1861, 8*. E. Bretschneider *Hist. European Bot. Discoveries in China* 1898, 374–79. H. Trimen *Hand-book Fl. Ceylon* v.5, 1900, 375. *D.N.B.* v.10, 33. E. M. H. Cox *Plant Hunting in China* 1945, 73–74. G. A. C. Herklots *Hong Kong Countryside* 1951, 164. *R.S.C.* v.1, 870.
Hong Kong plants at Kew, Oxford; Ceylon plants at Cambridge. MSS., letters and drawings at Kew.
Championia Gardner.

CHAMPNEYS May (*née* **Drummond**)
(*fl.* 1860s–1935)
Flower List of Hampstead Neighbourhood 1914.

CHANDLEE, Thomas (1824–1907)
b. Clogheen, County Tipperary 7 July 1824 *d.* Ballitore, County Kildare 12 April 1907
Studied flora of Kildare. 'Euphorbia cyparissias' (*Irish Nat.* 1893, 250).
N. Colgan and R. W. Scully *Cybele Hibernica* 1898, xxii.
Herb. of Kildare plants and portr. at Botanic Gardens, Glasnevin.

CHANDLER, Alfred (1804–1896)
b. Vauxhall, London 31 Jan. 1804 *d.* East Dulwich, London 10 Nov. 1896
Nurseryman, Vauxhall. Succeeded his father, Alfred Chandler, who had been in partnership with Napier, then Buckingham, 1827–33. Nursery renowned for camellias from about 1806 when A. Chandler senior started hybridising them. A. Chandler junior illustrated *Camellia Britannica* (text by E. B. Buckingham) 1825 and *Illustrations and Descriptions of...Camellieae* (text by W. B. Booth) 1831.
Gdnrs Mag. 1830, 291. *Gdnrs Chron.* 1896 ii 628. *J. Bot.* 1897, 32.

CHANDLER, Elizabeth (1818–1884)
b. Hinton-in-the-Hedges, Northants 29 April 1818 *d.* Isleworth, Middx 29 April 1884
'Plants of High Wycombe' (*Botanist's Chron.* 1864, 81–84).
G. C. Druce *Fl. Buckinghamshire* 1926, cii. H. J. Riddelsdell *Fl. Gloucestershire* 1948, cxxxi.
Bucks plants at BM(NH).

CHANDLER, J. (1700–1780)
b. Bath or Hungerford
Apothecary, Cheapside. Member of Botanical Society of London. Committee member of Chelsea Physic Garden, London.
Proc. Bot. Soc. Br. Isl. 1967, 315.

CHANDLER, Stafford Edwin (1880–1957)
b. London 17 March 1880 *d.* 1 Aug. 1957
DSc. Head of Plant and Animal Products Dept., Imperial Institute, London, 1936–45. Published papers on *Pyrethrum*. *World's Commercial Products* (with W. G. Freeman) 1907.
Proc. Bot. Soc. Br. Isl. 1958, 130–31.
Herb. at Croydon Natural History Society.

CHANTER, Mrs. Charlotte (1824–1882)
b. 24 Dec. 1824 *d.* Ilfracombe, Devon 20 March 1882
Sister of Charles Kingsley. *Ferny Combes* 1856.
Notes and Queries 1928 ii 213. W. K. Martin and G. T. Fraser *Fl. Devon* 1939, 774.

CHAPMAN, Mr. (*fl.* 1720s)
Nurseryman, near Pitfield Street, Hoxton, London.

CHAPMAN, Alfred Chaston (1869–1932)
d. 17 Oct. 1932
Educ. London University. Analytical chemist. President, Royal Microscopical Society, 1924. Wrote on yeasts and fungi imperfecti.
J. R. Microsc. Soc. 1932, 404–5 portr. *Nature* v.130, 1932, 654–55.

CHAPMAN, Arthur (–1920)
d. 9 Dec. 1920
Gardener to Sir G. Holford at Westonbirt, Glos for nearly 50 years.
Gdnrs Chron. 1920 ii 306 portr.

CHAPMAN, Frederick (1864–1943)
b. Camden Town, London 13 Feb. 1864 *d.* Stawell, Victoria 10 Dec. 1943
ALS 1896. Assistant, Geology Dept., Royal College of Mines, London, 1881–1902. Palaeontologist, National Museum, Melbourne, 1902–27. Commonwealth Palaeontologist, 1927–36. A keen botanist who encouraged cultivation of Australian plants in gardens.

E. J. Brady *Australia Unlimited* 1934, 128 portr. *Austral. J. Sci.* 1944, 122. *Proc. Linn. Soc.* 1943–44, 200–2. *Trans. R. Soc. S. Australia* 1944, 358. *Victorian Nat.* 1944, 152–54.

CHAPMAN, Henry J. (–1931)
d. Wylam-on-Tyne, Northumberland 30 Jan. 1931
VMM 1910. Gardener. Employed by R. H. Measures, Camberwell and N. C. Cookson, Wylam. Edited new edition of W. Watson's *Orchids* 1903.
Orchid Rev. 1931, 96.
Cypripedium chapmanii Cookson.

CHAPMAN, Henry L. R. (–1968)
d. Dec. 1968
Of East Lancing, Sussex. Kew gardener, 1919. Horticulturist to Egyptian Government, 1919. Botanist in charge of Beal Botanic Garden, Msu, 1926.
J. Kew Guild 1971, 65

CHAPMAN, Heywood (*fl.* 1870s)
Of Liverpool, later Birkenhead. Active member of Liverpool Naturalists' Field Club and contrib. to *Fl. Liverpool* 1872.
Trans. Liverpool Bot. Soc. 1909, 64.

CHAPMAN, James (1831–1872)
Commission agent. Explorer and adventurer. Collected plants in Damaraland, S.W. Africa, 1864; with Thomas Baines in S. Tropical Africa, 1864–65.
Fl. Zambesiaca v.1, 1960, 24.
Plants at Kew.

CHAPMAN, John (*fl.* 1510s–1530s)
Master-gardener to Cardinal Wolsey and Henry VIII at Hampton Court, Middx from 1515 to *c.* 1533. Nurseryman.

CHAPMAN, Thomas (*fl.* 1830s–1880s)
Herb. at Whitby Literary and Philosophical Society, Yorks.

CHARGE, Rev. John (*fl.* 1880s)
Friend of H. E. Fox. Plants from Devon and Yorks at Oxford.

CHARLES, Samuel George (1883–1960)
b. Monmouth 1883 *d.* Monmouth 1960
Collection of Monmouthshire plants at National Museum of Wales, Cardiff.
A. E. Wade *Fl. Monmouthshire* 1970, 16.

CHARLESWORTH, Benjamin (*fl.* 1760s)
MD Edinburgh 1769. Sent British plants to J. Hope (1725–1786).
Notes R. Bot. Gdn Edinburgh v.4, 1907, 125.

CHARLESWORTH, Joseph (*c.* 1851–1920)
d. Haywards Heath, Sussex 2 Aug. 1920
In 1880s at Heaton Bradford as orchid grower and importer. Collected plants in Andes, 1889. In partnership with E. Shuttleworth of Clapham. Nurseryman, Haywards Heath from 1908 specialising in orchids.
Gdnrs Mag. 1909, 627–28 portr. *J. Hort. Home Farmer* v.63, 1911, 538–39 portr. *Gdnrs Chron.* 1920 ii 78. *Orchid Rev.* 1920, 137–42 portr. *Amer. Orchid Soc. Bull.* 1958, 740.
Cypripedium charlesworthii Rolfe.

CHARLETON, William *see* Courten, W.

CHARLOTTE SOPHIA, H.M. Queen (1744–1818)
b. Mirow, Mecklenburg-Strelitz 19 May 1744
d. Kew, Surrey 17 Nov. 1818
Princess of Mecklenburg-Strelitz. Married George III in 1761. Pupil of Rev. J. Lightfoot whose herb. was bought for her by George III. Received instruction in flower painting from Francis Bauer. Kew Palace and its grounds (now Kew Gardens) was one of the royal residences. "Patroness of Botany and of the Fine Arts" (R. J. Thornton).
D.N.B. v.10, 123. *J. Bot.* 1915, 269–71. *Gdnrs Chron.* v.164 (18), 1968, 21–22. O. Hedley *Queen Charlotte* 1975 portr.
Strelitzia Aiton.

CHARLTON, Edward (1814–1874)
b. Hesleyside, Northumberland 23 July 1814
d. Newcastle, Northumberland 14 May 1874
MD Edinburgh 1836. DCL. Lecturer, Newcastle Medical School. Original member, Botanical Society of Edinburgh. Had herb.
Trans. Proc. Bot. Soc. Edinburgh v.12, 1876, 198–99. *Mon. Chron.* 1889, 443. G. Johnston *Correspondence* 1892, 29.

CHARLWOOD, George (*c.* 1784–1861)
d. Feltham, Middx 26 Aug. 1861
FLS 1824. Seedsman, 14 Tavistock Row (later Hart St., James St.), Covent Garden, London, 1820–39. In partnership with Cummings, 1839–71. Seed catalogues reprinted in J. C. Loudon's *Arboretum et Fruticetum Britannicum* v.4, 1838, 2618–20; *Hortus Lignosus Londinensis* 1838, 134–36. Helped in R. Sweet's *Hortus Britannicus* 1826.
R. Sweet *Fl. Australasica* 1827–28, 18. *Proc. Linn. Soc.* 1861–62, xc.
Charlwoodia Sweet.

CHARSLEY, Fanny Anne (1828–1915)
b. Beaconsfield, Bucks 23 July 1828 *d.* Hove, Sussex 21 Dec. 1915

In Melbourne, 1856. Drew plants which she published in *Wild Flowers around Melbourne* 1867. Returned to England, 1866. Correspondent of F. von Mueller.

Victorian Nat. 1908, 105; 1933, 261.

Hepipterum charsley F. Muell.

CHASE, Corrie Denew (1878–1965)

b. Keswick, Cumberland 1878 *d.* Belfast 15 Oct. 1965

MA Cantab 1900. Assistant master, Campbell College, Belfast, 1905. Botanised in N.E. Ireland. *Natural History of Campbell College and Cabin Hill* 1949. Contrib. to *Irish Nat. J.*

R. L. Praeger *Some Irish Nat.* 1949, 62. *Irish Nat. J.* 1966, 214–15.

Herb. at Ulster Museum.

CHASE, Joseph Smeaton (1864–1923)

b. London 8 April 1864 *d.* Bonning, California 29 March 1923

Natural history author. Emigrated to California, 1890. Resident of Palm Springs, 1915. *Cone-bearing Tress of California Mountains* 1911. *Yosemite Trails* 1911. *California Desert Trails* 1919. *California Coast Trails* 1913.

National Cyclop. Amer. Biogr. 1958, 317 portr.

CHATER, William (1802–1885)

b. Bumpstead, Essex 4 Feb. 1802 *d.* Saffron Walden, Essex 21 July 1885

Nurseryman, Saffron Walden. Specialised in hollyhocks; also landscaped gardens.

Garden v.28, 1885, 132. *Gdnrs Chron.* 1885 ii 155. *Victoria County Hist. Essex* v.2, 1907, 480.

CHATTERLEY, William Maddox (*fl.* 1830s)

First Secretary, Botanical Society of London. 'On Advantages of Botanical Statistics, illustrated by the Order Coniferae' (*Proc. Bot. Soc. London* 1839, 87–91). Contrib. to D. Cooper's *Fl. Metropolitana* 1836.

Bot. Soc. Exch. Club Br. Isl. Rep. 1932, 282.

CHAUNDY, Richard (*fl.* 1820s–1840s)

Nurseryman, St. Aldate, Oxford.

CHAVASSE, Mrs. Charles *see* Jackson, M. A.

CHEAL, John (1800–1896)

b. Crawley, Sussex 1800 *d.* 18 Feb. 1896

Nurseryman, Lowfield Nurseries, established at Crawley, 1871.

Gdnrs Chron. 1896 i 244. *J. Hort. Cottage Gdnr* v.32, 1896, 191–92 portr.

CHEAL, Joseph (*c.* 1848–1935)

d. 9 June 1935

VMH 1914. Nurseryman, Lowfield Nurseries, Crawley, Sussex. *Practical Fruit Culture* 1892.

Gdnrs Mag. 1894, 601 portr.; 1908, 957–58 portr. *Gdnrs Chron.* 1921 i 170 portr.; 1935 i 404. *Garden* 1923 ii–iii portr.

CHEALES, Rev. Alan (1828–1911)

b. Lincolnshire 1828 *d.* Edinburgh 3 June 1911

MA Cantab. Vicar, Brockham, Surrey, 1859–92. Rosarian. Established Brockham Rose Society, 1868; later joint founder of Reading Rose Society.

J. Hort. Cottage Gdnr v.59, 1909, 606–8 portr.; v.62, 1911, 543 portr.

CHEEL, Edwin (1872–1951)

b. Chartham, Kent 14 Jan. 1872 *d.* Sydney 19 Sept. 1951

Trained in horticulture and forestry. To Australia where he worked in private gardens and then at Sydney Botanic Gardens. Custodian of Lichens and Fungi, National Herb., Sydney, 1899–1908. Assistant Botanist, 1908–24; Curator, 1924–36; State Botanist, 1933–36. President, Linnean Society New South Wales, 1930. President, Royal Society New South Wales, 1931. Contrib. to *Proc. R. Soc. N.S.W.*

J. Proc. R. Soc. N.S.W. 1952, xiv. *Proc. Linn. Soc. N.S.W.* 1952, vi.

Plants at Sydney Botanic Gardens. Portr. at Hunt Library.

CHEESEMAN, Thomas Frederick (1846–1923)

b. Hull, Yorks 1846 *d.* Auckland, New Zealand 15 Oct. 1923

FLS 1873. To New Zealand with parents, 1854. Curator, Auckland Museum, 1873. Collected plants in Cook Islands, *c.* 1870–1923; Kermadec Islands, *c.* 1868–89, 1902–12; New Zealand. *Catalogue of Plants of New Zealand* 1906. *Manual of New Zealand Fl.* 1906. *Illustrations of New Zealand Fl.* (with W. B. Hemsley) 1914 2 vols.

Gdnrs Chron. 1923 ii 343. *Nature* v.112, 1923, 871. *N.Z. Herald* 16 Oct. 1923. *Proc. Linn. Soc.* 1923–24, 47–48. *Trans. Proc. N.Z. Inst.* 1923, xvii–xix portr. *J. Bot.* 1924, 60. *Kew Bull.* 1924, 27–28. R. Glenn *Bot. Explorers of N.Z.* 1950, 160–71. *R.S.C.* v.7, 381; v.9, 504; v.12, 153.

New Zealand plants at Kew.

Veronica cheesemanii Benth.

CHEESMAN, Lucy Evelyn (1881–1969)

b. Westwell, Kent 1881 *d.* 15 April 1969

FRES. Entomologist. Made a number of insect collecting expeditions on which she also collected plants. Pacific, 1924–25; New Hebrides, 1929–31; 1954–55; Papua, 1933–34; New Guinea, 1936, 1938–39; New Caledonia, 1949–50. *Things Worth While* 1957.

Fl. Malesiana v.1, 1950, 106. *Times* 17 April 1969. *Who was Who, 1961–1970* 199. *Entomol. Mon. Mag.* 1969, 217–19 portr.

Plants at BM(NH), Kew.

CHEESMAN, William Norwood (1847–1925)
b. Winterton, Lincs 1 Feb. 1847 *d.* Selby, Yorks 7 Nov. 1925

FLS 1903. Draper of Selby. Secretary, Selby Naturalists' Society. One of the founders of British Mycological Society, the formation of which was discussed at his home. Interested in Mycetozoa. President, Yorkshire Naturalist's Union, 1916. President, British Mycological Society, 1925. Contrib. to *Naturalist.*

Bot. Soc. Exch. Club Br. Isl. Rep. 1926, 88–89. *J. Bot.* 1926, 22–23. *Naturalist* 1926, 23–25 portr.; 1961, 55–56. *Trans. Br. Mycol. Soc.* 1926, 1–4 portr.

Fungi at Kew.

CHEETHAM, Christopher Arthington (1875–1954)
b. Horsforth, Yorks 1875

Textile designer. *Supplement to Yorkshire Fl.* (with W. A. Sledge) 1941. Bryologist. Contrib. to *Naturalist.*

Naturalist 1954, 159–63 portr.; 1961, 160; 1973, 7–8 portr.

CHEETHAM, James (–1890)
d. Rochdale, Lancs 1890

Nurseryman, Rochdale.

Garden v.37, 1890, 449. *Gdnrs Chron.* 1890 i 591. *J. Hort. Cottage Gdnr* v.20, 1890, 380.

CHEMYS, Charles (–1733)
d. Dublin 9 Sept. 1733

BA Dublin 1720. MB 1724. Professor of Botany, Trinity College, Dublin.

Gdnrs Chron. 1919 i 147.

CHERRY, John William (1846–1935)
d. Brimpton, Berks 8 Jan. 1935

Madras Forest Service. Assistant Conservator, 1869; Conservator, 1891.

Times 11 Jan. 1935.

CHESHIRE, William (–*c.* 1855)
d. Stratford-on-Avon, Warwickshire *c.* 1855

Working printer. 'Anacharis' (*Phytologist* 1855–56, 361–64).

J. E. Bagnall *Fl. Warwickshire* 1891, 502. *R.S.C.* v.1, 896.

Plants at Warwick Museum.

CHESNEY, Francis Rawdon (1789–1872)
b. Annalong, County Down 16 March 1789
d. Mourne, County Down 30 Jan. 1872

FRS 1834. DCL Oxon 1850. Explorer of Euphrates, 1835–37. *Expedition for Survey of Rivers Euphrates and Tigris 1835–37* 1850. 2 vols. *Narrative of Euphrates Expedition... 1835–1837* 1868.

Misc. Bot. v.1, 1842, 9. *J. Bot.* 1872, 96. *Proc. R. Geogr. Soc.* 1872, 301–4. L. Chesney and J. O'Donnell *Life of Late General F. R. Chesney* 1885 portr. *D.N.B.* v.10, 195–98. R. L. Praeger *Some Irish Nat.* 1949, 63. *Fl. Iraq* v.1, 1966, 112. *R.S.C.* v.1, 896; v.7, 382.

Iraq plants at BM(NH), Bologna. MS. list of plants by J. Lindley at Kew.

Chesneya Lindl.

CHESTER, Sir George (1886–1949)
b. 16 Jan. 1886 *d.* 21 April 1949

General Secretary, National Union of Boot and Shoe Operatives, 1930. Director of Bank of England, 1949. Secretary, Kettering and District Naturalists Society, 1908–30. Correspondent of G. C. Druce. Supplied plant records to G. C. Druce's *Fl. Northamptonshire* 1930.

Bot. Soc. Br. Isl. Yb. 1950, 73. *Who was Who, 1941–1950* 212. *First Fifty Years: Hist. Kettering District Nat. Soc. Field Club* 1956, 23–25.

Herb. at Kettering Museum; plants at Oxford.

CHESTER, William (–1906)
d. 28 July 1906

Head gardener to Duke of Devonshire at Chatsworth, Derbyshire 1891.

Garden v.70, 1906, 60 (vi).

CHESTERTON, J. Henry (–1883)
d. Puerto Berrio, Colombia 26 Jan. 1883

Collected orchids in S. America for Messrs J. Veitch, 1870–78; then for another 5 years on his own behalf. Introduced *Miltonia vexillaria.*

Florist Pomologist 1883, 64. *Garden* v.23, 1883, 258. *Gdnrs Chron.* 1883 i 352. J. H. Veitch *Hortus Veitchii* 1906, 59–60. *Orchid Rev.* 1913, 340; 1960, 166. *J. R. Hort. Soc.* 1948, 286–87. K. Lemmon *Covered Garden* 1962, 214–15.

CHILD, George (*fl.* 1870s)

Seedsman, London. Surviving partner of Beck, Henderson and Child; disposed of business to Waite, Burnell and Co.

CHILD, Sir Josiah (1630–1699)
b. London 1630 *d.* 22 June 1699

Naval store-dealer, Portsmouth, 1655. Baronet, 1678. Autocratic Director of East India Company. Had garden at Wanstead, Essex where he went to "prodigious cost in planting walnut-trees about his seate, and making fish-ponds, many miles in circuit" (J. Evelyn).

Archaeologia v.12, 1794, 186–87. *D.N.B.* v.10, 244–45.

CHILDREN, Anna *see* Atkins, A.

CHILDS, Archibald Prentice (–1881)
d. Bungay, Suffolk 14 March 1881

MRCS 1849. FRCS 1852. Lecturer on Materia Medica and Therapeutics, Royal School of Medicine and Surgery, Manchester. *British Botanist's Field Book* 1857.

CHILDS, Kathleen Amelia (1880–1952)

Plants at Alton Museum, Hants.

CHIPP, Thomas Ford (1886–1931)
b. Gloucester 1 Jan. 1886 *d.* Kew, Surrey 28 June 1931

BSc London 1909. FLS 1912. Gardener, Syon House, Middx, 1904–6. Kew gardener, 1906; Assistant, Kew Herb., 1906–8. Demonstrator in Botany, Birkbeck College, 1909–10. Assistant Conservator of Forests, Gold Coast, 1910–14. Assistant Director, Singapore Botanic Gardens, 1919–20. Deputy Conservator of Forests, Gold Coast, 1921–22. Assistant Director, Kew, 1922–31. Collected plants in Gold Coast, Malaya, and Sudan; Algeria, 1929–30. *List of Trees, Shrubs and Climbers of Gold Coast, Ashanti and Northern Territories* 1913. *List of Herbaceous Plants and Undershrubs* 1914. *Forest Officers' Handbook* 1922. *Gold Coast Forest* 1927. Contrib. to A. G. Tansley's *Aims and Methods in Study of Vegetation* 1926, *Kew Bull., Empire For. J.*

Ann. Applied Biol. 1931, 636–37. *Empire For. J.* 1931, 1–2. *Gdnrs Chron.* 1931 i 20. *J. Bot.* 1931, 214–16. *J. Kew Guild* 1931, 81–82; 1932, 191 portr. *Kew Bull.* 1931, 397–98, 433–40 portr. *Nature* v.128, 1931, 141–42. *Proc. Linn. Soc.* 1931–32, 169–74. *Pharm. J.* 1931, 18. *Times* 1 and 6 July 1931. *Scott. Geogr. Mag.* 1931, 286–87. *Fl. Malesiana* v.1, 1950, 107.

Plants, letters and MSS. at Kew. Portr. at Hunt Library.

CHISHOLM, John S. (–1942)
d. Edinburgh 27 Feb. 1942

Head, Dept. of Horticulture, Edinburgh and East of Scotland College of Agriculture. President, Horticultural Education Association, 1937–38.

Gdnrs Chron. 1942 i 129.

CHITEFIELD (*fl.* 1880s?)

Plants from United Provinces, India at Oxford.

CHITTENDEN, Frederick James (1873–1950)
b. West Ham, London 25 Oct. 1873 *d.* Dedham, Essex 31 July 1950

VMH 1917. VMM 1947. FLS 1908. Lecturer in Biology, East Anglia Institute of Agriculture, 1900–7. Director, Royal Horticultural Society's Laboratory, Wisley, 1907–19; Director of Gardens at Wisley, 1919–31. Editor, *Dictionary of Gardening* 1951. *Garden Doctor* 1920. Editor, *J. R. Hort. Soc.* 1908–39.

Bot. Mag. 1948 front. portr. *Gdnrs Chron.* 1950 ii 74. *J. R. Hort. Soc.* 1950, 424–26 portr. *Proc. Linn. Soc.* 1949–50, 229. *Times* 8 Aug. 1950. *Who was Who, 1940–1950* 214.

CHITTY, William (*c.* 1814–1894)
d. London 15 Nov. 1894

Nurseryman, Stamford Hill, London. "Did a large amount of literary work, which passed under the name of Shirley Hibberd, never seeking to push forward his own name."

Gdnrs Chron. 1894 ii 642. *Gdnrs Mag.* 1894, 737.

CHORLEY, John (*fl.* 1790s)

Had garden at Tottenham. "To whose Lady my collection stands indebted for several rare and valuable plants" (W. Curtis. *Bot. Mag.* 1791, t.157).

CHRISTIAN, Garth Hood (*c.* 1922–1967)
d. North Chailey, Sussex 26 Nov. 1967

Naturalist. Contrib. to *Country Life.*

Daily Telegraph 27 Nov. 1967. *Habitat* v.3 (12), 1968, 6.

CHRISTIE, Alexander Craig- *see* Craig-Christie, A.

CHRISTIE, Joseph (1838–1898)
b. Kilmarnock, Ayr 1838 *d.* Glasgow 8 July 1898

Foreman moulder with Messrs. Kesson and Campbell, Glasgow. Had good knowledge of fauna and flora of Glasgow district.

Trans. Nat. Hist. Soc. Glasgow v.5, 1897–98, 300–1.

CHRISTIE-MILLER, Charles Wakefield *see* Addendum

CHRISTISON, Sir Robert (1797–1882)
b. Edinburgh 18 July 1797 *d.* Edinburgh 27 Jan. 1882

MD Edinburgh 1819. LLD Edinburgh 1872. Baronet, 1871. Toxicologist. Professor of Medicine, Edinburgh, 1822–77. *Treatise on Poisons* 1829; ed 4 1845. 'On Poisonous Properties of Hemlock and its Alkaloid Conia' (*Trans. R. Soc. Edinburgh* v.13, 1836, 383–417). 'On Properties of Ordeal Bean of Old Calabar' (*Edinburgh Mon. J. Med. Sci.* v.21, 1855, 193–204). 'Exact Measurement of Trees' (*Trans. Bot. Soc. Edinburgh* v.13, v.14, 1879, 1883).

Calcutta J. Nat. Hist. v.8, 1847, 153–62. *Nature* v.25, 1881–82, 339–40. *Trans. Bot. Soc. Edinburgh* v.14, 1883, 266–77. *Life of Sir Robert Christison* 1885 2 vols. *D.N.B.* v.10, 290–91. *R.S.C.* v.1, 922; v.9, 516; v.14, 220.

Letters at Kew.

Christisonia Gardner.

CHRISTY, Robert Miller (1861–1928)
b. Chignal St. James, Essex 21 May 1861 *d.* London 25 Jan. 1928

FLS 1889. President, Essex Field Club, 1905–7, 1910. Botanised with E. T. Seton in Manitoba in 1880s. Studied *Ulmus.* It is said that he secured many of his *Ulmus* specimens with the use of a blunderbuss. '*Primula elator* in Britain' (*J. Linn. Soc.* 1897, 172–201). Edited *Essex Nat.* 1917–19. Contrib. to *Essex Nat., J. Bot., J. Ecol., J. Linn. Soc., New Phytol.*

Bot. Soc. Exch. Club Br. Isl. Rep. 1928, 700. *Essex Nat.* 1928, 110–12 portr.; 1954, 192. *J. Bot.* 1928, 115–17. *Proc. Linn. Soc.* 1927–28, 112–13. S. T. Jermyn *Fl. Essex* 1974, 19.

Canadian plants at BM(NH), Manchester Museum. *Ulmus* and *Primula* collections at Essex Field Club.

CHRISTY, Thomas (1832–1905)
b. 9 Dec. 1832 *d.* Wallington, Surrey 7 Sept. 1905

FLS 1876. In China, 1853–56. *Forage Plants and...the New System of Ensilage* 1877. *New Commercial Plants* 1878–97 3 vols. *Dictionary of Materia Medica* (with C. H. Leonard) 1892.

Proc. Linn. Soc. 1905–6, 36. *R.S.C.* v.14, 222.

MSS. at Kew.

CHRISTY, William (*c.* 1807–1839)
b. Kingston-upon-Thames, Surrey *c.* 1807 *d.* Clapham, London 24 July 1839

FLS 1828. Of Lambeth and Clapham. In Channel Islands, 1836. Collected plants in Norway and Madeira. Gave 16,000 plants to Botanical Society of Edinburgh, now at Royal Botanic Garden, Edinburgh. *Notes of Voyage to Hammerfest, Alten,...* 1837. *Recollections of Five Days in Tenerife c.* 1848. Contrib. to G. S. Gibson's *Fl. Essex* 1862.

Bot. Mag. 1831, t.3078–80. *Gdnrs Mag.* 1837, 184–85; 1839, 536. *Proc. Bot. Soc. Edinburgh* 1838–39, 119. *Proc. Linn. Soc.* 1838, 67. *J. Bot.* (Hooker) 1842, 133. A. M. Babington *Memorials...of C. C. Babington* 1897, 267.

Plants at Cambridge, Dublin, Edinburgh, Kew. Letters at Linnean Society.

Christya Ward & Harvey.

CHUBB and Co. (*fl.* 1830s)

Seedsmen near Mansion House, London. Trade card in Johnson Collection, Bodleian Library.

CHURCH, Arthur Henry (1865–1937)
b. Plymouth, Devon 28 March 1865 *d.* Oxford 24 April 1937

BSc London 1889. DSc 1904. FRS 1921. Lecturer and Reader in Botany, Oxford, 1903–30. Studied algae, phyllotaxis, evolution and floral morphology. *Relation of Phyllotaxis to Mechanical Laws* 1901–04. *Types of Floral Mechanism* 1908. *Thalassiophyta and Subaerial Transmigration* 1919. Contrib. to *Bot. Mem. Oxford.*

Nature v.139, 1937, 870–71; v.140, 1937, 268. *Times* 29 April 1937. *Chronica Botanica* 1938, 174 portr. *Obit. Notices Fellows R. Soc.* 1936–38, 433–43 portr.

MSS. and drawings at BM(NH).

CHURCH, Sir Arthur Herbert (1834–1915)
b. London 2 June 1834 *d.* Kew, Surrey 31 May 1915

MA Oxon 1894. FRS 1888. Professor of Chemistry, Cirencester, 1863–79. 'Plant-chemistry' (*J. Bot.* 1875–77). *Food-grains of India* 1886; Supplt. 1901. Revised (with W. T. Dyer) S. W. Johnson's *How Crops Grow* 1869. Contrib. to J. A. Grant's *Potato Disease* 1873. Contrib. to H. Trimen and W. T. Dyer's *Fl. Middlesex* 1869 and to *J. Bot.* Made collection of botanical drawings, now at Kew (*Kew Bull.* 1916, 162–68).

F. H. Davey *Fl. Cornwall* 1909, liv. *Gdnrs Chron.* 1915 i 335. *Kew Bull.* 1915, 263. *Nature* v.95, 1915, 399–400. *R.S.C.* v.1, 925; v.7, 389; v.9, 518; v.12, 158; v.14, 227.

CHURCHILL, Edward George Spencer (1876–1964)
b. 21 May 1876 *d.* 24 June 1964

BA Oxon. Capt. Grenadier Guards. *Herbal of Apuleius Barbarus.*

Who was Who, 1961–1970 205.

CHURCHILL, George Cheetham (1822–1906)
b. Nottingham 25 Sept. 1822 *d.* Clifton, Bristol 11 Oct. 1906

FGS. Solicitor. Of Manchester until 1863; later Clifton. Collected plants extensively in Europe. 'American *Woodsia glabella* in Tyrol and Carinthia' (*J. Bot.* 1864, 56–57). *Dolomite Mountains* (with J. Gilbert) 1864.

Kew Bull. 1906, 384–92. *J. Bot.* 1907, 40. *Trans. Liverpool Bot. Soc.* 1909, 64. H. N. Clokie *Account of Herb. Dept. Bot., Oxford* 1964, 146.

Plants at Kew, Oxford. Letters at Kew.

CHURCHILL, James Morss (*fl.* 1830s–1840s)

MRCS. FLS 1827–42. *Medical Botany* (with J. Stephenson) 1831. 4 vols.

CHURTON, Peter (*fl.* 1860s)

Herb. at Haslemere Museum, Surrey.

CLAPHAM, Abraham (*fl.* 1860s–1870s)

Of Scarborough, Yorks. Collected and grew ferns.

E. J. Lowe *Fern Growing* 1895, 182–83 portr. *Br. Fern Gaz.* 1909, 43.

CLAPPERTON, Hugh (1788–1827)
b. Annan, Dumfriesshire 1788 *d.* Chungary near Sokota, Abyssinia 13 April 1827

Capt. Royal Navy. African explorer. In East Indies, 1808–13; Canada, 1814–17; Nigeria, 1822–27. *Narrative of Travels...in Northern and Central Africa...1822–1824* (with D. Denham) 1826 (botany by R. Brown).

R. Lander *Records of Capt. Clapperton's Last Expedition to Africa* 1830 2 vols. *D.N.B.* v. 10, 272–74.

Plants at BM(NH).

Clappertonia Meissner.

CLAPTON, Edward (1830–1909)
b. Stamford, Lincs 28 Sept. 1830 *d.* Stamford 28 Sept. 1909

MD London 1857. FRCS. FLS 1861. Assistant Surgeon and Lecturer in Botany, St. Thomas's Hospital, London 1857.

Proc. Linn. Soc. 1909–10, 86.

CLARE, John (1793–1864)
b. Helpston, Northants 13 July 1793 *d.* Northampton 20 May 1864
Nature poet. "An accurate observer and possessed quite a good knowledge of plants" (G. C. Druce).
D.N.B. v.10, 384–86. *J. Northampton Nat. Hist. Soc.* 1912, 183–214. G. C. Druce *Fl. Northamptonshire* 1930, xci–cxv. J. W. and A. Tibble *John Clare* 1932. A. R. Horwood and C. W. F. Noel *Fl. Leicestershire* 1933, cclxxiii–cclxxiv portr. *Proc. Bot. Soc. Br. Isl.* 1955, 482–89.

CLARK, George (*fl.* 1830s)
Nurseryman, Strawberry Hall, Lincoln Road, Newark-on-Trent, Notts.

CLARK, Henry (–1782/3)
Nurseryman, Barnet, Herts.
Agric. Hist. Rev. 1974, 21, 34–35.

CLARK, Henry (*c.* 1702–1778)
d. 29 June 1778
Gardener and land agent to Earl of Gainsborough and Sir Dudley Ryder. Also part-time nurseryman at Chipping Campden, Glos.
C. Whitfield *Hist. Chipping Campden* 1958, 150, 172, 186–88, 190.

CLARK, James (*fl.* 1820s)
Nurseryman, East Retford, Notts.

CLARK, James Henry (1818–)
b. Gloucester 1818
Printer, Usk, Monmouthshire, 1834. Wrote and published *Cardiff and its Neighbourhood* 1853, *Usk and its Neighbourhood* 1856, *Sketches of Monmouthshire* 1868, all containing plant lists.
A. E. Wade *Fl. Monmouthshire* 1970, 13.
Herb. at Newport (Mon.) Museum.

CLARK, Jessie Jane (1881–1914)
b. 25 Aug. 1881 *d.* Westcliff, Essex 2 Feb. 1914
BSc London. Assistant, Kew Herb., 1909–13. 'Abnormal Flowers in *Amelanchier*' (*Ann. Bot.* 1912, 948–49). Contrib. to *Kew Bull.* 1913.
Kew Bull. 1914, 172. *J. Kew Guild* 1915, 240–41 portr.

CLARK, John (–1762)
d. March/July 1762
Nurseryman, West Bow, Edinburgh. His widow, Elizabeth Campbell-Clark, carried on the business and maintained another nursery at Pinkie.

CLARK, John (*fl.* 1820s)
Nurseryman, Silsden near Keighley, Yorks.

CLARK, John (–1929)
d. Selkirk 28 Sept. 1929
Gardener. Grew orchids for J. and F. S. Roberts at Bannerfield, Selkirk.
Orchid Rev. 1930, 22–23.

CLARK, John Aubrey (1826–1890)
b. 24 July 1826 *d.* 4 Aug. 1890
Of Street, Somerset. Mycologist. Correspondent of Rev. M. J. Berkeley and C E. Broome.
J. Bot. 1898, 313.
Fungi drawings at Kew.
Hygrophorus clarkii Berkeley & Broome.

CLARK, John Willis (1833–1910)
b. Cambridge 24 June 1833 *d.* Cambridge 10 Oct. 1910
BA Cantab 1856. Superintendent, Zoological Museum, Cambridge, 1866–91. Registrary of University, 1891–1910. 'Absorption of Nutriment by Leaves of Some Insectivorous Plants' (*J. Bot.* 1875, 268–74).
Nature v.84, 1910, 501–2. *R.S.C.* v.9, 525.

CLARK, Pooty (*fl.* 1790s).
Seedsman, Woodbridge, Suffolk.

CLARK, Thomas (*fl.* 1780s–1810s)
Nurseryman, Keswick, Cumberland. In partnership with William Clark and James Hanks of Keighley, Yorks, *c.* 1805.

CLARK, Thomas (*fl.* 1820s–1850s)
Nurseryman, East Retford, Notts.

CLARK, Thomas (1793–1864)
b. Greinton, Somerset 16 Nov. 1793 *d.* Wembdon, Somerset 26 May 1864
Contrib. to H. C. Watson's *Topographical Bot.* 1873–74, and to *Phytologist* v.4, 1852–54.
J. Bot. 1898, 311–13; 1905, 233–38. J. W. White *Fl. Bristol* 1912, 83–84. *Watsonia* 1950, 243. *R.S.C.* v.1, 933.
Herb. at Taunton Museum. Letters at Kew. Portr. at Hunt Library.

CLARK, Thomas Bennet (1854–1926)
b. 5 Feb. 1854 *d.* 16 Jan. 1926
Educ. Edinburgh University. Accountant. "An enthusiastic collector and cultivator." President, Botanical Society of Edinburgh, 1908–10. Contrib. to *Trans. Bot. Soc. Edinburgh.*
Trans. Bot. Soc. Edinburgh v.29, 1926, 308–9.

CLARK, William (*c.* 1763–1831)
d. Dulwich, London 25 July 1831
Amateur florist, Croydon. Raised tulips including 'Lawrance's La Joie'.
Gdnrs Mag. 1831, 639.

CLARK, William (*fl.* 1800s)
Nurseryman, Keighley, Yorks. In partnership with Thomas Clark of Keswick and James Hanks of Keighley, *c.* 1805.

CLARK, William (*fl.* 1820s–1830s)
Botanical artist to Horticultural Society of London. Contrib. 61 etchings to R. Morris's *Fl. Conspicua* 1825–30 and 52 etchings to J. Stephenson and J. M. Churchill's *Med. Bot.* 1827–31; also a few plates to *Pomological Mag.* and *Trans. Hort. Soc. London.*
R. Sweet *British Flower Gdn* v.2, 1838, t.142.

CLARK, William (*fl.* 1830s)
Nurseryman, Market Place, Newark-on-Trent, Notts.

CLARK, William (*c.* 1885–1963)
b. Bannerfield, Selkirk *d.* Kinver, Staffs 27 Nov. 1963
VMH 1948. Parks Superintendent, Southport, 1922.
Gdnrs Chron. 1948 i 42 portr.; 1963 ii 468.

CLARKE, Alfred (1848–1925)
b. Winchester, Hants 7 March 1848 *d.* Huddersfield, Yorks 20 Jan. 1925
In business in Huddersfield, 1885–1923. Mycologist. 'List of Fungi of Huddersfield District' (*Ann. Rep. Huddersfield Bot. Soc.* 1883).
Naturalist 1925, 79–83 portr.; 1961, 57–59. *Br. Mycol. Soc. News Bull.* 1965–66, 13–20 portr.
Herb. at Huddersfield Museum.

CLARKE, Benjamin (1813–1890)
b. Saffron Walden, Essex 5 Sept. 1813. *d.* Hampstead, London 4 Feb. 1890
MRCS. FLS 1845. *New Arrangement of Exogens* 1851. *New Arrangement of Phanerogamous Plants* 1866; ed. 3 1888. Contrib. to *J. Bot.*
J. Bot. 1890, 84–86. *Proc. Linn. Soc.* 1889–90, 94–95. *R.S.C.* v.1, 935; v.7, 395; v.12, 159; v.14, 246.
Letters at Kew.

CLARKE, Charles Baron (1832–1906)
b. Andover, Hants 17 June 1832 *d.* Kew, Surrey 25 Aug. 1906
MA Cantab 1859. FRS 1882. FLS 1867. Nephew of Benjamin and Joshua Clarke. Lecturer in Mathematics, Cambridge, 1857–65. Joined staff of Presidency College, Calcutta, 1865. Inspector of Schools, East Bengal, 1883–87. Superintendent, Calcutta Botanic Gardens, 1869–71. On retirement in 1887 settled at Kew where he continued to work on Indian botany, and contrib. largely to *Fl. British India.* Collected plants in India. President, Linnean Society, 1894–96. *List of Flowering Plants...of Andover* 1866. *Commelynacea et Cyrtandraceae Bengalenses* 1874. *Compositae Indicae* 1876. *New Genera and Species of Cyperaceae* 1908. *Illustrations of Cyperaceae* 1909.
Bot. Soc. Exch. Club Br. Isl. Rep. 1906, 202–3. *Gdnrs Chron.* 1906 ii 164. *J. Bot.* 1893, 137–38; 1906, 370–77 portr. *Kew Bull.* 1906, 271–81; 1908, 376–78. *Nature* v.74, 1906, 495. *Proc. Linn. Soc.* 1906–7, 38–42. *Proc. R. Soc. B.* v.79, 1907, l–lvi. *Times* 4 Sept. 1906. *D.N.B. Supplt 2* v.1, 366–67. *Who was Who, 1897–1916* 140. *Curtis's Bot. Mag. Dedications, 1827–1927* 243–44 portr. I. H. Burkill *Chapters on Hist. Bot. in India* 1965, 144–46. *R.S.C.* v.7, 395; v.9, 526; v.14, 246.
Plants at Kew, BM(NH), Calcutta, etc. MSS. and letters at Kew. Portr. at Hunt Library.
Clarkella Hook. f.

CLARKE, Edward (*fl.* 1750s)
Seedsman, The Naked Boy and Three Crowns, Strand, London. Successor to Edward Fuller.

CLARKE, Rev. Edward Daniel (1769–1822)
b. Willingdon, Sussex 5 June 1769 *d.* London 9 March 1822
BA Cantab 1790. LLD 1803. Travelled as tutor in Italy, 1792; Germany, 1794; N. Europe, 1799; S. Russia, 1800; Asia Minor, Palestine, Greece, 1801. Collected plants and minerals on his travels. Plants presented to A. B. Lambert. Rector, Harlton, Cambridgeshire, 1805; Yeldham, Essex, 1809–22. *Travels in Various Countries of Europe, Asia and Africa* 1810–23 6 vols portr.
J. Nichols *Lit. Anecdotes of Eighteenth Century* v.4, 389–91. W. Otter *Life and Remains of Rev. Edward Daniel Clarke* 1824 portr. P. Smith *Correspondence of Sir J. E. Smith* v.2, 1832, 159. *D.N.B.* v.10, 421–24. *Taxon* 1970, 519. *R.S.C.* v.1, 935.
Plants at BM(NH).

CLARKE, George (*fl.* 1800s)
Seedsman, Newark-on-Trent, Notts.

CLARKE, George (*fl.* 1840s)
Of Mahé, Seychelles. 'On Sea Cocoanut of the Seychelles' (*Proc. Linn. Soc.* 1842, 153–55). 'Some Further Particulars of Coco de Mer, *Lodoicea Sechellarum*' (*Ann. Nat. Hist.* v.5, 1840, 422–24; v.6, 1841, 408–10).

CLARKE, Henry (*fl.* 1820s–1830s)
Seedsman, 121 Leadenhall Street, London; *c.* 1830 at 315 Oxford Street, London.

CLARKE, Henry (1858–1920)
b. London 18 July 1858 *d.* 12 Oct. 1920
Medical Officer, Wakefield Prison, 1874–1908. Drew British plants.
Bot. Soc. Exch. Club Br. Isl. Rep. 1920, 101–2.

CLARKE, James (*fl.* 1630s–1650s)
Apothecary. Accompanied T. Johnson on simpling ride through Windsor Forest (Gerard *Herball* 1633, 30). Joined him again on excursion into Kent, 1632 and on tour of West of England, 1634.

CLARKE, James (*fl.* 1760s)
Nurseryman, Dorking, Surrey. Went bankrupt in 1767 and over 20,000 plants were auctioned.
Agric. Hist. Rev. 1974, 21, 25–33.

CLARKE, James (*fl.* 1770s)
Seedsman, Houghton-le-Spring, County Durham. Issued catalogue in 1779.
Gdnrs Chron. 1917 ii 135–36.

CLARKE, John (*fl.* 1760s)
Butcher, Barnes, Surrey. Raised seedling cedars of Lebanon; also successful with magnolias and other exotics. Supplied the nursery trade and landowners.
Trans. Linn. Soc. v.10, 1811, 274–75.

CLARKE, John (*fl.* 1770s–1780s)
Grocer and seedsman, 53 Castle Street, Liverpool.

CLARKE, John (*fl.* 1780s)
Nurseryman, County Roscommon, Ireland.

CLARKE, Joshua (1805–1890)
b. Saffron Walden, Essex 10 April 1805 *d.* Saffron Walden Feb. 1890
FLS 1853. Brother of Benjamin Clarke. Discovered *Lathyrus tuberosus*. Assisted G. S. Gibson with *Fl. Essex* 1862.
J. Bot. 1865, 164, 221; 1890, 192. *Proc. Linn. Soc.* 1887–88, 95. *Gdnrs Chron.* 1893 i 202. *R.S.C.* v.1, 936.
Plants formerly at Saffron Walden Museum, Kew.

CLARKE, Lilian Jane (1866–1934)
b. London 27 Jan. 1866 *d.* London 12 Feb. 1934
BSc London 1893. FLS 1905. Science mistress, James Allen's Girls School, Dulwich, 1896–1926. *Botany as Experimental Science in Laboratory and Garden* 1935.
J. Bot. 1934, 112–13. *Nature* v.133, 1934, 439–40. *Proc. Linn. Soc.* 1933–34, 150–51. *Chronica Botanica* 1935, 163.

CLARKE, Mrs. Louisa Lane (*née* **Lane**) (*c.* 1812–1883)
d. L'Hyereuse, Guernsey, Channel Islands 8 Nov. 1883
Married Rev. Thomas Clarke, Rector, Woodeaton, Oxfordshire. Contrib. to Redstone's *Guernsey and Jersey Guide* 1844. *Island of Alderney* 1851. *Microscope* 1858; ed. 2: *Objects for Microscope* 1863, 1870. '*Spiranthes autumnalis*' (*Intellectual Observer* 1863, 195–98). *Common Seaweeds of British Coast* 1865. Contrib. to *Hardwicke's Sci. Gossip*, *Recreative Sci. Mag.*, etc.
The Star (Guernsey) 13 Nov. 1883. *J. Bot.* 1928, 174–75.

CLARKE, Richard (*c.* 1757–1836)
Florist, Bridge Street, Cambridge. Issued catalogue in 1793.

CLARKE, Richard (*fl.* 1780s–1790s)
Nurseryman, Shangarry, County Galway, Ireland.
Irish For. 1967, 52.

CLARKE, Richard Trevor (1813–1897)
b. Welton Place, Daventry, Northants 29 Aug. 1813 *d.* Welton Place 11 April 1897
Colonel. Horticulturist. Hybridised cotton and begonias.
Gdnrs Chron. 1865, 366; 1872, 799; 1897 i 263 portr. *Garden* v.51, 1897, 308. *R.S.C.* v.14, 250.
Letters at Kew.

CLARKE, Robert (*fl.* 1840s–1860s)
Surgeon, Sierra Leone. Sent plants and information about economic products regarding Sierra Leone and Gold Coast to Kew. 'Congo Tobacco' (*Hooker's J. Bot.* 1851, 9–11). '*Paspalum exile*' (*Proc. Linn. Soc.* 1842, 155–57).
R.S.C. v.1, 937.

CLARKE, Stephen (*fl.* 1820s)
Of Ipswich, Suffolk. Wrote anonymously: *British Botanist* 1820; *Hortus Anglicus* 1822 2 vols.

CLARKE, Stephenson R. (*c.* 1862–1948)
VMH 1936. Had garden at Borde Hill, Sussex. Expert cultivator of trees, shrubs, greenhouse plants and bulbs.

CLARKE, T. H. (–1946)
d. 16 April 1946
Grew orchids.
Orchid Rev. 1946, 84.

CLARKE, Thomas (*fl.* 1760s–1780s)
Nurseryman, Shangarry, County Galway, Ireland.

CLARKE, Thomas (*fl.* 1780s)
Grocer and seedsman, 26 Fenwick Street, Liverpool.

CLARKE, Thomas (–1792)
d. Feb. 1792
MD. First Island Botanist and Curator of Bath Garden, Jamaica, 1777–87. Introduced many plants.
Hortus Eastensis ed. 5 1819, 367–407. F. Cundall *Historic Jamaica* 1915, 25–26. *Gdnrs Chron.* 1919 i 147.

CLARKE, William Ambrose (1841–1911)
b. Hinckley, Leics 6 Feb. 1841 *d.* Oxford 23 Feb. 1911
FLS 1890, 1909. Solicitor. Of Chippenham; to Oxford, 1892. 'First Records of British Flowering Plants' (*J. Bot.* 1892–96; reprinted 1897; ed. 2 1900; Supplt. in *J. Bot.* 1909, 413–16). 'British Botany in 19th Century' (*J. Bot.* 1901, 128–40).
Bot. Soc. Exch. Club Br. Isl. Rep. 1911, 51–52. *J. Bot.* 1911, 167–69 portr. *Proc. Linn. Soc.* 1910–11, 36. A. R. Horwood and C. W. F. Noel *Fl. Leicestershire* 1933, ccxxviii. D. Grose *Fl. Wiltshire* 1957, 40–41, *R.S.C.* v.14, 251.
Portr. at Hunt Library.

CLARKE, William Barnard (*fl.* 1840s)
b. Ipswich, Suffolk
MD. Fellow, Botanical Society of Edinburgh. 'Fl. Ipswich' (*Mag. Nat. Hist.* 1840, 124–30, 317–25).
R.S.C. v.1, 937.

CLARKE, Rev. William Branwhite (1798–1878)
b. East Bergholt, Suffolk 2 June 1798 *d.* Sydney 17 June 1878
BA Cantab 1821. FRS 1876. Curate, Ramsholt, Suffolk. Anglican clergyman, New South Wales, 1840–70. Discovered gold there in 1841, tin in 1849 and diamonds in 1859. Wrote papers on peat-bogs, submerged forests, carboniferous plants.
Nature v.18, 1878, 389–90. *Proc. Linn. Soc. N.S.W.* 1878–79, 429–41. *Sydney Morning Herald* 18 June 1878. *Proc. R. Soc.* v.28, 1879, i–iv. P. Mennell *Dict. Austral. Biogr.* 1892, 97–98. *D.N.B.* v.10, 450–52. *Hist. Geol. Soc.* 1907, 189. *J. Proc. Paramatta District Hist. Soc.* 1926, 45–47. *J. Proc. R. Soc. N.S.W.* 1942, 92–128. *J. Proc. R. Austral. Hist. Soc.* 1944, 345–58. P. Serle *Dict. Austral. Biogr.* v.1, 1949, 175–77. *R.S.C.* v.1, 937; v.7, 396; v.9, 528.

CLARKE, William George (1877–1925)
b. Stokesley, Yorks 7 Jan. 1877 *d.* 15 June 1925
Reporter on *Norwich Mercury*. Ornithologist and botanist. *In Breckland Wilds* 1925. Contrib. to *Trans. Norfolk Norwich Nat. Soc.*
Trans. Norfolk Norwich Nat Soc. v.12, 1924–25, 129–33 portr. C. P. Petch and E. L. Swann *Fl. Norfolk* 1968, 18.
Plants at Norwich Museum.

CLARKSON, George (*fl.* 1820s–1854)
Gardener and nurseryman, 174 Walmgate and Fulford Road, York.

CLAY, Charles (1801–1893)
b. Stockport, Cheshire 27 Dec. 1801 *d.* Poulton-le-Fylde, Lancs 19 Sept. 1893
LRCS Edinburgh 1823. Practised in Manchester. Palaeontologist and archaeologist. *Geological Sketches and Observations on Fossil Vegetable Remains, etc. from Great South Lancashire Coalfield* 1839.
D.N.B. Supplt. 2 30–32. *Trans. Liverpool Bot. Soc.* 1909, 65.

CLAY, John (*fl.* 1790s)
Gardener and seedsman, Mansfield, Notts.

CLAY, Samuel (*c.* 1816–1899)
d. Clacton, Essex 28 Oct. 1899
Nurseryman, Stratford, London.

CLAYTON, Henry James (–1914)
d. Kirkby Wharfe, Yorks Feb. 1914
Gardener at Grimston Park, Tadcaster, Yorks, 1872–1907. Contrib. to *Gdnrs Chron.*
Gdnrs Chron. 1914 i 120 portr., 139, 150–51, 206.

CLAYTON, Rev. John (*fl.* 1670s–1690s)
In Virginia before 1671. Rector, Crofton, Yorks, 1688. Dean of Kildare. Uncle to J. Clayton (1694–1773)? Letter on plants of Virginia in *Philos. Trans. R. Soc.* v.41, 1744, 143–62.
J. Bot. 1909, 297–301.
MS. on Virginia, 1671, at Royal Society.

CLAYTON, John (1694–1773)
d. Gloucester County, Virginia 15 Dec. 1773
Went to Virginia, 1705. *Fl. Virginica* 1739–43; ed. 2 1762. Herb. sent to J. F. Gronovius, now at BM(NH) (*Rhodora* 1915, 39–40; 1918, 21–28, 48–54, 65–73).
Gent. Mag. 1755, 407–8. *Philadelphia Med. Phys. J.* 1806, 139–45. W. Darlington *Reliquiae Baldwinianae* 1843, 165. *Appleton's Cyclop. Amer. Biogr.* v.1, 1887, 645–46. *D.N.B.* v.11, 13. *J. Bot.* 1909, 297–301. H. A. Kelly *Some Amer. Med. Botanists* 1915, 44–48. E. G. Swem *Brothers of the Spade* 1949, 175–76. *Rhodora* 1928, 232–37. *Gdn J. New York Bot. Gdn* 1951, 85–87. E. and D. S. Berkeley *John Clayton* 1963. *Virginia Mag. Hist. Biogr.* 1968, 415–36.
Claytonia Gron.

CLAYTON, John (*c.* 1846–1933)
d. Harrogate, Yorks 20 Dec. 1933
President, Bradford Naturalists and Microscopical Society, 1887. *Effects of Weather upon Vegetation* 1897. 'Cowthorpe Oak' (*Trans. Proc. Bot. Soc. Edinburgh* v.22, 1903, 396–414). *Sequoias* (MS.).
Naturalist 1934, 134–35.

CLAYTON, Sir Robert (1629–1707)
b. Bulwick, Northants 29 Sept. 1629 *d.* Marden, Surrey 16 July 1707
Merchant. Alderman of London, 1670–88; Lord Mayor, 1679–80. MP, London, 1679–81. Had garden at Marden.
Archaeologia v.12, 1794, 187. *D.N.B.* v.11, 17–19.

CLAYTON-BROWNE, Robert (1838–1906)
b. Newmount, Carlow 3 May 1838 *d.* Greenville, Carlow 15 Dec. 1906
Bryologist. Herb. and drawings at National Museum, Dublin.
Proc. R. Irish Acad. B. v.32, 1915, 75.

CLEEVE, Rev. Alexander (1747–1805)
b. Westminster, London 1747
BA 1770. Vicar, Wooler, Northumberland, 1780–1805. First Secretary, Horticultural Society of London, 1804–5.
J. R. Hort. Soc. 1933, 320–23 portr.

CLEGHORN, George (1716–1789)
b. Granton, Edinburgh 18 Dec. 1716 *d.* near Dublin Dec. 1789
MD Edinburgh 1736. Army Surgeon in Minorca, 1736–49. Lecturer and Professor of Anatomy in Dublin, 1751–89. Friend of J. Fothergill. *Observations on Epidemical Diseases in Minorca* 1751 (plants, pp. 12–45).
J. C. Lettsom *Mem. of John Fothergill* 1786, 227–37 portr. *D.N.B.* v.11, 25–26. R. H. Fox *Dr. John Fothergill and his Friends* 1919, 121–25.

CLEGHORN, Hugh Francis Clarke (1820–1895)
b. Madras, India 9 Aug. 1820 *d.* Stravithie, Fife 19 May 1895

MD Edinburgh 1841. LLD St. Andrews 1868. FLS 1851. Madras Medical Service, Mysore, 1842. Professor of Botany, Madras, 1852. Conservator of Forests, Madras, 1856; Inspector-General, 1867. *Hortus Madraspatensis* 1853. *General Index of Plants...in Dr. Wight's...Icones, etc.* 1856. *Forests and Gardens of South India* 1861.

Indian For. 1888, 395–401; 1895, 276–78; 1905, 227–34. *Trans. R. Scott. Arboricultural Soc.* 1888, 87–93. *Proc. R. Soc. Edinburgh* v.20, 1892–95, li–lx. *J. Bot.* 1895, 256. *Pharm. J.* 1894–95, 1085. *Trans. Bot. Soc. Edinburgh* v.20, 1895, 439–48. D. G. Crawford *Hist. Indian Med. Service, 1600–1913* v.2, 1914, 151. I. H. Burkill *Chapters on Hist. Bot. in India* 1965 *passim. R.S.C.* v.1, 948; v.7, 403; v.14, 265.

Herb. at Edinburgh Botanical Garden. Plants and letters at Kew.

Cleghornia Wight.

CLEMENS, Joseph (1862–1932)
b. St. Just, Cornwall 1862 *d.* Finschhafen, New Guinea 1936

Chaplain in U.S. Army, 1902. With his wife, Mary Knapp Clemens (*née* Strong), collected plants in Philippines, 1905–29; British N. Borneo, 1915–17, 1929, 1931–33; China, 1912–14; Indo-China, 1927; W. Java, 1932; N.E. New Guinea, 1935; W. U.S.A., 1912–19.

Chronica Botanica 1936, 89–90 portr. *J. New York Bot. Gdn* 1936, 117–18. *Kew Bull.* 1936, 287–88. J. Ewan *Rocky Mountain Nat.* 1950, 183. *Fl. Malesiana* v.1, 1950, 108–10 portr.

Plants at BM(NH), Kew, Chicago, Manila, etc.

CLEMENS, Mary Knapp *see* Clemens, J.

CLEMENT (or **Clements**), **John** (–1572)
d. Malines, Belgium 1572

MD Oxon. FRCP 1528. Professor of Greek. President, Royal College of Physicians, 1544. Tutor to Sir Thomas More's children. "The best botanist and herbalist of us all" (Sir T. More).

Gdnrs. Chron. 1872, 706. W. Munk *Roll of R. College of Physicians* v.1, 1878, 27. B. D. Jackson *Guide to Literature of Bot.* 1881, xxx. *D.N.B.* v.11, 33.

CLEMENTS, Frederick Moore (1857–1920)
d. Sydney 17 Aug. 1920

FLS 1917. Pharmaceutical chemist. Travelled in Central Africa, finally settling in New South Wales, 1881. Good knowledge of medicinal plants which he grew in his garden at Sydney. *Some Faces and Places of Clem. c.* 1918 (includes plants in his garden).

Proc. Linn. Soc. 1920–21, 45. *Proc. Linn. Soc. N.S.W.* 1921, 8. *Sydney Morning Herald* 18 and 20 Aug. 1920.

Library at Linnean Society, New South Wales.

CLEMINSHAW, Edward (1849–1922)
b. 1 June 1849 *d.* Birmingham 1922

BA Oxon 1873. Science master, Sherborne School. Analyst, Messrs. Chance's, Oldbury. Arranged moss collection at Birmingham University.

J. E. Bagnall *Fl. Warwickshire* 1891, 503. *J. Bot.* 1903, 366–71, 388–97; 1925, 25. *Rep. Br. Bryol. Soc.* 1924, 94. *Watson Bot. Exch. Club Rep.* 1924–25, 283.

Plants at Rugby School.

CLEPHANE, Arthur (*fl.* 1700s–1730s)

After retirement as ship's master in 1706 became merchant and seedsman, Edinburgh.

Agric. Hist. Rev. 1970, 151–60.

CLERK, Mr. (*fl.* 1730s–1740s)

Of Jersey, Channel Islands. Plants collected in Jersey at Oxford.

CLERK, William (*fl.* 1690s–1710s)

Surgeon. Brought plants from Turkey on J. Petiver.

J. E. Dandy *Sloane Herb.* 1958, 113–14.

CLERK, Rev. William (*fl.* 1710s–1730s)

Collected plants in Virginia, 1729, Carolina, Antigua, Montserrat, 1734, and Bermuda.

Proc. International Congress Plant Sci. 1926 v.2, 1929, 1526–27. J. E. Dandy *Sloane Herb.* 1958, 113.

Plants at BM(NH), Oxford.

CLEVELEY, John (1747–1786)
b. Deptford, London 25 Dec. 1747. *d.* Deptford? 25 June 1786

Marine painter. Draughtsman on Sir J. Banks's voyage to Iceland, 1772 and on Captain Phipps's Arctic voyage, 1774. Prepared finished drawings from sketches by S. Parkinson made during Capt. Cook's voyage, 1768–71.

D.N.B. v.11, 53.

Drawings at BM(NH).

CLEVERLY, John (*fl.* 1810s–1820s)

Nurseryman near the Bridge, Stoke Newington, London.

CLIFFE, Peter (–1885)

Gardener to Lord Egerton at Tatton Park, Cheshire.

Gdnrs Chron. 1885 i 124.

CLIFFORD, Dowager Lady (*fl.* 1820s)

Had garden at Paddington, London where *Artabotrys odoratissima* fruited in 1820 for first time in Europe.

E. Bretschneider *Hist. European Bot. Discoveries in China* 1898, 283.

CLIFFORD, J. R. S. (*c.* 1833–1910)
b. Pimlico, London *d.* 15 August 1910
Author. Contrib. to *J. Hort. Cottage Gdnr.*
J. Hort. Home Farmer v.61, 1910, 187 portr.

CLIFFORD, Thomas Hugh *see* Constable, *Sir* T. H. C.

CLIFTON, Edwin Samuel (1841–1929)
d. Ipswich, Suffolk 6 Feb. 1929
Homoeopathic chemist, Norwich, 1866. Cultivated medicinal plants.
Pharm. J. 1929, 167 portr.

CLIFTON, George (1823–1913)
b. 15 March 1823 *d.* Eastbourne, Sussex 12 Aug. 1913
Royal Navy. To W. Australia. Police Inspector, Freemantle, 1851. Sent Australian algae during 1851–64 to W. H. Harvey who dedicated vol. 1 of his *Phycologica Australica* to him.
W. H. Harvey *Phycologica Australica* v.5, 1863 preface v, t.279. *Trans. Linn. Soc.* v.3, 1891, 211. *J. W. Austral. Nat. Hist. Soc.* no. 6, 1909, 11–12. *J. Bot.* 1924, 328–30. *J. W. Austral. Hist. Soc.* 1936, 1–25; v.4, 1949–54, 67–68.
Algae at BM(NH). Letters at Kew. MSS. at Public Library, Perth.
Cliftonaea Harvey.

CLIFTON, William (*fl.* 1760s)
Attorney-General of Georgia in 1759. Chief Justice of West Florida. Sent Florida plants collected by negro servant in 1765 to J. Ellis.
Philos. Trans. R. Soc. v.60, 1771, 52–56. J. E. Smith *Selection of Correspondence of Linnaeus* v.1, 1821, 438, 571; v.2, 1821, 72. *J. Bot.* 1903, 87.
Cliftonia Banks.

CLINTON, P. (*fl.* 1820s)
MD. Professor of Medicinal Botany, Apothecaries Hall, Dublin. Translated and annotated A. Richards *Elémens de Botanique* 1829.
K. Baily *Irish Fl.* 1833, viii–ix.

CLINTON-BAKER, Henry William (1865–1935)
d. Bayfordbury, Herts 19 April 1935
Continued cultivation of pinetum established by his grandfather at Bayfordbury. *Illustrations of Conifers* 1909–13 3 vols. *Illustrations of New Conifers* (with A. B. Jackson) 1935.
Gdnrs Chron. 1935 i 295. *Kew Bull.* 1935, 336. *Nature* v.136, 1935, 13–14.

CLISBY, George (*fl.* 1830s)
Seedsman and herbalist, 449 West Strand; then 2 Hungerford Market, Strand. Successor to S. Maine in 1835. Issued catalogue in 1835.

CLISBY, J. (*fl.* 1830s)
Nurseryman, Thame, Oxfordshire.

CLITHEROE, William (1864–1944)
b. 21 May 1864 *d.* 21 May 1944
FLS 1903–16. Schoolteacher, Preston until 1929. Retired to Bowness-on-Windermere. Chairman, Botanical Section, Preston Scientific Society. Contrib. to *Fl. W. Lancashire* 1907.

CLIVE, Lady Charlotte Florentina, Duchess of Northumberland (–1866)
Granddaughter of Robert Clive of India. First to flower the Kaffir Lily, *Clivia*.

CLIVE, Edward, 1st Earl of Powis (1754–1839)
d. 16 May 1839
Son of Robert Clive of India. Governor of Madras, 1798–1803. Created Baron Clive of Walcot, 1794; Earl of Powis, 1804. Planted 2000 seedling Arolla pines at Walcot; also grew the rare exotic *Fourcroya*. "Remarkable for his physical vigour, which he retained to an advanced age, digging in his garden in his shirt-sleeves at six o'clock in the morning when in his eightieth year."
D.N.B. v.11, 108.

CLIVE, Lady Henrietta Antonia, Countess of Powis (1758–1830)
b. 3 Sept. 1758 *d.* Walcot, Shropshire 3 June 1830
Wife of Lord Clive (1754–1839). While in Mysore with her husband she discovered *Caralluma umbellata c.* 1800 and sent several specimens to Calcutta Botanic Gardens.
A. White and B. L. Sloane *Stapelieae* v.1, 1937, 88–91 portr.

CLOUGH, Edwin (–1883)
d. Ashton, Lancs 8 Feb. 1883
Workingman botanist of Ashton-under-Lyme. Active in local workingmen's botanical societies.
J. Bot. 1883, 192. *Trans. Liverpool Bot. Soc.* 1909, 65.

CLOUSTON, Rev. Charles (1800–1885)
b. Stromness, Orkney 1800 *d.* Stromness 1885
LLD St. Andrews. Minister of Sandwick, 1832. Orkney plants listed in Anderson's *Guide to Highlands* 1834.
T. Edmonston *Fl. Shetland* 1845, 54–55. *Scott. Nat.* 1885, 49–50. M. Spence *Fl. Orcadensis* 1914, xlvii–xlix. *R.S.C.* v.1, 960; v.14, 273.
Laminaria cloustonii Edmonston.

CLOWES, Frederic (*fl.* 1850s)
Of Bowness, Cumberland. Surgeon. Pteridologist. Found a marginate polypody in 1854. 'List for Windermere' (H. Martineau *Complete Guide to English Lakes* 1855). '*Lastrea remota*' (*Phytologist* v.4, 1860, 227–29).
Proc. Linn. Soc. 1855, 359–60. *Br. Fern Gaz.* 1909, 44.

CLOWES, G. (*fl.* 1850s)
Visited New South Wales for health reasons in mid 19th century. Collected plants which he sent to Kew.
J. D. Hooker *Fl. Tasmaniae* v.1, 1859, cxxvii. *J. R. Soc. N.S.W.* 1908, 97–98.

CLOWES, Rev. John (1777–1846)
b. Broughton Hall near Manchester 1 May 1777 *d.* Broughton Hall 28 Sept. 1846
BA Cantab 1799. MA 1805. Interested in botany and horticulture during last 10 years or so of his life. His outstanding collection of orchids went to Kew after his death. His gardener, W. Hammond, compiled *Catalogue of Orchidaceous Plants in Collection of Rev. John Clowes* 1842.
Curtis's Bot. Mag. Dedications, 1827–1927 75–76 portr. *Orchid Rev.* 1931, 65.
Portr. at Royal Horticultural Society.
Clowesia Lindl.

CLUNIES ROSS, William John (1850–1914)
b. London 1850 *d.* Nov. 1914
To Australia, 1864. Returned to England to study at King's College, London. Head, Technical College, Bathurst, New South Wales, 1884. Lecturer in Chemistry and Metallurgy, Sydney Technical College, 1903. 'Notes on Fl. Bathurst' (*Rep. Austral. Assoc. Advancement Sci.* no. 7, 1898, 467–81).
J. Proc. R. Soc. N.S.W. 1915, 8–9.

CLUTTON, Joseph (*fl.* 1720s)
Apothecary, Holborn, London. Botanist. Correspondent of R. Richardson.
D. Turner *Extracts from Lit. and Sci. Correspondence of Richard Richardson* 1835, 251.

COAKER, Jonas (*fl.* 1810s)
Nurseryman, Old Brompton, Kensington, London. Junior partner of Samuel Harrison.

COATES, Leonard (*fl.* 1870s)
To California, 1876, where he established a nursery.
J. Hort. Cottage Gdnr v. 50, 1905, 263–64.

COBB, Walter (*c.* 1836–1922)
d. 19 April 1922
Of Normanhurst, Rusper, Sussex. Cultivated orchids.
Orchid Rev. 1922, 153–54 portr.

COBBE, Amy Beresford (–1952)
d. 18 Feb. 1952
Sister of Miss Mabel Cobbe; both were keen botanists.
Daily Telegraph 20 Feb. 1952.

COBBE, Mabel (–1936)
d. Alassio, Italy 31 Dec. 1936
Daughter of Rev. Henry Cobbe. Collected British and continental plants.
Bot. Soc. Exch. Club Br. Isl. Rep. 1937, 435. H. J. Riddelsdell *Fl. Gloucestershire* 1948, cxlix.
Herb. at Bexhill Museum.

COBBETT, Henry (*fl.* 1820s–1850s)
Nurseryman, Horsell, Surrey. In partnership with Carmi.

COBBETT, William (1762–1835)
b. near Farnham, Surrey 9 March 1762 *d.* Kensington, London 18 June 1835
Journalist, social reformer and agriculturist. In U.S.A., 1792–1800, 1817–19. Nurseryman, Kensington, London, 1824–35. Issued in 1827 *Catalogue of American Trees, Shrubs and Seeds for Sale. American Gardener* 1821. *Cobbett's Cottage Economy* 1822. *Woodlands* 1825. *English Gardener* 1829. *Rural Rides* 1830.
D.N.B. v.11, 142–45. *Gdnrs Chron.* 1898 i 209–10; 1963 ii 158. *J. R. Agric. Soc. England* 1902, 1–26 portr. L. Melville *Life and Letters of William Cobbett* 1913. M. Bowen *Peter Porcupine* 1935. G. D. H. Cole *Life of William Cobbett* ed. 3 1947. W. Reitzel, *ed. Autobiography of William Cobbett* 1947. *Country Life* 1963, 464–65. *Huntia* 1965, 66–109 portr.
Portr. at National Portrait Gallery, Hunt Library.

COBBOLD, Thomas Spencer (1828–1886)
b. Ipswich, Suffolk 1828 *d.* London 20 March 1886
MD Edinburgh 1851. FRS 1864. FLS 1857. Helminthologist. Curator, Edinburgh Anatomical Museum, 1851–56. Professor of Botany, Royal Veterinary College, London, 1873. 'Embryogeny of *Orchis mascula*' (*Quart. J. Microsc. Sci.* 1853, 90–92).
Gdnrs Chron. 1886 i 406. *Proc. Linn. Soc.* 1885–86, 140–41. *Proc. R. Soc.* v.47, 1890, iv–v. *D.N.B.* v.11, 147. *R.S.C.* v.2, 2; v.7, 409; v.14, 277.

COCHRAN (Cockran?), James (*fl.* 1820s)
Seedsman and florist, 7 Duke Street, Grosvenor Square, London.

COCHRAN, John (1867–1961)
d. Kilmarnock 29 Nov. 1961
Joiner. Cultivated ferns.
Br. Fern Gaz. 1962, 95–96.

COCHRANE, Sir Alexander Forrester Inglis (1758–1832)
b. 22 April 1758 *d.* Paris 26 Jan. 1832
Officer, Royal Navy. Collected plants in Newfoundland.
W. J. Hooker *Fl. Boreali-americana* v.1, 1829, 172, 175, etc.

COCHRANE-BAILLIE, Charles Wallace Alexander Napier Ross, 2nd Baron Lamington (1860–1940)

Collected plants in India, *c.* 1890s and Persia, *c.* 1910s, now at Kew. Governor, Queensland, 1896–1901. Governor, Bombay, 1903–7. Collected plants in S.E. New Guinea, 1898.

Fl. Malesiana v.1, 1950, 310.

Plants at Brisbane.

COCKAYNE, Leonard (1855–1934)

b. Norton Lees, Derbyshire 7 April 1855 *d.* Wellington, N.Z. 8 July 1934

Educ. Owen College, Manchester. FRS 1912. FLS 1910. VMM 1932. To Australia, 1879. New Zealand, 1880. Teacher, 1881–85. Private experiment station, 1887–1904. Made botanical surveys for Lands and Survey Dept., 1906 onwards. Botanist, New Zealand State Forest Service. Founded Otari Open Air Plant Museum, Wellington. President, New Zealand Institute, 1918–19. *New Zealand Plants and their Story* 1910; ed. 3 1927. *Vegetation of New Zealand* 1921; ed. 2 1928. *Cultivation of New Zealand Plants* 1924. *Monograph on New Zealand Bush Forests* 1926–28. *Trees of New Zealand* (with E. P. Turner) 1928; ed. 5 1965.

Gdnrs Chron. 1926 ii 222 portr.; 1934 ii 111. *J. Bot.* 1934, 257–59. *J. N.Z. Inst. Hort.* 1934, 11–15 portr. *Kew Bull.* 1926, 428; 1929, 63–64; 1934, 313–17. *N.Z. Agric. J.* v.49, 1934, 161–63 portr. *Nature* v.134, 1934, 170. *Times* 9 July 1934. *Obit. Notices Fellows R. Soc.* 1935, 443–57 portr. *Proc. Linn. Soc.* 1934–35, 167–71. *Chronica Botanica* 1935, 231 portr. *Trans. Proc. R. Soc. N.Z.* v.65, 1936, 457–67 portr.; 1967, 1–18 portr. *Who was Who 1929–1940* 268. R. Glenn *Bot. Explorers of N.Z.* 1950, 148–59 portr. *Rec. Dominion Mus.* 1965, 265–76.

Plants at Wellington, Berlin, etc.

COCKAYNE, Rev. Thomas Oswald (1807–1873)

d. St. Ives, Cornwall 2 June 1873

BA Cantab 1828. Assistant master, King's College School, London until 1869. Philologist. *Leechdoms, Wortcunning, and Starcraft of Early England* 1864–66 3 vols.

D.N.B. v.11, 176.

COCKBURN, Sir James (–*c.* 1780)

d. Ireland *c.* 1780

Introduced *Gardenia thunbergia* in 1773. *Gent. Mag.* 1780, 153. *Bot. Mag.* 1807, t.1004. W. T. Aiton *Hortus Kewensis* v.1, 1810, 369.

COCKBURN, John (–1758)

Lord of the Admiralty. MP 1707–41. Had garden at Ormistoun, East Lothian.

J. Colville, *ed. Letters of John Cockburn of Ormistoun to his Gdnr, 1727–1744* 1904 portr. *J. Hort. Cottage Gdnr* v.48, 1904, 507–8.

COCKER, Alexander Morrison (1860–1920)

b. Aberdeen 1860 *d.* 19 Feb. 1920

Nurseryman, Sunnyside Nursery, Aberdeen. Grandson of founder of nursery.

Gdnrs Chron. 1920 i 110.

COCKER, James (*c.* 1806–1880)

d. Aberdeen 22 Oct. 1880

Established himself as a market gardener in Aberdeen in 1835. Later opened nurseries at Sunny Park, Froghall and Morningfield. Took two of his sons into partnership in 1874.

Florist Pomologist 1880, 188. *Garden* v.18, 1880, 472. *Gdnrs Chron.* 1880 ii 605.

COCKER, James (*c.* 1832–1897)

b. Corse *c.* 1832 *d.* Aberdeen 15 Sept. 1897

Nurseryman, Aberdeen. Son of J. Cocker (*c.* 1806–1880). Took his three sons, Alexander, James and William, into partnership.

Garden v.52, 1897, 232. *Gdnrs Chron.* 1897 ii 221. *Gdnrs Mag.* 1897, 588. *J. Hort. Cottage Gdnr* v.35, 1897, 290.

COCKER, James (1855–1894)

b. Aberdeen 1855 *d.* Aberdeen 21 Nov. 1894

Nurseryman, Aberdeen. Son of J. Cocker (*c.* 1832–1897).

Gdnrs Mag. 1894, 737.

COCKER, William (*c.* 1858–1913)

d. Aberdeen 27 Feb. 1913

Nurseryman, Springhill Nurseries, Aberdeen. Son of J. Cocker (*c.* 1832–1897).

J. Hort. Home Farmer v.63, 1911, 417–18 portr.; v.66, 1913, 242. *Garden* 1913, 128 (xvi). *Gdnrs Mag.* 1913, 179 portr.

COCKERELL, Theodore Dru Alison (1866–1948)

b. Norwood, London 22 Aug. 1866 *d.* San Diego, California, U.S.A. 26 Jan 1948

To Colorado, 1887. Held appointments in Public Museum, Kingston, Jamaica and New Mexico Agricultural College. Professor of Zoology, University of Colorado. 'Recollections of a Naturalist' (*Bios* 1935–40 portr.). Authority on flora and fauna of Colorado.

E. O. Essig *Hist. Entomology* 1931, 570–73 portr. *Amer. Philos. Yb.* 1948, 247–52. *Nat. Hist. Notes Nat. Hist. Soc. Jamaica* no. 33, 1948, 158. *Nature* v.161, 1948, 229–30. *Science* 1948, 295–96. J. Ewan *Rocky Mountain Nat.* 1950, 95–116 portr. *Univ. Colorado Studies Ser. Bibl.* no. 1, 1965, 1–124.

Plants at Kew. Portr. at Hunt Library.

Malvostrum cockerelli A. Nelson.

COCKFIELD, Joseph (*c.* 1740–1816)

d. March 1816

Of Upton, Essex. Friend of J. Fothergill and J. C. Lettsom. *Botanist's Guide* 1813. Letters, 1765–71 (J. Nichols *Illus. Lit. Hist. of Eighteenth Century* v.5, 753–808).

J. Smith *Cat. of Friends' Books* v.1, 1867, 438. H. Trimen and W. T. T. Dyer *Fl. Middlesex* 1869, 398.

COCKRAN (Cochran?), E. and S. (*fl.* 1820s)
Nurserymen and seedsmen, 7 Duke Street, Grosvenor Square, London.

COCKS, John (1787–1861)
b. Sussex 1787 *d.* Devonport 1861
MD. First to discover red seaweed *Stenogramme interrupta. Sea-weed Collector's Guide* 1853. *Algarum Fasciculi* 1855–60 (exsiccatae). 'Marine Algae' (*J. Linn. Soc.* v.4, 1859, 101–5).
R.S.C. v.2, 5.
Algae at BM(NH).

COCKS, Llewellyn J. (–1921)
d. Esher, Surrey 1921
Contrib. mosses to *Naturalist* 1897–98.
Naturalist 1973, 2.

COCKS, Reginald Wodehouse Somers (1863–1926)
b. Worcester 31 Aug. 1863 *d.* New Orleans, U.S.A. 21 Nov. 1926
MA Cantab 1889. Professor of Botany, Louisiana University, 1906–7. Professor of Botany, Tulane University, New Orleans, 1907.
Science v.64, 1926, 593. *Tulane News Bull.* 1925, 115–16. 'Letters from C. S. Sargent to R. C. Cocks, 1908–1926' (*J. Arnold Arb.* 1965, 1–44, 122–50, 324–61, 411–44 portr.)
Portr. at Hunt Library.
Quercus cocksii Sarg.

COCKS, William Pennington (1791–1878)
b. Devon Nov. 1791 *d.* Falmouth, Cornwall 10 July 1878
MRCS. Of Falmouth. Surgeon. Privately-printed papers on Falmouth flora and fauna from 1849 onwards. His collections lost.
F. H. Davey *Fl. Cornwall* 1909, li. *J. Conchology* v.27, 1971, 253–55 portr.
MS. 'Contributions to Fl. Falmouth', 1856 at BM(NH).

CODRINGTON, Rev. Robert Henry (1830–1922)
b. Wroughton, Wilts 15 Sept. 1830 *d.* Chichester, Sussex Sept. 1922
Educ. Oxford. Head of Melanasian Mission. Vicar, Wadhurst, Sussex. "A great lover of plants."
Bot. Soc. Exch. Club Br. Isl. Rep. 1922, 698. *Nature* v.110, 1922, 425.

COEL, James *see* Cole, J.

COEY, James (–1921)
d. Larne, County Antrim 8 Feb. 1921
Nurseryman, Donard Nursery, Newcastle, County Down, *c.* 1912. Hybridised daffodils.
Garden 1921, 108.

COFFIN, Albert Isaiah (*c.* 1791–1866)
d. London 1 Aug. 1866
MD Rostock. In London from 1850. *Botanic Guide to Health* 1845; ed. 36 1866. *Medical Botany* 1851.
F. Boase *Modern English Biogr. Supplt.* v.1, 1908, 701.
Portr. at Hunt Library.

COGGINS, George (*fl.* 1890s)
Of Clifton Hall Gardens, Cumberland. Contrib. to W. Hodgson's *Fl. Cumberland* 1898 (viii).

COLDEN, Cadwallader (1688–1776)
b. Ireland 7 Feb. 1688 *d.* Long Island, New York, U.S.A. 28 Sept. 1776
MD Edinburgh 1705. Practised medicine in Pennsylvania, 1708–18. Surveyor-General, New York Colony, 1719; Lieut.-Governor, 1761–75. Correspondent of Linneaus. 'Plantae Coldenghamiae' (*Act. Soc. R. Sci. Upsal.* 1743–50). Correspondence in *Amer. J. Sci. Arts* v.44, 1843, 85–113.
J. E. Smith *Selection of Correspondence of Linnaeus* v.1, 1821, 19, 286, 343; v.2, 451–58, 476. Thacker *Amer. Med. Biogr.* v.1, 1828, 234–38. A. L. A. Fée *Vie de Linné* 1832, 150. W. Darlington *Memorials of John Bartram and Humphry Marshall* 1849, 19–20, 326–33, 353. *Appleton's Cyclop. Amer. Biogr.* v.1, 1887, 683–84 portr. A. M. Keys *Cadwallader Colden* 1906. *Torreya* 1907, 21–34. *D.N.B.* v.11, 260–61. H. A. Kelly *Some American Med. Botanists* 1915, 38–43 portr. *Bull. Hist. Med.* 1958, 322–34. C. C. Gillispie *Dict. Sci. Biogr.* v.3, 1971, 343–45.
MSS. at Royal Society and New York Historical Society. Portr. at Hunt Library.
Coldenia L.

COLDEN, Jane *see* Farquhar, J.

COLDSTREAM, William (–1929)
Reported to Government of India on fruit culture in Himalayas and on forests of Simla Hill States. *Illustrations of Some of Grasses of Southern Punjab* 1889.

COLE, Miss Edith (*fl.* 1890s)
Collected plants with Mrs. Lort Phillips in British Somaliland, 1894–95. Discovered *Caralluma edithae.*
Kew Bull. 1895, 158, 211. A. White and B. L. Sloane *Stapelieae* v.1, 1937, 123.
Plants at Kew.

COLE, Edmund (*c.* 1840–1892)
d. Althorp, Northants 1892
Gardener to Earl Spencer at Althorp Park, 1878–92.
Gdnrs. Chron. 1892 i 702.

COLE, Sir Galbraith Lowry (1772–1842)
b. Dublin 1 May 1772 *d.* Hartford Bridge, Hants 4 Oct. 1842
Lieut.-General. Governor of Mauritius, 1823–28; of Cape Colony, 1828–33. Appointed W. Bojer as Professor of Botany, Mauritius.
Bot. Mag. 1828, t.2817. *D.N.B.* v.11, 264–66.
Bignonia colei Bojer ex Hook.

COLE, J. W. (*c*. 1863–1925)
d. 20 Feb. 1925
Nurseryman, Peterborough, Northants.

COLE, James (1563–1628)
b. Antwerp, Belgium 1563 *d*. May 1628
London merchant. Son-in-law of M. de Lobel (1538–1616). Had botanic garden at Highgate, London. Introduced *Cerasus laurocerasus*. Friend of J. Gerard.
M. de Lobel *Stirpium Illustrationes* 1655, 119. R. Pulteney *Hist. Biogr. Sketches of Progress of Bot. in England* v.1, 1790, 125. R. T. Gunther *Early Br. Botanists* 1922, 14 etc. R. H. Jeffers *Friends of John Gerard; Biogr. Appendix* 1969, 8–9.

COLE, Nathan (*c*. 1828–1909)
d. Exeter, Devon 2 Nov. 1909
Superintendant, Kensington Gardens. *Royal Parks and Gardens of London* 1877.
J. Hort. Cottage Gdnr v. 59, 1909, 476.

COLE, Rex Vicat (1870–1940)
d. 4 Feb. 1940
Artist; exhibited at Royal Academy. *British Trees* 1905–7 2 vols. *Artisitic Anatomy of Trees* 1916.
Who was Who, 1929–1940 271.

COLE, Richard (*fl*. 1730s)
Nurseryman, Battersea, London. Member of London Society of Gardeners.

COLE, Rev. Thomas (*fl*. 1720s)
Nonconformist minister at Gloucester. Correspondent of J. J. Dillenius (*Historia Muscorum* 1741, viii). Had herb. "which in a flight of religious zeal, and repentance, at having misspent his time in accumulating, he committed to the flames" (Pulteney).
R. Pulteney *Hist. Biogr. Sketches of Progress of Bot. in England* v.2, 1790, 191–92.

COLE, William (*c*. 1812–1864)
d. 28 Dec. 1864
Nurseryman, Fog Lane Nursery, Manchester, *c*. 1853.
Gdnrs Chron. 1865, 31–32. *J. Hort. Cottage Gdnr* v.33, 1865, 70–71.

COLE, William (1834–1904)
b. Overton, Hants 30 Nov. 1834 *d*. Feltham, Middx 16 Feb. 1904
Nurseryman, Vineyard Nurseries, Feltham, 1875. Grew strawberries and grapes.
Garden v.65, 1904, 158. *J. Hort. Cottage Gdnr* v.48, 1904, 163.

COLE, William *see* Coles, W.

COLEBROOK, Henry Thomas (1765–1837)
b. London 15 June 1765 *d*. London 10 March 1837
FRS 1816. FLS 1816. Sanskrit scholar. In India, 1783–1815. Chief Judge, Bengal. Furnished oriental names for W. Roxburgh's *Fl. Indica* 1820–32 (xi–xiv). Collected plants in Sylhet. Sent plants and drawings to W. J. Hooker and A. B. Lambert. 'On Frankincense (*Boswellia*)' (*Asiatick Researches* v.9, 1807, 377–82). 'On *Dryobalanops camphora*, or Camphor-tree of Sumatra' (*Asiatick Researches* v.12, 1816, 535–41).
J. E. Smith *Exotic Bot*. 1805–7, t.115. *Trans. Linn. Soc*. v.15, 1827, 355–70. T. E. Colebrooke, *ed. Misc. Essays with Life* 1873. *D.N.B*. v.11, 282–86. *Taxon* 1970, 519.
Plants at Kew.
Colebrookea Smith.

COLEMAN, Charles (*fl*. 1820s)
Nurseryman and land surveyor, Church Road, Lower Tottenham, Middx.

COLEMAN, Edith (–1951)
b. Surrey *d*. Sorrento, Victoria 1951
To Australia in early 20th century. Schoolteacher in Victoria. Interested in orchids. Contrib. to *Victorian Nat*.
Victorian Nat. 1950, 99–100; 1951, 46. *Austral. Orchid Rev*. v.16, 1951, 122.
Diuris colemanae Rupp.

COLEMAN, Sarah (*fl*. 1820s)
Nurseryman, Old Nursery, Marsh (now Park) Lane, Tottenham, Middx.

COLEMAN, William (*c*. 1743–1808)
Nurseryman, Tottenham, Middx.
Trans. London Middlesex. Archaeol. Soc. v.24, 1973, 193–94.

COLEMAN, William (1827–1908)
b. Rolleston, Leics 1827 *d*. Eastnor, Herefordshire 20 Feb. 1908
Gardener at Eastnor Castle, 1860. Wrote articles on conifers.
Gdnrs Chron. 1875 ii 517 portr; 1908 i 143 portr.

COLEMAN, Rev. William Higgins (*c*. 1816–1863)
b. Middx *c*. 1816 *d*. Burton-on-Trent, Staffs 12 Sept. 1863
BA Cantab 1836. Master at Christ's Hospital, Hertford; and, from 1847, at Ashby-de-la-Zouch. Described *Oenanthe fluviatilis* (J. Sowerby and J. E. Smith *English Bot*. Supplt. 2944). Discovered *Bunium bulbocastanum* 1839. *Fl. Hertfordiensis* (with R. H. Webb) 1849; supplts 1851 and 1859. First introduced the river basin delimitation into a county flora. MS. flora of East Grinstead, 1836, at Kew; of Dedham, 1838, at BM(NH) and Kew (*Essex Nat*. 1921, 303–7). Cambridge plants in H. C. Watson's *New Botanists' Guide* v.2, 1837, 598. Leicester plants in White's *Directory* 1863. *Fl. Leicestershire* (F. T. Mott *et al*.) 1886, based on MS. by Coleman.

Trans. Bot. Soc. Edinburgh v.8, 1863, 13. *J. Bot.* 1863, 318. H. C. Watson *Topographical Bot.* 1883, 540–41. R. A. Pryor *Fl. Hertfordshire* 1887, xlii–xliv. *D.N.B.* v.11, 290. *Proc. Linn. Soc.* 1933–34, 22. A. R. Horwood and C. W. F. Noel *Fl. Leicestershire* 1933, ccxi–ccxii. A. H. Wolley-Dod *Fl. Sussex* 1937, xlii. J. G. Dony *Fl. Hertfordshire* 1967, 14–15. S. T. Jermyn *Fl. Essex* 1974, 18. *R.S.C.* v.2, 13.

Herb. formerly at St. Albans Museum. Plants at Bolton Museum. MS. list of Herts plants at Linnean Society.

Rubus colemanni Bloxam.

COLEMAN, William Stephen (1829–1904)
b. Horsham, Sussex 1829 *d.* London 22 March 1904

Artist and book-illustrator. *Our Woodlands, Heaths and Hedges* 1859. Illustrated T. Moore's *British Ferns* 1861.

D.N.B. Supplt. 2 v.1, 382. *Entomol. Rec.* v.77, 1965, 47–48.

COLENSO, Rev. John William (1814–1883)
b. Cornwall 24 Jan. 1814 *d.* Durban, Natal 20 Jan. 1883

Bishop of Natal, 1853.

P. Ascherson and P. Graebner *Synopsis Mitteleuropaischen Fl.* v.3, 1906, 364.

Natal plants at Kew.

COLENSO, Rev. William (1811–1899)
b. 17 Nov. 1811 *d.* Napier, New Zealand 10 Feb. 1899

FRS 1886. FLS 1865. To New Zealand in 1833 as printer for Church Missionary Society. Helped J. D. Hooker with *Fl. Novae-Zelandiae. Excursion in Northern Island of New Zealand... 1841–42* 1844. *Essay on Botany of North Island of New Zealand* 1865.

Hooker's J. Bot. 1844, 3–4. *Nature* v.59, 1899, 420. *Proc. Linn. Soc.* 1898–99, 51–52. *Trans. N.Z. Inst.* 1898, 722–24 portr.; 1903, 350–54. *Yb. R. Soc.* 1901, 191–94. *Proc. R. Soc.* 1904, 57–60. T. F. Cheeseman *Manual of N.Z. Fl.* 1906, xxiv–xxv. L. Cockayne *N.Z. Plants* 1927, 9–12 portr. *Turnbull Library Rec.* no. 3, 1941, 5–10. A. G. Bagnall and G. C. Petersen *William Colenso* 1948 portr. A. M. Coats *Quest for Plants* 1969, 236–39. R. Glenn *Bot. Explorers of N.Z.* 1950, 86–99. *R.S.C.* v.2, 13; v.7, 415; v.9, 550; v.14, 302.

Plants at Kew, Wellington. Letters and MSS. at Kew (*J. Linn. Soc.* v.32, 1896, 197–208). MSS. at Mitchell Library, Sydney.

Colensoa Hook. f.

COLES, Benjamin (*fl.* 1820s)

Seedsman, King's Road, Chelsea, London.

COLES, J. (*fl.* 1900s)

In charge of orchid collection of Capt. Terry, Fulham; then Sander's firm at St. Albans. Foreman, Blenheim Palace. In charge of orchids at 'Woodlands', Streatham.

F. Boyle *Woodlands Orchids* 1901, 5–6 portr.

COLES (or Cole), William (1626–1662)
b. Adderbury, Oxford 1626 *d.* Winchester? 1662

BA Oxon 1651. Lived at Putney. *Art of Simpling* 1656. *Adam in Eden* 1657.

D.N.B. v.11, 277. G. C. Druce *Fl. Berkshire* 1897, cv. G. C. Druce *Fl. Buckinghamshire* 1926, lxxi–ii. G. C. Druce *Fl. Oxfordshire* 1927, lxvii–viii. *Trans. Herts. Nat. Hist. Soc.* v.12, 1904, 80–83. E. J. L. Scott *Index to Sloane Manuscripts* 1904, 116. *J. New York Bot. Gdn* v.41, 1940, 158–66.

COLEY, S. J. (–*c.* 1925)

Druggist of Stroud, Glos. Studied local flora. Had herb. consisting largely of contents of other herbaria.

H. J. Riddelsdell *Fl. Gloucestershire* 1948, cxxxix–cxl.

COLGAN, Nathaniel (1851–1919)
b. Dublin 28 May 1851 *d.* Dublin 20 Oct. 1919

Clerk, Metropolitan Police, Dublin. 'Henry Mundy and the Shamrock' (*J. Bot.* 1894, 109–11). 'Shamrock in Literature' (*J. R. Soc. Antiq. Ireland* v.6, 1896, 211–26, 349–61). *Fl. County Dublin* 1904. 'Clare Island Survey: Gaelic Plant and Animal Names' (*Proc. R. Irish Acad.* v.31(4), 1911, 1–30). With R. W. Scully saw through press ed. 2 of A. G. More's *Contributions towards Cybele Hibernica* 1898.

Irish Nat. 1919, 121–26 portr.; 1920, 23. *Bot. Soc. Exch. Club Br. Isl. Rep.* 1920, 102. *Irish Booklover* v.11, 1920, 66. *J. Bot.* 1920, 118. R. L. Praeger *Some Irish Nat.* 1949, 66 portr.

COLLENETTE, Cyril Leslie (1888–1959)
b. Woodford Green, Essex 13 Jan. 1888 *d.* 2 Nov. 1959

FRES. Entomologist, BM(NH) for whom he travelled extensively collecting insects. Collected plants in Malaya, 1913–23; Galapagos and Polynesia, 1924–25; French Guinea, Liberia, 1926; Matto Grosso, Brazil, 1927; Somaliland, 1929–30. *Sea-girt Jungles* 1926. *History of Richmond Park* 1937.

Kew Bull. 1926, 51; 1927, 121, 126; 1931, 401–14. *London Nat.* 1959, 136–38 portr. *Times* 26 Nov. 1959. *Nature* v.185, 1960, 211.

Plants at Kew.

COLLETT, Henry (*fl.* 1790s)

Seedsman, Tewkesbury, Glos.

COLLETT, Sir Henry (1836–1901)
b. Thetford, Norfolk 6 March 1836 *d.* Kew, Surrey 21 Dec. 1901

FLS 1879. KCB 1891. Joined Bengal Army, 1855; Colonel, 1884. Took up botany in 1877 and on retirement in 1893 worked at Kew on his *Fl. Simlensis* 1902 (biogr. account, xv–xxii). Collected plants in Afghanistan, Algeria, Burma, Canaries, Corsica, India, Java, Spain.

J. Linn. Soc. v.28, 1890, 1–150. *Garden* v.61, 1902, 16. *J. Bot.* 1902, 73–74. *Kew Bull.* 1902, 18–23. *Proc. Linn. Soc.* 1901–2, 28–30. *D.N.B. Supplt. 2* v.1, 384–85. *Fl. Malesiana* v.1, 1950, 602.

Plants and letters at Kew.

Neocollettia Hemsley.

COLLEY, Thomas (*fl.* 1820s–1830s)

Foreman at Fairburn's nursery, Oxford. Collected plants, especially orchids, in Demerara, British Guiana, 1833, for J. Bateman.

Gdnrs Mag. 1834, 571–72; 1835, 1–7. *Orchid Rev.* 1898, 11–12.

Collea Lindl.

COLLEY, William (*fl.* 1820s–1840s)

Nurseryman, King Street, Hammersmith, London. In partnership with Hill.

COLLIE, Alexander (1793–1835)

b. Wantonwells, Aberdeenshire June 1793 *d.* King George's Sound, W. Australia 8 Nov. 1835

FLS 1825. Surgeon, Royal Navy, 1813. On Beechey's voyage, 1825–28. Collected plants in California and Mexico, 1827; also in Chile and Australia. Explored country around Perth, 1829. Farmed on Swan River and Albany. Colonial Surgeon, Perth 1832. Collie's plant material worked over by W. J. Hooker and G. A. W. Arnott in *Botany of Captain Beechey's Voyage...1825–1828* 1830–41.

A. Lasègue *Musée Bot. de Benjamin Delessert* 1845, 497. W. H. Brewer and S. Watson *Bot. California* v.2, 1880, 554. *J. Nat. Hist. Soc. W. Austral.* no. 6, 1909, 12–13. L. Huxley *Life and Letters of Sir J. D. Hooker* v.1, 1918, 106. *Med. J. Austral.* 1935, 793–801. 1936, 537–40. S. D. McKelvey *Bot. Exploration of Trans-Mississippi West 1790–1850* 1955, 355.

Plants at BM(NH), Kew. Letters at Public Library, Perth, W. Australia.

COLLIE, Rev. Robert (1839–1892)

b. Aberdeenshire *d.* Sydney 18 April 1892

FLS 1882. Ordained Presbyterian minister, 1866. In charge of church at Newtown, Sydney, 1876–92. 'Specimens of Plants collected at King George's Sound by Rev. R. Collie' (*Proc. Linn. Soc. N.S.W.* 1889, 317–24). Herb. bequeathed to Linnean Society, New South Wales.

J. Proc. R. Soc. N.S.W. 1908, 98 portr. *R.S.C.* v.14, 308.

COLLIER, Ann (*fl.* 1810s)

Nurseryman, 2 Harrington Street, Liverpool.

COLLINGWOOD, Cuthbert (1826–1908)

b. Christchurch, Hants 25 Dec. 1826 *d.* Lewisham, London 20 Oct. 1908

BA Oxon 1849. MB Oxon 1854. FLS 1853. Lecturer in Botany, Royal Infirmary Medical School, Liverpool, 1858. Found *Elodea* pre-1861 at Rock Ferry, Cheshire (*Proc. Liverpool Lit and Philos. Soc.* v.15, 14). Surgeon and naturalist on HMS 'Rifleman' and 'Serpent' in China Seas, 1866–67. Collected plants on Pratas Islands, SW of Formosa. *On Scope and Tendency of Botanical Study* 1855. *Rambles of a Naturalist on Shores and Waters of the China Seas* 1867.

J. B. L. Warren *Fl. Cheshire* 1899, lxxx. *Nature* v.78, 1908, 673. *Proc. Linn. Soc.* 1908–9, 35. *Times* 22 Oct. 1908. *Liverpool Courier* 30 March 1909, 51. *Trans. Liverpool Bot. Soc.* v.1, 1909, 65. *D.N.B. Supplt. 2* v.1, 385. *Who was Who, 1897–1916* 148. *Fl. Malesiana* v.1, 1950, 113. *R.S.C.* v.2, 21; v.7, 417; v.9, 555; v.14, 312.

COLLINS, Dr. (*fl.* 1820s?)

MD. "Possessed of considerable botanical knowledge." MS. list of Barbados plants.

J. D. Maycock *Fl. Barbadensis* 1830, vii.

COLLINS, Elian Emily (*née* **Pemberton**) (*c* 1858–)

b. Burma *c.* 1858

Married D. J. Collins, a surveyor. Lived in Siam from 1877 until her death. Collected plants in Siam, now at BM(NH).

Bangkok Times Weekly Mail 15 Sept. 1936. *Blumea* v.11, 1961–62, 477–78.

COLLINS, Ernest Jacob (1877–1939)

d. 6 Feb. 1939

BSc London 1908. FLS 1903. Botanist, John Innes Institution, 1915–39. Contrib. to *Ann. Bot.*, 1918, 1935; *Gdnrs Chron.* 1921; *Proc. Linn. Soc.* 1924–25, 1928–29.

John Innes Inst. 1910–1935, 23, 41. *Proc. Linn. Soc.* 1938–39, 237–38.

COLLINS, James (*fl.* 1850s–1900s)

Curator, Pharmaceutical Society, 1868–72. Government Economic Botanist and Secretary and Librarian, Raffles Library and Museum, Singapore, 1873–77. 'India Rubber' (*J. Bot.* 1868, 2–22). 'Vernacular Names' (*J. Bot.* 1869, 360–63). 'On India Rubber' (*J. Soc. Arts.* v.18, 1870, 81–93). 'New or Little-known Vegetable Products' (*Pharm. J.* v.11, 1869–70, 66–70). 'Study of Economic Botany' (*Pharm. J.* 1871–72, 691–95, 713–15, 737–39). 'Materia Medica Papers' (*J. Bot.* 1872, 119–20). *Report on Caoutchouc of Commerce* 1872. Edited *J. Eastern Asia* 1875.

Collection of gums and resins at Singapore.

COLLINS, John (*fl.* 1820s)

Nurseryman, New Road, Marylebone, London.

COLLINS, Rev. John Coombes (1798–1867)
b. Bridgwater, Somerset 11 March 1798 *d.* Eastover, Somerset 30 Sept. 1867
Incumbent, St. John's, Bridgwater. List of Somerset plants in H. C. Watson's *New Botanist's Guide* 1835–37, 553–63.
J. Bot. 1898, 311–12; 1905, 233. J. W. White *Fl. Bristol* 1912, 84–5.

COLLINS, Robert (*fl.* 1790s)
Seedsman, Maidstone, Kent.

COLLINS, Samuel (*fl.* 1710s)
Of Irchester, Northants. *Paradise Retriev'd ...Method of Managing and Improving Fruit -trees* 1717.
Gdnrs Chron. 1909 i 113.

COLLINS, Thomas (*fl.* 1790s)
Gardener and seedsman, Havant, Hants.

COLLINS, William (*fl.* 1790s)
Gardener and seedsman, Dale Street, Liverpool.

COLLINSON, Michael (1727–1795)
b. Peckham, London 1727 *d.* 11 Aug. 1795
Only son of P. Collinson (1694–1768). Of Hendon, Middx and Chantry, Suffolk. 'British Orchids' (*Phytologist* 1861, 171–76).
J. Nichols *Lit. Anecdotes of Eighteenth Century* v.5, 315. W. Darlington *Mem. of John Bartram and Humphry Marshall* 1849, 446–60. N. G. Brett-James *Life of Peter Collinson* 1926, 206–16.

COLLINSON, Peter (1694–1768)
b. St. Clement's Lane, Lombard St., London 14 Jan. 1694 *d.* London 11 Aug. 1768
FRS 1728. Woollen-draper. Had garden at Peckham until 1749 when he settled at Mill Hill where he formed a botanic garden in which he grew many N. American plants. Had herb. Friend of Sir H. Sloane, J. Petiver, C. Linnaeus, J. Bartram, etc. Contrib. to *Gent. Mag.* 1751–66. MS. 'Account of Introduction of American Seeds into Great Britain' now at BM(NH) (*J. Bot.* 1925, 163–65).
Gent. Mag. 1770, 177–80. J. Fothergill *Some Account of Late Peter Collinson* 1770 portr. J. Fothergill *Some Anecdotes of Late Peter Collinson* 1785. J. C. Lettsom *Mem. of John Fothergill* 1786, 263–74 portr. R. Pulteney *Hist. Biogr. Sketches of Progress of Bot. in England* v.2, 1790, 275–77. *Trans. Linn. Soc.* v.10, 1811, 270–82. J. Nichols *Lit. Anecdotes of Eighteenth Century* v.5, 309–16; v.9, 609 portr. J. E. Smith *Selection of Correspondence of Linnaeus* v.1, 1821, 1–77. A. L. A. Fée *Vie de Linné* 1832, 142–48. J. C. Loudon *Arboretum et Fruticetum Britannicum* v.1, 1838, 42, 54–60, 81–82. L. W. Dillwyn *Hortus Collinsonianus* 1843. W. Darlington *Memorials of John Bartram and Humphry Marshall* 1849 *passim.* W. H. Dillingham *Tribute to Memory of Peter Collinson* 1851 portr. *Cottage Gdnr* v.7, 1852, 143. J. Smith *Cat. of Friends' Books* v.1, 1867, 443–44. *Gdnrs Chron.* 1895 ii 5–6 portr., 36; 1926 i 46–47 portr. E. J. L. Scott *Index to Sloane Manuscripts* 1904, 117. *A.L.A. Portrait Index* 1906, 329. R. H. Fox *Dr. John Fothergill and his Friends* 1919, 158–81. N. G. Brett-James *Life of Peter Collinson* 1926 portr. *Virginia Quart. Rev.* 1934, 218–33. N. G. Brett-James *Mill Hill* 1938. *Times* 13 May 1939 portr. *Bull. Friends Hist. Assoc.* v.36, 1947, 19–44. E. G. Swem *Brothers of the Spade* 1949. J. E. Dandy *Sloane Herb.* 1958, 114. J. Ewan *William Bartram* 1968 *passim. Friend* 1968, 962–64 portr. *J. R. Hort. Soc.* 1968, 329–33 portr. C. C. Gillispie *Dict. Sci. Biogr.* v.3, 1971, 349–51.
MSS. at Royal Society, BM, BM(NH), Linnean Society, Fitzwilliam Museum Cambridge, Society of Friends London. Portr. by Gainsborough at Mill Hill School. Portr. at Kew, Hunt Library.
Collinsonia L.

COLMAN, Sir Jeremiah (1859–1942)
b. 24 April 1859 *d.* Reigate, Surrey 16 Jan. 1942
MA Cantab. VMH 1908. Head of Colman and Sons, mustard firm. Cultivated orchids at Gatton Park, Reigate. *Hybridisation of Orchids* 1932.
Gdnrs Mag. 1907, 837–38 portr. *Orchid Rev.* 1914, 356 portr.; 1942, 49–50. *Austral. Orchid Rev.* v.7, 1942, 26 portr.; 39. *Gdnrs Chron.* 1942 i 42. *Times* 17 Jan. 1942. *Amer. Orchid Soc. Bull.* v.11, 1943, 339–42. *J. R. Hort. Soc.* 1943, 359–60. *Who was Who, 1941–1950* 239.

COLQUHOUN, Sir Robert (–1838)
Of Suez. Resident in Nepal, 1819. Collected plants in Kumaon. Patron of Calcutta Botanic Gardens.
Trans. Linn. Soc. v.13, 1822, 608–14. P. Smith *Correspondence of Sir J. E. Smith* v.2, 1832, 248.
Plants at Calcutta.
Colquhounia Wall.

COLTMAN-ROGERS, Charles (1854–1929)
d. Stanage Park, Radnorshire 20 May 1929
FLS 1921. MP 1884–85. Chairman, Radnorhsire County Council, 1896–1929. Lord Lieutenant, Radnorshire, 1922–29. Studied conifers. *Conifers and their Characteristics* 1920.
Proc. Linn. Soc. 1929–30, 195–96.

COLVILL, James (*c.* 1746–1822)
d. 28 Oct. 1822
Founded nursery in Kings Road, Chelsea, *c.* 1780s. In partnership with Buchanan, *c.* 1790. By 1811 nursery had between 30,000 and 40,000 square feet under glass.

Bot. Mag. 1789, t.73. T. Faulkner *Hist. and Topographical Description of Chelsea and its Environs* v.2, 1829, 162–63. E. Bretschneider *Hist. European Bot. Discoveries in China* 1898, 218.

COLVILL, James (*c.* 1777–1832)
d. 25 Jan. 1832
Son of J. Colvill (*c.* 1746–1822). Succeeded to his father's nursery in King's Road, Chelsea. Opened another nursery at Roehampton, Surrey by 1827. Had very large collection of Cape bulbs. Robert Sweet (1783–1835) worked for Colvills, 1819–26. Nursery occupied by Henry Adams and Durban, 1834–40.

COLVILL, William (*fl.* 1830s)
Of Arbroath, Forfarshire. 'On Circumstances under which Germs or Buds are produced in Trees and Woody Shrubs' (*Edinburgh J.* v.2, 1830, 421–26). 'Functions of Spiral Vessels' (*Edinburgh J.* v.3, 1830, 5–9).
Passiflora X *colvillii* Sweet.

COLVILL, William Henry (1838–1885)
b. 6 Sept. 1838 *d.* Laughton, Forfarshire 13 March 1885
Assistant Surgeon, Bombay, 1856; Surgeon, 1868; retired 1882. On Persian Gulf in 1865 with Lieut.-Colonel Pelly. Collected plants in Baghdad, 1873.
Fl. Iraq v.1, 1966, 112.
Plants at Kew.

COMBER, Harold Frederick (1897–1969)
b. Nymans, Sussex 1897 *d.* Gresham, Oregon, U.S.A. 23 April 1969
Gardener, Royal Botanic Garden, Edinburgh, 1920. Gardener to H. J. Elwes at Colesborne. Collected plants in Argentina, Chile and Peru, 1925–27; Tasmania, 1929–30. Hybridised lilies for Jan de Graff nursery, Oregon. *Andes Expeditions 1925–1926 and 1926–1927* 1928. *Field Notes of Tasmanian Plants Collected by H. F. Comber, 1929–30* 1930. 'New Classification of Genus Lilium' (*Lily Yb.* 1949, 86–105).
Gdnrs. Chron. 1931 ii 266 portr.; 13 June 1969, 33. *Lily Yb.* 1970, 224–25. *Quart. Bull. Alpine Gdn Soc.* 1973, 98–121 portr. *Int. Dendrological Soc. Yb.* 1975, 13–18. *See also* Addendum.
Plants, letters and MSS. at Kew.
Combera Sandwith.

COMBER, James (*c.* 1866–1953)
b. Ashdown Forest, Sussex *c.* 1866. *d.* Haywards Heath, Sussex 16 May 1953
VMH 1936. Gardener to Messel family at Nymans, Sussex where he created a pinetum, alpine garden and heath garden. Contrib. to *Gdnrs Chron.*
Gdnrs Chron. 1953 i 216.

COMBER, John (1869–1930)
d. 21 Aug. 1930
Of Guildford, Surrey. Archaeologist and botanist.
Bot. Soc. Exch. Club Br. Isl. Rep. 1930, 322.
Herb. at Sheffield University.

COMBER, Thomas Radcliffe (1837–1902)
b. Pernambuco, Brazil 14 Nov. 1837 *d.* Blackpool, Lancs 24 Jan. 1902
FLS 1878. Merchant, Manchester. 'Geographical Statistics of Extra-British European Fl.' (*J. Bot.* 1877 *passim*).
J. Bot. 1902, 386–88 portr. *J. R. Microsc. Soc.* 1902, 158. *Proc. Linn. Soc.* 1901–2, 30–31. *Trans. Liverpool Bot. Soc.* 1909, 65–66. *R.S.C.* v.1, 26; v.7, 419; v.9, 558; v.14, 320.
Diatoms and letters at BM(NH). Plants at Liverpool University. Portr. at Hunt Library.

COMINS, Rev. Richard Blundell (*fl.* 1880s–1890s)
Sent plants to Kew, 1882–1899, from Norfolk and Solomon Islands.
Kew Bull. 1892, 105.

COMPTON, Hon. and Rev. Henry (1632–1713)
b. Compton Wynyates, Warwick 1632 *d.* Fulham, London 7 July 1713
MA Cantab 1661. DD Oxon 1669. Bishop of Oxford, 1674; of London, 1675. Introduced many exotic trees into his garden at Fulham Palace. Friend of J. Ray, J. Bobart, etc.
J. Cockburn *Blessedness of Christians after Death* 1713. *Philos. Trans. R. Soc.* v.47, 1753, 242–47. R. Pulteney *Hist. Biogr. Sketches of Progress of Bot. in England* v.2, 1790, 105–7, 303. *Cottage Gdnr* v.7, 1850, 171. *D.N.B.* v.11, 443–47. E. J. L. Scott *Index to Sloane Manuscripts* 1904, 119. *J. Bot.* 1909, 45. A. White *et al. Succulent Euphorbieae* v.1, 1941, 35–36 portr. E. Carpenter *The Protestant Bishop* 1956. J. E. Dandy *Sloane Herb.* 1958, 114. *Gdnrs Chron.* 1964 ii 333 portr. *Huntia* 1965, 190–91 portr. *Arnoldia* 1972, 192–95.
Portr. by Kneller at National Portrait Gallery. Portr. at Hunt Library.
Comptonia Aiton.

CONINGSBY, Arthur (*c.* 1888–1966)
b. Sydenham, Kent *c.* 1888 *d.* Horton 11 March 1966
Cultivated orchids in Guernsey, Hampstead, Chichester, Southgate, etc. In employ of Messrs Sander, St. Albans. Raised hybrid, *Cymbidium coningsbyanum.*
Orchid Rev. 1957, 168–70 portr.; 1966, 157–58.

CONNELLY, Thomas (*fl.* 1820s)
Nurseryman, Hammersmith, London.

CONNOLD, Edward Thomas (1862–1910)
b. Hastings, Sussex 11 June 1862 *d.* Hastings 9 Jan. 1910
Studied British plant galls. *British Vegetable Galls* 1901. *British Oak Galls* 1908. *Plant Galls of Great Britain* 1909.
Gdnrs Chron. 1910 i 47. *Nature* v.82, 1910, 374.
Plant galls at Hastings Museum.

CONSIDEN, Dennis (*fl.* 1780s–1815)
d. 29 Dec. 1815
Surgeon, Scarborough, Yorks. In New South Wales, 1788–92, as First Assistant Surgeon. Sent plants to Sir Joseph Banks. Returned to England, 1793. Practised medicine in County Cork until 1805.
Hist. Rec. N.S.W. v.1 (2) 1892, 220–21. *Proc. Linn. Soc. N.S.W.* 1904, 477–78. *J. Proc. R. Soc. N.S.W.* 1908, 98–99. J. H. Maiden *Sir Joseph Banks* 1909, 155. *Med. J. Austral.* 1927, 770–73. *Austral. Dict. Biogr.* v.1, 1966, 242–43.
Eucalyptus consideniana Maiden.

CONSTABLE, Sir Thomas Hugh Clifford (*olim* Clifford) (1762–1823)
b. 4 Dec. 1762 *d.* Ghent, Belgium 25 Feb. 1823
Educ. Liège and Paris. Inherited Tixall, Staffs, 1786. Created baronet 1815. Took name of Constable on inheriting Burton Constable, Yorks, 1821. 'Fl. Tixalliana' (T. and A. Clifford *Topography and Historical Description of Parish of Tixall* 1817, 285–308; reissued separately 1818).
Gent. Mag. 1823 i 470–71. J. Nichols *Illus. Lit. Hist. of Eighteenth Century* v.5, 511–13. J. Gillow *Dict. English Catholics* v.1, 1885, 556–57. *D.N.B.* v.12, 45–46.

CONSTABLE, Timothy (*fl.* 1790s)
Nurseryman and gardener, Long Melford, Suffolk.

CONSTABLE, W. A. (1887–1954)
b. Colchester, Essex 1887 *d.* 1954
Nurseryman, Paddock Wood, Kent, 1929; transferred to Southborough near Tunbridge Wells, 1933. Amalgamated with Burnham Lily Nursery, 1934. Hybridised lilies.
Gdnrs Chron. 1950 ii 56 portr. *Lily Yb.* 1955. 9–10 portr.

CONSTABLE, William (1721–1791)
Of Burton Constable, Yorks. Dilettante. Had collection of plants.
Country Life 1970, 316–17 portr.

CONWAY, Charles (*fl.* 1790s–1870s)
Of Pontrhydyrun, Monmouth. Had herb. List of Monmouth plants in H. C. Watson's *New Botanists' Guide* v.2, 1837, 629–30.
H. C. Watson *Topographical Bot.* 1883, 542. H. A. Hyde and A. E. Wade *Welsh Flowering Plants* 1934, 1–2 portr. A. E. Wade *Fl. Monmouthshire* 1970, 11.
Herb. at National Museum of Wales.

CONWAY, John (*fl.* 1690s)
Friend of J. Petiver to whom he sent Madras and Cape plants.
J. E. Dandy *Sloane Herb.* 1958, 115.

CONWAY, Sir William Martin (1856–1937)
Collected plants in Balistan and Gilgit. *Climbing in Himalayas* 1894. *Climbing and Exploring Karakorum Himalayas* 1902 (includes plant list).
Kew Bull. 1893, 145; 1895, 20. *Pakistan J. For.* 1967, 342.
Plants at Kew.

COOK, Ernest Thomas (1870–1915)
d. 5 May 1915
On editorial staff of *Gdnrs Mag.* and *Gdning Illustrated.* Editor of *Garden.* Emigrated to Canada *c.* 1911.
Garden 1915, 240 (v). *Gdnrs Chron.* 1915 i 270 portr.

COOK, J. (*fl.* 1790s)
Gardener and seedsman, Halstead, Essex.

COOK, James (1728–1779)
b. Marton, Yorks 27 Oct. 1728 *d.* Hawaii 14 Feb. 1779
FRS 1776. Circumnavigator. First voyage, 1768–71, with J. Banks and D. Solander as naturalists; second voyage, 1772–75, with J. R. and G. Forster; third voyage, 1776–79, with W. Anderson and D. Nelson.
Companion Bot. Mag. v.2, 1837, 223–33. *Trans. Proc. N.Z. Inst.* 1900, 499–514; 1902, 24–45. *D.N.B.* v.12, 66–70. T. F. Cheeseman *Manual N.Z. Fl.* 1906, xi–xviii. E. D. Merrill 'Bot. of Cook's Voyages' (*Chronica Botanica* v.14, 1954, 164–383). *Fl. Malesiana* v.1, 1950, 114–15. J. C. Beaglehole, *ed. Journals of Captain James Cook* 1962 2 vols. *Austral. Nat. Hist.* 1969, 241–88 portr. *Notes Rec. R. Soc. London* v.24(1) 1969, 1–90 portr. G. M. Badger *Captain Cook, Navigator and Scientist* 1970 portr. J. C. Beaglehole *Life of Captain James Cook* 1974 portr. *Bull. Pacific Tropical Bot. Gdn* v.4 (4), 1974, 65–75.
Portr. at National Maritime Museum, National Gallery. MSS. at Royal Society.
Cookia Sonn.

COOK, John (*fl.* 1780s–1790s)
Nurseryman, Old Brompton, Kensington, London.

COOK, Moses (*fl.* 1660s–1715)
d. Little Hadham, Herts Feb. 1715
Laid out gardens at Cassiobury near Watford, Herts for Earl of Essex, 1660–81. Partner with London and Looker at Brompton Park Nursery, 1681–89. *Manner of Raising, Ordering and Improving Forrest Trees*: also

How to Plant, Make and Keep Woods, Walks, Avenues, Lawns, Hedges, etc. 1676; ed. 3 1724.

J. C. Loudon *Encyclop. Gdning* 1822, 1266. S. Felton *Portr. of English Authors on Gdning* 1830, 31–33. G. W. Johnson *Hist. of English Gdning* 1829, 116. *Cottage Gdnr* v.7, 1852, 891–92. *House and Gdn* Sept. 1973, 126–27.

COOK, Samuel Edward *see* Widdrington, S. E.

COOK, Thomas (*fl.* 1830s)

Nurseryman, Blue Anchor Road, Bermondsey, London.

COOK, Thomas H. (*c.* 1869–1947)

d. Kings Lynn, Norfolk 24 Sept. 1947

Gardener, Royal Gardens, Sandringham, Norfolk 1901.

Gdnrs Mag. 1901, 445 portr.; 1908, 117–18 portr. *J. Hort. Cottage Gdnr* v.43, 1901, 14–15 portr. *Gdnrs Chron.* 1947 ii 122.

COOK, Walter Robert Ivimey (1901–1952)

b. London 1901 *d.* Dinas Powis, Glam 1 Feb. 1952

BSc London 1923. FLS 1927. Demonstrator, King's College, London, 1923–28. Lecturer, Bristol University, 1928–31. Senior Lecturer in Botany, University of Wales, Cardiff, 1931–52. Mycologist. *Inter-relationships of Archimycetes* 1929. *Biology for Medical Students* (with C. C. Hentschel) 1932. *Plant Science Formulae* (with R. C. McLean) 1941. *Practical Field Ecology* (with R. C. McLean) 1946. *Textbook of Theoretical Botany* (with R. C. McLean) 1951. *Textbook of Practical Botany* (with R. C. McLean) 1952.

Nature v.169, 1952, 993. *Pharm. J.* 1952, 91. *Proc. Linn. Soc.* 1951–52, 283–84. *Times* 2 Feb. 1952.

Portr. at Hunt Library.

COOKE, Alice Sophia (*née* **Smart** *formerly* **Mrs. Bacon**) (1890–1957)

b. London 1 June 1890 *d.* Hassocks, Sussex 21 May 1957

BSc London 1912. FLS 1922. Lecturer, Huddersfield Technical College, 1914–20. Lecturer, Brighton Technical College, 1920–50. Studied Sussex flora.

Nature v.180, 1957, 411. *Proc. Linn. Soc.* 1956–57, 240–41.

COOKE, Ann (*fl.* 1790s)

Druggist and dealer in garden seeds, 27 Market Place, Manchester.

COOKE, Brian Kennedy- *see* Kennedy-Cooke, B.

COOKE, Edward (*fl.* 1630s–1640s)

Master of Apothecary's Company, 1640. Visited West of England with T. Johnson in 1634.

COOKE, Edward William (1811–1880)

b. Pentonville, London 27 March 1811 *d.* Tunbridge Wells, Kent 4 Jan. 1880

Royal Academician. Executed some of the drawings for Loddiges's *Bot. Cabinet* and possibly J. C. Loudon's *Encyclopaedia of Plants*. Cultivated ferns and contrib. accounts of ferns to L. H. Courtney's *Week in Isles of Scilly* 1867, 1882 and I. W. North's *Week in Isles of Scilly* 1850.

Art J. 1869, 253. *Gdnrs Chron.* 1880 i 41–42. *D.N.B.* v.12, 80–81.

Ferns at Croydon Natural History and Scientific Society.

COOKE, George (1781–1834)

b. London 22 Jan. 1781 *d.* Barnes, Surrey 27 Feb. 1834

Line-engraver. Engraved plates for many historical and topographical works and also G. Loddiges's *Bot. Cabinet* 1817–34.

D.N.B. v.12, 81–82.

COOKE, Mordecai Cubitt (1825–1914)

b. Horning, Norfolk 12 July 1825 *d.* Southsea, Hants 12 Nov. 1914

ALS 1877. VMH 1902. India Museum at India Office, 1861–80. In charge of Lower Cryptogams in Kew Herb., 1880–92. Extraordinarily industrious: had herb. of 46,000 specimens; made about 22,000 drawings, wrote over 300 articles. Edited *Grevillea*, 1872–92. *Manual of Botanic Terms* 1862. *Plain and Easy Account of British Fungi* 1862. *Index Fungorum Britannicorum* 1865. *Fungi Britannici Exsiccati* 1865–79. *Handbook of British Fungi* 1871 2 vols. *Fungi* 1875. *Mycographia, seu Icones Fungorum* 1875–79. *Illustrations of British Fungi* (*Hymenomycetes*) 1881–91 8 vols. *British Freshwater Algae* 1882–84 2 vols. *British Desmids* 1887. *British Edible Fungi* 1891. *Handbook of Australian Fungi* 1892. *Handbook of British Hepaticae* 1894.

J. Bot. 1887, 355; 1912, 296, 384; 1915, 58–66 portr. G. Lindau and P. Sydow *Thesaurus Litteraturae Mycologicae et Lichenologicae* v.1, 1907, 293–304. *Kew Bull.* 1912, 369; 1914, 392–93. *Gdnrs Chron.* 1914 ii 346 portr. *J. Hort.* v.69, 1914, 337. *J. Kew Guild* 1915, 243 portr. *Proc. Linn. Soc.* 1914–15, 23–24. *Trans. Brit. Mycol. Soc.* v.5, 1915, 169–85. *Bot. Soc. Exch. Club Br. Isl. Rep.* 1915, 250–51. *Phytopathology* 1916, 1–4 portr. *Who was Who, 1897–1916* 155. *Trans. Norfolk Norwich Nat. Soc.* 1937, 206–9 portr. *Taxon* 1974, 754 portr.

Herb. and drawings at Kew. Drawings and MSS. at Huddersfield Museum. Letters at BM(NH).

Cookella Saccado.

COOKE, Rev. Philip Henry (1859–1950)
b. Maidstone, Kent 5 Dec. 1859 *d.* Old Heathfield, Sussex 25 Oct. 1950

BA London 1881. Curate, Stepney, 1889–92; Hackney, 1892–98; Upper Clapton, 1903–13. Vicar, Ickleton, Cambridge, 1914–29. Studied Cambridgeshire flora.

A. H. Evans *Fl. Cambridgeshire* 1939, 15. *London Nat.* 1950, 84.

COOKE, Randle Blair (1880–1973)

Collected plants abroad and at home, particularly in Northumberland, Durham and Scotland.

Vasculum 1973, 51.

COOKE, Theodore (1836–1910)
b. Tramore, Waterford 6 Jan. 1836 *d.* Kew, Surrey 5 Nov. 1910

MA Dublin. LLD 1891. FLS 1892. In India, 1860–93. Principal, Poona College, 1865–93. Director, Botanical Survey W. India, 1891. Technical Sub-director, Research Dept., Imperial Institute, London, 1893–96. Collected plants in India. *Fl. Presidency Bombay* 1901–8 2 vols.

Gdnrs Chron. 1910 ii 402. *Kew Bull.* 1910, 350–52. *Proc. Linn. Soc.* 1910–11, 36–37. *J. Bot.* 1911, 64–66. R. L. Praeger *Some Irish Nat.* 1949, 67.

Plants at Dehra Dun, Kew. Letters and bust at Kew.

COOKE, Sir Thomas (*fl.* 1690s)

Had garden at Hackney, London in 1690s.

Archaeologia v.12, 1794, 186.

COOKE, Sir William (–1964)
d. Newbury, Berks 10 June 1964

Of Wyld Court, Newbury. Grew orchids, especially cymbidiums.

Orchid Rev. 1964, 272.

COOKSON, Norman C. (*c.* 1840–1909)
d. Wylam, Northumberland 15 May 1909

Raised many orchid varieties at Oakwood Grange, Wylam.

Gdnrs Mag. 1907, 223 portr., 224. *Gdnrs Chron.* 1909 i 334 portr. *Orchid Rev.* 1909, 166–67.

Dendrobium nobile cooksonianum Reichenb.

COOLING, Edwin (1808–1885)
b. Beeston Field, Notts 1808 *d.* 12 May 1885

Nurseryman, Derby, *c.* 1838.

Gdnrs Chron. 1885 i 679.

COOLING, George (*fl.* 1900s)

Nurseryman, Bath; specialised in roses.

J. Hort. Cottage Gdnr v. 45, 1902, 474 portr.

COOLING, W. F. (*fl.* 1900s)

Nurseryman, Bath; specialised in roses.

Gdnrs Mag. 1911, 179–80 portr.

COOMBER, Thomas (–1926)
b. East Grinstead, Sussex *d.* Llangattock, Mon 23 Dec. 1926

VMH 1911. Gardener to Lord Llangattock at The Hendre, Monmouth for about 47 years. Contrib. to *Gdnrs Chron.*, etc.

Gdnrs Mag. 1911, 199–200 portr. *J. Hort. Home Farmer* v.66, 1913, 581–82 portr. *Gdnrs Chron.* 1921 i 230; 1927 i 39.

COOMBER, William (1839–1923)
b. Tonbridge, Kent 1839 *d.* Teddington, Middx 21 Sept. 1923

Brother of T. Coomber (–1926). Kew gardener, 1860–61. Superintendent, Royal Botanic Gardens, Regent's Park, 1875–96.

Gdnrs Chron. 1923 ii 212. *J. Kew Guild* 1924, 255–56 portr.

COOMBS, Charles (*fl.* 1690s)

Surgeon. Sent plants from Calabar and Maryland to J. Petiver.

J. E. Dandy *Sloane Herb.* 1958, 115.

COOPER, Burton Frederick John (1872–1961)
d. Canterbury, Kent 1 Aug. 1961

Pharmaceutical chemist. Botanist.

Pharm. J. 1961, 174.

COOPER, Charlotte Angela (1871–1944)
b. Robin Hoods Bay, Yorks 29 Sept. 1871 *d.* 27 Feb. 1944

Educ. St. Andrews. Teacher, Newcastle, Rochester, Monmouth and Ascot. Secretary, Herts Natural History Society, 1924–34; President, 1934. Founder member of Whitby Naturalists Society. Mycologist.

Trans. Herts Nat. Hist. Soc. 1944, 5; 1945, 57. *N. Western Nat.* 1945, 279–80 portr. *Naturalist* 1961, 62.

COOPER, Daniel (*c.* 1817–1842)
d. Leeds, Yorks 24 Nov. 1842

ALS 1837. Curator, Botanical Society of London, 1837–38. Assistant, Zoological Dept., BM, 1839–41. Assistant-surgeon in Army, 1842. *Fl. Metropolitana* 1836; reissued with Supplt. 1837. *Catalogue of British Natural Orders and Genera* 1838. *Little Book of Botany* 1839.

Gdnrs Mag. 1837, 86–87. *Mag. Zool. Bot.* 1838, 163–70. *Proc. Bot. Soc. London* 1839, 1–6, etc. *Proc. Linn. Soc.* 1843, 173. *Gent. Mag.* 1843, 108. *Phytologist* v.2, 1847, 1008. J. Hardy, *ed. Selections from Correspondence of George Johnston* 1892, 71. *D.N.B.* v.12, 141. *Bot. Soc. Exch. Club Br. Isl. Rep.* 1932, 282–83. C. E. Salmon *Fl. Surrey* 1931, 51. *R.S.C.* v.2, 41.

COOPER, Edgar Franklin (1833–1916)
b. 24 Sept. 1833
Banker, Leicester. Botanist. Discovered *Potagometon cooperi* in River Soar.
A. R. Horwood and C. W. F. Noel *Fl. Leicestershire* 1937, ccxx.
Herb. at Leicester Literary and Philosophical Society.

COOPER, Edward William (1884–1950)
b. Salisbury Green near Southampton, Hants 1884 *d.* 24 April 1950
Kew gardener, 1893. With Messrs. Sander, orchid nursery, St. Albans, *c.* 1895. Contrib. to *Orchid Rev.*
Gdnrs Chron. 1930 i 166 portr.; 1950 i 192. *Orchid Rev.* 1950, 84.

COOPER, James Eddowes (1864–1952)
b. St. Helier, Jersey 5 Aug. 1864 *d.* Herne Bay, Kent 17 Aug. 1952
Employed by firm of tea-brokers, London. Botanist but primarily interested in conchology.
D. H. Kent *British Herb.* 1957, 49.
Herb. at BM(NH).

COOPER, John (1810s–1820s)
Gardener and seedsman, Standishgate, Wigan, Lancs.

COOPER, Joseph (*fl.* 1820s–1840s)
Gardener to Lord Fitzwilliam at Wentworth House, Yorks, 1820s–40. "Mr. Cooper is one of the most zealous and successful cultivators of rare plants in this kingdom." (*Bot. Register* 1836, t.1835).
Bot. Mag. 1836, t.3482.
Cooperia Herbert.

COOPER, Nathaniel (*fl.* 1790s)
Seed merchant, Dean Street, Canterbury Square, London.

COOPER, Neville Louis (*c.* 1865–1936)
d. Ramsgate, Kent 9 Jan. 1936
Picture-dealer, interested in conifers and rock plants. Had pinetum at Little Hall, St. Stephens, Canterbury; rock garden and conifers at Vernon Holme, Harbledown.
Kew Bull. 1936, 109–10. *Chronica Botanica* 1937, 154.

COOPER, Rev. Oliver St. John (*fl.* 1780s–1800s)
Vicar of Paddington, London. Floristic list in J. Nichols's *Collections towards History and Antiquities of Bedfordshire* 1783.

COOPER, Randle Blair (1880–1973)
d. Corbridge, Northumberland 13 Oct. 1973
MSc. Timber-broker. Had garden of rare plants at Kilbryde, Corbridge.
J. Scott. Rock Gdn Club v.14(1) 1974, 70–71.

COOPER, Robert (*fl.* 1840s)
Seedsman, Croydon, Surrey. Issued seed catalogue in 1846.

COOPER, Roland Edgar (1890–1962)
b. Kingston-upon-Thames, Surrey 1890 *d.* Southend-on-Sea, Essex 31 Jan. 1962
FRSE 1942. To India, 1907. Gardener, Royal Botanic Garden, Edinburgh, 1910. Collected plants for A. K. Bulley in Sikkim, 1913, Bhutan, 1914–15, W. Himalayas, 1916. Discovered many new species including *Lobelia nubigena*. Superintendent, Maymyo Botanic Garden, Burma, 1921. Assistant Curator, Royal Botanic Garden, Edinburgh, 1930; Curator, 1934–50.
W. J. Bean *Trees and Shrubs Hardy in British Isles* v.1, 1950, 18. *Gdnrs Chron.* 1962 i 128. *Yb. R. Soc. Edinburgh* 1963, 10–11. A. M. Coats *Quest for Plants* 1969, 169–72. H. R. Fletcher and W. H. Brown *R. Bot. Gdn, Edinburgh, 1670–1970* 1970, 247 portr.

COOPER, Thomas (1815–1913)
b. Dulwich, London 5 Sept. 1815 *d.* Kew, Surrey 16 May 1913
Curator of collection of exotic plants belonging to W. Wilson Saunders at Reigate. Collected for Saunders, Kew, Edinburgh and Dublin in S. Africa, 1859–62. Discovered new species including *Stapelia cooperi*. Father-in-law of N. E. Brown. Secretary, Holmesdale Natural History Society, Reigate.
Gdnrs Chron. 1913 i 360. *J. Hort. Home Farmer* v.66, 1913, 524. *J. Kew Guild* 1914, 146. *Cactus J.* v.5, 1937, 62–63 portr. A. White *et al. Succulent Euphorbieae* v.1, 1949, 49 portr. A. White and B. L. Sloane *Stapelieae* v.1, 1937, 103 portr. *Lesotho: Basutoland Notes Rec.* v.5, 1965–66, 22–24.
Plants at BM(NH), Kew, etc. Notebooks at Kew. Portr. at Hunt Library.
Aloe cooperi Baker.

COOPER, Thomas Henry (*fl.* 1750s–1840s)
FLS 1834. *Botany of County of Sussex* 1834. MS. flora of Notts. (H. C. Watson *New Botanists' Guide* v.1, 1835, 265).
J. Bot. 1875 Appendix, 6. *D.N.B.* v.12, 152–53. A. H. Wolley-Dod *Fl. Sussex* 1937, xlii.

COOPER, Rev. W. H. Windle (–1929)
d. 1 Jan. 1929
MA. Of Bournemouth. Had collection of orchids.
Orchid Rev. 1929, 61.

COOPER, William (*fl.* 1820s)
Gardener and seedsman, Standishgate, Wigan, Lancs. Probably successor to John Cooper.

COOPER, William Marsh (*fl.* 1850s–1890s)
Educ. London University, 1852. To China as interpreter, 1854. Consul, Ning-po, 1877–88. Sent plants to Kew, 1884–86.
Gdnrs Chron. 1886 i 753. E. Bretschneider *Hist. European Bot. Discoveries in China* 1898, 738–39. *Kew Bull.* 1901, 16.
Osmanthus cooperi Hemsley.

COOTE, William (*fl.* 1590s–1640s)
Fellow of Trinity College, Cambridge. Chaplain to Lord Herbert of Cherbury. Friend of John Parkinson and Thomas Johnson. With George Bowles discovered *Senecio saracenius.*
J. Parkinson *Theatrum Botanicum* 1640, 1053. C. E. Raven *English Nat.* 1947, 282–83.

COPE, Edward (*fl.* 1790s)
Gardener and seedsman, Chorley, Lancs.

COPE, Richard (*fl.* 1790s)
Gardener and seedsman, Chorley, Lancs.

COPELAND, Ralph (1837–1905)
b. Woodplumpton, Lancs 3 Sept. 1837 *d.* Edinburgh 27 Oct. 1905
Accompanied Lord Lindsay to Mauritius to observe transit of Venus, 1874. Collected plants on Island of Trinidad in S. Atlantic, 1874, and discovered there great tree fern, *Cyathea copelandii.* 'Ein Besuch auf der Insel Trinidad im südatlantischen Ocean' (*Abh. Nat. Ver. Bremen* v.7, 1882, 269–80). Astronomer Royal for Scotland, 1889.

COPLAND, William (*fl.* 1550s–1568/9)
d. London 1568/9
Printer. *Boke of Propreties of Herbes called an Herball c.* 1550. Authorship uncertain but the book is largely a copy of Rycharde Banckes's *Herball* 1525.
R. Pulteney *Hist. Biogr. Sketches of Progress of Bot. in England* v.1, 1790, 51–52. *D.N.B.* v.12, 174. E. S. Rohde *Old English Herbals* 1922, 58–59, 63. A. Arber *Herbals* 1938, 44, 274.

COPPINGER, Richard William (1847–1910)
b. Dublin 11 Oct. 1847 *d.* Farnham, Surrey 2 April 1910
MA Dublin 1870. Naval Surgeon and naturalist. Surgeon on HMS 'Alert' on voyage of polar exploration, 1875, and on exploring cruise in Patagonian and Polynesian waters, 1878–82. Whilst on 'Alert' collected plants in Magellen, Tahiti, Fiji, Australia, Torres Straits Islands, Singapore, Seychelles. Inspector-General of Hospitals and Fleets, 1901. *Cruise of the 'Alert', 1878–1882,* 1883.
D.N.B. Supplt 2 416. *Fl. Malesiana* v.1, 1950, 116.
Plants at Kew.

CORBET, Vincent *see* Pointer, V.

CORBETT, Herbert Henry (*fl.* 1890s)
Herb. at Doncaster Museum, Yorks.

CORBETT, Wilfred (*c.* 1902–1971)
d. 15 Feb. 1971
Kew gardener, 1923–25. Experimental Research Station, Cheshunt, 1925. Kent Advisory Officer on market gardening, 1931. In charge of County Demonstration Glasshouse Station, Swanley, 1935. Principal, Kent Horticultural Institute, Swanley, 1946.
J. Kew Guild 1972, 66–67 portr.

CORBYN, Samuel (*fl.* 1640s–1650s)
b. Worcestershire
Trinity College, Cambridge, 1648. MS. List of Cambridge plants at Oxford.
J. Bot. 1912, 76–79.

CORDER, Octavius (1828–1910)
b. Exeter? 1828 *d.* Brundall, Suffolk 5 Jan. 1910.
Pharmaceutical chemist. Of Norwich. President, Norfolk and Norwich Naturalists Society, 1880–81.
Pharm. J. v.24, 1893, 150–58 (Presidential address). *Trans. Norfolk Norwich Nat. Soc.* v.9, 1909–10, 138–39.

CORDER, Thomas (1812–1873)
b. Widford Hall, Essex 1812 *d.* Cranfield, Beds 10 Nov. 1873
ALS 1833. Discovered *Bupleurum falcatum* in Essex, 1831, and *Claytonia perfoliata* at Ampthill, 1852. In Australia, *c.* 1838–45. His herb. is supposed to have passed to his brother, Octavius Corder (1828–1910). 'On Prevalence of European Genera and Species of Plants in Hilly Parts of Province of S. Australia' (*Phytologist* v.2, 1845, 336–38).
J. G. Dony *Fl. Bedfordshire* 1953, 23–24.

CORMACK, Benjamin George (1866–1936)
b. 25 May 1866 *d.* 19 Aug. 1936
BSc MA Glasgow. Senior Assistant, Botany Dept., Glasgow University. Lecturer in Botany, Queen Margaret College. Professor of Botany, Anderson College, Glasgow, 1897–1933. Contrib. to *Ann. Bot., Trans. Linn. Soc.*
Who was Who, 1929–1940 289.

CORMACK, John (*fl.* 1800s)
Nurseryman, New Cross, Deptford, Kent. Formerly Crombie and Cormack, *c.* 1780; in 1830 Cormack and Sinclair; in 1843 Cormack and Oliver. Related to J. Cormack, seedsman, 8 Whitechapel (*fl.* 1780s)?
Trans. London Middlesex Archeol. Soc. v.24, 1973, 182–83.

CORMACK, William (*fl.* 1810s–1840s)
Nurseryman and market gardener, New Cross, Deptford, Kent.

CORMACK, William Epps (1796–1868)
b. St. John's, Newfoundland 5 May 1796 *d.* New Westminster, British Columbia May 1868
Explorer and naturalist. Founded Beothuck Institute, Newfoundland, 1827. Cultivated tobacco in Australia. Sent seeds to Kew from New Zealand. Founded agricultural society in British Columbia. List of Newfoundland plants in his *Narrative of a Journey across Island of Newfoundland in 1822* 1856; reprinted 1873 and 1928.
British Columbian 9 May 1868. *Canadian Rec. Sci.* v.7, 1896, 4. *Rhodora* 1911, 110–62. *J. Bot.* 1928, 175–76.

CORNISH, Charles John (1858–1906)
b. Salcombe, Devon 28 Sept. 1858 *d.* Worthing, Sussex 30 Jan. 1906
BA Oxon 1885. Classics master, St. Paul's School London, 1885–1906. *New Forest* 1894. *Naturalist on Thames* 1902. 'Surviving London Fl.' (*Essex Nat.* v. 13, 1904, 302). Contrib. to *Country Life.*
D.N.B. Supplt. 2 v.1, 420–21. *Who was Who, 1897–1916*, 159. *R.S.C.* v.14, 359.

CORNISH, Vaughan (1862–1948)
b. 22 Dec. 1862 *d.* 1 May 1948
MA Oxon. FRGS. Geographer. Concerned in preservation of rural England. *National Parks and Heritage of Scenery* 1930. *Scenery of England* 1932; ed.2 1937. *Historic Thorn Trees in British Isles* 1941. *Churchyard Yew* 1946.
Who was Who, 1941–1950 253.

CORNTHWAITE, Rev. Tullie (*c.* 1807–1879)
b. London *c.* 1807 *d.* Walthamstow? 1 May 1879
MA Oxon 1830. FLS 1864. Of Walthamstow, Essex. Had herb. and purchased some of E. Forster's plants. Herb. passed to Hildebrand Ramsden.
Herb. at Charterhouse School.

CORNWELL, George (*fl.* 1820s–1830s)
Nurseryman, gardener and greengrocer, Barnet, Herts.

CORNWELL, William (*fl.* 1790s–1820s)
Nurseryman, Barnet, Herts.

CORREIA DA SERRA, Rev. Jose Francisco (1750–1823)
b. Serpa, Portugal 1750 *d.* Caldas da Rainha, Portugal 1823
LLD. FRS 1796. In London 1794–1801 and 1821. Contrib. to *Trans. Linn. Soc.* v.5, 1800, 218–26; v.6, 1802, 211–13, *Philos. Trans. R. Soc.* v.86, 1796, 494–505; v.89, 1799, 145–56.
P. Smith *Correspondence of Sir J. E. Smith* v.2, 1832, 198.
MSS. at Royal Society. Letters at BM(NH).
Correa Andr.

CORRY, Thomas Hughes (1859–1883)
b. Belfast 19 Dec. 1859 *d.* drowned in Lough Gill 4 Aug. 1883
BA Cantab 1883. FLS 1882. Assistant Curator, Cambridge University Herb. Lecturer in Medical and Science Schools, Cambridge. Discovered *Hieracium hypocharoides* in Ireland, 1879. *Fl. N.E. Ireland* (with S. A. Stewart) 1888.
J. Bot. 1883, 313–14. *Proc. Linn. Soc.* 1882–83, 31–32. *Trans. Bot. Soc. Edinburgh* v.16, 1886, 60–61. *Irish Nat.* 1913, 29. *Proc. Belfast Nat. Field Club* v.6, 1913, 623–24. R. L. Praeger *Some Irish Nat.* 1949, 67. *R.S.C.* v.9, 583; v.14, 364.
MS. flora of N.E. Ireland, 1878 at Ulster Museum.

CORSON, James (1815–1841)
b. Dalscairth, Dumfriesshire 1815 *d.* at sea 16 June 1841; buried Timor
Surgeon on South Sea whaler. Collected plants and shells in South Sea Islands.
Gdnrs Mag. 1842, 369–71. *Fl. Malesiana* v.5, 1958, 36.
Scaevola corsoniana Don.

CORSTORPHINE, Margaret (*née* Buncle) (1863–1944)
b. Arbroath, Angus 2 April 1863 *d.* Arbroath 17 Sept. 1944
Wife of R. H. Corstorphine and collected plants with him. Proprietor of printers, T. Buncle and Co., Arbroath. Interested in flowers of Angus of which she prepared a MS. flora.
Bot. Soc. Exch. Club Br. Isl. Rep. 1943–44, 642–43. *N. Western Nat.* 1944–45, 305–10 portr.
Plants at St. Andrews and Dundee Universities, Edinburgh Botanic Garden.

CORSTORPHINE, Robert Henry (1874–1942)
b. Arbroath, Angus 13 Nov. 1874 *d.* Arbroath 25 March 1942
BSc St. Andrews. FLS 1933. Managing director, T. Buncle, printers. Had herb. of Angus flora.
Bot. Soc. Exch. Club Br. Isl. Rep. 1941–42, 451–53 portr. *N. Western Nat.* 1942, 152–53, 257–61 portr. *Proc. Linn. Soc.* 1941–42, 272.
Plants at Dundee and St. Andrews Universities.

CORT, James (*fl.* 1790s)
Ironmonger and seedsman, Market Place, Leicester.

CORY, Bessie Florence *see* Hassall, B. F.

CORY, Reginald (1871–1934)
d. Wareham, Dorset 12 May 1934
FLS 1922. Director of colliery, shipping and oil firms. Collected plants in S. Africa, 1927, West Indies, 1921, Atlas Mountains, 1932.

Interested in cultivation of dahlias. Financed expeditions to China and other countries to collect plants that could be grown in British gardens. Benefactor of Cambridge Botanic Garden and Library of Royal Horticultural Society.

Gdnrs Mag. 1913, 721–22 portr. *Gdnrs Chron.* 1934 i 337. *Proc. Linn. Soc.* 1933–34, 151–54.

COSBY, Pole (*fl.* 1740s–1760s)
Nurseryman, Stradbally, County Leix, Ireland.

Irish For. 1967, 54.

COSGRAVE, Ephraim Macdowel (1853–1925)
b. Dublin 17 July 1853 *d.* Dublin 16 Feb. 1925

BA 1875 Dublin. MD 1878. Lecturer in Botany and Zoology, Carmichael College of Medicine. President, Royal College of Physicians, Ireland, 1914–16. Writer on antiquarian subjects. *Student's Botany* 1885.

Irish Book-lover v.15, 29. *Who was Who, 1916–1928* 230.

COSSTICK, W. (*fl.* 1860s)
Amateur entomologist with superficial knowledge of botany. *Desiderata of Fl. etc. of Eastbourne* 1867.

A. H. Wolley-Dod *Fl. Sussex* 1937, xlv.

COSTA, Laurence (*fl.* 1750s)
Italian gardener in Dublin where he sold exotic plants.

Irish For. 1967, 47.

COSTELLO, Patrick (*fl.* 1790s)
Of Dublin. Had nursery in County Wicklow.

Irish For. 1967, 47.

COTTAM, Arthur (1838–1912)
d. Bridgwater, Somerset 23 Nov. 1912

Of Watford, Herts. Astronomer. Entomologist. Diatomist. 'Notes on Fl. Watford' (*Trans. Watford Nat. Hist. Soc.* v.1, 1875, 14–16).

R. A. Pryor *Fl. Hertfordshire* 1887, 1. *R.S.C.* v.14, 373.

COTTELL, J. (*fl.* 1830s)
Nurseryman, Westerham, Kent.

COTTINGHAM, George (*fl.* 1780s)
Nurseryman, County Dublin.

Irish For. 1967, 51–52.

COTTON, Arthur Disbrowe (1879–1962)
b. London 15 Jan. 1879 *d.* Hertford 27 Dec. 1962

FLS 1902. VMM 1935. VMH 1943. Assistant, Kew Herb., 1904–22 (mycologist, Plant Pathology Laboratory of Ministry of Agriculture, 1918–22); Keeper, Kew Herb., 1922–46. Collected plants on Mt. Kilimanjaro, Tanganyika, 1929. President, Linnean Society, 1943–46. President, British Mycological Society, 1913. Supplts to *Elwes Monograph of Genus Lilium* (with A. Groves) 1936–40. Contrib. to *Proc. Linn. Soc.*, *Trans. Br. Mycol. Soc.*, etc.

Mycol. Notes 1923, 1185 portr. *Gdnrs Chron.* 1924 i 170 portr. *J. Kew Guild* 1941, 4–6 portr.; 1963, 335–36 portr. *Lily Yb.* 1951–52, 9–10 portr.; 1964, 68–72 portr. *Bot. Mag.* v.172, 1958–59 portr. *Times* 29 Dec. 1962. *Br. Phycol. Bull.* 1963, 272. *Nature* v.197, 1963, 951. *Taxon* 1963, 129–37. *Proc. Linn. Soc.* 1964, 86–88. *Trans. Br. Mycol. Soc.* 1964, 141–42 portr.

Letters and African plants at Kew. Basidiomycete drawings at BM(NH). Portr. at Hunt Library.

COTTON, Barbara (*fl.* 1810s–1820s)
Of Newport Pagnell, Bucks. Artist. Exhibited at Royal Academy, 1815–22. Contrib. 2 plates to *Trans. Hort. Soc. London* v.5, 1822–25. 8 drawings of apples and peaches at Royal Horticultural Society.

Kew Bull. 1916, 165.

COTTON, Charles (1630–1686/7)
b. 28 April 1630 *d.* 16 Feb. 1686/7

Master in Ancient and Modern Languages, Cambridge. Friend of Isaac Walton to whose *Compleat Angler* he contributed. Interested in horticulture; had a notable garden at Beresford, Staffs. *Planter's Manual*... *for Raising*...*Fruit Trees* 1675.

G. W. Johnson *Hist. of English Gdning* 1829, 115. S. Felton *Portr. of English Authors on Gdning* 1830, 102–7. *J. Hort. Cottage Gdnr* v.54, 1875, 467–69 portr. *D.N.B.* v.12, 298–301. *Gdnrs. Chron.* 1918 i 174–75.

COTTON, Frederic (*fl.* 1850s)
Major, Madras Engineers. Discovered the teak forests of Anaimalai Hills. "A most indefatigable collector and successful cultivator of orchidaceous plants, and who has now a large, and for India, unique collection in his conservatory in Ootacamund" (R. Wight *Icones Plantarum Indiae Orientalis* v.5, 1852, 21).

Cottonia Wight.

COTTON, George (*fl.* 1790s)
Cornfactor and seedsman, Romford, Essex.

COTTON, Thomas Atkinson (1857–1925)
b. Driffield, Yorks 1857 *d.* Bournemouth, Hants 16 April 1925

FLS 1886. FZS. Of Bishopstoke, Hants. Secretary, Watson Botanical Exchange Club, 1889–99.

Watson Bot. Exch. Club Rep. 1924–25, 282–83.

COUCH, Jonathan (1789–1870)
b. Polperro, Cornwall 15 March 1789 *d.* Polperro 13 April 1870

FLS 1824. Practised medicine at Polperro, 1809–70. Naturalist. 'Potato Disease' (*Rep. Cornwall Polytechnic Soc.* 1845, 9–21; 1847, 1–11; 1848, 1–2).

Proc. Linn. Soc. 1869–70, xciv, c. G. C. Boase and W. P. Courtney *Bibliotheca Cornubiensis* 1874–82, 89–92. F. H. Davey *Fl. Cornwall* 1909, xliv. *D.N.B.* v.12, 323–24. *R.S.C.* v.2, 68; v.7, 446; v.12, 173.

COUCH, Thomas Quiller (1826–1884)
b. Polperro, Cornwall 28 May 1826 *d.* Bodmin, Cornwall 23 Oct. 1884

MRCS. Son of J. Couch (1789–1870). Phenologist. At Bodmin, 1855. Botany of Polperro in *Rep. Cornwall Polytechnic Soc.* 1848, 11–27; 1849, 29–30. 'Periodic Phenomena' (*J. R. Inst. Cornwall* 1864–78).

G. C. Boase and W. P. Courtney *Bibliotheca Cornubiensis* 1874–82, 94. F. H. Davey *Fl. Cornwall* 1909, xlvii. *D.N.B.* v.12, 324. *R.S.C.* v.2, 70; v.7, 446; v.12, 174.

COULDREY, Thomas (*fl.* 1790s)
Gardener and seedsman, Abingdon, Berks.

COULDRY, William (*fl.* 1800s–1830s)
Nurseryman, Old Kent Road, Camberwell, London.

COULTAS, Harland (–1877)
b. U.S.A.? *d.* London 2 Feb. 1877

Professor of Botany, Pennsylvania Medical University, Philadelphia. Lecturer, Charing Cross Hospital. *Principles of Botany as Exemplified in Cryptogamia* 1854. *What may be learnt from a Tree* 1860.

J. Bot. 1877, 192. C. S. Sargent *Silva of N. America* v.3, 1892, 84. M. Meisel *Biogr. Amer. Nat. Hist.* v.3, 1924, 557. H. N. Clokie *Account of Herb. Dept. Bot., Oxford* 1964, 149.

American plants at Manchester and Oxford.

COULTER, Thomas (1793–1843)
b. Dundalk, County Louth 1793 *d.* Dublin 1843

MD. Physician in Mexico, 1824. Collected plants in Mexico, 1825–28, 1834; California and Arizona, *c.* 1831–33. Collected with David Douglas in California. Curator, Herb., Trinity College, Dublin. 'Mémoire sur les Dipsacées' (*Mém. Soc. Phys. Genève* v.2, 1823).

J. R. Geogr. Soc. v.5, 1835, 59–70. *Proc. R. Irish Acad.* 1844, 553–57. *Bot. Gaz.* 1896, 519–31. *Notes Bot. School Dublin* v.1, 1896, 3–4. *J. Washington Acad. Sci.* 1943, 65–70. S. D. McKelvey *Bot. Exploration of Trans-Mississippi West, 1790–1850* 1955 *passim.* R. L. Praeger *Some Irish Nat.* 1949, 63. A. M. Coats *Quest for Plants* 1969, 341–44. *Contrib. Univ. Michigan Herb.* 1972, 226–27.

Plants and portr. at Trinity College, Dublin. Portr. at Hunt Library.

Coulteria H. B. K.

COURT, William (1843–1888)
b. Alphington, Devon 1843 *d.* Chelsea, London 17 Sept. 1888

Gardener, Lucombe and Prince, Exeter; Veitch nursery, Exeter and Chelsea. Hybridised nepenthes and sarracenias.

Garden v.34, 1888, 287. *Gdnrs Chron.* 1888 ii 338. *J. Hort. Cottage Gdnr* v.17, 1888, 289. J. H. Veitch *Hortus Veitchii* 1906, 102.

COURTAULD, Sydney (1840–1899)
b. Braintree, Essex 10 March 1840 *d.* Braintree 20 Oct. 1899

Cultivated orchids. *Ferns of British Isles* 1877.

Orchid Rev. 1899, 339.

Masdevallia X Courtauldiana Hort.

COURTEN, William (alias **Charleton**) (1642–1702)
b. London 28 March 1642 *d.* Kensington, London 26 March 1702

Friend of Tournefort, Sloane and Sherard. Established a museum of natural history and art objects in Middle Temple, 1684.

Philos. Trans. R. Soc. v.27, 1712, 485–500. S. Ayscough *Cat. of Manuscripts...in British Museum* 1782, 647–48. A. Kippis *Biographia Britannica* v.4, 1789, 334–53. R. Pulteney *Hist. Biogr. Sketches of Progress of Bot. in England* v.2, 1790, 74–75. T. Faulkner *Hist. and Antiquities of Kensington* 1820, 230. *D.N.B.* v.12, 335. E. J. L. Scott *Index to Sloane Manuscripts in British Museum* 1904, 126. C. E. Raven *John Ray* 1942, 211, 229, 353. J. E. Dandy *Sloane Herb.* 1958, 115–17.

Plants at BM(NH).

Courtenia R. Br.

COURTHOPE, Peter (1639–1724)
b. Cranbrook, Kent 1639 *d.* Danny, Sussex 1724

FRS. Cousin of John Ray.

J. Bot. 1934, 217–23.

COURTNEY, Daniel (*fl.* 1770s)
Nurseryman, Tralee, County Kerry.

COUTTS, John (1872–1952)
b. Lochnager, Aberdeenshire 17 Jan. 1872 *d.* Woking, Surrey 21 Dec. 1952

VMH 1933. VMM 1937. Kew gardener, 1896–1900. Gardener to Sir Thomas Acland, 1900–09. Foreman, then Assistant Curator, Kew, 1909–29; Deputy Curator, 1929–32; Curator, 1932–37. *Lilies* (with H. D. Woodcock) 1935. *Complete Book of Gardening* (with A. Edwards and A. Osborn) 1930.

Gdnrs Chron. 1923 ii 272 portr.; 1934 i 106 portr.; 1953 i 10. *J. Kew Guild* 1930, 749 portr.; 1938, 757–58; 1952–53, 132. *Kew Bull.* 1937, 396. *Times* 21 Dec. 1952. *Lily Yb.* 1953, 9–10 portr.; 1954, 130.

COVEL, Rev. John (1638–1722)
b. Horningsheath, Suffolk 2 April 1638 *d.* Cambridge 19 Dec. 1722

BA Cantab 1658. DD 1679. Chaplain to British Embassy, Constantinople, 1670–77. "Described and drawn himself many plants observed by him in Thrace, Greece and Asia the less" (*Further Correspondence of John Ray* 1928, 145). Master, Christ's College, Cambridge, 1688.

J. T. Bent, *ed. Extracts from Diaries of Dr. John Covel, 1670–1679* 1893 portr. *D.N.B.* v.12, 355–56. *J. Bot.* 1924, 351. C. E. Raven *John Ray* 1942, 110, 214–15, 226. J. E. Dandy *Sloane Herb.* 1958, 117.

COVENTRY, Bernard Okes (–1929)
Indian Forest Service. Collected plants in Kashmir, 1923–25. *Wild Flowers of Kashmir* 1923–30 3 vols.

Pakistan J. For. 1967, 342.

Plants at Kew.

COWAN, Alexander (1863–1943)
Paper maker at Penicuik; also a hill-farmer. President, Botanical Society of Edinburgh, 1937–39. President, British Pteridological Society.

Br. Fern Gaz. 1918, 273 portr. *Trans. Proc. Bot. Soc. Edinburgh* 1943–44, 211–12 portr.

COWAN, D. A. (*c.* 1874–1952)
d. Worthing, Sussex 18 June 1952

Son of John Cowan (1842–1929) who sent him to S. America *c.* 1896 to collect orchids. Employed by nursery firms of Charlesworth and Co. and Messrs J. and A. McBean. With his brother John, acquired nursery of Messrs Hassell, Southgate. After this nursery went bankrupt, he opened an orchid nursery in Surbiton, Surrey; later transferred to Blenheim and Worthing.

Gdnrs Chron. 1952 ii 9 portr. *Orchid Rev.* 1952, 133–34.

COWAN, James (–1823)
d. Lima, Peru 1823

Merchant. Travelled in Mexico and Peru. Sent plants to A. B. Lambert.

Trans. Linn. Soc. v.14, 1825, 573–77.

Cowania D. Don.

COWAN, John (1842–1929)
d. 1 Oct. 1929

Gardener at Niddrie House near Edinburgh. Attempted to promote his invention of using lime kilns to heat glass-houses. Acquired from Meredith the Vineyard Nurseries, Garston near Liverpool. Collected in Brazil for Liverpool Orchid Co. An important orchid importer.

J. Hort. Home Farmer v.63, 1911, 538–39 portr. *Gdnrs Chron.* 1929 ii 315. *Orchid Rev.* 1929, 324, 371–72.

Laelia cowanii Rolfe.

COWAN, John Macqueen (1892–1960)
b. Banchory, Kincardineshire 1892 *d.* Edinburgh 26 Oct. 1960

MA Edinburgh. BA Oxon. FRSE 1931. FLS 1928. VMM 1951. VMH 1958. Indian Forest Service, 1914. Superintendent, Royal Botanic Gardens, Calcutta, 1926–28. Collected plants in Sikkim, Bengal and Burma. Assistant to Regius Keeper, Royal Botanic Garden, Edinburgh, 1930–54. Collected plants with C. D. Darlington in Persia, 1930. Gardens Adviser, National Trust for Scotland. President, Botanical Society of Edinburgh, 1951–53. Authority on taxonomy of *Rhododendron. Trees of Northern Bengal* (with his wife, A. M. Cowan) 1929. Edited *Journeys and Plant Introductions of George Forrest* 1952. *Rhododendron Leaf* 1950. Contrib. to *Rhododendron Yb.*

Gdnrs Chron. 1952 i 102 portr.; 1956 i 7 portr.; 1960 ii 505. *Nature* v.188, 1960, 982–83. *Times* 31 Oct. 1960. *Forestry* 1961, 106–7. *Yb. R. Soc. Edinburgh* 1962, 28–30. *Rhododendron Camellia Yb.* 1962, 100–102 portr. H. R. Fletcher and W. H. Brown *R. Bot. Gdn, Edinburgh, 1670–1970* 1970, 244–45, 253 portr.

Plants at Edinburgh, Kew.

COWAN, Thomas (*fl.* 1830s)
Overseer, White River Estate, Jamaica. Sent plants to England.

Bot. Mag. 1840, t.3782.

COWAN, Rev. William Deans (1844–1924)
English missionary in Madagascar, 1874–82, where he collected plants.

Dansk Bot. Arkiv. v.7, 1932, x. *Comptes Rendus AETFAT 1960* 1962, 134.

Plants and drawings of Madagascar orchids at BM(NH).

COWBURN, Richard (*fl.* 1820s)
Nurseryman, Market Place, Otley, Yorks.

COWBURN, Thomas Brett (1839–1892)
b. Sydenham, Kent 16 Nov. 1839 *d.* Dermel Hill, 1892

Major, 52nd Light Infantry. Pteridologist. Collected ferns from 1888. Found *Scolopendrium vulgare cowburni.*

E. J. Lowe *Fern Growing* 1895, 180–81 portr.

COWDRY, Nathaniel Harrington (1849–1925)
b. Torrington, Devon 20 Sept. 1849 *d.* 25 Jan. 1925
To Canada with parents in 1854. Banker. Collected plants in N.W. Canada, 1882; in China, 1919–21.
Bull. Peking Soc. Nat. Hist. 1928, 47–194. *J. N. China Branch R. Asiatic Soc.* 1929, 120–26.
Plants at Arnold Arboretum, Kew. Portr. at Hunt Library.

COWDRY, Thomas (1812–)
b. Somerset 1812
Father of N. H. Cowdry (1849–1925). To Canada, 1854. Collected plants in England and Ontario.
Herb. at Agricultural College, Guelph, Ontario.

COWELL, Edward Byles (*fl.* 1870s–1900s)
Herb. at Ipswich Museum, Suffolk.

COWELL, John (*fl.* 1690s–1730)
Nurseryman. About 1718 acquired nursery of William Darby (–*c.* 1714) at Hoxton, London. *True Account of Aloë Americana* 1729. *Curious and Profitable Gardener* 1730; reissued as *Curious Fruit and Flower Gardener* 1732.
G. W. Johnson *Hist. of English Gdning* 1829, 198. *Cottage Gdnr* v.8, 1852, 121. *Gdnrs Chron.* 1909 ii 353–54. *Garden* 1925, 260. *Notes and Queries* v.157, 1929, 74, 146; v.158, 1930, 56; v.174, 235.

COWELL, Matthew Henry (*fl.* 1830s–1840s)
Of Faversham, Kent. Local Secretary, Botanical Society of London. *Floral Guide for E. Kent* 1839. *Series of Botanical Labels for Herbarium* 1841.

COWIE, John (*fl.* 1770s)
Seedsman, 21 Parliament Street, Westminster. In partnership with John Webb, *c.* 1775, and James Shiells, 1776–*c.* 1779.

COWLES, Mrs. Emily (*fl.* 1840s)
Wife of Charles Cowles, chemist of Stratford, Essex. Collected plants, chiefly in Edinburgh district, 1847–48.
Essex Nat. v.29, 1954, 189.
Herb. at Essex Field Museum.

COWLEY, Abraham (1618–1667)
b. London 1618 *d.* Chertsey, Surrey 28 July 1667
MA Cantab 1642. MD Oxon 1657. FRS 1662. "Botany, in the mind of Cowley, turned into poetry" (S. Johnson). In 1650s retired to "a fruitful part of Kent to pursue the study of simples." The result was a Latin poem, *Plantarum Libri Duo*, 1662.
R. Pulteney *Hist. Biogr. Sketches of Progress of Bot. in England* v.1, 1790, 282–89. *D.N.B.* v.12, 379–82.
Portr. at National Portrait Gallery.

COWLEY, Ebenezer (1848/9–1899)
b. Fairford, Glos 1848/9 *d.* Kamerunga, Queensland 8 Feb. 1899
Overseer, Kamerunga State Nursery, N. Queensland, 1889–99. Collected plants. Correspondent of F. M. Bailey. Contrib. to *Queensland Agric. J.*
Proc. R. Soc. Queensland 1890–91, 38–39.

COWLEY, Herbert (1885–1967)
b. Wantage, Berks 9 Jan. 1885 *d.* 1 Nov. 1967
Gardener, Messrs. Veitch and Kew. Edited *J. Kew Guild* 1909–14; *Garden* 1915–21; *Gdning Illustrated* 1923–36. Collected plants in Alps, Pyrenees, Majorca, Albania, Sicily, Elba, etc.
J. Kew Guild 1969, 930–31 portr. *Who was Who, 1961–1970* 249.

COWLISHAW, Henry (*fl.* 1740s–1777)
Nurseryman, Hodsock Park, Blyth, Notts.

COX, John (1814–1886)
b. Buckland, Berks Nov. 1814 *d.* 30 Aug. 1886
Gardener at Redleaf, Penhurst, Kent. Raised 'Redleaf' russet apple. Contrib. to *Gdnrs Chron.*, *J. Hort.*
Gdnrs Chron. 1875 ii 324–25 portr.; 1886 ii 378.

COX, Richard (*c.* 1776–1845)
d. Colnbrook, Bucks 20 May 1845
Brewer, Bermondsey, London. Retired to Colnbrook where he raised 'Cox's Orange Pippin' apple and 'Cox's Pomone' apple.
J. R. Hort. Soc. 1943, 347–49. A. Simmonds *Hort. Who was Who* 1948, 11–16.

COX, Richard (*fl.* 1830s–1840s)
Nurseryman, Ludlow, Shropshire.

COX, Thomas (*fl.* 1790s)
Seedsman and gardener, Eastbourne, Sussex.

COX, William (–1704)
d. Dec. 1704
Nurseryman, Kew Green, Surrey, probably as early as 1680.

COX, William (1680–1722)
d. March 1722
Son of W. Cox (–1704). Nurseryman, Kew Green, Surrey. Noted for his improvement of the 'Hotspur Pea'. Nursery acquired by Richard Butt.
Trans. London Middlesex Archaeol. Soc. v.24 1973, 192. *Agric. Hist. Rev.* 1974, 20, 25–33.

COX, William (*fl.* 1790s)
Grocer and seedsman, Gallow-tree-gate, Leicester.

COX, William (1822–1883)
b. Ludlow, Shropshire 1822 *d.* Madresfield Court, Worcs 8 May 1883
Gardener to Earl Beauchamp at Madresfield Court. Raised 'Madresfield Court Grape'. Revised *Paxton's Cottage Calendar*.
Gdnrs Chron. 1875 ii 785 portr.; 1883 i 642. *J. Hort. Cottage Gdnr* v.6 1883, 427.

COXE, Mr. (*fl.* 1810s)
Collected plants in New Holland, Australia, *c.* 1815. Plants were in A. B. Lambert's herb.
Taxon 1970, 520.

COXE, Daniel (*fl.* 1660s)
Educ. Cambridge. FRS 1664. Wrote on chemistry and medicine. *Enquiries concerning Vegetables* 1665.
C. E. Raven *John Ray* 1942, 187.

COYS, William (*c.* 1560–1627)
d. 27 March 1627
Of Stubbers, North Ockendon, Essex. *Yucca gloriosa* flowered in his garden in 1604 for the first time in England. Friend of M. de Lobel and J. Goodyer to whom he sent seeds. Introduced *Cymbalaria muralis*.
M. de Lobel *Rondelletii...Animadversiones* 1605, 469, 471, 498, 501. J. Gerard *Herball* 1633, 1626–29. M. de Lobel *Stirpium Illustrationes* 1655, 117, 120. J. Parkinson *Theatrum Botanicum* 1640, 84. R. T. Gunther *Early Br. Botanists* 1922, 128–30, 312–25. *Essex Nat.* 1938, 67–71; 1956, 306–11. S. T. Jermyn *Fl. Essex* 1974, 14.

COYTE, Rev. William Beeston (1740–1810)
d. Ipswich, Suffolk 3 March 1810
MB Cantab 1763. ALS 1788. FLS 1794. Of Yarmouth and Halesworth, Suffolk. Practised medicine at Ipswich. Published list of plants in his garden at Ipswich: *Hortus Botanicus Gippovicensis* 1796. *Index Plantarum* v.1, 1807. His father, William Coyte (1708–1775), had also interested himself in botany.
Gent. Mag. 1810 i 389. J. Nichols *Illus. Lit. Hist. of Eighteenth Century* v.6, 877–79. D. Turner *Extracts from Lit. and Sci Correspondence of Richard Richardson* 1835, 184. *D.N.B.* v.12, 424. *Suffolk Rev.* 1958, 201–3.

CRABBE, Rev. George (1754–1832)
b. Aldeborough, Suffolk 24 Dec. 1754 *d.* Trowbridge, Wilts 3 Feb. 1832
LLD Cantab 1784. Poet. Chaplain to Duke of Rutland at Belvoir, 1782–85. Curate, Stathern, Leics., 1785; Rector, Muston, Leics.; Rector, Trowbridge, 1813–32. Studied botany. Wrote and burnt 'Essay on Botany'. Plant list in J. Nichols' *Hist. of Leicestershire* v.1, 1795.
D. Turner and L. W. Dillwyn *Botanist's Guide* 1805, 537–65. W. M. Hind *Fl. Suffolk* 1889, 479. *D.N.B.* v.12, 428–31. *Proc. Suffolk Inst.* v.12, 1905, 223–32. *J. Bot.* 1907, 77–78. A. R. Horwood and C. W. F. Noel *Fl. Leicestershire* 1933, cxcviii–cc portr.

CRABTREE, John Henry (–1924)
b. Norden, Lancs *d.* 30 July 1924
Schoolmaster, Wardle, 1885. Factory inspector, 1894. *Woodland Trees and how to identify Them* 1915. *British Fungi and how to identify Them* 1916. *British Ferns and how to identify Them* 1919. *Grasses and Rushes and how to identify Them* 1920. *British Mosses and how to identify Them* 1924.
Rep. Br. Bryol. Soc. 1925, 182.

CRABTREE, Philippa (*fl.* 1764–1820s)
b. London 17 Nov. 1764
Flower painter of Bishopgate, London. Exhibited at Royal Academy, 1786–87.
Gdnrs Chron. 1920 i 278.
Drawings sold at Hodgson, 5 Feb. 1920; Sotheby, Nov. 1965.

CRACKNELL, John S. W. (*c.* 1895–1965)
b. Woodbridge, Suffolk *d.* Nov. 1965
VMH 1948. Managing Director, wholesale seed firm of Watkins and Simpson Ltd., 1949–60.
Gdnrs Chron. 1948 ii 2 portr.; 1966 i 31.

CRADDOCK, Dr. (*fl.* 1860s)
Bryophyta collected in Westmorland, Cumberland and Ireland at Oxford.

CRADDOCK, W. H. (*fl.* 1900s)
Burma Forest Service. In Malay Peninsula, 1902–3 and sent Pahang plants to Singapore.
Gdns Bull. Straits Settlements 1927, 119. *Fl. Malesiana* v.1, 1950, 119.
Impatiens craddockii Hook. f.

CRADWICK, William (*c.* 1862–1937)
d. Manderville, Jamaica 30 Aug. 1937
Kew gardener, 1885. Superintendent, Castleton Garden, Jamaica, 1888. Superintendent, Government Cinchona Plantations, 1889. Curator, Hope Gardens. Superintendent, Imperial Dept. of Agriculture, British West Indies. Contrib. to *Orchid Rev.* *Orchid Rev.* 1935 dedicated to him in recognition of his work in encouraging orchid cultivation in West Indies.
I. Urban *Symbolae Antillanae* v.3, 1902, 33. *Gdnrs Chron.* 1925 i 32 portr. *Orchid Rev.* 1935, 163 portr.; 1937, 323–24. *J. Kew Guild* 1938, 790 portr.
Plants at Kew.

CRAIB, William Grant (1882–1933)
b. Kirkside, Banff 10 March 1882 *d.* Kew, Surrey 1 Sept 1933

MA Aberdeen 1907. FRSE 1920. FLS 1920. Acting Curator, Herb., Royal Botanic Gardens, Calcutta, 1908. Assistant, Kew Herb., 1909–15. Lecturer in Forest Botany, Edinburgh, 1915–20. Regius Professor of Botany, Aberdeen, 1920–33. *Fl. Banffshire* 1912. 'Contributions to Fl. Siam' (*Kew Bull.* 1911–27). *Florae Siamensis Enumeratio* 1925–34.

Gdnrs Chron. 1933 ii 210. *J. Bot.* 1933, 299–300. *Kew Bull.* 1933, 409–12. *Nature* v.132, 1933, 471. *Times* 2 Sept. 1933. *J. Kew Guild* 1934, 371–72 portr. *Proc. Linn. Soc.* 1933–34, 154–55. *Proc. R. Soc. Edinburgh* v.53, 1934, 357–58. *Trans. Proc. Bot. Soc. Edinburgh* 1934, 471. *Chronica Botanica* 1935, 157.

Indian plants at Kew, Wroclaw Poland. Portr. at Hunt Library.

CRAIG, William (1832–1922)
b. near Strathaven, Lanarkshire 28 March 1832 *d.* Edinburgh 3 Feb. 1922

MD Edinburgh 1870. President, Botanical Society of Edinburgh, 1887–89; Secretary, 1900–12. Lecturer on Materia Medica, Royal College of Surgeons. Studied Scottish flora. *Manual of Materia Medica and Therapeutics* 1876.

Trans. Bot. Soc. Edinburgh v.28, 1920–21, 70–71. *Proc. R. Soc. Edinburgh* v.42, 1921–22, 390–91.

Craigia W. W. Smith & W. E. Evans.

CRAIG, William Nicol (*fl.* 1890s–1930s)
b. Milnthorpe, Westmorland

Gardener. To U.S.A., 1890. Nurseryman, Weymouth, Mass. *Lilies and their Culture in N. America* 1928. Contrib. to *Gdnrs. Chron.*

Gdnrs. Chron. 1931 i 194 portr.

CRAIG-CHRISTIE, Alexander (1843–1914)
b. Edinburgh 22 May 1843 *d.* Kincardine June 1914

FLS 1878. Shetland plants in *Trans. Bot. Soc. Edinburgh* v.10, 1870, 165–70. 'Stipules in Holly' (*J. Linn. Soc.* v.18, 1881, 467–68).

R.S.C. v.8, 387; v.12, 515; v.14, 217.

Plants at Kew. Portr. at Hunt Library.

CRAMB, Alexander (1810–1877)
b. Scone, Perth 1810 *d.* Tortworth, Glos 27 April 1877

Gardener to Earl of Ducie, Tortworth Court.

Gdnrs Chron. 1875 i 719–20 portr.; 1877 i 604.

CRAN, Marion (1875–1942)
b. S. Africa 1875 *d.* 2 Sept. 1942

Writer. Founder of Garden Club, Mayfair. Vice-President, National Gardens Guild. *Garden Register and Garden Talks* 1925. *Gardens of Character* 1930. *I Know a Garden* 1933.

Who was Who, 1941–1950 262.

CRAN, William (1854–1933)
d. 28 June 1933

MA Aberdeen. BD Edinburgh. Lecturer, Wesleyan Seminary, Antigua. Returned to Scotland, 1898. Collected Mycetozoa in West Indies, Aberdeen and Kincardine.

J. Bot. 1938, 319–27.

CRANE, Daniel Burton (–1938)
b. Highgate, London *d.* 16 July 1938

Gardener. *Pansies and Violas* 1908. *Book of Sweet Pea* 1910. *Chrysanthemums for Garden and Greenhouse* 1905; ed. 3 1918.

J. Hort. Cottage Gdnr v.52, 1906, 53, 55 portr. *Gdnrs Chron.* 1938 ii 72.

CRANFIELD, William Bathgate (1859–1948)
b. London 1 March 1859 *d.* 29 May 1948

VMH 1935. FLS 1927. Auctioneer in London. Collected ferns and grew daffodils, paeonies and irises in his garden at Enfield Chase. President, British Pteridological Society, 1920–48.

Br. Fern Gaz. 1922, 205 portr.; 1948, 200–2; 1949, 223. *Gdnrs Chron.* 1924 ii 142 portr.; 1948 i 196. *Daffodil Yb.* 1937, 1–2 portr. *Daffodil Tulip Yb.* 1949, 136–37. *Proc. Linn. Soc.* 1948–49, 85.

Fern collection went to Royal Horticultural Society Gardens, Wisley. Portr. at Hunt Library.

CRANSTON, John (*fl.* 1890s)

Nurseryman, Hereford; specialised in roses.

Gdnrs Mag. 1894, 396–97 portr.

CRASTIN, Cornelius (*c.* 1782–1849)
d. 10 Jan. 1849

Came to England from Holland, 1817. Took over Seven Sisters Nursery, Islington from Robert Enkel.

CRAWFORD, Alexander Robert (1840–1912)
b. Dublin 21 Feb. 1840 *d.* Moona Plains, N.S.W. 27 March 1912

Came to Australia to manage his uncle's farm in N.S.W. Collected plants in Australia for F. von Mueller.

J. Proc. R. Soc. N.S.W. 1921, 156.

Plants at Sydney Botanic Gardens.

CRAWFORD, Francis Chalmers (1851–1908)
b. N. Berwick 24 Aug. 1851 *d.* Edinburgh 9 Feb. 1908
Stockbroker. Fellow Botanical Society of Edinburgh, 1897. Demonstrator of Botany at Royal Botanic Garden, Edinburgh. *Anatomy of British Carices* 1910 (includes biogr. and portr.)
J. Bot. 1910, 339–41. *Trans. Bot. Soc. Edinburgh* v.24, 1912, 2–3. *R.S.C.* v.14, 396.
Herb. at Edinburgh.
Saxifraga X Crawfordii E. S. Marshall.

CRAWFURD, John (1783–1868)
Official of East India Company. In Burma in 1826 to negotiate treaty, accompanied by N. Wallich. Discovered *Amherstia nobilis*.
A. M. Coats *Quest for Plants* 1969, 151–53.
Crawfurdia Wall.

CRAWSHAY, De Barri (*c.* 1857–1924)
d. Sevenoaks, Kent 26 Dec. 1924
Of Sevenoaks. Cultivated orchids. 'Odontiodes' (*Orchid World* 1911, 75–84).
Orchid Rev. 1922, 293–99, 327–34; 1925, 55–56 portr.

CRAYCROFT (Creacroft), Capt. (*fl.* 1730s)
Collected plants in Greenland and Davis Straits, 1734–39.
J. E. Dandy *Sloane Herb.* 1958, 117.
Plants at BM(NH).

CREAGH, Charles Vandeleur (1842–1917)
d. 18 Sept. 1917
Governor, British North Borneo, 1888–95. Commander-in-Chief, Labuan, 1891–95. Gave his collection of Bornean plants to Kew.
Kew Bull. 1895, 272. *Who was Who, 1916–1928* 243. *Fl. Malesiana* v.1, 1950, 119.
Bauhinia creaghi Baker.

CREE, John (*c.* 1738–1816)
d. 27 Oct. 1816
Kew gardener. Collected plants in Carolina in 1760s. Founded nursery *c.* 1765 at Addlestone, Chertsey, Surrey. Supplied plants to Princess Augusta at Kew.
Bot. Repository v.2, 1801, 138. J. E. Smith *Selection of Correspondence of Linnaeus* v.1, 1821, 554. *Gdnrs Chron.* 1928 ii 30; 1950 i 21.
American plants at BM(NH).

CREE, John (*c.* 1800–1858)
b. Chertsey, Surrey *d.* 2 Jan. 1858
Son of J. Cree (*c.* 1738–1816) whose nursery at Addlestone he inherited. Issued large catalogues: *Hortus Addlestonensis* 1829 and *Catalogue of Herbaceous Plants, etc.*, 1837. Nursery closed *c.* 1838. J. Cree later appointed Surveyor of Highways and Assistant Overseer, Chertsey.
Gdnrs Mag. 1830, 381–82; 1831, 359–60. *Bot. Gdn.* v.8, 1839–40, 747.
Malva creeana R. Graham.

CREE, William (*fl.* 1780s–1810s)
Nurseryman, Pond Close Nursery, Addlestone, Chertsey, Surrey.

CRESPIGNY, Eyre Champion De *see* De Crespigny, E. C.

CRESSEY, William (*fl.* 1780s)
Seedsman and cork-cutter, High Street, Hull, Yorks.

CRESSWELL, Rev. Richard (1815–1882)
b. London 1 Dec. 1815 *d.* Teignmouth, Devon 10 April 1882
BA Oxon 1839. Curate, Salcombe Regis and Teignmouth. Studied algae and fungi. *Flowering Plants and Ferns of Sidmouth* 1846. Contrib. to *Phytologist*.
W. H. Harvey *Phycologia Britannica* 1846–51, t.160. W. K. Martin and G. T. Fraser *Fl. Devon* 1939, 773.
Plants and drawings at Exeter Museum. Letters at BM(NH).
Schizothrix cresswellii Harvey.

CRESSY (–before 1763)
Physician at Barbuda, West Indies whence he sent *Tetracera* seeds and plants to P. Miller (*Gardener's Dictionary* Abridged, ed. 5 1763).

CREWDSON, W. D. (–1966)
d. Kendal, Westmorland 24 June 1966
Amateur alpine plant grower. Contrib. to *Quart. Bull. Alpine Gdn Soc.*
Quart. Bull. Alpine Gdn Soc. 1966, 387.

CREWE, Rev. Henry Harpur– *see* Harpur-Crewe, *Rev.* H.

CRICHTON, Sir Alexander (1763–1856)
b. Edinburgh 2 Dec. 1763 *d.* Sevenoaks, Kent 4 June 1856
MD Leyden 1785. FRS 1800. FLS 1793. Knighted 1821. Physician, Westminster Hospital, 1794. Physician to Alexander I of Russia, 1804. 'Vegetable Remains found in Sandstone near Ballisadiere County Sligo' (*Proc. Geol. Soc.* v.2, 1838, 394–95).
Proc. R. Soc. v.8, 1856, 269–72. *Proc. Linn. Soc.* 1856–57, xxv–xxvi. W. Munk *Roll of R. College of Physicians* v.2, 1878, 416–18. *D.N.B.* v.13, 85–86. *R.S.C.* v.2, 93.

CRICHTON, James Smith (1841–1887)
b. Arbroath, Angus 2 April 1841 *d.* Arbroath 28 June 1887
MD Edinburgh 1864. FBS Edinburgh 1886. President, Natural History Association, Arbroath. Helped to compile Association's *Fl. Arbroath* 1882.
Scott. Nat. 1887, 175–76. *Trans. Bot. Soc. Edinburgh* v.17, 1889, 522–24.

CRICK, James (*fl.* 1790s)
Nurseryman, Rochford, Essex.

CRIDDLE, Norman (1875–1933)
b. Addlestone, Surrey 14 May 1875 *d.* Brandon, Manitoba, 4 May 1933
To Manitoba with parents in 1882. Entomologist, Canadian Dept. of Agriculture, 1913. Illustrated G. H. Clark and J. Fletcher's *Farm Weeds* 1906. Helped in preparation of check list of Manitoba flora in 1922.
Canadian Field Nat. 1933, 145–47 portr.
Plants at Ottawa.

CRIPPS, John Marten (1780–1853)
b. Sussex 1780 *d.* Novington 3 Jan. 1853
MA Cantab 1803. FLS 1803. Travelled in East with E. D. Clarke.
J. E. Smith *Exotic Bot.* v.1, 1805, t.120. H. C. Andrews *Bot. Repository* v.8, 1808, t.528. E. D. Clarke *Travels* v.1, 1810, additions, xxii. *Proc. Linn. Soc.* 1853, 231–32. M. A. Lower *Worthies of Sussex* 1865, 271–73. F. Boase *Modern English Biogr.* v.1, 1892, 781–82.
Plants at Cambridge.
Rubus crippsii Clarke.

CRIPPS, Thomas (*c.* 1809–1888)
d. Tunbridge Wells, Kent 17 April 1888
Nurseryman, Tunbridge Wells. First fuchsia with white sepals, 'Venus victrix', was sent out from this nursery in 1842.
Gdnrs Chron. 1888 i 504. *J. Hort. Cottage Gdnr* v.16, 1888, 339.

CRIPPS, William T. (*c.* 1840–1871)
d. 11 June 1871
Nurseryman, Tunbridge Wells, Kent.

CRISP, Sir Frank (1843–1919)
b. London 25 Oct. 1843 *d.* Henley, Oxfordshire 29 April 1919
BA London. LLB London. FLS 1870. VMH 1918. Solicitor. Had garden at Friar Park, Henley-on-Thames with celebrated rock garden. Secretary, Royal Microscopical Society, 1878–89. Treasurer, Linnean Society, 1881–1906. *Mediaeval Gardens* 1924 2 vols. Edited *J. R. Microsc. Soc.*
Gdnrs Chron. 1899 ii 321–24; 1919 i 232 portr.; 1925 i 139. *Proc. Linn. Soc.* 1904–5 portr.; 1918–19, 49–51. *Gdnrs Mag.* 1907, 857–58 portr. *Garden* 1916, iv portr. *J. Bot.* 1919, 200. *Curtis's Bot. Mag. Dedications, 1827–1927* 343–44 portr. *Who was Who, 1916–1928* 245. A. T. Gage *Hist. Linn. Soc. of London* 1938 portr. *R.S.C.* v.9, 603.
Portr. at Hunt Library.

CROALL, Alexander (1809–1885)
b. Brechin, Angus 1809 *d.* Stirling 19 May 1885
Teacher. Keeper of Museum and Herb., Derby, 1863. Curator, Smith Institute, Stirling, 1873. *Plants of Braemar* 1855 (exsiccatae). *Nature-printed British Seaweeds* (with W. G. Johnstone) 1859 4 vols.
Hooker's J. Bot. 1854, 284. S. Smiles *Men of Invention and Industry* 1884. *Scott. Nat.* 1885, 97–98, 148–53. *Reporter* (Stirling) 6 June 1885. *Trans. Bot. Soc. Edinburgh* v.16, 1886, 309–11. D. B. Morris *Alexander Croall, Nat.* 1933. *Mem. New York Bot. Gdn* v.19, 1969, 10–11. *R.S.C.* v.2, 95; v.7, 460; v.14, 406.
Plants at BM(NH), Kew, Stirling. Letters at Kew.

CROCKER, Charles William (1832–1868)
b. Chichester, Sussex 1832 *d.* Torquay, Devon 19 Feb. 1868
Foreman, Propagating Dept., Kew, 1857–63. 'Germination of *Cyrtandraceae*' (*J. Linn. Soc.* v.5, 1861, 65–67).
Gdnrs Chron. 1868, 242. *J. Kew Guild* 1895, 29. *R.S.C.* v.12, 177.

CROCKER, Emmeline (1858–1910)
b. Dulwich, London 1858 *d.* Funchal, Madeira 26 Feb. 1910
FLS 1907. Painted rhododendrons. Collected algae in Madeira. *Thirty-nine Articles on Gardening* 1908.
Proc. Linn. Soc. 1909–10, 87.

CROFTON, Robert (1603–1630s)
Nurseryman, Twickenham, Middx.

CROFTS, Daniel (*fl.* 1760s)
Gardener to Duke of Argyll at Whitton, Middx. *A Particular of Noble Large House, Gardens...of Late Duke of Argyle...at Whitton...* 1765 (includes list of trees and shrubs).

CROLL, David (*c.* 1841–1909)
d. Broughty Ferry near Dundee 7 Jan. 1909
Nurseryman, Broughty Ferry. Founded the nursery with his brother William in 1872. Specialised in roses.
J. Hort. Cottage Gdnr v. 58, 1909, 63.

CROLL, John (*fl.* 1910s)
Nurseryman, Broughty Ferry near Dundee. Brother of David Croll.
J. Hort. Home Farmer v.63, 1911, 418 portr.

CROMBIE, David (*fl.* 1900s–1920s)
Gardener to Marquis of Waterford at Curraghmore, County Waterford, 1900–23.
Gdnrs Chron. 1923 i 360 portr.

CROMBIE, Rev. James Morrison (1830–1906)
b. Aberdeen 20 April 1830 *d.* Ewhurst, Surrey 12 May 1906

MA Edinburgh. FLS 1869. Lecturer on Botany, St. Mary's Hospital, 1879–91. *Braemar* 1861. *Lichenes Britannici* 1870. *Monograph of Lichens found in Britain* 1894.
J. Bot. 1906, 248. *Kew Bull.* 1906, 225. *Proc. Linn. Soc.* 1905–6, 36–37. *Essex Nat.* v.29, 1954, 192. *R.S.C.* v.7, 461; v. 9, 605.
Herb. and MSS. at BM(NH). Plants at Kew.
Lecidea crombiei Jones.

CROMBIE, William (*fl.* 1700s)
Seedsman, London.
Agric. Hist. Rev. v.18, 1970, 153–54.

CROMBLEHOLME, Rev. John (*fl.* 1890s–1920s)
Rector, Clayton-le-Moors, Accrington, Lancs. Cultivated orchids. President, Manchester Orchid Society.
Gdnrs Chron. 1921 ii 228 portr.

CROOK, H. Clifford (*c.* 1882–*c.* 1974)
Vice-president, Alpine Garden Society. Authority on *Campanula*. *Campanulas; their Cultivation and Classification* 1951.
Quart. Bull. Alpine Gdn Soc. 1974, 212.

CROPPER, James (1773–1840)
Seedsman, Grappenhall, Cheshire and Latchford near Warrington, Lancs.
Trans. Hist. Soc. Lancs. Cheshire. 1960, 112.

CROSBY-SMITH, Joseph (1853–1930)
b. Keighley, Yorks 18 July 1853
FLS 1907. To New Zealand, 1876. Studied ferns and algae, Otago. Made botanical excursions in New Zealand, also to Campbell Islands, 1927.
Trans. N.Z. Inst. 1931, 174–76 portr.

CROSFIELD, Albert John (1852–1931)
b. Liverpool, 1852 *d.* Cambridge 6 Aug. 1931
Amateur botanist. Chairman, Friends' Foreign Mission Association, 1911–20.
Watson Bot. Exch. Club Rep. 1931–32, 101–2.
Herb. at BM(NH).

CROSFIELD, George (1754–1820)
b. Kendal, Westmorland 22 March 1754 *d.* Lancaster 10 Oct. 1820
Sugar dealer. Collected plants in Cheshire and Lancashire.
J. Bot. 1912, 369–71.
Herb. acquired by G. Crosfield (1785–1847).

CROSFIELD, George (1785–1847)
b. Warrington, Lancs 26 May 1785 *d.* Liverpool 15 Dec. 1847
Son of G. Crosfield (1754–1820). Grocer. Secretary, Botanical Society of Warrington. To Liverpool, 1819. *Calendar of Flora... during 1809 at Warrington* 1810. Herb. acquired by C. E. Salmon.
Annual Monitor 1849, 33–38. J. Smith *Cat. Friends' Books* v.1, 1867, 494. *Trans. Liverpool Bot. Soc.* 1909, 66. *D.N.B.* v.13, 213. *J. Bot.* 1912, 369–71.
Letters at Society of Friends, London. Silhouette in J. Kendrick's *Profiles of Warrington Worthies* 1853.

CROSS, Edward (*fl.* 1760s–1790s)
Seedsman, 152 Fleet Street, London.

CROSS, Robert Mackenzie (1836–1911)
b. Dunbarton 1836 *d.* Torrance, Stirlingshire 1 March 1911
Kew gardener, 1857. To Ecuador for *Cinchona*, 1859; Panama for *Castilloa*, 1875; Brazil for *Hevea*, 1876. *Report on Expedition to procure Seeds of Cinchona condaminea from...Ecuador* 1862. *Report on...Plants and Seeds of India-rubber Trees of Para...* 1877.
Gdnrs Chron. 1861, 735, 1047; 1911 ii 176. *J. Bot.* 1871, 319. *J. Kew Guild* 1911–12, 51–52. *Kew Bull.* 1911, 165–66.
Ecuador plants at Kew.

CROSS, William (*fl.* 1790s)
Gardener and seedsman, Chippenham, Wilts.

CROSS, William J. (1832–1885)
b. Whiteworth, Berks 10 Sept. 1832 *d.* Salisbury, Wilts 22 June 1885
Gardener to Lord Ashburton at Melchet, 1862–69. Partner with Steer in nursery at Ford and Salisbury, Wilts.
Gdnrs. Chron. 1877 ii 301–2 portr.; 1885 i 830.

CROSSE, Henry (*fl.* 1560s–1570s)
Registrar, Oxford University, 1566–70. Had herb garden.
R. T. Gunther *Early English Botanists* 1922, 304–5.

CROSSLAND Charles (1844–1916)
b. Halifax, Yorks 3 Sept. 1844 *d.* Halifax 9 Dec. 1916
FLS 1899. Butcher, Halifax. Mycologist. Authority on Discomycetes. Secretary, Mycological Committee of Yorkshire Naturalists' Union, 1893–1914. President, Yorkshire Naturalists' Union, 1907. *Fl. Parish of Halifax* (with W. B. Crump) 1904. *Fungus Fl. Yorkshire* (with G. E. Massee) 1905. *New and Rare British Fungi* (with G. E. Massee) 1906.
Naturalist 1910, 367–74 portr.; 1961, 57. *Kew Bull.* 1914, 173; 1917, 36–37. *Bot. Soc. Exch. Club Br. Isl. Rep.* 1916, 464. *J. Bot.* 1917, 62. *Trans. Br. Mycol. Soc.* 1917, 466–69.
Fungi and drawings at Kew.

CROSSLAND, Cyril (1878–1943)
b. Sheffield, Yorks 1878 *d*. Denmark 1943
PhD. Marine biologist. Director of pearl fisheries in Sudan until 1923. Wrote articles on botany, zoology and anthropology of Sudan. *Desert and Water Gardens of Red Sea* 1913.
R. Hill *Biogr. Dict. of Sudan* 1967, 106.

CROSSLAND, M. C. (*fl*. 1890s)
Plants collected in Togoland in 1891 now at Oxford.

CROSSLEY, Samuel (*fl*. 1810s)
Nurseryman, New Windsor, Manchester.

CROTCH, Rev. William Robert (1799–1877)
b. Oxford 1799 *d*. Catherington, Hants 8 May 1877
MA Oxon 1826. Master, Taunton Grammar School, 1854. Vicar, Catherington, 1872. 'List of Fungi' (*Somerset Archaeol. Soc. Proc*. v.5, 1852, 132–56).

CROUCH, Mr. (*fl*. 1670s)
Seedsman, Golden Ball, without Bishopsgate, London.
W. A. Shaw *Calendar of Treasury Books, 1676–79*.

CROUCH, Charles (1855–1944)
Farmer at Mead Hook near Pulloxhill, Beds. Nephew of Rev. J. F. Crouch (*c*. 1809–1889). Botanist.
J. G. Dony *Fl. Bedfordshire* 1953, 22, 29–30.
Plants at BM(NH), Luton Museum.

CROUCH, Rev. James Frederick (1809–1889)
b. Cainhoe, Beds 1809 *d*. Pembridge, Hereford 1889
BA Oxon 1830. BD 1841. Rector, Pembridge, 1849. '*Cuscuta hassiaca*' (*Trans Woolhope Nat. Field Club* 1868, 122).
J. Bot. 1889, 209. W. H. Purchas and A. Ley *Fl. Herefordshire* 1899, preface vi. J. G. Dony *Fl. Bedfordshire* 1953, 22. *Trans. Woolhope Nat. Field Club* 1954, 250.
Plants at Hereford Museum.

CROUCH, Rev. William (1818–1846)
b. Cainhoe, Beds 1818 *d*. Ridgmount near Woburn, Beds 1846
MA Cantab 1844. Brother of Rev. J. F. Crouch (1809–1889). Had livings at Lidlington and Cainhoe, Beds. Had herb. of Bedford plants.
J. Bot. 1889, 209. *Bot. Soc. Exch. Club Br. Isl. Rep*. 1935, 59. *J. Bedfordshire Nat. Hist. Soc. Field Club* 1946, 50–52. J. G. Dony *Fl. Bedfordshire* 1953, 21–22.
Herb. at Luton Museum.

CROUCHER, George (1833–1905)
b. Dunbar, E. Lothian 1833 *d*. Ochtertyre, Perthshire 27 June 1905
Gardener to Sir W. K. Murray at Ochtertyre, 1857–1905.

CROUCHER, J. (*fl*. 1860s)
Kew gardener, 1869. Contrib. articles on *Cactaceae* to *Garden*.
Bot. Mag. 1869, t.5812.
Gasteria croucheri Baker.

CROW, Francis (*fl*. 1790s–1810s)
Of Faversham, Kent. MS. catalogue of fossil fruits from Sheppey collected 1790–1810 with 831 drawings by himself at BM(NH).

CROWCROFT, Thomas (*fl*. 1810s–*c*. 1860)
Nurseryman, Doncaster, Yorks.

CROWDER, Abraham (*c*. 1734–1831)
Gardener to William Dixon. Nurseryman, Loversall near Doncaster, Yorks.
Naturalist 1972, 55.

CROWDER, Anderson (1792–1873)
Nurseryman, Cagthorpe in Horncastle, Lincs. Son of William Crowder (–1836). Succeeded by William Ashley Crowder (1859–1922).

CROWDER, Henry (*c*. 1784–1850s)
Nurseryman, 59 Christchurch Terrace, Doncaster, Yorks.

CROWDER, Michael (*c*. 1787–1850s)
Nurseryman, 59 Christchurch Terrace, Doncaster, Yorks.

CROWDER, Rowland Wood (*fl*. 1780s–1800s)
Nurseryman, Doncaster, Yorks.

CROWDER, William (–1831)
Nurseryman, Doncaster, Yorks.

CROWDER, William (–1836)
Nurseryman, Cagthorpe in Horncastle, Lincs.

CROWDER, William Law (*fl*. *c*. 1780–1850s)
Nurseryman, 15 Union Street, Doncaster, Yorks. Son of R. L. Crowder (*fl*. 1780s–1800s) and father-in-law of Samuel Appleby (1806–1870), nurseryman at Doncaster.

CROWE, James (1750–1807)
b. Norwich, Norfolk 1750 *d*. Lakenham, Norfolk 26 Jan. 1807
FLS 1788. Surgeon. Studied mosses, fungi and willows. Had a salicetum. "A most excellent British botanist" (Sir J. E. Smith). Contrib. to J. Sowerby and J. E. Smith's *English Bot*. W. Withering acknowledges his help in his *Bot. Arrangement* ed. 2 v.1, 1787, preface. Notes on Norfolk plants in copy of W. Hudson's *Fl. Anglica* at Linnean Society.
Trans Linn. Soc. v.4, 1798, 222. P. Smith *Correspondence of Sir J. E. Smith* v.1, 1832, 17, 535–37. *Trans. Norfolk Norwich Nat. Soc*. 1912–13, 663–71.
Crowea Smith.

CROWLEY, Philip (1837–1900)
b. Alton, Hants 28 Aug. 1837 *d.* Croydon, Surrey 20 Dec. 1900
FLS 1883. FZS. Brewer. Treasurer, Royal Horticultural Society, 1890–99. Master, Gardeners' Company. President, Croydon Natural History Society, 1881–82.
Gdnrs Chron. 1900 ii 481 portr. *Gdnrs Mag.* 1900, 847 portr. *J. Hort. Cottage Gdnr* v.41, 1900, 582 portr. *Proc. Linn. Soc.* 1900–1, 42. *Proc. Trans. Croydon Microsc. Nat. Hist. Club* 1900–1, xlvii–xlviii portr. *J. R. Hort. Soc.* v.25, 1900, 158 portr.

CROWTHER, Frank (1906–1946)
b. York 1906 *d.* 11 April 1946
Educ. Imperial College of Science, London. Plant physiologist, Gezira Research Farm, 1928. Investigated factors determining growth and yield of cotton under irrigation. Chief Plant Physiologist, Dept. of Agriculture, Sudan.
Nature v.157, 1946, 795–96. R. Hill *Biogr. Dict. of Sudan* 1967, 106.

CROWTHER, Henry (1848–1937)
b. Leeds, Yorks 1848 *d.* 29 Nov. 1937
Lecturer in Botany, Leeds Mechanics Institute. Curator, Leeds Museum. Secretary, Leeds Philosophical and Literary Society.
Who was Who, 1929–1940 313.

CROWTHER, James (1768–1847)
b. Manchester 24 June 1768 *d.* Manchester 6 Jan. 1847
Weaver and porter. Botanist and entomologist. Discovered *Cypripedium calceolus* at Malham, Yorks. Friend of J. Dewhurst, E. Hobson and W. Roscoe. Contrib. to J. B. Wood's *Fl. Mancuniensis* 1840. Assisted J. Hull in his *British Fl.* 1799.
Bot. Zeitung 1847, 158–59. *Chambers's J.* 1847, 215–18. *Manchester Guardian* 13 Jan. 1847. J. Cash *Where there's a Will there's a Way* 1873, 77–89. *Gdnrs Mag.* 1896, 365. *D.N.B.* v.13, 245. *Trans. Liverpool Bot. Soc.* 1909, 66.

CROWTHER, John (1859–1930)
b. Eccleshill 1859 *d.* Grassington, Yorks 14 Feb. 1930
Pharmacist, Grassington. Studied archaeology and botany of Upper Wharfedale, Yorks. *Rambles around Grassington* 1920. *Lady's Slipper Orchid; its History in Upper Wharfedale* 1930.
Naturalist 1930, 171–72 portr.
Helleborine crowtheri Druce.

CROWTHER, William (*c.* 1867–1895)
d. Aburi, Gold Coast 16 March 1895
Kew gardener, 1889. Curator, Botanical Station, Aburi, Gold Coast, 1890–95. Visited West Indies in 1893 to observe cultivation of economic crops.
Kew Bull. 1894, 227–28; 1895, 121; 1897, 307. *J. Kew Guild* 1895, 42–43 portr. *Gdnrs Mag.* 1895, 172 portr.
Plants and MSS. at Kew.

CROWTHER, William (*fl.* 1870s)
Nurseryman, Cemetery Road, Doncaster, Yorks.

CROWTHER, *see* Crowder

CROZIER, George (1792–1847)
b. Eccleston, Lancs 1792 *d.* Hulme, Manchester 16 April 1847
Saddler. Friend of J. Dewhurst, J. Crowther and R. Buxton. 'Note on Manchester Carex' (*Phytologist* 1844, 843–44).
Manchester Guardian 21 April 1847. J. Cash *Where there's a Will there's a Way* 1873, 119–29. L. H. Grindon *Country Rambles* 1882, 168–74. *Trans. Liverpool Bot. Soc.* 1909, 66.
Herb. at Manchester University.

CRUCKSHANKS, Alexander
(*fl.* 1820s–1850s)
Collected plants in Chile, 1825–28, and Europe, 1831–56. 'Excursion from Lima to Pasco' (*Bot. Misc.* 1830–31, 168–205).
Bot. Mag. 1827, t.2785–87 *passim. R.S.C.* v.2, 100.
Plants, letters and portr. by Sir D. Macnee at Kew. Portr. at Hunt Library.
Cruckshanksia Hook. & Arn.

CRUEGER, Hermann (1818–1864)
b. Hamburg, Germany 11 Feb. 1818 *d.* San Fernando, Trinidad 28 Feb. 1864
To Trinidad in 1841 as apothecary. Director, Botanic Garden and Government Botanist, 1857, in succession to W. Purdie. Collected plants in Jamaica, Trinidad and Venezuela. *Outline of Fl. Trinidad* 1858.
I. Urban *Symbolae Antillanae* v.3, 1902, 33–34. *Amer. Orchid Soc. Bull.* 1958, 79. *R.S.C.* v.2, 110; v.7, 470.
Herb. at Trinidad. Plants at Kew, Berlin, Göttingen, etc.
Ornithocephalus cruegeri Reichenb. f.

CRUICKSHANK, James (*c.* 1813–1847)
b. Montrose, Angus *c.* 1813 *d.* Dumfries 3 Dec. 1847
Cryptogamist. 'List of Jungermanniae, etc. observed...in Neighbourhood of Dumfries' (*Phytologist* 1842, 257–59). Had herb. of mosses and hepatics.
Phytologist 1848, 33. *J. Bot.* 1898, 293.
Letters at Kew.

CRUMP, Edward (–1927/8)
Kew gardener, 1871. Accompanied J. D. Hooker and J. Ball on botanical expedition to Atlas Mountains, Morocco, 1871. Later became market gardener at Whitnash near Leamington Spa, Warwickshire.

CRUMP, William (1843–1932)
b. Pontesbury, Shropshire 1843 *d.* Malvern Link, Worcs 30 Dec. 1932
VMH 1897. Gardener to Earls of Beauchamp at Madresfield Court near Malvern, 1883–1919.
Gdnrs Chron. 1899 i 50–51 portr.; 1919 ii 40 portr.; 1933 i 17. *Garden* v.61, 1902, 122 portr. *Gdnrs Mag.* 1909, 291–92 portr. *J. Hort. Home Farmer* v.66, 1913, 533–34 portr.

CRUMP, William Bunting (1868–1950)
b. Scarborough, Yorks 26 April 1868
Educ. Oxford. Science master, Heath Grammar School until 1914. Cinema proprietor, Brighouse until 1921. *Fl. Parish of Halifax* (with C. Crossland) 1904. Editor, *Halifax Nat.*
Naturalist 1950, 68–70 portr.

CRYER, John (1860–1926)
b. Baildon, Yorks 29 July 1860 *d.* Shipley, Yorks 7 May 1926
Schoolteacher, Bradford; later Inspector of Science and Superintendent of Gardening under local education authority. Botanised in Yorkshire. Found *Polygala amarella* (*J. Bot.* 1903, 114–15). Specialised in *Hieracia.*
Bot. Soc. Exch. Club Br. Isl. Rep. 1926, 89. *J. Bot.* 1926, 220.
Herb. at Leeds University.

CUBITT, George Eaton Stannard (*c.* 1875–1966)
d. 17 Sept. 1966
Assistant Inspector-General, India. Conservator of Forests, Malay Peninsula, 1914–29. Collected plants in Burma, 1910, and in Malaya.
Gdns Bull. Straits Settlements 1927, 119. *Fl. Malesiana* v.1, 1950, 120. *Malayan For.* 1967, 164.
Plants at Kew, Singapore.

CUBITT, Miss M. (*fl.* 1840s)
British herb. at Kew.

CUFF, Mr. (*fl.* 1780s)
Of Teddington, Middx. "A gentleman of great zeal and assiduity in cultivating plants and promoting the science of botany." Gave plants to William Curtis.
Bot. Mag. 1801, t.505.

CULBERT, R. C. (*fl.* 1950s)
d. in climbing accident in Karakorum
Collected plants around Mt. Haramosh, Kashmir in 1957.
Pakistan J. For. 1967, 342.
Plants at BM(NH).

CULHAM, Arthur Brook (–1948)
d. Aug. 1948
Kew gardener, 1909. Curator, Dept. of Agriculture, S. Nigeria, 1910. Curator, Dept. of Agriculture, Gold Coast, 1915–21.
Kew Bull. 1910, 197; 1914, 345. *J. Kew Guild* 1949, 786 portr.

CULHAM, William (*fl.* 1840s)
Nurseryman, Melton, Woodbridge, Suffolk.

CULLEN, Charles Sinclair (*fl.* 1800s)
Letter from Cullen in Sir J. E. Smith's correspondence at Linnean Society contains list of plants found on N.W. coast of Devon in 1806.
W. K. Martin and G. T. Fraser *Fl. Devon* 1939, 771.

CULLEN, William (1785–1862)
b. 17 May 1785 *d.* Allepey, Madras 1 Oct. 1862
Lieut.-General, Royal Artillery. Entered East India Company, 1804. Resident at Travancore. Meteorologist. Studied economic botany.
R. Wight *Icones Plantarum Indiae Orientalis* v.5, 1851, t.1761. E. G. Balfour *Cyclop. of India* ed. 2 v.1, 1871, 419.
Cullenia Wight.

CULLEN, William Henry (*fl.* 1830s–1840s)
MD St. Andrews 1837. Of Sidmouth, Devon. *Fl. Sidostiensis* 1849.

CULLIN, Daniel (*fl.* 1750s)
Nurseryman, Dublin.
Irish For. 1967, 47.

CULLINGFORD, William (*fl.* 1850s)
Nurseryman, Balls Pond Road, Islington, London.

CULLIS, John (*fl.* 1810s–1849)
Nurseryman, Ranelagh Gardens, Leamington Spa, Warwickshire.

CULLUM, Rev. Sir John (1733–1785)
b. Hawsted, Suffolk 21 June 1733 *d.* Hawsted 9 Oct. 1785
BA Cantab 1756. FRS 1775. Rector, Hawsted, 1772; Great Thurlow, 1774. Discovered *Veronica verna. History and Antiquities of Hawsted and Hardwick* 1774 (with list of plants). 'On Growth of Cedars in England' (*Gent. Mag.* 1779, 138). 'Yews in Churchyards' (*Gent. Mag.* 1779, 578). Projected a new 'Fl. Anglicana' which was never published.
D. Turner and L. W. Dillwyn *Botanist's Guide through England and Wales* 1805, 554. *Gent. Mag.* 1813 i 551. W. M. Hind *Fl. Suffolk* 1889, 105, 476–77. *D.N.B.* v.13, 283.
Plants at BM(NH). Portr. by Angelica Kauffmann at Hardwick.
Cullumia R. Br.

CULLUM, Sir Thomas Gery (1741–1831)
b. Hardwick House, Suffolk 30 Nov. 1741 *d.* Bury St. Edmunds, Suffolk 8 Sept. 1831

MRCS 1800. FRS 1787. FLS 1790. Brother of Rev. Sir J. Cullum (1733–1785). Practised medicine at Bury St. Edmunds. *Fl. Anglicae Specimen* 1774. Catalogue of Bury plants in E. Gillingwater's *Hist.... Account of St. Edmund's Bury...* 1804. Contrib. plant records for several counties to D. Turner and L. W. Dillwyn's *Botanist's Guide through England and Wales* 1805.

P. Smith *Correspondence of Sir J. E. Smith* v.2, 1832, 299–300. W. M. Hind *Fl. Suffolk* 1889, 477–78. F. H. Davey *Fl. Cornwall* 1909, xxxvi. *D.N.B.* v.13, 284.

Cullumia R. Br.

CULPEPER, Nicholas (1616–1654)
b. London 18 Oct. 1616 *d.* 10 Jan. 1654

Apothecary, herbalist and astrologer. Established medicinal garden in Red Lion Street, Spitalfields. *Physicall Directory* 1649 portr. *English Physitian* 1652 (frequently republished in new editions).

W. Cole *Art of Simpling* 1656, 76–78. R. Pulteney *Hist. Biogr. Sketches of Progress of Bot. in England* v.1, 1790, 180–81. *Gent. Mag.* 1797, 390–91, 472. *D.N.B.* v.13, 286–87. G. C. Druce *Fl. Buckinghamshire* 1926, lxxi. *Ann. Med. Hist.* 1931, 394–403. *Pharm. J.* v.72, 1931, 98–99; v.89, 1939, 113–14 portr. A. Arber *Herbals* 1938, 261–63 portr. *J. New York Bot. Gdn* v.41, 1940, 158–66. *J. Hist. Med.* v.11, 1956, 156–65; v.17, 1962, 152–67.

Portr. at Kew, Hunt Library.

CULVERWELL, William (–1910)
d. 19 June 1910

Hybridised vegetables and soft fruits.

Gdnrs Chron. 1899 ii 42 portr.; 1910 i 434 portr.

CUMING, Hugh (1791–1865)
b. West Alvington, Devon 14 Feb. 1791 *d.* London 10 Aug. 1865

FLS 1832. To Buenos Aires, 1819; in 1822 to Valparaiso, Chile where he became a dealer in natural history objects. Built himself a yacht, the 'Discover', sailing on his first expedition, 1827–28, via Juan Fernandez, Easter Island, etc. to Tahiti and Pitcairn. 2nd voyage, 1828–30, along W. coast of S. America. In England, 1831–35. 3rd voyage to Philippines, 1836–40. Notable shell and plant collector. Pioneer in shipping live orchids from Manila to England.

Gdnrs Mag. 1840, 116. *J. Bot.* (Hooker) 1841, 392–422. *London J. Bot.* 1844, 658–66; 1845, 3–11. A. Lasègue *Musée Botanique de B. Delessert* 1845, 258, 269–70. L. Reeve *Portr. of Men of Eminence in Literature, Sci. and Art* 1864, 41–46. *Gdnrs Chron.* 1865, 823–24. *J. Bot.* 1865, 325–26; 1866, 57–60, 347–49. *Bot. Zeitung* 1866, 31–32. *Proc. Linn. Soc.* 1865–66, lvii–lix. S. Vidal y Soler *Phanerogamae Cumingianae Philippinarum* 1885, xii–xv. *J. Conchology* v.8, 1895, 59–75 portr. *D.N.B.* v.13, 295–96. *Bull. Bureau Agric. Manila* no. 4, 1903, 23–26. I. Urban *Symbolae Antillanae* v.3, 1902, 34–35. *Bureau Govt. Laboratories, Manila* no. 35, 1905, 69–77. *Kew Bull.* 1908, 116–19. *Philippine J. Sci.* v. 30, 1926, 153–84 portr. *Occas. Papers Bernice P. Bishop Mus.* 1940, 81–90 portr. *Occas. Papers Mollusks Harvard* v.1 (3), 1945, 17–28 portr. *Fl. Malesiana* v.1, 1950, 120–22 portr. *Amer. Orchid Soc. Bull.* 1961, 357–59. S. P. Dance *Shell Collecting* 1966, 146–70 portr. *Taxon* 1970, 520. M. A. Reinikka *Hist. of the Orchid* 1972, 141–44 portr. *R.S.C.* v.2, 103; v.14, 426.

Plants at Kew, BM(NH), Oxford, etc. Letters at Kew. Portr. at Hunt Library.

Cumingia Vidal.

CUMMING, Linnaeus (1843–1927)
b. Cambridge 1 May 1843 *d.* Kilsby, Northants April 1927

MA Cantab. Mathematics and Science master, Rugby, 1875. Made large collection of *Rubi* which he gave to G. C. Druce.

Bot. Soc. Exch. Club Br. Isl. Rep. 1928, 701–2. H. J. Riddelsdell *Fl. Gloucestershire* 1948, cxxxi.

Plants at Rugby School (mostly destroyed) and Oxford.

CUMMINGS, Bruce Frederick (1889–1919)
d. Gerrards Cross, Bucks 22 Oct. 1919

'Rousseau as Botanist' (*J. Bot.* 1916, 80–84).

Selborne Mag. v.21, 1910, 2.

CUMMINS, Henry Alfred (1864–1939)
b. Cork 8 March 1864 *d.* Chelsea, London 1 Jan. 1939

MD Ireland. FLS 1893. Surgeon, Royal Army Medical Corps, 1888–1901, 1914–19. Collected plants in Sikkim and on borders of Bhutan, 1888, and on Ashanti expedition, 1895. Assistant, Kew Herb., 1906–9. Professor of Botany, University College, Cork, 1909.

Kew Bull. 1898, 65–82, 133–34; 1909, 391; 1939, 33–34. *Proc. Linn. Soc.* 1938–39, 238–39. *Who was Who, 1929–1940* 315–16.

Plants at Kew, Berlin.

CUNDALL, James Henry (1808–1884)
b. Trowbridge, Wilts 8 April 1808 *d.* Bristol 19 April 1884

Of Clifton, Bristol. Had herb. *Every-day Book of Natural History* 1866.

J. W. White *Fl. Bristol* 1912, 81–82, 109. H. J. Riddelsdell *Fl. Gloucestershire* 1948, cxxv.

CUNDY, Charles (1849–1933)
b. Raydon 1849 *d.* Sudbury, Suffolk 20 July 1933

Kew gardener, 1878–81. Nurseryman, Sudbury. President, Kew Guild, 1922–23.

J. Kew Guild 1922, 65 portr.; 1933, 285–86. *Gdnrs Chron.* 1933 ii 89.

CUNINGHAME, James *see* Cunningham, J.

CUNNACK, James (1831–1886)
b. Helston, Cornwall 27 Dec. 1831 *d.* Helston 11 May 1886

Bookseller. Correspondent of H. C. Watson. Found *Hypericum undulatum.* Had herb.

H. C. Watson *Topographical Bot.* 1883, 542. *J. Bot.* 1891, 98. F. H. Davey *Fl. Cornwall* 1909, liii portr.

Plants at Oxford.

CUNNINGHAM, Alexander (*fl.* 1780s)

Nurseryman, 4 Lisson Grove, Paddington, London.

CUNNINGHAM, Allan (1791–1839)
b. Wimbledon, Surrey 13 July 1791 *d.* Sydney 27 June 1839

Employed at Kew on Aiton's *Hortus Kewensis c.* 1808. Collected plants for Kew, 1814–31. To Brazil with James Bowie, 1814–16. At Sydney, 1816–26, from which he made voyages, 1817–21. In New Zealand, 1826. Returned to England, 1831. Succeeded his brother, Richard, as Superintendent of Botanic Garden, Sydney, 1836–38. 'Botany of Blue Mountains' (B. Field *New South Wales* 1825). 'Fl. Insularum Novae Zelandiae' (*Ann. Nat. Hist.* 1838, 210–16, 376–81, 455–62).

Bot. Mag. 1834, t.3313, 3323. *Bot. Register* 1840 Misc. Notes 1–3. *J. Bot.* (Hooker) 1842, 231–320 portr. *London J. Bot.* 1842, 107–28, 263–92; 1846, 661–62. *Gdnrs Mag.* 1843, 295. A. Lasègue *Musée Botanique de B. Delessert* 1845, 279, 284. J. D. Hooker *Fl. Tasmaniae* v.1, 1859, cxiv–cxvii. *Gdnrs Chron.* 1881 ii 440. *Proc. Linn. Soc.* 1840, 67–68; 1888–89, 34. *Revue Historique Litteraire de l'Isle Maurice* 1894, 113–18. J. H. Maiden *Allan Cunningham* 1903. *Kew Bull.* 1891, 309–10; 1906, 214–16. *Proc. Linn. Soc. N.S.W.* 1900, 770–71, 789. *D.N.B.* v.13, 308. C. F. P. Martius *Fl. Brasiliensis* v.1 (1), 1906, 14–15. *Rep. Austral. Assoc. Advancement Sci.* 1907, 194. *J. Proc. R. Soc. N.S.W.* 1909, 123–39. J. H. Maiden *Sir Joseph Banks* 1909, 141–55. *J. Linn. Soc.* v.45, 1920, 194–95. I. Lee *Early Explorers in Australia* 1925 (with lengthy extracts of his journal) portr. *Victorian Nat.* 1926, 163–69. *Austral. Pharm. Notes News* 1938, 266–67 portr. *Turnbull Library Rec.* no. 3, 1941, 5–10. *J. Proc. R. Austral. Hist. Soc.* v.21, 315–20; v.46, 1960, 323–42. *Fl. Malesiana* v.1, 1950, 122–23. *Austral. Genealogist* v.8, 1956, 72–79. *J. Scone Upper Hunter Hist. Soc.* 1959, 1–22; 1961, 203–9. H. N. Clokie *Account of Herb. Dept. Bot., Oxford* 1964, 151–52. *Austral. Dict. Biogr.* v.1, 1966, 265–67. A. M. Coats *Quest for Plants* 1969, 215–22 portr. W. G. McMinn *Allan Cunningham* 1970 portr. E. H. J. and G. E. E. Feeken *Discovery and Exploration of Australia* 1970, 86–90 portr.

Plants at Kew, BM(NH), Oxford. MS. journals at Kew, BM(NH), Mitchell Library Sydney. Portr. by D. Macnee at Kew, by J. E. H. Robinson at Linnean Society. Portr. at Hunt Library. Flower drawings from his expeditions in 1816 and 1818 at Herb. Sydney.

Alania Endl.

CUNNINGHAM, David Douglas (1843–1914)
b. Prestonpans, E. Lothian 29 Sept. 1843 *d.* Torquay, Devon 31 Dec. 1914

MD Edinburgh 1867. FRS 1889. FLS 1876. Pathologist. Bengal Medical Service, 1868–71. 'Mycoidea parasitica' (*Trans. Linn. Soc. Bot.* v.1, 1879, 301–16).

Nature v.94, 1915, 536–37. *Proc. Linn. Soc.* 1914–15, 24–26. *Proc. R. Soc. B.* v.89, 1916, xv–xx. D. G. Crawford *Hist. Indian Med. Service, 1600–1913* v.2, 1914, 160. *R.S.C.* v.7, 470; v.9, 617; v.12, 179; v.14, 428.

Cunninghamella Matruchot.

CUNNINGHAM, George (*fl.* 1810s–1830s)

Nurseryman, Liverpool. In partnership with Johnson at 71 Paradise Street and Wavertree, 1811–24. Later at 75 Paradise Street with his son, George.

CUNNINGHAM, George (*c.* 1800–1891)
d. 25 Feb. 1891

Nurseryman, Broad Green Road, Liverpool.

Gdnrs Chron. 1891 i 310–11.

CUNNINGHAM, J. F. (*fl.* 1890s)

Secretary, British Central Africa Administration. Plants collected in 1890s in British Central Africa at Kew.

Kew Bull. 1907, 234.

CUNNINGHAM, James (–*c.* 1709)

FRS 1699. Surgeon at East India Company Factory at Chusan and Amoy, 1698–1703. In Makassar, 1705. Imprisoned in Cochin-China for 2 years. Trader in Batavia, 1707. Sent plants from St. Helena, Ascension, Cape, Malacca and China to J. Ray, J. Petiver, L. Plukenet and Sir H. Sloane. "Incomparabilis botanicus et amicus noster" (L. Plukenet *Amaltheum Botanicum* 1705, 75). Contrib. to *Philos. Trans. R. Soc.*

Philos. Trans. R. Soc. v.21, 1699, 295–300. R. Pulteney *Hist. Biogr. Sketches of Progress of Bot. in England* v.2, 1790, 59–62. A. Hamilton *New Account of East Indies* v.2, 1744, 144–45, 205. *J. Bot.* 1882, 249–50; 1883, 12. E. Bretschneider *Hist. European Bot. Discoveries in China* 1898, 31–44. *D.N.B.* v.13, 312–13. E. J. L. Scott *Index to Sloane Manuscripts* 1904, 132. *J. Linn. Soc.* v.45, 1920, 34. *Hong Kong Nat.* v.5, 1934, 267–68. E. H. M. Cox *Plant-hunting in China* 1945, 40–42. *Fl. Malesiana* v.1, 1950, 123; v.5, 1958, 36. *Proc. Amer. Antiq. Soc.* 1952, 268–69. J. E. Dandy *Sloane Herb.* 1958, 117–22.

Letters to Petiver and plants at BM(NH).

Cunninghamia R. Br.

CUNNINGHAM, James (1784–1851)
b. Carluke, Lanarkshire 1784 *d.* Edinburgh 22 Oct. 1851

Nurseryman, Comley Bank near Edinburgh. First to take up hybridisation of rhododendrons in Scotland.

Gdnrs Chron. 1851, 695.

CUNNINGHAM, John (*c.* 1798–1878)
d. Paisley, Renfrew 28 March 1878

Raised many new auriculas.

Florist Pomologist 1878, 112.

CUNNINGHAM, Richard (1793–1835)
b. Wimbledon, Surrey 12 Feb. 1793 *d.* murdered by natives in Queensland *c.* 15 April 1835

Brother of A. Cunningham (1791–1839). Employed at Kew by W. T. Aiton as clerk and assisted in ed. 2 of *Hortus Kewensis* (*J. Bot.* 1912 Supplt 3, 12). Colonial Botanist and Superintendent, Botanic Garden, Sydney, 1833–35. Botanist on Darling River Expedition, 1835. (T. Mitchell *Three Expeditions into Interior of Eastern Australia* v.1, 1838, 147, 168–98).

Companion Bot. Mag. v.2, 1827, 210–21 portr. *Gdnrs Mag.* 1836, 326–28, 386–88; 1837, 619. *Mag. Zool. Bot.* 1837, 210–12. A. Lasègue *Musée Botanique de B. Delessert* 1845, 498. J. D. Hooker *Fl. Tasmaniae* v.1, 1859, cxvi, cxx, cxxiii. *Notes and Queries* v.4, 1863, 304. *Gdnrs Chron.* 1881 ii 440. *Proc. Linn. Soc. N.S.W.* 1900, 771–72, 789. J. H. Maiden *Richard Cunningham* 1903. *Kew Bull.* 1906, 212–14. *J. Proc. R. Soc. N.S.W.* 1908, 99–100. *D.N.B.* v.13, 317. *Austral. Dict. Biogr.* v.1, 1966, 268–69. A. M. Coats *Quest for Plants* 1969, 220–21, 232–33.

Plants, letters, and pencil portr. at Kew. Portr. at Hunt Library.

Cunninghamia R. Br.

CUNNINGHAM, Robert Oliver (1841–1918)
b. Prestonpans, E. Lothian 1841 *d.* Paignton, Devon 14 July 1918

MD Edinburgh. FLS 1870. Professor of Natural History, Belfast, 1871–1902. Collected plants in Magellan and Patagonia, 1866–69. *Notes on Natural History of Strait of Magellan...made during Voyage of H.M.S. 'Nassau'* 1871. 'Pleiotaxy in *Philesia*' (*J. Linn. Soc.* v.11, 1871, 477–79).

J. Bot. 1868, 60–63. *Irish Nat.* 1918, 128–29. *Nature* v.101, 1918, 390. *Tuatara* 1970, 71–72.

Plants at Kew.

CUPER, Boyder (*fl.* 1690s)

Gardener to Duke of Norfolk. Laid out pleasure ground at Lambeth, *c.* 1691, known as Cuper's Gardens.

W. Wroth *London Pleasure Gdns* 1896, 247–48.

CURDIE, Daniel (1810–1884)
b. Sliddery, Arran 9 Jan. 1810 *d.* near Camperdown, Victoria 22 Feb. 1884

MA Glasgow 1832. MD Edinburgh 1838. Pupil of W. J. Hooker. To Australia, 1839; returned to Scotland, 1851; back in Australia, 1854. Doctor at 'Tandarook' near Camperdown, Victoria. Sent Australian algae to W. H. Harvey and W. J. Hooker.

Ann. Mag. Nat. Hist. v.15, 1855, 333–34. W. H. Harvey *Phycologica Australica* v.1, 1858, xxxix. *Austral. Med. J.* 1884, 144. *Victorian Nat.* 1908, 105–6; 1949, 108. M. Kiddle *Men of Yesterday; Social Hist. of Western District of Victoria, 1834–1890* 1961, 92–95.

Plants at Kew.

Curdiaea Harvey.

CURL, S. M. (–1890)

MD. Of Rangitikei, New Zealand. Member of Wellington Philosophical Society, 1876–90. Notes on grasses, fodder plants, etc. in *Trans. N.Z. Inst.* v.9–v.13, 1876–81. "A good authority on botanical matters."

H. C. Field *Ferns of N.Z.* 1890, 145–46. *Trans. N.Z. Inst.* 1903, 355. *R.S.C.* v.9, 619; v.14, 433.

CURNOW, Richard (–1896)
d. at sea on voyage from Colombia 25 Aug. 1896

Collected orchids for Hugh Low and Co. in Philippines and Java.

Gdnrs Chron. 1896 ii 345. *Fl. Malesiana* v.1, 1950, 123.

CURNOW, William (*c.* 1809–1887)
d. Newlyn, Cornwall 24 Jan. 1887

Market-gardener. Cryptogamist. Supplied mosses to L. Rabenhorst for *Bryotheca Europaea*. Contrib. to *Phytologist* 1844. W. Cornwall mosses in *Penzance Nat. Hist. Soc. Rep.* 1862–65, 56–64; 1881–82, 117–20.

Notts. Midland Nat. 1887, 77. *J. Bot.* 1888, 128. *Penzance Nat. Hist. Soc. Rep.* 1887–88, 309. F. H. Davey *Fl. Cornwall* 1909, xliv–xlv portr. *R.S.C.* v.12, 179.

Herb. of *Hepaticae* at Mansfield. Letters at BM(NH).

Fissidens Curnowii Mitten.

CURREY, Frederick (1819–1881)
b. Norwood, Surrey 19 Aug. 1819 *d.* Blackheath, London 8 Sept. 1881

BA Cantab 1841. FRS 1858. FLS 1856. Mycologist. *Botany of the District...Rivers Cray, Ravensbourne and Thames...of Greenwich Natural History Club* 1858. Translated H. Schacht's *Microscope and its Application to Vegetable Anatomy and Physiology* 1855 and W. F. B. Hofmeister's *Germination...of Higher Cryptogamia* 1862. Edited ed. 2 of C. D. Badham's *Treatise on Esculent Funguses of England* 1864. 'Fungi of Greenwich' (*Phytologist* 1854, 121–23, 144–46). 'Nardoo Plants of Australia' (*J. Bot.* 1863, 161–67).

Gdnrs Chron. 1881 ii 412. *J. Bot.* 1881, 310–12. *Nature* v.24, 1881, 485–86. *Proc. Linn. Soc.* 1881–82, 59–60; 1888–89, 34–35.

Crayon portr. and 7 vols of MSS. on fungi at Linnean Society. Fungi at Kew. Letters at BM(NH).

CURROR, A. B. (*fl.* 1830s–1840s)

MD. Royal Navy. Collected plants in Angola, 1839–43.

Plants and letters at Kew.

Curroria Planch.

CURROR, John R. (*fl.* 1870s)

Gardener.

Gdnrs Chron. 1876 ii 328 portr.

CURTIS, Charles (1853–1928)
b. Barnstaple, Devon 1853 *d.* Barnstaple 16 Aug. 1928

Plant collector for Messrs J. Veitch and Sons, 1878–84. Superintendent, Botanic Gardens, Penang, 1884–1903. Assisted H. N. Ridley in some of the earliest experiments in tapping of Para rubber. Collected plants in Madagascar and Mauritius, 1878; Borneo, Sumatra, Java and Moluccas, 1880; Malay Peninsula, 1880–1903. 'Catalogue of Flowering Plants and Ferns...in...Penang' (*J. Asiatic Soc. Straits Branch* 1894, 67–163). Contrib. to *Agric. Bull. Straits Settlements.*

Gdnrs Mag. 1901, 795 portr. J. H. Veitch *Hortus Veitchii* 1906, 85–86. *Gdnrs Chron.* 1928 ii 159. *J. Bot.* 1881, 366–68; 1928, 332–33. *Kew Bull.* 1928, 383. *Orchid Rev.* 1928, 320. *Gdns Bull. Straits Settlements* 1927, 427. *J. Thailand Res. Soc. Nat. Hist. Supplt.* v.12 (1), 1939, 20–21. *J. R. Hort. Soc.* 1948, 288–89. *Fl. Malesiana* v.1, 1950, 124–25 portr. M. A. Reinikka *Hist. of the Orchid* 1972, 266–67.

Plants at Kew, BM(NH), Kuala Lumpur, Penang, Singapore.

Cirrhopetalum curtisii Hook. f.

CURTIS, Charles Henry (1869–1958)
b. Wimbledon, Surrey 3 June 1869 *d.* Brentford, Middx 24 March 1958

FLS 1925. VMH 1933. VMM 1957. Kew gardener 1889. Assistant Superintendent, Royal Horticultural Society Gardens, Chiswick. Sub-editor, then Editor, *Gdnrs Mag.* 1892–1917. Editor, *Gdnrs Chron.* 1918–50. Editor, *Orchid Rev.*, 1933–58. Secretary, National Chrysanthemum Society. *Sweet Peas and their Cultivation* 1908. *Book of the Flower Show* 1910. *Orchids for Everyone* 1910. *Book of Topiary* (with W. Gibson) 1904. *Annuals, Hardy and Half-hardy* 1912. *Orchids, their Distribution and Cultivation* 1950.

J. Kew Guild 1917, 335 portr.; 1934, 336; 1958, 583–85, 590–91 portr. *Gdnrs. Chron.* 1931 i 2 portr.; 1958 i 231. *J. R. Hort. Soc.* 1948, 192–93. *Daffodil Tulip Yb.* 1954, 9–10 portr. *Amer. Orchid Soc. Bull.* 1958, 371–72. *Orchid Rev.* 1958, 108–9. *Times* 25 March 1958.

Portr. at Hunt Library.

CURTIS, Charles M. (*c.* 1795–1839)
b. Norwich, Norfolk *c.* 1795 *d.* 16 Oct 1839

Brother of J. Curtis (1791–1862). Botanical artist. Contrib. plates to N. Wallich's *Plantae Asiaticae Rariores* 1830–32, J. F. Royle's *Illustrations of Botany...of Himalayan Mountains* 1839, J. Stephenson's *Medical Botany* 1827–31, R. Brown's *Miscellaneous Botanical Works* 1866–68, *Bot. Mag.*, *Pomological Mag.*

Art Union Nov. 1839, 167. *Gdnrs Chron.* 1933 ii 292.

CURTIS, Georgiana (*fl.* 1840s)

Drew t.3928 for *Bot. Mag.*

CURTIS, Henry (*c.* 1819–1889)
d. St. Helier, Jersey, Channel Islands 26 Nov. 1889

Grandson of W. Curtis (1746–1799). Head of Curtis, Sanford and Co., Devon Rosary, Torquay. *Beauties of the Rose* 1850–53 2 vols.

Gdnrs Chron. 1889 ii 639. *J. Hort. Cottage Gdnr* v.19, 1889, 486.

CURTIS, John (1791–1862)
b. Norwich, Norfolk 3 Dec. 1791 *d.* Islington, London 6 Oct. 1862

FLS 1822. Entomologist and artist. In youth did botanical drawing and engraving for Horticultural and Linnean Societies.

His *British Entomology* 1824–39 includes many plant drawings.

Gdnrs Chron. 1842, 774; 1862, 983; 1933 ii 292; v.164, 1968, 14–17. *Ipswich Mus. Portr. Ser.* 1852. *Proc. Linn. Soc.* 1862–63, xxxv–xli. *D.N.B. Supplt. 1* v.2, 99–100. W. Blunt *Art of Bot. Illus.* 1955, 186, 223, 238.

CURTIS, John Wright (1814–1864)

Of Alton, Hants. Had herb. which was acquired by C. E. Salmon.

Gdnrs Chron. 1921 i 150.

CURTIS, Maria (*fl.* 1830s)

Drew t.3412 and 3483 for *Bot. Mag.*

CURTIS, Sir Roger Colin Molyneux (1886–1954)

b. 12 Sept. 1886 *d.* Melbourne, Derby 11 Jan. 1954

BA Oxon 1910. H.M. Inspector of Schools. Treasurer, Botanical Society of British Isles, 1934–37. 'Adventive Fl. of Burton-upon-Trent' (*Bot. Soc. Exch. Club Br. Isl. Rep.* 1930, 465–69).

Proc. Bot. Soc. Br. Isl. 1954, 279.

CURTIS, Samuel (1779–1860)

b. Walworth, Surrey 29 Aug. 1779 *d.* 'La Chaire', Rozel, Jersey, Channel Islands 6 Jan. 1860

FLS 1810. First cousin and later son-in-law of W. Curtis (1746–1799). About 1800 acquired Walworth nursery belonging to Goring and Wright. Partner with Sturge to *c.* 1805; then Milliken to *c.* 1825. Also established nursery at Glazenwood, Coggeshall, Essex. Proprietor of *Bot. Mag.*, 1801–46. *Beauties of Flora* 1806–20. *Monograph on Genus Camellia* 1819.

Illus. London News 1 July 1843. *Cottage Gdnr* v.23, 1860, 335–36. *Proc. Linn. Soc.* 1860, xxii–xxiii. *Gdnrs Chron.* 1887 i 479, 671; 1933 ii 292, 417–18. *J. Bot.* 1899, 183–84. *D.N.B.* v.13. 349. *J. R. Hort. Soc.* 1933, 324–28 portr. W. H. Curtis *William Curtis, 1746–1799* 1941, 114–27 portr. *Trans. London Middlesex Archaeol. Soc.* v.24, 1973, 191.

Portr. at Kew.

CURTIS, Thomas (1749–)

Brother of William Curtis (1746–1799). Second proprietor of *Bot. Mag.*, 1799–1800.

CURTIS, William (1746–1799)

b. Alton, Hants 11 Jan. 1746 *d.* Brompton, London 7 July 1799

FLS 1788. Grew British plants on plot of ground at Grange Road, Bermondsey; established botanic garden at Higler's Lane, Lower Marsh, Lambeth, 1777; had nursery at Queen's Elm, Brompton, 1789–1811; Sloane Street, Chelsea, 1812–23. Praefectus Horti, Chelsea Physic Garden, 1772–77. *Fl. Londinensis* 1777–98 2 vols. *Enumeration of British Grasses* 1787; ed. 6 1824. *Hortus Siccus Gramineus* 1802 2 vols. *Lectures on Bot.* 1805 3 vols (includes 33p. sketch of life of Curtis by R. J. Thornton). Founded *Bot. Mag.*, 1787.

Gent. Mag. 1799 ii 628–29 635–39. *Ann. Bot.* v.1, 1805, 189–90. G. W. Johnson *Hist. of English Gdning* 1829, 244–46. S. Felton *Portr. of English Authors on Gdning* 1830, 184–85. H. Trimen and W. T. T. Dyer *Fl. Middlesex* 1869, 393–97. *J. Hort. Cottage Gdnr* v.31, 1876, 239–40 portr., v.46, 1903, 490–91 portr. R. H. Semple *Mem. Bot. Gdn at Chelsea* 1878, 104–9. *J. Bot.* 1881, 309–10; 1895, 112–14; 1899, 390–95; 1916, 153–54. *Gdnrs Chron.* 1887 i 479 *et seq.*; 1927 i 140; 1933 ii 292; 1946 i 19–20. Friends Institute *Biogr. Cat.* 1888, 156–60. *Nature* v.44, 1891, 86–87; v.157, 1946, 14–16. *D.N.B.* v.13, 349. *Bot. Soc. Exch. Club Br. Isl. Rep.* 1918, 412–14. *Curtis's Bot. Mag. Dedications, 1827–1927* xvii–xxi portr. W. H. Curtis *William Curtis, 1746–1799* 1941 portr. *Chronica Botanica* 1945, 75–76. *London Nat.* 1945, 3–12. *Proc. Linn. Soc.* 1945–46, 13–20 portr. *Endeavour* 1946, 13–17 portr. *J. R. Hort. Soc.* 1946, 98–100, 124–29 portr. W. Blunt *Art of Bot. Illus.* 1955, 183–85, 189–90. F. A. Stafleu *Taxonomic Literature* 1967, 90–97. W. Curtis *Short Hist. of Brown-tail Moth* 1782 (1969 reprint with biographical introduction).

Drawings at Kew. Portr. by Wright at Royal Horticultural Society. Portr. at Hunt Library. Letters at Society of Friends, London.

Curtisia Aiton.

CURTIS, William (*fl.* 1830s)

Drew t.3415 and 3530 for *Bot. Mag.*

CURWEN, Julia (*fl.* 1890s)

Of Roewath, Cumberland. Contrib. to W. Hodgson's *Fl. Cumberland* 1898.

CUSACK, Mrs. M. E. (*fl.* 1880s–1890s)

Collected plants in Colorado, 1888–90. T. D. A. Cockerell published several of her plant records in C. R. Orcutt's *West Amer. Scientist* (especially v.6, 1889, 134–36).

Bios v.8, 1937, 14–15. J. Ewan *Rocky Mountain Nat.* 1950, 192.

Herb. at Kew. Plants at Oxford.

CUSHING, John (–1819/20)

b. Ireland

Foreman gardener at nursery of Messrs. Lee and Kennedy, Hammersmith. *Exotic Gardener* 1812; ed. 3 1826.

Gdnrs Chron. 1963 i 87.

CUST, Lady Mary Anne (1800–1882)
d. 10 July 1882
Wife of Sir E. Cust (1794–1878), General and military historian. 205 drawings of plants and fishes made during voyage to West Indies in 1839, and to Madeira in 1866 now at BM(NH).
D.N.B. v.13, 356.

CUTBUSH, James (1827–1885)
b. Ashford, Kent 1827 *d.* Highgate, London 1 Aug. 1885
Nurseryman, Highgate, Barnet and Finchley, London.
Garden v.28, 1885, 160. *Gdnrs Chron.* 1885 ii 187.

CUTBUSH, William (*fl.* 1820s–1850s)
Nurseryman, West Hill, Highgate, London. Successor to nursery of William Bowstread.

CUTHBERT, George (1839–1914)
b. 21 July 1839 *d.* 8 July 1914
In 1875 joined the nursery at Southgate belonging to his uncle, Richard Cuthbert on whose death in 1903 the nursery passed to G. Cuthbert.
Gdnrs Chron. 1914 ii 43 portr., 63–64.

CUTHBERT, James (*fl.* 1790s)
Established nursery at Southgate, London in 1797.

CUTHBERT, John (–1830)
d. Rotherham, Yorks 7 Sept. 1830
Seedsman, Rotherham.
Gdnrs Mag. 1831, 512.

CUTHBERTSON, William (*c.* 1859–1934)
b. Penicuik, Midlothian *c.* 1859 *d.* Edinburgh 7 March 1934
VMH 1914. Senior partner, Messrs Dobbie and Co., Edinburgh. Director, Scottish Plant Breeding Station. President, National Sweet Pea Society, 1907. *Pansies, Violas and Violets* 1910. *Sweet Peas and Antirrhinums* 1915; ed. 2 1919. *Curtis's Bot. Mag. Dedications, 1827–1927* (with E. Nelmes). Contrib. to *Gdnrs Chron.*
Gdnrs Mag. 1907, 893–94 portr. *Garden* 1919, ii–iii portr. *Gdnrs Chron.* 1925 ii 402 portr.; 1934 i 169.

CUTLER, Catherine (–1866)
d. Exmouth, Devon 15 April 1866
Of Sidmouth, Devon. Algologist.
J. Bot. 1866, 238–39. *Bot. Zeitung* 1866, 268.
Algae at BM(NH).
Cutleria Grev.

CYPHER, James (1827–1901)
b. Tetbury, Glos 31 Dec. 1827 *d.* Cheltenham, Glos 1 Nov. 1901
Nurseryman, Exotic Nurseries, Queen's Road, Cheltenham. Specialised in orchids, particularly dendrobiums.
Garden v.60, 1901, 323–24 portr. *Gdnrs Chron.* 1901 ii 347–48 portr. *Gdnrs Mag.* 1901, 720 portr. *J. Hort. Cottage Gdnr* v.43, 1901, 430–31 portr. *Orchid Rev.* 1901, 356.

CYPHER, John James (*c.* 1854–1928)
d. Cheltenham, Glos 22 Jan. 1928
VMH 1910. Nurseryman, Cheltenham. Specialised in orchids and indoor plants. Introduced present display techniques for exhibiting large groups of plants.
Gdnrs Mag. 1910, 969–70 portr. *J. Hort. Cottage Gdnr* v.63, 1911, 539 portr. *Gdnrs Chron.* 1928 i 71 portr. *Orchid Rev.* 1928, 64.

CZAPLICKA, Miss (–1921)
d. 20 May 1921
Studied geography at Warsaw. Came to England in 1910. Lectured on Ethnology at Oxford and Bristol. Collected plants in Siberia, 1914–15, now at Oxford. *My Siberian Year* 1916.
Bot. Soc. Exch. Club Br. Isl. Rep. 1921, 356. *Nature* v.107, 1921, 466.

DADDS, John (–*c.* 1904)
Of Ilfracombe, Devon. Pteridologist. Found *Adiantum* c.v. *plumosa.* Raised *Polydactylous lastrea.*
Bot. Fern Gaz. 1909, 44.

DAFT, Richard (*fl.* 1830s)
Nurseryman, Beskwood Park, Nottingham.

DAGLISH, Eric Fitch (1892–1966)
b. London 29 Aug. 1892 *d.* 5 April 1966
Author and engraver. *Our Wild Flowers* 1923. *Marvels of Plant Life* 1924. *How to See Plants* 1936. 'The Gardener's Botany' (M. Hadfield, *ed. Gardener's Companion* 1936, 38–131).
Who was Who, 1961–1970 271.

DAINTREE, Richard (1831–1878)
b. Hemingford Abbots, Hunts Dec. 1831 *d.* 25 June 1878
Geologist. Collected plants in Kennedy district, N. Australia.
P. Mennell *Dict. Austral. Biogr.* 1892.
Plants at Kew.

DAINTREY, Edwin (1814–1887)
b. Petworth, Sussex 2 Sept. 1814 *d.* Randwick N.S.W. 3 Oct. 1887
To Sydney in early 1840s where he practised law. Took a keen interest in

Zoological Gardens. "An excellent botanist".
F. M. Bladen *Hist. Notes on Public Library of N.S.W.* 1906 portr. *J. Proc. Soc. N.S.W.* 1908, 100.
Pterostylis daintreana F. Muell.

DALBY, Robert (*né* **Blunt**) (1808–1884)
d. Castle Donington, Leics 1884
BA Cantab. Curate, Blaston, 1933; later Vicar of Medbourne, of Bisham, Berks, 1836 and of Belton, Leics 1840. Took name of Dalby by royal licence in 1853. Friend of Professor Henslow and C. C. Babington. Had herb.
A. R. Horwood and C. W. F. Noel *Fl. Leicestershire* 1933, ccix.
Herb. at Leicester Literary and Philosophical Society.

DALE, Francis (*fl.* 1730s)
b. Hoxton?
In Bahamas, 1730–32. Sent plants to his relative Samuel Dale (1659–1739) and to Peter Collinson.
J. Bot. 1883, 227. N. G. Brett-James *Life of Peter Collinson* 1926, 121. H. N. Clokie *Account of Herb. Dept. Bot., Oxford* 1964, 153.
Plants at BM(NH), Oxford.

DALE, George (–1781)
Nurseryman, Hebburn Quay, Gateshead, County Durham, *c.* 1734. Nursery acquired in 1782 by William Falla.
J. H. Harvey *Early Gdning Cat.* 1972, 32.

DALE, Ivan Robert (*c.* 1904–1963)
d. Johannesburg, S. Africa 9 Dec. 1963
BA Oxon 1928. FLS 1939 Forester, Kenya, 1928–38; Uganda, 1938–52. Settled in Kenya, 1952. *Descriptive List of Introduced Trees of Uganda* 1954. *Kenya Trees and Shrubs* (with P. J. Greenway) 1961.
Commonwealth For. Rev. 1964, 3. *Proc. Linn. Soc.* 1965, 99.
Plants at Kew.

DALE, John (–1662)
Doctor of Physick, St. Martin's-in-the-Field. Friend of John Goodyer. "Botanologus peritus" (C. Merrett *Pinax* 1667, A2 verso).
R. T. Gunther *Early Br. Botanists* 1922, 229–30, 294–98.

DALE, Joseph (1815–1878)
b. 29 June 1815 *d.* Leyton, London 31 Dec. 1878
Gardener at Middle Temple, London, 1843–78.
Gdnrs Chron. 1879 i 20, 21 portr. *Gdnrs' Yb. Almanack* 1880, 190.

DALE, Samuel (1659–1739)
b. Whitechapel, London 1659 *d.* Bocking, Essex 6 June 1739
ML 1730. Apothecary and physician. Practised at Braintree, Essex. Friend of John Ray (1627–1705) whose collections he catalogued. Collected in East Anglia and grew many plants from seed in his garden. *Pharmacologia* 1693; *Supplementum* 1705. Contrib. appendix to Taylor's *Hist. and Antiq. of Harwich and Dovercourt* 1730, 336–77. Contrib. to *Philos. Trans. R. Soc.*
Gent. Mag. 1739, 327. R. Pulteney *Hist. Biogr. Sketches of Progress of Bot. in England* v.2, 1790, 122–28. G. S. Gibson *Fl. Essex* 1862, 446–47. R. H. Semple *Mem. Bot. Gdn at Chelsea* 1878, 63–66. *J. Bot.* 1883, 193–97, 225–31 portr. *Notes and Queries* v.7, 1883, 408; v.8, 1883, 159, 217. E. J. L. Scott *Index to Sloane Manuscripts* 1904, 134. *D.N.B.* v.13, 385. *Essex Nat.* 1912–13, 134–36 portr.; 1918–19, 49–64; 1919–20, 65–69; 1956, 311–14. R. W. Gunther *Further Correspondence of John Ray* 1928 portr. C. E. Raven *John Ray* 1942 *passim.* J. E. Dandy *Sloane Herb.* 1958, 122. H. N. Clokie *Account of Herb. Dept. Bot., Oxford* 1964, 153. S. T. Jermyn *Fl. Essex* 1974, 15–16.
Plants at BM(NH), Oxford. Letters at Royal Society. Portr. at Apothecaries Hall. Portr. at Hunt Library.
Dalea L.

DALE, Thomas (1699/1700–1750)
MD Leyden. Nephew of Samuel Dale. Secretary, Botanical Society of London, 1726. In Charleston, Carolina, *c.* 1731. Sent plants to Samuel Dale. *De Pareira Brava et Serapia Officinarum* 1723.
W. Munk *Roll of R. College of Physicians* v.2, 1878, 362–63. *D.N.B.* v.13, 386. *Essex Nat.* 1919–20, 49–69. *Ann. Med. Hist.* 1931, 50–57. J. E. Dandy *Sloane Herb.* 1958, 122. H. N. Clokie *Account of Herb. Dept. Bot., Oxford* 1964, 153. J. I. Waring *Hist. Med. in S. Carolina* 1964, 204–6. *Proc. Bot. Soc. Br. Isl.* 1967, 310–11.
Plants at BM(NH).

DALE, Thomas (*fl.* 1800s)
Nurseryman, Balls Pond, Islington, London.

DALHOUSIE, Lady Christina *see* Ramsay, C.

DALLACHY, John (*c.* 1820–1871)
b. Scotland *c.* 1820 *d.* Rockingham Bay, Queensland 4 June 1871
Gardener at Kew and to Earl of Aberdeen. To Ceylon, 1847. Curator, Melbourne Botanic Garden. Collected plants in Queensland.

F. von Mueller *Plants Indigenous to... Victoria* v.1, 1860, 123, 159, 162, 167, 200, 229. F. von Mueller *Fragmenta Phytographiae Australiae* v.9, 1875, 140–41. *Victorian Nat.* 1908, 106–8. *Proc. R. Soc. Queensland* 1954, 12–14.

Plants at Melbourne Herb.

Dallachya F. Muell.

DALLAS, John E. S. (–1952)
b. London *d.* Leatherhead, Surrey 14 Nov. 1952

Civil servant. Ornithologist and botanist. *London Nat.* 1952, 114–15 portr.

DALLIMORE, William (1871–1959)
b. Tardebigge, Worcs 31 March 1871 *d.* Tonbridge, Kent 7 Nov. 1959

VMM 1924. VMH 1931. Kew gardener, 1891; foreman, Arboretum, 1901–8; Assistant, Museum, 1908; Keeper, Museum, 1926–36. Supervised development of National Pinetum at Bedgebury, Kent, 1925. *Pictorial Practical Tree and Shrub Culture* (with W. P. Wright) 1906. *Holly, Yew and Box* 1908. *Handbook of Coniferae* (with A. B. Jackson) 1923; ed. 4 1966.

Gdnrs Chron. 1925 i 406 portr.; 1932 i 2 portr.; 1936 i 210 portr. *J. Kew Guild* 1926, 365–66 portr.; 1959, 702–3 portr. *Kew Bull.* 1936, 189. *Nature* v.184, 1959, 1684. *Times* 10 Nov. 1959. *Empire For. Rev.* 1960, 1–2. *Quart. J. For.* 1960, 81–82.

Copy of unpublished autobiography at Kew. Portr. at Hunt Library.

DALLINGER, Rev. William Henry (1842–1909)
b. Devonport 5 July 1842 *d.* Lee, Kent 7 Nov. 1909

DSc Dublin 1892. FRS 1880. FLS 1882. Wesleyan minister, 1860–80. President, Wesley College, Sheffield, 1880–88. Microscopist. Contrib. to *Mon. Microsc. J.* 1873–76. President, Royal Miscroscopical Society, 1884–87.

J. R. Microsc. Soc. 1909, 699–702 portr. *Proc. Linn. Soc.* 1909–10, 87–89. *D.N.B. Supplt. 2* v.1, 462–63. *R.S.C.* v.7, 478; v.9, 627; v.14, 460.

DALLMAN, Arthur Augustine (1883–1963)
b. Cumberland 9 April 1883 *d.* Colwyn Bay, Denbighshire 20 March 1963

ALS 1935. Assistant Lecturer in Botany, Liverpool Technical School, 1903–5. Science master, Holt Secondary School, Liverpool, 1908–15; Hulme Grammar School, Manchester, 1917–20; Mexborough Secondary School, 1920–30. President, Liverpool Botanical Society, 1914; 1918 Doncaster Scientific Society, 1921–23, 1927. Editor, *N. Western Nat.*, 1926–55. Published papers on plant galls and flora of Flint and Denbigh.

Sorby Rec. v. 2 (3), 1966, 2–3.

MSS. including Flint and Denbigh floras at Liverpool Museums. Plants at Liverpool and Manchester Museums.

DALLY, Miss (*fl.* 1830s)

Drew t.3801 and 3808 for *Bot. Mag.*

DALLY, Thomas (*fl.* 1790s)

Seedsman, Bridgnorth, Shropshire.

DALMAN, R. (*fl.* 1790s)

Gardener and seedsman, Newark, Notts.

DALMAN, Thomas (*fl.* 1830s)

Gardener and seedsman, Bambygate, Newark, Notts.

DALTON, Rev. James (1764–1843)
b. York 14 Nov. 1764 *d.* Croft, Yorks 2 Jan. 1843

BA Cantab 1787. FLS 1803. Rector, Copgrove, 1789, Catterick, 1791, Croft, 1805–43. Collected and studied carices, lichens and mosses. Discovered *Scheuchzeria*, 1787. Contrib. to J. Sowerby and J. E. Smith's *English Bot.* (Not to be confused with his grandson, Rev. James Dalton (–1862) who was also active in N. Yorks.)

D. Turner and L. W. Dillwyn *Botanist's Guide through England and Wales* v.1, 1805, 663–64, 665–744. *Ann. Bot.* 1806, 197–98. W. J. Hooker and T. Taylor *Muscologia Britannica* 1818, 80–81. *Proc. Linn. Soc.* 1843, 172. *Rep. Yorks Philos. Soc.* 1897, xv–xvi; 1906, 48–53. *J. Bot.* 1919, 294. *Yorkshire Life* 1967, 47, 49 portr. *J. Soc. Bibl. Nat. Hist.* 1969, 117–20.

Herb. and Bryophyta drawings at York Museum. Letters at Kew, Linnean Society.

Daltonia Hook. & Taylor.

DALTON, John (1766–1844)
b. Eaglesfield, Cockermouth, Cumberland 6 Sept. 1766 *d.* Manchester 27 July 1844

DCL Oxon 1832. LLD Edinburgh 1834. FRS 1822. Chemist and physicist. Pupil of John Gough (1757–1825). Schoolmaster, Kendal, 1781–93. Professor of Natural Philosophy, New College, Manchester, 1793–99. President, Manchester Literary and Philosophical Society, 1817–44. Herb. in 11 vols. of plants collected near Kendal, 1790–93, was destroyed at Manchester Literary and Philosophical Society during 1940–45.

Annual Monitor 1845, 40–47. J. Smith *Cat. Friends' Books* v.1, 1867, 506–10. E. Baines *Hist. of County...of Lancaster* v.1, 1868, 413–15. H. Lonsdale *Worthies of Cumberland* 1874. H. Roscoe *John Dalton* 1895. *Trans. Liverpool Bot. Soc.* v.1, 1909, 66–67. *Mem. Proc. Manchester Lit. Philos. Soc.* v.63, 1918–19, 1–46. *D.N.B.* v.13, 428–35. *J. Soc. Bibl. Nat. Hist.* 1969, 117–20; 1970, 270–71. *R.S.C.* v.2, 22.

Plants at Royal Botanic Garden, Edinburgh. Portr. at National Portrait Gallery and Manchester Literary and Philosophical Society.

D'ALTON, St. Elroy (1847–1930)
b. Tipperary 1847 *d.* Melbourne 17 Dec. 1930
In Australia, *c.* 1875. Engineer at Dimboola, W. Victoria until *c.* 1918. Pioneer of West Wimmera flora. Collected plants which he sent to F. von Mueller.
Victorian Nat. 1949, 124 portr.
Trymalium D'Altoni F. Muell.

DALZELL, Nicol Alexander (1817–1878)
b. Edinburgh 21 April 1817 *d.* Edinburgh 18 Jan. 1878
MA Edinburgh 1837. In India, 1841–70. Conservator of Forests, Burma. Collected plants in India and Burma. *Catalogue of Indigenous Flowering Plants of Bombay Presidency* 1858. *Bombay Fl.* (with A. Gibson) 1861. Contrib. to *Hooker's J. Bot.*
R. Wight *Icones Plantarum Indiae Orientalis* v.5, 1851, 35. *D.N.B.* v.13, 448. *J. Bombay Nat. Hist. Soc.* 1939, 145. *Fl. Malesiana* v.1, 1950, 127. *R.S.C.* v.2, 135; v.7, 479; v.12, 182.
Plants and letters at Kew. Drawings at BM(NH).
Dalzellia Wight.

DALZIEL, John McEwen (1872–1948)
b. Nagpur, India 16 May 1872 *d.* Chiswick, Middx 1948
MB Edinburgh 1895. MD 1903. FLS 1917. Medical missionary, China, 1895–1902. West African Medical Service, 1905. Collected plants in China, Gold Coast, Nigeria, French Guinea, Gambia, Liberia, Sierra Leone. *Hausa Botanical Vocabulary* 1916. Assisted J. H. Holland with his *Useful Plants of Nigeria* 1922. Worked at Kew with J. Hutchinson during 1923–36 on *Fl. West Tropical Africa. Useful Plants of West Tropical Africa* 1937.
D. Fairchild *Exploring for Plants* 1930 portr. *J. Kew Guild* 1948, 693 portr. *Nature* v.161, 1948, 920–21.
Plants at Kew, BM(NH), Royal Botanic Garden, Edinburgh.

DAMPIER, William (1651–1715)
b. East Coker, Yeovil, Somerset bapt. 5 Sept. 1651. *d.* London March 1715
Circumnavigator and pirate. Collected plants in Java, Timor, New Guinea, Brazil and Australia on voyage to Australia in HMS 'Roebuck' in 1699. *New Voyage round the World* 1697. *Voyage to New Holland... 1699* 1703 (includes plant descriptions).
J. Ray *Historia Plantarum* v.3, 1704 Appendix 225–26. *Ann. Bot.* 1806, 531–32. *J. Bot.* 1873, 348. *Gdnrs Chron.* 1894 i 429–30, 464. *D.N.B.* v.14, 2–7. E. J. L. Scott *Index to Sloane Manuscripts* 1904, 134. *J. W. Austral. Nat. Hist. Soc.* 1909, 13–14. C. Wilkinson *William Dampier* 1922 portr. *Proc. Linn. Soc.* 1938–39, 44–50. *Fl. Malesiana* v.1, 1950, 128–29. *Nature* v.170, 1952, 408–9. J. E. Dandy *Sloane Herb.* 1958, 123. J. C. Shipman *William Dampier* 1962 portr. H. N. Clokie *Account of Herb. Dept. Bot., Oxford* 1964, 153. *W. Austral. Nat.* 1971, 173–78.
Plants at BM(NH), Oxford. Portr. by T. Murray at National Portrait Gallery. Portr. at Hunt Library.
Dampiera R. Br.

DANBY, Henry, Earl of *see* Danvers, H.

DANCER, Thomas (*c.* 1750–1811)
d. Kingston, Jamaica 1 Aug. 1811
MD Edinburgh 1771. To Jamaica, 1773. Curator, Botanic Garden, Bath, Jamaica, 1788. Island botanist, 1797. *Catalogue of Plants, Exotic and Indigenous, in the Botanical Garden* 1792.
Gent. Mag. 1811 ii 390. *Cottage Gdnr* v.8, 1852, 159. *J. Inst. Jamaica* 1892, 102–4, 141 portr. *Bot. Gaz.* 1897, 347–48. I. Urban *Symbolae Antillanae* v.1, 1898, 35; v.3, 1902, 35–36. *D.N.B.* v.14, 13–14. F. Cundall *Historic Jamaica* 1915, 25, 27, 72, 248–49. *Taxon* 1970, 520.
Plants at Kew. Letters at Royal Society of Arts.

DANDRIDGE, Joseph (*fl.* 1660s–1746)
b. Winslow, Bucks Jan. 1665 *d.* London 23 Dec. 1746
Of Stoke Newington. "A pattern-drawer in Moorfields" (Buddle MS.) Mycologist, ornithologist and lepidopterist. Had herb. Friend of W. Sherard and J. J. Dillenius. Corresponded with J. Petiver.
J. Nichols *Illus. Lit. Hist. of Eighteenth Century* v.1, 357; v.3, 782. D. Turner *Extracts from Lit. and Sci. Correspondence of Richard Richardson* 1835, 204. E. J. L. Scott *Index to Sloane Manuscripts* 1904, 1935. *Notes and Queries* v.165, 1933, 260. C. E. Raven *John Ray* 1942, 410–11. J. E. Dandy *Sloane Herb.* 1958, 123. *Entomol. Rec.* 1966, 89–94. *E. London Papers* 1966, 101–18. *Entomol. Gaz.* 1967, 73–89, 197–201.

DANIEL, Henry (*fl.* 1370s)
Dominican friar. 'Aaron Danielis, de re Herbaria, de Arboribus, Fruticibus' (MS. at Bodleian Library). Translated 'Book of the Virtues of Rosemary' (MS. at Trinity College, Cambridge).
R. Pulteney *Hist. Biogr. Sketches of Progress of Bot. in England* v.1, 1790, 23. *D.N.B.* v.14, 24.

DANIEL, Rev. Richard (–1864)
d. Combs, Suffolk 20 Feb. 1864
MA Cantab 1827. FLS 1862. Rector, Combs, 1836. Had large collection of mosses.
Proc. Linn. Soc. 1863–64, xxviii–xxix.

DANIEL, Samuel (–before 1707)
Surgeon. Sent plants from Greece and Levant to J. Petiver. 'Voyage to Levant' (*Memoirs for the Curious* 1707, 63–70).
J. Petiver *Museii Petiveriani* 1695, 211, 624. J. E. Dandy *Sloane Herb.* 1958, 123.

DANIELL, William Freeman (1818–1865)
b. Liverpool 1818 *d.* Southampton, Hants 26 June 1865
MRCS 1841. FLS 1855. Surgeon in army on coast of W. Africa, 1841–53, where he made a study of frankincense tree. Also in West Indies and in China in 1860. Contrib. to *Pharm. J.* v.9–v.18, 1849–63.
Ann. Mag. Nat. Hist. v.10, 1862, 195–202. *J. Bot.* 1865, 294–95. *Pharm. J.* v.7, 1865–66, 86. *Proc. Linn. Soc.* 1865–66, lix–lxi. E. Bretschneider *Hist. European Bot. Discoveries in China* 1898, 678–79. *D.N.B.* v.14, 35. I. Urban *Symbolae Antillanae* v.3, 1902, 36. *Trans. Liverpool Bot. Soc.* 1909, 67–68. *R.S.C.* v.2, 146, v.7, 483.
Plants at BM(NH), Kew.
Daniellia Benn.

DANVERS, Henry, 1st Earl of Danby (1573–1645)
b. Dauntsey, Wilts 28 June 1573 *d.* Cornbury Park, Oxford 20 Jan. 1645
Statesman. Founded Oxford Botanic Garden, 1621. Had over 1000 fruit trees in his gardens at Wimbledon House, Surrey.
R. Pulteney *Hist. Biogr. Sketches of Progress of Bot. in England* v.1, 1790, 165. F. N. Macnamara *Memorials of Danvers Family* 1895. *D.N.B.* v.14, 37–39. R. T. Gunther *Oxford Gdns* 1912, 1. S. H. Vines and G. C. Druce *Account of Morisonian Herb.* 1914. ix–xii portr.
Portr. at Hunt Library.

DARBISHIRE, Arthur Dukinfield (1880–1915)
d. Gailes, Ayr 26 Dec. 1915
Mendelism and Plant-breeding 1911. *Introduction to Biology* 1917 (with biogr. account). 'Crossing Peas' (*Proc. R. Soc. B.* v.80, 1908, 122–35). Contrib. to *New Phytol.*

DARBISHIRE, Otto Vernon (1870–1934)
b. Conway, Caernarvonshire 16 March 1870 *d.* Bristol 17 Oct. 1934
BA Oxon. PhD Kiel. FLS 1920. Lecturer in Botany, Kiel, 1897–98; Manchester, 1898–1909; Armstrong College, Newcastle, 1909–11; Bristol, 1911; Professor, 1919. Studied marine algae, lichens and ecology. 'Monographia Roccelleorum' (*Bibliotheca Bot.* Heft 45, 1898 102p.) 'Lichens of Swedish Antarctic Expedition' (*Wissensch. Ergebnisse Schwedischen Südpolar-Expedition, 1901–1903* v.4 (11), 1912 73p.) President, Bristol Naturalists' Society, 1932–34. President, British Mycological Society, 1923.
Gdnrs Chron. 1934 ii 327. *J. Bot.* 1934, 328; 1935, 21–22. *Kew Bull.* 1934, 451–52. *Nature* v.134, 1934, 726. *Proc. Bristol Nat. Soc.* v.7, 1934, 516. *Proc. Linn. Soc.* 1934–35, 171–72. *Times* 18 Oct. 1934. *Chronica Botanica* 1936, 176 portr. *Who was Who, 1929–1940* 329
Portr. at Hunt Library.

DARBY, Gerald (–1966)
d. 23 March 1966
Lily breeder.
Lily Yb. 1967, 127–28 portr.

DARBY, William (*fl.* 1670s–1710s)
Nurseryman at Hoxton, Shoreditch, London from about 1678 whence J. Petiver and others obtained plants. Propagated large numbers of *Liriodendron tulipifera*; also grew varigated hollies. "A famous gardener, noted for one of the first in England who chose the culture of exotic plants" (J. Cowell *Curious and Profitable Gardener* 1730).
Archaeologia v.12, 1794, 191–92. J. E. Dandy *Sloane Herb.* 1958, 123–24.

DARE, George (*fl.* 1680s–1690s)
Apothecary, London. Discovered *Hymenophyllum* at Tunbridge Wells. "Ad Herbariam Scientiam promovendam paratissimus" (Plukenet).
J. Petiver *Museii Petiveriani* 1695, 73. R. Morison *Plantarum Historiae Universalis Oxoniensis* v.3, 1699, 628. J. E. Dandy *Sloane Herb.* 1958, 124.
Darea Juss.

DARLINGTON, Hayward Radcliffe (*c.* 1863–1946)
VMH 1938. Cultivated roses. President, National Rose Society.
Gdnrs Mag. 1910, 483–84 portr.

DARNELL-SMITH, George Percy (1868–1942)
b. Chipping Norton, Oxford 15 Oct. 1868 *d.* Sydney, N.S.W. 10 April 1942
Microbiologist, N.S.W. Bureau of Microbiology, 1909–13. Biologist, N.S.W. Dept. Agriculture, 1913–24. Director, Botanic Gardens, Sydney, 1924–33.
Portr. at Hunt Library.

DARTMOUTH, 3rd Earl of *see* Legge, G.

DARWALL, Rev. Leicester (1813–1897)
d. Tenby, Pembrokeshire 22 July 1897

MA Cantab 1838. Incumbent, Criggion, Montgomeryshire. Had a salicetum. Contrib. to J. E. Leefe's *Salictum Britannicum* (exsiccatae).

Plants and letters at Kew.

DARWIN, Charles Robert (1809–1882)
b. Shrewsbury, Shropshire 12 Feb. 1809 *d.* Downe, Kent 19 April 1882

BA Cantab 1832. LLD 1878. FRS 1839. FLS 1854. Naturalist, HMS 'Beagle', 1831–36. Collected plants mainly in S. America, Galapagos Islands, Maldives Islands, Falkland Islands, Fernando de Noronha, Cocos Keeling Islands, Australia. *Origin of Species* 1859. *Fertilisation of Orchids* 1862. *Variation of Animals and Plants under Domestication* 1868. *Insectivorous Plants* 1875. *Movements and Habits of Climbing Plants* 1865. *Different Forms of Flowers on Plants of Same Species* 1877. *Power of Movements in Plants* 1880.

Ipswich Mus. Portr. Ser. 1852. *Nature* v.10, 1874, 79–81 portr.; v.25, 1882, 597; v.26, 1882, 49–51, 73–75, 97–100, 145–47, 169–71; v.60, 1899, 187–88; v.81, 1909, 7–14; v.149, 1942, 716–20; v.185, 1960, 216–17. *Garden* v.8, 1875, xi–xii portr. *Gdnrs Chron.* 1875 i 308–9 portr.; 1882 i 535–36 portr.; 1909 i 405–6, 412 portr.; 1927 ii 203. *J. Bot.* 1882, 165–68; 1928, 307–8; 1934, 177; 1935, 104–6. *Proc. Amer. Acad. Arts Sci.* 1882, 449–58. F. Darwin *Life and Letters of Charles Darwin* 1887 3 vols; *More Letters* 1903 2 vols. *Proc. Linn. Soc.* 1880–82, 60–62; 1887–88, 67–70; 1888–89, 35; 1957–58, 219–45. *Proc. R. Soc. Edinburgh* v.12, 1882–84, 1–6. *Proc. R. Soc.* v.44, 1888, i–xxv. *Trans. Herts. Nat. Hist. Soc.* 1893, 101–36. *D.N.B.* v.14, 72–84. N. L. Britton *Darwin and Bot.* 1909. J. R. Green *Hist. of Bot.* 1914, 451–56. *Kew Bull.* 1917, 212. *Nat. Hist.* v.23, 1923, 589–96. *Science* v.86, 1937, 468; v.87, 1938, 66. *Advancement Sci.* 1942, 262; 1945, 280–81; 1947, 91–96; 1960, 391–401. N. Barlow *ed. Charles Darwin and Voyage of Beagle* 1945 portr. V. W. von Hagen *South America called them* 1949, 213–88. N. Barlow, *ed. Autobiogr. of Charles Darwin* 1958 portr. *Darwinia* 1959, 563–83. *Discovery* 1959, 482–87 portr. *Notes Rec. R. Soc.* v.14 (1), 1959, 12–66 portr. *Handlist of Darwin Papers at...Cambridge* 1960. P. R. Bell *Darwin's Biological Work* 1960. *Proc. 10th Int. Congress Hist. Sci.* 1962, 971–74. G. de Beer *Charles Darwin* 1963 portr. R. B. Freeman *Works of Charles Darwin...Bibl. Handlist* 1965. J. Huxley and H. B. D. Kettlewell *Charles Darwin and his World* 1965. N. Barlow *ed. Darwin and Henslow... Letters, 1831–1860* 1967 portr. A. S. Gregor *Charles Darwin* 1967. W. Karp *Charles Darwin and Origin of Species* 1968 portr. C. C. Gillispie *Dict. Sci. Biogr.* v.3, 1971, 565–77. M. A. Reinikka *Hist. of the Orchid* 1972, 179–83 portr. D. L. Hull *Darwin and his Critics* 1973. *Br. J. Hist. Sci.* 1975, 62–69. *R.S.C.* v.9, 638; v.12, 184; v.14, 480.

'Beagle' plants at Cambridge, Kew. Letters at Cambridge, Kew, American Philosophical Society. Portr. by J. Collier at Linnean Society. Statue by Boehm at BM(NH).

Berberis darwinii Hook. f.

DARWIN, Erasmus (1731–1802)
b. Elston Hall, Notts 12 Dec. 1731 *d.* Derby 18 April 1802

MB Cantab 1755. MD Edinburgh. FRS 1761. FLS 1792. Grandfather of C. R. Darwin (1809–1882). Physician, scientist and poet. Doctor at Lichfield, 1756–81; moved to Derby. Member of Botanical Society of Lichfield whose other two members were Sir B. Boothby and J. Jackson. Edited translations of Linnaeus's *Systema Vegetabilium* and *Genera Plantarum* which were published as *A System of Vegetables* 1783 and *Families of plants* 1787. *The Botanic Garden* (poem) 1789, 1791. *Phytologia; or Philosophy of Agriculture and Gardening* 1800.

A. Seward *Mem. of Life of Dr. Darwin* 1804. S. Felton *Portr. of English Authors on Gdning* 1830, 164–73. *Gdnrs Mag.* 1838, 345–46. J. Dowson *Erasmus Darwin* 1861. *J. Hort. Cottage Gdnr* v.57, 1877, 196–97 portr. E. Krause *Erasmus Darwin* 1879 portr. *D.N.B.* v.14, 84–87. *J. Bot.* 1914, 322. H. Pearson *Doctor Darwin* 1930. *Bull. Hist. Med.* v.8, 1940, 844–47. *Isis* 1941, 315–25. *J. R. Hort. Soc.* 1941, 24–27. *Ann. Sci.* v.10, 1954, 314–20. *J. Soc. Ideas* 1955, 376–88. *Notes Rec. R. Soc.* 1959, 85–98 portr. D. King-Hale *Erasmus Darwin* 1963 portr. *Nature* v.200, 1963, 304–6; v.247, 1973, 87–91. R. E. Schofield *Lunar Society of Birmingham* 1963. *J. Soc. Bibl. Nat. Hist.* 1964, 210–13 portr. *Univ. Birmingham Hist. J.* v.11, 1967, 17–40. C. C. Gillispie *Dict. Sci. Biogr.* v.3, 1971, 577–81.

MSS. at Royal Society. Portr. by Joseph Wright at National Portrait Gallery. Portr. at Hunt Library.

Darwinia Rudge.

DARWIN, Francis (1848–1925)
b. Downe, Kent 16 Aug. 1848. *d.* Cambridge 19 Sept. 1925

BA Cantab 1870. FRS 1882. FLS 1875. Knighted 1913. Assistant to his father, C. R. Darwin, 1875–82. Lecturer in Botany,

Cambridge, 1884–88; Reader, 1888–1904. Founded School of Plant Physiology. President, British Association for Advancement of Science, 1908. *Power of Movement in Plants* (with C. R. Darwin) 1880. *Life and Letters of Charles Darwin* 1887 3 vols. *More Letters of Charles Darwin* 1903 2 vols. *Practical Physiology of Plants* (with E. H. Acton) 1894. *Elements of Botany* 1895.

Gdnrs Chron. 1925 ii 260. *J. Bot.* 1925, 333–34. *Nature* v.116, 1925, 583–84. *Proc. Linn. Soc.* 1925–26, 76–78. *Times* 21 Sept. 1925. F. O. Bower *Sixty Years of Bot. in Britain, 1875–1935* 1938, 78–79 portr. *Who was Who, 1916–1928* 264. *R.S.C.* v.7, 487; v.9, 638; v.12, 184; v.14, 480.

Portr. by W. Rothenstein at Botany School, Cambridge. Portr. at Hunt Library.

DARWIN, Robert Waring (1724–1816)
b. Newark, Notts 17 Oct. 1724 *d.* Elston, Notts 3/4 Nov. 1816

MD Leyden 1784. FRS 1788. Brother of Erasmus Darwin. *Principia Botanica* 1787.

Gent. Mag. 1816 ii 476.

MS. at Royal Society. Engraved portr. by Joseph Wright at Kew.

DASH, Thomas (*fl.* 1790s)
Gardener and seedsman, Gosport, Hants.

DAUBENY, Charles Giles Bridle (1795–1867)
b. Stratton, Glos 1795 *d.* Oxford 13 Dec. 1867

MA Oxon 1817. MD 1821. FRS 1822. FLS 1830. Professor of Chemistry, Oxford, 1822–55. Professor of Botany, Oxford, 1834. Collected plants in U.S.A., West Indies, Switzerland and Spain. *Oxford Botanic Garden* 1850. Edited E. M. Cox's *Popular Geography of Plants* 1855. *Lectures on Roman Husbandry* 1857. *Essay on Trees and Shrubs of Ancients* 1865. *Miscellanies* 1867.

Gdnrs Chron. 1867 1294–95; 1869, 1084; 1870, 1025; 1904 i 241–42. *J. Hort. Cottage Gdnr* v.13, 1867, 462–63. *Proc. Linn. Soc.* 1867–68, ci–civ. *Proc. R. Soc.* 1868–69, lxxiv–lxxx. *Quart. J. Geol. Soc. London* 1868, xxxii–xxxvi. *Trans. Bot. Soc. Edinburgh* v.9, 1868, 267–69. *J. Bot.* 1868, 32; 1869, 370. *Trans. Devon Assoc.* 1868, 303–8. W. Munk *Roll of R. College of Physicians* v.3, 1878, 254–58. G. C. Druce *Fl. Berkshire* 1897, clxxi–clxxii. *D.N.B.* v.14, 94–95. R. T. Gunther *Hist. of Daubeny Laboratory* 1904. W. Tuckwell *Reminiscences of Oxford* 1907, 33 portr. R. T. Gunther *Daubeny Laboratory Register, 1849–1923* 1924. G. C. Druce *Fl. Oxfordshire* 1927, cxv. *Berichte Schweizerischen Bot. Gesellschaft 50A* 1940, 202–4.

Plants, MSS. and portr. at Oxford. Letters at Kew. Portr. at Hunt Library.

Daubenya Lindl.

DAVALL, Edmund (1763–1798)
b. Holborn, London Aug. 1763 *d.* Orbe, Switzerland 26 Sept. 1798

FLS 1788. Correspondent of J. E. Smith. Left incomplete *Illustrations of Swiss Plants.* Sent seeds to W. Curtis.

Bot. Mag. 1795, t.351. *Ann. Bot.* 1805, 576–77. P. Smith *Correspondence of Sir J. E. Smith* v.2, 1832, 1–7, 70. *D.N.B.* v.14, 99. *Proc. Linn. Soc.* 1946–47, 42–65, 140; 1947–48, 179–84; 1948–49, 56–63; 1949–50, 185–88. *Taxon* 1973, 387–88.

Herb. formerly at Linnean Society.

Davallia Smith.

DAVENPORT, Arthur (*fl.* 1850s–1880s)
British Consular Service, China, 1857–85. Accompanied T. G. Grosvenor's mission to Yunnan, 1876. Sent plants and seeds to Kew.

E. Bretschneider *Hist. European Bot. Discoveries in China* 1898, 727. *Kew Bull.* 1901, 18.

DAVENPORT, James (*c.* 1799–1884)
d. Droylsden, Lancs 28 May 1884

Member of Manchester Botanists' Association. Known in Manchester as "father of fern-growers".

J. Hort. Cottage Gdnr v.8, 1884, 444.

DAVENPORT, Robert (*fl.* 1790s)
Seedsman, Ashby-de-la-Zouch, Leics.

DAVEY, Frederick Hamilton (1868–1915)
b. Ponsanooth, Cornwall 10 Sept. 1868 *d.* Perranwell, Cornwall 23 Sept. 1915

FLS 1903. *Fl. Cornwall* 1909. Contrib. Botany to *Victoria History of Cornwall* v.1, 1906, 49–111. Contrib. to *J. Bot.*

Bot. Soc. Exch. Club Br. Isl. Rep. 1915, 251–53. *Watson Bot. Exch. Club Rep.* 1914–15, 477–78 portr. *J. Bot.* 1916, 29–31 portr. E. Thurston and C. C. Vigurs *Fl. Cornwall. Supplement* 1922, ix–xv portr. J. E. Lousley *Fl. Isles Scilly* 1972, 84.

Herb. at Truro Museum. Portr. at Hunt Library.

Ulmus major var. *Daveyi* Henry.

DAVEY, James Thomas (1923–1959)
d. air crash, Bordeaux 24 Sept. 1959

Educ. Bristol University. Entomologist, Nigeria, 1947. Director, Research Service, O.I.C.M.A., Mali, 1954. Collected plants.

Nature v.184, 1959, 1104–5.

Plants at Kew, Ibadan.

DAVEY, John (1846–1923)
b. Stawley, Somerset 6 June 1846 *d.* Akron, Ohio, U.S.A. 8 Nov. 1923

"Father of tree surgery." Went to Warren, Ohio, 1873 as florist and landscape gardener. Established his own business at Kent, Ohio, 1881. Davey Institute of Tree Surgery, 1909.
Nat. Cyclopedia Amer. Biogr. v.22, 1932, 70 portr. *Amer. For.* v.65, 1959, 17–21 portr.

DAVEY, Thomas (*c.* 1758–1833)
Had small nursery at Camberwell until 1798 when he moved to King's Road, Chelsea. Noted for his carnations, pinks and tulips. His father was said to have been "an eminent florist".
J. Hort. Cottage Gdnr v.32, 1877, 316.

DAVIDSON, Alexander (*fl.* 1880s)
Collected plants with W. A. Sayer in N. Queensland, 1886–87.
Spiraeanthemum davidsonii F. Muell.

DAVIDSON, Alfred Augustus (–1885)
d. London June 1885
FLS 1881. Colonel, Madras Army, 1854. "Was for some time attached to the Nair Brigade in Travancore, in the mountains of which State he acquired a taste for Botany."
Proc. Linn. Soc. 1885–86, 141.
Plants at Kew.

DAVIDSON, Anstruther (1860–1932)
b. Caithness 19 Feb. 1860 *d.* Los Angeles, U.S.A. 3 April 1932
MB Glasgow. MD 1887. To U.S.A., 1889. Practised medicine in Los Angeles. Contrib. fauna and flora to J. Brown's *Hist. of Sanquhar* 1891. *Fl. Southern California* (with G. L. Moxley) 1923. Contrib. to *Bull. S. Calif. Acad.*
Amer. Men Sci. Who's Who in America ed. 4. *Madrono* v.2, 1934, 124–28 portr.
Herb. at Los Angeles Museum of Art, History and Science. Portr. at Hunt Library.

DAVIDSON, Sir Colin John (1878–1930)
b. 28 Oct. 1878 *d.* 12 June 1930
British consular service, Tokyo, 1903–27. 'Seaweed Industry of Japan' (*Bull. Imperial Inst.* 1906, 125–49).
Who was Who, 1929–1940 333.

DAVIDSON, George (1825–1911)
b. Nottingham 9 May 1825 *d.* San Francisco, U.S.A. 1 Dec. 1911
To U.S.A. in 1832. Astronomer and physiographer. Collected plants in California and Alaska.
Who's Who in America, 1903–1905 363. *National Acad. Sci. Biogr. Mem.* v.18, 1938, 189–217 portr. O. Lewis *George Davidson* 1954 portr.

DAVIDSON, Rev. George (*c.* 1854–1901)
d. Aberdeen 16 Sept. 1901
MA. LLD Aberdeen 1886. Minister, Logie-Coldstone, Aberdeenshire. Diatomist.
Gdnrs Chron. 1901 ii 233–34. *R.S.C.* v.9, 647.

DAVIDSON, George William (1836–1873)
b. Island of Sanday, Orkney 5 June 1836 *d.* Edinburgh 2 March 1873
MD Edinburgh. Professor of Anatomy, Veterinary College and Lecturer in Botany, School of Arts, Edinburgh.
Trans. Bot. Soc. Edinburgh v.12, 1876, 19.

DAVIDSON, John (*fl.* 1790s)
Seedsman, Tooley Street, Southwark, London.

DAVIDSON, John (1878–1970)
b. Aberdeen 6 Aug. 1878 *d.* 10 Feb. 1970
FLS 1912. Curator, Botany Museum, Aberdeen University, 1893–1911. Provincial botanist, British Columbia, 1911. Curator of Herb. and Botanic Garden, 1916. Professor of Botany, University of British Columbia. Founder and President, Vancouver Natural History Society. Contrib. to *Nat. Guide to the Americas. Conifers, Junipers and Yew* 1928.
Davidsonia 1970, 2–3 portr.

DAVIDSON, William (*fl.* 1820s–1830s)
b. Northumberland
First Superintendent, Botanic Garden, Hobart, 1828–34.
Bot. Mag. 1835, t.3415. *Papers Proc. R. Soc. Tasmania* 1913, 121–22. *J. Arnold Arb.* 1921, 51–52.

DAVIE, E. (*fl.* 1870s–1930s)
Herb. at Norwich Museum, Norfolk.

DAVIE, Robert Chapman (1887–1919)
b. Glasgow 1887 *d.* Largs, Edinburgh 4 Feb. 1919
MA Glasgow 1907. DSc 1915. Assistant, Botany Dept., Glasgow University. Lecturer in Botany, Edinburgh, 1913. Worked at ferns: *Penanema* and *Diacalype* (*Ann. Bot.* 1912, 245–68). 'Pinnar Trace in Ferns' (*Trans. R. Soc. Edinburgh* v.50, 1914, 349–78).
Bot. Soc. Exch. Club Br. Isl. Rep. 1919, 618–19. *Nature* v.103, 1919, 189–90. *Trans. Bot. Soc. Edinburgh* v.27, 1919, 342–44.

DAVIES, Arthur Elphinstone Sanger-*see* Sanger-Davies, A. E.

DAVIES, David (1870–1931)
b. Merthyr Tydfil, Glam 1870 *d.* Norway 15 Aug. 1931
Palaeobotanist. "Dafydd Ffossil." Collected fossil plants in Glamorgan. Contrib. to *Quart. J. Geol. Soc.* and *Trans. R. Soc.*
Nature v.128, 1931, 403.
Collections at National Museum of Wales.

DAVIES, George (1834–1892)
b. Brighton, Sussex 12 Feb. 1834 *d.* Brighton 6 April 1892
Bryologist. Mosses in Erridge's *Hist. Brighton* 1862. Contrib. to *Grevillea* v.2, 1873–74, 173–74; v.4, 1875–76, 76–77.
J. Bot. 1892, 288; 1893, 370. *Boissiera* v.5, 1941, 39–40. *R.S.C.* v.9, 648.
Letters and moss herb. at BM(NH).

DAVIES, Henry James (*c.* 1871–1948)
d. Bromley, Kent 21 Dec. 1948
Kew gardener 1889. Assistant Curator, Royal Botanic Gardens, Calcutta, 1894. Superintendent, Government Gardens, Allahabad, *c.* 1898. Superintendent, Lucknow Gardens, 1906–21.
J. Kew Guild 1939–40, 825–26 portr.; 1949, 788.

DAVIES, Rev. Hugh (1739–1821)
b. Llandyfrydog, Anglesey, 1739 *d.* Beaumaris, Anglesey 16 Feb. 1821
MA Oxon 1763. FLS 1790. Had living at Llandegfan, Anglesey, 1778–87. Rector, Aber, 1787–1816. Friend of W. Hudson. *Welsh Botanology* 1813. Contrib. to J. Sowerby and J. E. Smith's *English Bot.*
D. Turner and L. W. Dillwyn *Botanist's Guide through England and Wales* v.1, 1805 *passim. J. Bot.* 1898, 14; 1914, 318–19; 1927, 128. *D.N.B.* v.14, 138. *Dict. Welsh Biogr. down to 1940* 1959, 127–28. *Trans. Anglesey Antiq. Soc.* 1930 and 1931 *passim. Proc. Isle of Man Nat. Hist. Antiq. Soc.* v.6, 1960–63, 406–7. *R.S.C.* v.2, 166.
Herb. at BM(NH). MS. at Royal Society. Letters at BM(NH), Kew, Linnean Society. Fungi MSS. and drawings at BM(NH).
Daviesia Smith.

DAVIES, Isaac (1812–1897)
d. Ormskirk, Lancs 10 Oct. 1897
Nurseryman, Brook Lane Nurseries, Ormskirk.
J. Hort. Cottage Gdnr v.35, 1897, 387. *Garden* v.52, 1897, 354; v.53, 1898, 26–27 portr.

DAVIES, Rev. John (*c.* 1567–1644)
b. Llanrhaiadar, Denbighshire *c.* 1567 *d.* Mallwyd, Merioneth 15 May 1644
DD Oxon 1616. Rector, Mallwyd, 1604–8. Canon of St. Asaph. Welsh lexicographer. *Antiquae Linguae Britannicae Dictionarium* including *Botanologium* 1632.
H. Davies *Welsh Botanology* 1813, vi. *D.N.B.* v.14, 144–45. *Dict. Welsh. Biogr. down to 1940* 1959, 131–32.

DAVIES, John Henry (1838–1909)
b. Penketh, Warrington, Lancs 1838 *d.* Belfast 20 Aug. 1909
Manager, Glenmore bleach works, County Antrim. Collected mosses in Antrim. Contrib. papers on mosses to *Phytologist* 1857–58 and *Irish Nat.* 1900–7. Correspondent of W. H. Harvey and W. Wilson.
J. Bot. 1909, 451; 1910, 57, 79. *Irish Nat.* 1909, 235–36; 1913, 30–31. *Proc. Belfast Nat. Field Club* v.6, 1913, 625–26. *Proc. R. Irish Acad. B.* v.32, 1915, 76–77. *Bot. Soc. Exch. Club Br. Isl. Rep.* 1932, 290. R. L. Praeger *Some Irish Nat.* 1949, 70.
Mosses at Trinity College, Dublin.

Davies, Peter (*fl.* 1820s)
Seedsman and woollen draper, Bridge Street, Warrington, Lancs.

DAVIES, Richard (*fl.* 1690s)
Peterhouse College, 1693; Fellow, 1699. Assisted John Ray with plant records.
C. E. Raven *John Ray* 1942, 257.

DAVIES, Rev. Richard H. (*fl.* 1830s)
Collected plants on east coast of Tasmania in 1830s.
J. D. Hooker *Fl. Tasmaniae* v.1, 1859, cxxvii. *Papers Proc. R. Soc. Tasmania* 1909, 13.
Phebalium daviesi Hook. f.

DAVIES, Thomas (1829–1902)
b. Wavertree, Lancs 1829 *d.* Wavertree 6 May 1902
Nurseryman, Wavertree.
Garden v.61, 1902, 332.

DAVIES, William (*fl.* 1780s)
Seedsman, 61 Snow Hill, London.

DAVIES, William (1814–1891)
b. Holywell, Flintshire 1814
Palaeontologist. On staff of BM. Contrib. plant localities in Hampstead to H. Trimen and W. T. T. Dyer's *Fl. Middlesex* 1869.
H. Trimen and W. T. T. Dyer *Fl. Middlesex* 1869, 11.
Herb. at Queen Mary College, London.

DAVIES, William (*c.* 1854–1889)
d. 31 Oct. 1889
Joined his father's nursery at Brook Lane, Ormskirk, Lancs, 1875.
Garden v.36, 1889, 469.

DAVIES, William (1899–1968)
b. London 20 April 1899 *d.* 28 July 1968
BSc Wales 1923. DSc 1945. Grassland agronomist, Welsh Plant Breeding Station, 1923–28. Plant geneticist, Palmerston North, New Zealand, 1929–31. Head of Grassland Agronomy, Welsh Plant Breeding Station, 1933–38. Head of Grassland Survey of England, 1938–40. Assistant Director, Grassland Improvement Station, 1940–45; Director, 1945–49. Director, Grassland Research Institute, Hurley, 1949–64. President, British

Grassland Society, 1948, 1960. *Grasslands of Wales* 1937. *Grasslands of Falkland Islands* 1938. *Grass Crop* ed. 2 1960.
Who was Who, 1961–1970 285.

DAVIES, William Thomas (*c.* 1877–1951)
d. Port Dinorwic, Caernarvonshire 25 March 1951
Clerk at Port Dinorwic slate quarry for 50 years. Collected local ferns.
Br. Fern Gaz. 1951, 6–7.

DAVIS, George (*fl.* 1770s)
Seedsman and apothecary near Ball's Bridge, County Limerick, Ireland.
Irish For. 1967, 54.

DAVIS, James Richard Ainsworth-*see* Ainsworth-Davis, James Richard

DAVIS, John Ford (1773–1864)
b. Bath, Somerset 1773 *d.* Bath 1 Jan. 1864
MD Edinburgh 1797. Physician to Bath Hospital, 1817–34. 'Botany of Bath' (*Hist. Account of Bath* 1802).
W. Munk *Roll of R. College of Physicians* v.2, 1878, 67–68. *D.N.B.* v.14, 168.

DAVIS, Norman (*fl.* 1900s)
Nurseryman at Camberwell, London, then at Framfield, Sussex. Specialised in chrysanthemums.
Gdnrs Mag. 1911, 457–58 portr.

DAVIS, Rees Alfred (1855–1940)
b. Chepstow, Mon 1855
Helped to start Co-operative Fruit Growers' Association in California. Manager of orchards in S. Africa belonging to Cecil Rhodes, 1898. Government Horticulturist, Transvaal, 1902.

DAVIS, Richard (*fl.* 1830s)
Missionary in New Zealand. Sent seeds to Rev. John Noble Coleman at Ryde.
Bot. Mag. 1837, t.3584.

DAVIS, Walter (1847–1930)
b. Amport, Hants 14 Sept. 1847 *d.* Fulham, London 18 Nov. 1930
Trained as gardener on estates of Marquis of Winchester and Lady Herbert. Joined Messrs James Veitch at Chelsea, 1870; sent by Veitch to S. America in 1873 to collect plants, especially orchids. Wrote on plant propagation under pseudonym of Charles Bennett. Secretary to Geological and Royal Geographical Societies.
Gdnrs Chron. 1874 ii 710; 1930 ii 461. J. H. Veitch *Hortus Veitchii* 1906, 65–66. *J. Bot.* 1931, 24. *Orchid Rev.* 1931, 20. *J. R. Hort. Soc.* 1948, 287. M.A. Reinikka *Hist of the Orchid* 1972, 264–65.
Masdevallia davisii Reichenb. f.

DAVY, David Elisha (1769–1851)
d. 15 Aug. 1851
BA Cantab 1790. FLS 1793. Of Ufford and Yoxford, Suffolk. Contrib. to D. Turner and L. W. Dillwyn's *Botanist's Guide* 1805 and J. Sowerby and J. E. Smith's *English Bot.*, t.380, 381. MS. *Bibliotheca Botanica.*
W. M. Hind *Fl. Suffolk* 1889, 481.

DAVY, George Thomas (*fl.* 1840s)
Sent *Lardizabala* sp. from Chile to Messrs Veitch.
Bot. Mag. 1850, t.4501.

DAVY, Lady Joanna Charlotte (*née* **Flemmich**) (1865–1955)
b. London Feb. 1865 *d.* 28 Dec. 1955
Friend of G. C. Druce. Discovered *Carex microglochin* on Ben Lawers, 1928.
Proc. Bot. Soc. Br. Isl. 1956, 190–92.
Plants at Oxford and in collection of J. E. Lousley. Drawings of British orchids at BM(NH).

DAVY, Joseph Burtt *see* Burtt Davy, J.

DAVYES, Robert (1616–1666)
b. Gwysaney, Flintshire 19 Feb 1616
Catalogue of British names of plants in T. Johnson's ed. of Gerard's *Herball* 1633.
H. Davies *Welsh Botanology* 1813, vi.

DAWBER, Margaret (*fl.* 1880s–1890s)
D. H. Kent *British Herb.* 1957, 51.
Plants at Oxford, Nottingham.

DAWE, Morley Thomas (1880–1943)
b. Sticklepath, Devon 9 Sept. 1880 *d.* Kyrenia, Cyprus 14 July 1943
Kew gardener, 1900. Assistant, Botany, Forestry and Scientific Dept., Uganda, 1902; Head of Dept. 1903–10. Director, Botanic Gardens, Entebbe. Director of Agriculture, Cia de Mozambique, 1910–14. Agricultural adviser, Colombia. Investigated agriculture of Gambia, 1920 and later Angola. Commissioner of Lands and Forests, Sierra Leone, 1922. Director of Agriculture, Cyprus, 1928. Director of Agriculture and Fisheries, Palestine. Collected plants in Uganda, 1903–10; Mozambique, 1911–12; Brazil, 1914; Colombia, 1915–20; Angola, 1921–22; Sierra Leone, 1922–23. *Report on Botanical Mission...Buddu and Western and Nile Provinces of Uganda* 1905.
J. Linn. Soc. 1906, 495–544. *Kew Bull.* 1907, 234; 1939, 32. *Tropical Life* Oct. 1909, 166 portr. *Gdnrs Chron.* 1924 i 312 portr. *J. Kew Guild* 1925, 281–82 portr.; 1943, 301-2. *Mocambique* 1939, 65–66.
Plants at Kew. Portr. at Hunt Library.
Citharlxylum dawei Moldenke.

DAWES, John Samuel (1802–1878)
b. Birmingham 1802 *d.* Birmingham 20 Dec. 1878
FGS 1842. Ironmaster. 'Sternbergia' (*Quart. J. Geol. Soc.* 1845, 91–92). 'Halonia' (*Quart. J. Geol. Soc.* 1848, 289–91). 'Calamites' (*Quart. J. Geol. Soc.* 1849, 30–31; 1851, 196–99).
Proc. Geol. Soc. 1878–79, 54–55.

DAWES, William (*c.* 1758–1836)
d. Antigua 1836
Royal Marines. Arrived Botany Bay, 1788, where he built the first observatory in Australia. Collected plants during expeditions in New South Wales. Returned to England, 1791.
Hist. Rec. N.S.W. v.1–v.3, 1892–95 *passim. J. Proc. R. Austral. Hist. Soc.* 1924, 1–24; v.12, 227–30; v.13, 63–64. P. Serle *Dict. Austral. Biogr.* v.1, 1949, 225–27.

DAWNEY, James (*fl.* 1840s)
Nurseryman, Aylesbury, Bucks.

DAWODU, Thomas B. (–1920)
d. Ebute Metta, Nigeria 25 May 1920
Kew gardener, 1893–94. Assistant Curator, Lagos Botanic Station, 1894–1920. *Provisional List of...Flowering Plants of...Lagos and Ebute Metta* 1902.
Kew Bull. 1907, 234. *J. Kew Guild* 1921, 43 portr.

DAWSON, Jackson Thornton (1841–1916)
b. East Riding, Yorks 5 Oct. 1841 *d.* Jamaica Plain, Mass., U.S.A. 3 Aug. 1916
To U.S.A. with parents, 1849. First Superintendent, Arnold Arboretum.
Trans. Massachusetts Hort. Soc. 1916, 241–43. *J. Arnold Arb.* 1922, 168–69.
Portr. at Hunt Library.

DAWSON, Sir John William (1820–1899)
b. Pictou, Nova Scotia, Canada 13 Oct. 1820 *d.* Montreal 19 Nov. 1899
MA Edinburgh 1842. LLD 1884. FRS 1862. Knighted 1884. Professor of Geology and Principal, McGill College, Montreal, 1855–93. *Geological History of Plants* 1888.
Amer. Geol. v.26, 1900, 1–48 portr. *Bull. Geol. Soc. Amer.* 1900, 550–80. *D.N.B. Supplt. 1* v.2, 120–22. *Taxon* 1967, 81–85; 1974, 482 portr. *R.S.C.* v.12, 187; v.14, 506.
Portr. at Redpath Museum, Montreal.
Megalopteris dawsoni Hartt.

DAWSON, Warren Royal (1888–1968)
b. Ealing, Middx 13 Oct. 1888 *d.* 5 May 1968
ALS 1959. FRSE. Underwriter. *Leechbook* 1934. *J. E. Smith Papers* 1934. *T. H. Huxley Papers* 1946. *Banks Letters* 1958; Supplements 1962 and 1965.
Times 14 May 1968. *Who was Who, 1961–1970* 288.
MSS. and letters at BM. Letters at Kew.

DAWSON, William (*c.* 1714–*c.* 1776)
b. Leeds, Yorks *c.* 1714 *d.* Leeds *c.* 1776
Of Leeds. Surgeon and apothecary. Correspondent of J. Sherard, R.Richardson, etc. Contrib. to J. Blackstone's *Specimen Botanicum* 1746. List of Yorkshire plants in *Naturalist* 1855, 145–50.
R. Pulteney *Hist. Biogr. Sketches of Progress of Bot. in England* v.2, 1790, 272.
Herb. formerly at Ripon Mechanics' Institute.

DAWSON, William (*fl.* 1790s)
Seedsman, Whitby, Yorks.

DAY, E. (*fl.* 1830s)
Nurseryman, Shipston-on-Stour, Warwickshire

DAY, E. M. (1865–1934)
Of Minchinhampton, Glos. Specialised in Basidiomycetes.
H. J. Riddelsdell *Fl. Gloucestershire* 1948, cxlviii.
Herb. at Gloucester Museum. Fungi at Kew. Fungi drawings at BM(NH).

DAY, Francis Morland (1890–1962)
b. Burton-on-Trent, Staffs 1890 *d.* 16 Sept. 1962
Educ. Cambridge. Headmaster, Colwall near Malvern. Botanist. President, Worcestershire Nat. Club, 1958–59. Contrib. to *Trans. Worcs. Nat. Club.*
Proc. Bot. Soc. Br. Isl. 1963, 194.
Plants at Kew.

DAY, Gwendolen Helen (*c.* 1884–1967)
d. 20 June 1967
President, Bedford Natural History and Archaeological Society, 1935. Contrib. to J. G. Dony's *Fl. Bedfordshire* 1953.
Bedfordshire Nat. 1965–67, 46–47.

DAY, John (1824–1888)
b. London 3 Feb. 1824 *d.* Tottenham, London 15 Jan. 1888
FLS 1869. Amateur orchid grower of Tottenham. His orchid collection was sold in 1881 for £7000. Visited India, Ceylon, Malaya, Brazil and Jamaica and collected plants. Helped in J. Veitch's *Manual of Orchidaceous Plants* 1887–94.
Garden v.33, 1888, 83. *Gdnrs Chron.* 1888 i 88. *J. Hort. Cottage Gdnr* v.16, 1888, 60. *J. Bot.* 1888, 1–6. *Kew Bull.* 1906, 177–79.

Fl. Malesiana v.1, 1950, 131. *Orchid Rev.* 1972, 170–76; 1975, 19–21, 243–44.
Plants and 53 scrapbooks of orchid illustrations at Kew.

DAY, Robert Hague (1848–1928)
b. Liverpool 29 July 1848 *d.* 4 Jan. 1928
Bank manager, Bootle. Contrib. plant records to Appendix 3 of *Fl. Liverpool* 1887.
N. Western Nat. 1930, 236–41.
Herb. at National Museum of Wales.

DAY, William (*fl.* 1840s)
Nurseryman, St. Giles' Field, Oxford.

DEACON, Rev. Ernest (1872–1937)
b. Hethersett, Norfolk 10 March 1872 *d.* 24 Feb. 1937
Educ. Durham University. Held livings at Stanton, Croxden, Hartshill and Weeford. Botanised in N. Staffs. Edited *Trans. N. Staffs. Field Club.*
Trans. N. Staffs. Field Club v.71, 1936–37, 33–34 portr.

DEACON, John (–1912)
d. Highbury near Birmingham 11 Jan. 1912
Gardener to N. Chamberlain at Highbury, 1893.
Gdnrs Chron. 1912 i 46 portr.

DEAKIN, Richard (1808/9–1873)
d. Tunbridge Wells, Kent 18 Feb. 1873
MD Pisa 1838. Practised at Sheffield. *Florigraphia Britannica* (with R. Marnock) 1837. *Florigraphia Britannica* 1841–48 4 vols. *Fl. of Colosseum of Rome* 1855. *Flowering Plants of Tunbridge Wells* 1871. Drew plates for J. Bohler's *Lichenes Britannici* 1835.
J. Bot. 1873, 128. *Proc. Sheffield Nat. Club* 1914, 45–54 portr. *Sorby Rec.* v.2 (3), 1967, 5.
Lichen herb. and drawings at BM(NH). Letters at Kew.

DEAL, William (*c.* 1862–1912)
d. 2 Jan. 1912
Nurseryman, Kelvedon, Essex *c.* 1894. Specialised in sweet peas.
J. Hort. Home Farmer v.64, 1912, 45.

DE ALWIS, Harmanis (–1894)
b. Ceylon *d.* Peradeniya, Ceylon 10 June 1894
Draughtsman at Botanic Gardens, Peradeniya, 1823–61. Helped G. H. K. Thwaites with *Enumeratio Plantarum Zeylaniae* 1858–64.
J. Bot. 1894, 255–56. H. Trimen *Handbook Fl. Ceylon* v.5, 1900, 379.
Drawings at Peradeniya.
Alwisia Lindl.

DE ALWIS SENEVIRATNE, William (1843–1916)
b. Peradeniya, Ceylon 15 Sept. 1843 *d.* Peradeniya 30 Jan. 1916
Draughtsman at Botanic Gardens, Peradeniya, 1865–1902.
Ann. R. Bot. Gdns Peradeniya v.1, 1901–2, 6, 266.

DEAN, Alexander (1832–1912)
b. Hill near Southampton, Hants 22 March 1832 *d.* Kingston-upon-Thames, Surrey 20 Aug. 1912.
VMH 1904. Seed grower with his brother Richard at Bedfont, Middx. *Vegetable Culture* 1899. *Root and Stem Vegetables* 1910. Contrib. to horticultural periodicals and also weekly gardening article to *Reynolds News*.
Gdnrs Chron. 1904 ii 343 portr.; 1912 ii 160 portr. *J. Hort. Cottage Gdnr* v.49, 1904, 572–73 portr.; v.65, 1912, 192 portr. *Gdnrs Mag.* 1908, 257–58 portr.; 1912, 656 portr. *Garden* 1910, 263, 264–65 portr.

DEAN, James Godfrey (*c.* 1830–1900)
b. Slough, Bucks *d.* Titsey, Surrey 2 Aug. 1900
Gardener to Leveson-Gowers at Titsey Place, Surrey, 1855–1900.
Gdnrs Chron. 1900 ii 116–17.

DEAN, Richard (1830–1905)
b. Southampton, Hants 1 Feb. 1830 *d.* Ealing, Middx 21 Aug. 1905
VMH 1897. Secretary, National Floricultural Society. Secretary, National Chrysanthemum Society, 1890.
Gdnrs Chron. 1902 i 98; 1905 ii 169 portr. *Gdnrs Mag.* 1902, 79–80 portr.; 1905, 556 portr. *Garden* v.68, 1905, 134–35 portr. *J. Hort. Cottage Gdnr* v.51, 1905, 175 portr., 211.

DEAN, William (*c.* 1767–1881)
d. Croome Park, Worcs 11 May 1881
Gardener to Earl of Coventry at Croome d'Abitot, Worcs. *Account of Croome d'Abitot ...with...Hortus Croomensis* 1824.
Bot. Mag. 1800, t.466. *Gdnrs Mag.* 1831, 512.

DEAN, William (1825–1895)
b. Southampton, Hants 8 July 1825 *d.* 23 March 1895
Nurseryman, Shipley Nursery near Bradford, Yorks., 1857–*c.* 1876. Established Heath End Nursery, Farnham, Surrey. Florist and seedsman, Sparkhill, Birmingham. Specialised in cultivation of pansies and violas. Joint-editor with J. Sladden of *Gossip of the Garden* 1859. Editor of *Florists' Guide*.
Garden v.47, 1895, 227. *Gdnrs Chron.* 1895 i 404. *Gdnrs Mag.* 1895, 194, 208. *J. Hort. Cottage Gdnr* v.30, 1895, 270 portr.; 296.

DEANE, Henry (1847–1924)
b. Clapham, London 26 March 1847 *d.* Melbourne 12 March 1924

MA Galway. FLS 1886. Railway engineer. Joined N.S.W. Railway Construction Dept., 1880, eventually becoming Engineer-in-Chief. Authority on Tertiary flora of Australia. President, Royal Society, N.S.W. 1897 and 1907. President, Linnean Society N.S.W., 1895–97. Contrib. papers on *Eucalyptus* (with J. H. Maiden) to *Proc. Linn. Soc. N.S.W.* 1895–1901. Text for R. A. Fitzgerald's *Australian Orchids* v.2(5), 1894. *R.S.C.* v.14, 511.

Nature v.113, 1924, 865. *Proc. Linn. Soc.* 1923–24, 48–54. *Proc. Linn. Soc. N.S.W.* 1924, iv–v. *J. R. Soc. N.S.W.* 1924, 4–5. *Proc. R. Soc. Victoria* 1925, 255–56. *Austral. Encyclop.* 1965, 217–18.

MSS. at National Library, Canberra.

DEASY, Henry Hugh Peter (1866–1947)
b. Dublin 29 June 1866 *d.* 24 Jan. 1947

Captain, 16th Queen's Lancers. Travelled in Tibet with A. Pike, 1896; alone in 1897–98.

Geogr. J. v.9, 1897, 217–18. *Kew Bull.* 1897, 208; 1901, 18. E. Bretschneider *Hist. European Bot. Discoveries in China* 1898, 812. *J. Linn. Soc.* v.35, 1902, 158–61. *Who was Who, 1941–1950* 301. *Pakistan J. For.* 1967, 343.

Tibetan plants at Kew. Chinese Turkestan plants at BM(NH).

DEBBARMAN, Pyari Mohan (1887–1925)
b. Agartala, Bengal, India 6 Feb. 1887 *d.* Lucknow, India 8 Jan. 1925

BSc Calcutta. FLS 1920 Assistant, Botanical Survey of India, 1913. Contrib. to *J. Bombay Nat. Hist. Soc.* v.27, 1920, 179–81 and *J. Indian Bot. Soc.* v.3, 1922–23 *passim.*

J. Indian Bot. Soc. v.5, 1926, 47–48. *Proc. Linn. Soc.* 1925–26, 78.

DEBY, Julien (1826–1895)
b. Laeken, Belgium 10 March 1826 *d.* Sheffield, Yorks 14 April 1895

FRMS. Diatomist. Settled in London *c.* 1877.

Le Diatomiste v.2, 1895, 197–200 portr. *R.S.C.* v.14, 515.

Collection at BM(NH) (MS. Catalogue by J. Rattray).

Debya Pat.

DECKER, Sir Matthew (1679–1749)
b. Amsterdam, Holland 1679 *d.* 18 March 1749

London merchant. Director, East India Company. Sheriff of Surrey, 1729. In his garden on Richmond Green, Surrey the pineapple was grown in this country for the first time by his gardener Henry Telende. Painting by T. Netscher in Fitzwilliam Museum, Cambridge depicts the event.

R. Bradley *General Treatise of Husbandry and Gdning for...July 1725* 206–220. *D.N.B.* v.14, 266–67.

DE CRESPIGNY, Eyre Champion (1821–1895)
b. Vevey, Switzerland 5 May 1821 *d.* Beckenham, Kent 15 Feb. 1895

MD Heidelberg. In India, 1845–62. Conservator of Forests and Superintendent, Botanic Garden, Dapuri, Poona, 1859. *New London Fl.* 1877.

J. Bot. 1895, 127. D. H. Kent *Hist. Fl. Middlesex* 1975, 24.

Indian and Swiss plants at Manchester; S. African and Indian plants at Kew. Drawings of Indian plants at BM(NH).

DEERING, George Charles (*c.* 1690–1749)
b. Dresden, Saxony *c.* 1690 *d.* Nottingham 12 April 1749

MD Leyden and Rheims. Pupil of Boerhaave and B. de Jussieu. Secretary to Russian Envoy to Queen Anne, 1713. Returned to London, 1719. Member of Botanical Society of London, 1721–26. Physician at Nottingham from 1735. *Catalogus Stirpium... Nottingham* 1738. Assisted J. J. Dillenius in *Historia Muscorum* 1741.

Gent. Mag. 1783 ii 1014. R. Pulteney *Hist. Biogr. Sketches of Progress of Bot. in England* v.2, 1790 257–64. J. Nichols *Illus. Lit. Hist. of 18th Century* v.1, 211–20; v.3, 571 (contains fragments of Deering's autobiography). G. C. Gorham *Mem. of John and Thomas Martyn* 1830, 19. *Notes and Queries* v.1, 1850, 375. *D.N.B.* v.14, 279–80. *J. Bot.* 1909, 140. E. J. L. Scott *Index to Sloane Manuscripts* 1904, 138. *New Flora Sylva* v.10, 1938, 135–41. *Trans. Thoroton Soc.* v.45, 1941, 24. *Proc. Bot. Soc. Br. Isl.* 1967, 312.

Herb. purchased by Hon. Rothwell Willoughby not traced.

Deeringia R. Br.

DE FRAINE, Ethel Louise (1879–1918)
b. Aylesbury, Bucks 2 Nov. 1879 *d.* Falmouth, Cornwall 25 March 1918

DSc London. FLS 1908. Lecturer in Botany, Battersea Polytechnic, 1910–13; Westfield College, 1915. Contrib. to *Ann. Bot.*

Nature v.101, 1918, 150–51. *Proc. Linn. Soc.* 1917–18, 37–38.

DE LA CROIX *see* MacEncroe, D.

DELANY, Mary (*née* **Granville**) (1700–1788)
b. Coulston, Wilts 14 May 1700 *d.* Westminster, London 15 April 1788

Married Alexander Pendarves, 1718; Rev. Patrick Delany, 1743. In her 70s started her 'paper mosaicks' of flowers. *Autobiography and Correspondence* 1861–62 6 vols.

D.N.B. v.14, 308–10. *Gdnrs Chron.* 1896 ii 164, 248. G. Paston *Mrs. Delany* 1900. *Bot. Soc. Exch. Club Br. Isl. Rep.* 1924, 691–94. R. B. Johnston *Mrs. Delany* 1925. G. C. Druce *Fl. Buckinghamshire* 1926, lxxxix–xci. S. Dewes *Mrs. Delany* 1940 portr. W. Blunt *Art of Bot. Illus.* 1955, 154–55. *Country Life* 1963, 1273–74 portr.

Collection of 970 paper flower mosaics at BM. (*Connoisseur* v.78, 1927, 220–27). Portr. at National Portrait Gallery.

DELAP, Mrs. Drummond *see* Drummond-Delap, *Mrs.*

DELAP, Maude Jane (1866–1953)
b. County Kerry 7 Dec. 1866 *d.* Valentia Island, County Kerry 23 July 1953

ALS 1936. Contrib. to R. Scully's *Fl. County Kerry* 1916.

Irish Nat. J. 1958, 221–22 portr.

DE MOLE, Fanny Elizabeth (1835–1866)
b. 1 March 1835 *d.* Burnside, S. Australia 26 Dec. 1866

To Australia, 1856. *Wild Flowers of S. Australia* 1861.

Rep. Austral. Assoc. Advancement Sci. 1911, 226.

DENDY, Arthur (1865–1925)
b. Patricroft near Manchester 20 Jan. 1865 *d.* London 24 March 1925

DSc Manchester 1886. FRS 1908. FLS 1886. FZS. Demonstrator in Zoology, Melbourne University, 1884–94. Lecturer, later Professor of Biology, Canterbury College, N.Z., 1894–1903. Professor of Zoology, South African College, 1903–5; King's College, London, 1905–25. President, Quekett Microscopical Club, 1912–17. *Introduction to Study of Botany* (with A. H. S. Lucas) 1892. *Outlines in Evolutionary Biology* 1923.

Nature v.115, 1925 540–42. *Proc. Linn. Soc.* 1924–25, 67–71. *Proc. R. Soc.* v.99, 1926, xxxiii–xxxv portr. *Victorian Nat.* 1926, 243–44.

DENHAM, Dixon (1786–1828)
b. London 1 Jan. 1786 *d.* Freetown, Sierra Leone 8 May 1828

FRS 1826. Lieut-Colonel. African traveller. On expedition to N. Nigeria with H. Clapperton and W. Oudney. Superintendent of Liberated Africans, Sierra Leone, 1827. Governor, Sierra Leone, 1828. *Narrative of Travels...Northern and Central Africa...1822–1824* (with H. Clapperton) 1826.

D.N.B. v.14, 341–42.

Plants at BM (NH).

Denhamia Meissner.

DENISON, Henry (*fl.* 1790s)

Gardener and seedsman, Carlisle, Cumberland.

DENISON, N. (*fl.* 1860s–1880s)

Government officer, Sarawak, 1869. Secretary of Resident Magistrate, Perak, 1876. Superintendent, Lower Perak, 1881.

Fl. Malesiana v.1, 1950, 133.

Plants at Singapore.

DENISON, Sir William Thomas (1804–1871)
b. London 3 May 1804 *d.* East Sheen, Surrey 19 Jan. 1871

Lieut-General, Royal Engineers. Collected plants in Canada, 1827–31. Governor of Tasmania, 1846–55; New South Wales, 1855–61; Madras, 1861–66.

D.N.B. v.14, 355–57. M. Willis *By Their Fruits* 1949, 30, 78.

Plants at Melbourne.

Denisonia F. Muell.

DENNES, George Edgar (*fl.* 1817–1860s)
b. 18 March 1817 *d.* Australia?

FLS 1838. Solicitor. Secretary, Botanical Society of London, 1839–56. To Australia 1856? *London Cat. of British Plants* (edited with H. C. Watson) 1844. Notes on *London Cat. of British Plants* in *Phytologist* v.1, 1844, 1014–17, 1098–1100; v.2, 1847, 815.

F. Boase *Modern English Biogr.* v.2, 1912, 75. *J. Bot.* 1922, 364.

Plants at Kew, Glasgow University.

Vicia dennesiana Wats.

DENNING, William (1837–1910)
b. 1 April 1837 *d.* Hampton, Middx 2 April 1910

Kew gardener, 1855–56. Gardener to Lord Bolton and Lord Londesborough. Nurseryman, Hampton, 1887.

Garden 1910, 196. *Gdnrs Chron.* 1910 i 240 portr. *J. Hort. Home Farmer* v.60, 1910, 337. *Orchid Rev.* 1910, 135.

Lycaste denningiana Reichenb. f.

DENNIS, Norman (1912–1966)
d. Sept. 1966

FLS 1964. Jesuit priest. 'Records of non-vascular Cryptogams in Skye' (*Trans. Bot. Soc. Edinburgh* v.40, 1965–66, 204–31).

Proc. Linn. Soc. 1968, 145.

DENNIS, William (*fl.* 1820s)

Florist, 23 Grosvenor's Row, Chelsea, London.

DENNY, John (*c.* 1819–1881)
d. Stoke Newington, London 18 Nov. 1881
Medical Officer, Stoke Newington Dispensary for 34 years. Raised zonal pelargoniums. Founder member and Treasurer of Pelargonium Society. Contrib. to *Florist Pomologist*.
Gdnrs Chron. 1881 ii 702. *J. Hort. Cottage Gdnr* v.3, 1881, 474. *Gdnrs Yb. Almanack* 1882, 193.

DENNYER, Josiah (1869–1936)
Plants at Haslemere Museum, Surrey.

DENNYS, Nicholas Belfield (–1900)
d. Hong Kong 1900
Civil Dept., Royal Navy. Interpreter, Peking, 1863. Editor, *China Mail* (Hong Kong) and Curator of Museum, 1866–76. Assistant Protector of Chinese and Curator of Museum, Singapore, 1877. Magisterial duties at Singapore, 1879–88. Magistrate, British N. Borneo, 1894; Acting Judge, 1899. Editor, *British N. Borneo Herald.*
J. Straits Branch R. Asiatic Soc. 1901, 106–7. *Fl. Malesiana* v.1, 1950, 133.
Malayan and Bornean plants at Singapore.

DENSON, John (*fl.* 1820s–1870s)
ALS 1832. Gardener to N. S. Hodson at Bury St. Edmunds, Suffolk. Curator, Botanic Garden, Bury St. Edmunds, 1821–29. *Cat. of Hardy Trees...in Botanic Garden, Bury St. Edmunds* 1822. Botanical assistant to J. C. Loudon (*Arboretum et Fruticetum Britannicum* v.1, 1838, viii–ix). Edited Loudon's *Mag. Nat. Hist.* Contrib. to *Gdnrs Mag.*
Bot. Mag. 1823, t.2422; 1825, t.2581. *J. Bot.* 1900, 224–25. *R.S.C.* v.2, 240.

DENT, Mrs. Edith Vere (1863–1948)
d. 12 Oct. 1948
Founded Wild Flower Society.
Bot. Soc. Br. Isl. Yb. 1949, 62.

DENT, Hastings Charles (1855–1909)
b. London 23 June 1855 *d.* South Godstone, Surrey 6 March 1909
FLS 1885. Railway engineer. Entomologist. Collected plants in Thursday Islands, Vancouver Islands, China, Brazil. *A Year in Brazil* 1886.
J. Bot. 1909, 435–36. *Proc. Linn. Soc.* 1908–9, 35–36.
Plants at Manchester Museum.
Habenaria dentii Ridley.

DENT, Hilda Sophia Annesley (1903–1956)
b. 5 Oct. 1903 *d.* 19 Sept. 1956
Daughter of Mrs. E. V. Dent (1863–1948). President, Wild Flower Society.
Proc. Bot. Soc. Br. Isl. 1957, 325.

DENT, Peter (–1689)
d. Cambridge Oct. 1689
MB Lambeth 1678; Cantab 1680. "An eminent apothecary and botanist in the University [of Cambridge]" (J. Petiver *Philos. Trans. R. Soc.* v.27, 1712, 385). Friend of John Ray. Discovered *Papaver dubium.*
J. Ray *Historia Plantarum* v.1, 1686, 856. R. Pulteney *Hist. Biogr. Sketches of Progress of Bot. in England* v.1, 1790, 200. R. A. Pryor *Fl. Hertfordshire* 1887, 1. *D.N.B.* v.14, 378. C. E. Raven *John Ray* 1942, 54, 217, 227, 245.

DENTON, Sir George Chardin (1851–1928)
b. 22 June 1851 *d.* 9 Jan. 1928
Colonial Secretary, Lagos, 1888. Collected plants with his wife in S. Nigeria, 1895–1900. Governor of Gambia, 1900–11.
Who was Who, 1916–1928 281.
Plants and MSS. at Kew.

DENTON, William C. (1887–1953)
d. 1 April 1953
Grew *Euphorbia* and stemless *Mesembryanthemum.*
Cactus Succulent J. Great Britain v.15, 1953, 49.

DENYER, E. (*fl.* 1830s)
Nurseryman, Loughborough Road, Brixton, London.

DERMER, Thomas (*fl.* 1790s)
Seedsman, 25 Fenchurch Street, London. Partner of James Gordon.

DERRY, Robert (*fl.* 1880s–1910s)
Kew gardener. At Botanic Garden, Berbice, British Guiana. Assistant Superintendent of Forests, Malacca, 1885–88. In Malacca and Perak, 1889–1903. Assistant Superintendent, Botanic Gardens, Singapore, 1904–8. Superintendent of Gardens and Forests, Penang, 1908. Curator, Botanic Gardens, Singapore, 1909–13.
Kew Bull. 1896, 96; 1903, 31. H. N. Ridley *Fl. Malay Peninsula* v.1, 1922, xviii. *Gdns Bull. Straits Settlements* 1927, 120. *Fl. Malesiana* v.1, 1950, 133–34.
Plants at Kew, Singapore.

DES VANDES, Comtesse (–1832)
d. London 4 Feb. 1832
Had garden in Bayswater, London. Many of her plants were figured in botanical magazines.
Gdnrs Mag. 1832, 256.

DE TABLEY, 3rd Baron *see* Warren, J. B. L.

DEVONSHIRE, William George Spencer, 6th Duke *see* Cavendish, W.

DEWAR, A. (fl. 1840s)
MD. Of Dunfermline, Fife. Botanised in Clackmannan, Kinross, Perth and Fife.
Ann. Rep. Bot. Soc. Edinburgh v.3, 1850, 112–15. *Trans. Bot. Soc. Edinburgh* v.13, 1879, 214–15.
Plants at Kew.
Hieracium dewari Syme.

DEWAR, Daniel (*c.* 1860–1905)
b. Perthshire *c.* 1860 *d.* New York 7 May 1905
Kew gardener, 1880–93. Curator, Glasgow Botanic Garden, 1893–1902. *Synonymic List... of Genus Primula* 1886. With C. H. Wright revised G. W. Johnson's *Gdnrs Dict.* 1894. Contrib. to *Garden.*
Gdnrs Mag. 1893, 104 portr. *J. Kew Guild* 1905, 266–67 portr.

DEWHURST, John (1746–1818)
d. Southowram, Halifax, Yorks 22 Dec. 1818
Assisted James Bolton.
Gent. Mag. 1819 i 90. C. Crossland *Eighteenth Century Nat.: James Bolton* 1910, 28–29.

DEWHURST, John (*fl.* 1750s–*c.* 1835)
d. Salford, Manchester *c.* 1835
Fustian-cutter. President, Manchester Society of Botanists.
J. Cash *Where there's a Will there's a Way* 1873, 16–20. L. H. Grindon *Country Rambles* 1882, 207. *Trans. Liverpool Bot. Soc.* 1909, 68.

DICK, John Harrison (1877–1918)
b. Edinburgh 13 Oct. 1877 *d.* Brooklyn, New York 25 March 1918
Gardener. Reporter for *Gdning World* 1899. Sub-editor, *J. Hort.*, 1901; Editor, 1911. Editor, *Florists' Exchange* 1913. *Sweet Peas for Profit* 1914. *Commercial Carnation Culture* 1915.
Gdnrs Chron. 1918 i 182 portr.

DICK, R. K. (*fl.* 1800s)
Introduced *Wrightia coccinea* to Calcutta Botanic Gardens, 1805.
Bot. Mag. 1826, t.2696.

DICK, Robert (1811–1866)
b. Tullibody, Clackmannanshire Jan. 1811 *d.* Thurso, Caithness 24 Dec. 1866
Apprenticed to a baker. Self-taught botanist and geologist. Rediscovered *Hierochloe borealis*. Botanised in Caithness.
Geol. Mag. 1867, 142–44. *J. Hort. Cottage Gdnr* v.37, 1867, 130–31. J. Cash *Where there's a Will there's a Way* 1873, 170–72. S. Smiles *Robert Dick* 1878. *D.N.B.* v.15, 16. *Trans. Bot. Soc. Edinburgh* v.23, 1905, 44. *Trans. Inverness Sci. Soc. Field Club* 1912–18, 395–97. *Nature* v.210, 1966, 253. *Natur. Mus. Frank* 1966, 491–500 portr.
Herb. at Thurso Museum.

DICKENSON, Rev. Samuel (1730–1823)
d. Blymhill, Staffs 1823
LLB Cantab 1755. Rector, Blymhill. Contrib. notes on *Agrostis*, etc. to W. Withering's *Bot. Arrangement* ed. 3 1796 (see preface). Also contrib. to S. Shaw's *Hist. and Antiq. of Staffordshire* 1798.
J. Forbes *Salictum Woburnense* 1829 preface v–vi. *J. Bot.* 1901 Supplt. 71; 1914, 317. E. S. Edees *Fl. Staffordshire* 1972, 17.

DICKIE, George (1812–1882)
b. Aberdeen 23 Nov. 1812 *d.* Aberdeen 15 July 1882
MA Aberdeen 1830. MD 1842. MRCS London 1834. FRS 1881. ALS 1839. FLS 1863. Lectured in Botany and Zoology at Aberdeen. Professor of Natural History, Belfast, 1849–60. Professor of Botany, Aberdeen, 1860–77. *Fl. Aberdonensis* 1838. *Botanist's Guide to Counties of Aberdeen, Banff and Kincardine* 1860. *Fl. Ulster* 1864. Contrib. algae to G. Henderson and A. O. Hume's *Lahore to Yarkand* 1873.
H. C. Watson *New Botanists' Guide* 1835, 489–97. H. C. Watson *Topographical Bot.* 1873, 522. *J. Bot.* 1883, 30. *Proc. R. Soc.* v.34, 1882–83, xii–xiii. *Scott. Nat.* 1883, 3–8. *Trans. Bot. Soc. Edinburgh* v.16, 1886, 1–6. *Proc. Linn. Soc.* 1882–83, 40–41. S. A. Stewart and T. H. Corry *Fl. N.E. Ireland* 1888, xx–xxi. *D.N.B.* v.15, 32. *Proc. Belfast Nat. Field Club* v.6, 1913, 619. *Belfast Nat. Hist. Philos. Soc. Centenary Vol.* 1924, 71–72, 125–26. R. L. Praeger *Some Irish Nat.* 1949, 71. *R.S.C.* v.2, 283; v.7, 531; v.9, 696; v.14, 600.
Algae at BM(NH). Letters at BM(NH), Kew. Portr. at Hunt Library.

DICKINS, Frederick Victor (1838–1915)
b. 24 May 1838 *d.* Seend, Wilts 16 Aug. 1915
MB. FLS 1883. Surgeon on HMS 'Coromandel', 1863–65. Assistant Secretary, London University. Barrister, Yokohama. Studied Japanese ferns. Collected plants in Hong Kong and Japan. 'Japanese Bot.' (*Nature* v.69, 1904, 389). Sent plants and drawings from Hong Kong and Yokohama to J. D. Hooker at Kew.
Hooker's Icones Plantarum v.17, 1886, t.1659. *Nature* v.95, 1915, 708. *Who was Who, 1897–1916* 196–97. *R.S.C.* v.14, 600.
Plants and letters at Kew.
Nephrodium dickinsii Hook. f.

DICKINSON, Carola I. *see* Meikle, C. I.

DICKINSON, Francis (1816–1901)
b. Coalbrookdale, Shropshire 4 Jan. 1816 *d.* 24 Aug. 1901
Contrib. to W. A. Leighton's *Fl. Shropshire* 1841.
J. Bot. 1901, 434.

DICKINSON, John (*fl.* 1690s)
Sent plants from Bermuda to J. Petiver, 1692.
J. Petiver *Musei Petiveriani* 1695, 80. *J. Bot.* 1883, 258–61; 1919, 45–46. E. J. L. Scott *Index to Sloane Manuscripts* 1904, 143. J. E. Dandy *Sloane Herb.* 1958, 124–25.
Plants at BM(NH).

DICKINSON, Joseph (*c.* 1805–1865)
b. Lampleigh, Whitehaven, Cumberland *c.* 1805 *d.* Liverpool 21 July 1865
MD Dublin 1843. FRS 1854. FLS 1839. Lecturer in Botany, Liverpool School of Medicine, 1839. Secretary, Liverpool Botanic Garden. President, Liverpool Literary and Philosophical Society. *Fl. Liverpool* 1851; *Supplement* 1855.
Proc. Liverpool Lit. Philos. Soc. v.20, 1865–66, 3–4. J. B. L. Warren *Fl. Cheshire* 1899, lxxx. W. Munk *Roll of R. College of Physicians* v.4, 1878, 104–5. *D.N.B.* v.15, 36. *Trans. Liverpool Bot. Soc.* 1909, 68. *R.S.C.* v.2, 285.
Herb. at Liverpool University.

DICKINSON, Thomas (*fl.* 1850s)
Nurseryman, Guildford Nursery, London Road, and 34 High Street, Guildford, Surrey.

DICKINSON, William (*c.* 1799–1882)
b. Arlecdon, Cumberland *c.* 1799 *d.* Workington, Cumberland 22 June 1882
FLS 1855. Land-surveyor. Had herb., later in possession of family, now lost.
J. G. Baker *Fl. English Lake District* 1885, 12. W. Hodgson *Fl. Cumberland* 1898, xxix–xxx. *R.S.C.* v.2, 285.

DICKS, John (*fl.* 1770s)
Gardener to Duke of Kingston at Knightsbridge. *New Gardener's Dictionary* 1771.
J. C. Loudon *Encyclop. Gdning* 1822, 1275.

DICKS, S. B. (*c.* 1846–1926)
d. West Norwood, London 4 Jan. 1926
VMH 1925. Seedsman.
Gdnrs Chron. 1926 i 36.

DICKSON, Rev. Adam (–1776)
d. 25 March 1776
Minister, Duns, Berwickshire 1750–70; Whittingham, East Lothian, 1770–76. *Husbandry of the Ancients* 1778.
Cottage Gdnr v.7, 1851, 157.

DICKSON, Alexander (*c.* 1802–1880)
b. Hawthornden, Lothian *c.* 1802 *d.* Newtownards, County Down 11 Oct. 1880
Founded nursery at Newtownards 1836.
Garden v.18, 1880, 418. *Gdnrs Chron.* 1880 ii 541. *Florist and Pomologist* 1880, 176.

DICKSON, Alexander (1836–1887)
b. Edinburgh 21 Feb. 1836 *d.* Hartree, Peebles 30 Dec. 1887
MD Edinburgh 1860. MD Dublin. LLD Glasgow. FLS 1875. Professor of Botany, Trinity College, Dublin, 1866–68; Glasgow, 1868–79; Edinburgh and Regius Keeper, Botanic Garden, 1879–87. Contrib. to *Trans. Bot. Soc. Edinburgh*, *Trans. R. Soc. Edinburgh*.
Ann. Bot. v.1, 1888, 396–99. *Gdnrs Chron.* 1888 i 24. *J. Bot.* 1888, 63–64. *Nature* v.37, 1888, 229–30. *Proc. Linn. Soc.* 1887–88, 88–89. *Trans. Bot. Soc. Edinburgh* v.17, 1889, 508–16. *Scott. Nat.* 1888, 242–43. *D.N.B.* v.15, 41. F. W. Oliver *Makers of Br. Bot.* 1913, 300–1. H. R. Fletcher and W. H. Brown *R. Bot. Gdn Edinburgh, 1670–1970* 1970, 183–94 portr. *R.S.C.* v.2, 285; v.7, 532; v.9, 697; v.14, 600.
Letters at Kew.

DICKSON, Alexander (*c.* 1857–1949)
d. Newtownards, County Down 17 May 1949
VMH 1939. Nurseryman with his brother, Hugh at Royal Nurseries, Belmont, Belfast.
Gdnrs Mag. 1897, 410 portr. *J. Hort. Home Farmer* v.63, 1911, 418 portr. *Gdnrs Chron.* 1949 i 204; 1963 i 415 portr.

DICKSON, Archibald (*fl.* 1740s–1770s)
Nurseryman, Hassendeanburn near Hawick, Roxburghshire.
E. Hughes *North Country Life in 18th Century* v.2, 1965, 236.

DICKSON, Edward Dalzell (–1900)
d. Constantinople, Turkey 27 March 1900
MD. Collected oaks, etc. in Kurdistan. Physician at British Embassy, Constantinople.
Bot. Register 1840 Misc. Notes, 39–41; 1841, 24. *Ibis* 1900, 562.
Maltese plants at Cambridge.

DICKSON, Francis (1793–1866)
b. Edinburgh 25 Dec. 1793 *d.* Chester, Cheshire 3 March 1866
Nurseryman, Eastgate Street, Chester, 1820. Nurseries at Bache Pool, Piper's Ash, and Newton. In partnership with James Dickson, 1855. Succeeded by his son Francis Arthur Dickson. Friend of J. C. Loudon and T. A. Knight. The last of the great Dicksons of Edinburgh.
Gdnrs Chron. 1866, 272. *J. Hort. Cottage Gdnr* v.35, 1866, 241.

DICKSON, Francis Arthur (*c.* 1826–1888)
d. 27 Sept. 1888
Son of Francis Dickson (1793–1866). Nurseryman in partnership with James Dickson, Eastgate Street, Chester, 1866–88.
Gdnrs Chron. 1888 ii 392–93.

DICKSON, George (–1825)
Brother of James Dickson (1738–1822). Nurseryman, 10 St. Andrews Street, Edinburgh.

DICKSON, George (1832–1914)
b. 7 July 1832
VMH 1907. Head of Messrs Alexander Dickson and Sons, nurserymen, Newtownards, County Down. Specialised in roses.
J. Hort. Cottage Gdnr v.45, 1902, 473, 474 portr.; v.66, 1913, 228 portr.; v.69, 1914, 171. *Gdnrs Chron.* 1907 ii 201, 218 portr. *Gdnrs Mag.* 1907, 739–40 portr. *Garden* 1914, 440 (vi) portr.

DICKSON, George A. (*c.* 1835–1909)
b. Chester *c.* 1835 *d.* Newton, Chester, Cheshire 6 Feb. 1909
VMH 1897. Son of James Dickson (*c.* 1795–1867). Head of Dicksons Ltd., nurserymen, Chester.
Garden 1909, 96 (xii). *J. Hort. Cottage Gdnr* v.58, 1909, 129. *Gdnrs Chron.* 1909 i 110.

DICKSON, George Frederick (*fl.* 1830s)
Sent seeds from Mexico to Horticultural Society of London.
Bot. Mag. 1841, t.3878; 1843, t.4024.

DICKSON, Hugh (*c.* 1834–1904)
d. Belfast 5 May 1904
Nurseryman, Royal Nurseries, Belmont, Belfast, founded in 1869. Specialised in roses.
J. Hort. Cottage Gdnr v.45, 1902, 473, 475 portr. *Garden* v.65, 1904, 361. *Gdnrs Mag.* 1904, 338 portr.

DICKSON, James (1738–1822)
b. Traquhair, Peebles 1738 *d.* Croydon, Surrey 14 Aug. 1822
FLS 1788. Established herb and seed shop, Covent Garden, London, 1772. One of the foundersof Horticultural Society of London and Linnean Society. Botanised in N. Scotland and Hebrides. Discovered *Draba rupestris*, 1789. *Fasciculus...Plantarum Cryptogamicarum Britanniae* 1785–1801. *Collection of Dried Plants* 1789–91. *Hortus Siccus Britannicus* 1793–98.
A. H. Haworth *Miscellanae Naturalia* 1803, 4 (as 'Dixon'). *Gent. Mag.* 1822 ii 376–77. *Trans. Hort. Soc. London* v.5, 1824 Appendix 1–3. P. Smith *Correspondence of Sir J. E. Smith* v.2, 1832, 234. *J. Bot.* 1886, 103. *Proc. Linn. Soc.* 1888–89, 35. *D.N.B.* v.15, 44. *Gdnrs Chron.* 1904 i 147; 1964 ii 229 portr. *J. R. Hort. Soc.* 1943, 66–72 portr. *Huntia* 1965, 196–97 portr. *Mem. New York Bot. Gdn* v.19, 1969, 14–16. H. R. Fletcher *Story of R. Hort. Soc., 1804–1968* 1969, 33–34 portr. *R.S.C.* v.2, 285.
Herb. and mosses at BM(NH). Letters at BM(NH), Linnean Society. Portr. by H. P. Briggs at Royal Horticultural Society; by Wageman at Linnean Society. Portr. at Hunt Library.
Dicksonia L'Hérit.

DICKSON, James (*fl.* 1810s–1840s)
Nurseryman, Acre Lane, Clapham, Surrey.

DICKSON, James (*c.* 1839–)
Gardener to J. Jardine at Arkleton, Dumfriesshire, 1862.
Gdnrs Chron. 1876 ii 328–29 portr.

DICKSON, James Hill (*c.* 1795–1867)
d. 28 Dec. 1867
Nurseryman, Eastgate Street and Newton Nurseries, Chester, Cheshire.
Gdnrs Chron. 1868, 31.

DICKSON, John (1779–1847)
b. Dalkeith, Midlothian 5 April 1779 *d.* Tripoli, Libya 27 Feb. 1847
Physician in Tripoli, 1818–47, where he collected plants which were acquired by E. S.-C. Cosson and P. B. Webb.
E. Durand and G. Barratte *Florae Libycae Prodromus* 1910, xx.
Plants at Florence.

DICKSON, Joseph (–1874)
d. Berwick-on-Tweed, Northumberland 4 March 1874
MD Edinburgh 1842. Of Jersey. 'Notice of a Few Rare Plants collected...during...1839 in Jersey' (*Mag. Nat. Hist.* 1840, 226–30).

DICKSON, R. W. (*pseud.* **Alexander M'Donald**) (*fl.* 1790s–1810s)
MD. Of Hendon, Middx. Writer on agriculture. *Complete Dictionary of Practical Gardening* 1805–7 2 vols; recast and published anonymously both as *New Botanic Garden* and *New Flora Britannica* 1812.
G. W. Johnson *Hist. of English Gdning* 1829, 282. *Gdnrs Chron.* 1898 i 340.

DICKSON, Robert (*fl.* 1720s)
Founded nursery at Hassendeanburn near Hawick, Roxburghshire, 1728.

DICKSON, Robert (1804–1875)
b. Dumfries 1804 *d.* Harmondsworth, Slough? 13 Oct. 1875
MD Edinburgh 1826. FLS 1831. Practised medicine in London. Lectured on botany at St. George's Hospital. *Dry Rot* 1838. Wrote descriptions for B. Maund's *Botanist* v.1 and v.2.
Gdnrs Mag. 1839, 91. W. Munk *Roll of R. College of Physicians* v.4, 1878, 78–79. *D.N.B.* v.15, 44. *Notes and Queries* v.12, 1903, 149, 194, 236, 257.

DICKSON, T. A. (*c.* 1834–1899)
d. 9 May 1899
Florist, Centre Row, Covent Garden, London.
Gdnrs Chron. 1899 i 306.

DICKSON, Thomas (*c.* 1835–1877)
d. Chester, Cheshire 23 March 1877
Nurseryman, Upton, Cheshire.

DICKSON, W. (*fl.* 1860s)
MD. Physician in Canton, China. Collected *Ehretia dicksonii* in 1861.
E. Bretschneider *Hist. European Bot. Discoveries in China* 1898, 681.

DICKSON, William (–1835)
d. Bellwood, Perth 22 April 1835
Son of Robert Dickson of Hassendeanburn. Nurseryman of Dickson and Turnbull, Perth.
Gdnrs Mag. 1835, 328.

DICKSON, William Alfred (*c.* 1837–1891)
d. Chester, Cheshire 17 Dec. 1891
Nurseryman, Chester.
Gdnrs Chron. 1891 ii 769–70. *J. Hort. Cottage Gdnr* v.23, 1891, 536. *Garden* v.41, 1892, 18–19.

DIEFFENBACH, Ernest (*fl.* 1820s–1840s)
MD. Naturalist to New Zealand Company, 1839–41. Visited Chatham Islands, 1840. *Travels in New Zealand* 1843 2 vols. (botany in v.1, 419–31).
J. R. Geogr. Soc. v.11, 1841, 195–215. T. F. Cheeseman *Manual of New Zealand Fl.* 1906, xxi. A. M. Coats *Quest for Plants* 1969, 236.
Plants and letters at Kew.
Aciphylla dieffenbachii Kirk.

DIGBY, Edward Kenelm, 11th Baron (1894–1964)
b. 1 Aug. 1894 *d.* 29 Jan. 1964
VMH 1958. Had garden at Minterne, Dorset. Hybridised rhododendrons.
J. R. Hort. Soc. 1964, 161 portr. *Orchid Rev.* 1964, 102. *Rhododendron Camellia Yb.* 1965, 9–11. *Who was Who, 1961–1970* 304.

DIGBY, Sir Kenelm (1603–1665)
b. Gayhurst, Bucks 11 June/July 1603 *d.* London 11 June 1665
FRS 1663. Philosopher, writer and soldier. Knighted 1623. Lectured on botany at Gresham College, 1661. *Discourse concerning Vegetation of Plants* 1661.
D.N.B. v.15, 60–66. E. J. L. Scott *Index to Sloane Manuscripts* 1904, 144. J. F. Fulton *Sir Kenelm Digby* 1937. R. T. Petersson *Sir Kenelm Digby* 1956.
Portr. by Van Dyck at National Portrait Gallery. Portr. at Hunt Library.

DILL, Dr. (*fl.* 1840s)
MD. In Hong Kong, 1844, where he collected ferns which he sent to W. J. Hooker at Kew.
E. Bretschneider *Hist. European Bot. Discoveries in China* 1898, 373.

DILLENIUS, Johann Jacob (1684–1747)
b. Darmstadt, Germany 1684 *d.* Oxford 2 April 1747
MD Giessen. MD Oxon 1735. FRS 1724. Came to England 1721. First Sherardian Professor of Botany, Oxford, 1734–47. Edited J. Ray's *Synopsis* ed. 3 1724. *Catalogus Plantarum circa Gissam* 1718. *Hortus Elthamensis* 1732 2 vols. *Historia Muscorum* 1741. Dillenius drew and etched the plates for both these works. "*Dillenia* of all plants has the showiest flower and fruit, even as Dillenius made a brilliant show among botanists" (Linnaeus).
R. Pulteney *Hist. Biogr. Sketches of Progress of Bot. in England* v.2, 1790, 153–84. *Trans. Linn. Soc.* v.7, 1804, 101–15. J. E. Smith *Selection of Correspondence of Linnaeus* v.2, 1821, 82–160. A. L. A. Fée *Vie de Linné* 1832, 126–34. D. Turner *Extracts from Lit. and Sci. Correspondence of Richard Richardson* 1835, 209–12. *J. Bot.* (Hooker) 1834, 88–97. L. W. Dillwyn *Hortus Collinsonianus* 1843, 35. W. Darlington *Memorials of John Bartram* 1849, 309–12. *J. Bot.* 1875, 13–14. *J. Linn. Soc.* v.17, 1881, 553–81. G. C. Druce *Fl. Oxfordshire* 1886, 381–85. *Proc. Linn. Soc.* 1888–89, 35–36. *Sammlung gemeinverständlicher wissenschaftlicher Vorträge* Heft 49–72, 1889, 599–632. G. C. Druce *Fl. Berkshire* 1897, cxxxi–cxxxvi. *D.N.B.* v.15, 79. E. J. L. Scott *Index to Sloane Manuscripts* 1904, 144. G. C. Druce and S. H. Vines *Dillenian Herb.* 1907 portr. J. W. White *Fl. Bristol* 1912, 60–62. J. R. Green *Hist. Bot.* 1914, 163–73. *Bot. Soc. Exch. Club Br. Isl. Rep.* 1923, 351–55 portr. G. C. Druce *Fl. Buckinghamshire* 1926, lxxvi–lxxvii. *Svenska Linné-Sällsk. Arsskr.* v.21, 1938, 85–94. *Nature* v.145, 1940, 993–96. *Trans. Br. Mycol. Soc.* v.25, 1941, 220. J. E. Dandy *Sloane Herb.* 1958, 125. *Hunt Bot. Cat.* v.2 (1) 1961, lxvi–lxviii. H. N. Clokie *Account of Herb. Dept. Bot., Oxford* 1964, 30–36, 86–103. F. A. Stafleu *Taxonomic Literature* 1967, 107–8. C. C. Gillispie *Dict. Sci. Biogr.* v.4, 1971, 98–100.
Herb. and drawings of fungi at Oxford. Drawings of mosses and letters at BM(NH). Letters in Sherard correspondence at Royal Society. Portr. at Oxford Botanic Garden and Bodleian Library; copy at Linnean Society. Portr. at Hunt Library.
Dillenia L.

DILLISTONE, George (1877–1957)
b. Clare, Suffolk 1877 *d.* 30 Oct. 1957
Landscape architect. Editor, *Iris Yb.* Edited *Dykes on Irises* 1930.
Iris Yb. 1957, 26–27.

DILLMAN, John (*fl.* 1730s–1760s)
Gardener to Dame Elizabeth Molyneux at Kew House, Surrey before 1730. Gardener to Frederick Prince of Wales and later to his widow, Princess Augusta, at Kew, Carlton House and Leicester House, 1730–1750s.

DILLON-WESTON, William Alastair Royal (1899–1953)
b. Bristol 28 June 1899 *d.* Newmarket, Suffolk 20 Aug. 1953
MA Cantab 1925. PhD 1929. Mycologist, School of Agriculture, Cambridge, 1922–46. Plant pathologist, Eastern Province, Ministry of Agriculture, 1946–53. *Plant in Health and Disease* (with R. E. Taylor) 1948. *Diseases of Cereals* 1948. *Diseases of Potatoes, Sugar Beet and Legumes* 1948. *Diseases and Pests of Vegetable Crops* (with J. H. Stapley) 1948.
Grower v.40, 1953, 346. *Nature* v.172, 1953, 480.

DILLWYN, Lewis Weston (1778–1855)
b. Ipswich, Suffolk 21 Aug. 1778 *d.* Sketty Hall, Swansea, Glam 31 Aug. 1855
FRS 1804. FLS 1800. Of Walthamstow and (from 1803) Swansea. Porcelain manufacturer and naturalist. *British Confervae* 1802–9 (types at BM(NH)). *Botanist's Guide through England and Wales* (with D. Turner) 1805 2 vols. *Hortus Collinsonianus* 1843. *Materials for Fauna and Fl. of Swansea and Neighbourhood* 1848.
Proc. Linn. Soc. 1856, xxxvi–xxxix. *Quart. J. Geol. Soc. London* 1856, xl. Friend's Institute *Biogr. Cat.* 1888, 176–79. *D.N.B.* v.15, 90–91. G. C. Druce *Fl. Buckinghamshire* 1926, xcvii. *Bot. Soc. Exch. Club Br. Isl. Rep.* 1943–44, 655–56. *S. Wales and Monmouth Rec. Soc. Publ.* no. 5, 1963, 6–8. *Taxon* 1974, 548 portr.
Plants at BM(NH). Letters, lithographic and silhouette portr. at Kew. Portr. at Hunt Library. Letters at Society of Friends, London. Diary for 1817–52 in 36 vols at National Library of Wales. Diary of tour to Killarney, 1809, at Trinity College, Dublin.
Dillwynia Smith.

DILLWYN-LLEWELYN, Sir John Talbot (1836–1927)
b. 26 May 1836 *d.* 7 July 1927
MA Oxon. FLS 1859. VMH 1907. Had garden at Penllergaer, Glam. Vice-president, Royal Horticultural Society.
Garden v.67, 1905 frontispiece portr. *Gdnrs Chron.* 1927 ii 60, 200 portr. *Gdnrs Mag.* 1907 263 portr., 264. *Bot. Soc. Exch. Club Br. Isl. Rep.* 1927, 376. *Orchid Rev.* 1927, 246–47. *Who was Who, 1916–1928* 636.

DINGLEY, Henry (–1589)
d. Feb. 1589
Of Charlton, Worcs. Sheriff of Worcs. Listed plants in West Midlands.

DINGWALL, George (*fl.* 1850s–1900s)
b. Ross-shire
Gardener to Sir H. Campbell-Bannerman at Belmont Castle, Perthshire.
Garden v.66, 1904, 336 portr.

DINSLEY, William Featherstone (1854–1941)
b. Little Barugh, Yorks 3 Oct. 1854 *d.* Manchester 4 May 1941
Inspector, Great Central Railway Co. President, United Field Naturalists' Society, 1928–29. Lancashire botanist.
N. Western Nat. 1941, 210–11.

DISTIN, Henry (*fl.* 1810s–1840s)
MD. Of Westmorland, Jamaica. 'Dr. Doustan' who sent plants to W. Pamplin, now at Oxford?
I. Urban *Symbolae Antillanae* v.3, 1902, 37. H. N. Clokie *Account of Herb. Bot. Dept., Oxford* 1964, 157–58.
Jamaican plants at Kew (as from 'Distan').

DIVERS, John (–1916)
d. France 8/9 Oct. 1916
Kew gardener, 1912–14. Collected plants at Belvoir, Leics and Kew, Surrey.
Kew Bull. 1918, 157.
Plants at Kew.

DIVERS, W. H. (*c.* 1854–1942)
b. near Maidstone, Kent *d.* Surbiton, Surrey 5 July 1942
VMH 1912. Gardener to Duke of Rutland at Belvoir Castle. Lecturer on Horticulture, Surrey County Council. *Catalogue of Trees, Shrubs and Plants...of Belvoir Castle* 1904. *Spring Flowers at Belvoir Castle* 1909.
Gdnrs Chron. 1922 i 134 portr.; 1942 ii 40.

DIXON, Abraham (*c.* 1815–1907)
d. 30 April 1907
Had fine collection of yews and aquatic plants at Cherkley Court, Surrey.
Kew Bull. 1907, 246–47.

DIXON, Charles (1834–)
b. Rossie Priory, Perthshire
Gardener to Lady Holland at Holland House, Kensington, London.
Gdnrs Mag. 1910, 139–40 portr. *J. Hort. Home Farmer* v.60, 1910, 510–11 portr.

DIXON, Edmund Philip (1804–1887)
b. Donington, Lincs 4 July 1804 *d.* Hull, Yorks 2 Jan. 1887
Seedsman and florist, 67 Queen Street, Hull.
Gdnrs Chron. 1887 i 87.

DIXON, Frank (1871–1951)
b. Keighley, Yorks 7 May 1871 *d.* York 20 Dec. 1951

BSc London. Headmaster, Lydney Grammar School, Glos, 1903–32. Botanised around Lydney and Forest of Dean.

H. J. Riddelsdell *Fl. Gloucestershire* 1948, cxlviii. *N. Western Nat.* 1953, 454–56.

DIXON, George Brown (*fl.* 1850s–1910s)

Headmaster, Leicester. Entomologist and botanist.

A. R. Horwood and C. W. F. Noel *Fl. Leicestershire* 1933, ccxxx–ccxxxi.

DIXON, Henry Horatio (1869–1953)
b. Dublin 19 May 1869 *d.* Dublin 20 Dec. 1953

BSc Dublin 1892. FRS 1908. FLS 1925. Assistant, Botany Dept., Trinity College, Dublin, 1892–1904; Professor of Botany, 1904–50. Director, Trinity College Botanic Garden, 1906–51. Professor of Plant Biology, Trinity College, 1922. Studied cytology, transpiration and water relations of plants. *Transpiration and Ascent of Sap in Plants* 1914. *Practical Plant Biology* 1922; ed. 2 1943. *Transpiration Stream* 1924. Contrib. to *Proc. R. Dublin Soc.*, *Proc. R. Soc.*

R. L. Praeger *Some Irish Nat.* 1949, 72. *Nature* v.173, 1954, 239–40. *Times* 24 Dec. 1953. *Plant Physiol.* v.14, 615–19, portr. *Proc. Linn. Soc.* v.165, 1955, 213–16. *Obit. Notices Fellows R. Soc.* v.9, 1954, 79–97 portr. *D.N.B. 1951–1960* 1971, 302–3. *Who was Who, 1951–1960* 308. C. C. Gillispie *Dict. Sci. Biogr.* v.4, 1971, 130–31.

Portr. at Hunt Library.

DIXON, Hugh Neville (1861–1944)
b. Wickham Bishops, Essex 20 April 1861 *d.* Northampton 9 May 1944

BA Cantab 1883. MA 1886. FLS 1885. Headmaster at school for deaf, Northampton, 1884–1914. Secretary, Northamptonshire Naturalists' Society and Field Club, 1886–1931. President, British Bryological Society, 1923. *Student's Handbook of British Mosses* (with H. G. Jameson) 1896, 1904, 1924. *Studies in Bryology of New Zealand* 1913–29. Contrib. to *J. Bot.*, *Proc. Linn. Soc.*, *Bryologist.*

Bot. Soc. Exch. Club Br. Isl. Rep. 1943–44, 643–44; 1945, 17. *Bryologist* v.47, 1944, 135–46. *Nature* v.153, 1944, 705–6. *Proc. Linn. Soc.* 1943–44, 202–5. *Rep. Br. Bryological Soc.* 1944, 310–13. *Times* 11 May 1944. *Revue Bryologique Lichénologique* v.15, 1946, 117–19 portr.

Mosses at BM(NH). Plants at South London Botanical Institute. Portr. at Hunt Library.

DIXON, J. Dargue (*c.* 1877–1960)
d. Morecambe, Lancs 22 Feb. 1960

Cultivated ferns. On committee of British Pteridological Society from 1936.

Br. Fern Gaz. 1961, 59–60.

DIXON, James *see* Dickson, J.

DIXON, John H. (–1966)
d. 6 Jan. 1966

With his brother Cyril directed orchid business of Harry Dixon and Sons. Secretary, British Orchid Growers Association shows.

Orchid Rev. 1966, 62–63.

DIXON, William E. (*fl.* 1860s)

Nurseryman, Norwood Nursery, Beverley, Yorks.

DOBBIE, James (*c.* 1817–1905)
b. Gordon, Berwickshire *c.* 1817 *d.* Upper Craigmore, Rothesay 13 Oct. 1905

Nurseryman, Renfrew, 1866; moved to Rothesay, 1875. Sold business to William Cuthbertson in 1887.

Garden v.58, 1900, 199 portr. *Gdnrs Chron.* 1905 ii 297–98; 1965 ii 219 portr. *Gdnrs Mag.* 1905, 687. *J. Hort. Cottage Gdnr.* v.51, 1905, 358.

DOBBS, Arthur (1689–1765)
b. 2 April 1689 *d.* Town Creek, N. Carolina, U.S.A. 28 March 1765

Of Castle-Dobbs, County Antrim. Surveyor-General of Ireland, 1730. Governor, N. Carolina, 1754–65. Investigated insect pollination of flowers. 'Concerning Bees and their Method of Gathering Wax and Honey' (*Philos. Trans. R. Soc.* 1750, 536–49).

D.N.B. v.15, 130–32. *Bull. Torrey Bot. Club* 1949, 217–19.

DOBSON, John (*c.* 1832–1878)
d. Hounslow, Middx 3 May 1878

Nurseryman, Isleworth, Middx.

DOBSON, William (1820–1884)
b. Preston, Lancs 1820 *d.* Chester, Cheshire 8 Aug. 1884

Stationer, journalist and antiquary of Preston. Edited *Preston Chronicle*. *Rambles by the Ribble* 1864–83 3 series (contains floristic notes).

Preston Guardian 13 Aug. 1884. *D.N.B.* v.15, 138. *Trans. Liverpool Bot. Soc.* 1909, 68.

DOD, Anthony Hurt Wolley-*see* Wolley-Dod, A. H.

DOD, Rev. Charles Wolley-*see* Wolley-Dod, *Rev.* C.

DODD, Alfred John (*c.* 1882–1963)
b. Derbyshire *c.* 1882 *d.* 23 Dec. 1963

Collected fungi and mosses.

Somerset Archaeol. Nat. Hist. Soc. v.108, 1963–64, 176.

Mosses at BM(NH).

DODD, E. S. (–*c.* 1965)
Kew gardener, 1908–10. Calcutta Botanic Gardens, 1910–15. Tuxedo Park, New York, 1915; Superintendent, 1948–57.
J. Kew Guild 1966, 714 portr.

DODD, Harry (–1912)
d. Delhi 3 July 1912
Kew gardener, 1904–6. Curator, Onitsha Botanic Station, S. Nigeria, 1906–10. Royal Botanic Gardens, Calcutta, 1911.
Kew Bull. 1906, 224; 1911, 118; 1912, 300–1. *Gdnrs Chron.* 1912 ii 185.
MSS. and Nigerian plants at Kew.

DODD, Rev. John (*fl.* 1800s)
Vicar, Wigton, Cumberland, 1804. Contrib. new localities of Cumberland plants to D. Turner and L. W. Dillwyn's *Botanist's Guide* 1805, 143–65.

DODDS, William (*c.* 1808–1900)
d. Bristol Dec.? 1900
Gardener to Col. Baker at Salisbury, Wilts for 30 years. Specialised in dahlias.
Garden v.59, 1901, 16. *Gdnrs Chron.* 1901 i 15. *Gdnrs Mag.* 1901, 16.

DODGSON, David Scott (*fl.* 1870s)
Colonel. Collected a few plants, now at Kew, in Khasia and Sikkim, 1876.

DODS, George (*fl.* 1860s–1890s)
MD. In charge of missionary hospital, Canton, China, 1863–65. He and his wife collected plants in Hong Kong.
E. Bretschneider *Hist. European Bot. Discoveries in China* 1898, 697–98.

DODSWORTH, Rev. Joseph (1799–1877)
b. Denton, Lincs 1799 *d.* Bourne, Lincs 9 May 1877
BA Oxon 1819. Vicar, Bourne, 1842. Had herb. MS. notes in J. E. Smith's *Compendium Florae Britannicae* 1801 at BM(NH).
J. Bot. 1902, 102.
Plants at Lincoln Museum.

DODSWORTH, Rev. Matthew (1654–1697)
b. Badsworth, Yorks 1654 *d.* Sessay, Yorks 1697
BA Cantab 1674. Oxon 1675. MA 1678. Rector, Sessay, 1690–97. "Rei Herbariae Amantissimus" (L. Plukenet *Almagestum Botanicum* 1696, 201).
R. Morison *Plantarum Historiae Universalis Oxoniensis* v.3, 1699, 230. R. Pulteney *Hist. Biogr. Sketches of Progress of Bot. in England* v.2, 1790, 121. *J. Bot.* 1900, 337; 1909, 99–104. E. J. L. Scott *Index to Sloane Manuscripts* 1904, 146. G. C. Druce *Fl. Oxfordshire* 1927, lxxix–lxxx. C. E. Raven *John Ray* 1942, 245, 249. J. E. Dandy *Sloane Herb.* 1958, 126.
Plants at BM(NH).

DODWELL, Ephraim Syms (1819–1893)
b. Long Crendon, Bucks 28 Nov. 1819 *d.* Oxford 30 Nov. 1893
Cigar merchant. Grew *Dianthus.* Secretary, Oxford Carnation and Picotee Union. *Carnation and Picotee* 1887; ed. 3 1892. Co-founder of *Gossip of Gdn.*
Gdnrs Chron. 1883 ii 113, 117–18 portr.; 1893 ii 728–29 portr. *Florist and Pomologist* 1883, 134–37 portr. (reprinted from *Gdnrs Chron.* 1883). *Garden* v.44, 1893, 546. *J. Hort. Cottage Gdnr* v.27, 1893, 513. *Gdnrs Mag.* 1894, 235–36.

DOIG, D. (1821–1886)
b. Kirkinch, Angus 1821 *d.* Ross Priory, Dunbartonshire Feb. 1886
Gardener to Lord Kinnaird at Ross Priory, Contrib. to *Gdnrs Chron.*
Gdnrs Chron. 1886 i 248.

D'OMBRAIN, Rev. Henry Honywood (1818–1905)
b. Pimlico, London 10 May 1818 *d.* Ashford, Kent 23 Oct. 1905
VMH 1897. Vicar, Westwell, Kent. Founder member and Secretary, National Rose Society. Edited *Floral Mag.*, 1862–73, and *Rosarian's Yb.*
J. Hort. Cottage Gdnr v.21, 1890, 27–29 portr.; v.51, 1905, 408 portr. *Garden* v.57, 1900, 400 portr.; v.68, 1905, 282 portr. *Gdnrs Chron.* 1905 ii 319 portr. *Rose Ann.* 1907, 25–31 portr.

DOMINY, John (1816–1891)
b. Gittisham, Devon 1816 *d.* Chelsea, London 12 Feb. 1891
Gardener with Messrs Veitch at Exeter, later Chelsea, 1834–41, *c.* 1846–80. Produced the first known man-made orchid hybrid, *Calanthe dominii*, 1856. Hybridised nepenthes in addition to orchids.
Gdnrs Chron. 1870, 1181–82; 1891 i 277, 278 portr. *Garden* v.21, 1882, xi–xii portr.; v.39, 1891, 179. *J. Hort. Cottage Gdnr* v.1, 1880, 11–12 portr.; v.2, 1881, 439–40; v.22, 1891, 148–49 portr. J. H. Veitch *Hortus Veitchii* 1906, 99–101. *Amer. Orchid Soc. Bull.* 1956, 666 portr., 667, 679–87. M. A. Reinikka *Hist. of the Orchid* 1972, 196–98 portr.
Cypripedium dominianum Reichenb. f.

DON, David (1799–1841)
b. Doo Hillock, Forfarshire 21 Dec. 1799 *d.* London 8 Dec. 1841
ALS 1823. Librarian to A. B. Lambert and, from 1822, to Linnean Society. Professor of Botany, King's College, 1836–41. *Prodromus Florae Nepalensis* 1825 (types at BM(NH)). Edited, and wrote much of both series of R. Sweet's *British Flower Garden*

from *c*. 1830. 'Ericaceae' (*Edinburgh New Philos. J.* v.17, 1834, 150–60).

Bot. Misc. 1830–31, 61–64. *Proc. Linn. Soc.* 1842, 130, 145–49. *Ann. Mag. Nat. Hist.* 1842, 397–99. *Florist's J.* 1842, 15–19. *Gdnrs Mag.* 1842, 48. *Phytologist* v.1, 1844, 133–134. *Scott. Nat.* 1881, 115. E. Bretschneider *Hist. European Bot. Discoveries in China* 1898, 205–6. W. G. Don *Mem. of Don Family* 1897. *D.N.B.* v.15, 204–5. *J. Bot.* 1941, 171–72. *R.S.C.* v.2, 312.

Herb. formerly at Linnean Society, now at Botanic Garden, Brussels. Letters at Kew.

Donia R. Br.

DON, George (1764–1814)

b. Menmuir, Forfarshire Oct. 1764 *d*. Forfarshire 15 Jan. 1814

ALS 1803. Father of D. Don (1799–1841), G. Don (1798–1856) and P. H. Don (1806–1876). Superintendent, Edinburgh Botanic Garden, 1802. Nurseryman, Doo Hillock, Forfarshire. Found many rare Highland plants. *Herb. Britannicum* 1804–12. *Account of Plants of Forfarshire* 1813.

Bot. Gaz. v.3, 1851, 85–87. *Scott. Nat.* 1881, 62, 109, 149; 1884, 126–29, 176–78, 217–23, 258–69; 1885, 12. *Gdnrs Chron.* 1902 ii 151–52; 1910 ii 216–17. *Pharm. J.* 1902, 183–86. *Notes R. Bot. Gdn Edinburgh* v.3, 1904, 49–290; 1934, 167–72. *J. Bot.* 1906, 60–63, 137–38; 1936, 322; 1941, 171–72. *Mem. New York Bot. Gdn* v.19, 1969, 17–18. H. R. Fletcher and W. H. Brown *R. Bot. Gdn Edinburgh, 1670–1970.* 1970, 73–78. *R.S.C.* v.2, 314.

Herb. at Oxford. Letters at Linnean Society, National Library of Scotland. Monument at Forfar.

Donia G. & D. Don.

DON, George (1798–1856)

b. Doo Hillock, Forfarshire 17 May 1798 *d*. Kensington, London 25 Feb. 1856

ALS 1822. FLS 1831. Son of George Don (1764–1814). Foreman, Chelsea Physic Garden, 1816–21. Collected plants for Horticultural Society of London in Brazil, West Indies, Sao Tomé and Sierra Leone. "One of the most indefatigable, as well as accurate of botanists" (J. E. Smith). *General System of Gardening and Botany* 1831–37 4 vols. Edited ed. 3 of R. Sweet's *Hortus Britannicus* 1839. Prepared 1st Supplt. to J. C. Loudon's *Encyclopaedia of Plants* 1855. 'Botany' (*Encyclopaedia Metropolitana* v.7, 1845, 1–108).

Gdnrs Mag. 1829, 534; 1832, 203–4. *Cottage Gdnr* v.16, 1856, 152–53. *Proc. Linn. Soc.* 1855, xxxix–xli. *Bonplandia* v.5, 1857, 133–34. *D.N.B.* v.15, 206. C. F. P. von Martius *Fl. Brasiliensis* v.1 (1), 1906, 16–17. I. Urban *Symbolae Antillanae* v.3, 1902, 38. *J. Bot.* 1936, 322. K. Lemmon *Golden Age of Plant Hunters* 1968, 122–49. A. M. Coats *Quest for Plants* 1969, 248–49. *Comptes Rendus AETFAT 1960* 1962, 64, 96–97. *R.S.C.* v.2, 314.

Herb. at BM(NH). Plants at Cambridge, Kew. MS. African Journal at Royal Horticultural Society (*Trans. Hort. Soc.* v.5, v.6, 1824–26).

Memecylon donianum Planchon.

DON, Patrick Hall (1806–1876)

b. Edinburgh 1806 *d* Bedgebury, Kent 17 Aug. 1876

Gardener to James Bateman at Knypersley Hall, Cheshire until *c*. 1840. Foreman at Messrs Rollisson, Tooting, London. Co-editor of ed. 13 of J. Donn's *Hortus Cantabrigiensis* 1845.

W. G. Don *Mem. Don Family* 1897, 28.

DON, William (1879–1911)

d. Aberfeldy, Perthshire Nov. 1911

Kew gardener, 1902. Assistant Curator, Botanic Station, Tarkwa, Gold Coast, 1903. Curator, Botanic Station, Old Calabar, S. Nigeria, 1905.

Kew Bull. 1903, 31; 1905, 61; 1912, 56–57. *J. Kew Guild* 1911–12, 50 portr.

DONALD, James (1815–1872)

b. Forfarshire 1815 *d*. Hampton Court, Middx 13 Dec. 1872

Gardener at Chiswick, 1839–42; at Chatsworth House; at Hampton Court, 1856–72. 'Notes on Begonias' (*J. Hort. Soc.* 1846, 132–42).

Garden v.3, 1873, 40. *Gdnrs Chron.* 1873, 46. *R.S.C.* v.2, 314.

Herb. of British plants at BM(NH).

Donaldia Klotzsch.

DONALD, Robert (*fl*. 1810s–1850s)

Nurseryman, Goldworth Nursery, Woking, Surrey.

Gdnrs Mag. 1828, 122–24; 1829, 382. E. W. Brayley *Hist. Surrey* v.2, 1850, 27.

DONALDSON, G. (*fl*. 1860s)

Plants collected in County Down and County Antrim at Oxford.

DONCASTER, E. D. (–*c*. 1950)

Founder member of Alpine Garden Society. Collected plants in Bulgaria and Yugoslavia.

Quart. Bull. Alpine Gdn Soc. 1950, 341–42.

DONEGAN, Patrick (*fl*. 1790s–1810s)

Nurseryman, Dublin.

Irish For. 1967, 47.

DONKIN, Arthur Scott (*fl.* 1850s–1870s)
MD. *Natural History of British Diatomaceae* 1870–73.
G. B. De Toni *Sylloge Algarum* v.2, 1891, xxxvi. *R.S.C.* v.2, 318; v.7, 348.
Collection of microscope preparations at BM(NH).
Donkinia Ralfs.

DONN, James (1758–1813)
b. Monivaird, Perthshire 1758 *d.* Cambridge 14 June 1813
ALS 1795. FLS 1812. Under Aiton at Kew. Curator, Cambridge Botanic Garden, 1794–1813. *Hortus Cantabrigiensis* 1796.
Trans Hort. Soc. v.1, 1812, 262. *Gent. Mag.* 1813 i 663. *D.N.B.* v.15, 222. *J. Bot.* 1914, 319. *Bartonia* 1949, 9–11.

DONN, William (1787–1827)
b. Fifeshire 1787 *d.* Hull, Yorks 17 Dec. 1827
Gardener to I. Swainson at Twickenham; then Syon House, Middx. Curator, Botanic Garden, Hull, *c.* 1811.
Gdnrs Mag. 1828, 494.

DONOVAN, Edward *see* **O'Donovan, E.**

DOODY, Samuel (1656–1706)
b. Staffs 28 May 1656 *d.* London Nov. 1706
FRS 1695. Apothecary. Curator, Chelsea Physic Garden, 1692–1706. Friend of J. Petiver and L. Plukenet. Assisted J. Ray with his *Synopsis* and *Historia Plantarum.* "Rei herbariae peritissimus et maxima industria" (Ray).
L. Plunkenet *Almagestum Botanicum* 1696, 392. *Philos. Trans. R. Soc.* v.19, 1697, 390–1. R. Pulteney *Hist. Biogr. Sketches of Progress of Bot. in England* v.2, 1790, 108–9. R. Morison *Plantarum Historiae Universalis Oxoniensis* v.3, 1699, 214, 215. D. Turner *Extracts from Lit. and Sci. Correspondence of Richard Richardson* 1835, 11. R. H. Semple *Mem. Bot. Gdn Chelsea* 1878, 17. H. Trimen and W. T. T. Dyer *Fl. Middlesex* 1869, 376–78. G. C. Druce *Fl. Berkshire* 1897, cxxv–cxxvi. *D.N.B.* v.15, 236. E. J. L. Scott *Index to Sloane Manuscripts* 1904, 147. *J. Bot.* 1916, 113. C. E. Raven *John Ray* 1942, 232, 245, 249, 433. J. E. Dandy *Sloane Herb.* 1958, 126–27. D. H. Kent *Hist. Fl. Middlesex* 1975, 14.
Plants at BM(NH). MS. on mosses and funeral sermon by A. Buddle in Sloane MS. at BM.
Doodia R. Br.

DORRIEN-SMITH, Arthur Algernon (1876–1955)
b. 28 Jan. 1876 *d.* Tresco, Isles of Scilly 30 May 1955
VMH 1943. Continued development of gardens at Tresco Abbey from 1918. Collected plants in Australia, New Zealand and Chatham Islands, 1910.
Gdnrs Chron. 1955 i 234. *Gdning Illustrated* v.72, 1955, 195. *Times* 31 May 1955.

DORRIEN-SMITH, Thomas Algernon (*c.* 1845–1918)
d. 6 Aug. 1918
Nephew of Augustus Smith (1804–1872). Developed gardens at Tresco Abbey, Isles of Scilly.
Kew Bull. 1918, 242–43. *J. R. Hort. Soc.* 1947, 183–91.

DOUBLEDAY, Edward (1810–1849)
b. Epping, Essex 9 Oct. 1810 *d.* London 14 Dec. 1849
FLS 1843. Brother of Henry Doubleday (1808–1875). Entomologist. Assistant, Zoological Dept., BM. Vice-president, Botanical Society of London. '*Lilium martagon*' (*Phytologist* 1841, 62).
Proc. Linn. Soc. 1850, 84–87. *D.N.B.* v.15, 254.
N. American plants at BM(NH). Portr. at Entomological Society.

DOUBLEDAY, Henry (1808–1875)
b. Epping, Essex 1 July 1808 *d.* Epping 29 June 1875
Quaker entomologist and ornithologist. 'Bardfield Oxlips' (*Phytologist* v.1, 1844, 204, 975).
Entomologist 1877, 53–61, portr. Friend's Institute *Biogr. Cat.* 1888, 187–89. *D.N.B.* v.15, 254–55. H. W. Tompkins *Companion into Essex* 1938, 35–36. *Essex Nat.* 1961, 313–24 portr. *R.S.C.* v.1, 326; v.7, 551.

DOUGLAS, David (1799–1834)
b. Scone, Perth 25 June 1799 *d.* Hawaii 12 July 1834
ALS 1824. FLS 1828. At Glasgow Botanic Garden, 1820. Sent to N. America in 1823 by Horticultural Society of London to collect plants. At Rio, 1824; Columbia River, Oregon, 1825–27; California, 1830–32; Fraser River, 1832–33; Sandwich Islands, 1833–34. Introduced over 200 new species.
Gdnrs Mag. 1835, 271–72; 1836, 274–76, 384–86, 432–36, 602–9 portr.; 1840, 115; 1842, 289–301 portr. *Trans. Hort. Soc.* v.1, 1835, 403–14, 476–81. *Companion Bot. Mag.* 1836, 79–192 portr. *Cottage Gdnr* v.6, 1851, 263–64; v.12, 602 portr. *Canadian Nat. Geol.* 1860, 120–32, 200–8, 267–78, 329–49. *Overland Mon.* 1871, 105–13. *J. Hort. Cottage Gdnr* v.58, 1877, 96–97 portr. *Gdnrs Chron.* 1885 ii 173 portr; 1900 i 120; 1915 i 75 portr.; 1926 ii 250–51; 1928 i 37; 1955 i 228 portr.; 1964 i 67–68; 1965 i 576 portr. *Trans. Edinburgh Field Nat. Microsc. Soc.* 1893–94, 109–11. *Erythea* 1899, 174–75. *D.N.B.* v.15, 291. *Trans. Proc. Perthshire Soc. Nat. Sci.* v.5, 1909–10, 55–65 portr. *Quart. J. For.* 1915, 151–57. *Trans. R. Scott. Arboric. Soc.*

1915, 134–41. W. F. Wilson *David Douglas, Botanist at Hawaii* 1919 portr. *Forestry* 1931, 14–20, 154–58. *Leaflet Western Bot.* 1937, 59–62; 1938, 74–77, 94–98, 116–19; 1939, 170–74, 189–92; 1949, 160–62. *Br. Columbia Hist. Quart.* 1938, 89–94; 1940, 221–43. *California Hist. Soc. Quart.* 1939, 339–41. *J. R. Hort. Soc.* 1942, 121–28, 153–62; 1955, 272–74 portr. *Occas. Papers California Acad. Sci.* no. 20, 1943, 32–37 portr. A. G. Harvey *Douglas of the Fir* 1947 portr. *Gdn J. New York Bot. Gdn* v.2, 1952, 6–9 portr. S. D. McKelvey *Bot. Exploration of Trans-Mississippi West, 1790–1850* 1955, 299–341, 393–427. *Country Life* 1966, 1682–83 portr. R. and J. Young *Plant Detective: David Douglas* 1966 portr. *Times* 9 Sept. 1966. K. Lemmon *Golden Age of Plant Hunters* 1968, 150–81. A. M. Coats *Quest for Plants* 1969, 304–14. H. R. Fletcher *Story of R. Hort. Soc. 1804–1968* 1969 *passim* portr. *J. Scott. Rock Gdn Club* 1970, 147–52 portr. *Taxon* 1970, 521. W. Morwood *Traveller in a Vanished Landscape* 1973 portr. *Quart. Oregon Hist. Soc.* v.5, 215–71, 325–69; v.6, 76–97, 206–27, 288–309, 417–49. *R.S.C.* v.2, 327.

Plants at BM(NH), Cambridge, Kew. MS. Journal, 1823–27, at Royal Horticultural Society (published 1914 and 1959). Portr. by D. Macnee at Kew. Portr. at Hunt Library.

Douglasia Lindl.

DOUGLAS, Francis (1815–1886)
b. Kelso, Roxburgh 14 March 1815 *d.* Kelso 7 March 1886

MD Edinburgh 1836. Physician, Kelso. To India, 1845. President, Cuvierian Natural History Society, 1836–37. President, Berwickshire Naturalists' Club, 1841–42. Contrib. account of Berwick plants to *Hist. Berwick. Nat. Club* v.1, 1834–91, 132–33.

Hist. Berwick. Nat. Club 1886, 538–41. J. Hardy *Selections from Correspondence of G. Johnston* 1892, 73.

DOUGLAS, George Henry (1864–1933)
b. York 1864

On Sheffield Education Committee. Retired to Paignton where he studied botany. Secretary, Torquay Natural History Society, 1928–33.

W. K. Martin and G. T. Fraser *Fl. Devon* 1939, 778–79.

Herb. at Torquay Natural History Society.

DOUGLAS, James (1675–1742)
b. Scotland 1675 *d.* London April 1742

MD Rheims. FRS 1706. Physician to Queen Caroline. *Lilium Sarniense* 1725. *Arbor Yemensis Fructum Cofe Ferens* 1727. 'Crocus sativus' (*Philos. Trans. R. Soc.* v.32, 1723, 441–45).

R. Pulteney *Hist. Biogr. Sketches of Progress of Bot. in England* v.2, 1790, 234–36. W. Munk *Roll of R. College of Physicians* v.2, 1878, 77–79. *D.N.B.* v.15, 329–31. E. J. L. Scott *Index to Sloane Manuscripts* 1904, 148.

Bot. MSS. at Royal Society.

DOUGLAS, James (1837–1911)
b. Ednam near Kelso, Roxburgh 1837 *d.* Nov. 1911

VMH 1897. Nurseryman, Great Bookham, Surrey. *Hardy Florists' Flowers* 1880. *Hints on Culture of Carnation. Carnations* 1913.

J. Hort. Cottage Gdnr v.35, 1897, 145; v.58, 1909, 198–200 portr. *Garden* v.59, 1901, 150–51 portr.; 1911, 600 portr. *Gdnrs Mag.* 1910, 199–200 portr.; 1911, 908 portr. *Gdnrs Chron.* 1911 ii 404–5 portr. *Orchid Rev.* 1912, 10.

DOUGLAS, James (–1960)
d. Great Bookham, Surrey 11 May 1960

Son of James Douglas (1837–1911). Raised border carnations and auriculas.

Gdnrs Chron. 1960 i 446.

DOUGLAS, John (*fl.* 1710s–1743)

FRS 1720. Practised as surgeon in London before going to Antigua whence he sent J. Petiver plants in 1713. Returned to London and resumed medical practice.

E. J. L. Scott *Index to Sloane Manuscripts* 1904, 148. J. E. Dandy *Sloane Herb.* 1958, 127–28.

DOUGLAS, John (*fl.* 1860s)

Of Straffan, Kildare. Steward to Marquis of Kildare, Kilkeay Castle. 'Notes on Fl. Kildare' (*Gdnrs Chron.* 1864, 916).

Proc. R. Hort. Soc. v.5, 1865, 52. *Irish Nat.* 1905, 11–13.

Straffan plants at National Museum, Dublin.

DOUGLAS, Robert (1813–1897)
b. Gateshead, Durham 20 April 1813 *d.* Waukegan, Illinois, U.S.A. 1 June 1897

Emigrated to Canada, 1836; in Vermont, U.S.A., 1838 and Waukegan, 1844, where he established a nursery. Authority on conifers. Assisted C. S. Sargent with N. American dendrology. Contrib. to *Gdn Forest.*

Gdn Forest 1897, 230. L. H. Bailey *Standard Cyclop. Hort.* v.2, 1939, 1572 portr. U. P. Hedrick *Hist. Hort. in America to 1860* 1950, 325.

DOUGLAS, Rev. Robert C. (*fl.* 1850s)

Sent list of plants near Stafford to H. C. Watson.

J. Bot. 1901 Supplt., 73.

DOUGLAS, Sholto Charles John Hay, Earl of Morton *see* Addendum

DOUIE, Lady Frances Mary Elizabeth (*née* Roe) (1866–1965)
b. Amritsar, India 27 June 1866 *d.* Oxford 1965
Wife of Sir James Douie (1854–1935). Collected and painted flowers. Aurel Stein sent plants from Asia for her to paint. Friend of G. C. Druce to whose *Fl. Oxfordshire* she contributed.
Plant drawings at Kew, Oxford.

DOUSTON, Dr. *see* Distin, H.

DOVASTON, John Freeman Milward (1782–1854)
b. Westfelton, Shropshire 30 Dec. 1782 *d.* Westfelton 30 Aug. 1854
MA Oxon 1807. 'Annual Increase of Trunks of Trees' (*Gdnrs Mag.* 1836, 527–33).
D.N.B. v.15, 376–77. *R.S.C.* v.2, 329.
Plants at Liverpool Botanical Society.

DOVER, George (*fl.* 1830s–1850s)
Nurseryman, Magdalen Street, Norwich, Norfolk.

DOWDEN, Richard (1794–1861)
b. Bandon, County Cork 12 April 1794 *d.* Cork 5 Aug. 1861
Merchant. President, Cuvierian Society of Cork. *Walks after Wild Flowers* 1852.
Irish Nat. J. 1931, 237–38. R. L. Praeger *Some Irish Nat.* 1949, 73. *R.S.C.* v.2, 355; v.12, 203.

DOWKER, George (1828–1899)
b. Stourmouth, Kent 2 April 1828 *d.* Ramsgate, Kent 22 Sept. 1899
FGS. Studied Thanet plants. '*Falcaria rivini* in Kent' (*J. Bot.* 1889, 272). Contrib. plant records to *Fl. Kent* 1899.
F. J. Hanbury and E. S. Marshall *Fl. Kent* 1899, lxxvii. *J. Bot.* 1899, 496. G. M. Pittock *Fl. Thanet* 1903, 3. *R.S.C.* v.7, 554; v.14, 667.
Osmundites dowkeri Carruthers.

DOWN, St. Vincent B. (*fl.* 1900s–1910s)
Merchant, Singapore and Sarawak. Collected Malayan plants and gave them to Botanic Gardens, Singapore.
Gdns Bull. Straits Settlements 1927, 120.

DOWNES, Edward John (1893–1957)
b. Nenagh, Tipperary 7 June 1893 *d.* Manderville, Jamaica 1 Aug. 1957
Kew gardener, 1919. Assistant Superintendent, Hope Gardens, Jamaica, 1920; Superintendent, 1943–57.
J. Kew Guild 1962, 213 portr.

DOWNES, Harold (1867–1937)
b. Staverton, Devon 31 Jan. 1867 *d.* Ilminster, Somerset 9 Feb. 1937
MB. LRCP. LRCS. FLS 1917. Worked on flora of Somerset.
Proc. Linn. Soc. 1936–37, 191–92.
Plants at Yeovil Museum.

DOWNHAM, William (*fl.* 1790s)
Gardener and seedsman, 7 Cook Street, Liverpool.

DOWNIE, Dorothy G. (–1960)
b. Edinburgh *d.* 22 Aug. 1960
BSc Edinburgh 1917. Assistant to W. G. Craib at Aberdeen; later Lecturer and Reader. At Chicago University, 1925–28. Studied cycads and orchid mycorrhiza. Contrib. to *Trans. Proc. Bot. Soc. Edinburgh*.
Trans. Proc. Bot. Soc. Edinburgh v.39, 1959–60, 245–46.

DOWNIE, John (–1892)
d. Edinburgh 25 Nov. 1892
Nurseryman, 144 Princes Street, Edinburgh with nurseries at Beechill, Murrayfield. Originator of fancy pansies. For a while in partnership with Laird and Laing.
Gdnrs Chron. 1892 ii 681 portr., 771. *J. Hort. Cottage Gdnr* v.25, 1892, 506. A. Simmonds *Hort. Who was Who* 1948, 45.

DOWNTON, George (*fl.* 1870s)
Collected orchids for Veitch nursery in Central America, 1870–71 and in Chile, 1871–73.
J. H. Veitch *Hortus Veitchii* 1906, 57–58. *Orchid Rev.* 1913, 340–41.
Plants at Kew.

DOWSON, Walter John (1887–1963)
b. Bristol 22 May 1887 *d.* Cambridge 1 Sept. 1963
BA Cantab 1909. FLS 1917–28. Government mycologist, Kenya, 1913–19. Mycologist, Royal Horticultural Society Gardens, Wisley, 1920–28. Plant pathologist, Tasmania, 1928–32. Lecturer in Mycology, Cambridge, 1932–52. *Manual of Bacterial Plant Pathogens* 1949; ed. 2 1957.
Kew Bull. 1913, 90. *Nature* v.200, 1963, 630–31. *Times* 24 Sept. 1963. *B.M.S. News Bull.* v.21, 1964, 4–5.
Kenyan plants at Kew.

DOYLE, John (*fl.* 1750s)
Nurseryman, Stradbally, County Leix, Ireland.
Irish For. 1967, 54.

DOYLE, Martin *see* Hickey, *Rev.* W.

DOYLE, Percy W. (*fl.* 1850s)
Minister Plenipotentiary, Mexico. Sent plants to his brother and to Kew.
Bot. Mag. 1854, t.4824.

DRABBLE, Eric (1877–1933)
b. near Chesterfield, Derby 1877 *d.* Freshwater, Isle of Wight 3 Aug. 1933
DSc London 1903. FLS 1902. Lecturer in Botany, St. Thomas's Hospital Medical School, 1901–3. Lecturer, Royal College of Science, 1903–5. Lecturer, Liverpool University, 1905–6. Lecturer in Botany, Northern Polytechnic, 1906–24. Authority on British *Viola*. Contrib. to *J. Bot.*, *Bot. Soc. Exch. Club Br. Isl. Rep.*, *New Phytol.*
Bot. Soc. Exch. Club Br. Isl. Rep. 1933, 508–9. *Derbyshire Times* 12 Aug. 1933. *J. Bot.* 1933, 318–19. *Kew Bull.* 1933, 413. *Proc. Linn. Soc.* 1933–34, 156–57. *Watson Bot. Exch. Club Rep.* 1933–34, 206–7 portr. *Sorby Rec.* v.2 (3), 1967, 7.
Herb. at BM(NH). Portr. at Hunt Library.

DRAKE, Miss *see* Addendum

DRAKE, G. W. (*fl.* 1900s)
Nurseryman, Cathays Nurseries, Cardiff, Glam. Specialist in chrysanthemums.
J. Hort. Cottage Gdnr. v.55, 1907, 586–87 portr.

DRAKE, Rev. William Fitt (1786–1874)
b. Norwich, Norfolk 1786 *d.* West Halton, Lincs 5 May 1874
BA Cantab 1811. ALS 1810, 1834. FLS 1828. Vicar, St. Stephen, Norwich, 1811–31. Curate, St. Gregory, Norwich, 1831–36. Rector, West Halton, 1836–74. Friend of Sir J. E. Smith. Contrib. botanical articles to A. Rees's *Cyclopaedia* 1802.
P. Smith *Correspondence of Sir J. E. Smith* v.1, 1832, 489.

DRAKE-BROCKMAN, Ralph Evelyn (1875–)
Army medical officer in Somaliland and Abyssinia, 1904–15, where he collected plants. Discovered *Drakebrockmania crassa.*
A. White and B. L. Sloane *Stapelieae* v.1, 1937, 130 portr.
Plants at Kew.

DRAYTON, James (1681–1749)
d. 11 Sept. 1749
Apothecary, Maidstone, Kent. "A famous botanist of Maidstone" (epitaph on tombstone at Allington, Kent). Correspondent of J. Petiver.
W. Hudson *Fl. Anglica* 1762, 263. C. H. Fielding *Mem. of Malling and its Valley* 1893, 99, 142. E. J. L. Scott *Index to Sloane Manuscripts* 1904, 151. *J. Bot.* 1917, 55.

DRESSER, Christopher (1834–1904)
b. Glasgow 4 July 1834 *d.* Mulhouse, Alsace Nov. 1904
PhD Jena. FLS 1861. Lecturer in Botany, Dept. of Science and Art, London. Professor of Medical Botany, St. Mary's Hospital Schools, Hammersmith. Art Editor, *Furniture Gaz.*, 1880. Art Supervisor of Art Furnishers' Alliance, 1880–82. Designer in all media for leading manufacturers. 'Botany, as adapted to the Arts, and Art-manufacture' (*Art. J.* 1857–58). *Rudiments of Botany* 1859. *Unity in Variety, as deduced from the Vegetable Kingdom* 1859. *Popular Manual of Botany* 1860.
R.S.C. v.2, 342.
Plants at Kew.

DREW, F. G. (*c.* 1872–1916)
d. March 1916
Gardener. Lecturer and Superintendent of Horticulture, University College, Reading, 1909.
Gdnrs Mag. 1912, 353–54 portr. *Gdnrs Chron.* 1916 i 161.

DREW, Kathleen Mary (1901–1957)
b. Leigh, Lancs 1901 *d.* Manchester 14 Sept. 1957
BSc Manchester 1922. DSc 1939. Assistant lecturer in Botany, Manchester. Phycologist. Travelled in America and Hawaii. President, British Phycological Society, 1952. *Revision of Genera Chantransia, Rhodochorton and Acrochaetium* 1928. Contrib. to *Ann. Bot.*, *Bot. Rev.*, *Phytomorphology.*
Nature v.180, 1957, 889–90. *Phytomorphology* 1957, 407–8 portr. *Br. Phycol. Bull.* 1958, 1–12 portr. *Rev. Algologique* v.4, 1958, 3–6 portr.
Herb. at BM(NH). Portr. at Hunt Library.

DREWETT, Drewett O. (*c.* 1838–1910)
d. 9 March 1909
Of Riding Mill, Northumberland. Hybridised orchids, chiefly cypripediums.
Gdnrs Chron. 1910 i 191 portr. *Orchid Rev.* 1910, 104.

DREWETT, James (*c.* 1800–1885)
b. near Bristol *c.* 1800 *d.* Kingston, Surrey 14 May 1885
Gardener to Sir W. Heathcote at Hursley Park near Winchester; to Thomas Cubitt at Dorking, Surrey, 1850–75.
Gdnrs Chron. 1874 ii 399 portr.; 1885 i 708.

DREWITT, Frederic George Dawtrey (1848–1942)
b. Burpham, Surrey 29 Feb. 1848 *d.* Kensington, London 29 July 1942
Educ. Oxford. Surgeon, St. George's Hospital, London, 1876. Vice-chairman, Management Committee, Chelsea Physic Garden. *Romance of Apothecaries' Garden at Chelsea* 1922; ed. 3 1928. *Latin Names of Common Flowers* 1927.
Nature v.150, 1942, 228–29. G. H. Brown *Lives of Fellows of R. College of Physicians of London, 1826–1925* 1955, 327–28. *Who was Who, 1941–1950* 329.

DRINKWATER, Harry (1855–1925)
b. Northwich, Cheshire 1855 *d.* Wrexham, Denbighshire 11 July 1925

MD Edinburgh 1885. FRSE 1908. FLS 1910. Practised medicine at Sunderland and Wrexham. Student of genetics. Amateur botanical artist. Awarded J. G. Mendel medal at 4th International Conference on Genetics, Paris, 1911. *Lecture on Mendelism* 1910.

Bot. Soc. Exch. Club Br. Isl. Rep. 1926, 90. *N. Western Nat.* 1926, 40–41. *Proc. Linn. Soc.* 1925–26, 78–80. *Proc. R. Soc. Edinburgh* v.46, 385.

Drawings of British plants at National Museum of Wales.

DRIVER, Abraham Purshouse (*fl.* 1780s–1800s)

Nurseryman, Kent Road, Southwark, London. With W. Driver revised *Pomona Britanica* 1788.

Trans. London Middlesex Archaeol. Soc. v.24, 1973, 190–91.

DRIVER, Samuel (*fl.* 1720s–1779)

Nurseryman, Kent Road, Southwark, London.

Trans. London Middlesex Archaeol. Soc. v.24, 1973, 190–91.

DRIVER, William (*fl.* 1790s–1800s)

Nurseryman, Kent Road, Southwark, London. With A. Driver revised *Pomona Britanica* 1788.

Trans. London Middlesex Archaeol. Soc. v.24, 1973, 190–91.

DROPE, Francis (*c.* 1629–1671)
b. Cumnor, Berks *c.* 1629 *d.* Oxford 26 Sept. 1671

BA Oxon 1647. BD 1667. Prebendary, Lincoln Cathedral, 1669–70. *Short and Sure Guid in Practice of Raising and Ordering of Fruit-trees* 1672.

J. C. Loudon *Encyclop. Gdning* 1822, 1266. *D.N.B.* v.16, 21.

DROUGHT, Isabel (*fl.* 1840s–1850s)

Drawings of Ceylon plants, 1847–51, at BM(NH).

DROVER, William (*c.* 1841–)

Nurseryman, Fareham, Hants. Noted for his gardenias. *Chrysanthemum and its Growth* (with G. Drover and W. Adams).

J. Hort. v.68, 1914, 324 portr.

DRUCE, Francis (1873–1941)
b. Clapham, London 15 Jan. 1873 *d.* London 16 April 1941

BA Oxon 1894. FLS 1914. Treasurer, Linnean Society, 1931–40. Herb. and botanical library destroyed by fire in 1941.

Times Lit. Supplt. 31 Dec. 1938. *Chronica Botanica* 1940–41, 427. *J. Bot.* 1940, 287–88; 1941, 87. *Nature* v.147, 1941, 702. *Proc. Linn. Soc.* 1940–41, 293–94. *Times* 28 April 1941. *Bot. Soc. Exch. Club Br. Isl. Rep.* 1941–42, 453–56.

DRUCE, George Claridge (1850–1932)
b. Potter's Pury, Northants 23 May 1850 *d.* Oxford 29 Feb. 1932

MA Oxon 1899. FRS 1927. FLS 1879. Apprenticed to firm of retail chemists, Northampton, 1866. Established chemist's shop, Oxford, 1879. Topographical field botanist. Fielding Curator, Dept. of Botany, Oxford, 1895–1932. Secretary, Botanical Exchange Club of British Isles, 1904–32. *Fl. Oxfordshire* 1886; ed. 2 1927. *Fl. Berkshire* 1897. *Dillenian Herb.* (with S. H. Vines) 1907. *List of British Plants* 1908; ed. 2 1928. *Account of Morisonian Herb.* (with S. H. Vines) 1914. *Fl. Buckinghamshire* 1926. *Fl. Northamptonshire* 1930. *Comital Fl. British Isles* 1932.

Druggist and Chemist 1901, 158–62, 208. *Bot. Soc. Exch. Club Br. Isl. Rep.* 1925, 933–38; 1930, 480–96; 1931, 804–14. *Gdnrs Chron.* 1925 i 330 portr.; 1932 i 192. G. C. Druce *Fl. Buckinghamshire* 1926, cvi–cx. G. C. Druce *Fl. Northamptonshire* 1930, cxxi–cxxiii. G. C. Druce *Comital Fl. British Isles* 1932, x–xii. *Br. Fern Gaz.* 1932, 141–42. *J. Bot.* 1932, 141–44. *J. Northamptonshire Nat. Hist. Soc. Field Club* 1932, 124–26. *Kew Bull.* 1932, 157–58. *Nature* v.129, 1932, 426–27. *N. Western Nat.* 1932, 39–40 portr. *Obit. Notices Fellows R. Soc.* v.1, 1932, 12–14 portr. *Oxford Times* 4 March 1932. *Pharm. J.* 1932, 198 portr. *Proc. Linn. Soc.* 1931–32, 174–76. *Times* 1 and 5 March 1932. *D.N.B. 1931–1940* 239–41. *Who was Who, 1929–1940* 385–86. *Fl. Malesiana* v.1, 1950, 144. J. G. Dony *Fl. Bedfordshire* 1953, 30–31. H. N. Clokie *Account of Herb. Dept. Bot., Oxford* 1964, 49–52, 108–9.

Herb. and Library at Oxford. Portr. by de Laszlo at Radcliffe Science Library, Oxford. Portr. at Hunt Library.

DRUERY, Charles Thomas (1843–1917)
b. 25 May 1843 *d.* Acton, Middx 8 Aug. 1917

FLS 1885. VMH 1897. Pteridologist. Discovered apospory. President and Secretary, British Pteridological Society. *Choice British Ferns* 1888. *Book of British Ferns* 1903. *British Ferns and their Varieties* 1910. Editor, *Br. Fern Gaz.*, 1909–17.

J. Hort. Cottage Gdnr v.48, 1904, 367 portr. *Gdnrs Mag.* 1908, 389–90 portr. *Br. Fern Gaz.* 1910, 144; 1917, 201–4. *Garden* 1917, 351 portr. *Gdnrs Chron.* 1917 ii 63, 73 portr. *Bot. Soc. Exch. Club Br. Isl. Rep.* 1917, 87. *J. Bot.* 1917, 263–64. *Proc. Linn. Soc.* 1917–18, 38–39.

Portr. at Hunt Library.

DRUMMOND, David (*c.* 1813–1904)
b. Scotland *c.* 1813 *d.* 15 March 1904
Nurseryman, Dublin.
Garden v.65, 1904, 246.

DRUMMOND, Henry Andrews (*fl.* 1810s)
Captain of East Indiaman 'Castle Huntley'. Introduced two new Chinese chrysanthemums in 1819 for Horticultural Society of London.
E. Bretschneider *Hist. European Bot. Discoveries in China* 1898, 268.

DRUMMOND, James (*c.* 1784–1863)
b. Hawthornden, Midlothian *c.* 1784 *d.* Perth, W. Australia 27 March 1863
ALS 1810. Brother of Thomas Drummond (*c.* 1790–1835). Curator, Botanic Garden, Cork, 1809–29. Discovered *Spiranthes romanzoffiana.* To W. Australia, 1829. Government botanist and Superintendent of government gardens, W. Australia. Sent plants to Capt. Mangles and W. J. Hooker. In 1846 received honorarium of £200 for services rendered to botany. 'Swan River Orchids' (*Gdnrs Mag.* 1838, 425–29). Wrote accounts of W. Australia flora in *Hooker's J. Bot.* 1849, 247–51, 374–77; 1850, 30–32.
Gdnrs Mag. 1829, 328; 1840, 115. *Gdnrs Chron.* 1841, 341. A. Lasègue *Musée Botanique de B. Delessert* 1845, 282–83. T. Power *Bot. Guide County Cork* 1845 preface, iv–v. J. D. Hooker *Fl. Tasmaniae* v.1, 1859, cxxvi. *Proc. Linn. Soc.* 1863–64, xli–xlii. *Mem. of W. H. Harvey* 1869, 269–70. *D.N.B.* v.16, 33. *Proc. Linn. Soc. N.S.W.* v.25, 1900, 774. *J.Bot.* 1902, 29–30. *J. W. Austral. Nat. Hist. Soc.* no. 6, 1909, 14–16. R. L. Praeger *Some Irish Nat.* 1949, 74. *J. Proc. W. Austral. Hist. Soc.* 1953, 65–67. *Austral. Dict. Biogr.* v.1, 1966, 325–27. A. M. Coats *Quest for Plants* 1969, 223–27. R. Erickson *Drummonds of Hawthornden* 1969 portr. *Taxon* 1970, 521. *W. Austral. Nat.* 1971, 178–80. *R.S.C.* v.2, 346.
Plants at Cork, BM(NH), Kew, Paris, etc. Letters at Kew. Portr. at Hunt Library.
Drummondita Harvey (dedicated to the 2 Drummonds, with the termination *ita* "an I for James and a T for Thomas").

DRUMMOND, James (*fl.* 1800s)
President of English Factory, Canton in 1805; formerly in Macao. Sent a few plants to Kew.
E. Bretschneider *Hist. European Bot. Discoveries in China* 1898, 216–17.

DRUMMOND, James Lawson (1783–1853)
b. Larne, County Antrim 1783 *d.* Belfast 17 May 1853
MD Edinburgh 1814. Naval surgeon in Mediterranean, 1807–13. Professor of Anatomy, Academical Institution, Belfast, 1818–49. First President of Belfast Natural History Society. One of founders of Belfast Botanic Garden, 1820. *First Steps in Botany* 1823; ed. 2 1826. *Observations on Natural Systems of Botany* 1849.
Proc. Belfast Nat. Hist. Soc. 1882, 13. J. Hardy *Selections from Correspondence of George Johnston* 1892, 253. *D.N.B.* v.16, 33–34. *Proc. Belfast Nat. Field Club* 1913, 618. *Belfast Nat. Hist. Philos. Soc. Centenary Vol.* 1924, 72–73, 126. R. L. Praeger *Some Irish Nat.* 1949, 74. *Irish Nat. J.* 1958, 222–23.

DRUMMOND, James Montagu Frank (1881–1965)
b. India 1881 *d.* Exmouth, Devon 7 Feb. 1965
MA Cantab. FLS 1908. Lecturer in Botany, Armstrong College, Newcastle-upon-Tyne, 1906–9. Lecturer in Plant Physiology, Glasgow, 1909–21. Director, Scottish Plant Breeding Station, 1921–25. Professor of Botany, Glasgow, 1925–30. Professor of Botany, Manchester, 1930–46. Translated G. Haberlandt's *Physiological Plant Anatomy* 1914.
N. Br. Agriculturalist 17 March 1921 portr. *Nature* v.205, 1965, 1262. *Times* 11 Feb. 1965. *Who was Who, 1961–1970* 318.

DRUMMOND, James Ramsay (1851–1921)
b. May 1851 *d.* Acton, Middx 11 March 1921
BA Oxon 1872. FLS 1898. Indian Civil Service, 1874–1904. Collected plants in Punjab, Simla Hills, etc. Published list of Punjab plants in *J. Bombay Nat. Hist. Soc.* Curator of Calcutta Herb. during last few months in India. To Kew, 1905. '*Furcraea*' (*Rep. Missouri Bot. Gdn* 1906, 25–75). 'Grewias of Roxburgh' (*J. Bot.* 1911, 329–37, 357–63).
Bot. Soc. Exch. Club Br. Isl. Rep. 1921, 356–57. *J. Bot.* 1921, 174. *Kew Bull.* 1921, 123; 1922, 301–2. *Proc. Linn. Soc.* 1920–21, 47.
Plants at Dehra Dun, Kew. MSS. at Kew.

DRUMMOND, Patrick (*fl.* 1740s–1760s)
Merchant and seedsman with shop opposite Libberton's Wynd, Lawnmarket, Edinburgh.

DRUMMOND, Peter (*c.* 1798–1877)
d. 9 July 1877
Nurseryman, Stirling. Head of old established firm of William Drummond and Sons.
Gdnrs Yb. Almanack 1878, 192.

DRUMMOND, Thomas (*c.* 1790–1835)
b. Scotland *c.* 1790 *d.* Havana, Cuba March 1835
ALS 1830. Brother of James Drummond (*c.* 1784–1863). Succeeded G. Don in nursery at Forfar. First Curator, Belfast Botanic

Garden, 1828–31. Assistant naturalist to 2nd Land Arctic Expedition under Sir John Franklin, 1825–27. Collected plants in N. America, Canada and Texas for Glasgow Botanic Garden from 1831. Discovered *Orobus niger*. Arctic collections published in W. J. Hooker's *Fl. Boreali Americana* 1829–40. *Musci Scotici* 1828. *Musci Americani* 1841.

Edinburgh J. Sci. v.6, 1827, 110–13. J. Franklin *Narrative of Second Expedition... 1825–1827* 1828, 308–13. *Bot. Misc.* 1830, 95–96, 178–219. *Bot. Mag.* 1833, t.3287; 1835, t.3441; 1838, t.3626. *J. Bot.* (Hooker) 1834, 50–60, 183–202; 1843, 663–70. *Companion Bot. Mag.* 1835, 21–26, 39–49, 95–101; 1836, 170–77; v.2, 1836, 60–64. *Gdnrs Mag.* 1835, 608. A. Lasègue *Musée Botanique de B. Delessert* 1845, 196–98, 204. S. A. Stewart and T. H. Corry *Fl. N.E. Ireland* 1888, ix. *D.N.B.* v.16, 41. I. Urban *Symbolae Antillanae* v.3, 1902, 38–39. *Popular Sci. Mon.* 1909, 48–50 portr. *Irish Nat.* 1913, 28. *Southwest Rev.* 1930, 478–512. *Nature* v.135, 1935, 353. S. W. Geiser *Nat. of the Frontier* 1937, 73–105; ed. 2, 55–78. R. L. Praeger *Some Irish Nat.* 1949, 75. S. D. McKelvey *Bot. Exploration of Trans-Mississippi West 1790–1850* 1955, 486–507. H. N. Clokie *Account of Herb. Dept. Bot., Oxford* 1964, 159. A. M. Coats *Quest for Plants* 1969, 314–16 portr., 325–27. *R.S.C.* v.2, 347.

Plants at BM(NH), Kew, Oxford. Letters and portr. by D. Macnee at Kew. Portr. at Hunt Library.

Drummondia Hook.

DRUMMOND, W. (*fl.* 1830s)

Seedsman, 58 Dawson Street, Dublin. *Directions for Sowing and Cultivating Carrots* c. 1830–40.

DRUMMOND, W. C. (*c.* 1816–1893)
d. Bath, Somerset Dec. 1893

Nurseryman, Park Lane Nurseries, Bath. *Garden* v.44, 1893, 615.

DRUMMOND, William (1793–1868)
b. Bannockburn, Stirling 1793 *d.* Stirling 1868

Nurseryman, Agricultural Museum and Warehouse, Stirling.

Gdnrs Chron. 1868, 1293–94.

DRUMMOND, William Peter (1838–1906)
b. Stirling 1838 *d.* Edinburgh 18 Dec. 1906

Son of nurseryman, Peter Drummond. He and his brother, George, established business in George Street, Edinburgh with nurseries at Longfield.

Trans. Bot. Soc. Edinburgh v.23, 1908, 352.

DRUMMOND-DELAP, Mrs (*fl.* 1830s)

Plants from Pyrenees at Oxford.

DRUMMOND-HAY, Henry Maurice (*né* **Drummond**) (1814–1896)
b. Bath, Somerset 1814 *d.* Seggieden, Perth 3 Jan. 1896

Colonel. Black Watch Regiment, 1832–52. In command of Perthshire Militia, 1852–72. Hon. Curator, Perth Museum. Contrib. to *Scott. Nat.*, 1872–80.

Ann. Scott. Nat. Hist. 1896, 73–76 portr. *J. Bot.* 1896, 133. F. H. Davey *Fl. Cornwall* 1909, liv. *Trans. Proc. Perthshire Soc. Nat. Sci.* v.9, 1929–30, 31–33 portr. *R.S.C.* v.7, 927; v.10, 166; v.14, 686.

Herb. at Perth Museum.

Crista-galli var. *Drummond-Hayi* F. B. White.

DRURY, Heber (1819–1872)

Colonel, Madras Light Infantry. At Travancore, 1850. *Useful Plants of India* 1858; ed. 2 1873. *Handbook of Indian Fl.* 1864–69 3 vols.

R.S.C. v.2, 347.

DRURY, William D. (1857–1928)
b. Oxford 1857 *d.* 30 March 1928

On editorial staff of G. Nicholson's *Illustrated Dict. Gdning* 1884–88. Editor, *Bazaar, Exchange and Mart*, 1884–1926. *Home Gardening* 1898. *Book of Gardening* 1900. *Open-air Gardening* 1901. *Popular Bulb Culture* ed. 3 1910. *Hardy Perennials* 1920. *Fruit Culture for Amateurs* 1921.

Who was Who, 1916–1928 304.

DRYANDER, Jonas (1748–1810)
b. Göteborg, Sweden March 1748 *d.* London 19 Oct. 1810

MA Lund 1776. FLS 1788. Pupil of E. G. Lidbeck. Came to England, 1777. Succeeded D. C. Solander as librarian to Sir Joseph Banks, 1782. Librarian to Linnean Society, 1788. *Catalogus Bibliothecae Historico-naturalis Josephi Banks* 1798–1800 5 vols. 'Chloris Novae Hollandiae' (*Ann. Bot.* v.2, 1806, 504–32). Edited ed. 1 and part of ed. 2 of Aiton's *Hortus Kewensis* (*J. Bot.* 1912 Supplt 3, 1–16), and W. Roxburgh's *Plants of Coast of Coromandel* 1795–1819 (*Rees' Cyclop.* s.v. *Roxburghia*).

J. Nichols *Lit. Anecdotes of Eighteenth Century* v.9, 43–44. *Gent. Mag.* 1810 ii 398; 1811 i 158, 540. P. Smith *Correspondence of Sir J. E. Smith* v.1, 1832, 165, 591. R. A. Salisbury *Genera of Plants* 1866, 8. *Notes and Queries* v.12, 1879, 276. *D.N.B.* v.16, 64. *J. R. Soc. N.S.W.* 1908, 66–67. J. H. Maiden *Sir Joseph Banks* 1909, 96–100 portr. *Proc. Linn. Soc.* 1943–44, 99–102.

MSS. and letters at BM(NH). Letters at Fitzwilliam Museum, Cambridge. Portr. at Hunt Library.

Dryandra R. Br.

DU BOIS, Charles (1656–1740)
d. Mitcham, Surrey 20 Oct. 1740
FRS 1700. Treasurer, East India Company, 1702–37. Cultivated exotic plants in his garden at Mitcham. Sent plants by J. Petiver, L. Plukenet, I. Rand, J. Sherard, etc. His step-brother, Daniel, and sister sent him a large number of Indian plants. "Rei Herbariae cultor eximius" (L. Plukenet *Almagestum Botanicum* 1696, 4).
J. Ray *Synopsis* v.3, 1690, 364. *Gent Mag.* 1740, 525; 1747, 103; 1812 i 205. *Hooker's J. Bot.* 1854, 249. G. C. Druce *Fl. Berkshire* 1897, cxxvi. *D.N.B.* v.16, 77. E. J. L. Scott *Index to Sloane Manuscripts* 1904, 63. W. Foster *East India House* 1924, 113–24. *Bot. Soc. Exch. Club Br. Isl. Rep.* 1927, 463–93. J. E. Dandy *Sloane Herb.* 1958, 128–29. H. N. Clokie *Account of Herb. Dept. Bot., Oxford* 30, 81–86.
Plants at Oxford, BM(NH).
Duboisia R. Br.

DUCIE, 3rd Earl of *see* Moreton, H. J.

DUCK, John Nehemiah (*fl.* 1850s)
Natural History of Portishead 1852 contains local plant records.
J. W. White *Fl. Bristol* 1912, 87.

DUCKWORTH, William (*fl.* 1890s)
Of Carlisle and Ulverston, Cumberland. Contrib. plant records to W. Hodgson's *Fl. Cumberland* 1908, viii.

DUDGEON, Gerald Cecil (1867–1930)
b. 18 Oct. 1867 *d.* 4 May 1930
Superintendent of Agriculture, W. Africa. Collected plants in Gambia, Nigeria and Gold Coast, 1907–9. Managed tea, cocoa and coffee plantations in India, chiefly in Bengal and Punjab. Director-General, Egyptian Dept. Agriculture, 1910-13.
Who was Who, 1920–1940 388.
Some African plants at Kew.

DUDLEY, Arthur Horatio (1857–1921)
b. West Wycombe, Bucks 1857 *d.* Liverpool 2 Feb. 1921
Teacher of Botany, Liverpool Technical School, 1890–99. President, Liverpool Microscopical Society, 1908–9; Liverpool Botanical Society, 1920. Addresses in *Ann. Rep. Liverpool Microsc. Soc.* 1908, 10–27.
Proc. Liverpool Bot. Soc. 1912–22, 29–30 portr. *Lancashire and Cheshire Nat.* v.14, 1922, 199–200.

DUDLEY, Hon. Paul (*fl.* 1720s–1730s)
FRS 1721. Of Roxbury, New England. Correspondent of P. Collinson. 'Account of Poyson Wood Tree in New England' (*Philos. Trans. R. Soc.* v.31, 1720–21, 145–46). 'Observations on some of the Plants of New England...' (*Philos. Trans. R. Soc.* v.33, 1724–25, 194–200).
W. Darlington *Memorials of John Bartram and Humphry Marshall* 1849, 79.

DUDLEY-SMITH, Russell (1912–1967)
d. Winchcombe, Glos 6 Oct. 1967
Commander, Royal Navy. Civil servant. Botanised in Gloucestershire.
N. Gloucestershire Nat. Soc. J. v.18, 1967, 230–32. *Proc. Bot. Soc. Br. Isl.* 1968, 623–24.

DUFF, Sir Mountstuart Elphinstone Grant (1829–1906)
b. Eden, Aberdeen 21 Feb. 1829 *d.* Chelsea, London 12 Jan. 1906
MA Oxon 1853. FRS 1881. FLS 1872. Grandson of Sir W. Ainslie. Governor of Madras, 1881–86. Friend of J. B. L. Warren. Knew British plants well. *Notes from a Diary* 1897–1905 14 v. (contains much gossip about botanists and plants).
J. Bot. 1906, 79. *Proc. Linn. Soc.* 1905–6, 37–38. *Bot. Soc. Exch. Club Br. Isl. Rep.* 1906, 202. *D.N.B. Supplt 2* v.2, 150–51. *Curtis's Bot. Mag. Dedications, 1827-1927* 255–56 portr. *R.S.C.* v.7, 566.
Letters at Kew.
Iris grant-duffii Baker.

DUFF, Mungo Campbell (–1885)
b. Edinburgh *d.* Glasgow 28 Dec. 1885
Collected and cultivated British species of ferns.
Proc. Trans. Nat. Hist. Soc. Glasgow 1883–86, lxxv–lxxvi.

DUFFIELD, Mrs. Mary Elizabeth *see* Rosenberg, M. E.

DUGMORE, Horace Radclyffe (1845–1902)
d. June 1902
Of Parkstone, Dorset. Contrib. to *Garden.*
Garden v.61, 1902, 387.

DUGUID, Alexander Russell (1798–1872)
b. Bo'ness, Linlithgow 1798 *d.* 7 Oct. 1872
MD Edinburgh 1819. Practised medicine at Kirkwall, 1819–69. List of Orkney plants in W. J. Hooker's letters at Kew.
H. C. Watson *Topographical Bot.* 1883, 545. M. Spence *Fl. Orcadensis* 1914, xl, xli–xliii, 140.

DUKE, Oliver T. (*fl.* 1870s–1890s)
Medical officer. Collected plants with H. Hamilton in Baluchistan, 1876–94.
Pakistan J. For. 1967, 344.
Plants at Calcutta.

DULLEY, William (*fl.* 1830s–1840s)
Nurseryman, Lamb Farm, Hackney, London.

DÜMMER, Richard Arnold (*c.* 1887–1922)
b. Cape Town *c.* 1887 *d.* Uganda 21 Dec. 1922

Kew gardener, 1910–11. To Uganda, 1914. 'Enumeration of Bruniaceae' (*J. Bot.* 1912 Supplt., 1–37). 'Conifers of Lindley Herb., Cambridge' (*J. R. Hort. Soc.* v.39, 1913, 63–91).

J. Bot. 1923, 158. *J. Kew Guild* 1923, 175–76 portr. *Kew Bull.* 1923, 94.

Uganda and Kenya plants at Kew, BM(NH).

DUNBAR, George (1774–1851)
b. Coldingham, Berwick 1774 *d.* Rose Park, Edinburgh 6 Dec. 1851

MA Edinburgh 1807. Started as a gardener. Professor of Greek, Edinburgh, 1807–51. Grew ericas (catalogue in *Gdnrs Mag.* 1826, 131–35).

Cottage Gdnr v.7, 1851, 187. *D.N.B.* v.16, 153.

Dunbaria Wight.

DUNBAR, John (*fl.* 1840s)
Nurseryman, Park Nursery, Oxford.

DUNBAR, John (1859–1927)
b. Rafford, Morayshire 4 June 1859 *d.* Rochester, New York 13 June 1927

Gardener. To U.S.A., 1887. Assistant Superintendent, Rochester City Parks. Studied *Crataegus.*

C. S. Sargent *Silva of N. America* v.13, 1902, 121–22.

DUNCAN, Andrew (1744-1828)
b. Pinkerton, St. Andrew's 17 Oct. 1744 *d.* Edinburgh June 1828

MA St. Andrew's 1762. MD 1769. Professor of Physiology, Edinburgh, 1790–1821. Founder and President, Caledonian Horticultural Society, and established public experimental garden. Contrib. to *Mem. Caledonian Hort. Soc.*

J. C. Loudon *Encyclop. Gdning* 1822, 1285. S. Felton *Portr. of English Authors on Gdning* 1830, 190. *D.N.B.* v.16, 161–62. *R.S.C.* v.2, 401.

Portr. by Raeburn. Portr. at Hunt Library.

DUNCAN, Andrew (1773–1832)
b. Edinburgh 10 Aug. 1773 *d.* Edinburgh 13 May 1832

MA Edinburgh 1793. MD 1794. ALS 1795. Son of A. Duncan (1744–1828). Professor of Medical Jurisprudence, Edinburgh, 1807–19. Professor of Materia Medica, 1821–32. *Catalogue of Plants* 1826.

D.N.B. v.16, 163.

DUNCAN, Rev. James (*c.* 1802–1861)
b. Denholm, Roxburgh *c.* 1802 *d.* Denholm 30 Nov. 1861

Plant list in A. Jeffrey's *Hist. Roxburghshire* 1857. MS. flora of Jedburgh, *c.* 1830, at Kew.

H. C. Watson *New Botanists' Guide* 1835–37, 426–28. *Hist. Berwick Nat. Club* 1867, 322–24. H. C. Watson *Topographical Bot.* 1883, 543. J. Hardy *Selections from Correspondence of George Johnston* 1892, 62.

DUNCAN, James (1802–1876)
b. Aberdeen Oct. 1802 *d.* Calne, Wilts 11 Aug. 1876

Curator, Botanic Garden, Mauritius, 1849–65. *Catalogue of Plants in Royal Botanical Garden, Mauritius* 1863. His son, J. W. Duncan, presented drawings of Mauritius plants to Kew in 1894.

Kew Bull. 1894, 136; 1919, 284.

Mauritius and Rodriguez plants and letters at Kew.

DUNCAN, James Thompson (1884–1958)
b. Dublin 29 August, 1884 *d.* 3 June 1958

MD Dublin 1911. Lecturer in Clinical Medicine and Pathology, Edward VII College of Medicine, Singapore, 1914–19. Assistant Lecturer, School of Tropical Medicine, London, 1919. Reader in Medical Mycology, London University, 1947. Contrib. to *Rev. of Med. and Vet. Mycol.*

Trans. Br. Mycol. Soc. 1959, 121–22 portr.

DUNCAN, John (1794–1881)
b. Stonehaven, Kincardine 24 Dec. 1794 *d.* Alford, Aberdeenshire 9 Aug. 1881

Weaver.

J. Bot. 1881, 287–88. W. Jolly *Life of John Duncan* 1883 portr. (Appendix of Alford and Aberdeen plants). *Nature* v.23, 1880–81, 269–70; v.24, 1881, 6, 361. *Gdnrs Chron.* 1928 ii 22.

Herb. at Aberdeen.

DUNCAN, John Bishop (1869–1953)
b. Edinburgh 1869 *d.* Berwick-on-Tweed, Northumberland 4 Jan. 1953

Bank clerk, Moffat, Bewdley and Stratford-on-Avon. Treasurer, British Bryological Society, 1925–45; President, 1937–38. Edited *Moss Census Catalogue* ed. 2 1926. 'List of Bryophytes of Berwick' (*Trans. Bot. Soc. Edinburgh* v.34, 1946, 288–315). 'List of Bryophytes of Northumberland' (*Trans. Nat. Hist. Soc. Northumberland* v.10, 1950, 1–80).

Proc. Berwickshire Nat. Club 1952, 191–92 portr. *Trans. Br. Bryol. Soc.* 1953, 333–34.

Bryophytes at Hancock Museum, Newcastle-on-Tyne, Oxford.

DUNCAN, Rev. John Shute (1769–1844)
b. South Warnborough, Hants 1769 *d.* Bath, Somerset 14 May 1844

BA Oxon 1791. DCL 1830. FLS 1829. Keeper of Ashmolean Museum, 1823–26. *Botanical Theology* 1826.

Gent. Mag. v.22, 1844, 97–98.

DUNCAN, Peter Martin (1821–1891)
b. Twickenham, Middx 20 April 1821 *d.* Gunnersbury, Middx 28 May 1880

MB London 1846. FRS 1868. FLS 1880. Practised medicine at Colchester, 1848–60, and at Blackheath, 1860. Professor of Geology, King's College, London, 1870 and at Cooper's Hill College, *c.* 1871. President, Geological Society, 1876–78. *Micrographic Dictionary* (with J. W. Griffith) 1875 2 vols. 'Observations on Pollen-tube' (*Proc. Bot. Soc. Edinburgh* 1856, 10).

J. Bot. 1891, 224. *Geol. Mag.* 1891, 332–36. *Nature* v.44, 1891, 387–88. *Proc. Linn. Soc.* 1891–92, 65–68. *Proc. R. Soc.* 1891–92, iv–vii. *Quart. J. Geol. Soc.* 1892, 47–48. *J. R. Microsc. Soc.* 1895, 19–20. *D.N.B. Supplt. 1* v.2, 168–69. *R.S.C.* v.1, 402; v.7, 573; v.9, 750; v.14, 726.

DUNCANSON, Thomas (*fl.* 1820s)

Gardener, Botanic Garden, Edinburgh. At Kew, 1822–26, where he drew plants for W. T. Aiton.

J. Bot. 1912 Supplt. 3, 14–15. *Kew Bull.* 1916, 165.

c. 300 plant drawings with MS. list at Kew.

DUNCUMB, Rev. John (*fl.* 1800s)

Collections towards History and Antiquities of County of Hereford 1804 (includes plant lists).

Trans. Woolhope Nat. Field Club v.34, 1954, 237–38.

DUNDAS, James (1907–1966)
b. Edinburgh 22 July 1907 *d.* 29 May 1966

Forestry officer, 1929–51. Collected plants in N. and S. Nigeria and in W. Cameroons.

Plants at BM(NH), Kew.

DUNDAS, Maria *see* Callcott, *Lady* M.

DUNHILL, George (*fl.* 1780s)

Gardener and nurseryman, Pontefract, Yorks.

DUNLOP, Gavin Alfred (1868–1933)
b. Nottingham 1868 *d.* 3 April 1933

Curator, Warrington Museum, Lancs. Published articles on local flora.

N. Western Nat. 1933, 142–45 portr.

DUNLOP, John (1865–1895)
b. 28 June 1865 *d.* 2 July 1895

MB Owen College, Manchester, 1889. Physician. 'On some Rare Points in the Histology and Physiology of the Fruits and Seeds of Rhamnus' (with H. M. Ward) (*Ann. Bot.* v.1, 1887, 1–26).

DUNN, Edward John (1844–1937)
b. Bristol 1 Nov. 1844 *d.* Melbourne 20 April 1937

To Australia, 1849. On staff of Geological Survey, Victoria, until 1869. Geologist, Cape Town Administration, 1869. Returned permanently to Australia, 1883. Collected plants in N. Territory. Sent seeds of *Streptocarpus dunnii* to Kew in 1884.

Bot. Mag. 1886, t.6903. *Victorian Nat.* 1937, 20–22 portr. P. Serle *Dict. Austral. Biogr.* 1949, 259. *Austral. Encyclop.* v.3, 1965, 312–13.

DUNN, Malcolm (1837–1899)
b. Methven, Perthshire 1837 *d.* Dalkeith, Midlothian 11 May 1899

VMH 1897. Gardener to Duke of Buccleuch at Dalkeith Palace, 1871. Revised W. Williamson's *Horticultural Exhibitors' Handbook* 1892. Edited reports of Royal Caledonian Horticultural Society's Apple, Pear and Plum Congresses, 1885, 1889.

Gdnrs Chron. 1896 i 737 portr; 1899 i 306 portr., 318, 410. *Garden* v.55, 1899, 358. *Gdnrs Mag.* 1899, 293 portr. *J. Hort. Cottage Gdnr* v.38, 1899, 412–13 portr. *Trans. Bot. Soc. Edinburgh* v.21, 1899, 220–22.

DUNN, Stephen Troyte (1868–1938)
b. 26 Aug. 1868 *d.* Sheen, Surrey 18 April 1938

BA Oxon. FLS 1895. Private Secretary to Sir W. Thiselton-Dyer at Kew, 1898–1901. Assistant, Kew Herb., 1901–3. Superintendent, Botany and Forestry Dept., Hong Kong, 1903–10. Official guide, Kew, 1913–15. Assistant, Kew Herb., 1919–28. *Fl. Southwest Surrey* 1893. *Alien Fl. of Britain* 1905. *Fl. of Kwangtung and Hongkong* (with W. J. Tutcher) 1912.

Kew Bull. 1903, 30; 1938, 214–15. *J. Bot.* 1938, 183–84. G. A. C. Herklots *Hongkong Countryside* 1951, 167–68.

Plants collected in China, Korea, Taiwan and Japan at Hong Kong.

DUNSTALL, John (*fl.* 1640s–1670s)

Engraver of Blackfriars. *A Booke of Flowers, Fruicts, Beastes, Birds and Flies exactly Drawne* 1661.

D.N.B. v.16, 221.

DUNSTER, Rev. Henry Peter (1813–*c.* 1904)
b. Edmonton, Middx 1813 *d.* Woodbastwick, Norfolk *c.* 1904.

MA Oxon 1839. Vicar, Woodbastwick, 1848. *Young Collector's Handy Book of Botany* 1871.

DUPIN, Paul (*fl.* 1750s–1790s)

Apprenticed to Charles Minier, St. Martin-in-the-Field. Seedsman, 66 Strand, London.

DU PORT, Rev. James Mourant (1832–1899)
b. St. Peter Port, Guernsey, Channel Islands 14 April 1832 *d.* Denver, Norfolk 21 Feb. 1899

BA Cantab 1855. Rector, Denver, 1884. Had good knowledge of European flora. Mycologist. Contrib. to *Trans. Woolhope Club* and *Trans. Norfolk Norwich Nat. Soc.*

Trans. Br. Mycol. Soc. 1897–98, 82–83 portr. *Gdnrs Chron.* 1899 i 141. *J. Bot.* 1899, 192. *Trans. Norfolk Norwich Nat. Soc.* 1898–99, 426–27; 1937, 203. *Bull. Soc. Mycol. France* 1899, 150–51.

Letters at BM(NH).

Russula du porti Phillips.

DUPPA, Adeline Frances Mary
(*fl.* 1870s–1880s)

Drawings of Italian and Madeira plants and English fungi at BM(NH).

DUPPA, Richard (1770–1831)
b. Culmington, Shropshire 1770 *d.* London 11 July 1831

LLB Cantab 1814. High Sheriff of Radnor. *Illustrations of Lotus of Antiquity* 1813. *Illustrations of Lotus of the Ancients* 1816. *Classes and Orders of Linnean System of Botany* 1816 3 vols.

D.N.B. v.16, 243.

Sale of property by R. H. Evans, 3 Sept. 1831.

DUPUY, Edgar (*fl.* 1860s–1900s)

Dispensing chemist, St. Peter Port, Guernsey. Collected local plants, 1864–65. Acquired J. Gosselin's herb.

Hortus siccus in 1 vol. at La Société Guernesiaise Museum.

D'URBAN, William Stewart Mitchell
(1837–1934)
d. Topsham, Devon 1934

Curator, Royal Albert Memorial Museum, Exeter, 1865–85. Travelled widely in S. Africa and N. America. Collected ferns in Natal, 1860s, and in California, 1884. Contrib. to *Bot. Soc. Exch. Club Br. Isl. Rep.*

W. K. Martin and G. T. Fraser *Fl. Devon* 1939, 775. H. N. Clokie *Account of Herb. Dept. Bot., Oxford* 1964, 160.

Herb. at Exeter Museum. Ferns and MS. catalogue of ferns at Oxford.

DURHAM, Cornelius B. (*fl.* 1820s–1860s)

Of London. Miniature painter. Exhibited at Royal Academy, 1828–58. Made drawings of orchids for J. Day (1824–1888).

Kew Bull. 1906, 178.

DURHAM, Miss E. (*fl.* 1840s)

Drew t.3909 for *Bot. Mag.*

DURHAM, Frank Rogers (1872–1947)
b. 10 July 1872 *d.* Woolbrook near Sidmouth, Devon 1947

Secretary, Royal Horticultural Society, 1926–45.

J. R. Hort. Soc. 1947, 217–20 portr. *Times* 1 April 1947.

Plants collected in Gallipoli, Iraq, Tropical and S.W. Africa at Kew.

DURNFORD, J. (*fl.* 1880s)

Miner of Kuantan. Collected orchids in Malaya in 1880s and presented them to Botanic Gardens, Singapore.

Gdns Bull. Straits Settlements 1927, 120.

DURNFORD, Rev. Richard (1802–1895)
b. Sandleford, Berks 3 Nov. 1802 *d.* Basle, Switzerland 14 Oct. 1895

BA Oxon 1826. DD 1870. Rector, Middleton, Lancs, 1835–70. Bishop of Chichester, 1870–95. Had herb. and "a rare knowledge of botany and horticulture, and of natural history generally."

W. R. W. Stephens *Mem. R. Durnford* 1899 portr. *D.N.B. Supplt. 1* v.2, 170–71.

DUTHIE, James Allardyce (1868–1920)
d. Aberdeen 2 July 1920

Proprietor of Ben. Reid and Co., nurserymen, Aberdeen.

Gdnrs Chron. 1920 ii 28.

Mosses at BM(NH).

DUTHIE, John Firminger (1845–1922)
b. Sittingbourne, Kent 12 May 1845 *d.* Worthing, Sussex 23 Feb. 1922

BA Cantab 1867. FLS 1875. Professor of Natural History, Cirencester Agricultural College, 1875. Superintendent, Saharanpur Garden, 1876–1903. Kew Herb., 1903–7. Collected plants, principally in N.-W. Provinces. *List of North-west Indian Plants* 1881. *Field and Garden Crops of North-Western Provinces and Oudh* (with J. B. Fuller) 1882–93 3 vols. *Illustrations of Indigenous Fodder Grasses of the Plains of North-western India* 1886. *Fodder Grasses of Northern India* 1888. *Fl. Upper Gangetic Plain* 1903–29. Revised Sir R. Strachey's *Catalogue of Plants of Kumaon* 1906.

Bot. Soc. Exch. Club Br. Isl. Rep. 1922, 698–99. *J. Bot.* 1922, 151–53. *Kew Bull.* 1922, 125–28. *Proc. Linn. Soc.* 1921–22, 44–45. *Indian For. Bull.* (*Bot. Ser.*) no. 73, 1931, 25 portr. H. J. Riddelsdell *Fl. Gloucestershire* 1948, cxxxii. I. H. Burkill *Chapters on Hist. Bot. in India* 1965, 150–51. *R.S.C.* v.7, 584; v.9, 762.

Plants, letters and MS. catalogue of herb. at Kew. Portr. at Hunt Library.

Duthiea Hackel.

DUTHIE, William (*fl.* 1780s–1820s)
Nurseryman, Dog Row, Bethnal Green, London. In partnership with Alexander Duthie.

DUTTON, Hon. Francis Staker (1816–1877)
b. Cuxhaven, Germany
To Australia, *c.* 1840. Acquired extensive mining and farming interests. Premier, S. Australia, 1865. Collected plants. Agent-General for S. Australia in London, 1865–77.
Rep. Austral. Assoc. Advancement Sci. 1907, 174.
Algae at Manchester. Portr. at Hunt Library.
Duttonia F. Muell.

DYER, Harriet Ann Thiselton- *see* Thiselton-Dyer, H. A.

DYER, Richard (1651–1730)
BA Oxon and Fellow of Oriel College, 1673. "An excellent scholar and admirably well skilled in Botany" (Hearne). Contrib. preface to R. Morison's *Plantarum Historiae Universalis Oxoniensis* v.3, 1699.
R. Morison *Plantarum Historiae Universalis Oxoniensis* v.2, 1680, 628. *J. Bot.* 1926, 43–44. C. E. Raven *John Ray* 1942, 253, 435.

DYER, Thomas Webb (*fl.* 1780s–1830s)
MD. FLS 1799. Apothecary, Bristol Infirmary, 1789–1816. Contrib. Somerset plants to E. Shiercliff's *Bristol Guide*, 1789 and to D. Turner and L. W. Dillwyn's *Botanist's Guide* 1805, 519–28.
J. Sowerby and J. E. Smith *English Bot.* 1794, 614. Evans *Picture of Bristol* ed. 4, 1828, 71. J. W. White *Fl. Bristol* 1912, 76–77. H. J. Riddelsdell *Fl. Gloucestershire* 1948, cxvi.

DYER, Sir William Turner Thiselton- *see* Thiselton-Dyer, *Sir* W. T.

DYKES, Elsie Katherine (*née* **Kaye**) (–1933)
d. Raynes Park, Surrey 25 May 1933
Wife of W. R. Dykes whose *Notes on Tulip Species* 1930 she edited and illustrated. Grew irises.
Daily Mail 26 May 1933. *Gdnrs Chron.* 1933 i 394, 407. *Iris Yb.* 1933, 14–15 portr.

DYKES, Percy James (1909–1964)
Gardener, Royal Horticultural Society gardens, Wisley, 1929–1933, 1946 later becoming superintendent of the glasshouse dept.

DYKES, William Rickatson (1877–1925)
b. 4 Nov. 1877 *d.* Woking, Surrey 1 Dec. 1925
BA Oxon 1900. FLS 1920–25. VMH 1925. Master at Charterhouse School, 1903–19. Secretary, Royal Horticultural Society, 1920–25. Studied *Iris* and *Tulipa*. *Irises* 1912. *Genus Iris* 1913. *Handbook of Garden Irises* 1924. *Notes on Tulip Species* 1930. G. Dillistone, *ed. Dykes on Irises* 1932.
Bot. Soc. Exch. Club Br. Isl. Rep. 1925, 846. *Garden* 1925, 702–3 portr. *Gdnrs Chron.* 1925 ii 457 portr. *Nature* v.116, 1925, 908–9. *J. Bot.* 1926, 23–24. *J. R. Hort. Soc.* 1926, 177–82 portr.

DYMES, Thomas Alfred (1865–1944)
b. 30 Oct. 1865 *d.* 29 Oct. 1944
FLS 1912. ALS 1941. Bank clerk. Hon. Curator, Herb., Letchworth Museum. *Nature Study of Plants* 1920. Contrib. to *Proc. Linn. Soc.*, *J. Linn. Soc.*
Bot. Soc. Exch. Club Br. Isl. Rep. 1943–44, 644. *Proc. Linn. Soc.* 1943–44, 205.
Plants at Letchworth Museum, Kew, BM(NH), Oxford. Portr. at Hunt Library.

DYMOCK, William (1832–1892)
b. 10 Aug. 1832 *d.* Bombay 29 April 1892
Bombay Medical Staff, 1859. Surgeon-Major, 1873. Professor of Materia Medica, Grant College, Bombay, 1874–81. *Vegetable Materia Medica of Western India* 1883; ed. 2 1885. *Pharmacographia Indica* (with C. J. H. Warden and D. Hooper) 1889–93 3 vols. (biogr. in v.3, 1–3).
Pharm. J. v.22, 1892, 993. F. Boase *Modern English Biogr. Supplt. 2* 1912, 184.

DYSON, David (1823–1856)
b. Oldham, Lancs April 1823 *d.* Manchester 10 Dec. 1856
Mill-operative. Curator of Lord Derby's natural history collection at Knowsley Hall. Collected birds in Honduras and Venezuela, 1844–45, but did not send plants to BM as stated in *Proc. Manchester Field Club* 1900–1, 238–40 portr.
J. Bot. 1905, 134. *Lancs. Nat.* v.1, 1908, 167–170 portr., 192.

EAGLE, Archibald (*fl.* 1730s–1740s)
Merchant and seedsman, Smith's Land, Edinburgh. His widow, Elizabeth, carried on the business and had a nursery at Fountainbridge; later went into partnership with A. Henderson with nurseries at Meadowbank Road and Jocks Lodge.

EAGLE, Francis King (1769–1856)
b. Lakenheath, Suffolk 1769 *d.* Bury St. Edmunds, Suffolk 8 June 1856
LLB Cantab 1809. FLS 1807. Barrister. Bryologist. Contrib. to J. Sowerby and J. E. Smith's *English Bot.* 650, 2906, etc.

Proc. Linn. Soc. 1857, xxvii. *J. Bot.* 1888, 69. F. Boase *Modern English Biogr.* v.1. 1892, 950. K. M. Lyell *Life of Sir Charles J. F. Bunbury* v.2, 1906, 107, 111.
Herb. at Cambridge. Letters at Kew.

EALES, Luke (*fl.* 1660s–1690s)
MD. Of Welwyn, Herts. Sent plants to J. Ray. First recorded *Mentha piperita.*
R. Morison *Plantarum Historiae Universalis Oxoniensis* v.3, 1699, 495. R. A. Pryor *Fl. Hertfordshire* 1899, xxxix. C. E. Raven *John Ray* 1942, 263. J. E. Dandy *Sloane Herb.* 1958, 129.

EARDLEY, Sir Wilmot Sainthill (1852–1929)
b. 17 July 1852 *d.* 13 Nov. 1929
Indian Forest Service, 1873. Deputy Conservator of Forests, Garhwal District, United Provinces where he collected plants. Inspector-General of Forests, 1903–9.
Who was Who, 1929–1940 399.
Indian plants at Oxford.

EARLE, Rev. John (1824–1903)
b. Churchstowe, Devon 29 Jan. 1824 *d.* Oxford 31 Jan. 1903
BA Oxon 1845. Professor of Anglo-Saxon, Oxford, 1849–54, 1876–1903. Rector, Swanswick, Bath, 1857–1903. *English Plant Names from 10th to 15th Century* 1880.
D.N.B. Supplt. 2 v.1, 540–41. *Notes and Queries* v.11, 1903, 120. *Proc. Somerset Archaeol. Nat. Hist. Soc.* v.49, 1904, 193–195.

EARLE, Maria Theresa (*née* **Villiers**) (1836–1925)
b. 8 June 1836 *d.* Cobham, Surrey 27 Feb. 1925
Keen gardener. *Pot-pourri from a Surrey Garden* 1897. *More Pot-pourri from a Surrey Garden* 1899. *A Third Pot-pourri* 1903. *Memoirs and Memories* 1911. *Gardening for the Ignorant* (with E. Case) 1912. *Pot-pourri mixed by Two* (with E. Case) 1914.
Bot. Soc. Exch. Club Br. Isl. Rep. 1925, 846–47. *Gdnrs Chron.* 1925 i 174; 1961 ii 198–99. *Who was Who, 1916–1928* 317. *Lady* 23 March 1961, 475. D. Macleod *Gdnrs London* 1972, 151–56 portr.

EARLEY, William (*c.* 1835–1911)
d. Croydon, Surrey 17 Sept. 1911
Gardening editor of *Lloyd's News. How to Grow Mushrooms* 1869. *How to Grow Asparagus* 1873. *High-class Kitchen Gardening* 1875. *Garden Farmer* 1882. Edited *Villa Garden and Flower Show Manual* 1882.
Gdnrs Chron. 1911 ii 264–65 portr. *Gdnrs Mag.* 1911, 748. *J. Hort. Home Farmer* v.63, 1911, 333.

EARNSHAW, Frederick (1914–1952)
d. Edinburgh 27 April 1952
Educ. Cambridge, 1933–37. PhD Edinburgh. Assistant Lecturer in Botany, Edinburgh and East of Scotland College of Agriculture, 1937–40. Economic botanist, National Institute of Agricultural Botany, 1946. Genecologist, Scottish Plant Breeding Station, 1949.
Nature v.169, 1952, 993.

EASLEA, Walter (*c.* 1832–1919)
d. Feb. 1919
Gardener at Colchester, Stamford, Oxford and Waltham Cross. Grew roses.
Garden 1919, 96 portr.

EAST, Hinton (*fl.* 1770s–1792)
Of Kingston, Jamaica. Receiver-General. Had botanic garden at Liguanea from 1774; acquired, after his death in 1792, by the Government.
A. Broughton *Hortus Eastensis* (Appendix to B. Edwards' *Hist. Br. West Indies* v.1, 1792, 475). *Kew Bull* 1906, 64; *Additional Ser. 1* 1898, 139–40. F. Cundall *Historic Jamaica* 1915, 25, 228. A. Eyre *Bot. Gdns of Jamaica* 1966, 15–18.

EASTER, John (*c.* 1837–1912)
b. Kirby, Norfolk *c.* 1837 *d.* 25 April 1912
Gardener to Lord St. Oswald at Nostell Priory, Wakefield, 1890.
J. Hort. Home Farmer v.64, 1912, 401 portr.

EASTLAKE, Nathaniel (*fl.* 1820s)
Nurseryman, St. Mary Street, Bridgwater, Somerset.

EASTON, Nathaniel (*fl.* 1830s)
Of Plymouth, Devon. Had herb.
F. H. Davey *Fl. Cornwall* 1909, xli.

EASTWOOD, Charles (1839–1895)
b. Halifax, Yorks 1839 *d.* 21 Dec. 1895
Nephew of Samuel King (1810–1888) whom he assisted in preparing his plant specimens. Gardener at Kew, 1860–61; left to join his brother in nursery business at Lane House, Luddenden, Yorks.
W. B. Crump and C. Crossland *Fl. Parish of Halifax* 1904, lx. *N. Western Nat.* 1943, 277.

EATON, Rev. Alfred Edwin (1844–1929)
b. Little Bridge, Devon 25 April 1844
BA Cantab 1868. Curate, Ashbourne, Derby, 1869–71; Battlesden, Beds, 1872–73; Paddington, 1873–74; Thorncombe, Chard, Somerset, 1881–82. Vicar, Shepton Montague, Somerset, 1887–92. Entomologist. Naturalist on expedition to observe transit of Venus, 1874. Collected plants in British Isles, Spitzbergen, Magellen Straits, Cape of Good Hope, Kerguelen Island.
Bothalia 1950, 30. *Tuatara* 1970, 57–59.
Plants at Kew, BM(NH).

EATON, John (*fl.* 1810s)
Nurseryman, Horse Market, Warrington, Lancs.

EATON, Miss Mary Emily (1873–1961)
b. Coleford, Glos 27 Nov. 1873 *d.* North Newton, Somerset 4 Aug. 1961
Botanical artist at New York Botanical Gardens, 1911–32. Illustrated N. L. Britton and J. N. Rose's *Cactaceae* 1919–23 4 vols.
W. Blunt *Art of Bot. Illus.* 1955, 256.
Drawings at New York Botanical Garden, National Geographical Society, Washington, BM(NH). Portr. at Hunt Library.

EAVES, Daniel (1857–1936)
b. Liverpool 1857 *d.* May 1936
Headmaster, St. John's School, Fairfield, 1881. Writing master, Liverpool Institute, 1889–*c.* 1922. Collected plants in N. Wales, Lancashire and Cheshire.
Liverpool Public Museums. *Handbk Guide to Herb. Collections* 1935, 37.
Plants at Liverpool Museums destroyed in 1941.

ECKFORD, Henry (1823–1905)
b. Stenhouse, Midlothian 17 May 1823 *d.* Wem, Shropshire 5 Dec. 1905
VMH 1905. Gardener to Earl of Radnor at Coleshill, Berks, 1854–74. Nurseryman, Wem, 1888. Specialised in sweet peas.
Garden v.50, 1897, xi portr.; v.58, 1900, 69–70 portr.; v.68, 1905, 378 portr. *Gdnrs Chron.* 1905 ii 431–32 portr. *Gdnrs Mag.* 1905, 811 portr. *J. Hort. Cottage Gdnr* v.51, 1905, 547 portr.

EDDIE, Alexander (–1788)
d. Lambeth, London 3 May 1788
Seedsman, Woolpack and Crown, 68 Strand, London. Acquired business of Wilson and J. Sanders.
J. Harvey *Early Hort. Cat.* 1973, 11.

EDDIE, George (*fl.* 1780s)
Nurseryman, East Sheen, Surrey.

EDGAR, Patrick (*fl.* 1720s–1790s)
Nurseryman, Dublin. Joined by William Edgar, 1780s.

EDGERLEY, JOHN (*c.*1814–1849)
d. 9 June 1849
Arrived in Sydney, N.S.W., 1835. Collected plants in New Zealand for A. B. Lambert and Earl of Mountnorris, 1835–41. Nurseryman, Newmarket, N.Z., 1843–49.
Rec. Auckland Inst. Mus. 1970, 123–36. *Taxon* 1970, 521.

EDGEWORTH, Maria (1767–1849)
b. Black Bourton, Oxford 1 Jan. 1767 *d.* Edgeworthstown, Longford 24 May 1849
Novelist, *Dialogues on Botany for the Use of Young Persons* 1819.
A. J. C. Hare *Life and Letters of Maria Edgeworth* 1894. *D.N.B.* v.16, 380–82. R. L. Praeger *Some Irish Nat.* 1949, 77.

EDGEWORTH, Michael Pakenham (1812–1881)
b. Edgeworthstown, Longford, Ireland 24 May 1812 *d.* Eigg, Inverness 30 July 1881
FLS 1842. Half-brother of Maria Edgeworth. Pupil of R. Graham. Bengal Civil Service, 1831–81. Collected plants in Aden, 1846, India and Ceylon. *Pollen* 1877; ed. 2 1879. 'Plants from North-Western India' (*Trans. Linn. Soc.* v.20, 1851, 23–91). 'Florula Mallica' (*J. Linn. Soc.* v.6, 1862, 179–210). 'Florula of Banda' (*J. Linn. Soc.* v.9, 1866, 304–26). Contrib. *Caryophyllaceae* to *Fl. Br. India.*
A. Lasègue *Musée Botanique de B. Delessert* 1845; 433, 503. *J. Bot.* 1881, 288. *J. Linn.. Soc.* v.5, 1860, Supplt 1, iv. *Proc. Linn. Soc.* 1881–82, 63. *D.N.B.* v.16, 382–83. *Rec. Bot. Survey India* v.7, 1914, 5–6. H. N. Clokie *Account of Herb. Dept. Bot., Oxford* 1964, 162. I. H. Burkill *Chapters on Hist. Bot. in India* 1965, 74–75. *R.S.C.* v.2, 444; v.7, 594.
Plants at Kew, Oxford. Portr. at Hunt Library.
Edgeworthia Meissner.

EDMEADES, Robert (*fl.* 1770s–1780s)
Nurseryman, 11 (later 9) Fish Street Hill, London; nursery at Deptford, Kent. *Gentleman and Lady's Gardener* 1776.

EDMOND, James Williamson (–1875)
d. Inveresk, Midlothian 22 March 1875
MB Edinburgh 1870. Assisted B. Carrington with *British Hepaticae* 1874–75.
Trans. Proc. Bot. Soc. Edinburgh v.12, 1876, 409.

EDMONDS, Mr. (–1861)
d. 4 Feb. 1861
Gardener to Lady Lacon at Great Ormsby, Norfolk. Hybridised verbenas.
Florist 1861, 38.

EDMONDS, Charles (1811–1880)
b. North Aston, Oxford 7 Dec. 1811 *d.* Llandudno, Caernarvonshire 30 Dec. 1880
Gardener to Duke of Devonshire, etc. at Chiswick House, Surrey for more than 40 years.
Gdnrs Chron. 1875 ii 581 portr.; 1881 i 54 portr. *Gdnrs Yb. Almanack* 1882, 191.

EDMONDSON, John (1823–1894)
b. Penketh, Lancs 1823 *d.* Oct. 1894
Seedsman, 61 later 10 Dame Street, Dublin from 1851.
Gdnrs Chron. 1894 ii 479–80. *Gdnrs Mag.* 1894, 633.

EDMONDSTON, Thomas (1825–1846)
b. Buness, Shetland 20 Sept. 1825 *d.* Sua, Atacamas, Ecuador 24 Jan. 1846
Assistant Secretary, Botanical Society of Edinburgh. Discovered *Arenaria norvegica*. Naturalist on HMS 'Herald', 1845–46. *Fl. Shetland* 1845; ed. 2 1903 (biogr., 11–34 portr.). Contrib. to *Phytologist*.
Gdnrs Mag. 1840, 102. *Gdnrs Chron*. 1846, 411–12. *Phytologist* 1846, 580; 1851, 409. *Trans. Linn Soc*. v.20, 1851, 163–262. *Bonplandia* 1853, 4–5. B. C. Seeman *Narrative of Voyage of H.M.S. Herald* v.1, 1853, 67–69. *Mrs.* Edmonston *Young Shetlander* 1868. *D.N.B.* v.16, 397–98. *J. Bot*. 1912, 96. *New Shetlander* 1962, 7–9.
Plants from Falkland, Galapagos and Shetland Islands at Kew. MS. Fl. Shetland, 1837, at BM(NH). Letters at Kew.
Edmonstonia Seem.

EDMUNDS, Flavell (*fl.* 1850s)
Editor, *Hereford Times*. "A scholar of varied scientific attainments and a good botanist." Read botanical papers to Woolhope Naturalists' Field Club.
Trans. Woolhope Nat. Field Club v.34, 1954, 240.

EDWARD, George (*c.* 1816–1875)
d. 10 June 1875
Seedsman, York. Noted for florists' flowers and dahlias.
Gdnrs Yb. Almanack 1876, 171.

EDWARD, Thomas (1814–1886)
b. Gosport, Hants 25 Dec. 1814 *d.* Banff 27 April 1886
ALS 1866. Shoemaker. Curator, Banff Museum. Collected plants in Aberdeen and Banff.
J. Cash *Where there's a Will there's a Way* 1873, 172–214. S. Smiles *Life of a Scotch Nat*. 1876; ed. 2 1882. *J. Bot*. 1883, 61. *Proc. Linn. Soc*. 1885–86, 142–43. *Proc. Nat. Hist. Soc. Glasgow* 1886, lxxxiv–lxxxvi. *Scott. Nat*. 1886, 292–98. *D.N.B.* v.17, 106–7. *Trans. Inverness Sci. Soc. Field Club*. 1912–18, 438–40.

EDWARDS, David (*fl.* 1770s)
Seedsman, 5 Lower Thames Street, London. Successor to Ralph Prentice.

EDWARDS, Mrs. E. (–1881)
Contrib. botanical papers to *Sci. Gossip*.
Sci. Gossip v.17, 1881, 236.

EDWARDS, Edward (1812–1886)
d. Niton, Isle of Wight 10 Feb. 1886
Assistant, British Museum Library, 1839–46. First librarian, Manchester Free Library, 1850–58. Botanised in Herts. Contrib. to *Phytologist*.
R. A. Pryor *Fl. Hertfordshire* 1899, li. *D.N.B.* v.17, 115–17. W. A. Munford *Edward Edwards, 1812–1886* 1963 portr. *R.S.C.* v.7, 448.

EDWARDS, Eleanor Mary Wynne *see* Reid, E. M. W.

EDWARDS, Frank Charles (*c.* 1857–1934)
d. April 1934
Nurseryman, Headrow, Leeds, Yorks from 1897 with nurseries at Gledhow. *Cyclamen and how to grow Them* ed. 3 1897.
Gdnrs Chron. 1934 i 270.

EDWARDS, George (1694–1773)
b. Stratford, West Ham, Essex 7 April 1693 *d.* Plaistow, Essex 23 July 1773
FRS 1757. Artist. Ornithologist. Librarian, Royal College of Physicians. Correspondent of Linnaeus. Friend of Sir H. Sloane. *Gleanings of Natural History, exhibiting Figures of...Plants* 1758 portr. *Memoirs* 1776. Revised M. Catesby's *Natural History of Carolina...* 1754.
Gent. Mag. 1794 ii 902. J. Nichols *Lit. Anecdotes of Eighteenth Century* v.5, 317–26. W. Darlington *Memorials of John Bartram and Humphry Marshall* 1849, 419–20. *Essex Nat*. 1904, 343–49.
Portr. at Hunt Library. Library sold by James Robson, New Bond Street, 1774.

EDWARDS, Gerard (*fl.* 1830s–1860s)
Small collection of drawings of British plants, 1832–60, at BM(NH).

EDWARDS, Henry (1830–1891)
b. Ross, Hereford 17 Aug. 1830 *d.* New York 9 June 1891
Actor. In Victoria, Australia, 1853; Peru; Panama; U.S.A., 1867. Entomologist and botanist.
Entomol. News v.2, 1891, 129–30 portr. *Sci. Mon*. v.30, 240–49. *Victorian Nat*. 1891, 80. *Proc. Californian Acad*. 1893, 367. *J. New York Bot. Gdn* v.2, 1901, 125. E. O. Essig *Hist. Entomol*. 1931, 611–13 portr. J. Ewan *San Francisco* 1955, 48.
Plants at New York Botanical Garden.

EDWARDS, James (*fl.* 1830s–1850s)
Nurseryman, Bridge Street, York; nursery at Layerthorpe.

EDWARDS, John (*fl.* 1742–1790s)
Of Brentford, Middx. Artist. President, Society of Artists, 1776. *British Herbal* 1770. *Select Collection of...the most Beautiful Exotic and British Flowers which blow in our Gardens* 1755. *Collection of Flowers drawn after Nature* 1783–95.

EDWARDS, John (*fl.* 1810s–1820s)
Surgeon on 'Hecla' on W. E. Parry's Arctic expeditions, 1819–20, 1821–23. Had "extensive and well-preserved herbarium."
Plants collected listed in: W. E. Parry *Journal of Voyage for Discovery of North-west Passage...1819–1820; Supplt. to Appendix* 1824, cclxi–cccx; *Journal of Second Voyage...1821–1823; Appendix* 1825, 381–430.
Eutrema edwardsii R. Br.

EDWARDS, John (–1862)
d. 26 May 1862
Florist. Established *Garden Almanack* in 1853. Founder member of National Floricultural Society.
Florist and Pomologist 1862, 103–4.

EDWARDS, Richard (*fl.* 1630s)
Master of Apothecaries Company, 1634. Visited West of England with T. Johnson in 1634.

EDWARDS, Sydenham Teast (1768–1819)
b. Usk, Mon 5 Aug. 1768 *d.* Chelsea, London 8 Feb. 1819
FLS 1804. Botanical artist. Contrib. over 1200 plates to *Bot. Mag.*, 1787–1814. Founded *Bot. Register* in 1815 and contrib. many plates to it. Illustrated R. W. Dickson's *Complete Dictionary of Practical Gardening* 1805–7 2 vols. and W. Curtis's *Lectures on Botany* 1805 3 vols.
Bot. Mag. 1804, t.785. *Gent. Mag.* 1819 i 188. T. Faulkner *Hist. Topographical Description of Chelsea* v.2, 1829, 10–11. H. Trimen and W. T. T. Dyer *Fl. Middlesex* 1869, 396. *Gdnrs Chron.* 1887 i 479; 1898 i 340. *D.N.B.* v.17, 126. *Trans. Cardiff Nat. Soc.* 1910, 15–19. *J. Arnold Arb.* 1937, 183–84. W. Blunt *Art of Bot. Illus.* 1955, 192–93. F. A. Stafleu *Taxonomic Literature* 1967, 124–25.
Drawings at BM(NH).
Edwardsia Salisb.

EDWARDS, Thomas (*fl.* 1590s)
"Apothecarie in Excester, learned and skilfull...in the knowledge of plants." Took *Yucca gloriosa* to his friend, J. Gerard.
J. Gerard *Herball* 1597, 89, 143, 1359.

EDWARDS, Thomas (*fl.* 1800s–1840s)
FLS 1811. Contrib. article on botany to *Encylop. Metropolitana* (v.7, 1–108 and elsewhere).

EDWARDS, Wilfred Norman (1890–1956)
b. Peterborough, Northants 13 June 1890 *d.* 17 Dec. 1956
Educ. Cambridge. Palaeobotanist, BM(NH) 1913; Keeper of Dept. of Geology, 1938. Secretary, Geological Society, 1940–44; Vice-President, 1945–46.
J. Soc. Bibl. Nat. Hist. 1957, 231–37 portr.

EDWARDS, William Frederick (1776–1842)
b. Jamaica 6 April 1776 *d.* Versailles, France 24 Aug. 1842
MD. FRS 1829. Physiologist. To Paris in youth and became naturalised. *De l'influence de la Température sur la Germination* 1834. *Mémoire de Physiologie Agricole sur la Végétation des Céréales sous de Hautes Températures* 1835.
R.S.C. v.2, 453; v.7, 600 (4 papers erroneously assigned to his brother, Henri Milne-Edwards).

EDWARDS, Rev. Zachary James (1799–1880)
b. Wambrook, Chard, Dorset 1799 *d.* Misterton, Somerset 1880
BA Oxon 1821. MA 1827. Rector, Combe Pyne, Devon, 1840. *Ferns of the Axe* 1862.
Ferns at BM(NH).

EGERTON, Amelia *see* Hume, *Lady* A.

EGERTON, Mrs. Robert (*fl.* 1890s)
Collected plants in Baluchistan, 1890, now at Kew.
Pakistan J. For. 1967, 344.

EHRET, Georg Dionysius (1708–1770)
b. Heidelberg, Germany 30 Jan. 1708 (*b.* Erfurt 1710?) *d.* Chelsea, London 9 Sept. 1770
FRS 1757. Botanical artist. In 1736 settled in England where he was befriended by Duchess of Portland, R. Mead and Sir H. Sloane. Employed in Oxford Botanic Garden, 1750. Illustrated Linnaeus's *Hortus Cliffortianus* 1737; C. J. Trew's *Plantae Selectae* 1750–73. *Plantae et Papiliones Rariores* 1748–59. MS. autobiography translated in *Proc. Linn. Soc.* 1894–95, 41–58 portr. Contrib. papers to *Philos. Trans. R. Soc.*
R. Pulteney *Hist. Biogr. Sketches of Progress of Bot. in England* v.2, 1790, 284–93. *Gent. Mag.* 1803 i 199. J. E. Smith *Selection from Correspondence of Linnaeus* v.2, 1821, 480–81. *Notes and Queries* v.4, 1863, 432; v.5, 1864, 22, 35. *Proc. Linn. Soc.* 1883–84, 42–58. *J. Bot.* 1896, 316 portr. *D.N.B.* v.17, 167. W. Blunt *Georg Dionysius Ehret* 1953. W. Blunt *Art of Bot. Illus.* 1955, 143–49. P. Synge-Hutchinson *G. D. Ehret's Bot. Designs on Chelsea Porcelain* (*Concise Encyclop. of Antiq.* v.4, 1959, 73–77). C. Murdoch *G. D. Ehret* 1970 portr. *J. R. Hort. Soc.* 1970, 385–89. P. Mitchell *European Flower Painters* 1973, 103. *Taxon* 1973, 292 portr.
MS. biography by C. J. Trew and MS. autobiography at BM(NH). Drawings at Victoria and Albert Museum, BM(NH), Kew, Hunt Library. Portr. at Linnean Society. Portr. at Hunt Library.
Ehretia L.

ELDER, Mrs. Flora M. (–1964)
d. 14 Jan. 1964
Librarian. Leader of botanical excursions for Andersonian Naturalists of Glasgow.
Glasgow Nat. v.18, 1964, 386 portr.

ELEY, Charles Cuthbert (*c.*1873–1960)
d. East Bergholt, Suffolk 14 June 1960
VMH 1945. Founded Kynoch Press. Had garden at East Bergholt Place. Secretary, Rhododendron Society. *Gdning for Twentieth Century* 1923.
Gdnrs Chron. 1960 ii 51. *House and Gdn* June 1874, 118 portr.

ELGEE, Frank (1881–1944)
b. Middlesbrough, Yorks 8 Nov. 1880 *d.* Alton, Hants 7 Aug. 1944
Assistant Curator, Middlesbrough Museum, 1904; Curator, 1923–32. *Moorlands of North-eastern Yorkshire* 1912. 'List of Cleveland Lichens by Wm. Mudd' (*Proc. Cleveland Nat. Field Club* 1913, 34–52). 'Vegetation of Eastern Moorlands of Yorkshire' (*J. Ecol.* 1914, 1–18).
N. Western Nat. 1944, 264 portr.

ELLACOMBE, Rev. Henry Nicholson (1822–1916)
b. Bitton, Glos 18 Feb. 1822 *d.* Bitton 7 Feb. 1916
BA Oxon 1844. VMH 1897. Rector, Bitton, 1850, where he had a garden. Hon. Canon of Bristol, 1881. *Plant-lore and Garden-craft of Shakespeare* 1878; ed. 2 1884. *In a Gloucestershire Garden* 1895. *In my Vicarage Garden and Elsewhere* 1902. Contrib. to *Gdnrs Chron.*
Bot. Soc. Exch. Club Br. Isl. Rep., 1916, 464. *Garden* 1916, 107 portr., 1920, 32–34. *Gdnrs Chron.* 1916 i 107–8 portr.; 1965 i 129. *J. Bot.* 1916, 119; 1921, 50–52. *Kew Bull.* 1916, 51. *Times* 15 Feb. 1916. A. W. Hill *ed. Mem. H. N. Ellacombe* 1919 portr. *Curtis's Bot. Mag. Dedications, 1827–1927* 215–16 portr. *Country Life* 1967, 546–47 portr.
Letters at BM(NH), Kew. Portr. at Hunt Library.
Sedum ellacombianum Praeger.

ELLACOMBE, Rev. Henry Thomas (1790–1885)
b. Alphington, Devon 15 May 1790 *d.* Clyst St. George, Devon 30 July 1885
MA Oxon 1816. Father of H. N. Ellacombe (1822–1916). Vicar, Bitton, 1835–50; Clyst St. George, 1850–85. Archaeologist. Established a garden at Bitton which was developed by his son. *Mem. of Manor of Bitton* 1869 (includes flora).
Garden v.23, 1883, xii portr. *Gdnrs Chron.* 1885 ii 187. A. W. Hill *ed. Mem. H. N. Ellacombe* 1919, 12–31 portr.
Yucca ellacombei Baker.

ELLER, Rev. Charles Irvin (*fl.* 1810s–1870s)
BA Cantab 1867. Chaplain at Belvoir Castle, Leics. 'Fl. of Vale of Belvoir' (*Hist. of Belvoir Castle* 1841, 391–410).
A. R. Horwood and C. W. F. Noel *Fl. Leicestershire* 1933, ccix.

ELLINGHAM, J. (*fl.* 1820s)
Nurseryman, Imperial Institute Road, Queens Gate, London.
Trans. London Middlesex Archaeol. Soc. v.24, 1973, 186.

ELLIOT, Rev. Edward Arthur (1890–1960)
b. Kingsbridge, Devon 1890 *d.* 19 Feb. 1960
Educ. Oxford. Vicar, Dunstall near Burton-on-Trent, 1924–29. Rector, Cubley near Derby, 1932–36. Secretary, British Pteridological Society, 1950–59. Edited *Br. Fern. Gaz.* 1949–59.
Br. Fern Gaz. 1961, 58–59.

ELLIOT, George Francis Scott- *see* Scott-Elliot, G. F.

ELLIOT, John (1764–1828)
MA Oxon. FRS 1811. Managing Director, Elliot's Brewery, Westminster. Maintained garden of 20 acres near present Victoria Street where he grew grapes and other fruit. Treasurer and Vice-president, Royal Horticultural Society, 1809–29.
A. Simmonds *Hort. Who was Who* 1948, 68–70.

ELLIOT, Sir Walter (1803–1887)
b. Edinburgh 16 Jan. 1803 *d.* Wolfelee, Roxburghshire 1 March 1887
KCSI 1866. LLD Edinburgh 1878. FRS 1878. FLS 1859. Indian Civil Service, 1818–60. *Fl. Andhrica* 1859.
Proc. Linn. Soc. 1886–87, 39–40. *Proc. R. Soc.* v.42, 1887, viii–ix. *Trans. Bot. Soc. Edinburgh* v.17, 1889, 342–45. *Hist. Berwick Nat. Club* v.14, 1893, 365. R. Sewell *Sir Walter Elliot of Wolfelee* 1896 portr. *D.N.B.* v.17, 262–64. *R.S.C.* v.2, 481; v.7, 689; v.9, 789.

ELLIOTT, Mr. (*fl.* 1820s)
Staff surgeon. Sent seeds from St. Vincent to Liverpool Botanic Garden in 1825.
Bot. Mag. 1828, t.2813.

ELLIOTT, Clarence (1881–1969)
b. Hadley, Herts 3 Nov. 1881 *d.* Cheltenham, Glos 18 Feb. 1969
VMH 1951. Farmed in S. Africa for 3 years. Established Six Hills Nursery at Stevenage, Herts, 1907. Collected plants in Corsica, 1908; Falkland Islands, 1909; Chile and Andes, 1927–28. Edited 1907 and 1908 eds. of C. A. Johns's *Flowers of the Field. Rock Garden Plants* 1935.
Gdnrs Chron. 1922 i 176 portr.; 1952 i 30

portr.; 21 March 1969, 42–43 portr. *J. R. Hort. Soc.* 1969, 232. *Quart. Bull. Alpine Gdn Soc.* 1969, 159–66 portr. *Times* 20 Feb. 1969. A. M. Coats *Quest for Plants* 1969, 375–76. *Who was Who, 1961–1970*, 346.

Portr. at Hunt Library.

ELLIOTT, Charles Frederic (*fl.* 1890s)

Deputy Conservator of Forests, Baluchistan, where he collected plants, 1891–94, now at Kew and Calcutta.

Pakistan J. For. 1967, 344.

ELLIOTT, James (*fl.* 1830s)

Nurseryman, Lower Richmond Road, Putney, Surrey.

ELLIOTT, William (*fl.* 1800s)

Nurseryman, Newcastle, Northumberland.

ELLIOTT, William Robert (1860–1908)

b. 18 March 1860 *d.* Bedford 13 March 1908

Kew gardener. Superintendent, King's House Gardens, 1881–86. Curator, Botanic Garden, Grenada, 1886–89. Collector to West Indian Exploration Committee. Collected plants in Grenada, 1886–89; St. Vincent, 1891–92; Dominica, 1895–96. Forestry Officer, N. Nigeria, 1903–7, where he also collected plants.

Account of hepatics in *J. Linn. Soc.* v.30, 1895, 331–72; lichens in *J. Bot.* 1896 *passim. Kew Bull.* 1903, 31; 1908, 195–96; *Additional Ser. I* 1898, 74. I. Urban *Symbolae Antillanae* v.1, 1898, 101, 137, 157–58; v.3, 1902, 44. *J. Kew Guild* 1908, 429–30 portr.

Plants and MSS. at Kew.

Lejeunea elliottii Spruce.

ELLIOTT, William Thomas (1855–1938)

d. Thrandeston, Suffolk 23 April 1938

FLS 1916. Dentist in London, Birmingham and Stratford-on-Avon. Studied Myxomycetes.

Proc. Linn. Soc. 1937–38, 312–14.

ELLIS, Miss Alice B. (*fl.* 1870s–1880s)

Drew plates for R. Hogg and H. G. Bull's *Herefordshire Pomona* 1876–85 (v.1, ii).

ELLIS, Hon. Charles Arthur (1839–1906)

b. Lisbon 1839 *d.* 30 March 1906

FLS 1897. Had garden with fine collection of trees and shrubs at Frensham Hall near Haslemere, Surrey.

Proc. Linn. Soc. 1905–6, 38–40. *Curtis's Bot. Mag. Dedications 1827–1927* 295–96.

ELLIS, Daniel (*c.* 1772–1841)

b. Glos *c.* 1772 *d.* Edinburgh 17 Jan. 1841

MD Glasgow. FRSE. Army surgeon. *Inquiry into Changes induced on Atmospheric Air, by the Germination of Seeds...* 1807. *Further Enquiries* 1811. *Discourse on Subjects relating to Horticulture* 1829. Contrib. articles on 'Vegetable Anatomy' and 'Vegetable Physiology' to *Encyclop. Britannica* eds. 4–6, 1815–24.

Gdnrs Chron. 1841, 87. *Gdnrs Mag.* 1841, 188–90. *Scotsman* 27 Jan. 1841. *R.S.C.* v.2, 482.

ELLIS, David (1874–1937)

b. Aberystwyth, Cardiganshire 9 June 1874

d. Bearsden, Dunbartonshire 16 Jan. 1937

DSc London 1905. FRSE. Science master, County School, Aberystwyth, 1896–98; Wakefield, 1898–1900. Lecturer in Botany and Bacteriology, Glasgow and West of Scotland Technical College, 1904; Professor of Bacteriology, 1925. *Guide to Common Wild Flowers of West of Scotland. Guide to Common Wild Flowers in Wales* 1925. *Medicinal Herbs and Poisonous Plants* 1918.

Glasgow Nat. 1937, 26–28. *Pharm. J.* 1937, 89. *Proc. R. Soc. Edinburgh* 1936–37, 408–9. *Times* 19 Jan. 1937. *Chronica Botanica* 1938, 74 portr. *Who was Who, 1929–1940* 414.

ELLIS, Ernest Tetley (1893–1953)

b. 27 Nov. 1893 *d.* 30 May 1953

Horticultural writer. *Jottings of a Gentleman Gardener* 1917. *Jottings of an Allotment Gardener* ed. 3 1919. *Allotment Gardening for Profit* 1923. Edited ed. 2 of *Black's Gdning Dict.* 1921.

Who was Who, 1951–1960 344.

ELLIS, John (*c.* 1705–1776)

b. Ireland *c.* 1705 *d.* London 15 Oct. 1776

FRS 1754. London merchant. Agent for West Florida, 1764; for Dominica, 1770. Imported many American seeds and was particularly interested in the problems connected with the transportation of plants and seeds. Correspondent of Linnaeus who described him as "a bright star of natural history." *Directions for bringing over Seeds and Plants from East-Indies...to which is added Figure and Botanical Description of...Dionaea muscipula* 1770. *Historical Account of Coffee* 1774. *Description of Mangostan and Bread-fruit* 1775. *Natural History of Many Curious and Uncommon Zoophytes* (with D. Solander) 1786. Contrib. to *Philos. Trans. R. Soc.* 1759–70.

Gent. Mag. 1776, 483; 1794 ii 902. J. Nichols *Lit. Anecdotes of Eighteenth Century* v.9, 331–32. J. E. Smith *Selection of Correspondence of Linnaeus* v.1, 1821, 79–281. A. L. A. Fée *Vie de Linné* 1832, 169–85. *D.N.B.* v.17, 285–86. *Notes and Queries* v.6, 1889, 347. *J. R. Microsc. Soc.* 1901, 114–22. *Proc. Linn. Soc.* 1930–31, 114–16; 1933–34, 58–62. S. Savage *Calendar of Ellis Papers* 1948. F. A. Stafleu *Taxonomic Literature* 1967, 128–29.

Herb. at Lund Botanical Museum. MSS., drawings and letters at Linnean Society. Library sold by James Robson, 1786.

Ellisia L.

ELLIS, John William (1857–1916)
b. Doncaster, Yorks 24 Jan. 1857 *d.* Liverpool 20 Aug. 1916

MB Liverpool. Lieut.-Col., R.A.M.C. Secretary, Liverpool Naturalists' Field Club, 1882–89; President, 1899, 1910. 'Fungus Fl. of Hundred of Wirral' (*Proc. Liverpool Nat. Field Club* 1911–14). Contrib. new British fungi to *Trans. Br. Mycol. Soc.* from 1912.

J. Bot. 1917, 32. *Proc. Liverpool Nat. Field Club* 1916, 25–28 portr. *Trans. Br. Mycol. Soc.* 1916, 462–64. *Kew Bull.* 1917, 87. Liverpool Public Museums *Handbook and Guide to Herb. Collections* 1935, 17–18 portr. *R.S.C.* v.12, 218.

Herb. at Liverpool Museums. Fungi and slides at Kew.

ELLIS, Robert (*fl.* 1700s)

Of Charleston, Carolina. Collected plants in S. Carolina, 1700, and sent plants to J. Petiver.

J. Petiver *Musei Petiveriani* 1695, 79–80. E. J. L. Scott *Index to Sloane Manuscripts* 1904, 159. J. E. Dandy *Sloane Herb.* 1958, 129.

Plants at BM(NH).

ELLIS, Robert (*fl.* 1880s)

Collected plants in N.W. Himalayas, 1879–83, now at Oxford.

Erigeron ellisii Hook. f.

ELLIS, Thomas (*fl.* 1770s)

Gardener to Bishop of Lincoln. *Gardener's Pocket Kalendar* 1776.

ELLIS, Rev. William (1794–1872)
b. London 29 Aug. 1794 *d.* Hoddesdon, Herts 9 June 1872

Missionary. In Polynesia, 1817–25. In Madagascar, 1853–65, where he collected plants. Introduced *Ouvirandra.* Had garden at Hoddesdon. *Madagascar Revisited* 1867. *Three Visits to Madagascar during 1853–1854–1856* 1858.

Gdnrs Chron. 1872, 806. J. E. Ellis *Life of W. Ellis* 1873. *D.N.B.* v.17, 296–97. *R.S.C.* v.12, 218.

Letters and plants at Kew.

Grammangis ellisii Reichenb. f.

ELLIS, William (*fl.* 1730s–1758)
d. 1758

Farmer of Little Gaddesden, Herts. Writer on agriculture and botanist. *Timber-Tree Improved* 1738. *Compleat Cyderman* 1754.

J. Donaldson *Agric. Biogr.* 1854, 50–52. *D.N.B.* v.17, 295. J. G. Dony *Fl. Hertfordshire* 1967, 12.

ELLISON, Rev. Charles Christopher (1835–1912)

Vicar, Bracebridge, Lincs. Raised 'Ellison's' orange dessert apple.

A. Simmonds *Hort. Who was Who* 1948, 22–23.

ELLISON, George (1862–1941)
b. Warrington, Lancs 21 Oct. 1862 *d.* Graigfechan 28 May 1941

Authority on flora of Orkney.

Liverpool Public Museums *Handbook and Guide to Herb. Collections* 1935, 38. *N. Western Nat.* 1944, 23–26 portr.

Algae at Liverpool Museums.

ELLISTON-WRIGHT, Frederick Robert (–1966)
d. 6 Aug. 1966

MB. MRCS. Practised medicine at Braunton, Devon. All-round naturalist. Discovered Lundy Cabbage, *Brassica wrightii.* Contrib. to M. G. Palmer's *Fauna and Fl. of Ilfracombe District of North Devon* 1946. Contrib. articles on flora of Lundy to *J. Bot.*

Rep. Trans. Devonshire Assoc. 1967, 22.

ELLMAN, Rev. Ernest (1854–1929)
b. Berwick, Suffolk 1854 *d.* Bath, Somerset 30 Jan. 1929

Educ. Oxford. Botanised in S. England and on continent of Europe, especially Spain.

Bot. Soc. Exch. Club Br. Isl. Rep. 1929, 89–90. *J. Bot.* 1929, 154–55. *Kew Bull.* 1929, 140–41. A. H. Wolley-Dod *Fl. Sussex* 1937, xlvii.

Plants at Kew.

Teucrium ellmanii Hubbard & Sandwith.

ELLWOOD, A. G. (1870–1952)
b. Watford, Herts 1870 *d.* Burgess Hill, Sussex 1952

Cultivated orchids for Messrs B. S. Williams, Holloway, Messrs Sander, St. Albans. Director, Messrs Charlesworth and Co., 1934–50.

Orchid Rev. 1952, 114.

ELPHINSTONE, Sir Graeme Hepburn Dalrymple-Horn (1841–1900)
d. 23 May 1900

Coffee-grower in Ceylon. Planter in Perak. Collected plants on Taiping Hills, Malaya.

Gdns Bull. Straits Settlements 1927, 121. *Who was Who, 1897–1915* 227. *Fl. Malesiana* v.1, 1950, 152.

Plants at Singapore.

ELSE, Joseph (1874–1955)
d. May 1955

Principal, Nottingham School of Art, 1923–39. Did drawings of mosses.

Trans. Br. Bryol. Soc. 1956, 130–31.

ELSEY, Joseph Ravenscroft (1834–1857)
d. Springfield, St. Kitts 31 Dec. 1857

Surgeon and naturalist to N. Australian Expedition, 1856–57. Collected plants in St. Kitts for A. H. R. Grisebach.

Gdnrs Chron. 1858, 112. I. Urban *Symbolae Antillanae* v.3, 1902, 44.

Australian and St. Kitts plants at Kew.

ELWES, Henry John (1846–1922)
b. Colesbourne, Glos 16 May 1846 *d.* Colesbourne 26 Nov. 1922

FRS 1897. FLS 1874. VMH 1897. Sportsman, naturalist and arboriculturist. Travelled extensively in Asia where he collected plants. Founded *Quart. J. For.* during his Presidency of Royal English Arboricultural Society, 1907. *Monograph of Genus Lilium* 1877–80. *Trees of Great Britain and Ireland* (with A. Henry) 1906–13 7 vols. *Memoirs of Travel, Sport and Natural History* 1930.

Nature v.56, 1897, 55; v.110, 1922, 780–81. *Gdnrs Mag.* 1909, 235–36 portr. *Bot. Soc. Exch. Club Br. Isl. Rep.* 1922, 699–702. *Garden* 1922, 626, 638. *Gdnrs Chron.* 1922 ii 319–20, 334; 1930 i 308–9. *Bull. Soc. Dendrologique de France* 1923, 39–41. *J. Bot.* 1923, 30–31. *Kew Bull.* 1923, 36–43. *Proc. Linn. Soc.* 1922–23, 41–43. *Proc. R. Soc.* 1924, xlviii–liii. *J. R. Hort. Soc.* 1924, 40–46 portr. *Curtis's Bot. Mag. Dedications, 1827–1927*, 199–200 portr. *Who was Who, 1916–1928*, 329. *Lily Yb.* 1940, 4–6. A. M. Coats *Quest for Plants* 1969, 168–69. *Country Life* 1974, 1211–12 portr. *R.S.C.* v.7, 610; v.9, 792.

Plants at Kew.

Galanthus elwesii Hook. f.

ELWORTHY, Charles (1805–)
b. Bridgwater, Somerset 29 April 1805

Gardener to Sir J. Trevelyan at Nettlecombe, Somerset. Collected ferns.

Gdnrs Chron. 1877 ii 396 portr. *Br. Fern Gaz.* 1909, 44.

ELY, Benjamin (1779–1843)
b. 12 Jan. 1779 *d.* 26 March 1843

Began raising carnations in 1803 while still a blacksmith at Carlton. About 1827 became full-time florist and seedsman at Rothwell Haigh, Yorks. After his death his son, Benjamin (1810–) carried on the business.

R. Genders *Collecting Antique Plants* 1971, 226.

ELYOT, Sir Thomas (*c.* 1490–1546)
b. Wilts *c.* 1490 *d.* 20 March 1546

Diplomat and man of letters. Compiled first Latin–English dictionary, 1538, containing English renderings of the names of flora and fauna with occasional notes.

D.N.B. v.17, 347–50. C. E. Raven *English Nat.* 1947, 42–44, 73–74.

EMBLETON, Robert Castles (1806–1877)
b. Berwick-on-Tweed, Northumberland 14 Dec. 1806 *d.* Beadnell, Northumberland 6 June 1877

Surgeon. Fellow student of H. C. Watson. Had herb.

Hist. Berwick Field Club v.8, 1877, 229. H. C. Watson *Topographical Bot.* v.2, 1883, 543–44. *R.S.C.* v.2, 486; v.7, 610.

Herb. at Newcastle-on-Tyne Museum. Letters at Linnean Society.

EMERSON, Mr. (*fl.* 1830s?)

Some plants from India, S. Africa and Australia at Oxford.

EMMERTON, Isaac (*c.* 1736–1789)
d. 13 March 1789

Nurseryman and gardener, Monken Hadley, Herts and Barnet, Herts.

EMMERTON, Isaac (*c.* 1769–1823)
d. 28 Feb. 1823

Son of I. Emmerton (*c.* 1736–1789). Nurseryman, Monken Hadley and Barnet; also of Brook Street, Holborn, London. *Plain and Practical Treatise on Culture and Management of Auricula* 1816; ed. 2 1819.

J. C. Loudon *Encyclop. Gdning* 1822, 1288. *Garden* 1914, 288 (viii). R. Genders *Collecting Antique Plants* 1971, 18, 55–56.

EMMERTON, Thomas (*fl.* 1760s)

Nurseryman, Barnet, Herts.

EMMETT, Skelton Buckley (1818–1898)
b. Harrow, Middx 28 Sept. 1818 *d.* 15 Nov. 1898

To Hobart with his parents in 1819. Employed by Van Diemen's Land Company, 1835–53 in north of Tasmania where he collected plants. Correspondent of F. von Mueller who joined him on plant collecting trip in 1875.

Austral. Dict. Biogr. v.1, 1966, 356–57.

Plants at Melbourne.

EMPSON, James (–1765)

Keeper, Natural History Dept., BM. Edited J. Petiver's *Opera* 1767 2 vols.

Gent. Mag. 1765, 299. B. D. Jackson *Guide Lit. Bot.* 1881, 32.

ENGLEHEART, Rev. George Herbert (*c.* 1851–1936)
b. Channel Islands *c.* 1851 *d.* Dinton, Wilts 15 March 1936

MA. VMH 1900. Vicar, Chute Forest, Wilts, 1880–1902. Hybridised daffodils.

Garden v.51, 1897, xi portr. *Gdnrs Chron.* 1899 i 163 portr.; 1936 i 188. *Daffodil Yb.* 1933, 2–3, 29 portr. M. J. Jefferson-Brown *Daffodils and Narcissi* 1969, 23–25.

ENGLISH, Albert Charles (1861/3–1945)
d. Sydney, N.S.W. 1945
Went to Australia in 1881 as collector for British scientific institutions. Collected plants, etc. in New Guinea, 1887–1906. Government agent at Rigo, 1889–1907. Planter and trader at Rigo after 1907.
Pacific Islands Mon. v.4(1), 1933, 22; v.15 (10) 1945, 12. *Fl. Malesiana* v.1, 1950, 155.
Plants at Kew.

ENGLISH, James Lake (1820–1888)
b. Epping, Essex 21 Aug. 1820 *d.* Epping 12 Jan. 1888
Umbrella mender. Taxidermist, entomologist. *Manual for Preservation of Larger Fungi...also...Wild Flowers* 1882. Issued fascicles of Epping Forest Mosses, 1883–85. 'On Preparation of Fungi' (*Trans. Bot. Soc. Edinburgh* v.10, 1870, 28–29).
R. M. Christy *Birds of Essex* 1890, 19–21.

ENKEL, Robert (*fl.* 1830s)
Nurseryman, Islington, London. The site later became the Seven Sisters Nursery of Cornelius Crastin.

ENTWISTLE, Thomas (1851–1926)
b. Stockport, Lancs 1851 *d.* Sept. 1926
Kew gardener, 1874–75. In charge of herbaceous plants, Manchester Botanical Garden. President, Middleton Botanical Society for 15 years.
Heywood Advertiser 20 Nov. 1908. *N. Western Nat.* 1939, 267–68.

ENYS, John Davies (1837–1912)
b. Enys, Cornwall 11 Oct. 1837 *d.* Leeds, Yorks 7 Nov. 1912
Magistrate. In New Zealand, 1861–91, where he collected plants in the N.Z. Alps and Chatham Islands. 'Falmouth Algae' (*Rep. R. Cornwall Polytechnical Soc.* 1851, 24).
G. C. Boase and W. P. Courtney *Bibliotheca Cornubiensis* 1874–82, 1176. *Kew Bull.* 1893, 357; 1912, 393. *Trans. N.Z. Inst.* v.36, 1903, 355. T. F. Cheeseman *Manual N.Z. Fl.* 1906, xxxiii. *R.S.C.* v.7, 618; v.9, 802; v.14, 855.
Plants at Kew.
Ligusticum enysii T. Kirk.

EPPS, James (–1905)
d. West Indies 20 Feb. 1905
Proc. Croydon Nat. Hist. Soc. v.6, 1906, lxvi.
Plants at Croydon Nat. Hist. Soc. Museum.

EPPS, Wiliam James (*c.* 1817–1885)
d. Ringwood, Hants 18 May 1885
Nurseryman, Maidstone, Kent. Noted for fuchsias and Cape heaths.
J. Hort. Cottage Gdnr v.10, 1885, 440.

ERRINGTON, Robert (1799–1860)
b. Putney, Surrey 2 Nov. 1799
Gardener to Sir P. de M. G. Egerton at Oulton Park near Tarporley. *The Peach, its Culture, Uses and History* (with G. W. Johnson) 1847.
Cottage Gdnr v.16, 1856, 118–20 portr. *Gdnrs Chron.* 1860, 815. *J. Hort. Cottage Gdnr* v.38, 1899, 180 portr.

ERSKINE, Esme Nourse (1885–1962)
d. 7 July 1962
Major, King's African Rifles until 1924. Military Administration, Northern Frontier, Kenya, 1919–28. H. M.Consul, W. Abyssinia, 1928–37, where he collected plants.
Who was Who, 1961–1970 353.

ESCOMBE, Fergusson (1872–1935)
b. Bursledon, Hants 14 Sept. 1872 *d.* East Meon, Hants 12 Oct. 1935
BSc London. FLS 1896–1912. Collaborated with H. Brown at Jodrell Laboratory, Kew, 1897–1901, on germination and photosynthesis. At Research Laboratory, Guinness and Co., Dublin, 1901–4. Lecturer, S. Eastern Agricultural College, Wye, 1905–7. Contrib. to *Philos. Trans. R. Soc.* 1900–2.
Kew Bull. 1898, 62; 1901, 203–4; 1902, 24; 1905, 71; 1925, 96; 1935, 595. *Gdnrs Chron.* 1935 ii 406–7. *Nature* v.136, 1935, 900–1. *Times* 22 Oct. 1935. *Chronica Botanica* 1936, 185–86 portr. *J. Bot.* 1936, 19–21.

ESSEX, John (*fl.* 1770s–1800s)
Son of T. Essex (*c.* 1716–1799). Inherited his father's nursery at East Hill, Colchester, Essex.

ESSEX, Thomas (*c.* 1716–1799)
d. 6 April 1799
Nurseryman, Colchester, Essex.
Essex County Standard 15 April 1966.

EVANS, Alfred E. (*c.* 1880–1951)
d. Hove, Sussex 4 July 1951
Kew gardener, 1901. Assistant Curator, Botanic Station, Gold Coast, 1901. Inspector of Agriculture, 1907–20. Provincial Superintendent of Agriculture, Ashanti, 1920–23.
Kew Bull. 1901, 200. *Quart. J. Liverpool Univ. Inst. Comm. Res. Trop.* 1907, 9–17. *J. Kew Guild* 1951, 38–39 portr.
Plants and MSS. at Kew.

EVANS, Arthur Humble (1855–1943)
b. Scremerston, Northumberland 23 Feb. 1855 *d.* 28 March 1943
MA Cantab 1879. FRSE. Lecturer in History. Naturalist and botanist. Interested in *Arctium.* President, Berwickshire Natural History Club, 1900. 'Short Fl. Cambridge' (*Proc. Cambridge Philos. Soc.* 1911) revised

as *Fl. Cambridgeshire* 1939. Prepared part of A. Fryer's *Potamogetons of British Isles* 1898 for publication.

Bot. Soc. Exch. Club Br. Isl. Rep. 1943–44, 644–46. *N. Western Nat.* 1943, 224–26 portr.

EVANS, Edward (*c.* 1855–1933)
d. 23 Dec. 1933

Science master, Burnley Municipal College. *Botany for Beginners* 1899; 1906. *Plants and their Ways* 1908. *Intermediate Text-book of Botany* 1911.

Nature v.133, 1934, 17.

EVANS, Evan (*fl.* 1690s)

Surgeon. Friend of J. Petiver to whom he sent a few Icelandic plants.

J. E. Dandy *Sloane Herb.* 1958, 129.

EVANS, Evan Price (1882–1959)
b. Corris, Merioneth 19 Jan. 1882 *d.* Eastbourne, Sussex 27 April 1959

Educ. University College, Bangor, 1900. Teacher, Towyn and Ryhope, County Durham. Headmaster, Warrington Grammar School. Introduced plant ecology into school curriculum. *Plant Ecology and the School* (with A. G. Tansley) 1946. Contrib. to *J. Ecol.*

Nature v.185, 1960, 732.

EVANS, Frank James (–1928)
d. Perseverance, Trinidad 9 Aug. 1928

Kew gardener, 1902–3. Assistant Superintendent, Royal Botanic Gardens, Trinidad, 1903–12. Assistant Superintendent, Agricultural Dept., S. Nigeria, 1912–15. Managed German cacao estates, Cameroons.

Kew Bull. 1903, 31; 1912, 300; 1928, 336. *J. Kew Guild* 1929, 713.

Nigerian plants at Kew.

EVANS, Gareth Bevan (1935–1966)
b. Cadoxton, Glam 3 Sept. 1935 *d.* Kuala Lumpur, Malaya Sept. 1966

Educ. Cambridge. Member of Oxford expedition to Cameroons, 1962. Lecturer in Botany, University of Malaya, 1965. Collected ferns in British Guiana, 1961, and later in Africa and Malaya.

Times 10 Oct. 1966. *Br. Fern Gaz.* 1967, 364.

Plants at Kew.

EVANS, Sir Geoffrey (1883–1963)
b. Walmersley, Lancs 26 June 1883 *d.* Mayfield, Sussex 16 Aug. 1963

MA Cantab. Indian Agricultural Service, 1906–23. Director of Cotton Culture, Queensland, 1923–26. Principal, Imperial College of Tropical Agriculture, Trinidad, 1926–38. Economic Botanist, Kew Gardens, 1938–53; Acting Director, 1941–43.

J. Kew Guild 1943, 219 portr.; 1952–53, 122; 1963, 332. *Kew Bull.* 1954. 32–33. *Nature* v.200, 1963, 214–15. *Times* 19 Aug. 1963, portr. *Who was Who, 1961–1970* 356.

EVANS, H. A. (*fl.* 1880s)

Master, United Services College, Westward Ho. *Handlist of Plants occurring within Seven Miles of United Services College, Westward Ho* 1881.

W. K. Martin and G. T. Fraser *Fl. Devon* 1939, 775.

EVANS, Hugh (1874–1960)
b. Stamford, Lincs 14 Feb. 1874 *d.* Brentwood, Los Angeles, U.S.A. 10 Oct. 1960

To San Diego, California, 1892. In 1903 established a nursery at Los Angeles. A few years later sold it to Theodore Payne and moved to Santa Monica establishing in 1923 a garden and importing plants from S. Pacific, Australia and S. Africa. Established nursery with Reeves in West Los Angeles.

J. Californian Hort. Soc. v.12, 1951, 143–49 portr.

EVANS, Iltyd Buller Pole- *see* Pole-Evans, I. B.

EVANS, Rev. John (*fl.* 1768–1810s)
b. Lydney, Glos *c.* 1768

BA Oxon 1792. Master of Academy, Kingsdown. *Tour through Part of North Wales in...1798...with a View to Botanical Researches in that Alpine Country* 1800; ed. 2 1804.

D. Turner and L. W. Dillwyn *Botanist's Guide* 1805, 32. *J. Bot.* 1898, 14. *D.N.B.* v.18, 68.

EVANS, Joseph (1803–1874)
b. Tyldesley, Lancs 1803 *d.* Boothstown, Manchester 23 June 1874

Handloom weaver. Cultivated 300 species of medicinal plants. Founder and President of Boothstown Botanical Society. Contrib. to L. H. Grindon's *Manchester Fl.* 1859.

Gdnrs Chron. 1874 ii 614–15. L. H. Grindon *Country Rambles* 1882, 208. *Trans. Liverpool Bot. Soc.* 1909, 68.

EVANS, Thomas (*fl.* 1790s–1810s)

Of Stepney and East India Company House, London. Had a garden. Sent a plant collector to Pulo Penang. "Devoted almost his whole income to the acquirement of new and rare plants [from China and West Indies]."

H. C. Andrews *Bot. Repository* v.1, 1799, t.47; v.3, 1801, t.176. *Bot. Mag.* 1811, t.1416; 1813, t.1559; 1815, t.1783. E. Bretschneider *Hist. European Bot. Discoveries in China* 1898, 215–16. *Fl. Malesiana* v.1, 1950, 158.

Evansia Salisb.

EVANS, William (-1828)
Of Tyldesley, Lancs. Father of J. Evans (1803–1874). Botanist and correspondent of G. Caley, J. Hull and W. Withering.
J. Cash *Where there's a Will there's a Way* 1873, 132. L. H. Grindon *Country Rambles* 1882, 208. *Trans. Liverpool Bot. Soc.* 1909, 68–69.

EVANS, William (1851–1922)
b. Edinburgh 9 May 1851 *d.* Edinburgh 23 Oct. 1922
FRSE 1884. Field naturalist. 'Ricciae of Edinburgh' and 'Mosses of Isle of May' (*Trans. Bot. Soc. Edinburgh* v.23, 1908, 285–88; 348–51; v.28, 1923, 189–92).

EVANS, William Edgar (1882–1963)
b. Edinburgh 15 July 1882 *d.* 18 March 1963
BSc Edinburgh 1906. Assistant in Mycology, Heriot-Watt College, Edinburgh. Assistant in charge of Herb., Edinburgh Botanic Garden, 1919–44. Named G. Forrest's collections of Chinese and Tibetan plants. Collected plants in Iraq, 1917–19.
Yb. R. Soc. Edinburgh 1964, 19–20.
Plants at Edinburgh.

EVANS, William Wilson (1820–1885)
b. Dysart, Fifeshire 1 Jan. 1820 *d.* Edinburgh 5 May 1885
Gardener, Caledonian Horticultural Society Gardens; Curator, 1848. Assistant Secretary, Botanical Society of Edinburgh, 1843. Estate manager to Sir G. Clerk of Penicuik, 1864; and at Macbie Hill, Peebleshire, 1872. Had herb. of British plants. Discovered *Pallavicinia blyttii* in Scotland.
Trans. Bot. Soc. Edinburgh 1886, 311–12; 1910, 4.

EVELYN, John (1620–1706)
b. Wotton, Surrey 31 Oct. 1620 *d.* Wotton 27 Feb. 1706
MD Leyden 1641. DCL Oxon 1669. FRS 1663. Virtuoso. Diarist. Had garden at Sayes Court, Deptford and at Wotton. *Sylva* 1664. *Kalendarium Hortense* 1664. *Pomona* 1679. Translated *The French Gardiner* 1658 and *Compleat Gardener* 1693 by La Quintinye. *Directions for the Gardiner at Sayscourt* 1932.
W. Bray *Memoirs* 1819 2 vols. G. W. Johnson *Hist. English Gdning* 1829, 103–8. S. Felton *Portr. of English Authors on Gdning* 1830, 97–100. *Cottage Gdnr* v.5, 1850, 57. J. Donaldson *Agric. Biogr.* 1854, 26–29. *J. Hort. Cottage Gdnr* v.29, 1875, 249–50 portr.; v.46, 1903, 219 portr. *Gdnrs Chron.* 1895 ii 575–77, 772–73 portr.; 1964 i 85; 1965 i 70. *D.N.B.* v.18, 79–83. E. J. L. Scott *Index to Sloane Manuscripts* 1904, 183. E. Cecil *Hist. Gdning in England* 1910, 165–70. *Garden* 1915, 105. H. Evelyn *Hist. Evelyn Family* 1915. A. Ponsonby *John Evelyn* 1933. W. G. Hiscock *John Evelyn and his Family Circle* 1955. M. Hadfield *Gdning in Britain* 1960, 107–14. *Commonwealth For Rev.* 1965, 19–21. *Quart. J. For.* 1966, 40–43. *Country Life* 1967, 1006–7. H. Higham *John Evelyn Esquire* 1968. G. Keynes *John Evelyn* 1968. B. Saunders *John Evelyn and his Times* 1970. C. C. Gillispie *Dict. Sci. Biogr.* v.4, 1971, 494–97.
MS. and portr. by Kneller at Royal Society.
Evelyna Poeppig & Endl.

EVERARD, Charles Walter (1846–1890s)
b. Swaffham, Norfolk 1846
To China as student interpreter, 1867. Consul, Ichang, 1895. Sent plants to Kew, 1878–79.
E. Bretschneider *Hist. European Bot. Discoveries in China* 1898, 697. *Kew Bull.* 1901, 22.
Plants at Kew.
Nepeta everardi S. Moore.

EVERETT, Alfred Hart (1848–1898)
Zoologist. To Sarawak in 1869 to collect natural history specimens. Assistant Resident, Rejang, 1872. Resident of West Coast, British N. Borneo Co., 1883. Consul for Sarawak at Brunei, 1885. Resident, Trusan, 1885. Made several zoological collections; also collected plants in Borneo, 1888–94, Celebes, 1895, Lombok, 1896.
Sarawak Gaz. 1898, 136–37. *Fl. Malesiana* v.1, 1950, 158.
Plants at BM(NH), Kew, Singapore, etc.

EVERETT, Miss E. (*fl.* 1800s)
Of Isle of Wight where she collected algae. Correspondent of Dawson Turner and J. E. Smith.
D. Turner *Fuci* v.2, 1808, 110. J. Sowerby and J. E. Smith *English Bot.* 1970, 2116.

EVERSFIELD, William *see* Markwick, W.

EWART, Alfred James (1872–1937)
b. Liverpool 12 Feb. 1872 *d.* Melbourne 12 Sept. 1937
BSc Liverpool 1893. PhD Leipzig 1896. FRS 1922. FLS 1898. Science master, King Edward School, Bromsgrove. Lecturer, Birmingham University, 1897–1906. Professor of Botany and Government Botanist, Melbourne, 1906–37. Translated W. Pfeffer's *Physiology of Plants* 1900–6 3 vols. *On Physics and Physiology of Protoplasmic Streaming in Plants* 1903. *Recording Census of Victorian Fl.* 1908. *Fl. of N. Territory* (with O. B. Davies) 1917. *Handbook of Forest Trees* 1925. *Fl. Victoria* 1930. Contrib. to *Contributions to Fl. Australia* nos. 1–36.

Victorian Nat. 1905, 143; 1949, 88–89. *Times* 13 Sept. 1937. *Nature* v.141, 1938, 17. *Proc. Linn. Soc.* 1937–38, 314–17. *Obit. Notices Fellows R. Soc.* 1939, 465–69 port. *D.N.B. 1931–1940* 263–64. *Who was Who, 1929–1940* 425. *Fl. Malesiana* v.5, 1958, 41.

EWBANK, Rev. Henry (–1901)
d. Ryde, Isle of Wight 19 Oct. 1901
MA Oxon. Vicar, Ryde. Had a garden.
Garden v.58, 1900, iv, 459–61 portr.; v.60, 1901, 278 portr. *Gdnrs Chron.* 1901 ii 314, 327.

EWER, Samuel (1768–1815)
FLS 1789. Of Hackney, London. *Manuale sive Compendium Botanices* 1808.

EWER, Walter (*fl.* 1800s)
Sent plants from Sumatra, 1802–3, and Moluccas, 1809, to Botanic Gardens, Calcutta.
Fl. Malesiana v.5, 1958, 41–42.

EWING, Elizabeth Raymond (*née* **Burden**) (1860–1951)
b. Glasgow 25 Oct. 1860 *d.* 26 July 1951
President, Natural History Society, Glasgow, 1919. Contrib. to *Glasgow Nat.*
Glasgow Nat. 1951, 62–63.
Herb. with P. Ewing (1849–1913).

EWING, John (*c.* 1813–1896)
b. Fife *c.* 1813 *d.* 23 July 1896
Curator, Sheffield Botanical Gardens for more than 30 years.
Gdnrs Chron. 1896 ii 140. *J. Hort. Cottage Gdnr* v.33, 1896, 106. *J. Kew Guild* 1897, 10.

EWING, Rev. John Walter (*fl.* 1890s)
Some Materials for Fl. of Wrotham and Neighbourhood 1883.
Herb. at BM(NH) and in H. D. Geldart's collection at Norwich.

EWING, John William (*fl.* 1850s)
Nurseryman, 9 Exchange Street, Norwich and Eaton, Norfolk.

EWING, Juliana Horatia (*née* **Gatty**) (1841–1885)
d. Taunton, Somerset 13 May 1885
Author. Her keen interest in gardening is apparent in many of her children's stories.
Lady 23 March 1961, 474 portr. *House and Gdn* July/Aug. 1974, 123 portr.

EWING, Peter (1849–1913)
b. Kinross 13 July 1849 *d.* Glasgow 3 Aug. 1913
FLS 1894. Official of Phoenix Fire Assurance Association. President, Glasgow Natural History Society. *Glasgow Catalogue of Native and Established Plants* 1892; ed. 2 1899. Studied carices.
Bot. Soc. Exch. Club Br. Isl. Rep. 1913, 378–81. *Glasgow Nat.* v.5, 1913, 113–16. *J. Bot.* 1914, 296–98. *Proc. Linn. Soc.* 1913–14, 47–48. *R.S.C.* v.14, 902.
Herb. at Glasgow.

EWING, Rev. Thomas James (1813–1882)
b. Devon 1813 *d.* Feb. 1882
Educ. Cambridge. Arrived at Hobart, 1833. Ordained 1838. Headmaster, Queen's School for Orphans, 1839–47. Chaplain, St. John's School, 1839–63. Returned to England, 1863. Wrote on trees of Tasmania. Collected algae for W. H. Harvey.
Papers Proc. R. Soc. Tasmania 1909, 13. *Austral. Dict. Biogr.* v.1, 1966, 361. *R.C.S.* v.2, 535; v.7, 630.
Acanthococcus ewingii Harvey.

EYLES, Frederick (1864–1937)
b. Bristol 1864 *d.* Salisbury, Rhodesia 28 May 1937
Botanist, Dept. of Agriculture, Rhodesia where he collected plants, 1900–37. "Father of Rhodesian botany." 'Record of Plants collected in S. Rhodesia' (*Trans. R. Soc. S. Africa* 1916, 273-564).
Proc. Rhodesia Sci. Assoc. 1937, 1. *Rhodesia Herald* 29 May 1937. J. D. Clark ed. *Victoria Falls* 1952, 121–60. *Kirkia* 1967, 104 portr.
Herb. at Salisbury, Rhodesia.

EYLES, George (1815–1887)
d. Kew, Surrey 8 Dec. 1887
Gardener under J. Paxton at Chatsworth and at Crystal Palace, Sydenham. Superintendent, Royal Horticultural Society Gardens, South Kensington and Chiswick, 1859–71.
Garden v.32, 1887, 571. *Gdnrs Chron.* 1887 ii 754.

EYRE, Edward John (1815–1901)
b. Hornsea, Yorks 5 Aug. 1815 *d.* Tavistock, Devon 30 Nov. 1901
In Australia, 1833–45. Lieut.-Governor, New Zealand, 1846–53. Governor, St. Vincent, 1854–60; Jamaica, 1864–66. Collected plants in Australia. *Journals of Expeditions into Central Australia* 1845.
H. Hume *Life of Edward John Eyre* 1867. P. Mennell *Dict. Austral. Biogr.* 1892, 152. *Rep. Austral. Assoc. Advancement Sci.* 1907, 168. *J. W. Austral. Nat. Hist. Soc.* v.6, 1909, 16. *D.N.B. Supplt. 2* v.1, 641–44. *Austral. Encyclop.* v.3, 1965, 438–39. *Austral. Dict. Biogr.* v.1, 1966, 362–64.
MS. 'South Australia: Botanical Exploration (1854–1883)' at Kew. Plants at Melbourne. Portr. at Hunt Library.
Eyrea F. Muell.

EYRE, John (*fl.* 1850s)
General, Royal Artillery. Stationed in Hong Kong, 1849–51, where he collected plants and made drawings. Botanised with H. F. Hance. Discovered *Camellia hongkongensis.*
Hooker's J. Bot. 1851, 331. G. Bentham *Fl. Hongkongensis* 1861, 11*. E. Bretschneider *Hist. European Bot. Discoveries in China* 1898, 379–81. G. A. C. Herklots *Hongkong Countryside* 1951, 165.
Plants at Kew.
Eyrea vernalis Champ.

EYRE, Rev. William Leigh Williamson (1841–1914)
b. Padbury, Bucks 17 March 1841 *d.* Swarraton, Hants 25 Oct. 1914
Ordained priest, 1866. Rector, Swarraton, 1875. President, British Mycological Society, 1903. *List of Fungi of Grange Park and Neighbourhood, Hampshire* 1907. Contrib. lists of fungi to *Victoria County Hist. of Hampshire* v.2, 1900, 82–83.
J. Bot. 1915, 40. *Trans. Br. Mycol. Soc.* v.5, 1915, 185–86. *Proc. Hampshire Field Club Archaeol. Soc.* 1935, 75–77. *R.S.C.* v.14, 905.
Rubi at Haslemere Museum.
Chlorospora eyrei Massee.

FABER, John (*fl.* 1720s)
Introduced sweet-scented crab-tree, *Pyrus coronaria*, from N. America, 1724.
Bot. Mag. 1818, t.2009.

FACCIO, Nicholas (1664–1753)
b. Basle, Switzerland 16 Feb. 1664 *d.* Worcester 1753
FRS 1688. Mathematician. Came to England in 1687. Taught mathematics and was tutor to Marquis of Tavistock. *Fruit Walls Improved by inclining them to the Horizon* 1699.
J. C. Loudon *Encyclop. Gdning* 1822, 1266–67. *D.N.B.* v.18, 114–16.

FAGG, F. (*fl.* 1860s)
Master of 'South Western.' Collected a few Chinese plants for H. F. Hance in Hainan.
E. Bretschneider *Hist. European Bot. Discoveries in China* 1898, 690.

FAIR, Robert (*fl.* 1800s–1820s)
Seedsman and florist, 21 High Street, Borough, London.

FAIRBAIRN, George (*c.* 1846–)
b. Clapton, London *c.* 1846
Nurseryman, Carlisle, 1881. Specialised in carnations.
J. Hort. Home Farmer v.64, 1912, 537 portr.

FAIRBAIRN, James (*fl.* 1800s)
Nurseryman, Clapham, London.

FAIRBAIRN, John (*fl.* 1780s–1814)
d. Chelsea, London Dec. 1814
FLS 1788. Curator, Chelsea Physic Garden, 1784–1814. Correspondent of J. E. Smith. 'An Account of Several Plants Presented to Linnean Society, at Different Times by Mr. John Fairbairn and Mr. Thomas Hoy...by the President' (*Trans. Linn. Soc.* v.1, 1791, 249–54).
R. H. Semple *Mem. Bot. Gdn at Chelsea* 1878, 119.

FAIRBAIRN, Thomas (*fl.* 1820s–1840s)
Gardener to Sir J. Banks at Isleworth, 1805–16, to Prince Leopold at Claremont, Surrey, 1816–28. Later nurseryman, Broad Street, Oxford.
E. Bretschneider *Hist. European Bot. Discoveries in China* 1898, 286.

FAIRCHILD, Thomas (1667–1729)
b. Aldbourne, Wilts, *c.* May 1667 *d.* Hoxton, London 10 Oct. 1729
Nurseryman, Hoxton, *c.* 1690–1722 (list of plants in his garden in R. Bradley's *General Treatise of Husbandry and Gdning* v.3, 1724, 81–87). With William Darby "said to have been the first who raised tulip-trees in any quantity from seeds." His nephew, Stephen Bacon, acquired his nursery in 1729. *City Gardener* 1722. 'Some New Experiments concerning the Different and sometimes Contrary Motion of the Sap in Plants' (*Philos. Trans. R. Soc.* v.33, 1726, 127–29). Contrib. to *Catalogus Plantarum* 1730. First to raise a hybrid scientifically: *Dianthus caryophyllus x barbatus.* Correspondent of Linnaeus. Established by bequest the Fairchild sermon at Shoreditch Church.
R. Pulteney *Hist. Biogr. Sketches of Progress of Bot. in England* v.2, 1790, 238–39. J. Nichols *Illus. Lit. Hist. of Eighteenth Century* v.1, 371. J. C. Loudon *Encyclop. Gdning* 1822, 1268. G. W. Johnson *Hist. English Gdning* 1829, 191. S. Felton *Portr. English Authors on Gdning* 1830, 60–61. *Cottage Gdnr* v.6, 1851, 143; v.18, 1857, 195–97. *Gdnrs Chron.* 1881 i 48; 1893 i 546; 1912 i 65–66; 1927 i 366. *Gdnrs Mag.* 1896, 335. *D.N.B. Supplt 1* v.2, 198–99. *J. Heredity* 1932, 435–37. *Notes Rec. R. Soc.* 1940, 80–84. J. E. Dandy *Sloane Herb.* 1958, 129. *Huntia* 1965, 198–99.
MS. at Royal Society. Plants at BM(NH). Portr. by Van Blach at Oxford School of Botany. Portr. at Hunt Library.

FAIRCLOUGH, Benjamin (*fl.* 1810s)
Nurseryman, Fairclough Lane, Liverpool.

FAIRGRIEVE, Peter Walker (–1900)
b. Gallowayshire *d.* Dunkeld House, Perth 15 Feb. 1900

Gardener to Dowager Duchess of Athol at Dunkeld.

J. Hort. Cottage Gdnr v.40, 1900, 165 portr.

FAIRHAVEN, 2nd Baron *see* Broughton, H. R.

FALCONER, David (–1842)

Amateur botanist and gardener of Carlowrie near Edinburgh. Especially interested in growing irises.

Gdnrs Mag. 1842, 384.

FALCONER, Hugh (1808–1865)
b. Forres, Morayshire 29 Feb. 1808 *d.* London 31 Jan. 1865

MA Aberdeen 1826. MD Edinburgh 1829. FRS 1845. FLS 1844. Assistant-surgeon for East India Company in Bengal, 1830. Superintendent, Saharanpur Garden, 1832–41. Superintendent, Calcutta Botanic Gardens and Professor of Botany, Calcutta Medical College, 1848–55. *Palaeontological Mem. and Notes of Late H. Falconer* 1868 2 vols. *Rep. on Teak Forests of Tenasserim Provinces* 1852.

J. F. Royle *Essay on Productive Resources of India* 1840, 220–27. J. D. Hooker and T. Thomson *Fl. Indica* 1855, 67–68. *J. Bot.* 1865, 101. *Geol. Mag.* 1865, 142–44. *J. Hort. Cottage Gdnr* v.8, 1865, 234. *Photographic Portr. Men of Eminence in Lit. Sci. Art* v.3, 1865, 57–62 portr. *Proc. Linn. Soc.* 1864–65, xc–c. *Proc. R. Soc.* v.15, 1867, xiv–xx. *Proc. Geol. Soc.* 1865, xlv–xlix. *D.N.B.* v.18, 158–61. F. Darwin and A. C. Seward *More Letters of Charles Darwin* v.1, 1903, 252–58 portr. C. E. Buckland *Dict. Indian Biogr.* 1906, 142. D. G. Crawford *Hist. Indian Med. Service, 1600–1913* v.2, 1914, 147–48. I. H. Burkill *Chapters on Hist. Bot. in India* 1965 *passim.* C. C. Gillispie *Dict. Sci. Biogr.* v.4, 1971, 518–19. *R.S.C.* v.2, 551; v.7, 636.

Plants, MSS. and drawings at Kew. Portr. at Hunt Library.

Falconeria Royle.

FALCONER, John (–1547)
d. Ferrara, Italy 1547

Sent English plants to Amatus Lusitanus. Discovered *Anemone pulsatilla.* Studied with W. Turner at Bologna. 'Maister Falkonner's Boke' is the earliest English record of herb.

W. Turner *New Herball* 1551 sheet C, v (verso), vi. R. Pulteney *Hist. Biogr. Sketches of Progress of Bot. in England* v.1, 1790, 71–72. E. H. F. Meyer *Geschichte der Botanik* v.4, 1857, 270. *J. Bot.* 1863, 301–2. G. C. Druce *Fl. Oxfordshire* 1886, 372. *D.N.B.* v.18, 161. C. E. Raven *English Nat.* 1947, 69, 77–78, 82, 98.

Falconera Salisb.

FALCONER, Randle Wilbraham (1816–1881)
b. Bath? 1816 *d.* Bath, Somerset 6 May 1881

MD Edinburgh 1839. FRSE 1837. Practised medicine at Tenby, 1839–47. Physician to Bath General Hospital. *Contributions towards Catalogue of Plants...of Tenby* 1848 'Ancient History of the Rose' (*Gdn Mag.* 1839, 379–89).

Trans. Bot. Soc. Edinburgh v.14, 1883, 303. *D.N.B.* v.18, 162.

FALCONER, William (1744–1824)
b. Chester, Cheshire 23 Feb. 1744 *d.* Bath, Somerset 31 Aug. 1824

MD Edinburgh 1766. FRS 1773. Grandfather of R. W. Falconer (1816–1881). Physician, Chester Infirmary, 1767–70; Bath General Hospital, 1784–1819. *Historical View of Taste for Gardening and Laying Out Grounds among Nations of Antiquity* 1783. *Miscellaneous Tracts and Collections relating to Natural History* 1793 (includes list of plants known to Greeks). 'History of Sugar' (*Mem. Manchester Philos. Soc.* v.4, 1796, 291–301).

Gent. Mag. 1824, 374–75. S. Felton *Portr. of English Authors on Gdning* 1830, 183–84. *Cottage Gdnr* v.8, 1852, 299. W. Munk *Roll of R. College of Physicians* v.2, 1878, 278–80. *D.N.B.* v.18, 165–67. *R.S.C.* v.2, 552.

FALCONER, William (1850–1928)
b. Forres, Morayshire 2 Nov. 1850 *d.* Pittsburgh, U.S.A. 30 April 1928

Kew gardener, 1871. To USA 1872. Superintendent, Botanic Gardens, Harvard. Superintendent, Schenley Park, Pittsburgh, 1896. Superintendent, Allegheny Cemetery, 1903.

Trillia no. 9, 94–95. *J. Kew Guild* 1929, 711–12.

FALCONER, William (*fl.* 1890s)

British plants at Oxford.

FALKNER, Herbert John (1894–1951)
b. London 17 Aug. 1894 *d.* Torquay, Devon 10 March 1951

Studied Devon flora. Collected *Diatomaceae.*

Rep. Trans. Devonshire Assoc. 1951, 10–11.

FALKNER, Priscilla Susan *see* Bury, *Mrs.* E.

FALLA, John (*fl.* 1790s–1800s)

Son of W. Falla (*c.* 1739–1804). Nurseryman, Hebburn Quay, Jarrow, Durham.

FALLA, William (*c.* 1739–1804)
d. 20 May 1804

Foreman and bookkeeper to C. Thompson of Pickhill, Yorks. In 1782 purchased nursery of George Dale at Hebburn Quay, Jarrow, Durham.

FALLA, William (1761–1830)
b. 3 May 1761 *d.* 4 Aug. 1830
Son of W. Falla (*c.* 1739–1804). Partner in his father's nursery at Jarrow and Gateshead, Durham.

FALLA, William (1799–1836)
b. 28 Dec. 1799 *d.* 2 April 1836
Son of W. Falla (1761–1830). Nurseryman, Gateshead, Durham.

FANNIN, George (*fl.* 1850s)
Collected plants, especially orchids and asclepiads, in Natal in mid-19th century. Sent plants to W. H. Harvey in Dublin.
S. African J. Sci. 1971, 404–5.

FANNING, D. (*fl.* 1820s)
Plants from Caracas, Venezuela in A. B. Lambert's herb.
Taxon 1970, 522.

FARMER, Leo (*c.* 1875–1907)
d. Southsea, Hants 6 April 1907
Employed in Kew Herb., 1903–5, 1907. On Liverpool University W. African Expedition, 1906. Collected plants in Senegal, French Guinea, Liberia and Ivory Coast.
Gdnrs Chron. 1907 i 243. *J. Kew Guild* 1907, 381–82 portr. *Quart. J. Liverpool Univ. Inst. Comm. Res. in Trop.* 1907, 9–17.
Plants at BM(NH), Kew.

FARMER, Sir John Bretland (1865–1944)
b. Atherstone, Warwickshire 5 April 1865 *d.* Exmouth, Devon 26 Jan. 1944
MA Oxon. FRS 1900. FRSE 1927. FLS 1888. VMH 1933. Demonstrator in Botany, Oxford, 1887–92. Assistant Professor of Botany, Royal College of Science, London, 1892–95; Professor, 1895–1929. Edited *Gdnrs Chron.*, 1904–7; *Ann. Bot.; Sci. Progress. Practical Introduction to Study of Botany* 1899. *Plant Life* 1913. Co-editor with A. D. Darbishire of translation of H. de Vries's *Die Mutations Theorie* 1910–11 2 vols.
Nature v.62, 1900, 57; v.153, 1944, 397–98. *Gdnrs Mag.* 1912, 481–82 portr. *Gdnrs Chron.* 1934 i 54 portr.; 1944 i 64. *N. Western Nat.* 1944, 310–11 portr. *Times* 27 Jan. 1944. *Obit. Notices Fellows R. Soc.* 1945, 17–31. *Proc. Linn. Soc.* v.156, 1943–44, 205–7. *Yb. R. Soc. Edinburgh* 1945, 17. *D.N.B. 1941–1950* 245–46. *Who was Who, 1941–1950* 373–74. C. C. Gillispie *Dict. Sci. Biogr.* v.4, 1971, 545–46.
Portr. at Imperial College, London and Hunt Library.

FARNES, William W. (*fl.* 1820s)
Seedsman and florist, 52 West Smithfield, London.

FARQUHAR, Jane (*née* **Colden**) (1724–1766)
b. New York 27 March 1724 *d.* New York 10 March 1766
Daughter of Cadwallader Colden. Correspondent of J. Bartram. One of the first women to study the Linnean system.
J. E. Smith *Selection of Correspondence of Linnaeus* v.1, 1821, 40–45, 343. W. Darlington *Memorials of John Bartram* 1849, 202, 400–1. *J. Bot.* 1895, 12–15. *Torreya* 1907, 21–34. H. W. Rickett and E. C. Hall *Bot. Manuscript of Jane Colden, 1724–1766* 1963.
MS. 'Fl. Nov-Eboracensis' with drawings at BM(NH).

FARQUHAR, Robert (1821–1895)
b. Aberdeen 4 Jan. 1821 *d.* Aberdeen 12 Feb. 1896
Gardener at Fyvie Castle, Aberdeenshire.
Gdnrs Chron. 1875 ii 133 portr.; 1896 i 244 portr.

FARQUHAR, William (*c.* 1770–1839)
d. Perth 1839
Lieut.-Colonel of Engineers. Resident of Malacca, 1803–18. First Resident and Commandant of Singapore, 1819. Collected plants on Mount Ophir. Correspondent of N. Wallich. Employed Chinese artist to draw Malaccan plants and presented them to Royal Asiatic Society of Great Britain and Ireland in 1827.
C. H. Buckley *Anecdotal Hist. of Old Times in Singapore* 1902, 50, 105. *J. Straits Branch R. Asiatic Soc.* 1916, 153. *Gdns Bull. Straits Settlements* 1927, 121. *Gdns Bull. Singapore* 1949, 404–7; 1955, 530–33. *Fl. Malesiana* v.1, 1950, 163.
Plants in Wallich Herb. at Kew.
Myristica farquhariana Wallich.

FARQUHARSON, Charles Ogilvie (1888–1918)
b. Murtle, Aberdeenshire Nov. 1888 *d.* at sea 3 Oct. 1918
BSc Aberdeen 1911. Mycologist, Agricultural Dept., S. Nigeria 1912. Contrib. lists of fungi to *Kew Bull.* 1914, 253; 1917, 104. 'Notes on South Nigerian Mycetozoa' (with G. Lister) (*J. Bot.* 1916, 121–33).
Kew Bull. 1918, 353–61. *Nature* 1918, 192. *Trans. Br. Mycol. Soc.* v.6, 1919, 236–37.
Plants at Kew.

FARQUHARSON, Rev. James (1781–1843)
b. Coull, Aberdeenshire 1781 *d.* Alford, Aberdeenshire 3 Dec. 1843
MA Aberdeen 1798. LLD 1837. FRS 1830. Minister of Alford, 1813. *Agricultural Properties of Native Plants* 1835.
A. Murray *Northern Fl.* 1836 Appendix 2, v–xiv. *Gent. Mag.* v.21, 1844, 94–95. *D.N.B.* v.18, 224–25. *R.S.C.* v.2, 565.
Herb. at Carlisle Museum.

FARQUHARSON, Rev. James (-1906)
d. Selkirk 25 April 1906

LLD Aberdeen. Son of J. Farquharson (1781–1843). Minister of Selkirk. 'Plants of Selkirk' (*Hist. Berwickshire Nat. Club* v.8, 1876, 77–90).

Hist. Berwickshire Nat. Club v.19, 1905, 365–68. *Trans. Bot. Soc. Edinburgh* v.23, 1907, 216–17. *R.S.C.* v.7, 639; v.9, 830; v.14, 928.

FARQUHARSON, Marian Sarah (*née* **Ridley**) (1846–1912)
b. Privet, Hants 2 July 1846 *d.* Nice 20 April 1912

FLS 1908. Wife of R.F.O. Farquharson (1823–1890). *Pocket Guide to British Ferns* 1881.

J. Bot. 1903, 64. *Gdnrs Chron.* 1912 i 358. *Proc. Linn. Soc.* 1911–12, 45–46. *Who was Who, 1897–1916* 237. *Essex Nat.* v.29, 1954, 195.

British ferns at Essex Field Club.

FARQUHARSON, Robert F. Ogilvie (1823–1890)
d. Haughton, Aberdeenshire 3 May 1890

Diatomist. President, Alford Field Club.

J. Bot. 1890, 334–35. *Scott. Nat.* 1890, 289.

Docidium farquharsonii Roy.

FARRADY, James (*fl.* 1790s)

Seedsman, Settle, Yorks.

FARRAH, John (1849–1907)
b. Harrogate, Yorks 28 May 1849 *d.* Harrogate 13 Nov. 1907

FLS 1896. Contrib. botanical papers to *Naturalist*.

Naturalist 1907, 412–14 portr. *Proc. Linn. Soc.* 1907–8, 48.

FARRAR, Rev. Frederic William (1831–1903)
b. 7 Aug. 1831 *d.* 22 March 1903

Master at Harrow. Headmaster, Marlborough School. Dean of Canterbury, 1895. Wrote preface to J. C. Melvill's *Fl. Harrow* 1864. Contrib. to H. Trimen and W. T. T. Dyer's *Fl. Middlesex* 1869.

Who was Who, 1897–1916 238. D. H. Kent *Hist. Fl. Middlesex* 1975, 23.

FARRE, Frederick John (1804–1886)
b. London 16 Dec. 1804 *d.* Kensington, London 9 Nov. 1886

BA Cantab 1827. MD 1837. FLS 1835. Lecturer in Botany, St. Bartholomew's Hospital, 1831–54. Keeper of Herbaria, Medical Botanical Society, London.

Trans. Med. Bot. Soc. 1834–37, 206–18. W. Munk *Roll R. College of Physicians* v.4, 1878, 18–19. *Br. Med. J.* 1886 ii 1001–2. *Lancet* 1886 ii 1003–4. F. Boase *Modern English Biogr.* v.1, 1892, 1024.

FARRER, Reginald John (1880–1920)
b. Clapham, Yorks 17 Feb. 1880 *d.* Nyitada, Burma 16 Oct. 1920

Educ. Oxford. Plant collector, traveller, horticulturist, writer. Collected plants with W. Purdom in China, 1914–16; with E. H. M. Cox in Upper Burma, 1919. Influenced development of rock-gardening. *Garden of Asia* 1904. *Alpines and Bog Plants* 1908. *In a Yorkshire Garden* 1909. *My Rock-garden* 1909. *The Rock Garden* 1912. *On the Eaves of the World* 1917 2 vols. *English Rock-garden* 1919 2 vols. *Rainbow Bridge* 1921.

Bot. Soc. Exch. Club Br. Isl. Rep. 1920, 102–4. *Garden* 1920 588 portr. *Gdnrs Chron.* 1920 ii 247–48 portr.; 1921 i 31; 1956 i 677; v.167 (6) 1970, 6–7. *Kew Bull.* 1920, 370–71. *Nature* v.106, 1920, 413–14. *Geogr. J.* v.57, 1921, 69–70. *J. Bot.* 1921, 29. E. H. M. Cox *Farrer's Last Journey* 1926. *Who was Who, 1916–1928* 345. E. H. M. Cox *Plant Introductions of Reginald Farrer* 1930. *Quart. Bull. Alpine Gdn. Soc.* 1932, 176–213; 1961, 3–16; 1966, 127–36, 278–83; 1968, 29–47, 185–202. *J. R. Hort. Soc.* 1942, 287–93; 1970, 433–37; 1971, 234–36. E. H. M. Cox *Plant Hunting in China* 1945, 170–79. G. Taylor *Some Nineteenth Century Gdnrs* 1951, 116–60 portr. *Gdn J. New York Bot. Gdn* 1953, 179–80. M. Hadfield *Pioneers in Gdning* 1955, 227–31. *Lily Yb.* 1967, 10–12. A. M. Coats *Quest for Plants* 1969, 174–76.

Farreria Balf. f. & W. W. Smith.

FARRER, Thomas Henry, 1st Baron (1819–1899)
b. London 24 June 1819 *d.* Abinger Hall, Surrey 11 Oct. 1899

BA Oxon 1841. FLS 1869. Baronet, 1883. Baron, 1893. Secretary, Board of Trade. 'On Fertilisation of a Few Common Papilionaceous Flowers' (*Nature* 1872, 478–80, 498–501).

Proc. Linn. Soc. 1899–1900, 65–66. *D.N.B. Supplt. 2* 201–2. *Who was Who, 1897–1916* 238.

Portr. by Holl in possession of the family.

FARRER, William James (1845–1906)
b. Kendal, Westmorland 3 April 1845 *d.* Lambrigg, N.S.W. 16 April 1906

BA Cantab 1868. Surveyor, Lands Dept., N.S.W., 1870–86. Wheat experimentalist, 1898. Improved wheat by cross-breeding and selection.

Kew Bull. 1906, 226.

FAWCETT, Hugh Charles (1812–1890)
b. 16 May 1812 *d.* Stroud, N.S.W. 15 March 1890

Police Magistrate, N.S.W., 1862–70. Collected plants for F. von Mueller.

J. R. Soc. N.S.W. 1908, 100–1.

Cylicodaphne fawcettiana F. Muell.

FAWCETT, William (1851–1926)
b. Arklow, County Wicklow 13 Feb. 1851 *d.* Blackheath, London 14 Aug. 1926

BSc London 1879. FLS 1881. Assistant, Dept. of Botany, BM, 1880–86. Director, Public Gardens and Plantations, Jamaica, 1886–1908. Collected plants in Cayman Islands, 1888; Jamaica, 1887–1908. Initiated and edited *Bull. Bot. Dept. Jamaica* 1887–1902. *Guide to Botanic Gardens, Castleton, Jamaica* 1904. *Woods and Forests of Jamaica* 1909. *The Banana* 1913. *Fl. Jamaica* (with A. B. Rendle) 1910–36.

I. Urban *Symbolae Antillanae* v.1, 1898, 50–51; v.3, 1902, 46. *Bot. Soc. Exch. Club Br. Isl. Rep.* 1926, 90–91. *J. Bot.* 1926, 310–14. *Kew Bull.* 1908, 304; 1926, 427–28. *Nat. Hist. Mag.* (*BM*) v.1, 1927, 30–32 portr. *Proc. Linn. Soc.* 1926–27, 80–81. *West India Comm. Circ.* 1926, 322.

Plants at Kew. Portr. at Hunt Library.

FEILDEN *see* Tilden, R.

FEILDEN, Henry Wemyss (1838–1921)
b. 6 Oct. 1838 *d.* 18 June 1921

Lieut.-Colonel. Collected animals and plants in Greenland, *c.* 1876; Novaya Zemlya, 1882–98; Barbados, 1888–89; Kolgnev, *c.* 1896.

Who was Who, 1916–1928 347.

Plants at BM(NH).

FEILDEN, Rev. Oswald Mosley (1837–1924)
b. 16 Sept. 1837 *d.* Welsh Frankton, Shropshire 20 June 1924

BA Oxon. Rector, Welsh Frankton, 1865–1924. President, Offa Field Club, Oswestry. Contrib. to *Proc. Caradoc Field Club.*

J. Bot. 1924, 287.

FEILDING, J. B. (*fl.* 1890s)

In service of Sultan of Johore, 1892, where he collected plants.

Gdns Bull. Straits Settlements 1927, 121. *Fl. Malesiana* v.1, 1950, 163–64.

Plants at Singapore, BM(NH).

Ischaemum feildingianum Rendle.

FELGATE, George (*fl.* 1790s)

Nurseryman, Mansfield, Notts.

FELL, William (*c.* 1847–1903)
d. 22 March 1903

Acquired nursery of Rudolph Robson at Hexham, Northumberland *c.* 1879. In partnership with W. Milne, *c.* 1881.

Garden v.63, 1903, 252 portr.

FELLOWES, Rev. Charles (*c.* 1812–1896)
d. Shotesham, Norfolk 17 Dec. 1896

Vicar, Shotesham, 1838. Cultivated dahlias and picotees. President, National Dahlia Society.

Garden v.50, 1896, 524. *Gdnrs Chron.* 1896 ii 790. *Gdnrs Mag.* 1896, 896.

FELLOWS, Sir Charles (1799–1860)
b. Nottingham Aug. 1799 *d.* London 8 Nov. 1860

Traveller and archaeologist. Discovered ruins of Xanthus and of Tlos, 1838. Collected plants in Lycia, 1838–40. *Discoveries in Lycia* 1841 (plants, pp. 286–94).

D.N.B. v.18, 302–3.

FELTON, Robert Forester (*fl.* 1900s)

Florist, Hanover Square, London. *British Floral Decoration* 1910.

Gdnrs Mag. 1911, 81–82 portr.

FENN, Robert (1817–1912)
b. Rushbrooke, Suffolk 1817 *d.* Sulhamstead, Berks 20 March 1912

VMH 1903. Steward to Rev. G. St. John at Woodstock Rectory, Oxon. Smallholder, Sulhamstead, where he experimented in cross-breeding of plants, especially potatoes. Contrib. to *J. Hort.*

Gdnrs Mag. 1898, 296 portr. *J. Hort. Cottage Gdnr* v.56, 1908, 150 portr.; v.57, 1908, 6–8 portr.; v.64, 1912, 289 portr., 308.

FENNELL, James Hamilton (*fl.*1830s–1860s)

Antiquary. Zoologist. *Drawing-room Botany* 1840.

J. Timbs *Curiosities of London* 1868, 371. *Notes and Queries* v.4, 1851, 112; v.5, 1888, 169, 257, 404.

FENTON, Edward Wyllie (1889–1962)
b. Aberdeen 4 Nov. 1889 *d.* Milltimber, Aberdeenshire 16 Sept. 1962

MA Aberdeen 1912. DSc 1936. FLS 1923. FRSE. Lecturer in Botany, Aberdeen, 1913–18. Head, Biology Dept., Swindon and N. Wiltshire Technical Institute, 1918–20. Head, Dept. of Botany and Zoology, Seale Hayne Agricultural College, 1920–27. Head, Dept. of Botany, Edinburgh and East of Scotland College of Agriculture, 1927–54. President, Botanical Society of Edinburgh, 1943–45. Ecologist.

Yb. R. Soc. Edinburgh 1963, 16–17. *Proc. Linn. Soc.* 1964, 89–90.

FENWICK, Mark (*c.* 1861–1945)
d. Stow-on-the-Wold, Glos 28 Jan. 1945

VMH 1937. Created garden at Abbotswood, Stow-on-the-Wold.

Gdnrs Chron. 1945 i 64.

FENWICK, Roger (*or* **John**?) (*fl.* 1690s)

Of Spanish Town, Jamaica. Sent plants to J. Petiver. Had herb. of Jamaican plants.

Proc. Amer. Antiq. Soc. 1952, 361.

West Indian plants at Oxford.

FEREDAY, Rev. John (1813–1871)
b. Ellowes, Staffs 8 Nov. 1813 *d.* George Town, Tasmania 8 April 1871
MA Oxon. Episcopalian minister, George Town. He and his wife collected algae for W. H. Harvey.
W. H. Harvey *Phycologica Australica* v.1, 1858, 47; v.3, 1860, 173; v.4, 1862, Dedication; v.5, 1863, vi. S. Hannaford *Wild Flowers of Tasmania* 1866, 75, 85. W. H. Harvey *Memoir*, 1869, 282–83. *Proc. R. Soc. Tasmania* 1909, 14–15.
Algae at BM(NH), formerly at Paris Exhibition, 1855.
Cladophora feredayi Harvey.

FERGUSON, Daniel (–1864)
d. July 1864
At Glasgow Botanic Garden. Curator, Belfast Botanic Garden, 1836–64. *Popular Guide to Royal Botanic Garden of Belfast* 1851.

FERGUSON, David (*fl.* 1840s)
Nurseryman, Buckingham Road, Aylesbury, Bucks.

FERGUSON, Henry (*fl.* 1680s)
Seedsman, Blackfriars Wynd Head, Edinburgh.

FERGUSON, William (1820–1887)
b. July 1820 *d.* Colombo, Ceylon 31 July 1887
FLS 1862. Surveyor. Ceylon Civil Service, 1839–87. Collected plants, especially algae. *Palmyra Palm* 1850. *Scripture Botany of Ceylon c.* 1859. *Descriptive List of Ceylon Timber Trees* 1863. *Ceylon Ferns* 1872. *Ceylon Ferns and their Allies* 1880.
Ann. Bot. 1887, 403. *Ann. Mag. Nat. Hist.* 1887, 21–44. *Gdnrs Chron.* 1887 ii 312. *J. Bot.* 1887, 320. *Proc. Linn. Soc.* 1887–88, 89. *D.N.B.* v.18, 356–57. H. Trimen *Handbook Fl. Ceylon* v.5, 1900, 375. *R.S.C.* v.9, 848.
Algae at BM(NH). Letters at Kew.
Fergusonia Hook. f.

FERGUSON, William (*fl.* 1850s–1880s)
To Victoria, Australia, 1856. Inspector of State Forests, Victoria, 1869. Curator, Botanic Gardens, Melbourne, 1869–72. Quarrelled with F. von Mueller, Director of the Gardens at Melbourne. Inspector of Forests, 1889.
Rep. Austral. Assoc. Advancement Sci. 1911, 229. M. Willis *By their Fruits: Life of Ferdinand von Mueller* 1949, 92–97, 109, 126.

FERGUSON, William Hooker (*fl.* 1860s)
Son of D. Ferguson (–1864). Curator, Belfast Botanic Garden, 1864–68. Resigned after allegations of mishandling money.

FERGUSSON, Charles (–1904)
d. Nairn Feb. 1904
Nurseryman, Nairn.
Garden v.65, 1904, 141.

FERGUSSON, Rev. John (1834–1907)
b. Kerrow, Glen Shee, Forfarshire 1834 *d.* Edinburgh 6 Aug. 1907
LLD St. Andrews 1896. Minister, Glen Prosen, Forfarshire, 1867–68; New Pitsligo, Aberdeenshire, 1869; Fern, 1875. Bryologist. 'Forfarshire Mosses' (*Trans. Bot. Soc. Edinburgh* v.10, 1870, 245–51).
J. Bot. 1908, 31–32. *R.S.C.* v.7, 652; v.9, 348; v.12, 235.
Letters at BM(NH). Plants at Oxford.

FERNHALL, John (*fl.* 1790s)
Seedsman, Kidderminster, Worcs.

FIDLER, J. C. (–1903)
b. Reading, Berks *d.* Caversham, Berks 26 Dec. 1903
Seed and potato merchant, Friar Street, Reading.
J. Hort. Cottage Gdnr v.48, 1904, 16.

FIDLOR, Llewellyn L. (*fl.* 1820s)
Bombadier, Bombay Artillery. Draughtsman to Capt. W. H. Sykes.
M. Archer *Nat. Hist. Drawings in India Office Library* 1962, 40, 90.
Drawings of Deccan plants at India Office Library, BM(NH).

FIELD, Barron (1786–1846)
b. London 23 Oct. 1786 *d.* Torquay, Devon 11 April 1846
FLS 1825. Called to the Bar, 1814. Judge, Supreme Court, New South Wales, 1816–24. Resumed practice in England, 1827. Judge in Gibraltar, 1829. Sent plants and drawings to W. J. Hooker (Hooker *Exotic Fl.* v.3, 1827, t.232).
Proc. Linn. Soc. 1849, 298–99. *Bot. Zeitung* 1850, 392. *D.N.B.* v.18, 399–401. *J. Proc. R. Soc. N.S.W.* 1908, 101. *Taxon* 1970, 522.
Spanish plants at Kew. Portr. at Hunt Library. Sale at Sotheby 20 July, 1847.
Fieldia Cunn.

FIELD, Ernest (*c.* 1878–1970)
Gardener to Alfred de Rothschild at Halton House, Herts.
House and Garden Sept. 1974, 124 portr.

FIELD, Henry Claylands (1825–1911)
b. Holybourne, Hants 1825 *d.* Aramoko, Wanganui, New Zealand 1911
Civil engineer. To New Zealand, 1855. *Ferns of New Zealand* 1890. 'Notes on New Zealand Ferns' (*J. Bot.* 1878, 363–73).
Trans. Proc. N.Z. Inst. 1903, 355; 1911, iv–v. *R.S.C.* v.7, 656; v.9, 855; v.14, 989.
New Zealand plants at Kew.

FIELD, John (*fl.* 1760s–1800s)
Seedsman, 119 Lower Thames Street, London.

FIELD, Samuel (*fl.* 1820s–1840s)
Nurseryman, Boughton, Chester, Cheshire. Joined by John Field during 1840s.

FIELDER, C. R. (–1946)
d. Lyndhurst, Hants 17 Dec. 1946
VMH 1910. Gardener to Mrs. Burns at North Mimms, Herts. Horticultural adviser, Royal Horticultural Society.
Gdnrs Mag. 1910, 605–6 portr. *Gdnrs Chron.* 1923 ii 34 portr.; 1947 i 35.

FIELDING, Rev. Cecil Henry (1848–1918)
Vicar, Davington, Faversham, Kent. *Memories of Malling and its Valley: with a Fauna and Fl. of Kent* 1893.
J. Bot. 1893, 283–84.

FIELDING, Henry Borron (1805–1851)
b. Garstang, Lancs 17 Jan. 1805 *d.* Lancaster 21 Nov. 1851
FLS 1838. Added J. D. Prescott's and large part of A. B. Lambert's herb. to his own collection. Founded Fielding Curatorship at Oxford. *Sertum Plantarum* (with G. Gardner) 1844.
A. Lasègue *Musée Botanique de Benjamin Delessert* 1845, 330. *Cottage Gdnr* v.7, 1850, 188. *Proc. Linn. Soc.* 1852, 188. *Phytologist* 1852, 655–57. *Hooker's J. Bot.* 1854, 279, 284. G. C. Druce and S. H. Vines *Account of Herb. University of Oxford* 1897, 11–16. *Trans. Liverpool Bot. Soc.* 1909, 68–69. *D.N.B.* v.18, 424. H. N. Clokie *Account of Herb. Dept. Bot., Oxford* 1964, 43–44, 103–8, 164.
Letters at Kew.

FIELDING, Mary Maria (*fl.* 1830s–1880s)
Wife of H. B. Fielding (1805–1851). Illustrated her husband's *Sertum Plantarum* 1844.
Proc. Linn. Soc. 1852, 188.
6 vols. of drawings for sale by George's of Bristol, 1971.

FIFIELD, Samuel (*fl.* 1690s–1700s)
Surgeon. Sent plants from Campeachy, Mexico to J. Petiver.
J. Petiver *Musei Petiveriani* 1695, 94. E. J. L. Scott *Index to Sloane Manuscripts* 1904, 191. J. E. Dandy *Sloane Herb.* 1958, 130.

FIGGINS, Robert (*fl.* 1790s–1820s)
Nurseryman, Little Brittox, Devizes, Wilts.

FINCH, John Edward Montague (1842–1919)
Medical Officer, Mental Home, Humberstone, Leics for over 40 years. Had herb. which passed to W. Bell (1862–1925).
A. R. Horwood and C. W. F. Noel *Fl. Leicestershire* 1933, ccxxxiii.

FINCH, Louisa, Countess of Aylesford (*née* **Thynne**) (1760–1832)
b. 25 March 1760 *d.* Packington Hall, Coventry, Warwickshire 28 Dec. 1832
Married 4th Earl of Aylesford, 1781. Studied Warwickshire plants, 1784–1816. Correspondent of W. T. Bree, T. Purton, W. Withering and G. Don. Collection of 2830 drawings passed to Countess of Dartmouth.
J. E. Bagnall *Fl. Warwickshire* 1891, 493–94. *J. Bot.* 1908, 32. *Bot. Soc. Exch. Club Br. Isl. Rep.* 1914, 49. G. C. Druce *Fl. Buckinghamshire* 1926, xcv–xcvii portr. D. A. Cadbury *Computer-based Fl.... Warwickshire* 1971, 51.
Plants at Oxford. 69 flower drawings *c.* 1790 sold at Sotheby, 1 Nov. 1973.

FINDLAY, Bruce (1835–1896)
b. Streatham, Surrey 1835 *d.* Manchester 16 June 1896
VMH 1890. Kew gardener. Curator, Manchester Botanical Society, 1858; Secretary, 1875.
J. Hort. Cottage Gdnr v.3, 1881, 197 portr., 206–7, 497; v.20, 1890, 357 portr., 362–63; v.32, 1896, 582. *Gdnrs Chron.* 1881 ii 696. *Garden* v.49, 1896, 496. *Gdnrs Mag.* 1896, 415. *Sci. Gossip* 1896, 53. *J. Kew Guild* 1897, 34 portr.

FINDLAY, James (1821–1905)
b. Dumfriesshire 5 May 1821 *d.* 8 Nov. 1905
BA Edinburgh. Arrived in Melbourne, 1843. Accompanied F. von Mueller on plant collecting trip in New South Wales, 1874.
Plants at Melbourne.

FINDLAY, James (*fl.* 1860s)
Merchant in Burma. Found *Dendrobium findlayanum* while on journey to Chiengmai, 1866–68.
J. Thailand Res. Soc. Nat. Hist. Supplt. v.12, 1939, 17.

FINDLAY, Robert (*c.* 1882–1962)
d. Sunningdale, Berks 29 Dec. 1962
Curator, Royal Horticultural Society gardens, Wisley, Surrey, 1925–37.
Gdnrs Chron. 1963 i 34. *J. R. Hort. Soc.* 1963, 260.
British plants at Kew.

FINLAY, John (*c.* 1760–1802)
b. Glasgow 17 Feb. 1760? *d.* Glasgow 26 June 1802
FRS 1788. Army officer, Royal Engineers. In Guernsey, 1779–82. According to J. Gosselin he compiled a catalogue of Guernsey plants, now lost.

FINLAY, Kirkman (*fl.* 1820s–1880s)
d. San Fernando, Trinidad
MD. Collected plants at Gibraltar *c.* 1835, and in Canaries. Practised medicine in Trinidad, 1837. Collected plants in Antigua, Dominica and Grenada.
Ann. Rep. Trinidad Gdn 1887, 11. I. Urban *Symbolae Antillanae* v.3, 1902, 47–48; v.7, 1911, 74. *R.S.C.* v.2, 613.
Herb. at Trinidad; grasses at Kew. Letters at Kew.
Findlaya Hook. f.

FINLAY, William George Knox (*c.* 1895–1970)
d. 23 Feb. 1970
FLS 1959. VMH 1968. Army officer. Cotton grower in Brazil, 1922–28. Created garden at Keillour Castle near Perth. Noted for his cultivation of *Nomocharis*.
Lily Yb. 1969, 9 portr.; 1971, 154–55. *Gdnrs Chron.* v.167 (19), 1970, 9. *J. R. Hort. Soc.* 1970, 275–76. *J. Scott. Rock Gdn Club* 1970, 136–37 portr.

FINLAYSON, Daniel (1858–1939)
b. Wick, Caithness 1858 *d.* Wood Green, London 12 Feb. 1939
FLS 1898. Seed analyst, James Carter and Co., 1885–95. Founded own laboratory for seed testing, 1895.
Proc. Linn. Soc. 1938–39, 239.

FINLAYSON, George (*fl.* 1790s–1823)
b. Thurso, Caithness 1790 *d.* on passage from Calcutta to England, 1823
East India Company surgeon. Surgeon and naturalist to Crawfurd's mission to Siam and Cochinchina, 1821–23. Collected plants in Malaya and Siam which were distributed by N. Wallich, 1827–32. *Mission to Siam and Hue* 1826.
A. Lasègue *Musée Botanique de B. Delessert* 1845, 141–42. N. Wallich *Plantae Asiaticae Rariores* v.2, 1831, 48–50. *D.N.B.* v.19, 32. *Gdns Bull. Straits Settlements* 1927, 122. *J. Thailand Res. Soc. Nat. Hist. Supplt.* v.12, 1939, 8–9. *Fl. Malesiana* v.1, 1950, 165–66. M. Archer *Nat. Hist. Drawings in India Office Library* 1962, 48–51, 78. M. Archer *Br. Drawings in India Office Library* v.2, 1969, 419–20.
Plants at Kew. Drawings at Kew, India Office Library.
Finlaysonia Wall.

FINLAYSON, James (*c.* 1784–1864)
Florist of Paisley, Renfrewshire.
J. Hort. Cottage Gdnr v.32, 1864, 89–90.

FINNEY, Samuel (*fl.* 1830s–1840s)
Nurseryman, Windmill Hills, Gateshead, Durham. In partnership with Charles l'Anson. About 1841 took over the bankrupt business of William Falla and Co.

FIRMINGER, Rev. Thomas Augustus Charles (1812–1884)
b. London 13 March 1812 *d.* Edmonton, Middx 18 Jan. 1884
BA Cantab 1844. MA 1855. *Manual of Gardening for Bengal and Upper India* 1863; ed. 4 1890.
Gdnrs Chron. 1884 i 124. C. E. Buckland *Dict. Indian Biogr.* 1906, 146.

FISCHER, Cecil Ernest Claude (1874–1950)
b. Bombay 9 July 1874 *d.* 19 Oct. 1950
Studied forestry at Cooper's Hill College, 1892–95. Indian Forest Service, Madras, 1895–1926. Forest Entomologist, Dehra Dun, 1907–8. Principal, Madras Forest College, 1915–17. Conservator of Forests, Madras, 1920–23. Assistant for India, Kew Herb., 1926–40. 'Survey of Fl. Anaimalai Hills' (*Rec. Bot. Survey India* v.9, 1921, 1–218). Contrib. v.3 parts 8–11, 1928–36, to J. S. Gamble's *Fl. Presidency of Madras*.
Kew Bull. 1940, 204. *Empire For. Rev.* v.29, 1950, 292. *J. Kew Guild* 1950, 868 portr. *Nature* v.167, 1951, 16.
Indian plants at Calcutta, Kew.

FISH, David Sydney (*c.* 1881–1912)
d. Alexandria, Egypt 15 Nov. 1912
Gardener, Royal Botanic Garden, Edinburgh. Secretary and Garden Superintendent, Horticultural Society, Alexandria, 1906–12. Had orchid herb. *Book of the Winter Garden* 1906.
J. Hort. Home Farmer v.65, 1912, 503. *Trans. Edinburgh Field Nat. Microscop. Soc.* 1912–13, 48–49.

FISH, David Taylor (1824–1901)
b. Old Scone, Perthshire 25 Sept. 1824 *d.* Edinburgh 22 April 1901
Gardener to Sir T. Cullum at Hardwicke House. Horticultural journalist. *Cherry and Medlar* 1881. *Chrysanthemum* 1881. *Pear* *c.* 1881. *Plum* *c.* 1881. *Bulbs and Bulb Culture*. Edited *Cassell's Popular Gdning* 1884–86, 4 vols.
Gdnrs Chron. 1875 i 655–56 portr.; 1901 i 288–89 portr. *Garden* v.59, 1901, 308, 323 portr. *J. Hort. Cottage Gdnr* v.42, 1901, 374–75 portr.

FISH, Mrs. Margery (*c.* 1893–1969)
d. East Lambrook, Somerset March 1969
Created garden at Lambrook Manor, Somerset. Horticultural journalist. *We made a Garden* 1956. *An All the Year Garden* 1958.
Gdnrs Chron. 1961 i 383 portr.; v.166 (12), 1969, 13–15. *J. R. Hort. Soc.* 1969, 273–74. *Northern Gdnr* 1969, 97. *Times* 23 March 1969.

FISH, Robert (1808–1873)
b. New Scone, Perthshire 1808 *d*. Putteridgebury, Herts 23 Oct. 1873
Gardener to G. Sowerby at Putteridgebury; at Chiswick House, London. Contrib. to *Gdnrs Chron*., *Gdnrs Mag*. and *J. Hort*.
Florist and Pomologist 1873, 285–86. *Garden* v.4, 1873, 370. *Gdnrs Chron*. 1873, 1470–71.

FISHER, Alexander (*fl*. 1820s)
Assistant surgeon on 'Alexander' on John Ross's voyage to Baffin's Bay, 1818. Assistant surgeon on 'Hecla' on W. E. Parry's first Arctic voyage, 1819–20; surgeon on Parry's second Arctic voyage, 1821–23. Collected plants on these voyages.
W. E. Parry *Journal of Voyage...for Discovery of North-West Passage...1819–1820*. 1821. *Journal of Second Voyage...1821–1823* 1824.
Arctic plants at BM(NH).

FISHER, Charles (1823–1902)
b. Handsworth, Yorks 19 May 1823 *d*. Handsworth 21 March 1902
Nurseryman, Handsworth near Sheffield. Successor to his grandfather's nursery.
Garden v.61, 1902, 231. *Gdnrs Chron*. 1902 i 247 portr.

FISHER, Frederick (*fl*. 1810s)
Nurseryman, Edgware Road, Paddington, London.

FISHER, Harry (1860–1935)
b. Nottingham 3 June 1860 *d*. Grantham, Lincs 21 Jan. 1935
Chemist, Newick, 1886. Botanist to Jackson-Harmsworth Polar Expedition to Franz Joseph Land, 1894. Contrib. botanical sections to Victoria County Histories of Lancashire and Leicestershire. Authority on British *Rubi*. Director of *Grantham J*. 1911.
J. Bot. 1935, 166–67. R. C. L. and B. M. Howitt *Fl. Nottinghamshire* 1963, 18. E. J. Gibbons *Fl. Lincolnshire* 1975, 60.
Herb. at Nottingham Museum.

FISHER, Henry S. (–1881)
d. Liverpool 18 March 1881
Edited *Fl. Liverpool* (with F. M. Webb) 1872.
Trans. Liverpool Bot. Soc. 1909, 69. *R.S.C*. v.2, 627; v.7, 668.
Plants at Manchester University.

FISHER, James (*fl*. 1780s)
Seedsman, 130 Black Boy, Pall Mall, London. Successor to Samuel and F. Gray.

FISHER, Rev. Robert (1848–1933)
b. St. Bees, Cumberland 22 June 1848 *d*. Whitby, Yorks 11 May 1933
BA Cantab 1870. Held curacies at Hovingham and Manchester. Vicar, Sewerby, 1877–98; Beverley, 1898–1905. Rector, Stokesley, 1905–25. Canon of York, 1913. *Flower Land* 1890. *Flowering Plants of Whitby and District* 1928. *Flowers of Grass* 1931. *English Names of our Commonest Wild Flowers* 1932.
N. Western Nat. 1933, 334–36 portr. *Who was Who, 1929–1940* 450.

FISHER, St. John (*fl*. 1850s?)
Plants from Worcs, Cheshire and Lancs at Oxford.

FISHER, Thomas (*fl*. 1870s–1890s)
Herb. at Greenock Museum, Renfrewshire.

FISHER, William (*fl*. 1700s–1743)
d. c. Oct. 1743
Gardener to John Aislabie, Studley, Yorks. Helped to create the gardens at Studley Royal.

FISHER, William Rogers (1846–1910)
b. Sydney, N.S.W. 24 Feb. 1846 *d*. Oxford 13 Nov. 1910
BA Cantab 1867. MA Oxon 1905. Indian Forest Service, 1866. Director, Forest School, Dehra Dun. Assistant Professor of Forestry, Coopers Hill College, 1890. Professor of Forestry, Oxford, 1905. President, Royal English Arboricultural Society, 1904. Contrib. vols. 4 and 5 of *Schlich's Manual of Forestry*. *Manual of Indian Forest Botany* 1888. *Forest Protection* 1895; ed. 2 1907. Translated A. F. W. Schimper's *Plant Geography* 1903. Editor, *Indian Forester*, 1881–89.
Gdnrs Chron. 1910 ii 402 portr. *Nature* v.85, 1911, 113–114. *Quart. J. For*. 1911, 79–83. *Who was Who, 1897–1916* 245–46. *Proc. R. Soc. N.S.W*. 1921, 156–57.

FISHLOCK, Walter Charles (1875–1959)
b. Bathford, Somerset 14 Oct. 1875 *d*. Reading, Berks 20 Dec. 1959
Kew gardener, 1898. Imperial College of Tropical Agriculture, West Indies, 1902–20. Senior Curator, Dept. of Agriculture, Gold Coast, 1920–32. Collected plants in West Indies and Gold Coast.
J. Kew Guild 1959, 707.
Plants at Kew.

FISHWICK, John (*fl*. 1690s)
Sent plants from Andalusia, Africa and the Mediterranean to L. Plukenet. "Ornatissumus vir et nobis amicissimus" (Plukenet *Almagestum Botanicum* 1696, 85).
L. Plukenet *Almagestum Botanicum* 1696, 18, 54, 221. J. E. Dandy *Sloane Herb*. 1958, 130.

FITCH, Daniel (*c*. 1748–1818)
d. 4 Aug. 1818

Nurseryman, Fulham, Middx. Business carried on by widow and 3 sons; acquired in 1865 by James Veitch and Sons.

E. J. Willson *Hist. of Fulham* 1970, 245.

FITCH, John Nugent (1840–1927)
b. Glasgow 24 Oct. 1840 *d*. East Finchley, London 11 Jan. 1927

FLS 1877. Nephew of W. H. Fitch (1817–1892). Botanical artist and lithographer. Lithographed nearly 2500 plates for *Bot. Mag*. from 1878 when his uncle resigned as botanical artist. Illustrated *Floral Mag*. 1877–81, R. Warner and B. S. Williams's *Orchid Album* 1882–97 (original drawings at BM(NH)).

Bot. Soc. Exch. Club Br. Isl. Rep. 1927, 376. *J. Bot*. 1927, 118. *Proc. Linn. Soc*. 1926–27, 81–82.

Sale at Christie's on 12 April 1972 of 382 orchid drawings.

FITCH, Walter Hood (1817–1892)
b. Glasgow 28 Feb. 1817 *d*. Kew 14 Jan. 1892

FLS 1857. Botanical artist at Kew Gardens. Drew over 2700 plates for *Bot. Mag*., 1834–77 and nearly 500 plates for *Hooker's Icones Plantarum*, 1836–76. Produced over 10,000 published drawings. Among the works he illustrated were H. J. Elwes's *Monograph of Genus Lilium* 1877–80; J. D. Hooker's *Rhododendrons of Sikkim-Himalaya* 1849–51 and *Illustrations of Himalayan Plants* 1855. 'Botanical Drawing' (*Gdnrs Chron*. 1869, 7, 51, 110, 165, 221, 305, 389, 499).

London J. Bot. 1845, 641. *Gdnrs Chron*. 1879 ii 440; 1880 i 528; 1892 i 120. *Garden* v.41, 1892, 89. *J. Bot*. 1892, 100–2 portr. *Proc. Linn. Soc*. 1891–92, 68. *Nature* v.45, 1891–92, 302. *J. Kew Guild* 1895, 31. *Kew Bull*. 1915, 277–84, 392. L. Huxley *Life and Letters of Sir J. D. Hooker* v.2, 1918, 242–43. *Curtis's Bot. Mag. Dedications, 1827–1927* 167–68 portr. A. White and B. L. Sloane *Stapelieae* v.1, 1937, 87 portr. *Med. Biol. Illus*. 1969, 255–59 portr. W. Blunt *Art of Bot. Illus*. 1955, 223–30. *Gdn J*. 1975, 98–103 portr.

Drawings at Kew. Portr. at Hunt Library.

Fitchia Hook. f.

FITT, George (*fl*. 1840s)

Accountant of Great Yarmouth, Norfolk. Contrib. to *Phytologist* v.1–3, 1844–49.

London J. Bot. 1847, 287–89. *R.S.C*. v.2, 628.

Mosses at BM(NH). Letters at Kew.

FITTON, Sarah Mary (*fl*. 1810s–1860s)
b. Dublin

Conversations on Botany (with her sister Elizabeth) 1817; ed. 7 1831. *Four Seasons* 1865.

Proc. Geol. Soc. 1862, xxxiv. *Fl. de Serres* v.15, 1862–65, 185–86. *Gdnrs Chron*. 1951 ii 179–81.

Fittonia Coeman.

FITZALAN, Eugene F. Albini (1830–1911)
b. Londonderry 12 July 1830 *d*. Brisbane, Queensland June/July 1911

To Victoria, 1849. Seedsman, Brisbane, 1859. Plant collector on Burdekin River Expedition, 1860. Established nursery at Port Denison, Queensland, 1862. Moved to Cairns in 1887 whence he exported exotic plants. Collected at various times with J. Dallachy, ex Botanic Gardens, Melbourne.

F. von Mueller *Fragmenta Phytographiae Australiae* v.2, 1860–61, 108, 133, 167. *Gdnrs Chron*. 1861, 868. *J. Proc. R. Soc. N.S.W*. 1921, 157–59 portr. *Austral. Nat*. 1935, 120.

Fitzalania F. Muell.

FITZGERALD, Henry Purefoy (1867–1948)

FLS 1895. Science master, Wellington College, 1893–1908. Founder and first President, Botanical Essay Society. *Dictionary of Names of British Plants* 1885. *Concise Handbook of Climbers, Twiners and Wall Shrubs* 1906. *Wild Flowers*.

J. Bot. 1928, 150–52. *Proc. Linn. Soc*. 1948–49, 243–44.

FITZGERALD, Leslie Desmond Edward Foster Vesey- *see* Vezey-Fitzgerald, L. D. E. F.

FITZGERALD, Robert David (1830–1892)
b. Tralee, County Kerry 30 Nov. 1830 *d*. Sydney, N.S.W. 12 Aug. 1892

FLS 1874. In Sydney, 1856. In 1864 became interested in orchids after trip to Wallis Lake, N.S.W. Corresponded with C. Darwin about orchids. In 1869 collected plants with C. Moore in Lord Howe Island. Deputy Surveyor-General, 1873–87. *Australian Orchids* 1875–94 2 vols. Contrib. to *J. Bot*. 1883, 1885, 1891.

Gdnrs Chron. 1892 ii 404. *J. Bot*. 1892, 320. *Proc. Linn. Soc*. 1892–93, 23. *Sydney Mail* 3 Sept. 1892. *Victorian Nat*. v.9, 1892, 75–76; 1932, 233–42. *Proc. Linn. Soc. N.S.W*. v.21, 1896, 827; v.25, 1900, 784–85; v.69, 1944–45, 274–78. *J. R. Soc. N.S.W*. 1908, 102 portr. R. S. Rogers *Introduction to Study of South Australian Orchids* 1911, 51–52 portr. P. Serle *Dict. Austral. Biogr*. v.1, 1949, 298. *Austral. Orchid Rev*. v.4, 1939, 65; v.5, 1940, 111; v.17, 1952, 22–23. M. A. Reinikka *Hist. of the Orchid* 1972, 236–40 portr. *R.S.C*. v.9, 878.

Plants at Melbourne, BM(NH). Orchid drawings at Mitchell Library, Sydney. Portr. at Hunt Library.

Dracophyllum fitzgeraldi F. Muell.

FITZHERBERT, Sir Anthony (1470–1538)
b. Norbury, Derbyshire *d.* 27 May 1538
Judge. Sergeant-at-law, 1510. *Book of Husbandry* 1523. *Surveying and Book of Husbandry* 1547.
G. W. Johnson *Hist. English Gdning* 1829, 48–49. J. Donaldson *Agric. Biogr.* 1854, 4–7. *D.N.B.* v.19, 168–70. *Trans. Bibl. Soc.* 1896, 160–62.

FITZHERBERT, S. Wyndham (–1916)
d. Jan. 1916
Of Kingswear, Devon. *Book of Wild Garden* 1903. Contrib. to *Gdnrs Chron.*
Gdnrs Chron. 1916 i 54.

FITZ-ROBERTS, John *see* Robinson, J.

FITZWILLIAM, Charles William Wentworth, Viscount Milton (1786–1857)
b. London 4 May 1786 *d.* Wentworth Woodhouse, Yorks 4 Oct. 1857
Politician. Had notable garden at Wentworth. Patron of horticulture and of orchid-culture. "One of the oldest and most zealous friends of Natural Science in this country."
D.N.B. v.19, 224–25. *Curtis's Bot. Mag. Dedications, 1827–1927* 23–24 portr.
Miltonia Lindley.

FLANAGAN, Henry George (1861–1919)
b. Komgha, Cape Province 22 Jan. 1861 *d.* King Williamstown, S. Africa 23 Oct. 1919
FLS 1898. Farmer. Collected plants in Rhodesia.
Ann. Bolus Herb. v.3, 1924, 185–89 portr.
Herb. at National Herb., Pretoria.
Erica flanaganii Bolus.

FLATTERS, Abraham (*c.* 1848–1929)
b. West Stockwith, Notts *c.* 1848 *d.* Burnage, Lancs 2 March 1929
FRMS 1902. Director, Messrs Flatters and Garnett, Manchester. Lecturer in Microscopical Research, Manchester School of Technology. *Methods in Microscopical Research: Vegetable Histology* 1905. *The Cotton Plant* 1906; ed. 2 1912. Edited *Micrologist*.
N. Western Nat. 1929, 138–41.

FLEETWOOD, Rev. William (1656–1723)
b. London 1 Jan. 1656 *d.* Tottenham, Middx 4 Aug. 1723
MA Cantab 1683. DD 1705. Bishop of St. Asaph, 1708–14. Bishop of Ely, 1714–23. *Curiosities of Nature and Art in Husbandry and Gardening* 1707.
G. W. Johnson *Hist. English Gdning* 1829, 127–28. *D.N.B.* v.19, 269–71. D. McDonald *Agric. Writers* 1908, 160–63.

FLEMING, Andrew (*fl.* 1850s)
Geologist who mapped Salt Range, India and collected plants there and in Murree Hills, *c.* 1850.
Pakistan J. For. 1967, 345.
Plants at Kew, Edinburgh.
Astragalus flemingii Ali.

FLEMING, George (1809–1876)
b. Dunrobin, Sutherland 15 Jan. 1809 *d.* Hanchurch, Staffs 27 July 1876
Gardener to Duke of Sutherland at Trentham, Staffs.
Cottage Gdnr v.13, 1854, 33–37 portr. *Gdnrs Chron.* 1876 ii 180.

FLEMING, Gerald William Thomas Hunter (1895–1962)
d. Gloucester 19 June 1962
MRCS, LRCP 1920. DPM 1924. FLS 1941. Medical Superintendent, Barnwood House, Gloucester, 1937. Studied Gloucestershire flora and arranged botanical collections in Gloucester City Museum. Edited *Proc. Cotteswold Nat. Field Club* 1942–62.
Proc. Cotteswold Nat. Field Club 1960–61, 233–34. *Proc. Linn. Soc.* 1964, 90–91.
Herb. at BM(NH).

FLEMING, James (*fl.* 1800s)
Botanist on 'Cumberland' expedition under Lieut. Robbins in 1802 to explore islands off coast of Victoria and to visit Tasmania. Left New South Wales in 1804 for an appointment in a botanical garden in West Indies. "A good gardener and botanist." 'Journal of Explorations of Charles Grimes, kept by J. Fleming' (J. J. Shillinglaw *Hist. Rec. of Port Phillip* 1879, 15–30).
F. M. Bladen *Hist. Rec. of N.S.W.* v.5, 1897, 99, 136–37, 479. *J. Proc. R. Soc. N.S.W.* 1908, 102–3.
Journal of voyage to King Island and Port Phillip, 1802–3 at Sydney.

FLEMING, John (1747–1829)
d. London 17 May 1829
MD. FRS 1813. FLS 1816. Indian Medical Service, Bengal, 1768. In charge of Calcutta Botanic Garden during one of W. Roxburgh's leaves of absence. Returned to England, 1813. *Catalogue of Indian Medicinal Plants and Drugs* 1810.
W. Roxburgh *Plants of Coast of Coromandel* v.3, 1819, 44–45. *D.N.B.* v.19, 279 (corrected by *J. Bot.* 1916, 301–3). D. G. Crawford *Hist. Indian Med. Service, 1600–1913* v.2, 1914, 181–82. *Fl. Malesiana* v.5, 1958, 43.
Collection of Indian drawings at BM(NH).
Flemingia Roxb.

FLEMING, Rev. John (1785–1857)
b. Bathgate, Linlithgow 10 Jan. 1785 *d.* Edinburgh 18 Nov. 1857
DD St. Andrews 1814. FRSE 1814. Entered Presbyterian ministry and held charges at Bressay, Flisk and Clackmannan.

Lecturer in Natural History, Cork Institute, 1816. Professor of Natural Philosophy, Aberdeen, 1834. Professor of Natural Science, New College, Edinburgh, 1845. 'Outline Fl. West Lothian' (*Mem. Wernerian Soc.* v.2, 1814, 640). *Lithology of Edinburgh* 1859 i–civ, portr.

Trans. Bot. Soc. Edinburgh v.6, 1857, 16, 20–22. *D.N.B.* v.19, 279–80.

Flemingites Carruthers.

FLEMING, John (–1883)
d. Cliveden, Bucks 25 Nov. 1883

Gardener to Dukes of Sutherland and Westminster at Cliveden. *Spring and Winter Gardening* 1864.

Garden v.24, 1883, 498. *Gdnrs Chron.* 1883 ii 701. *J. Hort. Cottage Gdnr* v.6, 1883, 487.

FLEMMICH, Joanna Charlotte *see* Davy, *Lady* J. C.

FLEMWELL, George Jackson (1865–1928)
b. Mitcham, Surrey 29 May 1865 *d.* Lugano, Switzerland 6 March 1928

Artist, especially of Swiss Alps and their flora. Exhibited at Royal Academy. Wrote and illustrated *Alpine Flowers and Gardens* 1910, *Flower-fields of Alpine Switzerland* 1911. Illustrated H. S. Thompson's *Sub-Alpine Plants of Swiss Woods and Meadows* 1912.

Gdnrs Chron. 1928 i 201. *J. Bot.* 1928, 119; 1930, 192. *Who was Who, 1916–1928* 361.

Drawings at BM(NH).

FLETCHER, J. (–1916)

Of Cheltenham, Glos. Had herb. of local flora.

H. J. Riddelsdell *Fl. Gloucestershire* 1948, cxlii.

FLETCHER, James (1852–1908)
b. Ash, Kent 28 March 1852 *d.* Montreal 8 Nov. 1908

FLS 1886. Educ. King's School, Rochester. To Canada, 1874. Dominion Entomologist, 1884. Botanist and Entomologist, Experimental Farms, Canada, 1887. In charge of Dominion Arboretum and Botanic Garden, Ottawa, 1887–95. *Fl. Ottawaensis* 1879; 1888. *Farm Weeds of Canada* (with G. H. Clark) 1906; ed. 2 1909.

Proc. Linn. Soc. 1908–9, 37–38. *Science* v.28, 1908, 916–17. *Canadian Experimental Farms Rep.* 1909, 37–39 portr. *Ottawa Nat.* 1909, 189–234 portr.

Plants at Ottawa, Kew, etc.

FLETCHER, John (*c.* 1813–1882)
d. 24 May 1882

Florist, North Bierley, Bradford, Yorks. *J. Hort. Cottage Gdnr* v.4, 1882, 468.

FLIGHT, Frederick William (*c.* 1834–1910)
d. Winchester, Hants 15 Aug. 1910

Raised roses including 'Mrs. F. W. Flight'. *Gdnrs Mag.* 1910, 675.

FLINTOFF, Robert John (1873–1941)
b. Salford, Lancs 15 Aug. 1873 *d.* Goathland, Yorks 13 Sept. 1941

FLS 1932. FZS. Chemist. Partner, F. Scott and Co. Ltd., calico printers, Littleborough. Studied flora of East and North Ridings of Yorkshire. Founded Northern Ecological Association, 1936. Contrib. to *N. Western Nat.*

Naturalist 1941, 274. *N. Western Nat.* 1941, 332–44 portr. *Proc. Linn. Soc.* 1941–42, 275–78. *Whitby Gaz.* 19 Sept. 1941. *Chronica Botanica* 1942–43, 232. *J. Chem. Soc.* 1942, 64. H. N. Clokie *Account of Herb. Dept. Bot., Oxford* 1964, 166.

Plants at Oxford. Herb. at Keswick Museum. Botanical notebooks at BM(NH).

FLINTOFF, Thomas (*fl.* 1780s)

Of Knapton, Yorks. Surgeon. Sent fungi and notes to James Bolton.

J. Bolton *Hist. of Funguses about Halifax* 1788–91, 170.

FLINTOFT, Joseph (1796–1860)
b. Lastingham, Yorks 1796

Of Keswick, Cumberland.

N. Western Nat. 1943, 220–21; 1944, 277–80 portr.

Lakeland mosses at Carlisle Museum. Plants and MSS. at Keswick Museum.

FLOCKTON, Margaret Lillian (1862–1953)
d. Tennyson, N.S.W. 12 Aug. 1953

Went to Australia *c.* 1881 and for over 40 years was the artist at Sydney Botanic Gardens. Illustrated J. H. Maiden's books.

Sydney Morning Herald 15 Aug. 1953, 42. *Victorian Nat.* 1953, 116.

FLOOD, J. (*fl.* 1860s)

Member of A. C. Gregory's expedition to N.W. Australia, 1855–56, when he was botanical collector and assistant to F. von Mueller.

A. C. and F. T. Gregory *Journals of Australian Explorations* 1884, *passim. Proc. R. Soc. Queensland* v.8(2), 1890–91, 29.

Stylidium floodii F. Muell.

FLOOD, Margaret Greer (1896–1921)
b. Dublin 1896 *d.* Dublin 3 May 1921

Educ. Trinity College, Dublin. Technical Assistant, National Museum, Dublin, 1920. 'Exudation of Water by *Colocasia antiquorum*' (*Proc. R. Dublin Soc.* 1919). Published 3 papers with A. Henry on London Plane, Larch and Douglas Firs in *Proc. R. Irish Acad.*

Irish Nat. 1921, 65–67. *J. Bot.* 1921, 334.

FLORENCE, Ambrose *see* Lees, E.

FLORY, Sidney W. (–1954)
d. Slough, Bucks 22 Dec. 1954
Associated with his uncle, H. Tracy, in the management of orchid nursery, Twickenham. Acquired Messrs James Veitch's orchid nursery at Langley, Slough and went in partnership with J. M. Black.
Orchid Rev. 1955, 32.

FLOWER, Thomas Bruges (1817–1899)
d. Bath, Somerset 7 Oct. 1899
FRCS. FLS 1839. Had herb. *Fl. Thanetensis* 1847. Contrib. list of plants to W. Fletcher's *Tour Round Reading* 1840 and Robinson's *Environs of Reading* 1845. 'Fl. Wiltshire' (*Wiltshire Archaeol. Mag.* 1857–74). Contrib. to *J. Bot.*, *Phytologist*.
Proc. Linn. Soc. 1899–1900, 66. *J. Bot.* 1900, 32. G. C. Druce *Fl. Berkshire* 1897, clxix–clxx. J. W. White *Fl. Bristol* 1912, 88–91, 109. *Bot. Soc. Exch. Club Br. Isl. Rep.* 1941–42, 465. H. J. Riddelsdell *Fl. Gloucestershire* 1948, cxx. D. Grose *Fl. Wiltshire* 1957, 37–38. *R.S.C.* v.2, 646; v.7, 679; v.9, 887.

FODEN, William (*c.* 1841–1916)
d. Hemel Hempstead, Herts 16 May 1916
Kew gardener, 1861–63. Grew cucumbers at Flax Bourton, 1879. Nurseryman, Hemel Hempstead, 1882.
J. Kew Guild 1917, 374–75 portr.

FOGGITT, Gertrude (*née* **Bacon**) (1874–1949)
b. 19 April 1874 *d.* Sway, Hants 1949
Wife of T. J. Foggitt (1858–1934). Pioneer aviator and motorist. With Lady Davy made first discovery of *Carex microglochin* in Great Britain in 1923.
Bot. Soc. Br. Isl. Yb. 1951, 109–11.

FOGGITT, Thomas J. (1810–1895)
b. Durham 1810 *d.* Thirsk, Yorks 1895
Contrib. to J. G. Baker's *Fl. English Lake District* 1885 and other local floras.
Naturalist 1896, 202.

FOGGITT, Thomas Jackson (1858–1934)
b. Thirsk, Yorks 2 March 1858 *d.* Thirsk 30 Oct. 1934
Wholesale and retail chemist, Thirsk. Treasurer, Botanical Society and Exchange Club of British Isles, 1932. Studied flora of North Riding, Yorks.
Bot. Soc. Exch. Club Br. Isl. Rep. 1934, 808–10. *J. Bot.* 1934, 360. *N. Western Nat.* 1934, 391. *Times* 31 Oct. 1934. *Naturalist* 1935, 61–62.
Plants at BM(NH), Huddersfield Museum.

FOGGITT, William (1835–1917)
b. Yarm, Yorks 2 Feb. 1835 *d.* Thirsk, Yorks 10 May 1917
FLS 1903. Son of T. J. Foggitt (1810–1895). Original member of Thirsk Botanical Exchange Club; Curator, 1864–66. Good knowledge of plants of N.E. Yorks. Contrib. botanical account to E. Bogg's *Golden Vale of Mowbray* 1909, 249–72.
Bot. Soc. Exch. Club Br. Isl. Rep. 1917, 88; 1932, 290–97 portr. *J. Bot.* 1917, 200. *Naturalist* 1917, 205–6 portr. *Proc. Linn. Soc.* 1916–17, 44–45.
Plants at Kew.

FOORD-KELCEY, Mrs. Frances Louisa (*c.* 1862–1914)
Botanised in Leicestershire from 1896.
A. R. Horwood and C. W. F. Noel *Fl. Leicestershire* 1933, ccxxxiv.
Herb. at BM(NH). Plants at Oxford.

FOOT, Frederick James (*c.* 1831–1867)
b. Ireland *c.* 1831 *d.* drowned in Lough Kay, near Boyle 17 Jan. 1867
MA Dublin. Assistant geologist, Irish Geological Survey, 1856. 'Botany and Marine Zoology of Clare' (*Proc. Dublin Nat. Hist. Soc.* v.3, 1859–62). 'Plants in Burren' (*Trans. R. Irish Acad. Sci.* v.24, 1871, 143–60).
Geol. Mag. 1867, 95–96. R. L. Praeger *Some Irish Nat.* 1949, 81 portr. *R.S.C.* v.2, 653; v.7, 686.
Ferns at Kew.

FOOT, Harriet Sophia *see* Biddulph, H. S.

FORBES, Alexander (*fl.* 1790s)
In partnership with John Brunton, nurseryman, 83 High Street, Birmingham.

FORBES, Alexander (*fl.* 1810s–1860s)
Gardener to F. G. Howard at Levens Hall, Westmorland, 1810–62. *Short Hints on Ornamental Gardening* 1820.

FORBES, Arthur (1819–1879)
b. Douglas, Isle of Man 25 Jan. 1819 *d.* Aldershot, Hants 16 March 1879
Educ. Aberdeen and Cambridge. Ninth Laird of Culloden. Had herb. of local plants.
Trans. Proc. Bot. Soc. Edinburgh v.14, 1883, 20–22.

FORBES, Dorothy Hanbury- *see* Hanbury-Forbes, D.

FORBES, Edward (1815–1854)
b. Douglas, Isle of Man 12 Feb. 1815 *d.* Wardie, Edinburgh 18 Nov. 1854
FRS 1845. FLS 1843. Professor of Botany, King's College, London, 1842. Regius Professor of Natural History, Edinburgh, 1854. Contrib. Manx plants to H. C. Watson's *Topographical Bot.* 1883, 544. *Travels in Lycia, Milyas and the Cibyratis* (with T. Spratt) 1847 2 vols.

Ipswich Mus. Portr. Ser. 1852. *Gdnrs Chron.* 1854, 771–72. *Scotsman* 22 Nov. 1854. *Ann. Mag. Nat. Hist.* 1855, 35–52. *Edinburgh Mon. J.* 1855, 75–92. *Naturalist* 1855, 92–95. *Quart. J. Geol. Soc.* 1855, xxvii–xxxvi. *Trans. Bot. Soc. Edinburgh* v.5, 1858, 23–41. G. Wilson and A. Geikie *Mem. of Edward Forbes* 1861 portr. *Proc. Linn. Soc.* 1855, 408–12; 1888–89, 36. *D.N.B.* v.19, 388–92. F. Darwin and A. C. Seward *More Letters of Charles Darwin* v.1, 1903, 52–55. *Life and Letters and Journals of Sir C. J. F. Bunbury* v.2, 1906, 63–64. *J. Bot.* 1915, 286. *Centenary Rep., London Manx Soc.* 1915 portr. *Proc. Isle of Man Nat. Hist. Antiq. Soc.* v.2, 1923, 65–86; v.3, 1926–27, 167–70; v.6, 1960–63, 407–8. R. L. Praeger *Some Irish Nat.* 1949, 81–82. *J. Manx Mus.* 1965, 254–56. *R.S.C.* v.2, 654; v.12, 245.

Letters and Lycian plants at Kew. Isle of Man plants at BM(NH), King's College, London. MSS. at BM(NH). Bust by Sir J. Peel at Linnean Society. Portr. at Hunt Library. Sale at Sotheby, 11 Aug. 1857.

FORBES, Francis Blackwell (1839–1908)
b. U.S.A. 11 Aug. 1839 *d.* Boston, Mass., U.S.A. Nov. 1908

FLS 1875. In China, 1857–82. Friend of H. F. Hance. Account of his Chinese plants in *J. Bot.* 1876, 205–10. 'Enumeration of all Plants known from China' (with W. B. Hemsley) (*J. Linn. Soc.* v.23, 1886–88, 1–521; v.26, 1889–1902, 1–592; v.36, 1903–5, 1–686).

E. Bretschneider *Hist. European Bot. Discoveries in China* 1898, 720–23. *J. Bot.* 1905, 72; 1910, 19–22. *Proc. Linn. Soc.* 1908–9, 38–40. *R.S.C.* v.9, 896; v.15, 49.

Herb. and MSS. at BM(NH).

Euonymus forbesii Hance.

FORBES, George Ogilvie *see* Ogilvie-Forbes, G.

FORBES, Henry Ogg (1851–1932)
b. Drumblade, Aberdeenshire 30 Jan. 1851 *d.* Selsey, Sussex 27 Oct. 1932

Educ. Aberdeen and Edinburgh Universities. ALS 1879. Travelled in Portugal, 1875–77; in Dutch East Indies, 1878; New Guinea, 1885–86, 1889. Director, Canterbury Museum, N.Z., 1890–93. Director, Liverpool Museums, 1894–1911. Visited Socotra, 1898–99. Account of New Guinea plants collected by Forbes in *J. Bot.* 1886, 321–27, 353–60; *J. Bot.* 1923, Supplt. 1–64. Malayan plants in *J. Bot.* 1924, Supplt. 1–48; 1925, Supplt. 49–136; 1926, Supplt. 137–49. *Naturalist's Wanderings in the Eastern Archipelago* 1885.

Geogr. J. 1933, 93–94. *J. Bot.* 1933, 74–75. *Nature* v.131, 1933, 460–61. *Fl. Malesiana* v.1, 1950, 170.

Plants at BM(NH), Edinburgh, Kew. Drawings of Javan and Sumatran plants at BM(NH). MSS. at Kew.

Forbesina Ridley.

FORBES, James (1749–1819)
b. London 1749 *d.* Aix-la-Chapelle 1 Aug. 1819

FRS 1803. Official of East India Company in India, 1765–84. At Stanmore, Middx, 1804. *Oriental Memoirs* 1813–15 portr. with drawings of plants by himself; originals at Oscott College, Birmingham.

D.N.B. v.19, 397. *J. Bot.* 1917, 12–16.

FORBES, James (1773–1861)
b. Bridgend, Perthshire May 1773 *d.* Woburn Abbey, Beds 6 July 1861

ALS 1832. Gardener to Duke of Bedford at Woburn Abbey. Described *Epipactus purpurata* (J. Sowerby and J. E. Smith *English Bot. Supplt.* 2775). *Salictum Woburnense* 1829. *Hortus Woburnensis* 1833. *Journal of Horticultural Tour through Germany, Belgium and Part of France in...1835* 1837. *Pinetum Woburnense* 1839.

W. J. Hooker *Copy of Letter addressed to D. Turner...on...Death of Late Duke of Bedford* 1840, 11. *Proc. Linn. Soc.* 1861–62, civ–cv.

Letters and cultivated plants at Kew.

Oncidium forbesii Hook.

FORBES, John (*fl.* 1800s)
Nurseryman, 26 High Street, Birmingham.

FORBES, John (1798–1823)
d. Senna, Moçambique Aug. 1823

ALS 1822. Collected plants for Horticultural Society of London in Brazil, 1822, Madagascar, South and East Africa including Mozambique.

Trans. Hort. Soc. v.4, 1822, iii; v.5, 1824, iii. *Gdnrs Mag.* 1826, 360. W. J. Hooker *Exotic Fl.* v.2, 1827, 115. A. Lasègue *Musée Botanique de B. Delessert* 1845, 326, 327, 329, 376, 502. *D.N.B.* v.19, 405. A. de F. Gomes e Sousa *Exploradores e Naturalistas da Flora de Moçambique in Moçambique* Fasc. 20, 1939, 33–35. *J. R. Hort. Soc.* 1955, 271–72. *Kirkia* 1961, 133–37.

Plants at BM(NH), Kew. Journal for 1822–23 and letters at Royal Horticultural Society.

Forbesia Ecklon.

FORBES, John (–1909)
b. Aberfeldy, Perthshire *d.* 6 Sept. 1909

Nurseryman, Royal Nurseries, Hawick, Roxburghshire, 1870. Firm closed down *c.* 1969.

J. Hort. Cottage Gdnr v.59, 1909, 245, 293.

FORBES, W. (*fl.* 1820s)
Nurseryman, Gateshead, Durham.

FORBY, Rev. Joseph (*fl.* 1800s)
Brother of Rev. R. Forby (1759–1825). Discovered *Salix forbiana.*
J. Sowerby and J. E. Smith *English Bot.* 1344.

FORBY, Rev. Robert (1759–1825)
b. Stoke Ferry, Norfolk 1759 *d.* Fincham, Norfolk 20 Dec. 1825
BA Cantab 1781. MA 1784. FLS 1798. Rector, Fincham, 1789. "An able botanist" (J. E. Smith). *Vocabulary of East Anglia* 1830 2 vols (includes biogr. by D. Turner and portr.)
Gent. Mag. 1826 i 281. *D.N.B.* v. 19, 414. *Trans. Norfolk Norwich Nat. Soc.* 1912–13, 681–82 portr.
Portr. at Norwich Castle.
Salix forbiana Smith.

FORD, Charles (1844–1927)
b. 12 July 1844 *d.* Stanmore, Middx 14 July 1927
FLS 1885. ISO 1904. Superintendent, Botanical and Afforestation Dept., Hong Kong, 1871–1902. Collected plants in China and Formosa. *Catalogue of Plants in Government Gardens, Hong Kong* 1876.
E. Bretschneider *Hist. European Bot. Discoveries in China* 1898, 710–20. *Kew Bull.* 1927, 316–17. *Proc. Linn. Soc.* 1927–28, 114–15. *Curtis's Bot. Mag. Dedications, 1827–1927* 267–68 portr. E. H. M. Cox *Plant Hunting in China* 1945, 104–5, G. A. C. Herklots *Hong Kong Countryside* 1951, 115.
Plants at Kew.
Fordia Hemsley.

FORD, John (*fl.* 1760s–1780s)
MD. FLS 1789. Of Liverpool and Chester. Practised medicine at Leghorn and Rome and sent plants to John Ellis, his uncle.
J. Sowerby and J. E. Smith *English Bot.* 78. J. E. Smith *Selection of Correspondence of Linnaeus* v.2, 1821, 47–66.

FORD, Joseph (*c.* 1730s–1796)
d. 14 Feb. 1796
Nurseryman, New Bridge, Exeter, Devon.

FORD, Tom Henry (1849–1932)
b. Chewton Mendip, Somerset 10 Dec. 1849 *d.* Swansea, Glam 15 Sept. 1932
Kew gardener, 1875–76. Employed in private gardens in S. Wales.
J. Kew Guild 1933, 277 portr.

FORD, William (1760–1829)
b. Jan. 1760 *d.* Sept. 1829
Nurseryman, Exeter; first at St. Thomas, later at Longbrook Street. In partnership with Please. In 1831 the nursery was owned by Ann Ford.

FORDHAM, Henry (1803–1894)
Of Royston, Herts. Botanised locally.
J. G. Dony *Fl. Hertfordshire* 1967, 15.

FORDYCE, Mr. (*fl.* 1800s)
Gardener to Comtesse de Vandes at Bayswater, London.
Bot. Mag. 1813, t.1526.

FORDYCE, George (1736–1802)
b. Aberdeen 18 Nov. 1736 *d.* London 25 May 1802
MA Aberdeen 1750. MD Edinburgh 1758; Leyden 1759. FRS 1776. Grandfather of George Bentham. Physician, St. Thomas's Hospital, London. *Elements of Agriculture and Vegetation* 1765.
Gent. Mag. 1802 i 588. W. Munk *Roll of R. College of Physicians* v.2, 1878, 373–76. *D.N.B.* v.19, 432–33.
MSS. at Royal Society. Portr. at St. Thomas's Hospital. Wedgwood medallion.

FORDYCE, Mary Sophia *see* Bentham, *Lady* M. S.

FORMAN, Simon (*fl.* 1600s)
Had a garden at Lambeth.
R. T. Gunther *Early English Botanists* 1922, 309–10.

FORREST, George (1873–1932)
b. Falkirk, Stirlingshire 13 March 1873 *d.* Tengyueh, Yunnan, China 5 Jan. 1932
FLS 1924. VMH 1921. VMM 1927. Assistant, Herb., Royal Botanic Garden, Edinburgh, 1902. Made seven plant collecting expeditions in Yunnan, 1904–32. Introduced many new species of *Rhododendron* and *Primula*. 'Catalogue of Plants collected in Yunnan and Tibet' (*Notes R. Bot. Gdn Edinburgh* v.7, 1912–13, 1–411; v.14, 1924, 75–393; v.17, 1929–30, 1–406).
Gdnrs Mag. 1909, 445–46 portr. *Garden* 1924 ii–iii portr. *Curtis's Bot. Mag. Dedications, 1827–1927* 377–78 portr. *Gdnrs Chron.* 1932 i 53–54 portr.; v.174(17), 1973, 30–35. *J. Bot.* 1932, 79–81. *J. R. Hort. Soc.* 1932, 356–60 portr.; 1953, 448–54 portr.; 1973, 112–17 portr. *Kew Bull.* 1932, 106–7; 1948, 205–6. *Nature* v.129, 1932, 270. *New Fl. and Sylva* 1932, 180–86 portr. *Rhododendron Soc. Notes* 1932, 271–75. *Scott. Geogr. Mag.* v.48, 1932, 104–5. *Times* 16 and 19 Jan., 20 Feb. 1932. *Trans. Proc. Bot. Soc. Edinburgh* v.31, 1932, 239–43 portr. Scott. Rock Gdn Club *George Forrest VMH* 1935 portr. *Lily Yb.* 1940, 3 portr. E. H. M. Cox *Plant Hunting in China* 1945, 152–69 portr. A. W. Anderson *Coming of the Flowers* 1950, 241–48. J. M. Cowan *Journeys and Plant Introductions of George Forrest* 1952. M. Hadfield *Pioneers in Gdning* 1955, 204–13. A. M. Coats *Quest for*

Plants 1969, 123–27. H. R. Fletcher *Story of R. Hort. Soc. 1804–1968* 1969, 296–97 portr. T. Whittle *Plant Hunters* 1970, 211–18. *Country Life* 1973, 388–89 portr. *J. Scott. Rock Gdn Club* 1973, 169–75 portr.; 1974, 33–43. *Northern Gdnr* v.28(1), 1974, 20–23.

Plants at Edinburgh, Kew. Portr. at Hunt Library.

Rhododendron forrestii I. B. Balf.

FORREST, Sir John (1847–1918)
b. Bunbury, W. Australia 22 Aug. 1847 *d.* 3 Sept. 1918

Explorer and statesman. First Premier, W. Australia. Gave his plant collections to F. von Mueller. *Explorations in Australia* 1875 (list of plants on p.325–27).

Rep. Austral. Assoc. Advancement Sci. 1907, 170.

FORREST, Richard (*fl.* 1820s–1840s)

FLS 1829. Gardener to Duke of Westminster at Eaton Hall, Cheshire; to Duke of Northumberland in 1826 at Syon House, Middx where he designed the botanic garden. In 1835 took over the Kensington nursery of William Malcolm. *Alphabetical Catalogue of Plants of Syon Garden* 1831.

Gdnrs Mag. 1826, 349; 1827, 107; 1836, 696. J. C. Loudon *Arboretum et Fruticetum Britannicum* v.4, 1838, 2620–25. A. B. Jackson *Syon House Trees and Shrubs* 1910, viii.

FORREST, Thomas (*c.* 1729–*c.* 1802)

Midshipman, Royal Navy. Employed by East India Company to take nutmeg trees to Balambangan, north of Borneo. Deserted on arrival there. *Voyage to New Guinea and Moluccas from Balambangan* 1779 portr.

C. E. Buckland *Dict. Indian Biogr.* 1906, 151. *Fl. Malesiana* v.1, 1950, 178.

Forrestia A. Rich.

FORREST, William Hutton (1799–*c.* 1879)
b. Stirling 1799 *d.* Stirling *c.* 1879

MD 1818. 'Plants of Airthrey' (*Report of... Airthrey Mineral Springs* 1831).

Stirling Nat. Hist. Soc. Trans. 1907–8, 82–87.

FORSITT, Joseph (*fl.* 1700s–1760s)

Apothecary, Great Carter Lane, London, *c.* 1740s–60s. Member of Botanical Society of London.

Proc. Bot. Soc. Br. Isl. 1967, 315.

FORSTER, Benjamin Meggot (1764–1829)
b. 16 Jan. 1764 *d.* Hale End, Walthamstow, Essex 8 March 1829

Brother of E. Forster (1765–1849). *Introduction to Knowledge of Fungusses* 1820.

J. Nichols *Illus. Lit. Hist. of Eighteenth Century* v.8, 553–54. *Gent. Mag.* 1829 i 279. T. I. M. Forster *Recueil de ma Vie* ed. 3, 1837, 6. T. Forster *Epistolarium Forsterianum* v.2, 1850, xiii. *D.N.B.* v.20, 12. *Essex Nat.* 1920, 72–88, 221–37; 1934, 232–33, 268–73; 1954, 193; 1956, 321–22. *Notes and Queries* v.169, 1935, 387 S. T. Jermyn *Fl. Essex* 1974, 17. *R.S.C.* v.2, 669.

Herb. at Essex Field Club Museum. Letters at Linnean Society, Trinity College, Cambridge.

FORSTER, Edward (1765–1849)
b. Walthamstow, Essex 12 Oct. 1765 *d.* Woodford, Essex 23 Feb. 1849

FRS 1821. FLS 1800. Banker. Treasurer, Linnean Society, 1816. Contrib. to *Phytologist.*

J. Sowerby and J. E. Smith *English Bot.* 73, 1293, 2790. J. Nichols *Illus. Lit. Hist. of Eighteenth Century* v. 8, 554. *Ann. Mag. Nat. Hist.* v.2, 1839, 95–96; v.8, 1842, 433–35. *Proc. R. Soc.* v.5, 1849, 85–86. T. I. M. Forster *Epistolarium Forsterianum* v.2, 1850, xv. *Proc. Linn. Soc.* 1849, 39–40. *Ipswich Mus. Portr. Ser.* 1852. G. S. Gibson *Fl. Essex* 1862, 448–55. H. Trimen and W. T. T. Dyer *Fl. Middlesex* 1869, 397. *D.N.B.* v.20, 14. *Essex Nat.* 1920, 72–88, 221–37 portr.; 1934–35, 273–86; 1956, 322–24. A. H. Wolley-Dod *Fl. Sussex* 1937, xxxix–xl. S. T. Jermyn *Fl. Essex* 1974, 17. D. H. Kent *Hist. Fl. Middlesex* 1975, 18. *R.S.C.* v.2, 669.

Herb. and MS. notebooks at BM(NH). Letters at BM(NH), Kew, Linnean Society, Trinity College, Cambridge. Portr. by Eddis at Linnean Society. Portr. at Hunt Library. Library and herb. sold by Sotheby, 21–24 May 1849—herb. bought by R. Brown.

Luzula forsteri DC.

FORSTER, Johann Georg Adam (1754–1794)
b. Nassenhuben near Danzig, Prussia 26 Nov. 1754 *d.* Paris 10 Jan. 1794

FRS 1777. Son of J. R. Forster (1729–1798), whom he accompanied to Russia and England and on J. Cook's second voyage, 1772–75. Professor of Natural History, Cassel, 1779–84; Vilna. Librarian, University of Mainz, 1788–92. *Voyage Round the World in ...'Resolution'...1772–1775* 1777 2 vols. *De Plantis Esculentis Insularum Oceani Australis Commentatio Botanica* 1786. *Florulae Insularum Australium Prodromus* 1786. *De Plantis Magellanicis* 1787. *Herbarium Australe* 1797. *Characteres Generum Plantarum* (with J. R. Forster) 1776.

Gent. Mag. 1805 i 546. A. Lesson and A. Richard *Voyage de Découvertes de l'Astrolabe. Botanique* 1832 Introduction, iii–iv. A. Lasègue *Musée Botanique de B. Delessert*

1845, 365–66. *J. Bot.* 1863, 256; 1885, 360–68; 1902, 389–90. *Trans. Linn. Soc. Bot.* v.1, 1875, 58. *Nature* v.32, 1885, 501. *D.N.B.* v.20, 15–16. *J. R. Soc. N.S.W.* 1908, 68–69. *Emu* 1937, 95–99. G. Steiner and M. Häckel *Forster* 1952 portr. *Liverpool Libraries, Mus. Arts Committee Bull.* 1953, 5–25. *Chronica Botanica* 1954, 200–11. *Isis* v.46, 1955, 83–95. *Austral. J. Sci.* 1959, 118–19. *Hist. Studies, Austral. and N.Z.* v.8, 1959, 345–63. *Weimarer Beiträge* v.5, 1959, 527–61. *Trudy Inst. Istorii Estestvoznaniya Tekhniki* v.36, 1961, 176–201. *Proc. Linn. Soc. N.S.W.* 1963, 108–11. *Austral. Dict. Biogr.* v.1, 1966, 402–3. *Willdenowia* 1969, 279–94. *Taxon* 1970, 523. *Notulae Naturae* (Philadelphia) no. 437, 1971, 1–4.

Drawings (unpublished), engravings and letters at BM(NH). Set of the engravings at Leningrad Botanic Garden (*Acta Horti Petropolitani* v.9, 1884, 487–510). Plants at BM(NH), Kew, Paris.

Diforstera Baill.

FORSTER, Johann Reinhold (1729–1798)
b. Dirschau, Polish Prussia 22 Oct. 1729 *d.* Halle, Germany 9 Dec. 1798

DCL Oxon 1775. MD Halle 1781. FRS 1772. Came to England, 1766. Taught languages and natural history at Warrington Academy. Naturalist on J. Cook's second voyage, 1772–75. Left England in 1778. Professor of Natural History, Halle, 1780. *Fl. Americae Septentrionalis* 1771. *Characteres Generum Plantarum* (with J. G. A. Forster) 1776. *Beschreibungen der Gattungen von Pflanzen* (with J. G. A. Forster) 1779. *Observations made during a Voyage round the World* 1778. *Enchiridion* 1788.

Gent. Mag. 1798 i 357; 1805 i 546. A. Lasègue *Musée Botanique de B. Delessert* 1845, 365–66. *Nature* v.32, 1885, 501. *J. Bot.* 1885, 360–68; 1902, 389–90. *D.N.B.* v.20, 15. *J. Proc. R. Soc. N.S.W.* 1908, 67–69. *Emu* 1937, 95–99. *J. Proc. R. Austral. Hist. Soc.* 1926, 155–57. Liverpool Public Museums *Handbook and Guide to Herb. Collections* 1935, 50–51 portr. G. Steiner and M. Häckel *Forster* 1952 portr. *Liverpool Libraries, Mus. Arts Committee Bull.* 1953, 5–25. *Willdenowia* 1957, 778–80; 1969, 279–94. *Austral. J. Sci.* 1959, 118–19. *Proc. Linn. Soc. N.S.W.* 1963, 108–11. *Austral. Encyclop.* v.4, 1965, 163–64. *Austral. Dict. Biogr.* v.1, 1966, 403–4. *J. Pacific Hist.* 1967, 215–24. F. A. Stafleu *Taxonomic Literature* 1967, 156. *J. Soc. Bibl. Nat. Hist.* v.6, 1971, 1–8. *Notulae Naturae* (Philadelphia) no. 437, 1971, 1–4. M. E. Hoare *The Tactless Philosopher* 1976 portr.

Drawings and letters at BM(NH). Plants at BM(NH), Kew.

Forstera L. f. *Diforstera* Baill.

FORSTER, Richard (*fl.* 1590s–1610s)

MD Oxon. President, London College of Physicians, 1600–4, 1615. "This elegant plant was first found...when we accompanied Dr. Richard Forster." (L'Obel *Stirpium Illustrationes* 1655, 38).

C. E. Raven *English Nat.* 1947, 247.

FORSTER, Richard (*fl.* 1790s–1810s)

Medical Officer at Stowe Workhouse. "Surgeon of Stowe, who to the merit of a very skilful practitioner adds that of a scientific botanist." MS. list of Staffordshire plants at William Salt's library.

J. Bot. 1901 Supplt., 71. *Trans. N. Staffordshire Field Club* 1947–48, 98–99; 1948–49, 96–97.

FORSTER, Thomas Furly (1761–1825)
b. Walthamstow, Essex 5 Sept. 1761 *d.* Walthamstow 28 Oct. 1825

FLS 1800. Father of T. I. M. Forster (1789–1860). Discovered *Viola lactea*. *List of Tunbridge Wells Plants* 1801. *Fl. Tonbrigensis* 1816; ed. 2 1842 (biography, ix–xxiv). Joint author with his brothers B. M. and E. Forster of plant lists in Gough's *Camden*.

J. Sowerby and J. E. Smith *English Bot.* 240, 445, etc. G. F. W. Meyer *Primitiae Florae Essequeboënsis* 1818, 133–35. J. Nichols *Illus. Lit. Hist. of Eighteenth Century* v.8, 553. *Ann. Bot.* 1805, 59. T. I. M. Forster *Recueil de ma Vie* ed. 3, 1837, 5. T. I. M. Forster *Epistolarium Forsterianum* v.1, 1845, 33–41. *D.N.B.* v.20, 22. *Essex Nat.* 1924, 273; 1934, 230–31; 1956, 320–21. C. E. Salmon *Fl. Surrey* 1931, 49. A. H. Wolley-Dod *Fl. Sussex* 1937, xl–xli. S. T. Jermyn *Fl. Essex* 1974, 16. *R.S.C.* v.2, 671.

Letters at Linnean Society, Trinity College, Cambridge.

Forsteronia G. F. W. Meyer.

FORSTER, Thomas Ignatius Maria (1789–1860)
b. London 9 Nov. 1789 *d.* Brussels 2 Feb. 1860

MB Cantab 1818. FLS 1811. Naturalist and astronomer. *Index Fungorum* 1819. *Perennial Calendar* 1824. '*Papaver orientale*' (*Proc. Linn. Soc.* 1842, 158). Edited *Fl. Tonbrigensis* ed. 2 1842. *Recueil de ma Vie* 1837. *Epistolarium Forsterianum* 1845–50 2 vols.

Gdnrs Mag. 1828, 173–75. *Mag. Nat. Hist.* 1829, 64. *Gent. Mag.* 1860, 514. *Proc. Linn. Soc.* 1859–60, xxiii. J. Gillow *Dict. English Catholics* v.2, 1887, 318–24. *D.N.B.* v.20, 22–23.

Letters at BM(NH).

FORSTER, William (*fl.* 1840s–1860s)

Of Salford. Fern collector and cultivator.

Br. Fern Gaz. 1909, 44.

FORSYTH, Adam (*c.* 1830–1898)
b. Kelso, Roxburgh *c.* 1830 *d.* New Zealand 10 June 1898
Nurseryman, Stoke Newington, London. Specialised in chrysanthemums. Emigrated to New Zealand, 1874. Gardener at Awamoa.
Gdnrs Mag. 1898, 533.

FORSYTH, Alexander (*c.* 1809–1885)
d. 8 Nov. 1885
Gardener to Lord Stanley at Alderley Park and to Earl of Shrewsbury at Alton Towers. Contrib. to *Gdnrs Chron.* Possibly author of *Complete Practical Treatise on Culture and Economy of Potato* 1846.
Gdnrs Chron. 1885 ii 667.

FORSYTH, Gordon (*c.* 1903–1971)
b. Canada *c.* 1903 *d.* Haslemere, Surrey 22 May 1971
VMM. Gardener, Royal Horticultural Society's Garden, Wisley. Assistant Editor, *Gdnrs Chron.* Editor, *Popular Gdning.*
J. R. Hort. Soc. 1971, 366–67.

FORSYTH, William (1737–1804)
b. Old Meldrum, Aberdeen 1737 *d.* Kensington, London 25 July 1804
Gardener at Syon, 1763; at Physic Garden, Chelsea, 1771–84; then at St. James's and Kensington Palaces. One of the founders of Royal Horticultural Society. *Observations on Diseases, Defects and Injuries in all Kinds of Fruit and Forest Trees* 1791. *Treatise on Culture and Management of Fruit-trees* 1802; ed. 7 1824.
Gent. Mag. 1804 ii 787, 823; 1805 i 341. S. Felton *Portr. of English Authors on Gdning* 1830, 186. *Cottage Gdnr* v.4, 1850, 233. *J. Hort. Cottage Gdn* v.31, 1876, 147 portr. *Notes and Queries* v.2, 1874, 463; v.3, 1875, 15. R. H. Semple *Mem. Bot. Gdn at Chelsea* 1878, 112–13. *J. R. Hort. Soc.* 1941, 319–24, 374–81 portr. *Gdnrs Chron.* 1945 i 102–3; v.163(17), 1968, 15–16 portr. A. Simmonds *Hort. Who was Who* 1948, 52–62 portr. *Country Life* 6 July 1967, 9. R. Webber *Early Horticulturists* 1968, 101–14 portr. H. R. Fletcher *Story of R. Hort. Soc. 1804–1968* 1969, 25–27 portr.
Letters at Kew (reprinted in *Cottage Gdnr* v.7–9, 1852–53). Portr. at Hunt Library.
Forsythia Vahl.

FORSYTH, William (1772–1835)
b. Chelsea, London 26 Oct. 1772 *d.* London 28 July 1835
Son of W. Forsyth (1737–1804). Nurseryman in partnership with James Gordon at 25 Fenchurch Street, London with nursery at Mile End. *Botanical Nomenclator* 1794. Had excellent horticultural library. Prepared *Catalogue Raisonné of Gardening Works* and *Arboretum Britannicum*, both unpublished.
Gdnrs Mag. 1835, 496. *Gdnrs Chron.* 1904 i 147–48; 1930 i 235; 1945 i 102–3.
Library sold at Sotheby, 11 Nov. 1835.

FORSYTH, William (1864–1910)
b. Crieff, Perthshire 5 Oct. 1864 *d.* Sydney, N.S.W. 14 Sept. 1910
Gardener at Drummond Castle, Perthshire. Employed in Botanic Gardens, Sydney. Overseer of Centennial Park, Sydney, 1897. Collected Australian plants. 'Mosses of New South Wales' (*Proc. Linn. Soc. N.S.W.* 1899, 674–86).
Sydney Morning Herald 15 Sept. 1910. *Proc. Linn. Soc. N.S.W.* 1911, 10–11. *J. Proc. R. Austral. Hist. Soc.* 1932, 133.

FORSYTH-MAJOR, Charles Immanuel (1843–1923)
b. Basel, Switzerland 15 Aug. 1843 *d.* Kaufbeuren, Bavaria 25 March 1923
MD Basel 1868. FRS 1908. Of Scotch descent. Palaeontologist. Worked at BM(NH). Collected plants in Grecian Islands, 1886 and Madagascar, 1894–96. 'Etudes Botaniques' (with W. Barbey) (*Bull. Herb. Boissier* v.1–5, 1893–97). *Karpathos* (with W. Barbey) 1895.
J. Bot. 1923, 158. *Verhandlungen Naturforschenden Gesellschaft Basel* 1924–25, 1–23 portr. *R.S.C.* v.8, 310; v.10, 695; v.16, 1015.
Madagascar plants at BM(NH). Corsican plants at Kew.
Mimulopsis forsythii S. Moore.

FORTESCUE, Lady Eleanor (*fl.* 1830s)
MS. notes in her copy of *Fl. Devoniensis.*
Bot. Soc. Exch. Club Br. Isl. Rep. 1939–40, 281.

FORTESCUE, William Irvine (1851–1941)
b. Swanbister, Orkney 5 April 1851 *d.* Kincausie, Kincardine 10 May 1941
Farmer, Swanbister, Orkney. 'New List of Flowering Plants and Ferns of Orkney' (*Scott. Nat.* 1882, 318–26, 362–75).
M. Spence *Fl. Orcadensis* 1914, xlix–l.

FORTUNE, Robert (1812–1880)
b. Blackadder Town, Berwick 16 Sept. 1812 *d.* Brompton, London 13 April 1880
Gardener, Royal Botanic Garden, Edinburgh. At Horticultural Society of London's gardens at Chiswick by 1840. Collected plants in China, Luzon, Japan, etc. on 5 journeys between 1843 and 1861 for Horticultural Society of London, East India Company (twice), U.S. Patent Office and privately. On behalf of the East India Company he successfully introduced the tea plant from China into India. He was a pioneer in the use of the Wardian case for transporting plants over long distances. Curator, Chelsea

Physic Garden, 1846–48. *Wanderings in China* 1847. *Journey to Tea Countries of China* 1852. *Residence among the Chinese* 1857. *Yedo and Peking* 1863. Contrib. to *Athenaeum*, 1843–55, and *Gdnrs Chron.*

Gdnrs Chron. 1880 i 487–89; 1963 ii 185–86, 270, 331, 393, 397, 448; 1964 i 34–35 portr. *J. Hort. Cottage Gdnr* v.63, 1880, 337–39 portr.; v.21, 1890, 371 portr.; v.38, 1899, 411–12, 434–35 portr. *Rev. Hort. Belge* 1880, 272–74; 1893, 198–99. *J. Bot.* 1880, 160; 1894, 295–96. *Trans. Bot. Soc. Edinburgh* v.14, 1883, 161–63. E. Bretschneider *Hist. European Bot. Discoveries in China* 1898, 403–518. *D.N.B.* v.20, 50. *New Fl. and Sylva* v.8, 1936, 172–79. *J. R. Hort. Soc.* 1943, 161–71 portr.; 1955, 265–69, 276–80 portr.; 1972, 401–9. *Proc. Linn. Soc.* 1943–44, 8–16. E. H. M. Cox *Plant-hunting in China* 1945, 76–92 portr. *Fl. Malesiana* v.1, 1950, 179–80. G. A. C. Herklots *Hong Kong Countryside* 1951, 166–67. *Arb. Bull.* 1954, 84–85, 87. M. Hadfield *Pioneers in Gdning* 1955, 131–42. *Country Life* 1962, 358–59 portr. A. M. Coats *Quest for Plants* 1969, 71–75, 101–10. *Arnoldia* 1971, 1–18 portr. *R.S.C.* v.2, 672; v.12, 246.

Plants at BM(NH), Kew. Letters at Royal Horticultural Society, Kew, U.S. National Archives. Portr. at Hunt Library.

Fortunaea Lindl. *Fortunella* Swingle.

FOSS, Charles Calveley (1885–1953)
b. 9 March 1885 *d.* 9 April 1953

Brigadier. Deputy-Lieutenant, Bedfordshire. Amateur botanist.

Bedfordshire Nat. 1953, 45.

FOSTER, Capt. (*fl.* 1690s–1700s)

Had a garden at Lambeth, London, where he specialised in striped holly.

Archaeologia v.12, 1794, 190. J. E. Dandy *Sloane Herb.* 1958, 131.

FOSTER, Charles (1868–1911)
b. Talygarn, Glam 1868 *d.* Sutton Green, Surrey 16 Feb. 1911

Gardener to Lord Stradbroke at Henham, Suffolk. Director of *Times* Experimental Station near Guildford, 1909–11.

Gdnrs Mag. 1907, 673–74 portr.; 1911, 158 portr. *Garden* 1911, 96 portr. *Gdnrs Chron.* 1911 i 127–28. *J. Hort. Cottage Gdnr* v.62, 1911, 171 portr., 181.

FOSTER, Edgar William (1878–1921)
b. 25 May 1878 *d.* Netherne, Surrey 23 June 1921

Kew gardener, 1900. Curator, Botanic Station, Lagos, 1901. Assistant Conservator of Forests, Nigeria, 1906; later Senior Conservator until 1919. Collected plants in Nigeria.

Kew Bull. 1901, 81; 1921, 222. *J. Kew Guild* 1922, 117.

Plants and MSS. at Kew.

FOSTER, Fanny Isabel *see* Marshall, F. I.

FOSTER, J. (–1781)
d. Cheshunt, Herts 2 Jan. 1781

Seedsman, Strand, London.

FOSTER, James (*fl.* 1840s)

Nurseryman, Handsworth, near Sheffield, Yorks.

FOSTER, Sir Michael (1836–1907)
b. Huntingdon 8 March 1836 *d.* London 28 Jan. 1907

MD 1859. FRS 1872. FLS 1868. VMH 1897. KCB 1899. Professor of Practical Physiology, University College, London, 1869. Praelector of Physiology, Trinity College, Cambridge, 1870. Professor of Physiology, Cambridge, 1883–1903. Cultivated irises. *Textbook of Physiology* 1877; ed. 4 1883. *Bulbous Irises* 1892. Contrib. to *Gdnrs Chron., Garden.*

Garden v.53, 1898, xi portr. *Cambridge Rev.* 1907, 439–40. *Gdnrs Chron.* 1907 i 78–79 portr. *J. Physiol.* 1907, 233–46. *Kew Bull.* 1907, 65–66. *Mem. Proc. Manchester Lit. Philos. Soc.* 1906–7, xlv–xlviii. *Nature* v.75, 1907, 345–47. *Naturalist* 1907, 124–25 portr. *Proc. Linn. Soc.* 1906–7, 42–45. *Proc. R. Soc. B.* 1908, lxxi–lxxxi portr. *Times* 31 Jan. 1907. *D.N.B. Supplt 2* v.2, 44–46. *Iris Yb.* 1937, 50–54. *Notes Rec. R. Soc.* v.19, 1964, 10–32 portr. *Mercian* Jan. 1972, 3–5. *R.S.C.* v.2, 674; v.7, 602; v.9, 906; v.12, 246; v.15, 69.

MSS. at Linnean Society. Portr. by J. Collier at Royal Society.

Iris fosteriana Aitch. & Baker.

FOSTER, Nevin Harkness (1858–1927)
b. Brackaville, County Tyrone 18 March 1858 *d.* 23 Jan. 1927

FLS 1912. Hon. Secretary and Treasurer, Belfast Naturalists Field Club; President, 1909–10. Had collection of ferns in his garden at Hillsborough. Contrib. to *Irish Nat.*

Irish Nat. 1927, 197. *Proc. Rep. Belfast Nat. Hist. Philos. Soc.* 1926–27, 37–39.

FOSTER, W. E. (*c.* 1911–1954)
d. 3 Aug. 1954

Lecturer in Botany, Durham University, 1940–54.

Nature v.174, 1954, 586.

FOSTER, William (*c.* 1799–1877)
d. Stroud, Glos 31 March 1877

Nurseryman, Stroud.

Gdnrs Chron. 1877 i 508.

FOSTER-MELLIAR, Rev. Andrew (–1904)
d. Sproughton, Suffolk 14 Nov. 1904
Rectoř, Sproughton. Grew roses. *Book of the Rose* 1894; ed. 4 1910.
Gdnrs Mag. 1894, 397 portr. *J. Hort. Cottage Gdnr.* v.47, 1903, 397, 399 portr.; v.49, 1904, 441 portr. *Gdnrs Chron.* 1904 ii 352, 372.

FOTHERGILL, John (1712–1780)
b. Carr End, Wensleydale, Yorks 8 March 1712 *d.* London 26 Dec. 1780
MD Edinburgh 1736. FRS 1763. Travelled in Flanders before 1740. Practised medicine in Lombard Street from 1740. Had celebrated botanic garden at Upton, West Ham, 1762 (J. C. Lettsom *Hortus Uptonensis* 1783) and at Lea Hall, Cheshire, 1765. His large collection of natural history drawings including many flower paintings was purchased by Empress of Russia. *Some Account of Late Peter Collinson* 1770.
Gent. Mag. 1780 ii 592; 1781 i 165–67, 205–6; 1812 i 513. G. Thompson *Mem. of Life and View of Character of Late Dr. John Fothergill* 1782. J. C. Lettsom *Mem. of John Fothergill* 1783 portr. J. Nichols *Lit. Anecdotes of Eighteenth Century* v.9, 737–40 portr. W. Darlington *Memorials of John Bartram* 1849, 333–48, 495–515. *Cottage Gdnr* v.7, 1851, 327–28. J. Smith *Cat. Friends' Books* v.1, 1867, 628–29. Friends' Institute *Biogr. Cat.* 1888, 236–42. J. H. Tuke *Sketch of Life of John Fothergill* 1879. W. Munk *Roll of R. College of Physicians* v.2, 154–58. A. L. A. *Portr. Index* 1906, 530. H. A. Kelly *Some Amer. Med. Bot.* 1915, 18 portr. *J. Bot.* 1914, 319, 320, 323; 1920, 56–59. *D.N.B.* v.20, 66–68. R. H. Fox *John Fothergill and his Friends* 1919. *Essex Nat.* 1933, 86–92; 1938, 91–103. *N. Western Nat.* 1941, 1–3. *Trans. Med. Soc. London* v.64, 1943–46. 276–91. *Sci., Med. Hist.: Essays in Honour of Charles Singer* 1953, 173–78. *Proc. Amer. Philos. Soc.* v.98, 1954, 11–22; v.102, 1958, 413–19. B. C. Corner and C. C. Booth *Chain of Friendship; Selected Letters of Dr. John Fothergill of London, 1735–1780* 1971 portr. *House and Gdn* Oct. 1974, 151 portr.
Portr. by Hogarth at Royal College of Physicians. MSS. at Royal Society. Garden plants in Banks Herb. BM(NH).
Fothergilla L.

FOULIS, Robert (1799–1877)
b. Woodhouslee near Edinburgh 1799 *d.* Fordell, Fifeshire 21 Dec. 1877
Gardener at Fordell, 1827.
Gdnrs Chron. 1876 i 496 portr.; 1877 ii 817.

FOULKES, Rev. Robert (*c.* 1702–1729)
b. Llanfrothen, Merioneth *c.* 1702
BA Oxon 1725. MD. Of Llanbedr Duffryn Clwyd. Acquainted with Richard Richardson and Samuel Brewer. Had E. Lhuyd's MSS. and specimens. His name spelt 'Fowkes' in E. J. L. Scott's *Index to Sloane Manuscripts* 1904, 198 and 'Fowlkes' in R. Pulteney's *Hist. Biogr. Sketches of Progress of Bot. in England* v.2, 1790, 113.
J. E. Smith *Selection from Correspondence of Linnaeus* v.2, 1821, 171. D. Turner *Extracts from Lit. and Sci. Correspondence of Richard Richardson* 1835, 132, 167–68. *Nature in Wales* 1966, 14–16.
Plants at Cambridge, Oxford.

FOULKES, Rev. Thomas (*fl.* 1850s)
Missionary in India where he collected plants in Nilgiri Hills, *c.* 1855–60.
I. H. Burkill *Chapters on Hist. Bot. in India* 1965, 55.

FOUNTAIN, Robert (*fl.* 1790s)
Gardener and nurseryman, Waltham, Lincs.

FOUNTAINE, Rev. John (*c.* 1814–1877)
d. Southacre, Norfolk 28 Dec. 1877
Rector, Southacre. Keen gardener. Invented orchard-house railway. *Improved Method of Growing Fruit upon Orchard-house Principle* 1868.

FOWLDS, Allan (*c.* 1767–1842)
d. Kilmarnock 8 May 1842
He and his brother, Alexander, established a nursery at Kilmarnock in 1780s.
Gdnr's Mag. 1842, 336.

FOWLE, James (*fl.* 1820s)
Florist, 26 Vauxhall Walk, London.

FOWLER, Archibald (1816–1887)
b. Gilmore Hill near Glasgow 1816 *d.* 14 Aug. 1887
Gardener, Botanic Garden, Glasgow, 1832–36. Gardener to Earl of Stair at Castle Kennedy, 1840–87.
Gdnrs Chron. 1874 ii 582–84 portr.; 1887 ii 219 portr.

FOWLER, George (*fl.* 1820s)
Nurseryman, New Road, St. George's, London.

FOWLER, James (1830–)
b. Edinburgh 1830
Gardener to Earl of Harewood at Harewood House, Yorks, 1856.
Gdnrs Chron. 1875 ii 709 portr.

FOWLER, Joseph Gurney (1855–1916)
b. Woodford, Essex 5 Dec. 1855 *d.* Pembury, Kent 24 April 1916
Accountant. Treasurer, Royal Horticultural Society, 1899–1916. Cultivated and hybridised orchids.

Gdnrs Mag. 1907, 383 portr., 384; 1916, 219. *Gdnrs Chron*. 1916 i 240 portr. *Orchid Rev*. 1916, 121–22 portr. *J. R. Hort. Soc*. 1917, 209–12 portr.
Cattleya fowleri × Sander & Kraenzl.

FOWLER, Rev. William (1835–1912)
b. Winterton, Lincs 27 Feb. 1835 *d*. Winterton 7 March 1912
MA Cantab 1860. Vicar, Liversedge, Yorks, 1866–1910. Discovered *Selinum carvifolia*. Account of Lincs plants in *Phytologist* v.2, 1857–58, 331–33, 416 and *Naturalist* 1878–90.
Trans. Lincolnshire Nat. Union v.1, 1907, 219–21 portr. *Bot. Soc. Exch. Club Br. Isl. Rep*. 1912, 203–4. *J. Bot*. 1912, 320. *Naturalist* 1912, 121–23 portr. E. J. Gibbons *Fl. Lincolnshire* 1975, 56–57 portr. *R.S.C.* v.12, 248.

FOX, Alfred Russell (1853–1910)
b. Sheffield, Yorks 1853 *d*. Sheffield 5 Dec. 1910
FLS 1899. Pharmaceutical chemist. "An ardent field botanist".
Proc. Linn. Soc. 1910–11, 37.

FOX, Rev. Edward (*fl*. 1870s–1880s)
MA. Rector, Upper Heyford, Oxford. Made catalogue of Oxford plants.
G. C. Druce *Fl. Oxfordshire* 1927, cxx.

FOX, Edwin Fydell (1814–1891)
b. Brislington near Bristol 20 April 1814 *d*. Brislington 12 March 1891
Medical Officer, Brislington Lunatic Asylum. Collected and hybridised British ferns from 1869. Found a reflexed *Athyrium* in 1850.
Gdnrs Chron. 1891 i 374–75. E. J. Lowe *Fern Growing* 1895, 177–79 portr. *Br. Fern Gaz*. 1909, 44.

FOX, Rev. Henry Elliott (1841–1926)
b. Masulipatam, India 21 Oct. 1841 *d*. Putney, London 12 May 1926
BA Cantab 1864. Curate, Oxford, 1869. Vicar, Christchurch, Westminster, 1873; St. Nicholas, Durham, 1882. Hon. Secretary, Church Missionary Society, 1895–1910. Prebendary, London. Collected plants in Skye with M. A. Lawson and D. Oliver, 1868 (*J. Bot*. 1869, 108–14). With F. J. Hanbury in Caithness, 1885 (*J. Bot*. 1885, 333–38). In Palestine, 1890. Gave his vast herb. of British plants to G. C. Druce (*J. Bot*. 1876, 47).
Bot. Soc. Exch. Club Br. Isl. Rep. 1926, 91–92. *Who was Who, 1916–1928* 374. G. C. Druce *Comital Fl. Br. Isl*. 1932, xi. H. N. Clokie *Account of Herb. Dept. Bot., Oxford* 1964, 166–67.
Plants at Oxford, BM(NH), Kew.

FOX, Henry Stephen (1791–1846)
b. Chatham, Kent Sept. 1791 *d*. Washington, U.S.A. 13 Oct. 1846
British Minister, Buenos Aires, 1831; Rio de Janeiro, 1833; Washington, 1837. Uncle of C. J. F. Bunbury (1809–1886). Formed herb. at Rio de Janeiro, Monte Video, Porto Alegre, etc., 1831–33.
J. Bot. (Hooker) 1834, 178–79. C. J. F. Bunbury *Life and Letters and Journals* v.1, 1906, 358.
Plants at Cambridge (in Bunbury's Herb.), BM(NH), Kew.

FOX, J. Tregellis (*fl*. 1880s)
Of Leeds? Doctor and missionary who collected plants in Madagascar, 1885–87.
Orchids at Kew.

FOX, James Charles (*fl*. 1830s)
Nurseryman, Hawton Road, Newark, Notts.

FOX (or **Foxe**), **John** (*fl*. 1690s)
Surgeon. Sent plants to J. Petiver from Cape and Bengal.
J. Petiver *Musei Petiveriani* 1695, 39, 44, 80. *J. Linn. Soc*. v.45, 1920, 36. J. E. Dandy *Sloane Herb*. 1958, 131.

FOX, John (*fl*. 1770s)
Nurseryman in partnership with John Winter, Blyth, Notts.

FOX, Joseph (*fl*. 1770s–1800s)
Weaver, Norwich, Norfolk. Assisted J. E. Smith. Raised *Lycopodium* from spores.
Trans. Linn. Soc. v.2, 1794, 315; v.7, 1804, 297.

FOX, Louisa Emily *see* Bunbury, L. E.

FOX, Sir Stephen (1627–1716)
b. 27 March 1627 *d*. Chiswick, Middx 28 Oct. 1716
Statesman. Had garden at Chiswick, 1685-1716.
Archaeologia v.12, 1794, 185–86. *D.N.B.* v.20, 133–36.

FOX, Walter (1858–1934)
b. near Liverpool 1858 *d*. Shoreham, Sussex 26 July 1934
Kew gardener, 1876. Assistant Superintendent, Botanic Gardens, Singapore, 1879–1903. Superintendent, Garden and Forests, Penang, 1903–10. Collected plants in Malaya. Travelled in S. America where he reported on rubber. Collected plants in Loreto, Peru, 1911. *Guide to Botanic Gardens, Singapore* 1889.
Kew Bull. 1903, 31. *J. Kew Guild* 1905, 227 portr.; 1935, 462–66 portr. *Gdns Bull. Straits Settlements* 1927, 122; 1935, 164. *Fl. Malesiana* v.1, 1950, 180. *J. Thailand Res. Soc. Nat. Hist. Supplt*. 1939, 18.
Plants at Singapore, Kew.

FOX, Wilfrid Stephen (1875–1962)
b. Bromborough, Cheshire 21 March 1875 *d.* 22 May 1962

VMH 1948. Physician, St. George's Hospital, London. Created Winkworth Arboretum, Surrey. Founded Road Beautifying Association. Interested in *Sorbus.*

J. R. Hort. Soc. 1954, 80–92. *Times* 26 May 1962. R. R. Trail *Lives of Fellows of R. College of Physicians of London* v.5, 1968, 140–41.

FOX, William Tilbury (1836–1879)
b. Broughton, Winchester, Hants 1836 *d.* Paris 7 June 1879

MD London 1858. Physician at Charing Cross and University College Hospitals. 'Chignon Fungus' (*J. Bot.* 1867, 246–47; 1879, 224).

D.N.B. v.20, 139–40. *R.S.C.* v.9, 911.

FOX STRANGWAYS, William Thomas Horner *see* Strangways, W. T. H. F.

FOX TALBOT, William Henry *see* Talbot, W. H. F.

FOX WILSON, George (1896–1951)
b. 26 Jan. 1896 *d.* Weybridge, Surrey 9 Jan. 1951

Student, Royal Horticultural Society Gardens, Wisley, 1911. Entomologist, Wisley, 1919–51. President, Association of Applied Biologists, 1949, 1950. *Pests of Ornamental Garden Plants* 1937; ed. 2 1950. *Detection and Control of Garden Pests* 1947; ed. 2 1949. Contrib. to *J. R. Hort. Soc., Gdnrs Chron.*

Ann. Applied Biol. 1951, 311–17 portr. *Entomol. Mon. Mag.* 1951, 94. *Gdnrs Chron.* 1951 i 24. *Grower* v.35, 1951, 61. *J. R. Hort. Soc.* 1951, 117–18 portr. *Nature* v.167, 1951, 259. *Rose Annual* 1951, 145. *Times* 11, 16, 17 Jan. 1951.

FOXE, Simon (*fl.* 1630s)

Physician, London. President, College of Physicians, 1634–40. J. Parkinson in *Paradisus* 1629, 265, 941 refers to his 'Booke of dried herbes'.

FRAINE, Ethel Louise De (–1918)
d. Falmouth, Cornwall 25 March 1918

DSc London. FLS 1908. Lecturer in Botany, Whitelands Training College and Westfield College. Contrib. to *Ann. Bot.* 'Seedling Structure of Certain Cactaceae' (*Ann. Bot.* 1910, 125–75).

Nature v.101, 1918, 150–51.

FRAMPTON, Mary (1773–1846)
b. Moreton, Dorset 1773 *d.* Dorchester, Dorset 12 Nov. 1846

She produced 5 vols of drawings of Dorset plants which were in possession of the family.

J. C. Mansel-Pleydell *Fl. Dorsetshire* 1829, 39.

FRANCES, John (*fl.* 1790s)

Nurseryman, Castle Cary, Somerset.

FRANCIS, Edward Parke (*c.* 1802–1869)
d. Hertford 11 Jan. 1869

Rose-grower.

FRANCIS, George William (1800–1865)
b. London 1800 *d.* Adelaide 9 Aug. 1865

FLS 1839. Served apprenticeship with Loddiges nursery. Emigrated to Australia, 1849. First Director, Botanic Garden, Adelaide, 1855–65. *Catalogue of British Flowering Plants and Ferns* 1835; ed. 5 1840. *Analysis of British Ferns and their Allies* 1837; ed. 5 1860. *Grammar of Botany* 1840. *Little English Flora* 1840. *Illustrations of British Mosses* 1840. *Catalogue of the Vasculares... of Great Britain* 1852. *Catalogue of Plants... in...Botanic Garden, Adelaide* 1859. *Acclimatisation of...Animals and Plants* 1862.

Mag. Nat. Hist. 1835, 221–23. P. Mennell *Dict. Austral. Biogr.* 1892, 174. *D.N.B.* v.20, 167. *Rep. Austral. Assoc. Advancement Sci.* 1907, 174–75; 1911, 229–30. B. J. Best *Life and Works of George William Francis* 1965 portr. *R.S.C.* v.2, 696.

British plants at Kew. Letters at BM(NH). MSS. at Public Library, Adelaide.

Calocephalus francisii Benth.

FRANCIS, Rev. Robert Bransby (*c.* 1768–1850)
d. East Carleton, Norfolk 27 April 1850

MA Cantab 1794. FLS 1798. Vicar, Roughton, Norfolk, 1814. Studied *Jungermanniae.*

J. Sowerby and J. E. Smith *English Bot.* 605, 2569. *Proc. Linn. Soc.* 1851, 132. *Trans. Norfolk Norwich Nat. Soc.* 1937, 201.

Letters at Kew, Linnean Society. MSS. and drawings at Kew.

Jungermannia francisi Smith.

FRANKLAND, Sir Thomas (1750–1831)
b. Westminster, London 1750 *d.* 4 Jan. 1831

MA Oxon 1771. FLS 1796. 6th Baronet 1784. Of Thirkleby, Yorks. Drew algae at Scarborough. Had W. Hudson's marine plants.

J. Sowerby and J. E. Smith *English Bot.* 2340, etc. *Trans. Linn. Soc.* v.10, 1811, 157. P. Smith *Mem. and Correspondence of Sir J. E. Smith* v.1, 1832, 450–51; v.2, 167–69.

MS. at Royal Society. Letters at BM(NH) in Banks letters.

Franklandia R.Br.

FRANKLEN, Sir Thomas Mensel (1840–1928)

Educ. Oxford. Clerk of the Peace, Glamorgan. Had collection of British ferns.

Bot. Soc. Exch. Club Br. Isl. Rep. 1928, 702–3.

FRANKLIN, Mr. (*fl.* 1780s)
Of Lambeth. "An ingenious raiser" of carnations.
Bot. Mag. 1788, t.39.

FRANKLIN, J. (*fl.* 1820s)
Seedsman and florist, Cottage Place, City Road, London.

FRANKLYN, George (*fl.* 1700s)
Apothecary, Charlestown, Carolina, *c.* 1700. Sent plants to J. Petiver. Afterwards of Downton, Wilts.
J. Petiver *Musei Petiveriani* 1695, 80, 744. E. J. L. Scott *Index to Sloane Manuscripts* 1904, 202. J. E. Dandy *Sloane Herb.* 1958, 131.
Plants at BM(NH).

FRANKS, A. W. (*fl.* 1820s–1830s)
Plants at Luton Museum, Beds.

FRANKS, Frederica (*née* **Sebright**) (1796–)
Collected plants near Flamstead, Herts, *c.* 1820–23.
J. G. Dony *Fl. Hertfordshire* 1967, 13.
Herb. at Luton Museum.

FRANQUEVILLE, John De (*fl.* 1590s–1610s)
London merchant. Had a botanic garden where he grew exotic plants.
J. Gerard *Herball* 1597, 307. M. de Lobel *Adversaria* 1605, 486, 487. C. E. Raven *English Nat.* 1947, 241–42.

FRASER, Charles (*c.* 1788–1831)
b. Blair Athol, Perthshire *c.* 1788 *d.* Sydney 22 Dec. 1831
Served with 56th Regiment in East Indies, 1815. Arrived in Sydney with the 46th Regiment, 1816. Established Botanic Garden, Sydney, 1819. Plant collector on John Oxley's explorations in 1817, 1818, 1819. Colonial Botanist, New South Wales, 1821. At Swan River with James Stirling, 1826–27. Laid out public garden at Brisbane, 1828.
Bot. Misc. 1829–30, 221–69 (reprinted in *J. W. Austral. Nat. Hist. Soc.* no. 3, 1906, 16–35). R. K. Greville *Algae Britannicae* 1830, xii. *Companion Bot. Mag.* v.2, 1837, 230. A. Lasègue *Musée Botanique de B. Delessert* 1845, 498. J. D. Hooker *Fl. Tasmaniae* v.1, 1859, cxxiii. *Proc. Linn. Soc. N.S.W.* 1900, 770–72, 788–89, 791. J. H. Maiden *Charles Fraser* 1902. *Kew Bull.* 1906, 208–12. *Proc. R. Soc. N.S.W.* 1908, 103–4; 1921, xxx–xxxi. *J. Proc. R. Austral. Hist. Soc.* 1932, 103–8. *Austr. Dict. Biogr.* v.1, 1966, 416–17. *Tuatara* 1965, 160–61. *Taxon* 1970, 523. *R.S.C.* v.2, 702.
Plants at BM(NH), Kew. Letters at Kew. MSS. at Mitchell Library, Sydney.
Hakea fraseri R. Br.

FRASER, Donald (*fl.* 1820s)
Nurseryman, Inverness.

FRASER, Finlay (*c.* 1790–*c.* 1849)
Nurseryman, Lea Bridge Road, Leyton, Essex.

FRASER, George (1854–)
b. Draimie, Morayshire Oct. 1854
Established nursery at Ucluelet, British Columbia, 1894. Hybridised *Rhododendron* and *Rubus.*
Gdnrs Chron. 1928 i 348 portr.

FRASER, Gordon Travers (1882–1942)
b. Tralee, County Kerry 28 Aug. 1882 *d.* 1 Aug. 1942
BA Cantab. Civil engineer. Edited *Fl. Devon* (with W. K. Martin) 1939.
Bot. Soc. Exch. Club Br. Isl. Rep. 1941–42, 456–57.
Herb. at Torquay Natural History Society.

FRASER, Helen Charlotte *see* Gwynne-Vaughan, H. C.

FRASER, Hugh (*c.* 1834–1904)
d. Edinburgh 13 Jan. 1904
Nurseryman, Leith, Edinburgh. *Handbook of Ornamental Conifers, Rhododendrons...* 1875.
Garden v.65, 1904, 85. *Gdnrs Chron.* 1904 i 60.

FRASER, James (1820–*c.* 1863)
Son of F. Fraser (–*c.* 1849). Nurseryman, Lea Bridge Road, Leyton, Essex.

FRASER, Rev. James (1814–1902)
b. Granton-on-Spey, Morayshire 8 May 1814 *d.* Colvend, Kirkcudbrightshire 15 March 1902
DD Aberdeen. Incumbent, Colvend and Southwick. Found *Carex punctata* in Scotland (*J. Bot.* 1873, 47–48). Notes on botany of Colvend in local handbook (*J. Bot.* 1874, 63). Original member of Dumfriesshire and Galloway Natural History Society, 1862, and contrib. to its *Trans.*
G. F. S. Elliott *Fl. Dumfriesshire* 1891, 4.

FRASER, James Thomas (*fl.* 1820s)
Younger son of John Fraser (1750–1811). Partner with his elder brother, John Fraser (*fl.* 1790s–1860s) in continuing their father's nursery, King's Road, Chelsea, London, 1811–17, then by himself to 1827.

FRASER, John (1750–1811)
b. Tomnacloich? Inverness 1750 *d.* Chelsea, London 26 April 1811
FLS 1810. Hosier. Established nursery at Sloane Square, Chelsea in 1780s which was continued by his sons after his death. Collected plants in Newfoundland, 1780–84;

N. America 1785–1807. Collector for Czar of Russia, 1798. Issued nursery catalogues *c.* 1790 and 1796 (reprinted in *J. Bot.* 1899, 481–87; 1905, 329–31). Had herb. Published T. Walter's *Fl. Caroliniana* 1788. *Short History of Agrostis Cornucopiae* 1789 (*J. Bot.* 1921, 69–71).

Bot. Mag. 1802, t.563. *Gent. Mag.* 1811 i 596. T. Faulkner *Hist. Topographical Description of Chelsea* v.2, 1829, 41. *Companion Bot. Mag.* v.2, 1837, 300–5 portr. J. C. Loudon *Arboretum et Fruticetum Britannicum* v.1, 1838, 119–22. A. Lasègue *Musèe Botanique de B. Delessert* 1845, 199–200. *Cottage Gdnr* v.8, 1851, 250–52. *J. Hort. Cottage Gdnr* v.57, 1877, 238; v.58 1877, 156–58 portr. *D.N.B.* v.20, 213–14. *J. Bot.* 1907, 255; 1915, 271. *Proc. Linn. Soc.* 1906–7, 15. *Trans. Inverness Sci. Soc. Field Club* 1912–18, 416–17. I. Urban *Symbolae Antillanae* v.3, 1902, 48–49. F. A. Stafleu *Taxonomic Literature* 1967, 158. A. M. Coats *Quest for Plants* 1969, 281–85 portr. *Taxon* 1970, 523.

Letters at Kew, Royal Society of Arts. Portr. at Hunt Library.

Frasera Walter.

FRASER, John (*fl.* 1790s–1860s)

ALS 1848. Son of J. Fraser (1750–1811) with whom he travelled to N. America where he made at least 2 more collecting trips after his father's death. Issued nursery catalogue in 1813 (by T. Nuttall) (reprinted in *Pittonia* v.2, 1890, 116–19; discussed in *Rhodora* 1955, 290–93; 1956, 23–24). Had Hermitage Nursery at Ramsgate, 1817–35. Presented T. Walter's herb. to Linnean Society which they sold in 1863 (*Proc. Linn. Soc.* 1849, 53).

Companion Bot. Mag. v.2, 1837, 301–3. W. Darlington *Reliquiae Baldwinianae* 1843, 190, 234, 344. *J. Bot.* 1899, 482. I. Urban *Symbolae Antillanae* v.3, 1902, 49. *Rhodora* 1956, 23–24. A. M. Coats *Quest for Plants* 1969, 293.

Oenothera fraseri Pursh.

FRASER, John (1820–1909)

b. Glasgow 22 March 1820 *d.* Wolverhampton, Staffs 13 April 1909

MA Glasgow 1843. MD Glasgow 1852. To Wolverhampton, 1854. Added *Amblystegium confervoides* to British flora. 'Plants found in Staffordshire 1864' (*Trans. Dudley and Midland Geol. Sci. Soc. Field Club* 1865, 56–72).

J. Bot. 1901 Supplt., 73. *Trans. Bot. Soc. Edinburgh* v.24, 1912, 51–52. *Trans. N. Staffordshire Field Club* 1949–50, 92–93. *R.S.C.* v.7, 702.

Herb. at Hull University. Letters in Wilson correspondence at BM(NH).

FRASER, John (1821–1900)

d. 20 Jan. 1900

VMH 1897. Son of Finlay Fraser. Nurseryman, Lea Bridge Road, Leyton, Essex. For some years in partnership with his brother James. Business carried on by his son, John Finlay Fraser. Founder member of National Floricultural Society, 1851.

Garden v.57, 1900, 70, 91 portr. *Gdnrs Chron.* 1900 i 63 portr.

FRASER, John (1854–1935)

b. Newdeer, Aberdeenshire 31 Jan. 1854 *d.* London 24 Jan. 1935

FLS 1889. VMH 1922. VMM 1929. Gardener, Royal Horticultural Society Gardens, Chiswick, 1880–82; Kew, 1882–85. Worked with Sir John Lubbock on seedlings at Jodrell Laboratory, Kew, 1885–86, 1909–10. Assistant Editor, *Gdning World*, 1887; Editor, 1897–1909. Authority on *Mentha* and *Salix*. Editor of G. W. Johnson's *Gdnrs Dict.* (with A. Hemsley) 1917. *Fifty Best Roses for Amateur Growers* 1906. *Select Annuals and Biennials for Amateur Growers* 1906. *Select Dahlias and their Cultivation by Amateurs* 1906.

Gdnrs Mag. 1900, 57 portr. *Garden* 1922, 94 portr. *Gdnrs Chron.* 1922 i 110 portr.; 1935 i 85. *Bot. Soc. Exch. Club Br. Isl. Rep.* 1934, 811–12. *J. Bot.* 1935, 81–83. *J. Kew Guild* 1935, 467–68. *Kew Bull.* 1935, 97–98. *Nature* v.135, 1935, 422. *Proc. Linn. Soc.* 1934–35, 177–78. *Times* 13 Feb. 1935. *Chronica Botanica* 1936, 186 portr.

Herb. at Kew. Portr. at Hunt Library.

FRASER, M. (–1884/5)

d. Kawang, Borneo 1884/5

Medical Officer, Br. N. Borneo Company at Kudat, 1883.

Fl. Malesiana v.1, 1950, 182.

Some plants from Borneo at Kew.

FRASER, Margaret (*fl.* 1840s)

Nurseryman, Temple Street, Aylesbury, Bucks.

FRASER, Patrick Neill (1830–1905)

b. Edinburgh Aug. 1830 *d.* Edinburgh 27 Feb. 1905

Keen gardener with fine collection of ferns. *British Ferns and their Varieties* 1864.

Garden v.67, 1905, 154(ix). *Gdnrs Chron.* 1905 i 157. *Trans. Bot. Soc. Edinburgh* v.23, 1907, 208–9. *Mem. R. Caledonian Hort. Soc.* 1908, 83.

Fern herb. at Royal Botanic Garden, Edinburgh.

FREAM, William (1854–1906)

b. Gloucester 1854 *d.* Downton, Wilts 29 May 1906

BSc London 1877. LLD Montreal 1888. FLS 1882. Professor of Natural History, Royal Agricultural College, Cirencester, 1877–79. 'Fl. Water-meadows' (*J. Linn. Soc. Bot.* v.24, 1888, 454–64). *Elements of Agriculture* 1891; ed. 12 1944.

Br. Assoc. Advancement Sci. Rep. 1887, 767–68; 1889, 648–49. *D.N.B. Supplt 2* v.2, 54–55. *R.S.C.* v.15, 110.

FREE, Montague (1885–1965)
b. Cambridge 12 Dec. 1885 *d.* New York 27 Jan. 1965

Kew gardener, 1908–12. To U.S.A., 1912. Head gardener, Brooklyn Botanic Garden, 1914–45. *Gardening* 1937. *All about House Plants* 1946. *Plant Propagation in Pictures* 1957. *Plant Pruning in Pictures* 1961.

Gdnrs Chron. 1934 i 172 portr. *J. Kew Guild* 1965, 598–99 portr.

FREEMAN, Arthur (–1876)
d. Berber, Sudan 1 Oct. 1876

Kew gardener, 1873–75. Joined L. A. Lucas on expedition to Central Africa, 1875. Gardener in service of Egyptian government at Dufile where he collected plants, now at Kew.

Gdnrs Chron. 1933 i 117–18. R. Hill *Biogr. Dict. Sudan* 1967, 129–30.

FREEMAN, John (1784–1864)
b. Chipperfield, Herts 1784 *d.* Stratford, Essex 1864

Schoolmaster, Stratford. Friend of J. A. Brewer.

Essex Nat. 1918–19, 23; 1954, 193.

Herb. at Essex Field Club Museum.

FREEMAN, Joseph (1813–1907)
b. Bromley-by-Bow, Essex 1813.

LRCP. Son of J. Freeman (1784–1864). *Stratford Fl.* 1862. 'Hints on describing Species' (*Proc. Bot. Soc. London* 1839, 28). Contrib. to G. S. Gibson's *Fl. Essex* 1862.

Essex Nat. 1918–19, 23–25; 1954, 193.

Herb. at Essex Field Club Museum.

FREEMAN, Samuel (*fl.* 1840s)

Of Birmingham. Account of Birmingham and Bristol plants in *Phytologist* v.1, 1844, 261–62, 327–28. List in Morris's *Hist. Wye* 1842, 180.

J. E. Bagnall *Fl. Warwickshire* 1891, 500.

FREEMAN, William George (1874–)
b. Falmouth, Cornwall 9 Jan. 1874

BSc London 1898. FLS 1898. Assistant, Botanic Garden, Ceylon, 1896–97. Demonstrator in Botany, Royal College of Science, 1897–1900. Dept. of Agriculture, West Indies, 1900–3. Superintendent, Botanic Garden, Trinidad, 1911–17; Director of Agriculture, 1917–29.

Trinidad Tobago Yb. v.53, 161.

FREEMAN-MITFORD, Algernon Bertram
see Mitford, A. B. F.

FREER, Adam (*fl.* 1760s)

MD Edinburgh 1767. Sent British plants to J. Hope.

Notes from R. Bot. Gdn Edinburgh v.4, 1907, 125.

FRENCH, Alfred (1839–1879)
b. Banbury, Oxfordshire 1839 *d.* London 22 Oct. 1879

Journeyman baker, Banbury. From 1874 attendant in Botany Dept., BM. '*Salvia pratensis*' in Oxfordshire, (*J. Bot.* 1875, 292–94).

J. Bot. 1879, 352. G. C. Druce *Fl. Oxfordshire* 1927, cxix–cxx.

Herb. at BM(NH).

FRENCH, Charles (1840–1933)
b. Lewisham, Kent 10 Sept. 1840 *d.* Melbourne 21 May 1933

To Melbourne *c.* 1850. Apprenticed to nurseryman at Caulfield, Melbourne, 1858. Assistant in charge of glasshouses in Botanic Gardens, Melbourne, 1864. Assistant to F. von Mueller, 1865–81. Curator, Botanic Museum, Melbourne, 1881. Government Entomologist, 1889. Founded Field Naturalists' Club of Victoria, 1880. Contrib. articles on orchids to *Victorian Nat.*

J. Smith *Cyclop. Victoria* 1903, 275–76. *Victorian Nat.* 1933, 57–60 portr.; 1950, 146–47. P. Serle *Dict. Austral. Biogr.* v.1, 1949, 325–26. *Austral. Encyclop.* v.4, 1965, 213–14.

Prasophyllum frenchii F. Muell.

FRERE, Sir Bartle Henry Temple (1862–1953)
b. 26 Aug. 1862 *d.* 20 Feb. 1953

Educ. Cambridge. Police Magistrate, Gibraltar, 1902–11; Attorney-General, 1911–14; Chief Justice, 1914–22. *Guide to Fl. Gibraltar* 1910.

Who was Who, 1951–1960 398.

FRERE, Miss Catherine Frances (*fl.* 1870s)

Daughter of Sir H. B. E. Frere (1815–1884). Did drawing of Cape flowers, now at Kew.

Kew Bull. 1921, 223.

FRERE, Sir Henry Bartle Edward (1815–1884)
b. Clydach, Breconshire 29 March 1815 *d.* 29 May 1884

Entered Bombay Civil Service, 1834. Governor, Bombay, 1862–67. Governor, Cape, 1877–79. Collected plants in India and Zanzibar, now at Kew.

D.N.B. v.20, 257–66. *Pakistan J. For.* 1967, 345.

FRETTINGHAM, Henry (*c.* 1818–1884)
d. 30 Nov. 1884
Nurseryman, Rose Nurseries, Beeston, Notts. Specialised in roses and fruit.
Garden v.26, 1884, 490. *J. Hort. Cottage Gdnr* v.9, 1884, 509.

FRIEND, Rev. Hilderic (1852–1940)
b. High Wigsell, Kent 6 Nov. 1852 *d.* Birmingham 7 Feb. 1940
Methodist minister, Newton Abbot. *Glossary of Devonshire Plant Names* 1882. *Flowers and Flower Lore* 1884. *Ministry of Flowers* 1885. *Flowers and their Story* 1907. Contrib. to *Gdnrs Chron.* and *Naturalist.*
Gdnrs Chron. 1926 i 56 portr.; 1932 ii 369; 1940 i 157–58. *Nature* v.145, 1940, 414. *N. Western Nat.* 1943, 117–19. *Sorby Rec.* v.2(3), 1967, 10–11.
Herb. at Birmingham Museum.

FRITSCH, Felix Eugen (1879–1954)
b. London 26 April 1879 *d.* Cambridge 23 May 1954
BSc London 1898. DSc 1905. FRS 1932. FLS 1903. Assistant, Botanical Dept., Munich University, 1899–1901. Lecturer, Botany Dept., University College, London, 1902–6; Assistant Professor 1906–11. Head of Botany Dept., Queen Mary College, 1907; Professor, 1924–48. Algologist. President, Linnean Society, 1949–52. Translated H. Solereder's *Systematic Anatomy of Dicotyledons* (with L. A. Boodle) 1908. *Introduction to Study of Plants* 1914. *Elementary Studies in Plant Life* 1915. *Introduction to Structure and Reproduction of Plants* 1920. *Botany for Medical Students* 1921. *Plant, Form and Function* 1928. *Structure and Reproduction of Algae* 1935, 1945 2 vols.
Nature v.162, 1948, 562; v.174, 1954, 293–94. *Obit. Notices Fellows R. Soc.* 1954, 131–40 portr. *Proc. Linn. Soc.* 1953–54, 40–42. *Rev. Algologique* v.2, 1954, 131–40 portr. *Times* 24 May 1954. *J. Indian Bot. Soc.* 1956, 522–32 portr. *D.N.B. 1951–1960* 378–80. *Who was Who, 1951–1960* 399.
Portr. by F. M. Haines and plants at Queen Mary College.

FROMOW, James J. (*c.* 1855–1903)
d. 13 March 1903
Nurseryman, Chiswick, Middx.

FROMOW, William (1815–1886)
b. Attleborough, Norfolk, July 1815 *d.* 27 Nov. 1886
Nurseryman, Chiswick, Middx.
Gdnrs Chron. 1886 ii 728.

FROST, George (*c.* 1836–1912)
b. Brushford, Somerset *d.* Feb. 1912
Nurseryman, Bampton, Devon.
J. Hort. Home Farmer v.64, 1912, 147.

FROST, James (*c.* 1797–1835)
d. 3 May 1835
Nurseryman, Exotic Nursery, Lillington, Warwickshire.
Gdnrs Mag. 1835, 384.

FROST, John (1803–1840)
b. London 1803 *d.* Berlin 17 March 1840
FLS 1825. Knight of Brazilian Order of Southern Star. Lecturer, St. Thomas's Hospital. Founder in 1821 and Director of Medico-Botanical Society from which he was expelled for arrogant behaviour in 1830. Having incurred liabilities in respect of Millbank hospital-ship, fled to Paris in 1832. Later practised medicine in Berlin. *Science of Botany* 1827. Edited ed. 3 of W. Bingley's *Practical Introduction to Botany* 1831.
Gdnrs Mag. 1828, 324; 1831, 104. *D.N.B.* v.20, 286–87. *Practitioner* v.188, 262–66 portr.
Portr. at Hunt Library.

FROST, Philip (1804–1887)
b. Moreton Hampstead, Devon 10 July 1804 *d.* 10 May 1887
Foreman, Physic Garden, Chelsea, 1829–32. Gardener to Lady Grenville at Dropmore, Bucks, 1833. Found *Battarrea*, 1844 (*J. Bot.* 1916, 198–99).
Garden v.22, 1882, xii portr.; v.31, 1887, 478. *Gdnrs Chron.* 1887 i 649 portr. *J. Hort. Cottage Gdnr* v.14, 1887, 396.

FROST, Thomas (*c.* 1823–1882)
d. Aylesford, Kent 11 Nov. 1882
Nurseryman, Bower Nursery, Maidstone, Kent, 1863.
Gdnrs Chron. 1882 ii 695.

FRY, David (1834–1912)
b. Bristol 6 Jan. 1834 *d.* Bristol 24 Jan. 1912
Contrib. to J. W. White's *Fl. Bristol* 1912 and to *J. Bot.* from 1892.
J. Bot. 1912, 239–40. *Nature* v.102, 1918, 169. H. J. Riddelsdell *Fl. Gloucestershire* 1948, cxl. *R.S.C.* v.15, 158.
Plants at Bristol.

FRY, E. J. *see* Savage, E. J.

FRY, Sir Edward (1827–1918)
b. Bristol 4 Nov. 1827 *d.* Failand, Bristol 18 Oct. 1918
DCL Oxon. LLD Cantab. FRS 1883. FLS 1887. GCB 1907. Judge of High Court, 1877. Lord Justice, 1883. Brother of D. Fry (1834–1912). *Mycetozoa* (with his daughter Agnes) 1889. *British Mosses* 1892; ed. 2 1908. *Liverworts* (with Agnes) 1911.
Bot. Soc. Exch. Club Br. Isl. Rep. 1918, 354–55. *J. Bot.* 1918, 366. *Kew Bull.* 1919. 84–85. *Nature* v.102, 1918, 169–70. *Proc, Linn. Soc.* 1918–19, 53–54. *D.N.B. 1912–1921* 200–3. *Who was Who, 1916–1928* 383.
Portr. at Hunt Library.

FRY, George (*fl.* 1830s–1890s)
Nurseryman. Specialised in fuchsias. Patented design of garden pots.
Gdnrs Chron. 1898 i 87–88 portr.

FRY, George (1843–1934)
b. Arundel, Sussex 17 Jan. 1843 *d.* 29 June 1934
FLS 1882. Experimented on sowing of seeds, fertilisation of flowers, rise of sap and structure of woods.
Proc. Linn. Soc. v.147, 1934–35, 179.

FRYER, Alfred (1826–1912)
b. Chatteris, Cambridge 25 Dec. 1826 *d.* Chatteris 26 Feb. 1912
ALS 1897. Studied *Potamogeton. Potamogetons of the British Isles* (with A. Bennett) 1898–1915. Account of Hunts plants in *J. Bot.* 1884, 105–7. Contrib. to *J. Bot.*
Bot. Soc. Exch. Club Br. Isl. Rep. 1912, 195–201. *J. Bot.* 1912, 105–10 portr. *Proc. Linn. Soc.* 1911–12, 46–47. A. R. Horwood and C. W. F. Noel *Fl. Leicestershire* 1933, ccxix. *Hunts Fauna Fl. Soc. Annual Rep.* 1971, 12–13. *R.S.C.* v.15, 158, 176.
Plants at BM(NH), Manchester, Oxford. Portr. at Hunt Library.
Potamogeton fryeri Benn.

FRYER, Sir John Claud Fortescue (1886–1948)
b. Chatteris, Cambridge 13 Aug. 1886 *d.* 22 Nov. 1948
MA Cantab. FRS 1948. FRSE. Described vegetation of Aldabra Island during his stay there in 1908–9 (*Trans. Linn. Soc. Zool.* 1911, 397–442). Entomologist, Board of Agriculture, 1914. Director, Plant Pathology Laboratory, Harpenden, 1920. Secretary, Agricultural Research Council, 1944.
Kew Bull. 1919, 109, 113. *Ann. Applied Biol.* 1950, 136–37. *Who was Who, 1941–1950* 414.
Aldabra plants at Kew.

FULLER, Arabella (*later* **Morris**) (*fl.* 1720s–1740s)
Seedsman, The Naked Boy (and Three Crowns), Strand, London. The widow of Edward Fuller.

FULLER, E. H. (*fl.* 1830s)
Nurseryman, Worthing, Sussex.

FULLER, Edward (*fl.* 1670s–1720s)
Seedsman, The Naked Boy (and Three Crowns), Strand, London. Successor in business to William Lucas in 1679.

FULLER, James (*fl.* 1690s)
Seedsman, Ye Orange Tree, Strand, London.

FULLER, James (*c.* 1797–1879)
Nurseryman, Addlestone, Chertsey, Surrey. His son-in-law was William Fletcher, nurseryman of Ottershaw, Surrey.

FULTON, Thomas Alexander Wemyss (1855–1929)
d. 7 Oct. 1929
MB Edinburgh 1884. MD 1897. Superintendent of Scientific Investigations, Fisheries Board for Scotland, 1888–1921. 'Dispersion of Spores of Fungi by Agency of Insects' (*Ann. Bot.* v.3, 1889, 207–38).
Nature, v.124, 1929, 846–47. *Who was Who, 1929–1940* 485–86.

FURBER, Robert (*c.* 1674–1756)
d. Aug. 1756
Founded nursery at Kensington Gore, London soon after 1700. Acquired some of the exotic plants introduced by Bishop Compton to Fulham Palace. First known English source of the Moss rose. Firm continued by his assistant, John Williamson, 1756–83 when it was taken over by D. Grimwood. *Catalogue of English and Foreign Trees* 1727. *Twelve Months of Flowers* 1730. *Twelve Months of Fruit* 1732. *Flower Garden Displayed* 1732. *Short Introduction to Gardening* 1733.
Gdnrs Chron. 1910 i 33–34. G. Dunthorne *Flower and Fruit Prints of 18th and Early 19th Centuries* 1938, 12–15, 52. J. Harvey *Early Gdning Cat.* 1972, xi. 14–15. *Trans. London Middlesex Archaeol. Soc.* v.24, 1973, 185.

FURNASS, William (*fl.* 1780s–1790s)
Gardener and nurseryman, Kendal, Westmorland.

FYFE, William (–1912)
d. Wantage, Berks 24 June 1912
Gardener to Lord Wantage at Lockinge Park, Berks, *c.* 1890s
Gdnrs Mag. 1912, 498 portr.

FYFFE, Robert (*fl.* 1900s–1920s)
Kew gardener, 1908. Superintendent, Botanic, Forestry and Scientific Dept., Uganda, 1908. Plants collected in Uganda, 1909–25, at Kew.
Kew Bull. 1908, 195.

FYSHER, Robert (1698–1747)
BA Oxon 1718. MB 1725. Physician but never practised. Librarian, Bodleian Library, 1730. Member of John Martyn's Botanical Society of London.
Proc. Bot. Soc. Br. Isl. 1967, 315–16.

GAGE, Andrew Thomas (1871–1945)
b. Aberdeen 14 Dec. 1871 *d.* Strathpeffer, Ross 21 Jan. 1945

MA. LLD Aberdeen. FLS 1901. Assistant to Professor of Botany, Aberdeen, 1894–96. Indian Medical Service, 1897–98. Curator, Herb., Calcutta Botanic Gardens, 1898–1905; Director, 1905–25. Collected plants in Sikkim, Assam, Chittagong. Librarian and Assistant Secretary, Linnean Society, 1924–29. *Census of Indian Polygonums* 1903. *Vegetation of District of Minbu in Upper Burma* 1904. *History of Linnean Society of London* 1938. Contrib. to *Kew Bull.*, *Rec. Bot. Survey India.*

150th Anniversary Vol. R. Bot. Gdn Calcutta 1942, 6–7 portr. *Proc. Linn. Soc.* 1944–45, 105–9 portr. *Who was Who, 1940–1950* 416.

MSS. at Kew.

GAGE, Catherine (1816–1892)
b. Rathlin Island, County Antrim 1816 *d.* Rathlin Island 16 Feb. 1892

Account of Rathlin Island plants in *Ann. Mag. Nat. Hist.* v.5, 1850, 145–46.

Irish Nat. 1913, 26. *Proc. Belfast Nat. Field Club* 1913, 620.

GAGE, Sir Thomas (1781–1820)
d. Rome 27 Dec. 1820

FLS 1802. 7th Baronet. Of Hengrave Hall, Suffolk. Lichenologist. Collected plants in Suffolk, Ireland, Gibraltar and Portugal. Contrib. to J. Sowerby and J. E. Smith's *English Bot.* 1671, 2541, 2575, 2580, etc.

Ann. Bot. v.2, 1806, 555. *Bot. Mag.* 1806, t.935. *Trans. Hort. Soc. London* v.1, 1812, 328. J. G. Gage *Hist. Hengrave* 1822 portr. P. Smith *Mem. and Correspondence of Sir J. E. Smith* v.2, 1832, 235–37, 264–66. J. Gillow *Dict. English Catholics* v.2, 1887, 364–65.

Lichens at Cambridge. Letters at Cambridge University Library. Portr. at Hunt Library.

Gagea Salisb.

GAINES, N. (*fl.* 1830s–1840s)
Nurseryman, Surrey Lane Nursery, Battersea, London.

GAINSBOROUGH, Henry, 6th Earl of *see* Noel, H. *6th Earl of Gainsborough*

GAIRDNER, M. A. *see* Macaulay, M. A.

GAIRDNER, Meredith (1809–1837)
b. London 27 Nov. 1809 *d.* Honolulu, Hawaii 26 March 1837

MD Edinburgh. Surgeon to Hudson Bay Company, Columbia, 1832. Collected plants in N.W. America.

W. J. Hooker *Fl. Boreali-americana* v.2, 1840, 99. *Br. Columbia Hist. Quart.* 1945, 89–111. S. D. McKelvey *Bot. Exploration of Trans-Mississippi West, 1790–1850* 1955, 481–85.

Woods at BM(NH). Plants at Kew.

Pentstemon gairdneri Hook.

GALBRAITH, Miss (*fl.* 1870s)
Missionary in China. In 1875 discovered *Eriochrysis porphyrocoma* on Lien Chou River.

E. Bretschneider *Hist. European Bot. Discoveries in China* 1898, 707.

GALE, Rev. John Sadler (1835–1915)
MA. Rector, Cleeve, Somerset. Studied botany at Oxford. Presented his herb. of British and Swiss plants to Oxford.

H. N. Clokie *Account of Herb. Dept. Bot., Oxford* 1964, 169.

GALLOWAY, George (–1879)
d. Glasgow 1879

Seed merchant, Helensburgh, Dunbartonshire.

Gdnrs Chron. 1879 i 183.

GALLWEY, Richard (*fl.* 1720s)
Nurseryman, Dronwickbane, County Kerry.

GALPIN, Rev. Francis William (1858–1945)
b. Dorchester 25 Dec. 1858 *d.* 30 Dec. 1945

MA Cantab. FLS 1887. Vicar, Hatfield Broad Oak, 1891–1915; Witham, 1915–21; Faulkbourne, 1921–33. President, Essex Archaeological Society, 1921–26. *Account of Flowering Plants, Ferns and Allies of Harleston* 1888.

Proc. Linn. Soc. 1945–46, 74. *Who was Who, 1941–1950* 419.

GALPINE, John (*c.* 1769–1806)
d. Blandford Forum, Dorset 10 Jan. 1806

ALS 1798. *Synoptical Compend of British Botany* 1806; ed. 2 1820.

D.N.B. v.20, 388. *Essex Nat.* 1926, 232–33.

GALPINE, John Kingstone (*fl.* 1790s)
Nurseryman, Blandford Forum, Dorset.

GALVIN, John (*fl.* 1760s)
Nurseryman, County Dublin.

Irish For. 1967, 52.

GALVIN, William (1756–1832)
Nurseryman, Mount Talbot, County Roscommon, Ireland.

GAMBIER-PARRY, Thomas R. (1883–1935)
d. 15 Feb. 1935

Senior Assistant, Bodleian Library; Keeper, Oriental Dept. Friend of G. C. Druce.

Bot. Soc. Exch. Club Br. Isl. Rep. 1935, 19. H. J. Riddelsdell *Fl. Gloucestershire* 1948, cl.

Plants at Oxford.

GAMBLE, James Sykes (1847–1925)
b. London 2 July 1847 *d.* Liss, Hants 16 Oct. 1925

BA Oxon 1869. FRS 1899. FLS 1877. CIE 1899. Indian Forest Service, 1871–79. Director, Forestry School, Dehra Dun, 1890–99. *List of Trees, Shrubs and Large Climbers found in Darjeeling District, Bengal* 1878; ed. 2 1896. *Manual of Indian Timbers* 1881; ed. 2 1902. *Bambuseae of British India* 1896. *Materials for Fl. Malayan Peninsula* (with G. King) (*J. R. Asiatic Soc. Bengal* v.72–75, 1903–15). *Fl. Presidency Madras* 1915–25 (completed by C. E. C. Fischer in 1934). Edited *Indian Forester* 1878–82, 1891–99.

Indian Forester 1899, 162–63; 1926, 17–19, 209–12. C. E. Buckland *Dict. Indian Biogr.* 1906, 159. *Empire For. J.* 1925, 292–93. *Bot. Soc. Exch. Club Br. Isl. Rep.* 1925, 847–48. *Gdnrs Chron.* 1925 ii 359. *J. Bot.* 1925, 335. *Kew Bull.* 1925, 433–39; 1926, 12–17. *Nature* v.116, 1925, 684–85. *Proc. Linn. Soc.* 1925–26, 80–81. *Proc. R. Soc. B.* v.99, xxxviii–xliii portr. *Who was Who, 1916–1928* 391.

Herb., letters and MSS. at Kew.

Gamblea C. B. Clarke.

GAMMIE, George Alexander (1864–1935)

FLS 1899. Son of J. A. Gammie (1839–1924). Superintendent, Saharanpur Garden, Lloyd Botanic Garden, Darjeeling. Curator, Calcutta Botanic Garden. Economic botanist, Poona. Collected plants in Kashmir, 1891 and 1893; Sikkim, 1892; Brahmaputra Valley, 1894. 'Account of Botanical Tour in Sikkim' (*Kew Bull.* 1893, 297–315). 'Botanical Tour in Chamba and Kangra' (*Rec. Bot. Survey India* 1898, 183–214).

Proc. Linn. Soc. 1935–36, 207–8. I. H. Burkill. *Chapters on Hist. Bot. in India* 1965, 141. *Pakistan J. For.* 1967, 345.

Plants at Kew, Calcutta, Dehra Dun.

GAMMIE, James Alexander (1839–1924)
b. Kingcausie, Kincardineshire 12 Nov. 1839 *d.* Chiswick, Middx April 1924

Kew gardener, 1861. Manager, Cinchona plantation, Sikkim, 1865. Deputy Superintendent, Cinchona Dept. until retirement in 1897. Collected plants in Sikkim for Calcutta Botanic Gardens.

Kew Bull. 1898, 21–22. *J. Kew Guild* 1918, 399 portr.; 1925, 338–40 portr. *Gdnrs Chron.* 1924 i 246. I. H. Burkill *Chapters on Hist. Bot. in India* 1965, 139–40.

GAMMOCK, Alexander (*fl.* 1760s–1790s)

Nurseryman, Hoxton, Shoreditch, London.

GAPPER, Anthony *see* Southby, A.

GARDEN, Alexander (1730–1791)
b. Birse, Aberdeenshire 1730 *d.* London 15 April 1791

MD Glasgow 1785. FRS 1773. Practised medicine in Charleston, S. Carolina, 1752–83. Correspondent of P. Collinson, J. Ellis and Linnaeus.

J. E. Smith *Selection of Correspondence of Linnaeus* v.1, 1821, 282–605. Thacker *Amer. Med. Biogr.* v.1, 1828, 268–70. A. L. A. Fée *Vie de Linné* 1832, 162–66. Appleton's *Cyclop. Amer. Biogr.* v.2, 1887, 594–95. *D.N.B.* v.20, 406–7. W. Darlington *Reliquiae Baldwinianae* 1843, 390–400. H. A. Kelly *Cyclop. Amer. Med. Biogr.* v.1, 1915, 330–31. *Ann. Med. Hist.* 1928, 149–58. *Isis* v.38, 1948, 161–74. *Sci. Mon.* v.67, 1948, 17–22. *Recorder* v.23, 1959, 24–28. E. and D. S. Berkeley *Dr. Alexander Garden of Charles Town* 1970. *R.S.C.* v.2, 767.

Gardenia Ellis.

GARDEN, Robert Jones (*fl.* 1850s)

Army officer. Brought plants to W. J. Hooker at Kew from Natal, 1854; Asia Minor, 1857.

Bot. Mag. 1854, t.4817; 1855, t.4842. *Kew Bull.* 1901, 25.

GARDENER, John (*fl.* 1440s)

Possibly a gardener. Wrote verse treatise on gardening, *Feate of Gardening c.* 1440 (*Archaeologia* v.54, 1894)

E. Cecil *Hist. Gdning in England* 1910, 63–66.

MS. at Trinity College, Cambridge.

GARDINER, John (1861–1900)

BSc Manchester 1884. Dept. of Agriculture, Bahamas, 1885. Professor of Biology, University of Colorado, 1889–98. *Provisional List of Plants of Bahama Islands* (with L. J. K. Brace) 1889.

I. Urban *Symbolae Antillanae* v.1, 1898, 54–55; v.3, 1902, 51. J. Ewan *Rocky Mountain Nat.* 1950, 213–14.

GARDINER, Walter (1859–1941)
b. Burwell, Cambridge 1 Sept. 1859 *d.* 31 Aug. 1941

BA Cantab 1882. MA 1885. FRS 1890. Demonstrator in Botany, Cambridge, 1884–88; Lecturer, 1888–97. Researched on histology and published papers on protoplasmic continuity.

Cambridge Rev. 1941, 25–26. *Nature* v.148, 1941, 462–63. *Obit. Notices Fellows R. Soc.* 1941, 985–1004 portr. *Times* 3 Sept. 1941. *Who was Who, 1941–1950* 420.

GARDINER, William (1808–1852)
b. Dundee, Angus 13 July 1808 *d.* Dundee 21 June 1852

ALS 1849. Umbrella-maker. *Fl. of Ten Miles around Dundee* c. 1840. *Botanical Rambles in Braemar* 1845. *Twenty Lessons on British Mosses* 1846; 2nd series 1849 (with specimens). *Fl. Forfarshire* 1848. Contrib. to *Mag. Nat. Hist.* 1832–36.

London J. Bot. 1844, 138–39; 1845, 208–10. *Chambers's Edinburgh J.* 1847, 248–51. *Cottage Gdnr* v.8, 1851, 210. *Gdnrs Chron.* 1852, 406, 423–24. *Proc. Linn. Soc.* 1853, 244. *Gdnrs Mag.* 1896, 644. *Br. Assoc. Advancement Sci. Dundee Handbook* 1912, 447–50. *Trans. Bot. Soc. Edinburgh* v.26, 1917, 155–78. *R.S.C.* v.2, 767.

MSS. at Dundee Public Library. Letters at Kew. Plants at BM(NH), Kew, Dundee.

Sphoeria gardineri Berkeley.

GARDNER, Commodore (*fl.* 1780s)

Presented *Epidendrum* roots from Jamaica to Apothecaries Company, 1789.

Bot. Mag. 1791, t.152.

GARDNER, Charles Austin (1896–1970)

b. Lancaster 6 Jan. 1896 *d.* Subiaco, W. Australia 24 Feb. 1970

Emigrated with parents to W. Australia, 1909. Botanical collector, Forests Dept., W. Australia, 1920. Botanist to Kimberley Exploration Expedition, 1921. Government Botanist and Curator, W. Australia Herb., 1928–60. *Enumeratio Plantarum Australiae Occidentalis* 1930. *Flowers of Western Australia* 1935. *West Australian Wildflowers* ed. 8 1951. *Wild flowers of Western Australia* 1959. 'Contributions to Fl. of Western Australia' (*J. R. Soc. W. Austral.* 1922–64). 'Trees of Western Australia' (*J. Dept. Agric. W. Austral.* 1952–66).

Victorian Nat. 1970, 173–75. *W. Austral. Nat.* 1970, 168–72; 1971, 178–80. *J. R. Soc. W. Austral.* v.53, 1970, 63.

Plants at Canberra. Portr. at Hunt Library.

Rhizanthella gardneri R. S. Rogers.

GARDNER, Hon. Edward (1784–)

b. 9 March 1784

East India Company. Resident at Court of Rajah of Nepal at Katmandu, 1817. Sent mosses to W. J. Hooker and plants to N. Wallich.

W. J. Hooker *Musci Exotici* v.2, 1820, t.146. N. Wallich *Plantae Asiaticae Rariores* v.1, 1830, 33, etc.; v.2, 31.

Calymperes gardneri Hook.

GARDNER, George (1812–1849)

b. Glasgow May 1812 *d.* Neura Ellia, Ceylon 10 March 1849

MD Glasgow 1835. FLS 1842. Pupil of W. J. Hooker at Glasgow. Travelled and collected plants in Brazil, 1836–41. In 1843 came to Oxford to assist H. B. Fielding in arranging his herb. and wrote descriptions for *Sertum Plantarum* 1844–49. Superintendent, Botanic Gardens, Peradeniya, Ceylon, 1844. Collected plants in Mauritius on his way there and in Madras in 1845. *Musci Britannici* 1836 (exsiccatae) (*Companion Bot. Mag.* v.2, 1836, 3). *Travels in Interior of Brazil* 1846; ed. 2 1849.

Companion Bot. Mag. v.2, 1836, 3, 344–52. *J. Bot.* (Hooker) 1842, 199–206; 1849, 154–56; 1851, 188–89. *Gdnrs Chron.* 1849, 263; 1851, 343–44. *Proc. Linn. Soc.* 1849, 40–44. H. Trimen *Fl. Ceylon* v.5, 1900, 375–76. *D.N.B.* v.20, 431. C. F. P. von Martius *Fl. Brasiliensis* v.1 (1), 1906, 22–25. *Ann. R. Bot. Gdns Peradeniya* v.5, 1912, 265–301. H. N. Clokie *Account of Herb. Dept. Bot., Oxford* 1964, 169–70. I. H. Burkill *Chapters on Hist. Bot. in India* 1965, 51 *et seq.* A. M. Coats *Quest for Plants* 1969, 355. *R.S.C.* v.2, 768.

Brazilian plants at BM(NH). Ceylon plants and drawings at Kew. Plants and drawings at Oxford. Sale of herb. (*Gdnrs Chron.* 1851, 343–44). Library sold by Stevens, 11 July 1851.

Gardneria Wall.

GARDNER, John (*fl.* 1800s–1820s)

Seedsman and florist, 2 City Road, London.

GARDNER, John Starkie (1844–1930)

d. Dec. 1930

Founded metal works. 'On Coniferae' (*Proc. Geol. Assoc.* v.7, 1881). 'Fossil Plants' (*Proc. Geol. Assoc.* v.8, 1885). 'Fossil Grasses' (*Proc. Geol. Assoc.* v.9, 1887). *Monograph of British Eocene Fl.* (with C. von Ettingshausen) 1879–86 2 vols.

Who was Who, 1929–1940 493.

GARET, James *see* Garret J.

GARLAND, Lester Vallis Lester- *see* Lester-Garland, L. V.

GARLICK, Constance (–1934)

Taught botany at University College School, London. Contrib. botanical chapter to P. E. Vizard's *Guide to Hampstead* 1890.

D. H. Kent *Hist. Fl. Middlesex* 1975, 25.

GARNER, John (*fl.* 1830s)

Nurseryman, Church Street, Bawtry, Yorks.

GARNER, Robert (1808–1890)

b. Foley, Staffs *d.* Stoke, Staffs 16 Aug. 1890

FLS 1836. Surgeon at Stoke, 1834. 'Hybrid Vaccinium' (*Sci. Gossip.* 1872, 248–49). President, North Staffs Natural History Field Club.

Rep. N. Staffs Field Club 1886, 130–34 portr. *J. Bot.* 1901 Supplt., 72–73. F. Boase *Modern English Biogr.* v.2, 1912, 387. *N. Western Nat.* 1944, 143–46. *Trans. N. Staffs Field Club* 1950, 13–45. E. S. Edees *Fl. Staffordshire* 1972, 18–19.

GARNER, William James (*c.* 1873–1945)
d. Hale, Cheshire 21 Jan. 1945
Nurseryman, Langham Road Nurseries, Hale.
Gdnrs Chron. 1945 i 86.

GARNETT, Henry (1868–1931)
b. Waterford, Eire 1868 *d.* Manchester 3 Nov. 1931
Director, Flatters and Garnett, Manchester. Friend of J. H. Salter. Had herb.
Nature v.128, 1931, 930.

GARNETT, Rev. Philip Mauleverer (1906–1967)
d. 11 March 1967
BA Cantab 1927. MA 1931. Ordained 1952. Curate, Sholing, Hants, 1952–55; Kirkby Moorside, 1955–57. Vicar, Ledsham, 1957. Rural dean, Selby, 1964. Botanised in Yorkshire.
Naturalist 1967, 104.

GARNEYS, William *see* Hagger, J.

GARNIER, Rev. Thomas (1776–1873)
b. Wickham, Hants 26 Feb. 1776 *d.* Winchester, Hants 29 June 1873
DCL Oxon 1850. FLS 1798. Rector, Bishopstoke, 1807–68. Dean of Winchester, 1840. Contrib. Hants plants to *Hampshire Repository* v.1, 1798.
J. Sowerby and J. E. Smith *English Bot.* 1471. *Gdnrs Mag.* 1834, 124–31. *J. Bot.* 1873, 256. *D.N.B.* v.21, 10.
Herb. at BM(NH). Letters at Kew. Medallion, Winchester Cathedral.

GARNONS, Rev. William Lewes Pugh (1791–1863)
b. Wivenhoe, Essex 25 Sept. 1791 *d.* Ulting, Essex 5 March 1863
BA Cantab 1814. DD 1824. FLS 1825. Vicar, Ulting, 1848. Entomologist. Contrib. to G. S. Gibson's *Fl. Essex* 1862. Wrote text of J. W. Penfold and A. J. Robley's *Selection of Madeira Flowers* 1845.
Gent. Mag. 1863 i 526. *J. Bot.* 1863, 160; 1919, 98. C. C. Babington *Memorials, Journal and Bot. Correspondence* 1897, 267. F. H. Davey *Fl. Cornwall* 1909, xlix.
Herb. formerly at Saffron Walden Museum. Sale of property at Sotheby, 2 March 1864.

GARNSEY, Rev. Henry Edward Fowler (1826–1903)
b. Coleford, Glos 14 July 1826 *d.* Bath, Somerset 29 June 1903
BA Oxon 1846. Studied mosses. Helped in G. C. Druce's *Fl. Berks* 1897. Translated F. G. J. von Sachs's *Hist. Bot.* 1890, and other German botanical works.
J. Bot. 1903, 318. *Rep. Ashmolean Nat. Hist. Soc.* 1903, 26–30. G. C. Druce *Fl. Oxfordshire* 1927, cxxi.
Plants at Oxford. Portr. at Magdalen College.

GARRAWAY, James (*fl.* 1830s–1850s)
Nurseryman, Durdham Down Nurseries, Bristol. Firm founded by Miller and Sweet.

GARRAWAY, Stephen (*fl.* 1760s–1770s)
Nurseryman, The Rose, 139 Fleet Street, London.
J. Harvey *Early Gdning Cat.* 1972, 42.

GARRET, G. H. (*fl.* 1880s)
Travelling Commissioner, Sierra Leone. Sent plants to Kew, 1889.
Kew Bull. 1891, 245.

GARRETT (*or* GARET), James (*fl.* 1590s–1610)
b. Netherlands *d.* London 1610
London apothecary. "One of the earliest and most successful growers of the tulip" (J. Gerard). Friend of J. Gerard, Clusius and Rev. T. Penny. Translated Costa from the Spanish into English. "Honestissimus vir et idem rei herbariae studio valde se oblectans" (Clusius *Rariorum Plantarum Historia* 1601, v, cix).
M. De l'Obel *Stirpium Illustrationes* 1655, 3. *J. Bot.* 1899, 234–35. E. Cecil *Hist. Gdning in England* 1910, 148–49. R. H. Jeffers *Friends of John Gerard* 1967, 65–66; *Biogr. Appendix* 1969, 13.

GARRETT, Henry Burton Guest (*c.* 1871–1959)
b. Teddington, Middx *d.* Chieng Mai, Thailand 26 April 1959
To Siam, 1896. Assistant Conservator, Forestry Dept., *c.* 1899. Deputy Conservator on retirement in 1929 or 1932. Collected plants in N. Siam.
Blumea v.11, 1961–62, 478–79.
Plants at BM(NH), Kew, Leiden.
Garrettia Fletcher.

GARRETT, James R. (1820–1855)
Of Holywood, County Down. Solicitor. Botanised in N.E. Ireland. Fern fancier.
Irish Nat. 1913, 26. *Ann. Rep. Proc. Belfast Nat. Field Club* v.6, 1913, 620.

GARRETT, John (*c.* 1824–1893)
d. Gosforth, Northumberland 18 March 1893
Grew begonias and zonal pelargoniums. Vice-Chairman, Newcastle Botanical Society.
J. Hort. Cottage Gdnr v.26, 1893, 252.

GARRY, Robert (–1938)
d. 21 Jan. 1938
BSc Edinburgh. Science teacher, Glasgow High School for Girls. President, Andersonian Naturalists' Society, 1907–8. 'Some Recent Additions to Fresh-water Algae of Clyde Area' (*Glasgow Nat.* 1909, 13–14).
Glasgow Nat. 1940, 42.

GARSIDE, Sidney (1889–1961)
b. Ashton-under-Lyne, Lancs 28 Jan. 1889
d. Cape Town 3 Nov. 1961
BSc Manchester 1910. FLS 1922. Lecturer in Botany, Technical College, Salford. Lecturer in Botany, Victoria College, Stellenbosch, 1912. Returned to England, 1920. Lecturer in Botany, Bedford College, 1921–38. Returned to S. Africa, 1938. Assisted at Bolus Herb. Authority on *Proteaceae*.
J. S. African Bot. 1962, 231–35 portr.
Plants at Kew. Portr. at Hunt Library.

GARTH, Richard (*fl.* 1550s–1597)
d. Drayton, Hants 1597
In Diplomatic Service. Sent plants to Clusius. Made first record of *Polypogon monspeliensis*. "...Historiae Plantarum, cum Indicarum tum inquilinarum studiosissimi" (l'Obel).
J. Gerard *Herball* 1597, 757. M. de l'Obel *Stirpium Illustrationes* 1655, 85, 127. R. T. Gunther *Early English Botanists* 1922, 237–38. C. E. Raven *English Nat.* 1947, 243.

GARTSIDE, Miss (*fl.* 1800s)
Of Lancashire. "A lady eminently skilled in delineating botanical subjects." Exhibited flower paintings at the Royal Academy and elsewhere, 1781–1808.
Bot. Mag. 1803, t.699.

GATACRE, Sir William Forbes (1843–1906)
b. 3 Dec. 1843 *d.* 4 March 1906
Major-General. Commander of Chitral Relief Expedition, 1895. Collected plants in Chitral (described by J. F. Duthie in *Rec. Bot. Survey India* 1898).
Who was Who, 1897–1916 269. *Pakistan J. For.* 1967, 345.
Plants at Dehra Dun.

GATER, William Adam (*c.* 1835–1900)
b. Chestunt, Herts *c.* 1835 *d.* Slough, Bucks 15 Oct. 1900
Gardener at C. Turner's Royal Nursery, Slough, 1858. In charge of roses.
Gdnrs Chron. 1900 ii 296–97. *Gdnrs Mag.* 1900, 695.

GATES, Reginald Ruggles (1882–1962)
b. Middleton, Nova Scotia 1 May 1882 *d.* London 12 Aug. 1962
BSc McGill 1906. PhD Chicago 1908, FRS 1931. FLS 1912. Lecturer in Biology. St. Thomas's Hospital, 1912–14. Reader in Botany, King's College, London, 1919–21; Professor of Botany, 1921–42. Did research in cytology and genetics. Studied *Oenothera*. President, Royal Microscopical Society, 1930–31. *Mutation Factor in Evolution* 1915. *Mutations and Evolution* 1921. *Botanist in the Amazon Valley* 1927. *Taxonomy and Genetics of Oenothera* 1956.
Man v.62 art. 289, 1962 portr. *Mount Allison Rec.* 1962, 12–13 portr. *Nature* v.195, 1962, 1252–53. *Times* 13 Aug. 1962. *Indian J. Genetics Plant Breeding* 1963, 107–8. *Biogr. Mem. Fellows R. Soc.* 1964, 83–106 portr. *Proc. Linn. Soc.* v.175, 1964, 91–92. *Who was Who, 1961–1970* 416–17. C. C. Gillispie *Dict. Sci. Biogr.* v.5, 1972, 293–94.
Portr. at Hunt Library.

GATHORNE-HARDY, Hon. Robert (1902–1973)
b. 31 July 1902 *d.* Feb. 1973
Educ. Oxford. FLS 1960. Writer, bibliographer and botanist. *Wild Flowers in Britain* 1938. *Three Acres and a Mill* 1939 portr. *Garden Flowers* 1948. *Tranquil Gardener* 1958. *Native Garden* 1962.
Times 13 Feb. 1973.

GATTY, Juliana Horatia *see* Ewing, J. H.

GATTY, Margaret (*née* **Scott**) (1809–1873)
b. Burnham, Essex 3 June 1809 *d.* Ecclesfield, Yorks 4 Oct. 1873
Writer for children. Married Rev. Alfred Gatty, 1839. Studied algae from 1848–49. Correspondent of W. H. Harvey from 1850. *British Seaweeds* 1863. *Parables from Nature* 1885.
W. H. Harvey *Phycologica Australica* v.2, 1859, t.93. *J. Bot.* 1873, 352. *D.N.B.* v.21, 67–69. *N. Western Nat.* 1942, 150–52. *Sorby Rec.* v.2(3), 1967, 11. *Country Life* 1968, 625 portr.
Letters in G. A. Walker-Arnott correspondence at BM(NH). Portr. at Hunt Library.
Gattya Harvey.

GATTY, Stephen Herbert (1847–1922)
b. Ecclesfield, Yorks. 9 Oct. 1847
Herb. at Winchester College, Hants.

GAUKROGER, John (*fl.* 1830s–1840s)
Nurseryman, Bolton Brow in Skircoat, Sowerby Bridge, Halifax, Yorks.

GAUNT, Rev. C. (*fl.* 1850s)
Plants from Glos, Berks and Scotland at Oxford.

GAUT, Alfred (*c.* 1845–1928)
d. 27 Feb. 1928
Gardener. Lecturer in Horticulture, Shropshire, 1895; Leeds University, 1899–1917. *Seaside Planting of Trees and Shrubs* 1907. Contrib. to *Gdnrs Chron.*
Gdnrs Mag. 1911, 871–72 portr. *Gdnrs Chron.* 1928 i 182.

GAWLER, John *see* Ker, J. B.

GAYE, Caroline (1804–1883)
b. Sheffield, Yorks 1804 *d.* London 1883
Governess. Drew plants.
Bedfordshire Nat. 1955, 14–15.

GAYNER, Francis (1870–1933)
b. County Durham 1870 *d.* 23 Nov. 1933
MA. MB. FLS 1929. Physician, Redhill, Surrey. Collected and grew alpine plants.
Proc. Linn. Soc. 1933–34, 157–58.

GEAKE, Joseph John (1890–1917)
b. Guildford, Surrey 12 June 1890 *d.* Guildford 24 Sept. 1917
Chemist, Royal Veterinary College. Botanist.
Bot. Soc. Exch. Club Br. Isl. Rep. 1918, 355.

GEARY, Andrew Chapman (–1792)
Nurseryman, Milford, Salisbury, Wilts. In partnership with William Geary.

GEDDES, Sir Patrick (1854–1932)
b. Perth 2 Oct. 1854 *d.* Montpellier, France 17 April 1932
Educ. Edinburgh, Montpellier. Professor of Botany, University College, Dundee, 1888–1920. Professor of Civics and Sociology, Bombay, 1920–23. Planned garden at Dundee to illustrate an evolutionary classification. *Chapters in Modern Botany* 1893. *Life and Work of Sir Jagadis C. Bose* 1920. *Outlines of General Biology* 1931.
Nature v.129, 1932, 713–14. *Pharm. J.* 1932, 341. *Proc. R. Soc. Edinburgh* 1932, 452–54. *Times* 19 April 1932. *D.N.B. 1931–1940* 311–13. *Who was Who, 1929–1940* 502. P. Kitchen *A Most Unsettling Person* 1975 portr.

GEE, Edward Pritchard (1904–1968)
b. County Durham 1904 *d.* 22 Oct. 1968
Educ. Cambridge. Planter in Assam; retired in 1959 to Shillong. Amateur zoologist and conservationist. Collected orchids in Eastern Himalayas.
J. Bombay Nat. Hist. Soc. 1969, 361–64 portr.

GELDART, Alice Mary (1862–1942)
d. Thorpe, Norfolk 4 May 1942
FLS 1939. Daughter of H. D. Geldart (1831–1902). Added to her father's herb., now at Norwich Museum. President, Norfolk and Norwich Naturalists' Society, 1914, 1931. Authority on history of East Anglian botany. Edited and contrib. to *Trans. Norfolk Norwich Nat. Soc.*
Proc. Linn. Soc. 1942–43, 298–99. *Trans. Norfolk Norwich Nat. Soc.* 1943, 443–44.

GELDART, Herbert Decimus (1831–1902)
b. Feldthorpe Hall, Norwich 11 Feb. 1831 *d.* Thorpe Hamlet, Norwich 21 Sept. 1902
'Fl. Norwich' (*Trans. Norfolk Norwich Nat. Soc.* 1875–84). Account of Norfolk plants in *Victoria County Hist. Norfolk* v.1, 1901, 39–75.
J. Bot. 1902, 431–32. *Trans. Norfolk Norwich Nat. Soc.* 1902–3, 573–76 portr. *R.S.C.* v.12, 265; v.15, 253.
Herb. at Norwich Museum.

GELLAN, Thomas (*fl.* 1830s)
Nurseryman, Shacklewell, Hackney, London.

GENTLE, Alfred M. (*c.* 1874–1958)
d. 13 July 1958
Tradesman. Grew orchids. He and his wife collected orchids in Brazil, 1937.
Orchid Rev. 1958, 234–35.

GEOGHEGAN, Miss Frances (*fl.* 1890s)
Collected plants in Majorca, 1896 and 1900, and gave them to Botanic Garden, Glasnevin, Dublin.
Bot. Mag. 1903, t.7903; 1907, t.8161.

GEORGE, Edward (–1894)
d. 18 Jan. 1894
Gardener to Earl of Abingdon at Wytham Abbey, Berks.
J. Hort. Cottage Gdnr v.28, 1894, 89.

GEORGE, Edward (1830–1900)
b. Salisbury, Wilts 1830 *d.* Forest Hill, London 10 Oct. 1900
Bryologist and algologist.
J. Bot. 1900, 455.
Herb. at BM(NH).
Rhodophysema georgii Batters.

GEORGE, Edward Sanderson (*c.* 1801–1830)
d. Leeds, Yorks 9 Feb. 1830
Curator, Leeds Philosophical Hall. Had herb.
R. V. Taylor *Biographia Leodiensis* 1865–67, 320, 664.

GEORGE, James (1826–1911)
b. Farringdon, Berks 1826 *d.* Putney, London 10 March 1911
Gardener to Nicholson family, Putney Heath. Horticultural sundriesman, Putney.
J. Hort. Cottage Gdnr v.56, 1908, 468–70 portr.; v.62, 1911, 249 portr., 268. *Gdnrs Mag.* 1911, 231 portr.

GEPP, Antony (1862–1955)
b. High Easter, Essex 9 May 1862 *d*. Dec. 1955

BA Cantab 1885. Assistant, Botany Dept., BM(NH), 1886–1927. Algologist.

J. Bot. 1927, 184. *Nat. Hist. Mag*. 1927, 95–96. *Times* 19 Dec. 1955.

Portr. at Hunt Library.

GEPP, Ethel Sarel (*née* Barton) (1864–1922)
b. Hampton Court Green, Middx 21 Aug. 1864 *d*. Torquay, Devon 6 April 1922

Algologist. Worked in Dept. of Botany, BM(NH) and Kew. Contrib. to *Phycological Mem*., *J. Bot*., *J. Linn. Soc*., *Trans. Linn. Soc*.

J. Bot. 1922, 160, 193–95. *Nuova Notar*. ser. 34, 1923, 47–57 portr.

Plants at BM(NH).

Ethelia W. van Bosse.

GEPP, Maurice (1860–1947)
b. High Easter, Essex 1860 *d*. Shrewsbury, Shropshire 14 June 1947

Medical Officer of Health, South-west Shropshire, 1898–1941. President, Caradoc and Severn Valley Field Club, 1928, 1943. Botanist.

N. Western Nat. 1955, 258–59.

GÉRARD, — (–1840)

Gardener at Versailles. Came to England at outbreak of French Revolution. Collected herbs for Apothecaries Hall, etc. Lived in Marylebone. Knew London plants well. Had herb. (*Phytologist* v.6, 1862–63, 15–18). Said to have discovered *Wolffia arrhiza* at Putney *c*. 1816 (*J. Bot*. 1866, 263–64).

Part of his collection sold by Stevens, 7 Dec. 1861.

GERARD, Alexander (1792–1839)

Army surveyor. Collected plants in the Himalayas. 'Account of Journey through Himalaya' (*Edinburgh Philos. J*. 1824, 295).

I. H. Burkill *Chapters on Hist. Bot. in India* 1965, 31 *et seq*.

GERARD, James Gilbert (1794–1828)

Surgeon in Indian Army. Brother of A. Gerard (1792–1839). Collected plants in Himalayas. 'Account of Spiti Valley and Circumjacent Country' (*Asiatick Res*. 1833, 238).

I. H. Burkill *Chapters on Hist. Bot. in India* 1965, 31 *et seq*.

GERARD (*or* Gerarde), John (1545–1612)
b. Nantwich, Cheshire 1545 *d*. London Feb. 1612

Barber-surgeon. Had garden in Holborn; also supervised gardens of Lord Burghley in the Strand and at Theobalds, Herts. Travelled in Denmark, Russia, etc. (*Herball*, 1223). *Catalogus Arborum Fruticum ac Plantarum*...1596; ed. 2 1599 (reprinted with notes by B. D. Jackson as *Catalogue of Plants Cultivated in the Garden of John Gerard*...*1596–1599* 1876). *Herball* 1597; amended edition by T. Johnson in 1633; reprinted 1636; abridged version edited by M. Woodward published in 1927; reprinted 1964.

R. Pulteney *Hist. Biogr. Sketches of Progress of Bot. in England* v.1, 1790, 116–25. S. Felton *Portr. of English Authors on Gdning* 1830, 87–88. *Cottage Gdnr* v.6, 1851, 207. *Notes and Queries* v.11, 1855, 149, 204; v.3, 1899, 164; v.158, 1930, 242. H. Trimen and W. T. T. Dyer *Fl. Middlesex* 1869, 369. *J. Hort. Cottage Gdnr* v.28, 1875, 145–46. *Gdnrs Chron*. 1889 ii 219. G. C. Druce *Fl. Berkshire* 1897, xcvi–xcvii. *D.N.B*. v.21, 221–22. *Essex Nat*. 1899, 61–68, 169–73. *Gdnrs Mag*. 1897, 792–94 portr. *Garden* 1909, 613–14 portr.; 630–31. J. R. Green *Hist. Bot*. 1914, 34–43. G. C. Druce *Fl. Northamptonshire* 1930, xliv–xlv. C. E. Raven *English Nat*. 1947, 204–17. *Genealogists' Mag*. 1963, 137–45. R. H. Jeffers *Friends of John Gerard* 1967; *Biogr. Appendix* 1969. F. D. Hoeniger *Growth of Nat. Hist. in Stuart England* 1969, 6–11. C. C. Gillispie *Dict. Sci. Biogr*. v.5, 1972, 361–63. *House and Gdn* Nov. 1974, 144–46 portr.

Portr. in *Herball* and at National Portrait Gallery and Hunt Library.

Gerardia L.

GERARD, Rev. John (1840–1912)
b. Edinburgh 30 May 1840 *d*. London 13 Dec. 1912

BA London 1859. FLS 1900. Society of Jesus, 1856. *Fl. Stonyhurst* (anon.) 1886; ed. 2. 1891. *Essays on Un-natural History* 1910. Contrib. to *J. Bot*.

J. Bot. 1913, 59–60. *Proc. Linn. Soc*. 1912–13, 58.

GERARD, Patrick (1795–1835)

Army officer. Brother of A. Gerard (1792–1839). Sent Himalayan plants to N. Wallich.

I. H. Burkill *Chapters on Hist. Bot. in India* 1965, 31 *et seq*.

Pinus gerardiana Wall.

GERARD, Peter (*fl*. 1700s)

Virginian plants at Sloane herb.

J. E. Dandy *Sloane Herb*. 1958, 131–32.

GERRARD, William Tyrer (–1866)
d. Foul Point, Madagascar 1866

Of Natal. Collected plants in Natal, Zululand and Madagascar. *Synopsis Filicum Capensium* (with M. J. McKen) 1870.

W. H. Harvey *Genera of S. African Plants*

1868, 127–28. *Gdnrs Chron.* 1866, 1042. *J. Bot.* 1866, 367.
Plants at BM(NH), Kew. Letters at Kew. *Gerrardanthus* Harvey.

GERRISH, Richard (*c.* 1857–1929)
d. Salisbury, Wilts 16 Dec. 1929
Had large collection of orchids.
Orchid Rev. 1925, 327–34; 1930, 32.

GETTY, Edmund (1799–1857)
b. Belfast 1799 *d.* London Dec. 1857
Ballast master, Belfast. Wrote a number of botanical papers.
Belfast Nat. Hist. Philos. Soc. Centenary Vol. 1924, 76–78 portr., 129–30.

GIBBES, George *see* Gibbs, G.

GIBBES, Rev. Heneage (1802–1887)
b. Bath, Somerset 1802 *d.* Mutley, Plymouth, Devon 18 March 1887
MB Cantab 1826. Incumbent, All Saints, Sidmouth, 1847. Rector, Bradstone, Devon, 1870–83. Discovered *Euphorbia pilosa.*
C. C. Babington *Fl. Bathoniensis* 1834 preface, vi, 44. C. C. Babington *Memorials, Journal and Bot. Correspondence* 1897, xxii, xxxii. W. Munk *Roll of R. College of Physicians* v. 3, 1878, 14.

GIBBONS, H. J. J. F. (1856–1939)
Of Bristol where he botanised.
H. J. Riddelsdell *Fl. Gloucestershire* 1948, cxlix.
Herb. at Leicester University.

GIBBS, George (*fl.* 1640s)
Surgeon of Bath, Somerset. Catalogue of plants in his garden in T. Johnson's *Mercurius Botanicus* 1634–41. J. Parkinson in his *Theatrum Botanicum* 1640 reports that Gibbs had lately returned from Virginia "with a number of seeds and plants."
R. T. Gunther *Early English Botanists* 1922, 346–48.

GIBBS, John (–*c.* 1829)
b. Folkestone, Kent *d.* Twickenham, Middx *c.* 1829
Self-taught naturalist. In employ of the ornithologist, Colonel Montagu. Curator of Lord Waldegrave's museum, Strawberry Hill, Twickenham. Assistant, BM. Interested in London flora. MS. biogr. account by W. Pamplin in copy of H. Trimen and W. T. T. Dyer's *Fl. Middlesex* 1869 at Kew Gardens.

GIBBS, John (1822–1903)
b. Bermondsey, London 1822 *d.* Shelton, Beds 2 March 1903
Of Chelmsford, Essex. Wool-sorter. Taught botany at Mechanics Institute. Assistant Curator at Museum, 1868. *Lecture on Variations of Plants* 1861. *First Catechism of Botany* 1871; ed. 2 1878. *Symmetry of Flowers* 1890. Contrib. to *Proc. Essex Field Club* v.2–4, 1881–92.
Essex Nat. 1915–16, 89–96 portr., 203–5. *R.S.C.* v.15, 291.

GIBBS, Lilian Suzette (1870–1925)
b. London 10 Sept. 1870 *d.* Santa Cruz, Tenerife 30 Jan. 1925
FLS 1905. Collected plants in S. Rhodesia, 1905 (*J. Linn. Soc.* v.37, 1906, 425–94); Fiji, 1907 (*J. Linn. Soc.* v.39, 1909, 130–212); Mt. Kinabalu, 1910 (*J. Linn. Soc.* v.42, 1914, 1–240); Arfak Mts., 1913 (*Contribution to Phytogeography Fl. Arfak Mountains* 1917); Queensland and Tasmania, 1914.
Bol. Soc. Broteriana v.3, 1925, 239–41 portr. *Bot. Soc. Exch. Club Br. Isl. Rep.* 1925, 848. *Gdnrs Chron.* 1925 ii 222. *J. Bot.* 1925, 85, 116–17, 312. *Kew Bull.* 1925, 189. *Nature* v.115, 1925, 345. *Proc. Linn. Soc.* 1924–25, 72–74. *Fl. Malesiana* v.1, 1950, 190 portr.
Gibbsia Rendle.

GIBBS, Thomas (*fl.* 1790s–1820s)
Nurseryman, 90 Piccadilly, Old Brompton Road, London and Ampthill, Beds.
Bot. Mag. 1807, t.984, 1090, 1091; 1810, t.1299; 1818, t.1957. *Trans. London Middlesex Archaeol. Soc.* v.24, 1973, 187.

GIBBS, Thomas (1865–1919)
b. Burton-on-Trent, Staffs 1865 *d.* Lindfield, Sussex 8 Feb. 1919
Solicitor. Mycologist. To Sussex, 1916. Fl. Burton in *Trans. Burton Nat. Hist. Soc.* v.3–4, 1896–99. Contrib. to *Naturalist.*
Naturalist 1919, 177–80 portr. *Sorby Rec.* v.2(3), 1967, 11–12.
Coprinus gibbsii Mass. & Crossl.

GIBBS, Hon. Vicary (1853–1932)
b. Hampstead, London 12 May 1853 *d.* London 13 Jan. 1932
VMH 1916. MP, St. Albans, 1892–1904. Banker. Had garden at Aldenham House, Elstree with large collection of Michaelmas daisies and shrubs.
Gdnrs Mag. 1907, 545 portr.; 546. *Gdnrs Chron.* 1922 i 26 portr.; 1931 ii 22 portr.; 1932 i 71 portr. *J. R. Hort. Soc.* 1932, 361 portr. *Times* 14, 22 Jan. 1932. *Kew Bull.* 1932, 107. *D.N.B. 1931–1940* 336.

GIBSON, Alexander (1800–1867)
b. Laurencekirk, Kincardineshire 24 Oct. 1800 *d.* Bombay 16 Jan. 1867
MD Edinburgh. FLS 1853. Surgeon, East

India Company, 1825. Superintendent, Dapuri Garden, 1838–47. Conservator of Forests, Bombay, 1847–60. *Handbook to Forests of Bombay Presidency* 1863. *Bombay Fl.* (with N. A. Dalzell) 1861.

Proc. Linn. Soc. 1866–67, xxxiii. *D.N.B.* v.21, 272–73. D. G. Crawford *Hist. Indian Med. Service, 1600–1913* v.2, 1914, 151. *Indian For.* 1971, 152–61. *R.S.C.* v.2, 873.

Plants at Kew, Calcutta. Letters at Kew.

Gibsonia Stocks.

GIBSON, Alexander Henry (*fl.* 1870s–1880s)

Herb. at Royal Botanic Garden, Edinburgh.

GIBSON, George Stacey (1818–1883)

b. Saffron Walden, Essex 20 July 1818 *d.* London 5 April 1883

FLS 1847. Banker. Discovered *Galium vaillantii*, etc. *Fl. Essex* 1862. Contrib. to *Phytologist*.

J. Smith *Cat. Friends' Books* v.1, 1867, 838–39. *Herts and Essex Observer* 7, 14 and 21 April 1883. *J. Bot.* 1883, 161–65 portr. *Proc. Linn. Soc.* 1882–83, 41–42. *Trans. Essex Field Club* v.4, 1885, 1–8 portr. *D.N.B.* v.21, 276. H. C. Watson *Topographical Bot.* 1883, 545. F. H. Davey *Fl. Cornwall* 1909, xliv. *Essex Nat.* 1954, 189. S. T. Jermyn *Fl. Essex* 1974, 17. *R.S.C.* v.2, 874.

Plants at BM(NH), Essex Field Club. Letters at Kew, Society of Friends, London.

GIBSON, Jabez M. (1794–1838)

b. Saffron Walden, Essex 11 Dec. 1794 *d.* Saffron Walden 22 Feb. 1838

Uncle of G. S. Gibson (1818–1883). Banker. A founder of Saffron Walden Natural History Society and Museum, 1832. 'Rarer Plants observed near Coggeshall' (*Phytologist* 1844, 834–35).

Naturalist v.3, 1838, 283–84. R. M. Christy *Birds of Essex* 1890, 21–22.

GIBSON, John (*fl.* 1760s)

MD. Practised medicine in London. Surgeon, Royal Navy. *Fruit-gardener* 1768.

J. C. Loudon *Encyclop. Gdning* 1822, 1275. *Gdnrs Chron.* 1900 ii 301–2.

GIBSON, John (1815–1875)

b. Eaton Hall, Cheshire 18 June 1815 *d.* South Kensington, London 11 Jan. 1875

Gardener at Chatsworth House, Derbyshire. Sent by Duke of Devonshire to India in 1835 to collect *Amherstia nobilis* and orchids. Superintendent, Victoria Park, London, 1849; Superintendent, Greenwich Park, 1851; Battersea Park, 1858; Hyde Park, St. James's Park and Kensington Gardens, 1871–74.

Garden v.7, 1875, 60; v.19, 1881, xii portr. *Gdnrs Chron.* 1872, 865 portr.; 1875, 85–86 portr. G. F. Chadwick *Park and the Town* 1966, 123. K. Lemmon *Golden Age of Plant Hunters* 1968, 184–216. A. M. Coats *Quest for Plants* 1969, 154–55. M. A. Reinikka *Hist. of the Orchid* 1972, 193–95 portr.

Cymbidium gibsoni Paxton.

GIBSON, Robert John Harvey- *see* Harvey-Gibson, R. J.

GIBSON, Samuel (*c.* 1789/90–1894)

b. Hebden Bridge, Yorks 1789/90 *d.* Hebden Bridge 21 May 1849

Blacksmith. Entomologist, palaeontologist, conchologist. '*Carex pseudo-paradoxa*' (*Phytologist* v.1, 1844, 778–79; *J. Bot.* 1916, 17). Contrib. to E. Newman's *Hist. Br. Ferns* 1840 and H. Baines's *Fl. Yorkshire* 1840.

J. Cash *Where there's a Will there's a Way* 1873, 157–64. *Proc. Manchester Lit. Philos. Soc.* v.12, 1872–73, 45–47. *Leeds Mercury* 15 June 1889. J. B. L. Warren *Fl. Cheshire* 1899, lxxxi. W. B. Crump and C. Crossland *Fl. Parish of Halifax* 1904, lvi–lvii. *Trans. Liverpool Bot. Soc.* 1909, 70. *R.S.C.* v.2, 874.

Herb., formerly at Belle Vue Museum, Halifax, destroyed.

Hieracium gibsoni Backh.

GIBSON, Thomas (*fl.* 1810s–1870s)

Brother of S. Gibson (*c.* 1789/90–1849). Had herb. Contrib. to *Fl. Liverpool* 1872. Papers in *Br. Assoc. Advancement Sci. Rep.* 1870, Notices 115–17.

J. B. L. Warren *Fl. Cheshire* 1899, lxxxi. *Trans. Liverpool Bot. Soc.* 1909, 70.

GIDDINGS, Robert (*fl.* 1820s)

Nurseryman, New Street, Wells, Somerset.

GIDDY, Davies *see* Gilbert, D.

GIFFORD (*or* Gyfford), Mr. (*fl.* 1680s)

Gave plants from Madras and S. Africa to Bobart.

R. Morison *Plantarum Historiae Universalis Oxoniensis* v.3, 1699, 214, 236, 238, 347, 423, etc.

GIFFORD, Isabella (*c.* 1823–1891)

b. Swansea, Glam *c.* 1823 *d.* Minehead, Somerset 26 Dec. 1891

Phycologist. *Marine Botanist* 1848; ed. 3 1853. Contrib. Somerset plants to *Proc. Somerset Archaeol. Soc.* 1853, 116–23; 1855, 131–37. 'Tetraspores of Seirospora' (*J. Bot.* 1871, 113).

J. Bot. 1892, 81–83. *Notarisia* 1892, 1396–99.

Letters in Walker-Arnott correspondence at BM(NH). Plants at Somerset Archaeological Society.
Giffordia Batt.

GILBERT THE ENGLISHMAN (*fl.* 1250s)
Studied in Italy. Physician to Hubert, Archbishop of Canterbury. Following MSS. at Bodleian, New College, etc.: *De re Herbaria; De Viribus et Medicinis Herbarum, Arborum et Specierum; De Virtutibus Herbarum.*
Philos. Trans. R. Soc. v.63, 1773, 80–81. A. von Haller *Bibliotheca Botanica* v.1, 1771, 219; v.2, 1772, 658. R. Pulteney *Hist. Biogr. Sketches of Progress of Bot. in England* v.1, 1790, 22–23. *D.N.B.* v.21, 318. E. J. L. Scott *Index to Sloane Manuscripts* 1904, 215.

GILBERT, Davies (*né* Giddy) (1767–1839)
b. St. Erth, Cornwall 6 March 1767 *d.* Oxford 24 Dec. 1839
MA Oxon 1789. DCL 1832. LLD Cantab 1832. FRS 1791. FLS 1792. MP for Helston, then Bodmin, 1804–32. President, Royal Society, 1827–31. Married Mary Ann Gilbert, 1806, and took the name of Gilbert, 1817. Assisted W. Withering in his *Bot. Arrangement* ed. 3 1796.
Proc. Linn. Soc. 1840, 68–70. *D.N.B.* v.21, 323–24. F. H. Davey *Fl. Cornwall* 1909, xxxvii.

GILBERT, Edward Gillett (1849–1915)
b. Harleston, Norfolk 12 March 1849 *d.* Tunbridge Wells, Kent 17 Dec. 1915
Physician, London. Batologist. Contrib. to *J. Bot.*, 1903, 1907, 1912.
Bot. Soc. Exch. Club Br. Isl. Rep. 1916, 464–65. *J. Bot.* 1916, 70–71.
Rubi at Kew.

GILBERT, John (–1743)
Herb. at Truro Museum, Cornwall.

GILBERT, John (*c.* 1810–1845)
b. 14 March 1810? *d.* murdered by natives in Gulf of Carpentaria 28 June 1845
Collector on J. Gould's expedition to Australia, 1838–40. Returned in 1842 from visit to England. Explored in W. Australia. "Did good botanical collecting." With Leichhardt on expedition to Port Essington, 1844–45.
J. W. Austral. Nat. Hist. Soc. no. 6, 1909, 16–17; 1949, 23–53. *Rep. Austral. Assoc. Advancement Sci.* 1909, 377. *Emu* 1941, 112–29, 216–42; 1944, 131–50, 183–200; 1951, 17–31.
Plants at BM(NH), Kew, Vienna. Diary, 1844–45, and letters at Mitchell Library, Sydney.
Acacia gilberti Meissner.

GILBERT, Sir Joseph Henry (1817–1901)
b. Hull, Yorks 1 Aug. 1817 *d.* Rothamsted, Herts 23 Dec. 1901
PhD Giessen 1840. FRS 1860. FLS 1875. Knighted 1893. Conducted agricultural experiments with J. B. Lawes at Rothamsted from 1843. Professor of Rural Economy, Oxford, 1884–90. *Botanical Results of Experiments on Mixed Herbage of Permanent Meadow* 1882.
Gdnrs Chron. 1871, 1627 portr. *Nature* v.48, 1893, 327–29; v.65, 1901–2, 205–6. *J. R. Agric. Soc.* 1901, 347–55 portr. *Proc. Linn. Soc.* 1901–2, 34–35. A. D. Hall *Book of Rothamsted Experiments* 1905, xxxii–xl portr. F. W. Oliver *Makers of Br. Bot.* 1913, 233–42 portr. *D.N.B. 1901–1911* 106. *R.S.C.* v.2, 879; v.7, 773; v.9, 1007; v.15, 303.

GILBERT, Richard (1821–1895)
b. Worksop, Notts 6 Aug. 1821 *d.* Stamford, Lincs 22 Nov. 1895
Gardener to Marquis of Exeter at Burghley House, Stamford. Probably raised 'Richard Gilbert' apple.
Gdnrs Chron. 1875 ii 209–10 portr.; 1895 ii 658 portr.; v.160(25), 1966, 18 portr. *Garden* v.26, 1884, 332–33 portr.; v.48, 1895, 426, 427 portr.

GILBERT, Rev. Samuel (–*c.* 1692/4)
Chaplain to Jane, wife of Charles, 4th Baron Gerard. Rector, Quatt, Shropshire. Married Minerva, daughter of John Rea. *Florist's Vade-mecum* 1682; later eds. 1683, 1690, 1693, 1702.
G. W. Johnson *Hist. English Gdning* 1829, 119–20. *Cottage Gdnr* v.6, 1851, 107. *J. Hort. Cottage Gdnr* v.55, 1876, 172–73 portr. *D.N.B.* v.21, 334–35. *Garden* 1913, 5–6.

GILBERT, Thomas (–1891)
d. Hastings, Sussex 15 Feb. 1891
Gardener at Castledown. Succeeded to Springfield Nursery, Hastings on death of former proprietor, H. Barham.
J. Hort. Cottage Gdnr v.22, 1891, 167.

GILBERT-CARTER, Humphrey (1884–1969)
b. 19 Oct. 1884 *d.* Dawlish, Devon 4 Jan. 1969
Educ. Edinburgh and Cambridge. Economic Botanist, Botanical Survey of India, 1913–21. Director, Botanic Garden, Cambridge, 1921–50. University Lecturer in Botany, 1930–50. *Genera of British Plants* 1913. *Guide to Botanic Garden, Cambridge* 1922–47. *Descriptive Labels for Botanic Gardens* 1924. *Our Catkin-bearing Plants* 1930. *British Trees and Shrubs* 1936. *Glossary of British Fl.* 1950; ed. 3 1964. Translated Raunkiaer's *Plant Life Forms* 1937.

Nature v.221, 1969, 497–98. *Nature in Cambridgeshire* no. 12, 1969, 4–5. *Times* 6 Jan. 1969. *Watsonia* 1970, 71–74. *Who was Who, 1961–1970* 427. J. S. L. Gilmour and S. M. Walters *Humphrey Gilbert-Carter* 1975 portr.

Plants from Canaries, Madeira and Britain at Cambridge.

GILBY, William Hall (–*c.* 1821)

MD Edinburgh 1815. Geologist. President, Royal Society of Medicine. 'On Respiration of Plants' (*Edinburgh Philos. J.* v.4, 1821, 100–6).

D.N.B. v.21, 340. *R.S.C.* v.2, 884.

GILCHRIST, James (1813–1885)

b. Collin, Dumfriesshire 21 June 1813 *d.* Dumfries 7 Dec. 1885

MD 1850. President, Dumfries Field Club. 'Geological Relations of Alpine Plants' (*Proc. Bot. Soc. Edinburgh* 1855, 9–11).

Scott. Nat. 1886, 242–43. *Trans. Bot. Soc. Edinburgh* v.17, 1889, 2–11. *R.S.C.* v.2, 884.

Herb. at Royal Botanic Garden, Edinburgh.

GILES, George Michael James (*fl.* 1880s)

Surgeon in Indian Army. Collected plants on Gilgit-Chitral expedition, 1885–86. (*J. Linn. Soc.* v.35, 1902, 210–12).

Pakistan J. For. 1967, 346.

Plants at Kew, Calcutta.

GILES, John (*c.* 1725–1797)

Gardener to Lady Boyd at Lewisham, Kent. Foreman in nursery of Messrs. Russell, Lewisham, 1777. *Ananas, or Treatise on Pine Apple* 1767.

J. C. Loudon *Encyclop. Gdning* 1822, 1275. G. W. Johnson *Hist. English Gdning* 1829, 226–27.

GILES, John (*fl.* 1790s)

Nurseryman and hop-planter with Francis Giles, Farnham, Surrey.

GILES, W. F. (–1962)

VMH 1942. Authority on vegetables. Contrib. articles to *J. R. Hort. Soc.*

J. R. Hort. Soc. 1968, 47 portr.

GILES, William Ernest Powell (1847–1897)

b. Bristol 1847 *d.* 13 Nov. 1897

To Australia, 1850. Gold prospector in Victoria. Explored S. Australia, 1872 and Central Australia, 1873, collecting plants on both expeditions. 3rd expedition to Perth and back to Adelaide, 1875–76. Clerk, Warden's Office, Coolgardie, W. Australia. *Geographic Travels in Central Australia... 1872 to 1874* 1875 (with list of plants by F. von Mueller). *Journal of Forgotten Expedition* 1880. *Australia Twice Traversed...1872–1876* 1889 2 vols.

J. Bot. 1877, 269–81. *J. R. Geogr. Soc. Austral.* 1898, 138–40. *Rep. Austral. Assoc. Advancement Sci.* 1907, 169–70. E. Favene *Explorers of Australia and their Work* 1908 *passim.* G. Rawson *Desert Journeys* 1948, 13–18, 39–108. P. Serle *Dict. Austral. Biogr.* 1949, 343–44. L. F. Green *Ernest Giles* 1963. G. Dutton *Australia's Last Explorer: Ernest Giles* 1970.

MSS. at Mitchell Library, Sydney and South Australia Archives. Plants at Melbourne. Portr. at Hunt Library.

Cyperus gilesii Benth.

GILL, Charles Haughton (1841–1894)

b. Wells, Somerset 12 June 1841 *d.* 21 Feb. 1894

FRMS. Professor of Chemistry, London. Diatomist. Contrib. to *J. R. Microsc. Soc.* 1889–91.

Diatomiste v.2, 1894, 125–29. *J. R. Microsc. Soc.* 1894, 264–65, 284. *R.S.C.* v.7, 774; v.15, 306.

GILL, Edwin (*fl.* 1850s)

Nurseryman with John Gill, Blandford, Dorset.

GILL, Norman (*c.* 1878–1924)

d. Penryn, Cornwall 15 April 1924

Kew gardener, 1898. Assistant Curator, Royal Botanic Gardens, Calcutta, 1900–2. In charge of several government gardens. Superintendent, Government Gardens, Kumaon, 1909–24.

Gdnrs Chron. 1924 i 246. *J. Kew Guild* 1925, 338 portr.

Plants at Calcutta.

GILL, Richard Ernest (1875–1942)

b. Penryn, Cornwall 17 Aug. 1875 *d.* 17 Aug. 1942

Kew gardener, 1899–1901. Nurseryman, Kernick, Cornwall. Specialised in rhododendrons.

J. Kew Guild 1938, 733–34 portr.; 1942, 193–95.

GILL, Walter (1851–1929)

b. Welford, Northants 13 Oct. 1851 *d.* Malvern, S. Australia 17 July 1929

FLS 1890. Gardener, F. Gill and Co., nurserymen. To S. Australia, 1876. Sub-inspector, Crown Lands, S. Australia; Chief Forester, 1886. Conservator of Forests, S. Australia, 1890–1923. Collected plants for E. J. Brown and F. von Mueller.

H. T. Burgess *Cyclop. S. Australia* v.1, 1907, 350–51. *Fred John's Annual* 1914, 77–78. *Proc. Linn. Soc.* 1929–30, 196–97.

Acacia gillii Maiden.

GILL, William John (1843–1882)

b. Bangalore, India 1843 *d.* murdered by Bedouins in Sinai Desert, 11 Aug. 1882

Royal Engineers, 1864. Served in India, 1869–71. Travelled in China and Tibet,

1877. *River of Golden Sand* 1880 2 vols; ed. 2 1883 (contains references to plants).

E. Bretschneider *Hist. European Bot. Discoveries in China* 1898, 730–35. *J. Bot.* 1899, 68, 135–36. *D.N.B.* v.21, 355–57. E. H. M. Cox *Plant Hunting in China* 1945, 101–2.

Chinese plants at BM(NH).

GILL, Rev. William Wyatt (1828–1896)

Stationed in Pacific Islands; visited New Guinea, 1872–84. *Life in Southern Isles...in South Pacific and New Guinea* 1876. *Work and Adventure in New Guinea, 1877–1885* (with J. Chalmers) 1885.

Fl. Malesiana v.1, 1950, 191.

Plants at Kew.

GILLBANKS, Jackson (–1878)

d. Cumberland 19 May 1878

Amateur gardener. Contrib. to *Garden*.

Garden v.13, 1878, 534.

GILLETT, Gabriel (*fl.* 1800s)

Had a garden at Drayton Green, Middx.

Bot. Mag. 1807, t.1068.

GILLIES, John (1792–1834)

d. Edinburgh 24 Nov. 1834

MD 1817. Naval surgeon. Went to Buenos Aires, 1820. Resided at Mendoza, 1823–28. Returned to Scotland, 1829. Collected plants in Argentina and Chile. Correspondent of J. Miers and W. Jameson. MS. *Fl. Orcadensis* (with A. Duguid) 1832.

J. Miers *Travels in Chili and La Plata* v.1, 1826, 226–28. *Bot. Misc.* 1832–33, 129–30. A. Lasègue *Musée Botanique de B. Delessert* 1845, 486. H. C. Watson *Topographical Bot.* 1883, 545. *Notes Rec. R. Soc.* v.9, 1951, 115–36. H. N. Clokie *Account of Herb. Dept. Bot., Oxford* 1964, 171. *Taxon* 1970, 524. *Bol. Acad. Nac. Ciencias* (Cordoba) v.49, 1972, 71–75.

Plants at BM(NH), Kew, Oxford. Letters at BM(NH), Kew.

Crepis gillii S. Moore.

GILLILAND, Hamish Boyd (1911–1965)

b. Salisbury, S. Rhodesia 2 Oct. 1911 *d.* S. Africa 23 June 1965

BSc Edinburgh. FLS 1934. Lecturer in Botany, University of Witwatersrand, 1935–55. Professor of Botany, Singapore, 1955–65. Botany Dept., University of Natal, 1965. *Student's Key to Monocotyledons of Witwatersrand* 1952. *Common Malayan Plants* 1958; ed. 2 1962. *Grasses of Malaya* 1971. Contrib. to *S. African J. Bot., Gdns Bull., Singapore.*

S. African Forum Botanicum v.3(10), 1965, 1–3. *Nature* v.207, 1966, 808. *Gdns Bull., Singapore* 1967, 107–12 portr.

Rhodesian plants at Kew.

GILMAN, Edwin (*fl.* 1870s–1920s)

Gardener to Lord Shrewsbury at Ingestre Hall, Staffs, 1876 and at Alton Towers, Staffs.

Gdnrs Chron. 1922 ii 46 portr.

GILMOUR, Thomas (–1930)

MD. Member of Glasgow Natural History Society. Contrib. to P. Ewing's *Glasgow Catalogue of Native and Established Plants* 1892.

Glasgow Nat. v.9, 1930, 62.

GILRUTH, John Anderson (1871–1937)

d. 4 March 1937

Royal College of Veterinary Surgeons, 1891. Government Veterinary Surgeon, New Zealand, 1898. Professor, Veterinary Pathology, Melbourne, 1908–12. Administrator, Northern Territory, 1912–20. Collected plants in Northern Territory.

A. J. Ewart and O. B. Davies *Fl. Northern Territory* 1917 *passim. Who was Who, 1929–1940* 517.

GIMLETTE, John Desmond (1867–1934)

b. Southsea, Hants 28 Feb. 1867 *d.* Cheam, Surrey 24 April 1934

MRCS. LRCP. Surgeon Magistrate, Selinsing, Pahang, 1886. Joined Duff Co., at Kelantan, 1903. Collected plants. *Malay Poisons and Charm Cures* 1915. *Dictionary of Malayan Medicine* (with H. W. Thomson) 1939.

Gdns Bull. Straits Settlements 1927, 122. *J. Bot.* 1934, 176. *J. Malayan Branch R. Asiatic Soc.* 1934, 184. *Nature* v.133, 1934, 900–1. *Fl. Malesiana* v.1, 1950, 191–92.

Malayan plants at Singapore.

GIRAUD, Herbert John (1817–1888)

b. Faversham, Kent 14 April 1817 *d.* Shanklin, Isle of Wight 12 Jan. 1888

MD Edinburgh 1840. Physician, chemist, botanist. Went to India, 1842. Professor of Chemistry and, after 1845, of Botany, Grant Medical College, 1845. 'Vegetable Embryology' (*Ann. Nat. Hist.* v.5, 1840, 225–38). Drew *Myosurus* for W. Baxter's *Br. Phaenogamous Bot.* v.3, 1837, 204.

Trans. Linn. Soc. v.19, 1845, 161–70. *D.N.B.* v.21, 393–94. *R.S.C.* v.2, 902.

GIRDLESTONE, Theophilus William (–1899)

d. Sunningdale, Berks 25 June 1899

MA. FLS 1889. Proprietor, private school, Sunningdale. Grew dahlias. Secretary and President, National Dahlia Society.

Rosarian's Yb. 1892 portr. *Garden* v.56, 1899, 12. *Gdnrs Mag.* 1899, 387 portr. *Proc. Linn. Soc.* 1899–1900, 72.

GIRLING, Samuel (*fl.* 1830s)
Nurseryman, Stowmarket, Suffolk.

GIRTON, John (*fl.* 1830s)
Nurseryman, Stodman Street, Newark, Notts.

GISBORNE, John (1770–1850)
d. Twyford, Derbyshire June 1850
BA Cantab 1792. Brother of T. Gisborne (1758–1846). Of Wootton Hall and Orgreave Hall, Staffs. Poet. Had herb.
D.N.B. v.21, 400. *Trans. N. Staffordshire Field Club* 1947–48, 100. E. S. Edees *Fl. Staffordshire* 1972, 16.

GISBORNE, Rev. Thomas (1758–1846)
b. Yoxall, Barton-under-Needwood, Staffs 31 Oct. 1758 *d.* Durham 1846
BA Cantab 1780. FLS 1799. Uncle of C. C. Babington who called him "an ardent botanist." Perpetual Curate, Barton-under-Needwood, 1783. Prebendary, Durham, 1823 and 1826. Contrib. to J. Sowerby and J. E. Smith's *English Bot.* 438, etc.
Proc. Linn. Soc. 1846, 299–300. *D.N.B.* v.21, 401–2. *J. Bot.* 1901 Supplt., 71. *Trans. N. Staffordshire Field Club* 1947–48, 103–5. E. S. Edees *Fl. Staffordshire* 1972, 16.
Plants at BM(NH).

GISSING, Thomas Waller (1829–1870)
b. Halesworth, Suffolk 2 Aug. 1829 *d.* Wakefield, Yorks 28 Dec. 1870
Druggist. *Ferns and Fern-allies of Wakefield* 1862. *Materials for Fl. Wakefield* 1867.
J. Bot. 1871, 96. *Pharm. J.* v.1, 1871, 556–57. *R.S.C.* v.2, 907; v.7, 783; v.12, 276.
Herb. at Wakefield Museum.

GIUSEPPI, Paul Leon (–1947)
d. Felixstowe, Suffolk 13 Nov. 1947
MD London. FRCS. VMH 1947. Grew alpine plants in his garden at Felixstowe. Treasurer and later President, Alpine Garden Society. His wife discovered *Sempervivum andreanum* in Spain.
Quart. Bull. Alpine Gdn Soc. 1942, 247–48; 1948, 7–13.
Sempervivum giuseppi Wale.

GLADMAN, —
Sent plants to J. Petiver from Angola.
J. E. Dandy *Sloane Herb.* 1958, 132.

GLAISTER, Elsie J. (*fl.* 1870s–1890s)
Herb. at Carlisle Museum, Cumberland.

GLANVILLE, Bartholomaeus De (*alias* **Bartholomaeus Anglicus**) (*fl.* 1230s–1250s)
Franciscan friar. Professor of Theology, Paris. Went to Saxony, 1231. *De Proprietatibus Rerum* (dealing in part with plants); translated by John de Trevisa, printed *c.* 1495.
J. L. G. Mowat *Sinonoma Bartholomaei* 1882. *D.N.B.* v.21, 409–11. E. J. L. Scott *Index to Sloane Manuscripts* 1904, 41. C. E. Raven *English Nat.* 1947, 13–18.

GLASCOTT, Miss Louisa S. (*fl.* 1890s)
Of Alderton, New Ross, Ireland. Studied Irish Rotifera. Co-author with G. E. Barnett-Hamilton of 2 papers on Wexford plants (*J. Bot.* 1889, 4–8; 1890, 87–89).
R. L. Praeger *Some Irish Nat.* 1949, 86–87.

GLASGOW, Charles Ponsonby Robertson-
see Robertson-Glasgow, C. P.

GLASSCOCK, Henry (*c.* 1820–1891)
b. Bishop's Stortford, Herts *c.* 1820 *d.* Bishop's Stortford 1891
Grew dahlias.
Garden v.40, 1891, 433.

GLASSON, William Arthur (1828–1903)
b. Hayle, Cornwall 29 May 1828 *d.* Penzance, Cornwall 14 Jan. 1903
'Occurrence of Foreign Plants in W. Cornwall' (*Trans. Penzance Nat. Hist. Soc.* 1888–89, 62–69).
J. Bot. 1903, 111. F. H. Davey *Fl. Cornwall* 1909, lx.

GLASSPOOLE, Hampden Gledstanes (1825–1887)
b. Ormesby St. Michael, Norfolk 6 April 1825 *d.* Hammersmith, London 5 March 1887
Botanical Curator, Norwich Museum. Botanist, Alexandra Palace. Found *Carex trinervis*. '*Choetoceras armatum*' (*J. Bot.* 1878, 378). Contrib. to *Trans. Norfolk Norwich Nat. Soc.* and *Sci. Gossip.*
J. Bot. 1887, 382–83. *Trans. Norfolk Norwich Nat. Soc.* 1894–95, 78–79.
Herb. at Norwich Museum.

GLAZEBROOK, Thomas Kirkland (1780–1855)
b. Ashby-de-la-Zouch, Leics 4 June 1780 *d.* Southport, Lancs 17 Jan. 1855
FLS. Glass manufacturer, Warrington. Moved to Southport, 1835. List of plants in *Guide to Southport, North Meoles* 1809; ed. 2 1826.
D.N.B. v.21, 423. *Trans. Liverpool Bot. Soc.* 1909, 70.
Silhouette in J. Kendrick's *Profiles of Warrington Worthies* 1853.

GLEADOW, F. (*fl.* 1880s–1890s)
Deputy Conservator of Forests, India. Discovered *Gleadovia ruborum* at Jaunsar, 1898.
Bombay plants at Oxford.
Gleadovia Gamble & Prain.

GLEESON, John Matthew (–1899)
d. Madras, India 11 Aug. 1899
Kew gardener, 1868. To India in 1870 to take charge of cotton plantations. Superintendent of gardens of Agri-Horticultural Society of Madras, 1883–92. *Catalogue of Plants in Agri-Horticultural Society's Gardens, Madras* 1884.
Kew Bull. 1892, 286. *J. Kew Guild* 1900, 34–35.

GLEN, Rev. Andrew (*c.* 1666–1732)
d. Hathern, Leics 1 Sept. 1732
BA Cantab 1683. Rector, Hathern, 1694. Friend of J. Ray. Travelled in Sweden and Italy. Formed herb., 1685–92, of native and foreign plants.
R. Pulteney *Hist. Biogr. Sketches of Progress of Bot. in England* v.2, 1790, 63–64. *D.N.B.* v.21, 427. A. R. Horwood and C. W. F. Noel *Fl. Leicestershire* 1933, clxxxvii–viii.

GLENALMOND, Lord *see* Patton, G.

GLENDINNING, Robert (1805–1862)
b. Lanark 27 Sept. 1805 *d.* 9 Nov. 1862
Nurseryman, Turnham Green, Chiswick. R. Fortune sent most of his plants from his 3rd Chinese expedition, 1853–55, to this nursery. *Practical Hints on Culture of Pine Apple* 1839. *The Pinetum* (with G. Gordon) 1858. 'On Transplanting Large Evergreen Trees and Shrubs' (*J. Hort. Soc. London* 1849, 41–44).
Cottage Gdnr v.28, 1862, 655–56. E. Bretschneider *Hist. European Bot. Discoveries in China* 1898, 553. *Ann. Bot.* 1902, cxcvi.

GLENIE, Rev. Samuel Owen (1811–1875)
MA 1849. FLS 1863–73. To Colombo, Ceylon, 1834. Colonial chaplain, Trincomalee. Archdeacon, Kandy, 1870. Collected plants for G. H. K. Thwaites.
G. H. K. Thwaites *Enumeratio Plantarum Zeylaniae* 1864, vii–viii. H. Trimen *Handbook Fl. Ceylon* v.5, 1900, 380.
Glenniea Hook. f.

GLENNY, George (*c.* 1793–1874)
b. London *c.* 1793 *d.* Norwood, Middx 17 May 1874
Horticultural journalist. *Glenny's Handbook to Flower Garden and Greenhouse* 1851. *Hand-book to Fruit and Vegetable Garden* ed. 2 1853. *Culture of Fruits and Vegetables* 1860. *Culture of Flowers and Plants* ed. 2 1861. *Gardener's Every-day Book* 1863. *Handy Book on Gardening* 1863. Edited *Gdnrs Gaz.* and *Hort. J.*
Florist Pomologist 1874, 141. *J. Hort. Cottage Gdnr* v.51, 1874, 429. J. F. Wilson *Few Personal Recollections* 1896, 109 portr. *D.N.B.* v.21, 436. *Gdnrs Chron.* 1901 ii 425. *J. R. Hort. Soc.* 1944, 307–8. A. Simmonds *Hort. Who was Who* 1948, 63–64. *Country Life* 1971, 1149 portr.

GLOVER, C. *see* **Shinn, A.**

GLOVER, James (1844–1925)
d. Kircubbin, County Dublin 28 June 1925
Teacher in Dublin. Interested in natural history including botany.
Rep. Br. Bryol. Soc. 1925, 183.
Bryophyta at Oxford.

GLYN, Thomas (*fl.* 1630s)
Of Glynnllivon, Caernarvonshire. Friend of T. Johnson who dedicated *Mercurius Botanicus* 1634–41 to him. Found *Diotis maritima* in Wales.
J. Gerard *Herball* 1633, 63, 644, 996. R. Pulteney *Hist. Biogr. Sketches of Progress of Bot. in England* v.1, 1790, 136.

GLYNNE, Eryl *see* Smith, E.

GOADBY, Bede Theodoric (*c.* 1863–1944)
b. Kasauli, India *c.* 1863 *d.* Cottesloe, W. Australia 28 Sept. 1944
Lieut.-Col., Royal Engineers. To W. Australia, 1895. Collected plants there and also in New Britain, 1914. In later years collected orchids. President, Western Australian Naturalists Club.
Proc. Field Nat. Club Victoria 1945, 205. *Victoria Nat.* 1945, 30.
Herb. at Melbourne. Plants at Kew.
Goadbyella R. S. Rogers.

GODDARD, Rev. Edward Hungerford (1854–1947)
Vicar, Clyffe Pypard, Wilts, 1883–1935, where he created a garden with some 1300 species. Secretary, Wilts Archaeological and Natural History Society; Editor, 1890–1942.
G. Grigson *Gardenage* 1952, 75–81. D. Grose *Fl. Wiltshire* 1957, 41–42.

GODDARD, Harry James (*c.* 1864–1947)
d. Salisbury, Wilts 15 Aug. 1947
Botanist, Dunns Farm Seeds, Ltd., 1916. Contrib. to J. C. Mansel-Pleydell's *Fl. Dorsetshire* ed. 2 1895. *Goddard's Grasses of Great Britain* 1936.
Salisbury Times 22 Aug. 1947. *Salisbury and Winchester J.* 22 Aug. 1947. *Times* 15 Sept. 1947. *Watsonia* 1949–50, 186.
Grasses at Salisbury Museum.

GODDARD, Jonathan (*c.* 1617–1675)
b. Greenwich, London *c.* 1617 *d.* London 24 March 1675
MB Cantab 1638. MD 1643. FRS 1663. Warden, Merton College, Oxford, 1651. Professor of Physic, Gresham College, 1655. *Observations Concerning a Tree* 1664. *Fruit Trees' Secrets* 1664.

W. Munk *Roll of R. College of Physicians* v.1, 1878, 240–42. *D.N.B.* v.22, 24–26. E. J. L. Scott *Index to Sloane Manuscripts* 1904, 219.
MS. *Texture of Wood*, etc. at Royal Society.

GODFERY, Hilda Margaret (1871–1930)
d. Guildford, Surrey 17 Sept. 1930
Botanical artist who specialised in orchids. Drew plates for H. Correvon's *Album des Orchidées d'Europe* 1899, M. J. Godfery's *Monograph and Iconograph of Native British Orchidaceae* 1933 and *J. Bot.*
Bot. Soc. Exch. Club Br. Isl. Rep. 1930, 323. *J. Bot.* 1930, 343–44. *Orchid Rev.* 1930, 341.
Orchid drawings at BM(NH).

GODFERY, Masters John (1856–1945)
d. Torquay, Devon 9 April 1945
FLS 1915. Authority on British and European orchids, especially more critical genera such as *Epipactis* and *Ophrys*. *Monograph and Iconograph of Native British Orchidaceae* 1933. Contrib. to *J. Bot.*, *Orchid Rev.*, *Bull. Amer. Orchid Soc.*
Nature v.155, 1945, 627. *Orchid Rev.* 1945, 64. *Proc. Linn. Soc.* 1944–45, 110–11. *Times* 14 April 1945.

GODFREY, Robert (*c.* 1812–1874)
d. Ryde, Isle of Wight 20 Aug. ? 1874
After death of Hosea Waterer, joined A. Waterer as partner in Knapp Hill nursery, Woking, Surrey.

GODFREY, W. J. (*fl.* 1880s–1900s)
Nurseryman, Exmouth, Devon. Specialised in chrysanthemums.
Gdnrs Mag. 1911, 161–62 portr.

GODWIN, Fisher (*fl.* 1850s–1860s)
Nurseryman, 2–4 Norfolk Hall Market, Sheffield, Yorks.

GODWIN, William (*fl.* 1820s)
Nurseryman, Clifton near Ashbourne, Derbyshire.

GODWIN-AUSTEN, Henry Haversham (1834–1924)
Lieut.-Col., Survey of India. Zoologist. In Karakorum, 1857–60. Found *Pinus excelsa* at Rondu in Baltistan.
Pakistan J. For. 1967, 346.
Plants at Kew.

GOGARTY, Dr. (*fl.* 1840s)
Physician, Rio de Janeiro. Sent plants to Irish institutions including botanic garden at Glasnevin, Dublin.
Bot. Mag. 1842, t.3942.

GOLDHAM, C. (*fl.* 1890s)
Educational officer, Ipoh and Kuala Kangsor, Siam. Collected orchids which he sent to Singapore.
J. Thailand Res. Soc. Nat. Hist. Supplt. v.12, 1939, 23.

GOLDIE, Andrew (*fl.* 1870s–1880s)
Collected plants in New Guinea, 1876–82 for the Holloway nursery of B. S. Williams and possibly also for Sander and Sons.
Gdnrs Chron. 1879 i 597–98. *Fl. Malesiana* v.1, 1950, 195–96.
Plants at Melbourne.
Combretum goldieanum F. Muell.

GOLDIE, Rev. Hugh (*fl.* 1880s)
d. Creektown, Old Calabar, Nigeria
Scottish Presbyterian missionary, Old Calabar. Collected chiefly in Old Calabar and first brought *Aristolochia goldieana* to notice.
Kew Bull. Additional Series v.9, 1922, 23.
Plants at Kew.

GOLDIE, John (1793–1886)
b. Kirkoswald, Ayrshire 21 March 1793 *d.* Ayr, Ontario 23 July 1886
Gardener and plant collector. Discovered *Rumex aquaticus* (J. Sowerby and J. E. Smith *English Bot.* 2698). To Canada, 1817. *Diary of Journal through Upper Canada* 1819; reprinted 1897 portr.; 1967 portr.
Edinburgh Philos. J. v.6, 1822, 319–33. *Gdnrs Mag.* 1826, 85; 1827, 129–35. *Dominion Annual Register* 1886, 269–70. *Bot. Gaz.* 1886, 272–74. *Amer. J. Sci.* 1888, 260–61. *J. Bot.* 1888, 299–301. *Gdnrs Chron.* 1896 ii 531. *Trans. R. Soc. Canada* 1897, 125–30. *Fern Bull.* v.8, 1900, 73–75 portr. *Rhodora* 1968, 457–61. *R.S.C.* v.2, 929.
Aspidium goldianum Hook.

GOLDRING, William (1854–1919)
b. West Dean, Sussex May 1854 *d.* Kew, Surrey 26 Feb. 1919
Kew gardener, 1875–79. Assistant Editor, *Garden*, 1879. Editor, *Woods and Forests*. Landscape gardener.
J. Hort. Home Farmer v.66, 1913, 509 portr. *J. Kew Guild* 1913, 71 portr.; 1920, 503. *Garden* 1919, 110–11 portr. *Gdnrs Chron.* 1919 i 103.

GOLDSBURY, Joseph (*fl.* 1810s)
Nurseryman, 91 Whitechapel, London.

GOLLAN, William (–1905)
d. Naini Tal, United Provinces, India 12 Sept. 1905
Gardener, Edinburgh Botanic Garden. Superintendent, Saharanpur Botanic Garden, 1880s; Government Gardens, Lucknow, 1904. Collected plants for J. F. Duthie in Kashmir, 1893.
Pakistan J. For. 1967, 346.
Plants at Kew, Dehra Dun, New York, etc.

GOMME, Lady Elizabeth (*fl.* 1850s)
Wife of Commander-in-Chief, India. Collected a few plants, 1856.

GONNER, John (*fl.* 1770s–1790s)
Seedsman, adjoining White Hart Inn, Colchester, Essex.

GOOD, John Mason (1764–1827)
b. Epping, Essex 25 May 1764 *d.* Shepperton, Middx 2 Jan. 1827
MA Aberdeen 1820. FRS 1805. Practised medicine at Sudbury; came to London, 1793. *Structure and Physiology of Plants* 1808.
D.N.B. v.22, 110–11.

GOOD, Peter (–1803)
b. Scotland *d.* Sydney 11 June 1803
To Calcutta in 1796 to bring Christopher Smith's collection of plants back to England. Foreman, Kew Gardens until he was appointed as an assistant to Robert Brown on Flinders's voyage to Australia, 1801, on HMS 'Investigator'.
R. A. Salisbury *Paradisus Londinensis* v.1, 1805, t.41. *Bot. Mag.* 1806, t.958. J. D. Hooker *Fl. Tasmaniae* v.1, 1859, cxxiv. *Gdnrs Chron.* 1881 ii 568; 1952 i 216–17. *Kew Bull.* 1891, 301–2. F. M. Bladen *Hist. Rec. N.S.W.* v.5, 1897, 204. *J. Kew Guild* 1897, 28; 1946–47, 561–63. *J. Proc. R. Soc. N.S.W.* 1908, 105–6. J. H. Maiden *Sir Joseph Banks* 1909, 126–27. *Fl. Malesiana* v.1, 1950, 197. *J. S. African Bot.* 1974, 47–60.
MS. Journal, 1801–3, and lists at BM(NH).
Goodia Salisb.

GOODACRE, John Herbert (–1922)
d. 13 Aug. 1922
VMH 1907. Gardener to Lord Harrington at Elvaston Castle, Derbyshire, 1872–1919.
Gdnrs Chron. 1922 ii 130 portr.

GOODE, George (1858–1943)
b. Cambridge 26 Oct. 1858 *d.* Cambridge 21 Jan. 1943
MA. Assistant Librarian, Cambridge University. Secretary and Editor, Watson Botanical Exchange Club, 1905–20; Treasurer, 1920–34.
Bot. Soc. Exch. Club Br. Isl. Rep. 1943–44, 646.
Plants at Nottingham, Oxford.

GOODE, Henry (*fl.* 1860s–1880s)
Of Plymouth, Devon. Algologist. Had herb.
J. Bot. 1914, 107, 250.

GOODENOUGH, James Samuel (*fl.* 1880s–1900s)
Eurasian. Forest Service, Straits Settlements; Forest inspector, 1888–1901. Collected trees in Malacca and Selangor.
Gdns Bull. Straits Settlements 1927, 122.

GOODENOUGH, Rev. Samuel (1743–1827)
b. Kimpton, Weyhill, Hants 29 April 1743 *d.* Worthing, Sussex 12 Aug. 1827
MA Oxon 1767. DCL 1772. FRS 1789. Bishop of Carlisle, 1808–27. Founder member and first Treasurer, Linnean Society, 1788. 'Observations on British Species of Carex' (*Trans. Linn. Soc.* v.2, 1794, 126–211). 'Observations on British Fuci' (with T. J. Woodward) (*Trans. Linn. Soc.* v.3, 1797, 84–235).
Trans. Linn. Soc. v.2, 1794, 346–47. P. Smith *Mem. and Correspondence of Sir J. E. Smith* v.1, 1832, 289–90; v.2, 299. *J. Bot.* 1880, 256. *J. Hort. Cottage Gdnr* v.1, 1880, 29. G. C. Druce *Fl. Berkshire* 1897, cxliv–cxlvi. *D.N.B.* v.22, 124–25. *Annual Rep. Yorkshire Philos. Soc.* 1906, 54–55. H. J. Wilkinson *Hist. Account of Herb. Yorkshire Philos. Soc.* 1907, 10–11. G. C. Druce *Fl. Buckinghamshire* 1926, xci–xciii. G. C. Druce *Fl. Oxfordshire* 1927, cii–ciii. A. T. Gage *Hist. Linnean Soc. London* 1938 portr. *R.S.C.* v.2, 934.
Drawings of *Carex* at BM(NH). Herb. and letters, 1788–1810, at Kew. Portr. at Hunt Library.
Goodenia Smith.

GOODGER, William Frederick (*fl.* 1810s–1830s)
Physician at parochial infirmary, Marylebone, London, 1811–32. Had herb.
H. Trimen and W. T. T. Dyer *Fl. Middlesex* 1869, 398.

GOODIER, Joseph (1798–1885)
b. Manchester 24 April 1798 *d.* Castleton, Lancs 5 May 1885
Bleacher. "Came under the influence of Samuel Barlow and became an enthusiastic botanist."
Trans. Liverpool Bot. Soc. 1909, 70.

GOODRICH, Lawrence C. (*fl.* 1880s)
Collected ferns in Arabia in 1883, now at Kew.

GOODSIR, John (1814–1867)
b. Anstruther, Fife 14 March 1814 *d.* Wardie, Edinburgh 6 March 1867
MD Edinburgh. FRS 1846. Professor of Anatomy, Edinburgh, 1846–67. 'Conferva on Goldfish' (*Ann. Mag. Nat. Hist.* v.9, 1842, 333–37). 'Potato Disease' (*Phytologist* v.2, 1846, 469–71). *Anatomical Memoirs* 1868 2 vols.
Proc. R. Soc. v.16, 1868, xiv–xvi. *Trans. Bot. Soc. Edinburgh* v.9, 1868, 118–27. *D.N.B.* v.22, 137–39.

GOODWIN, William (*fl.* 1830s)
Nurseryman, Norwich Road, Ipswich, Suffolk

GOODYER, John (1592–1664)
b. Alton, Hants 1592 *d.* Petersfield, Hants 1664

Of Mapledurham, Hants. "A great lover and curious searcher of plants" (J. Parkinson). Critical on *Ulmus* (J. Gerard *Herball* 1633, 1479). Discovered *Frankenia laevis*, 1621.

J. Gerard *Herball* 1633, 560, 1625–29. J. Parkinson *Theatrum Botanicum* 1640 *passim*. T. Johnson *Mercurius Botanicus* 1634–41, alt. 2. C. Merrett *Pinax* 1666, A4 verso. R. Pulteney *Hist. Biogr. Sketches of Progress of Bot. in England* v.1, 1790, 135–36, 158. G. C. Druce *Fl. Berkshire* 1897, xcviii–xcix. *Cornhill Mag.* 1909 i 795–803. *Bot. Soc. Exch. Club Br. Isl. Rep.* 1916, 523–50. *J. Bot.* 1916, 375; 1917, 167; 1921, 119; 1938, 185–92. *Kew Bull.* 1919, 332. R. T. Gunther *Early Br. Botanists* 1922, 1–232, 372–401. G. C. Druce *Fl. Buckinghamshire* 1926, lxx–lxxi. C. E. Raven *English Nat.* 1947, 291–94.

MSS. at Magdalen College. Portr. at Hunt Library and Kew.

Goodyera R. Br.

GOOGE, Barnaby (*c.* 1540–1593/4)
b. Alvingham, Lincs 1540 *d.* Feb. 1593/4

Educ. Cambridge. Poet. Translated C. Heresbach's *Foure Bookes of Husbandrie* 1577.

G. W. Johnson *Hist. English Gdning* 1829, 71. *Cottage Gdnr* v.8, 1852, 157–58. J. Donaldson *Agric. Biogr.* 1854, 9–10. *D.N.B.* v.22, 151–52. D. McDonald *Agric. Writers* 1908, 46–55.

GORDON, Alexander (1813–*c.* 1873)

Gardener. Collected plants in Rocky Mountains and S. Carolina for G. Charlwood.

R. Sweet *Br. Flower Gdn* v.6, 1838, t.271. *London J. Bot.* 1845, 492–96. S. D. McKelvey *Bot. Exploration Trans-Mississippi West, 1790–1850* 1955, 818–23. J. Ewan *Rocky Mountain Nat.* 1950, 217.

Plants at Kew.

Penstemon gordoni Herb.

GORDON, Rev. George (1801–1893)
b. Urquhart 1801 *d.* Braebirnie, Elgin 12 Dec. 1893

LLD. Minister of Birnie near Elgin, 1832–89. *Collectanea for Fl. Moray* 1839 portr.

H. C. Watson *New Botanists' Guide* 1835–37, 498–502, 508–12. H. C. Watson *Topographical Bot.* 1883, 546. *Gdnrs Chron.* 1893 ii 809. *Ann. Scott. Nat. Hist.* 1894, 65–71 portr. *J. Bot.* 1894, 64, 160. *R.S.C.* v.2, 945; v.7, 800; v.10, 28; v.15, 382.

MS. *Fl. Moray* at BM(NH). Letters at Kew.

GORDON, George (1806–1879)
b. Lucan, County Dublin 25 Feb. 1806 *d.* Kew, Surrey 11 Oct. 1879

ALS 1841. At gardens of Horticultural Society of London at Chiswick, 1828; Superintendent of Hardy and Hothouse Departments. *The Pinetum* (with R. Glendinning) 1858; Supplt., 1862. Assisted J. C. Loudon in his *Arboretum et Fruticetum Britannicum* 1838.

Gdnrs Chron. 1879 ii 569. *Garden* v.13, 1878, 199; v.16, 1879, 382. *D.N.B.* v.22, 200. *R.S.C.* v.2, 945.

Herb. of conifers at Kew.

GORDON, George (1841–1914)
b. Frogmore Hall, Herts 1841 *d.* Kew, Surrey 14 June 1914

VMH 1897. Editor, *Gdnrs Mag.* 1890–1913, and *Gdning Yb.* President, National Sweet Pea Society, 1903; National Dahlia Society, 1913. *Books of Shrubs* 1903. *Dahlias* 1913. *Poppies* 1913. *Wasted Orchards of England.*

Gdnrs Mag. 1913, 875–76 portr.; 1914, 493–94 portr., 546. *Garden* 1914, 324 (viii) portr. *Gdnrs Chron.* 1914 i 448 portr., 460. *J. Hort.* v.68, 1914, 563 portr. *National Rose Soc. Annual* 1915, 123–25 portr. *Who was Who, 1897–1916* 283.

GORDON, James (*c.* 1708–1780)
d. Barking, Essex 20 Dec. 1780

Gardener to J. Sherard at Eltham, *c.* 1730–38 and to Lord Petre at Thorndon, Essex to 1742. Established nursery at Mile End in 1742; later at Bow. Seed shop at Thistle and Crown, 25 Fenchurch Street, London. Introduced many exotics including the camellia. His general catalogue *c.* 1770 was probably the first nurseryman's list in botanical form. Correspondent of Linnaeus. Nursery inherited by his sons, William, James and Alexander.

Philos. Trans. R. Soc. v.60, 1771, 520. R. Pulteney *Hist. Biogr. Sketches of Progress of Bot. in England* v.2, 1790, 241. J. E. Smith *Selection from Correspondence of Linnaeus* v.1, 1821, 93, 254, 500, 507; v.2, 73. D. Turner *Extracts from Lit. and Sci. Correspondence of Richard Richardson* 1835, 390, 394. L. W. Dillwyn *Hortus Collinsonianus* 1843, 4, 5. *Gdnrs Mag.* 1843, 637. E. Bretschneider *Hist. European Bot. Discoveries in China* 1898, 150–51. *J. Bot.* 1902, 389. *Gdnrs Chron.* 1905 i 201; 1966 i 488–89. *Essex Nat.* 1969, 203. J. Harvey *Early Gdning Cat.* 1972, 55–56. *Trans. London Middlesex Archaeol. Soc.* v.24, 1973, 193.

Plants from his garden in Banks herb. at BM(NH).

Gordonia Ellis.

GORDON, James (*fl.* 1750s)
MD Aberdeen. "A very ingenious and skilful physician and botanist who first initiated me into these studies" (Alexander Garden).
J. E. Smith *Correspondence of Linnaeus* v.1, 1821, 378.

GORDON, James (*fl.* 1750s–1770s)
Nurseryman, Fountainbridge near Edinburgh. Said by J. C. Loudon to have had one of the first tree nurseries in Scotland. *Catalogue of Shrubs and Flowers* 1758. *Planter's, Florist's and Gardener's Pocket Dictionary* 1774.
Gdnrs Chron. 1923 i 120.

GORDON, John (*fl.* 1800s)
Gardener, Chelsea Physic Garden. "Young man sent out here as a botanist, named Gordon. It appears that he is employed by [Col. J. A.] Woodford..." (letter of 10 March 1801 from Governor King of N.S.W.).
J. R. Soc. N.S.W. 1908, 106. J. E. B. Currey *Reflections on Colony of New South Wales* 1967, 47.

GORDON, Robert (*fl.* 1870s)
Of Ashton-under-Lyne and Dukinfield, Cheshire. Publican. "A good namer of plants." *Botanical Catalogue of British Phaenogamous Plants, Ferns, etc.* 1872.
Trans. Liverpool Bot. Soc. 1909, 70.

GORDON, Robert Jacob (1741–1795)
d. Cape of Good Hope 25 Oct. 1795
Dutchman of Scottish extraction. In S. Africa as Captain in Dutch East India Company. Travelled with W. Paterson, 1777–79. Made large collection of plants and animals. Collected and drew *Stapelias*.
F. Masson *Stapeliae Novae* 1796, preface, viii. *J. Bot.* 1884, 145; 1914, 75–77. *J. Linn. Soc.* v.45, 1920, 49–50. *S. African Biol. Soc. Pamphlet* 1949, 44–62.
Drawings at Rijks Museum.

GORDON, William (–1849)
d. Feb. 1849
FLS 1832. Surgeon. Of Welton near Hull. 'Analogy between Vegetables and Animals' (*Mag. Nat. Hist.* 1831, 385–93; 1832, 24–30, 118–28, 405–12, 507–12).
R.S.C. v.2, 945.

GORDON, William Thomas (1884–1950)
b. Glasgow 27 Jan. 1884 *d.* London 12 Dec. 1950
MA Edinburgh. BA Cantab 1910. Lecturer in Palaeontology, Edinburgh, 1910–13. Lecturer in Geology, King's College, London, 1914; Reader, 1919; Professor, 1920–49. Did research on Lower Carboniferous flora. Contrib. to *Ann. Bot.*, *Trans. Bot. Soc. Edinburgh*, *Trans. R. Soc. Edinburgh*.
Gdnrs Chron. 1950 ii 251. *Nature* v.167, 1951, 59–60. *Yb. R. Soc. Edinburgh* 1952, 16–18. *Who was Who, 1941–1950* 449.

GORE, John (*fl.* 1760s–1770s)
2nd Lieutenant on 'Endeavour', 1768–71. Helped J. Banks and D. Solander in their plant collecting on this voyage. Took command of 'Discovery' when Cook was killed in 1779.
J. E. T. Woods *Hist. Discovery and Exploration of Australia* 1865, 42–43. *J. Proc. R. Austral. Hist. Soc.* 1934, 293. J. C. Beaglehole *ed.* *'Endeavour' Journal of Joseph Banks, 1768–1771* 1962 2 vols *passim*.

GORHAM, George Cornelius (1787–1857)
Plants at BM(NH).

GORRIE, Archibald (1777–1857)
b. Logie Almond, Perthshire 1777 *d.* Errol, Perthshire 21 July 1857
Gardener at Logie House, Dupplin Castle and at Leith Walk nursery. Father of David and William Gorrie. Friend of G. Don. Contrib. to *Gdnrs Mag.*, *Mag. Nat. Hist.*, *Trans. Hort. Soc. London*.
Cottage Gdnr v.18, 1857, 291, 333–34. *Gdnrs Chron.* 1857, 582.

GORRIE, David (1822–1856)
b. Annat, Perthshire 1822 *d.* 12 March 1856
Horticultural journalist. *Illustrations of Scripture in Connection with Botanical Science* 1854.
Cottage Gdnr v.16, 1856, 98–99.

GORRIE, Robert Maclagen (*c.* 1897–1970)
d. Edinburgh 21 Dec. 1970
BSc Edinburgh 1922. DSc 1930. Indian Forest Service, 1930. Lecturer, Forest College, Dehra Dun. Edited *Indian Forester*, *Scottish For. J.*
Commonwealth For. Rev. 1971, 216.

GORRIE, William (*c.* 1811–1881)
b. Carse of Gowrie, Perthshire *c.* 1811 *d.* Newhaven near Edinburgh 6 Jan. 1881
Gardener to Messrs P. Lawson, nurseryman, Edinburgh. President, Botanical Society of Edinburgh. Contrib. to P. Lawson's *Agriculturalist's Manual* 1836 and *Treatise on Cultivated Grasses* 1843. Contrib. grasses to J. C. Morton's *Cyclop. Agric.* v.1, 1855, 996–1004.
Gdnrs Chron. 1881 i 89. *J. Hort. Cottage Gdnr* v.2, 1881, 32. *Gdnrs Yb. Almanack* 1882, 191. *Trans. Bot. Soc. Edinburgh* v.14, 1883, 298–300.
Plants at Glasgow.

GOSSE, Philip (1879–1959)
b. 18 Aug. 1879 *d.* 3 Oct. 1959
Grandson of P. H. Gosse. Naturalist on E. A. Fitzgerald's expedition to Andes, 1896–97. Collected plants, now at Kew. *Notes on Natural History of Aconcagua Valleys* 1899.
E. A. Fitzgerald *Highest Andes* 1899, 370 *et seq. Kew Bull.* 1920, 64. *Who was Who, 1951–1960* 438.

GOSSE, Philip Henry (1810–1888)
b. Worcester 6 April 1810 *d.* St. Marychurch, Torquay, Devon 23 Aug. 1888
FRS 1856. ALS 1849. Zoologist. In Canada, etc., 1827–44; in Jamaica, 1844. Collected plants in Jamaica. Devoted last years to study of Rotifera and orchids. *Canadian Naturalist* 1840. *Sojourn in Jamaica* 1846. *Wanderings through the Conservatories at Kew* 1856.
Gdnrs Chron. 1888 ii 250. *Proc. R. Soc.* v.44, 1888, xxvii–xxviii. *J. Inst. Jamaica* v.2, 1899, 574–81. *D.N.B.* v.22, 258–60. I. Urban *Symbolae Antillanae* v.3, 1902, 52. E. Gosse *Naturalist of the Sea Shore* 1890 portr. E. Gosse *Father and Son* 1907. F. Cundall *Historic Jamaica* 1915, 349–52. *J. Soc. Bibl. Nat. Hist.* v.3, 1955, 221–22.
'Orchids and their Culture', a vol. of cuttings and original drawings at BM(NH). Letters at Kew.

GOSSE, William Christie (1842–1881)
b. Hoddesdon, Herts 1842 *d.* Adelaide 12 Aug. 1881
To S. Australia with his parents in 1852. Joined S. Australia Survey Dept, *c.* 1859. Explored interior of Australia in 1873 and sent plants he collected to F. von Mueller. Deputy Surveyor-General, 1875.
E. Favenc *Explorers of Australia and their Life Work* 1908, 213–15. G. Rawson *Desert Journeys* 1948, 19–42. *Austral. Encyclop.* v.4, 1962, 341–42.

GOSSELIN, Joshua (1739–1813)
b. Guernsey, Channel Islands 6 Nov. 1739 *d.* Bengeo, Herts 27 May 1813
Had herb. 'Fl. Sarniensis' 1788 (in W. Berry's *Hist. Island of Guernsey* 1815). His grandson, Joshua Carteret Gosselin, also collected plants in Guernsey.
Rep. Trans. Guernsey Soc. Nat. Sci. Local Res. 1894, 344–46. E. D. Marquand *Fl. Guernsey* 1901, 22–24. *Bot. Soc. Br. Isl. Rep.* v.5, 1917, 141–42.

GOTOBED, Richard (–*c.* 1806)
FLS 1800. Of Eton. Contrib. Berks and Bucks lists to D. Turner and L. W. Dillwyn *Botanist's Guide* 1805.
J. Sowerby and J. E. Smith *English Bot.* 731, 1295, 1501. G. C. Druce *Fl. Berkshire* 1897, cxlvii. *J. Bot.* 1902, 322, 324.
Plants at Oxford.

GOTT, Sir Henry Thomas *see* Greening, *Sir* H. T.

GOTT, John (*c.* 1834–1931)
d. Kendal, Westmorland 19 April 1931
Collected British ferns. Founder member, British Pteridological Society.
Br. Fern Gaz. 1931, 84.

GOUGH, George Stevens, 2nd Viscount (1815–1895)
b. 13 Jan. 1815 *d.* Booterstown, Dublin 31 May 1895
BA Dublin 1836. FLS 1840. Captain, Grenadier Guards. Collected in Nilgiri Hills, S. India with W. Munro, 1842.
Illus. London News 1895, 734. *Proc. Linn. Soc.* 1895–96, 36–37.
Goughia Wight.

GOUGH, Henry (*fl.* 1700s)
Presented plants collected from East Indies, China, etc. to Sir Hans Sloane.
J. E. Dandy *Sloane Herb.* 1958, 132.

GOUGH, John (1757–1825)
b. Kendal, Westmorland 17 Jan. 1757. *d.* Kendal 28 July 1825
Mathematician. Blind from age of three but so developed his sense of touch that he became an accomplished botanist. Correspondent of W. Withering. 'Experiments and Observations on Vegetation of Seeds' (*Manchester Philos. Soc.* v.4, 1796, 310–23, 488–505).
J. Sowerby and J. E. Smith *English Bot.* 489. W. Withering *Bot. Arrangement* v.1, 1787, 455. Macpherson *Fauna of Lakeland* 1892, xxii–xxiv. *D.N.B.* v.22, 277–78. *R.S.C.* v.2, 959.

GOUGH, Thomas (1804–1880)
b. Middleshaw, Westmorland 30 Nov. 1804 *d.* Kendal? 17 July 1880
Son of J. Gough (1757–1825). Surgeon. Contrib. botanical notices to J. Hudson's *Guide to Lakes* 1843.
Naturalist 1894, 294. *Westmorland Notebook* 1889, 109 portr.

GOULD, Norman Kenneth (1897–1960)
b. Gravesend, Kent 23 Nov. 1897 *d.* 19 Sept. 1960
Gardener, Royal Horticultural Society gardens, Wisley, 1914; later plant recorder and lecturer in botany. Edited *Rhododendron Yb.*
J. R. Hort. Soc. 1960, 514–15 portr.

GOULD, William Buelow (*né* **Holland**) (*c.* 1804–1853)
b. Liverpool *c.* 1804 *d.* Hobart, Tasmania 11 Dec. 1853.

Studied at Royal Academy. Flower painter at Spode's pottery. Transported to Van Dieman's Land for theft, 1827. Drew Tasmanian plants.

H. Allport *Art in Tasmania* 1931, 20–25. *Papers Proc. R. Soc. Tasmania* 1959, 81–88 portr. W. Moore *Story of Austral. Art.* v.1, 1934, 31–32. *Austral. Encyclop.* v.4, 1965, 347–48.

Flower paintings at Queen Victoria Museum, Launceston. Portr. at Tasmanian Art Gallery.

GOULDING, Ernest (1872–1938)
d. Wood Green, London 15 Feb. 1938

BSc London 1898. FIC 1902. Imperial Institute, London, 1896–1935. Secretary to Advisory Committee on Vegetable Fibres, 1926. *Cotton and other Vegetable Fibres* 1917.

Kew Bull. 1938, 133–34. *Pharm. J.* 1938, 449.

GOULDING, Richard William (1868–1929)
b. Louth, Lincs 23 Nov. 1868 *d.* Louth 9 Nov. 1929

Librarian and private secretary to Duke of Portland at Welbeck Abbey. Antiquarian and naturalist. Wrote about early Louth botanists.

Bot. Soc. Exch. Club Br. Isl. Rep. 1929, 90–92.

GOURLAY, William Balfour- *see* Balfour-Gourlay, W.

GOURLIE, Robert (–1832)
d. Mendoza, Argentina 1832

Collected plants in Chile.

Bot. Misc. 1832–33, 208.

Gourliea Gillies.

GOURLIE, William (1815–1856)
b. Glasgow March 1815 *d.* Pollockshields, Glasgow 24 June 1856

FLS 1855. Pupil of W. J. Hooker and J. H. Balfour. Collected British plants, especially mosses, fossil plants. *Fl. Scotica Alpina* (exsiccatae).

Proc. Linn. Soc. 1856, xxvii–xxviii. *D.N.B.* v.22, 291.

Plants at Kew, Oxford. Letters at Kew. Sale of herb. by J. C. Stevens, 14 April 1858.

GOVAN, George (*fl.* 1820s–1830s)

MD. First Superintendent, Botanic Garden, Saharunpur, 1820–23. Correspondent of N. Wallich. First plant collector to penetrate Sabathu Hills near Simla in 1817. 'Natural History of Himalayan Mountains' (*Edinburgh J. Sci.* v.2, 1825, 17–38).

J. Bot. 1899, 462. *J. R. Central Asian Soc.* 1962, 47–57. I. H. Burkill *Chapters on Hist. Bot. in India* 1965, 33 *et seq.*

Govania Wall.

GOWEN, James Robert (–1862)

Of Highclere, Newbury. Secretary, Horticultural Society of London, 1845–50. Hybridised rhododendrons (*Bot. Register* 1831, t.1414, *Gdnrs Mag.* 1831, 62–63). 'Hybrid Amaryllis' (*Trans. Hort. Soc. London* 1822, 498–501; 1824, 361–63).

Bot. Cabinet 1831, t.1709. *Bot. Mag.* 1838, t.3676. *R.S.C.* v.2, 973.

Letters at Kew.

Govenia Lindl.

GOWER, William Hugh (1835–1894)
b. 6 Nov. 1835 *d.* Tooting, London 30 July 1894

Foreman, Kew Gardens until 1865. Employed by nursery firms of Jackson and Son, Kingston, Rollisson and B. S. Williams. *Orchids for Amateurs* (with J. Britten) *c.* 1879. Contrib. to *Garden* and *Orchid Rev.*

Garden v.46, 1894, 116. *J. Hort. Cottage Gdnr* v.29, 1894, 127. *Orchid Rev.* 1894, 258. *J. Kew Guild* 1896, 35 portr.

Herb. of garden plants at Kew, Edinburgh.

Coelogyne goweri Reichenb. f.

GRAEFER, John (–1802)
d. Brontë, Sicily 7 Aug. 1802

Of German origin? Came to England about middle of 18th century. Pupil of Philip Miller. Gardener to Earl of Coventry, then to James Vere at Kensington Gore. Joined nursery of Thompson and Gordon at Mile End, London about 1776. Laid out garden for Queen of Naples and Sicily at Caserta. Steward to Lord Nelson at Brontë in Sicily. *Descriptive Catalogue of...Herbaceous or Perennial Plants* 1789; ed. 4 1804.

J. C. Loudon *Encyclop. Gdning* 1822, 1279. G. W. Johnson *Hist. English Gdning* 1829, 248–49. *Gdn Hist. Soc. Newsletter* no. 16, 1972, 4–6.

GRAHAM, Alexander (–1878)

Herb. at Keighley Museum, Yorks.

GRAHAM, F. J. (–1902)
d. Cranford, Middx Feb. 1902

Fruit grower. Raised 'Graham's Yellow Perfection' wallflower.

Garden v.61, 1902, 116. *Gdnrs Mag.* 1902, 106.

GRAHAM, George John (1803–1878)
b. Brampton, Cumberland 1803 *d.* Ventnor, Isle of Wight 1 Jan. 1878

Collected plants in Mexico, 1827–29.

Bot. Register 1830, t.1370. G. Bentham *Plantae Hartwegianae* 1839–57, preface iv. *J. Bot.* 1905, 317–18.

Plants at BM(NH), Kew.

Salvia grahami Benth.

GRAHAM, Rev. Henry Longueville (1844–1921)
b. March 1844 *d.* Aug. 1921
Educ. Cambridge. Collected local plants while a pupil at Uppingham School. List of mosses in *Uppingham School Mag.* 1864.
A. R. Horwood and C. W. F. Noel *Fl. Leicestershire* 1933, cclxxviii.

GRAHAM, John (1805–1839)
b. Dumfriesshire 1805 *d.* Khandalla, Bombay 28 May 1839
To India 1828. Deputy post-master, Bombay. Superintendent, Botanic Garden, Bombay. *Catalogue of Plants growing in Bombay and its Vicinity* 1839 (posthumous biogr. iv).
W. Roxburgh *Fl. Indica* v.1, 1820, 52. *Gdnrs Chron.* 1841, 23. *J. Bot.* (Hooker) 1841, 300–1. A. Lasègue *Musée Botanique de B. Delessert* 1845, 433. *D.N.B.* v.22, 351. I. H. Burkill *Chapters on Hist. Bot. in India* 1965, 34 *et seq. R.S.C.* v.2, 977.

GRAHAM, Maria *see* Callcott, *Lady* M.

GRAHAM, Rev. Patrick (1756–1835)
d. Aberfoyle, Perthshire 4 Sept. 1835
DD. Minister of Aberfoyle. *Guide to Perthshire* 1810 (includes botany). Contrib. botanical lists to *Statistical Account of Scotland* v.17–18, 1796.
Proc. Perthshire Soc. Nat. Sci. v.4, 1907–8, cxc.

GRAHAM, Rex Alan Henry (*né* **Knowling**) (1915–1958)
b. Westminster, London 2 July 1915 *d* Richmond, Surrey 14 Dec. 1958
Educ. Oxford. Botanist in Kew Herb., 1956. Contrib. to *Fl. Tropical E. Africa* and *Watsonia.* Studied *Mentha.*
J. Kew Guild 1958, 594–95. *Proc. Bot. Soc. Br. Isl.* 1962, 505–7.
Plants at Oxford.

GRAHAM, Robert (1786–1845)
b. Stirling 7 Dec. 1786 *d.* Coldoch, Perthshire 7 Aug. 1845
MD Edinburgh 1808. FLS 1821. First Professor of Botany, Glasgow, 1818. Professor of Botany, Edinburgh, 1820–45. First President, Botanical Society of Edinburgh, 1836. Collected plants in Jersey, 1842, Ireland and Britain. Described N. Wallich's *Leguminosae.* Spent much time on 'Fl. Great Britain' which remained unfinished at his death. *Characters of Genera extracted from British Flora of W. J. Hooker* 1830.
W. J. Hooker *Exotic Fl.* v.3, 1827, t.189. *Edinburgh New Philos. J.* 1831, 1832. *Gdnrs Chron.* 1846, 390. *London J. Bot.* 1846, 11–12. *Phytologist* v.2, 1846, 572–73. *Proc. Linn. Soc.* 1846, 300–1. *Trans. Bot. Soc. Edinburgh* v.2, 1846, 59–63. C. Ransford *Biogr. Sketch of Robert Graham* 1846. *D.N.B.* v.22, 358. *Notes Edinburgh Bot. Gdn* v.3, 1904, 58. F. W. Oliver *Makers of Br. Bot.* 1913, 291–93. H. R. Fletcher and W. H. Brown *R. Bot. Gdn Edinburgh, 1670–1970* 99–112 portr. *R.S.C.* v.2, 977.
Letters at BM(NH), Kew. Crayon portr. by D. Macnee at Kew. Portr. at Hunt Library. Herb. and library sold by C. B. Tait and T. Nisbet, London, 6–7 April 1846.
Graemia Hook.

GRAHAM, Robert (*fl.* 1800s–1820s)
Nurseryman, 8 Camden Row, Camden Town, London.

GRAHAM, Robert James Douglas (1884–1950)
b. Perth 20 July 1884 *d.* St. Andrews, Fife 3 Sept. 1950
MA St. Andrews 1904. FRSE 1924. Economic botanist, Indian Agricultural Service, 1907; Pusa, 1908–17. Director of Agriculture, Mesopotamia, 1919–20. Lecturer in Botany, Edinburgh, 1921–34. Professor of Botany, St. Andrews, 1934. Secretary, Botanical Society of Edinburgh, 1930–41; President, 1942–43. 'Economic and Systematic Botany of Central Provinces of India' (DSc thesis 1917).
Gdnrs Chron. 1950 ii 154. *Nature* v.166, 1950, 976–77. *Yb. R. Soc. Edinburgh* 1951, 18–21.
Plants at Kew.

GRAHAM, S. J. (*c.* 1805–1882)
d. Cranford, Middx 17 Jan. 1882
Market gardener, Cranford.
Gdnrs Chron. 1882 i 122.

GRAHAM, Thomas (–1822)
d. on HMS 'Doris' off coast of Chile April 1822
Captain HM Packet Service. Brother of R. Graham (1786–1845) and husband of Maria Callcott (1785–1842). Brought plants to Edinburgh. Collected plants in Brazil. Described *Epidendrum ellipticum* (W. J. Hooker *Exotic Fl.* v.3, 1827, t.207).

GRAHAM, Thomas (1805–1869)
b. Glasgow 20 Dec. 1805 *d.* London 16 Sept. 1869
MA Glasgow 1826. DCL Oxon 1855. FRS 1836. Professor of Chemistry, Glasgow, 1830–37; University College, London, 1837–55. *Outlines of Botany* 1841; ed. 2 1848.
Archiv Pharmacie 1870, 85. L. Reeve *Photogr. Portr. of Men of Eminence in Literature, Sci. Art* v.3, *c.* 1865, 1–6 portr. *Proc. R. Soc.* v.18, 1870, xvii–xxvi. *Proc. Philos. Soc. Glasgow* 1883–84, 260–357; 1970, 116–27 portr.; 1972, 30–42. *D.N.B.* v.22, 361–63.

GRAHAM, Thomas (*fl.* 1820s)
Assistant Superintendent, Sydney Botanic Gardens until 31 March 1829 during the last years of Charles Fraser.
J. Proc. R. Austral. Hist. Soc. 1932, 108–9.

GRAHAM, Mrs. W. J. (*fl.* 1900s)
Contrib. to F. H. Davey's *Fl. Cornwall* 1909 lx.

GRAHAME, Charles J. (–1902)
d. Surbiton, Surrey 26 May 1902
Stock-broker. Rosarian.
Gdnrs Mag. 1902, 341.

GRAINGER, Jennie (*c.* 1891–1969)
d. Meltham, Yorks 15 May 1969
Schoolmistress, Helme, Yorks. Mycologist.
Naturalist 1969, 106.

GRAINGER, Rev. John (1830–1891)
b. Belfast 1830 *d.* Broughshane, County Antrim 1891
Rector, Skerry and Rathcavan. Had herb. Assisted in compilation of *Fl. Belfastiensis.*
Annual Rep. Proc. Belfast Nat. Field Club 1911–12, 625. *Irish Nat.* 1913, 30.

GRANGE, James (*fl.* 1810s–1820s)
Nurseryman, Kingsland, Dalston, London.

GRANT, Dr. (*fl.* 1900s)
Physician in Orkney. "A most enthusiastic botanist." Contrib. to M. Spence's *Fl. Orcadensis* 1914, li.

GRANT, Alexander (1848–1906)
b. Cullen, Banffshire 1848 *d.* Sydney 25 Dec. 1906
Employed in Royal Botanic Garden, Edinburgh. To Sydney, 1878. At Royal Botanic Garden, Sydney from 1882. Mycologist, Dept. of Agriculture. Vice-President, Horticultural Association of New South Wales.
Rep. Sydney Bot. Gdn 1906, 11. *Proc. Linn. Soc. N.S.W.* 1907, 5. *J. Proc. R. Austral. Hist. Soc.* 1932, 132.

GRANT, James Augustus (1827–1892)
b. Nairn 11 April 1827 *d.* Nairn 10 Feb. 1892
CB 1866. FRS 1873. FLS 1871. Lieut.-Colonel. African explorer. Accompanied J. H. Speke on exploration from Ukuni to Karague, 1861, and from Uganda to falls of Karuma, Faloro and Gondokoro, 1862–63. Collected plants which were listed in Speke's *Journal of Discovery of Source of Nile* 1863, 625–58 and in *Trans. Linn. Soc.* v.29, 1872–75, 1–190. *A Walk across Africa* 1864.
Blackwood's Mag. 1892, 573–81. *J. Bot.* 1892, 96. *Nature* v.45, 1892, 376. *Proc. Linn. Soc.* 1891–92, 68–69. *Proc. R. Soc.* v.50, 1892, xiv–xv. *D.N.B. Supplt 1* v.2, 339–41. *Uganda J.* 1953, 146–60. *Comptes Rendus AETFAT 1960* 1962, 207–8. R. Hill *Biogr. Dict. Sudan* 1967, 142. *R.S.C.* v.7, 816; v.10, 45.
Letters, MSS., sketches and plants at Kew.
Anthericum grantii Baker.

GRANT, James F. (–1930)
d. Inverurie, Aberdeen March 1930
MA Edinburgh. Editor, *Elgin Courant and Courier*. Studied flora of Caithness.
Gdnrs Chron. 1930 i 257.
Carex grantii Benn.

GRANT, James William (1788–1865)
b. 12 Aug. 1788 *d.* 17 Sept. 1865
In employ of East India Company, 1805–49. For a very short period in charge of Calcutta Botanical Garden. "One of the best microscopical observers of the present day" (J. O. Voight).
C. E. Buckland *Dict. Indian Biogr.* 1906, 176.
Plants at Kew.

GRANT, Mark Ogilvie- *see* Ogilvie-Grant, M.

GRANT, William (*fl.* 1820s)
Seedsman, victualler and spirit dealer, 25 High Street, Sheffield, Yorks.

GRANT DUFF, Sir Mountstuart Elphinstone *see* Duff, Sir M. E. G.

GRANVILLE, Castalia Rosalind, Countess of (*fl.* 1860s–1880s)
Large collection of her plants of Surrey and Walmer, Kent given to G. C. Druce by her daughter, Lady Victoria Russell.
G. C. Druce *Comital Fl. Br. Isl.* 1932, 173.
Plants at Oxford.

GRANVILLE, Mary *see* Delany, M.

GRATRIX, Samuel (–1929)
d. Manchester 1929
Amateur orchid grower, Manchester.
Orchid Rev. 1927, 84; 1929, 78.

GRAVES, George (1784–*c.* 1839)
b. Newington Butts, Surrey 23 May 1784 *d. c.* 2 June 1839
FLS 1812. Botanical colourist. Of Walworth, Peckham and Edinburgh. Had herb. Part owner and editor of *Fl. Londinensis* ed. 2, vols 1–3, 1817–26. *Monograph of British Grasses* 1822 (with plates reduced from *Fl. Londinensis*). *Hortus Medicus* 1834.
J. Smith *Cat. of Friends' Books* v.1, 1867, 862. *J. Bot.* 1916, 154. *Watsonia* 1951–53, 93–99.
Letters at Kew and in Winch correspondence at Linnean Society.

GRAVES, William (*c.* 1754–*post* 1827)
Of Newington Butts. Colourist of *Bot. Mag.* from its inception in 1787 until at least 1809, at which time he was credited with colouring more natural history plates than any other contemporary artist. Father of G. Graves (1784–*c.* 1839).
Watsonia 1951–53, 94–95.

GRAY, Alex. Hill (*fl.* 1850s–1900s)
Of Newbridge near Bath, Somerset. Rosarian; specialised in tea roses.
Gdnrs Mag. 1907, 707–8 portr.

GRAY, Christopher (1693/4–1764)
b. Jan. 1693/4 *d.* Fulham, London Nov. 1764
Nurseryman, King's Road, Fulham. His nursery was on site of garden of Mark Catesby whose *Hortus Britanno-Americanus* he published in 1763. Specialised in American plants. Bought part of Bishop Compton's collection of plants from his successor, *c.* 1713–14. *Catalogue of Trees, Shrubs, Plants and Flowers*...1755. On his death his nursery was acquired by William Burchell (*c.* 1725–1800).
G. W. Johnson *Hist. English Gdning* 1829, 202. J. C. Loudon *Arboretum et Fruticetum Britannicum* v.1, 1838, 76. C. S. Sargent *Silva of North America* v.4, 1892, 76. E. J. Willson *Hist. of Fulham to 1965* 1970, 239–40. J. Harvey *Early Hort. Cat.* 1973, 10.

GRAY, Edward Whitaker (1748–1806)
b. 21 March 1748 *d.* London 27 Dec. 1806
MD. FRS 1779. ALS 1788. Son of Samuel Gray (1694–1766). Librarian, Royal College of Physicians before 1773. Keeper of Natural History and Antiquities at BM, 1787–1806. Secretary, Royal Society, 1797. Sent plants from Oporto, Portugal to J. Banks.
J. Sowerby and J. E. Smith *English Bot.* 1631. *Gent. Mag.* 1807 i 90. W. Munk *Roll of R. College of Physicians* v.2, 1878, 298. *D.N.B.* v.23, 7.
MSS. and portr. by A. Callcott at Royal Society. Portr. at Hunt Library. Plants at BM(NH).

GRAY, George (1758–1819)
b. Newcastle, Northumberland 1758 *d.* Newcastle 9 Dec. 1819
Painter of fruit. "He was accounted one of the best botanists and chemists in this part of the country" (Bewick). To N. America on botanical expedition, 1787.
T. Bewick *Memoir* 1862, 63–64. *D.N.B.* v.23, 7.

GRAY, James (*fl.* 1780s–1810s)
On death of his partner, John Jeffreys, J. Gray assumed control of Brompton Park nursery in 1789. Thomas Wear and Robert Gray later became partners.
J. Harvey *Early Hort. Cat.* 1973, 14.

GRAY, James (1810–1883)
b. East Lothian 1810 *d.* Chelsea, London 24 Nov. 1883
Gardener. From 1846 in partnership with Brown and Henry Ormson as a builder of conservatories and plant houses.
Gdnrs Chron. 1883 ii 700–1; 1884 i 184 portr.

GRAY, John (*c.* 1835–1895)
d. St. Lucia 11 Jan. 1895
Gardener to Earl Brownlow and later to General Talbot at Worthy Park, Jamaica. First Curator, Botanic Station, St. Lucia, 1886–94.
Kew Bull. 1895, 39–40. *Kew Bull. Additional Ser. I* 1898, 83.
Plants at Kew.

GRAY, Rev. John Durbin (*c.* 1845–1925)
d. St. Leonards, Sussex 27 March 1925
MA Cantab. Vicar of Nayland, Suffolk, 1879–1909. Had herb. of British plants, now at Cambridge.
J. Bot. 1925, 182.

GRAY, John Edward (1800–1875)
b. Walsall, Staffs 12 Feb. 1800 *d.* London 7 March 1875
PhD Munich 1852. FRS 1832. FLS 1857. Son of S. F. Gray (1766–1828). Algologist and hepaticologist. Assistant, Zoology Dept., BM, 1824; Keeper, 1840–75. President, Botanical Society of London, 1836–59. *Natural Arrangement of British Plants* 1821 2 vols. *Handbook of British Waterweeds* 1864.
Ipswich Mus. Portr. Ser. 1852. *J. Bot.* 1865, 297–302; 1872, 374–75; 1875, 127; 1894, 96. *Trans. Bot. Soc. Edinburgh* v.10, 1869, 305–9; v.12, 1875, 409–10. J. Saunders *List of Books by J. E. Gray* 1872. *Ann. Mag. Nat. Hist.* 1875, 1–4, 281–85. *Gdnrs Chron.* 1875 i 334–35. *Proc. Linn. Soc.* 1874–75, xliii–xlvii. *Kew Bull.* 1894, 76–78. *Mem. Soc. Sci. Nat. Cherbourg* v.29, 1892–95, 1–36. *D.N.B.* v.23, 9–10. *J. Soc. Bibl. Nat. Hist.* 1974, 35–76 portr. A. E. Gunther *Century of Zoology in British Museum, 1815–1914* 1975. *R.S.C.* v.2, 998; v.7, 819; v.10, 49.
Herb. and MS. autobiogr. at BM(NH). Algae at Cambridge (*J. Bot.* 1891, 191). Letters and portr. by H. Phillips at Kew. Letters at American Philosophical Society. Portr. at Hunt Library.

GRAY, Maria Emma (*née* **Smith**) (1787–1876)
b. Greenwich Hospital, London 1787 *d.* London 9 Dec. 1876
Wife of J. E. Gray (1800–1875). Conchologist and algologist.
J. Bot. 1866, 45; 1877, 32; 1891, 191;

1892, 52. S. O. Gray *British Seaweeds* 1867, viii–ix. *D.N.B.* v.23, 11.
Algae at Cambridge.
Grayemma J. E. Gray.

GRAY, Peter (1818–1899)
b. Dumfries 18 Oct. 1818 *d.* Locharbriggs, Dumfries 3 June 1899
Journalist. *Lichens and Mosses* 1886. *Nithsdale Illustrated* 1894. Contrib. to *Phytologist* v.1, 1844; v.3, 1848–50; and to H. C. Watson *Topographical Bot.* 1883.
J. Bot. 1899, 336. *R.S.C.* v.15, 433.

GRAY, Robert (*fl.* 1760s–1840s)
Nurseryman, Brompton Park, Kensington, London. Brother of James Gray in whose business he was a partner after 1790. *Pigot's Directory* 1822 also records a Robert Gray, seedsman, at 39 Holborn Hill, London.

GRAY, Robert (*fl.* 1850s)
Of Alphington, Devon. Collected ferns.
Br. Fern Gaz. 1909, 44–45.

GRAY, Samuel (1694–1766)
Son of Samuel Gray senior whom he succeeded in the seedsman business at the Black Boy, Pall Mall, London.
J. Bot. 1872, 375. *Proc. Linn. Soc.* 1874–75, xliii.

GRAY, Samuel (1739–1771)
Son of S. Gray (1694–1766). Seedsman, Pall Mall. Translated parts of Linnaeus's *Philosophia Botanica* for J. Lee's *Introduction to Botany* 1760.

GRAY, Samuel Forfeit (1798–1872)
b. Walsall, Staffs 1798 *d.* Fulham, London 21 March 1872
FLS 1825. Eldest son of S. F. Gray (1766–1828). Contrib. to *Gdnrs Chron.*
Gdnrs Chron. 1872, 430.

GRAY, Samuel Frederick (1766–1828)
b. London 10 Dec.1766 *d.* Chelsea, London 2 April 1828
Son of S. Gray (1739–1771) and father of J. E. Gray (1800–1875). Druggist, Walsall, 1797–1800. Afterwards lectured on botany in London. *Supplement to Pharmacopoeia* 1818. *Natural Arrangement of British Plants* 1821 (mainly the work of J. E. Gray).
Athenaeum 1863, 368. J. Saunders *List of Books by J. E. Gray* 1872, 3. *Mem. Soc. Sci. Nat. Cherbourg* v.29, 1892–95, 1–36. *Bot. Notiser* 1893, 137–51. *J. Bot.* 1894, 96; 1922, 177. *Kew Bull.* 1894, 76–78. *D.N.B.* v.23, 20. *Chemist Druggist* v.58, 1901, 137. *Mycologia* 1941, 568–70; 1951, 376–78. *R.S.C.* v.2, 1012.
Portr. at Kew, Hunt Library.

GRAY, Samuel Octavus (1828–1902)
b. London 14 Oct. 1828 *d.* Rudgwick, Sussex 15 May 1902
Son of S. F. Gray (1798–1872). *British Seaweeds* 1867.
J. Bot. 1894, 96.
Algae at Manchester.

GRAY, William (–1745)
d. July 1745
Nurseryman, Parsons Green Lane, Fulham, Middx.
E. J. Willson *Hist. Fulham* 1970, 239–40.

GREATA, Louis Augustin (1857–1911)
b. London 13 March 1857 *d.* Los Angeles, U.S.A. 1 May 1911
Educ. in France and New York. Secretary, Pacific Coast Hardware and Metal Association. Collected plants in S. California, 1900–7, now in Parish herb.
Bull. S. California Acad. Sci. v.10, 1911, 67.
Salvia greatae Brandegee.

GREATOREX, Samuel (1804–1871)
Solicitor's clerk, Leicester. At Knighton, Leicester *c.* 1857 raised 'Annie Elizabeth' apple named after his daughter.
A. Simmonds *Hort. Who was Who* 1948, 24–25.

GREEN, Charles (–1886)
d. Reigate, Surrey Nov. 1886
Gardener to W. Borrer at Henfield, Sussex, to W. Saunders at Reigate, Surrey and to Sir G. Macleay at Bletchingley, Surrey.
Garden v.30, 1886, 530, 554.

GREEN, Charles Baylis (–1918)
d. Swanage, Dorset 6 Oct. 1918
Railway employee at Euston. Moved to Swanage, 1910. Pteridologist. Collected Middlesex and Dorset plants.
Br. Fern Gaz. 1911, 189. *Bot. Soc. Exch. Club Br. Isl. Rep.* 1918, 355–56, 418. *Proc. Bournemouth Nat. Sci. Soc.* v.10, 1917–18, 25–26 portr.
Plants at Bournemouth Nat. Sci. Society, South London Botanical Institute, Oxford.

GREEN, Conrad Theodore (1863–1940)
b. Kirkburton, Yorks 18 Nov. 1863 *d.* Birkenhead, Cheshire 17 April 1940
MB London 1887. MRCS. LRCP. FLS 1901–17, 1928. Physician, Birkenhead. President, Liverpool Botanical Society, 1934. Photographed plants. *Fl. Liverpool District* 1902; ed. 2 1933. *Preliminary Index of Local Fungi, mainly from Wirral* 1901–2.
Liverpool Public Museums *Handbook and Guide to Herb. Collections* 1935, 36. *N. Western Nat.* 1940, 170–72 portr. *Nature* v.145, 1940, 770. *Proc. Linn. Soc.* 1939–40, 364–66. *Proc. Liverpool Nat. Field Club* 1940, 8.
Plants at Birkenhead and Liverpool Museums. Photographic slides at BM(NH) and Liverpool Museums.

GREEN, Cyril (-1917)
d. Palestine Nov. 1917
BSc Aberystwyth 1911. In Dept. of Botany, National Museum of Wales 1914. Plant ecologist.
Nature v.100, 1917, 248.

GREEN, Donald Edwin (*c.* 1899–1968)
b. Caerleon, Mon *c.* 1899 *d.* Guildford, Surrey 31 Dec. 1968
MSc Bangor 1924. VMM 1961. Plant pathologist, Dept. Agriculture, Leeds University. Mycologist, Royal Horticultural Society gardens, Wisley, 1928–64. *Diseases of Vegetables* 1942.
Gdnrs Chron. 1939 i 160 portr.; 1954 ii 74 portr; 1969 i 35 portr. *J. R. Hort. Soc.* 1969, 138–39. *Times* 4 Jan. 1969.

GREEN, G. H. (1837–1918)
d. Stourbridge, Worcs 1918
Gardener to Countess of Stamford at Enville Hall, Worcs, 1861–1905.
Gdnrs Chron. 1918 i 162.

GREEN, Harold (1887–1941)
b. 28 June 1887 *d.* Basingstoke, Hants 14 Jan. 1941
Kew gardener, 1908. Assistant Superintendent, Botanic Garden, Hong Kong, 1911; Superintendent, 1920–22.
Kew Bull. 1911, 118. *J. Kew Guild* 1941, 87–88 portr.

GREEN, Henry Frederick (1868–*c.* 1945)
b. Ceylon 1868
Kew gardener, 1888–90. Curator, Botanic Station, Dominica, 1890–92. Superintendent, Agricultural Training School, Dominica, 1892–96. Surveyor for Gambia Development Syndicate, 1898–99. Gardener, Botanic Garden, Sibpur, 1900. At Government Cinchona Plantations, Munsong, Bengal. Left India *c.* 1919. Retired to Guernsey.
Kew Bull. Additional Ser. I 1898, 103–4. *J. Kew Guild* 1945, 468–69.

GREEN, Horace Edgar (1886–1973)
d. 15 Sept. 1973
Insurance official. Joined Liverpool Botanical Society, 1923; later became Secretary and President. Botanised in Merseyside and Deeside.
Watsonia 1974, 202.
Herb. at Liverpool Museums.

GREEN, Joseph Reynolds (1848–1914)
b. Stowmarket, Suffolk 3 Dec. 1848 *d.* Cambridge 3 June 1914
MA Cantab 1888. DSc 1894. FRS 1895. FLS 1889. Professor of Botany, Pharmaceutical Society, 1887–1907. Hartley Lecturer in Vegetable Physiology, Liverpool University, 1907. Edited R. Bentley's *Manual of Botany* 1895–96 2 vols; ed. 2 1897–1902. *Introduction to Vegetable Physiology* 1900. *History of Botany, 1860–1900* 1909. *History of Botany in United Kingdom* 1914 portr.
R.S.C. v.10, 52; v.15, 438.
Bot. Soc. Exch. Club Br. Isl. Rep. 1914, 52–53. *J. Bot.* 1914, 223. *Kew Bull.* 1914, 192–93. *Nature* v.93, 1914, 379–80. *Pharm. J.* v.38, 1914, 838 portr. *Who was Who, 1897–1916* 293–94.

GREEN, Rev. Thomas (-1788)
d. 7 June 1788
Professor of Botany and Woodwardian Professor of Fossils, Cambridge.
Gdnrs Chron. 1919 i 147.

GREEN, Rev. William (*fl.* 1720s–1740s)
Friend of S. Brewer and W. Jones.
H. N. Clokie *Account of Herb. Dept. Bot., Oxford* 1964, 173.
Welsh plants at Oxford.

GREENAWAY, William (*c.* 1834–1911)
d. Oxford May 1911
Secretary, Oxfordshire Chrysanthemum Society. Contrib. to *Gdnrs Mag.*
Gdnrs Mag. 1911, 352.

GREENER, John (*fl.* 1770s–1800s)
Nurseryman, Workington, Cumberland.
E. Hughes *North Country Life in 18th Century* v.2, 1965, 236.

GREENER, Peter (*fl.* 1820s–1830s)
Nurseryman, Workington, Cumberland.

GREENFIELD, Percy (1880–1970)
d. 8 Dec. 1970
Official in General Post Office. Secretary, British Pteridological Society, 1937–48.
Br. Fern Gaz. 1971, 221–22 portr.

GREENING (*alias* Gott), Sir Henry Thomas (*fl.* 1750s–1809)
Son of Thomas Greening (1684–1757). Gardener to King George III at Windsor, Berks. Changed name to Gott, 1769. Knighted 1774.

GREENING, John (-1770)
d. May 1770
Son of Thomas Greening (1684–1757). Gardener to King George II at Hampton Court Palace.

GREENING, Robert (-1758)
d. March 1758
Son of Thomas Greening (1684–1757). Gardener to Prince and Princess of Wales at Kew, Richmond, Surrey.

GREENING, Thomas (1684–1757)
Gardener to King George II.

GREENISH, Henry George (1855–1933)
b. London 1855 *d.* London 2 Aug. 1933
FLS 1884. Professor of Materia Medica, Pharmaceutical Society, 1890–1933. *Textbook of Materia Medica* 1899; ed. 6 1933.

Anatomical Atlas of Vegetable Powders (with E. Collin) 1904.
Kew Bull. 1933, 412–13. *Times* 3 Aug. 1933. *Proc. Linn. Soc.* 1933–34, 158–59.

GREENLEES, Thomas (1878–1950)
Herb. at Bolton Museum, Lancs.

GREENSHIELDS, Gavin (1822–1890)
b. Broughton, Peebleshire 1822 *d.* 28 March 1890
Blacksmith and later grocer. Raised pentstemons.
J. Hort. Cottage Gdnr v.24, 1892, 46–47.

GREENSHIELDS, John (*c.* 1802–1888)
b. Lanarkshire *c.* 1802 *d.* Sarsden, Oxford 12 May 1888
Gardener to Lord Ducie at Sarsden, 1834.
Gdnrs Chron. 1888 i 632.

GREENWAY, James (*c.* 1703–1794)
d. Dinwiddie County, Virginia 1794
MD. Had herb. of plants from Virginia and N. Carolina. "Misit ille aba. 1773–1775 plantas Virginicas siccatas, vivas a se collectas, ad 400, eo fine ad me, ut novam Floram Virginicam juncto labore concinnaremus, sed bello inter Anglos & colonias orto, conatus omnis profligatus est, ita ut nesciam num vivus adhuc supersit nec-ne?"
P. D. Giseke *Praelectiones* 1792, 226. *Rhodora* 1943, 301–3.

GREENWOOD, Alfred (1821–1862)
b. Springfield, Essex 8 March 1821
Of Chelmsford. At Penzance from 1845. 'Mosses of Chelmsford' (*Phytologist* v.2, 1846, 384–89). 'Mosses of Penzance' (*Penzance Nat. Hist. Soc. Rep.* 1846, 60–68).
Essex Nat. 1898, 336. *R.S.C.* v.3, 5.

GREENWOOD, W. J. H. (*fl.* 1890s–1910s)
Chemist at Cirencester, Glos. Discovered *Carex tomentosa* in Cirencester Park. *Fl. Cirencester and its Neighbourhood* 1914.
H. J. Riddelsdell *Fl. Gloucestershire* 1948, cxl.
Herb. at Cheltenham Museum.

GREGG, John (–1795)
FRS 1772. Of Charlestown, S. Carolina. Collected plants in West Indies, 1761–77, for Lord Hillsborough and J. Ellis.
J. E. Smith *Selection from Correspondence of Linnaeus* v.1, 1821, 189, 503, 509. I. Urban *Symbolae Antillanae* v.3, 1902, 53.
Greggia Solander.

GREGG, Mrs. Mary *see* Kirby, M.

GREGOR, Rev. Arthur George (1867–1954)
b. Retford, Notts 25 May 1867 *d.* West Worthing, Sussex 9 Nov. 1954
MA Durham. FLS 1948. Vicar of Firle, Sussex. Contrib. to A. H. Wolley-Dod's *Fl. Sussex* 1937.
Proc. Linn. Soc. 1953–54, 44–45. *Proc. Bot. Soc. Br. Isl.* 1955, 552–53.
Herb. at Kew.

GREGORSON, David (1836/7–1916)
b. Stewarton, Ayrshire 1836/7 *d.* California, U.S.A. 3 April 1916
Schoolmaster at Greenock and Kilsyth. To California, 1888. Inspector of Orchards. Papers on algae of Arran in *Proc. Trans. Nat. Hist. Soc. Glasgow* 1883–86.
Glasgow Nat. 1919, 77–80. *R.S.C.* v.15, 447.

GREGORY, Abraham (*fl.* 1780s)
Nurseryman, Faversham, Kent.

GREGORY, Sir Augustus Charles (1819–1905)
b. Farnsfield, Notts 1 Aug. 1819 *d.* Brisbane, Australia, 25 June 1905
To W. Australia with his parents in 1829. Survey Dept., W. Australia, 1841–54. Explored N. Territory and sought the lost explorer, F. W. L. Leichhardt, 1855–56. Renewed search for Leichhardt, 1858. Surveyor-General, Queensland, 1859–75. *Journals of Australian Exploration* (with F. T. Gregory) 1884.
J. D. Hooker *Fl. Tasmaniae* v.1, 1859, cxxiii. J. E. T. Woods *Hist. Discovery and Exploration of Australia* v.2, 1865, 280–91. P. Mennell *Dict. Austral. Biogr.* 1892, 196–97. E. Favenc *Explorers of Australia* 1908 *passim. D.N.B. 1901–1911* 160–61. *Austral. Encyclop.* v.3, 1962, 436–38; v.4, 384–85.
Plants at Melbourne. MSS. at Mitchell Library, Sydney. Portr. at Hunt Library.
Adonsonia gregorii F. Muell.

GREGORY, Eliza Standerwick (*née* Barnes) (1840–1932)
b. Thrapston, Northants 6 Dec. 1840 *d.* Weston-super-Mare, Somerset 22 March 1932
Studied *Viola. British Violets* 1912.
Bot. Soc. Exch. Club Br. Isl. Rep. 1932, 82. *J. Bot.* 1932, 144–45. *Watson Bot. Exch. Club Rep.* 1931–32, 102.
Herb. at BM(NH).

GREGORY, Francis Thomas (1821–1888)
b. Farnsfield, Notts 19 Oct. 1821 *d.* Toowoomba, Queensland 24 Oct. 1888
Younger brother of Sir A. C. Gregory (1819–1905). To W. Australia with his parents in 1829. Staff Surveyor, 1847. Explored W. Australia, 1846–61. Collected plants. Commissioner of Crown Lands, Queensland. *Journals of Australian Exploration* (with A. C. Gregory) 1884.

W. Howitt *Hist. Discovery of Australia, Tasmania and New Zealand* v.2, 1865, 343–51. J. E. T. Woods *Hist. Discovery and Exploration of Australia* v.2, 1865, 409–32. *Proc. R. Soc. Queensland* 1890–91, 29–30. E. Favenc *Explorers of Australia* 1908, 253–63. *Austral. Encyclop.* v.3, 1963, 449; v.4, 385.

MSS. at Mitchell Library, Sydney. Portr. at Hunt Library.

GREGORY, Frederick Gugenheim (1893–1961)
b. London 22 Dec. 1893 *d.* 27 Nov. 1961

FRS 1940. Assistant, Institute of Plant Physiology, Imperial College, London; Assistant Director, 1932; later Director and Professor, 1937–58. Researched on plant growth analysis and vernalization. Contrib. to *Ann. Bot.*

Times 30 Nov. 1961. *Nature* v.193, 1962, 118. *Plant Physiol.* 1962, 450. *Biogr. Mem. Fellows R. Soc.* v.9, 1963, 131–53. *Who was Who, 1961–1970* 457. C. C. Gillispie *Dict. Sci. Biogr.* v.5, 1972, 523–24.

GREGORY, John Walter (1864–1932)
b. Chelmsford, Essex 27 Jan. 1864 *d.* near Megantoni Falls, Peru 2 June 1932

Assistant, BM(NH), 1887–1900. Professor of Geology, Melbourne, 1900. Professor of Geology, Glasgow, 1905–1929. Collected plants for BM in Kenya, 1892–93 (*J. Linn. Soc.* 1895, 373–435). *Great Rift Valley* 1896.

Nature v.129, 1932, 930–31.

Plants from Tropical Africa and West Indies at BM(NH).

GREGORY, Reginald Philip (1879–1918)
b. Trowbridge, Wilts 7 June 1879 *d.* Cambridge 24 Nov. 1918

BA Cantab 1901. Son of Mrs. E. S. Gregory (1840–1932). Demonstrator in Botany, Cambridge, 1902–7; Lecturer, 1907. Geneticist. Contrib. papers on *Primula* to *J. Genetics B* 1911 and *Proc. R. Soc.* 1914.

Gdnrs Chron. 1918 ii 232. *Nature* v.102, 1918, 247–48, 284. *J. Bot.* 1919, 47.

GREGORY, William (1803–1858)
b. Edinburgh 25 Dec. 1803 *d.* Edinburgh 24 April 1851

MD Edinburgh 1828. Professor of Medicine and Chemistry, King's College, Aberdeen, 1839. Professor of Chemistry, Edinburgh, 1844–58. 'Marine Diatomaceae' (*Trans. R. Soc. Edinburgh* v.21, 1857, 473–542).

Trans. Bot. Soc. Edinburgh v.6, 1858, 75–79. *D.N.B.* v.23, 105. *R.S.C.* v.3, 9.

Diatoms at BM(NH). Letters in Walker-Arnott correspondence at BM(NH).

GREGORY, William (*fl.* 1830s–1840s)
Nurseryman, Cirencester, Glos.

GREGORY, William (*fl.* 1850s–1890s)
Interpreter, British Consular Service in China, 1854–90. Collected plants for H. F. Hance.

E. Bretschneider *Hist. European Bot. Discoveries in China* 1898, 534–35.

Plants at Kew.

Ixeris gregorii Hance.

GREGSON, Jesse (1837–1919)
b. Kent 4 Aug. 1837 *d.* Leura, N.S.W. 3 Aug. 1919

Arrived in Sydney, N.S.W., Jan. 1856. Farmed in Queensland. Assistant Superintendent, later Superintendent, Australian Agricultural Company, 1875–1905. Collected plants in Mount Wilson area of N.S.W.

Sydney Morning Herald 4 Aug. 1919. C. H. Currey *Mount Wilson, N.S.W.* 1968, 60–68.

Plants at Sydney. MSS. and letters at Mitchell Library, Sydney.

GRENFELL, Francis Wallace 1st Baron Grenfell (1841–1925)
b. 29 April 1841 *d.* Windlesham, Surrey 27 Jan. 1925

Field Marshal, 1908. President, Royal Horticultural Society, 1913–19. R.H.S. instituted Grenfell Medal in 1919 in commemoration of his services.

Garden 1915, iv portr. *Who was Who, 1916–1928* 436. *D.N.B. 1922–1930* 362–64. H. R. Fletcher *Story of R. Hort. Soc., 1804–1968* 1969, 266–67 portr.

GREVILLE, Hon. Charles Francis (1749–1809)
b. 12 May 1749 *d.* 23 April 1809

FRS 1772. FLS 1802. Of Paddington. One of the founders of Horticultural Society of London. Introduced and grew rare plants.

R. Brown *Prodromus Florae Novae Hollandiae* 1810, 375–80. E. Bretschneider *Hist. European Bot. Discoveries in China* 1898, 210–11. *Gdnrs Chron.* 1904 i 148. *J. R. Hort. Soc.* 1942, 219–32 portr. H. R. Fletcher *Story of R. Hort. Soc., 1804–1968* 1969, 30–33 portr.

Portr. at Hunt Library.

GREVILLE, Robert Kaye (1794–1866)
b. Bishop Auckland, Durham 13 Dec. 1794 *d.* Murrayfield, Edinburgh 4 June 1866

LLD Glasgow 1824. FRSE 1821. FLS 1827. *Scottish Cryptogamic Fl.* 1823–28 3 vols. *Fl. Edinensis* 1824. *Icones Filicum* (with W. J. Hooker) 1828–31 2 vols. *Algae Britannicae* 1830.

J. Sowerby and J. E. Smith *English Bot.* 2666, 2717. N. Wallich *Plantae Asiaticae Rariores* v.3, 1832, 5. J. G. Agardh *Species Genera et Ordines Algarum* v.1, 1848, xxxvi. *Gdnrs Chron.* 1866, 538. *J. Bot.* 1866, 238. *Trans. Bot. Soc. Edinburgh* v.8, 1866, 463–76. *D.N.B.* v.23, 164–66. *Meded. Bot. Mus. Herb. Rijksuniv. Utrecht* no. 283, 1968, 124–35. *R.S.C.* v.3, 12; v.7, 836.

Plants at BM(NH), Edinburgh, Glasgow. Letters at Kew, BM(NH). Crayon portr. by D. Macnee at Kew. Portr. at Hunt Library.

Grevillea R. Br. *Kayea* Wall.

GREW, Nehemiah (1641–1712)
b. Atherstone or Mancetter, Warwick 1641 *d.* London 25 March 1712

BA Cantab 1661. MD Leyden 1671. FRS 1671. Secretary, Royal Society, 1677–79. A pioneer of plant anatomy. *Anatomy of Vegetables Begun* 1672. *Idea of Phytological History Propounded* 1673. *Anatomy of Plants* 1682.

Gent. Mag. 1788 ii 1067. R. Pulteney *Hist. Biogr. Sketches of Progress of Bot. in England* v.1, 1790, 337–38. W. Munk *Roll of R. College of Physicians* v.1, 1878, 406–9. *D.N.B.* v.23, 166–68. *J. Bot.* 1902, 198–200. *J. R. Microsc. Soc.* 1902, 129–41. E. J. L. Scott *Index to Sloane Manuscripts* 1904, 228. F. W. Oliver *Makers of Br. Bot.* 1913, 44–64 portr. *Publ. Colonial Soc. Massachusetts* 1913, 142–86. J. R. Green *Hist. Bot.* 1914, 123–38. E. A. Johnson *Predecessors of Adam Smith* 1937, 115–38. *Chronica Botanica* v.6, 1941, 391–92; 1944, 73–84. *Nature* v.147, 1941, 630–32. *Proc. Linn. Soc.* v.153, 1940–41, 218–28; 1941–42, 166. *Isis* v.34, 1942, 7–16; v.51, 1960, 3–8. *Science* v.98, 1943, 13–14. J. S. L. Gilmour *Br. Botanists* 1944, 20–22. *Proc. Amer. Philos. Soc.* v.115, 1971, 502–5. C. C. Gillispie *Dict. Sci. Biogr.* v.5, 1972, 534–36.

Botanical MSS. at Royal Society.

Grewia L.

GREY, Charles Harvey (*né* Hoare) (1875–1955)
d. Malton, Yorks 24 July 1955

Educ. Oxford. FLS 1935. Adopted name of Grey in 1927. Banker. Founded Northern Horticultural Society's garden at Harlow Car near Harrogate. *Hardy Bulbs* 1938 3 vols.

Gdnrs Chron. 1955 ii 60. *Iris Yb.* 1955, 13–14. *Proc. Linn. Soc.* 1954–55, 135–36. *Times* 25 July 1955.

GREY, Eliza Lucy (*née* Spencer) (–1898)
d. London Sept. 1898

Collected plants in S. Australia, 1841–45, which she sent to Robert Brown at BM.

Letters in Brown correspondence at BM(NH).

GREY, Sir George (1812–1898)
b. Lisbon 12 April 1812 *d.* London 20 Sept. 1898

KCB. 83rd Foot, 1829–39. Explored west coast of Australia, 1837–39. Governor, S. Australia, 1841–45; New Zealand, 1845–53, 1861–67, 1875–84; Cape Colony, 1853–61. *Journals of Two Expeditions of Discovery in North-west and Western Australia... 1837–1839* 1841 2 vols (contains references to plants).

J. D. Hooker *Fl. Tasmaniae* v.1, 1859, cxix, cxx. *Proc. Dublin Univ. Zool. Bot. Assoc.* v.1, 1859, 137–40. W. L. and L. Rees *Life and Times of Sir George Grey* 1892. P. Mennell *Dict. Austral. Biogr.* 1892, 198–201. J. Milne *Romance of a Pro-consul... Life of Sir George Grey* 1899. *D.N.B. Supplt 1* v.2, 357–76. *J. W. Austral. Nat. Hist. Soc.* 1909, 17–18. *Emu* 1939, 216–26. P. Serle *Dict. Austral. Biogr.* v.1, 1949, 370–74. J. Rutherford *Sir George Grey, 1812–1898* 1961. *Austral. Encyclop.* v.4, 1965, 388–90. *Austral. Dict. Biogr.* v.1, 1966, 476–80. *R.S.C.* v.15, 456.

S. African plants at Kew. New Zealand ferns at Cambridge. MSS. at Cape Town and Auckland. Portr. at Hunt Library. Portr. by Herkomer at National Portrait Gallery.

Greyia Hook & Harvey.

GREY, J. (*fl.* 1850s?)

Some Mauritius plants at Oxford.

H. N. Clokie *Account of Herb. Dept. Bot., Oxford* 1964, 174.

GRIER, John (*c.* 1806–1879)
d. 20 Dec. 1879

Nurseryman, Ambleside, Westmorland.

GRIERSON, Robert (–1929)
b. Dublin *d.* Dublin 11 Sept. 1929

Solicitor, Glasgow. Secretary, Glasgow Natural History Society, 1922. Studied alien flora of Glasgow district. Contrib. to *Glasgow Nat.*

Bot. Soc. Exch. Club Br. Isl. Rep. 1930, 323–24. *Glasgow Nat.* v.9, 1930, 62.

Plants at Oxford and West of Scotland Agricultural College.

GRIERSON, Thomas Boyle (1818–1889)
b. Dumfries 19 Feb. 1818 *d.* 26 Sept. 1889

Practised medicine at Thornhill, Dumfriesshire. Local antiquarian and naturalist.

Trans. Bot. Soc. Edinburgh v.18, 1891, 479–80.

Herb. at Dumfries Museum.

GRIESSEN, Albert E. P. (1875–1935)
b. London 1875 *d.* London 6 Oct. 1935
Kew gardener, 1896. At Botanic Gardens, Calcutta, 1898. Superintendent, Taj Gardens, 1900. *Horticulture in France. Evolution of Moghul Gardens of Plains of India. Quelques Arbres à Fleurs de l'Inde. De la Distribution de certaines Espéces sur la Globe Terrestre.*
J. Kew Guild 1936, 582–84 portr.

GRIEVE, James (*c.* 1840–1924)
b. Peebles *c.* 1840 *d.* Edinburgh Sept. 1924
Hybridised pansies for Messrs Dickson and Co., Edinburgh, 1859–95. Nurseryman, Edinburgh, 1896.
J. Hort. Cottage Gdnr v.59, 1909, 486–87 portr. *Gdnrs Mag.* 1909, 547–48 portr.; 1914, 471 portr. *Garden* 1910, 29 portr. *Gdnrs Chron.* 1921 i 110 portr.; 1924 ii 226. A. Simmonds *Hort. Who was Who* 1948, 43–44.

GRIEVE, Peter (1812–1895)
b. Allanton, Berwickshire Dec. 1812 *d.* Bury St. Edmunds, Suffolk 26 Sept. 1895
Gardener to Earl of Lanesborough at Smithland Hall, Loughborough and to Rev. E. R. Benyon at Culford Hall, 1847.
Gdnrs Chron. 1875 ii 261 portr.; 1895 ii 405 portr. *Garden* v.48, 1895, 272. *Gdnrs Mag.* 1895, 625 portr. *J. Hort. Cottage Gdnr* v.31, 1895, 329 portr., 344.

GRIEVE, Symington (1850–1932)
d. 18 Feb. 1932
Vice-president, Botanical Society of Edinburgh, 1932. 'Physical Changes brought about by the Floating Power of Seaweed' (*Trans. Bot. Soc. Edinburgh* v.30, 1929, 72–103).
Times 20 Feb. 1932. *Trans. Bot. Soc. Edinburgh* v.31, 1932, 238.

GRIFFIN, James (*fl.* 1850s)
Nurseryman, Bath Nursery, Weston Road, Bath. Successor to J. Salter.

GRIFFIN, W. H. (1841–1921)
d. 30 Nov. 1921
S. Eastern Nat. 1922, xxxvi–xxxvii.
Plants at South London Botanical Institute.

GRIFFIN, William (–1827)
d. S. Lambeth, London 28 Jan. 1827
FLS 1817. Nurseryman, S. Lambeth, London. "Cultor felicissimus Bulborum" (R. A. Salisbury *Genera of Plants* 1866, 134).
Bot. Mag. 1812, t.1512; 1814, t.1618. *Bot. Register* 1820, t.511. *Gdnrs Mag.* 1827, 255. E. Bretschneider *Hist. of European Bot. Discoveries in China* 1898, 283.
Griffinia Ker.

GRIFFIN, William (*c.* 1752–1837)
d. Stapleford, Herts 28 Dec. 1837
Gardener at Kelham Hall, Notts and to S. Smith, Woodhall, Herts. *Treatise on Culture of Pineapple* ed. 2 1808. 'On Management of Grapes in Vineries' (*Trans. Hort. Soc.* v.4, 1822, 98–106).
J. C. Loudon *Encyclop. Gdning* 1822, 1284. *Gdnrs Mag.* 1838, 111–12.

GRIFFITH, John Edwards (1843–1933)
b. 18 June 1843 *d.* Bangor, Caernarvonshire 4 July 1933
Pharmaceutical chemist, Bangor. Naturalist and antiquary. *Fl. Anglesey and Caernarvonshire* 1895.
J. Bot. 1927, 127–28. *Dict. Welsh Biogr. down to 1940* 1954, 295–96.
Herb. at National Museum of Wales.

GRIFFITH, John Wynne (1763–1834)
b. Aberystwyth, Cardiganshire 1 April 1763 *d.* Garn, Heallan, Denbighshire 16 June 1834
FLS 1795. Bryologist and lichenologist. Discovered *Cotoneaster* 1783. Sent Welsh plants to J. E. Smith. Friend of the Witherings. Contrib. to W. Withering's *Botanical Arrangement* and to W. Bingley's *Tours round North Wales* 1800 (D. Turner and L. W. Dillwyn *Botanist's Guide* 1805, 166–80).
D. Turner *Extracts from Lit. Sci. Correspondence of Richard Richardson* 1835, vi. *J. Bot.* 1923, 225–29.
Oedipodium griffithianum Schwaegr.

GRIFFITH, William (–1742)
Nurseryman, Carlow, Ireland.
Irish For. 1967, 50.

GRIFFITH, William (1810–1845)
b. Ham, Surrey 4 March 1810 *d.* Malacca, Malaya 9 Feb. 1845
FLS 1840. Assistant-surgeon, Madras, 1832. In Assam with N. Wallich, 1835–36. In Bhutan, 1837–38; Afghanistan, 1839–41; Malaya, 1841–42, 1845. Superintendent, Calcutta Botanic Garden, 1842. *Journals of Travels in Assam, Burma, Bootan, Afghanistan and Neighbouring Countries* 1847. *Notulae ad Plantas Asiaticas* 1847–54 4 vols. *Palms of British East India* 1850.
N. Wallich *Plantae Asiaticae Rariores* v.3, 1832, 11. *Bot. Register* 1845 Misc. Notes, 36–37. *Gdnrs Chron.* 1845, 387; 1911 i 232. *J. Agric. Hort. Soc. India* 1845, 1–39. A. Lasègue *Musée Botanique de B. Delessert* 1845, 149–50, 432–33. *London J. Bot.* 1845, 371–75; 1848, 446–49. *Phytologist* v.2, 1845, 252–55. *Proc. Linn. Soc.* 1845, 239–44. *Calcutta J. Nat. Hist.* 1846, 294–306. J. D. Hooker and T. Thomson *Fl. Indica* 1855,

60–63. *J. Asiatic Soc. Bengal* 1856, 410–11. E. Bretschneider *Hist. European Bot. Discoveries in China* 1898, 359–60. *J. Bot.* 1899, 460–61. *D.N.B.* v.23, 240–41. D. G. Crawford *Hist. of Indian Med. Service, 1600–1913* v.2, 1914, 142, 144, 148. F. W. Oliver *Makers of Br. Bot.* 1913, 178–91 portr. L. Huxley *Life and Letters of Sir J. D. Hooker* v.2, 1918, 234, etc. *Fl. Malesiana* v.1, 1950, 201–2. R. H. Phillimore *Hist. Rec. of Survey of India* v.4, 1958, 446–47. M. Archer *Nat. Hist. Drawings in India Office Library* 1962, 41–46, 78–79. I. H. Burkill *Chapters on Hist. Bot. in India* 1965 *passim*. A. M. Coats *Quest for Plants* 1969, 153–54. *Notes R. Bot. Gdn Edinburgh* 1970, 159–75. C. C. Gillispie *Dict. Sci. Biogr.* v.5, 1972, 539–40. *Taxon* 1973, 424 portr. *R.S.C.* v.3, 18; v.12, 293.

Plants, MSS., Journals, letters and daguerreotype portr. at Kew. Portr. at Hunt Library.

Griffithia Wright & Arn.

GRIFFITHS, Miss Amelia Elizabeth (1802–1861)

Daughter of Mrs. A. W. Griffiths (1768–1858). Of Martinhoe, Devon. Collected plants in Devon.

Phytologist v.1, 1844, 521. W. K. Martin and G. T. Fraser *Fl. Devon* 1939, 773. Devonshire Assoc. *Fl. Devon* v.2(1), 1952, 70–71.

Plants at North Devon Athenaeum, Barnstaple and Torquay Museum. British fungi drawings at BM(NH).

GRIFFITHS, Amelia Warren (*née* Rogers) (1768–1858)

b. Pitton, Devon 14 Jan. 1768 *d.* Torquay, Devon 4 Jan. 1858

Algologist. Correspondent of W. H. Harvey and R. K. Greville. "The *facile regina* of British algologists" (W. H. Harvey). Contrib. to *Phytologist* v.1, 1844. Contrib. plants to O. Blewitt's *Panorama of Torquay* 1832.

J. Sowerby and J. E. Smith *English Bot.* 1926. D. Turner *Fuci* v.1, 1801, 80–81. P. Smith *Mem. and Correspondence of Sir J. E. Smith* v.1, 1832, 587. W. H. Harvey *Manual of Br. Algae* 1841, liv–lv. W. H. Harvey *Memoir* 1869, 149–50, 158, etc. *Trans. Penzance Nat. Hist. Soc.* 1890–91, 230. J. Hardy *Selections from Correspondence of George Johnston* 1892, 326. *J. Bot.* 1892, 51–52. K. M. Lyell *Life of Sir C. J. F. Bunbury* v.1, 1906, 256–57. *Kew Bull.* 1907, 19. *Trans. Proc. Torquay Nat. Hist. Soc.* v.2, 1953, 74–77.

Herb. at Torquay Museum. Algae at BM(NH). Letters at Kew and in Berkeley correspondence at BM(NH).

Griffitsia Agardh.

GRIFFITHS, Benjamin Millard (1886–1942)

b. Kidderminster, Worcs 1886 *d.* 25 March 1942

BSc Birmingham 1908. DSc 1923. FLS 1922. Demonstrator in Botany, Queen's University, Belfast, 1914. Lecturer in Botany, Reading, 1920; Newcastle, 1921. Head of Botany Dept., Durham, 1924–39. Studied ecology and taxonomy of fresh-water algae. President, Northern Naturalists' Union, 1927. Contrib. to *Vasculum*.

Nature v.149, 1942, 548. *Proc. Linn. Soc.* 1941–42, 278. *Vasculum* 27, 1942, 10–11.

Plants at Durham.

GRIFFITHS, David (1867–1935)

b. Aberystwyth, Cardiganshire 16 Aug. 1867 *d.* Takoma Park, D.C., U.S.A. 19 March 1935

To U.S.A. with his parents, *c.* 1870. MSc S. Dakota College of Agriculture. PhD Columbia 1900. Professor, Arizona, 1900–1. Mycologist and agronomist, USDA, 1901. Collected plants in Montana, New Mexico, Arizona and Wyoming.

J. New York Bot. Gdn 1908, 65–66; 1911, 129. *Lily Yb.* 1935, 132–33 portr. *Science* 1935, 426–27. *Yb. Amer. Amaryllis Soc.* 1935, 23–25 portr. *Chronica Botanica* 1936, 308 portr. H. B. Humphrey *Makers of North American Bot.* 1961, 101–103.

Plants at Washington. Portr. at Hunt Library.

Atriplex griffithsii Standley.

GRIFFITHS, Rev. Evan (1795–1873)

b. Gellibeblig near Bridgend, Glam 18 Jan. 1795 *d.* 31 Aug. 1873

Of Swansea, Glam. Preacher. *Y Hlysieulyft Tellluaidd* (Family Herbal) (with Rev. R. Price) 1849.

J. Bot. 1898, 18. *Dict. Welsh Biogr. down to 1940* 1959, 304.

GRIFFITHS, Griffiths Hooper (*c.* 1823–1872)

d. 19 Nov. 1872

MD. Muscologist and lichenologist. Secretary, Worcester Natural History Society. *Fl. Church Stretton c.* 1870.

GRIFFITHS, John (*fl.* 1800s)

In Sumatra *c.* 1800 whence he sent plants to Calcutta Botanic Garden.

Fl. Malesiana v.5, 1958, 47.

GRIFFITHS, Moses (1743–1819)

b. Carmarthenshire 1743

Servant of T. Pennant (1726–1798) for whom he drew and engraved illustrations for some of his works. 3 drawings for J. Lightfoot's *Fl. Scotica* 1777 at BM(NH).

GRIFFITHS, Waldron (1850–1936)

b. London 1850 *d.* 23 Feb. 1936

FRMS. Chemist. Perfected method of mounting freshwater algae.
Pharm. J. 1936, 328.

GRIGG, Thomas (*fl.* 1700s–1710s)
Of Parham Plantation, Antigua. Sent plants to J. Petiver.
Proc. Amer. Antiq. Soc. 1952, 320–21. J. E. Dandy *Sloane Herb.* 1958, 133.

GRIGOR, Alexander (*fl.* 1870s)
Nurseryman, Elgin, Morayshire.

GRIGOR, James (*c.* 1811–1848)
d. Norwich, Norfolk 22 April 1848
Nurseryman, Old Lakenham, Norwich. *Eastern Arboretum or Register of Remarkable Trees...in...Norfolk* 1841.
Notes and Queries v.7, 1889, 107, 257. *D.N.B.* v.22, 248.

GRIGOR, John (*c.* 1806–1881)
d. 19 May 1881
Nurseryman, Forres, Morayshire, 1826. *Arboriculture* 1868; ed. 2 1881.
Florist and Pomologist 1881, 112. *Gdnrs Chron.* 1881 i 704. *Gdnrs' Yb. Almanack* 1882, 192.

GRIMES, John (1859–1947)
b. Bubbenhall, Warwickshire 10 Nov. 1859 *d.* 6 Aug. 1947
President, Cardiff Naturalists Society, 1915–16. Amateur naturalist whose main interest was botany.
Trans. Cardiff Nat. Soc. 1950, 7.
Lantern slides at National Museum of Wales.

GRIMSHAW, James (1797–1857)
d. Jan. 1857
In family linen business. Married sister of J. Templeton (1766–1825). Wrote several botanical papers.
Belfast Nat. Hist. Philos. Soc. Centenary Vol. 1924, 132.

GRIMWOOD, Daniel (*c.* 1725–1796)
d. 6 Aug. 1796
Nurseryman, Kensington, London. In partnership with Samuel Hudson and Peter Barritt was successor to nursery of John Williamson and Co., 1783.
Gent. Mag. 1796 ii 706. J. Harvey *Early Hort. Cat.* 1973, 8.

GRIMWOOD, John (*fl.* 1790s)
Nurseryman, Charlemont Street, Dublin. Nursery at Rathgar, Dublin. Succeeded by his son, George.
Gdnrs Chron. 1928 ii 61–62. *Irish For.* 1967, 47.

GRINDLEY, Edith Isabel (–1948)
d. Brockenhurst, Hants 20 March 1948
FLS 1924. Bryologist.
Proc. Linn. Soc. 1948, 72.

GRINDON, Leopold Hartley (1818–1904)
b. Bristol 28 March 1818 *d.* Greenheys, Manchester 20 Nov. 1904
Founder member of Philo-botanical Society of Bristol. Cashier, Manchester, 1838–64. Founder member and President, Manchester Field Naturalists Society. Lectured in Manchester on botany and natural history. *Manchester Walks and Wild Flowers* 1858. *Manchester Fl.* 1859. *British and Garden Botany* 1864. *Trees of Old England* 1868. *Echoes in Plant and Flower Life* 1869. *Fairfield Orchids* 1872. *Pathway to Botany* 1872. *History of the Rhododendron* 1876. *Country Rambles* 1882. *Shakespeare Fl.* 1883. *Fruits and Fruit Trees* 1885.
J. Bot. 1865, 93–95; 1905, 30–31. *Gdnrs Chron.* 1904 ii 373, 393. *Trans. Liverpool Bot. Soc.* 1909, 70–71. J. W. White *Fl. Bristol* 1912, 87–88. *N. Western Nat.* 1930, 17–22 portr. *Notes from Manchester Mus.* no. 31, 1927, 17–22 portr. *Proc. Bristol Nat. Soc.* v.27, 1944, 27–36.
Herb. at Manchester.

GRISEBACH, Auguste Heinrich Rudolph (1814–1879)
b. Hanover, Germany 17 April 1814 *d.* Göttingen, Germany 9 May 1879
FLS 1859. Professor of Botany, Göttingen, 1841; Giessen, 1846. Employed by British Government on flora of West Indies, 1857. *Fl. Br. W. Indian Islands* 1864. *Vegetation der Erde* 1872.
Gdnrs Chron. 1879 ii 297. *J. Bot.* 1879, 191–92; 1880, 32. *J. Arnold Arb.* 1965, 243–85.
Grisebachia Klotzsch.

GRIST, William (*fl.* 1790s)
Seedsman, Maidstone, Kent.

GROOM, Charles Ottley (*called himself* **Napier** *and subsequently* **Prince of Mantua and Montserrat**) (*c.* 1840–1894)
b. Tobago, W. Indies *c.* 1840 *d.* London 17 Jan. 1894
Book of Nature 1870. Translated L. Figuier's *Vegetable Kingdom* 1866.
Cornhill Mag. v.33, 1912, 337–57.

GROOM, Henry (*fl.* 1820s–1850s)
Nurseryman and florist, Walworth, London. Acquired business of Curtis, Milliken and Co. Moved to Clapham Rise, London, 1842–43.

GROOM, Percy (1865–1931)
b. Wellington, Shropshire 12 Sept. 1865 *d.* Gerrards Cross, Bucks 16 Sept. 1931
BA Cantab 1887. FRS 1924. FLS 1889. Professor of Botany, Whampoa, China, 1889–92. Lecturer in Plant Physiology, Edinburgh, 1898–99. Head of Biological Dept., Royal Engineering College, Cooper's

Hill, 1899–1905. Lecturer in Botany, Northern Polytechnic, London, 1907–8. Assistant Professor of Botany, Imperial College of Science and Technology, 1908–11; Professor of Woods and Fibres, 1911–31. *Elementary Botany* 1898. *Trees and their Life Histories* 1907. Contrib. to *Ann. Bot.*

J. Bot. 1931, 266–67. *Nature* v.128, 1931, 695. *Proc. Linn. Soc.* 1931–32, 176–77. *Times* 18 Sept. 1931. *Obit. Notices Fellows R. Soc.* 1932, 63–64 portr. *Who was Who, 1929–1940* 562.

GROSE, Joseph Donald (*c.* 1901–1973)
d. Feb. 1973

Jeweller, Swindon, Wilts. *Fl. Wiltshire* 1957. Contrib. plant notes to *Wiltshire Archaeol. Nat. Hist. Mag.* 1937–69.

Wiltshire Archaeol. Nat. Hist. Mag. 1973, 35–36 portr. *Watsonia* 1974, 202–3.

GROULT, Rev. Philip (*fl.* 1800s)

FLS 1800. Of Walworth, London. "A very assiduous investigator of English plants" (J. E. Smith).

J. Sowerby and J. E. Smith *English Bot.* 919.

GROVE, Arthur Stanley (1865–1942)
d. Richmond, Surrey 2 Feb. 1942

MA. FLS 1903. VMH 1924. VMM 1934. Amateur gardener with interest in lilies. *Lilies* 1905. Contrib. two *Supplements* to H. J. Elwes's *Monograph of Genus Lilium* 1933–40 (1936–40 Supplt. with A. D. Cotton).

Lily Yb. 1935, 124 portr. *Gdnrs Chron.* 1942 i 73. *Proc. Linn. Soc.* 1941–42, 278–80.

GROVE, William Bywater (1848–1938)
b. Birmingham 24 Oct. 1848 *d.* Birmingham 6 Jan. 1938

BA Cantab 1871. Headmaster, Birmingham High School for Boys, 1887–1900. Lecturer in Botany, Studley Horticultural College, 1900–8. Lecturer in Botany, Birmingham Municipal Technical School, 1905–27. Hon Curator, Herb., Birmingham Univ. Mycologist. President, Birmingham Natural History Society. Contrib. 'New or Noteworthy Fungi' to *J. Bot.* 1884–1933. *Synopsis of Bacteria and Yeast Fungi* 1884. *British Rust Fungi* 1913. *British Stem- and Leaf-Fungi* 1935–37 2 vols. Contrib. *Pilobolidae* to A. H. R. Buller's *Researches on Fungi* v.6, 1934. Translated L. R. Tulasne's *Selecta Fungorum Carpologia* 1931 3 vols.

J. Bot. 1938, 86–87. *Kew Bull.* 1938, 83–84. *N. Western Nat.* 1938, 30–34 portr., 100.

Fungi at Kew, Birmingham University. Drawings and MSS. at BM(NH). Letters at Liverpool Museums.

GROVES, Elsie Margaret (*née* Reah) (1897–1956)
b. Sunderland, Durham 26 Aug. 1897 *d.* Ottawa 1 Dec. 1956

To Canada, 1922. Assisted her husband, Dr. W. Groves, mycologist in Canadian Dept. of Agriculture, in collecting fungi.

Canadian Field Nat. 1957, 9 portr.

Herb. at Dept. of Agriculture, Ottowa.

GROVES, Henry (1835–1891)
b. Weymouth, Dorset 1835 *d.* Florence, Italy 1 March 1891

FLS 1884. Pharmaceutical chemist in Florence, 1862. Botanised in Italy. Founder and President of Central Botanical Society of Tuscany to which he left his herb. 'Fl. Portland' (*Phytologist* v.2, 1857–58, 601–9). 'Coast Fl. Japygia' (*J. Linn. Soc.* v.21, 1886, 523–37).

J. Bot. 1891, 191–92. *Pharm J.* v.21, 1891, 894–95. *Proc. Linn. Soc.* 1890–91, 23–24. *R.S.C.* v.3, 331; v.7, 847; v.10, 71; v.15, 487.

Plants at BM(NH), Kew, Florence. Portr. at Hunt Library.

GROVES, Henry (1855–1912)
b. London 15 Oct. 1855 *d.* Clapham, London 2 Nov. 1912

FLS 1892. Brother of J. Groves (1858–1933) with whom he collaborated in 9th ed. of C. C. Babington's *Manual of British Botany* 1904. *Review of British Characeae* 1880. *Characeae Britannicae Exsiccatae* (with J. Groves) 1892, 1900 (*J. Bot.* 1892, 154; 1900, 453–54). Contrib. *Characeae* to I. Urban's *Symbolae Antillanae* v.7, 1911, 30–44. Contrib. to *J. Bot.*

Bot. Soc. Exch. Club Br. Isl. Rep. 1913, 376–78. *J. Bot.* 1913, 73–79 portr. *Nature* v.90, 1912, 284. *Proc. Linn. Soc.* 1912–13, 58–59. *Watson Bot. Exch. Club Rep.* 1932–33, 165–66.

Plants at Oxford.

Rosa hibernica var. *Grovesii* Baker.

GROVES, James (1858–1933)
b. London 19 Jan. 1858 *d.* Freshwater Bay, Isle of Wight 20 March 1933

FLS 1885. Worked in Army and Navy Stores. Co-editor with his brother Henry, of 9th ed. of C. C. Babington's *Manual of British Botany* 1904. *Monograph of British Charophyta* (with G. R. Bullock-Webster) 1920–24 2 vols. Contrib. *Characeae* to I. Urban's *Symbolae Antillanae* v.7, 1911, 30–44. *Characeae Britannicae Exsiccatae* (with H. Groves) 1892, 1900 (*J. Bot.* 1892, 154; 1900, 453–54). Contrib. to *J. Bot.*

Bryol. Soc. Rep. 1933, 172–73. *J. Bot.* 1933, 136–39 portr. *Kew Bull.* 1934, 274–75. *Proc. Linn. Soc.* 1932–33, 191–95. *Watson Bot. Exch. Club* 1932–33, 165–66 portr.

Plants at BM(NH), Oxford.

GROVES, Thomas Bennett (1829–1902)
d. July 1902
Pharmacist, Weymouth. Researched on aconite, aloin, kamala, senna, copaiba and opium. Contrib. to *Pharm. J.*
Pharm. J. v.69, 1902, 30, 34.

GRUBB, C. W. (–1954)
d. Lancaster 8 April 1954
Nurseryman, Bolton-le-Sands, Lancs. Cultivated ferns.
Br. Fern Gaz. 1954, 84.

GRUBBE, Anne Elizabeth Heath *see* Haviland, A. E. H.

GRUGEON, Alfred (1826–1913)
b. Spitalfields, London 17 July 1826 *d.* Walthamstow, Essex 14 Feb. 1913
Wood-turner. Taught botany at Working Men's College for 30 years. *Botany, Structural and physiological* 1873. Contrib. to H. Trimen and W. T. T. Dyer's *Fl. Middlesex* 1869.
J. Bot. 1917, 193–94.

GRUNDY, Clara (*fl.* 1870s)
Of Liverpool. Numerous records in *Fl. Liverpool* 1872 and 1902 and J. B. L. Warren's *Fl. Cheshire* 1899. "Undoubtedly the best lady botanist [in Liverpool] after Miss Potts" (J. B. L. Warren).
J. B. L. Warren *Fl. Cheshire* 1899, lxxxii. *Trans. Liverpool Bot. Soc.* 1909, 21. *Liverpool Bull.* v.14, 1967, 33–35.

GRUNDY, Sir Cuthbert Cartwright (1846–1946)
d. 1 Feb. 1946
FLS 1872. Artist. President, Royal Cambrian Academy for 21 years. *Notes on Food of Plants* 1871. *How does a Plant Grow* 1935.
Trans. Liverpool Bot. Soc. 1909, 71. *Times* 2 Feb. 1946. *Who was Who, 1941–1950* 477.

GRUNDY, Ellen (Mrs. Francis Boult) (1815–1894)
b. Liverpool 1815 *d.* 18 Dec. 1894
Began collecting plants *c.* 1861. Drew attention to adventive flora from ships' ballast used for road-making near Birkenhead. Contrib. to *Fl. Liverpool* 1872.
J. B. L. Warren *Fl. Cheshire* 1899, lxxviii. *Liverpool Bull.* v.14, 1967, 33–35.
Plants at Liverpool Museums.

GRUNDY, Maria Ann (Mrs. Swinton Boult) (1809–1871)
b. Liverpool 1809 *d.* 17 Aug. 1871
Contrib. to *Fl. Liverpool* 1872. Collected plants of Liverpool area.
J. B. L. Warren *Fl. Cheshire* 1899, lxxviii. *Liverpool Bull.* v.14, 1967, 33–35.
Plants at Liverpool Museums.

GUERMONPREZ, Henry Leopold Foster (1858–1924)
b. 5 July 1858 *d.* Bognor, Sussex 21 Dec. 1924
Of Bognor Regis, Sussex. Had herb. mainly of Sussex plants, now at Bognor Regis Museum.
A. H. Wolley-Dod *Fl. Sussex* 1937, xlvii–xlviii. *Bognor Post* 30 Nov. 1937, portr.

GUILDFORD, 5th Earl of *see* North, F.

GUILDING, Rev. Lansdown (1797–1831)
b. Kingstown, St. Vincent 29 May 1797 *d.* Bermuda 22 Oct. 1831
BA Oxon 1817. FLS 1817. Colonial Chaplain, St. Vincent. Zoologist. *Account of Bot. Gdn in Island of St. Vincent* 1825. Contrib. 17 plates to *Bot. Mag.*
Bot. Misc. v.1, 1829, 122–24. *Kew Bull.* 1899, 228–29. I. Urban *Symbolae Antillanae* v.3, 1902, 53–54; v.7, 1911, 74–75. *R.S.C.* v.3, 76.
Plants and drawings at Kew. Letters at Kew and in Swainson correspondence at Linnean Society.
Guildingia Hook.

GUILFOYLE, Michael (1809–1884)
b. Tipperary 1809 *d.* Paddington, N.S.W. 9 April 1884
Trained as gardener at Royal Exotic Nursery, Chelsea. Arrived in Sydney, 1851, where he established a nursery. Created new nursery at Double Bay where he did much to make Australian plants better known in horticulture.
J. Proc. R. Soc. N.S.W. 1921, 161. *Camellia News* no. 51, 1973, 31.

GUILFOYLE, William Robert (1840–1912)
b. Chelsea, London 8 Dec. 1840 *d.* Melbourne 26 June 1912
FLS 1869. Son of M. Guilfoyle (1809–1884) who took him to Sydney, 1851. Trained in his father's nursery. Botanist on 'Challenger' in South Seas, 1868 (*J. Bot.* 1869, 117–36). Farmed in Tweed River district, N.S.W. whence he sent plants to F. von Mueller. Director, Melbourne Botanic Gardens, 1873–1909. "He does with his trees what a pianist tries to do with his music" (Paderewski). 'Botanical Tour among the South Sea Islands' (*Sydney Mail* 1868, 1869). *Australian Botany* 1878. *A.B.C. of Botany* 1880. *Catalogue of Plants...in Melbourne Botanical Gardens* 1883.
F. von Mueller *Fragmenta* v.8, 1872, 33–34. T. W. H. Leavitt *Austral. Representative Men* ed. 2 1885, 410–12. P. Mennell *Dict. Austral. Biogr.* 1892, 205. *J. Proc. R. Soc. N.S.W.* 1921, 160–62. P. Serle *Dict. Austral. Biogr.* v.1, 1949, 383. *Camellia News* no. 51, 1973, 31–32. R. T. M. Pescott *W. R. Guilfoyle, 1840–1912* 1975 portr. *R.S.C.* v.7, 863; v.15, 521.
Guilfoylea F. Muell.

GUILLE, Miss Mary Elizabeth (–1903)
Daughter of Rector of St. Andrews. Collected plants in Guernsey, 1844–56.
Rep. Trans. Guernsey Soc. Nat. Sci. Local Res. 1903, 192–94.

GULLIVER, B. *and* **T. A.** (*fl.* 1860s)
Brothers who joined Capt. F. Cadell on HMS 'Eagle' in 1867 as botanical collectors. Collected plants in Gulf of Carpentaria and Arnhem Land. Specimens sent to F. von Mueller.
Austral. Encyclop. 1965, 446.
Plants at Melbourne.
Heterachne gulliveri Benth.

GULLIVER, George (1804–1882)
b. Banbury, Oxfordshire 4 June 1804 *d.* Canterbury, Kent 17 Nov. 1882
FRS 1838. Hunterian Professor of Comparative Anatomy and Physiology, 1861. *Catalogue of Plants collected in Neighbourhood of Banbury* 1841. *Notes on Researches in Botany* 1870; ed. 2 1880. Wrote on Raphides.
J. Bot. 1883, 31. *D.N.B.* v.23, 334–35. G. C. Druce *Fl. Oxfordshire* 1927, cxi. G. C. Druce *Fl. Northamptonshire* 1930, cxvi. *R.S.C.* v.3, 84; v.7, 865; v.10, 87; v.15, 533.
Herb. formerly at Chatham Literary Society.

GULSON, Mrs. (–1871)
Of Exmouth and Teignmouth, Devon. Collected Devon algae. Discovered *Atractophora hypnoides* near Exmouth.
Ann. Mag. Nat. Hist. v.15, 1855, 334.
Algae at Cambridge.
Gulsonia Harvey.

GUMBLETON, William Edward (1830–1911)
d. Queenstown, County Cork 4 April 1911
Grew rare plants at Belgrove, County Cork. Had notable botanical library. Contrib. to *Gdnrs Chron.*
Garden 1899 xii portr. *Gdnrs Mag.* 1907, 525 portr., 526. *Gdnrs Chron.* 1911 i 255, 266 portr. *Kew Bull.* 1911, 202–3.
Library at Glasnevin.

GUNN, Rev. George (1851–1900)
b. Edinburgh 3 June 1851 *d.* Peebles 12 Jan. 1900
MA Edinburgh. Minister, Stichill near Kelso, 1878–1900. Had herb. Cultivated alpines. Secretary, Berwickshire Naturalists Club.
Proc. Berwickshire Nat. Club 1899, 152–60. *Trans. Bot. Soc. Edinburgh* v.21, 1900, 277–80. *R.S.C.* v.15, 535.

GUNN, Ronald Campbell (1808–1881)
b. Cape Town 4 April 1808 *d.* Launceston, Tasmania 23 March 1881
FRS 1854. FLS 1850. Superintendent of convicts, Hobart, 1830. Police magistrate and Superintendent, Launceston, 1833. Became interested in botany through friendship with R. W. Lawrence (1807–1833). Sent plants to W. J. Hooker. Became large landowner in Launceston. Collected plants with J. D. Hooker during his visit to Tasmania. "There are few Tasmanian plants that Mr. Gunn has not seen alive" (J. D. Hooker). Secretary, Royal Society of Tasmania. *Observations on Fl. Geelong* 1842.
J. Bot. (Hooker) 1834, 241. A. Lasègue *Musée Botanique de B. Delessert* 1845, 283. J. D. Hooker *Fl. Tasmaniae* v.1, 1859, cxxv. W. H. Harvey *Phycologica Australica* v.5, 1863, v. *J. Bot.* 1881, 192. *Proc. Linn. Soc.* 1881–82, 63–64. *Proc. R. Soc.* v.34, 1883, xiii–xv. P. Mennell *Dict. Austral. Biogr.* 1892, 205–6. *D.N.B.* v.23, 342–43. *Papers Proc. R. Soc. Tasmania* 1909, 15–18. *Rep. Austral. Assoc. Advancement Sci.* 1911, 230 portr. *Tasmanian Nat.* v.1, 1926, 13–14. P. Serle *Dict. Austral. Biogr.* v.1, 1949, 383–84. *Victorian Nat.* 1949, 85–86; 1962, 197; 1965, 90–91. T. E. Burns and J. R. Skemp *Van Dieman's Land Correspondents* 1961 portr. *Austral. Encyclop.* 1962, 403–4. *Austral. Dict. Biogr.* v.1, 1966, 492–93. *R.S.C.* v.3, 87.
Plants at Sydney, BM(NH), Kew, Oxford. Letters at Kew, Mitchell Library Sydney. Crayon portr. at Kew.
Gunnia F. Muell.

GUNNIS, Francis G. (*fl.* 1890s)
Collected some plants in British Somaliland, 1894–95.
Kew Bull. 1895, 211.

GUNTHER, Robert William Theodore (1869–1940)
b. Surbiton, Surrey 23 Aug. 1869 *d.* South Stoke, Oxford 9 March 1940
MA Oxon. FLS 1900. Tutor, Magdalen College, Oxford, 1894–1928. Reader in History of Science, 1934–39. Founder and first Curator of Oxford Museum for History of Science. Curator, Oxford Botanic Garden, 1914–18. *Oxford Gardens* 1912. *Guide to Oxford Botanic Garden* 1914. *Early British Botanists and their Gardens* 1922. *Herbal of Apuleius Barbarus* 1925. *Further Correspondence of John Ray* 1928. *Life and Work of Robert Hooke* 1930–38 3 vols. *Greek Herbal of Dioscorides* 1934. *Correspondence of Dr. Plot and Oxford Philosophical Society* 1939. *Life and Letters of Edward Llwyd* 1940.
Nature v.145, 1940, 541–42. *Proc. Linn. Soc.* 1939–40, 366–68. *Times* 11 March 1940. *Chronica Botanica* 1942–43, 231–32. *D.N.B., 1931–1940* 381. *Who was Who, 1929–1940* 568. A. E. Gunther *Robert T. Gunther* 1967 portr.

GUPPY, Henry Brougham (1854–1926)
b. Falmouth, Cornwall Dec. 1854 *d.* Martinique, West Indies 23 April 1926

MB Edinburgh 1876. FRS 1918. FLS 1917. Surgeon, Royal Navy, 1876–85. Survey in Western Pacific on HMS 'Lark' 1881–84. Collected plants in Solomon Islands, Keeling Island, Java in 1880s. Botanical and geological exploration in Hawaiian and Fijian Islands, 1896–1900. Investigated littoral flora of Pacific side of S. America, 1903–4. Botanical work in West Indies, 1907–11. Studied plant distribution and dispersal. *Solomon Islands* 1887. *Observations of a Naturalist in the Pacific between 1896 and 1899* 1903–6. *Studies in Seeds and Fruits* 1912. *Plants, Seeds, and Currents in West Indies and Azores* 1917. Contrib. to J. *Linn. Soc. Bot.* v.29, 30, 44, 1892–1919.

Bot. Soc. Exch. Club Br. Isl. Rep. 1926, 92–93. *J. Bot.* 1926, 161–62. *Nature* v.38, 1888, 40; v.117, 1926, 699, 797–98. *Proc. Linn. Soc.* 1916–17, 40–41; 1926–27, 86–87. *Proc. R. Soc. Edinburgh* v.46, 1925–26, 362–64. *Proc. R. Soc.* v.101, 1927, xxviii–xxix portr. *Who was Who, 1916–1928* 444–45. *Fl. Malesiana* v.1, 1950, 205–6. *R.S.C.* v.10, 94; v.15, 537.

Plants and MSS. on Solomon Islands plants at Kew.

GURLE, Leonard (*c.* 1621–1685)

Had nursery between Spitalfields and Whitechapel, London. "A very eminent and ingenious nursery-man" (L. Meager). King's Gardener at St. James's Park, 1677.

Gdn Hist. v.3(3), 1975, 42–49.

GURNEY, Gerard Hudson (1893–1934)
d. 18 May 1934

Of Keswick Hall, Norwich. Ornithologist. Collected plants in East Africa, 1908. President, Norfolk and Norwich Naturalists Society, 1929–30. 'Natural History Experiences in British East Africa' (*Trans. Norfolk Norwich Nat. Soc.* v.8, 1907–8, 696–726).

Trans. Norfolk Norwich Nat. Soc. v.13, 1933–34, 509–10.

GUTCH, John Wheeley Gough (1809–1862)
b. Bristol 1809 *d.* London 30 April 1862

MRCS. FLS 1848. Of Swansea. Queen's Messenger, 1850. 'List of Plants met with in Neighbourhood of Swansea' (*Phytologist* v.1, 1844, 104–8, 118–21, 141–45, 180–87). Edited *Lit. Sci. Register* 1842–56.

D.N.B. v.23, 372–73. *Trans. Cardiff Nat. Soc.* 1952–53, 13. *R.S.C.* v.3, 95.

Plants at Charterhouse School.

GUTHRIE, Francis (1831–1899)
b. London 1831 *d.* Claremount, Cape Town 19 Oct. 1899

BA London 1850. LLB 1852. Professor of Mathematics, S. African College, 1876–98. 'Evolution illustrated by Distribution of Plants' (*Trans. Philos. Soc. S. Africa* 1888, 275–94).

J. Bot. 1899, 528. *Kew Bull.* 1899, 221. H. A. Baker and E. G. H. Oliver *Ericas in Southern Africa* 1967, x–xii portr. *R.S.C.* v.7, 876; v.10, 96; v.12, 300; v.15, 541.

Portr. at Hunt Library.

Guthriea Bolus.

GUTTRIDGE, James J. (*c.* 1866–1952)
b. Surrey *c.* 1866 *d.* 5 Dec. 1952

Kew gardener until 1891. Curator, Botanic Gardens, Glasgow. Curator, Botanic Garden and Superintendent, Public Parks, Liverpool.

Gdnrs Mag. 1909, 119–20 portr. *J. Kew Guild* 1952–53, 132.

GUYER, Richard Glode (*c.* 1870–1924)
b. Torquay? *c.* 1870 *d.* Edinburgh 21 April 1924

Pharmaceutical chemist. 'Cultivation of Medicinal Plants in Scotland' (*Pharm. J.* v.106, 1921, 146–49).

Trans. Bot. Soc. Edinburgh v.29, 1924, 117–18.

GWATKIN, Joshua Reynolds Gascoigne (1855–1939)
b. 24 March 1855 *d.* 12 Sept. 1939

MA Cantab. Compiled MSS. *List of Flowers near Potterne* 1921–26 and *List of Wiltshire Plants* 1929 which were used in D. Grose's *Fl. Wiltshire* 1957.

D. Grose *Fl. Wiltshire* 1957, 42.

Plant drawings at Linnean Society.

GWILLIM, Lady (*née* Symonds)
(–before 1809)

Wife of Sir Henry Gwillim, judge in Madras. Interested in botany and sent seeds of Indian plants to England.

Bot. Mag. 1806, t.977; 1809, t.1244.

GWINNELL, Wintour Frederic (1846–1921)
b. Wales 1846 *d.* Chiswick, Middx 26 Aug. 1921

BSc London. Lecturer, Regent Street Polytechnic. *Notes on Botany* 1876; ed. 2 1882.

GWYN, Nicholas (1710–1798)
b. Fakenham, Norfolk 14 July 1710 *d.* 20 Jan. 1798

MD Leyden. Studied under Boerhaave. Of Ipswich. "My very worthy and liberal friend, to whose penetrating genius, and learned researches, Botany owes much" (W. Curtis *Bot. Mag.* 1791, t.142). Friend of J. E. Smith.

J. E. Smith *Spicilegium Botanicum* 1791–92, 9, 13.

Letters in Smith correspondence at Linnean Society.

GWYNNE-VAUGHAN, David Thomas (1871–1915)
b. Llandovery, Carmarthenshire 12 March 1871 *d.* Reading, Berks 4 Sept. 1915

BA Cantab 1893. FLS 1907. Demonstrator in Botany, Glasgow, 1896–97. In Brazil, 1897, as botanist attached to rubber-prospecting expedition. Botanist on Skeat expedition to Malay Peninsula and Siam, 1899. Head of Botanical Dept., Birkbeck College, London, 1907. Professor of Botany, Belfast, 1909; Reading, 1914. *Practical Botany for Beginners* (with F. O. Bower) ed. 2 1902. 'Anatomy of Ferns' (*Ann. Bot.* 1901, 70–98; 1903, 689–742). 'Fossil Osmundaceae' (with R. Kidston) (*Trans R. Soc. Edinburgh* v.45, 1906–7, 759–80; v.46, 1908–9, 213–32, 651–67).

J. Bot. 1915, 342. *Kew Bull.* 1915, 389–90. *Nature* v.96, 1915, 61–62. *Proc. Linn. Soc.* 1915–16, 61–65. *Ann. Bot.* 1916, i–xxiv portr. *Proc. R. Soc. Edinburgh* v.36, 1915–16, 334–39. *Who was Who, 1897–1916* 302–3. *J. Thailand Res. Soc. Nat. Hist. Supplt.* v.12, 1939, 24–25. *Fl. Malesiana* v.1, 1950, 206–7. C. C. Gillispie *Dict. Sci. Biogr.* v.5, 1972, 604–5.

Plants at Kew, Cambridge.

Rhynia gwynne-vaughani Kidston.

GWYNNE-VAUGHAN, Helen Charlotte Isabella (*née* Fraser) (1879–1967)
b. London 21 Jan. 1879 *d.* 26 Aug. 1967

BSc London 1904. DSc 1907. FLS 1905. Lecturer in Botany, Royal Holloway College. Head of Botany Dept., Birkbeck College, 1909–17; Professor of Botany, 1921–44. President, British Mycological Society, 1928. *Fungi: Ascomycetes, Ustilaginales, Uredinales* 1922. *Structure and Development of Fungi* (with B. Barnes) 1930; ed. 2 1937.

Discovery 1944, 199–200, 219 portr. *Times* 22 Jan. 1962 portr.; 30 Aug. 1967 portr. *Trans. Br. Mycol. Soc.* 1968, 177–78 portr. M. Izzard *A Heroine in her Time* 1969. *Who was Who, 1961–1970* 470–71.

GWYTHER, Mr. (*fl.* 1810s)

Partner with Jenkins "who are preparing an extensive Botanic Garden [in New Road, London] for the use of subscribers; which, from its vicinity to the centre of the metropolis and the enjoyment of the pure air of the Regent's Park, promises to be a great acquisition to London Botanists."

Bot. Mag. 1816, t.1844.

GYFFORD, Mr. *see* Gifford, *Mr.*

HAARER, Alec Ernest (1894–1970)
b. Liverpool 10 Jan. 1894 *d.* Crawley Down, Sussex 20 Feb. 1970

FLS 1928. District Agricultural Officer, Tanganyika where he collected plants. *Jute Substitute Fibres* 1952. *Modern Coffee Production* 1956; ed. 2 1962. *Coffee Growing* 1963.

Plants at Kew.

HAAS, Paul (1877–1960)
b. London 1877 *d.* Cheam, Surrey 6 April 1960

Reader in Plant Chemistry, University College, London. *Introduction to Chemistry of Plant Products* (with T. G. Hill) 1913; ed. 2 1917; ed. 3 1921; ed. 4 1928.

Pharm. J. 1960, 346. *Nature* v.186, 1960, 595. *Who was Who, 1951–1960* 466.

HAAST, Sir Johann Franz Julius von (1824–1887)
b. Bonn, Germany 1 May 1824 *d.* Christchurch, New Zealand 16 Aug. 1887

PhD Tübingen 1862. FRS 1867. FLS 1864. KCMG 1886. To New Zealand, 1858. Government Geologist at Canterbury. Collected N.Z. plants.

J. D. Hooker *Handbook N.Z. Fl.* 1864, 12*. *N.Z. J. Sci.* v.2, 1884, 112–16 portr. *Nature* v.37, 1887–88, 87. *Proc. Linn. Soc.* 1887–88, 92–93. *Ann. Bot.* v.1, 1888, 403–4. *Proc. R. Soc.* v.46, 1889, xxiv–xxvi. P. Mennell *Dict. Austral. Biogr.* 1892, 206. *D.N.B.* v.23, 412–13. R. Speight *Nat. Hist. of Canterbury* 1927, 4–6 portr. H. F. von Haast *Life and Times of Sir Julius von Haast* 1948 portr. R. Glenn *Bot. Explorers of N.Z.* 1950, 111–27. *R.S.C.* v.7, 880; v.10, 103; v.15, 555.

Plants and letters at Kew. Portr. at Hunt Library.

Haastia Hook. f.

HACKER, John (*fl.* 1790s)

Seedsman, Basingstoke, Hants.

HACKETT, Walter (*c.* 1874–1957)
d. 21 May 1957

Kew gardener, 1897. Curator, Liverpool Botanic Garden, 1906–35.

Kew Bull. 1906, 173. *J. Kew Guild* 1957, 487–88 portr.

HADDEN, Norman G, (*c.* 1888–1971)
d. West Porlock, Somerset June 1971

VMH 1962. Grew irises in his garden at Porlock. Went plant collecting with W. R. Dykes in 1923.

Iris Yb. 1971, 18.

HADDINGTON, 6th Earl *see* Hamilton, T.

HAGGART, Donald Alexander (1850–1939)
d. Aberfeldy, Perthshire 15 Jan. 1939

Banker at Killin, Perthshire. Studied flora of Breadalbane mountains, Perthshire.

Bot. Soc. Exch. Club Br. Isl. Rep. 1938, 18–19.

HAGGER, John (-1895)
b. Cockermouth, Cumberland *d.* 1 March 1895

FLS 1891. Master, Repton School. Contrib. to *Fl. Repton* ed. 2 1881 with William Garneys (1832–1881), another master.

W. Wyatt and C. G. Thornton *Fl. Repandunensis* 1881 vii, viii.

Herb. at Nottingham.

HAILSTONE, Samuel (1768–1851)
b. Hoxton, London 1768 *d.* Bradford, Yorks 26 Dec. 1851

FLS 1801. Solicitor, Bradford. Authority on Yorkshire flora. Plant list in T. D. Whitaker's *Hist. and Antiq. of Deanery of Craven* ed. 2 1812.

J. Sowerby and J. E. Smith *English Bot.* 1035, 2737, etc. H. Baines *Fl. Yorkshire* 1840, preface. *Phytologist* v.1, 1844, 870. *J. Bot.* 1868, 65–66. J. James *Bradford* 1866, 316–18. *Proc. Linn. Soc.* 1852, 189. *D.N.B.* v.24, 2. *Annual Rep. Yorkshire Philos. Soc.* 1906, 58–59. *Trans. Liverpool Bot. Soc.* 1909, 71.

Herb. at York Museum.

Carex hailstoni S. Gibson.

HAINES, Frederick Merlin (1898–1964)
b. Winfrith, Dorset 8 July 1898 *d.* 29 Dec. 1964

BSc 1919. Lecturer in Plant Physiology, Queen Mary College, 1921; Professor of Botany, 1950–58. Studied transpiration in plants.

Nature v.181, 1958, 1242; v.201, 1964, 869–70. *Times* 4 Jan. 1964. *Who was Who, 1961–1970* 473–74.

HAINES, Henry Haselfoot (1867–1943)
b. London 12 April 1867 *d.* Berriew, Montgomery 6 Oct. 1945

FLS 1902. Indian Forest Service, 1888. Imperial Forest Botanist, 1906. Principal, Imperial Forest College, Dehra Dun, 1907. Conservator of Forests, 1909–19. *Forest Fl. Chota Nagpur...and Santhal Pargannahs* 1910. *Descriptive List of Trees...of Southern Circle, Central Provinces* 1916. *Botany of Bihar and Orissa* 1921 3 vols.

Indian For. 1946, 485. *Proc. Linn. Soc.* 1945–46, 68–70. *Pakistan J. For.* 1967, 361. *Who was Who, 1941–1950* 485–86. I. H. Burkill *Chapters on Hist. Bot. in India* 1965, 178–79.

Plants at Kew.

HAIRS, James (*fl.* 1770s–1810s)
Nurseryman, St. James's, Haymarket, London. Nurseries at Ranelagh, Chelsea and Ham Common, Surrey. In partnership with G. Smith and Ivie Hairs.

J. Harvey *Early Hort. Cat.* 1973, 10.

HALDIMAND, Jane *see* Marcet, J.

HALE, W. (*fl.* 1790s)
Of Alton, Hants. Lived in Halifax, Nova Scotia for several years. Brought seeds of fly-catching dogsbane to W. Curtis.

Bot. Mag. 1794, t.280.

HALES, Rev. Stephen (1677–1761)
b. Bekesbourne, Canterbury, Kent 17 Sept. 1677 *d.* Teddington, Middx 4 Jan. 1761

BA Cantab 1700. DD Oxon 1733. FRS 1718. Perpetual curate, Teddington, Middlesex 1709. Rector, Porlock, Somerset. Rector, Farringdon, Hants, 1722. Clerk of the Closet to Princess Dowager of Wales and chaplain to Prince of Wales. Plant physiologist. *Vegetable Staticks* 1727.

Gent. Mag. 1761, 32–33, 44; 1764, 273–78. *Hist. Acad. Paris* 1762, 213–30. *Annual Register of World Events* 1764, 42–49. J. E. Smith *Selection from Correspondence of Linnaeus* v.2, 1821, 25–43. J. von Sachs *Geschichte der Botanik* 1875, 514–21, 582–83. *Gdnrs Chron.* 1877 i 16 portr.; 1911 i 88–89 portr.; 1912 i 73. *D.N.B.* v.24, 32–36. R. Holt-White *Life and Letters of G. White* v.2, 1901, 230–31. E. J. L. Scott *Index to Sloane Manuscripts* 1904, 235. F. W. Oliver *Makers of Br. Bot.* 1913, 65–83. J. R. Green *Hist. Bot.* 1914, 198–206. F. Darwin *Rustic Sands* 1917, 115–39. *Ann. Med. Hist.* v.7, 1925, 109–16. *Nature* v.120, 1927, 228–31. E. Hawks and G. S. Boulger *Pioneers of Plant Study* 1928, 228–30. A. E. Clark-Kennedy *Stephen Hales* 1929 portr. *Isis* v.14, 1930, 422–23. *Notes Rec. R. Soc.* 1940, 53–63 portr. *Chronica Botanica* 1944, 86–92. J. S. L. Gilmour *Br. Botanists* 1944, 22–28 portr. *Act. VIe Congress Int. Hist. Sci.* 1950, 277–88. *Huntia* 1965, 200 portr. C. C. Gillispie *Dict. Sci. Biogr.* v.6, 1972, 35–48.

MSS. at Royal Society. Portr. by T. Hudson at National Portrait Gallery. Portr. at Hunt Library.

Halesia L.

HALES, William (1874–1937)
b. Leamington, Warwickshire 13 Feb. 1874 *d.* Chelsea, London 11 May 1937

ALS 1912. VMM 1930. VMH 1934. Kew gardener, 1894. Curator, Chelsea Physic Garden, 1899–1937.

Gdnrs Mag. 1899, 546 portr.; 1913, 489–90 portr. *Gdnrs Chron.* 1926 ii 302 portr.; 1935 i 20 portr.; 1937 i 354 portr. *J. Kew Guild* 1935, 405–7 portr.; 1938, 791–92 portr. *J. Bot.* 1937, 205–6. *Kew Bull.* 1937, 320–21. *Pharm. J.* 1937, 595. *Proc. Linn. Soc.* 1936–37, 192–93. *Times* 13 May 1937.

HALEY, Joseph (*fl.* 1810s)
Nurseryman, North Hall Nursery, Leeds, Yorks.

HALKET, Ann Cronin (–1965)
DSc. FLS 1911. Lecturer in Botany, Bedford College, London.
Proc. Linn. Soc. 1966, 119.

HALL, Dr. (*fl.* 1800s)
Member of Botanical Society of London.
Proc. Bot. Soc. Br. Isl. 1967, 319.

HALL, Abraham (*fl.* 1820s)
Nurseryman, Doncaster, Yorks. Grandson of Abraham Crowder.

HALL, Mrs. Agnes C. (1777–1846)
b. Roxburghshire 1777 *d.* London 1 Dec. 1846
Miscellaneous writer. Married Robert Hall, MD. *Elements of Botany* 1802.
Gent. Mag. 1847 i 97. *D.N.B.* v.24, 53–54.

HALL, Sir Alfred Daniel (1864–1942)
b. Rochdale, Lancs 22 June 1864 *d.* London 5 July 1942
FRS 1909. VMH 1935. Science master, King Edward VI School, Birmingham, 1884. Principal, Wye College, 1894–1902. Director, Rothamsted Experimental Station, 1902–12. Secretary, Board of Agriculture, 1917–19. Director, John Innes Horticultural Institute, 1926–39. *Book of the Tulip* 1929. *The Apple* (with W. B. Crane) 1933. *Genus Tulipa* 1940. *Reconstruction and the Land* 1941.
Gdnrs Chron. 1921 ii 243, 244 portr.; 1926 ii 121; 1942 ii 17. *Bot. Mag.* v.163, 1940–42 portr. *J. R. Hort. Soc.* 1942, 319–21 portr. *Nature* v.150, 1942, 114–15. *N. Western Nat.* 1942, 399–402 portr. *Times* 7, 17 July 1942. *Daffodil Tulip Yb.* 1946, 88. *Who was Who, 1941–1950* 487. H. E. Dale *Daniel Hall* 1956.

HALL, Rev. Charles Albert (1872–1965)
b. Eastfield, Northants 11 July 1872 *d.* Clynder, Dunbartonshire 27 Aug. 1965
Held pastorates at Hull, Bristol, Paisley, Southport and London, 1896–1935. *Wild Flowers and their Wonderful Ways* 1912. *Plant Life* 1915. *Pocket Book of British Wild Flowers* 1937. *Wild Flowers in their Haunts* 1944.
Who was Who, 1961–1970 475–76.

HALL, Enid Macalister (*née* **Phillips**) (1885–1941)
Of Clifton House, Beds.
Herb. at Royal Botanic Garden, Edinburgh.

HALL, (Francis?) (*fl.* 1720s)
Physician who treated slaves on their arrival at Buenos Aires. Sent plants to W. Sherard. Plants from Buenos Aires and Andes in Sloane Herb.
J. E. Dandy *Sloane Herb.* 1958, 133. H. N. Clokie *Account of Herb. Dept. Bot., Oxford* 1964, 176.

HALL, Francis (–1834)
d. killed in revolution at Quito, Ecuador 1834
In Canada and U.S.A., 1816–17; France, 1818; S. America, 1820. Colonel in Colombian army. Friend of W. Jameson (1796–1873). Sent plants to W. J. Hooker and Humboldt. *Travels in Canada and U.S.* 1818. *Travels in France* 1819. *Colombia* 1824–25. 'Travels in Ecuador' (*J. Bot.* (Hooker) 1834, 327–54; *Companion Bot. Mag.* 1835, 26–29, 52–65, 78–80).
A. Lasègue *Musée Botanique de B. Delessert* 1845, 472. *J. Bot.* 1941, 119–20.
Letters, MSS. and drawings at Kew.

HALL, Isaac (*fl.* 1780s)
Of Newton Cartmell near Ulverston, Lancs. Had herb.
J. Stokes *Bot. Commentaries* 1830, cxvi.

HALL, J. (*fl.* 1810s)
Nurseryman, New Road, St. Pancras, London.
Trans. London Middlesex Archaeol. Soc. v.24, 1973, 192.

HALL, John (*c.* 1529–*c.* 1566)
Physician, Maidstone, Kent. Poet and medical writer. *Poesie in Forme of a Vision* 1563 (contains references to plants).
D.N.B. v.24, 69–70. *Gdnrs Chron.* 1924 i 6, 40 portr. *J. Bot.* 1924, 63.

HALL, Kate Marion (1861–1918)
b. Newmarket, Suffolk Aug. 1861 *d.* Lingfield, Surrey 12 April 1918
FLS 1905. Curator, Stepney Museum, 1893–1909. 'Tmesipteris' (*Proc. R. Irish Acad.* v.2, 1891, 1–18). *Nature Rambles in London* 1908.
Proc. Linn. Soc. 1917–18, 61–63.

HALL, Leslie Beeching (1878–1945)
d. 28 Oct. 1945
Botanical curator, London Natural History Society for 6 years.
London Nat. 1945, 77. *Proc. Bournemouth Nat. Sci. Soc.* 1944–45, 22 portr.
Herb. at BM(NH).

HALL, Patrick Martin (1894–1941)
b. 14 March 1894 *d.* Fareham, Hants 5 Aug. 1941
FLS 1934. Surveyor. Botanised in Hampshire. Editor, *Bot. Soc. Exch. Club Br. Isl. Rep.*, 1936–38.
Bot. Soc. Exch. Club Br. Isl. Rep. 1941–42, 457–60. *Proc. Linn. Soc.* 1941–42, 280.
Plants at BM(NH), Oxford.

HALL, Robert (1763–1824)
b. Roxburghshire 1763 *d.* Chelsea, London 1824
MD Edinburgh. Naval surgeon on Jamaica station and medical officer to a Niger expedition. Practised in Jedburgh and London. *Elements of Botany* 1802.
D.N.B. v.24, 85. *R.S.C.* v.3, 139.

HALL, Thomas Batt (1814–1886)
b. Coggeshall, Essex 25 July 1814 *d.* Melbourne 26 Oct. 1886
At Liverpool, 1835–39. To Melbourne, 1852. *Fl. Liverpool* 1839. Contrib. to *Naturalist* 1837–39.
Essex Nat. v.4, 1890, 226. J. B. L. Warren *Fl. Cheshire* 1899, lxxxii. *Trans. Liverpool Bot. Soc.* 1909, 71. *J. Bot.* 1922, 279. *Lancashire and Cheshire Nat.* 1922, 244–62.
Herb. at Essex Field Club Museum. Letters at Liverpool Museums.

HALL, Thomas Kendrick (1848–1890)
d. drowned in SS 'Quetta' Feb. 1890
BA Cantab. Vicar, Whatton-in-the-Vale, Notts. Collected local plants while at school at Uppingham, Rutland.
A. R. Horwood and C. W. F. Noel *Fl. Leicestershire* 1933, cclxxvii–cclxxviii.

HALL, William (–1800)
d. London 3 April 1800
Of Whitehall, Chirnside, Berwickshire. "An enthusiastic botanist." Discovered and described *Rubus nessensis* (*suberectus* Anders).
Trans. R. Soc. Edinburgh v.3, 1794, 20–21. *J. Bot.* 1885, 372. *Hist. Berwickshire Nat. Club* 1889, 539–42.

HALLE, Hughes R. P. Fraser
(*fl.* 1840s–1860s)
Letters, Historical and Botanical 1851.
European plants and letters at Kew.

HALLEY, Edmund (1656–1742)
b. Haggerston, London 29 Oct. 1656 *d.* Greenwich, London 14 Jan. 1742
MA Oxon 1678. DCL 1710. FRS 1678. Astronomer and mathematician. Sent plants from Trinidad in 1700 to J. Petiver.
J. Petiver *Museii Petiveriani* 1695, 77, 80. *Gent. Mag.* 1742, 50; 1747, 455, 503. *D.N.B.* v.24, 104–9.

HALLIDAY, John (1806–)
b. Moffat, Dumfriesshire 27 Dec. 1806
Gardener to Earl of Mansfield at Scone Palace, Perth, 1852.
Gdnrs Chron. 1875 i 401 portr.

HALLOWELL, Ernest (1865–1936)
Foreman with firm of worsted spinners. Bryologist. 'Archegonia and Autherida in *Funaria hygrometrica*' (*Naturalist* 1934, 154).
Naturalist 1936, 207.

HALLOWES, Frederick (1907–1968)
b. 17 Dec. 1907 *d.* 12 Dec. 1968
Superintendent of Parks, Dukinfield, 1939; Nottingham, 1941–63. London County Council Dept. 1964. *Newstead Abbey; Collection of Trees and Shrubs* 1953.
Who was Who, 1961–1970 479.

HALLY, John (*c.* 1798–1879)
d. Arundel, Sussex 21 Dec. 1879
Nurseryman, Blackheath, London.
Florist and Pomologist 1880, 32.

HALSTEED (Halstead), William
(*fl.* 1700s)
Major. Of Charleston, S. Carolina. Sent plants to J. Petiver, 1702.
J. Petiver *Museii Petiveriani* 1695, 96. J. E. Dandy *Sloane Herb.* 1958, 133.

HAMBROUGH, Albert John (*c.* 1820–1861)
d. London 6 June 1861
FLS 1856. Of Steephill Castle, Isle of Wight. Found *Arum italicum.* Contrib. to W. A. Bromfield's *Fl. Vectensis* 1856. Seaweeds listed in E. Venables' *Isle of Wight* 1860.
Phytologist 1854, 194–95. *Trans. Bot. Soc. Edinburgh* v.7, 1861, 202. *Proc. Linn. Soc.* 1861–62, 90. *Bot. Soc. Exch. Club Br. Isl. Rep.* 1914, 50. *R.S.C.* v.3, 145.
Herb. acquired by C. E. Palmer (1830–1914).

HAMERTON, Philip Gilbert (1834–1894)
b. Laneside, Lancs 10 Sept. 1834 *d.* Paris 4 Nov. 1894
LLD Aberdeen 1894. Artist and essayist. Had herb. *c.* 1870. *Autobiography* 1897 portr.
D.N.B. Supplt 1 v.2, 380–81. *Trans. Liverpool Bot. Soc.* 1909, 71.

HAMILTON, Mr. (*fl.* 1830s–1840s)
His New Zealand plants were in A. B. Lambert's herb.
Taxon 1970, 525.

HAMILTON, Alexander Greenlaw
(1852–1941)
b. Bailieborough, County Cavan 14 April 1852 *d.* 21 Oct. 1941
To Australia with his parents in 1866. Lecturer in Biology, Teachers Training College, Sydney. Studied fertilisation of plants and xerophily in Australian plants. President, Linnean Society N.S.W., 1915–17. *Bush Rambles* 1937. Contrib. to *Proc. Linn. Soc. N.S.W.*, *J. R. Soc. N.S.W.*
Victorian Nat. 1941, 111. *Proc. Linn. Soc. N.S.W.* 1944, 176–84 portr.

HAMILTON, Arthur Andrew (1855–1929)
b. Liverpool 9 Sept. 1855 *d.* Croydon, N.S.W. 23 April 1929

To Australia in 1880, then to New Zealand. Employed in Botanic Gardens, Sydney, 1887. Assistant, Herb., Sydney, 1911–20. Contrib. to *Proc. Linn. Soc. N.S.W.*, *J. R. Soc. N.S.W.* and *Austral. Nat.*

Austral. Nat. 1929, 171. *Sydney Morning Herald* 25 April 1929. *Proc. Linn. Soc. N.S.W.* 1920, 260–64; 1930, v–vi. *J. Proc. R. Austral. Hist. Soc.* 1932, 132–33.

Plants at Sydney.

HAMILTON, Augustus (1854–1913)
b. Poole, Dorset 1854 *d.* Bay of Islands, New Zealand 12 Oct. 1913

Schoolmaster, scientist and ethnologist. To New Zealand, 1876. Registrar, University of Otago, 1890. Director, Dominion Museum, Wellington, 1903. Collected plants in Macquarie Islands, 1894 (*Trans. N.Z. Inst.* 1894, 354, 559–79).

Trans. N.Z. Inst. 1913, v–vii portr.

Plants at Kew.

Poa hamiltoni T. Kirk.

HAMILTON, Hon. Charles (1704–1786)
d. Bath, Somerset Sept. 1786

BA Oxon 1723. MP Truro, 1741–47. Leased Painshill, Surrey in 1738 where he created a garden. One of the originators of *ferme ornée*.

Gdn Hist. 1973, 39–68 portr.

Portr. by A. David at National Portrait Gallery.

HAMILTON, Charles (*c.* 1753–1792)
b. Belfast *c.* 1753 *d.* Hampstead, London 14 March 1792

Captain in military service of East India Company. Orientalist. 'Description of Mahwah Tree' (*Asiatick Researches* v.1, 1788, 300–8).

D.N.B. v.24, 140.

HAMILTON, Claudius (*fl.* 1690s)

Surgeon. Sent plants from Barbados to J. Petiver.

J. Petiver *Museii Petiveriani* 1695, 674. J. Dandy *Sloane Herb.* 1958, 133.

HAMILTON, Francis (*né* **Buchanan**) (1762–1829)
b. Branziet, Callander, Perthshire 15 Feb. 1762 *d.* Leny, Scotland 15 June 1829

MD Edinburgh 1783. FRS 1806. ALS 1788. FLS 1816. Bengal Medical Service, 1794–1815; Assistant Surgeon, 1794; Surgeon, 1807. Most of career spent on special missions and survey work. Superintendent, Botanic Garden, Calcutta, 1814–15. Superintendent, Institution for Promoting Natural History of India, 1803. 'Commentary on Hortus Malabaricus' (*Trans. Linn. Soc.* v.13–15, 17, 1822–27, 1837). *Journey from Madras through Countries of Mysore, Canara and Malabar* 1807 3 vols. *Account of Kingdom of Nepal* 1819. Contrib. mosses to J. Sowerby and J. E. Smith's *English Bot.* 1590, etc. D. Don's *Prodromus Florae Nepalensis* 1825 based on Hamilton's and Wallich's plants.

Bot. Mag. 1820, t.2170. *Bot. Misc.* 1830–31, 91. P. Smith *Mem. and Correspondence of Sir J. E. Smith* v.1, 1832, 555; v.2, 85–88. A. Lasègue *Musée Botanique de B. Delessert* 1845, 138–41. *J. Bot.* 1899, 458; 1902, 279–82. *Ann. Bot. Gdn Calcutta* v.10 (2), 1905, i–lxxv. *D.N.B.* v.7, 186. D. G. Crawford *Hist. Indian Med. Service, 1600–1913* v.2, 1914, 62, 140. R. H. Phillimore *Hist. Rec. Survey of India* v.2, 1950, 384. *Fl. Malesiana* v.1, 214; v.5, 1958, 49. M. Archer *Nat. Hist. Drawings in India Office Library* 1962, 29–33, 72–76. M. Archer *Br. Drawings in India Office Library* v.2, 1969, 397–400. I. H. Burkill *Chapters on Hist. Bot. in India* 1965 *passim. Taxon* 1970, 525.

Plants at BM(NH), Kew, Edinburgh. MSS. and drawings of Burmese plants at BM(NH). Drawings at India Office Library. Letters in Smith correspondence at Linnean Society.

Buchanania Sprengel.

HAMILTON, Gerald Edwin Hamilton Barrett- *see* Barrett-Hamilton, G. E. H.

HAMILTON, H. (*fl.* 1890s)

Army surgeon. Collected plants in Chitral in 1896. Probably also collected with O. T. Duke in 1876 in Quetta.

Pakistan J. For. 1967, 346.

Plants at Dehra Dun.

HAMILTON, Rev. James (1814–1867)
b. Paisley, Renfrewshire 27 Nov. 1814 *d.* London 24 Nov. 1867

MA Glasgow. DD Edinburgh. FLS 1848. Presbyterian minister, National Scottish Church, London, 1841–67. Contrib. botanical articles to P. Fairbairn's *Imperial Bible Dict.* 1866 2 vols.

Proc. Linn. Soc. 1867–68, civ–cv. *Trans. Bot. Soc. Edinburgh* v.9, 1868, 269–72. W. Arnot *Life of James Hamilton* 1870 portr. *D.N.B.* v.24, 188.

HAMILTON, Samuel (*fl.* 1900s)

BA. MB. Of Newport, Mon. Medical Officer of Health, Marshfield, Mon. *Fl. Monmouthshire* 1909.

A. E. Wade *Fl. Monmouthshire* 1970, 16.

HAMILTON, Thomas, 6th Earl of Haddington (1680–1735)
b. 29 Aug. 1680 *d.* New Hailes near Edinburgh 28 Nov. 1735
Planted more than 50 species of trees on his estate at Tynninghame near Dunbar. *Treatise on Manner of Raising Forest Trees...* 1761.
D.N.B. v.24, 212–13.

HAMILTON, Thomas (–1782)
d. 7 Jan. 1782
Professor of Anatomy and Botany, Glasgow, 1757.
Gent. Mag. 1782, 46. J. R. Green *Hist. Bot.* 1914, 283. *Gdnrs Chron.* 1919 i 147.

HAMILTON, William (*fl.* 1780s–1790s)
Professor of Anatomy and Botany, Glasgow, 1781–90.
Notes R. Bot. Gdn Edinburgh v.3, 1904, 57.

HAMILTON, William (1783–1856)
d. Plymouth, Devon 25 May 1856
MB. Collected plants in West Indies, 1814. *Prodromus Plantarum Indiae Occidentalis* 1825. 'Timber-trees of Choco, Mexico' (*Gdnrs Mag.* 1829, 44–47).
Bot. Mag. 1830, t.2996. I. Urban *Symbolae Antillanae* v.1, 1898, 64–65; v.3, 1902, 55–56. *R.S.C.* v.3, 147.

HAMILTON, William Phillips (*c.* 1842–1910)
d. Caversham, Oxford 1910
Nephew of William Phillips. Edited unpublished *Fl. Shropshire*. 'Shropshire Sphagna' (*J. Bot.* 1902, 416–19). 'Some Kirkcudbright Mosses' (*J. Bot.* 1901, 422–24).

HAMLIN, E. J. (*c.* 1878–1966)
d. Bridgwater, Somerset 3 March 1966
Collected Somerset plants now at Taunton Castle, Somerset.
Somerset Archaeol. Nat. Hist. Soc. v.110, 1965–66, 119.

HAMMOND, Henry (*fl.* 1780s–1810s)
Nurseryman, Rugeley, Staffs.

HAMMOND, Thomas (*fl.* 1790s)
Nurseryman, 2 Corn-Market, Worcester.

HAMOND, Charles Annesley (1856–1914)
b. London 1856 *d.* Twyford Hall, Norfolk 1 Feb. 1914
President, Norfolk and Norwich Naturalist Society, 1906–7. Botanist.
Trans. Norfolk Norwich Nat. Soc. 1912–13, 821–23 portr.

HAMPDEN, A. G. Hobart- *see* Hobart-Hampden, A. G.

HAMPTON, Frank Anthony (1888–1967)
b. Ewell, Surrey 2 Jan. 1888 *d.* Bampton, Oxfordshire 28 March 1967
Educ. Oxford. Psychiatrist. *Scent of Flowers and Leaves* 1925. *Herbs, Salads and Seasonings* 1930. *Wild Foods of Britain* 1939. Under pseudonym of Jason Hill wrote *Curious Gardener* 1932. *Contemplative Gardener* 1940.
Gdnrs Chron. v.161(18), 1967, 4. *Times* 8 April 1967.

HAMSON, John (1858–1930)
b. Naseby, Northants 11 Sept. 1858 *d.* Malvern, Worcs 4 June 1930
Reporter, *Leicester Post*, 1878. Assistant Editor, *Bedfordshire Times*, 1883. President, Bedfordshire Natural History Society. Lectured on botany at evening classes in Bedford. *Fl. Bedfordshire* 1906.
Bot. Soc. Exch. Club Br. Isl. Rep. 1935, 59–60. J. G. Dony *Fl. Bedfordshire* 1953, 28–29.
Plants at Bedfordshire Natural History Society.

HANBURY, Sir Cecil (*c.* 1871–1937)
d. 10 June 1937
FLS 1910. MP for North Dorset, 1924–37. Owner of gardens, herb. and library at La Mortola, Liguria, Italy, inherited from his father, Sir Thomas Hanbury. *La Mortola Garden* 1938.
Gdnrs Chron. 1937 i 433. *Kew Bull.* 1937, 356. *Proc. Linn. Soc.* 1937–38, 317–18. *Who was Who, 1929–1940* 587.
Portr. at Hunt Library.

HANBURY, Daniel (1825–1875)
b. London 11 Sept. 1825 *d.* Clapham, London 24 March 1875
FRS 1867. FLS 1855. Brother of Sir Thomas Hanbury. Pharmacologist in family firm of Allen, Hanbury and Barry. With J. D. Hooker in Syria, 1860. An early investigator of Chinese materia medica. Treasurer, Linnean Society, 1873–75. *Pharmacographia* (with F. A. Flueckiger) 1874; ed. 2 1879. *Science Papers* 1876.
Gdnrs Chron. 1875 i 429–30; 1875 ii 112 portr. *Nature* v.11, 1875, 428–29. *Pharm. J.* v.5, 1875, 797–99; v.61, 1925, 341–44 portr. *J. Bot.* 1875, 127–28, 192. *Proc. Linn. Soc.* 1874–75, xlvii–l. *Proc. R. Soc.* v.24, 1876, ii–iii. *Kew Bull.* 1893, 187. E. Bretschneider *Hist. European Bot. Discoveries in China* 1898, 815–16. F. B. Power *Influence and Development of some of the Researches of Daniel Hanbury* 1913. *D.N.B.* v.24, 270–71.
Herb. and library at Pharmaceutical Society. Letters at Kew. Portr. at Hunt Library.
Hanburia Seem.

HANBURY, F. A. (*fl.* 1860s)
'Note on Fl. Sarnicae' (*J. Bot.* 1863, 92). Plant specimens from Devon, Wilts, Suffolk, Jersey at Oxford.

HANBURY, Frederick Janson (1851–1938)
b. Stoke Newington, London 27 May 1851 *d.* East Grinstead, Sussex 1 March 1938
FLS 1873. VMH 1924. VMM 1927. Director, later Chairman of family firm, Allen and Hanbury, 1916–37. Had garden at East Grinstead where he grew orchid hybrids. *Fl. Kent* (with E. S. Marshall) 1897. *Illustrated Monograph of British Hieracia* 1889–98. Edited *London Catalogue of British Plants*, eds 8–11, 1886–1925. Contrib. to *J. Bot.*
Orchid Rev. 1934, 66–67 portr.; 1938, 128. *London Nat.* 1937, 24. *Bot. Soc. Exch. Club Br. Isl. Rep.* 1938, 19. *Chronica Botanica* 1938, 261. *Gdnrs Chron.* 1932 i 392 portr.; 1938 i 189. *J. Bot.* 1938, 211–12. *Nature* v.141, 1938, 544. *Proc. Linn. Soc.* 1937–38, 318–19. *Times* 3 March 1938. *Who was Who, 1929–1940* 587.
Herb. at BM(NH). Plants at Oxford.

HANBURY, Samuel (*fl.* 1890s)
Collected seeds in 'Rocky Mountains' U.S.A. from which *Yucca hanburii* Baker was grown in Thomas Hanbury's garden at La Mortola.
Kew Bull. 1892, 8.

HANBURY, Sir Thomas (1832–1907)
b. Clapham, London 21 June 1832 *d.* La Mortola, Liguria, Italy 9 March 1907
FLS 1878. VMH 1903. KCVO 1901. Merchant at Shanghai, 1853–71. Established botanic garden at La Mortola, 1867. Founded Botanical Institute of Genoa, 1892. Presented G. Wilson's garden at Wisley, Surrey to Royal Horticultural Society, 1903. *Letters of Sir Thomas Hanbury* 1913 portr.
Gdnrs Chron. 1907 i 172 portr. *J. Bot.* 1907, 216. *J. Hort. Cottage Gdnr* v.54, 1907, 244 portr., 374–75. *Kew Bull.* 1907, 132–36. *Proc. Linn. Soc.* 1906–7, 46–48. *Trans. Bot. Soc. Edinburgh* v.23, 1908, 355–58. *Who was Who, 1897–1916* 312. *Curtis's Bot. Mag. Dedications, 1827–1927* 263–64 portr. A. White and B. L. Sloane *Stapelieae* v.1, 1937, 133 portr. *Boissiera* fasc. 5, 1941, 52.
Letters at Kew. Portr. at Hunt Library.

HANBURY, Rev. William (1725–1778)
b. Bedworth, Warwickshire 1725 *d.* Church Langton, Leics 1 March 1778
BA Oxon 1748. MA St. Andrews 1769. Rector, Church Langton, 1753. In 1751 began sowing seeds "from distant countries, particularly North America." In 1758 set up a trust to run his nursery for the benefit of his church. *Essay on Planting* 1758. *Complete Body of Planting and Gardening* 1770 2 vols.
J. C. Loudon *Encyclop. Gdning* 1822, 1272. G. W. Johnson *Hist. English Gdning* 1829, 214. S. Felton *Portr. of English Authors on Gdning* 1830, 143–47. *Cottage Gdnr* v.7, 1852, 299. *J. Hort. Cottage Gdnr* v.55, 1876, 309–10 portr. *D.N.B.* v.24, 271–72. A. R. Horwood and C. W. F. Noel *Fl. Leicestershire* 1933, clxxxix. *Amateur Gdning* 2 May 1970, 34 portr.

HANBURY-FORBES, Mrs. Dorothy (–1972)
d. La Mortola, Liguria 29 Feb. 1972
FLS 1963. Restored the gardens at La Mortola, devastated after 1939–45 war. Presented them to Italian government in 1960.
J. R. Hort. Soc. 1972, 276. *Times* 7 March 1972.

HANCE, Henry Fletcher (1827–1886)
b. Brompton, London 4 Aug. 1827 *d.* Amoy, China 22 June 1886
PhD 1849. FLS 1878. In Hong Kong, 1844. Vice-consul at Whampoa, 1861–78. Consul at Canton, 1878–81 and 1883. Acting Consul at Amoy, 1886. 'Supplement to Bentham's Fl. Hongkongensis' (*J. Linn. Soc.* 1873, 95–144).
G. Bentham *Fl. Hongkongensis* 1861, 9*–10*. *J. Bot.* 1887, 1–11 portr. *Proc. Linn. Soc.* 1886–87, 40–41. *Gdnrs Chron.* 1886 ii 218–19. *Bot. Gaz.* 1887, 48. E. Bretschneider *Hist. European Bot. Discoveries in China* 1898, 365–70, 632–52. *D.N.B.* v.24, 272–73. *Acta Horti Bergiani* v.3(2), 1905, t.146 portr. *Tribuna Farmaceutica* v.9, 1941, 287. E. H. M Cox *Plant Hunting in China* 1945, 93–94. *Fl. Malesiana* v.1, 1950, 215. G. A. C. Herklots *Hong Kong Countryside* 1951, 163–64. *House and Gdn* March 1975, 133–34 portr. *R.S.C.* v.3, 156; v.7, 898; v.10, 127; v.15, 614.
Herb. at BM(NH). Letters at Kew. Portr. at Hunt Library. Sale at Sotheby 8 Aug. 1888.
Hancea Hemsley.

HANCOCK, Albany (1806–1873)
b. Newcastle, Northumberland 24 Dec. 1806 *d.* Newcastle 24 Oct. 1873
Zoologist and palaeontologist. 'On some Curious Fossil Fungi from the Black Shale of the Northumberland Coalfield' (with T. Attley) (*Ann. Mag. Nat. Hist.* v.4, 1869, 221–28; *Northumberland Nat. Hist. Trans.* v.3, 1870, 321–30).
Nature v.9, 1873, 43–44.

HANCOCK, John (*fl.* 1800s–1840s)
MD. Librarian, Medico-Botanical Society, London. Resided 25 years in British Guiana where he collected plants. 'Angostura Bark Tree' (*Trans. Med.-Bot. Soc.* v.1, 1829, 16–29, Appendix 11).
Taxon 1970, 525. *R.S.C.* v.3, 158.
Plants at Kew.

HANCOCK, Thomas (1783–1849)
b. Lisburn, County Antrim 1783 *d.* Lisburn 6 April 1849
MD Edinburgh 1809. Practised in London, 1809–30. 'Plants found near Bristol, 1836' (*Proc. Bot. Soc. London* 1839, 25–28). '*Lamium maculatum*' (*Proc. Bot. Soc. London* 1839, 32–34).
J. Smith *Cat. Friends' Books* v.1, 1867, 910–12. W. Munk *Roll of R. College of Physicians* v.3, 1878, 78. *D.N.B.* v.24, 275.

HANCOCK, William (*fl.* 1720s)
Collected plants at Pegu, Burma, now at Oxford.

HANCOCK, William (1847–1914)
b. Lurgan, Ulster 1847 *d.* Bristol 1914
Educ. Queen's College, Belfast. FLS 1884. In Chinese Imperial Customs, 1874. Collected plants in China, Formosa, Japan, Sumatra, Java, Jamaica, Guatemala, Mexico.
E. Bretschneider *Hist. European Bot. Discoveries in China* 1898, 747–50. *Kew Bull.* 1901, 29; 1922, 204. *Bot. Soc. Exch. Club Br. Isl. Rep.* 1922, 704. *Fl. Malesiana* v.1, 1950, 215.
Plants at Kew.
Hancockia Rolfe.

HANCORN, Philip (*fl.* 1790s)
In Portuguese navy. Chief of fleet in Brazil, 1797. "Rerum naturalium studiosus... etiam studiosarum fautor" (Gomes *Memorias dos Corresp.* 1812, 51).
J. Bot. 1896, 250.
Hancornia Gomes.

HANDASYDE, Thomas *and* **William** (*fl.* 1830s–1840s)
Nurserymen, Fisherrow, Musselburgh near Edinburgh.
Bot. Mag. 1838, t.3694.

HANDEY, John (*c.* 1836–1910)
d. Sedbergh, Yorks 13 Oct. 1910
JP. Member of Society of Friends. *Catalogue of Plants Growing in Sedbergh District* 1898.
Naturalist 1911, 17 portr.

HANDISYD, (Handyside) George (*fl.* 1690s)
Surgeon, Royal Navy. Collected plants in Juan Fernandez, Magellan and West Indies. Made first Fuegian collection.
E. J. L. Scott *Index to Sloane Manuscripts* 1904, 239. *J. Bot.* 1909, 207–8. J. E. Dandy *Sloane Herb.* 1958, 134.
Plants at BM(NH).

HANGER, Francis E. W. (1900–1961)
b. Dorset 1900 *d.* Wisley, Surrey 22 Oct. 1961
VMH 1953. Gardener to Lionel de Rothschild at Exbury, Hants, 1927. Curator, Royal Horticultural Society gardens, Wisley, 1946–61. Contrib. to *J. R. Hort. Soc.*
Gdnrs Chron. 1951 ii 128 portr.; 1961 ii 357. *J. R. Hort. Soc.* 1961, 513–16 portr. *Times* 24 Oct. 1961. *House and Gdn* March 1975, 134 portr.

HANHAM, Frederick H. (1806–1877)
b. Lansdown, Bath, Somerset 7 May 1806 *d.* Bath 26 May 1877
MRCS Edinburgh. In practice at Bath. *Natural Illustrations of British Grasses* (with dried specimens) 1846. *Manual for Victoria Park, Bath* 1857.

HANKEY, John Alexander (–1882)
FLS 1835. Discovered *Allium ambiguum* and *Polygonum dumetorum.*
J. Sowerby and J. E. Smith *English Bot.* 2803, 2811. H. N. Clokie *Account of Herb. Dept. Bot., Oxford* 1964, 177.
Plants at Oxford.

HANKEY, William Barnard (*fl.* 1830s–1860s)
Of Cranleigh, Surrey. Collected and cultivated ferns.
Br. Fern Gaz. 1909, 45. H. N. Clokie *Account of Herb. Dept. Bot., Oxford* 1964, 177.
Middle East plants at Oxford.

HANKIN, James (*fl.* 1790s–1820s)
Partner with James Blundell from 1790 in nursery at Ormskirk, Lancs.

HANKS, James (*fl.* 1790s–1830s)
Gardener with Messrs. Perfect of Pontefract, 1798. Nurseryman, Pontefract, 1802–9. Had nursery at Keighley, Yorks which he disposed of in 1811. Formed new nursery at Carlton near Pontefract. In partnership with Muscroft in 1830s.

HANLON, Cabel (*fl.* 1740s)
Nurseryman, Dublin.
Irish For. 1967, 47.

HANMER, Sir Thomas (1612–1678)
Had garden at Bettisfield, Flint. Friend of John Evelyn and John Rea. His MS. garden book, 1659, was edited by I. Elstob and published as *Garden Book of Sir Thomas Hanmer* 1933 portr.

J. Hanmer *Memorials of Parish and Family of Hanmer in Flintshire* 1877. E. Cecil *Hist. of Gdning in England* 1910, 164, 171–74. M. Hadfield *Gdning in Britain* 1960, 93–102. *House and Gdn* March 1975, 134 portr.
Portr. at Hunt Library.

HANNAFORD, Samuel (1828–1874)
b. Totnes, Devon 1828 *d.* Hobart, Tasmania 3 Jan. 1874
To Melbourne, 1853; to Hobart, 1858. Librarian, Public Library, Hobart. Sent plants to F. von Mueller and algae to W. H. Harvey. *Fl. Tottoniensis* 1851. *Jottings in Australia* 1856. *Wild Flowers of Tasmania* 1866. *Sea and Riverside Rambles in Victoria* 1860.
P. Mennell *Dict. Austral. Biogr.* 1892, 213. *Victorian Nat.* 1908, 108. *Papers Proc. R. Soc. Tasmania* 1909, 18–19. W. K. Martin and G. T. Fraser *Fl. Devon* 1939, 773.
Plants at Melbourne, Dublin. Letters at Kew.
Hannafordia F. Muell.

HANNAN, William Isaac (1847–1908)
b. Heywood, Lancs 1847 *d.* Ashton-under-Lyne, Lancs 28 June 1908
Cotton spinner and, later, school attendance officer. Friend of John Whitehead (1833–1896). Lectured on botany. Some botanical papers at Ashton-under-Lyne Library. *Textile Fibres of Commerce* 1903.
Heywood Advertiser 31 July 1908. *Trans. Liverpool Bot. Soc.* 1909, 72.

HANNINGTON, Rev. James (1847–1885)
b. Hurstpierpoint, Sussex 3 Sept. 1847 *d.* murdered in Uganda 30 Oct. 1885
BA 1873. DD Oxon 1884. FLS 1883. Curate, Hurstpierpoint, 1875–82. With Church Missionary Society in Uganda, 1882. Bishop of East Equatorial Africa, 1884–85. Led expedition which reached Lake Victoria Nyanza, 1885.
J. Bot. 1883, 245; 1886, 128; 1896, 55. *Proc. Linn. Soc.* 1885–86, 143. E. C. Dawson *James Hannington* 1887 portr. *D.N.B.* v.24, 307–8.
East African plants at Kew.
Asplenium hanningtoni Baker.

HANSARD, Providence (*fl.* 1780s)
Seedsman, Redcliffe Street, Bristol.

HANSFORD, Clifford George (1900–1966)
b. Wincanton, Somerset 1900 *d.* Port Shepstone, S. Africa 18 Feb. 1966
MA Cantab 1924. FLS 1931. Microbiologist, Dept. Agriculture, Jamaica, 1922–26. Mycologist, Dept. Agriculture, Uganda, 1926–46. Plant Pathologist, Tea Research Institute, Ceylon, 1946. Waite Institute, Adelaide, 1951–*c.* 1957. Cotton research in Bechuanaland, 1959–61.
S. African Forum Botanicum v.4(6), 1966, 2.
Fungi at Kew, Commonwealth Mycological Institute. Australian plants at Adelaide.

HANSON, Isaac (*fl.* 1780s)
Seedsman, Thames Street, London.

HANSON, Percy James (–1949)
Ornithologist. Had herb. and collection of botanical books which he bequeathed to London Natural History Society.
London Nat. 1948, 126.

HARBEN, Guy P. (–1949)
d. London 3 Dec. 1949
Director, Prudential Assurance Company. Grew orchids.
Orchid Rev. 1950, 23.

HARCOURT, Frederick George (1889–1970)
b. Sunninghill, Berks 23 March 1889 *d.* 18 Nov. 1970
Kew gardener, 1913–20. Agricultural Superintendent, Antigua, 1920. Curator and Agricultural Superintendent, Dominica, 1924.
J. Kew Guild 1962, 135–36 portr.; 1970, 1157–58.
Dryopteris harcourtii.

HARCOURT, Lewis Vernon, 1st Viscount Harcourt (1863–1922)
b. 31 Jan. 1863 *d.* 24 Feb. 1922
MP, Rossendale, Lancs. Colonial Secretary, 1910–15. Commissioner of Works, 1905–10, 1915–17. Had garden at Nuneham Park, Oxfordshire.
Bot. Soc. Exch. Club Br. Isl. Rep. 1922, 704–6.

HARDAKER, Walter Henry (1877–1970)
b. Yeadon, Yorks 6 July 1877 *d.* 19 Nov. 1970
BSc Leeds. MSc Birmingham 1910. Lecturer in Science, Handsworth Technical School, 1902–38. Surveyed local flora and added to records of *Fl. Warwickshire* 1971.
Proc. Birmingham Nat. Hist. Soc. 1972, 90–91. *Watsonia* 1972, 177.

HARDCASTLE, Emma Winifred *see* O'Malley, E. W.

HARDCASTLE, Lucy (*fl.* 1760s–1830s)
Of Derby. *Elements of Linnaean System of Botany* 1830. (MS. at BM(NH)).

HARDING, Rev. Michael (*c.* 1649–1690s)
b. Holywell, Oxford *c.* 1649
BA Oxon 1669. BD 1684. Annotated a copy of J. Ray's *Catalogus*, now at Oxford.
G. C. Druce *Fl. Oxfordshire* 1927, lxxxiii.

HARDMAN, Lawrence (1808–1896)
b. 23 July 1808 *d.* 15 May 1896
Of Liverpool. Diatomist.
Quart. J. Microsc. Sci. 1865, 55. G. B. De Toni *Sylloge Algarum* v.2, 1891, cxxvii.
Slides at BM(NH).
Triceratium hardmanianum Grev.

HARDWICKE, Robert (1824–1875)
d. 8 March 1875
FLS 1863. Of Dyke, Lincs. Publisher and founder of *Sci. Gossip* 1865.
Sci. Gossip 1875, 86.
Herb. sent to Bombay (D. H. Kent *Br. Herb.* 1957, 59).

HARDWICKE, Thomas (1755–1835)
d. Lambeth, London 3 March 1835
FRS 1813. FLS 1804. Entered Bengal Artillery, 1778; Major-General. Retired from India, 1823. Employed artists, both Indian and British, to make natural history drawings. Collected plants in S. Africa and St. Helena; in India, 1796; in Mauritius, 1811.
Bot. Misc. v.1, 1830–31, 89–91. P. Smith *Mem. and Correspondence of Sir J. E. Smith* v.2, 1832, 118–21. *Rep. Br. Assoc.* 1845, 188. *List of Books by J. E. Gray* 1872, 7. *J. Bot.* 1906, 235–41. *J. Soc. Bibl. Nat. Hist.* 1946, 55–69. M. Archer *Nat. Hist. Drawings in India Office Library* 1962, 8–10, 79. *Taxon* 1970, 525.
Plants and drawings at BM(NH). Plants at Kew. Letters at Kew, Linnean Society. Portr. by W. Hawkins at BM(NH).
Hardwickia Roxb.

HARDY, George (*c.* 1832–1894)
d. Timperley, Cheshire 26 March 1894
Of Pickering Lodge, Timperley. Grew orchids.
Orchid Rev. 1894, 130.
Cattleya × Hardyana.

HARDY, J. (*fl.* 1830s)
Nurseryman, Edmonton, Middx. Nursery opposite 'The Golden Fleece', Edmonton. Nursery acquired by T. Page in 1838.

HARDY, James (1815–1898)
b. Bilsdean, E. Lothian 1 June 1815 *d.* Old Cambus, Berwickshire 30 Sept. 1898
LLD Edinburgh 1890. Of Old Cambus. 'Lichen-flora of Eastern Borders' (*Hist. Berwickshire Nat. Club* v.4, 1863, 396–428). Contrib. other botanical papers to *Hist. Berwickshire Nat. Club.*
Ann. Scott. Nat. Hist. 1899, 1–6. *Hist. Berwickshire Nat. Club.* v.16, 1899, 341–72; v.17, 261–64. *Gdnrs Chron.* 1900 ii 72. *R.S.C.* v.3, 176; v.7, 907; v.10, 141; v.15, 639.
Letters in Wilson correspondence at BM(NH).

HARDY, John (1817–1884)
b. York 4 Nov. 1817 *d.* Manchester 15 Sept. 1884
Fellow Bot. Soc. Edinburgh, 1844. 'Curious Fern' (*Phytologist* v.1, 1844, 92–93). '*Leucojum vernum*, a probable British plant' (*J. Bot.* 1866, 88).
Rep. Proc. Manchester Sci. Students' Assoc. 1884, 71–79 portr. *Trans. Liverpool Bot. Soc.* 1909, 72. *R.S.C.* v.7, 907.
Herb. at Manchester.

HARDY, John (1834–1916)
b. Wooler, Northumberland *d.* Melbourne 1916
Village schoolteacher. Emigrated to Australia, 1853. Surveyor, Sydney. Geological Survey, Victoria, 1858. District Surveyor, Alexandra, Victoria. Sent plants to F. von Mueller.
Victorian Hist. Mag. 1943, 1–14.

HARDY, Robert Gathorne- *see* Gathorne-Hardy, R.

HARDY, Sir Thomas (*fl.* 1820s)
Sent seeds from S. America to Lady Campbell, *c.* 1820.
Bot. Mag. 1832, t.3150.

HARE, E. C. (*fl.* 1890s)
Collected plants in Samana Range and Safed Koh, N. West Frontier, 1898.
Pakistan J. For. 1967, 346.
Plants at Kew, Arnold Arboretum.

HARE, Richard (–*c.* 1826)
FLS 1810. Of Bath. Algologist. "Paid particular attention to the Algae of Devonshire" (D. Turner *Fuci* v.4, 1819, 4).

HARFORD, H. W. L. (–1921)
Of Horton Hall, Glos. Botanist.
H. J. Riddelsdell *Fl. Gloucestershire* 1948, cxlviii.

HARFORD, W. A. (–1925)
Of Badminton, Glos. Studied and drew local plants.
H. J. Riddelsdell *Fl. Gloucestershire* 1948, cxlix.

HARIOT, Thomas (1560–1621)
b. Oxford 1560 *d.* 2 July 1621
BA Oxon 1580. Mathematician, astronomer, surveyor, historian, mentor to Sir Walter Raleigh. Went to 'Virginia' (Roanoke River area, N.C.) 1585–86 and travelled inland 150 miles by his own reckoning. *Brief and True Report of New Found Land of Virginia* 1588 (includes description of vegetation).
D.N.B. v.24, 437–39. J. and N. Ewan *John Banister* 1970, 147, 361–62.

HARKER, James Allen (1847–1894)
b. 31 July 1847 *d.* Cirencester, Glos 19 Dec. 1894

FLS 1883. Professor of Natural History, Royal Agricultural College, Cirencester, 1881–94. Vice-president, Cotteswold Naturalists' Field Club which possesses his MS. account of *Botany of Environs of Cirencester*. Primarily a zoologist but interested in *Umbelliferae*, grasses and water-plants.

Proc. Linn. Soc. 1894–95, 32–33. H. J. Riddelsdell *Fl. Gloucestershire* 1948, cxxxviii. *R.S.C.* v.10, 142; v.12, 312; v.15, 642.

HARKNESS, John (*fl.* 1900s)

Nurseryman, Bedale, Yorks, 1880. Specialised in roses. Moved to Hitchin, Herts, 1899. *Practical Rose Growing* 1889.

Gdnrs Mag. 1895, 419 portr. *J. Hort. Cottage Gdnr* v.45, 1902, 473, 475 portr.

HARKNESS, Robert (1816–1878)
b. Ormskirk, Lancs 28 July 1816 *d.* Dublin 4 Oct. 1878

FRS 1856. Paleaontologist. Professor of Geology, Queen's College, Cork, 1853–78. 'Coal' (*Edinburgh New Philos. J.* v.57, 1854, 66–76). 'Subfossil Diatomaceae' (*Edinburgh New Philos. J.* v.2, 1855, 54–56).

Geol. Mag. 1878, 574–76 portr. *Mineralogical Mag.* 1879, 153–54. *Quart. J. Geol. Soc.* 1879, 41–44. *Proc. R. Soc. Edinburgh* v.10, 1880, 31–33. *D.N.B.* v.24, 390. *Trans. Liverpool Bot. Soc.* 1909, 72. *R.S.C.* v.3, 183; v.7, 908; v.10, 142.

HARKNESS, Robert (1855–1920)
d. Hitchin, Herts 10 Nov. 1920

Brother of J. Harkness with whom he established the Bedale nursery. Founded rose nursery at Hitchin, 1899 as Robert Harkness & Co.

Garden 1920, 591.

HARLAND, William Aurelius (–1857)
b. Scarborough, Yorks *d.* Hong Kong 12 Sept. 1857

MD Edinburgh 1845. Colonial surgeon at Hong Kong, 1848. Collected plants.

W. G. Walpers *Annales Botanicae Systematicae* v.2, 1852–53, 648. G. Bentham *Fl. Hongkongensis* 1861, 10*. *J. Bot.* 1887, 5. E. Bretschneider *Hist. European Bot. Discoveries in China* 1898, 371–73. G. A. C. Herklots *Hong Kong Countryside* 1951, 164.

Plants at Kew.

Harlandia Hance.

HARLEY, Andrew (*c.* 1872–1950)

VMH 1950. Had garden at Glendevon, Perthshire where he specialised in Sino-Himalayan plants.

J. R. Hort. Soc. 1951, 54 portr.

HARLEY, John (1833–1921)
b. Ludlow, Shropshire 1833 *d.* Pulborough, Sussex 9 Dec. 1921

MD. FLS 1863. Geologist. 'Parasitism of Mistle-toe' (*Trans. Linn. Soc.* v.24, 1863, 175–96).

Bot. Soc. Exch. Club Br. Isl. Rep. 1921, 357–58. *Nature* v.108, 1921, 575. *J. Bot.* 1922, 94. *Proc. Linn. Soc.* 1921–22, 45. *R.S.C.* v.3, 189; v.7, 909.

HARLEY, Margaret Cavendish *see* Bentinck, M.C. *Duchess of Portland*

HARLOW, James (*fl.* 1660s–1680s)

Gardener. Went to Virginia for John Watts of Chelsea Physic Garden, and to Jamaica for Sir Arthur Rawdon of Moira, County Down. Brought back plants, including many ferns, which were grown at Moira and then sent to H. Sloane.

L. Plukenet *Almagestum Botanicum* 1696, 34, 63, 260. R. Morison *Plantarum Historiae Universalis Oxoniensis* v.3, 1699, 563, 572, 579. H. Sloane *Nat. Hist. of Jamaica* v.1, 1707, preface, 3. R. Pulteney *Hist. Biogr. Sketches of Progress of Bot. in England* v.2, 1790, 81. *J. Bot.* 1886, 14. I. Urban *Symbolae Antillanae* v.3, 1902, 56. J. E. Dandy *Sloane Herb.* 1958, 134–35. H. N. Clokie *Account of Herb. Dept. Bot., Oxford* 1964, 177–78.

Plants at BM(NH), Oxford.

HARPER, Alan Gordon (1889–1917)
b. Dulwich, London 5 Jan. 1889 *d.* near Ypres, Belgium 1 June 1917

BA Oxon 1912. Deputy Professor of Botany, Presidency College, Madras, 1914.

Bot. Soc. Exch. Club Br. Isl. Rep. 1917, 88–90. *Nature* v.99, 1917, 312. *Oxford Mag.* 1917.

HARPUR-CREWE, Rev. Henry (1830–1883)
b. 30 Sept. 1830 *d.* Drayton Beauchamp, Bucks 7 Sept. 1883

BA Cantab 1851. Rector, Drayton Beauchamp, 1860. Entomologist. Botanised in Derbyshire, 1864. Cultivated crocuses.

Bot. Mag. 1875, t.6168. *Entomol. Mon. Mag.* 1883, 118–19. *Garden* v.24, 1883, 238; v.31, 1887, 371 portr. *Gdnrs Chron.* 1883 ii 347. *J. Bot.* 1883, 380–81. *Naturalist* 1883, 56–57. W. R. Linton *Fl. Derbyshire* 1903, 37. G. C. Druce *Fl. Buckinghamshire* 1926, c. A. Simmonds *Hort. Who was Who* 1948, 43. *R.S.C.* v.2, 92; v.7, 458; v.9, 602; v.12, 177.

Letters at Kew.

HARRADENCE, John Edward (1909–1970)
b. Reading, Berks 7 Aug. 1909 *d.* 11 March 1970

Manager, Indian branch of Sutton and Sons, seed merchants, 1939–55; Director, Reading office, 1964.

Bull. African Succulent Plant Soc. v.5(1), 1970, 28–29.

HARRIMAN, Rev. John (1760–1831)
b. Maryport, Cumberland 1760 *d.* Croft, Yorks 3 Dec. 1831

FLS 1798. Rector, Eglestone, Yorks and from 1801 of Gainford, Durham. Lichenologist. Discovered *Gentiana verna.* Contrib. to J. Sowerby and J. E. Smith's *English Bot.* 361, 2539, etc.

D. Turner and L. W. Dillwyn *Botanist's Guide* 1807, 143, 239. N. J. Winch *et al. Botanist's Guide* v.2, 1807, ii. *D.N.B.* v.24, 433–34. W. Hodgson *Fl. Cumberland* 1898, xxv. *Nat. Hist. Trans. Northumberland, Durham and Newcastle-upon-Tyne* v.14, 1903, 79.

Letters in Winch correspondence at Linnean Society and in Dawson Turner correspondence at Cambridge. Plants at Liverpool Museums.

Verrucaria harrimanni Ach.

HARRINGTON, Robert (*fl.* 1780s–1810s)

MD. Practised medicine in Carlisle, Cumberland. *General Principles of Vegetable Life* 1781.

D.N.B. v.24, 436–37.

HARRIS, George Francis Robert, 3rd Baron Harris (1810–1872)
b. Belmont, Kent 14 Aug. 1810 *d.* Belmont 23 Nov. 1872

BA Oxon 1832. Governor, Trinidad, 1846; Madras, 1854–59. Sent seeds of *Thunbergia harrisii* to W. J. Hooker.

Bot. Mag. 1857, t.4998. *D.N.B.* v.25, 5.

HARRIS, George Prideaux Robert (1775–1810)
d. Hobart, Tasmania 16 Oct. 1810

Of Exeter, Devon. To Port Phillip, Victoria as deputy surveyor, 1803. Farmed at Sandy Bay, Tasmania. Studied local flora and fauna.

Hist. Rec. Austral. 1916, 1921 *passim. J. Proc. R. Austral. Hist. Soc.* 1934, 302–3. R. and T. Rienits *Early Artists of Austral.* 1963 *passim. Austral. Dict. Biogr.* v.1, 1965, 516–17.

HARRIS, George St. Pierre (*c.* 1807–1901)
b. Chelsfield, Kent *c.* 1807 *d.* Orpington, Kent 26 Dec. 1901

Florist. Raised show and fancy dahlias.

Garden v.61, 1902, 30. *Gdnrs Mag.* 1902, 16. *J. Hort. Cottage Gdnr* v.44, 1902, 14.

HARRIS, George Thomas (*c.* 1856–1938)
b. Halesowen near Birmingham *d.* Devon 30 Nov. 1938

Retired to Devon in 1904. President, Devonshire Association, 1938. Specialised in fresh-water algae and diatom flora. Had collection of botanical photographs. Contrib. to *Trans. Devon Assoc. Advancement Sci.*, *J. Quekett Microsc. Club.*

J. Quekett Microsc. Club 1939, 128. W. K. Martin and G. T. Fraser *Fl. Devon* 1939, 779.

HARRIS, John (*fl.* 1780s–1810s)

Gardener and nurseryman, North Town, Taunton, Somerset. Nursery sold to James Poole (*c.* 1777–*c.* 1827).

HARRIS, Joseph (*fl.* 1700s–1720s)

Surgeon-apothecary. Member of Botanical Society of London.

H. N. Clokie *Account of Herb. Dept. Bot., Oxford* 1964, 178. *Proc. Bot. Soc. Br. Isl.* 1967, 316–17.

Plants at Oxford.

HARRIS, Moses (1730–*c.* 1788)

Artist and engraver. Secretary of Society of Aurelians, 1766. Accurately drew not only insects but also host plants. *Aurelian or Natural History of English Insects* 1766.

D.N.B. v.25, 20–21. *Popular Sci. Mon.* 1926, 560–64. A. A. Lisney *Bibl. Br. Lepidoptera 1608–1799* 1960, 156–75.

HARRIS, Richard (*fl.* 1530s–1540s)

Of Teynham, Kent. Fruiterer to Henry VIII.

Longman's Mag. 1904, 316–20. R. Webber *Early Horticulturists* 1968, 29–42.

HARRIS, Richard (*fl.* 1790s)

Seedsman, 5 Lower Thames Street, London.

HARRIS, Thomas (*fl.* 1800s)

Nurseryman, Battersea, London.

HARRIS, William (1860–1920)
b. Enniskillen, Fermanagh 15 Nov. 1860 *d.* Kansas City, U.S.A. 11 Oct. 1920

FLS 1899. Kew gardener, 1879. Superintendent, King's House Garden, Jamaica, 1881. Superintendent, Public Gardens and Plantations, 1908. Government botanist, 1917. Collected plants. 'Collecting Tour in Jamaica' (*Gdnrs Chron.* 1896 i 134–35, 197–98, 263–64).

I. Urban *Symbolae Antillanae* v.3, 1902, 56–57. *Gdnrs Chron.* 1920 ii 306. *J. Bot.* 1920, 298–99. *Kew Bull.* 1920, 218; 1921, 31–32. *Bot. Gaz.* v.71, 1921, 331–33 portr. *J. Kew Guild* 1921, 39. *Nature* v.106, 1921, 669. *Proc. Linn. Soc.* 1920–21, 49.

Plants at Kew.

Harrisia Britton. *Harrisella* Fawcett & Rendle.

HARRIS, Sir William Cornwallis (1807–1848)
b. Wittersham, Kent 1807 *d.* Surwar near Poona, India 9 Oct. 1848

In S. Africa, 1835. Major, East India Company, 1843. Knighted 1844. Superintending engineer of northern provinces of India, 1848. 'Trees producing Myrrh and Frankincense' (*Proc. Linn. Soc.* 1843, 181–83).
D.N.B. v.25, 28–29. *R.S.C.* v.3, 191.
MS. Catalogue by J. R. Roth of plants collected in 1842 by Harris at Kew.

HARRISON, Benjamin (1837–1921)
b. Ightham, Kent 14 Dec. 1837 *d.* Ightham 30 Sept. 1921
In his youth a keen botanist and archaeologist.
Nature v.108, 1921, 251, 286.

HARRISON, C. H. (*fl.* 1860s)
"Greatly interested in the introduction and cultivation of Indian orchids" (*Bot. Mag.* 1864, t.5433).
Saccolabium harrisonianum Hook.

HARRISON, C. W. (*fl.* 1830s)
Drew 17 plates for J. Harrison's *Floricultural Cabinet* 1833–38, and 18 for J. Harrison's *Gdnr's and Forester's Rec.* 1833–36. Nurseryman, Downham Market, Norfolk?

HARRISON, George (–1808)
Nurseryman, Leicester, 1788.

HARRISON, George (1816–1846)
b. Liverpool March 1816 *d.* London 20 Oct. 1846
Artist. His mother, M. Harrison, was a flower painter. *Weeds and Wild Flowers* 1847.

HARRISON, George (*fl.* 1830s)
Nurseryman, Downham Market, Norfolk.

HARRISON, Henry (*fl.* 1820s)
Of Rio de Janeiro, Brazil. Sent plants to Richard Harrison at Aigburgh near Liverpool.
Bot. Mag. 1827, t.2755.

HARRISON, John (–1788)
Nurseryman, East Bond Street, Leicester, 1764.

HARRISON, John (*fl.* 1790s)
Acquired the nursery in Brompton Road, Kensington, London belonging to his uncle, Henry Hewitt, on his death in 1791. His brother, Samuel, was a partner.
Trans. London Middlesex Archaeol. Soc. v.24, 1973, 186.

HARRISON, John (–1839)
Nurseryman, Leicester, 1808.

HARRISON, John (*fl.* 1850s)
Miner of St. Helens, Lancs. Botanised in Cheshire and Lancs. Contrib. to J. Dickinson's *Fl. Liverpool* 1851.
J. B. L. Warren *Fl. Cheshire* 1899, lxxxiii.

HARRISON, John (–1877)
d. Catterick Bridge, Yorks Feb. 1877
Nurseryman, North of England Rose Nurseries, Darlington, Durham.
Gdnrs Chron. 1877 i 208.

HARRISON, Sir John Burchmore (1856–1926)
b. Birmingham 29 May 1856 *d.* Georgetown, British Guiana 8 Feb. 1926
BA Cantab 1878. CMG 1901. Knighted 1921. Professor of Chemistry and Agricultural Science, Barbados, 1879–89. Raised seedling canes with T. R. Bovell (*Kew Bull.* 1888, 294; *J. Linn. Soc.* v.28, 1890, 199). Government Analyst and Professor of Chemistry, British Guiana, 1889; Director, Dept. of Science and Agriculture, 1904.
Kew Bull. 1888, 295–96; 1926, 191–92. *Bull. Geol. Soc. America* 1927, 45–52 portr.

HARRISON, John William Heslop- *see* Heslop-Harrison, J. W.

HARRISON, Joseph (–*c.* 1855)
Gardener to Lord Wharncliffe at Wortley Hall near Sheffield, Yorks. Nurseryman, Downham Market, Norfolk, 1834–48. Edited *Gardener's and Forester's Record* 1833, *Garden Almanack* 1842, *Gardener's and Naturalist's Almanack* 1852, *Floricultural Magazine and Florists' Magazine* 1833–54. *Horticultural Register* (with J. Paxton) 1832.
F. Boase *Modern English Biogr.* v.1, 1892, 1354. *Gdnrs Chronicle* 1950 ii 6.

HARRISON, Mary (*née* **Rossitter**) (1788–1875)
b. Liverpool 1788 *d.* 25 Nov. 1875
Painted wild flowers. Exhibited regularly at New Society of Painters in Water-Colours. Drew orchids for *Bot. Mag.* from plants sent from Rio de Janeiro by Henry and William Harrison (*Bot. Mag.* 1827, t.2755; 1828, t.2820; 1831, t.3109, etc.).
Bot. Mag. 1826, t.2699.
2 flower paintings at Walker Art Gallery, Liverpool.
Harrisonia Sims.

HARRISON, Samuel (*fl.* 1790s–1830s)
In partnership with his brother, John, in nursery in Brompton Road, Kensington, formerly belonging to their uncle, Henry Hewitt (–1791). In partnership with Jonas Coaker, 1812–16; with William Bristow, 1819–33. Firm went bankrupt in 1833.
Trans. London Middlesex Archaeol. Soc. v.24, 1973, 186.

HARRISON, Thomas (–1890)
Nurseryman, Leicester, 1839–82. In partnership with his sons, John (1844–1929) and Thomas (–1893). Moved shop to 33 Market Place, 1856.
Gdnrs Mag. 1914, 573 portr.

HARRISON, W. A. (–1916)
Nurseryman, Leicester.

HARRISON, Rev. William (–1593)
d. Windsor? 1593
Vicar, Radwinter, Essex, 1558. Canon, Windsor, 1586. Contrib. chapters on gardens and orchards in Holinshed's *Chronicle* 1577.
R. H. Jeffers *Friends of John Gerard* 1967, 21–23.

HARRISON, William (*fl*. 1720s)
Tradesman of Manchester. Correspondent of J. J. Dillenius to whom he sent plants. Had herb., particularly of exotic ferns.
L. Plukenet *Almagestum Botanicum* 1696, 26. J. J. Dillenius *Historia Muscorum* 1741, viii. R. Pulteney *Hist. Biogr. Sketches of Progress of Bot. in England* v.2, 1790, 190–91. J. Cash *Where there's a Will there's a Way* 1873, 6–7. *Trans. Liverpool Bot. Soc*. 1909, 22.
Herb. formerly at Manchester Library.

HARRISON, William (*fl*. 1820s)
Of Rio de Janeiro, Brazil. Sent plants to the Harrisons at Aigburgh near Liverpool.
Bot. Mag. 1826, t.2690; 1827, t.2776; 1828, t.2823, etc.

HARRISON, William (1821–)
At Liverpool Botanic Garden. Warder, Walton Gaol, Liverpool, 1855–84. Friend of W. Skellon. Botanised in Cheshire and Lancs. Some plant records in *Hist. Soc. Lancashire Cheshire Trans*. v.6, 1866, 254 *et seq*. Contrib. to J. Dickinson's *Fl. Liverpool* 1851.
J. B. L. Warren *Fl. Cheshire* 1899, lxxxiii. *Trans. Liverpool Bot. Soc*. 1909, 72.

HARRISS, S. A. (*fl*. 1890s)
Lieut.-Surgeon on Chitral Relief Expedition, 1895; collected plants in 1895 and 1899. (J. F. Duthie 'Bot. Chitral Relief Expedition, 1895' *Rec. Bot. Survey India* 1898, 139–81).
Pakistan J. For. 1967, 347.
Plants at Kew, Dehra Dun.

HARROLD, Thomas (*fl*. 1770s)
Nurseryman, Irishtown, County Limerick, Ireland.
Irish For. 1967, 54.

HARROW, Robert Lewis (1867–1954)
b. Kent, 1867 *d*. Godalming, Surrey 22 Dec. 1954
VMH 1926. Kew gardener, 1890. Royal Botanic Garden, Edinburgh, 1893; Curator, 1924–31. Director, Royal Horticultural Society gardens, Wisley, 1931–46.
Gdnrs Mag. 1909, 77–78 portr. *Gdnrs Chron*. 1927 i 42 portr.; 1931 ii 62 portr.; 1955 i 18. *J. Kew Guild* 1927, 455 portr.; 1955, 244. *Times* 28 Dec. 1954. H. R. Fletcher and W. H. Brown *R. Bot. Gdn Edinburgh, 1670–1970* 1970, 211–12 portr.

HART, Sir George Sankey (–1937)
b. 14 April 1866 *d*. 16 April 1937
Assistant Deputy Conservator of Forests, Punjab, 1887–1906. Conservator of Forests, Southern Circle, Central Provinces, 1906–8; Bengal, 1908–10. Chief Conservator of Forests, Central Provinces, 1910–13. Inspector-General of Forests, 1913–21.
Empire For. J. 1937, 6.

HART, Henry Chichester (1847–1908)
b. Raheny, County Dublin 29 July 1847 *d*. Carrablagh, Donegal 7 Aug. 1908
BA Dublin 1869. FLS 1875. Naturalist. Collected plants on British Polar Expedition (Arctic), 1875–76; Palestine Exploring Expedition, 1883–84. *List of Plants found in Islands of Aran* 1875. *Fl. Howth* 1887. *Some Account of Fauna and Fl. Sinai, Petra and Wady Arabah* 1891. *Fl. County Donegal* 1898. Contrib. to *J. Bot., Proc. R. Irish Acad., Irish Nat*.
Proc. R. Irish Acad. v.7, 1901, cxxii–cxxiii. *Irish Nat*. 1908, 249–54 portr. *J. Bot*. 1911, 121–22 portr. N. Colgan *Fl. County Dublin* 1904, xxx. *Proc. Belfast Nat. Field Club* v.6, 1913, 626. *Rec. Bot. Survey of India* v.8, 1933, 484. R. L. Praeger *Some Irish Nat*. 1949, 96–97 portr. *R.S.C*. v.7, 911; v.10, 147; v.15, 657.
Polar and Palestine plants at Kew, BM(NH). Letters at Kew.

HART, J. (*fl*. 1820s–1830s)
Drew 90 plates for P. W. Watson's *Dendrologia Britannica* 1825 and 175 plates for R. Sweet's *Br. Flower Gdn* 1832–37.

HART, John Hinchley (1847–1911)
b. Botesdale, Suffolk 1847 *d*. Port of Spain, Trinidad 20 Feb. 1911
FLS 1887. Landscape gardener, Nova Scotia, 1872. Superintendent, Cinchona plantations, Jamaica, 1881. Director, Botanic Garden, Trinidad, 1887–1908. Collected plants in Jamaica, 1881–87; Trinidad, 1887–97; Panama Canal, 1885. *Cacao* ed. 2 1900. Edited G. S. Jenman's *Ferns and Fern Allies of W. Indies* 1909. Started *Bull. Misc. Information R. Bot. Gdn Trinidad* and edited it for 11 years.

I. Urban *Symbolae Antillanae* v.3, 1902, 57. *J. Hort. Cottage Gdnr* v.53, 1906, 457 portr. *Proc. Agric. Soc. Trinidad* 1908, 217–19 portr.; 1911, 141–43. *Kew Bull.* 1908, 304; 1911, 162–63. *India-Rubber J.* 18 March 1911, 27 portr. *Gdnrs Chron.* 1911 i 184. *J. Bot.* 1911, 176. *Fern Bull.* 1911, 54. *Proc. Linn. Soc.* 1910–11, 39. *Contrib. U.S. National Herb.* v.27, 1928, 45. *R.S.C.* v.15, 658.

Plants at Kew, Trinidad. Letters at Kew.

Hemitelia hartii Baker.

HART, John William (1887–1916)
b. Wandsworth, London 28 April 1887 *d.* Somme, France 15 Sept. 1916

BSc London. Studied botany at Birkbeck College. Mycologist.

Trans. Br. Mycol. Soc. v.5, 1917, 464–66.

HART, M. (1820s–1830s)

Botanical artist. Contrib. over 800 plates to *Bot. Register* and a few to R. Sweet's *Cistineae* 1825–30 and *Geraniaceae* 1820–30.

HARTE, Walter (1709–1774)
b. Kentbury, Bucks *d.* Bath, Somerset March 1774

MA Oxon. Canon of Windsor, 1750. Vicar, St. Austell and St. Blaze, Cornwall. Poet and historian. *Essays on Husbandry* 1764.

J. C. Loudon *Encyclop. Gdning* 1822, 1273. G. W. Johnson *Hist. English Gdning* 1829, 218. *D.N.B.* v.25, 65–66.

HARTLAND, William Baylor (1836–1912)
b. Mallow, County Cork 1836 *d.* Cork 15 Sept. 1912

Nurseryman, Ballintemple, County Cork, 1878. Specialised in daffodils. *Little Book of Daffodils* 1884. *Original Little Book of Irish-grown Tulips* 1896. *Wayside Ireland, its Scenery, Botany, Agriculture and Peasantry* ed. 2 1895.

Garden 1908, 455 portr.; 1912, 496(x) portr. *Gdnrs Chron.* 1912 ii 255 portr.

HARTLESS, Amos C. (–*c.* 1941)

Kew gardener, 1888. At Royal Botanic Garden, Calcutta, 1889. To Mungpu, *c.* 1900. Superintendent, Government Gardens, Poona, 1903; Government Gardens, Saharanpur, *c.* 1906–20. Left India in 1923. 'Notes on Agriculture of Darjeeling District' (*Indian Gdning* 1899?).

J. Kew Guild 1941, 89. I. H. Burkill *Chapters on Hist. Bot. in India* 1965, 140.

HARTLEY, Isaac (1877–1941)
b. Kelbrook, Yorks 3 Aug. 1877 *d.* 13 Dec. 1941

Loom operator. Secretary, N.E. Lancashire Naturalists Union, 1923–41. Did research on parasitism of toothwort.

N. Western Nat. 1942, 261 portr.

HARTLEY, William (1829–1907)
d. Haslingden, Lancs 1907

Founder of Haslingden Natural History Society. Had good knowledge of local and British flora.

Lancashire Nat. 1908, 149. *Trans. Liverpool Bot. Soc.* 1909, 73.

HARTLIB, Samuel (*c.* 1600–*c.* 1670)

Came to England from Poland, *c.* 1628. Encouraged development of agriculture during Commonwealth. *Legacy...of Discourse of Husbandry* 1650. *Designe for Plentie* 1653.

J. C. Loudon *Encyclop. Gdning* 1822, 1264. G. W. Johnson *Hist. English Gdning* 1829, 96–97. S. Felton *Portr. English Authors on Gdning* 1830, 19–21. *Cottage Gdnr* v.7, 1852, 29. J. Donaldson *Agric. Biogr.* 1854, 21–24. H. Dircks *Biogr. Mem. of Samuel Hartlib* 1865. *D.N.B.* v.25, 72. D. McDonald *Agric. Writers* 1908, 68–78. G. H. Turnbull *Samuel Hartlib* 1920. *Notes Rec. R. Soc.* v.10, 1953, 101–30.

HARTOG, Marcus Manuel (1851–1924)
b. London 19 Aug. 1851 *d.* Paris 21 Jan. 1924

MA Cantab 1874. DSc London. FLS 1875. Assistant Director, Peradeniya Botanic Gardens, Ceylon, 1874–77. Demonstrator in Natural History, Owen's College, Manchester, 1879-82. Professor of Natural History, University College, Cork, 1882–1921. Translated H. E. Baillon's *Nat. Hist. Plants* 1871–88. 'Sapotaceae' (*J. Bot.* 1878, 65–71). 'Saproleguiaceae' (*Trans. R. Irish Acad. Sci.* v.30, 1895, 649–708. *Ann. Bot.* v.2, 1888–89, 201–16, 309–18; v.4, 1889–91, 299–300, 337–46; v.10, 1896, 98–100).

Bot. Soc. Exch. Club Br. Isl. Rep. 1924, 538–39. *Irish Nat.* 1924, 39–40. *J. Bot.* 1924, 148–49. *Nature* v.113, 1924, 243–44. *Times* 30 Jan. 1924. *Who was Who, 1916–1928* 472. *Irish Nat. J.* 1931, 240.

HARTWEG, Carl Theodor (1812–1871)
b. Carlsruhe, Germany 18 June 1812 *d.* Swetzingen, Baden, Germany 3 Feb. 1871

Collected plants for Horticultural Society of London. In Mexico, 1836–39, 1845–46; Guatemala, 1839–40; Ecuador, 1841–43; Jamaica, 1843; Madeira, 1854; Peru. 'Journal of Mission to California' (*J. Hort. Soc. London* 1846, 180–85; 1847, 121–25, 187–91; 1848, 217–28). 'Notes of Visit to Mexico, Guatemala and Equatorial America..., 1836 to 1843, in search of Plants and Seeds for Horticultural Society of London' (*Trans. Hort. Soc. London* v.3, 1848, 115–62).

G. Bentham *Plantae Hartwegianae* 1839–57. A. Lasègue *Musée Botanique de B. Delessert* 1845, 207–9. *Gdnrs Chron.* 1871, 313. *J. Bot.* 1871, 224. *J. Hort. Cottage Gdnr* v.45, 1871, 199–200; v.46, 1871, 214–15. *Erythea* 1897, 31–35, 51–56. W. B. Hemsley *Biologia Centrali-Americana* v.4, 1886, 126–28. N. Leon *Biblioteca Botanico-Mexicana* 1895, 352–53. *Leaflets W. Bot.* v.1, 1935, 180–81. *California Hist. Soc. Quart.* v.18, 1939, 342. *Occas. Papers California Acad. Sci.* no. 20, 1943, 47–48. *J. R. Hort. Soc.* 1955, 275–76. S. D. McKelvey *Bot. Exploration of Trans-Mississippi West, 1790-1850* 1955, 961–80. A. M. Coats *Quest for Plants* 1969, 317–19, 344–46. *Taxon* 1970, 526. M. A. Reinikka *Hist. of the Orchid* 1972, 187–88. *R.S.C.* v.3, 203.

Plants at Kew. Letters at Royal Horticultural Society.

Hartwegia Lindl.

HARVEY, Alexander (1811–1889)
b. Broomhill, Aberdeenshire 20 April 1811
d. London 25 April 1889

MA Aberdeen. MD Edinburgh 1835. Professor of Materia Medica, Aberdeen. Practised medicine at Southampton. *Trees and their Nature* 1856.

Scott. Nat. 1889, 97. F. Boase *Modern English Biogr.* v.1, 1892, 1364.

HARVEY, Enoch (*c.* 1826–1890)
d. Liverpool 1 Oct. 1890

Solicitor, Liverpool. Grew orchids.

Garden v.38, 1890, 357.

Dendrobium harveyanum Reichenb. f.

HARVEY, Frederick William (1882–1915)
b. Stebbing, Essex 1882 *d.* London 31 Aug. 1915

Kew gardener, 1903–5. Sub-editor, *Garden,* 1905; Editor, 1910–15. *Fruit-growing for Beginners* 1912. *Antirrhinums* 1911. *Nasturtiums* 1915.

Garden 1915, 431 portr., 444. *Gdnrs Chron.* 1915 ii 159 portr. *J. Hort.* v.71, 1915, 213, 223 portr. *J. Kew Guild* 1916, 311–12 portr. *Kew Bull.* 1915, 390. *Orchid Rev.* 1915, 305–6 portr. *Who was Who, 1897–1916* 320.

HARVEY, Joshua Reubens (*fl.* 1830s)

Physician, Cork. President, Medical Society of Cork. Local Secretary of Botanical Society of Edinburgh. Collected algae.

Irish Nat. J. 1972, 223–25.

Plants at University College, Cork.

HARVEY, William (*fl.* 1770s–1790s)

Nurseryman, Dublin.

Irish For. 1967, 47.

HARVEY, William Henry (1811–1866)
b. Summerville, Limerick 5 Feb. 1811 *d.* Torquay, Devon 15 May 1866

MD Dublin 1844. FRS 1858. FLS 1857. Colonial Treasurer at Cape, 1836–42. Professor of Botany, Royal Dublin Society, 1848; Trinity College, Dublin, 1856. Collected plants in S. Africa, 1835–42; Ceylon, 1854–56; W. Australia, 1854–56; Tasmania, 1854–56; Fiji; Friendly Islands; Florida. *Genera of S. African Plants* 1838. *Manual of British Algae* 1841. *Phycologia Britannica* 1846–51 4 vols. *Nereis Australis* 1847. *Nereis Boreali-Americana* 1851–58 3 vols. *Phycologica Australica* 1858–63 5 vols. *Thesaurus Capensis* 1859–63 2 vols. *Index Generum Algarum* 1860. *Fl. Capensis* (with O. W. Sonder) v.1–3, 1859–65.

Ipswich Mus. Portr. Ser. 1852. *Amer. J. Sci. Arts* 1866, 273–78. *Gdnrs Chron.* 1866, 537–38. *J. Bot.* 1866, 236–38. *Proc. Linn. Soc.* 1865–66, lxi–lxv. *J. Hort. Cottage Gdnr* v.38, 1867, 80–82. *Proc. R. Soc.* v.16, 1868, xxii–xxiii. L. J. Fisher *Mem. W. H. Harvey* 1869 portr. *Trans. S. African Philos. Soc.* v.4, 1887, li–liii. *Notes Bot. School Dublin* v.1, 1896, 5–8. *D.N.B.* v.25, 100. *J. Proc. R. Soc. N.S.W.* 1908, 69–71. R. Marloth *Fl. S. Africa* v.1, 1913, ix portr. F. W. Oliver *Makers of Br. Bot.* 1913, 204–24 portr. J. R. Green *Hist. Bot.* 1914, 437–42. *Irish Nat.* 1918, 162–63. R. L. Praeger *Some Irish Nat.* 1949, 98. *Hermanthena* v.103, 1966, 32–45 portr. H. A. Baker and E. G. H. Oliver *Ericas in S. Africa* 1967, ix–x. A. M. Coats *Quest for Plants* 1969, 264–65. C. C. Gillispie *Dict. Sci. Biogr.* v.6, 1972, 162–63. *R.S.C.* v.3, 205; v.7, 917.

Plants at BM(NH), Kew, Trinity College Dublin. Letters and crayon portr. by D. Macnee at Kew. Portr. at Hunt Library. Portr. at National Gallery Dublin. Letters at BM(NH).

Harveya Hook.

HARVEY-GIBSON, Robert John (1860–1929)
b. Helensburgh, Dunbartonshire 2 Nov. 1860
d. Glasgow 3 June 1929

DSc Aberdeen. FRSE. Demonstrator in Zoology, Edinburgh, 1882. Demonstrator in Biology, Liverpool, 1883. Lecturer in Botany, Liverpool, 1887; Professor of Botany, 1894–1921. Marine algologist. President, Liverpool Botanical Society, 1907–8. *Outlines of History of Botany* 1919. *British Plant Names and their Derivations* 1923. *Short History of Botany* 1926. *Elements of Botany* 1930. Translated L. Jost's *Lectures on Plant Physiology* 1907. Contrib. to *Ann. Bot., Trans. R. Soc. Edinburgh.*

Bot. Soc. Exch. Club Br. Isl. Rep. 1929, 92–93. *Gdnrs Chron.* 1929 i 453. *J. Bot.* 1929, 262. *Nature* v.124, 1929, 64–65. *N. Western Nat.* 1929, 141–42. *Pharm J.* 1929, 590. *Proc. R. Soc. Edinburgh* v.49, 1928–29, 384–85. *Times* 19 July 1929. *Who was Who 1929–1940* 604.

HARWARD, Simon (*fl.* 1570s–1620s)
Chaplain, New College, Oxford, 1577. Preacher, Warrington, Lancs; Banstead, Surrey. Schoolmaster, Tanridge, Surrey, *c.* 1604. *A Most Profitable New Treatise from Approved Experience of Art of Propagating Plants* 1623.
Cottage Gdnr v.6, 1851, 327.

HARWOOD, George (1842–1915)
b. near Taunton, Somerset 18 March 1842 *d.* 18 Jan. 1915
Trained as landscape gardener. To Sydney, *c.* 1862. At Sydney Botanic Gardens, 1873; Superintendent, 1891.
Sydney Morning Herald 19 Jan. 1915. *J. Proc. R. Austral. Hist. Soc.* 1931, 156; 1932, 130.

HASKELL, Gordon Mark Leo (1920–1967)
b. London 13 Feb. 1920 *d.* 29 June 1967
DSc London. FLS 1959. Geneticist, John Innes Horticultural Institute, 1948; Scottish Horticultural Research Institute, 1958. Head of Dept. of Biological Science, Portsmouth College of Technology, 1965. Published papers on taxonomy and cytology of *Rubus*.
Proc. Bot. Soc. Br. Isl. 1968, 496.

HASKET, Thomas *see* Hesketh, T.

HASLAM, Samuel Holker (–1856)
d. Milnthorpe, Westmorland 13 April 1856
FLS 1836. Presented his collections of insects and plants to Kendal Natural History Society. 'Remarks on Threatened Extermination of Rare Plants by Rapacity of Collector' (*Phytologist* 1843, 544–46).
Proc. Linn. Soc. 1856, xlii.

HASSALL, Abner (*c.* 1852–1922)
b. Manchester *c.* 1852 *d.* 15 June 1922
Manufacturing chemist. Cultivated orchids. Acquired nursery firm of Messrs Stanley Ashton and Co., Southgate, London in 1909.
Orchid Rev. 1922, 217–18 portr.

HASSALL, Arthur Hill (1817–1894)
b. Teddington, Middx 13 Dec. 1817 *d.* San Remo, Italy 9 April 1894
MD London 1851. FLS 1845. Public analyst. *History of British Freshwater Algae* 1845 2 vols. *Food and its Adulterations* 1855. *Adulterations Detected* 1857. *Narrative of a Busy Life* 1893.
J. Bot. 1894, 190–91. *Trans. Penzance Nat. Hist. Soc.* 1890–91, 231–32. *R.S.C.* v.3, 208; v.7, 918; v.15, 678.
Plants at BM(NH). Letters at Kew and in Berkeley correspondence at BM(NH).
Hassallia Berkeley.

HASSALL, Bessie Florence (*née* **Cory**) (1883–1954)
d. Oxford 15 Sept. 1954
Active member of Botanical Society of British Isles and sketched plants.
Proc. Bot. Soc. Br. Isl. 1955, 413.

HASSÉ, Alexander Cossart (1813–1894)
b. Leeds, Yorks Dec. 1813 *d.* Ockbrook, Derbyshire 12 Dec. 1894
Deacon, Moravian Church, 1844; Bishop, 1883. Collected plants in Silesia, now at Liverpool University.

HASTINGS, Somerville (1878–1967)
b. Warminster, Wilts 1878 *d.* Reading, Berks 7 July 1967
Educ. University College, London. FRCS 1904. LRCP 1912. MP, 1923–24, 1929–31, 1945–59. Surgeon and lecturer to Ear and Throat Dept., Middlesex Hospital. *Wild Flowers at Home. Alpine Plants at Home* 1905. *Toadstools at Home* 1905. *Summer Flowers of High Alps* 1910.
Times 8 July 1967 portr. *Reading Nat.* 1968, 52–54. *Who was Who, 1961–1970* 503.

HATCHER, Leonard W. (–1964)
d. 15 Jan. 1964
Managing director, Messrs Mansell and Hatcher, nurserymen, Rawdon, Leeds.
Orchid Rev. 1964, 102–3.

HATCHER, W. H. (*fl.* 1800s)
Manager, Price's Candleworks, Bromborough, Cheshire. Botanist.
J. B. L. Warren *Fl. Cheshire* 1899, lxxxiii.

HATCHER, William Henry (*c.* 1869–1951)
b. Alderbury, Wilts *c.* 1869 *d.* 29 May 1951
Employed in nurseries of W. Bull, W. Whiteley, Messrs Sander, Cowan and Co., Charlesworth and Co. With W. Mansell took over nursery firm of J. W. Moore at Rawdon, Leeds.
Orchid Rev. 1951, 112.

HATTON, Charles (1635–*c.* 1705)
Lieutenant-Governor of Guernsey. Keen gardener. Friend of J. Evelyn, R. Morison, Sir H. Sloane. Sent seeds from his garden in Guernsey to R. Morison. J. Ray's *Historia Plantarum* 1686 dedicated to Hatton. "The honorable and learned Charles Hatton, Esq. (to whom all our phytologists and lovers of horticulture are obliged, and myself in particular for many favours)" (J. Evelyn).
R. Morison *Plantarum Historiae Universalis Oxoniensis* v.2, 1680, 84. S. H. Vines and G. C. Druce *Morisonian Herb.* 1914, xxxii. C. E. Raven *John Ray* 1942, 66, 192. M. Hadfield *Gdning in Britain* 1960, 135–39.

HATTON, Rev. Charles Osborne Smeathman (1872–1932)
b. Scarborough, Yorks 1872 *d.* Hinton Admiral, Hants 19 Feb. 1932
BA Cantab 1894. FLS 1919. Vicar, Hinton Admiral, 1907–32. Chairman, Bournemouth Natural Science Society, 1927–32. Contrib. to *Proc. Bournemouth Nat. Sci. Soc.*
Proc. Linn. Soc. 1931–32, 177–78.

HATTON, Richard George (1864–1926)
d. 19 Feb. 1926
Educ. Birmingham School of Art. Professor of Fine Art, Armstrong College, Newcastle-upon-Tyne. *Craftsman's Plant-book* 1910.
Who was Who, 1916–1928 476.

HATTON, Sir Ronald George (1886–1965)
b. Yorks 6 July 1886 *d.* Benenden, Kent 11 Nov. 1965
Educ. Oxford. FRS 1944. VMH 1930. Director, Research Station, East Malling, 1919–49. Founder of Fruit Group, Royal Horticultural Society. Researched on fruit rootstocks. Contrib. to *J. R. Hort. Soc.*
Nature v.208, 1965, 1258. *Biogr. Mem. Fellows R. Soc.* 1966, 251–58 portr. R. Hort. Soc. *Fruit, Present and Future* 1966, 1–4 portr. *Who was Who, 1961–1970* 504.

HAUGHTON, John (1836–1889)
b. Carlow, Ireland 29 March 1836 *d.* Savernake, Wilts 26 Aug. 1889
Major-General, Royal Artillery. Collected plants in St. Helena, 1858–65.
Notes Bot. School Trinity College, Dublin v.1, 1898, 127. *R.S.C.* v.3, 220.
Plants at Kew, Dublin. Letters at Kew.

HAUGHTON, Rev. Samuel (1821–1897)
b. Carlow, Ireland 1821 *d.* Dublin 31 Oct. 1897
MD Dublin 1844. DCL Oxon 1860. FRS 1858. Cousin of J. Haughton (1836–1889). Professor of Geology, Dublin, 1851. Palaeobotanist.
Nature v.57, 1897, 55–56. *Irish Nat.* 1898, 1–4 portr. *Notes Bot. School Trinity College, Dublin* v.1, 1898, 126–27. *R.S.C.* v.3, 220; v.7, 923; v.10, 160; v.15, 687.

HAULKYARD, Dr. (*fl.* 1780s)
Of Oldham, Lancs. One of the founders of "a society of botanists" in Oldham *c.* 1785.
J. Holt *General View of Agriculture... Lancashire* 1795, 229. *Trans. Liverpool Bot. Soc.* 1909, 23.

HAVELAND, W. (1730–)
Apothecary, Bath, Somerset. Plants at Oxford.

HAVERFIELD, John (*c.* 1694–1784)
d. Kew, Surrey 21 Nov. 1784
Gardener at Kew to Augusta, Dowager Princess of Wales. Superintendent of Royal Gardens at Richmond, Surrey, *c.* 1760–84.
Leisure Hour Nov. 1862. *Kew Bull.* 1891, 289.
Portr. at Kew.

HAVERFIELD, John (*c.* 1741–1820)
Succeeded his father, J. Haverfield (*c.* 1694–1784), as Superintendent of Royal Gardens at Richmond, Surrey; retired in 1795.
Portr. at Kew, Hunt Library.

HAVERFIELD, Thomas (*fl.* 1760s–1800s)
Son of J. Haverfield (*c.* 1694–1784). Royal gardener at Hampton Court, Middx.

HAVILAND, Anne Elizabeth Heath (*née* **Grubbe**) (1818–)
To Prince Edward Island with parents in 1841 where she collected plants, 1849–54.
Kew Bull. 1948, 236.
Plants at Kew.

HAVILAND, Edwin (1823–1908)
b. Gloucester 20 July 1823 *d.* Sydney 22 May 1908
FLS 1885. Businessman. Interested in morphology and physiology of plants. Contrib. to *Proc. Linn. Soc. N.S.W.*, 1882–88.
J. R. Soc. N.S.W. 1908, 106–7 portr.

HAVILAND, George Darby (1857–*c.* 1901)
b. Warbleton, Sussex 19 Nov. 1857 *d.* Natal, S. Africa *c.* 1901
BA Cantab 1880. MB. FLS 1894. Surgeon and naturalist. Director, Raffles Museum, Singapore. Medical Officer, Sarawak, 1891–93. Curator, Government Museum, Kuching, Sarawak, 1893–95. Collected plants in Malaya, 1890; N. Borneo, 1891–95. 'Revision of Naucleae' (*J. Linn. Soc.* v.33, 1897, 1–94).
Kew Bull. 1891, 276; 1892, 249–50; 1894, 136; 1895, 42; 1907, 197–98. *Trans. Linn. Soc.* 1894, 69–263. *Gdns Bull. Straits Settlements* 1927, 123. *Fl. Malesiana* v.1, 1950, 222. *R.S.C.* v.15, 693.
Plants at Kew, Singapore.
Havilandia Stapf.

HAWES, E. F. (*fl.* 1900s)
Gardener, Royal Botanic Society, Regent's Park, London. Secretary, National Dahlia Society.
Gdnrs Mag. 1910, 641–42 portr.

HAWKER, Rev. William Henry (1828–1874)
b. Petersfield, Hants 13 Dec. 1828 *d.* Petersfield 26 May 1874
MA Cantab 1854. Helped H. Ardoino in his *Fl. Analytique du Département des Alpes-Maritimes* 1867 (preface xii). '*Asplenium fontanum* in Hampshire' (*Phytologist* v.4, 1853, 814–15).
Boissiera fasc. 5, 1941, 53.
Letters at Kew.

HAWKES, Abraham (*fl.* 1780s)
Nurseryman, Sandwich, Kent.

HAWKES, Rev. Henry (1805–1886)
b. Dukinfield, Cheshire 1 Feb. 1805 *d.* Liverpool 29 Jan. 1886
Educ. Glasgow. FLS 1842. Unitarian minister, Portsmouth, 1833–71. Botanist.
Proc. Linn. Soc. 1885–86, 143–44.
Portr. and library at Portsmouth Public Library.

HAWKINS, Miss Ellen (–1864)
Contrib. botanical appendix to Robertson's *Handbook to the Peak* 1854.
882 drawings of British plants at BM(NH).

HAWKINS, John (*fl.* 1730s–1790s)
Surgeon. At one time in employ of Sir H. Sloane. Later practised medicine at Dorchester, Dorset. 'Quinaquina' (*Trans. Linn. Soc.* v.3, 1797, 59–61).
J. Bot. 1909, 426–29. *Nat. Hist. Mag.* 1931, 21. J. E. Dandy *Sloane Herb.* 1958, 135–36.
Plate of Cinchona, 1739 with letter at BM(NH).

HAWKINS, John (1761–1841)
b. St. Erth, Cornwall 6 May 1761 *d.* Trewithian, Cornwall 4 July 1841
FRS 1791. Of Bignor, Sussex. Accompanied J. Sibthorp to Greece, 1786–87. Collected plants in Crete, 1794. One of founders of Horticultural Society of London.
J. Sibthorp *Fl. Graeca* v.1, 1806, vii. *Bot. Mag.* 1820, t.2146. P. Smith *Mem. and Correspondence of Sir J. E. Smith* v.1, 1832, 458, 468, 471, 568. *Gent. Mag.* 1841 ii 322–23. G. C. Boase and W. P. Courtney *Bibliotheca Cornubiensis* 1874–82. 222–23. *D.N.B.* v.25, 221. *J. R. Hort. Soc.* 1954, 461. *J. R. Inst. Cornwall* 1954, 98–106 portr. F. W. Steer *I am, My Dear Sir...Letters to and by John Hawkins* 1959. *Taxon* 1970, 526.
Letters at BM(NH), Linnean Society.

HAWKINS, Mary Esther (*née* **Sibthorp**) (*fl.* 1800s)
Daughter of H. W. Sibthorp (1712–1797). Married J. Hawkins (1761–1841). "This lady has all the botanical genius of her family. So enthusiastically fond of botany are Mr. and Mrs. Hawkins, that they were lately pursuing plants at the Lizard, in situations which to the dull, cautious visitor, would appear inaccessible" (R. Polwhele *Language, Literature and Lit. Characters of Cornwall* 1806, 121–23).

HAWKSHAW, Sir John (1811–1891)
b. Leeds, Yorks 1811 *d.* London 2 June 1891
FRS 1855. Knighted 1873. Civil engineer. President, British Association, 1875. 'Description of Fossil Trees found in Excavations for Manchester and Bolton Railway' (*Geol. Soc. Trans.* 1842, 173–80; *Geol. Soc. Proc.* 1842, 139–40, 269–70).
Proc. R. Soc. v.50, 1892, i–iv. *Quart. J. Geol. Soc.* 1892, 52–53. *D.N.B. Supplt* v.2, 402–4.

HAWLEY, Francis (*fl.* 1780s)
Nurseryman, Pontefract, Yorks.

HAWLEY, Sir Henry Cusack Wingfield, 6th Baronet (1876–1923)
b. Boston, Lincs 23 Dec. 1876 *d.* Bournemouth, Hants 18 Nov. 1923
BA Oxon. Mycologist who studied Pyrenomycetes. *Monograph on Fungi for Clare Island Survey* (with C. Rea) 1911.
Trans. Lincolnshire Nat. Union 1919, 9–11 portr. *Bot. Soc. Exch. Club Br. Isl. Rep.* 1923, 150. *Trans. Br. Mycol. Soc.* v.8, 1923, 226–30; v.9, 1924, 241–43. *J. Bot.* 1924, 60–61. *Naturalist* 1924, 186–88 portr.
Herb., MSS. and drawings at BM(NH).

HAWLEY, William (*fl.* 1820s)
Nurseryman, George Yard, Barnsley, Yorks.

HAWORTH, Adrian Hardy (1768–1833)
b. Hull, Yorks 19 April 1768 *d.* Chelsea, London 24 Aug. 1833
FLS 1798. Entomologist and botanist. In London, 1792; Cottingham near Hull, 1812–17. Intended to publish *Fl. Cottinghamensis* and formed herb. (*Observations on ...Mesembryanthemum* 479–80). A founder of Hull Botanic Garden, 1812. At Chelsea, 1817–33, where he grew succulents. Discovered *Cyperus fuscus* at Chelsea. Herb. bought by H. B. Fielding who destroyed most of the specimens after studying them. *Observations on Genus Mesembryanthemum* 1794. *Miscellanea Naturalia* 1803. *Synopsis Plantarum Succulentarum* 1812. *Supplementum Plantarum Succulentarum* 1819. *Saxifragearum Enumeratio* 1821. *Narcissinearum Monographia* ed. 2 1831. *Complete Works on Succulent Plants* 1965 (biogr., 9–57).
J. C. Loudon *Encyclop. Gdning.* 1822, 1280. T. Faulkner *Hist. and Topographical Description of Chelsea* v.2, 1829, 11–13. *Bot. Misc.* 1830, 69. *Gdnrs Mag.* 1833, 614, 635–40. *Gent. Mag.* 1833 ii 377–78. *Mag. Nat. Hist.* 1833, 562; 1836, 447. W. Herbert *Amaryllidaceae* 1837, 293–94. R. Sweet *Br. Flower Gdn* v.5, 1838, t.188, 194. *Cottage Gdnr* v.6, 1851, 157. *Ann. Mag. Nat. Hist.* v.7, 1871, 244–45. *J. Bot.* 1871, 148; 1916, 241–43. *D.N.B.* v.25, 246–47. *Trans. Hull Sci. Field Nat. Club* v.1, 1901, 229–32 portr. G. W. Reynolds *Aloes of S. Africa* 1950, 93, 227, etc. *Desert Plant Life* v.23, 1951, 41–43.

H. N. Clokie *Account of Herb. Dept. Bot., Oxford* 1964, 180. F. A. Stafleu *Taxonomic Literature* 1967, 194. C. C. Gillispie *Dict. Sci. Biogr.* v.6, 1972, 184. H. Herre *Genera of Mesembryanthemaceae* 1971, 42–43 portr. *R.S.C.* v.3, 235.

Plants at Oxford, Kew. Letters at Kew, in Swainson correspondence at Linnean Society and at Teyler Museum, Haarlam. Portr. at Hunt Library. Library sold by Sotheby, 11–13 March 1834.

Haworthia Duval.

HAWYS, John (*fl.* 1690s)

Doctor at Norwich, Norfolk. Correspondent of Sir H. Sloane and J. Petiver.

J. E. Dandy *Sloane Herb.* 1958, 136.

HAXTON, John (*fl.* 1790s–1800s)

ALS 1798. Gardener to Lord Macartney's Embassy to China, 1792–94. Collected plants. Brought back *Camellia* spp.

E. Bretschneider *Hist. European Bot. Discoveries in China* 1898, 217. *Fl. Malesiana* v.1, 1950, 222. A. M. Coats *Quest for Plants* 1969, 92–93.

Haxtonia D. Don.

Plants at BM(NH).

HAY, Mr. (*fl.* 1860s)

Sent small collection of plants from Silver Island, Yangtze River to H. F. Hance in 1863.

E. Bretschneider *Hist. European Bot. Discoveries in China* 1898, 689.

HAY, Alfred (*fl.* 1900s–1920s)

Educ. Royal Indian Engineering College, Coopers Hill. Professor of Electrical Technology, Indian Institute of Science, Bangalore.

Kew Bull. 1932, 350–51.

Paintings and wood specimens at Kew.

HAY, Edward (*fl.* 1760s–1780s)

Nurseryman, Dublin.

Irish For. 1967, 47.

HAY, Henry Maurice Drummond- *see* Drummond-Hay, H. M.

HAY, John (–1792)

d. Nov. 1792

Succeeded to nursery of Adam Holt at Leytonstone, Essex in 1759. Nursery acquired after 1792 by James Hill (*c.* 1761–1832) and later his widow Charlotte. Noted for collection of Red American oaks.

Trans. London Middlesex Archaeol. Soc. v.24, 1973, 190.

HAY, John (*fl.* 1810s–1820s)

Nurseryman, Lambeth, London.

HAY, Thomas (1874–1953)

b. Cullen, Banffshire 18 Aug. 1874 *d.* Haslemere, Surrey 22 Jan. 1953

VMH 1924. VMM 1940. Gardener to Duke of Fife and Marquess of Linlithgow. Superintendent, Greenwich Park, 1911; Regent's Park, 1919; Central Royal Parks, 1924–40. *Plants for Connoisseur* 1938.

Gdnrs Mag. 1909, 761–62 portr. *Bot. Mag.* 1929 portr. *Br. Fern Gaz.* 1953, 59. *Gdnrs Chron.* 1953 i 46. *Times* 23 and 26 Jan. 1953. *Lily Yb.* 1954, 128–30 portr.

HAY, Walter (*fl.* 1770s–1800s)

Nurseryman, St. Georges Field, London. In partnership with James and George Swinnerton.

Trans. London Middlesex Archaeol. Soc. v.24, 1973, 188.

HAY, William (*fl.* 1690s)

Surgeon. Sent plants from Newfoundland to J. Petiver.

J. Petiver *Museii Petiveriani* 1695, 653. J. E. Dandy *Sloane Herb.* 1958, 136.

HAY, William E. (*fl.* 1850s)

Superintendent, Simla Hill States. Collected plants at Karakorum, 1857 and in Lahul and Ladak, 1862. Sent Himalayan plants to M. P. Edgeworth.

Pakistan J. For. 1967, 347.

Plants at Calcutta, Kew, Oxford.

HAYDEN, Sir Henry Hubert (1869–1923)

d. Finsteraarhorn, Switzerland 13 Aug. 1923

Educ. Trinity College, Dublin. In Geological Survey of India, 1895; Director, 1910–21. 'Stratigraphical Position of the Gangamopteris Beds of Kashmir' (*Rec. Geol. Surv. India* v.36(1)).

Nature v.112, 1923, 450–51. *Alpine J.* v.35, 277–80 portr. *Who was Who, 1916–1928* 478.

HAYDON, Rev. George Philip (*c.* 1846–1913)

d. 21 Jan. 1913

Vicar, Hatfield, Yorks. Retired to Westbere, Canterbury, 1898. Hybridised narcissi.

Garden 1913, 64(xvi). *Daffodil Yb.* 1933, 30.

HAYDON, Walter (*c.* 1871–1925)

d. Brighton, Sussex 22 April 1925

Kew gardener, 1892. Curator, Botanical Station, Abuko, Gambia, 1894; Botanic Station, Sierra Leone, 1898–1901.

Kew Bull. 1898, 35–36. *J. Kew Guild* 1926, 427–28 portr.

HAYES, Sutton (–1863)

Correspondent of W. J. Hooker at Kew to whom he sent plants from Panama in 1860s.

Bot. Mag. 1862, t.5304, 5330.

Plants at Harvard. Letters at Kew.

HAYES, Thomas Richard (1864–1927)
b. Grasmere, Westmorland 1864 *d.* 12 June 1927
Nurseryman, Lake Road, Keswick, Cumberland. Specialised in design of rock gardens. Lectured in 1899 on 'Mountain Fl. of Wordsworth Country'.
Gdnrs Chron. 1924 ii 228 portr.

HAYHOW, Richard (*fl.* 1830s)
Nurseryman, St. Marks Road, Camberwell, London.

HAYNE, William Amherst (1847–1873)
b. Clifton, Bristol 4 Oct. 1847 *d.* Catania, Sicily 5 Jan. 1873
BA Cantab 1870. Collected plants in Palestine. 'On Fl. Moab' (*J. Bot.* 1872, 289–95). *Letters* 1873.
J. Bot. 1873, 96. *R.S.C.* v.7, 928.
Plants at Oxford, Kew.

HAYNES, Thomas (*fl.* 1790s–1810s)
Nurseryman, Oundle, Northants with shop at Stamford. *Improved System of Nursery Gardening* 1811. *Interesting Discoveries in Horticulture* 1811. *Treatise on Improved Culture of Strawberry, Raspberry and Gooseberry* 1812. *Essay on Soils and Composts* 1817.
J. C. Loudon *Encyclop. Gdning* 1822, 1285. G. W. Johnson *Hist. English Gdning* 1829, 283.

HAYWARD, Ida Margaret (1872–1949)
b. Trowbridge, Wilts 1872 *d.* Galashiels, Selkirk 2 Oct. 1949
FLS 1910. Studied natural history of border counties. *Adventive Fl. Tweedside* (with G. C. Druce) 1919.
Proc. Linn. Soc. 1949–50, 105–6. *Watsonia* 1949–50, 324.
Herb. of adventive plants at Edinburgh. Plants at Oxford.

HAYWARD, Joseph (*fl.* 1810s)
Yorkshire clothier. Moved to Plumstead, Kent. *Science of Horticulture* 1818. Contrib. to *Trans. Hort. Soc.*
J. C. Loudon *Encyclop. Gdning* 1822, 1290.

HAYWOOD, Daniel (*fl.* 1790s–1810s)
Nurseryman, Burslem, Staffs.

HAYWOOD, John (*fl.* 1790s)
Nurseryman, Leek, Staffs.

HAYWOOD, Thomas Burt (*c.* 1826–1900)
d. Reigate, Surrey 3 May 1900
Vice-President, Royal Horticultural Society. Treasurer, National Rose Society.
Gdnrs Chron. 1900 i 296 portr. *Gdnrs Mag.* 1900, 286 portr.

HEAD, William G. (1837–1897)
b. Worthing, Sussex 1837 *d.* 3 April 1897
Gardener, Royal Horticultural Society, Chiswick; Kew, 1872. In charge of Agri-Horticultural Society of Calcutta gardens at Alipore, 1872–98. Superintendent, Crystal Palace gardens, 1879–97.
Garden v.51, 1897, 272. *Gdnrs Chron.* 1897 i 241 portr. *Gdnrs Mag.* 1897, 207 portr. *J. Hort. Cottage Gdnr* v.34, 1897, 293. *J. Kew. Guild* 1897, 35.

HEAD, William (–1941)
b. West Indies *d.* 30 July 1941
Kew gardener, 1904–6. In India, 1906. Later Superintendent, Government gardens at Allahabad, Kumaon, Lucknow, etc. Left India, 1834.
J. Kew Guild 1942, 195–96.

HEADFORT, Geoffrey Thomas Taylour, 4th Marquess *see* Taylour, G. T.

HEADLEY, Rev. A. (–1899)
d. 15 Feb. 1899
Educ. Cambridge. Rector, Hardenhuish, Wilts, 1856–90. Regular contributor to *J. Hort.* under pen-name of 'Wiltshire Rector'.
J. Hort. Cottage Gdnr v.38, 1899, 143 portr.

HEADLY, C. B. (*c.* 1870–*c.* 1916)
Of Leicester. Assistant, Leicester Museum, *c.* 1898–1901. Botanist.
A. R. Horwood and C. W. F. Noel *Fl. Leicestershire* 1933, ccxxxiv.

HEADLY, Richard (*c.* 1795–1876)
d. 14 April 1876
Of Stapleford House near Cambridge. Cultivated auriculas, tulips, carnations and picotees.
Garden v.9, 1876, 420.

HEAL, John (*c.* 1841–1925)
b. Barnstaple, Devon *c.* 1841 *d.* Fulham, London 6 Nov. 1925
VMH 1897. VMM 1892. Gardener, Westacott nurseries near Barnstaple. Veitch nursery, 1863 for whom he hybridised rhododendrons, begonias, etc.
Gdnrs Chron. 1892 i 812–13 portr.; 1921 ii 204 portr.; 1925 ii 398. *Gdnrs Mag.* 1909, 643–44 portr. J. H. Veitch *Hortus Veitchii* 1906, 106–7.

HEALE, W. (*fl.* 1830s)
Nurseryman, Calne and Devizes, Wilts.

HEALEY, John Campbell (1843–1922)
b. Rochdale, Lancs June 1843 *d.* Rochdale 14 March 1922
In business in Rochdale. Interested in the local flora. 'Moss Fl. Rochdale' (*Proc. Liverpool Bot. Soc.* 1919–22, 16).
Lancashire and Cheshire Nat. v.14, 1922, 276. *Proc. Liverpool Bot. Soc.* 1919–22, 30.

HEAP, William (*fl.* 1780s–1790s)
Nurseryman and gardener, Cheadle, Staffs.

HEARSEY, Sir John Bennet (1793–1865)
b. Midnapore, India 21 Jan. 1793 *d.* Boulogne, France 21 Sept. 1865
In India, 1807. Indian Army, 1808; Lieut.-General, 1863. KCB 1856. Sent Punjab plants to Kew, 1850.
Times 27 Oct. 1865. V. C. P. Hodson *List of Officers of Bengal Army, 1758–1834* v.2, 1928, 423.

HEATH, Rev. Douglas Montagu (1881–1961)
Herb. at Birmingham.

HEATH, Francis George (1843–1913)
b. Totnes, Devon 15 Jan. 1843 *d.* Weymouth, Dorset 23 March 1913
Surveyor, Customs Dept., 1882–1904. Pioneer of open space movement. *Fern Paradise* 1875; ed. 7 1905. *Fern World* 1877; ed. 10 1910. *Our Woodland Trees* 1878. *My Garden Wild and what I grew There* 1881. *Where to find Ferns* 1881. *Fern Portfolio* 1885. *Our British Trees* 1908. *British Ferns* 1911. *Tree Lore* 1911. *Nervation of Plants* 1912. *British Fern Varieties* 1913.
Gdnrs Chron. 1913 i 212. *J. Hort. Home Farmer* v.66, 1913, 336. *Who was Who, 1897–1916* 327.

HEATH, William (*c.* 1810–1892)
d. 8 Oct. 1892
Nurseryman, Royal Exotic Nurseries, Cheltenham, Glos 1830.
Gdnrs Chron. 1892 ii 504.

HEATHCOTE, Rev. Evelyn Dawsonne (1844–1908)
b. London 11 Nov. 1844 *d.* Winchester, Hants 1 May 1908
BA Oxon 1867. Vicar, Sparsholt near Winchester, 1875. *Flowers of the Engadine* 1891 (drawings for this book at BM(NH)).

HEATHFIELD, Richard (1802–1848)
Hampstead plants at Kew.

HEATLEY, Margaret *see* Voss, M.

HEATON, John Deakin (–1880)
d. Claremont, Leeds, Yorks 1880
MD London 1843. Dean, Leeds Medical School. Herb. collected around Leeds, 1835–65, was in possession of F. Arnold Lees.
T. W. Reid *Memoir* 1883 portr. *Naturalist* 1900, 51–61.

HEATON, Rev. Richard (*fl.* 1620s–1660s)
DD Dublin 1661. Dean of Clonfert, 1662. Studied Irish flora.
J. Ray *Synopsis Methodica* 1724, 253. C. Threlkeld *Synopsis Stirpium Hibernicarum* 1727 preface. R. Pulteney *Hist. Biogr. Sketches of Progress of Bot. in England* v.2, 1790, 194–95. N. Colgan *Fl. County Dublin* 1904, xix. C. E. Raven *English Nat.* 1947, 302–4. R. L. Praeger *Some Irish Nat.* 1949, 100.

HEBDEN, Thomas (1849–1931)
d. Keighley, Yorks 3 Jan. 1931
Director, power-looms firm. Lichenologist and mycologist of Cullingworth. His correspondence with W. Nylander added several species of *Verrucaria* to the British flora.
J. Bot. 1931, 78–79. *Naturalist* 1931, 60–61; 1961, 56. *Nature* v.127, 1931, 346. *Trans. Br. Mycol. Soc.* v.18, 1933, 93. M. R. D. Seaward *Guide to Lichenological Collection of Thomas Hebden (1849–1931)* 1971 portr.
Herb., Library and MSS. at Keighley Museum.

HEBERDEN, Thomas (1703–1769)
MD. FRS 1761. Brother of W. Heberden (1710–1801). Of Funchal, Madeira, whence he sent plants to J. Banks.
J. Banks *Journal...during Captain Cook's First Voyage* 1896, 6–13. *J. Bot.* 1904, 175. *Pharm. J.* v.146, 1941, 68.

HEBERDEN, William (1710–1801)
b. London 6 Aug. 1710 *d.* London 17 May 1801
BA Cantab 1728. MD 1739. FRS 1749. Uncle of L. Blomefield (1800–1893). Practised medicine in Cambridge until 1748; afterwards in London. Lecturer in Materia Medica, Cambridge, 1745.
G. C. Gorham *Mem. of John and Thomas Martyn* 1830, 117. W. Munk *Roll of R. College of Physicians* v.2, 1878, 159–64. *D.N.B.* v.25, 359–60. A. C. Wootton *Chronicles of Pharmacy* v.1, 1910, 290 portr. *Pharm. J.* v.146, 1941, 68.
Herb. at St. John's College, Cambridge. MSS. at Royal Society.
Heberdenia Banks.

HECTOR, Sir James (1834–1907)
b. Edinburgh 16 March 1834 *d.* Wellington, New Zealand 5 Nov. 1907
MD Edinburgh 1856. FRS 1866. FLS 1875. KCMG 1887. Surgeon and geologist on exploring expedition of Capt. J. Palliser to western N. America, 1857–60. Government geologist, New Zealand, 1861. Director, Botanic Garden, Wellington, 1866. 'On Geographical Botany of New Zealand' (*Trans. N.Z. Inst.* 1868, 1–5).
P. Mennell *Dict. Austral. Biogr.* 1892, 225. *D.N.B. Supplt 2* v.2, 236–37. T. F. Cheeseman *Manual N.Z. Fl.* 1906, xxviii–ix. *Proc. Linn. Soc.* 1907–8, 50. *Quart. J. Geol. Soc.* 1908, lxi–lxii. *Trans. Bot. Soc. Edinburgh* v.23, 1908, 369–71. *Trans. N.Z. Inst.* 1907, 543–44, 554. J. Ewan *Rocky Mountains Nat.*

1950, 226. R. Glenn *Bot. Explorers of N.Z.* 1950, 111–27 portr. *R.S.C.* v.3, 246; v.7, 932; v.10, 174; v.12, 320; v.15, 714.
New Zealand plants at Kew, BM(NH). Letters at Kew.
Hectorella Hook. f.

HEDDERLEY, John S. (–1903)
d. Sneinton, Notts 1903
Florist. Raised carnations. Contrib. to *Midland Florist.*
Garden v.64, 1903, 124. *J. Hort. Cottage Gdnr* v.47, 1903, 156.

HEDDLE, Robert (*c.* 1827–1860)
d. Kirkwall, Orkney 28 Aug. 1860
Botanist and ornithologist in Orkney. *Historia Naturalis Orcadensis* (with W. B. Baikie) 1848.

HEDLEY, E. (*fl.* 1840s)
Nurseryman, Rose Hill Nursery, Yarm, Yorks.

HEDLEY, George Ward (1871–1941)
d. Cheltenham, Glos March 1941
BA Oxon 1894. Science master, Cheltenham College, 1894–1931. Organised much of the work on H. J. Riddelsdell's *Fl. Gloucestershire* 1948 (clv–clvi).
Cheltenham Chronicle and Gloucestershire Graphic 22 March 1941. *Proc. Cotteswold Nat. Field Club* 1941, 193–94.

HEDLEY, Robert (*fl.* 1830s)
Nurseryman, Spittall Hill, Yarm, Yorks.

HEDLEY, Mrs. Winifred (–1964)
d. 30 April 1964
Assisted her husband, G. W. Hedley (1871–1941), in his work on flora of Gloucestershire.
Proc. Cotteswold Nat. Field Club 1964, 113.

HEGINBOTHOM, Charles David (1874–1950)
Of Devizes, Wilts. Keen naturalist. Plant records in D. Grose's *Fl. Wiltshire* 1957, 44.

HELLON, Robert (1854–1924)
b. Workington, Cumberland 1854 *d.* Seascale, Cumberland 9 Nov. 1924
PhD Heidelberg. Analytical chemist, Whitehaven. County Analyst, Cumberland and Westmorland. President, Whitehaven Scientific Society. Botanised around Seascale.
Bot. Soc. Exch. Club Br. Isl. Rep. 1924, 539–40.

HELME, William (1785–1834)
b. Warrington, Lancs 27 March 1785 *d.* Preston, Lancs 11 April 1834
Warper. Entomologist. Member of botanical society meeting at 'Green Man', Lord Street, Preston. Collected plants with Mr. Tomlinson, surgeon.
P. Whittle *Topographical, Statistical and Hist. Account of...Preston* v.2, 1837, 291. *Trans. Liverpool Bot. Soc.* 1909, 73.

HELMS, Richard (1842–1914)
b. Altona, Germany 12 Dec. 1842 *d.* Sydney, N.S.W. 17 July 1914
Entomologist and conchologist. In Australia, 1858, New Zealand, 1862. On staff of Australian Museum, 1888. Government fruit inspector, W. Australia, 1896. Bacteriologist, Dept. of Agriculture, N.S.W., 1900. Botanist to Elder Expedition, W. Australia, 1891–92.
J. Bot. 1894, 78. *Rep. Austral. Assoc. Advancement Sci.* 1907, 171. *J. Proc. R. Soc. W. Australia* 1914, 28–29. *J. Proc. R. Soc. N.S.W.* 1915, 11–14; 1921, 162–63. *Austral. Encyclop.* 1965, 471. *R.S.C.* v.12, 324; v.15, 749.
Herb. at Manchester. Mosses at Oxford. MSS. at Mitchell Library, Sydney.
Helmsia H. Boswell.

HELYER, B. (*fl.* 1860s–1870s)
Of Ditchling, Sussex. Botanised in Sussex.
A. H. Wolley-Dod *Fl. Sussex* 1937, xliv–xlv.

HEMINGWAY, W. (*fl.* 1890s–1910s)
Amateur palaeobotanist of Derby and Barnsley. Collected coal measure plants, now at Sheffield Museum.
Naturalist 1892, 170. *Sorby Rec.* v.2(4), 1967, 3.
Equisetum hemingwayi Kidst.

HEMMINGS, Joseph (*fl.* 1820s)
Seedsman, Corn Market, Oxford.

HEMSLEY, Alfred (1851–1917)
b. 7 Jan. 1851 *d.* Lewisham, London 30 Jan. 1917
Gardener to H. B. May at Edmonton until 1899. *Book of Fern Culture* 1908. Co-editor with J. Fraser of G. W. Johnson's *Gdnrs Dict.* 1917. Contrib. to *Gdnrs Chron.* and *Amer. Florist.*
Garden 1917, 62. *Gdnrs Chron.* 1917 i 67.

HEMSLEY, Oliver Tietjens (1876–1906)
b. Richmond, Surrey 6 Feb. 1876 *d.* Lahore, India 6 Jan. 1906
Son of W. B. Hemsley (1843–1924). Kew gardener, 1893. At Government Cinchona Factory, Mungpoo, 1898. Assistant Curator, Calcutta, 1903. Superintendent, Agri-Horticultural Gardens, Lahore.
Garden v.69, 1906, 80. *Gdnrs Chron.* 1906 i 32 portr. *J. Kew Guild* 1906, 328–29 portr. *Kew Bull.* 1906, 393–94.

HEMSLEY, William Botting (1843–1924)
b. East Hoathly, Sussex 29 Dec. 1843 *d.* Broadstairs, Kent 7 Oct. 1924
FRS 1889. ALS 1875. FLS 1896. LLD

Aberdeen 1913. VMH 1909. Kew gardener, 1860. Assistant, Kew Herb., 1865–67, 1883–1908 (Keeper from 1899). *Handbook of Hardy Trees, Shrubs and Herbaceous Plants* 1877. *Diagnoses Plantarum Novarum* 1878–80. *Biologia Centrali-americana. Botany* 1879–88 5 vols. *Report on Scientific Results of Challenger Expedition. Botany* 1885–86. *Enumeration of all Plants known from China* (with F. B. Forbes) (*J. Linn. Soc.* v.23, v.26, v.36, 1886–1905). Contrib. to *Fl. Tropical Africa, Hooker's Icones Plantarum, Bot. Mag., Kew Bull., J. Bot., J. Linn. Soc.*

Nature v.39, 1889, 587; v.114, 1924, 616–17. E. Bretschneider *Hist. European Bot. Discoveries in China* 1898, 820–23. *Gdnrs Mag.* 1899, 10 portr.; 1907, 574–75 portr. *J. Kew Guild* 1899, 1 portr.; 1925, 331–37 portr.; 1943, 292–94. *J. Hort. Cottage Gdnr* v.58, 1909, 6–7 portr. *Kew Bull.* 1909, 22–23; 1924, 389–92. *Bot. Soc. Exch. Club Br. Isl. Rep.* 1924, 540–41. *Gdnrs Chron.* 1924 i 16 portr.; 1924 ii 275–76. *J. Bot.* 1925, 21–23. *Proc. Linn. Soc.* 1924–25, 76–78. *Proc. R. Soc. B.* v.98, 1925, i–ix. *Curtis's Bot. Mag. Dedications, 1827–1927* 283–84 portr. *Who was Who, 1916–1928* 484. A. H. Wolley-Dod *Fl. Sussex* 1937, xliv–xlv. F. A. Stafleu *Taxonomic Literature* 1965, 198–99. *R.S.C.* v.7, 947; v.10, 190; v.12, 324; v.15, 752.

Herb. in possession of F. C. S. Roper (1819–1896). Letters and MSS. at Kew.

Hemsleya Cogn.

HEMSTED, Rev. John (1746–1824)
b. Lynton, Cambridgeshire 11 June 1746 *d.* Bedford Feb. 1824

BA Cantab 1810. Correspondent of J. E. Smith. Contrib. to J. Sowerby and J. E. Smith's *English Bot.* 79, 201, etc.

Herb. at BM(NH) (*J. Bot.* 1918, 259–60).

HENBREY, R. (*fl.* 1840s)

Seedsman and florist, opposite New Inn, Croydon, Surrey.

HENCHMAN, Francis (*fl.* 1820s)

Introduced Australian plants to Clapton Nursery, London.

C. McIntosh *Fl. and Pomona* v.1, 1829, t.23. *Bot. Mag.* 1837, t.3607.

Chorizema henchmanni R. Br.

HENCHMAN, John (*fl.* 1830s)

Collected plants, especially orchids, in Demerara, 1834 and Cumana, Venezuela for the Low Nursery, Clapton, London. Acquired nursery and stock of T. Page, Edmonton, Middx *c.* 1840.

Gdnrs Mag. 1835, 113–18. *Bot. Mag.* 1837, t.3571.

HENDERSON, A. (*fl.* 1780s–1790s)

Seedsman, Kendal, Westmorland.

HENDERSON, Alexander (–1827)
d. Feb. 1827

Nurseryman, Edinburgh. Lord Provost, Edinburgh.

Gdnrs Mag. 1827, 255.

HENDERSON, Alexander (1780–1863)
b. Aberdeen

MD Edinburgh. Practised medicine in London, *c.* 1808. Secretary, Horticultural Society of London, 1841–45.

HENDERSON, Andrew (*fl.* 1790s–1840s)

Nurseryman, Pine Apple Place, Edgware Road, London. As John Andrew Henderson and Co. after 1840.

HENDERSON, Archibald (*c.* 1826–1879)
b. East Lothian *c.* 1826 *d.* 22 April 1879

Superintendent, Horticultural Society of London gardens at Chiswick, 1858. Gardener at Trentham, Staffs. Nurseryman, Thornton Heath, Croydon, Surrey.

Gdnrs Yb. Almanack 1880, 191.

HENDERSON, Archibald H. (–1921)

Collected plants in Guinea, 1880; Canada, 1898; Barbados.

Plants at Kew.

HENDERSON, Edward George
(*c.* 1782–1876)
d. 4 Nov. 1876

Nurseryman, Vine Place, Edgware Road, Paddington, London; Wellington Road Nurseries, St. John's Wood, *c.* 1836. Eldest son of Andrew Henderson (*fl.* 1790s–1840s), founder of Pine Apple Place Nursery. *Illustrated Bouquet* 1857–64 3 vols.

Florist and Pomologist 1876, 284. *Garden* v.10, 1876, 476. *Gdnrs Chron.* 1876 ii 630. *Gdnrs Yb. Almanack* 1877, 176.

HENDERSON, F. Y. (1894–1966)
d. 9 April 1966

DSc Glasgow. Plant physiologist, Imperial College, London, 1921; Professor of Timber Technology. Director, Forest Products Research Laboratory, 1945. *Handbook of Soft Woods* 1957.

Nature v.211, 1966, 129.

HENDERSON, Frederick (*c.* 1841–1895)
d. 24 Sept. 1895

FLS 1875. Lieut.-Colonel, 1880. Collected ferns in Nilghiris and at Simla. Had herb. Ferns of N. India in *Trans. Linn. Soc. Bot.* v.1, 1880, 425–26.

Proc. Linn. Soc. 1895–96, 37.

Plants and letters at Kew.

Polypodium hendersoni Atkinson.

HENDERSON, George (1836–1929)
b. Cullen 29 Nov. 1836 *d.* Inverness 23 June 1929

MD Aberdeen. FLS 1872. Professor of

Surgery, Lahore University. Medical officer and scientific collector to T. D. Forsyth's political mission to Yarkand, 1870. Plants collected in Tibet and Yarkand described in *Lahore to Yarkand* (with A. O. Hume) 1873, 308–46. Director, Royal Botanic Garden, Calcutta, 1872.

E. Bretschneider *Hist. European Bot. Discoveries in China* 1898, 796–98. *Gdnrs Chron.* 1929 i 60. *Geogr. J.* v.74, 1929, 196. *J. Bot.* 1929, 288. *Proc. Linn. Soc.* 1929–30, 197–98. *Pakistan J. For.* 1967, 347.

Apocynum hendersoni Hook. f.

HENDERSON, Hugh (–1778)

Nurseryman, Dublin.

Irish For. 1967, 47.

HENDERSON, John (*fl.* 1810s)

Gardener at Brechin Castle, Angus. Later nurseryman at Brechin. Contrib. to *Caledonian Mem.*

HENDERSON, John Andrew (*c.* 1795–1872)

Nurseryman, Pine Apple Place, Edgware Road, London.

HENDERSON, Joseph (*fl.* 1840s–1866)

d. Wentworth Woodhouse, Yorks 22 Nov. 1866

ALS 1842. Gardener to Lord Fitzwilliam at Wentworth House, Yorks, succeeding Joseph Cooper in 1840. 'Germination of Ferns' (*Mag. Zool. Bot.* 1837, 333–41). 'Equisetum' (*Trans. Linn. Soc.* v.18, 1841, 567–74).

Gdnrs Chron. 1866, ii 38; v.160(20), 1966, 22. *Proc. Linn. Soc.* 1866–67, xxxv. *R.S.C.* v.3, 273.

Letters in Berkeley correspondence at BM(NH).

Hendersonia Berkeley.

HENDERSON, Logan (*fl.* 1780s)

Botanist to Czar of Russia. In Crimea, 1787. Correspondent of W. Forsyth.

Cottage Gdnr v.8, 1852, 187–88.

HENDERSON, Montgomery (1808–1892)

b. Swanston near Edinburgh March 1808 *d.* Ashby, Leics 14 Feb. 1892

Gardener to Sir G. H. Beaumont at Orton Hall, Leics, 1838. Excelled in growing grapes and pine-apples.

Gdnrs Chron. 1874 ii 719 portr.; 1892 i 247–48 portr.

HENDERSON, Peter (1822–1890)

b. Edinburgh 9 June 1822 *d.* Jersey City, U.S.A. 17 Jan. 1890

Gardener and florist. Collected British plants when young. To New York, 1843 where he was employed by Thorburn and R. Buist. Established market garden in Jersey City, 1847. *Gardening for Profit* 1865. *Henderson's Practical Floriculture* 1868. *Gardening for Pleasure* 1875. *Handbook of Plants* 1881.

Bot. Gaz. 1890, 78. *Gdnrs Mag.* 1907, 48. *J. Hort. Cottage Gdnr* v.53, 1906, 574. *Bull. New Jersey Experiment Station* 1927, 239–40 portr. L. H. Bailey *Standard Cyclop. Hort.* v.2, 1939, 1578–79 portr. *Gdnrs Chron.* 1947 i 9.

HENFREY, Arthur (1819–1859)

b. Aberdeen 1 Nov. 1819 *d.* Turnham Green, Middx 7 Sept. 1859

FRS 1852. ALS 1843. FLS 1844. MRCS 1843. Lecturer in Botany, St. George's Hospital, London, 1847. Succeeded E. Forbes as Professor of Botany, King's College, 1854. *Outlines of Structural and Physiological Botany* 1847. *Rudiments of Botany* 1849; ed. 2 1858. *Vegetation of Europe* 1852.

Bonplandia 1859, 292–93. *Ann. Mag. Nat. Hist.* 1859, 311–12. *Cottage Gdnr* v.22, 1859, 385–86. *Proc. Linn. Soc.* 1859–60, xxiii–xxv. *Proc. R. Soc.* v.10, 1860, xviii–xix. *D.N.B.* v.25, 409–10. F. W. Oliver *Makers of Br. Bot.* 1913, 192–203. *Bot. Soc. Exch. Club Br. Isl. Rep.* 1933, 282–83. C. C. Gillispie *Dict. Sci. Biogr.* v.6, 1972, 266–67. *R.S.C.* v.3, 275; v.12, 324; v.15, 755.

Letters and engraved portr. at Kew.

Henfreya Lindl.

HENLEY, Robert, 1st Earl of Northington (*c.* 1708–1772)

d. Grainge, Hants 14 Jan. 1772

MA Oxon 1733. Lord High Chancellor, 1761. "Took great delight in the cultivation of plants...at Grainge" (J. E. Smith *Selection of Correspondence of Linnaeus* v.2, 1821, 66–71).

D.N.B. v.25, 417–19.

HENNEDY, Roger (1809–1877)

b. Carrickfergus, Belfast Aug. 1809 *d.* Bothwell, Lanarkshire 22 Oct. 1877

Professor of Botany, Andersonian Institution, Glasgow, 1863–77. *Clydesdale Fl.* 1865; ed. 4 1878 portr.

W. H. Harvey *Phycologica Australica* v.2, 1859, 75. *J. Bot.* 1877, 96. H. C. Watson *Topographical Bot.* 1883, 547. F. Boase *Modern English Biogr.* v.1, 1892, 1429–30. *D.N.B.* v.25, 422–23. *Br. Phycological Bull.* 1964, 385–86.

Herb. at Strathclyde University.

Hennedya Harvey.

HENRY, Augustine (1857–1930)

b. Cookstown, County Antrim 2 July 1857 *d.* Dublin 23 March 1930

MA Belfast 1878. LRCP Edinburgh 1880. FLS 1888. VMH 1906. VMM 1902. Medical officer, China, 1880–1900. Collected plants in Ichang, Hupeh, Szechuan, Yunnan,

Hainan, Formosa. Authority on Chinese materia medica. Reader in Forestry, Cambridge, 1907–13. Professor of Forestry, College of Science, Dublin, 1913–26. *Trees of Great Britain and Ireland* (with H. J. Elwes) 1906–13 7 vols. *Forests, Woods and Trees in Relation to Hygiene* 1919.

E. Bretschneider *Hist. European Bot. Discoveries in China* 1898, 774–94, 1092. *Garden* v.60, 1901, iv portr. *Gdnrs Chron.* 1901 ii 85–86 portr.; 1913 i 163–64 portr.; 1930 i 248–49; 1965 i 361. *Curtis's Bot. Mag. Dedications, 1827–1927* 299–300 portr. *Bot. Soc. Exch. Club Br. Isl. Rep.* 1930, 324. *Bull. Soc. Dendrologique France* 1930, 59–62 portr. *Empire For. J.* 1930, 129–31. *Indian For.* 1930, 323–24. *J. Bot.* 1930, 148–49. *Kew Bull.* 1930, 215–17. *Nature* v.125, 1930, 606–7. *Pharm. J.* 1930, 346. *Proc. Linn. Soc.* 1929–30, 198–200. *Quart. J. For.* 1930, 169–73 portr. *Times* 24 March 1930. *Who was Who, 1929–1940* 625. *J. R. Hort. Soc.* 1942, 10–15 portr. R. L. Praeger *Some Irish Nat.* 1949, 101. M. Hadfield *Gdning in Britain* 1960, 404–6. S. Pim *The Wood and the Trees* 1966 portr. A. M. Coats *Quest for Plants* 1969, 116–18 portr. *Lily Yb.* (U.S.A.) no. 27, 1974, 55–63.

Plants at Kew, Dublin. Portr. at Hunt Library.

HENRY, Caroline (*née* **Orridge**)
(–1894)

Wife of A. Henry (1857–1930). Collected plants in China, Japan and in Colorado, 1891–94.

Hooker's Icones Plantarum 1902, t.2726.

Plants at Kew.

Carolinella henryi Hemsley.

HENRY, Isaac Anderson- *see* Anderson-Henry, I.

HENRY, James (*c.* 1863–1936)
d. Dec. 1936

Teacher, Coleraine, N. Ireland. Botanised on north coast of Ireland.

Irish Nat. J. 1937, 154.

HENRY, John M. (1841–1937)
b. Inverkeilor, Forfarshire 16 Sept. 1841 *d.* 7 Oct. 1937

Kew gardener, 1865–67. To India, 1867. Superintendent, Baroda State Gardens, 1879–95. Nurseryman, Hartley Wintney, Hants.

Kew Bull. 1895, 318. *J. Kew Guild* 1935, 446; 1938, 792–93.

HENRY, Thomas (1734–1816)
b. Wrexham, Denbighshire 26 Oct. 1734 *d.* Manchester 18 June 1816

FRS 1775. Chemist. Surgeon-apothecary, Manchester. Secretary, Manchester Literary and Philosophical Society, 1781; President, 1807. First observed use of carbonic acid to plants. 'Influence of Fixed Air on Vegetation' (*Manchester Philos. Soc. Mem.* v.2, 1789, 357–65).

Mem. Manchester Lit. Philos. Soc. v.2, 1789, 357–65. J. E. Smith *Introduction to Physiol. and Systemat. Bot.* 1827, 169. *D.N.B.* v.26, 127–28. *R.S.C.* v.3, 292.

HENRY OF HUNTINGDON
(*c.* 1084–1155)

Archdeacon of Huntingdon, 1109. According to Bishop Tanner (Bibl. Bodley 6353) he wrote MS. *De Herbis, de Aromatibus et de Gemmis.*

R. Pulteney *Hist. Biogr. Sketches of Progress of Bot. in England* v.1, 1790, 21–22. *D.N.B.* v.26, 118–19.

HENSHALL, John (*fl.* 1850s)

Gardener to Baron J. H. Schröder. Possibly the Mr. Henshall who collected orchids in Java for Messrs Rollisson, Tooting, 1852. *Practical Treatise on Cultivation of Orchidaceous Plants* 1845.

Gdnrs Chron. 1852, 533–34, 581. *Bot. Mag.* 1854, t.4797; 1857, t.4972. *Fl. Malesiana* v.1, 1950, 226.

Plants at Kew.

Dendrobium henshallia Reichenb. f.

HENSHAW, John (*fl.* 1720s)

Superviser of nursery on Lord Kenmare's estate, County Cork. Later worked for Lord Barrymore.

Irish For. 1967, 50.

HENSHAW, Julia Wilmotte (*née* **Henderson**)
(1869–1937)
b. Shropshire 1869 *d.* Vancouver, Canada 20 Nov. 1937

Mountain Wild Flowers of America 1906. *Mountain Wild Flowers of Canada* 1906.

New York Times 21 Nov. 1937. J. Ewan *Rocky Mountains Nat.* 1950, 229.

HENSHAW, Nathaniel (–1673)
d. London Sept. 1673

MD Leyden 1653. FRS 1663. Practised medicine in Dublin. Discovered spiral vessels in *Juglans* 1661. *Aero-Chalinos* 1664; ed. 2 1677.

T. Birch *Hist. R. Soc.* v.1, 1756, 37. K. Sprengal *Historia Rei Herbariae* v.2, 1806, 9. *D.N.B.* v.26, 134.

HENSLOW, Anne *see* Barnard, A.

HENSLOW, Frances Harriet *see* Hooker, F. H.

HENSLOW, Rev. George (1835–1925)
b. Cambridge 23 March 1835 *d.* Bournemouth, Hants 30 Dec. 1925

BA Cantab 1858. FLS 1864. VMH 1897.

Son of J. S. Henslow (1796–1861). Curate, Steyning, Sussex, 1858. Headmaster, Hampton Lucy Grammar School, Warwick, 1861–65; Grammar School, Stove Street, London, 1865–72. Lecturer in Botany, St. Bartholomew's Medical School, 1886–90; also at Birkbeck and Queen's Colleges, London. Hon. Professor of Botany, Royal Horticultural Society. President, Ealing Natural History Society, 1882–1904. Popular lecturer and writer. *Origin of Floral Structures* 1888. *How to Study Wild Flowers* 1896. *Plants of the Bible* 1895; 1906. *South African Flowering Plants* 1903. *Heredity of Acquired Characters in Plants* 1908.

Garden v.59, 1901, 226 portr. *J. Hort. Cottage Gdnr* v.50, 1904, 30 portr. *Bot. Soc. Exch. Club Br. Isl. Rep.* 1925, 849. *Gdnrs Chron.* 1926, 36. *J. Bot.* 1926, 55–56. *Nature* v.117, 1926, 130. *Proc. Linn. Soc.* 1925–26, 82–83. *Who was Who, 1916–1928* 488. *R.S.C.* v.7, 954; v.10, 199; v.12, 326; v.15, 774.

HENSLOW, Rev. John Stevens (1796–1861)
b. Rochester, Kent 6 Feb. 1796 *d.* Hitcham, Suffolk 16 May 1861

BA Cantab 1818. FLS 1818. Professor of Botany, Cambridge, 1825–61. Vicar, Cholsey, Berks, 1832; Hitcham, 1839. Recommended his pupil, C. Darwin, as naturalist on HMS 'Beagle'. *Catalogue of British Plants* 1829; ed. 2 1835. *Principles of Descriptive and Physiological Botany* 1836. *Dictionary of Botanical Terms* 1858. Co-editor with B. Maund of *The Botanist* 1837–46 5 vols.

Cottage Gdnr v.26, 1861, 138. *Gdnrs Chron.* 1861, 505–6, 527–28, 551–52; 1911 ii 10–11. *J. Hort.* v.1, 1861, 138. *Proc. Linn. Soc.* 1861, xxv–xli. *Quart. J. Geol. Soc.* 1862, xxxv–xxxvii. L. Jenkyns *Mem. of Rev. J. S. Henslow* 1862 portr. *Trans. Bot. Soc. Edinburgh* v.7, 1863, 196–200. F. Darwin *Life of Charles Darwin* v.1, 1887, 168; *More Letters* (with A. C. Seward) v.1, 1903, 188–89 portr. G. C. Druce *Fl. Berkshire* 1897, clxiv–v. *D.N.B.* v.26, 135–36. *J. Hort. Soc.* 1912, 220–24 portr. F. W. Oliver *Makers of Br. Bot.* 1913, 151–63 portr. *Country Life* 1961, 1101–2 portr. N. Barlow *ed. Darwin and Henslow...Letters, 1831–1860* 1967 portr. *Trans. Suffolk Nat. Soc.* 1967, 406–8. C. C. Gillispie *Dict. Sci. Biogr.* v.6, 1972, 288–89. *E. Anglian Mag.* 1974, 632–34. *R.S.C.* v.3, 296; v.12, 326.

Plants at Cambridge, Kew. Letters at Kew, Fitzwilliam Museum, Cambridge and in Berkeley correspondence at BM(NH). Marble bust by T. Woolner at Kew. Portr. at Hunt Library. Sale at Stevens 1 and 23 July 1861.

Henslowia Wall.

HENWOOD, Thomas E. (–1937)

Pteridologist and cultivator of florists' flowers.

Br. Fern Gaz. 1937, 110 portr.

HEPBURN, Sir Archibald Buchan- *see* Buchan-Hepburn, *Sir* A.

HEPBURN, James Edward (1811–1869)
b. London 1811 *d.* Victoria, Canada 16 April 1869

Educ. Trinity College, Cambridge. Barrister. Ornithologist. Collected seeds for an English horticultural society. Collected in British Columbia, 1864.

Condar v.28, 1926, 249–53; v.33, 1931, 221.

Carex hepburnii Boott.

HEPWORTH, John (1802–1883)
b. Wyke near Halifax, Yorks 1802 *d.* 2 Jan. 1883

Florist. Raised auriculas and tulips.

Florist and Pomologist 1883, 32.

HERBERT, Hon. and Rev. William (1778–1847)
b. 12 Jan. 1778 *d.* London 28 May 1847

BA Oxon 1798. DCL 1808. BD 1840. MP Hants 1806; Cricklade, 1811. Rector, Spofforth, Yorks, 1814–40. Dean of Manchester, 1840–47. Cultivated bulbous plants and experimented in hybridisation. *Amaryllidaceae* 1837 (with 48 plates by author). Drew 61 plates for *Bot. Mag.* and 45 plates for *Bot. Register.*

Gdnrs Mag. 1830, 531–33. *Gent. Mag.* 1843 i 115–33. *Gdnrs Chron.* 1847, 372; 1886 i 23; 1950 i 178. *J. Hort. Soc. London* 1847, 249–93. *Garden* v.28, 1885, 400. *Proc. Manchester Lit. Philos. Soc.* v.25, 1885–86, 43–46. *D.N.B.* v.26, 234. *Trans. Liverpool Bot. Soc.* 1909, 73. *Curtis's Bot. Mag. Dedications, 1827–1927* 47–48 portr. H. F. Roberts *Plant Hybridization before Mendel* 1929, 94–102. *Herbertia* v.4, 1937, 3–4, 26–27, 63–69; 1947, 37–38. *J. Soc. Bibl. Nat. Hist.* 1952, 375–77. *Amer. Orchid Soc. Bull.* 1956, 686–87. W. Herbert *Amaryllidaceae* (1966 reprint with biogr.). F. A. Stafleu *Taxonomic Literature* 1967, 199. A. A. Guimond *Hon. and Very Rev W. Herbert* (unpublished thesis, Wisconsin Univ., 1966). C. C. Gillispie *Dict. Sci. Biogr.* v.6, 1972, 295–97.

Letters at Kew. Drawings at Royal Horticultural Society. Portr. by W. Beechey at Eton College.

Herbertia Sweet.

HERBST, Hermann Carl Cottlieb (*c.* 1830–1904)
d. Richmond, Surrey 18 March 1904
VMH 1897. Director, Botanic Gardens, Rio de Janeiro. Nurseryman, Richmond.
Garden v.65, 1904, 229 portr. *Gdnrs Chron.* 1904 i 204 portr. *Gdnrs Mag.* 1904, 222 portr. *J. Hort. Cottage Gdnr* v.48, 1904, 282.

HEREMAN, Samuel (*fl.* 1840s)
Associated with Joseph Paxton. *Blight on Flowers* 1840. Supervised 1868 ed. of Paxton's *Bot. Dict.*
Gdn J. New York Bot. Gdn 1965, 138–41.

HERLE, Thomas (*fl.* 1720s)
Sent plants from Lisbon to D. Du Bois at Oxford.

HERON, Andrew (–1729)
d. Bargaly, Kirkcudbright 1729
Lived at Bargaly from 1690. Botanist and horticulturist. "He is, in my opinion, the most learned and ingenious gentleman, in the article of gardening, I ever conversed with" (R. Bradley).
J. C. Loudon *Arboretum et Fruticetum Britannicum* v.1, 1838, 95–99. *Gdnrs Mag.* 1835, 718–20.

HERRGOTT, Joseph Franz Albert David (1823–1861)
b. Batavia, East Indies 1823 *d.* Melbourne 8 Oct. 1861
Botanist to B. H. Babbage's expedition to N.W. interior of S. Australia, 1858 (F. von Mueller *Report on Plants collected during Mr. Babbage's Expedition into N.W. Interior of S. Australia in 1858* 1859).
Adelaide Observer 2 Nov. 1861. J. D. Stuart *Explorations in Australia* 1865 *passim. Rep. Austral. Assoc. Advancement Sci.* 1907, 169.
Plants at Melbourne.

HERRING, Francis (*fl.* 1580s–1620s)
MA Cantab 1589. MD 1597. Physician, London, 1599–1628. Commendatory verses to J. Gerard published in his *Herball* 1597.
M. de Lobel *Stirpium Illustrationes* 1655, 38. *D.N.B.* v.26, 258. C. E. Raven *English Nat.* 1947, 247.

HERRINGTON, Arthur (1866–1950)
b. Coddenham, Sussex 30 March 1866 *d.* Madison, N. J., U.S.A. 29 March 1950
Gardener to W. Robinson at Gravetye. On staff of *Garden*. To U.S.A. as landscape gardener, 1895. *Chrysanthemum* 1905. Contrib. to *Gdns Mag.* and *Country Life.*
Gdnrs Chron. 1925 ii 82 portr.; 1935 ii 346 portr.
Portr. at Hunt Library.

HERVEY, Dudley Francis Amelius (1849–1911)
b. Great Chesterford, Essex 7 Jan. 1849 *d.* 1 June 1911
Chief clerk and interpreter, Penang, 1870–82. Resident Councillor, Malacca, 1882–93. Collected plants at Malacca and sent a few plants from Aden to Kew in 1892.
Kew Bull. 1891, 246; 1901, 31. *Gdns Bull. Straits Settlements* 1927, 123. *J. Malayan Branch R. Asiatic Soc.* 1927, 316. *Who was Who, 1897–1916* 335. *Fl. Malesiana* v.1, 1950, 227.
Plants at Kew.

HERVEY, Rev. George Aiden Kingsford (1893–1967)
b. 19 July 1893 *d.* Penrith, Cumberland 14 June 1967
BA Oxon 1916. MA 1920. Ordained priest 1923. Chaplain, Bryanston School, 1934–43. Vicar, Gilsland, 1943. Rector, Great Salkeld, 1946. Canon, Carlisle Cathedral, 1950. Botanist. Founder of Lake District Naturalists' Trust. Edited *Field Nat.*
Proc. Bot. Soc. Br. Isl. 1969, 623.

HERVEY, John (*c.* 1786–1829)
d. Comber, County Down 4 Sept. 1829
Nurseryman, Nurseryville near Comber.
Gdnrs Mag. 1829, 750–52.

HESKETH (*or* **Hasket**), **Thomas** (1561–1613)
b. Martholme Hall, Blackburn, Lancs 1561
d. Clitheroe, Lancs 7 Dec. 1613
Practised as physician and surgeon at Clitheroe. Correspondent of J. Gerard and J. Parkinson. Discovered *Andromeda* (J. Gerard *Herball* 1597, 1110).
J. Parkinson *Theatrum Botanicum* 1640, 740, 767, 1015, etc. M. de Lobel *Stirpium Illustrationes* 1655, 82, 93, 118. *Palatine Note-book* v.5, 1885, 7–9. *D.N.B.* v.26, 297. *Trans. Liverpool Bot. Soc.* 1909, 73. R. H. Jeffers *Friends of John Gerard* 1967, 61–62.

HESLOP-HARRISON, John William (1881–1967)
b. Birtley, County Durham 1881 *d.* Birtley 23 Jan. 1967
BSc Durham 1903. MSc 1916. DSc 1917. FRS 1928. FRSE. Science master, Middlesbrough High School, 1905–17. Lecturer in

Zoology, Armstrong College, 1919–27. Professor of Botany, King's College, Newcastle-upon-Tyne, 1927–46. President, Northern Naturalists' Union, 1946, 1957–58. Studied *Rosa*, *Salix* and *Rubus* and fauna and flora of Hebrides. Edited *Vasculum.*

Nature v.159, 1947, 260–61; v.213, 1967, 869–70. *Biogr. Mem. Fellows R. Soc.* 1968, 243–70 portr. *Times* 2 Feb. 1967. *Vasculum* v.52, 1967, 1–2. *Yb. R. Soc. Edinburgh* 1968, 15–19. *Watsonia* 1970, 181–82. *Who was Who, 1961–1970* 522–33.

HEWAN, D. Archibald (*fl.* 1860s)

Missionary doctor in Calabar, S. Nigeria. Collected plants in Fernando Po, 1862.

Plants at Edinburgh, Kew.

HEWARD, Robert (1791–1877)

d. Wokingham, Berks 24 Oct. 1877

FLS 1836. In Jamaica, 1823–26. 'Some Observations on Collection of Ferns from... Jamaica' (*Mag. Nat. Hist.* 1838, 453–67). 'Life of Allan Cunningham' (*J. Bot.* 1842, 231–320 portr.).

A. Lasègue *Musée Botanique de B. Delessert* 1845, 266. *Gdnrs Chron.* 1877 ii 571. *J. Bot.* 1877, 380. I. Urban *Symbolae Antillanae* v.3, 1902, 60. *J. Proc. R. Soc. N.S.W.* 1908, 71. H. N. Clokie *Account of Herb. Dept. Bot., Oxford* 1964, 183. *R.S.C.* v.3, 342.

Plants at Kew, Oxford. Letters at Kew.

Hewardia Hook.

HEWETT, Mr. (*fl.* 1840s)

Surgeon of East Ilsley, Berks. *History of Hundred of Compton* contains list of plants observed by Hewett and his son William. At BM(NH) is MS. by W. Hewett entitled *Account of Orchidaceous Plants found in...East Ilsley.*

G. C. Druce *Fl. Berkshire* 1897, clxx.

HEWGILL, Arthur (*fl.* 1840s–1860s)

MD. Of Repton, Derbyshire. Compiled list of Staffordshire plants (R. Garner *Nat. Hist. Staffordshire* 1844, 339).

W. Wyatt and C. G. Thornton *Fl. Repandunensis* 1866, vi.

HEWITT, H. Dixon (*fl.* 1890s–1920s)

Of Thetford, Norfolk. Analytical chemist. Herb. at Thetford Museum.

HEWITT, Henry (–1771)

d. 27 May 1771

Founded with his brother, Samuel (–1793), a nursery in Brompton Road, Kensington, London.

HEWITT, Henry (–1791)

Acquired his uncle's nursery in Brompton Road, Kensington, London in 1771. In partnership with Smith, Harrison and Cook. After his death nursery carried on by his nephews John and Samuel Harrison until bankruptcy in 1833.

Trans. London Middlesex Archaeol. Soc. v.24, 1973, 186.

HEWITT, John (1880–1961)

b. Dronfield, Derbyshire 23 Dec. 1880 *d.* Grahamstown, S. Africa 4 Oct. 1961

BA Cantab 1903. Zoologist. Curator, Sarawak Museum, 1905–8. Assistant, Transvaal Museum, 1909–10. Director, Albany Museum, Grahamstown, 1910–58.

Fl. Malesiana v.1, 1950, 229. *S. African J. Sci.* 1961, 312.

Sarawak plants at Kew, Singapore.

HEWLETT, James *see* Addendum

HEY, John (*c.* 1801–1837)

d. Leeds, Yorks 11 Dec. 1837

FLS 1837. FGS. Surgeon. Curator, Leeds Literary and Philosophical Society. Botanist and geologist. Mrs. Hey of Leeds who wrote *Moral of Flowers* 1833 was probably his wife.

R. V. Taylor *Biographia Leodiensis* 1865–67, 371–72, 665. *Leeds Mercury* 22 June 1889.

HEY, Thomas (*c.* 1840–1919)

b. Sheffield, Yorks *c.* 1840 *d.* Sheffield 15 Oct. 1919

Railwayman. Naturalist with particular interest in mycology. Founder member of British Mycological Society. President, Midland Railway Natural History Society.

Naturalist 1919, 404–5. *Sorby Rec.* v.2(4), 1967, 4.

HEYNE, Benjamin (1770–1819)

b. Döbra near Pirna, Germany 1 Jan. 1770 *d.* Vappera, Madras 6 Feb. 1819

MD. FLS 1813. Moravian missionary. Arrived at Danish Settlement of Tranquebar in 1792. Superintendent of Bangalore Gardens, 1802–8. Commissioned Indian artists to execute flower drawings. In England 1813 (preface to A. W. Roth's *Novae Plantarum Species* 1821).

Bot. Mag. 1815, t.1738. R. H. Phillimore *Hist. Rec. Survey of India* v.2, 1950, 113–15, 405–6. *Fl. Malesiana* v.5, 1958, 51. *Botanisk Tidsskrift* 1955, 56–57. M. Archer *Nat. Hist. Drawings in India Office Library* 1962, 27–28, 79–80. I. H. Burkill *Chapters on Hist. Bot. in India* 1965, 18 *et seq.* *R.S.C.* v.3, 345.

Drawings at Kew, India Office Library.

Heynea Roxb.

HEYNE, Ernest Bernhard (1825–1881)
b. Meissen, Saxony 15 Sept. 1825 *d.* Adelaide, Australia 16 Oct. 1881

To Victoria, 1849. At Botanic Garden, Melbourne, 1854–67. Nurseryman at Adelaide, 1868. Had herb. *Vines and their Synonyms* 1869. *Amateur Gardener* ed. 4 1886. *Fruit, Flower and Vegetable Garden* 1871.

P. Mennell *Dict. Austral. Biogr.* 1892, 231. *Vict. Nat.* 1908, 108–9. *Rep. Austral. Assoc. Advancement Sci.* 1911, 230.

Cyperus heynei Boeck.

HIBBERD, James Shirley (1825–1890)
b. Stepney, London 1825 *d.* Kew, Surrey 16 Nov. 1890

Horticultural journalist. Edited *Floral World* 1858–72 and *Gdnr's Mag.* 1861–90. *Rustic Adornments for Homes of Taste* 1856. *Garden Favourites* 1858. *Rose Book* 1864. *Fern Garden* 1869. *New and Rare Beautiful-leaved plants* 1870. *Field Flowers* 1870. *The Ivy* 1872. *Familiar Garden Flowers* 1879–97 5 vols.

Garden v.38, 1890, 495, 517, 541, 562, 592. *Gdnrs Chron.* 1883 ii 298–99, 305; 1890 ii 596–98 portr., 628, 660–61, 695, 752–53; v.164(19), 1968, 18. *J. Hort. Cottage Gdnr* v.21, 1890, 441–42 portr., 463. *J. Bot.* 1890, 382. *D.N.B.* v.26, 342–43. *R.S.C.* v.7, 975; v.12, 333; v.15, 829.

Portr. at Royal Horticultural Society. Portr. at Hunt Library. Library sold at Sotheby, 29 June 1891.

HIBBERT, George (1757–1837)
b. Manchester 1757 *d.* Watford, Herts 8 Oct. 1837

FRS 1811. FLS 1793. West Indian merchant. Alderman, London, 1798–1803. MP Seaford, 1806–12. Had botanic garden at Clapham with Joseph Knight as his gardener. Sent J. Niven to the Cape and J. Macfadyen to Jamaica for plants. Provided plants for H. C. Andrews's *Bot. Repository* 1797–1812 and *Geraniums* 1805.

J. Sowerby and J. E. Smith *English Bot.* 524. H. C. Andrews *Bot. Repository* v.2, 1800, 126. *Bot. Mag.* 1809, t.1218. *Gent. Mag.* 1838 i 96–99. *Proc. R. Soc.* v.4, 1838, 93–94. *J. Bot.* 1886, 296–97. E. Bretschneider *Hist. European Bot. Discoveries in China* 1898, 218. *D.N.B.* v.26, 343. *Trans. Liverpool Bot. Soc.* 1909, 74.

Herb. formerly at Linnean Society. Library sold by Evans, March–May 1829.

Hibbertia Andr.

HIBBERT, William (*fl.* 1830s)

MD. Army surgeon, Bombay. 'Mucor hyphaenes' (*Naturalist* v.4, 1838–39, 711).

HICK, Thomas (1840–1896)
b. Leeds, Yorks 5 May 1840 *d.* Bradford, Yorks 31 July 1896

BA, BSc London. ALS 1894. Assistant Lecturer in Botany, Owens College, Manchester, 1885. 'Protoplasmic Continuity in Algae' (*J. Bot.* 1884–85).

J. Bot. 1896, 488; 1897, 193–96 portr. *Naturalist* 1897, 81–84 portr. *Mem. Proc. Manchester Lit. Philos. Soc.* 1896–97, lxiii–lxv. *Proc. Linn. Soc.* 1896–97, 57. *Proc. Yorks Geol. Soc.* 1895, 43–57. *Trans. Liverpool Bot. Soc.* 1909, 74. *R.S.C.* v.10, 223; v.12, 333; v.15, 229.

Fossil plants at Manchester Museum. Letters at Kew.

HICKEY, Rev. William (*c.* 1787–1875)
d. Mulrankin, Wexford 24 Oct. 1875

BA Cantab. MA Dublin. Rector, Mulrankin, 1834–75. Wrote under pen-name of Martin Doyle. *Flower Garden* 1834. *Practical Gardening* 1836.

Garden v.8, 1875, 478. *J. Hort. Cottage Gdnr* v.54, 1875, 448.

HICKS, Henry (1837–1899)
b. St. Davids, Pembrokeshire 26 May 1837 *d.* Hendon, Middx 18 Nov. 1899

FRS 1885. Practised medicine at St. Davids and from 1871 at Hendon. Geologist, also interested in palaeobotany. President, Geological Society, 1896–98.

Proc. R. Soc. v.75, 1905, 106.

HICKS, John Braxton (1823–1897)
b. Lymington, Hants 1823 *d.* Lymington 28 Aug. 1897

MD London 1851. FRS 1862. FLS 1852. Obstetrician. Studied lichens and mosses. 'Gouidia of Mosses' (*Trans. Linn. Soc.* v.23, 1862, 567–88).

Proc. Linn. Soc. 1897–98, 37–38. *R.S.C.* v.3, 347; v.7, 976; v.10, 224; v.15, 831.

HICKS, Thomas (*fl.* 1630s)

Warden, Apothecaries Society, 1632. Accompanied T. Johnson on his botanical excursion in Kent in 1632, and his tour of west of England in 1634.

HICKSON, Sydney John (1859–1940)
b. Highgate, London 1859 *d.* Cambridge 6 Feb. 1940

DSc London 1883. FRS 1895. In East Indies, 1885–86. Collected plants in Indonesia, Mexico and Arizona. Professor of Zoology, Manchester, 1894–1926. Returned to Cambridge. *Naturalist in North Celebes* 1889.

Obit. Notices Fellows R. Soc. 1941, 383–94 portr. *Who was Who, 1929–1940* 636. *Fl. Malesiana* v.1, 1950, 231.

Plants at Kew.

HIERN, William Philip (1839–1925)
b. Stafford 19 Jan. 1839 *d.* Barnstaple, Devon 29 Nov. 1925

BA Cantab 1861. FRS 1903. FLS 1873. Studied Devon flora after he moved to Barnstaple in 1882. President, Devonshire Association, 1916. 'Monograph of Ebenaceae' (*Trans. Cambridge Philos. Soc.* v.12, 1873). *Catalogue of African Plants collected by F. Welwitsch in 1853–61* 1896–1900. Contrib. to D. Oliver's *Fl. Tropical Africa* v.2, 1871; v.3, 1877; *Fl. Capensis* v.4, 1904.

Bot. Soc. Exch. Club Br. Isl. Rep. 1925, 849–51. *J. Bot.* 1926, 51–54, 168. *Kew Bull.* 1926, 45–46. *Nature* v.117, 1926, 23. *Proc. Linn. Soc.* 1925–26, 83–84. *Who was Who, 1916–1928* 494. W. K. Martin and G. T. Fraser *Fl. Devon* 1939, 776–77. *R.S.C.* v.7, 977; v.10, 225; v.15, 833.

Plants at Exeter Museum, Kew, Cambridge. MSS. and portr. at BM(NH). Portr. at Hunt Library.

Hiernia S. Moore.

HIGGINS, Mrs. D. Martha (*c.* 1856–1920)

Of Luton. Interested in Mycetozoa. Her plants together with those of her sister, Emily Kate, at BM(NH).

J. G. Dony *Fl. Bedfordshire* 1953, 30.

HIGGINS, Rev. Henry Hugh (1814–1893)
b. Turvey Abbey, Bedfordshire 28 Jan. 1814 *d.* Liverpool 2 July 1893

BA Cantab 1836. Geologist. Chaplain, Rainhill Asylum, Liverpool, 1853–86. President, Liverpool Literary and Philosophical Society, 1859–62, 1889–90. Travelled in Palestine, Sinai and Egypt, 1848; West Indies, 1876. *On Some Fossil Ferns in Ravenhead Collection* (with F. P. Marrat) 1872. *Notes by Field Naturalist in Western Tropics* 1877. *Life in Lowest Organisms* 1880.

Research v.1, 1888, 8–9 portr. *Geol. Mag.* 1893, 380–84. *J. Bot.* 1893, 286–87. J. B. L. Warren *Fl. Cheshire* 1899, lxxxiii–lxxxiv. *Liverpool Lit. Philos. Soc. Proc.* v.68, 1893–94, 35–67 portr. *Trans. Liverpool Bot. Soc.* 1909, 74–75. *Lancashire Cheshire Nat.* v.14, 1922, 159–65. *R.S.C.* v.3, 348; v.7, 978; v.10, 226; v.15, 834.

West Indian cryptogams at Kew. Letters in Wilson correspondence at BM(NH). Portr. at Walker Art Gallery, Liverpool.

HIGGINS, Vera (*née* **Cockburn**) (1892–1968)
b. Jan. 1892 *d.* 17 Oct. 1968

MA Cantab. FLS 1945. VMH 1946. Scientific Officer, National Physical Laboratory. Authority on succulent plants and *Crassulaceae* in particular. *Naming of Plants* 1937. *Study of Cacti* 1933; ed. 2 1946. *Succulent Plants Illustrated* 1949. *Succulents in Cultivation* 1960. *Crassulas in Cultivation* 1964. Translated W. Kupper and P. Roshardt's *Cacti* 1961. Edited *J. Cactus Succulent Soc. Great Britain*, *Alpine Gdn Soc. Bull.*

Gdnrs Chron. 1937 i 180 portr.; 1961 ii 379 portr.; v.164, 1968, 27–28 portr. *National Cactus Succulent J.* 1968, 81 portr. *J. Cactus Succulent Soc. Great Britain* 1969, 1. *J. R. Hort. Soc.* 1969, 187–88.

HIGHLEY, Percy (1856–1929)
b. London 12 Aug. 1856 *d.* Wokingham, Berks 23 Jan. 1929

Natural history draughtsman and lithographer. Illustrated BM(NH) publications. Illustrated W. Fawcett and A. Rendle's *Fl. Jamaica* 1910, L. Newton's *Handbook of Br. Seaweeds* 1931, A. L. Smith's *Monograph of Br. Lichens* 1918 and articles in *J. Bot.*, *J. Linn. Soc.*

J. Bot. 1929, 86–87.

Drawings at BM(NH).

HIGSON, Thomas (1773–1836)
d. Kingston, Jamaica 21 Dec. 1836

Of Kingston. Merchant. Botanist and Curator, Bath garden, Jamaica, 1828–32, where he collected plants. Contrib. to A. H. R. Grisebach's *Fl. Br. W. Indian Islands* 1859–64.

Hooker's J. Bot. 1842, 138. *Bot. Gaz.* 1897, 348. I. Urban *Symbolae Antillanae* v.3, 1902, 61. F. Cundall *Historic Jamaica* 1915, 29, 175–76. *J. Bot.* 1922, 52.

Higsonia Robinson (ined.).

HILARY, Miss Daisy (1888–1959)

BSc Leeds. Biology teacher, Cockburn High School, Leeds. Mycologist and bryologist.

Naturalist 1959, 101.

HILDEBRAND, Arthur Hedding (1852–1918)
d. Puddletown, Dorset Jan. 1918

Civil servant in Burma for 30 years. Political officer, Southern Shan States, 1891–1902. Sent living plants to Kew Gardens.

Kew Bull. 1918, 32–33. *Orchid Rev.* 1918, 63. *Times* 7 Jan. 1918.

Dendrobium hildebrandii Rolfe.

HILL, Alexander (–1887)

Plants at Glasgow.

HILL, Sir Arthur William (1875–1941)
b. Watford, London 11 Oct. 1875 *d.* Kew, Surrey 3 Nov. 1941

BSc Cantab 1897. DSc 1919. FRS 1920. FLS 1908. KCMG 1931. VMH 1934. VMM 1936. Demonstrator, later Lecturer in Botany, Cambridge, 1899–1907. Assistant Director, Kew, 1907–22; Director, 1922–41. Expedition to Peru and Bolivia, 1903. Published papers on taxonomy of *Strychnos*, *Thesium*, *Malvastrum* and *Nototriche*.

J. Kew Guild 1920, 471 portr.; 1942, 129–39 portr. *Kew Bull.* 1921, 225–53; 1930, 63a. *Gdnrs Chron.* 1922 i 98 portr.; 1941 ii 188. *Nature* v.148, 1941, 619–23. *Pharm. J.* 1941, 164. *Times* 4 Nov. 1941 portr. *Chronica Botanica* 1942–43, 141–42. *Obit. Notices Fellows R. Soc.* 1942, 87–100 portr. *Proc. Linn. Soc.* 1941–42, 280–85. *Trans. R. Soc. N.Z.* 1942, xxxvi–xxxviii. *D.N.B., 1941–1950* 390–91. *Who was Who, 1941–1950* 541.

Plants and MSS. at Kew. Portr. at Hunt Library.

HILL, Charlotte (*fl.* 1830s)

Nurseryman, Leytonstone, Essex. Succeeded her husband, James Hill, in the nursery near Grove Green, later American Nursery. Succeeded by Alexander Protheroe and Thomas Morris.

HILL, Edward (1741–1830)

b. Ballyporeen, Tipperary 14 May 1814 *d.* Dublin 31 Oct. 1830

MD Dublin 1775. Lecturer in Botany, Dublin, 1773–85; Professor of Botany, 1785–1800. Acquired herb. from P. Browne.

Trans. Linn. Soc. v.4, 1798, 32. *Notes Bot. School, Dublin* v.1, 1896, 1–3. T. P. C. Kirkpatrick *Hist. Med. Teaching Trinity College, Dublin* 1912, 361.

HILL, Elizabeth (–1850)

Of Pilton, N. Devon. Algologist. "Good entomologist who has lately turned her very acute mind to the study of marine plants" (S. Goodenough). Ill-health prevented her working after about 1830. Correspondent of Dawson Turner.

J. Sowerby and J. E. Smith *English Bot.* 2084. D. Turner *Fuci* v.1, 1808, 60. R. K. Greville *Algae Britannicae* 1830, vi. P. Smith *Mem. Correspondence of Sir J. E. Smith* v.1, 1832, 544, 587.

Nitophyllum hilliae Grev.

Plants at Oxford.

HILL, Henry Charles (1852–1903)

Indian Forest Service, 1872; Inspector-General of Forests, 1900–3. Reported on forests of Malaya, 1899. Collected plants in India and Malaya.

Gdns Bull. Straits Settlements 1927, 124. *Fl. Malesiana* v.1, 1950, 231–32.

Plants at Kew, Singapore.

HILL, James (*c.* 1761–1832)

d. 17 Feb. 1832

Acquired nursery at Leytonstone, Essex belonging to John Hay. Succeeded by his widow, Charlotte Hill. When visited by Loudon in 1835 the nursery abounded in a "very great variety of Red American Oaks." From 1839–88 the nursery was carried on by Protheroe and Morris.

Trans. London Middlesex Archaeol. Soc. v.24, 1973, 190.

HILL, Jason *see* Hampton, F. A.

HILL, Sir John (1707–1775)

b. Peterborough, Northants 1707 *d.* London 21 Nov. 1775

MD St. Andrews 1750. Knight of Vasa 1774. Apothecary, James Street, Covent Garden. "First Superintendent of the Royal Gardens at Kew" (J. Thornton) but there is no positive evidence for this. Had botanic garden at Bayswater. *General Natural History* 1748–51 2 vols. *Useful Family Herbal* 1754. *British Herbal* 1756. *Eden; or, Compleat Body of Gardening* 1757. *Exotic Botany* 1759. *Hortus Kewensis* 1768. *Fl. Britanica* 1760. *Herbarium Britannicum* 1769. Under patronage of Lord Bute published *Vegetable System* 1759–75 26 vols.

Gent. Mag. 1771, 569; 1774, 282, 353. *Short Account of Life...of Late Sir John Hill* 1779. R. Pulteney *Hist. Biogr. Sketches of Progress of Bot. in England* v.2, 1790, 293–94. H. Field *Mem....of Bot. Gdn at Chelsea* 1820, 47. J. C. Loudon *Encyclop. Gdning* 1822, 1271. G. W. Johnson *Hist. English Gdning* 1829, 207–9. H. C. van Hall *Epistolae Ineditae C. Linnaei* 1830, 139. *Cottage Gdnr* v.5, 1850, 121–22. R. Chambers *Book of Days* v.2, 1864, 601–4. H. Trimen and W. T. T. Dyer *Fl. Middlesex* 1869, 391–92. *Kew Bull.* 1891, 294–95. *Bull. Torrey Bot. Club* 1899, 376–80. *D.N.B.* v.26, 397–401. *J. Bot.* 1908, 8–11; 1924, 353. *J. Northamptonshire Nat. Hist. Soc.* 1909, 134–51. F. W. Oliver *Makers Br. Bot.* 1913, 84–107 portr. *Gdnrs Chron.* 1915 ii 241. *Amer. Nat.* 1926, 417–42. G. C. Druce *Fl. Buckinghamshire* 1926, lxxvii–lxxxix portr. G. C. Druce *Fl. Northamptonshire* 1930, lxxv–lxxxvi. *Isis* v.34, 1942, 16–20. J. Wright *John Hill, Herbalist Extra-ordinary* 1957. *Herbarist* no. 21, 1955, 33–38. *Proc. R. Soc. Med.* v.53, 1960, 55–60. F. A. Stafleu *Taxonomic Literature* 1967, 202–3. *J. Amer. Med. Assoc.* 1970, 103–8 portr. C. C. Gillispie *Dict. Sci. Biogr.* v.6, 1972, 400–1.

British plants at BM(NH). MSS. at Royal Society. Original drawings of *Exotic Botany* at Alnwick Castle. Portr. at Hunt Library.

Hillia Jacq.

HILL, John Rutherford (1857–1941)
b. Windermere, Westmorland 1857 *d.* Balerno, Midlothian 17 July 1941

Lecturer in Pharmacy and Materia Medica, Edinburgh School of Medicine. Principal, School of Pharmacy. President, Botanical Society of Edinburgh, 1906–8, 1929–31.

Trans. Proc. Bot. Soc. Edinburgh 1940–41, 182.

HILL, Robert Southey (–1872)

MD. FLS 1856. Of Basingstoke, Hants. Prepared MS. flora of Hampshire. 'Notes on *Primula elatior*' (*Phytologist* v.1, 1842, 187–88).

J. Bot. 1872, 352. F. Townsend *Fl. Hampshire* 1883, xxix, xxxv.

Herb. and letters at BM(NH).

HILL, Thomas (*fl.* 1540s–1570s)

Of London. *A Most Briefe and Pleasante Treatyse Teachyng how to Dresse, Sowe and Set a Garden c.* 1558. *Gardener's Labyrinth* 1577 (under pen-name of Didymus Mountain).

G. W. Johnson *Hist. English Gdning* 1829, 66–68. *Cottage Gdnr* v.1, 1848, 16; v.6, 1851, 171. J. Donaldson *Agric. Biogr.* 1854, 12–15. *J. Hort. Cottage Gdnr* v.52, 1874, 448–50 portr. *Gdnrs Chron.* 1910 ii 209. *Huntington Library Quart.* v.7, 1944, 329–51. *J. Soc. Bibl. Nat. Hist.* v.2, 1952, 386–87.

HILL, Thomas George (1876–1954)
b. London 13 Feb. 1876 *d.* Hambledon, Surrey 25 June 1954

Educ. Royal College of Science. Demonstrator in Biology, St. Thomas's Hospital Medical School. Reader in Plant Physiology, University College, London, 1912; Professor of Plant Physiology, 1929–45. Researched with Ethel de Fraine on seedling structure. *Introduction to Chemistry of Plant Products* (with P. Haas) 1913; ed. 4 1928. *Essentials of Illustration* 1915.

Nature v.174, 1954, 159–60. *Times* 26 June 1954. *Who was Who, 1951–1960* 522.

HILL, Walter (1820–1904)
b. Scotsdyke, Dumfriesshire 31 Dec. 1820 *d.* Brisbane 4 Feb. 1904

At Botanic Garden, Edinburgh under W. McNab. At Kew Gardens, 1843–51. Arrived in Sydney, 1852. First Superintendent, Botanic Garden, Brisbane, 1855–81. Colonial botanist. Accompanied Sir G. Bowan on expedition to Cape York Peninsula, 1862. Botanist on G. E. Dalrymple's expedition to N. Queensland, 1873. *Report on Botany of Expedition...under G. E. Dalrymple* 1874.

Bot. Mag. 1861, t.5261. *Gdnrs Chron.* 1904 i 190. *J. Kew Guild* 1904, 206–7 portr. *J. Bot.* 1905, 280. *Rep. Austral. Assoc. Advancement Sci.* 1909, 377–78. *R.S.C.* v.12, 334.

Plants at Kew. Letters at Kew. Portr. at Hunt Library.

Musa hillii F. Muell.

HILL, William (*fl.* 1790s)

Gardener and seedsman, Merthyr Tydfil, Glam.

HILL, William (1824–1878)
b. Silsoe, Beds 1824

Gardener to J. Dempster at Keele Hall, Newcastle-under-Lyme, Staffs, 1850–78.

Gdnrs Chron. 1876 i 213 portr.; 1916 i 330–31 portr.

HILL, Wills, 1st Marquis of Downshire (1718–1793)
b. Fairford, Glos 30 May 1718 *d.* 7 Oct. 1793

FRS 1764. DCL Oxon 1771. 2nd Viscount Hillsborough, 1742. MP, Warwick, 1741–56. President, Board of Trade and Plantations, 1763–65, 1766. Secretary of State for Colonies, 1768–72. Numerous plants (including W. Indian plants from J. Gregg) in Banks herb. at BM(NH).

J. E. Smith *Selection from Correspondence of Linnaeus* v.1, 1821, 189, 509, 568–69. *D.N.B.* v.26, 427–29.

HILLAND, Miss (1857–1924)
b. 10 Feb. 1857 *d.* Westbourne, Hants 1 July 1924

Botanised on Hayling Island, Hants.

Bot. Soc. Exch. Club Br. Isl. Rep. 1924, 541–42.

HILLHOUSE, William (1850–1910)
b. Bedford 17 Dec. 1850 *d.* Malvern, Worcs 27 Jan. 1910

MA Cantab. FLS 1876. Assistant master, Bedford Modern School, 1867–77. Assistant Curator, Cambridge Herb., 1878–82. Professor of Botany, Mason College, Birmingham, 1882–1909. Secretary, Birmingham Botanical Horticultural Society, 1892–1905. *Bedfordshire Plant-list* 1875, 1876. 'Contribution towards a New Fl. Bedfordshire' (*Bedfordshire Mercury* 1876). Translated E. Strasburger's *Handbook of Practical Botany* 1887. Contrib. to *Trans. Bedfordshire Nat. Hist. Soc.*

Bot. Soc. Exch. Club Br. Isl. Rep. 1910, 534. *Gdnrs Chron.* 1910 i 96, 106. *J. Bot.* 1910, 105–6. *Nature* v.82, 1910, 405. *Proc. Linn. Soc.* 1909–10, 91–92. *Who was Who, 1897–1916* 339. *Bedfordshire Nat.* v.4, 1949, 40–42. J. G. Dony *Fl. Bedfordshire* 1953, 25–26. *R.S.C.* v.10, 233,; v.12, 334; v.15, 852.

Herb. at Birmingham.

Hillhousia G. S. West.

HILLIER, Edwin (1840–1929)
Founded nursery in Winchester, Hants in 1864. In 1870s purchased West Hill Nursery. Business continued by his sons, Edwin (1865–1944) and Arthur Richard (1877–1963).
Gdnrs Chron. 1964 i 329 portr.

HILLIER, Edwin Lawrence (1865–1944)
d. Winchester, Hants 8 Sept. 1944
VMH 1941. Partner in his father's (E. Hillier) nursery at Winchester.
Gdnrs Chron. 1938 i 86 portr.; 1944 ii 116; 1964 i 329 portr.

HILLIER, John Masters (*c.* 1861–1930)
b. Teddington, Middx *c.* 1861 *d.* Osterley, Middx 5 Oct. 1930
Assistant, Kew Gardens Museums, 1879; Keeper, 1901–26. Contrib. to *Kew Bull.*
Gdnrs Chron. 1923 i 296 portr.; 1930 ii 309. *J. Bot.* 1930, 344. *J. Kew Guild* 1924, 203 portr.; 1931, 77. *Kew Bull.* 1926, 220–21; 1930, 495.

HILLS, B. (*c.* 1884–1960)
d. 12 Nov. 1960
Gardener at Westonbirt, Glos. Orchid grower at Exbury to Lionel and Edmund de Rothschild. Contrib. to *Orchid Rev.*, *Gdnrs Chron.*
Orchid Rev. 1960, 375.

HILLSBOROUGH, 2nd Viscount *see* Hill, W.

HILTON, Thomas (1833–1912)
b. Brighton, Sussex 16 April 1833 *d.* Brighton 10 Feb. 1912
Hon. Curator, Brighton Museum. Had 3 herbs: one at BM(NH), one at Brighton Museum and one at Brighton and Hove Natural History Society. Contrib. to *Sci. Gossip* 1899, 1902 and *Proc. Brighton Nat. Hist. Soc.* 1892.
Bot. Soc. Exch. Club Br. Isl. Rep. 1912, 204–5. *Brighton Nat. Hist. Soc. Abstracts* 1912, 34 portr. *J. Bot.* 1912, 141. A. H. Wolley-Dod *Fl. Sussex* 1937, xlviii. *R.S.C.* v.15, 853.
Ranunculus × *hiltoni* H. & J. Groves.

HINCKS, Hannah (1798–1871)
Of Belfast. Eldest daughter of Rev. T. D. Hincks (1767–1857). Algologist. Formed herb. of Irish plants. Contrib. to G. Dickie's *Fl. Ulster* 1864.
W. H. Harvey *Phycologia Britannica* v.1, 1846, *t.*22.
Ectocarpus hincksiae Harvey.

HINCKS, T. C. (1840–1902)
d. Richmond, Yorks 13 March 1902
Cultivated orchids, particularly masdevallias.
Orchid Rev. 1903, 7.

HINCKS, Rev. Thomas Dix (1767–1857)
b. Dublin 24 June 1767 *d.* Belfast 24 Feb. 1857
LLD Glasgow 1834. Presbyterian minister, Cork, 1790. Founder member and Secretary of Royal Cork Institution. Corresponding Secretary, Belfast Botanical Society. Found *Hypericum linariifolium*, 1838. 'On Early Contributions to Fl. Ireland' (*Ann. Mag. Nat. Hist.* v.6, 1841, 1–12). Contrib. to J. Sowerby and J. E. Smith's *English Bot.* 2017, 2184.
D.N.B. v.26, 441–42. *Proc. Belfast Nat. Field Club* v.6, 1913, 617–18. H. F. Berry *Hist. R. Dublin Soc.* 1915, 192–93, 226–27. *Belfast Nat. Hist. Philos. Soc. Centenary Vol.* 1924, 83–85, 134. *Irish Nat. J.* 1931, 238–39. *R.S.C.* v.3, 355; v.7, 983; v.10, 234.; v.15, 854.
Herb. at Cork.

HINCKS, Rev. William (1794–1871)
b. Cork May 1794 *d.* Toronto, Canada 10 Sept. 1871
FLS 1826. Son of Rev. T. D. Hincks (1767–1857). Professor of Natural History, Cork, 1849–53; Toronto, 1854–71. 'Vegetable Monstrosities' (*Proc. Linn. Soc.* 1841, 118–19). Prepared monograph of *Oenothera* (*Gdnrs Mag.* 1838, 385–86).
H. Baines *Fl. Yorkshire* 1840, preface. *Canadian J.* v.13, 1872, 253–54. *Proc. Linn. Soc.* 1871–72, lxv–lxviii. J. Hardy *Selections from Correspondence of G. Johnston* 1892, 152. *Rep. Yorkshire Philos. Soc.* 1893, 36–38; 1906, 45–46. *D.N.B.* v.26, 441. *R.S.C.* v.3, 355; v.7, 983.
Plants at York Museum. Letters at Kew.

HIND, Rev. William Marsden (1815–1894)
b. near Belfast 21 Feb. 1815 *d.* Walsham-le-Willows, Suffolk 13 Sept. 1894
BA Dublin 1839. LLD 1870. Curate, Derriaghy, County Antrim, 1839. Perpetual curate, Pinner, 1861. Rector, Honnington, Suffolk, 1875. Contrib. to *Fl. Harrow* 1864; edited ed. 2 1876. *Fl. Suffolk* (with C. Babington) 1889. Contrib. Middlesex plants to *J. Bot.* 1871, 272; N. Cornish plants to *J. Bot.* 1873, 36–43, 99–101.
J. Bot. 1894, 352; 1907, 388–93. F. H. Davey *Fl. Cornwall* 1909, lvi. *Proc. Belfast Nat. Field Club* v.6, 1913, 619. R. L. Praeger *Some Irish Nat.* 1949, 103. *R.S.C.* v.3, 358; v.7, 984; v.15, 855.
Herb. at Trinity College, Dublin. Suffolk plants at Ipswich Museum.

HINDMARSH, William Thomas (1847–1913)
d. Alnwick, Northumberland 27 April 1913
FLS 1889. Solicitor. President, Berwickshire Naturalists' Club, 1905. Studied living

plants and Mendelism. Notes on *Shortia*, etc. in *J. R. Hort. Soc.* v.29, 1904, 32–37.
Proc. Berwickshire Nat. Club. 1912, 136–37. *Proc. Linn. Soc.* 1912–13, 59–60.

HINDON, Miss Fanny (–*c.* 1884)
Assisted Miss C. Cutler (–1866) in collecting and preserving seaweeds. Her own collections were acquired by Rev. R. Cresswell.

HINDS, Richard Brinsley (*c.* 1812–*c.* 1847)
d. Perth, W. Australia *c.* 1847
Surgeon, Royal Navy. Attached as surgeon/naturalist to HMS 'Sulphur', 1836–42. Collected plants during global voyage. *Botany of Voyage of HMS 'Sulphur'* 1844. *Regions of Vegetation* 1843. 'Sandwich Islands' from his unpublished journal on HMS 'Sulphur' (*Hawaiian J. Hist.* v.2, 1968, 102–35).
London J. Bot. 1843, 211–40. *Bot. Mag.* 1845, t.4135. E. Bretschneider *Hist. European Bot. Discoveries in China* 1898, 363–65. S. D. McKelvey *Bot. Exploration of Trans-Mississippi West, 1790–1850* 1955, 636–58. G. A. C. Herklots *Hong Kong Countryside* 1951, 163. *Fl. Malesiana* v.1, 1950, 233. *California Acad. Sci. Centennial Vol.* 1955, 52. *Contrib. Univ. Michigan Herb.* 1972, 245–46. *R.S.C.* v.3, 358.
Plants and letters at Kew. Plants and MS. diary at BM(NH).
Hindsia Benth.

HINDS, William (1811–1881)
b. Birmingham 1811 *d.* Birmingham 18 Oct. 1881
MRCS 1844. MD Aberdeen 1847. Professor of Botany, Queen's College, Birmingham, 1861–81. Lecturer in Botany, Midland Institute.
Gdnrs Chron. 1881 ii 695.

HINDSON, Isaac (*fl.* 1830s–1870s)
Studied flora of Kirkby Lonsdale, Westmorland, 1836–72. MS. notes of plant localities acquired by J. G. Baker.
J. G. Baker *Fl. English Lake District* 1885, 11.

HINTON, George Boole (1882–1943)
b. London 1882 *d.* Mexico Feb. or March 1943
Went to Japan with his parents, 1889; parents emigrated to U.S.A. 7 years later. In Mexico, 1911, where he worked as metallurgist, civil engineer and architect. Collected plants in Mexico, 1931–41.
Kew Bull. 1933, Appendix, 32; 1934, Appendix, 31; 1935, 632; 1936, 1, 346, 564; 1958, 155. *Bull. African Succulent Soc.* 1972, 158. *Contrib. Univ. Michigan Herb.* 1972, 246–47. *J. Arnold Arb.* 1972, 141–81 portr.
Plants at Kew, BM(NH), Washington, etc.
Hintonia Bullock.

HIRST, Harold Maude (1887–1956)
b. Leeds, Yorks 1887 *d.* Scarborough, Yorks 15 March 1956
Pharmacist, Scarborough, 1933. On management committee of Chelsea Physic Garden.
Pharm. J. 1956, 135 portr.

HIRST, Robert Michael (*fl.* 1820s)
Nurseryman, Handsworth, near Sheffield, Yorks.

HISLOP, Alexander (*c.* 1880–1945)
d. Bulawayo, Rhodesia 2 April 1945
Kew gardener. Assistant Superintendent, Municipal Gardens, Oudtshoorn, Cape, 1902. Curator, Pietermaritzburg Botanical Society's gardens, 1904–6. Curator, Agricultural Dept., S. Nigeria, 1908. Collected plants in Rhodesia.
Kew Bull. 1908, 376. *J. Kew Guild* 1945, 464–65.
Plants at Kew.

HITCHIN, Thomas (*fl.* 1810s–1830s)
d. Cambridge
Dyer of Norwich, later bank clerk at Cambridge. Had extensive collection of succulent plants at Norwich. Friend of A. H. Haworth.
Bot. Mag. 1821, t.2272 (as Kitchin); 1824, t.2517 (as Hitchen); 1830, t.3032. *Gdnrs Mag.* 1832, 244 (as Hitchen); 1833, 751. J. Forbes *J. Hort. Tour through Germany...in 1835* 1837, 147.
Hitchenia Wall.

HITT, Thomas (1690–*c.* 1770)
b. Aberdeenshire
Gardener to Lord Sutton at Kelham House, Notts and to Lord Manners at Bloxholme, Lincs. Afterwards nurseryman at Bromley, Kent and designer of gardens. *Treatise of Fruit Trees* 1755. *Modern Gardener* revised by J. Meader, 1771.
J. C. Loudon *Encyclop. Gdning* 1822, 1271–72. G. W. Johnson *Hist. English Gdning* 1829, 209–10.

HOARE, Charles Harvey *see* Grey, C. H.

HOARE, Clement (*c.* 1789–1849)
d. 18 Aug. 1849
Schoolmaster near Chichester. Moved to Shirley, 1841. Cultivated grape vine. *Practical Treatise on Cultivation of Grape Vine on Open Walls* 1839; ed. 3 1841. *Descriptive Account of Improved Method of Planting and Managing Roots of Grape-Vines* 1844.
Cottage Gdnr v.2, 1849, 306.

HOARE, Sarah (*c*. 1767–1855)
b. Bristol *c*. 1767 *d*. Bath? 14 April 1855
Pleasures of Botanical Pursuits 1818. *Poems on Conchology and Botany* 1831.
J. Smith *Cat. Friends' Books* v.1, 1867, 955–56.

HOBKIRK, Charles Codrington Pressick (1837–1902)
b. Huddersfield, Yorks 13 Jan. 1837 *d*. Ilkley, Yorks 29 July 1902
FLS 1878. Bank manager. Bryologist. President, Huddersfield Naturalists' Society. President, Yorkshire Naturalists' Union, 1892. *Huddersfield, its History and Natural History* 1859. *Synopsis of British Mosses* 1873; ed. 2 1884. *London Catalogue of British Mosses* (with H. Boswell) 1877. 'Mosses of West Riding' (*J. Bot*. 1873, 327–31, 358–63). 'Sur les formes du *Capsella*' (*Bull. Soc. R. Bot. Belg*. 1869, 449–58). Contrib. to *Naturalist*, *Phytologist*.
J. Bot. 1902, 431. *Naturalist* 1903, 105–8 portr. *Proc. Linn. Soc*. 1902–3, 30–31. *R.S.C*. v.3, 370; v.7, 992; v.10, 242; v.12, 336; v.15, 874.
Herb. at Dewsbury Museum. Portr. at Hunt Library.

HOBLYN, Rev. Richard Dennis (1803–1886)
b. Colchester, Essex 9 April 1803 *d*. London 22 Aug. 1886
BA Oxon 1824. *British Plants* 1851. *Botany* 1851.
D.N.B. v.27, 50.

HOBSON, Edward (1782–1830)
b. Manchester 1782 *d*. Bowdon, Cheshire 17 Sept. 1830
Grocer's assistant. Bryologist, entomologist, geologist. Friend of G. Caley and J. Horsefield. Correspondent of R. K. Greville and W. J. Hooker. First President, Banksian Society, Manchester, 1829. President, Lancashire Botanists. *Musci Britannici* (exsiccatae) 1818–22. Herb. purchased by Manchester Botanical Society.
Gdnrs Mag. 1830, 749; 1832, 94. *Mem. Lit. Philos. Soc. Manchester* 1842, 297–324. J. Cash *Where there's a Will there's a Way* 1873, 41–66. *D.N.B*. v.27, 51. *Victoria County Hist. Lancashire* v.1, 1906, 102. *Trans. Liverpool Bot. Soc*. 1909, 76.
Lejeunea hobsonniana.

HOBSON, H. Edgar (*fl*. 1890s)
In Chinese Customs Service. Collected plants in Chumbi Valley, Tibet, 1897, now at Kew.
E. Bretschneider *Hist. European Bot. Discoveries in China* 1898, 812. *Kew Bull*. 1898, 26; 1901, 32.
Impatiens hobsoni Hook. f.

HOBSON, William (*fl*. 1780s–1830s)
Brother of E. Hobson (1782–1830). Collected plants in California.
Gdnrs Mag. 1832, 94. *Bryologist* 1920, 36–37.

HOCKIN, John (–before 1885)
Of Dominica.
London J. Bot. 1843, 13. *J. R. Inst. Cornwall* v.8, 1885, 319.
2 vols of Dominica plants at Royal Institution, Cornwall. Letters at Kew.
Hockinia Gardner.

HODGES, Samuel (*fl*. 1830s)
Nurseryman, Cheltenham, Glos.

HODGINS, Edward (*fl*. 1780s)
Nurseryman, Dunganstown, County Wicklow.
Irish For. 1967, 57.

HODGKIN, Eliot (*c*. 1906–1973)
d. 6 March 1973
Educ. Oxford. VMH 1972. Overseas General Manager, I.C.I. Had fine collection of alpines and bulbous plants at Shelleys near Twyford, Berks. President, Alpine Garden Society, 1971–73.
Iris Yb. 1973, 19. *J. R. Hort. Soc*. 1973, 322–23. *Quart. Bull. Alpine Gdn Soc*. 1973, 156 portr. *Times* 8 March 1973.

HODGSON, Arthur Salisbury (–1934)
b. London *d*. Cardiff, Glam 9 March 1934
Headmaster, Pentre Secondary School, 1914–32. Mycologist with interest in Mycetozoa. President, Swansea Scientific and Field Naturalists' Society, 1933.
Proc. Swansea Sci. Field Nat. Soc. v.1, 1934, 213.

HODGSON, C. P. (*fl*. 1840s)
Consul, Japan where he discovered *Ligularia hodgsoni*.
Bot. Mag. 1863, t.5417.

HODGSON, Elizabeth (1814–1877)
d. Ulverstone, Lancs 26 Dec. 1877
Had herb. of Furness, Lancs mosses. 'Fl. Lake Lancashire' (*J. Bot*. 1874, 268–77, 296–305).
Gdnrs Chron. 1878 i 178. *J. Bot*. 1878, 64. *Trans. Liverpool Bot. Soc*. 1909, 76. *R.S.C*. v.3, 379; v.7, 994; v.10, 244; v.12, 337.
Plants and letters in Walker-Arnott correspondence at BM(NH).

HODGSON, J. K. (*fl*. 1870s)
Of Ulverstone, Lancs. With his wife, collected and grew fern varieties.
Br. Fern Gaz. 1909, 45.

HODGSON, William (1824–1901)
b. Raughtonhead Hill, Dalston, Cumberland 7 April 1824 *d*. Workington, Cumberland 27 March 1901

ALS 1884. Schoolmaster, Watermillock, later Aspatria. *Fl. Cumberland* 1898. Contrib. botany to *Victoria County Hist. Cumberland* v.1, 1901, 73–94. Contrib. to *Trans. Cumberland Assoc.*

J. Bot. 1901, 191. *Naturalist* 1901, 261–65 portr. *Proc. Linn. Soc*. 1900–1, 44. *R.S.C.* v.15, 379.

HODSON, Nathaniel Shirley
(*fl*. 1800s–1830s)

ALS 1823. Had garden in S. Lambeth. Moved to Abbey House, Bury St. Edmunds on retirement as War Office official. Established Bury St. Edmunds Botanic Garden, 1820 (*Gdnrs Mag*. 1835, 43). Sent crocuses to J. E. Smith (J. Sowerby and J. E. Smith *English Bot*. 2645) and plants to J. Sims for *Bot. Mag*. 1818, t.1955; 1821, 2276.

Gdnrs Mag. 1827, 236; 1837, 333.

HOFFMAN, Francis (*fl*. 1740s)

Specimen collected by him near Wantage in Sloane Herb., BM(NH).

J. E. Dandy *Sloane Herb*. 1958, 138.

HOFFMANN, George Christian
(1837–1917)
b. London 7 June 1837 *d*. Ottawa 8 March 1917

Educ. Royal School of Mines. Did research into tanning properties of bark, essential oils, etc. at Melbourne Botanic Garden. Geological Survey of Canada, 1872.

Nature v.99, 1917, 190. *Science* v.45, 1917, 330.

HOFFMAN, George Henry (1805–1882)
b. Margate, Kent 1805. *d*. Putney Heath, London 31 March 1882

Surgeon. Studied vine mildew and grew *Mucor* on onions.

Gdnrs Chron. 1848, 700; 1882 i 540.

HOGARTH, Andrew (*fl*. 1790s)

Nurseryman, Islington, London.

HOGG, Jabez (1817–1899)
b. Chatham, Kent 4 April 1817 *d*. Kensington, London 23 April 1899

FLS 1859. Ophthalmic surgeon. Surgeon, Royal Westminster Ophthalmic Hospital, 1871–78. *The Microscope* 1854; ed. 7 1869. *Vegetable Parasites* 1866.

D.N.B. Supplt I v.2, 432–33. *R.S.C.* v.3, 399; v.7, 1003; v.10, 255; v.15, 900.

HOGG, James (*fl*. 1830s)

Nurseryman, 31 Cloth Market, Newcastle-upon-Tyne and at Elswick, Northumberland.

HOGG, John (1800–1869)
b. Norton, Durham 21 March 1800 *d*. Norton 16 Sept. 1869

BA Cantab 1822. MA Oxon 1844. FRS 1839. FLS 1822. In Sicily, 1826. *On Natural History of Vicinity of Stockton-on-Tees* 1827. *Catalogue of Sicilian Plants* 1842. 'On Ballast Fl. of Coasts of Durham and Northumberland' (*J. Bot*. 1867, 47–48).

Proc. Linn. Soc. 1869–70, c–ci. H. C. Watson *Topographical Bot*. 1883, 548. J. Hardy *Selections from Correspondence of George Johnston* 1892, 186. *D.N.B.* v.27, 103–4. *R.S.C.* v.3, 399; v.7, 1004.

Letters at Kew and in Winch correspondence at Linnean Society.

HOGG, Mure (*c*. 1774–1832)
b. Bury St. Edmunds, Suffolk *c*. 1774 *d*. Bury St. Edmunds 9 Sept. 1832

Florist and market-gardener, Bury St. Edmunds.

Gdnrs Mag. 1833, 128.

HOGG, Robert (1818–1897)
b. Duns, Berwickshire 20 April 1818 *d*. Pimlico, London 14 March 1897

FLS 1861. LLD. After graduating at Edinburgh gained horticultural experience at Peter Lawson, Edinburgh and H. Ronalds, Brentford. Partner with Gray and Adams, nurserymen, Brompton Park, 1845–51. Co-editor of *Cottage Gdnr* 1848. *The Dahlia* 1853. *Vegetable Kingdom and its Products* 1858. *Apple and its Varieties* 1859. *Fruit Manual* 1860; ed. 5 1884. *Wild Flowers of Great Britain* (with G. W. Johnson) 1863–80 11 vols. *Herefordshire Pomona* (with H. G. Bull) 1876–85 2 vols.

Gdnrs Chron. 1871, 1292–93 portr.; 1897 i 188–89 portr.; 1897 ii 321–22; 1923 ii 218; v.164(26), 1968, 19–21 portr. *Garden* v.51, 1897, 217. *Gdnrs Mag*. 1897, 167 portr., 176. *J. Hort. Cottage Gdnr* v.34, 1897, 232–35 portr., 248–50, 268–69, 291, 314; v.57, 1908, 393–95 portr. *Proc. Linn. Soc*. 1896–97, 57–58. *Curtis's Bot. Mag. Dedications, 1827–1927* 203–4 portr.

Letters at Kew. Portr. at Hunt Library. Library sold at Sotheby, 3 Nov. 1897.

HOGG, Thomas (1771–1841)
b. Romaldkirk, Yorks 1771 *d*. Paddington, London 12 March 1841

Master of Academy, Paddington 1791. Florist, Paddington Green *c*. 1822. *Concise and Practical Treatise on Growth...of Carnation* 1820; ed. 6 1839. 'On Cultivation of Pinks' (*Trans. Hort. Soc*. v.4, 1821, 451–54).

J. C. Loudon *Encyclop. Gdning* 1822, 1286. *Cottage Gdnr* v.7, 1852, 185. *J. Hort. Cottage Gdnr* v.57, 1877, 347–48 portr.

HOGG, Thomas (1778–1855)
b. Polwarth, Berwickshire 20 Feb. 1778 *d.* New York 11 Oct. 1855
Gardener. To U.S.A., 1820. Nurseryman, New York.
J. Hort. Cottage Gdnr v.53, 1906, 546. U. P. Hedrick *Hist. Hort. in America to 1860* 1950, 213. A. M. Coats *Quest for Plants* 1969, 79.

HOGG, Thomas (1820–1892)
b. London 6 Feb. 1820 *d.* New York 30 Dec. 1892
To U.S.A. with his parents, 1820. Nurseryman, New York. Collected plants in N. Carolina and Virginia and in Japan.
Bull. Torrey Club 1893, 217–18. *Gdnrs Mag.* 1893, 56.
Portr. at Hunt Library.

HOGG, Thomas (–1908)
d. Crookston, Paisley, Renfrewshire 20 June 1908
Gardener to J. Gordon at Aitkenhead, Cathcart near Glasgow, and to A. Coates at Woodside, Paisley, 1889.
Gdnrs Chron. 1908 ii 32.

HOGGAN, Ismé Aldyth (1889–1936)
b. London 23 March 1889 *d.* Madison, Wisconsin, U.S.A. 28 Dec. 1936
BA Cantab 1922. MSc 1924. Fellow, Wisconsin Univ., 1924; Assistant Professor, 1933. Mycologist. Contrib. to *Phytopathology*.
Times 1 Jan. 1937. *Phytopathology* 1937, 1029–32 portr.

HOLBECH, Rev. Charles (1782–1837)
b. Farnborough, Warwickshire 14 May 1782 *d.* Farnborough 28 Nov. 1837
BA Oxon 1804. Vicar, 1812. Discovered *Linosyris* (J. Sowerby and J. E. Smith *English Bot.* 2505).
J. Bot. 1910, 232–33.

HOLBERT, Robert (*fl.* 1790s)
Nurseryman, Gloucester.

HOLCOMBE, Rev. John (1704–1770)
b. Mounton, Pembroke 1704 *d.* Tenby, Pembroke 1770
BA Cantab 1724. Rector, Tenby, 1730–70. Correspondent of Sir J. Cullum, Rev. J. Lightfoot and Sir J. Banks. Pembrokeshire botanist.
J. Bot. 1886, 22–23.
Cullum MSS. at Hardwick House, Bury St. Edmunds, Suffolk.

HOLDEN, Henry (*fl.* 1650s?)
Of Oxford? Collected Warwickshire plants.
D. A. Cadbury *et al. Computer-mapped Fl. Warwickshire* 1971, 47–48.
Plants at Birmingham Reference Library.

HOLDEN, Henry Smith (1887–1963)
b. Castleton, Lancs 30 Nov. 1887 *d.* 16 May 1963
DSc Manchester. Assistant Lecturer in Botany, Nottingham, 1909; Senior Lecturer, 1927; Professor of Dept. of Biology, 1928; Professor of Botany, 1932. Director, Forensic Science Laboratory, New Scotland Yard, 1936–51. Palaeobotanist.
Nature v.199, 1963, 330–31. *Times* 18 May 1963. *J. Soc. Bibl. Nat. Hist.* 1964, 230–34 portr.

HOLDEN, Samuel (*fl.* 1830s–1850s)
Botanical artist; specialised in orchids. Drew 12 plates for T. Moore and W. P. Ayres *Gdnrs Mag. Bot.* 1850–51 and 438 plates for *Paxton's Mag. Bot.* 1836–49.
Drawings at Victoria and Albert Museum, London.

HOLDICH, Benjamin (1770–1824)
b. Thorney, Cambridge Nov. 1770
Editor, *Farmer's J. Essay on Weeds of Agriculture* 1825.

HOLE, Robert Selby (1874–1938)
d. Belstone, Devon 28 June 1938
FLS 1904. CIE 1919. Indian Forest Service, 1896. Forest Botanist, Forest Research Institute, Dehra Dun, 1907–25. Conservator of Forests, 1920–25. *Manual of Botany for Indian Forest Students* 1909. *On Some Indian Forest Grasses and their Oecology* 1911. Contrib. to *Indian For.*, *J. Bombay Nat. Hist. Soc.*, *J. R. Asiatic Soc. Bengal.*
Indian Forester 1938, 625. *Kew Bull.* 1938, 305–6. *Proc. Linn. Soc.* 1938–39, 243–44.
Plants at Calcutta, Dehra Dun, Kew.

HOLE, Rev. Samuel Reynolds (1819–1904)
b. Ardwick, Manchester 5 Dec. 1819 *d.* Rochester, Kent 27 Aug. 1904
Educ. Oxford. VMH 1897. Curate and Vicar, Caunton near Newark, 1844–87. Dean of Rochester, 1887. First President, National Rose Society. *Book about Roses* 1869. *Book about the Garden and Gardeners* 1892. *Our Gardens* 1899. *Memories of Dean Hole* 1892.
Gdnrs Chron. 1871, 802–3 portr.; 1904 ii 170–71 portr., 207, 224, 226, 409; 1964 ii 253 portr. *Garden* v.2, 1872 portr.; v.58, 1900, 23–24 portr.; v.66, 1904, 4p. after 156 portr.; 1926, 371. *J. Hort. Cottage Gdnr* v.47, 1903, 397, 399 portr.; v.49, 1904, 201–2 portr. *Gdnrs Mag.* 1904, 589–90 portr. G. A. B. Dewar *Letters of Samuel Reynolds Hole* 1907. *Who was Who, 1897–1916* 346. *Country Life* 1967, 547–48 portr.; 1974, 451–52 portr. B. Massingham *Turn on the Fountains* 1974 portr.
Portr. at Hunt Library.

HOLFORD, Sir George Lindsay (1860–1926)
b. 2 June 1860 *d.* Westonbirt, Glos 11 Sept. 1926

Developed Westonbirt arboretum founded by his father, Robert Stayner Holford (1808–1892) in 1829. His celebrated orchid collection was acquired by H. G. Alexander Ltd. on his death.

Bot. Soc. Exch. Club Br. Isl. Rep. 1926, 97–100. *Gdnrs Chron.* 1926 ii 240. *Kew Bull.* 1926, 428. *Orchid Rev.* 1914, 201–2 portr.; 1926, 294–95, 373. *J. R. Hort. Soc.* 1927, 25–28 portr.

HOLL, Augustine (*fl.* 1780s)
Seedsman, Lower Close, Norwich, Norfolk.

HOLL, Harvey Buchanan (1820–1886)
b. Worcester? 28 Sept. 1820 *d.* Cheltenham, Glos 11 Sept. 1886

MD Aberdeen 1859. Civil surgeon in Crimea. On Geological Survey, Pennsylvania.

Geol. Mag. 1886, 526–28. *J. Bot.* 1886, 384. J. Amphlett and C. Rea *Bot. Worcestershire* 1909, xxviii. *R.S.C.* v.3, 404; v.7, 1005; v.10, 159; v.15, 907.

British lichens and mosses at BM(NH).

HOLLAND, Sir Henry (1788–1873)
b. Knutsford, Cheshire 27 Oct. 1788 *d.* London 27 Oct. 1873

MD Edinburgh 1811. FRS 1816. Practised medicine in London. President, Royal Institution. *General View of Agriculture of Cheshire* 1808 (includes references to plants).

Times 31 Oct. 1873. W. Munk *Roll of R. College of Physicians* v.3, 1878, 144–49. J. B. L. Warren *Fl. Cheshire* 1899, lxxxiv. *D.N.B.* v.27, 144–45.

HOLLAND, John Henry (1869–1950)
b. Chester, Cheshire 17 Oct. 1869 *d.* 6 Oct. 1950

Kew gardener, 1894. Assistant Curator, then Curator, Botanic Station, Old Calabar, Nigeria, 1896–1901. Collected plants in S. Africa and Nigeria. Assistant, Kew Museums, 1901–34. *Useful Plants of Nigeria* 1922. *Overseas Plant Products* 1937.

Kew Bull. 1896, 147; 1934, 397; 1951, 51. *J. Kew Guild* 1935, 428; 1950, 867 portr.

Plants at Kew. Portr. at Hunt Library.

HOLLAND, Robert (1829–1893)
b. Peckham, Surrey 2 Aug. 1829 *d.* near Acton Grange, Cheshire 16 July 1893

Contrib. to L. H. Grindon's *Manchester Fl.* 1859. *Dictionary of English Plant-names* (with J. Britten) 1878–86.

J. Bot. 1893, 241–43. *R.S.C.* v.3, 404; v.7, 1006; v.15, 907.

HOLLAND, William Buelow *see* Gould, W. B.

HOLLEY, Harry (*c.* 1875–1928)
d. Queenstown, S. Africa 15 Aug. 1928

Kew gardener, 1897. To S. Africa, 1898. Curator, Queenstown Public Gardens.

J. Kew Guild 1929, 713 portr.

HOLLIDAY, William H. (1834–1909)
b. Oxford 26 Oct. 1834 *d.* Oxford 12 May 1909

Bank clerk, Oxford. Botanised in Oxfordshire; interested in mosses.

G. C. Druce *Fl. Buckinghamshire* 1926, c–ci. G. C. Druce *Fl. Oxfordshire* 1927, cxvii.

Herb. at Oxford.

HOLLINGTON, Alfred Jordan (1845–1926)
d. Enfield, London 3 Dec. 1926

Hybridised orchids.

Orchid Rev. 1927, 23.

HOLLINGWORTH, John (1805–1888)
d. Maidstone, Kent 3 April 1888

Had paper-mill at Maidstone. Grew roses.

Gdnrs Chron. 1888 i 471.

HOLLOWAY, Robert (*fl.* 1820s)
Nurseryman, Queen Street, Wells, Somerset.

HOLME, Rev. John (–1829)
d. Freckenham, Suffolk 1829

MA Cantab 1818. FLS 1800. Vicar, Cherryhinton, Cambridgeshire; Freckenham, 1816. Contrib. to J. Sowerby and J. E. Smith's *English Bot.* 780, 947, 2266, etc.

R.S.C. v.3, 407.

HOLMES, Edward (*fl.* 1850s)
Nurseryman, Handsworth near Sheffield, Yorks.

HOLMES, Rev. Edward Adolphus (–1886)
d. South Elsham, Suffolk 3 June 1886

MA Cantab 1835. FLS 1834. Rector, South Elsham, 1833–86.

F. W. Galpin *Account of Flowering Plants...of Harleston* 1888, 23–24.

HOLMES, Edward Morell (1843–1930)
b. Wendover, Bucks 29 Jan. 1843 *d.* Sevenoaks, Kent 10 Sept. 1930

FLS 1875–1924. Lecturer in Botany, Westminster Hospital School, 1873–76. Lecturer in Materia Medica, Pharmaceutical Society, 1887–90. Curator of Museum at Pharmaceutical Society, 1872–1922; Emeritus Curator, 1922–30. Specialist on lichens and marine algae. *Catalogue of Collections in Museum of Pharmaceutical Society* 1878. *Holmes' Botanical Note-book* 1878. *Catalogue of Hanbury Herb. in Museum of Pharmaceutical Society* 1892. *Catalogue of Medicinal*

Plants in Museum of Pharmaceutical Society 1896. *Revised List of British Marine Algae* (with E. A. L. Batters) 1890. *Algae Britannicae Rariores Exsiccatae* 1883. Contrib. accounts of cryptogamic plants to several *Victoria County Histories*. Contrib. to *J. Bot.*, *J. Linn. Soc.*, *Pharm. J.*

Pharm. J. 1897, 214–15 portr.; 1930, 282–86 portr. *Gdnrs Chron.* 1915 ii 68 portr. *Bot. Soc. Exch. Club Br. Isl. Rep.* 1930, 325–26. *J. Bot.* 1930, 310–13. *Kew Bull.* 1930, 493–94. *Proc. Linn. Soc.* 1930–31, 177–79. *Who was Who, 1929–1940* 656–57. *Bot. J. Linn. Soc.* v.67, 1973, 1–2.

Algae at Birmingham. Mosses at Cambridge. Hepatics at National Museum of Wales. Lichens at Nottingham. Letters at Kew and Linnean Society.

HOLMES, George (1835–1910)
d. Stroud, Glos 17 Oct. 1909

Bryologist. Discovered *Eurhynchium rotundifolium*.

J. Bot. 1910, 64.

Herb. at Stroud Museum.

HOLMES, Robert (*fl.* 1820s)

Nurseryman, 3 Mount Street, Westminster Road, Lambeth, London.

HOLMES, Thomas Henry (1869–1944)
b. Skipton, Yorks 1869 *d.* Skipton 27 Jan. 1944

Employee of English Sewing Cotton Co., Skipton. All-round naturalist.

Naturalist 1944, 76–77.

Herb. at Skipton Museum.

HOLMES, William (–1830)
b. Lancs *d.* Moreton Bay, Queensland Aug. 1830

Emigrated to Australia, 1826. Colonial zoologist at Museum of New South Wales. Collected plants.

Sydney Gaz. 31 Aug. 1830. *Austral. Mus. Mag.* 1927, 3–4; 1961, 306.

HOLMES, William (1820–1878)
b. West Ham, London 26 Sept. 1820 *d.* 29 June 1878

Gardener to Dr. Frampton, Hackney, London, 1848. Proprietor, Frampton Park Nursery, Well Street, Hackney. Contrib. to *Gdnrs Chron.*, *Florist*.

Gdnrs Chron. 1878 ii 26. *Florist and Pomologist* 1878, 128.

HOLMES, William (*c.* 1851–1913)
d. 12 July 1913

Orchid grower to J. McCartney at Hey House, Bolton, Lancs.

J. Hort. Home Farmer v.67, 1913, 121.

HOLMES, William (*c.* 1852–1890)
d. 18 Sept. 1890

Son of W. Holmes (1820–1878). Succeeded to his father's nursery in Well Street, Hackney. Secretary, Stoke Newington Chrysanthemum Society.

Garden v.38, 1890, 309. *Gdnrs Chron.* 1890 ii 353, 357 portr. *J. Hort. Cottage Gdnr* v.21, 1890, 268, 296.

HOLT, Adam (*fl.* 1710s–1750)
d. Aug. 1750

Nurseryman, Leytonstone, Essex, 1710–29. Nursery owned by John Hay, 1759–92 and James Hill (*c.* 1761–1832) and his widow Charlotte.

HOLT, George Alfred (1852–1921)
b. Douglas, Isle of Man 18 May 1852 *d.* Sale, Cheshire 19 Dec. 1921

Druggist. Collected mosses and hepatics in N. Wales, Lake District, etc. Contrib. to B. Carrington's *Hepaticae Britannicae* 1878–90. Contrib. hepatics to *Fl. Ashton-under-Lyne* 1888, 75–80. 'Mosses of Isle of Man' (*Trans. Isle Man Nat. Hist. Soc.* v.1, 1888, 62–84).

J. Bot. 1922, 207–8. *Proc. Rep. Belfast Nat. Hist. Philos. Soc.* 1933–34, 26–27 portr. *J. Manx Mus.* 1934, 211–12 portr.

Plants at Manchester Natural History Society, Manx Museum. Portr. at Hunt Library.

Radula holtii Spruce.

HOLT, John (1743–1801)
b. Hattersley near Mottram-in-Longendale, Cheshire 1743 *d.* Walton, Lancs 21 March 1801

Schoolmaster, Walton grammar school near Liverpool. *General View of Agriculture of...Lancashire* 1795. 'On Submerged Forest north of Liverpool' (*Gent. Mag.* 1796)

Gent. Mag. 1801 i 285, 370; ii 193. *D.N.B.* v.27, 205–6. *Trans. Liverpool Bot. Soc.* 1909, 76.

HOLT, Simon (*fl.* 1780s)

Nurseryman, Newark-upon-Trent, Notts.

HOLTZE, Maurice William (1840–1923)
b. Hanover, Germany 8 July 1840 *d.* Kangaroo Island, S. Australia 12 Oct. 1923

FLS 1888. Studied botany at Royal Gardens at Hanover and St. Petersburg. Curator, Port Darwin Botanic Garden, 1872–91. Curator, Adelaide Botanic Gardens, 1891–1916. 'Narrative of Exploring Tour across Melville Island with Notes on its Botany' (*Trans. R. Soc. S. Australia* 1892, 114–20).

J. Proc. R. Soc. N.S.W. 1890, 73–79, 128. *Adelaide Advertiser* 15 Oct. 1923. *Who was Who, 1916–1928* 512. *Austral. Territories*

v.1, 1960, 37–45; v.2, 1961, 39–47. *Austral. Encyclop.* v.4, 1965, 525.
Plants at Pittsburgh.
Polyalthia holtzeana F. Muell.

HOLTZE, Nicholas (1868–1913)
b. Russia 1868 *d.* Port Darwin, Australia 26 May 1913
To Australia with his parents in 1872. Curator, Darwin Botanic Garden, 1891.
Kew Bull. 1913, 233. *Austral. Territories* v.1, 1960, 37–45; v.2, 1961, 39–47. *Austral. Encyclop.* v.4, 1965, 525.
Plants at Pittsburgh.

HOLWELL, John Zephaniah (1711–1798)
b. Dublin Sept. 1711 *d.* Pinner, Middx 5 Nov. 1798
FRS 1767. Indian Medical Service, 1732. Surgeon, Calcutta, 1749. 'A New Species of Oak' [Lucombe Oak] (*Philos. Trans. R. Soc.* v.62, 1772, 128–30).
D. G. Crawford *Hist. Indian Med. Service, 1600–1913* v.1, 1914, 154–77 portr.

HOMBERSLEY, Rev. Arthur (1866–1941)
Rector, All Saints Church, Port-of-Spain, Trinidad, 1894–1931. Botanised in Trinidad with W. E. Broadway (1863–1935), specialising in ferns.
Times 12 Sept. 1941. *Amer. Orchid Soc. Bull.* 1958, 82.
Ferns at BM(NH).
Epidendrum hombersleyi Summerh.

HOME, Francis (1719–1813)
b. 17 Nov. 1719 *d.* 15 Feb. 1813
MD Edinburgh 1750. Surgeon. Professor of Materia Medica, Edinburgh, 1768–98. Studied plant nutrition in relation to agriculture. *Principles of Agriculture and Vegetation* 1757; ed. 2 1759.
D.N.B. v.27, 228–29. *Chronica Botanica* 1944, 117–26.

HOME, Henry, Lord Kames (1696–1782)
b. Kames, Berwickshire 1696 *d.* 27 Dec. 1782
Scottish lawyer, philosopher and critic. Laid out a winter garden at Blair Drummond, Stirlingshire. *Essay on Gardening* 1762 3 vols.
J. C. Loudon *Encyclop. Gdning* 1822, 1273. *D.N.B.* v.27, 232–34.

HOME, Sir James Everard (1798–1853)
b. 25 Oct. 1798 *d.* Sydney 2 Nov. 1853
FRS 1825. Captain, Royal Navy. Collected plants in Australia, New Zealand, New Caledonia, China. Made observations on Norfolk Island pines (*Proc. Linn. Soc.* 1847, 321–22).
E. Bretschneider *Hist. European Bot. Discoveries in China* 1898, 362. *Curtis's Bot. Mag. Dedications, 1827–1927* 79–80 portr. *Fl. Malesiana* v.1, 1950, 240. *R.S.C.* v.3, 417.
Plants at Kew. Letters at Kew and in Brown correspondence at BM(NH).
Santalum homei Seem.

HONEY, Thomas (1872–1937)
b. Northumberland 1872
In service of Companhia de Moçambique where he collected plants, 1919–26.
Moçambique no. 23, 1940, 112–13 portr.

HONEYMAN, Alexander (1851–1884)
b. Ladybank, Fifeshire April 1851 *d.* Brighton, Sussex 10 Feb. 1884
Gardener on various estates in Scotland. Contrib. to *J. Hort. Cottage Gdnr.*
J. Hort. Cottage Gdnr v.8, 1884, 150–51 portr.

HOOD, Mr. (*fl.* 1820s)
Surgeon, S. Lambeth, London. Had extensive collection of succulent plants.
Bot. Mag. 1824, t.2518; 1826, t.2624.
Hoodia Sweet.

HOOD, W. (*c.* 1763–1828)
d. Dumfries 22 Aug. 1828
Nurseryman, Dumfries.
Gdnrs Mag. 1828, 384.

HOOD, William (*fl.* 1730s)
Seedsman, The Wheatsheaf, Hyde Park Corner, London.

HOOKE, Nathaniel (*c.* 1856–1935)
b. Knutsford, Cheshire *c.* 1856 *d.* Easton, Pa., U.S.A. 24 Dec. 1935
Gardener to G. Hardy at Pickering Lodge, Timperley, Cheshire. First to hybridise orchid, *Cattleya hardyana*. Settled in Easton, U.S.A., 1880s.
Orchid Rev. 1936, 63–64.

HOOKE, Robert (1635–1703)
b. Freshwater, Isle of Wight 18 July 1635 *d.* London 3 March 1703
MA Oxon 1663. MD Cantab 1691. FRS 1663. Experimental philosopher. Secretary, Royal Society, 1677–82. *Micrographia* 1665 (contains observations on plant anatomy).
D.N.B. v.27, 283–87. E. J. L. Scott *Index to Sloane Manuscripts* 1904, 261. L. C. Miall *Early Nat. (1530–1789)* 1912, 134–35. *Ann. Med. Hist.* v.9, 1927, 227–43. Pavlov *Palaeobiologica* v.1, 1928, 208–10. R. T. Gunther, *ed. Life and Work of Robert Hooke* 1930. R. T. Gunther, *ed. Diary of Robert Hooke* 1935. H. W. Robinson and W. Adams, *eds. Diary of Robert Hooke, F.R.S., 1672–1680* 1935. *Isis* 1941, 15–17 portr. *Chronica Botanica* 1944, 65–69. *Proc. R. Soc.* v.201A, 1950, 439–73. *Listener* 1951, 215–16. *Nature* v.136, 1935, 56–57, 358–61, 603–4; v.137, 1936, 702–3; v.171, 1953, 365–67. *Sci. Amer.* v.191, 1954, 94–98. *Endeavour* v.14, 1955, 12–18. M. Espinasse *Robert*

Hooke 1956. J. C. Crowther *Founders of Br. Sci.* 1960, 181–222. G. Keynes *Bibl. of Dr. Robert Hooke* 1960. *Notes Rec. R. Soc.* v.15, 1960, 137–45. A. R. Hall *Hooke's Micrographia 1665–1965* 1966. D. Hutchings, *ed. Late Seventeenth Century Scientists* 1969, 132–57. C. C. Gillispie *Dict. Sci. Biogr.* v.6, 1972, 481–88.

HOOKER, Frances Harriet (*née* **Henslow**) (1825–1874)
b. Cambridge 30 April 1825 *d.* Kew, Surrey 13 Nov. 1874

First wife of J. D. Hooker (1817–1911). Translated E. Le Maout and J. Decaisne's *Traité General de Botanique* in 1873.

Garden v.6, 1874, 486. *Gdnrs Chron.* 1874 ii 661. *J. Bot.* 1874, 383. M. Allan *Hookers of Kew, 1785–1911* 1967 *passim.*

HOOKER, Harriet Ann *see* Thiselton-Dyer, H. A.

HOOKER, Sir Joseph Dalton (1817–1911)
b. Halesworth, Suffolk 30 June 1817 *d.* Sunningdale, Berks 10 Dec. 1911

MD Glasgow 1839. FRS 1847. FLS 1842. CB 1869. KCSI 1877. GCSI 1897. OM 1907. Son of W. J. Hooker (1785–1865). Assistant Surgeon and naturalist on HMS 'Erebus', 1839–43. Botanist to Geological Survey, 1846. Explored and collected plants in Sikkim and Nepal, 1847–51. Assistant Director, Kew, 1855; Director, 1865–85. In Palestine and Syria, 1860. In Morocco with J. Ball, 1871. In Rocky Mountains with Asa Gray, 1877. Pioneer phytogeographer. Friend of C. Darwin with whom he collaborated in researches on evolutionary origin of species. *Botany of Antarctic Voyage* 1844–60 6 vols. *Rhododendrons of Sikkim-himalaya* 1849–51. *Himalayan Journals* 1854 2 vols. *Illustrations of Himalayan Plants* 1855. *Fl. Indica* (with T. Thomson) 1855. *Students' Fl. British Island* 1870; ed. 3 1884. *Genera Plantarum* (with G. Bentham) 1862–83 3 vols. *Fl. British India* 1872–97 7 vols. *Journal of Tour in Marocco and Great Atlas* (with J. Ball) 1878.

Ipswich Mus. Portr. Ser. 1852. L. Reeve, *ed. Portr. of Men of Eminence in Literature, Sci. and Art* v.2, 1864, 93–96 portr. *J. Hort. Cottage Gdnr* v.11, 1885, 496–97 portr. *Gdnrs Chron.* 1871, 8 portr.; 1911 ii 427–29, 436–37 portr., 448–50, 468–69; 1912 i 11–12, 26–27, 43. E. Bretschneider *Hist. European Bot. Discoveries in China* 1898, 813–14. *Nature* v.16, 1877, 537–39 portr.; v.88, 1911, 249–54. *J. Bot.* 1898, 487–89; 1911, 33–34; 1912, 1–9; 1919, 130–34. *D.N.B. Supplt 2* v.2, 294–99. *Proc. Linn. Soc.* 1897–98, 30–33; 1911–12, 26–39, 47–62; 1916–17, 93–94. *Gdnrs Mag.* 1911, 931–32 portr. *J. Kew Guild* 1911–12, 47–48 portr. *Kew Bull.* 1912, 1–34, 439–40 portr.; 1913, 91; 1918, 345–51. *Proc. R. Soc. B.* v.85, 1912, i–xxxv portr. *Bot. J. R. Bot. Soc. London* 1913, 103–10 portr. F. W. Oliver *Makers Br. Bot.* 1913, 302–23 portr. J. R. Green *Hist. Bot.* 1914, 467–91. Lord Redesdale *Memories* v.2, 1915, 734. *Who was Who, 1897–1916* 349–50. *Bengal Past and Present* 1917, 252–74. *Annual Rep. Smithsonian Inst.* 1918, 585–601. L. Huxley *Life and Letters of Sir Joseph Dalton Hooker* 1918 2 vols portr. F. O. Bower *Joseph Dalton Hooker* 1919 portr. *Proc. Amer. Acad. Arts Sci.* 1928, 257–66. *Curtis's Bot. Mag. Dedications, 1827–1927* 323–24 portr. *Rhododendron Yb.* 1950, 38–51. W. B. Turrill *Pioneer Plant Geography* 1953 portr. *Notes Rec. R. Soc.* v.14, 1959, 109–20 portr. *Proc. Wedgwood Soc.* 1959, 141–48. *Country Life* 1961, 1390–91. W. B. Turrill *Joseph Dalton Hooker* 1963 portr. I. H. Burkill *Chapters on Hist. Bot. in India* 1965 *passim. Progress* 1966, 190–97 portr. M. Allan *Hookers of Kew, 1785–1911* 1967 portr. F. A. Stafleu *Taxonomic Literature* 1967, 207–9. A. M. Coats *Quest for Plants* 1969, 157–64. C. C. Gillispie *Dict. Sci. Biogr.* v.6, 1972, 488–92. M. A. Reinikka *Hist. of the Orchid* 1972, 199–203 portr. *R.S.C.* v.3, 419; v.7, 1012; v.10, 267; v.12, 346; v.15, 930.

Plants, MSS., letters, drawings and portrs. at Kew. Letters in Berkeley correspondence at BM(NH) and T. H. Huxley correspondence at Imperial College, London. Portr. at Linnean Society and Hunt Library. Library sold at Sotheby, Wilkinson and Hodge, 17 May 1912.

Sirhookera O.K.

HOOKER, Joseph Symonds (1877–1940)
b. 18 Dec. 1877 *d.* 24 April 1940

Son of J. D. Hooker (1817–1911).

Plants at Royal Holloway College, London University.

HOOKER, Maria (*née* **Turner**) (1797–1872)
b. 29 May 1797 *d.* Torquay, Devon 16 Oct. 1872

Eldest daughter of D. Turner (1775–1858). Wife of W. J. Hooker (1785–1865). "For fully half a century Lady Hooker acted as her husband's helpmeet and amanuensis, and to her the successive series of the *Journal of Botany*, the *Botanical Miscellany* and others of Sir William's publications, owe much of their literary style and typographical accuracy" (*Gdnrs Chron.* 1872, 1427).

M. Allan *Hookers of Kew, 1785–1911* 1967 *passim* portr.

HOOKER, William (1779–1832)
b. London 1779 *d.* London 1832

Botanical artist. Employed by Horticultural Society of London to execute drawings of fruit in which he excelled. Engraved and coloured 30 plates in T. A. Knight's *Pomona Herefordiensis* 1811. Drew 119 plates for R. A. Salisbury's *Paradisus Londinensis* 1805–8, 24 plates for F. Pursh's *Florae Americae Septentrionalis* 1814 2 vols and 49 plates for *Pomona Londinensis* 1818.

Trans. Hort. Soc. London v.2, 1817, 62; v.3, 1822, v; v.5, 1824, i. J. C. Loudon *Encyclop. Gdning* 1822, 1285. *Gdnrs Chron.* 1886 ii 268. *J. Bot.* 1886, 51–52. *J. R. Hort. Soc.* 1927, 218–24. R. Hort. Soc. *Exhibition of MSS. Books* 1954, 12–13, 53.

Drawings at R. Hort. Soc.

Hookera Salisb.

HOOKER, William Dawson (1816–1840)
b. Glasgow 4 April 1816 *d.* Kingston, Jamaica 1 Jan. 1840

MD Glasgow 1839. Eldest son of W. J. Hooker (1785–1865). Ornithologist and entomologist. *Notes on Norway* 1837. *Inaugural Dissertation upon Cinchona* 1839.

A. Lasègue *Musée Botanique de B. Delessert* 1845, 395. *D.N.B.* v.27, 298.

HOOKER, Sir William Jackson (1785–1865)
b. Norwich, Norfolk 6 July 1785 *d.* Kew, Surrey 12 Aug. 1865

LLD Glasgow. DCL Oxon. FRS 1812. FLS 1806. Knight of Hanover, 1836. Collected plants in Iceland, 1809. Professor of Botany, Glasgow, 1820. Director, Kew, 1841–65. *Journal of Tour in Iceland* 1811. *British Jungermanniae* 1812–16. *Musci Exotici* 1818–20 2 vols. *Fl. Scotica* 1821. *Exotic Fl.* 1822–27 3 vols. *British Fl.* 1830. *Fl. Boreali-americana* 1829–40 2 vols. *Genera Filicum* 1842. *Species Filicum* 1846–64 5 vols. *Century of Orchidaceous Plants* 1849. *Filices Exoticae* 1859. *Niger Fl.* 1849. *British Ferns* 1861. *Second Century of Ferns* 1861. *Garden Ferns* 1862. *Botany of Captain Beechey's Voyage* (with G. A. W. Arnott) 1830–41. *Synopsis Filicum* (with J. G. Baker) 1868; ed. 2 1874. *Icones Filicum* (with R. K. Greville) 1828–31 2 vols. *Muscologia Britannica* (with T. Taylor) 1818. Edited and drew over 640 plates for *Bot. Mag.* from 1827. Founded *Botanical Miscellany* 1830–33; *Journal of Botany* 1834–42; *London Journal of Botany* 1842–48; *Hooker's Journal of Botany and Kew Garden Miscellany* 1849–57. *Companion to Botanical Magazine* 1835–36. *Icones Plantarum* 1836 to date.

Ipswich Mus. Portr. Ser. 1852. L. Reeve, *ed. Portr. of Men of Eminence in Literature, Sci. and Art.* v.1, 1863, 81–86 portr. *Gdnrs Chron.* 1865, 793, 818–19. *Illus. London News* 26 Aug. 1865, 193 portr. *J. Bot.* 1865, 326–28; 1903, 62; 1906, 176–77; 1909, 106–7. *J. Hort. Cottage Gdnr* v.9, 1865, 145. *Proc. Linn. Soc.* 1865–66, lxvi–lxxiii; 1888–89, 36. *Amer. J. Sci. Arts* v.41, 1866, 1–10. A. de Candolle *La Vie et les Écrites de W. Hooker* 1866. *Flora* 1866, 3–13. *Proc. R. Soc.* v.15, 1867, xxv–xxx. C. S. Sargent *Sci. Papers of Asa Gray* v.2, 1889, 321–32. E. Bretschneider *Hist. European Bot. Discoveries in China* 1898, 548–50. *D.N.B.* v.27, 296–99. *Ann. Bot.* 1902, ix–ccxxi portr. *J. R. Hort. Soc.* 1903, 908–31; 1941, 154–57. *A.L.A. Portr. Index* 1906, 714. F. W. Oliver *Makers Br. Bot.* 1913, 126–50 portr. *Trans. Norfolk Norwich Nat. Soc.* 1930–31, 87–105. Liverpool Public Museums *Handbook and Guide to Herb. Collections* 1935, 57–58 portr. J. S. L. Gilmour *Br. Botanists* 1944, 34–36 portr. M. Hadfield *Pioneers in Gdning* 1955, 90–95. *Proc. Wedgwood Soc.* 1959, 141–43. *Country Life* 1965, 600–2, 682–83 portr. M. Allan *Hookers of Kew, 1785–1911* 1967 portr. F. A. Stafleu *Taxonomic Literature* 1967, 209–18. *Meded. Bot. Mus. Herb. Rijksuniv. Utrecht* no. 283, 1968, 145–62. *Trans. Suffolk Nat. Soc.* v.14, 1969, 175–81. C. C. Gillispie *Dict. Sci. Biogr.* v.6, 1972, 492–95. *R.S.C.* v.3, 422; v.7, 1012; v.12, 346.

Herb., MSS., letters at Kew. Letters at BM(NH), Linnean Society. Portr. by D. Macnee and marble bust by T. Woolner at Kew. Portr. by Gambardella at Linnean Society. Portr. at Hunt Library.

Hookeria Smith.

HOOLEY, Henry (*fl.* 1790s)
Gardener and seedsman, Mansfield, Notts.

HOOLEY, Thomas (*fl.* 1790s)
Gardener and seedsman, Mansfield, Notts.

HOOPER, David (1858–1947)
b. Redhill, Surrey 1 May 1858 *d.* Bromley, Kent 31 Jan. 1947

Pharmacist and chemist. Quinologist to Government of Madras, 1884–96. Curator, Economic Section, Calcutta Museum, 1897–1912. Economic Botanist to Government of India, 1912–14. Collected plants in Sind. *Pharmacographia Indica* (with W. Dymock and C. J. H. Warden) 1889–93 3 vols. *Papers and Notes on Chemistry of Indian Plants from 1884 to 1913* 1913. 'Useful Drugs of Iran and Iraq' (with H. Field) (*Bot. Ser. Field Mus. Nat. Hist.* v.9, 1937, 73–241).

Pharm. J. v.158, 1947, 103 portr. *J. Chem. Soc.* 1948, 253–55. *Who was Who, 1941–1950* 559–60. *Pakistan J. For.* 1967, 348.

HOOPER, James (–1830/1)
d. at sea 1830/1

Kew gardener. Accompanied embassy of Lord Amherst to Peking, 1816, his task being to look after "a plant cabin, for the preservation of living specimens." Curator, Buitenzorg Botanic Gardens, 1817–30. Collected plants in Java, 1824.

Kew Bull. 1893, 174. E. H. M. Cox *Plant Hunting in China* 1945, 50. *Fl. Malesiana* v.1, 1950, 241.

Acacia hooperiana Zippel.

HOOPER, Mary Fawler *see* Maude, M. F.

HOOPER, Robert (1773–1835)
b. London 1773 *d.* London 6 May 1835

BA Oxon 1803. MB 1804. MD St. Andrews 1805. FLS 1796. Practised medicine in Savile Row, London. *Observations on Structure and Economy of Plants* 1797.

W. Munk *Roll of R. College of Physicians* v.3, 1878, 29–31. *D.N.B.* v.27, 306–7.

HOOPER, William (*fl.* 1640s–1660s)

Fellow of Magdalen, Oxford, 1643. Arboriculturist.

R. T. Gunther *Early Br. Botanists* 1922, 79–80.

HOPE, Charles William Webley (1832–1904)
b. Edinburgh 1832 *d.* Kew, Surrey 18 Feb. 1904

Civil engineer. Pteridologist. To India, 1859. Collected plants in Kumaon, 1861; Simla, 1871, and Western Himalayas. Returned to England, 1896. 'Ferns of N. Western India' (*J. Bombay Nat. Hist. Soc.* v.12–14, 1899–1903; *J. Bot.* 1896, 122–27).

J. Bot. 1904, 127–28. *R.S.C.* v.15, 933.

Ferns at BM(NH).

HOPE, Miss Frances Jane (–1880)
d. Wardie Lodge, Edinburgh 26 April 1880

Had garden at Wardie Lodge, noted for hellebores. *Notes and Thoughts on Gardens and Woodlands* 1881.

Garden v.17, 1880, 420. *Gdnrs Chron.* 1880 i 585–86; 1965 i 81, 237. *Trans. Bot. Soc. Edinburgh* v.14, 1883, 159.

HOPE, George (–1928)

Of Havering, Essex. Stock-broker. A naturalist in the Gilbert White tradition. Interested in plant collecting and horticulture.

Shooting Times and Country Mag. 25 July 1958 portr.

HOPE, J. (1875–1970)
b. Acomb, York 1875 *d.* Ness Gardens, Cheshire 26 Dec. 1970

Gardener at Cambridge Botanic Garden, 1893–96; Royal Botanic Garden, Edinburgh, 1908–13. Gardener to Mr. Bulley at Ness, 1913. Ness Gardens transferred to Liverpool University, 1948.

Gdnrs Chron. v.169(3), 1971, 18.

Primula sikkimensis subsp. *hopeana.*

HOPE, John (1725–1786)
b. Edinburgh 10 May 1725 *d.* Edinburgh 10 Nov. 1786

MD Glasgow 1750. FRS 1767. Pupil of C. Alston and B. de Jussieu. Professor of Botany and Regius Keeper, Edinburgh, 1761. First to teach Linnaen system in Scotland. Taught Smith, Salisbury, Pulteney, etc. Created botanic garden in Edinburgh. *Termini Botanici* 1778. *Genera Plantarum* 1780. 'On Rheum palmatum' (*Philos. Trans. R. Soc.* v.55, 1766, 290–93). Helped J. Lightfoot (*Fl. Scotica* v.1, 1777, preface xii). His Edinburgh plants, 1764–68 (*Notes Bot. Gdn Edinburgh* v.4, 1907, 123–92, 241–44).

Harveian Oration, Edinburgh 1789. R. Pulteney *Hist. Biogr. Sketches of Progress of Bot. in England* v.2, 1790, 17, 352. W. Darlington *Memorials of John Bartram and Humphry Marshall* 1849, 432–33. *D.N.B.* v.27, 321–22. *J. Bot.* 1907, 454. F. W. Oliver *Makers of Br. Bot.* 1913, 286–90. H. R. Fletcher and W. H. Brown *R. Bot. Gdn Edinburgh, 1670–1970* 1970, 57–67 portr.

MSS. at Royal Society. Letters at Kew. Portr. by J. Kay. Portr. at Hunt Library.

Hopea L.

HOPE, Thomas Charles (1766–1844)
b. Edinburgh 21 July 1766 *d.* Edinburgh 13 June 1844

MD Edinburgh 1787. FRS 1810. ALS 1788. Son of J. Hope (1725–1786). Professor of Chemistry, Glasgow, 1787–89; Edinburgh, 1796–1844. Had herb. *Tentamen Inaugurale, Quaedam de Plantarum Motibus et Vita Complectens* 1787. Helped in W. Withering's *Bot. Arrangement* ed. 3 1796.

Proc. Linn. Soc. 1845, 250–51. *D.N.B.* v.27, 329. *R.S.C.* v.3, 426; v.15, 933.

HOPKINS, Esther (*née* **Burton**) (1815–1897)
b. 18 Nov. 1815 *d.* Chester, Cheshire 27 May 1897

Of Bath. Discovered *Potamogeton decipiens.* Contrib. to H. C. Watson's *Topographical Bot.* 1883.

London Bot. Exch. Club Rep. 1866, 13. *J. Bot.* 1867, 71.

Herb. was in possession of D. M. Atkinson, Royal Infirmary, Glasgow.

Potamogeton burtoni Hopkins.

HOPKINS, George (*fl.* 1820s)

Collected *Crinum submersum* near Rio de Janeiro.

Bot. Mag. 1824, t.2463.

HOPKINS, W. Owen (*c.* 1884–1960)
d. London 7 July 1960

Cultivated orchids.

Orchid Rev. 1960, 279.

HOPKINSON, John (1844–1919)
b. Leeds, Yorks 15 Nov. 1844 *d.* Watford, Herts 5 July 1919
FLS 1875. Pianoforte maker, London. First Secretary, Watford Natural History Society. President, Hertfordshire Natural History Society, 1893. Secretary, Ray Society, 1902. Wrote introduction to A. R. Pryor's *Fl. Hertfordshire* 1887. 'Report on Phenological Observations in Hertfordshire, 1876–[82]' (*Trans. Watford Nat. Hist. Soc.* v.2, 1880, 37–40, 101–3, 229–36; v.1, 1882, 133–38, 257–63; v.2, 1884, 71–79, 181–86).
Bot. Soc. Exch. Club Br. Isl. Rep. 1919, 634. *Proc. Linn. Soc.* 1919–20, 43–45. *R.S.C.* v.7, 1013; v.10, 269; v.12, 347.

HOPKINSON, William (*fl.* 1790s)
Gardener and nurseryman, Derby.

HOPKIRK, Thomas (1785–1841)
b. Dalbeth, Glasgow 1785 *d.* Malone, Belfast 24 Aug. 1841
LLD Glasgow 1835. FLS 1812. Founded Botanical Institution, Glasgow. *Catalogue of Plants of Garden, Dalbeth* 1813. *Fl. Glottiana* 1813. *Fl. Anomoia* 1817.
J. Sowerby and J. E. Smith *English Bot.* 2532. *Trans. Nat. Hist. Soc. Glasgow* v.1, 1884–85, 196–259 portr. *D.N.B.* v.27, 341.
Portr. at Hunt Library.
Hopkirkia DC.

HOPPNER, H. P. (*fl.* 1820s)
Lieut., Royal Navy. Collected plants while on Arctic voyage with Captain Ross in 1820.
H. N. Clokie *Account of Herb. Dept. Bot., Oxford* 1964, 186.
Plants at Oxford.

HOPWOOD, Edmund (*fl.* 1830s)
Nurseryman, Richmond Road, Twickenham, Middx.

HORE, Rev. William Strong (1807–1882)
b. Stonehouse, Plymouth, Devon 29 March 1807 *d.* 19 Feb. 1882
BA Cantab 1830. MA Oxon 1851. FLS 1840. Vicar, Shebbear, Devon, 1855. Discovered *Trifolium molinerii.* Sent algae to W. H. Harvey. 'List of Plants found in Devonshire and Cornwall not mentioned by Jones in Fl. Devoniensis' (*Phytologist* 1841, 160–63).
Phytologist 1845, 239–40; 1851, 94–98. *Trans. R. Irish Acad. Sci.* v.22, 1854, 555–56. *J. Bot.* 1882, 288; 1906, 216. C. C. Babington *Memorials, J. and Bot. Correspondence* 1897, 268. F. H. Davey *Fl. Cornwall* 1909, xlii–xliii. W. K. Martin and G. T. Fraser *Fl. Devon* 1939, 772. *R.S.C.* v.3, 433.
Plants at Plymouth Institution, North Devon Athenaeum, Barnstaple. Letters at Kew.
Horea Harvey.

HORLICK, Sir James Nockells (1886–*c.* 1972)
b. Gloucestershire 22 March 1886
VMH 1963. Created garden on windswept island of Gigha, Argyll.
Who's Who 1972, 1552. *J. R. Hort. Soc.* 1973, 188–89.

HORMAN, Rev. William (*c.* 1458–1535)
b. Salisbury, Wilts *d.* Eton, Bucks 12 April 1535
DD Oxon. Fellow, New College, Oxford, 1477–85. Master of Eton, 1486; Vice-provost, 1503. Rector, East Wretham, Norfolk, 1494–1503. *Herbarum Synonyma.*
R. Pulteney *Hist. Biogr. Sketches of Progress of Bot. in England* v.1, 1790, 25. G. W. Johnson *Hist. English Gdning* 1829, 46–47. *D.N.B.* v.27, 352.
Brass effigy at Eton College Chapel.

HORNE, Edward (–1851)
d. Florence, Italy 18 March 1851
BCL Oxon. FLS 1812. Companion of Joseph Woods on botanical excursions.
Proc. Linn. Soc. 1851, 132.

HORNE, John (1835–1905)
b. Lethendy, Perthshire Jan. 1835 *d.* Jersey, Channel Islands 16 April 1905
FLS 1873. At Kew, 1859–60. At Botanic Garden, Mauritius, 1861–91 (Director from 1877). Collected plants in Mauritius and Seychelles. To Fiji Islands, 1876 where he also collected plants (ferns in *J. Bot.* 1879, 292–300). *A Year in Fiji* 1881. *Notes on Flat Island [Mauritius]* 1885. *Notes on Fl. Flat Island* 1886.
J. G. Baker *Fl. Mauritius and Seychelles* 1877, preface 11*. *Kew Bull.* 1892, 250–51. *Gdnrs Chron.* 1905 i 271. *J. Kew Guild* 1905, 266 portr. *J. Bot.* 1905, 192. *Proc. Linn. Soc.* 1904–5, 34. *Trans. Bot. Soc. Edinburgh* v.23, 1907, 212. *R.S.C.* v.7, 1017; v.10, 274.
Plants and letters at Kew.
Hornea Baker.

HORNER, Rev. Francis D. (–1912)
d. Burton-in-Lonsdale, Yorks 11 July 1912
Educ. Trinity College, Dublin. VMH 1897. Curate, Liverpool and Normanton, 1861–69; Kirkby Malzeard, Yorks, 1870. Grew auriculas and tulips. Secretary, North Auricula Society; Northern Carnation Society.
Florist and Pomologist 1883, 90–92 portr. *Gdnrs Chron.* 1883 i 530–31, 536 portr. *J. Hort. Cottage Gdnr* v.20, 1890, 343–44 portr.; v.62, 1911, 477 portr.; v.65, 1912, 74 portr. *Garden* v.57, 1900, 307–8 portr.; 1912, 372 (vi). *Gdnrs Mag.* 1907, 755–56 portr., 798; 1912, 566 portr.

HORNER, Katharine Murray *see* Lyell, K. M.

HORNER, Robert (*fl.* 1820s)
Nurseryman, Welburn in Bulmer near Malton, Yorks.

HORRELL, Ernest Charles (*fl.* 1870s–1930s)
Plants at BM(NH), Nottingham Museum, Oxford.

HORSEFIELD, John (1792–1854)
b. 18 July 1792 *d.* Prestwich, Manchester 6 March 1854
Handloom weaver of Bury and afterwards Prestwich. President, Prestwich Botanical Society, 1820–54. Raised *Narcissus bicolor horsefieldii*. 'Prestwich and Bury Societies' (*Gdnrs Mag.* 1830, 392–95).
R. Buxton *Bot. Guide to...Manchester* 1849, vi–vii. *Manchester Guardian* 2 March 1850. J. Cash *Where there's a Will there's a Way* 1873, 67–76. L. H. Grindon *Country Rambles* 1882, 192. *Gdnrs Chron.* 1894 ii 465–67; 1931 *i* 169. *Trans. Liverpool Bot. Soc.* 1909, 76. *Daffodil Yb.* 1933, 24–25.

HORSEFIELD, William (1816–1883)
b. Besses-o-the-Barn, Manchester 16 April 1816 *d.* Manchester 17 Jan. 1883
Son of J. Horsefield (1792–1854). Postman at Whitefield. Succeeded his father as President, Prestwich Botanical Society.
Gdnrs Chron. 1883, i 122. *J. Bot.* 1883, 192. *J. Hort. Cottage Gdnr.* v.6, 1883, 115. *Trans. Liverpool Bot. Soc.* 1909, 77.

HORSENELL, George (*c.* 1625–1697)
d. 7 April 1697
Surgeon, London. Correspondent of J. Ray (*Synopsis* 1690, 96). Brought plants from Antigua to L. Plukenet (*Almagestum Bot.* 1696, 15, 155, 240).
J. E. Dandy *Sloane Herb.* 1958, 138.

HORSFALL, Mrs. (*fl.* 1830s)
Wife of Charles Horsfall. Drew 5 plates for *Bot. Mag.*
Ipomoea horsfalliae Hook.

HORSFALL, Charles (*fl.* 1830s)
Had notable collection of plants at Everton near Liverpool.
Bot. Mag. 1835, t.3403.

HORSFALL, J. B. (*fl.* 1860s)
Son of Charles Horsfall. Of Bellamour Hall, Staffs. Had many exotic plants in his collection.
Bot. Mag. 1865, t.5486.

HORSFIELD, Thomas (1773–1859)
b. Bethlehem, Pa., U.S.A. 12 May 1773 *d.* London 14 July 1859
MD Pennsylvania 1798. FRS 1828. FLS 1820. Surgeon in service of Dutch in Java, 1801; transferred to East India Company, 1811. Collected plants in Java and Sumatra, 1802–18. East India Company Library and Museum, 1819; Keeper of Museum, 1836–59. Plants described with sketch of his travels in J. J. Bennett's *Plantae Javanicae Rariores* 1838–52.
A. Lasègue *Musée Botanique de B. Delessert* 1845, 494. *Proc. Linn. Soc.* 1860, xxv–xxvi. *Proc. R. Soc.* v.10, 1860, xix–xxi. *D.N.B.* v.27, 379–80. D. G. Crawford *Hist. Indian Med. Service, 1600–1913* v.2, 1914, 170–71. *Torreya* 1942, 1–9. M. Archer *Nat. Hist. Drawings in India Office Library* 1962, 46–48, 80–82. M. Archer *Br. Drawings in India Office Library* v.2, 1969, 446–65. *Straits Times Annual* 1972, 85–88. *Fl. Malesiana* v.8(1), Supplt. 2, 1974, 46 portr.
MSS., drawings and plants at BM(NH). Plants at Kew. Drawings at India Office Library.
Horsfieldia Blume.

HORSFIELD, Timothy (*fl.* 1830s)
Gardener and seedsman, 13 Bull Green, Halifax, Yorks.

HORSMAN, Frederick James Serle (*c.* 1842–1894)
d. 2 June 1894
Orchid grower and importer, Mark's Tey, Essex.
J. Hort. Cottage Gdnr v.28, 1894, 494. *Orchid Rev.* 1894, 194.

HORSMAN, Samuel (1698–1751)
MD Leyden 1721. MD Cantab 1728. Physician at Hatton Garden, London. Member of Botanical Society of London.
Proc. Bot. Soc. Br. Isl. 1967, 317.

HORT, Sir Arthur Fenton, 6th Baronet (1864–1935)
b. 15 Jan. 1864 *d.* Hurstbourne Tarrant, Hants 7 March 1935
MA Cantab. VMH 1930. Son of F. J. A. Hort (1828–1892). Master, Harrow School, 1888–1922. Raised irises. Edited Theophrastus's *Enquiry into Plants* 1916. *Unconventional Garden* 1928. *Garden Variety* 1935. Translated Linnaeus's *Critica Botanica* 1938.
Gdnrs Chron. 1922 ii 202 portr.; 1931 i 62 portr.; 1935 i 183. *Iris Yb.* 1935, 71–72. *Times* 9 and 12 March 1935. *Kew Bull.* 1936, 143–44. *Who was Who, 1929–1940* 665.

HORT, Rev. Fenton John Anthony (1828–1892)
b. Dublin 23 April 1828 *d.* Cambridge 30 Nov. 1892
BA Cantab 1850. MA Oxon 1856. DD 1876. LLD Dublin 1888. DCL Durham 1890. Rector, St. Ippolyts, Hitchin, 1857–72. Authority on *Rubi*. Correspondent of H. C. Watson. '*Rubus imbricatus* Hort' (*Ann. Mag. Nat. Hist.* v.7, 1851, 374–77). Contrib. to *Phytologist* v.2–4, 1845–53.

J. Bot. 1893, 63–64. A. F. Hort *Life and Letters of F. J. A. Hort* 1896 portr. 2 vols. *D.N.B. Supplt I* v.2, 443–47. J. W. White *Fl. Bristol* 1912, 86–87. R. L. Praeger *Some Irish Nat.* 1949, 105. *R.S.C.* v.3, 444.

Plants at BM(NH), Cambridge. Portr. at Hunt Library.

HORWOOD, Arthur Reginald (1879–1937)
b. Leicester 29 May 1879 *d.* Brentford, Middx 21 Feb. 1937

FLS 1913. Assistant then sub-curator, Leicester Museum, 1902–22. Assistant, Kew Herb., 1924–37. *Practical Field Botany* 1914. *Plant Life in British Isles* 1914–16 3 vols. *Outdoor Botanist* 1920. *New British Fl.* 1919 6 vols. Edited J. B. Hurry's *Woad Plant* 1930. *Fl. Leicestershire and Rutland* (with 3rd Earl of Gainsborough) 1933.

Bot. Soc. Exch. Club Br. Isl. Rep. 1937, 435–36. *J. Bot.* 1937, 113–14. *J. Kew Guild* 1937, 702–3. *Nature* v.139, 1937, 538. *Proc. Linn. Soc.* 1936–37, 195–96. *Times* 3 March 1937.

Plants at National Museum of Wales, Kew, Leicester Museum.

HOSACK, David (1769–1835)
b. New York 31 Aug. 1769 *d.* New York 22 Dec. 1835

MD Edinburgh. FRS 1817. FLS 1794. Pupil of J. E. Smith. Professor of Botany, Columbia College, 1795. Established New York Botanic Garden. *Hortus Elginensis* 1806.

Amer. J. Sci. Arts v.29, 1836, 395–96. *Gdnrs Mag.* 1836, 276. S. D. Gross *Lives of Eminent Amer. Physicians and Surgeons of 19th Century* 1861, 289–337. W. J. Youmans *Pioneers of Sci. in America* 1896, 100–10. *J. New York Bot. Gdn* 1900, 22–26. *Bull. New York Public Library* Sept. 1919, 7–9. *Columbia Univ. Quart.* v.31, 1939, 272–97. *Med. Recorder* 1948, 107–10. *Proc. Amer. Philos. Soc.* 1960, 293–313 portr. *Mem. Amer. Philos. Soc.* 1964, 1–246. C. C. Robbins *David Hosack* 1964 portr. C. C. Gillispie *Dict. Sci. Biogr.* v.6, 1972, 521.

Herb. probably destroyed by fire in 1866. MSS. at American Philosophical Society, Royal Society. Letters in Smith correspondence at Linnean Society. Portr. at New York Botanic Garden, Columbia and Princeton Universities.

Hosackia Douglas.

HOSE, Charles (1863–1929)
b. Hertfordshire 12 Oct. 1863 *d.* 14 Nov. 1929

Educ. Cambridge. Nephew of Rev. G. F. Hose (1838–1922). Entered services of Rajah of Sarawak 1884; retired as Judge of Supreme Court, 1907. Revisited Sarawak in 1909 and 1920. Collected plants and fauna in Borneo, N. Celebes and Malacca. *Pagan Tribes of Borneo* 1912 2 vols. *Fifty Years of Romance and Research* 1928. *Natural Man; a Record from Borneo* 1926 portr. *Field Book of a Jungle Wallah* 1929.

J. Bot. 1922, 273; 1929, 344. *Who was Who, 1929–1940* 666. *Fl. Malesiana* v.1, 1950, 245–46 portr.

Plants at BM(NH), Kew.

HOSE, Ernest Shaw (1871–1946)
b. 25 Nov. 1871 *d.* 12 Sept. 1946

Son of G. F. Hose (1838–1922). Malayan Civil Service, 1891. Colonial Secretary, Straits Settlements, 1924–25. In charge of Borneo Company's rubber plantations, Sarawak. Collected plants in Natuna Islands, 1894 and in Malaya, 1917.

Kew Bull. 1896, 40–41. *Gdns Bull. Straits Settlements* 1927, 124. *Fl. Malesiana* v.1, 1950, 246. *Who was Who, 1941–1950* 565.

Plants at Kew, Singapore.

Dolichos hosei Craib.

HOSE, Rev. George Frederick (1838–1922)
b. Cambridge 3 Sept. 1838 *d.* Guildford, Surrey 26 March 1922

Educ. Cambridge. DD. Curate, Marylebone, London, 1865–68. Chaplain, Malacca, 1868–73; Singapore, 1874–81. Bishop of Singapore and Sarawak, 1881–1908. President, Straits Branch, Royal Asiatic Society, 1877. Collected plants, chiefly ferns, in Borneo and Malaya. 'Catalogue of Ferns of Borneo and some Adjacent Islands' (*J. Straits Branch R. Asiatic Soc.* no. 32, 1899, 31–84).

J. Straits Branch R. Asiatic Soc. no. 57, 1911, 1–5 portr.; no. 86, 1922, 395. *J. Bot.* 1922, 272–73. *Gdns Bull. Straits Settlements* 1927, 124. *Fl. Malesiana* v.1, 1950, 246–47 portr.

Plants at Kew, Calcutta, Singapore.

Hosea Ridley.

HOSE, Gertrude (*fl.* 1900s)

Daughter of G. F. Hose (1838–1922). Collected grasses in Malaya in early 1900s.

Gdns Bull. Straits Settlements 1927, 124.

HOSIE, Sir Alexander (1853–1925)
b. 16 Jan. 1853 *d.* 10 March 1925

MA. FRGS. To China as student interpreter, 1876. Consul, Pagoda Island near Fu chow, 1895. Studied Chinese economic plants. Collected plants in China, Formosa, Tibet. *Three Years in Western China* 1890. *On Trail of Opium Poppy* 1914 2 vols.

E. Bretschneider *Hist. European Bot. Discoveries in China* 1898, 767–71. *Kew Bull.* 1901, 33; 1925, 346–47. *Who was Who, 1916–1928* 520. E. H. M. Cox *Plant Hunting in China* 1945, 106.

Hosiea Hemsley & E. H. Wilson.

HOSKINS, S. E. (1799–1888)
Physician. Presented his herb. of Guernsey plants (since lost) to La Société Guernesaise.

HOTHFIELD, Lord *see* Tufton, H. S. T.

HOTSON, John Ernest Buttery (1872–1944)
Captain. Entomologist. Collected plants in Persian Baluchistan and Makran coast, 1916–18, which were sent to Father Blatter in Bombay. 'Contributions towards Fl. Baluchistan from Materials supplied by Capt. J. E. B. Hotson' (*J. Indian Bot. Soc.* v.1, 1919, 84–91).
Plants at Bombay.

HOUBLON, Richard Archer (1884–1957)
b. London 13 Oct. 1884 *d.* Kilmurry 11 June 1957
Officer, Royal Horse Artillery. Naturalist. *Identification of Trees and Shrubs in Winter.*
G. Seaver *Richard Archer Houblon* 1970, portr.

HOUGHTON, Rev. William (1829–1895)
b. Liverpool 1829 *d.* 3 Sept. 1895
MA Oxon 1853. FLS 1859. Headmaster, Solihull grammar school, 1858. Rector, Preston, Wellington, Shropshire, 1860. *Natural History of Ancients* 1879. 'Notices of Fungi in Greek and Latin Authors' (*Ann. Mag. Nat. Hist.* 1885, 22–49, 153–54).
Proc. Linn. Soc. 1895–96, 37. *R.S.C.* v.3, 446; v.7, 1021; v.15, 954.
Letters at Kew and in Broome correspondence at BM(NH).

HOULSTON, John (*fl.* 1840s–1850s)
b. Scotland
At Birmingham Botanic Garden. At Kew, 1848–55, where he was foreman in orchid and fern houses. Contrib. papers on cultivated ferns to *Gdnrs Mag. Bot.* 1851 and *Gdn Companion and Florists' Guide* 1852.
J. Kew Guild 1894, 40–41.

HOULTON, Joseph (1788–1861)
b. Saffron Waldon, Essex 29 Feb. 1788 *d.* London 14 Jan. 1861
MRCS. MD Erlangen. FLS 1823. Surgeon to E. Norfolk militia till 1817. Practised at Saffron Walden and (from 1823) in London. Professor of Botany to Medico-botanical Society of London. Licensed lecturer by Society of Apothecaries. Contrib. to *Pharm. J.*
Med. Times and Gaz. 1861, 565. *R.S.C.* v.3, 446.

HOUSEMAN, James (*c.* 1796–1833)
d. Toft, Cheshire 23 Feb. 1833
Botanist. Contrib. to *Gdnrs Mag.*
Gdnrs Mag. 1834, 300.

HOUSTOUN, William (*c.* 1695–1733)
b. Scotland *c.* 1695 *d.* Jamaica 14 Aug. 1733
MD Leyden 1729. FRS 1733. Ship's surgeon in West Indies. Correspondent of P. Miller. Collected plants in Mexico and West Indies, 1729–33. *Reliquiae Houstounianae* 1781 with engravings by himself. Drawings in P. Miller's *Figures of...Plants* v.1, 1755, t.xliv and J. Martyn's *Historia Plantarum Rariorum* 1728, dec. 3, 3, 4; dec. 5, 3. 'Contrayerva' (*Philos. Trans. R. Soc.* v.37, 1733, 195–98).
Gent. Mag. 1733, 662. R. Pulteney *Hist. Biogr. Sketches of Progress of Bot. in England* v.2, 1790, 231–34. W. B. Hemsley *Biologia Centrali-Americana* v.4, 1886, 118–19. *J. Bot.* 1897, 225–34. I. Urban *Symbolae Antillanae* v.3, 1902, 62. *D.N.B.* v.27, 425–26. E. J. L. Scott *Index to Sloane Manuscripts* 1904, 264. *Osiris* 1948, 96. E. D. Johnston *Houstouns of Georgia* 1950. J. E. Dandy *Sloane Herb.* 1958, 139–40. F. A. Stafleu *Taxonomic Literature* 1967, 220. A. M. Coats *Quest for Plants* 1969, 332–33.
Houstonia L.

HOVE, Anton Pantaleon (*fl.* 1780s–1820s)
Of Warsaw. Kew plant collector on Guinea coast, 1785. Collected in India, 1787–88 and in Crimea, 1796. In England, 1829. Introduced geraniums (H. C. Andrews *Geraniums* sub *G. crassicaule*) and *Azalea pontica* (*Bot. Mag.* 1799, t.433; 1823, t.2383). *Tours...in Guzerat* 1855.
A. Afzelius and N. W. Elgenstierna *Genera Plantarum Guineensium* 1804, 25–26. J. C. Loudon *Encyclop. Plants* 1829, 610. *J. Linn. Soc. Bot.* v.45, 1920, 46–47. A. M. Coats *Quest for Plants* 1969, 57, 145–48.
Plants at BM(NH). MSS. at Wellcome Library.
Hovea Br.

HOW (or Howe), William (1620–1656)
b. London 1620 *d.* London 30 Aug. 1656
BA Oxon 1641. Physician. *Phytologia Britannica* 1650 (his inter-leaved copy at Magdalen College, Oxford). Edited M. de Lobel's *Stirpium Illustrationes* 1655.
R. Pulteney *Hist. Biogr. Sketches of Progress of Bot. in England* v.1, 1790, 169–74. J. R. Green *Hist. Bot.* 1914, 58–60. R. T. Gunther *Early Br. Botanists* 1922, 251–52, 276–94. *D.N.B.* v.28, 102. G. C. Druce: *Fl. Berks* 1897, ci–civ; *Fl. Buckinghamshire* 1926, lxx; *Fl. Oxford* 1927, lxvi–lxvii; *Fl. Northamptonshire* 1930, xlvi. C. E. Raven *English Nat.* 1947, 298–304. S. T. Jermyn *Fl. Essex* 1974, 14.

HOW, Rev. William Walsham (1823–1897)
b. Shrewsbury, Shropshire 13 Dec. 1823 *d.* Leenane, Connemara 10 Aug. 1897

BA Oxon 1845. DD 1886. Bishop of Bedford, 1879; of Wakefield, 1888. President, Yorkshire Naturalists Union, 1890. 'Botany of Great Orme's Head' (*Rep. Oswestry Field Club* 1865, 61–63). List of plants in Roberts's *Gossiping Guide to Wales*.

J. Bot. 1897, 464. *Naturalist* 1897, 299–308 portr. F. D. How *Bishop Walsham How* 1898. *D.N.B. Supplt I* v.3, 1–2.

HOWARD, Mr. (*fl.* 1770s)
Surgeon, Knutsford, Cheshire. About 1770 sent roots of *Saxifraga hirculus* to W. Curtis who figured it in *Fl. Londinensis* 1772–98.

J. B. L. Warren *Fl. Cheshire* 1899, lxxxiv.

HOWARD, Mrs. (*fl.* 1810s)
Nurserywoman, King's Road, Chelsea, London.

HOWARD, Sir Albert (1873–1947)
b. 8 Dec. 1873 *d.* London 20 Oct. 1947
BA Cantab 1899. MA 1902. FLS 1901. Mycologist and Agricultural Lecturer, Dept. of Agriculture for West Indies, 1899–1902. Botanist, S.E. Agricultural College, Wye, 1903–5. Economic Botanist, Indian Dept. of Agriculture, Pusa, 1905–24. Director, Institute of Plant Industry, Indore, 1925–31. *Wheat in India* (with G. L. C. Howard) 1909. *Crop Production in India* 1929. *Application of Science to Crop Production* 1929. *Agricultural Testament* 1940. *Farming and Gardening for Health and Disease* 1945. *Earth's Green Mantle* 1947.

Gdnrs Chron. 1947 ii 158. *Nature* v.160, 1947, 741–42. *Times* 21 Oct. 1947. *Who was Who, 1941–1950* 566–67.

HOWARD, John Eliot (1807–1883)
b. Plaistow, Essex 11 Dec. 1807 *d.* Tottenham, Middx 22 Nov. 1883
FRS 1874. FLS 1857. Quinologist. 'Pavon's Peruvian Barks' (*Pharm. J.* v.11, 12, 1852–53). *Illustrations of Nueva Quinologia of Pavon* 1862. *Quinology of East Indian Plantations* 1869–76.

Gdnrs Chron. 1883 ii 701. *Proc. Linn. Soc.* 1883–84, 35–36. *Trans. Essex Field Club* v.4, 1885, 8–11 portr. *D.N.B.* v.28, 48–49. *R.S.C.* v.3, 450; v.7, 1023; v.10, 279; v.12, 350; v.15, 960.

Letters at Kew.

Howardia Wedd.

HOWARD, M. (*fl.* 1820s)
Nurseryman, King's Road, Chelsea, London.

HOWARD, William (1877–1954)
d. 18 Jan. 1954
Member of Essex Field Club. Entered records of plants collected by Club members in inter-leaved copy of G. C. Druce's *Comital Fl. Br. Isl.* 1932.

Essex Nat. 1955, 266.

HOWARTH, Ralph (1889–1954)
b. Yorks 1889 *d.* 8 Feb. 1954
In textile trade. To Isle of Man, 1916, where he botanised. President, Isle of Man Natural History and Antiquarian Society.

Proc. Bot. Soc. Br. Isl. 1954, 280.

HOWARTH, Willis Openshaw (1890–1964)
b. Bury, Lancs 3 June 1890 *d.* Southport, Lancs 27 June 1964
BSc Manchester 1913. MSc 1918. DSc 1935. FLS 1922. Assistant Lecturer in Botany, University College, Cardiff and Reading. Lecturer in Botany, Manchester, 1919–56. Authority on flora of Lancashire. *Practical Botany for Tropics* (with R. G. G. Warne) 1959. Contrib. to *New Phytol.*, *J. Ecol.*, *J. Linn. Soc. Bot.*

Nature v.203, 1964, 695. *Proc. Linn. Soc.* 1966, 119–20.

HOWE, Alec C. (–1973)
d. Feb. 1973
Iris grower and hybridist.

Iris Yb. 1973, 19–23.

HOWE, W. E. (–1891)
Geologist and pteridologist of Matlock Bath, Derbyshire. *Ferns of Derbyshire* 1861; ed. 6 1867.

Sorby Rec. v.2(4), 1967, 6–7.

HOWES, Frank Norman (1901–1973)
b. S. Africa 2 Aug. 1901 *d.* 26 Feb. 1973
BA Durban 1922. DSc 1935. FLS 1925. On staff of Botanical Survey of S. Africa, 1922–24. Economic Botanist, Dept. of Agriculture, Gold Coast, 1924. Assistant, Kew Museums, 1926; Keeper, 1948–66. *Plants and Beekeeping* 1945. *Nuts* 1948. *Vegetable Gums and Resins* 1949. *Vegetable Tanning Materials* 1953. *Dictionary of Useful and Everyday Plants and their Common Names* 1975. Agricultural editor, *International Sugar J.*, 1967.

J. Kew Guild 1974, 340. *Econ. Bot.* 1975, 194–96.

Gold Coast plants at Kew.

HOWEY, William (*c.* 1729–1792)
d. 2 Dec. 1792
Purchased Putney Nursery from F. Hunt on his death in 1775. Bequeathed it to his sons John Howey (1762–1798) and Robert Howey (1764–1800). John's widow, Elizabeth, and his aunt Martha Howey carried on the business until John's infant son William attained his majority in 1819.

Surrey Archaeol. Collections v.69, 1973, 138–39. *Trans. London Middlesex Archaeol. Soc.* v.24, 1973, 191.

HOWIE, Charles (1811–1899)
d. St. Andrews, Fife 22 July 1899
Nurseryman, St. Andrews. Bryologist. Correspondent of W. Wilson, 1845–70. *Mosses of Fifeshire* (exsiccatae). *Moss Fl. Fife and Kinross* 1889. Contrib. to *Phytologist* 1857–59 and *Trans. Bot. Soc. Edinburgh* 1868–70.
Trans. Bot. Soc. Edinburgh v.9, 1867, 257–65. *R.S.C.* v.3, 451; v.7, 1024; v.10, 280; v.15, 967.
Herb. at St. Andrews.
Carduus carolorum Howie & Jenner.

HOWISON, William (*fl.* 1820s–1830s)
MD. Lecturer on Botany, Edinburgh. 'Turpentine' (*Trans. Highland Soc.* 1820, 495). 'Forest Trees of Russia' (*Edinburgh Philos. J.* v.12, 1825, 56–70).
Gdnrs Mag. 1838, 370–71. *R.S.C.* v.3, 451.

HOWITT, Alfred William (1830–1908)
b. Nottingham 17 April 1830 *d.* Bairnsdale, Victoria 7 March 1908
DSc Cantab 1904. FLS 1882. CMG 1906. Nephew of G. Howitt (1800–1873). To Australia, 1852. Farmer and expert bushman. Led expedition to sources of Mitchell River, 1860. Led expedition in search of Burke and Wills, 1861. Authority on Eucalypts. 'Eucalypts of Gippsland' (*Trans. R. Soc. Victoria* 1891, 81–200).
J. T. Woods *Hist. of Discovery and Exploration of Australia* v.2, 1865, 392–408. P. Mennell *Dict. Austral. Biogr.* 1892, 241. *Vict. Nat.* v.21, 1904, 4–5; v.24, 1908, 181–89 portr.; v.25, 1908, 71, 109. *Vict. Hist. Mag.* 1913, 1–24 portr. *Who was Who, 1897–1916* 356. P. Serle *Dict. Austral. Biogr.* v.1, 1949, 457. *Austral. Encyclop.* v.5, 1965, 21–22.
Plants at Melbourne. Portr. at Hunt Library.

HOWITT, Godfrey (1800–1873)
b. Heanor Wood, Derbyshire 10 Nov. 1800 *d.* Melbourne 1873
MD Edinburgh. Brother-in-law of M. Howitt (1797–1888). Practised medicine at Leicester and Nottingham. To Australia, 1839. *Muscologia Nottinghamensis* (with W. Valentine) (exsiccatae) 1833. *Nottinghamshire Fl.* 1839. Contrib. Stafford, Derby, Caernarvon and Nottingham plants to H. C. Watson's *New Botanists' Guide* 1835–37, 640–50.
Vict. Hist. Mag. 1913, 16–25. P. Serle *Dict. Austral. Biogr.* v.1, 1949, 457. R. C. L. and B. M. Howitt *Fl. Nottinghamshire* 1963, 16–17. *Sorby Rec.* v.2(4), 1967, 7.
Howittia F. Muell.

HOWITT, Mary (*née* **Botham**) (1797–1888)
b. Coleford, Glos 10 March 1797 *d.* Rome 30 Jan. 1888
Knew British plants well. *Sketches of Natural History* 1834 (verse). *Birds and Flowers and other Country Things* 1855 (verse). *Mary Howitt, an Autobiography* 1889 2 vols.
D.N.B. v.28, 122–23.

HOWITT, Richard (1799–1869)
b. Heanor Wood, Derbyshire 1799 *d.* near Sherwood Forest, Notts 1869
Emigrated with his brother, Godfrey, to Australia, 1839. Farmer. Botanist. Returned to England, 1844. *Impressions of Australia Felix* 1845.
Vict. Hist. Mag. 1913, 16–24.

HOWITT, William (*fl.* 1850s)
Father of A. W. Howitt (1830–1908). Returned to England from Australia, 1854. *Land, Labour and Gold* 1855 (includes an account of flora, fauna, geology, etc.).
Victorian Nat. 1973, 172–73.

HOWLAND, Joseph (*fl.* 1790s)
Seedsman, High Wycombe, Bucks.

HOWLETT, Charles J. (–1951)
d. Feltham, Middx 7 June 1951
Kew gardener, 1892. Uitenhage, S. Africa, 1894. Curator, Botanic Gardens, Graaff Reinet, 1899. Manager, Schaapkraal Nurseries, Tarkasted, Cape Province, 1915–32. Collected plants for Kew.
J. Kew Guild 1951, 41.

HOWSE, Richard (1821–1901)
Curator, Natural History Museum, Newcastle. Secretary, Tyneside Naturalists' Field Club. Published Catalogue of fossil plants from the Hutton Collection in *Northumberland, Durham and Newcastle-upon-Tyne Nat. Hist. Trans.*
Nature v.63, 1901, 499.

HOWSON, Rev. John (1817–1866)
b. Giggleswick, Yorks 2 Sept. 1817 *d.* Penrith, Cumberland 1 March 1866
Headmaster, Giggleswick Grammar School, Yorks. Fellow botanist of J. Windsor. Dean of Chester. *Illustrated Guide to... Craven with a Local Fl.* 1850.
J. Windsor *Fl. Cravoniensis* 1873, viii.

HOY, James Barlow (–1843)
d. 13 Aug. 1843
ALS 1788. FLS 1793. Brother of T. Hoy (*c.* 1750–1822). Gardener, Gordon Castle, Elgin.
J. Sowerby and J. E. Smith *English Bot.* 146, 289. *Trans. Linn. Soc.* v.2, 1794, 354–55.

HOY, Thomas (*c.* 1750–1822)
d. Isleworth, Middx 1 May 1822
FLS 1788. Gardener to Duke of Northumberland at Syon House, Middx. "An experienced botanist and able cultivator."

Trans. Linn. Soc. v.1, 1791, 249–54. *Mem. Wernerian Nat. Hist. Soc.* v.1, 1811, 26. G. J. Aungier *Hist. and Antiq. Syon Monastery* 1840, 170. *Gdnrs Chron.* v.163(17), 1968, 16.
Hoya R. Br.

HOYLE, G. W. (*c.* 1801–1872)
d. 26 May 1872
Raised pelargoniums. Secretary, Reading Horticultural Society.
Garden v.30, 1886, 415 portr.

HUBBARD, William (–1787)
Nurseryman, Market Harborough, Leics.

HUBERT, F. J. (*fl.* 1890s–1910s)
Nurseryman, Guernsey. In partnership with W. Mauger until 1897.

HUDSON, Alfred Wickens (1856–1928)
d. Cranbrook, Kent 11 Dec. 1928
Chemist and druggist. At Cranbrook, 1893–1928. Contrib. to F. J. Hanbury and E. S. Marshall's *Fl. Kent* 1899.
Pharm. J. 1928, 635.

HUDSON, Charles Edward (1896–1972)
b. London 1896 *d.* 6 March 1972
VMH 1953. Head of Dept. of Horticulture, Hertfordshire Institute of Agriculture, *c.* 1922; Vice-principal. Lecturer in Horticulture and Head of Horticultural Dept., Yorkshire Council for Agricultural Education, 1934. Head of Horticultural Dept., Midland Agricultural College, 1942. Entered N.A.A.S., 1946. Chief Horticultural Officer, Ministry of Agriculture. President, Horticultural Education Association, 1932.
Sci. Hort. v.24, 1972–73, 5–8 portr.

HUDSON, James (1846–1932)
b. Horsted Place, Sussex 29 May 1846 *d.* 28 May 1932
VMH 1897. Gardener at Gunnersbury House, Middx, 1876, first to H. J. Atkinson, then to Leopold de Rothschild.
Gdnrs Mag. 1907, 795–96 portr. *J. Hort. Home Farmer* v.62, 1911, 308–9 portr. *Gdnrs Chron.* 1919 i 194 portr.; 1932 i 432.

HUDSON, John (*fl.* 1880s–1900s)
Nurseryman, Leicester.
Gdnrs Mag. 1910, 663–64 portr.

HUDSON, Samuel H. (*c.* 1827–1904)
d. Epworth, Lincs 7 April 1904
Botanised in Lincolnshire.

HUDSON, William (1734–1793)
b. Kendal, Westmorland 1734 *d.* London 23 May 1793
FRS 1761. FLS 1791. Praefectus et praelector Chelsea Physic Garden, 1765–71. *Fl. Anglica* 1762; ed. 3 1798.
R. Pulteney *Hist. Biogr. Sketches of Progress of Bot. in England* v.2, 1790, 351–52. *Annual Register* 1793, 25–26. *Gent. Mag.* v.63, 1793, 485. *Trans. Linn. Soc.* v.4, 1798, 278. J. Nichols *Lit. Anecdotes of Eighteenth Century* v.9, 1815, 565–66. H. Davies *Welsh Botanology* 1813, iv. G. C. Gorham *Mem. of John and Thomas Martyn* 1830, 276. P. Smith *Mem. and Correspondence of Sir J. E. Smith* v.1, 1832, 161. H. Trimen and W. T. T. Dyer *Fl. Middx* 1869, 392. R. H. Semple *Mem. of Bot. Gdn at Chelsea* 1878, 88–93. G. C. Druce *Fl. Berkshire* 1897, cxxxvii–cxxxix. *D.N.B.* v.28, 155. F. H. Davey *Fl. Cornwall* 1909, xxiii–xxxiv. J. W. White *Fl. Bristol* 1912, 62–63. J. R. Green *Hist. Bot. in United Kingdom* 1914, 231–33. *J. Bot.* 1914, 320; 1927, 128. G. C. Druce *Fl. Oxfordshire* 1927, xcii. G. C. Druce *Fl. Northamptonshire* 1930, lxxv. C. E. Salmon *Fl. Surrey* 1931, 48. W. K. Martin and G. T. Fraser *Fl. Devon* 1939, 770. H. J. Riddelsdell *Fl. Gloucestershire* 1948, cxii. *Taxon* 1970, 526. C. C. Gillispie *Dict. Sci. Biogr.* v.6, 1972, 538–39.
Some plants sold with A. B. Lambert's herb. Herb. mostly burnt; remainder at BM(NH).
Hudsonia L.

HUGGAN, John (*fl.* 1770s)
MD Edinburgh 1771. Sent British plants to J. Hope (1725–1786).
Notes R. Bot. Gdn Edinburgh 1907, 125.

HUGHES, Dorothy *see* Poponoe, D. K.

HUGHES, E. Griffiths (*c.* 1835–1899)
d. Higher Broughton, Lancs 2 Dec. 1899
Chemist. *Sources of Plant Food* 1896.
Gdnrs Chron. 1899 ii 440.

HUGHES, Rev. Griffith (*fl.* 1707–1750s)
b. Towyn, Merioneth 1707
MA Oxon 1748. FRS 1750. Anglican mission, Pennsylvania, 1732–36. Rector, St. Lucy's, Barbados, 1736. *Natural History of Barbados* 1750 (plants, pp. 97–256; glossary mostly by P. Miller).
D.N.B. v.28, 175–76. I. Urban *Symbolae Antillanae* v.3, 1902, 62. *National Library of Wales J.* v.3, 1943, 19–22. *Hist. Mag. Protestant Episcopal Church* 1948, 151–63. *Osiris* 1948, 98–99. *Dict. Welsh Biogr. down to 1940* 1959, 374.

HUGHES, Joseph (*fl.* 1770s–1780s)
Nurseryman, Lea Bridge Road Nursery, Leyton, Essex, 1775–82.
Trans. London Middlesex Archaeol. Soc. v.24, 1973, 189–90.

HUGHES, William (*fl.* 1650s–1680s)
Lived in Jamaica. Visited Florida, *c.* 1652. Gardener to Viscountess Conway at Ragley. *Compleat Vineyard* 1665. *Flower-garden Enlarged* 1672; ed. 3 1683. *American Physitian; or a Treatise of Roots, Plants...in the English Plantations in America* 1672. *Flower-garden and Compleat Vineyard* 1683.
G. W. Johnson *Hist. English Gdning* 1829, 110–11. *D.N.B.* v.28, 190. *Notes and Queries* v.159, 1930, 358.

HUGHES-BULLER, R. (*fl.* 1900s)
Indian Forest Service. Collected plants in Baluchistan. Co-author with Rai Bahadur Diwan Jamiat Rai of botanical sections of Bolan, Makran and Las Bela Gazetteers, *c.* 1905.
Pakistan J. For. 1967, 348.

HULL, John (1761–1843)
b. Poulton-le-Fylde, Lancs 1761 *d.* London 17 March 1843
MD Leyden 1792. FLS 1810. LRCP 1819. Physician, Manchester. *British Fl.* 1799; ed. 2 1808. *Elements of Botany* 1800 2 vols.
Proc. Linn. Soc. 1839, 34. W. Munk *Roll of R. College of Physicians* v.3, 1878, 195. *D.N.B.* v.28, 195. *Trans. Liverpool Bot. Soc.* 1909, 77.
Plants at BM(NH).

HULLETT, Richmond William (1843–1914)
b. 15 Nov. 1843
FLS 1888. Schoolmaster. Principal, Raffles Institute, Singapore. Member of Gardens Committee, Singapore Botanic Gardens. Collected plants in Johore with Sir G. King in Java, Borneo, etc.
Gdns Bull. Straits Settlements 1927, 124.
Herb. at Singapore. Plants at Calcutta, Kew.
Hullettia King.

HULME, Frederick Edward (1841–1909)
b. Hanley, Staffs 29 March 1841 *d.* Kew, Surrey 11 April 1909
FLS 1869. Art master, Marlborough College, 1870. Professor of Drawing, King's College, 1885. *Series of Sketches from Nature of Plant Form* 1868. *Plants, their Natural Growth and Ornamental Treatment* 1874. *Suggestions in Floral Design* 1878–79. *Familiar Wild Flowers* 1878–1905 8 vols. *Flower Painting in Water Colours c.* 1883. *Wild Fruits of the Countryside* 1902. *Wild Flowers in the Seasons* 1907. *Familiar Swiss Flowers* 1908. *That Rock-garden of Ours* 1909. Illustrated J. S. Hibberd's *Familiar Garden Flowers* 1879–97 5 vols.
Proc. Linn. Soc. 1908–9, 41–42. *J. Bot.* 1909, 235. *D.N.B. Supplt 2* v.2, 321–22. *Who was Who, 1897–1916* 359–60.
Sale at Sotheby, 25 July 1899.

HULME, J. R. (*fl.* 1840s)
MD. Practised medicine at Scarborough, Yorks. *Scarborough Algae* 1842 (exsiccatae).

HUMBERSTON, Francis Mackenzie, Lord Seaforth and Mackenzie (1754–1815)
d. near Edinburgh 11 Jan. 1815
FRS 1794. FLS 1796. Lieut.-General. Governor of Barbados 1800–6. Sent algae to D. Turner (*Fuci* v.2, 1809, 130) and plants to A. B. Lambert. *List of West Indian Plants* (*B.M. Add. MS. 28610*).
R. Brown *Prodromus* 1810, 267. H. C. Andrews *Bot. Repository* 1807–8, t.502. *D.N.B.* v.28, 204–6. I. Urban *Symbolae Antillanae* v.3, 1902, 125. *J. Bot.* 1912, 171.
Letters in Smith correspondence at Linnean Society.
Seaforthia R. Br.

HUME, Sir Abraham (1748/9–1838)
b. London 20 Feb. 1748/9 *d.* Wormleybury, Herts 24 March 1838
FRS 1775. With assistance of his gardener, James Mean, cultivated many rare exotic plants at Wormleybury, Herts. Imported greenhouse plants and varieties of chrysanthemum from China.
E. Bretschneider *Hist. European Bot. Discoveries in China* 1898, 211–12. *D.N.B.* v.28, 208–9. *J. R. Hort. Soc.* 1941, 308–12. E. H. M. Cox *Plant Hunting in China* 1945, 47–48.

HUME, Allan Octavian (1829–1912)
b. St. Mary Cray, Kent 4 June 1829 *d.* Norwood, Surrey 31 July 1912
FLS 1901. FZS. CB 1860. Indian Civil Service, 1849–82; returned to England, 1890. Interested in Indian birds and big game. Founded South London Botanical Institute, 1907; his botanical books and residue of herb. there.
F. H. Davey *Fl. Cornwall* 1909, lxi. *Bot. Soc. Exch. Club Br. Isl. Rep.* 1912, 201–3. *J. Bot.* 1912, 347–48. *Nature* v.89, 1912, 584. *Proc. Linn. Soc.* 1912–13, 60–61. W. Wedderburn *A. O. Hume* 1913 portr. *D.N.B. 1912–1921* 277–78. *R.S.C.* v.7, 1036; v.10, 292; v.12, 353; v.15, 992.

HUME, Lady Amelia (*née* **Egerton**) (1751–1809)
b. 25 Nov. 1751 *d.* London 8 Aug. 1809
Wife of Sir A. Hume (1748/9–1838). Received plants from China and from W. Roxburgh for her garden. Pupil of J. E. Smith who dedicated his *Spicilegium Botanicum* 1791 to her.
J. E. Smith *Exotic Bot.* 1805, t.1. E. Bretschneider *Hist. European Bot. Discoveries in China* 1898, 212–13. *D.N.B.* v.28, 209. *J. R. Hort. Soc.* 1964, 497–99 portr.
Portr. by J. Reynolds. Portr. at Hunt Library.
Humea Smith.

HUMPHREY, William (-before 1792)
Of Norwich. Friend of J. E. Smith. "Senex optimus mihique olim familiarissimus" (J. E. Smith *Spicilegium Botanicum* 1791, t.12). Discovered *Battarea phalloides*. Contrib. to J. Sowerby and J. E. Smith's *English Bot.* 182, 805, 956.
Trans. Linn. Soc. v.7, 1804, 297. *Trans. Norfolk Norwich Nat. Soc.* 1937, 194–95.

HUMPHREYS, Rev. D. (*fl.* 1680s)
Of Anglesey. Sent specimens of marine algae to J. Bobart.
R. Morison *Plantarum Historiae Universalis Oxoniensis* v.3, 1699, 646, 647.

HUMPHREYS, Henry Noel (1810–1879)
b. Birmingham 4 Jan. 1810 *d.* London 10 June 1879
Artist and designer. Also interested in gardening design. *River Gardens* 1857. Contrib. flower drawings to Mrs. J. Loudon's *British Wild Flowers* 1846, etc., *Floral Cabinet and Mag. of Exotic Bot.* 1837–40, *Garden* 1872.
Garden v.15, 1879, 486–87; v.18, 1880, xii portr. *Gdnrs Chron.* 1879 i 766. *D.N.B.* v.28, 249.
Portr. at Hunt Library.

HUMPHREYS, John (*c.* 1850–1937)
b. Llanfyllin, Montgomeryshire *c.* 1850 *d.* Edgbaston, Birmingham June 1937
Dentist. Lecturer in Dental Anatomy, Birmingham until 1919. Lecturer in Medieval Archaeology, Birmingham, 1924.
Times 1 June 1937.
Worcestershire plants at Birmingham Museum.

HUMPHREYS, Thomas (*c.* 1867–1932)
b. Cheshire *c.* 1867 *d.* Birmingham 31 Oct. 1932
Kew gardener, 1887. Assistant Superintendent, Royal Horticultural Society gardens, Chiswick, 1892. Curator, Botanic Gardens, Birmingham, 1903.
Gdnrs Chron. 1903 ii 53–54 portr.; 1932 ii 347, 350. *Gdnrs. Mag.* 1903, 490 portr.; 1908, 809–10 portr. *J. Hort. Cottage Gdnr* v.47, 1903, 80 portr. *J. Kew Guild* 1933, 278–79 portr.

HUMPHRIES, Charles Henry (*fl.* 1890s)
Kew gardener, 1892. Curator, Botanic Station, Aburi, Gold Coast, 1895.
Kew Bull. 1895, 155.

HUMPIDGE, F. C. (1874–1944)
Of Stroud, Glos. Botanised in Forest of Dean.
H. J. Riddelsdell *Fl. Gloucestershire* 1948, clii.

HUNGERFORD, John (*fl.* 1656–1680s)
b. Reading, Berks *c.* 1656
BA Oxon 1677. FRCP 1687. MD. Of Oxford. Collected plants at Montpellier, France.
W. Munk *Roll of R. College of Physicians* v.1, 1878, 473. J. E. Dandy *Sloane Herb.* 1958, 141.
Plants at BM(NH).

HUNKIN, Rev. Joseph Wellington (1887–1950)
b. Truro, Cornwall 25 Sept. 1887 *d.* London 28 Oct. 1950
Educ. Cambridge. Rector, Rugby, 1927. Bishop of Truro, 1935–50. Had garden at Lis Escop, Truro. Contrib. to *J. R. Hort. Soc.*
Gdnrs Chron. 1950 i 122 portr.; 1950 ii 194. *Sunday Times* 29 Oct. 1950. *Times* 30 Oct. 1950. *J. R. Hort. Soc.* 1951, 264–66 portr. *Who was Who, 1940–1950* 577.

HUNNEMAN, John (*c.* 1760–1839)
d. London 5 March 1839
ALS 1831. London bookseller and agent for dried plants. Responsible for introduction of new plants (*Bot. Mag. passim*).
Gdnrs Mag. 1829, 162–63; 1839, 208, 302. *Bot. Mag.* 1831, t.3061. B. Maund *Bot. Gdn* v.5, 1833, 423. R. Sweet *Br. Flower Gdn* v.3, 1838, 276. *Proc. Linn. Soc.* 1839, 36.

HUNNYBUN, Edward Walter (1848–1918)
b. Norwich, Norfolk 21 Nov. 1848 *d.* Ventnor, Isle of Wight 3 July 1918
Solicitor. Illustrated C. E. Moss's *Cambridge British Fl.* 1914. 'Allium sphaerocephalum' (*J. Bot.* 1912, 64–65).
J. Bot. 1904, 318; 1914, 134; 1918, 248–50. *Bot. Soc. Exch. Club Br. Isl. Rep.* 1918, 356–57. *Watson Bot. Exch. Club Rep.* 1917–18, 48 portr. *Annual Rep. Huntingdonshire Fauna Fl. Soc.* 1969, 8–10.
Plants at Cambridge.
Ulmus nitens var. *Hunnybuni* Moss.

HUNT, Francis (-1662)
Gardener at Putney, Surrey. Founded Putney Nursery *c.* 1650. Business continued by his son, Francis.

HUNT, Francis (1652–1713)
d. June 1713
Acquired Putney Nursery, established by his father, Francis (-1662).

HUNT, Francis (1691–1763)
d. July 1763
Son of nurseryman, F. Hunt (1652–1713) whose business he acquired and developed. Member of London Society of Gardeners.
Surrey Archaeol. Collections v.69, 1973, 135–38. *Trans. London Middlesex Archaeol. Soc.* 1973, 191.

HUNT, Francis (*c*. 1729–1775)
d. Feb. 1775
Son of F. Hunt (1691–1763). Nurseryman, Putney. On his death the family nursery was sold to William Howey (*c*. 1729–1792).

HUNT, George Edward (*c*. 1841–1873)
d. Bowdon, Cheshire 26 April 1873
Bank clerk. Bryologist and conchologist. 'Botany of Mere, Cheshire' (*Manchester Lit. Philos. Soc. Proc*. 1871, 50–52). Contrib. to *Manchester Lit. Philos. Soc. Proc*. and *J. Bot*.
J. Bot. 1873, 191. L. H. Grindon *Manchester Banks and Bankers* 1877, 310. *Trans. Liverpool Bot. Soc*. 1909, 77. *R.S.C*. v.7, 1037; v.12, 293.
Mosses at BM(NH). Plants at Oxford. Letters at Kew and in Wilson correspondence at BM(NH).
Andreaea huntii Limpr.

HUNT, John (*fl*. 1790s–1820s)
Seedsman, 53 High Street, Borough, London.

HUNT, Samuel (–1763)
Nurseryman, Putney, Surrey. Brother of F. Hunt (1691–1763).

HUNT, Thomas Carew (–1886)
d. 12 Jan. 1886
Consul at Archangel, 1832; Azores, 1839–48; Bordeaux, 1866. Plants which he collected in Azores were distributed through Botanical Society of London.
London J. Bot. 1847, 381 *et seq*. F. D. C. Godman *Nat. Hist. Azores* 1870, 117. *Bot. Soc. Exch. Club Br. Isl. Rep*. 1921, 454–56. *Gdnrs Chron*. 1921 ii 119–20. *Kew Bull*. 1922, 47.
Plants at BM(NH), Kew.
Ammi huntii Watson.

HUNTBACK, James (*fl*. 1730s–1740s)
Gardener to Lord Petre at Ingatestone and Thorndon, Essex,1731–38.
Essex Nat. 1969, 202–3.

HUNTER, Alexander (1729–1809)
b. Edinburgh 1729 *d*. York 17 May 1809
MD Edinburgh 1753. FRS 1775. Practised medicine at York from 1763. Edited J. Evelyn's *Sylva* 1776, 1786, 1801, 1812 (with life of A. Hunter). *Illustration of Analogy between Vegetable and Animal Parturition* c. 1800. *Georgical Essays in which Food of Plants is particularly considered* 1770–72 4 vols. *New Method for Raising Wheat* 1796.
J. Donaldson *Agric. Biogr*. 1854, 57–58. *D.N.B*. v.28, 283–84.

HUNTER, Charles (1888–1926)
b. Great Aytoun, Yorks 23 July 1888 *d*. Bristol 23 Sept. 1926
MSc Durham. FLS 1923. Plant physiologist. Assistant Lecturer, Bristol University, 1912; Lecturer, 1919.
Proc. Linn. Soc. 1926–27, 88–89.

HUNTER, Edward (*fl*. 1790s–1820s)
ALS 1790. Steward at Caen Wood, Hampstead. Contrib. list of plants to J. J. Park's *Topography and Natural History of Hampstead* 1814.

HUNTER, Frederick Mercer (*fl*. 1870s–1880s)
Major. Collected plants in Aden. *Aden Handbook* 1873. *Account of British Settlement at Aden* 1877.
Rec. Bot. Survey of India v.7, 1914, 16–17.
Plants at Kew.

HUNTER, Herbert (*c*. 1883–1959)
d. Meldreth, Cambridge 21 Feb. 1959
DSc Leeds. Head, Plant Breeding Division, Dept. of Agriculture, Ireland, 1903–19. Senior Assistant, Plant Breeding Institute, Cambridge, 1924–36; Director, 1936–46. Researched on barley breeding. *Crop Varieties* 1954. *Barley Crop* 1956.
Chronica Botanica 1937, 157 portr. *Nature* v.183, 1959, 1016–17.

HUNTER, James (*fl*. 1860s–1900s)
Partner in nursery firm of Austin and McAslan, Glasgow.
Gdnrs Chron. 1917 ii 147–48 portr.

HUNTER, James Augustus (*fl*. 1780s–1820s)
Nurseryman, 25 later 18 High Street, Birmingham in partnership with Brunton and Forbes. In 1798 Hunter set up business on his own and was bankrupt by 1821.
Bot. Mag. 1818, t.2007.

HUNTER, James Henry (*c*. 1885–1924)
d. Njala, Sierra Leone 21 May 1924
MA Aberdeen 1907. Lecturer, Njala Agricultural College, 1923. Collected plants in Sierra Leone.
Gdnrs Chron. 1924 i 352.

HUNTER, John (1728–1793)
b. Kilbride, Lanark 14 Feb. 1728 *d*. London 16 Oct. 1793
FRS 1767. Surgeon and anatomist. *Memoranda on Vegetation* 1860.
Proc. Linn. Soc. 1888–89, 37. *Gdnrs Chron*. 1895 i 236. *D.N.B*. v.28, 287–93. G. C. Peachey *Mem. of William and John Hunter* 1924 portr.
MSS. at Royal Society. Portr. at Apothecaries Hall.

HUNTER, John (1737–1821)
b. Leith, Midlothian 29 Aug. 1737 *d.* London 13 March 1821
Educ. Edinburgh University. Captain, Royal Navy. Arrived in New South Wales, 1788; Governor, 1795–1800. Vice-Admiral, 1810. Accomplished natural history artist. Sent natural history specimens to Sir J. Banks. *Historical Journal of Transactions at Port Jackson and Norfolk Island* 1793.
Hist. Rec. N.S.W. v.3, 1895, *passim. D.N.B.* v.28, 294. *J. Proc. R. Austral. Hist. Soc.* 1901, 21; v.14, 344–62; v.50, 32–57. *Austral. Mus. Mag.* v.6, 291–304. *Austral. Dict. Biogr.* v.1, 1965, 566. *Austral. Encyclop.* v.5, 1965, 32. *Emu* 1965, 83–95.
Drawings entitled, 'Birds and Flowers of New South Wales' 1788–90 at National Library, Canberra.

HUNTER, Robert (*fl.* 1790s)
Nurseryman, Hexham, Northumberland.

HUNTER, Robert (*fl.* 1850)
Merchant at Bangkok. Sent specimens of cardamons to D. Hanbury, now at Kew.
J. Thailand Res. Soc. Nat. Hist. Supplt v.12, 1939, 11.

HUNTER, Rev. Robert (1823–1897)
b. Newburgh, Fife 3 Sept. 1823 *d.* Epping, Essex 25 Feb. 1897
MA Aberdeen 1840. LLD 1883. Lexicographer and theologian. Missionary, Free Church of Scotland in India, 1846–55. Collected plants in Bermuda, 1844. 'Bermuda Ferns' (*J. Bot.* 1877, 367).
J. Bot. 1897, 158–59. *D.N.B. Supplt. 1* v.3, 14–15. *R.S.C.* v.3, 476; v.10, 295; v.15, 998.
Plants at BM(NH).

HUNTER, Robert E. (–before 1847)
MD. Of Margate, Kent. *Description of Isle of Thanet* 1796 (includes list of plants). MS. Fl. of Thanet.
T. B. Flower *Fl. Thanetensis* 1847, vi.

HUNTER, Rev. Sylvester Joseph (1829–1896)
b. Bath, Somerset 13 Sept. 1829 *d.* Blackburn, Lancs 20 June 1896
MA Cantab. S.J. 'Conjugation in Spirogyra' (*J. Bot.* 1885, 185).
Stonyhurst Mag. July 1896. *Trans. Liverpool Bot. Soc.* 1909, 77. *J. Bot.* 1908, 187.
Plants at St. Benno's College, St. Asaph.

HUNTER, Thomas (–1965)
d. April 1965
Kew gardener, 1908. Curator, Botanic Gardens, Aburi, Gold Coast, 1911; Kusami Agricultural Station, 1913–31.
Kew Bull. 1911, 348. *J. Kew Guild* 1965, 603 portr.

HUNTER, William (*fl.* 1790s)
Gardener and seedsman, Mansfield, Notts.

HUNTER, William (1755–1812)
b. Montrose, Angus 1755 *d.* Batavia, East Indies Dec. 1812
MA Aberdeen 1777. MD 1808. Ship's surgeon in East, 1781. Medical service of East India Company, 1783–1812; Chief of Medical Service, Java, 1811. 'Nauclea Gambir' (*Trans. Linn. Soc.* v.9, 1808, 218–24). MS. 'Fl. of Prince of Wales's Island, Penang' 1803 at BM(NH) (published in *J. R. Asiatic Soc., Straits Branch* no. 53, 1909, 49–127).
W. Roxburgh *Fl. Indica* v.2, 1824, 532–33. *D.N.B.* v.28, 305. *J. Bot.* 1906, 235; 1916, 143–44. D. G. Crawford *Hist. Indian Med. Service, 1600–1913* v.1, 1914, 207. *Gdns Bull. Straits Settlements* 1927, 125, 146. *Fl. Malesiana* v.1, 1950, 252–53. *R.S.C.* v.3, 476.
Plants at BM(NH).
Hunteria Roxb.

HUNTER-WESTON, A. Gould (*fl.* 1890s)
Lieut.-Colonel, Royal Engineers. Collected plants in Kashmir and Baltistan, 1889–90, now at BM(NH) and Kew.
Pakistan J. For. 1967, 348.

HUNTINGDON, Henry of *see* Henry of Huntingdon

HUNTINGDON, Rev. Robert (*fl.* 1680s)
MA Oxon. Fellow of Merton College. Chaplain at Aleppo, 1671–81. Provost of Trinity College, Dublin and Bishop of Raphoe. Sent plants from Aleppo to Bobart at Oxford.
R. Morison *Plantarum Historiae Universalis Oxoniensis* v.3, 1699, 82, 87, 105, 144, 166, 604, etc. G. C. Druce and S. H. Vines *Account of Herb. Univ. Oxford* 1897, 49. C. E. Raven *John Ray* 1942, 276.
Plants at Oxford.

HUNTLEY, Rev. J. T. (*fl.* 1820s)
Of Kimbolton, Hunts. Had garden of rare plants, especially orchids.
Bot. Mag. 1828, t.2799. *Bot. Register* v.23, 1837, t.1991.
Huntleya Bateman ex Lindl.

HUNTLY, Mariette Antoinetta Pegus, Marchioness (1821–1893)
b. Uffington 1821 *d.* Orton, Northants 1893
Made collection of central European plants and flower drawings.
G. C. Druce *Fl. Northamptonshire* 1930, cxx.

HURRY, Jamieson Boyd (1857–1930)
b. Liverpool 8 June 1857 *d.* Bournemouth, Hants 13 Feb. 1930

BA Cantab 1879. MD 1885. Medical Officer of Health, Reading. *Woad Plant and its Dye* 1930.
Bot. Soc. Exch. Club Br. Isl. Rep. 1930, 325.

HURST, Cecil Prescott (–1956)
FLS 1928. Contrib. plant records to D. Grose's *Fl. Wiltshire* 1957.
Proc. Linn. Soc. 1956–57, 242–43.

HURST, Charles Chamberlain (1870–1947)
d. near Horsham, Sussex 17 Dec. 1947
DSc Cantab. Hybridised orchids. Researched on Mendelian inheritance in animals and plants, particularly *Rosa*. Unpublished monograph on *Rosa*. Contrib. chapters on evolution of garden roses to G. S. Thomas's *Old Shrub Roses* 1955. *Orchid Stud Book* (with R. A. Rolfe) 1909. *Genetics* 1925. *Mechanics of Creative Evolution* 1932. Contrib. to *Orchid Rev.*
Gdnrs Chron. 1921 ii 296 portr.; 1948 i 16. *Orchid Rev.* 1934, 226–28 portr. *Times* 24 Dec. 1947. *Nature* v.161, 1948, 46–47. *Who was Who, 1941–1950* 580.

HURST, Henry Alexander (*c.* 1825–1882)
d. Liverpool 1882
Merchant. Of Knutsford and Liverpool. Collected with Letourneaux in Egypt; also collected plants in India, 1862; Madeira, 1868. Contrib. to *Mem. Manchester Lit. Philos. Soc.*
J. Bot. 1867, 63. *Proc. Manchester Lit. Philos. Soc.* v.18, 1878–79, 133–36. *Trans. Liverpool Bot. Soc.* 1909, 77–78. *R.S.C.* v.7, 1041.
Herb. at Manchester.

HURST, Samuel John (1864–1936)
d. Boston, Lincs 22 Oct. 1936
Druggist. Botanist.
Pharm. J. 1936, 471.

HURST, William (*c.* 1799–1868)
d. 24 Dec. 1868
Seedsman and florist, 6 Leadenhall Street, London.

HURST, William (*c.* 1831–1882)
d. London 11 Feb. 1882
Seedsman, Houndsditch, London.
J. Hort. Cottage Gdnr v.4, 1882, 132.

HUSKINS, Charles Leonard (*c.* 1898–1953)
b. Walsall, Staffs *d.* Madison, U.S.A. 26 July 1953
MS Alberta. Family emigrated to Alberta when he was 9. Studied for PhD under R. R. Gates at King's College, London. Research Fellow, John Innes Horticultural Institution, 1927–30. Associate Professor of Botany, McGill University, 1930. Professor of Botany, Wisconsin, 1945. Geneticist.
Nature v.173, 1954, 59–60.

HUSSEY, A. M. (*née* **Reed**)
(–before 1859)
Of Hayes, Kent. *Illustrations of British Mycology* 1847–55.
Letters in Berkeley correspondence at BM(NH).
Husseia Berkeley.

HUSSEY, Benjamin (*fl.* 1760s)
Collected plants in Falkland Islands, 1767.
R. A. Salisbury *Prodromus Stirpium* 1796, 89, 90.

HUSSEY, Jessie L. (1862–1899)
b. Port Eliot, S. Australia 5 June 1862 *d.* Port Eliot 16 March 1899
Collected algae in S. Australia. Correspondent of F. von Mueller and J. G. Agardh.
Rep. Austral. Assoc. Advancement Sci. 1911, 230.
Herb. at Adelaide Museum. Plants at BM(NH).
Crysymenia husseyana Agardh.

HUSSEY, William (*fl.* 1850s)
Nurseryman, Norwich, Norfolk.

HUTCHINGS, James Mason (1818–1902)
b. Towcester, Northants 1818 *d.* Yosemite Valley, U.S.A. 2 Nov. 1902
Went to California, 1849. Gold miner turned writer. Established first hotel in Yosemite Valley. Interested in natural history. Collected *Lathyrus splendens* near Campo, California. *Scenes of Wonder and Curiosity in California* 1860. Edited *Hutchings California Magazine* 1856–61.
F. Walker *San Francisco's Lit. Frontier* 1939 *passim*. F. P. Farquhar *Yosemite* 1948, 18–21, 76–77.
Plant collection destroyed in San Francisco earthquake, 1906.

HUTCHINS, Sir David Ernest (1850–1920)
b. 22 Sept. 1850 *d.* New Zealand 11 Nov. 1920
Indian Forest Service, 1870–85. Forest Service, Cape Colony, 1885; later Director of Forest Operations, British East Africa. Collected plants in Kenya, 1903–7. In W. Australia, 1914 and New Zealand, 1916. Knighted 1920. *Descriptive Catalogue of Best Trees to Plant in Cape Colony* 1899. *Report on Forests of Kenia* 1907. *Report on Cyprus Forestry* 1909.
Nature v.106, 1920, 540–41. *Indian Forester* 1921, 125–27. *J. Bot.* 1921, 29–30. *Kew Bull.* 1921, 32–33. *Trans. Proc. N.Z. Inst.* 1921, vii–viii portr. *Who was Who, 1916–1928* 534.
Brachylaena hutchinsii Hutchinson.

HUTCHINS, Miss Ellen (1785–1815)
b. Ballylickey, County Cork 1785 *d.* Bantry, County Cork 1815

Algologist and bryologist. "She could find almost anything" (J. E. Smith). Contrib. to J. Sowerby and J. E. Smith's *English Bot.* 1915, 2480, 2523, 2652, etc., L. W. Dillwyn's *Br. Confervae* 1802 and D. Turner's *Fuci* 1802 for which she drew some plates.

W. H. Harvey *Phycologia Britannica* 1846, t.124. *J. Bot.* 1912, 63. *Proc. R. Irish Acad. B.* v.32, 1915, 70–71; v.38, 1929, 182. *Bryologist* 1918, 78–80. R. L. Praeger *Some Irish Nat.* 1949, 107.

Drawings of algae and letters at Kew. Algae at BM(NH).

Hutchinsia R. Br.

HUTCHINS, John (*fl.* 1830s)
Nurseryman, Tilbrook, Hunts.

HUTCHINS, Thomas (*fl.* 1770s)
Chief factor in Hudson Bay Company. MS. 'Observations on Hudson's Bay' at Library of Hudson Bay Company. Sent plants to Sir J. Banks, now at BM(NH).

J. Bot. 1922, 239, 336–37.

HUTCHINSON, Claude Mackenzie (1869–1941)
b. 29 April 1869 *d.* 2 Aug. 1941

Educ. Cambridge. Lecturer in Agricultural Chemistry, Colonial College, 1898–1904. Mycologist, Indian Tea Association, Assam, 1904–9. Bacteriologist, Agricultural Institute, Pusa, 1909–26. Chief Scientific Adviser, Imperial Chemical Industries (India), 1926–31. Worked on influence of bacteria on soil fertility, nitrogen fixation, green manures, etc.

Nature v.148, 1941, 367. *Who was Who, 1941–1950* 581. *Chronica Botanica* 1942–43, 183.

HUTCHINSON, John (1884–1972)
b. Wark-on-Tyne, Northumberland 7 April 1884 *d.* Kew, Surrey 2 Sept. 1972

FRS 1947. FLS 1918. VMH 1944. VMM 1945. Kew gardener, 1904. Assistant, Kew Herb., 1905; Keeper of Kew Museums, 1936–48. Plant taxonomist. Botanised in S. Africa, 1928–29 and 1930. Illustrated many of his books. *Families of Flowering Plants* 1926–34 2 vols; ed. 3 1973. *Fl. W. Tropical Africa* (with J. M. Dalziel) 1927–36 2 vols. *Common Wild Flowers* 1945. *More Common Wild Flowers* 1948. *Uncommon Wild Flowers* 1950. *Botanist in Southern Africa* 1946. *British Flowering Plants* 1948. *Story of Plants and their Uses to Man* (with R. Melville) 1948. *Genera of Flowering Plants* 1964–67 2 vols. *Evolution and Phylogeny of Flowering Plants* 1969. Contrib. to *Kew Bull.*

Kew Bull. 1934, 344; 1949, 18; 1974, 1–14 portr. *Herbertia* 1939, 27–30 portr. *Gdnrs Chron.* 1940 i 171–72; 1948 i 180 portr.; 1964 ii 564. *J. Kew Guild* 1948, 656; 1959, 650–51. *Nature* v.161, 1948, 799–800; v.240, 1972, 367–68 portr. *Proc. Linn. Soc.* 1966, 104–5. *Modern Men of Sci.* Ser. 2, 1968, 253–54 portr. *Bull. Bot. Survey India* 1971, 167–68 portr. Hunt Library *Cat. 3rd International Exhibition of Bot. Art* 1972, 79 portr. *Times* 9 Sept. 1972. *Bothalia* 1973, 1–3 portr. *J. R. Hort. Soc.* 1973, 30–31. *Taxon* 1973, 705–6. *Watsonia* 1973, 408. *Biogr. Mem. Fellows R. Soc.* 1975, 345–57 portr.

Plants, drawings and MS. autobiography at Kew. Portr. at Kew, Hunt Library.

HUTCHINSON, Peter Orlando (1810–1897)
b. Winchester, Hants 17 Nov. 1810 *d.* Sidmouth, Devon 1 Oct. 1897

Antiquarian and botanist. *Ferns of Sidmouth* 1862.

Rep. Trans. Devon Assoc. 1903, 338–52. *J. Bot.* 1931, 137.

HUTCHINSON, Robert Russell (1870–1951)
b. 28 Aug. 1870 *d.* 12 June 1951

Bank clerk, Tunbridge Wells and Wallingford.

Yb. Bot. Soc. Br. Isl. 1953, 92.

Herb. at Croydon Natural History and Scientific Society.

HUTCHINSON, Rev. Thomas (1846–1916)
MA. Vicar, Kimbolton. Botanised in Herefordshire.

Trans. Woolhope Nat. Field Club v.34, 1954, 254.

HUTCHINSON, Rev. Thomas Neville (*fl.* 1890s)
Participated in botanical work of Rugby School Natural History Society.

J. E. Bagnall *Fl. Warwickshire* 1891, 503.

HUTCHISON, William (1815–)
b. Caputh, Perthshire 8 Feb. 1815

Gardener to A. L. Gower at Castle Malgwyn, Cardiganshire, 1838.

Gdnrs Chron. 1875 ii 646–47 portr.

HUTTON, Frederick Wollaston (1836–1905)
b. Gate Burton, Gainsborough, Lincs 16 Nov. 1836 *d.* at sea 27 Oct. 1905

FRS 1892. 23rd Royal Welch Fusiliers, 1855. Captain, 1862. To New Zealand, 1865. Provincial Geologist, Otago, 1873. Professor of Natural Science, 1877; of Biology, Canterbury, 1879. Wrote papers on *Capsella*, distribution of New Zealand flora, etc.

P. Mennell *Dict. Austral. Biogr.* 1892, 243. *Trans. N.Z. Inst.* v.36, 1903, 357. *Proc. Linn. Soc. N.S.W.* v.30, 1905, 606–7. *Proc. R. Soc.* v.79, 1907, xli–xliv. *Proc. R. Irish Acad.* B.

v.32, 1915, 74. R. Speight *Nat. Hist. of Canterbury* 1927, 6–9 portr. *R.S.C.* v.3, 480; v.7, 1043; v.10, 300; v.12, 357; v.15, 1008.

Herb. of mosses and drawings at National Museum, Dublin. Letters in Wilson correspondence at BM(NH).

HUTTON, Henry (–1868)
d. Java, East Indies 1868

Messrs Veitch plant collector. Sent to Java and Malaya in 1866 to collect orchids. In Moluccas and Timor, 1868.

J. H. Veitch *Hortus Veitchii* 1906, 54, 130. *J. R. Hort. Soc.* 1948, 286. *Fl. Malesiana* v.1, 1950, 253. K. Lemmon *Covered Gdn* 1962, 214.

HUTTON, Henry (*fl.* 1850s–1896)
d. Rondebosch, S. Africa Nov. 1896

Of South Molton, Devon. Official in S. Africa previous to 1857. He and his wife sent plants to W. H. Harvey and J. D. Hooker.

W. H. Harvey *Thesaurus Capensis* v.2, 1863, 1–2.

Letters at Kew.

Massonia huttoni Baker.

HUTTON, Janet (*née* **Robertson**)
(*fl.* 1800s–1820s)

Wife of East India merchant. At Penang, 1802–8; at Calcutta, 1817–23. Drew plants.

Kew Bull. 1894, 135–36.

Drawings of Indian and Malayan plants at Kew.

HUTTON, Thomas (*fl.* 1780s–1820s)

Of Keswick, Cumberland. Botanist and mineralogist. Established museum of plant and mineral specimens, 1786. Guide for tourists in Lake District. Gave much erroneous information on Lake plants.

D. Turner and L. W. Dillwyn *Botanist's Guide* 1805, 143. H. C. Watson *New Botanists' Guide* 1835, 310. D. Turner *Extracts from Lit. and Sci. Correspondence of Richard Richardson* 1835, 279. *Phytologist* v.2, 1845, 74. J. G. Baker *Fl. English Lake District* 1885, 10. W. Hodgson *Fl. Cumberland* 1898, xxvi.

HUTTON, William (1797–1860)
b. Sunderland, Durham 21 March 1797 *d.* West Hartlepool, Durham 21 Nov. 1860

FRS 1840. Geologist. *Fossil Fl. Great Britain* (with J. Lindley) 1831–37 3 vols.

Gent. Mag. 1861 i 111. *Trans. Northumberland Durham Newcastle Nat. Hist. Soc.* v.10, 1888, 19–151. J. Hardy *Selections from Correspondence of George Johnston* 1892, 46–47. *D.N.B.* v.28, 363.

Fossil plants at Newcastle Museum. Letters at Kew.

Huttonia Sternb.

HUXLEY, Thomas Henry (1825–1895)
b. Ealing, Middx 4 May 1825 *d.* Eastbourne, Sussex 29 June 1895

DCL Oxon 1885. FRS 1851. FLS 1858. Assistant-surgeon, Royal Navy, 1846; on HMS 'Rattlesnake', 1846–50. Professor of Natural History, Royal School of Mines, 1854–85. President, Royal Society, 1883. 'Gentians' (*J. Linn. Soc.* v.24, 1887, 101–24).

L. Reeve *Portr. of Men of Eminence in Lit., Sci. and Arts* v.1, 1863, 127–34 portr. *Nature* v.9, 1873–74, 257–58 portr.; v.52, 1895, 226–29, 240–49, 318–20; v.62, 1900, 10–12; v.63, 1900–1, 92–96, 116–19, 184–85; v.64, 1901, 145–51; v.115, 1925, 697–752 portr. *J. Bot.* 1895, 312. *Nat. Sci.* 1895, 119–28 portr. *Proc. Linn. Soc.* 1895–96, 38–40. *Proc. R. Soc.* v.59, 1895–96, xlvi–lxvi. *Trans. New York Acad. Sci.* 1895, 40–50. L. Huxley *Life and Letters of T. H. Huxley* 1900 2 vols portr. P. C. Mitchell *T. H. Huxley*, 1900. *D.N.B. Supplt. 1* v.3, 22–31. E. Clodd *T. H. Huxley* 1902. J. R. A. Davis *T. H. Huxley* 1907. L. Huxley *T. H. Huxley* 1920. H. F. Osborn *Impressions of Great Nat.* 1924, 73–98 portr. C. Ayres *T. H. Huxley* 1932. E. W. MacBride *Huxley* 1934. J. Huxley *T. H. Huxley's Diary of Voyage of HMS Rattlesnake* 1935. F. O. Bower *Sixty Years of Bot. in Britain* 1938, 45–47 portr. W. R. Dawson *Huxley Papers* 1946. C. Bibby *Scientist Extraordinary* 1959. R. W. Clark *The Huxleys* 1968 portr. J. Pingree *T. H. Huxley* 1968, 1969 2 vols. C. C. Gillispie *Dict. Sci. Biogr.* v.6, 1972, 589–97. *R.S.C.* v.3, 482; v.7, 1045; v.10, 302; v.12, 357; v.15, 1010.

MSS. at Imperial College of Science, London. Portr. at Hunt Library.

HYAM, George Neville (1886–1958)
b. Willesden, Middx 16 June 1886 *d.* Melbourne 28 Aug. 1958

Educ. Birmingham University. Farmer, 1904–10. To Victoria in 1910 as food preserving technician. Horticultural work in New Zealand, 1916–23. Nurseryman and garden designer, Melbourne, 1923. Horticultural supervisor, Victoria Dept. of Agriculture, 1939. President, Victorian Field Naturalists Club, 1935–36.

Proc. Field Nat. Club Victoria 1958, 87. *Victorian Nat.* 1959, 48–49.

HYAMS, Edward (1910–1975)
b. London 30 Sept. 1910 *d.* Besançon, France 25 Nov. 1975

Novelist and garden historian. *Grape Vine in England* 1949. *Vineyards in England* 1953. *From the Waste Land* 1950. *Soil and Civilisation* 1952. *Orchard and Fruit Garden* (with A. A. Jackson) 1961. *Strawberry Growing Complete* 1962. *English Garden* 1964. *An*

Englishman's Garden 1965. *Ornamental Shrubs for Temperate Zone Gardens* 1965–66 3 vols. *Dionysus* 1966. *Irish Gardens* 1967. *Gardener's Bedside Book* 1968. *Great Botanical Gardens of the World* 1969. *English Cottage Gardens* 1970. *Capability Brown and Humphry Repton* 1971. *History of Gardens and Gardening* 1971. *Plants in the Service of Man* 1971. *Survival Gardening* 1975.
Who's Who 1974, 1665. *Times* 28 Nov. 1975.

HYDE, George (*fl.* 1780s)
Helped to found a "society of botanists" in Oldham, *c.* 1785.
J. Holt *General View of Agriculture... Lancashire* 1795, 229. *Trans. Liverpool Bot. Soc.* 1909, 78.

HYDE, Harold Augustus (*c.* 1892–1973)
d. Cardiff, Glam 19 March 1973
BA Cantab. FLS 1926–39, 1946. Teacher, Birmingham, Stamford, Tonbridge, 1914–17, 1919–22. National Museum of Wales, 1922; Keeper of Dept. of Botany until 1962. *Welsh Timber Trees* 1931; ed. 3 1961. *Welsh Flowering Plants* (with A. B. Wade) 1934; ed. 2 1957. *Welsh Ferns* (with A. B. Wade) 1940; ed. 5 1969. *Atlas of Airborne Pollen Grains* (with K. F. Adams) 1958.
Mus. Assoc. Mon. Bull. v.13, 1973, 19–20. *Watsonia* 1974, 113–14.

HYDE, Rev. Thomas (1636–1703)
b. Billingley, Yorks 16 May 1636 *d.* Oxford 18 Feb. 1703
MA Oxon 1659. DD 1682. Keeper of Bodleian, 1665–1701. Professor of Arabic, Oxford, 1691; of Hebrew, 1697. Orientalist. *Epistola de Mensuris Sinensium necnon de Herbae Cha* 1688.
D.N.B. v.28, 401.

HYDE, William Wellington (1812–1880)
Surgeon, Bloxham. Botanised in Oxfordshire.
G. C. Druce *Fl. Oxfordshire* 1927, cxi.
Plants at BM(NH).

HYLL, Thomas *see* Hill, T.

HYNDMAN, Mrs. E. N. *see* Miles-Thomas, E. N.

HYNDMAN, George Crawford (1796–1867)
b. Belfast 14 Oct. 1796 *d.* Belfast 18 Nov. 1867
Auctioneer, Belfast. Collected plants in N. Ireland. Contrib. to S. A. Stewart and T. H. Corry's *Fl. N.E. Ireland* 1888, xv. Had herb.
J. Bot. 1879, 264–65. *Irish Nat.* 1913, 27. *Belfast Nat. Hist. Philos. Soc. Centenary Vol.* 1924, 86–87 portr. R. L. Praeger *Some Irish Nat.* 1949, 107.

HYNES, Sarah (*c.* 1860–1938)
d. Randwick, N.S.W. 28 May 1938
BA Sydney 1891. On staff of Sydney Botanic Gardens.
Sydney Morning Herald 30 May 1938. *Proc. Linn. Soc. N.S.W.* v.64, 1939, 1.

HYSLOP, James Macadam (*fl.* 1850s)
Collected plants in N. Iraq, 1852–54, now at Kew.

I'ANSON, Charles (*fl.* 1840s)
Nurseryman, High Street, Gateshead and 46 Groat Market, Newcastle, Northumberland. In partnership with Samuel Finney.

I'ANSON, James (1784–1821)
b. Darlington, Durham 7 April 1784 *d.* 10 June 1821
Linen weaver of Darlington. Had herb. of local plants.
Proc. Bot. Soc. Br. Isl. 1858, 39–40.
Herb. at BM(NH).

IBBETSON, Agnes (*née* **Thomson**) (1757–1823)
b. London 1757 *d.* Exmouth, Devon Feb. 1823
Contrib. physiological papers to *J. Nat. Philos, Chemistry and the Arts* and *Philos. Mag.* Had herb.
Bot. Mag. 1810, t.1259. *Mon. Mag.* v.31, 1811, 601. *D.N.B.* v.28, 409. B. D. Jackson *George Bentham* 1906, 39.
Collection of woods at BM(NH).
Ibbetsonia Sims.

IBBETT, Thomas (*fl.* 1830s–1840s)
Florist, Mount Pleasant, Bull's Field, Woolwich, Kent. Specialist in pinks, carnations. 'On the Origin of the Pink' (*Floricultural Cabinet* v.9, 1841, 106–9).

IBBOTSON, Henry (1814–1886)
d. York 12 Feb. 1886
Schoolmaster. Of Ganthorpe near Malton, Yorks. Sold dried plants. *Catalogue of Phaenogamous Plants of Great Britain* 1848. *Ferns of York* 1884. 'Rarer Plants found near Castle-Howard, Yorkshire' (*Phytologist* 1844, 577–79). 'List of Mosses found near Castle-Howard, Yorkshire' (*Phytologist* 1844, 781–82).
London J. Bot. 1845, 496–97. *Nat. Hist. J. School Rep.* 15 March 1886, 42. *D.N.B.* v.28, 410–11. *Annual Rep. Yorkshire Philos Soc.* 1906, 67–68. *J. Bot.* 1935, 107.
Herb. with F. Townsend (1822–1905). Letters at Kew.

ICK, William (1800–1844)
b. Newport, Shropshire 1800 *d.* Birmingham 28 Sept. 1844

PhD 1843. Curator, Birmingham Philosophical Institute. List of Birmingham plants in *Analyst* v.6, 1837, 20–28. Fossil trees in *J. Geol. Soc.* 1845, 43–46.

J. E. Bagnall *Fl. Warwickshire* 1891, 499–500. D. A. Cadbury *et al. Computer-mapped Fl.... Warwickshire* 1971, 55. *R.S.C.* v.3, 489.

Herb. at Birmingham.

ILCHESTER, 4th Earl of *see* Strangways, W. T. H. F.

ILIFF, William Tiffin (–1876)
d. Epsom, Surrey 17 Feb. 1876

MD London 1856. FLS 1833. Member of Medico-botanical Society. Collected plants in Syria, West Indies. 'Experiments on Roots of *Canna indica*' (*Br. Assoc. Rep.* 1847, 85).

Gdnrs Mag. 1839, 678–79. *R.S.C.* v.3, 490.

Plants at Kew.

IM THURN, Sir Everard Ferdinand
see Thurn, *Sir* E. F. I.

IMRAY, John (1811–1880)
b. N. Scotland 11 Jan. 1811 *d.* St. Aroment, Dominica 22 Aug. 1880

MD. To Dominica, 1832, where he botanised.

Gdnrs Chron. 1880 ii 361. *J. Bot.* 1880, 320. *Lancet* 1880 ii 559–60. I. Urban *Symbolae Antillanae* v.3, 1902, 67–68.

Plants and letters at Kew.

Vaccinium Imrayi Hook.

INCHBALD, Peter (1816–1896)
b. Doncaster, Yorks 1816 *d.* Hornsea, Yorks 13 June 1896

FLS 1880. Of Storthes Hall, Huddersfield. Knew British and South European plants. *Llandudno Botany, Entomology, Ornithology* 1864. Yorkshire plants in *Phytologist* v.3, 1849, 445–49.

Proc. Linn. Soc. 1896–97, 58–59. *Sorby Rec.* v.2(4), 1967, 7–8. *R.S.C.* v.3, 493; v.8, 3; v.10, 306; v.12, 359; v.16, 13.

Plants at Kew.

INDER, Robert W. (*c.* 1885–1949)
d. 20 Nov. 1949

FLS 1932. Indian Forest Service, 1911–39. Studied flora of Sind.

Proc. Linn. Soc. 1951, 230–31. *Empire For. J.* 1950, 93.

Field note-books at Kew.

INGAMELLS, David (*c.* 1863–1946)
b. Lincs *c.* 1863 *d.* March 1946

VMH 1945. Gardener. Salesman, Covent Garden, 1894. Nursery at Maidenhead. Specialised in chrysanthemums.

Gdnrs Mag. 1910, 119–20 portr. *Gdnrs Chron.* 1946 i 156.

INGEN-HOUSZ, Jan (1730–1799)
b. Breda, Netherlands 8 Dec. 1730 *d.* Bowood, Wilts 7 Sept. 1799

MD Vienna. FRS 1779. Came to England, 1764–65. *Experiments upon Vegetables* 1779 portr.

Gent. Mag. 1799 ii 900–1. S. H. Vines *Lectures on Physiology of Plants* 1886, 75. *D.N.B.* v.28, 433–34. J. Wiesner *Jan Ingen-Housz* 1905 portr. *Chronica Botanica* 1949, 285–393. C. C. Gillispie *Dict. Sci. Biogr.* v.7, 1973, 11–16.

MSS. at Royal Society.

Ingenhouzia Moc. & Sess.

INGHAM, William (1854–1923)
b. Manchester 1854 *d.* York 25 May 1923

BA London. Bryologist. On staff of York Education Office. Secretary, Moss Exchange Club, 1903–22. *Handbook Cryptogamous Fl. York District* 1906. Edited *Census Catalogue British Mosses* 1907 and *Hepatics* 1913. Contrib. to *J. Bot.* 1907–9.

Br. Bryological Soc. Rep. 1923, 40–41; 1926, 253. *J. Bot.* 1923, 318–19. *Naturalist* 1923, 238–39 portr.; 1973, 2–3. *New Phytol.* 1925, 312.

Bryophytes at Leeds.

INGLIS, Andrew (1837–1875)
b. Edinburgh 1837 *d.* Aberdeen? 13 March 1875

MD Edinburgh 1859. Professor of Midwifery, Aberdeen, 1869. Active field botanist.

Trans. Proc. Bot. Soc. Edinburgh v.12, 1875, 410–11.

INGLIS, David (*c.* 1843–1921)
b. Fifeshire *c.* 1843 *d.* 28 Aug. 1921

Gardener to Duke of Buccleuch at Drumlanrig, Dumfriesshire.

Gdnrs Chron. 1921 ii 142.

INGLIS, R. (*fl.* 1830s)

Of Canton. Visited Shipke on India–Tibet Frontier *c.* 1830. Sent plants to R. Brown.

I. H. Burkill *Chapters on Hist. Bot. in India* 1965, 80.

Plants at Liverpool.

INGLIS, Robert Alexander (1869–)
b. Calcutta 14 March 1869

BA Cantab. Electrical engineer, 1891–1913. To Canada, 1914. Assistant botanist and librarian, Central Experimental Farm, Ottawa, 1915.

INGRAM, Alexander (1821–1881)
b. Chaple of Garioch, Aberdeen 10 Dec. 1821 *d.* 5 Nov. 1881

Gardener to J. J. Blandy at Reading, 1854 and to Duke of Northumberland at Alnwick, 1867.

Gdnrs Chron. 1876 i 657 portr.; 1881 ii 637 portr. *Florist and Pomologist* 1881, 188.

INGRAM, J. (*fl.* 1830s–1840s)

Nurseryman, Southampton, Hants.

INGRAM, John (*c.* 1822–1876)
d. 10 Dec. 1876
Acquired nursery at Huntingdon, formerly belonging to his uncle, Mr. Wood.
Gdnrs Yb. and Almanack 1878, 190.

INGRAM, Thomas (*c.* 1796–1872)
d. Slough, Bucks 9 March 1872
Head gardener to Queen Charlotte, 1816. Superintendent, Royal Gardens at Windsor, 1833.
Florist and Pomologist 1872, 96. *Garden* 1872, 387. *Gdnrs Chron.* 1872, 364. *J. Hort. Cottage Gdnr* v.47, 1872, 230–31.

INGRAM, William (1820–1894)
b. Frogmore, Berks 1820 *d.* 9 Jan. 1894
Son of T. Ingram (*c.* 1796–1872). Gardener to Duke of Rutland at Belvoir Castle, Grantham, 1853–94.
Gdnrs Chron. 1875 i 336 portr.; 1894 i 50 portr. *Gdnrs Mag.* 1894, 31 portr.

INGWERSEN, Walter Edward Theodore (1883–1960)
b. Hamburg, Germany 1883 *d.* East Grinstead, Sussex 14 Feb. 1960
VMH 1944. VMM 1950. To England, *c.* 1902. Nurseryman, East Grinstead, 1925; specialised in alpines. Collected plants in Balkans and Caucasus. *Wild Flowers in the Garden* 1951.
Gdnrs Chron. 1935 i 234 portr.; 1953 ii 42 portr.; 1960 i 126. *Quart. Bull. Alpine Gdn Soc.* v.28, 1960, 147. *Times* 17 Feb. 1960.

INNES, John (1829–1904)
London businessman. Part of his estate, including his house and two acres of ground at Merton, Surrey bequeathed "for the promotion of horticultural instruction, experiment and research." The John Innes Institution moved to Bayfordbury, Herts in 1946 and to Norwich in 1967.

INNIS, James (*fl.* 1790s)
Nurseryman, Bridlington, Yorks.

INSCH, James (1877–1951)
b. Morayshire 1877
FLS 1940. Tea planter in India, 1902–33. Bequeathed his library of tea culture to Linnean Society.
Proc. Linn. Soc. 1950–51, 253.

INSTONE, Henry (*fl.* 1830s–1840s)
Nurseryman, Shrewsbury, Shropshire.

IRVINE, Alexander (1793–1873)
b. Daviot, Aberdeenshire 1793 *d.* Chelsea, London 13 May 1873
Educ. Aberdeen University. Schoolmaster at Albury, Bristol, Guildford and from 1851 at Chelsea. *London Fl.* 1838 (author's copy, with additions at BM(NH)—*J. Bot.* 1921, 178–79). *Illustrated Handbook of British Plants* 1858. Edited *Phytologist* 1855–63 and *Botanists Chron.* 1863–65.
Gdnrs Mag. 1837, 184. *J. Bot.* 1872, 222; 1873, 222–23; 1921, 178–79. *Gdnrs Chron.* 1873, 1017–18. *D.N.B.* v.29, 48. J. B. L. Warren *Fl. Cheshire* 1899, lxxxv. G. C. Druce *Fl. Buckinghamshire* 1926, xcix. G. C. Druce *Fl. Northamptonshire* 1930, cxv–cxvi. C. E. Salmon *Fl. Surrey* 1931, 51–52. D. H. Kent *Hist. Fl. Middlesex* 1975, 20. *R.S.C.* v.3, 498.
Plants at Manchester, Kew.

IRVINE, Frederick Robert (1898–1962)
b. Newcastle, Northumberland 30 April 1898 *d.* Accra, Ghana 19 Aug. 1962
DSc Newcastle. FLS 1927. Lecturer, Achimota College, Gold Coast, 1924–40. Administrative officer, Edinburgh University, 1940. Returned to Ghana, 1961. Collected plants in W. Africa, 1924–39. *Plants of Gold Coast* 1930. *West African Botany* 1931. *West African Agriculture* 1934. *Woody Plants of Ghana* 1961.
Nature v.196, 1962, 319. *Proc. Linn. Soc.* 1961–62, 155. *Times* 22 Aug. 1962.
Herb. at Lagos, Ghana. Plants at BM(NH), Kew, Edinburgh. MSS. at Edinburgh. Portr. at Hunt Library.

IRVING, Christopher (–1856)
b. Dalton, Dumfriesshire *d.* Lea, Glos 27 Feb. 1856
LLD Aberdeen 1818. Schoolmaster. *Catechism of Botany* 1821. *Outline of Kingdoms of Nature for Use of Schools* 1841.

IRVING, Edward George (1816–1855)
b. Hoddom, Dumfriesshire 1 April 1816 *d.* Abbeokuta, Lagos, Nigeria 1855
MD Edinburgh. Entered Navy, 1840. Surgeon, Royal Navy. Collected plants in S. Africa, 1843 and S. Nigeria. 'Cultivation of Cotton in Western Africa' (*Hooker's J. Bot.* 1855, 297–302).
J. R. Geogr. Soc. v.26, 1856, clxxvii–clxxviii. *Trans. Linn. Soc.* v.23, 1862, 167.
Plants and letters at Kew.
Irvingia Hook. f.

IRVING, Walter (1867–1934)
b. Wickham Market, Suffolk 3 Aug. 1867 *d.* Lightwater, Surrey 23 April 1934
Kew gardener, 1890; Assistant Curator, 1922. *Everyman's Book of the Greenhouse* 1907. *Rock Gardening* 1925. *Saxifrages* (with R. A. Malby) 1914.
Gdnrs Mag. 1912, 791–92 portr. *Gdnrs Chron.* 1923 i 330 portr.; 1934 i 337. *J. Kew Guild* 1928, 555 portr.; 1934, 374 portr.

IRWIN, George (*fl.* 1740s–1770s)
Nurseryman, Southampton, Hants.

IRWIN, Rev. J. J. (*fl.* 1860s)
Colonial chaplain, Hong Kong. Collected some plants for H. F. Hance.
E. Bretschneider *Hist. European Bot. Discoveries in China* 1898, 689.
Quercus irwinii Hance.

IRWYN (*or* Irvine), John
MD. Plants in J. Petiver's herb.
J. E. Dandy *Sloane Herb.* 1958, 141.

ISAAC, Japhet S. (–1918)
Clerk, Botanic Gardens, Singapore. Collected plants under H. N. Ridley.
Gdns Bull. Straits Settlements 1927, 125. *Fl. Malesiana* v.1, 1950, 255.
Plants at Singapore.

IVENS, Arthur J. (1897–1954)
d. London 26 April 1954
FLS 1950. Nurseryman, with his father at Harrietsham, Kent until 1927, then with Hillier's at Winchester, 1927–37. Nursery manager, Messrs Clibrams Ltd. at Altrincham, then back to Messrs Hillier's as general manager, 1946–54. Studied *Lapponicum* series of *Rhododendron* and shrubby *Potentilla*.
Gdnrs Chron. 1954 i 184. *Proc. Linn. Soc.* 1952–53, 216–17.

IVERY, James (*c.* 1823–1872)
d. 2 Aug. 1872
Nurseryman, Dorking, Surrey. Specialised in Indian azaleas and British ferns.
Gdnrs Chron. 1872, 1075.

IVIMEY COOK, Walter Robert *see* Cook, W. R. I.

JACK, James (1863–1955)
b. Calf of Man, Isle of Man 1863 *d.* Arbroath, Angus 26 March 1955
FLS 1890. Pharmaceutical chemist. Collected algae of Forfarshire coast.
Pharm. J. 1955, 255.

JACK, William (1795–1822)
b. Aberdeen 29 Jan. 1795 *d.* Bencoolen, Sumatra 15 Sept. 1822
MA Aberdeen 1811. In Bengal Medical Service, 1813. To Sumatra with S. Raffles in 1818. Collected plants in India, Malaya, Sumatra. Greater part of his collections and MSS. lost by fire on the way to Europe. His Malayan plants in *J. Arnold Arb.* v.33, 1952, 199–251. *Malayan Miscellanies* 1820–21 2 vols. 'Malayan Plants' (*Calcutta J. Nat. Hist.* 1843, 1–62, 160–231, 305–74). 'William Jack's Letters to Nathaniel Wallich, 1819–1821' (*J. Straits Branch R. Asiatic Soc.* no. 73, 1916, 147–268).
W. Roxburgh *Fl. Indica* v.2, 1824, 321–24. *Companion Bot. Mag.* 1835, 121–47. A. Lasègue *Musée Botanique de B. Delessert* 1845, 145–47. *J. Straits Branch R. Asiatic Soc.* no. 25, 164; no. 65, 1913, 43. *D.N.B.* v.29, 86. *Gdns Bull. Straits Settlements* 1927, 125. *Fl. Malesiana* v.1, 1950, 256–57; v.5, 1958, 55. *Notes R. Bot. Gdn Edinburgh* 1954, 219–27. F. A. Stafleu *Taxonomic Literature* 1967, 227–28. *Taxon* 1970, 526. *R.S.C.* v.3, 506.
Plants at Geneva, BM(NH), Edinburgh, Kew.
Jackia Wall.

JACKMAN, Arthur George (1866–1926)
b. 16 July 1866 *d.* 16 March 1926
Nurseryman, Woking, Surrey.
Gdnrs Chron. 1960 ii 594 portr.

JACKMAN, George (1801–1869)
b. 25 March 1801 *d.* 12 Feb. 1869
Son of W. Jackman (1763–). Nurseryman, St. John's, Woking, Surrey, 1830.
Florist and Pomologist 1869, 72. *Gdnrs Chron.* 1869, 197–98; 1960 ii 594–95 portr.

JACKMAN, George (1837–1887)
b. Woking, Surrey 13 March 1837 *d.* Woking 29 May 1887
Son of G. Jackman (1801–1869). Nurseryman, St. John's, Woking. Nursery moved to site between Woking and Mayford in 1885. *Clematis as a Garden Flower* (with T. Moore) 1872.
Gdnrs Chron. 1887 i 748; 1960 ii 594–95 portr.

JACKMAN, William (1763–)
b. 13 March 1763
Founded nursery at St. John's, Woking, Surrey, 1810.
Gdnrs Chron. 1960 ii 594.

JACKSON, Albert Bruce (1876–1947)
b. Newbury, Berks 14 Feb. 1876 *d.* Kew, Surrey 14 Jan. 1947
FLS 1917. VMM 1925. Assistant, Kew Herb. 1907–10. Technical Assistant, Imperial Institute, London, 1910–32. Part-time specialist in *Coniferae*, BM(NH), 1932–47. Assisted R. J. Elwes and A. Henry in preparation of *Trees of Great Britain and Ireland* 1906–13. Wrote accounts of tree collections at Syon House, 1910, Yattenden Court, 1911, Albury Park, 1913, Westonbirt, 1927, Borde Hill, 1935. *Handbook of Coniferae* (with W. Dallimore) 1923; ed. 4 1966. *Illustrations of New Conifers* (with H. C. Baker) 1935. *Identification of Conifers* 1946.
Gdnrs Chron. 1924 i 44 portr.; 1947 i 46. *Empire For. Rev.* 1947, 7. *Kew Bull.* 1947, 63–64. *Nature* v.159, 1947, 156. *Proc. Linn. Soc.* 1946–47, 132–33. *Quart. J. For.* 1947, 63. *Times* 16 Jan. 1947. *Trans. Br. Bryol. Soc.* 1948, 132–33. *Watsonia* 1949–50, 123–24.
Plants at BM(NH), Kew. Bryophytes at South London Botanical Institute.

JACKSON, Benjamin Daydon (1846–1927)
b. Stockwell, London 3 April 1846 *d.* London 12 Oct. 1927

Hon. AM and PhD Uppsala. FLS 1868. Botanical Secretary, Linnean Society, 1880–1902; General Secretary, 1902–26. Botanical bibliographer. Edited *Index Kewensis* 1893–95; Supplt I with T. Durand, 1901–6. Edited J. Gerard's *Catalogue of Plants* 1876; W. Turner's *Libellus* 1877; A. R. Pryor's *Fl. Hertfordshire* 1887. *Guide to Literature of Botany* 1881. *Vegetable Technology* 1882. *Glossary of Botanic Terms* 1900; ed. 4 1928. *George Bentham* 1906. *Linnaeus* 1923. *Catalogue of Library of Linnaean Society* 1925.

E. Bretschneider *Hist. European Bot. Discoveries in China* 1898, 823–24. *Gdnrs Chron.* 1923 ii 244 portr.; 1926 i 396 portr.; 1927 ii 336. *Bot. Soc. Exch. Club Br. Isl. Rep.* 1925, 756; 1927, 376–78. *J. Bot.* 1927, 314–17. *Kew Bull.* 1927, 421–23. *Nature* v.117, 1926, 502; v.120, 1927, 665–66. *Proc. Linn. Soc.* 1927–28, 119–23. *Who was Who, 1916–1928* 545. *R.S.C.* v.10, 313; v.16, 38.

Plants at Kew.

JACKSON, Charles

Of Barnstaple, Devon. Collected British ferns.

Br. Fern Gaz. 1909, 45.

JACKSON, David (*c.* 1805–1889)
d. 6 June 1889

Hand-loom silk weaver of Middleton, Lancs. Raised florists' flowers.

Garden v.35, 1889, 591.

JACKSON, George (*c.* 1780–1811)
b. Aberdeen *c.* 1780 *d.* London 12 Jan. 1811

FLS 1808. Had charge of A. B. Lambert's herb. Identified Crimean plants with Lambert in E. D. Clarke's *Travels* v.1, 1810, 738–47; v.2, 1812, xvii. 'Account of *Ormosia*' (*Trans. Linn. Soc.* v.10, 1811, 358–64). Contrib. to J. Sowerby and J. E. Smith's *English Bot.* 1251, 2459. Edited *Bot. Repository*, 1807–11.

J. Bot. 1916, 242–43.

Jacksonia R. Br.

JACKSON, George (*fl.* 1840s–1860s)

Nurseryman, Kingston-upon-Thames, Surrey.

JACKSON, John (*fl.* 1740s–1790s)
d. Lichfield, Staffs

Proctor, Lichfield Cathedral. Formed with E. Darwin and B. Boothby the Botanical Society at Lichfield. Translated and printed Linnaeus's *System of Vegetables* 1783 and *Families of Plants* 1787.

A. Seward *Mem. of Dr. Darwin* 1804, 98–100. R. Garner *Nat. Hist. County of Stafford* 1844, 7. *Gdnrs Mag.* 1838, 345. *J. Bot.* 1914, 322.

JACKSON, John Reader (1837–1920)
b. Chelsea, London 26 May 1837 *d.* Lympstone, Devon 28 Oct. 1920

ALS 1868. Keeper, Kew Museums, 1858–1901. Edited new ed. of B. H. Barton and T. Castle's *British Fl. Medica* 1877. *Commercial Botany of 19th Century* 1890.

Kew Bull. 1901, 201; 1920, 368–69. *J. Kew Guild* 1902, 55 portr.; 1921, 37–38. *Garden* 1920, 567. *Gdnrs Chron.* 1920 ii 234. *J. Bot.* 1920, 298. *Nature* v.106, 1920, 511. *Proc. Linn. Soc.* 1920–21, 49–50.

JACKSON, Mrs. Maria Elizabeth
(*fl.* 1790s–1820s)

Of Somersal Hall, Uttoxeter, Stafford. *Botanical Dialogues* 1797. *Botanical Lectures* 1804. *Sketches of Physiology of Vegetable Life* 1811. *Florist's Manual* 1816; ed. 2 1822.

JACKSON, Mary Ann (*afterwards* **Mrs. Charles Chavasse**) (*fl.* 1830s–1840s)

Of Lichfield. Probably daughter of John Jackson. *Pictorial Fl.* 1840, illustrated from plants collected in Wales, etc., 1834–38. 'Catalogue of some of the Rarer Species of Plants found in the Neighbourhood of Lichfield' (*Analyst* v.6, 1837, 297–98).

JACKSON, Thomas (*c.* 1851–1888)
d. Kingston-upon-Thames, Surrey 7 June 1888

Nurseryman, Kingston-upon-Thames. Secretary, Kingston and Surbiton Chrysanthemum Society.

J. Hort. Cottage Gdnr v.16, 1888, 487.

JACKSON, Thomas Herbert Elliot
(1903–1968)
b. Dorset 12 Jan. 1903 *d.* Kitale, Kenya 22 May 1968

Coffee grower in Kenya. Entomologist and botanist.

East African Standard 14 June 1968. *Times* 11 June 1968. *J. Lepidopterists' Soc.* 1969, 131–34.

Kenya plants at Kew.

JACKSON, William (*fl.* 1630s)

Of Oxford. Accompanied T. Johnson on his tour of west of England in 1634.

JACKSON, William (*fl.* 1840s)

Nurseryman, Cross Lane, Bedale, Yorks. Specialised in rhododendrons.

M. Hadfield *Gdning in Britain* 1960, 333.

JACKSON, William (1820–1848)
b. Dundee, Angus 10 Oct. 1820 *d.* Dundee 12 March 1848

Tailor. Botanist and friend of W. Gardiner (1808–1852). Founder member of Dundee Naturalists' Association.

Phytologist 1848, 109. *Chambers's Edinburgh J.* 1849, 165.

JACOB, Edward (*c.* 1710–1788)
b. Canterbury, Kent *c.* 1710 *d.* Faversham, Kent 26 Nov. 1788

Surgeon. *Plantae Favershamiensis* 1777 portr.

J. Nichols *Lit. Anecdotes of Eighteenth Century* v.7, 194, 601. R. Pulteney *Hist. Biogr. Sketches of Progress of Bot. in England* v.2, 1790, 272–73. *Gent. Mag.* 1788 ii 1127. *D.N.B.* v.29, 114.

Portr. at Hunt Library. Sale at Sotheby, 13 Feb. 1789.

JACOB, Eustace (*fl.* 1860s)

Captain. Sent plants from Chusan and Hong Kong to Kew, 1863–64.

E. Bretschneider *Hist. European Bot. Discoveries in China* 1898, 689. *Kew Bull.* 1901, 31.

JACOB, Rev. John (1796–1849)
b. London 19 Jan. 1796 *d.* London 29 Aug. 1849

LLD. Master, Devonport Grammar School. Minister, St. Aubyn Chapel. *West Devon and Cornwall Fl.* 1835–37.

T. R. A. Briggs *Fl. Plymouth* 1880, xxx. F. H. Davey *Fl. Cornwall* 1909, xli. W. K. Martin and G. T. Fraser *Fl. Devon* 1939, 772.

Herb. at Plymouth Institution, destroyed in 1941.

JACOB, Rev. Joseph (1858–1926)
d. Whitewell, Flintshire Feb. 1926

BA Cantab. Collected plants while a pupil at Uppingham School. Vicar, Whitewell, 1884–1926. *Daffodils* 1910. *Tulips* 1912. *Hardy Bulbs for Amateurs* 1924. Edited *Daffodil Yb.* Contrib. to *Garden.*

J. Hort. Cottage Gdnr v.64, 1912, 348–49 portr. *Garden* 1912, 104 portr.; 1926, 109 portr. *Gdnrs Chron.* 1926 i 125. A. R. Horwood and C. W. F. Noel *Fl. Leicestershire* 1933, cclxxxi. *Daffodil Yb.* 1949, 9–10 portr.

Plants at Uppingham School.

JAGO, Rev. George (–1726)

Of Looe, Cornwall and Harberton, Devon. Correspondent of J. Petiver.

C. E. Raven *John Ray* 1942, 367. J. E. Dandy *Sloane Herb.* 1958, 141–42.

JAMES, F. N. (*née* Vobes) (–1934)
d. 29 Oct. 1934

At Aberystwyth University College. Investigated peat of submerged forest of Borth, Cardiganshire.

Chronica Botanica 1935, 157–58 portr.

JAMES, Frank Linsly (1851–1890)
b. Liverpool 21 April 1851 *d.* San Benito, Spanish Guinea 21 April 1890

MA Cantab 1881. Explorer. Collected plants with J. G. Thrupp in Somaliland, 1884–85. Plants listed in his *Unknown Horn of Africa* 1888 portr.; ed. 2 1890.

D.N.B. v.29, 208–9.

Plants at Kew.

JAMES, Sir Henry Evan Murchison (1846–1923)
b. 20 Jan. 1846 *d.* 20 Aug. 1923

Under-Secretary, Government of Bombay, 1873. Postmaster-General, Bombay, 1875; Bengal, 1880. Commissioner, Ahmedabad, 1889–91; Sind, 1891–1900. Travelled with Younghusband and H. Fulford in Manchuria, 1886 when he collected plants which he sent to Kew. Collected plants in Aberdare Mountains, Kenya, 1905 and Uganda. *Long White Mountain or a Journey in Manchuria* 1888.

E. Bretschneider *Hist. European Bot. Discoveries in China* 1898, 766. *Kew Bull.* 1901, 35. *Who was Who, 1911–1928* 550.

Senecio jamesii Hemsley.

JAMES, J. (*fl.* 1850s)

Of Vauvert, Guernsey, Channel Islands. Collected ferns.

Br. Fern Gaz. 1909, 45.

JAMES, J. Thomas (–1791)
d. 2 Jan. 1791

Nurseryman, Cuper's Bridge, Lambeth, London. Sheriff of County of Surrey, 1774.

J. R. Hort. Soc. 1934, 242.

JAMES, John (*fl.* 1680s)

Ship's surgeon. Slave in Barbary for nearly 20 years. Drawings of plants made there with descriptions in Sloane MS. 4009; some reproduced by J. Petiver, *Gazophylacii* 1705, 57–58, 66, t.37, 38, 40.

E. J. L. Scott *Index to Sloane Manuscripts* 1904, 278. J. E. Dandy *Sloane Herb.* 1958, 142.

JAMES, Moses (*fl.* 1720s–1730s)

Seedsman, Stangate Street, Lambeth, London.

JAMES, Robert (1801–1871)
b. Downhampton, Glos 1801 *d.* 28 Nov. 1871

Publican, Stoke Newington, London. Founded Stoke Newington Chrysanthemum Society, *c.* 1846.

Gdnrs Mag. 1871, 557. *Floral World and Gdn Guide* 1872, 60–61.

JAMES, Hon. Robert (*c.* 1873–1960)
d. 13 Dec. 1960

VMH 1953. Had garden at Richmond, Yorks. Authority on lilies.

Gdnrs Chron. 1961 ii 80–81 portr.

JAMES, Thomas (*c.* 1720–1782)
d. Blackheath, Kent 1782
Lieut.-Colonel, Royal Artillery. At Gibraltar, 1748–55. List of Gibraltar plants in *History of Herculean Straits* v.2, 1771, 338–43, copied in Sir J. Talbot Dillon's *Travels through Spain* (*J. Bot.* 1914 Supplt iv–v).
M. Colmeiro *La Botanica...de la Peninsula Hispano-Lusitana* 1858, 75, 137.

JAMESON, Hampden Gurney (1852–1939)
b. London 1852 *d.* Leatherhead, Surrey 26 Dec. 1939
BA Oxford 1877. Had curacies at Norlands, Lincoln and Eastbourne. Vicar, Eastbourne, 1896–1917. *Illustrated Guide to British Mosses* 1893. *Student's Handbook of British Mosses* (with H. N. Dixon) 1896. *Illustrated Guide to Trees and Flowers of England and Wales* 1909; ed. 4 1942. Illustrated S. M. Macvicar's *Student's Handbook of British Hepatics* 1912.
Br. Bryol. Soc. Rep. 1944–45, 315–16.
Moss herb. at BM(NH).

JAMESON, James Sligo (1856–1888)
b. Alloa, Clackmannan 17 Aug. 1856 *d.* Bangala, Congo 17 Aug. 1888
Travelled in S. Africa, 1878–81 returning to England with zoological and botanical specimens. Naturalist to H. M. Stanley's expedition to relieve Emin Pasha, 1887. *Story of Rear Column of Emin Pasha Relief Expedition* 1890.
D.N.B. v.29, 232–34. R. Hill *Biogr. Dict. of Sudan* ed. 2 1967, 192.

JAMESON, Mrs. Rachel (–1893)
Travelled with her husband, Hugh Jameson, Deputy Inspector, Royal Navy. Collected plants in Cape and Falkland Islands, *c.* 1841.
Kew Bull. 1907, 70.
Plants at Kew.

JAMESON, William (1796–1873)
b. Edinburgh 3 Oct. 1796 *d.* Quito, Ecuador 23 June 1873
MD Edinburgh 1818. To Greenland, 1818 (plants in *Mem. Wernerian Nat. Hist. Soc.* v.3, 1821, 416); S. America, 1820; at Quito, 1826. Professor of Chemistry and Botany, Quito, 1827. Assayer at Quito mint, 1832; Director, 1861. Collected plants in Argentina, Brazil, Colombia, Ecuador, Peru, Venezuela. *Herb. Edinense* (with J. R. Scott) 1820 (with specimens). *Synopsis Plantarum Aequatoriensium* 1865 3 vols. *Journal of Voyage from Rio to Peru* 1822.
Mem. Wernerian Nat. Hist. Soc. v.5, 1824, 187–205. *Bot. Misc.* 1832–33, 212. *Companion to Bot. Mag.* 1835, 111–16. *London J. Bot.* 1843, 643–61; 1845, 378–85. *Gdnrs Chron.* 1872, 1622; 1873, 1151; 1892 ii 218. *J. Bot.* 1873, 318–19; 1909, 151. *Trans. Bot. Soc. Edinburgh* v.12, 1873, 19–28. *D.N.B.* v.29, 236. *Bibliotheca Botanica* Heft 116, 1937, 47–48. *R.S.C.* v.3, 532; v.8, 12.
Plants at Kew, BM(NH). Letters and portr. by A. Salas at Kew. Portr. at Hunt Library.
Jamesonia Hook. & Grev.

JAMESON, William (1815–1882)
b. Leith, Midlothian 1815 *d.* Dehra Dun, India 18 March 1882
MD Edinburgh. FLS 1864. FRSE. Bengal Medical Service, 1838. Curator, Museum, Asiatic Society, Bengal, 1838. Superintendent, Saharunpur Garden, 1842–75. Collected plants in India and Burma.
J. D. Hooker and T. Thomson *Fl. Indica* 1855, 73. *Proc. Linn. Soc.* 1882–83, 42. *Trans. Bot. Soc. Edinburgh* v.14, 1883, 288–95. *D.N.B.* v.29, 236. D. G. Crawford *Hist. Indian Med. Service, 1600–1913* v.2, 1914, 152. I. H. Burkill *Chapters on Hist. Bot. in India* 1965, 149–50. *R.S.C.* v.3, 533; v.7, 12; v.14, 288.
Plants at Kew.

JAMIESON, Andrew (*c.* 1842–1895)
d. Madras, India 17 Aug. 1895
Kew gardener, 1866. Curator, Gardens and Parks, Ootacamund, Nilgiris, 1868–94.
Gdnrs Chron. 1895 ii 429. *Kew Bull.* 1895, 231. *J. Kew Guild* 1896, 34 portr.

JAMIESON, Thomas Francis (1829–1924)
b. Aberdeen 26 April 1829 *d.* Milltimber near Aberdeen 3 May 1924
Director, North of Scotland Agricultural Research Association for over 20 years. Studied assimilation of nitrogen by plants.
Gdnrs Chron. 1924 i 296.

JANE, Frank William (1901–1963)
b. London April 1901 *d.* Ibadan, Nigeria 6 May 1963
DSc. Lecturer in Botany, University College, London, 1932–45; Reader, 1945–49. Professor of Botany, Royal Holloway College, 1949–63. Temporary Professor and Head of Dept. of Botany, Ibadan. *Structure of Wood* 1956. Contrib. to *J. R. Microsc. Soc.*, *New Phytol.*, *Trans. Hertfordshire Nat. Hist. Soc.*
Br. Phycol. Bull. 1963, 272–73. *Nature* v.199, 1963, 329–30. *Times* 13 and 20 May 1963. *Trans. Hertfordshire Nat. Hist. Soc.* v.26(1), 1964, 1. *Essex Nat.* 1965, 333. *Who was Who, 1961–1970* 592.

JANES, Edwin Ridgeway (*c.* 1885–1958)
VMH 1949. Horticulturist. Introduced new colour forms of *Primula malacoides* and *P. sinensis*. *Flower Garden* 1952. *My Early-flowering Chrysanthemums* 1952. *My Onions* 1952. *My Sweet Peas* 1952. *My Tomatoes* 1952. *Sweet Peas* 1953. *Vegetable Garden* 1954.

JANNOCH, Theodor Carl W. (1850–1925)
b. Pomerania, Prussia 1850 *d.* Dersingham Hall, Norfolk 29 Oct. 1925
Kew gardener, 1874. Had nursery at Dersingham where he specialised in lilies of the valley and lilacs.
J. Kew Guild 1926, 426–27 portr.

JANSON, Joseph (1789–1846)
b. Tottenham, Middx 12 July 1789 *d.* Stoke Newington, Middx 30 April 1846
FLS 1831. Discovered *Spiranthes aestivalis* in England.
Proc. Linn. Soc. 1840, 80; 1846, 301.
Jansonia Kippist.

JANSON, Thomas Corbyn (1809–1863)
b. 1 July 1809 *d.* Stamford Hill, London 23 June 1863
FLS 1843. Banker. Formed collection of plants of Tunbridge Wells neighbourhood. Friend of J. Woods and E. Forster.
Proc. Linn. Soc. 1863, xxix–xxx.

JARVIS, Claude Scudamore (1879–1953)
b. Forest Gate, London 20 July 1879 *d.* Ringwood, Hants 8 Dec. 1953
Soldier, administrator and orientalist. Governor, Sinai, 1922–36. Keen naturalist and botanist. *Half a Life* 1943.
Times 10 Dec. 1953. *D.N.B. 1951–1960* 542.

JAY, Bernard Alwyn (1911–1961)
d. 25 July 1961
MA Cantab. FLS 1937. Deputy Director, Timber Development Association, 1938–61. *British Timbers* (with E. R. B. Boulton) 1944. *Conifers in Britain* 1952.
Forestry v.34, 1961, 218. *Proc. Linn. Soc.* 1960–61, 139.

JEANNERETT, Dr. (*fl.* 1840s)
Of Tasmania. Sent algae from Port Arthur to W. H. Harvey.
W. H. Harvey *Nereis Australis* 1847, 20. *Proc. R. Soc. Tasmania* 1909, 20.
Jeannerettia Hook. f. & Harvey.

JEANS, Rev. George (–1863)
Vicar, Alford. Contrib. to *J. Hort. Cottage Gdnr*, etc.
J. Hort. Cottage Gdnr v.29, 1863, 272–73.

JEBB, George Robert (1838–1927)
b. Baschurch, Shropshire 30 Nov. 1838
Railway engineer. Authority on flora of N. Wales where he recorded *Lastrea rigida*.
N. Western Nat. 1927, 183–85.
Letters at Liverpool Museums.

JEEVES, John (*fl.* 1790s)
Seedsman, Hitchin, Herts.

JEFFERIES, John (*c.* 1818–1904)
b. Somerford Keynes, Glos *c.* 1818 *d.* Cirencester, Glos July 1904
Acquired nursery in Cirencester belonging to Mr. Gregory, 1850.
Garden v.66, 1904, 68.

JEFFERIES, William J. (1844–)
Nurseryman, Royal Nurseries, Cirencester, Glos.
Gdnrs Chron. 1925 ii 282 portr.

JEFFERS, Robert Haynes (–1973)
d. Chelmsford, Essex 11 Dec. 1973
FLS 1937. 'Edward Morgan and the Westminster Physic Garden' (*Proc. Linn. Soc.* 1953, 102–33). 'Richard Pulteney MD, FRS (1730–1801) and his Correspondents' (*Proc. Linn. Soc.* 1959–60, 15–26). *Friends of John Gerard* 1967; *Biogr. Appendix* 1969.
MSS. at Linnean Society.

JEFFERY, Henry John (1885–1950)
d. Peacehaven, Sussex 24 July 1950
FLS 1909. Lecturer, School of Pharmacy, Pharmaceutical Society, 1907–11. Assistant, Imperial Institute, 1911–44. *British Plants; their Biology and Ecology* (with J. F. Bevis) 1911; ed. 2 1920.
Proc. Linn. Soc. 1949–50, 231.

JEFFERY, Stephen (*fl.* 1840s)
Nurseryman, St. Giles Road, Oxford.

JEFFREY, James (*c.* 1850–1916)
b. Berwick, Northumberland *c.* 1850 *d.* 6 Jan. 1916
Gardener to Hope family at St. Mary's Isle, Kirkcudbright.
Garden 1916, 48.

JEFFREY, John (1826–1854)
b. Forneth, Perthshire 14 Nov. 1826 *d.* Arizona, U.S.A. Feb. 1854
Gardener, Royal Botanic Garden, Edinburgh, 1849. Collected in California, Oregon and Vancouver for Oregon Association, 1850–54 (*see* A. Murray *Bot. Expedition to Oregon* 1853).
Gdnrs Chron. 1852, 551; 1853, 243, 663; 1871, 1423; 1872, 464–65, 573–74, 636–37; 1914 i 13. *Hooker's J. Bot.* 1853, 315–17. *Trans. Bot. Soc. Edinburgh* v.6, 1860, 350–51; v.11, 1872, 322–38. E. Ravenscroft *Pinetum Britannicum* 1863, 45–46. *Proc. Biol. Soc. Washington* v.11, 1897, 57–60. *Forestry* v.6, 1932, 5–8. *California Hist. Soc. Quart.* v.18,

1939, 343. *Notes R. Bot. Gdn Edinburgh* v.20, 1939, 1–53. *Br. Columbia Hist. Quart.* 1946, 281–90. J. Ewan *Rocky Mountain Nat.* 1950, 239. *Gdn J. New York Bot. Gdn* 1954, 118–19. *J. R. Hort. Soc.* 1955, 521–22. *Oregon Hist. Quart.* 1967, 111–24. A. M. Coats *Quest for Plants* 1969, 319–20.

Plants at Edinburgh.

Pinus jeffreyi A. Murray.

JEFFREY, John (–1886)
Of Balsusney, Kirkcaldy, Fife. Arboriculturist. *Trees and Shrubs of Fife and Kinross* (with C. Howie) 1879.

Trans. Bot. Soc. Edinburgh v.17, 1889, 19–20.

JEFFREY, John Frederick (1866–1943)
b. Saffron Walden, Essex 1866 *d.* 1 March 1943

Son of W. R. Jeffrey. Assistant, Kew Herb., *c.* 1882–94. Curator, Herb., Royal Botanic Garden, Edinburgh, 1894–1917.

J. Kew Guild 1943, 304.

JEFFREY, William Rickman
Of Ashford, Kent. Discovered *Vicia lutea* in Kent.

F. J. Hanbury and E. S. Marshall *Fl. Kent* 1899, lxxv.

JEFFREYS, John (–1789)
Nurseryman, Brompton Park, Kensington, London. Successor to G. London and H. Wise. In partnership with James Gray who carried on the business after Jeffreys's death.

JEFFRIES, R. (*fl.* 1810s–1830s)
Founded Ipswich Nursery, St. Matthew Street, Ipswich, Suffolk, *c.* 1810. James and W. B. Jeffries also partners.

JEKYLL, Miss Gertrude (1843–1932)
b. London 29 Nov. 1843 *d.* Godalming, Surrey 8 Dec. 1932

VMH 1897. VMM 1922. Pioneer of modern informal style of gardening. Collaborated with Sir E. Lutyens. *Wood and Gardening* 1899. *Home and Garden* 1900. *Lilies for English Gardens* 1901. *Wall and Water Gardens* 1901. *Roses for English Gardens* (with E. Mawley) 1902. *Some English Gardens* 1904. *Colour in Flower Garden* 1908. *Children and Gardens* 1908. *Gardens for Small Country Houses* (with L. Weaver) 1912. *Annuals and Biennials* 1916. *Garden Ornament* 1918; ed. 2 1927.

Country Life v.8, 1900, 730–39. *Garden* 1926 ii–iii portr. *Gdnrs Chron.* 1932 ii 439, 440 portr.; 1960 ii 165, 180, 201; v.162(13), 1967, 7; v.162(14), 1967, 7. *Curtis's Bot. Mag. Dedications, 1827–1927* 393–94 portr. *Times* 10 Dec. 1932. *J. Bot.* 1933, 24. F. Jekyll *Gertrude Jekyll* 1934 portr. *Who was Who, 1929–1940* 700. *Saturday Book* 1956, 197–200. M. Hadfield *Gdning in Britain* 1960, 363–70. *Huntia* 1965, 201–2 portr. B. Massingham *Miss Jekyll* 1966 portr. *Garden* 1974, 310–12 portr.

Drawings and planting plans at California University. Portr. by W. Nicholson at National Portrait Gallery. Portr. at Hunt Library.

JELLY, Mr. (–before 1849)
Wrote *Fl. Bathonica* (unpublished).

Phytologist v.3, 1849, 581.

JENKIN, Nelson West (1882–*c.* 1852)
Physician. Collected plants in Spain, Italy, Balkans, Scandinavia. Cultivated alpine plants.

Quart. Bull. Alpine Gdn Soc. 1952, 54–56.

JENKIN, Thomas James (1885–1965)
b. Pembrokeshire 8 Jan. 1885 *d.* 7 Nov. 1965

Adviser on agricultural botany, University College, Bangor, 1915–19; University College, Aberystwyth, 1919–20. Senior research officer, Welsh Plant Breeding Station, 1920–40; Assistant Director, 1940–42; Director and Professor of Agricultural Botany, 1942–50. Contrib. to *Handbuch der Pflanzenzüchtung* ed. 2, v.4 1959.

Rep. Welsh Plant Breeding Station 1965, 7 portr. *Nature* v.210, 1966, 12–13. *Who was Who, 1961–1970* 596.

JENKINS, Edmund Howard (*c.* 1855–1921)
b. Cheltenham, Glos *c.* 1855 *d.* Surbiton, Surrey 9 Nov. 1921

Nurseryman, Queen's Road Nursery, Hampton Hill, Middx. *Rock Gardens and Alpine Plants* 1911. *Hardy Flower Book* 1913. *Small Rock Garden* 1913. *Rock Garden* 1920.

Garden 1921, 585, 601 portr. *Gdnrs Chron.* 1921 ii 266 portr.

JENKINS, Francis (1793–1866)
b. St. Clement, Cornwall 4 Aug. 1793 *d.* Gowhatty, Assam 28 Aug. 1866

In Bengal Army, 1809–61. Major-General. Commissioner of Assam.

W. J. Hooker *Genera Filicum* 1842, t.lxxv, B. *J. Asiatic Soc. Bengal* 1856, 410. G. C. Boase and W. P. Courtney *Bibliotheca Cornubiensis* v.1, 1874, 273, 1247. *Rep. Penzance Nat. Hist. Soc.* 1884–85, 17. I. H. Burkill *Chapters on Hist. Bot. in India* 1965 *passim.*

Plants at Kew.

Jenkinsia Hook.

JENKINS, Thomas (1800–1832)
d. London 7 April 1832
Nurseryman, Gloucester Place, New Road, Marylebone, London. In partnership with Mr. Gwyther, 1810–16. Nursery in Regent's Park became gardens of Royal Botanic Society *c.* 1838. *Hortus Marybonensis* 1819.
Bot. Mag. 1816, t.1844. J. C. Loudon *Encyclop. Gdning* 1822, 1291. *Gdnrs Mag.* 1832, 384.

JENKINSON, James (*c.* 1739–1808)
d. 15 Oct. 1808
Of Yealand, Lancs. *Generic and Specific Description of British Plants* 1775.
Naturalist 1902, 34.

JENKINSON, John Wilfred (1871–1915)
d. Gallipoli 4 June 1915
Educ. Oxford. DSc 1905. Lecturer on Embryology, Oxford, 1906. Collected plants when a boy at Bradfield College. Several of his plant records are in G. C. Druce's *Fl. Berkshire* 1897.
Nature v.95, 1915, 456.

JENKINSON, Robert C. (*c.* 1900–1970)
d. 22 Aug. 1970
Had garden at Knaphill Manor. Chairman of Knaphill Nursery.
J. R. Hort. Soc. 1970, 534–36.

JENMAN, George Samuel (1845–1902)
b. Plymouth, Devon 24 Aug. 1845 *d.* Georgetown, British Guiana 28 Feb. 1902
FLS 1881. Kew gardener, 1871–73. Curator, Castleton, Jamaica, 1873. Superintendent, Botanic Garden, Georgetown, 1879. Collected plants in N. America and West Indies. 'Synoptical List...of Ferns and Fern Allies of Jamaica' (*Bull. Bot. Dept. Jamaica* nos. 18–49, 1890–98). *Ferns and Fern Allies of British West Indies* 1898–1900. 'On Jamaica Ferns of Sloane's Herb.' (*J. Bot.* 1886, 14–17, 33–43).
Gdnrs Chron. 1902, 234. *J. Bot.* 1902, 237. *J. Kew Guild* 1902, 92, 150 portr. *Proc. Linn. Soc.* 1901–2, 237. *Kew Bull.* 1903, 29. I. Urban *Symbolae Antillanae* v.1, 1898, 80–82; v.3, 1902, 67. *R.S.C.* v.12, 367; v.16, 97.
Plants at Georgetown Botanic Garden, Washington, New York.
Jenmania Rolfe.

JENNER, Charles (1810–1893)
b. Chatham, Kent 1 Sept. 1810 *d.* Portobello, Edinburgh 27 Oct. 1893
Draper in Edinburgh from 1830. President, Botanic Society of Edinburgh, 1867. Algologist. Had botanic garden at Portobello. Contrib. to *Trans. Bot. Soc. Edinburgh* v.5–9, 1858–68.
Trans. Bot. Soc. Edinburgh v.20, 1894, 23–29. *R.S.C.* v.3, 544; v.8, 23; v.16, 97.
Algae at BM(NH).
Didymodon jennerii Schimper.

JENNER, Edward (1803–1872)
b. 13 March 1803 *d.* Lewes, Sussex 13 March 1872
ALS 1838. Studied microscopic algae. Had herb., now lost. *Fl. Tunbridge Wells* 1845. Drew figures in J. Ralfs's *Br. Desmideae* 1848.
J. Sowerby and J. E. Smith *English Bot.* 2925. *Gdnrs Chron.* 1872, 398. *Proc. Linn. Soc.* 1871–72, lxix–lxx. *D.N.B.* v.29, 324–25. A. H. Wolley-Dod *Fl. Sussex* 1937, xliii.
Algae and letters in Broome correspondence at BM(NH). Portr. at Hunt Library.

JENNER, J. H. A. (*fl.* 1850s–1900s)
Of Battle, Sussex. Had herb. Contrib. to *Proc. Eastbourne Nat. Hist. Soc.* 1887 and *Proc. Brighton Nat. Hist. Soc.* 1889.
A. H. Wolley-Dod *Fl. Sussex* 1937, xlvii.
Plants at Lewes Museum, BM(NH), Kew, Oxford.

JENNINGS, Alfred Vaughan (1864–1903)
b. Hampstead, London 1864 *d.* Christiania 11 Jan. 1903
FLS 1888. Demonstrator in Botany, Dublin Royal College of Science, 1895–98. Geologist. Collected plants in New Zealand. 'Tmesipteris' (*Proc. R. Irish Acad.* v.2, 1891, 1–18).
New Phytol. 1903, 65–66. *Proc. Linn. Soc.* 1902–3, 31–32. *Quart. J. Geol. Soc.* 1903, lv–lvi. *R.S.C.* v.16, 98.
Plants at Glasnevin, Kew.

JENNINGS, John (*fl.* 1820s)
Nurseryman, Mornington Nursery, Hampstead Road, London.

JENNINGS, Robert Henry (*fl.* 1880s)
Captain. Plants collected in Persia, 1886, now at Kew.

JENYNS, Leonard *see* Blomefield, L.

JEPSON, William (1812–1897)
b. Altrincham, Cheshire 1812 *d.* Bolton, Lancs 29 July 1897
MD Edinburgh. Lectured on botany at Pine Street Medical School, Manchester. Assisted R. Buxton with his *Bot. Guide to... Manchester* 1849. Friend of J. Martin.
Trans. Liverpool Bot. Soc. 1909, 78.
Plants at Tyldesley Natural History Society.

JERDON, Archibald (1819–1874)
b. Bonjedward, Roxburghshire 21 Sept. 1819 *d.* Jedburgh, Roxburgh 28 Jan. 1874
Cryptogamist. Contrib. to *Phytologist* v.2–6, 1857–63.
Hist. Berwickshire Field Club v.7, 1874, 338–46. *Trans. Bot. Soc. Edinburgh* v.12, 1874, 201–2. *R.S.C.* v.3, 547; v.8, 25.
MS. of Hymenomycetes and letters at BM(NH). Plants at Royal Botanic Garden, Edinburgh.

JERDON, Thomas Caverhill (1811–1872)
b. Biddick House, County Durham 12 Oct. 1811 *d.* Upper Norwood, London 12 June 1872
FLS 1864. Brother of A. Jerdon (1819–1874). Assistant surgeon, East India Company in Madras, 1835. Zoologist. Collected plants in various parts of India, especially in the south. *Impatiens jerdoniae* named after his wife who drew plates for R. Wight's *Icones Plantarum Indiae Orientalis* 1840–53.
Hist. Berwickshire Field Club v.7, 1870, 143–57. *Proc. Linn. Soc.* 1872–73, xxxii–xxxiii. *D.N.B.* v.29, 338. D. G. Crawford *Hist. of Indian Med. Service, 1600–1913* v.2, 1914, 146–47.
Indian plants at Kew.
Jerdonia Wight.

JERMYN, Stanley Thomas (1909–1973)
b. Benfleet, Essex 1909 *d.* 23 Sept. 1973
FLS 1962. Secretary, Essex Naturalists' Trust, 1966. His *Fl. Essex* in course of publication on his death.
B.S.B.I. News v.3(2), 1974, 15. *London Nat.* 1974, 105. *Watsonia* 1974, 203–5. *Essex Nat.* v.33, 1973–74, 109–10.
Herb. at Smithsonian Institution, Washington. MSS. at Essex Record Office.

JERRAM, Martyn Ralph Knight (1890s–1945)
d. Bangor, Caernarvonshire 10 May 1945
Assistant Conservator of forests, Punjab. Divisional forest officer, Rawalpindi. Lecturer, Forest School, University of Wales, 1932. *Textbook on Forest Management* 1935. *Outline of Forestry* 1938. *Elementary Forest Mensuration* 1939.
Empire For. J. 1945, 2–3.

JERRETT, W. (*fl.* 1840s)
Florist, Cambridge Heath Nursery, Hackney, London.

JESSON, Enid Mary *see* Addendum

JESSOP, C. Hale (*fl.* 1820s)
Nurseryman, Somerset House Nursery, St. James's Square, Cheltenham, Glos.

JEUNE, Bertram Hanmer Bunbury Symons- *see* Symons-Jeune, B. H. B.

JEWITT, Thomas Orlando Sheldon (1799–1869)
b. Derbyshire 1799 *d.* London 30 May 1869
Self-taught wood-engraver. Botanist. Illustrated with wood-engravings and lithographs W. H. Harvey's *Phycologia Britannica* 1846–51, A. Catlow's *Popular Greenhouse Botany* 1857, J. Harrison's *Floricultural Cabinet* 1833.

JOAD, George Curling (1837–1881)
b. Walmer, Kent 30 Nov. 1837 *d.* Wimbledon, Surrey 24 Oct. 1881
FLS 1871. Travelled in S. Africa. Collected alpine flora which was bequeathed to Kew in 1881 and formed basis of the rock garden.
Proc. Linn. Soc. 1881–82, 64. *J. Bot.* 1883, 53. *Curtis's Bot. Mag. Dedications, 1827–1927* 219–20 portr.
Plants at Kew, Marlborough College. Letters at Kew.

JOB, William (*fl.* 1770s)
Gardener, Clifton, Bristol. *Poems on Various Subjects* 1785.
Gdnrs Chron. 1937 i 226.

JOHNS, Rev. Charles Alexander (1811–1874)
b. Plymouth, Devon 31 Dec. 1811 *d.* Winchester, Hants 28 June 1874
BA Dublin 1841. FLS 1836. Master, Helston Grammar School; headmaster, 1843–47. *Fl. Sacra* 1840. *Botanical Rambles* 1847–52. *Flowers of the Field* 1851; rev. ed. 1949. *Forest Trees of Britain* 1854 2 vols.
J. Sowerby and J. E. Smith *English Bot.* 2792, 2949–50. *J. Bot.* 1874, 256. *Proc. Linn. Soc.* 1874–75, lii–liii. *D.N.B.* v.30, 3. F. H. Davey *Fl. Cornwall* 1909, xlvi–xlvii portr. *J. Soc. Bibl. Nat. Hist.* 1970, 259–69. *R.S.C.* v.3, 555.
Plants and letters at Kew. MS. diary at Record Office, Truro.

JOHNS, William (1771–1845)
b. Kilmanllwyd, Pembroke 1771 *d.* Higher Broughton, Lancs 27 Nov. 1845
FLS 1824–37. Unitarian minister, Gloucester, Totnes, Nantwich. Opened private school in Manchester, 1804. Friend of John Dalton who resided with him for 26 years. Secretary and Vice-president, Manchester Literary and Philosophical Society. *Practical Botany* 1826.
D.N.B. v.30, 4. *Trans. Liverpool Bot. Soc.* 1909, 78.

JOHNSON, Alfred (*c.* 1840–1899)
d. Boston, Lincs 2 Sept. 1899
Managing director of seed firm in Boston, established by his father, William Wade Johnson in 1820.
Gdnrs Chron. 1899 ii 212.

JOHNSON, Rev. Andrew (1830–1893)
b. Holborn, London 16 June 1830 *d*. Lee, Kent 7 April 1893

MA Cantab. FLS 1871. Headmaster, St. Olave's School, Southwark, London. *Botany Reading Books* 1881.

JOHNSON, Arthur Tysilo (1873–1956)
d. Bulkeley Mill near Conway, Caernarvonshire 20 Sept. 1956

VMM 1942. Schoolmaster, St. Asaph Grammar School. Market gardener at Bulkeley Mill. Collected plants in California, Canada and Europe. *Garden in Wales* 1927. *Hardy Heaths* 1928. *Woodland Garden* 1937. *Garden Today* 1938. *Mill Garden* 1950; rev. ed. 1956.

Gdnrs Chron. 1952 ii 132 portr.; 1956 ii 358.

JOHNSON, Charles (1791–1880)
b. London 5 Oct. 1791 *d*. Camberwell, London 21 Sept. 1880

FLS 1824. Lecturer, Guy's Hospital, 1830–73 and to Medical Botanical Society, London. Edited J. Sowerby and J. E. Smith's *English Bot*. ed. 2 1832–46. *Ferns of Great Britain* 1855. *Fern Allies* 1856. *British Poisonous Plants* 1856. *Grasses of Great Britain* 1861.

J. Bot. 1880, 351–52. *J. Hort. Cottage Gdnr* v.1, 1880, 306. *D.N.B.* v.30, 7. *Guy's Hospital Rep*. 1971, 67–68.

JOHNSON, Charles Pierpoint (–1893)
d. Camberwell, London 6 March 1893

Son of C. Johnson (1791–1880). *British Wild Flowers* 1858–60; ed. 3 1876. *British Poisonous Plants* (with C. Johnson) ed. 2 1861. *Useful Plants of Great Britain* 1861–62; ed. 2 1863.

J. Bot. 1893, 128.

JOHNSON, Christopher (1782–1866)
b. Lancaster 23 July 1782 *d*. Lancaster 21 June 1866

Educ. at Edinburgh. Surgeon. Friend of W. Smith (1808–1857) whose collection of diatoms he acquired. Correspondent of J. Ralfs, R. K. Greville and T. G. Rylands. 'On Animal Nature of Diatomaceae' (translated from Italian article by G. Meneghini) in *Botanical and Physiological Memoirs* 1853, edited by A. Henfrey.

J. Bot. 1907, 455. *Trans. Liverpool Bot. Soc*. 1909, 78–79. *Lancashire Nat*. v.8, 1916, 395–402 portr.

Slides at BM(NH). Letters in Walker-Arnott correspondence at BM(NH).

JOHNSON, Rev. Edward (*fl*. 1850s?)

Of Travancore, India. List of orchids he collected in *Madras J. Literature and Sci*. 1858, 215.

Plants at Kew.

JOHNSON, G. F. (*c*. 1875–1954)
d. 26 April 1954

Gardener to Rothschild family at Waddesdon Manor, Bucks, 1892 for 60 years.

Gdnrs Chron. 1952 i 70 portr.; 1954 i 176.

JOHNSON, George William (1802–1886)
b. Blackheath, Kent 5 Nov. 1802 *d*. Croydon, Surrey 29 Oct. 1886

FLS 1830. Barrister, Gray's Inn, 1836. Professor of Moral and Political Economy, Calcutta, 1836–42. *Outlines of Botany* 1827 (*see Gdnrs Mag*. 1829, 80). *History of English Gardening* 1829. *Principles of Practical Gardening* 1845. *Dictionary of Modern Gardening* 1846; revised as *Cottage Gardener's Dictionary* 1857; ed. 5 1860. *Science and Practice of Gardening* 1862. *Wild Flowers of Great Britain* (with R. Hogg) 1863–80. 11 vols. *Gardener's Almanack* 1844–66. Founder and editor of *Cottage Gdnr* 1849.

J. Hort. Cottage Gdnr v.3, 1881, 11–14 portr.; v.13, 1886, 401–4 portr.; v.57, 1908, 391–93 portr. *Gdnrs Chron*. 1886 ii 592; 1950 i 106–7. *D.N.B.* v.30, 12–13.

JOHNSON, Henry (*fl*. 1880s–1918)

Chemist, Barnsley, Yorks. Assisted F. A. Lees in collecting plants for his *Fl. West Yorkshire* 1888.

Sorby Rec. v.2(4), 1967, 9.

Herb. formerly at Barnsley Naturalists and Scientific Society; now lost.

JOHNSON, James Yate (1820–1900)
d. Funchal, Madeira 3 Feb. 1900

Zoologist. 'Plants of Madeira' (*J. Bot*. 1857, 161–65). '*Helichrysum devium*' (*Gdnrs Chron*. 1888 ii 62).

C. A. de Menezes *Fl. Archipelago da Madeira* 1914, 225–26. *R.S.C.* v.3, 556; v.8, 29.

Plants at Kew.

JOHNSON, John Charles Sperrin- *see* Sperrin-Johnson, J. C.

JOHNSON, Rev. Ralph (–1689)

Vicar, Brignall, Yorks, 1662–89. Friend and correspondent of John Ray. "A person of the greatest curiosity in botany, ornithology, antiquities, etc." (Thoresby).

C. E. Raven *John Ray* 1942 *passim*.

JOHNSON, Robert (–1919)

Orchid grower to Thomas Statter at Stand Hall, Whitefield, Manchester.

Orchid Rev. 1919, 76–78.

JOHNSON, Thomas (*c*. 1597–1644)
b. Selby, Yorks *c*. 1597 *d*. Basing House, Hants Sept. 1644

MD Oxon 1643. Apothecary. Lived in Lincolnshire and London. Had physic garden on Snow Hill, 1633. Killed in Civil War. *Iter...Cantianum* 1629 (MS. copy by

S. Dale in BM(NH)). *Enumeratio Plantarum in Ericeto Hampstediano* 1632. Revised J. Gerard's *Herball* 1633; reprinted 1636. *Mercurius Botanicus* 1634–41. *Opuscula Omnia Botanica* 1847.

R. Pulteney *Hist. Biogr. Sketches of Progress of Bot. in England* v.1, 1790, 126–34. *Gdnrs Chron.* 1848, 831. *Cottage Gdnr* v.6, 1851, 313. *Notes and Queries* v.11, 1855, 149, 204. H. Trimen and W. T. T. Dyer *Fl. Middlesex* 1869, 369–72. G. C. Druce *Fl. Berkshire* 1897, xcix–c. *Essex Nat.* 1899, 174–75. F. J. Hanbury and E. S. Marshall *Fl. Kent* 1899, lvii–lx. *D.N.B.* v.30, 47–48. E. J. L. Scott *Index to Sloane Manuscripts* 1904, 239, 282. J. W. White *Fl. Bristol* 1912, 53–56. J. R. Green *Hist. Bot.* 1914, 43–47. R. T. Gunther *Early Br. Botanists* 1922, 273–77. G. C. Druce *Fl. Buckinghamshire* 1926, lxx. *Br. J. Surgery* 1928, 181–87. H. W. Kew and H. G. Powell *Thomas Johnson* 1932. *J. Bot.* 1933, 51–52. C. E. Raven *English Nat.* 1947, 273–97. R. H. Jeffers *Friends of John Gerard* 1967, 87–95. W. S. C. Copeman *Worshipful Soc. of Apothecaries of London* 1968, 75–76. F. D. and J. F. M. Hoeniger *Growth of Nat. Hist. in Stuart England from Gerard to R. Soc.* 1969, 20–22. J. S. L. Gilmour, *ed. Thomas Johnson; Bot. Journeys in Kent and Hampstead* 1972. C. C. Gillispie *Dict. Sci. Biogr.* v.7, 1973, 146–48. D. H. Kent *Hist. Fl. Middlesex* 1975, 12–13.

Johnsonia R. Br.

JOHNSON, Thomas (1863–1954)
b. Barton upon Humber, Lincs 1863 *d.* 9 Sept. 1954

Educ. Royal College of Science, London. FLS 1890. Demonstrator in Botany, Royal College of Science, 1885–90. Professor of Botany, Royal College of Science, Dublin, 1890–1928. Director, Seed Testing and Plant Diseases Station, Ireland, 1900–10. Keeper of botanical collections in National Museum, Dublin. Studied algae and fossil plants.

R. L. Praeger *Some Irish Nat.* 1949, 109. *Proc. Linn. Soc.* 1955–56, 47–48. *Who was Who, 1951–1960* 589.

Plants at National Museum Dublin, Kew.

JOHNSON, W. B. (*c.* 1764–1830)
d. Coxbench, Derby 13 Jan. 1830

MB. Of Coxbench. Collected plants in Switzerland and with G. H. E. Muhlenberg in U.S.A. Contrib. to J. Pilkington's *View of Present State of Derbyshire* 1789 2 vols.

J. Stokes *Bot. Commentaries* v.1, 1830, Dedication. *J. Bot.* 1914, 320.

JOHNSON, Rev. William (1844–1919)
b. Halifax, Yorks 11 Feb. 1844 *d.* Harrogate, Yorks 20 July 1919

FLS 1888. Primitive Methodist minister in Yorkshire, 1864–1908. Lichenologist. *North of England Lichen Herb.* 1894–1918 13 fasc.

Naturalist 1917, 88; 1918, 103; 1929, 285–86; 1972, 13–14. *Essex Nat.* v.29, 1954, 193.

Lichens at Leeds, Essex Field Club Museum.

JOHNSON, William Henry (1875–)
b. East Grinstead, Sussex 30 Aug. 1875

FLS 1901–20. Kew gardener, 1898. Curator, Botanic Station, Gold Coast, 1898. Director of Agriculture, Gold Coast, 1904–6. Companhia de Moçambique, 1906. Collected plants. Director of Agriculture, S. Nigeria until *c.* 1920. *Hints on Cultivation and Preparation of Cocoa* 1899. *Cultivation and Preparation of Para Rubber* 1904. *Cocoa* 1912. *Elementary Tropical Agriculture* 1913.

Kew Bull. 1904, 13; 1906, 383. *Moçambique* 1939, 65.

Plants at Kew.

JOHNSON, Rev. William James Percival (1854–1928)
d. Nyasaland Oct. 1928

To Nyasaland as missionary, 1876. Archdeacon of Nyasaland. Collected plants in East Africa; now at Kew. *My African Reminiscences, 1875–1895* 1924.

Proc. R. Geogr. Soc. 1884, 512–36. *Kew Bull.* 1929, 32. B. H. Barnes *Johnson of Nyasaland* 1933. *Moçambique* 1939, 45–47 portr. *Comptes Rendus AETFAT 1960* 1962, 165–66.

JOHNSTON, George (1773–1835)
b. Old Rayne, Aberdeenshire 5 Jan. 1773 *d.* 13 June 1835

Gardener to Earl of Aberdeen at Haddo House, Aberdeen, 1805–35.

Gdnrs Mag. 1835, 552.

JOHNSTON, George (1797–1855)
b. Simprin, Berwickshire 27 July 1797 *d.* Berwick 30 July 1855

MD Edinburgh 1819. LLD Aberdeen. Surgeon at Berwick from 1819. *Fl. Berwick-on-Tweed* 1829–30 2 vols. (illustrated by his wife). *Bot. Eastern Borders* 1853. Edited *Mag. Zool. Bot.* 1838. Contrib. to J. Sowerby and J. E. Smith's *English Bot.* 2776, 2866.

Cottage Gdnr v.14, 1855, 350–51. *Gdnrs Chron.* 1855, 597. *Phytologist* 1855, 140–42. *Archiv Pharmacie* 1856, 338. *Hist. Berwickshire Field Club* v.3, 1856, 202–8. J. Hardy *ed. Selections from Correspondence of Dr. George Johnston* 1892 portr. *D.N.B.* v.30, 61. *Endeavour* 1955, 136–39. *R.S.C.* v.3, 563; v.8, 30.

Herb. formerly at Tweedside Physical and Antiquarian Society Museum not traced. Letters at Kew, BM(NH), Linnean Society. Portr. at Kew, Hunt Library.

JOHNSTON, George (1837–1887)
b. Fingask, Inverness-shire 31 Oct. 1837 *d*. Edinburgh 30 Sept. 1887

Gardener to Earl of Strathmore at Glamis Castle, Angus, 1866–87.

Gdnrs Chron. 1875 i 239 portr.; 1887 ii 473 portr.

JOHNSTON, Sir Harry Hamilton (1858–1927)
b. Kennington, London 12 June 1858 *d*. Worksop, Notts 31 July 1927

DSc Cantab 1902. GCMG 1901. Artist, naturalist and government official. Held various government appointments in Africa. Special Commissioner, Uganda Protectorate, 1899–1901. In Angola, 1882, Congo, 1884. Led scientific expedition to Mt. Kilimanjaro, 1885 (plants enumerated in *Trans. Linn. Soc. Bot*. v.2, 1887, 327–55). In Central Africa, 1888–96. *River Congo* 1884. *Kilimanjaro Expedition* 1886. *British Central Africa* 1897. *Uganda Protectorate* 1902 2 vols. *Liberia* 1906 2 vols (plants described by O. Stapf). *Story of My Life* 1923. A. Johnston *Life and Letters* 1929.

Scott. Geogr. Mag. 1888, 513–36. *Bot. Soc. Exch. Club Br. Isl. Rep*. 1927, 378–79. *J. Bot*. 1927, 258–60. *Kew Bull*. 1927, 315–16. *Times* 1 Aug. 1927 portr. R. Oliver *Sir Harry Johnston and the Scramble for Africa* 1957. *R.S.C*. v.10, 344; v.16, 123.

Plants at Kew.

JOHNSTON, Henry Halcro (1856–1939)
b. 13 Sept. 1856 *d*. Kirkwall, Orkney 18 Oct. 1939

MD. DSc Edinburgh 1894. FLS 1895. FRSE 1895. Army Medical Dept., 1881–1919. Collected plants in Afghanistan, Mauritius, Canaries, Madeira, Egypt, Gambia, Natal, India, Sierra Leone. Collected in Orkney Islands from 1919. Contrib. papers, particularly on flora of Mauritius and Orkney and Shetland Islands to *Trans. Bot. Soc. Edinburgh*.

Kew Bull. 1939, 670. *Proc. Linn. Soc*. 1939–40, 369–71. *Proc. R. Soc. Edinburgh* 1938–39, 271–72. *Who was Who, 1929–1940* 719.

Herb. at Royal Botanic Garden, Edinburgh, Orkney Museum. Plants at BM(NH), Kew, Oxford. Portr. at Hunt Library.

JOHNSTON, James (*fl*. 1790s)
Seedsman, Wolverhampton, Staffs.

JOHNSTON, James Finlay Weir (1796–1855)
b. Paisley, Renfrewshire 13 Sept. 1796 *d*. Durham 18 Sept. 1855

MA Glasgow 1796. FRS 1837. Reader in Chemistry, Durham, 1833–55. *Potato Disease in Scotland* 1845–46. Contrib. notes to G. J. Mulder's *Chem. of Vegetable and Animal Physiol*. 1845. *Chemistry of Common Life* 1855 2 vols.

D.N.B. v.30, 65–66. *R.S.C*. v.3, 562.

JOHNSTON, Lawrence Waterbury (–1958)
d. Menton, France 27 April 1958

Created notable gardens at Hidcote, Glos and Serre de la Madone, Menton.

Gdnrs Chron. 1958 i 363. M. Hadfield *Gdning in Britain* 1960, 427–28.

JOHNSTON, Pelham (*fl*. 1700s)
Collected plants in Spain.

J. E. Dandy *Sloane Herb*. 1958, 142.

JOHNSTON, Robert Mackenzie (1844–1918)
b. Petty, Inverness-shire 27 Nov. 1844 *d*. Hobart, Tasmania 20 April 1918

FLS 1879. To Victoria, 1870. Railway clerk, Tasmania. Registrar-General and Government Statistician of Tasmania, 1881. Botanised in Tasmania with R. C. Gunn (1808–1881). *Field Memoranda for Tasmanian Botanists* 1874.

P. Mennell *Dict. Austral. Biogr*. 1892, 252. *Papers Proc. R. Soc. Tasmania* 1918, 136–42; 1923, 110–13. *Mercury* (Hobart) 22 and 23 April 1918. *J. R. Soc. N.S.W*. v.55, 1921, 163. P. Serle *Dict. Austral. Biogr*. v.1, 1949, 478–79.

JOHNSTON, Miss S. D. (*fl*. 1890s)
Collected plants in Cowhill, Dumfriesshire.

Proc. Trans. Dumfriesshire Galloway Nat. Hist. Antiq. Soc. 1891–92, 12–13.

JOHNSTONE, C. E. (*fl*. 1870s–1900s)
Of Sussex. Plants from Devon, Sussex, Kent, Surrey, London area, France, Germany, Switzerland at Oxford.

H. N. Clokie *Account of Herb. Dept. Bot., Oxford* 1964, 190.

JOHNSTONE, George Horace (1882–1960)
VMH 1951. Created garden at Trewithen, Cornwall, 1904. Authority on rhododendrons, magnolias, camellias, daffodils. *Asiatic Magnolias in Cultivation* 1955.

Daffodil Tulip Yb. 1960, 9–10 portr. *J. R. Hort. Soc*. 1960, 267–71 portr.

JOHNSTONE, James (*c*. 1643–1737)
d. 11 May 1737

Secretary of State for Scotland, 1692–96. Had garden at Orleans House, Twickenham, Middx. "Secretary Johnstone had the best collection of fruit of most gentlemen in England...Dr. Bradley, in his *Treatise on Gardening*, ranked him among the finest gardeners in the kingdom" (J. Macky *Journey through England* 1720).

D.N.B. v.30, 64–65.

JOHNSTONE, R. B. (1856–1934)
b. Australia 1856
To England, 1863. Secretary, Andersonian Naturalists' Society, 1899–1913; President, 1917–18. Mycologist. Collected material for Gastromycetes in *Fauna, Fl. and Geology of Clyde Area* 1901 ed. by G. F. Scott Eliot *et al.*
Glasgow Nat. 1936, 134–35.

JOHNSTONE, Thomas Scott (*fl.* 1870s–1917)
d. Carlisle, Cumberland 5 Feb. 1917
President, Carlisle Natural History Society. Had herb. of Carlisle flora. Contrib. to *Trans. Carlisle Nat. Hist. Soc.*
Naturalist 1917, 110–11.
Plants at Carlisle Museum.

JOHNSTONE, William Grosart (–*c.* 1860)
d. London *c.* 1860
Nature-printed British Sea-weeds (with A. Croall) 1859–70 4 vols.
Scott. Nat. 1885, 150.

JOINVILLE (Jonville), Joseph (*fl.* 1800s)
F. North, Governor of Ceylon, brought Joinville to Ceylon as "clerk for natural history and agriculture." In charge of Governor's garden at Peliyagoda. Collected plants at Kandy, 1880. Drew plates for Cordiner's *Description of Ceylon* 1807.
Ann. R. Bot. Gdns Peradeniya v.1, 1901, 3.
Plants at BM(NH).

JOLLIFFE, John (1822–1887)
b. Niton, Isle of Wight? 1822 *d.* Isle of Wight 1887
Assistant surgeon, Royal Navy, 1845. Staff surgeon, survey ship, 'Pandora', surveying New Zealand, 1850–56. Collected New Zealand plants.
Diary, 1851–56 at Mitchell Library, Sydney.

JOLY, John (1857–1933)
b. Hollywood, King's County, Ireland 1 Nov. 1857 *d.* Dublin 8 Dec. 1933
Educ. Trinity College, Dublin. FRS 1892. Professor of Geology, Dublin, 1897–1933. Studied ascent of sap in plants with H. H. Dixon (1869–1953).
Obit. Notices Fellows R. Soc. London 1934, 259–86 portr. H. H. Dixon *John Joly* 1941 portr.

JONES, Alan Philip Dalby (1918–1946)
d. Nigeria 8 Oct. 1946
BSc Cantab 1940. Forestry officer, Nigeria, 1941–46. Botanist, Forest Herb., Ibadan.
Kew Bull. 1947, 37.
Nigerian plants at BM(NH), Kew.

JONES, Arthur Coppen (1866–1901)
b. London 1866 *d.* Davos, Switzerland 8 March 1901
FLS 1891. Bacteriologist. Pupil of T. H. Huxley. Translated A. Fischer's *Structure and Functions of Bacteria* 1900.
J. Bot. 1901, 191–92. *R.S.C.* v.16, 133.

JONES, Arthur Mowbray (1826–1889)
b. Ringwood, Hants 8 Jan. 1826 *d.* Clifton, Bristol 28 Feb. 1889
Entered Army, 1849; retired as Colonel, West York Militia, 1882. Pteridologist and lichenologist. One of founders of Pteridologist Society. 'Abnormal Ferns' (*Ann. Bot.* v.3, 1889, 27–31). Supervised publication of *Nature-printed Impressions of Varieties of British Species of Ferns* 1876–80.
Gdnrs Chron. 1889 i 310. E. J. Lowe *Fern Growing* 1895, 172–76 portr. *Br. Fern Gaz.* v.1, 1910, 65–70. *R.S.C.* v.16, 133.
Lichens and fungi at National Museum, Dublin.

JONES, Daniel Angell (1861–1936)
b. Liverpool 14 July 1861 *d.* Bristol 6 Oct. 1936
MSc Wales 1918. ALS 1925. Schoolmaster, Machynlleth, 1886–92 and Harlech, 1892–1924. Secretary, British Bryological Society; President, 1935–36. Contrib. flora to T. P. Ellis's *Story of Two Parishes, Dolgelly and Llanelltyd* 1928. Contrib. to *J. Bot.*, *Naturalist*.
Annales Bryologici 1936, 155 portr. *Bot. Soc. Exch. Club Br. Isl. Rep.* 1936, 212. *J. Bot.* 1936, 351–52. *Nature* v.138, 1936, 871. *N. Western Nat.* 1936, 374–75. *Proc. Linn. Soc.* 1936–37, 196–97. *Chronica Botanica* 1937, 154. H. J. Riddelsdell *Fl. Gloucestershire* 1948, cl.
Plants at BM(NH). MS. flora of Merioneth and plants at National Museum of Wales.

JONES, David Thomas (*fl.* 1810s)
Of Llanllyfni, Caernarvonshire. *Herbal, neu Lysieu-Lyfr* 1817; ed. 3 *c.* 1862.
J. Bot. 1898, 18.

JONES, Edward (*fl.* 1840s)
Nurseryman, Holywell, Flintshire.

JONES, Eric Marsden Marsden- *see* Marsden-Jones, E. M.

JONES, Fred Ronald (1925–1967)
b. 27 March 1925
FLS 1955. Lecturer, Chelsea Polytechnic. 'Notes on Alga Compsopogon Mont.' (*J. Linn. Soc.* 1955, 261–70).

JONES, George Howard (–1945)
MA Cantab. Mycologist, Dept. of Agriculture, Nigeria, 1924–29. Chief mycologist, Ministry of Agriculture, Egypt, 1929–39.

JONES, George Neville (1904–1970)
b. Boston, Lincs 26 Jan. 1904 *d.* Urbana, U.S.A. 25 June 1970
BSc State College, Washington 1930. PhD 1937. Emigrated to Canada with parents. Instructor in Biology, Harvard, 1937–39. Professor of Botany, Illinois. Authority on flora of Illinois. *Vascular Plants of Illinois* (with G. D. Fuller) 1955. *Taxonomy of American Species of Linden* 1968.
Plant Sci. Bull. v.16(3), 1970, 10. *Taxon* 1971, 597–602 portr.
Portr. at Hunt Library.
Rosa jonesii St. John.

JONES, H. J. (*c.* 1856–1928)
d. Plymouth, Devon July 1928
VMH 1925. In partnership with Norman Davis as chrysanthemum grower at Camberwell, London. Nurseryman, Ryecroft Nursery, Hither Green, Lewisham, 1890. *Chrysanthemum Guide* 1895–97 3 vols. *Chrysanthemum Album* 1896. *Chrysanthemum* 1898. *Portfolio of New Chrysanthemums* 1901.
Gdnrs Chron. 1928 ii 60 portr.

JONES, Harry Richard Briton- *see* Briton-Jones, H. R.

JONES, Rev. Hugh (*fl.* 1690s–1701)
d. Jan. 1701
Went to Maryland, 1696. Minister of Christ Church parish, County Calvert, Virginia. Sent "several volumes of plants" from Maryland to J. Petiver. "A very curious person in all parts of Natural History" (Petiver). 'Remarks by Mr. James Petiver... on some Animals, Plants, etc. sent him from Maryland by the Reverend Mr. Hugh Jones' (*Philos. Trans. R. Soc.* 1698, 393–406).
E. J. L. Scott *Index to Sloane Manuscripts* 1904. *Proc. International Congress Plant Sci. 1926* v.2, 1929, 1529. *Proc. Amer. Antiq. Soc.* 1952, 292–303. J. E. Dandy *Sloane Herb.* 1958, 142–43.
Plants at BM(NH).

JONES, James (*fl.* 1760s–1770s)
Nurseryman, Dublin.
Irish For. 1967, 47.

JONES, James (*fl.* 1830s)
Nurseryman, Pontefract, Yorks. 'On Culture of Dahlia' (*Floricultural Cabinet* 1834, 198).

JONES, Jezreel (–1731)
d. London 21 May 1731
Traveller. Clerk to Royal Society, 1698. Secretary, Portuguese Embassy. Consul at Algiers. In Barbary, 1698–99 and 1701–4. British envoy to Morocco, 1704. Sent plants from Portugal to J. Petiver and from Spain to H. Sloane. Collected and drew plants in Barbary.
J. Petiver *Museii Petiveriani* 1695, 45. *Gent. Mag.* 1731, 221. S. Ayscough *Cat. of Manuscripts...in British Museum* 1782, 648. *D.N.B.* v.30, 122. E. J. L. Scott *Index to Sloane Manuscripts* 1904, 283. J. E. Dandy *Sloane Herb.* 1958, 143–44.
Plants at BM(NH).

JONES, John (*fl.* 1770s)
Nurseryman, Dublin. Kaven as partner, 1776–78. Connected with Dickson, Scottish nurseryman.
Irish For. 1967, 47.

JONES, Rev. John Evans (*c.* 1858–1937)
d. St. Asaph, Flintshire 16 July 1937
Educ. Oxford 1882. Vicar, Brymbo, 1897–1907; Dyserth, Flintshire, 1907–34. Canon of St. Asaph, 1928. President, Dyserth and District Field Club. Botanist.
N. Western Nat. 1937, 410–12 portr.

JONES, John Matthew (1828–1888)
b. Frontfaith Hall, Montgomery 7 Oct. 1828 *d.* Halifax, Nova Scotia 7 Oct. 1888
FLS 1859. Zoologist. To America, *c.* 1854. 'On the Vegetation of the Bermudas' (*Trans. Nova Scotia Inst. Sci.* v.3, 1874, 237–80).
Proc. Nova Scotia Inst. Sci. v.10, 1898–1902, lxxx–lxxxii. *R.S.C.* v.3, 573; v.8, 34; v.10, 349; v.12, 371.

JONES, Rev. John Pike (1790–1857)
b. Chudleigh, Devon 1790 *d.* Cheadle, Staffs 4 Feb. 1857
BA Cantab 1813. Curate, North Bovey, Devon. *Botanical Tour through Various Parts of the Counties of Devon and Cornwall* 1820; ed. 2 1821. *Fl. Devoniensis* (with J. F. Kingston) 1829.
T. R. A. Briggs *Fl. Plymouth* 1880, xxix. *J. Bot.* 1882, 74. *Notes and Queries* v.12, 1855, 29. *D.N.B.* v.30, 141. F. H. Davey *Fl. Cornwall* 1909, xxxviii–xxxix. W. K. Martin and G. T. Fraser *Fl. Devon* 1939, 771.
Herb. at Exeter Museum.

JONES, John Wynne (1859–1923)
b. Caernarvon 18 Feb. 1859 *d.* Pittsburgh, U.S.A. 17 Aug. 1923
Gardener, Penrhyn Castle. To U.S.A. as young man. Foreman, Phipps Conservatory, Schenley Park, Pittsburgh. President, Botanical Society of Western Pennsylvania, 1912–13.
Trillia no. 7, 1921–23, 1–2 portr.

JONES, Joseph (1867–1934)
b. Cheshire 27 Feb. 1867 *d.* Dominica 1 May 1934
Kew gardener, 1889. Curator, Botanic Garden, Dominica, 1892–1925.
J. Kew Guild 1921, 3 portr.; 1934, 375; 1935, 468–69 portr.

JONES, Richard (1740–1831)
b. Morville, Shropshire 2 May 1740 *d.* Packington, Warwickshire 5 Aug. 1831
Gardener to Earl of Aylesford at Packington, 1781.
Gdnrs Mag. 1831, 639–40.

JONES, Robert C. Fowler (1865–1952)
d. Ilkley, Yorks 19 April 1952
Mycologist. Financed publication of *Cat. Yorkshire Fungi* 1937.
Naturalist 1952, 129; 1961, 64–65.

JONES, Theobald (1790–1868)
b. Dublin 1790 *d.* London 7 Feb. 1868
FLS 1842. In Royal Navy, 1803–65. Admiral. Lichenologist. Contrib. papers on Irish lichens to *Proc. Dublin Nat. Hist. Soc.* v.3, 1863–65, 114–50, 280–90.
J. Bot. 1866, 158. *Proc. Linn. Soc.* 1867–68, cv–cvii. F. Boase *Modern English Biogr.* v.2, 1897, 142. *Proc. Irish Acad.* v.38B, 1929, 185–86. R. L. Praeger *Some Irish Nat.* 1949, 110.

JONES, William (*fl.* 1730s)
Of Anglesey. Plants from him in Dillenius's herb., Oxford.

JONES, Sir William (1746–1794)
b. London 28 Sept. 1746 *d.* Calcutta 27 April 1794
MA Oxon 1773. FRS 1772. FLS 1791. Orientalist. Judge of Supreme Court, Calcutta, 1783–94. Knighted, 1793. 'Catalogue of Indian Plants comprehending their Sanscrit Names...' (*Asiatick Researches* v.4, 1792, 229–36). 'Botanical Observations on Select Indian Plants' (*Asiatick Researches* v.4, 1792, 237–312).
Works of Sir William Jones 1799 6 vols. *D.N.B.* v.30, 174–77. A. J. Arberry *Asiatic Jones* 1946 portr. Asiatic Society of Bengal *Sir William Jones* 1948 portr. *Hist. Today* 1971, 57–64 portr. *Notes Rec. R. Soc. London* 1975, 205–30. *R.S.C.* v.3, 576.
Jonesia Roxb.

JONES, William Neilson (1883–1974)
d. 8 Oct. 1974
MA Cantab. Lecturer in Botany, University College, Reading, 1908. Assistant Lecturer in Botany, Bedford College, London, 1913; Professor of Botany, 1920–48. *Textbook of Plant Biology* (with M. C. Rayner) 1920. *Plant Chimeras and Graft Hybrids* 1934; ed. 2 1969. *Problems in Tree Nutrition* (with M. C. Rayner) 1944.
Times 31 Oct. 1974.

JONVILLE, Joseph *see* Joinville, J.

JORDAN, A. J. (*c.* 1873–1906)
d. 6 Aug. 1906
Kew gardener, 1898. Agricultural Instructor, Montserrat, West Indies, 1899. Curator, Botanic Station, Montserrat; Antigua, 1905. Government House Gardens, Trinidad, 1905.
J. Kew Guild 1906, 329. *Kew Bull.* 1906, 395–96.

JORDAN, Charles (–1907)
d. 8 July 1907
Nurseryman, Silverhall Nursery, Isleworth, Middx. *c.* 1871. Superintendent, Victoria, Greenwich and Hyde Parks.
J. Hort. Cottage Gdnr v.55, 1907, 53.

JORDAN, F. (*c.* 1865–1958)
b. Yorks *c.* 1865 *d.* Edenbridge, Sussex 4 Aug. 1958
VMH 1930. Gardener at Ford Manor, Lingfield, Surrey 1916–35.
Gdnrs Chron. 1931 i 80 portr.; 1956 ii 401 portr.; 1958 ii 102.

JORDAN, Robert C. R. (*fl.* 1840s)
Of Teignmouth and Lympstone, Devon. Contrib. plant records for Teignmouth district to *Phytologist*.
W. K. Martin and G. T. Fraser *Fl. Devon* 1939, 772.
Plants at Birmingham.

JORDEN, George (1783–1871)
b. Clee Hills, Farlow, Shropshire 1783 *d.* Bewdley, Worcs 1871
Butler to J. Fryer of Bewdley. Self-taught botanist. Studied *Rubi*. Discriminated *Thymus serpyllum* and *T. chamaedrys*. Contrib. to *Phytologist* v.1–6, 1855–63.
F. Boase *Modern English Biogr.* v.2, 1897, 148. J. Amphlett and C. Rea *Bot. Worcestershire* 1909, xxiv. *R.S.C.* v.3, 580.
MS. *Fl. Bellus Locus* [*Bewdley*] and herb., mainly *Rubi*, at Worcester Museum.

JOSHUA, William (1828–1898)
b. London 13 Aug. 1828 *d.* Cheltenham, Glos 18 Jan. 1898
FLS 1877. Lichenologist and algologist. *Microscopical Slides of British Lichens* nos. 1–48, 1879. *Microscopical Slides of British Freshwater Algae* (with E. M. Holmes) nos. 1–48, 1880. Papers on *Desmidieae* in *J. Bot.*, 1882–83.
H. J. Riddelsdell *Fl. Gloucestershire* 1948, cxxxii. *R.S.C.* v.10, 358.
Herb. and microscopic preparations at BM(NH). Drawings of *Desmidieae* at Kew.

JOSSELYN, John (*fl.* 1630s–1670s)
b. Willingdale Doe, Essex?
In Massachusetts, 1633–74. *New-England's Rarities discovered in Birds, Beasts, Fishes, Serpents and Plants of that Country* 1672 (plants, pp. 41–91). *Two Voyages to New England* 1674.
D.N.B. v.30, 208.

JOWETT, Thomas (*c*. 1801–1832)
b. Colwick, Notts *c*. 1801 *b*. Morton, Notts 1832

Surgeon, Nottingham. Published a series of 'Botanical Calendars' of Nottingham plants in *Nottingham J*. 1826 under name of Il Rosajo.

Annual Rep. Trans. Nottingham Nat. Soc. 1906–7, 59–72. *J. Bot*. 1909, 134, 139. R. C. L. Howitt and B. M. Howitt *Fl. Nottinghamshire* 1963, 16.

Herb. formerly at Bromley House Library, Nottingham.

JOWITT, John Fort (1846–1915)
b. 16 Sept. 1846 *d*. London 1915

BA Oxon 1869. Ceylon planter. '*Apluda varia*' (*Ann. R. Bot. Gdns Peradeniya* 1907, 85–88). '*Cymbopogon nardus*' (*Ann. R. Bot. Gdns Peradeniya* 1908, 185–93).

JOYCE, John (*fl*. 1790s)

Nurseryman, 87 High Street, Gateshead, County Durham.

JOYCE, Stanley (*fl*. 1750s–1770s)

Son of W. Joyce (–1767). Nurseryman, Gateshead, County Durham.

JOYCE, William (–1767)

Nurseryman, Gateshead, County Durham.

Gdn Hist. v.2(2), 1974, 34–44.

JUDD, Daniel (1815–1884)
b. Edmonton, Middx Oct. 1815 *d*. Shefford, Beds Dec. 1884

Gardener to Earl Spencer at Althorpe, 1848, Lord Hill at Hawkstone, 1864–75, Earl of Warwick at Warwick Castle.

Gdnrs Chron. 1875 i 785 portr.; 1884 ii 761.

JUDD, William H. (1888–1946)
b. Preston Brook, Cheshire 14 July 1888 *d*. Boston, U.S.A. 23 May 1946

Kew gardener, 1910–13. To Arnold Arboretum, U.S.A., 1913 where he remained as plant propagator.

Gdnrs Chron. 1932 i 232 portr.; 1946 i 274. *Arnoldia* 1946, 25–28 portr. *J. Kew Guild* 1946–47, 595–98.

Portr. at Hunt Library.

JULIUS, Sir George Alfred (1873–1946)
b. Norwich, Norfolk 29 April 1873 *d*. Sydney, N.S.W. 28 June 1946

BSc Canterbury, N.Z. 1896. To Victoria with his parents, 1884. In New Zealand, 1890. Wood technologist. *Physical Characteristics of Hardwoods of Western Australia* 1906. *Economic Use of Australian Hardwoods* 1907. *Physical Characteristics of Hardwoods of Australia* 1907.

Who's Who in Australia 1938, 281–82. *Austral. J. Sci*. 1946, 14–15. *Sydney Morning Herald* 29 June 1946 portr. *Who was Who, 1941–1950* 620. *Austral. Encyclop*. v.1, 1965, 361; v.5, 149.

Portr. at Hunt Library.

JURIN, James (1684–1750)
b. London? Dec. 1684 *d*. London 29 March 1750

BA Cantab 1705. MD Cantab 1716. FRS 1717 or 1718. Secretary, Royal Society, 1721–27. Correspondent of Sir H. Sloane and J. Petiver.

D.N.B. v.30, 229–30. J. E. Dandy *Sloane Herb*. 1958, 144.

JUST, John (1797–1852)
b. Natland, Kendal, Westmorland 3 Dec. 1797 *d*. Bury, Lancs 14 Oct. 1852

Assistant schoolteacher, Kirkby Lonsdale, 1817 and Bury, 1832. Cryptogamic botanist and geologist. Lecturer in Botany, Manchester School of Medicine, 1833–52. Professor of Botany, Manchester Institution, 1848.

Phytologist v.1, 1844, 396. *Gent. Mag*. 1852 ii 652–53. *Mem. Manchester Lit. Philos. Soc*. v.11, 1854, 91–121. J. Cash *Where there's a Will there's a Way* 1873, 136–45. *D.N.B*. v.30, 230–31. *Trans. Liverpool Bot. Soc*. 1909, 79.

JUSTEN, Frederick (1832–1906)
b. Bonn, Germany 29 Feb. 1832 *d*. Soho, London 20 Nov. 1906

FLS 1886. Bookseller (Dulau and Co.) Executor of F. Welwitsch. Well acquainted with botanical literature.

Notes and Queries v.6, 1906, 458. *St. Annes Mon. Paper* 1906, 534 portr. *J. Bot*. 1907, 62 portr. *Proc. Linn. Soc*. 1906–7, 48.

Justenia Hiern.

JUSTICE, James (1698–1763)
b. 25 Sept. 1698 *d*. Leith, Midlothian 2 Aug. 1763

FRS 1730. Principal Clerk to Court of Sessions, 1727. Spent the greater part of his fortune gardening on his estates at Crichton, Midlothian and Justicehall, Berwickshire. A 'tulip maniac'. The first to fruit pine-apples in Scotland. *Directions for Propagating Hyacinths* 1743. *Scots Gardiners Director* 1754; reissued as *British Gardener's Director* 1764. *British Gardener's Calendar* 1759; ed. 4 1765.

J. C. Loudon *Encyclop. Gdning* 1822, 1271. G. W. Johnson *Hist. English Gdning* 1829, 206–7. *Gdnrs Chron*. 1899 i 309–10. *Gdn Hist*. v.2(1), 1973, 41–62; v.2(2), 1974, 51–74; v.3(2), 1975, 37–67; v.4(2), 1976, 53–91.

Justicia Houston.

KANE, Lady Katherine Sophia (*née* Baily) (1811–1886)
b. 11 March 1811 *d*. Dublin 25 Feb. 1886
Married Sir John Kane, 1838. *Irish Fl.* 1833.
D.N.B. v.30, 239. N. Colgan *Fl. County Dublin* 1904, xxvii. *Annual Rep. Proc. Belfast Nat. Field Club* 1913, 623. R. W. Scully *Fl. Kerry* 1916, xiii. R. L. Praeger *Some Irish Nat.* 1949, 111.
Herb. at University College, Cork.

KAY, P. Crichton (–1954)
d. Cowley, Middx 16 Oct. 1954
VMH 1951. Authority on production of flowers for market. President, British Flower Industry Association.
Gdnrs Chron. 1952 i 46 portr.; 1954 ii 180.
Portr. at Hunt Library.

KAY, Peter E. (*c*. 1853–1909)
d. Aug. 1909
VMH 1897. Claigmar Vineries, Church End, Finchley, London.
Gdnrs Chron. 1909 ii 160.

KAYE, Sir Richard (1736–1809)
b. Kirkheaton, Yorks 11 Aug. 1736 *d*. Lincoln 25 Dec. 1809
BA Oxon 1757. DCL. FRS 1765. Dean of Lincoln. List of plants in flower at Welbeck, 29 Aug. 1777, and at Kirkby, 16 Aug. 1774, containing first records for Nottingham (BM Add. MSS. 18, 565 in R. W. Goulding's *Sir Richard Kaye, Bart.* 1925 portr.).
Bot. Soc. Exch. Club Br. Isl. Rep. 1925, 813.

KAYE SMITH, Miss A. Dulcie (–1955)
d. Sept. 1955
Secretary, Hastings Natural History Society, 1926–50. Contrib. to A. H. Wolley-Dod's *Fl. Sussex* 1937.
Hastings and East Sussex Nat. v.8, 1956, 157.

KEARSE, Mrs. Mary *see* Lawrance, M.

KEARSLEY, John (*fl*. 1860s)
Nurseryman, Woodhouse Hill, Hunslet and 44 Vicar Lane, Leeds, Yorks.

KEARTLAND, George Arthur (1848–1926)
b. Wellingborough, Northants 11 June 1848
Australian plant collector and ornithologist. Zoologist and botanist on Calvert expedition to N.W. Australia, 1896.
Victorian Nat. 1926, 48–52 portr.
Gardenia keartlandi Tate.

KEBLE MARTIN, Rev. W. *see* Martin, Rev. W. K.

KEDIE, William (*fl*. 1830s)
Nurseryman, Mundford Nursery near Brandon, Suffolk.

KEEBLE, Sir Frederick William (1870–1952)
b. London 2 March 1870 *d*. London 19 Oct. 1952
Educ. Cambridge. FRS 1913. Knighted, 1922. Plant physiologist, Ceylon. Assistant Lecturer in Botany, Owens College, Manchester, 1897–98. Lecturer in Botany, Reading, 1902; Professor, 1907–14. Director, Royal Horticultural Society gardens, Wisley, 1914–19. Professor of Botany, Oxford, 1920–27. Agricultural Adviser to I.C.I., 1927–32; established Jealott's Hill Research Station. Fullerian Professor, Royal Institution, 1938–41. Scientific Editor, *Gdnrs Chron.* 1908–19. *Practical Plant Physiology* 1911. *Life of Plants* 1926. *Polly and Freddie* 1936 (autobiography). *Science lends a Hand in the Garden* 1939.
Gdnrs Chron. 1922 i 298 portr.; 1952 ii 170. *Pharm. J.* 1952, 297. *Times* 21 Oct. 1952. *Nature* v.171, 1953, 63. *D.N.B. 1951–1960* 564–65. *Who was Who, 1951–1960* 603.

KEEGAN, Peter Quinn (–1916)
d. Patterdale, Westmorland 10 Aug. 1916
LLD Dublin. 'Experiments in Floral Colours' (*Nature* v.61, 1899, 105–6). 'Leaf Decay and Autumn Tints' (*Nature* v.69, 1903, 30). Contrib. articles on plant chemistry and colour to *Naturalist* and *Knowledge* 1910–11.
Nature v.98, 1916, 296. *Times* 28 Nov. 1916.

KEELING, A. J. (*c*. 1858–1920)
d. 13 Sept. 1920
Nurseryman, Grange Nurseries, Westgate Hill, Bradford, Yorks. Hybridised orchids.
Orchid Rev. 1920, 163–64.

KEELING, A. J. (–1962)
Son of A. J. Keeling (*c*. 1858–1920). Acquired his father's orchid nursery at Westgate Hill, Bradford, Yorks.
Orchid Rev. 1962, 326–27.

KEEN, Isaac (*fl*. 1780s–1810s)
Nurseryman, East Street, Southampton, Hants.

KEENAN, Richard Lee (*fl*. 1900s)
Kew gardener, 1865–67. Tea planter at Cachar, India, 1867. Sent plants to Kew.

KEENS, Michael (*c*. 1762–1835)
b. Isleworth, Middx *c*. 1762
Market gardener, Isleworth. 'Account of New Strawberry' (*Trans. Hort. Soc.* 1814, 101–2). 'On Cultivation of Strawberries in the Open Ground' (*Trans. Hort. Soc.* 1817, 392–97).
J. C. Loudon *Encyclop. Gdning* 1822, 1287. *Gdnrs Chron.* 1950 ii 30. G. M. Darrow *The Strawberry* 1966, 79–80.

KEILL, James (1673–1719)
b. Scotland 27 March 1673 *d.* Northampton 16 July 1719
MD Cantab. FRS 1711. Physician, Northampton, 1703–19. Collected plants in Cyprus and Rhodes which he sent to J. Petiver.
J. Petiver *Museii Petiveriani* 1695, 219. *D.N.B.* v.30, 309–10. J. E. Dandy *Sloane Herb.* 1958, 148.

KEIR, Walter (*fl.* 1690s)
Surgeon. Sent plants from Johore and China to J. Petiver.
J. Petiver *Museii Petiveriani* 1695, 44, 45, 80. *Fl. Malesiana* v.1, 1950, 276. J. E. Dandy *Sloane Herb.* 1958, 148.
Plants at BM(NH).

KEITH, Sir Arthur (1866–1955)
b. Persley, Aberdeen 5 Feb. 1866 *d.* Downe, Kent 7 Jan. 1955
MB Aberdeen 1888. FRS 1913. Anatomist and anthropologist. To Siam as medical officer to Gold Fields of Siam Ltd., 1889. Collected plants in Bangtapan, now at Singapore Herb. Senior Demonstrator in Anatomy, London Hospital, 1895; Head of Dept., 1899. Conservator, Royal College of Surgeons Museums, 1908.
J. Thailand Res. Soc. Nat. Hist. Supplt. v.12, 1939, 21. *Biogr. Mem. Fellows R. Soc.* 1955, 145–62. *Who was Who, 1951–1960* 805. C. C. Gillispie *Dict. Sci. Biogr.* v.7, 1973, 278–79.

KEITH, Rev. George Skene (1752–1823)
b. Mar near Aberdeen 6 Nov. 1752 *d.* Tulliallan, Perthshire 7 March 1823
MA Aberdeen 1770. Minister, Keith-Hall and Kinkell, 1778–1822; Tulliallan, 1822–23. 'Observations on British Grasses' in his *General View of Agriculture of Aberdeenshire* 1811.
D.N.B. v.30, 322–24.

KEITH, Rev. James (1825–1905)
b. Keith, Banffshire 23 Dec. 1825 *d.* Forres, Morayshire 11 Aug. 1905
MA Aberdeen 1845. LLD 1882. Minister at Forres. Mycologist. Papers on fungi and mosses in *Scott. Nat.*
Ann. Scott. Nat. Hist. 1905, 193–95 portr. *J. Bot.* 1905, 334–35.
Plants at Forres Museum.
Peziza keithia Phillips.

KEITH, Rev. Patrick (1769–1840)
b. Scotland 1769 *d.* Stalisfield, Kent 25 Jan. 1840
MA Glasgow. FLS 1805. Vicar, Stalisfield. *System of Physiological Botany* 1816 2 vols. *Botanical Lexicon* 1837.
Proc. Linn. Soc. 1840, 70–71. *R.S.C.* v.3, 628.
Keithia Benth.

KELAART, Edward Frederick (*c.* 1818–1860)
b. Ceylon *c.* 1818 *d.* at sea 31 Aug. 1860
MD 1841. FLS 1846. Returned to Ceylon, 1841 and 1849. At Gibraltar, 1843–45. *Fl. Calpensis* 1846.
Proc. Linn. Soc. 1861, xli–xlii. *R.S.C.* v.3, 630.
Letters at Kew.

KELCEY, Mrs. Frances Louisa Foord- *see* Foord-Kelcey, F. L.

KELLERMAN, August Emil (1813–1847)
d. Edinburgh 27 June 1847
Assistant Curator, Edinburgh Botanical Society, 1837–39.
Annual Rep. Bot. Soc. Edinburgh 1837–38, 20, 79. *Botanische Zeitung* 1847, 692–93.

KELSALL, Miss E. J. (1832–1897)
d. Blackrock, County Dublin 28 June 1897
Field botanist.
Irish Nat. 1897, 233.

KELSALL, Harry Joseph (*fl.* 1890s–1920s)
Lieut.-Colonel, Royal Artillery. Stationed at Singapore. Made expedition with Harry Lake from east to west of Johor, 1892. Ornithologist 'St. George' Pacific Expedition, 1924–25. Collected plants for Singapore Botanic Gardens. 'Notes on Trip to Bukit Etam, Selangor' (*J. Straits Branch R. Asiatic Soc.* no. 23, 1891, 67–75). 'Account of Trip up to Pahang, Tembeling and Tahan Rivers' (*J. Straits Branch R. Asiatic Soc.* no. 25, 1894, 33–56).
Kew Bull. 1926, 51. *Gdns Bull. Straits Settlements* 1927, 125. *Fl. Malesiana* v.1, 1950, 276.

KELSALL, Rev. John Edward (–1924)
Botanised in Milton district, Hants.
J. F. Rayner *Supplt. to Townsend's Fl. Hampshire* 1926, xiv.

KELWAY, James (1815–1899)
b. Westholme, Somerset 2 Nov. 1815 *d.* Langport, Somerset 17 May 1899
Nurseryman, Langport, 1851. Specialised in gladioli.
Garden v.36, 1889 portr.; v.55, 1899, 378. *Gdnrs Mag.* 1899, 305–6 portr. *Gdnrs Chron.* 1899 i 343–44 portr.

KELWAY, William (1839–)
Son of J. Kelway (1815–1899), nurseryman, Langport, Somerset, who took him into partnership in 1864.
Garden v.58, 1900, 148–49 portr.; 1909, 503–4 portr. *J. Hort. Home Farmer* v.64, 1912, 54–55 portr.

KEMP, George (*fl.* 1820s–1840s)
Nurseryman, Barnet, Herts.

KEMPE, Rev. Hermann (1844–*c.* 1907)
b. Deuben, Saxony 26 March 1844
Lutheran missionary in S. Australia, 1875–93. Sent plants to F. von Mueller. Central Australian plants in *Trans. Proc. R. Soc. S. Australia.*
Rep. Austral. Assoc. Advancement Sci. 1907, 187.
Plants at Adelaide.
Acacia kempeana F. Muell.

KEMPLIN, Hugh (*fl.* 1740s)
Nurseryman, Cork, Ireland.
Irish For. 1967, 51.

KENDRICK, James (1771–1847)
b. Warrington, Lancs 14 Jan. 1771 *d.* Warrington 30 Nov. 1847
MD Edinburgh 1833. FLS 1802. Practised medicine in Warrington from 1793. President, Warrington Natural History Society. Friend of T. Nuttall. Contrib. notes to G. Crosfield's *Calendar of Flora* 1810. Memoir with portr. in his *Profiles of Warrington Worthies* ed. 2 1854.
Palatine Notebook v.2, 1882, 113–16 portr. *D.N.B.* v.30, 410. *Trans. Liverpool Bot. Soc.* 1909, 79.
Rhododendron kendrickii Nutt.

KENNEDY, Edmund Besley Court (1818–1848)
b. Guernsey, Channel Islands 5 Sept. 1818 *d. c.* 11 Dec. 1848
Assistant surveyor, Sydney, N.S.W., 1840. Led expedition along Victoria River, 1847, and collected plants. Killed by aborigines on expedition to Cape York, 1848.
J. Macgillivray *Voyage of HMS Rattlesnake* 1852, Addendum. W. Carron *Narrative of Expedition of...E. B. Kennedy...* 1849. R. L. Jack *Northmost Australia* 1925 *passim. J. Proc. R. Austral. Hist. Soc.* 1949, 1–25. *Austral. Encyclop.* v.5, 1965, 175. *Austral. Dict. Biogr.* v.2, 1967, 43–44.
MSS. at Mitchell Library, Sydney.

KENNEDY, Georgiana *see* Molloy, G.

KENNEDY, John (–1790)
d. Jan. 1790
Gardener to Sir T. Gascoigne at Parlington House, Yorks. *Treatise upon Planting, Gardening...*1776.
Gdnrs Chron. 1912 ii 187–88.

KENNEDY, John (1759–1842)
b. Hammersmith, Middx 30 Oct. 1759 *d.* Eltham, Kent 18 Feb. 1842
Son of Lewis Kennedy (*c.* 1721–1782). Entered into partnership with James Lee (1715–1795) at Vineyard Nursery, Hammersmith. Sold his share in the nursery in 1818 and retired to Eltham. Advised Empress Josephine on her garden at Malmaison for which he also provided plants. Author of *Page's Prodromus* 1817. Wrote many plant descriptions for *Bot. Repository* v.1–5, 1797–1804.
E. P. Ventenat *Jardin de la Malmaison* 1803, t.104. J. C. Loudon *Encyclop. Gdning* 1822, 1291. G. W. Johnson *Hist. English Gdning* 1829, 301. *J. Bot.* 1904, 296–97; 1916, 241. E. J. Willson *James Lee and the Vineyard Nursery, Hammersmith* 1961, 52–55.
Kennedia Vent.

KENNEDY, Lewis (–1743)
d. 6 Oct. 1743
Gardener to Duke of Bedford.
Gent. Mag. 1743, 543–44.

KENNEDY, Lewis (*c.* 1721–1782)
Partner of James Lee (1715–1795) at Vineyard Nursery, Hammersmith, Middx *c.* 1745.

KENNEDY, Patrick Beveridge (1874–1930)
b. Mt. Vernon, Scotland 17 June 1874 *d.* Berkeley, U.S.A. 18 Jan. 1930
AB Toronto, 1894. PhD Cornell 1899. Associate Professor of Botany, Nevada, 1900; Professor of Botany, 1904–13. Associate Professor of Agronomy, Berkeley, 1913; Professor. Professor, California Botanical Society, 1915.
Bot. Soc. Amer. Publ. v.105, 1931, 19–20. *Madrono* 1931, 34–35 portr. *Torreya* v.45(3), 1945, 93–96. *J. Range Management* v.4, 1951, 107–11.
Nevada plants at Kew. Portr. at Hunt Library.

KENNEDY, Richard (1785–1810)
b. Kilmore, County Down 1785 *d.* Kilmore 15 June 1810
MD. First discovered *Hottonia palustris* in Ireland.
Belfast Mon. Mag. June 1810. *Irish Nat. J.* 1935, 306–7.

KENNEDY, William A. (–1922)
d. Finchley, London 12 July 1922
Kew gardener, 1880. Government Cinchona plantations, Darjeeling, 1880. Curator, Botanic Garden, Darjeeling. Returned to England on retirement in 1911.
J. Kew Guild 1923, 177.

KENNEDY-COOKE, Brian (1894–1963)
b. 22 Oct. 1894 *d.* 13 June 1963
Educ. Oxford. Sudan Political Service, 1920. Governor, Kassala Province, 1935–41. *Trees of Kassala Province* 1944.
Who was Who, 1961–1970 624.

KENNION, Edward (1744–1809)
b. Liverpool 15 Jan. 1744 *d.* London 14 April 1809

FSA. Teacher of Drawing, London, 1789. *Essay on Trees in Landscape* 1815 (with memoir).

D.N.B. v.31, 12–13. *Trans. Liverpool Bot. Soc.* 1909, 79.

KENNON, Joseph (*fl.* 1830s)
Nurseryman, Bagnigge Wells Road, London.

KENRICK, George Cranmer (1806–1869)
d. 13 Nov. 1869

FLS 1832. MRCS. In practice at Melksham, Wilts. "He took a great interest in botany."

Proc. Linn. Soc. 1869–70, cii.

KENT, Adolphus Henry (1828–1913)
b. Bletchingley, Surrey 1828 *d.* Fulham, London 12 Sept. 1913

BA London 1871. ALS 1889. Employed by Messrs Veitch for 35 years; private secretary to Sir H. J. Veitch. *Manual of Coniferae* 1881; ed. 2 1900. *Manual of Orchidaceae* 1887–94.

Gdnrs Chron. 1913 ii 211. *J. Bot.* 1913, 304–5. *Orchid Rev.* 1913, 302–3. *Proc. Linn. Soc.* 1913–14, 52–53.

KENT, Elizabeth (*fl.* 1820s)
Of London. Gave lessons in botany. *Fl. Domestica* 1823. *Sylvan Sketches* 1825. 'Linnean System of Plants' (*Mag. Nat. Hist.* v.1–2, 1829). 'Notes on British Plants' (*Mag. Nat. Hist.* v.1, 1829, 83, 378–79). Prepared ed. 3 of J. Galpine's *Synoptical Compend of Br. Bot.* 1829.

Gdnrs Mag. 1828, 104; 1830, 487. *Gdnrs Chron.* 1916 i 27. A. M. Coats *Flowers and their Histories* 1968, 327–28. *R.S.C.* v.3, 638.

KENT, William (–1840)
FLS 1813. Had garden at Upper Clapton, London. Moved to Bath where he grew many rare plants.

Bot. Mag. 1814, t.1666; 1815, t.1831; 1826, t.2630. *Proc. Linn. Soc.* 1840, 71.

KENT, William (1789–1850)
d. Chadderton, Lancs 9 Dec. 1850

"A noted botanist and herbalist in humble life."

Trans. Liverpool Bot. Soc. 1909, 79.

KENT, William (*c.* 1842–1913)
d. Royton, Lancs 28 Nov. 1913

Medical botanist. President, Rochdale Botanical Society.

Lancashire Nat. v.6, 1913, 337.

KENTISH, Richard (1731–1792)
b. Yorks 1731 *d.* Bridlington, Yorks 5 April 1792

MD Edinburgh 1784. Practised medicine at Huntingdon. President, Society of Naturalists, Edinburgh, 1782. *Experiments and Observations on New Species of Bark...being...a...History of Genus of Cinchona* 1784.

Gent. Mag. 1792, 388. W. Munk *Roll of R. College of Physicians* v.2, 1878, 413–14. *J. Bot.* 1915, 137–38. *Gdnrs Chron.* 1919 i 147.

KENYON, John (*fl.* 1780s)
Nurseryman, Windsor, Berks.

KENYON, William (–before 1868)
Nailmaker. Of Settle, Yorks. "His delight was in Botany. He was well acquainted with most of the flowering plants and ferns growing near Settle."

J. Windsor *Fl. Cravoniensis* 1873, dedication.

KEOGH, Rev. John (*c.* 1681–1754)
DD. Chaplain to Lord Kingston. Incumbent of Mitchelstown, County Cork. *Botanologia Universalis Hibernica* 1735.

R. Pulteney *Hist. Biogr. Sketches of Progress of Bot. in England* v.2, 1790, 201–2. *D.N.B.* v.31, 33.

KER, Charles Henry Bellenden (*c.* 1785–1871)
d. Cannes, France 2 Nov. 1871

FRS 1819. Son of J. B. Ker (1764–1842). Barrister, Lincoln's Inn, 1814. Legal reformer. *Icones Plantarum sponte China Nascentium* 1821.

Gdnrs Mag. 1839, 429–30. *Gdnrs Chron.* 1871, 1589. *J. Bot.* 1872, 32. E. Bretschneider *Hist. European Bot. Discoveries in China* 1898, 186–89. *D.N.B.* v.31, 47. F. A. Stafleu *Taxonomic Literature* 1967, 239–40.

Letters at Kew.

KER, John Bellenden (*olim* Gawler) (1764–1842)
b. Ramridge, Andover, Hants? 1764 *d.* Ramridge June 1842

Name changed by Royal permission to Ker Bellenden, 5 Nov. 1804, but always used as Bellenden Ker. *Recensio Plantarum* 1801. *Select Orchideae c.* 1816. *Iridearum Genera* 1827. Edited *Bot. Register* 1815–24. Contrib. largely on *Irideae* to *Bot. Mag.*

Trans. Linn. Soc. v.10, 1810, 166. W. Herbert *Amaryllidaceae* 1837, 269, etc. *J. Bot.* 1884, 146; 1902, 419–22. *D.N.B.* v.31, 52–53. *R.S.C.* v.3, 638; v.4, 279 (papers attributed to Masson).

Bellendena R. Br.

KER, Robert Preston (1816–1886)
b. Hawick, Roxburgh 21 July 1816 *d.* 4 March 1886

Nurseryman, Barsett Street, Liverpool, 1860 and Aigburth nurseries, 1870.

Garden v.29, 1886, 248. *Gdnrs Chron.* 1886 i 344. *J. Hort. Cottage Gdnr* v.12, 1886, 192.

KER, Robert Wilson (1839–1910)
d. Liverpool 3 May 1910
VMH 1909. Son of R. P. Ker (1816–1886). Nurseryman, Aigburth Nurseries, Liverpool.
Gdnrs Chron. 1910 i 306 portr. *Gdnrs Mag.* 1910, 366. *J. Hort. Home Farmer* v.60, 1910, 430.

KERMODE, Sydney Alfred Pizey
(*c.* 1862–1925)
MA Cantab. Vicar, Kirk Onchan, 1890–1904. Vicar, Bowdon, Haddenham. 'Fl. Isle of Man' (*Yn Lioar Manninagh* v.2, 1900, 273–91).
J. Bot. 1901, 212. *Proc. Isle of Man Nat. Hist. Antiq. Soc.* v.6, 1960–63, 413–14.

KERNAN, John (*fl.* 1830s)
Florist, 4 Great Russell Street, Covent Garden, London.

KERR, Arthur Francis George (1877–1942)
b. Kinlough, County Leitrim 7 Feb. 1877 *d.* Hayes, Kent 21 Jan. 1942
MD Dublin. FLS 1923. Medical Officer of Health, Siam, 1902–20. Government Botanist, Ministry of Commerce, 1920–32. Collected plants for description in W. G. Craib's *Enumeratio Fl. Siamensis* and continued the work after Craib's death in 1933. Contrib. to *J. Siam Soc. Nat. Hist.*
J. Siam Soc., Nat. Hist. Supplt. v.8, 1932, 344–45. *J. Thailand Res. Soc. Nat. Hist. Supplt.* v.12, 1939 portr. *Proc. Linn. Soc.* 1941–42, 285–86. *Fl. Malesiana* v.1, 1950, 277–78. *Blumea* v.11, 1962, 427–93 portr.
Plants at Kew, BM(NH). Sketches and photographs at Kew. Portr. at Hunt Library.

KERR, Rev. Frederick Hugh Woodhams
(1885–1958)
b. Cerne Abbas, Dorset 10 Aug. 1885 *d.* Hazaribegh, India May 1958
DD Dublin 1907. Youngest brother of A. F. G. Kerr (1877–1942). Missionary, Hazaribegh. Curate, Arboe, Ireland. Returned to India after death of his wife. Collected plants in India.
Irish Times 13 June 1958. *Blumea* v.11, 1962, 479–80.
Plants at BM(NH).

KERR, George W. (1865–1930)
b. Dumfries 1865 *d.* Doylestown, U.S.A. 14 Jan. 1930
Gardener. To U.S.A. *c.* 1908. Hybridised sweet peas and dahlias. President, American Sweet Pea Society.
Gdn Fl. Annual 1918, 195–96. *Gdnrs Chron.* 1930 i 95.

KERR, James (1738–1782)
Ship's surgeon, 1763–72. Surgeon to East India Company at Dacca, 1774. *Account of Tree producing Terra Japonica* [*Mimosa catechu*] 1779.
W. Woodville *Med. Bot.* v.2, 1792, 183–86. M. Archer *Nat. Hist. Drawings in India Office Library* 1962, 6, 83–84.
MS. on Jacca [*Artocarpus integrifolia*] at BM(NH). MSS. and drawings at India Office Library.

KERR, John Graham (1869–1957)
b. Arkley, Herts 18 Sept. 1869 *d.* 21 April 1957
Educ. Cambridge. FRS 1909. Naturalist, Pilcomayo Expedition, Argentina, 1889–91. 'Botany of Pilcomayo Expedition' (*Trans. Proc. Bot. Soc. Edinburgh* 1894, 44–78). Professor of Natural History, Glasgow, 1902–35. Zoologist. *Naturalist in the Gran Chaco* 1950.
Kew Bull. 1891, 276. *Biogr. Mem. Fellows R. Soc.* 1958, 155–66 portr. *Who was Who, 1951–1960* 614.
Argentinian plants at Kew.

KERR, Mark E. (1883–1950)
b. Ireland 1883 *d.* Pearl Harbour, Hawaii 29 April 1950
Of Independence, Inyo County, California.
Leaflet Western Bot. 1957, 94.
Herb. at California Academy.
Lupinus kerrii Eastw.

KERR, William (–1814)
d. Ceylon 1814
Kew gardener *c.* 1800. Collected plants for Kew in Canton, 1803; Java and Philippines. Superintendent of gardens on Slave Island and at King's House, Colombo.
Trans. Linn. Soc. v.12, 1818, 154–57. *Trans. Hort. Soc.* v.3, 1822, 424–25. A. B. Lambert *Description of Genus Pinus* ed. 2, 1832, 111. *Gdnrs Chron.* 1881 ii 570. *Kew Bull.* 1891, 304. *J. Kew Guild* 1897, 29. E. Bretschneider *Hist. European Bot. Discoveries in China* 1898, 189–91. H. Trimen *Handbook Fl. Ceylon* v.5, 1900, 373. *Ann. R. Bot. Gdns, Peradeniya* v.1, 1901, 3. E. H. M. Cox *Plant Hunting in China* 1945, 49. *Fl. Malesiana* v.1, 1950, 278. K. Lemmon *Golden Age of Plant Hunters* 1968, 109–19. A. M. Coats *Quest for Plants* 1969, 98–100.
MS. journal of mission to Luzon, 1805 at BM(NH).
Kerria D.C.

KETT, Mrs. Hannah (*fl.* 1790s)
Of Seething, Norfolk. Plant records in J. Sowerby and J. E. Smith's *English Bot.* 69, 318, 514, 691, etc.

KEY, John *see* Caius, J.

KEYNES, John (*c.* 1806–1878)
d. Salisbury, Wilts 17 Feb. 1878
Nurseryman, Salisbury, *c.* 1836. Acquired Castle Street Nurseries, *c.* 1845. Specialised in dahlias.
Florist and Pomologist 1878, 48. *Garden* v.13, 1878, 176, 417 portr. *Gdnrs Chron.* 1878 i 248–49. *J. Hort. Cottage Gdnr* v.34, 1897, 577.

KEYS, Alfred (–1958)
d. West Palm Beach, U.S.A. Aug. 1958
Kew gardener, 1919. Assistant Curator, Botanic Station, Dominica, 1919. To U.S.A. 1923. Employed on rubber investigations at Plant Introduction Garden, Miami, Florida.
Kew Bull. 1919, 237. *J. Kew Guild* 1958, 595.

KEYS, Isaiah Waterloo Nicholson (1818–1890)
b. Devonport, Devon 12 March 1818 *d.* Plymouth, Devon 4 Nov. 1890
Bookseller and printer, Devonport. 'Fl. Devon and Cornwall' (*Trans. Plymouth Inst.* 1865–66, 21–69; 1867–68, 45–83; 1868–69, 107–44, 181–304). Contrib. to *Phytologist* 1847–50.
T. R. A. Briggs *Fl. Plymouth* 1880, xxx, xxxiii. *J. Bot.* 1890, 382. F. H. Davey *Fl. Cornwall* 1909, xlv–xlvi. W. K. Martin and G. T. Fraser *Fl. Devon* 1939, 773. *R.S.C.* v.3, 646; v.8, 71.
Plants at Plymouth Institution destroyed in 1941. Plants at Cambridge, Oxford.
Rhododendron keysii Nutt.

KIDD, Herbert Henry (*c.* 1883–1936)
d. Pietermaritzburg, Natal 24 May 1936
Kew gardener, 1904. Assistant Curator, Pietermaritzburg, 1909; Curator, 1917.
J. Kew Guild 1937, 699.

KIDD, James (1800–1867)
b. Edinburgh 1 Aug. 1800 *d.* Sydney, N.S.W. 15 Feb. 1867
Assistant Overseer, Botanic Gardens, Sydney, 1830; Overseer, 1838. Superintendent, 1844–47; reverted to Overseer, 1847–67, on appointment of J. C. Bidwill. Collected plants in Blue Mountains. Sent plants to W. J. Hooker at Kew.
Sydney Morning Herald 16 Feb. 1867. J. H. Maiden *Sydney Bot. Gdns Biogr. Notes* no. 7, 1903, 3–4. *Kew Bull.* 1906, 217. *J. Proc. R. Soc. N.S.W.* 1908, 107 portr.
Portr. at Hunt Library.

KIDDER, Jerome H. (*fl.* 1870s)
Plants collected on Kerguelen Islands, 1874–75, at Kew.

KIDSTON, Robert (1852–1924)
b. Bishopton House, Renfrewshire 29 June 1852 *d.* Gilfachdach, Glam 13 July 1924
LLD. DSc Edinburgh. FRS 1902. FRSE 1886. Of Stirling. Palaeobotanist. *Catalogue of Palaeozoic Plants in...British Museum* 1886. *Fossil Plants of Carboniferous Rocks of Great Britain* (*Mem. Geol. Survey Mus.* pts. 1–6, 1923–24). Contrib. papers to *Trans. R. Soc. Edinburgh*, *Philos. Trans. R. Soc.*
Bot. Soc. Exch. Club Br. Isl. Rep. 1924, 542. *Gdnrs Chron.* 1924 i 67. *J. Bot.* 1924, 255–56. *Naturalist* 1924, 364–66 portr. *Nature* v.114, 1924, 321–22. *Proc. R. Soc. B.* v.98, 1925, xiv–xxii portr. *Proc. R. Soc. Edinburgh* v.44, 1923–24, 248–52. *Glasgow Nat.* 1930, 56. *Who was Who, 1916–1928* 584–85. F. O. Bower *Sixty Years of Bot. in Britain, 1875–1935* 1938, 71–72 portr. *Trans. Proc. Perthshire Soc. Nat. Sci.* 1938, xviii–xix. *R.S.C.* v.10, 393; v.12, 385; v.16, 262.
Plants and books at Glasgow. Plants at Stirling Museum.
Kidstonia Zeill.

KILBURN, William (1745–1818)
b. Dublin 1745 *d.* Wallington, Surrey 23 Dec. 1818
Apprenticed to calico-printer near Dublin; later became proprietor of calico-printing firm at Wallington. Drew and engraved plates for W. Curtis's *Fl. Londinensis* 1777–98.
D.N.B. v.31, 101. W. Blunt *Art. of Bot. Illus.* 1955, 189–90.

KINAHAN, George Henry (1829–1908)
b. Dublin 19 Dec. 1829 *d.* Clontarf, Dublin 5 Dec. 1908
District surveyor, Geological Survey of Ireland, 1869. President, Royal Geological Society of Ireland, 1880–81. 'On New Locality for Polypodium Phegopteris' (*Proc. Dublin Nat. Hist. Soc.* v.3, 1859–62, 90–91).
Irish Nat. 1909, 29–31 portr.

KINAHAN, John Robert (1828–1863)
d. Dublin 2 Feb. 1863
MD Dublin. FLS 1858. Geologist. Lecturer in Botany, Carmichael School, Dublin. Secretary, Dublin Natural History Society. 'List of Ferns and their Allies found in County Dublin' (*Phytologist* v.5, 1854, 196–201). Contrib. papers on ferns to *Proc. Dublin Nat. Hist. Soc.*

KING, Mrs. (*fl.* 1880s–1910s)
Wife of W. G. King, Indian Civil Service. Drew Burmese and Madras plants, 1886–1910, now at Kew.
Kew Bull. 1923, 404.

KING, Bolton (1860–1937)
Nephew of C. E. Palmer (1830–1914). Collection of British plants at Oxford.

KING, Ernest William (–1930)
d. Coggeshall, Essex March 1930
Seed grower, Coggeshall. Specialised in sweet peas. President, Sweet Pea Society.
Gdnrs Chron. 1927 ii 300 portr.; 1930 i 257 portr.

KING, Sir George (1840–1909)
b. Peterhead, Aberdeen 12 April 1840 *d.* San Remo, Italy 12 Feb. 1909
MB Aberdeen 1865. FRS 1887. FLS 1870. KCIE 1905. VMH 1901. Bengal Medical Service, 1865. In charge of gardens at Saharanpur, 1868. Superintendent, Botanic Garden, Calcutta and Professor of Botany, 1871. Director, Botanical Survey of India, 1891–98. Founded *Ann. R. Bot. Gdn Calcutta* 1887. *Manual of Cinchona Cultivation in India* 1876. *Materials for Fl. Malayan Peninsula* (with J. S. Gamble) 1889–1909. Collected plants in Andaman Islands, India, Burma, Philippines. 'Orchids of Sikkim-Himalaya' (*Ann. R. Bot. Gdn Calcutta* v.8, 1898.
Kew Bull. 1898, 54–56; 1909, 68–72, 193–97. *Calcutta J. Med.* 1898, 297–300. *Gdnrs Chron.* 1901 i 293 portr.; 1909 i 138 portr. *Agric. Bull. Straits Settlements* 1909, 169. *J. Bot.* 1909, 120–22 portr. *Nature* v.79, 1909, 493–94. *Proc. Linn. Soc.* 1908–9, 42–45. *Proc. R. Soc. B.* v.81, 1909, xi–xxviii. *Trans. Bot. Soc. Edinburgh* v.24, 1910, 46–48. D. G. Crawford *Hist. Indian Med. Service, 1600–1913* v.2, 1914, 145. *Curtis's Bot. Mag. Dedications 1827–1927* 207–8 portr. *Who was Who, 1897–1916* 398. *150th Anniversary Vol. R. Bot. Gdn Calcutta* 1942, 5–6 portr. *Fl. Malesiana* v.1, 1950, 282 portr. I. H. Burkill *Chapters Hist. Bot. in India* 1965, 167–73 portr. *R.S.C.* v.8, 75; v.10, 398; v.16, 277.
Plants at Calcutta, Kew. Portr. at Hunt Library.
Indokingia Hemsley.

KING, James (*fl.* 1820s)
Seedsman and florist, 3 St. James's Place, Hackney Road, London.

KING, John (*fl.* 1820s)
Nurseryman, St. James's Place, Hackney Road, London.

KING, L. G. (*c.* 1868–1903)
d. Matlock, Derbyshire 30 Jan. 1903
Seedsman, Coggeshall, Essex and Reading, Berks.
J. Hort. Cottage Gdnr v.46, 1903, 127.

KING, Mark (1828–1901)
b. Lilliesleaf, Roxburghshire 28 Dec. 1828 *d.* 9 May 1901
Gardener to W. Kinghorn of Leith. Botanist. Contrib. to *Trans. Edinburgh Field Nat. Soc., Proc. Scott. Hort. Assoc.*
Trans. Edinburgh Field Nat. Microsc. Soc. 1900–1, 231–33.

KING, Philip Parker (1791–1856)
b. Norfolk Island 13 Dec. 1791 *d.* Sydney, N.S.W. 25 Feb. 1856
FRS 1824. FLS 1824. Educated in England. Joined Royal Navy, 1807; Rear-Admiral, 1855. Surveyed Australian coast, 1817–22. P. P. King, A. Cunningham and J. S. Roe collected plants on these expeditions. Charted coast of Peru, Chile and Patagonia, 1826. *Narrative of Survey of Intertropical and Western Coasts of Australia, 1818–1822* 1826. *Narrative of Surveying Voyages of...Adventure and Beagle* (with R. Fitzroy) 1839 3 vols.
Gent. Mag. 1856 i 426. *Proc. Linn. Soc.* 1856–57, xxviii–xxxi. P. Mennell *Dict. Austral. Biogr.* 1892, 260–1. *D.N.B.* v.31, 149–50. J. Gregson *Austral. Agric. Company, 1824–1875* 1907, 100–35. *J. R. Soc. N.S.W.* 1908, 107–8. *J. Proc. R. Soc. W. Australia* 1915, 115–24. I. Lee *Early Explorers in Australia* 1925, 310–400, 427–88. *J. Proc. R. Austral. Hist. Soc.* 1934, 308–10. *Austral. Encyclop.* v.5, 1965, 189–90. *Austral. Dict. Biogr.* v.2, 1967, 61–64. *Taxon* 1970, 527–28. *R.S.C.* v.3, 655.
Plants at BM(NH), Edinburgh, Kew. Letters at BM(NH), Kew. MSS. at Mitchell Library, Sydney. Portr. at Hunt Library.
Kingia R. Br.

KING, Richard (*c.* 1811–1876)
d. London 4 Feb. 1876
MRCS 1832. Arctic traveller and ethnologist. Surgeon and naturalist to expedition of Sir G. Back to Great Fish River, 1833–35. *Narrative of Journey to Shore of Arctic Ocean* 1836 (includes list of plants by W. J. Hooker).
D.N.B. v.31, 152. *R.S.C.* v.3, 656; v.8, 75.

KING, Samuel (*fl.* 1780s)
Nurseryman, Daventry, Northants.

KING, Rev. Samuel (1810–1888)
b. Midgley, Yorks 12 June 1810 *d.* Luddenden, Yorks 10 Jan. 1888

Nurseryman. Afterwards Baptist minister. Contrib. to *Phytologist* and L. C. Miall's *Fl. West Riding* 1862.

W. B. Crump and C. Crossland *Fl. Parish of Halifax* 1904, lviii–lx portr. *N. Western Nat.* 1943, 275–84.

Herb. formerly at Halifax Museum.

KING, Thomas (1834–1896)
b. Lochwinnoch, Renfrewshire 14 April 1834
d. Fochabers, Morayshire 14 Sept. 1896

Mycologist. Professor of Botany, Anderson's College, Glasgow, 1889. Collected plants at Valparaiso, Chile, 1864–73. Edited R. Hennedy's *Clydesdale Fl.* ed. 5 1891.

Gdnrs Chron. 1896 ii 375, 405. *Gdnrs Mag.* 1896, 665. *J. Bot.* 1896, 487–88. *Ann. Scott. Nat. Hist.* 1897, 1–4. *Trans. Nat. Hist. Soc. Glasgow* 1896–97, 1–12 portr. *R.S.C.* v.16, 278.

Plants at Glasgow. Portr. at Hunt Library. *Alstroemeria kingii* Philippi.

KING, Thomas (1835–1902)
b. Devizes, Wilts 1835 *d.* 17 April 1902

Gardener at Devizes Castle, 1860.

Garden v.61, 1902, 296. *Gdnrs Mag.* 1902, 282. *J. Hort. Cottage Gdnr* v.44, 1902, 394–95.

KING, William (*fl.* 1700s)

Botanical artist of Totteridge, Herts. Drew flowers in Peter Collinson's garden.

W. Blunt *Art of Bot. Illus.* 1950, 151.

Drawings at BM(NH).

KINGDON, Boughton (1816–1896)
b. Plymouth, Devon 9 April 1816 *d.* Sydney, N.S.W. 1896

Of Ryde, Isle of Wight. Apothecary. Practised at Exeter, Croydon and Sydney. Translated A. P. de Candolle's *Vegetable Organography* 1839–40.

KINGDON WARD, Francis (1885–1958)
b. Manchester 6 Nov. 1885 *d.* London 8 April 1958

MA Cantab. FLS 1920. VMH 1932. VMM 1934. Son of H. M. Ward (1854–1906). Schoolmaster, Shanghai, 1907. Collected in China, Burma, Tibet, Thailand, 1909–56. Introduced rhododendrons, primulas, meconopsis, gentians and lilies. Best-known introduction is *Meconopsis betonicifolia*. *Land of Blue Poppy* 1913. *In Farthest Burma* 1921. *Mystery Rivers of Tibet* 1923. *Romance of Plant Hunting* 1924. *Plant Hunting on the Edge of the World* 1930. *Plant Hunting in the Wilds* 1931. *Plant Hunter in Tibet* 1934. *Plant Hunter's Paradise* 1937. *Return to the Irrawaddy* 1956. *Pilgrimage for Plants* 1960.

J. Bot. 1928, 155. *Gdnrs Chron.* 1932 i 322 portr.; 1958 i 278. E. H. M. Cox *Plant Hunting in China* 1945, 180–91. *Fl. Malesiana* v.1, 1950, 561. *Geogr. J.* 1958, 422. *Lily Yb.* 1958, 9–11 portr.; 1966, 11–14. *Nature* v.181, 1958, 1505–6. *Times* 10 April 1958 portr. *J. R. Hort. Soc.* 1959, 206–12 portr. *D.N.B. 1951–1960* 587–88. *Who was Who, 1951–1960* 620–21. A. M. Coats *Quest for Plants* 1969, 128–31. U. Schweinfurth *Exploration...F. K. Ward* 1975.

Letters at Kew. Portr. at Hunt Library. *Kingdon-Wardia* Marquand.

KINGHORN, Francis Rodney (1813–1887)
b. Lennoxlove, Haddington 13 Feb. 1813
d. Richmond, Surrey 11 June 1887

Gardener, Orleans House, Twickenham, *c.* 1837. Nurseryman, Sheen Nurseries, Richmond, 1855.

Garden v.31, 1887, 569. *Gdnrs Chron.* 1887 i 817. *J. Hort. Cottage Gdnr* v.14, 1887, 513.

KINGSBURY, James (*c.* 1821–1884)
d. Southampton, Hants 27 Dec. 1884

Nurseryman, Bevois Valley Nursery Southampton.

Gdnrs Chron. 1885 i 61.

KINGSLEY, Rev. Charles (1819–1875)
b. Holne, Devon 12 June 1819 *d.* Eversley, Hants 23 Jan. 1875

MA Cantab. FLS 1856. Rector, Eversley, 1844. Canon of Chester, 1869. Author. *At Last: a Christmas in the West Indies* 1872 (with a chapter on Botanic Gardens, Trinidad). 'Bio-geology' (*J. Bot.* 1872, 53–57). *Letters and Memories* 1877 portr. 2 vols.

Proc. Linn. Soc. 1874–75, lvi–lx. *J. Bot.* 1875, 64. *Garden* v.11, 1877, 95. *J. Kew Guild* 1925, 318–22. *D.N.B.* v.31, 175–81. K. M. Lyell *Life...of Sir C. J. F. Bunbury* v.2, 1906, 195–96. *R.S.C.* v.8, 76.

Plants at Cambridge, Kew.

KINGSLEY, Henry (1830–1876)
b. Barnack, Northants 2 Jan. 1830 *d* Cuckfield, Sussex 24 May 1876

Younger brother of C. Kingsley (1819–1875). In Australia, 1853–57, where he was a gold miner, agricultural labourer and policeman. His early novels, such as *Recollections of Geoffrey Hamlyn* 1859, abound with references to flora and fauna of N.S.W.

D.N.B. v.31, 181–82. *Victorian Nat.* 1958, 123–36.

Portr. at Hunt Library.

KINGSLEY, Henry (*fl.* 1830s–1860s)

MD. FLS 1852. Of Uxbridge and Stratford-on-Avon. Corresponding member, Botanical Society of London. Correspondent of H. C. Watson.

Herb. at Butler School, Harrow School.

KINGSLEY, Mary Henrietta (1862–1900)
b. Islington, London 13 Oct. 1862 *d.* Simon's Town, S. Africa 3 June 1900
Niece of C. Kingsley (1819–1875). Author, traveller and ethnologist. Collected plants in Cameroons, 1896. *Ascent of Great Peak of Cameroon* 1895. *Travels in West Africa* 1897.
S. Gwynn *Life of Mary Kingsley* 1932. C. Howard *Mary Kingsley* 1957 portr. O. Campbell *Mary Kingsley* 1957 portr. D. Middleton *Victorian Lady Travellers* 1965, 149–76 portr.
Plants at Kew.

KINGSLEY, Rev. W. T. (*fl.* 1860s)
Cousin of C. Kingsley (1819–1875). Rector of Kilvington for 57 years. Died 101. Botanist and gardener. Outside his Rectory he put up a notice: "Trespassers beware! Polypodiums and Scolopendriums set in these grounds."
Bot. Soc. Exch. Club Br. Isl. Rep. 1932, 295.

KINGSTON, John Filmore
(–before 1850)
Fl. Devoniensis (with Rev. J. P. Jones) 1829. 'A Sketch of Distribution of Animals and Plants in South-western Extremity of Great Britain' (*Edinburgh J. Nat. Geogr. Soc.* v.3, 1831, 340–52).
Notes and Queries v.162, 1932, 302. W. K. Martin and G. T. Fraser *Fl. Devon* 1939, 771.

KINGSTON, Robert Creaser (*c.* 1846–1872)
d. Kew, Surrey 21 June 1872
Assistant, Kew Herb.
Gdnrs Chron. 1872, 876. *J. Bot.* 1872, 224.
Kingstonia Hook. f.

KINSEY, W. E. (–1943)
Inspector of Mines, Malaya, 1902. Assistant, later Deputy Conservator of Forests, Negri Sembilan, Malaya, 1905–27. Collected plants, now at Kuala Lumpur.
Gdns Bull. Straits Settlements 1927, 126. *Malayan Forester* 1948, 58. *Fl. Malesiana* v.1, 1950, 282.

KIPPIST, Richard (1812–1882)
b. Stoke Newington, Middx 11 June 1812 *d.* Chelsea, London 14 Jan. 1882
ALS 1842. Assisted J. Woods in *Tourists' Fl.* 1850. Librarian, Linnean Society, 1842–80. Discovered *Clathrus cancellatus* (*Phytologist* v.3, 1850, 1070–71).
Trans. Linn. Soc. v.28, 1873, 416–19. *Gdnrs Chron.* 1882 i 91. *J. Bot.* 1882, 63–64. *J. Hort. Cottage Gdnr* v.4, 1882, 73. *Nature* v.25, 1882, 275. *Proc. Linn. Soc.* 1881–82, 64–65. *D.N.B.* v.31, 197–98. *J. R. Soc. N.S.W.* 1908, 72. A. T. Gage *Hist. Linnean Soc. London* 1938 portr.
Kippistia Miers.

KIRBY, Elizabeth (1823–1873)
b. Leicester 15 Dec. 1823 *d.* Melton Mowbray, Leics June 1873
Co-author with her sister, Mary Kirby, of *Fl. Leicestershire* 1848 and ed. 2 1850, *Plants of Land and Water* 1857, *Chapters on Trees* 1873.
D.N.B. v.31, 198. A. R. Horwood and C. W. F. Noel *Fl. Leicestershire* 1933, ccxii–ccxiii.
Plants at Leicester Literary and Philosophical Society.

KIRBY, Mary (*afterwards* Gregg)
(1817–1893)
b. Leicester 27 April 1817 *d.* Brooksby, Leics 15 Oct. 1893
Co-author with her sister, Elizabeth Kirby, of *Fl. Leicestershire* 1848; ed. 2 1850, *Plants of Land and Water* 1857, *Chapters on Trees* 1873. *Letters from my Life* 1887.
Notes and Queries v.150, 1926, 430. A. R. Horwood and C. W. F. Noel *Fl. Leicestershire* 1933, ccxii–ccxiii.

KIRBY, Sarah *see* Trimmer, S.

KIRBY, Rev. William (1759–1850)
b. Witnesham, Suffolk 19 Sept. 1759 *d.* Barham, Suffolk 4 July 1850
BA Cantab 1781. FRS 1818. FLS 1796. Rector, Barham, 1796. Entomologist. 'Fungi parasitic on Wheat' (*Trans. Linn. Soc.* v.5, 1800, 112–15).
Ipswich Mus. Portr. Ser. 1852. J. Freeman *Life of Rev. W. Kirby* 1852 portr. *Proc. Linn. Soc.* 1851, 133–35; 1888–89, 37. *D.N.B.* v.31, 199–200. E. O. Essig *Hist. Entomology* 1931, 670–72.
Letters in Smith correspondence and portr. at Linnean Society. Portr. at Hunt Library, Kew.

KIRBY, William Forsell (1844–1912)
b. Leicester 14 Jan. 1844 *d.* Chiswick, Middx 20 Nov. 1912
FLS 1890. Nephew of Elizabeth Kirby. Entomologist. Assistant, Zoological Dept., BM(NH). *British Flowering Plants* 1906.
Proc. Linn. Soc. 1912–13, 61–62. *Who was Who, 1897–1916* 400. *R.S.C.* v.3, 658; v.8, 77; v.10, 401; v.12, 387; v.16, 284.

KIRCKWOOD, John (*fl.* 1690s)
Surgeon. Sent plants to J. Petiver from Angola and Old Calabar.
J. Petiver *Museii Petiveriani* 1695, 155, 167. J. Petiver *Gazophylacii* 1705, t.9. J. E. Dandy *Sloane Herb.* 1958, 151.

KIRK, Sir John (1832–1922)
b. Barry, Arbroath 19 Dec. 1832 *d.* Sevenoaks, Kent 15 Jan. 1922
MD Edinburgh 1854. FRS 1887. FLS 1864. GCMG 1886. KCB 1890. Administrator,

explorer and naturalist. On D. Livingstone's expedition, 1858. Surgeon, Zanzibar, 1866; Vice-Consul, 1867; Agent and Consul, 1873–86. Collected plants at Kilimanjaro, Malawi, Mozambique, Rhodesia, Seychelles, Somaliland, Zambia, Zanzibar. Contrib. to *J. Bot.*

Kew Bull. 1896, 80–82; 1922, 49–63. *Outward Bound* Sept. 1921, 7–16 portr. *J. Bot.* 1922, 96; 1933, 206. *Nature* v.109, 1922, 114–15. *Proc. Linn. Soc.* 1921–22, 45–47. *Times* 16 Jan. 1922 portr. *Proc. R. Soc. B.* v.94, 1923, xi–xxx. *Curtis's Bot. Mag. Dedications 1827–1927* 239–40 portr. Sir R. Coupland *Kirk on the Zambesi* 1928. A. White and B. L. Sloane *Stapelieae* v.1, 1937, 107 portr. *Moçambique* 1939, 69–73 portr. Sir R. Coupland *Exploitation of East Africa, 1856–1890* 1939. *Fl. Zambesiaca* v.1(1), 1960, 35–37. *Comptes Rendus AETFAT 1960* 1962, 178–79. *Kirkia* 1960, 5–10 portr.; 1967, 104 portr. R. Foskett *ed. Zambesi Doctors: David Livingstone's Letters to John Kirk, 1859–1872* 1964. *R.S.C.* v.3, 662; v.8, 78; v.10, 403.

Plants at Kew. Portr. at Hunt Library. MSS. sold at Sotheby, 21 March 1966.

Kirkia Oliver.

KIRK, John William Carnegie (1878–1962)
b. Zanzibar 28 Jan. 1878 *d.* 7 March 1962

Son of Sir J. Kirk. Officer in British Army; retired in 1921. Interested in botany and gardening. *British Garden Fl.* 1927.

King's College, Cambridge Annual Rep. 1962, 43.

KIRK, Thomas (1828–1898)
b. Coventry, Warwickshire 18 Jan. 1828 *d.* Wellington, New Zealand 8 March 1898

FLS 1871. To New Zealand, 1863. Curator, New Zealand Institute. Lecturer in Natural Science, Wellington College, N.Z. Chief Conservator of State Forests, N.Z. *Forest Fl. N.Z.* 1889. *Students' Fl. N.Z.* 1899. Contrib. to *Phytologist* 1847–60, *Trans. N.Z. Inst.* 1868–97. Collected plants in Auckland, Campbell Islands, N.Z. 1863–96.

J. Bot. 1898, 489–90; 1900, 144 portr. J. E. Bagnall *Fl. Warwickshire* 1891, 500, 502. H. C. Watson *Topographical Bot.* 1883, 549. *Proc. Linn. Soc.* 1897–98, 39. *Kew Bull.* 1898, 57. *Trans. N.Z. Inst.* 1903, 358–64. T. F. Cheeseman *Manual N.Z. Fl.* 1906, xxx–xxxi. L. Cockayne *Bot. Survey Stewart Island* 1909, 12. *Rec. Dominion Mus.* 1965, 93–100. R. Glenn *Bot. Explorers of N.Z.* 1950, 141–47 portr. *Tuatara* 1970, 68; 1973, 51–56. *R.S.C.* v.3, 662; v.8, 79; v.10, 403; v.12, 386; v.16, 289.

Plants at BM(NH), Kew, Auckland Museum.

Dacrydium kirkii F. Muell.

KIRKE, Joseph (*fl.* 1760s–1820s)

Cromwell's Garden Nursery, Cromwell Road, Kensington, founded soon after 1700 by John Kirke, occupied by Joseph Kirke in 1766. Until after 1824 the firm was in the hands of Joseph Kirke senior and his sons William and Joseph; in 1836 of John Kirke.

Trans. London Middlesex Archaeol. Soc. v.24, 1973, 186.

KIRKHAM, Thomas (*c.* 1709–1785)
d. 13 April 1785

Seedsman, Wool Pack and Crown near Durham Yard, Strand, London.

KIRKWOOD, Esther Judith Grant (1887–1969)
b. 8 Dec. 1887 *d.* Jan./Feb. 1969

FLS 1956. Teacher, Tiffins Girls School, Kingston-upon-Thames, Surrey. *Plant and Flower Forms* 1923.

KIRTIKAR, Kanoba Ranchhoddas (1850–1917)
d. 9 May 1917

FLS 1893. Surgeon-Major I.M.S. Professor of Materia Medica, Bombay, 1887. 'Poisonous Plants of Bombay Presidency' (*J. Bombay Nat. Hist. Soc.* 1892–1904). *Indian Medicinal Plants* (with Basu Baman Das) 1918 6 vols. Collected Indian mosses.

Proc. Linn. Soc. 1916–17, 48. *R.S.C.* v.16, 294.

Bryosedgwickia kirtikarii Card. & Dixon.

KITCHENER, Francis Elliott (1838–1915)
b. Newcastle, Northumberland 30 Dec. 1838 *d.* 6 July 1915

BA Cantab 1861. FLS 1867. Fellow of Trinity College, 1863. Headmaster, High School, Newcastle-under-Lyme, 1873–91. President, Rugby School Natural History Society, 1867–74. *Naked-eye Botany* 1892. Contrib. to *J. Bot.* Assisted his wife, Frances Anna, in *A Year's Botany* 1874.

J. E. Bagnall *Fl. Warwickshire* 1891, 502–3. *Who was Who, 1897–1915* 401. *R.S.C.* v.3, 667; v.8, 82.

KITCHIN, Thomas *see* Hitchin, T.

KITCHING, Langley (1835–1910)
b. Leeds, Yorks 7 July 1835 *d.* Bewdley, Worcs 9 Jan. 1910

Of Bewdley. Member, Society of Friends. Collected plants in Madagascar *c.* 1880. Articles by J. G. Baker: 'On a Collection of Ferns made by Langley Kitching in Madagascar' (*J. Bot.* 1880, 326–30, 369–73); 'Notes on Collection of Flowering Plants made by L. Kitching in Madagascar' (*J. Linn. Soc.* v.18, 1881, 264–81).

Herb. at Birmingham.

Kitchingia Baker.

KITSON, Sir Albert Ernest (1868–1937)
d. 8 March 1937
Senior geologist, Victoria, 1903. Principal, Mineral Survey, S. Nigeria, 1906–11. Director, Geological Survey, Gold Coast, 1913–30. Collected plants in Australia, Gold Coast, Togoland.
Who was Who, 1929–1940 761.
Plants at Missouri Botanic Garden.

KITSON, Fanny
Of Torquay, Devon. Clergyman's daughter. Collector and cultivator of ferns.
Br. Fern Gaz. 1909, 45.

KITTON, Frederic (1827–1895)
b. Cambridge 24 April 1827 *d.* London 22 July 1895
Diatomist and microscopist. President, Norwich Naturalists Society, 1873. Contrib. information on diatoms to M. C. Cooke's *Ponds and Ditches* 1880 and H. Mason's *Hist. Norfolk* 1884. *Norfolk Diatoms Ser. I–IV nos. 1–100* 1885. Contrib. to *Sci. Gossip* 1865–85.
F. G. Kitton *Frederic Kitton* 1895 portr. *Diatomiste* v.2, 1895, 201–4 portr. *J. Bot.* 1895, 312. *J. Quekett Microsc. Club* v.6, 1895, 152–53. *Sci. Gossip* 1895, 221. *Trans. Norfolk Norwich Nat. Soc.* v.6, 1895–96, 201–3. *R.S.C.* v.8, 83; v.10, 407; v.12, 387; v.16, 299.
Diatoms at BM(NH). Letters in Walker-Arnott correspondence at BM(NH).
Kittonia Grove & Sturt.

KLAASSEN, Henderina (Rina) Victoria *see* Scott, H. V.

KLOSS, Cecil Boden (1877–)
b. Warwickshire 1877
Zoologist. Employed in Botanic Gardens, Singapore 1903 and 1907. Sub-director, Kuala Lumpur Museum. Director, Raffles Museum, Singapore, 1923–31. Collected plants with H. C. Robinson in Perak, Selangar and Kedah.
Gdns Bull. Straits Settlements 1927, 126. *Fl. Malesiana* v.1, 1950, 285–86.
Plants at BM(NH), Kew, Singapore.
Klossia Ridley.

KNAPP, John Leonard (1767–1845)
b. Shenley, Bucks 9 May 1767 *d.* Alveston, Glos 29 April 1845
FLS 1796. *Gramina Britannica* 1804; ed. 2 1842. *Journal of a Naturalist* 1829.
J. Sowerby and J. E. Smith *English Bot.* 1127. *Proc. Linn. Soc.* 1845, 244–45. *D.N.B.* v.31, 235–36. J. W. White *Fl. Bristol* 1912, 85. G. C. Druce *Fl. Buckinghamshire* 1926, c. H. J. Riddelsdell *Fl. Gloucestershire* 1948, cxvii–cxviii.
Plants at BM(NH), Edinburgh. Letters in Smith correspondence at Linnean Society. Wax bust by Parker at Kew. Portr. at Hunt Library.
Knappia Smith.

KNAPP, Lydia Margaret (*fl.* 1820s–1830s)
Herb. at Whitehaven Museum, Cumberland.

KNIGHT, Charles (*c.* 1818–1895)
d. New Zealand 1895
Educ. University College, London. MRCS 1840. FRCS 1869. FLS 1857. To New Zealand before 1852. Lichenologist. Collected in Australia and New Zealand. 'Notes on the *Stictei* in the Kew Museum' (*J. Linn. Soc.* v.11, 243–46). 'Contribution to Lichenographia of N.S.W.' (*Trans. Linn. Soc. Bot.* v.2, 1881, 37–51). Papers on N.Z. lichens in *Trans. Linn. Soc.*, 1860, 1878; *Trans. N.Z. Inst.* 1875–84.
R.S.C. v.3, 686; v.8, 91; v.10, 419; v.16, 336.
N.Z. lichens at BM(NH). N.Z. plants at Kew.

KNIGHT, Frances *see* Acton, F.

KNIGHT, Henry (*fl.* 1830s)
Collected plants in Florida for his uncle, Joseph Knight (*c.* 1777–1855).
Floral Cabinet v.2, 1838, 47–48, 51–52.
Ismene knightii Knowles & Westc.

KNIGHT, Henry (1834–1896)
b. Taunton, Somerset 14 Dec. 1834 *d.* Brussels 9 Sept. 1896
Gardener at Dalkeith Palace, 1853–59 and Royal Gardens, Laeken, Brussels.
Gdnrs Chron. 1896 ii 377 portr. *J. Hort. Cottage Gdnr* v.33, 1896, 268–69.

KNIGHT, Henry Herbert (1862–1944)
b. Sutton Maddock, Shropshire 24 May 1862 *d.* Cheltenham, Glos 4 Jan. 1944
MA Cantab. Mathematics master, Llandovery School, 1888–1907. Retired to Cheltenham. Cryptogamist. President, British Bryological Society, 1933–34. Authority on Gloucestershire flora. Contrib. to *Proc. Cotteswold Nat. Field Club.*
Proc. Cotteswold Nat. Field Club v.28, 1943, 67–68. *Rep. Br. Bryol. Soc.* 1940–43, 228–29. *Bot. Soc. Exch. Club Br. Isl. Rep.* 1943–44, 647. *Times* 7 Jan. 1944. H. J. Riddelsdell *Fl. Gloucestershire* 1948, clvi–clviii.
Bryophyta and lichens at National Museum of Wales. Portr. at Hunt Library.

KNIGHT, Joseph (*c.* 1777–1855)
b. Brindle, Walton-le-Dale, Lancs *c.* 1777 *d.* Banbury, Oxfordshire 27 July 1855

Gardener to G. Hibbert (1757–1837) whose collections he obtained. Nurseryman, Exotic Nursery, King's Road, Chelsea, 1808; partner with Thomas A. Perry in 1853. *On the Cultivation of the Plants belonging to the Natural Order of Proteaceae* 1809. *Synopsis of Coniferous Plants grown in Great Britain and Sold by Knight and Perry, Chelsea* 1850.

Bot. Mag. 1809, t.1218. *Trans. Hort. Soc.* v.1, 1812, 262. T. Faulkner *Hist. and Topographical Description of Chelsea* v.1, 1829, 61–62. J. Gillow *Dict. English Catholics* v.4, 1887, 74–75. *J. Bot.* 1886, 296–300. E. Bretschneider *Hist. European Bot. Discoveries in China* 1898, 220. M. Hadfield *Gdning in Britain* 1960, 338–39. F. A. Stafleu *Taxonomic Literature* 1967, 241. M. Allan *Tom's Weeds* 1970, 15–19.

KNIGHT, Robert Cedric (1891–1935)
d. Torquay, Devon 28 Jan. 1935

Plant Physiology Dept., Imperial College of Science, 1919. Head of Physiology Section, East Malling Research Station, 1920; afterwards Assistant Director.

Gdnrs Chron. 1935 i 101. *Nature* v.135, 1935, 363–64. *Chronica Botanica* 1936, 179 portr.

KNIGHT, Robert Lanier (–1972)
d. 15 Feb. 1972

On staff of Empire Cotton Growing Corporation; in Sudan researching on disease resistant cottons, 1929. Chief geneticist of Sudan Government. Returned to England in 1954 when he joined staff of East Malling Research Station.

Ann. Applied Biol. 1972, 187 portr. *Chronica Horticulturae* v.12, 1972, 24 portr. *Nature* v.236, 1972, 316.

KNIGHT, Thomas Andrew (1759–1838)
b. Weobley, Herefordshire 12 Aug. 1759 *d.* London 11 May 1838

Educ. Balliol College, Oxford. FRS 1805. FLS 1807. Plant physiologist and pomologist. Friend of Sir J. Banks. President, Horticultural Society of London, 1811–38. *Treatise on Culture of Apple and Pear* 1797. *Pomona Herefordiensis* 1811. *Selection from Physiological and Horticultural Papers* 1841.

J. C. Loudon *Encyclop. Gdning* 1822, 1280–81. *Gdnrs Mag.* 1829, 87–88; 1838, 303. G. W. Johnson *Hist. English Gdning* 1829, 271–72. *Gdnrs Chron.* 1841, 351–52; 1877 i 169–70, 177 portr.; 1881 i 182–83; 1915 ii 145–46 portr.; 1922 ii 192; 1933 ii 457–58. *Cottage Gdnr* v.6, 1851 43–44. R. Hogg and H. G. Bull *Herefordshire Pomona* v.1, 1876, 29–47 portr. *J. Hort. Cottage Gdnr* v.56, 1876, 428–29 portr. *Flora* 1893, 38–42. *D.N.B.* v.31, 263–64. *Gdning Mag.* 1900, 783–85 portr. *J. R. Hort. Soc.* 1901, 198 portr.; 1938, 319–24 portr. J. R. Green *Hist. Bot.* 1914, 295–305. *J. Bot.* 1931, 77. *Plant Physiol.* v.14, 1939, 1–8 portr. *Fruit Yb.* 1948, 10–13. M. Hadfield *Pioneers in Gdning* 1955, 143–49. M. Hadfield *Gdning in Britain* 1960, 270–73. G. M. Darrow *The Strawberry* 1966, 75–78 portr. R. Webber *Early Horticulturists* 1968, 115–24 portr. H. R. Fletcher *Story of R. Hort. Soc. 1804–1968* 1969 *passim*, portr. C. C. Gillispie *Dict. Sci. Biogr.* v.7, 1973, 409–10.

Letters at BM(NH), Linnean Society. Portr. by S. Cole at Kew. Portr. at Hunt Library.

Knightia R. Br.

KNIGHT, William (1786–1844)
b. Aberdeen 17 Sept. 1786 *d.* Aberdeen 3 Dec. 1844

MA Aberdeen 1802. LLD 1817. Lecturer in Botany, Aberdeen University. Professor of Natural Philosophy, Academical Institution, Belfast, 1816–22. Friend of R. Brown. Had herb. *Outlines of Botany* 1813; ed. 2 1828.

D.N.B. v.31, 266–67.

KNIGHTS, Arthur Ivor James (*c.* 1900–1948)
d. Newtown, Montgomery 25 Aug. 1948

Officer in Trinidad Constabulary. Collected and grew orchids. Contrib. to *Orchid. Rev.*

Orchid Rev. 1948, 159.

KNOWLDIN, Edward (1850–1936)
b. Lambeth, London 1850 *d.* Dublin 16 March 1936

Gardener. Secretary, Royal Horticultural and Arboricultural Society of Ireland, 1908–31. Contrib. to *Gdnrs Chron.*, *Gdnrs Mag.*, *Garden.*

Gdnrs Chron. 1936 i 329–30.

KNOWLES, George Beauchamp (*fl.* 1820s–1850s)

ALS 1834. FLS 1834. Surgeon. Professor of Botany, Birmingham School of Medicine, 1829–52. Co-editor with F. Westcott of *Floral Cabinet and Magazine of Exotic Botany* 1837–40.

Knowlesia Hassk.

KNOWLES, Gilbert (1674–after 1725)

MD. *Materia Medica Botanica* 1723 portr. (in verse). Correspondent of Sir H. Sloane.

R. Pulteney *Hist. Biogr. Sketches of Progress of Bot. in England* v.1, 1790, 282. *D.N.B.* v.31, 296. E. J. L. Scott *Index to Sloane Manuscripts* 1904.

KNOWLES, Matilda Cullen (1864–1933)
b. Ballymena, County Antrim 31 Jan. 1864 *d.* Dublin 27 April 1933

Assistant, Science and Art Museum, Dublin, 1902. 'Lichens of Ireland' (*Proc. R. Irish Acad. B.* 1929, 179–434).

Irish Nat. J. 1933, 191–93 portr. *J. Bot.* 1933, 230–31. R. L. Praeger *Some Irish Nat.* 1949, 116–17 portr.

Herb. at Dublin Museum.

KNOWLING, Rex Alan Henry *see* Graham, R. A. H.

KNOWLTON, Thomas (1691–1781)
b. Chislehurst, Kent 1691 *d.* Londesborough, Yorks 29 Nov. 1781

Gardener to J. Sherard at Eltham until 1725 and to Earl of Burlington at Londesborough. Letter to M. Catesby in *Philos. Trans. R. Soc.* v.44, 1748, 100–2, 124–27.

W. Aiton *Hortus Kewensis* v.1, 1789, x. J. Bolton *Filices Britannicae* 1785–90, 77. R. Pulteney *Hist. Biogr. Sketches of Progress of Bot. in England* v.2, 1790, 239–41. J. E. Smith *Selection from Correspondence of Linnaeus* v.2, 1821, 78–80. R. Richardson *Extracts from Lit. and Sci. Correspondence* 1835, 301–3. *D.N.B.* v.31, 303–4. *J. Bot.* 1914, 318.

MS. at Royal Society. Letters to Brewer at BM(NH).

Knowltonia Salisb.

KNOWLTON, Thomas (1757–1837)
b. Keighley, Yorks 1757 *d.* Darley Dale, Derbyshire 11 Sept. 1837

FLS 1795. Grandson of T. Knowlton (1691–1781). Steward to Duke of Devonshire; had gardens at Londesborough, Edensor and Darley Dale, and was "a skilful botanist."

Gent. Mag. 1837 ii 435; 1838 i 544–45. *Proc. Linn. Soc.* 1846, 302. *Leeds Mercury* 29 June 1889.

Library sold by Lewis, 12–14 Feb. 1846.

KNOX, George (*fl.* 1790s)

Gardener and seedsman, Alnwick, Northumberland.

KNOX, John (1831–1914)
b. Kirkcudbright 1831 *d.* Forfar 8 July 1914

Schoolmaster at Crieff, Forfar, 1866. Studied flora of Forfar.

Bot. Soc. Exch. Club Br. Isl. Rep. 1914, 56–58.

KNOX, Margaret (–1952)
b. London *d.* 29 April 1952

Studied painting at Slade School of Art. Drew British plants.

Proc. Bot. Soc. Brit. Isl. 1954, 105–6.

Drawings at BM(NH).

KNOX, Robert (1640/1–1720)
d. London July 1720

To Madras with his father in 1657. Captive in Ceylon, 1659–79. Commander, East India Company. *Historical Relation of Island of Ceylon in East Indies* 1681 (contains account of trees of Ceylon).

H. Trimen *Handbook to Fl. Ceylon* v.5, 1900, 372–73. *D.N.B.* v.31, 330.

Knoxia L.

KOENIG, Carl Dietrich Eberhard (1774–1851)
b. Brunswick, Germany 1774 *d.* London 6 Sept. 1851

FRS 1810. FLS 1802. Keeper of Mineralogy, BM 1813. *Tracts relative to Botany, translated from Different Languages* 1805. Joint editor with J. Sims of *Ann. Bot.* 1804–6; also drew plates. Translated K. Sprengel's *Introduction to Study of Cryptogamous Plants* 1807.

Gent. Mag. v.36, 1851, 435–36. *D.N.B.* v.31, 343–44.

Portr. at Hunt Library, Kew.

Koniga R. Br.

KOENIG, Johann Gerhard (*c.* 1728–1785)
b. Courland, Latvia *c.* 1728 *d.* Jagrenathporum, India 26 June 1785

Pupil of Linnaeus. In Ireland, 1765. In India, 1768. Physician to Danish settlement in Carnatic. Naturalist to Nabob of Arcot. On Madras establishment, East India Company, 1778. To Siam and Malacca, 1778–79. Collected plants in Bornholm, Ceylon, India, Malacca, Siam. MS. journal at BM(NH) (*J. R. Asiatic Soc. Straits Branch* no. 26, 1894, 58–201; no. 27, 57–133). Bequeathed plants and MSS. to Sir J. Banks.

W. Roxburgh *Plants of Coast of Coromandel* v.1, 1795, preface. A. J. Retzius *Observationes* v.3, 1791, 6. N. Wallich *Plantae Asiaticae Rariores* v.3, 1832, 50. A. Lasègue *Musée Botanique de B. Delessert* 1845, 557–58. H. Trimen *Handbook to Fl. Ceylon* v.5, 1900, 373. D. G. Crawford *Hist. Indian Med. Service 1600–1913* v.2, 1914, 142–43. *J. Bot.* 1933, 143–53, 175–87. *Proc. Linn. Soc.* 1932–33, 46–47. *Kew Bull.* 1932, 49, 256; 1934, 221. *J. Thailand Res. Soc. Nat. Hist. Supplt.* 1939, 3–9. I. H. Burkill *Chapters on Hist. Bot. in India* 1965 *passim.*

Letters in Banks correspondence at BM(NH).

Koenigia L.

KOETTLITZ, Reginald (1861–1916)
b. Dover, Kent 1861 *d.* Somerset, S. Africa Jan. 1916

LRCP. Naturalist and explorer. With Jackson–Harmsworth Polar expedition,

1894–97. To Somaliland and Abyssinia, 1898; also Brazil. Medical officer to Scott's 1st Antarctic expedition, 1902.
Nature v.96, 1916, 600.
MSS. on phytoplankton at BM(NH).

KOPSCH, Henry C. T. (*fl.* 1860s–1890s)
Chinese Customs Service, 1862. Superintendent, Chinese Postal Service, 1897. Sent seedlings of *Illicium verum* to C. Ford at Hong Kong. Contrib. articles on Chinese plants to *Notes and Queries on China and Japan* 1868–69.
E. Bretschneider *Hist. European Bot. Discoveries in China* 1898, 693–95.

KOSTER, Henry (1793–1820)
b. Liverpool 1793 *d.* Pernambuco, Brazil 1820
Of Portuguese descent. In Brazil, 1809–15. *Travels in Brazil* 1816 (with botanical appendix).
J. Bot. 1896, 242.

KRICHAUFF, Friedrich Eduard Heinrich Wulf (1824–1904)
b. Schleswig, Germany 18 Dec. 1824 *d.* Adelaide, Australia 29 Oct. 1904
Gardener, Botanic Gardens, Kiel. Emigrated to S. Australia, 1848. Friend of F. von Mueller with whom he collected plants on Bugle Ranges. Chairman, Central Agricultural Bureau, S. Australia, 1888–1902. 'Notes on Nicobar Islands' (*Allgemeine Gartenzeit* 1847, 25–30). 'On *Cytisus adami*' (*Allgemeine Gartenzeit* 1848, 25–28). 'Timber Supply of Australia' (*Trans. Scott. Arb. Soc.* v.8, 1877, 110–35).
F. von Mueller *Rep. on Plants collected during Mr. Babbage's Expedition* 1858, 7–8. P. Mennell *Dict. Austral. Biogr.* 1892, 263–64. *S. Austral. J. Agric.* 1904, 137. *Rep. Austral. Assoc. Advancement Sci.* 1907, 175.
Hibiscus krichauffianus F. Muell.

KRIEG, David (–1713)
b. Saxony, Germany
FRS 1698. Physician. To London in 1697 when he became a friend of J. Petiver. Collected plants in Maryland with W. Vernon, 1698. Sent plants to J. Petiver (*Musei Petiveriani* 1695, 45, 95) and to Bobart. Correspondent of Dale. "Medicus ornatissimus" (Plukenet).
J. Ray *Historia Plantarum* v.3, 1704, iii. L. Plukenet *Almagesti Botanici Mantissa Plantarum* 1700, 80. R. Pulteney *Hist. Biogr. Sketches of Progress of Bot. in England* v.2, 1790, 57–58. *Philadelphia Med. Physiol. J.* v.2(2), 1806, 139–42. E. J. L. Scott *Index to Sloane Manuscripts* 1904, 294. *Proc. International Congress Plant Sci.* v.2, 1926, 1929, 1527–28. C. E. Raven *John Ray* 1942, 301. *Proc. Amer. Antiq. Soc.* 1952, 307–10. J. E. Dandy *Sloane Herb.* 1958, 151–52.
Plants at BM(NH).

KUHN, Adam (1741–1817)
b. Philadelphia, U.S.A. 17 Nov. 1741 *d.* Philadelphia, U.S.A. 5 July 1817
MD Edinburgh 1767. Pupil of Linnaeus, 1762–64. Professor of Botany, Philadelphia, 1768.
J. W. Harshberger *Botanists of Philadelphia* 1899, 88–91. H. A. Kelly *Some Amer. Med. Botanists* 1915, 69–74 portr.
Kuhnia L.

KURZ, Wilhelm Sulpiz (*c.* 1833–1878)
b. Munich, Germany *c.* 1833 *d.* Pulo Penang 15 Jan. 1878
Pupil of Martius. At Botanic Garden, Buitenzorg, Dutch East Indies. Curator, Calcutta Herb., 1864. Collected plants in Andaman Islands, India, Burma, Malaya, Java. *Report on Vegetation of Andaman Islands* 1867; reprinted 1870. *Preliminary Report on Forest and other Vegetation of Pegu* 1875. *Bamboo and its Use* 1876. *Forest Fl. of British Burma* 1877 2 vols. Contrib. to *J. Bot.*
J. Asiatic Soc. Bengal v.40, 1871, 461. *Trans. Linn. Soc. Bot.* v.1, 1876, 119–31. *Flora* 1878, 113–14. *J. Bot.* 1878, 127–28. *D.N.B.* v.31, 346. *R.S.C.* v.8, 138; v.10, 480; v.12, 419.
Plants at Calcutta, Kew.
Kurzianda Kuntze.

KYD, Robert (1746–1793)
b. Forfarshire 1746 *d.* Calcutta 26 May 1793
Bengal Infantry, 1765. Lieut.-Colonel, 1782. Established Calcutta Botanic Garden, 1787.
Ann. Bot. Gdn Calcutta v.4, 1893, i–xi portr. *J. Bot.* 1899, 456–57. *D.N.B.* v.31, 348. R. H. Phillimore *Hist. Rec. Survey of India* v.1, 1945, 347.
Letters in Banks correspondence at BM(NH). MS. on tea at Kew.
Kydia Roxb.

KYLE, Thomas (*fl.* 1780s)
Gardener to David Stewart Moncrieff of Moredun near Edinburgh. *Treatise on Management of Peach and Nectarine Trees* 1783.
J. C. Loudon *Encyclop. Gdning* 1822, 1278.

LACAITA, Charles Carmichael (1853–1933)
b. Edinburgh 5 April 1853 *d.* Selham, Sussex 17 July 1933
BA Oxon. FLS 1882. Barrister. MP for Dundee, 1885–87. Collected plants in Sikkim, 1913 and in Spain, 1925–28. Owned estates in Italy where he botanised. Contrib. etymological appendix to G. Maw's *Monograph of Genus Crocus* 1886. 'Piante Italiane Critiche o Rare' (*Nuovo Giornale Botanico Italiano* 1910–27). 'Plants collected in

Sikkim...1913' (*J. Linn. Soc.* 1917, 457–92). Contrib. to *J. Linn. Soc.*, *J. Bot.*, *Cavanillesia.*

J. Bot. 1933, 259–62. *Nuovo Giorn. Bot. Ital.* 1933, 447–52 portr. *Proc. Linn. Soc.* 1933–34, 160–62. *Who was Who, 1929–1940* 769.

Herb. at BM(NH).

LACE, John Henry (1857–1918)
d. Exmouth, Devon 9 June 1918

FLS 1888. Indian Forest Service, 1881; Inspector-General of Forests. Chief Conservator of Forests, Burma, 1908–13. Collected plants in Afghanistan, Baluchistan, India, Burma. 'Sketch of Vegetation of British Baluchistan' (*J. Linn. Soc.* 1891, 288–327). *List of Trees, Shrubs and Principal Climbers, etc. recorded from Burma* 1913.

Indian For. 1918, 369–70; 1919, 448. *Kew Bull.* 1918, 341. *Proc. Linn. Soc.* 1918–19, 56–57.

Plants at Edinburgh, Kew, Oxford.

LACE, Margaret *see* Roscoe, M.

LAFFAN, J. de C. (–1930)
d. 4 July 1930

Army major. President, Reading Natural History Society. Botanist.

S. Eastern Nat. 1930, xliv.

LAFFERTY, Henry Aloysius (1891–1954)
b. Ardstraw, County Tyrone 10 May 1891
d. 19 July 1954

Educ. Royal College of Science, Dublin. Assistant, Seeds and Plant Disease Division, Dept. of Agriculture, Dublin until 1921; Head of Division, 1923–54. Investigated flax diseases.

Nature v.174, 1954, 585–86.

LAFLIN, Tom (1914–1972)
d. 10 Dec. 1972

Horticultural adviser, Ministry of Agriculture. Contrib. records of mosses to J. G. Dony's *Fl. Bedfordshire* 1953 and D. A. Cadbury's *et al. Computer-mapped Fl.... Warwickshire* 1971.

J. Bryol. 1973, 468–69.

Bryophyte collection at Cambridge. Warwickshire plants at Birmingham.

LA GASCA Y SEGURA, Mariano (1776–1839)
b. Encinacorva, Aragon, Spain 4 Oct. 1776
d. Barcelona, Spain 23 June 1839

FLS 1831. Professor of Botany and Director, Royal Garden, Madrid, 1807. To England, 1822. In Jersey, 1831–34. *Amenidades Naturales de las Espanas* 1811. *Genera et Species Plantarum* 1816. *Hortus Siccus Londinensis* 1826–27. Papers on Spanish plants in *Gdnrs Mag.* 1826–28. Catalogue of Jersey plants in *5th Rep. Jersey Agric. Soc.* 1839.

Gdnrs Mag. 1827, 220–21. *Bot. Misc.* 1829–30, 64. *Proc. Linn. Soc.* 1840, 71–72. M. Colmeiro *La Botanica* 1858, 191–95. L. V. Lester-Garland *Fl. Island of Jersey* 1903, x. *J. Bot.* 1908, 163–70; 1924, 347–50. *Michigan Bot.* 1968, 11–13. *R.S.C.* v.3, 801.

Plants at BM(NH), Madrid.

LAGG, John *see* Legg, Thomas

LAIDLAW, Charles Glass Playfair (1887–1915)
b. London 1887 *d.* killed in action Béthune 2 April 1915

BA Cantab 1910. Engaged in plant physiological research at Imperial College of Science, 1912–14.

J. Ecol. 1915, 242. *New Phytol.* 1915, 210–11.

LAING, James (*c.* 1857–1917)
d. 9 Sept. 1917

Nurseryman, Kelso, Roxburghshire. In partnership with R. V. Mather.

Garden 1917, 402(v).

LAING, John (1823–1900)
b. Carriston near Brechin, Angus Oct. 1823
d. 8 Aug. 1900

VMH 1897. Gardener to Earl of Rosslyn at Dysart House. Partner with Laird and Downie, nurserymen, Edinburgh, 1860. Partnership dissolved in 1874. J. Laing took over London branch at Stanstead Park Nursery, London, 1875. Hybridised tuberous begonias.

Tuberous Begonia 1888, 24–26 portr. *Garden* v.58, 1900, 119–20 portr. *Gdnrs Chron.* 1900 ii 117 portr. *Gdnrs Mag.* 1900, 504. *J. Hort. Cottage Gdnr* v.41, 1900, 157 portr.

LAING, R. (*c.* 1809–1887)
d. 12 Jan. 1887

Nurseryman, Twickenham, Middx.

Gdnrs Chron. 1887 i 120.

LAIRD, David Pringle (*c.* 1853–1905)
d. Loch Awe, Argyllshire Sept. 1905

Nurseryman, Edinburgh.

J. Hort. Cottage Gdnr v.51, 1905, 260.

LAIRD, R. B. (1823–1895)
b. Balgone, E. Lothian 16 May 1823 *d.* Dundee, Angus 4 March 1895

Brother of W. P. Laird (–1872), becoming senior partner in his nursery at Dundee on his death in 1872. In partnership with J. Downie, 17 South Frederick Street, Edinburgh, 1848. J. Laing joined partnership in 1860 which was dissolved in 1874.

Garden v.47, 1895, 174. *Gdnrs Chron.* 1895 i 307. *Gdnrs Mag.* 1895, 137–38 portr.

LAIRD, William Pringle (–1872)
b. Balgone, E. Lothian *d.* Fountainbrae, Monifieth, Angus 14 Aug. 1872
Nurseryman, Dundee, 1833; nurseries at Blackness until 1868; also at Fountainbrae, 1858. In partnership with Andrew Sinclair, 1859.
Gdnrs Chron. 1872, 1170–71.

LAKE, Harry W. (*fl.* 1890s)
Miner and surveyor. Explored Johore and with H. J. Kelsall made a crossing of it from east to west in 1892, collecting plants on the way for the Singapore Botanic Gardens. 'A Journey on the Sembrong River' (*J. Straits Branch R. Asiatic Soc.* no. 26, 1894, 1–24).
Gdns Bull. Straits Settlements 1927, 126. *Fl. Malesiana* v.1, 1950, 307–8.

LAKE, Henry (*fl.* 1820s)
Nurseryman, Matthews Field, Bridgwater, Somerset.

LAKER, William (*fl.* 1820s)
Nurseryman, West Street, Horsham, Sussex.

LAKIN, Charles (*c.* 1840–1916)
Doctor in Leicester. Interested in pharmaceutical properties of British plants. Discovered *Polygonum bistorta* at Bradgate, Leics.
A. R. Horwood and C. W. F. Noel *Fl. Leicestershire* 1933, ccxxxiii.

LAKIN, Joseph (1828–1895)
b. Derby 1828 *d.* Temple Cowley, Oxford 4 March 1895
Police Superintendent, Chipping Norton. Raised new varieties of carnations and picotees.
Gdnrs Mag. 1895, 138 portr. *J. Hort. Cottage Gdnr* v.30, 1895, 204, 251, 293.

LAMB, Henry (1858–1905)
b. Maidstone, Kent April 1858 *d.* Maidstone 15 July 1905
Fl. Maidstone 1899.
J. Bot. 1905, 280.
Plants at Maidstone and Haslemere Museums.

LAMB, Joshua (1856–1943)
b. 3 March 1856 *d.* 7 Jan. 1943
Of Sibford, Oxfordshire. Had a garden of British wild flowers.
Countryside v.12, 1943, 157.
Herb. at Haslemere Museum.

LAMB, Percy Hutchinson (*c.* 1883–1937)
d. Bedford Oct.? 1937
Educ. Leeds University. Cotton planter, Egypt. Director of Agriculture, Uganda. Director of Agriculture, N. Nigeria, 1912–30. Introduced plants from West Indies for cultivation in Nigeria.
Times 8 Oct. 1937.

LAMB, Thomas (*fl.* 1790s)
MD. ALS 1789. FLS 1794. Of Newbury, Berks. Plant records in J. Sowerby and J. E. Smith's *English Bot.* 1931, 2031.

LAMBERT, Aylmer Bourke (1761–1842)
b. Bath, Somerset 2 Feb. 1761 *d.* Kew, Surrey 10 Jan. 1842
Educ. Oxford 1779. FRS 1791. FLS 1788. Built up large herb. and library. Discovered *Carduus tuberosus* 1813. Vice-president, Linnean Society, 1796–1842. *Description of Genus Cinchona* 1797. *Description of Genus Pinus* 1803 2 vols.; ed. 5 1842 2 vols. Contrib. to J. Sowerby and J. E. Smith's *English Bot.* 1359, 2562, etc.
Bot. Misc. 1829–30, 62. *Gdnrs Mag.* 1838, 58. *Bot. Mag.* 1842, t.3922. *Gdnrs Chron.* 1842, 271, 439. *Gent. Mag.* 1842, 667–68. *London J. Bot.* 1842, 394–96. *Proc. Linn. Soc.* 1842, 137–39; 1929–30, 32–33. W. Darlington *Reliquiae Baldwinianae* 1843, 196–97. A. Lasègue *Musée Botanique de B. Delessert* 1845, 75. *Kew Bull.* 1891, 526–27. E. Bretschneider *Hist. European Bot. Discoveries in China* 1898, 205–7. *D.N.B.* v.32, 6–7. *J. Bot.* 1905, 219; 1930, 94–95. *J. Linn. Soc.* 1930, 439–66. *Madrono* v.10, 1949, 33–47. *Proc. Amer. Philos. Soc.* 1952, 599–628. D. Grose *Fl. Wiltshire* 1957, 33–34. F. A. Stafleu *Taxonomic Literature* 1967, 256–58. *Taxon* 1970, 489–553 portr.; 1972, 614 portr. *R.S.C.* v.3, 812.
Plants at BM(NH). Portr. by Russell at Linnean Society. Portr. at Hunt Library. Letters at Kew, BM, Linnean Society. Library sold at Sotheby 18 April 1842. Herb. sold by Sotheby 27 June 1842; annotated sale catalogue at BM(NH).
Aylmeria Martius. *Lambertia* Smith.

LAMBERT, John (1619–1683)
Major-General. Notable gardener. Lived at Wimbledon Manor, 1652; Guernsey, 1662; St. Nicholas Island, Plymouth, 1670. Credited with the introduction of Guernsey lily, *Nerine sarniensis.*
D.N.B. v.32, 11–18. W. H. Dawson *Cromwell's Understudy; Life and Times of General John Lambert* 1938 portr. M. Hadfield *Gdning in Britain* 1960, 91–92. *Amateur Gdning* v.87, no. 4459, 1970, 34 portr.

LAMBOURNE, Amelius Richard Mark Lockwood, 1st Baron (1847–1928)
b. 17 Aug. 1847 *d.* Romford, Essex 26 Dec. 1928
VMH 1922. President, Royal Horticultural Society, 1917–28.

Garden 1920, ii–iii portr. *Bot. Soc. Exch. Club Br. Isl. Rep.* 1928, 704–5. *Gdnrs Chron.* 1929 i 18 portr. *J. R. Hort. Soc.* 1929, 253–55 portr. *Who was Who, 1916–1928* 602.

LAMBOURNE, John (–1965)
b. Sussex *d.* Worthing, Sussex 9 Nov. 1965
Kew gardener, 1909. Manager, Sapintas Coconut Estate, Malayan Agricultural Service, 1912.
Kew Bull. 1912, 155. *J. Kew Guild* 1965, 600.

LAMINGTON, Lord *see* Cochrane-Baillie, C. W. A.

LAMONT, Charles Peter (*c.* 1892–1949)
b. Upper Deeside, Aberdeenshire *c.* 1892 *d.* 26 May 1949
Gardener, Royal Botanic Garden, Edinburgh, 1914; Assistant Curator, 1921.
Gdnrs Chron. 1949 ii 8.

LAMONT, Rev. James (1844–1928)
b. Crathie 11 July 1844 *d.* Edinburgh 28 Aug. 1928
FLS 1893. Ordained Presbyterian minister at Portsmouth, 1871. Pastor, Hong Kong, 1873. Moderator of Presbyterian Church of N.S.W., 1895. Collected plants in Hong Kong and N.S.W. for BM(NH).
E. Bretschneider *Hist. European Bot. Discoveries in China* 1898, 706. *J. Bot.* 1928, 301. *Proc. Linn. Soc.* 1928–29, 143.
Cardamine lamontii Hance.

LAMPARD, E. R. J. (–1961)
d. 7 Sept. 1961
Secretary and Treasurer, Amateur Orchid Growers' Society, 1951–61.
Orchid Rev. 1961, 332.

LANCASHIRE, Robert (*c.* 1798–1875)
d. 12 March 1875
Of Middleton. One of leading northern florists, specialising in auriculas; raised 'Lancashire Lass'.
Gdnrs Yb. Almanack 1876, 171.

LANCASTER, Sydney Percy (1886–)
b. Meerut, India 19 July 1886
FLS 1920. Assistant, Agricultural and Horticultural Society of India, 1904; Assistant Secretary, 1910; Secretary, 1914–53. Senior Technical Assistant, National Botanic Gardens, Lucknow, 1953–59, 1961. *Amateur in an Indian Garden* 1929. Wrote 55 *Bulletins* for National Botanic Gardens.
Herbertia 1939, 40–43 portr. *Indian Hort.* April/June 1966 portr.

LANCE, Christopher (*fl.* 1790s)
Gardener and seedsman, Blandford Forum, Dorset.

LANCE, Edward Jarman (1788–1863)
b. Lewisham, Kent 8 Aug. 1788 *d.* Reading, Berks 25 Oct. 1863
Golden Farmer 1831. *Hop Farmer* 1838. *Food of Plants* 1842.
Notes and Queries 1928, 389, 430; 1929, 13.

LANCE, John Henry (1793–1878)
d. Holmwood, Dorking 12 Jan. 1878
FLS 1828. Barrister. Commissary Judge in Surinam, 1828–34. Discovered *Cycnoches loddigesii* in Surinam.
Bot. Register v.9, 1836, t.1887. J. Lindley *Sertum Orchidaceum* 1838, t.13. F. Boase *Modern English Biogr.* v.2, 1897, 289. *Orchid Rev.* 1898, 56–57.
Oncidium lanceanum Lindl.

LANDON, F. H. (1909–1956)
d. Sept. 1956
BA Oxon. Malayan Forest Service, 1932. Chief Research Officer, Forest Research Institute, Kepong, Selangor. Collected plants.
Fl. Malesiana v.5, 1958, 61.

LANDON, Sylvanus (*fl.* 1670s–1700s)
Surgeon. Brought plants to J. Petiver from Spain, Azores and Borneo.
J. Petiver *Museii Petiveriani* 1695, 45. J. E. Dandy *Sloane Herb.* 1958, 153. *Fl. Malesiana* v.5, 1958, 61.
Plants at BM(NH).

LANDRETH, David (1752–1836)
b. Haggerston, Northumberland 1752
Emigrated to Canada in 1781, then to Philadelphia, where *c.* 1786 with his brother Cuthbert, he started a nursery and seed business.
J. W. Harshberger *Botanists of Philadelphia and their Work* 1899, 91. U. P. Hedrick *Hist. of Hort. in America to 1860* 1950, 203. *Trans. Amer. Philos. Soc.* v.53(2), 1963, 64 and *passim.*

LANDSBOROUGH, Rev. David (1779–1854)
b. Dalry, Glen Kens, Galloway 11 Aug. 1779 *d.* Saltcoats, Ayrshire 12 Sept. 1854
ALS 1849. Minister, Stevenston, Ayrshire, 1811. Minister, Saltcoats, 1843. Discovered *Ectocarpus landsburgii. Popular History of British Seaweeds* 1849; ed. 3 1857. Contrib. to W. H. Harvey's *Phycologia Britannica* 1846–51. *Excursions to Arran* 1847; ed. 2 1875 includes memoir.
Gent. Mag. 1854 ii 402–3. *Proc. Linn. Soc.* 1855, 426. *D.N.B.* v.32, 62–63.
Plants at Edinburgh, Kew.
Landsburgia Harvey.

LANDSBOROUGH, Rev. David (*c.* 1826–1912)
d. Kilmarnock 22 Nov. 1912
Studied flora of Ayrshire and Island of Arran. Contrib. to *Trans. Edinburgh Bot. Soc.*
Garden 1912, 620(x). *Gdnrs Chron.* 1912 ii 418.

LANDSBOROUGH, William (1825–1886)
b. Ayrshire 21 Feb. 1825 *d.* Caloundra, Queensland 16 March 1886
Son of Rev. D. Landsborough (1779–1854). Emigrated to N.S.W. Farmed in Queensland. Explored Queensland and collected plants which he sent to F. von Mueller. Inspector of Brands, Moreton District, 1872–86. *Journal of Expedition from Carpentaria in search of Burke and Wills* 1862. *Exploration of Australia from Carpentaria to Melbourne* 1866.
W. Howitt *Hist. of Discovery in Australia, Tasmania and New Zealand* v.2, 1865, 284–96. J. E. T. Woods *Hist. of Discovery and Exploration of Australia* v.2, 1865, 390–91, 435–44. *D.N.B.* v.32, 63. P. Serle *Dict. Austral. Biogr.* v.2, 1949, 5–6. *Austral. Encyclop.* v.5, 1965, 236.
Portr. at Hunt Library.

LANE, A. W. (*fl.* 1840s)
Surgeon, HMS 'Illustrious'. Collected plants in West Indies, *c.* 1843.
J. Bot. 1884, 226. I. Urban *Symbolae Antillanae* v.3, 1902, 71.
List of Bermuda plants at Bermuda Library. Plants at Edinburgh, Kew.

LANE, Frederick Q. (–1907)
d. 2 June 1907
Nurseryman, Berkhamsted, Herts.
Garden 1903, 322–23 portr. *Gdnrs Mag.* 1907, 419.

LANE, George (*c.* 1810–1885)
d. St. Mary's Cray, Kent Nov. 1885
Nurseryman, St. Mary's Cray.
Garden, v.28, 1885, 597.

LANE, George Thomas (1867–1936)
b. Long Bredy, Dorset 5 Oct. 1867 *d.* London 23 April 1936
Kew gardener, 1889. Assistant Curator, Royal Botanic Garden, Calcutta, 1891; Curator, 1896–1923.
Kew Bull. 1923, 403. *J. Kew Guild* 1924, 217; 1933, 225–27 portr.; 1936, 584–85 portr.

LANE, James (*fl.* 1750s–1790s)
Nurseryman near Sadler's Wells, Clerkenwell, London.

LANE, John Edward (*c.* 1807–1889)
b. Berkhamsted, Herts *c.* 1807 *d.* Berkhamsted 17 July 1889
Son of Harvey Lane (*fl.* 1830s), nurseryman. Nurseryman, Berkhamsted. Rose and fruit grower.
Garden v.27, 1885, xii portr.; v.36, 1889, 65. *J. Hort. Cottage Gdnr* v.19, 1889, 60.

LANE, Louisa *see* Clarke, L. L.

LANE-POOLE, Charles Edward (1885–1970)
b. Sussex 1885 *d.* Sydney, N.S.W. 22 Nov. 1970
S. African Forest Dept., 1906. Conservator of Forests, Sierra Leone, 1910. W. Australia, 1917. Commonwealth Inspector-General of Forests, 1927–45. Collected plants in Sierra Leone, W. Australia, New Guinea. *Report on Forests of Sierra Leone* 1911. *List of Trees, Shrubs, Herbs and Climbers of Sierra Leone* 1916. *Forest Resources of Territories of Papua and New Guinea* 1925.
Fl. Malesiana v.1, 1950, 311; v.5, 1958, 61 portr. *Commonwealth For. Rev.* 1971, 1–3 portr. *Who was Who, 1961–1970* xxxvii.
Plants at Kew.

LANG, William Henry (1874–1960)
b. Withyham, Groombridge, Sussex 12 May 1874 *d.* Storth near Milnthorpe, Westmorland 29 Aug. 1960
BSc Glasgow 1894. MB 1895. FRS 1911. Studied tropical cryptogams with A. G. Tansley in Ceylon and Malaya, 1900. Lecturer in Botany, Glasgow, 1902. Professor of Cryptogamic Botany, Manchester, 1909–40. 'Studies in Morphology of Isoetes' (*Mem. Manchester Lit. Philos. Soc.* 1914–15). Contrib. to *Ann. Bot., Philos. Trans. R. Soc., Trans R. Soc. Edinburgh.*
Nature v.188, 1960, 102–3. *Times* 31 Aug. 1960. *Biogr. Mem. Fellows R. Soc.* v.7, 1961, 147–60. *Yb. R. Soc. Edinburgh* 1961, 18–21. *D.N.B. 1951–1960* 1971, 606–8. C. C. Gillispie *Dict. Sci. Biogr.* v.8, 1973, 4–7.
Portr. at Hunt Library.

LANGABEER, Francis G. (1898–1964)
Treasurer, Quekett Microscopical Club, 1952. Microscopist and botanist.
J. Quekett Microsc. Club 1965, 14–15 portr.

LANGDON, Allan George (1893–1972)
d. 4 Feb. 1972
FLS 1961. VMH 1955. Nurseryman, Bath, Somerset. Specialised in delphiniums, begonias and gloxinias.
National Begonia Soc. Bull. June 1972, 4.

LANGDON, C. F. (*c.* 1868–1947)
b. Newton St. Loe, Somerset *c.* 1868 *d.* Bath, Somerset 4 March 1947.
VMH 1935. Nurseryman, Newton St. Loe. In partnership with J. B. Blackmore, 1901.
Gdnrs Chron. 1947 i 96.

LANGELIER, Réné (*fl.* 1840s)
Nurseryman, St. Helier, Jersey, Channel Islands.

LANGLANDS, Robert (*fl.* 1750s)
MD Edinburgh 1750. Sent British plants to J. Hope (1725–1786).
Notes R. Bot. Gdn Edinburgh 1907, 125.

LANGLEY, Batty (1696–1751)
b. Twickenham, Middx 1696 *d.* Soho, London 3 March 1751
Architect and landscape gardener. *Sure Method of Improving Estates, by Plantations fo Oak, Elm*...1728. *New Principles of Gardening* 1728. *Pomona* 1729. *Landed Gentleman's Useful Companion* 1741.
Gent. Mag. 1751, 139. G. W. Johnson *Hist. English Gdning* 1829, 197–98. *Cottage Gdnr* v.8, 1852, 93–94. *D.N.B.* v.32, 108–9. *Gdnrs Chron.* 1913 i 143; 1922 ii 180. E. J. L. Scott *Index to Sloane Manuscripts* 1904, 298. *J. Land Agents' Soc.* 1948, 315–17.
Mezzotint portr. by J. Carwithan.

LANGLEY, Larret (*fl.* 1830s)
FLS 1827. Teacher at Brompton Academy, Rotherham, Yorks. 'Fl. Rotherham' (*Mag. Nat. Hist.* 1829, 269–70). Co-author with Rev. W. Moorhouse of local plant list in *Village Magazine or Wath Repository* 1831.
Sorby Rec. v.2(4), 1967, 12.

LANKESTER, Charles Herbert (1879–1969)
b. Southampton, Hants 14 June 1879 *d.* San José, Costa Rica 8 July 1969
To Costa Rica, 1900. Coffee planter. Created outstanding garden of native flora. Collected plants in Costa Rica, Brazil, Mexico, Guatemala, Canaries, Uganda, Kenya, Rhodesia, Sudan.
A. S. and P. P. Calvert *Year of Costa Rican Nat. Hist.* 1917 *passim. Amer. Orchid Soc. Bull.* 1964, 43–50 portr.; 1969, 860–62 portr. *Orchid Rev.* 1970, 8–10 portr. *Orquidea* (Mexico) v.3, 1973, 216–19 portr.
Portr. at Hunt Library. Plants at Kew.
Lankesterella Ames.

LANKESTER, Edwin (1814–1874)
b. Melton, Suffolk 23 April 1814 *d.* Margate, Kent 30 Oct. 1874
MD Heidelberg 1839. FRS 1845. FLS 1840. Professor of Natural History, New College, London, 1850. Secretary, Ray Society, 1844. President, Microscopical Society of London, 1859. *Vegetable Substances used for Food of Man c.* 1840. *Natural History of Plant Yielding Food* 1845. 'Setae of *Funaria*' (*Ann. Nat. Hist.* v.4, 1840, 361–64). Contrib. botanical articles to *Penny Cyclopaedia*, 1833–41, from letter R. Translated M. J. Schleiden's *Principles of Sci. Bot.* 1849. Contrib. to *Phytologist* 1844.
Ipswich Mus. Portr. Ser. 1852. *J. Bot.* 1874, 383. *Nature* v.11, 1875, 15–16. *Trans. Bot. Soc. Edinburgh* v.12, 1876, 202–3. *J. R. Microsc. Soc.* 1895, 16–17. *D.N.B.* v.32, 137–39.
Portr. at Hunt Library. Sale at Sotheby, 16 Feb. 1875.
Lankesteria Lindl.

LANKESTER, Phoebe (*née* Pope) (1825–1900)
b. 10 April 1825 *d.* London 9 April 1900
Wife of E. Lankester (1814–1874). *Plain and Easy Account of British Ferns* 1860; 1881. *Wild Flowers worth Notice* 1861. *Talks about Plants* 1879. Contrib. popular portion of J. T. I. Boswell's *English Bot.* ed. 3 1863–72.
Times 14 April 1900.

LAPORTE *see* Castelnau, F. L. de L. de *Comte*

LARBALESTIER, Charles Du Bois (1838–1911)
b. St. Brelade's, Jersey, Channel Islands 29 Oct. 1838 *d.* St. Helier, Jersey 4 April 1911
BA Cantab 1863. FLS 1882. Lichenologist. Collected in Ireland, Channel Islands and at Cambridge. *Lichenes Caesarienses et Sargienses Exsiccati* fasc. 1–6, nos. 1–280, 1867–72. *Lichenes Exsiccati circa Cantabrigiam* nos. 1–35, 1896. *Lichen Herb.* fasc. 1–9, nos. 1–360, 1879–81.
Times 11 April 1911. *J. Bot.* 1912, 69–70. *Proc. Irish Acad. B.* 1929, 186–87.
Herb. at Birmingham.
Microglaena larbalestieri A. L. Sm.

LARDER, Joseph (–1923)
d. 28 Nov. 1923
Botanist, mainly of cryptogams, of Louth, Lincs. Secretary, Louth Antiquarian and Naturalists' Society.
Trans. Lincoln Naturalists' Union 1923, 20.

LARPENT, Sir George (*fl.* 1840s)
Had garden at Roehampton, Surrey. *Aerides larpentae* Reichenb. f. named after his wife who imported orchids.
Gdnrs Chron. 1847, 732.

LARTER, Miss Clara Ethelinda (1847–1936)
b. Leeds, Yorks 27 June 1847 *d.* Torquay, Devon 13 May 1936
FLS 1912. Botanised in Devon. *Notes on Botany of North Devon* 1897. *Manual of Fl. Torquay* 1900. Editor-in-chief from 1930 of *Fl. Devon* 1939.
Bot. Soc. Exch. Club Br. Isl. Rep. 1936, 212–13. *Proc. Linn. Soc.* 1936–37, 200–2. W. K. Martin and G. T. Fraser *Fl. Devon* 1939, 777–78.
Herb. at Torquay Natural History Museum. Plants at Oxford.

LASCELLES, Rev. Edwin (-1923)
d. Midhurst, Sussex 1923
Rector, Newton St. Loe, Somerset, 1878–1904. Grew zonal pelargoniums, begonias and delphiniums.
A. Simmonds *Hort. Who was Who* 1948, 41–42.

LASLETT, Thomas (1811–1887)
b. Chatham, Kent 18 June 1811 *d.* Old Charlton, Kent 6 April 1887
In Admiralty timber-yard at Chatham. *Timber and Timber-trees, Native and Foreign* 1875; ed. 2 by H. M. Ward, 1894.

LAST, George Valentine Chapman (1875–1945)
b. Liverpool 14 Feb. 1875 *d.* Hilbre Island, Cheshire 9 June 1945
MRCS. LRCP 1919. FLS 1922. Pharmaceutical chemist, Liverpool. District Medical Officer. Helped to form herb. of Liverpool Botanical Society. Interested in *Primulaceae* and *Gentianaceae*.
Bot. Soc. Exch. Club Br. Isl. Rep. 1945, 13–14. *N. Western Nat.* 1945–46, 283–85.
Plants at Liverpool, Kew.

LAST, Joseph Thomas (1847/8–1933)
b. Tuddenham, Suffolk 1847/8 *d.* Shortlands, Kent 13 Dec. 1933
Member of Church Missionary Society in British East Africa. Collected plants in Madagascar, 1885–99, Zanzibar, *c.* 1899–1901, Arabia, 1908–10.
Times 16 Dec. 1933. *Geogr. J.* 1934, 352. *Moçambique* 1939, 61–63; 1940, 113–14 portr.
Plants at Kew.

LATHAM, Robert Gordon (1812–1888)
b. Billingborough, Lincs 24 March 1812 *d.* Putney, Surrey 9 March 1888
BA Cantab 1832. MD London. Ethnologist and philologist. Sent plants to C. C. Babington. List of Peterborough plants in *J. Bot.* 1902, 102.
Br. Med. J. 1888 i 672. *D.N.B.* v.32, 168–69.

LATHAM, Samuel (*fl.* 1710s–1760s)
Apothecary, Newgate Street, London. Member of Botanical Society of London, founded in 1721.
Proc. Bot. Soc. Br. Isl. 1967, 317.

LATHAM, William Bradbury (1835–1914)
b. Bicknacre, Essex 13 Feb. 1835 *d.* Leighton Buzzard, Beds 17 Dec. 1914
VMM 1901. Kew gardener, 1855–57. Curator, Birmingham Botanic Garden, 1868–1903.
Gdnrs Chron. 1901 i 294 portr., 296; 1903 ii 53 portr.; 1914 ii 418 portr. *Garden* v.59, 1901, 182–83 portr.; v.63, 1903, 386 portr.; 1915, 12(vi). *J. Kew Guild* 1903, 111 portr.; 1915, 241–43. *Orchid Rev.* 1915, 46.
Cypripedium lathamianum Reichenb. f.

LATHBURY, Rev. Peter (1760–1820)
b. Westerfield, Suffolk 1760 *d.* 3 Aug. 1820
Rector, Great and Little Livermere, Suffolk, 1804–20. Had herb. which was sold at public auction, Livermere, 22 Sept. 1820.

LATIMER, William (-1929)
d. 7 Aug. 1929
Orchid grower.
Orchid Rev. 1929, 276.

LATROBE, Charles Joseph (1801–1875)
b. London 20 March 1801 *d.* Litlington, Sussex 2 Dec. 1875
Travelled in Europe and later in N. America with Washington Irving. Superintendent, Port Phillip district, Victoria, 1839–54. Established Melbourne Botanic Gardens. Had extensive knowledge of Australian plants. Returned to England, 1854.
P. Mennell *Dict. Austral. Biogr.* 1892, 269. *D.N.B.* v.32, 182–83. *Hist. Rec. Australia* v.20–26, 1924–25 *passim*. *Victorian Nat.* 1908, 109–10. A. Pratt *Centenary Hist. Victoria* 1934, 100–6. P. Serle *Dict. Austral. Biogr.* v.2, 1949, 13–15. *Austral. Encyclop.* v.5, 1965, 246–48. A. Gross *Charles Joseph Latrobe* 1956.
Plants and letters at Kew. MSS. at Public Library of Victoria.
Latrobea Meissner.

LAUDER, Alexander (*fl.* 1790s)
Nurseryman, 33 Stokes Croft, Bristol.

LAUDER, Peter (*fl.* 1790s–1810s)
Nurseryman, Lawrence Hill, Bristol. Business taken over by William Spring in 1814.

LAURENCE, Rev. John (1668–1732)
b. Stamford, Northants 1668 *d.* Bishop's Wearmouth, Durham 17 May 1732
BA Cantab 1689. Rector, Yelvertoft, Northants, 1700. Rector, Bishop's Wearmouth, 1722–32. Gardener. *Clergy-man's Recreation* 1714 portr.; ed. 6 1726. *Gentleman's Recreation* 1716; ed. 3 1723. *Fruit-garden Kalendar* 1718; ed. 2 1726. *New System of Agriculture* 1726; ed. 2 1727.
J. C. Loudon *Encyclop. Gdning* 1822, 1267. G. W. Johnson *Hist. English Gdning* 1829, 155–58. *Cottage Gdnr* v.6, 1851, 79. *J. Hort. Cottage Gdnr* v.55, 1876, 270–71 portr.; v.55, 1907, 153 portr. *D.N.B.* v.32, 206. *Gdnrs Chron.* 1900 i 129–30, 413–14; 1900 ii 42. G. O. Bellawes *Notes on Life and Works of John Laurence* 1905. D. McDonald *Agric. Writers* 1908, 167–70. *Fruit Yb.* 1958, 63–69. *Huntia* 1965, 117–37 portr.

LAURENCE, John William (1828–)
b. St. John's, Isle of Wight 1828
Kew gardener, 1844. Gardener to Bishop of Winchester at Farnham Castle, 1852–75. Specialised in orchids.
Gdnrs Chron. 1877 i 785 portr.

LAURIE, Arthur Pillans (1861–1949)
b. 6 Nov. 1861 *d.* Haslemere, Surrey 7 Oct. 1949
MA Cantab. FRSE 1885. Principal, Heriot-Watt College, Edinburgh, 1900–28. Professor of Chemistry, Royal Academy of Arts, 1912–36. *Food of Plants* 1893.
Yb. R. Soc. Edinburgh 1950, 24–25. *Who was Who, 1941–1950* 665.

LAURIE, Charlotte Louisa (–1933)
b. West Indies
On teaching staff of Cheltenham Ladies College, 1880–1910. Secretary, Cheltenham Science Society. *Flowering Plants* 1903. *Introduction to Elementary Botany* 1907. *Textbook of Elementary Botany* 1910.
H. J. Riddelsdell *Fl. Gloucestershire* 1948, cxxxv–cxxxvi.

LAURIE, Malcolm Vyvyan (1901–1973)
b. 30 Aug. 1901 *d.* 30 Nov. 1973
BSc Cantab 1923. Silviculturist, Indian Forest Service, Madras, 1925–35. Forest Research Institute, Dehra Dun, 1936–40. Deputy Director, Timber Supplies, New Delhi, 1940–43; Director, 1943–46. Director, Alice Holt Station, Forestry Commission, 1946–59. Professor of Forestry, Oxford, 1959–68.
Times 3 Dec. 1973.

LAUTERER, Joseph (1848–1911)
b. Freiburg, Germany 18 Nov. 1848 *d.* Brisbane, Australia 29 July 1911
MD. Surgeon in Franco-Prussian war. To New South Wales, 1885 and settled in Brisbane a few months later. President, Royal Society of Queensland, 1896. Interested in flora of Moreton Bay. *Excursionsflora v. Freiburg* 1874. 'Gums and Resins of Queensland Plants' (*Bot. Bull. Dept. Agric. Queensland no. 13*, 1896, 35–80). Papers, mainly on plant-chemistry, in *Proc. R. Soc. Queensland* v.11–14, 1895–99.
J. Proc. R. Soc. N.S.W. 1921, 163–64.

LAVER, Henry (1829–1917)
b. Paglesham, Essex 12 Oct. 1829 *d.* Colchester, Essex 31 Aug. 1917
MRCS. Surgeon, Essex County Hospital, 1877–97. Hon. Curator, Colchester Castle.
Essex Rev. v.26, 1917, 204. *Essex Nat.* v.29, 1954, 189.
Essex plants and letters at Essex Field Club.

LAVEROCK, William Shepherd (1865–1947)
b. Aberdeen 13 July 1865 *d.* Aughton Park near Ormskirk, Lancs 4 June 1947
MA Aberdeen 1888. Schoolteacher, Aberdeen, 1888–93. Assistant, Liverpool Museums, 1896–1928. Collected natural history specimens in Malaya in 1914 for Liverpool Museums. President, Liverpool Naturalists' Field Club, 1905, 1918, 1923–25, 1929–30. President, Liverpool Biological Society, 1924–26. Helped in compilation of C. T. Green's *Fl. Liverpool District* 1902.
N. Western Nat. 1947, 285–87 portr. *Proc. Liverpool Nat. Field Club* 1947, 5–7.

LAVINGTON, Robert (*fl.* 1800s)
Nurseryman, Devizes, Wilts.

LAW, Benjamin (*fl.* 1810s–1830s)
Nurseryman, College Lane, Northampton.

LAW, Ernest (1854–1930)
b. 26 Aug. 1854 *d.* 25 Feb. 1930
BA London. Barrister. Designed gardens at Brompton Hospital Sanatorium, Frimby and Esher Place, and knot gardens at Stratford-upon-Avon and Hampton Court. *Shakespeare's Garden, Stratford-upon-Avon* 1922. *Flower-lover's Guide to Gardens of Hampton Court Palace* 1923. *Hampton Court Gardens, Old and New* 1926.

LAW, Henry (*fl.* 1820s–1830s)
Gardener and seedsman, Drapery, Northampton. Successor to John Law.

LAW, John (*fl.* 1810s)
Nurseryman, Drapery, Northampton.

LAW, John Sutherland (1810–1885)
b. India 31 May 1810 *d.* 10 July 1885
FLS 1856. Bombay Civil Service, 1826. In India until 1854. Collected plants in Deccan, Concan, etc.
J. Graham *Cat. of Plants growing in Bombay* 1839, iii. A. Lasègue *Musée Botanique de B. Delessert* 1845, 156, 158, 433. R. Wight *Icones Plantarum Indiae* v.3, 1846, t.1070. *Taxon* 1970, 530.
Plants at Kew, Oxford. Letters at Kew.
Lawia Wight.

LAW, Wilby (*c.* 1838–1916)
d. 8 Dec. 1916
Member of Northampton Natural History Society; contrib. botanical papers to its *Journal.*
J. Northants Nat. Hist. Soc. Field Club 1916, 239.

LAWES, Sir John Bennet (1814–1900)
b. Rothamsted, Herts 28 Dec. 1814 *d.* Rothamsted 31 Aug. 1900

Educ. Oxford. FRS 1854. Agricultural chemist. Founded Rothamsted Experimental Station on his estate, 1843. Published independently and with his technical adviser, J. H. Gilbert.

Gdnrs Chron. 1871, 917–18 portr.; 1900 ii 190. *Nature* v.48, 1893, 327–29. *Gdnrs Mag.* 1900, 570 portr. A. D. Hall *Book of Rothamsted Experiments* 1905, xxi–xxxii portr. *D.N.B. Supplt I* v.3, 79–82. C. C. Gillispie *Dict. Sci. Biogr.* v.8, 1973, 92–93.

Portr. at Hunt Library.

LAWES, William George (1839–1907)
b. Aldermaston, Berks 1839 *d.* Sydney 1907

Missionary, London Missionary Society at Niué, Savage Islands, 1861–74. New Guinea Mission, 1874–1906.

J. King *W. G. Lawes of Savage Islands and New Guinea* 1909. *Fl. Malesiana* v.1, 1950, 315.

Plants at Melbourne.

LAWFIELD, Wilfred Norman (1913–1962)
b. May 1913 *d.* 30 Aug. 1962

Kew gardener, 1937–40. Assistant Editor, *Amateur Gdning* 1950. *Encyclopaedia of Garden Pests and Diseases* (with J. van Konynburg) 1958. *Lawns and Sportsgreens* 1959.

J. Kew Guild 1962, 206.

LAWRANCE, Mary (*afterwards* **Kearse**) (*fl.* 1790s–1830s)

Of London. Artist. Exhibited at Royal Academy. Taught botanical drawing. Friend of R. Sweet. *Collection of Roses from Nature* 1796–99. *Sketches of Flowers from Nature* 1801. *Collection of Passion Flowers coloured from Nature* 1802.

J. für Botanik. v.3, 1800, 211. K. G. Rössig *Die Rosen* 1802–20, preface. *Ann. Bot.* v.1, 1805, 25. *D.N.B.* v.32, 248. E. A. Willmott *Genus Rosa* v.1, 1910, 92. W. Blunt *Art of Bot. Illus.* 1955, 211.

Rosa lawranceana Sweet.

LAWRANCE, William (*fl.* 1820s)

Nurseryman, Diana Place, New Road, London.

LAWRENCE, George (*fl.* 1840s)

Gardener at Hendon Vicarage, Middx. 'Catalogue of Cacti at Hendon Vicarage' (*Gdnrs Mag.* 1841, 313–21).

J. Bot. 1916, 338. N. L. Britton and J. N. Rose *Cactaceae* v.3, 1922, 109.

LAWRENCE, H. C. (–1917)
b. Cambridge *d.* 14 Oct. 1917

Nurseryman, Chatham, Kent, 1868.

Garden 1917, 488(v).

LAWRENCE, Lady Iris (*c.* 1888–1955)
d. 16 June 1955

VMH 1942. Widow of Sir William Lawrence (1870–1934). President, Alpine Garden Society.

Quart. Bull. Alpine Gdn Soc. 1955, 269–70.

LAWRENCE, Sir James John Trevor (1831–1913)
b. London 30 Dec. 1831 *d.* Burford, Surrey 22 Dec. 1913

KCVO 1902. VMH 1900. VMM 1913. Indian Medical Service, 1854–64. MP Surrey, 1875–92. President, Royal Horticultural Society, 1885–1913. Cultivated orchids.

Garden v.69, 1906 portr.; 1914, 16 portr. *Orchid Rev.* 1911, 49–50 portr.; 1914, 7–8, 92–93, 176; 1921, 10–12, 37–39. *Gdnrs Chron.* 1885 i 605, 607 portr.; 1913 i 220–22 portr.; 1914 i 8, 15 portr. *Gdnrs Mag.* 1913, 986. *J. R. Hort. Soc.* 1913, 513–22 portr. Indian Hort. v.68, 1914, 23 portr. D. G. Crawford *Hist. Indian Med. Service, 1600–1913* v.2, 1914, 182. *Curtis's Bot. Mag. Dedications, 1827–1927* 235–36 portr.

Plants at Kew.

Trevoria F. C. Lehmann.

LAWRENCE, John (1753–1839)
b. Colchester, Essex 22 Jan. 1753 *d.* Peckham, London 17 Jan. 1839

Contrib. additions to W. Curtis's *Practical Observations on Br. Grasses,* ed. 5–7 1812–34.

D.N.B. v.32, 265–67.

LAWRENCE, Rev. John *see* Laurence, *Rev.* J.

LAWRENCE, Louisa (*née* **Senior**) (*c.* 1803–1855)
b. Broughton House near Aylesbury, Bucks *c.* 1803 *d.* 14 Aug. 1855

Married William Lawrence. FRS 1828. Created notable garden at Drayton Green, Middx in 1830s and at Ealing Park in 1840. Famous for her hothouse plants, particularly orchids. First to flower *Amherstia nobilis* in England in 1849. Her son, J. J. T. Lawrence, became President of Royal Horticultural Society.

Gdnrs Mag. 1838, 305–22 (description of garden at Drayton Green). *Curtis's Bot. Mag. Dedications, 1827–1927* 59–60 portr. *J. R. Hort. Soc.* 1955, 423–28. *Country Life* 1973, 580–81 portr.

LAWRENCE, Robert William (1807–1833)
b. 18 Oct. 1807 *d.* Formosa, Tasmania 18 Oct. 1833

Arrived in Tasmania, 1825. Correspondent of W. J. Hooker whom he introduced to R. C. Gunn. Collected plants. 'Notes on Excursion up the Western Mountains of

Van Dieman's Land' (*J. Bot.* (Hooker) 1834, 235–41).

Companion Bot. Mag. v.1, 1836, 272–77. A. Lasègue *Musée Botanique de B. Delessert* 1845, 328. *Hooker's Icones Plantarum* 1840, t.261. J. D. Hooker *Fl. Tasmaniae* v.1, 1859, cxxv. G. Bentham *Fl. Australiensis* 1863, 189–90. *Proc. Linn. Soc. N.S.W.* 1900, 773. *Proc. R. Soc. Tasmania* 1909, 20–21. *Rep. Austral. Assoc. Advancement Sci.* 1911, 231. T. E. Burns and J. R. Skemp *Van Dieman's Land Correspondents* 1961 *passim. R.S.C.* v.3, 894.

Plants and letters at Kew.

Lawrencia Hook.

LAWRENCE, Sir William Matthew Trevor (1870–1934)

b. 17 Sept. 1870 *d.* 4 Jan. 1934

BA Oxon. PhD Heidelberg. VMH 1929. Son of Sir J. J. T. Lawrence (1831–1913). Treasurer, Royal Horticultural Society, 1924–28. President, Cactus and Succulent Society, 1932–34, Iris Society, Alpine Garden Society. Contrib. to *Gdnrs Chron.*

Br. Fern Gaz. 1934, 251–52. *Gdnrs Chron.* 1934 i 33. *Cactus J.* v.2, 1934, 47 portr. *Quart. Bull. Alpine Gdn Soc.* 1934, 222–23. *Times* 5 Jan. 1934 portr. *Who was Who, 1929–1940* 788.

Portr. at Hunt Library.

LAWSON, Abercrombie Anstruther (1874–1927)

b. Pittinwem, Fife 1874 *d.* Sydney, N.S.W. 26 March 1927

Berkeley, 1893. FRSE 1898. FLS 1909. Instructor in Botany, Stanford University, 1893. Lecturer in Botany, Glasgow University, 1907–12. Professor of Botany, Sydney, 1912. Accompanied Setchell on algologic expedition to the Aleutian Islands. '*Bowenia*' (*Trans. R. Soc. Edinburgh* v.54, 1926, 357–94). Contrib. papers on Gymnosperms and *Psilotaceae* to *Trans. R. Soc. Edinburgh, Bot. Gaz., Ann. Bot., Proc. Linn. Soc. N.S.W.*

G. C. Druce *Fl. Oxfordshire* 1927, cxx–cxxi. *J. Bot.* 1927, 203–4. *Kew Bull.* 1927, 223–24. *Nature* v.119, 1927, 753–54. *Proc. R. Soc. Edinburgh* v.47, 1926–27, 374–77. *Sydney Univ. Sci. J.* v.2, 17–20. P. Serle *Dict. Austral. Biogr.* v.2, 1949, 17. *R.S.C.* v.16, 641.

Portr. at Hunt Library.

LAWSON, Charles (1794–1873)

b. Edinburgh 1794 *d.* Edinburgh 21 Dec. 1873

Son of Peter Lawson who founded nursery in Edinburgh in 1770. Succeeded to business on his father's death in 1821. Raised Lawson's Cypress from seeds sent from Sacramento River in 1854. *Agriculturist's Manual* 1836. *Agrostographia* 1842. Began publication of E. J. Ravenscroft's *Pinetum Britannicum* 1863–84.

Garden v.11, 1877, xv–xvi portr. C. S. Sargent *Silva of N. America* v.10, 1896, 120–21. *Gdnrs Chron.* 1904 ii 36–37.

Cupressus lawsoniana A. Murray.

LAWSON, George (1827–1895)

b. Newport, Fife 12 Oct. 1827 *d.* Halifax, Nova Scotia 10 Nov. 1895

Instructor in Botany, Edinburgh University, 1848–58. Professor of Chemistry and Natural History, Kingston, Ontario, 1858–63. Professor of Chemistry and Mineralogy, Halifax, 1863–95. *Royal Water-lily of South America* 1851. *Synopsis of Ferns and Filicoid Plants of Canada* 1864. *Wild Flowers of N. America* 1867. *Fern Fl. of Canada* 1889. Contrib. to *Phytologist* 1847–48.

Proc. R. Soc. Canada v.12, 1894, 49–52. *Proc. Nova Scotia Inst.* 1895–96, xxiii–xxxi portr.; v.13, lxxxvii–lxxxviii. *Amer. Fern J.* 1941, 59–62. G. F. G. Stanley *Pioneers of Canadian Sci.* 54–80 portr. *R.S.C.* v.3, 895; v.8, 177; v.10, 532; v.12, 434; v.16, 641.

Herb. at Ottawa.

LAWSON, Henry P. (*c.* 1863–1956)

d. Sept. 1956

Cultivated orchids. Contrib. preface to *Addendum to Sander's List of Orchid Hybrids, 1949–51* 1952.

Orchid. Rev. 1956, 233.

LAWSON, Isaac (–*c.* 1747)

b. Scotland *d.* Oosterhout, Netherlands *c.* 1747

MD Leyden 1737. Physician in the Army. Friend of Linnaeus and Gronovius. Printed (with Gronovius) Linnaeus's *Systema Naturae*. Plants from Padua, with list, in Sloane Herb. at BM(NH).

W. G. Maton *General View of Writings of Linnaeus* 1805, 49, 530. J. E. Smith *Selection from Correspondence of Linnaeus* v.1, 1821, 18; v.2, 173, 175. D. Turner *Extracts from Lit. Sci. Correspondence of Richard Richardson* 1835, 345. *Naturalist* 1894, 243. E. J. L. Scott *Index to Sloane Manuscripts* 1904, 302. *D.N.B.* v.32, 291. *J. Bot.* 1912, 262. J. E. Dandy *Sloane Herb.* 1958, 153–54.

Plants at BM(NH), Linnean Society.

Lawsonia L.

LAWSON, John (–1711)

b. Scotland *d.* burnt by Indians in N. Carolina 1711.

Surveyor-General N. Carolina from 1700. *New Voyage to Carolina* 1709; plants listed on pp. 89–114 (reprint 1967). Sent plants to J. Petiver.

J. Nichols *Illus. Lit. Hist. Eighteenth Century* v.4, 489. *D.N.B.* v.32, 294–95. E. J. L. Scott *Index to Sloane Manuscripts* 1904, 302. *Proc. International Congress Plant Sci.* v.2, 1929, 1530–31. *J. Heredity* v.49, 1958, 137–38. *Proc. Amer. Antiq. Soc.* 1952, 335–42. J. E. Dandy *Sloane Herb.* 1958, 154. H. Savage *Lost Heritage* 1970, 27–56.

Plants at BM(NH).

LAWSON, John (*fl.* 1870s)

Collected plants in S. Africa, Australia and New Zealand.

H. N. Clokie *Account of Herb. Dept. Bot., Oxford* 1964, 197.

Plants at Oxford.

LAWSON, Marmaduke Alexander (1840–1896)

b. Seaton-Carew, Durham 20 Jan. 1840 *d.* Madras, India 14 Feb. 1896

MA Cantab 1864. FLS 1869. Professor of Botany, Oxford, 1868–82. Director, Botanic Dept., Ootacamund, 1882. Government Botanist and Director of Cinchona Plantations, Nilgiris, 1885. Collected plants. 'On Fl. of Skye' (*J. Bot.* 1869, 108–14). 'Oxford Botanists' (*Gdnrs Chron.* 1870, 1024). Contrib. to *Fl. Tropical Africa* and *Fl. Br. India*.

J. Bot. 1896, 191, 239. *Kew Bull.* 1896, 185. *Proc. Linn. Soc.* 1895–96, 40–41. G. C. Druce *Fl. Berkshire* 1897, clxxvii. *Bot. Soc. Exch. Club Br. Isl. Rep.* 1924, 699–70. H. N. Clokie *Account of Herb. Dept. Bot., Oxford* 1964, 48–49. *R.S.C.* v.7, 178; v.10, 532; v.16, 640.

Plants and MSS. at Oxford. Letters and plants at Kew. Portr. at Hunt Library.

LAWSON, Peter (*fl.* 1770s–1821)

Founded nursery in Edinburgh, 1770. Son acquired business.

J. Harvey *Early Hort. Cat.* 1973, 12–13.

LAWSON, Rev. Thomas (1630–1691)

b. near Settle, Yorks 10 Oct. 1630 *d.* Great Strickland, Westmorland 12 Nov. 1691

"Father of Lakeland Botany." Vicar, Rampside, Low Furness. Schoolmaster, Great Strickland near Lowther, Westmorland. Sent plants to R. Morison (*Plantarum Historiae Universalis Oxoniensis* v.3, 1699, 450, etc.). List of plants in *Correspondence of John Ray* 1848, 197–210.

J. Nichols *Lit. Anecdotes of Eighteenth Century* v.1, 233. J. Ray *Synopsis Methodica* 1690, 43, etc. J. Ray *Historia Plantarum* v.2, 1688, preface. L. Plukenet *Almagestum Botanicum* 1696, 8. R. Pulteney *Hist. Biogr. Sketches of Progress of Bot. in England* v.2, 1790, 116–18. D. Turner *Extracts from Lit. Sci. Correspondence of Richard Richardson* 1835, 5–7. J. Smith *Cat. Friends' Books* v.2, 1867, 88–92. J. G. Baker *Fl. English Lake District* 1885, 7–8. W. Hodgson *Fl. Cumberland* 1898, xxiii–xxiv. *D.N.B.* v.32, 297–98. *Trans. Nat. Hist. Northumberland, Durham and Newcastle-upon-Tyne* v.14, 1903, 74–75. *Trans. Liverpool Bot. Soc.* 1909, 80. A. Wilson *Fl. Westmorland* 1938, 70–71. *Proc. Linn. Soc.* 1947–48, 3–12. C. E. Raven *John Ray* 1942, 233. J. E. Dandy *Sloane Herb.* 1958, 154.

MS. note-book at Linnean Society. MSS. at Friends' Meeting House, Devonshire Street, London. Plants at Oxford.

Hieracium lawsonia Vill.

LAWSON, William (*fl.* 1570s–1610s)

Of Teesmouth, Yorks. Horticultural writer. *New Orchard and Garden* 1618; 1623, 1626. *Country Housewife's Garden* 1618.

G. W. Johnson *Hist. English Gdning* 1829, 76–78. *Cottage Gdnr* v.6, 1851, 327. *Gdnrs Chron.* 1898 ii 413, 434–35, 450–51. *D.N.B.* v.32, 298.

LAWTON, John Roland Sylvester (–1970)

d. Australia 1970

Dept. of Botany, Ibadan, S. Nigeria, *c.* 1957. Collected *Dioscorea*.

Plants at Ibadan, Kew.

LAXTON, Edward Augustine Lowe (*c.* 1869–1951)

d. Bedford 22 Feb. 1951

VMH 1932. Son of T. Laxton (*c.* 1830–1893). Fruit and vegetable nurseryman, Bedford.

Gdnrs Mag. 1912, 715–16 portr. *Fruit Yb.* 1949, 9–10 portr. *Gdnrs Chron.* 1951 i 80.

LAXTON, Thomas (*c.* 1830–1893)

b. Tinwell, Rutland *c.* 1830 *d.* 6 Aug. 1893

Nurseryman, Sandy and Bedford. Raised new varieties of peas, beans and strawberries. Conducted some breeding experiments on *Pisum* for C. Darwin. 'Notes on some Changes and Variations in the Offspring of Cross-fertilised Peas' (*J. R. Hort. Soc.* 1872, 10–14).

Garden v.44, 1893, 152, 175–76. *Gdnrs Chron.* 1893 ii 195; 1930 ii 454–56. *J. Hort. Cottage Gdnr* v.27, 1893, 121, 151 portr. G. M. Darrow *The Strawberry* 1966, 81–83.

LAY, George Tradescant (*fl.* 1820s–1845)

d. Amoy, China 1845

Naturalist on Beechey's voyage, 1825–28. Agent in China of British and Foreign Bible Society, 1836. British Consul at Amoy, 1843. Collected with A. Collie in California and Mexico in 1827, and in Macao and the Philippines. *Natural History Calendar at Foo-*

chow-foo 1844–45. Philippine plants in *Chinese Repository* v.7, 1838, 422–37.

J. Hort. Soc. 1846, 119–26. *Trans. Hort. Soc.* v.3, 1848, 237–45. A. Lasègue *Musée Botanique de B. Delessert* 1845, 84. *Notes and Queries* v.5, 1852, 386. E. Bretschneider *Hist. European Bot. Discoveries in China* 1898, 290–94. *Fl. Malesiana* v.1, 1950, 315–16. S. D. McKelvey *Bot. Exploration of Trans-Mississippi West, 1790–1850* 1955, 343–44, 355, 357. *R.S.C.* v.3, 896.

Letters and Californian plants at Kew. Macao plants at BM(NH).

Layia Hook. & Arn.

LAYARD, Mr. (*fl.* 1850s)

Of Melbourne, Victoria. Collected algae for W. H. Harvey.

J. D. Hooker *Fl. Tasmaniae* v.1, 1859, cxxvi.

LAYARD, Sir Austen Henry (1817–1894)
b. Paris 5 March 1817 *d.* London 5 July 1894

Excavator of Ninevah and politician. Collected plants in N. Iraq *c.* 1845.

D.N.B. Supplt. v.3, 82–85.

Plants at BM(NH), Edinburgh.

LAYCOCK, J. (*c.* 1887–1960)
d. 3 Dec. 1960

Friend of Botanic Gardens, Singapore. Collected orchids. Collaborated with R. E. Holttum in hybridising orchids.

Fl. Malesiana Bull. 1961, 795.

LAYCOCK, John (*fl.* 1820s)

Nurseryman, Sowerby Bridge, Halifax, Yorks.

LEA, John (*fl.* 1820s)

Nurseryman, Sowerby Bridge, Halifax, Yorks.

LEA, Rev. Thomas Simcox (1857–1939)
b. Hereford 21 April 1857 *d.* Exmouth, Devon 2 Aug. 1939

Educ. Oxford. Ordained 1881. Collected plants in Australia, 1885–86; and in Pernambuco and Fernando de Noronha with H. N. Ridley, 1887.

C. F. P. von Martius *Fl. Brasiliensis* v.1(1), 1906, 39. *J. Linn. Soc.* v.45, 1920, 196. *J. Bot.* 1939, 275. *Trans. Woolhope Nat. Field Club* v.34, 1955, 260–61.

Plants at BM(NH), Kidderminster Museum.

LEAK, George William (1868–1963)
b. Selby, Yorks June 1868 *d.* 20 July 1963

VMH 1930. VMM 1959. Director, Messrs R. H. Bath, Wisbech. Authority on narcissi and tulips. Vice-president, Royal Horticultural Society.

Gdnrs Chron. 1931 i 40 portr. *Daffodil Tulip Yb.* 1953, 9–10 portr.; 1965, 176–77. *J. R. Hort. Soc.* 1963, 442–43 portr.

LEAR, J. G. (*fl.* 1830s)

To Ceylon in 1837 or earlier as a plant collector for J. Knight of Exotic Nursery, Chelsea. Acting Superintendent, Botanic Gardens, Peradeniya, Ceylon, 1838–40. Collected and described orchids. One of the earliest tea planters in Ceylon.

Paxton's Mag. Bot. 1839, 267–68. *Ann. R. Bot. Gdns Peradeniya* v.1, 1901, 6.

Ipomoea learii Paxton.

LEARED, Arthur (1822–1879)
b. Wexford 1822 *d.* London 17 Oct. 1879

BA Dublin 1845. MD Dublin 1860, Oxon 1861. To Morocco 1872, 1877, 1879. 'Morocco Drugs' (*Pharm. J.* v.3, 1872, 621–25; v.5, 1874, 521–23; v.6, 141–42).

Proc. R. Geogr. Soc. v.1, 1879, 802. *D.N.B.* v.32, 326–27. *R.S.C.* v.3, 903; v.8, 180.

Letters at Kew.

LEATE, (*or* Lete) Nicholas (–1631)
d. London 1631

Levantine merchant. Introduced many rare plants.

J. Gerard *Herball* 1597, 246, 804. M. de Lobel *Animadversiones* 1605, 490. J. Parkinson *Paradisus Terrestris* 1629, 420. *D.N.B.* v.32, 327–28.

LEATHER, John Walter (1860–1934)
b. Rainhill, Lancs 26 Dec. 1860 *d.* 14 Nov. 1934

PhD Bonn. Agricultural chemist, Indian Agricultural Service, 1892–1916. Researched on plant chemistry. Contrib. to *Agric. Ledger*.

Nature v.135, 1934, 58. *Who was Who, 1929–1940* 790.

LEATHES, Rev. George Reading (1779–1836)
b. Rudham, Norfolk 1779 *d.* Shropham, Norfolk 1 Jan. 1836

BA Cantab 1803. FLS 1805. Vicar, Limpenhoe, 1803–36. Plant records in J. Sowerby and J. E. Smith's *English Bot.* 1823, etc. *Gent. Mag.* 1836, 439–40.

Plants at Ipswich Museum.

Leathesia S. F. Gray.

LEBOUR, George Alexander Louis (1847–1918)
d. Newcastle, Northumberland 7 Feb. 1918

Educ. Royal School of Mines. Geological Survey, 1867–73. Lecturer, Geological Surveying, Durham College of Science, 1873–79; Professor of Geology. *Illustrations of Fossil Plants* 1877.

Nature v.100, 1918, 487. *Who was Who, 1916–1928* 614.

LEBROCQ, Philip *see* Brocq, P. le

LECAAN, Jean Polus (*fl.* 1690s–1710s)
MD. Physician to English Hospital at Malines, Belgium. Sent plants to J. Petiver. *Advice to Gentlemen in the Army...in Spain and Portugal* 1708 (descriptions of 'Medicinal Spanish Plants', pp. 59–92).
E. J. L. Scott *Index to Sloane Manuscripts* 1904, 303. J. E. Dandy *Sloane Herb.* 1958, 154–55.

LECHMERE, Arthur Eckley (1885–1919)
b. Fownhope, Hereford 1885 *d.* Long Ashton, Bristol 14 Feb. 1919
BSc London 1909. BSc Bristol 1910. DSc Paris 1911. Mycologist, Long Ashton Research Station, 1914. 'Bacterial Disease of Swedes' (with J. H. Priestley) (*J. Agric. Sci.* 1910, 390–97). 'Two Embryo-sac Mother-cells in the Ovule of *Fritillaria*' (*New Phytol.* 1910, 257–59). 'Investigation of a Species of Saprolegnia' (*New Phytol.* 1910, 305–19). 'Ivory Coast Fungi' (*Bull. Soc. Mycol. France* 1913, 303–31).
Kew Bull. 1919, 164–68.

LECKY, Miss Susan (1837–1896)
b. Cork 27 Sept. 1837 *d.* London 22 Oct. 1896
Drew landscapes and flowers.
Kew Bull. 1917, 211–12.
Drawings at Kew.

LE COUTEUR, Sir John (*fl.* 1830s–1860s)
FRS 1843. Colonel, Jersey Militia. President, Jersey Agricultural Society, 1834–39. *On the Varieties, Properties, etc. of Wheat* 1836 (drew plates).
Gdnrs Mag. 1837, 231. *R.S.C.* v.3, 921.

LEDGER, Charles (1818–1905)
Spent many years in S. America. Living in Tucuman in 1880. Collected seed of *Cinchona calisaya* in Bolivia.
C. R. Markham *Peruvian Bark* 1880, 280. 'Ledger Bark or Red Bark' by J. H. Holland (*Kew Bull.* 1932, 1–17).
Cinchona ledgeriana Moeus.

LEDGERWOOD, James (*fl.* 1790s)
Seedsman, Ulverston, Lancs.

LE DOUX, Jacques Alphonse (*c.* 1881–1961)
b. Liverpool *c.* 1881 *d.* Johore, Malaya 1 April 1961
Rubber-planter in Kota Tinggi district, Malaya. Sent plants to H. N. Ridley.
Gdns Bull., Singapore 1961, 328–29 portr.

LEE, Ann (1753–1790)
d. Hammersmith, Middx 1790
Daughter of James Lee (1715–1795). "Drew and painted plants...with grace and accuracy." A pupil of Sydney Parkinson.
P. Smith *Correspondence of Sir J. E. Smith* v.2, 1832, 25. *J. Bot.* 1914, 323; 1917, 65–66. E. J. Willson *James Lee and the Vineyard Nursery, Hammersmith* 1961, 56, 60–62. A. M. Lysaght *Joseph Banks in Newfoundland and Labrador, 1766* 1971, 102–3.
Drawings at National Library Canberra, Kew, BM(NH).

LEE, Miss Annie (–1930)
d. 30 Nov. 1930
Co-Secretary, Liverpool Botanical Society, 1917–21. Contrib. to *Lancashire and Cheshire Nat.*
N. Western Nat. 1931, 33–34.

LEE, Charles (1808–1881)
b. Hammersmith, Middx 8 Feb. 1808 *d.* Hounslow, Middx 2 Sept. 1881
Son of J. Lee (1754–1824). In 1827 joined his brother John as a partner in nursery at Hammersmith.
Garden v.20, 1881, 281. *Gdnrs Chron.* 1881 ii 330–31. *Gdnrs Yb. and Almanack* 1882, 192–93. *J. Hort. Cottage Gdnr* v.3, 1881, 247–48 portr. *Florist and Pomologist* 1881, 160.

LEE, Ernest (1886–1915)
b. Stanley-Lane End, Yorks 11 April 1886 *d.* killed in action, Flanders 10 July 1915
ARCS 1909. FLS 1911. Assistant Lecturer, Birkbeck College, 1910–13. Lecturer in Agricultural Botany, Leeds, 1913. 'Morphology of Leaf-fall' (*Ann. Bot.* 1911, 51–106). 'Seedling Anatomy of Sympetalae' (*Ann. Bot.* 1912, 727–46; 1914, 303–39).
Ann. Bot. 1915, 639–41. *J. Ecol.* 1915, 243. *New Phytol.* 1915, 300–1. *Proc. Linn. Soc.* 1915–16, 65–66.

LEE, Henry (1826–1888)
d. Brixton, London 31 Oct. 1888
FLS 1866. Surgeon. Naturalist to Brighton Aquarium, 1872. Founded Croydon Microscopical Club. *Vegetable Lamb of Tartary* 1887.
D.N.B. v.32, 357. *R.S.C.* v.8, 188; v.10, 547; v.12, 436.

LEE, James (1715–1795)
b. Selkirk 1715 *d.* Hammersmith, Middx 25 July 1795
Gardener at Syon House, Middx and to Duke of Argyll at Whitton, Middx. Partner with J. Kennedy in Vineyard Nursery, Hammersmith, *c.* 1745. Had plant collectors in America and at the Cape. Introduced *Fuchsia coccinea.* Correspondent of Linnaeus. *Introduction to Botany* 1760; ed. 4 1810 portr. Contrib. to W. Aiton's *Hortus Kewensis* (v.1, 1789, x).
Bot. Mag. 1788, t.56. R. Pulteney *Hist. Biogr. Sketches of Progress of Bot. in England* v.2, 1790, 349. A. H. Haworth *Observations on Genus Mesembryanthemum*

1794, 22, 25–28. M. Tenore *Viaggio* v.3, 1828, 138–46. P. Smith *Mem. and Correspondence of Sir J. E. Smith* v.2, 1832, 117, 183. J. C. Loudon *Arboretum et Fruticetum Britannicum* v.1, 1838, 78–79. T. Faulkner *Hist. Hammersmith* 1839, 42. *J. Hort. Cottage Gdnr* v.57, 1877, 64–66 portr. *Gdnrs Chron.* 1881 ii 330–31; 1899 i 56; 1952 i 114 portr. *D.N.B.* v.32, 357–58. *J. Bot.* 1915, 66–67, 112; 1917, 65–66. E. J. Willson *James Lee and the Vineyard Nursery, Hammersmith* 1961 portr. R. Webber *Early Horticulturists* 1968, 91–100 portr.

Letters in Smith correspondence at Linnean Society. Portr. at Hunt Library.

Leea L.

LEE, James (1754–1824)
d. 10 June 1824

Son of J. Lee (1715–1795). Partner in Lee and Kennedy's Vineyard Nursery, Hammersmith until 1818; alone 1818–24; left property to his 2nd wife Eliza and his sons John, Charles and James Lee.

LEE, John (*c.* 1805–1899)
d. Kensington, London 20 Jan. 1899

Son of J. Lee (1754–1824). In partnership with his brother Charles in nursery in Hammersmith. Retired in 1877 when his nephew William joined the firm; returned in 1881 after the death of Charles.

Garden v.26, 1884, xii portr.; v.55, 1899, 60 portr. *Gdnrs Chron.* 1893 i 259 portr.; 1899 i 56. *Gdnrs Mag.* 1899, 51. *J. Hort. Cottage Gdnr* v.38, 1899, 63 portr.

LEE, John Edward (1808–1887)
b. Hull, Yorks 21 Dec. 1808 *d.* Torquay, Devon 18 Aug. 1887

FGS 1859. Palaeontologist and antiquarian. 'On the Dispersion of Plants' (*Mag. Nat. Hist.* 1832, 522–32).

D.N.B. v.32, 363. *R.S.C.* v.3, 924; v.8, 188; v.10, 547.

LEE, John Ramsay (1868–1959)
b. Helensburgh, Dunbartonshire 26 May 1868 *d.* Glasgow 14 Jan. 1959

Hon. MA Glasgow 1950. Cashier, Glasgow. Hon. Curator, Herb., Glasgow University, 1938. President, Andersonian Naturalists' Society, 1903–4. President, Natural History Society, Glasgow, 1911–14; President of combined societies, 1931–33. *Fl. Clyde Area* 1933. Contrib. plant records to *Census Catalogue of British Mosses* 1907.

Glasgow Nat. 1951, 33 portr.; 1958, 31–36 portr.; 1959, 111–14. *Trans. Br. Bryol. Soc.* 1959, 640. *Proc. Bot. Soc. Br. Isl.* 1960, 103–5.

Herb. at Glasgow.

LEE, Phineas Fox (–1912)
b. Dewsbury, Yorks *d.* Dewsbury 12 March 1912

Botanical Secretary, Yorkshire Naturalists' Union. Had herb. 'Fl. Dewsbury and Neighbourhood' (*Yorkshire Nat. Union Trans.* 1885, 1888, 1889; *Bot. Trans. Yorkshire Nat. Union* 1891; *Naturalist* 1912).

Naturalist 1912, 123–25 portr.

Herb. acquired by C. C. P. Hobkirk (1837–1902).

LEE, Sarah (*née* **Wallis**) (1791–1856)
b. Colchester, Essex 10 Sept. 1791 *d.* 23 Sept. 1856

First woman to collect plants systematically in tropical West Africa. Accompanied her first husband, T. E. Bowdich (1791–1824) to West Africa, 1814, 1815 and 1823. After her husband's death she continued to collect and draw plants; the major part of her collection was lost at sea. Contrib. botanical appendix, pp. 244–67, to T. E. Bowdich's *Excursions in Madeira and Porto Santo* 1825. Married Robert Lee in 1829. *Trees, Plants and Flowers* 1854.

Gent. Mag. 1856, 653–54. J. Hardy *Selections from Correspondence of George Johnston* 1892, 177. *D.N.B.* v.6, 43; v.32, 379. *R.S.C.* v.1, 550.

Plants at BM(NH).

LEE, William Arthur (1870–1931)
b. Hindley, Lancs 1870 *d.* Birkenhead, Cheshire 17 July 1931

BA 1902. MA Dublin 1906. PhD 1909. Chief Superintendent of Telegraphs, Liverpool General Post Office. President, Liverpool Botanical Society. 'Muscineae of the Wirral' (*Lancashire and Cheshire Nat.* 1921, 1928).

N. Western Nat. 1931, 235–37 portr.

LEEBODY, Mrs. Mary Isabella (–1911)
b. Portaferry, County Down *d.* Londonderry 19 Sept. 1911

Of Londonderry. Discovered *Teesdalia medicaulis* in Ireland, 1896.

N. Colgan and R. W. Scully *Cybele Hibernica* 1898, 38. *Proc. R. Irish Acad.* 1901, cxxvii; 1915, 77. *Irish Nat.* 1911, 218. *Proc. Belfast Nat. Field Club* v.6, 1913, 626. R. L. Praeger *Some Irish Nat.* 1949, 118. *R.S.C.* v.16, 672.

Herb. at Dublin Museum.

LEECH, William (*fl.* 1780s)

Seedsman, High Wycombe, Bucks.

LEECHMAN, Alleyne *see* Addendum.

LEEDS, Edward (1802–1877)
b. Pendleton, Lancs 9 Sept. 1802 *d.* Bowdon, Cheshire 1877

Nurseryman, Manchester. Hybridised narcissi. Joined his son's stockbroker's business.

Bot. Mag. 1834, t.3295. R. Sweet *Br. Flower Gdn* v.4, 1838, t.65. *Gdnrs Mag. Bot.* v.3, 1851, 169. *Gdnrs Chron.* 1894 ii 561–62, 625–26; 1931 i 169. *Trans. Liverpool Bot. Soc.* 1909, 80. *Daffodil Yb.* 1933, 23–24.

Letters and plants at Kew.

Narcissus leedsii Moore.

LEEFE, Rev. John Ewbank (–1889)
b. Richmond, Yorks *d.* Redcar, Yorks 1889

BA Cantab 1835. FLS 1868. Vicar, Audley End, Essex, 1841; Richmond, Yorks, 1842–44; Cresswell, Northumberland, 1849–87. Arranged *Salix* in *London Cat. of Plants* 1844 and in W. E. Steele's *Handbook of Field Botany* ed. 2 1851. Papers on *Salix* in *Trans. Bot. Soc. Edinburgh* 1840, 155–63 and *J. Bot.* 1870–72. *Salictum Britannicum* (exsiccatae) 1842–44, 1870–72.

Nature v.36, 1887, 159. *R.S.C.* v.3, 925; v.8, 188.

Letters, MS. on *Salix* and plants at Kew. *Salix* at Edinburgh.

LEEMAN, John (*c.* 1843–1918)
d. Heaton Mersey, Lancs 14 Jan. 1918

Of West Bank House, Heaton Mersey. Cultivated orchids.

Orchid Rev. 1918, 39.

LEES, Edwin (1800–1887)
b. Worcester 12 May 1800 *d.* Worcester 21 Oct. 1887

FLS 1835. Printer and stationer. Worcester plants in *Guide to City and Cathedral* (Ambrose Florence, *pseud.*). *Affinities of Plants with Man* 1834. *Botanical Looker-out among the Wild Flowers* 1842; ed. 2 1851. *Botany of Worcestershire* 1867. *Plants of Worcestershire Arranged...in Four Divisions* 1867. *Botany of Malvern Hills* 1843; ed. 3 1868. Arranged *Rubi* in W. E. Steele's *Handbook of Field Botany* 1847. Contrib. to *Phytologist.*

J. Sowerby and J. E. Smith *English Bot.* t.2981. *Ann. Bot.* 1887–88, 406–8. *J. Bot.* 1887, 384. *Proc. Linn. Soc.* 1887–88, 93. J. Amphlett and C. Rea *Bot. Worcestershire* 1909, xxi–xxv. *D.N.B.* v.32, 394. H. J. Riddelsdell *Fl. Gloucestershire* 1948, cxxii–cxxiii. *Trans. Woolhope Nat. Field Club* v.34, 1954, 242–43. A. E. Wade *Fl. Monmouthshire* 1970, 11. *R.S.C.* v.3, 925; v.8, 189; v.16, 674.

Letters in Broome correspondence at BM(NH).

Rubus leesii Bab.

LEES, Frederick Arnold (1847–1921)
b. Leeds, Yorks 20 Jan. 1847 *d.* Leeds 17 Sept. 1921

MRCS 1871. FLS 1872. Founder and editor, 1873–86, Botanical Record Club. *Fl. West Yorkshire* 1888. *W. Yorkshire* (with J. W. Davis) 1878. 'Botany and Outline Fl. Lincolnshire' (*W. White's Directory* 1892). 'Vegetation of Yorkshire and Supplement to Fl. of County' (*Naturalist* 1937 *passim*). Contrib. to *J. Bot.*, *Naturalist.*

J. Bot. 1922, 97–100 portr. *Bot. Soc. Exch. Club Br. Isl. Rep.* 1922, 358–63. *Trans. Lincolnshire Nat. Union* 1966, 153–59. *Sorby Rec.* v.2(4), 1967, 12–13. *Naturalist* 1921, 372–73 portr.; 1967, 77–80; 1968, 133–35; 1970, 125–29; 1973, 35–36. E. J. Gibbons *Fl. Lincolnshire* 1975, 57. *R.S.C.* v.8, 189; v.10, 540; v.12, 436.

Herb. at Cartwright Hall, Bradford. Library at Bradford Public Library.

Carex pilulifera var. *Leesii* Ridley.

LEES, W. H. (*fl.* 1890s)

Gardener to Duchess of Montrose at Sefton Lodge, Newmarket; to F. A. Bevan at Trent Park, 1892.

Gdnrs Mag. 1895, 761 portr. *J. Hort. Cottage Gdnr* v.31, 1895, 533 portr.

LEESON, William (–1722)

Gardener and nurseryman, Hodsock Woodhouse, Blyth, Notts.

LEFROY, Helena (*née* **Trench**) (1820–1908)
b. Queen's County, Ireland 1820 *d.* Aghaderg, County Down 1908

Discovered *Euphorbia peplis* at Waterford, Ireland.

N. Colgan and R. W. Scully *Cybele Hibernica* 1898, 520. *Annual Rep. Proc. Belfast Nat. Field Club* 1911–12, 621. *Irish Nat.* 1913, 27. R. L. Praeger *Some Irish Nat.* 1949, 118.

Herb. formerly at Y.M.C.A., Banbridge, County Down.

LEFROY, Sir John Henry (1817–1890)
b. Ashe, Hants 28 Jan. 1817 *d.* Lewarne, Cornwall 11 April 1890

FRS 1848. KCMG 1877. General, 1870. Governor of Bermuda, 1871–77; Tasmania, 1880–82. Collected plants in Bermuda and Tasmania. 'Botany of Bermuda' (*Bull. U.S. National Mus.* no. 25, 1884, 35–141).

J. Bot. 1883, 105. *Proc. Geogr. Soc.* 1891, 115–22. *D.N.B.* v.32, 399–404. I. Urban *Symbolae Antillanae* v.1, 1898, 95; v.3, 1902, 73. *R.S.C.* v.3, 930; v.10, 552; v.16, 678.

Letters and Bermuda plants at Kew. Tasmanian plants at Leningrad.

Statice lefroyi Hemsley.

LEGG, Thomas (*fl.* 1760s)

Ostler, Crown Inn, Alton, Hants. In his youth a friend of William Curtis. Knew the local flora.

R. H. Semple *Mem. Bot. Gdn, Chelsea* 1878, 104–5. W. H. Curtis *William Curtis* 1941, 5.

LEGGE, George, 3rd Earl of Dartmouth (1755–1810)
b. 3 Oct. 1755 *d.* Cornwall 1 Nov. 1810
MA Oxon 1775. DCL 1778. FLS 1790. MP. Lord Chamberlain, 1804. Had garden at Sandwell Hall, Staffs. Contrib. to J. Sowerby and J. E. Smith's *English Bot.* First President, Horticultural Society of London, 1804–10. President, Society for Promoting Natural History.
Gent. Mag. 1810 ii 500. *D.N.B.* v.32, 410. H. R. Fletcher *Story of R. Hort. Soc., 1804–1968*, 1969, 40 portr.

LEGGE, Lady Joan (–1939)
d. Bhyundar, Himalayas 4 July 1939
Youngest daughter of 6th Earl of Dartmouth. Collected plants in Kumaon, India.
Gdnrs Chron. 1939 ii 89. *Kew Bull.* 1939, 522. *Times* 7 July 1939.
Plants at Kew.

LE HUNTE, Sir George Ruthven (1852–1929)
b. 20 Aug. 1852 *d.* 29 Jan. 1929
BA Cantab 1873. In government service, Fiji, 1875–87; Dominica, Mauritius, Barbados, 1887–97. Lieut.-Governor, New Guinea, 1898–1903. Governor, S. Australia, 1903–8; Trinidad and Tobago, 1908–15. Sent New Guinea plants to F. M. Bailey.
Who was Who, 1916–1928 619. *Fl. Malesiana* v.1, 1950, 251–52.
Plants at Kew.

LEICHHARDT, Friedrich Wilhelm Ludwig (1813–1848)
b. Trebatsch, Prussia 23 Oct. 1813 *d.* lost in Australia 1848
Arrived in Sydney, 1842. Lectured on botany and geology. Made several expeditions on which he collected plants. *Journal of Overland Expedition in Australia from Moreton Bay to Port Essington* 1847. Contrib. to *Tasmanian J. Nat. Sci.* v.3, 1847, 18–51, 81–113.
London J. Bot. 1845, 278–91; 1847, 342–64. J. D. Hooker *Fl. Tasmaniae* v.1, 1859, cxxi. F. von Mueller *Fragmenta Phytographiae Australiae* v.10, 1876, 67–68. *Gdnrs Chron.* 1885, 656. *D.N.B.* v.32, 426–27. *J. R. Soc. N.S.W.* 1908, 108–10. C. D Cotton *Ludwig Leichhardt* 1938. *Proc. R. Soc. Queensland* 1954, 10–12. W. Beard *Journey Triumphant* 1955. R. Erdos *Ludwig Leichhardt* 1963. *Austral. Encyclop.* v.5, 1965, 281–85. *Austral. Dict. Biogr.* v.2, 1967, 102–4. M. Aurousseau *Letters of F. W. L. Leichhardt* 1968 portr. 3 vols. *J. R. Austral. Hist. Soc.* 1971, 74–78. *R.S.C.* v.3, 939.
Letters at Kew. Herb. at Sydney. MSS. at Mitchell Library, Sydney. Portr. at Hunt Library.
Leichhardtia F. Muell.

LEIGH, Charles (1662–1700s)
b. Singleton, Lancs 1662
BA Oxon 1683. MD Cantab 1689. FRS 1685. Practised medicine in London and Manchester. *Natural History of Lancashire, Cheshire, and the Peak in Derbyshire* 1700 portr.
R. Pulteney *Hist. Biogr. Sketches of Progress of Bot. in England* v.1, 1790, 353–54. *D.N.B.* v.32, 431–32. E. J. L. Scott *Index to Sloane Manuscripts* 1904, 305. *Trans. Liverpool Bot. Soc.* 1909, 80.
Leighia Cass.

LEIGH, Clara Maria *see* Pope, C. M.

LEIGHTON, Darwin (–1943)
d. July 1943
Collected ferns in Great Britain and Switzerland.
Br. Fern Gaz. 1948, 202–3.

LEIGHTON, James (1855–1930)
b. Kincardine O'Neil, Aberdeenshire 19 Jan. 1855 *d.* King William's Town, S. Africa 22 Jan. 1930
Kew gardener, 1878–80. To S. Africa, 1881. Curator, King William's Town Botanic Garden until 1887. Established his own nursery, specialising in roses.
Gdnrs Chron. 1930 i 217. *J. Kew Guild* 1930, 833–34 portr.

LEIGHTON, Rev. William Allport (1805–1889)
b. Shrewsbury, Shropshire 7 May 1805 *d.* Shrewsbury 25 Feb. 1889
BA Cantab 1833. FLS 1865. *Catalogue of Cellulares or Flowerless Plants of Great Britain* 1837. *Fl. Shropshire* 1841. *British Species of Angiocarpous Lichens* 1851. *Lichenes Britannici Exsiccati* fasc. I–XIII, nos. 1–410, 1851–67. *Lichen-fl. of Great Britain, Ireland and the Channel Islands* 1871; ed. 3 1879.
Trans. Shropshire Archaeol. Soc. 1886 portr. *Ann. Bot.* 1889–90, 465–67. *J. Bot.* 1889, 111–13. *D.N.B.* v.33, 8–9. B. Lynge *Index...Lichenum Exsiccatorum* 1915–19, 323–32. *Essex Nat.* v.29, 1954, 193. *R.S.C.* v.3, 943; v.8, 197.
Letters at Kew, BM(NH). Herb. at Kew. Lichens at Essex Field Club. Portr. at Kew, Hunt Library.
Leightonia Trev.

LEIPNER, Adolph (1827–1894)
b. Dresden, Germany 13 Aug. 1827 *d.* Clifton, Glos 1 April 1894
To England in 1848. Teacher of German and Natural Science at Clifton, 1854. Lecturer in Botany and Vegetable Physiology, Bristol Medical School. Professor of Botany, University College, Bristol, 1886; formed botanic garden there. Secretary, Bristol

Naturalists' Society. 'Silica in Rubiaceae' (*Quart. J. Microsc. Sci.* 1857, 134–37). 'List of Mosses of Bristol District' (*Proc. Bristol Nat. Soc.* v.3, 1868, 21–22).

J. Bot. 1894, 224. *Proc. Bristol Nat. Soc.* v.9, 1899, 81–83 portr. J. W. White *Fl. Bristol* 1912, 99–100. H. J. Riddelsdell *Fl. Gloucestershire* 1948, cxxv–cxxvi. *R.S.C.* v.3, 944; v.8, 198.

Moss herb. at Bristol Museum. Letters in Broome correspondence at BM(NH).

LEITCH, John (*c.* 1849–1896)
b. Monimail, Fife *c.* 1849 *d.* Silloth, Cumberland 22 Dec. 1896

MB Edinburgh 1871. Son of W. Leitch (1814–1864). Contrib. to W. Hodgson's *Fl. Cumberland* 1898 (xxxii).

J. Bot. 1897, 112. *R.S.C.* v.16, 693.

Herb. at Carlisle Museum.

LEITCH, William (1814–1864)
b. Rothesay, Isle of Bute 1814 *d.* 9 May 1864

President, Botanical Society of Canada, 1860–62.

Amer. Annual Cyclop. 1864, 625–26.

LEITH, Andrew Henderson (1807–1875)
b. Edinburgh 23 Nov. 1807 *d.* Tunbridge Wells, Kent 28 Nov. 1875

MD. Surgeon in East India Company, Bombay. Collected plants, especially ferns, in N. and S. India. Correspondent of E. J. Lowe.

Pakistan J. For. 1967, 351.

Plants at Manchester, Oxford.

LE LACHEUR, Mrs. Cecilia Mary Nutter (*c.* 1883–1968)
d. Littlehampton, Sussex 24 Sept. 1968

FLS 1925. "Attached to study of Botany."

Daily Telegraph 26 Sept. 1968.

LEMANN, Charles Morgan (1806–1852)
b. London 1806 *d.* Bath, Somerset 26 Aug. 1852

MD Cantab 1833. FLS 1831. Collected plants in Madeira, 1837–38, and Gibraltar, 1840–41. MS. Fl. of Madeira (*see* R. T. Lowe *Manual Fl. Madeira* v.1, 1868 preface).

Companion Bot. Mag. 1837, 340. *Gdnrs Chron,* 1842, 127. *Proc. Linn. Soc.* 1853, 234–35. *Hooker's J. Bot.* 1853, 307–9.

Herb. at Cambridge. Plants and letters at Kew.

Carlemannia Benth.

LEMON, Sir Charles (*c.* 1784–1868)

MP. Had garden at Carclew, Cornwall noted for its rhododendrons.

J. R. Hort. Soc. 1943, 262–63.

LENDY, Auguste Frederic (*c.* 1825–1889)
d. Sunbury-on-Thames, Middx 10 Oct. 1889

FLS 1861. Major, 4th Middx Regiment. Cultivated orchids.

Gdnrs Chron. 1889 ii 450.

LEONARD, H. Selfe (–1902)
d. Rome Feb. 1902

Established Guildford Hardy Plant Co. at Guildford, Surrey.

Garden v.61, 1902, 151.

LESCHALLAS, Henry Pigé (–1947)

Army Captain. *Small Alpine Garden* 1937.

Quart. Bull. Alpine Gdn Soc. 1947, 74.

LESLIE, John Erskine (*c.* 1877–1962)
b. Aberlour, Banffshire *c.* 1877 *d.* 2 Nov. 1962

Kew gardener, 1902. Assistant Curator, Royal Botanic Garden, Calcutta, 1902–4. Held appointments in Bengal, Nagpur, New Delhi, Darjeeling. Curator, Calcutta Botanic Garden, 1931–32. Collected plants in Sikkim and Tibet.

J. Kew Guild 1962, 205 portr.

LESTER, J. Brown (*fl.* 1890s)

MD Edinburgh. Collected plants during Anglo-French Gambia Delimitation Commission, 1890–1.

Kew Bull. 1891, 268–75. *Bull. Herb. Boissier* 1907, 83.

Plants at Kew.

LESTER-GARLAND, Lester Vallis (*né* Lester) (1860–1944)
b. 28 July 1860 *d.* Bath, Somerset 23 March 1944

Educ. Oxford. FLS 1899. Assistant master, St. Edwards School, Oxford, 1883–86. Lecturer, St. John's College, Oxford, 1886–96. Principal, Victoria College, Jersey, 1896–1911. *Fl. of Island of Jersey* 1903 (the first British flora to adopt Engler classification).

Bot. Soc. Exch. Club Br. Isl. Rep. 1944, 647–48. *Nature* v.153, 1944, 613. *Times* 27 March 1944.

Herb. at Kew.

LESUEUR, A. Denis C. (–*c.* 1969)

Forestry consultant, 1924. Head, Forestry Dept., Royal Agricultural College, Cirencester, 1925–46. *Care and Repair of Ornamental Trees* 1934; ed. 2 1949.

Arboricultural Assoc. J. v.1, 1969, 247–48 portr.

LE TALL, Benjamin Bower (1858–1906)
b. Woodhouse, Yorks 1858 *d.* Hobart, Tasmania 1906

MA London. Master at Bootham School, 1883; Friends' School at Hobart, 1893. Field botanist. Had herb. Edited H. Ibbotson's *Ferns of York* 1884. Joint editor of *Nat. Hist. J.*, 1877–93.

Bootham Register 1914, 145.

LETE, Nicholas *see* Leate, N.

LETHERLAND, J. (1699–1764)
b. Stratford-on-Avon, Warwickshire 1699
MD Leyden 1724. MD Cantab 1736. Physician, St. Thomas's Hospital, 1736; to the Queen, 1761. Member of Botanical Society of London.
Proc. Bot. Soc. Br. Isl. 1967, 317–18.

LETT, Rev. Henry William (1838–1920)
b. Hillsborough, County Down 4 Dec. 1838 *d.* Aghaderg, Loughbrickland, County Down 26 Dec. 1920
TCD. Rector, Aghaderg, 1886. Canon of Dromore. Discovered *Rubus lettii* in Ireland, 1901. *List...of all Species of Hepatics hitherto found in British Islands* 1902. *Census Report on Mosses of Ireland* 1915. Contrib. to *J. Bot.*
Bot. Soc. Exch. Club Br. Isl. Rep. 1920, 363. *Irish Nat.* v.30, 1921, 41–43. *J. Bot.* 1921, 75–76. R. L. Praeger *Some Irish Nat.* 1949, 118–19. *Who was Who, 1916–1928* 625.
Plants at National Museum, Dublin. Portr. at Hunt Library.

LETTS, Arthur (1862–1923)
b. Holdenby, Northants 17 June 1862 *d.* Los Angeles, U.S.A. 18 May 1923
Merchant. To Canada, 1883; to Seattle, then Los Angeles, 1893. His garden of succulents and cycads formed basis of Huntington and University of California, Los Angeles collections.
W. H. B. Kilner *Arthur Letts* 1927, 255–60 portr.

LETTS, Edmund Albert (1852–1918)
b. Sydenham, Kent 27 Aug. 1852 *d.* Isle of Wight 19 Feb. 1918
Professor of Chemistry, Bristol, 1876; Belfast, 1879. Wrote papers on the relation of *Ulva latissima* to the nitrogen content of the water in which it grows.
Nature v.101, 1918, 7–8. *Who was Who, 1916–1928* 625.

LETTSOM, John Coakley (1744–1815)
b. Tortola, Virgin Islands 22 Nov. 1744 *d.* London 1 Nov. 1815
MD Leyden 1769. FRS 1773. FLS 1797. Physician in London, 1770. President, Medical Society of London, 1775, 1809, 1813. Had a botanic garden arranged on Linnaean system at Camberwell. *Natural History of Tea-tree* 1772. *Hortus Uptonensis* 1783.
J. Nichols *Illus. Lit. Hist. of Eighteenth Century* v.2, 657–88 portr. *Gent. Mag.* 1801 ii 761; 1804 i 473; 1815, 2, 469–73, 577. T. J. Pettigrew *Mem. of Life and Writings of J. C. Lettsom* 1817. *Cottage Gdnr* v.5, 1850, 79. W. Darlington *Memorials of John Bartram and Humphry Marshall* 1849, 541–49. *J. Hort. Cottage Gdnr* v.57, 1877, 385–86 portr. W. Munk *Roll of R. College of Physicians* v.2, 1878, 287–90. Friends' Inst. *Biogr. Cat.* 1888, 423–29. *D.N.B.* v.33, 134–36. *Notes and Queries* v.5, 1906, 393, 514. *J. Bot.* 1914, 320. H. A. Kelly *Some Amer. Med. Botanists* 1915, 19. St. Clair Thomson *J. C. Lettsom* 1918. R. H. Fox *Dr. John Fothergill and his Friends* 1919, 99–117. *Lancet* 1929, 855. J. J. Abraham *Lettsom* 1933. *Nature* v.132, 1933, 948–49. *Agriculture* 1948, 17–20. *Osiris* 1948, 115. T. Hunt *ed. Med. Soc. of London, 1773–1973* 1–21 portr., 101–4. *R.S.C.* v.3, 979.
Sale at Leigh and Sotheby 26 March 1811 and 3 April 1816.
Lettsomia Roxb.

LEVINGE, Harry Corbyn (*c.* 1831–1896)
d. Knockdrin Castle, Mullingar, Westmeath 11 March 1896
Secretary, Bengal Public Works Dept. Collected ferns in Sikkim, Kashmir and Nilgiris. Discovered *Chara denudata* in Ireland. Contrib. to *Irish Nat., J. Bot.*
Irish Nat. 1896, 107. *J. Bot.* 1896, 240. *Nature* v.53, 1896, 583–84. *Sci. Proc. R. Dublin Soc.* 1903, 122–32. R. L. Praeger *Some Irish Nat.* 1949, 119–20. *R.S.C.* v.16, 753.
Herb. at Royal Dublin Soc.
Adiantum levingei Baker.

LEWIN, John William (*c.* 1770–1819)
d. Sydney, N.S.W. 27 Aug. 1819
ALS 1801. Ornithologist, entomologist and "painter and drawer in natural history." In Australia from 1798. Drew plants collected on John Oxley's expedition, 1817.
Proc. Linn. Soc. 1846, 299. *D.N.B.* v.33, 170. *J. Bot.* 1902, 303. *Proc. Linn. Soc. N.S.W.* 1902, 747–49. *J. R. Soc. N.S.W.* 1908, 110–11; 1921, 151. *J. Proc. R. Austral. Hist. Soc.* 1920, 236–40; 1956, 153–86. *Austral. Nat.* 1930, 1–3. G. P. Whitley *Some Early Nat. and Collectors in Australia* 1933, 9–10. P. Serle *Dict. Austral. Biogr.* v.2, 1949, 31–33. *Austral. Encyclop.* v.5, 1965, 293–94. *Austral. Dict. Biogr.* v.2, 1967, 111–12.
Plant drawings at Mitchell Library and Botanic Gardens, Sydney.

LEWIS, David (*fl.* 1790s)
Seedsman, 28 Cornhill, London. Business acquired by Warner and Seaman, 1795.

LEWIS, Francis John (1875–1955)
b. London 1875 *d.* London 24 May 1955
MSc 1908 DSc 1912 Liverpool. FLS 1900. FRSE. Demonstrator in Botany, Liverpool

University, 1900–5; Lecturer in Phytogeography, 1905–12. Professor of Botany, Alberta, 1912–35. Professor of Botany, Fouad I University of Cairo, 1935–46. Visiting lecturer in Plant Physiology, Royal Holloway College, 1947–48. 'Geographical Distribution of Vegetation of Basins of River Eden, Tees, Wear and Tyne' (*Geogr. J.* v.23, 1904, 313–31; v.24, 1904, 267–85). Contrib. to *Ann. Bot.*, *J. Ecol.*, *J. Linn. Soc.*, etc.

Nature v.176, 1955, 237–38. *Proc. Linn. Soc.* 1954–55, 136–37. *Times* 28 May 1955. *Who was Who, 1951–1960* 658.

LEWIS, Frederick (1857–1930)
b. Ceylon 18 July 1857 *d.* Ceylon 19 July 1930

FLS 1894. Forest Dept., Ceylon for 18 years. *Vegetable Products of Ceylon* 1934.

Proc. Linn. Soc. 1930–31, 180.

Vateria lewisiana Trimen.

LEWIS, Rev. George (*fl.* 1690s–1700s)

Of Madras. Sent Cape of Good Hope plants to J. Petiver.

J. Petiver *Museii Petiveriani* 1695, 261, 784. J. E. Dandy *Sloane Herb.* 1958, 155.

Plants at BM(NH).

LEWIS, John Harbord (–before 1890)

FLS 1874–90. Of Liverpool. Friend of A. E. Lomas and F. M. Webb. Contrib. to *Fl. Liverpool* 1872. Collected plants in S. Africa and Europe.

J. Bot. 1874, 160; 1875, 199. *Fl. Liverpool* 1872, ii. J. B. L. Warren *Fl. Cheshire* 1899, xciii. *Trans. Liverpool Bot. Soc.* 1909, 80.

Herb. at Manchester. Plants at Liverpool, Oxford.

Polypodium lewisii Baker.

LEWIS, John Spedan (1885–1963)
d. Stockbridge, Hants 21 Feb. 1963

FLS 1933. Founded John Lewis Partnership which established the John Spedan Lewis Trust for the Advancement of Natural Sciences in 1955. Botanist and zoologist. Supported expeditions; gave plants to Kew. Had garden at Longstock House near Stockbridge.

Proc. Linn. Soc. 1964, 94–95.

LEWIS, W. T. (*fl.* 1800s–1860s)

Uncovenanted official, Bencoolen, 1806–24. Head, Land Dept., Malacca. Assistant Resident Councillor, Penang, 1840. Resident, Malacca. Councillor, Penang, 1855–60. Friend of W. Griffith (1810–45) to whom he sent Malay plants.

J. Malay Branch R. Asiatic Soc. v.3(2), 1925, 96. *Fl. Malesiana* v.1, 1950, 323.

LEWTON-BRAIN, Lawrence (1879–1922)
b. Swanton Morley, Norfolk 26 June 1879 *d.* Kuala Lumpur, Malaya 24 June 1922

BA Cantab 1898. FLS 1903. Demonstrator in Botany, Cambridge, 1900. Mycologist, Dept. Agriculture, W. Indies, 1902–5. Assistant Director, Pathology Division, Hawaiian Sugar Planters' Association Experiment Station, 1905; Director, 1907. Director of Agriculture, Federated Malay States, 1910–22. Technical Adviser, 1922. 'Anatomy of Leaves of British Grasses' (*Trans. Linn. Soc. Bot.* 1904, 315–59).

Kew Bull. 1903, 30; 1910, 253; 1922, 199. *Malay Agric. J.* 1922, 150. *Who was Who, 1916–1928* 629.

LEY, Rev. Augustin (1842–1911)
b. Hereford 3 April 1842 *d.* Ross, Herefordshire 23 April 1911

MA Oxon. Vicar, Sellack, 1878–1908. Studied *Rosa*, *Rubus*, *Hieracium*, *Ulmus*, mosses, etc. *Fl. Herefordshire* (with W. H. Purchas) 1889. Contrib. botany to *Victoria County Hist. Hereford* v.1, 1908, 39–54. 'British Roses of Mollis–Tomentosa Group' (*J. Bot.* 1907, 200–10). Contrib. to *Trans. Woolhope Nat. Field Club.*

Trans. Woolhope Nat. Field Club 1908–11, 195–204 portr.; 1954, 250–53. *Rep. Watson Bot. Exch. Club* 1910–11, 274–76 portr. *Bot. Soc. Exch. Club Br. Isl. Rep.* 1911, 46–50. *J. Bot.* 1911, 201–6 portr. H. J. Riddelsdell *Fl. Gloucestershire* 1948, cxxxvi. A. E. Wade *Fl. Monmouthshire* 1970, 14.

Plants at Birmingham, Oxford. Portr. at Hunt Library.

Rubus leyanus Rogers.

LEYCESTER, Augustus Adolphus (–1892)
b. Whitehead near Maidenhead, Berks

Farmer in New South Wales. Keen naturalist; collected plants.

Daily Mirror (Sydney) 2 Aug. 1961, 22.

Australian plants at Kew.

LEYCESTER, William (1775–1831)
b. Surrey 1775 *d.* Puri, India 24 May 1831

Judge in Bengal Civil Service, 1804. President, Agricultural and Horticultural Society of India, 1827.

D. McClintock *Companion to Flowers* 1966, 46.

Plants at Kew.

Leycesteria Wall.

LEYEL, Hilda Winifred Ivy (*née* **Wauton**) (1880–1957)
b. London 6 Dec. 1880 *d.* London 15 April 1957

Sold herbal preparations from Culpeper House, Baker Street, London from 1927. Founded Society of Herbalists. *Magic of Herbs* 1926. *Herbal Delights* 1937. *Compassionate Herbs* 1946. *Elixirs of Life* 1948. *Hearts-ease* 1949. *Green Medicine* 1952.

Cinquefoil 1957. Edited M. Grieve's *Modern Herbal* 1931 2 vols.

D.N.B. 1951–1960 1971, 631–32. *Who was Who, 1951–1960* 659–60.

LEYLAND, Roberts (1784–1847)
b. Halifax, Yorks 1784 *d.* Halifax 15 Nov. 1847

Printer. Botanical Curator, Halifax Literary and Philosophical Society, 1830. Collected lichens. Contrib. to H. C. Watson's *New Botanists' Guide* 1835–37, 659, etc.

W. B. Crump and C. Crossland *Fl. Parish of Halifax* 1904, liv–lvi.

Herb., formerly at Belle Vue Museum, Halifax, now destroyed. Plants at BM(NH). Letters at Kew.

Sticta leylandi Taylor.

LHOTSKY, Johann (*fl.* 1800–1860s)
b. Lemberg, Galicia 27 June 1800

MD Vienna. To Brazil, 1830 where he collected plants. Australia, 1832. Collected plants in New South Wales, and also around Hobart, Tasmania, in 1836. In London, 1839–40. *Journey from Sydney to Australian Alps* 1835. *Botanical Geography of New Holland* 1843. 'On New Method of introducing Palms of Large Size into Hothouses' (*Gdnrs Mag*, 1840, 596–97).

Proc. Linn. Soc. Zool. 1839, 4, 11–13, 39–40. A. Lasègue *Musée Botanique de B. Delessert* 1845, 281–82. C. F. P. von Martius *Fl. Brasiliensis* v.1(1), 1906, 42. *Proc. R. Soc. Queensland* 1890–91, 125. *J. Proc. R. Soc. N.S.W.* 1908, 72–74. *Austral. Zool.* 1922–24, 223–26. *J. Proc. R. Austral. Hist. Soc.* 1934, 318–21. *Med. J. Australia* 1938, 661–67. G. P. Whitley *Some Early Nat. and Collectors in Australia* 1933, 30–33. *Austral. Encyclop.* v.5, 1965, 296–97. *Austral. Dict. Biogr.* v.2, 1967, 114–15. *Taxon* 1970, 530–31. *R.S.C.* v.4, 2; v.6, 713; v.12, 445.

Plants at Berlin. Letters at Kew.

Lhotskya Schauer.

LHUYD (or Lloyd), Edward (1660–1709)
b. Llanfihangel, Cardigan 1660 *d.* Oxford 30 June 1709

MA Oxon 1701. FRS 1708. Geologist, botanist, philologist, antiquary. Keeper, Ashmolean Museum, 1690–1709. "Plantarum Britannicarum sagacissimus investigator" (Plukenet *Almagestum Botanicum* 1696, 261). Discovered *Lloydia*, *Daboecia*, etc. First to record (in E. Gibson's edition of Camden's *Britannia* 1695) that the mountains of Britain have a distinctive alpine flora. His list of Snowdon's plants included in J. Ray's *Synopsis Methodica* 1690.

R. Morison *Plantarum Historiae Universalis Oxoniensis* v.3, 1699, 478, 499, 573, 633, etc. *Philos. Trans. R. Soc.* v.27, 1712, 524–26; v.28, 1714, 93–101. N. Owen *Mem. Edward Llwyd* (*British Remains* 1777, 129–84). R. Pulteney *Hist. Biogr. Sketches of Progress of Bot. in England* v.2, 1790, 110–16. *Gent. Mag.* 1807 i 419, 421. H. Davies *Welsh Botanology* 1813, ix–xi. *Trans. Hort. Soc. London* v.1, 1812, 328–29. D. Turner *Extracts from Lit. and Sci. Correspondence of Richard Richardson* 1835, 12–14, 62–64, etc. E. Lankester *Correspondence of John Ray* 1848, 482–84. *D.N.B.* v.33, 217–19. *N. Western Nat.* 1934, 224–29. C. E. Raven *John Ray* 1942 *passim.* R. T. Gunther *Edward Llwyd; Life and Letters* 1945. *Nature* v.159, 1947, 154–55. J. E. Dandy *Sloane Herb.* 1958, 155–57. *Dict. Welsh Biogr. down to 1940* 1959, 565–67. *Country Life* 1960, 428–29 portr. H. N. Clokie *Account of Herb. Dept. Bot., Oxford* 1964, 199. *Trans. Hon. Soc. Cymmrodorion* 1965(1), 59–114. C. C. Gillispie *Dict. Sci. Biogr.* v.8, 1973, 307–8. *B.S.B.I. Welsh Region Bull.* no. 23, 1975, 3–8.

Plants at BM(NH), Oxford. Letters in Sherard correspondence at Royal Society. Letters at Trinity College, Dublin. Letters sold at Sotheby in 1807, believed destroyed in fire. Portr. at Hunt Library.

Lloydia Salisb.

LIGHTBODY, George (–1872)
d. 9 June 1872

Of Falkirk? Grew auriculas; raised numerous varieties.

Garden v.1, 1872, 714.

LIGHTFOOT, Catherine Anne (1807–1898)

Sister of Rector of Lincoln College, Oxford. Made a collection of plants at Wootton, Stockleigh and Pomeroy, Northants.

G. C. Druce *Fl. Buckinghamshire* 1926, c. G. C. Druce *Fl. Northamptonshire* 1930, cxv. G. C. Druce *Comital Fl. of Br. Isl.* 1932, xii.

Plants at Oxford.

LIGHTFOOT, Rev. John (1735–1788)
b. Newent, Glos 9 Dec. 1735 *d.* Uxbridge, Middx 20 Feb. 1788

MA Oxon 1766. FRS 1781. FLS 1788. Rector, Gotham. Chaplain to Dowager Duchess of Portland. Travelled through Scotland with T. Pennant, 1772. *Fl. Scotica* 1777 2 vols.; ed. 2 1789. Transcript of MS. Journal of Excursion in Wales, 1773 at BM(NH), published in *J. Bot.* 1905, 290–307. Herb. bought for Queen Charlotte (*J. Bot.* 1915, 269–71).

Gent. Mag. 1780, 542; 1788, 183, 269. *Trans. Linn. Soc.* v.4, 1798, 280. P. Smith *Mem. and Correspondence of Sir J. E. Smith* v.1, 1832, 289. *Notes and Queries* v.8, 1877, 129, 275. *D.N.B.* v.33, 231–32. *Lancashire Nat.* v.1, 1907, 102–3. C. Crossland *Eighteenth Century Naturalist: James Bolton* 1910, 7. G. C. Druce *Fl. Berks.* 1897, cxxxix–cxl. G. C. Druce *Fl. Buckinghamshire* 1926, xciv–xcv. G. C. Druce *Fl. Oxfordshire* 1927, xci–xcii. H. J. Riddelsdell *Fl. Gloucestershire* 1948, cxiii–cxiv. F. A. Stafleu *Taxonomic Literature* 1967, 268.

Plants at Kew, BM(NH). Letters at BM(NH), Linnean Society, Hardwick Hall. Herb. sold at Christie's 2 June 1821.

Lightfootia L'Hérit.

LIGHTFOOT, John Prideaux Weston (1871–1919)
b. Crosby Ravensworth, Westmorland 4 Aug. 1871 *d.* Ketton, Rutland 26 April 1919

Employed by English Church Union Office. Assisted in preparation of a united flora of Leicestershire and Rutland. Secretary, Rutland Archaeological and Natural History Society, 1912–17.

A. R. Horwood and C. W. F. Noel *Fl. Leicestershire* 1933, cclxxxi.

LIND, Mr.

Collected plants in Lord Howe Island for F. von Mueller before 1875.

J. R. Soc. N.S.W. 1908, 111.

LIND, James (1736–1812)
b. Scotland 17 May 1736 *d.* London 17 Oct. 1812

MD Edinburgh 1768. FRS 1777. Visited China in 1766. Subsequently became physician to Royal Household at Windsor.

Gent. Mag. 1812 ii 405. E. Bretschneider *Hist. European Bot. Discoveries in China* 1898, 152.

Plants at BM(NH).

LINDLEY, George (*c.* 1769–1835)
d. Catton, Norfolk 20 May 1835

Nurseryman, Catton near Norwich. *Plan of an Orchard* 1796. 'Account of...Best Varieties of Apples...in...Norfolk' (*Trans. Hort. Soc.* v.4, 1820, 65–71.) *Guide to Orchard and Kitchen Garden* 1831, edited by his son, John Lindley.

LINDLEY, John (1799–1865)
b. Catton, Norfolk 5 Feb. 1799 *d.* Turnham Green, Middx 1 Nov. 1865

PhD Munich 1832. FRS 1828. FLS 1820. Son of George Lindley. Assistant in J. Banks's library, 1819. Garden clerk, Horticultural Society of London, Chiswick, 1822. Assistant Secretary, Horticultural Society, 1827; Secretary, 1858–63. Professor of Botany, University College, London, 1829–60. Praefectus Hort., Chelsea. In 1838 prepared a report on royal gardens at Kew which led to creation of Royal Botanic Gardens, Kew. *Rosarum Monographia* 1820. *Digitalium Monographia* 1821. *Collectanea Botanica* 1821 (drawings at BM(NH). *Synopsis of British Fl.* 1829; ed. 2 1835. *Introduction to Natural System of Botany* 1830; ed. 2 1835. *Genera and Species of Orchidaceous Plants* 1830–40. *Introduction to Botany* 1832; ed. 4 1848. *Ladies' Botany* 1834; ed. 5 1856. *Victoria Regia* 1837. *Fl. Medica* 1838. *Sertum Orchidaceum* 1838. *Vegetable Kingdom* 1846; ed. 3 1853. *Folia Orchadicea* 1852–59. *Fossil Fl. of Great Britain* (with W. Hutton) 1831–37 3 vols. *Treasury of Botany* (*with* T. Moore) 1866 2 vols. *Paxton's Flower Garden* (with J. Paxton) 1850–53 3 vols. Edited *Bot. Register* from 1826; *Gdnrs Chron.* from 1841.

N. Wallich *Plantae Asiaticae Rariores* v.1, 1830, 25. A. Lasègue *Musée Botanique de B. Delessert* 1845, 328–29, etc. *Ipswich Mus. Portr. Ser.* 1852. *Cottage Gdnr* v.15, 1856, 307–10 portr. *Bull. Soc. Bot. Belgique* v.4, 1865, 411–23. *Gdnrs Chron.* 1860, 703; 1865, 1057–58, 1082–83; 1877 ii 19; 1891 i 17 portr.; 1904 i 148; 1965 ii 362, 386, 406, 409, 430, 434, 451 portr., 457, 478, 481, 502, 507, 526. *J. Bot.* 1865, 384–88; 1891, 158; 1934, 177. *J. Hort. Cottage Gdnr* v.34, 1865, 381–82; v.45, 1902, 353–54 portr. *Proc. Linn. Soc.* 1865–66, lxxiii–lxxvii. *Portr. Men of Eminence in Literature, Sci and Art* v.4, 1866, 23–26 portr. *Proc. R. Soc.* v.15, 1867, xxx–xxxii. E. Bretschneider *Hist. European Bot. Discoveries in China* 1898, 186–89. *D.N.B.* v.33, 277–79. *J. R. Soc. N.S.W.* 1908, 74–75. *J. W. Austral. Nat. Hist. Soc.* no. 6, 1909, 31–32. *J. R. Hort. Soc.* 1913, 63–69; 1955, 557 portr.; 1965, 457–62. F. W. Oliver *Makers of Br. Bot.* 1913, 164–77 portr. J. R. Green *Hist. Bot. in U.K.* 1914, 336–53. *Orchid Rev.* 1917, 75–79. *Amer. Orchid Soc. Bull.* 1932, 35–36 portr.; 1964, 564–68 portr. *Country Life* 1965, 293–98. *Listener* 1965, 847–49 portr. *Taxon* 1965, 293–98; 1973, 230 portr. F. A. Stafleu *Taxonomic Literature* 1967, 269–72. M. A. Reinikka *Hist. of the Orchid* 1972, 153–59 portr. C. C. Gillispie *Dict. Sci. Biogr.* v.8, 1973, 371–73. *R.S.C.* v.4, 31; v.12, 448.

Herb. at Cambridge. Orchid herb. and drawings at Kew. Letters at Kew, BM(NH). MSS. and portr. by E. V. Eddis at Royal Horticultural Society. Portr. at Kew, Hunt Library.

Lindleya H.B.K. *Neolindleya* Kränzl.

LINDSAY, Lady Anne *see* Barnard, *Lady* A.

LINDSAY, Rev. John (*fl.* 1750s–1788)
d. Spanish Town, Jamaica 2 Nov. 1788.

DD Edinburgh 1773. Rector, St. Thomas-ye-Vale, Jamaica, 1768; St. Catherine, Spanish Town, 1773. Executed drawings of Jamaican plants, 1761–69.

W. Fawcett and A. B. Rendle *Fl. Jamaica* v.3, 1914, v. *J. Bot.* 1915, 105–7.

Drawings and MSS. at Bristol Museum.

LINDSAY, John (*fl.* 1780s–1803)
d. 1803
FRSE 1793. Of Westmorland, Jamaica. Surgeon. Correspondent of J. Banks and J. Hope. Discovered *Cinchona brachycarpa* 1785. 'Germination of Ferns' (*Trans. Linn. Soc.* v.2, 1794, 93–100).
J. Bot. 1915, 106–7. *R.S.C.* v.4, 34.
Plants at BM(NH). MSS. on *Mimosa* at Royal Society.
Lindsaea Dryander.

LINDSAY, Nancy (–1973)
d. Jan. 1973
Had garden at Sutton Courtenay, Berks. L. Johnson left her his French garden at Le Serre de la Madoune from which she sent Cambridge a great collection of his plants.
Times 4 Jan. 1973.

LINDSAY, Robert (1846–1913)
b. Edinburgh 7 May 1846 *d.* Edinburgh 24 Sept. 1913
Curator, Royal Botanic Garden, Edinburgh, 1883–96. President, Botanic Society of Edinburgh, 1889. 'Nepenthes' (*Trans. Bot. Soc. Edinburgh* v.18, 1891, 229–40).
Gdnrs Chron. 1896 i 709 portr.; 1913 ii 265–66. *Garden* 1913, 504(x). *Notes Edinburgh Bot. Gdn* v.14, 1923–24, x. H. R. Fletcher and W. H. Brown *R. Bot. Gdn Edinburgh, 1670–1970* 1970, 191–92 portr. *R.S.C.* v.16, 795.

LINDSAY, William Lauder (1829–1880)
b. Edinburgh 19 Dec. 1829 *d.* Edinburgh 24 Nov. 1880
MD Edinburgh 1852. FLS 1858. FRSE. Combined geological and botanical studies with his practice of medicine. Physician, Murray's Royal Asylum, Perth, 1854–79. Collected plants in Great Britain, Iceland, New Zealand. *Popular History of British Lichens* 1856. *Contributions to New Zealand Botany* 1868. Contrib. papers on lichens to *J. Bot.* 1886–89.
Gdnrs Chron. 1880 ii 734. *Proc. Linn. Soc.* 1880–82, xviii–xix. *J. Bot.* 1881, 64. *Proc. R. Soc. Edinburgh* v.11, 1881–82, 736–39. *Trans. Bot. Soc. Edinburgh* v.14, 1883, 163–64. *D.N.B.* v.33, 216. T. F. Cheeseman *Manual of New Zealand Fl.* 1906, xxvii.
Plants and letters at Kew. Lichens at Edinburgh. Portr. at Hunt Library.

LINDSELL, E. B. (*fl.* 1900s)
Cultivated roses. President, National Rose Society, 1906.
Gdnrs Mag. 1908, 527–28 portr.

LINEHAM, Patrick Aloysius (1904–1973)
b. 7 March 1904 *d.* 6 March 1973
DSc. Research Demonstrator, Plant Breeding, University College, Dublin, 1927. Assistant, Dept. of Agricultural Botany, Belfast, 1930; Lecturer, 1944; Reader, 1950; Professor, 1951–69. Contrib. papers on grassland research to *J. Br. Grassland Soc.*

LINES, Oliver (1884–1965)
b. Kings Heath, Birmingham 28 Jan. 1884 *d.* Signal Mount, Tennessee, U.S.A. 1 April 1965
Gardener at Westonbirt, Messrs. Charlesworth and Co. Ltd. At John Sloan, Lenox, Mass., U.S.A., 1910. Started own business as Lines Orchids, Signal Mount, Tennessee with his son, 1947.
Orchid Rev. 1965, 210.

LINGWOOD, Mrs. Edmund (*fl.* 1820s)
Plants at Ipswich Museum, Suffolk.

LINGWOOD, Robert Maulkin (–1887)
d. 2 June 1887
BA Cantab 1836. FLS 1839. Botanist and entomologist. In Channel Islands and in Ireland with C. C. Babington, 1837. Found *Fumaria densiflora*.
W. A. Clarke *First Rec. Br. Flowering Plants* 1897, 8.
Herb. at Exeter Museum.

LINNELL, John (1822–1906)
Son of the artist, John Linnell (1792–1882). Botanist.
C. E. Salmon *Fl. Surrey* 1931, 53.
Herb. of British plants at BM(NH).

LINNELL, Mary (*c.* 1828–*c.* 1881)
Daughter of the artist, John Linnell (1792–1882). Made a number of flower studies, possibly under the influence of John Ruskin.

LINTON, Rev. Edward Francis (1848–1928)
b. Diddington, Hunts 16 March 1848 *d.* Southbourne, Hants 9 Jan. 1928
MA Oxon. FLS 1914. Curate, St. Paul's, Preston. Rector, Sprowston, Norfolk, 1878–88. Moved to Bournemouth, 1900. Rector, Edmondsham, Dorset, 1901–20. With his brother, W. R. Linton, published sets of *Hieracia*, *Rubi*, and *Salices*. *Fl. Bournemouth* 1900; ed. 2 1919. Contrib. to *J. Bot.*
Bot. Soc. Exch. Club Br. Isl. Rep. 1928, 708–10. *J. Bot.* 1928, 81–86 portr. *Watson Bot. Exch. Club Rep.* 1927–28, 416–17. *Proc. Bournemouth Nat. Sci. Soc.* v.20, 1927–28, 105–6 portr. H. N. Clokie *Account of Herb. Dept. Bot., Oxford* 1964, 201.
Plants at BM(NH), Cambridge, Bournemouth Natural Science Society.

LINTON, William James (1812–1897)
b. Mile End, London 7 Dec. 1812 *d.* New Haven, U.S.A. 1 Jan. 1898
Wood-engraver. *Ferns of the English Lake Country* 1865; ed. 2 1878. *Memories* 1895.
D.N.B. Supplt 1 v.3, 100–2. F. B. Smith *Radical Artisan: William James Linton, 1812–1897* 1973 portr.

LINTON, Rev. William Richardson (1850–1908)
b. Diddington, Hunts 2 April 1850 *d.* Shirley, Derbyshire 4 Jan. 1908

BA Oxon 1873. Tutor, C.M.S. College, Islington, 1876. In Palestine, 1881. Vicar, Shirley, 1886. Discovered *Rubus durescens* in Britain. *Fl. Derbyshire* 1903. *Account of British Hieracia* 1905. Contrib. account of botany to *Victoria County Hist. Derbyshire* v.1, 1905, 39–50. Published sets of *Hieracia, Rubi,* and *Salices* with his brother, E. F. Linton.

Bot. Soc. Exch. Club Br. Isl. Rep. 1908, 344. *J. Bot.* 1908, 65–71 portr. *Naturalist* 1908, 51. *Watson Bot. Exch. Club Rep.* 1907–8, 128–29 portr. *Trans. Liverpool Bot. Soc.* 1909, 80. *Sorby Rec.* v.2(4), 1967, 13.

Plants at BM(NH), Liverpool.

LIPSCOMBE, Rev. Christopher (1781–1843)

Bishop of Jamaica. Supposed to have collected fungi and lichens.

H. N. Clokie *Account of Herb. Dept. Bot., Oxford* 1964, 201.

LISBOA, José Camillo (*c.* 1822–1897)
d. Poona, India 1 May 1897

FLS 1888. Practised medicine in Bombay. Papers on Bombay plants in *J. Bombay Branch R. Asiatic Soc.* v.14, 1879, 264–66; v.15, 1882, 203–24, and on grasses in *J. Bombay Nat. Hist. Soc.*

Proc. Linn. Soc. 1897–98, 41. *R.S.C.* v.10, 610; v.16, 810.

Letters at Kew.

LISLE, Edward (*c.* 1666–1722)
d. June 1722

Of Crux Easton, Hants. *Observations in Husbandry* 1757 which contains a chapter on 'the orchard or fruit garden'.

J. Hort. Cottage Gdnr v.55, 1876, 430–31 portr.

LISLE, Mrs. Mary (*fl.* 1700s)
Of Crux Easton, Hants. Wife of E. Lisle (*c.* 1666–1722). Specimens of cultivated and wild flowers in Sloane Herb.

J. E. Dandy *Sloane Herb.* 1958, 157.

LISTER, Arthur (1830–1908)
b. Upton House, West Ham, Essex 17 April 1830 *d.* Lyme Regis, Dorset 19 July 1908

FRS 1898. FLS 1873. Of Leytonstone, Essex. Wine merchant. Worked at Mycetozoa from 1884. President, British Mycological Society, 1906. *Monograph of Mycetozoa* 1894 (plates by himself and his daughter, Gulielma). *Guide to British Mycetozoa* 1895; ed. 2 1905.

Nature v.58, 1898, 33; v.78, 1908, 325. *Gdnrs Chron.* 1908 ii 71. *J. Bot.* 1908, 331–34. *Proc. Linn. Soc.* 1908–9, 46–47. *D.N.B. Supplt 2*, 469. *Essex Nat.* v.18, 1917, 216–20. *Trans. Br. Mycol. Soc.* v.35, 1952, 188–89 portr. *Proc. Dorset Nat. Hist. Archaeol. Soc.* v.83, 1961, 79–81 portr. *R.S.C.* v.16, 812.

Drawings and slides at BM(NH).

LISTER, Gulielma (1860–1949)
b. Leytonstone, Essex 28 Oct. 1860 *d.* Leytonstone 18 May 1949

FLS 1904. Daughter of A. Lister (1830–1908). President, British Mycological Society, 1912, 1932. President, Essex Field Club, 1916–19. Revised A. Lister's *Monograph of the Mycetozoa* ed. 2 1911; ed. 3 1926. Provided drawings for her cousin, F. J. Hanbury's *Illustrated Monograph of British Hieracia* 1889–98 and A. Dallimore and A. B. Jackson's *Handbook of Coniferae* 1923.

Nature v.164, 1949, 94; v.188, 1960, 362–63. *London Nat.* 1949, 141–42. *Times* 6 June 1949. *Essex Nat.* 1950, 214 portr. *Trans. Br. Mycol. Soc.* 1950, 165–66 portr. *School Nat. Study* 1961, 6–7. *J. Quekett Microsc. Club* 1949, 60. *Proc. Dorset Nat. Hist. Archaeol. Soc.* v.83, 1961, 79–81 portr. S. T. Jermyn *Fl. Essex* 1974, 19.

Herb. at Bedford College, London, Essex Field Club. MSS. and drawings at BM(NH). Letters and MSS. at Kew.

LISTER, Joseph, 1st Baron Lister (1827–1912)
b. Upton, Essex 5 April 1827 *d.* Walmer, Kent 10 Feb. 1912.

BA London 1847. FRCS 1852. Professor of Surgery, Glasgow, 1860. Collected plants while living in Essex, 1844–48; some plants also collected by his brother William Henry and his sister Mary.

Who was Who, 1897–1916 431. *D.N.B. 1912–1921* 339–43. *Essex Nat.* 1925, 104–7; 1933, 92–102 portr.; 1954, 194.

Plants at Essex Field Club. Portr. at Royal College of Surgeons.

LISTER, Joseph Jackson (1857–1927)
b. Leytonstone, Essex 3 Aug. 1957 *d.* Grantchester, Cambridge 5 Feb. 1927

BA Cantab 1880. FRS 1900. FLS 1906. Son of A. Lister (1830–1908). Zoologist. Collected plants in Algeria, Canary Islands, Australasia. Plants collected in Christmas Island, 1888 described in *J. Linn. Soc.* v.25, 1890, 351–52.

Kew Bull. 1892, 151. *J. Bot.* 1927, 83. *Nature* v.119, 1927, 360. *Proc. Linn. Soc.* 1926–27, 90–92. *Proc. R. Soc. B.* v.102, 1928, i–v portr. *Times* 7 Feb. 1927. *R.S.C.* v.16, 812.

Acrostichum listeri Baker.

LISTER, Martin (*c.* 1638–1712)
b. Radcliffe, Notts *c.* 1638 *d.* London or Epsom 2 Feb. 1712

MA Cantab 1662. MD Oxon 1684. FRS 1671. FRCP., later Censor. Zoologist. Friend of J. Ray. At York, 1670; to London, 1681. Physician to Queen Anne, 1709. Contrib. articles on natural history, including botany, to *Philos. Trans. R. Soc.*

R. Morison *Plantarum Historiae Universalis Oxoniensis* v.3, 1699, 368. E. Lankester *Correspondence of John Ray* 1848, 111–25. E. Lankester *Memorials of John Ray* 1846, 17. W. Munk *Roll of R. College of Physicians* v.1, 1878, 442–45. *Yorkshire Archaeol. J.* 1873, 297–320. *Notes and Queries* v.2, 1875, 208, 433; v.4, 1875, 16, 177, 236; v.7, 1883, 137; v.3, 1893, 286, 337, 391; v.4, 1893, 276. *D.N.B.* v.33, 350–51. *Rep. Archit. and Archaeol. Soc. Lincolnshire and Nottinghamshire* v.25, 1900, 329–70. E. J. L. Scott *Index to Sloane Manuscripts* 1904, 312. L. C. Miall *Early Nat.* 1912, 130–34. *Proc. Linn. Soc.* 1921–22, Supplt. 18. *N. Western Nat.* 1941, 241–45. C. E. Raven *John Ray* 1942, 138. J. E. Dandy *Sloane Herb.* 1958, 190. H. N. Clokie *Account of Herb. Dept. Bot., Oxford* 1964, 201.

Plants at Oxford, BM(NH).

Listera R. Br.

LISTER, Robert (*fl.* 1810s)

Nurseryman, High Street, Bradford, Yorks.

LISTER, Thomas (*fl.* 1880s)

Of Flimby, Cumberland. Contrib. to W. Hodgson's *Fl. Cumberland* 1898, viii.

LISTON, Henrietta, Lady *see* Addendum

LISTON, Rev. William (1781–1864)

b. Aberdour, Perthshire 1781 *d.* Redgorton, Perthshire 1864

Minister, Redgorton. List of plants for Scone parish in *New Statistical Account Scotland* 1844.

Proc. Perthshire Soc. Nat. Sci. v.4, 1907–8, clxcii.

LITTLE, Henry (*c.* 1833–1914)

d. Twickenham, Middx 14 April 1914

Cultivated orchids.

Orchid Rev. 1914, 157.

Cypripedium littleanum × Hort.

LITTLE, Joseph Edward (1861–1935)

b. Tonbridge, Kent 12 Feb. 1861 *d.* 18 Jan. 1935

MA Oxon. Headmaster, Hitchin Grammar School, 1885–97. President, Hitchin Natural History Club, 1891. Contrib. chapter on botany in R. Hine's *Nat. Hist. of Hitchin Region* 1934. Contrib. articles, particularly on trees and shrubs, to *J. Bot.*

J. Bot. 1933, 265; 1935, 232–33. *Countryside* v.10, 1935, 204–5 portr. J. G. Dony *Fl. Bedfordshire* 1953, 31–32. J. G. Dony *Fl. Hertfordshire* 1967, 17.

Herb. at Cambridge. Plants at BM(NH), Kew, Oxford. MSS. at Hitchin Museum.

LITTLEDALE, St. George R.

(*c.* 1851–1931)

d. Bracknell, Berks 16 April 1931

Accompanied by his wife, travelled through East Turkestan and Tibet, 1893–95; small collection of plants from Tibet at Kew. In 1898 Mrs. Littledale collected plants in Mongolia, also at Kew. Collected in Kamchatka, 1900, Thian Shan Mountains, 1901, Persia, 1902, E. Caucasus, 1908.

E. Bretschneider *Hist. European Bot. Discoveries in China* 1898, 809–11. *Kew Bull.* 1896, 99–100, 207–16; 1898, 26; 1901, 41. *J. Linn. Soc.* 1902, 151–52. *J. R. Geogr. Soc.* v.78, 1931, 95–96.

Littledalea Hemsley.

LITTLEJOHN, James (1818–1893)

b. Crieff, Perth 1818

Kew gardener, 1844–46. Rose-grower, Madison, New Jersey, U.S.A.

J. Hort. Cottage Gdnr v.53, 1906, 574.

LITTLEJOHN, Robert (1756–1818)

b. Scotland 1756 *d.* Hobart 26 Oct. 1818

Settled in Tasmania, 1803–4. "A man of wealth of learning and a naturalist of repute." Sent plants to A. B. Lambert.

H. C. Andrews *Botanists Repository* 1797–1815, 531, 574. R. Brown *Prodromus Florae Novae Hollandiae* v.1, 1810, 434. *Hobart Town Gazette* 28 Nov. 1818. *Austral. Dict. Biogr.* v.2, 1967, 120.

LITTLEWOOD, John (*fl.* 1770s–1825)

d. 13 May 1825

Nurseryman, Handsworth near Sheffield, Yorks.

LITTON, Samuel (1781–1847)

d. Dublin 4 June 1847

MA Dublin 1804. MD Edinburgh 1806. Librarian, Royal Dublin Society, 1815–25; Professor of Botany, 1826.

Dublin Univ. Mag. Feb. 1848 portr. *Bot. Mag.* 1853 t. 4723. H. F. Berry *Hist. R. Dublin Soc.* 1915, 446. *Irish Booklore* 1971, 181–82 portr.

Herb. at Glasnevin, Dublin.

Littonia Hook.

LIVETT, Miss M. A. G. (*fl.* 1910s–1930s)

Watsonia 1950, 242–43.

Somerset plants at Taunton Museum, Somerset.

LIVINGSTONE, David (1813–1873)

b. Blantyre, Lanarkshire 19 March 1813 *d.* Lake Bangweulu, Rhodesia 1 May 1873

African explorer. Zambesi, 1858–63. *Missionary Travels and Researches in South Africa* 1857 (contains botanical observations). *Narrative of Expedition to the Zambesi and its Tributaries* 1865. *Last Journals of David Livingstone* 1874.

D.N.B. v.33, 384–96. *Comptes Rendus AETFAT, 1960* 1962 175–76. M. Gelfand *Livingstone the Doctor* 1957.

LIVINGSTONE, John (–1829)
d. at sea on voyage to China 8 July 1829

Surgeon, East India Company. In China in 1793 and from 1803; in Macao. Introduced rice paper, 1805 (*Bot. Misc.* 1829, 88–91). 'Observations on Difficulties which have existed in Transportation of Plants from China to England, and suggestions for obviating them' (*Trans. Hort. Soc*, v.3, 1822, 421–29). 'Account of Method of dwarfing Trees and Shrubs as practised by the Chinese' (*Trans. Hort. Soc.* v.4, 1822, 224–31).

H. C. Andrews *Botanist's Repository* v.10, 1811, t.612. *Bot. Mag.* 1828, t.2802; 1829, t.2908. E. Bretschneider *Hist. European Bot. Discoveries in China* 1898, 266–68. E. H. M. Cox *Plant Hunting in China* 1945, 52. *R.S.C.* v.4, 61.

Plants at Kew.

LLEWELYN, Sir John Talbot Dillwyn- *see* Dillwyn-Llewelyn, *Sir* J. T.

LLOYD, Edward *see* Lhuyd, E.

LLOYD, Francis Ernest (1868–1947)
b. Manchester 4 Oct. 1868 *d.* Carmel, California, U.S.A. 17 Oct. 1947

AB Princeton 1891. FLS 1925. Professor of Botany, McGill University, 1912–34. Member of expeditions to Mexico, 1890; Alaska, 1896; Dominica, 1903. *Comparative Embryology of Rubiaceae* 1902. *Physiology of Stomata* 1908. *Guayule* 1911. *Carnivorous Plants* 1942.

J. New York Bot. Gdn 1947, 292. *Who was Who, 1941–1950* 692.

Herb. at New York Botanic Garden. Portr. at Hunt Library.

LLOYD, George (1815–1843)
b. Wales 17 Oct. 1815 *d.* Gurneh 31 Oct. 1843

Member of Cairo Literary Society. Excavated Thebes, 1839–43.

E. Prisse *Oriental Album* 1851 portr. W. R. Dawson *Who was Who in Egyptology* 1951.

Herb. and drawings at Botanic Garden, Montpellier.

LLOYD, George N. (1804–1889)
b. Albrighton, Shropshire 19 March 1804 *d.* Berkhamstead, Herts 5 July 1889

MD Edinburgh 1826. Lectured on botany in Edinburgh. Collected plants, 1825–43, in British Isles with W. J. Hooker and G. A. Arnott. *Botanical Terminology* 1826. *Fasciculus Gramineae Britannicae* 1840.

Gdnrs Mag. 1828, 405. *Gdnrs Chron.* 1923 i 329. *Kew Bull.* 1923, 189.

LLOYD, Harold Buchan (*fl.* 1890s–1900s)

Kew gardener, 1898. Assistant Curator, Botanic Gardens, Old Calabar, Nigeria, 1898. Collected plants. In partnership with his brother in Osney Nurseries, Paignton, Devon, *c.* 1900. Emigrated to Canada.

Kew Bull. 1898, 136. *J. Kew Guild* 1906, 308.

Plants at Kew.

LLOYD, James (1810–1896)
b. London 17 March 1810 *d.* Nantes, France 10 May 1896

B.-ès-L. Lorient 1829. *Fl. Loire-inférieure* 1844. *Fl. de l'ouest de la France* 1854; ed. 5 1897.

Companion Bot. Mag. v.2, 1837, 265. *Bull. Soc. Bot. Deux-Sevres* 1896, 118–24. *Bull. Soc. Sci. Nat. Ouest France* v.6, 1896, 137–58 portr. *J. Bot.* 1896, 328.

Herb. at Nantes. Portr. at Hunt Library. *Arenaria lloydii* Jord.

LLOYD, John (*c.* 1791–1870)
b. Herefordshire *c.*1791 *d.* London 24 Jan. 1870

Gardener. Of Wandsworth, London. Contrib. many papers including some on *Lastrea* to *Phytologist* 1845–63.

Gdnrs Chron. 1870, 180. *R.S.C.* v.4, 64.

LLOYD, Joseph (–1942)
d. Dec. 1942

Farmer of Birkdale, Southport, Lancs. Fern cultivator and exhibitor at Southport.

Br. Fern Gaz. 1948, 203.

LLOYD, Llewelyn Cyril (*c.* 1905–1968)
b. Nottingham *c.* 1905 *d.* Cardeston, Shropshire 23 Aug. 1968

FLS 1944. Master printer. Settled at Shrewsbury in 1931. Held various offices in Shropshire Archaelogical and Natural History Society. *Handbook of Shropshire Fl.* (with E. M. Rutter) 1957.

Trans. Shropshire Archaeol. Nat. Hist. Soc. v.58, 1968, 186.

LLOYD, Morris (*fl.* 1640s)

Of Prislierworth (Treiorworth), Anglesey. Discovered *Oxyria* in Wales.

J. Parkinson *Theatrum Botanicum* 1640,745.

LOBB, Thomas (1820–1894)
b. Cornwall 1820 *d.* Devoran, Cornwall 30 April 1894

Collected plants for Veitch nursery in India, Burma, Malaya, Java, Borneo and Philippines from 1840s. His plants listed in *London J. Bot.* 1847–48.

London J. Bot. 1847, 145–46. *Cottage Gdnr* v.13, 1855, 273–74. *J. Bot.* 1894, 191. *Gdnrs Chron.* 1894 i 636; 1942 ii 137. J. H. Veitch *Hortus Veitchii* 1906, 41–44. *Philippine J. Sci. Bot.* 1915, 171–94. *J. R. Hort.*

Soc. 1942, 48–50; 1948, 285–86. *Nature* v.149, 1942, 438. *Fl. Malesiana* v.1, 1950, 325–26. A. M. Coats *Quest for Plants* 1969, 205–6. M. A. Reinikka *Hist. of the Orchid* 1972, 208–10.

Plants at Kew, BM(NH), Oxford.

Lobbia Planchon.

LOBB, William (1809–1864)
b. Perran-ar-worthal, Cornwall 1809 *d.* San Fransisco, U.S.A. 1863

Brother of T. Lobb. Collected plants for Veitch nursery in Brazil, Chile, Patagonia, Peru, Ecuador, Colombia, 1840–48, and in California and Oregon, 1849–57.

Gdnrs Mag. 1837, 554. *London J. Bot.* 1947, 145–46. J. Veitch *Manual of Coniferae* 1881, 258. C. S. Sargent *Silva of N. America* v.10, 1896, 60. J. H. Veitch *Hortus Veitchii* 1906, 37–40. F. H. Davey *Fl. Cornwall* 1909, xlviii–xlix. *Muhlenbergia* v.7, 1911, 100–3. *Kew Bull.* 1920, 61–62. *Forestry* 1932, 5–8. *Smithsonian Misc. Collections* v.87, 1932, 1–13. *Gdnrs Chron.* 1938 ii 388–89; 1963 ii 406–7. *J. R. Hort. Soc.* 1942, 48–50. *Leaflet Western Bot.* v.5, 1949, 155–56. S. D. McKelvey *Bot. Exploration of Trans-Mississippi West, 1790–1850* 1955, 1092–94. A. M. Coats *Quest for Plants* 1969, 321–22, 373–74. *University California Publ. in Bot.* v.67, 1973, 1–36.

Plants at Kew, BM(NH). Letters at Kew.

Lobbia Planchon.

LOBEL, Matthias de (1538–1616)
b. Lille, Flanders 1538 *d.* Highgate, London 3 March 1616

To England, 1584. Botanographer to James I. Pupil of Rondelet at Montpellier. To Denmark with Lord Zouch, 1592. Superintendent of Zouch's garden. Hackney. *Stirpium Adversaria Nova* (with P. Pena) 1570. *Plantarum seu Stirpium Historia* 1576. *Icones Stirpium* 1591. *Stirpium Illustrationes* 1655.

R. Pulteney *Hist. Biogr. Sketches of Progress of Bot. in England* v.1, 1790, 96–109. *Bull. Soc. Bot. Belgique* 1862, 16–19. H. Trimen and W. T. T. Dyer *Fl. Middlesex* 1869, 369. E. H. F. Meyer *Geschichte der Botanik* v.4, 1875, 358–66. *Bull. Fed. Soc. Hort. Belgique* 1857, 1–25. *Bull. Soc. Bot. France* 1897, xi–xlvii. L. Legré *P. Pena et M. de Lobel* 1898. *J. Bot.* 1899, 88–92. J. W. White *Fl. Bristol* 1912, 48–50. E. J. L. Scott *Index to Sloane Manuscripts* 1904, 314. R. T. Gunther *Early English Botanists* 1922, 245–53 portr. C. E. Raven *English Nat.* 1947, 235–40. H. J. Riddelsdell *Fl. Gloucestershire* 1948, cix. *Gdnrs Chron.* 1964 ii 432 portr. R. H. Jeffers *Friends of John Gerard* 1967 *passim*. C. C. Gillispie *Dict. Sci. Biogr.* v.8, 1973, 435–36. S. T. Jermyn *Fl. Essex* 1974, 14.

Lobelia L.

LOBJOIT, Sir William George (1860–1939)
b. Putney, Surrey 1860 *d.* Woburn, Bucks 28 May 1939

VMH 1932. Chairman, market garden firm of W. J. Lobjoit and Son. Hon. Controller of Horticulture under Ministry of Agriculture, 1920–27.

Gdnrs Chron. 1939 i 359–60. *Pharm. J.* 1939, 578. *Times* 30 May 1939.

LOCHHEAD, William (–1815)
d. St. Vincent 22 March 1815

Surgeon. Of Antigua. Curator, Botanic Garden, St. Vincent, 1811–15.

L. Guilding *Account of Bot. Gdn Island of St. Vincent* 1825, 18, 22. *Kew Bull.* 1892, 95, 97. *R.S.C.* v.4, 67.

Drawings at BM(NH).

LOCK, Robert Heath (1879–1915)
b. Eton, Bucks 1879 *d.* Eastbourne, Sussex 26 June 1915

BA Cantab 1900. DSc 1910. FLS 1912. To Ceylon, 1902. Curator, Cambridge Herb., 1903–8. Assistant Director, Peradeniya Gardens, Ceylon, 1908–12. Engaged in plant breeding. *Recent Progress in Study of Variation, Heredity and Evolution* 1906; ed. 3 1911. *Rubber and Rubber Planting* 1913.

Ann. R. Bot. Gdns Peradeniya 1912, 257–64. *Gdnrs Chron.* 1915 ii 32. *Kew Bull.* 1915, 307. *Nature* v.95, 1915, 515. *Who was Who, 1897–1916* 435. *Proc. Linn. Soc.* 1915–16, 66–67.

LOCKE, John (1632–1704)
b. Wrington, Somerset 29 Aug. 1632 *d.* Oates, High Laver, Essex 28 Oct. 1704

MA Oxon. Philosopher. *Observations upon Growth and Culture of Vines and Olives* 1766.

G. W. Johnson *Hist. English Gdning* 1829, 225–26. *D.N.B.* v.34, 27–36. *Times* 8 March 1960. *Bodleian Library Rec.* 1962, 42–46. K. Dewhurst *John Locke* 1963 portr. R. I. Aaron *John Locke* 1965. C. C. Gillispie *Dict. Sci. Biogr.* v.8, 1973, 436–40.

Herb. in 2 vols. and MSS. at Bodleian Library.

LOCKHART, David (–1846)
b. Cumberland *d.* Trinidad 1846

Kew gardener. Assistant to C. Smith on Congo Expedition, 1816. First Superintendent, Botanic Garden, Trinidad, 1818–46. Introduced spice trees to the island.

R. Brown *Obervations...on Herb. collected by Prof. C. Smith* 1818, lxiii. *Bot. Mag.* 1827, t.2715. *London J. Bot.* 1847, 40. J. Smith *Rec. R. Bot. Gdns, Kew* 1880, 230. *Gdnrs Chron.* 1885 ii 236–37. *Kew Bull.* 1891, 310–11; 1901, 41. *Kew Bull. Additional Ser. I* 1898, 55. *D.N.B.* v.34, 14. I. Urban *Symbolae Antillanae* v.3, 1902, 78.

Letters, Congo and Trinidad plants at Kew. Bahia plants at BM(NH).

Lockhartia Hook.

LOCKHART, Theodore *and* Charles (*fl.* 1840s)

Seedsmen, 156 Cheapside, London and Haarlem, Holland.

LODDIGES, Conrad (*c.* 1739–1826)
b. Herzberg? *c.* 1739 *d.* Hackney, London 13 March 1826

Came to England as gardener *c.* 1761. In 1771 took over nursery in Hackney founded by J. Busch. Introduced plants from Michaux and Bartram.

Bot. Mag. 1806, t.965. J. C. Loudon *Encyclop. Gdning* 1822, 84. P. W. Watson *Dendrologia Britannica* v.1, 1825, ix–x. *Gdnrs Mag.* 1826, 229; 1829, 379. *Bot. Misc.* 1829–30, 74–77. W. Robinson *Hist. Hackney* v.2, 1843, 133. *Athenaeum* 1899 i 214, 245, 311.

Cultivated plants at Kew.

Loddigesia Sims.

LODDIGES, Conrad (–1865)
d. Hackney, London 20 Jan. 1865

Nurseryman.

Gdnrs Chron. 1865, 74.

LODDIGES, George (1784–1846)
b. Hackney, London 12 March 1784 *d.* Hackney 5 June 1846

FLS 1821. Son of C. Loddiges (*c.* 1739–1826). Nurseryman, Hackney. Conducted *Bot. Cabinet* 1817–34; plates mostly drawn by himself.

Proc. Linn. Soc. 1839–40, 334–35. *J. Hort. Soc.* 1846, 224–25. *Gdnrs Chron.* 1852, 616. *J. R. Microsc. Soc.* 1895, 4. E. Bretschneider *Hist. European Bot. Discoveries in China* 1898, 281–82. *Athenaeum* 1899 i 214, 245, 311. *J. R. Hort. Soc.* 1938, 428 portr. *Huntia* 1965, 203 portr.

Letters and cultivated plants at Kew. Portr. by J. Renton at Royal Horticultural Society. Portr. at Hunt Library.

Acropera loddigesii Lindl.

LODER, Sir Edmund Giles (1849–1920)
b. London 7 Aug. 1849 *d.* Leonardslee, Sussex 14 April 1920

Created garden at Leonardslee; established collection of conifers and rhododendrons.

Gdnrs Chron. 1920 i 210. *Kew Bull.* 1920, 175. *Nature* v.105, 1920, 301–2. *Quart. J. For.* v.14, 201–2. A. E. Pease *Edmund Loder* 1923 portr. *Shooting Times and Country Mag.* 1952, 218–19 portr.

LODER, Gerald Walter Erskine, 1st Baron Wakehurst (1861–1936)
b. 25 Oct. 1861 *d.* Wakehurst, Sussex 30 April 1936

MA Cantab. LLD. FLS 1914. VMH 1936. MP for Brighton, 1889–1905. President, Royal Arboricultutal Society, 1926–27. President, Royal Horticultural Society, 1929–31. Acquired Wakehurst Place, 1903 where he established an important collection of trees and shrubs.

Bot. Mag. 1933 portr. *Br. Fern Gaz.* v.7, 1936, 49–50. *Gdnrs Chron.* 1936 i 303. *J. R. Hort. Soc.* 1936, 253–54 portr. *Kew Bull.* 1936, 286–87. *Proc. Linn. Soc.* 1935–36, 213–15. *Quart. J. For.* 1936, 195. *Times* 1 May 1936. *Who was Who, 1929–1940* 1398.

LODGE, F. A. (*fl.* 1890s)

Deputy Conservator, Madras Forest Dept. Sent to Trinidad in 1899 to report on forest conservancy.

Kew Bull. 1899, 220. I. Urban *Symbolae Antillanae* v.3, 1902, 78.

Plants at Trinidad.

LODGE, William (*fl.* 1830s)

Nurseryman, Broughton Lane, Strangeways, Manchester, Lancs.

LOFFT *see* Casborne, *Mrs.*

LOFTHOUSE, Thomas Ashton (1868–1944)
d. 1 Sept. 1944

FLS 1936. Architect. Keen gardener. 'Plants in E. Pyrenees' (*J. R. Hort. Soc.* 1927, 153–57). 'Plant Hunting in Sierra Nevada' (*J. R. Hort. Soc.* 1931, 225–33.) 'Further Notes on Plants seen in Sierra Nevada of Spain' (*J. R. Hort. Soc.* 1933, 307–13).

Quart. Bull. Alpine Gdn Soc. v.12, 1944, 223–24.

LOFTUS, William Kennett (*c.* 1821–1858)
b. Rye, Sussex *c.* 1821 *d.* at sea Nov. 1858

Geologist on Turco-Persian Frontier Commision, 1849–52. Conductor of Assyrian Excavation Expedition, 1853–55. Collected plants in N. Iraq and Persia. *Travels in Chaldaea and Susiana* 1859.

Gent. Mag. v.6, 1859. 435. *Trans. Tyneside Nat. Club* v.4, 1859, 98–99. *D.N.B.* v.34, 80–81.

Plants at BM(NH), Kew. Letters at Kew.

LOGAN, James (1674–1751)
b. Lurgan, County Armagh 20 Oct. 1674 *d.* Stenton, U.S.A. 31 Oct. 1751

To N. America with Penn, 1699. Secretary of Pennsylvania, 1701; Governor, 1736–38. American correspondent of P. Collinson. *Experimenta et Meletemata de Plantarum Generatione* Leyden, 1739; in English, London, 1747.

R. Pulteney *Hist. Biogr. Sketches of Progress of Bot. in England* v.2, 1790, 277–78. J. Armistead *Mem. of J. Logan* 1851. J. Smith *Cat. of Friends' Books* v.2, 1867, 129–30. W. C. Armor *Lives of Governors of State of Pennsylvania* 1873, 136–41 portr. Friends' Inst. *Biogr. Cat.* 1888, 439–40. Appleton's *Cyclop. of Amer. Biogr.* v.4, 3–4 portr. *Bot. Gaz.* 1894, 307–12. J. W. Harshberger *Botanists of Philadelphia and their Work* 1899, 41–42. *D.N.B.* v.34, 81–83. N. G. Brett-

James *Life of Peter Collinson* 1926, 155–57. *J. Heredity* 1932, 440–43 portr.
Logania R. Br.

LOMAX, Alban Edward (1861–1894)
d. Liverpool 4 May 1894
Nephew of E. A. Lomax (1810–1895). Pharmaceutical chemist, Liverpool. '*Cerastium carpetanum*' (*J. Bot.* 1893, 331).
J. Bot. 1894, 384. *Trans. Liverpool Bot. Soc.* 1909, 80.
Herb. at Liverpool.

LOMAX, Elizabeth Anne (*née* **Smithson**) (1810–1895)
b. Pontefract, Yorks 22 Feb. 1810 *d.* Torquay, Devon 16 March 1895
Married Robert Lomax, 1842. Member of Botanical Exchange Club.
J. Bot. 1895, 160.
Herb. at Manchester.

LOMAX, James (1857–1934)
b. Elton, Lancs 1857 *d.* Bolton, Lancs 15 Oct. 1934
ALS 1913. Palaeobotanist. Founded Lomax Palaeobotanical Laboratories which became part of the Coal Research Association. Distributed slides of fossil plants.
N. Western Nat. 1934, 392–93. *Proc. Linn. Soc.* 1934–35, 182–83.

LONDON, George (–1714)
d. Edgar, Herts 12 Jan. 1714
Pupil of John Rose. Gardener to Bishop Compton and William and Mary. In partnership with R. Looker, M. Cook and J. Field in nursery at Brompton Park, Kensington, 1681; H. Wise as sole partner in 1694. His daughter, Henrietta, who married J. Peachy in 1706, made botanical drawings some of which are at Badminton. "De rebus Botanicus optima meriti" (Plukenet). *Compleat Gard'ner* (with H. Wise) 1699. *Retir'd Gard'ner* (with H. Wise) 1706.
J. Petiver *Museii Petiveriani* 1695, 45–46. S. Switzer *Ichnographia* v.1, 1718, 68–93. G. W. Johnson *Hist. English Gdning* 1829, 123–24. S. Felton *Portr. of English Authors on Gdning* 1830, 35–43. *Cottage Gdnr* v.5, 1851, 171. *J. Hort. Cottage Gdnr* v.45, 1871, 195–97. *Gdnrs Chron.* 1892 i 361–62, 621–22; 1913 ii 181–82; 1923 ii 218. E. J. L. Scott *Index to Sloane Manuscripts* 1904, 167. *Garden* 1920, 428–29. *J. Linn. Soc.* v.45, 1920, 32–33. D. Green *Gdnr to Queen Anne* 1956 *passim.* J. E. Dandy *Sloane Herb.* 1958, 157–59. J. and N. Ewan *John Banister* 1970, 24 and *passim.*
Plants in Sloane Herb., BM(NH). Library sold 22 March 1713/4.

LONG, Annie Doris (*formerly* **Clement Shorter**) (*née* **Banfield**) (1895–1964)
b. Penzance, Cornwall 10 Sept. 1895 *d.* Isles of Scilly 17 July 1964
Founded Trenoweth Valley Flower Farm Ltd, St Keverne, Cornwall; also established trial grounds at Walton-on-Thames, Surrey in 1937. Cultivated narcissi.
Daffodil Tulip Yb. 1966, 176–77.

LONG, Bertram Raymond (1890–1962)
b. County Wicklow 1890 *d.* Jan. 1962
VMM 1927. Hybridised irises.
Iris Yb. 1962, 15–16.

LONG, Edward (1734–1813)
b. St. Blazey, Cornwall 1734 *d.* Arundel Park, Sussex 13 March 1813
Secretary to Governor of Jamaica, 1756–62. *History of Jamaica* 1774 3 vols. ('Synopsis of Vegetable and other Productions' v.2, 674–864).
D.N.B. v.34, 100–1. I. Urban *Symbolae Antillanae* v.1, 1898, 97; v.3, 1902, 79. F. Cundall *Hist. Jamaica* 1915 *passim.*

LONG, Ernest Philip (1878–1947)
b. Wilton, Wilts. 1878 *d.* 25 Nov. 1947
Kew gardener, 1900. Superintendent, Government Gardens, Simla and Delhi, 1902–33. Collected Indian mosses and lichens.
J. Kew Guild 1946–47, 601.

LONG, Frank Reginald (1884–1961)
b. Winchester, Hants 20 Oct. 1884 *d.* 5 Dec. 1961
Kew gardener, 1905. In charge of Hill Garden, Taiping, Malaya, 1908–10. Superintendent, Government Plantations, Perak. Rubber planter, 1912–21. Port Elizabeth Parks Dept., S. Africa, 1922; Superintendent, 1929–44. Collected plants in Malaya and S. Africa.
Gdnrs Chron. 1936 i 82 portr. *Fl. Malesiana* v.1, 1950, 328. *J. Kew Guild* 1962, 208 portr.
Plants at Kew, Singapore, Port Elizabeth.

LONG, Frederick (1840–1927)
d. 23 Feb. 1927
MRCS. LRCP. Doctor at Wells-next-the-Sea, Norfolk until 1899; retired to Norwich. Collected grasses in Madeira and Zululand, 1916. Discovered *Sonchus arvensis* L. var. *angustifolius* Meyer in 1922. Compiled catalogue to herb. of J. D. Salmon.
Trans. Norfolk Norwich Nat. Soc. 1926–27, 375–76. *Watson Bot. Exch. Club Rep.* 1926–27, 369–70.
British plants at Norwich Museum.

LONG, James (*fl.* 1650s)
Seedsman, The Barge, Billingsgate, London.

LONG, James Walter (1864–1948)
d. Newport, Isle of Wight July 1948
Civil servant. Interested in alien flora of Britain.
Watsonia 1949–50, 187.
Herb. at BM(NH).

LONG, John Albert (1863–1944)
b. Yeadon, Yorks 1863 *d.* Menston-in Wharfedale, Yorks 5 June 1944
FRMS 1927. Teacher at Keighley and Burley-in-Wharfedale; headmaster, Bradford 1898–1923. Algologist.
J. R. Microsc. Soc. 1944, 168.
Collection of diatom slides at Royal Microscopical Society.

LONGLEY, Charles (*c.* 1830–1915)
d. 1 Jan. 1915
Acquired nursery founded by his father at Rainham, Kent.
Garden 1915, 48 (vi).

LOOKER, Roger (–1685)
d. 3 March 1685
Gardener to Earl of Sandwich and to Queen Catherine of Braganza. Nurseryman, St. Martin in the Fields, London and partner with M. Cook, J. Field and G. London in Brompton Park Nursery. Bequeathed to his son, William, his share in Brompton Park Nursery.
Cottage Gdnr v.7, 1852, 891.

LORD, John Keast (1818–1872)
b. Cornwall 1818 *d.* Brighton, Sussex 9 Dec. 1872
Veterinary surgeon. British Army, 1855–58. In Canada in 1858 as naturalist to Commision to establish boundary line on 49th Parallel. Collected ferns in Sinai, 1868. *Naturalist in Vancouver's Island* 1866.
D.N.B. v.34, 136. J. Ewan *Rocky Mountain Nat.* 1950, 254.

LORD, Joseph (*fl.* 1700s)
Correspondent of J. Petiver to whom he sent plants from Carolina.
Proc. International Congress Plant Sci. 1926 v.2, 1929, 1527. J. E. Dandy *Sloane Herb.* 1958, 159.
Plants at BM(NH).

LORD, Robert (1818–1886)
b. Todmorden, Yorks 2 Jan. 1818 *d.* Todmorden 18 Aug. 1886
Nurseryman, Hole Bottom, Todmorden. Specialised in carnations and picotees.
Gdnrs Chron. 1886 ii 283.

LORKIN, Robert (*fl.* 1620s–1630s)
Apothecary. Accompanied T. Johnson (1597–1644) on his visits to Kent in 1629 and 1632, and on his tour of West of England in 1634.

LORT-PHILLIPS, Mrs. E. (*fl.* 1890s)
Wife of explorer, E. Lort-Phillips. Collected about 350 species with Miss Edith Cole on expedition from Berbera to Golis Mountains, 1895. Discovered *Echidnopsis somalensis* and *Huernia somalica.*
Kew Bull. 1895, 158, 211. A. White and B. L. Sloane *Stapelieae* v.1, 1937, 123.
Plants from Somaliland at Kew.

LOTHIAN, Thomas (–1863)
d. 23 May 1863
Nurseryman, Campbelltown, Argyllshire.
Phytologist 1863, 605.

LOUDON, Jane Wells (*née* **Webb**) (1807–1858)
b. Birmingham 19 Aug. 1807 *d.* Bayswater, London 13 July 1858
Married J. C. Loudon (1783–1843) in 1830. Horticultural writer. *Young Lady's Book of Botany* 1838. *Gardening for Ladies* 1840. *Ladies' Flower-garden* 1840–48 4 vols. *Ladies' Companion to Flower-garden* 1841. *Botany for Ladies* 1842. *British Wild Flowers* 1845. *Tales about Plants* 1853. *My Own Garden* 1855. Founded and edited *Ladies Mag. of Gdning* 1842.
Cottage Gdnr v.20, 1858, 248, 255–59. *D.N.B.* v.34, 148–49. *Gdnrs Chron.* 1922 ii 368; 1923 i 77, 110–11; 1946 ii 102–3; 1955 ii 192, 222–23, 265; 1961 ii 364–65. B. Howe *Lady with Green Fingers* 1961 portr.
Pelargonium × Webbianum Penny.

LOUDON, John Claudius (1783–1843)
b. Cambuslang, Lanarkshire 8 April 1783 *d.* Bayswater, London 14 Dec. 1843
FLS 1806. To London, 1803. Travelled in N. Europe, 1813–15; in Italy, 1819. Landscape gardener and horticultural writer. Founded and edited *Gdnrs Mag.* from 1826 and *Mag. of Nat. Hist.* from 1828. *Treatise on...Hot Houses* 1805. *Greenhouse Companion* 1824. *Encyclopaedia of Gardening* 1822. *Encyclopaedia of Plants* 1829. *Hortus Britannicus* 1830. *Arboretum et Fruticetum Britannicum* 1838 8 vols. *Hortus Lignosus Londinensis* 1838. *Suburban Gardener* 1838. *Suburban Horticulturist* 1842. *Self-instruction for Young Gardeners* 1845.
G. W. Johnson *Hist. English Gdning* 1829, 277–81. *Gdnrs Mag.* 1843, 673–74, 679–81. *Proc. Linn. Soc.* 1844, 204–6; 1888–89, 37. *Gdnrs Chron.* 1844, 7; 1845, 754–55; 1921 i 246; 1923 ii 172; 1938 ii 456–57; 1949 i 132–33; 1963 ii 324 portr. Memoir by his widow with portr. prefixed to his *Self-instruction for Young Gardeners* 1845. *Cottage Gdnr* v.5, 1850, 143; v.20, 1858, 255–59. J. Donaldson *Agric. Biogr.* 1854, 87–89. R. Chambers's *Book of Days* v.2, 1864, 693–94. *Garden* v.1, 1872, 697–98; v.2, 1872, 1–2, 47–48, 488–90, 532–34; v.3, 1873, 47–49. *J. Hort. Cottage Gdnr* v.57, 1877, 292–96 portr., 333–35; 1902 i 238–40 portr. *D.N.B.* v.34, 149–50. *Gdnrs Mag.* 1915, 580, 598. *J. R. Hort. Soc.* 1936, 277–84. *Official Architect and Planning Rev.* 1951, 37–40 portr.,

88–91, 155–56, 265–68. G. Taylor *Some Nineteenth Century Gdnrs* 1951, 17–67 portr. M. Hadfield *Pioneers in Gdning* 1955, 157–70. F. A. Stafleu *Taxonomic Literature* 1967, 293. J. Gloag *Mr. Loudon's England* 1970 portr. P. Willis *Furor Hortensis* 1974, 76–88 portr.

Portr. by J. Linnell at Linnean Society. Portr. at Kew, Hunt Library.

LOUDON, William (1830–1907)
b. Musselburgh, Midlothian 1830 *d.* N. Berwick 27 Jan. 1907

Educ. Edinburgh University. In Indian Revenue Survey, 1854. Administrator-General of Bombay until 1879. Had keen interest in horticulture and a collection of drawings of Indian plants.

Trans. Bot. Soc. Edinburgh v.23, 1908, 353.

LOUGHNANE, James B. (1905–1970)
b. Crinkle, Ireland 1905 *d.* Dublin 30 Aug. 1970

DSc 1956. Lecturer, University College, Dublin, 1956; Professor of Plant Pathology, 1966–70.

Br. Mycol. Soc. Bull. v.5, 1971, 29. *International Newsletter on Plant Pathology* v.1(2), 1 portr.

LOUSLEY, Job (1790–1855)
b. Newbury, Berks 20 Nov. 1790 *d.* Hampstead Norris, Berks 8 July 1855

Farmer of Blewbury and Hampstead Norris. Contrib. plant localities to W. Hewett's *Hist. and Antiq. of Hundred of Compton, Berks* and to A. Russell's *Hist. of Newbury* 1839.

Phytologist 1849, 716. G. C. Druce *Fl. Berks* 1897, clxviii–clxix. *Notes and Queries* v.10, 1963, 429–30. *Proc. Bot. Soc. Br. Isl.* 1964, 203–9. *Berkshire Archaeol. J.* 1967–68, 57–65.

LOUSLEY, Job Edward (1907–1976)
b. Berks 18 Sept. 1907 *d.* 6 Jan. 1976

Employee of Barclays Bank. Authority on docks and sorrels. Had herb. of alien plants. President, South London Botanical Institute; Botanical Society of British Isles, 1961–65; London Natural History Society, 1963–64. *Wild Flowers of Chalk and Limestone* 1950; ed. 2 1969. *Fl. Isles of Scilly* 1971. At time of his death had completed *Fl. Surrey* started by D. P. Young. Edited *Studies of Distribution of British Plants* 1951; *Changing Face of Britain* 1953; *Progress in Study of British Fl.* 1957. Contrib. to *J. Bot.*

Times 9 Jan. 1976.

Herb. at Reading.

LOVE, John (1806–)
b. Kilbarchan, Renfrewshire 10 April 1806

Handloom weaver, Kilbarchan. Florist specialising in pinks.

J. Hort. Cottage Gdnr v.23, 1891, 284–86 portr.

LOVELADY, John (*fl.* 1890s–1948)

Of Haslingden, Lancs. Collected British ferns.

Br. Fern Gaz. 1948, 203.

LOVELL, Robert (*c.* 1630–1690)
b. Lapworth, Warwickshire *c.* 1630 *d.* Coventry, Warwickshire Nov. 1690

BA Oxon 1650. Practised medicine at Coventry. *Pambotanologia sive Enchiridion Botanicum* 1659; ed. 2 1665.

R. Pulteney *Hist. Biogr. Sketches of Progress of Bot. in England* v.1, 1790, 181–83. *D.N.B.* v.34, 174–75. C. E. Raven *English Nat.* 1947, 46–47.

LOW, Edward Valentine (1866–1931)
d. 27 March 1931

In family nursery firm of Messrs Hugh Low and Co., Clapton and Enfield. Set up new business at Vale Bridge, Wivelsfield, Sussex where he grew orchids.

Orchid Rev. 1931, 150.

LOW, Rev. George (1746–1795)
b. Edzell, Forfar 1746 *d.* Bissay, Orkney 13 March 1795

MA St. Andrews 1771. Minister, Bissay, 1774. Contrib. list of Orkney plants (unacknowledged) to G. Barry's *Hist. of Orkney Islands* 1805, 266–85. *Fauna Orcadensis* 1813. *Tour through Orkney and Shetland in 1774* 1879.

M. Spence *Fl. Orcadensis* 1914, xxxvii–xxxviii.

LOW, Hugh (1793–1863)

Came from Scotland and entered J. L. Mackay's nursery at Clapton, London, 1823; acquired ownership in 1831.

J. Soc. Bibl. Nat. Hist. 1968, 328.

LOW, Sir Hugh (1824–1905)
b. Clapton, London 10 May 1824 *d.* Alassio, Italy 18 April 1905

FLS 1894. FZS 1893. KCMG 1883. GCMG 1889. Son of H. Low (1793–1863). To Borneo 1844. Colonial Secretary, Labuan. Ascended Kina Balu, 1851. Resident in Perak, 1877–89 whence he sent plants to Kew. Authority on orchids and other tropical plants. *Sarawak* 1848. *Journal 1877* (*J. Malay Br. R. Asiatic Soc.* 1954, 1–108).

Gdnrs Mag. 1826–35 *passim*. *Proc. Zool. Soc.* 1873, 337–61. *Agric. Bull. Straits and Federated Malay States* 1905, 239–41. *Gdnrs Chron.* 1905 i 264–65. *J. Bot.* 1905, 192. *Orchid Rev.* 1905, 182–83; 1932, 163–66, 208–11, 236–38, 267–70, 292–94. *Proc. Linn. Soc.* 1904–5, 39–42. *Times* 22 April 1905. *Who was Who, 1897–1915*, 439. *Fl. Malesiana* v.1, 1950, 332. *J. Soc. Bibl. Nat. Hist.* 1968, 327–43 portr. A. M. Coats *Quest for Plants*

1969, 206–7. M. A. Reinikka *Hist. of the Orchid* 1972, 226–29 portr.
Plants and letters at Kew.
Nepenthes lowii Hook. f.

LOW, Hugh (*c.* 1861–1893)
d. Clapton, London 17 Sept. 1893
Nurseryman, Clapton. Son of S. H. Low (1826–1890).
J. Hort. Cottage Gdnr v.27, 1893, 265.

LOW, James (–1852)
Lieut.-Colonel, Madras Army. In civil charge of Province Wellesley, 1823–40. Assistant Resident, Singapore. Collected a few plants at Penang, now at Kew. *Dissertation on Soil and Agriculture of Penang* 1828.
D.N.B. v.34, 183–84. *Gdns Bull. Straits Settlements* 1927, 127. *Fl. Malesiana* v.1, 1950, 332.

LOW, Stuart Henry (1826–1890)
b. Clapton, London 4 Jan. 1826 *d.* Clapton 1890
Son of H. Low (1793–1863). Nurseryman, Clapton. Established Bush Hill Park Nursery, Enfield, 1882.
Garden v.37, 1890, xii portr. *Gdnrs Chron.* 1890 i 560–61. *Fl. Malesiana* v.1, 1950, 331.

LOW, Stuart Henry (*c.* 1863–1952)
d. 29 Sept. 1952
Son of S. H. Low (1826–1890). Nurseryman, Clapton, London. Developed Bush Hill Park Nursery, Enfield, Middx. Specialised in orchids.
J. Hort. Cottage Gdnr v.63, 1911, 539 portr. *Gdnrs Chron.* 1937 i 82 portr.; 1952 ii 150. *Amer. Orchid Soc. Bull.* 1952, 856 portr. *Orchid Rev.* 1952, 182–83 portr.

LOWE, Charles (*fl.* 1790s)
Seedsman, Market Deeping, Lincs.

LOWE, Edward Joseph (1825–1900)
b. Nottingham 11 Nov. 1825 *d.* Shirenewton, Mon 10 March 1900
FRS 1867. FLS 1857. Cultivated British ferns. Hybridist. *Ferns, British and Exotic* 1856–60 8 vols. *Natural History of British Grasses* 1858. *Natural History of New and Rare Ferns* 1862. *Our Native Ferns* 1865–67 2 vols. *Fern Growing* 1895.
Gdnrs Chron. 1871, 803 portr.; 1900 i 173. E. J. Lowe *Fern Growing* 1895, 187 portr. *Garden* v.57, 1900, 215–16. *Gdnrs Mag.* 1900, 183 portr. *J. Bot.* 1900, 152. *Proc. Linn. Soc.* 1899–1900, 74–75. *Proc. R. Soc.* v.75, 1904, 101–3. *Br. Fern Gaz.* 1909, 46; 1967, 301–8. *R.S.C.* v.4, 95; v.8, 266, v.10, 640; v.12, 460; v.16, 886.
Letters at Kew.

LOWE, John (1830–1902)
b. Cheadle, Staffs 14 May 1830 *d.* Weybridge, Surrey 12 Dec. 1902
MD Edinburgh 1857. FLS 1890. Practised medicine in London and King's Lynn, Norfolk. *Yew-trees of Great Britain and Ireland* 1897. Contrib. to *Trans. Bot. Soc. Edinburgh*, *Trans. Norfolk Norwich Nat. Soc.*
Br. Med. J. v.2, 1902, 1974–75. *Gdnrs Chron.* 1902 ii 459. *Trans. Norfolk Norwich Nat. Soc.* 1902–3, 578–79. *R.S.C.* v.4, 97; v.8, 266; v.10, 640; v.12, 460; v.16, 887.

LOWE, Joseph (1843–1929)
b. Cookham, Berks 1843 *d.* 4 Dec. 1929
VMH 1923. Jobbing gardener, Uxbridge, 1864. Established nursery at Uxbridge, 1867; took George Shawyer into partnership *c.* 1897. At one time had the largest cut-flower establishment in the world. Noted rose-grower.
Gdnrs Chron. 1923 i 100 portr.; 1929 ii 476.

LOWE, Obadiah (*fl.* 1720s)
Nurseryman, Battersea, London.

LOWE, Richard (*c.* 1827–1908)
d. Tettenhall, Staffs 6 July 1908
Nurseryman, North Road and Penn Road, Wolverhampton. In partnership with Mr. Mowbray until 1862; then with his brother William Lowe until 1871.
Gdnrs Chron. 1908 ii 78 portr.

LOWE, Rev. Richard Thomas (1802–1874)
b. 4 Dec. 1802 *d.* drowned off Scilly Isles 13 April 1874
BA Cantab 1825. MA Oxon 1843. English chaplain in Madeira, 1832–54. Rector, Lea, Lincs, *c.* 1854. Botanised in Orkney Islands (*J. Bot.* 1864, 12), Canary Islands, Cape Verde Islands, 1862–64, Madeira, *c.* 1829–63, Morocco. 'Primitiae Faunae et Florae Maderae et Portus Sancti' (*Trans. Cambridge Philos. Soc.* 1831, 1–66). 'Novitiae Florae Maderensis' (*Trans. Cambridge Philos. Soc.* 1838, 523–51.) *Manual Fl. Madeira* 1857–72 2 vols. *Florulae Salvagicae Tentamen* 1869.
Bot. Mag. 1830, t.2988; 1833, t.3227, 3234. *J. Bot.* 1866, 157–58; 1874, 287–88; 1875, 192. *D.N.B.* v.34, 196–97. C. A. de Menezes *Fl. Archipelago da Madeira* 1914, 226–27.
Herb., Madeira MSS. and letters at Kew. Plants at BM(NH). Sale at Sotheby 21 June 1875.
Lowea Lindl.

LOWE, William Henry (*c.* 1815–1900)
d. Wimbledon, Surrey 26 Aug. 1900
MD Edinburgh 1840. FRCP Edinburgh 1846. Practised in Edinburgh until 1875 when he moved to Wimbledon. President, Royal College of Physicians of Edinburgh, 1873–75. President, Royal Botanical Society of London.

LOWER, N. Y. (–1926)
d. St. Blazey, Cornwall 20 Nov. 1926

Daffodil grower and hybridist. President, Midland Daffodil Society.
Garden 1926, 738 (ix).

LOWNDES, Donald George (1899–1956)
b. Ootacamund, India 1899 *d.* 28 Sept. 1956
Colonel in Royal Garhwal Rifles. Collected plants in Himalayas, etc. Botanist on H. W. Tilman's expedition to Nepal, 1950.
Lily Yb. 1958, 90–91 portr. *Quart. Bull. Alpine Gdn Soc.* 1956, 358–60; 1970, 170–73.
Gentiana lowndesii Blatter.

LOWNE, Benjamin Thompson (*fl.* 1860s)
Collected plants in Palestine and Syria, 1863–64. 'Vegetation of Western and Southern Shores of Dead Sea' (*J. Linn. Soc.* v.9, 1866, 201–8).
Plants at Kew.

LOWNE, Benjamin Thompson (1878–1956)
b. Finchley, London 1878 *d.* Worthing, Sussex March 1956
Electrical engineer. Studied Sussex flora. Contrib. localities to A. H. Wolley-Dod's *Fl. Sussex* 1937.
Proc. Bot. Soc. Br. Isl. 1957, 328.
Herb. at Kew.

LOYDELL, Alfred (1849–1910)
b. Woodend, Northants 1849 *d.* Acton, Middx 1 Jan. 1910
Studied Middlesex plants. Herb. purchased by G. C. Druce.
J. Bot. 1910, 88, 269; 1911, 66. G. C. Druce *Comital Fl. Br. Isles* 1932, xii. *R.S.C.* v.16, 888.

LUBBOCK, Sir John, 1st Baron Avebury (1834–1913)
b. London 30 April 1834 *d.* Kingsgate Castle, Kent 28 May 1913
DCL Oxon. LLD Cantab. FRS 1858. FLS 1858. Succeeded to baronetcy, 1865; created Baron, 1900. Banker. Entomologist, anthropologist, botanist. President, Linnean Society, 1881–86. *On British Wild Flowers considered in Relation to Insects* 1875. *Contribution to our Knowledge of Seedlings* 1892 2 vols. *Buds and Stipules* 1899. *Flowers, Fruit and Leaves* 1903. *Notes on Life History of British Flowering Plants* 1905.
Gdnrs Chron. 1913 i 375 portr. *J. Bot.* 1913, 222–23. *Nature* v.91, 1913, 350–51; v.133, 1934, 632–34. *Proc. Linn. Soc.* 1913–14, 53–59. *Selborne Mag.* 1913, 125–32 portr. H. G. Hutchinson *Life of Sir John Lubbock* 1914 2 vols. *D.N.B. Supplt. 1912–1921* 345–47. *Who was Who, 1897–1916* 29–30. U. G. Duff *Life-work of Lord Avebury* 1924. *Notes Rec. R. Soc.* v.13, 1958, 49–58; v.17, 1962, 183–91. *Science* 1959, 1087–92 portr. C. C. Gillispie *Dict. Sci. Biogr.* v.8, 1973, 527–29. *R.S.C.* v.4, 104; v.8, 267; v.10, 643; v.13, 210.
Portr. at Linnean Society, Hunt Library.

LUCAS, Arthur Henry Shakespeare (1853–1936)
b. Stratford-on-Avon, Warwickshire 7 May 1853 *d.* Albury, N.S.W. 10 June 1936
MA Oxon. BSc London. Science master, Wesley College, Melbourne, 1883. Headmaster, Newington College, Stanmore, N.S.W., 1892–98. Senior science master, Sydney Grammar School, 1899; Headmaster, 1920–23. President, Victorian Field Naturalists Club, 1887–89. President, Linnean Society N.S.W., 1907–9. Curator of Algae, Sydney Botanic Gardens. Friend of F. von Mueller. Collected plants. *Introduction to Study of Botany* (with A. Dendy) 1892; ed. 3, 1915. *A. H. S. Lucas, Scientist; his own Story* 1937. Contrib. to *Victorian Nat.*, *Proc. R. Soc. Tasmania.*
J. Bot. 1936, 296–97. *Nature* v.138, 1936, 234. *Victorian Nat.* 1936, 54–55. *Chronica Botanica* 1937, 57–58 portr. *Proc. Linn. Soc. N.S.W.* 1937, 243–52. P. Serle *Dict. Austral. Biogr.* v.2, 1949, 49–51. *Austral. Encyclop.* v.5, 1965, 382–83.
Portr. at Sydney Grammar School.

LUCAS, Caroline Catherine *see* Wilkinson, C. C.

LUCAS, Charles James (*c.* 1853–1928)
d. London 17 April 1928
Had large collection of orchids at Warnham Court, Horsham.
Orchid World 1910–11, 198–201 portr. *Orchid Rev.* 1922, 206–10; 1928, 149.

LUCAS, Edward (*fl.* 1780s)
Cornfactor and seedsman, 183 High Holborn, London.

LUCAS, Robert (–1733/4)
Nurseryman, Cheshunt, Herts.

LUCAS, William (–1679)
Seedsman, The Naked Boy, Strand, London.
J. Harvey *Early Gdning Cat.* 1972, 15–23, 65–74.

LUCK, Peter (*fl.* 1790s)
Seedsman, Louth, Lincs.

LUCKLEY, John Lamb (1822–1899)
b. Alnwick, Northumberland 1822 *d.* Clayport 1 March 1899
Cabinet-maker. Poet, writer and journalist. Studied Alnwick flora. *Botany of Alnwick* 1860. 'Botanical Rambles' (*Alnwick and County Gaz.* 1893).
Hist. Berwick Nat. Club v.24, 1920, 232–33. *Vasculum* v.15, 1929, 20. *J. Bot.* 1931, 77–78.

LUCOMBE, William (*c.* 1696–1794)
d. Sept. 1794
Gardener to Thomas Ball of Mamhead, Devon. Founded nursery at St. Thomas's,

Exeter, 1720; shop at New Bridge, Exeter. Raised hybrid Lucombe oak *c.* 1765. Son, William, joined business. R. T. Pince (*c.* 1804–1871) was partner by 1828.

Quart. J. For. 1970, 199–202.

LUCY, W. C. (*fl.* 1840s–1890s)

Merchant of Gloucester. Geologist. In 1864 formed collection of 200 species of plants for "British Botanical Competition."

H. J. Riddelsdell *Fl. Gloucestershire* 1948, cxxviii.

Plants at Gloucester Museum.

LUDGATER, James (*fl.* 1820s)

Seedsman and florist, 111 Church Street, Shoreditch, London.

LUDLOW, Frank (1885–1972)

b. Chelsea, London 10 Aug. 1885 *d.* 25 March 1972

BA Cantab 1908. Vice-principal, Sind College, Karachi. Opened school at Gyantse, Tibet, 1923–26. Retired to Srinagar. Collected birds and plants for BM(NH). Met G. Sherriff in 1928 with whom he made a number of collecting expeditions in E. Himalayas and S. E. Tibet. In charge of British Mission in Lhasa, 1942–43.

Bot. Mag. 1962–63 portr. A. M. Coats *Quest for Plants* 1969, 198–201. *J. R. Hort. Soc.* 1972, 416–17. *Times* 27 March 1972. H. R. Fletcher *A Quest of Flowers* 1975 portr. *Bull. BM (NH) Bot.* v.5(5), 1976, 243–88.

Portr. at Hunt Library.

Rhododendron ludlowii Cowan.

LUEHMANN, Johann Georg (1843–1904)

b. Buxtende, Germany May 1843 *d.* Victoria 18 Nov. 1904

FLS 1885. To Victoria, 1862. Assistant to F. von Mueller, 1869, whom he helped in *Key to System of Victorian Plants* 1885–88. Curator, Melbourne Herb. and Government Botanist, 1896. Authority on eucalypts and acacias.

Proc. Linn. Soc. 1904–5, 42. *Victorian Nat.* 1904–5, 108; 1908, 110–11. *R.S.C.* v.16, 903.

Plants at Kew. Portr. at Hunt Library.

Eugenia luehmanni F. Muell.

LUFKIN, John (*fl.* 1690s–1710s)

Pharmacist, Colchester, Essex. Sent specimen of marine alga to Herb. Sherard.

J. E. Dandy *Sloane Herb.* 1958, 159.

Plants at BM(NH).

LUGARD, Mrs. Charlotte Eleanor (1859–1939)

Miniature painter who exhibited at Royal Academy. Drew many of the plants collected by her husband, E. J. Lugard. Accompanied brother-in-law, Lord Lugard, on expedition in Bechuanaland, 1896–97. Collected plants for Kew (*Kew Bull.* 1909, 81–146).

Kew Bull. 1925, 348–49. A. White and B. L. Sloane *Stapelieae* v.1, 1937, 124 portr. *J. Cactus Succulent Soc. Amer.* 1941, 89.

Drawings of Ngamiland plants at Kew.

Monadenium lugardae N. E. Brown.

LUGARD, Edward James (1865–1944)

b. Worcester 23 March 1865

Major, British Army. Collected plants in Bechuanaland with his brother, F. D. Lugard, 1896; with his wife in Bechuanaland, Basutoland and Ngamiland, 1897–98. Collected in Nigeria, 1905–7, Kenya, 1930–31. Plants identified by Kew botanist, N. E. Brown (*Kew Bull.* 1909, 81–146).

A. White and B. L. Sloane *Stapelieae* v.1, 1937, 123–24 portr. *J. E. Africa Nat. Hist. Soc. National Mus.* v.28, 1970, 53–55.

Caralluma lugardi N. E. Brown.

LUGARD, Sir Frederick Dealtry, 1st Baron Lugard (1858–1945)

b. 22 Jan. 1858 *d.* 11 April 1945

Collected plants in Bechuanaland with his brother, E. J. Lugard, 1896. Governor, Hong Kong, 1907–12, Nigeria, 1912–19.

Who was Who, 1941–1950 705.

LUKAR, Roger *see* Looker, R.

LUKER, Warren (–1784)

d. 5 Nov. 1784

Founded nursery, City Road, London, *c.* 1760. From 1780 carried on as Luker, Smith and Lewis and from 1785–1849 as Smith.

LUKIS, Catherine Rabey *see* Mansell, C. R.

LUMB, Dennis (1871–1951)

b. 6 Feb. 1871 *d.* Dalton-in-Furness, Yorks 26 Aug. 1951

Headmaster, Broughton Road School, Dalton-in-Furness. Studied *Euphrasia.*

Watsonia 1953, 360.

Herb. at Bradford Natural History Society.

Euphrasia lumbii Druce.

LUMSDAINE, John (*fl.* 1790s–1810s)

Eurasian, probably the first 'native' to be appointed to East India Company's service. Assistant surgeon, 1797; Surgeon, 1812. Botanist and Superintendent, Spice Plantations, Fort Marlborough, Sumatra, 1815. Had good knowledge of Sumatran flora.

D. G. Crawford *Hist. Indian Med. Service, 1600–1913* v.1, 1914, 502–3. *Fl. Malesiana* v.1, 1950, 54.

LUNAM, George (–1947)

d. 1 June 1947

Science teacher, Whitehall Secondary School, Glasgow. Vice-president, Natural History Society of Glasgow, 1925, 1939. Specialised in fresh-water algae. Contrib. to *Glasgow Nat., Ann. Andersonian Nat. Soc.*

Glasgow Nat. v.16, 1947, 35–36.

Plants at Glasgow.

LUNAN, John (*fl.* 1810s)
Hortus Jamaicensis 1814.
I. Urban *Symbolae Antillanae* v.1, 1898, 97–98 (plants not at Kew as stated).
Lunania Hook.

LUNT, William (1805–1883)
Market gardener, Mitcham, Surrey.
Country Life 1967, 1648–49 portr.

LUNT, William (1871–1904)
b. Ashton-under-Lyne, Lancs 16 Dec. 1871 *d.* St. Kitts, West Indies 3 Jan. 1904
Kew gardener, 1892. Collected plants on Theodore Bent's expedition to Hadramaut Valley, S. Arabia, 1893–94. Assistant Superintendent, Trinidad Botanic Garden, 1894. Curator, Botanic Station, St. Kitts-Nevis, 1898. *Visitors' Guide to Royal Botanic Gardens, Trinidad* 1895.
Kew Bull. 1893, 366; 1894, 328–43. I. Urban *Symbolae Antillanae* v.3, 1903, 79. *Gdnrs Chron.* 1904 ii 221. *J. Kew Guild* 1904, 208–9 portr. *Rec. Bot. Survey India* v.7, 1914, 19–20.
Plants at Kew, Trinidad.
Verbascum luntii Baker.

LUPTON, John Watts (*c.* 1793–1877)
Nurseryman, 2 Bootham Terrace and The Nurseries, Clifton, York. Succeeded by his son, James Killingbeck Lupton.

LUSH, CHARLES (1797–1845)
b. 6 Nov. 1797 *d.* Hyderabad, India 4 July 1845
MD. FLS 1820. Lecturer in botany, St. Thomas's Hospital, 1825. Superintendent, Botanic Garden, Dapooree, Poona, 1827–37. 'Acacia' (*Trans. Linn. Soc.* v.18, 1841, 217).
Proc. Linn. Soc. 1846, 302–3. *R.S.C.* v.4, 131.

LUSHINGTON, Alfred Wyndham (*c.* 1860–1920)
Indian Forest Service, Madras. *Vernacular List of Trees, Shrubs and Woody Climbers in Madras Presidency* 1915 3 vols. *Nature and Uses of Madras Timbers* 1919.
Indian Forester 1920, 421–23.
Plants at Kew.

LUXFORD, George (1807–1854)
b. Sutton, Surrey 7 April 1807 *d.* Walworth, London 12 June 1854
ALS 1836. Printer. Lecturer in Botany, St. Thomas's Hospital, 1846–51. *Fl. of Neighbourhood of Reigate, Surrey* 1838. Edited *Phytologist* 1841–54. Contrib. to *Mag. Nat. Hist.*
Proc. Linn. Soc. 1855, 426–27. *D.N.B.* v.34, 302–3. C. E. Salmon *Fl. Surrey* 1931, 51. *Bot. Soc. Exch. Club Br. Isl. Rep.* 1932, 285.

LYALL, David (1817–1895)
b. Auchinblae, Kincardineshire 1 June 1817 *d.* Cheltenham, Glos 2 March 1895
MD Aberdeen. FLS 1862. Assistant surgeon on HMS 'Terror' in Ross's Antarctic voyage, 1839–42. Surgeon and naturalist on HMS 'Acheron' in New Zealand. Discovered *Ranunculus lyalii* in New Zealand during this voyage. Surgeon and naturalist on HMS 'Assistance' in Arctic, 1852. On British Columbia Boundary Commission, 1858–61. Collected plants in Antarctic, Australia, New Zealand, Norfolk Island, N.W. U.S.A., S.W. Canada.
J. D. Hooker *Fl. Tasmaniae* v.1, 1859, 548–49. *J. Linn. Soc.* v.7, 1863, 124–47. *Ann. Scott. Nat. Hist.* 1895, 263. *J. Bot.* 1895, 127, 209–11. *Proc. Linn. Soc.* 1894–95, 33–35. C. S. Sargent *Silva of N. America* v.12, 1898, 16. J. Veitch and Sons *Manual of Coniferae* 1900, 400. *Contrib. U.S. National Herb.* 1906, 16–17. L. Huxley *Life and Letters of Sir J. D. Hooker* 1918 2 vols *passim*. J. Ewan *Rocky Mountain Nat.* 1950, 255. R. Glenn *Bot. Explorers of N. Z.* 1950, 115–20. *R.S.C.* v.4, 137; v.16, 927.
Letters and plants at Kew.
Lyallia Hook. f.

LYALL, John B. (*c.* 1827–1912)
d. Drumlithie, Kincardineshire 12 April 1912
Schoolmaster, Peebles. Local botanist. Contrib. chapters on flora to Peebles *Guide*.
Gdnrs Chron. 1912 i 269.
Herb. formerly at Peebles Museum, now lost.

LYALL, Robert (*fl.* 1780s–1831)
b. Scotland *d.* Port Louis, Mauritius Sept. 1831
MD Edinburgh. FLS 1824. In Russia, 1815–23; Mauritius, 1827; Madagascar, 1828; Antananarivo, 1828–29. Contrib. papers on irritability of plants to *Nicholson's J.* v.24–28, 1809–11.
Gdnrs Chron. 1892 ii 519–21. *D.N.B.* v.34, 304–5. *J. Bot.* 1906, 35. *Comptes Rendus AETFAT 1960* 1962, 131.
Plants at Kew.
Vernonia lyallii Baker.

LYE, James (–1906)
d. Easterton, Wilts Feb. 1906
Gardener at Cliffe Hall, Wilts. Raised fuchsias.
J. Hort. Home Farmer v.63, 1911, 198.

LYELL, Charles (1767–1849)
b. Kinnordy, Forfarshire 7 March 1767 *d.* Kinnordy 8 Nov. 1849
MA Cantab 1794. FLS 1813. Contrib. lichens to J. Sowerby and J. E. Smith's *English Bot.* MS. flora of Kirriemuir. Studied roses. Sent *Jungermanniae* to W. J. Hooker (Hooker *Br. Jungermanniae* 1816, t.77).
Gdnrs Chron. 1849, 727. *Proc. Linn. Soc.* 1850, 87–88. *Proc. Geol. Soc.* 1876, 53–69. *D.N.B.* v.34, 319. *J. Bot.* 1899, 143–44.

Silhouette portr. on porcelain at Kew. Letters at BM(NH), Kew, Linnean Society. Herb. at BM(NH).
Lyellia R. Br.

LYELL, Sir Charles (1797–1875)
b. Kinnordy, Forfarshire 14 Nov. 1797 *d.* London 22 Feb. 1875
BA Oxon 1819. DCL 1854. FRS 1826. FLS 1819. Son of C. Lyell (1767–1849). Baronet, 1864. Geologist and palaeontologist. *Principles of Geology* 1828.
Ipswich Mus. Portr. Ser. 1852. L. Reeve *Portr. of Men of Eminence in Lit., Sci. and Art* v.1, 1863, 9–12 portr. *Nature* v.11, 1874, 341–42; v.12, 1875, 325–27. *Proc. Linn. Soc.* 1874–75, liii–lvi. *Proc. R. Soc.* v.25, 1877, xi–xiv. K. M. Lyell *Life, Letters and J. of Sir C. Lyell* 1881 3 vols. *D.N.B.* v.34, 319–24. L. Huxley *Life and Letters of Sir J. D. Hooker* 1918 2 vols *passim. Notes Rec. R. Soc.* v.14, 1959, 121–38 portr. E. Bailey *Charles Lyell* 1962 portr. R. C. Olby *Early Nineteenth Century European Scientists* 1967, 119–52. L. G. Wilson *Sir Charles Lyell's Sci. Js. on Species Question* 1970 portr. C. C. Gillispie *Dict. Sci. Biogr.* v.8, 1973, 563–76. *R.S.C.* v.4, 138; v.8, 284; v.10, 665; v.16, 980.
Letters at American Philosophical Society. Portr. at Hunt Library.

LYELL, Mrs. Katharine Murray (*née* **Horner**) (1817–1915)
d. 19 Feb. 1915
Sister-in-law of Sir C. Lyell. Collected plants in Ganges Delta, Assam and Khasia Hills. *Geographical Handbook of all the Known Ferns* 1870. Edited *Life of Sir Charles J. F. Bunbury* 1906 2 vols.
L. Huxley *Life and Letters of Sir J. D. Hooker* 1918 2 vols *passim.*
Ferns at Kew.

LYLE, Peter (*fl.* 1820s–1890s)
Scottish gardener. Raised show pansies.
J. Hort. Cottage Gdnr v.26, 1893, 423–24 portr.

LYLE, Thomas (–1859)
d. Glasgow 20 April 1859
MD. Practised medicine in Glasgow. Bryologist.
Mosses and letters in Wilson correspondence at BM(NH). Plants at Paisley Museum.

LYMBURN, Robert (*c.* 1793–1843)
b. Scotland *c.* 1793 *d.* Kilmarnock 31 Oct. 1843
Nurseryman at Kilmarnock in partnership with Mr. Dreghorn. Contrib. articles on plant physiology to *Gdnrs Chron.* and *Gdnrs Mag.*
Gdnrs Mag. 1843, 677.

LYNAM, James (1812–1885)
b. Ballybrummel, Carlow 1812 *d.* Raheen, Galway Oct. 1885
Corresponding Secretary, Botanical Society of London, 1852. Found *Sisyrinchium angustifolium* in Galway, 1845. *Climates of the Earth* 1857 (botanical chart).

LYNCH, Richard Irwin (1850–1924)
b. St. Germans, Cornwall 1 June 1850 *d.* Torquay, Devon 7 Dec. 1924
ALS 1881. VMH 1906. VMM 1923. Kew gardener, 1867. Curator, Botanic Garden, Cambridge, 1879–1919. *List of Fern and Fern Allies in Botanic Garden, Cambridge* 1897. *Book of the Iris* 1904.
Garden v.59, 1901, 110 portr. *Gdnrs Chron.* 1901 i 294 portr., 295–96; 1924 ii 414. *Gdnrs Mag.* 1906, 179 portr.; 1908, 175–76 portr. *J. Hort. Cottage Gdnr* v.52, 1906, 277 portr. *J. Kew Guild* 1907, 347 portr.; 1925, 341–42 portr. *Bot. Soc. Exch. Club Br. Isl. Rep.* 1924, 543–44. *Nature* v.114, 1924, 942. *J. Bot.* 1925, 32. *Proc. Linn. Soc.* 1924–25, 80–81. *Curtis's Bot. Mag. Dedications, 1827–1927* 359–60 portr. *Who was Who, 1916–1928* 654.

LYNCH, Thomas Kerr (1818–1891)
b. Partry, Ballinrobe, County Mayo 1818 *d.* London 27 Dec. 1891
On second Euphrates expedition, 1837–42. Collected plants in N. Persia, 1849. *Visit to the Suez Canal* 1866.
D.N.B. v.34, 338.
Plants at BM(NH), Glasgow.

LYNE, Robert Nunez (1864–1961)
b. 8 Aug. 1864 *d.* 21 June 1961
Director of Agriculture, Zanzibar, 1896. Collected plants in Zanzibar and Pemba, 1901–2. Director of Agriculture, Ceylon, 1912. Editor, *Tropical Agric.* 1912–16.
Who was Who, 1961–1970 700.
Plants at Kew.

LYNES, Hubert (1874–1942)
b. 27 Nov. 1874 *d.* 10 Nov. 1942
Entered Royal Navy, 1887; Rear-Admiral; retired 1922; settled in Tanganyika. Collected plants in Darfur, Sudan, 1920–22; Morocco, 1923–25; Portuguese Guinea and Belgian Congo, 1930–34; Tanganyika, 1931–32. 'Notes on Natural History of Jebel Marra' (*Sudan Notes and Rec.* 1921, 119–37).
Rev. Zoo. Bot. Afr. 1938, 1–129. *Who was Who, 1941–1950* 711. *Comptes Rendus AETFAT 1960* 1962, 223, 234, 244.
Plants at BM(NH), Brussels, Kew.

LYNN, Mary Johnstone (1891–)
b. Carrickfergus, Antrim 1891
BSc Belfast. DSc 1937. Senior Lecturer in Botany, Belfast. Studied phyto-ecology of

tidal zone in N. Ireland. Contrib. to *Irish Nat. J.*

R. L. Praeger *Some Irish Nat.* 1949, 121–22.

LYNN, W. (*fl.* 1840s)

Nurseryman, Henley-on-Thames, Oxfordshire.

LYNN, William (*fl.* 1840s)

Seedsman and florist, 3 Providence Place, High Street, Kingsland, Hackney, London.

LYON, Sir David Bowes- *see* Bowes-Lyon, *Sir* D.

LYON, John (*c.* 1765–1814)

b. Gillogie, Forfarshire *c.* 1765 *d.* Ashville, N. Carolina, U.S.A. 13 Sept. 1814

Gardener to William Hamilton in Philadelphia, *c.* 1796–1803, 1805. To mountains of W. Pennsylvania, 1799; retraced trail of John and William Bartram into Georgia and E. Florida, 1802–4; Southern States and Tennessee, 1807–8; New York, 1810; Tennessee and N. Carolina, 1814. To England with cargoes of plants, 1806, 1812. 'Journal, 1799–1814' (*Trans. Amer. Philos. Soc.* 1963, 1–69).

H. C. Andrews *Botanist's Repository* v.1, 1798, t.42. T. Nuttall *Genera of N. American Plants* v.1, 1818, 266–67. *Bot. Mag.* 1813, t.1566. *London J. Bot.* 1842, 11. C. S. Sargent *Silva of N. America* v.5, 1893, 80. J. W. Harshberger *Botanists of Philadelphia* 1899, 133. A. M. Coats *Quest for Plants* 1969, 289–91.

Had herb., since lost. MS. Journal at American Philosophical Society.

Lyonia Nuttall.

LYON, Peter (*fl.* 1810s)

Apothecary and physic gardener at Comely-garden, Edinburgh. *Observations on Barrenness of Fruit Trees* 1813. *Treatise on Physiology and Pathology of Trees* 1816.

J. C. Loudon *Encyclop. Gdning* 1822, 1287.

LYONS, Caroline (*fl.* 1830s)

Plants collected in Pyrenees and Nice at Oxford.

LYONS, Israel (1739–1775)

b. Cambridge 1739 *d.* London 1 May 1775

Astronomer. Taught botany to J. Banks. On Arctic Expedition, 1773. *Fasciculus Plantarum circa Cantabrigiam Nascentium* 1763.

Gent. Mag. 1775, 254. J. Sowerby and J. E. Smith *English Bot.* 459. J. Nichols *Illus. Lit. Hist. of Eighteenth Century* v.4, 475. *Mem. Wernerian Nat. Hist. Soc.* v.1, 1811, 66. G. C. Gorham *Mem. John and Thomas Martyn* 1830, 122. *D.N.B.* v.34, 357–58.

Lyonsia R. Br.

LYONS, John Charles (1792–1874)

b. Ladiston, County Westmeath 22 Aug. 1792 *d.* Ladiston 3 Sept. 1874

Educ. Oxford. Antiquary. Imported and grew orchids. *Remarks on Management of Orchidaceous Plants* 1843; ed. 2 1845. *Catalogue of Orchidaceous Plants at Ladiston* 1845. Contrib. to W. J. Hooker's *Century of Orchidaceous Plants* 1849.

D.N.B. v.34, 358. *J. R. Hort. Soc.* 1943, 175–78 (reprinted in *Orchid Rev.* 1943, 102–6).

Letters at Kew.

LYSONS, Rev. Daniel (1762–1834)

b. 28 April 1762 *d.* Hempstead Court, Glos 3 Jan. 1834

BA Oxon 1782. FRS 1797. FLS 1798. Topographer. Rector, Rodmarton, 1804–33. Friend of A. B. Lambert. Cornish plants listed in his *Magna Britannia* 1814, cxcviii–cc. "An Excellent botanist" (*Gent. Mag.* v.17, 1842, 668).

D.N.B. v.34, 361–62. G. C. Druce *Fl. Buckinghamshire* 1926, xcviii.

Letters at Kew, BM(NH). Sale at R. H. Evans, 17 March 1828 and 30 Nov. 1835. Portr. at Hunt Library.

LYTE, Henry (*c.* 1529–1607)

b. Lytes Cary, Somerset *c.* 1529 *d.* Lytes Cary 15 Oct. 1607

Antiquary. *A Nievve Herball or Historie of Plantes* 1578 (translated from R. Dodoens).

R. Pulteney *Hist. Biogr. Sketches of Progress of Bot. in England* v.1, 1790, 88–95. *J. Bot.* 1875, 349. *Proc. Somerset Archaeol. Nat. Hist. Soc.* 1892, 1–100. *D.N.B.* v.34, 364–65. J. W. White *Fl. Bristol* 1912, 50. *Somerset Dorset Notes and Queries* March 1917, 1–4. *Proc. Linn. Soc.* 1921–22, 19–20. C. E. Raven *English Nat.* 1947, 199–204. R. H. Jeffers *Friends of John Gerard* 1967, 23–25.

MSS. at Oxford.

LYTTEL, Rev. Edward Shefford (1868–1944)

d. 20 Aug. 1944

FLS 1929. Professor of History, Southampton, 1912–34. President, Alpine Garden Society, 1944. Grew lilies.

Proc. Linn. Soc. 1943–44, 212–13. *Quart. Bull. Alpine Gdn Soc.* v.12, 1944, 126–27, 220–23 portr. *Lily Yb.* 1946, 1–2 portr.

MacALLA, William *see* McCalla, W.

McALPINE, Archibald Nichol (1855–1924)

b. Saltcoats, Ayrshire 1 June 1855 *d.* Glasgow 2 Dec. 1924

BSc London 1876. Studied in Germany. Professor of Botany, New Veterinary College, Edinburgh and afterwards (for 21 years) at W. of Scotland Agricultural College. *Biological Atlas* (with D. McAlpine) 1880. *How to know Grasses by their Leaves* 1890.

Translated F. G. Stebler and C. Schroeter's *Best Forage Plants* 1889.

Gdnrs Chron. 1924 ii 414. *J. Bot.* 1925, 32.

Plants at West of Scotland Agricultural College.

McALPINE, Daniel (1848–1932)
b. Ayrshire 21 Jan. 1848 *d.* Leitchville, Victoria 12 Oct. 1932

Studied biology under T. H. Huxley and botany under W. T. T. Dyer at Royal College of Science. Lecturer, Heriot Watt College, Edinburgh, 1877. Lecturer in Biology, Melbourne University, 1884; then Lecturer in Botany, College of Pharmacy, Melbourne. Plant Pathologist, Dept. of Agriculture, Victoria, 1890–1915. *Botanical Atlas* (with A. N. McAlpine) 1880. *Systematic Arrangement of Australian Fungi* 1895. *Fungus Diseases of Citrus Trees in Australia* 1899. *Fungus Diseases of Stone-fruit Trees in Australia* 1902. *Rusts of Australia* 1906. *Handbook of Fungus Diseases of the Potato in Australia* 1911.

F. John's Annual 1914, 125–26. *Proc. Linn. Soc. N.S.W.* 1933, iv. *Victorian Nat.* 1949, 107–8.

Portr. at Hunt Library.

McANDREW, James (1836–1917)
b. New Spymie, Morayshire 29 Jan. 1836 *d.* Edinburgh 4 July 1917

Headmaster, Kells parish school. *List of Flowering Plants of Dumfriesshire and Kirkcudbrightshire* 1882. Contrib. to G. F. S. Elliot's *Fl. Dumfriesshire and Dumfries District* 1891. Contrib. notes on flora of Galloway to *Trans. Dumfriesshire and Galloway Nat. Hist. Antiq. Soc.*

Bot. Soc. Exch. Club Br. Isl. Rep. 1919, 619. *Trans. Dumfriesshire and Galloway Nat. Hist. Antiq. Soc.* 1916–18, 264–65.

Mosses at Edinburgh. Plants at Dumfriesshire and Galloway Natural History Society.

McARDLE, David (1849–1934)
b. Dublin 28 Nov. 1849 *d.* Ilford, Essex 2 June 1934

Son of P. McArdle (1808–1883), foreman, Glasnevin, 1831–68. Plant collector and clerk, Botanic Gardens, Glasnevin, 1869–1923. Studied Irish Bryophyta. 'List of Irish Hepaticae' (*Proc. R. Irish Acad. B.* 1904, 387–502). Contrib. to *Irish Nat.*

J. Bot. 1934, 292. *Chronica Botanica* 1935, 195. *Irish Nat.* 1935, 182. R. L. Praeger *Some Irish Nat.* 1949, 123.

Herb. at BM(NH). Plants and slides at Dublin Museum.

MACARTHUR, Elizabeth (1769–1850)
b. Devon? 1769 *d.* Sydney, N.S.W. 9 Feb. 1850

Emigrated to Sydney with her husband, John Macarthur. Studied the local flora.

S. M. Onslow *Some Early Rec. of Macarthurs of Camden* 1914 *passim. Austral. Encyclop.* v.5, 1965, 400–1.

McARTHUR, Peter (*fl.* 1830s)

Nurseryman, Burwood Place, Edgware Road, London.

MACARTHUR, Sir William (1800–1882)
b. Parramatta, N.S.W. 16 Dec. 1800 *d.* Camden Park, N.S.W. 29 Oct. 1882

Knighted, 1856. Sent his gardener, P. Reedy, to New Guinea with W. Macleay's expedition for new plants. *Catalogue of Plants cultivated at Camden Park, N.S.W.* 1850. Commissioner from N.S.W. to Paris Exhibition, 1855. *Catalogue des Collections de Bois indigènes* (with C. Moore) 1855. List of S. Australian woods sent to Edinburgh reported in *Proc. Bot. Soc. Edinburgh* 1856, 61–70.

J. D. Hooker *Fl. Tasmaniae* v.1, 1859, cxxvi. P. Mennell *Dict. Austral. Biogr.* 1892, 287. *J. Proc. R. Soc. N.S.W.* 1908, 111–12. *Rep. Austral. Assoc. Advancement Sci.* 1911, 231 portr. S. M. Onslow *Some Early Rec. of Macarthurs of Camden* 1914 *passim. Fl. Malesiana* v.1, 1950, 335. *Austral. Encyclop.* v.5, 1965, 406.

Plants at BM(NH), Kew. Letters at Kew.

Macarthuria Endl.

McASLAN, Alexander (–1841)

Partner in nursery firm of Robert Austin and McAslan, 114 Tron Gate, Glasgow.

Gdnrs Chron. 1917 ii 148.

McASLAN, Duncan (–1741)

Nurseryman, Glasgow; succeeded by his widow, then his son, John.

Gdnrs Chron. 1917 ii 147.

McASLAN, John (*fl.* 1710s)

Founded nursery near Hutcheson's Hospital, Glasgow, 1717. His brother, Duncan, was associated with him.

Gdnrs Chron. 1917 ii 147.

McASLAN, John (*fl.* 1750s–1815)
d. 1815

Son of D. McAslan (–1741). Removed family nursery from Hutcheson's Hospital to The Hill, Glasgow.

Gdnrs Chron. 1917 ii 147 portr.

MACAULAY, M. A. (*née* **Gairdner**) (*fl.*1910s)

Of Mumbwa, N. Rhodesia where she collected plants, 1911–12, now at Kew.

Comptes Rendus AETFAT 1960 1962, 185.

Crotalaria macaulayae Baker f.

MACAULAY, William (1810–1900)
b. St. Ninians, Stirlingshire 7 Feb. 1810 *d.* Selkirk 24 Nov. 1900
Gardener at Haining near Selkirk. Seedsman, Selkirk for about 27 years.
Gdnrs Chron. 1900 ii 423.

McBAIN, Alan Matthew (*c.* 1902–1936)
d. 10 Jan. 1936
BSc Glasgow. Assistant, Virus Disease Investigations, Scottish Society for Research in Plant Breeding, 1930–36. Investigated physiological aspects of potato virus.
Nature v.137, 1936, 217.

McBEAN, Albert Alexander (–1942)
d. Haywards Heath, Sussex 31 July 1942
Acquired his father's (J. McBean) nursery at Cooksbridge, Sussex in 1910. Specialised in cymbidiums.
Amer. Orchid Soc. Bull. 1942, 199; 1958, 738–40; 1969, 985. *Orchid Rev.* 1942, 150–51.

McBEAN, James (1840–1910)
b. Marlborough, Scotland 22 Oct. 1840 *d.* Cooksbridge, Sussex 21 Aug. 1910
Started orchid nursery at Cooksbridge near Lewes, Sussex in 1879.
Gdnrs Chron. 1910 ii 170 portr. *Amer. Orchid Soc. Bull.* 1969, 985.

MacBETH, Charles (*fl.* 1830s)
Nurseryman, Stone Bridge, Goole, Yorks.

McCALLA (*or* **MacALLA**), **William** (*c.* 1814–1849)
b. Roundstone, Connemara *c.* 1814 *d.* Connemara May 1849
Schoolmaster. Algologist. Discovered *Erica mackaiana*. Sent algae to W. J. Hooker, W. H. Harvey, etc. *Algae Hibernicae* 1845 (exsiccatae).
Companion Bot. Mag. v.1, 1836, 158. *J. Bot.* (Hooker) 1842, 71. W. H. Harvey *Phycologia Britannica* v.1, 1846, 84, 293. *Phytologist* v.2, 1847, 742–46. C. C. Babington *Mem, J. and Bot. Correspondence* 1897, 45, 274–75. *Notes R. Bot. Gdn Edinburgh* v.2, 1902, 155. R. L. Praeger *Some Irish Nat.* 1949, 122.
Plants at Trinity College, Dublin. Letters at Kew.
Cladophora macallana Harvey.

M'CALLAN, David (*fl.* 1790s)
Seedsman, Gainsborough, Lincs.

MacCALLUM, *Mrs.* **Bella Dytes MacIntosh** (–1927)
b. New Zealand
BA Canterbury, N.Z. 1908. FLS 1921. To England, 1919. Assistant, Dept. of Botany, Edinburgh University, 1920–21.
Proc. Linn. Soc. 1930–31, 180.

McCARTHY, Rev. J. (*fl.* 1860s–1870s)
To China in service of China Island Mission, 1867. Travelled from the Yangtze to Burma and collected a few plants which he sent to Kew.
Kew Bull. 1901, 42. E. H. M. Cox *Plant Hunting in China* 1945, 101.
Nephrodium mcCarthyi Baker.

McCLELLAND, John (1800–1883)
d. St. Leonards, Sussex 11 July 1883
FLS 1841. Bengal Medical Service, 1830–65. Zoologist. Superintendent, Botanic Garden, Calcutta, 1846–48, 1858. *Plan...with List of Plants in Natural and Medicinal Gardens Calcutta* 1847. Founded *Calcutta Mag. Nat. Hist.* Edited *Calcutta J. Nat. Hist.* 1841–47. Edited W. Griffith's works, 1847–54.
London J. Bot. 1848, 446–49. F. Boase *Modern English Biogr.* v.2, 1897, 371–72. D. G. Crawford *Hist. of Indian Med. Service, 1600–1913* v.2, 1914, 148. M. Archer *Nat. Hist. Drawings in India Office Library* 1962, 41–46, 87.
Plants at Kew.

McCLOUNIE, John (*fl.* 1890s)
Forester, British Central African Administration, Nyasaland, 1893. Surveyed forests on Mlanje Mountain, 1895. Collected plants in Nyasaland and N. Rhodesia, 1903. 'Journey across Nyika Plateau' (*Geogr. J.* 1903, 423–37).
Kew Bull. 1895, 158. *Comptes Rendus AETFAT 1960* 1962, 167, 186.
Plants at Kew.

McCOIG, Malcolm (–1789)
d. Edinburgh 25 Feb. 1789
Principal gardener, Royal Botanic Garden, Edinburgh, 1782–89. *Fl. Edinburgensis* (unpublished).
Notes R. Bot. Gdn Edinburgh v.3, 1903–4, 5, 20.

McCONACHIE, Rev. George (–1901)
b. Glenrinnes, Banffshire *d.* Rerrick, Kirkcudbrightshire May 1901
Educ. Aberdeen University. Minister, Rerrick from 1877. Contrib. papers on cryptogams to *Trans. Dumfriesshire and Galloway Nat. Hist. Antiq. Soc.*
Ann. Scott. Nat. Hist. 1901, 193.

McCORMICK, Robert (1800–1890)
b. Runham, Norfolk 22 July 1800 *d.* Wimbledon, Surrey 28 Oct. 1890
Assistant surgeon, Royal Navy, 1823. Pupil of J. Lindley. Collected plants on Parry's Arctic and Ross's Antarctic expeditions. *Voyages of Discovery* 1884 2 vols. portr.
D.N.B. v.35, 11.
Plants at BM(NH).
Quercus mcCormickii Carruth.

M'CORQUODALE, William (*c.* 1810–1891)
b. Argyll *d.* 17 April 1891
Forester to Earl of Mansfield, 1838. Contrib. to *Trans. Scott. Arboricultural Soc.*
Garden v.39, 1891, 401.

McCOSH, James (1811–1894)
b. Carskeoch, Ayrshire 1 April 1811 *d.* Princeton, U.S.A. 16 Nov. 1894
Theologian and philosopher. Professor of Logic, Queen's College, Belfast, 1850–68. Botanised in Belfast. President, College of New Jersey, Princeton, 1868–88.
Belfast Nat. Hist. Philos. Soc. Centenary Vol. 1924, 91, 144.

McCOY, Sir Frederick (1817 or 1823–1899)
b. Dublin 1817 or 1823 *d.* Melbourne 13 May 1899
FRS 1880. DSc Cantab 1886. KCMG 1891. Professor of Mineralogy and Geology, Queen's College, Belfast, 1852. Professor of Natural Sciences, 1854. Founded National Museum of Natural History and Geology, Melbourne. President, Field Naturalists' Club of Victoria, 1880–82. 'Fossil Botany of Coal of Australia' (*Ann. Nat. Hist.* v.20, 1847, 145–57). *Palaeontology of Victoria* 1874.
P. Mennell *Dict. Austral. Biogr.* 1892, 289–90. *Geol. Mag.* 1899, 283–87. *D.N.B. Supplt 1* v.3, 119. *Proc. R. Soc.* v.75, 1904, 43–45. *Vict. Nat.* 1899, 19; 1906, 79–80; 1940, 5–6 portr. P. Serle *Dict. Austral. Biogr.* v.2, 1949, 68–69. *Austral. Encyclop.* v.5, 1965, 414. *R.S.C.* v.4, 151; v.8, 289; v.10, 672; v.16, 948.

McCREA, Mrs. Rawdon (–1949)
d. Feb. 1949
Secretary, Botanical Section, Société Guernesiaise. Contrib. botanical records to the Society's *Trans.*
Rep. Trans. La Société Guernesiaise 1949, 354–55.
Herb. at Société Guernesiaise.

MacCULLOCH, John (1773–1835)
b. Guernsey, Channel Islands 6 Oct. 1773 *d.* Cornwall 21 Aug. 1835
MD Edinburgh 1793. LRCP 1808. FRS 1820. FLS 1801. Geologist to Trigonometrical Survey, 1814. President, Geological Society, 1816–17. 'Fl. Islands of Jersey and Guernsey' (T. Quayle *General View of Agric....of Islands on Coast of Normandy* 1815, 330–37). 'Naturalization of Plants' (*Quart. J. Sci.* 1826, 200–15).

McDAKIN, S. Gordon (–1918)
d. Dover, Kent 6 Feb. 1918
Captain in Black Watch. Contrib. papers on 'Verification of Records of Fl.' and 'Verification of Botanical Records' to Dover Sciences Society of which he became President.
Nature v.101, 1918, 28.

M'DONALD, Alexander *see* Dickson, R. W.

MACDONALD, Charlotte *see* Smith, C.

McDONALD, Donald (*c.* 1857–1931)
d. 27 Feb. 1931
FLS 1898. Seedsman and journalist. *English Vegetables and Flowers in India and Ceylon* 1890. *Sweet-scented Flowers and Fragrant Leaves* 1895. *My Garden Companion* 1905. *Agricultural Writers...1200–1800* 1908. *Fruit Culture and Utility* 1924.
Proc. Linn. Soc. 1930–31, 181.
Part of library at Royal Horticultural Society.

MacDONALD, James (*fl.* 1810s)
Gardener to Duke of Buccleugh at Dalkeith Park near Edinburgh. Contrib. to *Mem. Caledonian Hort. Soc.*
J. C. Loudon *Encyclop. Gdning* 1822, 1285.

MacDONALD, James (1855–1930)
b. Huntingtower, Perthshire 1855 *d.* Harpenden, Herts 14 Nov. 1930
FLS 1923. Founder of firm of grass specialists at Harpenden. *Lawns, Links and Sportsfields* 1923.
Gdnrs Chron. 1925 ii 382 portr.; 1930 ii 439. *Proc. Linn. Soc.* 1930–31, 181–83.

McDONALD, James Edwin (1871–1938)
b. Stockport, Cheshire 11 Jan. 1871 *d.* Stockport, Cheshire 10 Oct. 1938
Hatter. Journalist on *Stockport Advertiser* and *Echo*. Botanical referee for Manchester Field Naturalists' Society.
N. Western Nat. 1938, 234–35. *Stockport Advertiser* 14 Oct. 1938.

MacDONALD, John (*fl.* 1890s)
Superintendent of Works, Port Moresby, New Guinea, 1897. Collected mosses on Owen Stanley Range.
Annual Rep. Br. New Guinea 1897–98, 98; 1898–99, 41–45. *Fl. Malesiana* v.1, 1950, 335.
Plants at BM(NH).

McDONELL, J. C. (*fl.* 1890s)
Indian Forest Service. Collected plants in Kashmir and Punjab, *c.* 1890–95. Interested in ferns.
Pakistan J. For. 1967, 351.

McDOUALL, Kenneth (*c.* 1870–1945)
d. 17 May 1945
Created garden of interesting plants at Logan, Wigtownshire.
Gdnrs Chron. 1945 ii 34.

McEACHARN, Neil Boyd (1885–1964)
b. 28 Oct. 1885 *d.* 1964
FLS 1953. Created gardens at Villa Taranto, Lake Maggiore, now the property

of the Italian government. Grew plants particularly from N. America and Asia. *Villa Taranto* 1954. *Catalogue of Plants in Gardens of Villa Taranto* 1963.

Kew Bull. 1955, 319–20. *Proc. Linn. Soc.* 1966, 120–21.

McELROY, John F. (*c.* 1818–1887)
d. 9 Jan. 1887

Gardener to A. J. Lewis of Camden Hill, Kensington, London. Secretary, United Horticultural Benevolent and Provident Society. Contrib. to *Gdnrs Mag.*

Gdnrs Chron. 1887 i 87. *J. Hort. Cottage Gdnr* v.14, 1887, 29.

MacELWEE, Alexander (1869–1923)
b. Glasgow 28 Jan. 1869 *d.* 23 Jan. 1923

Emigrated with parents to U.S.A., 1883. Curator of botanical collections, Philadelphia College of Pharmacy, 1894. Curator, Herb., Philadelphia Botanical Club; President, 1918–23. Landscape gardener, 1904.

J. W. Harshberger *Botanists of Philadelphia and their Work* 1899, 399–401. *Bartonia* v.9, 1926, 1–6 portr.

MacENCROE, Demetrius (*alias* De la Croix) (*fl.* 1720s)
b. Ireland

MD. *Connubia Floram Latino Carmine Demonstrata* prefixed to S. Vaillant's *Botanicon Parisiense* 1727; published separately in Paris, 1728; ed. 2 Bath, 1791.

F. Atterbury *Epistolary Correspondence and Miscellanies* v.4, 1783–98, 167.

M'EVOY, John (*fl.* 1780s–1790s)

Nurseryman, Collon, County Louth, Ireland.

Irish For. 1967, 55.

MACFADYEN, Allan (1860–1907)
b. Glasgow 26 May 1860 *d.* Hampstead, London 1 March 1907

MD Edinburgh 1886. Lectured on bacteriology, Lister Institute, London; Director, 1891. Fullerian Professor of Physiology, Royal Institution, 1901–4. *The Cell as the Unit of Life* 1908 portr. (memoir, x–xvi).

Nature v.75, 1907, 443. *D.N.B. Supplt 2* v.2, 519–20. *Who was Who, 1897–1916* 451. *R.S.C.* v.16, 955.

MACFADYEN, James (1800–1850)
b. Glasgow 1800 *d.* Jamaica 1850

MD Glasgow 1821–22. FLS 1838. Island Botanist, Jamaica, 1826–28. Established Jamaica Botanic Garden. *Fl. Jamaica* 1837 (v.2 never published but incomplete copies at BM(NH) and Kew). 'List of Plants growing in Plain of Liguanea' (*Jamaica Almanack 1842* part 1, 1–18). *Description of Nelumbium jamaicense* 1847.

Proc. Linn. Soc. 1851, 135–36. *J. Inst. Jamaica* v.1, 1892, 141. C. S. Sargent *Silva of N. America* v.2, 1892, 73. I. Urban *Symbolae Antillanae* v.1, 1898, 99, 100; v.3, 1902, 79–80. W. Fawcett and A. B. Rendle *Fl. Jamaica* v.5, 1926, xiv. F. A. Stafleu *Taxonomic Literature* 1967, 294–95. *R.S.C.* v.4, 157.

Plants, letters and portr. by J. J. Macfadyen at Kew.

Macfadyena A. DC. *Fadyenia* Endl.

MACFARLAN, A. J. (–1868/9)

MD. Of Edinburgh. Curator, Botanic Society of Edinburgh. 'Nectary of Ranunculus' (*Trans. Bot. Soc. Edinburgh* v.5, 1858, 169–70).

R.S.C. v.4, 157.

MACFARLANE, Rev. George (–1884)
d. Coldingham, Berwick 1884

Berwickshire plants listed in *Trans. Bot. Soc. Edinburgh* v.16, 1886, 26–28, 192.

MACFARLANE, J. L. (1836–*c.* 1913)

Artist and lithographer. Drew floral plates for nurserymen such as W. Bull, T. Cripps, W. Paul and B. S. Williams. Contrib. to horticultural periodicals such as *Florist and Pomologist.*

MACFARLANE, John Muirhead (1855–1943)
b. Kirkcaldy, Fife 28 Sept. 1855 *d.* Lancaster, New Hampshire, U.S.A. 16 Sept. 1943

BSc Edinburgh 1880. FRSE 1885. Lecturer in Botany, Royal Veterinary College, 1881. To U.S.A., 1891. Professor of Botany, University of Pennsylvania, *c.* 1893–1920.

J. W. Harshberger *Botanists of Philadelphia and their Work* 1899, 367–72. *Times* 20 Sept. 1943. *Yb. R. Soc. Edinburgh* 1945, 19–20.

Portr. at Hunt Library.

MacFARLANE, Rev. Samuel (1837–1911)
d. Southport, Lancs 1911

Missionary of London Missionary Society in Lifu, Loyalty Islands, Somerset, New Guinea Mission, 1874. Murray Island, 1877. *Among the Cannibals of New Guinea* 1888.

Fl. Malesiana v.1, 1950, 335–36.

Plants at Melbourne.

Dendrobium macfarlanii F. Muell.

MacGARROCH, James Burgess (1765–1782)
b. Kirkmichael 30 Nov. 1765 *d.* 23 July 1782

Only son of Rev. J. Burgess (1725–1795); took his mother's family name. A lichenologist like his father.

M'GAVEN, D. (*fl.* 1790s)

Nurseryman, Andover, Hants.

McGIBBON, James (*fl.* 1840s–1860s)
b. Inveresk, Midlothian

Kew gardener, 1848. Superintendent, Botanic Garden, Cape Town, 1850. *Catalogue of Plants in Botanic Garden, Cape Town* 1858.

Plants at Kew.

McGILL, Hilton (*fl.* 1900s–1910s)
Sea captain. Planter in Kelantau, 1904–13 where he collected plants for H. N. Ridley at Singapore Botanic Gardens.
Gdns Bull. Straits Settlements 1927, 127.

MacGILLIVRAY, John (1822–1867)
b. Aberdeen 18 Dec. 1822 *d.* Sydney, N.S.W. 6 June 1867
Son of W. MacGillivray (1796–1852). Zoologist. Professional collector of natural history objects on HMS 'Fly', 'Rattlesnake' and 'Herald', 1842–54. *Narrative of Voyage of HMS Rattlesnake* 1852 2 vols. Letters published in *Hooker's J. Bot.* 1853, 279–82; 1854, 353–63.
J. D. Hooker *Fl. Tasmaniae* v.1, 1859, cxvii. B. C. Seemann *Fl. Vitiensis* 1865, vii. *Gdnrs Chron.* 1867, 1027. *J. Bot.* 1867, 316. *D.N.B.* v.35, 91. *Rep. Austral. Assoc. Advancement Sci.* 1909, 379–81 portr. W. MacGillivray *Life of William MacGillivray* 1910, 111. *Austral. Zoologist* 1937, 40–63. *Fl. Malesiana* v.1, 1950, 336–37. *Victorian Nat.* 1950, 79. *Austral. Encyclop.* v.1, 3, 5. *Austral. Dict. Biogr.* v.2, 1967, 167–68. *R.S.C.* v.4, 158.
Plants at BM(NH), Kew. Letters at Kew. MSS. at Mitchell Library, Sydney. MS. Journal of voyage of HMS 'Herald' at Admiralty Library. Portr. at Hunt Library.
Nothopanax macgillivrayi Seem.

MacGILLIVRAY, Paul Howard (1834–1895)
b. Aberdeen 1834 *d.* Bendigo, Victoria 9 July 1895
MA Aberdeen 1854. LLD 1889. FLS 1880. Son of W. MacGillivray (1796–1852). In Australia, 1855. Surgeon, Bendigo Hospital, 1857. *Catalogue of Flowering Plants and Ferns growing in Neighbourhood of Aberdeen* 1853.
Ann. Scott. Nat. Hist. 1895, 262. *J. Bot.* 1895, 383. *Papers Proc. Linn. Soc. N.S.W.* 1895, 624. *Proc. Linn. Soc.* 1895–96, 42–43. *Victorian Nat.* 1895, 48. *R.S.C.* v.4, 158; v.8, 292; v.10, 675; v.16, 958.

MacGILLIVRAY, William (1796–1852)
b. Aberdeen 25 Jan. 1796 *d.* Aberdeen 5 Sept. 1852
MA Aberdeen 1815. LLD 1844. FRSE. Zoologist. Conservator, Royal College of Surgeons Museum, Edinburgh, 1831–41. Professor of Natural History, Aberdeen, 1841. Abridged W. Withering's *Systematic Arrangement of British Plants* 1830. Translated A. Richard's *Nouveau Élémens de Botanique* 1831. *Natural History of Dee-side and Braemar* 1855.
D.N.B. v.35, 90–92. *Gdnrs Chron.* 1900 ii 399–400. W. MacGillivray *Life of W. MacGillivray* 1910. *R.S.C.* v.4, 159; v.8, 292.
Selaginella macgillivrayi Baker.

McGREDY, Samuel (*c.* 1861–1926)
d. Portadown, County Armagh 26 April 1926
Nurseryman, Portadown, 1880. Specialised in roses.
J. Hort. Home Farmer v.63, 1911, 419 portr. *Garden* 1925, ii–iii portr.; 1926, 290 (vi). *Gdnrs Chron.* 1963 ii 196–97.

MACGREGOR, Ann *see* Watt, A.

MacGREGOR, Donald (1877–1933)
b. 24 June 1877 *d.* Havant, Hants 6 Aug. 1933
Kew gardener, 1902. Superintendent, Parks and Gardens, Shanghai, 1904–29. Collected plants.
Gdnrs Chron. 1924 ii 106 portr.; 1929 ii 178 portr.; 1933 ii 134. *J. Kew Guild* 1933, 284.

MacGREGOR, Sir William (1846–1919)
b. Towie, Upper Donside, Aberdeenshire 20 Oct. 1846 *d.* Aberdeen 3 July 1919
MB Aberdeen 1872. MD 1874. KCMG 1889. Medical Officer in Colonial Service in Seychelles, 1873; Mauritius, 1874; Fiji, 1875. Administrator, British New Guinea, 1888–89. Governor, Lagos, 1901–4; Newfoundland, 1904–9; Queensland, 1909–14. Collected plants in New Guinea, Lagos and Labrador, now at Kew. *British New Guinea, Country and People* 1897.
Kew Bull. 1907, 76–78; 1908, 135–37; 1920, 31–32. *Trans. R. Geogr. Soc. Austral.* v.8, 1890, 45–62. *Aberdeen Grammar School Mag.* Oct. 1919, 13–16 portr. *D.N.B. 1912–1921* 357–58. *Who was Who, 1916–1928* 671. *Fl. Malesiana* v.1, 1950, 337–42 portr.; v.5, 1958, 64.
Portr. at Hunt Library.

McGRIGOR, Sir James (1771–1858)
b. Cromdale, Inverness 9 April 1771 *d.* London 2 April 1858
MA Aberdeen 1788. MD Edinburgh 1804. LLD 1826. FRS 1816. Baronet, 1830. KCB 1850. Army surgeon, 1793–1814; Director-General, 1815–51. President, Medical Botanical Society, 1828. Collected plants in Jersey, 1799; Mauritius, *c.* 1820. *Autobiography and Services of Sir J. McGrigor* 1861.
D.N.B. v.35, 102–5. D. G. Crawford *Hist. of Indian Med. Service, 1600–1913* v.1, 1914, 319–20.
Plants at BM(NH).

McGROUTHER, Thomas (1858–1941)
b. Grahamston, Falkirk, Stirling Aug. 1858 *d.* Larbert, Stirling 2 July 1941
Solicitor, Glasgow. President, Falkirk Natural History and Archaeological Society for 12 years. Authority on Falkirk flora.
Glasgow Nat. 1943, 101–2.
Herb. at Glasgow University.

MACHADO, Alfred Dent (–1910)
d. 12 June 1910

Police officer. Miner at Tomo Gold Mines, 1892–93. On staff of Botanic Gardens, Singapore, 1902–3. Planter in Perak, later in charge of United Rubber Estates, Singapore. Collected plants in Perak.

Agric. Bull. Straits and Federated Malay States 1910, 328–29. *Gdns Bull. Straits Settlements* 1927, 127. *Fl. Malesiana* v.1, 1950, 342.

Plants at Singapore.

Borassus machadonis Ridley.

M'HATTIE, John W. (*c.* 1859–1923)
b. Morayshire *d.* 29 April 1923

VMH 1920. Gardener. Superintendent of Parks and Gardens, Edinburgh, 1901.

Gdnrs Mag. 1907, 281 portr., 282. *Gdnrs Chron.* 1923 i 252 portr.

MACHELL, Christopher (1747–1827)
d. Beverley, Yorks 24 Sept. 1827

Lieut.-Colonel, 15th Regiment of Foot. Retired to Beverley. *c.* 1789. Friend of R. Teesdale (*c.* 1740–1804). Studied botany.

Annual Rep. Yorks Philos. Soc. 1906, 54.

Plants at York Museum.

McHUTCHEON, John (1819–1887)
b. Ayrshire 1819 *d.* 26 March 1887

Gardener. Assistant Editor of *Garden.*

Garden v.31, 1887, 294 portr.

McINDOE, James (1836–1910)
b. Renfrewshire 1836 *d.* Dartford, Kent 16 March 1910

VMH 1897. Gardener to Sir J. Pease at Hutton Hall, Guisborough, Yorks, 1874.

Gdnrs Chron. 1903 ii 257–58; 1910 i 191, 203, 206, 207 portr. *J. Hort. Cottage Gdnr* v.47, 1903, 438 portr.; v.60, 1910, 269–70 portr.

M'INTOSH, Charles (1794–1864)
b. Abercairny, Perthshire Aug. 1794 *d.* Murrayfield, Edinburgh 9 Jan. 1864

ALS 1854. Gardener to King of Belgians at Claremont, Esher, Surrey; and at Dalkeith. *Practical Gardener and Modern Horticulturist* 1828–29 2 vols. *Flora and Pomona* 1829. *Flower Garden* 1838. *Greenhouse, Hothouse and Stove* 1838. *Orchard* 1839. *Book of the Garden* 1853–55 2 vols. *Larch Disease* 1860.

Gdnrs Chron. 1864, 50. *Proc. Linn. Soc.* 1863–64, xlii–xliii.

Portr. at Hunt Library.

MacINTOSH, Charles (1839–1922)
b. Inver, Perthshire 27 March 1839 *d.* Inver 5 Jan. 1922

Postman. Studied mosses and fungi. Contrib. to F. B. W. White's *Fl. Perthshire* 1898.

Bot. Soc. Exch. Club Br. Isl. Rep. 1921, 364–65. H. Coates *A Perthshire Nat.* 1923. *J. Bot.* 1922, 188–89. *Trans. Proc. Perth Soc. Nat. Sci.* v.7, 1921–22, 174–78.

McINTOSH, James (1814–1890)
b. London 1814 *d.* Weybridge, Surrey 5 Nov. 1890

Engineer. Had garden at Duneevan, Oatlands Park, Weybridge, Surrey.

Garden v.38, 1890, 473. *J. Hort. Cottage Gdnr* v.21, 1890, 421–22 portr.

McINTOSH, William Carmichael (1838–1931)
d. 1 April 1931

MB, ChB Edinburgh. Professor of Natural History, St. Andrews, 1882. Marine zoologist; also interested in plants.

Trans. Proc. Bot. Soc. Edinburgh 1930–31, 351.

MacINTYRE, Aeneas *see* Addendum

McINTYRE, Archibald (1828–1887)
b. Netherby near Carlisle 19 Oct. 1828 *d.* 4 June 1887

Gardener to Earl of Clare at Mount Shannon near Limerick. Foreman, Kew Gardens. Superintendent, Victoria and Greenwich Parks.

Gdnrs Chron. 1887 i 779–80. *J. Hort. Cottage Gdnr* v.14, 1887, 489.

MacIVOR, William Fordyce *see* Mavor, W. F.

McIVOR, William Graham (–1876)
b. Dollar, Clackmannan *d.* Ootacamund, India 8 June 1876

Kew gardener, 1845. To Madras, 1858. Superintendent, Botanic Garden, Ootacamund, 1848. Acclimatised cinchona in Nilgiris, India. *Hepaticae Britannicae* 1847 (exsiccatae). *Notes on Propagation and Cultivation of Medicinal Cinchonas or Peruvian Bark Trees* 1863.

J. Sowerby and J. E. Smith *English Bot.* 2948. *Gdnrs Chron.* 1876 ii 150. *J. Bot.* 1876, 224. *Trans. Bot. Soc. Edinburgh* v.13, 1879, 11–12.

Plants at Kew.

MACK, John (*fl.* 1820s)

Missionary teacher, Serampore, India. He and his wife presented plants collected in Khasia Hills in 1826 to W. J. Hooker.

I. H. Burkill. *Chapters on Hist. Bot. in India* 1964, 45, 197.

MacKAY, James Townsend (1775–1862)
b. Kirkcaldy, Fife 29 Jan. 1775 *d.* Dublin 25 Feb. 1862

LLD Dublin 1850. ALS 1806. To Dublin, 1804. Founder and Curator, Botanic Garden, Trinity College, 1806–62. *Systematic Catalogue of Rare Plants found in Ireland* 1806. *Catalogue of Plants found in Ireland* 1825. *Fl. Hibernica* 1836. Contrib. to J. Sowerby and J. E. Smith's *English Bot.*

D. Turner *Fuci* v.1, 1801, 116. *Gdnrs Mag.* 1831, 229–30. *Mag. Nat. Hist.* 1831, 167. *J. Hort.* v.2. 1862, 457–58. *Proc. Linn. Soc.*

1861–62, cv. *Notes Bot. School Dublin* no. 1, 1896, 2. *D.N.B.* v.35, 127. N. Colgan *Fl. County Dublin* 1904, xxvi–xxvii. *Notes from R. Bot. Gdn, Edinburgh* v.3, 1904, 94–95. R. W. Scully *Fl. County Kerry* 1916, xii–xiii. *Curtis's Bot. Mag. Dedications, 1827–1927* 119–20 portr. R. L. Praeger *Some Irish Nat.* 1949, 125 portr. *Gdnrs Chron.* v.162(19), 1967, 18 portr.

Herb. at Trinity College, Dublin (N. Colgan and R. W. Scully *Cybele Hibernica* 1898, xxix). Letters at Kew and Winch correspondence at Linnean Society. Portr. at Hunt Library.

Mackaya Harvey.

MacKAY, John (1772–1802)
b. Kirkcaldy, Fife 25 Dec. 1772 *d.* Edinburgh 14 April 1802

ALS 1796. Superintendent, Royal Botanic Garden, Edinburgh, 1800. Collected plants with G. Don in Scotland, 1792. Contrib. to J. Sowerby and J. E. Smith's *English Bot.*

Scots Mag. Feb. 1804. *Notes from R. Bot. Gdn, Edinburgh* 1904, 21–49, 95. H. R. Fletcher and W. H. Brown *R. Bot. Gdn Edinburgh, 1670–1970* 1970, 71–72.

Letters in Winch correspondence at Linnean Society.

MacKAY, John (*fl.* 1820s–1830s)

Nurseryman, Clapton Nursery, Upper Clapton Road, Hackney, London.

Gdnrs Mag. 1829, 379.

MacKAY, John Bain (1795–1888)
b. Echt, Aberdeenshire 5 Feb. 1795 *d.* Totteridge, Herts 9 Aug. 1888

FLS 1824. Nurseryman, Clapton, London.

Garden v.34, 1888, 287.

McKAY, Robert (1889–1964)
b. County Antrim 1889 *d.* 4 May 1914

BSc Royal College of Science, Ireland. Assistant, Plant Disease Division, Dept. of Agriculture, Ireland, 1920; Head of Plant Pathology Dept., 1938. Lecturer, University College, Dublin, 1940; Professor of Plant Pathology, 1945.

Nature v.203, 1964, 124.

MacKELLAR, Archibald Campbell (*c.* 1854–1931)
d. Frogmore, Berks 22 May 1931

VMH 1909. Gardener at Sandringham and Windsor Castle.

Gdnrs Mag. 1907, 723–24 portr. *Gdnrs Chron.* 1923 ii 346 portr.; 1931 i 442 portr.

MacKELLAR, Robert (–1903)
d. 27 Jan. 1903

Gardener, Abney Hall, Cheadle, Staffs.

Garden v.63, 1903, 151 portr.

M'KEN, Mark John (1823–1872)
b. Maxwelltown, Dumfries 1823 *d.* Pietermaritzburg, S. Africa 20 April 1872

Collected plants in Jamaica, 1847 and in Natal. Curator, Natal Botanic Garden, 1851–53, 1860–72. *Ferns of Natal* 1869. *Synopsis Filicum Capensium* (with W. T. Gerrard) 1870.

Gdnrs Chron. 1872, 806. *J. Bot.* 1872, 223–24. A. White and B. L. Sloane *Stapelieae* v.1, 1937, 103. *S. African J. Sci.* 1971, 403–4.

Ferns and letters at Kew.

McKENZIE, Alexander (*fl.* 1840s–1910s)

Manager, Warriston Nursery. Treasurer, Scottish Horticultural Association.

Gdnrs Chron. 1899 i 60 portr. *J. Hort. Cottage Gdnr* v.59, 1909, 271–72 portr.

MACKENZIE, Charles (–*c.* 1841)

MD Edinburgh. FRS 1815. FLS 1812. Contributed article on Botany to Brewster's ed. of *Edinburgh Encyclopaedia* 1830 (signed C.M.).

R.S.C. v.4, 162.

MACKENZIE, Charles (*fl.* 1820s–1830s)

Consul-General, Haiti, 1826. Resident at Havana. Collected plants for W. J. Hooker in Mexico, 1824 and Haiti, 1828. *Notes on Haiti* 1830.

I. Urban *Symbolae Antillanae* v.3, 1902, 80.

Plants at Kew, Göttingen. Letters at Kew.

MacKENZIE, Colin (1754–1821)
b. Island of Lewis 1754 *d.* Calcutta 8 May 1821

2nd Lieutenant, Madras Engineers, 1783; Colonel, 1819. Surveyor-General of India, 1819. Indian antiquary and topographer. His Indian artist drew examples of flora of Mysore.

D.N.B. v.35, 138–39. M. Archer *Nat. Hist. Drawings in India Office Library* 1962, 38–39, 84–86.

Botanical drawings at India Office Library.

MacKENZIE, Daniel (*fl.* 1790s)

One of the first English engravers to devote himself to botanical illustration. Employed by Sir J. Banks on plates of J. Cook's first voyage. Engraved plates in F. Bauer's *Delineations of Exotick Plants...at Kew* 1796–1803, W. Roxburgh's *Plants of Coast of Coromandel* 1795–1819, A. B. Lambert's *Genus Pinus*, etc.

J. für die Botanik 1800, 426–27. *J. Bot.* 1899, 181–82.

MACKENZIE, Francis Humberston *see* Humberston, F. M.

MacKENZIE, Sir George Steuart (1780–1848)
b. 22 June 1780 *d.* Oct. 1848

Of Coul, Ross-shire. Mineralogist. Introduced new varieties of apples. *Choice of Wheat for Seed* 1844. *Short Plea for Advancement of Scottish Husbandry by Science* 1848.
D.N.B. v.35, 149–50.

MACKENZIE, Osgood Hanbury (1842–1922)
d. Inverewe, Wester Ross 15 April 1922
Established notable garden on inhospitable site at Inverewe. *Hundred Years in the Highlands* 1921 contains many references to Ross-shire plants.
Bot. Soc. Exch. Club Br. Isl. Rep. 1922, 707. *J. R. Hort. Soc.* 1950, 436 portr.

McKIBBEN, J. N. (*fl.* **1880s**)
b. N. Ireland?
School teacher, Maryborough, Victoria. Collected plants for F. von Mueller. On botanical expedition to King Island, Bass Strait with C. French and W. A. Sayer in 1887. 'Orchids of Loddon Valley' (*Southern Sci. Rec.* 1883, 100–7).
Victorian Nat. 1888, 140–47; 1949, 104.
Plants at Herb., Melbourne.
Thelymitra mackibbinii F. Muell.

MACKIE, Frederick (*fl.* **1830s–1840s**)
Nurseryman, 10 Exchange Street, Norwich; nursery at Lakenham, Norfolk.
Bot. Mag. 1837, t.3569.

MACKIE, John (*fl.* **1770s–1797**)
Nurseryman, St. Stephen's Road, later 10 Exchange Street, Norwich, *c.* 1773–1797.
J. Harvey *Early Gdning Cat.* 1972, 142–51.

MACKIE, John (–1818)
Son of J. Mackie (–1797). In partnership with his brother, William Aram Mackie, in the family nursery at Norwich.

MACKIE, Robert (*fl.* **1780s–1790s**)
Occupied Kingsland Nursery, Islington from 1787. Later in partnership with Lewis. Owned by T. Bassington from 1800.

MACKIE, Sarah (*fl.* **1770s–1833**)
d. Norwich, Norfolk 4 Sept. 1833
Wife of W. A. Mackie and proprietor of the Norwich nursery after his death in 1817 and that of his brother J. Mackie in 1818.
Gdnrs Mag. 1833, 751.

MACKIE, William Aram (–1817)
Nurseryman in partnership with his brother, John, at Norwich, Norfolk.

McKIMM, Charles (–1908)
Foreman, Belfast Botanic Garden; Curator, 1877–1908.

McLACHLAN, Robert (1837–1904)
b. Upper East Smithfield 10 April 1837 *d.* Lewisham, London 23 May 1904
FRS 1877. FLS 1862. FZS 1881. Primarily a zoologist but also a botanist. Had herb. and collected plants in Australia and China.
Proc. Linn. Soc. 1904–5, 42–43. E. O. Essig *Hist. of Entomol.* 1931, 707–8 portr.

MACLAGAN, Philip Whiteside (1818–1892)
b. Edinburgh 1818 *d.* Berwick, Northumberland 25 May 1892
MD 1840. Son-in-law of G. Johnston (1797–1855). Army surgeon in Canada, 1841–53. 'Plant Collecting in W. Canada' (*Ann. Mag. Nat. Hist.* v.20, 1847, 11–14).
Athenaeum 4 June 1892, 731. *R.S.C.* v.4, 165; v.16, 975. F. Boase *Modern English Biogr.* v.2, 1897, 644.
Plants at Cambridge, Edinburgh, Kew.

McLAREN, Henry Duncan, 2nd Baron Aberconway (1879–1953)
b. 16 April 1879 *d.* Bodnant, Denbighshire 23 May 1953
FLS 1927. VMH 1934. MP 1910–22. President, Royal Horticultural Society, 1931–53. Created gardens at Bodnant, growing especially rhododendrons, camellias and magnolias.
Gdnrs Chron. 1953 i 206. *J. R. Hort. Soc.* 1953, 230–32 portr. *Orchid Rev.* 1953, 105–6. *Proc. Linn. Soc.* 1952–53, 217–18. *Times* 25 May 1953 portr. *Rhododendron Camellia Yb.* 1954, 7–11 portr. *Who was Who, 1951–1960* 2.
Portr. by Sir O. Birley at Royal Horticultural Society.

McLAREN, John (1815–1888)
b. Methven, Perthshire 1815
Gardener to Samuel Whitbread at Cardington, Beds, 1846. Made collection of Bedfordshire plants in 1864.
J. G. Dony *Fl. Bedfordshire* 1953, 23.
Plants at Luton, BM(NH).

McLAREN, John (1846–1943)
b. Stirling 20 Dec. 1846 *d.* San Francisco, U.S.A. 12 Jan. 1943
Gardener on estates in Scotland. To California, 1869 where he continued as gardener. Superintendent of Parks, San Francisco, 1887. *Gardening in California: Landscape and Flower* 1909; ed. 3 1924.
Dict. Amer. Biogr. Supplt. 3, 1973, 490–91.
Portr. at Hunt Library.

McLAREN, Laura, Lady Aberconway (*née* **Pochin**) (–1933)
d. Antibes, France 4 Jan. 1933
VMH 1931. Had garden at Bodnant, Denbighshire where she cultivated new and rare plants. Mother of H. D. McLaren, Lord Aberconway (1879–1953).
Gdnrs Chron. 1933 i 34.

McLAREN, Malcolm Shaw (*c.* 1892–*c.* 1947)

Teacher of French and music, Burford. Entered into partnership in Oxonian Nurseries, Wheatley, Oxfordshire, 1944.
Quart. Bull. Alpine Gdn Soc. 1948. 340.

McLEAN, John (*fl.* 1820s–1830s)
Assistant Superintendent, Botanic Gardens, Sydney, 1829; Acting Superintendent, 1832–33, 1835–36.
Sydney Herald 18 July 1832. *Sydney Gaz.* 28 Feb. 1833. *Proc. Linn. Soc. N.S.W.* v.25, 1900, 789. *J. R. Soc. N.S.W.* 1908, 112. *J. Proc. R. Austral. Hist. Soc.* 1932, 109–10.

MACLEAN, John (*fl.* 1830s–1850s)
Merchant, Lima, 1832–54. Sent plants to W. J. Hooker and W. Herbert. Employed A. Mathews to collect plants.
Hooker's Icones Plantarum 1837, t.109. *Bot. Mag.* 1842, t.3979.
Letters and plants at Kew.
Macleania Hook.

MACLEAR, John Fiot Lee Pearse (*fl.* 1870s)
Captain, Royal Navy. With expedition of surveying ship 'Alert', 1878–82. Collected some plants on Christmas Island, Indian Ocean, now at Kew.
Fl. Malesiana v.1, 1950, 343.
Dicliptera maclearii Hemsley.

MACLEAY, Alexander (1767–1848)
b. Ross-shire 24 June 1767 *d.* Sydney, N.S.W. 18 July 1848
FRS 1809. FLS 1794. Secretary, Linnean Society, 1798–1825. As Colonial Secretary, N.S.W., Botanic Gardens at Sydney came under his supervision. First President, Australian Museum, Sydney, 1836.
Bot. Mag. 1820, t.2153; 1823, t.2377. *Sydney Herald* 22 July 1848. *Proc. Linn. Soc.* 1849, 45–47; 1888–89, 37. J. Hardy *Selections from Correspondence of G. Johnston* 1892, 133. *Macleay Mem. Vol.* 1893. *D.N.B.* v.35, 205. *J. Proc. Linn. Soc. N.S.W.* 1908, 112–13; 1920, 567–635; 1942, 5–7. A. T. Gage *Hist. Linn. Soc. London* 1938 portr. P. Serle *Dict. Austral. Biogr.* v.2, 1949, 93. *Austral. Encyclop.* v.2, 1965, 5, 7.
Letters and portr. at Linnean Society. Portr. at Kew, Hunt Library.
Macleaya R. Br.

MACLEAY, Sir George (1809–1891)
b. London 1809 *d.* Mentone, France 26 June 1891
FLS 1860. CMG 1869. KCMG 1875. To Australia to join his father, A. Macleay, in 1826. Accompanied C. Sturt on expedition in S. Australia, 1829. Had garden of native plants at Brownlow Hill near Sydney. Returned to England in 1859 and settled at Pendell Court, Surrey.
Macleay Mem. Vol. 1893. P. Mennell *Dict. Austral. Biogr.* 1892, 304. *D.N.B.* v.35, 205. *Proc. Linn. Soc. N.S.W.* 1920, 630–35. P. Serle *Dict. Austral. Biogr.* v.2, 1949, 94. *Austral. Encyclop.* v.2, 1965, 3, 5. *J. Proc. R. Austral. Hist. Soc.* v.27, 291–92.

MACLEAY, Sir William John (1820–1891)
b. Wick, Caithness 13 June 1820 *d.* Sydney, N.S.W. 7 Dec. 1891
Nephew of A. Macleay. To Australia as farmer 1839. Conducted expedition to New Guinea, 1875, which yielded botanical and zoological specimens. First President, Linnean Society, N.S.W. Knighted in 1889 for services to science and politics. Contrib. to *Proc. Linn. Soc., N.S.W.*
Nature v.13, 1875, 153–54. *Gdnrs Chron.* 1876 i 52. *Proc. Linn. Soc., N.S.W.* 1891, 707–16; 1904, 5–23; 1925, 1–5, 185–272; 1942–43, 9–15. *Macleay Mem. Vol.* 1893, xii–li. *D.N.B.* v.35, 1893, 206. P. Serle *Dict. Austral. Biogr.* v.2, 1949, 94–96. S. D. Macmillan *Squatter went to Sea; Story of Sir W. Macleay's New Guinea Expedition 1875 and his Life in Sydney* 1957. *Fl. Malesiana* v.1, 1950, 343–44.
Plants at Melbourne. MSS. at Linnean Society, N.S.W. and University.

MACLEAY, William Sharp (1792–1865)
b. London 30 July 1792 *d.* Sydney, N.S.W. 26 Jan. 1865
BA Cantab 1814. FLS 1821. Joined his father, A. Macleay (1767–1848), in N.S.W., 1839. Established gardens at Elizabeth Bay, Sydney. 'On Identity of certain General Laws regulating Distribution of Insects and Fungi' (*Trans. Linn. Soc.* v.14, 1825, 46–68).
Proc. Linn. Soc. 1864–65, c–ciii. *D.N.B.* v.35, 206–7. *J. Proc. R. Soc. N.S.W.* 1908, 113–14. *Proc. Linn. Soc. N.S.W.* 1920, 591–629; 1942, 7–9. S. Serle *Dict. Austral. Biogr.* v.2, 1949, 96–97. *Austral. Encyclop.* v.2, 1965, 25.
Letters at Linnean Society and at Mitchell Library, Sydney.

M'LEISH, Alexander (–1828)
d. Dublin 15 Dec. 1828
Landscape gardener and nurseryman, Dublin, *c.* 1814.
Gdnrs Mag. 1829, 112.

McLEOD, Daniel (–1866)
Assistant Curator, Kew Gardens, 1858–64. Went to Cachar, Assam to work on tea plantations of T. McMeekin.
J. Kew Guild 1895, 29.

McLEOD, David (*fl.* 1850s–1932)
b. Berwickshire
Gardener to S. Gratrix at Manchester.

Set up business as commercial grower.
Orchid Rev. 1932, 310.

McLEOD, James Findlay (1863–1950)
b. Inverness-shire 1863
VMH 1929. Gardener at Dover House, Roehampton, Surrey, 1889.
Gdnrs Chron. 1910 i 365 portr.

McLEOD, John (1788–1849)
b. Stornoway, Isle of Lewis 1788 *d.* Hochelaga, Lower Canada 24 July 1849
Fur-trader for North West Company before 1821 and then Hudson's Bay Company; retired in 1842. Friend of W. F. Tolmie and sent him plants from Snake River and Jackson Hole area, many of which are listed in the Supplt. to W. J. Hooker and G. A. W. Arnott's *Bot. of Captain Beechey's Voyage* 1830–41.
Encyclop. Canada v.4, 1936, 209. S. D. McKelvey *Bot. Exploration of Trans-Mississippi West, 1790–1850* 1955, 627–635.
Letters in British Columbia archives at Victoria.

McLEOD, Roderick G. (*fl.* 1870s)
Of Liverpool. Cryptogamist. Had herb.
Trans. Liverpool Bot. Soc. 1909, 81.

MACLOSKIE, Rev. George (1834–1920)
b. Castledawson, Tyrone 14 Sept. 1834 *d.* 4 Jan. 1920
LLD London 1871. DSc Belfast 1874. Minister, Ballygoney, 1861–74. Professor of Natural History, Princeton, New Jersey, 1874–1906. Studied flora of Patagonia. *Elementary Botany* 1893.
J. W. Harshberger *Botanists of Philadelphia and their Work* 1899, 293–94. *Nature* v.104, 1920, 540–41. *Science* 1920, 180–81. *R.S.C.* v.8, 300; v.10, 683; v.12, 471.
Viola macloskeyi F. E. Lloyd.

McLOUGHLIN, John (–1857)
d. 3 Sept. 1857
Chief Factor, Hudson's Bay Company, 1824–46. Awarded silver medal of Horticultural Society of London, 1826 for assistance to botanical collectors.
Madrono 1958, 268–72 portr.

McLUCKIE, John (1890–1956)
b. Killermont, Dunbartonshire 12 Aug. 1890 *d.* Sydney, N.S.W. 27 Sept. 1956
MA Glasgow. DSc Sydney 1923. Dept. of Botany, Sydney, 1915. One of the founders of plant ecology in Australia. *Australia and New Zealand Botany* (with H. S. McKee) 1954. Contrib. to *Proc. Linn. Soc. N.S.W.*
Who's Who in Australia 1950, 471. *Austral. J. Sci.* 1956, 109–10. *Nature* v.178, 1956, 1148–49.

MACLURE, William (1763–1840)
b. Ayr 1763 *d.* San Angel, Mexico 23 March 1840
To U.S.A., 1796. Agriculturist, geologist. President, Philadelphia Academy of Natural Sciences, 1817–39.
S. G. Martin *Mem. William Maclure* 1841. *Rep. U.S. National Mus.* 1904, 217–21, 705 portr.
Maclura Nutt.

McMAHON, Bernard (*c.* 1775–1816)
b. Ireland *c.* 1775 *d.* Philadelphia, U.S.A. 16 Sept. 1816
Went to U.S.A., 1796. Nurseryman, Philadelphia, 1809. Main distributor of seeds to Europe. *American Gardeners' Calendar* 1806; ed. 11 1857.
T. Nuttall *Genera of N. American Plants* v.1, 1818, 211–12. C. S. Sargent *Silva of N. America* v.7, 1895, 86–87. J. W. Harshberger *Botanists of Philadelphia and their Work* 1899, 117–19. L. H. Bailey *Standard Cyclop. Hort.* v.2, 1939, 1586. U. P. Hedrick *Hist. Hort. in America to 1860* 1950, 197–203. *J. Soc. Bibl. Nat. Hist.* 1960, 363–80.
Mahonia Nuttall.

MacMAHON, Philip (1857–1911)
b. Dublin 13 Dec. 1857 *d.* Fraser Island, Queensland 14 April 1911
Kew gardener, 1881. Curator, Botanic Garden, Hull, 1882. On botanical expedition to Central America. In India, 1887. To Victoria, 1888. Curator, Botanic Garden, Brisbane, 1889. Director of Forests, Queensland, 1905.
P. Mennell *Dict. Austral. Biogr.* 1892, 306. *Gdnrs Chron.* 1911 i 358. *J. Kew Guild* 1911–12, 49 portr.

M'MILLAN, Alexander (*c.* 1835–1924)
d. 7 April 1924
Gardener to Sir W. Jardine at Jardine Hall, Lockesbie; at Trinity Cottage, Edinburgh. Hybridised rhododendrons.
Gdnrs Mag. 1910, 463–64 portr. *Gdnrs Chron.* 1924 i 230.

MACMILLAN, Rev. Hugh (1833–1903)
b. Aberfeldy, Perthshire 17 Sept. 1833 *d.* Edinburgh 24 May 1903
LLD St. Andrews 1871. DD Edinburgh 1879. FRSE 1871. Minister, Free Church, Glasgow, 1864–78; Greenock, 1878–1901. Moderator of General Assembly of Free Church, 1897. *First Forms of Vegetation* 1861. *Bible Teachings in Nature* 1867. *Rambles in search of Alpine Plants* 1869. *Ministry of Nature* 1871. *Poetry of Plants* 1902.
D.N.B. Supplt 2 v.2, 543–44. *Who was Who, 1897–1916* 461. *R.S.C.* v.4, 169; v.10, 684; v.16, 982.

MACMILLAN, Hugh Fraser (1869–1948)
b. Glen Urquhart, Inverness 4 June 1869 *d.* Ealing, Middx 19 Nov. 1948
Kew gardener, 1893. At Royal Botanic, Gardens, Peradeniya, Ceylon, 1895; Curator 1912–25. *Handbook of Tropical Gardening and Planting with Special Reference to Ceylon* 1910; ed. 5 1962. Contrib. to *J. Tropical Agric.*
Kew Bull. 1895, 155. *J. Kew Guild* 1948, 697.

McNAB, Catherine Mary (1809–1857)
b. Richmond, Surrey 13 Feb. 1809 *d.* Dailly, Ayrshire 1857
Eldest daughter of W. McNab (1780–1848). *Botany of the Bible* 1850–51. Prepared sheets of *Object Lessons in Botany.*
Notes from R. Bot. Gdn, Edinburgh v.3, 1908, 323.

McNAB, Gilbert (1815–1859)
b. Edinburgh 26 Nov. 1815 *d.* St. Ann's, Jamaica 21 Jan. 1859
MD Edinburgh 1836. Son of W. McNab (1780–1848). In Shetland, 1837. To Jamaica, 1838. Assisted J. Macfadyen in his *Fl. Jamaica.*
Trans. Bot. Soc. Edinburgh v.6, 1860, 354–55. I. Urban *Symbolae Antillanae* v.3, 1902, 80. *Notes from R. Bot. Gdn, Edinburgh* v.3, 1908, 321–22.
Plants at Kew, Edinburgh, Glasgow, Oxford. Letters at Kew.

McNAB, James (1810–1878)
b. Richmond, Surrey 25 April 1810 *d.* Edinburgh 19 Nov. 1878
Son of W. McNab (1780–1848) whom he succeeded as Curator of Royal Botanic Garden, Edinburgh, 1849. Collected plants in N. America, 1834 (*Edinburgh New Philos. J.* 1835, 56–64). Superintendent, Caledonian Horticultural Society, 1835. President, Edinburgh Botanical Society, 1872. Drew plates for *Bot. Mag.* (t.2930, 3025, 3190, 3746, 3888, 3903, 3964).
Gdnrs Chron. 1871, 1033 portr.; 1878 ii 661 portr. *Garden* v.12, 1877, 16 portr.; 1878, 459 portr. *J. Bot.* 1878, 382. *Trans. Bot. Soc. Edinburgh* 1879, 381–83. C. S. Sargent *Silva of N. America* v.10, 1896, 110. S. D. McKelvey *Bot. Exploration of Trans-Mississippi West, 1790–1850* 1955, 1094–95. H. R. Fletcher and W. H. Brown *R. Bot. Gdn Edinburgh, 1670–1970* 1970, 138–48 portr. *R.S.C.* v.4, 170; v.8, 300; v.10, 685; v.12, 471.
Plants at Edinburgh. Portr. at Hunt Library.
Cupressus macnabiana A. Murray.

MACNAB, Robert (*fl.* 1830s–1840s)
Of Perth. Gardener at Kinfauns Castle, 1836 and Ruthven House, 1841. Contrib. notes on botany of Perthshire in *Perthshire Courier* 1836–40 (reprinted in *Trans. Proc. Perthshire Soc. Nat. Sci.* v.7, 1920, 71–74). *North British Cultivator* 1842.

McNAB, William (1780–1848)
b. Dailly, Ayrshire 12 Aug. 1780 *d.* Edinburgh 1 Dec. 1848
ALS 1825. Kew gardener, 1801–10. Superintendent, Royal Botanic Garden, Edinburgh, 1810–48. *Hints on Planting and General Treatment of Hardy Evergreens in the Climate of Scotland* 1830. *Treatise on Propagation, Cultivation and General Treatment of Cape Heaths* 1832.
Gdnrs Mag. 1832, 210–11. *Gdnrs Chron.* 1848, 812. *Cottage Gdnr* v.1, 1849, 165–66. *Bot. Gaz.* 1849, 53–56. *Proc. Linn. Soc.* 1849, 52. *Chambers' J.* v.13, 1850, 60–61. *Notes from R. Bot. Gdn Edinburgh* 1908, 293–324 portr. H. R. Fletcher and W. H. Brown *R. Bot. Gdn Edinburgh, 1670–1970* 1970, 80–90 portr. *R.S.C.* v.4, 170.
Herb. passed to W. R. McNab. Portr. at Kew, Hunt Library.
McNabia Benth.

McNAB, William Ramsay (1844–1889)
b. Edinburgh 9 Nov. 1844 *d.* Dublin 3 Dec. 1889
MD Edinburgh 1866. FLS 1877. Son of J. McNab (1810–1878). Professor of Natural History, Royal Agricultural College, Cirencester. Professor of Botany, Royal College of Science, Dublin, 1872–89. Scientific Superintendent of Botanic Gardens, Glasnevin. *Botany: Outlines of Classification of Plants* 1878. *Botany: Outlines of Morphology and Physiology* 1878.
Gdnrs Chron. 1889 ii 670. *Ann. Bot.* 1888–89, 477–79. *J. Bot.* 1890, 51–52. *Nature* v.41, 1889–90, 159–60. *Proc. Linn. Soc.* 1889–90, 97–98. *Scott. Nat.* 1890, 283–84. *D.N.B.* v.35, 238. *R.S.C.* v.8, 301; v.10, 685; v.12, 471; v.16, 983.
Herb. at National Museum, Dublin. Letters at Kew.

McNAIR, James (*fl.* 1880s–1890s)
At Hope Nurseries, Jamaica. Superintendent, Botanic Garden, Lagos, Nigeria, 1887–90. Curator, Botanic Station, British Honduras, 1892.
Kew Bull. 1896, 101–3. *Kew Bull. Additional Ser. 9* 1922, 25–26.

McNAIR, John Frederick Adolphus (–1910)
Colonial Engineer and Surveyor-General, Straits Settlements. H.M. Commissioner, Perak. Lieut.-Governor, Penang, 1881–82. Collected specimens of timber trees for N. Cantley. *Perak and the Malays* 1878.
Gdns Bull. Straits Settlements 1927, 127. *Fl. Malesiana* v.1, 1950, 344.

McNIVEN, Charles (*fl.* 1770s–1815)
d. 8 May 1815
Nurseryman in partnership with his brother, Peter, at Alport Lane, Manchester.

McNIVEN, Peter (*fl.* 1780s–1818)
d. 9 Jan. 1818
Nurseryman in partnership with his brother, Charles, at Alport Lane, Manchester.

MACOUN, James Melville (1862–1920)
b. Belleville, Ontario 1862 *d.* Ottawa 8 Jan. 1920
Educ. Albert University, Belleville. FLS 1914. Son of J. Macoun (1831–1920). Chief of Biological Division, Geological Survey of Canada, 1888–1917. Assisted his father with *Catalogue of Canadian Plants* 1883–1902. *Check list of Canadian Plants* 1889. *List of Plants of Pribilof Islands* 1899. Contrib. to *Ottawa Nat.*
Bot. Gaz. v.70, 1920, 240–42 portr. *Canadian Field Nat.* 1920, 38–40 portr. *J. Bot.* 1921, 149. *Bot. Notiser* 1940, 308.
Plants at BM(NH), Kew.
Papaver macounii Greene.

MACOUN, John (1831–1920)
b. Maralin, County Down 17 April 1831 *d.* Sidney, British Columbia 18 July 1920
MA Syracuse. FRS Canada. FLS 1886. To Ontario, *c.* 1850. Farmed until 1857. Schoolteacher until 1874. Professor of Botany, Albert College, Belleville, Ontario, 1874. Professor of Natural History, 1875–79. Botanist to Dominion Government, 1881. Assistant Director and Naturalist to Geological Survey, 1887. *Catalogue of Canadian Plants* 1883–1902. *Autobiography* 1922 portr.
Canadian Field Nat. 1920, 110–14 portr. *Bot. Gaz.* v.71, 1921, 236–37 portr. *Bryologist* 1921, 39–41. *J. Bot.* 1921, 149. N. Polunin *Bot. Canadian Eastern Arctic* 1940, 14. *Bot. Notiser* 1940, 319–20. *Who was Who, 1916–1928* 687. J. Ewan *Rocky Mountain Nat.* 1950, 257–58. *R.S.C.* v.8, 302; v.10, 685; v.12, 471.
Plants at BM(NH), Kew, Ottawa. Portr. at Hunt Library.
Oreocarya macounii Eastwood.

MacOWAN, Peter (1830–1909)
b. Hull, Yorks 13 Nov. 1830 *d.* Uitenhage, S. Africa 1 Dec. 1909
BA London 1857. DSc Cape 1901. FLS 1885. To Grahamstown, 1861. Director, Botanic Garden, Cape Town, 1881–91. Government Botanist, 1892–95. *Herb. Norm. Austro-Africanum* (with H. Bolus) 1884 (exsiccatae). *Plants that furnish Stock Food at the Cape* 1887. *Collecting and Preserving of Botanical Specimens* 1893. *Manual of Practical Orchard-work at the Cape* 1896.
J. Bot. 1872, 159; 1910, 64. *Gdnrs Chron.* 1910 i 57. *Kew Bull.* 1910, 84–90. *Proc. Linn. Soc.* 1909–10, 92–93. *S. African J. Sci.* 1910, 71–79 portr. R. Marloth *Fl. S. Africa* v.1, 1913, ix–x portr. A. White and B. L. Sloane *Stapelieae* v.1, 1937, 112 portr. J. Hutchinson *Botanist in S. Africa* 1946, 643–44. *Bothalia* 1950, 36–37. *R.S.C.* v.10, 685; v.12, 472; v.16, 985.
Plants at Albany Museum Grahamstown, Kew. Portr. at Hunt Library.
Macowania Oliver. *Macowanites* Kalchbr.

McPHAIL, James (*fl.* 1770s–1800s)
Gardener to Earl of Liverpool at Addiscombe Place near Croydon. *Treatise on Culture of Cucumber* 1803; ed. 2 1795. *Gardener's Remembrancer* 1803; new ed. 1819.
J. C. Loudon *Encyclop. Gdning* 1822, 1280. G. W. Johnson *Hist. English Gdning* 1829, 253–54.

MacPHERSON, Alexander
(*fl.* 1870s–1890s)
Gardener, Queensland Board of Enquiry into Causes of Disease in Livestock and Plants, 1877–80. Attendant, Brisbane Museum, 1885–91. Collected plants near Stanthorpe, Queensland.
Pugh's Almanac, Queensland Directory and Men of the Time 1886, 1888–91. *Rep. Austral. Assoc. Advancement Sci.* 1909, 381.
Saccolabium macphersonii F. Muell.

MacPHERSON, Ann (*fl.* 1820s)
Seedsman, 166 Oxford Street, London.

McPHERSON, Donald (1886–1917)
d. Leith, Midlothian 11 Nov. 1917
BSc Edinburgh. Had herb. while at University.
J. Ecol. 1918, 93.

MACRAE, James (–1830)
d. Ceylon June 1830
Gardener. At Botanic Garden, St. Vincent, *c.* 1823. Collected plants for Horticultural Society of London in Sandwich and Galapagos Islands, Chile and Brazil. Superintendent, Botanic Gardens, Ceylon, 1827–30.
Trans. Hort. Soc. London v.5, 1824, v–vi; v.6, 1826, iii–iv. A. Lasègue *Musée Botanique de B. Delessert*, 1845. 455. I. Urban *Symbolae Antillanae* v.3, 1902, 80. H. Trimen *Handbook Fl. Ceylon* v.5, 1900, 374. C. F. P. von Martius *Fl. Brasiliensis* v.1(1), 1906, 49. *J. R. Hort. Soc.* 1955, 275.
Plants at Cambridge. Journal of Sandwich Islands, 1824–26, at Royal Horticultural Society.
Macraea Lindl.

McRAE, William (1878–1952)
b. Scotland 26 May 1878 *d.* Edinburgh 8 July 1952
DSc Edinburgh. FRSE 1936. Mycologist, Imperial Institute of Agricultural Research, Pusa, India. Contrib. to *Mem. Dept. Agric. India*, *Indian J. Agric. Sci.*
Nature v.170, 1952, 561. *Yb. R. Soc. Edinburgh* 1953, 37. *Who was Who, 1951–1960* 721.

MACREE, John (–1804)
Paisley muslin worker; raised the laced pink, 'Paisley Gem'.
C. O. Moreton *Old Carnations and Pinks* 1955, 33–34.

MACREIGHT, Daniel Chambers (1799–1857)
b. Armagh, Ireland 1799 *d.* Jersey, Channel Islands 10 Dec. 1857
BA Dublin 1820. MD 1827. FLS 1833. Lecturer, Middlesex Hospital. Founder member of Botanical Society of London, 1836. Collected plants in Ireland (J. Sowerby and J. E. Smith *English Bot.* 2770), in Pyrenees (R. Sweet *Br. Flower Gdn* 1838, t.202). *Manual of British Botany* 1837.
G. H. Brown *Lives of Fellows R. College of Physicians 1826–1925* 1955, 4. *Berichte Schweizerischen Bot. Gesellschaft* v.50a, 1940, 303.
Macreightia A. DC.

MacRITCHIE, Rev. William (1754–1837)
b. Clunie, Perthshire 1754 *d.* Clunie 6 Dec. 1837
Correspondent of R. Brown. Found *Hutchinsia alpina* at Ingleboro'. *Diary of Tour through Great Britain* 1795 (publ. 1897; biogr.).
J. Bot. 1863, 359; 1876, 52. *Proc. Perthshire Soc. Nat. Sci.* v.4, 1907–8, cxc–cxci. *R.S.C.* v.4, 172.
Plants at BM(NH).

MacSELF, Albert James (*c.* 1879–1952)
d. Reading, Berks 6 March 1952
VMH 1950. Editor, *Amateur Gdning* 1926–46.
Br. Fern Gaz. 1952, 6–7. *Gdnrs Chron.* 1951 ii 60 portr.; 1952 i 92.

MACVICAR, Rev. John Gibson (1801–1884)
b. Dundee, Angus 16 March 1801 *d.* Moffat, Dumfries 12 Feb. 1884
MA St. Andrews. Lecturer in Natural History, St. Andrews, 1827. Pastor, Scottish Church in Ceylon, 1839–52. Minister, Moffat, 1853. 'Vegetable Morphology' (*Trans. Bot. Soc. Edinburgh* v.6, 1860, 401–18).
Trans. Bot. Soc. Edinburgh v.16, 1886, 95–98. *D.N.B.* v.35, 285–86. *R.S.C.* v.4, 172; v.8, 302.

MACVICAR, Symers Macdonald (1857–1932)
b. Moffat, Dumfries 27 Dec. 1857 *d.* Invermoidart, Isle of Shona 27 Feb. 1932
LRCP. LRCS Edinburgh. Authority on hepatics. *Census Catalogue of British Hepatics* 1905. *Revised Key of Hepatics of British Islands* 1906. *Student's Handbook of British Hepatics* 1912; ed. 2 1926. Contrib. to *Ann. Scott. Nat. Hist.*
Bot. Soc. Exch. Club Br. Isl. Rep. 1932, 83. *J. Bot.* 1932, 258–61 portr. *Rep. Br. Bryol. Soc.* 1932, 82–83. *Rev. Bryologique* 1932, 165–68.
Herb. at BM(NH).
Macvicaria Nicholson.

McWEENEY, Edmund Joseph (1864–1925)
b. Dublin 1864 *d.* Dublin 20 June 1925
MD 1891. Professor of Pathology, Catholic University, Dublin, 1891. Studied fungi of Dublin area. Discovered *Stysanus ulmariae.* Contrib. to *Irish Nat.*
Br. Med. J. 1925 ii 41. *Lancet* 1925 i 1371. R. L. Praeger *Some Irish Nat.* 1949, 127.

McWILLIAM, James Ormiston (1808–1862)
d. London 4 May 1862
Assistant Surgeon, Royal Navy 1829. Surgeon, SS 'Scout', W. coast of Africa, 1836. Surgeon, SS 'Albert', Niger Expedition, 1841. Mission to Cape Verde Islands, 1845. Collected plants in Australia when the ships he served in were in Australian waters. *Medical History of Niger Expedition* 1843.
D.N.B. v.35, 287–88. *J. Proc. R. Soc. N.S.W.* 1908, 114. *Rep. Austral. Assoc. Advancement Sci.* 1911, 231–32.
Plants at BM(NH), Botanic Gardens, Sydney.

MADDEN, Bartholomew (*fl.* 1770s)
Nurseryman, Dublin.
Irish For. 1967, 48.

MADDEN, Edward (1805–1856)
b. Ireland 1805 *d.* Edinburgh June 1856
FRSE. President, Botanical Society of Edinburgh, 1853. Officer in Bengal Artillery, 1830–49. Sent seeds to Botanic Gardens, Glasnevin, 1841–49. Collected plants in Aden, Suez, Cairo and Malta from 1849. 'Diary of Excursion to the Shatool and Boorum Passes over the Himalayas, 1845' (*J. Asiatic Soc. Bengal* v.15, 1846, 79–135). 'Notes of Excursion to Pindree Glacier' (*J. Asiatic Soc. Bengal* v.16, 1847, 226–66).
Trans. Bot. Soc. Edinburgh v.5, 1856, 116–40. *Rec. Bot. Survey India* v.7, 1914, 7. *J. R. Hort. Soc.* 1972, 203–6. *R.S.C.* v.4, 173.
Plants, letters and MSS. at Kew.
Maddenia Hook. f. & Thomson.

MADDEN, Francis (*fl.* 1780s–1790s)
Nurseryman, Ballinasloe, County Galway.
Irish For. 1967, 53.

MADDEN, Michael (*fl.* 1760s–1790s)
Nurseryman, Ballinasloe, County Galway.
Irish For. 1967, 53.

MADDOCK, James (*c.* 1715–1786)
d. 24 Sept 1786
Quaker said to have moved from Warrington, Lancs, *c.* 1765–75. Founded Walworth Nursery, London. *Florist's Directory or a Treatise on Culture of Flowers* 1792. Nursery acquired by his son, James (1763–1825).
J. Harvey *Early Gdning Cat.* 1972, 35–36. *Trans. London Middlesex Archaeol. Soc.* v.24, 1973, 191.

MADDOCK, James (1763–1825)
d. April 1825
Son of J. Maddock (*c.* 1715–1786) whose nursery at Walworth he inherited. In 1792 disposed of nursery to Goring and Wright.
Bot. Mag. 1797, t.379. *Trans. London Middlesex Archaeol. Soc.* v.24, 1973, 191.

MADDOCK(S), John (*fl.* 1790s–1830s)
Nurseryman at Lampblack Hill, Bristol, 1795–1816; moved to site near Stokes Croft.

MADDOX, John (*c.* 1749–1828)
d. Oxford 8 April 1828
Gardener at Christ Church, Oxford. Assisted W. Baxter in *Br. Phaenogamous Bot.* 1834–43 (v.6, 1843, 415).

MADDOX, Richard Leach (1816–1902)
b. Bath, Somerset 4 Aug. 1816
MD Aberdeen 1851. Photomicrographer. Practised medicine in Constantinople, 1847–55, at Ryde and Southampton. 'Mucor' (*Monthly Microsc. J.* 1869, 140–47). 'Cultivation of Micro-fungi' (*Mon. Microsc. J.* 1870, 14–24).
J. R. Microsc. Soc. 1903, 530–32. *R.S.C.* v.16, 990.

MAFFETT, Isabella (1842–1907)
b. Glasslough, County Monaghan 1842 *d.* Dublin 23 April 1907
Mother Superior, Church House, Clyde Road, Dublin. Had herb. (*see* S. A. Stewart and T. H. Corry *Fl. N.E. Ireland* 1888, xv).
Annual Rep. Proc. Belfast Nat. Field Club 1911–12, 625; 1913, 625.

MAHER, John (*fl.* 1810s–1830s)
Gardener to D. Beale at Edmonton, Middx, Duke of Norfolk, etc. Contrib. to *Trans. Hort. Soc. London* v.2–4, New Ser. v.1, 1812–35.
J. C. Loudon *Encyclop. Gdning* 1822, 1288.

MAHON, John (1870–1906)
b. Dublin 12 May 1870 *d.* London 6 April 1906
Kew gardener, 1891–97; to Zomba as forester, 1897–99. Curator, Botanic Garden, Uganda until 1903. Collected plants in Uganda. Contrib. to *J. Kew Guild.*
Kew Bull. 1901, 200; 1906, 394–95. *Gdnrs Chron.* 1906 i 256 portr. *J. Kew Guild* 1906, 327–28 portr.
Plants and letters at Kew.
Dissotis mahoni Hook. f.

MAHOOD, Ronald William (*c.* 1882–1970)
d. 20 March 1970
Founded nursery in 1910 at Burscough, Ormskirk, Lancs with his brother who left the business in the 1930s. Mr. Cross and his son, Christopher, became partners.
Gdnrs Chron. 167(15), 1970, 7.

MAIDEN, Joseph Henry (1859–1925)
b. London 25 April 1859 *d.* Sydney 16 Nov. 1925
FRS 1916. FLS 1888. To Australia, 1880. Curator, Technological Museum, Sydney, 1881–96. Director, Botanic Garden and Government Botanist, 1896–1924. Studied Australian flora, especially *Acacia* and *Eucalyptus* and formed National Herb. at Botanic Garden, Sydney. Secretary, Royal Society N.S.W. for 22 years; President, 1896, 1911. President, Linnean Society N.S.W., 1901–2. *Useful Native Plants of Australia* 1889. *Wattles and Wattle-barks* 1890; ed. 3 1906. *Bibliography of Australian Economic Botany* 1892. *Flowering Plants and Ferns of N.S.W.* 1895–98. *Manual of Grasses of N.S.W.* 1898. *Forest Fl. of N.S.W.* 1904–25 8 vols. *Critical Revision of Genus Eucalyptus* 1903–33 8 vols. *Illustrations of N.S.W. Plants* 1907–11. *Sir Joseph Banks* 1909. *Native Fl. N.S.W.* 1911.
Proc. Linn. Soc. 1914–15, 22–23; 1925–26, 84–90. *Austral. Pharm. Notes and News* 1924, 147–48 portr. *Bot. Soc. Exch. Club Br. Isl. Rep.* 1925, 853–54. *Gdnrs Chron.* 1924 ii 400 portr. *J. Bot.* 1926, 138–41. *Victorian Nat.* 1925, 192–94. *Kew Bull.* 1926, 107–10. *Nature* v.117, 1926, 57–58. *Proc. Linn. Soc. N.S.W.* 1926, 4–5; 1930, 355–70 portr. *Proc. R. Soc. B.* 1926, viii portr. *Who was Who, 1916–1928* 693. *J. Proc. R. Austral. Hist. Soc.* v.18(3), 128–29. *Austral Encyclop.* v.1, 1965, 3–6. *R.S.C.* v.16, 1009.
Maidenia Rendle.

MAIDSTONE, Nathanael (*fl.* 1690s–1720s)
Of Barbourne, Worcs. Sent plants from China and India to J. Petiver and seeds to H. Sloane.
E. J. L. Scott *Index to Sloane Manuscripts* 1904, 132. J. E. Dandy *Sloane Herb.* 1958, 160.
Plants at BM(NH).

MAIN, James (*c.* 1775–1846)
b. Edinburgh? *c.* 1775 *d.* Chelsea, London 1846
ALS 1829. Gardener. Collected plants in China for Gilbert Slater, 1792–94 (*Hort. Register* v.5, 1836, 62 *et seq.*); afterwards employed by J. S. Hibbert. Edited *Hort. Register* 1835–36. *Villa and Cottage Florist's Directory* 1830; ed. 2 1835. *Illustrations of Vegetable Physiology* 1833. *Popular Botany* 1835. *Hortus Dietetica* 1845. Contrib. to *Gdnrs Mag.*
Proc. Linn. Soc. 1846, 303–4. A. M. Coats *Quest for Plants* 1969, 96–98. *R.S.C.* v.4, 192.

MAIN, Thomas Wilson (*c.* 1879–1944)
d. 15 April 1944
Kew gardener, 1901. Superintendent, Batu Tiga, Selangor Rubber Centre, 1906. In charge of Hill Garden, Taiping, 1907. Assistant Curator, Botanic Gardens, Singapore, 1908–10. Manager, Charg rubber estates, Malacca, 1910–20. Returned to U.K., 1920. Estates bailiff, Tregenna Park Hotel, St. Ives, 1922–31. Park Superintendent, Finsbury, London, 1935.
Kew Bull. 1906, 88, 383. *Gdns Bull. Straits Settlements* 1927, 127. *J. Kew Guild* 1945, 471. *Fl. Malesiana* v.1, 1950, 345.
Plants at Singapore.

MAINE, S. (*fl.* 1830s)
Seedsman, 449 West Strand, London.

MAINGAY, Alexander Carroll (1836–1869)
b. Great Ayton, Yorks 25 Oct. 1836 *d.* murdered Rangoon 14 Nov. 1869
MD Edinburgh 1858. Indian Medical Service, 1859; to China, 1860. In charge of jail, Malacca, 1862–67. Collected plants in N. China, Burma, Malaysia, etc. Part author of W. Mudd's *Manual of British Lichens* 1861 (*see Trans. Bot. Soc. Edinburgh* 1873, 36–40).
J. Bot. 1870, 63. E. Bretschneider *Hist. European Bot. Discoveries in China* 1898, 679. *Gdns Bull. Straits Settlements* 1927, 108. *Fl. Malesiana* v.1, 1950, 345. *R.S.C.* v.8, 309.
MS. flora of Malacca, drawings, letters and plants at Kew.
Maingaya Oliver.

MAIR, Robert Young (*fl.* 1900s)
Became partner in Prestwick nursery belonging to his father, George Mair, in 1902. Acquired additional land at Glenburn.
Gdnrs Chron. 1934 i 72 portr.

MAITLAND, Thomas Douglas (1885–)
b. Perth 8 May 1885
Kew gardener, 1909. Curator, Botanic Garden, Calabar, S. Nigeria, 1910. District agricultural officer, Uganda, 1913. Chief of Economic Plants Division, Kenya, 1919. Government Botanist, Uganda, 1922. Superintendent, Botanic Garden, W. Cameroons, 1927–31. Collected plants in Uganda. 'Grassland Vegetation of Cameroon Mountain' (*Kew Bull.* 1932, 417–25).
Kew Bull. 1910, 64; 1913, 125; 1921, 171. *J. Kew Guild* 1954, 176–77 portr.
Plants at Kew.
Digitaria maitlandii Hubbard.

MAJOR, Charles Immanuel Forsyth- *see* Forsyth-Major, C. I.

MAJOR, George (*fl.* 1830s)
Nurseryman, Preston Cottage, Leeds, Yorks.

MAJOR, H. (*fl.* 1830s)
Nurseryman, Knosthorpe, Leeds, Yorks.

MAJOR, James (*c.* 1737–1831)
d. Lewisham, London 18 March 1831
Gardener. Raised the earliest laced pinks.
R. Genders *Collecting Antique Plants* 1971, 211.

MAJOR, Joshua (*c.* 1787–1866)
d. Knowsthorpe, Leeds Yorks 26 Jan. 1866
Landscape gardener and nurseryman, Knowsthorpe. *Treatise on Insects most prevalent on Fruit Trees* 1829. *Theory and Practice of Landscape Gardening* 1852. *Ladies' Assistant in the Formation of their Flower Gardens* 1861. Contrib. to *Gdnrs Mag.*
Gdnrs Chron. 1866, 128. *J. Hort. Cottage Gdnr* v.35, 1866, 101–2.

MAKINS, Frederick Kirkwood (–1956)
d. Bruton, Somerset 7 Feb. 1956
FLS 1935. Indian Forest Service, 1912–28. *Identification of Trees and Shrubs* 1936; ed. 2 1948. *Concise Fl. Britain* 1939. *British Trees in Winter* 1945.
Proc. Linn. Soc. 1955–56, 51. *Forestry* 1956, 136.
Herb. at Sexey's Boys School, Bruton.

MALBY, Reginald A. (–1924)
d. 7 Aug. 1924
Photographed flowers and gardens. *Story of my Rock Garden* 1912. *With Camera and Rucksack in the Oberland and Valais* 1913. *Saxifrages or Rockfoils* (with W. Irving) 1914.
Garden 1924, 576(viii).

MALCOLM, Alexander (*fl.* 1800s)
Nurseryman, Stockwell, London.

MALCOLM, Alexander (–1945)
VMH 1928. Hybridised sweet peas.
Gdnrs Chron. 1929 i 112 portr.

MALCOLM, Neill (*fl.* 1890s)
Lieutenant. Collected plants with Capt. M. S. Wellby in Tibet, 1896.
Geogr. J. v.9, 1897, 215–17. *Kew Bull.* 1897, 208. M. S. Wellby *Through Unknown Tibet* 1898. *J. Linn. Soc.* v.35, 1902, 152–55.

MALCOLM, William (*fl.* 1750s–1798)
d. 5 Nov. 1798
Nurseryman, Kennington, near London in 1750s. Moved to larger nursery at Stockwell, 1788. *Catalogue of Hot-house and Green-house...Herbaceous Plants* 1771.
Bot. Mag. 1790, t.129; 1803, t.616. *Gent. Mag.* 1798 ii 1083.

MALCOLM, William (*c.* 1768–1835)
d. Kemnay, Aberdeenshire 13 Sept. 1835
Son? of W. Malcolm (*fl.* 1750s–1798). Nurseryman, Stockwell. Firm became Malcolm and Doughty, 1805–10 and Malcolm and Sweet, 1811; closed in 1815.
Gdnrs Mag. 1835, 720. J. C. Loudon *Arboretum et Fruticetum Britannicum* v.1, 1838, 77, 79. *Trans. London Middlesex Archaeol. Soc.* v.24, 1973, 188–89.
Malcomia R. Br. (possibly refers to his father).

MALLARD, Mrs. (*fl.* 1840s)
As wife of Capt. Mallard visited Port Phillip, Victoria before 1844. Collected algae which she gave to W. H. Harvey.
W. H. Harvey *Nereis Australis* 1847, 40–41. *Victorian Nat.* 1908, 111.
Algae at Trinity College, Dublin.
Polysiphonia mallardiae Harvey.

MALLER, Benjamin (–1884)
b. Ashling, Sussex *d.* Lee, Kent 1 Jan. 1884
Nurseryman, Burnt Ash Lane Nurseries, Lee.
Gdnrs Chron. 1884 i 26.

MALLESON, Rev. Frederic Amadeus (1819–1897)
b. London 19 June 1819 *d.* Broughton-in-Furness, Lancs 14 Nov. 1897
BA Dublin 1853. At Pulborough, Sussex, 1843–46. Vicar, Broughton-in-Furness, 1870. Friend of W. Borrer and J. Ruskin. 'Notes on a Collection of Woods' (*Gdnrs Chron.* 1845, 56–57). MS. flora of Cumberland.
W. Hodgson *Fl. Cumberland* 1898, xxxii–xxxiii. *Naturalist* 1898, 32. *Nature Notes* 1898, 54–55. *Trans. Liverpool Bot. Soc.* 1909, 81. A. H. Wolley-Dod *Fl. Sussex* 1937, xliii. *R.S.C.* v.4, 204.
Letters and MS. flora of Sussex at Kew.

MALONE, Michael (1875–1943)
d. 15 Feb. 1943
Gardener, Botanic Garden, Lister Park, Bradford, Yorks. Studied Cryptogams.
Naturalist 1943, 99–100 portr.

MANDER, Thomas (*fl.* 1840s–1890s)
Nurseryman, Ranelagh Gardens, Leamington Spa, Warwickshire. Successor to I. Cullis. Succeeded by Mr. Greenfield in 1891.

MANDEVILLE, Henry John (1773–1861)
b. Suffolk 1773 *d.* Buenos Aires, Argentine 16 March 1861
H.B.M. Minister at Buenos Aires "...to whom we are indebted for the introduction of this [*Mandevilla*] and many other interesting plants" (J. Lindley).
Bot. Register 1840, t.7. *Bot. Mag.* 1840, t.3797.
Mandevilla Lindl.

MANGHAM, Sydney (1886–1962)
b. London 15 May 1886 *d.* 30 July 1962
BSc Cantab 1908. Demonstrator, Cambridge Botany School, 1908–11. Lecturer in Botany, Armstrong College, Newcastle-upon-Tyne, 1911–20. Professor of Botany, Southampton, 1920–51. *Introduction to Botany* 1926. *Earth's Green Mantle* 1939. Contrib. to *Ann. Bot.*, *New Phytol.*
Nature v.196, 1962, 111–12. *Times* 3 Aug. 1962. *Who was Who, 1961–1970* 746.

MANGLES, George (*fl.* 1830s)
Elder brother of J. Mangles (1786–1867). Arrived in W. Australia, 1829. Superintendent of Government Stock, Swan River Settlement. Collected plants with J. Drummond (*c.* 1784–1863).
J. W. Austral. Nat. Hist. 1909, 19. *Rep. Austral. Assoc. Advancement Sci.* 1926, 43. M. Uren *Land looking West* 1948, 71, 250, 301.

MANGLES, James (1786–1867)
d. Fairfield, Exeter, Devon 18 Nov. 1867
FRS 1825. Entered Royal Navy, 1800; retired as Commander. Visited Swan River Settlement, W. Australia, 1831. James Drummond, Mrs. Molloy and others collected W. Australian plants for him. *Floral Calendar* 1839.
Gdnrs Mag. 1839, 702. *Bot. Register* 1835, t.1703. S. L. Endlicher *Novarum Stirpium Decas* 1839, 25–26. *Proc. R. Geogr. Soc.* v.12, 1868, 229. *Gdnrs Chron.* 1884 ii 311; 1931 i 133. *D.N.B.* v.36, 33–34. *J. W. Austral. Nat. Hist. Soc.* v.2, 1909, 19–20. M. Uren *Land looking West* 1948, 251–52. A. Hasluck *Portr. with Background* 1955 *passim*. *Austral. Encyclop.* v.6, 1965, 117. R. Erikson *Drummonds of Hawthornden* 1969, 17–23, 25–29, 35–36. *Gdn Hist.* v.1(3), 1973, 42–46.
MSS. at W. Australian Archives, Perth.
Manglesia Endl.

MANGLES, James Henry (*c.* 1832–1884)
d. Haslemere, Surrey 24 Aug. 1884
FLS 1874. Son of J. Mangles (1786–1867). Chairman, London and South Western Railway. Had collection of rhododendrons at Valewood near Haslemere. Contrib. to *Gdnrs Chron.* 1881 and *Garden*.
Garden v.26, 1884, 219. *Gdnrs Chron.* 1884 ii 311. *Proc. Linn. Soc.* 1883–86, 106. *Rhododendron Soc. Notes* v.1 (2), 1917, 45–116.
Letters at Kew.

MANGLES, Robert (–1860)
Brother of J. Mangles (1786–1867). Laid out garden at Sunninghill near Ascot. Introduced many W. Australian plants.
Gdn Hist. v.1(3), 1973, 42.

MANN, Gustav (1836–1916)
b. Hanover, Germany 20 Jan. 1836 *d.* Munich, Germany 22 June 1916
Kew gardener, 1859. Collected on Niger Expedition, 1859–62 (*see J. Linn. Soc.* v.6, 1861, 2, 27–30). Indian Forest Service, 1863–91. 'Palms of W. Tropical Africa' (*Trans. Linn. Soc.* v.24, 1864, 421–39).
J. Bot. 1863, 190, 224, 384. *Bull. Soc. Bot. France* v.29, 1882, 184. *Gdnrs Chron.* 1916 i 176 portr. *Kew Bull.* 1907, 247–48; 1916, 237. *J. Kew Guild* 1917, 373–74 portr. *Nature* v.98, 1916, 74. *Curtis's Bot. Mag. Dedications, 1827–1927* 275–76 portr. *Comptes Rendus AETFAT 1960* 1962, 98. *R.S.C.* v.4, 215; v.16, 1042.
MS. journals, letters, portr. and plants at Kew.
Mannia Hook. f. *Manniella* Reichenb.

MANN, Harold Hart (1872–1961)
b. 16 Oct. 1872 *d.* 2 Dec. 1961
BSc Leeds. FLS 1902. Scientific Officer, Indian Tea Association, 1900–7. Principal, Agricultural College, Poona, 1907–18. Director of Agriculture, Bombay, 1918–27. Assistant Director, Woburn Experimental Station, 1928–56. *Pests and Blights of Tea Plants* (with G. Watt) ed. 2 1903. *Tea Soils of Assam and Tea Manuring* 1901.
Nature v.193, 1962, 321. *Who was Who, 1961–1970* 747.

MANN, Robert James *see* Addendum

MANNINGHAM, Rev. Thomas (1684–1750)
b. East Tisted, Hants 1684 *d.* Slinfold, Sussex May 1750
MA Oxon. DD Cantab 1724. Rector, Slinfold, 1711. Prebendary, Westminster, 1720. Friend of J. J. Dillenius. Sent mosses to Linnaeus (Dillenius *Historia Muscorum* 1741, viii, ix). "A very nice botanist" (Petiver in *Philos. Trans. R. Soc.* v.27, 1712, 375–76). Introduced rare plants at Slinfold.
J. Ray *Synopsis Methodica* 1724, preface. D. Turner *Extracts from Lit. Sci. Correspondence of Richard Richardson* 1835, 180. H. Trimen and W. T. T. Dyer *Fl. Middlesex* 1869, 389. *Sussex Archaeol. Coll.* v.33, 1883, 198. G. C. Druce *Fl. Berkshire* 1897, cxxxvi. J. E. Dandy *Sloane Herb.* 1958, 161.
Plants at Oxford.

MANSEL-PLEYDELL, John Clavell (*né* **Mansel**) (1817–1902)
b. Smednore, Dorset 4 Dec. 1817 *d.* Whatcombe, Dorset 3 May 1902
BA Cantab 1839. FLS 1870. Served 30 years in Queen's Own Yeomanry Cavalry. President, Dorset Natural History and Antiquarian Field Club, 1875–1902. *Fl. Dorset* 1874; ed. 2 1895.
J. Bot. 1902, 260–63 portr. *Proc. Linn. Soc.* 1901–2, 39–40. *D.N.B. Supplt. 2* v.2, 562–63. *Who was Who, 1897–1916* 471. *R.S.C.* v.8, 321; v.12, 481; v.16, 1046.
Herb. at County Museum, Dorchester. Plants at Kew. Portr. at Hunt Library.

MANSELL, Lady Catherine Rabey (*née* **Lukis**) (1781–1841)
b. Guernsey, Channel Islands 1781 *d.* Guernsey 1841
Studied conchology and marine botany.
Algae at BM(NH).

MANSELL, William (*c.* 1871–1948)
d. 14 Feb. 1948
With W. H. Hatcher as partner, Mansell took over the nursery of J. W. Moore, Rawden, Leeds, Yorks. Specialised in orchids.
Orchid Rev. 1948, 64.

MANSFIELD, Brendan P. (–1949)
b. Dublin *d.* Christchurch, New Zealand 28 March 1949
Kew gardener, 1925. To New Zealand, 1929. Director, Botanic Gardens, Dunedin, 1932. Director, Botanic Gardens, Christchurch, 1945.
Gdnrs Chron. 1949 i 174. *J. Kew Guild* 1949, 784.

MANT, Rev. Richard (1776–1848)
b. Southampton, Hants 12 Feb. 1776 *d.* Ballymoney, County Antrim 2 Nov. 1848
BA Oxon 1797. Bishop of Killaloe, 1820; of Down, 1823. Familiar with Irish flora.
D.N.B. v.36, 96–98. *Ann. Rep. Proc. Belfast Nat. Field Club* 1911–12, 618. *Irish Nat.* 1913, 24–25.

MANTELL, Gideon Algernon (1790–1852)
b. Lewes, Sussex 3 Feb. 1790 *d.* London 11 Nov. 1852
MRCS 1811. FRS 1825. FLS 1813. Practised medicine in Lewes and London. Palaeontologist.
Proc. Linn. Soc. 1853, 235–37. *Quart. J. Geol. Soc.* 1853, xxii–xxv. B. Burke *Colonial Gentry* v.1, 1891, 176. S. Spokes *Gideon Algernon Mantell* 1927. E. C. Curwen *J. of Gideon Mantell...1818–1852* 1940. *Proc. R. Soc. Med.* 1972, 215–25.
Fossils at BM(NH).

MANTELL, Joshua (*fl.* 1830s)
Surgeon, Newick, Sussex. Raised florists' flowers. *Floriculture* 1832.
Gdnrs Chron. 1955 ii 24.

MAPLET, Rev. John (–1592)
BA Cantab 1563. MA 1567. Rector, Great Leighs, Essex, 1568; Northolt, Middx, 1576. *A Green Forest* 1567 (includes "herbes, trees and shrubs").
R. Pulteney *Hist. and Biogr. Sketches of Progress of Bot. in England* v.1, 1790, 86. *D.N.B.* v.36, 112–13.

MAPPLEBECK, John E. (1842–*c.* 1905)
b. Birmingham 23 June 1842
FLS 1873–1904. Pteridologist. Raised numerous hybrid fern varieties. Found *Asplenium filix-foemina* var. *Mapplebeckii* Lowe. In New Zealand, 1863–66.
E. J. Lowe *Fern Growing* 1895, 184–86 portr. *Br. Fern Gaz.* 1909, 46.
Ferns at Birmingham Museum.

MAPSON, Leslie William (1907–1970)
b. Cambridge 17 Nov. 1907 *d.* 3 Dec. 1970
Educ. Cambridge. FRS 1969. Lecturer in Biochemistry, Portsmouth College of Technology, 1936. Scientific Officer, Food Investigation Board, D.S.I.R., 1938. Head of Plant Biochemistry Division, Food Research Institute. Contrib. to *Biochem. J.*
Who was Who, 1961–1970 750. *Biogr. Mem. Fellows R. Soc. 1972* 427–44 portr.

MARCAN, Alexander (1883–1953)
b. Leeds, Yorks 16 Oct. 1883 *d.* London 12 Aug. 1953
To Bangkok as Chief Assayer to Royal Mint, 1909. Government Chemist, Siam, 1917. Collected plants in Siam.
Bangkok Times 23 April 1932. *J. Siam Soc. Nat. Hist. Supplt.* v.8, 1932, 345–46. *Times* 26 Aug. 1953. *Blumea* v.11, 1961–62, 480–81.
Herb. at BM(NH). Plants at Kew, Singapore.
Marcania Imlay.

MARCET, Jane (*née* **Haldimand**) (1769–1858)
b. Geneva 1769 *d.* London 28 June 1858
Married A. J. G. Marcet, Swiss doctor, in 1790. *Conversations on Vegetable Physiology* 1829 2 vols.; ed. 3 1839.
H. Martineau *Biogr. Sketches* ed. 4 1876, 368. *D.N.B.* v.36, 122–23. *Berichte Schweizerischen Bot. Gesellschaft* v.50A, 1940, 305–6. *Gdnrs Chron.* 1951 ii 238.

MARCH, George (*fl.* 1830s)
"...to whom I am indebted for the plant [*Gesneria marchii*], which was collected on his estate in the Organ Mountains of Brazil, and whose kindness to the various botanical collectors in that distant country, entitles him to the compliment" (*Bot. Mag.* 1839, t.3744).
Gesneria marchii Wailes.

MARCH, William Thomas (*c.* 1795–*c.* 1872)
d. Jamaica *c.* 1872
Of Spanish Town. Lawyer. Secretary to Governor, Jamaica, 1868. Had herb. Sent plants to Kew, etc.
I. Urban *Symbolae Antillanae* v.3, 1902, 81. *R.S.C.* v.4, 227; v.8, 323.

MARCHANT, Angela (–1970)
Knowledgeable on irises and bulbs of Middle East.
Iris Yb. 1970, 9.

MARGARY, Augustus Raymond (*fl.* 1860s)
d. murdered near Manwyne, Burma
To China in 1867 as student interpreter. Consul in Formosa in 1870 where he became interested in botany.
E. H. M. Cox *Plant Hunting in China* 1945, 100. T. Whittle *Plant Hunters* 1970, 177.

MARGERISON, Samuel (1857–1917)
b. Calverley, Yorks 1857 *d.* Leeds, Yorks 7 June 1917
Timber merchant. All-round naturalist. Studied plant ecology of Calverley Woods, Yorks.
Bot. Soc. Exch. Club Br. Isl. Rep. 1917, 90. *Nature* v.99, 1917, 329. *Naturalist* 1917, 235–36, 270 portr.
Rosa involuta Smith var. *Margerisoni* Wolley-Dod.

MARIES, Charles (*c.* 1851–1902)
b. Stratford-on-Avon, Warwickshire *c.* 1851 *d.* Gwalior, India 11 Oct. 1902
FLS 1887. VMH 1897. Employed by James Veitch and Sons to collect in Japan and China, 1877–79. Superintendent of Gardens to Maharajah of Durbhungah, 1882; later to Maharajah of Gwalior. Contrib. to *Garden*.
Garden v.24, 1883, 444 portr.; v.62, 1902, 318. E. Bretschneider *Hist. European Bot. Discoveries in China* 1898, 741–44. *Gdnrs Mag.* 1901, 795, 796 portr.; 1902, 738 portr. *Gdnrs Chron.* 1902 ii 360–61 portr. *Proc. Linn. Soc.* 1902–3, 34–35. *J. Hort. Cottage Gdnr* v.45, 1902, 452. J. H. Veitch *Hortus Veitchii* 1906, 79–84, 96. E. H. M. Cox *Plant Hunting in China* 1945, 103–4. A. M. Coats *Quest for Plants* 1969, 79–82.
Plants at Kew, BM(NH). Unpublished drawings and MS. of *Cultivated Mangoes of India* at Kew. Portr. at Hunt Library.
Fraxinus mariesii Hook. f.

MARKHAM, Sir Clements Robert (1830–1916)
b. Stillingfleet, Yorks 20 July 1830 *d.* London 30 Jan. 1916
LLD Cantab. DSc Leeds. FRS 1873. FLS 1864–81. KCB 1896. Cadet, Royal Navy, 1844. In Peru, 1852–53 studying Inca ruins. Later employed by India Office until 1877.

Planned and executed a project for acclimatisation of Peruvian cinchona in India. Secretary, Royal Geographical Society, 1863–88; President, 1893–1905. *Notes on Culture of Cinchonas* 1859. *Travels in Peru and India while superintending the Collection of Cinchona Plants and Seeds in South America* 1862. *Peruvian Bark* 1880.

Gdnrs Chron. 1860, 732. *Geogr. J.* 1905, 121–23; 1916, 165–72; 1968, 343–52 portr. *Kew Bull.* 1916, 50–51. A. H. Markham *Life of Sir Clements R. Markham* 1917 portr. C. C. Gillispie *Dict. Sci. Biogr.* v.9, 1974, 123–24. *R.S.C.* v.4, 245; v.8, 232; v.10, 724; v.17, 27.

Andes plants at Kew. Portr. at Hunt Library.

Markhamia Seem.

MARKHAM, Ernest (–1937)
d. East Grinstead, Sussex 6 Dec. 1937

Gardener to W. Robinson at Gravetye, Sussex. *Clematis* 1935; ed. 3 1951. *Raspberries and Kindred Fruits* 1936.

Gdnrs Chron. 1937 ii 439.

MARKHAM, Gervase (*c.* 1568–1637)
b. Cottam near Newark, Notts? *c.* 1568

Hackwriter on agriculture and horticulture. *English Husbandman* 1613. *Farewell to Husbandry* 1620.

G. W. Johnson *Hist. English Gdning* 1829, 78–80. S. Felton *Portr. English Authors on Gdning* 1830, 88–89. *Cottage Gdnr* v.6, 1851, 391. J. Donaldson *Agric. Biogr.* 1854, 18–20. *J. Hort. Cottage Gdnr* v.53, 1875, 370–72 portr.; N.S. v.55, 1907, 369–70 portr. *D.N.B.* v.36, 166–68. D. McDonald *Agric. Writers* 1908, 84–96. *Isis* v.35, 1944, 106–18.

MARKHAM, Lady Mary (*née* **Thynne**) (–1814)

Daughter of Marquis of Bath and sister of Countess of Aylesford (1760–1832). Botanised in Buckinghamshire. Had herb., formerly in possession of family.

G. C. Druce *Fl. Buckinghamshire* 1926, xcvii.

MARKHAM, Richard Anthony *see* Salisbury, R. A.

MARKHAM, William (*fl.* 1830s)

Nurseryman, Stockwell Green, London.

MARKWICK, William (*afterwards* **Eversfield**) (1739–1813)
b. Battle, Sussex 1739 *d.* 6 April 1813

ALS 1788. FLS 1792. Country gentleman. Correspondent of G. White. Kept a natural history diary *c.* 1768–76 entitled *Calendar of Flora.* Communicated a paper on *Plantae Sussexiensis* to Linnean Society 16 Nov. 1802.

Notes and Queries v.8, 1889, 287. *Zoologist* 1890, 335–45. *Hastings and East Sussex Nat.* v.3, 1922, 179–200 portr. *Canadian Field Nat.* 1932, 32–33. *J. Bot.* 1933, 348–51. A. H. Wolley-Dod *Fl. Sussex* 1937, xxxix.

Drawings, MSS. and *Florula Canadensis* 3 vols at Linnean Society. MSS. at Hastings Museum.

MARLAY, Charles Brinsley (1831–1912)
d. 18 June 1912

BA Cantab 1853. MA 1858. Of Belvedere House, Mullingar. Vice-President, Royal Botanical Society. Contrib. to *Gdnrs. Chron.* 1879, 449, 609.

Who was Who, 1897–1916 473.

MARLBOROUGH, 5th Duke of *see* Spencer-Churchill, G.

MARNOCK, Robert (1800–1889)
b. Kintore, Aberdeenshire 12 March 1800 *d.* London 15 Nov. 1889

FLS 1846. Gardener, Bretton Hall, Wakefield, 1830. Curator, Sheffield Botanic Garden, 1834 and Royal Botanic Society's Garden, Regents Park, 1840–69. During 1840s in partnership with Manley in nursery at Sheffield. *Florigraphia Britannica* (with R. Deakin) 1837. *The Plantation, Leighton Buzzard* 1872. Edited *Floricultural Mag.* 1836–42.

Gdnrs Mag. 1840, 603. *Garden* v.6, 1874 portr.; v.36, 1889, 489–90; v.38, 1890, 140 portr. *Gdnrs Chron.* 1882 i 565 portr., 567–68; 1889 ii 588–89 portr. *D.N.B.* v.36, 192–93.

MARQUAND, Cecil Victor Boley (1897–1943)
b. Richmond, Surrey 7 June 1897 *d.* Isle of Skye 1 July 1943

BA Cantab 1919. MA 1922. FLS 1922. Son of E. D. Marquand (1848–1918). Assistant, Welsh Plant Breeding Station, 1919–22, working on *Avena.* Assistant, Kew Herb., 1923–39. Contrib. papers on E. Asian flora to *Kew Bull.*

Bot. Soc. Exch. Club Br. Isl. Rep. 1943–44, 648–49. *J. Kew Guild* 1943, 304–5 portr. *Kew Bull.* 1923, 128. *Nature* v.152, 1943, 322–23. *Proc. Linn. Soc.* 1943–44, 213–14.

Plants at Kew. Portr. at Hunt Library.

MARQUAND, Ernest David (1848–1918)
b. Guernsey, Channel Islands 8 Feb. 1848 *d.* Totnes, Devon 16 Feb. 1918

ALS 1902. *Fl. Guernsey and Lesser Channel Islands* 1901. Contrib. botanical papers to *Trans. Guernsey Soc. Nat. Sci.*

F. H. Davey *Fl. Cornwall* 1909, lix. *Bot. Soc. Exch. Club Br. Isl. Rep.* 1918, 357–60. *J. Bot.* 1918, 187–89. *Proc. Linn. Soc.* 1917–18, 39–40. *Trans. Guernsey Soc. Nat. Sci.* 1918, 83–90. *R.S.C.* v.10, 725; v.12, 486.

Herb. at La Société Guernesiaise Museum and with C. R. P. Andrews (1870–1951).
Salvia marquandii Druce.

MARRAT, Frederick Price (1820–1904)
b. Broadway, Yorks 16 March 1820 *d.* Liverpool 5 Nov. 1904
Bryologist, geologist and conchologist. Assistant, Liverpool Museums. Friend of W. Wilson, J. Dickinson and H. H. Higgins. Pioneer worker on Cryptogams in Liverpool area. 'On Fossil Ferns in Ravenhead Collection' (with H. H. Higgins) (*Proc. Liverpool Geol. Soc.* 1872, 97–109). Had moss herb. (*J. Bot.* 1910, 102–5). Contrib. to *Proc. Lit. Philos. Soc.*
J. B. L. Warren *Fl. Cheshire* 1899, lxxxvi. *Proc. Liverpool Nat. Field Club* 1904, 8. *Trans. Liverpool Bot. Soc.* 1909, 81. *Lancashire Nat.* v.13, 1920–21, 47–56, 76–82, 127–29. Liverpool Museums *Handbook and Guide to Herb. Collections* 1935, 24–25. *R.S.C.* v.4, 248; v.8, 335; v.10, 726.
Mosses at Liverpool Museums. Letters in Wilson correspondence at BM(NH).
Bryum marratii Hook & Wilson.

MARRIOTT, St. John (1870–1927)
b. Sandbach, Cheshire 1 Sept. 1870 *d.* London 7 Oct. 1927
Of Woolwich. *British Woodlands as illustrated by Lessness Abbey Woods* 1925. 'Notes on Bryophyta of Essex' (*S.E. Union Sci. Soc. Essex Handbook* 1926, 56–62).
Bot. Soc. Exch. Club Br. Isl. Rep. 1927, 380–81. *J. Bot.* 1927, 317. F. O. Whitaker *Mem. of St. John Marriott* 1928 portr.
Herb. at Woolwich Museum.

MARRIOTT, William E. (1880–1965)
d. Natal, S. Africa 3 July 1965
Kew gardener, 1904. To S. Africa, 1904. Curator, Pietermaritzburg Botanic Gardens; retired 1950.
J. Kew Guild 1966, 706 portr.

MARSDEN, William (1754–1836)
b. Verval, Wicklow 16 Nov. 1754 *d.* Edgegrove, Herts 6 Oct. 1836
FRS 1783. DCL Oxon 1786. Orientalist and numismatist. Writer, East India Company, Bencoolen, Sumatra, 1771–79. Secretary to Admiralty. Collected plants in Bencoolen. *History of Sumatra* 1783; ed. 3 1811.
Mem. Wernerian Soc. v.1, 1811, 28–31. *Proc. R. Soc.* v.3, 1836, 436. *Brief Mem. of Life and Writings of Late William Marsden* 1838 2 vols. *D.N.B.* v.36, 206–7. C. E. Buckland *Dict. Indian Biogr.* 1906, 275. *Fl. Malesiana* v.1, 1950, 348. M. Archer *Nat. Hist. Drawings in India Office Library* 1962, 17–19, 86.
Plants at BM(NH), Kew.
Marsdenia R. Br.

MARSDEN-JONES, Eric Marsden (1887–1960)
b. Tilston, Cheshire 8 May 1887 *d.* Bath, Somerset 26 Aug. 1960
FLS 1914. Established Potterne Biological Station, 1923. Collaborated with W. B. Turrill on experimental taxonomy of *Silene, Centaurea, Ranunculus, Anthyllis* and *Saxifraga*. *British Knapweeds* (with W. B. Turrill) 1954. *Bladder Campions* (with W. B. Turrill) 1957. Contrib. to *J. Ecol., J. Linn. Soc., Kew Bull., Proc. Linn. Soc.*
J. Kew Guild 1960, 812. *Proc. Bot. Soc. Br. Isl.* 1961, 353–54. *Proc. Linn. Soc.* 1959–60, 132–33.

MARSH, Albert Stanley (1892–1916)
b. Crewkerne, Somerset 1 Feb. 1892 *d.* killed near Armentières 6 Jan. 1916
BA Cantab 1912. Demonstrator, Cambridge, 1913. 'Azolla' (*J. Bot.* 1914, 209–13). 'Anatomy of *Stangeria*' (*New Phytol.* 1914, 18–30).
Ann. Bot. v.33, 1916, 25–27. *Bot. Soc. Exch. Club Br. Isl. Rep.* 1916, 465. *J. Bot.* 1916, 71. *Nature* v.96, 1916, 683. *New Phytol.* 1916, 81–85.

MARSH, Florence Hannah Bacon (1881–1948)
b. Warwickshire 1881
FLS 1935. Studied flora of Herefordshire.
Proc. Linn. Soc. 1948–49, 85.

MARSH, Henry (1665–1741)
Gentleman gardener with garden in Hammersmith, Middx, perhaps close to his mansion in Frog Lane. "A curious collector of rare and uncommon trees." Supplied 150 oaks for Canons, Middx in 1719.
P. Miller *Gdnrs Dict.* abridg. ed. 5 1763, Sorbus no. 2. *London Topographical Rec.* 1965, 56.

MARSH, Maria (*née* **Ransom**) (*fl.* 1830s–1840s)
Of Hitchin and Luton. Had herb. of Herts plants.
Herts Nat. Hist. Soc. Trans. 1897, 167–68.

MARSH, Rev. Thomas Orlebar (1749–1831)
b. Felmersham, Beds 1749 *d.* Felmersham 25 Dec. 1831
FLS 1797. Vicar, Stevington, Beds. Contrib. to J. Sowerby and J. E. Smith's *English Bot.* 499 and C. Abbot's *Fl. Bedfordiensis* 1798.
Gent. Mag. 1832 i 281.
Letters at BM (Add. MS. 23205).

MARSH, William Thomas (*c.* 1795–*c.* 1872)
Clerk of Supreme Court, Jamaica. Sent plants to Kew, 1854–62.
Kew Bull. 1901, 44. I. Urban *Symbolae Antillanae* v.3, 1902, 81.

MARSHALL, Alexander (*c.* 1639–1682)
Botanical artist. "Painted on velom a book of Mr. Tradescant's choicest flowers and plants" (H. Walpole *Anecdotes*).
C. E. Raven *English Nat.* 1947, 306. National Book League *Flower Books* 1950, 24. A. P. Oppé *English Drawings; Stuart and Georgian Periods...at Windsor Castle* 1950, 74–75. W. Blunt *Art of Bot. Illus.* 1955, 125–26. M. Allan *Tradescants* 1964, 55. *Proc. Bot. Soc. Br. Isl.* 1964, 230.
Two volumes of drawings at Windsor Castle.

MARSHALL, Rev. Charles (*c.* 1747–1818)
d. 12 March 1818
Vicar, Brixworth, Northants. *Introduction to...Gardening* 1796.
J. C. Loudon *Encyclop. Gdning* 1822, 1282. *Cottage Gdnr* v.5, 1851, 349.

MARSHALL, Rev. Edward Shearburn (1858–1919)
b. London 7 March 1858 *d.* Tidenham, Glos 25 Nov. 1919
BA Oxon 1881. FLS 1887. Vicar, Milford, Surrey, 1890–1900. Rector, West Monkton, Somerset, 1904–19. *Fl. Kent* (with F. J. Hanbury) 1899. *Supplt. to Fl. Somerset* 1914. Contrib. *Betula* to C. E. Moss's *Cambridge Br. Fl.* v.2, 1914. Contrib. to *J. Bot.*
Bot. Soc. Exch. Club Br. Isl. Rep. 1919, 619–22. *J. Bot.* 1920, 1–6 portr. *Nature* v.104, 1919, 377. *Proc. Linn. Soc.* 1919–20, 45–46. *Watson Bot. Exch. Club Rep.* 1918–20, 89–90 portr. W. K. Martin and G. T. Fraser *Fl. Devon* 1939, 777. H. J. Riddelsdell *Fl. Gloucestershire* 1948, cxli. D. Grose *Fl. Wiltshire* 1957, 42.
Plants at Cambridge, Oxford. Portr. at Hunt Library.
Hieracium marshalli E. F. Linton.

MARSHALL, Henry (1775–1851)
b. Kilsyth, Stirlingshire 1775 *d.* Edinburgh 5 May 1851
Educ. Edinburgh Univ. FRSE. Army surgeon in Ceylon, 1809–21. Inspector-General of army hospitals. 'Coco-nut Tree' (*Mem. Wernerian Soc.* v.5, 1824, 107–48; reprinted 1836).
D.N.B. v.36, 237–38. *R.S.C.* v.4, 250.

MARSHALL, Henry Samuel Amos (1892–1974)
b. Chiswick, Middx 29 May 1892 *d.* Orpington, Kent 31 March 1974
FLS 1948. Assistant, Library, Kew, 1932; Librarian, 1948–61. Contrib. papers on botanical bibliography to *J. Soc. Bibl. Nat. Hist.*, *Kew Bull.*
Gdnrs Chron. 1961 i 39 portr. *J. Kew Guild* 1974, 343–44 portr.

MARSHALL, Humphry (1722–1801)
b. West Bradford, U.S.A. 10 Oct. 1722 *d.* West Bradford 5 Nov. 1801
Cousin of J. Bartram (1699–1777). Correspondent of Sir J. Banks. Founded Marshallton Botanic Garden. *Arbustum Americanum* 1785.
F. Pursh *Fl. Americae Septentrionalis* v.1, 1814, vi–vii. W. Darlington *Mem. of John Bartram and Humphry Marshall* 1849, 485–585. J. W. Harshberger *Botanists of Philadelphia and their Work* 1899, 77–85. *Bull. Chester County Hist. Soc.* 1913, 1–34. *Kew Bull.* 1915, 309–10. R. H. Fox *Dr. John Fothergill and his Friends* 1919, 191–93. H. B. Humphrey *Makers of N. American Bot.* 1961, 166–68.
Plants at BM(NH).

MARSHALL, James (*fl.* 1690s–1700s)
Surgeon. Sent plants from Virginia to J. Petiver (*Museii Petiveriani* 1695, 178) and L. Plukenet (*Amaltheum Botanicum* 1705, 102).

MARSHALL, James (*fl.* 1790s)
Gardener and seedsman, Romsey, Hants.

MARSHALL, Joseph Jewison (1860–1934)
d. Lyndhurst, Hants 12 April 1934
Pharmaceutical chemist, Beverley. Grimsby, 1909–16. Studied Bryophyta of East Riding, Yorks and Lincolnshire. Contrib. mosses to J. F. Robinson's *Fl. East Riding of Yorkshire* 1902.
Trans. Lincolnshire Nat. Union 1933, 180. *Naturalist* 1934, 209; 1962, 133–36; 1964, 67–68; 1973, 3–4.
Plants at Hull Museum destroyed in 1939–45 war. Bryophytes at Lincolnshire Naturalists' Union.

MARSHALL, Moses (1758–1813)
b. W. Bradford, U.S.A. 30 Nov. 1758 *d.* W. Bradford 1 Oct. 1813
Nephew of Humphry Marshall whom he helped in preparing *Arbustum Americanum* 1785. Sent plants to Europe.
W. Darlington *Mem. of John Bartram and Humphry Marshall* 1849, 545–48. J. W. Harshberger *Botanists of Philadelphia and their Work* 1899, 97–107. H. A. Kelly *Some American Medical Botanists* 1929, 75–81.
Marshallia Schreber.

MARSHALL, William (*fl.* 1560s–1580s)
Apprentice surgeon to J. Gerard (1545–1612) who sent him to the Mediterranean region to collect plants.
E. Cecil *Hist. Gdning in England* 1910, 149. R. H. Jeffers *Friends of John Gerard; Biogr. Appendix* 1969, 22–23.

MARSHALL, William (1745–1818)
b. Yorks *d.* Cleveland, Yorks 1818

Planter in West Indies. Returned to England *c.* 1774 and farmed in Surrey. Agent to Sir H. Harbord in Norfolk, 1780–84. Settled at Stafford. *Planting and Ornamental Gardening* 1785; ed. 2 1796.

G. W. Johnson *Hist. English Gdning* 1829, 246–47. J. Donaldson *Agric. Biogr.* 1854, 63–64.

MARSHALL, William (1815–1890)
b. Ely, Cambridge 1815 *d.* Ely 4 Feb. 1890

Solicitor. *New Water Weed; Anacharis alsinastrum* 1852. Contrib. botany to S. H. Miller and S. B. J. Skertchley's *Fenland* 1878, 294–320.

Phytologist v.2, 1845, 284–85; v.4, 1852, 705–15. *R.S.C.* v.4, 251; v.8, 339.

Letters at Kew.

MARSHALL, William (1835–1917)
b. Hackney, London 20 Nov. 1835 *d.* Bexley, Kent 11 Nov. 1917

VMH 1906. VMM 1909. Cultivated orchids.

Gdnrs Mag. 1897, 92 portr.; 1906, 162 portr., 172, 179; 1909, 525–26 portr. *J. Hort. Cottage Gdnr* v.52, 1906, 277 portr.; v.58, 1909, 116–18 portr. *Gdnrs Chron.* 1917 ii 202–3 portr. *Orchid Rev.* 1917, 252–53 portr.

Portr. at Royal Horticultural Society.

Oncidium marshallianum Reichenb. f.

MARSHALL, William Emerson (1873–1937)
b. Edinburgh 1873 *d.* Califon, New Jersey, U.S.A. 25 May 1937

Emigrated to U.S.A. Nurseryman, New York. *Consider the Lilies* 1927.

Gdnrs Chron. 1937 i 433.

MARSHAM, Robert (1707/8–1797)
b. 27 Jan. 1707/8 *d.* Stratton Strawless, Norfolk 4 Sept. 1797

FRS 1780. Kept calendar of natural phenomena for more than 50 years. Contrib. papers on growth of trees to *Trans. Philos. Soc.* v.51–71, 1751–97.

J. E. Smith *Selection from Correspondence of Linnaeus* v.2, 1821, 42–43. *Trans. Bot. Soc. Edinburgh* v.19, 1893, 587–90. J. Grigor *Eastern Arboretum* 1841, 85. *Trans. Norfolk Norwich Nat. Soc.* v.2, 133–95. *Quart. J. For.* 1914, 63–66. *Country Life* 1975, 935 portr.

Letters in Banks correspondence at BM(NH).

MARTIN, Abraham (*fl.* 1780s–1810s)

Founded nursery at Northgate, Cottingham, Yorks in 1788. Later a shop was opened in Hull, perhaps by his son, Samuel F. Martin.

MARTIN, Claude (1731–1800)
b. Lyons, France 4 Jan. 1731 *d.* Lucknow, India 13 Sept. 1800

Went to India, 1751. Officer in East India Company; Major-General, Bengal Army. To Lucknow, 1776.

Bengal: Past and Present v.2, 1908, 277–87; 1909, 102–9. *J. Bot.* 1919, 263–64. *Kew Bull.* 1919, 207–8.

Plants and botanical drawings at Kew.

Andropogon martini Roxb.

MARTIN, Dorothy (1882–1949)

Art teacher, Roedean School, Brighton, 1916–46. Collection of over 300 flower drawings at Royal Horticultural Society.

MARTIN, George Anne (*c.* 1807–1867)
d. Ventnor, Isle of Wight 7 Jan. 1867

MD Edinburgh 1837. Of Ventnor. *The 'Undercliff' of Isle of Wight* 1849 ('Botany of the Undercliff', 297–351).

Trans. Bot. Soc. Edinburgh v.9, 1868, 90–91.

MARTIN, J. D. (–*c.* 1941)
d. S. Rhodesia *c.* 1941

In Forest Dept., N. Rhodesia, 1930. Collected plants in Barotseland. 'Baikiaea Forests of N. Rhodesia' (*Empire For. J.* 1940, 8–18). *Report on Forestry in Barotseland* 1941.

Comptes Rendus AETFAT 1960 1962, 188.

Plants at Kew.

MARTIN, James (–1899)
d. 27 Sept. 1899

VMM 1894. Hybridiser for nursery firm of Sutton and Sons, Reading, Berks.

Gdnrs Chron. 1899 ii 285 portr.

MARTIN, John (*c.* 1783–1855).
b. Tyldesley, Lancs *c.* 1783 *d.* Tyldesley 13 Aug. 1855

Hand-loom weaver. "An accurate botanist" (W. J. Hooker).

Phytologist v.1, 1842, 199. R. Buxton *Bot. Guide to...Manchester* 1849, xxiv. J. Cash *Where there's a Will there's a Way* 1873, 108–18. L. H. Grindon *Country Rambles* 1882, 211. *Trans. Liverpool Bot. Soc.* 1909, 81.

MARTIN, Martin (*c.* 1660–1719)
d. London 1719

Educ. Edinburgh University 1681. Governor to Laird of Macdonald in Skye, 1686–92. Toured Inner and Outer Hebrides, *c.* 1695. *Late Voyage to St. Kilda* 1698. *Description of Western Islands of Scotland* 1703 (both books include references to native flora).

D.N.B. v.36, 288–89. *Watsonia* 1953–56, 36–40. *Notes and Queries* v.203, 1958, 109–11.

MARTIN, William (*fl.* 1590s–1600s)

Barber-surgeon, London. Warden, Barber-Surgeons Company, 1599–1601. Sent one of his apprentices to collect plants in Mediterranean region.

R. H. Jeffers *Friends of John Gerard; Biogr. Appendix* 1969, 23.

MARTIN, William (1767–1810)
b. Mansfield, Notts 1767 *d.* 31 May 1810
FLS 1796. Friend of J. Bolton (*fl.* 1750s–1799). Correspondent of J. Banks and A. B. Lambert.
Mon. Mag. v.32, 1811, 556–65.

MARTIN, Rev. William Keble (1877–1969)
b. Radley, Berks 9 July 1877 *d.* Woodbury, Devon 26 Nov. 1969
BA Oxon 1899. MA 1907. FLS 1928. Curate, Beeston, Ashbourne and Lancaster, 1902–9. Vicar, Wath-upon-Dearne, Rotherham, 1909–21. Rector, Haccombe, Devon, 1921–34; Great Torrington, 1934–43; Combe-in-Teignhead, 1943–49. Botanist and botanical artist. *Fl. Devon* (with G. T. Fraser) 1939. *Concise British Fl. in Colour* 1965. *Over the Hills* 1968 portr. (autobiography). *Sketches for the Fl.* 1972.
Sunday Times 14 June 1965 portr. *Sunday Times Mag.* 23 June 1968 portr. *Times* 28 Nov. 1969 portr. *Rep. Trans. Devonshire Assoc.* 1970, 284–86.

MARTINDALE, Joseph Anthony (1837–1914)
b. Stanhope, County Durham 19 July 1837
d. Staveley, Westmorland 3 April 1914
Schoolmaster. Lichenologist. President, Kendal Natural History Society, 1912. Studied flora of Westmorland. Contrib. to *J. Bot.*
Bot. Soc. Exch. Club Br. Isl. Rep. 1914, 51. *J. Bot.* 1914, 241–45 portr. *Naturalist* 1914, 157–59 portr.
Herb. at Kendal Museum.
Ephebeia martindalei Crombie.

MARTINEAU, Lady Alice (*c.* 1866–1956)
d. 7 Feb. 1956
Herbaceous Garden 1913. *Gardening in Sunny Lands* 1924. *Secrets of Many Gardens* 1924.
Gdnrs Chron. 1956 i 163.

MARTYN, John (1699–1768)
b. London 12 Sept. 1699 *d.* Chelsea, London 29 Jan. 1768
FRS 1727. Practised medicine in London, 1727–52. Professor of Botany, Cambridge, 1732–62. Friend of P. Blair and W. Sherard. In 1721 founded Botanical Society of London which met weekly at Rainbow Coffee House, Watling Street. *Methodus Plantarum circa Cantabrigiam Nascentium* 1727. *Historia Plantarum Rariorum* 1728–37 (58 original drawings at Royal Society). Translated J. P. de Tournefort's *History of Plants* 1732 2 vols. Translated Virgil's *Georgicks* 1746.
R. Pulteney *Hist. Biogr. Sketches of Progress of Bot. in England* v.2, 1790, 205–18. *Gent. Mag.* 1794 ii 898. J. Nichols *Lit. Anecdotes of Eighteenth Century* v.3, 638. G. C. Gorham *Mem. John and Thomas Martyn* 1830, 278. A. L. A. Fée *Vie de Linné* 1832, 249. T. Faulkner *Hist. and Topographical Description of Chelsea* v.1, 1829, 161–66. D. Turner *Extract from Lit. and Sci. Correspondence of Richard Richardson* 1835, 282, 287. *Gdnrs Chron.* 1887 i 381. *D.N.B.* v.36, 317–19. J. R. Green *Hist. Bot.* 1914, 176–84. *Med. Hist.* 1961, 361–74. *Proc. Bot. Soc. Br. Isl.* 1967, 305–24. F. A. Stafleu *Taxonomic Literature* 1967, 300. D. H. Kent *Hist. Fl. Middlesex* 1975, 16.
Herb. at Cambridge. MSS. at Royal Society. Correspondence at BM(NH). Sale at L. D. Lockyer and C. Reymers, 21 July 1768.
Martynia L.

MARTYN, Rev. Thomas (1735–1825)
b. Chelsea, London 23 Sept. 1735 *d.* Pertenhall, Beds 3 June 1825
BA Cantab 1756. BD 1766. FRS 1786. FLS 1788. Son of J. Martyn (1699–1768). Professor of Botany, Cambridge, 1762. One of the earliest exponents of Linnaean classification. *Plantae Cantabrigiensis* 1763. *Catalogus Horti Botanici Cantabrigiensis* 1771. *Thirty-eight Plates with Explanations* 1788; new ed. 1817. *Fl. Rustica* 1792–95 4 vols. *Language of Botany* 1793; ed. 2 1796. Edited P. Miller's *Gdnr's Dict.* 1803–7. Translated J. J. Rousseau's *Letters on Elements of Botany* 1785.
R. J. Thornton *New Illus. of Sexual System of Linnaeus* 1799 portr. J. Nichols *Lit. Anecdotes of Eighteenth Century* v.3, 156. J. Nichols *Illus. of Lit. Hist. of Eighteenth Century* v.5, 752–53. *Gent. Mag.* 1825 ii 85–87. G. C. Gorham *Mem. of John and Thomas Martyn* 1830, 278–80. H. Trimen and W. T. T. Dyer *Fl. Middlesex* 1869, 392–93. *J. Hort. Cottage Gdnr* v.56, 1876, 76–78 portr. *D.N.B.* v.36, 321–23. F. H. Davey *Fl. Cornwall* 1909, xxxiv. J. R. Green *Hist. Bot.* 1914, 229–40. C. E. Salmon *Fl. Surrey* 1931, 47. F. A. Stafleu *Taxonomic Literature* 1967, 300–2. *Taxon* 1972, 630 portr. *R.S.C.* v.4, 270.
Letters at BM(NH) and in Smith correspondence at Linnean Society. Portr. at Hunt Library. Sale at Stewart, Wheatley and Adlaid, 6 Dec. 1827.

MARTYN, Rev. Thomas Waddon (–1918)
d. Oxford 1918
MA Oxon. Curate, Plumstead, 1874–76; Stoke d'Abernon, 1876–79; Buckingham, 1879–81. Rector, Heathe, Oxford, 1881–87; Stoke Abbots, Bucks, 1887–1918. Botanised in Oxford and Bucks.
Bot. Soc. Exch. Club Br. Isl. 1918, 360–61.

MASCALL, Leonard (–1589)
d. Farnham Royal, Bucks 10 May 1589

Clerk of the kitchen to Archbishop Parker. Of Plumstead, Sussex. *Booke of Arte and Maner, Howe to Plant and Graffe all Sortes of Trees* 1569; later eds up to 1640.

Cottage Gdnr v.1, 1849, 50; v.6, 1851, 285. *J. Hort. Cottage Gdnr* v.53, 1875, 78–79 portr. *D.N.B.* v.36, 404–5. D. McDonald *Agric. Writers* 1908, 37–46.

MASEFIELD, John Richard Beech (1850–1932)
b. Stone, Staffs 11 March 1850 *d.* 16 Feb. 1932

BA Cantab 1872. Solicitor, Cheadle, Staffs. Collected plants in vicinity of Cheadle, now in possession of E. S. Edees. President, North Staffordshire Field Club. 'Staffordshire Ferns' (*Trans. N. Staffordshire Field Club* 1908–9, 102–7).

Trans. N. Staffordshire Field Club 1931–32, 144–47 portr. E. S. Edees *Fl. Staffordshire* 1972, 20.

MASKELL, Ernest John (1895–1958)
b. Cambridge 1 Feb. 1895 *d.* 20 Dec. 1958

Educ. Cambridge. FRS 1939. Plant physiologist, Horticultural Research Station, Cambridge, 1922–24. Rothamsted Experimental Station, 1924–26. Cotton Research Station, Trinidad, 1926–30. Lecturer, Cambridge, 1930; later Reader in Plant Physiology. Professor of Botany, Birmingham, 1951–58.

Nature v.168, 1951, 497; v.183, 1959, 219–20. *Times* 24 Dec. 1958. *Who was Who, 1951–1960* 742.

MASLEN, Arthur J. (–1954)
d. 5 Feb. 1954

Lecturer, Geology Dept., Chelsea Polytechnic. Carried out research on fossil plants at Jodrell Laboratory, Kew. *Class Book of Botany* (with G. P. Mudge) 1903.

MASON – (*fl.* 1690s)

Surgeon. Sent plants from Angola to his friend, J. Petiver (*Museii Petiveriani* 1695, 176).

J. E. Dandy *Sloane Herb.* 1958, 161.

Plants at BM(NH).

MASON, A. (1826–1888)
d. Grange-over-Sands, Lancs 1888

Formerly in army. Veterinary surgeon of Grange. List of plants in Aspland's *Guide to Grange* 1869.

Naturalist 1894, 123. *Trans. Liverpool Bot. Soc.* 1909, 81.

MASON, Frances Mary (*née* **Young**) (1882–1932)
b. 12 Nov. 1882 *d.* London 8 Aug. 1932

Kew gardener, 1917–18. To New Zealand. Employed by Messrs Duncan and Davies Nurseries, New Plymouth, N.Z. Sent plants to T. F. Cheeseman.

J. Kew Guild 1933, 279–80.

MASON, Rev. Francis (1799–1874)
b. York 2 April 1799 *d.* Rangoon 2 March 1874

DD Brown University. In U.S.A., 1818–30. Missionary in Burma from 1830. *Fl. Burmanica* 1851. *Burma, its People and Productions* 1860. *Story of a Working-man's Life* 1870 (autobiography).

Ripley and Dana *American Cyclop.* v.11, 1883, 241. C. E. Buckland *Dict. Indian Biogr.* 1906, 278.

MASON, Francis Archibald (*c.* 1878–1936)
b. Leicester *c.* 1878 *d.* Leeds, Yorks 4 March 1936

Brewer's chemist, Leeds. President, Leeds Naturalists' Club, 1924; Yorkshire Naturalists' Union, 1934. Mycologist.

Kew Bull. 1921, 172. *Naturalist* 1936, 133–34 portr.; 1961, 64. *N. Western Nat.* 1936, 58.

MASON, John (*fl.* 1780s–1810s)

Seedsman and florist, Orange Tree, 152 Fleet Street, London, 1781. His son succeeded to the business in 1820. Principal importers of Dutch bulbs.

MASON, Joseph (*fl.* 1820s)

Nurseryman, Derby.

MASON, Marianne Harriet (1845–1932)
b. Morton Hall, Notts Feb. 1845 *d.* Rondebosch, S. Africa 7 April 1932

First official woman inspector of boarded-out children. Plant collector and botanical artist. Made journeys in Cape, Rhodesia, Uganda, etc.

Kew Bull. 1932, 203–4. *Times* 9 April 1932. *Who was Who, 1929–1940* 911.

Drawings, plants and letters at Kew.

MASON, Nathaniel Haslope (*fl.* 1850s–1860s)

FLS 1856. Collected plants in Madeira, 1855–57. Published sets of Madeira ferns and woods.

Hooker's J. Bot. 1855, 317–18. C. A. de Menezes *Fl. do Archipelago da Madeira* 1914, 227–28.

Plants at Kew, Oxford. Letters at Kew.

MASON, Peter (–1719)
d. Aug. 1719

Nurseryman, Isleworth, Middx.

Agric. Hist. Rev. 1974, 20, 25–34.

MASON, Peter (1680–1730)

Son of P. Mason (–1719). Nurseryman, Isleworth, Middx. "...has one of the best collections of English fruits of any nursery-man in England" (Batty Langley, 1728).

MASON, Philip B. (–1905)
Herb. at Bolton Museum, Lancs.

MASON, Samuel (*fl.* 1800s)
Of Yarmouth, Norfolk. Collected and drew seaweeds. "A most indefatigable collector, as well as a most accurate observer of these plants" (D. Turner).
C. J. and J. Paget *Sketch of Nat. Hist. Yarmouth* 1834, xxix. *J. Bot.* 1893, 281.
3 vols. of drawings at Kew.

MASON, Thomas (1818–1903)
Arrived in Wellington, N.Z. in 1841. Made a reputation as a horticulturist, having "the finest botanical collection in the colony." Contrib. to *Trans. Proc. N.Z. Inst.*
Trans. Proc. N.Z. Inst. 1902, xx portr.

MASON, Thomas Godfrey (1890–1959)
b. Ireland 24 July 1890 *d.* 22 Oct. 1959
Educ. Trinity College, Dublin. Lecturer in Botany, Alberta, 1919–20. Economic Botanist, Barbados, 1920–22. Professor of Botany, Imperial College of Tropical Agriculture, Trinidad, 1922–23. Senior Botanist, Nigeria, 1924–25. Contrib. to *Ann. Bot.*
Kew Bull. 1920, 218. *Who was Who, 1951–1960* xxix.

MASON, Rev. William Wright (1853–1932)
b. Wainfleet St. Mary, Lincs 31 Oct. 1853 *d.* Louth, Lincs 11 Aug. 1932
BA 1876. Rector, Leverton, Lincs, 1878–94; Bootle, 1894; Melmert, 1913; Salmonby, 1924–29. "The most consistent of Lincolnshire botanists" (E. A. Woodruffe-Peacock). Sent part of his plant collection to G. C. Druce. Added *Myosotis brevifolia* to British flora.
Trans. Lincolnshire Nat. Union 1932, 69–70 portr. G. C. Druce *Comital Fl. Br. Isl.* 1932, xii. *N. Western Nat.* 1937, 316–21 portr. E. J. Gibbons *Fl. Lincolnshire* 1975, 59 portr.
MSS. at Liverpool Botanical Society, Bootle Museum, Lincoln Museum. Plants at Oxford.
Crataegus monogyna var. *Masonii.*

MASSEE, George Edward (1850–1917)
b. Scampton, Yorks 20 Dec. 1850 *d.* Sevenoaks, Kent 17 Feb. 1917
FLS 1895–1915. ALS 1916. VMH 1902. Mycologist. To S. America with R. Spruce. Illustrated Spruce's *Hepaticae Amazonicae et Andinae* 1884. Principal Assistant, Kew Herb., 1893–1915. First President, British Mycological Society, 1896–98. *British Fungi* 1891. *British Fungus-fl.* 1892–95 4 vols. *Monograph of Myxogastres* 1892. *Text-book of Plant Diseases* 1899; ed. 3 1907. *European Fungus Fl.: Agaricaceae* 1902. *Textbook of Fungi* 1906. *Diseases of Cultivated Plants and Trees* 1910. *British Fungi and Lichens* 1911. *Fungus Fl. Yorkshire* (with C. Crossland) 1905. *Enemies of the Rose* (with F. V. Theobald) 1908.
J. Hort. Cottage Gdnr v.45, 1902, 311 portr. *Gdnrs Mag.* 1908, 739–40 portr. *J. Kew Guild* 1908, 399 portr.; 1918, 426. *Mycol. Notes* 1910, 445–46 portr. *Naturalist* 1913, 291–93 portr.; 1917, 139–42; 1961, 60–62. *Bot. Soc. Exch. Club Br. Isl. Rep.* 1917, 90–91. *Gdnrs Chron.* 1917 i 91 portr. *Garden* 1917, 78 portr. *J. Bot.* 1917, 223–27 portr. *Kew Bull.* 1915, 118–20; 1917, 84–85; 1922, 335–48. *Nature* v.99, 1917, 9–10. *Orchid Rev.* 1917, 58–59. *Proc. Linn. Soc.* 1916–17, 49–51. *Trans. Br. Mycol. Soc.* v.5, 1917, 469–73.
Herb. at New York Botanic Garden. Drawings at BM(NH). Portr. at Hunt Library.
Massea Sacc.

MASSEE, Ivy (*fl.* 1910s)
Daughter of G. E. Massee (1850–1917). She "was most active in the field collecting and when indoors painting." Mycologist. *Mildews, Rusts and Smuts* (with G. E. Massee) 1913.
Naturalist 1961, 62.

MASSEY, John Dickinson (1870–1943)
b. Aughton, Lancs 1870 *d.* Liverpool 19 Dec. 1943
President, Liverpool Naturalists' Field Club, 1932–33. Contrib. plant records to C. T. Green's *Fl. Liverpool District* 1933.
Proc. Liverpool Nat. Field Club 1943, 5. *N. Western Nat.* 1944–45, 69–71.
Herb. dispersed among members of Liverpool Botanical Society.

MASSEY, Richard Middleton
(*c.* 1678–1743)
b. Cheshire *c.* 1678 *d.* Rostherne, Cheshire 27 March 1743
MD Aberdeen 1720. FRS 1712. Practised medicine in Wisbech, 1705–25; in London, 1725–39. Correspondent of H. Sloane and J. Petiver. Plate of fungi in J. Martyn's *Historia Plantarum Rariorum* dec. 3, 1728, 9 drawn by him.
W. Munk *Roll of R. College of Physicians* v.2, 1878, 93–94. E. J. L. Scott *Index to Sloane Manuscripts* 1904, 346. *J. Bot.* 1915, 243–49. J. E. Dandy *Sloane Herb.* 1958, 161.

MASSIE, Barbara (*née* **Townshend**)
(1781–1816)
b. 11 Aug. 1781 *d.* Chester, Cheshire 9 July 1816
Of Wincham Hall, Cheshire.
Lancashire and Cheshire Nat. v.10, 1917, 166–68.
Herb. at Chester Natural Science Society.

MASSON, Francis (1741–1805)
b. Aberdeen Aug. 1741 *d.* Montreal 23 Dec. 1805

FLS 1796. Gardener. First plant collector sent out from Kew. To the Cape, S. Africa, 1772–74; Canaries and Azores, 1776–82; Spain and Portugal, 1783–85; Cape and interior with Thunberg, 1786–95; New York and Montreal, 1798. Sent plants to A. B. Lambert. *Stapeliae Novae* 1796.

Philos. Trans. R. Soc. v.66, 1776, 268–317. H. C. Andrews *Bot. Repository* 1799, t.46. *Ann. Bot.* 1806, 592–93. *Gent. Mag.* 1806 ii 1077. J. Nichols *Lit. Anecdotes of Eighteenth Century* v.8, 620. J. E. Smith *Selection from Correspondence of Linnaeus* v.2, 1821, 559–65. A. L. A. Fée *Vie de Linné* 1832, 233. P. Smith *Mem. and Correspondence of Sir J. E. Smith* v.2, 1832, 117, 183. *Hort. Register* 1836, 67. A. Lasègue *Musée Botanique de B. Delessert* 1845, 178–79. *Cottage Gdnr* v.8, 1852, 286–87. *Gdnrs Chron.* 1881 ii 335; 1945 i 26. *J. Bot.* 1884, 114–23, 144–48; 1885, 227; 1886, 335–36; 1904, 2–3; 1917, 70–74. *Trans. S. African Philos. Soc.* v.4, 1887, xxxix–xlii. *Proc. Linn. Soc.* 1888–89, 38. *Kew Bull.* 1891, 295–96. *J. Kew Guild* 1897, 26–27. *D.N.B.* v.37, 16. I. Urban *Symbolae Antillanae* v.3, 1902, 82–83. C. A. de Menezes *Fl. Archipelago da Madeira* 1914, 228. *J. Linn. Soc.* v.45, 1920, 41–43, 50–51 portr. A. White and B. L. Sloane *Stapelieae* v.1, 1937, 82–85 portr. J. Hutchinson *Botanist in S. Africa* 1946, 613, 617. *S. African Geogr. J.* 1947, 16–31. *Proc. Amer. Philos. Soc.* 1950, 150. *J. S. African Bot.* 1958, 203–18 portr.; 1959, 167–88, 283–310; 1960, 9–15; 1961, 15–45, 167–68. C. Lighton *Cape Floral Kingdom* 1960, 17–27. V. S. Forbes *Pioneer Travellers of S. Africa* 1965, 37–45 portr. *Huntia* 1965, 204 portr. F. A. Stafleu *Taxonomic Literature* 1967, 301–2. K. Lemmon *Golden Age of Plant Hunters* 1968, 43–73 portr. *Dict. S. African Biogr.* v.1, 1968. A. M. Coats *Quest for Plants* 1969, 252–59 portr., 323–24. *Taxon* 1970, 532.

Letters, plants and drawings at BM(NH). Portr. by G. Garrard at Linnean Society. Portr. at Hunt Library. Plants at Academy of Natural Sciences, Philadelphia.

Massonia L.

MASTERS, George (*fl.* 1720s)

Nurseryman, Strand-on-the-Green, Chiswick, Middx.

MASTERS, John (*fl.* 1800s)

Nurseryman, Dover Street, Canterbury, Kent.

MASTERS, John W. (*c.* 1792–1873)
d. Faversham, Kent 15 Feb. 1873

Gardener, Calcutta Botanic Garden to 1838. 'Fl. Naga Hills' (*J. Asiatic Soc. Bengal* v.13, 1844; v.25, 1856, 410).

Plants at Kew, Calcutta.

Mastersia Benth.

MASTERS, Maxwell Tylden (1833–1907)
b. Canterbury, Kent 15 April 1833 *d.* Ealing, Middx 30 May 1907

MRCS 1856. MD St. Andrews 1862. FRS 1870. FLS 1860. Son of W. Masters (1796–1874). Lecturer, St. George's Hospital, 1855–68. Editor, *Gdnrs Chron.* 1865–1907. *Vegetable Morphology* 1862. *Vegetable Teratology* 1869. *Botany for Beginners* 1872. Papers on *Passifloraceae* and *Coniferae* in *J. Linn. Soc.* and *Trans. Linn. Soc.* Contrib. to *J. Bot.*, *Fl. Tropical Africa* and *Fl. Br. India.*

J. Hort. Cottage Gdnr v.22, 1891, 109–10 portr.; v.54, 1907, 509–10 portr. *Bot. Soc. Exch. Club Br. Isl. Rep.* 1907, 263–64. C. F. P. von Martius *Fl. Brasiliensis* v.1(1), 1906, 185–86. *Gartenflora* 1907, 377–80 portr. *Gdnrs Chron.* 1907 i 368–69 portr., 377, 398, 418. *Gdnrs Mag.* 1907, 419 portr. *J. Bot.* 1907, 257–58 portr. *Kew Bull.* 1907, 325–34. *Nature* v.76, 1907, 157. *Orchid Rev.* 1907, 201–3 portr. *Proc. Linn. Soc.* 1907–8, 54–56. *J. R. Hort. Soc.* 1909, 153 portr. *Who was Who, 1897–1916* 481. L. Huxley *Life and Letters of Sir J. D. Hooker* v.1, 1918, 383. G. C. Druce *Fl. Oxfordshire* 1927, cxiv–cxv. *D.N.B. Supplt. 2* v.2, 586–87. *Curtis's Bot. Mag. Dedications, 1827–1927* 191–92 portr. H. N. Clokie *Account of Herb. Dept. Bot., Oxford* 1964, 208. *R.S.C.* v.4, 280; v.8, 382; v.10, 743.

Plants, MSS and letters at Kew. Plants at Canterbury. Portr. at Hunt Library.

Maxwellia Baillon.

MASTERS, William (1796–1874)
b. Canterbury, Kent 7 July 1796 *d.* Canterbury 26 Sept. 1874

Nurseryman, Exotic Nursery, 25 St. Peter's Street, Canterbury. Founded Canterbury Museum, 1823; Hon. Curator, 1823–46. Had garden arranged on natural system. Hybridised passion-flowers, etc. *Hortus Duroverni* ed. 3 1831 (*see Gdnrs Mag.* 1831, 609–10).

Florist and Pomologist 1874, 261. *Gdnrs Chron.* 1874 ii 437. *J. Hort. Cottage Gdnr* v.52, 1874, 297.

Bust at Canterbury Museum.

MATEER, Rev. Samuel (1835–1893)
b. Belfast 1835 *d.* Trevandrum, India 24 Dec. 1893

FLS 1870. To India, 1859. Wesleyan Missionary in Travancore. Anthropologist. Contrib. Tamil plant-names to *J. Linn. Soc.* v.13, 1871, 25–30.

Congregational Yb. 1895, 221 portr. F. Boase *Modern English Biogr.* Supplt. v.2, 1921, 178–79.

MATEER, William (*fl.* 1830s–1840s)
MD. Professor of Botany, Belfast Academy Institution. Collected plants in N. Ireland.
S. A. Stewart and T. H. Corry *Fl. N.E. Ireland* 1888, xv.

MATHER, Rev. Cotton (1663–1728)
FRS 1714. Sent plants from Boston, Massachusetts to J. Petiver.
J. E. Dandy *Sloane Herb.* 1958, 161.

MATHESON, Thomas (*fl.* 1870s)
Nurseryman, 35 Oldgate, Morpeth, Northumberland.

MATHEW, William (*fl.* 1790s–1820s)
FLS 1793. Of Bury St. Edmunds. Contrib. largely to J. Sowerby and J. E. Smith's *English Bot.* 1793–99.

MATHEWS, Andrew (–1841)
d. Chachapoyas, Peru 24 Nov. 1841
ALS 1825. Gardener for Horticultural Society of London at Chiswick. Collected plants in Peru and Chile, 1830–41.
J. Bot. (Hooker) 1834, 176–77. *Hooker's J. Bot.* 1842, 392–94. *Companion Bot. Mag.* v.1, 1835–36, 17–19, 305. *Proc. Linn. Soc.* 1843, 173. A. Lasègue *Musée Botanique de B. Delessert* 1845, 255–57. H. N. Clokie *Account of Herb. Dept. Bot., Oxford* 1964, 209. *Gdn J. New York Bot. Gdn* 1964, 7–10. A. M. Coats *Quest for Plants* 1969, 373. *Taxon* 1970, 532.
Plants at BM(NH), Kew, Oxford. MS. Fl. Peruana at BM(NH). Letters at Kew.
Mathewsia Hook. & Arn.

MATHEWS, Miss Elizabeth (–*c.* 1811)
Of Belmont, Herefordshire. Provided most of the drawings for T. A. Knight's *Pomona Herefordiensis* 1811.
J. Bot. 1931, 77.

MATHEWS, John (*fl.* 1780s)
Collected plants in Sierra Leone. *Voyage to River Leone* 1788.
J. Vallot *Études sur la Fl. du Sénégal* 1883, 24.

MATHEWS, Joseph William (–1949)
b. Cheshire *d.* Durban, S. Africa 23 Sept. 1949
Kew gardener, 1895. Nurseryman, Cape Town, 1895. Curator, Botanic Garden, Kirstenbosch, 1913–36. *Cultivation of Nonsucculent S. African Plants* 1938. Contrib. to *J. Bot. Soc. S. Africa.*
Gdnrs Chron. 1949 ii 152. *J. Kew Guild* 1949, 781–82. C. Lighton *Cape Floral Kingdom* 1960, 40–41.

MATHEWS, William (1828–1901)
b. Hagley, Worcs 10 Sept. 1828 *d.* Tunbridge Wells, Kent 5 Sept. 1901
MA Cantab 1856. Of Birmingham. President, Alpine Club, 1868–70. Friend of C. C. Babington and W. W. Newbould. *Fl. Algeria* 1880. *Fl. Clent and Lickey Hills* 1881. 'History of County Botany of Worcester' (*Midland Nat.* 1887–93). Contrib. to *Phytologist, J. Bot.*
Alpine J. v.20, 1901, 521–26. *J. Bot.* 1901, 352, 428–29. *Trans. Surveyors' Inst.* 1901–2, 525–26. *Kew Bull.* 1906, 173–74. J. Amphlett and C. Rea *Bot. Worcestershire* 1909, xxv–xxvi. *R.S.C.* v.8, 353; v.10, 744.
Plants at Glasgow University, Kew, Worcester Museum.

MATHIAS, Hayward (–1912)
d. 10 Feb. 1912
Nurseryman, Medstead, Hants. In partnership with P. Smith. Specialised in carnations. Secretary, Perpetual-flowering Carnation Society.
J. Hort. Home Farmer v.64, 1912, 172–73 portr.

MATHIAS, William Thomas (1900–1954)
b. Cardigan 1900 *d.* Liverpool 19 May 1954
Educ. Univ. College, Aberystwyth. FLS 1928. Assistant Lecturer in Botany, Bangor, 1923. Lecturer in Botany, Liverpool, 1924–54. Algologist.
Nature v.173, 1954, 1211. *Proc. Linn. Soc.* 1952–53, 218.

MATON, William George (1774–1835)
b. Salisbury, Wilts 31 Jan. 1774 *d.* London 30 March 1835
BA Oxon 1794. MD 1801. FRS 1800. FLS 1794. Physician at Westminster Hospital, London, 1800–8. 'Some Account of Medicinal and other Uses of Various Substances prepared from Trees of Genus Pinus' (A. B. Lambert *Genus Pinus* 1803, 65–82). *Natural History of Western Counties* 1797. *Natural History of Wiltshire* 1843. Plants of Salisbury in R. C. Hoare's *Hist. Modern Wiltshire* 1843, 654.
J. Nichols *Illus. of Lit. Hist. of Eighteenth Century* v.8, xlv–xlvi. P. Smith *Mem. and Correspondence of Sir J. E. Smith* v.2, 1832, 121–22. *Gent. Mag.* 1837 i 175–76. J. A. Paris *Biogr. Sketch of Late W. G. Maton* 1838. R. C. Hoare *Hist. Modern Wiltshire* v.6, 1843, 654–57. W. Munk *Roll of R. College of Physicians* v.3, 1878, 6–11. *Proc. Linn. Soc.* 1888–89, 38. F. H. Davey *Fl. Cornwall* 1909, xxxvi. *D.N.B.* v.37, 60. *Tribuna Farmaceutica* v.10, 1942, 64. D. Grose *Fl. Wiltshire* 1957, 34–35. *R.S.C.* v.4, 285.
Plants at BM(NH). Bust by W. Behnes at Linnean Society.
Matonia R. Br.

MATTERSON, William (1815–1890)
d. 23 April 1890
MD. Botanical Curator, Yorkshire Philosophical Society.

Yorkshire Herald 24 April 1890. *Annual Rep. Yorkshire Philos. Soc.* 1906, 46.
Plants at York Museum.

MATTHEW, Charles Geikie (1862–1936)
b. Newmilne, Perthshire 30 July 1862 *d.* Guildtown, Perthshire 11 Jan. 1936
MB Edinburgh 1885. FLS 1909. Ship's surgeon, P. & O. Shipping Company, 1885–89; Royal Navy, 1889–1909, 1914–18. Collected ferns in Japan, China, Malaya, Philippines and Hawaii, 1903–13; Malaya, 1911–13. *Notes on Ferns of Hong Kong and the adjacent Mainland* 1908. 'Enumeration of Chinese Ferns' (*J. Linn. Soc.* 1911, 339–93).
Kew Bull. 1936, 189. *Times* 13 Jan. 1936. *Chronica Botanica* 1937, 154. *Proc. Linn. Soc.* 1936–37, 203–4. *Trans. Proc. Perthshire Soc. Nat. Sci.* 1938, xci–xcii portr. *Fl. Malesiana* v.1, 1950, 351.
Ferns at Kew.

MATTHEW, Patrick (1790–1874)
b. near Scone, Perth 20 Oct. 1790 *d.* Gourdie Hill, Perthshire 8 June 1874
'Nature's Law of Selection' (*Gdnrs Chron.* 1860, 312–13). "I freely acknowledge that Mr. Matthew has anticipated by many years the explanation which I have offered of the origin of species under the name of natural selection" (C. Darwin, *Gdnrs Chron.* 1860, 362–63). *On Naval Timber and Arboriculture* 1831 (with appendix on natural selection).
Gdnrs Mag. 1832, 702–3. *Br. Assoc. Handbook Dundee* 451–57 portr. *J. Bot.* 1912, 193–94. *R.S.C.* v.4, 294; v.12, 493.

MATTHEWS, Sydney (*c.* 1887–1954)
d. Swansea, Glam 28 June 1954
Nurseryman, Port Talbot, Glam, 1913. Nursery later extended to Sketty and West Cross.
Orchid Rev. 1954, 148.

MATTHEWS, Washington (1843–1905)
b. Killiney, County Dublin 17 July 1843 *d.* Washington, U.S.A. 29 April 1905
MD Iowa 1864. Assistant surgeon, U.S. Army. Ethnobotanist. Collected plants about Fort Wingate, New Mexico, 1882. *Navajo Names for Plants* 1886.
S. Watson *Bot. California* v.2, 1880, 559.
Gilia matthewsii Gray.

MATTOCK, John R. (*fl.* 1910s)
Nurseryman, New Headington, Oxford. Specialised in roses.
J. Hort. Home Farmer v.63, 1911, 513 portr.

MAUDE, Ashley Henry (1850–1933)
d. 3 Oct. 1933
FLS 1908. Barrister. Mayor of Eastbourne, Sussex, 1903–4. Collected plants in Britain, Europe, Algeria, Canary Islands and the Cape.
Proc. Linn. Soc. 1933–34, 164–65. *Times* 5 Oct. 1933.
Herb. at BM(NH).

MAUDE, Mary Fawler (*née* **Hooper**) (1819–1913)
b. London 25 Oct. 1819 *d.* Overton, Cheshire 30 July 1913
Married Rev. Joseph Maude, 1841. *Scripture Natural History* 1848.

MAUGER, W. (*fl.* 1890s–1915)
Nurseryman, Guernsey, Channel Islands. In partnership with F. J. Hubert until 1897.

MAUGER, Mrs. William P. (*fl.* 1860s)
Helped S. O. Gray in *British Sea-Weeds* 1867.
Maugeria S. O. Gray.

MAUGHAN, Edward James (1790–1868)
b. Edinburgh 1790 *d.* Edinburgh 1868
Son of R. Maughan (1769–1844). Inspector of Taxes, Perth and Edinburgh. "A keen botanist." Contrib. localities to various floras.
Notes R. Bot. Gdn Edinburgh 1908, 292.

MAUGHAN, Robert (1769–1844)
b. Edinburgh 1769 *d.* London 1844
FLS 1809. Original member of Botanical Society of Edinburgh, 1836. To London, 1840. 'List of Rarer Plants of Edinburgh' (*Mem. Wernerian Nat. Hist. Soc.* v.1, 1811, 215–48, 626–28).
R. K. Greville *Fl. Edinensis* 1824, vi. *Notes R. Bot. Gdn Edinburgh* 1908, 292. *R.S.C.* v.4, 298.

MAULE, Alexander James (*c.* 1821–1884)
d. Bristol 5 May 1884
Nurseryman, Stapleton Road Nurseries, Bristol. Youngest son of William Maule, founder of the nurseries.
Gdnrs Chron. 1884 i 653. *J. Hort. Cottage Gdnr* v.8, 1884, 383.

MAULE, William (*fl.* 1810s–1860s)
Founded nursery at Stapleton Road, Bristol. Specialised in conifers. Introduced *Cydonia maulei* from Japan in 1869.
M. Hadfield *Gdning in Britain* 1960, 336.

MAUND, Benjamin (1790–1863)
d. Sandown, Isle of Wight 21 April 1863
FLS 1827. Druggist and bookseller, Bromsgrove, Worcs. *Botanic Garden* 1825–51 13 vols (original drawings at BM(NH)); re-issued in 1878. *Botanist* 1836–41 5 vols. *Botanical Souvenir* (with J. S. Henslow).
Gdnrs Mag. 1839, 90–91. *Proc. Linn. Soc.* 1863–64, xxx–xxxi. *D.N.B.* v.37, 91. *J. Bot.* 1918, 235–43; 1928, 184. F. A. Stafleu *Taxonomic Literature* 1967, 302–3.
Letters in Bentham correspondence at Kew. Window and tablet in Bromsgrove Church (*Gdnrs Chron.* 1928 i 290). Library sold by Palmer of Bromsgrove, 13 May 1858.

MAVOR, Alexander (*fl.* 1830s–1840s)
Nurseryman, Aylesbury, Bucks.

MAVOR, Rev. William Fordyce (*olim* **MacIvor**) (1758–1837)
b. New Deer, Aberdeen 1 Aug. 1758 *d.* Woodstock, Oxfordshire 29 Dec. 1837
LLD Aberdeen 1789. Schoolmaster, Burford, Oxford, 1775. Vicar, Hurley, Berks, 1789. Rector, Woodstock. Sent notes on plants to W. Baxter. *Lady's and Gentleman's Botanical Pocket Book...of Indigenous Botany* 1800. *General View of Agriculture of Berkshire* 1809 (contains first lengthy list of Berks plants).
Cottage Gdnr v.8, 1852, 221–22. *D.N.B.* v.37, 108–9. G. C. Druce *Fl. Berkshire* 1897, cxlix–clii. G. C. Druce *Fl. Oxfordshire* 1927, ciii–civ.

MAW, George (1832–1912)
b. London 10 Dec. 1832 *d.* Kenley, Surrey 7 Feb. 1912
FLS 1860. Of Benthall Hall, Broseley, Shropshire. Botanised with J. D. Hooker and J. Ball in Morocco, 1871. Studied *Crocus* which he collected in native localities. *Monograph of Genus Crocus* 1886.
Garden v.14, 1878, xii portr. *Gdnrs Chron.* 1881 i 205–6; 1912 i 111. *Gdnrs Mag.* 1912, 147. *Nature* v.88, 1912, 561. *Kew Bull.* 1912, 155–56. *Proc. Linn. Soc.* 1911–12, 62–63. *Times* 8 Feb. 1912. *Curtis's Bot. Mag. Dedications, 1827–1927* 187–88 portr. W. K. Martin and G. T. Fraser *Fl. Devon* 1939, 773. W. Blunt *Art. Bot. Illus.* 1955, 238. A. M. Coats *Quest for Plants* 1969, 33–35. *R.S.C.* v.4, 303; v.8, 360; v.10, 751.
Maw's *Crocus* drawings at Kew; types at BM(NH).
Draba mawii Hook. f.

MAWDESLEY, John (*fl.* 1790s)
Nurseryman, Enfield, Middx.

MAWE, Thomas (1760s–1770s)
Gardener to Duke of Leeds at Kiveton Park, Yorks. *Every Man his own Gardener* 1767 and *Universal Gardener and Botanist* 1778 (with J. Abercrombie but actually written by the latter).
J. C. Loudon *Encyclop. Gdning* 1822, 1275. G. W. Johnson *Hist. English Gdning*, 1829, 220.

MAWLEY, Edward (*c.* 1842–1916)
d. Berkhamsted, Herts 15 Sept. 1916
VMH 1904. Secretary, National Rose Society, 1877–1914; President, 1915. Also interested in dahlia cultivation.
J. Hort. Cottage Gdnr v.23, 1891, 69–70 portr.; v.47, 1903, 397, 399 portr.; v.49, 1904, 573 portr. *Garden* v.58, 1900, 26–27 portr.; 1914, 605 portr.; 1916, 471 portr. *Gdnrs Mag.* 1903, 443–44 portr.; 1907, 503 portr., 504; 1916, 399 portr. *Gdnrs Chron.* 1916 ii 148–49 portr. *National Rose Soc. Annual* 1917, 17–27 portr. *Rose Bull.* 1972, 37–38 portr.

MAWSON, Robert R. (*c.* 1864–1910)
d. Windermere, Westmorland 15 Dec. 1910
Nurseryman, Lakeland Nurseries, Windermere.
Gdnrs Mag. 1910, 1024. *J. Hort. Home Farmer* v.61, 1910, 593–94 portr.

MAWSON, Thomas Hayton (1861–1933)
d. Lancaster 14 Nov. 1933
FLS 1921. Landscape architect. *Art and Craft of Garden Making* 1901; 1927. *Civic Art* 1911. *Life and Work of an English Landscape Artist* 1927.
Gdnrs Mag. 1910, 403–4 portr. *Proc. Linn. Soc.* 1933–34, 165–66.

MAWSON, Thomas William (*c.* 1850–1876)
d. Burringham, Yorks 16 Sept. 1876
MD Edinburgh. In medical practice at Burringham. To Surinam. 'Ferns of Derwent Valley' (*Trans. Bot. Soc. Edinburgh* v.11, 1873, 499–502).
Trans. Bot. Soc. Edinburgh v.13, 1879, 10.

MAXWELL, E. (–before 1839)
Lieut., 11th Dragoons. Collected plants at Kunawar in Himalayas. Sent plants to J. F. Royle.
J. F. Royle *Illus. of Bot....of Himalayan Mountains* 1833–40, 52.
Thalictrum maxwellii Royle.

MAXWELL, George (1804–1880)
b. April 1804 *d.* near Albany, W. Australia Jan. 1880
Settled in Albany. Collected Australian plants for 30 years. Accompanied James Drummond, 1846–47, and also F. von Mueller.
G. Bentham *Fl. Australiensis* v.1, 1863, 14*. *Rep. Austral. Assoc. Advancement Sci.* 1911, 232. *J. W. Austral. Nat. Hist. Soc.* 1909, 20. *J. Linn. Soc.* v.45, 1920, 196. *W. Austral. Nat.* 1948, 118.
Plants at BM(NH), Kew, Melbourne. Letters at Kew.
Eriostemon maxwelli F. Muell.

MAXWELL, Sir Herbert Eustace (1845–1937)
b. Edinburgh 8 Jan. 1845 *d.* Monreith, Wigtownshire 30 Oct. 1937
FRS 1898. VMH 1917. MP for Wigtown, 1880–1906. Had garden of rare trees and shrubs at Monreith. Flower illustrations in *New Fl. and Silva* and *Gdnrs Chron. Memories of the Months* 1897. *Scottish Gardens* 1908. *Trees* 1915. *Flowers* 1923.

Garden 1917 iii–iv portr. *Gdnrs Chron.* 1921 ii 308 portr.; 1937 ii 349 portr. *Curtis's Bot. Mag. Dedications, 1827–1927* 381–82 portr. *Kew Bull.* 1937, 513–14. *Nature* v.140, 1937, 959–60. *Times* 1 Nov. 1937 portr. *Obit. Notices R. Soc. London* 1938, 387–93 portr. *Scott. For. J.* 1938, 87–91. *Who was Who, 1929–1940* 920.
Portr. at Hunt Library.

MAY, H. (*c.* 1827–1880)
d. 3 Nov. 1880
Nurseryman, Leeming Lane, Bedale, Yorks.

MAY, Henry Benjamin (*c.* 1845–1936)
d. Chingford, Essex 1 July 1936
VMH 1910. VMM 1927. Nurseryman, Edmonton, Middx, 1870. Specialised in ferns. *Seventy Years in Horticulture* 1928 portr.
Gdnrs Mag. 1908, 139–40 portr. *J. Hort. Home Farmer* v.66, 1913, 123–25 portr. *Gdnrs Chron.* 1925 i 176 portr., 445–47 portr; 1936 ii 17.

MAY, William (*fl.* 1840s–1870s)
Nurseryman, Hope Nursery, Leeming Lane, Bedale, Yorks.

MAYCOCK, James Dottin (–1837)
d. Barbados 1837
MD. FLS 1829. In Barbados for many years. *Fl. Barbadensis* 1830. List of plants in A. Halliday's *West Indies* 1837, 389–408.
Proc. Linn. Soc. 1840, 72. I. Urban *Symbolae Antillanae* v.3, 1902, 83. *R.S.C.* v.4, 305.
Maycockia A. DC.

MAYERS, William Frederick (–1878)
d. Shanghai, China March 1878
To China as student interpreter, 1858. Chinese Secretary, British Legation, Peking. Contrib. articles on Chinese plants to *Notes and Queries on China and Japan* 1867–69.
E. Bretschneider *Hist. European Bot. Discoveries in China* 1898, 695–96.

MAYES, Martin (*c.* 1801–1858)
d. 4 April 1858
Nurseryman, Durdham Down Nurseries, Bristol. In partnership with James Garraway.
Cottage Gdnr v.20, 1858, 70.

MAYFIELD, Arthur (1868–1956)
b. Norwich, Norfolk 1868 *d.* Mendlesham, Suffolk 5 Sept. 1956
FLS 1921. Teacher at Norwich, Great Yarmouth, Mendlesham, 1883–1931. Bryologist. Contrib. to *J. Bot.*, *Trans. Norfolk Norwich Nat. Soc.*
Proc. Linn. Soc. 1956–57, 40.
Herb. at Norwich Museum. Micro-fungi at Kew and Commonwealth Mycological Institute.

MAYNARD, Arthur William (1890–1944)
b. Chislehurst, Kent 11 March 1890 *d.* Grahamstown, S. Africa 22 Oct. 1944
Kew gardener, 1912. At Botanic Gardens, Kirstenbosch, 1914. Curator, Botanic Gardens, Grahamstown, 1936–44.
J. Kew Guild 1944, 406–7.

MAYNARD, Frederick P. (*fl.* 1890s)
Surgeon-captain. Served with David Prain on Baluchistan-Afghan Boundary Commission, 1896. Prain named his plant collections (*Rec. Bot. Survey India* 1896, 125–37).
Pakistan J. For. 1967, 351.

MAYO, Herbert (1796–1852)
b. London 3 April 1796 *d.* Badweilbach, Germany 15 May 1852
MRCS 1819. MD Leyden. FRS 1828. Surgeon, Middlesex Hospital, 1827–42. Professor of Anatomy, King's College, London, 1830–36. 'Motion of *Mimosa*' (*Quart. J. Sci.* 1827 pt 2, 76–83).
J. von Sachs *Hist. Bot.* 1890, 550. *D.N.B.* v.37, 172–73. *R.S.C.* v.4, 313.
Portr. in Hope Collections, Oxford.

MAYOW, John (1643–1679)
b. London May 1643 *d.* London Oct. 1679
BCL Oxon 1665. FRS 1678. Plant physiologist and chemist. Wrote on agricultural chemistry. *Five Medico-physical Treatises* 1674.
D.N.B. v.37, 175–77. *Isis* 1931, 47–96, 504–43. *Chronica Botanica* 1944, 69–73.
Portr. at Hunt Library.

MEAD, John Phillips (1886–1951)
b. London 10 Aug. 1886 *d.* 2 Jan. 1951
Assistant Conservator of Forests, Malaya, 1907–16. Conservator of Forests, Sarawak, 1919–28. Temporary Forest Adviser, Fiji, 1926. Director of Forests, Malaya, 1930–40. *Mangrove Forests of West Coast of Federated Malay States* 1912. *Report on Forests of Fiji Islands* 1927. Contrib. to *Indian Forester*, *Malayan Forester*.
Malayan Forester 1940, 101–2; 1951, 58–59. *Fl. Malesiana* v.1, 1950, 353. *Who was Who, 1951–1960* 751–52.
Plants at Kuala Lumpur.

MEAD, Richard (1673–1754)
b. Stepney, London 11 Aug. 1673 *d.* London 16 Feb. 1754
MD Padua 1695. FRS 1703. FRCP 1716. Practised medicine at Stepney, 1696. Physician to George II, 1727. Patron of Mark Catesby. Engaged G. Ehret to paint plants.
G. Edwards *Essays* 1770, 124–25. J. Nichols *Lit. Anecdotes of Eighteenth Century* v.1, 269; v.6, 212–23. J. E. Smith *Correspondence of Linnaeus* v.2, 1821, 481. *D.N.B.* v.37, 181–86. *Proc. Linn. Soc.* 1894–95, 54.

Portr. at Hunt Library.
Dodecatheon meadia L.

MEADE, Richard John (*fl.* 1870s)
Colonel. Collected grasses in north-west India in 1870s, now at Kew.
Kew Bull. 1901, 45.

MEADER, James (*fl.* 1770s–1780s)
Gardener to Duke of Northumberland at Syon House, Isleworth, Middx; to Empress Catherine at St. Petersburg, Russia, 1779–87. *Planter's Guide* 1779. Edited T. Hitt's *Modern Gardener* 1771.
J. C. Loudon *Encyclop. Gdning* 1822, 1276.

MEAGER, Leonard (*c.* 1620–)
Gardener. Foreman, Brompton Park Nursery, London in 1690s. *English Gardener* 1670; ed. 9 1699. *Mystery of Husbandry* 1697.
J. Donaldson *Agric. Biogr.* 1854, 38–40. D. McDonald *Agric. Writers* 1908, 148–51. *J. Ministry of Agric.* 1930, 879–85.

MEAN, James (*fl.* 1810s–1820s)
Gardener to Sir Abraham Hume at Wormleybury, Herts. Edited ed. 2 of J. Abercrombie's *Practical Gdnr* 1817 and his *Gdnr's Companion* 1818. Contrib. *to Trans. Hort. Soc. London.*
J. C. Loudon *Encyclop. Gdning* 1822, 1289. *J. R. Hort. Soc.* 1941, 311.

MEARES, Richard Goldsmith (1780–1862)
d. York, W. Australia 1862
Captain. Served in Napoleonic Wars. Emigrated to Swan River, W. Australia, *c.* 1831. Resident Magistrate at York near Perth. Collected plants for J. Mangles (1786–1867) *c.* 1835–42.
M. Uren *Land looking West* 1948, 227. A. Hasluck *Portr. with Background* 1955, 176 portr. R. Erickson *Drummonds of Hawthornden* 1969, 25–26.

MEARNS, John (*fl.* 1780s–1790s)
Nurseryman, Springfield, Chelmsford, Essex. In partnership with Thomas Sorrell.

MEASURES, Richard Isaac (*c.* 1833–1907)
d. Camberwell, London 8 Aug. 1907
Had a large orchid collection at Cambridge Lodge, Camberwell.
Orchid Rev. 1907, 239–40, 282–83.
Masdevallia × *Measuresiana*

MECKLENBURG-STRELITZ, Charlotte
see Charlotte Sophia, *H. M. Queen*

MEDHURST, Walter Henry (1796–1857)
b. London 29 April 1796 *d.* London 24 Jan. 1857
Protestant missionary, Batavia; Shanghai, 1843–55. *Silk Manufacture and Cultivation of Mulberry* 1849.
E. Bretschneider *Hist. European Bot. Discoveries in China* 1898, 558–59.

MEDHURST, Sir Walter Henry (1822–1885)
b. China 1822 *d.* Torquay, Devon 26 Dec. 1885
Son of W. H. Medhurst (1796–1857). To China as interpreter, 1842; later British Consul. Collected some plants for H. F. Hance.
E. Bretschneider *Hist. European Bot. Discoveries in China* 1898, 535. *D.N.B.* v.37, 203.

MEDLAND, George (1808–1894)
b. 18 Feb. 1808 *d.* 3 Aug. 1894
Gardener with Summerlands Nursery, Exeter for more than 20 years. Hybridised plants.
Gdnrs Chron. 1894 ii 164–65.

MEEHAN, Edward (*c.* 1798–1882)
d. Nov. 1882
Gardener to Harcourt family at St. Clare, Ryde, Isle of Wight.
Garden v.22, 1882, 434.

MEEHAN, Joseph (1840–)
b. Ryde, Isle of Wight 9 Nov. 1840
Brother of T. Meehan (1826–1901) whom he joined at Germantown, Pennsylvania. Contrib. to *Florist's Exchange.*
Florist's Exchange 6 Aug. 1904.

MEEHAN, Thomas (1826–1901)
b. Potter's Bar, Herts March 1826 *d.* Philadelphia, U.S.A. 19 Nov. 1901
Kew gardener, 1846–48. Established nursery at Germantown, Pennsylvania *c.* 1853. Edited *Gdnrs Mon.*, 1859–89. Founded *Meehan's Mon.*, 1891. State Botanist. Collected plants in New England, Colorado, Utah, Alaska. *Native Flowers and Ferns of United States* 1878–79 2 vols. Contrib. to *Proc. Acad. Nat. Sci. Philadelphia.*
Appleton's Cyclop. Amer. Biogr. v.4, 1888, 285–86. *Bull. Torrey Bot. Club* 1894, 33–34. *J. Kew Guild* 1894, 38–43; 1901, 36–37 portr. C. S. Sargent *Silva of N. America* v.9, 1896, 82. J. W. Harshberger *Botanists of Philadelphia and their Work* 1899, 249–56 portr. *Fern Bull.* v.9, 1901, 87–88 portr. *Gdnrs Chron.* 1901 i 383–84 portr.; 1901 ii 383 portr. *Meehan's Mon.* 1901, 187 portr.; 1902, 13–19. *Nature* v.65, 1901, 132. *J. Bot.* 1902, 38–41. L. H. Bailey *Standard Cyclop. Hort.* v.2, 1939, 1587–88 portr. J. Ewan *Rocky Mountain Nat.* 1950, 196, 259, 263, 266, 288. *R.S.C.* v.4, 319; v.8, 368; v.10, 761; v.12, 496.
Portr. at Hunt Library.
Meehania Britton.

MEEK, Michael (*fl.* 1820s)
Nurseryman, Sober Low, Ainderby Steeple near Northallerton, Yorks.

MEEN, Margaret (*fl.* 1770s–1820s)
Of Bungay, Suffolk. Settled in London. Botanical artist. Exhibited at Royal Academy, 1775–85. *Exotic Plants from Royal Gardens at Kew* 1790.
Kew Bull. 1893, 147; 1925, Appendix 2, 65; 1933, 2–3. *Gdnrs Chron.* 1894 i 197–98, 241.
Drawings at Kew.

MEGAW, Rev. William Rutledge (1885–1953)
b. Carrowdore, County Down 1885 *d.* Dec. 1953
BA. MRIA. Minister, Ahoghill, 1910–19; Newtownbreda, 1919–50. President, Belfast Naturalists Field Club, 1922–23. Revised bryological section of S. A. Stewart and T. H. Corry's *Fl. N.E. Ireland* ed.2 1938. Contrib. to *Irish Nat. J.*
R. L. Praeger *Some Irish Nat.* 1949, 129–30. *Irish Nat. J.* 1954, 181–83 portr.
Herb. at Queen's University Belfast, Ulster Museum.

MEIKLE, Carola Ivena (*née* **Dickinson**) (1900–1970)
b. Alston, Cumberland 27 April 1900 *d.* Wootton Courtney, Somerset 27 March 1970
Algologist, Kew Herb., 1929–59. *British Seaweeds* 1963.

MEIKLEJOHN, Alexander (*c.* 1798–1885)
d. Stirling 18 Feb. 1885
Specialised in growing auriculas.
Gdnrs Chron. 1885 i 321.

MEINERTZHAGEN, Richard (1878–1967)
b. London 3 March 1878 *d.* 17 June 1967
Chief political officer, Palestine and Syria, 1919–20; military adviser, Middle East Dept., Colonial Office, 1921–24. Ornithologist. Collected plants in Palestine and Egypt, 1929; Algeria, 1931; central Sahara, 1931; Pyrenees, 1933; Syria, 1934. *Diary of a Black Sheep* 1964.
Times 19 June 1967. J. Lord *Duty, Honour, Empire: Life and Times of Colonel Richard Meinertzhagen* 1971 portr.
Plants at BM(NH), Kew.

MELDRUM, Robert Hunt (1858–1933)
d. June 1933
Teacher, Perth. Headmaster, Tibbermore, Perth. Collected local mosses and made herb. of European specimens. Contrib. to *Trans. Proc. Perthshire Soc. Nat. Sci.*
Trans. Proc. Perthshire Soc. Nat. Sci. 1938, xci–xcii portr.
Herb at Royal Botanic Garden, Edinburgh, Perthshire Society of Natural Sciences.

MELLER, Charles James (*c.* 1836–1869)
b. Surrey *c.* 1836 *d.* Berrima, Sydney 26 Feb. 1869
MD. FLS 1867. Surgeon and naturalist on D. Livingstone's African expedition, 1860–63. Superintendent, Botanic Garden, Mauritius, 1865. Collected plants on Zambesi, and in Madagascar and Mozambique. 'On Botany, Geology...of...Madagascar' (*J. Asiatic Soc.* v.20, 1863, 388–96).
J. Bot. 1869, 212. *Proc. Linn. Soc.* 1869–70, cii–ciii. *Moçambique* 1939, 38–39. *R.S.C.* v.4, 330.
Plants, drawings and letters at Kew.
Mellera S. Moore.

MELLERSH, William Lock (1872–1941)
Botanist. Formed 'Gloucestershire Wild Garden' in Cheltenham, Glos in 1896.
H. J. Riddelsdell *Fl. Gloucestershire* 1948, cxlv.

MELLIAR, Rev. Andrew Foster *see* Foster-Melliar, A.

MELLOR, John (1767–1848)
b. Oldham, Lancs 1767 *d.* Oldham 5 Oct. 1848
Hand-loom weaver and cotton-spinner; afterwards nurseryman. Collected plants chiefly in N. England and Scotland. Friend of G. Caley. "Father of working men botanists of Lancashire."
W. Johns *Practical Bot.* 1826, vi. *Cottage Gdnr* v.1, 1848, 74. R. Buxton *Bot. Guide to Flowering Plants...of Manchester* 1849, ix. J. Cash *Where there's a Will there's a Way* 1873, 90–93. *Trans. Liverpool Bot. Soc.* 1909, 82. *Weekly Courier* 2 Jan. 1909 portr. *Lancashire Nat.* v.11, 1918, 5–7.

MELLOR, Thomas (*c.* 1822–1882)
d. Ashton-under-Lyne, Lancs 1 May 1882
Shoemaker. Grew florist flowers, particularly auriculas.
Gdnrs Chron. 1882 i 646.

MELVILL, James Cosmo (1845–1929)
b. Hampstead, London 1 July 1845 *d.* Meole Brace, Shropshire 4 Nov. 1929
BA Cantab 1868. DSc Manchester 1904. FLS 1870. MP, S. Salford. Conchologist and botanist. President, Manchester Literary and Philosophical Society, 1897–99; Caradoc and Severn Valley Field Club, 1904–14. Collected plants in N. America, 1871–72. Had herb. to which he added collections made by Boswell Syme, C. G. Pringle, R. Tate, T. Kirk, E. L. Layard, etc. *Fl. Harrow* 1864. Contrib. to *Mem. Proc. Manchester Lit. Philos. Soc.*, *J. Bot.*
I. Urban *Symbolae Antillanae* v.3, 1902, 84. *Lancashire Nat.* 1912, 323–29 portr. *Bot. Soc.*

Exch. Club Br. Isl. Rep. 1929, 94–96. *Nature* v.124, 1929, 921. *J. Bot.* 1930, 20–22. *Mem. Proc. Manchester Lit. Philos. Soc.* v.74, 1930, ii–iv. *Notes Manchester Mus.* no. 33, 1930, 150–56 portr. *N. Western Nat.* 1929, 195–97; 1930, 150–56 portr. *Proc. Linn. Soc.* 1929–30, 211–13. *Trans. Caradoc Severn Valley Field Club* v.8, 1930, 165–73 portr.

Plants at Harrow School, Manchester University.

Habenaria melvillii Ridley.

MELVILLE, Andrew Smith (–1876)
Of Galway. Lecturer on Botany and Geology, Edinburgh School of Arts.

Athenaeum 22 July 1876, 119.

MELVILLE, Robert (1723–1809)
b. Monimail, Fife 12 Oct. 1723 *d.* 20 Aug. 1809

FRS. Brigadier-General. Governor of West Indies, 1763. Founded St. Vincent Botanic Garden, 1765.

D.N.B. v.37, 246–47.

MENNELL, Henry Tuke (1835–1923)
b. Scarborough, Yorks 1835 *d.* Croydon, Surrey 9 Dec. 1923

FLS 1863. Alpinist. Secretary, Northumberland and Durham Natural History Society. To Croydon as tea merchant, 1861. 'Notes on Botany of Swanage' (*J. Bot.* 1882, 51–53).

J. Bot. 1924, 81–82. *Watson Bot. Exch. Club Rep.* 1922–23, 243–44.

Herb. in possession of C. E. Salmon (1872–1930). Plants at Kew.

MENZIES, Archibald (1754–1842)
b. Stix House, Aberfeldy, Perthshire 15 March 1754 *d.* Notting Hill, London 15 Feb. 1842

FLS 1790. Gardener, Royal Botanic Garden, Edinburgh. Pupil of J. Hope. On botanical tour of Highlands, 1778. Collected Scottish plants for J. Fothergill and W. Pitcairn. Assistant-surgeon, Royal Navy, 1782. Surgeon and naturalist under Capt. Vancouver, 1791–95 (C. F. Newcombe, *ed. Menzies' Journal of Vancouver's Voyage... 1792* 1923, vii–xx portr.) Introduced *Araucaria imbricata*, 1796.

J. E. Smith *Plantarum Icones* 1789–91, 56. *Bot. Misc.* 1829–30, 69–70. *Gent. Mag.* 1842 i 139–41. *Proc. Linn. Soc.* 1842, 139–41; 1888–89, 38; 1943–44, 170–83. W. H. Harvey *Nereis Boreali-americana* v.1, 1851, 42–43. J. D. Hooker *Fl. Tasmaniae* v.1, 1859, cxiv. *J. Bot.* 1886, 101; 1924, 119–23. *Kew Bull.* 1891, 299–300. *D.N.B.* v.37, 258. *J. W. Austral. Nat. Hist. Soc.* 1909, 20–21. *Gdnrs Chron.* 1921 ii 324 portr.; 1931 ii 32–33. *Quart. California Hist. Soc.* 1924, 265–340. *Trans. Scott. Arboricultural Soc.* 1924, 16–25. *Madrono* 1929, 262–66 portr. *Forestry* 1931, 5–8. *Condar* 1932, 243–52. *Occas. Papers California Acad. Sci.* no. 20, 1943, 14–18 portr. *Bull. Hist. Med.* 1948, 796–811. *Br. Columbia Hist. Quart.* 1951, 151–59. *Fl. Malesiana* v.5, 1958, 67–68. *Notes R. Bot. Gdn Edinburgh* 1954, 219–27. S. D. McKelvey *Bot. Exploration of Trans-Mississippi West, 1790–1850* 1955, 26–60. W. R. Dawson *Banks Letters* 1958, 604–7. *Trans. R. Soc. N.Z.* 1960, 63–64. H. N. Clokie *Account of Herb. Dept. Bot., Oxford* 1964, 209. *Taxon* 1970, 532–33. *J. Scott Rock Gdn Club* 1972, 91–97. *R.S.C.* v.4, 345.

Herb. at Edinburgh; plants at BM(NH), Kew. MS. journal at BM, Bloomsbury and Royal Society. Portr. by E. Eddis at Kew. Portr. at Hunt Library.

Menziesia Smith.

MENZIES, James (1854–1945)
b. Ruthvenfield, Perth 1854 *d.* 2 Dec. 1945

Dyer's finisher, Perth, 1895–1943. President, Perthshire Society of Natural Sciences, 1926–28. In 1912 discovered fungus *Calycella menziesi*. Contrib. to *Trans. Proc. Perthshire Soc. Nat. Sci.*

Trans. Proc. Perthshire Soc. Nat. Sci. 1949, xxi–xxii.

MENZIES, John (*fl.* 1840s–1850s)
Nurseryman, Haughshaw Lane, Halifax, Yorks.

MENZIES, Robert (*fl.* 1790s)
Principal Gardener, Royal Botanic Garden, Edinburgh. Elder brother of A. Menzies (1754–1842).

Notes R. Bot. Gdn Edinburgh v.3, 1904, 21.

MERCER, George Enos (1896–1918)
b. 30 Dec. 1896 *d.* 3 Oct. 1918

Botanised in Leicestershire.

A. R. Horwood and C. W. F. Noel *Fl. Leicestershire* 1933, ccxxxiv.

MERCER, John (*fl.* 1810s)
Nurseryman, 122 Whitechapel, Liverpool.

MERCER, Stephen Pascal (1891–1944)
b. Abbots Bromley, Staffs 1891 *d.* 18 Aug. 1944

Lecturer in Agricultural Botany, Armstrong College, Durham, 1916. Assistant Director, Seed Testing Station for England and Wales, 1919–22. Head, Seed Testing Division, N. Ireland, 1922. Professor of Agricultural Botany, Belfast, 1924. *Farm and Garden Seeds* 1938; ed. 2 1948.

Nature v.154, 1944, 389–90. *Who was Who, 1941–1950* 787.

MEREDITH, Louisa Anne (*née* **Twamley**) (1812–1895)

b. Birmingham 20 July 1812 *d.* Hobart, Tasmania 21 Oct. 1895

To Australia, then Tasmania, 1830. Illustrated her own works. *Romance of Nature* 1836; ed. 3 1839. *Wild Flowers* 1838. *My Home in Tasmania* 1852–53 2 vols. *Some of my Bush Friends in Tasmania* 1860; 1881.

P. Mennell *Dict. Austral. Biogr.* 1892, 320. *Papers Proc. R. Soc. Tasmania* 1909, 21–22. *J. Proc. R. Austral. Hist. Soc.* 1921, 215; 1929, 3–14. *Austral. Nat.* 1924, 182. W. Blunt *Art of Bot. Illus.* 1955, 220. *Austral. Encyclop.* v.6, 1965, 57.

MERRETT, Christopher (1614–1695)
b. Winchcombe, Glos 16 Feb. 1614 *d.* London 19 Aug. 1695

BA Oxon 1635. MD 1643. FRS 1663. Glass maker. First Keeper of Library and Museum of College of Physicians, 1654. *Pinax Rerum Naturalium Britannicarum* 1666.

R. Pulteney *Hist. Biogr. Sketches of Progress of Bot. in England* v.1, 1790, 290–97. J. E. Smith *English Fl.* v.1, 1824, vii–viii. W. Munk *Roll R. College of Physicians* v.1, 1878, 258–64. H. Trimen and W. T. T. Dyer *Fl. Middlesex* 1869, 372–73. *D.N.B.* v.37, 288–89. *Essex Nat.* 1900, 231–36. J. W. White *Fl. Bristol* 1912, 56–57. C. E. Raven *English Nat.* 1947, 305–20. J. E. Dandy *Sloane Herb.* 1958, 162–63. C. C. Gillispie. *Dict. Sci. Biogr.* v.9, 1974, 312–13. S. T. Jermyn *Fl. Essex* 1974, 15. D. H. Kent *Hist. Fl. Middlesex* 1975, 13.

Plants at BM(NH).

Merrettia Gray.

MERRIFIELD, Mary Philadelphia (*née* **Watkins**) (1804–1889)
b. Brompton, London 15 April 1804 *d.* Stapleford, Cambridgeshire 4 Jan. 1889

Algologist. *Natural History of Brighton* 1860 (includes account of botany). 'Marine Algae at Brighton' (*Phytologist* v.6, 1862–68, 513–23). 'Additions to British Marine Fl.' (*J. Bot.* 1876, 147–48). 'Nitophyllum versicolor' (*J. Linn. Soc.* v.14, 1875, 421–23).

J. Linn. Soc. v.14, 1875, 421–23. *Ann. Bot.* v.3, 1889, 484. *J. Bot.* 1889, 160. A. H. Wolley-Dod *Fl. Sussex* 1937, xlv. *R.S.C.* v.4, 351; v.10, 783.

Herb. at BM(NH). Letters at Kew.

Merrifieldia Ag.

MERRYWEATHER, Ernest Arthur (1872–1924)

Nurseryman, Southwell, Notts.

Garden 1924, 411 portr.

MERRYWEATHER, Henry (1839–)
b. 24 Jan. 1839

Nurseryman, Southwell, Notts. Introduced 'Bramley's Seedling' apple.

Gdnrs Mag. 1894, 397 portr.; 1912, 733–34 portr. *J. Hort. Home Farmer* v.61, 1910, 370 portr. *Garden* 1914, iv portr.

MERTON, Lionel Francis Herbert (1919–1974)
b. 8 June 1919 *d.* 29 Sept. 1974

Educ. Cambridge. Assistant Lecturer in Botany, Leeds University, 1947–50. Plant Ecologist, Moroccan Locust Research Team, 1950–55. Plant Ecologist, Dept. of Agriculture, Cyprus, 1955–57. Special Adviser, British Middle East Development Division, Beirut, 1957–59. Lecturer in Botany, University of Malaya, Kuala Lumpur, 1960–64. Lecturer in Botany, Sheffield University, 1965. Collected plants in Malaya, Cyprus, Aldabra. 'History and Status of Woodlands of Derbyshire Limestone' (*J. Ecol.* 1970, 723–43).

MESNY, William (1842–1919)

Of Jersey, Channel Islands. Major-General in Chinese Imperial Army. Travelled with Capt. William John Gill from Chengtu to Bhano, 1877. Collected plants for H. F. Hance in Kwangsi, 1879; Kweichow, 1880; Szechuen, 1880; Turkestan, 1881.

E. Bretschneider *Hist. European Bot. Discoveries in China* 1898, 744–47. E. H. M. Cox *Plant Hunting in China* 1945, 101–2.

Plants at Kew.

Salix mesnyi Hance.

MESSEL, Leonard Charles Rudolf (1873–1953)
b. 19 Feb. 1873 *d.* Cuckfield, Sussex 2 Feb. 1953

MA Oxon. FLS 1931. VMH 1945. Continued development of garden at Nymans, Sussex created by his father, L. Messel (–1915). Grew rhododendrons.

Kew Bull. 1919, 240. *Gdnrs Chron.* 1953 i 64. *Proc. Linn. Soc.* 1952–53, 88. *Times* 5 Feb. 1953.

METHVEN, John (–1913)
d. Edinburgh 24 May 1913

Nurseryman, East Princes Street, Edinburgh. Son of T. Methven (*c.* 1819–1879), founder of the nursery.

Garden 1913, 296(viii). *J. Hort. Home Farmer* v.66, 1913, 524.

METHVEN, Thomas (*c.* 1819–1879)
b. Kennoway, Fifeshire *d.* 13 Jan. 1879

Nurseryman, Edinburgh; nurseries at Stanwell, Leith Walk and Warriston.

Gdnrs Chron. 1879 i 88.

MEWSE, Benjamin (*fl.* 1690s)

Sent plants from Surat, India to J. Petiver.

J. E. Dandy *Sloane Herb.* 1958, 163.

MEYER, Canon Horace Rollo (1868–1953)
b. July 1868 *d.* Little Gaddesden, Herts 6 March 1953

MA Cantab. VMH 1949. Rector, Clophill, Beds; Walton-at-Stone, Herts. Canon, St.

Albans, 1934. Raised varieties of daffodils and irises. Founder of London Gardens Guild which in 1927 became the National Gardens Guild.

Gdnrs Chron. 1928 ii 182 portr.; 1950 i 40 portr. *Iris Yb.* 1953, 15–16.

MEYRICK, Edward (1854–1938)
b. Ramsbury, Wilts 24 Nov. 1854 *d.* Ramsbury 30 March 1938

Educ. Cambridge. FRS 1904. Schoolmaster, Australia and New Zealand, 1877–86; Marlborough College, 1887–1914. All-round naturalist, particularly interested in Lepidoptera. Contrib. botanical papers to Marlborough *Nat. Hist. Soc. Rep.*

Nature v.141, 1938, 776–77. *Proc. Entomol. Soc. Washington* v.40, 1938, 177–79. *Trans. R. Soc. N.Z.* v.68, 1938, 141–42. *Obit. Notices Fellows R. Soc.* 1939, 531–48 portr.

MEYRICK, William (*fl.* 1750s–1800s)

Of Birmingham. Surgeon, West Bromwich. *New Family Herbal* 1789. *Miscellaneous Botany* 1794.

MIALL, Louis Compton (1842–1921)
b. Bradford, Yorks 1842 *d.* Leeds, Yorks 21 Feb. 1921

FRS 1892. Professor of Botany, Yorkshire College of Science (Leeds University), 1876–1907. *Fl. West Riding...of Yorkshire* (with B. Carrington) 1862. *Early Naturalists* 1912.

Bot. Soc. Exch. Club Br. Isl. Rep. 1921, 363–64. *J. Bot.* 1921, 117–18. *Naturalist* 1921, 183–84 portr.; 1923, 116 portr. *Nature* v.107, 1921, 16–18. *Times* 22 Feb. 1921. *Who was Who, 1916–1928* 727–28. *R.S.C.* v.8, 396; v.10, 799; v.12, 506; v.17, 216.

MIDDLEMIST, John (*fl.* 1800s–1830s)

Nurseryman, in partnership with Alexander Wood, Cape Nursery, Hammersmith, 1802.

Bot. Mag. 1809, t.1224.

MIDDLETON, Mr.

Surgeon on merchant ship. Collected seeds at Port Desire, Patagonia in late 18th century.

J. Bot. 1924, 351–52.

MIDDLETON, Cecil Henry (1885–1945)
b. Northants 1885 *d.* Surbiton, Surrey 18 Sept. 1945

Kew gardener, 1906–8. Worked in a number of nurseries. Horticultural journalist and broadcaster. *Talks about Gardening* 1935. *More Gardening Talks* 1936. *Outlines of a Small Garden* 1937. *Winter Flowering Plants* 1957.

Gdnrs Chron. 1938 i 68 portr.; 1945 ii 144. *J. Kew Guild* 1945, 471–73. *Times* 19 Sept. 1945.

MIDDLETON, Robert Morton (1846–1909)
b. Sowerby, Yorks 25 Jan. 1846 *d.* Wallington, Surrey 8 Aug. 1909

FLS 1880. Spent 2 years in Chile on behalf of Church Missionary Society where he collected plants, now at BM(NH). 'First Fuegian Collection' (*J. Bot.* 1909, 207–12).

J. Bot. 1909, 396. *Proc. Linn. Soc.* 1909–10, 94.

British plants at McGill University.

MIDDLEWOOD, William (*fl.* 1770s)

Seedsman, Market Street Lane, Manchester, Lancs.

MIDDLEWOOD, William (*fl.* 1780s–1790s)

Nurseryman, Dublin.

Irish For. 1967, 48.

MIERS, John (1789–1879)
b. London 25 Aug. 1789 *d.* Kensington, London 17 Oct. 1879

FLS 1839. FRS 1843. Engineer. In S. America, 1819–38. *Travels in Chile and La Plata* 1826 2 vols. *Illustrations of South American Plants* 1849–57 2 vols. *Contributions to Botany* 1851–71 3 vols. *On Apocynaceae of S. America* 1878. Contrib. to *Trans. Linn. Soc.*

Gdnrs Chron. 1879 ii 522. *Nature* v.20, 1879, 614. *Proc. R. Soc.* v.29, 1879, xxii–xxiii. *J. Bot.* 1880, 33–36 portr., 219–20. *Proc. Linn. Soc.* 1888–89, 38–39. *D.N.B.* v.37, 369. C. F. P. von Martius *Fl. Brasiliensis* v.1(1), 1906, 63–64. *Tribuna Farmaceutica* v.10, 1942, 84. *Taxon* 1970, 533. *R.S.C.* v.4, 382; v.8, 402; v.10, 807.

Herb. and MSS. at BM(NH). Letters at Kew. Portr. at Hunt Library.

Miersia Lindl.

MILAM, Mr. *see* Mylam, Mr.

MILEHAM, George (*fl.* 1880s–1900s)

Gardener to A. Miller at Emlyn House, Leatherhead, Surrey. Specialist in chrysanthemums.

J. Hort. Cottage Gdnr v.55, 1907, 587–88 portr. *Gdnrs Mag.* 1909, 1001–2 portr.

MILES, Mrs. (–1884)
d. Shirehampton near Bristol 15 Nov. 1884

Flower painter. Contrib. plates to *Garden*.

Garden v.26, 1884, 450.

MILES, Archibald Clarence (1887–1969)
b. 2 July 1887 *d.* 14 March 1969

Kew gardener, 1908. Curator, Botanic Station, Assuontsi, Gold Coast, 1909–19. Gold Coast Dept. of Agriculture, 1921–47. Collected plants in Ivory Coast, Gold Coast, S. Nigeria.

Kew Bull. 1909, 22. *J. Kew Guild* 1969, 1031.

Plants at BM(NH), Kew.

MILES, Beverley Alan (1937–1970)
b. Porthcawl, Glam 17 Nov. 1937 *d.* 26 Jan. 1970
BA Cantab 1959. MA 1963. Biology master, St. Mary's Grammar School, Sidcup, Kent, 1959. Curator, South London Botanical Institute, 1969. Studied *Rosa.*
Watsonia 1971, 423–24.
Herb. and MSS at Cambridge.

MILES, Douglas Frank Streatfield (1908–*c.* 1965)
b. Eastbourne, Sussex 30 May 1908
Pharmacist, Croydon, Surrey. President, Croydon Natural History and Scientific Society, 1949. Amassed a large collection of colour transparencies of plants.
Proc. Croydon Nat. Hist. Sci. Soc. 1965, 272.

MILES, George Thomas (1831–1904)
b. Clewer, Berks 10 Jan. 1831 *d.* 17 Nov. 1904
Gardener to Lord Carrington at Wycombe Abbey, Bucks, 1858. Contrib. to *Gdnrs Chron.*
Gdnrs Chron. 1875 ii 72 portr.; 1904 ii 373. *Garden* v.66, 1904, 371 portr.

MILES, Rev. Henry (1698–1763)
b. Stroud, Glos 2 June 1698 *d.* Tooting, Surrey 10 Feb. 1763
DD Aberdeen 1744. FRS 1743. 'Seed of Fern' (*Philos. Trans. R. Soc.* v.41, 1744, 770–75).
D.N.B. v.37, 378.

MILES, William (*c.* 1834–1883)
b. Ashtead, Surrey *c.* 1834 *d.* 4 Feb. 1883
Nurseryman, Brighton, Sussex.

MILES-THOMAS, Miss E. N. (*afterwards* **Hyndman**) (–1944)
d. Aug. 1944
Educ. London University. FLS 1908. On staff of Botanical Depts., Bedford College, 1908–16; University College, Leicester, 1923–37. 'Theory of Double Leaf-trace founded on Seedling Structure' (*New Phytol.* 1907, 77–91).
Proc. Linn. Soc. 1943–44, 235–36.

MILFORD, Mrs. Helen A. (–1940)
d. 11 Sept. 1940
Collected plants in Basutoland. Of Chedworth, Glos.
Quart. Bull. Alpine Gdn Soc. 1940, 334–35.

MILL, George Grote (–1853)
d. Madeira 15 July 1853
Younger brother of J. S. Mill (1806–73). 'List of Flowering Plants...in Neighbourhood of Great Marlow, Bucks' (*Phytologist* 1844, 983–95).
G. C. Druce *Fl. Buckinghamshire* 1926, xcix–c.

MILL, John Stuart (1806–1873)
b. London 20 May 1806 *d.* Avignon, France 8 May 1873
Logician and economist. Official in East India Company, London. "I make a flora of every district in which I settle. I made a flora of Surrey" (J. S. Mill). Found *Impatiens biflora* at Albury, Surrey, 1822. Collected plants in Anatolia, 1862 (*J. Bot.* 1875, 236–37). Left MS. notes for flora of Avignon. Contrib. short notes to *Phytologist* 1841–63 and to J. A. Brewer's *Fl. Surrey* 1863.
Nature v.8, 1873, 47. *Annual Rep. Kew Gdns* 1875, 14. *J. Bot.* 1873, 191–92; 1904, 297. *Garden* v.3, 1873, 402. *Gdnrs Chron.* 1890 i 234. *D.N.B.* v.37, 390–99. G. C. Druce *Fl. Berkshire* 1897, clxix. G. C. Druce *Fl. Buckinghamshire* 1926, c. C. E. Salmon *Fl. Surrey* 1931, 50. *Notes and Queries* v.176, 1939, 8. M. St. John Packe *Life of John Stuart Mill* 1954. *Taxon* 1973, 392. C. C. Gillispie *Dict. Sci. Biogr.* v.9, 1974, 383–86. D. McClintock *Companion to Flowers* 1966, 77–78. *R.S.C.* v.4, 387.
Plants at Kew, Harvard University, National Arboretum Washington. Letters at Kew.
Sedum millii Baker.

MILLAIS, John Guille (*c.* 1865–1931)
b. London 24 March 1865 *d.* Horsham, Sussex March 1931
VMH 1927. Naturalist, botanical artist and gardener. *Rhododendrons and the various Hybrids* 1917, 1924 2 vols. *Magnolias* 1927.
Gdnrs Chron. 1931 i 269. *J. Bot.* 1931, 144. *Times* 27 March 1931. *Trans. Proc. Perthshire Soc. Nat. Sci.* v.9, 1938, xlii–xliii. *Who was Who, 1929–1940* 940.

MILLAR, James (1762–1827)
b. Ayr 1762 *d.* Edinburgh 13 July 1827
MD Edinburgh. FRCP. Lecturer in Natural History, Edinburgh. *Guide to Botany* 1818.
D.N.B. v.37, 401. *R.S.C.* v.4, 387.

MILLAR, Robert (*fl.* 1730s–1740s)
MD Edinburgh. Surgeon. Employed by H. Sloane and others to collect plants in W. Indies and Central America, 1734–40.
E. J. L. Scott *Index to Sloane Manuscripts* 1904, 368. J. E. Dandy *Sloane Herb.* 1958, 165. A. M. Coats *Quest for Plants* 1969, 333.

MILLAR, Robert Cockburn (1853–1929)
d. Edinburgh 19 April 1929
Accountant, Edinburgh. Fellow of Botanical Society of Edinburgh, 1890.
Trans. Proc. Bot. Soc. Edinburgh 1928–29, 178–80.

MILLARD, Miss (*fl.* 1840s–1870s)
H. J. Riddelsdell *Fl. Gloucestershire* 1948, cxxv.
Plants at Gloucester Museum.

MILLARD, Frederick William (*c.* 1862–1944)
d. East Grinstead, Sussex 18 April 1944
VMH 1932. Had garden at Felbridge, Sussex, specialising in alpine plants. Contrib. to *Field, Game Keeper*.
Gdnrs Chron. 1935 ii 220 portr.; 1944 i 194. *Quart. Bull. Alpine Gdn Soc.* 1944, 88–90.

MILLARD, Walter Samuel (*c.* 1867–1952)
d. Tunbridge Wells, Kent 24 March 1952
Joined firm of Phipson and Co., Bombay in 1887. Secretary and Editor, Bombay Natural History Society. *Some Beautiful Indian Trees* (with E. Blatter) 1927.
Nature v.169, 1952, 690.

MILLEN, Henry (1871–1908)
b. East Woodhay, Hants 1871 *d.* Tobago, Windward Islands 15 Nov. 1908
Kew gardener, 1889. Curator, Botanic Station, Lagos, Nigeria, 1890–98. Curator, Botanic Station, Tobago, 1898–1907. Collected plants.
Kew Bull. 1892, 72. *J. Kew Guild* 1908, 429 portr. *Nigerian Field* v.22, 1957, 130–35.
Plants at Berlin, Kew.

MILLEN, John (–1635)
d. Oct. 1635
Nurseryman, Old Street, Cripplegate, Middx. His nursery had "the choisest fruits this kingdom yeelds" (J. Parkinson *Theatrum Botanicum* 1640, 1456).

MILLER, Charles (1739–1817)
b. Chelsea, London 27 Aug. 1739 *d.* London 6 Oct. 1817
Younger son of P. Miller (1691–1771). First Curator of Cambridge Botanic Garden, 1762–70. Went to India, Sumatra, etc., 1770–72. Experimented with cultivation of wheat. Account of Islands of Sumatra in *Philos. Trans. R. Soc.* v.68, 1779, 160–79.
G. C. Gorham *Mem. John and Thomas Martyn* 1830, 114. *Fl. Malesiana* v.1, 1950, 362.
Letters in Martyn correspondence at BM(NH).

MILLER, Hugh (1802–1856)
b. Cromarty, Ross 10 Oct. 1802 *d.* Edinburgh 23 Dec. 1856
Stone mason. Self-educated geologist; also interested in palaeobotany. *Testimony of the Rocks* 1857.
D.N.B. v.37, 408–10.

MILLER, Hugh Francis Ridley (–1962)
d. Sevenoaks, Kent 12 Nov. 1962
Founder member of Alpine Garden Society. Collected sempervivums. Hybridised irises. President, British Iris Society, 1959.
Iris Yb. 1963, 14–15.

MILLER, John (Johann Sebastian Müller) (1715–*c.* 1790)
b. Nürnberg, Germany 1715 *d.* London *c.* 1790
Draughtsman and engraver. Settled in London, 1744. *Illustratio Systematis Sexualis Linnaei* 1770–77. Executed plates for Lord Bute's *Botanical Tables* 1785 (unused plates at BM(NH); *see J. Bot.* 1916, 84–87). His drawings or engravings also appear in P. Miller's *Figures of Plants* 1755–60; J. Evelyn's *Silva* 1776; J. C. Lettsom's *Tea-tree* 1772, etc.
Gdnrs Chron. 1887 i 451; 1890 i 255–56. *D.N.B.* v.37, 412–14. *J. Bot.* 1913, 255–57; 1919, 353; 1936, 208–9. *Country Life* 1965, 561 portr. F. A. Stafleu *Taxonomic Literature* 1967, 331.
Drawings at BM(NH), Windsor Castle.

MILLER, John (*fl.* 1740s)
Gardener to Lord Petre at Thorndon Hall, Essex.
Essex Nat. 1969, 204–5.

MILLER, John (*fl.* 1820s–1830s)
Nurseryman, Whiteladies Road, Bristol. Successor to Miller and James Sweet, 1786–1808; Sweets and Miller, 1824–37. Went bankrupt in 1837. Business acquired by James Garraway.
J. Harvey *Early Hort. Cat.* 1973, 16.

MILLER, John Frederick (*fl.* 1770s–1790s)
Son of J. S. Miller (1715–*c.* 1790). Went to Iceland with J. Banks in 1772 as draughtsman (*J. Bot.* 1913, 255–57; 1919, 353). *Cimelia Physica* 1776–94. Illustrated R. Weston's *Universal Botanist* 1770–77. His brother James also drew plants for Banks.
E. Smith *Life of Sir Joseph Banks* 1911, 181. *Ibis* 1921, 302–9. F. A. Stafleu *Taxonomic Literature* 1967, 313.
Drawings of Iceland plants at BM(NH) (*J. Bot.* 1907, 314).

MILLER, John W. (*c.* 1822–1902)
Gardener to Lord Foley at Worksop Manor, Notts, 1864.
Gdnrs Chron. 1902 i 381–82.

MILLER, Joseph (–1748)
d. Chelsea, London 29 March 1748
Member of Society of Apothecaries, 1693; Warden, 1737; Master, 1738. Demonstrator at Chelsea Physic Garden, 1740–48. Friend of J. Martyn. *Botanicum Officinale* 1722. List of plants sent to Royal Society from Chelsea (*Philos. Trans. R. Soc.* v.42–46, 1744–52).

Gent. Mag. 1748, 187. D. Turner *Extracts from Lit. and Sci. Correspondence of Richard Richardson* 1835, 188. R. H. Semple *Mem. Bot. Gdn at Chelsea* 1878, 67–72. G. C. Druce and S. H. Vines *Dillenian Herb.* 1907, 3–19.

Icones Plantarum (unpublished) and herb. formerly at Apothecaries Hall.

MILLER, Rev. Joseph Kirkman
(1786–1850s)
b. Bockleton, Worcs 1786

BA Cantab 1808. Vicar, Walkeringham, Notts, 1819–55. MS 'Fl. Walkeringhamensis' in *Naturalist* 1895, 159–71. Contrib. to C. H. J. Anderson's *Short Guide to County of Lincoln* 1847.

MILLER, Oliphant Bell (1882–1966)
b. Scotland 27 May 1882 *d.* Bulawayo, Rhodesia 4 Aug. 1966

FLS 1949. Forest officer, S. Africa. Farmer, N. Rhodesia. Head of Forestry, Bechuanaland. Collected plants in Rhodesia.

Comptes Rendus AETFAT 1960 1962, 188. *Proc. Linn. Soc.* 1968, 145.

MILLER, Philip (1691–1771)
b. Bromley, Greenwich or Deptford 1691 *d.* Chelsea, London 18 Dec. 1771

FRS 1729. His father had a market garden near Deptford. Philip was a commercial florist at St. George's Fields, near Southwark, London. Gardener at Physic Garden, Chelsea, 1722–70. "Hortulanorum princeps" (Linnaeus). *Gardener's and Florist's Dictionary* 1724. *Catalogue of Trees, Shrubs, Plants and Flowers...in Gardens near London* 1730 (original drawings at BM). *Catalogus Plantarum Officinalium quae in Horto Botanico Chelseyano Aluntus* 1730. *Gardener's Dictionary* 8 eds, 1731–68 (Linnean nomenclature adopted in ed. 8 1768). *Gardener's Kalendar* 1732. *Figures of most Beautiful, Useful and Uncommon Plants* 1755–60 2 vols. Lists of plants in R. Pococke's *Description of the East* v.1, 1743, 284; v.2, 1745, 187–96 and notes at end of G. Hughes *Nat. Hist. Barbados* 1750.

Gent. Mag. 1755, 395, 513; 1771, 571; 1815 ii 80. R. Pulteney *Hist. Biogr. Sketches of Progress of Bot. in England* v.2, 1790, 241–50. F. Ehrhart *Beiträge zur Naturkunde* v.6, 1791, 158–78. T. Martyn's ed. of P. Miller's *Gdnrs and Botanists Dict.* 1807, i–x, xxxv. J. Nichols *Illus. of Lit. Hist. of Eighteenth Century* v.1, 321–24. J. E. Smith *Selection from Correspondence of Linnaeus* v.1, 1821, 84, 255. G. W. Johnson *Hist. English Gdning* 1829, 192–97. S. Felton *Portr. English Authors on Gdning* 1830, 138–41. T. Faulkner *Hist. and Topographical Description of Chelsea* v.1, 1829, 250; v.2, 184–86. *Bot. Misc.* 1829–30, 67–69. D. Turner *Extracts from Lit. and Sci. Correspondence of Richard Richardson* 1835, 273–75. J. Rogers *Vegetable Cultivator* 1839, 335–43. *Cottage Gdnr* v.5, 1849, 157; v.7, 1850, 109–10. *J. Hort. Cottage Gdnr* v.56, 1876, 76. R. H. Semple *Mem. Bot. Gdn at Chelsea* 1878, 79–87. *Gdnrs Chron.* 1887 i 451; 1889 ii 219. P. Kalm *Visit to England* 1892, 108–11. *J. Bot.* 1898, 51–54; 1913, 132–35; 1923, 136–37. E. Bretschneider *Hist. European Bot. Discoveries in China* 1898, 107–11. *D.N.B.* v.37, 420–22. E. J. L. Scott *Index to Sloane Manuscripts* 1904, 368. *Garden* 1912, 71–72. *Rhodora* v.29, 1927, 17–19. *J. Heredity* 1932, 437–39. *Svenska Linné-Sällskapets Arsskrift* v.21, 1938, 85–94. J. E. Dandy *Sloane Herb.* 1958, 165–68. *Cat. Bot. Books in Collection of Rachel McMasters Miller Hunt* v.2, 1961, lxxv–lxxvii. C. Wall *Hist. Worshipful Soc. of Apothecaries of London* v.1, 1963, 169–77, 409–12 portr. F. A. Stafleu *Taxonomic Literature* 1967, 313–15. P. Miller *Abridgement to Gdnrs Dict.* 1969 facsimile, v–xv. *J. R. Hort. Soc.* 1971, 556–63. *Taxon* 1972, 646 portr. *Trans. Bot. Soc. Edinburgh* v.41, 1972, 293–307. C. C. Gillispie *Dict. Sci. Biogr.* v.9, 1974, 390–91. *J. Soc. Bibl. Nat. Hist.* 1974, 125–41.

Herb. at BM(NH). Portr. at Kew, Hunt Library. Library sold by Baker and Leigh, 12–15 April 1774.

Milleria L.

MILLER, Miss Rebecca (*fl.* 1810s)

Botanical artist at Liverpool Botanic Garden. Daughter of a "military man." Drew for W. Roscoe. Sketch-book of S. African flowers (including *Strelitzia*) at Liverpool Museums attributed to her.

MILLER, Thomas (1807–1874)
b. Gainsborough, Lincs 31 Aug. 1807 *d.* London 24 Oct. 1874

Poet and novelist. Apprenticed to basket-maker. *Common Wayside Flowers* 1860.

D.N.B. v.37, 424–25.

MILLER, William (*fl.* 1710s–1760s)

Royal gardener, Abbey of Holyrood House, Edinburgh, 1719. Quaker seedsman and nurseryman "at the Abbey."

MILLER, William (1828–1909)
b. Knockdow, Argyllshire 29 Nov. 1828 *d.* Berkswell, Warwickshire 16 April 1909

Gardener to Earl of Craven at Combe Abbey, Warwickshire, 1861.

Gdnrs Chron. 1875 i 465–66 portr.; 1909 i 271–72 portr.

MILLER, William (*c.* 1831–1898)
d. Ballycanew, County Wexford June? 1898

Educ. Trinity College, Dublin. Classical master, Stackpoole School, Kingstown and

Ennis College. *Dictionary of English Names of Plants* 1884.
Garden v.54, 1898, 32.

MILLER, William Duppa (1868–1933)
b. Tupsley, Hereford 5 July 1868 *d.* 7 Nov. 1933
Shipyard engineer, etc. Botanised with Rev. E. S. Marshall. Knew the flora of Devon and Somerset.
Bot. Soc. Exch. Club Br. Isl. Rep. 1933, 510–11. W. K. Martin and G. T. Fraser *Fl. Devon* 1939, 779.

MILLER, William Frederick (1834–1918)
b. Edinburgh 18 Sept. 1834 *d.* Winscombe, Somerset 28 April 1918
Engraver. Discovered *Carex buxbaumii* at Arisaig, Scotland, 1895. Contrib. to *J. Bot.* 1882–1910. 'Sydney Parkinson' (*J. Friends' Hist. Soc.* v.8, 1911, 123–27).
J. Bot. 1918, 221. *Bot. Soc. Exch. Club Br. Isl. Rep.* 1918, 361.
Herb. at Bristol Museum.

MILLETT, Charles (*fl.* 1820s–1830s)
Official of East India Company. At Canton, Ceylon, Malabar and Macao.
Bot. Mag. 1831, t.3058; 1832, t.3148. E. Bretschneider *Hist. European Bot. Discoveries in China* 1898, 298–301.
Letters and Chinese plants at Kew.
Millettia W. & A.

MILLETT, Miss Louisa (1801–1871)
Collected plants in the Isles of Scilly; recorded in *Rep. Penzance Nat. Hist. Antiq. Soc.* 1852.
J. E. Lousley *Fl. Isles of Scilly* 1972, 81.

MILLETT, Miss Matilda (1805–1855)
Collected plants in the Isles of Scilly; recorded in *Rep. Penzance Nat. Hist. Antiq. Soc.* 1852.
J. E. Lousley *Fl. Isles of Scilly* 1972, 81.

MILLIGAN, Joseph (1807–*c.* 1883)
b. Dumfriesshire 1807
MRCS 1829. FLS 1850. Surgeon. Medical Superintendent of convict discipline, Van Dieman's Land Company, 1830. Superintendent of aborigines, Flinders Island, 1843–48. Secretary, Royal Society of Tasmania. Collected plants.
J. D. Hooker *Fl. Tasmaniae* v.1, 1859, cxxvii. *Proc. Linn. Soc.* 1883–84, 36–37. *Papers Proc. R. Soc. Tasmania* 1909, 22–23; 1913, t.xx portr.; 1943, 212–18. *Tasmanian Nat.* 1926, 6–8. T. E. Burns and J. R. Skemp *Van Dieman's Land Correspondents* 1961 *passim. Victorian Nat.* 1970, 145. *J. Proc. R. Austral. Hist. Soc.* v.19, 318.
Plants at BM(NH), Kew. Letters at Kew. Portr. at Hunt Library.
Milligania Hook. f.

MILLINGTON, Sir Thomas (1628–1704)
b. Newbury, Berks 1628 *d.* London 5 Jan. 1704
BA Cantab 1649. MA 1657. MD Oxon 1659. FRCP 1672. FRS. Sedleian Professor of Natural Philosophy, 1675. Royal physician. Alleged discoverer of sexuality in plants. "Sir Thomas Millington, he told me, that he conceived that the entire [stamens] doth serve as the male for the generation of the seed" (N. Grew).
R. Pulteney *Hist. Biogr. Sketches of Progress of Bot. in England* v.1, 1790, 336–37. W. Munk *Roll of R. College of Physicians* v.1, 1878, 363. B. Stillingfleet *Misc. Tracts relating to Nat. Hist.* 1775, xi. *D.N.B.* v.37, 442.
Portr. at Royal College of Physicians; Hunt Library.

MILLS, Charles (*fl.* 1830s)
Nurseryman, Blyth, Notts.

MILLS, Frederick William (1868–1949)
b. Huddersfield, Yorks 26 March 1868 *d.* Monkleigh, Devon 5 Oct. 1949
FRMS 1912. FLS 1918. Solicitor until 1922. Microscopist, especially of *Diatomaceae. Photography applied to the Microscope* 1891. *Introduction to Study of Diatomaceae* 1893. *Index to Genera and Species of Diatomaceae and their Synonyms* 1933–35.
J. R. Microsc. Soc. 1950, 292–93. *J. Quekett Microsc. Club* 1950, 205. *Proc. Linn. Soc.* 1949–50, 117–18.

MILLS, George (*c.* 1787–1871)
b. Hants *c.* 1787 *d.* Ealing, Middx 30 Sept. 1871
Gardener at Gunnersbury Park, Middx, *c.* 1833–53; then rose grower. *Improved Mode of Cultivating the Cucumber and Melon* 1841. *Some Additional Observations on Forcing Cucumbers c.* 1843. *Culture of the Pineapple* 1845.
Florist and Pomologist 1871, 264.

MILLS, William Hobson (1873–1959)
b. 6 July 1873 *d.* 22 Feb. 1959
MA Cantab. FRS 1923. Emeritus Reader in Stereochemistry, Cambridge. Botanised in East Anglia.
Nature v.183, 1959, 929–30. *Proc. Bot. Soc. Br. Isl.* 1959, 365. *Times* 23 Feb. 1959. *Biogr. Mem. Fellows R. Soc.* 1960, 201–25 portr. *Who was Who, 1951–1960* 768.

MILLSON, Alvan (–1896)
Assistant Colonial Secretary, Lagos. Collected, plants in Yoruba, Nigeria, 1890.
Kew Bull. 1891, 206–7.

MILN, George Peddie (1861–1928)
b. near Dundee, Angus 30 Nov. 1861 *d.* Warrington, Lancs 14 Feb. 1928

Managing Director, Messrs Garton Ltd., seed merchants, Warrington. President, Agricultural Seed Trade Association.
N. Western Nat. 1929, 25–27.

MILNE, Rev. Colin (*c.* 1743–1815)
b. Aberdeen *c.* 1743 *d.* Deptford, Kent 2 Oct. 1815
LLD Aberdeen. Rector (non-resident), North Chapel, Petworth, Sussex. *Botanical Dictionary* 1770; ed. 3 1805. *Institutes of Botany* 1771. *Indigenous Botany* (with A. Gordon) 1793.
R. J. Thornton *New Illus. of Sexual System of Linnaeus* 1807 portr. G. W. Johnson *Hist. English Gdning* 1829, 232. *Cottage Gdnr* v.8, 1852, 185. *Notes and Queries* v.4, 1881, 334. G. C. Druce *Fl. Berkshire* 1897, cxliii–cxliv. *D.N.B.* v.38, 6.
Portr. at Hunt Library.
Milnea Roxb.

MILNE, David (1876–1954)
b. Scotland 9 May 1876 *d.* Brechin, Angus 11 Feb. 1954
BSc Aberdeen. Economic botanist, Punjab, 1907–21. Dean of Faculty of Agriculture, Punjab University, 1923–33. Director of Agriculture, Punjab, 1922–33. Selected new types of long staple cotton in Punjab. *Date Palm and its Cultivation in Punjab* 1918. *Handbook on Field and Garden Crops in the Punjab* (with Ali Muhammad) 1931.
Nature v.173, 1954, 425. *Who was Who, 1951–1960* 769.

MILNE, J. (*fl.* 1800s–1810s)
Gardener, Fonthill, Wilts, later nurseryman of same place.

MILNE, Joshua (1776–1851)
d. Upper Clapton, London 4 Jan. 1851
FLS 1834. Actuary. Studied mosses and hepatics.
Proc. Linn. Soc. 1851, 136–37. *D.N.B.* v.38, 8–9.

MILNE, Thomas (*c.* 1767–1838)
d. 29 Jan. 1838
Nurseryman, Fulham, Middx in partnership with Whitley and Brames, 1810–33.
Gdnrs Mag. 1835, 160.

MILNE, Thomas (*fl.* 1790s–1840s)
ALS 1795. Curator, Botanic Garden, Oxford before 1796. Contrib. to W. Withering's *Arrangement* ed. 3 1796 (v.1, xii).
L. W. Dillwyn *Materials for Fauna and Fl. Swansea* 1848, 31.

MILNE, Thomas (*c.* 1823–1910)
d. Aberdeen Sept. 1910
Nurseryman, Aberdeen.
Gdnrs Mag. 1910, 757.

MILNE, William Grant (–1866)
b. Scotland *d.* Creek Town, Old Calabar 3 May 1866
Gardener, Royal Botanic Garden, Edinburgh. Botanist on 'Herald' Expedition to Fiji, 1852–56. Collected plants in West Africa, 1862–66. Letter from Fiji in *Hooker's J. Bot.* 1857, 106–15; from Gabon in *J. Bot.* 1865, 193–95.
B. C. Seemann *Fl. Vitiensis* 1865, vii. *Gdnrs Chron.* 1866, 731. *J. Bot.* 1866, 272. *Trans. Bot. Soc. Edinburgh* v.8, 1866, 71–73, 485–86. *Rep. Austral. Assoc. Advancement Sci.* 1909, 381. *Fl. Malesiana* v.1, 1950, 363. *R.S.C.* v.4, 396; v.8, 408.
Plants at BM(NH). Letters, MSS. and Lord Howe Island plants at Kew.
Polypodium milnei Hook.

MILNE-REDHEAD, George Bertram (1866–1951)
b. Manchester 1866 *d.* Cheltenham, Glos 27 Oct. 1951
MA Cantab. Son of R. Milne-Redhead (1828–1900). Barrister. Contrib. botanical notes to *Gdnrs Chron.* Supplied records to H. J. Riddelsdell's *Fl. Gloucestershire* 1948 and D. Grose's *Fl. Wiltshire* 1957.
Proc. Cotteswold Nat. Field Club 1951, 64.

MILNE-REDHEAD, Humphrey (1906–1974)
b. Batcombe, Somerset 27 May 1906 *d.* Dumfries 16 March 1974
MB, ChB Edinburgh 1937. Rubber planter in Malaya for 5 years. Physician, Darlington, 1940–45; Mainsriddle, Kirkcudbrightshire, 1947–73. Botanised in Scotland. 'Checklist of Flowering Plants of Dumfries, Kirkcudbright and Wigtown' (*Trans. Dumfriesshire Galloway Nat. Hist. Soc.* v.49, 1972, 1–19).
B.S.B.I. News v.3(2), 1974, 15–16. *B.S.E. News* no. 15, 1975, 9–11. *Watsonia* 1975, 449–50.
Herb. at Dumfries Museum. Bryophytes at University College Bangor.

MILNE-REDHEAD, Richard (1828–1900)
b. Manchester 16 Jan. 1828 *d.* Clitheroe, Lancs 24 Feb. 1900
FLS 1865. Barrister. Collected plants in India, West Indies, Brazil, etc. 'Desert Fl. of Sinai' (*J. Linn. Soc.* 1866, 208–29). Contrib. to *Phytologist*, *Garden*, *Gdnrs Chron.*
Proc. Linn. Soc. 1900–1, 47–48. *Trans. Liverpool Bot. Soc.* 1909, 82. *Dalesman* 1975, 796–97. *R.S.C.* v.8, 713.
Letters and Syrian plants at Kew.

MILNER, Edward (–1884)
b. Darley, Derbyshire *d.* Norwood, Middx 26 March 1884
FLS 1878. Horticulturist and landscape gardener. Landscaped gardens at Crystal Palace, Sydenham and many municipal parks.

Gdnrs Chron. 1884 i 459. *Proc. Linn. Soc.* 1883–84, 37.

MILNER, Henry Ernest (*c.* 1845–1906)
b. Derbyshire *c.* 1845 *d.* 10 March 1906
VMH 1897. Son of E. Milner (–1884). Landscape gardener; was closely connected with Earl's Court Exhibition, 1892.
Gdnrs Chron. 1906 i 175.

MILNER, James Donald (1874–1927)
b. 20 Nov. 1874 *d.* 15 Aug. 1927
Keeper and Secretary, National Portrait Gallery, 1916. *Catalogue of Portraits of Botanists in Museums of Royal Botanic Gardens, Kew* 1906.
Who was Who, 1916–1928 735.

MILROY, A. J. W. (–1936)
b. 9 Oct. 1883 *d.* Shillong, Assam 26 Sept. 1936
To Assam, 1908. Conservator of Forests.
Indian Forester 1936, 761–62.
Pasania milroyii Purkay Das.

MILSOM, Francis Eric (1886–1945)
b. London 24 May 1886 *d.* Kirkburton, Yorks 5 Dec. 1945
BSc London 1908. Pharmaceutical chemist, I.C.I., Huddersfield. Nephew of W. H. Burrell. Bryologist. Compiled *Yorkshire Liverworts* 1946.
Naturalist 1946, 83–84 portr.; 1961, 160. *Rep. Br. Bryol. Soc.* 1946, 319.
Herb. at Leicester University.

MILSUM, John Noel (*c.* 1890–1945)
b. Hants *c.* 1890 *d.* 4 Jan. 1970
FLS 1921. Kew gardener, 1913. Assistant Superintendent, Government Plantations, Federated Malay States, 1913. State Agricultural Officer, Perak, 1937. Chief Field Officer, Malayan Agricultural Service, 1946–48. Director of Agriculture, Seychelles, 1948–50. Discovered the grass, *Eulalia milsumi*.
Kew Bull. 1913, 314. *Fl. Malesiana* v.1, 1950, 363. *J. Kew Guild* 1970, 1161–62 portr.
Plants at Kew, Kuala Lumpur.

MINIER, Charles (*c.* 1710–1790)
d. Kew, Surrey 23 March 1790
Seedsman, The Orange Tree, Strand, London. In partnership with John Mason and Robert Teesdale.
J. Harvey *Early Hort. Cat.* 1973, 21.

MINIER, Charles (*c.* 1738–1808)
d. London Dec. 1808
Son of C. Minier (*c.* 1710–1790). Seedsman, The Orange Tree, Strand, London. In partnership with his brother William.
J. Harvey *Early Hort. Cat.* 1973, 21.

MINIER, William (*fl.* 1800s–1830s)
Son of C. Minier (*c.* 1710–1790). Seedsman, The Orange Tree, Strand, London. In partnership with his brother Charles.
J. Harvey *Early Hort. Cat.* 1973, 21.

MITCHELL, Dr. (*fl.* 1690s)
Irish physician and botanist. "Dr. Wood, Dr. Mitchell and I have resolved to be as curious as our leisures will permit in making a collection of what plants this kingdom affords" (J. Ray).
J. Bot. 1911, 125.

MITCHELL, Anna Helena (1794–1882)
b. Gothenburg, Sweden 22 May 1794 *d.* Montrose, Angus 14 Jan. 1882
Lichenologist and algologist. Worked with A. Croall and J. Gilchrist.
Plants at Montrose Museum.

MITCHELL, George (*fl.* 1790s)
Seedsman and florist, 19 New Bond Street, London.
Trans. London Middlesex Archaeol. Soc. 1973, 187.

MITCHELL, James (*fl.* 1810s–1850s)
b. Wooler, Northumberland
Surgeon, Royal Navy. Friend of G. Johnston (1797–1855). Described *Mentha crispa* in J. Sowerby and J. E. Smith's *English Bot.* 2785.
G. Johnston *Bot. Eastern Borders* 1853, 161. *R.S.C.* v.4, 408.

MITCHELL, James (*c.* 1808–1873)
d. 10 May 1873
Nurseryman, Piltdown Nursery, Maresfield, Sussex.

MITCHELL, John (1711–1768)
b. Lancaster County, Virginia 1711 *d.* 29 Feb. 1768
Educ. Edinburgh University. MD Leyden 1719. FRS 1748. Returned to Virginia by 1735 where he was a physician at Urbanna. In London, 1746. Botanised with Duke of Argyll in N. Scotland. *Nova Plantarum Genera* 1741. *Dissertatio Brevis de Principiis Botanicorum* 1769.
R. Pulteney *Hist. Biogr. Sketches of Progress of Bot. in England* v.2, 1790, 278–81. J. E. Smith *Selection from Correspondence of Linnaeus* v.1, 1821, 34; v.2, 399, 442–51, etc. A. L. A. Fée *Vie de Linné* 1832, 149–50. W. Darlington *Reliquiae Baldwinianae* 1843, 363–67. P. Kalm *Visit to England* 1892 ed. *passim*. H. A. Kelly *Some Amer. Med. Bot.* 1915, 33–37. *D.N.B.* v.38, 70. *Annual Rep. Amer. Hist. Assoc.* 1918, 199–219. *Virginia Mag. Hist.* 1931, 126–35, 206–20; 1932, 48–62, 97–110, 268–79, 335–46; 1933, 59–70, 144–56; 1962, 43–48; 1968, 437–43. *Library of Congress J. of Current Acquisitions* v.1(4), 1944, 36–38. *Osiris* 1948, 99–100. *Bull. Med.*

Hist. 1957, 132–36; 1964, 241–59. *Virginia Cavalcade* no. 4, 1963, 32–39. E. and D. S. Berkeley *Dr. John Mitchell* 1974.
Mitchella L.

MITCHELL, John (*c.* 1761–1838)
d. Moncrieffe House, Perthshire 7 Feb. 1838
Gardener at Moncrieffe House.
Gdnrs Mag. 1838, 640.

MITCHELL, John (1762–)
b. London 1762
Land steward. Of Stanstead, Sussex and Keighley, Yorks. *Dendrologia* 1827, 15–18.

MITCHELL, John (1814–1904)
b. near Southampton, Hants 1814 *d.* Escrick, Yorks 20 Feb. 1904
Gardener to Lord Wenlock at Escrick, 1854.
Gdnrs Chron. 1904 i 157.

MITCHELL, M. D. *see* Stelfox, M. D.

MITCHELL, Richard (*fl.* 1830s)
Seedsman, Mitchells Yard, Lower Clapton, Hackney, London.

MITCHELL, Sir Thomas Livingstone (1792–1855)
b. Craigend, Stirlingshire 15 May 1792 *d.* Darling Point, N.S.W. 5 Oct. 1855
DCL Oxon 1839. FRS 1839. Lieut.-Colonel, Royal Artillery. Deputy Surveyor-General, N.S.W., 1827. Led expedition in N.S.W. 1831 and collected plants. On expeditions to find Murray River, 1835, 1836. In N. Australia, 1845–47. *Three Expeditions into Interior of Eastern Australia* 1838 2 vols. *Journal of Expedition into Interior of Tropical Australia* 1848.
London J. Bot. 1847, 364–72. J. D. Hooker *Fl. Tasmaniae* v.1, 1859, cxx–cxxi. J. E. T. Woods *Hist. Discovery and Exploration of Australia* v.1, 1865, 366–95. W. Woolls *Lectures on Vegetable Kingdom* 1879, 43–45. E. Favenc *Hist. Austral. Exploration from 1788–1888* 1888, 103–15. *D.N.B.* v.38, 74–76. *Victorian Nat.* 1904, 19; 1936, 113–19; 1948, 199–201, 217; 1949, 86–87. J. H. L. Cumpston *Thomas Mitchell* 1954. L. B. Gardiner *Thomas Mitchell* 1962. F. A. Stafleu *Taxonomic Literature* 1967, 325.
Plants at BM(NH), Kew. Letters at Kew. Portr. at Hunt Library.
Capparis mitchellii Lindl.

MITCHELL, W. F. (*c.* 1875–1949)
b. Birmingham *c.* 1875 *d.* Leek Wootton, Warwickshire 6 Dec. 1949
Raised narcissi.
Daffodil Tulip Yb. 1950, 144 portr.

MITCHELL, W. J. W. (*c.* 1874–1965)
VMH 1947. Gardener at Westonbirt, Glos.
Gdnrs Chron. 1948 i 138 portr.

MITCHELL, William (*fl.* 1820s)
Nurseryman, 1 Borough Road, London.

MITCHELL, William (–1873)
d. Edinburgh 10 April 1873
Associate, Botanical Society of Edinburgh, 1858. Contrib. papers on internodes, etc. to *Trans. Bot. Soc. Edinburgh* v.6, 1860, 31–35; v.10, 1870, 288–95.
Trans. Bot. Soc. Edinburgh v.12, 1876, 29–31.

MITCHELL, William Stephen (1840–1892)
b. Bath, Somerset 25 March 1840 *d.* London 9 Nov. 1892
BA Cantab 1866. FLS 1867. 'Alum Bay Leaf-bed' (*Rep. Br. Assoc. Advancement Sci.* 1866, 146–48 (specimens at BM(NH)). 'Introduction of Potato' (*Antiquary* v.13, 1886, 146–50).

MITCHELSON, George (*fl.* 1780s–1800s)
Nurseryman, The Oval, Kennington, Surrey. In partnership with James Mitchelson. Continued as Mitchelsons' until about 1835. In 1803 the business was in the hands of J. Mitchelson who is said to have lived to 100. Also had nursery at Kingston, Surrey.
Trans. London Middlesex Archaeol. Soc. v.24, 1973, 188.

MITCHINSON, John (–1901)
d. Feb. ? 1901
Nurseryman, Quay Street, Truro, Cornwall. Acquired St. Austell Nursery belonging to his grandfather. Opened branch at Penzance.
Gdnrs Chron. 1901 i 163.

MITCHINSON, Rev. John (1833–1918)
b. Durham 23 Sept. 1833 *d.* Gloucester 25 Sept. 1918
Educ. Oxford. Ordained 1858. Headmaster, King's School, Canterbury, Kent, 1859–73. Bishop of Barbados and Windward Islands, 1873–81. Master, Pembroke College, Oxford.
Bot. Soc. Exch. Club Br. Isl. Rep. 1918, 261–64. *Nature* v.102, 1918, 93. *Who was Who, 1916–1928* 737. A. R. Horwood and C. W. F. Noel *Fl. Leicestershire* 1933, ccxx.
Herb. of British plants at Oxford.

MITFORD, Algernon Bertram Freeman-, 1st Baron Redesdale (1837–1916)
b. London 24 Feb. 1837 *d.* Batsford Park, Glos 17 Aug. 1916
FLS 1896. VMH 1904. Diplomatic service, 1858. In Japan, 1866–70. Secretary, Office of Works, 1874–86. Responsible for improvements in gardens of Buckingham Palace, Hyde Park, St. James's Park. Always sympathetic to improvements at Kew Gardens. Cultivated hardy bamboos at Batsford. *Bamboo Garden* 1896. *Memories* 1915 portr. *Further Memories* 1917.

Garden v.62, 1902, iv portr.; 1916, 423 portr. *Gdnrs Chron.* 1916 ii 100 portr. *J. Bot.* 1916, 375. *Kew Bull.* 1916, 237. *Proc. Linn. Soc.* 1916–17, 45–46. *Times* 18 Aug. 1916. *D.N.B. 1912–1921* 380–82. *Curtis's Bot. Mag. Dedications, 1827–1927* 279–80 portr. *Who was Who, 1916–1928* 874–75. *R.S.C.* v.15, 113.

MITTEN, William (1819–1906)
b. Hurstpierpoint, Sussex 30 Nov. 1819 *d.* Hurstpierpoint 27 July 1906

ALS 1847. Bryologist. Sussex mosses in *Ann. Mag. Nat. Hist.* v.8, 1851, 51–59, 305–24, 362–70. *Catalogue of British Mosses* 1866. *Musci Austro-Americani* 1869.

Bot. Soc. Exch. Club Br. Isl. Rep. 1906, 203. *J. Bot.* 1906, 329–32 portr. *J. New York Bot. Gdn* Feb. 1907, 28–32. *Gdnrs Chron.* 1907 i 89. *Bryologist* 1907, 1–5. *Kew Bull.* 1906, 283–84. *Proc. Linn. Soc.* 1906–7, 49–54. A. H. Wolley-Dod *Fl. Sussex* 1937, xliii. *Gdn J. New York Bot. Gdn* 1964, 146–48 portr. *R.S.C.* v.4, 416; v.8, 412; v.10, 823.

Mosses and hepatics at New York Botanical Garden. Letters and note-books at Kew. Portr. at Hunt Library.

Mittenia Lindb.

MITTON, Richard (*fl.* 1830s)
Nurseryman, Pontefract, Yorks.

MOBBS, Eric Charles (–*c.* 1972)
BA Cantab. Indian Forest Service, 1922. Silviculturist, United Provinces. Principal and Professor of Forestry, Dehra Dun, 1938. Director of Forest Education, 1945–47. Professor, Forestry School, Bangor, 1947–67. President, Society of Foresters of Great Britain, 1960, 1961.

Commonwealth For. Rev. 1973, 12–13.

MOCKERIDGE, Florence Annie
(*c.* 1889–1958)
d. 18 Dec. 1958

BSc London 1909. DSc 1917. FLS 1919. Did research for Professor W. B. Bottomley, 1911–17. Demonstrator and Lecturer in Botany, King's College, London, 1917–22. Lecturer in Botany, University College, Swansea, 1922; Professor, 1936–54. Edited *Proc. Swansea Field Nat. Soc.*

Nature v.183, 1959, 150. *Proc. Linn. Soc.* 1958–59, 135.

MOFFAT, Charles Bethune (1859–1945)
b. Isle of Man 16 Jan. 1859 *d.* Dublin 14 Oct. 1945

BA Dublin 1879. Called to the Bar, 1881. President, Dublin Naturalists' Field Club, 1906; Secretary, 1914–15; Treasurer, 1922–35. Contrib. to N. Colgan and R. W. Scully's *Contributions towards a Cybele Hibernica* 1898, R. L. Praeger's *Irish Topographical Botany* 1901. Also contrib. to *Irish Nat.*

Irish Nat. 1946, 349–70 portr.

MOGGRIDGE, John Traherne (1842–1874)
b. Woodfield, Mon 8 March 1842 *d.* Mentone, France 24 Nov. 1874

FLS 1869. Son of M. Moggridge (1803–1882). Studied *Ophrys* (*J. Bot.* 1866, 167–68; 1867, 317–18). *Contributions to Fl. Mentone* 1864–68.

Gdnrs Chron. 1874 ii 723. *J. Bot.* 1875, 63–64. *Proc. Linn. Soc.* 1874–78, lxi. *Bull. Soc. Bot. France* 1883, cxxiii–cxxiv. *Boissiera* fasc. 5, 1941, 64. *R.S.C.* v.8, 415.

Plants, drawings and letters at Kew. Portr. at Hunt Library.

MOGGRIDGE, Matthew (1803–1882)
b. 16 July 1803 *d.* Kensington, London 14 July 1882

FLS 1877. Collected plants figured in his son's *Fl. Mentone*. 'On the Zones of the Coniferae from the Mediterranean to the Crest of the Maritime Alps' (*J. Bot.* 1867, 48–50).

Proc. Linn. Soc. 1882–83, 42–43. *Trans. Cardiff Nat. Soc.* 1952–53, 13. *R.S.C.* v.4, 421.

Plants at Kew. Letters in Broome correspondence at BM(NH).

MOLE, Fanny Elizabeth De *see* De Mole, F. E.

MOLESWORTH, Caroline (1794–1872)
b. Pencarrow, Cornwall 4 Nov. 1794 *d.* Cobham, Surrey 29 Dec. 1872

Cobham Journals...Phenological Observations...1825 to 1850 1880.

J. Sowerby and J. E. Smith *English Bot.* 2613. *J. Bot.* 1874, 209; 1882, 28.

Plants and letters at Kew.

MOLESWORTH, Richard (*fl.* 1780s)
Of Peckham, London. "A gentleman... anxious to promote the cause of Botany."

Bot. Mag. 1793, t.225.

MOLINEUX, Rev. James (1791–1873)
b. Leigh, Lancs 1791 *d.* Rochdale, Lancs 13 Nov. 1873

Warehouseman, Manchester. Schoolmaster, Wilmslow. Nonconformist itinerant preacher. Methodist minister, Rochdale, 1837. Lectured on botany. *Botany made Easy* 1867. Herb. in possession of Rochdale Pioneers.

Rochdale Observer Nov. 1873. *Trans. Liverpool Bot. Soc.* 1909, 82.

MOLLOY, Georgiana (*née* **Kennedy**)
(1805–1843)
b. Carlisle, Cumberland 23 May 1805 *d.* Vasse, W. Australia 8 April 1843

Emigrated with her husband to Augusta, W. Australia, 1830. Collected plants and seeds for Capt. J. Mangles. "British Herbaria are indebted [to her] for many valuable communications" (Meisner in *Hooker's J. Bot.* 1855, 382).

J. Proc. W. Austral. Hist. Soc. v.1, 1927–31, 30–84; v.4, 1949–54, 65. A. Hasluck *Portr. with Background: a Life of Georgiana Molloy* 1955. *Austral. Encyclop.* v.6, 1965, 117. *Austral. Dict. Biogr.* v.2, 1967, 244–45. R. Erickson *Drummonds of Hawthornden* 1969, 26–29, 86–87.

Plants at Kew.

Molloya Meisner.

MOLONEY, Sir Cornelius Alfred (1848–1913)
d. Fiesole, Italy 13 Aug. 1913

KCMG 1890. In W. Africa from 1867. Governor of Lagos, Nigeria, 1887–91. In West Indies, 1891–1904. Founded Botanic Station, Lagos, 1887. *Sketch of Forestry of West Africa* 1887.

Kew Bull. 1888, 149–56. *Geogr. J.* v.42, 1913, 501–2.

Plants and economic products at Kew.

MOLONEY, Hon. Evelyn R. (*née* **Napier**) (–1952)

Botanist at Coryndon Museum, Nairobi, Kenya. Collected plants for Kew.

Kew Bull. 1952, 439.

Dicliptera napierae E. A. Bruce.

MOLY, James (–1910)
d. Charmouth, Dorset 15 April 1910

Collected and grew British ferns. Of Axminster, Devon and Charmouth.

Br. Fern Gaz. 1909, 46; 1910, 86–87.

MOLYNEUX, Edwin (1851–1921)
b. Sproxton, Yorks 1851 *d.* Bishop's Waltham, Hants 12 Nov. 1921

VMH 1897. Gardener to W. H. Myers at Swanmore Park, Bishop's Waltham, 1878. *Chrysanthemums and their Culture* 1886. *Grape Growing for Amateurs* 1891.

J. Hort. Cottage Gdnr v.11, 1885, 450–51 portr.; v.62, 1911, 100 portr. *Gdnrs Chron.* 1903 ii 219 portr.; 1921 ii 281–82 portr. *Gdnrs Mag.* 1908, 156–57 portr. *Garden* 1921, 611 portr.

MOLYNEUX, Herbert Ernest (1868–1916)
d. Southampton, Hants 22 Nov. 1916

Treasurer, National Rose Society, 1904–8. Co-editor of ed. 4 of A. Forster-Melliar's *Book of the Rose* 1910.

J. Hort. Cottage Gdnr v.47, 1903, 398 portr. *Garden* 1916, 589–90 portr. *Gdnrs Chron.* 1916 ii 273. *Gdnrs Mag.* 1916, 480 portr. *National Rose Soc. Annual* 1917, 133.

MOLYNEUX, N. (*fl.* 1880s–1900s)

Gardener at Rookesbury Park, Wickham, Hants. Raised chrysanthemums.

Gdnrs Mag. 1909, 271–72 portr.

MOLYNEUX, Sir Thomas (1661–1733)
b. Dublin 14 April 1661 *d.* Dublin? 19 Oct. 1733

BA Dublin 1680. MD 1687. FRS 1686. Baronet 1730. Physician to the Forces. Professor of Medicine, Dublin, 1717. Friend of W. Sherard. Discovered *Saxifraga umbrosa* 1697.

C. Threlkeld *Synopsis Stirpium Hibernicarum* 1727, Appendix. *Gent. Mag.* 1733, 607. R. Pulteney *Hist. Biogr. Sketches of Progress of Bot. in England* v.2, 1790, 196. *D.N.B.* v.38, 137–38. W. A. Clarke *First Rec. of Br. Flowering Plants* 1897, 52. T. P. C. Kirkpatrick *Hist. Med. Teaching in Trinity College, Dublin* 1912, 362. H. F. Berry *Hist. R. Dublin Soc.* 1915, 7–8. N. Colgan *Fl. County Dublin* 1904, xxi–xxii. H. N. Clokie *Account of Herb. Dept. Bot., Oxford* 1964, 213.

MONCKTON, Horace Woollaston (1857–1930)
d. 14 Jan. 1930

FLS 1892. Barrister. Treasurer, Linnean Society, 1905–30. President, Geologists' Association, 1902–4. *Fl. Bagshot District* 1916. *Fl. District of Thames Valley Drift* 1919.

Bot. Soc. Exch. Club Br. Isl. Rep. 1931, 629. *J. Bot.* 1931, 55–56. *Nature* v.127, 1931, 206. *Proc. Linn. Soc.* 1930–31, 186–88.

Herb. at Reading.

MONEY, Daniel (*c.* 1763–1833)
d. 4 April 1833

Nurseryman, Haverstock Nursery, Hampstead Road, London.

Gdnrs Mag. 1829, 737–40; 1833, 384.

MONGREDIEN, Augustus (*c.* 1806–1888)
d. Forest Hill, London 30 March 1888

Nurseryman, Heatherside Nurseries, Farnham, Surrey. *Trees and Shrubs for English Plantations* 1870. *Heatherside Manual of Hardy Trees and Shrubs* 1874–75.

J. Hort. Cottage Gdnr v.16, 1888, 299; v.30, 1895, 296.

MONINGTON, Harold Warren (1867–1924)
b. Plumstead, Kent 11 Aug. 1867 *d.* Swanage, Dorset 5 Sept. 1924

FLS 1898–1923. Studied British botany, especially mosses. Contrib. mosses to *Victoria County History of Surrey* v.1, 1902, 51–56. Contrib. to *J. Bot.*

J. Bot. 1925, 275.

Mosses at BM(NH).

MONKMAN, Charles (*fl.* 1860s)
Of Malton, Yorks. Fern collector.
Br. Fern Gaz. 1909, 46.

MONRO, Claude Frederick Hugh (1863–1918)
b. London 29 April 1863. *d.* Weybridge, Surrey 14 Aug. 1918
Collected plants in S. Rhodesia, 1900–16. 'Grasses in Rhodesia' (*Proc. Rhodesia Sci. Assoc.* v.6, 1906, 5–67). Sent seeds to Kew.
J. Bot. 1918, 335.
Plants at BM(NH), Cape Town.
Fockea monroi S. Moore.

MONRO, Sir David (1813–1877)
b. Edinburgh 1813 *d.* Newstead near Nelson, New Zealand 15 Feb. 1877
Speaker, New Zealand Parliament, 1861–62. Sent New Zealand plants to Kew. 'Botany of Nelson and Marlborough' (*Trans. N.Z. Inst.* 1868 Essays, 6–21).
Nature v.16, 1877, 15–16. *D.N.B.* v.38, 182. T. F. Cheeseman *Manual of New Zealand Fl.* 1925, xxvi.
Senecio monroi Hook. f.

MONRO, George (*c.* 1847–1920)
d. 30 May 1920
VMH 1897. Founder of wholesale horticultural firm at Covent Garden, London.
Gdnrs Mag. 1907, 483 portr. *Garden* 1920, 305. *Gdnrs Chron.* 1920 i 287 portr.

MONRO, George (1876–1951)
b. 24 March 1876 *d.* Finchley, London 14 Nov. 1951
VMH 1934. Son of G. Monro (*c.* 1847–1920). Covent Garden merchant. Founder and President, British Florists' Federation. Treasurer, Royal Horticultural Society, 1938–42.
Gdnrs Chron. 1935 i 88; 1951 ii 201. *Orchid Rev.* 1952, 16. *Daffodil Tulip Yb.* 1953, 83–89. *Who was Who, 1951–1960* 778.

MONSON, Lady Anne (*née* **Vane**) (*c.* 1714–1776)
d. Calcutta 18 Feb. 1776
Botanised at the Cape with C. P. Thunberg and F. Masson, 1774, and also in India with her husband, George Monson, an officer in the Army. Suggested and helped in J. Lee's *Introduction to Botany* 1760.
H. C. Andrews *Botanist's Repository* 1803, t.276. C. P. Thunberg *Fl. Capensis* v.1, 1807, 12. *D.N.B.* v.38, 196. *J. Bot.* 1918, 147–49. E. J. Willson *James Lee and the Vineyard Nursery, Hammersmith* 1961, 39–40.
Monsonia L.

MONTGOMERY, Archibald Sim (1844–1922)
b. Brentford, Middx 1844 *d.* Hilperton, Wilts 12 April 1922
Timber merchant, Brentford. Came to Cheltenham, Glos, 1910.
Bot. Soc. Exch. Club Br. Isl. Rep. 1922, 707. H. J. Riddelsdell *Fl. Gloucestershire* 1948, cxlviii.
Herb. at Cheltenham and Gloucester Museums.

MONTGOMERY, Duncan (*c.* 1778–1857)
b. Comrie, Perthshire *c.* 1778 *d.* Buchanan House, Stirlingshire 15 Sept. ? 1857
Gardener at Cambridge, Kew and to Duke of Montrose at Buchanan House.
Gdnrs Chron. 1857, 680.

MOODY, Robert (*fl.* 1730s)
Gardener to City of Dublin; also had a seed shop.
Irish For. 1967, 48.

MOODY, Thomas (*fl.* 1830s–1840s)
Nurseryman, Castle Street, Salisbury, Wilts. In partnership with Mr. Geary.

MOON, Alexander (–1825)
b. Scotland *d.* Ceylon 1825
Kew gardener, 1815. Collected plants at Gibraltar and in Barbary. Superintendent, Botanic Garden, Peradeniya, Ceylon 1817–25. Formed herb. and collection of flower drawings. *Catalogue of Indigenous and Exotic Plants...in Ceylon* 1824.
Kew Bull. 1891, 304. *Ann. R. Bot. Gdns, Peradeniya* v.1, 1901, 3–6. H. Trimen *Handbook to Fl. Ceylon* v.5, 1900, 373–74. *J. Bot.* 1906, 281.
Plants at BM(NH), Kew. MS. (1823) with native drawings at BM(NH).
Moonia Arn.

MOON, Henry George (1857–1905)
b. London 18 Feb. 1857 *d.* St. Albans, Herts 6 Oct. 1905
Artist. In 1880 joined art dept. of *Garden* which he illustrated for many years. Illustrated F. Sander's *Reichenbachia* 1886–94 and W. Robinson's *English Flower Garden* 1883 and *Fl. and Sylva* 1903–5.
Fl. and Sylva v.3, 1905, 346–48 portr. *Garden* v.68, 1905, 234, 267–68. *Gdnrs Chron.* 1905 ii 287 portr. *Orchid Rev.* 1905, 374. *J. Bot.* 1906, 182–83. W. Blunt *Art. of Bot. Illus.* 1955, 239–40. *Amer. Orchid Soc. Bull.* 1965, 329–30 portr. M. A. Reinikka *Hist. of the Orchid* 1972, 283–86.

MOON, John McKay (1901–1960)
b. Portrush, County Antrim 1901
Teacher. President, Belfast Naturalists' Field Club, 1948–49. Assisted in *Fl. Fermanagh* (unpublished). Contrib. to *Irish Nat. J.*
Irish Nat. J. 1961, 217–20 portr.

MOONEY, Herbert Francis (1897–1964)
b. 6 Nov. 1897 *d.* 20 Aug. 1964

ScD Dublin. FLS 1946. Indian Forest Service, 1921–47. Conservator of Forests, Orissa. British Middle East Office, Beirut, 1947–62. Collected plants in Middle East, especially Ethiopia. *Supplement* to H. H. Haines' *Botany of Bihar and Orissa* 1950. *Glossary of Ethiopian Plant Names* 1963. Contrib. to *Indian Forest Rec.*, *J. Indian Bot. Soc.*

Indian Forester 1965, 204. *Commonwealth For. Rev.* 1964, 271–72. *Proc. Linn. Soc.* 1965, 224–25. *Irish Nat. J.* 1967, 316–18. *Who was Who, 1961–1970* 797–98.

Plants at Kew, National Herb. Addis Ababa, Ethiopia.

MOOR, Samuel Albert (–1944)

MA Cantab. Lecturer in Botany, University College, Aberystwyth. Principal, Grasia College, India. Headmaster, Kendal Grammar School, 1908–24. Translated W. Detmer's *Das Pflanzenphysiologische Praktikum.*

Who was Who, 1941–1950 820–21.

MOORCROFT, William (*c.* 1765–1825)
b. Lancs *c.* 1765 *d.* Andekhui, Afghanistan 15 Aug. 1825

Veterinary surgeon in East India Company; in Bengal, 1808. Collected plants with N. Wallich in Nepal. Travelled in N.W. India, etc., 1819–25. *Travels in Himalayan Province of Hindustan and the Panjab* (with G. Trebeck) 1841 (with biogr.).

N. Wallich *Plantae Asiaticae Rariores* v.3, 1832, 7–9. J. F. Royle *Illus. of Bot....of the Himalayan Mountains* 1839, 2. *J. R. Geogr. Soc.* v.1, 1832, 233–47. *D.N.B.* v.38, 337–38. *Trans. Liverpool Bot. Soc.* 1909, 82. D. G. Crawford *Hist. Indian Med. Service, 1600–1913* v.2, 1914, 141. *Pakistan J. For.* 1967, 352. *R.S.C.* v.4, 455.

Plants at BM(NH).

Moorcroftia Chois.

MOORE, Alexander (*c.* 1855–1884)
d. Campbelltown, N.S.W. 14 March 1884

Kew gardener. Superintendent, State Nursery, N.S.W.

Garden v.25, 1884, 378.

MOORE, Charles (1820–1905)
b. Dundee, Angus 10 May 1820 *d.* Sydney, N.S.W. 30 April 1905

FLS 1863. Brother of D. Moore (1807–1879). Kew gardener, 1847. Arrived in Sydney, 1848. Director, Botanic Gardens, Sydney, 1848–96. Toured S. Pacific in HMS 'Havannah' in 1850 and collected plants. Also collected in Lord Howe Island, 1869. *Lord Howe's Island* 1869. *Census of Plants of N.S.W.* 1884. *Handbook of Fl. N.S.W.* (with E. Betche) 1893. *Catalogue of Plants in Government Botanic Gardens, Sydney* 1895. Contrib. to *J. Proc. R. Soc. N.S.W.*

J. D. Hooker *Fl. Tasmaniae* v.1, 1859, cxxiii–cxxiv. *Garden* v.32, 1887 portr. *Gdnrs Chron.* 1894 ii 185 portr.; 1905 i 299 portr. *J. Bot.* 1905, 280. *J. Kew Guild* 1905, 264–66 portr. *Rep. Sydney Bot. Gdn* 1905, 4–5. *Trans. Proc. Bot. Soc. Edinburgh* 1907, 212–13. *J. R. Soc. N.S.W.* 1908, 114–15. *J. Proc. R. Austral. Hist. Soc.* 1932, 126–28. *Fl. Malesiana* v.1, 1950, 369. *Austral. Encyclop.* v.6, 1965, 143. *R.S.C.* v.8, 430.

Plants at Kew, Sydney. Letters at Kew. Portr. at Hunt Library.

Eucryphia moorei F. Muell.

MOORE, Clifford (1870–1944)

Chemist, Rugeley, Staffs. Botanised in Staffs.

E. S. Edees *Fl. Staffordshire* 1972, 21.

Plants at Mid-Staffs Field Club.

MOORE, David (*olim* **Muir**) (1807–1879)
b. Dundee, Angus 23 April 1807 *d.* Glasnevin, Dublin 9 June 1879

PhD Zurich 1863. ALS 1840. FLS 1861. Changed his name to conceal his Scottish origin. Foreman, Trinity College Garden, Dublin, 1829–34. Botanist to Ordnance Survey of Ireland, 1834. Curator, Botanic Garden, Glasnevin, 1838. Found *Inula salicina* 1843. *Concise Notices of Indigenous Grasses of Ireland best suited for Agriculture* 1843, 1850 (with specimens). *Contributions towards a Cybele Hibernica* (with A. G. More) 1866; ed. 2 1898. *Synopsis of Irish Mosses* 1873. *Design in Structure and Fertilisation of Plants as Proof of Existence of God* 1875.

Gdnrs Chron. 1871, 739 portr.; 1879 i 756–57 portr. *Garden* v.13, 1878, xii portr.; v.15, 1879, 486. *Nature* v.20, 1879, 198. *Trans. Bot. Soc. Edinburgh* v.14, 1883, 36–39. S. A. Stewart and T. H. Cory *Fl. N.E. Ireland* 1888, xviii–xx. *D.N.B.* v.38, 345. N. Colgan *Fl. County Dublin* 1904, xxvii–xxix. *Curtis's Bot. Mag. Dedications, 1827–1927* 183–84 portr. *Proc. Irish Acad. B.* 1928, 183–84. *Irish Nat. J.* 1935, 302–6 portr. R. L. Praeger *Some Irish Nat.* 1949, 133–34 portr. *R.S.C.* v.4, 456; v.8, 430; v.10, 840; v.12, 518.

Plants at National Museum, Dublin. Letters at Kew and in Berkeley correspondence at BM(NH). Portr. at Hunt Library.

Rosa moorei Baker.

MOORE, Frederick (1835–1916)
b. Chelsea, London 1835 *d.* Ilford, Essex 4 Aug. 1916

Gardener at Eisgrub, Moravia, 1873–79. On editorial staff of *Gdnrs Chron.*

Gdnrs Chron. 1916 ii 76–77 portr.

MOORE, Sir Frederick William (1857–1950)
b. Dublin 3 Sept. 1857 *d.* Dublin 23 Aug. 1950

ALS 1893. FLS 1911. VMH 1897. VMM 1932. Curator, Trinity College Garden, Dublin, 1877–79. Succeeded his father, David Moore, as Curator of Botanic Garden, Glasnevin, 1879–1922.

Gdnrs Mag. 1907, 463 portr., 464; 1911, 529–30 portr. *Gdnrs Chron.* 1909 ii 329–30 portr.; 1921 i 158 portr.; 1922 i 252 portr.; 1949 ii 90. *Garden* 1911 iv portr. *Orchid Rev.* 1911, 241–42 portr. *Curtis's Bot. Mag. Dedications, 1827–1927* 327–28 portr. *J. R. Hort. Soc.* 1949, 473–75 portr. *Nature* v.164, 1949, 559. *Proc. Linn. Soc.* 1949–50, 108–9. R. L. Praeger *Some Irish Nat.* 1949, 134. *Irish Nat. J.* 1950, 23–24 portr.

Portr. at Hunt Library.

MOORE, George (–1729)

Nurseryman, Twickenham, Middx.

MOORE, George F. (*c.* 1855–1927)
d. Bourton-on-the-Water, Glos 19 June 1927

VMH 1925. Cultivated orchids.

Gdnrs Chron. 1927 i 454. *Orchid Rev.* 1927, 218–19.

MOORE, George Fletcher (1798–1886)
b. Donemana, County Tyrone 10 Dec. 1798

LLB Dublin 1820. Arrived in Freemantle, W. Australia, 1830. Advocate-General, 1834. Collected plants for J. Mangles. Returned to Ireland, 1852.

M. Uren *Land Looking West* 1948 *passim.* A. Hasluck *Portr. with a Background* 1960, 61, etc. *Austral. Dict. Biogr.* v.2, 252–54.

Letters and journals at Battye Library, Perth.

MOORE, H. Cecil (1836–1908)
b. India 1836

Medical Officer of Health for rural Hereford. President, Woolhope Naturalists' Field Club, 1896. Contrib. plant records to W. H. Purchas and A. Ley's *Fl. Herefordshire* 1889.

Trans. Woolhope Nat. Field Club v.34, 1954, 255–56.

MOORE, Rev. Henry Kingsmill (1853–1943)
b. Dublin 1853 *d.* Dundrum, County Down 1 Dec. 1843

FLS 1930. Principal, Church of Ireland Training College, Dublin. Had collection of ferns. Contrib. to *Bull. R. Hort. Soc. Ireland.*

Proc. Linn. Soc. 1943–44, 214.

MOORE, Hugh Armytage- *see* Armytage-Moore, H.

MOORE, John (–1857)

FLS 1826. President, Botanical and Horticultural Society, Manchester.

W. C. Williamson *Reminiscences of a Yorkshire Nat.* 1896, 81. *Trans. Liverpool Bot. Soc.* 1909, 82–83.

MOORE, John Chisnall (1871–1920)
b. Grantham, Lincs 28 Aug. 1871 *d.* Woodstock, Oxford 5 May 1920

Kew gardener, 1893–95. Curator, Botanic Station, St. Lucia, 1895; Agricultural Superintendent, 1898–1914. Superintendent of Agriculture, Grenada, 1914–19.

Kew Bull. 1895, 155; 1914, 345; 1919, 447; 1920, 250. *J. Kew Guild* 1921, 40 portr.

MOORE, Oswald Allen (–1862)

Of York. Botanical Curator, Yorkshire Museum, 1840–62. MS. *Fl. Yorkshire* (H. C. Watson *Topographical Bot.* 1883, 550).

J. G. Baker *N. Yorkshire* 1863, 343. *Rep. Yorkshire Philos. Soc.* 1893, 36–37; 1906, 46. H. J. Wilkinson *Hist. Account of Herb. of Yorkshire Philos. Soc.* 1907, 2.

Plants at York Museum.

MOORE, Richard (1860–1899)
d. Rangoon, Burma 18 Aug. 1899

Schoolmaster at Rangoon from about 1880. Assistant Superintendent, Southern Shan States, 1888. 'Orchids of Shan States' (*Orchid Rev.* 1895, 169–72).

Gdnrs Chron. 1899 ii 349. *Orchid Rev.* 1899, 355–56.

MOORE, Robert Reginald Heber (1858–1942)
d. 27 Sept. 1942

MD Dublin. *Pruning in Summer: an Unorthodox Method of growing Apples and Pears for Amateurs* 1933.

Who was Who, 1941–1950 811.

MOORE, Spencer Le Marchant (1851–1931)
b. Hampstead, London 1 Nov. 1851 *d.* Streatham, London 14 March 1931

BSc London 1872. FLS 1875. In Kew Herb., 1872–80. Botanist on prospecting expedition, Matto Grosso, Brazil, 1891–92. Visited W. Australia, 1894–96. Part-time unestablished assistant in BM(NH). Studied taxonomy of some groups of *Compositae.* Edited J. B. L. Warren's *Fl. Cheshire* 1899. Assistant Editor, *J. Bot.*, 1876–79. Contrib. to *J. Bot.*, *J. Linn. Soc.*, *Trans. Linn. Soc.*

E. Bretschneider *Hist. European Bot. Discoveries in China* 1898, 819–20. C. F. P. von Martius *Fl. Brasiliensis* v.1, 1906, 65–66. *J. Bot.* 1931, 134–36 portr. *Nature* v.127, 1931, 528. *Kew Bull.* 1931, 428. *Proc. Linn. Soc.* 1930–31, 191–92. *Orchid Rev.* 1931, 178. *Times* 18 March 1931.

MOORE, Thomas (1821–1887)
b. Stoke, Guildford, Surrey 21 May 1821 *d.* Chelsea, London 1 Jan. 1887

FLS 1851. Gardener, Regent's Park, London, 1844–47. Curator, Physic Garden, Chelsea, 1848. *Handbook of British Ferns*

1848; ed. 3 1857. *Ferns of Great Britain and Ireland* (ed. by J. Lindley) 1855. *Illustrations of Orchidaceous Plants* 1857. *Index Filicum* 1857–62. *Account of Exotic Cultivated Ferns* (with J. Smith) 1858. *Florists' Guide* (with W. P. Ayres) 1850. *Treasury of Botany* (with J. Lindley) 1866 2 vols. Edited *Floral Mag.* 1861.

Gdnrs Chron. 1882 i 709 portr.; 1887 i 48 portr., 81, 86. *Garden* v.31, 1887, 36–37 portr., 43. *J. Bot.* 1887, 63–64. *J. Hort. Cottage Gdnr* v.14, 1887, 3; v.34, 1897, 577. *Proc. Linn. Soc.* 1886–87, 41–42. *Ann. Bot.* 1888, 409–10. *D.N.B.* v.37, 385–86. *R.S.C.* v.4, 458; v.8, 432; v.10, 842; v.12, 518.

Herb. at BM(NH). Ferns and letters at Kew.

Davallia moorei Hook.

MOORE, Walter Cecil (1900–1967)
b. Frome, Somerset 21 June 1900 *d.* 18 Nov. 1967

MA Cantab 1926. Mycological Assistant, Plant Pathology Laboratory, Harpenden, 1925; Deputy Director, 1948; Director, 1949–62. President, British Mycological Society, 1941; Association of Applied Biologists, 1947–48. Editor, *Trans. Br. Mycol. Soc.* 1946–51. *Diseases of Bulbs* 1939. *Diseases of Crop Plants* 1943. *Diseases of Cereals* 1945. *British Parasitic Fungi* 1959.

Times 27 Nov. 1967. *Ann. Applied Biol.* 1968, 167 portr. *Trans. Br. Mycol. Soc.* 1969, 353–54 portr.

MOORE, William (*fl.* 1880s–1890s)

On staff of Felsted School, Essex? Collected plants near Felsted and Rochester, Kent, 1880–85.

Essex Nat. v.29, 1954, 189.

Herb. at Essex Field Club.

MOORMAN, J. W. (1843–)
b. St. Brelade, Jersey 31 May 1843

Gardener at Coombe Bank, Kingston-on-Thames, Surrey. Superintendent, Victoria Park, London.

J. Hort. Cottage Gdnr v.58, 1909, 418–20 portr.

MORE, Alexander Goodman (1830–1895)
b. London 5 Sept. 1830 *d.* Dublin 22 March 1895

FLS 1856. FRSE. Curator, Natural History Museum, Dublin, 1881–87. 'Supplement to 'Fl. Vectensis'' (*J. Bot.* 1871, 72–76, 135–45, 167–72, 202–11). *Contributions towards a Cybele Hibernica* (with D. Moore) 1866; ed. 2 1898. Contrib. to *Phytologist*, *J. Bot.*

Garden v.47, 1895, 227. F. M. More *Life and Letters of A. G. More* 1898 portr. *Gdnrs Chron.* 1887 ii 195. *Irish Nat.* 1895, 109–16 portr. *J. Bot.* 1895, 225–27 portr. *Proc. Linn. Soc.* 1894–95, 36. N. Colgan *Fl. County Dublin* 1904, xxviii–xxx. R. W. Scully *Fl. County Kerry* 1916, xv. R. L. Praeger *Some Irish Nat.* 1949, 134–35 portr. *R.S.C.* v.4, 416; v.8, 435; v.10, 845.

Herb. and portr. at National Museum, Dublin. Letters at Kew. Portr. at Hunt Library.

MORE, Frances Margaret (–1909)
d. Dublin 7 Feb. 1909

Sister of A. G. More (1830–1895). Discovered *Neotinea intacta* in Ireland, 1864.

N. Colgan *Fl. Dublin* 1904, xxvii–xxix. *Irish Nat.* 1909, 132.

MORE, John (*fl.* 1820s)

Nurseryman, King's Road, Chelsea, London.

MORE, Robert (1703–1780)
d. Jan. 1780

FRS 1729. Of Shrewsbury. Travelled in Europe. Friend of Linnaeus. "An excellent botanist" (Watson *Philos. Trans. R. Soc.* v.44, 1748, 236). 'Note on Manna of Naples' (*Philos. Trans. R. Soc.* v.46, 1749, 470–71).

D.N.B. v.38, 428.

Morea Miller.

MORE, Thomas (*fl.* 1670s–1720s)

Known as "the Pilgrim botanist." Collected plants for J. Sherard and other subscribers in New England, New York and Pennsylvania, 1722–24.

D. Turner *Extracts from Lit. and Sci. Correspondence of Richard Richardson* 1835, 181, 206. H. N. Clokie *Account of Herb. Dept. Bot., Oxford* 1964, 214. A. M. Coats *Quest for Plants* 1969, 271–73. G. F. Frick and R. F. Stearns *Mark Catesby* 1961, 114–27. R. F. Stearns *Sci. in Br. Colonies of America* 1970, 473–77.

Plants at Oxford. Letters at Royal Society.

MOREAU, Reginald E. (*c.* 1897–1970)
d. 30 May 1970

Ornithologist. To Egypt, 1920. Secretary, E. African Agricultural Research Station, Tanganyika, 1928. Collected plants in E. Africa especially orchids. At Edward Grey Institute, Oxford, 1947–68.

Times 3 June 1970.

Orchids, drawings and photographs of orchids at Kew.

MORETON, Henry Haughton Reynolds- *see* Reynolds-Morton, H. H.

MORETON, Henry John, 3rd Earl of Ducie (1827–1921)
b. London 25 June 1827 *d.* Tortworth, Glos 28 Oct. 1921

FRS 1855. FLS 1889. Established a notable arboretum at Tortworth House.

Bot. Soc. Exch. Club Br. Isl. Rep. 1921, 357. *Kew Bull.* 1921, 316–17. *Proc. Linn. Soc.*

1921–22, 43–44. *Proc. R. Soc. B.* 1922, i–ii portr.

MORETON, Rev. John (*fl.* 1710s)

Rector, Oxendon, Northants. *Natural History of Northamptonshire* 1712 (includes list of plants).

MOREY, Frank (1858–1925)

b. Newport, Isle of Wight 4 March 1858 *d.* Newport 29 Dec. 1925

FLS 1906. Founded Isle of Wight Natural History Society, 1919; Secretary and Editor. *Guide to Natural History of Isle of Wight* 1909.

Bot. Soc. Exch. Club Br. Isl. Rep. 1925, 855–56. *Proc. Linn. Soc.* 1925–26, 91–92.

MORGAN, Edward (*fl.* 1610s–1680s)

Gardener in charge of medical garden in Westminster. Grew plants from seeds sent to him from Tangiers. Accompanied T. Johnson to Wales, 1639. "A very skilful botanist" (J. Evelyn). "Vir in rebus botanicis haud infimae notae" (L. Plukenet *Almagestum Botanicum* 1696, 191, 224, 329).

R. Morison *Plantarum umbelliferarum* 1672, 2. J. C. Loudon *Arboretum et Fruticetum Britannicum* v.1, 1838, 50, 74. *J. Bot.* 1903, 152. G. C. Druce and S. H. Vines *Account of Herb., Oxford* 1919, 51–53. *Bot. Soc. Exch. Club Br. Isl. Rep.* 1919, 722–24. R. T. Gunther *Early English Botanists* 1922, 351–54. C. E. Raven *English Nat.* 1947, 317–18. *Proc. Linn. Soc.* 1951–52, 102–33; 1955–56, 96–101. J. E. Dandy *Sloane Herb.* 1958, 168–70.

Hortus siccus at Bodleian Library.

MORGAN, Hugh (*fl.* 1540s–1613)

d. Battersea, London Sept. 1613

Apothecary to Queen Elizabeth. Had a botanic garden near Coleman Street, London and at Battersea. Introduced *Clematis viticella*, etc. "A curious conserver of simples" (J. Gerard *Herball* 1597, 1308).

W. Turner *New Herball* v.3, 1568, 33. M. de Lobel *Adversaria* 1605, 294, 343, 373. J. Parkinson *Paradisus* 1629, 437. R. Brown *Prodromus* 1810, 441. *Gent. Mag.* 1814 ii 3. R. T. Gunther *Early English Botanists* 1922, 415. C. E. Raven *English Nat.* 1947, 116–17. J. E. Dandy *Sloane Herb.* 1958, 169. L. G. Matthews *Royal Apothecaries* 1967, 74–78. R. H. Jeffers *Friends of John Gerard; Biogr. Appendix* 1969, 23–24.

Morgania R. Br.

MORGAN, Robert (1863–1900)

b. Norwood, Surrey 9 May 1863 *d.* London 6 Nov. 1900

FLS 1887. Artist and lithographer. Illustrated A. Fryer's *Potamogeton* (drawings at BM(NH)); plates in *J. Bot.* 1882–1900.

Gdnrs Chron. 1900 ii 361. *J. Bot.* 1900, 489–92 portr. *Proc. Linn. Soc.* 1900–1, 46–47.

MORGAN, Thomas Owen (*c.* 1801–1878)

b. Cardiganshire *c.* 1801 *d.* Aberystwyth, Cardiganshire 5 Dec. 1878

Barrister. *New Guide to Aberystwyth* 1848, 1858 (with list of plants). *Fl. Ceretícae Superioris...Plants...of Aberystwyth* 1849.

MORGAN, William (*fl.* 1800s)

Gardener to Henry Brown at North Mimms Place, Herts. Contrib. to *Trans. Hort. Soc. London.*

J. C. Loudon *Encyclop. Gdning* 1822, 1289.

MORIARTY, Mrs. Henrietta Maria (*fl.* 1800s)

Novelist. *Viridarium* 1803; ed. 2 as *Fifty Plates of Greenhouse Plants* 1807.

J. Bot. 1917, 52–54.

MORING, Percy (1865–1930)

d. 26 Dec. 1930

Secretary, Dover Hospital, Kent. Secretary, Dover Scientific Society. Collected plants in neighbourhood of Dover.

S.-Eastern Nat. 1931, xxxiv.

Plants at BM(NH).

MORISON, Robert (1620–1683)

b. Dundee, Angus 1620 *d.* London 11 Nov. 1683

MA Aberdeen 1638. MD Angers 1648. MD Oxon 1669. Gardener to Gaston d'Orléans at Blois, 1650–60. Royal Physician and Botanist on Restoration in 1660. First Professor of Botany, Oxford, 1669. Discovered *Carex vesicaria* 1699. *Phytologia Britannica* 1650. *Praeludia Botanica* 1669. *Plantarum Umbelliferarum* 1672. *Plantarum Historiae Universalis Oxoniensis* 1680–99.

A. Wood *Athenae Oxonienses* v.2, 1692, 851–52. R. Pulteney *Hist. Biogr. Sketches of Progress of Bot. in England* v.1, 1790, 298–312. A. L. A. Fée *Vie de Linné* 1832, 245. *Hooker's J. Bot.* 1854, 248. A. Franchet *Fl. Loir-et-Cher* 1885, xiii–xvi. G. C. Druce *Fl. Berkshire* 1897, cxiii–cxvi. *D.N.B.* v.39, 61–63. E. J. L. Scott *Index to Sloane Manuscripts* 1904, 379. F. W. Oliver *Makers of Br. Bot.* 1913, 16–43 portr. J. R. Green *Hist. Bot.* 1914, 98–110. S. H. Vines and G. C. Druce *Account of Morisonian Herb.* 1914 portr. R. T. Gunther *Early English Botanists* 1922, 355–57. *Bot. Soc. Exch. Club Br. Isl. Rep.* 1923, 345–47 portr. G. C. Druce *Fl. Oxfordshire* 1927, lxxix. C. E. Raven *John Ray* 1942, 183–86. H. N. Clokie *Account of Herb. Dept. Bot., Oxford* 1964, 10–13. *Notes R. Bot. Gdn Edinburgh* 1973, 151–58. C. C. Gillispie *Dict. Sci. Biogr.* v.9, 1974, 528–29.

MSS. and portr. at Oxford. Portr. at Hunt Library.

Morisonia L.

MORLAND, Sir Samuel (1625–1695)
b. Sulhamstead, Reading, Berks 1625 *d.* Hammersmith, Middx 30 Dec. 1695
Baronet, 1660. Diplomat, mathematician, inventor. Master of Mechanics to Charles II. 'Parts and Use of the Flower' (*Philos. Trans. R. Soc.* v.23, 1704, 1474–79).
R. Pulteney *Hist. Biogr. Sketches of Progress of Bot. in England* v.1, 1790, 339–40. *D.N.B.* v.39, 68–73. E. J. L. Scott *Index to Sloane Manuscripts* 1904, 379.
MS. autobiography at Lambeth Palace.

MORLEY, Christopher Love (*c.* 1646–1702)
MD Leyden 1679. FRCP 1680. Plants from Leyden and Paris gardens in Sloane Herb. at BM(NH). 'Catalogus Planta Villa Nostra Hamstiadensis' in Sloane MSS. 1394, 16.
W. Munk *Roll of R. College of Physicians* v.1, 1878, 450. *D.N.B.* v.39, 73–74. E. J. L. Scott *Index to Sloane Manuscripts* 1904, 379. J. E. Dandy *Sloane Herb.* 1958, 170.

MORLEY, John (1829–1886)
d. Birmingham? 10 Dec. 1886
Pteridologist. Secretary, Birmingham Natural History Society, 1876–86.
Midland Nat. 1887, 24.

MORREL, Robert (*fl.* 1810s)
Nurseryman, Westgate, Bradford, Yorks.

MORRIS, —
Plant collector employed by H. F. Hance (1827–1886)
G. A. C. Herklots *Hong Kong Countryside* 1951, 164.
Diospyros morrisiana Hance.

MORRIS, Allister G. (*c.* 1887–1967)
d. Germany 24 Aug. 1967
Partner in firm of Protheroe and Morris, horticultural auctioneers, 1907–62. Conducted orchid sales for Messrs Sander, Stuart Low and other growers.
Orchid Rev. 1967, 379–80.

MORRIS, Arabella *see* Fuller, A.

MORRIS, Daniel (1844–1933)
b. Loughor, Glam 26 May 1844 *d.* Boscombe, Hants 9 Feb. 1933
MA Dublin 1876. DSc Dublin. CMG 1893. FLS 1903. VMH 1897. Assistant, Botanic Gardens, Peredeniya, Ceylon, 1877–79, where he studied coffee diseases. Director, Public Gardens, Jamaica, 1879–86. Associated with development of economic resources of West Indies. Assistant Director, Kew, 1886–98. Imperial Commissioner of Agriculture, West Indies, 1898–1908. *Cacao* 1882. *Planting Enterprise in West Indies* 1883. *Native and other Fibre Plants* 1884. *Vegetable Resources of West Indies* 1888. *Sisal Industry in the Bahamas* 1896. *Forest Resources of Newfoundland* 1916. Contrib. to *Kew Bull.*
J. Kew Guild 1896, 1 portr.; 1933, 277–78. *Bot. Gaz.* 1897, 354–55. *Gdnrs Mag.* 1897, 31 portr. *Kew Bull.* 1898, 234; 1933, 110–11. I. Urban *Symbolae Antillanae* v.3, 1902, 89–90. *Proc. Bournemouth Nat. Sci. Soc.* v.25, 56 portr. *Gdnrs Chron.* 1933 i 126. *J. Bot.* 1933, 101–2. *Nature* v.131, 1933, 231, 266. *Proc. Linn. Soc.* 1932–33, 197–99. *Times* 10 Feb. 1933. *Who was Who, 1929–1940* 970.

MORRIS, George Field (1831-1909)
b. 2 Sept. 1831 *d.* Wanstead, Essex 2 Jan. 1909
Entered his father's nursery at Leytonstone, Essex, 1845. In partnership with Protheroe as orchid auctioneers, Cheapside, London.
Orchid Rev. 1909, 58.

MORRIS, Harold G. (–1928)
d. at sea 1928
Member of the firm of orchid auctioneers, Protheroe and Morris, Cheapside, London.
Orchid Rev. 1928, 149.

MORRIS, James (*fl.* 1840s)
MD. Made collection of Middlesex plants during 1845–50.
H. Trimen and W. T. T. Dyer *Fl. Middlesex* 1869, 11.

MORRIS, John (1810–1886)
b. Homerton, London 19 Feb. 1810 *d.* St. John's Wood, London 7 Jan. 1886
FGS 1845. Professor of Geology, University College, London, 1855–77. *Catalogue of British Fossils* 1843; ed. 2 1854. *Fossil Fl. Rajmahal Series, Rajmahal Hills, Bengal* (with T. Oldham) 1862–63.
Geol. Mag. 1878, 481–87 portr.; 1886, 95–96. *J. Bot.* 1886, 64. *Proc. Geol. Assoc.* 1886, 386–410. *Proc. Geol. Soc.* 1885–86, 44–47. *D.N.B.* v.39, 98.
Volkmannia morrisii Hook. f.

MORRIS, Richard (*fl.* 1820s–1830s)
FLS 1825. Surveyor and landscape gardener. *Botanist's Manual* 1824. *Fl. Conspicua* 1825–30.

MORRIS, Sydney (*c.* 1851–1924)
d. St. Jean de Luz, France 24 March 1924
VMH 1924. VMM 1924. Created garden at Earlham Hall, Norfolk. Raised cultivars of montbretia.
Garden 1924, 234(ix).

MORRIS, Valentine (1727–1789)
b. Antigua, West Indies 27 Oct. 1727 *d.* London 26 Aug. 1789
West Indian planter and Governor of St. Vincent. Created garden at Piercefield, Mon. Interested in introduction of breadfruit trees to West Indies.
I. Waters *The Unfortunate Valentine Morris* 1964 portr.

MORRIS, William (1705–1763)
b. Anglesey 6 May 1705 *d.* Holyhead, Anglesey 29 Dec 1763
Comptroller of Customs at Holyhead, 1758. MS. 'Collection of plants gathered in Anglesey'. MS. notes in a copy of J. Ray's *Synopsis* 1724 at BM Library. Contrib. to H. Davies's *Welsh Botanology* 1813; "a good practical botanist" (Davies).
H. J. Rose *Biogr. Dict.* v.10, 1857, 235. *J. Bot.* 1898, 13. *Dict. Welsh Biogr. down to 1940* 1959, 666–67.

MORRISON, Alexander (1849–1913)
b. Dalmeny, Edinburgh 15 March 1849 *d.* Cheltenham, Melbourne 7 Dec. 1913
MD Edinburgh. To Australia, 1877; practised medicine in Victoria. Forced by ill-health to abandon medicine, *c.* 1892. Botanist, Bureau of Agriculture, W. Australia, 1897–1906. Travelled in Pacific and collected plants in New Hebrides which he sent to F. von Mueller. Contrib. to *J. Nat. Hist. Soc. W. Australia, Victorian Nat.*, *J. Bot.* Articles on *Drosera* in *Trans. Bot. Soc. Edinburgh* 1905, 1907.
J. Nat. Hist. Sci. Soc. W. Australia v.5, 1914, 108. *J. R. Soc. N.S.W.* 1921, 164.
Herb. at Edinburgh.

MORRISON, John (–1781)
d. Stratford, Essex March 1781
Gardener to J. Fothergill (1712–1780) at Upton House, West Ham. "An ingenious botanist."
Gdnrs Chron. 1917 ii 236.

MORRISON, William (*fl.* 1820s)
b. Scotland
Kew gardener. Manager of sugar plantation, Barbados. Collected plants in Barbados and Trinidad, 1824–28. About 1828 to Swan River colony, Australia where he also collected plants.
Bot. Mag. 1841, t.3893. *Kew Bull.* 1891, 317. *J. W. Austral. Hist. Soc.* v.6, 1909, 21. R. Erickson *Drummonds of Hawthornden* 1969, 43.

MORSE, Ernest William (1887–1917)
b. 30 June 1887 *d.* Cape Coast, Gold Coast 30 Aug. 1917
Kew gardener, 1908. Curator, Botany and Agricultural Dept., Gold Coast, 1911.
J. Kew Guild 1918, 427.

MORT, James (–1907)
b. Kenyon, Lancs *d.* Lymm, Cheshire 1907
Surveyor and sanitary inspector. Active member of Warrington Field Club. MS. *Fl. Lymm.*
Trans. Liverpool Bot. Soc. 1909, 83.

MORTIMER, John (*c.* 1656–1736)
b. London *c.* 1656
FRS. Merchant, Tower Hill, London. Had estate at Toppingo Hall, Essex in which he planted many trees. *Whole Art of Husbandry* 1707; ed. 6 1761.
G. W. Johnson *Hist. Gdning* 1829, 128–30. *Cottage Gdnr* v.7, 1852, 207. *D.N.B.* v.39, 128–29. D. McDonald *Agric. Writers* 1908, 158–60. *Gdnrs Chron.* 1909 ii 257–58.

MORTIMER, S. (–1923)
d. 22 July 1923
Nurseryman, Farnham, Surrey.
Gdnrs Chron. 1923 ii 95 portr.

MORTON, Rev. John (1670/1–1726)
b. Lincs 18 July 1670/1 *d.* Great Oxendon, Northants 18 July 1726
BA Cantab 1691. BA Oxon 1694. FRS 1703. Rector, Great Oxendon, 1706. Correspondent of H. Sloane and E. Lhuyd. Attempted systematic arrangement of the flora in *Natural History of Northamptonshire* 1712, 360–407 (copy with MS. notes at BM Library).
R. Pulteney *Hist. Biogr. Sketches of Progress of Bot. in England* v.1, 1790, 354–55. J. Nichols *Illus. Lit. Hist. of Eighteenth Century* v.1, 326–27. D. Turner *Extracts from Lit. Sci. Correspondence of Richard Richardson* 1835, 85–86. *D.N.B.* v.13, 1050–51. *J. Northampton Nat. Hist. Soc.* v.14, 1908, 259–72, 293–324. E. J. L. Scott *Index to Sloane Manuscripts* 1904, 380. G. C. Druce *Fl. Northamptonshire* 1930, xlviii–lxxiv. C. C. Gillispie *Dict. Sci. Biogr.* v.9, 1974, 539–40.

MORTON, William Lockhart (1824–1898)
In Australia *c.* 1841. Botanised in the Mallee, Victoria and also Mackay district of Queensland. Contrib. to *Trans. R. Soc. Victoria.*
Victorian Nat. 1908, 111; 1970, 146. *J. Proc. R. Austral. Hist. Soc.* 1951, 277–79.
Plants at Herb., Melbourne.
Rhysotoechia mortoniana F. Muell.

MOSELEY, Harriet (*fl.* 1830s–1860s)
Of Malvern, Worcs. Contrib. to E. Lees's *Bot. Malvern Hills* 1843 and to W. A. Leighton's *Fl. Shropshire* 1841.
E. Lees *Bot. Worcestershire* 1867, xc.
Plants and drawings at BM(NH).

MOSELEY, Henry Nottidge (1844–1891)
b. Wandsworth, London 14 Nov. 1844 *d.* Parkstone, Dorset 10 Nov. 1891
MA Oxon 1872. FRS 1877. FLS 1880. Naturalist on 'Challenger' Expedition, 1872–76. Linacre Professor of Human Anatomy, Oxford, 1881. *Notes by a Naturalist...during the Voyage of H.M.S. Challenger*. ed. 2 1892.
Nature v.45, 1891, 79–80. *Proc. Linn. Soc.* 1890–91, 72–73. I. Urban *Symbolae Antillanae* v.1, 1898, 68–69; v.3, 1902, 90–91.

D.N.B. v.39, 176–77. *Fl. Malesiana* v.1, 1950, 371–72. *Tuatura* 1970, 51–55. *R.S.C.* v.8, 445; v.10, 859.

Plants and letters at Kew. 'Challenger' journal at BM(NH).

Moseleya Hemsley.

MOSELEY, Walter Michael (*fl.* 1790s–1800s)

Of Glashampton, Worcs. "A skilful and most assiduous botanist" (T. Butt). Plants recorded in J. Sowerby and J. E. Smith's *English Bot.* 494, 1005.

P. Smith *Correspondence of Sir J. E. Smith* v.1, 1832, 440.

MOSES, John (*fl.* 1820s–1830s)

Nurseryman, Netherend, Penrith, Cumberland.

MOSLEY, Charles (1875–1933)

b. Huddersfield, Yorks 27 Dec. 1875 *d.* 11 Dec. 1933

Printer and stationer, Lockwood, Yorks. Curator, Wakefield Museum, 1930. Studied local flora. *The Oak: its Natural History, Antiquity and Folklore* 1910.

Naturalist 1934, 29–30.

MOSS, Miss (*fl.* 1830s)

Contrib. drawing to *Bot. Mag.* 1837, t.3573. Possibly daughter or sister of John Moss of Otterspool near Liverpool.

MOSS, Charles Edward (1870–1930)

b. Hyde, Cheshire 7 Feb. 1870 *d.* Johannesburg, S. Africa 11 Nov. 1930

DSc Manchester 1907. FLS 1912. Teacher in Somerset and Manchester. Curator, Cambridge Herb., 1907–16. Professor of Botany, Witwatersrand University, 1916; Johannesburg, 1917–30. Collected plants in Lourenço Marques. *Vegetation of Peak District* 1913. *Geographical Distribution of Vegetation in Somerset* 1907. *Cambridge British Fl.* 1914–20. Contrib. to *Halifax Nat.*, *Naturalist*, *J. Bot.*, *Gdnrs Chron.*

Bot. Soc. Exch. Club Br. Isl. Rep. 1930, 326–27. *Times* 4 and 9 Dec. 1930. *J. Bot.* 1931, 20–23. *J. Ecol.* 1931, 209–14. *Kew Bull.* 1931, 106–7. *Naturalist* 1931, 55–59 portr. *Nature* v.127, 1931, 98–99. *N. Western Nat.* 1931, 34–36. *Proc. Linn. Soc.* 1930–31, 193–96. *Who was Who, 1929–1940* 974.

Plants at Cambridge, BM(NH), Johannesburg.

MOSS, John (*fl.* 1830s)

Had outstanding collection of orchids at Otterspool near Liverpool.

Bot. Mag. 1839, t.3722.

MOSS, John Snow (*c.* 1859–1913)

d. Bishops Waltham, Hants 11 Sept. 1913

Cultivated orchids. *Coelogyne mossiae* named after his wife.

Kew Bull. 1894, 156. *Orchid Rev.* 1913, 301–2.

MOSSMAN, Samuel (*fl.* 1850s)

Collected plants in Australia and New Zealand, 1850. Mosses described by C. Müller in *Botanische Zeitung* 1851, 545–52, 561–67.

Hooker's J. Bot. 1851, 31.

Plants at BM(NH), Liverpool Museums.

MOTLEY, James (–1859)

b. Isle of Man? *d.* murdered by natives at Bangkal, Borneo 1859

Civil engineer in Labuan, 1851–52. Superintendent of coal mining operations, S.E. Borneo, 1854. Collected plants in Malaysia, 1852–53. Contrib. to *Phytologist* and *J. Bot.* Contrib. Carmarthen plants to H. C. Watson's *Topographical Bot.*

L. W. Dillwyn *Materials for Fauna and Fl. Swansea* 1848 *passim*. *Trans. Linn. Soc.* v.23, 1862, 157. *Naturalist* 1902, 343–51; 1903, 4. *J. Straits Branch R. Asiatic Soc.* no. 79, 1918, 37–38. *Gdns Bull. Straits Settlements* 1927, 128. *Fl. Malesiana* v.1, 1950, 373. *R.S.C.* v.4, 495.

Plants and letters at Kew. Plants were at Royal Institution S. Wales, Swansea.

Barclaya motleyi Hook. f.

MOTT, Frederick Thompson (1825–1908)

b. Loughborough, Leics 24 March 1825 *d.* Birstall, Yorks 14 March 1908

Wine merchant. *Fl. Odorata* 1843. *Charnwood Forest* ed. 3 1868 (with list of plants). *Fruits of all Countries* 1883. *Fl. Leicestershire* (with E. F. Cooper and others) 1886.

A. R. Horwood and C. W. F. Noel *Fl. Leicestershire* 1933, ccxviii. *R.S.C.* v.10, 863; v.12, 522.

Plants at Kew, Manchester. Portr. at Hunt Library.

MOTT, L. (–1940)

Of Cheltenham, Glos. Botanised in Gloucestershire and neighbouring counties.

H. J. Riddelsdell *Fl. Gloucestershire* 1948, cli.

Herb. formerly at Cheltenham College.

MOTT, William (*fl.* 1830s)

Nurseryman, Basingstoke, Hants.

MOULE, Rev. George E. (*fl.* 1850s–1890s)

Church of England Missionary Society. To China, 1858. At Hang chou fu where he collected some plants for H. F. Hance.

E. Bretschneider *Hist. European Bot. Discoveries in China* 1898, 705.

MOULT, William (1819–1896)

b. Wollaton, Notts 15 Feb. 1819 *d.* Newcastle, Northumberland 15 Oct. 1896

Gardener to Earl of Ravensworth at Ravensworth Castle, County Durham, 1848–83.
J. Hort. Cottage Gdnr v.33, 1896, 397.

MOULTON, John Coney (1886–1926)
b. St. Leonards, Sussex 1886 *d.* London 6 June 1926
Educ. Oxford. Army major. Curator, Sarawak Museum, 1905–15; founded *Sarawak Museum J.* 1911. Director, Raffles Museum and Library, Singapore, 1919–23. Chief Secretary, Sarawak. Collected plants in Borneo. 'Some Plants collected on Mr. Moulton's Expedition to Batu Lawi (Borneo)' (*J. Straits Branch R. Asiatic Soc.* no. 63, 1912, 59–63).
J. Malay Branch R. Asiatic Soc. 1926, 264–65. *Fl. Malesiana* v.1, 1950, 373–74.
Plants at Sarawak.
Moultonianthus Merr.

MOUNT, George (*c.* 1844–1927)
d. 22 April 1927
VMH 1925. Nurseryman, St. Dunstan's, Canterbury, 1884; acquired Exotic Nursery of late Mr. Masters, 1888. Established nursery at Folkestone, 1893.
Gdnrs Mag. 1895, 419–20 portr. *J. Hort. Home Farmer* v.67, 1913, 525–26 portr. *Gdnrs Chron.* 1927 i 296 portr., 311.

MOUNT, Henry (*c.* 1847–1934)
d. Canterbury, Kent 2 Feb. 1934
Gardener. Employed at Danesbury House, Welwyn and at Penzance. Cultivated ferns.
Br. Fern Gaz. 1934, 252–53.

MOUNT, Rev. William (1545–1602)
b. Mortlake, Surrey 1545 *d.* London Dec. 1602
Fellow, King's College, Cambridge, 1566. Master of Savoy, 1593. Botanised in Kent.
R. T. Gunther *Early Br. Botanists* 1922, 253–63.

MOUNTAIN, Didymus *see* Hill, T.

MOUNTFORT, Charles Clayton (–1960)
d. 11 May 1960
MA. FLS 1950. Teacher. Founder member of Alpine Garden Society. Edited *Quart. Bull. Alpine Gdn Soc.* Travelled in Dolomites.
Quart. Bull. Alpine Gdn Soc. 1960, 93–94.

MOUNTNORRIS, Lord *see* Annesley, G.

MOWBRAY, William (*c.* 1792–1832)
b. Hitchin, Herts *c.* 1792 *d.* Hitchin 10 July 1832
Curator, Manchester Botanical and Horticultural Garden.
Hort. Register 1832, 670.

MOXON, Margaret Louisa (1863–1920)
b. London 30 Sept. 1863 *d.* New Milton, Hants 11 July 1920
Drawings of British and Swiss plants at Kew.
Kew Bull. 1920, 371–72.

MOYLE, Walter (1672–1721)
b. Bake, St. Germans, Cornwall 3 Nov. 1672 *d.* Bake 9 June 1721
Exeter College, Oxford, 1689. MP Saltash, 1695–98. Ornithologist and botanist. Correspondent of E. Lhuyd. Contrib. to J. Ray's *Synopsis* ed. 2 1696.
J. Nichols *Illus. Lit. Hist. Eighteenth Century* v.1, 375, 389. D. Turner *Extracts from Lit. Sci. Correspondence of Richard Richardson* 1835, 249. *D.N.B.* v.39, 246–48.

MOYSEY, Lewis (1863–1918)
d. at sea 20 Jan. 1918
Doctor in Nottingham. Palaeontologist.
Nature v.101, 1918, 28.
Fossil plant collection at Cambridge.

MOZE, Dr. (–1733)
d. 21 Jan. 1733
"A learned antiquary and botanist" (*Gent. Mag.* 1738).
Gdnrs Chron. 1919 i 147.

MUDD, William (1830–1879)
b. Bedale, Yorks 1830 *d.* Cambridge April 1879
ALS 1868. Gardener at Great Ayton, Yorks. Curator, Botanic Garden, Cambridge. *Manual of British Lichens* 1861. *Herb. Lichenum Britannicorum* 1861 (*J. Bot.* 1863, 152–56). *British Cladoniae* 1865 (with specimens).
J. Bot. 1879, 160. *Gdnrs Chron.* 1864, 1010; 1879 i 558–59. *Trans. Bot. Soc. Edinburgh* v.14, 1883, 40–41. B. Lynge *Index...Lichenum Exsiccatorum* 1915–19, 391–98. *R.S.C.* v.4, 502.

MUDIE, Robert (1777–1842)
b. Forfarshire 28 June 1777 *d.* London 29 April 1842
Journalist. *Vegetable Substances* 1828. *Botanic Annual* 1832. *Popular Guide to Observation of Nature* 1832.
Gdnrs Mag. 1840, 602–3. *D.N.B.* v.39, 263–64. *R.S.C.* v.4, 502.

MUEHLENBERG, Rev. Gotthilf Heinrich Ernst (1753–1815)
b. New Providence, U.S.A. 17 Nov. 1753 *d.* Lancaster, Pa., U.S.A. 23 May 1815
DD Princeton 1787. To Halle, 1763; returned to America, 1770. *Catalogus Plantarum Americae Septentrionalis* 1813. *Descriptio uberior Graminum et Plantarum Calamariarum Americae Septentrionalis* 1817.
W. Darlington *Reliquiae Baldwinianae* 1843, 17–179. W. Darlington *Mem. of John Bartram and Humphry Marshall* 1849, 466–74. J. W. Harshberger *Botanists of Philadelphia and their Work* 1899, 92–97 portr.

Herb. at American Philosophical Society.
Muehlenbergia Schreber.

MUELLER, Sir Ferdinand Jakob Heinrich von (1825–1896)
b. Rostock, Germany 30 June 1825 *d.* Melbourne 9 Oct. 1896

MD. PhD Kiel. FRS 1861. FLS 1859. KCMG 1879. To Australia, 1847. Government Botanist, Victoria, 1852. Botanist to A. C. Gregory's expedition in search of Leichhardt, 1855. Director, Botanic Garden, Melbourne, 1857–73. President, Australian Association for Advancement of Science, 1890. Collected plants in many parts of Australia. *Fragmenta Phytographiae Australiae* 1858–82 12 vols. *Plants Indigenous to Colony of Victoria* 1860–65 2 vols. *Eucalyptographia* 1879–84. *Key to System of Victorian Plants* 1885–88 2 vols. Assisted G. Bentham in *Fl. Australiensis* 1863–78 7 vols.

J. D. Hooker *Fl. Tasmaniae* v.1, 1859, cxxii–cxxiii. P. Mennell *Dict. Austral. Biogr.* 1892, 335–36. *Gartenflora* 1895, 454–56 portr. *Gdnrs Chron.* 1873, 743–44 portr.; 1896 ii 464–66 portr.; 1897 i 110–11; 1960 i 482. *Times* 12 Oct. 1896. *Austral. J. Pharmacy* 1896, 298 portr. *Garden* v.50, 1896, 322. *Kew Bull.* 1896, 218–19. *Victorian Nat.* 1892, 101; 1896, 87–92 portr.; 1897, 94–97; 1904, 17–28; 1922, 98–102; 1935, 79–82; 1948, 132–37; 1950, 169–76; 1952, 179–80; 1957, 110–14. *J. Bot.* 1897, 272–78 portr. *Berichten der Deutschen Bot. Gesellschaft* 1897, 56–70. *Proc. Linn. Soc.* 1896–97, 60–64. *Proc. R. Soc.* v.63, 1898, xxxii–xxxvi. *Proc. Linn. Soc. N.S.W.* 1900, 778–80. *Rep. Austral. Assoc. Advancement Sci.* 1907, 195–96. *J. W. Austral. Nat. Hist. Soc.* no. 6, 1909, 32–33. *J. R. Soc. N.S.W.* 1908, 77–80; 1912, 8–11. *S. Austral. Nat.* v.2, 1921, 76–82. *Victorian Hist. Mag.* v.10, 1924, 1–55 portr. *Curtis's Bot. Mag. Dedications, 1827–1927* 135–36 portr. P. Serle *Dict. Austral. Biogr.* v.2, 1949, 167–70. M. Willis *By their Fruits: Life of Ferdinand von Mueller* 1949 portr. A. H. Chisholm *Ferdinand von Mueller* ed. 2 1963. *Austral. Encyclop.* v.6, 1965, 190–91. A. M. Coats *Quest for Plants* 1969, 227. M. A. Reinikka *Hist. of the Orchid* 1972, 230–31 portr. *S. Austral. Nat.* v.49, 1975, 51–64. *R.S.C.* v.4, 515; v.8, 459; v.10, 874.

Herb. at Melbourne. Letters, plants and portr. at Kew. Letters in Berkeley correspondence at BM(NH). Portr. at Hunt Library.
Sirmuellera Kuntze.

MUELLER, Hugo (1834/5–1915)
b. Tirschenreuth, Germany 1834/5 *d.* Camberley, Surrey 23 May 1915

PhD Göttingen. LLD St. Andrews. FRS 1866. To England *c.* 1855. Chemist. Economic botanist. 'Resin of *Ficus rubiginosa*' (*Philos. Trans. R. Soc.* v.150, 1860, 43–56). 'Fibres Végétales' (*Moniteur Scientifique* 1878, 554).

Kew Bull. 1915, 261–63. *Nature* v.95, 1915, 376–77. *R.S.C.* v.4, 521; v.8, 462; v.10, 876; v.17, 404.

MUELLER, Johann Sebastian *see* Miller, John

MUIR, David *see* Moore, D.

MUIR, John (1838–1914)
b. Dunbar, East Lothian 21 April 1838 *d.* Los Angeles, U.S.A. 24 Dec. 1914

Educ. Wisconsin University. Naturalist and conservationist. Influential in establishment of U.S. National Parks, especially Yosemite. Collected plants. *Mountains of California* 1894.

W. F. Bade *Life and Letters of John Muir* 1923–24 2 vols. *Who was Who, 1897–1916* 512.
Senecio muiri Greenman.

MUIR, Thomas (1833–1926)
b. 1 May 1833 *d.* Oct. 1926

Emigrated to W. Australia with his parents in 1844. Farmer. Collected plants for F. von Mueller.

J. Proc. W. Austral. Hist. Soc. v.4, 1926, 69–70. *Austral. Encyclop.* v.5, 1965, 478.
Plants at Melbourne.

MULDER, John Frederick (1841–1921)
b. Kent 3 Dec. 1841 *d.* Geelong, Australia 28 Dec. 1921

Emigrated to Victoria with his parents *c.* 1848. Taxidermist. Naturalist. Collected in Geelong and Cape Otway districts. Sent plants to F. von Mueller. Contrib. to *Geelong Nat.*

Geelong Nat. 1922, 3–10. *Victorian Nat.* 1922, 138.
Plants at Melbourne.

MULGRAVE, 2nd Baron *see* Phipps, C. J.

MULLOCK, Catharine (*fl.* 1820s–1830s)
Nurseryman, Nursery Grounds, High Street, Nantwich, Cheshire.

MULLOCK, Isaac (*fl.* 1790s)
Nurseryman, Houghton Moss near Nantwich, 1791; at Shrewsbury, 1799–1805.

MULLOCK, Peter (*fl.* 1800s–1820s)
Nurseryman, Nantwich, Cheshire.

MUNBY, Giles (1813–1876)
b. York 1813 *d.* Farnham, Surrey 12 April 1876

Studied medicine and botany at Edinburgh and in Paris under A. de Jussieu. MD Montpellier. In Algeria, 1839–59, where he collected plants. 'Natural History of Dijon' (*Mag. Nat. Hist.* 1836, 113–19). *Fl. de l'Algérie* 1847. *Catalogus Plantarum in Algeria* 1859; ed. 2 1866.

Gdnrs Chron. 1876 i 539; 1876 ii 260–62 portr. *J. Bot.* 1876, 160. *Trans. Bot. Soc. Edinburgh* v.13, 1879, 12–14. E. Cosson *Compendium Florae Atlanticae* v.1, 1881, 73–74. *Kew Bull.* 1894, 194. *D.N.B.* v.39, 289–90. *Annual Rep. Yorkshire Philos. Soc.* 1906, 70–71. *R.S.C.* v.4, 542; v.8, 470.

Plants and drawings at Kew, York Museum. Letters at Kew. Portr. at Kew, Hunt Library.

Munbya Pomel.

MUNDAY, Alfred John (1883–1947)
b. 13 Aug. 1883 *d.* 4 April 1947

Employed by Messrs Sander, St. Albans, Herts. In charge of orchid collection, Royal Botanic Garden, Edinburgh, 1911–13, 1919–47. Raised many new orchid hybrids.

Gdnrs Chron. 1947 i 152. *Orchid Rev.* 1947, 83–84.

MUNDY, Henry (1627–1682)
b. Henley, Oxfordshire 1627 *d.* Henley 28 June 1682

BA Oxon 1647. Master, Henley Grammar School, 1656. *Commentarii de aere vitali, esculentis ac potulentis* 1680.

J. Bot. 1889, 262–63; 1894, 109–11; 1918, 56.

MUNFORD, Rev. George (*c.* 1794–1871)
b. Great Yarmouth, Norfolk *c.* 1794 *d.* East Winch, Norfolk 17 May 1871

Of Magdalen Hall, Oxford. Vicar, East Winch, 1849. 'Flowering Plants of W. Norfolk' (*Ann. Mag. Nat. Hist.* 1841, 171–91). Contrib. botany to W. White's *Hist. Norfolk* 1863.

Trans. Norfolk Norwich Nat. Soc. 1871–72, 12. *R.S.C.* v.4, 544.

MUNRO, Donald (–1853)
b. Scotland *d.* 9 April 1853

FLS 1821. Gardener to George Don at Forfar. Gardener to Horticultural Society of London, 1820–50. Contrib. to *Trans. Hort. Soc. London.*

Proc. Linn. Soc. 1853, 237. H. R. Fletcher *Story of R. Hort. Soc., 1804–1968* 1969, 87.

MUNRO, James (1807–1831)
b. 4 May 1807 *d.* 2 Dec. 1831

British Resident, Sikkim. Collected ferns and orchids.

Cymbidium munronianum King & Pantl.

MUNRO, James (*fl.* 1810s)

Nurseryman, Bridge Road, Lambeth, London.

MUNRO, William (1818–1880)
b. Druids Stoke, Glos 1818 *d.* Taunton, Somerset 29 Jan. 1880

FLS 1840. CB 1857. Entered army, 1834. General, 39th Regiment. Collected plants in India in 1840s and in Barbados, 1870–75. *Hortus Agrensis* 1844. *Timber Trees of Bengal* 1847. 'Monograph of Bambusaceae' (*Trans. Linn. Soc.* v.26, 1868, 1–157). 'Gramineae' (W. H. Harvey *Genera of S. African Plants* ed. 2 1868, 427–58).

Garden v.17, 1880, 143. *Gdnrs Chron.* 1880 i 169. *J. Bot.* 1880, 96. *Trans. Bot. Soc. Edinburgh* v.14, 1883, 158–59. E. Bretschneider *Hist. European Bot. Discoveries in China* 1898, 814–15. I. Urban *Symbolae Antillanae* v.3, 1902, 91. *D.N.B.* v.39, 313. *R.S.C.* v.4, 545; v.8, 471; v.12, 572.

Plants, letters and MSS. including' Hortus Bangalorensis' 1837 at Kew. Portr. at Hunt Library.

Munroa Torrey. *Munronia* Wight.

MURCHISON, Charles (1830–1879)
b. Springfield, Vere, Jamaica 26 July 1830 *d.* London 23 April 1879

MD Edinburgh 1851. LLD 1870. FRS 1866. In Bengal Medical Service, 1853–55. Lecturer in Botany, St. Mary's Hospital, London, 1856.

Proc. R. Soc. v.29, 1879, xxiii–xxv. *Trans. Bot. Soc. Edinburgh* v.14, 1883, 33–36. *D.N.B.* v.39, 316–17. D. G. Crawford *Hist. Indian Med. Service, 1600–1913* v.2, 1914, 178. G. H. Brown *Lives of Fellows of R. College of Physicians of London, 1826–1925* 1955, 113–14.

Indian plants at Kew.

MURDOCH, Alfred M. Burn- *see* Burn-Murdoch, A. M.

MURE, — (*fl.* 1820s)

MD. Of Ayrshire. Grandfather of J. F. Cathcart (1802–1851). Had herb. and botanical library. "A botanist of considerable attainments" (J. D. Hooker *Illus. of Himalayan Plants* 1855, ii).

MURIE, James (1832–1925)
b. Glasgow 30 March 1832 *d.* Leigh-on-Sea, Essex 25 Dec. 1925

FLS 1868. Pathologist, Glasgow Royal Infirmary. Naturalist and Medical Officer to J. Petherick's journey to the Upper White Nile to meet J. H. Speke and J. A. Grant, 1861–63.

Proc. Linn. Soc. 1925–26, 92–94. R. Hill *Biogr. Dict. of Sudan* ed. 2 1967, 282–83.

Plants at Kew.

MURPHY, Edmund (–1866)

At Trinity College, Dublin Botanic Garden. Professor of Agriculture, Queen's College, Cork, 1849. Secretary, Dublin Arboricultural Society, 1831. List of Irish plants in *Mag. Nat. Hist.* 1829, 436–38.

Treatise on Agricultural Grasses 1844 (with specimens). Contrib. to *Hort. Register*. *R.S.C.* v.4, 554.
Gdnrs Chron. 1921 ii 196.

MURPHY, Paul Aloysius (1887–1938)
b. County Kilkenny 1887 *d.* 27 Sept. 1938
BA Dublin 1913. Plant pathologist, Prince Edward Island. Plant pathologist, Dept. of Agriculture, Ireland, 1921. Professor of Plant Pathology, University College, Dublin, 1927–38. Did research on potato diseases.
Nature v.142, 1938, 986.

MURRAY, Alexander (*c.* 1798–1838)
d. Aberdeen 1838
MD. Of Aberdeen. 'Connection of Rocks with Plants' (*Mag. Nat. Hist.* 1833, 335–44). *Northern Fl.…belonging to North and East of Scotland* 1836.
Mag. Nat. Hist. 1838, 160. W. Jolly *Life of John Duncan* 1883, 417. *R.S.C.* v.4, 554; v.12, 528.

MURRAY, Alister (–1904)
d. Feb. 1904
Gardener at Dabton, Thornhill, Dumfriesshire. Collected plants in the Lothians. Made a special study of mosses and liverworts.
Garden v.65, 1904, 141.

MURRAY, Amelia Matilda (1795–1884)
d. Glenberrow, Hereford 7 June 1884
"Excellent botanist and artist." Maid of honour to Queen Victoria, 1837–56. Collected *Asplenium verecundum* in Florida, 1855. *Letters from United States, Cuba and Canada* 1856.
D.N.B. v.39, 347–48.

MURRAY, Andrew (before 1810–1850)
d. Cambridge 4 July 1850
Curator, Botanic Garden, Cambridge, 1845–50.

MURRAY, Andrew (1812–1878)
b. Edinburgh 19 Feb. 1812 *d.* Kensington, London 10 Jan. 1878
FLS 1861. FRSE 1857. Entomologist. President, Botanical Society of Edinburgh, 1858–59. Assistant Secretary, Royal Horticultural Society, 1860–64. *Botanical Expedition to Oregon* 1853. *Pines and Firs of Japan* 1863. *Book of Royal Horticultural Society* 1863.
Entom. Mon. Mag. v.14, 1878, 215–16. *Garden* v.13, 1878, 69. *Gdnrs Chron.* 1878 i 86. *J. Bot.* 1878, 63–64. *Hardwicke's Sci. Gossip* 1878, 50–51. *J. R. Hort. Soc.* 1879, 46–48. *Trans. Bot. Soc. Edinburgh* v.13, 1879, 379–81. *D.N.B.* v.39, 349–50. H. R. Fletcher *Story of R. Hort. Soc. 1804–1968* 1969 *passim. R.S.C.* v.4, 555; v.8, 475; v.10, 891; v.12, 529.
Letters at Kew and in Berkeley correspondence at BM(NH). Portr. at Hunt Library.
Pinus murrayana Balf.

MURRAY, Lady Charlotte (1754–1808)
d. Bath, Somerset 1808
Of Athol House, Scotland. *The British Garden* 1799 2 vols (anonymous); ed. 3 1808.
J. Sowerby and J. E. Smith *English Bot.* 404. *Gdnrs Chron.* 1950 ii 238–39.

MURRAY, Denis (*fl.* 1840s)
Of Cork. Gardener. Contrib. list of fungi to T. Power's *Botanist's Guide for County of Cork* 1845.
Letters in Berkeley correspondence at BM(NH).

MURRAY, Rev. Desmond P. (1887–1967)
b. London 21 Aug. 1887 *d.* 2 March 1967
Ordained priest, St. Edmunds College, 1911. Dominican Mission, S. Africa, 1924–28. Priest, Holy Cross, Leicester, 1928–50. Collected plants in S. Africa and British Isles. *Species Revalued* 1955.
Irish Nat. J. v.18(5), 1975, 160–61.
Plants at University College Dublin and Cork.

MURRAY, George Robert Milne (1858–1911)
b. Arbroath, Angus 11 Nov. 1858 *d.* Stonehaven, Kincardine 16 Dec. 1911
FLS 1878. FRS 1897. Pupil of de Bary at Strassburg. Assistant, Botany Dept., BM, 1876; Keeper, 1895–1905. Lecturer on Botany, St. George's Hospital, London, 1882–86. Naturalist attached to solar eclipse expedition to West Indies, 1886. Diatomist and algologist. *Handbook of Cryptogamic Botany* (with A. W. Bennett) 1889. *Introduction to Study of Seaweeds* 1895. Edited *Phycological Memoirs* 1892–95.
Nature v.56, 1897, 56; v.88, 1911, 287. I. Urban *Symbolae Antillanae* v.3, 1902, 91. *Gdnrs Chron.* 1911 ii 466. *J. Bot.* 1912, 73–75 portr. *Proc. R. Soc. B.* v.86, 1913, xxi–xxiii. *D.N.B. Supplt 2* v.2, 667–68. *Who was Who, 1897–1916* 516. C. C. Gillispie *Dict. Sci. Biogr.* v.9, 1974, 587–88. *R.S.C.* v.10, 891.
Portr. at Hunt Library.
Schizophyllum murrayi Massee.

MURRAY, Sir Hugh (1861–1941)
b. 27 April 1861 *d.* Salisbury, Wilts 9 Feb. 1941
Indian Forest Service, Bombay Presidency, 1882–1911. Forestry Commissioner, England, 1924–34.
Nature v.147, 1941, 474. *Who was Who, 1941–1950* 830.

MURRAY, James (1872–1942)
b. Carlisle, Cumberland 9 June 1872 *d.* 7 March 1942

Entomologist and botanist. Made collection of mosses.
N. Western Nat. 1942–43, 115–16 portr.

MURRAY, James A. (*fl.* 1880s)
Plants and Drugs of Sind 1881.
Pakistan J. For. 1967, 352.
Plants at Kew.

MURRAY, James Patrick (*fl.* 1860s)
Surgeon and plant collector with A. W. Howitt on his second expedition in S. Australia, 1861–62. On staff of Melbourne Hospital. 'Enumeration of Plants collected by Dr. J. Murray...in 1862' (*Annual Rep. Government Botanist, Victoria* 15 April 1863).
Austral. Encyclop. v.2, 1965, 21.
Plants at Melbourne.
Acacia murrayana F. Muell.

MURRAY, John (*c.* 1786–1851)
b. Stranraer, Wigtonshire *c.* 1786 *d.* Stranraer 28 June 1851
FLS 1819. *Syllabus of Six Lectures on Physiology of Plants* 1833 (anonymous). *Descriptive Account of Palo de Vaca or Cow-Tree of the Caracas* 1837; ed. 2 1838. *Economy of Vegetation or Phœnomena of Plants* 1838 (anonymous). *Strictures on Morphology* 1845. Contrib. to *Naturalist* v.1, 1837 and Thomson's *Ann. Philos.* v.16, 1820.
D.N.B. v.39, 394. *R.S.C.* v.4, 557.
Letters at Kew.

MURRAY, John Graham (*c.* 1879–1943)
b. Kippin, Stirlingshire *c.* 1879 *d.* Reepham, Lincolnshire 8 March 1943
FLS 1911. Gardener, Botanic Garden, Cambridge, 1900–3. Kew gardener, 1903–4. Chief Horticultural Officer, Lindsey County, Lincs.
Proc. Linn. Soc. 1942–43, 305–6.

MURRAY, Patrick (–1671)
d. Avignon, France 1671
Laird of Livingston, West Lothian. Pupil of Andrew Balfour. Had a collection of plants at Livingston which formed the main source of supply for the Edinburgh Botanic Garden, established in 1670.
A. B. Balfour *Letters to a Friend* 1700. R. Brown *Prodromus Fl. Novae Hollandiae* 1810, 267–68. J. C. Loudon *Encyclop. Gdning* 1860, 280. H. R. Fletcher and W. H. Brown *R. Bot. Gdn, Edinburgh, 1670–1970* 1970, 6–7.
Livistona R. Br.

MURRAY, Rev. Richard Paget (1842–1908)
b. Thornton, Isle of Man 26 Dec. 1842 *d.* Shapwick, Dorset 29 Oct. 1908
MA Cantab 1867. FLS 1882. Ordained 1868. Vicar, Shapwick, 1882–1908. Botanised in Ireland, 1885; in Portugal, 1888; and in Canary Islands (of which he was preparing a flora). 'Fl. Somerset' (*Proc. Somerset Archaeol. Nat. Hist. Soc.* v.39–42, 1893–96). 'Notes on Species of *Lotus-Pedrosia*' (*J. Bot.* 1897, 381–87). 'Canarian and Madeiran Crassulaceae' (*J. Bot.* 1899, 201–4). Published with E. F. and W. R. Linton and W. M. Rogers set of British *Rubi*, 1892–95.
Bot. Soc. Exch. Club Br. Isl. Rep. 1908, 346–47. *J. Bot.* 1909, 1–2 portr. J. W. White *Fl. Bristol* 1912, 97–98. *R.S.C.* v.8, 476; v.10, 892.
Plants at BM(NH), Kew.

MURRAY, Robert (*fl.* 1800s–1820s)
Nurseryman, Hertford Nursery, Hertford.

MURRAY, Stewart (*c.* 1789–1858)
Superintendent, Botanic Garden, Glasgow until 1852. *Glasgow Botanic Garden* 1849.
J. Sowerby and J. E. Smith *English Bot.* 2684. *Bot. Mag.* 1838, t.3674.
Letters at Kew.
Zygopetalum murrayanum Gardn.

MURRAY, William (*fl.* 1850s)
Brother of A. Murray (1812–1878) for whom he collected the material for his 'Notes upon Californian Trees' (*Trans. Bot. Soc. Edinburgh* v.6, 1859, 210–17). Sent seeds of *Chamaecyparis lawsoniana* from California in 1854 to Peter Lawson and Son.
E. J. Ravenscroft *Pinetum Britannicum* 1863–84 3 vols *passim.*

MURRELL, Olive (–1957)
d. 16 Aug. 1957
Had nursery at Orpington, Kent.
Iris Yb. 1957, 24–25.

MURRELL, Percy (*fl.* 1910s)
Nurseryman, Portland Nurseries, Shrewsbury, Shropshire. Rose specialist.
J. Hort. Home Farmer v.63, 1911, 419 portr.

MURTON, Henry James (1853–1881)
b. Cornwall 1853 *d.* Bangkok, Siam 20 Sept. 1882
Kew gardener, 1872–73. Superintendent, Singapore Botanic Gardens, 1875–80. In charge of royal gardens, Bangkok, 1881. *Catalogue of Plants under Cultivation in Botanical Gardens, Singapore* 1879.
J. Kew Guild 1899, 32–33 portr. *Gdns Bull. Straits Settlements* 1918, 93–108; 1927, 128. *Thailand Res. Soc. Nat. Hist. Supplt* 1939, 18–19. *Fl. Malesiana* v.1, 1950, 377–78. *R.S.C.* v.10, 892; v.12, 529.
Plants at Kew, Singapore. MS. *Fl. Singapore* lost.
Murtonia Craib.

MUSGRAVE, Mrs. (*fl.* 1890s)
Possibly wife of A. Musgrave, Government Secretary, Port Moresby, New Guinea. Collected mosses in New Guinea, 1897, now at BM(NH).
J. Linn. Soc. 1922, 502, 504. *Fl. Malesiana* v.1, 1950, 378.

MUSGRAVE, Charles Thomas (1862–1949)
b. 15 April 1862 *d.* 30 Jan. 1949
VMH 1926. Registrar, Land Registry. Treasurer, Royal Horticultural Society, 1921–24. Created gardens at Hascombe Place near Godalming, Surrey.
Bot. Mag. 1936 portr. *Gdnrs Chron.* 1949 i 58. *Times* 2 Feb. 1949.

MUSK, Harold (1899–1935)
b. Haveringland, Norfolk 2 April 1899 *d.* Dar-es-Salaam, Tanganyika 20 April 1935
Kew gardener, 1924–25. At Wye Agricultural College, 1925–26; Imperial College of Tropical Agriculture, Trinidad, 1926–27. District Agricultural Officer, Tanganyika, 1928–35.
Kew Bull. 1928, 112; 1935, 219. *J. Kew Guild* 1936, 581–82 portr.

MUSSON, Charles Tucker (1856–1928)
b. Nottingham 14 Dec. 1856 *d.* Gordon, N.S.W. 9 Dec. 1928
Emigrated to Australia, 1887. Science master, Hawkesbury Agricultural College, N.S.W., 1891–1919. Contrib. botanical papers to *Proc. Linn. Soc. N.S.W.*
Sydney Morning Herald 11 Dec. 1928. *Proc. Linn. Soc. N.S.W.* 1929, vii.

MUSTOE, William Robert (1878–1942)
b. Leckhampton, Glos June 1878 *d.* Jodhpur, India 22 July 1942
Kew gardener, 1904. Superintendent, Lawrence Gardens, Lahore, India, 1905. Superintendent of Horticultural Operations, New Delhi.
Gdnrs Chron. 1942 ii 148. *J. Kew Guild* 1942, 199–201 portr.

MYERS, John Golding (1897–1942)
b. near Rugby, Warwickshire 22 Oct. 1897 *d.* Amadi, Sudan 3 Feb. 1942
To New Zealand, 1911. Entomologist, N.Z. Dept. of Agriculture, 1919–24. On staff of Imperial Institute of Entomology, 1926; Imperial College of Tropical Agriculture, Trinidad, 1934. Economic botanist to Government of Anglo-Egyptian Sudan, 1937.
Nature v.149, 1942, 406.
Plants from British Guiana and Sudan at Kew; main collection at Wad Medani, Sudan.

MYLAM (*or* **MILAM**), **Mr.** (*fl.* 1720s)
Physician employed to treat slaves on their arrival in Brazil.
Plants from Buenos Aires at Oxford.

MYLES, Rev. Percy Watkins Fenton (1849–1891)
b. Kilmoe, County Cork 27 Feb. 1849 *d.* Ealing, Middx 7 Oct. 1891
BA Dublin 1867. FLS 1887. Ordained 1873; had living at Ealing, 1874. Contrib. to *J. Bot.* and G. Nicholson's *Illus. Dict. Gdning* 1884–88.
J. Bot. 1891, 349–50 portr. *Proc. Linn. Soc.* 1891–92, 73.

MYLNE, — (*fl.* 1850s)
Collected flowering plants and algae near Swan River, W. Australia.
W. H. Harvey *Nereis Australis* 1847, 21. G. Bentham *Fl. Australiensis* v.1, 1863, 39. *J. W. Austral. Nat. Hist. Soc.* no. 6, 1909, 21–22.
Plants at Cambridge, Kew.
Hibbertia mylnei Benth.

NAIRNE, Alexander Kyd (*fl.* 1880s)
Collected plants in Bombay, 1888. *Flowering Plants of Western India* 1894.

NAPIER, Mr. (*fl.* 1800s)
Nurseryman near Vauxhall. Firm became Napier and Chandler in early 1800s.
Bot. Mag. 1803, t.697. *Trans. London Middlesex Archaeol. Soc.* v.24, 1973, 189.

NAPIER, Charles Ottley *see* Groom, C. O.

NAPIER, E. R. *see* Moloney, E. R.

NAPIER, Walter G. (–1934)
d. Edinburgh 1 Aug. 1934
MA. BSc Edinburgh. Partner in firm of D. Napier and Sons, herbalists and chemists. Botanist.
Pharm. J. 1934, 195.

NAPIER, Sir Walter John (1857–1945)
b. 10 July 1857 *d.* 14 Feb. 1945
Lawyer, Singapore. Attorney-General, Straits Settlements, 1907–9. Collected plants now at Herb., Singapore.
Gdns Bull. Straits Settlements 1927, 129. *Fl. Malesiana* v.1, 1950, 380. *Who was Who, 1941–1950* 834.

NAPPER, Diana Margaret (1930–1972)
b. Woking, Surrey 23 Aug. 1930 *d.* London 31 March 1972
BSc Exeter 1951. FLS 1967. Laboratory Assistant, Coffee Research Station, Kenya, 1954. Assistant, East African Herb., Nairobi, 1955. On staff of Kew Herb., 1965. Studied *Acanthaceae*, *Cyperaceae* and *Gramineae* of E. Africa. *Illustrated Guide to Grasses of Uganda* (with K. W. Harker) 1961. *Grasses of Tanganyika* 1965. 'Cyperaceae of East Africa' (*J. East Africa Nat. Hist. Soc.* 1963–71). Contrib. to *Fl. Tropical East Africa* 1971, *J. East Africa Nat. Hist. Soc.*, *Kew Bull.*
East Africa Nat. Hist. Soc. Bull. 1972, 75. *J. Kew Guild* 1972, 165–66 portr. *Kew Bull.* 1973, 1–4 portr.

NAPPER, Matthew (*fl.* 1820s)
Nurseryman, Common, Horsham, Sussex.

NASH, C. Anthony M. (–1972)
d. British Solomon Islands 25 Aug. 1972

Aberdeen University, 1946–49. Forester, Nigeria, 1949. Deputy Conservator of Forests, British Solomon Islands, 1968.
Commonwealth For. Rev. 1972, 288.

NASH, Daniel (*c.* 1806–1874)
d. 28 Nov. 1874
Senior partner in Minier, Nash and Adams, 63 Strand, London.

NASMITH (*or* **Naysmith), John** (–*c.* 1619)
b. Posso, Peeblesshire *d.* Earlston, Berwick? *c.* 1619
Surgeon and botanist to James VI (James I of England). Had a garden in London. Friend of Lobel.
M. de Lobel *Animadversiones* 1605, 487, 489, 496. *D.N.B.* v.40, 112–13.

NASMYTH, Sir James (–1779)
d. Philiphaugh, Selkirkshire 4 Feb. 1779
Studied under Linnaeus in Sweden. Introduced many fine conifers and other trees on family estate at Dawyck, Peebles.
D.N.B. v.40, 116. *J. R. Hort. Soc.* 1947, 5–12.
Nasmythia Huds.

NATION, William (1826–1907)
b. Staplegrove, Somerset 1826 *d.* Clapham, London 18 Oct. 1907
Kew gardener, 1848. To Peru, 1850. Professor at Guadeloupe College, Lima.
Bot. Mag. 1864, t.5432. *Gdnrs Chron.* 1907 ii 330. *J. Bot.* 1907, 455. *J. Kew Guild* 1907, 379–80. *Kew Bull.* 1908, 46.
Letters and Peruvian plants at Kew.
Quamoclit nationis Hook.

NAUEN, John Charles (1903–1943)
b. Coventry, Warwickshire 1903 *d.* Siam–Burma Railway Oct. 1943
Kew gardener, 1928. Horticulturist, Dept. of Agriculture, Bermuda, 1928–35. Assistant Curator, Botanic Gardens, Singapore, 1935–39. Botanic Garden, Penang, 1939–41.
Kew Bull. 1935, 336. *J. Kew Guild* 1945, 477–78. *Gdns Bull., Singapore* 1947, 266. *Fl. Malesiana* v.1, 1950, 380.
Plants at Singapore.

NAYLOR, Frederick (1811–1882)
d. Kew, Surrey 21 Dec. 1882
Had large collection of autographs of botanists. 'On *Asplenium petrarchae* DC as an Irish Plant' (*Trans. Bot. Soc. Edinburgh* v.8, 1865, 365–66). 'Notice of Rare Plants collected in South-west of England' (*Trans. Bot. Soc. Edinburgh* v.8, 1866, 262–64).
J. Bot. 1883, 192.
Plants at BM(NH), Kew, Oxford.

NAYLOR, Hannah Elizabeth *see* Wilkinson, H. E.

NAYLOR, James (*c.* 1827–1901)
b. near Bakewell, Derbyshire *c.* 1827 *d.* Harrow, Middx 19 June 1901
Nurseryman, Roxeth, Harrow, *c.* 1867.
Gdnrs Chron. 1901 i 425.

NAYSMITH, John *see* Nasmith, J.

NEAL, G. (*fl.* 1780s)
Nurseryman, Old Kent Road, Camberwell, Surrey. Specialised in pine-apples.
Trans. London Middlesex Archaeol. Soc. v.24, 1973, 181.

NEAL, Robert (–1915)
d. Wandsworth, London 26 Oct. 1915
Nurseryman, Trinity Road, Wandsworth.
J. Hort. v.71, 1915, 433 portr.

NEALE, Adam (*fl.* 1770s)
Gardener to J. Blackbourne at Warrington, Lancs. *Catalogue of Plants in Garden of John Blackbourne* 1779.
G. W. Johnson *Hist. English Gdning* 1829, 238.

NEALE, J. J. (*c.* 1854–1919)
d. 29 Dec. 1919
Of Cardiff, Glam and Kenton, Devon. Cultivated orchids and insectivorous plants.
Orchid Rev. 1920, 7–8.

NEALE, S. (*fl.* 1820s)
Seedsman, 74 Hackney Road, London.

NEAME, Sir Thomas (*c.* 1885–1973)
b. 23 Dec. 1885 *d.* 28 Aug. 1973
VMH 1958. Farmer and fruit grower, Faversham, Kent. Chairman, East Malling Research Station, 1945–60.
J. R. Hort. Soc. 1973, 551–52.
Portr. at Hunt Library. Library sold at Sotheby, 5 Feb. 1974.

NECK, Rev. Aaron (1769–1852)
b. St. Marychurch, Devon 1769 *d.* Kingskerswell, Devon 4 Oct. 1852
BA Oxon 1791. Incumbent, Kingskerswell, 1832. Discovered *Bupleurum aristatum*, 1802. Contrib. to J. P. Jones and J. F. Kingston's *Fl. Devoniensis* 1829.
W. K. Martin and G. T. Fraser *Fl. Devon* 1939, 771.
Letters at Kew.

NECKAM (*or* **Necham), Rev. Alexander** (1157–1217)
b. St. Albans, Herts 6 Sept. 1157 *d.* Kempsey, Worcs 1217
Master of Dunstable. Taught in Paris, 1180. Abbot of Augustinians, Cirencester, 1213. *De Naturis Rerum* in Rolls Series (including plants).
Gdnrs Chron. 1881 i 361–62; 1910 i 273. *D.N.B.* v.40, 154–55. E. Cecil *Hist. Gdning in England* 1910, 59–61. C. E. Raven *English Nat.* 1947, 4–9.

NEEDHAM, James (1849–1913)
b. Hebden Bridge, Yorks March 1849 *d.* Hebden Bridge 14 July 1913

Ironmoulder. Took up botany, 1885. Mycologist and bryologist. Contrib. fungi records to W. B. Crump and C. Crossland's *Fl. Parish of Halifax* 1904.

Naturalist 1913, 294–98 portr.; 1961, 56–57; 1973, 4.

Gnomonia needhamii Mass. & Crossl.

NEEDHAM, Rev. John Turberville (1713–1781)
b. London 10 Sept. 1713 *d.* Brussels 30 Dec. 1781

FRS 1747. Roman catholic priest and scientist. Director of Imperial Academy, Brussels, 1768–80. *Account of some New Microscopical Discoveries* 1745. *Observations upon the Generation, Composition and Decomposition of Animal and Vegetable Substances* 1749. Contrib. to *Philos. Trans. R. Soc.*

J. Nichols *Lit. Anecdotes of Eighteenth Century* v.7, 283, 635. J. Gillow *Dict. English Catholics* v.5, 157–60. *D.N.B.* v.40, 157–59.

Needhamia R. Br.

NEEDHAM, Walter (*c.* 1631–*c.* 1691)
d. 5 April *c.* 1691

FRS 1667. Physician and anatomist. Helped John Ray with his *Historia Plantarum* 1686–88 and supplied details of medical properties of plants to his *Catalogus Plantarum* 1660.

D.N.B. v.40, 164–65.

NEILL, Patrick (1776–1851)
b. Edinburgh 25 Oct. 1776 *d.* Canonmills, Edinburgh 5 Sept. 1851

MA Edinburgh. LLD 1834. ALS 1807. FLS 1813. FRSE. Printer. Vice-President, Botanical Society of Edinburgh, 1836. Secretary, Caledonian Horticultural Society and also Wernerian Society. Friend of G. Don. Collected Scottish plants. *Tour through some of the Islands of Orkney and Shetland* 1806 (includes plant lists). *Journal of Horticultural Tour through some Parts of Flanders, Holland and North of France in 1817* 1823. *Fruit, Flower and Kitchen Garden* 1845. 'Fuci' (Brewster's *Edinburgh Encyclop.* 1830).

Gdnrs Mag. 1835, 673–74; 1836, 333–41; 1843, 87–88, 455–59. R. Sweet *Br. Flower Gdn* v.6, 1838, 233. *Gdnrs Chron.* 1851, 567, 663–64; 1902 ii 297–98; 1923 i 320–21. *Cottage Gdnr* v.7, 1851, 121. *Proc. Linn. Soc.* 1852, 191–92. *J. Bot.* 1863, 96. *D.N.B.* v.40, 178–79. *Notes Bot. Gdn Edinburgh* v.3, 1904, 94–96. M. Spence *Fl. Orcadensis* 1914, xlv–xlvii.

Letters at Kew and in Smith correspondence at Linnean Society.

Neillia D. Don.

NELMES, Ernest (1895–1959)
b. Bevington, Glos 18 April 1895 *d.* Kew, Surrey 5 Feb. 1959

Kew gardener, 1921. Assistant, Kew Herb., 1923; later Librarian. Authority on *Carex*.

J. Kew Guild 1959, 709 portr.

Plants at Kew. Portr. at Hunt Library.

NELSON, David (–1789)
d. Koepang, Timor 20 July 1789

Kew gardener. Collected plants for J. Banks on Cook's third voyage, 1776–80. Collected breadfruit trees on HMS 'Bounty', 1787.

R. Brown *Prodromus* 1810, 480–81. *Gdnrs Chron.* 1881 ii 267. *Kew Bull.* 1891, 297. E. Bretschneider *Hist. European Bot. Discoveries in China* 1898, 153–54. *J. R. Soc. Tasmania* 1909, 23–24. J. H. Maiden *Sir Joseph Banks* 1909, 124–25. *J. Bot.* 1916, 351–52. *J. Linn. Soc.* v.45, 1920, 48. I. Lee *Early Explorers in Australia* 1925, 76–77. A. W. Anderson *Coming of the Flowers* 1950, 143–49. *Fl. Malesiana* v.1, 1950, 383. K. Lemmon *Golden Age of Plant Hunters* 1968, 79–106.

Australian, Cape, Macao and Timor plants at BM(NH).

Nelsonia R. Br.

NELSON, Edward Milles (1851–1938)
d. 20 July 1938

President, Quekett Microscopical Club, 1893–95. Authority on theory of microscope. Examined structure of diatoms.

J. Bot. 1938, 280. *Nature* v.142, 1938, 385–86.

NELSON, John (*fl.* 1860s)

Of Lymington, Hants. Arboriculturist. *Pinaceae* 1866 (under pseudonym of 'Senilis').

NELSON, Rev. John Gudgeon (1818–1882)
b. Winterton, Norfolk 14 July 1818 *d.* Aldborough, Norfolk 14 April 1882

Rector, Aldborough, *c.* 1860. Raised *Phlox nelsonii* and *Lachenalia nelsoni*.

Garden v.21, 1882, 284. *Gdnrs Chron.* 1882 i 540.

NELSON, William (1852–1922)
b. Sheffield, Yorks 5 March 1852

Son of John Nelson, nurseryman, Rotherham, Yorks. Nurseryman, Bradway near Sheffield, Yorks. To Transvaal, 1876. Collected plants in S. Africa. Nurseryman, Johannesburg, 1890s. Published account of his travels in *Masbro Advertiser* 19 May 1877 onwards.

Gdnrs Chron. 1880 ii 198.

S. African plants at Kew. MSS. at Africana Museum, Johannesburg.

Albuca nelsoni N. E. Brown.

NESFIELD, William Andrews (1793–1881)
b. Chester-le-Street, Durham 19 Feb. 1793
d. Regent's Park, London 2 March 1881
Educ. Cambridge. Artist and landscape gardener. Designed Kew Gardens and Royal Horticultural Society's gardens at South Kensington.
Garden v.19, 1881, 296. *Gdnrs Chron.* 1881 i 342. *J. Hort. Cottage Gdnr* v.2, 1881, 192–93. *D.N.B.* v.40, 226–27.

NEVE, Arthur (*fl.* 1890s)
Kashmir Medical Mission. Collected plants in Baltistan, Ladak and Kashmir, 1895–98.
Plants at Kew.

NEVE, J. R. (*fl.* 1900s)
Of Chipping Camden, Glos. Left his herb. to J. Morris of Broadway, Glos. Contrib. list of Chipping Campden plants to P. C. Rushen's *Handbook to Chipping Campden* *c.* 1905.
H. J. Riddelsdell *Fl. Gloucestershire* 1948, cxlvii.

NEVILL, Lady Dorothy Fanny (*née* Walpole) (1826–1913)
b. London 1 April 1826 *d.* 24 March 1913
Had notable garden at Dangstein near Midhurst, Sussex where she cultivated orchids, nepenthes and other tropical plants.
Curtis's Bot. Mag. Dedications, 1827–1927 115–16 portr.

NEVINS, John Birkbeck (–1903)
b. Shetland? *d.* Liverpool June 1903
MD London 1846. Lecturer in Botany at Liverpool Medical School. 'Archetype of Flowering Plants' (*Trans. Bot. Soc. Edinburgh* v.6, 1860, 355–57). 'Ripening of Seeds' (*Trans. Bot. Soc. Edinburgh* v.8, 1866, 166–69).

NEW, Rev. Charles (1840–1875)
b. Fulham Jan. 1840 *d.* Mombasa, Kenya 13 Feb. 1875
Methodist missionary. To Zanzibar, 1863. Collected plants on Kilimanjaro, Tanganyika, 1870s. *Life, Wanderings and Labours in East Africa* 1874.
J. Bot. 1872, 235–36; 1875, 160. *J. Linn. Soc.* v.14, 1873, 141–46. *Proc. R. Geogr. Soc.* 1874–75, 387–89.
Plants at Kew.
Helichrysum newii Oliver & Hiern.

NEW, William (–1873)
d. Aug. 1873
Gardener, Botanic Garden, Belfast. Kew gardener, *c.* 1855. Superintendent, Botanic Garden, Bangalore, 1858–63; reappointed 1865–73. *Catalogue of Plants in Public Garden, Bangalore, June 1861.*
Strobilanthes newii Bedd.

NEWBERRY, Percy Edward *see* Addendum

NEWBERRY, W. J. (*c.* 1880–1954)
d. Pietermaritzburg, Natal 24 July 1954
Kew gardener, 1903. Botanic Gardens, Pietermaritzburg, 1906; Curator, 1910–17. Superintendent of Parks, Pietermaritzburg, 1917–40.
J. Kew Guild 1955, 248.

NEWBERY, William (–before 1797)
Of Stockland, Devon. "A noted herbalist" (Z. J. Edwards). Discovered *Lobelia urens* and *Coronopus didymus.*
W. Hudson *Fl. Anglica* 1778, 280, 378. Z. J. Edwards *Ferns of the Axe* 1862, 115–16.

NEWBIGGING, Alexander Tweedie (*c.* 1815–1885)
b. Broughton, Peeblesshire *d.* 19 Aug. 1885
Married daughter of T. Kennedy, nurseryman of Dumfries. When Kennedy retired in 1856 carried on business with R. Cowan.
Gdnrs Chron. 1885 ii 252.

NEWBOLD, Sir Douglas (1894–1945)
b. 13 Aug. 1894 *d.* Khartoum, Sudan 23 March 1945
Educ. Oxford. Sudan Political Service, 1920; Civil Secretary, 1929. Governor, Kardofan Province, 1932–38. Made explorations 1923, 1927, 1929–30. Collected plants. Contrib. to *Geogr. J.*, *Sudan Notes and Rec.*
D.N.B. 1941–1950 623–24. *Who was Who, 1941–1950* 842. R. Hill *Biogr. Dict. of Sudan* ed. 2 1967, 294.
Plants at Kew. Portr. at University, Khartoum.

NEWBOLD, Thomas John (1807–1850)
b. Macclesfield, Cheshire 8 Feb. 1807 *d.* Mahabuleshwar, India 29 May 1850
Madras Light Infantry, 1828. Sent plants from Mount Ophir, Sumatra to N. Wallich in Calcutta. About 1838 prepared zoological and botanical catalogue of Straits Settlements which was never published.
J. Straits Branch Asiatic Soc. 1887, 143–48. *D.N.B.* v.40, 314–15. C. E. Buckland *Dict. Indian Biogr.* 1906, 315. *Gdns Bull. Straits Settlements* 1927, 129. *Fl. Malesiana* v.1, 1950, 384.

NEWBOULD, Rev. William Williamson (1819–1886)
b. Sheffield, Yorks 20 Jan. 1819 *d.* Kew, Surrey 16 April 1886
BA Cantab 1842. FLS 1863. Curate, Bluntisham, Hunts, 1845. Comberton, Cambridgeshire, 1846. In Scotland, 1845; Ireland, 1852, 1858; N. Wales, 1862. Friend of Babington and H. C. Watson. 'Notes on Fl. Matlock' (*J. Bot.* 1884, 334–44). Edited ed. 2 of H. C. Watson's *Topographical Bot.* (with J. G. Baker) 1883.

J. Bot. 1881, 89; 1886, 159–60, 161–74 portr.; 1905, 218. *Gdnrs Chron.* 1886 i 569. H. C. Watson *Topographical Bot.* 1883, 551–52. *Proc. Linn. Soc.* 1885–86, 145–46. *Trans. Bot. Soc. Edinburgh* v.17, 1889, 15–17. G. C. Druce *Fl. Berkshire* 1897, clxxiv–clxxvi. *D.N.B.* v.40, 315–16. G. C. Druce *Fl. Oxfordshire* 1927, cxvii–cxviii. *N. Western Nat.* 1947, 223–25 portr.

Plants and letters at Oxford. MSS. at BM(NH).

Newbouldia Seem.

NEWINGTON, Samuel (*c.* 1822–1882)
d. Ticehurst, Sussex 3 July 1882

Superintendent of a mental home. Invented horticultural equipment such as a tree-lifter, cylinder vinery, etc. Contrib. to *Gdnrs Chron.*

Gdnrs Chron. 1882 ii 90.

NEWMAN, Edward (1801–1876)
b. Hampstead, London 13 May 1801 *d.* Peckham, London 12 June 1876

FLS 1833. At Godalming, Surrey, 1817–26. *History of British Ferns and Allied Plants* 1840; ed. 4 1865 (illustrated by himself). *Letters of Rusticus* 1849. Edited *Phytologist* 1841–54.

T. P. Newman *Mem. of Life...of Edward Newman* 1876. *Entomologist* 1876, i–xxiv portr. *Gdnrs Chron.* 1876 i 823. *J. Bot.* 1876, 223–24. *Zoologist* 1876, iii–xxii portr. *Trans. Bot. Soc. Edinburgh* v.13, 1879, 14–16. Friends Institute *Biogr. Cat.* 1888, 467–72. *D.N.B.* v.40, 338. *Amer. Fern. J.* 1931, 144–45. C. E. Salmon *Fl. Surrey* 1931, 52. *R.S.C.* v.4, 600; v.8, 494.

Ferns and drawings at BM(NH). Letters at Society of Friends, Royal Entomological Society. Portr. at Hunt Library.

NEWMAN, Francis William (*c.* 1796–1859)
d. Hobart, Tasmania 23 Aug. 1859

Superintendent, Botanic Garden, Hobart, 1845–59. *Catalogue of Plants in Royal Society's Gardens, Hobart Town, Tasmania* 1857.

Rep. R. Soc. Van Dieman's Land for Hort. 1846, 1; 1848, 8–9; 1859, 22. *Papers Proc. R. Soc. Tasmania* 1909, 24. *J. Arnold Arb.* 1921, 52. *Austral. Encyclop.* v.2, 1965, 58.

NEWMAN, John (*fl.* 1820s–1840s)
b. Kew, Surrey *d.* Mauritius

Kew gardener. Collected plants in Brazil for Lee and Kennedy's nursery, 1822. Curator, Botanic Garden, Mauritius, 1825–48.

Bot. Misc. 1829–30, 291–92.

Letters at Kew.

NEWMAN, William (–1789)

Nurseryman, Chichester, Sussex. Succeeded by his son, William Newman.

NEWSHAM, John Clark (1873–1927)
d. Usk, Mon 30 Oct. 1927

FLS 1906. Kew gardener, 1896. Principal, Hampshire Agricultural Institute, 1900. Principal, Monmouthshire Agricultural Institution, 1914. *Horticultural Notebook* 1906. *Propagation and Pruning of Hardy Trees, Shrubs and Miscellaneous Plants* 1913. *Potato Book* 1917. *Crops and Tillage* 1921. *Fruit Growing for Profit* 1921.

Gdnrs Chron. 1927 ii 436. *J. Kew Guild* 1928, 626.

NEWTON, Henry (*fl.* 1850s)

Plumber, Ashton-under-Lyne, Lancs. Discovered *Carex ornithopoda* with J. Whitehead at Millersdale, Derbyshire.

Trans. Liverpool Bot. Soc. 1909, 83.

NEWTON, James (1639–1718)

MD. Friend of J. Ray and Commelin. Found *Arabis stricta* (J. Ray *Historia Plantarum* v.1, 1686, 817). *Enchiridion Universale Plantarum* 1689 (incomplete). *Compleat Herbal* (begun 1680; edited by his son) 1752 portr.; ed. 6 1802.

J. Ray *Synopsis* ed. 2 1696; ed. 3 1724 preface. *Philos. Trans. R. Soc.* v.20, 1698, 263–64. R. Morison *Plantarum Historiae Universalis Oxoniensis* v.3, 1699, 202, 232, 237, 628, 647. H. Trimen and W. T. T. Dyer *Fl. Middlesex* 1869, 389. E. J. L. Scott *Index to Sloane Manuscripts* 1904, 393. C. E. Raven *John Ray* 1942, 218, 259. J. E. Dandy *Sloane Herb.* 1958, 170.

Plants at BM(NH).

NEWTON, James (–1750)
d. Islington, London 5 Nov. 1750

MD. Botanist. Kept private asylum near Islington turnpike.

D.N.B. v.40, 393. *Gdnrs Chron.* 1919 i 147.

Portr. at Hunt Library.

NEWTON, John (*fl.* 1600s)

Surgeon. Of Colyton, Devon. Brought an *Avens* and *Lobelia cardinalis* from New England for John Parkinson.

J. Parkinson *Theatrum Botanicum* 1640, 596. C. E. Raven *English Nat.* 1947, 263.

NEWTON, John (*fl.* 1780s)

Helped to found a "society of botanists" in Oldham, Lancs, *c.* 1785.

J. Holt *General View of Agric....Lancashire* 1795, 229. *Trans. Liverpool Bot. Soc.* 1909, 83.

NEWTON, John (*c.* 1837–1907)
d. Streatham, London 7 Nov. 1907

Gardener, Inner Temple Gardens. Chrysanthemum grower. *Culture of Chrysanthemum* ed. 7 1881; ed. 18 1893.

J. Hort. Cottage Gdnr v.55, 1907, 458.

NEWTON, Rev. Thomas (*c.* 1542–1607)
b. Prestbury, Cheshire *c.* 1542 *d.* Little Ilford, East Ham, Essex May 1607

Rector, Little Ilford, 1583. *Herbal for the Bible* 1587, from Lemnius.

R. Pulteney *Hist. Biogr. Sketches of Progress of Bot. in England* v.1, 1790, 108–9. *D.N.B.* v.40, 402–3. *Gdnrs Chron.* 1935 ii 9–10, 81–82.

NEWTON, William Charles Frank (1895–1927)
b. Thetford, Norfolk 16 Feb. 1895 *d.* Merton, Surrey 22 Dec. 1927

BSc London 1921. FLS 1925. Cytologist, John Innes Horticultural Institute, 1922. Interested in application of cytology to taxonomy. 'Studies in Somatic Chromosomes' (*Ann. Bot.* 1924, 197–206). 'Chromosome Studies in *Tulipa* and some Related Genera' (*J. Linn. Soc.* v.47, 1926, 339–54).

J. Bot. 1928, 51. *Nature* v.121, 1928, 27–28. *Proc. Linn. Soc.* 1927–28, 126–27.

NICHOL, William (1836–1859)
b. Edinburgh March 1836 *d.* Alexandria, Egypt 7 May 1859

MD Edinburgh 1857. Pupil of J. H. Balfour. Bryologist.

Trans. Bot. Soc. Edinburgh v.6, 1859, 290–92.

Plants at Glasgow.

NICHOLLS, Amelia (*fl.* 1840s)

Nurseryman, Marsh End, Newport Pagnell, Bucks. Widow of Joseph Nicholls?

NICHOLLS, Charles Henry (1866–1938)
b. Kirdford, Sussex 30 Dec. 1866 *d.* Watlington, Oxford 15 Jan. 1938

BA Oxon 1890. FLS 1893. Worked at Kew, Cinchona (Jamaica), Botanic Gardens, Peradeniya, Ceylon and Edinburgh. On staff of Imperial Institute, London, 1903–9. Economic botanist. Contrib. to *Bull. Imperial Institute*.

Proc. Linn. Soc. 1937–38, 323–24.

NICHOLLS, Frank (1699–1778)
b. London 1699 *d.* Epsom, Surrey 7 Jan. 1778

MD Oxon 1729. FRS 1728. Lecturer, Royal College of Physicians, 1734, 1736, 1748–49. Physician to George II, 1753. Correspondent of William Curtis, 1775.

Proc. Linn. Soc. 1946–47, 75–76.

MSS. at Royal Society. Botanical MSS. at Linnean Society.

NICHOLLS, Sir Henry Alfred Alford (1851–1926)
b. London 27 Sept. 1851 *d.* Dominica 9 Feb. 1926

MD 1875. MRCS 1878. FLS 1908. Knighted 1926. Medical Officer, Dominica, 1875–1925. *On Cultivation of Liberian Coffee in West Indies* 1881. *Elementary Textbook of Tropical Agriculture* 1891. *Cultivation of Banana in Dominica* 1890.

I. Urban *Symbolae Antillanae* v.3, 1902, 92. *Kew Bull.* 1926, 192. *Proc. Linn. Soc.* 1925–26, 94–95. *West India Committee Circ.* 1925, 370; 1926, 72. *Who was Who, 1916–1928* 776.

Plants at Kew.

NICHOLLS, Joseph (*fl.* 1830s)

Nurseryman, Marsh End, Newport Pagnell, Bucks. Succeeded by Amelia Nicholls.

NICHOLLS, Rebecca (*fl.* 1790s)

Nurseryman, Bodmin, Cornwall.

NICHOLLS, Robert (*fl.* 1710s–1740s)

Apothecary. Master, Apothecary's Company, 1741. Correspondent of J. Blackstone. Sent fungi to J. J. Dillenius at Oxford.

H. Trimen and W. T. T. Dyer *Fl. Middlesex* 1869, 391. E. J. L. Scott *Index to Sloane Manuscripts* 1904, 394. G. E. Dandy *Sloane Herb.* 1958, 172. H. N. Clokie *Account of Herb. Dept. Bot., Oxford* 1964, 217.

Plants at BM(NH).

NICHOLLS, Thomas (*fl.* 1790s)

Nurseryman, 86 High Street, Birmingham.

NICHOLSON, Charles (1868–1940)
b. Ringwood, Hants Sept. 1868 *d.* Penzance, Cornwall 21 March 1940

Primarily an entomologist but also interested in botany. President, London Natural History Society, 1897.

Gdnrs Chron. 1935 ii 56 portr. *London Nat.* 1939, 35.

Plants at London Natural History Society.

NICHOLSON, George (1847–1908)
b. Ripon, Yorks 7 Dec. 1847 *d.* Richmond, Surrey 20 Sept. 1908

ALS 1886. FLS 1898. VMH 1897. Kew gardener, 1873; Curator, 1886–1901. *Illustrated Dictionary of Gardening* 1884–88 4 vols.; *Supplement* 1900 2 vols.

Gdnrs Mag. 1894, 325 portr.; 1907, 323 portr., 324. *Garden* v.48, 1895, xi portr. *J. Bot.* 1896, 81–82 portr.; 1908, 337–39 portr. *Gdnrs Chron.* 1908 ii 239 portr. *Bot. Soc. Exch. Club Br. Isl. Rep.* 1908, 314–15. *Kew Bull.* 1908, 422–27. *J. Kew Guild* 1908, 428 portr. *D.N.B. Supplt 2* v.3, 12–13. *Curtis's Bot. Mag. 1827–1927* 303–4 portr. *R.S.C.* v.10, 521; v.12, 538.

Herb. at Aberdeen.

Neonicholsonia Dammer.

NICHOLSON, Henry (*c.* 1660–1732/3)
d. Dublin 1732/3

MD Leyden 1709. FRS 1716. Lecturer in Botany, Dublin, 1711–32. *Methodus Plantarum in Horto Medico Collegii Dublinensis* 1712.

E. J. L. Scott *Index to Sloane Manuscripts* 1904, 394. T. P. C. Kirkpatrick *Hist. Med. Teaching in Trinity College, Dublin* 1912, 362.

NICHOLSON, Henry Alleyne (1844–1899)
b. Penrith, Cumberland 11 Sept. 1844 *d.* Aberdeen 19 Jan. 1899
PhD Göttingen 1862. MD Edinburgh 1869. FRS 1897. FLS 1876. Professor of Natural History, Toronto, 1871; Durham, 1874; St. Andrews, 1875; Aberdeen, 1882. *Manual of Palaeontology* 1872 (palaeobotany, 473–503); ed. 2 1879.
Geol. Mag. 1899, 138–44. *Nature* v.59, 1899, 298–99. *Proc. Linn. Soc.* 1898–99, 54–56. *Quart. J. Geol. Soc.* 1899, lxiv–lxvi. *Yb. R. Soc.* 1899, 189. *Who was Who, 1897–1916* 525. *R.S.C.* v.8, 502; v.10, 921.

NICHOLSON, John (*fl.* 1820s)
Nurseryman, Sheffield, Yorks.

NICHOLSON, Thomas (1799–1877)
b. Dumfriesshire 1799 *d.* Antigua, West Indies 8 July 1877
Ship's surgeon. In Antigua, 1819, 1822–77. Sent plants and drawings to W. J. Hooker (*Bot. Mag.* 1831, t.3071, t.3098).
Bot. Misc. 1830, 271–73. I. Urban *Symbolae Antillanae* v.1, 1898, 116; v.3, 1902, 92. V. L. Oliver *Caribbeana* v.2, 1912, 287.
Letters and plants at Kew. MS. *Hortus Antiguensis* at Dept. Agriculture, Jamaica.
Castela nicholsoni Hook.

NICHOLSON, William A. (1858–1935)
b. Birkenhead, Cheshire 1858 *d.* Norwich, Norfolk 28 May 1935
Bank clerk. Secretary and Editor, Norfolk and Norwich Naturalists Society, 1891–1912. *Fl. Norfolk* 1914.
Trans. Norfolk Norwich Nat. Soc. 1935, 110–11 portr.
Plants at Norwich Museum.

NICHOLSON, William Edward (1866–1945)
b. Lewes, Sussex 24 March 1866 *d.* Mullion, Cornwall 13 Feb. 1945
FLS 1919. Solicitor, Lewes. President, British Bryological Society, 1929–30. Friend of H. N. Dixon with whom he made a number of plant collecting expeditions in Europe. 'Hepatics of Sussex' (*Hastings and E. Sussex Nat.* 1911, 243–92). Described hepatics in H. Handel-Mazzetti's *Symbolae Sinicae* v.5, 1930, 7–15, 21–35.
Nature v.155, 1945, 507. *Proc. Linn. Soc.* 1944–45, 68–69. *Rep. Br. Bryol. Soc.* 1945, 313–15. *Times* 19 Feb. 1945. *Bot. Soc. Exch. Club Br. Isl. Rep.* 1946–47, 233. *Rev. Bryol. Lichénol.* v.15, 1946, 120–22 portr.
Herb. at Cambridge.
Barbula nicholsoni Culmann.

NICKSON, John (–1809)
Nurseryman, Nether Knutsford, Cheshire. Took John Carr into partnership *c.* 1780. His widow, Margaret (–1823) continued the business.

NICOL, Walter (–1811)
d. March 1811
Gardener to Marquis of Townsend at Raynham, Norfolk; and later at Wemyss Castle. Landscape gardener in Edinburgh, *c.* 1797. Joint Secretary, Caledonian Horticultural Society. *Scotch Forcing and Kitchen Gardener* 1798. *Practical Planter* 1799. *Villa Garden Directory* 1809. *Gardener's Kalendar* 1810. *Planter's Kalendar* 1812.
J. C. Loudon *Encyclop. Gdning* 1822, 1282. S. Felton *Portr. of English Authors on Gdning* 1830, 82–83. *Cottage Gdnr* v.6, 1851, 1. *Gdnrs Chron.* 1902 ii 298.

NICOL, William (1768–1851)
d. Edinburgh 2 Sept. 1851
Invented section-cutting of fossil wood in 1827. 'Observations on Recent and Fossil Coniferae' (*Edinburgh Philos. J.* v.16, 1834, 137–58, 310–14).
J. Bot. 1867, 90. *R.S.C.* v.4, 615.
Fossils at BM(NH).
Nicolia Ung.

NICOLL, David (*fl.* 1880s)
Gardener, Rossie House near Perth, 1889. Raised chrysanthemums.
J. Hort. Cottage Gdnr v.52, 1906, 54, 55 portr.

NICOLLES, J. (*c.* 1802–1832)
d. Wandsworth, London 1832
Gardener to R. Pettiward at Great Finborough Hall, Suffolk.
Gdnrs Mag. 1833, 128.

NICOLSON, Rev. William (1655–1727)
b. Plumbland? Cumberland 1655 *d.* Londonderry 14 Feb. 1727
BA Oxon 1676. FRS 1705. Rector, Salkeld, Cumberland, 1682. Bishop of Carlisle, 1702. Bishop of Londonderry, 1718–26. Archbishop of Cashel, 1726–27. MS. *Catalogus Plantarum Angliae* 1690 (with N. England localities). Sent plants to A. Buddle and J. Dillenius. Buddle's MS. *Methodus* dedicated to him and others.
J. Ray *Synopsis* 1724, 172. W. Hodgson *Fl. Cumberland* 1898, xxiv–xxv. *Proc. Belfast Nat. Field Club* v.6, 1913, 615. *Trans. Cumberland Westmorland Antiq. Archaeol. Soc.* 1901–5, 1935, 1947 *passim*.
18 vols of MSS. transcriptions from diaries at Carlisle Public Library.

NID, John (*fl.* 1660s)
Senior Fellow, Trinity College, Cambridge. "My intimate friend Mr. Nid, a notable botanist" (J. Ray).
E. Lankester *Memorials of John Ray* 1846, 11. *J. Bot.* 1934, 217, 219. C. E. Raven *John Ray* 1942, *passim*.

NIELD, James (1825–1895)
b. Oldham, Lancs 17 Jan. 1825 *d.* Oldham 7 April 1895
Printer. Bryologist and geologist. President, United Field Naturalists. *Botanical Excursion to Grampian Mountains* (with T. Rogers) 1877.
Heywood Advertiser 11 Dec. 1908. *Trans. Liverpool Bot. Soc.* 1909, 83.
Herb. at Oldham Museum.

NIGHTINGALE, Thomas (1810–1865)
Royal Navy. To Brazil and Polynesia, 1833. *Oceanic Sketches, with Botanical Appendix by Dr. Hooker* 1835 portr. Plants also described in B. C. Seemann's *Fl. Vitiensis* 1865–73.
Ferns at Kew; mosses at BM(NH).

NIMMO, Joseph (*fl.* 1830s–1854)
b. India? *d.* 1854
In Government service at Surat, 1819. "The acknowledged head of the corps botanique of Bombay—a gentleman whose diligence in collecting is only equalled by his liberality in distributing the proceeds" (R. Wight). Collected plants in Socotra, 1834–39. Sent plants to Wight. Completed J. Graham's *Catalogue of Plants growing in Bombay and its Vicinity* 1839.
Madras J. Sci. v.5, 1837, 311–13. *J. Bot.* (Hooker) 1841, 300–1. I. B. Balfour *Bot. Socotra* 1888, xvi.
Plants at Kew.
Nimmoia Wight.

NISBET, John (1853–1914)
b. Edinburgh 2 Oct. 1853 *d.* Exmouth, Devon 30 Nov. 1914
Indian Forest Service, 1875. Conservator of Forests, Burma, 1895–1900. Professor of Forestry, West of Scotland Agricultural College, 1908–12. *British Forest Trees* 1893. *Essays on Silviculture* 1893. *Studies in Forestry* 1894. *The Forester* 1905 2 vols. *Our Forests and Woodlands* ed. 2 1908. *Elements of British Forestry* 1911.
Gdnrs Chron. 1915 i 50. *Indian Forester* 1915, 95–96, 134–36. *Who was Who, 1897–1916* 527.
Burmese plants at Kew.

NISBET, Thomas (–1946)
b. Glasgow *d.* Helensburgh, Dunbartonshire 31 Dec. 1946
MA Glasgow 1895. Headmaster, Dennistoun, Glasgow. President, Andersonian Society, 1919–20. Contrib. on flora of South Ardgoil to *Ann. Andersonian Soc.* 'Plant Geography of Ardgoil' (*Scott. Geogr. Mag.* 1911, 449–66).
Glasgow Nat. 1947, 33–35.

NIVEN, James (*c.* 1774–1827)
b. Pennicuik, Edinburgh *c.* 1774 *d.* Pennicuik 9 Jan. 1827
Father of N. Niven (1799–1879). Gardener, Royal Botanic Garden, Edinburgh. Gardener at Syon House, Middx *c.* 1796. Collected plants in Western Cape, S. Africa for George Hibbert, 1798–1803. Made another collecting trip there, 1803–12, for syndicate which included Lee and Kennedy.
H. C. Andrews *Botanist's Repository* 1801, t.193. *Trans. Linn. Soc.* v.10, 1810, 46, 134. *Gdnrs Mag.* 1827, 255. A. Lasègue *Musée Botanique de B. Delessert* 1845, 447. *Bot. Gaz.* 1849, 106. *J. Linn. Soc.* v.45, 1920, 45. *J. S. African Bot.* 1960, 12. A. M. Coats *Quest for Plants* 1969, 260. *Taxon* 1970, 534.
Plants at BM(NH). MS. on heaths at Kew.

NIVEN, James Craig (1828–1881)
b. Dublin 1828. *d.* Hull, Yorks 16 Oct. 1881
Son of N. Niven (1799–1879). Kew gardener, 1846. Curator, Botanic Garden, Hull, 1853. *Catalogue of Hardy Herbaceous Plants in Royal Gardens of Kew* 1853. Edited re-issue of B. Maund's *Bot. Gdn* 1878.
Florist and Pomologist 1881, 176. *Garden* v.20, 1881, 432. *Gdnrs Chron.* 1881 ii 541–42, 589 portr. *J. Bot.* 1881, 352. *J. Kew Guild* 1898, 37–38 portr. *Naturalist* 1903, 310.

NIVEN, Lawrence (–1876)
Superintendent, Botanic Gardens, Singapore, 1861 whilst also Superintendent of adjoining nutmeg plantation. Manager, Botanic Gardens, 1874–75.
Gdns Bull. Straits Settlements 1918, 58–63, 93, 177 portr.

NIVEN, Ninian (1799–1879)
d. Dublin 18 Feb. 1879
Son of J. Niven (*c.* 1774–1827). Curator, Botanic Garden, Glasnevin, Dublin, 1834–38. Nurseryman, Drumcondra, Dublin, 1838. *Visitor's Companion to Botanic Garden, Glasnevin* 1838. 'Detail of Experiments on Vegetable Physiology' (*Gdnrs Mag.* 1838, 161).
Garden v.7, 1875, xi–xii portr.; v.15, 1879, 186–87. *Gdnrs Chron.* 1879 i 277. H. F. Berry *Hist. of R. Dublin Soc.* 1915, 195–96. *Irish Booklore* 1971, 183 portr. *R.S.C.* v.4, 627.
Letters and plants at Kew.

NIX, Charles G. A. (*c.* 1874–1956)
d. Warninglid, Sussex 8 March 1956
VMH 1923. Authority on fruits, especially apples and pears.
Gdnrs Chron. 1956 i 279.

NOBLE, Charles (*fl.* 1840s–1880s)
Partner with John Standish in nursery at Bagshot Park, Surrey. Established Sunningdale Nursery, Berks in 1847.

Bot. Mag. 1860, t.5169. M. Hadfield *Gdning in Britain* 1960, 329–30. *Gdnrs Chron.* v.162(5), 1967, 16.

NOBLE, Francis (–1756)
Nurseryman, Newark-upon-Trent, Notts.

NOBLE, William (*fl.* 1820s–1830s)
Seedsman with John Noble at 152 Fleet Street, London. Succeeded to the business and premises of Edward Cross and John Mason and Son *c.* 1827; *c.* 1850 firm had become Noble, Cooper and Bolton.

NOCK, William (*fl.* 1870s–1900s)
Kew gardener, 1874. On Cinchona Plantations, Jamaica, *c.* 1875–80 where he also collected plants. In Ceylon, 1881–1904 where he was in charge of the garden at Hakgala. Acting Director at Botanic Gardens at Peradeniya after H. Trimen's death in 1896. Succeeded at Hakgala by John Knighton Nock (1880–1909) who in turn was followed by J. J. Nock in 1911.
I. Urban *Symbolae Antillanae* v.3, 1902, 92. *J. Kew Guild* 1904, 176. *Gdnrs Chron.* 1910 ii 317. *Kew Bull.* 1911, 319.
Strobilanthes nockii Trimen.

NODDER, Frederick Polydore (*fl.* 1770s–1800s)
Appointed botanical painter to Her Majesty, 1785. Drew and/or engraved plates for T. Martyn's *Fl. Rustica* 1792–95, J. White's *J. of Voyage to N.S.W.* 1790, J. L. Knapp's *Gramina Britannica* 1804. Drawings in J. Banks's and D. Solander's collections at BM(NH), some reproduced in *Illus. of Austral. Plants* 1904.
D.N.B. v.41, 86–87. *J. Bot.* 1916, 280.
Some drawings at Kew.

NOEHDEN, Georg Heinrich (1770–1826)
b. Göttingen, Germany 23 June 1770 *d.* London 14 March 1826
LLD Cantab 1796. FRS 1820. FLS 1800. Came to England, 1793. Assistant Librarian, BM, 1820. 'Varieties of Citrus' (*Trans. Hort. Soc. London* v.3, 1822, 1–19). Contrib. to J. Sowerby and J. E. Smith's *English Bot.* 738, 858. Secretary, Royal Asiatic Society, 1823. His brother, Hans Adolphus (1775–1804), came to England, 1799–1800.
Gent. Mag. 1826, 466. J. C. F. Hoefer *Nouvelle Biogr. Générale* v.38, 173–74. H. J. Rose *Biogr. Dict.* v.10, 1857, 317. G. C. Druce *Fl. Berkshire* 1897, clv–clvii. *R.S.C.* v.4, 631.
Letters in Smith correspondence at Linnean Society. Bust at Royal Asiatic Society. Sale at Sotheby, 21 Feb. 1827.

NOEL, Charles William Francis, 3rd Earl of Gainsborough (1850–1926)
b. 20 Oct. 1850 *d.* 17 April 1926
Landowner. Botanist. *Fl. Leicestershire and Rutland* (with A. R. Horwood) 1933.
A. R. Horwood and C. W. F. Noel *Fl. Leicestershire* 1933, cclxxix–cclxxxi portr. *Who was Who, 1916–1928* 388.

NOEL, Lady Elizabeth (1731–1801)
b. London? 1731 *d.* Bath, Somerset 1801
Eldest sister of H. Noel, 6th Earl of Gainsborough. Contrib. to W. Withering's *Bot. Arrangement* ed. 3, v.1, 1796, ix; v.4, 145. MS. Fl. Rutland was in the possession of the family.
A. R. Horwood and C. W. F. Noel *Fl. Leicestershire* 1933, cclxxii.
Letters in Smith correspondence at Linnean Society.

NOEL, Miss Emilia Frances (–1950)
d. 19 March 1950
FLS 1905. Collected plants in Kashmir. *Some Wild Flowers of Kashmir* 1903.
Proc. Linn. Soc. 1949–50, 232.
Kashmir plants at Liverpool University.

NOEL, Henry, 6th Earl of Gainsborough (1743–1798)
b. Exton, Rutland 19 April 1743 *d.* 8 April 1798
Hon. member of Linnean Society, 1789. Contrib. to J. Sowerby and J. E. Smith's *English Bot.* 50, 1188. Dedication in J. Bolton's *Hist. Fungusses* 1788.
C. Crossland *Eighteenth Century Nat.: James Bolton* 1910, 11, 19. A. R. Horwood and C. W. F. Noel *Fl. Leicestershire* 1933, cclxxiii.
Herb. formerly at Exton. Portr. in possession of family.

NORMAN, Cecil (1872–1947)
b. London 12 Sept. 1872 *d.* Ickham, Kent 2 April 1947
FLS 1923. Studied *Umbelliferae*. Collected plants in Jamaica, 1924. Contrib. to W. Fawcett and A. B. Rendle's *Fl. Jamaica.*
Proc. Linn. Soc. 1946–47, 153–54.
Plants at BM(NH).

NORMAN, George (1824–1882)
b. Hull, Yorks 1824 *d.* Peebles 5 July 1882
Studied Yorks plants. 'List of Diatomaceae...in...Hull' (*Trans. Microsc. Soc.* v.8, 1860, 59–71).
Trans. Hull Field Nat. Club v.1, 1900, 105–12; v.2, 1902, 14. *Naturalist* 1903, 307–8. *R.S.C.* v.4, 643; v.8, 517; v.10, 939; v.13, 541.
Diatoms at Hull Museum. Letters in Walker-Arnott correspondence at BM(NH).
Pleurosigma normani Ralfs.

NORMAN, George (–1906)
d. 28 Feb. 1906

VMH 1901. Gardener to Marquis of Salisbury at Hatfield, Herts, 1876.

Gdnrs Chron. 1901 i 294 portr.; 1906 i 142 portr. *Garden* v.69, 1906, 155 portr. *J. Hort. Cottage Gdnr* v.52, 1906, 218 portr.

NORMANSELL, Harry Thomas (–1843)
d. Peradeniya, Ceylon 7 June 1843

Surgeon. Director, Botanic Gardens, Peradeniya, 1840–43.

Ann. R. Bot. Gdns Peradeniya 1901, 6.

NORRIS, Sir William (1793–1859)
b. London? 7 Nov. 1793 *d.* Sunningdale, Berks 7 Sept. 1859

Barrister. To India, 1829. Recorder, Penang and Malacca, 1836–47. Chief Justice of Ceylon, 1847. Friend of W. Griffith (1810–1845) to whom he sent plants. Sent plants from Mount Ophir, Sumatra to G. Gardner. Particularly interested in ferns.

Hooker's J. Bot. 1849, 326. F. Boase *Modern English Biogr.* v.2, 1897, 1170. *Gdns Bull. Straits Settlements* 1927, 129. *Fl. Malesiana* v.1, 1950, 389.

Ferns and letters at Kew.

NORTH, Frederick, 5th Earl of Guildford (*c.* 1766–1827)
b. 7 Feb. 1766 *d.* London 14 Oct. 1827

DCL Oxon 1793. LLD Cantab 1821. FRS 1794. MP, Banbury, 1792. Secretary of State, Corsica, 1795. Governor of Ceylon, 1798–1805. Sent Ceylon plants to J. Banks.

D.N.B. v.41, 164–66.

NORTH, Rev. Isaac William (1810–)
b. London 28 July 1810

Chaplain, Isles of Scilly, 1841–51. *Week in the Isles of Scilly* 1850 contains some plant records.

F. H. Davey *Fl. Cornwall* 1909, xlix.

NORTH, Miss Marianne (1830–1890)
b. Hastings, Sussex 1830 *d.* Alderley, Glos 30 Aug. 1890

Travelled around the world and made over 800 paintings of plants, now in the North Gallery at Kew. *Recollections of a Happy Life* 1892 2 vols. *Some Further Recollections of a Happy Life* 1893.

Nature v.26, 1882, 155–56; v.45, 1891–92, 602–3; v.48, 1893, 291–92. *Gdnrs Chron.* 1885 i 77–78. *Icones Plantarum* 1884, t.1473. *J. Bot.* 1890, 329–34. *J. Hort. Cottage Gdnr* v.21, 1890, 271–72; v.51, 1905, 364–65. *Garden* v.58, 1900, 300–1 portr. *D.N.B.* v.41, 168–69. *Trees* 1953, 77–79. *Cornhill* 1962, 319–29 portr. *Geogr. Mag.* 1962, 445–62 portr. *J. R. Hort. Soc.* 1964, 231–40. *Lady* 1969, 506–7 portr.

Letters and marble bust by C. Dressler at Kew.

Northea Hook. f.

NORTH, Richard (–1766)
d. Jan. 1766

Nurseryman north of Westminster Bridge Road and in Lambeth, Surrey. Disposed of nursery to James Sheilds (or Shiells). *Gardener's Catalogue of Hardy Trees, Shrubs, Flowers, Seeds* 1759. *Treatise on Grasses and the Norfolk Willow* 1760.

J. C. Loudon *Encyclop. Gdning* 1822, 1272. G. W. Johnson *Hist. English Gdning* 1829, 215. *Trans. London Middlesex Archaeol. Soc.* v.24, 1973, 187.

NORTH, William (*fl.* 1790s–1800s)

Nurseryman, Asylum Road, Lambeth, Surrey.

Bot. Mag. 1803, t.633. *Trans. London Middlesex Archaeol. Soc.* v.24, 1973, 188.

NORTHCOTE, Lady Rosalind Lucy Stafford (1873–1950)
d. 31 Dec. 1950

Authority on herbs. *Books of Herbs* 1912.

P. Coats *Flowers in Hist.* 1970, 240 portr.

NORTHINGTON, 1st Earl of *see* Henley, R.

NORTHUMBERLAND, Dukes of *see* Percy

NORTON, James (1824–1906)
b. Sydney 5 Dec. 1824 *d.* Sydney 18 July 1906

Solicitor, Sydney. One of the founders and Treasurer, Linnean Society New South Wales. Keen gardener and studied flora of New South Wales.

J. Proc. R. Soc. N.S.W. 1908, 115–16.

Adenochilus nortoni Fitz.

NOTCUTT, Roger Crompton (1869–1938)
d. Woodbridge, Suffolk 30 Jan. 1938

FLS 1927. Nurseryman, Woodbridge, 1896. *Handbook of Flowering Trees and Shrubs for Gardeners* 1926.

Proc. Linn. Soc. 1937–38, 324–25. *Gdnrs Chron.* 1938 i 99.

NOTCUTT, William Lowndes (1819–1868)
b. Wilbarston, Notts 19 April 1819 *d.* Cheltenham, Glos 15 Sept. 1868

Pharmacist. Of Fareham, Kettering and Cheltenham. Correspondent of Robert Dick (1811–1866). *Handbook of British Plants* 1865. 'Catalogue of Plants observed in Neighbourhood of Daventry, Northamptonshire' (*Phytologist* 1843, 500–8).

H. C. Watson *Topographical Bot.* 1883, 552. G. C. Druce *Fl. Northamptonshire* 1930, cxvii. H. J. Riddelsdell *Fl. Gloucestershire* 1948, cxxviii. *R.S.C.* v.4, 646.

Plants at BM(NH), Carlisle Museum, Manchester.

NOTCUTT, William Russel (–before 1802)
d. Surinam before 1802

FLS 1796. Of Ipswich. Lectured on Chemistry, New College, Hackney, 1796. "An ardent naturalist."

J. Sowerby and J. E. Smith *English Bot.* 1049.

NOWELL, John (1802–1867)
b. Todmorden, Yorks 1802 *d.* Todmorden 28 Oct. 1867

Handloom weaver. Bryologist. Discovered *Cinclidium stygium*, 1836. Contrib. mosses to *Supplement*, 1854 to H. Baines's *Fl. Yorkshire*. MS. Fl. Todmorden partly published in *Lancashire Nat.* 1907, 52, 54, 76–77, 100–1, 115–18.

R. Buxton *Bot. Guide to...Manchester* 1849, x, xiv. J. Cash *Where there's a Will there's a Way* 1873, 102–3. *Manchester Quart.* 1882 i 205. W. B. Crump and C. Crossland *Fl. Parish of Halifax* 1904, lx–lxi. *Trans. Liverpool Bot. Soc.* 1909, 83. *Naturalist* 1961, 155–56; 1973, 5–6.

Herb. and portr. formerly at Todmorden Public Library. Letters in Wilson correspondence at BM(NH).

Nowellia Mitt.

NOWELL, William (1880–1968)
b. Heptonstall, Yorks 9 May 1880 *d.* 1 Oct. 1968

Educ. Royal College of Science, London. Assistant Superintendent, Dept. of Agriculture, Barbados. Mycologist, Imperial Dept. of Agriculture for West Indies. Assistant Director, Dept. of Agriculture, Trinidad and Tobago. Director of Agriculture, British Guiana. Director, East African Agricultural Research Station, Amani; retired 1936. *Diseases of Crop Plants in Lesser Antilles* 1923. Contrib. to *Ann. Applied Biol.*, *Ann. Bot.*

Kew Bull. 1913, 359; 1920, 218. *Who was Who, 1961–1970* 844–45.

NUTT, Rev. W. Harwood (*fl.* 1890s)

Missionary at Fwambo, N. Rhodesia, 1890s. 'Account of Journey to Lake Rukwas' (*Br. Central African Gaz.* 15 Oct., 1 Nov. 1895).

Comptes Rendus AETFAT 1960 1962, 167–68.

NUTTALL, Mrs. Gertrude Clarke (–1929)
d. St. Albans, Herts 4 May 1929

BSc. One of the first women to take a degree in botany. *Wild Flowers as they Grow* 1911–14. *Trees and how they Grow* 1913. *Beautiful Flowering Shrubs* 1920.

J. Bot. 1929, 183. A. R. Horwood and C. W. F. Noel *Fl. Leicestershire* 1933, ccxxx.

NUTTALL, John (–1849/50)

Of Tithewar, County Wicklow. Arboriculturist. Knew Irish plants well.

Hooker's J. Bot. 1850, 94.

NUTTALL, Thomas (1786–1859)
b. Long Preston, Settle, Yorks 5 Jan. 1786
d. Nutgrove near Wigan, Lancs 10 Sept. 1859

FLS 1813. Printer. To Philadelphia, 1808. Travelled in U.S.A., 1811–34; to Sandwich Islands, 1836. Curator, Botanic Garden, Harvard, 1822–34. Returned to England, 1842. *Cat. of New and Interesting Plants...for Sale at Messrs Fraser's Nursery* 1813 (*Pittonia* 1890, 114–19; *J. Bot.* 1899, 481–87). *Genera of N. American Plants* 1818 2 vols (biogr. in 1971 reprint, ix–xxxvii). *Journal of Travels into Arkansas Territory* 1821. *Introduction to Systematic and Physiological Botany* 1827. *N. American Sylva* 1842 2 vols. 'Descriptions of New Species and Genera of Plants in...Compositae' (*Trans. Amer. Philos. Soc.* 1841, 283–453).

J. Bot. (Hooker) 1841, 108–10. *Gdnrs Chron.* 1860, 556; 1964 ii 529 portr. *Proc. Linn. Soc.* 1859, xxvi–xxix. *Cottage Gdnr* v.23, 1860, 349–50. *Proc. Amer. Philos. Soc.* 1860, 297–315; 1942, 65–67; 1950, 145–47; 1960, 86–100 portr. A. Gray *Letters* v.1, 1893, 326. *Popular Sci. Mon.* v.46, 1895, 577 portr., 689–96. *Proc. Biol. Soc. Washington* 1899, 109–21. J. W. Harshberger *Botanists of Philadelphia and their Work* 1899, 112. 151–59 portr., 176. *D.N.B.* v.41, 276–77, *Trans. Liverpool Bot. Soc.* 1909, 84–85. *J. Bot.* 1922, 57. *J. Arnold Arboretum* 1927, 24–55; 1928, 33–34. *Dict. Amer. Biogr.* v.13, 1934, 596–97. *Madrono* 1934, 143–47. F. W. Pennell *Travels and Sci. Collections of Thomas Nuttall* 1936, 1–51. *Bartonia* no. 18, 1936, 1–51 portr.; no. 19, 1937, 50–53; no. 20, 1938–39, 1–6; no. 29, 1957–58, 10. *Little Gdns* (Seattle) 1937, 6–25. *Occas. Papers California Acad. Sci.* no. 20, 1943, 42–46 portr. *Chronica Botanica* 1951, 1–88 portr. *Sci. Mon.* 1952, 84–90. *Rhodora* 1952, 293–303; 1954, 253–57; 1956, 20–24; 1968, 429–38. S. D. McKelvey *Bot. Exploration of Trans-Mississippi West 1790–1850* 1955, 139–49, 164–87, 586–626. *Missouri Hist. Soc. Bull.* 1956, 249–52. *Leaflets Western Bot.* no. 3, 1959, 33–42. *Cambridge Hist. Soc. Proc.* 1961, 69–86. *Pennsylvania Mag. Hist. Biogr.* 1961, 423–38. *Appalachia* 1964, 44–63. *Castanea* 1966, 187–98. *Michigan Bot.* 1967, 81–94. J. E. Graustein *Thomas Nuttall* 1967 portr. F. A. Stafleu *Taxonomic Literature* 1967, 339–41. A. M. Coats *Quest for Plants* 1969, 295–304. *Taxon* 1970, 534; 1973, 92 portr. *R.S.C.* v.4, 650; v.8, 521.

Plants at BM(NH), Kew, Harvard, Philadelphia. Portr. at Hunt Library.

Nuttallia Torrey & Gray.

NUTTING, William James (*c.* 1827–1910)
d. Bromley, Kent 12 June 1910
Seedsman, Southwark Street, London.
Gdnrs Chron. 1910 i 414 portr.

OATES, Frank (1840–1875)
b. Meanwoodside near Leeds, Yorks 6 April 1840 *d.* Jantje's Kraal, S. Rhodesia 5 Feb. 1875
To Durban, S. Africa, 1873 and undertook a natural history exploring journey to Matabeleland. Also collected plants in Central America and California. *Matabeleland and the Victoria Falls* 1881 (botany by D. Oliver); ed. 2 1889.
Plant Life v.2, 1946, 74.
S. African plants at Kew.
Lippia oatesii Rolfe.

O'BRIEN, Charlotte Grace (1845–1909)
b. 23 Nov. 1845 *d.* Foynes, County Limerick 3 June 1909
Of Foynes. *Wild Flowers of the Undercliff, Isle of Wight* 1881. 'Fl. of Barony of Shanid' (with M. C. Knowles) (*Irish Nat.* 1907, 185–201).
S. Gwynn *Charlotte Grace O'Brien* 1909.

O'BRIEN, James (1842–1930)
b. Llanelly, Glam 28 Jan. 1842 *d.* Harrow, Middx 30 Dec. 1930
VMH 1897. Gardener to R. S. Holford at Westonbirt, Glos. In charge of R. Hanbury's orchid collection at Ware, Herts. General Manager, Messrs E. G. Henderson and Son, St. John's Wood. Secretary, Orchid Committee, Royal Horticultural Society, 1889–1923. Contrib. on orchid hybrids to *Gdnrs Chron.*
Gdnrs Chron. 1891 ii 225 portr.; 1899 ii 25–26 portr.; 1922 ii 118 portr.; 1924 i 30 portr.; 1931 i 19 portr. *Gdnrs Mag.* 1908, 909–10 portr. *Bot. Soc. Exch. Club Br. Isl. Rep.* 1931, 630–31. *J. Bot.* 1931, 56. *Orchid Rev.* 1931, 52.
Portr. at Hunt Library.

O'BRIEN, Robert Donough (1847–1917)
d. Limerick 9 April 1917
Brother of C. G. O'Brien (1845–1909). Botanised in Limerick area. Discovered *Scirpus triqueter* in Ireland. Contrib. to *Irish Nat.*, *J. Limerick Field Club.*
Irish Nat. 1917, 113. R. L. Praeger *Some Irish Nat.* 1949, 136.

O'CONNOR, Charles Anthony (1883–1963)
b. 20 Aug. 1883 *d.* Mauritius 22 March 1963
Superintendent of Gardens of Curepipe and Réduit, Mauritius, 1903; later Senior Agricultural Officer until his retirement in 1944. At Kew, 1918–19. Inspector of Agriculture, Zanzibar, *c.* 1919–22. Introduced economic plants to Mauritius. Interested in flora of Mauritius. He saved the endemic palm, *Hyophorbe vaughanii*, from extinction.
Proc. R. Soc. Arts Sci. Mauritius v.4(2), 1968–72, 61–64 portr.

O'CONNOR, G. M. (–1897)
d. 20 Nov. 1897
Of Ballycastle, County Antrim. Had good knowledge of local flora.
Irish Nat. 1898, 15.

O'CONNOR, Patrick (1889–1969)
PhD Dublin 1929. Keeper, Natural History Division, National Museum of Ireland, 1929. *Handlist of Irish Plants* 1934. Contrib. to *Irish Nat. J.*
R. L. Praeger *Some Irish Nat.* 1949, 137.
Mycological herb. at Botanic Gardens, Glasnevin, Dublin.

O'DONOVAN, Edward (*afterwards* **Donovan**) (1768–1837)
d. Lambeth, London 1 Feb. 1837
FLS 1799. Zoologist. Founded London Museum and Institute of Natural History, 1807. *Essay on Minute Parts of Plants in General* 1789–90. *Botanical Review, or Beauties of Flora* 1790; also illustrated both books.
D.N.B. v.15, 235–36.

O'DONOVAN, John Emmet (1898–1966)
b. Skibbereen, County Cork 1898 *d.* Skibbereen 15 Feb. 1966
Teacher. Secretary, West Cork Field Club. Botanised in West Cork. Contrib. natural history column to *Skibbereen Star* and botanical papers with B. O'Regan to *Irish Nat. J.*
Irish Nat. J. 1966, 185–86.

OGBORNE, Thomas (*fl.* 1780s)
Seedsman, Wootton Bassett, Wilts.

OGDEN *see* Ugden

OGILBY, Alan (*fl.* 1470s)
Of Scotland. Resided in Constantinople and Venice. *De Virtutibus Herbarum.*
T. Tanner *Bibliotheca Britannico-Hibernica* 1748, 560. A. von Haller *Bibliotheca Botanica* v.1, 1771, 245. R. Pulteney *Hist. Biogr. Sketches of Progress of Bot. in England* v.2, 1790, 2.

OGILVIE-FORBES, George (*olim* **Ogilvie**) (1820–1886)
b. Aberdeen 1820 *d.* Boyndie near Banff 25 June 1886
MA Aberdeen 1839. MD Edinburgh 1842. Lecturer in Physiology, Aberdeen, 1860–77. One of the founders of Scottish Cryptogamic Society. Studied morphology of ferns. Contrib. to *Ann. Mag. Nat. Hist.*
Scott. Nat. 1887, 1–2. *R.S.C.* v.4, 664.

OGILVIE-GRANT, Mark (*c.* 1906–1969)
d. London 13 Feb. 1969
Interested in Greek flora.
J. R. Hort. Soc. 1969, 232–33.

OGLANDER, John (*c.* 1778–1825)
b. Brading, Isle of Wight? *c.* 1778 *d.* Oxford 30 Oct. 1825
MA Oxon 1804. Sub-warden, Merton College, 1824. Bryologist. Contrib. mosses to T. Purton's *Bot. Description of Br. Plants in Midland Counties* 1817, appendix, 95, etc.

OGLE, John Joseph (1857–1909)
b. Lincoln 4 Feb. 1857 *d.* Bootle, Lancs 19 Dec. 1909
Librarian and Curator, Bootle Public Library and Museum, 1887–1900. Director of Higher Education, Bootle. 'Fertilisation of *Saxifraga*' (*Midland Nat.* 1883, 73–75).
Trans. Liverpool Bot. Soc. 1909, 62–63 portr. *R.S.C.* v.12, 545.

OGLE, William (1827–1905)
b. Oxford 1827 *d.* London 16 May 1905
MA Oxon 1852. MD 1861. Lecturer in Physiology, St. George's Hospital. Translated K. von Marilaun's *Flowers and their Unbidden Guests* 1878. Contrib. papers on fertilisation to *Popular Sci. Rev.* 1869–70.
Who was Who, 1897–1916 536. *R.S.C.* v.8, 527; v.12, 545.

OGLETHORPE, James Edward (1696–1785)
b. London 22 Dec. 1696 *d.* Cranham 1 July 1785
FRS 1749. Founder of Colony of Georgia, 1732.
D.N.B. v.42, 43–47. J. E. Dandy *Sloane Herb.* 1958, 172.
American plants at BM(NH).

OKELL, Mr. (*fl.* 1800s)
Of Chester, Cheshire. Contrib. Cheshire plant localities to *Magna Britannia*.
J. B. L. Warren *Fl. Cheshire* 1899, lxxxvi.

O'KELLY, Patrick B. (*fl.* 1890s–1930s)
Of Ballyvaughan, County Clare. Discovered *Potamogeton pygmaeus*, 1891 and *Limosella aquatica*, 1893 in Ireland. Had intimate knowledge of flora of Burren, County Carlow.
R. L. Praeger *Some Irish Nat.* 1949, 137.
Plants at Oxford.

OLDACRE, (*or* Oldaker) Isaac
(*fl.* 1810s–1820s)
Gardener to Czar of Russia at St. Petersburg. Gardener to Lady Banks at Spring Grove, Isleworth, Middx, 1816. Contrib. to *Gdnrs Mag.*, *Trans. Hort. Soc. London.*
J. C. Loudon *Encyclop. Gdning* 1822, 1289.

OLDFIELD, Augustus Frederick (1820–1887)
b. London 12 Jan. 1820 *d.* London 22 May 1887
Collected plants in Tasmania, N.S.W. and W. Australia, 1850s. In England by 1863. Sent plants to F. von Mueller.
J. D. Hooker *Fl. Tasmaniae* v.1, 1859, cxxvii. G. Bentham *Fl. Australiensis* v.1, 1863, 14*. *J. W. Austral. Nat. Hist. Soc.* no. 6, 1909, 22. *Papers Proc. R. Soc. Tasmania* 1909, 24–26. *Rep. Austral. Assoc. Advancement Sci.* 1911, 232–34.
Herb. and letters at Kew. MSS. at Royal Botanic Gardens, Sydney.
Lasiopetalum oldfieldii F. Muell.

OLDFIELD, Richard Albert K.
(*fl.* 1830s–1850s)
Physician to Laird and Lander Niger expedition, 1832–34. Collected some plants in Nigeria, 1832–34; Sierra Leone, 1851–57. *Narrative of Expedition into Interior of Africa by River Niger...in 1832–34* (with M. Laird) 1837 2 vols.
Comptes Rendus, AETFAT 1960 1962, 65.
Plants at Kew.

OLDHAM, Richard (1837–1864)
d. Amoy, China 13 Nov. 1864
Kew gardener, 1859. Collected plants for Kew in Eastern Asia, 1861, and in Khasia Hills, India, 1861–62.
Bot. Zeitung 1866, 260. *J. Linn. Soc.* 1865, 163–70. *J. Bot.* 1866, 239–40. *J. Kew Guild* 1895, 32; 1897, 29. E. Bretschneider *Hist. European Bot. Discoveries in China* 1898, 682–88. *Fl. Malesiana* v.1, 1950, 393. A. M. Coats *Quest for Plants* 1969, 76–78.
Plants at BM(NH), Kew. Letters at Kew.
Desmodium oldhami Oliver.

OLDHAM, Thomas (1816–1878)
b. Dublin 4 May 1816 *d.* Rugby, Warwickshire 20 July 1878
BA Dublin 1836. LLD 1874. FRS 1848. Director, Geological Survey, Ireland, 1840–50; Indian Geological Survey, 1850–76. *Fossil Fl. of Rajmahal Series, Rajmahal Hills, Bengal* (with J. Morris) 1862–63.
Geol. Mag. 1878, 382–84. *Quart. J. Geol. Soc.* 1878–79, 46–48. *D.N.B.* v.42, 111–12. *R.S.C.* v.4, 672; v.8, 528; v.10, 953; v.12, 545.

OLDHAM, W. R. (*c.* 1870–1949)
d. Gatcombe, Isle of Wight
VMH 1937. Nurseryman, Windlesham, Surrey. President, Horticultural Trades Association.
Gdnrs Chron. 1949 i 48.

OLIVER, Daniel (1830–1916)
b. Newcastle, Northumberland 6 Feb. 1830
d. Kew, Surrey 21 Dec. 1916

LLD Aberdeen 1891. FRS 1863. FLS 1853. Assistant, Kew Herb., 1858; Keeper, 1864–90. Professor of Botany, University College, London, 1861–88. *Lessons in Elementary Botany* 1864. *First Book of Indian Botany* 1869. *Illustrations of Principal Natural Orders of the Vegetable Kingdom* 1874. Edited *Fl. Tropical Africa* vols. 1–3, 1868–77. Edited *Icones Plantarum* 1890–95.

Proc. Linn. Soc. 1892–93, 19–20; 1916–17, 53–57; 1924–25, 59–61. *Kew Bull.* 1893, 188–89; 1917, 31–36. *J. Kew Guild* 1898, 1 portr. *Gdnrs Chron.* 1916 ii 314. *Bot. Soc. Exch. Club Br. Isl. Rep.* 1916, 466–67. *J. Bot.* 1917, 89–95 portr. *Nature* v.98, 1916, 331. *Proc. R. Soc. B* 1917, xi–xv portr. *Curtis's Bot. Mag. Dedications, 1827–1927* 159–60 portr. *R.S.C.* v.4, 674; v.8, 528; v.10, 953; v.12, 546.

Portr. by J. W. Forster at Kew. Portr. at Hunt Library.

Pavetta oliveriana Hiern.

OLIVER, Francis Wall (1864–1951)
b. Richmond, Surrey 10 May 1864 *d.* Limpsfield, Surrey 14 Sept. 1951

MA Cantab. FRS 1905. FLS 1886. Son of D. Oliver (1830–1916). Lecturer in Botany, University College, London, 1888; Professor of Botany, 1890–1929. Professor of Botany, Cairo University, 1929–35. President, British Ecological Society, 1915–16. Palaeobotanist and ecologist. Translated Kerner von Marilaun's *Nat. Hist. Plants* 1894–95. Edited *Makers of British Botany* 1913. *Tidal Lands* (with A. E. Carey) 1918.

Gdnrs Chron. 1925 i 388 portr. *Nature* v.168, 1951, 809–11. *Times* 18 Sept. 1951. *Pharm. J.* 1951, 217. *Obit. Notices Fellows R. Soc.* 1952, 229–40 portr. *Trans. Herts. Nat. Hist. Soc.* 1953, 1. M. A. Oliver *Letters from Egypt* 1932. *D.N.B. 1951–1960* 1971, 779–80. *Who was Who, 1951–1960* 832.

Portr. by F. A. Biden Footner at University College, London.

OLIVER, John William (1851–1914)
d. Ireland 24 Dec. 1914

Indian Forest Service, 1874–1905; in Burma, 1874–84; at Dehra Dun, 1886. One of the pioneers of systematic forestry in Burma.

Indian Forester 1915, 133.

Dysoxylum oliveri Brandis.

OLIVER, Joseph William (1833–1907)
d. Harborne, Birmingham 9 Jan. 1907

Taught botany in Birmingham for 30 years. *Elementary Botany* 1890. *Systematic Botany* 1894.

Gdnrs Chron. 1907 i 48.

OLIVER, Samuel Pasfield (1838–1907)
b. Bovinger, Essex 30 Oct. 1838 *d.* Worthing, Sussex 31 July 1907

Captain, Royal Artillery. Geographer and antiquary. Studied fauna and flora of Mascarene Islands. Collected ferns on St. Helena, 1876 (plants at Kew). *Life of Philibert Commerson* 1909 (biogr. of Oliver, v–viii portr.).

D.N.B. 1901–1911 44–45.

O'MAHONEY, Rev. Thaddeus (1823–1879)
d. Dublin July 1879

Discovered *Simethis planifolia*, 1848 and *Epipactis atropurpurea*, 1851 in Ireland. 'Botanical Excursion in Clare' (*Proc. Dublin Nat. Hist. Soc.* v.1, 1860, 30–34).

H. C. Watson *Cybele Britannica* ed. 2, xxxii, 354. *Notes Bot. School Dublin* v.2, 1909, 46.

O'MALLEY, Lady Emma Winifred (*née* **Hardcastle**) (1847–1927)
b. Essex 1847 *d.* Cuddesdon, Oxon 25 June 1927

Wife of Sir E. L. O'Malley, Attorney-General for Jamaica, 1876–80; Hong Kong, 1880–89. Studied ferns of Jamaica and Hong Kong. Contrib. account of Chinese plants to *Sci. Gossip*.

Bot. Soc. Exch. Club Br. Isl. Rep. 1927, 379–80.

Plants at BM(NH).

O'MEARA, Rev. Eugene (*c.* 1815–1880)
d. Newcastle Lyons, County Dublin 20 Jan. 1880

MA Dublin 1858. A founder of Dublin Microscopical Club. 'Diatomaceae' (*J. Bot.* 1872–73).

J. Bot. 1880, 128. *Nature* v.21, 1880, 423. *Acta Horti Bergiani* v.3(1), 1897–1903, t.35 portr. *R.S.C.* v.4, 684; v.8, 530.

Diatoms at BM(NH).

ONDAATJE, William Charles (–1888)
d. Eastbourne, Sussex Oct. 1888

FLS 1882. Surgeon. Superintendent, Botanic Gardens, Peradeniya, Ceylon, 1843–44. 'A Few Remarks on Poisonous Properties of the *Calotropis gigantea*' (*Ceylon J. Asiatic Soc.* 1866, 157–59).

ONSLOW, Hon. Muriel (*née* **Wheldale**) (–1932)
d. 19 May 1932

Educ. Cambridge. Assistant in Plant Biochemistry, Cambridge, 1915–26; Lecturer in Plant Biochemistry, 1926. Studied plant oxidase systems. *Anthocyanin Pigments of Plants* 1916; ed. 2 1925. *Practical Plant Biochemistry* 1920; ed. 3 1929. *Principles of Plant Biochemistry* 1931. Contrib. to *Biochem. J.*, *J. Genetics*, *Proc. R. Soc.*

Nature v.129, 1932, 859. *Times* 20 May 1932. *Who was Who, 1929–1940* 1025.

ORCHARD, Charles (*c.* 1846–)
b. Ventnor, Isle of Wight *c.* 1846
Gardener, Coombe House, Croydon, Surrey.
J. Hort. Cottage Gdnr v.60, 1910, 54–56 portr.

ORD, George Walker (1871–1899)
b. King Edward, Aberdeenshire 1871 *d.* Glasgow 9 Aug. 1899
Assistant, Kelvingrove Museum, Glasgow, 1886. Keen naturalist with interest in botany. Secretary, Clydesdale Naturalists' Society.
Ann. Scott. Nat. Hist. 1899, 193–96.

ORDOYNO, Garrett (*c.* 1723–1795)
d. 29 Nov. 1795
Nurseryman, Newark-upon-Trent, Notts. In partnership with his brother, Jacob Ordoyno (*c.* 1734–1812).

ORDOYNO, Thomas (*fl.* 1790s–1810s)
Of Newark-upon-Trent, Notts. *Fl. Nottinghamiensis* 1807. Nephew of Garrett and Jacob Ordoyno. Probably responsible for the botanical parts of the nursery catalogue issued by G. and J. Ordoyno, *c.* 1795.
J. Bot. 1914, 321.

ORME, Robert (1865–1934)
b. Dublin 1865 *d.* Aug. 1934
Studied flora of Devon.
Bot. Soc. Exch. Club Br. Isl. Rep. 1934, 810–11.

ORMEROD, Miss Eleanor A. (1828–1901)
b. Sedbury Park, Glos 11 May 1828 *d.* St. Albans, Herts 20 July 1901
VMH 1901. Horticultural entomologist. *Manual of Injurious Insects* 1881. *Handbook of Insects injurious to Orchard and Bush-fruit* 1898.
Garden v.60, 1901, 58, 86 portr., 103–4. *Gdnrs Mag.* 1901, 478. *J. Hort. Cottage Gdnr* v.42, 1901, 257 portr., 260. *Who was Who, 1897–1916* 539. H. J. Riddelsdell *Fl. Gloucestershire* 1948, cxxxv.

ORR, David (–1892)
b. Belfast? *d.* Dublin 1892
Assistant to Dr. Moore at Botanic Garden, Glasnevin, Dublin. Collected mosses in Antrim, Dublin and Wicklow. 'Some Mosses collected in Ireland' (*J. Bot.* 1881, 83–84).
Irish Nat. 1913, 29. *Proc. Belfast Nat. Field Club* 1913, 623. *Proc. Irish Acad.* v.32, 1915, 74. R. L. Praeger *Some Irish Nat.* 1949, 139.
Plants at BM(NH), National Museum, Dublin. Letters in Wilson correspondence at BM(NH).

ORR, Matthew Young (*c.* 1883–1953)
d. 9 Sept. 1953
Educ. University College, Cardiff. Botanist, Royal Botanic Garden, Edinburgh, 1913–47. Authority on conifers. Contrib. to *Trans. Bot. Soc. Edinburgh.*
Yb. R. Soc. Edinburgh 1954, 31–32.

OSBORN, Arthur (1878–1964)
b. Sonning, Berks 16 Dec. 1878 *d.* Kew, Surrey 24 Feb. 1964
Kew gardener, 1899; Assistant Curator; Deputy Curator, Arboretum. *Shrubs and Trees for the Garden* 1933. *Complete Book of Gardening* (with J. Coutts and A. Edwards) 1930.
J. Kew Guild 1934, 313–14 portr.; 1964, 463.

OSBORN, Emma (1822–1877)
Daughter of R. Osborn (*c.* 1780–1866). Amateur flower painter. Paintings in possession of M. Walpole, Loughborough, Leics.

OSBORN, Mary (1814–1898)
Daughter of R. Osborn (*c.* 1780–1866). Amateur flower painter. Paintings in possession of M. Walpole, Loughborough, Leics.

OSBORN, Robert (*c.* 1780–1866)
Nurseryman, Fulham, Middx. Partner of Reginald Whitley, 1833. The business continued by his sons William and Thomas. *Catalogue of Hardy Trees and Shrubs cultivated by Whitley and Osborn* 1840.

OSBORN, Theodore George Bentley (1887–1973)
b. Clacton, Essex 2 Oct. 1887 *d.* 3 June 1973
MSc Manchester 1911. Lecturer in Economic Botany, Manchester, 1908–12. Professor of Botany, Adelaide and Consulting Botanist to Government of S. Australia, 1912–27. Professor of Botany, Sydney, 1928–37. Professor of Botany, Oxford, 1937–53. Returned to Adelaide, 1953. President, Linnean Society N.S.W., 1932. Contrib. to *J. Dept. Agric. S. Australia.*, *Trans. R. Soc. S. Australia.*
Nature v.139, 1937, 746; v.171, 1953, 373; v.244, 1973, 377. *Chronica Botanica* 1938, 72 portr. *S. Austral. Nat.* v.48, 1973, 9. *Trans. R. Soc. S. Australia* 1973, 316–20 portr. *Times* 6 June 1973. *Who's Who* 1972, 2411.

OSBORN, Thomas (*c.* 1819–1872)
d. Fulham, London 28 Jan. 1872
Nurseryman, Fulham.
Garden v.1, 1872, 250.

OSBORN, William (–1872)
d. 7 March 1872
Brother of Thomas Osborn. Nurseryman, Fulham Nursery, London.
Garden v.1, 1872, 387. *J. Hort. Cottage Gdnr* v.47, 1872, 235.

OSBORNE, Philip Valentine (1891–1943)
b. Feb. 1891 *d.* Kalimoong, India 16 June 1943
Kew gardener, 1912–13. At Royal Botanic Garden, Calcutta, 1913; later Curator. Served in Bengal Cinchona Dept.
J. Kew Guild 1943, 308–9.

OSBOURNE, Edith *see* Blake, *Lady* E.

O'SHANESY, John (1834–1899)
b. Ballybunion, County Kerry July 1834 *d.* Rockhampton, Queensland July 1899
Gardener. To Queensland, 1861. Gardener, Botanic Gardens, Brisbane. Established nursery at Rockhampton, 1866. Collected grasses, etc. for F. von Mueller.
Rep. Austral. Assoc. Advancement Sci. 1909, 382.

O'SHANESY, Patrick Adams (1837–1884)
b. Ratton, County Kerry 1837 *d.* Rockhampton, Queensland Dec. 1884
FLS 1879–84. Gardener. To Brisbane, Queensland, 1864. Employed by his brother John in his Rockhampton nursery until 1876 when he established his own business. Correspondent of F. von Mueller. *Contributions to Fl. Queensland* 1880.
Rep. Austral. Assoc. Advancement Sci. 1909, 382.
Solanum shanesii F. Muell.

OSMASTON, Bertram Beresford (1868–1961)
b. Yeldersley Hall, Derbyshire 1868 *d.* 6 Sept. 1961
Indian Forest Service, 1888. Retired as Chief Conservator of Central Provinces.
Nature v.192, 1961, 705–6. *J. Bombay Nat. Hist. Soc.* v.60, 1963, 709–10. *Who was Who, 1961–1970* 860.
Kashmir plants at Kew.

OTLEY, Jonathan (1766–1856)
b. Scroggs? Cumberland Jan. 1766 *d.* Keswick, Cumberland 7 Dec. 1856
Of Keswick. Watchmaker, naturalist and guide. *Guide to the Lakes* ed. 4 1834 (including plant list); ed. 6 1838.
Trans. Cumberland Assoc. 1876–77, 125–69. W. Hodgson *Fl. Cumberland* 1898, xxv.

OUDNEY, Walter (1790–1824)
b. Edinburgh Dec. 1790 *d.* Katagum, Nigeria 12 Jan. 1824
MD Edinburgh 1817. Naval Surgeon, 1810. Surgeon to D. Denham and H. Clapperton's expedition to N. Nigeria, 1822–24.
D. Denham and H. Clapperton *Narrative of Travels and Discoveries in Northern and Central Africa...1822–24* 1826. T. Nelson *Biogr. Mem. of late Dr. W. Oudney* 1830. *J. Linn. Soc.* v.17, 1879, 328–29. *D.N.B.* v.42, 354.
Plants at BM(NH), Kew.
Oudneya R. Br.

OULTON, Rev. Richard (1812–1880)
b. Cooldagh, County Antrim 1812 *d.* Holywood, County Down 1880
Curate, St. Anne's, Belfast. Registrar, Queen's College, Belfast. Botanised in Counties Antrim, Armagh and Down. His herb. of Irish plants was sold *c.* 1900.
Proc. Belfast Nat. Field Club 1913, 620. *Irish Nat.* 1913, 26.

OUTRAM, Alfred (1847–1899)
b. Tooting, London 1847 *d.* London 8 Dec. 1899
Gardener. Traveller for nursery firm of B. S. Williams, Holloway, London.
Gdnrs Chron. 1899 ii 457 portr. *Orchid Rev.* 1900, 5.

OWEN, Miss (*fl.* 1830s)
Accompanied her brother, Rev. Francis Owen, on a mission in S. Africa, 1837–40. Plants collected in Natal now at Trinity College, Dublin.
S. African J. Sci. 1959, 319–20; 1971, 402.

OWEN, G. D. (*c.* 1842–1894)
d. 22 Feb. 1894
Of Rotherham, Yorks. Cultivated orchids.
Orchid Rev. 1895, 100.
Phaius × *Owenianus*.

OWEN, George (*c.* 1500–1558)
d. 18 Oct. 1558
MD Oxon 1527. Physician to Henry VIII, Edward VI, Queen Mary. W. Turner in preface to *New Herball* 1551 acknowledges his assistance.
D.N.B. v.42, 407–8. C. E. Raven *English Nat.* 1947, 69.

OWEN, Jeremiah (*fl.* 1830s)
Nurseryman, Blue Anchor Road, Bermondsey, London.

OWEN, Robert (*c.* 1839–1897)
d. Maidenhead, Berks 8 May 1897
Nurseryman, Maidenhead.
Garden v.51, 1897, 362. *Gdnrs Mag.* 1897, 281 portr.

OWEN, William Fitzwilliam (*fl.* 1820s)
Captain, Royal Navy. "His botanical knowledge secured him against the mistakes committed by others" and enabled him to collect roots of *Cocculus palmatus* in E. Africa, 1825.
Bot. Mag. 1830, t.2970–71.

OXLEY, Thomas (*fl.* 1800s–1820s)
Nurseryman, Beast Fair, Pontefract, Yorks. In partnership with Mr. Scholey.

OXLEY, Thomas (1805–1886)
b. Dublin 7 June 1805 *d.* Southampton, Hants 6 March 1886

BA. MD. Surgeon, East India Company. At Penang, 1831. Senior Surgeon, Straits Settlements until 1857. Entomologist. Interested in economic plants, especially gutta-percha trees. 'Botany of Singapore' (*J. Ind. Archipel.* 1850, 436–40).

Calcutta J. Nat. Hist. v.5, 1845, 115–16. *J. R. Asiatic Soc. Bengal* v.33, 1900, 35–36. D. G. Crawford *Hist. Indian Med. Service, 1600–1913* v.2, 1914, 90. *Gdns Bull. Straits Settlements* 1927, 129. *Fl. Malesiana* v.1, 1950, 397. *R.S.C.* v.4, 730.

Plants at Calcutta, Kew.

Durio oxleyanus Griff.

PACKE, Charles (1826–1896)
b. Prestwold, Leics 1826 *d.* Stretton Hall, Leics 16 July 1896

BA Oxon 1849. FLS 1870. *Guide to Pyrenees* 1862; 1867 with botanical notes.

Alpine J. v.18, 1896, 236–42. *J. Bot.* 1897, 415–16. *Proc. Linn. Soc.* 1896–97, 66. A. R. Horwood and C. W. F. Noel *Fl. Leicestershire* 1933, ccxix.

Herb. at Cambridge.

PACKER, J. J. (*fl.* 1850s)

Stationer of Thirsk, Yorks. First librarian of Thirsk Botanical Society. Married one of J. G. Baker's sisters.

Bot. Soc. Exch. Club Br. Isl. Rep. 1932, 289.

PAGE, John Courtney (–1947)
d. Haywards Heath, Sussex 8 Sept. 1947

Secretary, National Rose Society. Edited *Rose Annual.*

Gdnrs Chron. 1947 ii 104. *Times* 9 Sept. 1947.

PAGE, Mary Maud (1867–1925)
b. London 21 Sept. 1867 *d.* Cape Town 8 Feb. 1925

To S. Africa, 1911. Botanical artist, Bolus Herb., Cape Town, 1915–25.

Ann. Bolus Herb. v.4, 1926, 56–61 portr.

PAGE, Thomas (*fl.* 1830s)

Nurseryman, Edmonton, Middx.

PAGE, W. H. (–1928)
d. 9 Sept. 1928

Nurseryman, Tangley Nurseries, Hampton, Middx.

Gdnrs Chron. 1923 i 72 portr.; 1928 ii 239 portr. *J. Hort. Home Farmer* v.64, 1912, 30–31 portr.

PAGE, William Bridgewater (1790–1871)
d. 12 April 1871

Employed at Vineyard Nursery of Lee and Kennedy, Hammersmith. Married Amelia Emily Kennedy. Nurseryman, Old Spa Gardens, Southampton, Hants. In partnership with W. Toogood *c.* 1861. John Kennedy is considered to be the author of *Page's Prodromus* 1817.

G. W. Johnson *Hist. English Gdning* 1829, 301. *Gdnrs Mag.* 1835, 60–61. *J. Bot.* 1916, 241.

PAGE, Winifred Mary (1887–1965)
b. 11 Nov. 1887 *d.* 9 March 1965

FLS 1951. Contrib. to *Trans. Br. Mycol. Soc.*

Proc. Linn. Soc. 1966, 121.

PAGET, Sir James (1814–1899)
b. Great Yarmouth, Norfolk 11 Jan. 1814 *d.* Regent's Park, London 30 Dec. 1899

DCL Oxon 1868. LLD Cantab 1874. FRS 1851. FLS 1872. Baronet, 1871. Surgeon. *Sketch of Natural History of Yarmouth* (with C. J. Paget) 1834 (includes plants). Contrib. to H. C. Watson's *Topographical Bot.* 1883.

J. Bot. 1900, 62; 1904, 298–99. *Proc. Linn. Soc.* 1899–1900, 79–80. *Proc. R. Soc.* v.75, 1904, 136–40. S. Paget *Mem. and Letters of Sir J. Paget* 1901 portr. *D.N.B. Supplt 3* 240–42. *R.S.C.* v.4, 739; v.10, 981.

Plants at Norwich Museum, Kew.

Pagetia F. Muell.

PAINE, William (*fl.* 1730s)

Made plant collections in West and Eastern England. Sent algae to Joseph Andrews.

J. Bot. 1872, 174–75; 1904, 299–300. *Notes and Queries* v.11, 1880, 367. J. E. Dandy *Sloane Herb.* 1958, 174.

Plants, chiefly algae, with MS. list at BM(NH). Plants at Wadham College, Oxford.

PAINTER, Rev. William Hunt (1835–1910)
b. Birmingham 16 July 1835 *d.* Shrewsbury, Shropshire 12 Oct. 1910

At Edgbaston, Derby and Bristol; Rector of Stinchley, 1894–1909. *Contribution to Fl. Derbyshire* 1889. Contrib. to *J. Bot.*

J. Bot. 1901 Supplt, 73–74; 1909, 451; 1911, 125–26 portr. *Bot. Soc. Exch. Club Br. Isl. Rep.* 1910, 533–34. A. R. Horwood and C. W. F. Noel *Fl. Leicestershire* 1933, ccxxvi–ccxxvii. *R.S.C.* v.10, 982.

Herb. at University College, Aberystwyth. Plants at Kew, Oxford.

Fumaria × *Painteri* Pugsley.

PALETHORPE, Joseph (*fl.* 1830s)

Nurseryman, Chain Lane, Newark, Notts.

PALETHORPE, Thomas (*fl.* 1790s–1800s)

Nurseryman, Appletongate, Newark, Notts.

PALEY, Frederick Apthorp (1815–1888)
b. Easingwold, Yorks 14 Jan. 1815 *d.* Boscombe, Hants 11 Dec. 1888

BA Cantab 1838. LLD Aberdeen 1883. Greek scholar. *Wild Flowers of Dover* 1850. *Wild Flowering Plants of Peterborough* 1860.

J. Gillow *Dict. of English Catholics* v.4, 1885–87, 234–37. *D.N.B.* v.43, 99–101. G. C. Druce *Fl. Northamptonshire* 1930, cxxi.

PALGRAVE, Mary *see* Turner, M.

PALGRAVE, Thomas (1804–1891)
b. Great Yarmouth, Norfolk 27 Aug. 1804 *d.* Llansaintffraid, Montgomeryshire Jan. 1891

Cousin of Sir W. J. Hooker. Solicitor, Liverpool. Bryologist.

J. Bot. 1904, 300. *Trans. Liverpool Bot. Soc.* 1909, 85. *Lancashire Nat.* v.13, 1920–21, 76–82, 127–29. Liverpool Museums *Handbook and Guide to Herb. Collections* 1935, 26–27.

Herb. and letters at Liverpool Museums. Letters in Wilson correspondence at BM(NH).

PALMER, Charlotte Ellen (1830–1914)
b. Ladbroke, Warwickshire 1830 *d.* Odiham, Hants 27 Feb. 1914

Contrib. to F. Townsend's *Fl. Hampshire* 1883 and J. E. Bagnall's *Fl. Warwickshire* 1891. Also painted flowers.

J. Bot. 1908, 32. *Bot. Soc. Exch. Club Br. Isl. Rep.* 1914, 48–51. G. C. Druce *Comital Fl. Br. Isl.* 1932, xii. H. N. Clokie *Account of Herb. Dept. Bot., Oxford* 1964, 220. D. A. Cadbury *Computer-based Fl.... Warwickshire* 1971, 51–52.

Herb. at Oxford. Plants at Basingstoke Museum.

PALMER, Edward (1831–1911)
b. Wilton, Norfolk 12 Jan. 1831 *d.* Washington, U.S.A. 10 April 1911

To U.S.A., 1849. Surgeon at various army posts in Colorado, Kansas, Arizona, 1862–67. Collected plants in Arizona, California, Mexico, 1862–1910. *List of Plants collected in S.W. Chihuahua...1885.*

Proc. Amer. Acad. v.11, 1876, 80–81; 1879, 213–88; 1886, 414–45; 1887, 396–465; 1889, 36–82. D. C. Eaton *List of Marine Algae collected by Dr. Edward Palmer* 1875. *Contrib. U.S. National Herb.* v.1, 1890–95 *passim*. *Amer. Fern J.* 1911, 143–47. *Bot. Gaz.* v.52, 1911, 61–63. portr. *Popular Sci. Mon.* 1911, 341–54. *Brittonia* v.5, 1943, 64–79. *Amer. Midland Nat.* 1943, 768–78. R. McVaugh *Edward Palmer* 1956. *Madrono* v.16, 1961, 1–4. *Contrib. Univ. Michigan Herb.* 1972, 288–89. C. C. Gillispie *Dict. Sci. Biogr.* v.10, 1974, 285. *R.S.C.* v.10, 984.

Plants collected in Mexico at Kew. Field notes at Washington. Portr. at Hunt Library.

Palmerella A. Gray.

PALMER, Edward Henry (1840–1882)
d. Wady Sudr, Sinai Aug. 1882

Orientalist. Surveyed Sinai, 1868–70. *Desert of the Exodus* 1871 2 vols (with plant lists).

Sir W. Besant *Life and Achievements of E. H. Palmer* 1883. *Rec. Bot. Survey of India* v.8, 1933, 478.

PALMER, Ernest Jesse (1875–1962)
b. Leicester 8 April 1875 *d.* Webb City, Missouri 25 Feb. 1962

To U.S.A. 1878. Collected plants *c.* 1900 under joint auspices of Arnold Arboretum and Missouri Botanic Garden. Assistant, Arnold Arboretum, 1921.

J. Arnold Arb. v.43, 1962, 351–58 portr.

Portr. at Hunt Library.

PALMER, James (*fl.* 1820s)

Nurseryman, Irongate, Derby.

PALMER, John (*fl.* 1820s)

Nurseryman, Queen Street, Derby.

PALMER, John Adnah Jordan (1883–1951)
b. Horetown, County Wexford 5 April 1883 *d.* Greystones, County Wicklow 22 Dec. 1951

Member of Dublin Naturalists' Field Club, 1926. Had particular interest in ferns. Contrib. to *Irish Nat.*

Irish Nat. J. 1952, 305–6.

PALMER, Thomas Carey (*fl.* 1810s)

Of Bromley, Kent. Received Chinese plants via his relative, Richard Rawes, Captain of 'Warren Hastings', 1816–20.

E. Bretschneider *Hist. European Bot. Discoveries in China* 1898, 282–83.

PALMER, William (1856–1921)
b. Penge, Kent 1 Aug. 1856 *d.* New York 8 April 1921

To U.S.A., 1868. Taxidermist, etc. U.S. National Museum, 1874. Made collections, mainly birds on O. Bryant's expedition to Java, 1909. Interested in ferns. 'Ferns of the Dismal Swamp, Virginia' (*Proc. Biol. Soc. Washington* v.13, 1899, 61–70). President, American Fern Society, 1917–18.

Amer. Fern. J. 1917, 20–21; 1923, 23–24. *Auk* 1922, 305–21 portr. *Fl. Malesiana* v.1, 1950, 399.

Java plants at Washington.

PALMER, Hon. William Jocelyn Lewis (1894–1971)
b. 15 Sept. 1894 *d.* Guernsey, Channel Islands 6 June 1971

Educ. Oxford. FLS 1937. VMH 1954. Treasurer, Royal Horticultural Society, 1953–65. Keen gardener; hybridised *Agapanthus*.

J. R. Hort. Soc. 1971, 458–60. *Lilies 1972 and Allied Genera* 50–51. *Watsonia* 1972, 178–79.

PAM, Albert (1875–1955)
b. London 26 June 1875 *d.* Broxbourne, Herts 2 Sept. 1955

MA Oxon. FLS 1939. VMH 1943. Financier. Authority on S. American plants which he grew in his garden. Introduced *Pamianthe*. *Adventures and Recollections* 1945.

Herbertia 1938, 75–76 portr. *Bot. Mag.* v.170, 1954 portr. *Proc. Linn. Soc.* 1954–55, 137–38. *Times* 5 Sept. 1955. *Who was Who, 1951–1960* 846.

Library sold at Sotheby, 7 May 1956.

Pamianthe Stapf.

PAMPLIN, James (1785–1865)

Nurseryman, Lea Bridge Road, Leyton, Essex, 1838–60; followed by William Pamplin, 1861–69. This nursery is not to be confused with the Lea Bridge Road Nursery founded by Finlay Fraser in 1823.

Trans. London Middlesex Archaeol. Soc. 1973, 189–90.

PAMPLIN, William (1741–1805)

Father of William (1768–1844) and James Pamplin (1785–1865). Nurseryman, Walthamstow, Essex.

PAMPLIN, William (1768–1844)
b. Walthamstow, Essex 1768 *d.* Lavender Hill, Battersea 25 Sept. 1844

Employed in his father's nursery at Walthamstow. Acquired nursery in King's Road, Chelsea; later moved to Battersea. Plants grown by him are figured in R. Sweet's publications. Father of W. Pamplin (1806–1899).

Y Seren 16 Ionawr 1937, 6; 23 Ionawr 1937, 7. *Yr Haul* 1939, 326–27, 354–55.

PAMPLIN, William (1806–1899)
b. Chelsea, London 5 Aug. 1806 *d.* Llandderfel, Merionethshire 9 Aug. 1899

ALS 1830. Son of W. Pamplin (1768–1844). Bookseller (with J. Hunneman), publisher and distributor of exsiccatae. Retired to Llandderfel where he tried to establish the North Wales Central Botanic Gardens. *Catalogue of Rarer Species of Indigenous Plants...in Vicinity of Battersea and Clapham* 1827. Published *Phytologist* 1855–63 to which he also contributed. Contrib. to H. Trimen and W. T. T. Dyer's *Fl. Middlesex* 1869 and other local floras.

Gdnrs Mag. 1839, 303–4. *J. Bot.* 1864, 391; 1899, 521–24 portr. G. C. Druce *Fl. Berkshire* 1897, clxx–clxxi. *Proc. Linn. Soc.* 1899–1900, 80–81. *Essex Nat.* v.19, 1920, 76. G. C. Druce *Fl. Oxfordshire* 1927, cxiii–cxiv. *Y Seren* 16 Ionawr 1937, 6; 23 Ionawr 1937, 7. *Yr Haul* 1939, 326–27, 354–55. *Gdnrs Chron.* 1942 i 175. W. Matthews *Br. Diaries, 1442–1929* 1950, 214. H. N. Clokie *Account of Herb. Dept. Bot., Oxford* 1964, 220–21. *R.S.C.* v.4, 748.

Plants at Oxford. Letters at Kew, University College, Bangor, National Library of Wales.

Gentiana pamplinii Druce.

PANTLING, Robert (1857–1910)
d. Suez, Egypt 6 Feb. 1910

Kew gardener, 1875. Curator, Botanic Gardens, Calcutta, 1879. Deputy Superintendent, Cinchona Plantation, Bengal, 1897–1910. Studied and collected Indian orchids. Illustrated G. King's 'Orchids of Sikkim-Himalaya' (*Ann. R. Bot. Gdn Calcutta* v.8, 1898).

J. Kew Guild 1909–10, 491–92 portr. *Trans. Bot. Soc. Edinburgh* v.24, 1912, 101–3.

Pantlingia Prain.

PANTON, Joseph Anderson (1831–1913)
b. Knockiemil, Aberdeenshire 2 June 1831 *d.* Melbourne 25 Oct. 1913

Educ. Edinburgh University. To Australia, 1851. Commissioner of Crown Lands, Victoria. Police Magistrate, Victoria, 1862. Collected plants in W. Australia which he sent to F. von Mueller.

Austral. Encyclop. v.6, 1965, 443.

Plants at Melbourne.

Eremophila pantonii F. Muell.

PAPPE, Carl Wilhelm Ludwig (1802–1862)
b. Hamburg, Germany 1802 *d.* Cape Town 14 Oct. 1862

MD Leipzig 1827. Cape botanist, 1835. *List of S. African Indigenous Plants* 1847. *Fl. Capensis Medicae Prodromus* 1850; ed. 2 1857. *Silva Capensis* 1854; ed. 2 1862. *Synopsis Filicum Africae Australis* (with R. W. Rawson) 1858.

W. H. Harvey and O. W. Sonder *Fl. Capensis* 1859, 11*. *J. Bot.* 1896, 117–19.

Herb. at Cape Town. Plants and letters at Kew.

Pappea Ecklon & Zeyher.

PARFITT, Edward (1820–1893)
b. East Tuddenham, Norfolk 17 Oct. 1820 *d.* Exeter, Devon 15 Jan. 1893

Gardener. Librarian at Devon and Exeter Institution, 1861–93. Contrib. to *Trans. Devonshire Assoc.*

J. Bot. 1893, 160. *D.N.B.* v.43, 205. W. K. Martin and G. T. Fraser *Fl. Devon* 1939, 774–75. *R.S.C.* v.4, 756; v.8, 561; v.10, 989.

Herb. and 'Devon Fungi' in 12 vols with coloured drawings (unpublished) at Torquay Natural History Society. Letters in Broome correspondence at BM(NH).

PARIS, John Ayrton (1785–1856)
b. Cambridge 7 Aug. 1785 *d.* London 24 Dec. 1856

MD Cantab 1813. President, Royal College of Physicians, 1844–56. *Guide to Mount's Bay and the Land's End* 1824 (contains a number of plant records).

D.N.B. v.43, 206–7. F. H. Davey *Fl. Cornwall* 1909, xxxix.

PARISH, Rev. Charles Samuel Pollock (1822–1897)
b. Calcutta 26 Jan. 1822 *d.* Roughmoor, Somerset 18 Oct. 1897

BA Oxon 1841. Chaplain to forces at Moulmein, 1852–78. Collected orchids and ferns. Retired to Somerset, 1878. Contrib. to *J. Asiatic Soc., Bengal*. Contrib. account of orchids to F. Mason's *Burmah* v.2, 1883, 148–202.

Trans. Linn. Soc. v.23, 1862, 170. *Gdnrs Chron.* 1897 ii 295. *J. Bot.* 1897, 464. *Orchid Rev.* 1897, 322–23. *Kew Bull.* 1898, 313. *Curtis's Bot. Mag. Dedications, 1827–1927* 171–72 portr. *J. Thailand Res. Soc. Nat. Hist. Supplt.* 1939, 13. *Amer. Orchid Soc. Bull.* 1962, 206–7. M. A. Reinikka *Hist. of the Orchid* 1972, 211–14 portr. *R.S.C.* v.4, 757; v.8, 562; v.12, 557.

Herb. at Taunton Museum. Drawings at Kew.

Parishia Hook. f.

PARISH, William Hawtayne (*fl.* 1840s)

Lieut. Sent Himalayan plants, in particular ferns, from Mandi and Kulu to J. D. Hooker and T. Thomson in 1847.

I. H. Burkill *Chapters on Hist. Bot. in India* 1965, 81–82.

PARK, Kenneth John Frederick (*c.* 1928–1960)
d. 2 March 1960

Educ. King's College, Newcastle. Officer-in-charge, Nature Conservancy's Moor House Station in the Pennines, 1954. Botanist.

Nature v.186, 1960, 595.

Herb. at BM(NH).

PARK, Mungo (1771–1806)
b. Foulshiels, Selkirk 10 Sept. 1771 *d.* Niger 1806

LRCP Edinburgh 1791. ALS 1793. Surgeon and traveller. Brother-in-law of James Dickson. Protégé of J. Banks. To India and Sumatra, 1793; to Africa, 1795–97, 1805–6. Practised medicine at Peebles, 1799–1804. *Travels in Interior Districts of Africa* 1799 portr. *Journal of Mission to Interior of Africa, 1805* 1815.

D. Denham and H. Clapperton *Narrative of Travels...in Northern and Central Africa* 1826, 234–35. *Hist. Berwick Nat. Club* v.10, 1883, 300–4. *Trans. Bot. Soc. Edinburgh* v.19, 1893, 593. D. G. Crawford *Hist. Indian Med. Service, 1600–1913* v.2, 1914, 57–58. *D.N.B.* v.43, 218–21. S. Gwynn *Mungo Park and the Quest of the Niger* 1934. *Fl. Malesiana* v.1, 1950, 400. *Comptes Rendus AETFAT, 1960* 1962, 61–63.

Plants at BM(NH). Drawings at Royal Botanic Garden, Edinburgh.

Parkia R. Br.

PARKE, Thomas Heazle (1857–1893)
b. Drumsna, Country Roscommon 27 Nov. 1857 *d.* Alt na Craig, Argyll 10 Sept. 1893

Surgeon and African traveller. With H. M. Stanley on Emin Pasha expedition, 1887–89. Botanised. *My Experiences in Equatorial Africa...* 1891. *Guide to Health in Africa* 1893 (contains floristic information).

D.N.B. v.43, 228–30.

PARKER, Charles Eyre (1822–1895)

Collected plants of Torquay and neighbourhood from 1845. Contrib. to R. Stewart's *Handbook of Torquay Fl.* 1860.

W. K. Martin and G. T. Fraser *Fl. Devon* 1939, 773.

Plants at Torquay Natural History Society.

PARKER, Charles Sandbach (–1869)
b. Glasgow *d.* 1869

Of Aigburth, Liverpool and Blochairn, Glasgow. Studied under De Candolle at Geneva. Assisted W. Roscoe in his *Monandrian Plants* 1824–29. Collected plants in N. America, Guiana and West Indies from 1824.

W. J. Hooker *Exotic Fl.* v.2, 1824, t.147. *Bot. Mag.* 1837, t.3595. A. Lasègue *Musée Botanique de B. Delessert* 1845, 492. *Annual Rep. Kew Gdns* 1869, 8. I. Urban *Symbolae Antillanae* v.3, 1902, 98. *Trans. Liverpool Bot. Soc.* 1909, 86. *R.S.C.* v.4, 758.

Plants at Kew, Liverpool. Letters at Kew.

Parkeria Hook.

PARKER, Edward Harper (*fl.* 1860s–1890s)

British Consular Service, China, 1869–95. Consul, Chungking, 1880–81. Collected plants in neighbourhood. Later Consul at Wenchow and in Korea. Travelled in Annam, 1892. *Up the Yangtze* 1891 (account of botanical explorations, pp. 301–8). 'Notice of Fl. of Hainan' (*China Rev.* 1892, 167–68).

E. Bretschneider *Hist. European Bot. Discoveries in China* 1898, 750–55. E. H. M. Cox *Plant Hunting in China* 1945, 106.

Rubus parkeri Hance.

PARKER, George Williams (–1904)
d. Potaro, British Guiana June 1904
Physician to Queen Ranavalona in Madagascar where he collected plants. Went to British Guiana to collect plants for Kew.
Plants and MS. *Guiana Plant List 1897–1900* at Kew.

PARKER, John Cowham (*c.* 1774–1841)
b. c. 1774 *d.* Hull, Yorks 28 Jan. 1841
Wine-merchant, Hull. One of the founders of Hull Botanic Garden. Had a garden.
Gdnrs Mag. 1841, 190. R. W. Corlass *Sketches of Hull Authors* 1879, 108.

PARKER, Nicholas (*fl.* 1720s)
Nurseryman, Strand-on-the-Green, Chiswick, Middx.

PARKER, Richard (1856–)
Gardener at Impney, Worcs.
J. Hort. Cottage Gdnr v.19, 1889, 511–12 portr.

PARKER, Richard Neville (1884–1958)
b. 4 Dec. 1884 *d.* 12 April 1958
Indian Forest Service, 1905–39. Chief Conservator of Forests, Punjab, 1932. Forest Botanist, Forest Research Institute, Dehra Dun, 1922–32. Collected plants in Burma, India. *Forest Fl. of Punjab with Hazara and Delhi* 1918; ed. 2 1924. *Common Indian Trees and how to know them* 1933.
Indian Forester 1959, 270 portr. *Pakistan J. For.* 1967, 353.
Plants at Dehra Dun, Kew.

PARKER, Robert (*fl.* 1850s–1880s)
In partnership with B. S. Williams in nursery in Holloway, London, 1854; partnership dissolved, 1861. Had nursery at Tooting, Surrey.

PARKER, Thomas (*c.* 1821–1880)
d. 27 Feb. 1880
Nurseryman, St. Michael's Hill and Stapleton Nurseries, Bristol.
Gdnrs Chron. 1880 i 311.

PARKER, Wilfred Henry (1888–1938)
b. Christchurch, New Zealand 29 Sept. 1888 *d.* 11 Jan. 1938
To England, 1897. Educ. Cambridge. Assistant to Sir Rowland Biffen, School of Agriculture, Cambridge, 1913. Director, National Institute of Agricultural Botany, Cambridge, 1919.
Who was Who, 1929–1940 1044.

PARKIN, John (1873–1964)
b. Wigton, Cumberland 25 Jan. 1873 *d.* Wigton 29 March 1964
MA Cantab 1897. FLS 1902. Scientific Assistant, Peradeniya Botanic Gardens, Ceylon, 1898–99. Lecturer, Cambridge University, 1899–1911. Contrib. to *J. Linn. Soc., Proc. Linn. Soc., J. Indian Bot. Soc.*
Nature v.202, 1964, 953. *Proc. Linn. Soc.* 1965, 100–1.

PARKINSON, Charles Edward (1890–1945)
b. 7 May 1890 *d.* Edinburgh 25 March 1945
Assistant Conservator of Forests, Andaman Islands, 1912. Indian Forest Service, Burma, 1916. Forest Botanist, Burma, 1925. Forest Research Institute, Dehra Dun, 1929. Collected plants in Andaman Islands, 1915; Kulu, 1934; Kumaun, 1935; Jaunsar, 1936. *Forest Fl. of Andaman Islands* 1923. Responsible for series of botanical bulletins on Burmese trees.
Empire For. J. 1945, 5. *Indian Forester* 1945, 306.
MSS. at Kew.

PARKINSON, James (1755–1824)
d. Hoxton, London 21 Dec. 1824
FGS. Surgeon. Palaeontologist. Practised medicine at Hoxton, 1785. *Organic Remains of a Former World* 1804–11 3 vols. (v.1 contains vegetable kingdom).
Nature v.51, 1894, 31–32. *D.N.B.* v.43, 314–15. M. Critchley *James Parkinson* 1955. *J. Soc. Bibl. Nat. Hist.* 1976, 451–66. *R.S.C.* v.4, 760.
Trigonocarpus parkinsoni Brongn.

PARKINSON, John (1567–1650)
b. Notts 1567 *d.* London Aug. 1650
Apothecary. King's Herbarist ('Botanicus Regius Primarius'). Had garden in Long Acre, London. *Paradisi in Sole Paradisus Terrestris* 1629. *Theatrum Botanicum* 1640 portr.
R. Pulteney *Hist. Biogr. Sketches of Progress of Bot. in England* v.1, 1790, 138–54. G. W. Johnson *Hist. English Gdning* 1829, 82–93. S. Felton *Portr. of English Authors on Gdning* 1830, 90–91 portr. *Cottage Gdnr* v.6, 1851, 249–50. H. Trimen and W. T. T. Dyer *Fl. Middlesex* 1869, 372. *J. Hort. Cottage Gdnr* v.53, 1875, 493 portr. *Garden* v.3, 1873, 151 portr.; 1910, 288–89, 300–1. G. C. Druce *Fl. Berkshire* 1897, c. *Essex Nat.* 1899, 176–79. *Gdnrs Mag.* 1899, 834–35. *D.N.B.* v.43, 35. *Gdnrs Chron.* 1902 i 317–18 portr. J. W. White *Fl. Bristol* 1912, 52–53. J. R. Green *Hist. Bot.* 1914, 48–53. R. T. Gunther *Early Br. Botanists* 1922, 265–71. G. C. Druce *Fl. Buckinghamshire* 1926, lxix–lxx. G. C. Druce *Fl. Northamptonshire* 1930, xlv–xlvi. *Daffodil Yb.* 1937, 15–18. C. E. Raven *English Nat.* 1947, 248–73. W. S. C. Copeman *Worshipful Society of Apothecaries in London* 1968, 74–75. F. D. and J. F. M. Hoeniger *Growth of Nat. Hist. in Stuart England* 1969, 17–20. *Missouri Bot. Gdn Bull.* 1970, 4–10.
Parkinsonia L.

PARKINSON, John (-1719)
d. 14 Nov. 1719
Nurseryman, Lambeth, London.

PARKINSON, John (*c.* 1772–1847)
d. Paris 3 April 1847
FRS 1840. FLS 1795. Consul-General in Mexico, 1838. Sent plants to Kew.
Bot. Mag. 1840, t.3778. *Proc. Linn. Soc.* 1847, 336–37. F. D. Goodman and O. Salvin *Biologia Centrali-Americana* v.4, 1886, 128–29. *Curtis's Bot. Mag. Dedications, 1827–1927* 51–52 portr. *Taxon* 1970, 523–24.
Epidendrum parkinsonianum Hook.

PARKINSON, Sydney (*c.* 1745–1771)
b. Edinburgh *c.* 1745 *d.* Indian Ocean 26 Jan. 1771
Woollen draper. Protégé of J. Banks, who sent him to draw at Kew, 1767, and with whom he went as draughtsman on J. Cook's voyage, 1768. The first artist to draw the Australian landscape. *Journal of a Voyage to the South Seas* 1773; ed. 2 1784.
Trans. Proc. N.Z. Inst. 1877, 108–34. *D.N.B.* v.43, 317–18. J. H. Maiden *Sir Joseph Banks* 1909, 62–64 portr. *J. Friends Hist. Soc.* 1911, 123–27. *J. Bot.* 1912, 72. *J. Proc. R. Austral. Hist. Soc.* 1921, 100–1. *J. Soc. Bibl. Nat. Hist.* 1950, 190–93 portr. *Chronica Botanica* 1954, 326–63. *Gdnrs Chron.* 1954 i 24–25 portr. E. J. Willson *James Lee and the Vineyard Nursery, Hammersmith* 1961, 56–60. R. and T. Rienits *Early Artists of Australia* 1963. J. C. Beaglehole, *ed. Endeavour Journal of Joseph Banks, 1768–1771* 1962 2 vols *passim*. *Australian Dict. Biogr.* v.2, 1967, 314. *Endeavour* 1968, 5–7. F. A. Stafleu *Taxonomic Literature* 1967, 349. A. M. Lysaght *Joseph Banks in Newfoundland and Labrador, 1766* 1971, 102–3 portr. *Captain Cook's Florilegium* 1973 (engravings from drawings by Parkinson).
Drawings and portr. at BM(NH).
Ficus parkinsoni Hiern.

PARKS, John Damper (*c.* 1792–1866)
d. 11 Jan. 1866
Collected plants for Horticultural Society of London in China and Java, 1823–24. Gardener to Earl of Arran at Bognor, Sussex. Nurseryman, Dartford, Kent. Contrib. to *Gdnrs Mag.*
Trans. Hort. Soc. London v.5, 1824, 427–28, preface, v. *Gdnrs Mag.* 1829, 572–73. E. Bretschneider *Hist. European Bot. Discoveries in China* 1898, 271–74. E. H. M. Cox *Plant Hunting in China* 1945, 58–59. *Fl. Malesiana* v.1, 1950, 401. *J. R. Hort. Soc.* 1955, 270.
MS. journal at Royal Horticultural Society. Plants at Kew.

PARNELL, Frederick Richard (*c.* 1886–1971)
d. 9 Jan. 1971
Economic botanist, Madras Government, 1913. Bred rice at Coimbatore. Joined staff of Empire Cotton Growing Corporation, 1924. Bred U4 strain of Upland cotton at Cotton Experimental Station, Barberton, Transvaal. Developed Cotton Research Station, Uganda.
Times 13 Jan. 1971.

PARNELL, Richard (1810–1882)
b. Bramford Speke, Devon 1810 *d.* Edinburgh 28 Oct. 1882
MD. FRSE 1837. Ink manufacturer. *Grasses of Scotland* 1842. *Grasses of Britain* 1845.
J. Bot. 1883, 30–31. *Scott. Nat.* 1883, 43. *Trans. Bot. Soc. Edinburgh* v.16, 1886, 6–8. *R.S.C.* v.4, 763.
Plants at Kew, Edinburgh.
Poa parnellii Bab.

PARNELL, William (1833–1906)
b. Ireland 1833 *d.* Glasnevin, Dublin 28 Nov. 1906
Employed in Kew Herb., 1852 and assisted G. Bentham with his *British Fl.* Foreman, Botanic Garden, Glasnevin, Dublin, 1868–1906. Knew British plants well.
Gdnrs Chron. 1906 ii 400. *J. Kew Guild* 1907, 382.

PARRY, Charles Christopher (1823–1890)
b. Admington, Glos 28 Aug. 1823 *d.* Davenport, Iowa 20 Feb. 1890
MD Columbia College 1846. To America, 1832. Settled in Davenport, 1846. Botanist to Mexican Boundary Survey, 1849–52. In Rocky Mountains, 1861. Botanist to Agricultural Dept., Washington, 1869–71; to San Domingo Commission, 1871. 'Enumeration of Plants of Rocky Mountains' (with G. Engelmann and A. Gray) (*Amer. J. Sci.* 1862, 249–53, 331–41).
J. Bot. 1872, 64. *Bull. Torrey Bot. Club* 1878, 280; 1890, 74–75. *Proc. Davenport Acad. Sci.* 1878, 279–82; 1893, 35–52 portr. W. B. Hemsley *Biologia Centrali-Americana. Bot.* v.4, 1887, 131. *Garden* v.37, 1890, 285. *J. Hort. Cottage Gdnr* v.20, 1890, 244–45. C. S. Sargent *Silva of N. America* v.7, 1895, 130. I. Urban *Symbolae Antillanae* v.3, 1902, 98. H. A. Kelly *Some Amer. Med. Botanists* 1914, 180–86 portr. M. Meisel *Biogr. Amer. Nat. Hist.* v.1, 1924, 217; v.3, 635. *Dict. Amer. Biogr.* v.14, 1934, 261–62. J. Ewan *Rocky Mountain Nat.* 1950, 278. S. D. McKelvey *Bot. Exploration of Trans-Mississippi West, 1790–1850* 1955, 1031–38. *R.S.C.* v.4, 767; v.8, 565; v.12, 358.
Herb. at Davenport Academy of Natural Sciences. Portr. at Kew and Hunt Library. Letters at Kew.
Parryella Torrey & Gray.

PARRY, Francis (*fl.* 1860s)
Merchant at Canton and Hong Kong. Collected plants for H. F. Hance.
E. Bretschneider *Hist. European Bot. Discoveries in China* 1898, 682.
Hedyotis parryi Hance.

PARRY, Richard (*fl.* 1810s)
Served in India. Resident of Sumatra, 1807–11. Commissioned Indian artists to draw local flora and fauna.
M. Archer *Nat. Hist. Drawings in India Office Library* 1962, 19, 87–88.
Drawings at Kew.

PARRY, Thomas R. Gambier- *see* Gambier-Parry, T. R.

PARRY, Sir William Edward (1790–1855)
b. Bath, Somerset 19 Dec. 1790 *d.* Ems, Germany 8 July 1855
FRS 1821. DCL Oxon 1829. Knighted 1829. Made three expeditions to Arctic, 1819–20, 1821–23, 1824–25. Extensive botanical collections made by Parry and his officers. *Journal[s of Voyages] for Discovery of a North-West Passage* 1821–1826 3 vols.
J. R. Geogr. Soc. 1856, clxxxii–clxxxv. E. Parry *Mem. of Rear Admiral Sir W. E. Parry* 1857 portr. *D.N.B.* v.43, 392–93. N. Polunin *Bot. of Canadian Eastern Arctic* 1940, 11, 20. F. A. Stafleu *Taxonomic Literature* 1967, 350.
Plants at BM(NH), Kew. Letters at Kew. Portr. at National Portrait Gallery.

PARSONS, Alfred (–*c.* 1923)
Kew gardener, 1865. Went to India in 1860s. In charge of Ajmer Garden and later Simla Gardens.
J. Kew Guild 1924, 256 portr.

PARSONS, Alfred William (1847–1920)
b. Beckington, Somerset 2 Dec. 1847 *d.* Broadway, Worcs 16 Jan. 1920
RA 1911. President, Society of Painters in Watercolour, 1905. Illustrated A. B. Freeman-Mitford's *Bamboo Gdn* 1896 and Miss E. A. Willmott's *Genus Rosa* 1910–14.
Gdnrs Chron. 1891 i 439; 1920 i 70. *D.N.B. 1912–1921* 426–27. *Who was Who, 1916–1928* 817.
Genus Rosa drawings at Royal Horticultural Society.

PARSONS, Anthony (1810–1880)
b. Merton, Surrey 1810 *d.* 25 Dec. 1880
Gardener to Capt. Blake, Danesbury, Welwyn, Herts, 1851.
Gdnrs Chron. 1875 i 601 portr.; 1881 i 22 portr.

PARSONS, Christopher (1807–1882)
b. Southchurch, Essex 1807 *d.* Southchurch 23 Sept. 1882
FLS 1839. MS. 'Fl. of Rochford Hundred'.
Essex Nat. v.3, 1889, 55–56.
Diaries and herb. at Southend.

PARSONS, Henry Franklin (1846–1913)
b. Frome, Somerset 27 Feb. 1846 *d.* Croydon, Surrey Oct. 1913
MD London 1870. In medical practice, 1867–73. Medical Officer of Health, Goole, Yorks, 1874–79. Medical Inspector, Local Government Board, 1879–92. Assistant Medical Officer, Local Government Board, 1892–1911. President, Croydon Natural History and Scientific Society, 1893, 1912. 'Maritime Plants and Tidal Rivers of the West Riding' (*Naturalist* 1876, 113–20). Contrib. to *Trans. Croydon Nat. Hist. Soc.*
F. A. Lees *Fl. W. Yorkshire* 1888, 97–98. R. P. Murray *Fl. Somerset* 1896, xvii. *Bot. Soc. Exch. Club Br. Isl. Rep.* 1914, 51–52. *J. Bot.* 1914, 24, 280. *Naturalist* 1914, 8–9 portr.; 1961, 54. *Proc. Trans. Croydon Nat. Hist. Sci. Soc.* 1914, clxxiv–clxxvi portr.; 1948–51, 76–77. *Who was Who, 1897–1915* 548–49. *Watsonia* 1951, 100-1.
Plants at Croydon and Taunton Museums.

PARSONS, James (1705–1770)
b. Barnstaple, Devon March 1705 *d.* London 4 April 1770
MD Rheims 1736. FRS 1741. Practised medicine in London. *Microscopical Theatre of Seeds* 1745. *Pharmacopoeia Edinburgensis* 1752.
Gent. Mag. 1812 i 206. W. Munk *Roll of R. College of Physicians* v.2, 1878, 175–77. *D.N.B.* v.43, 403–4.
Parsonsia R. Br.

PARSONS, John (1742–1785)
b. York 1742 *d.* Oxford 9 April 1785
MA Oxon 1766. MD 1772. Pupil of Hope. Professor of Anatomy, Oxford, 1766. Had herb. Contrib. to J. Lightfoot's *Fl. Scotica* v.1, 1777, xiv. Discovered *Ranunculus reptans* (*ibid*, 289).
D.N.B. v.43, 405.

PARSONS, Philip (*fl.* 1630s)
Principal, Hart Hall, Oxford. Accompanied T. Johnson on his tour of West of England in 1634.

PARSONS, Thomas (*fl.* 1780s)
Nurseryman, Fareham, Hants.

PARTINGTON, Charles Frederick (–*c.*1857)
On staff of London Institution. Scientific writer. *Introduction to Science of Botany* 1835.
Hort. Register v.5, 1836, 18. *D.N.B.* v.43, 427.

PASCOE, Francis Polkinghorne (1813–1893)
b. Penzance, Cornwall 1 Sept. 1813 *d.* Brighton, Sussex 20 June 1893

FLS 1852. Surgeon in Royal Navy until 1843. Entomologist. Correspondent of H. C. Watson. 'Cornish Plants' (*Bot. Gaz.* 1850, 37–39).

G. C. Boase and W. P. Courtney *Bibliotheca Cornubiensis* 1874, 427–29. *J. Bot.* 1893, 287. *Proc. Linn. Soc.* 1893–94, 33–34. *D.N.B.* v.43, 435. F. H. Davey *Fl. Cornwall* 1909, xlvii–xlviii. *R.S.C.* v.4, 769; v.8, 566; v.10, 996; v.12, 558.

New Zealand plants at Kew.

PASMORE, Rev. Henry (–before 1699)
d. Jamaica before 1699

Sent plants to J. Petiver.

J. Petiver *Museii Petiveriani* 1695, 46. J. E. Dandy *Sloane Herb.* 1958, 174.

PATERSON, Alexander (1822–1898)
d. Bridge of Alan, Stirlingshire 1898

MD Edinburgh 1843. Physician. Had garden at Bridge of Alan.

Garden v.53, 1898, 398. *Gdnrs Chron.* 1898 i 301.

PATERSON, John Ligertwood (1820–1882)
b. Midmar, Aberdeenshire 1820 *d.* Brazil 9 Dec. 1882

MD. Fellow of Botanical Society of Edinburgh, 1872. Practised medicine in Brazil. Sent living plants to Royal Botanic Garden, Edinburgh.

Trans. Proc. Bot. Soc. Edinburgh 1886, 9–11.

PATERSON, Robert (–1933)
d. Auchendrane, Ayr 4 Aug. 1933

Of Stamperland, Glasgow and Stonehurst, Ardingly, Sussex. Cultivated orchids.

Orchid Rev. 1933, 284–85.

PATERSON, William (1755–1810)
b. Montrose, Angus 10 Aug. 1755 *d.* on voyage from Australia 21 June 1810

FRS 1798. FLS 1797. Colonel. Collected plants in Cape *c.* 1778. "He professed to travel at the expense of certain individuals, and possessed some small knowledge of Botany, but was in fact a mere gardener" (C. P. Thunberg *Travels in Europe, Africa and Asia* v.4, 1796, 271). To India, 1781–85. To Australia, 1791. Collected plants for J. Banks in Norfolk Island and N.S.W. Lieut.-Governor, N.S.W., 1800–10. *Narrative of Four Journeys into the Country of the Hottentots and Caffraria in...1777–1779* 1789.

Bot. Mag. 1795, t.300. R. Brown *Prodromus Florae Novae Hollandiae* v.1, 1810, 303–4. A. Lasègue *Musée Botanique de B. Delessert* 1845, 278, 446. *Cottage Gdnr* v.8, 1851, 328–29, 351, 364–65, 378–79; v.9. 1851, 3. J. D. Hooker *Fl. Tasmaniae* v.1, 1859, cxxiv. *D.N.B.* v.44, 26–28. *Hist. Rec, N.S.W.* v.1–8, 1895–1901 *passim. J. R. Soc. N.S.W.* v.42, 1908, 116. J. H. Maiden *Sir Joseph Banks* 1909, 197–99. *J. Linn. Soc.* v.45, 1920, 45–46. *Times* 24 Nov. 1956. *Australian Encyclop.* v.2–8, 1965 *passim.* V. S. Forbes *Pioneer Travellers of South Africa* 1965, 81–92 portr. *Taxon* 1970, 535.

Tasmanian plants at BM(NH). Letters to W. Forsyth at Kew. Letters at Mitchell Library, Sydney. MS. journal at Swedenborg House, London.

Patersonia R. Br.

PATERSON, William (*c.* 1824–1896)
b. Mortlach, Banffshire *c.* 1824 *d.* Balmoral, Aberdeenshire 5 Oct. 1896

Gardener to the Queen at Balmoral, 1847–92.

Gdnrs Chron. 1876 ii 528 portr.; 1896 ii 439. *Gdnrs Mag.* 1896, 701.

PATEY, G. S. (*fl.* 1860s–1880s)

Of East Hendred, Berks and Newton Abbott, Devon. Collected ferns.

Br. Fern Gaz. 1909, 47.

PATON, A. W. *see* Walker, A. W.

PATON, Cyril Ingram (1874–1949)

Authority on flora of Isle of Man. 'List of Flowering Plants...of Isle of Man' (*N. Western Nat.* v.8, 1933 Supplt. 67p).

Proc. Isl. of Man Nat. Hist. Antiq. Soc. v.6, 1960–63, 415.

Herb. at Ramsey Library.

PATTISON, Ernest (*c.* 1840–1916)

Of Leicester. Interested in ferns.

A. R. Horwood and C. W. F. Noel *Fl. Leicestershire* 1933, ccxxvii–ccxxviii.

PATTISON, Samuel Rowles (1809–1901)
b. Stroud, Glos 27 Oct. 1809 *d.* 27 Nov. 1901

FGS 1839. Of Launceston, Cornwall. *Chapters on Fossil Botany* 1849.

G. C. Boase and W. P. Courtney *Bibliotheca Cornubiensis* 1874, 430–31. *Nature* v.65, 1901, 109. *Quart. J. Geol. Soc.* 1902, lxii.

PATTON, Donald (1884–1959)
d. Glasgow 14 July 1959

MA Glasgow 1906. BSc 1911. PhD 1923. FRSE 1926. Science teacher, Bellahouston Academy, 1909–19. Assistant, then Senior Lecturer in Botany, Glasgow University, 1919–23. Lecturer in Botany, West of Scotland Training College for Teachers, 1923–40. President, Andersonian Naturalists Society, 1925–26. President, Natural History Society, Glasgow, 1926–29, 1952–54. Authority on mountain floras. Studied flora of

Upper Clydesdale. Curator, Glasgow University Herb., 1949. Edited *Glasgow Nat.* 1940–47. *Nature Study for Beginners* 1928.
Yb. R. Soc. Edinburgh 1959–60, 97–98. *Glasgow Nat.* 1961, 169–70. *Proc. Bot. Soc. Br. Isl.* 1961, 215–16.

PATTON, George, Lord Glenalmond (1803–1869)
b. The Cairnies, Perthshire 1803 *d.* Glenalmond, Perthshire 20 Sept. 1869
BA Cantab 1826. Lord Justice-Clerk of Scotland, 1867. Chairman, Oregon Botanical Association. Introduced conifers, etc. to his estate at Cairnies.
Gdnrs Chron. 1869, 1043–44. *D.N.B.* v.44, 64–65.
Abies pattoniana A. Murray.

PAUL, Arthur William (*fl.* 1870s–1910s)
Nurseryman, Waltham Cross, Herts. Specialised in roses.
Gdnrs Mag. 1897, 411–12 portr.; 1909, 739–40 portr. *J. Hort. Cottage Gdnr* v.45, 1902, 474 portr. *Garden* 1910, 324–25 portr.

PAUL, Rev. David (1845–1929)
b. Banchory, Kincardineshire 28 Aug. 1845 *d.* Edinburgh 12 July 1929
MA Aberdeen. DD Edinburgh 1915. FLS 1926. Minister, Roxburgh, 1876–96. Clerk of General Assembly of Church of Scotland, 1912–28. President, British Mycological Society, 1918. President, Cryptogamic Society of Scotland, 1922. President, Botanical Society of Edinburgh, 1899–1901. President, Scottish Alpine Botanical Club, 1923. Contrib. to *Trans. Bot. Soc. Edinburgh*, *Trans. Berwickshire Nat. Club.*
J. Bot. 1929, 236. *Times* 13 July 1929. *Proc. Berwickshire Nat. Club* 1929, 143–45. *Proc. Linn. Soc.* 1929–30, 214. *Bot. Soc. Exch. Club Br. Isl. Rep.* 1930, 96. *Trans. Proc. Bot. Soc. Edinburgh* v.30, 1931, 183–86. *Who was Who, 1929–1940* 1054.
Primula pauliana W. W. Smith & Forrest.

PAUL, George (1841–1921)
b. Dec. 1841 *d.* Cheshunt, Herts 18 Sept. 1921
VMH 1897. Nephew of W. Paul (1822–1905). Nurseryman, Old Nurseries, Cheshunt. Specialised in roses.
Garden v.62, 1902, 8–9 portr.; 1910, iv portr. *Gdnrs Mag.* 1910, 1–2 portr. *J. Hort. Home Farmer* v.62, 1911, 194–96 portr. *Gdnrs Chron.* 1921 i 242 portr.; 1921 ii 166.

PAUL, George Laing (*fl.* 1910s)
Son of G. Paul (1841–1921). Nurseryman, Cheshunt, Herts.
Gdnrs Mag. 1897, 412; 1911, 139–40 portr. *J. Hort. Home Farmer* v.63, 1911, 419–20 portr.

PAUL, William (1822–1905)
b. Cheshunt, Herts 16 June 1822 *d.* Waltham Cross, Herts 31 March 1905
FLS 1875. VMH 1897. Nurseryman, Cheshunt. Founded Royal Nurseries, Waltham Cross, 1860. Specialised in roses. *Rose Garden* 1848. *Contributions to Horticultural Literature* 1892. Co-editor of *Florist and Pomologist*.
Garden v.57, 1900, 166 portr.; v.63, 1903, iv portr.; v.67, 1905, 213–14 portr. *Gdnrs Chron.* 1905 i 216–17 portr., 233; 1905 ii 405–6. *Gdnrs Mag.* 1905, 231 portr. *J. Hort. Cottage Gdnr* v.50, 1905, 305 portr., 325. *Proc. Linn. Soc.* 1904–5, 46–47. *Rose Annual* 1971, 36–39.
Library sold at Sotheby, 27–29 Nov. 1905.

PAUL, William (1825–1880)
d. Paisley, Renfrewshire 12 Dec. 1880
Nurseryman, Crossflat Nursery, Paisley. Specialised in florists' flowers.
Gdnrs Chron. 1880 ii 821–22.

PAULSON, Robert (1857–1935)
b. Hendon, Middx 24 April 1857 *d.* 1 March 1935
FLS 1913. Schoolteacher. President, Essex Field Club, 1920–23. Mycologist. Contrib. to *Essex Nat.*
Essex Nat. 1935, 62–63 portr. *J. Bot.* 1935, 232. *J. Quekett Microsc. Club* 1935, 180. *Proc. Linn. Soc.* 1934–35, 186.
Herb. at BM(NH).

PAXTON, Sir Joseph (1803–1865)
b. Milton Bryant, Beds 1803 *d.* Sydenham, Kent 8 June 1865
FLS 1831. Knighted 1851. Gardener, Horticultural Society of London, Chiswick, 1823. Head gardener to Duke of Devonshire at Chatsworth, 1826, where he achieved the first blooming of *Victoria amazonica*. Designed the Crystal Palace. Edited *Hort. Register* 1831–34; *Paxton's Mag. Bot.* 1834–49. One of founders of *Gdnrs Chron.* 1841. *Practical Treatise on Cultivation of Dahlia* 1838. *Pocket Botanical Dictionary* 1840. *Paxton's Flower Garden* (with J. Lindley) 1850–53 3 vols.
Gdnrs Mag. 1834, 230–32. *Bot. Register, 1838* Misc. Notes., 61. *Cottage Gdnr.* v.12, 1854, 409–11 portr. *Gdnrs Chron.* 1865, 554–55; 1924 i 120; 1935 i 331, 353, 371; 1962 i 368–69; 1965 i 19 portr. *J. Hort.* v.33, 1865, 446–47 portr.; v.34, 1865, 12–13; 1891 i 9 portr.; v.43, 1901, 325–26, 335 portr. *J. Bot.* 1865, 231–32. *Proc. Linn. Soc.* 1865–66, lxxxi–lxxxiii. *D.N.B.* v.44, 103–4. *J. R. Hort. Soc.* 1934, 477–81. V. G. Markham *Paxton and the Bachelor Duke* 1935 portr. A. Simmonds *Hort. Who was Who* 1948, 46–52. M. Hadfield *Pioneers in Gdning* 1955,

170–82. *Fl. Malesiana Bull.* 1956, 489–90. G. F. Chadwick *Works of Sir Joseph Paxton, 1803–1865* 1961 portr. F. A. Stafleu *Taxonomic Literature* 1967, 352. *Sorby Rec.* v.2(3), 1967, 73–77. *Morris Arb. Bull.* 1968, v.19(1), 8–13. M. A. Reinikka *Hist. of the Orchid* 1972, 164–68 portr. J. Anthony *Joseph Paxton* 1973 portr.

Letters at Kew. Portr at Hunt Library.

Paxtonia Lindl.

PAYNE, Miss Bellamira F.
(*fl.* 1900s–1910s)

Of Chester, Cheshire. 'Fl. of Chester Abbey' (*Lancs Cheshire Nat.* v.11, 1919, 273–80). 'List of Plants growing within a Two Mile Radius of Chester...' (with A. Payne and Miss Cummings) (*29th Rep. Chester Soc. Nat. Sci.* 1900, 31–34).

PAYNE, Charles Harman (*c.* 1854–1925)
d. London 23 Feb. 1925

Horticultural journalist. *Short History of Chrysanthemum* 1885. *Florist's Bibliography* 1908; *Supplement* 1912.

Gdnrs Mag. 1908, 885–86 portr. *J. Hort.* v.63, 1911, 294–95 portr. *Garden* 1925, 134 (xii). *Gdnrs Chron.* 1921 ii 70 portr.; 1925 i 140 portr., 156, 159, 173.

PAYNE, Frederick William (1852–1927)
d. London 17 April 1927

BA London 1871. Schoolmaster, City of London School, 1876–1919. Diatomist. *Liostephania and Allies* 1922. 'Notes on Diatoms' (*Nuova Notarisia* 1925, 29–30; *J. Bot.* 1925, 256–62).

J. Bot. 1927, 177–78, 208.

Plants at BM(NH).

PAYNE, John (*fl.* 1700s–1710s)

Apothecary, London. Member of Botanical Society of London.

Proc. Bot. Soc. Br. Isl. 1967, 318.

PAYNE, John Henry (1857–1931)
b. Barnsley, Yorks July 1857 *d.* Newhill, Yorks 30 May 1931

Analytical chemist. Contrib. articles on fungi and mosses to *Naturalist*.

Naturalist 1931, 252–54 portr. *N. Western Nat.* 1931, 167–69.

PAYNE, Laurence Gilbert (1893–1949)
b. Salford, Lancs 1893 *d.* 10 March 1949

Bank clerk. President, London Natural History Society, 1946–48. Collected and cultivated ferns. Discovered *Dryopteris cristata* in Surrey.

London Nat. 1946, 16; 1948, 124–26 portr. *N. Western Nat.* 1953, 105–7 portr.

PAYNE, Theodore (1872–1963)
b. Church Brompton, Northants 19 June 1872 *d.* Los Angeles, California 6 May 1963

To Los Angeles, 1893. Purchased Hugh Evans's nursery, 1900s.

Lasca Leaves Autumn 1954, 74–77. *Bull. Soc. California Acad. Sci.* v.62, 1963, 221–24 portr.

Portr. at Hunt Library.

Lupinus paynei Davidson.

PEACH, Rev. Charles Pierrepont
(*c.* 1829–1886)
d. 17 Sept. 1886

Vicar, Appleton-le-Street, Yorks. Keen gardener. Contrib. to *J. Hort. Cottage Gdnr.*

J. Hort. Cottage Gdnr v.13, 1886, 272, 290–91; v.38, 1899, 181.

PEACH, Charles William (1800–1886)
b. Wansford, Northants 30 Sept. 1800 *d.* Edinburgh 28 Feb. 1886

ALS 1868. Coastguardsman and naturalist. Accounts of fossil plants in *Trans. Bot. Soc. Edinburgh* 1876–79.

S. Smiles *Robert Dick* 1878, 238–58 portr. *Proc. Linn. Soc.* 1883–86, 146–47. *J. Northampton Nat. Hist. Soc.* 1886, 33–40. *Geol. Mag.* 1886, 190–92. *Nature* v.33, 1886, 446–47. *Scott. Nat.* 1886, 289–92. *Trans. Edinburgh Geol. Soc.* 1887, 327–29. *Trans. Bot. Soc. Edinburgh* v.17, 1889, 11–13. *R.S.C.* v.4, 791; v.8, 576; v.10, 1009; v.12, 563.

Fossil plants at BM(NH).

PEACOCK, Edward Adrian Woodruffe- *see* Woodruffe-Peacock, *Rev.* E. A.

PEACOCK, John T. (*fl.* 1870s)

Of Sudbury House, Hammersmith, London. Had extensive collection of succulent plants.

Annual Rep. Kew Gdns 1878, 6. *Kew Bull.* 1897, 232.

PEAL, Samuel Edward (1834–1897)
b. 31 Dec. 1834 *d.* Moran, Sibsagur, Assam 29 July 1897

Tea-planter, India, 1862. Studied grasses and trees of Assam.

Nature v.56, 1897, 421.

PEARCE, Horace (1838–1900)
b. Hadley Lodge, Shropshire 21 Nov. 1838 *d.* Stourbridge, Worcs 19 Feb. 1900

FLS 1876. President, Worcestershire Field Club.

Proc. Linn. Soc. 1899–1900, 81.

Plants at BM(NH).

PEARCE, Nathaniel (1779–1820)
b. Acton, Middx 14 Feb. 1779 *d.* Cairo, Egypt June 1820

To Abyssinia with H. Salt, 1805; stayed there until 1818. Sent plants to R. Brown. *Life and Adventures* 1831.

Gent. Mag. 1819 ii 40; 1820 ii 568. *D.N.B.* v.44, 149–50.

Plants at BM(NH).

PEARCE, Richard (-1868)
b. Devonport *d.* Panama 17 July 1868

Gardener with Veitch nursery, Exeter, *c.* 1858. Collected plants in Chile and Ecuador, 1859 and elsewhere in S. America. In 1867 went to Panama to collect plants for W. Bull.

Gdnrs Chron. 1868, 874, 893. *J. Bot.* 1868, 320. *J. Hort.* v.15, 1868, 134. J. H. Veitch *Hortus Veitchii* 1906, 45–48. *Kew Bull.* 1920, 62. A. M. Coats *Quest for Plants* 1969, 374.

Plants at Kew, BM(NH).
Pearcea Regel.

PEARCE, Sydney Albert (1906–1972)
b. near Romsey, Hants 16 June 1906 *d.* Rowde, Wilts 28 March 1972

Kew gardener, 1928; Assistant Curator, 1937; Deputy Curator, 1967–68. President, Kew Guild, 1964–65. *Flowering Shrubs* 1953. *Climbing and Trailing Plants* 1957. *Ornamental Trees for Garden and Roadside Planting* 1961. *Flowering Trees and Shrubs* 1965.

J. Kew Guild 1964, 386–87 portr.; 1972, 150–51. *Gdnrs Chron.* v.164(18), 1968, 28 portr. *Arboricultural Assoc. J.* 1972, 48–51 portr.

PEARLESS, Anne *see* Pratt, A.

PEARSALL, William Harold (1891–1964)
b. Stourbridge, Worcs 23 July 1891 *d.* Morecambe, Lancs 14 Oct 1964

Educ. Manchester University. FRS 1940. FLS 1920. Son of W. H. Pearsall (1860–1936). Lecturer in Botany, Leeds University, 1919; Reader, 1922–38. Professor of Botany, Sheffield, 1938–44; University College, London, 1944–57. Plant ecologist. Edited *J. Ecol.* and *Ann. Bot. Mountains and Moorlands* 1949. *Report on Ecological Survey of Serengeti National Park* 1957.

Naturalist 1937, 1–2 portr.; 1965, 102. *Times* 15 Oct. 1964 portr. *Nature* v.205, 1965, 21. *Ann. Bot.* 1965 front. portr. *J. Indian Bot. Soc.* 1965, 145–46. *Tropical Ecol.* 1965, 1–2 portr. *Proc. Linn. Soc.* 1966, 121–22. *Who was Who, 1961–1970* 879. *Biogr. Mem. Fellows R. Soc.* v.17, 1971, 511–40 portr.

PEARSALL, William Harrison (1860–1936)
b. 6 June 1860 *d.* Matfield, Kent 12 Aug. 1936

Schoolteacher, Dalton-in-Furness. Secretary, Botanical Society and Exchange Club of British Isles, 1931–36. Botanised in Lake District. Authority on pondweeds. Contrib. to *Lancashire Nat.* Edited C. E. Salmon's *Fl. Surrey* 1931.

Bot. Soc. Exch. Club Br. Isl. Rep. 1936, 211–12. *J. Bot.* 1936, 352–53. *N. Western Nat.* 1936, 375–76. W. K. Martin and G. T. Fraser *Fl. Devon* 1939, 779.

Plants at Oxford, South London Botanical Institute.

PEARSON, Mr. (*fl.* 1690s)

Nurseryman, Hoxton, London.

Archaeologia v.12, 1794, 191.

PEARSON, Alfred Hetley (-1930)

VMH 1911. Nurseryman, Chilwell and later at Lowdham, Notts. Acquired business on death of his father, J. R. Pearson, in 1876.

J. Hort. Home Farmer v.61, 1910, 370–71 portr. *Gdnrs Chron.* 1921 ii 14 portr.; 1930 ii 203–4.

PEARSON, Arthur Anselm (1874–1954)
b. London 12 April 1874 *d.* Hindhead, Surrey 13 March 1954

FLS 1917. Chairman, British Belting and Asbestos Ltd. President, Yorkshire Naturalists Union, 1946. President, British Mycological Society, 1931, 1952. Authority on Agarics. Contrib. to *Trans. Br. Mycol. Soc.*, *Yorkshire Nat.*

Proc. Linn. Soc. v.165, 1955, 218–19. *Nature* v.173, 1954, 709. *Naturalist* 1954, 163–65 portr.; 1961, 66. *Times* 16 March 1954. *Trans. Br. Mycol. Soc.* v.37, 1954, 321–23 portr. *Bull. Soc. Mycol. France* 1957, 13–17 portr.

Letters, MSS. and plants at Kew.

PEARSON, Charles Edward (1856–1929)
d. Lowdham, Notts 14 Nov. 1929

FLS 1904. VMH 1924. Nurseryman, Lowdham. With his brother, A. H. Pearson, founded *Horticultural Advertiser*. Secretary, Horticultural Trades Association.

Gdnrs Chron. 1925 i 246 portr. *Br. Fern Gaz.* 1929, 25. *Proc. Linn. Soc.* 1929–30, 214–15.

PEARSON, Henry Harold Welch (1870–1916)
b. Long Sutton, Lincs 28 Jan. 1870 *d.* Wynberg, S. Africa 3 Nov. 1916

BA Cantab 1896. ScD 1907. FRS 1916. FLS 1901. At Kew Herb., 1899. Professor of Botany, South African College, Cape Town, 1903. Director of Cape Botanical Garden, 1913. Founder of Kirstenbosch Botanical Garden. Studied *Welwitschia* and *Gnetum*. Collected plants in S.W. Africa. Contrib. to *Philos. Trans. R. Soc.*, *Ann. Bot.*

Bot. Soc. Exch. Club Br. Isl. Rep. 1916, 467–68. *Gdnrs Chron.* 1916 ii 238, 250 portr.; v.160(18), 1966, 20 portr. *Kew Bull.* 1916, 277–81; 1917, 85. *Nature* v.98, 1916, 211. *Ann. Bolus Herb.* v.2, 1917, 131–47 portr. *Ann. Bot.* 1917, ii–xviii. *J. Kew Guild* 1917, 377–78 portr. *J. Bot.* 1917, 62. *Proc. Linn.*

Soc. 1916–17, 57–60. *Proc. R. Soc. B.* 1917, lx–lxvii. *Trans. R. Soc. S. Africa* 1918, 139–45. *Bot. Gaz.* v.63, 1917, 150–51 portr. *Curtis's Bot. Mag. Dedications, 1827–1927* 347–48 portr. *Who was Who, 1916–1928* 823.

Plants at Cape Town, Kew. Portr. at Hunt Library.

Pearsonia Dümmer.

PEARSON, J. Duncan (*fl.* 1880s)
Nurseryman, Lowdham, Notts. Hybridised daffodils.

Garden v.66, 1904, 415 portr. *Gdnrs Mag.* 1912, 679–80 portr. *J. Hort. Home Farmer* v.64, 1912, 348–49 portr.

PEARSON, John *see* Addendum

PEARSON, John Royston (1819–1876)
b. Chilwell, Notts 29 Jan. 1819 *d.* Chilwell 14 Aug. 1876

Nurseryman, Chilwell. Raised new varieties of pelargoniums and grapes. Specialised in tulips and carnations. *Vine Culture under Glass*, 1866. *Hints on Construction and Management of Orchard Houses* 1861. Contrib. to *J. Hort. Cottage Gdnr.*

Garden v.10, 1876, 228; v.29, 1886, xii portr. *Gdnrs Chron.* 1876 ii 278–79, 290–91. *J. Hort. Cottage Gdnr* v.56, 1876, 168–70; v.34, 1897, 577; v.38, 1899, 181, 182 portr.; v.59, 1909, 230–31. *Gdnrs Yb. and Almanack* 1877, 176.

PEARSON, Sir Ralph Sneyd (1874–1958)
b. Dec. 1874 *d.* Thame, Oxfordshire 8 Dec. 1958

FLS 1907. Knighted 1933. Indian Forest Service, 1898. Forest Research Institute, Dehra Dun, 1909–25. Director, Forest Products Research Laboratory, Princes Risborough, 1925–33. *Note on Uses of Rosha Grass* 1916. *Commercial Timbers of India* (with H. R. Brown) 1932 2 vols.

Indian Forester 1959, 629 portr. *Nature* v.183, 1959, 150. *Proc. Linn. Soc.* 1958–59, 135–36. *Who was Who, 1951–1960* 859.

PEARSON, Robert Hooper (1866–1918)
b. Brewood, Staffs 18 July 1866 *d.* 11 June 1918

Kew gardener, 1889–90. On staff of *Gdnrs Chron.* 1892; Editor, 1907. President, Kew Guild, 1911. Edited *Present Day Gdning* series. *Book of Garden Pests* 1908.

J. Kew Guild 1910, 449 portr.; 1919, 459–62. *Garden* 1918, 246 portr. *Gdnrs Chron.* 1918 i 246–47 portr. *Kew Bull.* 1918, 213. *Nature* v.101, 1918, 310. *Who was Who, 1916–1928* 824.

PEARSON, William Henry (1849–1923)
b. Pendleton, Lancs 22 July 1849 *d.* Manchester 19 April 1923

ALS 1907. Yarn agent. Hepaticologist. *Hepaticae of British Isles* 1899–1902 2 vols.

Mem. Torrey Bot. Club v.4, 1893, 62–64. *Lancashire Nat.* v.2, 1909, 182; v.15, 1923, 197–98 portr. *J. Bot.* 1923, 184, 194–97. *Proc. Linn. Soc.* 1922–23, 45. *Rep. Br. Bryol. Soc.* 1923, 39. *Bryologist* 1924, 12–14 portr.; 96–101.

Herb. at BM(NH).

Jungermannia pearsoni Spruce.

PEATFIELD, William (–*c.* 1954)
Physicist. Contrib. to A. H. Wolley-Dod's *Fl. Sussex* 1937. Contrib. botanical section to *Hastings, a Survey of Times Past and Present.*

Hastings and East Sussex Nat. v.8, 1955, 107.

PECHEY, John (1654–1717)
b. Chichester, Sussex Dec. 1654 *d.* Chichester June 1717

BA Oxon 1675. FRCP 1684. Practised medicine in London. *Compleat Herbal of Physical Plants* 1694; ed. 2 1707.

R. Pulteney *Hist. Biogr. Sketches of Progress of Bot. in England* 1790 i 184–85. W. Munk *Roll of R. College of Physicians* v.1, 1878, 433. *D.N.B.* v.44, 184. *Notes and Queries* v.11, 1891, 366. C. J. S. Thompson *Quacks of Old London* 1928, 132.

Pecheya Scop.

PECK, A. E. (*fl.* 1910s)
Yorkshire mycologist.

Naturalist 1961, 63.

PEDDIE, John (–1840)
d. Ceylon Aug. 1840

Lieut.-Colonel 72nd Regiment, 1832. Collected plants in Natal, 1839–40.

J. Bot. (Hooker) 1840, 124, 265. *D.N.B.* v.44, 204–5.

Peddiea Harvey.

PEED, John (*c.* 1830–1902)
d. Streatham Park, Wandsworth, London 24 Dec. 1902

Nurseryman at Roupell Park and Streatham Park, Wandsworth.

Gdnrs Mag. 1903, 14. *J. Hort. Cottage Gdnr* v.46, 1903, 7.

PEED, Thomas (1857–1926)
d. 6 May 1926

Son of J. Peed (*c.* 1830–1902) whose nursery he acquired. Also had nurseries at West Norwood and Morden, Surrey.

Gdnrs Chron. 1925 ii 122 portr.; 1926 i 362 portr.

PEERS, John (1838–1902)
b. Warrington, Lancs 4 Nov. 1838 *d.* Warrington 11 Jan. 1902
Founder and Secretary, Warrington Field Naturalists' Society. Studied *Carex*. 'List of Less Common Plants found about Warrington' (*Phytologist* v.6, 1863, 444–58).
Trans. Liverpool Bot. Soc. 1909, 86.

PEETE, William (1771–1848)
b. 27 June 1771 *d.* Bromley, Kent 4 Feb. 1848
FLS 1794. Surgeon, Dartford, Kent, 1795–1833. Described *Silene patens* in J. Sowerby and J. E. Smith *English Bot.* 2748. Contrib. to *Phytologist* v.1, 1844, 587.
Proc. Linn. Soc. 1848, 377.
Herb. in possession of S. H. Bickham (*J. Bot.* 1916, 139–40).

PEGLER, Louis Wellesley Hemington (1852–1927)
b. Colchester, Essex 18 Nov. 1852 *d.* Exeter, Devon 26 Feb. 1927
Surgeon. Presented his collection of fossil trees and ferns to BM(NH).
Bot. Soc. Exch. Club Br. Isl. Rep. 1927, 380. *Who was Who, 1916–1928* 825.

PEGUS, Mariette Antoinetta, Marchioness of Huntley (1821–1893)
Of Orton, Northants. Made herb. of central European plants, 1876–80.
G. C. Druce *Fl. Northamptonshire* 1930, cxx.

PEIRSON, Daniel (1819–1899)
Schoolmaster, Hitchin, Reading and Hertford. Contrib. plant records to A. R. Pryor *Fl. Hertfordshire* 1887.
J. Bot. 1915, 343.

PEIRSON, Henry (1852–1915)
b. Hertford 23 Oct. 1852 *d.* Hertford 4 June 1915
Son of D. Peirson (1819–1899). Hybrid *Gymnadenia* (*J. Bot.* 1899, 360; 1907, 278). *Azolla* (*J. Bot.* 1915, 308–9).
J. Bot. 1915, 343.

PELLY, Sir Lewis (1825–1892)
b. Stroud, Glos 14 Nov. 1825 *d.* Falmouth, Cornwall 22 April 1892
Political Resident on Persian Gulf, 1862–67. Collected plants in Central Arabia, 1865, now at Kew.
D.N.B. v.44, 275–77. *Rec. Bot. Survey India* v.8, 1933, 476.

PEMBERTON, Elian Emily *see* Collins, E. E.

PEMBERTON, Rev. Joseph Hardwick (–1926)
d. Romford, Essex 23 July 1926
Curate, Romford, 1880–1903. Diocesan inspector, St. Albans, 1891–1914. Cultivated roses. President, National Rose Society, 1911. *Roses, their History, Development and Cultivation* 1908.
Gdnrs Mag. 1907, 657–58 portr. *J. Hort. Cottage Gdnr* v.47, 1903, 398, 399 portr. *Garden* 1926, 454. *Gdnrs Chron.* 1910 ii 457 portr.; 1926 ii 99 portr.

PENA, Pierre (*fl.* 1530–1600s)
b. Jouques, Aix, France *c.* 1530
At Montpellier under Rondelet with M. de Lobel, 1565. In England with Lobel, 1566–72; afterwards practised in France. *Stirpium Adversaria Nova* (with M. de Lobel) 1570.
J. Bot. 1899, 88–92. L. Legré *Pierre Pena et Mathias de Lobel* 1899. *Notes and Queries* v.10, 1908, 365, 435. A. C. Wootton *Chronicles of Pharmacy* v.1, 1910, 256. *Bot. Soc. Iber. Cienc. Nat.* 1928, 128. *Berichte der Schweizerischen Botanischen Gesellschaft* v.50A, 1940, 362–64.

PENDAR, William (*fl.* 1760s–1770s)
Nurseryman, Woolhampton, Berks.

PENFOLD, Mrs. Jane Wallas (*fl.* 1820s–1840s)
Lived in Madeira. *Madeira Flowers, Fruits and Ferns* 1845 (plates by her).
J. Bot. 1919, 97–99.

PENFORD, Charles (1834–1908)
d. Havant, Hants 2 July 1908
Gardener, Leigh Park, Havant. Raised many chrysanthemum varieties.
Gdnrs Chron. 1908 ii 32.

PENNANT, Thomas (1726–1798)
b. Downing, Whiteford, Flintshire 14 June 1726 *d.* Downing 16 Dec. 1798
DCL Oxon 1771. FRS 1767. Zoologist and antiquary. With J. Lightfoot in Scotland, 1772. Correspondent of Linnaeus. *History of Whiteford and Holywell* 1796 (plant list, 152–55). *Literary Life* 1793 portr.
P. Smith *Mem. and Correspondence of Sir J. E. Smith* v.1, 1832, 300, 448–49. *D.N.B.* v.44, 320–23. *J. Arnold Arb.* 1948, 186–92. *Ann. Sci.* 1954, 258–71 portr. *Dict. Welsh Biogr. down to 1940* 1959, 745. C. Urness *Nat. in Russia: Letters from P. S. Pallas to Thomas Pennant* 1967. *Library* 1970, 136–49.
Letters at Chester Public Library, Fitzwilliam Museum Cambridge, Queen's College Oxford. Portr. at Hunt Library.
Pennantia Forster.

PENNECK, Henry (*c.* 1762–1834)
b. Paul, Cornwall *c.* 1762 *d.* Penzance, Cornwall 31 March 1834
MD. ALS 1799. Contrib. to J. Sowerby and J. E. Smith's *English Bot.* 850, 1316.
J. P. Jones *Bot. Tour through...Devon and Cornwall* 1820, 29.

PENNECK, Rev. Henry (1801–1862)
b. Penzance, Cornwall 7 Aug. 1801 *d.* Penzance 24 April 1862

BA Cantab 1826; Oxon 1847. Son of H. Penneck (*c.* 1762–1834). Curate, Morvah, 1826. Friend of J. Ralfs. Had herb. Contrib, to J. Sowerby and J. E. Smith's *English Bot.* 2818, 2845, etc.

Gent. Mag. 1862 ii 106–7. C. C. Babington *Memorials, J. and Bot. Correspondence* 1897, 268. F. H. Davey *Fl. Cornwall* 1909, xlii.

PENNELL, Charles (*c.* 1821–1891)
d. 21 June 1891

Nurseryman, Lincoln.

J. Hort. Cottage Gdnr v.23, 1891, 5.

PENNELL, Richard (*fl.* 1820s–1869)

Nurseryman, Gowt's Bridge, Lincoln.

PENNEY, F. Gordon (*fl.* 1870s–1910s)

Malayan Civil Service, 1876–1907. Sent plants collected in Pahang to Botanical Gardens, Singapore.

Gdns Bull. Straits Settlements 1927, 130.

PENNEY, William (–1903)

ALS 1856. Of Poole, Dorset. 'Similarity in the Medical Properties of Two Species of Cotyledon' (*Pharm. J.* 1852, 66–67).

Proc. Linn. Soc. 1903–4, 35.

PENNINGTON, Fred (*c.* 1801–1850s)

Nurseryman, Market Place, East Retford, Notts.

PENNY, Rev. Charles William (1837–1898)
b. West Ilsley, Berks 1837 *d.* Wokingham, Berks 30 March 1898

MA Oxon 1863. FLS 1872. Bursar and assistant master, Wellington College, 1861–91. Contrib. local plant records to *Rep. Wellington College Nat. Hist. Soc.*, 1869–74. Contrib. to *J. Bot.*

G. C. Druce *Fl. Berkshire* 1897, clxxx–clxxxi. *J. Bot.* 1898, 208. *Proc. Linn. Soc.* 1897–98, 46–47.

Plants at Wellington College.

PENNY, George (–1838)
d. Milford, Surrey 22 Dec. 1838

ALS 1829. Nurseryman, Milford. *Hortus Epsomensis.* Contrib. to *Gdnrs Mag.* 1829–32 as "Alpha".

Gdnrs Mag. 1834, 170; 1839, 96.

PENNY, Rev. Thomas (*c.* 1530–1589)
b. Gressingham, Lancs *c.* 1530 *d.* London? 1589

BA Cantab 1551, MA 1557. MD. DD. Prebendary of St. Paul's, 1560. Entomologist. "A second Dioscorides" (J. Gerard). Sent drawings and Balearic plants to Clusius, 1580–81. Discovered *Cornus suecica.* Correspondent of C. Gesner, M. de Lobel, J. Gerard.

Clusius *Rariorum Plantarum Historia* v.1, 1601, 68. M. de Lobel *Stirpium Illustrationes* 1655, 162; *Adversaria* 1570, 358, 394, 397. W. Munk *Roll of R. College of Physicians* v.1, 1878, 82–83. *D.N.B.* v.44, 337. R. T. Gunther *Early Br. Botanists* 1922, 234–35. *Annual Rep. Lancs Cheshire Entomol. Soc.* 1928–30, 31–52. C. E. Raven *English Nat.* 1947, 153–71.

PENNY, Thomas (*fl.* 1790s)

Nurseryman, Castle Cary, Somerset.

PENSTON, Miss Norah Lillian (1903–1974)
b. 20 Aug. 1903 *d.* 1 Feb. 1974

BA Oxon 1927. PhD 1930. FLS 1933. Did research in plant physiology under W. O. James. Demonstrator in Botany, King's College, London, 1929–33; Assistant Lecturer, 1933–36; Lecturer, 1936–45; Head of Dept., 1940–44. Vice-Principal, Wye College, 1945–51. Principal, Bedford College, 1951–64.

Times 5 Feb. 1974.

PENTLAND, Joseph Barclay (1797–1873)
b. Ireland 1797 *d.* London 12 July 1873

Collected plants in Bolivia and Peru, 1826–28. Consul-General in Bolivia, 1836–39.

Bot. Register 1839, t.68. *D.N.B.* v.44, 350–51. *Taxon* 1970, 535–36. *R.S.C.* v.4, 820.

Letters at Kew.

Pentlandia Herbert.

PENZANCE, Lord *see* Wilde, J. P.

PEPLOW, Mrs. (*fl.* 1710s?)

Barbados plants in Sherard Herb. at Oxford.

PEPPARD, Luke (*fl.* 1760s–1790s)

Nurseryman, Dublin. Connected with Anderson of Edinburgh.

Irish For. 1967, 48.

PERCEVAL, Cecil Henry Spencer (1849–1920)
b. 8 May 1849 *d.* Morpeth, Northumberland May 1920

Of Longwitton Hall, Morpeth. Studied fungi and British orchids. Discovered *Orchicoeloglossum mixtum* at Longwitton, 1891. '*Battarea phalloides*' (*Rep. Br. Assoc. Advancement Sci.* 1875 ii 158–59). 'Notes on Uncommon Fungi' (*Trans. Br. Mycol. Soc.* v.2, 1904, 91–92). Contrib. to *J. Bot.*

Orchid Rev. 1919, 144; 1920, 164. *J. Bot.* 1921, 76–77.

PERCIVAL, James (1828–1902)
b. Prestwick, Ayr 1828 *d.* Rochdale, Lancs 17 Aug. 1902

Spindle and fly maker; later a hide dealer. President, Manchester Botanical Association. Bryologist. Contrib. to L. H. Grindon's *Manchester Fl.* 1859.

L. H. Grindon *Country Rambles* 1882, 199, 202–3. *Gdnrs Chron.* 1902 ii 182. *J. Hort. Cottage Gdnr* v.45, 1902, 226. *Manchester City News* 23 Aug. 1902. *Trans. Liverpool Bot. Soc.* 1909, 86. *Lancashire Nat.* v.11, 1918, 53.

PERCIVAL, John (1863–1949)
b. Wensleydale, Yorks 3 April 1863 *d.* Mortimer, Berks 26 Jan. 1949

BA Cantab 1888. FLS 1893. Professor of Botany, South Eastern Agricultural College, Wye, 1894–1902. Professor of Agriculture, Reading, 1907; Professor of Agricultural Botany, 1912–32. *Fl. Wensleydale* 1888. *Agricultural Botany* 1900; ed. 8 1936. *Common Weeds of Farm and Garden* (with H. C. Long) 1910. *Wheat Plant* 1921. *Wheat in Great Britain* 1934; ed. 2 1948.

Gdnrs Chron. 1949 i 58. *Nature* v.163, 1949, 275. *Proc. Linn. Soc.* 1948–49, 248–51. *Times* 29 Jan. 1949. *Watsonia* 1949–50, 264. *Who was Who, 1941–1950* 904.

Herb. at Reading University.

PERCIVAL, R. P. (*c.* 1835–1885)
d. Southport, Lancs Dec. 1885

Cultivated orchids.

Garden v.28, 1885, 641.

PERCIVAL, Thomas (1740–1804)
b. Warrington, Lancs 29 Sept. 1740 *d.* Manchester, Lancs 30 Aug. 1804

MA Leyden 1765. FRS 1765. Practised medicine in Manchester from 1767. President, Manchester Literary and Philosophical Society. 'Speculations on Perceptive Power of Vegetables' (*Mem. Manchester Philos. Soc.* v.2, 1789, 122–46). *Works: Literary, Moral and Medical* 1807 4 vols. (memoir by E. Percival).

D.N.B. v.44, 383–84. *Trans. Liverpool Bot. Soc.* 1909, 86.

Silhouette in J. Kendrick *Profiles of Warrington Worthies* 1853.

PERCY, Algernon, 4th Duke of Northumberland (1792–1865)
b. Syon House, Isleworth, Middx 15 Dec. 1792 *d.* Alnwick, Northumberland 12 Feb. 1865

Developed gardens at Syon House.

D.N.B. v.44, 390–91. *Gdnrs Chron.* 1865, 148.

PERCY, Hugh (*né* Smithson), 1st Duke of Northumberland (1715–1786)
b. Newby Wiske, Yorks 1715 *d.* Syon House, Isleworth 6 June 1786

FRS 1736. Developed gardens at Syon House. "Not only a great encourager of botanical studies, but greatly skilled in the science himself" (P. Miller). Miller dedicated 8th ed. of *Gdnrs Dict.* 1768 to him.

D.N.B. v.44, 418–20.

Portr. by Reynolds, Hamilton and Sharples.

Piercea Miller.

PERCY, Hugh, 2nd Duke of Northumberland (1742–1817)
b. Aug. 1742 *d.* 10 July 1817

Introduced new plants for garden at Syon House.

E. Bretschneider *Hist. European Bot. Discoveries in China* 1898, 150. *D.N.B.* v.44, 420–22.

PERCY, Hugh, 3rd Duke of Northumberland (1785–1847)
b. 20 April 1785 *d.* Alnwick Castle, Northumberland 11 Feb. 1847

MA Cantab 1805. FLS 1833. Expended large sums on his garden at Syon House, Isleworth. *Clivia* Lindl. named after his wife.

E. Bretschneider *Hist. European Bot. Discoveries in China* 1898, 287. *D.N.B.* v.44, 422–23.

PERCY, John (1817–1889)
b. Nottingham 23 March 1817 *d.* London 19 June 1889

MD Edinburgh 1838. FRS 1847. Pupil of A. de Jussieu. Metallurgist. Collected plants in S.E. France and Switzerland, 1836–37.

First Rep. Bot. Soc. Edinburgh 1836, 43–45. *Nature* v.40, 1889, 206. *Proc. R. Soc.* 1889, xxxv–xl. *Trans. Bot. Soc. Edinburgh* v.17, 1889, 521–22. *Quart. J. Geol. Soc. London* 1890, 45–48. *D.N.B.* v.44, 425–27.

Plants at Edinburgh.

PEREIRA, Jonathan (1804–1853)
b. Shoreditch, London 22 May 1804 *d.* London 20 Jan. 1853

MD Erlangen 1840. FRS 1838. FLS 1828. Apothecary at Aldersgate dispensary where he became Professor of Materia Medica of New Medical School which took its place. Professor of Materia Medica, Pharmaceutical Society, 1843. *Elements of Materia Medica and Therapeutics* 1839–40.

Pharm. J. 1852–53, 409–16 portr.; 1927, 554–55. *Proc. Linn. Soc.* 1853, 237–39; 1888–89, 39. *D.N.B.* v.45, 1–2. R. R. Trail *Lives of Fellows of R. College of Physicians of London* v.5, 1965, 42–43.

Letters at Kew. Library sold at Sotheby, 7–8 June 1853.

PERFECT, Bertram F. (–1953)
d. Smallfields Hospital, Surrey 3 April 1953
Gardener to Sir J. Colman, Gatton Park, Reigate, Surrey. In charge of orchid collection, Duckylls, East Grinstead, Surrey, 1942–51.
Orchid Rev. 1953, 106.

PERFECT, Grosvenor (1768–1822)
Nurseryman, Pontefract, Yorks.

PERFECT, John (1717–1762)
Nurseryman, Pontefract, Yorks.

PERFECT, John (1749–1800)
Nurseryman, Pontefract, Yorks.

PERFECT, William (1718–1785)
Nurseryman, Pontefract, Yorks.

PERIN, Dr. (*fl.* 1640s)
Book of plants collected at Padua, Italy in Sloane Herb. BM(NH).
J. E. Dandy *Sloane Herb.* 1958, 175.

PERKINS, Arthur (*c.* 1833–1905)
d. 18 June 1905
Nurseryman, Coventry, Warwickshire.
J. Hort. Cottage Gdnr v.50, 1905, 562.

PERKINS, Mrs. E. E. (*fl.* 1830s)
Of Chelsea. "Professor of Botanical Flower Painting." Lectured on botany. *Elements of Botany* 1837. Illustrated J. H. Fennell's *Drawing Room Botany* 1840.

PERKINS, Edward (*fl.* 1830s–1860s)
Nurseryman, Holloway Down Nursery, Langthorne Road, Leyton and Wanstead, Essex.

PERKINS, John (*fl.* 1820s–1840s)
Nurseryman, Bath Street, Northampton.

PERKINS, Thomas (*c.* 1822–1880)
d. 3 April 1880
Nurseryman, Northampton; in partnership with John Marsh, 1857–67. Business carried on by his son John.
Gdnrs Chron. 1880 i 471.

PERKINS, William (*fl.* 1760s–*c.* 1825)
Acquired Holloway Down Nursery, Langthorne Road, Leyton on the death of its owner, Spencer Turner.
Trans. London Middlesex Archaeol. Soc. v.24, 1973, 190.

PERMAN, Edgar Philip (1866–1947)
d. Llanafan Fawr, Brecon 27 May 1947
DSc London. Lecturer in Chemistry, University College, Cardiff, 1892; Assistant Professor, 1904. President, Cardiff Naturalists Society, 1913. Collected plants for National Museum of Wales.
Trans. Cardiff Nat. Soc. 1950, 6–7.

PERRIN, George Samuel (1849–1900)
d. Ballarat, Victoria 24 Dec. 1900
FLS 1885. To Victoria with his parents, 1853. Forester, S. Australia, 1880, Tasmania, Victoria. Had knowledge of Australian plants. Friend of F. von Mueller.
Proc. Linn. Soc. 1900–1, 47.

PERRY, Amos (1841–1913)
b. 22 Sept. 1841 *d.* Enfield, Middx 10 June 1913
Founded nursery at Winchmore Hill, London, 1889; later moved to Enfield.
Garden 1913, 320(x) portr. *Gdnrs Chron.* 1913 i 410 portr.; v.168, (14) 1970, 18. *Gdnrs Mag.* 1913, 446. *J. Hort. Home Farmer* v.66, 1913, 572 portr.

PERRY, Amos (1871–1953)
d. Kirby-le-Soken, Essex 21 Aug. 1953
VMH 1935. VMM 1950. Nurseryman, Enfield, Middx. Hybridised irises. *Water, Bog and Moisture Loving Plants c.* 1930; ed. 3 1936.
Garden 1910, 224–25 portr. *Gdnrs Chron.* 1923 i 44 portr.; 1935 i 104 portr.; 1953 ii 86. *Br. Fern Gaz.* 1953, 59. *Iris Yb.* 1953, 16–18.

PERRY, Charles James (*c.* 1822–1873)
d. Castle Bromwich, Birmingham 11 April 1873
Florist who specialised in dahlias, verbenas and roses. Contrib. to *Gdnrs Chron.*
Gdnrs Chron. 1873, 547–48. *J. Hort. Cottage Gdnr* v.49, 1873, 337–38.

PERRY, Frederick (1860–1935)
b. Lympstone, Devon 1860 *d.* 23 Oct. 1935
Gardener to Viscount Ridley, Blagdon Hall, Northumberland, 1902–22.
Orchid Rev. 1935, 375.

PERRY, Peter James (*c.* 1836–1881)
d. 23 Jan. 1881
Nurseryman, Banbury, Oxford.
Gdnrs Chron. 1881 i 155.

PERRY, William Groves (1796–1863)
b. Warwick 1796 *d.* Warwick 25 March 1863
Bookseller. Curator, Warwickshire Museum, 1840. Secretary, Warwickshire Natural History and Archaeological Society. *Plantae Varvicenses Selectae* 1820.
Mag. Nat. Hist. v.2, 1829, 268–69. *Phytologist* 1844, 700–2; 1863, 605. *Trans. Bot. Soc. Edinburgh* v.8, 1863, 14. J. E. Bagnall *Fl. Warwickshire* 1891, 494–97. J. Amphlett and C. Rea *Bot. of Worcestershire* 1909, xxii. D. A. Cadbury *Computer-mapped Fl.... Warwickshire* 1971, 53–54.
Herb. at Warwick Museum.

PERRY, William Wykeham (*c.* 1846–1894)
d. 14 June 1894
Fleet-Paymaster, Royal Navy. Collected plants at Amsterdam Island, 1873; in Socotra, 1876; at Aden, 1878; at Chefoo, 1881; in Korea, 1883.
Kew Bull. 1894, 397–98. E. Bretschneider *Hist. European Bot. Discoveries in China* 1898, 756. *Rec. Bot. Survey India* v.7, 1914, 16.
Plants and letters at Kew.
Aloe perryi Baker.

PERRYCOSTE, Frank H. (–1938)
Herb. at Truro Museum.

PERTZ, Dorothea Frances Matilda (1859–1939)
b. London March 1859 *d.* Cambridge 6 March 1939
BA Cantab. FLS 1905. Plant physiologist. Contrib. to *Ann. Bot.*
Nature v.143, 1939, 590–91. *Proc. Linn. Soc.* 1938–39, 245–47.

PETCH, Thomas (1870–1948)
b. Hornsea, Yorks 1870 *d.* King's Lynn, Norfolk 24 Dec. 1948
BA, BSc London. Mycologist, Royal Botanic Gardens, Peradeniya, Ceylon, 1904–25, where he studied fungal diseases of the rubber tree. Director, Tea Research Institute of Ceylon, 1925–28. President, British Mycological Society, 1920. President, Yorkshire Naturalists' Union, 1931. Edited and contrib. to *Ann. R. Bot. Gdns, Peradeniya. Fungi of Certain Termite Nests* 1906. *Physiology and Diseases of Hevea Brasiliensis* 1911. *Diseases and Pests of Hevea Rubber Tree* 1921. *Diseases of the Tea Bush* 1923.
Trans. Hull Sci. Field Nat. Club 1904, 182–83 portr. *Naturalist* 1949, 50; 1961, 65. *Nature* v.163, 1949, 202–3.
Mycological collections at Kew.
Petchia Livera.

PETER, Rev. John (1833–1877)
Of Bala Merionethshire. "An enthusiastic botanist." 'William Salesbury fel Llysisnwr' (*Traethodydd* 1873, 156–81).
J. Bot. 1898, 13.

PETERKEN, Joseph Henry Garfield (1893–1973)
b. 24 Aug. 1893 *d.* Amersham, Bucks 13 Sept. 1973
Accountant, London. Knew flora of Epping Forest, Essex. Treasurer, British Bryological Society, 1946–65. 'Handlist of Bryophytes of London Area' (*London Nat.* no. 44, 1965, 43–71).
J. Bryol. 1974, 139–40. *London Nat.* 1974, 107. *Watsonia* 1975, 320.
Bryophytes at British Bryological Society.

PETERS, Charles Thomas (*fl.* 1880s)
Collected plants in Afghanistan, now at Kew.

PETHERICK, John (1813–1882)
Mining engineer. Entered service of Muhammad Ali Pasha in Egypt, 1845. Gum arabic trader, Kordofan, 1848–53. British vice-consul, Khartoum, 1858. Led expedition to rescue J. H. Speke and J. A. Grant, 1862. Left Sudan in 1865. Collected plants. *Egypt, the Soudan and Central Africa* 1861. *Travels in Central Africa* (with his wife) 1869.
R. Hill *Biogr. Dict. of Sudan* ed. 2 1967, 305–6.
Plants at Kew.

PETHYBRIDGE, George Herbert (1871–1948)
b. Bodmin, Cornwall 1 Oct. 1871 *d.* Bodmin 23 May 1948
BSc London 1892. PhD Göttingen. FLS 1925. Botanist, Department of Agriculture, Dublin, 1900; Economic Botanist, 1908–23. Investigated potato diseases. Mycologist, Plant Pathology Laboratory, Harpenden, 1923–36. President, British Mycological Society, 1926. 'Vegetation of District lying South of Dublin' (with R. L. Praeger) (*Proc. R. Irish Acad. B.* v.25, 1905, 124–80). 'Census Catalogue of Irish Fungi' (with J. Adams). (*Proc. R. Irish Acad. B.* 1910, 120–66). Edited *J. Pomology* 1936–48.
Nature v.161, 1948, 1002. *Proc. Linn. Soc.* 1947–48, 186–87. *J. Hort. Sci.* 1948, 69–71. *Times* 26 May 1948. *Ann. Applied Biol.* 1949, 414–17. R. L. Praeger *Some Irish Nat.* 1949, 143. *Trans. Br. Mycol. Soc.* 1950, 161–65 portr.

PETIVER, James (1663/4–1718)
b. Hillmorton, Warwickshire 1663/4 *d.* London 2 April 1718
FRS 1695. Apothecary to the Charterhouse. Demonstrator at Chelsea, 1709. Contrib. list of Middx plants to Gibson's *Camden*, to Ray's *Historia* v.2, v.3 and *Synopsis* ed. 2 and to *Philos. Trans. R. Soc.*, 1697–1717. *Museii Petiveriani* 1695–1703. *Opera* 1764 2 vols. 'Journal of Botanical Tour from London to Dover...1714' (*Phytologist* 1862, 114–20).
C. Plumier *Nova Plantarum Americanum Genera* 1703, 50. R. Pulteney *Hist. Biogr. Sketches of Progress of Bot. in England* v.2, 1790, 31–43. J. E. Smith *Selection from Correspondence of Linnaeus* v.2, 1821, 161–70. H. Trimen and W. T. T. Dyer *Fl. Middlesex*

1869, 379–86. E. Bretschneider *Hist. European Bot. Discoveries in China* 1898, 33. *Antiquary* v.35, 1899, 118–20. *J. Bot.* 1899, 227. *D.N.B.* v.40, 85–86. E. J. L. Scott *Index to Sloane Manuscripts* 1904, 417. C. E. Salmon *Fl. Surrey* 1931, 45–46. A. H. Wolley-Dod *Fl. Sussex* 1937, xxxvii–xxxviii. *Proc. Amer. Antiq. Soc.* 1952, 243–379. J. E. Dandy *Sloane Herb.* 1958, 175–82. H. N. Clokie *Account of Herb. Dept. Bot., Oxford* 1964, 223–24. W. S. C. Copeman *Worshipful Soc. of Apothecaries of London* 1968, 76–77. D. H. Kent *Hist. Fl. Middlesex* 1975, 14.
Herb. at BM(NH).
Petiveria Plumier.

PETO, Harold Ainsworth (1854–1933)
Landscape gardener. Created a notable garden at Iford Manor, near Bradford-on-Avon, Wilts.
M. Hadfield *Gdning in Britain* 1960, 366–67.

PETRE, Robert James, 8th Baron Petre (1713–1742)
b. Ingatestone, Essex 3 June 1713 *d.* 2 July 1742
FRS 1731. Had garden at Thorndon, Essex. Introduced *Camellia japonica*. "He spared no pains or expense to procure seeds and plants from all parts of the world and was as ambitious to preserve them" (P. Collinson).
Trans. Linn. Soc. v.10, 1811, 273–74. D. Turner *Extracts from Lit. and Sci. Correspondence of Richard Richardson* 1835, 315, 340–41, 389, 392. W. Darlington *Memorials of John Bartram and Humphry Marshall* 1849, 145, 157–58. E. J. L. Scott *Index to Sloane Manuscripts* 1904, 418. *Dublin Rev.* Oct. 1914, 307–21. R. H. Fox *Dr. John Fothergill and his Friends* 1919, 166. *Essex Nat.* 1938, 73–87; 1969, 201–6. *Gdn J. New York Bot. Gdn* 1957, 141–43, 152, 189–91; 1958, 6, 21. *Essex J.* 1970, 56–61. *Gdn Hist. Soc. Occas. Paper* no. 2, 1970, 27–37.
Herb. at Sutro Library, California.
Petrea L.

PETRIE, Donald (1846–1925)
b. Morayshire 7 Sept. 1846 *d.* 1 Sept. 1925
MA Aberdeen 1867. Inspector of Schools, Otago, N.Z., 1874. Chief Inspector, Auckland Board of Education, 1894. Contrib. papers on New Zealand flora to *Trans. New Zealand Inst.*
T. F. Cheeseman *Manual of New Zealand Fl.* 1906, xxxi–xxxii. *Trans. New Zealand Inst.* v.56, 1926, vii–ix portr. *Rec. Dominion Mus.* 1958, 89–99.
Plants at Dominion Museum, Wellington.

PETRIE, Robert (*fl.* 1790s)
Nurseryman, Kendal, Westmorland.

PETTIGREW, Andrew (*c.* 1830–1903)
d. Cardiff, Glam 26 April 1903
Gardener to Lord Bute at Cardiff Castle.
J. Hort. Cottage Gdnr v.23, 1891, 137 portr.; v.46, 1903, 389 portr., 390. *Garden* v.63, 1903, 305 portr. *Gdnrs Chron.* 1903 i 287 portr. *Gdnrs Mag.* 1903, 296 portr.

PETTIGREW, Hugh A. (1871–1947)
b. Ayr 24 Jan. 1871 *d.* France 25 Jan. 1947
Kew gardener, 1891–93. Gardener to Earl of Plymouth at St. Fagans Castle, Glam, 1901–35.
Gdnrs Mag. 1910, 739–40 portr. *J. Kew Guild* 1945, 591.

PETTIGREW, William Wallace (*c.* 1867–1947)
d. Charmandean, Sussex 9 Feb. 1947
VMH 1926. Son of A. Pettigrew (*c.* 1830–1903). Kew gardener, 1890. Parks Superintendent, Cardiff and Manchester. *Commonsense Gardening* 1925. *Municipal Parks* 1937.
Gdnrs Mag. 1907, 625–26 portr. *Gdnrs Chron.* 1947 i 86.

PETTY, Samuel Lister (–1919)
d. Ulverston, Lancs 10 May 1919
President of N. Lonsdale Field Club. Botanised in Lake District. Contrib. papers on N. Lancashire flowers to *Naturalist*.
Bot. Soc. Exch. Club Br. Isl. Rep. 1919, 625. *Lancashire Nat.* v.12, 1919, 68–69 (as Perry). *Naturalist* 1919, 248.

PHELAN, John (*fl.* 1740s–1780s)
Nurseryman, Dublin. Also designed gardens.
Irish For. 1967, 48.

PHELPS, Rev. William (1776–1856)
b. Flax Bourton, Somerset 1776 *d.* Oxcombe, Lincs 17 Aug. 1856
BA Oxon 1797. Rector, Oxcombe, 1851. Topographer. *Calendarium Botanicum* 1810.
D.N.B. v.45, 150.

PHILBRICK, Frederick Adolphus (*c.* 1836–1910)
Recorder of Colchester, 1874. Judge, Dorset County Court, 1895. Cultivated orchids.
Orchid Rev. 1911, 30.
Laelia philbrickiana × Reichenb. f.

PHILIP, Robert Harris (1852–1912)
d. Hull, Yorks 15 April 1912
Diatomist. Catalogued G. Norman's diatoms. *Diatomaceae of Hull District.*
Trans. Hull Sci. Field Nat. Club 1911, 219–22. *Naturalist* 1912, 150–51 portr., 327.

PHILIPS, John (1676–1708/9)
b. Bampton, Oxfordshire 30 Dec. 1676 *d.* 15 Feb. 1708/9
Educ. Oxford. Poet. His poem *Cyder* 1706 contains much practical horticultural advice. "There were many books written on the same subject in prose which do not contain so much truth as that poem" (P. Miller).
Cottage Gdnr v.5, 1851, 299–300. *D.N.B.* v.45, 175–77.

PHILLIP, Arthur (1738–1814)
b. London 11 Oct. 1738 *d.* Bath, Somerset 31 May 1814
First Governor of N.S.W. Plants collected in Australia and New Zealand, 1788–92, formerly in A. B. Lambert's herb., now at Cambridge.
D.N.B. v.45, 188–89. *Taxon* 1970, 536.

PHILLIPPS, Leonard (*fl.* 1810s–1830s)
Nurseryman, Fruit Tree Nursery, Vauxhall, London. *Catalogue of Fruit Trees for Sale* 1814.
J. R. Hort. Soc. 1954, 363–64.

PHILLIPS, Mrs. E. Lort- *see* Lort-Phillips, Mrs. E.

PHILLIPS, Henry (1779–1840)
b. Henfield, Sussex 1779 *d.* Brighton, Sussex 8 March 1840
FLS 1825. *Pomarium Britannicum* 1820; ed. 2 1821. *History of Cultivated Vegetables* 1822 2 vols. *Sylva Florifera* 1823 2 vols. *Flora Domestica* 1823. *Flora Historica* 1824. *Floral Emblems* 1825.
G. W. Johnson *Hist. English Gdning* 1829, 304–5. *D.N.B.* v.45, 201–2. *Gdnrs Chron.* 1950 ii 178. *Gdn Hist. Soc. Newsletter* no. 14, 1971, 2–4.

PHILLIPS, John (1800–1874)
b. Marden, Wilts 25 Dec. 1800 *d.* Oxford 24 April 1874
DCL 1866. FRS 1834. Curator, York Museum, 1824–40. Professor of Geology, Trinity College, Dublin, 1844–53. Keeper, Ashmolean Museum, 1854–70. *Geology of Yorkshire* 1829–36 (includes fossil plants).
Geol. Mag. 1870, 301–6. *Proc. Geol. Polytechnic Soc. West Riding Yorks* 1883, 3–20. *Quart. J. Geol. Soc. London* 1875, xxxvii–xliii. *D.N.B.* v.45, 207–8. *Naturalist* 1923, 100 portr. *Proc. Yorkshire Geol. Soc.* 1933, 153–87. *Advancement Sci.* v.12, 1955, 97–104. *R.S.C.* v.4, 888; v.8, 617.
Sale at Sotheby, 27 May 1875.
Caulopteris phillipsii Brongn.

PHILLIPS, John (1859–1917)
b. Ross Priory, Dunbartonshire 1859 *d.* Edinburgh 13 Sept. 1917
Nurseryman, Granton Road Nurseries, Edinburgh. President, Scottish Horticultural Association, 1917.
Garden 1917, 402(v). *Gdnrs Chron.* 1917 ii 123 portr.

PHILLIPS, Joseph (–*c.* 1747)
Nurseryman, Devizes, Wilts.

PHILLIPS, Mrs. Margaret Sturrock (1857–1937)
Herb. at Hitchin Museum.

PHILLIPS, Philip (1873–)
b. India 1873
Indian Forest Service, 1897–1901. Malayan Forest Service, 1901–26. Pioneer of forest conservation in Malaya. Collected plants in Pahang, now at Kuala Lumpur.
Gdns Bull. Straits Settlements 1927, 130. *Fl. Malesiana* v.1, 1950, 406.

PHILLIPS, Reginald William (1854–1926)
b. Talgarth, Brecon 15 Oct. 1854 *d.* Leominster, Hereford 2 Dec. 1926
BA Cantab 1884. DSc London 1898. FLS 1890. Head of Dept. Biology, University College, Bangor, 1884; Professor of Botany until 1922. Studied marine algae. 'Algae' in *Encycl. Britannica* eds. 10 and 11. Contrib. papers on red seaweeds to *Ann. Bot.* and *New Phytol.*
Bot. Soc. Exch. Club Br. Isl. Rep. 1926, 101. *J. Bot.* 1927, 80–83. *Nature* v.119, 1927, 92–93. *N. Western Nat.* 1927, 185–87 portr. *Proc. Linn. Soc.* 1927–28, 128–29. *Who was Who, 1916–1928* 836.

PHILLIPS, Robert Albert (1866–1945)
b. Courtmacsherry, County Cork 29 April 1866 *d.* Cork 20 Nov. 1945
Discovered *Ranunculus lutarius* in Ireland, 1894. Contrib. to *Irish Nat.*, *Proc. R. Irish Acad.*
Irish Nat. J. 1946, 391–94 portr. R. L. Praeger *Some Irish Nat.* 1949, 143–44 portr.
Herb. at Dublin Museum.

PHILLIPS, Sarah (*fl.* 1740s)
Nurseryman, The George, Devizes, Wilts.

PHILLIPS, William (1803–1871)
b. Norwich, Norfolk April 1803 *d.* Sydney, N.S.W. June 1871
To Sydney, 1842. Schoolmaster. Friend of F. Leichhardt. Collected plants in Sydney and Blue Mountains.
Proc. R. Soc. N.S.W. 1921, 165.
Hypocalymma phillipsii Harvey.

PHILLIPS, William (1822–1905)
b. Presteigne, Radnorshire 4 May 1822 *d.* Shrewsbury, Shropshire 22 Oct. 1905
FLS 1875. Mycologist and antiquary. *Guide to Botany of Shrewsbury* 1878. *Manual of British Discomycetes* 1887. Contrib. botany to *Victoria County History of Shropshire.*
D.N.B. Supplt 2 v.3, 115–16. *Gdnrs Chron.* 1905 ii 331–32 portr. *J. Bot.* 1905, 361–62 portr.; 1906, 184. *Proc. Linn. Soc.* 1905–6, 44–45. *Trans. Caradoc and Severn Valley Field Club* 1906, 121–23. *Trans. Soc. Cymmrodorion* 1932–33, 172–74. *Trans. Woolhope Nat. Field Club* 1955, 232–67. *R.S.C.* v.8, 618; v.11, 10; v.12, 575.
Herb., drawings and letters at BM(NH).
Phillipsia Berkeley.

PHILLIPS, William Edwards (*fl.*1800s–1850s)
To Penang, 1800, in service of East India Company; Governor, 1819–26. Founded Ayer Hitam Garden, Penang, 1823. Gave plants to N. Wallich, also to Horticultural Society of London, afterwards sold to Planchon.
C. Curtis *Cat. of Flowering Plants and Ferns...in...Penang* 1894, 99–100. *J. Malayan Branch R. Asiatic Soc.* 1923, 8–9. *Gdns Bull. Straits Settlements* 1927, 130. *Fl. Malesiana* v.1, 1950, 406.
Plants at Kew.

PHILLIPS, William Henry (1830–1923)
d. Holywood, County Down 13 March 1923
Had large collection of British ferns. President, British Pteridological Society, 1904–5. President, Belfast Naturalists' Field Club, 1905–7. 'Ferns of Ulster' (with R. L. Praeger) (*Proc. Belfast Nat. Field Club* v.2, 1887, Appendix 1, 1–26).
Proc. R. Irish Acad. v.7, 1901, cxxxvii. *Br. Fern Gaz.* 1909, 47; 1923, 3. *Irish Nat.* 1923, 48. R. L. Praeger *Some Irish Nat.* 1949, 144.

PHILLIPSON, Henry (*fl.* 1810s)
Nurseryman, Cottingham and Beverley, Yorks.

PHIPPEN, George (1836–1885)
b. Horton, Bucks 1836 *d.* Reading, Berks 1 May 1885
Nurseryman, Reading.
Garden v.27, 1885, 430.

PHIPPS, Constantine John, 2nd Baron Mulgrave (1744–1792)
b. 30 May 1744 *d.* 10 Oct. 1792
Captain, Royal Navy. MP Newark. Arctic navigator, 1773 in 'Racecourse' and 'Carcass' with Israel Lyons. *Voyage towards North Pole* 1774 (plants described by Solander, 200–4).
A. Lasègue *Musée Botanique de B. Delessert* 1845, 395. *J. Soc Bibliogr. Nat. Hist.* 1964, 208–9.
Phippsia R. Br.

PICKARD, Joseph Fry (1876–1943)
b. Silverdale, Lancs 3 April 1876 *d.* Leeds, Yorks 18 Feb. 1943
Draper. Had herb. of British plants. Assisted A. Wilson in his *Fl. Westmorland* 1938. Contrib. to *Naturalist*, *N. Western Nat.*
Bot. Soc. Exch. Club Br. Isl. Rep. 1943–44, 649–50. *N. Western Nat.* 1945–46, 285 portr.

PICKEN, Joseph (*c.* 1806–1835)
d. Liverpool 16 Oct. 1835
Nurseryman in partnership with W. Caldwell at High Street, Knutsford, Cheshire and at Liverpool. "A scientific practical botanist" (Loudon).
Gdnrs Mag. 1836, 164.

PICKERING, Percival Spencer Umfreville (1858–1920)
d. Harpenden, Herts 5 Dec. 1920
Educ. Oxford. FRS 1890. Lecturer in Chemistry, Bedford College, 1881–88. Director, Woburn Experimental Fruit Farm. *Science and Fruit Growing* (with H. A. Russell) 1919.
Gdnrs Mag. 1907, 909 portr., 910. *Gdnrs Chron.* 1920 ii 283 portr. T. M. Lowry and Sir J. Russell *Scientific Work of Late Spencer Pickering* 1927. *Who was Who, 1916–1928* 838.

PICKSTONE, William (1823–)
b. Whitefield, Lancs 1823
Cultivated 'Sir Watkin' daffodil.
Gdnrs Chron. 1894 ii 773–74.

PICOT, Henry Philip (*fl.* 1890s)
Captain, Indian Staff Corps. Collected plants on Kuen Luen Plains, Kashmir, 1892 (listed in *J. Linnean Soc. Bot.* v.30, 1894, 107, 123–24; v.35, 1902, 212).
E. Bretschneider *Hist. European Bot. Discoveries in China* 1898, 812. *Kew Bull.* 1901, 52.

PIDDINGTON, Henry (1797–1858)
b. Uckfield, Sussex 1797 *d.* Calcutta 7 April 1858
Meteorologist. Commander in mercantile marine from which he retired *c.* 1830. Curator, Museum of Economic Geology, Calcutta. *English Index to the Plants of India* 1832. *Tabular View of the Generic Characters in Roxburgh's Fl. India* 1836.
E. G. Balfour *Cyclopaedia of India* v.4, 1871–73, 567. *D.N.B.* v.45, 256–57.

PIERARD, Francis (*fl.* 1830s)
Indian civil servant, afterwards resided at Kew. Sent plants to Royal Botanic Garden, Calcutta.

N. Wallich *Plantae Asiaticae Rariores* v.2, 1831, 37.
Pierardia Roxb.

PIERCE, Edward (*fl.* 1830s–1840s)
Nurseryman, Yeovil, Somerset.

PIERCE, Edwin (*fl.* 1880s)
Collected plants on coast of Baluchistan, now at Kew.
Pakistan J. For. 1967, 353.
Eragrostis piercei Benth.

PIERCE, S. (*c.* 1902–1963)
d. Ebford, Devon 30 Dec. 1963
Orchid grower at Ebford Nurseries near Exeter.
Orchid Rev. 1964, 47.

PIERCY, W. (1825–1897)
b. 27 June 1825
Of Forest Hill, London. Cultivated chrysanthemums.
Gdnrs Mag. 1897, 54.

PIERS, John (*fl.* 1790s)
Nurseryman, Raheny, Dublin, Ireland.

PIESSE, George William Septimus (1820–1882)
b. 30 May 1820 *d.* Chiswick, Middx 23 Oct. 1882
Analytical chemist, London. *Art of Perfumery* 1855; ed. 5 1891. *Lectures on Perfumes, Flower Farming, and of obtaining the Odours of Plants* 1865.
G. L. M. Strauss *England's Workshops* 1864, 170–78. *Chemist and Druggist* 15 Nov. 1882, 496 portr. *J. Chem. Soc.* v.43, 1883, 255. F. Boase *Modern English Biogr.* v.2, 1897, 1530–31.

PIGGOTT, Horatio (1821–1913)
d. Tunbridge Wells, Kent 7 Dec. 1913
Lichen herb., drawings and correspondence with M. J. Berkeley at BM(NH).

PIGOTT, Arabella Elizabeth *see* Roupell, A. E.

PIGOTT, Edward (*fl.* 1790s–1800s)
Algologist. Friend of J. Stackhouse and correspondent of D. Turner to whom he sent specimens. Imprisoned in France.
J. Stackhouse *Nereis Britannica* 1801, xxvi. D. Turner *Fuci* v.1, 1801, 130.
Letters at Trinity College, Cambridge.

PIGOTT, Luke (*fl.* 1820s)
Florist, Sherdington, Glos. *Treatise on Culture and Management of the Carnation, Picotee and Pink* 1820.

PIKE, Arnold (*fl.* 1890s)
Collected plants in Spitzbergen, 1894. Collected with Capt. H. H. P. Deasy in Tibet, 1896.
E. Bretschneider *Hist. European Bot Discoveries in China* 1898, 812. *Kew Bull.* 1901, 52. *J. Linn. Soc. Bot.* v.35, 1902, 158–61.
Plants at Kew.

PIKE, Edward L. (–*c.* 1772)
d. Chelsea Hospital, London *c.* 1772
Of Bideford, Devon. Lived many years in Eddystone Lighthouse. Contrib. list of Bideford plants to *General Mag. of Arts and Sci.* 1755 and to Martin's *Nat. Hist. of England* v.1, 1759, 35–36.
Watkins *Hist. Bideford* 1792, 279. W. K. Martin and G. T. Fraser *Fl. Devon* 1939, 770.

PILCHER, Charles (–1900)
d. Wandsworth, London 29 Dec. 1900
Gardener to S. Rucker, West Hill, Wandsworth and in charge of orchid collection.
Orchid Rev. 1901, 47–48.
Laelia × *Pilcheri*.

PILKINGTON, Geoffrey Langton (*c.* 1885–1971)
d. 8 Jan. 1971
Educ. Oxford. VMH 1960. VMM 1965. Chairman, Pilkington Bros Ltd. Had garden at Grayswood Hill near Haslemere, Surrey. Hybridised irises. President, Iris Society, 1925–26, 1937–39.
Bot. Mag. 1970 dedicated to him. *J. R. Hort. Soc.* 1972, 109–10. *Times* 10 Jan. 1972.

PILKINGTON, William (1758–1848)
b. Hatfield, Yorks 7 Sept. 1758 *d.* Hatfield 13 Aug. 1848
FLS 1795. Architect. Had herb. Contrib. to J. Sowerby and J. E. Smith's *English Bot.* 1276, 2029.
Proc. Linn. Soc. 1849, 47–49; 1888–89, 39. *D.N.B.* v.45, 302. *R.S.C.* v.4, 912.
MSS. at BM(NH). Pencil drawing at Linnean Society.

PIM, Greenwood (1851–1906)
b. Monkstown, County Dublin 4 May 1851 *d.* Monkstown 14 Nov. 1906
BA Dublin 1872. FLS 1876. Mycologist. President, Dublin Naturalists Field Club, 1888–89. Contrib. to *Sci. Proc. R. Dublin Soc.*
Gdnrs Chron. 1906 ii 364. *Irish Nat.* 1907, 169. *Proc. Irish Acad. B.* 1929, 187. *R.S.C.* v.11, 23; v.12, 577.

PINCE, Robert Taylor (*c.* 1804–1871)
d. 9 Oct. 1871

Partner with W. Lucombe in nursery, St. Thomas, Exeter, Devon. Specialised in fuchsias. Corresponded with J. C. Loudon and other leading horticulturists.
Gdnrs Chron. 1871, 1330.

PINK, William (*fl*. 1790s)
Seedsman, Winchester, Hants.

PINKERTON, William (*fl*. 1780s–1790s)
Nurseryman, Wigan, Lancs.

PINWELL, William Stackhouse Church (1831–1926)
b. 1 Jan. 1831 *d*. Trehane, Cornwall 30 May 1926
VMH 1914. Captain. Collected sedges in Sind, India before 1857 and plants in Malacca. Had garden at Trehane.
Gdnrs Chron. 1926 i 430. *J. R. Hort. Soc*. 1950, 326–31. *Pakistan J. For*. 1967, 353.
Plants at Kew.

PIPER, George Wren (*c*. 1838–1912)
d. Uckfield, Sussex 9 April 1912
Nurseryman, Uckfield; specialised in roses. After his death his son, Thomas Wren Piper, carried on the business.
Gdnrs Chron. 1912 i 269. *Gdnrs Mag*. 1912, 329.

PIQUET, John (1825–1912)
b. St. Hélier, Jersey 16 March 1825 *d*. St. Hélier 5 Sept. 1912
Druggist. Presented a collection of algae to H. van Heurck for his *Prodrome de la Fl. des Alques Marines des Iles Anglo-Normandes* 1908. *Phanerogamous Plants and Ferns of Jersey* 1896.
Bot. Soc. Exch. Club Br. Isl. Rep. 1912, 205–7; 1920, 177. *J. Bot*. 1912, 371–74 portr. G. R. Balleine *Biogr. Dict. Jersey* 1848, 551.

PIRIE, Mary (1821–1885)
b. Aberdeen 1821 *d*. Portsey, Banffshire 8 Feb. 1885
Flowers, Grasses and Shrubs 1860.
Aberdeen J., Notes and Queries v.1, 15.

PITCAIRN, William (1711–1791)
b. Dysart, Fife 1711 *d*. Islington, London 25 Nov. 1791
MD Rheims. MD Oxon 1749. FRS 1770. Physician, St. Bartholomew's Hospital, London, 1750–80. President, Royal College of Physicians, 1775–85. Had botanic garden in Islington from 1775 (plants in Banks's Herb.).
Gent. Mag. 1794 ii 903. W. Munk *Roll of R. Coll. of Physicians* v.2, 1878, 172–74. *D.N.B*. v.45, 334–35. *J. Bot*. 1902, 389. *Huntia* 1965, 207 portr.
Portr. by Sir J. Reynolds at Royal College of Physicians. Part of library sold by Benjamin White, 1792.
Pitcairnia L'Hérit.

PITCHFORD, John (*c*. 1737–1803)
d. Norwich, Norfolk 22 Dec. 1803
ALS 1788. FLS 1796. Surgeon, Norwich from 1769. Friend of J. E. Smith. Discovered *Holosteum umbellatum*. Contrib. to J. Sowerby and J. E. Smith's *English Bot*. 27, 229, etc.
W. Withering *Bot. Arrangement* v.1, 1787, 989. *Trans. Linn. Soc*. v.7, 1804, 295–96. P. Smith *Mem. and Correspondence of Sir J. E. Smith* v.1, 1832, 41–44, 107–8, 128–31, 274. W. M. Hind *Fl. Suffolk* 1889, 477. *Trans. Norfolk Norwich Nat. Soc*. 1874–75, 28–30; 1912–13, 656–61. *J. Bot*. 1902, 321.
Letters at Kew.

PITT, Edmund (*fl*. 1650s–1670s)
Mayor of Worcester, 1656. "A very knowing botanist." Discovered *Sorbus domestica* in Wyre Forest (*Philos. Trans. R. Soc*. v.12, 1678, 978–79).
E. Lees *Bot. Worcestershire* 1867, lxxxviii. *Berrow's Worcester J*. 29 July 1911.

PITT, William (–1823)
b. Tettenhall, Staffs *d*. Tettenhall 18 Sept. 1823
Engaged by Board of Agriculture to prepare reports on the state of agriculture of several counties: Stafford, 1794; Worcester, 1810; Northampton, 1813; Leicester, 1819. List of plants in *Agric. Survey of Stafford* 1796, *Agric. Survey of Worcester* 1810, *Agric. Survey of Northamptonshire* 1813.
D.N.B. v.45, 386. *J. Bot*. 1901, Supplt., 72. J. Amphlett and C. Rea *Bot. Worcestershire* 1909, xxi. *J. Northamptonshire Nat. Hist. Soc*. 1910, 222–23. G. C. Druce *Fl. Northamptonshire* 1930, lxxxvii–lxxxviii. E. S. Eedes *Fl. Staffordshire* 1972, 16–17.

PITTOCK, George Mayris (1832–1916)
b. Deal, Kent 1832 *d*. 5 March 1916
MB London 1855. *Fl. Thanet* 1903. '*Ambrosia trifida*' (*J. Bot*. 1903, 379).
J. Bot. 1903, 224.

PITTS, John (1864–1944)
b. Dunster, Somerset 1864 *d*. 8 Jan. 1944
Gardener, Grittleton Park, Wilts, 1894–1942.
Orchid Rev. 1944, 38–40.

PLACE, Francis (*fl*. 1800s–1820s)
Of Charing Cross, London. Introduced plants from Chile.
Bot. Mag. 1823, t.2399; 1824, t.2523.

PLANCHON, Jules Émile (1823–1888)
b. Ganges, Hérault, France 21 March 1823
d. Montpellier, France 1 April 1888
DSc Paris 1844. FLS 1855. Studied under A. St. Hilaire. Assistant in W. J. Hooker's herb. at Kew, 1844–48. Professor of Botany, Ghent, 1849; Montpellier, 1881. Distinguished *Ulex gallii* 1849.
G. Bentham *Fl. Australiensis* v.1, 1863, 8.* *Proc. Linn. Soc.* 1887–88, 95–96. *Ann. Bot.* 1888–89, 423–28. *Gdnrs Chron.* 1895 i 461. *J. R. Soc. N.S.W.* 1910, 152–58 portr. *R.S.C.* v.4, 932; v.8, 631; v.12, 579.
Letters at Kew.
Planchonia Bl.

PLANER, Richard (*fl*. 1690s–1700s)
Surgeon. Sent plants to J. Petiver from Guinea Coast.
J. Petiver *Musei Petiveriani* 1695, 46, 95. J. E. Dandy *Sloane Herb.* 1958, 182.
Plants at BM(NH).

PLANT, Andrew (*fl*. 1790s)
Nurseryman, Stockport, Cheshire.

PLANT, Joseph (*fl*. 1820s–1840s)
Nurseryman, Cheadle, Staffs. Contrib. to *Floricultural Cabinet*.

PLANT, Robert W. (–*c*. 1858)
d. St. Lucia Lake, S. Africa *c*. 1858
Kew gardener. Nurseryman, Cheadle, Staffs. Arrived in Natal, 1852. Discovered *Stapelia gigantea*. Sent plants to Kew. Curator, Botanic Gardens, Durban, 1854–56. *New Gardener's Dictionary c*. 1840. 'Notes of Excursion in Zulu Country' (*Hooker's J. Bot*. 1852, 222–23, 257–65).
Hooker's J. Bot. 1853, 225–29; 1857, 190. A. White and B. L. Sloane *Stapelieae* v.1, 1937, 102. J. Hutchinson *Botanist in Southern Africa* 1946, 643. *S. African J. Sci.* 1971, 404.
Stapelia plantii MacKen.

PLANT, Thomas (*fl*. 1810s–1820s)
Nurseryman, Cheadle, Staffs.

PLAT (*or* **Platt**), **Sir Hugh** (1552–1608)
b. London 1552 *d*. Stepney, London 1608
BA Cantab 1571. Knighted 1605. Writer on agriculture and inventor. Had gardens in Bethnal Green and St. Martin's Lane, London. *Jewell House of Art and Nature* 1594. *Floreas Paradise* 1608; ed. 5 1659. *Garden of Eden* 1653.
G. W. Johnson *Hist. English Gdning* 1829, 69–70. S. Felton *Portr. English Authors on Gdning* 1830, 13–16. *Cottage Gdnr* v.6, 1851, 185. J. Donaldson *Agric. Biogr.* 1854, 11–12. *J. Hort. Cottage Gdnr* v.53, 1875, 252–53. *D.N.B.* v.45, 407–9. E. J. L. Scott *Index to Sloane Manuscripts* 1904, 425. *Gdnrs Chron.* 1936 i 352–53; 1936 ii 62–63. *Univ. Missouri Studies* v.21, 1946, 93–118. R. H. Jeffers *Friends of John Gerard* 1967, 81–83.

PLATT, Thomas (–1842)
d. London 8 Oct. 1842
One of J. Sibthorp's executors who supervised the publication of *Fl. Graeca* 1806–40.
J. Sibthorp *Fl. Graeca* v.1, 1806, iv. *Linnaea* v.19, 1847, 459.

PLATTEN, Edward William (*c*. 1881–1951)
d. April 1951
Bookseller and stationer. Member of British Pteridological Society, 1929–51. *History of Needham Market* 1926.
Br. Fern Gaz. 1952, 7.

PLATTES, Gabriel (*c*. 1600–1655)
Writer on agriculture. *Treatise on Husbandry* 1638. *Discovery of Infinite Treasure hidden since the World's Beginning* 1639. *Practical Husbandry Improved* 1639. *Countryman's Recreation, or Art of Planting, Graffeing and Gardening* 1654.
G. W. Johnson *Hist. English Gdning* 1829, 93–95. *D.N.B.* v.45, 410. D. McDonald *Agric. Writers* 1908, 78–84. *Chronica Bot.* 1944, 51–53.

PLAYFAIR, David Thomson (1855–1904)
b. March 1855 *d*. Bournemouth, Hants 1 Feb. 1904
MD Edinburgh. FLS 1888. Had herb.
J. Bot. 1904, 96.

PLAYFAIR, George MacDonald Home (*fl*. 1870s–1880s)
MA Dublin. To China as student interpreter, 1872. Consul, Ning-po, 1895. Sent plants from China and Formosa to Kew, 1872–89.
E. Bretschneider *Hist. European Bot Discoveries in China* 1898, 709–10. *Kew Bull.* 1901, 53.
Rubus playfairii Hemsley.

PLAYFAIR, Rev. Patrick M. (1858–1924)
b. 5 Nov. 1858 *d*. 6 Oct. 1924
MA Edinburgh. DD St. Andrews. Minister. Botanist.
Who was Who, 1916–1928 842.

PLAYFAIR, Sir Robert Lambert (1828–1899)
b. St. Andrews, Fife 21 March 1828 *d*. St. Andrews 18 Feb. 1899
Madras Artillery, 1846. Assistant Political Resident, Aden, 1854–62. Consul-General, Algeria, 1867. Ichthyologist. *History of Arabia Felix* 1859.
D.N.B. Supplt 1 v.3, 272–73. *Rec. Bot. Survey India* v.7, 1914, 9–10. *R.S.C.* v.8, 635.
Plants at Kew.
Balsamodendron playfairii Hook. f.

PLEYDEL, John Clavell Mansel- *see* Mansel-Pleydel, J. C.

PLOMLEY, Francis (*c.* 1805–1860)
d. Maidstone, Kent 9 Jan. 1860
MD. FLS 1845. Studied hop-mildew. Lectured on botany to Weald of Kent Farmers' Club, 1849–51.
Proc. Linn. Soc. 1860–61, xlii.

PLOT, Robert (1640–1696)
b. Sutton Baron, Borden, Kent 1640 *d.* Sutton Baron 30 April 1696
BA Oxon 1661. DCL 1671. FRS 1677. Secretary, Royal Society, 1682. First Keeper of Ashmolean Museum, Oxford, 1683. *Natural History of Oxfordshire* 1677. *Natural History of Staffordshire* 1679.
Gent. Mag. 1795, 1089. J. Nichols *Lit. Anecdotes of Eighteenth Century* v.9, 547–48. R. Pulteney *Hist. Biogr. Sketches of Progress of Bot. in England* v.1, 1790, 350–52. G. C. Druce *Fl. Berkshire* 1897, cxii–cxiii. E. J. L. Scott *Index to Sloane Manuscripts* 1904, 426. *D.N.B.* v.45, 424–26. G. C. Druce *Fl. Oxfordshire* 1927, lxxv–lxxix. R. T. Gunther *Robert Plot and Correspondence of Philos. Soc. of Oxford* 1939. J. E. Dandy *Sloane Herb.* 1958, 182–83. *Hist. Today* 1970, 112–17 portr. *Trans. Kent Field Club* 1971, 213–24.
Plants at BM(NH).
Plotia Adans.

PLOWRIGHT, Charles Bagge (1849–1910)
b. King's Lynn, Norfolk 3 April 1849 *d.* North Wootton, Norfolk 24 April 1910
MD Durham 1890. FLS 1884. Medical Officer of Health, Freebridge, Lynn Rural District Council for over 32 years. President, British Mycological Society, 1899. *Sphaeriacei Britannici* 1873–78 (exsiccatae). *Monograph of British Uredineae and Ustilagineae* 1889. Contrib. to *Gdnrs Chron.*, *Grevillea*, *J. Bot.*, *Trans. Br. Mycol. Soc.*, *Trans. Norfolk Norwich Nat. Soc.*
P. A. Saccardo *Sylloge Fungorum* v.2, 1883, 635–39. *Gdnrs Chron.* 1910 i 286. *Nature* v.83, 1910, 287. *Trans. Br. Mycol. Soc.* v.3, 1910, 231–32. *Trans. Norfolk Norwich Nat. Soc.* 1910–11, 275–82; 1937, 203–6 portr. *R.S.C.* v.8, 635; v.11, 35; v.12, 580; v.17, 927.
Fungi and letters at Kew. Herb. at Birmingham University. Drawings, MSS., letters at BM(NH). Portr. at Hunt Library.
Plowrightia Sacc.

PLUES, Margaret (*c.* 1840–*c.* 1903)
Of Ripon, Yorks. In middle-age entered Roman Catholic convent at Weybridge, Surrey where she became Mother Superior and died. *Rambles in Search of Ferns* 1861. *Rambles in Search of Mosses* 1861. *Rambles in Search of Flowerless Plants* 1864. *British Grasses* 1867.
Country Life 1968, 624.
Letters in Broome correspondence at BM(NH).

PLUKENET, Leonard (1642–1706)
b. 4 Jan. 1642 *d.* Westminster, London 6 July 1706
MD. Physician at Westminster where he had a small botanic garden. Superintendent, Hampton Court Garden and Queen's Botanist. *Phytographia* 1691–96. *Almagestum Botanicum* 1696. *Almagesti Botanici Mantissa* 1700. *Amaltheum Botanicum* 1705. (His own copies of his books with MS. notes at BM(NH).)
P. D. Giseke *Index Linneanus in L. Plukenetii Opera Botanica* 1779 (preface). R. Pulteney *Hist. Biogr. Sketches of Progress of Bot. in England* v.2, 1790, 18–29. H. Trimen and W. T. T. Dyer *Fl. Middlesex* 1869, 374–76. *Trans. Watford Soc.* v.1, 1875, 23. *J. Bot.* 1882, 338–42; 1883, 213; 1894, 247–48; 1900, 336–38. E. Bretschneider *Hist. European Bot. Discoveries in China* 1898, 33–44. *D.N.B.* v.45, 432. E. J. L. Scott *Index to Sloane Manuscripts* 1904, 426. A. White and others *Succulent Euphorbieae* v.1 1941, 36 portr. C. E. Raven *John Ray*, 1942, 232. I. H. Burkill *Chapters on Hist. Bot. in India* 1965, 9–10. J. E. Dandy *Sloane Herb.* 1958, 183–88. F. A. Stafleu *Taxonomic Literature* 1967, 359–60.
Plants at BM(NH). Portr. at Hunt Library.
Plukenetia Plumier.

PLUNKETT, Hon. Katherine (1820–)
b. Kilsaran, County Louth 1820
Volume of her flower paintings at Botanic Gardens, Glasnevin, Dublin.

POCKETT, Thomas William (1857–1952)
b. Cheltenham, Glos 11 Feb. 1857 *d.* Healesville, Victoria 1 Nov. 1952
To Australia, 1880. Curator, Malvern Public Gardens, Melbourne. Hybridised roses, carnations and chrysanthemums.
The Age (Melbourne) 3 Nov. 1952. *Australian Encyclop.* v.7, 1965, 156.

POCOCK, Robert (1760–1830)
b. Gravesend, Kent 21 Feb. 1760 *d.* Dartford, Kent 26 Oct. 1830
Printer. Antiquary. Founded Natural History Society of Kent, 1812.
G. M. Arnold *Robert Pocock* 1883 portr. *J. Bot.* 1884, 53–55 portr. *D.N.B.* v.46, 6–7.
Plants at BM(NH).

POCOCKE, Rev. Richard (1704–1765)
b. Southampton, Hants 1704 *d.* Tullamore, Charleville, County Cork 25 Sept. 1765

BA Oxon 1725. DCL 1733. FRS 1741. Bishop of Ossory, 1756. Bishop of Meath, 1765. Travelled in East, 1737–40. *Description of the East* 1743 (plants listed by P. Miller and illustrated by G. D. Ehret).

Gent. Mag. 1765, 444. *Bot. Mag.* 1788, t.61. J. Nichols *Lit. Anecdotes of Eighteenth Century* v.2, 157–58. *Trans. Linn. Soc.* v.12, 1817, 1–5. A. Lasègue *Musée Botanique de B. Delessert* 1845, 409. *D.N.B.* v.46, 12–14.

MSS. at BM(NH).

Pocockia Ser.

POË, John T. Bennett- *see* Bennett-Poë, J. T.

POHLMANN, Edward (1825–1886)
b. Halifax, Yorks 1825 *d.* Halifax 27 Nov. 1886

Grew auriculas.

Gdnrs Chron. 1886 ii 761.

POINTER, Vincent (*alias* **Corbet**)
(–1619)

Nurseryman, Twickenham, Middx. "A most curious planter and improver of all manner of rare fruites" (Gerard).

J. Gerard *Herball* 1597, 1269. *Gdnrs Chron.* 1917 ii 235. C. E. Raven *English Nat.* 1947, 214.

POLE-EVANS, Iltyd Buller (1879–1968)
b. Llanmaes, Glam 3 Sept. 1879 *d.* Umtali, Rhodesia 16 Oct. 1968

MA Cantab. FLS 1907. Mycologist, Transvaal Government, 1905. Chief, Division of Plant Pathology and Mycology, Dept. of Agriculture, S. Africa, 1911. Director, Botanical Survey S. Africa, 1918–39. *Roadside Observations on Vegetation of East and Central Africa* 1948. Edited *Flowering Plants of S. Africa*, *Bothalia*.

Kew Bull. 1939, 669. A. White *et al. Succulent Euphorbieae* v.1, 1941, 54–55 portr. *Bothalia* 1971, 131–35 portr. *Who was Who, 1961–1970* 902.

Plants at Kew. Portr. at Hunt Library.

Polevansia De Winter.

POLLARD, Joseph (1825–1909)
b. Ware, Herts 1825

Of High Down near Pirton, Herts from 1841.

J. G. Dony *Fl. Bedfordshire* 1953, 25.

Herb. at Hitchin Museum.

POLLEXFEN, Rev. John Hutton (*recte* **Pollexsen**) (1813–1899)
b. Kirkwall, Orkney 1813 *d.* Middleton Tyas, Yorks 5 June 1899

MD Edinburgh 1835. BA Cantab 1843. Algologist.

W. H. Harvey *Nereis Australis* 1847, 22. *J. Bot.* 1899, 438–39.

Algae at BM(NH).

Pollexfenia Harvey.

POLWHELE, Rev. Richard (1760–1838)
b. Truro, Cornwall 6 Jan. 1760 *d.* Truro 12 March 1838

Of Polwhele, Cornwall. Vicar, Manaccan, Cornwall 1794–1816. *Historical Views of Devonshire* 1793–97. *History of Cornwall* 1803–8. (Both books contain plant lists).

D. Turner and L. W. Dillwyn *Botanist's Guide through England and Wales* 1805, 194–95. J. Nichols *Illus. of Lit. Hist. of Eighteenth Century* v.8, 646–48 portr. *Gent. Mag.* 1838 i 545–49. G. C. Boase and W. P. Courtney *Bibliotheca Cornubiensis* 1874, 506–17. *D.N.B.* v.46, 71–73. F. H. Davey *Fl. Cornwall* 1909, xxxvi–xxxviii. W. K. Martin and G. T. Fraser *Fl. Devon* 1939, 770–71.

PONSFORD, Samuel (*fl.* 1840s)

Established nursery at Loughborough Park, Brixton, London, 1843.

PONTEY, Alexander (*fl.* 1790s–1821)

Nurseryman, Upperhead Row, Leeds, Yorks.

PONTEY, Alexander (*fl.* 1830s–1840s)

Nurseryman, 21 Cornwall Street, Plymouth, Devon.

PONTEY, Charles (*fl.* 1830s)

Nurseryman, 29 Upperhead Row, Leeds, Yorks.

PONTEY, Francis (*fl.* 1790s–1810s)

Nurseryman, Huddersfield and Kirkheaton, Yorks.

PONTEY, Henry P. (*fl.* 1830s–1850s)

Nurseryman, Lepton in Kirkheaton, Yorks.

PONTEY, John (*fl.* 1820s–1840s)

Nurseryman, 8 Kirkgate, Huddersfield, Yorks.

PONTEY, Martha (*fl.* 1820s)

Nurseryman, 30 Upperhead Row, Leeds, Yorks.

PONTEY, William (*fl.* 1780s–1831)

Planter and tree pruner to Duke of Bedford. Nurseryman, Lepton in Kirkheaton, Yorks. *Profitable Planter* 1800. *Forest Pruner* 1805. *Rural Improver* 1823.

J. C. Loudon *Encyclop. Gdning* 1822, 1283. G. W. Johnson *Hist. English Gdning* 1829, 276. *Gdnrs Chron.* 1963 i 193. J. H. Harvey *Early Gdning Cat.* 1972, 38, 41.

PONTEY, William (*fl.* 1830s–1860s)

Nurseryman, Lepton in Kirkheaton, Yorks in partnership with his uncle, W. Pontey (*fl.* 1780s–1831).

POOL, Thomas (*fl.* 1690s)
Of Nottingham. "A diligent enquirer into Natural History" (J. Petiver *Museii Petiveriani* 1695, 89). Plants in Sloane Herb. at BM(NH).
J. E. Dandy *Sloane Herb.* 1958, 188.

POOLE, Charles Edward Lane- *see* Lane-Poole, C. E.

POOLE, Henry (*fl.* 1790s)
Nurseryman, Reading, Berks.

POOLE, James (*c.* 1777–*c.* 1827)
Nurseryman, North Town, Taunton, Somerset in partnership with his brother William; *c.* 1827 nursery passed to John Young (*c.* 1790–1862).

POOLE, Rev. John (*c.* 1771–1857)
b. Over Stowey, Somerset *c.* 1771 *d.* 16 May 1857
BA Oxon 1792. Rector, Enmore, Somerset, 1796; Swainswick, 1811. Contrib. plant records to H. C. Watson's *New Botanists' Guide* v.2, 1837, 553.

POOLE, Joseph (*fl.* 1810s–1820s)
Brother-in-law of Dr. Clarke Abel whom he accompanied to China in 1816. Collected plants in Canton for nursery firm of Barr and Brookes, Newington Green, London, 1818–19.
Trans. Hort. Soc. London v.4, 1822, 333. E. Bretschneider *Hist. European Bot. Discoveries in China* 1898, 221. E. H. M. Cox *Plant Hunting in China* 1945, 50, 62.

POPE, Alexander (–*c.* 1853)
Partner in John Pope and Sons, nurserymen, Handsworth, Birmingham.
Gdnrs Mag. 1831, 237–38, 410. *Gdnrs Chron.* 1962 i 63.

POPE, Clara Maria (*née* **Leigh**) (*fl.* 1760s–1838)
d. London 24 Dec. 1838
Married Francis Wheatley, R.A., then Alexander Pope, the actor. Exhibited flower paintings at Royal Academy, 1796–1838. Contrib. plates to S. Curtis's *Beauties of Flora* 1806–20 and *Monograph on Genus Camellia* 1819.
Art-Union March 1839, 23. *D.N.B.* v.46, 130. A. Graves *R. Acad. of Arts* v.6, 1906, 181. *J. Bot.* 1918, 126–27. *J. R. Hort. Soc.* 1933, 325. W. Blunt *Art. Bot. Illus.* 1955, 211–12. *Country Life* 1967, 1246–48 portr.
Drawings of *Paeonia* at BM(NH).

POPE, Henry (*fl.* 1850s)
Nurseryman, Cowick Street, Exeter, Devon.

POPE, John (*fl.* 1810s–1830s)
Son of Luke Pope (*c.* 1740–1825). Nurseryman, Handsworth, Birmingham. His son, Alexander, was a partner in the firm.

POPE, John (*c.* 1847–1918)
d. 26 Jan. 1918
Nurseryman, Birmingham.
Gdnrs Chron. 1918 i 62.

POPE, Luke (*c.* 1740–1825)
Nurseryman, Smethwick and Handsworth, Birmingham.
Gdnrs Chron. 1962 i 63 portr.

POPE, Luke Linnaeus (*fl.* 1800s)
Son of John Pope. Botanical artist. Drew plants in his father's nursery, John Pope and Sons, Handsworth, Birmingham.
M. Hadfield *Gdning in Britain* 1960, 293. *Gdnrs Chron.* 1962 i 63.

POPE, Phoebe *see* Lankester, P.

POPENOE, Dorothy Kate (*née* **Hughes**) (1899–1932)
b. Ashford, Kent 19 June 1899 *d.* Guatemala City 31 Dec. 1932
Assistant, Kew Herb., 1918–23. Worked with O. Stapf on *Gramineae.* Contrib. to *Kew Bull.* Bureau of Plant Industry, Washington, 1923.
J. Bot. 1933, 320. *J. Kew Guild* 1933, 285. *Kew Bull.* 1933, 304.
Panama plants at Kew. Portr. at Hunt Library.

PORTER, George (*fl.* 1800s–1830s)
Overseer, Botanic Garden, Calcutta until 1822. Accompanied N. Wallich to Singapore, 1822. Remained in Penang as schoolteacher; put in charge of the Botanic Garden. Sent plants to Wallich.
C. Curtis *Cat. of Flowering Plants and Ferns...in...Penang* 1894, 99. *J. Straits Branch R. Asiatic Soc.* no. 65, 1913, 40. *Gdns Bull. Straits Settlements* 1927, 130; 1929, 426. *Fl. Malesiana* v.1, 1950, 412.
Plants at Kew.

PORTER, John Scott (1801–1880)
b. Limavady, Londonderry 31 Dec. 1801 *d.* 5 July 1880
Read *Botany of New Testament* (unpublished) to Belfast Natural History Society, 1834.
Belfast Nat. Hist. Philos Soc. Centenary Vol. 1924, 97–98 portr.

PORTER, Sir Robert Ker (1777–1842)
b. Durham 1777 *d.* Petrograd 4 May 1842
Knighted 1813. Historical painter. Zoologist. Consul at Caracas, Venezuela, 1826–41. *Travels in Georgia, Persia* 1821.

Hooker's Icones Plantarum 1852, t.864. *Bot. Mag.* 1839, t.3723. *Gent. Mag.* 1850 ii 364. *D.N.B.* v.46, 190–92.

Plants at Kew, Oxford. MSS. sold at Sotheby, 27 June 1966.

Porteria Hook.

PORTLAND, Margaret, Duchess of *see* Bentinck, M. C.

POTTER, Beatrix (1866–1943)
b. S. Kensington, London 6 July 1866 *d.* Sawrey, Lancs 22 Dec. 1943

Writer and illustrator of children's books. Went to Kew Herb. in 1896 to discuss her fungi drawings some of which appear in W. P. K. Findlay's *Wayside and Woodland Fungi* 1967.

D.N.B. 1941–1950 686–87. A. C. Moore *Art of Beatrix Potter* 1955. L. Linder *Journal of Beatrix Potter* 1966.

Fungi drawings at Armitt Library, Ambleside, Westmorland.

POTTER, Henry Arthur (1874–1949)
b. Peckham, Camberwell, London 16 July 1874 *d.* 3 Feb. 1949

FLS 1930. Pharmaceutical chemist. Studied *Vinca rosea*, *Plantago psyllium* and Cape aloes while travelling in S. Africa.

Proc. Linn. Soc. 1948–49, 251.

POTTER, Rev. Michael Cressé (1858–1948)
b. Corlby, Warwickshire 7 Sept. 1858 *d.* New Milton, Hants 9 March 1948

BA Cantab 1881. Assistant Curator, Cambridge Herb., 1884–91. In Portugal, 1886; Ceylon, 1888–89. Head of Botany Dept. (later Professor), Armstrong College, Newcastle, 1889–1925. President, British Mycological Society, 1909. President, Bournemouth Natural Science Society, 1940–41. *Elementary Textbook of Agricultural Botany* 1893. Translated E. Warming's *Handbook of Systematic Botany* 1894.

Nature v.161, 1948, 590–91, 673–74. *Proc. Bournemouth Nat. Sci. Soc.* 1947–48, 78. *Times* 16 March 1948. *Who was Who, 1941–1950* 929–30.

POTTS, Eliza (1809–1873)
b. Chester, Cheshire 11 March 1809 *d.* Funchal, Madeira 6 Dec. 1873

Of Chester and Glanyr Afon, Denbigh. Friend of W. Wilson. Contrib. to T. B. Hall's *Fl. Liverpool* 1839.

J. B. L. Warren *Fl. Cheshire* 1899, lxxxvii. *J. Bot.* 1910, 41; 1911, Supplt 1, 4.

Plants at BM(NH), Chester Museum.

POTTS, George (1877–1948)
b. Northallerton, Yorks 1877 *d.* 11 July 1948

Educ. Armstrong College, Newcastle-upon-Tyne. Professor of Botany, University College, Bloemfontein, 1905–38. Botanist in charge, Botanical Survey, Central Area of Union of S. Africa, 1918–38.

Who was Who, 1941–1950 930.

POTTS, George Honington (1830–1907)
b. London 1830 *d.* Lasswade, Midlothian 6 June 1907

Painter and decorator, Edinburgh. Grew saxifrages, sedums and sempervivums in his garden at Lasswade.

Trans. Bot. Soc. Edinburgh v.23, 1908, 368–69.

POTTS, John (–1822)
d. Chiswick, Middx 5 Oct. 1822

Gardener for Horticultural Society of London for which he collected plants on round trip to China and Bengal, 1821–22.

Trans. Hort. Soc. London v.5, 1824, 427; v.7, 1830, 25. A. Murray *Book of R. Hort. Soc.* 1863, 16. E. Bretschneider *Hist. European Bot. Discoveries in China* 1898, 269–71. *Fl. Malesiana* v.1, 1950, 414. G. A. C. Herklots *Hong Kong Countryside* 1951, 162. *J. R. Hort. Soc.* 1955, 269–70.

MS. Journal 1821–22 at Royal Horticultural Society.

Pottsia Hook. & Arn.

POTTS, John (*fl.* 1840s–1850s)

Manager of Mint at Chihuahua, Mexico. Sent cacti to F. Scheer.

B. Seemann *Bot. Voyage of 'Herald'* 1852, 285–86. *Taxon* 1970, 536.

Mammillaria pottsii Scheer.

POTTS, Thomas Henry (1824–1888)
d. Christchurch, New Zealand 27 July 1888

To New Zealand, *c.* 1853. Naturalist, especially ornithology. Pioneer of forest conservation. *Out in the Open* 1882 (list of N.Z. ferns, 237–80).

R. Speight *Nat. Hist. of Canterbury* 1927, 10–11 portr. *R.S.C.* v.8, 650; v.11, 53.

Plants at Kew.

POUPART, John (*c.* 1876–1945)
d. Weybridge, Surrey 26 Jan. 1945

Son of W. Poupart (*c.* 1846–1936). Horticultural salesman, Covent Garden, London.

Gdnrs Chron. 1945 i 52.

POUPART, William (*c.* 1846–1936)
d. Walton-on-Thames, Surrey 24 Dec. 1936

VMH 1922. Nurseryman, Marsh Farm, Twickenham. President, Market Gardeners', Nurserymen and Farmers' Association.

Gdnrs Mag. 1906, 746 portr. *Gdnrs Chron.* 1937 i 16. *Daffodil Yb.* 1939, 1–4 portr.

POWELL, Miss F. S. (*fl.* 1820s–1860s)

Of Henbury, Glos.

J. W. White *Fl. Bristol* 1912, 97. H. J. Riddelsdell *Fl. Gloucestershire* 1948, cxviii.

Plants at Bristol Museum.

POWELL, Henry (1864–1920)
b. Goodrich, Hereford 13 March 1864 *d.* Mazeras, Kenya 5 June 1920
Kew gardener, 1888. Curator, Botanic Station, St. Vincent, 1890–93. Assistant Director of Agriculture, British East Africa, 1903. Chief of Economic Division, 1907.
I. Urban *Symbolae Antillanae* v.3, 1902, 106. *Kew Bull.* 1903, 31; 1920, 220–21. *J. Kew Guild* 1921, 40.
Kenyan plants at Kew.
Boscia powellii Sprague & M. L. Greene.

POWELL, James Thomas (1833–1904)
b. Daventry, Northants 3 April 1833 *d.* Parkstone, Dorset 14 Jan. 1904
Schoolmaster. Member of Watson Exchange Club, 1885–1900. Collected British *Rubi.* Discovered *Rubus powellii* Rogers at High Beech, Essex.
J. Bot. 1904, 95. *Essex Nat.* v.29, 1954, 189–90, 195. S. T. Jermyn *Fl. Essex* 1974, 19.
Rubi collection at Essex Field Club.

POWELL, Nathaniel (*fl.* 1740s–1773)
Seedsman, Kings Head near Fetter Lane, Holborn, London.
Gdnrs Chron. 1918 i 223.

POWELL, Rev. Thomas (1809–1887)
d. Penzance, Cornwall 6 April 1887
FLS 1867. Missionary, Upolu, Samoa, 1860–85. Mosses described by W. Mitten in *J. Linn. Soc. Bot.* v.10, 1868, 166–95. Papers on Samoan ferns and plant-names in *J. Bot.* 1868.
R.S.C. v.8, 654; v.11, 57.
Plants at BM(NH), Kew. Letters at Kew.
Powellia Mitt.

POWER, A. (*fl.* 1780s–1800s)
Botanical artist, Maidstone. Exhibited at Royal Academy, 1800.
Kew Bull. 1916, 167. W. Blunt *Art of Bot. Illus.* 1955, 194.
Drawings at Victoria and Albert Museum, Kew.

POWER, John (1758–1847)
b. Polesworth, Warwickshire 15 April 1758 *d.* Atherstone, Warwickshire 1847
MD. Studied lichens. *Calendar of Fl. at Market Bosworth* 1807.
J. Bot. 1901, Supplt., 72. A. R. Horwood and C. W. F. Noel *Fl. Leicestershire* 1933, cc.
Herb. formerly at Holmesdale Natural History Society.

POWER, John Arthur (1810–1886)
d. Bedford 9 June 1886
BA Cantab 1832. Entomologist. Medical tutor in London. Friend of C. C. Babington. Studied *Atriplex.*
Trans. Bot. Soc. Edinburgh v.1, 1840, 5–6. C. C. Babington *Memorials, J. and Bot. Correspondence* 1897, 266. A. R. Horwood and C. W. F. Noel *Fl. Leicestershire* 1933, ccx–ccxi.
Herb. formerly at Holmesdale Natural History Society.

POWER, Robert (*fl.* 1780s–1790s)
Nurseryman, Galway, Ireland.
Irish For. 1967, 53.

POWER, Thomas (*fl.* 1840s)
MD. Lecturer in Botany, Cork School of Medicine. 'Botanist's Guide for County of Cork' (J. R. Harvey, J. D. Humphreys, T. Power *Contributions towards Fauna and Fl. of County of Cork* 1845).
Proc. Irish Acad. B. 1915, 71–72; 1929, 183. *Irish Nat. J.* 1931, 238.

POYNTER, Vincent *see* Pointer, V.

PRAEGER, Robert Lloyd (1865–1953)
b. Holywood, County Down 25 Aug. 1865 *d.* Belfast 5 May 1953
BA Belfast 1885. ALS 1949. Assistant Librarian, National Library of Ireland, 1898; Librarian, 1920–24. President, Royal Irish Academy, 1931–34. Edited *Irish Nat.*, *Bull. R. Hort. Soc. Ireland. Open-air Studies in Botany* 1897. 'Ferns of Ulster' (with W. H. Phillips) 1887. *Irish Topographical Botany* 1901 (in this work he divided Ireland into its 40 vice-counties). *Tourist's Fl. West of Ireland* 1909. 'Account of Genus Sedum as found in Cultivation' (*J. R. Hort. Soc.* 1921, 1–314). *Aspects of Plant Life* 1921. *Account of Sempervivum Group* 1932. *Botanist in Ireland* 1934. *The Way that I Went* 1937. *Natural History of Ireland* 1950. *Some Irish Naturalists* 1949 (146–47 biogr., portr.).
Geogr. J. 1953, 368–69. *Proc. Bot. Soc. Br. Isl.* 1954, 106–10. *Irish Nat.* 1954, 141–76. *Who was Who, 1951–1960* 887–88.
Herb. at Dublin Museum.

PRAEGER, William Emilius (1863–1936)
b. Belfast 10 Sept. 1863 *d.* 13 Aug. 1936
Brother of R. L. Praeger (1865–1953). Professor of Biology, Kalamazoo College, Michigan, 1905–34. Studied plant physiology and ecology.
Chronica Botanica 1937, 296 portr.

PRAIN, Sir David (1857–1944)
b. Fettercairn, Kincardineshire 11 July 1857 *d.* Whyteleafe, Surrey 16 March 1944
MA Aberdeen 1878. MB Edinburgh 1882. LRCS Edinburgh 1883. FRS 1905. FLS 1888. VMH 1912. VMM 1933. CMG 1912. Indian Medical Service, 1884–87. Curator of Herb. and Librarian, Royal Botanic Garden, Calcutta, 1887; Superintendent, 1898–1905. Professor of Botany, Medical College, Calcutta, 1898–1905. Director, Royal Botanic Gardens, Kew, 1905–22. President,

Linnean Society, 1916–19. *Bengal Plants* 1903. *Account of Genus Dioscorea in the East* (with I. H. Burkill) 1936–38. Contrib. to *J. Asiatic Soc. Bengal*, *J. Linn. Soc.*, *Proc. Linn. Soc.*, *Kew Bull.*

J. Bot. 1906, 21–22 portr. *J. Kew Guild* 1906, 287 portr.; 1922, 82; 1944, 345–46 portr. *Bull. Acad. Géogr. Bot.* 1912, 21–22 portr. *Gdnrs Chron.* 1922 i 85, 86 portr.; 1944 i 144. *Curtis's Bot. Mag. Dedications, 1827–1927* 291–92 portr. *Kew Bull.* 1930, 96; 1935, 97. *150th Anniversary Vol. R. Bot. Gdn Calcutta* 1942, 6–7 portr. *Nature* v.153, 1944, 426–27; v.180, 1957, 162–63. *Ann. Applied Biol.* 1944, 169. *Obit. Notices Fellows R. Soc.* 1944, 747–70 portr. *Proc. Linn. Soc.* 1943–44, 223–29. *Times* 18 March 1944. *Yb. Amer. Philos. Soc.* 1944, 379–83. *Yb. R. Soc. Edinburgh* 1945, 22–24. *Chronica Botanica* 1946, 374–76. *D.N.B. 1941–1950* 695–96. *Who was Who, 1941–1950* 934.

Portr. by F. A. de B. Footner at Kew. Portr. at Hunt Library.

PRANKERD, Theodora Lisle (1878–1939)
b. London 21 June 1878 *d.* Reading, Berks 11 Nov. 1939

BSc London 1903. DSc 1929. FLS 1919. Lecturer, Bedford College, 1912–17. Lecturer in Botany, University College, Reading, 1917–39. Palaeobotanist. *Studies in Geotropism of Pteridophyta* 1922–36. Contrib. to *Ann. Bot.*, *Bot. Gaz.*, *J. Linn. Soc.*, *New Phytol.*

Nature v.144, 1939, 932–33. *Proc. Linn. Soc.* 1939–40, 371–73.

PRATT, Anne (*afterwards* **Pearless**) (1806–1893)
b. Strood, Kent 5 Dec. 1806 *d.* Shepherd's Bush, London 27 July 1893

Married John Pearless, 1866. *Flowers and their Associations* 1828. *Field, Garden, and Woodland* 1838. *Wild Flowers* 1852 2 vols. *Flowering Plants of Great Britain* 1855 5 vols.

Women's Penny Paper 9 Nov. 1889 portr. *J. Bot.* 1893, 288; 1894, 205–7. *J. Hort. Cottage Gdnr* v.27, 1893, 102. *D.N.B.* v.46, 284–85.

Portr. at Hunt Library.

PRATT, Anne Maria *see* Barkly, *Lady* A. M.

PRATT, Antwerp E. (*fl.* 1880s–1910s)

Explorer and professional zoological collector. Collected insects and some plants in China, 1887–90; New Guinea, 1902–3; S. America, 1912. *To the Snows of Tibet through China* 1891. *Two Years among New Guinea Cannibals* 1906.

J. Linn. Soc. Bot. v.29, 1892, 298–322. E. Bretschneider *Hist. European Bot. Discoveries in China* 1898, 802–5. *Kew Bull.* 1901, 53. E. H. M. Cox *Plant Hunting in China* 1945, 132–35. *Fl. Malesiana* v.1, 1950, 415–16.

Chinese plants at Kew.
Clematis prattii Hemsley.

PRATT, John (–*c.* 1663)

MD Cantab 1645. Fellow Trinity College Cambridge. *Catalogue of Plants of England* (Sloane MSS. 591).

J. Bot. 1871, 15, 175.

PRENTICE, Charles (–1894)
d. Woolloongabba, Queensland 20 April 1894

Of Cheltenham, Glos. Contrib. papers on *Lindsaea* to *J. Bot.* 1872–73.

Phytologist 1854, 127. *Proc. R. Soc. Queensland* 1894, 50. *Rep. Austral. Assoc. Advancement Sci.* 1909, 381. H. J. Riddelsdell *Fl. Gloucestershire* 1948, cxxv. *R.S.C.* v.8, 657.

Letters at Kew. Australian plants at Kew.
Cheilanthes prenticei Luerssen.

PRESCOTT, John D. (–1837)
d. Petrograd, Russia 21 Feb. 1837

Correspondent of W. J. Hooker and J. Lindley. Collected plants in Russia. Undertook *Cyperaceae* for N. Wallich's *List*, 114.

W. J. Hooker *Exotic Fl.* v.2, 1824, t.115. *Companion Bot. Mag.* v.2, 1836, 342–43. A. Lasègue *Musée Botanique de B. Delessert* 1845, 132, 330. H. N. Clokie *Account of Herb. Dept. Bot., Oxford* 1964, 228.

Herb. at Oxford. Letters at Kew. Library sold at Sotheby, 10–13 Aug. 1842.

PRESS, George (*fl.* 1820s–1830s)

Gardener to Edward Gray, Harringay House, Hornsey, London. Cultivated several new varieties of *Camellia japonica* (*C. japonica* var. *punctata*, 'Gray's Invincible Camellia').

E. Bretschneider *Hist. European Bot. Discoveries in China* 1898, 286.

PRESSLY, D. (1828–)
b. Stuartfield, Aberdeen Nov. 1828

Gardener at Knockmannon Lodge, 1845.
Gdnrs Chron. 1876 i 757 portr.

PRESTOE, Henry (1842–1923)
b. Dummer near Andover, Hants 6 Jan. 1842 *d.* Brighton, Sussex 24 Sept. 1923

Kew gardener, 1865. Government Botanist and Superintendent, Botanic Gardens, Trinidad after H. Crueger, 1864–86. *Catalogue of Plants cultivated in Royal Botanical Gardens, Trinidad from 1865 to 1870* 1870.

Kew Bull. Additional Ser. I 1898, 56. I. Urban *Symbolae Antillanae* v.1, 1898, 131; v.3, 1902, 106. *R.S.C.* v.11, 63.

Plants and letters at Kew. Plants at Trinidad Herb.
Prestoëa Hook. f.

PRESTON, Arthur Waters (1855–1931)
b. Norwich, Norfolk 1855 *d.* Norwich 1931
Solicitor, Norwich. Meteorologist and phenologist. Secretary, Norfolk and Norwich Horticultural Society, 1881–88; President. President, Norfolk and Norwich Naturalists Society, 1897–98.
Trans. Norfolk Norwich Nat. Soc. v.13, 1930–31, 187–89.
Herb. at Norwich Museum.

PRESTON, Charles (1660–1711)
b. Lasswade, Midlothian 12 July 1660 *d.* Dec. 1711
MD Edinburgh 1694. Professor of Botany and Master of Physick Garden, Edinburgh, 1706. Correspondent of J. P. de Tournefort, J. Ray, L. Plukenet and J. Petiver. Sent plants from Scotland to H. Sloane.
J. Petiver *Musei Petiveriani* 1695, 266. L. Plukenet *Almagesti Botanici Mantissa Plantarum* 1700, 12. *Mem. Wernerian Nat. Hist. Soc.* v.1, 1811, 69–70. E. Lankester *Correspondence of John Ray* 1848, 380–88. E. J. L. Scott *Index to Sloane Manuscripts* 1904, 436. *Notes R. Bot. Gdn Edinburgh* 1935, 64–102. J. E. Dandy *Sloane Herb.* 1958, 188. H. R. Fletcher and W. H. Brown *R. Bot. Gdn Edinburgh, 1670–1970* 1970, 26–30.
Plants at BM(NH). MSS. at Royal Society.
Prestonia R. Br.

PRESTON, Frederick George (1882–1964)
b. Warborough, Oxfordshire 23 July 1882 *d.* Cambridge 8 Jan. 1964
VMH 1938. Kew gardener, 1904. Foreman, University Botanic Garden, Cambridge, 1909; Superintendent, 1919–47. *Greenhouse* 1951. Contrib. to *R. Hort. Soc. Dict. Gdning* 1951, *Gdnrs Chron.*
J. Kew Guild 1949, 738–39 portr.; 1964, 470–71 portr. *Br. Fern Gaz.* 1964, 195. *Gdnrs Chron.* 1964 i 72 portr.

PRESTON, George (1665/6–1749)
d. Gorton, Lasswade, Midlothian 16 Feb. 1749
Brother of C. Preston (1660–1711). Apothecary. Master of Physick Garden and Professor of Botany, Edinburgh, 1712–38. *Catalogus Omnium Plantarum quas in Seminario Medicinae* 1712; ed. 2 1716. "An indefatigable botanist" (P. Blair *Bot. Essays* 1720 preface).
P. Blair *Misc. Observations...with Remarks in Bot.* 1718, 101. R. Pulteney *Hist. Biogr. Sketches of Progress of Bot. in England* v.2, 1790, 8–9. J. Nichols *Illus. of Lit. Hist. of Eighteenth Century* v.1, 323. E. J. L. Scott *Index of Sloane Manuscripts* 1904, 437. *Notes R. Bot. Gdn Edinburgh* 1935, 103–34. H. R. Fletcher and W. H. Brown *R. Bot. Gdn Edinburgh, 1670–1970* 1970, 30–36.

PRESTON, Isabella (1881–1965)
b. Lancaster 1881 *d.* Ottawa, Canada 31 Dec. 1965
VMM 1935. To Ontario, 1912. Joined staff of Central Experimental Farm, Ottawa, 1920. Specialised in plant hybridisation.
Lily Yb. 1967, 129–31.
Portr. at Hunt Library.

PRESTON, Sir Robert (*fl.* 1790s)
Of Woodford, Essex. Introduced varieties of camellias from China.
E. Bretschneider *Hist. European Bot. Discoveries in China* 1898, 213.

PRESTON, Rev. Thomas Arthur (1838–1905)
b. Westminster, London 10 Oct. 1838 *d.* Thurcaston, Leics 6 Feb. 1905
BA Cantab 1856. FLS 1872. Master, Marlborough College, 1858–85; founder of College Natural History Society. Rector, Thurcaston, 1885. Phenologist. *Fl. Marlborough* 1863. *Flowering Plants of Wiltshire* 1888. 'Phenological Observations around Marlborough' (*J. Bot.* 1865, 203–9).
J. Bot. 1905, 362–64. *Proc. Linn. Soc.* 1904–5, 49–51. A. R. Horwood and C. W. F. Noel *Fl. Leicestershire* 1933, ccxxvi. D. Grose *Fl. Wiltshire* 1957, 38–39. *R.S.C.* v.8, 658; v.11, 64.
Plants at BM(NH), Kew. Letters at Kew.

PRICE, Edward (–1784)
d. Kinnlgwyd, Breconshire 14 Sept. 1784
"Well-known for his researches into the vegetable creation" (*Gent. Mag.* 1784 ii 718).

PRICE, John (1803–1887)
b. Pwll-y-Crochen, Colwyn Bay, Denbighshire 3 June 1803 *d.* Chester, Cheshire 14 Oct. 1887
Educated at Chester, then Shrewsbury. MA Cantab 1828. Settled at Chester. *Old Price's Remains* 1863–64. Papers on proliferous leaves in *Liverpool Nat. J.* 1867, 163–64 and *Proc. Chester Soc. Nat. Sci.* no. 2, 1878, 63–66.
J. Bot. 1888, 32. *R.S.C.* v.8, 662.

PRICE, Rev. Rees (1807–1869)
Of Cwmllynfell. *Y hlysieu-lyft Telluaidd* (Family Herbal) (with Rev. E. Griffiths) 1849.
J. Bot. 1898, 18.

PRICE, Robert (–1761)
Of Foxley, Hereford. Friend of B. Stillingfleet for whose *Observations on Grasses* 1762 he drew plates (Stillingfleet *Misc. Tracts* 1762, 372).
W. Coxe *Lit. Life and Select Works of Benjamin Stillingfleet* v.1, 1811, 160–61 portr.; v.2, 169–82. *D.N.B.* v.46, 341.

PRICE, Thomas ('Carnhuanawe') (1787–1848)
Of Cwmdu, Breconshire. Cultivated British plants.
J. Williams *Lit. Remains of Rev. T. Price* 1854, 280. *J. Bot.* 1898, 16.

PRICE, William Robert (1886–1975)
d. 30/31 July 1975
BA Cantab 1908. FLS 1909. Temporary Assistant, Kew Herb., 1909. Collected plants in Formosa, 1912. Contrib. to *Proc. Cotteswold Nat. Field Club.*
MSS. *Plant Collecting in Formosa* and *Life of a Botanist* at Kew.

PRICHARD, Hesketh Vernon Hesketh- (1876–1922)
b. India 17 Nov. 1876 *d.* Gorhambury, St. Albans, Herts 14 June 1922
Traveller, big-game hunter, naturalist. *Through the Heart of Patagonia* 1902 contains list of plants he collected in 1900. 'Mr. Hesketh Prichard's Patagonian Plants' (*J. Bot.* 1904, 321–34, 367–78).
J. Bot. 1922, 214–15. *Times* 15 June 1922. E. Parker *Hesketh Prichard...a Mem.* 1924.
Fagelia prichardi Rendle.

PRICKETT, George (*c.* 1844–1941)
d. Enfield, Middx 16 Jan. 1941
Nurseryman specialised in chrysanthemums. President, National Chrysanthemum Society, 1933.
Gdnrs Chron. 1941 i 38.

PRIEST, Myles (*fl.* 1830s)
Nurseryman, Reading, Berks.

PRIEST, Robert (*c.* 1560–*c.* 1596)
b. Middx *c.* 1560 *d.* London *c.* 1596
MD Cantab 1580. FRCP 1583–89? Began translation of R. Dodoen's *Pemptades*, the basis of J. Gerard's *Herball* 1597.
M. de Lobel *Animadversiones* 1605, 59; *Stirpium Illustrationes* 1655, 13. J. Gerard *Herball* 1633 preface. R. Pulteney *Hist. Biogr. Sketches of Progress of Bot. in England* v.1, 1790, 119. B. D. Jackson *Cat. of Plants...in Garden of J. Gerard* 1876, xiii. W. Munk *Roll of R. College of Physicians* v.1, 1878, 98. R. H. Jeffers *Friends of John Gerard; Biogr. Appendix* 1969, 26–27.

PRIESTLEY, Joseph (1733–1804)
b. Birstall, Yorks 13 March 1733 *d.* Northumberland, Pa., U.S.A. 6 Feb. 1804
LLD Edinburgh 1764. FRS 1766. To N. America, 1794. *Experiments relating to Natural Philosophy* 1781 (dealing with chlorophyllian action). *Memoirs* 1810.
J. Corry *Life of J. Priestley* 1804. H. C. Bolton *Sci. Correspondence of Joseph Priestley* 1892. T. E. Thorpe *Joseph Priestley* 1906. *Trans. Liverpool Bot. Soc.* 1909, 86–87. *D.N.B.* v.46, 357–76. A. Holt *Life of Joseph Priestley* 1931. *Proc. R. Inst.* 1931, 395–430. *Isis* 1934, 81–97 portr. *Notes Rec. R. Soc.* 1939, 32–33 portr. *Proc. Amer. Philos. Soc.* 1942, 103–7. *Chronica Botanica* 1944, 139–45.
Portr. by J. Millar at Royal Society. Portr. at Hunt Library.
Priestleya DC.

PRIESTLEY, Joseph Hubert (1883–1944)
b. Tewkesbury, Glos 5 Oct. 1883 *d.* Leeds, Yorks 31 Oct. 1944
BSc London 1905. FLS 1908. Head of Dept. of Botany, Bristol University, 1905–11. Professor of Botany, Leeds, 1911–44. President, Yorkshire Naturalists' Union, 1925. Authority on developmental morphology and anatomy. *Introduction to Botany* (with L. Scott) 1938.
Nature v.154, 1944, 694–95. *Proc. Linn. Soc.* 1943–44, 231–32. *Times* 4 Nov. 1944. *Bot. Soc. Exch. Club Br. Isl. Rep.* 1945, 15. *Naturalist* 1945, 17–18 portr. *New Phytol.* 1946, 3–4.

PRIESTLEY, Sir William Overend (1829–1900)
b. Leeds, Yorks 24 June 1829 *d.* London 11 April 1900
MD Edinburgh 1853. LLD 1884. FLS 1888. KCB 1893. Physician, King's College Hospital, London. 'British Species of *Carex*' (*Trans. Bot. Soc. Edinburgh* v.4, 1853, 71–74).
Proc. Linn. Soc. 1899–1900, 81–82. *D.N.B. Supplt. 1* v.3, 287–88. *R.S.C.* v.5, 20.
Plants and drawings at BM(NH).

PRINCE, Edith (1877–1953)
b. Sept. 1877 *d.* 9 Sept. 1953
Teacher. Organised herb. of Essex Field Club.
Essex Nat. v.29, 1954, 195, 212–13.
Herb. at Essex Field Club.

PRINCE, George (*c.* 1831–1896)
d. Oxford 3 March 1896
Rose-grower, Longworth, Oxford, 1866.
Gdnrs Mag. 1896, 155 portr. *J. Hort. Cottage Gdnr* v.32, 1896, 258.

PRINCE, John (*fl.* 1770s–1820s)
To Fort Marlborough, Bencoolen, 1787. Assistant to Resident of Natal, 1790; Resident, 1794–98. Resident of Sumatra. Sent plants from Sumatra to W. Roxburgh and N. Wallich. To Straits as Resident Councillor, taking an active interest in the attempts to keep going the Botanical Garden started by Raffles.
Fl. Malesiana v.1, 1950, 416–17; v.8(1), Supplt 2, 1974, 77.
Drawings at Kew.

PRINCE, John Critchley (1808–1866)
b. Wigan, Lancs 21 June 1808 *d.* Hyde, Cheshire 5 May 1866
Of Ashton-under-Lyne, Lancs. Working-man naturalist and poet.
Heywood Advertiser 30 Nov. 1917; 11 Jan. 1918.

PRING, George Henry (1885–1974)
b. Exmouth, Devon 2 Dec. 1885 *d.* West Chester, Pa., U.S.A. 9 May 1974
Kew gardener, 1899. Foreman, Missouri Botanical Garden, 1906; Superintendent, 1928–63. Collected orchids in Andes, 1923; further collecting in Panama, West Indies and Costa Rica.
Gdnrs Chron. 1934 i 208 portr.; 1948 ii 148 portr. *J. Kew Guild* 1936, 535–37; 1950, 827–30 portr; 1974, 347–49. *Amer. Orchid Soc. Bull.* 1963, 282–86 portr.; 1974, xxiv. *Missouri Bot. Gdn Bull.* v.51, 1963, 1–11 portr.; v.63(5), 1974, 2p.
Portr. at Hunt Library.

PRINGLE, William (*c.* 1742–1813)
Nurseryman, Sydenham, Kent; later Chelsea.
Bot. Mag. 1814, t.1617; 1821, t.2286.

PRINSEP, H. C. (–*c.* 1930)
Gardener to Lord Portman at Buxted Park, Sussex. Cultivated orchids.
Gdnrs Mag. 1908, 485–86 portr. *Orchid Rev.* 1930, 315.

PRIOR, David (*c.* 1832–1916)
d. May 1916
Nurseryman, The Nurseries, Colchester, Essex until 1901.
J. Hort. Home Farmer v.65, 1912, 299–300 portr. *Gdnrs Mag.* 1916, 232.

PRIOR, Richard Chandler Alexander (*olim* Alexander) (1809–1902)
b. Corsham, Wilts 6 March 1809 *d.* London 6 Dec. 1902
BA Oxon 1830. MD 1837. FLS 1851. Took name of Prior in 1859. One time curator of Fielding Herb., Oxford. Collected plants in S. Africa, 1846–48; West Indies, 1849; Canada, 1850; S. Europe, etc. *Popular Names of British Plants* 1863; ed. 3 1879.
I. Urban *Symbolae Antillanae* v.3, 1902, 107. *Gdnrs Chron.* 1902 ii 460–61; 1903 i 137. *J. Bot.* 1903, 108–9. *Kew Bull.* 1903, 32; 1909, 317–18. *Lancet* 1903 i 339. *Proc. Linn. Soc.* 1902–3, 35–37. D. Grose *Fl. Wiltshire* 1957, 36–37.
Herb. and letters at Kew. Plants at BM(NH). Portr. at Hunt Library.
Prioria Griseb.

PRIOR, William D. (*fl.* 1910s)
Nurseryman, Colchester, Essex.

PROSSER, L. N. (1910–1970)
b. Worthing, Sussex 20 March 1910 *d.* 22 March 1970
Kew gardener, 1934–38. Head gardener, Salford Corporation. Assistant District Superintendent, Johannesburg, 1948. Superintendent of Parks, Port Elizabeth, 1958. Had herb. Sent S. African plants to Kew.
J. Kew Guild 1970, 1159–60 portr.

PROTHEROE, Alexander (*c.* 1803–1885)
d. 9 Dec. 1885
With Thomas Morris founded firm of horticultural auctioneers and nurserymen at Highbury, London, *c.* 1830. Nursery at Leytonstone, Essex, formerly of J. Hay and J. and R. Hill.
Gdnrs Chron. 1885 ii 818.

PROTHEROE, William Henry (1846–1899)
b. Leytonstone, Essex 1846 *d.* Leytonstone 2 Dec. 1899
Son of A. Protheroe (*c.* 1803–1885). Horticultural auctioneer, Cheapside, London, 1883.
J. Hort. Cottage Gdnr v.26, 1893, 431–32 portr. *Gdnrs Chron.* 1899 ii 440 portr. *Gdnrs Mag.* 1899, 776 portr.

PROUDLOCK, Robert Louis (1862–1948)
b. Hepscott Moor, Northumberland 10 Nov. 1862 *d.* Grouville, Jersey 27 Sept. 1948
Gardener, Royal Botanic Garden, Edinburgh, 1883. Kew gardener, 1886. Assistant Curator, Royal Botanic Garden, Calcutta, 1889; Curator, 1891–96. Collected plants in Lower Burma, 1892. Curator, Botanic Gardens, Ootacamund, 1896. Collected plants in Nilgiris for Botanical Survey of India. Retired from Indian Government service, 1918. Collected plants in Iceland, 1934.
J. Kew Guild 1936, 501–3 portr.; 1948, 690–91 portr.
Iceland and Jersey plants at Kew.

PRYER, William Burgess (–1899)
Member of A. Dent's expedition to N. Borneo, 1877. Agent and Resident of British North Borneo Co. at Sandakan. Collected plants, chiefly orchids.
Fl. Malesiana v.1, 1950, 417–18.
Begonia pryeriana Ridley.

PRYOR, Alfred Reginald (1839–1881)
b. Hatfield, Herts 24 April 1839 *d.* Baldock, Herts 18 Feb. 1881
BA Oxon 1862. FLS 1874. *Fl. Hertfordshire* 1887. Contrib. to *J. Bot.*
J. Bot. 1881, 276–78. *Proc. Linn. Soc.* 1880–82, 19. *D.N.B.* v.46, 437–38. G. C. Druce *Fl. Buckinghamshire* 1926, civ. G. C. Druce *Fl. Oxfordshire* 1927, cxxi. J. G. Dony *Fl. Bedfordshire* 1953, 25. J. G. Dony *Fl. Hertfordshire* 1967, 16.

Herb. formerly at St. Albans Museum. Portr. at Hunt Library.
Papaver rhoeus var. *Pryorii* Druce.

PUDDLE, Frederick Charles (*c.* 1877–1952)
d. Haslemere, Surrey 23 Aug. 1952
VMH 1937. VMM 1946. Gardener to Lord Aberconway at Bodnant, Denbighshire.
Gdnrs Chron. 1952 ii 90. *J. R. Hort. Soc.* 1952, 360–61. *Orchid Rev.* 1952, 147–48.

PUGSLEY, Herbert William (1868–1947)
b. Bristol 24 Jan. 1868 *d.* Wimbledon, Surrey 18 Nov. 1947
BA London 1889. FLS 1920. Civil servant in Admiralty, 1896–1928. Contrib. papers on *Fumaria*, *Rupicapnos*, *Euphrasia*, *Narcissus* and *Hieracium* to *J. Bot.*, *J. Linn. Soc.*, *J. R. Hort. Soc.*
Gdnrs Chron. 1947 ii 200. *Times* 22 Nov. 1947. *Daffodil Tulip Yb.* 1948, 90–91. *Naturalist* 1948, 13–15. *Nature* v.161, 1948, 121–22. *Proc. Linn. Soc.* 1947–48, 187–93. *Watsonia* 1949–50, 124–30.
Herb. at BM(NH).

PULHAM, James Robert (1873–1957)
d. 20 April 1957
Specialist in construction of rock gardens. Secretary, British Pteridological Society, 1948–50.
Gdnrs Mag. 1912, 105–6 portr. *Br. Fern Gaz.* 1958, 203–4.

PULLEN, John (*fl.* 1740s)
Nurseryman, Bedminster, Somerset.

PULTENEY, Richard (1730–1801)
b. Mountsorrel Leics 17 Feb. 1730 *d.* Blandford, Dorset 13 Oct. 1801
MD Edinburgh 1764. FRS 1762. FLS 1790. Surgeon at Leicester and from 1765 at Blandford. Contrib. plant lists to *Philos. Trans. R. Soc.* v.49, 1757, 803–66, J. Hutchin *Hist. Antiq. Dorset* 1774 2 vols, J. Nichols *Hist. Antiq. County of Leicester* 1795–1811 4 vols. *General View of Writings of Linnaeus* 1781. *Historical and Biographical Sketches of Progress of Botany in England* 1790 2 vols.
J. Nichols *Lit. Anecdotes of Eighteenth Century* v.8, 196 portr. *Gent. Mag.* 1801, 1058, 1207. *Cat. Library of Nat. Hist., Medicine, etc. of R. Pulteney* 1802. M. Kirby *Fl. Leicestershire* 1850, xi, xv. *Cottage Gdnr* v.8, 1852, 314–15. G. C. Gorham *Mem. John and Thomas Martyn* 1830, 102–3. *Proc. Linn. Soc.* 1888–89, 39; 1958–59, 15–26. W. Munk *Roll of R. College of Physicians* v.2, 1878, 264–65. *D.N.B.* v.47, 26–28. *J. Bot.* 1919, 100. A. R. Horwood and C. W. F. Noel *Fl. Leicestershire* 1933, clxxxix–cxciv portr. F. A. Stafleu *Taxonomic Literature* 1967, 368–69.
Plants at BM(NH). MS. *Fl. Anglica* and Catalogue of English plants at BM(NH), MS. *Catalogus Stirpium* 1747 at Leicester Museum, MS. *Opusculum Botanicum* 1749 at Linnean Society. Letters at BM(NH), Linnean Society, Fitzwilliam Museum Cambridge. Portr. by S. Beach at Linnean Society. Sale at Leigh Sotheby 26 April 1802.
Pultenaea Smith.

PURCHAS, Rev. William Henry (1823–1903)
b. Ross, Herefordshire 12 Dec. 1823 *d.* Alstonfield, Staffs 16 Dec. 1903
BA Durham 1857. Held livings of Ticknall, Derbyshire, 1857–65, Lydney, Glos, 1865–70, Alstonfield, 1870–1903. Studied *Rubus*, *Rosa*, *Hieracium*, etc. *Fl. Herefordshire* (with A. Ley) 1889. Contrib. to *Bot. Gaz.*, *J. Bot.*
J. Bot. 1904, 80–82 portr. *Trans. Woolhope Nat. Field Club* 1902–4, 341–44 portr.; 1954, 248–49. A. R. Horwood and C. W. F. Noel *Fl. Leicestershire* 1933, ccxviii. H. J. Riddelsdell *Fl. Gloucestershire* 1948, cxxxi. *R.S.C.* v.5, 43; v.8, 673.
Plants at BM(NH), Kew, Oxford. MS. list of Aberystwyth plants at Kew. Portr. at Hunt Library.
Rubus purchasianus Rogers.

PURDIE, Alexander (1859–1905)
b. Edinburgh 23 Oct. 1859 *d.* Perth, W. Australia 17 July 1905
MA. Professor of Geology, Ballarat. Director of Education, W. Australia. Orchidologist. Contrib. to *J. Proc. Mueller Bot. Soc. W. Australia.*
J. W. Austral. Nat. Hist. Soc. no. 6, 1909, 24.
Plants at Melbourne.
Boronia purdieana Diels.

PURDIE, William (*c.* 1817–1857)
b. Scotland *c.* 1817 *d.* Trinidad 10 Oct. 1857
Gardener, Royal Botanic Garden, Edinburgh. Superintendent, Botanic Gardens, Trinidad, 1846–57. Collected plants in Trinidad, Jamaica, Colombia, Venezuela and other parts of tropical S. America. 'Journal of Botanical Mission to West Indies in 1843–4' (*London J. Bot.* 1844, 501–33; 1845, 14–27).
London J. Bot. 1847, 40–41. *Gdnrs Chron.* 1857, 792. *Kew Bull. Additional Ser.* 1898, 55–56. I. Urban *Symbolae Antillanae* v.1, 1898, 132; v.3, 1902, 107–8. *Caldasia* v.5(21), 1948, 90–95. *J. Arnold Arb.* 1965, 269–70. A. M. Coats *Quest for Plants* 1969, 348. *R.S.C.* v.5, 43.
Plants at Kew, Trinidad. Letters at Kew.
Purdiaea Planchon.

PURDOM, William (1880–1921)
b. Heversham, Westmorland 10 April 1880 *d.* Peking 7 Nov. 1921

Kew gardener, 1902–8. Plant collector for Messrs Veitch and Arnold Arboretum in China, 1909–12. Travelled in Kansu region with Reginald Farrer. Forestry adviser to Chinese Government, 1915–21.

Gdnrs Chron. 1921 ii 294. *J. Arnold Arb.* v.3, 1921, 55–56. *Kew Bull.* 1921, 408. *J. Kew Guild* 1922, 115–16 portr. E. H. M. Cox *Plant Hunting in China* 1945, 174. C. H. Hough *Westmorland Rock Gdn* 1948, 51–54. A. M. Coats *Quest for Plants* 1969, 132–36.

Plants at Kew.

Allium purdomii W. W. Smith.

PURSH (*originally* Pursch), Friedrich Traugott (1774–1820)
b. Grossenhain, Saxony 4 Feb. 1774 *d.* Montreal, Canada 11 July 1820

Travelled in U.S.A., 1799–1811, and afterwards in Canada. In England, 1811–15. *Fl. Americae Septentrionalis* 1814 2 vols. *Journal of Botanical Excursion in...1807* 1869 (original MS. at American Philosophical Society). Edited eds. 8 and 9 and joint editor of later eds. of J. Donn's *Hortus Cantabrigiensis.*

Amer. J. Sci. Arts 1825, 269–74; 1882, 323–26. *J. Bot.* (Hooker) 1841, 107–8; 1857, 256–57. A. Lasègue *Musée Botanique de B. Delessert* 1845, 460–61. *Bonplandia* 1857, 149–50. *J. Bot.* 1870, 63–64. *Bot. Gaz.* v.7, 1882, 141–43. *Proc. Trans. R. Soc. Canada* 1897, 3–6. *Proc. Acad. Nat. Sci. Philadelphia* 1898, 13. J. W. Harshberger *Botanists of Philadelphia and their Work* 1899, 113–17. I. Urban *Symbolae Antillanae* v.3, 1902, 108–9. *Torreya* 1904, 132–36. *J. Acad. Sci. Washington* 1927, 351. *Rhodora* 1943, 415–16. *Bartonia* 1935, 24–32; 1949, 27–29. *Proc. Amer. Philos. Soc.* 1950, 142–44; 1952, 599–628. *Yb. Amer. Philos. Soc.* 1955, 160–63. H. B. Humphrey *Makers of N. Amer. Bot.* 1961, 202–4. F. A. Stafleu *Taxonomic Literature* 1967, 369. *Taxon* 1970, 536–37. C. C. Gillispie *Dict. Sci. Biogr.* v.11, 1975, 217–19.

Plants at BM(NH).

Purshia DC.

PURTON, Thomas (1768–1833)
b. Endon Burnell, Bridgnorth, Shropshire 10 May 1768 *d.* Alcester, Warwickshire 29 April 1833

FLS 1821. Surgeon; practised in London, 1791–95 and in Alcester. Mycologist. *Botanical Description of British Plants in Midland Counties* 1817–21 2 vols.

Mag. Nat. Hist. 1836, 606–70. E. Lees *Bot. Worcestershire* 1867, lxxxix. J. E. Bagnall *Fl. Warwickshire* 1891, 495–96. G. C. Druce *Fl. Berkshire* 1897, clvii. *J. Bot.* 1901 Supplt. 72. *R.S.C.* v.5, 46.

Fungi, MSS. and letters at Kew. Herb. at Worcester Museum.

PURVES, John McLennan (1876–)
b. 11 Dec. 1876

Kew gardener, 1899–1900. Government forester, Nyasaland, 1900; Head of Botany and Forestry Dept., 1906–23. Collected plants in Nyasaland, now at Kew.

Comptes Rendus AETFAT 1960 1962, 168.

QUEKETT, Edwin John (1808–1847)
b. Langport, Somerset Sept. 1808 *d.* 28 June 1847

FLS 1836. Microscopist and surgeon. Lecturer in Botany, London Hospital, 1835. Contrib. papers on ergot to *Trans. Linn. Soc.* v.18, 1841, 453–73; v.19, 1845, 137–42.

Bot. Register 1839, Misc. Notes, 4. *Proc. Linn. Soc.* 1848, 378–79. *Phytologist* 1848, 110. *D.N.B.* v.47, 98. *London Hospital Gaz.* 1969, 15, 19.

Quekettia Lindl.

QUEKETT, Eliza Catherine *see* White, E. C.

QUEKETT, J. F. (*fl.* 1870s)

Resident at Shanghai. Made small collection of Chinese plants which he presented to Kew.

E. Bretschneider *Hist. European Bot. Discoveries in China* 1898, 701.

QUEKETT, John Thomas (1815–1861)
b. Langport, Somerset 11 Aug. 1815 *d.* Pangbourne, Berks 20 Aug. 1861

FRS 1860. FLS 1857. Brother of E. J. Quekett (1808–1847). Microscopist. Professor of Histology, Royal College of Surgeons, 1852. *Lectures on Histology* 1852. *Catalogue of Fossil Plants in Museum of Royal College of Surgeons* 1855.

Proc. Linn. Soc. 1861–62, xciii–xciv. *J. R. Microsc. Soc.* 1895, 6–7. *D.N.B.* v.47, 97–99. *R.S.C.* v.5, 53.

QUILTER, H. E. (*c.* 1850–1915)

Amateur geologist and botanist. Contrib. paper on microfungi to *Trans. Leics. Lit. Philos. Soc.*

A. R. Horwood and C. W. F. Noel *Fl. Leicestershire* 1933, ccxxxiii.

RABITTI, Anna, Contessi di San Giorgio (*née* Harley) (1803–1874)
b. Florence, Italy 31 July 1803 *d.* Florence 18 May 1874

Catalogo poliglotto delle piante 1870.

RABONE, Thomas H. (1833–1895)
b. Wellesbourne, Warwickshire 16 April 1833 *d.* 20 July 1895

Gardener to Lord Shrewsbury at Alton Towers, Staffs.

Gdnrs Chron. 1876 ii 489–90 portr.; 1895 ii 105 portr., 137–38. *J. Hort. Cottage Gdnr* v.31, 1895, 104.

RADCLIFFE, Rev. John (1749–1828)
b. Oxford 6 July 1749 *d.* Much Hadham, Herts July 1828

Professor of Divinity, Oxford. Bishop of Oxford, 1799; of Bangor, 1807; of London, 1809. Botanist.

Bot. Soc. Exch. Club. Br. Isl. Rep. 1917, 142.

RADCLYFFE, Rev. William Frederick (*c.* 1801–1880)
d. Okeford Fitzpaine, Dorset 8 July 1880

BA Oxon 1929. Rector, Rushton; Okeford Fitzpaine, 1866–80. Wrote horticultural articles.

Gdnrs Chron. 1880 ii 78.

RAE, John (1813–1893)
b. Clestrain, Stromness, Orkney 30 Sept. 1813 *d.* Kensington, London 22 July 1893

MD Edinburgh 1833. LLD. FRS 1880. Arctic explorer. *Narrative of Arctic Expedition in 1846–47* 1850.

Canadian Rec. Sci. v.5, 1893, 484–87. *Proc. R. Soc.* v.60, 1897, v–vii. *D.N.B.* v.47, 151–53.

Portr. at National Portrait Gallery.

RAFFILL, Charles P. (1876–1951)
b. Tredegar, Mon April 1876 *d.* Coventry, Warwickshire 27 March 1951

VMH 1939. Kew gardener; Assistant Curator. Authority on rhododendrons, lilies, irises, fuchsias.

Gdnrs Chron. 1951 i 66 portr., 112. *Iris Yb.* 1951, 14–15. *J. Kew Guild* 1951, 43 portr. *Lily Yb.* 1951–52, 134–35 portr. *Times* 29 March 1951.

RAFFLES, Sir Thomas Stamford Bingley (1781–1826)
b. at sea off Jamaica 5 July 1781 *d.* Highwood Hill, Middx 4 July 1826

FRS 1817. FLS 1825. East India Company official. Under-Secretary, Pulo Penang, 1805. Lieut.-Governor of Java, 1811–16. Governor, Bencoolen, 1818. Founder and first President of Zoological Society of London. *History of Java* 1817.

S. Raffles *Mem. of the Life...of T. S. Raffles* 1830 portr. *D.N.B.* v.47, 161–65. D. C. Boulger *Life of Sir Stamford Raffles* 1897 portr.; new ed. 1973. R. Coupland *Raffles, 1781–1826* 1926 portr. J. A. B. Cook *Sir Stamford Raffles* 1918. E. Hahn *Raffles of Singapore* 1948. *Fl. Malesiana* v.1, 1950, 425; v.5, 1958, 79. C. E. Wurtzburg *Raffles of the Eastern Isles* 1954 portr. M. Archer *Nat. Hist. Drawings in India Office Library* 1962, 15–17, 88. J. S. Bastin *Raffles in Java and Sumatra* 1957 portr. M. Collis *Raffles* 1966 portr. *Straits Times Annual* 1969, 35–37 portr.; 1971, 58–63. C. C. Gillispie *Dict. Sci. Biogr.* v.11, 1975, 261–62.

Plants at Kew. MSS. and drawings at India Office Library.

Rafflesia R. Br.

RAGG, Lonsdale (1866–1945)
b. Wellington, Shropshire 23 Oct. 1866 *d.* 31 July 1945

Educ. Oxford. Prebendary of Buckden, Lincoln Cathedral. Archdeacon of Gibraltar, 1934. Founder and editor of the *Tree Lover* from 1932 and known for his drawings of trees. *Some of my Tree Friends* 1931. *Trees I have met* 1933. *Tree Lore in the Bible* 1935.

Nature v.156, 1945, 289. *Who was Who, 1941–1950* 950.

RAINE, Frederick (1851–1919)
b. Durham 16 Sept. 1851 *d.* Hyères, France 24 April 1919

His plant records are in E. Jahandiez's *Les Iles d'Hyères* 1905.

J. Bot. 1920, 30–31.

Herb. at Museum, Newcastle-upon-Tyne.

RALFS, John (1807–1890)
b. Millbrook, Southampton, Hants 13 Sept. 1807 *d.* Penzance, Cornwall 14 July 1890

MRCS 1832. Practised briefly as a surgeon but settled in Penzance, 1837 and devoted himself to botany. President, Penzance Natural History and Antiquarian Society, 1883–84. *British Phaenogamous Plants and Ferns* 1839. *British Desmidieae* 1848. Joint editor of A. Pritchard's *History of Infusoria* ed. 4 1861.

J. Bot. 1890, 289–93 portr. *J. Hort. Cottage Gdnr* v.21, 1890, 71–72. *Trans. Penzance Nat. Hist. Soc.* 1890–91, 225–40. *D.N.B.* v.47, 209–10. A. G. Lewis *John Ralfs, an Old Cornish Botanist* 1907 portr. F. H. Davey *Fl. Cornwall* 1909, lvii–lviii portr. J. E. Lousley *Fl. Isles of Scilly* 1972, 83. *R.S.C.* v.5, 80; v.12, 597.

Published set of British algae at BM(NH). MS. Fl. of W. Cornwall and Scilly at Penzance Library. Plants at Kew, South London Botanical Institute. Letters at BM(NH), Kew. Portr. at Hunt Library.

Ralfsia Berkeley.

RALPH, Thomas Shearman (1813–1891)
d. Melbourne 22 Dec. 1891

MRCS. ALS 1842. To Australia, 1851. Practised medicine in Melbourne. *Elementary Botany* 1849. *Icones Carpologicae* 1849. Edited T. Johnson's *Opuscula Omnia Botanica* 1847.

Argus (Melbourne) 23 Dec. 1891. *Austral. Med. Gaz.* 1892, 117. *Victorian Nat.* 1908, 111. *R.S.C.* v.5, 81; v.8, 689; v.11, 96.

New Zealand plants at Kew.

RAM, William (*c.* 1535–1602)
b. Pleshey, Essex *c.* 1535 *d.* Colchester, Essex 1602

Educ. Cambridge. Deputy Town Clerk, Colchester, 1573–78. *Rams Little Dodeon* 1606 (epitome of H. Lyte *Nievve Herball* 1578).

J. Gerard *Herball* 1597, 278. R. H. Jeffers *Friends of John Gerard; Biogr. Appendix* 1969, 28.

RAMAGE, George A. (1864–1933)
b. Edinburgh 1864 *d.* Dominica 11 Jan. 1933

Educ. Edinburgh. Collected plants on H. N. Ridley's expedition to Fernando de Noronha, 1887. Collected for West Indian Commission to Dominica.

I. Urban *Symbolae Antillanae* v.3, 1902, 109. *Rep. Br. Assoc. Advancement Sci.* 1888, 437–38; 1889, 93–94; 1891, 448. *J. Bot.* 1933, 75.

Plant drawings at BM(NH).

RAMSAY, Christina, Countess of Dalhousie (*née* Broun) (1786–1839)
b. 28 Feb. 1786 *d.* Edinburgh 22 Jan. 1839

Wife of 9th Earl of Dalhousie, Commander-in-Chief, East Indies, 1829–32. Collected plants in Nova Scotia, 1816–28; India and Penang.

Gdnrs Mag. 1826, 255. *Trans. Bot. Soc. Edinburgh* 1836–37, 50; 1837–38, 40; 1838–39, 52. J. D. Hooker and T. Thomson *Fl. Indica* 1855, 70. *Curtis's Bot. Mag. Dedications, 1827–1927* 27–28 portr.

Herb. at Edinburgh, Kew. Letters at Kew. *Dalhousiea* Graham.

RAMSAY, James (1812–1888)
b. Kilwinning, Ayrshire 1812 *d.* Glasgow 10 Sept. 1888

Lecturer in Botany, Glasgow Mechanics Institute, 1867–68. Contrib. papers on Scottish plants to *Proc. Trans. Nat. Hist. Soc. Glasgow*.

Proc. Trans. Nat. Hist. Soc. Glasgow v.3, 1888–89, vii–viii.

RAMSAY, Robert (*fl.* 1750s)

MD Edinburgh 1757. Sent British plants to J. Hope (1725–1786).

Notes from R. Bot. Gdn Edinburgh v.4, 1907, 126.

RAMSBOTTOM, James Kirkham (1891–1925)
b. Manchester 11 Oct. 1891 *d.* New York 9 Feb. 1925

Gardener, Chelsea Physic Garden; Royal Horticultural Society, Wisley, 1911. Investigated diseases of daffodils. Assistant Editor, *Gdnrs Mag.*, 1914–16. Assistant Editor, *Gdnrs Chron.*, 1924. 'Iris Leaf-blotch Disease' (*J. R. Hort. Soc.* 1915, 481–92). 'Investigations on Narcissus Disease' (*J. R. Hort. Soc.* 1918, 51–64).

Gdnrs Chron. 1925 i 68 portr., 120. *J. Bot.* 1925, 85. *Daffodil Yb.* 1940, 2–4 portr.

RAMSBOTTOM, John (1885–1974)
b. Manchester 25 Oct. 1885 *d.* 14 Dec. 1974

BA Cantab 1909. FLS 1914. VMH 1950. VMM 1944. Mycologist, BM(NH), 1910; Deputy Keeper of Botany, 1928; Keeper, 1930–51. President, British Mycological Society, 1924, 1946. President, Linnean Society, 1937–41. President, Essex Field Club, 1937. President, Society for Bibliography of Natural History, 1944–72. Secretary, British Mycological Society, 1921–45; Editor, 1919–42. Secretary, Linnean Society, 1923–35. *Handbook of Larger British Fungi* 1923. *Book of Roses* 1941. *Edible Fungi* 1943. *Poisonous Fungi* 1945. *Mushrooms and Toadstools* 1953. Contrib. to *Proc. Linn. Soc.*, *Trans. Br. Mycol. Soc.*, *Essex Nat.*

J. Bot. 1910, 216; 1917, 167; 1928, 32; 1929, 344. *Times* 1 June 1938; 17 Dec. 1974. *Gdnrs Chron.* 1922 ii 246 portr.; 1930 i 38 portr.; 1950 i 60 portr. *Nature* v.166, 1950, 387–88. *Discovery* 1950, 371 portr. *Essex Nat.* 1951, 271. *Trans. Br. Mycol. Soc.* 1966, 1–2; 1975, 1–6 portr. *Proc. Linn. Soc.* 1966, 105–6. *Essex Nat.* v.33, 1973–74, 110. *B.S.E. News* no. 15, 1975, 11–13.

Portr. at Hunt Library.

RAMSDEN, Hildebrand (1836–1899)

Herb. at Charterhouse School.

RANADE, N. B. (–1897)
d. Poona, India 15 Oct. 1897

Curator, Herb., Poona.

Rep. Bot. Survey India 1897–98, 10.

RAND, Isaac (–1743)
d. London 1743

FRS 1719. Apothecary. Praefectus Horti Chelsiani, 1724–43. Helped E. Blackwell with her *Curious Herbal* 1737. *Index Plantarum Officinalium...in Horto Chelseiano* 1730. *Horti Medici Chelseiani Index Compendiarius...* 1739.

J. Ray *Synopsis Methodica* 1724, preface. L. Plukenet *Almagesti Botanici Mantissa Plantarum* 1700, 112. J. Nichols *Illus. Lit. Hist. of Eighteenth Century* v.1, 338–39. D. Turner *Extracts from Lit. and Sci.*

Correspondence of R. Richardson 1835, 125. H. Trimen and W. T. T. Dyer *Fl. Middlesex* 1869, 388–89. R. H. Semple *Mem. Bot. Gdn at Chelsea* 1878, 41–63. *Notes and Queries* v.10, 1896, 193. E. J. L. Scott *Index to Sloane Manuscripts* 1904, 443. *D.N.B.* v.47, 268–69. G. C. Druce *Fl. Berkshire* 1897, cxxxvi–cxxxvii. C. E. Salmon *Fl. Surrey* 1931, 46–47. J. E. Dandy *Sloane Herb.* 1958, 229. C. Wall *Hist. Worshipful Soc. of Apothecaries of London* v.1, 1963, 169–71 portr. W. S. C. Copeman *Worshipful Soc. of Apothecaries of London* 1968, 78–79.

Plants at BM(NH).

Randia L.

RAND, Richard Frank (1856–1937)
b. Plaistow, Essex 12 Oct. 1856 *d.* Brightlingsea, Essex 3 Jan. 1937

MD Edinburgh 1889. FRCS 1883. FLS 1898. Medical officer, British South Africa Company. Collected plants in Mashonaland, Rhodesia, 1890. Contrib. 'Wayfaring Notes' to *J. Bot.*, 1898–1926.

J. Bot. 1937, 79–80. *Times* 6 Jan. 1937. *Comptes Rendus AETFAT 1960* 1962, 168–69.

Plants at BM(NH).

RANDAL, Mr. (*fl.* 1700s)

Gardener at Fort St. George, Madras. *Eriocaulon quinquangulare* collected by him and sent to C. Du Bois.

J. E. Dandy *Sloane Herb.* 1958, 189.

RANDALL, Mr. (–1896)
d. 16 Dec. 1896

Nurseryman, Exeter, Devon.

J. Hort. Cottage Gdnr v.33, 1896, 629.

RANDALL, Harry J. (–1967)
d. May 1967

Breeder of irises, narcissi and hemerocallis.

Iris Yb. 1967, 13–15 portr. *Daffodil Tulip Yb.* 1968, 179 portr.

RANDOLPH, Rev. Francis (1713–1797)
d. Oxford 18 Feb. 1797

MA Oxon 1736. DD 1763. Principal, Alban Hall, Oxford, 1759. "A skilful botanist, and well acquainted with most branches of natural history."

Gent. Mag. 1797 i 254.

RANDOLPH, Rev. John (1749–1813)
b. Oxford 6 July 1749 *d.* Much Hadham, Herts 28 July 1813

BA Oxon 1771. DD 1783. FRS 1811. Rector, Ewelme, Oxford, 1796–99. Bishop of Oxford, 1799; Bangor, 1807; London, 1809. List of Ewelme plants in *Rep. Ashmolean Nat. Hist. Soc.* 1917, 23–42.

D.N.B. v.47, 274–75. *Bot. Soc. Exch. Club Br. Isl. Rep.* 1917, 142 (as Radcliffe). *J. Bot.* 1918, 335. G. C. Druce *Fl. Oxfordshire* 1927, cv–cvi.

RANGER, Hugh (*c.* 1850–1912)
d. Liverpool 27 May 1912

Manager, Aigburth Nurseries, Liverpool. Plant hybridist.

J. Hort. Home Farmer v.64, 1912, 527.

RANKIN, William Munn (1878–1951)
b. Skipton, Yorks 1878 *d.* Lyme Regis, Dorset 13 Jan. 1951

BSc Manchester. Principal, Technical College, Bournemouth, 1911. Awarded hon. MSc by Leeds University for research on plant ecology in the Pennines. Revised ed. 13 of C. L. Laurie's *Elementary Bot.* 1950.

Proc. Bournemouth Nat. Sci. Soc. v.41, 1950–51, 61–62.

RANSOM, Arthur (1832–1912)

Editor, *Bedfordshire Times.* Bedfordshire botanist.

J. G. Dony *Fl. Bedfordshire* 1953, 26.

RANSOM, Francis (*fl.* 1890s–1935)
d. Hitchin, Herts 19 Dec. 1935

Chairman, William Ransom and Son Ltd., manufacturing chemists, Hitchin. Secretary, British Pharmaceutical Conference, 1893–1903; President, 1909–10. President, Hitchin Natural History Society. *Medicinal Plant Names* 1899.

Chemist and Druggist 1893, 479–81 portr.; 1910, 185–89 portr.; 1935, 753 portr. *Br. Fern Gaz.* 1936, 43–48.

RANSOM, Maria *see* Marsh, M.

RANSOM, William (1826–1914)
b. Hitchin, Herts 1826 *d.* Hitchin 1 Dec. 1914

FLS 1885. "Grower of herbs, distiller of lavender and peppermint" at Hitchin.

Proc. Linn. Soc. 1914–15, 32.

RANSON, Mrs. Florence Mary (*c.* 1884–1970)
b. Essex *c.* 1884 *d.* 27 July 1970

Secretary, Essex County Herbs Committee, 1940–46. *British Herbs* 1949.

Times 28 July 1970.

RAPER, George (*c.* 1760s–1790s)

Royal Navy, 1783. In Australia, 1790–92, where he did many water colour drawings of the fauna and flora, now at BM(NH) and Mitchell Library, Sydney.

Austral. Mus. Mag. 1938, 296. *J. Proc. R. Austral. Hist. Soc.* 1934, 295; 1964, 32–57. *Austral. Dict. Biogr.* v.2, 1967, 363–64.

RASHLEIGH, William (1777–1855)
b. 11 Jan. 1777 *d.* Kilmarsh, Cornwall 14 May 1855

FRS 1814. Algologist. Correspondent of D. Turner. Had notable collection of exotic plants.

D. Turner *Fuci* v.2, 1809, 43. G. C. Boase and W. P. Courtney *Bibliotheca Cornubiensis* 1874, 547–48.

RATCLIFFE, Ronnie (–1965)
d. 17 April 1965
Founded nursery at Chilton, Didcot, Berks, *c.* 1930. Specialised in carnations, chrysanthemums and cypripediums.
Orchid Rev. 1965, 182.

RATHBONE, Mary May (1866–1960)
d. Chipping Campden, Glos 15 Nov. 1960
FLS 1908. One of the founders of Botanical Research Fund, 1913–14. 'Notes on *Myriactis areschougii* and *Coilodesme californica*' (*J. Linn. Soc.* v.35, 1904, 670–75).
Proc. Linn. Soc. 1960–61, 66.

RATTRAY, George (1872–1941)
b. Aberdeenshire 1872 *d.* 13 May 1941
MA Aberdeen 1894. Taught for a few years in S. Africa returning to Aberdeen to take BSc. Returned to S. Africa in 1903. Principal, Selborne College, East London, 1904–*c.* 1931. Collaborated with J. Hutchinson in account of cycads in *Fl. Capensis* 1933, 24–44.
J. S. African Bot. 1941, 213 portr.

RATTRAY, J. (*fl.* 1830s–1880s)
Surgeon on 'Buccaneer' making soundings off west coast of Africa. Collected plants in S. Tomé and Principe, 1886.
A. W. Exell *Cat. Vascular Plants of S. Tomé* 1944, 12.
Plants at Edinburgh.

RATTRAY, John (1858–1900)
b. Dunkeld, Perthshire 29 June 1858 *d.* Perth 9 Dec. 1900
MA Aberdeen 1880. FRSE. FLS 1892. Diatomist. 'Aulacodiscus' (*J. Bot.* 1888, 97–102). Diatoms of Noronha (*J. Linn. Soc.* v.27, 1890, 81–86). 'Distribution of Marine Algae of Firth of Forth' (*Trans. Bot. Soc. Edinburgh* v.16, 1886, 420–66). Contrib. to *J. R. Microsc. Soc.*, *J. Quekett Microsc. Club*, *Proc. R. Soc. Edinburgh*.
MS. catalogue of J. Deby collection at BM(NH).
Rattrayella De Toni.

RAUTHMELL, Rev. Richard (*fl.* 1720s–1740s)
b. Little Bowland, Lancs
BA Cantab 1713. Curate, Whitewell, Bowland. Had herb.
D. Turner *Extracts from Lit. and Sci. Correspondence of R. Richardson* 1835, 314, 355. *Trans. Liverpool Bot. Soc.* 1909, 87.

RAVEN, Rev. Charles Earle (1885–1964)
b. 4 July 1885 *d.* Cambridge 8 July 1964
DD Cantab. FLS 1942. Lecturer in Theology, Emmanuel College, Cambridge, 1909–20. Rector, Bletchingley, 1920–24. Canon of Liverpool, 1924–32. Regius Professor of Divinity, Cambridge, 1932–50. President, Botanical Society of British Isles, 1951–55. *A Wanderer's Way* 1928 (autobiography). *John Ray, Naturalist* 1942. *English Naturalists from Neckham to Ray* 1947. *Teilhard de Chardin* 1962. Contrib. to *Nature*.
Times 10 July 1964 portr. *Br. J. Hist. Sci.* 1965, 254–56. *Proc. Linn. Soc.* 1966, 122–24. *Who was Who, 1961–1970* 935. F. W. Dillistone *Charles Raven* 1975 portr.

RAVENSCROFT, Edward James (1816–1890)
b. Edinburgh 1816 *d.* London 15 Nov. 1890
Printer and publisher, Edinburgh. Compiled text and completed publication (parts 34–52) of C. Lawson's *Pinetum Britannicum* 1863–84 3 vols.
Gdnrs Chron. 1890 ii 605–6; 1904 ii 36–37. F. A. Stafleu *Taxonomic Literature* 1967, 375.

RAVENSHAW, Rev. Thomas Fitzarthur Torin (1829–1882)
b. London 1829 *d.* London 26 Sept. 1882
MA Oxon 1854. Curate, Ilfracombe, Devon. Rector, Pewsey, Wilts, 1857. *New List of Flowering Plants and Ferns growing Wild in County of Devon* 1860; Supplt., 1872. Contrib. 'Botanical Notes from Devon' to *Phytologist*, 1855–56. Contrib. 'Botany of North Devon' to Stewart's *North Devon Handbook* ed. 4 1877.
J. Bot. 1882, 352. W. K. Martin and G. T. Fraser *Fl. Devon* 1939, 773. F. Boase *Modern English Biogr.* v.3, 1901, 49. *R.S.C.* v.5, 110.

RAWDON, Sir Arthur (1660s–1695)
Of Moira, County Down. Friend of W. Sherard with whom he botanised in Ireland and of H. Sloane. Sent James Harlow to Jamaica.
Annual Rep. Proc. Belfast Nat. Field Club 1911–12, 615–16. *Irish Nat.* 1913, 21–22. J. E. Dandy *Sloane Herb.* 1958, 189.
Plants at BM(NH).

RAWDON, Patricia (–1954)
d. 7 Feb. 1954
Assistant, Herb., Kew in charge of National *Dianthus* Collection, 1952. 'An Interpretation of *Dianthus viscidus* Borv. et Chaubard' (*Kew Bull.* 1953, 543–44).
Kew Bull. 1954, 34.

RAWLINGS, Arthur (*c.* 1857–1911)
d. Grand Forks, British Columbia, Canada 14 Feb. 1911
Nurseryman, Dahlia Nurseries, Romford, Essex. To British Columbia as fruit farmer, 1900.
Gdnrs Mag. 1911, 198.

RAWLINGS, George (*c*. 1820–1903)
d. Whitebrook, Mon 17 Jan. 1903
Nurseryman, Romford, Essex. Cultivated dahlias.
Garden v.63, 1903, 113–14. *Gdnrs Chron.* 1903 i 79–80. *J. Hort. Cottage Gdnr* v.46, 1903, 105–6.

RAWLINGS, John (*c*. 1855–1890)
d. 4 June 1890
Son of G. Rawlings (*c*. 1820–1903). Nurseryman, Romford, Essex.
Garden v.37, 1890, 563. *J. Hort. Cottage Gdnr* v.20, 1890, 508.

RAWLINSON, Thomas E. (–1882)
d. Lancs 3 Feb. 1882
In Victoria, Australia; returned to England, 1880. Possibly the Rawlinson acknowledged as a collector of algae by W. H. Harvey in his *Phycologia Australica* v.5, 1863, 6.
Trans. Proc. R. Soc. Victoria 1883, 294–95. *Victorian Nat.* 1908, 111.

RAWSON, Rev. A. (*c*. 1819–1891)
d. Windermere, Westmorland 18 May 1891
Vicar, Bromley Common, Kent. Bred florists' flowers, particularly pelargoniums.
Garden v.39, 1891, 535.

RAWSON, Sir Rawson William (1812–1899)
b. London 8 Sept. 1812 *d*. S. Kensington, London 20 Nov. 1899
KCMG 1875. Colonial Secretary, Cape of Good Hope, 1854–64. Governor, Windward Islands, 1869–75. Pteridologist. *Synopsis Filicum Africae Australis* (with C. W. L. Pappe) 1858.
Phytologist v.4, 1852, 695–96. W. H. Harvey and O. W. Sonder *Fl. Capensis* v.1, 1859, 67. *J. Bot.* 1896, 118; 1900, 63. *Kew Bull.* 1899, 221–22. I. Urban *Symbolae Antillanae* v.3, 1902, 109. *Who was Who, 1897–1916* 588. *R.S.C.* v.8, 708; v.11, 116.
Ferns at BM(NH), Kew. Letters at Kew.
Rawsonia Harvey & Sonder.

RAY, Elizabeth (1625–)
Sister of J. Ray (1627–1705). "Withering's story drawn from William Atkinson that Ray's sister collected plants for him (*Arrangement* ed. 3, vol. 2, 286) is quite unsupported" (C. E. Raven *John Ray* 1942, 4).

RAY, Rev. John (1627–1705)
b. Black Notley, Essex 29 Nov. 1627 *d*. Black Notley 17 Jan. 1705
MA Cantab 1651. FRS 1667. Father of English natural history. *Catalogus Plantarum circa Cantabrigiam* 1660; ed. 2 1685 (translation by A. H. Ewen and C. T. Prime in 1975). *Catalogus Plantarum Angliae* 1670; ed. 2 1677. *Historia Plantarum* 1686–1704 3 vols. *Synopsis Methodica Stirpium Britannicarum* 1690; ed. 3 1724.
R. Pulteney *Hist. Biogr. Sketches of Progress of Bot. in England* v.1, 1790, 189–281. *Mag. Nat. Hist.* v.2, 1829, 84–86. E. Lankester *Memorials of J. Ray* 1846. E. Lankester *Correspondence of J. Ray* 1846. R. W. T. Gunther *Further Correspondence* 1928. *Cottage Gdnr* v.5, 1851, 221. G. S. Gibson *Fl. Essex* 1862, 444–46. *J. Bot.* 1863, 32; 1870, 82–84; 1893, 107–9; 1929, 116–19; 1934, 217–23. H. Trimen and W. T. T. Dyer *Fl. Middlesex* 1869, 373–74. *J. Hort. Cottage Gdnr* v.31, 1876, 512–16 portr. *Trans. Essex Field Club* v.4, 1886, 171–88. J. E. Bagnall *Fl. Warwickshire* 1891, 490–92. *Proc. Essex Field Club* v.4, 1892, clix–clxxiii. *Proc. Linn. Soc.* 1888–89, 40; 1935–36, 71–73; 1941–42, 3–10. G. C. Druce *Fl. Berkshire* 1897, cxvii–cxxi. *D.N.B.* v.47, 339–44. *Essex Nat.* 1900, 331–33; 1904, 219–29; 1912–13, 129–32 portr., 145–56; 1954, 190. *A.L.A. Portr. Index* 1906, 1207–8. F. H. Davey *Fl. Cornwall* 1909, xxxi–xxxiii. J. W. White *Fl. Bristol* 1912, 57–60. F. W. Oliver *Makers of English Bot.* 1913, 28–43 portr. J. R. Green *Hist. Bot.* 1914, 67–97 portr. *Essex Rev.* 1917, 57–71, 129–40. G. C. Druce *Fl. Buckinghamshire* 1926, lxxii–lxxiv. E. Hawks *Pioneers of Plant Study* 1928, 204–12 portr. G. C. Druce *Fl. Northamptonshire* 1930, xlvi–xlviii. C. E. Salmon *Fl. Surrey* 1931, 44–45. D. C. Gunawardena *Studies in Biol. Works of J. Ray* (London University thesis 1933). A. C. Seward *John Ray* 1937 portr. A. H. Wolley-Dod *Fl. Sussex* 1937, xxxvi–xxxvii. C. E. Raven *John Ray* 1942; ed. 2 1950 portr. *Isis* 1943, 319–24. *Naturalist* 1943, 79–85. *Nature* v.158, 1946, 345; v.175, 1955, 103–5. *J. Hist. Medicine* 1947, 250–61. G. Keynes *John Ray; a Bibliogr.* 1951. J. E. Dandy *Sloane Herb.* 1958, 189–92. J. C. Crowther *Founders of Br. Sci.* 1960, 94–130 portr. *J. Soc. Bibliogr. Nat. Hist.* 1963, 97–99 portr.; 1974, 111–23. F. A. Stafleu *Taxonomic Literature* 1967, 375–76. D. A. Cadbury *Computer-mapped Fl.... Warwickshire* 1971, 46–47. J. Ray *Synopsis Methodica* (1973 reprint, 5–18). *Taxon* 1973, 204 portr. C. C. Gillispie *Dict. Sci. Biogr.* v.11, 1975, 313–18.
Herb. at BM(NH). Letters and portr. attributed to Mrs. Beale at BM(NH). Letters at Fitzwilliam Museum. Portr. at National Portrait Gallery. Copy of bust by Roubilliac at Linnean Society. Portr. at Hunt Library.
Rajania L.

RAYER, Jacob (1735–1797)
b. Winchcombe, Glos March 1735 *d*. London 16 March 1797

Messenger to Medical Society of London and "dayman" to East India Company. Friend of J. E. Smith. Botanised in home counties. Discovered *Althaea hirsuta* at Cobham, 1792. Herb. formerly at Medical Society of London. Contrib. to J. Sowerby and J. E. Smith's *English Bot.* 65, 71, etc.

Gent. Mag. 1797 i 436. J. Symons *Synopsis Plantarum* 1798, 200. *Phytologist* 1862–63, 181–82. W. A. Clarke *First Rec. of Br. Flowering Plants* 1897, 8, 30.

RAYNER, John (*fl.* 1790s–1800s)

Nurseryman, near the Blackmoor's Head, High Street, Nottingham.

RAYNER, John Frederick (1854–1947)

d. Westerham, Kent 19 Jan. 1947

Florist, Swaythling, Southampton, Hants. Studied flora of Hampshire. President, Southampton Natural History Society. *Standard Catalogue of English Names of Our Wild Flowers* 1927. *Supplement to F. Townsend's Fl. Hampshire* 1929. *In Search of Wild Flowers* 1933. Contrib. fungi to F. Morey's *Guide to Nat. Hist. Isle of Wight* 1909.

Bot. Soc. Exch. Club Br. Isl. Rep. 1947, 233–34.

Plants at Bournemouth Natural History Society.

RAYNER, Mabel Mary Cheveley (–1948)

d. Wareham, Dorset 17 Dec. 1948

BSc London 1908. DSc 1915. Married W. Neilson Jones, 1912. Head of Botany Dept., University College, Reading. Authority on mycotrophy in plants. *Mycorrhiza* 1927. *Trees and Toadstools* 1945. *Practical Plant Physiology* (with Sir F. Keeble) 1911. *Textbook of Plant Biology* (with W. N. Jones) 1920.

Forestry v.22, 1948, 241–44. *Times* 31 Dec. 1948. *Empire For. J.* 1949, 5. *Nature* v.163, 1949, 275–76. *Who was Who, 1941–1950* 958.

RAYNTON, Mr.

Had garden at Enfield, Middx in 1690s.

Archaeologia v.12, 1794, 189.

REA, Carlton (1861–1946)

b. Worcester 7 May 1861 *d.* Worcester June 1946

MA. BCL Oxon. Barrister. President, British Mycological Society, 1908, 1921; Secretary, 1896–1920. President, Worcestershire Naturalists' Club, 1902–6, 1934–38, 1946. *Botany of Worcestershire* (with J. Amphlett) 1909. *British Basidiomycetes* 1922. Contrib. to *Trans. Br. Mycol. Soc.*

Trans. Worcestershire Nat. Club 1945–47, 79–81 portr. *Trans. Br. Mycol. Soc.* 1948, 180–85 portr.

Fungi drawings and MSS. at BM(NH).

Hygrophorus reai Maire.

REA, John (*fl.* 1620s–1677)

d. Kinlet, Shropshire Oct. 1677

Nurseryman, Kinlet, Shropshire. Introduced *Corylus colurna* 1665. *Flora: seu de Florum Cultura* 1665; ed. 2 1676 (also titled *Flora, Ceres et Pomona*).

G. W. Johnson *Hist. English Gdning* 1829, 111. S. Felton *Portr. English Authors on Gdning* 1830, 23–28. *J. Hort.* v.30, 1876, 172–73. *D.N.B.* v.47, 349. *Garden* 1912, 652. M. Hadfield *Gdning in Britain* 1960, 114–21.

REA, Margaret Williamson (1875–)

b. Belfast 1875

BSc Belfast 1919. Collected Mycetozoa. Contrib. to *New Phytol.*, *Irish Nat. J.*

R. L. Praeger *Irish Nat.* 1949, 147.

READ, John (1809–1880)

b. Great Grimsby, Lincs 21 Oct. 1809

Brewer, Market Rasen, Lincoln. Florist specialising in auriculas. Contrib. to *Florists' Guide.*

Florist and Pomologist 1880, 160. *Gdnrs Chron.* 1880 ii 314.

READE, Rev. Joseph Bancroft (1801–1870)

b. Leeds, Yorks 5 April 1801 *d.* Bishopsbourne, Canterbury, Kent 12 Dec. 1870

BA Cantab 1825. FRS 1838. Rector, Stone, 1839–59; Ellesborough, 1859–63; Bishopsbourne, 1863–70. Microscopist and photographer. President, Microscopical Society. 'Spiral Vessels in Roots' (*Ann. Nat. Hist.* 1838, 111–13).

Mon. Microsc. J. 1871, 92–96. *D.N.B.* v.47, 360–61. *R.S.C.* v.5, 114; v.8, 710.

READE, Oswald Alan (1848–1929)

d. Lowestoft, Suffolk 14 April 1929

FLS 1898. Pharmaceutical chemist in Royal Navy, 1873–1908. *Plants of the Bermudas, or Somers' Islands* 1885.

I. Urban *Symbolae Antillanae* v.3, 1902, 100. *Pharm. J.* 1929, 393. *J. Bot.* 1930, 22.

Bermuda plants at Kew. Maltese plants and notes at BM(NH).

READE, William Winwood *see* Addendum

READER, Henry Charles Lyon (1840–)

b. 12 Aug. 1840

Did botanical work for Rugby School Natural History Society.

J. E. Bagnall *Fl. Warwickshire* 1891, 503.

READER, Henry Peter (1850–1929)

BA Oxon. Dominican priest. Studied flora of Rugeley and Cannock Chase, Staffs. Contrib. to *Fl. Leicestershire* 1933 and *Fl. Gloucestershire* 1948.

A. R. Horwood and C. W. F. Noel *Fl. Leicestershire* 1933, ccxxviii–ccxxix. H. J. Riddelsdell *Fl. Gloucestershire* 1948, cxxxviii–cxxxix. E. S. Edees *Fl. Staffordshire* 1972, 21–22.

Plants at BM(NH), Bristol University, Oxford, Stoke on Trent Museum.

REAH, Elsie Margaret *see* Groves, E. M.

REDESDALE, 1st Baron *see* Mitford, A. B. F.-

REDGROVE, Herbert Stanley (1887–1943)
b. London Feb. 1887 *d.* Pangbourne, Berks 13 March 1943

BSc London 1907. Analytical chemist. Interested in natural and synthetic scents. Had herb. *Scent and all about it* 1928. *Spices and Condiments* 1933.

Gdnrs Chron. 1939 i 194 portr.; 1943 i 130.

Plants at BM(NH), Kew.

REDHEAD *see* Milne-Redhead

REDMOND, David (–1905)
b. Connor, County Antrim, Ireland

Supplied plant records to S. A. Stewart and T. H. Corry's *Fl. N.E. Ireland* 1888.

Annual Rep. Proc. Belfast Nat. Field Club 1911–12, 624–25.

REED, A. M. *see* Hussey, A. M.

REED, James (*fl.* 1690s)

"Plants from Barbados by James Reid the quaker sent thither on King W^ms^ account, 1692" (Herb. Sir Hans Sloane, 55). Sent plants to J. Petiver and W. Courten.

J. Petiver *Musei Petiveriani* 1695, 31 (Rheed). J. Petiver *Opera Historiam Naturalem Spectantia* v.2, 1764: *Petiveriana* nos. 161–269. L. Plukenet *Almagestum Botanicum* 1696, 15 (Reede). D. Turner *Extracts from Lit. and Sci. Correspondence of Richard Richardson* 1835, 11. I. Urban *Symbolae Antillanae* v.3, 1902, 110. *Proc. Amer. Antiq. Soc.* 1952, 291–92. J. E. Dandy *Sloane Herb.* 1958, 192–93. A. M. Coats *Quest for Plants* 1969, 332.

REEKS, Henry (1838–1882)
b. Standen, Berks 15 March 1838 *d.* Thruxton, Hants 20 Feb. 1882

FLS 1866. 'Newfoundland Plants' (*J. Bot.* 1871, 16). 'Plants of East Woodhay' (*Rep. Newbury Field Club* 1870–71, 60–70).

J. Bot. 1882, 352. *Proc. Linn. Soc.* 1880–82, 65. G. C. Druce *Fl. Berkshire* 1897, clxxii–clxxiii.

REES, John (1890–1955)
b. 14 Sept. 1890 *d.* 24 Aug. 1955

Assistant Lecturer in Agriculture, University College, Bangor, 1915. Teacher, Caernarvon, Merioneth, 1919. Adviser in Agricultural Botany, University College, Cardiff, 1946. *Preliminary Survey of Rough Pastures of Lleyn Peninsula* (MSc thesis, Bangor, 1928).

REES, Kingsley (–1931)

Secretary of Société Guernesiaise. Studied flora of Guernsey.

Rep. Trans. Société Guernesiaise 1931, 136, 145.

REEVES, John (1774–1856)
b. West Ham, Essex 1 May 1774 *d.* Clapham, Surrey 22 March 1856

FRS 1817. FLS 1817. Inspector of Tea for East India Company. To China, 1812; at Canton and Macao. Correspondent of Sir J. Banks. Early collector of Chinese plants and minerals. Commissioned and supervised Chinese artists to depict native plants and animals. Sent plants to Horticultural Society of London.

Bot. Register v.15, 1829, no. 1236. *Gdnrs Mag.* 1835, 112. *Cottage Gdnr* v.16, 1856, 21. *Gdnrs Chron.* 1856, 212. *Proc. Linn. Soc.* 1856, xliii–xlv. *J. Bot.* 1894, 293, 298; 1897, 427. E. Bretschneider *Hist. European Bot. Discoveries in China* 1898, 256–63. *D.N.B.* v.47, 416. E. H. M. Cox *Plant Hunting in China* 1945, 52–57. Royal Horticultural Society *Exhibition of MSS., Books...* 1954, 14, 53. A. M. Coats *Gdn Shrubs and their Histories* 1963, 393–95. P. J. P. Whitehead and P. I. Edwards *Chinese Nat. Hist. Drawings* 1974 portr.

Chinese drawings at BM(NH), Royal Horticultural Society.

REEVES, John Russell (1804–1877)
d. Wimbledon, Surrey 1 May 1877

FRS 1834. FLS 1832. Son of J. Reeves (1774–1856) whom he joined in China in 1827. Lived 30 years in Canton. Had herb.

Gdnrs Chron. 1877 i 604. *J. Bot.* 1877, 192; 1894, 293. E. Bretschneider *Hist. European Bot. Discoveries in China* 1898, 263–66.

Spiraea reevesiana Lindl.

REEVES, Rev. John William (1816–1862)
b. King's Somborne, Hants 22 Oct. 1816 *d.* King's Somborne 6 Jan. 1862

BA Cantab 1840. Had herb. of Hampshire plants.

F. Townsend *Fl. Hampshire* 1904, xxxv (where he is erroneously described as Garnier's nephew).

Plants at BM(NH).

REEVES, Walter Waters (1819–1892)
b. Beckley, Sussex 14 Feb. 1819 *d.* Pickering, Yorks 18 May 1892

Assistant Secretary, Royal Microscopical Society, 1868–84. One of the founders of Quekett Club. Had herb. 'Plants of Farnham' (*Bot. Gaz.* 1850, 76–79). Contrib to. *J. Bot.* and J. A. Brewer's *Fl. Surrey* 1863.

J. Bot. 1892, 212–13. C. E. Salmon *Fl. Surrey* 1931, 53. *R.S.C.* v.5, 127.

Herb. at Scarborough Museum. Plants and MSS. at Kew.

REID, Clement (1853–1916)
b. London 6 Jan. 1853 *d.* Milford-on-Sea, Hants 10 Dec. 1916

FRS 1899. FLS 1888. On staff of Geological Survey, 1874–1913. Studied fossil *Characeae* and Tertiary seeds. Collected plants in Cyprus, 1908. *Origin of British Fl.* 1899. *Submerged Forests* 1913. Contrib. to *Trans. Norfolk Norwich Nat. Soc.*

Nature v.98, 1916, 312. *Bot. Soc. Exch. Club Br. Isl. Rep.* 1916, 465–66. *J. Bot.* 1917, 145–51 portr. *Proc. Linn. Soc.* 1916–17, 61–64. *Proc. R. Soc.* 1917, viii–x portr. *Trans. Norfolk Norwich Nat. Soc.* 1916–17, 292. *Who was Who, 1916–1928* 878.

Plants at Kew.

REID, Eleanor Mary Wynne (*née* **Edwards**) (1860–1953)
b. Denbigh 1860 *d.* Milford-on-Sea, Hants 28 Sept. 1953

Mathematics teacher, Westfield College. Took up botany after marriage to C. Reid (1853–1916). *Pliocene Fl. of Dutch–Prussian Border* (with C. Reid) 1915. *Bembridge Fl.* (with M. E. J. Chandler) 1926. *London Clay Fl.* (with M. E. J. Chandler) 1933.

Times 9 Oct. 1953. *Nature* v.173, 1954, 190.

REID, Francis Alexander (–1862)
d. Beauly, Inverness 10 Nov. 1862

Madras army, 1819. Lieut.-Colonel, 1860. Director, Madras Horticultural Society Garden.

F. Boase *Modern English Biogr.* v.3, 1901, 99.

Reidia Wight.

REID, George (*c.* 1826–1881)
d. 18 July 1881

Nurseryman, Aberdeen.

Florist and Pomologist 1881, 144. *Gdnrs Chron.* 1881 ii 122–23.

REID, Hugo (1809–1872)
b. Edinburgh 21 June 1809 *d.* London 13 June 1872

To U.S.A., 1858. Principal, Dalhousie College, Halifax, Nova Scotia. *Outlines of Medical Botany* 1832. *Botanical Classification* 1838.

D.N.B. v.47, 428–29.

REID, James *see* Reed, J.

REID, John (*fl.* 1680s)

Gardener to Sir George Mackenzie at Rosehaugh, Aberdeenshire. *Scots Gardener* 1683; later eds up to 1766.

G. W. Johnson *Hist. English Gdning* 1829, 120.

REID, Moses (*fl.* 1790s–1830s)

Nurseryman, Pepper Street, Middlewich, Cheshire.

REID, Robert (–1863)
b. Langholm, Dumfriesshire

Florist, New York. Contrib. to *Gdnrs Chron.*

J. Hort. Cottage Gdnr v.53, 1906, 547.

REILLY, John (*c.* 1793–1876)

Irish schoolmaster, afterwards coastguardsman. Collected plants in various parts of Ireland.

J. Bot. 1877, 179.

Plants at National Museum, Dublin.

REILLY, Terence (*fl.* 1780s–1800s)

Nurseryman, Ballybeg, County Meath, Ireland.

Irish For. 1967, 56.

RELHAN, Rev. Richard (1754–1823)
b. Dublin 1754 *d.* 28 March 1823

MA Cantab 1779. FRS 1787. FLS 1789. ALS 1798. Rector, Hemingby, Lincoln, 1791. *Fl. Cantabrigiensis* 1785–93; ed. 3 1820. Contrib. to J. Sowerby and J. E. Smith's *English Bot.*

G. C. Gorham *Mem. John and Thomas Martyn* 1830, 126–27. C. C. Babington *Fl. Cambridgeshire* 1860, x–xi. *D.N.B.* v.48, 6–7.

Letters in J. E. Smith correspondence at Linnean Society.

Relhania L'Hérit.

RENCH *see* Wrench

RENDLE, Alfred Barton (1865–1938)
b. London 19 Jan. 1865 *d.* Leatherhead, Surrey 11 Jan. 1938

BA Cantab 1887. DSc London 1899. FRS 1909. FLS 1888. VMH 1917. VMM 1929. Assistant, Dept. of Botany, BM(NH), 1888; Keeper, 1906–30. President, Linnean Society, 1923–27. Edited *J. Bot.*, 1924–38. *Catalogue of African Plants collected by Dr. F. Welwitsch, 1853–61* 1899. *Classification of Flowering Plants* 1904–25 2 vols. *Fl. Jamaica* (with W. Fawcett) 1910–36 7 vols. Edited J. Britten and G. S. Boulger's *Biogr. Index of Deceased Br. and Irish Botanists* 1931.

Gdnrs Chron. 1921 ii 256 portr.; 1930 i 2 portr.; 1938 i 65. *J. Bot.* 1930, 64; 1938, 64–68 portr. *Chronica Botanica* 1938, 268 portr. *J. R. Microsc. Soc.* 1938, 86–87. *Kew Bull.* 1938, 81–82. *Nature* v.141, 1938, 400–1. *Obit. Notices Fellows R. Soc.* 1939, 511–17 portr. *Proc. Linn. Soc.* 1937–38, 327–33 portr. *Times* 13 Jan. 1938 portr. *D.N.B. 1931–1940* 730–31.

Herb. at BM(NH). Portr. at Hunt Library.

RENDLE, William Edgcumbe (1820–1881)
b. Compton Giffard, Devon 10 Feb. 1820 *d.* Eastbourne, Sussex 3 Sept. 1881
Nurseryman, Plymouth, Devon. Invented glazing system for greenhouses.
Garden v.20, 1881, 281; v.22, 1882, 217 portr. *Gdnrs Chron.* 1881 ii 349.

RENNIE, Rev. James (1787–1867)
b. 26 Feb. 1787 *d.* Adelaide, S. Australia 25 Aug. 1867
MA Glasgow 1815. ALS 1829. Professor of Zoology, King's College, London, 1830–34. To Australia, 1840. Edited *Mag. Bot.*, 1833–34. *Alphabet of Botany* 1833. *Handbook of Plain Botany* 1835. *Familiar Introduction to Botany* 1849. Contrib. to *Hort. Register.*
D.N.B. v.48, 18–19. *R.S.C.* v.5, 162; v.8, 731.

RENTON, Dorothy G. (–1966)
VMM 1954. Had garden at Branklyn, Perth.
Quart. Bull Alpine Gdn Soc. 1966, 151.

RENTON, John (*c.* 1747–1810)
d. 18 July 1810
Nurseryman, Hoxton Field Nursery, Hoxton, Shoreditch, London.
J. R. Hort. Soc. 1938, 422–28.

RENTON, Robert (*c.* 1840–*c.* 1915)
Of Greenlaw, Berwickshire. Collected British mosses.
Garden 1915, 36(vi). *Gdnrs Chron.* 1915 i 36.

RENWICK, John (–1918)
d. 23 July 1918
Treasurer, Geological Society, Glasgow, 1885–1917. Contrib. papers on trees of the Clyde area to *Glasgow Nat.* 'Measurement of Notable Trees' (with R. M'Kay) (*Handbook Nat. Hist. of Glasgow and W. Scotland* 1901, 131–47).
Glasgow Nat. v.8, 1926, 198. *R.S.C.* v.18, 142.

REUTHE, Gustavus (–1942)
d. Keston, Kent 28 Oct. 1942
VMH 1938. Nurseryman, Keston, Kent.
Gdnrs Chron. 1942 ii 190. *Quart. Bull. Alpine Gdn Soc.* 1942, 249; 1943, 45–46.

REVELL, John (*fl.* 1830s)
Of Pitsmoor, Sheffield, Yorks. Raised florists' flowers, e.g. the pink, 'Revell's Lady Wharncliffe'. Contrib. to *Floricultural Cabinet.*

REYNARDSON, Samuel (–1721)
d. Hillingdon, Middx 1721
Nurseryman, Hillingdon. Collection sold to Robert Walpole. Sloane MS. 4015, ff. 19–33 contains rough drawings of fungi mainly from his orchard.
L. Plukenet *Amaltheum Botanicum* 1705, 63. L. Plukenet *Almagesti Botanici Mantissa Plantarum* 1700, 51, 85, 147.
E. J. L. Scott *Index to Sloane Manuscripts* 1904, 450. J. E. Dandy *Sloane Herb.* 1958, 193.

REYNOLDS, William (*fl.* 1830s–1850s)
Nurseryman, Lakenham Hall Road, Norwich, Norfolk.

REYNOLDS-MORETON, Henry Haughton, Lord Moreton (1857–1920)
d. London 27 Feb. 1920
MP for W. Gloucestershire, 1880–85. Studied Oxford flora.
Bot. Soc. Exch. Club Br. Isl. Rep. 1920, 104–5.

RHIND, William (*fl.* 1830s–1860s)
MRCS. Lecturer in Botany, Marischal College, Aberdeen. *History of the Vegetable Kingdom* 1840–41. *Catechism of Botany* 1833.

RHODES, Osman (*c.* 1829–1869)
d. 8 July 1869
Nurseryman and hot-water engineer, Sydenham Park, London.
Gdnrs Chron. 1869, 767.

RHODES, Rev. Philip Grafton Mole (1885–1934)
b. Birmingham 1885 *d.* Evesham, Worcs 16 Dec. 1934
BA Cantab. DD Fribourg. FLS 1925. Curate, Kidderminster, Worcs. Professor, Ascott College, Birmingham. Minister, Evesham. Mycologist and bryologist. Contrib. to A. H. Evans's *Short Fl. Cambridgeshire* 1911, *Trans. Worcestershire Nat. Club, Rep. Soc. Guernésiase.*
Rep. Br. Bryol. Soc. 1934, 246–47. *Proc. Linn. Soc.* 1934–35, 187.
Lichens and hepatics at Birmingham Museum; fungi at Kew; mosses at BM(NH).

RHYDDERCH, Sion (*alias* Roderick), John (*fl.* 1730s)
Printer. Of Shrewsbury, Shropshire. *Y geir Cyfr Saesneg a Chymraeg* (*English–Welsh Herbal*) 1737.
H. Davies *Welsh Botanology* 1813, xi.

RICH (*or* Ryche), John (*fl.* 1580s–1593)
Apothecary to Queen Elizabeth I. Master of the Grocer's Company, 1580. "Maister Riche, apothecary, whose garden where I saw many other good and strange herbes which I never saw anywhere elles in all England" (W. Turner).
C. E. Raven *English Nat.* 1947, 117. L. G. Mathews *R. Apothecaries* 1967, 78–79.

RICH, Mary Florence (1865–1939)
b. Weston-super-Mare, Somerset 1865 *d.* Ruislip, Middx 20 April 1939
Educ. Oxford. FLS 1926. Taught at Roedean, Sussex. Established Granville School, Leicester. Hon. Research Assistant, Dept. of Botany, Queen Mary College, London. Studied freshwater algae. Published papers in collaboration with F. E. Fritsch. Contrib. to *Ann. Bot.*, *J. Bot.*, *Trans. R. Soc. S. Africa.*
J. Bot. 1939, 184. *Nature* v.143, 1939, 845. *Proc. Linn. Soc.* 1938–39, 251–52. *Times* 22 April 1939.

RICHARDS, Edward Alfred (1880–1927)
b. Winwick, Lancs 22 Oct. 1880 *d.* Wallasey, Cheshire 1 Feb. 1927
Salesman to Flour Millers, Liverpool. Bryologist.
Rep. Br. Bryol. Soc. 1927, 322.
Herb. at BM(NH).

RICHARDS, Francis John (1901–1965)
b. Burton-on-Trent, Staffs 1 Oct. 1901 *d.* Wye, Kent 2 Jan. 1965
MSc Birmingham. FRS 1954. Research Officer, Institute of Plant Physiology, Imperial College, London, 1926–39. Head, Imperial College Laboratory, Rothamsted, 1939–59. Director, Plant Morphogenesis and Nutrition, Wye, 1958. Editor of *Plant and Soil.* Contrib. to *Ann. Bot.*
Nature v.205, 1965, 853–54. *Times* 4 Jan. 1965. *Biogr. Mem. Fellows R. Soc.* 1966, 423–36 portr. *Who was Who, 1961–1970* 952–53.

RICHARDS, Franklin T. (1847–1905)
b. Kensington, London 18 March 1847 *d.* London 14 April 1905
Fellow, Trinity College, Oxford. Botanised in Oxfordshire.
G. C. Druce *Fl. Oxfordshire* 1927, cxxii.

RICHARDS, John (*c.* 1826–1877)
d. Gunnersbury, Middx 18 July 1877
Gardener to Baron L. de Rothschild at Gunnersbury, Middx.
Gdnrs Chron. 1877 ii 153.

RICHARDS, Rev. Thomas (*c.* 1710–1790)
b. Glam *c.* 1710 *d.* Coychurch, Glam 20 March 1790
Welsh lexicographer. Botanology in *Antiquae Linguae Britannicae Thesaurus* 1753.
D.N.B. v.48, 219.

RICHARDS, Thomas (*fl.* 1780s–1800s)
Nurseryman, Kingsland, Hackney, London.

RICHARDS, Thomas John (*c.* 1860–1937)
d. 5 Dec. 1937
MA Oxon. President, Eastbourne Natural History Society, 1927. Assisted in compilation of A. H. Wolley-Dod's *Fl. Sussex* 1937.
Trans. J. Eastbourne Nat. Hist., Photogr. Archaeol. Soc. 1938, 25.

RICHARDSON, Mr. (*fl.* 1690s)
Had garden at East Barnet, Herts in 1690s.
Archaeologia v.12, 1794, 189–90.

RICHARDSON, Adam Dewar (1857–1930)
b. Garvald, East Lothian 1857 *d.* Jan. 1930
Royal Botanic Garden, Edinburgh, 1880; Curator, 1896–1902. Secretary, Scottish Horticultural Society. Edited *Trans. Scott. Hort. Soc.*, 1906–21.
Gdnrs Chron. 1929 i 196 portr.; 1930 i 54–55. H. R. Fletcher and W. H. Brown *R. Bot. Gdn Edinburgh, 1670–1970* 204–5 portr.

RICHARDSON, David Lester (1800–1865)
In India, 1819–27, 1829–61. Major in Bengal Army. Principal, Hindoo College, Calcutta. Editor from 1861 of Allen's *Indian Mail. Flowers and Flower-gardens* 1855.
Memoir by J. W. Kaye in *Calcutta Rev.* v.16, 1851, 289–320.

RICHARDSON, J. Lionel (1890–1961)
b. Tramore, County Waterford, Ireland 1890 *d.* Dublin 14 Oct. 1961
VMM 1952. Raised narcissi.
J. R. Hort. Soc. 1961, 532–34 portr. *Daffodil Tulip Yb.* 1963, 16–19. M. J. Jefferson-Brown *Daffodils and Narcissi* 1969, 29–31.

RICHARDSON, James (1806–1851)
b. Lincs 1806 *d.* Ungouratoua, Egypt 4 March 1851
African traveller. 'Dates of Fezzan' (*Hooker's J. Bot.* 1850, 233–36). *Travels in Desert of Sahara* 1845–46 2 vols portr. *Mission to Central Africa* 1850–51.
D.N.B. v.48, 226–27. *R.S.C.* v.5, 188.

RICHARDSON, Sir John (1787–1865)
b. Dumfries 5 Nov. 1787 *d.* Grasmere, Westmorland 5 June 1865
MD Edinburgh 1816. LLD Dublin 1857. FRS 1825. FLS 1825. Surgeon and Naturalist on J. Franklin's north Polar expeditions, 1818–22, 1825. Botanical appendix in Franklin's *Narrative of Journey to Shores of Polar Sea* 1823. Plants collected by Richardson recorded in W. J. Hooker's *Fl. Boreali-Americana* 1840.
Proc. Linn. Soc. 1865–66, lxxxiv–lxxxvi. *J. R. Geogr. Soc.* 1866, cxxxii–cxxxiv. *Proc. R. Soc.* 1867, xxxvii–xliii. J. McIlraith *Life of Sir John Richardson* 1868 portr.

D.N.B. v.48, 233–35. *J. R. Naval Med. Service* 1924, 160–72 portr.; 1936, 180–87 portr. *Rhodora* 1924, 199–200; 1936, 412–13. *New England J. Med.* 1955, 26–27 portr. *J. Soc. Bibl. Nat. Hist.* 1969, 202–17; 1972, 98–117 portr. *Mem. New York Bot. Gdn* 1975, 386–87. R. E. Johnson *Sir John Richardson* 1976 portr. *R.S.C.* v.5, 188.

Plants at BM(NH), Kew, Oxford. Letters at Kew, Scott Polar Research Institute, Cambridge. Portr. by D. Macnee at Kew, by S. Pearce at National Portrait Gallery. Portr. at Hunt Library.

Heuchera richardsonii R. Br.

RICHARDSON, John Matthew (*c.* 1797–1882)
d. Newcastle, N.S.W. 28 July 1882

Transported to Australia, 1822. Employed in Botanic Gardens, Sydney. Accompanied John Oxley as plant collector in Queensland, 1823 and 1824. Settled on Melville Island near Darwin, 1825–28. Collected plants in Timor, 1826. Botanist and collector on T. L. Mitchell's expedition along Murray River, 1836.

J. Proc. R. Soc. N.S.W. 1908, 117. *Victorian Nat.* 1908, 112; 1948, 200. *Austral. Dict. Biogr.* v.2, 1967, 377.

Hibiscus richardsonii Sweet.

RICHARDSON, Richard (1663–1741)
b. North Bierley, Yorks 6 Sept. 1663 *d.* North Bierley 21 April 1741

MB Oxon. MD Leyden 1690. FRS 1712. Physician at North Bierley where he created a garden. "He was a highly accomplished and admired botanist having the best collection of native and foreign plants in the north of England" (J. Nichols). Found *Trichomanes radicans* in Yorkshire. 'Subterraneous Trees' (*Philos. Trans. R. Soc.* v.19, 1697, 526–28). *De Cultu Hortorum* 1699. *Index Hort. Bierleiensis* 1737. D. Turner *Extracts from Lit. Sci. Correspondence of Richard Richardson* 1835. MS. *Deliciae Hortenses* 1896.

J. Petiver *Musei Petiveriani* 1695, 95. R. Morison *Plantarum Historiae Universalis Oxoniensis* v.3, 1699, 386, 488, 626. R. Pulteney *Hist. Biogr. Sketches of Progress of Bot. in England* v.2, 1790, 185–88. J. Nichols *Illus. Lit. Hist. of Eighteenth Century* v.1, 225–52 portr. *Gdnrs Mag.* 1828, 127–28. R. A. Salisbury *Genera of Plants* 1866, 114. *D.N.B.* v.48, 240–41. *Naturalist* 1906, 257–60. *Bradford Sci. J.* 1912, 234–41 portr. E. J. L. Scott *Index to Sloane Manuscripts* 1904, 452. *Notes and Queries* v.2, 1916, 403, 447. C. E. Salmon *Fl. Surrey* 1931, 46. J. E. Dandy *Sloane Herb.* 1958, 194.

Plants at Oxford, BM(NH). Letters at Royal Society and Bodleian Library. Portr. at Hunt Library.

Richardia L. *Richardsonia* Kunth.

RICHARDSON, Rev. William (–1768)
b. Ullswater, Cumberland

Rector, Dacre, Cumberland, 1742. Contrib. botany to W. Hutchinson's *Hist. County Cumberland* 1794.

N. J. Winch *Contrib. Fl. Cumberland* 1833, 4.

RICHARDSON, Rev. William (1740–1820)
d. Clonfele, County Antrim, Ireland 1820

BA Dublin 1763. DD 1778. Rector of Moy and Clonfeacle. Agriculturist. 'Useful Grasses' (*Trans. Irish Acad. Sci.* v.11, 1810, 87–119). Letters on Fiorin Grass in *Gent. Mag.* 1809–16.

D.N.B. v.48, 253. *R.S.C.* v.5, 190.

RICHARDSON, William (1797–1879)
b. Hebburn, Northumberland 31 Aug. 1797
d. Alnwick, Northumberland 18 April 1879

Saddler, Alnwick. Discovered *Psamma baltica.* 'Northumbrian Botany' (*Phytologist* v.5, 1861, 97–100). 'Plants of Holy Island' (*Phytologist* v.6, 1862–63, 10–15).

J. Bot. 1872, 21; 1879, 192. *Proc. Berwickshire Field Club* v.9, 1879, 184–91. *Nat. Hist. Trans. Northumberland, Durham and Newcastle-upon-Tyne* v.14, 1903, 85.

RICHE John *see* Rich, J.

RICHMOND, John (*fl.* 1760s–1770s)
Nurseryman, Edinburgh.

RICHMOND, Sir Robert Daniel (*c.* 1879–1948)
d. 1 May 1948

Indian Forest Service, 1898. Chief Conservator of Forests, Madras, 1927–32.

Who was Who, 1941–1950 973.

RICHMOND, Thomas (1695–1758)

BA Oxon 1715. Physician? Member of Botanical Society of London.

Proc. Bot. Soc. Br. Isl. 1967, 318.

RICKETTS, George (*fl.* 1660s–1706)
d. Hoxton, London July 1706

Nurseryman Hoxton, Shoreditch, London. "The best and most faithful florist now about London" (Rea).

Archaeologia v.12, 1794, 190–91. S. Felton *Portr. English Authors on Gdning* 1830, 61–62. *J. R. Hort. Soc.* 1938, 425. *Gdnrs Chron.* 1943 i 206. J. Harvey *Early Gdning Cat.* 1972, 10.

RICKETTS, James (*fl.* 1680s–1710s)

Son of G. Ricketts (*fl.* 1660s–1690s). Nurseryman, Hoxton, Shoreditch, London.

RICKETTS, O. F. (*c.* 1867–)
Resident for many years at Kuching, Sarawak. Collected living plants for Singapore Botanic Gardens. 'Orchids of Sarawak' (*Orchid Rev.* 1936, 77–83).
Fl. Malesiana v.1, 1950, 435.

RIDDELL, Maria (*née* **Woodley**) (*fl.* 1772–1800s)
b. St. Kitts? *c.* 1772
Visited Madeira and St. Kitts, 1788; Antigua and Barbuda, 1790. *Voyages to Madeira and Leeward Isles* 1792 (contains list of Antigua plants).
D.N.B. v.48, 272. *J. New York Bot. Gdn* 1906, 275–79. *J. Bot.* 1907, 118–19.

RIDDELSDELL, Rev. Harry Joseph (1866–1941)
d. 17 Oct. 1941
BA Oxon, London. ALS 1925. Sub-warden, St. Michael's Theological College, Aberdare and Llandaff, 1897–1914. Rector, Wigginton, Oxford, 1914–18; Bloxham, Oxford, 1918–36. Authority on *Rubus*. *Fl. Glamorgan* 1907. Co-editor of *Fl. Gloucestershire* 1948. Contrib. to *J. Bot.*
Bot. Soc. Exch. Club Br. Isl. Rep. 1941–42, 460–62 portr. *Proc. Cotteswold Nat. Field Club* 1941, 194–96 portr. *Proc. Linn. Soc.* 1941–42, 294–95. *Chronica Botanica* 1942–43, 354. *Nature* v.149, 1942, 376. H. J. Riddelsdell *Fl. Gloucestershire* 1948, clii–clv.
Herb. at BM(NH). MS. *Fl. Glamorgan* at National Museum of Wales.

RIDER, William Lake (*fl.* 1830s–1850s)
Nurseryman, Heath Nursery, Moortown, Leeds, Yorks. Specialised in *Ericaceae*.

RIDGE, William Till Boydon (1872–1943)
b. Stoke-on-Trent, Staffs 18 Feb. 1872 *d.* Stoke-on-Trent 2 Feb. 1943
BSc London 1910. Schoolmaster, Hanley High School, 1894–1932. President, North Staffordshire Field Club, 1918, 1930. 'Fl. N. Staffordshire' (*Trans. N. Staffordshire Field Club* 1922–29).
Trans. N. Staffordshire Field Club 1942–43, 26–28 portr. *Bot. Soc. Exch. Club Br. Isl. Rep.* 1943–44, 651. *N. Western Nat.* 1944, 73–74 portr. E. S. Edees *Fl. Staffordshire* 1972, 22.

RIDING, J. B. (*fl.* 1910s)
Nurseryman, Chingford, Essex, 1890.
Gdnrs Mag. 1912, 189–90 portr.

RIDLEY, Henry Nicholas (1855–1956)
b. West Harling, Norfolk 10 Dec. 1855 *d.* Kew, Surrey 24 Oct. 1956
BA Oxon. FRS 1907. FLS 1881. Assistant, Botany Dept., BM(NH), 1880–88. On expedition to Fernando de Noronha, 1887. Director, Botanic Gardens, Singapore, 1888–1911. Helped to introduce rubber into Malaya. Collected plants and made numerous observations on relations between animals and plants. *Spices* 1912. *Fl. Malay Peninsula* 1922–25 5 vols. *Dispersal of Plants throughout the World* 1930. Founded and edited *Agric. Bull. Straits Settlements*. Contrib. to *J. Bot.*, *J. Straits Branch Asiatic Soc.*
C. F. P. von Martius *Fl. Brasiliensis* v.1 (1), 1906, 88–89. *Gdnrs Chron.* 1928 i 330 portr. *J. Bot.* 1928, 128. *Kew Bull.* 1928, 158–59. *Curtis's Bot. Mag. Dedications, 1827–1927* 315–16 portr. *Gdns Bull. Straits Settlements* 1935, 1–48 portr. *Fl. Malesiana* v.1, 1950, 435–37 portr. *M.A.H.A. Mag.* 1955, 3–6. *Times* 12 Dec. 1955; 25 Oct. 1956. *Nature* v.176, 1956, 1092–93. *Pharm. J.* 1956, 357. *Proc. Linn. Soc.* 1956–57, 35–38. *Biogr. Mem. Fellows R. Soc.* 1957, 141–59 portr. *Proc. Bot. Soc. Br. Isl.* 1957, 328–31. *Taxon* 1957, 1–6 portr. *J. Malayan Branch R. Asiatic Soc.* v.33, 1960, 104–9. *D.N.B. 1951–1960* 841–42. M. A. Reinikka *Hist. of the Orchid* 1972, 279–82 portr.
Plants at Kew, BM(NH), Singapore. MSS. at Kew. Portr. at National Portrait Gallery. Portr. at Hunt Library.
Ridleyella Schlechter. *Ridleyinda* Kuntze.

RIDLEY, Marian Sarah *see* Farquharson, M. S.

RIDLEY, Mathew (*c.* 1847–1904)
d. Bulrampur, India 17 Sept. 1904
Kew gardener, 1870. To India in 1870s to supervise cotton plantations. Superintendent, Government Gardens, Lucknow.
Gdnrs Chron. 1904 ii 295. *J. Kew Guild* 1904, 209.

RIGG, John (1777–1833)
Son of T. Rigg (*c.* 1746–1835). Nurseryman, Fishergate, York.

RIGG, Thomas (*c.* 1746–1835)
Nurseryman, Fishergate, York.

RILEY, John (–1846)
d. York 1846
Of Papplewick, Notts. Had fern herb. 'Hybridity in Ferns' (*Proc. Bot. Soc. London* 1839, 60–62). *Catalogue of Ferns* 1841.
Phytologist 1847, 779–80.

RILEY, Laurence Athelstan Molesworth (1888–1928)
b. Jersey 1888 *d.* Warnborough, Hants 13 March 1928
MA Oxon 1927. Voluntary worker in Kew Herb., 1920. Botanist on 'St. George' Pacific Expedition, 1924–25. Contrib. papers on flora of Tropical America to *Kew Bull.*, *J. Bot.*
J. Bot. 1928, 118–19. *Kew Bull.* 1928, 157–58.
Herb. at Kew.

RILSTONE, Francis (1881–1953)
b. Penhallow, Cornwall 5 Nov. 1881 *d.* Truro, Cornwall 22 Jan. 1953
ALS 1937. Headmaster, Polperro Primary School. Authority on Cornish *Rubus. Bryophyte Fl. Cornwall* 1948.
Trans. Br. Bryol. Soc. 1954, 501. *Proc. Bot. Soc. Br. Isl.* 1954, 110–13 portr.
Plants at BM(NH), Oxford.
Rubus rilstonei Barton & Riddelsdell.

RITCHIE, David (1809–1866)
b. Tibbermore, Perth 19 July 1809 *d.* 28 May 1866
MD Edinburgh 1830. Indian Medical Service, 1831–66. Deputy Surgeon-General, 1866. Collected chiefly in Bombay but also in Sind and W. Punjab. First found *Nannorhops ritchiena* in Khyber Pass.
Pakistan J. For. 1967, 354.
Plants at Edinburgh, Calcutta.

RITCHIE, James (*c.* 1808–1885)
b. Scotland *c.* 1808 *d.* 11 March 1885
Nurseryman, Philadelphia, U.S.A.
J. Hort. Cottage Gdnr v.53, 1906, 547.

RITCHIE, Joseph (*c.* 1788–1819)
b. Otley, Yorks *c.* 1788 *d.* Murzuk, Fezzan 20 Nov. 1819
Surgeon and African traveller. Collected plants near Tripoli, etc.
G. F. Lyon *Narrative of Travels in N. Africa* 1821 *passim.* D. Denham and H. Clapperton *Narrative of Travels...in Northern and Central Africa* 1826, 209, 225. *D.N.B.* v.48, 323–24. *J. Bot.* 1917, 278–79.
Ritchiea R. Br.

RITTERSHAUSEN, P. R. C. (*c.* 1895–1965)
d. 9 Sept. 1965
Nurseryman, Burnham Nurseries Ltd., Croydon, Surrey; moved to Kingsteignton, Newton Abbot, Devon. *Successful Orchid Cultivation* 1953. Contrib. to *Orchid Rev.*
Orchid Rev. 1965, 323.

RIVERS, T. Francis (1831–1899)
d. Sawbridgeworth, Herts 17 Aug. 1899
VMH 1897. Succeeded his father, T. Rivers (1798–1877), as nurseryman at Sawbridgeworth. Specialised in fruit.
J. Hort. Cottage Gdnr v.23, 1891, 499 portr.; v.39, 1899, 161–62 portr. *Gdnrs Mag.* 1894, 581 portr.; 1899, 531 portr. *Garden* v.56, 1899, 172. *Gdnrs Chron.* 1899 ii 50 portr., 179.

RIVERS, Thomas (1798–1877)
b. Sawbridgeworth, Herts 27 Dec. 1798 *d.* Sawbridgeworth 17 Oct. 1877
Nurseryman, Sawbridgeworth. Specialised in roses and fruit. *Rose Amateur's Guide* 1837. *The Orchard House* 1859. *Miniature Fruit Garden* ed. 6 1854.
Garden v.3, 1873 portr.; v.12, 1877, 411–12. *Florist and Pomologist* 1877, 264. *Gdnrs Chron.* 1877 ii 522–24. *J. Hort. Cottage Gdnr* v.33, 1877, 327–29 portr., 342–44; v.34, 1897, 576–77; v.38, 1899, 181, 183 portr. *D.N.B.* v.48, 333. M. Hadfield *Gdning in Britain* 1960, 290–92.
Portr. at Royal Horticultural Society.

RIVERS, Thomas Alfred Hewitt (1863–1915)
b. 5 Aug. 1863 *d.* 6 Aug. 1915
Nurseryman, Sawbridgeworth, Herts.
Garden v.62, 1902, 408–9. *J. Hort. Home Farmer* v.61, 1910, 371 portr. *Gdnrs Mag.* 1911, 601–2 portr. *Gdnrs Chron.* 1915 ii 109 portr.

ROAD, John (*fl.* 1690s)
Gardener at Hampton Court, Middx. Collected plants in Virginia *c.* 1690.
M. Hadfield *Gdning in Britain* 1960, 142.

ROB, Catherine Muriel (1906–1975)
b. Thirsk, Yorks 21 Feb 1906 *d.* Thirsk 6 Feb. 1975
Recorder of flowering plants in N. Riding for Yorkshire Naturalists' Union for 37 years. Secretary, Yorkshire Naturalists' Union, 1958; President, 1969.
Naturalist 1975, 67–68 portr.

ROBB, Mary Anne (*née* Boulton) (1829–1912)
Of Tew Park until her marriage in 1856. Had a garden at Liphook, Hants where to deter trespassers she displayed a notice: "Beware of the Lycopodium." Friend of W. Robinson and E. A. Bowles. Collected plants in Greece, 1891. Introduced *Euphorbia robbiae*, 'Mrs. Robb's Bonnet', to British gardens.
Kew Bull. 1912, 203. B. D. Morley *Wild Flowers of the World* 1970, 35. *J. R. Hort. Soc.* 1973, 306–10 portr.

ROBBIE, James (1904–1972)
b. Aberdeen 28 May 1904 *d.* 23 Feb. 1972
Kew gardener, 1925. To Ministry of Agriculture, Sudan, 1927; Chief Horticulturist until retirement in 1952.
J. Kew Guild 1972, 155 portr.

ROBBINS, Randolph William (1871–1941)
b. Hackney, London 11 March 1871 *d.* Torquay, Devon 1 Aug. 1941
Warden, Guy's Hospital, Southwark, London. President, London Natural History Society, 1920. Contrib. flora to L. G. Fry's *Oxted, Limpsfield and Neighbourhood* 1932. Joint editor of 'Botanical Records of London Area' (*London Naturalist*, 1929–36).
Bot. Soc. Exch. Club Br. Isl. Rep. 1941–42, 462–63. *London Nat.* 1941, 2–11 portr.
Herb. at London Natural History Society.

ROBERTS, David (*fl.* 1820s)
Surgeon. Of Melin-y-coed, Llanrwst, Denbighshire. Son of Richard Roberts with whom he annotated W. Withering's *Bot. Arrangement*.
J. Williams *Faunula Grustensis* 1830, 4.

ROBERTS, Harry (1871–1946)
b. Bishop's-Lydeard, Somerset 1871 *d.* 12 Nov. 1946
Educ. Bristol University. Science master, Leek Grammar School. Practised medicine in Cornwall and Stepney, London. *Book of Old Fashioned Flowers* 1901. *Chronicles of a Cornish Garden* 1901. *Beginner's Book of Gardening* 1911. *Book of Rarer Vegetables* (with G. Wythes) 1906.
Who was Who, 1941–1950 982.

ROBERTS, Jack (–1960)
d. 8 April 1960
BSc. FLS 1950. Lecturer in Botany, Edinburgh University.
Pharm. J. 1960, 346.

ROBERTS, James Ernest Helme (–1948)
d. 25 Aug. 1948
MD London. FRCS. President, Medical Society of London. Founder member of Alpine Garden Society. His collection of *Sempervivum* acquired by Royal Horticultural Society's Gardens, Wisley.
Quart. Bull. Alpine Gdn Soc. 1948, 339–40.

ROBERTS, Joel (*fl.* 1850s)
Nurseryman, Pennycomequick, Plymouth, Devon.

ROBERTS, John (–1828)
d. Bangor, Caernarvon 1828
Of Bangor. Had herb.
J. E. Griffith *Fl. Anglesey and Caernarvonshire* 1895, iii.

ROBERTS, John (–before 1843)
ALS. Of Salisbury and Norwich. Miller. Helped G. Maton with botany of Salisbury.
G. Maton *Nat. Hist. of...Wiltshire* 1843, 10.

ROBERTS, Mary (1788–1864)
b. Homerton, London 18 March 1788 *d.* Brompton, London 13 Jan. 1864
At Painswick, Glos from 1790. Granddaughter of T. Lawson (1630–1691). *Wonders of the Vegetable Kingdom* 1822 (anon). *Seaside Companion* 1835. *Plants and Animals of America* 1839. *Flowers of Matin and Evensong* 1845. *Voices from the Woodlands* 1850. Her brother Oade (1786–1821) of the Inner Temple, contrib. to W. Withering's *Arrangement* ed. 6 1818.
J. Smith *Cat. Friends' Books* v.2, 1867, 500–1. *Notes on Painswick* 1881, 12. *D.N.B.* v.48, 388–89. H. J. Riddelsdell *Fl. Gloucestershire* 1948, cxviii.

ROBERTS, Nellie (–1959)
VMM 1953. Appointed by Royal Horticultural Society in 1897 to paint orchids.
Garden v.58, 1900, 112. *Orchid Rev.* 1959, 150.

ROBERTS, Richard (–before 1828)
Of Melin-y-coed, Llanwrst, Denbighshire. Knew British plants well. Annotated W. Withering's *Bot. Arrangement*.
J. Williams *Faunula Grustensis* 1830, 4.

ROBERTS, Robert Alun (1894–1969)
b. Nantlle Valley, Caernarvon 10 March 1894 *d.* 15 May 1969
BSc Bangor. PhD 1928. Lecturer in Agricultural Botany, University College, Bangor, 1919–26; Head of Dept., 1926–40; Professor, 1945–60. Chairman, Welsh Committee of Nature Conservancy until 1955. Carried out ecological survey of Lleyn Peninsula.
Nature v.223, 1969, 109. *Times* 21 May 1969. *Who was Who, 1961–1970* 20.

ROBERTS, William (1862–1940)
b. Madron, Cornwall 15 Dec. 1862 *d.* 9 April 1940
Journalist and author from 1882. Suggested to publisher, L. Upcott Gill, the publication of a new dictionary of gardening. His text was revised and extended by G. Nicholson and others and published as *Illustrated Dict. Gdning* 1884–88. Contrib. articles on bibliography and horticultural history to *Gdnrs Chron.*, *J. R. Hort. Soc.*
Notes and Queries v.178, 1940, 288. *Times* 11 April 1940. *Who was Who, 1929–1940* 1152. J. Soc. Bibl. Nat. Hist. 1974, 139–40.

ROBERTSON, Mr. (*fl.* 1750s)
Established nursery in County Kilkenny *c.* 1756. Possibly connected with John Robertson of Kilkenny (–1839) who bequeathed his books to the Botanic Gardens, Glasnevin, Dublin.
Irish For. 1967, 53.

ROBERTSON, Agnes *see* Arber, A.

ROBERTSON, Rev. Andrew (*fl.* 1780s–1840s)
Minister, Inverkeithing, Fife, 1792–1845. Contrib. botany of his parish to *New Statistical Account of Scotland* v.9, 1845, 230. List of plants in *Trans. Bot. Soc. Edinburgh* v.20, 1894, 85.

ROBERTSON, Archibald (1789–1864)
b. Cockburnspath, Dunbar 3 Dec. 1789 *d.* Clifton, Bristol 19 Oct. 1864
MD Edinburgh 1817. FRS 1836. Practised medicine at Northampton, 1818–53. *De Rebus Physiologiae Vegetabilium atque Botanices* 1822.
Proc. R. Soc. v.14, 1865, xvii. *D.N.B.* v.48, 402–3. *R.S.C.* v.5, 230.

ROBERTSON, Rev. Archibald (1853–1931)
b. Sywell, Northants 29 June 1853 *d.* Oxford March 1931
Dean, Trinity College, Oxford, 1879–83. Principal, Bishop Hatfield's Hall, Durham, 1883–97. Vice-Chancellor, London University, 1897–1903. Bishop of Exeter, 1903–16. Botanist.
Bot. Soc. Exch. Club Br. Isl. Rep. 1930, 327–28. *Times* 31 Jan. 1931. *D.N.B. 1931–1940* 734–35.

ROBERTSON, Benjamin (*c.* 1732–1800)
Of Stockwell, London where he established a botanic garden. Friend of A. H. Haworth (1768–1833).
Gent. Mag. 1800 ii 1294; 1801 i 274; 1801 ii 762, 936. A. H. Haworth *Miscellanea Naturalia* 1803, 190. *Bot. Mag.* 1804, t.760. *Gdnrs Mag.* 1835, 108. *J. Bot.* 1924, 352.
Robertsonia Haw.

ROBERTSON, Daniel (*fl.* 1790s–1800s)
Nurseryman, Belfast, Ireland.

ROBERTSON, David (1806–1896)
b. Glasgow 28 Nov. 1806 *d.* Millport, Cumbrae 20 Nov. 1896
LLD Glasgow 1894. FLS 1876. Algologist. 'Botany of Loch Ryan' (*Proc. Nat. Hist. Soc. Glasgow* v.1, 1868, 21–36). His wife had a collection of algae (*J. Bot.* 1891, 212, 229).
T. R. R. Stebbing *Nat. of Cumbrae* 1891. *J. Bot.* 1897, 32. *Proc. Linn. Soc.* 1896–97, 66–67. *Trans. Nat. Hist. Soc. Glasgow* v.5, 1896–97, 18–42. *R.S.C.* v.5, 230; v.8, 760; v.11, 194.

ROBERTSON, James (*fl.* 1760s–1770s)
Pupil of John Hope. Made botanical survey of the "distant parts of Scotland" 1768 (*Philos. Trans. R. Soc.* v.59, 1769, 241–42). Found *Eriocaulon* in Skye. Collected plants at St. Helena, Cape and in China, 1772; at St. Iago, Johanna Island, Bombay, and Madras, 1775.
E. Bretschneider *Hist. European Bot. Discoveries in China* 1898, 154. *J. Bot.* 1899, 87–88.
Plants at BM(NH).

ROBERTSON, Janet *see* Hutton, J.

ROBERTSON, John (–1839)
d. Kilkenny 27 Aug. 1839
Nurseryman, Kilkenny. Contrib. to *Gdnrs Mag., Trans. Hort. Soc. London.*
Gdnrs Mag. 1839, 584.

ROBERTSON, John (–1865)
b. Perthshire *d.* Glasgow 24 March 1865
Gardener, Kew and at Kinfauns Castle, Perthshire. Prepared *Fl. Perthensis* (unpublished) (*J. Bot.* 1873, 48–49).
Trans. Bot. Soc. Edinburgh v.8, 1865, 337. *R.S.C.* v.5, 230.
MSS. at Perth Museum.

ROBERTSON, Rev. John (*fl.* 1880s)
Collected plants in British Honduras, 1889–90.
Kew Bull. 1924, 8.
Plants at BM(NH).

ROBERTSON, John George (1803–1862)
b. Glasgow 15 Oct. 1803 *d.* Baronald, Lanark 1862
To India as botanist and naturalist on scientific expedition, *c.* 1828. Emigrated to Tasmania, 1831. Farmer in Victoria, 1840. Sent plants to W. J. Hooker at Kew. Returned to Scotland, 1854.
Victorian Nat. 1908, 112–13; 1941, 30–31. *J. Proc. R. Austral. Hist. Soc.* 1951, 257–59. T. E. Burns and J. R. Skemp *Van Diemen's Land Correspondents, 1827–1849* 1961, 67, 69, 99, 139.
Plants at Melbourne.
Ranunculus robertsonii Benth.

ROBERTSON, Peter S. (1818–1879)
b. Comrie, Perthshire 4 Nov. 1818 *d.* 16 Sept. 1879
Gardener, Royal Botanic Garden, Edinburgh, 1837–43. Manager, Peter Lawson and Son, nurserymen, Edinburgh.
Florist and Pomologist 1879, 176. *Gdnrs Chron.* 1879 ii 409, *Gdnrs Yb. Almanack* 1880, 192. *Trans. Bot. Soc. Edinburgh* v.14, 1883, 39–40.

ROBERTSON, Robert Alexander (1873–1935)
b. Rattray, Perthshire 1873 *d.* 22 Jan. 1935
MA, BSc Edinburgh. FLS 1903. Lecturer in Botany, St. Andrews University, 1891; Reader, 1915; Professor, 1929–34. President, Botanical Society of Edinburgh, 1915. *Plant Growth Rhythms* 1913.
Nature v.135, 1935, 364. *Proc. Linn. Soc.* 1934–35, 188–89. *Proc. R. Soc. Edinburgh* v.45, 1934–35, 164–65. *Trans. Proc. Bot. Soc. Edinburgh* 1935, 533–34.
Herb. at St. Andrews University.

ROBERTSON, William (*fl.* 1760s)
Nurseryman, Thomastown, County Kilkenny, Ireland.
Irish For. 1967, 57.

ROBERTSON, William (–1846/7)
Of Newcastle, Northumberland. "A very accurate investigator of Lichens" (J. Sowerby and J. E. Smith *English Bot.* 2602).
Trans. Nat. Hist. Northumberland, Durham v.14, 1903, 83–84.
Letters at Kew. Plants at Newcastle Museum.

ROBERTSON, William Naismith (–1844)
b. Kilmany, Fifeshire *d.* July 1844
In Sydney, 1829. Gardener to W. Macarthur at Camden Park, N.S.W. Acting Superintendent, Botanic Gardens, Sydney, 1842–44.
Sydney Herald 12 July 1844. *J. Proc. Soc. N.S.W.* 1908, 117. *J. Proc. R. Austral. Hist. Soc.* 1932, 122–23.

ROBERTSON-GLASGOW, Charles Ponsonby (*fl.* 1900s)
Collected fungi in Singapore and Perak, 1898 which were sent to Kew.
Gdns Bull. Straits Settlements 1927, 130.

ROBINS, Thomas (1743–1806)
b. Bath, Somerset 1743
Son of Thomas Robins (1715–1770), landscape painter and occasional painter of flowers. Flower painter. "Mr. Robins begs to announce that his pictures of exotic plants and insects may be seen at his apartments in Chandos Buildings and at Mr. Richard's Print Shop, Broad Street" (*Bath Chronicle* 29 Nov. 1787).
W. Blunt *Art. of Bot. Illus.* 1955, 151–52. *Country Life* 1967, 654; 1975, 1800–1.
Drawings sold at Sotheby Sept. 1967, June 1969.

ROBINSON, Mrs. (–1847)
Of Fareham, Hants. Friend of G. E. Smith.
G. C. Druce *Fl. Buckinghamshire* 1926, c.
Plants at BM(NH).

ROBINSON, Anthony (–1768)
b. Sunderland, Durham *d.* Jamaica July 1768
"Practitioner in Physic and Surgery" in Jamaica. Drawings, MSS., and portr. at Institute of Jamaica, used by J. Lunan in *Hortus Jamaicensis* 1814.
Amer. Nat. 1894, 775–80. I. Urban *Symbolae Antillanae* v.3, 1902, 114. *J. Bot.* 1922, 49–52.
Ampelocissus robinsonii Planchon.

ROBINSON, Charles Budd (1871–1913)
b. Pictou, Nova Scotia 26 Oct. 1871 *d.* Amboyna 5 Dec. 1913
MA Dalhousie 1891. PhD Columbia 1906. Assistant Curator, Herb., New York Botanic Garden, 1906. Economic Botanist, Manila Bureau of Science, 1908. Contrib. papers on Philippine botany to *Philippine J. Sci.* 1908–14.
Kew Bull. 1914, 192. *Philippine J. Sci. Bot.* 1914, 191–97.

ROBINSON, Fred (*fl.* 1910s–1920s)
Lawyer, Watton, Norfolk. Friend of G. C. Druce. Norfolk botanist.
C. P. Petch and E. L. Swann *Fl. Norfolk* 1968, 18.

ROBINSON, Rev. George (*c.* 1824–1893)
d. Beech Hill, Armagh, Ireland 5 Sept. 1893
MA. Rector, Tartaraghan, County Armagh until 1892. Contrib. plant records to G. Dickie's *Fl. Ulster* 1864, S. A. Stewart and T. H. Corry's *Fl. N.E. Ireland* 1888.
Irish Nat. 1893, 296; 1913, 31. *Proc. Belfast Nat. Field Club* v.6, 1913, 626. R. L. Praeger *Some Irish Nat.* 1949, 149.

ROBINSON, Gilbert Wooding (–1942)
Kew gardener, 1924. Assistant Curator, Kew, 1930. Curator, Chelsea Physic Gardens, 1937.
Gdnrs Chron. 1937 ii 56 portr. *Richmond and Twickenham Times* 6 June 1942.

ROBINSON, Henry (1866–1932)
b. Colne, Lancs 12 July 1866 *d.* 14 May 1932
Founder member of Colne Naturalists' Society. From 1907 member of Liverpool Botanical Society to which he sent plant records of the Colne district.
N. Western Nat. 1932, 236.

ROBINSON, Herbert Christopher (1874–1929)
b. Liverpool 4 Nov. 1874 *d.* Oxford 30 May 1929
Zoologist. Visited Queensland, 1896; Malay Peninsula, 1901–2. Director, Federated Malay States Museums, 1903–26. Inspector of Fisheries, 1906. First man to ascend Gunong Tahan in Pahang, Malaya collecting plants above 6000 ft. Collected plants in Malaya and Siam.
Gdn Bull. Straits Settlements 1927, 130. *J. Bot.* 1929, 261. *J. Siam Nat. Hist. Soc. Nat. Hist. Supplt* v.8, 1929, 57–58. *Nature* v.124, 1929, 239–40. *Orchid Rev.* 1929, 303. *J. Federated Malay States Mus.* 1930, 1–12. *J. Malay Branch R. Asiatic Soc.* 1930, 361–63. *Fl. Malesiana* v.1, 1950, 441–42 portr.
Plants at BM(NH), Kew.
Eugenia robinsoniana Ridley.

ROBINSON, Hugh (1845–1890)
b. Belfast 1845

Draper. President, Belfast Naturalists' Field Club, 1887–89.

Belfast Nat. Hist. Philos. Soc. Centenary Vol. 1924, 100.

Herb. at Ulster Museum.

ROBINSON, James Fraser (1857–1927)

Schoolmaster, Hull, Yorks. President, Hull Scientific and Field Naturalists' Club. 'Fl. East Riding of Yorkshire' (*Trans. Hull Sci. Field Nat. Club* v.2, 1902).

Naturalist 1903, 310; 1923, 104 portr.; 1927, 126–28 portr. *J. Bot.* 1931, 88.

Herb. formerly at Hull Museum destroyed in 1943.

ROBINSON, James Frodsham (1838–1884)
b. Netherton, Frodsham, Cheshire 16 July 1838 *d.* Frodsham 4 Nov. 1884

Druggist. Curator, Owen's College Museum, Manchester, *c.* 1879–82. Sent lists of Caernarvon, Flint and Anglesey plants to H. C. Watson's *Topographical Bot.* not to be accepted as accurate (*Topographical Bot.* ed. 1 pt. 2, 618; ed. 2, 554–55). Contrib. Frodsham plants to *J. Bot.* 1868, 95–96.

Trans. Bot. Soc. Edinburgh v.16, 1886, 313. J. B. L. Warren *Fl. Cheshire* 1899, lxxxvii. *J. Bot.* 1904, 300–1; 1907, 138; 1912, 239. *R.S.C.* v.5, 238; v.8, 764.

Plants at Royal Botanic Garden, Edinburgh.

ROBINSON, John (*alias* **Fitzroberts**) (*fl.* 1690s–1710s)

Of the Gill, Kendal, Westmorland. Correspondent of J. Petiver and J. Ray. "An expert botanist" (*Philos. Trans. R. Soc.* v.27, 1711, 376).

J. Ray *Synopsis Methodica* 1696, 325. L. Plukenet *Almagestum Botanicum* 1696, 173. Nicholson *Ann. of Kendal* 1832, 244. E. J. L. Scott *Index to Sloane Manuscripts* 1904, 193. J. E. Dandy *Sloane Herb.* 1958, 195–96.

ROBINSON, Joseph (1811–1890)
b. Chelmsford, Essex Feb. 1811 *d.* Slough, Bucks 1890

Gardener to James Simpson at Thames Bank, Pimlico until 1858. Contrib. to *Florist, Gdn Misc.*

Garden v.37, 1890, 449. *Gdnrs Chron.* 1890 i 591–92.

ROBINSON, Sir Tancred (–1748)
b. Yorks *d.* 29 March 1748

MB 1679. MD 1685. FRS 1684. Knighted 1714. Physician to George I. Studied with H. Sloane under Tournefort. Friend of J. Ray. "Insignis Botanicus" (L. Plukenet *Almagestum Botanicum* 1696, 374). 'Tubera Terrae' (*Philos. Trans. R. Soc.* v.17, 1693, 824–26).

Gent. Mag. 1748, 187. R. Pulteney *Hist. Biogr. Sketches of Progress of Bot. in England* v.2, 1790, 118–21. E. Lankester *Memorials of J. Ray* 1846, 10. W. Munk *Roll of R. College of Physicians* v.1, 1878, 469–70. *D.N.B.* v.49, 45–46. E. J. L. Scott *Index to Sloane Manuscripts* 1904, 456. C. E. Raven *John Ray* 1942 *passim.* J. E. Dandy *Sloane Herb.* 1958, 196.

Robinsonia Scop.

ROBINSON, Rev. Thomas (–1719)

Rector, Ousby, Cumberland, 1672–1719. Correspondent of J. Ray. *Natural History of Westmorland and Cumberland* 1709.

R. Pulteney *Hist. Biogr. Sketches of Progress of Bot. in England* v.1, 1790, 354.

ROBINSON, Wilfrid (1885–1930)
b. Hull, Yorks 1885 *d.* Aberystwyth, Cardiganshire 7 March 1930

MSc Manchester 1914. DSc 1919. FLS 1927. Son of J. F. Robinson (1857–1927). Lecturer in Botany, Manchester University, 1914–26. Professor of Botany, University College, Aberystwyth, 1926–30. Contrib. to *Ann. Bot.*, *Philos. Trans. R. Soc.*, *Trans. Br. Mycol. Soc.*

Bot. Soc. Exch. Club Br. Isl. Rep. 1930, 329. *J. Bot.* 1930, 119–20. *Mem. Proc. Manchester Lit. Philos. Soc.* 1930–31, v–x. *Naturalist* 1930, 173–75 portr. *Nature* v.125, 1930, 57. *N. Western Nat.* 1930, 116–17 portr. *Proc. Linn. Soc.* 1929–30, 216–18.

ROBINSON, Sir William (1836–1912)
d. 1 Dec. 1912

Governor, Bahamas, 1874–80; Windward Islands, 1881–84; Barbados, 1884; Trinidad, 1885; Hong Kong, 1891–98.

Kew Rep. 1880, 30. I. Urban *Symbolae Antillanae* v.3, 1902, 114. *Who was Who, 1897–1916* 606.

Bahamas plants at Kew.

ROBINSON, William (1838–1935)
b. Ireland 15 July 1838 *d.* East Grinstead, Sussex 12 May 1935

FLS 1866. Under-gardener, Royal Botanic Society, Regent's Park, 1861–67. Pioneer of natural gardening. Friend of G. Jekyll. Created garden at Gravetye, East Grinstead. Founded *Garden* 1871, *Gardening Illustrated* 1879, *Fl. and Sylva* 1903. *Gleanings from French Gardens* 1868. *Parks, Promenades and Gardens of Paris* 1869. *Alpine Flowers for English Gardens* 1870. *Wild Garden* 1870. *Hardy Flowers* 1871. *English Flower Garden* 1883. *Garden Design and Architects' Gardens* 1892. *Gravetye Manor* 1911. *Home Landscapes* 1914.

Gdn Mag. 1920, 253–57 portr. *J. Bot.* 1935, 167–68. *Morning Post* 13 May 1935. *Proc. Linn. Soc.* 1934–35, 189–90. *Times* 13 May 1935. *Gdnrs Chron.* 1935 i 323–24 portr.,

378–79; 1936 i 24–25; 1964 ii 581. G. Taylor *Some Nineteenth Century Gdnrs* 1951, 68–115 portr. M. Hadfield *Pioneers in Gdning* 1955, 215–21. *J. R. Hort. Soc.* 1956, 506–7. M. Hadfield *Gdning in Britain* 1960, 360–70. *Northern Gdnr* 1969, 121–22. *Gdn Hist.* v.2(3), 1974, 12–21. *J. Scott. Rock Gdn Club* 1974, 45–49.

ROBLEY, Mrs. Augusta J. (*fl.* 1840s)
Daughter of Mrs. J. W. Penfold. *Selection of Madeira Flowers drawn and coloured from Nature* 1845.
J. Bot. 1919, 97–99.

ROBSON, Edward (1763–1813)
b. Darlington, Durham 17 Oct. 1763 *d.* Tottenham, Middx 21 May 1813
ALS 1790. Nephew of Stephen Robson (1741–1779). "A very assiduous and accurate botanist" (J. Sowerby and J. E. Smith *English Bot.* 70). Correspondent of W. Withering and J. E. Smith. Described *Ribes spicatum* (*Trans. Linn. Soc.* v.3, 1797, 240–41). Contrib. to *English Bot.* 92, 611, 1290, etc. Plant lists in Brewster's *Stockton* and Hutchinson's *Durham.* Drawing of *Lycoperdon* in *Gent. Mag.* 1792 i 113.
J. Bolton *Hist. Fungusses* 1788, 170. *Gent. Mag.* 1813 i 666. *Nat. Hist. Trans. Northumberland, Durham...* 1903, 78. *D.N.B.* v.49, 62. *Mem. Proc. Lit. Philos. Soc. Manchester* 1918–19, 1–46. *J. Bot.* 1922, 278.
Robsonia Spach.

ROBSON, John (–1886)
d. Hunton, Kent 1 Feb. 1886
Gardener at Linton Park, Kent until 1876. Contrib. to *J. Hort.*
Gdnrs Chron. 1886 i 183.

ROBSON, Joseph (1817–1884)
b. Cockermouth, Cumberland 13 April 1817 *d.* Cleator, Cumberland 19 April 1884
ALS 1854. Mining engineer. 'Catalogue of Wild Plants of Gosforth (Cumberland) in 1853' (*Phytologist* v.5, 1854, 1–9).
W. Hodgson *Fl. Cumberland* 1898, xxxi–xxxii. *R.S.C.* v.5, 243.

ROBSON, Stephen (1741–1779)
b. Darlington, Durham 24 June 1741 *d.* Darlington 16 May 1779
Linen manufacturer and grocer, Darlington. Correspondent of W. Curtis. *British Fl.* 1777 (MS. Supplt by E. Robson at BM(NH)). *Plantae Rariores Agro Dunelmensi* (privately printed).
D. Turner and L. W. Dillwyn *Botanist's Guide through England and Wales* 1805, 241. *D.N.B.* v.49, 62. *Nat. Hist. Trans. Northumberland, Durham* v.14, 1903, 77–78. *Friends' Quart. Examiner* 1917, 14–31, 265–82. *J. Bot.* 1922, 278.
Plants at York Museum.

ROBSON, William (1882–1923)
b. Wark, Northumberland 6 May 1882 *d.* Montserrat 22 April 1923
Kew gardener, 1903. Curator, Botanic Station, Montserrat, 1905–23.
Kew Bull. 1905, 60; 1923, 238–39. *J. Kew Guild* 1924, 253–54 portr.

ROBY, John (1793–1850)
b. Wigan, Lancs 5 Jan. 1793 *d.* drowned near Portpatrick, Wigtonshire 18 June 1850
Antiquary. Banker, Rochdale, Lancs, 1819–47. Lectured on botany at Rochdale, 1838 (*Naturalist* v.4, 1838–39, 55). *Seven Weeks in Belgium, Switzerland* 1838 (contains plant localities).
Legendary and Poetical Remains 1854 (with memoir). *D.N.B.* v.49, 65. *Trans. Liverpool Bot. Soc.* 1909, 87.
Portr. at Rochdale Library.

ROCHFORD, Edmund (1854–1919)
b. 16 Nov. 1854 *d.* March 1919
Nurseryman, Cheshunt, Herts.
M. Allan *Tom's Weeds* 1970 portr.

ROCHFORD, John (1851–1919)
b. 14 Feb. 1851 *d.* 15 Dec. 1919
Nurseryman, Enfield, Middx.
M. Allan *Tom's Weeds* 1970 portr.

ROCHFORD, Joseph (1856–1932)
b. 3 Dec. 1856 *d.* 29 June 1932
VMH 1925. Nurseryman, Broxbourne, Herts.

ROCHFORD, Joseph Patrick (*c.* 1882–1965)
d. c. April 1965
VMH 1942. Nurseryman, Tottenham, Middx.
Gdnrs Chron. 1965 i 445.

ROCHFORD, Michael (*c.* 1819–1883)
d. 29 Jan. 1883
Nurseryman, Page Green, Tottenham, Middx.
Florist and Pomologist 1883, 48. *Gdnrs Chron.* 1883 i 223. M. Allan *Tom's Weeds* 1970, 20 portr.

ROCHFORD, Thomas (1849–1901)
b. 15 April 1849 *d.* Cheshunt, Herts 12 Oct. 1901
Founded nursery at Tottenham, Middx in 1877. Also had Turnford Hall Nurseries, Cheshunt, 1883.
Garden v.60, 1901, 258–59 portr. *Gdnrs Chron.* 1901 ii 295 portr. *Gdnrs Mag.* 1901, 682. R. Webber *Early Horticulturists* 1968, 127–36 portr. M. Allan *Tom's Weeds* 1970 portr.

ROCHFORD, Thomas Samuel (1877–1918)
b. 8 June 1877 *d.* 27 Sept. 1918
Son of T. Rochford (1849–1901). Nurseryman, Turnford Hall Nurseries, Cheshunt.
M. Allan *Tom's Weeds* 1970 portr.

ROCKLEY, Lady Alicia Margaret (*formerly* **Cecil**) (*née* **Amherst**) (1865–1941)
d. Poole, Dorset 14 Sept. 1941
Married Sir Evelyn Cecil (Baron Rockley) 1898. Collected plants in Mozambique, 1899; Rhodesia, 1900; Ceylon, Australia, New Zealand; Canada, 1927. *History of Gardening* 1895; ed. 3 1910. *Children's Gardens* 1902. *London Parks and Gardens* 1907. *Wild Flowers of Great Dominions of British Empire* 1935. *Some Canadian Wild Flowers* 1937. *Historic Gardens of England* 1938.
Kew Bull. 1907, 234; 1935, 338. *Moçambique* 1940, 114–17 portr. *Times* 15 Sept. 1941. *Chronica Botanica* 1942–43, 354. *Who was Who, 1941–1950* 992.
MSS. at Chelsea Physic Garden. Plants at Kew.
Kaempferia cecilae N. E. Brown.

ROCQUE, Bartholomew (*fl.* 1730s–1760s)
Brother of John Rocque, the cartographer. Nurseryman, Walham Green, Fulham, Middx. *Treatise on Hyacinth* 1755. *Practical Treatise of Cultivating Lucern* 1761.
J. C. Loudon *Encyclop. Gdning* 1822, 1271.

RODERICK, John *see* Rhydderch, S.

RODGER, Sir Alexander (*c.* 1875–1950)
b. Greenock, Renfrewshire *c.* 1875 *d.* 30 Sept. 1950
Indian Forest Service, Burma, 1898–1911. Instructor, Forest Research Institute, Dehra Dun, 1911–13. Conservator of Forests, 1920–22; Inspector-General, 1926. President, Forest Research Institute, Dehra Dun, 1925–30. *Handbook of Forest Products of Burma* 1920. Edited J. H. Lace's *List of Trees and Shrubs of Burma* ed. 2 1922.
Indian Forester 1931, 197–99. *Empire For. Rev.* 1950, 291. *Nature* v.166, 1950, 764. *Quart. J. For.* 1951, 125. *Who was Who, 1941–1950* 992.

RODWAY, James (1848–1926)
b. Trowbridge, Wilts 1848 *d.* Georgetown, British Guiana 19 Nov. 1926
FLS 1886. Curator, Museum of Natural History and Anthropology, British Guiana. *History of Guiana* 1891–94 3 vols. *In the Guiana Forest* 1894.
Proc. Linn. Soc. 1926–27, 94.

RODWAY, Leonard (1853–1936)
b. Torquay, Devon 5 Oct. 1853 *d.* Kingston, Tasmania 9 March 1936
Dental surgeon. To Tasmania, 1890. Hon. Government Botanist, 1896–1932. Lecturer in Botany, Tasmania University, 1922–29. Director, Herb. and Botanic Garden, Hobart, 1928–32. *Tasmanian Fl.* 1903. *Tasmanian Bryophyta* 1914–16 2 vols. *Some Wild Flowers of Tasmania* 1910; ed. 2 1922.
Kew Bull. 1932, 459–60; 1936, 287. *Papers Proc. R. Soc. Tasmania* 1928, 163–65; 1936, 94–96 portr. *Trans. Proc. Bot. Soc. Edinburgh* 1936, 250. *Chronica Botanica* 1937, 60 portr. *Who was Who, 1929–1940* 1162. P. Serle *Dict. Austral. Biogr.* 1949, 285.
Plants at Melbourne Botanic Garden. Library at Royal Society, Tasmania. Portr. at Hunt Library.

ROE, Jeremiah Freeman (*fl.* 1830s)
Nurseryman, 47 London Street, Norwich, Norfolk.

ROE, John Septimus (1797–1878)
b. Newbury, Berks 8 May 1797 *d.* Perth, W. Australia 23 May 1878
FLS 1828. Lieut., Royal Navy, 1813–27. Surveyor-General, W. Australia, 1828–71. Collected plants in W. Australia, Persian Gulf, Ceylon. Sent plants to C. von Huegel, A. B. Lambert, W. J. Hooker. 'Journey of Discovery into Interior of Western Australia' (*Hooker's J. Bot.* 1854–55).
J. D. Hooker *Fl. Tasmaniae* v.1, 1859, cxxi–cxxii. *Proc. R. Geogr. Soc.* v.1, 1879, 277–78. P. Mennell *Dict. Austral. Biogr.* 1892, 394. *J. W. Austral. Nat. Hist. Soc.* no. 6, 1908, 24–25. *D.N.B.* v.49, 88. I. Lee *Early Explorers in Australia* 1925 *passim. W. Austral. Nat.* 1948, 115. *J. Proc. W. Austral. Hist. Soc.* 1954, 67. *Austral. Dict. Biogr.* v.2, 1967, 390. *Taxon* 1970, 537–38. *R.S.C.* v.4, 251.
MSS. at Perth Public Library. Portr. at Hunt Library.
Roea Huegel.

ROEBUCK, William Denison (1851–1919)
b. Leeds, Yorks 5 Jan. 1851 *d.* 15 Feb. 1919
FLS 1884. President, Yorkshire Naturalists' Union, 1903. Instituted fungus forays for Y.N.U.
Proc. Linn. Soc. 1918–19, 64–65.

ROFFEY, Rev. John (1860–1927)
b. 25 Aug. 1860 *d.* Riva, Lake Garda, Italy 3 July 1927
BA Oxon 1884. Held curacies in Notts and London. Studied British flora especially *Hieracia. Hieracia* in *London Cat.* 1925, 26–30 and *J. Bot.* 1925, 315–22.
Bot. Soc. Exch. Club Br. Isl. Rep. 1927, 380. *J. Bot.* 1927, 229–30. *Kew Bull.* 1927, 317.
Plants at BM(NH).

ROGER-SMITH, Hugh (1877–1955)
d. Shrewsbury, Shropshire Sept. 1955
Travelled and collected plants in the Alps, Greece and Albania. Contrib. to *Quart. Bull. Alpine Gdn Soc.*
Quart. Bull. Alpine Gdn Soc. 1955, 383–86 portr.

ROGERS, Amelia Warren *see* Griffiths, A. W.

ROGERS, Charles Coltman- *see* Coltman-Rogers, C.

ROGERS, Charles Gilbert (1864–1937)
b. Plymouth, Devon 1864 *d.* Woking, Surrey 18 Nov. 1937

FLS 1902. Indian Forest Service. Assistant Conservator, Bengal, 1888–90. Instructor, Indian Forestry School, Dehra Dun, 1890–1906. Conservator, Central Provinces, 1906–10. Chief Conservator, Burma, 1913–19. Collected plants in India, Burma, and Kenya (with his brother F. A. Rogers).

Kew Bull. 1938, 84. *Proc. Linn. Soc.* 1937–38, 333–34. *Who was Who, 1929–1940* 1162.

ROGERS, Rev. Frederick Arundel (1876–1944)

Educ. Oxford. Chaplain, South African Church Railway Mission; Head of Mission, 1911. Archdeacon, Pietersburg, Transvaal, 1914. Collected plants in Rhodesia, Belgian Congo, 1904–25; Iraq, 1928–29; Syria, Cyprus, Greece, Switzerland, 1930.

Kew Bull. 1921, 289–301. *Comptes Rendus AETFAT 1960* 1962, 169–70, 185–86.

Plants at BM(NH), Kew.

Caralluma rogersii E. A. Bruce & R. A. Dyer.

ROGERS, George (*fl.* 1780s–1799)

Nurseryman, St. Peter's Churchyard, and Nuns Gardens, Chester, Cheshire.

ROGERS, John (1752–1842)
b. Richmond, Surrey 10 Feb. 1752 *d.* Southampton, Hants 9 Nov. 1842

Gardener, Royal Gardens at Richmond and Kew. At Southampton where he assisted his son John to found a nursery in 1812. Had herb. *Fruit Cultivator* 1834. *Vegetable Cultivator...Life of P. Miller* 1839. H. M. Gilbert *Sketch of Life and Reminiscences of John Rogers* 1889 portr.

Gdnrs Mag. 1835, 99; 1842, 667. *J. Kew Guild* 1902, 90–92 portr.

ROGERS, Patrick Kerr (1776–1828)
b. Newtonstewart? Ireland 1776 *d.* Williamsburg, U.S.A. 1 Aug. 1828

MD Pennsylvania 1802. Student of B. S. Barton. Professor of Natural History and Chemistry, William and Mary College, 1819–28. *Investigation of Properties of Liriodendron Tulipifera* 1802.

J. Bot. 1900, 231.

ROGERS, Thomas (1827–1901)
b. St. Helens, Lancs 1827 *d.* Patterdale, Westmorland 30 May 1901

Of Manchester. Cryptogamist. Herb. rich in Australian mosses. *Botanical Excursion to Breadalbane Mountains* 1875. *Botanical Excursion to Grampian Mountains* (with J. Neild) 1877.

J. Bot. 1901, 395–96. *Trans. Liverpool Bot. Soc.* 1909, 87. W. K. Martin and G. T. Fraser *Fl. Devon* 1939, 775–76.

ROGERS, William Henry (1818–1898)
b. 13 Feb. 1818 *d.* 30 Nov. 1898

Nurseryman, Red Lodge Nursery, Southampton, Hants. Son of John Rogers, the founder and grandson of J. Rogers (1752–1842)

Gdnrs Chron. 1898 ii 427 portr. *J. Hort. Cottage Gdnr* v.37, 1898, 436.

ROGERS, Rev. William Moyle (1835–1920)
b. Helston, Cornwall 12 July 1835 *d.* Bournemouth, Hants 26 May 1920

FLS 1881. Vice-Principal, Theological College, Capetown, 1860. To England, 1862. Curate, Trusham, Devon, 1876. Vicar, Bridgerule, 1882–85. *Handbook of British Rubi* 1900. Contrib. to *J. Bot.*

Bot. Soc. Exch. Club. Br. Isl. Rep. 1920, 105–6. *J. Bot.* 1920, 161–64. F. H. Davey *Fl. Cornwall* 1909, lviii. A. R. Horwood and C. W. F. Noel *Fl. Leicestershire* 1933, ccxxvii. W. K. Martin and G. T. Fraser *Fl. Devon* 1939, 775–76. D. Grose *Fl. Wiltshire* 1957, 39–40.

Plants at BM(NH), Kew, Queen Ethelburga's School, Harrogate, Yorks. Portr. at Hunt Library.

Rubus rogersii F. Linton.

ROGET, Peter Mark (1779–1869)
b. London 18 Jan. 1779 *d.* West Malvern, Worcs 13 Sept. 1869

MD Edinburgh 1798. FRS 1815. Physician in Manchester and London. Professor of Physiology, Royal Institution, 1833–36. Secretary, Royal Society, 1827. 'Animal and Vegetable Physiology' (*Bridgewater Treatise* 1834).

Proc. R. Soc. v.18, 1870, xxviii–xl. W. Munk *Roll of R. College of Physicians* v.3, 1878, 71–74. *D.N.B.* v.49, 149–51.

ROHDE, Eleanor Sinclair (1880s–1950)
d. Reigate, Surrey 23 June 1950

Educ. Oxford. *Garden of Herbs* 1920. *Old English Herbals* 1922. *Old English Gardening Books* 1924. *Scented Garden* 1931. *Oxford College Gardens* 1932. *Story of the Garden* 1932. *Gardens of Delight* 1934. *Shakespeare's Wild Flowers* 1935. *Herbs and Herb Gardening* 1936. Contrib. to *Country Life*, *The Field*, *Sphere*.

Gdnrs Chron. 1950 ii 12; v.164(21), 1968, 18–19. *Times* 24 June 1950. *Who was Who, 1941–1950* 994. D. Macleod *Gardener's London* 1972, 157–64 portr.

ROLAND, Thomas (*c.* 1861–1929)
b. Birkenhead, Cheshire *c.* 1861 *d.* Lynn, Mass., U.S.A. 11 Dec. 1929
To U.S.A., 1884. Nurseryman, Nahant, Mass., 1894. Specialised in orchids.
Gdnrs Chron. 1930 i 18. *Orchid Rev.* 1930, 55.

ROLFE, Robert Allen (1855–1921)
b. Ruddington, Notts 12 May 1855 *d.* Kew, Surrey 13 April 1921
ALS 1885. VMH 1921. Assistant, Kew Herb., 1880. Founder and editor of *Orchid Rev.* 1893–1920. *Selagineae* in *J. Linn. Soc.* v.20, 1883, 338–58; Philippine plants v.21, 1884, 283–316. Accounts of orchids in *Fl. Tropical Africa* and *Fl. Capensis.*
E. D. Merrill *Bot. Work in Philippines* 1903, 27–28. *Gdnrs Mag.* 1909, 157–58 portr. *J. Kew Guild* 1911–12, 1 portr. *J. Hort. Home Farmer* v.65, 1912, 467 portr. *Garden* 1921, 207 portr. *Gdnrs Chron.* 1921 i 74, 204 portr. *Bot. Soc. Exch. Club. Br. Isl. Rep.* 1921, 365–67. *J. Bot.* 1921, 182–83. *Kew Bull.* 1921, 123–27. *Orchid Rev.* 1921, 5–8 portr.; 1933, 19–22 portr. *Proc. Linn. Soc.* 1920–21, 52–53. *Nature* v.107, 1921, 276–77. *Amer. Orchid Soc. Bull.* 1933, 38–39 portr. M. A. Reinikka *Hist. of the Orchid* 1972, 275–78 portr. *Taxon* 1973, 234 portr.
Plants at Kew.
Rolfea Zahlbr.

ROLLISSON, George (*c.* 1799–1879)
d. Balham, London 15 Dec. 1879
Eldest son of founder of nursery at Upper Tooting, Surrey.
Gdnrs Chron. 1880 i 23. *Gdnrs Yb. Almanack* 1881, 190. *Gdnrs Mag.* 1893, 591–92.

ROLLISSON, William (*c.* 1765–1842)
d. 27 June 1842
Founded Springfield Nursery, Upper Tooting, Surrey. Famous for its orchids which were eventually purchased by Messrs Veitch.
Gdnrs Mag. 1842, 336. *Gdnrs Mag.* 1893, 591–92. *Trans. London Middlesex Archaeol. Soc.* v.24, 1973, 193.

ROLLISSON, William (*c.* 1802–1875)
d. 18 June 1875
Nurseryman, Upper Tooting, Surrey.
Gdnrs Chron. 1875 i 825–26.

ROLLITT, Sir Albert Kaye, (*c.* 1842–1922)
VMH 1917. Chairman, Hull Botanic Garden. President, National Chrysanthemum Society. Largely responsible for National Diploma in Horticulture.
Gdnrs Mag. 1907, 775–76 portr.

ROMANES, George John (1848–1894)
b. Kingston, Canada 20 May 1848 *d.* Oxford 23 May 1894
BA Cantab 1870. LLD Aberdeen. FRS 1879. FLS 1875. Zoologist. Friend of C. Darwin. Wrote on evolution. Founded Romanes Lectureship at Oxford, 1891. Experimented on graft-hybrids. *Darwin and after Darwin* 1892–97 3 vols.
Nature v.50, 1894, 168–69. *Proc. Linn. Soc.* 1893–94, 34. *Proc. R. Soc.* v.57, 1895, vii–xiv. *Life and Letters of George John Romanes* 1896 portr. *D.N.B.* v.49, 177–80. *R.S.C.* v.8, 772; v.11, 211; v.12, 627; v.18, 281.

ROMANS, Bernard (*c.* 1720–1783)
b. Holland *c.* 1720 *d.* at sea between Jamaica and U.S.A. 1783
Educated as engineer and surveyor in England. King's botanist in Florida, 1763–71. *Natural History of Florida* 1775.
J. E. Smith *Selection from Correspondence of Linnaeus* v.1, 1821, 596. C. S. Sargent *Silva of North America* v.4, 1892, 5. *Appleton's Cyclop. Amer. Biogr.* v.5, 1888, 313–14. *D.N.B.* v.49, 180–81.

RONALDS, Hugh (*c.* 1726–1788)
d. 7 Jan. 1788
Founded nursery at Isleworth, Middx.
Trans. London Middlesex Archaeol. Soc. v.24, 1973, 184.

RONALDS, Hugh (1759–1833)
b. Brentford, Middx 4? March 1759 *d.* Brentford 22 Nov. 1833
Son of H. Ronalds (*c.* 1726–1788). Nurseryman, Brentford, Middx. Formed herb. of Kew plants. *Pyrus Malus Brentfordiensis* 1831; illustrated by his daughter Elizabeth.
Gdnrs Mag. 1829, 736–37; 1834, 96. *Gent. Mag.* 1834 i 337–38. *Hort. Register* 1834, 92. *Gdnrs Chron.* 1950 ii 140. *J. Kew Guild* 1972, 127–29.

RONALDS, Robert (*c.* 1799–1880)
d. 29 Aug. 1880
Son of H. Ronalds (*c.* 1759–1833). Nurseryman, Brentford, Middx.
Florist and Pomologist 1880, 160. *Gdnrs Chron.* 1880 ii 346. *Gdnrs Yb. Almanack* 1881, 191.

ROOKE, Hayman (*c.* 1722–1806)
d. 18 Sept. 1806
Major, R.A. Of Whitehaven, Cumberland. Botanical artist. *Oaks... at Welbeck* 1790.
Gent. Mag. 1806 ii 886.

ROOKE, John (1807–1872)
b. North Shields, Northumberland 1807 *d.* Whitehaven, Cumberland 1872
Art teacher at Whitehaven. Botanical artist.
W. Hodgson *Fl. Cumberland* 1898, xxvi–xxvii.

ROOTSEY, Samuel (1788–1855)
b. Colchester, Essex 12 Feb. 1788 *d.* Bristol 4 Sept. 1855

FLS 1811. Druggist, Bristol, 1812. Lectured on botany. Had herb. *Syllabus of Botanical Lectures* 1818. 'Observations upon some of the Medicinal Plants mentioned by Shakespeare.' (*Trans. Med.-bot. Soc. London* 1832–33, 83–96; 1834–37, 195).

Proc. Linn. Soc. 1856, xlv–xlvii. J. W. White *Fl. Bristol* 1912, 77–78. H. J. Riddelsdell *Fl. Gloucestershire* 1948, cxvii.

ROPER, Freeman Clarke Samuel (1819–1896)
b. Hackney, London 23 Sept. 1819 *d.* Eastbourne, Sussex 28 July 1896

FLS 1857. President, Eastbourne Natural History Society. *Catalogue of Works on the Microscope* 1865. *Fl. Eastbourne* 1875. Contrib. to *J. Bot., Proc. Eastbourne Nat. Hist. Soc.*

J. Bot. 1896, 430–31 portr. *Proc. Linn. Soc.* 1896–97, 67. A. H. Wolley-Dod *Fl. Sussex* 1937, xlvii.

Herb. at Brighton Museum. Diatoms at BM(NH). Portr. at Hunt Library.

ROPER, Ida Mary (1865–1935)
b. Bristol 25 Aug. 1865 *d.* Bristol 8 June 1935

FLS 1909. President, Bristol Naturalists' Society, 1913–16. Studied *Viola* and flora of Bristol area. Discovered *Nitella mucronata* var. *gracillima* 1917.

Bot. Soc. Exch. Club Br. Isl. Rep. 1935, 19. *J. Bot.* 1935, 233–34; 1936, 78–79. *Nature* v.136, 1935, 134–35. *Proc. Bristol Nat. Soc.* 1935, 26–27 portr. *Proc. Linn. Soc.* 1935–36, 212–13. *Rep. Br. Bryol. Soc.* 1935, 310. *Naturalist* 1936, 35. H. J. Riddelsdell *Fl. Gloucestershire* 1948, cxliii.

Herb. and books at Leeds University. Plants at Oxford.

ROSCOE, Margaret (Mrs. Edward) (*née* **Lace**) (*fl.* 1820s–1830s)

Daughter-in-law of W. Roscoe (1753–1831). Botanical artist. *Floral Illustrations of the Seasons* 1829. Contrib. plates to W. Roscoe's *Monandrian Plants* 1824–29. Was to have illustrated W. Hincks's *Monograph on Genus Oenothera* which was never published.

Gdnrs Mag. 1830, 76–77; 1838, 385–86. *Trans. Liverpool Bot. Soc.* 1909, 87.

ROSCOE, William (1753–1831)
b. Liverpool, Lancs 8 March 1753 *d.* Liverpool 30 June 1831

FLS 1804. Banker, Liverpool. MP, Liverpool, 1806–7. Founded Liverpool Botanic Garden, 1802. Correspondent of J. E. Smith. *Monandrian Plants* 1824–29. Contrib. to *Trans. Linn. Soc.*

H. Roscoe *Life of William Roscoe* 1833 portr. P. Smith *Mem. and Correspondence of J. E. Smith* v.2, 1832, 301. *Gent. Mag.* 1831 ii 179–81. E. Baines *Hist. Lancashire* v.2, 1870, 377–79. *D.N.B.* v.49, 222–25. *Trans. Liverpool Bot. Soc.* 1909, 87–88. Liverpool Public Museums *Handbook and Guide to Herb. Collections* 1935, 59 portr. *Liverpool Libraries, Mus. and Arts Committee Bull.* 1955, 19–61 portr.; 1962–63, 9–13. G. Chandler *William Roscoe of Liverpool* 1953 portr. *Gdnrs Chron.* 1964 ii 529 portr. F. A. Stafleu *Taxonomic Literature* 1967, 399–405. *Huntia* 1965, 206 portr.

Plants, letters and original drawings for *Monandrian Plants* at Liverpool City Libraries and Museums. Letters at Kew. Portr. at National Portrait Gallery; plaster bust at Kew.

Roscoea Smith.

ROSE, Frederick J. (1885–1956)
d. Southampton, Hants 5 May 1956

VMH 1943. Gardener to Lord Swaythling at Townhill Park, South Stoneham, Hants, 1919.

Gdnrs Chron. 1951 i 202 portr.; 1956 i 590. *Lily Yb.* 1957, 130–31.

ROSE, Hugh (*c.* 1717–1792)
d. Norwich, Norfolk 18 April 1792

Apothecary, Norwich. Taught J. E. Smith. Correspondent of W. Hudson. *Elements of Botany* 1775.

Trans. Linn. Soc. v.7, 1804, 297–99. *Notes and Queries* v.4, 1863, 395. *Trans. Norfolk Norwich Nat. Soc.* 1874–75, 27–28; 1912–13, 648–50, 655.

Herb. passed to J. E. Smith.

ROSE, John (*c.* 1621/2–1677)
b. Amesbury, Wilts Oct. *c.* 1621/2 *d.* 10 Sept. 1677

Gardener to Duchess of Somerset, Earl of Essex and Duchess of Cleveland. Keeper of St. James's Garden, *c.* 1661. *English Vineyard Vindicated* 1666 (biogr. account in 1966 facsimile).

G. W. Johnson *Hist. English Gdning* 1829, 112–13. *J. Hort. Cottage Gdnr* v.54, 1875, 114–16 portr. *Antiq. Chronicle Lit. Advertiser* July 1882, 27. *Gdnrs Chron.* 1938 ii 420–21. M. Hadfield *Gdning in Britain* 1960, 125–26.

ROSENBERG, Mary Elizabeth (*afterwards* **Duffield**) (1820–1914)
d. 13 Jan. 1914

Of Bath, Somerset. Member of Royal Institute of Painters in Water Colours, 1861. *Corona Amaryllidacea* 1839. *Museum of Flowers* 1845. *Art of Flower-painting* 1856; ed. 19 1882.

D.N.B. v.49, 248.

ROSENHEIM, Paul (1878–1939)
d. London Oct. 1939
Stockbroker. Founder member of Alpine Garden Society. Contrib. to *Quart. Bull. Alpine Gdn Soc.*
Quart. Bull. Alpine Gdn Soc. 1940, 71–72.

ROSS, Charles (*c.* 1825–1917)
b. Dalmeny, Midlothian *c.* 1825 *d.* Westgate-on-Sea, Essex 4 Feb. 1917
VMH 1908. Gardener to Col. Houblin at Welford Park, Newbury, Berks, 1860–1908. Raised new apple cultivars including 'Charles Ross'.
Gdnrs Mag. 1899, 673 portr.; 1908, 789–90 portr. *Garden* 1908, 584–85 portr.; 1917, 62 portr. *J. Hort. Cottage Gdnr* v.43, 1901, 329 portr.; v.64, 1912, 346. *Gdnrs Chron.* 1908 ii 299 portr.; 1917 i 67 portr. A. Simmonds *Hort. Who was Who* 1948, 44 portr.

ROSS, David (*c.* 1810–1881)
d. 21 Feb. 1881
Classics teacher in Edinburgh. *Some Account of a Botanical Tour in Mountains of Auvergne and Switzerland* 1861 portr. *Account of Botanical Excursions round...Paris* 1862. *Account of Botanical Rambles in the Pyrenees* 1863.

ROSS, Henry James (1820–1902)
b. Malta 1820 *d.* Florence, Italy 17 July 1902
Had large collection of orchids near Florence; introduced many new species. Correspondent of Reichenbach. His wife, Janet (1842–1927), made water-colour drawings of the orchids, now at Kew.
Orchid Rev. 1902, 282–84; 1911, 202–5; 1917, 213–14; 1927, 283. *Kew Bull.* 1917, 85–86.
Coelogyne rossiana Reichenb. f.

ROSS, John (*fl.* 1800s–1830s)
Nurseryman, Caledonian Nursery, Stoke Newington Road, Hackney, London.

ROSS, John (*fl.* 1830s–1840s)
Orchid collector in Mexico *c.* 1836 for G. Barker of Birmingham.
G. B. Knowles and F. Westcott *Floral Cabinet* v.3, 1840, 180. *Bot. Mag.* 1846, t.4203.

ROSS, Rev. John (*fl.* 1870s–1890s)
Of United Presbyterian Church of Scotland. In Manchuria, 1872. Sent plants to Kew, 1877–90.
E. Bretschneider *Hist. European Bot. Discoveries in China* 1898, 703–4. *Kew Bull.* 1901, 56.
Iris rossii Baker.

ROSS, Sir John Foster George (1848–1926)
b. Rostrevor, County Down 27 July 1848 *d.* Rostrevor 10 July 1926
Had garden of trees and shrubs at Rostrevor.
Kew Bull. 1926, 368. *Who was Who, 1916–1928* 910. *Curtis's Bot. Mag. Dedications, 1827–1927* 385–86 portr.

ROSS, Joseph (1873–1962)
b. Banbury, Oxford 8 May 1873 *d.* Chingford, Essex Nov. 1962
Journalist. Curator, Essex Field Club Museum. Studied flora of Epping Forest. Contrib. to *Essex Nat.*, *London Nat.*
London Nat. 1962, 112–13. *Essex Nat.* 1963, 158–60. S. T. Jermyn *Fl. Essex* 1974, 19–20.

ROSS, William (*c.* 1763–1825)
d. 14 Nov. 1825
Possibly John Ross intended. Nurseryman, Stoke Newington, London. Raised camellia, 'Ross's Seedling'.
Gdnrs Mag. 1826, 95–96.

ROSS, William John Clunies *see* Clunies Ross, W. J.

ROSSITER, William (–1897)
d. 18 Oct. 1897
Teacher. Founder of South London Working Men's College, 1868. *First Book of Botany* 1866.
Times 22 Oct. 1897. F. Boase *Modern English Biogr.* Supplt 3, 1921, 501.

ROSSITTER, Mary *see* Harrison, M.

ROTHERAM, John (*c.* 1750–1804)
b. Hexham, Northumberland *c.* 1750 *d.* St. Andrews, Fife 6 Nov. 1804
MD Uppsala. FLS 1788. FRSE 1792. Pupil of Linnaeus. Assistant Professor of Chemistry, Edinburgh, 1793. Professor of Natural Philosophy, St. Andrews, 1795–1804. *Sexes of Plants Vindicated* 1790.
Gent. Mag. 1804 ii 1079; 1830 ii 565. *D.N.B.* v.49, 300.

ROTHERY, Henry Cadogan (1817–1888)
b. London 1817 *d.* Bagshot, Surrey 2 Aug. 1888
BA Cantab 1840. Lawyer. From 1842 was employed in ecclesiastical and Admiralty courts. Wreck Commissioner, 1876. Collected plants in British Guiana, Dominica, Madagascar.
Plants at Kew, BM(NH). Letters at Kew.
D.N.B. v.49, 303–4.

ROTHSCHILD, Ferdinand James de (1839–1898)
b. Paris 1839 *d.* 17 Dec. 1898
Settled in England, 1860. Created garden at Waddesdon, Bucks.
Gdnrs Chron. 1898 ii 457–58. *D.N.B. Supplt. 3* 304–5.

ROTHSCHILD, Leopold de (1845–1917)
b. 22 Nov. 1845 *d.* 29 May 1917
VMM. Had garden at Ascott, Leighton Buzzard, Bedfordshire.
Garden 1913, iv portr.; 1917, 216 portr.

ROTHSCHILD, Lionel Nathan de (1882–1942)
b. Ascot, Berks 25 Jan. 1882 *d.* London 28 Jan. 1942
VMH 1929. Created garden at Exbury, Hants; specialised in orchids, rhododendrons, azaleas.
Gdnrs Mag. 1909, 507–8 portr. *Gdnrs Chron.* 1942 i 62, 73. *Orchid Rev.* 1942, 65–67. *Rhododendron Yb.* 1946, 1–2 portr. C. E. Lucas Phillips and P. N. Barber *Rothschild Rhododendrons* 1967, 4–21 portr.

ROTHSCHILD, Lionel Walter, 2nd Baron (1868–1937)
b. London 8 Feb. 1868 *d.* Tring, Herts 27 Aug. 1937
FRS 1911. VMH 1897. Banker. Created garden and museum of natural history at Tring Park.
Gdnrs Mag. 1907, 609–10 portr. *Nature* v.140, 1937, 574. *Times* 28 Aug. 1937. *Obit. Notices Fellows R. Soc.* 1938, 385–86 portr. *D.N.B. 1931–1940* 754–55.
Gloriosa rothschildiana O'Brien.

ROTHSCHILD, Nathan Meyer, 1st Baron (1840–1915)
b. London 8 Nov. 1840 *d.* 31 March 1915
Had garden at Tring Park, Herts.
Gdnrs Chron. 1915 i 204. *Times* 1 April 1915. *Who was Who, 1897–1916* 613–14. *D.N.B. 1912–1921* 480–81.
Phalaenopsis rothschildiana Day ex Rolfe.

ROTHSCHILD, Nathaniel Charles (1877–1923)
b. 9 May 1877 *d.* Ashton Wold, Northants 12 Oct. 1923
MA Cantab 1901. Founded Society for Promotion of Nature Reserves, 1912. Interested in orchids.
Bot. Soc. Exch. Club Br. Isl. Rep. 1923, 151–55. *Outlook* Summer 1972, 19–20 portr.

ROTTLER, John Peter (1749–1836)
b. Strasbourg June 1749 *d.* Vepery, Madras, India 27 Jan. 1836
PhD Vienna 1795. In Ceylon, 1788. Danish missionary at Madras. Orientalist. Collected plants in Coromandel, 1795–96.
Annual Register 1837, 172. *Bot. Gaz.* 1851, 55. *Hooker's J. Bot.* 1851, 67. *Madras J. Lit. Sci.* v.22, 1861, 1–17. *J. Bot.* 1873, 210. *Bull. Liverpool Mus.* 1957, 27–33. I. H. Burkill *Chapters on Hist. Bot. in India* 1965, 17. *R.S.C.* v.5, 304.
Herb. and MS. Catalogue at Kew. Plants at Liverpool. Letters in Smith correspondence at Linnean Society.
Rottlera Roxb.

ROUND, Frank Harold (*c.* 1878–1958)
d. 7 Feb. 1958
Drawing master at Charterhouse School. Contrib. flower drawings to W R. Dykes's *Genus Iris* 1913.
W. Blunt *Art of Bot. Illus.* 1955, 250–51. *Iris Yb.* 1958, 19.
Iris drawings at BM(NH).

ROUPELL, Arabella Elizabeth (*née* **Pigott**) (1817–1914)
b. Edgmond, Shropshire 23 March 1817 *d.* Loddon, Berks 31 July 1914
Married T. B. Roupell of East India Company. Travelled in Cape, 1843–45 where she painted the flora. Returned to England, 1858. *Specimens of Fl. of S. Africa* 1849. *More Cape Flowers* 1964.
Bot. Mag. 1849, t.4466. *Hooker's J. Bot.* 1850, 127–28. *Quart. Bull. S. African Library* v.27(1), 1972, 4–14.
Paintings at Cape Town University.
Roupellia Hook.

ROURKE, James (–1919)
Superintendent, Glasgow Botanic Gardens. Assistant Superintendent of Parks, Glasgow, 1915.
Glasgow Nat. 1930, 52–53.

ROUSE, William (*fl.* 1690s–1700s)
"An ingenious Botanist and eminent Apothecary in London." Found *Orchis hircina* at Dartford, Kent.
J. Ray *Synopsis Methodica* 1696, 235–36. J. E. Dandy *Sloane Herb.* 1958, 196.

ROWDEN, Frances Arabella (*afterwards* **Countess St. Quentin**) (*fl.* 1800s–1840)
d. near Paris *c.* 1840
Schoolmistress at Chelsea. *Poetical Introduction to Botany* 1801; ed. 3 1818.
J. Bot. 1921, 329–31.

ROWLAND, John William (1852–1925)
b. 30 Dec. 1852 *d.* 18 Oct. 1925
MB Edinburgh. In India, 1877–78. Assistant Colonial Surgeon, Gold Coast, 1880. Chief Medical Officer, S. Nigeria. Botanist on Sir Gilbert Carter's expedition to Yoruba country, S. Nigeria, 1896. Collected plants in Gold Coast and Nigeria.
Kew Bull. 1893, 146, 369. *Who was Who, 1916–1928* 913.
Plants at Kew, Berlin.

ROWLANDS, Samuel Pryce (1887–1957)
d. Doncaster, Yorks 7 Dec. 1957
Educ. King's College, London. Collected ferns. Contrib. to *Br. Fern Gaz.*
Br. Fern Gaz. 1958, 200–1.

ROXBURGH, John (*fl.* 1770s–1820s)
Son of W. Roxburgh (1751–1815). Collected plants at Cape, *c.* 1799–1806 and in Chittagong, 1810–11. Overseer, Botanic Garden, Calcutta, 1813–17. Sent plants to A. B. Lambert.
J. Bot. 1918, 202–3; 1919, 28–34.
Plants at BM(NH).

ROXBURGH, William (1751–1815)
b. Craigie, Ayrshire 3 June 1751 *d.* Edinburgh 18 Feb. 1815
MD Edinburgh. FLS 1799. Madras Medical Service, 1776–80. Superintendent, Samalkot Botanic Garden, 1781–93. Superintendent, Calcutta Botanic Garden and Chief Botanist of East India Company, 1793–1813. At Cape, 1798, 1799, 1814; St. Helena, 1814; plant list in A. Beatson's *Tracts relative to Island of St. Helena* 1816. *Plants of Coast of Coromandel* 1795–1819 3 vols. *Hortus Bengalensis* 1814. *Fl. Indica* 1820–24 2 vols; ed. 2 1832 3 vols.
Trans. Linn. Soc. v.10, 1811, 46. *Trans. Soc. Arts* v.33, 1815 portr. *Cottage Gdnr* v.6, 1851, 65. *Bot. Misc.* 1830–31, 90–91. *Gdnrs Chron.* 1896 i 781–82. *Ann. R. Bot. Gdn Calcutta* v.5, 1895, 1–9 portr. E. Bretschneider *Hist. European Bot. Discoveries in China* 1898, 237–46. *D.N.B.* v.49, 368–70. *J. Bot.* 1899, 457–58; 1919, 28. Liverpool Public Museums *Handbook Guide to Herb. Collections* 1935, 44–45 portr. *Kew Bull.* 1956–57, 297–399. *Nature* v.207, 1965, 1234–35. F. A. Stafleu *Taxonomic Literature* 1967, 403. M. Archer *Nat. Hist. Drawings in India Office Library* 1962, 20–23, 65–66, 102. I. H. Burkill *Chapters on Hist. Bot. in India* 1965 *passim* portr. A. M. Coats *Quest for Plants* 1969, 148–49 portr. *Taxon* 1970, 538; 1974, 52 portr. *Endeavour* 1975, 84–89 portr.
MS. *Fl. Indica* at BM(NH), Kew, India Office Library. Drawings at Kew, BM(NH), Calcutta. Plants at BM(NH), Liverpool Museums. Letters at BM(NH), Linnean Society. Portr. at Hunt Library.
Roxburghia Dryander.

ROXBURGH, William (*fl.* 1780s–1810)
d. Padang, Sumatra 21 Sept. 1810
Son of W. Roxburgh (1751–1815). Collected plants in Rajmahal, 1800; Chittagong, 1801–2; Penang, 1802; Sumatra, 1803. Superintendent of East India Company's Spice Plantations, Fort Marlborough.
W. Roxburgh *Fl. Indica* v.1, 1832, 554; v.2, 51; v.3, 456. *J. Bot.* 1919, 28–34. *Fl. Malesiana* v.1, 1950, 450–51.

ROY, John (1826–1893)
b. Fowlis Webster, Perth 24 Feb. 1826 *d.* Aberdeen 18 Dec. 1893
LLD Aberdeen 1889. Desmidologist. Correspondent of H. C. Watson. Contrib. to J. Sowerby and J. E. Smith's *English Bot.* ed. 3. 'Desmidieae of County of Leicester' (*Fl. Leicestershire* 1886, 327–36). 'Notes on Japanese Desmids' (with J. P. Bisset) (*J. Bot.* 1886, 193–96, 237–42). 'Freshwater Algae of Embridge Lake and Vicinity, Hampshire' (*J. Bot.* 1890, 334–38).
Ann. Scott. Nat. Hist. 1894, 72–75 portr. *J. Bot.* 1894, 159–60.
Herb. acquired by Rev. J. Keith (1825–1905).

ROYDS, Sir John (*c.* 1750–1817)
b. Halifax, Yorks *c.* 1750 *d.* India 24 Sept. 1817
MA Oxon 1774. Knighted 1801. Puisne Judge, Supreme Court of Bengal. "A zealous botanist and an eminent benefactor of the science" (W. Roxburgh *Plants of Coast of Coromandel* v.3, 1819, 87).
Roydsia Roxb.

ROYLE, John Forbes (1798–1858)
b. Cawnpore, India 20 May 1798 *d.* Acton, Middx 2 Jan. 1858
Educ. Edinburgh. MD Munich 1833. FRS 1837. FLS 1833. Surgeon, East India Company in Bengal, 1819. Curator, Saharunpur Garden, 1823; retired to England, 1831. Professor of Materia Medica, King's College, London, 1837–56. Secretary, Horticultural Society of London, 1851–58. *Illus. Bot.…of Himalayan Mountains* 1833–40 2 vols. *Manual of Materia Medica* 1847; ed. 2 1852–53. *Fibrous Plants of India* 1855.
Cottage Gdnr v.19, 1857, 225, 249–50. *Gdnrs Chron.* 1858, 20–21. *Proc. Linn. Soc.* 1858, xxxi–xxxvii. *Proc. R. Soc.* 1859, 547–49. *J. Indian Art* 1888, 65–66 portr. *D.N.B.* v.49, 375–76. D. G. Crawford *Hist. Indian Med. Service, 1600–1913* v.2, 1914, 149–50, 171. *Kew Bull.* 1933, 378–90. *J. Arnold Arb.* 1943, 484–87. *N. Western Nat.* 1953, 250–65. *Pharm. J.* 1953, 74–75, 78–79. *Liverpool Libraries, Mus. Arts Committee Bull.* v.3(3), 1954, 5–38 portr. I. H. Burkill *Chapters on Hist. Bot. in India* 1965 *passim.* M. Archer *Nat. Hist. Drawings in India Office Library* 1962, 25–27, 88–89. F. A. Stafleu *Taxonomic Literature* 1967, 403–4. A. M. Coats *Quest for Plants* 1969, 150–51. *Bull. Inst. Hist. Med.* (*Hyderabad*) v.3, 1973, 79–87. *R.S.C.* v.5, 316.
Plants at Kew, Liverpool Museums. Letters at Kew, Royal Society of Arts.
Roylea Wall.

ROZEA, Richard (–*c.* 1829)
Surgeon. Of Marylebone, London. Herb. collected in 1815–23 with W. F. Goodyer, afterwards in possession of E. G. Varenne.
H. Trimen and W. T. T. Dyer *Fl. Middlesex* 1869, 398.

RUCK, Albert John (–1965)
Gardener to Sir Joseph Chamberlain at Birmingham. Orchid grower for Messrs Sander, Bruges; later at St. Albans branch of the firm.
Orchid Rev. 1966, 28–29.

RUDDER, Augustus (1828–1904)
b. Birmingham 10 Nov. 1828 *d*. Cabramatta, N.S.W. 11 Dec. 1904
Emigrated to Australia. Forester, 1884–96. Contrib. to *Agric. Gaz. N.S.W.*
Proc. Linn. Soc. N.S.W. 1904, 778–79. *J. R. Soc. N.S.W*. 1908, 117–18 portr.
Plants at Botanic Gardens, Sydney.
Eucalyptus rudderi Maiden.

RUDGE, Edward (1763–1846)
b. Evesham, Worcs 27 June 1763 *d*. Evesham 3 Sept. 1846
FRS 1805. FLS 1802. *Plantarum Guianae Rariorum Icones et Descriptiones* 1805. 'New Holland Plants' (*Trans. Linn. Soc*. v.10, 1811, 283–303). 'On History...of Carnation' (*Gdnrs Mag*. 1832, 428–32). His wife, Anne Rudge, drew plates for *Plantarum Guianae* and 'New Holland Plants'.
Bot. Mag. 1806, t.935; 1824, t.2465. *Trans. Linn. Soc*. v.8, 1807, 325–29. *Proc. Linn. Soc*. 1847, 315–17, 337–38. *Gent. Mag*. 1846 ii 652–53. G. F. W. Meyer *Primitiae Fl. Essequeboënsis* 1818, 198–201. *D.N.B*. v.49, 383–84. *J. Bot*. 1912, 63–64; 1917, 344–45. *Bull. Jardin Bot. Bruxelles* 1957, 243–65. F. A. Stafleu *Taxonomic Literature* 1967, 405.
Plants at BM(NH).
Nymphaea rudgea Meyer.

RUDGE, Samuel (1727–1817)
b. Thornhaugh, Northants 1727 *d*. Watlington, Oxon 24 Jan. 1817
Uncle of E. Rudge (1763–1846). Of Elstree, Herts. Made "innumerable MS. notes in almost every botanical work that he possessed" (*Gent. Mag*. 1817 i 181–82).
A. R. Pryor *Fl. Hertfordshire* 1887, xlii. G. C. Druce *Fl. Berkshire* 1897, cxlvi–cxlvii. *J. Bot*. 1917, 345.
Berks plants at BM(NH).

RUFFORD, Rev. William Squire (*c*. 1786–1836)
d. Lower Sapey, Worcs 17 April 1836
BA Oxon 1808. Rector, Badsey, Worcs, 1817; Lower Sapey, 1831. Helped T. Purton in *Midland Fl*. 1817–21, especially with lichens and fungi (preface iv; appendix, vii); his wife, A. Rufford, drew some of the plates. Had herb.
Mag. Nat. Hist. 1836, 607. J. E. Bagnall *Fl. Warwickshire* 1891, 495–96.

RUSKIN, John (1819–1900)
b. London 8 Feb. 1819 *d*. Brantwood, Lancs 20 Jan. 1900
MA Oxon 1843. Slade Professor of Art, Oxford, 1870–79. *Frondes Agrestes* 1875. *Proserpina: Studies of Wayside Flowers* 1875–86. *Hortus Inclusus* 1887.
J. Bot. 1900, 199. *D.N.B. Supplt 3* 305–27.

RUSSELL, Alexander (*c*. 1715–1768)
b. Edinburgh *c*. 1715 *d*. London 28 Nov. 1768
MD Glasgow. FRS 1756. Physician to English factory at Aleppo, Syria, *c*. 1740–53. Physician, St. Thomas's Hospital, 1759–68. Sent seeds from Aleppo to P. Collinson. *Natural History of Aleppo and Parts adjacent* 1756 (contains long lists of plants).
Gent. Mag. 1771, 109–11. J. C. Lettsom *Mem. of John Fothergill* 1786, 241–59 portr. W. Munk *Roll of R. College of Physicians* v.2, 1878, 230–31. *D.N.B*. v.49, 426–27. R. H. Fox *Dr. John Fothergill and his Friends* 1919, 118–20. *J. Bot*. 1937, 185.
Plants at BM(NH). Portr. at Hunt Library.
Russelia Jacq.

RUSSELL, Anna (*née* Worsley) (1807–1876)
b. Arno's Vale, Bristol Nov. 1807 *d*. Kenilworth, Warwick 11 Nov. 1876
Of Kenilworth. Studied and drew fungi drawings at BM(NH). Correspondent of H. C. Watson. 'Note on a List of Newbury Plants' (*Phytologist* v.3, 1849, 716). 'List of some of the Rarer Fungi found near Kenilworth' (*J. Bot*. 1868, 90–91).
H. C. Watson *Topographical Bot*. 1883, 555. J. E. Bagnall *Fl. Warwickshire* 1891, 505. G. C. Druce *Fl. Berkshire* 1897, clxvii–clxviii. J. W. White *Fl. Bristol* 1912, 78–79. *Bot. Soc. Exch. Club Br. Isl. Rep*. 1932, 287. H. J. Riddelsdell *Fl. Gloucestershire* 1948, cxxi.

RUSSELL, Sir Edward John (1872–1965)
b. Frampton, Glos 31 Oct. 1872 *d*. Goring, Oxford 12 July 1965
Educ. University College, Aberystwyth. FRS 1917. Lecturer in Chemistry, Manchester University, 1898–1901. Head of Chemical Dept., Wye College, 1901–7. Soil chemist, Rothamsted Experiment Station, 1907–12; Director, 1912–42. *Soil Conditions and Plant Growth* 1912; ed. 7 1942. *Plant Nutrition and Crop Production* 1926. *The Land called Me* (autobiography) 1956 portr.
Nature v.150, 1942, 732. *Times* 13 July 1965 portr. *Biogr. Mem. Fellows R. Soc*. 1966, 457–77 portr. *Who was Who, 1961–1970* 988.

RUSSELL, George (1857–1951)
b. Stillington, Yorks 1857 *d*. 15 Oct. 1951
VMM 1937. Gardener in York. Bred lupins.
Gdnrs Chron. 1951 ii 153. *Times* 17 Oct. 1951. R. Parrett *Russell Lupin* 1959, 1–13 portr.

RUSSELL, Isaac (*fl.* 1820s–1840s)
Of Oxford. "Botanical draughtsman and glass painter." Drew over 200 plates for W. Baxter's *Br. Phaenogamous Bot.* 1834–43 (see t.505).
J. Bot. 1919, 60.

RUSSELL, John (*c.* 1731–1794)
Nurseryman, Lewisham, Kent. His nursery was "the largest concern of the kind in the neighbourhood of London and one of the largest in the kingdom." Business continued by his sons John (1766–1808) and Thomas (*c.* 1773–1810) and son-in-law, John Willmott (1775–1834). Closed as Willmott and Chaundy in 1860.
Trans. London Middlesex Archaeol. Soc. v.24, 1973, 189.

RUSSELL, John, 6th Duke of Bedford (1766–1839)
b. 6 July 1766 *d.* Rothiemurchus, Perthshire 20 Oct. 1839
FLS 1816. Patron of G. Gardner (1812–1849). Played a major role in the discussion on the future of Kew as the national botanic garden in 1830s. Wrote introductions to botanical works issued in connection with his estate at Woburn Abbey: *Hortus Gramineus Woburnensis* 1816; *Hortus Ericaeus Woburnensis* 1825; *Salictum Woburnense* 1829; *Pinetum Woburnense* 1839.
Bot. Mag. 1839, t.3717. W. J. Hooker *Copy of Letter...to D. Turner...on the...Death of the Late Duke of Bedford* 1840. *Memorials of Sixth Duke of Bedford* 1842. *D.N.B.* v.49, 454. *Curtis's Bot. Mag. Dedications, 1827–1927* 19–20 portr.
Letters at Kew, Linnean Society. Portr. at Hunt Library.
Epiphyllum russellianum Hook.

RUSSELL, John Louis (1897–1976)
b. Richmond, Surrey, 1897 *d.* Chertsey, Surrey 25 Feb. 1976
VMH 1957. Acquired Richmond Nurseries, Windlesham, Surrey (founded by his father L. R. Russell) in 1939; retired 1966.
Times 4 March 1976.

RUSSELL, Louis R. (1863–1942)
b. Hampstead, London 18 Oct. 1863 *d.* Windlesham, Surrey 25 Feb. 1942
VMH 1929. Nurseryman, Richmond Nurseries, Windlesham, Surrey.
Gdnrs Chron. 1942 ii 107, 117.

RUSSELL, Patrick (1727–1805)
b. Edinburgh 6 Feb. 1727 *d.* London 2 July 1805
MD Edinburgh. FRS 1777. Succeeded his half-brother, Alexander Russell, as physician at Aleppo, 1753. Botanist to the East India Company at Madras, 1785–89. Revised *Nat. Hist. of Aleppo* 1794.
W. Roxburgh *Plants of Coast of Coromandel* 1795 preface. *Gent. Mag.* 1805 ii 683. Cunningham *Lives of Illustrious and Eminent Englishmen* v.8, 1856, 118–20. *D.N.B.* v.49, 469–70. *J. Bot.* 1900, 81–82; 1937, 185–86. *Taxon* 1970, 540. *R.S.C.* v.5, 345.
Aleppo plants and drawings at BM(NH). Indian plants at BM(NH), Kew. Portr. at Hunt Library.

RUSSELL, Thomas Hawkes (1851–1913)
b. 30 March 1851 *d.* Edgbaston, Birmingham 31 July 1913
FLS 1906. Solicitor. *Mosses and Liverworts* 1908; ed. 2 1910 (illustrated by himself).
Proc. Linn. Soc. 1913–14, 61.

RUSSON, Jamas Baron (*c.* 1903–1965)
d. 13 Nov. 1965
Acquired McBean's Orchids Ltd., 1964.
Orchid Rev. 1966, 28.

RUST, Joseph (1826–1894)
b. Strachan, Kincardineshire 26 March 1826 *d.* Eridge, Surrey 26 April 1894
Gardener to Marquess of Abergavenny at Eridge. Contrib. to *Gdnrs Chron.*
Gdnrs Chron. 1894 i 573, 729 portr.

RUTHERFORD, Daniel (1749–1819)
b. Edinburgh 3 Nov. 1749 *d.* Edinburgh 15 Dec. 1819
MD Edinburgh 1772. FRSE 1788. FLS 1796. Chemist. Professor of Botany and Regius Keeper, Royal Botanic Garden, Edinburgh, 1786–1819. *Characteres Generum Plantarum* 1793.
W. Wadd *Nugae Chirurgieae* 1824, 136. *D.N.B.* v.50, 5–6. F. W. Oliver *Makers of Br. Bot.* 1913, 290–91. H. R. Fletcher and W. H. Brown *R. Bot. Gdn Edinburgh, 1670–1970* 1970, 68–79. *R.S.C.* v.5, 347.
Portr. at Hunt Library.

RUTHERFORD, Sir John (*c.* 1855–1932)
d. London 26 Feb. 1932
Of Beardwood, Blackburn, Lancs. Exhibited orchids at meetings of Manchester and North of England Society. *Orchid Rev.* v.38, 1930 dedicated to him.
Orchid Rev. 1932, 118.

RUTHERFORD, Robert (*fl.* 1690s–1700s)
"This hearty person and kind friend gathered me several plants in Carolina" (J. Petiver).
J. E. Dandy *Sloane Herb.* 1958, 196–97.

RUTTER, Edward Montagu (*c.* 1890–1963)
d. Shrewsbury, Shropshire 5 Oct. 1963
Superintendent, L.N.E. Railway. Botanical recorder for Caradoc and Severn Valley Field Club, 1954–63. *Handlist of Shropshire Fl.* (with L. C. Lloyd) 1957.
Trans. Shropshire Archaeol. Soc. v.57(2), 1962–63, 170.

RUTTER, John (*c.* 1711–1772)
d. Feb. 1772
Gardener at Wandsworth, London. *Modern Eden* (with D. Carter) 1767.
J. C. Loudon *Encyclop. Gdning* 1822, 1275.

RUTTY, John (1697–1775)
b. Wilts 25 Dec. 1697 *d.* Dublin 26 April 1775
MD Leyden 1723. Physician, Dublin, 1724–75. *Natural History of County of Dublin* 1772. *Materia Medica* 1777.
Notes and Queries v.7, 1859, 147, 264, 324, 423. J. Smith *Cat. of Friends' Books* v.2, 1867, 520–24. N. Colgan *Fl. County Dublin* 1904, xxii–xxiii. *D.N.B.* v.50, 31–32.
Ruttya Harvey.

RYAN, George Michael (*c.* 1859–1932)
d. London 15 Jan. 1932
FLS 1902. Indian Forest Service, 1882. Deputy Conservator of Forests, Bombay Presidency, 1892–1914. Official guide, Kew Gardens, 1919.
J. Kew Guild 1932, 194–95. *Kew Bull.* 1932, 48. *Proc. Linn. Soc.* 1931–32, 185–86. *Times* 12 Jan. 1932.

RYAN, John (*fl.* 1770s–1790s)
MD. FRS 1798. FLS 1798. Collected plants in Santa Cruz, Montserrat and Trinidad. Sent seeds to Chelsea Physic Garden, 1778. Furnished M. Vahl with most of the plants figured in his *Eclogae Americanae* 1796 (preface and dedication).
R. H. Semple *Mem. Bot. Gdn at Chelsea* 1878, 109. I. Urban *Symbolae Antillanae* v.3, 1902, 117.
Plants at BM(NH).
Ryania Vahl.

RYAN, Thomas (–1910)
d. 22 March 1910
Gardener to Lord Annesley at Castlewellan, County Down.
Gdnrs Chron. 1910 i 240.

RYCHE, John *see* Rich, J.

RYLANCE, Charles (*c.* 1811–1884)
d. Ormskirk, Lancs 20 Aug. 1884
Nurseryman, Bold Lane Nursery, Ormskirk.
Florist and Pomologist 1884, 160. *Gdnrs Chron.* 1884 ii 283.

RYLANDS, Thomas Glazebrook (1818–1900)
b. Warrington, Lancs 24 May 1818 *d.* Warrington 14 Feb. 1900
FLS 1862. Wire manufacturer. Diatomist. 'Varieties of British Ferns' (*Naturalist* 1839, 283–88, 412–18).
Proc. Linn. Soc. 1901–2, 41–42. R. D. Radcliffe *Mem. Thomas Glazebrook Rylands* 1901 portr. *J. Bot.* 1907, 455–56. *Trans. Liverpool Bot. Soc.* 1909, 88. *R.S.C.* v.5, 349.
Diatomaceae and letters at BM(NH). Plants at Glasgow University, Warrington Museum.

RYLE, Reginald John (1854–1922)
d. Guy's Hospital, London 4 Dec. 1922
Educ. Oxford. MRCS 1883. In medical practice at Barnet and Brighton. Botanist.
Bot. Soc. Exch. Club Br. Isl. Rep. 1922, 707–8.

SABINE, Sir Edward (1788–1883)
b. Dublin 14 Oct. 1788 *d.* Richmond, Surrey 26 June 1883
FRS 1818. FLS 1817. DCL Oxon 1855. KCB 1869. Brother of J. Sabine (1770–1837). General, Royal Artillery. President, Royal Society, 1861–71. Astronomer to Arctic expeditions under J. Ross, 1818 and W. E. Parry, 1819–20; collected plants in Melville Island and in Greenland (W. E. Parry *Voyage* 1824, cclxi–cccx; *Second Voyage* 1825, 382).
Proc. R. Soc. 1892, xliii–li. E. Bretschneider *Hist. European Bot. Discoveries in China* 1898, 251–52. *D.N.B.* v.50, 74–78. *R.S.C.* v.5, 351; v.8, 805; v.11, 251.
Plants at BM(NH). Letters at Kew. MSS. at Wellcome Historical Medical Library. Portr. at Hunt Library. Sale at Sotheby, 10 March 1884.
Pleuropogon sabinii R. Br.

SABINE, Joseph (1770–1837)
b. Tewin, Herts 6 June 1770 *d.* London 24 Jan. 1837
FRS 1799. FLS 1798. Brother of Sir E. Sabine (1788–1883). Barrister until 1808. Inspector-General of assessed taxes, 1808–35. Secretary, Horticultural Society of London, 1816–30. Plant list in R. Clutterbuck's *Hist...Hertfordshire* 1815. Contrib. to *Trans. Hort. Soc. London.*
J. C. Loudon *Encyclop. Gdning* 1822, 1289. *Gdnrs Mag.* 1837, 144. *Hort. J.* 1837, 2–6 portr. *Mag. Nat. Hist.* 1837, 390–92. *Cottage Gdnr* v.6, 1851, 363. R. A. Pryor *Fl. Hertfordshire* 1887, xli–xlii. *D.N.B.* v.50, 79. *Gdnrs Chron.* 1924 i 106. *Curtis's Bot. Mag. Dedications, 1827–1927* 11–12 portr. *Condar* v.30, 1928, 293. H. R. Fletcher *Story of R. Hort. Soc. 1804–1968* 1969 *passim.*
MSS. and letters at Linnean Society. Letters at Kew.
Sabinea DC.

SACKVILLE-WEST, Victoria May (Vita) (1892–1962)
b. Knole, Kent March 1892 *d.* Sissinghurst, Kent 2 June 1962

VMM 1955. Married Harold Nicholson. Author and horticultural journalist. Created garden at Sissinghurst. *Some Flowers* 1937. *In Your Garden* 1951. *Hidcote Manor Garden* 1952. *In Your Garden Again* 1953. *More for Your Garden* 1955. *Even More for Your Garden* 1958. *V. Sackville-West's Garden Book* 1968.

Who was Who, 1961–1970 992. D. Macleod *Gdnr's London* 1972, 176–85 portr. M. Steven *V. Sackville-West* 1973 portr.

SADLER, John (1837–1882)
b. Gibbleston, Fife 3 Feb. 1837 *d.* Edinburgh 9 Dec. 1882

Assistant to J. H. Balfour, 1854. Assistant Secretary, Botanical Society of Edinburgh, 1858–79. Curator, Royal Botanic Garden, Edinburgh, 1879. *Flowers on Moffat Hills* 1858. Assisted in J. H. Balfour's *Fl. Edinburgh* 1863. Contrib. to *Trans. Bot. Soc. Edinburgh.*

Gdnrs Chron. 1879 i 76, 81 portr.; 1882 ii 793. *Garden* v.22, 1882, 542. *Hist. Berwick Nat. Club* v.10, 1882, 72–83. *J. Bot.* 1883, 31–32. *J. Hort. Cottage Gdnr* v.6, 1883, 55–56. *Scott. Nat.* 1883, 43–44. *Trans. Bot. Soc. Edinburgh* v.16, 1886, 11–15. C. S. Sargent *Silva N. America* v.8, 1895, 62. H. R. Fletcher and W. H. Brown *R. Bot. Gdn Edinburgh 1670–1970* 1970, 146, 189–91 portr.

Mosses at BM(NH). Letters at Kew. Portr. at Hunt Library.

Salix sadleri Syme.

SAGE, George (1824–1889)
b. Hillingdon, Middx 1824 *d.* 17 July 1889

Gardener to Lord Brownlow at Ashridge Park, Herts, 1858.

Gdnrs Chron. 1876 i 45 portr.; 1889 ii 109 portr.

ST. BRODY, Gustavus A. Ornano (1828–1901)
b. France 1828 *d.* Wallingford, Berks 22 Nov. 1901

BSc Paris. PhD Göttingen. FLS 1863. *Fl. Weston-super-Mare* 1856. 'New Gloucestershire Plants' (*J. Bot.* 1866, 121–23).

J. Bot. 1898, 291–92; 1902, 127–28; 1907, 407–8; 1911, 224. J. W. White *Fl. Bristol* 1912, 93–95. H. J. Riddelsdell *Fl. Gloucestershire* 1948, cxxix–cxxxi.

Herb at Gloucester Museum.

ST. JOHN, Sir Oliver Beauchamp Coventry (1837–1891)
b. Ryde, Isle of Wight 21 March 1837 *d.* Quetta, India 3 June 1891

In Indian Army, 1858. Officiating agent to Governor-General of India in Baluchistan, 1861. President, Maisur and Kurg, 1889. Collected plants in Persia, now at Kew.

Kew Bull. 1901, 57. *D.N.B.* v.50, 157–58.

ST. JOHN, Spenser (*fl.* 1840s–1860s)

Secretary to J. Brooke whom he accompanied to Borneo, 1848. Consul-General, British N. Borneo, 1855. Deputy, Republic of Haiti, 1861. Collected some *Nepenthaceae*. *Life in Forests of Far East* 1862.

Fl. Malesiana v.1, 1950, 265.

SAINTLOO, Edward (–1578)

Of Somerset. Correspondent of M. de Lobel. Discovered *Cnicus eriophorus*.

M. de Lobel *Adversaria* 1605, 370.

ST. QUENTIN, Countess *see* Rowden, F. A.

SALAMAN, Redcliffe Nathan (1874–1955)
b. London 12 Sept. 1874 *d.* Barley, Herts 12 June 1955

BA Cantab 1896. MRCS 1900. MD 1904. FRS 1935. Director, Pathology Institute, London Hospital, 1901–26. Director, Potato Virus Research Institute, 1926–39. Carried out research on virus-free potatoes. *Potato Varieties* 1926. *History and Social Influence of Potato* 1949.

Gdnrs Chron. 1926 ii 41, 162 portr. *Palestine J. Bot. Rehovot Ser.* 1949, 187–91. *Biogr. Mem. Fellows R. Soc.* 1955, 239–45 portr. *Nature* v.176, 1955, 97. *D.N.B. 1951–1960* 1971, 860–61. *Who was Who, 1951–1960* 960.

Portr. by C. Salaman at National Institute of Agricultural Botany.

SALISBURY, Richard Anthony (*né* Markham) (1761–1829)
b. Leeds, Yorks 2 May 1761 *d.* London 23 March 1829

Educ. Edinburgh University. FRS 1787. FLS 1788. Son of Richard Markham, Leeds clothmaker; took name of Salisbury as a condition of a legacy. Descended from H. Lyte. Pupil of John Aikin (1747–1822). Had garden at Chapel Allerton, Yorks and later acquired P. Collinson's garden at Mill Hill, London. Secretary, Horticultural Society of London, 1805–16. *Icones Stirpium Rariorum Descriptionibus Illustratae* 1791. *Prodromus Stirpium in Horto ad Chapel Allerton Vigentium* 1796. *Generic Characters in the English Botany*... 1806. *Paradisus Londinensis* 1805–8 2 vols. *Genera of Plants* 1866. 'On the Cultivation of Rare Plants...' (*Trans. Hort. Soc. London* v.1, 1812, 261–366).

P. Smith *Mem. and Correspondence of Sir J. E. Smith* v.2, 1832, 329. *Gdnrs Mag.* 1835,

380. W. Herbert *Amaryllidaceae* 1837, 295–96. W. Darlington *Memorials of J. Bartram and H. Marshall* 1849, 474. G. Bentham *Fl. Australiensis* v.6, 1873, 350, 352. *J. Bot.* 1886, 51–53, 296–300; 1916, 57–65; 1917, 345. J. Gillow *Dict. English Catholics* v.4, 1887, 468–70. E. Bretschneider *Hist. European Bot. Discoveries in China* 1898, 209–10. *D.N.B.* v.50, 192–94. *Gdnrs Chron.* 1904 i 148. *J. R. Hort. Soc.* 1944, 58–65, 95–100 portr. F. A. Stafleu *Taxonomic Literature* 1967, 413–14. H. R. Fletcher *Story of R. Hort. Soc., 1804–1968* 1969, 34–36 portr. *Contrib. N.S.W. National Herb.* v.4(6), 1973, 369–72.

Plants and portr. by W. J. Burchell at Kew. Drawings and notes at BM(NH). Letters at Kew, BM(NH). Portr. at Hunt Library.

Salisburia Smith.

SALISBURY, Rev. William *see* Salusbury, Rev. W.

SALISBURY, William (–1823)

Nurseryman, Old Brompton Road, Chelsea, in partnership with William Curtis, 1798. Disposed of this nursery to W. Malcolm. Acquired another nursery in Sloane Street *c.* 1809. *Hortus Paddingtonensis* 1797. *Hortus Siccus Gramineus* 1802–6 (exsiccatae). *Botanist's Companion* 1816 2 vols. *Cottager's Agricultural Companion* 1822.

Gent. Mag. 1810 ii 113. J. C. Loudon *Arboretum et Fruticetum Britannicum* 1838, 75. H. Trimen and W. T. T. Dyer *Fl. Middlesex* 1869, 395. *D.N.B.* v.50, 201. *Gdnrs Chron.* 1900 i 65–66; 1949 ii 172. *J. Bot.* 1902, 324. *Trans. London Middlesex Archaeol. Soc.* v.24, 1973, 186.

SALMON, Charles Edgar (1872–1930)

b. 22 Nov. 1872 *d.* Reigate, Surrey 1 Jan. 1930

FLS 1902. Architect. President, Holmesdale Natural History Club, 1929. Authority on *Limonium.* Co-author of 2nd Supplt to H. C. Watson's *Topographical Bot.* 1930. *Fl. Surrey* 1931 portr. Contrib. to *J. Bot.*, *Bot. Soc. Exch. Club Br. Isl. Rep.*

Bot. Soc. Exch. Club Br. Isl. Rep. 1929, 96–98. *Proc. Linn. Soc.* 1929–30, 218–21. *Watson Bot. Exch. Club Rep.* 1929–30, 7–11 portr. *J. Bot.* 1930, 50–53 portr.; 1932, 53–56. *Kew Bull.* 1930, 126. *Nature* v.125, 1930, 643. *S. Eastern Nat.* 1930, xliii. A. H. Wolley-Dod *Fl. Sussex* 1937, xlviii. H. J. Riddelsdell *Fl. Gloucestershire* 1948, cxlix–cl.

Herb. at BM(NH). Portr. at Hunt Library.

SALMON, Ernest Stanley (1872–1959)

d. 12 Oct. 1959

Did research on fungi at Jodrell Laboratory, Kew, 1899–1906. Mycologist, S.E. Agricultural College, Wye, 1906–37. Reader in Mycology, London University, 1912; Professor, 1925. President, British Mycological Society, 1911. Authority on hop diseases. *Monograph of Erysiphaceae* 1900.

Nature v.184, 1959, 1188. *Times* 18 Oct. 1959 portr.

Portr. at Hunt Library.

SALMON, John Drew (*c.* 1802–1859)

d. London 5 Aug. 1859

FLS 1852. Manager, Wenham Lake Ice Company. MS. notes incorporated in J. Brewer's *Fl. Surrey* 1863. 'Outline of Fl. Neighbourhood of Godalming' (*Phytologist* v.2, 1846, 447–57). 'On Division of County of Surrey into Botanical Districts' (*Phytologist* v.4, 1852, 558–66, 719–20).

Phytologist v.5, 1861, 350. *Proc. Linn. Soc.* 1860, xxiv–xxv. *Trans. Norfolk Norwich Nat. Soc.* v.2, 1877–78, 420. *D.N.B.* v.50, 209–10. C. E. Salmon *Fl. Surrey* 1931, 52. *R.S.C.* v.5, 382.

Herb. and diaries at Norwich Museum. Plants at Holmesdale Natural History Club.

SALMON, Thomas (*fl.* 1810s)

Nurseryman, Latchford, Warrington, Lancs.

SALMON, William (1644–1712)

b. 2 June 1644 *d.* Dec. 1712

MD. Professor of Medicine. *Botanologia* 1710.

R. Pulteney *Hist. Biogr. Sketches of Progress of Bot. in England* v.1, 1790, 185–88. *Notes and Queries* v.8, 1889, 92. *D.N.B.* v.50, 209–10. C. E. Salmon *Fl. Surrey* 1931, 46.

Portr. at Hunt Library. Sale at T. Ballard, 16 Nov. 1713 and 10 March 1714.

SALT, Henry (1780–1827)

b. Lichfield, Staffs 14 June 1780 *d.* Dessuke, Alexandria 30 Oct. 1827

FRS 1812. FLS 1812. Secretary to Lord Valentia in India and Africa, 1802–5. Travelled in Abyssinia, 1805 and 1810. British Consul-General, Egypt, 1815–27. Collector of antiquities. *Voyage to Abyssinia* (with 'List of New and Rare Plants collected ...1805–10') 1814. Sent algae to D. Turner (*Fuci* v.4, 1819, 38).

Gent. Mag. 1828 i 374. J. J. Halls *Life...of Henry Salt* 1834 portr. A. Lasègue *Musée Botanique de B. Delessert* 1845, 167, 301, 323, 504. *D.N.B.* v.50, 212–13. *Rec. Bot. Survey India* v.7, 1914, 2–5. *Kirkia* 1961, 130–32. R. Hill *Biogr. Dict. Sudan* 1967, 331.

Plants at BM(NH).

Saltia R. Br.

SALT, Henry Stephen (1851–1939)
b. India 1851 *d.* 19 April 1939
Educ. Cambridge University. Assistant master, Eton College, 1875–84. Secretary, Humanitarian League, 1891–1920. *Call of the Wild Flower* 1922. *Our Vanishing Wild Flowers* 1928.
Who was Who, 1929–1940 1190.

SALT, Jonathan (1759–1810)
b. Sheffield, Yorks 3 March 1759 *d.* Sheffield 2 Aug. 1810
FLS 1797. Cutler. Discovered *Carex elongata* 1807. Contrib. to J. Sowerby and J. E. Smith's *English Bot.* 358, 598, 648, 1920, 2018.
R. E. Leader *Reminiscences of Old Sheffield c.* 1875, 312. F. A. Lees *Fl. W. Yorkshire* 1888, 106. *List of Plants collected... by Jonathan Salt and now in Sheffield Public Museum* 1889. *Naturalist* 1967, 47–50.
Herb. and MS. *Fl. Sheffield* at Sheffield Museum.

SALTER, Anthony (*fl.* 1600–1630s)
MA Oxon 1624. MA Cantab 1626. MD 1633. Of Exeter. Physician. Correspondent of T. Johnson.
J. Parkinson *Theatrum Botanicum* 1640, 1219. R. Pulteney *Hist. Biogr. Sketches of Progress of Bot. in England* v.1, 1790, 154 (Sadler by mistake).

SALTER, J. (*fl.* 1800s)
Nurseryman, Bath, Somerset. Later in partnership with Wheeler? *Treatise upon Bulbous Roots...* 1816.
J. C. Loudon *Encyclop. Gdning* 1822, 1289.

SALTER, John (1798–1874)
b. Hammersmith, London 27 Jan. 1798 *d.* 10 May 1874
Cheesemonger. Established nursery for sale of English flowers at Versailles, 1838–48. Founded nursery at Hammersmith where he specialised in chrysanthemums. *Chrysanthemum: its History and Culture* 1865.
Florist and Pomologist 1874, 140–41. *Gdnrs Chron.* 1874, 643. *J. Hort. Cottage Gdnr* v.51, 1874, 402.

SALTER, John Henry (1862–1942)
b. Westleton, Suffolk 5 June 1862 *d.* Aberystwyth, Cardigan 5 Aug. 1942
BSc London. Studied at Bonn under Strasburger, 1896. Taught at various schools, 1881–88. First Professor of Botany, University College, Aberystwyth, 1891–1908. Lived in Canary Islands and S. France, 1908–16. Studied *Mentha* and *Salix*. *Flowering Plants and Ferns of Cardiganshire* 1935.
Chronica Botanica 1942–43, 354. *Nature* v.150, 314–15. *N. Western Nat.* 1942, 265–67 portr.
Herb. at National Museum of Wales. Natural history diary at National Library of Wales. Botanical drawings at University College, Aberystwyth.

SALTER, John William (1820–1869)
b. 15 Dec. 1820 *d.* Gravesend, Kent 2 Aug. 1869
ALS 1842. Apprenticed as artist and engraver in 1835 to J. de C. Sowerby whose daughter, Sally, he married in 1846. Chief Assistant to E. Forbes in Geological Survey, 1846; Palaeontologist, 1854–63. Contrib. plates to *Supplement to English Bot.* vols 4 and 5, 1849–63; also publisher of v.5. Illustrated C. P. Johnson's *British Wild Flowers* 1858.
Geol. Mag. 1869, 432, 477–80. *J. Bot.* 1869, 280. *Proc. Linn. Soc.* 1869–70, cvii–cviii. *Quart. J. Geol. Soc.* 1870, xxxvi–xxxix. *D.N.B.* v.50, 217. *J. Soc. Bibl. Nat. Hist.* 1974, 538–45. *R.S.C.* v.5, 382; v.8, 819; v.12, 646.
Drawings at BM(NH), Oxford Botanic Garden.

SALTER, Samuel James Augustus (1825–1897)
b. Poole, Dorset 10 Aug. 1825 *d.* Basingstoke, Hants 28 Feb. 1897
MB London. FRS 1863. FLS 1853. Dental surgeon. 'Vitality of Seeds' (*J. Linn. Soc.* v.1, 1856, 140–42). 'Polliniferous Ovules in *Passiflora*' (*Trans. Linn. Soc.* v.24, 1863, 143–50).
Proc. Linn. Soc. 1896–97, 68–69. *Gdnrs Chron.* 1897 i 163. *Proc. R. Soc.* v.61, 1897, iii–iv. *R.S.C.* v.5, 384; v.8, 820; v.11, 268; v.12, 646.
Letters at Kew.

SALTER, Terence Macleane (1883–1969)
b. Cheltenham, Glos 5 Feb. 1883 *d.* Cape Town, S. Africa 30 March 1969
Royal Navy paymaster until 1931. Emigrated to Cape Town, 1935. Collected plants in Cape. *Genus Oxalis in South Africa* 1944. *Fl. Cape Peninsula* (co-editor with R. S. Adamson) 1950.
Gdnrs Chron. v.165(23), 1969, 40. *J. S. African Bot.* 1971, 177–82 portr.
Plants at Cape Town, BM(NH), Kew. MSS. and Library at Kirstenbosch. Portr. at Hunt Library.
Oxalis salteri L. Bolus.

SALTER, Thomas Bell (1814–1858)
d. Southampton, Hants 30 Sept. 1858
MD Edinburgh. FLS 1837. Practised medicine at Ryde. 'Short Account of Botany of Poole' (J. Sydenham *Hist. Poole* 1839, 465–91). Co-editor of W. A. Bromfield's *Fl. Vectensis* 1856. Papers on *Rubi* in *Bot. Gaz.* 1850, 113–31, 147–56. Contrib. to *Phytologist*.

Proc. Linn. Soc. 1859, xxxiv–xxxv. *R.S.C.* v.5, 385.
Plants at Kew, Ryde Museum.
Rubus salteri Bab.

SALTMARSH, Joseph (–1872)
d. 14 June 1872
Nurseryman, Moulsham Nurseries, Chelmsford, Essex.
Gdnrs Chron. 1872, 838.

SALTMARSH, Thomas Joseph (*c.* 1827–1899)
d. Chelmsford, Essex 2 June 1899
Nurseryman, Chelmsford.
Garden v.55, 1899, 438. *Gdnrs Chron.* 1899 i 401.

SALTONS, John (–1794)
d. 4 July 1794
Curator, Botanic Gardens, Cambridge.
Gdnrs Chron. 1919 i 147.

SALUSBURY, Sir John (1567–1612)
d. 24 July 1612
Had garden at Lleweni near Denbigh. Compiled garden lists for 1596, 1607, 1608.
R. T. Gunther *Early Br. Botanists* 1922, 238–45, 306–9. *J. Bot.* 1923, 117–18.

SALUSBURY, Rev. William (*c.* 1520–1600)
b. Llansannan, Denbighshire *c.* 1520
Lexicographer. Compiled a Welsh botanology; "an original work, showing close observation of plant life."
J. Bot. 1898, 12–13; 1917, 259–69. *D.N.B.* v.50, 196–200. E. S. Roberts *ed. Llysieulyfr Meddvginiaethol* 1916 (preface). R. T. Gunther *Early Br. Botanists* 1922, 238, 242.

SALVIN, Osbert (1835–1898)
b. Finchley, Middx 25 Feb. 1835 *d.* Haslemere, Surrey 1 June 1898
Educ. Cambridge. FLS 1864. Zoologist. Accompanied G. U. Skinner on journey in Guatemala, 1857–58; collected plants.
Nature v.58, 1898, 129. *Proc. Linn. Soc.* 1898–99, 59–61. *Auk* v.15, 286, 343–45.

SALWEY, Rev. Thomas (1791–1877)
b. Ludlow, Shropshire 19 Oct. 1791 *d.* Worthing, Sussex 3 Dec. 1877
BD Cantab 1824. FLS 1824. Rector, Oswestry, Shropshire, 1833–72. Lichenologist. Issued centuries of lichens. Botanised in Guernsey and found *Allium triquetrum*, 1847. List of plants of Barmouth in *Visitor's Guide to Merioneth* 1863. Contrib. papers on lichens to *Ann. Nat. Hist.*, *Trans. Bot. Soc. Edinburgh*. Contrib. to J. Sowerby and J. E. Smith's *English Bot.* 2667, 2796, 2861, 2963.
J. Bot. 1878, 63. B. Lynge *Index... Lichenum Exsiccatorum* 1915–19, 471–75. *R.S.C.* v.5, 387; v.12, 647.
Plants at Kew. MS. botanical journal, 1836–44 and lichens at Ludlow Museum. Letters at Kew, BM(NH).
Lecidea salweii Borrer.

SAMBACH, Richard (*fl.* 1690s–1720s)
MD. Of Worcester. Sent plants to J. Petiver from East Indies. "Humanissimus vir ac chirurgus peritissimus" (L. Plukenet).
Phil. Trans. R. Soc. v.20, 1698, 317. J. E. Dandy *Sloane Herb.* 1958, 198.
Plants at BM(NH).

SAMPSON, G. Theophilus (1831–1897)
b. Hull, Yorks 3 Dec. 1831 *d.* London 29 Dec. 1897
In government service at Canton, 1858–89; returned to England, 1889. Correspondent of H. F. Hance.
Ann. Sci. Nat. Bot. v.5, 1866, 202. E. Bretschneider *Hist. European Bot. Discoveries in China* 1898, 652–61, 1091. *Lingnan Sci. J.* v.7, 1931, 158–59.
Chinese plants at BM(NH), Kew.
Vitex sampsoni Hance.

SAMPSON, Hugh Charles (1878–1953)
b. Simla, India 2 May 1878 *d.* 29 Nov. 1953
Educ. Edinburgh University. Transvaal Agricultural Dept., 1903. Indian Agricultural Service, 1906–23. Economic botanist, Kew, 1927–38. Authority on cotton growing. Collected plants in Gambia, 1929; Gold Coast and Nigeria, 1929 and 1938. *Cultivated Crop Plants of British Empire and Anglo-Egyptian Sudan* 1936.
J. Kew Guild 1952–53, 133. *Nature* v.171, 1953, 16.
Plants at Kew.

SAMSON, F. (–1929)
Of Redland, Bristol. Studied local flora with his half-sister, I. M. Roper (1865–1935).
H. J. Riddelsdell *Fl. Gloucestershire* 1948, cxliii–cxliv.

SANCROFT, Rev. William (1616/7–1693)
b. Fressingfield, Suffolk 30 Jan. 1616/7
MA Cantab 1641. Archbishop of Canterbury, 1678. Possibly collected plants at Padua, Italy.
D.N.B. v.50, 244–50. J. E. Dandy *Sloane Herb.* 1958, 199–200.

SANDEMAN, Christopher Albert Walter (1882–1951)
b. London 25 Nov. 1882 *d.* 20 April 1951
MA Oxon. FLS 1949. Travelled extensively in Tropical S. America where he collected plants. *A Forgotten River* 1939. *Wanderer in Inca Land* 1948. *No Music in Particular* 1943. *Thyme and Bergamot* 1947. 'Richard Spruce' (*J. R. Hort. Soc.* 1949, 531–44).
Proc. Linn. Soc. 1951–52, 280–82. *Who was Who, 1951–1960* 963.

Plants at Kew, Oxford, Colombian National Herb.
Sandemania Gleason.

SANDER, Charles Fearnley (1874–1957)
b. Lewisham, Kent 24 Dec. 1874 *d.* Bedford 10 Jan. 1957
Associated with his father, H. F. C. Sander, in founding and maintaining Messrs Sander and Sons, Bruges, 1894. Largely responsible for the Bruges establishment, 1920–34.
Orchid Rev. 1957, 55–56. A. Swinson *Frederick Sander: the Orchid King* 1970 *passim.*

SANDER, Frederick K. (1876–1951)
d. St. Albans, Herts 9 Jan. 1951
Son of H. F. C. Sander (1847–1920). Kew gardener, 1896. Director, Messrs Sander's orchid nurseries, St. Albans. *Sanders Complete List of Orchid Hybrids* 1946.
J. Kew Guild 1950, 870–71 portr. *Gdnrs Chron.* 1951 i 24.

SANDER, Henry Frederick Conrad (1847–1920)
b. Bremen, Germany 1847 *d.* Bruges, Belgium 23 Dec. 1920
FLS 1886. VMH 1897. To England 1865 where he worked in the nurseries of Messrs Carter, Forest Hill. Began as seedsman in George Street, St. Albans, Herts; moved to camp district of St. Albans, 1881. Established orchid nursery at Summit, New Jersey in 1880s; sold it to J. Lager and H. Hurrell in 1896. Established orchid nursery at St. André, Bruges, 1894. *Reichenbachia* 1886–94.
J. Hort. Cottage Gdnr v.22, 1891, 449, 451 portr. *Garden* 1921, 25. *Gdnrs Chron.* 1921 i 12 portr. *Kew Bull.* 1921, 33. *Orchid Rev.* 1921, 45–50 portr. *Proc. Linn. Soc.* 1920–21, 54–55. A. Swinson *Frederick Sander: the Orchid King* 1970 portr. M. A. Reinikka *Hist. of the Orchid* 1972, 260–63 portr.
Letters at Kew.
Sanderella O. Kuntze.

SANDER, John (*fl.* 1790s–1810s)
Nurseryman, Keswick, Cumberland. Introduced Keswick Codlin apple.

SANDER, Louis L. (1878–1936)
b. St. Albans, Herts 10 April 1878 *d.* Bruges, Belgium 27 Aug. 1936
Son of H. F. C. Sander (1847–1920) Director at Bruges, St. Albans and Southgate branches of Sander's nurseries.
Gdnrs Chron. 1936 ii 188. *Orchid Rev.* 1936, 290–92. A. Swinson *Frederick Sander: the Orchid King* 1970 *passim.*

SANDERS, Edgar (1827–1907)
b. nr East Grinstead, Sussex 10 Oct. 1827 *d.* 29 Sept. 1907
Gardener. To New York, 1853. Nurseryman, Chicago, 1857. President, American Association of Nurserymen, 1876.
J. Hort. Cottage Gdnr v.55, 1907, 365.

SANDERS, Gilbert (*fl.* 1840s–1860s)
Algologist. Collected Irish algae. 'On the Fructification of the Genus *Desmarestia*' (*Dublin Nat. Hist. Soc. Proc.* 1849–55, 34–36).
Notes Bot. School Dublin Jan. 1901, 150. *R.S.C.* v.5, 392; v.8, 826.

SANDERS, Thomas William (1855–1926)
b. Martley, Worcs 6 Nov. 1855 *d.* 13 Oct. 1926
FLS 1902. Horticultural journalist. Edited *Amateur Gdning* 1887–1926. *Sanders' Encyclopaedia of Gardening* 1895. Wrote many books on gardening.
Gdnrs Chron. 1926 ii 339 portr. *Proc. Linn. Soc.* 1926–27, 95–96. *Who was Who, 1916–1928* 929–30.

SANDERSON, Arthur Rufus (1877–1932)
b. Bradford, Yorks 1877 *d.* Austwick, Yorks 25 Dec. 1932
Pathologist with Rubber Growers' Association, Malaya, 1918–31; and later Rubber Research Institute. Studied Mycetozoa. Sent plants to Singapore Botanic Gardens, 1920. Contrib. to *Naturalist.*
Gdns Bull. Straits Settlements 1927, 131. *Naturalist* 1933, 127–28. *Proc. Linn. Soc.* 1932–33, 200–1.

SANDERSON, Harry (1871–1917)
b. Galashiels, Selkirkshire 2 March 1871 *d.* Western Front 23 April 1917
Partner in firm of tweed manufacturers. Had garden at Galashiels where he grew alpines.
Bot. Soc. Exch. Club Br. Isl. Rep. 1917, 91–92. *Gdnrs Chron.* 1917 i 197, 207 portr.

SANDERSON, Henry (*c.* 1852–1881)
Florist, Whalton, Newcastle-on-Tyne, Northumberland. Noted for his polyanthus.
Gdnrs Chron. 1881 ii 797. *Florist and Pomologist* 1882, 16.

SANDERSON, John (1820/1–1881)
b. Greenock, Renfrewshire 1820 or 1821 *d.* Durban, S. Africa March 1881
To Durban, 1850. Secretary, Horticultural Society of Natal. Collected plants in S. Africa. Sent plants and drawings to W. H. Harvey and Kew.
Bot. Mag. 1853, t.4716. W. H. Harvey *Fl. Capensis* v.1, 1859, 9*.
Sandersonia Hook.

SANDERSON, Sir John Scott Burdon- *see* Burdon-Sanderson, Sir J. S.

SANDFORD, Major (*fl.* 1860s)
"A gentleman who has done much during a long sojourn in W. Australia to explore the natural history products of that colony" (*Bot. Mag.* 1862, t.5350). Sent seeds to Thompson's nursery, Ipswich.

SANDFORD, Joseph (*fl.* 1780s)
Nurseryman, Chipping Camden, Glos.

SANDS, William Norman (1875–1943)
b. Littlebourne, Kent 21 July 1875 *d.* Richmond, Surrey 17 Jan. 1943
FLS 1912. Kew gardener, 1897. Curator, Botanic Station, Antigua, 1899–1904. Superintendent of Agriculture, St. Vincent, 1904–19. Joined Royal Society expedition to West Indies, 1907 to study vegetation after volcanic eruptions, 1904–5. Assistant Economic Botanist, Malaya, 1920–30. Principal Agricultural Officer, Kedah, 1930–35.
Kew Bull. 1899, 133. *J. Kew Guild* 1929, 655–56 portr.; 1943, 309–10.

SANDWITH, Mrs. Cecil Ivry (1871–1961)
d. Clifton, Bristol 6 Feb. 1961
FLS 1949. Collected plants in Chilanga district, N. Rhodesia. Made collection of local grasses for Home Office Laboratory at Bristol. Contrib. notes to J. W. White's *Fl. Bristol* 1912. Contrib. 'Bristol Botany' to *Proc. Bristol Nat. Soc.* 1935–47.
Proc. Bristol Nat. Soc. 1961, 103–6.
Rhodesian plants at Kew, Salisbury.

SANDWITH, Noel Yvri (1901–1965)
b. Harworth, Notts 8 Sept. 1901 *d.* Kew, Surrey 7 May 1965
MA Oxon. FLS 1936. Son of C. I. Sandwith (1871–1961). Collected plants in British Guiana, 1929 and 1937. Travelled in Europe, particularly Albania, and N. Africa and Tropical America. Authority on *Bignoniaceae*. 'Contributions to Fl. Tropical America' (*Kew Bull.* 1930–68).
H. J. Riddelsdell *Fl. Gloucestershire* 1948, clx–clxi. *Proc. Bristol Nat. Soc.* 1966, 128–30. *Kurtziana* 1966, 247–50 portr. *Proc. Linn. Soc.* 1966, 124. *Proc. Bot. Soc. Br. Isl.* 1967, 418–22. *J. Kew Guild* 1966, 710–11 portr. *Taxon* 1966, 245–55 portr.
Plants at Bristol University, Kew, Rijksherb. Leiden. Letters at Kew. Portr. at Hunt Library.
Sandwithia Laujouw.

SANDYS, Edwin (*c.* 1689–1724)
b. Petherton, Somerset *c.* 1689 *d.* Oxford? 1724
MA Oxon 1718. Professor of Botany, Oxford, 1719–24.
J. Nichols *Illus. Lit. Hist. of Eighteenth Century* v.1, 359. D. Turner *Extracts from Lit. Sci. Correspondence of R. Richardson* 1835, 11. *Gdnrs Chron.* 1919 i 147.

SANDYS, Rev. G. W. (–*c.* 1840)
Of Miserden Park, Glos. Botanised in Stroud area. Some of his specimens quoted in R. Walker's *Fl. Oxfordshire* 1833. Had herb., since lost.
H. J. Riddelsdell *Fl. Gloucestershire* 1948, cxxi.

SANDYS, Letitia Hannah Damer (1840–1911)
b. Isle of Wight 1840
Wife of Rev. B. N. White-Spunner, Rector of Donaghmore, County Tyrone. Had herb. of Irish plants.
Annual Rep. Proc. Belfast Nat. Field Club 1911–12, 627–28. *Irish Nat.* 1913, 32.

SANFORD, William Ashford (1818–1902)
b. Nynhead Court, Somerset 1818 *d.* 28 Oct. 1902
Educ. Cambridge. Colonial Secretary, W. Australia 1852. Collected local flora, especially algae, which he gave to W. H. Harvey. Returned to England, 1871. President, Somerset Archaeological and Natural History Society, 1872 and 1892.
Proc. Somerset Archaeol. Nat. Hist. Soc. 1903, 122–25. *J. W. Austral. Nat. Hist. Soc.* 1909, 25.
Asparagopsis sanfordiana Harvey.

SANG, Edward (*fl.* 1810s)
Nurseryman, Kirkcaldy, Fifeshire. Edited W. Nicol's *Planter's Kalendar* 1820. Contrib. to *Mem. Caledonian Hort. Soc.*
J. C. Loudon *Encyclop. Gdning* 1822, 1285.

SANGER-DAVIES, Arthur Elphinstone (1885–1945)
b. Tunbridge Wells, Kent 1885 *d.* Sumatra 22 March 1945
Forest Dept., Malaya, 1907. State Forest Officer, Negri Sembilan and Malacca.
Malayan Forester 1937, 188–89 portr.; 1948, 58. *Fl. Malesiana* v.1, 1950, 459.
Plants at Kuala Lumpur.

SANGSTER, John (*c.* 1796–1881)
d. Romford, Essex 27 May 1881
Nurseryman in partnership with Hay and Anderson at Newington Butts, Surrey. Raised peas.
Gdnrs Chron. 1881 i 737–38. *J. Hort. Cottage Gdnr* v.2, 1881, 464–65. *Gdnrs Yb. Almanack* 1882, 192.

SANSOM, F. (*fl.* 1780s–1810s)
Artist. In Rotterdam, 1788–90; later in London. Engraved plates in W. Curtis's *Fl. Londinensis*, *Bot. Mag.* 1800–15, R. W. Dickson's *Complete Dict. Practical Gdning* 1807, W. Roxburgh's *Plants of Coast of Coromandel* 1795–1819, W. Curtis's *Lectures on Bot.* 1805, L. W. Dillwyn's *Br. Confervae* 1809.

Index to Bot. Mag., 1787–1904 xvii. G. Dunthorne *Flower and Fruit Prints of 18th and Early 19th Centuries* 1938. C. Nissen *Botanische Buchillustration* 1951.

SANSOM, Thomas (1816/7–1872)
d. Liverpool 22 March 1872
ALS 1843. Friend of W. Bean (1817–1864) and F. Brent (1816–1903) with whom he formed a herb. and presented to Historic Society of Lancashire and Cheshire, 1855, now at Liverpool Museums. Contrib. to J. Dickinson's *Fl. Liverpool* 1851. Contrib. papers on mosses to *Proc. Lit. Philos. Soc. Liverpool* 1849–55.
Trans. Liverpool Bot. Soc. 1909, 88–89. *R.S.C.* v.5, 397.

SAREL, Henry Andrew (*c.* 1825–1886)
d. Rollesby Hall, Norfolk 1886
General, 17th Lancers. In China, 1860. Lieut.-Governor, Guernsey, Alderney and Sark. Collected ferns on Yang-tze, 1860; ferns named by W. J. Hooker in T. W. Blakiston's *Five Months on the Yang-tze* 1862.
E. Bretschneider *Hist. European Bot. Discoveries in China* 1898, 688–89.
Asplenium sarelii Hook.

SARGANT, Ethel (1863–1918)
b. London 28 Oct. 1863 *d.* Sidmouth, Devon 16 Jan. 1918
Educ. Cambridge. FLS 1904. Worked under D. H. Scott at Jodrell Laboratory, Kew, 1892–93. Had private laboratory at Reigate where she studied plant anatomy, especially anatomy of seedlings. Contrib. to *Ann. Bot.*
Ann. Bot. 1918, i–v. *Bot. Soc. Exch. Club Br. Isl. Rep.* 1918, 364–65. *Cambridge Mag.* 1918, 361. *J. Bot.* 1918, 115–16. *Kew Bull.* 1918, 125–26. *Nature* v.100, 1918, 428–29. *New Phytol.* 1918, 120–28. *Proc. Linn. Soc.* 1918, 41–42. *Times* 23 Jan. 1918. *Girton Rev.* 1927, 17–26. *Who was Who, 1916–1928* 933. *J. Soc. Bibl. Nat. Hist.* 1968, 370–71, 883–84 portr.

SARGEAUNT, John (1857–1922)
b. Irthlingborough, Northants 12 Aug. 1857 *d.* Hove, Sussex 20 March 1922
Classical master, Westminster School, 1890–1918. *Trees, Shrubs and Plants of Virgil* 1920. *Westminster Verses* 1922 (with memoir).
Who was Who, 1916–1928 933.

SARGESON, William (*c.* 1806–1886)
d. Trafford Park, Lancs 24 June 1886
Gardener to Sir H. F. de Trafford at Trafford Park.
J. Hort. Cottage Gdnr v.13, 1886, 158–59.

SAUL, John (1819–1897)
b. Castle Martyr, County Cork 25 Dec. 1819 *d.* Washington D.C., U.S.A. 11 May 1897
Manager of nurseries at Bristol. To U.S.A., 1851. Landscape gardener, nurseryman and plant importer.
Gdnrs Chron. 1897 i 392. L. H. Bailey *Standard Cyclop. Hort.* v.2, 1939, 1594.

SAUL, Michael (1817–1892)
b. Castle Martyr, County Cork 29 June 1817 *d.* Manchester 27 Aug. 1892
Gardener to Lord Stourton at Stourton Castle, Yorks, 1848–78. Contrib. to *Gdnrs Chron.*
Gdnrs Chron. 1876 i 108–9 portr.; 1892 ii 349–50 portr.

SAUNDERS, Bernard (*fl.* 1840s)
Nurseryman, St. Helier, Jersey, Channel Islands.

SAUNDERS, Charles B. (1824–1893)
b. 4 Jan. 1824 *d.* Jersey, Channel Islands 1 Aug. 1893
Nurseryman, Jersey. On his death the business was acquired by Mr. Becker.
Gdnrs Chron. 1893 ii 249.

SAUNDERS, Edith Rebecca (1865–1945)
b. Brighton, Sussex 14 Oct. 1865 *d.* Cambridge 6 June 1945
Educ. Cambridge. Lecturer, Cambridge University, 1889–1925; Director of Balfour Laboratory, Newham College. President, Genetical Society, 1936–38. Wrote papers on floral morphology and genetics. Contrib. to *Ann. Bot.*, *J. Bot.*, *J. Genetics*, *New Phytol.*, *Proc. Linn. Soc.*
Nature v.156, 1945, 198–99, 385. *Proc. Linn. Soc.* 1945–46, 75–76. *Times* 8 June 1945.

SAUNDERS, Sir Edwin (*c.* 1814–1901)
d. Wimbledon, Surrey 15 March 1901
Dental surgeon. President, National Chrysanthemum Society, 1891.
Garden v.59, 1901, 213–14. *Gdnrs Mag.* 1901, 185 portr.

SAUNDERS, Miss Elsie (*fl.* 1900s–1920s)
Nurse at St. Stephens Mission Hospital, Delhi. Collected plants in Murree and Kashmir.
Plants at Kew.
Rosa saundersiae Rolfe.

SAUNDERS, Eric (1912–1966)
b. 1 Nov. 1912
Commercial artist. Secretary, Essex Field Club, 1962–66. Interested in bryophytes, charophytes and marine algae. Made new plant records for Essex.
Essex Nat. 1966, 339–42 portr. S. T. Jermyn *Fl. Essex* 1974, 21.

SAUNDERS, George Sharp (–1910)
b. Wandsworth, London *d.* Burgh Heath, Surrey 10 April 1910

FLS 1899. Son of W. W. Saunders (1809–1879). Made drawings of teratological plants. Edited *J. R. Hort. Soc.*, 1906–8. Contrib. to *Garden.*

Garden v.68, 1905, 380 portr.; 1910, 208. *Gdnrs Chron.* 1910 i 272.

SAUNDERS, Miss Helen (1830–1914)
b. South Molton, Devon 1830

Botanised in S. Molton neighbourhood. Contrib. to *Trans. Devonshire Assoc.*

W. K. Martin and G. T. Fraser *Fl. Devon* 1939, 777.

Herb. at Seale-Hayne Agricultural College.

SAUNDERS, James (1839–1925)
b. Salisbury, Wilts 30 March 1839 *d.* Luton, Beds 17 April 1925

ALS 1900. Straw hat manufacturer, Luton. Studied British Mycetozoa. *Wild Flowers of Bedfordshire* 1897. *Field Flowers of Bedfordshire* 1911. Contrib. to *J. Bot.*, *Midland Nat.*, *Trans. Hertfordshire Nat. Hist. Soc.*

J. Bot. 1925, 180–82. *Bot. Soc. Exch. Club Br. Isl. Rep.* 1925, 856–57. *Proc. Linn. Soc.* 1924–25, 81–82. G. C. Druce *Fl. Buckinghamshire* 1926, civ. *J. Bedfordshire Nat. Hist. Soc. Field Club* 1947, 58–61. J. G. Dony *Fl. Bedfordshire* 1953, 27–28.

Herb. at Luton Museum. Portr. at Hunt Library.

SAUNDERS, Katherine (*fl.* 1880s–1890s)

Wife of J. R. Saunders, pioneer of Natal sugar industry. Sent Natal plants to Kew.

S. African J. Sci. 1971, 405.

Flower drawings at Tongaat and Natal Museums.

SAUNDERS, M. (–1877)
d. Cork, Ireland 3 April 1877

Nurseryman, Friars' Walk, Cork. Succeeded by his son, David.

Gdnrs Chron. 1877 i 478.

SAUNDERS, Mary Anne *see* Stebbing, M. A.

SAUNDERS, Samuel (*fl.* 1780s–1790s)

Of Leatherhead, Surrey. Friend of J. E. Smith. *Introduction to... Botany* 1792.

SAUNDERS, William (1822–1900)
b. St. Andrews, Fifeshire 7 Dec. 1822 *d.* Washington DC, U.S.A. 11 Sept. 1900

Kew gardener. To U.S.A., 1848. In partnership with T. Meehan, 1854. Botanist and Superintendent of Horticulture, U.S. Dept. of Agriculture, 1862. Introduced many economic plants to U.S.A.

L. H. Bailey *Standard Cyclop. Hort.* v.2, 1939, 1594–95. E. C. Jellett *Gdns and Gdnrs of Germantown* 1914, 249–343.

Portr. at Hunt Library.

SAUNDERS, William (1836–1914)
b. Crediton, Devon 16 June 1836 *d.* London, Ontario 13 Sept. 1914

FLS 1886. To Canada, *c.* 1847. Pharmacist, entomologist, plant-breeder. First Director, Experimental Farms, Canada, 1886–1911. Produced hybrid cultivars in *Ribes*, *Rubus*, *Vitis*, *Rosa*, *Pyrus*, *Triticum*. *Useful Trees and Shrubs of Northwest Plains of Canada* 1889.

Agric. Gaz. Canada 1914, 766–70 portr. L. H. Bailey *Standard Cyclop. Hort.* v.2, 1939, 1595 portr.

SAUNDERS, William Frederick (1834–1901)
b. Wandsworth, London 7 April 1834 *d.* Clapham, London 26 Dec. 1901

FLS 1858. Son of W. W. Saunders (1809–1879). Had herb. Contrib. to J. A. Brewer's *Fl. Surrey* 1863.

Proc. Linn. Soc. 1901–2, 42.

SAUNDERS, William Wilson (1809–1879)
b. Little London, Wendover, Bucks 4 June 1809 *d.* Worthing, Sussex 13 Sept. 1879

FRS 1853. FLS 1833. Underwriter at Lloyds. Treasurer, Linnean Society, 1861–73. Secretary, Royal Horticultural Society, 1863–66. *Refugium Botanicum* 1869–73 5 vols. *Mycological Illustrations* 1871–72. Oxfordshire plants in *Mag. Nat. Hist.* 1839, 239–42.

Gdnrs Chron. 1871, 136–37 portr.; 1879 ii 368. *J. Bot.* 1879, 320. *Nature* v.20, 1879, 836–37. *Proc. Entomol. Soc.* 1879, lxvi–lxvii. *D.N.B.* v.50, 331–32. G. C. Druce *Fl. Buckinghamshire* 1926, xcviii. G. C. Druce *Fl. Oxfordshire* 1927, cx–cxi. *Curtis's Bot. Mag. Dedications, 1827–1927* 155–56 portr. *R.S.C.* v.5, 412; v.8, 837; v.11, 289.

Herb. at Oxford. Letters at Kew. Library sold at Sotheby, Aug. 1873. Fungi drawings at BM(NH).

Saundersiana Royle.

SAVAGE, E. J. (*née* Fry) (–1948)

MSc Aberystwyth 1919. Assistant Lecturer in Botany, Aberystwyth, 1919. Lecturer in Botany, Westfield College, London, 1925. Married S. Savage (1886–1966) in 1932.

Nature v.162, 1948, 766.

SAVAGE, Spencer (1886–1966)
b. 23 June 1886 *d.* Southampton, Hants 3 Nov. 1966

FLS 1929. Clerk, Linnean Society, 1909; Librarian and Assistant Secretary, 1929–51. Translated Book 2 of Crispin van de Passe's *Hortus Floridus* 1929. *Catalogue of Manuscripts of Linnean Society* 1935–48. *Catalogue of Linnaean Herb.* 1945. Catalogued J. E. Smith's herb. in 1963; reproduced on microfilm, 1966.

Svenska Linnésällsk. Årsskr. 1949, 5. *Times* 8 Nov. 1966. *Taxon* 1967, 67. *Proc. Linn. Soc.* 1968, 145.

Portr. at Hunt Library.

SAVERY, George Brooke (1874–1937)
d. Silverton, Devon 23 Nov. 1937
Civil engineer. Studied *Rubus* and *Rosa* of Silverton area. 'Mosses of Silverton' (*Trans. Devon Assoc.* 1910).
Rep. Br. Bryol. Soc. 1937, 63. W. K. Martin and G. T. Fraser *Fl. Devon* 1939, 778.
Plants at Exeter Museum.

SAVILLE, Rev. John (1736–1803)
d. 2 Aug. 1803
Vicar Choral, Lichfield Cathedral. Botanist and friend of Jonathan Stokes.
J. Bot. 1914, 320–21.

SAWER, John Charles (–1904)
d. Brighton, Sussex 23 Aug. 1904
FLS 1881. Interested in perfumes and perfume-bearing plants. *Rhodologia; a Discourse on Roses* 1894.
Proc. Linn. Soc. 1904–5, 54.

SAWERS, William (*fl.* 1850s)
Of Londonderry, N. Ireland. Algologist.
Irish Nat. J. v.15, 1967, 346–47.
Marine algae at Queen's University, Belfast.

SAWYER, Sir James (1844–1919)
b. Carlisle, Cumberland 11 Aug. 1844 *d.* 27 Jan. 1919
Professor of Materia Medica, Queen's College, Birmingham. Studied the medicinal properties of plants.
Bot. Soc. Exch. Club Br. Isl. Rep. 1919, 634.

SAWYER, Robert (*c.* 1812–1907)
d. Bartley, Hants Oct? 1907
Nurseryman, High Street, Southampton, Hants until 1875.
J. Hort. Cottage Gdnr v.55, 1907, 340.

SAXTON, Walter Theodore (1882–1973)
b. Cleeve, Somerset 16 Oct. 1882 *d.* 27 Feb. 1973
FLS 1909. Professor, Gujarat College, Ahmedabad, India; later at Depts. of Botany, Reading and Cape Town.
Biol. J. Linn. Soc. 1973, 392.

SAYERS, Willielma Jane (–1959)
d. June 1959
BA Belfast 1893. Assisted in field work for R. L. Praeger and W. R. Megaw's *Fl. North E. Ireland* ed. 2 1938.
R. L. Praeger *Some Irish Nat.* 1949, 151. *Irish Nat. J.* 1960, 173–74.

SCAMPTON, John (*fl.* 1690s–1710s)
"That ingenious botanist" (Petiver). Sent *Calamagrostis lanceolata* to J. Petiver from Leicestershire (Petiver *Gram. Musc. Fung. Br. Conc.* 1716 no. 69); also botanised in Derbyshire (Petiver *Museii Petiveriani* 1695, 74).
J. Bot. 1915, 175–76. A. R. Horwood and C. W. F. Noel *Fl. Leicestershire* 1933, clxxxvi–clxxxvii. J. E. Dandy *Sloane Herb.* 1958, 200.

SCHAW, Frances Sara *see* Sharland, F. S.

SCHEER, Frederick (*c.* 1792–1868)
b. Rügen, Germany *c.* 1792 *d.* Northfleet, Kent 30 Dec. 1868
Cultivated cacti at Kew Green. Contrib. cacti to B. Seemann's *Bot.... Herald* 1852–57. *Kew and its Gardens* 1840. 'New *Mamillaria*' (*London J. Bot.* 1845, 136–37).
Bot. Mag. 1853, t.4743. *Gdnrs Chron.* 1869, 964–65. *J. Bot.* 1869, 268–70. *Kew Bull.* 1891, 324–25. *R.S.C.* v.5, 447.
Letters at Kew. Portr. at Hunt Library.
Scheeria Seem.

SCHLICH, Sir William (1840–1925)
b. Hesse-Darmstadt, Germany 25 Feb. 1840
d. Oxford 28 Sept. 1925
PhD Giessen 1867. FRS 1901. FLS 1885. KCIE 1909. Indian Forest Service, 1866. In Burma until 1870. Conservator of Forests, Bengal, 1872–79. Conservator of Forests, Punjab, 1880. Inspector-General of Forests until 1885. First Editor of *Indian Forester*, 1875–79. Royal Indian Engineering College, Cooper's Hill, Windsor, 1885–1906. Professor of Forestry, Oxford, 1905–19. *Manual of Forestry* (with W. R. Fisher) 1889–96 5 vols; ed. 5 1925.
Indian Forester 1889, 45–51; 1925, 625–32 portr. *Bot. Soc. Exch. Club Br. Isl. Rep.* 1925, 857. *Gdnrs Chron.* 1925 ii 298; 1929 i 306. *Empire For. J.* 1925, 160–67. *Nature* v.116, 1925, 617–18. *Proc. Linn. Soc.* 1925–26, 95–96. *Times* 1 Oct. 1925. *Proc. R. Soc. B.* v.101, 1927, vi–xi portr. *Who was Who, 1916–1928* 937. I. H. Burkill *Chapters on Hist. Bot. in India* 1965, 163–64. *R.S.C.* v.18, 529.
Indian plants at Oxford.

SCHMID, Bernhard (1787–1857)
Missionary. To India, 1817. Collected plants in Nilgiri Hills.
J. K. Zenker *Plantae Indicae, quas... collegit B. Schmid* 1835–37. I. H. Burkill *Chapters on Hist. Bot. in India* 1965 *passim.*

SCHNEIDER, George (1848–1917)
b. Paris 1848 *d.* Fulham, London 2 Jan. 1917
Came to England, 1870. Employed by Veitch and Sons, Chelsea for 30 years. *Book of Choice Ferns* 1892–94 3 vols. *Choice Ferns for Amateurs* 1905.
Gdnrs Chron. 1917 i 11.

SCHNEIDER, Henry *see* Carlton, H.

SCHOLES, John (-1943)
Schoolmaster, Archimota School, Gold Coast, 1939–43. Collected plants in Gold Coast and Togoland, now at BM(NH), Kew.

SCHOLEY, John (*fl*. 1850s–1860s)
Nurseryman, Ropergate, Pontefract, Yorks.

SCHOMBURGK, Sir Moritz Richard (1811–1891)
b. Frihault, Saxony, Germany 5 Oct. 1811 *d*. Adelaide, Australia 24 March 1891
PhD. Trained in Royal Gardens, Potsdam and San Souci Gardens. Accompanied his brother Robert to British Guiana, 1840. Director, Adelaide Botanic Garden, 1865. *Reisen in British-Guiana* 1847–49 3 vols. *Botanical Reminiscences of British Guiana* 1876. 'Fl. South Australia' (W. Harcus *S. Australia* 1876, 205–80).
J. Bot. 1891, 224. *Proc. Linn. Soc. N.S.W*. 1900, 793–94. *Rep. Austral. Assoc. Advancement Sci*. 1907, 176. *Proc. R. Geog. Soc. A/Asia S. Austral. Branch* 1933–34, 446–47. *Austral. Encyclop*. v.8, 1965, 16. *R.S.C*. v.5, 520; v.8, 879; v.18, 571.
Plants and letters at Kew. Portr. at Adelaide Botanic Garden.

SCHOMBURGK, Sir Robert Hermann (1804–1865)
b. Freiburg, Germany 5 June 1804 *d*. Berlin 11 March 1865
PhD Königsberg. FRS 1859. Knighted 1844. Brother of Sir M. R. Schomburgk. In West Indies, 1830; British Guiana, 1835–39, 1840–44. Discovered *Victoria amazonica*. Consul, St. Domingo, 1848–57; Bangkok, 1857–64. Collected plants in Siam. *A Description of British Guiana* 1840. *History of Barbados* 1848.
A. Lasègue *Musée Botanique de B. Delessert* 1845, 216–19. *Leopoldina* Hft 1, 1859, 34–39. *J. Asiatic Soc. Bengal* v.32, 1864, 387–99. *Bot. Zeitung* 1865, 131–32. *J. Bot*. 1865, 136; 1903, 307. *D.N.B*. v.50, 437–39. I. Urban *Symbolae Antillanae* v.3, 1902, 121–23. *J. Thailand Res. Soc., Nat. Hist. Supplt* v.12, 1939, 11–13. *Fl. Malesiana* v.1, 1950, 475. *Bol. Archivo Generalde Nacion* v.12, 1949, 267–76 portr. *R.S.C*. v.5, 520; v.8, 879; v.18, 571.
Guiana plants and drawings at BM(NH). Malayan plants and lithograph portr. after E. Eddis at Kew.
Schomburgkia Lindl.

SCHRÖDER, Sir Henry (1824–1910)
d. Sidmouth, Devon 20 April 1910
VMH 1897. Merchant banker. Had garden at Egham, Surrey. Collected orchids. Benefactor of Royal Horticultural Society.
Gdnrs Chron. 1910 i 280–81 portr.; 1963 i 350–51, 355 portr. *Gdnrs Mag*. 1910, 333. *J. Hort. Cottage Gdnr* v.60, 1910, 381–82 portr. *Orchid Rev*. 1910, 133–35.
Miltonia schroederiana × O'Brien.

SCHUNCK, Henry Edward (1820–1903)
b. Manchester 16 Aug. 1820 *d*. Manchester 13 Jan. 1903
FRS 1850. Studied colouring matters of vegetable substances, especially *Rubia tinctorum*, 1846–55.
Nature v.67, 1903, 275. *Times* 14 Jan. 1903. *Proc. R. Soc*. v.75, 1904, 261–65. *D.N.B. 1901–1911* 274–75.

SCHWARTZ, Ernest Justus (1870–1939)
b. London 11 Nov. 1870 *d*. 29 Dec. 1939
MA Cantab 1891. BSc London 1900. DSc 1914. FLS 1906. Lecturer, King's College, London, 1897–1935. Mycologist. Contrib. to *Ann. Bot*.
Proc. Linn. Soc. 1939–40, 374–75.

SCOFFERN, John (1814–1882)
b. Dutson, Cornwall 9 Oct. 1814 *d*. Wimbledon, Surrey 14 Feb. 1882
MB London 1843. Professor of Chemistry, Aldergate Street College, London, 1840. *Outlines of Botany* 1857.
F. Boase *Modern English Biogr*. v.3, 1901, 445. *R.S.C*. v.5, 602.

SCORESBY, William (1790–1857)
b. Cropton, Yorks 5 Oct. 1790 *d*. Torquay, Devon 21 March 1857
FRS 1824. FRSE 1819. Captain in Greenland fishery where he collected plants. Correspondent of J. Banks who probably encouraged him to study natural history of Arctic.
D.N.B. v.51, 6–8.

SCORTECHINI, Rev. Benedetto (1845–1886)
b. Cupramontana, Italy 1845 *d*. Calcutta 4 Nov. 1886
LLB. FLS 1881. In Queensland, 1871–84; Straits Settlements, 1884. Collaborated with F. M. Bailey, F. von Mueller, G. King. Queensland fungi described in *Revue Mycologique* 1885, 93–94, *Atti R. Istituto Veneto Scienze* v.3, 1885, 711–43; v.6, 1887–88, 388–89. Contrib. to *Proc. Linn. Soc. N.S.W., J. Bot*.
J. Bot. 1887, 321–25; 1893, 225–26. *Proc. R. Soc. Queensland* 1887, 2–8. *Revue Mycologique* 1887, 123. *Rep. Austral. Assoc. Advancement Sci*. 1909, 381. *R.S.C*. v.11, 370.
Plants at BM(NH), Kew, Calcutta, Perak.
Scortechinia Hook. f.

SCOT, Reynold (*c*. 1539–*c*. 1599)
Educ. Oxford. Of Smeeth, Kent. *Perfite Platforme of a Hoppe Garden* 1574.
Cottage Gdnr v.6, 1851, 171. D. McDonald *Agric. Writers* 1908, 34–36.

SCOTT, Andrew (*fl.* 1730s)
Maryland plants at BM(NH).
E. J. L. Scott *Index to Sloane Manuscripts* 1904, 316. J. E. Dandy *Sloane Herb.* 1958, 202.

SCOTT, Benjamin Charles George (*fl.* 1860s–1890s)
Interpreter, China Consular Establishment, 1867. Collected specimens of *Fraxinus chinensis* and *Ligustrum incidum* for Kew.
Kew Bull. 1893, 108–11; 1901, 59. E. Bretschneider *Hist. European Bot. Discoveries in China* 1898, 739.

SCOTT, Charles (1864–1907)
Collected in South of Scotland and in Argyllshire. Discovered *Marsupella pearsoni* in Scotland.
Trans. Bot. Soc. Edinburgh v.25, 1910, 5.

SCOTT, Dukinfield Henry (1854–1934)
b. London 28 Nov. 1854 *d.* Oakley, Hants 29 Jan. 1934
BA Oxon 1876. PhD Würzburg 1881. FRS 1894. FLS 1880. Lecturer, University College, London, 1882. Lecturer, Royal College of Science, London, 1884–92. Hon. Keeper, Jodrell Laboratory, Kew, 1892–1906. Palaeobotanist. President, Linnean Society, 1908–12. President, Royal Microscopical Society, 1904–6. *Introduction to Structural Botany* 1894–96. *Studies in Fossil Botany* 1900; ed. 3 1920–23. *Evolution of Plants* 1911. *Extinct Plants and Problems of Evolution* 1924. Contrib. to *Ann. Bot.*, *New Phytol.*, *J. Linn. Soc.*, *Philos. Trans. R. Soc.*
Nature v.50, 1894, 56; 1934, v.133, 317–19; v.174, 1954, 992–93. *New Phytol.* 1925, 9–16; 1934, 73–76. *J. Bot.* 1928, 56; 1934, 83–88, 95. *Current Sci.* 1934, 392–95 portr. *J. Indian Bot. Soc.* 1934, 305–9 portr. *J. Kew Guild* 1934, 376–77. *J. R. Microsc. Soc.* 1934, 32–34. *Kew Bull.* 1934, 128–33. *N. Western Nat.* 1934, 59–61. *Obit. Notices Fellows R. Soc.* 1934, 205–27. *Proc. Bournemouth Nat. Sci.* v.26, 1933–34, 73 portr. *Proc. Linn. Soc.* 1933–34, 166–69. *Ann. Bot.* 1935, 823–40 portr. *Chronica Botanica* 1935, 157, 167 portr. *Proc. R. Soc. Edinburgh* 1935, 224. *Who was Who, 1929–1940* 1207. *D.N.B. 1931–1940* 796–97. F. O. Bower *Sixty Years of Bot. in Britain 1875–1935* 1938, 66–68 portr. C. C. Gillispie *Dict. Sci. Biogr.* v.12, 1975, 258–60.
Fossil plants at BM(NH). Portr. at Hunt Library.

SCOTT, Henderina Victoria (*née* **Klaassen**) (–1929)
d. Oakley, Hants 18 Jan. 1929
FLS 1905. Married D. H. Scott, 1887. Studied fossil botany and plant physiology. Prepared illustrations for her husband's *Introduction to Structural Botany* 1894–96, *Studies in Fossil Botany* 1900. Contrib. to *Ann. Bot.*, *New Phytol.*
Bot. Soc. Exch. Club Br. Isl. Rep. 1929, 98. *J. Bot.* 1929, 37. *Nature* v.123, 1929, 287. *Proc. Linn. Soc.* 1928–29, 146–47.

SCOTT, Henry (*fl.* 1750s–1760s)
Nurseryman, Weybridge, Surrey.

SCOTT, Hercules R. (*fl.* 1830s)
Of Edinburgh. Advocate. Sent list of plants to A. Murray.
A. Murray *Northern Fl.* 1836, 21.

SCOTT, Hugh (1885–1960)
b. Lee, London 16 Sept. 1885 *d.* Henley, Oxfordshire 1 Nov. 1960
FRS 1941. FLS 1927. Curator in Entomology, Museum of Zoology, Cambridge, 1909–28. Made entomological expeditions to N.E. Africa and S.W. Arabia, 1926–27, 1937–38, 1948–49, 1952–53. Assistant Keeper, BM(NH), 1930–48. Also collected plants on these expeditions. *In the High Yemen* 1942.
Biogr. Mem. Fellows R. Soc. 1961, 229–42 portr. *Geogr. J.* 1961, 142. *Proc. Linn. Soc.* 1960–61, 68.
Notebooks and letters at BM(NH). Plants at Kew.

SCOTT, James (*c.* 1740–1770)
d. 1770
Nurseryman, Turnham Green, Middx.
Trans. London Middlesex Archaeol. Soc. v.24, 1973, 182. *Gdn Hist.* 1974, 58.

SCOTT, James Reid (1839–1877)
b. Earlston 1839 *d.* 25 Aug. 1877
Emigrated to Tasmania; elected to Legislative Assembly there, 1867. Botanist.
Rep. R. Soc. Tasmania 1887, 11. *Austral. Encyclop.* v.3, 1965, 460.

SCOTT, James Robinson (*c.* 1789–1821)
d. 30 Aug. 1821
FLS 1817. FRSE. Of Edinburgh. Surgeon, Royal Navy. Lecturer on Botany, Edinburgh. *Herbarium Edinense* (with W. Jameson) 1820 (exsiccatae).

SCOTT, John (1836–1880)
b. Denholm, Roxburghshire 1836 *d.* Garvald, E. Lothian 10 June 1880
FLS 1873. Gardener, Chatsworth. Curator, Botanic Garden, Calcutta. Correspondent of C. Darwin. 'Tree-ferns of British Sikkim' (*Trans. Linn. Soc.* v.30, 1875, 1–44). Contrib. to *J. Linn. Soc.*
Gdnrs Chron. 1880 i 794. *J. Bot.* 1880, 224. *Trans. Bot. Soc. Edinburgh* v.14, 1883, 160–61. F. Darwin and A. C. Seward *More Letters of Charles Darwin* v.1, 1903, 217–22; v.2, 302–32. *Fl. Malesiana* v.5, 1958, 85.
Plants at Calcutta. Letters at Kew.

SCOTT, Margaret *see* Gatty, M.

SCOTT, Munro Briggs (1889–1917)
b. Fifeshire 29 April 1889 *d.* killed near Arras, France 12 April 1917
MA, BSc Edinburgh. Assistant, Herb., Kew, 1914. Contrib. to *Kew Bull.*
J. Bot. 1917, 263. *Kew Bull.* 1917, 210–11. *J. Kew Guild* 1918, 420 portr.

SCOTT, Robert (1757–1808)
d. Dublin 18 Sept. 1808
MD. Bryologist. Professor of Botany, Dublin, 1785–1808. Friend of Dawson Turner. Contrib. to J. Sowerby and J. E. Smith's *English Bot.* 1181, 1391, 1564, 2489.
Trans. Dublin Soc. v.3, 1803, 157–61; v.4, 1804, 199. P. Smith *Mem. Correspondence of Sir J. E. Smith* v.2, 1832, 165. *J. Bot.* 1907, 305. N. Colgan *Fl. County Dublin* 1904, xxv. T. P. C. Kirkpatrick *Hist. of Med. Teaching in Trinity College* 1912, 208–10. *R.S.C.* v.5, 606.
Algae at National Museum, Dublin. Miniature portr. at Kew. Portr. at Hunt Library.
Scottia R. Br.

SCOTT, Robert (*fl.* 1850s–1860s)
Horticultural Curator, Botanic Garden, Calcutta until 1865. Collected plants in Burma, 1855, 1860–61.

SCOTT, Robert Robinson (1827–1877)
b. Belfast 1827 *d.* Harrisburg, Pennsylvania, U.S.A. 24 June 1877
Gardener, Belfast and Kew Botanic Gardens. To U.S.A. where he founded *Philadelphia Florist* in 1852. Discovered *Asplenium ebenoides.*
J. Kew Guild 1894, 41–42. *Fern Bull.* 1903, 50–51 portr.

SCOTT, Thomas (*fl.* 1820s–1830s)
Of Glasgow. Merchant, Launceston, Tasmania. Correspondent of W. J. Hooker to whom he sent seeds.
J. D. Hooker *Fl. Tasmaniae* v.1, 1859, cxxvii. *Proc. R. Soc. Tasmania* 1909, 26. *Rep. Austral. Assoc. Advancement Sci.* v.13, 1912, 234.

SCOTT, Walter Francis Montague Douglas, 5th Duke of Buccleugh (1806–1884)
b. Dalkeith, Midlothian 25 Nov. 1806 *d.* Bowhill, Selkirkshire 16 April 1884
MA Cantab 1827. FLS 1833. Had gardens at Dalkeith Palace and Drumlanrig Castle, Dumfries. President, Highland Agricultural Society, 1831. President, Royal Horticultural Society, 1862–83.
Proc. Linn. Soc. 1883–86, 31. *D.N.B.* v.51, 25–26.

SCOTT, William (–1827)
d. 3 Oct. 1827
Student gardener under W. Aiton at Royal Gardens, Kew. Nurseryman, Dorking, Surrey.
Gdnrs Mag. 1828, 256.

SCOTT, William (1859–1897)
b. Lomnay, Aberdeenshire 12 Sept. 1859 *d.* Stirling 3 Oct. 1897
FLS 1895. Employed in private gardens, nurseries and at Kew. Assistant Director of Forests and Gardens, Mauritius, 1881; Director, 1893–97. Advised on formation of Botanic Garden in Seychelles.
Kew Bull. 1897, 403; 1919, 285. *Proc. Linn. Soc.* 1897–98, 49–50. *J. Kew Guild* 1898, 35–36 portr.

SCOTT-ELLIOT, George Francis (1862–1934)
b. Calcutta 6 Jan. 1862 *d.* Dumfries 20 June 1934
BA Cantab 1882. BSc Edinburgh. FLS 1890. Botanist, French and English Delimination Commission of Sierra Leone Boundary, 1891–92. On British East Africa expedition, 1893–94. Lecturer in Botany, Royal Technical College, Glasgow, 1896–1904. President, Dumfries and Galloway Natural History and Antiquarian Society. *Fl. Dumfriesshire* 1891–96. *Naturalist in Mid-Africa* 1896. 'Expedition to Ruwenzori and Tanganyika' (*Geogr. J.* 1893, 301–24). 'Botanical Results of Sierra Leone Boundary Commission' (*J. Linn. Soc.* v.30, 1894, 64–100). *Nature Studies* 1903. *Romance of Plant Life* 1906. *First Course in Practical Botany* 1906. *Botany Today* 1910.
Geog. J. 1894, 349–52; 1895, 301–24; 1934, 280. *Nature* v.50, 1894, 549–50. *Commerce* 1895, 693–96 portr. *Kew Bull.* 1895, 77–78; 1897, 304. *J. Bot.* 1934, 233–35. *N. Western Nat.* 1934, 282–89 portr., 390–91. *Proc. Linn. Soc.* 1934–35, 174–75. *Glasgow Nat.* 1936, 128–30. *Trans. J. Proc. Dumfriesshire Galloway Nat. Hist. Antiq. Soc.* 1936, 351–52. *Moçambique* 1941, 26–31. *Comptes Rendus AETFAT 1960* 1962, 181–82. *Taxon* 1969, 425–28.
Plants at Kew, BM(NH).
Bulbophyllum elliotii Rolfe.

SCOULER, John (1804–1871)
b. Glasgow 31 Dec. 1804 *d.* Glasgow 13 Nov. 1871
MD Glasgow 1827. LLD 1850. FLS 1829. Surgeon and naturalist on voyage to Columbia River, 1824–25. Professor of Geology, Glasgow, 1829. Professor of Geology, Zoology and Botany to Royal Dublin Society, 1833–54. Journals in *Edinburgh J. Sci.* v.5, 1826, 195–214; v.6, 1827, 228–36;

Quart. Oregon Hist. Soc. v.6, 1905, 54–75, 159–205, 276–87.
Bot. Misc. 1829–30, 34. *Glasgow Herald* 18 Nov. 1871. *D.N.B.* v.51, 122–23. *Occas. Papers California Acad. Sci.* no. 20, 1943, 37–38. S. D. McKelvey *Bot. Exploration of Trans-Mississippi West, 1790–1850* 1955, 284–98. *Glasgow Nat.* 1962, 210–12. *Br. Phycol. Bull.* 1964, 385–86. *Taxon* 1970, 541. *R.S.C.* v.5, 607.
Plants at Kew, Dublin, Strathclyde University.
Scouleria Hook.

SCRASE-DICKENS, Charles R. (–1947)
VMH 1934. Amateur gardener and successful grower of difficult plants.

SCUDAMORE, John, 1st Viscount (1601–1671)
b. Holme Lacy, Herefordshire 16 Feb. 1601 *d.* 8 June 1671
Educ. Oxford. MP Herefordshire. Ambassador to Paris, 1635–36. Had garden at Holme Lacy where he took a great interest in the planting and grafting of his orchard trees.
R. Hogg and H. G. Bull *Herefordshire Pomona* v.1, 1876, 63–92 portr.
MSS. at BM.

SCULLY, Reginald William (1858–1935)
d. Cork, Ireland May 1935
FLS 1889. Studied medicine but never practised. Discovered *Hieracium scullyi* in Ireland, 1894. Co-editor with N. Colgan of A. G. More *Contributions towards a Cybele Hibernica* 1898. *Fl. County Kerry* 1916. Contrib. to *J. Bot.*, *Irish Nat.*
Irish Nat. J. v.5, 1935, 283–84 portr. *Proc. Linn. Soc.* 1935–36, 213. R. L. Praeger *Some Irish Nat.* 1949, 153 portr.
Herb. at National Museum, Dublin.

SCULTHORPE, Cyril Duncan (1939–1969)
b. 4 Aug. 1939 *d.* 5 April 1969
MA Cantab. FLS 1965. Lecturer in Biology, Salford Technical College; Biology Dept., West Ham College of Technology. Interested in temperate and tropical aquatic vascular plants.

SCUPHAM, John Robertson (1840–1927)
b. Edinburgh 5 April 1840 *d.* San Francisco, U.S.A. 30 May 1927
Civil engineer. To California at close of American Civil War. Collected Californian plants.
Leaflet W. Bot. v.8(5), 1957, 98.
Herb. at California University.
Ribes scuphami Eastw.

SEABROOK, W. (*fl.* 1900s)
Nurseryman, Chelmsford, Essex.
Gdnrs Mag. 1911, 751–52 portr.

SEAFORTH, Lord *see* Humberston, F. M.

SEALE, William (*c.* 1820–1885)
d. 6 Aug. 1885
Nurseryman, The Gardens, Wildernesse, Sevenoaks, Kent.
Gdnrs Chron. 1885 ii 219.

SEALY, James (–before 1834)
Friend of A. H. Haworth. Collected plants in County Cork, Ireland. Herb. sent to W. J. Hooker.
Letters in Winch correspondence at Linnean Society.

SEARLE, Henry (Harry) (1844–1935)
b. West Fen, Cambridge 19/22 Dec. 1844 *d.* Oldham, Lancs 26 Jan. 1935
Secretary, Ashton-under-Lyne Linnean Botanical Society. President, Middleton Botanical Society. Had herb.
Heywood Advertiser 13 Nov. 1909. *N. Western Nat.* 1939, 262–69 portr.
Letters at Liverpool Museums.

SEBBORN, John (–1780)
Seedsman, Colchester, Essex.

SEBRIGHT, Frederica *see* Franks, F.

SECRETT, Frederick Augustus (1886–1964)
b. Ealing, Middx 11 Feb. 1886 *d.* Milford, Surrey 18 July 1964
FLS 1927. VMH 1936. VMM 1963. Vice-President, Royal Horticultural Society, 1963. Market gardener, Milford.
Gdnrs Chron. 1964 ii 131. *J. R. Hort. Soc.* 1964, 423–24 portr. *J. R. Soc. Arts* 1964, 785–86. *Proc. Linn. Soc.* 1966, 124.

SEDEN, John (1840–1921)
b. Dedham, Essex 10 July 1840 *d.* Worthing, Sussex 24 Feb. 1921
VMH 1897. Gardener at Veitch nursery, Chelsea, London, 1861. Taught how to hybridise orchids by John Dominey.
Gdnrs Chron. 1899 ii 41 portr.; 1904 ii 464–65; 1921 i 120 portr. *Garden* v.67, 1905, 136 portr. J. H. Veitch *Hortus Veitchii* 1906, 103–5. M. A. Reinikka *Hist. of the Orchid* 1972, 244–45.

SEDGWICK, Rev. John
Of Lincoln. Few local plants in Sir Hans Sloane's herb.
J. E. Dandy *Sloane Herb.* 1958, 202.

SEDGWICK, Leonard John (1883–1925)
b. Bristol 27 April 1883 *d.* Bombay, India 27 June 1925

BA Cantab 1905. FLS 1916. Indian Civil Service. Studied flora of W. India. 'Plants of N. Gujarat' (with W. T. Saxton) (*Rec. Bot. Survey India* v.6, 1918, 209–323). Contrib. to *J. Bombay Nat. Hist. Soc.*, *J. Indian Bot.*

J. Indian Bot. v.5, 1926, 48–49. *Proc. Linn. Soc.* 1925–26, 98–99.

Mosses at BM(NH).

SEEMANN, Berthold Carl (1825–1871)
b. Hanover, Germany 25 Feb. 1825 *d.* Javali Mine, Nicaragua 10 Oct. 1871

PhD Göttingen. FLS 1852. Naturalist on HMS 'Herald' on voyages on west coast of America and in Arctic seas, 1846–51. In Venezuela, 1864. Founded and edited *Bonplandia* 1853–63 and *J. Bot.* 1863–69. *Narrative of Voyage of HMS 'Herald'... 1845–1851* 1853 2 vols. *Botany of Voyage of HMS 'Herald'...1845–1851* 1852–57. *Popular History of the Palms* 1856. *Fl. Vitiensis* 1865–73. *Revision of Natural Order Hederaceae* 1868.

E. Walford, *ed. Portr. of Men of Eminence in Literature, Sci. and Art* v.5, 1866, 53–58 portr. *Gdnrs Chron.* 1871, 1678–79 portr. *Bot. Zeitung* 1872, 503–9. *J. Bot.* 1872, 1–7 portr.; 1889, 102; 1921, 22–24. *Proc. Linn. Soc.* 1871–72, lxxiv–lxxix. *J. Kew Guild* 1894, 41; 1895, 31–32. E. Bretschneider *Hist. European Bot. Discoveries in China* 1898, 384–86. *D.N.B.* v.51, 194–95. *Fl. Malesiana* v.1, 1950, 481. G. A. C. Herklots *Hong Kong Countryside* 1951, 165–66. J. W. Parham *Plants of Fiji Islands* 1964, xvi–xviii portr. *J. Soc. Bibl. Nat. Hist.* 1963, 151–52. F. A. Stafleu *Taxonomic Literature* 1967, 441–42. *Contrib. Univ. Michigan Herb.* 1972, 303–4. *R.S.C.* v.5, 622; v.8, 926; v.12, 671.

Plants at BM(NH), Kew. Letters and MS. Journal at Kew. Portr. at Hunt Library.

Seemannia Regel.

SELBY, Prideaux John (1788–1867)
b. Alnwick, Northumberland 23 July 1788 *d.* Twizell, Northumberland 27 March 1867

MA Durham 1839. FLS 1826. High Sheriff, Northumberland, 1823. Joint-founder of *Mag. Zool. Bot.* 1837. *British Forest Trees* 1842.

Ipswich Mus. Portr. Ser. 1852. *Proc. Linn. Soc.* 1866–67, xxxvii–xxxviii. L. Blomefield *Reminiscences of Prideaux John Selby* 1885. *D.N.B.* v.51, 210–11.

Letters in Winch correspondence at Linnean Society. Sale at Sotheby, 27 June 1894.

SELIGMAN, Charles Gabriel (1873–1940)
b. London 24 Dec. 1873 *d.* Oxford 19 Sept. 1940

MD London 1905. FRS 1919. Visited New Guinea, 1898–99, 1904; Ceylon, 1907–8 and Africa. Professor of Ethnology, London, 1913–34.

Obit. Fellows R. Soc. 1941, 627–46 portr. *Who was Who, 1929–1940* 1215. *D.N.B. 1931–1940* 802. *Fl. Malesiana* v.1, 1950, 482; v.5, 1958, 86.

New Guinea plants at Kew.

SELLER, William (1798–1869)
b. Peterhead, Aberdeenshire 1798 *d.* Edinburgh 11 April 1869

MD Edinburgh 1821. President, Botanical Society of Edinburgh, 1857. 'Nutrition of Plants' (*Edinburgh New Philos. J.* v.39, 1845, 50–68). 'Plants from Davis Strait' (*Trans. Bot. Soc. Edinburgh* v.2, 1846, 215–23).

Trans. Bot. Soc. Edinburgh v.10, 1869, 202–3. *Proc. R. Soc. Edinburgh* v.7, 1869–72, 26–30. *R.S.C.* v.5, 634; v.8, 931.

Portr. at Royal College of Physicians, Edinburgh.

SEMPLE, Charles Edward Armand (1845–1895)
d. 5 March 1895

BA Cantab 1867. MB 1873. *Aids to Botany* 1877.

"SENILIS" *see* Nelson, J.

SENIOR, Louisa *see* Lawrence, L.

SERVICE, James (*c.* 1824–1901)
b. Ayrshire *c.* 1824 *d.* 15 Oct. 1901

Nurseryman, Corberry and Janefield Nurseries, Maxwelltown, Dumfries.

Garden v.60, 1901, 278. *Gdnrs Chron.* 1901 ii 314. *Gdnrs Mag.* 1901, 715. *J. Hort. Cottage Gdnr* v.43, 1901, 376.

SERVICE, Robert (–1911)
d. 8 May 1911

Nurseryman, Corberry and Janefield Nurseries, Maxwelltown, Dumfries.

J. Hort. Home Farmer v.62, 1911, 426.

SESSIONS, F. (1836–1920)

In business in Gloucester. Had herb., now at Gloucester Museum.

H. J. Riddelsdell *Fl. Gloucestershire* 1948, cxxiii.

SEWARD, Sir Albert Charles (1863–1941)
b. Lancaster, Lancs 9 Oct. 1863 *d.* Oxford 11 April 1941

Educ. Cambridge. FRS 1898. FLS 1901. Knighted 1936. Lecturer in Botany, Cambridge, 1890; Professor, 1906–36. Palaeobotanist. President, Yorkshire Naturalists Union, 1910. *Fossil Plants as Tests of Climate* 1892. *Wealden Fl.* 1894–95 2 vols. *Fossil Plants* 1898–1919 4 vols. *A Summer in Greenland* 1921. *Plant Life through the Ages* 1931; ed. 2 1933.

Nature v.58, 1898, 34; v.142, 1938, 386–87; v.147, 1941, 667–68. *New Phytol.* 1941, 161–64. *Obit. Notices Fellows R. Soc.* 1941, 867–80 portr. *N. Western Nat.* 1941, 209–10 portr. *Pharm. J.* 1941, 186. *Proc. Geol. Soc. London* 1941, lxxviii–lxxxi. *Proc. Linn. Soc.* 1940–41, 299–300. *Times* 12 April 1941. *Chronica Botanica* 1942–43, 40–41 portr. *Berichten Deutschen Bot. Gesellschaft* 1955, 101–4 portr. *D.N.B. 1941–1950* 771–73. *Who was Who, 1941–1950* 1040. C. C. Gillispie *Dict. Sci. Biogr.* v.12, 1975, 339–40.

Portr. by J. Gunn at Downing College, Cambridge. Portr. at Hunt Library.

SEWARD, Ebenezer (–1893)
d. 21 July 1893

Florist of Godmanchester, Hunts. Raised pyrethrum 'Golden Feather'.

Gdnrs Chron. 1893 ii 137.

SEWARD, John (*fl.* 1790s)

MD. ALS 1796. Of Worcester. Discovered *Hypericum dubium.*

J. Sowerby and J. E. Smith *English Bot.* 296.

SEWELL, Philip (1865–1928)
b. Malton, Yorks 2 Feb. 1865 *d.* York 29 May 1928

Tutor to Hanbury children at La Mortala. 'Fl. Coasts of Lapland and of Yugor Straits (N.W. Siberia), as observed during Voyage of the 'Labrador' in 1888' (*Trans. Proc. Bot. Soc. Edinburgh* v.17, 1889, 444–81).

Herb. at BM(NH).

SEWELL, S. Arthur (*fl.* 1900s)

Of Buckhurst Hill, Essex.

Essex Nat. v.13, 1903, 90; v.29, 1954, 190.

Plants at Essex Field Club.

SEXBY, John James (*fl.* 1900s)

Superintendent of Parks and Gardens, London County Council. *Municipal Parks, Gardens and Open Spaces of London* 1898.

Gdnrs Mag. 1907, 403 portr., 404.

SEYMER, Henry (1745–1800)
b. Hanford, Dorset 1745 *d.* 3 Dec. 1800

BCL Oxon 1771. DCL 1777. Of Hanford. Step-uncle of A. B. Lambert. Friend of R. Pulteney and D. C. Solander. Had garden of exotics.

J. Nichols *Lit. Anecdotes of Eighteenth Century* v.8, 201–2. J. Hutchin *Hist. Dorset* v.4, 1873, 66–67. *Proc. Linn. Soc.* 1888–89, 40.

Portr. at Linnean Society.

Seymeria Pursh.

SEYMOUR, Capt. (*fl.* 1830s)

"Capt. Seymour of the Royal Navy, to whose kindness and zeal for horticulture we are much indebted for many new and good plants" (G. B. Knowles and F. Westcott *Floral Cabinet* v.2, 1838, 66).

SHACKLETON, Lydia (1828–1914)
b. Ballitore, Kildare 22 Nov. 1828 *d.* Rathgar, County Dublin 10 Dec. 1914

Botanical artist. Drawings at BM(NH).

SHADWELL, Charles Lancelot (1840–1919)
b. 16 Dec. 1840 *d.* Oxford Feb. 1919

Provost, Oriel College, Oxford, 1905. Curator, Botanic Garden, Oxford. Botanised in Berkshire.

Bot. Soc. Exch. Club Br. Isl. Rep. 1919, 625.

SHAILER, Anthony (*fl.* 1780s)

Nurseryman, Little Chelsea, Kensington, London.

SHAILER, Francis (*fl.* 1800s)

Nurseryman, Little Chelsea, Kensington, London. Brother of Henry Shailer senior.

SHAILER, Henry (*fl.* 1770s–1820s)

Nurseryman, Little Chelsea, Kensington, London. Specialised in roses.

Trans. London Middlesex Archaeol. Soc. v.24, 1973, 182. E. J. Willson *Hist. Hammersmith* 1965, 94.

SHAILER, James (*fl.* 1780s–1810s)

Nurseryman, Little Chelsea, Kensington, London.

SHAKESPEAR, Roger (*fl.* 1770s–1780s)

Sent plants from Jamaica and N. and S. America to J. Banks.

I. Urban *Symbolae Antillanae* v.3, 1902, 126.

SHAKESPEAR, William Henry (–1915)

Indian Army, 1898. Political service, 1903. Political agent at Kuwait where he collected plants now at BM(NH).

SHAND, William (*c.* 1838–1903)
b. Pittenkerrie, Kincardineshire *c.* 1838 *d.* Lancaster 23 Sept. 1903

Nurseryman, New Street, Lancaster, 1885.

Gdnrs Chron. 1903 ii 247.

SHANKS, Archibald (1870–1951)
b. Gourock, Renfrewshire Sept. 1870 *d.* Dalry, Ayrshire 1 May 1951

Analytical chemist. Discovered *Senecio ernisfolius* in Clyde area.

Glasgow Nat. v.17, 1951, 63–64.

SHANKS, James (1854–1912)
b. 4 Nov. 1854 *d.* Ballyfounder, N. Ireland 2 Nov. 1912

Farmer. Studied botany and geology of Little Ards, County Down.

Annual Rep. Proc. Belfast Nat. Field Club 1911–12, 628. *Irish Nat.* 1913, 33.

SHANNE, Richard (1561–1627)
Of Woodrowe near Methley, Yorks where he created a notable garden.
R. T. Gunther *Early English Botanists* 1922, 264–65, 310–12.

SHARLAND, Arthur (1849–1917)
MD. Spent many holidays plant collecting in N. Devon.
W. K. Martin and G. T. Fraser *Fl. Devon* 1939, 778.
Herb. at N. Devon Athenaeum, Barnstaple.

SHARLAND, Frances Sara (*née* **Schaw**) (1813–1859)
b. Jamaica 1813 *d.* George Town, Tasmania 1859
To Tasmania with her parents, 1833. Collected algae at mouth of River Tamar.
Papers Proc. R. Soc. Tasmania 1909, 26. *Austral. Dict. Biogr.* v.2, 1967, 421, 437. P. L. Brown, *ed. Clyde Company Papers* v.1, 1941, 209–10; v.5, 1963, 402–3.

SHARPE, Sir Alfred (1853–1935)
b. Lancaster 19 May 1853 *d.* 10 Dec. 1935
Solicitor. Colonial appointment, Fiji. Joined African Lake Corporation's Defence Force. Vice-Consul, British Central Africa, 1891; Consul, 1894; Deputy Commissioner, 1896; Commissioner, 1897. Governor, Nyasaland, 1907–10. Collected plants in Nyasaland, now at Kew.
Who was Who, 1929–1940 1223. *Comptes Rendus AETFAT 1960* 1962, 170–71.

SHARPE, Charles (*c.* 1830–1897)
d. Sleaford, Lincs 8 March 1897
Seed merchant, Sleaford.
Gdnrs Mag. 1897, 161 portr.

SHARPE, Daniel (1806–1856)
b. Marylebone, London 6 April 1806 *d.* London 31 May 1856
FRS 1850. FLS 1828. Geologist. Merchant in Portugal where he resided, 1835–38; collected plants.
Proc. R. Soc. v.8, 1856–57, 275–79. *Quart. J. Geol. Soc. London* 1857, xlv–lxiv. *D.N.B.* v.51, 421–22. F. Darwin and A. C. Seward *More Letters of Charles Darwin* v.2, 1903, 131.
Portuguese plants at BM(NH).

SHARPLES, Arnold (1887–1937)
b. Great Harwood, Burnley, Lancs 25 Nov. 1887 *d.* St. Anne's-on-Sea, Lancs 6 Aug. 1937
Educ. Royal College of Science, London, 1908–12. Assistant Mycologist, Dept. of Agriculture, Federated Malay States, 1913–16; Mycologist, 1916–30. Head of Pathological Division, Rubber Research Institute of Malaya, 1930–34. *Diseases and Pests of Rubber Tree* 1936. Contrib. to *Ann. Bot.*, *Ann. Applied Biol.*
Kew Bull. 1916, 107; 1937, 443–44. *Malayan Agric. J.* 1937, 390. *Nature* v.140, 1937, 494–95.

SHARROCK, Rev. Robert (1630–1684)
b. Drayton Parslow, Bucks June? 1630 *d.* Bishop Waltham, Hants 11 July 1684
BCL Oxon 1654. DCL 1661. Prebendary of Winchester, 1665; Archdeacon, 1684. Rector, Bishop Waltham. *History of Propagation and Improvement of Vegetables...* 1660 [1659].
J. C. Loudon *Encyclop. Gdning* 1822, 1265. G. W. Johnson *Hist. English Gdning* 1829, 110. *D.N.B.* v.51, 430–31. *Isis* v.51, 1960, 3–8. C. C. Gillispie *Dict. Sci. Biogr.* v.12, 1975, 357.

SHAW, C. W. (1849–1884)
d. 22 Dec. 1884
On editorial staff of *Garden*. Editor of *Gdning Illus.* 1879–84.
Garden v.26, 1884, 550.

SHAW, Frederick John Freshwater (1885–1936)
b. 16 Dec. 1885 *d.* Pusa, India 29 July 1936
BSc London 1909. DSc 1916. FLS 1911. Mycologist, Agricultural Research Institute, Pusa, 1910–29; Economic Botanist and joint Director, 1929–34; Director, 1934–36. Contrib. to *Mem. Dept. Agric. India*.
Kew Bull. 1936, 393. *Nature* v.138, 1936, 317–18. *Proc. Linn. Soc.* 1936–37, 211. *Who was Who, 1929–1940* 1225.

SHAW, George (1751–1813)
b. Bierton, Bucks 10 Dec. 1751 *d.* London 22 July 1813
BA Oxon 1769. MD 1787. FRS 1789. FLS 1788. Assistant Lecturer in Botany, Oxford, 1786. Keeper of natural history collections at BM, 1807–13. Vice-president, Linnean Society, 1788. *Cimelia Physica* 1796. Contrib. to J. Sowerby and J. E. Smith's *English Bot.* v.16–18.
R. J. Thornton *New Illustration of Sexual System of...Linnaeus* 1803 portr. *Gent. Mag.* 1813 ii 290. *Gent. Mag.* 1813 ii 290–92. G. Lipscomb *Hist. Antiq. County of Buckinghamshire* v.2, 1831, 102–3, 273. *D.N.B.* v.51, 436. *Austral. Dict. Biogr.* v.2. 1967, 437.
Portr. at Hunt Library.

SHAW, Henry (1800–1889)
b. Sheffield, Yorks 24 July 1800 *d.* St. Louis, Missouri, U.S.A. 25 Aug. 1889
Merchant. To U.S.A., 1818. Founded Missouri Botanic Garden, 1848 and Shaw School of Botany, Washington. Published G. Engelmann's *Bot. Works* 1887.

Gdnrs Chron. 1890 i 46. *Missouri Bot. Gdn Rep.* 1890, 1–90 portr. *J. Bot.* 1891, 190. A. Gray *Letters* v.2, 1893, 752–53. *Amer. Orchid Soc. Bull.* 1932, 78–81 portr. L. H. Bailey *Standard Cyclop. Hort.* v.2, 1939, 1595–96 portr. *Missouri Bot. Gdn Bull.* v.30, 1942, 100–12; v.55(6), 1967, 1–16 portr.
Letters at Kew. Portr. at Hunt Library.

SHAW, James (*fl.* 1780s–1790s)
Gardener to Lord Mulgrave at Mulgrave Castle, Yorks. *Plans...of Forcing Houses in Gardening* 1794.
Gdnrs Chron. 1963 ii 358; 1965 i 6.

SHAW, John (–1891)
d. S. Africa 1891
PhD. MD. FLS 1873. Of Glasgow. Bryologist. To Colesberg, S. Africa, 1867 as Principal, Dutch Church Gymnasium. Contrib. papers on mosses to *J. Bot.* 1865–66.
Proc. Nat. Hist. Soc. Glasgow v.3, 1891, lxxx. *R.S.C.* v.8, 244; v.12, 677.
S. African plants at Kew. Letters in Wilson correspondence at BM(NH).
Campylopus shawii Wilson.

SHAW, Rev. Thomas (1693–1751)
b. Kendal, Westmorland 4 June 1693 *d.* Oxford 15 Aug. 1751
BA Oxon 1716. DD 1734. FRS 1734. Chaplain at English factory, Algiers, 1720–33. Principal, St. Edmund Hall, 1740. Vicar, Bramley, 1742–51. *Travels...to Several Parts of Barbary and the Levant* 1738 (with plants described by J. J. Dillenius).
R. Pulteney *Hist. Biogr. Sketches of Progress of Bot. in England* v.2, 1790, 173–74. D. Turner *Extracts from Lit. Sci. Correspondence* 1835, 348. *Hooker's J. Bot.* 1854, 250–51. *J. Bot.* 1880, 256. E. S.-C. Cosson *Compendium Fl. Atlanticae* v.1, 1881, 7–8. *D.N.B.* v.51, 446–47. E. J. L. Scott *Index to Sloane Manuscripts* 1904, 491.
Plants at Kew, Oxford. Portr. at Hunt Library.
Shawia Forster.

SHAWE, Rev. J. (*fl.* 1880s)
Correspondent of H. C. Watson. Botanised in Herefordshire.
Trans. Woolhope Nat. Field Club v.34, 1954, 254–55.

SHAWYER, George E. (*c.* 1864–1943)
d. 11 Aug. 1943
VMH 1933. Had nursery in partnership with E. J. Lowe at Uxbridge, Middx.
Gdnrs Chron. 1937 i 282 portr.; 1943 i 72.

SHEA, Charles E. (*fl.* 1860s–1900s)
Of Foots Cray, Kent. Raised chrysanthemums. President, National Chrysanthemum Society.
J. Hort. Cottage Gdnr v.30, 1895, 140 portr. *Garden* v.68, 1905, 74–75 portr.

SHEARER, George (*fl.* 1860s–1870s)
MD. Of Liverpool. Physician, Han-Kow Hospital, China, 1868. Collected plants at Kiu-Kiang, 1873.
J. Bot. 1875, 199–202, 225–31. E. Bretschneider *Hist. European Bot. Discoveries in China* 1898, 699–701.
Plants at Kew.
Sheareria S. Moore.

SHEARER, Johnstone (1827–1916)
b. Aberdeen 1827 *d.* Glasgow 11 April 1916
Photographer. 'Fl. Stirling' (*Ann. Andersonian Nat. Soc. Glasgow* 1893, 66–70). Contrib. to *Trans. Nat. Hist. Soc. Glasgow*.
Glasgow Nat. v.8, 1916, 33–35.

SHEARMAN, Edward James (1798–1878)
b. Wrington, Somerset 1798 *d.* Rotherham, Yorks 2 Oct. 1878
MD Jena. FRSE. Physician. *Essay on Properties of Animal and Vegetable Life* 1845.
J. R. Microsc. Soc. 1879, 334.

SHEBBEARE, Edward Oswald (1884–1964)
b. 8 March 1884 *d.* 11 Aug. 1964
Indian Forest Service, 1906. Chief Conservator of Forests, United Provinces until retirement in 1938. *Trees of the Duars and Terai* 1957.
Indian Forester 1964, 792–93 portr. *Who was Who, 1961–1970* 1025.

SHEETING, Richard (*fl.* 1780s)
Nurseryman, Windsor, Berks.

SHEFFIELD, Rev. William (*c.* 1732–1795)
b. Henley, Warwickshire *c.* 1732 *d.* Oxford 23 June 1795
BA Oxon 1754. DD 1778. Keeper, Ashmolean Museum, Oxford, 1772–95. Provost, Worcester College. Friend of J. Banks and G. White.
J. Nichols *Illus. Lit. Hist. of Eighteenth Century* v.5, 517. G. C. Druce *Fl. Berkshire* 1897, cxxxvii. G. C. Druce *Fl. Oxfordshire* 1927, xc. R. Holt-White *Life and Letters of G. White* v.1, 1901, 187–89, 210–12.
Sheffieldia Forster.

SHEILDS (*recte*) **(Shiells), James**
(*fl.* 1760s–1780s)
Acquired nursery of R. North, Lambeth. In 1776 in partnership with John Cowie, 21 Parliament Street, Westminster; with Walter Hay by 1780.
Trans. London Middlesex Archaeol. Soc. v.24, 1973, 187–88.

SHELDON, John Joseph (1858–1942)
d. 14 Feb. 1942
FLS 1931. Collected and grew British ferns. Treasurer, British Pteridological Society.
Proc. Linn. Soc. 1941–42, 295.

SHELDRAKE, Timothy (*fl.* 1730s–1750s)
MD. Of Norwich. Moved to London. *Gardener's Best Companion in the Greenhouse* 1756. *Botanicum Medicinale* 1759.
D.N.B. v.52, 27. E. J. L. Scott *Index to Sloane Manuscripts* 1904, 492.

SHELFORD, Robert Walter Campbell (1872–1912)
b. Singapore 1872 *d.* Oxford 1912
Educ. Cambridge. Zoologist and anthropologist. Curator, Sarawak Museum, 1897. Assistant Curator, Oxford Museum, 1905. *Naturalist in Borneo* 1916. 'Trip to Mount Penrissen' (*J. Straits Branch R. Asiatic Soc.* 1900, 1–26).
Fl. Malesiana v.1, 1950, 484.
Plants at Sarawak Museum.

SHENSTONE, James Chapman (1854–1935)
d. 8 May 1935
Had herb. now at Essex Field Club. Contrib. to *Essex Nat.*
Essex Nat. v.11, 1900, 223–24; v.25, 1935, 64; v.29, 1954, 190. S. T. Jermyn *Fl. Essex* 1974, 18–19. D. H. Kent *Hist. Fl. Middlesex* 1975, 26.

SHEPHERD, Henry (*c.* 1783–1858)
d. 14 Jan. 1858
FLS 1827. Nephew of J. Shepherd (*c.* 1764–1836) whom he succeeded in 1836 as Curator, Liverpool Botanic Garden. First to grow ferns from spores (*Trans. Hort. Soc. London* v.3, 1822, 338–41). Prepared dissections for W. Roscoe's *Monandrian Plants* 1824–29.
P. W. Watson *Dendrologia Britannica* v.1, 1825, lxvii. *Naturalist* 1838–39, 398–99. T. B. Hall *Fl. Liverpool* 1839, vii–viii. J. B. L. Warren *Fl. Cheshire* 1899, lxxxviii. Liverpool Public Museums *Handbook and Guide to Herb. Collections* 1935, 32. *Proc. Bot. Soc. Br. Isles* 1968, 170–71.
Plants at Liverpool Museums. Letters at Kew.

SHEPHERD, John (*c.* 1764–1836)
b. Gosforth, Cumberland *c.* 1764 *d.* Liverpool 27 Sept. 1836
First Curator, Liverpool Botanic Garden, 1803–36. Friend of W. Roscoe, J. E. Smith, T. Nuttall. Discovered *Erythraea latifolia* 1803. Contrib. plant records to W. Withering's *Systematic Arrangement of Br. Plants* ed. 5 1812. *Catalogue of Plants in Botanic Garden at Liverpool* 1808.
T. Nuttall *Genera of N. American Plants* v.2, 1818, 240–41. *Gdnrs Mag.* 1836, 724. *Naturalist* v.4, 1838–39, 395. E. Bretschneider *Hist. European Bot. Discoveries in China* 1898, 223. J. B. L. Warren *Fl. Cheshire* 1899, lxxxviii. *Trans. Liverpool Soc. Bot.* 1909, 89. *Lancs and Cheshire Nat.* v.12, 1919–20, 171–73, 217–23; v.17, 1924–25, 126–29, 157–60, 198–200. Liverpool Public Museums *Handbook and Guide to Herb. Collections* 1935, 33–35 portr. *N. Western Nat.* 1938, 10–15 portr. *Huntia* 1965, 208 portr.
Plants at Liverpool Museums. Portr. at Walker Art Gallery, Liverpool and Hunt Library.
Shepherdia Nuttall.

SHEPHERD, Joseph (1807–1859)
d. Sowerby, Yorks 7 June 1859
A founder and President of Todmorden Botanical Society. *List of Plants of Halifax* 1836. Contrib. to J. B. Wood's *Fl. Mancuniensis* 1840 and to completion of H. Baines's *Fl. Yorkshire* 1840.

SHEPHERD, Thomas (1780–1835)
b. Balcarres, Colinsburg, Fifeshire 1780 *d.* Sydney, N.S.W. 30 Aug. 1835
Nurseryman, Hackney, London. In Sydney, 1827 where he established Darling Nursery which was carried on by his sons. *Lectures on Horticulture of New South Wales* 1835. *Lectures on Landscape Gardening in Australia* 1836.
Gdnrs Mag. 1837, 584. *J. Proc. R. Soc. N.S.W.* 1908, 118. *Camellia News* no. 51, 1973, 28–31.

SHEPHERD, Thomas William (1824–1884)
b. Hackney, London 11 March 1824 *d.* Ashfield, Sydney, N.S.W. 27 Aug. 1884
Emigrated with his father, T. Shepherd (1780–1835) to Australia. Employed in his father's nursery in Sydney. Sent plants which he collected in N.S.W. to Rev. W. Woolls and F. von Mueller.
F. von Mueller *Fragmenta Phytographiae Australiae* v.1, 1858, 190. *Sydney Morning Herald* 29 Aug. 1884. *J. R. Soc. N.S.W.* 1908, 119.
Bulbophyllum shepherdi F. Muell.

SHEPPARD, Alfred William (1861–1938)
b. Camberwell, London 26 Nov. 1861 *d.* Dorking, Surrey 8 Oct. 1938
FLS 1919. Clerk with London publishers. Mainly interested in microscopic botany. Editor, *J. Quekett Microsc. Club*, 1910–24.
Times 11 Oct. 1938. *J. Quekett Microsc. Club* 1939, 128–30 portr. *Proc. Linn. Soc.* 1938–39, 256.

SHEPPARD, John (1785–1879)
b. Frome, Somerset 16 Oct. 1785 *d.* Frome 30 April 1879

Woollen trader. *On Trees, their Uses and Biography* 1848.

SHEPPARD, S. (*fl.* 1830s)

Nurseryman, Union Street, Bedford.

SHERARD, James (1666–1738)
b. Bushby, Leics 1 July 1666 *d.* 12 Feb. 1738

MD Oxon 1731. FRCP 1732. FRS 1706. Younger brother of W. Sherard (1658/9–1728). Apothecary. Had garden at Eltham, Kent noted for rare plants. Persuaded J. J. Dillenius from 1724 to devote himself to *Hortus Elthamensis* 1732.

Gent. Mag. 1738, 109; 1796 ii 810–11. R. Pulteney *Hist. Biogr. Sketches of Progress of Bot. in England* v.2, 1790, 150–52. J. Nichols *Illus. Lit. Hist. of Eighteenth Century* v.1, 403–16. J. Nichols *Lit. Anecdotes of Eighteenth Century* v.3, 651–52. W. Munk *Roll of R. College of Physicians* v.2, 1878, 127–28. G. C. Druce *Fl. Berkshire* 1897, cxxxi. *D.N.B.* v.52, 66–67. E. J. L. Scott *Index to Sloane Manuscripts* 1904, 492. G. C. Druce *Fl. Oxfordshire* 1927, lxxxvi. J. E. Dandy *Sloane Herb.* 1958, 202.

Portr. at Hunt Library. Library sold by T. Osborne, 1766.

SHERARD, William (*né* **Sherwood**) (1658/9–1728)
b. Bushby, Leics 27 Feb. 1658/9 *d.* London 11 Aug. 1728

BCL Oxon 1683. DCL 1694. FRS 1720. Pupil of Tournefort. Consul at Smyrna, 1703–16. Founded Sherardian Chair of Botany at Oxford, bequeathing library, herb. and MS. *Pinax* in 5 vols. Brought J. J. Dillenius to England. Visited Ireland, Cornwall and Jersey (J. Ray *Synopsis Methodica* 1690 appendix 237–39; *J. Bot.* 1916, 335–36). *Schola Botanica* (under pseudonym S.W.A.) 1689. Edited P. Hermann's *Paradisus Batavus* 1698.

R. Pulteney *Hist. Biogr. Sketches of Progress of Bot. in England* v.2, 1790, 141–50. J. Nichols *Illus. Lit. Hist. of Eighteenth Century* v.1, 339–403. *Gent. Mag.* 1796 ii 811. G. C. Gorham *Mem. of John and Thomas Martyn* 1830, 11–12. *Hooker's J. Bot.* 1854, 248–49. *J. Bot.* 1874, 129–38; 1914, 322; 1916, 335–36. R. H. Semple *Mem. of Bot. Gdn at Chelsea* 1878, 48–50. W. Munk *Roll of R. College of Physicians* v.2, 1878, 127. *J. Linn. Soc.* v.24, 1887, 129–55. G. C. Druce *Fl. Berkshire* 1897, cxxvii–cxxx. *D.N.B.* v.52, 67–68. G. C. Druce and S. H. Vines *Dillenian Herb.* 1907, xxix–xxxi. *Bot. Soc. Exch. Club Br. Isl. Rep.* 1923, 350–51. G. C. Druce *Fl. Oxfordshire* 1927, lxxxiii–lxxxvi. G. Pasti *Consul Sherard* 1950 (unpublished thesis, University of Illinois). J. E. Dandy *Sloane Herb.* 1958, 202–3. G. A. Lindeboom *Boerhaave's Correspondence* Pt. 1, 1962, 38–162. H. N. Clokie *Account of Herb. Dept. Bot., Oxford* 1964, 17–30, 58–81. J. and N. Ewan *John Banister* 1970, 11–17. *Janus* v.58, 1971, 37–61. C. C. Gillispie *Dict. Sci. Biogr.* v.12, 1975, 394.

Plants at BM(NH), Oxford. Letters at Royal Society, Zurich Central Library. Portr. at Hunt Library. Library at Oxford.

SHERARE, James (*c.* 1809–1834)
d. Hampstead, London 3 Nov. 1834

Gardener to Sir J. Hay at Kings Meadows, Notts. Contrib. to *Gdnrs Mag.*

Gdnrs Mag. 1835, 56.

SHERBOURNE, Margaret Dorothea (*née* **Willis**) (1791–1846)
b. Prescot, Lancs 3 Oct. 1791 *d.* Prescot 6 Nov. 1846

Of Hurst House, Prescot. "A zealous and successful collector of plants" (*Bot. Mag.* v.69, 1843, Dedication). First to flower *Sherbournia foliosa* in England.

Curtis's Bot. Mag. Dedications, 1827–1927 63–64 portr.

Letters at Kew.

SHERRARD, G. O. (*c.* 1883–1959)
d. Sept. 1959

Educ. Royal College of Science, Dublin. Studied plant breeding under W. Bateson at John Innes Horticultural Institution. Lecturer in Horticulture, Royal College of Science, Dublin. University College, Dublin, 1926; Professor of Horticulture, 1940. President, Horticultural Educational Association, 1953. President, Royal Horticultural Society of Ireland, 1955–58.

Nature v.184, 1959, 1531.

SHERRATT, John (*c.* 1829–1882)
d. Biddulph, Staffs 20 June 1882

Gardener to J. Bateman at Biddulph Grange. Partner in Knypersley Nursery of Sherratt and Pointon, Biddulph.

Gdnrs Chron. 1882 ii 25.

SHERRIFF, George (1898–1967)
b. Larbert, Stirlingshire 3 May 1898 *d.* Kirriemuir, Angus 19 Sept. 1967

VMH 1953. Explorer and plant collector. British Vice-Consul, Kashgar, Chinese Turkestan, 1928–32. With F. Ludlow collected plants mainly in Bhutan and S.E. Tibet, 1933–49.

Bot. Mag. 1962–63 portr. *Taxon* 1967, 573. *Times* 22 Sept. 1967. *Geogr. J.* 1968, 175–76. *J. R. Hort. Soc.* 1968, 11–19 portr. A. M. Coats *Quest for Plants* 1969, 198–201. H. R. Fletcher *A Quest of Flowers* 1975 portr.

Portr. at Hunt Library.

SHERRIN, William Robert (1871–1955)
b. Twickenham, Middx 20 May 1871 *d.* E. Dulwich, London 22 March 1955

ALS 1919. Had taxidermist's shop at Ramsgate, Kent. Articulator, Zoology Dept., BM(NH), 1895–1928; in charge of moss collection, Botany Dept., 1928–47. Curator, South London Botanical Institute, 1919. President, British Bryological Society, 1945–47. *Illustrated Handbook of British Sphagna* 1927.

Kew Bull. 1948, 57. *Nature* v.175, 1955, 750–51. *Proc. Bot. Soc. Br. Isl.* 1955, 553–56. *Proc. Linn. Soc.* 1954–55, 139–41. *Br. Fern Gaz.* 1955, 134. *Times* 23 March 1955. *Trans. Br. Bryol. Soc.* 1956, 132–35.

Plants at BM(NH), South London Botanical Institute.

SHERRING, Richard Vowell (1847–1931)
b. Bristol 23 Dec. 1847 *d.* Hallatrow near Bristol 4 Feb. 1931

FLS 1886. Collected plants in West Indies and British Guiana, 1873, 1890–91. Vice-President, Bournemouth Natural Science Society. Recorded *Spartina townsendii* in Poole Harbour. Friend of G. S. Jenman.

Rep. Br. Assoc. Advancement Sci. 1890, 447; 1891, 356–57; 1892, 354. *Ann. Bot.* 1892, 95–102. I. Urban *Symbolae Antillanae* v.3, 1902, 126. *J. Bot.* 1931, 104–5. *Proc. Bournemouth Nat. Sci. Soc.* v.23, 1930–31, 64–65 portr. *Proc. Linn. Soc.* 1930–31, 196–98.

Plants at BM(NH), Kew, Cambridge, Oxford.

Hemitelia sherringi Jenman.

SHERWOOD, John (*c.* 1806–1883)
b. Scotland *c.* 1806

Nurseryman, Philadelphia, U.S.A.; specialised in roses and camellias.

J. Hort. Cottage Gdnr v.53, 1906, 546.

SHERWOOD, Nathaniel Newman (1846–1916)
b. France April 1846 *d.* Kelvedon, Essex 20 July 1916

VMH 1897. Proprietor, Hurst and McMullen, seed merchants, 1890. Master, Gardeners' Company, 1896–98. President, National Sweet Pea Society, 1910.

J. Hort. Cottage Gdnr v.26, 1893, 7 portr.; v.61, 1910, 54–55 portr. *Gdnrs Mag.* 1910, 61–62 portr.; 1916, 335 portr. *Garden* 1912, iv portr.; 1916, 376 portr. *Gdnrs Chron.* 1916 ii 54 portr.

SHIELDS, Andrew (*fl.* 1780s–1805)

Seedsman, Brentford, Middx.

SHIELLS, James *see* Sheilds, J.

SHILL, J. E. (–1940)
d. Sept. 1940

Gardener to Baron B. Schröder at Dell Park, Englefield Green, Surrey. Grew orchids.

Orchid Rev. 1941, 136.

SHILLING, John (*fl.* 1830s)

Nurseryman, North Warnborough, Odiham, Hants.

SHILLITO, James (*fl.* 1850s–1899)
d. Australia 1899

Of Prestwick, Ayr. Assistant, Botanic Garden, Liverpool, *c.* 1850–57. Contrib. to J. Dickinson's *Fl. Liverpool* 1851 and *Supplement* 1855. Bought Henry Shepherd's herb. Emigrated to Australia, 1857.

J. B. L. Warren *Fl. Cheshire* 1899, lxxxviii. *Trans. Liverpool Bot. Soc.* 1909, 89. Liverpool Public Museums *Handbook and Guide to Herb. Collections* 1935, 30–31. *Proc. Bot. Soc. Br. Isl.* 1968, 170.

Plants at Liverpool Museums.

SHINGLER, William (*c.* 1842–1925)
b. Shropshire *c.* 1842 *d.* 29 June 1925

Gardener to Lord Hastings at Melton Constable Park, Norfolk.

Gdnrs Chron. 1925 ii 60.

SHINN, Mrs. Arthur (*née* Glover) (*c.* 1893–1938)
d. Mlanje, Nyasaland Aug. 1938

Nursing sister, Universities Mission. In Nyasaland, *c.* 1894. Collected plants in Mlanje district, 1913.

Comptes Rendus AETFAT 1960 1962, 171–72.

Plants at BM(NH).

SHIRLEY, John F. (1849–1922)
b. Dorchester 11 Aug. 1849 *d.* 5 April 1922

BSc London. Teacher in Birmingham. Headmaster, Roma, Queensland, 1878. Inspector of Schools, 1879; Senior Inspector, 1879. Principal Teacher, Training College, Brisbane, 1914–19. President, Royal Society of Queensland. Collected plants extensively in Queensland. Contrib. additions to lichen fl. of Queensland to *Proc. R. Soc. Queensland. Additions to Fossil Fl. Queensland* 1898.

Proc. Linn. Soc. N.S.W. 1900–1, 785; 1923, vii–viii. *Queensland Nat.* v.3, 1922, 87–88. *Proc. R. Soc. Queensland* 1922, 2–3; 1959, 94–95. *Austral. Encyclop.* v.8, 1965, 123.

SHOESMITH, H. (–1967)

VMH 1953. Raised chrysanthemums in his garden at Woking, Surrey.

Gdnrs Chron. 1954 ii 32 portr.

SHOOLBRED, William Andrew (1852–1928)
b. Wolverhampton, Staffs 11 March 1852 *d.* Chepstow, Mon 25 Jan. 1928

MRCS. FLS 1902. Surgeon, Chepstow. Botanised with Rev. E. S. Marshall in Scotland and Ireland, 1894–1916. *Fl. Chepstow* 1920.

Gdnrs Chron. 1928 i 91. *J. Bot.* 1928, 118. *Proc. Linn. Soc.* 1927–28, 137. *Watson Bot. Exch. Club Rep.* 1927–28, 416. *Bot. Soc. Exch. Club Br. Isl. Rep.* 1928, 710–11. *N. Western Nat.* 1931, 137. H. A. Hyde and A. B. Wade *Welsh Flowering Plants* 1934, 3 portr. H. J. Riddelsdell *Fl. Gloucestershire* 1948, cxl–cxlii. R. E. Wade *Fl. Monmouthshire* 1970, 15.

Herb. at National Museum of Wales.

Hieracium shoolbredii E. S. Marshall.

SHORE, Harriet (–1828)

Had garden at Norton between Chesterfield and Sheffield. "An investigator and collector of plants" (J. Stokes *Commentaries* 1830).

SHORE, Margaret Emily (1819–1839)

b. Bury St. Edmunds, Suffolk 25 Dec. 1819 *d.* Madeira 7 July 1839

Numerous notes on plants in *Journal of Emily Shore* 1891 portr.

SHORE, Rev. T. W. (1756–1822)

Of Otterton, Devon. Probably assisted in preparation of J. P. Jones and J. F. Kingston's *Fl. Devoniensis* 1829.

Rep. Trans. Devonshire Assoc. Advancement Sci., Literature and Art 1954, 284.

SHORE, Thomas William (1861–1947)

b. Churcham, Glos 5 Nov. 1861 *d.* 19 Feb. 1947

BSc London 1880. MD London 1883. Lecturer in Biology at St. Bartholomew's Hospital, 1885–1925; Dean, 1905. *Elementary Vegetable Biology* 1887.

Who was Who, 1941–1950 1052.

SHORT, Thomas (*c.* 1690–1772)

b. Edinburgh *c.* 1690 *d.* Rotherham, Yorks 28 Nov. 1772

MD Edinburgh. Of Sheffield and, from 1762, of Rotherham. *Technical Words in Botany* 1731. *Medicina Britannica* 1746.

J. Nichols *Lit. Anecdotes of Eighteenth Century* v.1, 451. *Gent. Mag.* 1807 i 404; 1807 ii 823; 1808 ii 773. E. J. L. Scott *Index to Sloane Manuscripts* 1904, 494. *D.N.B.* v.52, 154–55.

SHORT, Thomas Kier (*fl.* 1830s)

Of Martin Hall, Notts. Collected plants in New Zealand from 1836 and "in all quarters of the world". Contrib. to *Naturalist* 1838.

SHORTER, Mrs. Clement *see* Long, A. D.

SHOVE, Rosamund F. (*c.* 1878–1954)

d. Richmond, Surrey 17 Oct. 1954

Educ. Cambridge 1896. Lecturer in Biology, Maria Grey Training College, Isleworth, Middx, 1921–38. Edited *School Nat. Study*, 1940–53.

Nature v.174, 1954, 995–96. *School Nat. Study* v.50, 1955, 1–2.

SHRIMPTON, John (1840–1899)

b. Norwood, Middx 1840 *d.* Hanwell, Middx 23 Oct. 1899

Gardener to W. Seward at Hanwell. Raised chrysanthemums.

Garden v.56, 1899, 408. *Gdnrs Chron.* 1899 ii 381. *Gdnrs Mag.* 1899, 740.

SHRUBBS, Arthur Sidney (1858–1922)

b. Cambridge 8 Oct. 1858 *d.* Hinxton, Cambridge 28 Oct. 1922

Museum Assistant, Botany School, Cambridge, 1870–1922. 'Records of Autumnal or Second Flowering of Plants' (with F. Darwin) (*New Phytol.* 1922, 48).

Cambridge Univ. Rep. 1922–23, 432.

Plants at Cambridge.

SHURLY, Ernest William (1888–1963)

b. Hampstead, London 17 Aug. 1888 *d.* St. Albans, Herts 11 Nov. 1963

FLS 1955. Studied *Mammillaria*. Founder member and President, Cactus and Succulent Society of Great Britain, 1956. President, Mammillaria Society.

Cactus Succulent J. Great Britain 1964, 1.

Portr. at Hunt Library.

SHUTER, James (–*c.* 1827)

MD. FLS 1819. Government naturalist, Madras.

Bot. Mag. 1834, t.3302. R. Wight and G. A. W. Arnott *Prodromus* v.1, 1834, 207. *J. Bot.* (Hooker) 1841, 157–58.

Plants at Kew.

Shuteria Wight.

SHUTTLEWORTH, Edward (1829–1909)

d. Putney, London 5 March 1909

Collected orchids in Colombia for William Bull, 1873. In partnership with J. Carder, afterwards with J. Charlesworth.

Gdnrs Chron. 1909 i 176. *Orchid Rev.* 1909, 125.

Masdevallia shuttleworthii Reichenb. f.

SHUTTLEWORTH, Robert James (1810–1874)

b. Dawlish, Devon Feb. 1810 *d.* Hyères, France 19 April 1874

Educ. Geneva and Edinburgh where he studied medicine, 1830–32. FLS 1856. Conchologist and botanist. In Berne, Switzerland, 1833–66; Hyères, 1866–74. Correspondent of Meisner. Collected plants in Ireland, 1831. *Notitiae Malacologicae* ed. 2 1878 (with biogr.).

C. H. Godet *Fl. Jura* v.1, 1852, iv. *Verh. Schweiz. Nat. Ges.* v.57, 1873–74, 5–6. *Trans. Bot. Soc. Edinburgh* v.12, 1874, 203–4. *J. Bot.* 1878, 179–80. *Bull. Soc. Bot. France* v.30, 1883, cxxxi. *D.N.B.* v.52, 176–77. *Trans. Liverpool Bot. Soc.* 1909, 89–90. *Boissiera* fasc. 5, 1941, 87–88.

Plants at BM(NH). Letters at Kew.

Shuttleworthia Meisner.

SIBBALD, Sir John (1833–1905)
b. Edinburgh 24 June 1833 *d.* 20 April 1905

Commissioner of Lunacy in Scotland, 1878–99. Fellow, Botanical Society of Edinburgh, 1851–1905. Wrote on *Volvox.*

Trans. Bot. Soc. Edinburgh v.23, 1906, 211. *Who was Who, 1897–1916* 648.

SIBBALD, Sir Robert (1641–1722)
b. Edinburgh 15 April 1641 *d.* Edinburgh Aug. 1722

MD Leyden 1661. Knighted 1682. First Professor of Medicine, Edinburgh, 1685. With A. Balfour instituted Botanical Garden at Edinburgh. *Scotia Illustrata* pt. 2, 1684.

Gent. Mag. 1788 ii 1067. Sir W. Jardine *Ornithology* (Nat. Library) 1841, 1–69. *Remains* 1837 portr. W. Munk *Roll of R. College of Physicians* v.1, 1878, 439–41. *D.N.B.* v.52, 179–81. E. J. L. Scott *Index to Sloane Manuscripts* 1904, 495. *Notes R. Bot. Gdn Edinburgh* v.19, 1933, 6–13. F. P. Hett *Mem. Sir Robert Sibbald* 1932. H. R. Fletcher and W. H. Brown *R. Bot. Gdn Edinburgh, 1670–1970* 1970, 5–10 portr.

MS. life at Advocate's Library, Edinburgh. Portr. at College of Physicians, Edinburgh and Hunt Library. Library sold at Edinburgh, 1722.

SIBLEY, Joseph

Bedfordshire botanist of late 18th century. "A gentleman who has paid much attention to the orchis tribe."

J. G. Dony *Fl. Bedfordshire* 1953, 20.

SIBORNE, Richard (*fl.* 1750s–1770s)

Nurseryman, Lea Bridge Road Nursery, Leyton, Essex.

Trans. London Middlesex Archaeol. Soc. v.24, 1973, 189.

SIBRAY, Henry (*c.* 1837–1894)
d. Handsworth, Sheffield 24 Nov. 1894

Nurseryman, Handsworth; in partnership with Messrs Fisher and Son.

J. Hort. Cottage Gdnr v.29, 1894, 24–25.

SIBREE, Rev. James (1836–1929)
b. Hull, Yorks 14 April 1836 *d.* 6 Sept. 1929

Educ. Hull Collegiate College. Missionary with London Missionary Society in Madagascar until 1915. *Madagascar and its People with...Botany* 1870. *Naturalist in Madagascar* 1915. 'Madagascar Wild Flowers' (*Antananarivo Annual* 1888, 511–12). Edited *Antananarivo Annual.*

Who was Who, 1929–1940 1234.

SIBSON, Rev. Edmund (1796–1847)
d. Ashton-in-Makerfield, Lancs 1847

Vicar, Ashton-in-Makerfield. "A good botanist."

Trans. Liverpool Bot. Soc. 1909, 90.

SIBTHORP, Humphrey Waldo (*c.* 1713–1797)
b. Canwick, Lincoln *c.* 1713 *d.* Instow, Devon 12 Aug. 1797

MA Oxon 1737. MD 1745. Professor of Botany, Oxford, 1747–83. Correspondent of Linnaeus.

W. Darlington *Memorials of J. Bartram and H. Marshall* 1849, 428–30. *D.N.B.* v.52, 189. *J. Bot.* 1910, 28. G. C. Druce *Fl. Oxfordshire* 1927, xc. H. N. Clokie *Account of Herb. Dept. Bot., Oxford* 1964, 36–37.

MSS. at Oxford. Portr. at Hunt Library.

Sibthorpia L.

SIBTHORP, John (1758–1796)
b. Oxford 28 Oct. 1758 *d.* Bath, Somerset 8 Feb. 1796

MA Oxon 1780. MD 1784. FRS 1788. FLS 1788. Succeeded his father, Humphrey, as Professor of Botany, Oxford, 1784–96. Travelled in Greece with Ferdinand Bauer, 1786–87 and in Cyprus, Asia Minor, etc., 1794–95. *Fl. Oxoniensis* 1794. *Fl. Graecae Prodromus* 1806–13 2 vols. *Fl. Graeca* 1806–40 10 vols. (biogr. in Preface).

Hooker's J. Bot. 1854, 250. *D.N.B.* v.52, 189–90. *J. Bot.* 1910, 257; 1911, 66–67. *Bot. Soc. Exch. Club Br. Isl. Rep.* 1917, 143; 1923, 355–63 portr. *Kew Bull.* 1926, 111. G. C. Druce *Fl. Oxfordshire* 1927, xciv–cii. H. N. Clokie *Account of Herb. Dept. Bot., Oxford* 1964, 38–42. *Taxon* 1967, 168–78; 1970, 353–61 portr. *Huntia* 1965, 210–11 portr. F. A. Stafleu *Taxonomic Literature* 1967, 445–46. A. M. Coats *Quest for Plants* 1969, 25–28.

Greek plants and Bauer drawings at Oxford. Letters at BM(NH), Linnean Society. MSS. and portr. at Oxford. Portr. at Hunt Library.

SIBTHORP, Mary Esther *see* Hawkins, M. E.

SIDDALL, John Davies (1844–1914)
d. 25 April 1914

'The American Waterweed *Anacheris elsinastrum* Bab., its Structure and Habits' (*Proc. Chester Soc. Nat. Sci.* v.3, 1885, 125–33).

SIDEBOTHAM, Joseph (1824–1885)
b. Apethorne, Hyde, Cheshire 1824 *d.* Bowdon, Cheshire 30 May 1885
FLS 1878. Calico printer, Manchester. One of the founders of Manchester Field Naturalists' Society. Collected diatoms, etc. Contrib. to *Phytologist*.
Proc. Linn. Soc. 1883–86, 107. *J. Bot.* 1885, 319–20. *Trans. Liverpool Bot. Soc.* 1909, 90. *R.S.C.* v.5, 683; v.8, 948.
Plants at Manchester University.

SIDNEY, Rev. Edwin (1797–1872)
d. Cornard Parva, Suffolk 22 Oct. 1872
BA Cantab 1841. Rector, Cornard Parva. *Blights of Wheat* 1846.
Ipswich Mus. Portr. Ser. 1852. S. A. Allibone *Critical Dict. of English Literature* 2095–96. *Gdnrs Chron.* 1872, 1460. *R.S.C.* v.5, 684.

SIGERSON, George (1838–1925)
b. Holyhill, Strabane, County Tyrone 1838 *d.* Dublin 17 Feb. 1925
MD Queen's University, Cork, 1859. "Physician, scientist and publicist." Contrib. botanical papers to *Proc. R. Irish Acad.*
Irish Book Lover v.15, 1925, 7, 22. *R.S.C.* v.8, 952; v.11, 414.

SIGSTON, Benjamin (*fl.* 1780s–1790s)
Nurseryman, Beverley, Yorks.

SILLIARD, Zancke (*fl.* 1640s)
Apothecary, Dublin. Sent *Drosera anglica* to J. Parkinson.
J. Parkinson *Theatrum Botanicum* 1640, 1053.

SILLITOE, Frederick Sampson (1877–1957)
b. Redhill, Surrey 1877 *d.* 4 Dec. 1957
Kew gardener, 1900. Superintendent of Gardens, Khartoum, Sudan, 1903. Collected plants in Yei River Valley, Congo border, 1919.
Gdnrs Chron. 1927 ii 162 portr.; 1957 ii 441. *J. Kew Guild* 1928, 577; 1931, 3–5 portr.; 1957, 485 portr.
Plants at Kew.

SILSBURY, Martin (*fl.* 1900s)
Printer, Shanklin, Isle of Wight. Hybridised chrysanthemums.
Gdnrs Mag. 1911, 23–24 portr.

SILVER, Alexander (1841–1882)
b. Forfarshire 1841 *d.* London 16 July 1882
MA Aberdeen 1862. MD 1863. Assistant Professor, Aberdeen. Lecturer in Botany, London Hospital, 1867; later at Charing Cross Hospital. *Outlines of Elementary Botany* 1866; ed. 2 1877.
J. Bot. 1866, 30–31. *Med. Times Gaz.* 1882 ii 113.

SILVERLOCK, Henry (*fl.* 1800s–1840s)
Nurseryman, North Street, Chichester, Sussex.

SILVERTHORN, G. W. (–1962)
Secretary, Bristol and West of England Orchid Society, 1949–62. Contrib. to Society's *Bulletin*.
Orchid Rev. 1962, 327.

SIM, D. (*fl.* 1900s)
Employed *c.* 1900 by Monrovian Rubber Co. and contributed to A. Whyte's collection. Collected plants in Liberia *c.* 1904, now at Kew.

SIM, John (*c.* 1812–1893)
b. Aberdeenshire *c.* 1812 *d.* Dunfermline, Fife 1893
Herd-boy; afterwards in 92nd Regiment, 1832–55; then sergeant-instructor of Militia in Perth. In West Indies, 1841. Contrib. to *Phytologist*, *Naturalist*.
C. C. Babington *Memorials, Journal* 1897, 340. *R.S.C.* v.5, 699; v.12, 682.
Herb. at Perthshire Society of Natural Science. Letters at Kew and in Wilson correspondence at BM(NH).

SIM, John (1824–1901)
b. Aberdeen 1824 *d.* West Cults, Aberdeen 24 June 1901
Market gardener. Had herb. of *Hepaticae*. *On Botany of Scotston Moor* 1868.
Gdnrs Chron. 1901 ii 13–14. *Ann. Scott. Nat. Hist.* 1901, 193; 1902, 179–82. *Deeside Field* 1933, 54.

SIM, R. (*c.* 1828–1882)
d. Foots Cray, Kent 1882
Son of R. Sim (1791–1878). Nurseryman, Foots Cray. Keen pteridologist.
Gdnrs Chron. 1882 ii 472.

SIM, Robert (1791–1878)
b. Belhelvie, Aberdeen 26 Aug. 1791 *d.* Foots Cray, Kent 3 Aug. 1878
Established nursery at Foots Cray, 1830; specialised in ferns. *Catalogue of British and Exotic Ferns* 1863. Published set of mosses of W. Kent.
Gdnrs Chron. 1878 ii 223. *Florist and Pomologist* 1878, 144.

SIM, Thomas (–1912)
Of Fyvie, Aberdeenshire. Had good knowledge of flora and fauna of Buchan district, Aberdeenshire. Hybridised garden plants.
Gdnrs Mag. 1912, 188(xiv).

SIM, Thomas Robertson (1856–1938)
b. near Aberdeen 25 July 1856 *d.* Durban, S. Africa 23 July 1938

FLS 1894. Kew gardener, 1878. Horticulturist, S. Africa, 1879. Curator, Botanic Garden, King Williamstown, S. Africa, 1888. Conservator of Forests, Natal, 1902. Established nursery, 1907. Explored forests of Portuguese East Africa. *Handbook of Ferns of Kaffraria* 1891. *Forests and Forest Fl. of Colony of Cape of Good Hope* 1907. *Forest Fl. and Forest Resources of Portuguese East Africa* 1909. *Ferns of S. Africa* 1915. *Flowering Trees and Shrubs for use in S. Africa* 1919.

J. Kew Guild 1939–40, 920–21. *Moçambique* 1939, 51–53 portr. *Bothalia* 1964, 174. *S. African J. Sci.* 1971, 409.

SIMMONDS, Arthur (1892–1968)
b. 25 Feb. 1892 *d.* 16 March 1968

VMH 1947. VMM 1962. Assistant Secretary, Royal Horticultural Society, 1925–56; Secretary, 1956–62. Contrib. to *J. R. Hort. Soc. Horticultural Who was Who* 1948.

Gdnrs Chron. 1948 i 74 portr.; 1962 i 149. *J. R. Hort. Soc.* 1968, 238–42 portr. H. R. Fletcher *Story of R. Hort Soc., 1804–1968* 1969 *passim* portr. *Who was Who, 1961–1970* 1033.

SIMMONDS, Peter Lund (*né* Lund) (1814–1897)
b. Aarhuus, Denmark 1814 *d.* London Oct. 1897

FLS 1886. Exhibition Commissioner. *Commercial Products of Vegetable Kingdom* 1854. *Coffee and Chicory* 1864. *Hops* 1877. *Tropical Agriculture* 1877. Edited *J. Applied Sci.*

Br. Roll of Honour 1887, 479. *Athenaeum* 1897 ii 493. *R.S.C.* v.5, 700; v.8, 957.

SIMMONDS, Thomas Williams (–1804)
b. Dartford? Kent *d.* Trinidad 1804

Of Settle, Yorks. Surgeon. Naturalist to Lord Seaforth, Governor of Barbados, 1803.

London J. Bot. 1844, 400. J. Windsor *Fl. Cravoniensis* 1873, Dedication, viii. I. Urban *Symbolae Antillanae* v.3, 1902, 127. *R.S.C.* v.5, 700.

Simmondsia Nutt.

SIMMONITE, William Joseph (*fl.* 1840s–1850s)

Of Sheffield. *Medical Botany* 1848.

S. A. Allibone *Critical Dict. of English Literature* 2104. *R.S.C.* v.5, 700.

SIMMONS, Winifred *see* Bussey, W.

SIMONITE, Benjamin (1834–1909)
b. Sheffield, Yorks 25 June 1834 *d.* Sheffield April 1909

Cutler-smith. Workingman florist, Sheffield.

J. Hort. Cottage Gdnr v.57, 1877, 463–64 portr.; v.58, 1909, 312.

SIMONS, Charles J. (*fl.* 1820s–1850s)

Government apothecary, Assam, India. Collected plants in Assam, Khasia and Mikir Hills. Sent plants to W. J. Hooker and Botanic Garden, Calcutta. Left India, 1858.

Mahonia simonsii Takeda.

SIMPSON, A. (*c.* 1830–1898)
d. 4 May 1898

Nurseryman, Heworth Nurseries, York.

Gdnrs Chron. 1898 i 300–1. *Gdnrs Mag.* 1898, 319. *J. Hort. Cottage Gdnr* v.36, 1898, 406.

SIMPSON, Benjamin (*fl.* 1780s–1800s)

Nurseryman, Dublin. Supplied Botanic Garden, Glasnevin.

Irish For. 1967, 48.

SIMPSON, John (–1740)

Nurseryman, Hoxton, Shoreditch, London. Succeeded to the nursery formerly of Thomas Fairchild (1667–1729) and later of Fairchild's nephew Stephen Bacon (1709–1734).

J. R. Hort. Soc. 1938, 426.

SIMPSON, John (*c.* 1837–1910)
b. Thornhill, Dumfriesshire *c.* 1837 *d.* Wadsley, York 19 Jan. 1910

Gardener and forester to Lord Wharncliffe at Wortley Hall, Yorks, 1864. *New Forestry* 1900. *Quick Fruit Culture* 1900.

Gdnrs Chron. 1910 i 80 portr.

SIMPSON, John Baird (1894–1960)
b. Ardclach, Nairnshire 14 Jan. 1894 *d.* 28 June 1960

BSc Aberdeen 1914. FRSE 1932. Geologist, Edinburgh office of Geological Survey of Great Britain, 1920–45. President, Edinburgh Geological Society, 1950–52. Authority on fossil pollen.

Yb. R. Soc. Edinburgh 1961, 29–31.

SIMPSON, Norman Douglas (1890–1974)
b. Carlton Miniott, Yorks 23 Sept. 1890 *d.* Bournemouth, Hants 29 Aug. 1974

BA Cantab 1911. FLS 1922. Cotton Research Board, Giza, 1921–26. Irrigation Services, Ministry of Public Works, Egypt, 1926–29. Taxonomist, Agricultural Dept., Ceylon, 1931–32. Studied British flora. *Bibliographical Index of British Fl.* 1960.

Times 2 Oct. 1974. *Watsonia* 1975, 403–10.

Plants at BM(NH), Kew. Library at Cambridge.

SIMPSON, R. S. (*fl.* 1840s)

Captain. Collected plants in Simla, Khasia Hills, Bolan Pass, Baluchistan.

A. Lasègue *Musée Botanique de B. Delessert* 1845, 331. I. H. Burkill *Chapters on Hist. Bot. in India* 1965, 81, 129.

Plants at Oxford.

SIMPSON, Richard (*-c.* 1783)
Nurseryman, Tentergate, Knaresborough, Yorks.

SIMS, John (1749–1831)
b. Canterbury, Kent 13 Oct. 1749 *d.* Dorking, Surrey 26 Feb. 1831
MD Edinburgh 1774. FRS 1814. FLS 1788. Studied at Leiden, 1773–74. Settled in London, 1779. Physician to Princess Charlotte. Edited *Ann. Bot.* (with C. D. E. Koenig) 1804–6; *Bot. Mag.* 1801–26 (Preface to v.15, 1801). Herb. purchased by G. Bentham at auction in 1829.
Bot. Misc. 1829–30, 69. W. Munk *Roll of R. College of Physicians* v.2, 1878, 322. *Gdnrs Chron.* 1887 i 641. *D.N.B.* v.52, 281–82. *Friends Quart. Examiner* v.47, 1913. *Practitioner* 1950, 156–70 portr. *R.S.C.* v.5, 707.
Plants and letters at Kew. Portr. at Kew, Hunt Library. Library sold by Thomas, 26–27 May 1829.
Simsia R. Br.

SIMSON, Augustus (1836–1918)
b. London 1836 *d.* Launceston, Tasmania 21 May 1918
In Queensland, 1863; Tasmania about 1873. Had wide knowledge of Tasmanian flora. Secretary, Northern Tasmanian Natural Science Association.
Examiner (Launceston) 22 May 1918. *Papers Proc. R. Soc. Tasmania* 1918, 143–44. *Trans. R. Soc. S. Australia* 1918, 299. *Proc. Linn. Soc. N.S.W.* 1919, 18.

SINCLAIR, Andrew (*c.* 1796–1861)
b. Paisley, Renfrewshire *c.* 1796 *d.* drowned in R. Rangitata, N.Z. 25 March 1861
MD. FLS 1857. Surgeon, Royal Navy, *c.* 1824. Surgeon on HMS 'Sulphur', 1834–42. Surgeon, convict ship in Australia, 1842. Colonial Secretary, New Zealand, 1844–56. Collected plants in Australia, New Zealand and Mexico. 'Vegetation of Auckland' (*Hooker's J. Bot.* 1851, 212–17).
Gdnrs Chron. 1861, 773; 1961 ii 8–9. *Phytologist* 1861, 284–86. *Proc. Linn. Soc.* 1861–62, xcv–xcvi. *D.N.B.* v.52, 289–90. *Fl. Malesiana* v.1, 1950, 485. R. Glenn *Bot. Explorers of N.Z.* 1950, 107–14. *Contrib. Univ. Michigan Herb.* 1972, 311–12. *R.S.C.* v.5, 707.
Plants at Kew, BM(NH). Letters at Kew.

SINCLAIR, Archibald (*c.* 1731–1795)
d. 7 Oct. 1795
Buried in Penarth churchyard, Glam, where his epitaph describes him as "a justly celebrated and scientific botanist".
J. Bot. 1907, 381.

SINCLAIR, George (*c.* 1750–1833)
d. April 1833
Father of G. Sinclair (1786–1834). Gardener to G. Baillie at Mellerstain, Berwickshire, 1778–1833.
Gdnrs Mag. 1833, 512.

SINCLAIR, George (1786–1834)
b. Mellerstain, Berwickshire 1786 *d.* Deptford, Kent 13 March 1834
Gardener to Duke of Bedford at Woburn Abbey, Beds. Nurseryman, New Cross, Kent in partnership with Cormack. *Hortus Gramineus Woburnensis* (with specimens) 1816; ed. 2 1825; ed. 4 1838. *Hortus Ericaeus Woburnensis* 1825. Made additions to J. Donn's *Hortus Cantabrigiensis* ed. 12 1831. 'On cultivating a Collection of Grasses' (*Gdnrs Mag.* 1826, 26–29, 112–16).
Gdnrs Mag. 1834, 192. *D.N.B.* v.52, 294–95.
Letters in Smith correspondence at Linnean Society.

SINCLAIR, James (1809–1881)
b. Morayshire 1809 *d.* Melbourne 29 April 1881
Trained in London in drawing and landscape gardening. Drew plants for Thomas Knight, nurseryman at Kew. Landscaped gardens for Prince Woronzorf in Russia, 1838–51. Emigrated to Australia, 1854. Landscaped Fitzroy Gardens, Melbourne. Published *Gdnrs Mag.* June 1855. Contrib. to Loudon's *Gdnrs Mag.*
Victorian Hist. Mag. 1940, 81–87.
Drawings at Melbourne Herb.

SINCLAIR, James (1913–1968)
b. Bu of Hoy, Orkney 29 Nov. 1913 *d.* Kirkwall, Orkney 15 Feb. 1968
BSc Edinburgh 1936. Teacher in Orkney where he studied algae. Botanist, Royal Botanic Garden, Edinburgh, 1946–48. Curator, Herb., Botanic Gardens, Singapore, 1948–63. Collected plants in Malaya, Borneo, Indonesia, Philippines. 'Notes on New Guinea Annonaceae' (*Gdnrs Bull. Singapore* v.15, 1956, 4–373). 'Genus Myristica in Malesia and outside Malesia' (*Gdnrs Bull. Singapore* v.23, 1968, 1–540).
Gdns Bull. Singapore v.23, 1968, i–xxiii portr. *Fl. Malesiana Bull.* no. 23, 1969, 1671–72.
Portr. at Hunt Library.

SINCLAIR, James A. (*c.* 1864–1940)
d. Aug. 1940
Photographic apparatus dealer. Interested in ferns. Vice-President, British Pteridological Society, 1936–38. Travelled in France, N. Africa, India.
Br. Fern Gaz. 1948, 204.

SINCLAIR, Sir John (1754–1835)
b. Thurso Castle, Caithness 10 May 1754 *d.* Edinburgh 21 Dec. 1835

DCL Glasgow 1788. FRS 1784. FLS 1810. Baronet 1786. President, Board of Agriculture, 1793–98, 1806–13. *Hints on Vegetation* 1796. *Inquiry into Blight* 1809.

Gent. Mag. 1836 i 431–33. *Cottage Gdnr* v.9, 1853, 77, 98. J. Sinclair *Mem. of Life...of Sir J. Sinclair* 1838. *D.N.B.* v.52, 301–5. R. Mitchison *Agricultural Sir John* 1962.

Portr. at Hunt Library.

SKAN, Sidney Alfred (1870–1939)
b. Kineton Green, Warwickshire 13 Aug. 1870 *d.* Kew, Surrey 19 Dec. 1939

Kew gardener, 1892. Assistant, Kew Herb. 1894; Librarian, 1899–1933. Contrib. to *Index Florae Sinensis* 1899. Co-author with W. B. Hemsley of *Scrophulariaceae* in *Fl. Tropical Africa* 1906.

Kew Bull. 1907, Appendix 5, 136; 1940, 46–47. *Gdnrs Chron.* 1940 i 86. *J. Kew Guild* 1939–40, 918–19 portr.

SKEEN, James (*fl.* 1700s)

Surgeon. Sent plants from Guinea to J. Petiver.

J. Petiver *Museii Petiveriani* 1695, 95. J. E. Dandy *Sloane Herb.* 1958, 203.

Plants at BM(NH).

SKELLON, William (*fl.* 1850s)

Employed in Liverpool Botanic Garden. Friend of William Harrison and contrib. with him to J. Dickinson's *Fl. Liverpool* 1851. Emigrated to Australia *c.* 1851 where he died.

J. B. L. Warren *Fl. Cheshire* 1899, lxxxviii–lxxxix. *Trans. Liverpool Bot. Soc.* 1909, 90.

SKENE, David (1731–1770)
b. Aberdeen 13 Aug. 1731 *d.* Dec. 1770

MD Edinburgh 1753. Settled in Aberdeen as assistant to his father, also a physician. Correspondent of Linnaeus, J. Ellis and T. Pennant. "Contemplated the preparation of a complete fauna and flora of his own neighbourhood." Had herb.

A. Murray *Northern Fl.* 1836, ix–xi. *Proc. R. Soc. Edinburgh* v.4, 1857–62, 164–67. *Ann. Sci. Nat. Hist.* 1911, 178.

SKEPPER, Edmund (1825–1867)
b. Oulton, Suffolk 20 Oct. 1825 *d.* Bury St. Edmunds, Suffolk 2 June 1867

Druggist, Harwich and Bury. *Fl. Suffolk* (with J. S. Henslow) 1860.

W. M. Hind *Fl. Suffolk* 1889, 485–88.

Plants and letters at BM(NH).

SKERTCHLY, Sydney Barber Josiah (1850–1926)
b. Anstey, Leics 14 Dec. 1850 *d.* Molendinar, Queensland 2 Feb. 1926

Educ. Royal School of Mines, London. Geologist. Professor of Botany, Hong Kong. Geological Survey Dept., Queensland, 1891. President, Queensland Naturalists' Society. *Queensland Nat.* 1926, 70–72 portr. *Proc. R. Soc. Queensland* 1926, 1–2. *Austral. Encyclop.* v.8, 1965, 138–39.

SKEY, William (1835–1900)
b. London 8 April 1835 *d.* 4 Oct. 1900

To New Zealand, 1860. Government analyst, Mines Dept. Did work on phytochemistry; isolated poisons of many New Zealand shrubs.

Trans. N.Z. Inst. 1901, 554.

SKINNER, George Ure (1804–1867)
b. Newcastle-upon-Tyne, Northumberland 18 March 1804 *d.* Aspinwall, Panama 9 Jan. 1867

FLS 1866. Merchant in Leeds and Guatemala. To Guatemala, 1831. Collected orchids for J. Bateman and sent plants to W. J. Hooker.

J. Bateman *Orchidaceae of Mexico* 1837, t.xiii. *Bot. Mag.* 1838, t.3638; 1857, t.5009. *Gdnrs Chron.* 1867, 180–81. *J. Bot.* 1867, 91. *Proc. Linn. Soc.* 1866–67, xxxviii–xl. *Trans. Bot. Soc. Edinburgh* v.9, 1868, 91–94. W. B. Hemsley *Biologia Centrali-Americana Bot.* v.4, 1886, 124–25. *Amer. Orchid Soc. Bull.* 1964, 403–5. M. A. Reinikka *Hist. of the Orchid* 1972, 169–73 portr.

Portr. at Hunt Library.

Uroskinnera Lindl. *Cattleya skinneri* Batem.

SKINNER, Rev. Richard (*c.* 1729–1795)
b. Didmarton, Glos *c.* 1729 *d.* 27 Nov. 1795

MA Oxon 1753. Rector, Bassingham, Lincolnshire, 1774. Knew Welsh plants.

J. R. Forster *Characteres Generum Plantarum* 1776, 58. R. Holt-White *Life and Letters of G. White* v.1, 1901, 187–89, 210–12. *J. Bot.* 1905, 303.

SKIPPON, Sir Philip (*fl.* 1640s–1670s)

BA Cantab 1660. Knighted 1674. Pupil of J. Ray whom he accompanied to Wales in 1658 and on the Continent in 1663. Had herb.

J. Ray *Historia Plantarum* v.2, 1688, 1310. E. Lankester *Memorials of J. Ray* 1846, 12. *D.N.B.* v.52, 356. *J. Bot.* 1934, 222.

SKIRVING, William (*fl.* 1830s)

Nurseryman, 17 Queen's Square, Liverpool; also Walton Nursery.

SKRIMSHIRE, William (1766–1830)

Of Wisbech, Cambridgeshire. Entomologist. His brother, Dr. Fenwick Shrimshire, of Blair, discovered *Campanula rapunculoides* in 1800 (J. E. Smith *Fl. Britannica* v.1, 1800, 237–38). Contrib. paper on seeds of *Iris* as substitute for Coffee to *Nicholson's J. Nat. Philos.* v.22, 1809, 70–73. Contrib. to J. Sowerby and J. E. Smith's *English Bot.*

J. Bot. 1918, 260–61. *Proc. Bot. Soc. Br. Isl.* 1956, 133. *R.S.C.* v.5, 712.

SLACK, Henry (-before 1845)
Of London and Epsom, Surrey. Received medal of Society of Arts for dissecting microscope, 1831. 'Elementary Tissue of Plants' (*Trans. Soc. Arts* v.49, 1833, 127–57). 'Motion of Fluids in Plants' (*Trans. Soc. Arts* v.49, 1833, 177–79).
W. Griffith *Palms of Br. East India* 1850, 161–62.
Slackia Griff.

SLACK, R. (*fl.* 1830s)
Botanised in Cheshire *c.* 1837–39.
J. B. L. Warren *Fl. Cheshire* 1899, lxxxix.

SLADE, Sir James Benjamin (1861–1950)
b. Hampstead, London 12 Dec. 1861 *d.* Southborough, Kent 15 April 1950
VMH 1943. Head of Messrs Protheroe and Morris, nursery stock auctioneers.
Gdnrs Chron. 1950 i 172. *Orchid Rev.* 1950, 84. *Who was Who, 1941–1950* 1061.

SLATER, Gilbert (*c.* 1753–1793)
d. Leyton, Essex 30 Oct. 1793
Manager of several East India Company ships. Imported plants for his garden at Leyton, including *Hydrangea hortensis c.* 1790 and is said to have flowered it for the first time in this country, "having two persons collecting for him in the East Indies, at the expense of £500 a year."
Gent. Mag. 1793 ii 1054. *Bot. Mag.* 1794, t.284; 1799, t.438. H. C. Andrews *Botanist's Rep.* 1798,t.25. *J. Botanique* (Desvaux) v.1, 1808, 243–45. *Bot. Cabinet* 1821, 513. *Gdnrs Mag.* 1827, 128. *Hort. Reg.* 1836, 63–64, 336. E. Bretschneider *Hist. European Bot. Discoveries in China* 1798, 213–15. *Essex Nat.* 1938–39, 162–68, 198–204.
Slateria Desvaux.

SLATER, John (*fl.* 1820s–1840s)
Nurseryman, Yorkersgate, Malton, Yorks.

SLATER, John Samuel (1850–1911)
b. Calcutta 1850 *d.* Ealing, Middx 7 April 1911
Civil engineer, Indian Public Works Dept. Inspector of Schools, 1897–1904. Photographed pollen.
Kew Bull. 1912, 56.
Specimens and photos at Kew.

SLATER, Matthew B. (*c.* 1829–1918)
b. Malton, Yorks *c.* 1829 *d.* Malton Feb. 1918
FLS 1889. Bryologist. Friend of R. Spruce. 'Muscineae of N. Yorkshire' (*Trans. Yorkshire Nat. Union* pt. 33, 1906, 417–645).
Naturalist 1903, 361–62 portr.; 1918, 108–9 portr. *J. Bot.* 1918, 191.

SLATTER, John Whewell (1829–1896)
d. 6 Jan. 1896
BA Cantab. Collected plants in Australia, 1870–80, now at National Herb.

SLEEMAN, Sir William Henry (1788–1856)
b. Stratton, Cornwall 8 Aug. 1788 *d.* at sea 10 Feb. 1856
Major-General. Served in Nepal war, 1814–16. Resident, Gwalior, 1843–49; Lucknow, 1849–54. 'Age and Flowering of Bamboos' (*Trans. Agric. Hort. Soc. India* v.3, 1839, 139–41; v.4, 189–91). *Rambles and Recollections of an Indian Official* 1844.
Gent. Mag. 1856 ii 243–44. G. C. Boase and W. P. Courtney *Bibliotheca Cornubiensis* 1874, 656–57. *D.N.B.* v.52, 373–74. *R.S.C.* v.5, 714.

SLOANE, Sir Hans (1660–1753)
b. Killyleagh, County Down 16 April 1660 *d.* Chelsea, London 11 Jan. 1753
MD Orange 1684; Oxon 1701. FRS 1685. Baronet 1716. Pupil of Tournefort. Physician to Governor of Jamaica, 1687–89. Secretary, Royal Society, 1693–1712; President, 1727–41. Gave Chelsea Physic Garden to Apothecaries' Company. *Catalogus Plantarum quae in Insula Jamaica...* 1696. *Voyage to... Madeira, Barbados,...and Jamaica* 1707–25 2 vols.
Hist. Acad. Paris 1753, 305–20. *Lit. Mag.* 1790, 313–18 portr. R. Pulteney *Hist. Biogr. Sketches of Progress of Bot. in England* v.2, 1790, 65–96. J. Nichols *Illus. Lit. Hist. of Eighteenth Century* v.1, 269–89. T. Faulkner *Hist. and Topographical Description of Chelsea* 1829 *passim.* D. Turner *Extracts from Lit. Sci. Correspondence of R. Richardson* 1835, 43–46. E. Lankester *Memorials of J. Ray* 1846, 40–42. E. Edwards *Lives of Founders of Br. Mus.* 1870, 274–312. W. Munk *Roll of R. College of Physicians* v.1, 1878, 460–67. R. H. Semple *Mem. Bot. Gdn at Chelsea* 1878, 26–31. I. Urban *Symbolae Antillanae* v.1, 1898, 154–57; v.3, 1902, 130–31. *D.N.B.* v.52, 379–80. *Contrib. U.S. National Herb.* 1908, 113–58. *Nat. Hist. Mag.* 1935, 49–64, 97–116, 145–64 portr. *Ann. Med. Hist.* 1938, 390–404 portr. C. E. Raven *John Ray* 1942 *passim. Hist. Rev.* 1945, 66–75 portr. *Br. Mus. Quart.* v.18(1), 1953, 1–26 portr. *Notes R. Soc. London* v.10, 1953, 81–100 portr. G. R. de Beer *Sir Hans Sloane and the Br. Mus.* 1953. *Hermathena* 1953, 3–13. *Nature* v.171, 1953, 64–65. W. E. St. John Brooks *Sir Hans Sloane* 1954 portr. J. E. Dandy *Sloane Herb.* 1958 portr. *Gdnrs Chron.* 1960 i 163. *Irish J. Med. Sci.* 1960, 201–12. *Irish Nat.* 1961, 197–200. *Med. Hist.* 1961, 154–56. *Proc. Bot. Soc. Br. Isl.* 1961, 136–41. C. Wall *Hist. Worshipful Society of Apothecaries of London* v.1, 1963 *passim* portr. F. A. Stafleu *Taxonomic Literature* 1967, 447–48. A. M. Coats *Quest for Plants* 1969, 330–33 portr. C. C. Gillispie *Dict. Sci. Biogr.* v.12, 1975, 456–59.

Letters, MSS. and Library at BM. Herb. at BM(NH). Portr. by Kneller at National Portr. Gallery. Statue by Rysbrach at Chelsea Physic Garden. Portr. at Hunt Library.
Sloanea L.

SLOCOCK, Oliver Charles Ashley (*c*. 1907–1970)
d. 26 March 1970
VMH 1964. Nurseryman, Goldsworth Nursery, Woking, Surrey; specialised in rhododendrons.
Gdnrs Chron. v.167(15), 1970, 7. *J. R. Hort. Soc*. 1970, 276–78. *Times* 1 April 1970.

SLOCOCK, Walter Charles (*c*. 1853–1926)
b. Wraysbury, Bucks *c*. 1853
VMH 1916. Nurseryman, Goldsworth Nursery, Woking, Surrey; noted for conifers and rhododendron hybrids.
Gdnrs Chron. 1926 i 2 portr.; 1960 i 378–79 portr.

SMALES, Charles Bertram (–1927)
d. France 5 March 1927
Burma Forest Service, 1893; Chief Conservator, 1924. Had knowledge of forest flora of Burma.
Indian Forester 1927, 376–77.

SMALL, James (1889–1955)
b. Brechin, Angus 1889 *d*. Lisburn, County Antrim 28 Nov. 1955
BSc London 1913. MSc 1916. FLS 1918. Lecturer in Botany, Bedford College, London, 1916–20. Professor of Botany, Belfast, 1920–54. *Origin and Development of Compositae* 1919. *Textbook of Botany* 1921; ed. 4 1937. *Practical Botany* 1931. *Pocket-lens Plantlore* 1931. *Modern Aspects of pH* 1934.
R. L. Praeger *Some Irish Nat*. 1949, 155–56. *Nature* v.177, 1956, 258–59. *Irish Nat*. 1956, 49–50. *Pharm. J*. 1955, 505. *Proc. Bot. Soc. Br. Isl*. 1956, 192–93. *Yb. R. Soc. Edinburgh* 1957, 48–50. *Who was Who, 1951–1960* 1010.
Herb. at Queen's University, Belfast.

SMALL, Moses (*fl*. 1760s–1798)
Nurseryman, Chichester, Sussex.

SMART, Alice Sophia *see* Cooke, A. S.

SMART, John (*fl*. 1700s)
Surgeon. Of Sassafras Creek, Maryland. Sent Hudson's Bay and Maryland plants to J. Petiver.
Mon. Misc: or, Mem. for Curious v.3, 1709, 133. E. J. L. Scott *Index to Sloane Manuscripts* 1904, 499. J. E. Dandy *Sloane Herb*. 1958, 209.
Plants at BM(NH).

SMEATHMAN, Henry (–1786)
d. 1 July 1786
To Sierra Leone in 1771 to collect plants for J. Fothergill, etc. Collected also in Madagascar and West Indies. Friend of J. C. Lettsom.
Gent. Mag. 1786 ii 620. R. H. Fox *Dr. John Fothergill* 1919, 213–14. *J. Washington Acad. Sci*. 1940, 296–97. *Comptes Rendus AETFAT, 1960* 1962, 56.
Plant at BM(NH), Washington.
Smeathmannia R. Br.

SMEATON, Thomas Drury (*c*. 1832–1908)
b. London *c*. 1832 *d*. Mount Lofty, S. Australia 18 Feb. 1908
To Australia in 1853 where he joined the staff of Bank of Australia. Collected plants.
Trans. Proc. R. Soc. S. Australia 1908, 395.

SMEE, Alfred (1818–1877)
b. Camberwell, London 18 June 1818 *d*. Wallington, Croydon 11 Jan. 1877
MRCS 1840. FRS 1841. FLS 1875. Surgeon, Bank of England, 1841. *The Potato Plant* 1846. *Report on Vegetable Parchment* 1858. *My Garden* 1872.
Gdnrs Chron. 1877 i 88, 108–10 portr. Mrs. Odling *Mem. of A. Smee* 1878. *D.N.B*. v.52, 398–99. *R.S.C*. v.5, 715.

SMEE, Alfred Hutchison (1841–1901)
b. Finsbury, London 1841 *d*. Carshalton, Surrey 8 Nov. 1901
Son of A. Smee (1818–1877). Medical Officer, Gresham Life Assurance Company. Cultivated orchids. Experimented on *Cytisus adami* (*J. Bot*. 1901, 436).
Gdnrs Chron. 1901 ii 358–59. *Gdnrs Mag*. 1901, 751. *Orchid Rev*. 1901, 356–57.
Saccolabium smeeanum Reichenb. f.

SMELLIE, William (1740–1795)
b. Edinburgh 1740 *d*. Edinburgh 24 June 1795
FRSE. Printer. Zoologist. Pupil and deputy lecturer of J. Hope. Had herb. of Scottish plants. *Dissertation on Sexes of Plants* 1765.
R. Kerr *Mem.…of W. Smellie* 1811 portr. *D.N.B*. v.52, 400–1.

SMILES, Frederick Henry (–1895)
d. Korat, Siam May 1895
Joined Royal Survey Dept., Siam, 1891. Collected plants, now at Kew.
Kew Bull. 1895, 38–39, 198. *J. Thailand Res. Soc. Nat. Hist. Supplt* v.12, 1939, 22.

SMITH, Albert Malins (1879–1962)
b. Waltham, Lincs 14 April 1879 *d*. 26 Nov. 1962

Educ. Cambridge. Assistant, Royal Botanic Gardens, Peradeniya, Ceylon. Demonstrator, Botany School, Cambridge. Lecturer in Plant Physiology, Manchester University. Principal, East Anglian Institute of Agriculture. Head, Biology Dept., Bradford Technical College, 1919–45. Studied algae and plant ecology. Contrib. to *Naturalist*.

Naturalist 1963, 69–70.

Herb. at Bradford University.

SMITH, Alexander (*fl.* 1750s)

Surgeon and apothecary. Of Croydon. Discovered *Scrophularia vernalis* 1759.

G. C. Gorham *Mem. of John and Thomas Martyn* 1830, 106.

SMITH, Alexander (1832–1865)

b. Kew, Surrey 17 Dec. 1832 *d.* Kew 15 May 1865

Son of John Smith (1798–1888). Curator, Kew Museum, 1856–58. Clerk, Kew Herb., 1863. Contrib. to J. Lindley and T. Moore's *Treasury of Botany* 1866.

Gdnrs Chron. 1865, 464, 986. *J. Bot.* 1865, 199–200. *J. Kew Guild* 1895, 30–31. *Kew Bull.* 1914, 85–87. *R.S.C.* v.5, 716.

MSS. and portr. at Kew.

SMITH, Alexander (1894–1948)

b. Peterhead, Aberdeenshire 8 April 1894 *d.* 23 July 1948

BSc Aberdeen 1921. PhD Cantab 1927. Assistant Lecturer in Botany, Aberdeen University, 1926–27. Assistant Mycologist, Plant Pathology Laboratory, Harpenden, 1928–48.

Trans. Hertfordshire Nat. Hist. Soc. 1950, 85; 1951, 136.

SMITH, Alfred William (*c.* 1855–1927)

Market gardener in Middx: Bedfont, Feltham, and Ashford. "The cabbage king."

R. Webber *Early Horticulturists* 1968, 172–83. D. Macleod *Gdnrs London* 1972, 202–4.

SMITH, Annie Lorrain (1854–1937)

b. Halfmorton, Dumfriesshire 25 Oct. 1854 *d.* London 7 Sept. 1937

FLS 1904. Worked at BM(NH) on fungi and lichens, 1892–1933. Assistant, Royal Agricultural Society. President, British Mycological Society, 1907, 1917. Edited and completed J. Crombie's *Monograph of British Lichens* 1911–18 2 vols. *Handbook of British Lichens* 1921. *Studies of Trees and Flowers* 1911. Prepared lichen and fungi abstracts for *J. R. Microsc. Soc.*

J. Bot. 1937, 329–30. *Kew Bull.* 1937, 442–43. *Proc. Linn. Soc.* 1937–38, 337–39. *Times* 14 Sept. 1937.

MSS. at BM(NH). Portr. at Hunt Library.

SMITH, Arthur Algernon Dorrien- *see* Dorrien-Smith, A. A.

SMITH, Augustus (1804–1872)

b. London Sept. 1804 *d.* Plymouth, Devon 31 July 1872

Lord Proprietor of Isles of Scilly, 1834. Established garden at Tresco Abbey. Pioneered commercial production of early potatoes.

J. R. Hort. Soc. 1947, 177–83. M. Hadfield *Gdning in Britain* 1960, 357–58. E. Inglis-Jones *Augustus Smith of Scilly* 1969 portr.

SMITH, Charles (*c.* 1715–1762)

b. Waterford, Ireland *c.* 1715 *d.* Bristol July 1762

MD Dublin 1738. Of Dublin. Plant lists in his histories of Waterford, 1746, Cork, 1750, Kerry, 1756.

R. Pulteney *Hist. Biogr. Sketches of Progress of Bot. in England* v.2, 1790, 202–4. N. Colgan and R. W. Scully *Cybele Hibernica* 1898, 353–54. R. W. Scully *Fl. County Kerry* 1916, xi–xii. *D.N.B.* v.53, 20.

SMITH, Charles (*c.* 1830–1910)

d. Guernsey, Channel Islands 1910

Nurseryman, Caledonia Nurseries, Guernsey.

Gdnrs Chron. 1910 i 158 portr.

SMITH, Charles (–1912)

d. 10 May 1912

Nurseryman, Caledonia Nurseries, Guernsey.

Gdnrs Chron. 1912 i 358.

SMITH, Charlotte (*née* Macdonald) (*fl.* 1830s–1850s)

One of R. C. Gunn's collectors and also looked after his herb. Collected algae in Tasmania for W. H. Harvey.

Bot. Register 1839 Appendix 50. J. D. Hooker *Fl. Tasmaniae* v.1, 1859, cxxvi. *Gdnrs Chron.* 1878 ii 345–46. *Papers Proc. R. Soc. Tasmania* 1909, 27. *J. R. Soc. N.S.W.* 1921, 166. T. E. Burns and J. R. Skemp *Van Diemen's Land Correspondents* 1961, 63, 75, 82.

Polyphacum smithiae Hook. f. & Harvey.

SMITH, Christen (1785–1816)

b. Drammen, Norway 17 Oct. 1785 *d.* Congo 21 Sept. 1816

Travelled in British Isles, 1814. Professor of Botany, University of Christiania. To Madeira, Tenerife, etc. 1815; Congo Expedition, 1816. Journal in J. K. Tuckey's *Narrative of Expedition to...the Congo...in 1816* 1818, 229–336; biogr. lxiii–lxxi; his plants described by R. Brown.

Edinburgh New Philos. J. 1826, 209–16. F. C. Kiaer *Dagbog paa Reisen til de Canariske* 1815. T. Durand and H. Schinz *Études Fl. Congo* pt. 1, 1896, 6. *Comptes Rendus AETFAT 1960* 1962, 82–83, 114–15.
Plants and MS. autobiography at BM(NH). Plants at Kew.
Christiana DC.

SMITH, Christopher (–1807)
d. Penang, Malaya 1807
FLS 1793. Assistant to J. Wiles on voyage of 'Providence', 1791–93. Botanist to East India Company at Calcutta, 1794. To Moluccas, 1796. Superintendent, Botanic Garden, Penang, 1805–6.
Ann. Bot. 1805, 569–73. *Gdnrs Chron.* 1881 ii 267. *Kew Bull.* 1891, 300. I. Lee *Captain Bligh's Second Voyage to the South Sea* 1920, 18–19, 81, 210. *Papers Proc. R. Soc. Tasmania* 1920, 119–21; 1922, 10–12. *J. Bot.* 1922, 23–25. *Singapore Nat.* 1923, 37–38. *Taxon* 1970, 543.
Plant drawings and plants at BM(NH).

SMITH, Christopher Parker (1835–1892)
b. Brighton, Sussex 13 Oct. 1835 *d.* Hassocks, Sussex 15 Nov. 1892
Bryologist. Collected plants in Highlands of Scotland. Acquired E. Jenner's herb. *Moss-Fl. Sussex* 1870.
J. Bot. 1893, 31–32.

SMITH, Rev. Colin (*fl.* 1820s–1850s)
Minister, Inveraray, Argyll. Sent mosses and lichens to W. J. Hooker and W. Wilson. 'Biographical Notice of Late Capt. D. Carmichael' (*Bot. Misc.* 1830–31, 1–59, 258–343; 1832–33, 23–76).
Letters at Kew and in Wilson correspondence at BM(NH).

SMITH, Denys D. Munro (1890–1971)
b. Bristol 1890 *d.* Bristol 1 Dec. 1971
Physician, Bristol. Interested in bryophytes and flora of Bristol Gorge. Discovered *Pohlia lutescens* at Oldbury Park, Bristol, 1963. Contrib. to *Proc. Bristol Nat. Soc.*
Watsonia 1973, 408–9.

SMITH, Edward (*fl.* 1780s–1840s)
Nurseryman in partnership with Samuel Smith at Dalston, Hackney, London.
Trans. London Middlesex Archaeol. Soc. v.24, 1973, 184.

SMITH, Edward (*c.* 1818–1874)
b. Heanor, Derbyshire *c.* 1818 *d.* London 16 Nov. 1874
MD London 1843. FRS 1860. Visited Texas, 1849. Lecturer in Botany, Charing Cross Hospital, 1851. *Structural and Systematic Botany* 1854. *Foods* 1872.
D.N.B. v.53, 31–32.

SMITH, Edwin Dalton (*fl.* 1820s–1840s)
FLS 1823. Of Chelsea. Botanical artist. Illustrated B. Maund's *Bot. Gdn* v.1–6, 1825–36 (original drawings at BM(NH)); R. Sweet's *Fl. Australasica* 1827–28 (original drawings at BM(NH)); R. Sweet's *Br. Flower Gdn* 1823–34; R. Sweet's *Florists' Guide* 1827–32; R. Sweet's *Geraniaceae* 1820–30; P. W. Watson's *Dendrologia Britannica* 1823–25.
J. Bot. 1918, 238.

SMITH, Mrs. Elizabeth (*fl.* 1790s)
Of Bownham House, Minchinhampton Common, Gloucestershire; botanised locally.
J. Sowerby and J. E. Smith *English Bot.* 437. H. J. Riddelsdell *Fl. Gloucestershire* 1948, cxvi.

SMITH, Eryl (*née* **Glynne**) (*c.* 1893–1930)
b. Bangor, Caernarvonshire *c.* 1893 *d.* London 25 Jan. 1930
MB London 1918. Wife of Dr. Malcolm A. Smith who was in medical practice in Siam. Collected plants, especially ferns, in Siam, 1922; Hainan, 1923; Cambodia, 1924; Celebes, 1925 and Malaya. Worked on ferns of Malaya at Kew Herb., 1928–30.
Gdns Bull. Straits Settlements 1927, 132. *Kew Bull.* 1928, 141; 1930, 175, 398. *Gdns Chron.* 1930 ii 268. *J. Siam Nat. Hist. Soc. Supplt* v.8, 1930, 128–29. *Fl. Malesiana* v.1, 1950, 489. *Blumea* v.11, 1961–62, 483–84.
Plants at BM(NH), Kew.
Rinorea smithiae Craib.

SMITH, F. Percy (–1945)
d. 24 March 1945
Pioneer of British film production, specialising in the interpretation of nature and microscopic work. Some of his best documentaries were *Plants of the Underworld* 1930, *Gathering Moss* 1933 and *Life of a Plant.*
Nature v.155, 1945, 538.

SMITH, Frederick John (1853–1919)
b. Castle Donnington, Derby 1853 *d.* Colyton, Devon 30 April 1919
MB Oxon 1885; London 1891. Physician, London Hospital, 1912.
Bot. Soc. Exch. Club Br. Isl. Rep. 1919, 625–26. *Nature* v.103, 1919, 191.
Herb. at Exeter Museum.

SMITH, Frederick Porter (1833–1888)
d. Shepton Mallet, Somerset 29 March 1888
MB London 1855. Medical missionary in Central China. *Contributions towards Materia Medica and Natural History of China* 1871.
Pharm. J. v.18, 1888, 859.

SMITH, Frederick William (1797–1835)
b. London 25 Aug. 1797 *d.* Shrewsbury, Shropshire 18 Jan. 1835
Son of miniature painter, Anker Smith. Brother of E. D. Smith. Botanical artist. Illustrated *Paxton's Mag. Bot.* 1834–37; *Florists' Mag.* 1835–36 of which he was also the editor; R. Sweet's *Br. Flower Gdn* 1834–35.
J. Bot. 1918, 238.

SMITH, G. Campbell (*fl.* 1830s)
Land surveyor, Banff. Discovered *Pinguicula alpina* in Ross, 1831.
A. Murray *Northern Fl.* 1836, 17.

SMITH, George (*c.* 1812–1883)
d. 26 March 1883
Nurseryman, Tollington Nursery, Hornsey Road, London.
Gdnrs Chron. 1883 i 446. *J. Hort. Cottage Gdnr* v.6, 1883, 279.

SMITH, George (*fl.* 1820s–1850s)
Nurseryman, Liverpool Road, Islington, London.

SMITH, George (1825–1899)
b. Perth 1825 *d.* 19 Feb. 1899
Gardener, Vice-Regal Lodge, Phoenix Park, Dublin.
Gdnrs Chron. 1875 i 45–46 portr. *Garden* v.55, 1899, 138.

SMITH, George (*fl.* 1830s)
Nurseryman, Farnham, Surrey.

SMITH, George (1895–1967)
d. 29 March 1967
BSc Manchester 1916. MSc 1918. Chemist to firm of cotton manufacturers, 1919. Assistant, London School of Hygiene and Tropical Medicine, 1930–61, where he worked on the taxonomy and care of fungal cultures. President, British Mycological Society, 1945. *Introduction to Industrial Mycology* 1939; ed. 5 1960. Published monographs on *Paecilomyces* and *Scopulariopsis*.
Nature v.215, 1967, 109. *Trans. Br. Mycol. Soc.* 1967, 339–40 portr.

SMITH, George Percy Darnell- *see* Darnell-Smith, G. P.

SMITH, George Whitfield (*fl.* 1860s–1900s)
Botanical collector, St. Vincent. Curator, Botanic Garden, Grenada, 1890–94. Travelling Superintendent, Imperial Dept. of Agriculture for West Indies.
Kew Bull. Additional Ser. v.1, 1898, 74. I. Urban *Symbolae Antillanae* v.3, 1902, 131.
Plants at Kew.

SMITH, Rev. Gerard Edwards (1804–1881)
b. Camberwell, London 1804 *d.* Ockbrock, Derby 21 Dec. 1881
BA Oxon 1829. Curate, Sellinge, Kent, 1830–32. Vicar, Cantley, Yorks, 1844–46; Osmaston, Derby, 1854–71. Described *Ophrys arachnites* and *Statice binervosa* in J. Sowerby and J. E. Smith's *English Bot.* 2596, 2683. *Catalogue of...Plants, collected in South Kent* 1829.
J. Bot. 1882, 63. H. C. Watson *Topographical Bot.* 1883, 556. W. H. Painter *Contrib. Fl. Derbyshire* 1889, 7. *D.N.B.* v.53, 43.
Sussex and Kent plants at Oxford. Letters at Kew.

SMITH, Gordon Levesley (1937–1963)
b. Stamford, Lincs 8 April 1937 *d.* 8 Dec. 1963
BA Cantab 1961. 'Studies in Potentilla' (*New Phytol.* 1963, 264–300).
Proc. Bot. Soc. Br. Isl. 1964, 417–18.

SMITH, H. B. Willoughby (1879–)
b. York 1879
MB London 1902. FRCS 1909. Physician, Gainsborough, Lincs, 1910. President, Lincolnshire Naturalists' Union, 1924. Collected Yorkshire plants.
Trans. Lincolnshire Nat. Union 1927, 33–34 portr.
Plants at Beverley Public Library.

SMITH, Henry (1786–1868)
b. Shanklin, Isle of Wight 3 June 1786 *d.* Southsea, Hants 14 June 1868
Royal Marines, 1809–48. Had good botanical library. Made MS. lists of Hants, Wilts and Hull plants. Contrib. plant list to D. T. Ansted's *Channel Islands* ed. 2 1865.
E. D. Marquand *Fl. Guernsey* 1891, 26. *J. Bot.* 1928, 274. D. Grose *Fl. Wiltshire* 1957, 35. D. McClintock *Wild Flowers of Guernsey* 1975, 27–28.

SMITH, Henry (*fl.* 1810s)
MD. FLS 1816. Physician, Salisbury Infirmary. *Fl. Sarisburiensis* 1817.

SMITH, Henry Ecroyd (1823–1889)
b. Doncaster, Yorks 28 Aug. 1823 *d.* Middleham, Yorks 25 Jan. 1889
Draper. Curator, 'Egyptian Museum', Liverpool; Curator, Meyer Collection, Liverpool Museum *c.* 1876. Naturalist and archaeologist. Contrib. to *Fl. Liverpool* 1872, *Liverpool Nat. J.*, *Phytologist*.
J. B. L. Warren *Fl. Cheshire* 1899, lxxxix. *Trans. Liverpool Bot. Soc.* 1909, 90–91. *R.S.C.* v.5, 724.

SMITH, Henry George (1852–1924)
b. Littlebourne, Kent 1852 *d.* Sydney, N.S.W. 19 Sept. 1924

To New South Wales, 1883. Employed in Sydney Technological Museum where he later collaborated with J. H. Maiden in papers on plant chemistry. Studied essential oils of Australian plants. President, Royal Society N.S.W., 1913–14. *Research on Eucalypts especially in regard to their Essential Oils* (with R. T. Baker) 1902. *Research on Pines of Australia* (with R. T. Baker) 1910. *Research on Eucalypts of Tasmania and their Essential Oils* (with R. T. Baker) 1912.

J. R. Soc. N.S.W. 1925, 11–12. *Proc. Linn. Soc. N.S.W.* 1925, 8. *Rep. Austral. Assoc. Advancement Sci.* 1926, 2–3. P. Serle *Dict. Austral. Biogr.* 1949, 332–33. *Proc. R. Austral. Chem. Inst.* 1960, 309–16 portr. *Austral. Encyclop.* v.8, 1965, 149–50.

SMITH, Henry Harwood (1867–1935)
d. Haywards Heath, Sussex 13 Jan. 1935

FLS 1923. Managing Director, Messrs Charlesworth and Co. nursery. Grew orchids.

Gdnrs Chron. 1925 i 50 portr.; 1935 i 53. *Orchid Rev.* 1935, 35 portr., 63. *Proc. Linn. Soc.* 1934–35, 191.

SMITH, Hugh Roger- *see* Roger-Smith, H.

SMITH, James (*c.* 1760–1848)
b. Ayrshire *c.* 1760 *d.* Ayr 1 Jan. 1848

Superintendent, W. Curtis's botanic garden, Lambeth, London, until 1784. Nurseryman, Monkwood, Ayr, 1784. Discovered *Veronica hirsuta*. "This simple monument to the father of Scottish Botany" (gravestone in Ayr churchyard). 'Varieties of British Plants cultivated and sold by James Smith and Son...' (*Gdnrs Mag.* 1830, 713–18).

W. J. Hooker *Fl. Scotica* 1821, 6. *Gdnrs Mag.* 1840, 46–47. *Gdnrs Chron.* 1932 ii 36.

SMITH, James (1837–1903)
b. Belhelvie near Aberdeen 1837 *d.* Cullen, Aberdeenshire 17 Sept. 1903

VMH 1897. Gardener to Earl of Roseberry at Mentmore, Leighton Buzzard, Beds, 1875–1903. Contrib. to *Gdnrs Chron.*

Garden v.64, 1903, 228; v.65, 1904, 32. *Gdnrs Chron.* 1903 ii 231 portr. *Gdnrs Mag.* 1903, 651 portr. *J. Hort. Cottage Gdnr* v.47, 1903, 296.

SMITH, Sir James Edward (1759–1828)
b. Norwich, Norfolk 2 Dec. 1759 *d.* Norwich 17 March 1828

MD Leyden 1786. FRS 1785. Knighted 1814. Purchased Linnaeus's collections, 1784. Founded Linnean Society, 1788; President, 1788–1828. *Plantarum Icones Hactenus Ineditae* 1789–91. *Reliquiae Rudbeckianae* 1789. *Icones Pictae Plantarum Rariorum* 1790–93. *Spicilegium Botanicum* 1791–92. *Fl. Britannica* 1800–4 3 vols. *Exotic Botany* 1804–5 2 vols. *English Fl.* 1824–28 4 vols. *English Fl.* (with J. Sowerby) 1790–1814 36 vols. Contrib. botanical articles to *Rees' Cyclop.* 1802–20.

J. Nichols *Illus. Lit. Hist. Eighteenth Century* v.6, 830–50 portr. *Gent. Mag.* 1828 i 297–300 portr. *Mag. Nat. Hist.* 1829 i 91–93. P. Smith *Mem. Correspondence of...Sir J. E. Smith* 1832 2 vols. *Cottage Gdnr* v.5, 1850, 185; v.9, 1853, 36. *Proc. Linn. Soc.* 1887–88, 22–28; 1931–32, 83–84; 1933–34, 10–21; 1936–37, 6–11. *J. Bot.* 1896, 308; 1898, 297–302; 1902, 321–22; 1915, 34–36; 1918, 276–77. G. C. Druce *Fl. Berkshire* 1897, clviii–clxii. E. Bretschneider *Hist. European Bot. Discoveries in China* 1898, 207–8. *D.N.B.* v.53, 61–64. *Trans. Norfolk Norwich Nat. Soc.* 1912–13, 645–92. W. R. Dawson *Smith Papers* 1934. A. T. Gage *Hist. Linn. Soc. of London* 1938 portr. *Ann. Sci.* 1949, 105–14. *J. Soc. Bibl. Nat. Hist.* 1957, 281–90; 1956, 133–34. F. A. Stafleu *Taxonomic Literature* 1967, 449–51. C. C. Gillispie *Dict. Sci. Biogr.* v.12, 1975, 471–72. *R.S.C.* v.5, 725.

Herb., MSS. and Library at Linnean Society. Letters at BM(NH). Portr. and bust by F. Chantry at Linnean Society. Portr. at Kew, Hunt Library.

Smithia Aiton.

SMITH, Rev. John *see* Smyth, J.

SMITH, John (*fl.* 1830s)

Gardener to Dykes Alexander, Ipswich, Suffolk. *Treatise on Artificial Growth of Cucumbers and Melons* 1833.

SMITH, John (1798–1888)
b. Aberdour, Fife 5 Oct. 1798 *d.* Kew, Surrey 14 Feb. 1888

ALS 1837. Gardener, Royal Botanic Garden, Edinburgh, 1818; Kew, 1822; Curator, 1842–64. *Cultivated Ferns* 1857. *Ferns: British and Foreign* 1866. *Domestic Botany* 1871. *Historia Filicum* 1875. *Bible Plants* 1878. *Adam Spade* 1879. *Records of Royal Botanic Gardens, Kew* 1880. *Dictionary of Popular Names of Plants* 1882.

Gdnrs Mag. 1840, 590–91; 1842, 189. *Cottage Gdnr* v.16, 1856, 15–17 portr. *J. Bot.* 1864, 191–92; 1865, 184–85; 1888, 102–3. *Gdnrs Chron.* 1865, 986; 1876 i 363–65 portr.; 1888 i 216. *J. Hort. Cottage Gdnr* v.12, 1886, 287. *Proc. Linn. Soc.* 1887–88, 96–98. *Ann. Bot.* 1888–89, 429–30. *Garden* v.33, 1888, 399. *J. Kew Guild* 1895, 27–28; 1965, 576–87 portr. *Kew Bull.* 1914, 85–87; 1920, 71. *Br. Fern Gaz.* 1967, 330–34 portr.

Ferns, etc. at BM(NH). MSS., unpublished autobiography, letters and portr. at Kew. Portr. at Hunt Library.

SMITH, John (1821–1888)
b. Kelso, Roxburghshire 1821 *d.* Kew, Surrey 11 May 1888
Gardener to Duke of Roxburgh; to Duke of Northumberland at Syon House, Middx, 1859–64. Curator, Kew, 1864–86.
Gdnrs Chron. 1873, 1701–2 portr.; 1888 i 624. *J. Hort. Cottage Gdnr* v.16, 1888, 410. *J. Kew Guild* 1897, 32–33 portr.

SMITH, John (*fl.* 1830s–1840s)
Nurseryman, Dalston, Hackney, London.

SMITH, John (1845–1930)
b. Clarkston, Lanarkshire 1845 *d.* Kilwinning, Ayrshire 30 Nov. 1930
Engineer. *Botany of Ayrshire* 1896. Contrib. to *Glasgow Nat., Ann. Andersonian Nat. Soc.*
Glasgow Nat. 1920, 20 portr.; 1930, 100.

SMITH, John Henderson (1875–1952)
d. 28 Nov. 1952
Educ. Oxford. MB, ChB Edinburgh 1903. Assistant bacteriologist, Lister Institute, Oxford until 1916. Mycologist, Rothamsted Experiment Station, 1919; Head, Dept. of Plant Pathology, 1933–40. President, Association of Applied Biologists, 1936–37.
Gdnrs Chron. 1952 ii 238. *Times* 2 Dec. 1952. *Nature* v.171, 1953, 16–17.

SMITH, Joseph (*fl.* 1830s–1840s)
Nurseryman, Westerham, Kent.

SMITH, Joseph (*c.* 1774–1857)
d. London 26 May 1857
FRS 1819. FLS 1811. Barrister. "Well acquainted with British plants, and wrote a memoir on the Guernsey Lily, which however has not been published."
Proc. Linn. Soc. 1857–58, xxxvii.

SMITH, Joseph Crosby- *see* Crosby-Smith, J.

SMITH, Mathew Richard (*c.* 1749–1819)
b Henton St. George, Somerset 22 June 1749 *d.* Sylhet, India 14 July 1819
Magistrate, Pundua, Sylhet where he had a garden of native plants. Sent plants to Botanic Garden, Calcutta. Collected in Bengal, 1810–16.
N. Wallich *Plantae Asiaticae Rariores* v.3, 1832 *passim. Bot. Mag.* 1831, t.3049. I. H. Burkill *Chapters on Hist. Bot. in India* 1965 *passim.*

SMITH, Maria Emma *see* Gray, M. E.

SMITH, Martin Ridley (–1908)
d. Hayes, Kent 8 Nov. 1908
VMH 1897. Grew carnations. President, National Carnation and Picotee Society.
Gdnrs Mag. 1907, 205 portr., 206. *Gdnrs Chron.* 1908 ii 354 portr.

SMITH, Mary Playne (*fl.* 1840s–1850s)
Of Nailsworth, Glos. Had herb. of local plants.
H. J. Riddelsdell *Fl. Gloucestershire* 1948, cxxi.
Herb. at Gloucester Museum.

SMITH, Matilda (1854–1926)
b. Bombay 30 July 1854 *d.* Kew, Surrey 29 Dec. 1926
ALS 1921. Botanical artist, Kew. Contrib. plates to *Bot. Mag.* 1878–1921, *Hooker's Icones Plantarum* 1881–1921, W. B. Hemsley's *Rep. on... Voyage of HMS Challenger* 1885, I. B. Balfour's *Bot. of Socotra* 1888.
J. Kew Guild 1916, 265 portr.; 1922, 83; 1927, 527–28 portr. *Kew Bull.* 1920, 210; 1921, 317–18; 1927, 135–39. *Bot. Soc. Exch. Club Br. Isl. Rep.* 1926, 383. *Gdnrs Chron.* 1927 i 40. *J. Bot.* 1927, 57–58. *Proc. Linn. Soc.* 1926–27, 100–1. *Gesneriad Saintpaulia News* v.1(2), 1964, 34 portr. *Gloxinian* v.20(1), 1970, 11 portr.
Smithiantha Kuntze. *Smithiella* S. T. Dunn.

SMITH, Noel James Gillies (1899–)
b. 25 Dec. 1899
PhD Cantab 1926. FLS 1934. Assistant in Botany, Aberdeen University, 1925–26. Professor of Botany, Rhodes University College, S. Africa, 1926–48. Botanised in S. Africa, particularly Namib Desert, 1933. Contrib. to *Ann. Applied Biol.*

SMITH, Pleasance (*née* **Reeve**) (1773–1877)
b. Lowestoft, Suffolk 11 May 1773 *d.* Lowestoft 3 Feb. 1877
Wife of Sir J. E. Smith and benefactor of Linnean Society. *Mem. and Correspondence of Sir J. E. Smith* 1832 2 vols.
J. Bot. 1877, 95–96.

SMITH, Richard (1748–1810)
d. Beeston 1810
Nurseryman, St. John's Nurseries, Worcester.

SMITH, Richard (1780–1848)
Son of R. Smith (1748–1810). Nurseryman, Worcester.

SMITH, Robert (1873–1900)
b. Dundee, Angus 11 Dec. 1873 *d.* Edinburgh 28 Aug. 1900
BSc Dundee 1896. Demonstrator in Botany, Dundee. Ecologist. Made botanical survey of Scotland. 'Plant Associations of Tay Basin' (*Trans. Perthshire Soc. Nat. Sci.* v.2, 1898, 200–17; v.3, 1900, 69–87 portr.). Contrib. to *Scott. Geogr. Mag.*
Ann. Scott. Nat. Hist. 1901, 1–2. *Irish Nat.* 1901, 48. *J. Bot.* 1901, 30–33 portr. *Scott. Geogr. Mag.* 1909, 597–600 portr. *Naturalist* 1923, 107 portr. *J. Ecol.* 1929, 2–3 portr. *Advancement of Sci.* 1957, 253.

SMITH, Rose Carr (*fl.* 1900s–1920s)
Contributed plant records to Devonshire Association of which she was a member.
W. K. Martin and G. T. Fraser *Fl. Devon* 1939, 778.
Plants at Torquay Natural History Society Museum.

SMITH, Roy Leslie (1892–1973)
b. Cardiff, Glam 26 June 1892 *d.* Cardiff 6 April 1973
Electrician, Cardiff. Studied adventive flora of Cardiff and Barry docks.
Watsonia 1974, 205.
Herb. at National Museum of Wales.

SMITH, Russell Dudley- *see* Dudley-Smith, R.

SMITH, Samuel (*fl.* 1780s–1830s)
Nurseryman, Dalston, Hackney, London.

SMITH, Thomas (*fl.* 1620s)
Apothecary. Accompanied T. Johnson (1597–1644) on simpling ride through Windsor Forest.
J. Gerard *Herball* 1633, 30.

SMITH, Thomas (–*c.* 1825)
FRS 1816. FLS 1799. Of London. Microscopist. Friend of R. Brown. 'Carduus and Cnicus' (*Trans. Linn. Soc.* v.13, 1822, 592–603).
Trans. Linn. Soc. v.19, 1845, 341–45. *R.S.C.* v.5, 732.
Sale of property at Sotheby, 9 May 1825.
Thismia Griff.

SMITH, Thomas (1820–1904)
b. 31 Aug. 1820 *d.* Stranraer, Wigtownshire 18 June 1904
Nurseryman, Stranraer, 1861.
Garden v.65, 1904, 378. *Gdnrs Chron.* 1904 i 367.

SMITH, Thomas (*c.* 1840–1919)
d. Newry, N. Ireland 23 May 1919
VMH 1906. Nurseryman, Daisy Hill Nurseries, Newry, 1887.
Gdnrs Chron. 1919 i 287 portr.

SMITH, Thomas (1857–1955)
VMH 1950. Horticulturist, Mayland, Essex. *French Gardening* 1909. *Profitable Culture of Vegetables* 1911.
R. Webber *Early Horticulturists* 1968, 162–71.

SMITH, Thomas Algernon Dorrien- *see* Dorrien-Smith, T. A.

SMITH, Mrs. W. Anderton (*fl.* 1840s)
Discovered *Epipogium aphyllum* in Herefordshire, 1842.
Trans. Woolhope Nat. Field Club v.34, 1954, 247–48.

SMITH, William (*fl.* 1770s–1789)
Nurseryman, Kensington, London. Junior partner of Henry Hewitt.

SMITH, William (*fl.* 1790s–1840s)
Nurseryman, High Street, Burton-on-Trent, Staffs.

SMITH, William (*c.* 1804–1828)
b. Hopetoun, Linlithgowshire *c.* 1804 *d.* London 15 Nov. 1828
ALS 1828. Gardener at Chiswick gardens of Horticultural Society of London. Studied British roses.
Gdnrs Mag. 1829, 495–96.

SMITH, Rev. William (1808–1857)
b. Ballymoney, County Antrim 12 Jan. 1808 *d.* Cork 6 Oct. 1857
FLS 1847. Professor of Natural History, Cork, 1854. *Synopsis of British Diatomaceae* 1853–56 2 vols.
Proc. Linn. Soc. 1857–58, xxxvii–xxxviii. *R.S.C.* v.5, 733.
Diatomaceae at BM(NH) (list published 1859). Plants at University College, Cork.

SMITH, William (*fl.* 1810s–1840s)
Nurseryman, Norbiton Common, Surrey.

SMITH, William Arthur Hans Bernhard- *see* Bernhard-Smith, W. A. H.

SMITH, William Gardner (1866–1928)
b. Dundee, Angus 20 March 1866 *d.* Edinburgh 8 Dec. 1928
BSc Dundee 1890. PhD Munich 1894. Lecturer in Botany, Yorkshire College, Leeds. Head, Biological Dept., East of Scotland College of Agriculture, 1908. President, British Ecological Society, 1917–18. *Diseases of Plants induced by Cryptogamic Parasites* 1897 (translation of work by K. Tubeuf). 'Geographical Distribution of Vegetation in Yorkshire' (with C. E. Moss and W. M. Rankin) (*Geogr. J.* v.21, 1903, 375–401; v.22, 1903, 149–78).
Bot. Soc. Exch. Club Br. Isl. Rep. 1928, 711–12. *Pharm. J.* 1928, 591. *J. Bot.* 1929, 53–56 portr. *J. Ecol.* 1929, 170–73. *Kew Bull.* 1929, 90. *Naturalist* 1929, 153–56 portr. *Nature* v.123, 1929, 101. *Trans. Proc. Bot. Soc. Edinburgh* 1929, 175–78.

SMITH, William Robertson (1828–1912)
b. Athelstaneford, East Lothian 21 March 1828 *d.* Washington, D.C. 7 July 1912
Kew gardener. To Philadelphia, 1853. Superintendent, Botanic Garden, Washington, 1854–1912. Grew plants from Wilkes Expedition.
Gdnrs Mag. 1907, 92. *Torreya* 1912, 199. *J. Kew Guild* 1913, 112–13 portr. L. H. Bailey *Standard Cyclop. Hort.* v.2, 1939, 1597.

SMITH, Rev. William Sunderland
(–1912)
d. Antrim, N. Ireland 1912
Studied littoral flora of Lough Neagh. *Gossip about Lough Neagh* 1885 (includes plant list).
Irish Nat. 1913, 32. *Proc. Belfast Nat. Field Club* v.6, 1913, 627.

SMITH, Sir William Wright (1875–1956)
b. Lochmaben, Dumfriesshire 2 Feb. 1875
d. Edinburgh 15 Dec. 1956
MA Edinburgh 1896. FRS 1945. FLS 1918. VMH 1925. VMM 1930. Lecturer in Botany, Edinburgh University, 1902–7. Keeper, Herb., Royal Botanic Garden, Calcutta, 1907–11. Director, Botanical Survey of India. Botanised in Himalayas. Deputy Keeper, Royal Botanic Garden, Edinburgh, 1911; Keeper, 1922–56. Published papers on *Primula* in *Trans. R. Soc. Edinburgh*, *Trans. Bot. Soc. Edinburgh*.
Gdnrs Chron. 1932 i 434 portr.; 1956 ii 676. *Pharm. J.* 1956, 477. *Times* 18 Dec. 1956. *J. R. Hort. Soc.* 1957, 103–6 portr. *Biogr. Mem. Fellows R. Soc.* 1957, 193–202 portr. *Empire For. Rev.* 1957, 3–4. *Forestry* 1957, 45. *Nature* v. 179, 1957, 126–27. *Proc. Bot. Soc. Br. Isl.* 1957, 427–28. *Taxon* 1957, 55–56. *Trans. Proc. Bot. Soc. Edinburgh* 1957, 142–45 portr. *Yb. R. Soc. Edinburgh* 1958, 47–50. H. R. Fletcher and W. H. Brown *R. Bot. Gdn Edinburgh, 1670–1970* 1970, 225–26, 233–55 portr.
Portr. at Hunt Library.

SMITH, Winifred (1858–1925)
b. Mortlake, Surrey 5 Nov. 1858 *d.* London 24 Dec. 1925
BSc London 1904. FLS 1908–20. Lecturer in Botany, University College, London. '*Macaranga triloba*' (*New Phytol.* 1903, 79–82). 'Anatomy of Sapotaceous Seedlings' (*Trans. Linn. Soc. Bot.* v.7, 1909, 189–200).
J. Bot. 1926, 56.

SMITH, Winifred Lily Boys- *see* Boys-Smith, W. L.

SMITH, Worthington George (1835–1917)
b. London 23 March 1835 *d.* Dunstable, Beds 27 Oct. 1917
FLS 1868. VMM 1907. Botanical artist and mycologist. Artist for *Gdnrs Chron.* 1869–1910. President, British Mycological Society, 1904. *Mushrooms and Toadstools* 1867. 'Clavis Agaricinorum' (*J. Bot.* 1870, 137–45, 176–82, 213–23, 246–52). *Diseases of Field and Garden Crops* 1884. *Guide to Sowerby's Models of British Fungi in... BM(NH)* 1893. *Synopsis of Br. Basidiomycetes* 1908. Illustrated *Floral Mag.* 1869–76; A. Smee's *My Gdn* 1872; *Illus. Br. Fl.* (with W. H. Fitch) 1887.
Gdnrs Chron. 1907 i 188 portr.; 1917 ii 180–81 portr. *J. Bot.* 1909, 35–36; 1918, 243–47. *J. Hort. Home Farmer* v.64, 1912, 214 portr. *Nature* v.100, 1917, 170, 191, 209. *Bot. Soc. Exch. Club Br. Isl. Rep.* 1917, 92–93. *Kew Bull.* 1918, 31–32. *Proc. Linn. Soc.* 1917–18, 42–43. *Trans. Br. Mycol. Soc.* 1917–19, 65–67. *Naturalist* 1923, 109 portr. W. Blunt *Art of Bot. Illus.* 1955, 242–44. *Bedfordshire Mag.* 1967, 73–79 portr.

SMITH-PEARSE, Rev. Thomas Northmore Hart (1854–1943)
b. 9 June 1854 *d.* 11 Jan. 1943
MA Oxon. Master, Marlborough College, 1879–89. Headmaster, Epsom College, 1889–1914. *Fl. Epsom and its Neighbourhood* 1917.
Who was Who, 1941–1950 1075.

SMITHAM, Richard Willyams (1858–1928)
d. Fowey, Cornwall 23 Nov. 1928
Headmaster, Fowey Council Boys' School. Cryptogamist and algologist.
Rep. Br. Bryol. Soc. 1928, 141.
Herb. at Museum of Cornwall Institution, Truro.

SMITHIES, James John (*c.* 1850–1931)
d. Kendal, Westmorland 8 June 1931
Of Manchester and Kendal. Collected ferns in Lake District and Ireland.
Br. Fern Gaz. 1931, 84–85.

SMITHSON, Elizabeth Anne *see* Lomax, E. A.

SMITTEN, George (*fl.* 1780s–1800s)
Nurseryman, Dublin.
Irish For. 1967, 48.

SMYTH (*or* Smith), Rev. John (*fl.* 1690s)
"Minister to Royal African Company in the English Factory at Cabo Corso, vulgarly called Cape Coast, in Guinea." Sent Guinea plants to J. Petiver.
J. Petiver *Museii Petiveriani* 1695, 21. *Philos. Trans. R. Soc.* v.19, 1697, 677–86. J. E. Dandy *Sloane Herb.* 1958, 209.
Plants at BM(NH).

SMYTHE, Crosby Wilson (1879–1909)
d. London 15 Oct. 1909
Kew gardener, 1900–4. Curator, Botanic Station, Sierra Leone, 1904. Agricultural Superintendent, Sierra Leone. Collected plants in Sierra Leone and Gold Coast.
Kew Bull. 1904, 13; 1909, 391. *J. Kew Guild* 1910, 490. A. H. Unwin *West African Forests and Forestry* 1920 portr.

SMYTHE, Francis Sydney (*c.* 1901–1949)
d. 27 June 1949
Climbed in the Alps and Himalayas where he collected plants for Kew and Edinburgh. *Valley of the Flowers* 1938.
Gdnrs Chron. 1949 ii 16. *Nature* v.164, 1949, 266.

SMYTHE, William (*fl.* 1870s)
Gardener to W. Nicholson at Basing Park, Hants, *c.*1876. Hybridised kidney beans.
Gdnrs Mag. 1898, 168 portr. *Gdnrs Chron.* 1899 ii 42 portr.

SNEATH, J. S. (1841–1924)
E. J. Gibbons *Fl. Lincolnshire* 1975, 60.
Herb. at Lincoln Museum.

SNELGROVE, Edward (1859–1934)
b. near Leek, Staffs 1859 *d.* Sheffield, Yorks 9 April 1934
Headmaster, Sheffield schools. President, Sheffield Naturalists' Club. *Object Lessons in Botany from Forestry, Field and Garden* 1894–96 3 vols. Botany of Sheffield District (*Br. Assoc. Handbook* 1910, 405–33).
N. Western Nat. 1934, 162–64.

SNELL, John (1879–1920)
b. Cornwall 1879 *d.* Preston, Lancs 19 April 1920
BSc London. Lecturer in Botany at Birkbeck College. '*Vicia Faba*' (*Ann. Bot.* 1911, 845–55).
J. Bot. 1920, 158.

SNELLING, Lilian (1879–1972)
b. St. Mary Cray, Kent 8 June 1879 *d.* St. Mary Cray 12 Oct. 1972
VMH 1955. Botanical artist. At Royal Botanic Garden, Edinburgh, 1916–21. Illustrated *Bot. Mag.* 1922–52; *Supplement to Elwes' Monograph of Genus Lilium* 1934–40; F. Stoker's *Book of Lilies* 1943; F. C. Stern's *Study of Genus Paeonia* 1946.
Bot. Mag. 1952–53 portr. W. Blunt *Art of Bot. Illus.* 1955, 251. *Gdnrs Chron.* 1956 i 7 portr. Hunt Library *Cat. 3rd International Exhibition of Bot. Art and Illus.* 1972, 159 portr. *Times* 17 Oct. 1972. *J. R. Hort. Soc.* 1973, 139. Hunt Library *Artists from R. Bot. Gdns, Kew* 1974, 56–57 portr.
Portr. at Hunt Library. Drawings at Kew.

SNEYD, John (1734–1809)
Of Bishton, Staffs. Friend of J. Banks. Had a garden. Contrib. to W. Withering's *Systematic Arrangement* 1796 and assisted Botanical Society of Lichfield in their production of *Families of Plants* 1787.
J. Bot. 1914, 322.

SNOOKE, William Drew (1787–1857)
b. Wool, Dorset 6 Nov. 1787 *d.* Ryde, Isle of Wight 5 Sept. 1857
Fl. Vectiana 1823.

SNOW, George Robert Sabine (1897–1969)
b. 19 Jan. 1897 *d.* Vernet-les-Bains, France 24 July 1969
Educ. Oxford. FRS 1948. Fellow, Magdalen College, 1922–60. Botanist with special interest in geotropism and phyllotaxis. Contrib. to *Ann. Bot.*, *Proc. R. Soc.*, *New Phytol.*
Times 4 Aug. 1969. *Biogr. Mem. Fellows R. Soc.* 1970, 499–522 portr.

SNOW, Seward (*c.* 1793–1869)
d. Wrest Park, Beds 10 March 1869
Vegetable grower. Re-introduced 'Muscat Hamburgh' grape.
Florist and Pomologist 1869, 96.

SNOWDEN, Joseph Davenport (1886–1973)
b. Silverdale, Staffs 31 May 1886 *d.* 9 May 1973
Kew gardener, 1909. Assistant Agricultural Officer, Uganda, 1911; District Agricultural Officer, 1918; Senior District Agricultural Officer, 1928. Economic Botanist, 1930–31. Collected plants, especially grasses, for Kew. *Cultivated Races of Sorghum* 1936. *Grass Communities and Mountain Vegetation of Uganda* 1953.
Nature v.134, 1934, 844. *J. Kew Guild* 1960, 761–62 portr.; 1974, 345–46.
Herb. at BM(NH). Plants and *Sorghum* note-books at Kew.
Snowdenia C. E. Hubbard.

SOLANDER, Daniel Carl (1733–1782)
b. Piteå, Norrland, Sweden 19 Feb. 1733 *d.* London 13 May 1782
MD Uppsala. DCL Oxon 1771. FRS 1764. To England, 1760. Assistant Librarian, BM, 1763; Keeper, Natural History Dept., 1773. Accompanied J. Banks on J. Cook's first voyage, 1768–71, and to Iceland, 1772. Librarian to Banks, 1771. Edited Linnaeus's *Elementa Botanica* 1756; J. Ellis's *Nat. Hist. of...Zoophytes* 1786. 'Gardenia' (*Philos. Trans. R. Soc.* v.52, 1762, 654–61). Assisted W. Aiton in early planning of *Hortus Kewensis* 1789.
Gent. Mag. 1782, 263; 1784 ii 886. *Berlinische Monatsschrift* v.6, 1785, 240–49. R. Pulteney *Hist. Biogr. Sketches of Progress of Bot. in England* v.2, 1790, 350–51. J. E. Smith *Selection of Correspondence of Linnaeus* 1821 *passim.* A. L. A. Fée *Vie de Linné* 1832, 180–84. *Proc. Linn. Soc.* 1888–89, 40. J. Banks *Journal...1768–1771* 1896, xxxviii–xlii portr. *J. Bot.* 1897, 481–82; 1912 Supplt, 3. *D.N.B.* v.53, 212–13. T. M. Fries *Linné* v.2, 1903, 62–64 portr. *J. Proc. R. Soc. N.S.W.* 1905, 34–39; 1908, 81–82; 1921, 166–67. J. H. Maiden *Sir Joseph Banks* 1909, 73–76 portr. E. Smith *Life of Sir Joseph Banks* 1911 *passim. J. Linn. Soc.* v.45, 1920, 47, 51. *Kungl. Svenska Vetenskapsakad. Arsbok* 1940, 279–301. P. Serle *Dict. Austral. Biogr.* v.2, 1949, 340–41. *Fl. Malesiana* v.1, 1950, 493. *Gdnrs Chron.* 1953 ii 140–41 portr. *Chronica Botanica* 1954, 172–75.

W. R. Dawson *Banks Letters* 1958 *passim. Svenska Linné-Sällsk. Arsskr.* v.43, 1960, 53–71, 146–47; v.45, 1962, 128–37; 1954–55, 23–64. *J. Soc. Bibl. Nat. Hist.* 1962, 57–62. *Isis* 1964, 62–67; 1967, 367–74. *Austral. Dict. Biogr.* v.2, 1967, 456–57. *Austral. Encyclop.* v.8, 1965, 195. *Trans. Amer. Philos. Soc.* v.58(8), 1968, 1–66. C. C. Gillispie *Dict. Sci. Biogr.* v.12, 1975, 515–16.

Plants and MSS. at BM(NH). Portr., medallion and MSS. at Linnean Society. MSS. at Wedgwood Museum, Etruria. Letters at Fitzwilliam Museum Cambridge, Mitchell Library Sydney. Portr. at Hunt Library.

Solandra L.

SOLE, William (1741–1802)
b. Thetford, Cambridgeshire June 1741 *d.* Bath, Somerset 7 Feb. 1802

ALS 1788. Apothecary, Bath. MS. *Fl. Bathonica* 1782. *Menthae Britannicae* 1798.

D. Turner and L. W. Dillwyn *Botanist's Guide through England and Wales* 1805, 747. *Phytologist* v.3, 1849, 581. *D.N.B.* v.53, 213. J. W. White *Fl. Bristol* 1912, 71–73. *J. Bot.* 1914, 317. *Bot. Soc. Exch. Club Br. Isl. Rep.* 1914, 50. G. C. Druce *Comital Fl. Br. Isl.* 1932, xii. *Proc. Linn. Soc.* 1937–38, 52–58. H. J. Riddelsdell *Fl. Gloucestershire* 1948, cxiii. D. Grose *Fl. Wilts* 1957, 33.

Mentha and letters to Lambert at BM(NH).

SOLLY, Edward (1819–1886)
b. London 11 Oct. 1819 *d.* Sutton, Surrey 2 April 1886

FRS 1843. FLS 1842. Brother-in-law of J. Royle. Hon. Professor of Chemistry, Horticultural Society of London, 1842. Professor of Chemistry, Addiscombe, Surrey, 1845–49. *Rural Chemistry* 1843. 'Experiments on the Inorganic Constituents of Plants' (*Trans. Hort. Soc. London* v.3, 1848, 35–92). 'On Seed-steeping' (*Trans. Hort. Soc. London* v.3, 1848, 197–210). 'On the Exhaustion of Soils' (*Trans. Hort. Soc. London* v.3, 1848, 189–95).

D.N.B. v.53, 214–15. *R.S.C.* v.5, 745–46.

SOLLY, Richard Horsman (1778–1858)
b. London 29 April 1778 *d.* London 31 March 1858

MA Cantab 1803. FRS 1807. FLS 1826. Studied plant physiology and anatomy.

Bot. Register 1831, t.1466. *Proc. Linn. Soc.* 1857, xxxviii–xl. *Proc. R. Soc.* v.9, 1859, 549–50.

Sollya Lindl.

SOMERSET, Mary (*née* Capel), Duchess of Beaufort (*c.* 1630–1714)
d. Badminton, Glos 7 Jan. 1714

Had botanic gardens at Badminton and Chelsea and a large collection of flower drawings.

D. Turner *Extracts from Lit. and Sci. Correspondence of R. Richardson* 1835, 33. J. C. Loudon *Arboretum et Fruticetum Britannicum* 1838, 61. E. J. L. Scott *Index to Sloane Manuscripts* 1904, 35. A. W. Hill *Henry Nicholson Ellacombe* 1919, 186–87. *Garden* 1920, 428–29. J. E. Dandy *Sloane Herb.* 1958, 209–15. A. M. Coats *Gdn Shrubs and their Histories* 1963, 375–76. *Gdnrs Chron.* 1964 ii 333 portr. *House and Gdn* 1972, 136–37 portr.

Plants at BM(NH).

Beaufortia R. Br.

SOMERVILLE, Alexander (1842–1907)
b. Glasgow 25 March 1842 *d.* Glasgow 5 June 1907

BSc Glasgow. FLS 1881. Merchant, Calcutta for 15 years. 'Additional Records for Scilly Isles' (*J. Bot.* 1893, 118–20). Contrib. to *Trans. Glasgow Nat. Hist. Soc.*, *Trans. Bot. Soc. Edinburgh.*

Ann. Scott. Nat. Hist. 1907, 193–95 portr. *J. Bot.* 1907, 288. *Bot. Soc. Exch. Club Br. Isl. Rep.* 1907, 262–63. *Proc. Linn. Soc.* 1907–8, 61. *Trans. Bot. Soc. Edinburgh* v.23, 1908, 365–67. *Trans. Nat. Hist. Soc. Glasgow* v.8, 1906–8, 227–30. *Watson Bot. Exch. Club Rep.* 1906–7, 74–75 portr. F. H. Davey *Fl. Cornwall* 1909, lx. J. E. Lousley *Fl. Isles of Scilly* 1972, 83.

Herb. at Glasgow University.

SOMERVILLE, Sir William (1860–1932)
b. Cormiston, Lanarkshire 30 May 1860 *d.* Boar's Hill, Oxford 17 Feb. 1932

BSc Edinburgh 1887. FLS 1891. KBE 1926. Lecturer in Forestry, Edinburgh University, 1889–91. Professor of Agriculture, Newcastle-upon-Tyne, 1891–99. Professor of Agriculture, Cambridge, 1899–1901. Assistant Secretary, Board of Agriculture, 1901–6. Professor of Rural Economy, Oxford, 1906–25. President, Royal English Arboricultural Society, 1900–1, 1922–24. Edited *J. Arboricultural Soc.* 1910–23. *Textbook of Diseases of Trees* 1894. *How a Tree Grows* 1927.

Nature v.129, 1932, 389–90. *Times* 18 Feb. 1932. *Proc. Linn. Soc.* 1931–32, 187–89. *Proc. R. Soc. Edinburgh* v.52, 1931–32, 479–80. *Who was Who, 1929–1940* 1266. *D.N.B. 1931–1940* 826–27.

SOMMERVILLE, Thomas (*c.* 1783–1810)
d. Edinburgh 17 March 1810

Superintendent, Royal Botanic Garden, Edinburgh.

Mem. Wernerian Nat. Hist. Soc. v.1, 1811, 246. *Notes from R. Bot. Gdn Edinburgh* 1908, 291–92.

SOPPITT, Henry Thomas (1858–1899)
b. Bradford, Yorks 21 June 1858 *d.* Halifax, Yorks 1 April 1899

Mycologist. '*Alcidium leucospermum*' (*J. Bot.* 1893, 273–74). Contrib. botany to J. Gray's *Through Airedale from Goole* 1891.

Gdnrs Chron. 1899 i 239–40 portr. *J. Bot.* 1899, 240. *Naturalist* 1899, 157–60 portr.; 1961, 54–55. *Trans. Br. Mycol. Soc.* 1897–98, 83–85 portr. W. B. Crump and C. Crossland *Fl. of Parish of Halifax* 1904, lxiii.

SORBY, Henry Clifton (1826–1908)
b. Woodbourne, Sheffield, Yorks 10 May 1826 *d.* Sheffield 8 March 1908

LLD Cantab 1879. FRS 1857. FLS 1875. Chemist. Mineralogist. 'Analysis of Colouring Matters' (*Proc. R. Soc.* v.15, 1867, 433–35).

J. Bot. 1876, 16–18. *Naturalist* 1906, 137–44 portr. *Geol. Mag.* 1908, 193–204 portr. *Proc. Linn. Soc.* 1907–8, 61–62. *Proc. R. Soc.* v.80, 1908, lvi–lxvi. *D.N.B. Supplt 2* v.3, 355–56. *R.S.C.* v.5, 752; v.8, 983; v.11, 452; v.18, 854.

Letters at Kew. Portr. at Hunt Library.

SORRELL, John (–1811)

Nurseryman, Springfield near Chelmsford Essex.

SORRELL, Thomas (–1797)
d. 8 May 1797

Nurseryman, Springfield near Chelmsford, Essex.

SORRELL, Thomas (–1843)
d. 22 April 1843

Nurseryman, Chelmsford, Essex.

SORRELL, Thomas (–1849)
d. 10 Dec. 1849

Nurseryman, Chelmsford, Essex.

SOULSBY, Basil Harrington (1864–1933)
b. Christchurch, N.Z. 3 Nov. 1864 *d.* Reading, Berks 14 Jan. 1933.

BA Oxon 1897. FLS 1930. Library, BM, 1892–1909. Library, BM(NH), 1909; Librarian, 1921–30. *Catalogue of Works of Linnaeus* ed. 2 1933.

J. Bot. 1933, 73–74. *Nature* v.131, 1933, 230. *Proc. Linn. Soc.* 1932–33, 203–5. *Times* 16 Jan. 1933.

SOUTELLINHO, Baron de *see* Tait, A. W.

SOUTHALL, Henry (1826–1916)
b. Leominster, Herefordshire 16 July 1826 *d.* 27 Jan. 1916

Draper, Ross-on-Wye, Hereford. Grew alpine plants. Botanist. President, Woolhope Naturalists Field Club, 1889–91, 1903. Contrib. to *Trans. Woolhope Nat. Field Club.*

Trans. Woolhope Nat. Field Club 1914–17, 296–98; 1954, 246–47.

SOUTHBY, Anthony (*olim* **Gapper**)
(*fl.* 1830s–1840s)

MD. Of Bridgwater, Somerset. Sent catalogue of Somerset and Wiltshire plants to H. C. Watson. With R. Spruce in Pyrenees, 1845.

Ann. Mag. Nat. Hist. v.3, 1849, 501–3. H. C. Watson *Topographical Bot.* 1883, 556. *Essex Nat.* v.29, 1954, 195. *R.S.C.* v.5, 762.

Plants at Taunton.

Southbya Spruce.

SOUTTER, J. P. (*fl.* 1880s)

Of Bishop Auckland, Durham. British plants at Oxford.

SOWDEN, Harry (1870–1936)
b. Leeds, Yorks 28 June 1870 *d.* York 30 Aug. 1936

Policeman, York. President, Yorkshire Conchological Society, 1928. Bryologist.

N. Western Nat. 1936, 377–79 portr.

SOWERBY, Arthur De Carle (1885–1954)
b. Taiyuan-Fu, Shanghai 8 July 1885 *d.* 16 Aug. 1954

Pioneer naturalist in China where he collected for BM(NH) and U.S. National Museum. Zoologist. *Through Shen-kan* (with R. S. Clark) 1912. Founded and edited *China J. Sci. Arts* 1923–41 to which he contributed some botanical articles: 'Animals and Plants of China's Trade' (v.14, 1931, 256–62); 'Economic Animals and Plants of China' (v.18, 1933, 289–99); 'Fl. of Chinese Art' (v.26, 1937, 310–20). *Nature Notes: a Guide to Fauna and Fl. of a Shanghai Garden* 1939.

A. De C. Sowerby *Sowerby Saga* 1952. *Nature* v.174, 1954, 723. R. R. Sowerby *Sowerby of China* 1956 portr. *J. Soc. Bibl. Nat. Hist.* 1974, 529–34.

SOWERBY, Charles Edward (1795–1842)
b. London 1 Feb. 1795 *d.* London 7 May 1842

ALS 1827. Assisted his father, James, and his brother, James De Carle, with their natural history publications. Superintended smaller cheaper edition of *English Bot. Illustrated Catalogue of British Plants* 1842. Contrib. to J. Sowerby and J. E. Smith's *English Bot.* 2446.

Proc. Linn. Soc. 1842, 149. A. De C. Sowerby *Sowerby Saga* 1952. J. Collins *Sowerby Family* 1973. *J. Soc. Bibl. Nat. Hist.* 1974, 534.

SOWERBY, Charlotte Caroline (1820–1865)
Eldest daughter of George Brettingham Sowerby. Contrib. plates to E. G. Henderson's *Illustrated Bouquet* 1857–64 3 vols. Lithographed 12 plates in E. Hamilton's *Fl. Homoeopathica* 1852–53 2 vols.
A. De C. Sowerby *Sowerby Saga* 1952.

SOWERBY, Henry (1825–1891)
b. 28 March 1825 *d.* Doylesford, Victoria 15 Sept. 1891
Son of George Brettingham Sowerby. Assistant Librarian, Linnean Society, 1842–53. To Australia, 1854. Draughtsman, Melbourne University. Gold prospector. Lithographed 47 plates in E. Hamilton's *Fl. Homoeopathica* 1852–53 2 vols.
D.N.B. v.53, 305. *J. Soc. Bibl. Nat. Hist.* 1974, 534.

SOWERBY, James (1757–1822)
b. London 21 March 1757 *d.* Lambeth 25 Oct. 1822
ALS 1789. FLS 1793. Botanical artist. *Easy Introduction to Drawing Flowers according to Nature* 1788. *English Botany* (text by J. E. Smith) 1790–1814 36 vols. *Coloured Figures of English Fungi or Mushrooms* 1795–1815 4 vols. Illustrated J. E. Smith's *Exotic Botany* 1804–5. Engraved plates in J. Sibthorp's *Fl. Graeca* 1806–40 10 vols.
Gent. Mag. 1822 ii 568. *Cottage Gdnr* v.5, 1850, 29. *J. Bot.* 1872, 231–35, 356–60; 1888, 231–35, 268–69; 1903–4, Supplts; 1905 *passim*; 1918, 276–77. *D.N.B.* v.53, 305–7. *Trans. Br. Mycol. Soc.* 1933, 167–70. *Zeitschr. für Pilzkunde* 1929, 102–8. H. J. Riddelsdell *Fl. Gloucestershire* 1948, cxv–cxvi. W. Blunt *Art of Bot. Illus.* 1955, 190–92. A. De C. Sowerby *Sowerby Saga* 1952. *Sterkiana* no. 23, 1966, 1–6 portr. F. A. Stafleu *Taxonomic Literature* 1967, 452. J. Collins *Sowerby Family* 1973. *J. Soc. Bibl. Nat. Hist.* 1974, 386–89, 402–17, 484–92 portr. C. C. Gillispie *Dict. Sci. Biogr.* v.12, 1975, 552.
Herb., few models of fungi, MSS., letters and drawings at BM(NH). Portr. at Kew, Hunt Library.
Sowerbaea Smith.

SOWERBY, James (1815–1834)
b. 18 Nov. 1815 *d.* 1 Feb. 1834
Son of J. De Carle Sowerby. Published *Mushrooms and Champignon Illustrated* 1832 (plates adapted from J. Sowerby's *Coloured Figures of English Fungi*).
J. Soc. Bibl. Nat. Hist. 1974, 535.

SOWERBY, James Bryant (1855–1934)
b. 15 April 1855 *d.* 21 March 1934
Son of W. Sowerby (1827–1906) whom he succeeded as Secretary, Royal Botanic Society, Regent's Park, London, 1906–13.

SOWERBY, James De Carle (1787–1871)
b. Lambeth 5 June 1787 *d.* London 26 Aug. 1871
FLS 1823. FZS 1826. Eldest son of James Sowerby with whom he collaborated. Illustrated D. Turner's *Muscologiae Hibernicae Spicilegium* 1804 (original drawings at BM(NH)). With his brother, Charles Edward, produced *Supplt. to English Bot.* 1831–40 (original drawings at BM(NH)). Founder member of Royal Botanic Society and Gardens, Regent's Park, London, 1838; Secretary, 1839–69; his son, William, succeeded him as Secretary.
Gdnrs Chron. 1871, 1260 (reprint of *Lancet* 1871, 451–52). *Geol. Mag.* 1871, 478–79. *J. Bot.* 1871, 319–20; 1903 Supplt., 1–4. *Proc. Linn. Soc.* 1871–72, lxxix–lxxx. *Gdnrs Mag.* 1888, 494–95 portr. *D.N.B.* v.53, 307–8. A. De C. Sowerby *Sowerby Saga* 1952. *Sterkiana* no. 23, 1966, 1–6. J. Collins *Sowerby Family* 1973. *J. Soc. Bibl. Nat. Hist.* 1974, 389–95, 493–509.
Letters at BM(NH), Kew. Portr. at Hunt Library.

SOWERBY, John Edward (1825–1870)
b. Lambeth, London 17 Jan. 1825 *d.* Clapham, London 28 Jan. 1870
Eldest son of C. E. Sowerby (1795–1842). Botanical artist and publisher. *Ferns of Great Britain* (text by C. Johnson) 1855. *British Poisonous Plants* (text by C. Johnson) 1856; ed. 2 1861. *Fern Allies* (text by C. Johnson) 1856. *British Wild Flowers* (text by C. P. Johnson) 1860; 1863. *Useful Plants of Great Britain* (text by C. P. Johnson) 1862. *Grasses of Great Britain* (text by C. Johnson) 1857–61. *English Botany* ed. 3 1863–86. *Illustrated Key to Natural Orders of British Plants* 1865.
Gdnrs Chron. 1870, 559. *D.N.B.* v.53, 308. A. De C. Sowerby *Sowerby Saga* 1952. *Sterkiana* no. 23, 1966, 1–6. *J. Soc. Bibl. Nat. Hist.* 1974, 535–36.

SOWERBY, William (1827–1906)
b. 12 Feb. 1827 *d.* Ware, Herts 9 March 1906
FLS 1872. Son of J. De Carle Sowerby. Succeeded his father in 1869 as Secretary of Royal Botanic Society; succeeded by his son, James Bryant, 1906. Engraved plates in M. J. Berkeley and C. E. Brown's 'Notices of British Fungi' (*Ann. Mag. Nat. Hist.* 1864–65). *Catalogue of Medicinal and Economic Plants in Gardens, Regent's Park, Royal Botanic Society of London* 1882.
Gdnrs Mag. 1888, 494–95 portr. *Gdnrs Chron.* 1906 i 175. *Proc. Linn. Soc.* 1905–6, 46–47. A. De C. Sowerby *Sowerby Saga* 1952. *J. Soc. Bibl. Nat. Hist.* 1974, 396–98 536–37.
Portr. at Hunt Library.

SOWTER, Frederick Archibald (1899–1972)
b. Leicester 30 Aug. 1899 *d.* Leicester 16 Nov. 1972.
FLS 1944. Sales manager, Courtauld's Ltd., Leicester. Bryologist and lichenologist, particularly for Leicestershire and Rutland. President, British Bryological Society, 1958. Edited *Trans. Br. Bryol. Soc.* 1947–55. *Cryptogamic Fl. Leicestershire and Rutland: Bryophytes* 1941; *Lichens* 1950. Contrib. to *Trans. Br. Bryol. Soc., Trans. Leicester Lit. Philos. Soc.*
J. Bryol. 1973, 465–69 portr. *Lichenologist* 1973, 345–48 portr. *Trans. Leicester Lit. Philos. Soc.* 1973, 20–24 portr. *Watsonia* 1974, 114.
Herb. at Leicester Museum. Portr. at Hunt Library.

SPALDING, John (*c.* 1814–1905)
b. Woodside, Perthshire *c.* 1814 *d.* Feb. 1905
Nurseryman, New London, Conn., U.S.A.
J. Hort. Cottage Gdnr v.53, 1906, 547.

SPARE, Gordon H. (–1940/5)
d. Prisoner-of-war Thailand 1940/5
Kew gardener, 1929. Rubber planter, Johore, 1929–32; later in Perak and Kedah. Collected plants in Malaya.
Fl. Malesiana v.1, 1950, 496.
Plants at Kew, Singapore.
Fagraea sparei Henderson.

SPARROW, Harry Davis (1890–1959)
b. Rochford, Essex 1890 *d.* Rochford 1959
Veterinary surgeon, Rochford. Botanised in S.E. Essex.
S. T. Jermyn *Fl. Essex* 1974, 20.

SPEECHLEY, William (–1804)
d. 7 June 1804
Nurseryman, Newark-on-Trent, Notts.

SPEECHLEY, William (*c.* 1733–1819)
b. near Peterborough, Northants *d.* Great Milton, Oxford 1 Oct. 1819
Gardener to Duke of Portland at Welbeck Abbey, Notts, 1767. Skilled in growing pine-apples and grapes. "The Moses of modern British vine dressers" (Loudon). *Treatise on Culture of the Pine Apple* 1779; ed. 2 1796. *Treatise on Culture of the Vine* 1790. *Practical Hints in Domestic Rural Economy* 1820.
J. C. Loudon *Encyclop. Gdning* 1822, 1277. *Gdnrs Mag.* 1828, 383–84. G. W. Johnson *Hist. English Gdning* 1829, 238–40. S. Felton *Portr. of English Authors on Gdning* 1830, 81–82. *Gdnrs Chron.* 1910 i 193, 211–12.

SPEED, Thomas (1832–1883)
b. Abington, Cambridgeshire 19 Dec. 1832 *d.* Chatsworth, Derbyshire Dec. 1883
Gardener to Sir Edward Walker at Berry Hill, Mansfield, 1859 and Duke of Devonshire at Chatsworth, 1868.
Gdnrs Chron. 1874, 783 portr.; 1884 i 26. *J. Hort. Cottage Gdnr* v.8, 1884, 9.

SPEED, Walter (*c.* 1835–1921)
d. Bangor, Caernarvonshire 8 Oct. 1921
VMH 1897. Gardener at Penrhyn Castle, Bangor.
Gdnrs Mag. 1909, 943 portr. *Garden* 1921, 531. *Gdnrs Chron.* 1921 ii 202.

SPENCE, John (1848–)
b. Little Gransdem, Cambridgeshire 14 Jan. 1848
To U.S.A. *c.*1866. Settled in Santa Barbara, California where he discovered *Papaver californicum.*
Bot. Gaz. v.11, 1886, 181.

SPENCE, Magnus (1853–1919)
b. Birsay, Orkney 1 Jan. 1853 *d.* St. Ola, Orkney 20 Aug. 1919
Schoolmaster. *Fl. Orcadensis* 1914 portr. Contrib. Orkney algae to *J. Bot.* 1918, 281–85, 337–40.
Bot. Soc. Exch. Club Br. Isl. Rep. 1919, 626; 1926, 206. *J. Bot.* 1919, 293; 1927, 78–79. *Orchadian* 8 July 1926.
Herb. at Orkney Natural History Society.

SPENCER, Eliza Lucy *see* Grey, E. L.

SPENCER, Herbert (1820–1903)
b. Derby 27 April 1820 *d.* Brighton, Sussex 8 Dec. 1903
Philosopher. *Principles of Biology* 1864–67.
Nature v.69, 1903–4, 155–56. *D.N.B. Supplt 2* v.3, 360–69. D. Duncan *Life and Letters of H. Spencer* 1908 portr.

SPENCER, James (1834–1898)
b. Luddenden, Yorks 1834 *d.* Akroyden, Yorks 9 July 1898
Geologist. Interested in fossil flora of Halifax Hard Bed coal. Discovered club moss, *Lepidodendron spenceri.* Chairman, Yorkshire Fossil Fl. Committee, 1896. Contrib. articles on 'Recreations in Fossil Botany' to *Sci. Gossip* 1881–83.
Proc. Yorkshire Geol. and Polytechnic Soc. 1898, 473–76.

SPENCER, John (1809–1881)
b. Langley, Derbyshire 27 June 1809 *d.* Calne, Wilts 10 Jan. 1881
Gardener to Lord Lansdowne at Bowood, Wilts. Founder member of British Pomological Society, 1854. Joint proprietor of *Florist*, 1854–62. Contrib. to *Gdnrs Chron.*
Florist and Pomologist 1881, 32. *Gdnrs Chron.* 1881 i 89–90. *J. Hort. Cottage Gdnr* v.2, 1881, 31–32, 134–35 portr.

SPENCER, Sir Walter Baldwin (1860–1929)
b. Stretford, Lancs 23 June 1860 *d.* Ushuaia, Patagonia 14 July 1929

BA Oxon 1884. KCMG 1916. Professor of Biology, Melbourne, 1887–1919. Secretary, Royal Society of Victoria, 1889–98; President, 1904. President, Victorian Field Naturalists Club, 1891–93, 1895–97. Collected plants on King Island, 1887, in Victoria, and Central Australia, 1894.

Nature v.124, 1929, 347. *J. Proc. R. Soc. N.S.W.* 1930, 7–8. *Proc. Linn. Soc. N.S.W.* v.55, 1930, v. *Victorian Nat.* 1929, 102–7 portr. *Trans. R. Soc. S. Australia* 1929, 391–92. R. R. Maratt and T. K. Penniman *Spencer's Last Journey* 1931; *Spencer's Scientific Correspondence with Sir J. G. Frazer and Others* 1932. P. Serle *Dict. Austral. Biogr.* 1949, 347–50. R. T. M. Pescott *Collections of a Century; Hist. of...National Mus. of Victoria* 1954, 89–133. *Austral. Encyclop.* v.8, 1965, 230–31.

Portr. and MSS. at National Museum of Victoria.

SPENCER-CHURCHILL, George, 5th Duke of Marlborough (1766–1840)
b. 6 March 1766 *d.* Blenheim, Oxfordshire 5 March 1840

MA Oxon 1786. DCL 1792. MP 1790. Had garden at Whiteknights, Reading, Berks. "Few at present patronize the science [of horticulture], through all its branches, with so much vigour and liberality, or have equal knowledge in its theory and practice" (H. C. Andrews *Botanist's Repository* v.5, 1804, 343).

P. Smith *Mem. and Correspondence of Sir J. E. Smith* v.1, 1832, 434. *J. Bot.* 1916, 245. M. Hadfield *Gdng in Britain* 1960, 275–76.

Blandfordia Andr.

SPENDER, Reginald Edward Sydney
(*c.* 1878–1965)
d. Sherborne, Dorset 13 Oct. 1965

Hybridised irises. *Iris Culture for Amateurs* (with L. Pesel) 1937.

Iris Yb. 1966, 17–18.

SPERRIN-JOHNSON, John Charles
(1885–1948)
b. 2 Oct. 1885 *d.* 19 May 1948

MA, MB University of Ireland. Professor of Biology, Auckland, N.Z., 1914–31. Professor of Botany, Cork, 1932.

Who was Who, 1941–1950 1086.

SPICER, Rev. William Webb (*c.* 1820–1879)
b. Westminster *c.* 1820 *d.* Itchen Abbas, Hants 28 April 1879

BA Oxon 1843. MA 1848. Rector, Itchen Abbas, 1850–74. Travelled and collected plants in Tasmania. *Handy Book to Collection...of Freshwater and Marine Algae* 1867. *Handbook of Plants of Tasmania* 1878. Contrib. to *Papers Proc. R. Soc. Tasmania.*

Papers Proc. R. Soc. Tasmania 1909, 27. *R.S.C.* v.8, 989.

Tasmanian plants at Oxford. Plants at Kew, Winchester College.

Helichrysum spiceri F. Muell.

SPITTALL, Robert (*fl.* 1820s–1840s)

Surgeon, Edinburgh. President, Plinian Natural History Society, 1829. 'Repetition of M. Dutrochet's Experiments on Mimosa pudica' (*Edinburgh New Philos. J.* v.8, 1830, 60–64).

SPON, Mr. (*fl.* 1790s–1800s)

Nurseryman, Egham, Surrey.

Bot. Mag. 1806, t.963.

SPOTTISWOODE, Lady John Scott
(1810–1901)
d. March 1901

Of Spottiswoode, Berwickshire. Amateur botanist.

Proc. Berwickshire Nat. Club 1902, 306–8.

SPOTTSWOOD, Robert (*fl.* 1670s)

Surgeon. Lived at Tangier. Sent plants to A. Balam and R. Morison. 'Catalogue of Tangier Plants' (*Philos. Trans. R. Soc.* v.19, 1696, 239–49).

R. Morison *Plantarum Historiae Universalis Oxoniensis* v.2, 1680, 69. E. S.-C. Cosson *Compendium Fl. Atlanticae* v.1, 1881, 7. E. J. L. Scott *Index to Sloane Manuscripts* 1904, 508. J. E. Dandy *Sloane Herb.* 1958, 84.

SPRAGUE, Thomas Archibald (1877–1958)
b. Edinburgh 7 Oct. 1877 *d.* Cheltenham, Glos 22 Oct. 1958

BSc Edinburgh 1898. FLS 1903. Collected plants on Capt. H. W. Dowding's expedition to Venezuela and Colombia, 1898. Assistant, Kew Herb., 1900; Deputy Keeper, 1930–45. Visited Canary Islands with John Hutchinson, 1913. Authority on *Loranthaceae*, plant nomenclature and old herbals.

H. J. Riddelsdell *Fl. Gloucestershire* 1948, clxi. *J. Kew Guild* 1958, 594 portr. *Pharm. J.* 1958, 336. *Nature* v.182, 1958, 1483–84. *Times* 31 Oct. 1958. *Proc. Cotteswold Nat. Field Club* 1957–58, 80. *Proc. Linn. Soc.* 1959–60, 134–35. *Taxon* 1960, 93–102 portr.

Plants at Kew. Portr. at Hunt Library.

SPRATT, George (*fl.* 1820s–1840s)

MRCS. Edited and drew plates for *Fl. Medica* 1828–30. *Medico-botanical Pocketbook* 1836. Contrib. to and drew for v.5 of W. Woodville's *Medical Bot.* 1832.

SPRUCE, Richard (1817–1893)
b. Ganthorpe, Yorks 10 Sept. 1817 *d.* Coneysthorpe, Castle Howard, Yorks 28 Dec. 1893

PhD Berlin 1864. ALS 1893. Collected plants in S. America, 1849–64. 'Mosses of Eskdale' (*Phytologist* 1844, 540–44). *Report on Expedition to procure Seeds...of Cinchona succirubra* 1861. 'Palmae Amazonicae' (*J. Linn. Soc.* v.11, 1869, 65–185). 'Musci and Hepaticae of Pyrenees' (*Ann. Mag. Nat. Hist.* v.3–4, 1849). 'Hepatics of Amazons and Andes' (*Trans. Bot. Soc. Edinburgh* v.15, 1884, 1–588). *Notes of a Botanist on Amazons and Andes*, ed. by A. R. Wallace, 1908.

J. Bot. 1864, 199–201; 1894, 50–53. *Nature* v.49, 1894, 317–19. *Revue Bryol.* 1896, 61–79. *Kew Bull.* 1894, 32–33; 1908, 464. *Proc. Linn. Soc.* 1893–94, 35–37. *Trans. Bot. Soc. Edinburgh* v.20, 1894, 99–109. *D.N.B.* v.53, 431–32. *Ann. Bot.* 1900, xi–xiv portr. C. F. P. von Martius *Fl. Brasiliensis* v.1, 1906, 113–16. *Annual Rep. Yorkshire Philos. Soc.* 1906, 59–67. *Naturalist* 1909, 45–48; 1961, 156–57; 1971, 129–31. B. Lynge *Index...Lichenum Exsiccatorum* 1915, 494–501. *Bibliotheca Botanica* Heft 116, 1937, 52–53. *J. New York Bot. Gdn* 1944, 73–80 portr. *Geogr. Mag.* 1949, 481–88. *J. R. Hort. Soc.* 1949, 531–44 portr. V. W. von Hagen *South America called Them* 1949, 291–374. *Bot. Mus. Leaflet Harvard Univ.* 1951, 29–78 portr.; 1969, 121–32 portr. *Northern Gdnr* 1953, 20–27, 55–61, 87–93, 121–25; 1971, 15–16, 21. *Ciencia e Cultura* v.20, 1968, 37–49. *Rhodora* 1968, 313–39 portr. C. C. Gillispie *Dict. Sci. Biogr.* v.12, 1975, 594. *R.S.C.* v.5, 785; v.8, 993; v.11, 469; v.12, 697.

Plants at Kew, York Museum, Harvard. Letters at Kew, BM(NH). MSS. at Kew. Diary, 1841–63 at Linnean Society.

Sprucea Benth. *Sprucella* Stephani.

SPRY, Mrs. Constance (1886–1960)
b. Derby 5 Dec. 1886 *d.* Winkfield Place near Windsor, Berks 3 Jan. 1960

Flower arranger. *Flower Decoration* 1934. *Flowers in House and Garden* 1937. *Constance Spry's Garden Book* 1940. *Summer and Autumn Flowers* 1951. *Winter and Spring Flowers* 1951. *Favourite Flowers* 1959.

D.N.B. 1951–1960 915–16. *Who was Who, 1951–1960* 1032–33. E. Coxhead *Constance Spry* 1976.

SQUIBB, R. W. (*fl.* 1840s)
Nurseryman, Fisherton Nursery, Salisbury, Wilts.

SQUIBB, Robert (–before 1817)
New York 1781. Acquired John Watson's nursery in Charleston, S.C., 1785. Collected plants for European gardens. *Gardener's Kalender for S. Carolina and N. Carolina* 1787; ed. 3? 1827. Assisted T. Walter with *Fl. Caroliniana* 1788.

Bot. Mag. 1789, t.104; 1794, t.259. S. Elliott *Sketch of Bot. of S. Carolina and Georgia* v.1, 1821, 556. U. P. Hedrick *Hist. of Hort. in America to 1860* 1950, 141, 471.

SQUIRE, Peter (*fl.* 1860s)
Pharmaceutical chemist, Oxford Street, London. MS. *List of Indigenous Plants found at Basmead Manor c.* 1860 at Bedfordshire Record Office.

J. G. Dony *Fl. Bedfordshire* 1953, 22.

STABLER, George (1839–1910)
b. Wellburn, Yorks 3 Sept. 1839 *d.* Levens, Westmorland 4 Jan. 1910

Schoolfellow of R. Spruce. Schoolmaster. Bryologist and hepaticologist. 'Hepaticae of Balmoral' (*Trans. Bot. Soc. Edinburgh* v.22, 1902, 249–54; v.24, 1911, 101). Hepatics and mosses of Westmorland in *Naturalist* 1888–98; added many new species to British flora.

J. Bot. 1910, 160–61. *Naturalist* 1910, 97. *Trans. Bot. Soc. Edinburgh* v.24, 1912, 101.

Letters in Wilson correspondence at BM(NH).

Stableria Lindb.

STABLES, W. J. (*c.* 1878–1957)
d. Wark, Northumberland 29 July 1957

Orchid grower to de Barri Crawshay at Sevenoaks, 1901; later to C. Cookson at Hexham. Established own nursery at Houxby Gardens, Wark.

Orchid Rev. 1957, 188.

STABLES, William Alexander (1810–1890)
b. Cullen, Banff 1810 *d.* Calcots, Elgin 21 June 1890

Factor for Cawdor. Pupil of R. Graham. Contrib. to Rev. G. Gordon's *Collectanea for Fl. Moray* 1839 and A. Murray's *Northern Fl.* 1836. Herb. acquired by H. C. Watson.

H. C. Watson *Topographical Bot.* 1883, 556. *Ann. Scott. Nat. Hist.* 1894, 66–67.

Plants at Glasgow University.

STACEY, W. (–1888)
d. April 1888

Nurseryman, Dunmow, Essex; noted for his verbenas.

J. Hort. Cottage Gdnr v.16, 1888, 321.

STACKHOUSE, Emily (1811–1870)
b. Modbury, Devon 1811 *d.* Truro, Cornwall March 1870

Contrib. Truro plants and Cornwall mosses to *J. R. Inst. Cornwall* 1865–67. Drew figures in C. A. Johns's *Week at the Lizard* 1848.

F. H. Davey *Fl. Cornwall* 1909, liii. *R.S.C.* v.8, 994.

Letters in Wilson correspondence at BM(NH).

STACKHOUSE, John (1742–1819)
b. Trehane, Cornwall 1742 *d.* Bath, Somerset 22 Nov. 1819

FLS 1795. Discovered *Viola lactea*, 1796. Herb. in possession of J. E. Smith. *Nereis Britannica* 1795–1801; ed. 2 1816. *Illustrationes Theophrasti* 1811 portr. Edited Theophrastus's *De Historia Plantarum* 1813–14 2 vols. *Extracts from Bruce's Travels* 1815. Contrib. to W. Withering's *Arrangement* 1796.

Gent. Mag. 1819 ii 569; 1820 i 88. P. Smith *Mem. and Correspondence of Sir J. E. Smith* v.1, 1832, 415. *Gdnrs Chron.* 1899 ii 237. *D.N.B.* v.53, 441–42. *Trans. Woolhope Nat. Field Club* v.34, 1954, 236–37.

Plants at BM(NH). Two vols of drawings at Linnean Society. Portr. at Hunt Library.

Stackhousia Smith.

STACKHOUSE, Thomas (–1886)
d. Rocky Mouth, Clarence River, N.S.W. 1886

Commander, Royal Navy. Founder and first Secretary, Linnean Society of New South Wales. Collected plants at Yamba, Clarence River.

Proc. Linn. Soc. N.S.W. 1886, 1221. *J. R. Soc. N.S.W.* 1908, 121.

STACY, Theophilus (*fl.* 1680s)

Seedsman, Rose and Crown without Bishopsgate, London.

STAIR, John James Hamilton Dalrymple, 12th Earl of (1879–1961)
b. 1 Feb. 1879 *d.* 4 Nov. 1961

VMH 1953. MP, Wigtownshire, 1906. Had garden at Lochinch Castle, Wigtownshire. Authority on rhododendrons.

Gdnrs Chron. 1954 ii 116 portr. *Who was Who, 1961–1970* 1067.

STANDEN, Richard Spiers (1835–1917)
b. Oxford 11 Oct. 1835 *d.* Romsey, Hants 29 July 1917

FLS 1893. Tailor, London and Oxford.

Entomologist 1917, 263–64. *Proc. Linn. Soc.* 1917–18, 43–44. *Rep. Watson Bot. Exch. Club* 1916–17, 5–7 portr.

British plants at BM(NH).

STANDISH, John (1814–1875)
b. Yorks 25 March 1814 *d.* Ascot, Berks 24 July 1875

Gardener to Duchess of Gloucester at Bagshot Park, Surrey. Nurseryman in partnership with C. Noble at Bagshot, 1847–56. Nurseryman at Ascot where he received many of R. Fortune's Chinese plants. *Practical Hints on Planting Ornamental Trees* 1852.

Florist and Pomologist 1875, 216. *Garden* v.8, 1875, 98. *Gdnrs Chron.* 1875 ii 139, 229 portr.; v.162(5), 1967, 16–17. *J. Hort. Cottage Gdnr* v.54, 1875, 97–98, 136. *Gdnrs Yb. Almanack* 1876, 171–72. E. Bretschneider *Hist. European Bot. Discoveries in China* 1898, 552–53.

STANDLEY, James (*fl.* 1810s)

Nurseryman, 18 Market Place, Manchester, Lancs.

STANFIELD, Dennis Percival (–1971)
d. 13 May 1971

FLS 1959. Agricultural Dept., Nigeria, 1926; transferred to administration, 1930; became District Officer, retiring in 1949. Collected plants in Nigeria, 1956. Contrib. to *Nigerian Trees* 1960–64 2 vols. *Fl. Nigeria :Grasses* (with J. Lowe) 1970.

AETFAT Bull. no. 22, 1972, 34.

Stanfieldiella Brenan.

STANFORD, Edward Charles Curtis (–1899)
d. Glenwood, Dunbartonshire Dec. 1899

Pharmacist. Wrote on economic uses of algae and founded Scottish seaweed industry. 'Remarks on a Specimen of Sea-weed Char, from Laminaria digitata' (*Br. Pharm. Conference Proc.* 1867, 40–41). 'On Algin—a New Substance obtained from some of the Commoner Species of Marine Algae' (*Glasgow Philos. Soc. Proc.* 1883, 241–56).

Pharm. J. v.9, 1899, 591. *R.S.C.* v.5, 796; v.8, 998; v.11, 475.

STANFORD, H. (*fl.* 1830s)

Nurseryman, St. Leonards, Sussex.

STANGER, William (1812–1854)
b. Wisbech, Cambridgeshire 1812 *d.* Durban, S. Africa 14 March 1854

MD Edinburgh. Practised in London. Visited Australia. Naturalist to Niger Expedition, 1841. Surveyor-General, Natal, 1845.

Hooker's J. Bot. 1853, 228. *Gent. Mag.* 1854 ii 84–85. *Quart. J. Geol. Soc.* 1855, xlii–xliii. F. Boase *Modern English Biogr.* v.3, 1901, 704.

Plants at BM(NH), Kew.

Stangeria Moore.

STANHOPE, Philip Henry, 4th Earl (1781–1855)
b. London 7 Dec. 1781 *d.* Sevenoaks, Kent 2 March 1855

FRS 1807. President, Medico-Botanical Society, London, 1829–37.

Gent. Mag. 1855 ii 89.

Stanhopea Frost ex Hook.

STANLEY, Lady Beatrix (1877–1944)
d. Market Harborough, Leics 3 May 1944

Wife of Governor of Madras, Sir George Frederick Stanley, 1929–34. Made drawings of flora of Madras now at library of Royal Horticultural Society. Had garden at Sibbertoft Manor. For a few years edited *New Fl. and Sylva.*

Gdnrs Chron. 1944 i 202. *Times* 4 May 1944. *Daffodil Tulip Yb.* 1946, 92.

STANSFIELD, Abraham (1802–1880)
b. Kebcote-in-Stansfield, Yorks 12 Jan. 1802 *d.* Todmorden, Yorks 15 Aug. 1880

Nurseryman, Todmorden. President, Todmorden Botanical Society, 1852. Collected ferns. 'Botany of Forest of Rossendale' (T. Newbiggin *History of Forest of Rossendale* 1868). 'Fl. Todmorden' (with J. Nowell) partly published in *Lancashire Nat.* 1907–8.

Gdnrs Chron. 1880 ii 283–84. *J. Hort. Cottage Gdnr* v.1, 1880, 187–88. *Gdnrs Yb. Almanack* 1881, 191. W. B. Crump and C. Crossland *Fl. Parish of Halifax* 1904, lxi–lxii. *Lancashire Nat.* 1909, 257 portr.; 1913, 4–5. *Trans. Liverpool Bot. Soc.* 1909, 91.

Herb. formerly at Todmorden Public Library, now missing.

STANSFIELD, Frederick Wilson (1854–1937)
b. Todmorden, Yorks 1854 *d.* Reading, Berks 1 March 1937

MD Manchester 1900. FLS 1927. Public vaccinator, Reading. Collected and grew ferns. President, British Pteridological Society. Edited *Br. Fern Gaz.* 1917–37.

Gdnrs Chron. 1930 i 280 portr. *Br. Fern Gaz.* 1937, 107–9; 1938, 149 portr., 176–80. *J. Bot.* 1937, 114–15. *Proc. Linn. Soc.* 1936–37, 212.

Portr. at Hunt Library.

STANSFIELD, Herbert (*c.* 1856–1928)
d. Sale, Cheshire 28 April 1928

Grew ferns. Contrib. to *Br. Fern Gaz.*

Br. Fern Gaz. 1928, 218–20.

STANSFIELD, Thomas (*c.* 1826–1879)
d. 30 Dec. 1879

Nurseryman, Todmorden, Yorks, specialising in ferns. Secretary, Todmorden Botanical Society.

Florist and Pomologist 1880, 32.

STANSFIELD, Tom (–*c.* 1943)

FLS 1938. Collected and grew ferns.

Proc. Linn. Soc. 1943–44, 233.

STANSFIELD, William Henry (1850–1934)
b. Todmorden, Yorks 25 May 1850 *d.* Tangier 7 March 1934

Employed in nursery at Todmorden. Established nursery at Southport, Lancs, 1875 noted for alpine plants. Collected ferns in British Isles, Pyrenees, Alps and N. Africa.

Br. Fern Gaz. 1934, 253–54. *Gdnrs Chron.* 1934 i 205. *N. Western Nat.* 1934, 164–67. Liverpool Museums *Handbook and Guide to Herb. Collections* 1935, 28 portr.

Fern herb. at Liverpool Museums.

STANTON, George (1840–1920)
b. Bramley, Surrey 10 Dec. 1840 *d.* Henley, Oxfordshire 14 March 1920

Kew gardener, 1862–64. Gardener to J. Noble at Park Place, Henley. 'Kew and Kew Men Fifty Years Ago' (*J. Kew Guild* 1915, 206–9).

J. Kew Guild 1919, 431–32 portr. *Bot. Soc. Exch. Club Br. Isl. Rep.* 1920, 107. *Garden* 1920, 162 portr.

STAPF, Otto (1857–1933)
b. Ischl, Austria 23 March 1857 *d.* Innsbruck, Austria 3 Aug. 1933

PhD Vienna. FRS 1908. ALS 1898. FLS 1908. VMH 1928. VMM 1931. Assistant to Professor Kerner von Marilaun, 1882–89. Travelled in Persia, 1885. Assistant, Kew Herb., 1891; Keeper, 1909–22. *Botanische Ergebnisse der Polak'schen Expedition nach Persian* 1882. *Die Arten der Gattung Ephedra* 1889. Edited *Index Londinensis* 1929–31. Contrib. *Gramineae* to *Fl. Capensis* and *Fl. Tropical Africa.* Contrib. to *Ann. Bot., J. Linn. Soc., Kew Bull.*

J. Kew Guild 1922, 83. *Proc. Linn. Soc.* 1926–27, 53–54; 1933, 369–71. *J. Bot.* 1928, 56; 1933, 296–99. *J. Kew Guild* 1933, 283–84 portr. *Kew Bull.* 1933, 366, 369–90. *Bot. Soc. Exch. Club Br. Isl. Rep.* 1932, 511–14. *Gdnrs Chron.* 1933 ii 134. *Nature* v.132, 1933, 305. *Obit. Notices Fellows of R. Soc.* 1933, 115–18 portr. *Times* 8 Aug. 1933; 9 Sept. 1933. *Bot. Mag.* 1934 portr. *Berichten Deutschen Bot. Gesellschaft* 1934, 210–22 portr. *J. R. Hort. Soc.* 1934, 127–30 portr. *Chronica Botanica* 1935, 29 portr. *Who was Who, 1929–1940* 1281–82.

Letters and MSS. at Kew.

STAPLEDON, Sir Reginald George (1882–1960)
b. Northam, Devon 22 Sept. 1882 *d.* Bath, Somerset 16 Sept. 1960

Educ. Cambridge. FRS 1939. Pioneer of grassland science. On staff of Royal Agricultural College, Cirencester, 1910. Advisory officer in Agricultural Botany, University College, Aberystwyth, 1912. Director, Welsh Plant Breeding Station, 1919–42. Director, Grassland Improvement Station, Stratford-on-Avon. President, British Grassland Society, 1945. *The Land: Now and Tomorrow* 1935. *The Way of the Land* 1943.

Nature v.149, 1942, 549; v.188, 1960, 363–64. *Listener* 1943, 407 portr. *Times* 17 Sept. 1960 portr. *Who was Who, 1951–1960*

1037. *D.N.B. 1951–1960* 920–21. R. Waller *Prophet of the New Age* 1962 portr.

STAPLES, R. P. (*fl.* 1820s)
Possibly Robert Ponsonby Staples, British Consul in Latin America, *c.* 1820–35. Collected plants for A. B. Lambert in Mexico, *c.* 1833.
Bot. Mag. 1826, t.2620. *Taxon* 1970, 543.

STARK, Robert Mackenzie (1815–1873)
b. Dirleton, E. Lothian 1815 *d.* London 29 Sept. 1873
Nurseryman, Edinburgh; specialised in alpines. 'Muscology of Cirencester' (*Ann. Nat. Hist.* v.4, 1840, 211–12). *Popular History of British Mosses* 1854; ed. 2 1860. *Marine Aquarium* 1857.
Trans. Bot. Soc. Edinburgh v.8, 1866, 414; v.12, 1873, 29. *Garden* v.4, 1873, 310. *J. Bot.* 1873, 352.

STARKER, Thomas (*fl.* 1790s)
Nurseryman, Gateshead, Durham.

STATTER, John Whewell (1829–1896)
d. 6 Jan. 1896
BA Cantab. Collected plants in Australia, 1870–80.
Plants at BM(NH).

STAUNTON, Sir George Leonard (1737–1801)
b. Cargin, Galway 19 April 1737 *d.* London 14 Jan. 1801
MD Montpellier 1758. DCL Oxon 1790. FRS 1787. FLS 1789. Baronet 1785. Physician in West Indies, 1762–79. Visited Brazil. Secretary to Lord Macartney in Madras, 1781–84 and in China, 1792–94. Collected plants in China. *Authentic Account of Embassy...to Emperor of China* 1797–98.
Gent. Mag. 1801 i 89–90. *J. Bot.* 1884, 81. G. T. Staunton *Mem. Life and Family of Sir G. L. Staunton* 1823. E. Bretschneider *Hist. European Bot. Discoveries in China* 1898, 156–63. *D.N.B.* v.54, 113–14. *Taxon* 1970, 543–44. *J. R. Hort. Soc.* 1974, 339–47.
Plants at BM(NH). Letters in Smith correspondence at Linnean Society.
Stauntonia D.C.

STAUNTON, John (*fl.* 1850s–1870s)
Of Longbridge, Warwick. Diatomist.
Slides and letters at BM(NH).
Alloioneis stauntoni Grunow.

STAWARD, Richard (*c.* 1873–1961)
b. Kirriemuir, Fifeshire *c.*1873 *d.* Ware, Herts 22 May 1961
Superintendent of gardens at Ware Park Hospital, 1921–48. *Practical Hardy Fruit Culture* 1920.
Gdnrs Chron. 1961 i 519.

STEBBING, Edward Percy (1870–1960)
d. 21 March 1960
FLS 1902. FRSE 1923. Indian Forest Service, 1893. Forest Zoologist, Forest Research Institute, Dehra Dun, 1906–10. Lecturer in Forestry, Edinburgh University, 1910–51. *Forests of India* 1921–26 3 vols. *Forests and Erosion* 1941.
Nature v.186, 1960, 515–16. *Yb. R. Soc. Edinburgh* 1961, 32–33. *Who was Who, 1951–1960* 1038.

STEBBING, Mary Anne (*née* Saunders) (–1927)
d. 21 Jan. 1927
FLS 1904. Wife of Rev. T. R. R. Stebbing FRS. Daughter of W. W. Saunders. Drew plants now at Kew.
Proc. Linn. Soc. 1926–27, 103–4.

STEDMAN, Frank William (–1920)
d. Ashford, Kent 12 Nov. 1920
Chemist, Ashford. Botanised in Kent.
Bot. Soc. Exch. Club Br. Isl. Rep. 1920, 107.

STEEL, George (1809–1891)
b. Richmond, Surrey 1809 *d.* 23 July 1891
Established in 1841 with his brother William a nursery in Kew Road, Richmond; specialised in rhododendrons and lilies.
Gdnrs Chron. 1891 ii 144.

STEEL, Thomas (1858–1925)
b. Glasgow 8 Sept. 1858 *d.* 17 Aug. 1925
Chemist. On staff of Colonial Refining Co., Sydney, 1882–1918. Collected fungi in New South Wales. Contrib. to *Victorian Nat.*
Rep. Bot. Gdn Sydney 1919, 8. *Proc. Linn. Soc. N.S.W.* 1926, vii.

STEELE, Matthew (*fl.* 1830s)
Introduced plants from Demerara, British Guiana.
Bot. Mag. 1837, t.3573.
Maxillaria steelii Hook.

STEELE, William Edward (1816–1883)
b. Belfast 15 June 1816 *d.* Bray, County Wicklow 6 May 1883
BA Dublin 1857. MD 1856. Director, Science and Art Museum, Dublin. *Handbook of Field Botany* 1847; ed. 2 1851.
J. Bot. 1883, 192.
Herb. and bust at National Museum, Dublin.

STEGGALL, John (*fl.* 1820s–1860s)
MRCS 1825. MD Bologna and Pisa. *Pupil's Introduction to Botany* 1829.

STEGGALL, Rev. William (*fl.* 1820s–1850s)
BA Cantab 1826. MA 1829. Vicar, Hunston, Suffolk, 1846.
W. M. Hind *Fl. Suffolk* 1889, 489.
Herb. of 30 vols, 1830–34, formerly at Stowlangtoft Hall, now missing.

STEINHAUER, Daniel (1785–1852)
b. Wales 1785 *d.* Bethlehem, Pa., U.S.A. 1852
Teacher in Moravian schools in Lancaster, Nazareth, Bethlehem, Zoneville and Chilicothe. Sent plants to Schweinitz in Philadelphia and C. W. Short in Cheneyville, La.
Bartonia no. 36, 1966, 1–24.

STELFOX, Arthur Wilson (1883–1972)
b. Belfast 1883 *d.* 19 May 1972
ARIBA 1909. ALS 1947. Architect. Joined staff of National Museum, Dublin, 1920–48. Grew Irish plants in his garden.
Nature v.238, 1972, 175–76 portr. *Irish Nat. J.* 1973, 285–302 portr.
Plants at BM(NH).

STELFOX, Margarita Dawson (*née* **Mitchell**) (1886–1971)
b. Lisburn, County Antrim 1886 *d.* Aug. 1971
Wife of A. W. Stelfox. Contrib. papers on Mycetozoa to *Irish Nat.*
R. L. Praeger *Some Irish Nat.* 1949, 158–59. *Irish Nat. J.* 1973, 296.

STENHOUSE, John (1809–1880)
b. Glasgow 21 Oct. 1809 *d.* Glasgow 31 Dec. 1880
LLD Aberdeen 1850. FRS 1848. Lecturer in Chemistry at St. Bartholomew's Hospital, London, 1851–57. Assayer to the Mint, 1865–70. Wrote numerous papers on chemistry of lichens.
Proc. R. Soc. v.31, 1881, xix–xxi. *D.N.B.* v.54, 149. *R.S.C.* v.5, 819; v.8, 1010; v.11, 489.

STENNETT, Ralph (*fl.* 1800s)
Of Bath, Somerset. Natural history draughtsman. Exhibited at Royal Academy, 1803.
Flower drawings at BM(NH). Flower drawings for sale in Marlborough Rare Books Cat. 45, 1961 and Quarich Cat. 851, 1964.

STENNING, Lewis (1901–1965)
b. Sorn, Ayrshire 17 Nov. 1901 *d.* London 4 March 1965
VMH 1964. Kew gardener, 1925; Assistant Curator, 1929; Curator, 1960–65.
Gdnrs Chron. 1939 i 226 portr.; 1965 i 295 portr. *J. Kew Guild* 1965, 601–2 portr. *Orchid Rev.* 1965, 109. *Times* 6 March 1965.

STEP, Edward (1855–1931)
b. London 11 Nov. 1855 *d.* Wimbledon, Surrey 8 Nov. 1931
FLS 1896. President, South London Entomological and Natural History Society, 1894, 1903. President, British Empire Naturalists' Association. *Plant Life* 1880. *Wayside and Woodland Blossoms* 1895–96. *Favourite Flowers of Garden and Greenhouse* 1895–97. *Romance of Wild Flowers* 1899. *Wayside and Woodland Trees* 1904. *Wild Flowers Month by Month* 1905. *Wayside and Woodland Ferns* 1908. *Toadstools and Mushrooms of the Countryside* 1913. *Spring Flowers of the Wild* 1927. *Summer Flowers of the Wild* 1927.
Nature v.128, 1931, 863. *Observer* 22 Nov. 1931. *J. Bot.* 1932, 18–19. *Naturalist* 1932, 15–16 portr. *N. Western Nat.* 1932, 41–42. *Proc. Linn. Soc.* 1931–32, 189–90. *Who was Who, 1929–1940* 1285.

STEPHEN, John Horne (–1915)
b. Broughty Ferry, Forfarshire *d.* Calcutta, India 29 Dec. 1915
Kew gardener, 1888. Gardener, Lal Bagh Gardens, Bangalore, 1891. Superintendent, Government Gardens, Nagpur, 1895.
Kew Bull. 1895, 231. *J. Kew Guild* 1917, 378 portr.

STEPHENS, A. J. (1861–1941)
Of Cheltenham, Glos. Botanised in Gloucestershire.
H. J. Riddelsdell *Fl. Gloucestershire* 1948, cl.

STEPHENS, Arthur Bligh (1855–1908)
d. Taiping, Malaya Jan. 1908
Planter, Malaya, 1873–92. Assistant Indian Immigration Agent, Malay States, 1892–1903. Deputy Conservator of Forests, Perak. Collected orchids and forest trees.
Agric. Bull. Straits and Federated Malay States 1908, 66–67. *Gdns Bull. Straits Settlements* 1927, 132. *Fl. Malesiana* v.1, 1950, 505.
Plants at Kuala Lumpur.

STEPHENS, Henry Oxley (1816–1881)
Surgeon, Bristol. Contrib. to *Phytologist* and to E. H. Swete's *Fl. Bristoliensis* 1854. 'Mycology of Bristol' (*Ann. Nat. Hist.* v.4, 1840, 246–53).
J. W. White *Fl. Bristol Coal-field* 1887, ii. J. W. White *Fl. Bristol* 1912, 80–81. H. J. Riddelsdell *Fl. Gloucestershire* 1948, cxix.
Herb. formerly at Bristol Naturalists' Society. Letters at Kew.
Stephensia Tul.

STEPHENS (Stevens), Rev. Lewis (1654–1724/5)
b. Braunton, Devon 1654 *d.* Menheniot, Cornwall 1 Jan. 1724/5
BA Cantab 1677. MA Oxon 1678. Vicar, Treneglos and Warbstow, 1678–85; Menheniot, 1685–1724. Marine algologist. Correspondent of A. Buddle and W. Sherard to whom he sent seaweeds and *Physospermum* which he added to the British flora. "A learned clergyman and skilful in Botaniks."
R. Morison *Plantarum Historiae Universalis Oxoniensis* v.3, 1699, 627, 646, 647, 648.

C. E. Raven *John Ray* 1942, 257. J. E. Dandy *Sloane Herb.* 1958, 216. H. N. Clokie *Account of Herb. Dept. Bot., Oxford* 1964, 248.

STEPHENS (Steephens), Philip (1619/20–1679)
b. Devizes, Wilts 1619/20 *d.* London 4 Feb. 1679

MA Cantab; Oxon 1645. MD Oxon 1656. Principal, Hart Hall, 1653–60. *Catalogus Horti Botanici Oxoniensis* (with W. Browne and the Bobarts) 1658.

R. Pulteney *Hist. Biogr. Sketches of Progress of Bot. in England* v.1, 1790, 166–67. W. Munk *Roll of R. College of Physicians* v.1, 1878, 296. G. C. Druce *Fl. Oxfordshire* 1886, 374.

STEPHENS, Thomas (*fl.* 1830s–1840s)

Nurseryman, South Road, Taunton, Somerset.

STEPHENS, William (–1760)
d. Dublin 28 June? 1760

MD Leyden 1718. MD Dublin 1724. FRS 1718. Lecturer in Chemistry, Dublin, 1733–60. Sent plants to J. Petiver. *Botanical Elements* 1727.

J. J. Dillenius *Hortus Elthamensis* 1732, 388. T. P. C. Kirkpatrick *Hist. of Med. Teaching in Trinity College, Dublin* 1912, 363. H. F. Berry *Hist. of R. Dublin Soc.* 1915, 456. E. J. L. Scott *Index to Sloane Manuscripts* 1904, 512. J. E. Dandy *Sloane Herb.* 1958, 216.

MS. *Catalogue Hort. Dublin,* 1726 at BM(NH).

STEPHENS, William (–1866)
d. near Mooloolah, Queensland

Of Richmond, Surrey. Murdered by natives while collecting plants for Botanic Gardens at Brisbane.

Gdnrs Chron. 1866, 520.

STEPHENSON, John (*fl.* 1820s–1850s)

MD Edinburgh. FLS 1829. *Medical Botany* (with J. M. Churchill) 1827–31 4 vols.

STEPHENSON, Rev. Thomas (1855–1948)
b. Brackley, Northants 1855 *d.* Hindhead, Surrey 15 April 1948

BA. DD. Methodist minister. Studied *Dactylorchis* with his son, T. A. Stephenson. Contrib. papers, mainly on British orchids, to *J. Bot.* Contrib. to W. K. Martin and G. T. Fraser's *Fl. Devon* 1939.

Nature v.161, 1948, 799. *Watsonia* 1949–50, 187–89.

Herb. at Torquay Natural History Society.

STEPHENSON, Thomas Alan (1898–1961)
b. Burnham-on-Sea, Somerset 19 Jan. 1898 *d.* 3 April 1961

Educ. University College, Aberystwyth. FRS 1951. FLS 1937. His first scientific work was on orchids in collaboration with his father, Rev. T. Stephenson. Professor of Zoology, University of Cape Town, 1931–40. Professor, University College, Aberystwyth, 1940–61. Taught flower painting by H. Drinkwater. Exhibition of his paintings held in London, 1964.

Br. Phycol. Bull. 1961, 94. *Biogr. Mem. Fellows R. Soc.* 1962, 137–48 portr. *Proc. Linn. Soc.* 1961–62, 153–55.

STEPHENSON, William (*fl.* 1810s–*c.* 1863)
d. Taree, N.S.W. *c.* 1863

MRCS 1814. Army surgeon in India and China. Surgeon and collector to T. L. Mitchell's expedition in Australia, 1845–46.

J. Proc. R. Soc. N.S.W. 1908, 121–22.

Plants at BM(NH).

Siebera stephensonii Benth.

STERLING, George (*c.* 1806–1885)
d. Edinburgh 1885

Gardener at Melville Castle, Midlothian. Botanist.

Garden v.27, 1885, 76.

STERN, Sir Frederick Claude (1884–1967)
b. London 8 April 1884 *d.* London 10 July 1967

FLS 1925. VMH 1940. Banker. Treasurer, Linnean Society, 1941–58. Chairman, John Innes Horticultural Institution, 1947–61. Created chalk garden at Highdown, Sussex. *Study of Genus Paeonia* 1946. *Snowdrops and Snowflakes* 1956. *A Chalk Garden* 1960.

Kew Bull. 1925, 1–6. *Gdnrs Chron.* 1932 i 340 portr.; 1956 i 6 portr.; v.162(7), 1967, 4. *Bot. Mag.* 1951 portr. *Lily Yb.* 1960, 11–12 portr.; 1968, 9–10. *Iris Yb.* 1967, 15–17. *J. R. Hort. Soc.* 1967, 379–81 portr. *Times* 11 July 1967. *Who was Who, 1961–1970* 1076.

Letters and MSS. at Kew. Portr. at Hunt Library.

STEUART, James Henry Augustus (1834–1895)
b. Ewhurst, Surrey 1834 *d.* Ventnor, Isle of Wight 26 Feb. 1895

Botanised on Isle of Wight. '*Gentiana amarella* var. *præcox*' (*J. Bot.* 1889, 217).

J. Bot. 1894, 181 (as S. H. Stewart); 1895, 128.

Herb. at Liverpool University.

STEVEN, Henry Marshall (1893–1969)
b. 24 June 1893 *d.* 15 Feb. 1969

BSc. PhD Edinburgh. Lecturer in Silviculture, Imperial Forestry Institute, Oxford, 1924–30. Professor of Forestry, Aberdeen, 1938–63. President, Society of Foresters of Great Britain, 1950–51. Edited *Forestry,*

1926–45. *Native Pinewoods of Scotland* (with A. Carlisle) 1959.
Who was Who, 1961–1970 1076.

STEVENS, Eliza *see* Allen, E.

STEVENS, George (–1902)
d. Putney, London 27 March 1902
Nurseryman, St. John's Nurseries, Putney; specialised in chrysanthemums.
Gdnrs Mag. 1902, 221. *J. Hort. Cottage Gdnr* v.44, 1902, 304.

STEVENS, Rev. Lewis *see* Stephens, *Rev.* L.

STEVENS, Zadok (*c.* 1833–1886)
d. Trentham, Staffs 20 Oct. 1886
Gardener to Duke of Sutherland at Trentham. Contrib. to *Gdnrs Chron.*
Gdnrs Chron. 1886 ii 537.

STEVENSON, Rev. Henry (*fl.* 1710s–1760s)
Headmaster, Free School, Retford, Notts. *Young Gard'ner's Director* 1716. *Gentleman Gard'ner Instructed* 1716.
Gdnrs Chron. 1923 ii 56; v.164(22), 1968, 17–18.

STEVENSON, Rev. John (1836–1903)
b. Coupar Angus, Perthshire 1836 *d.* Glamis, Angus 27 Nov. 1903
DD St. Andrews 1888. Mycologist. A founder of Scottish Cryptogamic Society. *Mycologia Scotica* 1879. *British Fungi: Hymenomycetes* 1886 2 vols.
Ann. Scott. Nat. Hist. 1904, 1–3. *J. Bot.* 1904, 64. *R.S.C.* v.11, 495.
Portr. at Hunt Library.

STEVENSON, John Barr (*c.* 1881–1950)
d. Ascot, Berks 27 May 1950
VMH 1939. Had celebrated collection of rhododendrons at Tower Court, Ascot. Edited *Species of Rhododendron* 1930.
Bot. Mag. 1950 portr. *Gdnrs Chron.* 1950 i 232. *Rhododendron Yb.* 1950, 5–6 portr.

STEVENSON, Robert (*fl.* 1730s–1750s)
Nurseryman, Dublin.
Irish For. 1967, 49.

STEVENSON, Thomas (–1938)
d. Cowley, Middx 22 Feb. 1938
Nurseryman, Colham Green Nurseries, Hillingdon, Middx.
Gdnrs Chron. 1938 i 154.

STEWART, Archibald (*fl.* 1690s–1700s)
Surgeon. Sent ferns from Darien to J. Petiver.
J. Petiver *Museii Petiveriani* 1695, 553. E. J. L. Scott *Index to Sloane Manuscripts* 1904, 513. J. E. Dandy *Sloane Herb.* 1958, 216.
Plants at BM(NH).

STEWART, Charles (*fl.* 1790s–1820s)
ALS 1791. Printer. Secretary, Natural History Society of Edinburgh. Edited J. Lee's *Introduction to Botany* 1806 and J. J. Dillenius's *Historia Muscorum* 1811, which he printed.

STEWART, Gilbert A. C. (–1876)
d. Melrose, Roxburghshire 12 Jan. 1876
Investigated adventive flora of Gala and Tweed.
Trans. Bot. Soc. Edinburgh 1870, 20, 170; 1879, 16.

STEWART, Rev. James (1831–1905)
b. Edinburgh 14 Feb. 1831 *d.* Lovedale, Cape Colony 21 Dec. 1905
MD Glasgow 1866. DD. Missionary in Central Africa from 1862. Collected plants in Zambesi, 1862–73. *Botanical Diagrams* 1857.
J. Bot. 1906, 144. J. Wells *Life of J. Stewart* 1909. *D.N.B. Supplt 2* v.3, 416–19. *Moçambique* 1939, 35–38. *Comptes Rendus AETFAT 1960* 1962, 172. J. P. R. Wallis *Zambesi Journal of James Stewart, 1862–1863* 1952.
Plants at BM(NH), Kew. Portr. at United Free Church Hall, Edinburgh.

STEWART, John (–1820)
d. Edinburgh 3 Nov. 1820
Lecturer in Botany, Edinburgh. *Hortus Cryptogamicus Edinensis* 1819 (exsiccatae). Wrote article 'Musci' in Brewster's *Encyclopaedia* 1830.
W. J. Hooker *Fl. Scotica* 1821, 139. *Mem. Wernerian Nat. Hist. Soc.* v.3, 1821, 444.

STEWART, John Lindsay (*c.* 1832–1873)
b. Fettercairn, Kincardineshire *c.* 1832 *d.* Dalhousie, Lahore 5 July 1873
MD Edinburgh 1856. FLS 1865. Assistant surgeon, Bengal, 1856. Superintendent, Botanic Garden, Saharanpur. Conservator of Forests, Punjab, 1864. *Notes of Botanical Tour in Ladak or Western Tibet* 1869. *Forest Fl. N. India* 1869. *Punjab Plants* 1869.
J. Bot. 1873, 319–20. *Proc. Linn. Soc.* 1873–74, lvii. *Proc. R. Soc. Edinburgh* v.8, 1872–75, 321–22. *Trans. Bot. Soc. Edinburgh* v.12, 1876, 31–33. D. Brandis *Forest Fl. N.W. and Central India* 1874, xiii–xv, xx.
Plants at Kew, Edinburgh, Dehra Dun. Letters at Kew.

STEWART, Lawrence Baxter (1877–1934)
b. Kenordy, Kirriemuir, Fifeshire 1877 *d.* 30 Jan. 1934
Gardener, Royal Botanic Garden, Edinburgh, 1901; Curator, 1932–34.
Gdnrs Chron. 1934 i 85, 103. H. R. Fletcher and W. H. Brown *R. Bot. Gdn Edinburgh, 1670–1970* 1970, 228 portr.

STEWART, Neil (*c.* 1814–1875)
d. Edinburgh 8 Dec. 1875
Botanical artist. 'Colour and Fertilisation' (*Trans. Bot. Soc. Edinburgh* v.11, 1871, 190; v.13, 1879, 16).
Trans. Bot. Soc. Edinburgh v.13, 1879, 16.

STEWART, R. B. (*fl.* 1830s)
Lectured on botany. *Outlines of Botany* 1835 (*Hort. Register* v.4, 1835, 352, 353, 390).

STEWART, Robert (1811–1865)
MRCS. Practised medicine in Torquay, Devon. *Handbook of Torquay Fl.* 1860.
W. K. Martin and G. T. Fraser *Fl. Devon* 1939, 774.

STEWART, Samuel Alexander (1826–1910)
b. Philadelphia 5 Feb. 1826 *d.* Belfast 15 June 1910
ALS 1904. To Belfast, 1836. Pupil of R. Tate. Curator, Museum of Belfast Natural History and Philosophical Society, 1891–1901. Discovered *Hieracium stewartii* in Ireland, 1891. *List of Mosses of N.E. Ireland* 1874. *Fl. N.E. Ireland* (with T. H. Corry) 1888.
Bot. Soc. Exch. Club Br. Isl. Rep. 1910, 532–33. *Irish Nat.* 1910, 201–9; 1913, 29–30. *Nature* v.83, 1910, 530. *J. Bot.* 1911, 122–23 portr. *Rep. Proc. Belfast Nat. Field Club* 1910–11, 356, 410–34 portr.; 1913, 623–24; 1933–34, 2–26 portr. *Proc. Linn. Soc.* 1910–11, 40–41. *Belfast Nat. Hist. Soc. Centenary Vol.* 1924, 102–4 portr., 155–56. R. L. Praeger *Some Irish Nat.* 1949, 159–60 portr. *R.S.C.* v.11, 498; v.12, 704.
Plants at National Museum Dublin, Ulster Museum Belfast. Letters at Ulster Museum. Portr. at National Museum Dublin.

STEWART, William (–1886)
d. 12 Sept. 1886
Acquired nursery at Dundee and Broughty Ferry, established by his father, John, in 1809.
Gdnrs Chron. 1886 ii 377.

STIDOLPH, Davey (1793–1840s)
Son of G. Stidolph (1760–1848). Nurseryman, Bromley, Kent.

STIDOLPH, Godfrey (1733/4–1818)
b. 21 Feb. 1733/4 *d.* June 1818
Nurseryman, Bromley, Kent.

STIDOLPH, Godfrey (1760–1848)
d. Feb. 1848
Son of G. Stidolph (1733/4–1818). Nurseryman, Bromley, Kent.

STIDOLPH, Henry (*fl.* 1780s–1800s)
Nurseryman, Bromley, Kent.

STIDOLPH, John (*fl.* 1840s)
Son of G. Stidolph (1733/4–1818). Acquired Bromley nursery, Kent, 1848.

STIDOLPH, William (1772–1855)
d. Jan. 1855
Son of G. Stidolph (1733/4–1818). Took over Bromley 'Workhouse' nursery, 1818 and Bromley 'College' nursery, 1848, Kent.

STILES, Matthew Henry (1846–1935)
b. Lutterworth, Leics 22 May 1846 *d.* Doncaster, Yorks 9 May 1935
Pharmaceutical chemist, Doncaster. Microscopist. Contrib. papers on *Diatomaceae* to *Naturalist* and *N. Western Nat.* Supplied material to W. and G. S. West's *Alga-Fl. Yorkshire* 1901. Contrib. article on algae and *Diatomaceae* to British Association's *Sheffield Handbook*, 1910, 443–47.
N. Western Nat. 1935, 270–74.

STILES, Walter (1886–1966)
b. Shepherd's Bush, London 23 Aug. 1886 *d.* 19 April 1966
MA Cantab 1912. FRS 1928. FLS 1927. Assistant Lecturer in Botany, Leeds, 1910–19. Professor of Botany, Reading, 1919–29; Birmingham, 1929–51. Plant physiologist. *Permeability* 1924. *Photosynthesis* 1925. *Introduction to Principles of Plant Physiology* 1936; ed. 2 1950. *Trace Elements in Plants and Animals* 1946; ed. 3 1961. *Respiration of Plants* (with W. Leach) 1932; ed. 4 1960. Contrib. to *Ann. Bot.*, *New Phytol.*, *Proc. R. Soc.*
Nature v.168, 1951, 497. *Times* 22 April 1966. *Biogr. Mem. Fellows R. Soc.* 1967, 342–57 portr. *Who was Who, 1961–1970* 1081.

STILL, Arthur Langford (1872–1944)
b. Croydon, Surrey 3 April 1872
MA Oxon. Analytical chemist. Interested in *Carex* and *Mentha.* Contrib. to *J. Bot.*
Bot. Soc. Exch. Club Br. Isl. Rep. 1943/4, 651–53.
Plants at BM(NH).

STILLINGFLEET, Benjamin (1702–1771)
b. Wood Norton, Norfolk *d.* London 15 Dec. 1771
BA Cantab 1723. Botanical adviser to W. Hudson (P. Smith *Mem. Correspondence of J. E. Smith* v.2, 1832, 473). *Miscellaneous Tracts relating to Natural History* 1759; ed. 3 1775.
R. Pulteney *Hist. Biogr. Sketches of Progress of Bot. in England* v.2, 1790, 349. J. Nichols *Lit. Anecdotes of Eighteenth Century* v.2, 336–37, 719 portr.; v.7, 399, 682. J. Nichols *Illus. Lit. Hist. of Eighteenth Century* v.8, 103. *Gent. Mag.* 1776, 162; 1777, 440. W. Coxe *Lit. Life and Select Works of*

B. Stillingfleet 1811 portr. 2 vols. *Cottage Gdnr* v.7, 1851, 79 portr. G. C. Druce *Fl. Berkshire* 1897, cxliii. *D.N.B.* v.54, 373–75. *Notes and Queries* v.193, 1948, 224–26.

Portr. at Kew, Hunt Library. Sale at Baker and Leigh, 3 Feb. 1772.

Stillingia L.

STIMSON, Ada (1842–1915)

Of Marston Moretaine, Beds. Married Sir John Thomas, a prominent Congregationalist, 1879. Botanised in her youth in Beds where she first found *Luzula forsteri*.

J. G. Dony *Fl. Bedfordshire* 1953, 26.

STIRLING, Sir James (1791–1865)
d. 22 April 1865

Admiral, 1862. First Lieut.-Governor, W. Australia, 1828–39. "He advanced the study of Western Australian plants by every means in his power" (J. H. Maiden).

J. W. Austral. Nat. Hist. Soc. 1909, 25–26.

Physolobium stirlingii Benth.

STIRLING, Sir James (1836–1916)
b. Aberdeen 3 May 1836 *d.* Goudhurst, Kent 27 June 1916

MA Cantab. Lord Justice of Appeal, 1900–5. Collected mosses from round his house at Goudhurst.

J. Bot. 1901, 179. *Times* 28 June 1916. *D.N.B. 1912–1921* 510–11. *Who was Who, 1916–1928* 1001.

Herb. at Tunbridge Wells Museum.

STIRLING, James (1852–1909)
b. Geelong, Victoria 9 Jan. 1852 *d.* Riverside, California 1909

FLS 1883. Land officer, Omes, Victoria, 1878. Assistant geologist, Mining Dept., Victoria. Plants collected in Victorian Alps determined by F. von Mueller. 'Census of Plants of Australian Alps' (*Trans. Bot. Soc. Edinburgh* v.22, 1901–4, 319–95).

Victorian Nat. v.6, 1890, 166–67; 1949,126. *Melbourne Age* (newspaper) 5 Aug. 1909. *Rep. Austral. Assoc. Advancement Sci.* v.13, 1911, 234.

Helichrysum stirlingi F. Muell.

STIRLING, John Stirling (–1900)
d. Gargunnock, Stirling 18 May 1900

Lieut.-Colonel, Royal Artillery. Contrib. papers on flora of Stirlingshire to *Trans. Stirling Nat. Hist. Soc.* 1890–91, 88–102; 1891–92, 74–102.

Herb. at Stirling.

STIRLING, William (–1967)
d. 15 Oct. 1967

Ophthalmic surgeon. Grew orchids.

Orchid Rev. 1967, 418–19.

STIRRAT, Alexander H. (*c.* 1877–1910)
b. Falkirk *c.* 1877 *d.* Johannesburg, S. Africa 6 Feb. 1910

Gardener, Glasgow Botanic Gardens. To S. Africa, 1900. Superintendent, Parks and Zoological Gardens, Johannesburg.

Gdnrs Chron. 1910 i 128.

STIRTON, James (1833–1917)
b. Scotland 1833 *d.* Glasgow 14 Jan. 1917

MD Edinburgh 1858. FLS 1875. Professor of Midwifery, St. Mungo's College, Glasgow, 1889. 'Dr. Stirton's New British Mosses Revised' by H. N. Dixon (*J. Bot.* 1923, 10–17, 46–52, 69–75). Contrib. to W. A. Leighton's *Lichen Fl. of Great Britain* 1879, *Glasgow Nat.*, *Grevillea*, *Scott. Nat.*

Proc. Linn. Soc. 1916–17, 71–75. *Glasgow Nat.* 1919, 142–44.

Plants at BM(NH), Glasgow.

Stirtonia R. Br. (of N.Z.).

STOCK, Daniel (*fl.* 1820s–1860s)

Of Bungay, Suffolk and Stoke Newington, London. Botanist and entomologist. Contrib. to *Mag. Nat. Hist.* from 1828 (often as "D.S."). Local Secretary, Botanical Society of London, 1839.

H. C. Watson *New Botanists' Guide* 1835, 112. H. C. Watson *Topographical Bot.* 1883, 556–57. *Trans. Suffolk Nat. Soc.* 1946, 8–12.

Plants at BM(NH), Kew. Letters and MS. list of Bungay plants at Kew.

STOCKDALE, Frank Arthur (1883–1949)
b. 24 June 1883 *d.* 13 Aug. 1949

BA Cantab. FLS 1907. Mycologist, Imperial Dept. of Agriculture for West Indies, 1905. Assistant Director, Dept. of Agriculture, British Guiana and Government Botanist, 1909. Director of Agriculture, Mauritius, 1912; Ceylon, 1916. Agricultural Adviser for Colonies, 1929–40.

Kew Bull. 1909, 150; 1916, 277. *Nature* v.164, 1949, 397–98. *Who was Who, 1941–1950* 1110.

STOCKEN, Chris. M. (*c.* 1922–1966)
d. East Greenland 1966

Lieut. Commander, Royal Navy. Explored the flora of Andalucia while stationed at Gibraltar. Interested in *Narcissus*. *Andalusian Flowers and Countryside* 1969.

Gdnrs Chron. v.160(17), 1966, 18. *Northern Gdnr* 1966, 190.

STOCKS, John Ellerton (1822–1854)
b. Cottingham, Hull, Yorks 1822 *d.* Cottingham 30 Aug. 1854

MD London. FLS 1848. Pupil of J. Lindley. On Bombay medical staff from 1847. Collected plants in Scinde, Baluchistan, etc. Acting Conservator of Forests and Superintendent of Botanic Gardens, Bombay during A. Gibson's absence. Brought plants and economic products to Kew, 1853.

Hooker's J. Bot. 1853, 304–6; 1854, 308–10. *Gdnrs Chron.* 1854, 580. *Gent. Mag.* 1854 ii 401–2. *Proc. Linn. Soc.* 1855, 416–17. J. D. Hooker and T. Thomson *Fl. Indica* 1855, 152. *R.S.C.* v.5, 836.
Letters at Kew.
Ellertonia Wight.

STODDART, James (*fl.* 1810s)
Nurseryman, College Lane, Northampton.

STOKER, Fred (*c.* 1879–1943)
d. Loughton, Essex 20 July 1943
MB. FRCS. FLS 1918. VMH 1937. *Shrubs for the Rock Garden* 1934. *Gardener's Progress* 1938. Contrib. to *Gdnrs Chron.*
Gdnrs Chron. 1934 i 36 portr.; 1943 i 48. *Proc. Linn. Soc.* 1942–43, 207–8. *Quart. Bull. Alpine Gdn Soc.* 1943, 245–47 portr. *Times* 23 July 1943. *J. R. Hort. Soc.* 1965, 397–400.
Library at Royal Horticultural Society.

STOKES, Charles (1783–1853)
d. Gray's Inn, London 28 Dec. 1853
FRS 1821. FLS 1808. 'Recent Wood Petrified' (*Trans. Geol. Soc.* v.5, 1836, 207–14). Had collection of fossil woods.
Proc. Linn. Soc. 1854, 312–13. C. J. F. Bunbury *Life and Letters and Journals* v.1, 1906, 200–1. *R.S.C.* v.5, 838.
Letters at Kew and in Brown correspondence at BM(NH). Library sold at Sotheby, 30–31 May 1854.

STOKES, John Lort (1812–1885)
d. Wales 11 June 1885
Entered Royal Navy, 1824; Admiral, 1877. Assistant surveyor on HMS 'Beagle' 1831; in command, 1841. Keen naturalist and encouraged plant collectors such as B. Bynoe.
J. D. Hooker *Fl. Tasmaniae* v.1, 1859, 117. *Times* 13 June 1885. *D.N.B.* v.54, 400–1. A. H. Chisholm *Strange New World* 1941, 38–40, 285–87. G. C. Ingleton *Charting a Continent* 1944, 58–60, 72–73. P. Serle *Dict. Austral. Biogr.* 1949, 368–69. *Austral. Encyclop.* v.8, 1965, 300.

STOKES, Jonathan (1755–1831)
b. Chesterfield, Derbyshire 1755 *d.* Chesterfield 30 April 1831
MD Edinburgh 1782. ALS 1790. Friend of the younger Linnaeus. Had herb. Contrib. references to figures and drew plates for Withering's *Bot. Arrangement* 1787. *Botanical Materia Medica* 1812 4 vols. *Botanical Commentaries* 1830.
P. Smith *Mem. and Correspondence of J. E. Smith* v.1, 1832, 118–20. E. Lees *Bot. Worcestershire* 1867, lxxxviii. *J. Bot.* 1901, Supplt., 70–71; 1914, 299–306, 317–23. G. C. Druce *Fl. Northamptonshire* 1930, lxxxvi.
Letters in Winch correspondence at Linnean Society. Sale at Sotheby, 23 Nov. 1856.
Stokesia L'Hérit.

STOKES, Whitley (1763–1845)
d. Dublin 13 April 1845
MD Dublin 1793. Lecturer in Natural History, Trinity College, Dublin, 1816. Professor of Medicine, Dublin, 1830–43. Friend of Dawson Turner and also of J. Templeton with whom he collected plants in Ulster. Bryologist. Contrib. mosses to J. Sowerby and J. E. Smith's *English Bot.* 1273, 2403, etc.
D. Turner *Muscologiae Hibernicae Spicilegium* 1804, vi. *Proc. R. Irish Acad.* v.3, 1847, 198; v.32, 1915, 69. *J. Bot.* 1898, 360. *D.N.B.* v.54, 401. *Proc. Belfast Nat. Field Club.* v.6, 1913, 627.
Portr. at Kew.
Hypnum stokesii Smith.

STOLTERFOTH, Henry (1836–1907)
b. 27 Sept. 1836 *d.* Chester 6 Oct. 1907
Diatomist. 'List of Diatomaceae found in Chester and District and Cwm Bychan' (*Proc. Chester Soc. Nat. Sci.* v.2, 1878, 28–41). 'Surface Dredging in the Dee' (*Proc. Chester Soc. Nat. Sci.* v.3, 1885, 93–97).
37th Rep. Chester Soc. Nat. Soc. 12 portr.
Micro slides at Chester Museum.
Rhizosolenia stolterfothii H. Perak.

STONE, Robert (*c.* 1751–1829)
d. 6 Jan. 1829
FLS 1790. Of Bedingham Hall, Bungay, Suffolk. Found *Hydnum imbricatum.* Had herb. Contrib. to J. Sowerby and J. E. Smith's *English Bot.* 458, 1467, etc. and to W. Withering's *Bot. Arrangement.*
Mag. Nat. Hist. v.2, 1829, 120. P. Smith *Mem. and Correspondence of J. E. Smith* v.1, 1832, 43. *J. Bot.* 1902, 321. *Trans. Norfolk Norwich Nat. Soc.* 1914, 672; 1937, 199.

STONEHOUSE, Mr.
Surgeon. Sent St. Helena plants to J. Petiver.
J. E. Dandy *Sloane Herb.* 1958, 216.

STONEHOUSE, Rev. Walter (1597–1655)
b. London 1597 *d.* London? 1655
BA Oxon 1617. BD 1629. Rector, Darfield, Yorks. Friend of T. Johnson. Travelled much in England and Wales. Discovered *Viola palustris.* MS. catalogue of his garden at Magdalen College in *Gdnrs Chron.* 1920 i 240–41, 256, 268, 296.
J. Bot. 1920, 170–73. R. T. Gunther *Early Br. Botanists* 1922, 271–73, 348–51, 416. C. E. Raven *English Nat.* 1947, 300–2.

STONESTREET, George (*fl.* 1690s)
Brother of Rev. W. Stonestreet. Collected plants at the Cape, Ascension Island and St. Helena. Sent plants to L. Plukenet.
J. Petiver *Musei Petiveriani* 1695, 143, 149. *J. Linn. Soc.* v.45, 1920, 36. J. E. Dandy *Sloane Herb.* 1958, 216. H. N. Clokie *Account of Herb. Dept. Bot., Oxford* 1964, 249.
Plants at BM(NH).

STONESTREET, Rev. William (–1716)
MA Cantab 1681. Rector, St. Stephen Walbrook, London, 1689. Correspondent of J. Ray, A. Buddle, J. Petiver and L. Plukenet. Discovered *Euphorbia portlandica*, etc.
R. Morison *Plantarum Historiae Universalis Oxoniensis* v.3, 1699, 185. J. Nichols *Illus. of Lit. Hist. of Eighteenth Century* v.3, 341–42. H. Trimen and W. T. T. Dyer *Fl. Middlesex* 1869, 389. G. C. Druce *Fl. Berkshire* 1897, cxxvii. G. C. Druce *Fl. Buckinghamshire* 1926, lxxiv–lxxv. E. J. L. Scott *Index to Sloane Manuscripts* 1904, 514. C. E. Raven *John Ray* 1942, 263, 402. J. E. Dandy *Sloane Herb.* 1958, 217.
Plants at BM(NH), Oxford.

STONHAM, William Burne (–1896)
d. Maidstone, Kent 6 Dec. 1896
FLS 1895. Botanised in Kent when young.
Proc. Linn. Soc. 1896–97, 68.

STOPES, Marie Charlotte Carmichael (1880–1958)
b. Edinburgh 15 Oct. 1880 *d.* Norbury Park, Surrey 2 Oct. 1958
DSc London 1905. PhD Munich. FLS 1909. Lecturer in Palaeobotany, Manchester University, 1904–14; University College, London, 1914–20. Marriage to R. Ruggles Gates annulled in 1916; married H. V. Roe, 1918. Pioneer of birth control and sex education. *Study of Plant Life* 1907; ed. 2 1910. *Ancient Plants* 1910. *Catalogue of Cretaceous Fl.* 1913–15 2 vols.
A. Maude *Authorised Life of Marie C. Stopes* 1924. K. Briant *Marie Stopes* 1962. *Nature* v.182, 1958, 1201–2. *Times* 3 and 6 Oct. 1958. *Proc. Linn. Soc.* 1959–60, 135–36.
Portr. at National Portrait Gallery, London, National Gallery of Edinburgh, Hunt Library.

STOPPS, Arthur James (1833–1931)
b. Devon 11 Nov. 1833 *d.* Hunter's Hill, Victoria Aug. 1931
To Melbourne, 1856. Acting Surveyor-General, N.S.W. Botanical artist and lithographer. Drew most of the plates for R. D. Fitzgerald's *Australian Orchids* 1875–94 2 vols.
Victorian Nat. v.48, 1932, 242–45 portr.

STOREY, Harold Haydon (1894–1969)
b. Manchester 10 June 1894 *d.* Nairobi, Kenya 5 April 1969
Educ. Cambridge. FRS 1946. Mycologist, Natal Herb., S. Africa, 1923. Plant Pathologist, E. African Agricultural Research Station, Amani, Tanganyika, 1928. Deputy Director, E. African Agricultural and Forestry Research Organisation, 1948. Published papers on virus diseases of tobacco, groundnuts, maize, cassava. Contrib. to *Ann. Applied Biol.*
Ann. Applied Biol. 1969, 188 portr. *Biogr. Mem. Fellows R. Soc.* 1969, 239–46 portr. *Nature* v.222, 1969, 905.

STOREY, John (*c.* 1801–1859)
d. Newcastle, Northumberland 8 Oct. 1859
Sent Newcastle list and plants to H. C. Watson's *Topographical Bot.* 1883, 557–58. Was preparing flora of Northumberland and Durham. Contrib. to *Trans. Tyneside Nat. Club* v.1–3.
Nat. Hist. Trans. Northumberland, Durham and Newcastle-upon-Tyne 1903, 85.
Plants at Hancock Museum, Newcastle.

STOREY, Thomas (–*c.* 1742)
Quaker preacher. Friend of P. Collinson who sent him N. American seeds and plants for his garden.
N. G. Brett-James *Life of Peter Collinson* 1926, 114–17.

STORRIE, John (1843–1901)
b. Muiryett, Cambusnethan, Lanark 2 June 1843 *d.* Cardiff, Glam 2 May 1901
ALS 1899. Curator, Cardiff Museum, 1877–92. *Fl. Cardiff* 1886.
J. Bot. 1901, 434. *Proc. Linn. Soc.* 1900–1, 50–51. H. A. Hyde and A. E. Wade *Welsh Flowering Plants* 1934, 2 portr.
Herb. at National Museum of Wales.

STORY, George Fordyce (1800–1885)
b. 4 June 1800 *d.* Kelvedon, Tasmania 7 June 1885
MA Aberdeen 1820. MD Edinburgh 1824. In London, 1825–28. To Tasmania, 1829. Assistant surgeon, Waterloo Point Military Station, 1829. Secretary, Royal Society of Tasmania, 1844. Supervised development of botanic garden, Hobart for one year. Collected plants for F. von Mueller and J. Backhouse.
Papers Proc. R. Soc. Tasmania 1909, 27–28. *Austral. Dict. Biogr.* v.2, 1967, 490.
Plants at Melbourne.

STOVIN, Mrs. Margaret (*fl.* 1820s–1830s)
Of Newbold, Chesterfield, Derbyshire. Had herb. "An investigator and collector of plants" (J. Stokes *Commentaries* 1830).
J. Bot. 1914, 321.
Herb. at Middlesborough Museum.

STOWELL, Rev. Hugh Ashworth (1830–1886)
b. Pendleton, Lancs 1830 *d.* Breadsall, Derbyshire 16 March 1886
BA Oxon 1852. MA 1855. Rector, Breadsall, 1865. 'Fl. Faversham' (*Phytologist* v.1–2, 1855–58). 'Fl. Isle of Man' (*Phytologist* v.4, 1860, 161–69).
Trans. Liverpool Bot. Soc. 1909, 91. *R.S.C.* v.5, 846.
Portr. at Hunt Library.

STRACHEY, Henry (1816–1912)
Elder brother of Sir R. Strachey. Member of Tibetan Boundary Commission, 1848. Surveyed Upper Indus Valley. Collected plants at Pangong Lake in Eastern Ladak.
I. H. Burkill *Chapters on Hist. Bot. in India* 1965, 83.

STRACHEY, Sir Richard (1817–1908)
b. Sutton Court, Somerset 24 July 1817 *d.* Hampstead, London 12 Feb. 1908
LLD Cantab 1892. FRS 1854. FLS 1859. GCSI 1897. Entered Bombay Engineers, 1836. Lieut.-General, 1875. With J. E. Winterbottom in N.W. Himalaya and adjacent Tibet, collecting in Punch, Kashmir, 1847 and in Garhwal, Kumaun, 1840–49. *Catalogue of Plants of Kumaun and of the adjacent Portions of Garhwal and Tibet* 1906.
Hooker's J. Bot. 1853, 306–7. *J. Linn. Soc.* v.25, 1902, 136–40. *J. Bot.* 1908, 95. *Kew Bull.* 1908, 127–29. *Proc. Linn. Soc.* 1907–8, 63–64. *Quart. J. Geol. Soc.* 1908, lix–lxi. *Times* 13 Feb. 1908. *D.N.B. Supplt 2* v.3, 439–42. I. H. Burkill *Chapters on Hist. Bot. in India* 1965, 31.
Plants at Kew, Gray Herb. Harvard.
Stracheya Benth.

STRANGE, Frederick (–1854)
d. murdered Percy Island, Queensland 15 Oct. 1854
In S. Australia, *c.* 1836. Set up as collector of natural history specimens in Sydney, 1840. Collected in New Zealand, 1848–49. Explored central Queensland coast with W. Hill.
Hooker's J. Bot. 1857, 189. *J. R. Soc. N.S.W.* 1908, 122–24 portr. *Victorian Nat.* 1948, 220–21. *Austral. Zoologist* 1947, 96–114 portr. *Austral. Encyclop.* v.1, 1958, 102; v.7, 54; v.8, 315.
Plants at BM(NH). Portr. at Hunt Library.
Strangea Meissner.

STRANGE, John (1732–1799)
b. Barnet, Herts 1732 *d.* Ridge, Herts 19 March 1799
MA Cantab 1755. FRS 1766. DCL Oxon 1793. British Resident, Venice, 1773–88. *Lettera sopra...Conferva Plinii* 1764.
Gent. Mag. 1799 i 348. J. Nichols *Lit. Anecdotes of Eighteenth Century* v.8, 9–10. *D.N.B.* v.55, 23.

STRANGWAYS, William Thomas Horner Fox, 4th Earl of Ilchester (1795–1865)
b. 7 May 1795 *d.* Melbury, Dorset 10 Jan. 1865
BA Oxon 1816. MA 1820. FRS 1821. FLS 1821. Secretary, English Legation, Vienna, 1834. Minister at Berlin, 1840–49. Had collection of rare and interesting plants in his garden at Abbotsbury, Dorset. Contrib. to *Gdnrs Mag.*
Bot. Register v.10, 1837, 1956. *Quart J. Geol. Soc. London* 1865, xlix–l. E. Bretschneider *Hist. European Bot. Discoveries in China* 1898, 287. *Nature* v.206, 1965, 10–11.
Plants and letters at Kew. Portr. at Hunt Library.

STRATTON, Frederic (1840–1916)
b. Newport, Isle of Wight 16 Nov. 1840 *d.* Newport 5 Dec. 1916
FLS 1869. Solicitor. *Wild Flowers of Isle of Wight* 1900. 'Flowering Plants and their Allies' (F. Morey *Guide to Nat. Hist. of Isle of Wight* 1909, 126–99). Contrib. to *J. Bot.*
Bot. Soc. Exch. Club Br. Isl. Rep. 1916, 468–69. *J. Bot.* 1917, 20–22. *Proc. Linn. Soc.* 1916–17, 66–67. *R.S.C.* v.8, 103; v.11, 515.
Plants at Oxford.

STREDWICK, Harry (*c.* 1877–1966)
d. St. Leonards-on-Sea, Sussex 1966
VMH 1943. Cultivated dahlias and raised several new cultivars.
Gdnrs Chron. v.161(6), 1967, 4.

STREDWICK, James (–1917)
Nurseryman, St. Leonards-on-Sea, Sussex. Hybridised dahlias.
Gdnrs Mag. 1910, 801–2 portr. *Garden* 1917, 135 portr.

STREET, Charles (1859–1946)
b. Wookey, Somerset 27 June 1859 *d.* Kelso, Roxburghshire 11 March 1946
Gardener to Duke of Roxburghe at Floors Castle, Kelso, *c.* 1889.
Gdnrs Chron. 1946 i 156.

STREETEN, Robert James Nicholl (1800–1849)
b. London 28 June 1800 *d.* Worcester 10 May 1849
MD Edinburgh 1824. FLS 1846. Practised medicine at Worcester. 'Myosotis' (*Naturalist* v.1, 1837, 169–75). Contrib. to *Phytologist*.
Proc. Linn. Soc. v.2, 1849, 48. *R.S.C.* v.5, 853.

STREETER, Fred. (1877–1975)
b. Pulborough, Sussex June 1877 *d.* 1 Nov. 1975

VMH 1945. Gardener with J. Veitch and Sons, Chelsea; with A. Pam at Wormleybury, Herts; with Lord Leconfield at Petworth Park, Sussex. Well-known BBC broadcaster on gardening.

Times 3 Nov. 1975 portr.

STRETCH, Richard Harper (1837–1926)
b. Nantwich, Cheshire 25 Nov. 1837 *d.* Seattle, U.S.A. 22 March 1926

To U.S.A. 1861, again in 1863. Collected plants in W. Nevada, 1865.

E. O. Essig *Hist. of Entomol.* 1931, 767–70 portr. *Madrono* v.19, 1967, 32.

STRICKLAND, Agnes (1796–1874)
b. London 19 Aug. 1796 *d.* Southwold, Suffolk 13 July 1874

Historian. *Floral Sketches, Fables and other Poems* 1836; ed. 2 1861. *Sketches from Nature* 1830. *Narratives of Nature c.* 1845.

J. M. Strickland *Life of A. Strickland* 1887. *D.N.B.* v.55, 48–50.

Portr. at National Portrait Gallery.

STRICKLAND, Catharine Parr *see* Traill, C. P.

STRICKLAND, Sir Charles William (1819–1909)
b. 6 Feb. 1819 *d.* 31 Dec. 1909

FLS 1877. Grew orchids, bulbous plants and hardy fruits at Hildenley Hall, Malton, Yorks.

Gdnrs Chron. 1910 i 59 portr. *Orchid Rev.* 1910, 38–39. *Proc. Linn. Soc.* 1909–10, 101–2.

Dendrobium stricklandianum Reichenb. f.

STRICKLAND, Charlotte (*c.* 1759–1833)
d. Apperley, Glos 2 June 1833

She and her sister, Julia Sabina (*c.* 1765–1849), drew plates for *Select Specimens of British Plants* 1797–1809 edited by their brother-in-law, Strickland Freeman. "Certainly rank as artists in the first line" (James Sowerby and J. E. Smith *English Bot.* 637).

STRICKLAND, Hugh Edwin (1811–1853)
b. Rigton, Yorks 2 March 1811 *d.* Clarborough, Lincs 14 Sept. 1853

BA Oxon 1832. FRS 1852. Geologist and zoologist. 'Natural System in Zoology and Botany' (*Ann. Mag. Nat. Hist.* v.6, 1841, 184–94). 'Report on Vitality of Seeds' (with others) (*Br. Assoc. Rep.* 1845, 337–39).

Quart. J. Geol. Soc. London 1854, xxv–xxvi. W. Jardine *Mem. of H. E. Strickland* 1858 portr. *D.N.B.* v.55, 50–52. *R.S.C.* v.5, 855.

STRONACH, David (*fl.* 1790s)

"Gardener and botanist" on Lord Macartney's embassy to China, 1793–94. Compiled list of plants in the Camoens gardens in Macao, 1794 (list now at Cornell University).

J. L. Cranmer Byng *Embassy to China* 1962, 317–18.

STRONACH, James (*fl.* 1820s)

Nurseryman, 57 High Street, Lewes, Sussex.

STRONACH, William Gavin (*fl.* 1850s–1870s)

BA London 1856. To China as student interpreter, 1861. Sent plants to H. F. Hance and BM(NH).

E. Bretschneider *Hist. European Bot. Discoveries in China* 1898, 708–9.

Clematis stronachii Hance.

STRONG, Mary Knapp *see* Clemens, J.

STROUD, T. B. (*fl.* 1820s)

Gardener to Duke of Northumberland at Syon House, Middx, 1822. *Elements of Botany* 1821.

STRUTT, Jacob George (*fl.* 1810s–1850s)

Landscape painter and etcher. At Lausanne and Rome, 1830–51. *Sylva Britannica* 1822; ed. 2 1831–36. *Deliciae Sylvarum* 1828–29. *Sylva Italica* 1844. 'Forest Trees of Europe' (*Mag. Nat. Hist.* v.1, 1829, 37–42, 242–48).

D.N.B. v.55, 64.

STRZELECKI, Count Paul Edmund De (1796–1873)
b. Poland 1796 *d.* London 6 Oct. 1873

Educ. Edinburgh. FRS 1853. DCL Oxon 1860. KCMG 1869. To New South Wales. *Physical Description of New South Wales and Van Dieman's Land* 1845.

Hooker's J. Bot. 1857, 308–9. P. Mennell *Dict. Austral. Biogr.* 1892, 444. F. Boase *Modern English Biogr.* v.3, 1901, 805.

Fossils at BM(NH).

Strzeleckya F. Muell.

STUART, Alexander (*fl.* 1700s–1740s)

MD. Surgeon in merchant service, travelling in Far East. Sent plants to Sir H. Sloane.

J. E. Dandy *Sloane Herb.* 1958, 218.

STUART, Charles (1802–1877)
d. Parramatta, N.S.W. Sept. 1877

Friend of F. von Mueller for whom he collected plants in Tasmania, 1842–52. Also collected for R. Gunn and algae for W. H. Harvey. In last years was gardener.

J. D. Hooker *Fl. Tasmaniae* v.1, 1859, cxxvii. W. H. Harvey *Phycologica Australica* v.5, 1863, t.ccxciv. *Rep. Austral. Assoc. Advancement Sci.* 1907, 28. *J. R. Soc. N.S.W.* 1908, 124–26. *Papers Proc. R. Soc. Tasmania* 1909, 28–29. *Victorian Nat.* v.52, 1935, 106–10, 132–37, 154–57.

Plants at Herb., Melbourne. Letters at Herb., Melbourne, Mitchell Library, Sydney.

Areschougia stuartii Harvey.

STUART, Charles (1825–1902)
b. Woodhall, Edinburgh 30 March 1825 *d.* Chirnside, Berwickshire 12 Feb. 1902

MD Edinburgh 1846. FRSE 1884. President, Berwickshire Naturalists' Club, 1873. Hybridised pansies. Contrib. to *Proc. Berwickshire Nat. Club.*

Trans. Bot. Soc. Edinburgh v.19, 1893, 63; v.22, 1902, 191–94. *Garden* v.45, 1894, xii portr.; v.61, 1902, 132. *Ann. Scott. Nat. Hist.* 1902, 65–66, 126. *Gdnrs Chron.* 1902 i 133–34 portr. *Gdnrs Mag.* 1902, 122. *J. Hort. Cottage Gdnr* v.44, 1902, 171. *Proc. Berwickshire Nat. Club* v.18, 1902, 171–75.

Erica tetralix subsp. *Stuarti* Macfarlane.

STUART, James (1802–1842)
b. Ireland 1802 *d.* Port Macquarie, N.S.W. May 1842

Surgeon, Dublin, 1829; Liverpool, 1833. Arrived in Sydney, N.S.W., 1834. Assistant Surgeon to the Colony, 1836. Transferred to Norfolk Island, 1838–40, where he painted the local fauna and flora.

J. Proc. R. Austral. Hist. Soc. v.19, 323. *Austral. Zool.* 1955, 120–31. *Austral. Encyclop.* v.8, 1965, 331.

Paintings at Mitchell Library, Sydney.

STUART, John, 3rd Earl of Bute (1713–1792)
b. Edinburgh 25 May 1713 *d.* London 10 March 1792

Prime Minister, 1762–63. Had botanic gardens at Luton Hoo, Beds and Highcliffe, Hants. Advised Princess Augusta on the development of her garden at Kew. *Botanical Tables* 1785 (*J. Bot.* 1916, 84–87).

J. E. Smith *Selection from Correspondence of Linnaeus* v.1, 1821, 26–31, 33. D. Turner *Extracts from Lit. and Sci. Correspondence of R. Richardson* 1835, 407. W. Darlington *Memorials of J. Bartram and H. Marshall* 1849, 296. *Kew Bull.* 1891, 290–92; 1892, 306–8. *D.N.B.* v.55, 92–98. J. A. Lovat-Fraser *John Stuart Earl of Bute* 1912. *Huntia* 1965, 212–13 portr. A. M. Coats *Lord Bute* 1975 portr.

Plants at BM(NH), Kew. MSS. and drawings of *Tabular Distribution of Vegetable Kingdom* 1783 at BM(NH). Portr. by Sir J. Reynolds at National Portrait Gallery. Portr. at Kew, Hunt Library. Sale at Leigh and Sotheby 5 May 1794 and 21 April 1809. Herb. (lot 148 in A. B. Lambert's sale) bought by Gibson.

Stewartia L. (*Stuartia* L.) *Butea* Roxb.

STUART, Rev. John (1743–1821)
b. Killin, Perth 1743 *d.* Luss, Dunbartonshire 24 May 1821

DD Glasgow 1795. ALS 1793. Of Luss from 1777. Travelled in Highlands and Hebrides with T. Pennant and J. Lightfoot, 1772, and assisted the latter with his *Fl. Scotica* (v.1, 1777, xii–xiii). Discovered *Juncus biglumis.* Contrib. to J. Sowerby and J. E. Smith's *English Bot.* 898, 2586.

W. Wright *Memoir* 1828, 145–46. P. Smith *Mem. and Correspondence of J. E. Smith* v.1, 1832, 55. *D.N.B.* v.55, 101–2. *Proc. Perthshire Soc. Nat. Sci.* v.4, 1907–8, clxxxix–cxc.

Letters in Smith correspondence at Linnean Society.

Salix stuartiana Smith.

STUART, John McDouall (1815–1866)
b. Fifeshire 7 Sept. 1815 *d.* London 5 June 1866

To Australia, 1838, joining Government Survey Service. Made several expeditions into Central Australia, 1858–62, and collected plants for F. von Mueller. *Exploration across Continent of Australia, 1860–1862* 1863. *Explorations in Australia: Journals* 1864.

G. Bentham *Fl. Australiensis* v.1, 1863, 13, 456. B. Threadgill *South Australian Land Exploration, 1856–1888* 42–73. N. G. Stuart *Life of Charles Stuart* 1899. *D.N.B.* v.55, 103–4. *Rep. Austral. Assoc. Advancement Sci.* 1907, 169. J. H. L. Cumpston *Charles Stuart* 1951. M. S. Webster *J. McD. Stuart* 1958. *Austral. Dict. Biogr.* v.2, 1967, 495–98.

MS. journals at Mitchell Library, Sydney. Portr. at Hunt Library.

Diplopeltis stuartii F. Muell.

STUKELEY, William (1687–1765)
b. Holbeach, Lincs 7 Nov. 1687 *d.* London 3 March 1765

MB Cantab 1708. MD 1719. FRS 1718. Antiquary. Rector, St. George-the-Martyr London, 1747. Collected additions to J. Ray's *Catalogus Plantarum circa Cantabrigiam.* Correspondent of P. Collinson.

D.N.B. v.55, 127–29. S. Piggott *William Stukeley* 1950 portr.

STURDY, Herbert Hastings (–1933)

Botanised in Settle, Yorks.

Naturalist 1933, 128–29.

STURROCK, Abram (1843–1886)
b. Padanaram, Forfarshire Sept. 1843 *d.* Rattray, Perthshire 13 March 1886

Schoolmaster. Studied water plants.
Scott. Nat. 1886, 298–99.
Herb. at Perthshire Society of Natural Sciences Museum.
Potamogeton pusillus var. *Sturrockii* A. Benn.

STURT, Charles (1795–1869)
b. Bengal 28 April 1795 *d.* Cheltenham, Glos 16 June 1869
FLS 1833. Capt. 39th Regt.; left Army, 1833. Friend of Robert Brown. Colonial Secretary, 1849–51. Explored Southern Australia, 1828–31; Central Australia, 1844–46. Collected plants. Returned to England, 1853. *Two Expeditions into Interior of Southern Australia...1828, 1829, 1830 and 1831* 1833. *Narrative of Expedition into Central Australia...1844, 1845, 1846* 1849 2 vols (botanical appendix by R. Brown).
Proc. Linn. Soc. 1869–70, cx–cxii. *J. R. Geogr. Soc.* v.14, 1870, cxxxiii–cxxxvii. N. G. Sturt *Life of Charles Sturt* 1899 portr. *D.N.B.* v.55, 136–38. *Rep. Austral. Assoc. Advancement Sci.* 1907, 167–68. P. Serle *Dict. Austral. Biogr.* v.2, 1949, 384. J. H. L. Cumpston *Charles Sturt* 1951. D. Kennedy *Charles Sturt* 1958. G. Farwell *Rider to an Unknown Sea* 1963. *Austral. Encyclop.* v.8, 1965, 335. *Austral. Dict. Biogr.* v.2, 1967, 495–98. *Hist. Today* 1967, 735–42. *J. Proc. R. Austral. Hist. Soc.* v.15, 49–92. *R.S.C.* v.5, 880.
Plants and letters at Kew. Portr. at Hunt Library. MSS. at Mitchell Library, Sydney.
Sturtia R. Br.

STURT, Gerald (1860–1947)
d. 11 Jan. 1947
Amateur naturalist. 'On a Fossil Marine Diatomaceous Deposit from Oamaru, Otago, New Zealand' (with E. Grove) (*J. Quekett Microsc. Club* 1886, 321–30; 1889, 7–12, 63–78, 131–48).
J. Quekett Microsc. Club v.2, 1948, 228.

SUDELL, Richard (–1968)
d. Kuwait 18 Nov. 1968
Gardener, Kew, 1915. President, Institute of Landscape Architects, 1955. *Landscape Gardening* 1933. *New Illustrated Gardening Encyclopaedia* 1937. *Herbaceous Borders and the Waterside* 1938. *Teach Yourself Gardening* 1946. *Town and Suburban Garden* 1950. *Labour-saving Garden* 1952. *Practical Gardening...in Pictures* 1952. *Garden Planning* 1952.
J. Kew Guild 1972, 64–65.

SULLIVAN, David (1836–1895)
d. Moyston, Ararat, Victoria, Australia 2 June 1895
FLS 1884. Headmaster, Moyston. Correspondent of F. von Mueller. Contrib. to *Victorian Nat.*
Proc. Linn. Soc. 1895–96, 47. *Victorian Nat.* 1895, 36; 1908, 113. *Austral. Encyclop.* v.6, 1965, 42. *R.S.C.* v.18, 1036.
Caleya sullivani F. Muell.

SULLIVAN, Dennis (*fl.* 1740s–1760s)
Nurseryman, Cork.
Irish For. 1967, 51.

SUMMERHAYES, Victor Samuel (1897–1974)
b. Street, Somerset 21 Feb. 1897 *d.* 27 Dec. 1974
BSc London 1920. Spitzbergen, 1921. In charge of orchid herb., Kew, 1924–64. *Wild Orchids of Britain* 1951; ed. 2 1968. Revised orchids for ed. 2 of *Fl. West Tropical Africa* 1968. Contrib. orchids to *Fl. Tropical East Africa* 1968. Contrib. to *Kew Bull.*
Amer. Orchid Soc. Bull. 1975, 200 portr. *Bull. Br. Ecol. Soc.* v.6(2), 1975, 5–6. *Gdnrs Chron.* v.177(4), 1975, 7. *Orchid Rev.* 1975, 77–78. *Times* 7 Jan. 1975.
Letters at Kew.

SUMMERS, Rev. William Henry (1850–1906)
b. Dorking, Surrey 27 June 1850 *d.* Hungerford, Berks 30 April 1906
Congregational Minister, Beaconsfield, Bucks; later, Hungerford. *Notes on Buckinghamshire Botany* 1894.
Congregational Yb. 1907, 176. G. C. Druce *Fl. Buckinghamshire* 1926, civ–cv.

SUNDERLAND, John Edward (1839–1903)
b. Ashton-under-Lyne, Lancs 7 May 1839 *d.* Hatherlow, Cheshire 21 Feb. 1903
"An enthusiastic botanist."
Trans. Liverpool Bot. Soc. 1909, 91.
Herb. formerly at Stalybridge Field Naturalists Society.

SUTCLIFFE, Herbert (1880–1930)
b. Bradford, Yorks 1880 *d.* Malaya 26 Nov. 1930
FLS 1921. Mycologist, Malay Peninsula Agricultural Association, 1913–17. Mycologist and pathologist, Rubber Growers' Association, 1917–26. Pathologist, Rubber Research Institute of Malaya, 1926–30.
Proc. Linn. Soc. 1930–31, 198.

SUTHERLAND, James (*c.* 1639–1719)
d. Edinburgh 24 June 1719
King's Botanist for Scotland, 1699. Professor of Botany, Edinburgh, 1695–1706. *Hortus Medicus Edinburgensis* 1683. Sent Scottish plants to J. Petiver.

R. Morison *Plantarum Historiae Universalis Oxoniensis* v.3, 1699, 114, 342, etc. J. Petiver *Museii Petiveriani* 1695, 70, 95. R. Pulteney *Hist. Biogr. Sketches of Progress of Bot. in England* v.2, 1790, 4. J. Newton *Compleat Herbal* 1752, 5. D. Turner *Extracts from Lit. and Sci. Correspondence of R. Richardson* 1835, 27, 68–71. J. C. Loudon *Arboretum et Fruticetum Britannicum* 1838, 50–51. *Notes R. Bot. Gdn, Edinburgh* v.1, 1900–1, vi; v.12, 1919–21, v–vi; 1933, 13–61. E. J. L. Scott *Index to Sloane Manuscripts* 1904, 517. F. W. Oliver *Makers of Br. Bot.* 1913, 281–82. J. E. Dandy *Sloane Herb.* 1958, 218–19. H. R. Fletcher and W. H. Brown *R. Bot. Gdn Edinburgh, 1670–1970* 1970, 11–19.

Sutherlandia R. Br.

SUTHERLAND, Peter Cormack (1822–1900)
b. Latheron, Caithness 1822 *d.* Durban, Natal 30 Nov. 1900

MD Aberdeen 1847. To Davis Strait as surgeon to whaler, 1844 (*Trans. Bot. Soc. Edinburgh* v.2, 1846, 215). Surgeon on HMS 'Sophia' in search of missing ships, 'Erebus' and 'Terror'. Surveyor-General, Natal, 1855–87. Sent plants to W. H. Harvey and Kew. *Journal of Voyage in Baffin's Bay and Barrow Straits...1850–51* (list of plants in Appendix) 1852 2 vols.

W. H. Harvey *Fl. Capensis* v.1, 1859, 9*–10*. *Gdnrs Chron.* 1901 i 195. *J. Bot.* 1901, 191. *Kew Bull.* 1901, 170–1. *Rhodora* 1936, 411–12. *S. African J. Sci.* 1971, 405. *R.S.C.* v.5, 889; v.8, 1047.

S. African plants and letters at Kew.

Greyia sutherlandi Harvey.

SUTHERLAND, William (*c.* 1833–1920)
d. Catford, London 12 Feb. 1920

Kew gardener, 1856–60. Gardener at Minto, Roxburghshire. *Handbook of Hardy Herbaceous and Alpine Flowers* 1871. Contrib. to *Gdnrs Chron.*

Gdnrs Chron. 1920 i 110.

SUTTON, Alfred (*c.* 1818–1897)
d. Reading, Berks 7 Aug. 1897

Partner in seed firm, Reading. Son of J. Sutton (1777–1863).

Gdnrs Mag. 1897, 498–99 portr.; 1906, 39 portr. *J. Hort. Cottage Gdnr* v.35, 1897, 138–39.

SUTTON, Alfred (1859–1931)
d. Dec. 6 1931

Employed at Patent Office. Bryologist. Contrib. to *Trans. Herts. Nat. Hist. Soc.*

Rep. Br. Bryol. Soc. 1931, 389. *Trans. Herts. Nat. Hist. Soc.* 1932, xxxvii.

SUTTON, Arthur Warwick (1854–1925)
b. Reading, Berks 5 July 1854 *d.* Bournemouth, Hants 15 April 1925

FLS 1886. VMH 1897. Son of M. H. Sutton (1815–1901). Senior partner in firm of Messrs Sutton and Sons, seed merchants, Reading. Investigated production of various food-crops. '*Brassica* Crosses' (*J. Linn. Soc.* v.38, 1908, 337–49). 'Notes on some Wild Forms and Species of Tuber-bearing Solanums' (*J. Linn. Soc.* v.38, 1908, 446–53).

Garden 1909, iv portr. *Bot. Soc. Exch. Club Br. Isl. Rep.* 1925, 858. *Gdnrs Chron.* 1925 i 292 portr. *J. Bot.* 1925, 151. *Proc. Linn. Soc.* 1924–25, 82–83. *Who was Who, 1916–1928* 1015.

SUTTON, Rev. Charles (1756–1846)
b. Norwich, Norfolk 6 March 1756 *d.* Tombland, Norwich 28 May 1846

BA Cantab 1779. DD 1806. ALS 1791. Pupil of John Pitchford. Distinguished *Orobanche elatior*. 'British *Orobanche*' (*Trans. Linn. Soc.* v.4, 1798, 173–88). Contrib. to J. Sowerby and J. E. Smith's *English Bot.* 20, 568, etc.

Proc. Linn. Soc. 1847, 341–42. *Trans. Norfolk Norwich Nat. Soc.* 1902–3, 453–66; 1912–13, 682. *R.S.C.* v.5, 889.

Letters in Smith correspondence at Linnean Society.

Suttonia A. Rich.

SUTTON, Ernest Phillips Foquett
(–1972)
d. Aug. 1972

VMH 1952. Director, Sutton Seeds Ltd., Reading.

J. R. Hort. Soc. 1972, 496.

SUTTON, John (1777–1863)

Founded Sutton and Sons, seed merchants, Reading, 1806.

J. R. Hort. Soc. 1956, 234 portr. M. Hadfield *Gdning in Britain* 1960, 331–32.

SUTTON, Leonard Goodhart (1863–1932)
d. Reading, Berks 13 June 1932

FLS 1899. Son of M. H. Sutton (1815–1901). Senior partner in firm of Messrs Sutton and Sons, seed merchants, Reading.

Garden 1922, ii–iii portr. *Gdnrs Chron.* 1932 i 470 portr. *Nature* v.130, 1932, 85–86. *J. R. Hort. Soc.* 1932, 361 portr. *Proc. Linn. Soc.* 1932–33, 208–9. *Times* 14 June 1932. *Who was Who, 1929–1940* 1314.

SUTTON, Leonard Noel (–1965)
d. Reading, Berks 8 Oct. 1965

Director, Sutton and Sons, seed merchants, Reading. Son of L. G. Sutton (1863–1932). *Cool Greenhouse* 1935.

Gdnrs Chron. 1965 ii 504.

SUTTON, Martin Hope (1815–1901)
b. Reading, Berks 14 March 1815 *d.* Reading 4 Oct. 1901

Son of J. Sutton (1777–1863) and partner in family firm of seed merchants, Reading. *Sutton's Amateur's Guide in Horticulture* 1856. *Laying Down Land to Permanent Pasture* ed. 2 1863; ed. 11 1875.

Garden v.28, 1885, xii portr.; v.60, 1901, 255–56 portr. *Gdnrs Chron.* 1901 ii 279–80 portr. *Gdnrs Mag.* 1901, 655–56 portr.; 1906, 39 portr. *J. Hort. Cottage Gdnr* v.43, 1901, 341, 363 portr.

SUTTON, Martin Hubert Foquett (1875–1930)
b. 28 Nov. 1875 *d.* 27 March 1930

BA Oxon 1895. FLS 1909. Partner in seed firm of Sutton and Sons, Reading, Berks. *Book of the Links* (and others) 1912. *Effect of Radioactive Ores and Residues on Plant-life* 1915. *Electrification of Seeds by Wolfryn Process* 1920.

Gdnrs Chron. 1930 i 277. *Proc. Linn. Soc.* 1929–30, 223–24. *Who was Who, 1929–1940* 1314.

SUTTON, Martin John (1850–1913)
b. Reading, Berks 1850 *d.* London 14 Dec. 1913

FLS 1886. Senior partner in seed firm of Sutton and Sons, Reading. Experimented on improvement of agricultural grasses, etc. *Permanent and Temporary Pastures* 1886; ed. 9 1929 (based on Martin Hope Sutton's *Laying down Land to Permanent Pasture*).

Garden 1913, 639 portr. *Gdnrs Chron.* 1913 ii 450 portr. *Gdnrs Mag.* 1913, 968 portr. *J. Hort. Home Farmer* v.67, 1913, 599 portr. *Proc. Linn Soc.* 1913–14, 63. *D.N.B. 1912–1921* 517–18. *Who was Who, 1897–1916* 691.

SUTTOR, George (1774–1859)
b. Chelsea, London 11 June 1774 *d.* Alloway Bank, Bathurst, N.S.W. 5 May 1859

FLS 1843. Worked in his father's market garden. Took charge of plants on voyage to New South Wales, 1800. Farmed there. 'Forest-trees of Australia' (*Proc. Linn. Soc.* 1843, 177–78; 1860, xxxiii–xxxiv). *Culture of Grape Vine and Orange in Australia and New Zealand* 1843. *Memoirs, Historical and Scientific of Right Honourable Sir Joseph Banks* 1855.

J. H. Maiden *Sir Joseph Banks* 1909, 209–12 portr. F. M. Bladen *Hist. Rec. N.S.W.* v.3, 1895 *passim.* G. Mackaness *Mem. of G. Suttor* 1948. *Austral. Dict. Biogr.* v.2, 1967, 498–500. *Camellia News* no. 51, 1973, 27. *R.S.C.* v.5, 890.

MS. memoir at Mitchell Library, Sydney.

SWABEY, Christopher (1906–1972)
b. Constantinople, Turkey *d.* Asmara, Ethiopia 4 April 1972

Educ. Edinburgh University. Assistant Conservator, Forest Service, Trinidad, 1928. Jamaica Forest Service, 1937. British Guiana, 1946. Uganda, 1951–57. Forestry Adviser to Secretary of State until 1965. Director, Commonwealth Forestry Bureau, 1965–71.

Commonwealth For. Rev. 1972, 104–5.

SWAFFER, Arnold (1926–1951)
b. 16 March 1926

BA Cantab 1947. ALS 1947. Research Assistant in Cotton Physiology, Manchester University; Assistant Lecturer in Botany and Zoology, 1949.

Proc. Linn. Soc. v.164, 1953, 284.

SWAILES, John (*fl.* 1800s–1840s)
Gardener and nurseryman, Beverley, Yorks.

SWAILES, Robert (*fl.* 1810s–1820s)
Nurseryman, Beverley, Yorks.

SWAILES, Thomas (*fl.* 1840s)
Nurseryman, Beverley, Yorks.

SWAINSON, Isaac (1746–1812)
b. Hawkshead, Lancs 1746 *d.* Twickenham, Middx 7 March 1812

MD 1785. Of Frith Street, Soho, London. Cousin of W. Swainson (1789–1855). Formed private botanic garden at Twickenham (afterwards managed by Robert Castles) and a collection of 11,000 botanical plates, now at BM(NH).

H. C. Andrews *Botanist's Repository* 1804, 348. *Gent. Mag.* 1812 i 300. J. C. Loudon *Arboretum et Fruticetum Britannicum* 1838, 75, 2533. H. Swainson Cowper *Hawkshead* 1899, 406 portr.

Library sold by Rushworth and Jervis 29 May 1841.

Swainsona Salisb.

SWAINSON, William (1789–1855)
b. Newington Butts, London 8 Oct. 1789 *d.* Fern Grove, Hutt Valley, N.Z. 7 Dec. 1855

FRS 1820. FLS 1816. Clerk, H.M. Customs, London, *c.* 1803; in Malta and Sicily, 1807–15. To Brazil with Koster, 1816–18. Applied unsuccessfully for post in BM, 1822 and 1837. Arrived in Wellington, N.Z., 1841; to Sydney, 1851. Engaged by Tasmanian and Victorian Governments to report on timber trees, 1853: *Botanical Report on Victoria* 1853. *Naturalists' Guide* 1822; ed. 2 1824. *Taxidermy* 1840 (includes autobiography).

Edinburgh Philos. J. 1819, 369–73. *Gdnrs Mag.* 1828, 377–78; 1833, 521–23. *Naturalist* v.4, 1839, 397. *Hooker's J. Bot.* 1854, 30, 186–90. *Gent. Mag.* 1856 i 522–23. *Proc. Linn. Soc.* 1855–56, xlix–liii; 1899–1900, 14–61. P. Mennell *Dict. Austral. Biogr.* 1892, 451. *D.N.B.* v.55, 192–93. *Proc. Linn. Soc. N.S.W.* 1901, 796–98. C. F. P. von Martius *Fl. Brasiliensis* v.1(1), 1906, 117. *Victorian Nat.* 1908, 113–14. *Trans. Liverpool Bot. Soc.* 1909, 91–92. *Rep. Austral. Assoc. Advancement Sci.* 1911, 235. *J. Proc. R. Austral. Hist. Soc.* 1923, 43–44. Liverpool Public Museums *Handbook and Guide to Herb. Collections* 1935, 63. *Emu* 1950, 208–10. S. P. Dance *Shell Collecting* 1966, 124–26 portr. *Turnbull Library Rec.* v.1, 1967, 6–19. C. C. Gillispie *Dict. Sci. Biogr.* v. 13, 1976, 167–68.

Sicilian plants at Cambridge. Greek plants at Liverpool. Brazilian plants and drawings of N.Z. trees at Kew. Letters at BM(NH), Kew, Linnean Society. MSS. at Mitchell Library Sydney, National Library Canberra, Dominion Museum Wellington.

Swainsonia Salisb.

SWALES, John (*c.* 1864–1908)
b. Egton Bridge, Yorks *c.* 1864 *d.* Middlesbrough, Yorks April 1908

Of Whitby, Yorks. 'List of Whitby Plants' (*Whitby Official Guide*). Contrib. to B. Reynolds's *Whitby Wild Flowers* 1915, 25.

SWALLOW, James (*fl.* 1760s–1810s)

Nurseryman, Oxford Road, Reading, Berks.

SWAN, Joseph (–1872)

Engraver, Trongate, Glasgow. Engraved plates for *Bot. Mag.* 1826–45; W. J. Hooker's *Exotic Fl.* 1822–27; W. J. Hooker and R. K. Greville's *Icones Filicum* 1827–31; W. J. Hooker's *Fl. Boreali-Americana* 1829–40, etc.

SWAN, William (*c.* 1841–1919)
d. Staines, Middx 3 Oct. 1919

Gardener to John Day of Tottenham; to S. Mendel of Manley Hall, Manchester, 1870; to W. Leech of Fallowfield, Manchester *c.* 1873–*c.* 1886. Hybridised orchids.

Gdnrs Chron. 1919 ii 195. *Orchid Rev.* 1919, 178–80.

SWANTON, Ernest William (1870–1958)
b. Dibden, Hants 28 June 1870 *d.* Twickenham, Middx 21 Oct. 1958

ALS 1920. Schoolmaster and private tutor for some years. Curator, Haslemere Museum, 1897–1948. President, British Mycological Society, 1916. President, British Conchological Society, 1921. *Annotated Catalogue of Edible British Fungi* 1900. *Fungi and how to know Them* 1909; ed. 2 1922. *British Plant Galls* 1912. *Yew Trees of England* 1958.

Nature v.182, 1958, 1412. *Times* 23 Oct. 1958. *Proc. Bot. Soc. Br. Isl.* 1960, 105–6.

SWARBRICK, Thomas (1900–1965)
b. 8 Jan. 1900 *d.* 26 Nov. 1965

BSc. PhD Leeds. Physiologist and pomologist, Long Ashton Agricultural and Horticultural Research Institute, 1926–45. Research and Development Dept., Royal Dutch Shell, 1945–51. Director, Scottish Horticultural Research Institute, 1951. *Harnessing the Hormone* 1933.

Who was Who, 1961–1970 1095.

SWATMAN, Cecil Charles (1884–1958)
d. 24 Sept. 1958

Diatomist. Contrib. to *J. Quekett Microsc. Club.*

J. Quekett Microsc. Club 1958, 144 portr.

SWAYNE, Rev. George (*c.* 1746–1827)
b. Evilton, Somerset *c.* 1746 *d.* Dyrham, Glos 24 Oct. 1827

BA Oxon 1766. Vicar, Pucklechurch, Glos, 1772. Rector, Dyrham, 1806. Correspondent of W. Withering. *Gramina pascua* 1790 (exsiccatae).

J. W. White *Fl. Bristol* 1912, 65–66. H. J. Riddelsdell *Fl. Gloucestershire* 1948, cxvi.

SWEET, James (*c.* 1839–1924)
b. London *c.* 1839

VMH 1901. Nurseryman, Whetstone, Leics.

Gdnrs Chron. 1901 i 294–95 portr. *J. Hort. Cottage Gdnr* v.42, 1901, 259 portr. M. Allan *Tom's Weeds* 1970, 46–51 portr.

SWEET, James (*fl.* 1790s)

Nurseryman, Bristol with John Sweet and John Miller.

SWEET, Robert (1783–1835)
b. Cockington, Devon 1783 *d.* Chelsea, London 20 Jan. 1835

FLS 1812. Nurseryman, Stockwell, 1810–15; Fulham, 1815–19; Chelsea, 1819–26; moved residence to Chelsea, 1830. *Hortus Suburbanus Londinensis* 1818. *Geraniaceae* 1820–28 4 vols; Supplement, 1828–30. *Cistineae* 1825–30. *Sweet's Hortus Britannicus* 1826. *Fl. Australasica* 1827–28 (drawings by E. D. Smith at BM(NH)). *Florist's Guide* 1827–32 2 vols. *British Flower Garden* 1823–38 7 vols. Contrib. to *Mag. Nat. Hist.*

Trial of Robert Sweet at Old Bailey 1824. *Gdnrs Mag.* 1830, 487; 1835, 159–60. *Mag. Nat. Hist.* 1835, 410–11. *Hort. J. R. Lady's Mag.* v.3, 1835, 79. *J. Kew Guild* 1894, 34. *D.N.B.* v.55, 197. *New Fl. and Silva* v.8, 1935, 16–21. *J. Land Agents' Soc.* 1949, 187–89. F. A. Stafleu *Taxonomic Literature* 1967, 463–64.

Letters at Kew.

Sweetia DC.

SWEETENHAM, George (–1746)
Nurseryman, Dublin.
Irish For. 1967, 49.

SWETE, Edward Horace (1827–1912)
d. Seaton, Devon 4 Dec. 1912
MD. Surgeon. Of Clifton, Bristol. First Lecturer in Botany, Bristol Medical School. *Fl. Bristoliensis* 1854.
J. W. White *Fl. Bristol* 1912, 95–96. *J. Bot.* 1913, 69–70. H. J. Riddelsdell *Fl. Gloucestershire* 1948, cxxvi.

SWINDEN, Nathaniel (*fl.* 1740s–1800s)
Nurseryman, Brentford, Middx. *Beauties of Flora Display'd* 1778.
J. C. Loudon *Encyclop. Gdning* 1822, 1277. *Gdnrs Chron.* 1942 ii 4. *Trans. London Middlesex Archaeol. Soc.* v.24, 1973, 184.

SWINHOE, John (*fl.* 1730s–1750s)
Nurseryman, Brompton Park, London.

SWINHOE, R. C. G. (–1927)
d. Mandalay, Burma Aug. 1927
Lawyer, Mandalay. Collected orchids. 'Fertilisation of *Vanda coerulii*' (*Orchid Rev.* 1926, 336).
Orchid Rev. 1927, 289.

SWINHOE, Robert (1836–1877)
b. Calcutta, India 1 Sept. 1836 *d.* London 28 Oct. 1877
Educ. London University. FRS 1876. At Amoy, 1861. Consul, Taiwan, 1865; Ning-po, 1873–75. Collected plants and animals in Formosa. *List of Plants from Island of Formosa* 1863.
Ibis 1878, 126–28. *J. Bot.* 1878, 96. E. Bretschneider *Hist. European Bot. Discoveries in China* 1898, 661–78. *Quart. J. Taiwan Mus.* 1965, 335–39 portr. *R.S.C.* v.5, 898; v.8, 1048.
Chinese plants at BM(NH), Kew. Letters at Kew.
Rubus swinhoei Hance.

SWINTON, George (1780–1854)
b. 5 Nov. 1780 *d.* 17 June 1854
Writer, East India Company, 1802. Persian Secretary to Governor-General of India, 1814. Acting Chief Secretary to Government and officiating Superintendent of the Botanical Garden, 1826. Chief Secretary, Bengal, 1827. Retired 1833. Collected some plants for N. Wallich. "Has always been ready to promote the interests of science and the welfare of the Tenasserim Provinces."
Swintonia Griff.

SWITZER, Stephen (1682–1745)
b. Micheldever, Hants Feb. 1682 *d.* 8 June 1745
Gardener at Blenheim, 1706; with Lord Orrery, 1724–31. Nurseryman, Millbank, London with a market stall in Westminster Hall. *Ichnographia Rustica* 1718 3 vols. *Practical Fruit Gardener* 1724. *Practical Kitchen Gardiner* 1727. *Dissertation on True Cythisus of the Ancients* 1731. Edited *Practical Husbandman and Planter* 1733–34.
J. C. Loudon *Encyclop. Gdning* 1822, 1267. G. W. Johnson *Hist. English Gdning* 1829, 158–82. S. Felton *Portr. English Authors on Gdning* 1830, 45–53. J. Donaldson *Agric. Biogr.* 1854, 44. *Cottage Gdnr* v.6, 1852, 93; v.13, 1854, 53–54. *D.N.B.* v.55, 141–42. D. McDonald *Agric. Writers* 1908, 177–81. *Gdnrs Chron.* 1923 i 230.

SWYNNERTON, Charles Francis Massey (1877–1938)
b. India 3 Dec. 1877 *d.* Tanganyika 8 June 1938
FLS 1907. Farmer, S. Rhodesia, *c.* 1896. Game warden, Tanganyika, 1919. Director, Tsetse Research Dept., 1929. Collected plants in Rhodesia and Mozambique. 'Contribution to our Knowledge of Fl. Gazaland' (*J. Linn. Soc.* 1911–12, 1–245).
J. Bot. 1938, 212–13. *Nature* v.142, 1938, 198–99. *Proc. Linn. Soc.* 1938–39, 254–56. *Sisal Rev.* July 1938, 11–12. *Tanganyika Notes and Rec.* Dec. 1938, 3–4. *Moçambique* 1939, 47–51 portr. *Excelsa* no. 4, 1974, 2–12 portr.
Plants at BM(NH), Kew.
Aloe swynnertonii Rendle.

SYDENHAM, Robert (1848–1913)
b. Salisbury, Wilts 1848 *d.* Birmingham 19 July 1913
Nurseryman, Tenby Street, Birmingham. President, National Sweet Pea Society, 1912. *All about Sweet Peas* ed. 4 1908.
Gdnrs Mag. 1912, 65 portr.; 1913, 573 portr. *Garden* 1913, 379 portr. *Gdnrs Chron.* 1913 ii 75 portr. *J. Hort. Home Farmer* v.64, 1912, 349 portr.; v.67, 1913, 100 portr.

SYKES, Mary Gladys *see* Thoday, M. G.

SYKES, William Henry (1790–1872)
b. Friezing Hall, Yorks 25 Jan. 1790 *d.* London 16 June 1872
FRS 1832. Bombay Army, 1804–31; Lieut.-Colonel, 1831. MP Aberdeen, 1857–72. Collected plants in Bombay, 1826–30.
Proc. R. Soc. 1871–72, xxxiii–xxxiv. *D.N.B.* v.55, 258. M. Archer *Nat. Hist. Drawings in India Office Library* 1962, 40, 89–90. *R.S.C.* v.5, 899.
MSS. at BM(NH). Drawings at BM(NH), India Office Library.

SYME, George (1844–)
b. Perthshire 11 July 1844

Gardener at Dickson and Turnbull, Perth. Superintendent, Botanic Garden, Jamaica, 1879–84.
I. Urban *Symbolae Antillanae* v.3, 1902, 135–36.
Jamaican plants at Kew.

SYME, John Thomas Irvine *see* Boswell, J. T. I.

SYMINGTON, Colin Fraser (1905–1943)
b. Edinburgh 3 Nov. 1905 *d.* Nigeria 9 Sept. 1943
BSc Edinburgh. FLS 1940. Malayan Forestry Service, 1927. Botanical Assistant to F. W. Foxworthy, Forest Research Institute, Kepong, 1929; in charge 1933. Forest Botanist, S. Nigeria, 1942. Contrib. to *Gdns Bull., Singapore.*
Empire For. J. 1943, 98. *Proc. Linn. Soc.* 1943–44, 233–35. *Fl. Malesiana* v.1, 1950, 515 portr.
Plants at Kuala Lumpur, Ibadan, Oxford. Portr. at Hunt Library.
Symingtonia van Steenis.

SYMMONS, John (*fl.* 1790s)
Of Paddington, London. Had "a collection of hardy herbaceous plants superior to most in this country."
Bot. Mag. 1796, t.328.

SYMONDS, Rev. William Samuel (1818–1887)
b. Hereford 1818
Father-in-law of J. D. Hooker. Rector, Pendock, Worcs, 1845–83. "Gifted and interested in all nature and natural science." President, Woolhope Club, 1854. President, Malvern Field Club, 1854–71.
Trans. Woolhope Nat. Field Club v.34, 1954, 240–41.

SYMONS, Rev. Jelinger (1778–1851)
b. Leyton, Essex 1778 *d.* Radnage, Bucks 20 May 1851
MA Cantab 1797. FLS 1798. Curate, Whitburn, Durham. Rector, Radnage, 1833–51. *Synopsis Plantarum Insulis Britannicis Indigenarum* 1798.
Proc. Linn. Soc. 1852, 192–93. *D.N.B.* v.55, 280. *Nat. Hist. Trans. Northumberland, Durham and Newcastle-upon-Tyne* v.14, 1903, 79.
List of Durham plants, 1805 in Winch letters at Linnean Society.

SYMONS-JEUNE, Bertram Hanmer Bunbury (1886–1959)
d. Jamaica Jan. 1959
FLS 1921. Specialist in rock gardening. *Natural Rock Gardening* 1932; ed. 3 1955. *Phlox* 1953.
Proc. Linn. Soc. 1958–59, 136.

SYMPSON, Alexander (*fl.* 1690s)
Surgeon. Presented J. Petiver with plants from Gallipoli.
J. E. Dandy *Sloane Herb.* 1958, 219.

SYNNET, W. (*fl.* 1820s)
Resided several years at the Cape and brought home many bulbous plants.
R. Sweet *Br. Flower Gdn* 1826, t.138, 150.
Synnetia Sweet.

TABER, George (*c.* 1819–1895)
d. Feb. 1895
Seedsman, Rivenhall, Essex. Later in partnership with Robert Cooper at Southwark, London and Witham, Essex, 1887.
J. Hort. Cottage Gdnr v.30, 1895, 164–65 portr.

TAGG, Elizabeth (*c.* 1696–1779)
d. Jan. 1779
Apparently widow of Thomas Tagg (1695–1750s) and mother of J. Tagg. Had nursery at Paradise Gardens, Oxford.

TAGG, Harry Frank (*c.* 1873–1933)
b. Maidstone, Kent *c.* 1873 *d.* Edinburgh 1933
FLS 1899. Assistant, Royal Botanic Garden, Edinburgh, 1894; Keeper of Museum. Contrib. to *Species of Rhododendron* 1930, *Notes R. Bot. Gdn Edinburgh, Trans. Proc. Bot. Soc. Edinburgh.*
Proc. Linn. Soc. 1933–34, 171–72. *Times* 12 Aug. 1933. *Gdnrs Chron.* 1933 ii 170. *Nature* v.132, 1933, 342. *Trans. Proc. Bot. Soc. Edinburgh* 1934, 470. *Yb. Rhododendron Assoc.* 1934, 24. H. R. Fletcher and W. H. Brown *R. Bot. Gdn Edinburgh, 1670–1970* 1970, 220–21.

TAGG, James (*fl.* 1750s–1790s)
Son of Thomas Tagg (1695–1750s). Nurseryman, Paradise Gardens, Oxford.

TAGG, Thomas (1695–1750s)
b. 4 Feb. 1695
Nurseryman, Paradise Gardens, Oxford. "He succeeded one [Thomas] Wrench, the best Kitchen Gardiner in England, to whom he had been servant, and whose widow he married" (T. Hearne *Remarks and Collections* 1921, 355).

TAGG, Thomas (*fl.* 1800s–1837)
Probably son of James Tagg. Nurseryman, Paradise Gardens, Oxford.

TAHOURDIN, Charles Baynard (–1942)
d. Wallington, Surrey 5 July 1942
Solicitor. Studied and painted orchids. *Native Orchids of Britain* 1925. Contrib. to *Orchid Rev.*
Bot. Soc. Exch. Club Br. Isl. Rep. 1943–44, 653.

TAIT, Alfred Wilby, Baron de Soutellinho (1847–1917)
b. Oporto, Portugal 25 Oct. 1847 *d.* Oporto 15 March 1917
FLS 1887. Wine exporter. Rediscovered *Narcissus cyclamineus* near Oporto. *Notes on Narcissus of Portugal* 1886.
Proc. Linn. Soc. 1916–17, 75–76. *Gdnrs Chron.* 1962 i 288.
Iris taitii Foster.

TAIT, Henry (–1869)
d. 11 Nov. 1869
Nurseryman, Sydenham Nursery, Kelso, Roxburghshire for 30 years.
Gdnrs Chron. 1869, 1212.

TAIT, William (1833–1904)
d. 17 May 1904
LLD Aberdeen 1895. Of Kintore, Aberdeenshire whose flora he studied.
Ann. Scott. Nat. Hist. 1904, 137–38.

TAIT, William Chester (1844–1928)
b. Oporto, Portugal June 1844 *d.* Oporto 7 April 1928
Brother of A. W. Tait. Correspondent of C. Darwin. Pioneer in introduction of *Eucalyptus* as a forest tree for commercial purposes into Portugal.
Gdnrs Chron. 1928 i 327.

TALBOT, Dorothy Amaury (1871–1916)
b. 20 Dec. 1871 *d.* Degama, Nigeria 28 Dec. 1916
Wife of P. A. Talbot with whom she collected plants in S. Nigeria from 1909.
A. B. Rendle *Cat. of Plants collected by Mr. and Mrs. P. A. Talbot in Oban District, S. Nigeria* 1913. *J. Bot.* 1917, 85–86.
Plants at BM(NH).
Talbotiella Bak. f. *Amauriella* Rendle.

TALBOT, Percy Amaury (1877–1945)
b. 26 June 1877 *d.* 28 Dec. 1945
Surveyor. District officer, Nigeria. Collected plants in Nigeria, Liberia and Sierra Leone.
O. Macleod *Cities and Chiefs of Central Africa* 1912 (plant list, 301–8). A. B. Rendle *Cat. of Plants collected by Mr. and Mrs. P. A. Talbot in Oban District, S. Nigeria* 1913. *Comptes Rendus AETFAT, 1960* 1962, 72. *Nigerian Field* 1967, 191–92.
Plants at BM(NH), Kew. Drawings of Nigerian plants at BM(NH).
Talbotia S. Moore.

TALBOT, William Alexander (1847–1917)
b. Ireland 1847 *d.* Chateau d'Oex, Switzerland 23 July 1917
FLS 1884. Indian Forest Service, 1875; Conservator of Forests, Bombay, 1901; retired 1909. Collected plants in Switzerland, 1911–17. *Systematic list of Trees, Shrubs and Woody-climbers of Bombay Presidency* 1894; ed. 2 1902. *Forest Fl. Bombay Presidency and Sind* 1909–11 2 vols.
Kew Bull. 1921, 93–94. *Proc. Linn. Soc.* 1920–21, 56–58. *Indian Forester* 1925, 577–79.
Herb. at College of Agriculture, Poona. Plants at Kew.
Impatiens talboti Hook. f.

TALBOT, William Henry Fox (1800–1877)
b. Melbury, Dorset 11 Feb. 1800 *d.* Lacock Abbey, Wilts 17 Sept. 1877
MA Cantab 1825. FRS 1831. FLS 1831. Pioneer of photography. Collected plants in Ionian Islands, now at Kew. MS. Fl. Corfu (G. Bentham *Labiatarum Genera et Species* 1832–36, 742). 'Note on Vellozia elegans' (*Trans. Bot. Soc. Edinburgh* v.9, 1868, 79).
P. Smith *Mem. and Correspondence of J. E. Smith* v.2, 1832, 293–98. *Nature* v.16, 1877, 464, 523–25. *Trans. Bot. Soc. Edinburgh* v.9, 189–90, 192). *D.N.B.* v.55, 339–41.
Letters at Kew.

TALBOT DE MALAHIDE, Milo John Reginald Talbot, 7th Baron (1912–1973)
b. 1 Dec. 1912 *d.* at sea 14 April 1973
FLS 1968. British Diplomatic Service, 1937–58. Had garden at Malahide Castle near Dublin. Sponsored *Endemic Fl. of Tasmania* (illus. by Miss M. Stones) 1967-75 5 vols. Contrib. to *J. R. Hort. Soc.*
Times 15 May 1973. *Endemic Fl.* v.5, 1975, 309–11.

TALLACK, John C. (–1909)
b. Cornwall *d.* 12 Nov. 1909
Gardener at Shipley Hall, Derbyshire. *Book of the Greenhouse* 1901. Contrib. to *Gdnrs Chron.*
Gdnrs Chron. 1909 ii 351.

TANNER, Henry Charles Baskerville (*fl.* 1880s)
Colonel. Collected plants in Astor and Gilgit, India.
Pakistan J. For. 1967, 357.
Plants at Kew.

TANNOCK, David (–1952)
d. Richmond, Surrey 3 June 1952
Kew gardener, 1898. In charge of Agricultural School, Dominica. Superintendent, Botanic Garden, Dunedin, New Zealand, 1903.
Kew Bull. 1903, 30. *J. Kew Guild* 1952–53, 134–35.

TANSLEY, Sir Arthur George (1871–1955)
b. London 15 Aug. 1871 *d.* Grantchester, Cambridge 25 Nov. 1955
Educ. Cambridge. FRS 1915. Knighted 1950. Lecturer, University College, London, 1893–1906. Lecturer in Botany, Cambridge,

1906–23. Professor of Botany, Oxford, 1927–37. Plant ecologist. Founded and edited *New Phytologist*, 1902–31. Edited *J. Ecology*, 1917–37. President, British Ecological Society, 1913. Chairman, Nature Conservancy, 1949–53. Edited *Types of British Vegetation* 1911. *Elements of Plant Biology* 1922. *Practical Plant Ecology* 1923. *Aims and Methods in Study of Vegetation* (with T. F. Chipp) 1926. *British Isles and their Vegetation* 1939. *Our Heritage of Wild Nature* 1945. *Britain's Green Mantle* 1949.

Gdnrs Chron. 1927 ii 182 portr. J. B. Baker *Sir Arthur Tansley* 1955. *Nature* v.176, 1955, 1245–46. *Times* 28 Nov. 1955. *Forestry* 1956, 90. *New Phytol.* 1956, 145–46 portr. *Proc. Bot. Soc. Br. Isl.* 1956, 99–100. *Biogr. Mem. Fellows R. Soc.* 1957, 227–46 portr. *Ecology* 1957, 658–59. *J. Ecol.* 1958, 1–8. *D.N.B. 1951–1960* 953–54. *Who was Who, 1951–1960* 1067.

Portr. at Hunt Library.

TANSLEY, Joseph (*fl.* 1820s)

Nurseryman, 19 Beckford Row, Walworth, Surrey.

TANTON, Ransley (*c.* 1835–1878)
d. 26 April 1878

Nurseryman, Epsom, Surrey.

Garden v.13, 1878, 429.

TARRANT, William (–1872)
d. Shanghai, China 1872

Collected some plants near Ning-po, China for H. F. Hance. Edited *Friend of China.*

E. Bretschneider *Hist. European Bot. Discoveries in China* 1898, 535.

TATE, Alfred (*fl.* 1900s)

Had garden at Downside, Leatherhead, Surrey. Vice-President, National Rose Society.

Gdnrs Mag. 1908, 197–98 portr.

TATE, Alexander Norman (1837–1892)
b. Wells, Somerset 24 Feb. 1837 *d.* Oxton, Cheshire 22 July 1892

Analytical chemist and science teacher, Liverpool. Conducted botanical classes in Liverpool for many years. Founded and edited *Research.*

Naturalist 1892, 305–8 portr. *D.N.B.* v.55, 375. *Trans. Liverpool Bot. Soc.* 1909, 92.

TATE, George (1805–1871)
b. Alnwick, Northumberland 21 May 1805 *d.* Alnwick 7 June 1871

Linen draper. Postmaster, Alnwick, 1848–71. *History of Alnwick* 1865–69. Contrib. fossil flora to G. Johnston's *Bot. Eastern Borders* 1853, 289–317. Had museum of fossils.

Proc. Berwickshire Field Club v.6, 1871, 269–80. *D.N.B.* v.55, 377–78. *R.S.C.* v.5, 915.

Beyrichia tatei R. Jones.

TATE, George Henry Hamilton (1894–1953)
b. London 30 April 1894 *d.* Morristown, N.J., U.S.A. 24 Dec. 1953

Field Assistant, American Museum of Natural History, 1922. Collected plants in S. America.

Plants at New York Botanic Garden.

TATE, George Ralph (1835–1874)
b. Alnwick, Northumberland 27 March 1835 *d.* Fareham, Hants 23 Sept. 1874

MD Edinburgh. FLS 1869. Son of G. Tate (1805–1871). Assistant-surgeon, Royal Artillery. Collected plants in China, now at Kew. Had herb. *New Fl. of Northumberland and Durham* (with J. G. Baker) 1868. Contrib. to A. G. More's *Supplement to Fl. Vectensis* (*J. Bot.* 1871, 1).

Proc. Linn. Soc. 1874–75, lxiv–lxvi. *Hist. Berwickshire Nat. Club* v.7, 1876, 334–37. E. Bretschneider *Hist. European Bot. Discoveries in China* 1898, 531. *R.S.C.* v.8, 1861.

Plectranthus tatei Hemsley.

TATE, James Charles (*fl.* 1820s–1830s)

Nurseryman, Sloane Street, Chelsea, London.

T. Faulkner *Hist. and Topographical Description of Chelsea and its Environs* v.2, 1829, 346–47. *J. Hort. Cottage Gdnr* v.56, 1876, 145. E. Bretschneider *Hist. European Bot. Discoveries in China* 1898, 284. G. A. C. Herklots *Hong Kong Countryside* 1951, 162. *Taxon* 1970, 544.

TATE, Ralph (1840–1901)
b. Alnwick, Northumberland March 1840 *d.* Adelaide, Australia 20 Sept. 1901

ALS 1867. FLS 1883. Nephew of G. Tate (1805–1871). Science master, Trade and Mining School, Bristol. Geologist. Founder of Belfast Naturalists Field Club. Assistant Curator, Geological Society of London, 1864. Professor, Natural Science, Adelaide, 1875. Collected plants in Shetlands, 1865; Chontales, Nicaragua, 1867 (plants at Kew); Arnhem Land, Australia, 1882. *Fl. Belfastiensis* 1863. *Handbook of Fl. Extra-tropical S. Australia* 1890. *Botany of Horn Expedition* 1896.

Annual Rep. Proc. Belfast Nat. Field Club 1901–2, 31–35. S. A. Stewart and T. H. Corry *Fl. N. East of Ireland* 1888, xxi–xxii. *Rep. Austral. Assoc. Advancement Sci.* 1901, 741; 1907, 177–78, 181–87; 1911, 235–36. *Victorian Nat.* 1901, 88–89. *Irish Nat.* 1902, 36–39. *J. Bot.* 1902, 75–76. R. S. Rogers *Introduction to Study of S. Austral. Orchids*

1911, 48–50 portr. R. L. Praeger *Some Irish Nat.* 1949, 161–62 portr. P. Serle *Dict. Austral. Biogr.* v.2, 1949, 408. *Austral. Encyclop.* v.8, 1965, 439. *R.S.C.* v.8, 1061; v.11, 555; v.12, 721; v.18, 32.
Plants at BM(NH). Letters at Herb., Melbourne.
Tatea F. Muell.

TATE, Thomas (*c.* 1842–1934)
b. Alnwick, Northumberland *c.* 1842 *d.* Rockhampton, Queensland 21 Jan. 1934
Emigrated to Australia. Assistant-surgeon on 'Maria' which was wrecked on Great Barrier Reef, 1872. Botanist to William Hann's expedition in N. Queensland; collected plants which were sent to Kew.
Proc. R. Soc. Queensland 1890–91, xxxiii. R. L. Jack *Northmost Australia* v.2, 1921. *Austral. Encyclop.* v.4, 1965, 423.
Premna tateana Bailey.

TATHAM, John (1793–1875)
b. Settle, Yorks 20 Sept. 1793 *d.* Settle 12 Jan. 1875
Druggist. Fellow of Botanical Society of Edinburgh, 1841. Assisted J. Windsor in *Fl. Cravoniensis* 1873 and H. Baines in *Fl. Yorkshire* 1840.
J. Sowerby and J. E. Smith *English Bot.* 2890, 2905. *J. Bot.* 1875, 64. *Naturalist* 1893, 25–40. *Kew Bull.* 1916, 31.
Herb. and MSS. at Kew. Plants at BM(NH).

TATTERSHALL, George (*fl.* 1820s)
Nurseryman, Silkstone, Yorks.

TATUM, Edward John (1851–1929)
b. Salisbury, Wilts 1851 *d.* Salisbury 1929
Solicitor, Salisbury. Contrib. records of Wilts and Hants plants to *J. Bot.* 1893–97.
D. Grose *Fl. Wilts* 1957, 41.

TAUTZ, Frederick George (*c.* 1845–1894)
d. 1 Feb. 1894
Of Dublin House, Ealing, Middx. Cultivated orchids.
Orchid Rev. 1894, 66.
Cypripedium × *Tautzianum.*

TAYLOR, Adam (*fl.* 1760s)
Gardener to J. Sutton, New Park near Devizes, Wilts. *Treatise on Ananas or Pineapple* 1769.
J. C. Loudon *Encyclop. Gdning* 1822, 1275.

TAYLOR, C. (1762–1818)
b. Youlgrave, Derbyshire 1762 *d.* London 28 Nov. 1818
Surgeon, Royal Navy. Surgeon and botanist to Sierra Leone Co., 1791–92. Collected plants at Verdun, 1796 whilst a prisoner of the French. Sent plants to Stokes.
J. Bot. 1914, 322–23.

TAYLOR, Fanny Hope (–before 1886)
Collected ferns in Jamaica, 1852–54.
J. Bot. 1886, 269–70. I. Urban *Symbolae Antillanae* v.3, 1902, 136.
Plants at Kew.
Athyrium taylorianum Jenman.

TAYLOR, Frederick Beatson (1851–1931)
b. 29 Nov. 1851 *d.* Bournemouth, Hants 22 Jan. 1931
BA Cantab. Judge, Indian Civil Service until 1901. Diatomist. *Notes on Diatoms* 1929.
J. Bot. 1931, 103–4. *Proc. Bournemouth Nat. Sci. Soc.* v.23, 1930–31, 65.

TAYLOR, George (*c.* 1857–1941)
d. Gerrard's Cross, Bucks 4 April 1941
Gardener to Ramsden family at Byram, Yorks and Bulstrode, Gerrard's Cross, 1919.
Gdnrs Chron. 1941 i 162.

TAYLOR, George Crosbie (1901–1962)
b. Edinburgh 1901 *d.* 20 April 1962
BSc Edinburgh. FLS 1925. Gardening editor, *Country Life* and *Garden*. Garden correspondent, *Observer*. *Propagation of Hardy Trees and Shrubs* (with F. P. Knight) 1927. *Garden Making by Example* 1932. *Modern Garden* 1936; ed. 5 1952.
Quart. J. For. 1963, 238–39. *Proc. Linn. Soc.* 1964, 96–97.

TAYLOR, George M. (*c.* 1875–1955)
d. 1 Jan. 1955
Horticulturist. *Roses* 1945. *Old Fashioned Flowers* 1946. *British Garden Flowers* 1946. *Lilies for the Beginner* 1947. *British Herbs and Vegetables* 1947. *Little Garden* 1948. *Book of the Rose* 1949.
Gdnrs Chron. 1955 i 28.

TAYLOR, Harold Victor (1887–1965)
b. Taunton, Somerset 6 May 1887 *d.* 14 Nov. 1965
BSc London. VMH 1937. VMM. Science master, Taunton, 1906–9. Inspector, Board of Agriculture, 1912; Deputy Controller of Horticulture, 1920; Commissioner of Horticulture, 1927. Senior Advisory Officer, National Agricultural Advisory Service, 1945–48. Chairman, Horticultural Advisory Council. Chairman, Fruit Group, Royal Horticultural Society. *Asparagus* 1932. *Salad Crops* 1932. *Apples of England* 1936; ed. 3 1946. *Plums of England* 1949.
Gdnrs Chron. 1965 ii 553. Royal Horticultural Society *Fruit, Present and Future* 1966, 5–9 portr. *Who was Who, 1961–1970* 1102.

TAYLOR, Humphrey (*fl.* 1770s)
Of Chelsea, London. Sold essences and cordials prepared from the products of his herb garden.

TAYLOR, James (1823–1913)
d. Clashfarquhar, Aberdeen 30 Jan. 1913
Surgeon on whalers, 1856–61. Collected plants in Davis Strait and Baffin Bay (plant list in *Trans. Bot. Soc. Edinburgh* v.7, 1862, 323–33).
W. Jolly *Life of John Duncan* 1883, 373–81. *R.S.C.* v.5, 920.
Plants at BM(NH), Kew.

TAYLOR, John (*fl.* 1790s–1820s)
Nurseryman, Bagshot, Surrey.

TAYLOR, John Ellor (1837–1895)
b. Levenshulme, Lancs 21 Sept. 1837 *d.* Ipswich, Suffolk 28 Sept. 1895
FLS 1873. Geologist. Sub-editor, *Norwich Mercury* 1863. Editor, *Hardwicke's Science Gossip* 1872–93. Curator, Ipswich Museum, 1872–93. *Flowers, their Origin, Shapes, Perfumes and Colours* 1878. *Sagacity and Morality of Plants* 1884; 1891.
Geol. Mag. 1895, 528. *J. Bot.* 1895, 352. *Proc. Linn. Soc.* 1895–96, 47. *Science Gossip* 1895, 210 portr. *Times* 1 Oct. 1895. *Quart. J. Geol. Soc.* 1896, lxxv–lxxvi. *D.N.B.* v.55, 450. *Trans. Liverpool Bot. Soc.* 1909, 92.

TAYLOR, John Martin (1886–1947)
b. Kirkintilloch, Dunbartonshire 12 March 1886 *d.* Lake District 8 June 1947
MD 1912. Medical Officer of Health, Thorne near Doncaster, Yorks. Studied local flora.
Naturalist 1947, 131–32. *N. Western Nat.* 1947, 289–91 portr.

TAYLOR, Joseph (*fl.* 1810s)
Nurseryman, 19 Lowerhead Row, Leeds, Yorks.

TAYLOR, Joseph (*fl.* 1810s–1830s)
Of Newington Butts, Surrey. *Arbores Mirabiles* 1812. *Complete Weather Guide... with...a Curious Botanical Clock* 1812. *Bible Garden* 1836.

TAYLOR, Joseph (*fl.* 1830s–1840s)
Nurseryman, Aylesbury Road, Wendover, Bucks.

TAYLOR, Norman (1834–1894)
b. Surrey 3 Oct. 1834
Emigrated to Australia, 1854. Joined Victorian Geological Survey, 1856. Geologist to William Hann's expedition in N. Queensland. Collected plants which he sent to F. von Mueller. Contrib. Appendix to R. W. E. M'Ivor's *Chemistry of Agriculture* 1879.
Bull. Geol. Survey of Victoria no. 23, 1910, 23–25. *Austral. Encyclop.* v.5, 1965, 423.
Bulbophyllum taylori F. Muell.

TAYLOR, Robert Alexander (1898–1956)
b. Strathkinnes, Fife 11 Dec. 1898 *d.* 15 May 1956
BSc St. Andrews Univ. 1921. PhD 1936. FRSE 1951. Rubber planter, Ceylon. Demonstrator, Botany Dept., Dundee, 1933; Lecturer, 1945.
Yb. R. Soc. Edinburgh 1957, 52–54.

TAYLOR, Samuel (*fl.* 1800s–1820s)
Of Moston, Manchester, after of Bungay, Suffolk. 'Experiments on Growth of Whitethorn' (*Tilloch. Philos. Mag.* 1806, 39–43). 'Experiments on Smut in Wheat' (*Tilloch. Philos. Mag.* 1822, 350–53).
Trans. Liverpool Bot. Soc. 1909, 92. *R.S.C.* v.5, 923.

TAYLOR, Simon (1742–*c.* 1796)
b. Oct. 1742
Trained in drawing at William Shipley's School, London. Botanical artist. Painted Kew Garden plants for Lord Bute and J. Fothergill. Paintings bought by Empress of Russia.
J. E. Smith *Selection from Correspondence of Linnaeus* v.1, 1821, 255. *D.N.B.* v.55, 465. *Kew Bull.* 1933, 43; 1971, 167–69. *J. R. Soc. Arts* 1966, 525–30.
Drawings at BM(NH), Kew.

TAYLOR, Thomas (–1848)
b. East Indies *d.* Dunkerron, Kerry Feb. 1848
BA Dublin 1807. MD Dublin 1814. FLS 1814. Professor of Botany, Cork Scientific Institution. *Muscologia Britannica* (with W. J. Hooker) 1818; ed. 2 1827. Contrib. Musci to part 2 of J. T. Mackay's *Fl. Hibernica* 1836. 'North Ireland Fungi' (*Ann. Nat. Hist.* v.5, 1840, 3–6). 'Australian Mosses' (*Phytologist* v.1, 1844, 1093–94).
London J. Bot. 1848, 162–63. *Hooker's J. Bot.* 1849, 63. *Proc. Linn. Soc.* 1848, 379–80. *D.N.B.* v.55, 470–71. *Ann. Bot.* 1902, xxi–xxii. *Proc. Irish Acad.* v.32, 1915, 72. R. W. Scully *Fl. Kerry* 1916, xiii. *Irish Nat. J.* 1931, 238. *Meded. Bot. Mus. Herb. Rijksunivers. Utrecht* no. 283, 1968, 235.
Lichens and drawings at Boston Society of Natural History. Mosses and hepatics at Harvard. Letters at Kew.
Tayloria Hook.

TAYLOR, Thomas William (1878–1932)
b. Glos June 1878 *d.* London 4 March 1932
Kew gardener, 1902; Assistant Curator, 1922; Curator, 1929–32.
Gdnrs Chron. 1929 i 356 portr.; 1932 i 134 portr., 212. *J. Kew Guild* 1930, 772–73 portr.; 1931, 191–92 portr. *Kew Bull.* 1932, 112, 156–57.

TAYLOR, William (*fl.* 1830s)
Nurseryman, Millbrook, Southampton, Hants.

TAYLOR, William (*fl.* 1850s–1910s)
Gardener at Shrubland Park, Suffolk and Royal Horticultural Society gardens, Chiswick. Specialised in grapes.
J. Hort. Home Farmer v.61, 1910, 436–37 portr.

TAYLOR, Rev. William Ernest (1856–1927)
b. Worcester 25 Jan. 1856 *d.* Bath, Somerset 2 Oct. 1927
BA Oxon 1878. Ordained, 1880. With Church Missionary Society in British East Africa, Cairo and Sudan, 1880–1904. Rector, Halton Holgate, Lincs, 1921. Collected plants in East Tropical Africa, 1885–88.
J. Linn. Soc. v.30, 1895, 373–435. *J. Bot.* 1927, 317.
Plants at BM(NH).
Lissochilus taylorii Ridley.

TAYLOUR, Geoffrey Thomas, 4th Marquess of Headfort (1878–1943)
b. 12 June 1878 *d.* Cahir Park, County Tipperary 29 Jan. 1943
FLS 1930. VMH 1939. President, Royal Horticultural and Arboricultural Society of Ireland, 1920. Created garden at Headfort House, Kells, County Meath, noted for its conifers, rhododendrons and lilies.
Gdnrs Chron. 1943 i 70. *Proc. Linn. Soc.* 1942–43, 299–300. *Who was Who, 1941–1950* 520.

TEDLIE, Henry (*c.* 1792–*c.* 1818)
b. Ireland *c.* 1792 *d.* Cape Coast Castle *c.* 1818
Assistant-surgeon to T. E. Bowdich's Mission to Ashantee (T. E. Bowdich *Mission from Cape Coast Castle to Ashantee* 1849; medicinal plants, 370–74).
Plants at BM(NH), Kew. MSS. at BM(NH).
Senecio tedliei Oliver & Hiern.

TEESDALE, Robert (*c.* 1740–1804)
d. Hammersmith, London 25 Dec. 1804
FLS 1788. Gardener at Castle Howard, Yorks where his father, Robert Teesdale (–1773), had also been gardener. Junior partner in Minier, Mason and Teesdale, seedsmen, 60 Strand, London, 1775. Friend of J. E. Smith. Discovered *Carex tomentosa*, 1799. *Catalogue of the More Rare Plants...in the Neighbourhood of Castle Howard* 1792 (*Annual Rep. Yorkshire Philos. Soc.* 1962, 20–31). 'Plantae Eboracenses' (*Trans. Linn. Soc.* v.2, 1794, 103–25; v.5, 1800, 36–95). Contrib. to J. Sowerby and J. E. Smith's *English Bot.* 202, 2046, 2517, etc.
D. Turner and L. W. Dillwyn *Botanist's Guide* v.1, 1805, 663–65. *Trans. Linn. Soc.* v.11, 1815, 283–87. *Bot. Trans. Yorkshire Nat. Union* 1885, 198–99. *Annual Rep. Yorkshire Philos. Soc.* 1893, 45; 1906, 53. *Naturalist* 1967, 37–47. *R.S.C.* v.5, 927.
Plants at York Museum. Herb. sold in sale of W. T. Aiton's effects by Foster and Son, 54 Pall Mall, London, 4 Sept. 1851.
Teesdalia R. Br.

TEGG, James (1832–1902)
b. Midgham, Berks 29 March 1832 *d.* Wokingham, Berks 5 March 1902
Gardener to John Walter at Bearwood, Wokingham.
Garden v.61, 1902, 170. *Gdnrs Chron.* 1902 i 184. *J. Hort. Cottage Gdnr* v.44, 1902, 259.

TELFAIR, Annabella (*née* Chamberlain) (–1832)
d. Port Louis, Mauritius 23 May 1832
Wife of C. Telfair (1778–1833). Sent algae from Mauritius to W. J. Hooker (*J. Bot.* (Hooker) 1834, 147–57). Also sent drawings to *Bot. Mag.* t.2751, 2817, 2970.
Bot. Mag. 1830, t.2976.
Letters at Kew.
Bignonia telfairiae Bojer.

TELFAIR, Charles (1778–1833)
b. Belfast, Ireland 1778 *d.* Port Louis, Mauritius 14 July 1833
Surgeon, Mauritius. Supervisor, Botanic Garden, Mauritius, 1826–29.
Bot. Mag. 1826, t.2681; 1827, t.2751; 1830, t.2970. *Bot. Misc.* 1830–31, 123–24. N. Wallich *Plantae Asiaticae Rariores* v.2, 1832, 79. *J. Bot.* (Hooker) 1834, 150–51. J. Desjardins *Notice Historique sur Charles Telfair* 1836. *D.N.B.* v.56, 8. *Kew Bull.* 1919, 284. *Curtis's Bot. Mag. Dedications, 1827–1927* 15–16 portr. *R.S.C.* v.5, 929.
Plants and letters at Kew.
Telfairia Hooker.

TELFORD, George (–1704)
d. York Nov. 1704
Founder of a famous firm of nurserymen which took a lease of the Friars Gardens, York and was stated to have lasted 150 years when it was sold to Thomas and James Backhouse in 1816.
Yorkshire Archaeol. J. 1969, 352–57. J. Harvey *Early Gdning Cat.* 1972, 30–31.

TELFORD, George (1687–1711)
b. York 5 March 1687 *d.* York 1711
Son of G. Telford (–1704). Nurseryman, York.

TELFORD, George (1749–1834)
b. York June 1749 *d.* Widmore, Kent 27 Dec. 1834
Younger son of J. Telford (1716–1770). Nurseryman with his brother John (1744–1830) at Tanner Row, York. *Catalogue of Forest-trees, Fruit-trees, and Ever-green and Flowering Shrubs* 1775.

TELFORD, John (1689–1771)
b. York 1689 *d.* 12 Nov. 1771
Son of G. Telford (–1704). Nurseryman, York. "One of the first that brought our northern gentry into the method of planting and raising all kinds of forest trees for use and ornament" (F. Drake *Eboracum* 1736).

TELFORD, John (1716–1770)
b. York 13 Sept. 1716 *d.* 17 Dec. 1770
Eldest son of J. Telford (1689–1771). Took over the York nursery on his father's retirement in 1762. Succeeded by his sons John (1744–1830) and George (1749–1834).
York Courant 18 Dec. 1770.

TELFORD, John (1744–1830)
b. York May 1744 *d.* York 12 Oct. 1830
Son of J. Telford (1716–1770). Partner in family nursery at York with his brother George (1749–1834) from 1770 until the business was acquired by T. and J. Backhouse, 1816. *Catalogue of Forest-trees, Fruit-trees and Ever-green and Flowering Shrubs* 1775.

TELLAM, Richard Vercoe (1826–1908)
b. Tregustick, Withiel, Cornwall 9 Feb. 1826 *d.* Wadebridge, Cornwall 18 Sept. 1908
Farmer. Cryptogamist and entomologist. Contrib. to I. W. N. Keys's *Fl. Devon and Cornwall* 1865–69.
Bot. Soc. Exch. Club Br. Isl. Rep. 1908, 345. *J. Bot.* 1908, 360–63. F. H. Davey *Fl. Cornwall* 1909, liv–lv portr.
Herb. at Truro Museum. Algae at Bodmin Museum.
Tellamia Betters.

TEMPANY, Sir Harold Augustin (1881–1955)
b. 23 July 1881 *d.* London 2 July 1955
Educ. University College, London. Agricultural chemist, Leeward Islands, 1903. Director of Agriculture, Mauritius, 1916–28. Director of Agriculture, Federated Malay States, 1929–36. Agricultural Adviser to Colonial Office, 1940. *Principles of Tropical Agriculture* (with G. E. Mann) 1930. *Agriculture in West Indies* 1942. *Introduction to Tropical Agriculture* (with D. H. Grist) 1958.
Kew Bull. 1916, 277; 1928, 301. *Who was Who, 1951–1960* 1073.

TEMPERLEY, George William (*c.* 1875–1967)
d. 30 Nov. 1967
Of Newcastle-upon-Tyne, Northumberland. Knew plants of N.E. England.
Vasculum 1968, 3–4.

TEMPERLEY, Nicholas (1844–1923)
b. Hexham, Northumberland 1844 *d.* Gateshead, Durham 30 Sept. 1923
In business, Newcastle-upon-Tyne, 1863–1908. Botanist.
Trans. Nat. Hist. Soc. Northumberland 1923–26, 108–9. *Rep. Br. Bryol. Soc.* 1924, 95.

TEMPLE, Mungo (1834–1902)
b. Fifeshire Oct. 1834 *d.* Falkirk 16 April 1902
Gardener to Sir T. D. Brodie at Carron House, Falkirk.
Gdnrs Chron. 1902 i 284 portr.

TEMPLE, Sir William (1628–1699)
b. Blackfriars, London 1628 *d.* 27 Jan. 1699
Statesman and writer. Had garden at West Sheen, Richmond, Surrey (mentioned by J. Evelyn in his *Diary*) and also at Moor Park, Surrey. *Upon the Gardens of Epicurus* 1685.
Archaeologia v.12, 1794, 184–85. G. W. Johnson *Hist. of English Gdning* 1829, 120–21. *Cottage Gdnr* v.5, 1851, 235. *J. Hort. Cottage Gdnr* v.56, 1876, 467–68 portr.; 1903, 267 portr. W. C. Hazlitt *Gleanings in Old Gdn Literature* 1887, 182–87. *D.N.B.* v.56, 42–51. *Country Life* 1974, 76–77 portr.
Portr. by P. Lely at National Portrait Gallery.

TEMPLEMAN, Andrew (1887–1945)
b. Aberdour, Fifeshire 22 April 1887 *d.* Seaton, Cumberland 16 Dec. 1945
Geologist. Fossil collector.
Bot. Soc. Exch. Club Br. Isl. Rep. 1945, 15–16.
Plants at BM(NH), Oxford.

TEMPLEMAN, William Gladstone (–1970)
d. Dec. 1970
MSc. PhD. FLS 1944. Deputy Director, Jealott's Hill Research Station. 'Experiments with Plant Growth Substances for the Rooting of Cuttings' (with C. R. Metcalfe) (*Kew Bull.* 1939, 441–56).

TEMPLETON, John (1766–1825)
b. Belfast, Ireland 1766 *d.* Malone, Belfast 15 Dec. 1825
ALS 1794. Founder member of Belfast Natural History Society, 1821. Found *Rosa hibernica, Entosthodon templetoni*, etc. Contrib. to J. Sowerby and J. E. Smith's *English Bot.* 508, 2196, etc., L. W. Dillwyn's *Br. Confervae* 1802–9, D. Turner's *Fuci* 1808–19 and *Muscologiae Hibernicae Spicilegium* 1804. MS. *Catalogue Plants of Ireland* 1793–1811.
Mag. Nat. Hist. v.1, 1829, 403–6; v.2, 1829, 305–10. J. C. Loudon *Arboretum et Fruticetum Britannicum* 1838, 111–12. *Ann. Nat. Hist.* v.5, 1840, 3–6. J. Hardy *Selections from Correspondence of G. Johnston*

1892, 279. *D.N.B.* v.56, 54–55. N. Colgan *Fl. Dublin* 1904, xxiv–xxv. *J. Bot.* 1907, 304–5. *Ann. Rep. Proc. Belfast Nat. Field Club* 1913, 616–17. *Irish Nat.* 1913, 22–24. *Proc. Linn. Soc.* 1924–25, 21–22. S. A. Stewart and T. H. Corry *Fl. N.E. Ireland* ed. 2 1938, xlvii–liii. R. L. Praeger *Some Irish Nat.* 1949, 163–65. *Irish Nat. J.* 1966, 229–30, 318–22; 1967, 266–67, 350–53.

Fl. Hibernica 5 vols. and MSS. at Ulster Museum. Plants at Queen's University, Belfast. MSS. and drawings of Irish Cryptogams at BM(NH).

Templetonia R. Br.

TENISON-WOODS, Rev. Julian Edmund (1832–1889)
b. Southwark, London 15 Nov. 1832 *d.* Sydney, N.S.W. 7 Oct. 1889

MA Oxon. FLS 1863. To Tasmania, 1854. Ordained Catholic priest, 1857. Parish priest, Penola, S. Australia, 1857–67. Studied local flora and sent plants to F. von Mueller. Director of Catholic Education in Australia, 1867. Missionary work in N.S.W. and Queensland, 1871–72. President, Linnean Society of N.S.W., 1880–81. *History of Discovery and Exploration of Australia* 1865 2 vols. Contrib. to *Proc. Linn. Soc. N.S.W.*

Ann. Bot. 1889–90, 494–95. *J. Proc. R. Soc. N.S.W.* 1890, 2–10; 1908, 82. *Proc. Linn. Soc.* 1889–90, 98–99. *Proc. Linn. Soc. N.S.W.* v.4, 1889, 1301–9. *Geol. Mag.* 1890, 288. P. Mennell *Dict. Austral. Biogr.* 1892, 521–22. *D.N.B.* v.62, 410. *Rep. Austral. Assoc. Advancement Sci.* 1907, 179. *Catholic Encyclop.* v.15, 1912, 702–3 portr. *Victorian Nat.* 1928, 194–95. G. O'Neill *Life of Rev. Julian Tenison Woods* 1929. *Fl. Malesiania* v.1, 1950, 519. *J. Catholic Hist. Soc.* v.1, 1960, 26–39. *Austral. Encyclop.* v.8, 1965, 454–55. *R.S.C.* v.6, 436; v.8, 1270; v.12, 724.

Drawings at Mitchell Library, Sydney.

Leucopogon woodsii F. Muell.

TENNENT, John (*fl.* 1720s–1760s)

MD. FRS 1765. Of Virginia and New York. Corresponded with Sir H. Sloane and other Fellows of Royal Society about the medicinal qualities of American plants.

Osiris 1948, 106.

TENTERDEN *see* Abbott, Charles, *1st Baron Tenterden*

TEPPER, Johann Gottlieb Otto (1841–1923)
b. Neutomisch, Poland 1841 *d.* Adelaide, Australia 16 Feb. 1923

FLS 1879. To S. Australia, 1847. Entomologist, S. Australia Museum, Adelaide, 1897–1910. 'Fl. von Clarendon (S.A.)' (*Bot. Centralblatt* v.63, 1895, 1–9).

R.S.C. v.11, 567; v.12, 724; v.19, 56.

Portr. at Hunt Library.

TETLEY, William N. (1861–1928)
d. Hoylake, Cheshire 13 March 1928

Schoolteacher at Enniskillen, N. Ireland for nearly 35 years. Bryologist.

Rep. Br. Bryol. Soc. 1927, 68–69.

TEWE, Robert (*fl.* 1630s)

Sent plants from Russia, 1631.

R. T. Gunther *Early English Botanists* 1922, 63, 362–63.

THATCHER, Mrs. E. J. (*c.* 1853–1934)
d. Chew Magna, Somerset 21 Oct. 1934

Cultivated ferns.

Br. Fern Gaz. 1934, 253.

THATCHER, Samuel (*fl.* 1780s–1790s)

Seedsman and netmaker, The Raven, 147 Fleet Street, London.

THEOBALD, Frederick Vincent (1868–1930)
d. Wye, Kent 6 March 1930

MA Cantab. VMH 1926. In charge of Economic Zoology Section, BM(NH), 1900–3. Professor of Agricultural Zoology, London University. *Enemies of the Rose* 1908. *Insect Pests of Fruits* 1909. *Insect Enemies of the Allotment Holder* 1918. *Plant Lice or Aphididae of Great Britain* 1926–29 3 vols.

Gdnrs Chron. 1930 i 237. *Nature* v.125, 1930, 607–8. *Who was Who, 1929–1940* 1336–37.

THEOBALD, James (*fl.* 1720s)

A few foreign plants in Sloane herb.

J. E. Dandy *Sloane Herb.* 1958, 219.

THICKENS, Rev. W. (*fl.* 1840s)

Vicar, Kerseley, near Coventry, Warwickshire. 'Localities for Botrychium lunaria' (*Phytologist* v.3, 1848, 222–23).

J. E. Bagnall *Fl. Warwickshire* 1891, 501.

THICKNESSE, Ralph (1719–1790)
b. Barthomley, Cheshire 1719 *d.* Wigan, Lancs 12 Feb. 1790

MD Cantab. BA Oxon 1730. *Treatise on Foreign Vegetables* 1749.

Gent. Mag. 1790 i 185, 272–73, 399–400. *D.N.B.* v.56, 132. *Trans. Liverpool Bot. Soc.* 1909, 93.

THISELTON-DYER, Lady Harriet Ann (*née* Hooker) (1854–1945)
b. Hitcham, Suffolk 23 June 1854 *d.* Weir Quay, Devon 16 Dec. 1945

Eldest daughter of Sir J. D. Hooker. Married W. T. Thiselton-Dyer. Received tuition in botanical drawing from W. H. Fitch. Contrib. drawings to *Bot. Mag.* 1878–1906, I. B. Balfour's *Bot. of Socotra* 1888, *Gdnrs Chron.*

Curtis's Bot. Mag. Dedications, 1827–1927 311–12 portr. A. White and B. L. Sloane *Stapelieae* v.1, 1937, 113–14 portr. *Gdnrs Chron.* 1946 i 12. *Nature* v.157, 1946, 186. Hunt Library *Artists from R. Bot. Gdns, Kew* 1974, 62–63 portr.

Drawings at Kew.

THISELTON-DYER, Sir William Turner (1843–1928)
b. Westminster, London 28 July 1843 *d.* Witcombe, Glos 23 Dec. 1928

BA Oxon 1867. BSc London 1870. FRS 1880. FLS 1872. KCMG 1899. Professor of Natural History, Royal Agricultural College, Cirencester, 1868. Professor of Botany, Royal College of Science, Dublin, 1870. Professor of Botany, Royal Horticultural Society, 1872. Assistant Director, Kew, 1875; Director, 1885–1905. *Fl. Middlesex* (with H. Trimen) 1869. Edited *Fl. Capensis, Fl. Tropical Africa.* Contrib. to *Nature, Ann. Bot.*

J. Hort. Cottage Gdnr v.11, 1885, 562–63 portr. *Gdnrs Mag.* 1899, 4 portr. *Garden* v.57, 1900, iv portr.; v.69, 1906, 3–4 portr. *Kew Bull.* 1905, 62–63; 1929, 32, 65–75 portr. *Gdnrs Chron.* 1923 ii 52 portr.; 1929 i 1–2. *Nature* v.112, 1923, 182; v.116, 1925, 474–75; v.123, 1929, 212–15. *Curtis's Bot. Mag. Dedications, 1827–1927* 355–56 portr. *Times* 27 Dec. 1928. *Who was Who, 1916–1928* 1033. *Bot. Soc. Exch. Club Br. Isl. Rep.* 1929, 719–20. *J. Bot.* 1929, 54–57. *J. Kew Guild* 1929, 715–19 portr. *Proc. Linn. Soc.* 1928–29, 147–49. *Pharm. J.* 1929, 19. *D.N.B. 1922–1930* 830–32. *Proc. R. Soc. B.* v.106, 1930, xxiii–xxix. *Proc. R. Soc. Edinburgh* 1930, 377. A. White and B. L. Sloane *Stapelieae* v.1, 1937, 113 portr. F. O. Bower *Sixty Years of Bot. in Britain* 1938, 48–51 portr. C. C. Gillispie *Dict. Sci. Biogr.* v.13, 1976, 341–44.

Letters, MSS. and projected life at Kew. Portr. at Hunt Library.

THOBURN, Peter? (*fl.* 1780s)

Founded nursery at Brompton, Kensington, London in 1784 and from 1788 to 1790 run in partnership with Reginald Whitley (*c.* 1754–1835).

Trans. London Middlesex Archaeol. Soc. v.24, 1973, 186.

THODAY, David (1883–1964)
b. Honiton, Devon 5 May 1883 *d.* Llanfairfechan, Caernarvonshire 30 March 1964

Educ. Cambridge. FRS 1942. Demonstrator in Botany, Cambridge, 1909–11. Lecturer in Plant Physiology, Manchester University, 1911–18. Professor of Botany, Cape Town, 1918–22. Professor of Botany, Bangor, 1923–49. Professor of Plant Physiology, Alexandria University, 1950–51. *Botany: a Senior Book for Schools* 1915; ed. 5 1935.

Nature v.202, 1964, 1161–62. *Times* 1 April 1964. *Biogr. Mem. Fellows R. Soc.* v.11, 1965, 177–85 portr. *Who was Who, 1961–1970* 1111.

THODAY, Mary Gladys (*née* Sykes) (*c.* 1884–1943)
d. 9 Aug. 1943

Educ. Cambridge. Married D. Thoday (1883–1964), 1910. To Cape Town in 1918 where she studied flora of S. Africa. Contrib. to *Ann. Bot.*

Nature v.152, 1943, 406.

THOMAS, Edward John Haynes (–1930)
b. Chester *d.* Chester 3 Jan. 1930

LRCP. LRCS. Practised medicine in Chester. Had good knowledge of flora of Flintshire and Denbighshire.

N. Western Nat. 1930, 187–89. *Observer* 11 Jan. 1930.

THOMAS, Edward Joseph (1869–1958)
b. 30 July 1869 *d.* 11 Feb. 1958

MA St. Andrews 1900. BA Cantab 1905. Kew gardener, 1894. Oriental Librarian, Cambridge. "He knew the British flora by heart."

Emmanuel College Mag. v.40, 1958, 106–8.

THOMAS, Ethel Nancy Miles (–1944)
d. Aug. 1944

DSc. FLS 1908. Research Assistant to Ethel Sargent, 1897–1901. Lecturer in Botany, Bedford College, London, 1908. Head, Botanical Dept., University College, Cardiff, 1918–19. Keeper, Dept. of Botany, National Museum of Wales, 1919–21. Lecturer in Biology, University College, Leicester, 1923–37. 'Some Points in Anatomy of *Acrostichum aureum*' (*New Phytol.* v.4, 1905, 175–89). 'Theory of Double Leaf-trace founded on Seedling Structure' (*New Phytol.* v.6, 1907, 77–91). 'Seedling Anatomy of Ranales, Rhoeadales and Rosales' (*Ann. Bot.* v.28, 1914, 695–733).

Nature v.154, 1944, 481–82. *Proc. Linn. Soc.* 1943–44, 235–36. *Who was Who, 1941–1950* 1142.

THOMAS, Harry Higgott (1876–1956)
b. 2 July 1876

VMH 1948. Kew gardener, 1897. Joined staff of *Gardener*, 1900. Edited *Popular Gardening*, 1907–47. Prolific horticultural journalist. *Rose Book* 1913. *Bulb Growing for Amateurs* 1915. *Complete Amateur Gardener* 1924. *Dahlias, Gladioli and Begonias* 1926. *The Greenhouse* 1953, etc.

J. Kew Guild 1952–53, 85–86 portr. *Gdnrs Chron.* 1956 i 250.

THOMAS, Hugh Hamshaw (1885–1962)
b. Wrexham, Denbighshire 29 May 1885 *d.* Cambridge 30 June 1962
BA Cantab 1907. FRS 1934. FLS 1925. Curator, Botany Museum, Cambridge, 1909–23; Lecturer, 1923–37; Reader in Plant Morphology, 1937–50. President, Linnean Society, 1955–58. Palaeobotanist.
Nature v.167, 1951, 176; v.195, 1962, 754–55. *Times* 2 July 1962. *Proc. Linn. Soc.* 1961–62, 156–58. *Biogr. Mem. Fellows R. Soc.* 1963, 287–99 portr. *Br. J. Hist. Sci.* 1963, 280–83. *Who was Who, 1961–1970* 1112. C. C. Gillispie *Dict. Sci. Biogr.* v.13, 1976, 345–46.
Portr. at Hunt Library.

THOMAS, L. D. (*fl.* 1840s)
Of Tinnevelly, India. Sent some plants to W. J. Hooker in 1844.

THOMAS, Owen (1843–1923)
b. Hermon, Anglesey 1843 *d.* Hanwell, Middx 27 May 1923
VMH 1897. Gardener to Sir R. Peel at Drayton, to Duke of Devonshire at Chatsworth, at Windsor Castle, 1891.
J. Hort. Cottage Gdnr v.8, 1884, 210–11 portr. *Garden* v.58, 1900, 258–59 portr.; 1923, 299 portr. *Gdnrs Mag.* 1910, 331–32 portr. *J. Hort. Home Farmer* v.61, 1910, 342–44 portr. *Gdnrs Chron.* 1922 ii 160 portr.; 1923 i 312 portr.

THOMAS, Thomas Henry (1839–1915)
b. Pontypool, Mon 1839
Painter and artist to *Graphic* and *Daily Graphic*. Formed collection of Welsh plants. 'Notes on some Fine Specimens of Oak, Yew, Elm and Beech...' (*Trans. Cardiff Nat. Soc.* 1880, 15–24).
A. E. Wade *Fl. Monmouthshire* 1970, 13.
Herb. and MSS. at National Museum of Wales.

THOMAS, William (1823–1895)
b. Swansea, Glam 1823 *d.* 24 May 1895
Nurseryman, Wolverhampton, Staffs, 1857–80. Park Superintendent, Wolverhampton, 1881–95.
Gdnrs Chron. 1895 i 692.

THOMPSON, Mrs. Agar (*fl.* 1850s)
Collected ferns.
Br. Fern Gaz. 1909, 48.

THOMPSON, Arnold (1876–1959)
b. Rawdon, Yorks 15 June 1876 *d.* Skipton, Yorks 27 Jan. 1959
BSc London. Teacher, King Edward VII School, Sheffield until 1936. Interested in bryology, particularly *Sphagnaceae*. Secretary, British Bryological Society, 1936–47; President, 1948 and 1949. President, Craven Naturalists' Society, 1944–45. Helped to revise *Census Catalogue of British Sphagna* 1946. Contrib. to *J. Bot.*, *Naturalist*.
Naturalist 1959, 100; 1961, 158–59. *Trans. Br. Bryol. Soc.* 1960, 783.

THOMPSON, Charles John S. (–1943)
b. Liverpool *d.* 14 July 1943
Educ. Liverpool University. Hon. Curator, Museum, Royal College of Surgeons of England. *Mystery and Lore of Perfume* 1927. *Mystery and Art of Apothecary* 1939. *Mystic Mandrake in Lore and Legend* 1934.
Who was Who, 1941–1950 1144–45.

THOMPSON, Charles Norval (*c.* 1834–1874)
d. 29 Dec. 1874
Son of R. Thompson (1798–1869). Subeditor, *J. Hort.*
Gdnrs Yb. and Almanack 1875, 170.

THOMPSON, Christopher (*fl.* 1750s–1782)
Son of W. Thompson (1759–1811). Nurseryman, Pickhill near Thirsk, Yorks.
J. Harvey *Early Gdning Cat.* 1972, 33.

THOMPSON, H. P. (–1946)
d. Oct. 1946
Collected alpine plants in Yugoslavia and Balkans.
Quart. Bull. Alpine Gdn Soc. 1940, 213; 1946, 223.
Sempervivum thompsonianum Wale.

THOMPSON, Harold Stuart (1870–1940)
b. Bridgwater, Somerset March 1870 *d.* Bristol 3 March 1940
Educ. Cambridge. FLS 1901–7. ALS 1930. Secretary, Watson Botanical Exchange Club, 1900–4, 1920–34. *Alpine Plants of Europe* 1911. *Sub-alpine Plants of the Swiss Woods and Meadows* 1912. *Flowering Plants of Riviera* 1914. Contrib. to *J. Bot.*, *J. Ecol.*
Bot. Soc. Exch. Club Br. Isl. Rep. 1939–40, 231. *J. Bot.* 1940, 102–3. *Proc. Bristol Nat. Soc.* v.9(2), 1940, 88 portr. *Proc. Linn. Soc.* 1939–40, 377–78. H. J. Riddelsdell *Fl. Gloucestershire* 1948, cxliv.
Plants at Birmingham, Bristol, Reading. Library at Bristol.

THOMPSON, Henry Nilus (–1938)
b. Burma *d.* 9 July 1938
Forestry Dept., Burma, 1889. Nigeria, 1902. Director of Forests and Agriculture, S. Nigeria, 1906; Director of Forests for whole of Nigeria, 1914–29. Re-organised Victoria Botanic Garden, W. Cameroons. *Report on Forests, Gold Coast* 1910. Edited J. H. J. Farquhar's *Oil Palm* 1913.
Kew Bull. 1938, 303–4.
Plants at Kew.

THOMPSON, John *see* Thomson, J. W.

THOMPSON, James (*fl.* 1750s)
Florist, Newcastle-upon-Tyne, Northumberland. *Distinguishing Properties of a Fine Auricula* 1757. *Dutch Florist* 1758.
C. O. Moreton *The Auricula* 1964, 39, 90.

THOMPSON, John (*fl.* 1720s–*c.* 1758)
Nurseryman, The Rose, Kings Road, Chelsea, London. Member of London Society of Gardeners.

THOMPSON, John (*c.* 1778–1866)
d. Gateshead, Durham 26 March 1866
Of Crowhall Mill, Northumberland. Miller. Discovered *Carex irrigua*, 1841.
J. Sowerby and J. E. Smith *English Bot. Supplt.* 2895. *Trans. Northumberland Nat. Hist. Soc.* v.1, 1866–67, 257–58.
Letters in Winch correspondence at Linnean Society.

THOMPSON, Rev. John Thomas (1752–1811)
b. Mevagissey, Cornwall 1752 *d.* Penzance, Cornwall April 1811
Curate, Zennor, Cornwall. Contrib. notes on *Daucus maritimus* to W. Withering's *Bot. Arrangement* v.2, 1796, 290–91.
J. P. Jones *Bot. Tour through...Devon and Cornwall* 1820, 33. G. C. Boase and W. P. Courtney *Bibliotheca Cornubiensis* 1874–82, 718–19.

THOMPSON, John Vaughan (1779–1847)
b. Berwick-on-Tweed, Northumberland 19 Nov. 1779 *d.* Sydney, N.S.W. 21 Jan. 1847
ALS 1807. FLS 1810. Zoologist. Surgeon, 37th Regiment. In West Indies, 1800–9; Madagascar and Mauritius, 1812–16. District medical inspector, Cork, Ireland, 1816–34. In charge of convict medical dept., Sydney, 1835. *Catalogue of Plants growing in Vicinity of Berwick-upon-Tweed* 1807. 'Piper' (*Trans. Linn. Soc.* v.9, 1808, 200–3).
D.N.B. v.56, 218–20. *J. Bot.* 1912, 169–71. *Proc. R. Soc. Arts and Sci. Mauritius* 1953, 241–48. C. C. Gillispie *Dict. Sci. Biogr.* v.13, 353–56.
Thompsonia R. Br. *Vaughania* S. Moore.

THOMPSON, Percy George (1866–1953)
b. Rotherhithe, London 1866 *d.* 7 April 1953
Architect. Secretary, Essex Field Club. Editor, *Essex Nat.* Interested in lichens and mosses.
Essex Nat. 1954, 211–12 portr.

THOMPSON, Rachel Ford (1856–1906)
b. York 31 Aug. 1856 *d.* Southport, Lancs 9 Dec. 1906
Daughter of Silvanus Thompson. Studied flora of Settle, Yorks, 1882–93. Assisted F. J. Hanbury in his work on *Hieracia.* Contrib. *Hieracia* to C. C. Babington's *Manual of Br. Bot.* 1904, 232–70. Contrib. to F. A. Lees's *Fl. West Yorkshire* 1888.
Wings (Women's Temperance Assoc.) v.25, 17 portr. *Naturalist* 1893, 25–27. *J. Bot.* 1907, 78. *Trans. Liverpool Bot. Soc.* 1909, 93.

THOMPSON, Richard Horatio Ely (*fl.* 1870s)
Forest Conservator, Oudh, India. Sent collection of plants in 1870 to D. Brandis who forwarded them to Kew.

THOMPSON, Robert (1798–1869)
b. Echt, Aberdeenshire Sept. 1798 *d.* Chiswick, Middx 7 Sept. 1869
At Horticultural Society of London's Chiswick gardens, 1824–69 where he became Superintendent of the Fruit Dept. *Gardener's Assistant* 1859. Contrib. to *Gdnrs Chron., Gdnrs Mag., Trans. Hort. Soc.*
Gdnrs Chron. 1869, 963, 989–90; 1918 i 121–22 portr. *J. Hort. Cottage Gdnr* v.42, 1869, 209–10; v.58, 1877, 54–56 portr. *R.S.C.* v.5, 959.
MSS. at Royal Horticultural Society.

THOMPSON, Silvanus (1818–1881)
b. Liverpool 20 March 1818 *d.* Settle, Yorks 3 Feb. 1881
Son-in-law of J. Tatham (1793–1875). Schoolmaster at Friends' School, York. Contrib. to H. Baines's *Fl. Yorkshire* 1840. Contrib. to *Phytologist* v.1, 1844.
J. Sowerby and J. E. Smith *English Bot.* 2890. *Naturalist* 1893, 26–27. *Trans. Liverpool Bot. Soc.* 1909, 93.
Plants at BM(NH), Wellington N.Z.

THOMPSON, William (1759–1811)
Nurseryman, Pickhill near Thirsk, Yorks. His sons, Christopher and William, inherited the nursery.

THOMPSON, William (1805–1852)
b. Belfast 2 Nov. 1805 *d.* London 17 Feb. 1852
Began as linen-draper. Zoologist and algologist. President, Belfast Natural History Society, 1843–52. Discovered *Elatine hydropiper* in Ireland, 1836. *Natural History of Ireland* 1849–56 (biogr. and portr. in v.4, 1856, x–xxx). Contrib. to W. H. Harvey's *Phycologia Britannica* 1846–51. Plant records in G. Dickie's *Fl. Ulster* 1864.
S. A. Stewart and T. H. Corry *Fl. N.E. Ireland* 1888, xv. J. Hardy *Selections from Correspondence of G. Johnston* 1892, 458–59. *D.N.B.* v.56, 227–28. *Annual Rep. Belfast Nat. Field Club* 1913, 621–22. *Irish Nat.* 1913, 27–28. *Belfast Nat. Hist. Philos. Soc. Centenary Vol.* 1924, 106–7 portr. R. L.

Praeger *Some Irish Nat.* 1949, 166–67 portr. *R.S.C.* v.5, 960.
Algae at Belfast Museum.

THOMPSON, William (1823–1903)
b. Ipswich, Suffolk 18 May 1823 *d.* Ipswich 3 July 1903
VMH 1897. Founded nursery at Ipswich; later known as Thompson and Morgan. *English Flower Garden* 1852–53. *Gardening Book of Annuals* 1855.
Garden v.64, 1903, 36–37 portr. *Gdnrs Chron.* 1903 ii 30, 44. *J. Hort. Cottage Gdnr* v.47, 1903, 25. *Curtis's Bot. Mag. Dedications, 1827–1927* 195–96 portr.
Letters at Kew. Portr. at Hunt Library.

THOMPSON, William (*c.* 1832–1916)
d. 22 Dec. 1916
Of Stone, Staffs. Cultivated orchids, particularly *Odontoglossum.*
Orchid Rev. 1917, 31.
Odontoglossum thompsonianum × Garnier.

THOMPSON, William J. (*fl.* 1880s–1890s)
Superintendent, Castleton Gardens, Jamaica, 1889–1890; Kings House Garden, 1890–92.
I. Urban *Symbolae Antillanae* v.3, 1902, 136.

THOMSON, Agnes *see* Ibbetson, A.

THOMSON, Agnes C. (*fl.* 1880s)
Niece of T. Thomson (1817–1878) and correspondent of G. C. Druce (1850–1932).
H. N. Clokie *Account of Herb. Dept. Bot., Oxford* 1964, 253.
British and Indian plants at Oxford.

THOMSON, Anthony Todd (1778–1849)
b. Edinburgh Jan. 1778 *d.* Ealing, Middx 3 July 1849
MD Edinburgh 1799. FLS 1812. Professor of Materia Medica, University College, London, 1828. *Lectures on Elements of Botany* 1822. *Vegetable Physiology* 1827. Edited J. Thomson's *The Seasons* with natural history notes, 1847.
Pharm. J. v.9, 1849, 90–95. *Proc. Linn. Soc.* 1850, 91. *D.N.B.* v.56, 235–36.
Plants and drawings at Cork University.
Thomsonia Wall.

THOMSON, Archibald (*c.* 1753–1832)
d. London 5 Jan. 1832
Gardener to Lord Bute at Luton Hoo, Beds. In partnership with Messrs. Gordon and Dermer, nurserymen, Mile End Road, London.
Bot. Mag. 1820, t.2143. *Gdnrs Mag.* 1832, 256.
Magnolia thomsoniana Hort.

THOMSON, Sir Charles Wyville (1830–1882)
b. Bonsyde, Linlithgow 5 March 1830 *d.* Bonsyde 10 March 1882
LLD Aberdeen 1853; Dublin 1878. FRS 1869. FLS 1872. Knighted 1876. Zoologist. Lecturer in Botany, Aberdeen, 1850; Professor of Botany, 1851. Professor of Natural History, Cork, 1853. Professor of Mineralogy and Geology, Belfast, 1854; Professor of Natural History, 1860. Professor of Natural History, Edinburgh, 1870–81. Chief of 'Challenger' staff, 1872–76.
Nature v.14, 1876, 85–87 portr. *Proc. Linn. Soc.* 1881–82, 67–68. *Proc. R. Soc. Edinburgh* v.12, 1882–84, 58–80. *Scott. Nat.* 1883, 44–46. *Trans. Bot. Soc. Edinburgh* v.14, 1883, 278–82. *D.N.B.* v.56, 237–38. W. A. Herdman *Founders of Oceanography* 1923, 27–68 portr. R. L. Praeger *Some Irish Nat.* 1949, 167–68. R. Harre *Some Nineteenth Century Br. Scientists* 1969, 1–30.
Letters at Kew.

THOMSON, David (1823–1909)
b. Torloisk, Isle of Mull 5 March 1823 *d.* Esk Bank, Dalkeith, Midlothian 22 Oct. 1909
VMH 1897. Gardener to Duke of Buccleuch at Drumlanrig Castle, 1868–97. Edited *Scott. Gdnr*, 1854–82. *Practical Treatise on...Pine-apple* 1866. *Handy Book of Flower Garden* 1868. *Handy Book of Fruit Culture under Glass* 1873.
J. Hort. Cottage Gdnr v.40, 1868, 358–59; v.20, 1890, 363–64 portr.; v.34, 1897, 204–5 portr.; v.57, 1908, 294–96 portr.; v.59, 1909, 427–28 portr. *Gdnrs Chron.* 1903 ii 241 portr.; 1909 ii 296–97 portr.

THOMSON, David Stewart (*c.* 1818–1905)
b. near Stirling *c.* 1818 *d.* Wimbledon, Surrey 6 June 1905
Nurseryman and landscape gardener, Wimbledon.
Gdnrs Chron. 1905 i 368 portr. *J. Hort. Cottage Gdnr* v.50, 1905, 518.

THOMSON, David William (*c.* 1855–1915)
d. Edinburgh 11 March 1915
Nurseryman, Edinburgh, 1877.
Gdnrs Chron. 1915 i 159–60 portr. *Garden* 1915, 114(x).

THOMSON, George (*fl.* 1700s–1767)
d. 18 May 1767
MD Aberdeen. LRCP 1742. Practised medicine at Maidstone, Kent. *Virtues of Plants* 1734.
W. Munk *Roll of R. College of Physicians* v.2, 1878, 149.

THOMSON, George (1819–1878)
b. Balfron near Glasgow 26 May 1819 *d.* Victoria, W. Cameroons 14 Dec. 1878
Architect, Glasgow. Baptist missionary, Victoria, 1871. Contrib. to R. Hennedy's *Clydesdale Fl.* 1878.
Proc. Nat. Hist. Soc. Glasgow v.4, 1878–79, 51–52. W. C. Thomson *Mem. of George Thomson* 1881.
Plants at Kew.

THOMSON, George Malcolm (1848–1933)
b. Calcutta, India 2 Oct. 1848 *d.* 25 Aug. 1933
FLS 1879. In New Zealand, 1868. Science master, Otago High School, 1877–1903. Analytical chemist, 1906–14. Founded Portobello Marine Biological Station, 1902. *Ferns and Fern Allies of N.Z.* 1882. *Introductory Class-book of Botany* 1891. *N.Z. Naturalist's Calendar* 1909. *Naturalisation of Plants and Animals in N.Z.* 1922.
T. F. Cheeseman *Manual of N.Z. Fl.* 1906, xxxii. *Nature* v.132, 1933, 539. *Proc. Linn. Soc.* 1933–34, 172–73. *Who was Who, 1929–1940* 1344–45. R. Glenn *Bot. Explorers of N.Z.* 1950, 138–40.

THOMSON, Gideon (–before 1855)
Brother of T. Thomson (1817–1878). Collected plants in Madras, etc., now at Kew.
J. D. Hooker and T. Thomson *Fl. Indica* v.1, 1855, 73. *Fl. Malesiana* v.5, 1958, 92.

THOMSON, James (*fl.* 1830s)
Son of A. Thomson (*c.* 1753–1832). Nurseryman, Mile End Road, London.

THOMSON, James (–1925)
d. 28 Nov. 1925
Nurseryman, Kelso, Roxburghshire.
Gdnrs Chron. 1925 ii 457.

THOMSON, Miss Jane Smithson (*c.* 1876–1972)
d. Dublin 29 Jan. 1972
Member of Dublin Naturalists Field Club and British Bryological Society.
Herb. at University College, Galway.

THOMSON, John (*c.* 1852–1901)
d. Galashiels, Selkirkshire 27 April 1901
Nurseryman, Galashiels.
Garden v.59, 1901, 342. *Gdnrs Mag.* 1901, 280 portr.

THOMSON (Thompson), John W. (*c.* 1805–1895)
d. Hayward's Heath, Sussex 25 March 1895
Kew gardener, 1819. Gardener to Duke of Northumberland at Syon House, Middx. Nurseryman, Hammersmith, London, 1835–60. Nurseryman, Haywards Heath, 1876.
J. Kew Guild 1894, 14, 32 portr.; 1895, 44. *Gdnrs Chron.* 1895 i 498. *J. Hort. Cottage Gdnr* v.30, 1895, 313. *Kew Bull.* 1895, 120.

THOMSON, Joseph (1858–1895)
b. Thornhill, Dumfries 14 Feb. 1858 *d.* London 2 Aug. 1895
Pupil of J. H. Balfour (1808–1884). African traveller. Naturalist to A. K. Johnston's expedition to Central Africa, 1878–80. Collected plants in E. Equatorial Africa, 1879–85, now at Kew. *To Central African Lakes and Back* 1882. *Through Masi Land* 1885. *Travels in the Atlas and Southern Morocco* 1889.
J. Linn. Soc. v.21, 1885, 397–406. J. B. Thomson *Joseph Thomson, African Explorer* 1896 portr. *Geogr. J.* v.6, 1895, 289–91 portr. *D.N.B.* v.56, 262–65. *J. Bot.* 1895, 313. *Comptes Rendus AETFAT 1960* 1962, 212.
Letters at Kew.
Impatiens thomsoni Hook. f.

THOMSON, Robert (1840–1908)
d. Thornton Heath, Surrey 28 Dec. 1908
Kew gardener, 1862. Botanic Gardens, Castleton, Jamaica, 1862–79. Wrote articles on rubber trees.
J. Kew Guild 1911–12, 48–49 portr.
Polypodium thomsonii Jenman.

THOMSON, Spencer (*c.* 1817–1886)
d. Torquay, Devon 12 Aug. 1886
MD St. Andrews 1840. FRSE 1836. Practised medicine at Burton-on-Trent and Torquay. *Wanderings among the Wild Flowers* 1854. *Wild Flowers worth Notice* 1858. *Wayside Weeds* 1864.
F. Boase *Modern English Biogr.* Supplt. 3, 1921, 684. *R.S.C.* v.5, 970.

THOMSON, Thomas (1773–1852)
b. Crieff, Perthshire 12 April 1773 *d.* Kilmun, Argyllshire 2 July 1852
MD Edinburgh 1799. FRS 1811. FLS 1812. Professor of Chemistry, Glasgow, 1818. *Chemistry of Organic Bodies: Vegetables* 1838.
Gent. Mag. 1852 ii 202–6. *Pharm. J.* v.12, 1852–53, 95. *Proc. Linn. Soc.* 1853, 240–41. *D.N.B.* v.56, 271–72. C. C. Gillispie *Dict. Sci. Biogr.* v.13, 1976, 372–75. *R.S.C.* v.5, 202.

THOMSON, Thomas (1817–1878)
b. Glasgow 4 Dec. 1817 *d.* London 18 April 1878
MD Glasgow. FRS 1855. FLS 1852. Son of T. Thomson (1773–1852). Pupil of W. J. Hooker. Surgeon, Bengal army, 1839. Joined J. D. Hooker in Himalayan expedition, 1849. Superintendent, Botanic Garden, Calcutta, 1854–61. Collected plants in Kashmir, etc. *Western Himalaya and Tibet; a Narrative of a Journey...1847–1848* 1852. *Fl. Indica* (with J. D. Hooker) 1855. 'Account of Calcutta Herbarium' (*J. Asiatic Soc. Bengal* 1856, 405–18).
Bonplandia 1853, 181–82. *Gdnrs Chron.* 1878 i 529–30. *J. Bot.* 1878, 160; 1899, 461–

62. *Nature* v.18, 1878, 15–16. *Proc. R. Geogr. Soc.* v.22, 1878, 309–15. *Kew Bull.* 1895, 236. *D.N.B.* v.56, 272–73. D. G. Crawford *Hist. Indian Med. Service, 1600–1913* v.2, 1914, 144–45. *Curtis's Bot. Mag. Dedications, 1827–1927* 111–12 portr. I. H. Burkill *Chapters on Hist. Bot. in India* 1965 *passim.* A. M. Coats *Quest for Plants* 1969, 164. *R.S.C.* v.5, 976; v.8, 1080.

Plants at Kew, BM(NH), Oxford, etc. Letters and crayon portr. by G. Richmond at Kew. Portr. at Hunt Library.

Hedyotis thomsoni Hook. f.

THOMSON, William (1814–1895)
b. Bowden, Roxburghshire 27 March 1814 *d.* 12 Jan. 1895

Gardener to Duke of Buccleuch at Dalkeith. Grape grower at Clovenfords, 1871. *Practical Treatise on Cultivation of Grape Vine* 1862; ed. 10 1890. Edited *Gardener* 1869–82.

J. Hort. Cottage Gdnr v.21, 1890, 263, 273 portr.; v.30, 1895, 53–54 portr. *Gdnrs Mag.* 1894, 581, 585 portr.; 1895, 40. *Gdnrs Chron.* 1895 i 82–83, 115 portr.

THOMSON, William (1825–1899)
b. 19 July 1825 *d.* Teignmouth, Devon 16 Sept. 1899

Curator, Museum of Human and Comparative Anatomy, King's College, London, 1846–51. Superintendent, Natural History Dept., Crystal Palace, 1852–53. Secretary, City of London Club, 1856–94. Contrib. to *Gdnrs Chron., Garden.*

Gdnrs Chron. 1899 ii 250.

THOMSON, William (*c.* 1836–1912)
d. 25 March 1912

Seedsman, Melbourne Place, Edinburgh.

J. Hort. Home Farmer v.64, 1912, 318.

THOMSON, William (1849–1893)
b. Barnet, Herts 20 Dec. 1849 *d.* Galashiels, Selkirkshire 30 July 1893

Nurseryman, Tweed Nurseries, Clovenfords, Galashiels.

Gdnrs Chron. 1893 ii 167, 219 portr.

THOMSON, Rev. William Cooper (*fl.* 1820s–1870s)

Nephew of G. Thomson (1819–1878). Missionary in Calabar, 1849–65. Afterwards practised medicine in Liverpool. 'Notice of Ferns from Old Calabar' (*Trans. Proc. Bot. Soc. Edinburgh* 1860, 357–58). *Mem. of George Thomson* 1881.

J. Linn. Soc. v.8, 1864, 158–62. *Trans. Liverpool Bot. Soc.* 1909, 93.

Plants at Edinburgh, Kew.

THOMSON, Mrs. William Cooper (*fl.* 1860s)

Wife of Rev. W. C. Thomson. Went out as a bride to West Africa, botanised and corresponded with Hooker, and died young.

Bot. Mag. 1862, t.5313.

Clerodendron thomsonae Balfour.

THORBURN, Grant (1773–1863)
b. Dalkeith, Midlothian 18 Feb. 1773 *d.* New Haven, Conn., U.S.A. 21 Jan. 1863

Emigrated to U.S.A., 1794. Established first seedhouse in New York City, 1802. *Forty Years' Residence in America* 1834 portr. *Fifty Years' Reminiscences of New York* 1845. *Life and Writings* 1852.

Gdnrs Mag. 1832, 278–79. *J. Hort. Cottage Gdnr* v.29, 1863, 127–29 portr.; v.55, 1907, 327–28 portr. *Gdnrs Mag.* 1907, 16. *Gdnrs Chron.* 1929 i 320 portr., 328–29. L. H. Bailey *Standard Cyclop. Hort.* v.2, 1939, 1600 portr. U. P. Hedrick *Hist. Hort. in America to 1860* 1950, 204.

Portr. at Hunt Library.

THORLEY, Samuel (*fl.* 1830s)

Nurseryman, Market Place, Northwich, Cheshire.

THORLEY, William (*fl.* 1800s–1810s)

Nurseryman, Hulme, Manchester, Lancs.

THORNBECK, Mr. (*fl.* 1740s)

Surgeon, Ingleton, Yorks. Supplied plant records to J. Blackstone's *Specimen Botanicum* 1746.

R. Pulteney *Hist. Biogr. Sketches of Progress of Bot. in England* v.2, 1790, 272.

THORNCROFT, George (1857–1934)
b. Kent 1857 *d.* Barberton, Transvaal 19 July 1934

To Natal, S. Africa, 1882, thence to Transvaal where he botanised. Sent plants to J. Medley Wood. Contrib. to *Gdnrs Chron.* under pen-name 'Kof Kof'.

Bot. Mag. 1919, t.8824. *Gdnrs Chron.* 1934 ii 201. *Kew Bull.* 1934, 338. *Lantern* v.13(1), 1963, 84 portr.

Plants at Kew and various S. African herbaria.

Thorncroftia longiflora N. E. Brown.

THORNHILL, John (*fl.* 1800s–1850s)

Of Gateshead, Durham. *Botanist's Guide through Counties of Northumberland and Durham* (with N. J. Winch and R. Waugh) 1805–7. *Fasciculus of Thirty-five Dried Specimens of Grasses* 1806. Contrib. to J. Sowerby and J. E. Smith's *English Bot.* 1163, 2807, etc. His son, John (–1882), lectured on Botany at Newcastle Medical College.

Nat. Hist. Trans. Northumberland, Durham and Newcastle-upon-Tyne v.14, 1903, 81–82.

Plants at BM(NH). Letters at Kew.
Phascum thornhillii Wils.

THORNS, Frank William (*c.* 1904–1971)
d. 19 July 1971
Kew gardener, 1926–28. To Sudan. Curator, Botanic Gardens, Durban, 1931–36. Curator, Botanic Gardens, Kirstenbosch. Director of Parks, Durban, 1947.
J. Kew Guild v.9(76), 1972, 63 portr.

THORNTON, Archie J. (1894–1957)
b. Chester 6 July 1894 *d.* 19 Oct. 1957
Kew gardener, 1919–22. Assistant Superintendent, Botanical and Forestry Dept., Hong Kong, 1922. To Canada, 1925. Nurseryman, Poughkeepsie, New York, U.S.A. *Rock Garden Primer* 1929.
Kew Bull. 1922, 125. *J. Kew Guild* 1957, 489.

THORNTON, Robert John (*c.* 1768–1837)
d. London 21 Jan. 1837
MB Cantab 1793. MD St. Andrews 1805. Pupil of Thomas Martyn. Succeeded J. E. Smith as Lecturer in Botany, Guy's Hospital, London. *New Illustration of Sexual System of Linnaeus* (including *Temple of Flora*) 1799–1807. *Botanical Extracts* 1810 3 vols. *New Family Herbal* 1810. *British Flora* 1812 5 vols. *Elements of Botany* 1812 2 vols.
Gent. Mag. 1837 ii 93–95. W. Munk *Roll of R. College of Physicians* v.3, 1878, 98–100. *Gdnrs Chron.* 1894 ii 89–90, 154, 276–78. *D.N.B.* v.56, 304–6. *J. R. Hort. Soc.* 1947, 281–85 portr., 450–53. *Country Life* 1951, 733–35 portr. *Times Literary Supplement* 1 June 1951, 348. G. Grigson and H. Buchanan *Thornton's Temple of Flora* 1952 portr. W. Blunt *Art. of Bot. Illus.* 1955, 203–8. F. A. Stafleu *Taxonomic Literature* 1967, 466. *Guy's Hospital Reports* 1971, 64–67. *R.S.C.* v.5, 982.
Portr. at Hunt Library.
Thorntonia Reichenb.

THOROLD, William Grant (*fl.* 1890s)
Surgeon Capt., Indian Medical Service. Collected plants on journey with Capt. H. Bower across Tibet, 1891–92.
J. Linn. Soc. v.30, 1894, 101–40; v.35, 1902, 142–48. E. Bretschneider *Hist. European Bot. Discoveries in China* 1898, 806–7. *Kew Bull.* 1901, 64.
Plants at Kew.
Iris thoroldi Baker.

THORPE, John (1682–1750)
b. Penshurst, Kent 12 March 1682 *d.* Rochester, Kent 30 Nov. 1750
BA Oxon 1701. MD 1710. FRS 1705. Under-Secretary, 1713. Assisted Sir H. Sloane in *Philos. Trans. R. Soc.* Practised medicine at Rochester. Sent plants to Rev. A. Buddle. Edited J. J. Scheuchzer's *Helveticus sive Itineris Alpina* 1708.
J. Nichols *Lit. Anecdotes of Eighteenth Century* v.3, 509–22. D. Turner *Extracts from Lit. and Sci. Correspondence of R. Richardson* 1835, 94. *D.N.B.* v.56, 320–21. E. J. L. Scott *Index to Sloane Manuscripts* 1904, 530. J. E. Dandy *Sloane Herb.* 1958, 219.

THOZET, Anthelme (*c.* 1826–1878)
b. near Lyons, France *c.* 1826 *d.* Rockhampton, Queensland 31 May 1878
FLS 1867. Gardener, Botanic Gardens, Sydney, 1856–58. To Port Curtis goldfields, Queensland, 1858. Collected plants in Queensland which he sent to F. von Mueller. Farmed in Rockhampton.
J. Bot. 1878, 320. *Proc. R. Soc. Queensland* v.8(2), 1890–91, 32. *Rep. Austral. Assoc. Advancement Sci.* 1909, 382–83; 1911, 236. *J. Proc. R. Soc. N.S.W.* 1910, 153–54. *J. Proc. R. Austral. Hist. Soc.* 1932, 129–30.
Thozetia F. Muell.

THRELFALL, William (1862–1888)
b. Hollowfork, Preston, Lancs 1862 *d.* drowned in River Dryala, Kurdistan March 1888
BA Oxon 1885. FLS 1887. Botanised between Caucasus and Persia. Died on plant collecting expedition.
Gdnrs Chron. 1888 i 632. *Proc. Linn. Soc.* 1887–88, 98. *Trans. Liverpool Bot. Soc.* 1909, 95.
Persian plants at Kew.

THRELKELD, Rev. Caleb (1676–1728)
b. Keibergh, Kirk Oswald, Cumberland 31 May 1676 *d.* Dublin 28 April 1728
MA Glasgow 1698. MD Edinburgh 1713. Dissenting minister and physician. Settled in Dublin, 1713. Had a botanic garden. *Synopsis Stirpium Hibernicarum* 1727.
Gent. Mag. 1777, 63–64. R. Pulteney *Hist. Biogr. Sketches of Progress of Bot. in England* v.2, 1790, 196–201. J. C. Loudon *Encyclop. Gdning* 1878, 282. S. A. Stewart and T. H. Corry *Fl. N.E. Ireland* 1888, xiv. N. Colgan *Fl. County Dublin* 1904, xix–xx. *D.N.B.* v.56, 325. *J. Bot.* 1924, 353–54. R. L. Praeger *Some Irish Nat.* 1949, 168–69. *Proc. R. Irish Acad. B.* v.74(1), 1974, 1–6.
Threlkeldia R. Br.

THRING, Mrs. Lydia Eliza (1830–1925)
b. Brockhurst, Hants 4 Aug. 1830
Her list of Rutland plants was published in *Uppingham School Magazine* 1864.
A. R. Horwood and C. W. F. Noel *Fl. Leicestershire* 1933, cclxxvi–cclxxvii portr.

THROP, William (*fl.* 1820s–1840s)
Nurseryman, Shaw Syke, Halifax, Yorks.

THURN, Sir Everard Ferdinand Im (*c.* 1852–1932)
d. Prestonpans, E. Lothian 8 Oct. 1932
MA Oxon. LLD Edinburgh. KCMG 1905. KBE 1918. Curator of Museum, Georgetown, British Guiana, 1877–82. Magistrate, Pomeroon, 1882–91. Government Agent, N.W. District, 1891–99. Collected plants. 'Botany of Roraima Expedition' (*Trans. Linn. Soc.* v.2, 1887, 249–70). 'Sketches of Wild Orchids in Guiana' (*J. R. Hort. Soc.* 1898, 40–52). Founded journal, *Timehri*, in 1882. Colonial Secretary in Ceylon, 1901. Governor, Fiji Islands, 1904–10. Fiji plants determined in *J. Linn. Soc.* 1915, 15–39.
Geogr. J. 1932, 556–57. *J. Bot.* 1932, 341–42. *Kew Bull.* 1932, 461–62. *Marlburian* 1932, 206–7. *Nature* v.130, 1932, 602–3. *Orchid Rev.* 1932, 352. *Times* 11 Oct. 1932. *Who was Who, 1929–1940* 690.
Plants at Kew.
Thurnia Hook. f. *Everardia* Ridley.

THURSTON, Edgar (1855–1935)
b. London 14 July 1855 *d.* Penzance, Cornwall 5 Oct. 1935
LRCP 1877. Superintendent, Government Museum, Madras, 1885–1909. After retirement settled in Cornwall where he studied the flora and horticulture. Supplement to F. Hamilton Davey's *Fl. Cornwall* 1922. *British and Foreign Trees and Shrubs in Cornwall* 1930.
Bot. Soc. Exch. Club Br. Isl. Rep. 1935, 19–20. *Kew Bull.* 1935, 594–95. *Times* 8 Oct. 1935. *J. Bot.* 1936, 21–22. *Who was Who, 1929–1940* 1251.
Plants at Kew, Nottingham Museum.

THURSTON, Sir John Bates (1836–1897)
b. London 31 January 1836 *d.* Suva, Fiji Feb. 1897
British Consul, Levuka, 1866; Fiji, 1869. Deputy Governor, Fiji, 1882; Governor, 1887–97. Sent plants from Solomon Islands and Fiji to Kew.
Colonies and India 13 Feb. 1897, 14 portr. *Kew Bull.* 1897, 169. *D.N.B.* v.56, 357–58.

THURTLE, George (*fl.* 1840s)
Nurseryman, Mile End Nursery, Cross Road, Mile End, Norwich, Norfolk.

THWAITES, George Henry Kendrick (1812–1882)
b. Bristol 9 July 1812 *d.* Kandy, Ceylon 11 Sept. 1882
PhD. FRS 1865. FLS 1854. CMG 1878. Secretary, Botanical Society of London, 1839. Superintendent, Botanic Gardens, Peradeniya, Ceylon, 1849; Director, 1857–80. Introduced cultivation of cinchona to Ceylon. Contrib. records of Bristol plants to H. C. Watson's *Topographical Bot.* and to *Phytologist* from 1841. *Enumeratio Plantarum Zeylaniae* 1858–64.
Gdnrs Chron. 1874 i 438 portr.; 1882 ii 505. *J. Bot.* 1882, 351–52. *Nature* v.26, 1882, 632–33. *Proc. Linn. Soc.* 1882–83, 43–47. *Trans. Bot. Soc. Edinburgh* v.16, 1886, 8–9. *Tropical Agric.* v.14, 1894, 75–79. *D.N.B.* v.56, 361–62. H. Trimen *Handbook to Fl. Ceylon* v.5, 1900, 376–79. *Ann. R. Bot. Gdns Peradeniya* v.1, 1901, 7–10. J. W. White *Fl. Bristol* 1912, 79–80. *Curtis's Bot. Mag. Dedications, 1827–1927* 131–32 portr. H. J. Riddelsdell *Fl. Gloucestershire* 1948, cxxi–cxxii. *R.S.C.* v.5, 989.
Ceylon plants at Kew. Ferns at Liverpool Museums. Bristol plants at Bristol Museum destroyed in 1940. Letters at Kew, BM(NH). Portr. at Hunt Library.
Thwaitesia Montague=*Kendrickia* Hook. f.

THWAITES, Richard G. (–1943)
d. 15 June 1943
Of Streatham, London. Hybridised orchids including *Dendrobium thwaitesiae.*
Gdnrs Mag. 1910, 159–60 portr. *Orchid Rev.* 1943, 163–64.

THWAITES, W. A. (*fl.* 1900s)
b. Marsham, Yorks?
Gamekeeper, Marsham. Collected fungi for C. Crossland (1844–1916).
Naturalist 1961, 65–66.

THYNNE, Louisa *see* Finch, L. *Countess of Aylesford*

THYNNE, Mary *see* Markham, M.

THYNNE, Thomas, 1st Viscount Weymouth (1640–1714)
d. 28 July 1714
Reputed to have introduced the Weymouth pine, *Pinus strobus*, in 1705 and planted it extensively on his estate at Longleat, Wilts.
D.N.B. v.56, 368–69. D. McClintock *Companion to Flowers* 1966, 208–9.

TIERNAN, John (*fl.* 1790s–1800s)
Nurseryman, Dublin.
Irish For. 1967, 49.

TIETKINS, William Henry (1844–1933)
b. London 20 Aug. 1844 *d.* Lithgow, N.S.W. 19 April 1933
Educ. Christ's Hospital School. To S. Australia, 1859 where he worked in the goldfields there and in N.S.W. Joined W. Giles's trans-Australian expeditions, 1873 and 1875 and collected plants. Led expedition into central Australia, 1889. *Journal of Central Australian Exploring Expedition* 1891. 'List of Plants Collected during Mr. Tietkins

Expedition into Central Australia' (*Proc. R. Soc. S. Australia* v.13, 94–109, 170–71). Surveyor, N.S.W. Dept. of Lands, 1891–1909.

Rep. Austral. Assoc. Advancement Sci. 1907, 171. *Austral. Encyclop.* v.8, 1965, 501.

MS. diary of explorations in S. Australia at Public Library of S. Australia.

TIGHE, Right Hon. W. F. (*c.* 1793–1878)
d. Woodstock, Ireland 11 June 1878

Created arboretum at Woodstock Park.

Garden v.14, 1878, 24.

TIGHE, William (1766–1816)
b. Rossana, County Wicklow 1766

'Plants of Coast of Wexford' (*Trans. R. Dublin Soc.* v.3, 1802, 147–56). *The Plants* 1808–11.

N. Colgan and R. W. Scully *Cybele Hibernica* 1898, xxxv.

TILDEN, Richard (*fl.* 1700s)

Sent *Gentiana germanica* from St. Albans to J. Petiver and I. Rand.

F. Pursh *Fl. Americae Septentrionalis* v.1, 1814, xviii. A. R. Pryor *Fl. Hertfordshire* 1887, li, 274 (misprinted Feilden). J. E. Dandy *Sloane Herb.* 1958, 219.

Plants at BM(NH). Hudson Bay plants at Oxford.

TILEY, Edward (*fl.* 1850s)

Nurseryman, 14 Abbey Churchyard, Bath, Somerset; rose grounds at back of Sydney Gardens, Bathwick.

TILLERY, William (1808–1879)
b. Kilmarnock, Ayrshire 1808 *d.* 25 April 1879

Gardener to Duke of Portland at Welbeck, Notts. Contrib. to *Gdnrs Chron.*

Gdnrs Chron. 1874, 463 portr.; 1879 i 593 portr.

TILLOTSON, Dr. (*fl.* 1690s)

Had garden at Enfield, Middx.

Archaeologia v.12, 1794, 188.

TILLSON, Arthur G. (*fl.* 1890s)

Kew gardener, 1889. Curator, Botanic Station, Antigua, 1890–97. Collected plants.

I. Urban *Symbolae Antillanae* v.3, 1902, 136.

Plants at Kew.

TILLYARD, George B. (1819–1889)
b. Hendon, Middx 1819 *d.* 6 Sept. 1889

Gardener to Earl of Yarborough at Brocklesby Park, Lincs.

Gdnrs Chron. 1889 ii 304–5 portr.

TIMINS, Elizabeth Helen *see* Barkly, *Lady* E. H.

TINDAL, Miss M. I. (*fl.* 1830s)

Collected British plants, now at Essex Field Club.

Essex Nat. v.29, 1954, 191.

TINDALL, George (*fl.* 1800s–1830s)

Partner with his brother, William, in four nurseries at Beverley, Yorks.

TINDALL, Mrs. Isabella Mary (1850–1928)

Contrib. to *J. Bot.* on hepatics. Herb. acquired by F. J. Smith (1853–1919).

Hepatics at BM(NH).

TINDALL, William (*fl.* 1820s–1830s)

Nurseryman, Beverley, Yorks.

TINKER, Jethro (1788–1871)
b. Stalybridge, Cheshire 25 Sept. 1788 *d.* Stalybridge 10 March 1871

Weaver. Gardener, botanist and entomologist. Correspondent of E. Hobson (1782–1830).

R. Buxton *Botanical Guide to...Manchester* 1849, ix–x. J. Cash *Where There's a Will There's a Way* 1873, 135. J. B. L. Warren *Fl. Cheshire* 1899, xc. *Victoria County Hist. Lancashire* v.1, 1906, 102. *Trans. Liverpool Bot. Soc.* 1909, 93.

Herb. formerly at Stalybridge and monument inscribed "Our local Linnaeus."

TINLEY, George F. (–1931)
d. 1 Aug. 1931

Kew gardener, 1899. Assistant editor, *Gdnrs Chron.* Editor, *Fruit, Flower and Vegetable Trades J.*

J. Kew Guild 1932, 192–93.

TISDALL, Henry Thomas (*c.* 1836–1905)
b. Waterford *c.* 1836 *d.* Melbourne, Victoria 10 July 1905

FLS 1883–90. Arrived in Melbourne, 1858. Schoolmaster in Victoria until 1894. Lecturer in Botany, Teacher's College of Victoria, 1894. Cryptogamist. *Botany Notes* 1894. Contrib. to *Victorian Nat.*

Australasian (newspaper) 22 July 1905. *Victorian Nat.* 1905, 56–58; 1908, 114; 1949, 107.

Fungi at Kew. Plant drawings at Melbourne Herb.

TITFORD, William Jowit (*fl.* 1784–1820s)
b. Jamaica 13 May 1784

MD. Brought up in London by his uncle and aunt. *Sketches towards Hortus Botanicus Americanus* 1811–12.

Central Rev. or Ann. of Literature v.24, Nov. 1811. *Mon. Mag.* v.33, 1812, 94.

TOBIN, George (1768–1838)
b. Salisbury, Wilts 13 Dec. 1768 *d.* Teignmouth, Devon April 1838
Royal Navy, 1780. Artist–naturalist on HMS 'Providence' under Capt. Bligh, 1791. Made drawings of plants collected by J. Wiles and C. Smith on this voyage. Appointed Capt., 1802; Rear-Admiral, 1837.
D.N.B. v.56, 422. I. Lee *Captain Bligh's Second Voyage to South Sea* 1920 *passim.* G. Mackaness *Life of Vice-Admiral William Bligh* v.1, 1931 *passim. Austral. Mus. Mag.* 1933, 44–50. *J. Proc. R. Austral Hist. Soc.* 1934, 297. *J. Proc. R. Austral. Hist. Soc.* 1949, 54–55.
MS. Journal and drawings at Mitchell Library, Sydney.

TOBIN, James Webbe (–1814)
d. Nevis, West Indies 30 Oct. 1814
Sent seeds to A. B. Lambert.
Bot. Mag. 1815, t.1779. W. Hamilton *Prodromus Plantarum Occidentalis* 1825, 56–57. *D.N.B.* v.56, 423.
Tobinia Desv.

TODD, Emily Sophia (1859–1949)
b. London 19 May 1859 *d.* Wantage, Berks 16 April 1949
Of Aldbourne, Wilts. Amateur botanist.
Watsonia 1949–50, 325. D. Grose *Fl. Wiltshire* 1957, 43–44.
Herb. at Swindon Museum. Plants at Oxford.
×*Rosa toddii* Wolley-Dod.

TOFIELD, Thomas (1730–1779)
b. Wilsic, Yorks 1730 *d.* Wilsic 1779
MA. Of Doncaster, Yorks. Correspondent of W. Hudson. Had herb.
Trans. Linn. Soc. v.12, 1818, 237. *J. Bot.* 1924, 306.
Tofieldia Hudson.

TOLL, George (*c.* 1835–1884)
b. Dunsford, Devon *c.* 1835 *d.* Manchester 23 June 1884
Nurseryman, Hullard Hall Nursery, Manchester.
Florist and Pomologist 1884, 128. *Gdnrs Chron.* 1884 ii 27.

TOLMIE, William Fraser (1812–1886)
b. Inverness 3 Feb. 1812 *d.* Victoria, B.C., Canada 8 Dec. 1886
MD. Pupil of W. J. Hooker. To Fort Vancouver as medical officer, 1832. Canadian Geological Survey. Made first recorded trip on Mt. Rainier, Aug. 1833. Contrib. to W. J. Hooker's *Fl. Boreali-Americana. Journal of William Fraser Tolmie* 1963 portr. *Journal* (*Washington Hist. Quart.* 1906, 77–81; 1912, 229–41; 1932, 205–77).
Companion Bot. Mag. v.2, 1836, 159. *Dominion Annual Register* 1886, 290–91. *Amer. J. Sci.* v.33, 1887, 244–45. Kerr *Biogr. Dict. Br. Columbia* 1890, 307–9. *Beaver* 1937, 29–32 portr. *Br. Columbia Hist. Quart.* 1937, 227–40. S. D. McKelvey *Bot. Exploration of Trans-Mississippi West, 1790–1850* 1955, 464–80.
Plants and letters at Kew. Portr. at Hunt Library.
Tolmiea Hook.

TOMALIN, E. F. J. (*c.* 1916–1973)
BSc London. Chemical engineer. Grew irises. Secretary, Iris Society, 1937–46.

TOMKINS, Charles (*c.* 1796–1864)
d. 17 May 1864
FLS 1823. Physician at Abingdon, Berks and Weston-super-Mare, Somerset. Amateur botanist.
Proc. Linn. Soc. 1863–64, xxxii.

TOMLINSON, George (1696–1760)
b. Aug. 1696 *d.* Hathern, Leics 10 Feb. 1760
Uncle and teacher of R. Pulteney. Had herb.
J. Nichols *Hist.…Leicester* v.3, 1804, 846–48.

TOMLINSON, William James Coleman (1863–1921)
d. Belfast June 1921
Of Belfast. On staff of Midland Railway. Botanised locally. Contrib. to *Irish Nat.*
Irish Nat. 1921, 108. R. L. Praeger *Some Irish Nat.* 1949, 169.
Herb. at Queen's University, Belfast.

TONGE (Tongue), Rev. Israel (Ezereel) (1621–1680)
b. Tickhill, Yorks 11 Nov. 1621 *d.* London 18 Dec. 1680
BA Oxon 1643. DD 1656. Rector, St. Michael's, Wood Street, London; St. Mary Stayning, London; Aston, Herefordshire. Communications relating to vegetation in *Philos. Trans. R. Soc.* v.3–5, 1668–71.
D.N.B. v.57, 32.

TOOGOOD, William (*c.* 1827–1892)
d. Southampton, Hants July 1892
Formerly partner and then proprietor of nursery firm of W. B. Page and Sons, Southampton.
Gdnrs Chron. 1892 ii 79.

TOOHEY, Matthew (1854–1926)
b. County Clare 21 April 1854 *d.* St. Helens, Lancs 26 April 1926
Entered Society of Jesus, 1877. Chaplain, Whiston Institution, Prescot, 1906–26. Botanised in Lancashire, Denbighshire and Flintshire.
N. Western Nat. 1926, 153–54.

TOPPIN, Sidney Miles (1878–1917)
b. Clonmel, County Tipperary 12 June 1878 *d.* Ypres, Belgium 24 Sept. 1917

Major, Royal Artillery. FLS 1912. Collected plants in Chitral and N. Burma, 1900–12. 'Balsams of Chitral and the Kachin Hills' (*Kew Bull.* 1920, 345–67).

Kew Bull. 1918, 156–57. *Proc. Linn. Soc.* 1917–18, 45.

Plants at Kew.

Impatiens toppinii Dunn.

TOPPING, Timothy (*fl.* 1750s–1770s)

Seedsman, 'at the Sun on London Bridge'. Succeeded by Topping and Fenton, 121 Upper Thames Street, London.

TOWARD, Andrew (*c.* 1796–1881)
d. 7 May 1881

Gardener to Duke of Gloucester at Bagshot Park, Surrey. Land Steward to H.M. Queen at Osborne, Isle of Wight.

Florist and Pomologist 1881, 96.

TOWERS, George John (*fl.* 1830s–1840s)

"Horticultural chemist." *Domestic Gardener's Manual* 1830. 'Investigation of Structure of the Balsam' (*Gdnrs Mag.* 1832, 403–7).

Gdnrs Mag. 1831, 57–60. *Gdnrs Chron.* 1950 i 206–7.

TOWNDROW, Richard Francis (1845–1937)
b. Malvern Link, Worcs 29 Oct. 1845 *d.* Malvern Wells 25 Dec. 1937

ALS 1915. Grocer. Botanised in Malvern district. Contrib. records to W. H. Purchas and A. Ley's *Fl. Herefordshire* 1899 and J. Amphlett and C. Rea's *Bot. Worcestershire* 1909. Contrib. to *J. Bot.*

Bot. Soc. Exch. Club Br. Isl. Rep. 1937, 436–38. *Proc. Linn. Soc.* 1937–38, 341–42. *Trans. Worcestershire Nat. Club* 1958–59, 146–50; 1961, 146–50.

Herb. at Malvern Public Library.

TOWNLEY, E. W. (*fl.* 1780s–1790s)

Nurseryman and land surveyor, 27 Crosby Row, Walworth, Surrey.

TOWNLEY, John (*fl.* 1830s–1840s)

Of Preston, Lancs. Agricultural writer. *Diseases, Regeneration and Culture of Potato* 1847.

Trans. Liverpool Bot. Soc. 1909, 94.

TOWNLEY, Thomas (–1857)
d. Manchester 9 Sept. 1857

Shoe-maker, Blackburn, Lancs, then Liverpool and Manchester. Taught G. Crozier (1792–1847). Contrib. to R. Buxton's *Bot. Guide* 1849.

J. Cash *Where there's a Will there's a Way* 1873, 129–31. L. H. Grindon *Country Rambles* 1882, 172–74. *Trans. Liverpool Bot. Soc.* 1909, 94.

TOWNSEND, Frederick (1822–1905)
b. Rawmarsh, Yorks 5 Dec. 1822 *d.* Cimiez, Alpes Maritimes 16 Dec. 1905

BA Cantab 1850. FLS 1878. Friend of C. C. Babington and W. W. Newbould. *Notes on Fl. Hampshire* 1879. *Fl. Hampshire* 1883; ed. 2 1904. 'Contributions to Fl. Scilly Isles' (*J. Bot.* 1864, 102–20). '*Euphrasia*' (*J. Bot.* 1897–98). Contrib. botany (with W. M. Rogers) to *Victoria County Hist. Hampshire* v.1, 1900, 47–87. The plants collected by his sister, Elizabeth (*fl.* 1840s–1850s) of Honington Park, Warwick, were acquired by C. E. Palmer (1830–1914).

Bot. Soc. Exch. Club Br. Isl. Rep. 1905, 8–9. *J. Bot.* 1906, 113–15 portr. *Proc. Linn. Soc.* 1905–6, 47–49. *Trans. Bot. Soc. Edinburgh* v.23, 1906, 217–18. F. H. Davey *Fl. Cornwall* 1909, lii. H. J. Riddelsdell *Fl. Gloucestershire* 1948, cxxxii. J. E. Lousley *Fl. Isles of Scilly* 1972, 82. *R.S.C.* v.6, 17; v.8, 1105; v.11, 630; v.12, 739.

Plants at Kew, Oxford. Herb. at S. London Botanical Institute. Portr. at Hunt Library.

Spartina townsendii Groves.

TOWNSEND, Rev. Joseph (1739–1816)
b. London 4 April 1739 *d.* Pewsey, Wilts 9 Nov. 1816

MA Cantab 1765. Rector, Pewsey. Geologist. 'Food of Plants' (*Nicholson J. Nat. Philos.* v.23, 1809, 5–15).

D.N.B. v.57, 106–7. *R.S.C.* v.6, 17.

Portr. in R. J. Thornton's *New Illus. Sexual System* 1807. Portr. at Hunt Library.

TOWNSHEND, Barbara *see* Massie, B.

TOWNSON, Robert (*fl.* 1790s–1800s)
b. Shropshire *d.* Australia

LLD. MD Göttingen 1795. FRSE 1791. Correspondent of R. A. Salisbury. Discovered *Saxifraga rivularis* in Britain, 1790. *Travels in Hungary* 1797 (with botanical appendix). 'Perceptivity of Plants' (*Trans. Linn. Soc.* v.2, 1794, 267–72).

C. Linnaeus *Fl. Lapponica* 1792, 143. *D.N.B.* v.57, 133. *J. Bot.* 1914, 323. *R.S.C.* v.6, 17.

Letters in Winch correspondence at Linnean Society.

TOWNSON, William (1850–1926)
b. Liverpool 1850 *d.* Thames, New Zealand 11 Aug. 1926

To N.Z. as a young man. Pharmaceutical chemist. Collected plants on Tararua Range, Mounts Ruapehu and Egmont, etc.

T. F. Cheeseman *Manual of N.Z. Fl.* 1906, xxx. *Trans. Proc. N.Z. Inst.* v.58, 1927, 186–88 portr.

TOYNBEE, Mrs. Henry (*fl.* 1850s)
Drawings of marine animals and plants made during voyages between England and India, via the Cape, 1856–58, at BM(NH).

TOZER, Rev. Henry Fanshawe (1829–1916)
b. Plymouth, Devon 18 May 1829 *d.* Oxford 2 June 1916
MA Oxon. Travelled in Europe and the East. *Highlands of Turkey* 1869. His wife's specimens from Devon, Wilts, Berks and Worcs were presented to G. C. Druce, now at Oxford.
Geogr. J. v.48, 1916, 176–77. *Kew Bull.* 1920, 29–30.
Plants at Kew, Oxford.

TOZER, Rev. John Savery (*c.* 1790–1836)
d. drowned near Shrewsbury, Shropshire March 1836
Curate, St. Petrock, Exeter, Devon. Discovered *Erica ciliaris* 1828. Contrib. to J. P. Jones and J. F. Kingston's *Fl. Devoniensis* 1829 and to W. J. Hooker's *Br. Fl.* 1830.
J. Sowerby and J. E. Smith *English Bot.* 2618, 2628. *Gent. Mag.* 1836, 438. *Mag. Zool. Bot.* 1837, 112. F. H. Davey *Fl. Cornwall* 1909, xxxix.
Plants at Kew, Glasgow. Letters at Kew.
Bryum tozeri Grev.

TRACY, Henry Amos (*c.* 1850–1910)
b. Colchester, Essex *c.* 1850 *d.* 18 Aug. 1910
Nurseryman, Orchid and Bulb Nursery, Amyand Park Road, Twickenham, Middx. One of the first to import *Odontoglossum crispum* and popular orchids and to sell them at low prices.
Gdnrs Chron. 1910 ii 169 portr. *J. Hort. Home Farmer* v.61, 1910, 187–88. *Orchid Rev.* 1910, 272.
Cymbidium tracyanum Hort.

TRADESCANT, John (–1638)
b. London? *d.* Lambeth April 1638
Gardener to Charles I, 1629. Went to Russia, 1618 (J. Hamel *Tradescant der aeltere 1618 in Russland* 1847 portr.); Algiers, 1620; Egypt, etc. Friend of J. Parkinson. Had garden and museum at Lambeth.
J. Parkinson *Paradisus* 1629, 152, 346; *Theatrum Botanicum* 1640, 218, 343. R. Lovell *Herball* 1659, 551. *Philos. Trans. R. Soc.* v.46, 1752, 160–61; v.63, 1773, 79–88. R. Pulteney *Hist. Biogr. Sketches of Progress of Bot. in England* v.1, 1790, 175. G. W. Johnson *Hist. English Gdning* 1829, 98–100. *Gdnrs Chron.* 1852, 163–64, 294; 1881 i 87–88; 1926 ii 442 portr.; 1928 ii 201; 1947 ii 324 portr. *Cottage Gdnr* v.24, 1860, 344–45 portr. *D.N.B.* v.57, 143–45. *J. Bot.* 1895, 33–38; 1918, 197, 202; 1920, 171, 248. *J. Pomology* 1920, 188–96. R. W. T. Gunther *Early Br. Botanists* 1922, 328–46. *Bot. Soc. Exch. Club Br. Isl. Rep.* 1927, 555. *J. R. Hort. Soc.* 1928, 308–17. M. Hadfield *Pioneers in Gdning* 1955, 21–25. M. Allan *The Tradescants...1570–1662* 1964 portr. R. Webber *Early Horticulturists* 1968, 66–77. A. M. Coats *Quest for Plants* 1969, 45–46 portr. C. C. Gillispie *Dict. Sci. Biogr.* v.13, 1976, 449–50.
Portr. at Ashmolean Museum, Hunt Library.
Tradescantia L.

TRADESCANT, John (1608–1662)
b. Meopham, Kent 4 Aug. 1608 *d.* Lambeth 22 April 1662
Son of J. Tradescant (–1638) whom he succeeded as Gardener to Charles I. Collected *Tradescantia*, *Liriodendron*, etc. in Virginia. *Musaeum Tradescantianum* 1656 (includes catalogue of his garden).
R. Pulteney *Hist. Biogr. Sketches of Progress of Bot. in England* v.1, 1790, 178–79. *Philos. Trans. R. Soc.* v.46, 1752, 160–61. R. Weston *Cat. of English Authors on Agric., Bot.* 1773, 28–30. S. Felton *Portr. of English Authors on Gdning* 1830, 92–93. *Cottage Gdnr* v.4, 1849, 269; v.8, 1851, 3; v.25, 1861, 304–5 portr., 335–37. *Gdnrs Chron.* 1881 i 87–88; 1928 ii 201. *D.N.B.* v.57, 145–47. *A.L.A. Portr. Index* 1906, 1456. *J. Bot.* 1918, 200–1. *Torreya* 1927, 43. *J. R. Hort. Soc.* 1928, 308–17. R. W. T. Gunther *Early Br. Botanists* 1922, 272–73. M. Hadfield *Pioneers in Gdning* 1955, 21–25. M. Allan *The Tradescants...1570–1662* 1964 portr. *Huntia* 1965, 214–15 portr. C. C. Gillispie *Dict. Sci. Biogr.* v.13, 1976, 451.
Museum collections at Ashmolean Museum. Portr. at Ashmolean Museum, National Portr. Gallery, Hunt Library.

TRAIL (*olim* Traill), James William Helenus (1851–1919)
b. Birsay, Orkney 4 March 1851 *d.* Aberdeen 18 Sept. 1919
MA Aberdeen 1870. MB 1876. FRS 1893. FLS 1875. Professor of Botany, Aberdeen, 1877–1919. President, British Mycological Society, 1902. Botanist on expedition to Brazil, 1873. *Fl. Buchan* 1902–4. 'Palms of Amazon' (*J. Bot.* 1876–77). Edited F. B. W. White's *Fl. Perthshire* 1898. Edited *Scott. Nat.*, 1883–92. Botanical Editor of *Ann. Scott Nat. Hist.*, 1892–1911.
C. F. P. von Martius *Fl. Brasiliensis* v.1(1), 1906, 121–23. *Nature* v.48, 1893, 10; v.104, 1919, 76–77. *Bot. Soc. Exch. Club Br. Isl. Rep.* 1919, 626–28. *Gdnrs Chron.* 1919 ii 172; 1923 ii 360. *J. Bot.* 1919, 318–21; 1924, 150–51. *Kew Bull.* 1919, 378–88; 1920, 32–33. *Proc. Linn. Soc.* 1919–20, 49–51. *New Phytol.* 1920, 46–48. *Proc. R. Soc. B.* v.91, 1920, vii–xi. *Scott. Nat.* 1920, 1–5 portr. *Trans. Br.*

Mycol. Soc. 1920. 297–98. *J. W. H. Trail: a Memorial Vol.* 1923 portr. *Who was Who, 1916–1928* 1050. *R.S.C.* v.8, 1106; v.11, 631; v.12, 739; v.18, 182.

Herb. at Aberdeen University. Portr. at Hunt Library.

TRAILL, Catharine Parr (*née* Strickland) (1802–1899)

b. London 9 Jan. 1802 *d.* Lakefield, Ontario 29 Aug. 1899

Of Reydon Hall, Suffolk. Sister of Agnes Strickland; married Lieut. T. Traill and emigrated to Canada, 1832. *Canadian Wild Flowers* 1868. *Studies of Plant Life in Canada* 1885. *Pearls and Pebbles* 1894 portr.

Canadian Men and Women of the Time 1898, 1017. *J. Bot.* 1899, 448. *Appleton's Cyclop. Amer. Biogr.* v.6, 1889, 153. *Canadian Field Nat.* v.60, 1946, 97–101. *Encyclop. Canadiana* v.10, 1962, 119–20. *Ontario Nat.* 1966, 17–21. *Country Life* 1970, 182–83 portr.

Herb. destroyed by fire, 1857; some specimens at Herb. Ottawa and Herb. Queen's University, Kingston.

Aspidium marginale var. *Traillae* Lawson.

TRAILL, Charles (–1898)

d. Ulva, New Zealand 1898

Sent extensive collections of Stewart Island plants to T. Kirk.

T. Kirk *Students' Fl. N.Z.* 1899, 265.

Olearia Traillii Kirk.

TRAILL, George William (1836–1897)

b. Kirkwall, Orkney 26 Oct. 1836 *d.* Joppa, Edinburgh 7 April 1897

Clerk in Standard Life Assurance Co. Studied algae of East Coast and N. Scotland. *Monograph of Algae of Firth of Forth* 1885. Contrib. to *Trans. Bot. Soc. Edinburgh.*

J. Bot. 1897, 440. *Ann. Scott. Nat. Hist.* 1898, 7–8.

Herb. at Botanical Society of Edinburgh.

Trailliella Batters.

TRAILL, James (*fl.* 1820s–1853)

d. Cairo 11 Feb. 1853

ALS 1827. Gardener at Chiswick, Middx and to Ibrahim Pasha at Cairo, 1834. '*Hoya*' (*Trans. Hort. Soc. London* v.7, 1830, 16–30).

Companion Bot. Mag. v.1, 1836, 319. *Flora* 1841, 16. *R.S.C.* v.6, 18.

TRAILL, William (1818–1886)

b. Kirkwall, Orkney 8 Sept. 1818 *d.* St. Andrews, Fife 10 Dec. 1886

MD Edinburgh 1841. Conchologist. Employed by East India Company. Collected plants in India, China, Singapore, Malacca for nearly 20 years. 'Submarine Forests in Orkney' (*J. Bot.* 1867, 174–82).

Nature v.35, 1886, 419. *Scott. Nat.* 1887, 50–51. *Trans. Bot. Soc. Edinburgh* v.17, 1889, 17–19. *R.S.C.* v.6, 21; v.8, 1107.

TRANTER, J. R. (–1911)

d. Henley, Oxfordshire 20 April 1911

Builder. Raised dahlia varieties: 'Mrs. Tranter', etc.

Gdnrs Mag. 1911, 334.

TRAVERS, Henry (*fl.* 1830s–1870s)

Son of W. T. L. Travers with whom he studied the New Zealand flora.

R. Glenn *Bot. Explorers of N.Z.* 1950, 131–34.

TRAVERS, William Thomas Locke (1819–1903)

b. near Newcastle, County Limerick 9 Jan. 1819 *d.* Wellington, New Zealand 26 April 1903

FLS 1863. In N.Z. from 1849. Studied alpine plants of South Island. Ornithologist. Contrib. to *Trans. Proc. N.Z. Inst.*

J. Bot. 1864, 324. *Trans. Proc. N.Z. Inst.* v.35, 1902, xviii–xix portr. *Proc. Linn. Soc.* 1907–8, 64–65. R. Glenn *Bot. Explorers of N.Z.* 1950, 128–34. *R.S.C.* v.8, 1110; v.11, 635; v.19, 191.

Plants and letters at Kew.

Traversia Hook. f.

TRAVIS, William Gladstone (1877–1958)

b. Liverpool, Lancs 29 April 1877 *d.* 29 March 1958

Employed by firm of patent agents. Secretary of South Lancashire Flora Committee, 1906–57. *Travis's Fl. South Lancashire* 1963 was dedicated to him in acknowledgement of his work on the flora. Contrib. to *J. Bot.*

Proc. Bot. Soc. Br. Isl. 1960, 470–73.

Herb. at Liverpool Museums.

TREEN, W. H. (*fl.* 1860s)

Nurseryman, Victoria Nursery, Rugby, Warwickshire.

TREGELLES, George Fox (1859–1943)

b. Tottenham, Middx March 1859 *d.* Barnstaple, Devon 25 July 1943

Bank clerk, Truro, Falmouth, Penzance and Barnstaple until 1920. Secretary, Penzance Natural History and Antiquarian Society. Contrib. articles on algae of Devon and of Lundy Island to *Trans. Devon Assoc.* 1931, 1932, 1937.

Trans. Devon Assoc. 1943, 22–23.

Plants formerly at Ilfracombe Museum. MS. list of Devon algae at N. Devon Athenaeum.

TRELAWNEY, Henry (*fl.* 1700s)

Of Butshead, Plymouth, Devon. Received seeds from Peter Collinson. Particularly successful in raising pines.

N. G. Brett-James *Life of Peter Collinson* 1926, 113.

TRENCH, Helena *see* Lefroy, H.

TRESEDER, William (*c.* 1829–1893)
d. Cardiff, Glam 22 March 1893
Nurseryman, Cardiff.
Gdnrs Chron. 1893 i 397.

TREUTLER, William John (1841–1915)
b. Dinapore, India 23 Oct. 1841 *d.* Hove, Sussex 20 March 1915
MD Edinburgh. FLS 1868. At Kew, 1869–74. Physician at Fletching, Sussex, 1875 and Hove, 1890. Collected in Sikkim, 1874.
Proc. Linn. Soc. 1914–15, 33. *S.E. Nat.* 1915, xlvi–xlvii portr.
Herb. and MS. catalogue of Sikkim plants at Kew.
Treutlera Hook. f.

TREVELYAN, Sir Walter Calverley, 6th Baronet (1797–1879)
b. Newcastle-upon-Tyne, Northumberland 31 March 1797 *d.* Wallington, Northumberland 23 March 1879
BA Oxon 1820. FRSE 1822. Discovered *Romulea parviflora*. *On Vegetation and Temperature of Faroe Islands* 1837. Contrib. to J. Sowerby and J. E. Smith's *English Bot.* 2798 and *Mag. Nat. Hist.* 1830.
Gdnrs Chron. 1879 i 412. *J. Bot.* 1879, 160. *Proc. R. Soc. Edinburgh* 1879–80, 354–56. *Trans. Bot. Soc. Edinburgh* v.14, 1883, 8–11. *D.N.B.* v.57, 210. *Nat. Hist. Trans. Northumberland, Durham and Newcastle-upon-Tyne* v.14, 1903, 84.
Plants at Hancock Museum Newcastle, Edinburgh, Kew, Glasgow. Letters at Kew, Linnean Society.
Diderma trevelyani Fries.

TREVITHICK, W. E. (1858–1929)
b. Helston, Cornwall 1858 *d.* Kells, County Meath 10 Feb. 1929
Gardener to Lord Headfort at Headfort House, County Meath, 1912.
Gdnrs Chron. 1929 i 151.

TREVITHICK, William Edward (1900–1958)
b. Kells, County Meath 1900 *d.* Ruislip, Middx 1958
Kew gardener, 1920. Assistant, Kew Herb., 1923. Contrib. drawings to *Fl. West Tropical Africa* 1927–29 and J. Hutchinson's *Families of Flowering Plants* 1926, 1934. Left Kew and became commercial advertiser.
J. Kew Guild 1958, 592 portr. Hunt Library *Artists from R. Bot. Gdns Kew* 1974, 64–65 portr.

TREVOR, Sir Charles Gerald (1882–1959)
b. 28 Dec. 1882 *d.* 20 May 1959
Assistant Conservator, Indian Forest Service, Punjab, 1908. Conservator, United Provinces, 1920. Professor of Forestry, Dehra Dun, 1926. Chief Conservator of Forests, Punjab, 1930–33. Inspector-General of Forests, 1933–37. *Revised Working Plan for Kulu Forests* 1920. *Manual of Indian Silviculture* (with H. G. Champion) 1938.
Who was Who, 1951–1960 1099–1100.

TREVOR-BATTYE, Aubyn Bernard Rochfort (1855–1922)
b. 17 July 1855 *d.* Las Palmas, Canary Islands 20 Dec. 1922
Educ. Oxford. Of Ashford Chace, Petersfield. Explored Arctic Russia, N.W. America, Africa, Nepal, etc. Collected plants on Kolguev Island, Barents Sea, 1894. *Icebound on Kolguev* 1895. *Camping in Crete* 1909.
Times 22 Dec. 1922. *Geogr. J.* 1923, 229–30.

TREVOR-JONES, Henry Tudor (1855–1901)
Collected plants while a pupil at Uppingham School, Rutland. Plant lists published in *Uppingham School Mag.* 1873–74.
A. R. Horwood and C. W. F. Noel *Fl. Leicestershire* 1933, cclxxviii.

TRIMEN, Henry (1843–1896)
b. Paddington, London 26 Oct. 1843 *d.* Peradeniya, Ceylon 16 Oct. 1896
MB London 1865. FRS 1888. FLS 1866. Assistant, Botany Dept., BM, 1869–79. Director, Botanic Gardens, Peradeniya, 1879. *Fl. Middlesex* (with W. T. T. Dyer) 1869. *Medicinal Plants* (with R. Bentley) 1875–80 4 vols. *Hortus Zeylanicus* 1888. *Handbook to Fl. Ceylon* 1893–1900 (biogr. in vol. 5, 1900, 380–81).
Nature v.38, 1888, 12–13; v.152, 1943, 470. *J. Bot.* 1896, 489–94 portr. *Kew Bull.* 1896, 147–48, 219–20. *D.N.B.* v.57, 230–31. *Ann. R. Bot. Gdns Peradeniya* v.1, 1901, 10–12. *Tropical Agriculturist* 1900, 1–6 portr. *Proc. R. Soc.* v.75, 1904, 161–65. *Curtis's Bot. Mag. Dedications, 1827–1927* 259–60 portr. *R.S.C.* v.6, 40; v.8, 1115; v.11, 644; v.19, 203.
British plants at BM(NH). Ceylon plants at Kew. Letters at Kew. Portr. at Hunt Library.
Trimenia Seem.

TRIMEN, Roland (1840–1916)
b. London 29 Oct. 1840 *d.* Epsom, Surrey 25 July 1916
FRS 1883. FLS 1871. Brother of Henry Trimen. Entomologist. To Cape, 1860. Curator, South African Museum, 1873–95. Correspondent of C. Darwin. Studied fertilisation of orchids (*Disa* and *Bonatea* in *J. Linn. Soc.* v.7, 1863, 144–47; v.9, 1865, 156–60).
J. Bot. 1873, 354; 1916, 279. *Proc. Linn. Soc.* 1916–17, 76–78. *R.S.C.* v.6, 40; v.8, 1116; v.11, 645; v.12, 740; v.19, 204.
Cape plants at BM(NH).
Melianthus trimenianus Hook. f.

TRIMMER, Rev. Kirby (1804–1887)
b. Poplar, London 22 Dec. 1804 *d.* Norwich, Norfolk 9 Oct. 1887
BA Oxon 1828. Vicar, St. George Tombland, Norwich, 1842. *Fl. Norfolk* 1866; Supplt 1884.
Ann. Bot. 1887–88, 412. *J. Bot.* 1887, 383–84.
Mentha at BM(NH), Oxford. Herb. acquired by H. D. Geldart (1831–1902).

TRIMMER, Sarah (*née* Kirby) (1741–1810)
b. Ipswich, Suffolk 1741 *d.* Brentford, Middx 15 Dec. 1810
Aunt of William Kirby. *Easy Introduction to the Knowledge of Nature* 1782.
Gent. Mag. 1811 i 86. C. L. Balfour *Sketch of Mrs. Trimmer* 1854. *D.N.B.* v.57, 231–32.
Portr. at National Portrait Gallery.

TRISTRAM, Rev. Henry Baker (1822–1906)
b. Eglingham, Northumberland 11 May 1822 *d.* Durham 8 March 1906
BA Oxon 1844. DD Durham 1882. LLD Edinburgh 1868. FRS 1868. FLS 1857. Rector, Castle Eden, 1849–60. Master, Greatham Hospital, 1860–74. Canon of Durham, 1870. President, Tyneside Naturalists' Club. Travelled in Palestine and Egypt. '*Cyperus papyrus*' (*J. Linn. Soc.* v.9, 1866, 329–30). *Natural History of the Bible* 1867; ed. 3 1873. *Fauna and Fl. Palestine* 1884.
J. Bot. 1906, 144. *Proc. R. Soc.* v.80, 1908, xlii–xliv. *D.N.B. Supplt 2* v.3, 535–36. *Who was Who, 1897–1916* 715–16. *R.S.C.* v.6, 44; v.8, 12; v.11, 647.
Palestine plants at Cambridge. Letters at Kew.

TRISTRAM, Ruth Mary (*née* Cardew) (1886–1950)
b. 25 April 1886 *d.* 22 Oct. 1950
FLS 1911. Amateur botanist. Studied *Plantago*.
Watsonia 1951–53, 139.

TROTT, Henry William (1857–)
b. 17 Jan. 1857
Member of Rugby School Natural History Society. Edited *Register of Plants found within Ten Miles of Rugby* (with L. Cumming) 1876.
J. E. Bagnall *Fl. Warwickshire* 1891, 503.

TROTTER, E. W. (*fl.* 1880s)
Director, Punjab Post and Telegraph Dept. Collected ferns and grasses, 1885–90.
Pakistan J. For. 1967, 358.
Plants at Gordon College Rawalpindi, Kew.

TROTTER, James (*fl.* 1780s–1790s)
Seedsman and fruiterer, Newcastle-upon-Tyne, Northumberland.

TROTTER, Leslie Batten Currie (1882–1964)
b. Coleford, Glos 4 Aug. 1882 *d.* 2 Oct. 1964
MB. DCL. MD. FLS 1953. Physician, Ledbury, 1913–39. President, British Bryological Society, 1956. Did histological illustrations for *Census Catalogue of British Mosses*.
Proc. Linn. Soc. 1965, 225–26. *Trans. Br. Bryol. Soc.* 1965, 836–37 portr.
Mosses at National Museum of Wales. Drawings at Linnean Society.

TROTTER, Richard Durant (1887–1968)
d. 20 March 1968
Educ. Cambridge. VMH 1952. Chairman, Alliance Assurance Co. Friend of E. A. Bowles. Treasurer, Royal Horticultural Society, 1929–38, 1943–48.
J. R. Hort. Soc. 1968, 276–77. *Who was Who, 1961–1970* 1132.

TROUP, Robert Scott (1874–1939)
b. Banbury, Oxfordshire 13 Dec. 1874 *d.* Oxford 1 Oct. 1939
BSc Aberdeen. FRS 1926. FLS 1923. Indian Forest Service, Burma, 1897. Economist, Forest Research Institute, Dehra Dun, 1905. Silviculturalist, 1909–15. Assistant Inspector-General of Forests, 1915–19. Professor of Forestry, Oxford, 1920–39. *Silviculture of Indian Trees* 1921 3 vols. *Silvicultural Systems* 1928. *Exotic Forest Trees in British Empire* 1932. *Forests and State Control* 1938.
Empire For. J. 1939, 187–89. *Kew Bull.* 1939, 521–22. *Nature* v.144, 1939, 699–700. *Times* 3 Oct. 1939. *Proc. Linn. Soc.* 1939–40, 378–81. *Chronica Botanica* 1940–41, 187–88 portr. *Quart. J. For.* 1940, 7–8. *Who was Who, 1929–1940* 1366. *Obit. Notices Fellows R. Soc.* v.3, 1940, 217–19 portr.

TROW, Albert Howard (1863–1939)
b. Newtown, Montgomeryshire 28 Feb. 1863 *d.* Penarth, Glam 26 Aug. 1939
DSc London. FLS 1898. Professor of Botany, University College, Cardiff, 1892–1919; Principal, 1919–29. President, Cardiff Naturalists Society, 1908, 1920. Edited *Fl. Glamorgan* 1911. Contrib. to *Ann. Bot.*, *J. Genetics*.
Proc. Linn. Soc. 1939–40, 381–82. *Trans. Cardiff Nat. Soc.* 1939, 9–10. *Who was Who, 1929–1940* 1366.
Herb. and portr. at University College, Cardiff.

TROWE, Gilbert (*c.* 1685–1734)
b. Abingdon, Berks *c.* 1685 *d.* Oxford ? 1734
BA Oxon 1704. MD Oxon 1723. Professor of Botany, Oxford, 1724–28.

TROWELL, Samuel (–*c.* 1747)
Steward of Estates of Benchers of Inner

Temple. "A very ingenious gentleman and celebrated garden artist" (G. W. Johnson). *New Treatise of Husbandry, Gardening and other Curious Matters*...1738.

G. W. Johnson *Hist. English Gdning* 1829, 202. J. Donaldson *Agric. Biogr.* 1854, 53.

TROWER, Charlotte Georgiana (1855–1928)
b. Stanstead Bury, Herts 1855 *d.* Stanstead Bury 8 Nov. 1928

Botanist and botanical artist. Illustrated *British Brambles* (*Bot. Soc. Exch. Club Br. Isl. Rep.* 1928) and M. Skene's *Flower Book for the Pocket* 1935. Her sister, Alice (1853–1929), collected plants for her to paint.

Bot. Soc. Exch. Club Br. Isl. Rep. 1928, 851–55 portr.; 1929, 98–99. *J. Bot.* 1929, 22.

Drawings of British plants at Oxford.

TRUELOVE, William (–1894)
d. Brixton, London 16 Jan. 1894

Foreman of Arboretum, Kew; retired 1892.

Kew Bull. 1892, 185–86; 1894, 74. *J. Hort. Cottage Gdnr* v.28, 1894, 70.

TRUMP, C. (*fl.* 1900s–1910s)

Pharmacist. Son of T. W. Trump with whom he collected Gloucestershire plants.

H. J. Riddelsdell *Fl. Gloucestershire* 1948, cxlvii.

Herb. formerly at Cheltenham College. Plants acquired by Rev. W. Butt (1850–1917).

TRUMP, T. W. (*fl.* 1900s–1910s)

Bandmaster to Cheltenham College Cadet Corps Band. Collected Gloucestershire plants with his son.

H. J. Riddelsdell *Fl. Gloucestershire* 1948, cxlvii.

Herb. formerly at Cheltenham College. Plants acquired by Rev. W. Butt (1850–1917).

TRUSLER, Rev. John (1735–1820)
b. London July 1735 *d.* Bathwick, Somerset 1820

BA Cantab 1757. Curate, printer, bookseller. *Elements of Modern Gardening* 1784. *Art of Gardening* 1793. *Lady's Gardener's Companion* 1816.

J. C. Loudon *Encyclop. Gdning* 1822, 1820. *D.N.B.* v.57, 268–69.

TRUSTED, George (–1837)
b. Herefordshire *d.* 14 May 1837

Showed "zeal in his contributions to our Herbarium" (*Annual Rep. Bot. Soc. Edinburgh* 1837–38, 39).

TRYON, Henry (1856–1943)
b. Buckfastleigh, Devon 20 Dec. 1856 *d.* Brisbane, Queensland 15 Nov. 1943

Medical student, London Hospital. Farmed in New Zealand. Assistant, Queensland Museum, 1883; Assistant Curator. Government Entomologist, Queensland, 1894; Government Pathologist, 1901–29. President, Natural History Society of Queensland. Collected sugar-cane varieties in Papua, 1895.

Proc. R. Soc. Queensland 1890–91, 38; 1945, 77–80. *Queensland Agric. J.* 1929, 178–83. *Victorian Nat.* 1943, 128. *Fl. Malesiana* v.1, 1950, 533. *Austral. Encyclop.* v.9, 1965, 54.

Plants in Queensland Herb.

Bryum tryonii Broth.

TUBB, William (*fl.* 1800s–1820s)

Nurseryman, King's Road, Chelsea, London.

TUCK, James H. (*fl.* 1830s–1840s)

Had nursery in Eaton Square, Pimlico, London; later moved to Sloane Street.

J. Hort. Cottage Gdnr v.56, 1876, 145.

TUCKER, E. (–1868)
d. Margate, Kent 9 March 1868

Gardener, Margate. Discovered grape mildew, *Oidium tuckeri.*

Gdnrs Chron. 1868, 298.

TUCKER, J. W. (*c.* 1876–1951)

Printer, Exeter. Cultivated ferns. 'Variegated Hartstongues' (*Br. Fern Gaz.* 1911, 160–61).

Br. Fern Gaz. 1952, 7.

Scolopendrium ramocristatum tuckeri Druery.

TUCKER, John (*fl.* 1790s)

Nurseryman, Wells, Somerset.

TUCKER, Robert (1832–1905)
b. Walworth, London 26 April 1832 *d.* Worthing, Sussex 29 Jan. 1905

MA Cantab. Mathematician. Contrib. plant records to H. Trimen and W. T. T. Dyer's *Fl. Middlesex* 1869 and F. Townsend's *Fl. Hampshire* 1883. Isle of Wight plants in *J. Bot.* 1870–74 and A. G. More's 'Supplement to "Fl. Vectensis"' 1871.

J. Bot. 1905, 168. F. H. Davey *Fl. Cornwall* 1909, lvi. *R.S.C.* v.8, 1126.

TUCKWELL, Rev. William (1829–1919)
b. Oxford 27 Nov. 1829 *d.* Pyrford, Surrey 1 Feb. 1919

MA Oxon 1852. Headmaster, Taunton School, 1864–77. Rector, Stockton, Warwickshire, 1878–93; Waltham, Lincs, 1893–1905. Contrib. to R. P. Murray's *Fl. Somerset* 1896. *Tongues in Trees* 1891. Contrib. to *Gdnrs Chron.* as "Corycius Senex". 'Fl. Quantocks' (*Nature* v.7, 1872, 48–49).

T. F. Kirby *Winchester Scholars* 1888, 319. *Bot. Soc. Exch. Club Br. Isl. Rep.* 1919, 628–29.

TUDOR, Richard (1798–1886)
b. Kirby Stephen, Westmorland 1798 *d.* Bootle, Lancs 26 July 1886
MRCS. To Liverpool, 1811. Spent some time in Greenland. Practised medicine at Shrewsbury and Bootle. Contrib. to T. B. Hall's *Fl. Liverpool* 1839.
Bootle Times 7 Nov. 1885. *Trans. Liverpool Bot. Soc.* 1909, 94.

TUFNAIL, Frank (1861–1899)
b. Reading, Berks 18 Feb. 1861 *d.* Reading 3 June 1899
FLS 1897. Seed-grower. Contrib. to G. C. Druce's *Fl. Berkshire* 1897 (clxxxiii–clxxxv).
Proc. Linn. Soc. 1899–1900, 83.
Herb. at Reading University.

TUFTON, Henry Sackville Thanet, 3rd Baron Hothfield (*c.* 1897–1961)
d. Englefield Green, Egham, Surrey 20 Aug. 1961
Grew and judged cymbidiums.
Orchid Rev. 1961, 296–97.

TUGGY, Ralph (–1632/3)
d. March 1632/3
Florist, Westminster, London. His garden was famous for its pinks, carnations and auriculas. Friend of T. Johnson (1597–1644).
E. Cecil *Hist. Gdning in England* 1910, 150.

TUGWELL, James (*fl.* 1820s)
Nurseryman, East Parade, Horsham, Sussex.

TULK, John Augustus (1814–1896)
b. Middx 1814 *d.* Chertsey, Surrey 12 March 1896
MA Cantab 1847. LRCP 1849. FRMS 1877. Diatomist. 'On Preparing Diatoms' (*Trans. Microsc. Soc.* 1863, 4–8).
Diatoms at BM(NH).
Rutilaria tulkii Castr.

TULL, Jethro (1674–1740)
b. Basildon, Berks 1674
Barrister, Gray's Inn, 1699; bencher, 1724. Farmed at Hungerford, Berks. Agricultural writer. Advocate of theory that earth was the chief factor in nutrition of crops. *Horse-hoing Husbandry* 1731.
D.N.B. v.57, 304–6. *Plant Physiol.* 1941, 223–26. *Chronica Botanica* 1944, 92–96.

TULLIDEPH, Walter (–1794)
Amanuensis to James Douglas. Planter and medical practitioner in Antigua from 1727. Returned to Great Britain in 1758 and settled in Dundee. Member of John Martyn's Botanical Society of London.
G. C. Gorham *Mem. of J. and T. Martyn* 1830, 19–20. V. L. Oliver *Hist. of Island of Antigua* v.3, 1899, 156–62, 408–10. E. J. L. Scott *Index to Sloane Manuscripts* 1904, 538. *Proc. Bot. Soc. Br. Isl.* 1967, 312–13.

TUNSTAL, Mrs. Thomazin (*fl.* 1620s)
Of Bull-banke near Hornby Castle, Yorks. Sent many plants to J. Parkinson and botanised at Ingleborough, Yorks.
J. Parkinson *Paradisi in Sole* 1629, 348; *Theatrum Botanicum* 1640, 286. R. Pulteney *Hist. Biogr. Sketches of Progress of Bot. in England* v.1, 1790, 154. *Trans. Liverpool Bot. Soc.* 1909, 94.

TUOMEY, Michael (1805–1857)
b. Cork 29 Sept. 1805 *d.* Tuscaloosa, Ala., U.S.A. 30 March 1857
Collected marine algae on Florida Keys, *c.* 1852.
Appleton's Cyclop. Amer. Biogr. v.6, 1889, 180. *J. Elisha Mitchell Soc.* v.7, 1891, 98–103, 115–16. *Rep. U.S. National Mus.* 1904, 387 portr., 713.
Portr. at Hunt Library.

TUPPER, James Perchard (*fl.* 1790s–1831)
d. 1831
MD. FLS 1797. Of London. Pupil of J. E. Smith. To Paris, 1817 (Surgeon Extraordinary to Prince Regent). *Essay on Probability of Sensation in Vegetables* 1811; ed. 2 1817.

TURNBULL, Andrew (1804–1886)
b. Legerwick, Berwickshire 18 Jan. 1804 *d.* Bothwell Castle, Lanarkshire 18 April 1886
Gardener at Bothwell Castle, 1828.
Gdnrs Chron. 1874 ii 329 portr.; 1880 i 179; 1886 i 603. *Garden* v.29, 1886, 412. *J. Hort. Cottage Gdnr* v.12, 1886, 318–19.

TURNBULL, Archibald (*c.* 1790–1875)
b. Hawick, Roxburghshire *d.* 19 Jan. 1875
Partner with his uncle, William Dickson, in a nursery at Perth. Founder member of Perthshire Horticultural Society.
Florist and Pomologist 1875, 48. *Gdnrs Yb. and Almanack* 1876, 170.

TURNBULL, Robert (*fl.* 1810s–1820s)
Nurseryman, Daventry, Northants.

TURNBULL, Robert (*c.* 1813–1891)
b. Knaresborough, Yorks *c.* 1813 *d.* Scarborough, Yorks 19 Jan. 1891
Taught geology by William Smith. *Index of British Plants* 1889.
J. Bot. 1890, 28–29.

TURNER, Miss (*fl.* 1820s)
Collected plants in Sierra Leone, 1826. Most of her collection lost but some plants at Kew.

TURNER, Arthur (-1930)
d. Dec. 1930

Son of C. Turner (1818–1885). Nurseryman, Royal Nurseries, Slough, Bucks.

Gdnrs Mag. 1894, 397 portr. *J. Hort. Cottage Gdnr* v.45, 1902, 474, 475 portr. *Gdnrs Chron.* 1921 i 270 portr.; 1931 i 19 portr.

TURNER, Charles (1818–1885)
b. Wilton, Wilts 3 May 1818 *d.* Slough, Bucks 9 May 1885

Acquired Royal Nurseries, Slough, from W. Cutter *c.* 1845. *Culture of the Pansy* 1850. Edited *Florist*, 1851–60. Promoted 'Cox's Orange Pippin' apple.

Gdnrs Chron. 1883 ii 134–35, 145 portr.; 1885 i 643. *Garden* v.25, 1884, xi–xii portr.; v.27, 1885, 458. *J. Hort. Cottage Gdnr* v.10, 1885, 401 portr.; v.34, 1897, 577.

TURNER, Charles (1864–1926)
b. Hingham, Norfolk 1864 *d.* Wilmslow, Cheshire 10 Sept. 1926

FLS 1922. Lecturer, Manchester School of Pharmacy, 1890; Principal, 1895–1912. Vice-President, Manchester Microscopical Society, 1899–1914. Authority on freshwater algae and desmids.

J. Bot. 1926, 288. *Pharm. J.* v.63, 1926, 388. *Proc. Linn. Soc.* 1926–27, 104–5. *N. Western Nat.* 1927, 31–32.

TURNER, Charles Hampden
(*fl.* 1810s–1820s)

Of Rook's Nest, Surrey. Introduced from China varieties of camellia through Capt. R. Wellbank in East-Indiaman 'Cuffnels', 1810–20.

E. Bretschneider *Hist. European Bot. Discoveries in China* 1898, 218.

TURNER, Dawson (1775–1858)
b. Great Yarmouth, Norfolk 18 Oct. 1775 *d.* London 20 June 1858

FRS 1802. FLS 1797. Banker, Yarmouth. *Synopsis of British Fuci* 1802 2 vols. *Muscologiae Hibernicae Spicilegium* 1804. *Fuci* 1808–19 4 vols. *Specimen of Lichenographia Britannica* (with W. Borrer) 1839. *Botanist's Guide through England and Wales* (with L. W. Dillwyn) 1805 2 vols. Contrib. to J. Sowerby and J. E. Smith's *English Bot.*

Trans. Linn. Soc. v.10, 1811, 318–19. *Proc. Linn. Soc.* 1858, xl–xliv. *Trans. Norfolk Norwich Nat. Soc.* 1894–95, 76–77; 1920–21, 179–93. *D.N.B.* v.57, 334–35. *J. Bot.* 1902, 320–21; 1912, 64. H. Turner *Turner Family of Mulbarton and Yarmouth*, rev. by F. Johnson 1907. F. H. Davey *Fl. Cornwall* 1909, xxxvii. G. C. Druce *Fl. Buckinghamshire* 1926, xcvii. *Original Papers of Norfolk Norwich Archaeol. Soc.* v.26, 1936, 59–72. *J. Soc. Bibl. Nat. Hist.* 1950, 218–22; 1958, 303–10 portr. *Trans. Cambridge Bibl. Soc.* 1961, 232–56. A. N. L. Munby *Cult of the Autograph Letter in England* 1962, 33–60 portr. F. A. Stafleu *Taxonomic Literature* 1967, 477. *R.S.C.* v.6, 67.

Herb., MS. *Fl. Norfolciensis* (incomplete), 2 vols of botanical memoranda, drawings of algae, and photostats of correspondence with W. Borrer at Kew. 82 vols of letters at Trinity College, Cambridge. Letters also at BM, Linnean Society, Liverpool and Norwich Public Libraries. Portr. at Hunt Library. Library sold at Sotheby, 7 March 1853; Puttick and Simpson, 16–23 May, 6 June 1859.

Dawsonia R. Br.

TURNER, Edward Phillips (1865–1937)
b. Havant, Hants 1865 *d.* Hamilton, New Zealand 20 May 1937

FLS 1935. To New Zealand with his parents, 1870. Director of Forestry, N.Z. Collected plants which he sent to T. F. Cheeseman and L. Cockayne. *Trees of New Zealand* (with L. Cockayne) 1928.

Proc. Linn. Soc. 1937–38, 342–43.

Pittosporum turneri Petrie.

TURNER, Frederick (1852–1939)
b. Pontefract, Yorks 17 April 1852 *d.* Sydney, N.S.W. 17 Oct. 1939

FLS 1892. Curator, Queensland Acclimatisation Society's gardens, 1874–90. Economic botanist, Dept. of Agriculture, N.S.W., 1890. Consultant botanist to Government of W. Australia. *Census of Grasses of N.S.W.* 1890. *Forage Plants of Australia* 1891. *Australian Grasses* 1895. *Australian Grasses and Pasture Plants* 1921. *Botanical Surveys of New England, the Darling and S. West and N. West N.S.W.* 1914.

F. Turner *Australian Grasses* 1895, 3–6. J. A. Alexander *Who's Who in Australia* 1938, 500–1. *Proc. Linn. Soc.* 1939–40, 382. *Sydney Morning Herald* 19 Oct. 1939. *Who was Who, 1929–1940* 1371–72.

TURNER, George Creswell (1858–1940)
b. Leicester 13 Aug. 1858 *d.* Leicester 17 Oct. 1940

FLS 1897. Leicester businessman. President, Leicester Literary and Philosophical Society, 1915–16; Chairman, Botanical Section, 1910–40. Field botanist. Helped finance publication of A. R. Horwood and C. W. F. Noel's *Fl. Leicestershire* 1933.

Proc. Linn. Soc. 1940–41, 301–2.

TURNER, Rev. George Edward (1810–1869)
b. Corsham, Wilts 1810 *d.* Ryde near Sydney, N.S.W. 10 Jan. 1869

Curate, Monkton Farleigh and South Wraxhall, Wilts. Chaplain, Tasmania, *c.* 1838. Hon. Secretary, Sydney Botanic Gardens Committee. Interested in horticulture and botany, particularly microscopy.

J. Proc. R. Soc. N.S.W. 1908, 126.

TURNER, Harry (*c.* 1848–1906)
d. Langley, Bucks 14 Sept. 1906

VMH 1897. Son of C. Turner (1818–1885). Nurseryman, Royal Nurseries, Slough, Bucks.

Gdnrs Chron. 1906 ii 218 portr. *Gdnrs Mag.* 1906, 637 portr. *J. Hort. Cottage Gdnr* v.53, 1906, 278.

TURNER, Henry (*c.* 1810–1876)
d. Bury St. Edmunds, Suffolk 22 Oct. 1876

Curator, Botanic Garden, Bury St. Edmunds until 1857.

Gdnrs Chron. 1876 ii 630. *Gdnrs Yb. and Almanack* 1877, 176.

TURNER, James (1786–1820)
b. Great Yarmouth, Norfolk 17 July 1786
d. London 2 Jan. 1820

FLS 1806. Brother of D. Turner (1775–1858). Banker, Halesworth, Norfolk. Lichenologist. Discovered *Lepidium draba*. Contrib. to J. Sowerby and J. E. Smith's *English Bot.* 1499–1501, 1892.

W. J. Hooker *Br. Fl.* 1830, 297.

TURNER, John (*fl.* 1690s–1730s)

Nurseryman, The Orange Tree, Strand, London.

J. Harvey *Early Gdning Cat.* 1972, 28.

TURNER, John (*fl.* 1800s–1820s)

Nurseryman, 99 New Bond Street, London.

TURNER, John (*fl.* 1820s)

FLS 1821–26. Assistant Secretary, Horticultural Society of London, 1818; dismissed for embezzlement in 1826. 'Some Account of *Ipomoea tuberosa*' (*Trans. Hort. Soc. London* v.1, 1810, 184–86).

TURNER, Magdalene (*fl.* 1840s–1880s)

Of Oxford. Sent Jersey algae to W. H. Harvey for *Phycologia Britannica* 1846–51 (t.315).

Cladophora magdalenae Harvey.

TURNER, Maria *see* Hooker, M.

TURNER, Mary (*née* **Palgrave**) (1774–1850)
b. 16 Jan. 1774 *d.* Great Yarmouth, Norfolk 17 March 1850

Married Dawson Turner (1775–1858) for whose *Fuci* she drew and engraved plates. Her daughter Maria (1797–1872) collected mosses, and, together with her sister, Elizabeth (1799–1852), drew and engraved them for W. J. Hooker.

J. Bot. 1912, 64.

TURNER, Peter (*c.* 1542–1614)
d. 27 May 1614

Doctor. MP for Bridport, Dorset. Friend of Thomas Penny. J. Bauhin records that he was presented with a hortus siccus of English plants by P. Turner (*Historia Plantarum Universalis* v.1, 1650, 225).

C. E. Raven *English Nat.* 1947, 165, 171.

TURNER, R. Lister (*fl.* 1875–1930s)
b. Apia, Samoa 1875

MA Glasgow 1900. Missionary, Papua, 1902–39. Collected plants in New Guinea.

Fl. Malesiana v.1, 1950, 534.

Plants at BM(NH), Brisbane.

Jasminum turneri C. T. White.

TURNER, Robert (*fl.* 1626–1680s)
b. Reading, Berks 30 July 1626

Of Holdshott, Hants; afterwards of Wokingham, Berks and London. "Astrological botanist." *Botanologia: the British Physician* 1664; reissued 1687 portr.

R. Pulteney *Hist. Biogr. Sketches of Progress of Bot. in England* v.1, 1790, 180. *D.N.B.* v.57, 354. R. T. Gunther *Early Br. Botanists* 1922, 234. G. C. Druce *Fl. Buckinghamshire* 1926, lxxii. A. H. Wolley-Dod *Fl. Sussex* 1937, xxxvi.

TURNER, Robert (*fl.* 1830s)

Nurseryman, Sheffield, Yorks.

TURNER, Robert (1848–1894)
b. Strathaven, Glasgow 29 Dec. 1848 *d.* Glasgow 20 March 1894

Member of Cryptogamic Society of Scotland. Had herb.

Proc. Trans. Nat. Hist. Soc. Glasgow v.4, 1892–94, 73–78.

TURNER, Spencer (*c.* 1728–1776)
d. 7 Jan. 1776

Nurseryman, Holloway Down Nursery, Langthorne Road, Leyton, Essex. His name is commemorated by the hybrid semi-evergreen oak raised in the nursery, *Quercus* × *Turneri*. Turner's will offered the nursery at a valuation to his servant William Perkins, who carried it on to 1825.

Essex Nat. 1940, 45–48. *Trans. London Middlesex Archaeol. Soc.* v.24, 1973, 190.

TURNER, Thomas Pike (1843–1908)
b. Andover, Hants 1843 *d.* Hammersmith, London 19 March 1908

Nurseryman, King Street and Bridge Road, Hammersmith. Contrib. to *Gdnrs Chron.*

Gdnrs Chron. 1908 i 223.

TURNER, Rev. W. Y. (*fl.* 1870s)
Missionary doctor. In Port Moresby, New Guinea, 1876–77. Possibly the plant collector cited by F. von Mueller.
Fl. Malesiana v.1, 1950, 534.
Plants at Melbourne.

TURNER, Rev. William (*c.* 1508–1568)
b. Morpeth, Northumberland *c.* 1508 *d.* London 7 July 1568
BA Cantab 1529–30. MD Oxon. Physician to Duke of Somerset, Syon, Middx. Dean of Wells, 1550. "Father of English Botany." Had gardens at Kew. *Libellus de re Herbaria* 1538 (reprinted 1965; biogr. 3–37). *Names of Herbes* 1548 (reprinted 1882; 1965). *New Herball* 1551–68.
R. Pulteney *Hist. Biogr. Sketches of Progress of Bot. in England* v.1, 1790, 56–76. *Cottage Gdnr* v.7, 1851, 107. H. Trimen and W. T. T. Dyer *Fl. Middlesex* 1869, 364–69. *Notes and Queries* v.5, 1894, 146; v.2, 1916, 507. G. C. Druce *Fl. Berkshire* 1897, xciv–xcvi. *Kew Bull.* 1891, 280–81. *D.N.B.* v.57, 363–66. *Nat. Hist. Trans. Northumberland, Durham and Newcastle-upon-Tyne* v.14, 1903, 69–71. J. W. White *Fl. Bristol* 1912, 46–48. R. T. Gunther *Early Br. Botanists* 1922, 416. G. C. Druce *Fl. Oxfordshire* 1927, lxi–lxiii. *Estates Mag.* 1937, 367–72. *J. Bot.* 1941, 132–33. C. E. Raven *English Nat.* 1947, 48–137. *Proc. Leeds Philos. Soc. Sci. Sect.* 1959, 109–38. S. T. Jermyn *Fl. Essex* 1974, 13. C. C. Gillispie *Dict. Sci. Biogr.* v.13, 1976, 501–3.
Turnera L.

TURNER, William Barwell (1845–1917)
b. Birmingham, Warwickshire 9 June 1845 *d.* Leeds, Yorks 11 May 1917
Consulting brewer and analytical chemist. President, Leeds Naturalists' Club, 1881. Algologist. *Fresh Water Algae of East India* 1892. Contrib. to *Naturalist.*
Acta Horti Bergiani v.3(1), 1897–1903, t.35 portr. *Naturalist* 1917, 202–5 portr.
Algae slides at Leeds University.

TURRILL, William Bertram (1890–1961)
b. Woodstock, Oxfordshire 14 June 1890 *d.* Kew, Surrey 15 Dec. 1961
BSc London 1915; DSc 1928. FRS 1958. FLS 1925. VMH 1956. VMM 1953. Assistant, Kew Herb., 1909; Keeper, 1946–57. Studied flora of Balkans and phytogeography. Collaborated with E. M. Marsden-Jones in genetical, transplant and other experiments at Potterne Biological Station. *Plant Life of Balkan Peninsula* 1929. *British Plant Life* 1948. *Pioneer Plant Geography* 1953. *British Knapweeds* (with E. M. Marsden-Jones) 1954. *Bladder Campions* (with E. M. Marsden-Jones) 1957. *Royal Botanic Gardens, Kew* 1959. *J. D. Hooker* 1963. Edited *Vistas in Botany* 1959–64.
Kew Bull. 1950, 136; 1958, 29–30; 1960, 218–19, 467; 1965, 298. *J. Kew Guild* 1948, 639–40 portr.; 1962, 203–4 portr. *Times* 16 Dec. 1961 portr. *Gdnrs Chron.* 1961 ii 500. *Lily Yb.* 1962, 9–10 portr. *Nature* v.193, 1962, 826–27. *Proc. Bot. Soc. Br. Isl.* 1963, 194–96. *Who was Who, 1961–1970* 1138.
Plants and letters at Kew. Portr. at Hunt Library.

TURTON, T. (*fl.* 1870s–1910s)
Kew gardener, 1872. Gardener at Maiden Erleigh, Reading and at Sherborne Castle, Dorset.
Gdnrs Mag. 1911, 509–10 portr.

TURTON, William (1762–1835)
b. Olveston, Glos 21 May 1762 *d.* Bideford, Devon 28 Dec. 1835
BA Oxon 1785. MB 1791. FLS 1804. Practised medicine at Swansea, Glam. Discovered *Draba aizoides* (J. Sowerby and J. E. Smith *English Bot.* 1271, 1338). Edited Linnaeus's *General System of Nature* 1806. Prepared a pocket flora.
D.N.B. v.57, 377–78.

TUSSER, Thomas (*c.* 1524–1580)
b. Essex *c.* 1524 *d.* London 1580
Farmed at Cattiwade, Suffolk. Agricultural writer and poet. *One Hundred Pointes of Good Husbandrie* 1557. *Five Hundred Pointes of Good Husbandrie* 1573.
G. W. Johnson *Hist. English Gdning* 1829, 49–51. *Cottage Gdnr* v.5, 1851, 285–86. J. Donaldson *Agric. Biogr.* 1854, 7–9. *J. Hort. Cottage Gdnr* v.51, 1874, 330–32, 384–85, 425–26; v.52, 1874, 189. *D.N.B.* v.57, 379–81. D. McDonald *Agric. Writers* 1908, 23–30. *Gdnrs Chron.* 1935 ii 300–1. *Nature* v.120, 1932, 348–49.

TUTCHER, Frederick George (*c.* 1888–1920)
b. Kingsweston, Glos *c.* 1888
Brother of W. J. Tutcher (1867–1920). Kew gardener, 1902–5. Inspector, Horticultural Division, Ministry of Agriculture, 1911. Collected fungi.
Gdnrs Chron. 1920 ii 314. *J. Kew Guild* 1921, 40.

TUTCHER, William (*fl.* 1840s)
Nurseryman, Bridport, Dorset.

TUTCHER, William James (1867–1920)
b. Bristol 22 Nov. 1867 *d.* Hong Kong 5 April 1920
FLS 1904. Kew gardener, 1888. To Hong Kong, 1891; Superintendent, Botanical and Forestry Dept., 1910. *Fl. Kwangtung and Hongkong* (with S. T. Dunn) 1912. *Gardening for Hong Kong* ed. 2 1913. Chinese plants listed in *J. Linn. Soc.* v.37, 1905, 58–70.
Gdnrs Chron. 1912 i 202–4 portr.; 1920 i

208. *Kew Bull.* 1920, 136–38. *Proc. Linn. Soc.* 1919–20, 51–52. *J. Kew Guild* 1921, 39. *Sunyatsenia* v.1, 1933, 157–87.
Tutcheria Dunn.

TUXWORTH, G. (*fl.* 1830s)
Nurseryman, Louth, Lincs.

TWAMLEY, Louisa Anne *see* Meredith, L. A.

TWEDDLE, David (–1875)
b. Dappley Moor, Cumberland
Had a school in Workington, Cumberland. Manager, Carlisle and District Branch Bank from 1865. Assisted W. Dickinson (*c.* 1799–1882) who acquired his herb.
W. Hodgson *Fl. Cumberland* 1898, xxviii–xxix.

TWEEDIE, John (1775–1862)
b. Lanarkshire 10/12 April 1775 *d.* Santa Catalina, Buenos Aires, Argentine 1 April 1862
Gardener at Eglinton Castle, Ayrshire; also at Royal Botanic Garden, Edinburgh. To Buenos Aires, 1825. Corresponded with W. J. Hooker. 'Buenos Ayres' (*Ann. Nat. Hist.* v.4, 1840, 8–15).
Companion Bot. Mag. v.1, 1836, 235–36. *J. Bot.* (Hooker) 1834, 178–79. A. Lasègue *Musée Botanique de B. Delessert* 1845, 486–87. *Gdnrs Chron.* 1862, 597. C. F. P. von Martius *Fl. Brasiliensis* v.1(1), 1906, 123–24. *Holmbergia* 1945, 3–14 portr. A. M. Coats *Quest for Plants* 1969, 357–59. H. R. Fletcher and W. H. Brown *R. Bot. Gdn Edinburgh, 1670–1970* 1970, 70.
Plants at BM(NH), Kew, Oxford. Letters at Kew.
Tweedia Hook.

TWINING, Elizabeth (1805–1889)
d. Twickenham, Middx 24 Dec. 1889
Illustrations of Natural Orders of Plants 1849–55 2 vols; ed. 2 1868.
D.N.B. v.57, 388. *R.S.C.* v.6, 74.
Plant drawings at BM(NH).

TWINING, Thomas (1806–1895)
d. Twickenham, Middx 16 Feb. 1895
Of Twickenham whose flora he studied. Authority on technical education. *Botanic Stand* 1883.
D.N.B. v.57, 388–89. S. H. Twining *Two Hundred and Twenty Five Years in the Strand...Embodying an Account of the Descendants of Thomas Twining* 1931. D. H. Kent *Hist. Fl. Middlesex* 1975, 22.
Plants at Bangor, Kew, Oxford.

TYACKE, Nicholas (1812–1900)
b. Godolphin, Cornwall 1 Oct. 1812 *d.* Chichester, Sussex 7 May 1900
MD Edinburgh 1836. Physician at Chichester from 1840. Had herb. of Sussex plants, now at Oxford. '*Lamium intermedium*' (*Rep. Bot. Soc. Edinburgh* 1836–37, 33).
J. Sowerby and J. E. Smith *English Bot.* 2914, 2983.
A. H. Wolley-Dod *Fl. Sussex* 1937, xlii.

TYAS, Rev. Robert (1811–1879)
b. 4 Nov. 1811 *d.* East Tilbury, Essex 11 April 1879
BA Cantab 1848. Vicar, East Tilbury, 1872. *Popular Flowers* 1844. *Choice Garden Flowers* 1848. *Favourite Field Flowers* 1848–50. *Flowers from the Holy Land* 1851. *Sentiment of Flowers* 1853. *Language of Flowers* 1869.
F. Boase *Modern English Biogr.* v.3, 1901, 1058–59. *Gdnrs Chron.* 1935 ii 178–79.

TYERMAN, John Simpson (*c.* 1830–1889)
d. Tregony, Cornwall 24 Nov. 1889
Curator, Liverpool Botanic Garden. Collected and grew ferns. Contrib. to *Gdnrs Chron.*
Gdnrs Chron. 1889 ii 639. *Br. Fern Gaz.* 1923, 53–54.

TYLER, Edward (*fl.* 1830s)
Nurseryman, West Street, Marlow, Bucks.

TYLOR, Alfred (1824–1884)
b. 26 Jan. 1824 *d.* Carshalton, Surrey 31 Dec. 1884
FLS 1849. Brass founder. Geologist. *On Growth of Trees* 1886.
Geol. Mag. 1885, 142–44. *Proc. Geol. Soc.* 1884–85, 42–43. *D.N.B.* v.57, 422. H. F. Jones *Samuel Butler* v.1, 1920, 410–11.

TYSO, Carey (*fl.* 1840s–1850s)
Nurseryman, Wallingford, Berks.

TYSO, Rev. Joseph (*fl.* 1830s–1840s)
Of Wallingford, Berks. *Catalogue of Choice Ranunculuses...grown and sold by Rev. Joseph Tyso* 1833. *The Ranunculus; How to grow it* 1847.
R. Genders *Collecting Antique Plants* 1971, 181.

TYSON, Rev. Michael (1740–1780)
b. Stamford, Lincs 19 Nov. 1740 *d.* Lambourne, Essex 3 May 1780
BA Cantab 1764. BD 1775. FRS 1779. Rector, Lambourne, 1778. Friend of Israel Lyons and Gray.
J. Nichols *Lit. Anecdotes of Eighteenth Century* v.8, 204–10. *D.N.B.* v.57, 449–50. A. R. Horwood and C. W. F. Noel *Fl. Leicestershire* 1933, cclxxii–cclxxiii.

TYSON, William (1851–1920)
b. Port Royal, Jamaica March 1851 *d.* Port Alfred, Grahamstown, S. Africa 13 April 1920

FLS 1896. To S. Africa in early life. Teacher in Cape Colony. In Forest Dept., Kimberley, 1888–93; in Agricultural Dept., 1893–1904. Collected plants in S. Africa from 1877.

Kew Bull. 1920, 176. *Ann. Bolus Herb.* v.3, 1921, 89–121 portr.

Plants at Bolus Herb., BM(NH), Kew.

Tysonia Bolus.

UDALE, James (1851–1927)
b. Uttoxeter, Staffs 1851 *d.* Droitwich, Worcs 28 Jan. 1927

FLS 1917. Kew gardener, 1875. Gardener to Sir H. Watson at Shirecliffe Hall, Sheffield. Technical instructor in horticulture, Worcs, 1891. Contrib. to *Gdnrs Chron.*

Gdnrs Chron. 1927 i 106. *Proc. Linn. Soc.* 1926–27, 105. *J. Kew Guild* 1928, 625–26 portr.

UGDEN (Ogden), Mr. (*fl.* 1690s)
Surgeon. Plants from Alicante in Spain and elsewhere at BM(NH).

J. Petiver *Museii Petiveriani* 1698, no. 226. J. E. Dandy *Sloane Herb.* 1958, 222.

ULLYETT, Henry (*fl.* 1860s–1880s)
BSc. 'Ferns of High Wycombe' (*Botanists Chron.* 1864, 99). 'Ferns of Wycombe District' (*High Wycombe Nat. Hist. Mag.* v.1, 1866–68, 156). *Rambles of a Naturalist round Folkestone* 1880 (list of plants, 129–38).

R.S.C. v.8, 1135.

UNDERWOOD, John (*fl.* 1780s–1834)
b. Scotland *d.* Aug. 1834

ALS 1797. Superintendent, Botanic Garden, Glasnevin, Dublin, 1798–1833. *Catalogue of Plants...at Glasnevin* 1800–4.

J. White *Grasses of Ireland* 1808, xvi. N. Colgan *Fl. Dublin* 1904, xxv–xxvi. H. F. Berry *Hist. R. Dublin Soc.* 1915, 100.

UNWIN, W. J. (*fl.* 1900s)
Of Histon, Cambridge. Raised varieties of sweet peas. *Sweet Pea and its Cultivation* ed. 10 1936.

Gdnrs Mag. 1910, 565–66 portr.

UNWIN, William Charles (1811–1887)
d. Lewes, Sussex 23 April 1887

Ornithologist and entomologist. *Illustrations of British Mosses* 1878 (plates by Unwin). Plant lists in M. P. Merrifield's *Sketch of Nat. Hist. Brighton* 1864, iv–v. Contrib. to B. R. Morris's *Naturalist* v.3, 1853.

Entomol. Mon. Mag. v.24, 1887, 47.

UPJOHN, Joseph (*c.* 1799–1883)
d. Rondebosch, S. Africa

Collected bulbous plants in S. Africa.

Garden v.24, 1883, 216.

UPJOHN, W. B. (1843–)
b. Cley, Norfolk May 1843

Gardener to Lord Ellesmere at Worsley Hall, Lancs.

Gdnrs Chron. 1923 ii 318 portr.

UPTON, C. (–1927)
Geologist and zoologist. Curator, Gloucester Museum, 1920. Made collection of diatoms in canal between Gloucester and Sharpness.

H. J. Riddelsdell *Fl. Gloucestershire* 1948, cxlix.

URQUHART, Daniel (–1880)
d. Broughty Ferry, Dundee 30 Aug. 1880

Nurseryman, Dundee, Angus.

Gdnrs Chron. 1880 ii 314.

URQUHART, Francis Gregor (1813–*c.* 1890)
b. Craigston, Aberdeenshire 1813

Lieut.-Colonel. Collected plants in N. China and Hong Kong, 1860.

E. Bretschneider *Hist. European Bot. Discoveries in China* 1898, 401.

Ferns at Kew.

URWIN, Joseph (*fl.* 1820s)
Nurseryman, Richmond, Yorks.

USHER, Rev. Robert (–1943)
d. 22 June 1943

MA Cantab. FLS 1897. Held several livings in diocese of Sarum, Wilts and Dorset. All-round naturalist; interested in cryptogamic botany.

Proc. Linn. Soc. 1943–44, 236–37.

UVEDALE, Rev. Robert (1642–1722)
b. Westminster, London 25 May 1642 *d.* Enfield, Middx 17 Aug. 1722

BA Cantab 1662. LLD 1682. Master at Enfield Grammar School *c.* 1670 and proprietor of boarding school. Non-resident Rector, Orpington, Kent, 1696. Correspondent of many contemporary botanists. Had a large garden of exotic plants. "Cl. vir olim condiscipulus noster" (L. Plukenet *Almagestum Botanicum* 1696, 36).

R. Morison *Plantarum Historiae Universalis Oxoniensis* v.3, 1699, 530, 610, 613. R. Pulteney *Hist. Biogr. Sketches of Progress of Bot. in England* v.2, 1790, 30. *Gent. Mag.* 1804 ii 612; 1814 ii 206; 1815 ii 6. *Archaeologia* v.12, 1794, 188. D. Turner *Extracts from Lit. Sci. Correspondence of R. Richardson* 1835, 15–17, 24–25. *J. Bot.* 1891, 9–18. *D.N.B.* v.58, 76–77. *Gdnrs Chron.* 1890 ii 505; 1902 ii 31, 62–63; 1927 ii 181–82. *Geneal. Mag.* 1902, 109–11. E. J. L. Scott *Index to Sloane Manuscripts* 1904, 541. *Notes and Queries* 1916 ii 361–63 etc. J. E. Dandy *Sloane Herb.* 1958, 223, 225.

Herb. at BM(NH).

VACHELL, Charles Tanfield (1848–1914)
b. Cardiff, Glam 1848 *d.* Cardiff 1914
MD. Physician, Cardiff. Secretary and President of Cardiff Naturalists' Society. With his daughter, Eleanor, formed herb. of British plants.
Trans. Cardiff Nat. Soc. 1914, 1–6 portr.; 1917, 17–19. *Bot. Soc. Exch. Club Br. Isl. Rep.* 1915, 253–54.
Herb. at National Museum of Wales.

VACHELL, Eleanor (1879–1948)
b. Cardiff, Glam 1879 *d.* 6 Dec. 1948
FLS 1917. Daughter of C. T. Vachell (1848–1914). With her father was recording secretary of *Fl. Glamorgan* (*Trans. Cardiff Nat. Soc.* 1907–11). President, Cardiff Naturalists' Society, 1936–37. Contrib. weekly article on wild flowers to *Western Mail.*
Proc. Linn. Soc. 1948–49, 252. *Trans. Cardiff Nat. Soc.* 1948, 1. *Watsonia* 1949–50, 325–27.
Herb. at National Museum of Wales.

VACHELL, Rev. George Harvey (1799–)
b. Littleport, Cambridgeshire 1799
BA Cantab 1821. Chaplain at East India Company Factory at Macao, 1828–36.
E. Bretschneider *Hist. European Bot. Discoveries in China* 1898, 294–98. *Fl. Malesiana* v.1, 1950, 536. G. A. C. Herklots *Hong Kong Countryside* 1951, 163.
Plants at Kew, Cambridge, Oxford.
Vachellia Wight & Arn.

VAIR, James (*c.* 1825–1887)
b. Faldonside, Roxburghshire *c.* 1825 *d.* Heathfield 24 Feb. 1887
Gardener to Lady D. Nevill at Dangstein, Sussex.
J. Hort. Cottage Gdnr v.14, 1887, 170, 193.

VAIZEY, John Reynolds (1862–1889)
b. London 10 Sept. 1862 *d.* Cambridge 24 Feb. 1889
BA Cantab 1884. Visited Norway for mosses. Contrib. papers on mosses to *J. Linn. Soc.* 1887 and *Ann. Bot.* 1888–90.

VALENTIA, Lord *see* Annesley, G.

VALENTINE, William (*fl.* 1810s–1884)
d. Tasmania 1884
FLS 1831. Of Nottingham. Microscopist. Friend of J. T. Quekett and J. Lindley. To Tasmania, *c.* 1839? *Muscologia Nottinghamensis* (with G. Howitt) 1833 (with specimens). Contrib. to *Trans. Linn. Soc.*
A. Gray *Letters* v.1, 1893, 144. *R.S.C.* v.6, 101.

VALLANCE, G. D. (–1889)
d. Exeter, Devon 17 Aug. 1889
Gardener at Tresco Abbey, Scilly Isles.
J. Hort. Cottage Gdnr v.19, 1889, 157.

VALLENTIN, Elinor Frances (*née* Bertrand) (1873–1924)
b. Falkland Islands 14 Jan. 1873 *d.* Plymouth, Devon 12 March 1924
Wife of R. Vallentin (1859–1934). *Illustrations of Flowering Plants and Ferns of Falkland Islands* 1921 (original drawings at Kew).
J. Linn. Soc. v.39, 1911, 313–39; v.43, 1915, 137–43. *Kew Bull.* 1924, 283–87. *Proc. Linn. Soc.* 1934–35, 192–93.

VALLENTIN, Rupert (1859–1934)
b. Walthamstow, Essex 15 Nov. 1859 *d.* Nov. 1934
FLS 1889. Zoologist. Collected plants in Falkland Islands and described algae in V. F. Boyson's *Falkland Islands* 1924.
Proc. Linn. Soc. 1934–35, 191–94.

VANE, Anne *see* Monson, A.

VARENNE, Ezekiel George (1811–1887)
b. Marylebone, London 6 May 1811 *d.* Kelvedon, Essex 22 April 1887
MRCS 1833. Surgeon, Kelvedon, *c.* 1847. Lichenologist. Contrib. to *Phytologist* and G. S. Gibson's *Fl. Essex* 1862.
Essex Nat. v.5, 1891, 1–30, 42–44 portr.; v.12, 1901, 167–68; v.18, 1917–18, 133–34, 292–300 portr.; v.29, 1954, 195. S. T. Jermyn *Fl. Essex* 1974, 18. *R.S.C.* v.6, 110.
Lichens, hepatics, fungi at Essex Field Club. MS. list of Essex plants at Kew.

VARLEY, Cornelius (1781–1873)
b. Hackney, London 21 Nov. 1781 *d.* Stoke Newington, London 21 Oct. 1873
Water-colour artist. Microscopist. '*Chara vulgaris*' (*Trans. Microsc. Soc.* v.2, 1849, 93–104).
D.N.B. v.58, 148–49. *J. Bot.* 1920, 50–53. *R.S.C.* v.6, 111.

VASEY, George (1822–1893)
b. Scarborough, Yorks Feb 1822 *d.* Washington, U.S.A. 4 March 1893
To U.S.A. with parents, 1823. Practised medicine at Ringwood, Illinois, 1848–66. Botanist in charge of U.S. National Herb., 1872–93. Botanised in the Rockies. *Grasses of the U.S.* 1883. *Illustrations of N. American Grasses* 1891–93 2 vols.
Bot. Gaz. 1893, 170–83 portr. *Bull. Torrey Club* 1893, 170, 218–20. *Science* v.21, 1893, 145. J. Ewan *Rocky Mountain Nat.* 1950, 326–27.
Vaseya Thurb. *Vaseyochloa* Hitchc.

VAUGHAN, David Thomas Gwynne- *see* Gwynne-Vaughan, D. T.

VAUGHAN, Francis (*fl.* 1690s)
MD. Of Clonmel, County Tipperary. Made catalogue of Wexford plants.
E. Lankester *Correspondence of J. Ray* 1848, 304–5, 313, 319–21. *J. Bot.* 1911, 125.

VAUGHAN, George (*fl.* 1800s–1830s)
Nurseryman, 22 Market Place, Manchester, Lancs.

VAUGHAN, Helen C. Gwynne- *see* Gwynne-Vaughan, H. C.

VAUGHAN, James (1792–1863)
b. Killerton, Devon 25 Jan. 1792 *d.* Exeter, Devon 14 May 1863
MRCS. Assistant Surgeon in Bombay Army and Port Surgeon at Aden. Returned to England, 1853. 'Notes upon Drugs observed at Aden, Arabia' (*Pharm. J.* 1852, 226–29, 268–71; 1853, 385–88).
Rec. Bot. Survey India v.7, 1914, 8–9. *R.S.C.* v.6, 114.

VAUGHAN, Rev. John (1855–1922)
b. Finchingfield, Essex 22 Jan. 1855 *d.* Winchester, Hants 10 July 1922
MA Cantab 1876. Vicar, Porchester, 1800; Langrish, 1897; Droxford, 1902–10. Canon of Winchester Cathedral. *Wild Flowers of Selborne* 1906. *Music of Wild Flowers* 1920.
Bot. Soc. Exch. Club Br. Isl. Rep. 1922, 708–12; 1924, 661–66. *J. Bot.* 1922, 243–44. *Watson Bot. Exch. Club Rep.* 1922–23, 204. *Papers Proc. Hampshire Field Club Archaeol. Soc.* v.9, 1925, 430–31. *Who was Who, 1916–1928* 1069.
Plants at Oxford.

VAUX, Thomas
Of Bedford. Friend of Charles Abbot (*c.* 1761–1817). Amateur botanist.
J. G. Dony *Fl. Bedfordshire* 1953, 20.

VEITCH, Anna Mildred (1889–1949)
Daughter of Peter C. M. Veitch (1850–1929). Managing Director, Robert Veitch and Son Ltd.
International Dendrology Soc. Yb. 1972, 66.

VEITCH, Arthur (1844–1880)
b. Exeter, Devon 24 Feb. 1844 *d.* 25 Sept. 1880
Brother of J. G. Veitch (1839–1870). Partner in family nursery firm.
Gdnrs Chron. 1880 ii 440. *J. Hort. Cottage Gdnr* v.1, 1880, 300, 327, 597–98 portr.
Portr. at Hunt Library.

VEITCH, Sir Harry James (1840–1924)
b. Exeter, Devon 29 June 1840 *d.* Slough, Bucks 6 July 1924
FLS 1886. VMH 1906. Knighted 1912. Son of James Veitch (1815–1869). Admitted into partnership in Veitch firm in 1865; at Chelsea, Coombe Wood and Langley. Hybridised orchids. 'Fertilisation of *Cattleya labiata*' (*J. Linn. Soc.* v.24, 1888, 395–406). Contrib. to *J. R. Hort. Soc.*
Garden v.59, 1901, v portr.; 1909, 288 portr.; 1912, 291 portr. *Gdnrs Chron.* 1899 i 40 portr.; 1910 i 72–74 portr.; 1912 i 364 portr.; 1924 ii 19, 20 portr.; 1921 ii 192 portr.; 1925 i 67; 1964 ii 139–40. *Gdnrs Mag.* 1906, 179 portr.; 1908, 843–44 portr.; 1910, 81–82 portr.; 1912, 413–14 portr. *J. Hort. Cottage Gdnr* v.52, 1906, 277 portr.; v.56, 1908, 268–70 portr. *Orchid Rev.* 1912, 41–43 portr. *Bot. Soc. Exch. Club Br. Isl. Rep.* 1924, 545–46. *J. Bot.* 1924, 253. *Kew Bull.* 1924, 300–1. *Nature* v.114, 1924, 95–96. *Proc. Linn. Soc.* 1924–25, 83–84. *Times* 7 July 1924. *Curtis's Bot. Mag. Dedications, 1827–1927* 331–32 portr. *Who was Who, 1916–1928* 1070. *D.N.B. 1922–1930* 368–69. *J. R. Hort. Soc.* 1948, 244–47 portr. *Fl. Malesiana* v.1, 1950, 538. *International Dendrology Soc. Yb.* 1972, 63–64 portr. M. A. Reinikka *Hist. of the Orchid* 1972, 246–49 portr.
Portr. by H. G. Rivière at Royal Horticultural Society. Portr. at Hunt Library.
Masdevallia harryana Reichenb. f.

VEITCH, James (1792–1863)
b. Killerton, Devon 25 Jan. 1792 *d.* Exeter, Devon 14 May 1863
Nurseryman, Mount Radford, Exeter. With his son, James (1815–1869), as partner acquired Knight and Perry's nursery, Chelsea, 1853.
Cottage Gdnr v.13, 1855, 273–75 portr. J. H. Veitch *Hortus Veitchii* 1906, 9–12, 27 portr. *International Dendrology Soc. Yb.* 1972, 63–69 portr.
Portr. at Hunt Library.

VEITCH, James (1815–1869)
b. Exeter, Devon 24 May 1815 *d.* Chelsea, London 10 Sept. 1869
FLS 1862. Son of J. Veitch (1792–1863) with whom he was in partnership. Moved to Chelsea in 1853 and ceased all interest in Exeter firm in 1864. VMM instituted in his honour.
Gdnrs Chron. 1869, 990; 1923 i 197. *J. Hort. Cottage Gdnr* v.42, 1869, 230–31. *Proc. Linn. Soc.* 1869–70, cxiv–cxv. *Garden* v.10, 1876, xv–xvi portr. J. H. Veitch *Hortus Veitchii* 1906, 12–18, 27 portr. *International Dendrology Soc. Yb.* 1972, 63–69 portr.
Portr. at Hunt Library.
Veitchia Wendl.

VEITCH, James Herbert (1868–1907)
b. Chelsea, London 1 May 1868 *d.* Exeter, Devon 13 Nov. 1907
FLS 1889. Son of J. G. Veitch (1839–1870). Nurseryman at Chelsea, Coombe Wood, Langley and Feltham. Travelled in India, Japan, Australia, etc. collecting plants. *A Traveller's Notes* 1896. *Hortus Veitchii* 1906. *Veitchian Nurseries* 1903.

E. Bretschneider *Hist. European Bot. Discoveries in China* 1898, 767. J. H. Veitch *Hortus Veitchii* 1906, 27 portr., 89–91. *Garden* 1907, 562. *Gdnrs Chron.* 1907 ii 360 portr.; v.160(20), 1966, 12–13. *Gdnrs Mag.* 1907, 856 portr. *J. Hort. Cottage Gdnr* v.55, 1907, 497 portr., 511. *Nature* v.77, 1907, 86–87. *Orchid Rev.* 1907, 371. *Proc. Linn. Soc.* 1907–8, 65–66. *D.N.B. Supplt. 2* v.3, 555. *Fl. Malesiana* v.1, 1950, 539. A. M. Coats *Quest for Plants* 1967, 229–30.

Plants at Kew. Portr. at Hunt Library.

VEITCH, John (1752–1839)
b. Jedburgh, Roxburghshire 1752 *d.* Mount Radford, Exeter, Devon 1839

Brought from Scotland by Sir Thomas Dyke Acland to lay out gardens at Killerton near Exeter. Founded nursery at Budlake near Killerton, 1808. Moved to Mount Radford, 1832. Succeeded by J. Veitch (1792–1863).

J. H. Veitch *Hortus Veitchii* 1906, 8–9, 27 portr. *International Dendrology Soc. Yb.* 1972, 63–69 portr.

Portr. at Hunt Library.

VEITCH, John Gould (1839–1870)
b. Exeter, Devon April 1839 *d.* Coombe Wood, Surrey 13 Aug. 1870

FLS 1866. Son of J. Veitch (1815–1869). Visited Japan, China and Philippines, 1860; Australia and Pacific, 1864. Reported his travels in *Gdnrs Chron.* 1866, 1867, etc.

Gdnrs Chron. 1870, 1117. *J. Hort. Cottage Gdnr* v.44, 1870, 132. *Proc. Linn. Soc.* 1870–71, xc–xci. E. Bretschneider *Hist. European Bot. Discoveries in China* 1898, 551–52. J. H. Veitch *Hortus Veitchii* 1906, 18–21, 27, 49–52. *Fl. Malesiana* v.1, 1950, 539. A. M. Coats *Quest for Plants* 1969, 69–71.

Plants at Kew. Portr. at Hunt Library.

Veitchia Wendl.

VEITCH, John Gould (*c.* 1869–1914)

MA Cantab. Son of J. G. Veitch (1839–1870). Partner in family nursery firm.

J. H. Veitch *Hortus Veitchii* 1906, 27. *Gdnrs Chron.* 1914 ii 256 portr.

VEITCH, Peter Christian Massyn (1850–1929)
b. Cape of Good Hope 1850 *d.* Exeter, Devon 9 March 1929

VMH 1916. Son of R. T. Veitch (1823–1885). Collected plants for Veitch firm in Fiji, Australia, New Zealand, New Guinea, Borneo, 1875–78. Succeeded his father in charge of Exeter branch of the firm.

J. H. Veitch *Hortus Veitchii* 1906, 27, 67–69 portr. *Gdnrs Mag.* 1909, 663–64 portr. *Garden* 1910, 632–33 portr. *Gdnrs Chron.* 1922 ii 62 portr.; 1929 i 214 portr. *Orchid Rev.* 1929, 114. *J. R. Hort. Soc.* 1948, 287–88. *Fl. Malesiana* v.1, 1950, 539.

Plants at BM(NH), Kew.

Spathoglottis petri Reichenb. f.

VEITCH, Robert Toswill (1823–1885)
d. Torquay, Devon 18 Jan. 1885

Son of J. Veitch (1792–1863). Farmer, Cape of Good Hope. Returned to Exeter branch of Veitch firm in 1856–57. Nurseryman of New North Road and High Street, Exeter in 1864 as Robert Veitch and Son.

Garden v.27, 1885, 76. *Gdnrs Chron.* 1885 i 124. J. H. Veitch *Hortus Veitchii* 1906, 27.

VELLEY, Thomas (1748–1806)
b. Chipping Ongar, Essex 1748 *d.* Reading, Berks 6 June 1806

DCL Oxon 1787. FLS 1792. Of Bath and Liverpool. Lieut.-Colonel, Oxford Militia. Algologist. Friend of D. Turner and J. E. Smith. *Coloured Figures of Marine Plants found on Southern Coast of England* 1795.

J. Sowerby and J. E. Smith *English Bot.* 1690. *Ann. Bot.* 1806, 593. *Trans. Linn. Soc.* v.5, 1800, 145–58. P. Smith *Mem. and Correspondence of J. E. Smith* v.2, 1832, 343–44. *Gent. Mag.* 1806 i 588. *Naturalist* v.4, 1839, 398. *D.N.B.* v.58, 202. *Trans. Liverpool Bot. Soc.* 1909, 94. Liverpool Museums *Handbook and Guide to Herb. Collections* 1935, 54. *N. Western Nat.* 1938, 72–78. *R.S.C.* v.6, 131.

Herb. at Liverpool Museums. Sale at Leigh and Sotheby, 15 June 1807.

Velleia Smith.

VENNING, Alfred Reid (–1927)

Malayan Civil Service, 1893–1908. Collected plants when Secretary to the Resident at Perak, 1900–3.

Gdns Bull. Straits Settlements 1927, 133. *Fl. Malesiana* v.1, 1950, 540.

Plants at Singapore.

VENNING, Francis Esmond (1882–1970)
b. Opalgalla, Ceylon 26 Jan. 1882 *d.* Hythe, Kent 28 Aug. 1970

Brigadier, Indian Army until retirement in 1933. President, Southampton Natural History Society, 1947. Studied ferns.

Southern Evening Echo (Southampton) 31 Aug. 1970. *Who was Who, 1961–1970* 1149.

VERE, James (*fl.* 1790s–1800s)

Had garden at Kensington Gore, London. "A great encourager of botanical science" (*Bot. Mag.* 1812, t.1436). Friend of W. Curtis.

Bot. Mag. passim. H. C. Andrews *Botanist's Repository* v.1, 1798, t.21.

Garden plants at BM(NH).

Verea Andr.

VERNEY, John (*fl.* 1660s–1670s)
Sent seeds from Middle East to his father, Sir R. Verney, at Claydon, Bucks.
F. P. Verney *Mem. of Verney Family* 1892. M. Hadfield *Gdning in Britain* 1960, 142.

VERNEY, Sir Ralph (1612–1696)
Politician. MP, Aylesbury, 1640; Buckingham, 1680, 1685, 1689. Had garden at Claydon, Bucks.
F. P. Verney *Mem. of Verney Family* 1892. *D.N.B.* v.58, 264–65. M. Hadfield *Gdning in Britain* 1960, 93–94.

VERNON, Mr. (*fl.* 1730s)
Merchant at Aleppo, Turkey. "Transplanted weeping willow from river Euphrates ...and planted it in his seat at Twickenham-Park" *c.* 1730 (P. Collinson).
Trans. Linn. Soc. v.10, 1811, 275.

VERNON, William (*fl.* 1680s–1710s)
BA Cantab 1688. FRS 1702. Collected plants with D. Krieg in Maryland, 1698, and later in England. Corresponded with H. Sloane, J. Petiver, R. Uvedale, etc.
J. Petiver *Museii Petiveriani* 1695, no. 89. R. Pulteney *Hist. Biogr. Sketches of Progress of Bot. in England* v.2, 1790, 57–58. *Philadelphia Med. Physical J.* v.2(2), 1806, 139–42. D. Turner *Extracts from Lit. Sci. Correspondence of R. Richardson* 1835, 37–40, 73–75, 79–80. H. Trimen and W. T. T. Dyer *Fl. Middlesex* 1869, 389. J. B. L. Warren *Fl. Cheshire* 1899, xc. E. J. L. Scott *Index to Sloane Manuscripts* 1904, 546. *Proc. International Congress Plant Sci. 1926* v.2, 1929, 1528–29. C. E. Raven *John Ray* 1942, 257. *Proc. Amer. Antiq. Soc.* 1952, 303–7. *Fl. Malesiana* v.1, 1950, 541. J. E. Dandy *Sloane Herb.* 1958, 226–28.
Plants at BM(NH).
Vernonia Schreber.

VERNON, William (1811–1890)
b. Epsom, Surrey 1811 *d.* Sydney, N.S.W. 6 Jan. 1890
Gardener, Botanic Gardens, Sydney; in charge of Herb., 1857. Gardener to T. S. Mort at Darling Point, Sydney. Sent plants to F. von Mueller.
G. Bentham *Fl. Australiensis* v.1, 1863, 14. *J. R. Soc. N.S.W.* 1908, 126–27 portr.
Ionidium vernonii F. Muell.

VERT, James (–1929)
d. 10 April 1929
Acquired nursery of Webb and Brand, Saffron Walden, Essex.
Gdnrs Chron. 1929 i 302.

VESEY-FITZGERALD, Leslie Desmond Edward Foster (*c.* 1910–1974)
d. Nairobi, Kenya 3 May 1974
Entomologist. Ecologist to Tanzania National Parks. Pioneer conservationist in E. Africa where he also collected plants. 'Central African Grasslands' (*J. Ecol.* 1963, 243–74).
Bull. Br. Ecol. Soc. v.5, 1974, 16. *EANHS Bull.* June 1974, 87.

VICARY, Nathaniel (*fl.* 1830s–1850s)
Major, 2nd European Regiment, Bengal Army. Collected plants at Peshawar, *c.* 1838. 'Botany of Sinde' (*Ann. Mag. Nat. Hist.* 1848, 420–34). 'Some Notes on Botany of Sinde' (*J. Asiatic Soc. Bengal* v.16, 1847, 1152–68).
J. D. Hooker and T. Thomson *Fl. Indica* v.1, 1855, 70. *J. Asiatic Soc. Bengal* v.25, 1856, 410. J. D. Hooker *Fl. Tasmaniae* v.1, 1859, cxxvii. *J. R. Soc. N.S.W.* 1908, 127. *R.S.C.* v.6, 149.
Indian and N.S.W. plants at Kew, Calcutta.
Vicarya Wall.

VICK, James (1818–1882)
b. Portsmouth, Hants 23 Nov. 1818 *d.* Rochester, New York 16 May 1882
To U.S.A., *c.* 1830. Printer. Purchased A. J. Downing's *Horticulturalist*, 1850. Established seed business, 1860.
Garden v.21, 1882, 66 portr. L. H. Bailey *Standard Cyclop. Hort.* v.2, 1939, 1601 portr.

VICKERS, Anna (1852–1906)
b. Bordeaux, France 28 June 1852 *d.* Roscoff, Finisterre, Spain 1 Aug. 1906
Algologist. Travelled in Canaries, 1895–96; in Antilles, 1898–99. *Phycologia Barbadensis* 1908. 'Liste des Algues de la Barbade' (*Ann. Sci. Nat. Bot.* 1905, 45–66).
Vickersia Karsakoff.

VIDAL, Capt. (*fl.* 1840s ?)
Captain, Royal Navy. Discovered *Campanula vidalii* in Azores.
Hooker's Icones Plantarum 1844, t.684. *Bot. Mag.* 1853, t.4748.

VIDLER, Edward Alexander (1863–1942)
b. London 13 Aug. 1863 *d.* Melbourne 28 Oct. 1942
Employed by Cassells, publisher. To Australia, 1888. Editor and proprietor of *Evening News* (Geelong), *Tatler* (Melbourne) and *The Spinner* (Melbourne). Joint Hon. Curator, Maranoa Native Gardens, Victoria. *An Australian Flower Painter: A. E. Oakley* 1923. *Our Own Trees: a First Book on Australian Forests* 1930.
J. A. Alexander *Who's Who in Australia* 1938, 504. *The Age* (Melbourne) 30 Oct. 1942. *Victorian Nat.* 1943, 180.

VIGNE, Godfrey Thomas (1801–1863)
b. 1 Sept. 1801 *d.* Woodford, Essex 12 July 1863

Called to Bar, Lincoln's Inn, 1824. In India 1833–39 and collected plants in Kashmir which he presented to J. F. Royle. *Travels in Kashmir, Ladak, Iskardo* 1842 2 vols (botanical appendix by Royle, 440–62).

Gent. Mag. 1863 ii 250. *D.N.B.* v.58, 309. *Pakistan J. For.* 1967, 359. I. H. Burkill *Chapters on Hist. Bot. in India* 1965, 75–76. R. H. Phillimore *Hist. Records of Survey of India* v.4, 1958, 472.

VILLIERS, Maria Theresa *see* Earle, M. T.

VINES, Sydney Howard (1849–1934)
b. Ealing, Middx 31 Dec. 1849 *d.* Exmouth, Devon 4 April 1934

BSc London 1873. BA Cantab 1875. FRS 1885. FLS 1878. Fellow and Lecturer, Christ's College, Cambridge, 1876–88; Reader in Botany, 1883–88. Professor of Botany, Oxford, 1888–1919. President, Linnean Society, 1900–4. Did pioneer work on investigation of proteolytic enzymes of plants. *Lectures on Physiology of Plants* 1886. *Student's Text-book of Botany* 1895. *Dillenian Herb.* (with G. C. Druce) 1907. *Account of Morisonian Herb.* (with G. C. Druce) 1914. Translated K. Prantl's *Elementary Text-book of Botany* 1880 and F. Sachs's *Text-book of Botany* 1882.

J. Bot. 1934, 139–41. *Kew Bull.* 1934, 134–35. *Nature* v.133, 1934, 675–77. *Proc. Linn. Soc.* 1933–34, 173–79 portr. *Obit. Notices Fellows R. Soc.* 1934, 185–88 portr. *Times* 6 April 1934. *Chronica Botanica* 1935, 176 portr. F. O. Bower *Sixty Years of Bot. in Britain* 1938, 51–54 portr. *D.N.B. 1931–1940* 881–82. *Who was Who, 1929–1940* 1392.

Portr. by Collier at Linnean Society.

VINTEN, Albert George (*c.* 1871–1962)
d. 9 Feb. 1962

VMH 1958. Chrysanthemum grower, Oldland Nursery, Balcombe, Sussex. Vice-President, National Chrysanthemum Society.

Gdnrs Chron. 1962 i 146.

VIPAN, John Alexander Maylin (1849–1939)
b. Stibbington Hall, Hunts 24 May 1849 *d.* 20 March 1939

Army officer, serving mainly in India and Burma. Cultivated plants, chiefly orchids and ferns, at Stibbington Hall.

Gdnrs Chron. 1882 ii 134; 1965 ii 22–23 portr.

Vanda vipanii Reichenb. f.

VIZE, Rev. John Edward (1831–1916)
b. 7 March 1831 *d.* Bristol March 1916

MA Dublin and Oxon. Priest, 1860. Vicar, Forden, Welshpool, 1873–1910. Mycologist. *Fungi Britannici* 1873–75. *Micro-fungi Britannici* 1878–88. *Micro-fungi Exotici* 1883 (exsiccatae). '*Aecidium depauperans*' (*Gdnrs Chron.* 1876 ii 361). Contrib. to *Woolhope Club Trans.*

R.S.C. v.8, 1160; v.11, 711; v.19, 381.

Slides, specimens and letters at BM(NH). Portr. at Hunt Library.

VOBES, F. N. *see* James, F. N.

VOELCKER, John Augustus (1854–1937)
b. Cirencester, Glos 24 June 1854 *d.* 6 Nov. 1937

BA, BSc London. PhD Giessen. FLS 1925. Chemist, Royal Agricultural Society of England, 1884–1937. Director, Woburn Experimental Station, 1885–1936. Agricultural Advisor, Government of India, 1889–90. Researched on agricultural chemistry, green manuring. *Improvement of Indian Agriculture* 1892. *Woburn Experiment* (with E. J. Russell) 1936. Contrib. articles on manuring to *Encyclop. Britannica* and *Standard Cyclop. of Agric.*

Nature v.140, 1937, 1001. *Chronica Botanica* 1938, 178 portr. *Proc. Linn. Soc.* 1937–38, 344–45. *Who was Who, 1929–1940* 1393.

VOGAN, James (*fl.* 1820s)

Seedsman, 17 Tooley Street, Southwark, London.

VOIGT, Joachim Otto (1798–1843)
b. Nordborg, Denmark 22 March 1798 *d.* London 22 June 1843

Surgeon, Serampore, India, 1827. Superintendent, Botanic Garden, Serampore, 1834; Botanic Garden, Calcutta, 1842. *Hortus Suburbanus Calcuttensis* 1845.

Calcutta J. Nat. Hist. v.4, 1844, 388–90. C. Christensen *Danske Bot. Hist.* 1924–26, 199–200. *Fl. Malesiana* v.1, 1950, 546.

Plants at Copenhagen.

Givotia Griff.

VOSS (Vossius), Rev. Isaac (1618–1689)
b. Leyden, Holland 1618 *d.* Windsor, Berks 21 Feb. 1689

DCL Oxon 1670. Canon of Windsor, 1673–89. Edited Pliny's *Natural History* 1669. Owned Rauwolf's herb.

L. Plukenet *Almagestum Botanicum* 1696, 76, 141. *D.N.B.* v.58, 392–96.

VOSS, Margaret (*née* **Heatley**) (1885–1953)
b. Blackburn, Lancs 1885 *d.* Johannesburg, S. Africa 1953

Instructor in Botany, Wellesley College, 1910–19. Curator, Moss Herb., University of Witwatersrand, 1930s–50. Interested in *Gnidia, Hasiosiphon* and marine angiosperms of E. Africa.

Kew Bull. 1954, 34.

VOWELL, Richard Prendergast (-1911)
d. 30 Oct. 1911
With R. M. Barrington discovered *Epilobium alsinefolium* in Ireland, 1884.
Irish Nat. 1911, 218. R. L. Praeger *Some Irish Nat.* 1949, 172.
Herb. at Dublin National Museum.

VYVYAN, Lady Clara (*c.* 1886–1976)
b. Australia *c.* 1886 *d.* 1 March 1976
Had garden at Trelowarren, Cornwall. *Letters from a Cornish Garden* 1972.
Times 4 March 1976.

WABY, John Frederick (1848–1923)
b. London 26 Oct. 1848 *d.* Georgetown, British Guiana 28 Dec. 1923
FLS 1901. Gardener, Botanic Garden, Trinidad, 1873; Botanic Garden, British Guiana, 1878–1914. Collected plants in Barbados, 1895. *Tropical Gardening in British Guiana* 1893.
I. Urban *Symbolae Antillanae* v.3, 1902, 138. *Kew Bull.* 1914, 85. *J. Board Agric. Br. Guiana* v.17, 1924, 26–27.
Plants at Kew.

WADDELL, Rev. Coslett Herbert (1858–1919)
b. Maralin, County Down 6 March 1858 *d.* Grey Abbey, County Down 8 June 1919
MA, BD Trinity College, Dublin. Rector, Saintfield and Grey Abbey. Secretary, Moss Exchange Club, 1896–1903. *Catalogue of British Hepaticae* 1897. Contrib. to *J. Bot.*
Bot. Soc. Exch. Club Br. Isl. Rep. 1919, 629–30. *Irish Nat.* 1919, 108. *J. Bot.* 1919, 358–59. *R.S.C.* v.19, 423.
Plants at National Museum Dublin, Ulster Museum.

WADDELL, Laurence Austine (1854–1938)
b. Cumbernauld, Dunbartonshire 29 May 1854 *d.* Craigmore, Rothesay 19 Sept. 1938
MB Glasgow 1878. FLS 1891. Indian Medical Service, 1880. Collected plants in India, Sikkim, Tibet.
Proc. Linn. Soc. 1938–39, 257–60.
Plants at Kew.
Primula waddellii Balf. & Smith.

WADDINGTON, William (1866–1945)
b. Lower Broughton, Lancs 16 Aug. 1866 *d.* Southport, Lancs 26 Nov. 1945
Monumental mason, N. Wingfield, Derbyshire. President, Southport Scientific Society, 1939, 1940. Contrib. botanical papers to *Southport Sci. Soc. Rep.*
N. Western Nat. 1947, 291–93 portr.
Herb. and MSS. at Southport Museum.

WADE, Walter (1760–1825)
d. Dublin July 1825
MD. FRS 1811. ALS 1792. Professor of Botany, Dublin Society and established their garden in 1796. Discovered *Eriocaulon* in Ireland. *Catalogus Systematicus Plantarum Indigenarum in Comitatu Dublinensi Inventarum* 1794. *Plantae Rariores in Hibernia Inventae* 1804. *Sketch of Lectures on Meadow and Pasture Grasses* 1808. *Salices* 1811. *Prospectus of Arrangements...at Glasnevin* 1818.
Mag. Nat. Hist. v.2, 1829, 305. P. Smith *Mem. and Correspondence of J. E. Smith* v.2, 1832, 127, 147–50, 160–63. K. Baily *Irish Fl.* 1833, vii. *D.N.B.* v.58, 421–22. N. Colgan *Fl. County Dublin* 1904, xxiii–xxiv. *Proc. Irish Acad.* v.32, 1915, 68–69. R. L. Praeger *Some Irish Nat.* 1949, 173. *R.S.C.* v.6, 221.

WADLOW, Henry John (1858–1950)
b. Shifnal, Shropshire Nov. 1858 *d.* 17 Sept. 1950
Headmaster, Stoney Middleton, Derbyshire. Headmaster, Frenchay near Bristol until retirement in 1918. Botanist.
Proc. Bournemouth Nat. Sci. Soc. v.40, 1949–50, 92.

WAGER, Sir Charles (1666–1743)
d. Fulham, London 24 May 1743
Admiral, 1731. Knighted, 1708. MP, Westminster, 1732–42. Grew exotic plants at Fulham. Friend of Peter Collinson, Lord Petre, etc. Introduced *Acer saccharinum* from N. America in 1725.
Bot. Mag. 1789, t.74. *Trans. Linn. Soc.* v.10, 1811, 282. *D.N.B.* v.58, 428–30. J. E. Dandy *Sloane Herb.* 1958, 228. *J. Soc. Bibl. Nat. Hist.* 1970, 289–90.
Gibraltar plants at BM(NH).

WAGER, Harold William Taylor (1862–1929)
b. Stroud, Glos 11 March 1862 *d.* Hawkswick, Yorks 17 Nov. 1929
FRS 1904. FLS 1893. Lecturer in Botany, Yorkshire College of Science, 1888–94. H.M. Inspector of Schools, 1894–1926. President, British Mycological Society, 1910. President, Yorkshire Naturalists Union, 1913. Studied cytology of fungi. Contrib. to *Naturalist*.
Naturalist 1915, 246; 1930, 141–43 portr.; 1961, 63. *Bot. Soc. Exch. Club Br. Isl. Rep.* 1929, 99. *Nature* v.124, 1929, 953–54. *Times* 20 Nov. 1929. *J. Bot.* 1930, 18–20. *Proc. Linn. Soc.* 1929–30, 224–26. *Proc. R. Soc. B.* 1930, xix–xxii portr. *Who was Who, 1929–1940* 1396.

WAGHORNE, Rev. Arthur Charles (1851–1900)
b. London 1851 *d.* Garden Town, Jamaica 11 April 1900

Missionary, Newfoundland. *Newfoundland and Labrador Plants* 1893. Fl. of Newfoundland and Labrador in *Trans. Nova Scotia Inst. Sci.* v.8, 1892–93, 359–73; v.9, 1894–95, 83–100, 361–401.

WAILES, George (*c.* 1802–1882)
d. Gateshead, Durham 1882
Naturalist; interested in alpines and orchids. "A gentleman who has for many years occupied himself with the cultivation and scientific study of orchids" (*J. Hort. Soc. London* v.4, 1849, 263).
Gdnrs Chron. 1882 ii 727.
Wailesia Lindley.

WAINWRIGHT, Joseph (*c.* 1813–1884)
d. Wakefield, Yorks 10 April 1884
FLS 1856. Solicitor, Wakefield. President, Yorkshire Naturalists' Union. President, Wakefield Naturalists' Society, 1871–84. Keen gardener.
Proc. Linn. Soc. 1883–86, 42.

WAINWRIGHT, R. (*fl.* 1790s)
Surgeon, Dudley, Worcs. Contrib. plant records to S. Shaw's *Hist. and Antiq. of Staffordshire* v.2, 1801, 6–7.

WAINWRIGHT, Thomas (1826–1916)
b. Leeds, Yorks 7 April 1826 *d.* Barnstaple, Devon 29 April 1916
Schoolmaster and antiquarian. Phenologist. Contrib. to *J. Bot.*
Bot. Soc. Exch. Club Br. Isl. Rep. 1916, 469. *J. Bot.* 1916, 208–10. W. K. Martin and G. T. Fraser *Fl. Devon* 1939, 778.

WAIT, Rev. Walter Oswald (1852–1936)
d. 25 Dec. 1936
Educ. Oxford. Vicar, Titley, Herefordshire, 1908–36. Studied local flora. Botanical editor of Woolhope Naturalists Field Club, 1917–49.
Trans. Woolhope Nat. Field Club 1936–38, 63; 1954, 258.

WAITE, Percival Colin (1859–1907)
b. London 1859 *d.* Edinburgh 13 Feb. 1907
Studied at Montpellier. Demonstrator, University College, Dundee.
Trans. Bot. Soc. Edinburgh v.23, 1908, 353–55.

WAKEFIELD, Elsie Maud (1886–1972)
b. Birmingham 3 July 1886 *d.* Richmond, Surrey 17 June 1972
MA Oxon. FLS 1911. Assistant, Kew Herb., 1910; Deputy Keeper, 1945–51. President, British Mycological Society, 1929. President, Kew Guild, 1944–45. Specialist in Basidiomycetes. *Edible and Poisonous Fungi* 1910 (which she also illustrated). *Common British Fungi* (with R. W. G. Dennis) 1950. *Observer's Book of Common Fungi* 1954. Contrib. to *Trans. Br. Mycol. Soc.*
J. Kew Guild 1944, 343–44 portr.; 1972, 163–64. *Taxon* 1952, 113–14 portr. *Trans. Br. Mycol. Soc.* 1966, 355 portr.; 1973, 167–74 portr. *Bull. Br. Mycol. Soc.* 1972, 81–82. *Times* 5 July 1972. Hunt Library *Artists from R. Bot. Gdns Kew* 1974, 66–67 portr.
Plants at Kew. Portr. at Hunt Library.
Wakefieldia Corner & Hawker.

WAKEFIELD, Priscilla (*née* Bell) (1751–1832)
b. Tottenham, Middx 31 Jan. 1751 *d.* Ipswich, Suffolk 12 Sept. 1832
Author and philanthropist. *Introduction to Botany* 1796; ed. 11 1841.
Gent. Mag. 1832 ii 650. J. Smith *Cat. of Friends' Books* v.2, 1867, 848–51. *D.N.B.* v.58, 455–56. *Gdnrs Chron.* 1950 ii 130–31.
Portr. by Gainsborough.

WAKEFIELD, Rev. Thomas (1836–1901)
b. Derby 23 June 1836 *d.* Southport, Lancs 15 Dec. 1901
Methodist missionary in East Equatorial Africa.
J. Bot. 1904, 95–96. E. S. Wakefield *Thomas Wakefield* 1904 portr. *Trans. Liverpool Bot. Soc.* 1909, 94.
Plants and letters at Kew.
Turraea wakefieldii Oliver.

WAKEHURST, 1st Baron *see* Loder, G. W. E.

WAKELING, J. (*fl.* 1840s)
Botanical artist and teacher of flower painting, Walworth, London. Contrib. plates to *Hort. J. Florists' Register*, W. Paul's *Rose Gdn* 1848.

WAKELY, Charles (1870–1932)
b. Thornford, Dorset 1870 *d.* Chelmsford, Essex 20 Aug. 1932
Kew gardener, 1890. Horticultural instructor, East Anglian Institute of Agriculture, 1895; later Superintendent.
Gdnrs Chron. 1931 i 272 portr.; 1932 ii 168. *J. Kew Guild* 1933, 280–81 portr.

WALCOTT, John (1754/5–1831)
b. Ireland 1754/5 *d.* Bath, Somerset 5 Feb. 1831
Lived mainly at Bath and Highnam Court, Gloucester. *Fl. Britannica Indigena* 1778.

WALCOTT, Pemberton (*fl.* 1860s)
"Joined as a volunteer for the collection of specimens of natural history and botany" on F. Gregory's expedition to N.W. Australia, 1861.
J. W. Austral. Nat. Hist. Soc. no. 6, 1909, 26.
Corchorus walcottii F. Muell.

WALDUCK, Thomas (*fl.* 1710s)
Captain. Correspondent of J. Petiver to whom he sent Barbados plants.
J. E. Dandy *Sloane Herb.* 1958, 228–29.

WALE, Roydon Samuel (–1952)
d. Leicester Nov. 1952
FLS 1938. Pathologist attached to East Grinstead Hospital, later transferred to Leicester. Cultivated and studied rock-plants. Collected plants in Yugoslavia and Albania. *Genus Sempervivum* 1943. Contrib. to *Quart. Bull. Alpine Gdn Soc.*
Proc. Linn. Soc. 1952–53, 221–22. *Quart. Bull. Alpine Gdn Soc.* 1953, 64–65.

WALFORD, Thomas (1752–1833)
b. Whitley 14 Sept. 1752 *d.* Whitley 6 Aug. 1833
FLS 1797. Antiquary. Of Birdbrook, Essex. Correspondent of J. Sowerby. Contrib. to J. Sowerby and J. E. Smith's *English Bot.* 259, 446, 630. *Scientific Tourist* 1818 (botany, v.1, 69–76 and county lists).
Gent. Mag. 1833 ii 469. *D.N.B.* v.59, 40. *R.S.C.* v.6, 239.
Letters at BM(NH).

WALKDEN, Harold (1885–1949)
d. Sale, Manchester 15 April 1949
FLS 1939. Mycologist. 'Isolation of Organism causing Crown Gall on Chrysanthemum frutescens in Britain' (*Ann. Bot.* v.35, 1921, 137–38). 'Critical Study of Crown Gall' (with W. Robinson) (*Ann. Bot.* v.37, 1923, 299–321).
Proc. Linn. Soc. 1949–50, 233–34.

WALKER, A. L. (*fl.* 1870s)
Captain in Army stationed in Hong Kong in 1870s.
Bot. Mag. 1876, t.6225. E. Bretschneider *Hist. European Bot. Discoveries in China* 1898, 698.

WALKER, A. W. (*née* Paton) (*fl.* 1820s–1840s)
Wife of G. W. Walker (–1844) with whom she collected plants in Ceylon, 1820–40. *Patonia* was dedicated to her by R. Wight "in return for many contributions from her accomplished pencil illustrations of the Flora of Ceylon."
Companion to Bot. Mag. v.2, 1837, 194–200. *Bot. Mag.* 1839, t.3770; 1846, t.4327. R. Wight *Icones Plantarum Indiae Orientalis* v.5, 1852, 20, t.1750. *R.S.C.* v.6, 239. H. Trimen *Handbook to Fl. Ceylon* v.5, 1900, 374.
Plants at BM(NH). Letters at Kew.
Liparis walkeriae Hook.

WALKER, Andrew (*fl.* 1820s)
Acting Superintendent, Botanic Gardens, Peradeniya, Ceylon, 1825–27.
Ann. R. Bot. Gdns, Peradeniya v.1, 1901, 6.

WALKER, Rev. Francis Augustus (1841–1905)
b. Southgate, Middx 1841 *d.* Cricklewood, Middx 31 Jan. 1905
BA Oxon 1864. DD 1883. FLS 1871. Collected plants in Iceland, 1889. Had herb.
J. Bot. 1890, 79. *Proc. Linn. Soc.* 1904–5, 55–56. *R.S.C.* v.8, 1183; v.11, 738; v.19, 446.

WALKER, Frederick (1829–1889)
b. Southgate, Middx 4 Dec. 1829 *d.* Southgate 20 Dec. 1889
Collected plants near Abingdon, Berks.
G. C. Druce *Fl. Berkshire* 1897, clxxx.

WALKER, George Warren (–1844)
In India, 1803–14. Adjutant-General, Ceylon, 1830–37. General, 21st Foot, 1840. Collected plants in Ceylon, Penang and Singapore.
Bot. Mag. 1828, t.2826. *J. Bot.* (Hooker) 1834, 180; 1841, 189. *Gdn Bull. Straits Settlements* 1927, 133. *Fl. Malesiana* v.1, 1950, 555; v.5, 1958, 97. *J. Bombay Nat. Hist. Soc.* 1956, 50. I. H. Burkill *Chapters on Hist. Bot. in India* 1965, 50.
Plants at Kew, Oxford. Letters at Kew.
Impatiens walkeri Hook.

WALKER, Rev. James (1794–1854)
d. Liverpool, N.S.W. 27 Oct. 1854
MA Oxon. Headmaster, King's School, Parramatta. Botanised in New South Wales. Gave plants to W. Woolls (1814–1893) who recorded them in his *Species Plantarum Parramettensium.* Rector, Liverpool, 1846.
W. Woolls *Plants...of Sydney* 1880, 1. W. W. Armstrong *Some Early Recollections of Rylstone* 1905. *J. R. Soc. N.S.W.* 1908, 127–28. S. M. Johnstone *Hist. King's School [Parramatta]* 1932, 97–98, 100–1, 140.

WALKER, James (–1875)
d. Nov. 1875
Of Mossley. Botanist.
Hardwicke's Sci.-Gossip 1876, 18–19.

WALKER, James (1837–1911)
b. Rutherglen, Lanarkshire 1837 *d.* Ham, Surrey 12 Feb. 1911
Nurseryman, Ham Common, Surrey.
Garden 1911, 96 portr. *Gdnrs Chron.* 1911 i 111 portr. *Gdnrs Mag.* 1911, 136.

WALKER, Rev. John (1731–1803)
b. Edinburgh 1731 *d.* Colinton, Edinburgh 31 Dec. 1803
DD Edinburgh. FRS 1794. Minister, Glencorse, Midlothian, 1758–62; Moffat, 1762–83; Colinton, 1783. Professor of Natural History, Edinburgh, 1779–1803. Teacher of Robert Brown (1773–1858). Discovered *Veronica fruticulosa* in Britain, 1782. *Experiments on Sap* 1785. *Essays on Natural History* 1808; ed. 2 1812. MS. *Adversaria*, 1771

dividing 4 Linnean genera of algae into 14 (Brewster's *Edinburgh Encyclop*. v.10, 1830, 3–4). Contrib. to J. Sowerby and J. E. Smith's *English Bot.*

J. C. Loudon *Arboretum et Fruticetum Britannicum* v.1, 1838, 87–91. *Jardine's Nat. Library: Ornithology* v.3, 17–50 portr. *Ann. Scott. Nat. Hist.* 1895, 257. *D.N.B.* v.59, 74. *Trans. Bot. Soc. Edinburgh* 1959, 180–203 portr. C. C. Gillispie *Dict. Sci. Biogr.* v.14, 1976, 131–33. *R.S.C.* v.6, 245.

Portr. at Hunt Library.

WALKER, John (1839–1895)
b. Boothtown, Halifax, Yorks 24 June 1839
d. Halifax 16 May 1895

Worsted manufacturer. Local botanist. List of plants in *Circulator* (Halifax, 1866–68).

W. B. Crump and C. Crossland *Fl. Parish of Halifax* 1904, lxii–lxiii.

Plants at Halifax Museum.

WALKER, John (–1895)
d. 8 May 1895

Nurseryman, Thame, Oxfordshire, 1846.

Gdnrs Chron. 1895 i 625–26.

WALKER, Joseph (–before 1810)

Had "a very copious collection of Stapelias, among other rare plants, in his extensive collection at Stockwell" (*Bot. Mag.* 1804, t.786; 1810, t.1326).

WALKER, Joseph (*c.* 1815–1877)
d. Eccleshall, Yorks 1 June 1877

Florist; raised new varieties of auriculas.

Florist and Pomologist 1877, 192. *Gdnrs Yb. and Almanack* 1878, 192.

WALKER, Rev. Richard (1679–1764)
d. Cambridge 15 Dec. 1764

BA Cantab 1706. DD 1728. Professor of Moral Philosophy, Cambridge, 1744. Founded Cambridge Botanic Garden, 1761. *Short Account of Late Donation of a Botanic Garden to...Cambridge* 1763.

Gent. Mag. 1765, 212. *Philos. Trans. R. Soc.* v.53, 1764, 131. *D.N.B.* v.59, 81–82.

Walkeria Miller.

WALKER, Rev. Richard (1791–1870)
b. Norwich, Norfolk 17 March 1791 *d.* Olveston, Glos 31 Dec. 1870

BA Oxon 1814. BD 1824. FLS 1829. Vice-President, Magdalene College, Oxford. *Fl. Oxfordshire* 1833.

G. C. Druce *Fl. Berkshire* 1897, clxv–clxvi. G. C. Druce *Fl. Oxfordshire* 1927, cix–cx.

Portr. at Hunt Library.

WALKER, Robert (–1890)
d. March 1890

Of Richmond, Aberdeenshire. Botanist.

Gdnrs Chron. 1890 i 400.

WALKER, Thomas (*fl.* 1700s)

Judge of Vice-Admiralty Court, Bahamas. Sent plants from New Providence, Bahamas to J. Petiver.

J. Petiver *Museii Petiveriani* 1695, no. 96. E. J. L. Scott *Index to Sloane Manuscripts* 1904, 557. J. E. Dandy *Sloane Herb.* 1958, 229.

WALKER, William (1821–)
b. Manchester, Lancs 30 Aug. 1821.

Professor of Drawing. *Forest Trees* 1876.

Trans. Liverpool Bot. Soc. 1909, 95.

WALKER ARNOTT, George Arnold *see* Arnott, G. A. W.

WALL, George (*c.* 1821–1894)
d. London 18 Dec. 1894

FLS 1872. Pteridologist. Friend of G. H. K. Thwaites (1812–1882). In Ceylon from 1846. *Catalogue of Ferns Indigenous to Ceylon* 1873. *Check List* 1879.

J. Bot. 1895, 63. H. Trimen *Handbook to Fl. Ceylon* v.5, 1900, 379–80. *Ceylon Observer* 21 Dec. 1894.

Ferns at Kew. Plants at Liverpool, Peradeniya Ceylon.

Trichomanes wallii Thwaites.

WALLACE, Alexander (1829–1899)
b. London 1829 *d.* Colchester, Essex 7 Oct. 1899

MA Oxon 1858. MD 1861. Physician, Metropolitan Free Hospital, St. Pancras. Had garden at Colchester. Agent for new introductions of lilies and orchids. *Notes on Lilies and their Culture* 1873; ed. 2 1879.

Garden v.56, 1899, 290. *Gdnrs Chron.* 1899 ii 303. *J. Bot.* 1899, 496. *J. Hort. Cottage Gdnr* v.39, 1899, 311. *J. R. Hort. Soc.* 1952, 274–80 portr. *R.S.C.* v.8, 1186.

WALLACE, Alfred Russel (1823–1913)
b. Usk, Mon 8 Jan. 1823 *d.* Broadstone, Dorset 7 Nov. 1913

LLD Dublin 1882. DCL Oxon 1889. FRS 1893. FLS 1872. OM 1910. Collected with H. W. Bates in S. America, 1848–52. In Malay Archipelago, 1854–62; in 1858 while in Moluccas independently reached the same conclusions as Darwin regarding natural selection. *Palm Trees of the Amazon and their Uses* 1853. *Narrative of Travels on Amazon* 1853. *Malay Archipelago* 1869 2 vols. *Island Life* 1880. *Darwinism* 1889. *My Life* 1905 2 vols portr.

Garden v.64, 1903, iv portr. *Nature* v.73, 1905, 145–46; v.183, 1959, 723–25. C. F. P. von Martius *Fl. Brasiliensis* v.1(1), 1906, 130–31. *Gdnrs Chron.* 1913 ii 242. *Popular Sci. Mon.* v.83, 1913, 523–37. *Zoologist* 1913, 468–71. *Auk* 1914, 138–41. *J. Bot.* 1914, 15–18. *Proc. Dorset Nat. Hist. Antiq. Field Club*

1914, lxxxiv–lxxxvi portr. *Proc. Linn. Soc.* 1913–14, 63–65; 1957–58, 219–26; 1958–59, 139–54. *Orchid Rev.* 1914, 5–7. *Who was Who, 1897–1916* 737. *D.N.B. 1912–1921* 546–49. J. Marchant *A. R. Wallace* 1916 2 vols. L. T. Hogben *A. R. Wallace* 1920. H. F. Osborn *Impressions of Great Naturalists* 1924, 1–32 portr. *Proc. R. Soc. B.* v.95, 1924, i–xxxv portr. *Isis* v.7, 1925, 25–57; v.14, 1930, 133–54. *Fl. Malesiana* v.1, 1950, 555–57. *Notes R. Soc. London* v.13, 1958, 73–74; v.14, 1959, 67–84 portr. B. J. Loewenberg *Darwin, Wallace and Theory of Natural Selection* 1959. W. George *A. R. Wallace: Biologist Philosopher* 1964 portr. A. Williams-Ellis *Darwin's Moon* 1966 portr. *J. Hist. Medicine and Allied Sci.* v.21, 1966, 333–57. H. L. McKinney *Wallace and Natural Selection* 1972 portr. *Br. J. Hist. Sci.* 1972, 177–99. *Proc. Croydon Nat. Hist. Sci. Soc.* v.15, 1973, 81–100 portr. C. C. Gillispie *Dict. Sci. Biogr.* v.14, 1976, 133–40. *R.S.C.* v.6, 247; v.8, 1186; v.11, 739; v.19, 450.

MSS. at Linn. Soc. Some S. American plants at Kew. Portr. at National Portrait Gallery, BM(NH).

Wallacea Spruce.

WALLACE, James (*fl.* 1680s–1720s)

MD. In service of East India Company. Edited his father's *Description of Orkney* 1693 (plants in ed. 2 1700). Visited New Caledonia in Darien; gave plants to J. Petiver and H. Sloane.

J. Petiver *Museii Petiveriani* 1695, no. 53. *Philos. Trans. R. Soc.* v.22, 1700, 536–43. R. Pulteney *Hist. Biogr. Sketches of Progress of Bot. in England* v.2, 1790, 8. E. J. L. Scott *Index to Sloane Manuscripts* 1904, 558. M. Spence *Fl. Orcadensis* 1914, xxxv–xxxvi. *D.N.B.* v.59, 100. *Contrib. from U.S. National Herb.* v.27, 1928, 41–42. J. E. Dandy *Sloane Herb.* 1958, 229–30.

Plants at BM(NH).

WALLACE, Robert (–1838)

d. Columbia River, British Columbia 22 Oct. 1838

Gardener to Duke of Devonshire at Chatsworth, Derbyshire. Sent with Peter Banks by J. Paxton to collect plants on N.W. coast of N. America. Both were drowned in a boating accident.

Gdnrs Mag. 1839, 479–80. V. R. Markham *Paxton and the Bachelor Duke* 1935, 68–72. *Beaver, Outfit* Sept. 1942, 19–21. S. D. McKelvey *Bot. Exploration of Trans-Mississippi West, 1790–1850* 1955, 797–98.

WALLACE, Robert Whistler (1867–1955)

d. London 16 Nov. 1955

VMH 1923. VMM 1936. Nurseryman at Kilnfield Gardens, Colchester, Essex and then The Old Gardens, Tunbridge Wells, Kent. Specialised in lilies.

Gdnrs Mag. 1911, 217–18 portr. *Lily Yb.* 1949, 9–10 portr.; 1957, 131–32. *Gdnrs Chron.* 1955 ii 216. *Iris Yb.* 1956, 14.

WALLACE, Thomas (1891–1965)

b. 5 Sept. 1891 *d.* 2 Feb. 1965

MSc Durham 1919. FRS 1953. VMH 1952. Chemist, Long Ashton Research Station, 1919–23; Deputy Director, 1924–43; Director and Professor of Horticultural Chemistry, 1943–57. *Diagnosis of Mineral Deficiencies in Plants by Visual Symptoms* 1943; ed. 3 1961. *Science and Fruit* (with R. W. Marsh) 1953. *Modern Commercial Fruit Growing* (with R. G. W. Bush) 1956. Edited *J. Hort. Sci.*, 1943–58.

Fruit Yb. 1951–52, 9 portr. *Gdnrs Chron.* 1953 i 20 portr. *J. Hort. Sci.* v.40, 1965, 175–76. *Nature* v.206, 1965, 558–59. *Who was Who, 1961–1970* 1164.

WALLACE, W. E. (1851–1941)

b. Inkberrow, Worcs 1851 *d.* Dunstable, Beds 6 May 1941

VMH 1926. Nurseryman, Eaton Bray, Dunstable, 1886. Pioneer in cultivation of perpetual flowering carnations.

Gdnrs Chron. 1924 i 142 portr.; 1941 i 198.

WALLACE, William (1844–1897)

b. Cupar, Fife 11 May 1844 *d.* Oxford 18 Feb. 1897

BA Oxon 1868. Professor of Moral Philosophy, Oxford, 1882–97.

D.N.B. v.59, 116.

Herb. at Stowe School, Bucks.

WALLEN, Matthew (*fl.* 1770s–1780s)

b. Ireland

Resident in Jamaica. Cultivated plants. Assisted P. Browne (*c.* 1720–1790). "Botanices praecellens cultor et promotor" (O. Swartz *Fl. Indiae Occidentalis* v.1, 1797, 247).

J. E. Smith *Icones Pictae Plantarum Rariorum* 1790, t.3, 10. *Kew Bull. Additional Ser. I* 1898, 139. F. Cundall *Historic Jamaica* 1915, 25.

Wallenia Swartz.

WALLER, Alfred Rayney (1867–1922)

b. York 1867 *d.* 19 July 1922

MA Cantab 1905. Secretary, Cambridge University Press. First Secretary of Watson Botanical Exchange Club, 1884.

Watson Bot. Exch. Club Rep. 1922–23, 203.

WALLER, Rev. Horace (1833–1896)

b. London 1833 *d.* East Liss, Hants 22 Feb. 1896

Missionary in Central Africa, 1861–62. Rector, Twywell, Northants, 1874.

J. Bot. 1896, 190. *D.N.B.* v.59, 129. *Moçambique* 1939, 61–62 portr.

Letters and Mozambique plants at Kew.

Walleria J. Kirk.

WALLER, Richard (*c.* 1650–1715)
d. Northaw, Herts Jan. 1715

FRS 1681. Secretary, Royal Society of London, 1687–1715. Album of paintings by Waller of British flora at Royal Society.

Notes Rec. R. Soc. London 1940, 92–94 portr. W. Blunt *Art of Bot. Illus.* 1955, 130.

Portr. by T. Murray after Kneller at Royal Society.

WALLICH, George Charles (1815–1899)
b. Calcutta 15 Nov. 1815 *d.* London 31 March 1899

MD Edinburgh 1836. FLS 1860. Son of N. Wallich (1786–1854). Indian Medical Service, 1838–56. Contrib. papers on diatoms to *Ann. Mag. Nat. Hist.* 1860, 1863.

Proc. Linn. Soc. 1897–98, 29–30. *Trans. Bot. Soc. Edinburgh* v.21, 1899, 222–24. D. G. Crawford *Hist. Indian Med. Service, 1600–1913* v.2, 1914, 147. *D.N.B.* v.59, 136. *R.S.C.* v.6, 252; v.8, 1188; v.11, 743.

MS. catalogue of Bengal diatoms with drawings at BM(NH).

WALLICH, Nathaniel (*olim* Nathan Wolff) (1786–1854)
b. Copenhagen 28 Jan. 1786 *d.* London 28 April 1854

MD Copenhagen 1821. FRS 1829. FLS 1818. Studied under Vahl. Surgeon to Danish settlement, Serampore, 1807. In East India Company service, 1813. Superintendent, Botanic Garden, Calcutta, 1815–41. Collected plants at Cape, 1842–43. *Tentamen Florae Napalensis Illustratae* 1824. *Numerical List of Dried Specimens of Plants in East India Company's Museum* 1828. *Plantae Asiaticae Rariores* 1830–32 (drawings at BM(NH)).

Bot. Misc. 1830–31, 92–95. *Edinburgh New Philos. J.* v.9, 1830, 125–29. P. Smith *Mem. and Correspondence of J. E. Smith* v.2, 1832, 246–64. *Ipswich Mus. Portr. Ser.* 1852. *Bonplandia* 1854, 139–40. *Gdnrs Chron.* 1854, 284; 1964 ii 29 portr. *Gent. Mag.* 1854, 84. *Proc. Linn. Soc.* 1854, 314–18; 1888–89, 41. *Bot. Tidsskrift* v.12, 1880–81, 105–6. E. Bretschneider *Hist. European Bot. Discoveries in China* 1898, 246–47. *D.N.B.* v.59, 135–36. *J. Bot.* 1899, 458–59. *J. Straits Branch R. Asiatic Soc.* no. 65, 1913, 39–48. D. G. Crawford *Hist. Indian Med. Service, 1600–1913* v.2, 1914, 143–44. *Kew Bull.* 1913, 255–63; 1925, 312–14. C. Christensen *Danske Bot. Historie* v.1, 1924–26, 251–53 portr.; v.2, 157–66. Liverpool Museums *Handbook and Guide to Herb. Collections* 1935, 46–47 portr. *150th Anniversary Vol. R. Bot. Gdn Calcutta* 1942, 3–4 portr. *Fl. Malesiana* v.1, 1950, 557–58. M. Archer *Nat. Hist. Drawings in India Office Library* 1962, 23–25, 90–91. I. H. Burkill *Chapters on Hist. Bot. in India* 1965 *passim* portr. F. A. Stafleu *Taxonomic Literature* 1967, 489–90. D. E. Allen *Victorian Fern Craze* 1969, 8–15. A. M. Coats *Quest for Plants* 1969, 149–53 portr. *Taxon* 1970, 544–45. C. C. Gillispie *Dict. Sci. Biogr.* v.14, 1976, 145–46. *R.S.C.* v.6, 252.

Herb, letters and portr. by D. Macnee at Kew. Portr. at Hunt Library, Linnean Society.

Wallichia Roxb.

WALLIS, Anthony (1879–1919)
b. Reading, Berks 14 July 1879 *d.* Penrith, Cumberland 28 Aug. 1919

MA Cantab. School inspector. 'Pembroke and Carmarthen Plants' (*J. Bot.* 1919, 347–50) biogr.

Bot. Soc. Exch. Club Br. Isl. Rep. 1919, 630.

WALLIS, Arthur (1816–1856)
d. 24 April 1856

Of Chelmsford, Essex and Brighton, Sussex. 'Plants of Chelmsford' (*Proc. Bot. Soc. London* 1839, 34–41). 'Orchideae of Essex' (*Ann. Nat. Hist.* v.4, 1840, 270–71).

WALLIS, Edward Jonathan (–1928)
d. Kew, Surrey 23 Feb. 1928

Horticultural and botanical photographer. Trained as a wood-engraver. Photographs published in W. Watson's *Gdnrs Assistant* 1905, W. J. Bean's *Trees and Shrubs Hardy in British Isles* 1914, *Illus. Guide to R. Bot. Gdns.*

Kew Bull. 1928, 254–55.

Photographic negatives at Kew.

WALLIS, Rev. John (1714–1793)
b. South Tindal, Northumberland 1714 *d.* Norton near Stockton 19 July 1793

MA Oxon 1740. Curate, Simonburn, *c.* 1746–72; Billingham, 1776–92. *Natural History and Antiquities of Northumberland* 1769 2 vols (*see* letters of 19 April, 7 June 1831 in N. J. Winch correspondence at Linnean Society).

R. Pulteney *Hist. Biogr. Sketches of Progress of Bot. in England* v.1, 1790, 356. *Gent. Mag.* 1793 ii 769–70. J. Nichols *Lit. Anecdotes of Eighteenth Century* v.7, 704; v.8, 758–60. *D.N.B.* v.59, 145. *Nat. Hist. Trans. Northumberland, Durham and Newcastle-upon-Tyne* v.14, 1903, 77.

WALLIS, John (*fl.* 1780s–1830s)
b. Sussex

Timber surveyor. Of Lambeth. *Dendrology* 1833; ed. 2 1835.

Gdnrs Mag. 1834, 51.

WALLIS, John (*c.* 1830–1907)
d. Woore, Staffs 4 Feb. 1907

Gardener to Rev. W. Sneyd at Keele, Staffs, 1878.

Gdnrs Chron. 1907 i 111 portr.

WALLIS, John Richard (1917–1944)
b. Lamberhurst, Kent 26 March 1917 *d.* 23 Nov. 1944
Accountant. Botanised in Kent.
Bot. Soc. Exch. Club Br. Isl. Rep. 1943–44, 654.
Herb. at South London Botanical Institute.

WALLIS, Sarah *see* Lee, S.

WALPOLE, Edward Horace (1880–1964)
d. 12 Dec. 1964
Had garden at Mount Usher near Dublin.
J. R. Hort. Soc. 1965, 194–98 portr.

WALSH, Harold (1881–1962)
Of Luddendenfoot, Yorks. Workingman naturalist. Compiled list of bryophytes in Halifax parish, Yorks.
Naturalist 1973, 8.

WALSH, James Joseph (*fl.* 1880s)
Royal Navy. Collected plants in West Indies, 1889.
Rep. Br. Assoc. Advancement Sci. 1890, 448. I. Urban *Symbolae Antillanae* v.3, 1902, 138.
Plants at Kew.

WALSH, Rev. Robert (1772–1852)
b. Waterford, Ireland 1772 *d.* Finglas, Dublin 30 June 1852
BA Dublin 1796. Vicar, Finglas, 1839. Chaplain to British Embassy at Constantinople, 1820–27, 1831–35. *History of Dublin* 1815 (includes catalogue of plants). 'Plants of Constantinople' (*Trans. Hort. Soc.* v.6, 1826, 32–33).
Dublin Univ. Mag. v.15, 172 portr. N. Colgan *Fl. County Dublin* 1904, xxvi. F. Boase *Modern English Biogr.* v.3, 1901, 1178–79. *R.S.C.* v.6, 256.

WALSHE, Rev. Thomas J. (1861–1938)
b. County Waterford, Ireland 21 Oct. 1861 *d.* Liverpool 16 Aug. 1938.
MA Liverpool. Chaplain, Notre Dame Training College, 1899–*c.*1919. Founder member of Liverpool Botanical Society.
N. Western Nat. 1938, 169–71. *Who was Who, 1929–1940* 1409.

WALSINGHAM, Frank Gordon (*c.* 1890–1964)
d. 14 Feb. 1964
Kew gardener, 1912–13. Assistant Director of Horticulture, Egyptian Dept. of Agriculture, 1913. Collected plants in Egypt now at Kew. Superintendent, Harvard University Gardens in Cienfuegas, 1932–56.
Kew Bull. 1913, 359. *J. Kew Guild* 1964, 466 portr.

WALTER, Carl (*c.* 1831–1907)
b. Mecklenburg, Germany *c.* 1831 *d.* Melbourne 11 Oct. 1907
In Victoria from *c.* 1857. Collected plants for F. von Mueller. Arranged vegetable products in Technological Museum, Melbourne. 'Plants new to Victoria' (*Victorian Nat.* v.16, 1899, 98–101).
Victorian Nat. v.24, 1907, 110; v.25, 1908, 114. *Austral. Encyclop.* v.1, 1965, 285.
Prostanthera walteri F. Muell.

WALTER, Thomas (*c.* 1740–1789)
b. Hampshire *c.* 1740 *d.* St. John's, S. Carolina 17/18 Jan. 1789
Planter. *Fl. Caroliniana* 1788 (MS. brought to England in 1785 by J. Fraser).
Garden and Forest 1897, 301–2. *Ann. Rep. Missouri Bot. Gdn* 1905, 31–56. *Bull. Charleston Mus.* v.3, 1907, 33–37; v.7, 1911, 10–13. *J. Elisha Mitchell Sci. Soc.* 1910, 31–42. *Rhodora* 1914, 117–19; 1915, 129–37. *J. Bot.* 1921, 69–74. *Smithsonian Misc. Coll.* 1936, 1–6. *Kew Bull.* 1939, 331–34. F. A. Stafleu *Taxonomic Literature* 1967, 493.
Herb. at BM(NH). Letters at Kew.
Walteriana Fraser.

WALTERS, Samuel (–1885)
d. Hilperton, Wilts 11 Nov. 1885
Florist of Hilperton.
Garden v.28, 1885, 548. *Gdnrs Chron.* 1885 ii 667.

WALTHAM, Thomas Ernest (*c.* 1869–1950)
d. Christchurch, Hants 28 Sept. 1950
Flower photographer. *Common British Wild Flowers Easily Named* 1927.
Proc. Bournemouth Nat. Sci. Soc. 1949–50, 92.

WALTHAM, William (*fl.* 1810s)
Nurseryman, 199 Deansgate, Manchester, Lancs.

WALTON, George Chapman (1845–1934)
b. Scaldwell, Northants 3 Sept. 1845 *d.* Folkestone, Kent 6 Nov. 1934
FLS 1881. Botanical Curator, Folkestone Museum, 1888–1934. President, Folkestone Natural History Society, 1905–34. *List of Flowering Plants Found in Neighbourhood of Folkestone* 1893.
Proc. Linn. Soc. 1934–35, 194–95.
Herb. at Folkestone Museum.

WALTON, H. J. (*fl.* 1900s)
Surgeon on Tibet Frontier Commission of 1904. Collected plants on way from Sikkim to Lhasa.
Primula waltoni Watt.

WALTON, John (1834–1914)
b. Newcastle-upon-Tyne, Northumberland 14 Jan. 1834 *d.* Rochester, U.S.A. 13 May 1914
To N. America, 1860. Settled in Rochester. Sometime Curator of Conchology, Rochester

Academy of Science. Illustrated *Portfolio of Rare and Beautiful Flowers* 1885.

Proc. Rochester Acad. v.5, 1919, 273–74. *Huntia* 1965, 171–79.

Drawings at Monroe County Dept. of Parks.

WALTON, John (1895–1971)
b. London 14 May 1895 *d.* Dundee, Angus 13 Feb. 1971

MA Cantab. FRSE 1931. Botanist on Oxford expedition to Spitzbergen, 1921. Lecturer in Botany, Manchester University, 1924–30. Professor of Botany, Glasgow, 1930–62. President, Botanical Society of Edinburgh, 1962–64. 'Fossil Fl. of Karroo System in Wankie District, S. Rhodesia' (*Geology of Central Part of Wankie Coalfield* 1929, 62–75). *Introduction to Study of Fossil Plants* 1940; ed. 2 1953.

Nature v.196, 1962, 112. *Plant Sci. Bull.* June 1971, 19. *Times* 16 Feb. 1971. *Palaeobotanist* v.21, 1972, 127–28 portr. *Glasgow Nat.* v.19, 1973, 70–71 portr.

WANT, Thomas (–1937)
d. Stoke d'Abernon, Surrey 20 March 1937

Founder and Editor of *Nurseryman and Seedsman* 1894, *Garden Work for Amateurs* 1912, *Market Grower and Salesman* 1923.

Gdnrs Chron. 1937 i 213.

WARBURG, Edmund Frederic (1908–1966)
b. London 22 March 1908 *d.* Oxford 9 June 1966

MA Cantab. FLS 1934. Son of O. E. Warburg (1876–1937). Assistant Lecturer, Bedford College, London 1938–41. Curator, Druce Herb., Oxford, 1948. President, British Bryological Society, 1962–63. President, Botanical Society of British Isles, 1965. *Fl. British Isles* (with A. R. Clapham and T. G. Tutin) 1952; ed. 2 1962. Edited *Census Catalogue of British Mosses* 1963. Edited *Watsonia*, 1949–60.

Nature v.212, 1966, 240. *Proc. Bot. Soc. Br. Isl.* 1966, 207–8; 1967, 67–69 portr. *Times* 11, 14 and 24 June 1966. *Br. Fern Gaz.* 1967, 363–64. *Trans. Br. Bryol. Soc.* 1967, 375–77 portr. *B.S.B.I. News* v.2(1), 1973, 16.

Plants at BM(NH), Oxford; *Euphrasia* collection at Cambridge.

WARBURG, Sir Oscar Emanuel (1876–1937)
b. London 6 Feb. 1876 *d.* 1 July 1937

FLS 1929. Chairman, London County Council, 1925–26. Collected and cultivated *Quercus*, *Cistus* and *Sorbus*. 'Preliminary Study of Genus Cistus' (with E. F. Warburg) (*J. R. Hort. Soc.* 1930, 1–52). 'Oaks in Cultivation in British Isles' (with E. F. Warburg) (*J. R. Hort. Soc.* 1933, 176–89).

Kew Bull. 1937, 357. *Proc. Linn. Soc.* 1937–38, 346. *Who was Who, 1929–1940* 1411.

WARBURTON, Peter Egerton (1813–1889)
d. Adelaide, S. Australia 5 Nov. 1889

Entered Royal Navy, *c.* 1825–29. Joined Bombay Army. Emigrated to S. Australia, 1853 where he became Commissioner of Police. Collected plants in S. and W. Australia, sending them to F. von Mueller. Led expedition from Alice Springs, 1872. *Journey across Western Interior of Australia* 1875.

E. Favenc *Hist. Austral. Exploration, 1788–1888* 1888, 190–93, 255–62. B. Threadgill *South Austral. Land Exploration, 1855–1880* 1922 *passim*. G. Rawson *Desert Journeys* 1948, 183–259. *Austral. Encyclop.* v.9, 1965, 160–61.

Plants at Herb., Melbourne.

WARD, Francis Kingdon *see* Kingdon Ward, F.

WARD, George (1791–1880)
b. Witton, Blackburn, Lancs 1791 *d.* Blackburn 23 July 1880

Handloom-weaver. Blackburn plants listed in *Lancashire Nat.* v.6, 1913, 228–31, 235 portr.

WARD, Harry Marshall (1854–1906)
b. Hereford 21 March 1854 *d.* Torquay, Devon 26 Aug. 1906

BA Cantab 1879. Dsc 1892. FRS 1888. FLS 1886. Investigated coffee leaf disease in Ceylon, 1880–82. Assistant Lecturer, Owens College, Manchester, 1883–85. Professor of Botany, Royal Indian Engineering College, Coopers Hill, 1885–95. Professor of Botany, Cambridge, 1895–1906. President, British Mycological Society, 1900–1. *Diseases of Plants* 1889. *Timber and Some of its Diseases* 1889. *The Oak* 1892. *Disease in Plants* 1901. *Grasses* 1901. *Trees* 1904–9 4 vols.

Bot. Centralblatt v.102, 1906, 367–68. *Gdnrs Chron.* 1906 ii 164. *J. Bot.* 1906, 422–25. *Kew Bull.* 1906, 281–82. *Nature* v.74, 1906, 493–95. *Bot. Soc. Exch. Club Br. Isl. Rep.* 1906, 203–4. *Proc. Linn. Soc.* 1906–7, 54–57. *Ann. Bot.* 1907, ix–xiii. *Trans. Proc. Bot. Soc. Edinburgh* v.23, 1907, 218–32. *New Phytol.* 1907, 1–9. *Trans. Liverpool Bot. Soc.* 1909, 95–98. *Phytopathology* 1913, 1–2. *Proc. R. Soc. London B.* v.83, 1911, i–xiv. F. W. Oliver *Makers of Br. Bot.* 1913, 262–79. *D.N.B. Supplt. 2* v.3, 589–91. *Who was Who, 1897–1916* 741. F. O. Bower *Sixty Years of Bot. in Britain* 1938, 54–56 portr. J. S. L. Gilmour *Br. Botanists* 1944, 43–47. *R.S.C.* v.11, 747.

Plants and letters at Kew. Portr. at Hunt Library.

Wardomyces Brooks & Hansford.

WARD, Henry William (1840–1916)
b. Portarlington, County Leix 1840 *d.* Rayleigh, Essex 8 Feb. 1916

Gardener to Lord Radnor at Longford Castle, Wilts. *Potato Culture for the Million* 1891. *My Garden* 1900. *Book of the Grape* 1901. *Book of the Peach* 1903. *Flowers and Flower Culture*.

Gdnrs Chron. 1916 i 97 portr., 108–9.

WARD, James (1802–1873)
b. Wensley, Yorks 25 Dec. 1802 *d.* Barton-on-Irwell, Manchester 6 March 1873

Of Richmond, Yorks until 1871; later Barton-on-Irwell. Helped to form herb. for Botanical Society of Edinburgh. Correspondent of W. J. Hooker, H. C. Watson, J. S. Henslow, J. H. Balfour and C. C. Babington. Contrib. Richmond plants to H. C. Watson's *New Botanists' Guide* 1835–37, 274–99. *Salictum Britannicum* (with J. E. Leefe) 1842–43 (exsiccatae).

J. Sowerby and J. E. Smith *English Bot*. 2737, 2955. *J. Bot*. 1873, 222; 1898, 271–73. *Trans. Bot. Soc. Edinburgh* 1876, 35–36. *Trans. Liverpool Bot. Soc*. 1909, 98.

Herb. at BM(NH). Letters in Winch correspondence at Linnean Society.

WARD, John (1831–1903)
b. 29 Jan. 1831 *d.* 19 Nov. 1903

Market gardener, Leytonstone, Essex.

Gdnrs Chron. 1903 ii 374 portr.

WARD, John R. (–1895)
d. Nagpur, India 15 Jan. 1895

Kew gardener. Superintendent, Public Gardens, Nagpur, *c.* 1894.

J. Kew Guild 1894, 16–17; 1895, 43 portr.

WARD, Nathaniel Bagshaw (1791–1868)
b. London 1791 *d.* St. Leonards, Sussex 4 June 1868

MRCS. FRS 1852. FLS 1817. Physician, London. Inventor of Wardian case for transporting plants. Took an interest in Chelsea Physic Garden. *On Growth of Plants in Closely Glazed Cases* 1842; ed. 2 1852. *Aspects of Nature* 1864.

Gdnrs Chron. 1853, 647; 1868, 655–56. *Gent. Mag*. 1868 ii 271. *J. Bot*. 1868, 223; 1926, 200; 1934, 359. *Trans. Bot. Soc. Edinburgh* v.9, 1868, 426–30. *Proc. Linn. Soc*. 1868–69, cxii–cxix; 1888–89, 41; 1934–35, 2. *Proc. R. Soc. Edinburgh* v.18, ii–iv. *J. R. Microsc. Soc*. 1895, 2–3. *D.N.B*. v.59, 328–29. R. H. Semple *Mem. of Bot. Gdn at Chelsea* 1878, 219–30. *Curtis's Bot. Mag. Dedications, 1827–1927* 99–100 portr. *Morton Arb. Quart*. v.9(4), 1973, 49–55.

Plants at BM(NH), Kew, Oxford. Portr. by J. P. Knight at Linnean Society. Portr. at Apothecaries Hall London, Hunt Library. Letters at Kew.

Wardia Harvey & Hook.

WARD, Robert (1860–*c.* 1934)
b. Perthshire 23 April 1860

Kew gardener, 1883. Botanic Station, British Guiana, 1886; later Superintendent, Botanic Gardens and Agricultural Stations, British Guiana; retired 1926. Collected plants.

J. Board of Agric. Br. Guiana 1926, 176–79. *J. Kew Guild* 1927, 469–71; 1935, 471.

WARD, William (*c.* 1826–1883)

Gardener to Lady Foley at Stoke Edith Park, Herefordshire. Provided information for R. Hogg and G. Bull's *Herefordshire Pomona* 1876–85.

J. Hort. Cottage Gdnr v.6, 1883, 73.

WARE, Thomas Softly (*c.* 1824–1901)
d. Barnard Castle, Durham 30 May 1901

Nurseryman, Hale Farm Nurseries, Tottenham, Middx, 1857.

Gdnrs Chron. 1901 i 392. *Gdnrs Mag*. 1901, 379.

WARE, Walter T. (–1917)
d. 16 Dec. 1917

Nurseryman, Inglescombe near Bath with further nurseries in Wilts and Lincs. Specialised in daffodils.

Garden 1918, 11.

WARE, William Melville (1892–1955)
b. 19 Dec. 1892 *d.* 15 Feb. 1955

MSc London 1923. DSc 1935. Lecturer in Botany, Wye College *c.* 1919; Assistant Mycologist, 1924–37; Head of Mycology Dept., 1937–46. Did research on diseases of fruit, hops and mushrooms.

J. Hort. Sci. v.30, 1955, 149–50. *Nature* v.175, 1955, 494–95.

WAREDRAPER, Thomas (*fl.* 1830s)

Nurseryman, Well Street, Hackney, London.

WARING, Edward John (1819–1891)
d. London 22 Jan. 1891

MD. FLS 1863. Jamaica Medical Service. Madras Medical Service, 1849–65. *Remarks on Uses of Some Bazaar Medicines* 1860. *Pharmacopoeia of India* 1868.

D. G. Crawford *Hist. Indian Med. Service, 1600–1913* v.2, 1914, 150. *R.S.C.* v.6, 266; v.8, 1196.

WARING, Rev. Holt (1766–1850)

Rector, Shankill and Lurgan. Dean of Dromore. His garden at Waringstown House, County Down contained a large collection of hardy ferns.

Irish Nat. 1913, 25. *Proc. Belfast Nat. Field Club* v.6, 1913, 619.

WARING, Rev. Richard Hill (*c.* 1720–1794)
b. Shrewsbury, Shropshire *c.* 1720 *d.* Berwick near Shrewsbury 11 Sept. 1794

FRS 1769. Of St. James's, Westminster and Ince, Cheshire. 'Plants found in Several Parts of England' (*Philos. Trans. R. Soc.* v.61, 1772, 359–89).

Gent. Mag. 1794 ii 966, 1051. J. B. L. Warren *Fl. Cheshire* 1899, xci, cxii.

WARING, Rupert Thomas Tremayne (1899–1968)
b. 25 May 1899 *d.* 16 May 1968

FLS 1962. Interested in culture and hybridisation of orchids.

WARNE, Leslie Gordon Glynne (1908–1962)
b. Faringdon, Berks 25 June 1908 *d.* Jodrell Bank, Cheshire 3 May 1962

BSc Bristol 1932. MSc 1933. PhD 1936. FLS 1953. Lecturer in Botany, Manchester University, 1934; Senior Lecturer in Horticulture, 1948–62. Professor of Botany, Karachi University, 1953–56. Revised J. M. Lowson's *Textbook of Botany* 1945; ed. 4 1963. *Practical Botany for the Tropics* (with W. O. Howarth) 1959. Contrib. to *Ann. Bot.*, *New Phytol.*, *J. Pomology*.

Nature v.195, 1962, 120–21. *Proc. Linn. Soc.* 1964, 97–99.

WARNER, Charles B. (–1869)
d. Hoddesdon, Herts 27 July 1869

Of Hoddesdon. Cultivated orchids.

Gdnrs Chron. 1869, 819.

WARNER, Frederick Isaac (1841–1896)
d. Winchester, Hants 8 Nov. 1896

FLS 1872. Had herb. Winchester plants in *Proc. Winchester Sci. Soc.* v.1, 1871, 37–52. Contrib. to F. Townsend's *Fl. Hampshire* 1883.

J. Bot. 1897, 32. *Proc. Linn. Soc.* 1896–97, 71–72. *R.S.C.* v.8, 1197.

WARNER, John (*c.* 1674–1760)
b. Whitechapel, London 22 Dec. 1674 *d.* 24 Feb. 1760

Merchant, Rotherhithe, London. "A gentleman eminent for his skill in the most curious articles of horticulture." Introduced Black Hamburg vine in 1720.

Gdnrs Chron. 1917 ii 235–36. *J. R. Hort. Soc.* 1971, 371–72.

WARNER, Richard (*c.* 1713–1775)
b. London *c.* 1713 *d.* Woodford, Essex 11 April 1775

BA Oxon 1734. Correspondent of Linnaeus. *Plantae Woodfordienses* 1771–84.

R. Pulteney *Hist. Biogr. Sketches of Progress of Bot. in England* v.2, 1790, 281–83. J. Nichols *Lit. Anecdotes of Eighteenth Century* v.3, 74–75; v.8, 596; v.9, 642. *Gent. Mag.* 1816 ii 104. *D.N.B.* v.59, 398–99. G. S. Gibson *Fl. Essex* 1862, 447–48. G. C. Druce and S. H. Vines *Account of Herb.... Oxford* 1897, 49–51. *Essex Nat.* 1920–21, 72–88, 221–37; 1923–24, 206–17, 245–51, 268–76; 1938, 87–91; 1954, 195; 1956, 314–19. S. T. Jermyn *Fl. Essex* 1974, 16.

Herb. at Oxford, Essex Field Club. Library at Wadham College, Oxford.

Warneria Miller.

WARNER, Robert (*c.* 1814–1896)
d. Chelmsford, Essex 17 Dec. 1896

FLS 1874. *Select Orchidaceous Plants* 1862–91. *Orchid Album* (with B. S. Williams) 1882–97 11 vols.

Gdnrs Chron. 1896 ii 790. *Orchid Rev.* 1897, 7. *Proc. Linn. Soc.* 1897–98, 52. *Orchid Album* v.11, 1897, t.521.

Cattleya warneri T. Moore.

WARNER, Thomas (*fl.* 1590s–1600s)

Of Horsleydown, Bermondsey. Friend of J. Gerard who called him "a diligent and most effectionate lover of plants." 'Warner's Rose' named after him.

C. E. Raven *English Nat.* 1947, 214.

WARNER, Thomas (*fl.* 1840s)

Nurseryman, Leicester Abbey.

WARRE, Edmond (1837–1920)
b. London 12 Feb. 1837 *d.* 22 Jan. 1920

Headmaster and Provost, Eton College. Late in life became keen gardener and acquainted with H. N. Ellacombe and W. T. Thiselton-Dyer.

C. R. L. Fletcher *Edmond Warre* 1922 **portr.**

WARRE, Frederick (*fl.* 1820s)

Collected orchids in Brazil, 1829.

Bot. Register v.29, 1843, Misc. Notes 14.

Warrea Lindley.

WARREN, Cyril N. (–1970)

Kew gardener, 1912. Assistant orchid collector with Eugene Andre in S. America. Curator, Agricultural Dept., S. Nigeria, 1915. To Canada, 1923. Nurseryman, Berkeley, California. Founder of *Orchid Digest*.

Kew Bull. 1915, 411. *J. Kew Guild* 1963, 298; 1969, 988–92; 1970, 1158 portr.

WARREN, Elizabeth Andrew (1786–1864)
b. Truro, Cornwall 28 April 1786 *d.* Flushing, Cornwall 5 May 1864

Algologist. *Botanical Chart for Schools* 1839. Falmouth algae in *Rep. Cornwall Polytechnic Soc.* 1849, 31–37. Herb. of Cornish plants presented to Cornwall Horticultural Society, 1834.

W. H. Harvey *Phycologia Britannica* v.4, 1851, no. 354. *Rep. Cornwall Polytechnic Soc.* 1864, 11–14. *J. Bot.* 1865, 101–3. F. H. Davey *Fl. Cornwall* 1909, xliii. *R.S.C.* v.6, 269.

Schizosiphon warreniae Casp.

WARREN, John Byrne Leicester, 3rd Baron de Tabley (1835–1895)
b. Tabley Hall, Knutsford, Cheshire 26 April 1835 *d.* Ryde, Isle of Wight 22 Nov. 1895
MA Oxon 1856. FLS 1864. Poet and numismatist. Correspondent of H. C. Watson. Studied *Rubi*, *Rumex*. 'Fl. Hyde Park and Kensington Gardens' (*J. Bot.* 1871, 227–38; 1875, 336). *Fl. Cheshire* 1899 portr.
Contemporary Rev. Jan. 1896, 84–99. *J. Bot.* 1896, 77–80; 1900, 74–76. *Naturalist* 1896, 49–50. *D.N.B.* v.59, 415–16. *Gdnrs Chron.* 1899 ii 332. H. Walker *J. B. L. Warren, Lord de Tabley* 1903. *Watsonia* 1973, 431. *R.S.C.* v.8, 1198; v.11, 752; v.14, 587.
Herb. at Botanic Garden Glasnevin.
Rumex warrenii Trimen.

WARRY, William A. (*c.* 1874–1958)
d. 26 April 1958
Kew gardener, 1896. Gardener to Messrs Hubert and Manger, Guernsey.
J. Kew Guild 1958, 595; 1959, 711.
Herb. at La Société Guernesiaise Museum, Channel Islands.

WASHINGTON, Thomas (*fl.* 1830s–1840s)
Nurseryman, Hopwood Lane, Halifax, Yorks.

WATERER, Anthony (1822–1896)
d. Knaphill, Surrey 16 Nov. 1896
Nurseryman, Knaphill, Woking.
Garden v.34, 1888 portr.; v.50, 1896, 420, 421 portr. *Gdnrs Chron.* 1896 ii 628, 757; 1897 i 307 portr. *J. Hort. Cottage Gdnr* v.34, 1897, 576. *Baileya* 1953, 37–41.

WATERER, Anthony (*c.* 1850–1924)
d. 24 July 1924
Nurseryman, Knaphill, Woking, Surrey.
Garden 1924, 540(v), 576.

WATERER, Frank Gomer (*c.* 1868–1945)
Nurseryman, Bagshot, Surrey. Specialised in rhododendrons.
Garden 1921, ii–iii. *Times* 2 March 1945.

WATERER, Frederick (1822–1871)
d. 4 Oct. 1871
Eldest son of J. Waterer (*c.* 1783–1868). With his brothers John and Michael succeeded to family nursery at Bagshot, Surrey on the death of his father. Nursery specialised in American plants.
Gdnrs Chron. 1871, 1330.

WATERER, John (*c.* 1783–1868)
d. 2 Nov. 1868
Nurseryman, Bagshot, Surrey.
Gdnrs Chron. 1868, 1168.

WATERER, Michael (1745–1827)
Founded nursery at Knaphill near Woking, Surrey, 1790. Specialised in American plants.
Gdnrs Mag. 1829, 382. E. W. Brayley *Hist. of Surrey* v.2, 1850, 27.

WATERER, Ralph Ronald (–1971)
d. 8 Jan. 1971
Educ. Cambridge. Colonial Forest Service, 1923; in Cyprus for 23 years. Conservator of Forests, Kenya, 1951–57.
Commonwealth For. Rev. 1971, 216–17.

WATERFALL, Charles (1851–1938)
b. Leeds, Yorks 7 Jan. 1851 *d.* Chester, Cheshire 26 Jan. 1938
FLS 1911. Of Hull and Chester. Knowledgeable on flora of East Riding of Yorks, Derbyshire and Cheshire. Contrib. to J. F. Robinson's *Fl. East Riding of Yorkshire* 1902.
Bot. Soc. Exch. Club Br. Isl. Rep. 1938, 19–20. *Naturalist* 1938, 130. *N. Western Nat.* 1938, 104–6. *Proc. Linn. Soc.* 1937–38, 346. H. J. Riddelsdell *Fl. Gloucestershire* 1948, cxliv.
Herb. at Sheffield University. Plants at Oxford.
Epilobium waterfallii E. S. Marshall.

WATERFALL, William Booth (1850–1915)
b. Tyneside 1850 *d.* Bristol 4 Oct. 1915
FLS 1908. To Bristol, 1878. Formed moss herb. Contrib. to *Census Cat. of Br. Mosses* 1907.
Proc. Linn. Soc. 1915–16, 15. *Proc. Bristol Nat. Soc.* 1941, 290. H. J. Riddelsdell *Fl. Gloucestershire* 1948, cxlii.
Plants at BM(NH).

WATERFIELD, William (1832–1907)
b. Westminster, London 14 Aug. 1832 *d.* 24 Jan. 1907
FLS 1876. East India Company, 1852. Accountant-General, Allahabad, 1874. Had interest in Indian flora but never published anything.
Proc. Linn. Soc. 1906–7, 57–58.

WATERHOUSE, Benjamin (1754–1846)
b. Newport, Rhode Island 4 March 1754 *d.* Cambridge, Mass. 2 Oct. 1846
MD Leyden, 1780. Pupil of J. Fothergill. Professor of Natural History, Brown University, 1784. Founded Harvard Botanic Garden. *The Botanist* 1811.
P. Smith *Mem. and Correspondence of J. E. Smith* v.2, 1832, 173–75. *Proc. Cambridge Hist. Soc.* 1909, 5–22.

WATERHOUSE, Frederick George (1815–1898)
d. Adelaide, S. Australia 1898
In Australia, 1852. Naturalist on J. M. Stuart's expedition across Australia, 1861–62. Collected plants on Kangaroo Island, 1861. Curator, Adelaide Museum, 1862. *Features and Productions of Country on Stuart's Track across Australia* 1863.

G. Bentham *Fl. Australiensis* v.1, 1863, 14. *Trans. Proc. R. Soc. S. Australia* v.6, 1882, 133. *Rep. Austral. Assoc. Advancement Sci.* 1907, 178. *Austral. Encyclop.* v.2, 1965, 58.

MS. notebooks at S. Australian Archives and Mitchell Library.

Stenanthemum waterhousii Benth.

WATERMAN, William (*fl.* 1840s–1850s)

Overseer, Botanic Gardens, Sydney, 1846; transferred to Inner Domain section of the Gardens, 1848; resigned in 1852 to go gold-mining.

J. Proc. R. Soc. N.S.W. 1908, 128.

WATERS, Rev. James (*fl.* 1820s–1850s)

In Jamaica from 1826. Curate, St. Elizabeth, Jamaica, 1837. Collected plants, now at Kew.

Kew Bull. 1901, 68. I. Urban *Symbolae Antillanae* v.3, 1902, 139.

WATKINS, Alfred (*c.* 1846–1937)

d. Broadstairs, Kent 15 Jan. 1937

VMH 1927. In partnership with James Simpson opened wholesale seed firm in Savoy Street, London, 1876; moved to Essex Street, 1882; later had trial grounds at Twickenham and Feltham, etc.

Gdnrs Chron. 1926 ii 362 portr.; 1937 i 63.

WATKINS, Burton Mounsher (1816–1892)

b. Liverpool Dec. 1816 *d.* Treaddow, Herefordshire 30 July 1892

Relieving Officer at Ross, Herefordshire for 50 years. 'Florula of Dowards' (*Trans. Woolhope Field Nat. Soc.* 1881–82, 53–85). Contrib. to W. H. Purchas and A. Ley's *Fl. Herefordshire* 1889.

J. Bot. 1892, 319–20. *Trans. Liverpool Bot. Soc.* 1909, 98. *Trans. Woolhope Field Nat. Club* 1954, 241.

WATKINS, Charles R. Waddle (1800–)

b. Mangalore, India 19 Nov. 1800

Captain in Bombay Army; dismissed the services, 1829. *Principles and Rudiments of Botany* 1858.

WATKINS, Mary Philadelphia *see* Merrifield, M. P.

WATLING, Thomas (1762–1806)

b. Dumfries 19 Sept. 1762

Taught drawing. Transported to Australia for forgery, 1792. Assigned to J. White, Surgeon-General, a keen naturalist who made use of Watling to draw fauna and flora. Returned to England. In Calcutta, 1801–3. *Letters from an Exile at Botany Bay to his Aunt in Dumfries* 1794; reprinted 1945.

J. Bot. 1902, 302–3. *J. Proc. R. Austral. Hist. Soc.* 1919, 227–29. *J. Proc. R. Soc. N.S.W.* 1908, 128; 1921, 169. *Emu* 1931, 101; 1938, 331–32; 1950, 66–67. H. S. Gladstone *Thomas Watling* 1938. *Trans. Dumfries Galloway Nat. Hist. Antiq. Soc.* v.20, 1938, 70–133. *Austral. Mus. Mag.* 1938, 302. *Austral. Avian Rec.* 1922, 22–32. *Proc. R. Zool. Soc. N.S.W.* 1956–57, 162–65. B. Smith *European Vision and S. Pacific, 1768–1850* 1960 *passim.* R. and T. Rienits *Early Artists of Australia* 1963 *passim. Austral. Encyclop.* v.9, 1965, 214. *Austral. Dict. Biogr.* v.2, 1967, 574–75.

Drawings at BM(NH), Mitchell Library Sydney.

WATLINGTON, John (–1659)

d. Sept.? 1659

Friend of E. Ashmole. "Apothecary in Reading, and a very good botanist" (Ashmole). Contrib. to W. How's *Phytologia Britannica* 1650.

G. C. Druce *Fl. Berkshire* 1897, cxi–cxii. G. C. Druce *Fl. Oxfordshire* 1927, lxix.

WATSON, Forbes (1840–1869)

b. Mansfield, Notts 7 Feb. 1840 *d.* Nottingham 28 Aug. 1869

MRCS 1861. Surgeon to Nottingham Union. *Flowers and Gardens* 1872 portr.

WATSON, Gavin (1795–1858)

b. Lanarkshire 1795 *d.* Philadelphia, U.S.A. 1 Nov. 1858

MA Glasgow. To U.S.A., 1828. Physician in Philadelphia. Sent plants to Botanical Society of Edinburgh.

Trans. Bot. Soc. Edinburgh v.6, 1860, 159. J. W. Harshberger *Botanists of Philadelphia and their Work* 1899, 246–47.

WATSON, Hewett Cottrell (1804–1881)

b. Firbeck, Yorks 9 May 1804 *d.* Thames Ditton, Surrey 27 July 1881

FLS 1834. Lived at Thames Ditton from 1833. Botanist on HMS 'Styx' in survey of Azores, 1842. "Father of British topographical botany." *Outlines of Geographical Distribution of British Plants* 1832; ed. 3 1843. *New Botanist's Guide* 1835–37 2 vols. *Cybele Britannica* 1847–59 4 vols.; Supplt 1860. *Topographical Botany* 1873–74 2 vols. *London Catalogue of British Plants* 1844–74. Contrib. botany to F. D. C. Godman's *Nat. Hist. of Azores* 1870, 113–288.

Naturalist 1839, 264–69 portr.; 1864, 42. *Bot. Centralblatt* v.7, 1881, 254–55. *Gdnrs Chron.* 1881 ii 177. *J. Bot.* 1881, 257–65; 1883, 343–46. *Trans. Bot. Soc. Edinburgh* v.14, 1883, 300–3. G. C. Druce *Fl. Berkshire* 1897, clxxiii–clxxiv. *D.N.B.* v.60, 7–9. *Acta Horti Bergiani* v.3(2), 1905, t.145 portr. F. H. Davey *Fl. Cornwall* 1909, xl–xli. *Trans. Liverpool Bot. Soc.* 1909, 99. J. W. White *Fl. Bristol* 1912, 78–79. G. C. Druce *Fl. Oxfordshire* 1927, cxviii–cxix. G. C. Druce *Fl.*

Northamptonshire 1930, cxx–cxxi. G. C. Druce *Comital Fl. Br. Isl.* 1932, xii–xviii portr. *Watsonia* 1949–50, 3–5. *Proc. Bot. Soc. Br. Isl.* 1965, 110–12. D. A. Cadbury *Computer-mapped Fl.... Warwickshire* 1971, 56–57. D. H. Kent *Hist. Fl. Middlesex* 1975, 21. C. C. Gillispie *Dict. Sci. Biogr.* v.14, 1976, 189–90.

MSS. at BM(NH), Kew. Herb., letters and portr. by M. Carpenter at Kew. Letters in Winch correspondence at Linnean Society. Library sold by Quaritch, 1882.

Eleocharis watsoni Bab.

WATSON, James (*fl.* 1770s–1790s)
Brother of W. Watson (*c.* 1718–1793) with whom he was in partnership in a nursery in Islington, London, *c.* 1776.

WATSON, James (1845–1932)
d. Irvine, Ayr 31 Jan. 1932
Hosiery manufacturer, Glasgow and Irvine. Grew orchids.
Orchid Rev. 1932, 80.

WATSON, James George (–1838)
d. July 1838
Superintendent of Peradeniya Botanic Gardens, Ceylon, 1832–38. Sent plants to N. Wallich and J. Lindley. Described by General G. W. Walker as "an ignoramus who could not read the language of Botany."
Ann. Bot. R. Bot. Gdns Peradeniya v.1, 1901, 6. I. H. Burkill *Chapters on Hist. Bot. in India* 1965, 50, 97, 106.

WATSON, James Gilbert (1889–1950)
d. 5 Jan. 1950
Agricultural and Forestry Depts, Malaya, 1913. Forest Economist, Kepong, 1926. Director of Forestry, Malaya, 1940. *Malayan Plant-names* (*Malay For. Rec.* no. 5, 1928). *Mangrove Forests of Malay Peninsula* (*Malay For. Rec.* no. 6, 1928). Contrib. to *Imperial For. J.*, *Malayan Forester*.
Gdns Bull. Straits Settlements 1927, 133. *Empire For Rev.* 1950, 2–3. *Fl. Malesiana* v.1, 1950, 561–62.
Plants at Kuala Lumpur.

WATSON, John (1520–1584)
MD Oxon. Dean of Winchester, 1572; Bishop, 1580. Gunther suggests he was the author of some plant records.
R. T. Gunther *Early Br. Botanists* 1922, 305–6.

WATSON, John Forbes (1827–1892)
b. Aberdeenshire 1827 *d.* Norwood, Surrey 29 July 1892
MD Aberdeen 1848. FLS 1859. Bombay Medical Service, 1850; taught physiology at Grant Medical College. Reporter on Economic Products of India and Keeper of the Museum, India Office, London, 1858–79. *Growth of Cotton in India* 1859. *Index to Native and Scientific Names of Indian and other Eastern Economic Plants and Products* 1868. *Report on Cultivation and Preparation of Tobacco in India* 1871. *Plants of Kumaon c.* 1873. *Report on Indian Wheat* 1879.
J. Bot. 1864, 390. *D.N.B.* v.60, 15–16. F. Boase *Modern English Bot.* v.3, 1901, 1226–27. C. E. Buckland *Dict. Indian Biogr.* 1906, 442–43. D. G. Crawford *Hist. Indian Med. Service, 1600–1913* v.2, 1914, 171–72.
Letters at Kew.
Origanum watsoni Schmidt.

WATSON, Miss Mary Winifred Emia Cradock (1887–1969)
b. London 1887 *d.* 16 Jan. 1969
Kew gardener, 1915–19. Established nursery at Ditchling, Sussex in 1930s.
J. Kew Guild 1970, 1164 portr.

WATSON, Peter William (1761–1830)
b. Hull, Yorks Aug. 1761 *d.* Cottingham, Hull 1 Sept. 1830
FLS 1824. Tradesman of Hull. Founder member of Hull Botanic Garden. *Dendrologia Britannica* 1825 2 vols.
Gdnrs Mag. 1831, 512. J. C. Loudon *Arboretum et Fruticetum Britannicum* v.1, 1838, 188. *D.N.B.* v.60, 22. *Naturalist* 1903, 240.

WATSON, Samuel Harding (1837–1928)
b. Mildenhall, Suffolk 1 Dec. 1837 *d.* Waunakee, Wisc., U.S.A. 12 Dec. 1928
To Dunkirk, Wisconsin, 1847. University of Wisconsin, 1858–61. Lawyer. Collected plants in Wisconsin.
Michigan Botanist v.8, 1969, 35–37.
Herb. at Milton College.

WATSON, Thomas (*fl.* 1790s–1800s)
Nurseryman, Colebrooke Row, Islington, London. Succeeded his brothers William and James Watson in 1792. In 1798 the first to flower the Pontic azalea in Britain.

WATSON, Walter (1872–1960)
b. Wetherby, Yorks 30 March 1872 *d.* Taunton, Somerset 10 Jan. 1960
BA London. DSc. ALS 1918. Biology master, Sexeys, Bruton, Somerset, 1902; Taunton School, 1908–39. President, Somerset Natural History and Archaeological Society. *Readable School Botany* 1923. *Elementary Botany* 1926. *Census Catalogue of British Lichens* 1953. Contrib. to *J. Bot.*, *J. Ecol.*, *Proc. Somerset Archaeol. Nat. Hist. Soc.*, *Trans. Br. Mycol. Soc.*
Trans. Br. Bryol. Soc. 1960, 783–84. *Trans. Br. Mycol. Soc.* 1960, 581–82 portr. *Lichenologist* 1960, 207–8.
Flowering plants at Taunton Museum; lichens at Kew.

WATSON, Sir William (1715–1787)
b. London 3 April 1715 *d.* London 10 May 1787

MD Halle and Wittenberg 1757. LRCP 1759. FRCP 1784. FRS 1741. Knighted 1786. Physician to Foundling Hospital, London, 1762–87. Did much to introduce Linnean system into this country. Contrib. papers to *Philos. Trans. R. Soc.* 1744–63, e.g. v.42 'On Culture of Mushrooms'; v.46 'Account of... John Tradescant's Botanic Garden at Lambeth'; v.47 'Account of Bishop of London's Garden at Fulham'. Grasses lent to B. Stillingfleet (*Miscellaneous Tracts...to Natural History* 1775, xxviii).

R. Pulteney *Hist. Biogr. Sketches of Progress of Bot. in England* v.2, 1790, 295–340. J. E. Smith *Selection from the Correspondence of Linnaeus* v.2, 1821, 481. J. C. Loudon *Encyclop. Gdning* 1822, 1270. G. W. Johnson *Hist. English Gdning* 1829, 203–4. S. Felton *Portr. of English Authors on Gdning* 1830, 142. W. Munk *Roll of R. College of Physicians* v.2, 1878, 348–50. *D.N.B.* v.60, 45–47. C. C. Gillispie *Dict. Sci. Biogr.* v.14, 1976, 193–97.

Portr. by L. F. Abbot at Royal Society. Portr. at Hunt Library.

Watsonia Miller.

WATSON, William (*c.* 1718–1793)
d. 29 Jan. 1793

Before 1769 founded a nursery near Colebrooke Row, Islington. His brother James became a partner *c.* 1776 and in 1792 his brother Thomas succeeded to the business.

Trans. London Middlesex Archaeol. Soc. v.24, 1973, 185.

WATSON, William (1832–1912)
b. Aberdeen 19 March 1832 *d.* 16 June 1912

MD Aberdeen. Assistant surgeon, East India Company, 1853. Sanitary Commissioner, North West Provinces, 1859. Deputy Surgeon-General. Commissioned by Government to prepare account of flora of Kumaon. President, Botanical Society of Edinburgh, 1897–99.

Trans. Edinburgh Field Nat. Microsc. Soc. 1911–12, 447–52. *Trans. Bot. Soc. Edinburgh* v.27, 1919, 339–42.

WATSON, William (1858–1925)
b. Garston, Liverpool 14 March 1858 *d.* St. Albans, Herts 30 Jan. 1925

ALS 1904. VMH 1916. VMM 1892. Kew gardener, 1879; Assistant Curator, 1886; Curator, 1901–22. Contrib. articles on palms to *Gdnrs Chron.* 1884–93. Revised R. Thompson's *Gardener's Assistant* 1902, 1905. *Cactus Culture for Amateurs* 1889; ed. 2 1903. *Orchids* 1890. *Rhododendrons and Azaleas* 1911. *Climbing Plants* 1915.

Gdnrs Chron. 1887 ii 137; 1892 i 812–13 portr.; 1922 ii 30 portr.; 1925 i 102 portr., 135. *Garden* v.89, 1925, 96. *J. Bot.* 1925, 85. *J. Kew Guild* 1923, 160; 1925, 342; 1926, 394, 422–25 portr. *Kew Bull.* 1925, 94–96. *Nature* v.115, 1925, 271–72. *Proc. Linn. Soc.* 1924–25, 84–85. *Curtis's Bot. Mag. Dedications, 1827–1927* 307–8 portr.

Streptocarpus watsoni Hort.

WATSON, William Charles Richard (1885–1954)
b. Chislehurst, Kent 1885

Post Office clerk. Studied *Rubus*. *Handbook of Rubi of Great Britain and Ireland* 1958.

Proc. Bot. Soc. Br. Isl. 1955, 556–61.

Rubus at BM(NH), South London Botanical Institute.

Rubus watsonii W. H. Mills.

WATT, Ann (*née* **Macgregor**) (–1832)

Wife of J. Watt (1736–1819). Sent algae from Cornwall to W. Withering and J. Stackhouse.

J. Stackhouse *Nereis Britannica* 1801, xxix. *D.N.B.* v.60, 52. *J. Bot.* 1914, 322.

WATT, Sir George (1851–1930)
b. Old Meldrum, Aberdeenshire 24 April 1851 *d.* Lockerbie, Dumfriesshire 2 April 1930

MB Glasgow. FLS 1874. Professor of Botany, Calcutta, 1873. Secretary, Indian Revenue and Agricultural Dept., 1884. Formed herb. of Bengal plants. Reporter on Economic Products to Government of India, 1887–1903. Lectured on Indian botany at Edinburgh University. *First Step in Botany* 1876. *Lessons in Elementary Botany* 1877. *Dictionary of Economic Products of India* 1889–96 9 vols. *Pests and Blights of the Tea Plant* 1898; ed. 2 1903. *Wild and Cultivated Cotton Plants of the World* 1907. *Commercial Products of India* 1908.

C. E. Buckland *Dict. Indian Biogr.* 1906, 443. *Bot. Soc. Exch. Club Br. Isl. Rep.* 1930, 329. *J. Bot.* 1930, 149–50. *Kew Bull.* 1930, 331. *Nature* v.125, 1930, 677–78. *Pharm. J.* 1930, 402. *Proc. Linn. Soc.* 1929–30, 226–29. *Times* 5 April 1930. *Who was Who, 1929–1940* 1423–24. I. H. Burkill *Chapters on Hist. Bot. in India* 1965, 147–49 portr.

Herb. at Edinburgh. Portr. at Hunt Library.

WATT, Helen Winifred Boyd (*née* **De Lisle**) (1879–1968)
d. 8 Jan. 1968

Studied flora of Bournemouth where she lived. President, Bournemouth Natural History Society, 1949–50.

Proc. Bournemouth Nat. Hist. Soc. 1966–67, 23–24. *Proc. Bot. Soc. Br. Isl.* 1968, 624–25.

WATT, Hugh Boyd (1858–1941)
b. Glasgow 1 Dec. 1858 *d.* Bournemouth, Hants 17 Feb. 1941
Under-writer. 'Early Tree-planting in Scotland' (*Glasgow Nat.* 1911). 'Observations on some Land Trees' (*Glasgow Nat.* 1914).
Glasgow Nat. 1943, 99–100.

WATT, James (1736–1819)
b. Greenock, Renfrewshire 19 Jan. 1736 *d.* Birmingham 25 Aug. 1819
LLD Glasgow 1806. FRS 1785. Engineer and pioneer of the steam-engine. First found *Erythraea pulchella* in Cornwall, 1796.
W. Withering *Bot. Arrangement* v.2, 1796, 255. W. A. Clarke *First Rec. of Br. Flowering Plants* 1897, 96. *D.N.B.* v.60, 51–62. *J. Bot.* 1914, 322.
Portr. at National Portrait Gallery.

WATT, James Cromar (1862–1940)
b. Aberdeen 1862 *d.* 19 Nov. 1940
Architect and enameller. Made botanical expedition to Sikkim in 1925.
Who was Who, 1929–1940 xxxii. *Who was Who, 1941–1950* 1211.

WATT, John (–before 1703)
Surgeon. Collected plants with J. Skeen in Guinea.
J. E. Dandy *Sloane Herb.* 1958, 230.
Plants at BM(NH).

WATT, Lawrence Alexander (1849–1939)
b. Banff 1849 *d.* 20 Jan. 1939
Employed in John Brown shipyard. Authority on flora of Clydesdale and Dunbartonshire. Had herb. Contrib. to R. Hennedy's *Clydesdale Fl.* 1891.
Glasgow Nat. 1940, 44–48.
Herb. at Glasgow University. Plants at Oxford.

WATTAM, William Edward Locking (1872–1953)
b. Oldham, Lancs 1872 *d.* Huddersfield, Yorks
Solicitor's clerk, Huddersfield. Contrib. papers on lichens to *Naturalist*, 1911–53.
Naturalist 1953, 141–42 portr.; 1961, 159.

WATTERS, Thomas (*c.* 1840–1904)
d. 10 Jan. 1904
In China, 1863–94; Consul, Swatow, 1875; Ichang, 1878; Canton, 1893. Collected plants for H. F. Hance in China and Formosa.
E. Bretschneider *Hist. European Bot. Discoveries in China* 1898, 739–40. *Kew Bull.* 1901, 68.
Plants from Formosa and Korea at Kew.
Polygala wattersii Hance.

WATTS, Mr. (*fl.* 1690s)
Had garden near Enfield, Middx.
Archaeologia v.12, 1794, 189.

WATTS, David (*fl.* 1780s)
Nurseryman, 83 St. James's Street, London.

WATTS, Sir Francis (1859–1930)
b. 1 Nov. 1859 *d.* Trinidad 26 Sept. 1930
Government analytical chemist, Antigua, 1889–98; Jamaica, 1898–99. Government Chemist and Superintendent of Agriculture, Leeward Islands, 1899–1909. Principal, Imperial College of Tropical Agriculture. *Nature Teaching* 1901.
J. Bot. 1930, 344. *Kew Bull.* 1930, 494–95. *Nature* v.126, 1930, 656. *Who was Who, 1929–1940* 1424–25.

WATTS, G. A. R. (1873–1949)
d. Fleet, Hants 9 Nov. 1949
Indian army officer. Contrib. plant records to W. K. Martin and G. T. Fraser's *Fl. Devon* 1939.
Bot. Soc. Br. Isl. Yb. 1950, 73.
Notebooks at BM(NH).

WATTS, Henry (1828–1889)
d. Melbourne, Victoria 16 Dec. 1889
Sent numerous algae to W. H. Harvey. First Librarian of Victorian Field Naturalists' Club, 1881–82. Contrib. to *Victorian Nat.*
W. H. Harvey *Phycologica Australica* v.4, 1862, t.233. *Victorian Nat.* v.6, 1890, 138–39; v.25, 1908, 115; 1949, 109.
Wrangelia wattsii Harvey.

WATTS, John (*fl.* 1670s–1690s)
Apothecary. Curator, Chelsea Physic Garden, 1680–93. Sent James Harlow to collect plants in Virginia.
E. Lankester *Correspondence of John Ray* 1846, 158, 159. R. H. Semple *Mem. of Bot. Gdn at Chelsea* 1878, 14. E. J. L. Scott *Index to Sloane Manuscripts* 1904, 561. C. Wall *Hist. of Worshipful Soc. of Apothecaries of London* v.1, 1963, 165–67.

WATTS, Rev. John Stanhawe (*fl.* 1750s–1800s)
BA Cantab 1772. FLS 1798. Of Ashill, Norfolk. Contrib. to J. Sowerby and J. E. Smith's *English Bot.* 544, 556, 667.

WATTS, William (*fl.* 1800s)
Nurseryman, Walcot Place, Lambeth, London.

WATTS, William Marshall (1844–1919)
d. 13 Jan. 1919
Educ. at Owen's College, Manchester. Science master, Manchester Grammar School and Giggleswick School, 1868–1904. *School Fl.* 1878.
Who was Who, 1916–1928 1102.

WATTS, Rev. William Walter (1856–1920)
b. Ivybridge, Devon 5 Oct. 1856 *d.* Canterbury, Victoria 20 Sept. 1920
FLS 1919. To Australia, 1887; settled in Milton, Queensland. To New Zealand, 1893 where he began his study of ferns and mosses. Settled in N.S.W., eventually becoming Presbyterian minister at Gladesville, Sydney. Hon. Custodian in National Herb., N.S.W., 1909–16. 'Census Muscorum Australiensium' (with T. Whitelegge) 1902–5. Contrib. to *Proc. Linn. Soc. N.S.W., J. Proc. R. Soc. N.S.W.*
Proc. Linn. Soc. N.S.W. Supplt. 1902, 1–90; 1906, 91–163. *Handbook Br. Assoc. Advancement Sci.* 1914, 446–52. *J. Proc. R. Soc. N.S.W.* 1921, 3–4, 169. *Victorian Nat.* 1949, 105–6.
Plants at Herb., Sydney. Portr. at Hunt Library.

WAUGH, Richard (–1806)
Joint editor of *Botanist's Guide through the Counties of Northumberland and Durham* 1805–7 2 vols.
Nat. Hist. Trans. Northumberland, Durham and Newcastle-upon-Tyne 1903, 82.

WAUTON, Hilda Winifred Ivy *see* Leyel, H. W. I.

WAVELL, William (*fl.* 1780s–1820s)
MD Edinburgh. FLS 1823. Practised medicine at Barnstaple, Devon. Mineralogist. Devon plant records in R. Polwhele's *Hist. Devonshire* 1793–97 3 vols.
R. J. Thornton *Sketch of Life and Writings of... W. Curtis* 1805, 4–5. D. Turner and L. W. Dillwyn *Botanist's Guide through England and Wales* 1805, 195. H. Trimen and W. T. T. Dyer *Fl. Middlesex* 1869, 393. W. K. Martin and G. T. Fraser *Fl. Devon* 1939, 771. *Rep. Trans. Devonshire Assoc.* 1962, 623–29.

WAYNMAN, Thomas (–1795/6)
Seedsman, opposite Town Hall, High Street, Colchester, Essex.

WEAR, Sylvanus (1858–1920)
b. Felton, Northumberland 21 May 1858 *d.* Belfast 13 Nov. 1920
Flour-miller. Field naturalist.
Annual Rep. Proc. Belfast Nat. Field Club 1920–21, 97 portr. *Irish Nat.* 1921, 23. R. L. Praeger *Some Irish Nat.* 1949, 174.
Herb. at Belfast Museum.

WEARE, James (*fl.* 1790s–1830s)
Nurseryman, Coventry, Warwickshire.

WEARE, Thomas (–1829)
d. Kensington, London 28 Feb. 1829
Nurseryman, Brompton Park, Kensington in partnership with J. Gray.
Gdnrs Mag. 1829, 240.

WEATHERILL, Richard (*c.* 1825–1883)
Nurseryman, Woodside Nursery, Finchley, London; specialised in camellias.
Gdnrs Chron. 1883 i 512.

WEATHERLEY, Charles (*fl.* 1850s)
Nurseryman, Market Square, Aylesbury, Bucks.

WEATHERS, John (1867–1928)
b. Newmarket, County Cork 17 June 1867 *d.* Isleworth, Middx 10 March 1928
Kew gardener, 1888. Assistant Secretary, Royal Horticultural Society. Market gardener, Isleworth. *Practical Guide to Garden Plants* 1901. *French Market Gardening* 1909. *Bulb Book* 1911. *Commercial Gardening* 1913. *Twentieth Century Gardening* 1913. *My Garden Book* 1925.
Gdnrs Chron. 1893 i 637 portr.; 1928 i 201. *J. Hort. Cottage Gdnr* 1903, 299 portr. *J. Bot.* 1928, 119. *J. Kew Guild* 1928, 627–28 portr. *Orchid Rev.* 1928, 149.

WEATHERS, Patrick (1869–1933)
d. Isleworth, Middx 1933
Kew gardener, 1889. Curator, Royal Botanic Gardens, Old Trafford, Manchester, 1897. Secretary, Manchester Royal Botanic Society. Garden editor of *Field.*
Gdnrs Mag. 1897, 17 portr. *Gdnrs Chron.* 1933 i 108. *J. Kew Guild* 1933, 279.

WEAVER, Mary (*fl.* 1830s)
Nurseryman, Handbridge, Chester.

WEAVER, Thomas (*fl.* 1820s–1840s)
Nurseryman, Handbridge, Chester.

WEAVER, Thomas (1803–1875)
b. Dymock, Glos 28 June 1803 *d.* Winchester, Hants 21 Jan. 1875
Gardener under W. Baxter at Botanic Garden, Oxford. Gardener to Wardens of Winchester College, 1835. Contrib. to *J. Hort.*
Gdnrs Chron. 1875 i 149. *Gdnrs Yb. Almanack* 1876, 171.

WEBB, Edward (*c.* 1844–1913)
d. Paris 21 Jan. 1913
Seed merchant, Stourbridge, Worcs, 1873.
Garden 1913, 64(xvi).

WEBB, Frederick Morgan (1841–1880)
b. Stafford 1841 *d.* Edinburgh Oct. 1880
Curator, Herb., Royal Botanic Garden, Edinburgh, 1876. Studied *Rosa* and *Rubus*. Edited *Fl. Liverpool* (with H. S. Fisher) 1872. Helped J. B. L. Warren in his *Fl. Cheshire* 1899. Contrib. to F. J. Hanbury and E. S. Marshall's *Fl. Kent* 1899. '*Utricularia*' (*J. Bot.* 1876, 142–47).
J. Bot. 1880, 382–83. *Trans. Bot. Soc. Edinburgh* v.13, 1879, 88–114, cviii, cx; v.14, 1883, 163. J. B. L. Warren *Fl. Cheshire* 1899, xciv. *Trans. Liverpool Bot. Soc.* 1909, 99. *R.S.C.* v.11, 761.
Herb. at Edinburgh.
Rosa involuta var. *Webbii* Baker.

WEBB, Jane Wells *see* Loudon, J. W.

WEBB, John (*fl.* 1750s–1760s)
Seedsman at the Acorn near Westminster Bridge, London.

WEBB, Philip Barker (1793–1854)
b. Milford, Surrey 10 July 1793 *d.* Paris 31 Aug. 1854
BA Oxon 1815. FRS 1824. FLS 1818. In the East, 1818; in Spain, 1826; Canaries, 1828–30; Ireland, 1851. *Iter Hispaniense,... Plants collected in...Spain and in Portugal* 1838. *Otia Hispanica* 1839. *Histoire Naturelle des Iles Canaries* (with S. Berthelot) 1835–60 3 vols.
Gdnrs Chron. 1854, 580, 598. *Hooker's J. Bot.* 1854, 310–15. *Bonplandia* 1855, 260–61; 1857, 97–100. *Bull. Soc. Bot. France* 1856, 51–52. F. Parlatore *Elogio di Filippo Barker Webb* 1856 portr. *D.N.B.* v.60, 105–9. *Webbia* 1905, i–xi, 1–11. *Comptes Rendus AETFAT 1960* 1962, 84. *J. Soc. Bibl. Nat. Hist.* 1937, 60–63. F. A. Stafleu *Taxonomic Literature* 1967, 494–95. A. M. Coats *Quest for Plants* 1969, 32–33 portr. *Monographiae Biologicae Canarienses* no. 4, 1973, 30–48.
Herb. at Florence. MS. biogr. by Jacques Gay, 1856 at BM(NH). Letters at Kew. Portr. at Hunt Library. Library sold at Hodgsons 24 June 1927.
Webbia DC.

WEBB, Rev. Robert Holden (*c.* 1806–1880)
d. Essendon, Herts March 1880
BA Cantab 1829. Rector, Essendon, 1843. *Fl. Hertfordiensis* (with W. H. Coleman) 1849; Supplts 1851, 1859; additions in *J. Bot.* 1872, 182–84.
J. Bot. 1880, 128. A. R. Pryor *Fl. Hertfordshire* 1887, xliv. J. G. Dony *Fl. Hertfordshire* 1967, 14–15. *R.S.C.* v.6, 287; v.8, 1204.

WEBB, Wilfred Mark (1868–1952)
b. Primrose Hill, London 28 May 1868 *d.* 7 Jan. 1952
FLS 1890. Lecturer in Biology for Essex County Council, 1893–98 and Surrey County Council, 1901–8. *Principles of Horticulture* 1907.
Gdnrs Mag. 1911, 313–14 portr. *Who was Who, 1951–1960* 1146.

WEBB, William Spencer (1784–1865)
b. 2 Dec. 1784 *d.* 4 Feb. 1865
Army engineer and surveyor in India from 1802. Explored Kumaun, 1815–22. Collected plants in Himalayas and sent them to N. Wallich whose collectors R. Blinkworth and Kamrup he assisted.
R. H. Phillimore *Hist. Rec. of Survey of India* v.2, 1950, 453–54; v.3, 1954, 512–13. I. H. Burkill *Chapters on Hist. Bot. in India* 1965, 30, 31, 77, 78, 82.

WEBBER, John (*c.* 1750–1793)
b. London *c.* 1750 *d.* London 29 April 1793
RA 1791. Draughtsman on J. Cook's 3rd voyage, 1776–79. Executed at least one flower drawing.
D.N.B. v.60, 112–13. *J. Bot.* 1916, 346; 1917, 54.

WEBSTER, Angus Duncan (*fl.* 1890s–1920s)
b. Balmoral, Aberdeenshire
Manager of woodlands to Lord Penrhyn. Park superintendent, Regents Park, 1896–1920. *British Orchids* 1890; ed. 2 1898. *Fl. Kent* 1893. *Hardy Ornamental Flowering Trees and Shrubs* 1893; ed. 3 1908. *Greenwich Park* 1902. *Coniferous Trees* 1918. *London Trees* 1920.
Gdnrs Mag. 1909, 40–41 portr. *Gdnrs Chron.* 1920 i 14.

WEBSTER, George (1851–1924)
b. Aldborough, Yorks 22 June 1851 *d.* York Aug. 1924
Traveller for Messrs Backhouse and Sons, nurserymen, York. Field botanist.
F. A. Lees *Fl. West Yorkshire* 1888, 100. *Bot. Soc. Exch. Club Br. Isl. Rep.* 1924, 547–48. *J. Bot.* 1924, 341–42.
Rosa at Oxford.

WEBSTER, Rev. George Russell Bullock- *see* Bullock-Webster, *Rev.* G. R.

WEBSTER, Henry (*fl.* 1860s)
Seedsman, 29 Upperhead Row, Leeds, Yorks.

WEBSTER, Rev. James (*fl.* 1880s)
Presbyterian missionary at Mukden, Manchuria. Collected plants now at Kew.
E. Bretschneider *Hist. European Bot. Discoveries in China* 1898, 765.
Viola websteri Hemsley.

WEBSTER, John (1814–1890)
b. Blanerne, Berwickshire 8 Dec. 1814 *d.* Gordon Castle, Banffshire 11 March 1890
Gardener to Duke of Richmond at Gordon Castle, 1850.
Gdnrs Chron. 1875 i 528 portr.; 1890 i 333 portr. *J. Hort. Cottage Gdnr* v.20, 1890, 259.

WEBSTER, Joshua (*fl.* 1780s–1790s)
d. Chelsea, London
Practised medicine at St. Albans and Chigwell. Herbalist with sound knowledge of botany. *True and Brief Account...of the Cerevisia Anglicana or Syrupated English Diet Drink...* 1799.

WEBSTER, William Henry Bayley (*fl.* 1820s)
Surgeon on HM sloop 'Chanticleer' on survey work in S. Atlantic, 1828–30. Collected plants on Staten and Deception Islands. *Narrative of Voyage to Southern Atlantic Ocean* 1834 2 vols.
Tuatara 1965, 158.

WEDDELL, Hugh Algernon (1819–1877)
b. Painswick, Glos 22 June 1819 *d.* Poitiers, France 22 July 1877
MD Paris 1841. FLS 1859. Pupil of A. de Jussieu. In S. America, 1843–48 and 1851. Worked at *Cinchona, Balanophoreae* and *Urticaceae. Histoire Naturelle des Quinquinas* 1849. *Chloris Andina* 1855–57 2 vols. *Memoire sur le Cynomorium coccineum* 1860. *Monographie de la Famille des Urticées* 1856.
Gdnrs Chron. 1877 ii 217. *J. Bot.* 1877, 288. *Trans. Bot. Soc. Edinburgh* v.13, 1879, 122. P. N. E. Fournier *Notice Biographique sur H. A. Weddell* 1880. C. F. P. von Martius *Fl. Brasiliensis* v.1(1), 1906, 136–39. *Philippine J. Sci. C. Bot.* 1910, 471–73. F. A. Stafleu *Taxonomic Literature* 1967, 496–97. *R.S.C.* v.6, 296; v.8, 1209; v.12, 772.
Plants at Paris. Letters at Kew. Portr. at Hunt Library.
Algernonia Baillon. *Weddellina* Tulasne.

WEDGWOOD, Allen (1893–1915)
d. Gallipoli 1915
Son of Mrs. M. L. Wedgwood with whom he went botanising.
Catalogue of Wedgwood Herb. 1920. *Proc. Bot. Soc. Br. Isl.* 1954, 114–15.
Herb. at Marlborough College.

WEDGWOOD, John (1766–1844)
b. Etruria, Staffs March 1766 *d.* Tenby, Pembrokeshire 24 Jan. 1844
FLS 1794. Eldest son of Josiah Wedgwood. A founder of Horticultural Society of London, 1804; Treasurer, 1804–6. 'Culture of Dahlia' (*Trans. Hort. Soc. London* v.1, 1812, 113–15).
Proc. Linn. Soc. 1845, 245. *Cottage Gdnr* v.9, 1853, 357–59; 377–78. *Gdnrs Chron.* 1904 i 145–46. *J. Bot.* 1931, 143–44. *J. R. Hort. Soc.* 1931, 65–68 portr.; 196–200. H. R. Fletcher *Story of R. Hort. Soc., 1804–1968* 1969, 23–25 portr.
Letters at Kew.

WEDGWOOD, Mary Louisa (*née* Bell) (1854–1953)
b. 23 Nov. 1854 *d.* Slough, Bucks 17 April 1953
Created a herb. of British plants at Marlborough College as a memorial to her son. *Wedgwood Catalogue* 1945.
Proc. Bot. Soc. Br. Isl. 1954, 114–15. D. Grose *Fl. Wilts* 1957, 44–45.
Plants at Marlborough, Oxford.
Rubus wedgwoodiae Barton & Riddelsd.

WEEKS, Edward (*fl.* 1810s–1820s)
Gardener to Viscount Kirkwall in Wales. Hot-house builder, Horticultural Repository, King's Road, London. *Forcer's Assistant* 1814.
J. C. Loudon *Encyclop. Gdning* 1822, 1288.

WEEKS, Henry (1866–)
b. Foots Cray, Kent 1866
Gardener to Lady Byron at Thrumpton Hall, Notts.
Gdnrs Chron. 1899 ii 23 portr.; 42–43. *Gdnrs Mag.* 1899, 72 portr.

WEIGHELL, William (–1802/3)
ALS 1799. Of Sunderland, Durham. First collector of ballast plants. Collected *Fuci.* Herb. in N. J. Winch's *Botanist's Guide through...Northumberland and Durham* v.1, 1805, vi.
Nat. Hist. Trans. Northumberland, Durham and Newcastle-upon-Tyne v.14, 1903, 79–80.
Letters in Winch correspondence at Linnean Society.

WEIGHTMAN, Miss Mary (1883–1941)
b. Litherland, Lancs 30 July 1883 *d.* 7 April 1941
BSc Liverpool 1904. Teacher, Birkenhead High School for Girls. Secretary, Liverpool Botanical Society.
N. Western Nat. 1941, 344–45.

WEIR, John (–1898)
d. East Barnet, Herts 28 April 1898
To Peru with Clements Markham, 1859. Plant collector for Royal Horticultural Society in Brazil and N. Granada, 1861–64; returned to England, 1865. Lists and journal in *Proc. R. Hort. Soc.* 1861–65. 'Musci Austro-Americani' by W. Mitten (*J. Linn. Soc.* v.12, 1869) includes Weir's collection.

J. Hort. Cottage Gdnr v.34, 1865, 481. *J. Bot.* 1871, 383. *Gdnrs Chron.* 1898 i 301. *Kew Bull.* 1898, 175. *J. R. Hort. Soc.* 1898–99, 115–16. C. F. P. von Martius *Fl. Brasiliensis* v.1(1), 1906, 139–40.
Plants at BM(NH), Kew.

WEIR, John Jenner (1822–1894)
b. Lewes, Sussex 9 Aug. 1822 *d.* Beckenham, Kent 23 March 1894
FLS 1865. All-round naturalist.
Sci. Gossip v.1, 1894, 49–50 portr.

WEISS, Frederick Ernest (1865–1953)
b. Huddersfield, Yorks 2 Nov. 1865 *d.* Sydenham, Kent 7 Jan. 1953
BSc London. FRS 1917. FLS 1888. VMH 1946. Professor of Botany, Manchester, 1892–1930. President, Linnean Society, 1931–34. President, Manchester Literary and Philosophical Society, 1911–13. *Plant Life and its Romance* 1928. Translated P. Sorauer's *Popular Treatise on Physiology of Plants for Use of Gardeners* 1891.
N. Western Nat. 1930, 73–77; 1953, 603–8 portr. *Br. Fern Gaz.* 1953, 58. *Gdnrs Chron.* 1953 i 28. *Mem. Proc. Manchester Lit. Philos. Soc.* 1953, i–iii. *Nature* v.171, 1953, 285–86. *Pharm. J.* 1953, 43. *Times* 9 Jan. 1953. *Who was Who, 1951–1960* 1149–50.
Herb. at University College, Cardiff.

WELDON, Walter Frank Raphael (1860–1906)
b. Highgate, London 15 March 1860 *d.* London 13 April 1906
BA Cantab 1881. FRS 1890. FLS 1891. Professor of Zoology, University College, London, 1890–99. Professor of Comparative Anatomy, Oxford, 1899–1906. Papers on inheritance in *Biometrika* 1901–4.
Proc. R. Soc. B. v.80, 1908, xxv–xl. *Proc. Linn. Soc.* 1905–6, 109–14. *D.N.B. Supplt 2* v.3, 629–31.
Bust at Oxford Museum.

WELLBY, Montague Sinclair (1866–1900)
d. Paardekop, S. Africa 5 Aug. 1900
Commissioned into cavalry, 1886. Capt., 18th Hussars, 1894. Explored interior of Somaliland, 1894–95. Collected plants with Lieut. N. Malcolm in journey across N. Tibet, 1896. In Abyssinia, 1898. *Through Unknown Tibet* 1898 (plant list, 423).
Kew Bull. 1897, 208; 1901, 68. *Geogr. J.* v.9, 1897, 215–17. E. Bretschneider *Hist. European Bot. Discoveries in China* 1898, 812. *J. Linn. Soc.* v.35, 152–55. *Comptes Rendus AETFAT 1960* 1962 214. R. Hill *Biogr. Dict. of Sudan* 1967, 376.
Abyssinian and Tibetan plants at Kew.
Saussurea wellbyii Hemsley.

WELLESLEY, Richard Colley, Marquess Wellesley (1760–1842)
b. Dangan, County Meath 20 June 1760 *d.* Brompton, London 26 Sept. 1842
Governor-General, Fort William, India, 1798–1805. Keen naturalist. Collection of natural history drawings by Indian artists, including plants, at India Office Library.
D.N.B. v.60, 211–23. M. Archer *Nat. Hist. Drawings in India Office Library* 1962, 6–8, 91–98.

WELLS, Arthur Quinton (1896–1956)
b. 22 June 1896 *d.* Inverness 9 Oct. 1956
BCh Oxon. Practised medicine at Eyam, Derbyshire, 1923–25. Lecturer in Bacteriology, St. Bartholomew's Hospital, 1930–36. Pathologist, Bureau of Animal Population, Oxford, 1937–39. Authority on alpine plants. Chairman, Oxford Botanic Garden.
Times 11 Oct. 1956. *Who was Who, 1951–1960* 1151.

WELLS, Benjamin (–1904)
d. Crawley, Sussex Jan. 1904
Fruit nurseryman, Crawley.
J. Hort. Cottage Gdnr v.48, 1904, 73.

WELLS, Ben (1900–*c.* 1942)
b. 15 March 1900
FLS 1936. Nurseryman, Merstham, Surrey; introduced Korean chrysanthemums.
Proc. Linn. Soc. 1947–48, 72.

WELLS, William (1848–1916)
Nurseryman, Earlswood, Surrey, then Merstham, Surrey; specialised in chrysanthemums.
Gdnrs Mag. 1899, 687 portr. *Garden* 1916, 132(viii) portr.

WELLSTED, James Raymond (1805–1842)
d. France 1842
Lieut., Royal Navy. On East India Company's ship 'Palinurus', 1830 surveying Gulf of Aqaba and northern part of Red Sea; southern coast of Arabia, 1833 and Oman, 1835. Collected plants in Sinai Peninsula and Socotra. Sent plants to A. B. Lambert and J. Lindley. *Travels in Arabia* 1838 2 vols. *Travels to City of the Caliphs* 1840.
Rec. of Bot. Survey of India v.8, 1933, 467–68. *Liverpool Bull.* 1958, 4–20. *Taxon* 1970, 545.
Some plants at Liverpool.
Wellstedia Balf. f.

WELLWOOD, James (*fl.* 1820s–1870s)
Handloom weaver of Parkhead near Glasgow. Botanised in W. Scotland.
Gdnrs Chron. 1874 ii 121.

WELSH, James (*c.* 1833–1885)
d. 18 June 1885

Gardener to Dicksons and Co., nurserymen, Edinburgh, 1862.

Trans. Bot. Soc. Edinburgh v.16, 1886, 312–13.

WELTON, Thomas (*fl.* 1820s)
Nurseryman, Checkley near Cheadle, Staffs.

WELWITSCH, Friedrich Martin Josef (1806–1872)
b. Maria-Saal near Klagenfurt, Austria 5 Feb. 1806 *d.* London 20 Oct. 1872

MD Vienna 1836. ALS 1858. FLS 1865. In Lisbon, 1839–53; Director of Botanic Garden; in Angola, 1853–61; in London, 1861–72. 'Sertum Angolense' (*Trans. Linn. Soc.* v.27, 1869, 1–94).

J. Bot. 1864, 255, 326–39; 1873, 1–11 portr.; 1875, 380–82; 1897, 1, 369–74. *Gdnrs Chron.* 1872, 1426–27, 1586. *Proc. Linn. Soc.* 1872–73, xxxvii–xliv. *Bull. Soc. Bot. France* 1882, 195–96. W. P. Hiern *Cat. of African Plants collected by Dr. Welwitsch in 1853–1861* 1896–1901 2 vols. *D.N.B.* v.60, 243–45. *Curtis's Bot. Mag. Dedications, 1827–1927* 147–48 portr. *Portugaliae Acta Biologica* 1959, 257–323; 1961, 324–551. *Comptes Rendus AETFAT 1960* 1962, 97, 115. F. A. Stafleu *Taxonomic Literature* 1967, 497–98. A. M. Coats *Quest for Plants* 1969, 249. *Biol. J. Linn. Soc.* 1972, 269–303 portr. M. A. Reinikka *Hist. of the Orchid* 1972, 174–78 portr. *Garcia de Orta Bot.* v.1(1–2), 1973, 1–2, 101–4. H. Dolezal *Friedrich Welwitsch vide e obra* 1974 portr. *R.S.C.* v.6, 310; v.8, 1218.

Plants at BM(NH), Kew, Lisbon. Letters at Kew. MSS. at BM(NH). Portr. at Hunt Library.

Welwitschia Hook. f.

WENDY, Thomas (*fl.* 1500s–1560s)
BA 1518–19. Fellow, Gonville Hall College, 1519–24. Physician to Henry VIII and his successors. Mentioned by W. Turner in Dedication to his *New Herball* 1551 as having "much knowledge in herbes."

C. E. Raven *English Nat.* 1947, 69.

WENHAM, Francis Herbert (1823–1908)
d. 11 Aug. 1908

'Circulation in *Anacharis*' (*J. Quart. Microsc. Sci.* 1855, 277–83). 'Potato Blight' (*Mon. Microsc. J.* 1874, 35–36).

J. R. Microsc. Soc. 1908, 693–97. *R.S.C.* v.6, 320; v.8, 1218; v.11, 783; v.12, 775.

WENYON, Rev. Charles (*fl.* 1880s–1890s)
MD. In charge of Wesleyan Missionary Society's Hospital, Fat Shan near Canton, China. Sent plants to Kew.

E. Bretschneider *Hist. European Bot. Discoveries in China* 1898, 765.

WEST, Daniel (*fl.* 1770s)
Nurseryman, New Road, St. Pancras, London.

Gdnrs Mag. 1829, 737. *Trans. London Middlesex Archaeol. Soc.* v.24, 1973, 192.

WEST, Edward (1797–)
Of Warrington, Lancs. Agricultural writer. *Essay on Potato Disease* 1876.

Trans. Liverpool Bot. Soc. 1909, 99.

WEST, George Stephen (1876–1919)
b. Bradford, Yorks 20 April 1876 *d.* Birmingham 7 Aug. 1919

BA Cantab 1898. DSc Birmingham 1908. FLS 1901. Son of W. West (1848–1914). Professor of Natural Science, Royal Agricultural College, Cirencester, 1899–1906. Lecturer in Botany, Birmingham University, 1906; Professor of Botany, 1909. 'Alga-flora of Cambridgeshire' (*J. Bot.* 1899). *Treatise on British Freshwater Algae* 1904–11. *Monograph of British Desmidiaceae* (with W. West) 1904–11. *Algae* 1916.

Bot. Soc. Exch. Club Br. Isl. Rep. 1919, 630–31. *J. Bot.* 1919, 283–84. *Kew Bull.* 1919, 314–15. *Nature* v.103, 1919, 470–71. *Proc. Linn. Soc.* 1919–20, 52–53. *Proc. Birmingham Nat. Hist. Philos. Soc.* v.14, 1920–21, 139–46. *Who was Who, 1916–1928* 1110. R. L. Praeger *Some Irish Nat.* 1949, 176.

Plants at Birmingham. Drawings at BM(NH).

WEST, Tuffen (1823–1891)
b. Leeds, Yorks 1823 *d.* Frensham, Surrey 19 March 1891

FLS 1861. Botanical artist. Illustrated W. Smith's *Synopsis of British Diatomaceae* 1853–56 2 vols.

J. Bot. 1891, 224.

WEST, Vita Sackville- *see* Sackville-West. V.

WEST, William (1848–1914)
b. Leeds, Yorks 22 Feb. 1848 *d.* Bradford, Yorks 14 May 1914

FLS 1887. Pharmaceutical chemist. Lecturer in Botany, Technical College, Bradford, 1886. *Monograph of British Desmidiaceae* (with G. S. West) 1904–11. 'Notes on Fl. of Shetland' (*J. Bot.* 1912, 265–75). Contrib. to F. A. Lees's *Fl. West Yorkshire* 1888 and *J. Bot.*

Bot. Soc. Exch. Club Br. Isl. Rep. v.4, 1914, 53–56. *J. Bot.* 1914, 161–64 portr. *Naturalist* 1914, 227–30, 257–60 portr. *Nature* v.93, 1914, 327–28. *Proc. Linn. Soc.* 1913–14, 65–67. R. L. Praeger *Some Irish Nat.* 1949, 176.

WEST, William (1875–1901)
b. Bradford, Yorks 11 Feb. 1875 *d.* Mozufferpore, India 14 Sept. 1901

BA Cantab 1896. Son of W. West (1848–1914). 'Notes on Cambridgeshire Plants' (*J. Bot.* 1898, 246–59, 491–92).

J. Bot. 1901, 353. *Naturalist* 1901, 303–4. *Nature* v.93, 1914, 327–28.

Microscopic preparations at BM(NH).

WESTCOMBE, Thomas (1815–1893)
d. Worcester 9 May 1893

Collected British plants. Grew stapelias; collection and drawings (by his sister) at Kew. Helped E. Lees in his *Bot. of Malvern Hills* 1852.

J. Bot. 1893, 192. *Kew Bull.* 1893, 186; 1916, 168. F. H. Davey *Fl. Cornwall* 1909, xlvii.

Letters at Kew.

WESTCOTT, Frederic (–1861)

ALS 1841. Of Erdington, Birmingham. Had herb. Joint editor with G. B. Knowles of *Floral Cabinet*, 1837–40. Described *Cibotium barometz* (*Ann. Nat. Hist.* v.5, 1840, 130–31). Contrib. to *Phytologist*.

Botanic Gdn v.6, 1835–36, 498. *R.S.C.* v.6, 329.

WESTERN, William Henry (1871–1948)
b. Southport, Lancs 1871 *d.* Darwen, Lancs 27 June 1948

For more than 40 years contrib. weekly nature article to *Blackburn Times*. Authority on local flora. Founded *Lancashire Nat.*

N. Western Nat. 1948, 169–70 portr.

WESTLAND, Alexander B. (*fl.* 1880s–1890s)

Kew gardener, 1883. Assistant Superintendent, Botanical and Afforestation Dept., Hong Kong, 1883–90. Collected *Rhododendron westlandii* on Lan Tan Island and other plants on adjacent islands. To Taj Mahal Gardens, India. In U.S.A. in 1890s.

E. Bretschneider *Hist. European Bot. Discoveries in China* 1898, 764–65. G. A. C. Herklots *Hong Kong Countryside* 1951, 167.

Aristolochia westlandi Hemsley.

WESTON, A. Gould Hunter- *see* Hunter-Weston, A. G.

WESTON, Francis (*fl.* 1680s–1710s)

Seedsman "at the Flower-de-Luce, over against the May-Pole in the Strand."

WESTON, Richard (*c.* 1733–1806)
d. Leicester 20 Oct. 1806

Thread-hosier, Leicester. *Tracts on Practical Agriculture and Gardening* 1769. *Universal Botanist and Nurseryman* 1770–77 4 vols. *Gardener's and Planter's Calendar* 1773. *Fl. Anglicana* 1775.

Gent. Mag. 1806 ii 1080–81. S. Felton *Portr. of English Authors on Gdning* 1830, 66–70. D. Turner and L. W. Dillwyn *Botanist's Guide through England and Wales* 1805, 195. *D.N.B.* v.60, 369–70. *Gdnrs Chron.* 1899 ii 353; 1953 i 221–22.

WESTON, William Alastair Royal Dillon- *see* Dillon-Weston, W. A. R.

WETHERELL, James (*fl.* 1850s)

H.M. Vice-Consul, Bahia, Brazil. Sent some Brazilian plants to England.

Bot. Mag. 1855, t.4835.

Billbergia wetherelli Hook.

WEYMOUTH, Thomas, Viscount *see* Thynne, T.

WHALLEY, Thomas (*fl.* 1780s–1810s)

Seedsman, Castle Street, Liverpool; also (1814) at Maghull Nurseries near Ormskirk, Lancs.

WHAN, Rev. William Taylor (1829–1901)
b. Moneymore, Londonderry 30 Oct. 1829
d. Skipton, Victoria 2 April 1901

MA Belfast. To Australia, 1860. Presbyterian minister, Skipton until 1884; Minister, Port Fairy, Victoria. Collected plants for F. von Mueller.

G. Bentham *Fl. Australiensis* v.1, 1863, 14. *Victorian Nat.* 1908, 115–16. *Proc. Belfast Nat. Field Club* v.6, 1913, 623.

Plants at Melbourne.

Acacia whanii F. Muell.

WHARTON, Henry Thornton (1846–1895)
b. Mitcham, Surrey 1846 *d.* Hampstead, London 22 Aug. 1895

MA Oxon 1874. MRCS. Mycologist and ornithologist. Contrib. chapter on flora to J. L. Lobley's *Hampstead Hill* 1889, 73–80.

Ibis 1896, 159. *D.N.B.* v.60, 402. *R.S.C.* v.11, 792; v.19, 570.

WHARTON, Mrs. Elizabeth
(*fl.* 1790s–1810s)

Large collection of her watercolour drawings of flowers and fungi with MS. biogr. sketch sold at Christie's, 17 May 1967.

WHATELEY, Henry (–1915)
d. 18 Dec. 1915

Nurseryman, Spring Lane, Kenilworth, Warwickshire, 1875.

Orchid Rev. 1916, 18.

WHATELY, Thomas (–1772)
d. 26 May 1772

Of Nonsuch Park, Surrey. Politician. Secretary to Earl of Surrey. *Observations on Modern Gardening* 1770.

J. C. Loudon *Encyclop. Gdning* 1822, 1276. G. W. Johnson *Hist. English Gdning* 1829, 233. S. Felton *Portr. of English Authors on Gdning* 1830, 72–78. *D.N.B.* v.60, 429–30.

WHATELY, Thomas (-1821)
d. Isleworth, Middx 16 Nov. 1821
Surgeon, Old Jewry, London. Contrib. to W. Withering's *Bot. Arrangement* 1787 and J. Sowerby and J. E. Smith's *English Bot.* 442.

WHEATCROFT, Alfred (1895–1965)
b. Nottingham 1895 *d.* 23 Jan. 1965
In partnership with his brother Harry established nursery at Nottingham, 1919; specialised in roses.
Gdnrs Chron. 1965 i 152 portr.

WHEELER, Daniel (*c.* 1819–1894)
d. Chelmsford, Essex 26 Feb. 1894
MRCS 1841. Of Reigate, Surrey and later of Chelmsford. '*Cuscuta trifolii*' (*Phytologist* v.1, 1844, 753–55).

WHEELER, Edwin (1833–1909)
b. Clifton, Bristol 8 Feb. 1833 *d.* Bristol 28 April 1909
Homoeopathic druggist, Bristol. Had herb. 2449 drawings of British fungi, 1880–95, presented to BM(NH), 1895.
J. W. White *Fl. Bristol* 1912, 98–99. H. J. Riddelsdell *Fl. Gloucestershire* 1948, cxxxii.

WHEELER, Elizabeth (*fl.* 1810s–1820s)
Nurseryman, Northgate Street, Gloucester.

WHEELER, George (*c.* 1791–1878)
d. Warminster, Wilts 10 June 1878
Succeeded to his father's nursery which was founded in Warminster in 1773.
Florist and Pomologist 1878, 112. *Garden* v.13, 1878, 586. *Gdnrs Chron.* 1878 i 805–6.

WHEELER, James (*fl.* 1750s–1760s)
Nurseryman, Gloucester. *Botanist's and Gardener's New Dictionary* 1763.
J. C. Loudon *Encyclop. Gdning* 1822, 1273. G. W. Johnson *Hist. English Gdning* 1829, 217.

WHEELER, James (*fl.* 1840s)
Nurseryman, 2 New Bond Street, Bath. In partnership with John Salter.

WHEELER, James Cheslin (-1860)
Nurseryman, 99 Northgate Street, Gloucester.

WHEELER, James Daniel (*fl.* 1820s–1840s)
Nurseryman, Gloucester.

WHEELER, James Lowe (*fl.* 1820s–1870)
d. 1870
FLS 1823. Son of T. Wheeler (1754–1847). Botanical Demonstrator, Chelsea Physic Garden, 1821–34. *Catalogus Rationalis Plantarum Medicinalium in Horto Societatis Pharmaceuticae Londinensis...Cultarum* 1830.
R. H. Semple *Mem. of Bot. Gdn Chelsea* 1878, 164–68.

WHEELER, Leonard Richmond (1888–1948)
b. Highgate, London 23 July 1888 *d.* 25 Sept. 1948
BA London. FLS 1939. 'Botany of Antigua' (*J. Bot.* 1916, 41–52). *Vitalism* 1939. *Harmony of Nature* 1948.
Proc. Linn. Soc. 1948–49, 253–54.
Plants at BM(NH).

WHEELER, Richard (*fl.* 1690s)
Sent plants from Norway "for many years" to J. Petiver (*Museii Petiveriani* 1695, 47).

WHEELER, Thomas (1754–1847)
b. London 24 June 1754 *d.* London 10 Aug. 1847
FLS 1799. Botanical Demonstrator, Chelsea Physic Garden, 1778–1820. Pupil of W. Hudson. Wrote Latin texts for H. C. Andrews's *Heathery* 1804–12.
Proc. Linn. Soc. 1848, 380–81. R. H. Semple *Mem. of Bot. Gdn Chelsea* 1878, 152–64.
Portr. at Apothecaries' Hall.

WHELDALE, Muriel *see* Onslow, M.

WHELDON, James Alfred (1862–1924)
b. Northallerton, Yorks 26 May 1862 *d.* Liverpool 28 Nov. 1924
FLS 1901. ALS 1923. Pharmacist to H.M. Prison, Walton, Liverpool, 1891–1922. "An excellent all-round botanist and an expert at mosses." President, Liverpool Botanical Society. *York Catalogue of British Mosses* 1888. *Fl. West Lancashire* (with A. Wilson) 1907. 'Lichens of South Lancashire' (with W. G. Travis) (*J. Linn. Soc.* v.43, 1915, 87–136).
Proc. Liverpool Bot. Soc. 1908–9, portr. *Lancashire Nat.* v.4, 1911, 265–69 portr. *Nature* v.114, 1924, 904. *Bot. Soc. Exch. Club Br. Isl. Rep.* 1924, 548–52 portr. *Bryologist* 1925, 28. *J. Bot.* 1925, 52–54; 1926, 80–81. *Naturalist* 1925, 29–31 portr. *Proc. Linn. Soc.* 1924–25, 85–87. *Rep. Br. Bryol. Soc.* 1925, 178–80. *Revue Bryol.* v.53, 1926, 10–13. *N. Western Nat.* 1953, 394–98. *R.S.C.* v.19, 573.
Herb. at National Museum of Wales Cardiff. Portr. at Hunt Library.
Drepanocladus aduncus var. *Wheldoni* Renault.

WHELER, Rev. Sir George (1650–1724)
b. Breda, Holland 1650 *d.* Durham 15 July 1724
MA Oxon 1683. DD 1702. FRS 1677. Knighted 1682. Prebendary of Durham, 1684. Vicar, Basingstoke, Hants, 1685. Rector, Houghton-le-Spring, 1709–23. Travelled in France and Italy, 1673–75 and in Greece and Levant, 1675–76 collecting plants, coins, classical MSS. and antique

marbles. Brought plants to L. Plukenet, R. Morison and J. Ray. List of plants in Ray's *Collected Travels* v.2, 1693, 30–34. Introduced *Hypericum calycinum*. *Journey into Greece* 1682.

L. Plukenet *Almagestum Botanicum* 1696, 49, 72, 190, etc. R. Morison *Plantarum Historiae Universalis Oxoniensis* v.3, 1699, 362, 376, 385, etc. R. Pulteney *Hist. Biogr. Sketches of Progress of Bot. in England* v.1, 1790, 357–58. A. à Wood *Athenae Oxonienses* v.2, 1813–20, 388. *Gent. Mag.* 1833 ii 107–12 portr. *J. Bot.* 1894, 170–1. *D.N.B.* v.60, 445–46. E. J. L. Scott *Index to Sloane Manuscripts* 1904, 566. G. C. Druce and S. H. Vines *Account of Herb. of University of Oxford* 1897, 53–55. J. E. Dandy *Sloane Herb.* 1958, 230. A. M. Coats *Quest for Plants* 1969, 15–17.

Herb. at Oxford. Plants at BM(NH).
Convolvulus wheleri Vahl.

WHITE, Rev. (*fl.* 1840s)
Chaplain, Singapore. Gave plants to T. E. Cantor which went to Kew and Cambridge.
J. Asiatic Soc. Bengal v.23, 1854, 623–50. *Fl. Malesiana* v.1, 1950, 570.

WHITE, Adam (1817–1878)
b. Edinburgh 29 April 1817 *d.* Glasgow 30 Dec. 1878
FLS 1846. In Zoological Dept., BM, 1835–63. 'Peloria' (*Ann. Nat. Hist.* v.4, 1840, 286–87).
Entomol. Mon. Mag. 1879, 210–11. *J. Bot.* 1879, 96. *D.N.B.* v.61, 31. *Rhodora* 1936, 410. *R.S.C.* v.6, 347.
Scrapbook of plants at Toronto University.

WHITE, Charles Frederick (1818–1896)
b. Poplar, London 12 Feb. 1818 *d.* Clapton, London 20 Nov. 1896
FLS 1876. Drew mosses, microscopic fungi and pollen; drawings presented to Kew Herb. President, Ealing Microscopical and Natural History Club. 'Pollen from Egyptian Funereal Garlands' (*J. Linn. Soc.* v.21, 1884, 251, t.6).
Proc. Linn. Soc. 1896–97, 72–73.

WHITE, Mrs. Claude (*fl.* 1890s)
Wife of Political Officer in Sikkim before 1898. Helped G. King and R. Pantling. Discovered *Cymbidium whiteae* in Sikkim.

WHITE, David (1767–1818)
d. 'Apollo' in Bombay harbour 6 Jan. 1818
Surgeon on Bombay Establishment. 'Malabar Cardamom' (*Trans. Linn. Soc.* v.10, 1811, 229–55).
W. R. Dawson *Cat. of Smith Papers* 1934, 96. *R.S.C.* v.6, 349.
Letters in J. E. Smith correspondence at Linnean Society.

WHITE, Edward (*c.* 1873–1952)
b. Worthing, Sussex *c.* 1873 *d.* Woking, Surrey 6 Jan. 1952
VMH 1920. Landscape gardener; planned the first Royal Horticultural Society Chelsea Flower Show. President, Institute of Landscape Architects, 1931–33.
Gdnrs Chron. 1922 i 146 portr.; 1931 ii 42 portr.; 1952 i 28. *Who was Who, 1951–1960* 1158.

WHITE, Eliza Catherine (*née* Quekett) (1812–1875)
b. Langport, Somerset 1812 *d.* Ealing, Middx 14 Nov. 1875
Wife of C. F. White (1818–1896). "A good British botanist, a keen collector of mosses, micro-fungi, Bryozoa, etc."
Proc. Linn. Soc. 1896–97, 73.

WHITE, Ernest William (1858–1884)
b. Eythorne, Kent 20 June 1858 *d.* Philadelphia, U.S.A. 29 Nov. 1884
With parents *c.* 1864 to Argentine where he later collected plants.
Ibis 1885, 335–36.
Plants at BM(NH).

WHITE, Francis Buchanan White (1842–1894)
b. Perth 20 March 1842 *d.* Perth 3 Dec. 1894
MD Edinburgh 1864. FLS 1873. Son of F. I. White. President, Perthshire Society of Natural Science, 1867–72, 1884–92. Edited *Scott. Nat.*, 1871–82. 'Revision of British Willows' (*J. Linn. Soc.* v.27, 1890, 333–457). *Fl. Perthshire* 1898 (biogr. and portr. xxviii–lix).
Trans. Proc. Perthshire Soc. Nat. Sci. v.1, 1889–90, 155–206; v.2, 1897–98, xlv–xlvii. *Ann. Scott. Nat. Hist.* 1895, 73–91 portr. *J. Bot.* 1895, 49–52. *Proc. Linn. Soc.* 1894–95, 38–39. *Sci. Gossip* v.1, 1894, 241–42 portr. *D.N.B.* v.61, 35–36. *R.S.C.* v.8, 1229; v.11, 795; v.12, 780.
Herb. at Perth Museum. Letters in M. J. Berkeley and W. Wilson correspondence at BM(NH).

WHITE, Francis Isaiah (1815–1898)
d. Perth 8 Oct. 1898
MD Edinburgh 1838. *Inaugural Dissertation on Geography of Plants* 1838.
Trans. Proc. Perthshire Soc. Nat. Sci. v.3, 1903, ii.

WHITE, Rev. Gilbert (1720–1793)
b. Selborne, Hants 18 July 1720 *d.* Selborne 26 June 1793
MA Oxon 1746. Curate, Selborne. Marked Selborne plants in a copy of W. Hudson's *Fl. Anglica* (*J. Bot.* 1893, 289–94). *Natural History of Selborne* 1789. *Naturalist's Calendar* 1795. *Garden Kalendar, 1751–1771* (facsimile of MS. in British Library) 1975.

Nature v.12, 1875, 481–82. *Gdnrs Chron.* 1897 ii 366. E. A. Martin *Bibliography of Gilbert White* 1897 portr. *D.N.B.* v.61, 36–48. R. Holt-White *Life and Letters of Gilbert White of Selborne* 1901 portr. 2 vols. W. H. Mullens *Gilbert White of Selborne* 1907. *Hastings and East Sussex Nat.* 1909, 153–73. *Selborne Mag.* 1913, 65–67; 1914, 126–31 portr. D. Prain *Gilbert White as a Botanist* 1923. W. Johnson *Gilbert White* 1928. W. Johnson, *ed. Journals of Gilbert White* 1931. A. H. Wolley-Dod *Fl. Sussex* 1937, xxxviii–xxxix. W. S. Scott *White of Selborne* 1950 portr. R. M. Lockley *Gilbert White* 1954 portr. C. S. Emden *Gilbert White in his Village* 1956. R. Holt-White *Letters to Gilbert White* 1960. *J. Hist. Biol.* 1969, 363–90. A. Rye *Gilbert White and his Selborne* 1970. *Country Life* 1970, 247–51. *New Scientist* 1970, 126–28. C. C. Gillispie *Dict. Sci. Biogr.* v.14, 1976, 299–300.

MSS. at Gilbert White Museum, Selborne.

WHITE, Harry (*c.* 1857–1938)
d. 27 May 1938

VMH 1927. Nurseryman, Sunningdale Nurseries, Windlesham, Surrey. One of the pioneers of Lincolnshire bulb industry.

Gdnrs Chron. 1938 i 403.

WHITE, J. T. (*c.* 1848–1930)

VMH 1927. Nurseryman, Daffodil Nurseries, Spalding, Lincs.

Gdnrs Chron. 1925 i 230 portr.

WHITE, James (*fl.* 1790s)

Seedsman, 91 Whitechapel, London.

WHITE, James Walter (1846–1932)
b. London 8 Aug. 1846 *d.* Bristol 26 Oct. 1932

FLS 1889. Pharmaceutical chemist, Clifton, Bristol, 1874. President, Bristol Naturalists' Society, 1907–9. Lectured in Systematic Botany at Bristol University. *Fl. Bristol Coalfield* 1886. *Fl. Bristol* 1912. Contrib. to *J. Bot.*, *Pharm. J.*, *Proc. Bristol Nat. Soc.*

Bot. Soc. Exch. Club Br. Isl. Rep. 1932, 83–86. *J. Bot.* 1931, 248; 1933, 47–49 portr. *Proc. Bristol Nat. Soc.* 1932, 341–42. *Proc. Cotteswold Nat. Field Club* 1931–32, 230–33. *Proc. Linn. Soc.* 1932–33, 210–12. *Annual Rep. Watson Bot. Exch. Club* 1932–33, 160–63 portr. H. J. Riddelsdell *Fl. Gloucestershire* 1948, cxxxvi. J. E. Lousley *Fl. Isles of Scilly* 1971, 85.

Herb. at Bristol University. Plants at Oxford.

WHITE, John (*fl.* 1540s–1590s)

Cartographer and artist to W. Raleigh's Virginia expeditions, 1585, 1587.

Gdnrs Chron. 1923 i 200–1. *Virginia Mag. Hist. Biogr.* v.35, 1927, 419–30; v.36, 1928, 17–26, 124–29. *Hist. Today* 1963, 310–20. P. Hulton and D. B. Quinn *American Drawings of John White, 1577–1590* 1964 2 vols. R. H. Jeffers *Friends of John Gerard; Biogr. Appendix* 1969, 33.

Drawings at BM.

WHITE, John (*c.* 1756–1832)
d. Worthing, Sussex 20 Feb. 1832

FLS 1796. Joined Royal Navy as surgeon's mate, 1778; promoted surgeon, 1780; Chief Surgeon, N.S.W., 1786. Collected plants which he sent to J. E. Smith and A. B. Lambert. Returned to England, 1794. *Journal of Voyage to N.S.W.* 1790 (with botanical appendix by J. E. Smith) (reprinted with biogr., 1962).

J. D. Hooker *Fl. Tasmaniae* v.1, 1859, cxxiv. G. B. Barton *Hist. N.S.W.* 1889 *passim. J. R. Soc. N.S.W.* v.43, 1908, 128–29. *Emu* 1924, 209–15. *Univ. Sydney Med. J.* 1928, 115. G. Mackaness *Admiral Arthur Phillip* 1937 *passim. Med. J. Australia* 1933, 183–87. J. B. Cleland *Nat. in Medicine with particular Reference to Australia* 1950, 549–63. E. Ford *Med. Practice in Early Sydney* 1955. *Austral. Mus. Mag.* 1936–38, 298–301. *Austral. Dict. Biogr.* v.2, 1967, 594–95. R. and T. Rienits *Early Artists of Australia* 1963 *passim. Taxon* 1970, 545.

Letters in J. E. Smith correspondence at Linnean Society.

WHITE, John (–1837)
d. Dec. 1837

Gardener, Botanic Garden, Glasnevin, Dublin, 1797–1834. Collected plants in Ireland for Glasnevin. *Essay on Indigenous Grasses of Ireland* 1808. Contrib. most of localities to K. Baily's *Irish Fl.* 1833.

Phytologist v.2, 1845, 345–46. N. Colgan *Fl. County Dublin* 1904, xxvii. H. F. Berry *Hist. of R. Dublin Soc.* 1915, 191.

WHITE, Joseph Hill (*c.* 1848–1907)
d. Worcester 12 April 1907

Succeeded his uncle, Mr. Haywood, as proprietor of St. John's Nursery, Worcester.

J. Hort. Cottage Gdnr v.54, 1907, 340.

WHITE, Miss Matilda (*fl.* 1830s)

Discovered *Ornithopus pinnatus* on Tresco, Isles of Scilly, 1838.

J. E. Lousley *Fl. Isles of Scilly* 1972, 81.

WHITE, Robert (*c.* 1810–1877)
d. Parkstone, Dorset 28 May 1877

Nurseryman, Poole and Parkstone Nurseries.

WHITE, Thomas (from 1776 **Thomas Holt White**) (1724–1797)
b. Compton, Surrey 19 Oct. 1724 *d.* Feb. 1797

FRS 1777. Brother of Gilbert White. Articles on British trees signed T.H.W. in *Gent. Mag.*

WHITE, W. H. (*c.* 1859–1942)
b. Exeter, Devon *c.* 1859 *d.* Leatherhead, Surrey 14 July 1942
In charge of Sir Trevor Lawrence's orchid collection at Burford Lodge, Dorking until 1914. *Book of Orchids* 1902. Contrib. to *Gdnrs Chron., Orchid Rev.*
Gdnrs Chron. 1942 ii 48. *Orchid Rev.* 1942, 151–52.

WHITE, William (*fl.* 1760s)
Nurseryman, Mile End, London.

WHITE-SPUNNER, Letitia Hannah Damer (*née* **Sandys**) (1840–1911)
b. Isle of Wight 1840 *d.* Danaghmore, County Tyrone 1911
Had herb. of Irish flora.
Proc. Belfast Nat. Field Club v.6, 1913, 627–28.

WHITEAVES, Joseph Frederick (1835–1909)
To Canada, 1862. Curator, Montreal Natural History Society. Palaeontologist, Geological Survey of Canada.
J. Ewan *Rocky Mountain Nat.* 1950, 335.
Scottish and Canadian plants at Oxford.

WHITEHEAD, Sir Charles (1834–1912)
b. Kent 7 May 1834 *d.* 29 Nov. 1912
FLS 1871. Farmed until 1879. Chairman, Botanical and Zoological Committee, Royal Agricultural Society of England. *Market Gardening for Farmers* 1880. *Fruit Growing in Kent* 1881. *Hop Cultivation* 1893. *Hints on Vegetable and Fruit Farming* 1890; ed. 4 1893.
Nature v.90, 1912, 390. *Proc. Linn. Soc.* 1912–13, 64–65. *Who was Who, 1897–1916* 759.

WHITEHEAD, Rev. Edward (1789–1827)
b. Bolton-le-Moors, Lancs 1789 *d.* Eastham, Worcs 4 June 1827
BA Oxon 1808. BD 1820. Rector, Eastham, 1805. Discovered *Aconitum* in British Isles, 1819.
T. Purton *Midland Fl.* v.3, 1821, 47. *Trans. Liverpool Bot. Soc.* 1909, 99.

WHITEHEAD, John (1833–1896)
b. Dukinfield, Cheshire 1833 *d.* Oldham, Lancs 6 May 1896
Cotton operative. Bryologist. First President, Manchester Cryptogamic Society. President, Ashton Linnean Botanical Society. Discovered *Chara braunii*. Contrib. to R. Braithwaite's 'Sphagnaceae Exsiccatae'. Mosses in *Fl. Ashton-under-Lyne* 1888, *Naturalist* 1886, 85–100, *J. Bot.* 1894, 193–201.
Cottage Gdnr v.3, 1861, 585. *J. Bot.* 1897, 89–91 portr. *Heywood Advertiser* 14 Aug. 1908. *Trans. Liverpool Bot. Soc.* 1909, 100.
Herb. at Manchester University. Letters in W. Wilson correspondence at BM(NH).
Amblystegium filicinum var. *Whiteheadii* Wheldon.

WHITEHEAD, John (1860–1899)
b. Muswell Hill, London 30 June 1860 *d.* Hainan, China 2 June 1899
Ornithologist who also collected plants. Collected in Corsica, 1882–83, in Borneo, 1884–88 and in Philippines, 1893–96. *Exploration of Mount Kina Balu* 1893.
J. Bot. 1896, 355; 1899, 526. *D.N.B.* v.61, 104. E. D. Merrill *Bot. Work in Philippines* 1903, 29. *Fl. Malesiana* v.1, 1950, 571–72.
Plants from Mt. Kinabalu and Philippines at BM(NH).
Rhododendron whiteheadii Rendle.

WHITEHEAD, Robert Bovill (1867–1946)
b. Fiume 1867
Had garden at Piddletrenthide, Dorset.
Quart. Bull. Alpine Gdn Soc. 1946, 108–9.

WHITEHEAD, Tatham (1889–1964)
b. Burnley, Lancs 23 June 1889 *d.* 3 Feb. 1964
Assistant Lecturer in Agriculture and Forest Botany, Armstrong College, Newcastle-upon-Tyne, 1920–46. Mycologist, University College, Bangor, 1946–55. *Potato in Health and Disease* (with T. P. McIntosh and W. M. Findlay) ed. 2 1945; ed. 3 1953.
B.M.S. News Bull. no. 23, 1964–65, 31.

WHITELEGG, George G. (*c.* 1877–1957)
d. Knockholt, Kent 19 Oct. 1957
VMH 1952. Landscape gardener noted for his rock and water gardens at the Chelsea Flower Shows. In partnership with Page at Chislehurst until 1915; partner with Major Murrell at Orpington until 1923.
Gdnrs Chron. 1953 i 38 portr.; 1957 ii 329. *Iris Yb.* 1957, 26.

WHITELEGGE, Thomas (1850–1927)
b. Stockport, Cheshire 17 May/Aug. 1850 *d.* Sydney, N.S.W. 4 Aug. 1927
Workingman naturalist. Secretary and President, Ashton Linnean Botanical Society. Had herb. Corresponded with C. Darwin. To Australia, 1883. On staff of Australian Museum, Sydney until 1908. Authority on mosses and ferns. 'Census Muscorum Australiensium' (with W. W. Watts) (*Proc. Linn. Soc. N.S.W.* 1902 Supplt. 1–90; 1905 Supplt. 91–163).
Austral. Mus. Mag. 1927, 133. *Sydney Morning Herald* 5 Aug. 1927. *Proc. Linn. Soc. N.S.W.* 1928, 3. *Rec. Austral. Mus.* 1929, 265–77 portr. *Austral. Encyclop.* v.9, 1965, 297–98.

WHITESIDE, Robert (*c.* 1866–1960)
d. Nov. 1960
Had laundry at Lancaster. Founder member of N. British Pteridological Society, 1891.
Br. Fern Gaz. 1963, 133–34.
Herb. at BM(NH).

WHITFIELD (Withfield), Thomas (*fl.* 1840s)
Plant collector for Lord Derby in Sierra Leone and Gambia.
Bot. Mag. 1845, t.4119, 4155. E. Jardin *Herborisations sur la Côte Occidentale d'Afrique 1845–48* 1851, 15. *Bull. Herb. Boissier* 1907, 83.
Plants at BM(NH), Kew. Letters at Kew.
Whitfieldia Hook.

WHITING, James Edward (*c.* 1850–1927)
Taxidermist, Hampstead, London. Contrib. botanical notes to *Hampstead Annual* 1901 and T. W. Barratt's *Annals of Hampstead* 1912.
D. H. Kent *Hist. Fl. Middlesex* 1975, 25.

WHITLA, Francis (*fl.* 1830s–1850s)
Of Belfast, later of Dublin. Knew Irish plants well. Distinguished *Equisetum trachyodon*. 'Fossil *Equiseta*' (*J. Geol. Soc. Ireland* 1838, 79–81). Contrib. to J. T. Mackay's *Fl. Hibernica* 1836 and G. Dickie's *Fl. Ulster* 1864.
London J. Bot. 1846, 311. *Proc. Belfast Nat. Field Club* v.6, 1913, 627. R. L. Praeger *Some Irish Nat.* 1949, 177. *R.S.C.* v.6, 351.
Letters at Kew and in W. Wilson correspondence at BM(NH).
Whitlavia Harvey.

WHITLAW, Charles (1776–1829)
b. East Lothian 1776
Quack doctor. In Botanic Garden, Edinburgh, 1794–96. To New York, 1796; travelled in N. and S. America and West Indies, 1803–16. To London, 1826. *New Medical Discoveries* (with translation of Linnaeus's *Materia Medica*) 1829 (biogr., 42–112).
Gent. Mag. 1820 i 31–32. W. Darlington *Reliquiae Baldwinianae* 1843, 114, 123, 206.

WHITLEY, Eva (–1926)
d. 22 Feb. 1926
BSc London 1891. FLS 1907. Teacher, Maida Vale High School, London. Botanist.
Proc. Linn. Soc. 1925–26, 99.

WHITLEY, Reginald (*c.* 1754–1835)
d. 28 Jan. 1835
In partnership with Peter? Thoburn in nursery in Cromwell Road, Kensington, 1788–90; Whitley and Barrit, 1796–1801; Whitley and Brames until 1810. Whitley, Brames and Milne acquired Burchell's Fulham Nursery in 1810; Whitley and Osborn from 1833. In 1808 Whitley obtained seed of Chinese white paeony, *Paeonia lactiflora Whitleyi*. A. H. Haworth's *Miscellanea Naturalia* pt. 3, 1803 dedicated to him.
Bot. Mag. 1792, t.180. *Hort. J. and R. Lady's Mag.* v.3, 1835, 79–80. *Gdnrs Mag.* 1835, 160. E. Bretschneider *Hist. European Bot. Discoveries in China* 1898, 222–23. *Trans. London Middlesex Archaeol. Soc.* v.24, 1973, 183, 186.

WHITLOCK, John (*c.* 1740–1827)
d. Coventry, Warwickshire Jan. 1827
Gardener to Lord Kilmorey. In partnership with Mr. Bagley, nurseryman, Chelsea.
Gdnrs Mag. 1827, 488.

WHITMEE, Rev. Samuel James (1838–1925)
b. Stagsden, Beds 26 May 1838 *d.* London 10 Dec. 1925
Missionary of London Missionary Society in Samoa, Loyalty Islands, etc., 1863–77. Pastor, Dublin and Bristol. Returned to Samoa, 1891–94. Collected plants in Samoa (ferns described by J. G. Baker in *J. Bot.* 1876, 9–13, 342–45).
Times 14 Dec. 1925. *J. Bot.* 1926, 24. *Kew Bull.* 1926, 46. *Nature* v.117, 1926, 351. *R.S.C.* v.11, 798.
Plants at BM(NH), Kew, Oxford.
Cyathea whitmeei Baker.

WHITMILL, Benjamin (*fl.* 1720s–1730s)
Gardener at Hoxton, Shoreditch, London. *Kalendarium Universale* 1726; ed. 7 1765.

WHITTAKER, Mr. (*fl.* 1840s)
Collected plants in Adelaide district and also Encounter Bay of S. Australia. Sent plants to Kew.
Rep. Austral. Assoc. Advancment Sci. 1907, 179.
Drosera whittakeri Planchon.

WHITTAKER, Joseph (–1894)
d. Morley, Derbyshire 29 March 1894
Gardener and botanist.
Gdnrs Chron. 1894 i 379. *J. Hort. Cottage Gdnr* v.28, 1894, 239.
Herb. formerly at Derby Museum destroyed by floodwater.

WHITTALL, Edward (1851–1917)
Merchant at Smyrna, Turkey. Collected plants which he sent to Kew.
Kew Bull. 1893, 147; 1899, 81–82. *Garden* v.46, 1894, xii portr. *Lloyd's Log* June 1972, 18–19 portr. *Quart. Bull. Alpine Gdn Soc.* 1975, 240, 245–46.
Letters at Kew.
Fritillaria whittallii Baker.

WHITTINGHAM, Charles (*fl.* 1780s–1790s)
Nurseryman, Much Park Street, Coventry, Warwickshire.

WHITTLE, Peter Armstrong (1789–1867)
b. Inglewhite, Goosnargh, Lancs 9 July 1789 *d.* Liverpool 7 Jan. 1867

Bookseller and printer, Preston, Lancs, 1810–51. Lived in Bolton, Lancs then Liverpool. List of plants in his histories of Southport, Lytham and Blackpool. *Topographical ...Account of...Preston* (v.2, 1837, 336, autobiogr.). *Fl. Prestoniensis* 1837 (unpublished).

Men of the Time 1865, 825. *D.N.B.* v.61, 158. F. Boase *Modern English Biogr.* v.3, 1901, 1329. *Trans. Liverpool Bot. Soc.* 1909, 100.

WHITTON, James (1851–1925)
b. Methven Castle, Perthshire 1851 *d.* Glasgow 30 Oct. 1925

VMH 1912. Gardener at Glamis Castle. Superintendent, Glasgow Public Parks, 1893 and Curator, Botanic Gardens, 1902.

Gdnrs Mag. 1907, 591 portr., 592. *Garden* 1925, 650. *Gdnrs Chron.* 1923 i 314 portr.; 1925 ii 378.

WHITWELL, George (*c.* 1840–1924)
d. Kendal, Westmorland 17 June 1924

Collected ferns. Secretary, British Pteridological Society. 'Personal Finds [of ferns in Lake District]' (*Br. Fern Gaz.* 1909, 29–34).

Br. Fern Gaz. 1909, 48; 1924, 73–75.

WHITWELL, William (1839–1920)
b. Manchester 30 Oct. 1839 *d.* Knowle, Warwickshire 16 Dec. 1920

FLS 1892. Civil servant. Contrib. to *J. Bot.* and F. A. Lees's *Fl. West Yorkshire* 1888.

Bot. Soc. Exch. Club Br. Isl. Rep. 1921, 367–69. *J. Bot.* 1921, 84–85. A. R. Horwood and C. W. F. Noel *Fl. Leicestershire* 1933, ccxxvii.

Herb. at Birmingham Museum.

WHYMPER, Edward (1840–1911)
b. London 27 April 1840 *d.* Chamonix, France 16 Sept. 1911

Wood-engraver and mountaineer. In Greenland, 1867 and 1872 (plants described by O. Heer in *Philos. Trans. R. Soc.* v.159, 1869, 445–88). *Travels among the Great Andes of the Equator* 1892 (with plant lists).

J. Bot. 1890, 161–62. *D.N.B. Supplt. 2* v.3, 656–58. *Kew Bull.* 1915, 64. *Bibliotheca Botanica* Heft 116, 1937, 55. F. S. Smythe *Edward Whymper* 1940 portr. *R.S.C.* v.8, 1233; v.11, 799.

Greenland algae and Andean plants at BM(NH).

Helosis whymperi Baker f.

WHYTE, Alexander (1834–1908)
b. Fettercairn, Kincardineshire 5 March 1834 *d.* High Barnet, Herts 21 Dec. 1908

FLS 1894. To British Central Africa, 1891. Curator, Botanic Garden, Uganda, 1898. Director of Agriculture, British East Africa, 1902. West African Gold Concessions Company, Liberia, 1903. Collected plants on M'lanje Mountains, 1891; Uganda and Liberia, 1904.

J. Bot. 1892, 244–45; 1909, 155. *Trans. Linn. Soc.* v.4, 1894, 1–67. *Gdnrs Chron.* 1909 i 16. *Kew Bull.* 1909, 24. *Proc. Linn. Soc.* 1908–9, 51–52. *Comptes Rendus AETFAT 1960* 1962, 45.

Plants at BM(NH), Kew.

Widdringtonia whytei Rendle.

WHYTEHEAD, Rev. William (1757–1817)

Vicar, Atwick, Yorks. Herb. of plants collected at Hornsea, Yorks now at Hull University.

WHYTOCK, James (*c.* 1845–1926)
d. 31 Jan. 1926

VMH 1914. Gardener to Duke of Buccleugh at Dalkeith until 1921. President, Scottish Horticultural Association, 1908–10. President, Botanical Society of Edinburgh, 1917–20.

Gdnrs Chron. 1921 i 282 portr.; 1926 i 107. *Trans. Bot. Soc. Edinburgh* v.29, 1927, 309–10.

Whytockia W. W. Smith.

WICKHAM, Sir Henry Alexander (1846–1928)
b. London 29 May 1846 *d.* 28 Sept. 1928

Inspector of Forests, India. Acting District Commissioner in Colon. Collected seeds of Para rubber in Brazil which were sent to Kew in 1876 where they were germinated and sent to the Far East to form the beginnings of the rubber industry. *Journey through the Wilderness* 1872. *Introduction, Plantation and Cultivation of Para (Hevea) Rubber* 1908.

Pharm. J. 1928, 342. *Who was Who, 1916–1928* 1123.

Portr. at Hunt Library.

WICKHAM, John Clements (1798–1864)
b. Leith, Midlothian 21 Nov. 1798 *d.* France 6 Jan. 1864

Entered Royal Navy, 1812. Second in command on HMS 'Beagle'; in command of 'Beagle', 1837–41 surveying Australian coast. Keen naturalist and assisted B. Bynoe with his collections. Resigned from Royal Navy, 1841. Police magistrate, Moreton Bay, 1842; Government Resident there, 1853.

J. D. Hooker *Fl. Tasmaniae* v.1, 1859, 126. *Victorian Nat.* 1927, 187. *Austral. Encyclop.* v.9, 1965, 300–1. *Austral. Dict. Biogr.* v.2, 1967, 597.

WICKHAM, William (1831–1897)
b. London 1831 *d.* Binsted Wyck, Hants 16 May 1897
MA Oxon 1857. FLS 1879. MP for Petersfield, Hants. Studied the natural history, and especially the botany, of East Hampshire. Interested in the state of flora of Selborne since Gilbert White's time.
Proc. Linn. Soc. 1896–97, 73–74.

WIDDRINGTON, Samuel Edward (*olim* Cook) (1787–1856)
d. Felton, Northumberland 11 Jan. 1856
Entered Royal Navy, 1802; retired as Commander, *c.* 1824. Took name of Widdrington, 1840. In Spain, 1829–32, 1843. 'European Pines' (*Ann. Nat. Hist.* v.2, 1839, 163–78; v.3, 1839, 296–302; v.8, 1842, 87–90). 'Vegetation of Spain' (*Rep. Br. Assoc. Advancement Sci. 1847* v.2, 88–89).
D.N.B. v.61, 182. F. Boase *Modern English Biogr.* v.3, 1901, 1335.
Widdringtonia Endl.

WIDNALL, Samuel (*fl.* 1830s)
Nurseryman, Grantchester, Cambridgeshire.

WIGAN, Sir Frederick (*c.* 1828–1907)
d. 2 March 1907
Had orchid collection at Clare Lawn, East Sheen, Surrey.
Orchid Rev. 1907, 112.
Odontoglossum × *Wiganianum.*

WIGG, Lilly (1749–1829)
b. Smallburgh, Norfolk 25 Dec. 1749 *d.* Yarmouth, Norfolk 29 March 1829
ALS 1790. Of Yarmouth: shoemaker, schoolmaster and clerk in Dawson Turner's bank. Instructed D. Turner in algae. Contrib. to J. Sowerby and J. E. Smith's *English Bot.* 205, 419, 571, 847, 2247 and to W. Withering's *Bot. Arrangement* 1787–92.
Trans. Linn. Soc. v.6, 1802, 126, 136. *Gent. Mag.* 1830 i 184–85. C. J. and J. Paget *Sketch of Nat. Hist. Yarmouth* 1834, xxix. *Trans. Norfolk Norwich Nat. Soc.* v.2, 1876–77, 269–74; 1912–13, 687–88. W. M. Hind *Fl. Suffolk* 1889, 480. *D.N.B.* v.61, 192–93. *J. Bot.* 1902, 321.
MS. on esculant plants 1810 at BM(NH). MS. *Fl. Cibaria* and paper silhouette portr. at Kew. Portr. by C. J. Paget at Linnean Society. Portr. at Hunt Library.
Fucus wigghii D. Turner.

WIGHAM, Robert (1785–1855)
b. Tanfield, Durham 6 Jan. 1785 *d.* Norwich, Norfolk 15 Feb. 1855
Tobacco manufacturer. Diatomist.
K. Trimmer *Fl. Norfolk* 1866, viii. F. G. Kitton *Mem. of F. Kitton* 1895, 7–8. *Trans. Norfolk Norwich Nat. Soc.* v.7, 1901–2, 298–303 portr.
Herb. acquired by H. D. Geldart.
Choetocerus wighamii Brightw.

WIGHT, Robert (1796–1872)
b. Milton, Duncra Hill, East Lothian 6 July 1796 *d.* Grazeley, Berks 26 May 1872
MD Edinburgh 1818. FRS 1855. FLS 1832. To India as assistant-surgeon in East India Company, 1819; stationed at Madras. Superintendent, Botanic Garden, Madras, 1826. Surgeon, 1831. *Contributions to Botany of India* 1834. *Icones Plantarum Indiae Orientalis* 1840–53 6 vols. *Illustrations of Indian Botany* 1840–50 2 vols. *Spicilegium Neilgherrense* 1846–51 2 vols. *Prodromus Florae Peninsulae Indiae Orientalis* (with G. A. W. Arnott) v.1, 1834.
Bot. Misc. 1830–31, 95–97. N. Wallich *Plantae Asiaticae Rariores* v.1, 1830, 72; v.2, 55. *J. Bot.* (Hooker) 1841, 156–201 portr. *Gdnrs Chron.* 1872, 731–32. *Proc. Linn. Soc.* 1872–73, xliv–xlvii. *Trans. Bot. Soc. Edinburgh* v.11, 1873, 363–88 portr. *J. Bot.* 1872, 180, 223; 1899, 459–60. *D.N.B.* v.61, 194–95. H. Trimen *Handbook to Fl. Ceylon* v.5, 1900, 374–75. *Curtis's Bot. Mag. Dedications, 1827–1927* 143–44 portr. A. White and B. L. Sloane *Stapelieae* v.1, 1937, 97–98 portr. *Nature* v.154, 1944, 566–69. I. H. Burkill *Chapters on Hist. Bot. in India* 1965 *passim.* F. A. Stafleu *Taxonomic Literature* 1967, 501–2. *Taxon* 1970, 545. M. A. Reinikka *Hist. of the Orchid* 1972, 149–52 portr. *R.S.C.* v.6, 364.
Herb. and letters at Kew. Drawings and MSS. at BM(NH). Portr. by D. Macnee at Kew. Portr. at Hunt Library.
Wightia Wall.

WIGHTON, John (1803–1878)
b. Craiglockhart, Edinburgh 25 May 1803 *d.* Costessey Park, Norfolk 23 Nov. 1878
Gardener to Earl of Stafford at Costessey Park near Norwich. Contrib. to *Florist and Pomologist.*
Gdnrs Chron. 1877 i 401–2 portr.; 1878 ii 733.

WILCOX, James Fowler (1823–1881)
b. Pilton, Somerset 1823 *d.* South Grafton, N.S.W. 11 July 1881
Zoologist, primarily ornithologist. Joined HMS 'Rattlesnake' 1846 and left ship at Sydney, 1851. Established himself as a dealer in natural history specimens. With his son, J. C. Wilcox, collected plants in New Guinea which he sent to F. von Mueller. Introduced Jacaranda tree to streets of Grafton where he settled *c.* 1855.

J. Macgillivray *Narrative of Voyage of HMS Rattlesnake* 1852, 231, 318. *J. R. Soc. N.S.W.* 1908, 129–30. *Fl. Malesiana* v.1, 1950, 575. *Austral. Dict. Biogr.* v.2, 1967, 167.
MS. journal at Mitchell Library, Sydney.
Pleiococca wilcoxiana F. Muell.

WILD, Charles James (*fl.* 1870s–1890s)
FLS 1889. Of Manchester, *c.* 1878–83. Collected *Musci* in Queensland, 1887.
British cryptogams at Oxford.

WILD, Thomas (*fl.* 1830s)
Nurseryman, Ipswich, Suffolk.

WILDE, James Plaisted, 1st Baron Penzance (1816–1899)
b. London 12 July 1816 *d.* 9 Dec. 1899
Judge of Provincial Courts of Canterbury and York. Had garden at Eashing Park, Surrey; interested in hybridising roses.
Gdnrs Mag. 1899, 792 portr. *Garden* v.57, 1900, 80–81 portr. *Who was Who, 1897–1916* 558.

WILDEGOSE, Robert (*fl.* 1820s–1830s)
Of Daventry, Northants. List of plants by Wildegose and author's sister in G. Baker's *Hist. and Antiq. of Northamptonshire* 1822–41.

WILDSMITH, William (*c.* 1838–1890)
b. Bradford, Yorks *c.* 1838 *d.* Jan. 1890
Gardener to Lord Eversley at Heckfield, Hants.
Garden v.37, 1890, 117, 141. *Gdnrs Chron.* 1890 i 171.

WILES, James (*fl.* 1790s–1800s)
Gardener to R. A. Salisbury. Botanist on HMS 'Providence' under Capt. W. Bligh, 1791–93. At Botanic Garden, Liguanea, Jamaica, 1793–1805. Edited *Hortus Eastensis* 1806. Sent plants to A. B. Lambert.
Kew Bull. 1891, 300–1. I. Urban *Symbolae Antillanae* v.3, 1902, 140. *J. Linn. Soc.* v.45, 1920, 48. F. Watson *Hist. Rec. of Australia* ser. 1, 1916, 5–30. I. Lee *Captain Bligh's Second Voyage to South Sea* 1920, 18–19. *J. Bot.* 1922, 24. *Papers Proc. R. Soc. Tasmania* 1922, 10–12. *Fl. Malesiana* v.1, 1950, 575. *Taxon* 1970, 545.
Plants at BM(NH). Letters at Kew.

WILFORD, Charles (–1893)
d. Wimbledon, Surrey 1893
Assistant, Kew Herb., *c.* 1854–57. Collected plants in Hong Kong, 1857–58; Formosa, 1858; Korea and Japan, 1859. On return to England looked after herb. and collection of living plants of W. Wilson Saunders.
Hooker's J. Bot. 1857, 273–74. E. Bretschneider *Hist. European Bot. Discoveries in China* 1898, 400, 539–44. *J. Kew Guild* 1901, 38. E. H. M. Cox *Plant Hunting in China* 1945, 94–95.
Plants and letters at Kew.
Tripterygium wilfordii Hook. f.

WILKIE, David (–1961)
Assistant Curator, Royal Botanic Garden, Edinburgh. *Gentians* 1936. Contrib. to *Quart. Bull. Alpine Gdn Soc.*
Quart. Bull. Alpine Gdn Soc. 1961, 132.

WILKIN, Simon (1790–1862)
b. Costessey, Norfolk 27 July 1790 *d.* Hampstead, London 28 July 1862
FLS 1811. Printer and publisher, Norwich. Entomologist. Friend of J. E. Smith. Had private botanic garden of hardy plants at Costessey. Sent *Conferva capillaris* to J. Sowerby and J. E. Smith's *English Bot.* 2364.
Proc. Linn. Soc. 1862–63, xlvi–xlix. *D.N.B.* v.61, 259.
Letters at Kew.

WILKINS, Charlotte *see* Wilson, C.

WILKINS, William Henry (–1966)
b. Norfolk *d.* 25 April 1966
MA 1924. FLS 1922. Demonstrator in Botany, Oxford, 1924–47; Reader, 1947.
Times 27 April 1966.

WILKINSON, Caroline Catherine, Lady (*née* Lucas) (1822–1881)
b. Llandebie, Carmarthen 10 May 1822 *d.* Llandovery, Carmarthen 2 Oct. 1881
Married Sir J. G. Wilkinson in 1856. *Weeds and Wild Flowers* 1858. Made drawings of fungi.
J. Bot. 1880, 224; 1882, 159–60.

WILKINSON, Hannah Elizabeth (*née* Naylor) (1810–1892)
b. Batley Carr, Yorks 6 May 1810 *d.* Anerley Surrey 24 Jan. 1892
Wife of E. S. Wilkinson of Enfield in 1849. Herb. formerly at London School of Medicine for Women.

WILKINSON, Henry John (1859–1934)
b. Ogleforth, Yorks 28 Dec. 1859 *d.* York 6 Dec. 1934
Director, Terry's Confectionery Works, York. Curator, Herb., Yorkshire Philosophical Society, 1892. 'Historical Account of Herb.' (*Annual Rep. Yorkshire Philos. Soc.* 1906, 45–71). President, York and District Field Naturalists' Society, 1901, 1909–10, 1931–33. President, York Florists' Society, 1923. 'Phanerogamic Fl. and Vascular Cryptogams' (*York Handbook Br. Assoc. 1906* 275–93).
J. Bot. 1935, 106–8. *Naturalist* 1935, 60–61 portr. *N. Western Nat.* 1935, 149–53 portr.
Herb. at Yorkshire Philosophical Society. Plants at Nottingham University.

WILKINSON, Sir John Gardner (1797–1875)
b. Hardendale, Westmorland? 5 Oct.? 1797 *d.* Llandovery, Carmarthen 29 Oct. 1875
DCL Oxon 1852. FRS 1834. Knighted 1839. Explorer and Egyptologist. Collected plants in Egypt, 1821–30, and made drawings.
J. Bot. 1880, 224. *D.N.B.* v.61, 274–76. E. Durand and G. Barratte *Florae Libycae Prodromus* 1910, xxvi.
Plants at BM(NH).

WILKINSON, John Grimshaw (1856–1937)
b. Leeds, Yorks 6 Jan. 1856 *d.* Leeds 28 Feb. 1937
A blind botanist who could distinguish many plants by touch, taste and smell. President, Leeds Naturalists Club.
Naturalist 1915, 249. *Bot. Soc. Exch. Club Br. Isl. Rep.* 1937, 437–38. *Gdnrs Chron.* 1937 i 299. *J. Bot.* 1937, 142–43. *N. Western Nat.* 1937, 182–83.
Herb. at Leeds University.

WILKINSON, Thomas (*fl.* 1790s–1810s)
Nurseryman, Barton-on-Irwell, Lancs.

WILKINSON, William Henry (–1918)
FLS 1893. FRMS. Of Sutton Coldfield, Warwickshire. Lichenologist.
Proc. Linn. Soc. 1918–19, 16.
Plants at Birmingham University.

WILKS, Rev. William (1843–1923)
b. Ashford, Kent 19 Oct. 1843 *d.* Shirley, Croydon 2 March 1923
MA Cantab. VMH 1912. Curate, Croydon, 1866. Vicar, Shirley, 1879–1912. Secretary, Royal Horticultural Society, 1888–1920. Created a garden where he bred the Shirley poppy. *Elementary Handbook of Fruit Culture* (with G. Bunyard). *Selected List of Hardy Fruits* (with G. Bunyard) 1914.
J. Hort. Cottage Gdnr v.20, 1890, 91–92 portr. *Gdnrs Chron.* 1892, 690 portr.; 1913 i 34–35 portr.; 1919 ii 271 portr.; 1923 i 127–28 portr. *Gdnrs Mag.* 1894, 291 portr. *Garden* v.57, 1900, 384–86 portr.; v.72, 1908, iv portr.; 1919, 563 portr.; 1923, 128, 138 portr. *Bot. Soc. Exch. Club Br. Isl. Rep.* 1923, 155–56. *J. R. Hort. Soc.* 1923, 157–60 portr. *Nature* v.111, 1923, 403–4. *Curtis's Bot. Mag. Dedications, 1827–1927* 371–72 portr. *Who was Who, 1916–1928* 1126.
Portr. at Hunt Library.

WILLARD, Jesse (*fl.* 1900s)
b. Hawkhurst, Kent
Gardener to Baroness Burdett-Coutts at Holly Lodge, Highgate.
Gdnrs Chron. 1909 ii 314–15 portr.

WILLEY, Frederick Enos (1871–1898)
b. Exeter, Devon 1871 *d.* Freetown, Sierra Leone 19 Jan. 1898
Kew gardener, 1892. Curator, Aburi Garden, Gold Coast, 1894; Botanic Station, Freetown, Sierra Leone, 1895.
Kew Bull. 1895, 318; 1897, 303–17; 1898, 57–60. *Gdnrs Mag.* 1898, 73. *J. Kew Guild* 1898, 36–37 portr.

WILLIAMS, Benjamin Samuel (1824–1890)
b. Hoddesdon, Herts 2 March 1824 *d.* Holloway, London 24 June 1890
FLS 1879. Son of J. Williams (1797–1891). Gardener to the Warner family at Hoddesdon. Commenced nursery with Robert Parker at Seven Sisters Road, Holloway, 1854–61; moved to Victoria and Paradise Nurseries, Upper Holloway. *Hints on Cultivation of British and Exotic Ferns and Lycopodiums* 1852. *Orchid-growers' Manual* 1852; ed. 6 1885. *Select Ferns and Lycopods* 1868. *Choice Stove and Greenhouse Ornamental-leaved Plants* 1870; ed. 3 1873. *Orchid Album* (with R. Warner) 1882–97 11 vols. Contrib. to R. Warner's *Select Orchidaceous Plants* 1862–97.
J. Hort. Cottage Gdnr v.62, 1879, 468. *Orchid Album* v.9, 1891, [1–3] portr. *Proc. Linn. Soc.* 1890–92, 27. *Gdnrs Chron.* 1890 i 801; ii 19 portr. M. A. Reinikka *Hist. of the Orchid* 1972, 223–25 portr.
Dendrobium williamsianum Reichenb. f.

WILLIAMS, C. H. (*fl.* 1860s)
Correspondent of Sir W. J. Hooker to whom he sent plants from Bahia, Brazil.
Bot. Mag. 1864, t.5485.
Epistephium williamsii Hook. f.

WILLIAMS, Rev. Charles (1796–1866)
b. London 18 July 1796 *d.* Sibbertoft, Northants 16 June 1866
Congregational minister at Newark, Salisbury, London and Sibbertoft. *The Vegetable World* 1833.
Gdnrs Mag. 1833, 352–53. *D.N.B.* v.61, 398.

WILLIAMS, Rev. Edward (1762–1833)
b. Eaton Mascott, Shropshire 1762 *d.* Shrewsbury, Shropshire 3 Jan. 1833
BA Oxon 1783. Antiquary. Rector, Chelsfield, Kent, 1817–33. Discovered *Elatine hexandra* in Britain, 1798. Had herb. MS. catalogue of Shropshire plants (W. A. Leighton *Fl. Shropshire* 1841, ix). 'Shropshire Lichens' (*Ann. Mag. Nat. Hist.* v.1, 1868, 183–88). Contrib. to J. Sowerby and J. E. Smith's *English Bot.* 904, 955, 2360.
Gent. Mag. 1833 i 182–83; ii 155. *D.N.B.* v.61, 395–96.

WILLIAMS, Frederick Newton (1862–1923)
b. Brentford, Middx 19 March 1862 *d.* Brentford 6 May 1923

LRCP. FLS 1884. Practised medicine at Brentford. Authority on *Caryophyllaceae*. *Enumeratio Specierum Varietatumque Generis Dianthus* 1885. *Notes on Pinks of Western Europe* 1889. *Provisional and Tentative List of Orders and Families of British Flowering Plants* 1895; ed. 2 1898. *Prodromus Florae Britannicae* 1901–12. Contrib. to *J. Bot.*, *Proc. Linn. Soc.*, *Bull. Herb. Boissier*.

Bot. Soc. Exch. Club Br. Isl. Rep. 1923, 156–64. *J. Bot.* 1923, 249–52. *Proc. Linn. Soc.* 1922–23, 45–46. D. H. Kent *Hist. Fl. Middlesex* 1975, 26. *R.S.C.* v.19, 632.

MS. fragment of *Fl. Middlesex* at BM(NH). Plants at Kew. Portr. at Hunt Library.

WILLIAMS, George (1762–1834)
b. Catherington, Hants 1762 *d.* Oxford 17 Jan. 1834

BA Oxon 1781. MD 1788. FLS 1798. Professor of Botany, Oxford, 1795–1834. Sent *Mesembryanthema* to A. H. Haworth (*Saxifragearum Enumeratio* 1821, 98, etc.)

F. Pursh *Fl. Americae Septentrionalis* 1814, xviii. *Bot. Misc.* 1829–30, 57–61. *Gent. Mag.* 1834 i 334–36. D. Turner *Extracts from Lit. and Sci. Correspondence of R. Richardson* 1835, ix–x. W. Munk *Roll of R. College of Physicians* v.2, 1878, 467–69. G. C. Druce *Fl. Berkshire* 1897, clviii. *D.N.B.* v.61, 399. G. C. Druce *Fl. Oxfordshire* 1927, cv.

Letters in J. E. Smith and N. J. Winch correspondence at Linnean Society. Plants and MSS. at Oxford.

WILLIAMS, Henry (*fl.* 1890s)

FLS 1890. Son of B. S. Williams (1824–1890). Nurseryman, Upper Holloway, London. Co-author with R. Warner of *Orchid Album* from v.9, 1891–v.11, 1897. Revised ed. 7 1894 of B. S. Williams's *Orchid-Grower's Manual.*

WILLIAMS, Iolo Aneurin (1890–1962)
b. Middlesbrough, Yorks 18 June 1890 *d.* Kew, Surrey 17 Jan. 1962

FLS 1926. Journalist, bibliographer, botanist. *Where the Bee Sucks* 1929. *Flowers of Marsh and Stream* 1946. Contrib. to *J. Bot.*

Proc. Bot. Soc. Br. Isl. 1962, 507–8. *Proc. Linn. Soc.* 1961–62, 106–7. *Times* 19 Jan. portr.; 23 Feb. 1962. *Illus. London News* 12 Jan. 1963, 58. *Who was Who, 1961–1970* 1207.

Plants at BM(NH).

WILLIAMS, Rev. J. (*fl.* 1880s)

Collected plants at Dera Ismail Khan and Wajuristan, 1888.

Pakistan J. For. 1967, 360.

Plants at Dehra Dun India, Edinburgh.

WILLIAMS, James (1797–1891)
b. Jan. 1797 *d.* 24 Dec. 1891

Gardener to James Warner of Hoddesdon, Herts.

Gdnrs Chron. 1892 i 25, 53 portr.

WILLIAMS, John (1801–1859)
b. Llansantffraid, Glan Conway, Denbighshire 1 March 1801 *d.* Mold, Flint 1 Nov. 1859

MRCS Dublin 1832. MD St. Andrews 1858. Gardener at Kew and Chelsea. *Faunula Grustensis* 1830.

J. Bot. 1910, 232.

Herb. at University College, Bangor.

WILLIAMS, John (*fl.* 1830s)

Nurseryman, Burnham, Bucks.

WILLIAMS, John Charles (1861–1939)
b. Caerhays Castle, Cornwall 30 Sept. 1861 *d.* Caerhays 29 March 1939

Had garden at Caerhays where he grew many new plant introductions by George Forrest from W. China; noted for rhododendrons and magnolias.

Gdnrs Chron. 1939 i 223. *Kew Bull.* 1939, 252–53. *Curtis's Bot. Mag. Dedications, 1827–1927* 363–64 portr. *J. R. Hort. Soc.* 1943, 9–18, 43–48; 1966, 280.

WILLIAMS, John Lloyd (1854–1945)
b. Llanrwst, Denbighshire 1854 *d.* Bath, Somerset 15 Nov. 1945

DSc 1906. FLS 1916. Schoolmaster, Garn Dolbenmaen Board School. Lecturer in Botany, University College, Bangor. Professor of Botany, University College, Aberystwyth, 1914–26. Discovered *Juncus macer* in Wales. Contrib. to J. E. Griffith's *Fl. Anglesey and Caernarvonshire* 1895. *Flowers of Wayside and Meadow* 1927. *Atgofion tri Chwarter Canrif* (autobiography).

Nature v.157, 1946, 399. *N. Western Nat.* 1946, 110–13 portr. *Proc. Linn. Soc.* 1945–46, 72–74.

Letters at Liverpool Museums.

WILLIAMS, Joseph (*fl.* 1850s)

MD. Lecturer in Botany, Dublin. Revised ed. 3 of J. H. Balfour's *Manual of Botany* 1855. *Botanists' Vade-mecum* 1855; ed. 2 1856.

WILLIAMS, Percival Dacres (1865–1935)
d. Lanarth, Cornwall 6 Nov. 1935

VMH 1927. High Sheriff of Cornwall, 1903. Developed garden at Lanarth on the Lizard; especially interested in daffodils.

Br. Fern Gaz. 1935, 13. *Gdnrs Chron.* 1935 ii 361. *Daffodil Yb.* 1936, 1–2 portr.

WILLIAMS, R. Dorrington (*c.* 1889–1943)
d. Carmarthenshire 7 Oct. 1943

In charge of clover breeding at Welsh Plant Breeding Station. Published papers on genetics of *Trifolium pratense*.

Nature v.152, 1943, 471.

WILLIAMS, Richard (*fl.* 1780s–*c.* 1826)
Acquired Turnham Green Nursery, Chiswick, Middx, 1785; specialised in heathers and introduced plants from Australia and the Cape. 'Williams Bon Chrétien' pear.
Bot. Mag. 1793, t.240; 1795, t.303. *Trans. London Middlesex Archaeol. Soc.* v.24, 1973, 182.

WILLIAMS, Robert Orchard (1891–1967)
b. West Lulworth, Dorset 24 Jan. 1891 *d.* Cape Town 28 March 1967
ALS 1954. Kew gardener, 1916. Curator, Trinidad Botanic Garden, 1916. Deputy Director of Agriculture, Trinidad, 1939. Director of Agriculture, Zanzibar, 1945–48. *School Gardening in the Tropics* ed. 3 1949. Contrib. several families to *Fl. Trinidad and Tobago* 1928–47. Co-author with his son of ed. 3 of *Useful and Ornamental Plants of Trinidad and Tobago* 1941.
Kew Bull. 1916, 23. *Proc. Linn. Soc.* 1968, 145. *J. Kew Guild* 1969, 927–28 portr.

WILLIAMS, Samuel (*c.* 1898–1965)
d. 5 April 1965
BSc Manchester 1920. FRSE 1926. Assistant Lecturer in Botany, Manchester University, 1921. Lecturer in Botany, Glasgow University, 1923–64. Published papers on morphology of ferns. Contrib. to *Trans. R. Soc. Edinburgh*.
Nature v.194, 1962, 728; v.206, 1965, 1195. *Yb. R. Soc. Edinburgh* 1966, 39–40.

WILLIAMS, Rev. Theodore (1785–1875)
Vicar, Hendon, Middx, 1812–75. Created a garden of 1½ acres which was highly praised by J. C. Loudon.
Country Life 1961, 931, 933 portr.

WILLIAMS, Rev. Thomas (*c.* 1550–*c.* 1620)
b. Arddu r'Mynaich, Trefriw, Caernarvon *c.* 1550
Physician and lexicographer. *Llyfr Llysiau*.
W. Rowland *Cambrian Bibliogr.* 1869, 113–16. *J. Bot.* 1898, 13. *D.N.B.* v.61, 454–55.

WILLIAMS, Walter Henry (–1894)
d. Salisbury, Wilts 14 Jan. 1894
Nurseryman in partnership with J. Keynes and John Wyatt, Castle Street, Salisbury. Secretary, Wiltshire Horticultural Society.
Gdnrs Chron. 1894 i 84. *Gdnrs Mag.* 1894, 45–46 portr. *J. Hort. Cottage Gdnr* v.28, 1894, 52.

WILLIAMS, William (1805–1861)
b. Llanfwrog, Denbighshire *d.* 13 June 1861
Of Llanberis, Caernarvonshire. Groom and driver. Local botanist.
Phytologist 1861, 223, 307–9. *Times* 25 June 1861.

WILLIAMS, Rev. William Leonard (1829–1916)
d. Napier, New Zealand 1 Sept. 1916
BA Oxon 1852. DD 1897. Archdeacon of Waiapu, N.Z., 1862; Bishop, 1894–1909. Collected plants *c.* 1875–1906 in East Cape and Hawke's Bay, N.Z. Contrib. Maori plant-names to T. F. Cheeseman's *Manual of Botany* 1906, 1094–1111.
Trans. Proc. N.Z. Inst. 1896, 509–32, 554.

WILLIAMSON, Alexander (1819–1870)
b. Dunkeld, Perthshire 1819 *d.* Kingston, Surrey 1870
Kew gardener, 1843; foreman of Herbaceous Dept; Curator of Pleasure Grounds, 1848–66.
J. Kew Guild 1895, 28–29.

WILLIAMSON, Rev. Alexander (1829–1890)
b. Falkirk 5 Dec. 1829 *d.* Chefoo, China 28 Aug. 1890
BA Glasgow. LLD. In China for the London Missionary Society, 1855–58; agent in China for National Bible Society of Scotland, 1863–90. Collected plants in N. China. *Elements of Botany* 1858 (in Chinese). *Journeys in North China, Manchuria and Eastern Mongolia* 1870 (plant list in vol. 2).
E. Bretschneider *Hist. European Bot. Discoveries in China* 1898, 690–91. *D.N.B.* v.62, 2. *R.S.C.* v.8, 1244.

WILLIAMSON, Charles (*c.* 1748–1835)
b. Aberdeen *c.* 1748 *d.* Oxford 31 Dec. 1835
Gardener at Botanic Garden, Oxford, *c.* 1790–*c.* 1832.
Gdnrs Mag. 1836, 108.

WILLIAMSON, Helen Stuart (*née* Chambers) (1884–1934)
d. 4 Dec. 1934
Research Assistant, Botany Dept., Birkbeck College, 1926. Joint author of series of papers on fungi.
Nature v.134, 1934, 998.

WILLIAMSON, Isaac (–*c.* 1860)
Of Stockport, Cheshire. Botanised in Cheshire.
J. B. L. Warren *Fl. Cheshire* 1899, xci.

WILLIAMSON, John (*fl.* 1750s–1780s)
Acquired in 1756 Kensington Nursery founded by Robert Furber; taken over by Daniel Grimwood in 1783.
Trans. London Middlesex Archaeol. Soc. v.24, 1973, 185.

WILLIAMSON, John (–1780)
d. Edinburgh Sept. 1780
Gardener, Royal Botanic Garden, Edinburgh, *c.* 1756–80. MS. *Narrative of Experiments on Trees* 1769 in Garden Library.
Notes R. Bot. Gdn Edinburgh 1904, 18–20.

WILLIAMSON, John (1774–1877)
Curator, Scarborough Museum. Discovered many plants in Yorkshire Oolite.
Fossil plants at Scarborough, BM(NH).
Williamsonia Brongn.

WILLIAMSON, John (1839–1884)
b. Abernethy, Perthshire 23 July 1839 *d*. Alderson, Virginia, U.S.A. 17 June 1884
Self-taught botanist. To U.S.A. *c*. 1866; settled in Louisville. Established a brass foundry and etched, especially ferns, on metal. Friend of G. E. Davenport. *Ferns of Kentucky* 1878. *Fern etchings* ed. 2 1879.
Bot. Gaz. 1884, 122–26. *Bull. Torrey Bot. Club* 1884, 104–5. *Fern Bull*. 1900, 1–5 portr. G. H. Tilton *Fern Lover's Companion* 1922, 212, 215 portr.
Portr. at Hunt Library.

WILLIAMSON, William (1843–)
Gardener at Tarvit, Fifeshire. Lecturer in Horticulture, Edinburgh, 1902. Nurseryman, Logie Green, Edinburgh. *Horticultural Exhibitors' Handbook* 1892. *British Gardener* 1901. *Smallholder's Handbook* 1912.
Gdnrs Mag. 1902, 581 portr. *Gdnrs Chron*. 1930 i 380 portr.

WILLIAMSON, William Crawford (1816–1895)
b. Scarborough, Yorks 24 Nov. 1816 *d*. Clapham, Surrey 23 June 1895
MRCS 1840. LLD Edinburgh 1883. FRS 1854. Son of J. Williamson (1774–1877). Curator, Museum of Manchester Natural History Society, 1835–38. Surgeon, Chorlton-on-Medlock Dispensary, 1841–68. First Professor of Natural History, Owens College, Manchester, 1851; of Botany, 1880–92. Founder of modern palaeobotany. Contrib. to J. Lindley and W. Hutton's *Fossil Fl. Great Britain* 1831–37. 'On Organization of Fossil Plants of the Coal Measures' (*Philos. Trans. R. Soc*. 1871–93). Contrib. to *Manchester Lit. Philos. Soc. Mem. Reminiscences of a Yorkshire Naturalist* 1896.
Geol. Mag. 1895, 383–84. *J. Bot*. 1895, 298–300. *Nature* v.52, 1895, 441–43. *Naturalist* 1896, 25–33 portr. *Proc. R. Soc*. v.60, 1897, xxvii–xxxii. *D.N.B*. v.62, 9–12. *Trans. Liverpool Bot. Soc*. 1909, 100–5. *Gdnrs Chron*. 1911 i 328. F. W. Oliver *Makers of Br. Bot*. 1913, 243–60 portr. F. C. Bower *Sixty Years of Bot. in Britain, 1875–1935* 1938, 68–70 portr. C. C. Gillispie *Dict. Sci. Biogr*. v.14, 1976, 396–99. *R.S.C*. v.6, 380; v.8, 1245; v.11, 817; v.19, 638.
Fossil plant slides at BM(NH). Letters at Kew. Portr. at Manchester University. Portr. at Hunt Library.

WILLIS, James (*fl*. 1790s)
Nurseryman, Burford, Oxfordshire.

WILLIS, John Christopher (1868–1958)
b. Birkenhead, Cheshire 20 Feb. 1868 *d*. Montreux, Switzerland 21 March 1958
BA Cantab 1891. DSc 1905. FRS 1919. FLS 1897. Assistant, Botany Dept., Glasgow University, 1894–96. Director, Botanic Gardens, Peradeniya, Ceylon, 1896–1911. Director, Botanic Garden, Rio de Janeiro, Brazil, 1912–15. *Manual and Dictionary of Flowering Plants and Ferns* 1897; ed. 7 1966. *Agriculture in the Tropics* 1909. *Age and Area* 1922. *Course of Evolution* 1940. *Birth and Spread of Plants* 1949. Edited *Tropical Agriculturist, Empire Cotton Growing Review*. Contrib. to *Ann. Bot., J. Linn. Soc., Proc. R. Soc., Philos. Trans. R. Soc., Ann. R. Bot. Gdns, Peradeniya*.
Tropical Life 1906, 120 portr. *Nature* v.151, 1943, 247; v.181, 1958, 1103–4. *Times* 26 March 1958. *Proc. Linn. Soc*. 1956–57 245–50. *Biogr. Mem. Fellows R. Soc*. 1958, 353–59 portr. *Archivio Botanico* 1958, 90–116. *Who was Who, 1951–1960* 1176. I. H. Burkill *Chapters on Hist. Bot. in India* 1965, 203–5.

WILLIS, Margaret Dorothea *see* Sherbourne, M. D.

WILLISEL (Willisell), Thomas (–*c*. 1675)
d. Jamaica *c*. 1675
Collected botanical and zoological specimens for Royal Society in England and Scotland; also collected for R. Morison, C. Merrett, J. Ray, Sherard. Gardener to John Vaughan, 3rd Earl of Carbery in Jamaica, 1674.
J. Ray *Catalogus Plantarum Angliae* 1670, 340. J. Ray *Historia Plantarum* 1686, 1488. J. Petiver *Museii Petiveriani* 1695, 742. R. Pulteney *Hist. Biogr. Sketches of Progress of Bot. in England* v.1, 1790, 347–49. J. Cash *Where there's a Will there's a Way* 1873, 2–3. *D.N.B*. v.62, 26–27. *J. Bot*. 1909, 101. J. E. Dandy *Sloane Herb*. 1958, 230–31. A. M. Coats *Quest for Plants* 1969, 330.

WILLISON, Alexander (*fl*. 1820s)
Gardener and nurseryman, Bridge Street, Whitby, Yorks.

WILLISON, William (*c*. 1806–1875)
d. 16 Oct. 1875
Nurseryman, Rose Nursery, Whitby, Yorks. Hybridised *Clematis willisoni*.
Gdnrs Chron. 1875 ii 533. *Gdnrs Yb. and Almanack* 1876, 172.

WILLMER, John Thomas (*c*. 1786–1879)
d. 20 Nov. 1879
Nurseryman, Sunbury, Middx.
Garden v.16, 1879, 546. *Gdnrs Chron*. 1879 ii 733. *Gdnrs Yb. and Almanack* 1881, 190.

WILLMOTT, Ellen Ann (1858–1934)
b. Isleworth, Middx 19 Aug. 1858 *d.* Brentwood, Essex 27 Sept. 1934

FLS 1904. VMH 1897. Created garden at Warley Place, Essex. Also had gardens at Aix-les-Bains and on Riviera. *Warley Garden in Spring and Summer* 1909; ed. 2 1924. Commissioned Alfred Parsons to illustrate *Genus Rosa* (text by J. G. Baker) 1910–14 2 vols.

Garden 1907, iv portr. *Essex Nat.* 1912, 40–60; 1935, 311–12; 1966, 370–75. *Gdnrs Chron.* 1934 ii 255, 365; 1964 ii 385 portr. *Kew Bull.* 1934, 397–98. *Br. Fern Gaz.* 1934, 273–74. *Nature* v.134, 1934, 726. *Proc. Linn. Soc.* 1934–35, 195–97. *Times* 28 Sept. 1934. *Curtis's Bot. Mag. Dedications, 1827–1927* 319–20 portr. *J. R. Hort. Soc.* 1948, 180–82. *Garden* 1975, 586–87 portr.

Herb. Warleyense at Kew.

Ceratostigma willmottianum Stapf.

WILLMOTT, John (1775–1834)

Son-in-law of John Russell (*c.* 1731–1794) in whose Lewisham Nursery in Kent he was a partner. The firm later became John Willmott and Co., and finally Willmott and Chaundy, closing in 1860.

Trans. London Middlesex Archaeol. Soc. v.24, 1973, 189.

WILLOUGHBY, Francis *see* Willughby, F.

WILLS, John (1832–1895)
b. Chard, Somerset 1832 *d.* S. Kensington, London 9 July 1895

Gardener to Sir P. Egerton at Oulton Park, Tarporley, Cheshire. Floral decorator, Onslow Crescent, Kensington with a nursery at Annerley, Fulham. In 1877 suggested enclosing the Albert Memorial in glass to preserve it from "the vitiated atmosphere of London."

Garden v.48, 1895, 52. *Gdnrs Chron.* 1895 ii 78. *Gdnrs Mag.* 1895, 436, 440 portr. *J. Hort. Cottage Gdnr* v.31, 1895, 53–54 portr. *Br. Fern Gaz.* 1909, 48. R. Webber *Early Horticulturists* 1968, 156.

WILLSHIRE, William Hughes (1816–1899)
d. London 1899

MD Edinburgh 1836. Lecturer in Botany, Charing Cross Hospital, 1838–47. President, Medical Society of London, 1855. *Principles of Botany* 1840.

Br. Med. J. 1899 i 703. F. Boase *Modern English Biogr.* v.3, 1901, 1398.

WILLUGHBY, Francis (1635–1672)
b. Middleton, Warwickshire 22 Nov. 1635 *d.* Middleton 3 July 1672

BA Cantab 1656. FRS 1663. Naturalist. Botanised with J. Ray in England and on continent. Brought Italian plants to J. Petiver. Contrib. to *Philos. Trans. R. Soc.* 1669–70.

J. Martyn *Plantae Cantabrigiensis* 1763, preface, 13. J. F. Denham *Mem. of F. Willughby* 1838. *D.N.B.* v.62, 54–57. E. J. L. Scott *Index to Sloane Manuscripts* 1904, 571. C. E. Raven *John Ray* 1942 *passim.* J. E. Dandy *Sloane Herb.* 1958, 231. *J. Soc. Bibl. Nat. Hist.* 1972, 71–85. C. C. Gillispie *Dict. Sci. Biogr.* v.14, 1976, 412–14.

Plants and MSS. at Nottingham University. MSS. at Royal Society. Bust at Trinity College Cambridge.

Willughbeia Roxb.

WILMER, John (1697–1769)
b. 27 June 1697 *d.* Westminster, London 1769

MD. Apothecary. Demonstrator, Chelsea Physic Garden, 1748–64. Had herb. Member of John Martyn's Botanical Society.

G. C. Gorham *Mem. of John and Thomas Martyn* 1830, 7. R. H. Semple *Mem. of Bot. Gdn at Chelsea* 1878, 71, 74–75. C. W. Foster and J. J. Green *Hist. of Wilmer Family* 1888. *Proc. Bot. Soc. Br. Isl.* 1967, 313–14.

WILMOT, John (*fl.* 1810s–1830s)

Fruit gardener, Isleworth, Middx. Contrib. to *Trans. Hort. Soc. London.*

J. C. Loudon *Encyclop. Gdning* 1822, 1288.

WILMOTT, A. T. (*fl.* 1840s)

Surgeon, Ross, Herefordshire. Contrib. plant records to H. C. Watson's *Topographical Bot.* 1883.

Trans. Woolhope Nat. Field Club v.34, 1954, 255.

WILMOTT, Alfred James (1888–1950)
b. Tottenham, Middx 31 Dec. 1888 *d.* S. Kensington, London 26 Jan. 1950

MA Cantab. FLS 1911. Assistant, Botany Dept., BM(NH), 1911; Deputy Keeper, 1931–50. Edited ed. 10 of C. C. Babington's *Manual of Br. Bot.* 1922. Secretary, Botanical Society of British Isles, 1941–48. Contrib. to *J. Bot.*

Gdnrs Chron. 1950 i 48. *Nature* v.165, 1950, 629–30. *Proc. Linn. Soc.* 1949–50, 234–36. *Pharm. J.* 1950, 93. *Times* 28 Jan., 4 Feb. 1950. *Watsonia* 1951–53, 63–70 portr. *Who was Who, 1941–1950* 1246.

Plants at BM(NH).

WILSON, Mr. (*fl.* 1740s)

Tailor, Norwich. Had herb.

Trans. Linn. Soc. v.7, 1804, 296–97. *Trans. Norfolk Norwich Nat. Soc.* 1874–75, 26.

WILSON, Albert (1862–1949)
b. Calder Mount, Lancs 12 Oct. 1862 *d.* Priest Hutton, Lancs 15 May 1949

FLS 1900. Pharmaceutical chemist, Bradford, Yorks. *Fl. West Lancashire* (with J. A. Wheldon) 1907. *Fl. Westmorland* 1938. Compiled ed. 3 of *Census Catalogue of British Hepatics* 1930. Contrib. to *J. Bot.*, *N. Western Nat.*

Naturalist 1949, 153–54. *Proc. Linn. Soc.* 1949–50, 116–17. *Trans. Br. Bryol. Soc.* 1950, 402–3. *Watsonia* 1949–50, 327–31. *N. Western Nat.* 1953, 391–99 portr.
Herb. at York Museum.

WILSON, Alexander M. (1868–1953)
d. Presteigne, Radnorshire Aug. 1953
Raised narcissi. Contrib. to A. F. Calvert's *Daffodil Growing for Pleasure and Profit* 1929.
Daffodil Tulip Yb. 1951–52, 9–10 portr.; 1955, 97.

WILSON, Alexander Stephen (1827–1893)
b. Rayne, Aberdeenshire 1827 *d.* Aberdeen 16 Nov. 1893
Civil engineer. *Ergot* 1876. *Botany of Three Records* 1878. *Bushel of Corn* 1883.
Gdnrs Chron. 1893 ii 665. *Ann. Scott. Nat. Hist.* 1894, 52–54. *J. Bot.* 1894, 31. *R.S.C.* v.8, 1248; v.11, 820; v.12, 785.
Letters in Berkeley correspondence at BM(NH).

WILSON, Rev. Alexander Stoddart (1854–1909)
d. Inverkeithing, Fife 8 Feb. 1909
MA Glasgow. Lecturer, Anderson's College, Glasgow. 'Dispersion of Seeds' (*Proc. Trans. Nat. Hist. Soc. Glasgow* v.3, 1888–89, 50–81, 101–26).
Glasgow Nat. v.1, 1909, 61–62.

WILSON, Alfred Gurney (*c.* 1878–1957)
d. Hove, Sussex 19 Jan. 1957
FLS 1909. VMH 1943. VMM 1952. Orchid grower. Founded and edited *Orchid World* 1910–16. Edited *Orchid Rev.* 1921–32. Benefactor of Royal Horticultural Society.
J. Hort. Home Farmer v.65, 1912, 466–67 portr. *J. R. Hort. Soc.* 1957, 178–79 portr. *Orchid Rev.* 1957, 55.
Orchid drawings at Royal Horticultural Society.

WILSON, Andrew (*fl.* 1790s)
Seedsman, Market Harborough, Leics.

WILSON, Rev. Charles Thomas (*fl.* 1880s)
Missionary in Uganda. Sent plants to Kew, 1880.
H. H. Johnson *Uganda Protectorate* v.1, 1902, 329.

WILSON, Charlotte (*fl.* 1840s)
Discovered *Simethis* at Bournemouth, 1847.
Gdnrs Chron. 1847, 467 (misprinted Wilkins).

WILSON, E. Kenneth (*c.* 1870–1947)
Had garden at Cannizaro, Wimbledon where he cultivated orchids.
Orchid Rev. 1947, 84.

WILSON, Edward Arthur (1858–1924)
d. Bristol, Glos 2 June 1924
FLS 1928. Interested in botany, gardening and microscopy.
Proc. Linn. Soc. 1924–25, 87. *Watson Bot. Exch. Club Rep.* 1923–24, 244–45.

WILSON, Edward S. (–1846)
Of Buglawton, Cheshire. 'Cowslip and Primrose' (*Phytologist* v.2, 1846, 377–79, 550–52).
J. B. L. Warren *Fl. Cheshire* 1899, xci. *R.S.C.* v.6, 385.

WILSON, Ernest Henry (1876–1930)
b. Chipping Campden, Glos 15 Feb. 1876
d. Worcester, Mass., U.S.A. 15 Oct. 1930
VMH 1912. VMM 1906. Kew gardener, 1897–99. Collected plants in China for Veitch and Sons, 1899–1902, 1903–5. Returned to China on behalf of Arnold Arboretum, 1906–9, 1910–11; Japan, 1914, 1917–19; Far East, India and Africa, 1920–21. Assistant Director, Arnold Arboretum, 1919; Keeper, 1927–30. *Naturalist in Western China* 1913 2 vols. *Cherries of Japan* 1916. *Conifers and Taxads of Japan* 1916. *Romance of our Trees* 1920. *Monograph of Azaleas* (with A. Rehder) 1921. *Lilies of Eastern Asia* 1925. *Aristocrats of the Garden* 1926. *Plant Hunting* 1927 2 vols. Contrib. to *J. Arnold Arb.*
J. Kew Guild 1899, 30; 1900, 24–25; 1923, 139 portr.; 1931, 67–73 portr. J. H. Veitch *Hortus Veitchii* 1906, 92–96. *Gdnrs Chron.* 1930 ii 333–34 portr., 394; 1950 i 198. C. S. Sargent *Plantae Wilsonianae* 1911–17 3 vols. *World's Work* v.27, 1913, 41–52. *Rhododendron Soc. Notes* 1925, 61–62; 1932, 289–92. *Curtis's Bot. Mag. Dedications, 1827–1927* 367–68 portr. *Bot. Soc. Exch. Club Br. Isl. Rep.* 1930, 329–30. *J. Arnold Arb.* 1930, 181–92 portr. *Kew Bull.* 1930, 498–99. *Nature* v.126, 1930, 693–94. *Orchid Rev.* 1930, 341–42. *J. Bot.* 1931, 18–20. *J. R. Hort. Soc.* 1931, 79 portr.; 1942, 351–59. *Quart. J. For.* 1931, 9–11. *Scott. For.* 1931, 116–21. E. Farrington *Ernest H. Wilson* 1931. *Proc. Amer. Acad. Arts Sci.* 1936, 602–4. *Lily Yb.* 1939, 3–6 portr.; 1968, 17–25. *Who was Who, 1929–1940* 1469. E. H. M. Cox *Plant Hunting in China* 1945, 136–51 portr. M. Hadfield *Pioneers in Gdning* 1955, 193–203. A. M. Coats *Gdn Shrubs and their Histories* 1963, 396–98. A. M. Coats *Quest for Plants* 1969, 118–22. D. J. Foley *Flowering World of 'Chinese' Wilson* 1969 portr.
Portr. at Hunt Library.
Magnolia wilsonii Rehder.

WILSON, Rev. Francis Robert Muter (1832–1903)
d. Canterbury, Victoria June 1903

Presbyterian minister, Kew, Melbourne. Lichenologist. Contrib. to *Victorian Nat.*
Victorian Nat. 1908, 116; 1949, 106–7. *Rep. Austral. Assoc. Advancement Sci.* 1911, 255.
Plants at Herb., Sydney.

WILSON, George Fergusson (1822–1902)
b. Wandsworth, London 25 March 1822 *d.* Weybridge, Surrey 28 March 1902
FRS. FLS 1875. VMH 1897. Chemist and Managing Director of Messrs Price, candle maker. His garden at Wisley, Surrey was purchased by Sir T. Hanbury for the Royal Horticultural Society.
Garden v.57, 1900, 17 portr.; v.61, 1902, 231. *Gdnrs Chron.* 1902 i 228. *Curtis's Bot. Mag. Dedications, 1827–1927* 251–52 portr. *Lily Yb.* 1934, 4 portr. H. R. Fletcher *Story of R. Hort. Soc., 1804–1968* 1969 *passim.*

WILSON, George Fox *see* Fox Wilson, G.

WILSON, Guy Livingstone (1885–1962)
b. Broughshane, County Antrim 1885 *d.* 5 Feb. 1962
VMH 1950. Hybridised daffodils. Contrib. to *Daffodil Tulip Yb.*
Gdnrs Chron. 1951 ii 22 portr.; 1964 i 121. *Daffodil Tulip Yb.* 1957, 9–10 portr.; 1963, 20–26 portr.; 1966, 14–19. M. J. Jefferson-Brown *Daffodils and Narcissi* 1969, 27–29.

WILSON, Miss Henrietta (–1863)
d. Sept. 1863
Of Woodville. Gardener, botanist and entomologist. *Chronicles of a Garden* 1863.
Lady 23 March 1961, 475.

WILSON, Henry (*c.* 1822–1892)
d. 6 May 1892
Journalist, Halifax, Yorks. Grew auriculas.
J. Hort. Cottage Gdnr v.24, 1892, 398–99.

WILSON, Herbert Ward (1877–1955)
b. Bradford, Yorks 29 Sept. 1877 *d.* Melbourne 1 Oct. 1955
BSc Melbourne 1919. To Australia *c.* 1883. Teacher, Victoria. Lecturer, Teacher's College, Melbourne 1924–42. Botanist.
Melbourne Univ. Gaz. 1955, 84–85. *Education Mag.* (Melbourne) 1956, 97–108. *Victorian Nat.* 1956, 148–58 portr. *Austral. Encyclop.* v.9, 1965, 321–22.

WILSON, Hugh (*fl.* 1790s)
Nurseryman, Hersham, Surrey.

WILSON, James (*fl.* 1780s–1790s)
Seedsman, 68 Smithfield, London.

WILSON, Rev. James (1855–1937)
b. Essex 1855 *d.* Cheltenham, Victoria 21 June 1937
To Australia in 1880s as Congregational minister, Beechworth, Victoria. Collected plants which he sent to F. von Mueller. Transferred to church at Beaconsfield, Victoria. Later specialised in fungi, collecting for Lloyd Herb., Cincinnati.
Victorian Nat. 1937, 50; 1949, 108.
Melanogarter wilsonii.

WILSON, James Hewetson (*c.* 1827–1850)
b. Chorlton, Manchester *c.* 1827 *d.* Worth, Sussex 12 Nov. 1850
BA Oxon 1850. FLS 1847. Translated A. de Jussieu's *Elements of Botany* 1849.
Proc. Linn. Soc. 1851, 139. *Trans. Liverpool Bot. Soc.* 1909, 105.
Herb. at St. Andrews University.

WILSON, John (1696–1751)
b. Longsleddale near Kendal, Westmorland 1696 *d.* Kendal 15 July 1751
Shoemaker and baker. *Synopsis of British Plants* 1744.
R. Pulteney *Hist. Biogr. Sketches of Progress of Bot. in England* v.2, 1790, 264–69. *Gent. Mag.* 1791 i 804. J. Cash *Where there's a Will there's a Way* 1873, 7–8. *D.N.B.* v.62, 106. *Nat. Hist. Trans. Northumberland, Durham and Newcastle-upon-Tyne* 1903, 75–76. F. H. Davey *Fl. Cornwall* 1909, xxxii. A. Wilson *Fl. Westmorland* 1938, 71.
Wilsonia R. Br.

WILSON, John A. (*fl.* 1860s–1914)
d. Bowness, Cumberland 8 May 1914
Collected ferns together with his wife.
Br. Fern Gaz. 1909, 48; 1914, 210.

WILSON, John Bracebridge (1828–1895)
b. Topcroft, Norfolk 1828 *d.* Geelong, Victoria 22 Oct. 1895
BA Cantab 1852. FLS 1882. To Australia in 1850s. Headmaster, Geelong Grammar School, 1863. Phycologist and marine zoologist.
J. Bot. 1896, 48. *Proc. Linn. Soc.* 1895–96, 48–49. *Victorian Nat.* 1895, 81; 1908, 116–17 portr. *Proc. Linn. Soc. N.S.W.* 1896, 624.
Algae at BM(NH).

WILSON, John Charles (1850/1–1925)
d. Manchester 11 Sept. 1925
FLS 1910. Solicitor, Manchester. Bryologist.
Br. Bryol. Soc. Rep. 1926, 255.

WILSON, John Hardie (1858–1920)
b. St. Andrews, Fife 1858 *d.* St. Andrews 13 Jan. 1920
DSc 1889. Lecturer in Botany, St. Andrews University. Planned the first botanic garden there. Lecturer in Agriculture, 1900. Produced strain of disease-resistant potatoes. *Rambles round St. Andrews.*
J. Hort. Cottage Gdnr v.46, 1903, 159 portr.; v.56, 1908, 585 portr. *Gdnrs Chron.* 1920 i 59–60. *Kew Bull.* 1920, 71. *Nature* v.104, 1920, 539–40.

WILSON, Joseph (*fl.* 1790s–1820s)
Nurseryman, Derby.

WILSON, Joseph (1853–1934)
b. Castle Bellingham, County Louth 13 Aug. 1853 *d.* London 25 July 1934
FRMS 1908. Excise and Customs Service, 1874–1920. Member of Essex Field Club and Quekett Microscopical Club. Diatomist.
J. Quekett Microsc. Club 1934, 86–87. *J. R. Microsc. Soc.* 1935, 32–33.

WILSON, Joshua (*fl.* 1820s–1840s)
Nurseryman, Cheapside, Derby.

WILSON, Leonard W. (1887–1951)
Inspector of Taxes, Luton and Bedford. Botanist.
Bot. Soc. Br. Isl. 1952, 87.

WILSON, Lucy Sarah (*née* Atkins) (*fl.* 1820s)
Of Chipping Norton, Oxfordshire. Married Rev. Daniel Wilson. *Botanical Rambles* 1822; ed. 2 1826.
J. Smith *Cat. of Friends' Books* v.1, 1867, 141–42.

WILSON, Malcolm (1882–1960)
d. 8 July 1960
BSc London 1905. DSc 1911. FRSE 1923. FLS 1910. Demonstrator in Botany, Imperial College London, 1909. Lecturer in Mycology, Edinburgh University; Reader, 1925–51. President, British Mycological Society, 1934. President, Botanical Society of Edinburgh, 1933–34. *British Rust Fungi* (with D. M. Henderson) 1966.
Ann. Appl. Biol. 1961, 386. *Yb. R. Soc. Edinburgh* 1961, 39–41.

WILSON, Nathaniel (1809–1874)
b. Scotland 18 April 1809 *d.* Clarendon, Jamaica 2 May 1874
Kew gardener, 1834. Botanist, Jamaica, 1846–67. 'Outline Fl. Jamaica' (*Rep. Geol. Jamaica* 1869, 263–91).
Kew Bull. 1891, 321; Additional Ser. v.1, 1898, 143. *Bull. Bot. Dept. Jamaica* v.8, 1901, 182–84. I. Urban *Symbolae Antillanae* v.3, 1902, 140–41. *R.S.C.* v.6, 388.
Ferns and letters at Kew.

WILSON, R. (*fl.* 1760s)
Seedsman, Woolpack and Crown, Strand, London.

WILSON, R. (*fl.* 1800s)
Of Medomsley, Durham. Sent fungi drawings to N. J. Winch.
N. J. Winch *Botanist's Guide through... Northumberland and Durham* v.2, 1807, i.

WILSON, Robert (1855–1927)
b. Austwick, Yorks 1855
Wheelwright, Hellifield, Yorks. Studied local flora, especially ferns.
Naturalist 1929, 61–62.

WILSON, Thomas Braidwood (1792–1843)
b. Braidwood, Lanarkshire 30 April 1792 *d.* Braidwood, N.S.W. 11 Nov. 1843
Surgeon, Royal Navy, 1815. Surgeon-Superintendent of convict transport ships to Australia. Friend of A. Cunningham of Sydney Botanic Gardens to whom he sent plants and seeds of W. Australian plants. Settled at Braidwood, 1836. *Narrative of Voyage round the World* 1835.
Sydney Morning Herald 20 Nov. 1843. *J. W. Austral. Nat. Hist. Soc.* 1909, 26–27. *Hist. Rec. of Australia* Ser. 1, v.17, 1923, 264–65. A. Hogg *Back to Braidwood Celebrations* 1925, 17–21. M. Uren *Land Looking West* 1948 *passim. J. W. Austral. Hist. Soc.* 1965, 7–28. *Austral. Dict. Biogr.* v.2, 1967, 612.
Grevillea wilsoni Cunn.

WILSON, William (*fl.* 1770s)
Worked under P. Miller. Gardener to Earl of Glasgow near Paisley, Renfrewshire. *Treatise on Forcing of Early Fruits and Management of Hot-walls* 1777.
J. C. Loudon *Encyclop. Gdning* 1822, 2283.

WILSON, William (1799–1871)
b. Warrington, Lancs 7 June 1799 *d.* Paddington near Warrington 3 April 1871
Solicitor. Described mosses for J. D. Hooker's *Fl. Antarctica*, B. C. Seemann's *Bot....Herald*, etc. *Bryologia Britannica* 1855. Contrib. to J. Sowerby and J. E. Smith's *English Bot.* 2664, 2686, 2723.
Gdnrs Chron. 1871, 554. *Gdnrs Mag.* 1871, 183. *Warrington Examiner* April 1871. *J. Bot.* 1871, 159–60; 1873, 128; 1875, 180. *Trans. Bot. Soc. Edinburgh* v.11, 1873, 170–74. J. Cash *Where there's a Will there's a Way* 1873, 145–56. J. Cash *Late William Wilson* 1886. *Naturalist* 1887, 181–90. *Bot. Gaz.* 1897, 348–49. J. B. L. Warren *Fl. Cheshire* 1899, xci. *D.N.B.* v.62, 147. *Trans. Liverpool Bot. Soc.* 1909, 105–7.
Plants and drawings at BM(NH). Plants, MSS. and medallion portr. at Warrington Museum. Letters at BM(NH), Kew. Portr. at Hunt Library.
Wilsonia Gill. & Hook.

WILSON, William (1803–1876)
b. 3 March 1803 *d.* 18 June 1876
Gardener to Duke of Devonshire at Holker, Lancs. Plant list for Cartmel, Lancs in C. M. Jopling's *Sketch of Furness and Cartmel* 1843, 257–72.
Naturalist 1894, 124. *Trans. Liverpool Bot. Soc.* 1909, 107.

WILSON, William (*c.* 1805–1896)
d. 18 Oct. 1896
Farmer, Alford, Aberdeenshire. Botanist.
Sci. Gossip 1896, 195.

WILSON, William (*c.* 1867–*c.* 1948)
Founder member of British Pteridological Society, 1891; Treasurer, 1903–10.
Br. Fern Gaz. 1948, 204–5.

WILTSHEAR, Felix Gilbert (1882–1917)
b. Kensington, London 15 Jan. 1882 *d.* France 23 Nov. 1917
Librarian, Botany Dept., BM(NH). Contrib. bibliographical notes to *J. Bot.* 1909, 1912–15.
J. Bot. 1918, 117–18.

WILTSHIRE, Samuel Paul (1891–1967)
b. Burnham-on-Sea, Somerset 13 March 1891 *d.* 13 May 1967
MA Cantab. Mycologist, Long Ashton Research Station, 1919–22. Assistant Director, Imperial Bureau of Mycology, Kew, 1924–39; Director, 1940–56. President, British Mycological Society, 1943. Published papers on *Alternaria* 1933 and *Stemphylium* 1938.
Nature v.215, 1967, 221. *Times* 16 May 1967. *Trans. Br. Mycol. Soc.* 1967, 513–14. *Who was Who, 1961–1970* 1217.
Portr. at Hunt Library.

WINCH, Joseph (*fl.* 1830s)
Nurseryman, Burnham, Bucks.

WINCH, Nathaniel John (1768–1838)
b. Hampton, Middx 26 Dec. 1768 *d.* Newcastle-upon-Tyne, Northumberland 5 May 1838
FLS 1803. ALS 1821. Secretary, Newcastle Infirmary. Discovered *Pyrola media. Botanist's Guide through...Northumberland and Durham* (with J. Thornhill and R. Waugh) 1805–7 2 vols. *Essay on Geographical Distribution of Plants...of Northumberland, Cumberland and Durham* 1819; ed. 2 1825. *Remarks on Fl. Cumberland* 1825. Welsh plants in *Mag. Nat. Hist.* v.2, 1829, 278–82.
Ann. Nat. Hist. v.1, 1838, 415. *J. Linn. Soc.* v.4, 1860, 195–96. *Northumberland Nat. Hist. Trans.* v.8, 1889, 307–25. G. C. Druce *Fl. Berkshire* 1897, clxvi–clxvii. W. Hodgson *Fl. Cumberland* 1898, xxvi. *J. Bot.* 1903, 380. *D.N.B.* v.62, 154. *R.S.C.* v.6, 392.
Letters and plants at BM(NH). Herb. at Hancock Museum, Newcastle. Portr. miniature at Kew. Portr. at Hunt Library.
Winchia A. DC.

WINDEBANK, W. (–1878)
d. Salisbury, Wilts 10 Oct. 1878
Nurseryman, Bevois Mount Nursery, Southampton, Hants. At one time in partnership with Mr. Kingsbury.
Gdnrs Chron. 1878 ii 509.

WINDSOR, John (1787–1868)
b. Settle, Yorks 1787 *d.* Manchester 1 Sept. 1868
FRCS. FLS 1814. Surgeon, Manchester from 1815. 'Settle Plants' (*Phytologist* 1855–58). *Fl. Cravoniensis* 1873.
Proc. Linn. Soc. 1868–69, cxx. *Naturalist* 1894, 289. *Trans. Liverpool Bot. Soc.* 1909, 107–8. F. Boase *Modern English Biogr. Supplt.* v.3, 1921, 926–27.
Herb. at Manchester University.
Windsoria Mitt.

WINGATE, George (1852–1936)
b. 21 Nov. 1852 *d.* Godalming, Surrey 21 Aug. 1936
FLS 1887. Colonel in Indian Army. In charge of grass farms for supplying fodder for Army horses. Collected grasses in N.W. Frontier Province and Baluchistan.
Proc. Linn. Soc. 1936–37, 221–22. *Pakistan J. For.* 1967, 360.
Plants and drawings of grasses at Kew.

WINGFIELD, Anne Frances (1823–1914)
b. 21 April 1823
Of Tickencote Hall, Rutland. Active botanist in Rutland.
A. R. Horwood and C. W. F. Noel *Fl. Leicestershire* 1933, cclxxvi portr.

WINGFIELD, George (*fl.* 1690s)
Surgeon. Sent African plants to J. Petiver.
J. E. Dandy *Sloane Herb.* 1958, 231.

WINN, Charles (*c.* 1829–1917)
d. March 1917
Of Selly Hill, Birmingham. Cultivated orchids.
Orchid Rev. 1917, 80–82.
Cypripedium winnianum Reichenb. f.

WINNECKE, Charles George Alexander (1856–1902)
b. Norwood, S. Australia *d.* Adelaide, S. Australia 1902
Explorer and geologist. Collected plants in Central Australia, 1883 and near Stuart's Range, 1885. Leader of Horn expedition, 1894. 'List of Plants collected by C. Winnecke...' (*Trans. Proc. R. Soc. S. Australia* v.8, 1886, 160).
Rep. Austral. Assoc. Advancement Sci. 1907, 170–71.
Triumfetta winneckeana F. Muell.

WINTER, John (*fl.* 1770s)
Nurseryman, Blyth, Notts.

WINTER, John Newnham (*c.* 1831–1907)
d. Kew, Surrey 18 Jan. 1907
MRCS. Pupil of J. Lindley. Collector and grower of filmy ferns.
Gdnrs Chron. 1907 i 64. *Kew Bull.* 1907, 68–69.

WINTER, Walter Percy (1867–1950)
b. Cheltenham, Glos 1867
BSc. Master, Cheltenham Grammar School. Master, Shipley, Yorks. President, Yorkshire Naturalists Union. *Classification Chart of Commoner British Orders of Flowering Plants* 1865. Contrib. to *Bradford Sci. J.*
Naturalist 1951, 8–9 portr.

WINTERBOTTOM, James Edward (1803–1854)
b. Reading, Berks 7 April 1803 *d.* Rhodes, Greece 4 July 1854
MA Oxon 1828. MB 1833. FLS 1830. Accompanied Sir Richard Strachey on Tibetan Boundary Commission and collected plants with him in Himalayas and Tibet, 1848–49. 'Kumaon...Plants' (with R. Strachey) (E. I. Atkinson *Economic Products of N.W. Provinces* 1876, 403–670).
Hooker's J. Bot. 1854, 307–8, 345–49. *Proc. Linn. Soc.* 1855, 418–19. J. D. Hooker and T. Thomson *Fl. Indica* v.1, 1855, 65–66. *Kew Bull.* 1900, 19–20. *J. Linn. Soc.* v.35, 1902, 136–40. *Fl. Malesiana* v.1, 1950, 581; v.5, 1958, 99. *J. Bombay Nat. Hist. Soc.* 1956, 77–78.
Plants at Calcutta, BM(NH), Kew.

WINTHROP, John (1681–1747)
b. Boston, U.S.A. 1681 *d.* London 1747
FRS 1734. Spent later years of his life in London. All-round naturalist.
Sci. Mon. 1931, 354–55. J. E. Dandy *Sloane Herb.* 1958, 231.

WINTLE, G. S. (*fl.* 1880s?)
Of Gloucester. Botanised locally. Supplied plants to G. E. S. Boulger and helped to compile his list of Painswick plants.
H. J. Riddelsdell *Fl. Gloucestershire* 1948, cxxviii.

WIPER, Joseph (*c.* 1847–1930)
d. Victoria, British Columbia 1930
Founder member of British Pteridological Society; later Treasurer. Emigrated to Canada.
Br. Fern Gaz. 1930, 33.

WISE, Frederick Clunie (1884–1962)
b. 4 Oct. 1884 *d.* 12 Aug. 1962
FRMS. President, Quekett Microscopical Society, 1939. Diatomist. Contrib. to *J. Quekett Microsc. Club.*
J. Quekett Microsc. Club 1962, 76–77.

WISE, Henry (1653–1738)
d. Warwick 15 Dec. 1738
Partner with George London in nursery at Brompton, London. Superintendent of royal gardens at Hampton Court, Kensington, etc. *Compleat Gard'ner* (with G. London) 1699; ed. 6 1717. *Retir'd Gard'ner* (with G. London) 1706.
Gent. Mag. 1738, 666. G. W. Johnson *Hist. of English Gdning* 1829, 124–27. *Gdnrs Chron.* 1892 i 361–62, 621–22. *J. R. Hort. Soc.* 1939, 464–79. D. Green *Gardener to Queen Anne* 1956 portr. *Northern Gdnr* 1972, 86–87.

WISE, Thomas (1854–1932)
b. Glasgow 1854 *d.* Dec. 1932
Founded Hamilton Field Club. President, Andersonian Society. Knew flora of Clyde area.
Glasgow Nat. v.12, 1934, 32.
Herb. at Glasgow University.

WISE, William (*c.* 1843–1935)
b. Launceston, Cornwall *c.* 1843 *d.* Helston, Cornwall 15 Jan. 1935
Chemist and druggist. Contrib. to *J. Bot.*
Pharm. J. 1935, 103.
Herb. at Launceston Museum and College.

WISEMAN, Percy (*c.* 1890–1970)
d. 21 Aug. 1970
Manager, The Nurseries, Bagshot, Surrey, 1925–63.
J. R. Hort. Soc. 1971, 39–40.

WITHAM, Gilbert (*fl.* 1610s–1680s)
Educ. Trinity College, Cambridge. Rector, Garforth, Yorks, 1644–84. Sent local plants to C. Merrett. In 1668 showed *Actaea spicata* and *Pyrola minor* to J. Ray.
C. Merrett *Pinax* 1666 *passim.* C. E. Raven *English Nat.* 1947, 318–19.

WITHAM, Henry Thomas Maire (*né* **Silvertop**) (1779–1844)
b. Minster Acres, Northumberland 28 May 1779 *d.* Lartington Hall, Yorks 28 Nov. 1844
Observations on Fossil Vegetables 1831. *Internal Structure of Fossil Vegetables* 1833.
F. W. Oliver *Makers of Br. Bot.* 1913, 243–45 portr. *R.S.C.* v.6, 404.
Pitys withami Scott.

WITHERING, William (1741–1799)
b. Wellington, Shropshire 28 March 1741 *d.* Birmingham 6 Oct. 1799
MD Edinburgh 1766. FRS 1785. FLS 1791. Practised medicine at Stafford and from 1775 at Birmingham. Chief Physician to Birmingham General Hospital. *Botanical Arrangement of all the Vegetables...in Great Britain* 1776 2 vols; ed. 7 1830 4 vols. *Miscellaneous Tracts* 1822 2 vols.
Gdnrs Mag. 1828, 536–37. *Cottage Gdnr* v.7, 1850, 43. *J. Linn. Soc.* v.4, 1860, 196. J. E. Bagnall *Fl. Warwickshire* 1891, 492–93. G. C. Druce *Fl. Berkshire* 1897, cxlii.

D.N.B. v.62, 268–70. F. H. Davey *Fl. Cornwall* 1909, xxxv. J. W. White *Fl. Bristol* 1912, 63. *J. Bot.* 1914, 300–1, 322. *Nature* v.147, 1941, 325. *Trans. N. Staffordshire Field Club* 1947–48, 90–95. T. W. Peck and D. Wilkinson *William Withering of Birmingham* 1950 portr. F. A. Stafleu *Taxonomic Literature* 1967, 508–9. D. A. Cadbury *Computer-mapped Fl.... Warwickshire* 1971, 49–51. C. C. Gillispie *Dict. Sci. Biogr.* v.14, 1976, 463–65.

Lichens at Kew (*Grevillea* v.12, 1883–84, 56–62, 70–76). Letters and R. Brown's MS. notes on his herb. (Cryptogams) at BM(NH). Portr. at Hunt Library.

Witheringia L'Hérit.

WITHERING, William (1775–1832)
b. Birmingham 1775 *d.* London 1832

LLD. FLS 1801. Son of W. Withering (1741–1799). Edited and partly illustrated 4th, 5th and 6th editions of his father's *Botanical Arrangement* and *Miscellaneous Tracts.*

WITHERS, Mr. (*fl.* 1700s–1730s)

Apothecary. Member of John Martyn's Botanical Society.

Proc. Bot. Soc. Br. Isl. 1967, 319.

WITHERS, Mrs. Augusta Innes (*née* **Baker**) (*c.* 1793–1860s)

Of Lisson Grove, London. Appointed painter of flowers to Queen Adelaide, 1830. Flower and Fruit Painter in Ordinary to Queen Victoria, 1864. Drew plants for *Trans. Hort. Soc. London, Bot. Mag., Floral Cabinet*, J. Bateman's *Orchidaceae of Mexico*, R. Thompson's *Gdnrs Assistant*, etc.

Gdnrs Mag. 1831, 95; 1834, 452. *J. Bot.* 1918, 242. W. Blunt *Art. of Bot Illus.* 1955, 216. A. Graves *R. Soc. of Arts* v.8, 1906, 326–27.

Drawings at BM(NH), Royal Horticultural Society. A few letters at Windsor Castle.

WITHERS, George (*fl.* 1830s)

Nurseryman together with Thomas Withers in partnership with Speechley and Ordoyno, Newark, Notts.

WITHERS, Robert (–1856)

Of Bath, Somerset. Had botanical library and herb., especially of Bath plants. 'Figures and Description of *Scirpus savii*...' (*Phytologist* v.3, 1850, 865–67).

R.S.C. v.6, 404.

MS. on Bath plants at Kew.

WITHERS, William (*fl.* 1700s–1730s)

Apothecary. Member of Botanical Society of London.

Proc. Bot. Soc. Br. Isl. 1967, 319.

WITHFIELD, Thomas *see* Whitfield, T.

WITHYCOMBE, Cyril Luckes (1898–1926)
b. Walthamstow? Essex 27 Oct. 1898 *d.* Cambridge 5 Dec. 1926

PhD London. Entomologist. Lecturer, Imperial College, Trinidad, 1923–26; Cambridge University, 1926. 'Function of Bladders in *Utricularia*' (*J. Linn. Soc.* v.46, 1924 401–13).

Entomol. 1927, 25–26 portr.

WITT, Christopher (1675–1765)
b. Wilts 1675 *d.* Germantown, Philadelphia Jan. 1765

Physician. To U.S.A., 1704; joined theosophical colonists on the Wissahickon. Settled in Germantown where he formed a large garden, one of the first botanical gardens in N. America. Corresponded with J. Bartram and P. Collinson to whom he sent plants.

L. W. Dillwyn *Hortus Collinsonianus* 1843, 37, 39, etc. W. Darlington *Memorials of J. Bartram and H. Marshall* 1849, 86. J. W. Harshberger *Botanists of Philadelphia* 1899, 42–46. *Amer. J. Pharm.* v.77, 1905, 311–23; v.80, 1908, 414–15.

WITTS, Rev. Edward Francis (1813–1886)

Of Upper Slaughter, Glos. Had herb.

H. J. Riddelsdell *Fl. Gloucestershire* 1948, cxix.

Herb. was in possession of Miss A. Broome-Witts.

WODROW (Woodrow), John (–1768)
d. 19 Dec. 1768

Doctor of Glasgow. "A celebrated botanist."

Gent. Mag. 1768, 591. *Gdnrs Chron.* 1919 i 147.

WOLFE, Thomas Birch (*né* **Birch**) (1801–1880)
d. Steyning, Sussex 1880

Assumed name of Wolfe before 1869 on succeeding to estate of Wood Hall, Arkesden, Essex. Collected and purchased plants—many from central Europe and N. Africa—during 1840s and 1860s. Collection presented by Ipswich Museum to Kew in 1952.

Sci. Gossip 1880, 235. *Trans. Suffolk Nat. Soc.* v.10, 1957, 246–47.

WOLFERSTAN, Littleton Edward Pipe (*fl.* 1860s–1920s)

Malayan Civil Service, 1889–1922. Collected plants in the Dindings, 1900.

Gdns Bull. Straits Settlements 1927, 134.

WOLLASTON, Alexander Frederick Richmond (1875–1930)
b. Clifton, Bristol 1875 *d.* Cambridge 3 June 1930

BA Cantab 1846. LRCP. Traveller and naturalist. Collected plants on BM expedition to Ruwenzori, 1905–7. In Dutch New Guinea, 1909–11, 1912–13; Mount Everest expedition, 1921; Colombia, 1923. *From Ruwenzori to the Congo* 1908. 'Report on Botany of Wollaston Expedition to Dutch New Guinea' by H. N. Ridley (*Trans. Linn. Soc.* 1916, 1–284).

Nature v.125, 1930, 944, 981–82. *Who was Who, 1929–1940* 1479–80. *Fl. Malesiana* v.1, 1950, 584–85.

WOLLASTON, George Buchanan (1814–1899)
b. Clapton, London 26 April 1814 *d.* Chislehurst, Kent 26 March 1899

Collected ferns. Contrib. to *Phytologist*.

Gdnrs Mag. 1899, 211. *J. Bot.* 1899, 447–48. *Br. Fern Gaz.* 1909, 48. *R.S.C.* v.6, 428; v.19, 695.

WOLLASTON, T. Vernon (*fl.* 1870s)

Plants collected on St. Helena in 1877 at Oxford.

WOLLEY-DOD, Anthony Hurt (1861–1948)
b. Eton, Bucks 17 Nov. 1861 *d.* 21 June 1948

VMH 1897. Son of C. Wolley-Dod (1826–1904). Lieut.-Colonel, Royal Artillery; at Woolwich Arsenal. Collected plants in S. Africa, 1900–1, Gibraltar, 1913–14, California, 1921. Authority on British roses. *Fl. Gibraltar* 1914. *Roses of Britain* 1924. *Revision of British Roses* 1931. Edited *Fl. Sussex* 1937. Contrib. to *J. Bot.*

Quart. Bull. Alpine Gdn Soc. 1948, 239–40. *Times* 23 June 1948. *Wild Flower Mag.* 1948, 220. *Watsonia* 1949–50, 331–34.

Herb. at BM(NH). Plants at Kew, Oxford. Portr. at Hunt Library.

WOLLEY-DOD, Rev. Charles (*né* **Wolley**) (1826–1904)
b. Wirksworth, Derbyshire 21 March 1826 *d.* Malpas, Cheshire 14 June 1904

BA Cantab 1849. VMH 1897. Assistant master, Eton College, 1850–78. Contrib. notes to *Fl. and Sylva*, *Gdnrs Chron.*, etc.

Garden v.54, 1898, xii portr.; v.65, 1904, 455 portr. *Gdnrs Chron.* 1904 i 392–93; v.164(15), 1968, 14. *Gdnrs Mag.* 1904, 414 portr. *Country Life* 1967, 548.

Letters at Kew and in C. E. Broome correspondence at BM(NH).

WOLSELEY, Frances Garnet, Viscountess (1872–1936)
b. 15 Sept. 1872 *d.* Ardingly, Sussex 24 Dec. 1936

Founded College for Lady Gardeners, Glynde, 1901. *Gardening for Women* 1908. *In a College Garden* 1915. *Gardens, their Form and Design* 1919.

Who was Who, 1929–1940 1480.

WOLSELEY, Rev. R. (1772–1815)

Recorded Staffordshire plants.

J. Bot. 1901 Supplt., 71.

WOLSEY, George (*c.* 1814–1870)
d. Guernsey, Channel Islands 25 Sept. 1870

"An exceedingly intelligent nurseryman" of Guernsey who also collected local plants. Discovered *Isoetes hystrix* in Guernsey. Secretary, Guernsey Horticultural Society.

E. D. Marquand *Fl. Guernsey* 1901, 26. *R.S.C.* v.6, 432.

WOLSTENHOLME, George Edwin (1907–1962)
b. 9 July 1907 *d.* Ypres, Belgium 22 July 1962

Kew gardener, 1929. Horticulturist, Bolivia, 1934–39. Curator, Botanic Gardens, Georgetown, British Guiana, 1950–62.

J. Kew Guild 1962, 202 portr.

WONFOR, Desmond John (1910–1971)
b. Dublin 30 June 1910 *d.* 23 Nov. 1971

To S. Africa in 1921 to join his father. Chief horticulturist, Joubert Park, Cape Town. Took charge of family nursery in Cape Town.

S. African Orchid J. March 1972, 8–9.

WONFOR, Thomas William (*c.* 1827–1878)
d. Brighton, Sussex 20 Oct. 1878

FLS 1877. Diatomist and microscopist. Secretary, Brighton and Sussex Natural History Society, 1853–78.

Naturalist v.4, 1878, 75–76. *R.S.C.* v.8, 1266.

WOOD, Alex (*fl.* 1870s–1910s)

Pharmacist, Brentford, Middx. Sent Middlesex plant records to H. Trimen. Had herb.

WOOD, Charles (*fl.* 1830s–1840s)

Nurseryman, Woodlands Nursery, Maresfield, Uckfield, Sussex. Contrib. to *Floricultural Cabinet*.

WOOD, Emily Margaret (1865–1907)
b. Calcutta, India 23 Aug. 1865 *d.* Birkenhead, Cheshire 28 Oct. 1907

To England, 1871. Lectured on botany at Liverpool. Made more than 800 plant drawings and contrib. to C. T. Green's *Fl. Liverpool District* 1902. Revised G. F. Atkinson's *First Studies of Plant Life* 1905. 'Llandudno and its Fl.' (*Research* 1888, 58).

J. Bot. 1907, 454–55. *Proc. Liverpool Nat. Field Club* 1907, 11–14 portr. *Weekly Courier* 25 April 1908. *Trans. Liverpool Bot. Soc.* 1909, 108.

Plants formerly at Grosvenor Museum, Chester.

WOOD, Geoffrey Howarth Spencer (1927–1957)
b. 12 Aug. 1927 *d.* Kuala Belait, Borneo 5 May 1957

MA Oxon 1948. FLS 1954. Uganda Forest Service, 1949–54. Forest Botanist, N. Borneo, 1954–57 where he collected bryophyta. *Check List of Forest Fl. of N. Borneo* (with J. Agama) 1956. Contrib. to *Malayan Forester*.

Empire For. Rev. 1957, 229–30. *Malayan Forester* 1957, 121. *Nature* v.179, 1957, 1332. *Proc. Linn. Soc.* 1956–57, 38–40. *Fl. Malesiana* v.5, 1958, 100. *Gdns Bull., Singapore* 1960, 498–501.

Plants at Kepong, Kew, Leiden, Oxford.

WOOD, Rev. Henry Hayton (1825–1882)
b. Westward, Cumberland 28 Sept. 1825 *d.* Westward 3 Nov. 1882

MA Oxon 1851. Son of Rev. R. Wood (1796–1883). Rector, Holwell, Dorset, 1857. Bryologist. One of founders of Dorset Field Club.

J. Bot. 1883, 380–81.

Moss herb. at BM(NH).

WOOD, James (*c.* 1792–1830)
d. Huntingdon 18 Nov. 1830

Nurseryman, Huntingdon; business founded by his father John Wood.

Gdnrs Mag. 1831, 384.

WOOD, John Bland (1813–1890)
b. Pontefract, Yorks 3 Dec. 1813 *d.* Manchester 11 Feb. 1890

MD. Bryologist. Employed R. Buxton as collector. *Fl. Mancuniensis* 1840. Contrib. to J. Dickinson's *Fl. Liverpool* 1851 and Supplt., 1855. Contrib. to *Phytologist*.

J. Bot. 1890, 86–87. J. B. L. Warren *Fl. Cheshire* 1899, xcii. *Trans. Liverpool Bot. Soc.* 1909, 108. *Lancashire Nat.* v.13, 1920, 41–44. *R.S.C.* v.6, 433.

Herb. at Manchester University. Plants at Kew. Letters at Kew and in W. Wilson correspondence at BM(NH).

WOOD, John Frederick (*fl.* 1840s)

Nurseryman, The Coppice, near Nottingham.

WOOD, John Joseph (1828–1867)
b. Arcot, India 1828 *d.* Madras, India 23 June 1867

Surgeon, Indian Medical Service, 1847. Assistant to Professor of Botanical and Medical College, Madras, 1859. Contrib. Supplt., 245–85, to R. N. Brown's *Hand Book of Trees, Shrubs and Herbaceous Plants ...in...Madras*...ed. 2 1866.

WOOD, John Medley (1827–1915)
b. Mansfield, Notts 1 Dec. 1827 *d.* Durban, S. Africa 26 Aug. 1915

DSc Cape Univ. 1913. ALS 1887. To Durban, 1852. Traded in Zululand then became a farmer. Curator, Durban Garden, 1882; Director of Natal Herb. *Guide to Trees and Shrubs in Natal Botanic Gardens, Durban* 1897. *Natal Plants* v.2–6, 1899–1912. *Handbook to Fl. Natal* 1907. *Revised List of Fl. Natal* 1908.

Gdnrs Chron. 1915 ii 268. *Kew Bull.* 1915, 417–19. *Nature* v.96, 1915, 174–75. *Proc. Linn. Soc.* 1915–16, 73–74. *Ann. Bolus Herb.* v.2, 1916, 33–36 portr. *Curtis's Bot. Mag. Dedications, 1827–1927* 339–40 portr. A. White and B. L. Sloane *Stapelieae* v.1, 1937, 120 portr. *S. African J. Sci.* 1971, 406–7.

Plants at Kew. Portr. at Hunt Library.

Woodia Schlechter.

WOOD, N. (*fl.* 1690s)

Irish physician and botanist. Correspondent of J. Ray.

J. Bot. 1911, 125.

WOOD, Robert (–1728)

Curator, Edinburgh Town's Physic Garden, 1712–23. Made collection of seeds.

J. Nichols *Illus. of Lit. Hist. of Eighteenth Century* v.1, 389, 391–97, 400–1. D. Turner *Extracts from Lit. and Sci. Correspondence of R. Richardson* 1835, 201–3. G. C. Druce and S. H. Vines *Dillenian Herb.* 1907, lxxix.

WOOD, Rev. Robert (1796–1883)
b. Tallentire, Cockermouth 18 Dec. 1796 *d.* Westward, Cumberland 15 March 1883

Rector, Westward, 1822–83. Had herb. '*Alchemilla conjuncta* in Cumberland' (*J. Bot.* 1872, 308).

J. Bot. 1883, 380. W. Hodgson *Fl. Cumberland* 1898, xxvii–xxviii. *R.S.C.* v.8, 1268.

WOOD, William (*fl.* 1740s–1780s)

Nurseryman, Woodbridge, Suffolk.

WOOD, Rev. William (1745–1808)
b. Collingtree, Northants 29 May 1745 *d.* Leeds, Yorks 1 April 1808

BD. FLS 1791. Nonconformist minister at Debenham, Suffolk, Stamford, Lincs, Ipswich, Suffolk, 1770–72; Mill Hill Chapel, Leeds, 1773–1808. Correspondent of W. Withering and contrib. to ed. 2 of *Bot. Arrangement* 1787–92. Contrib. to *Rees' Cyclopaedia*, letters B and C and to J. Sowerby and J. E. Smith's *English Bot.* 57, 775.

C. Wellbeloved *Mem. of...W. Wood* 1809. P. Smith *Mem. and Correspondence of J. E. Smith* v.1, 488. R. V. Taylor *Biographia Leodiensis* 1865, 232–37. J. B. L. Warren *Fl. Cheshire* 1899, xcii. *D.N.B.* v.62, 379–80.

WOOD, William (*fl.* 1780s–1790s)
Nurseryman, Grantham, Lincs.

WOOD, William (*fl.* 1810s)
Nurseryman, Westgate, Bradford, Yorks.

WOOD, William (*fl.* 1830s–1840s)
Nurseryman, Woodlands Nursery, Maresfield, Uckfield, Sussex.

WOOD, William Leslie (1885–1951)
b. Stepney, London 23 July 1885 *d.* Duxford, Cambridgeshire 8 Feb. 1951
Kew gardener, 1908. Superintendent, Government Gardens and Plantations, Perak, 1910–14. Manager, Jinjang Rubber Estate. Manager, Pontian Rubber Estate, S. Johore, 1919–21. Assistant, Kew Herb., 1921–23. Returned to Malaya, 1923. Superintendent, Johore Palace Gardens, 1928–34. Nurseryman, Duxford, Cambridgeshire.
Kew Bull. 1910, 253. *J. Kew Guild* 1950, 873.

WOODBRIDGE, John (*c.* 1832–1888)
b. Amersham, Bucks *c.* 1832 *d.* 16 April 1888
Kew gardener. Gardener to Duke of Northumberland at Syon House, Middx, 1870.
Garden v.33, 1888, 375. *Gdnrs Chron.* 1888 i 503–4. *J. Hort. Cottage Gdnr* v.16, 1888, 319 portr.

WOODCOCK, Hubert Bayley Drysdale (1867–1957)
b. Antigua 1867 *d.* Bisley, Surrey 12 Feb. 1957
FLS 1941. County Courts Judge. Cultivated garden at Lypiatt Park, Glos. *Lilies* (with J. Coutts) 1935. *Lilies of the World* (with W. T. Stearn) 1950.
Kew Bull. 1935, 582. *Lily Yb.* 1950, 9–10 portr.; 1958, 88–90 portr. *Who was Who 1951–1960* 1189.
Portr. at Hunt Library.

WOODCOCK, W. K. (*c.* 1841–1897)
Nurseryman, Victoria Nursery, Humberstone, Leics.
J. Hort. Cottage Gdnr v.35, 1897, 34 portr., 49.

WOODFORD, Charles Morris (1852–1927)
d. Steyning, Sussex 4 Oct. 1927
Collector for Tring Museum in S. Pacific, 1871. Settled in Suva, Fiji, 1882. Acting Consul, Samoa, 1895. Resident Commissioner, British Solomon Islands, 1897–1915. Between 1907 and 1913 sent herb. specimens to Kew.
Kew Bull. 1927, 421. *Geogr. J.* 1928, 206–7.
Livistona woodfordi Ridley.

WOODFORD, E. John Alexander (*fl.* 1790s)
Had garden at Belmont House, Vauxhall, London where he was "particularly successful in cultivating plants from the hottest climates" (Salisbury).
Bot. Mag. 1801, t.506, t.520; 1817, t.1906. R. A. Salisbury *Paradisus Londinensis* 1805, t.42.

WOODFORD, Sir Ralph (–1828)
d. at sea near Haiti May 1828
Governor, Trinidad, 1813–28. Interested in flora of Trinidad and sent plants to W. J. Hooker.
Bot. Mag. 1827, t.2719. I. Urban *Symbolae Antillanae* v.3, 1902, 141.
Bletia woodfordii Hook.

WOODFORDE, James (1771–1837)
b. Ansford, Somerset 1771 *d.* Castle Cary, Somerset 6 July 1837
MD Edinburgh 1825. FLS 1826. *Catalogue of...Plants Growing in Neighbourhood of Edinburgh* 1824.

WOODHAM, H. J. (1881–1957)
d. 4 June 1957
Cultivated orchids for Messrs. Armstrong and Brown, Orchidhurst, Tunbridge Wells, Kent, 1901–57.
Orchid Rev. 1957, 166.

WOODHEAD, John Ezra (1883–1967)
b. Wyke, Yorks 3 April 1883 *d.* 22 Sept. 1967
BSc London 1922. Pharmaceutical chemist. Studied *Rubus*. Assisted P. D. Sell in editing W. C. R. Watson's *Handbook of Rubi of Great Britain and Ireland* 1958.
Proc. Bot. Soc. Br. Isl. 1968, 497–98.
Plants at Cambridge, Lancaster University.

WOODHEAD, Thomas (*c.* 1832–1882)
d. Halifax, Yorks 30 April 1882
Manager, Shibden Head Brewery. Florist of Halifax; specialised in auriculas.
Gdnrs Chron. 1882 i 646.

WOODHEAD, Thomas William (1863–1940)
b. Huddersfield, Yorks 1863 *d.* Huddersfield 1940
MSC Leeds 1915. PhD Zurich. FLS 1899. Lecturer, Huddersfield Technical College, 1891; Head, Biological Dept. President, Yorkshire Naturalists' Union, 1922; British Ecological Society, 1926–27. *Ecology of Woodland Plants in Neighbourhood of Huddersfield* 1906. *Study of Plants* 1915. *Junior Botany* 1922. Edited *Naturalist* 1903–32.
Naturalist 1940, 157–58 portr. *Proc. Linn. Soc.* 1939–40, 383–84.

WOODHOUSE, Edward John (1884–1917)
d. France 18 Dec. 1917
BA Cantab 1906. FLS 1909. Economic Botanist to Government of Bengal, 1907. Principal, Sabour Agricultural College, 1911. Captain, Indian Army.
Agric. J. India 1918, 242 portr. *Kew Bull.* 1918, 33. *Nature* v.100, 1918, 429–30. *Proc. Linn. Soc.* 1917–18, 46.

WOODHOUSE, Rev. Thomas
(*fl.* 1860s–1870s)
Vicar, Ropley, Hants. Contrib. to W. H. Purchas and A. Ley's *Fl. Herefordshire* 1899. Contrib. 'Notes on Natural History of Aymestry' and papers on yew and beech trees to *Trans. Woolhope Nat. Field Club.*
Trans. Woolhope Nat. Field Club v.34, 1954, 250.

WOODLEY, Maria *see* Riddell, M.

WOODMAN, Henry (*c.* 1698–1758)
d. Jan. 1758
Nurseryman, Strand-on-the-Green, Middx; business passed to his widow.
Trans. London Middlesex Archaeol. Soc. v.24, 1973, 182.

WOODMAN, William Robert (*c.* 1829–1891)
d. Brondesbury, Middx 20 Dec. 1891
Succeeded his uncle, R. T. Prince, in management of Exeter nursery of Lucombe, Prince and Company; retired 1883.
Garden v.41, 1892, 19. *Gdnrs Chron.* 1892 i 25, 42 portr.

WOODROW, George Marshall (1846–1911)
b. 14 Feb. 1846 *d.* Lanarkshire 8 June 1911
Kew gardener, 1865. To India to be in charge of Ganeshkhind Experimental Garden; also responsible for official gardens at Poona, 1872. Lecturer, Royal College of Science, Poona, 1879. Director, Botanic Survey, Western India, 1893–99. *Hints on Gardening in India* ed. 2 1877. *The Mango* 1904.
J. Kew Guild 1916, 309–10 portr.
Plants at Kew.

WOODRUFF, George (1864–1891)
d. Asaba, Nigeria 2 Jan. 1891
Kew gardener, 1887. Curator, Botanical Plantation, Abutsi, Nigeria, 1889. Sent seeds to Kew.
J. Kew Guild 1895, 39–42 portr.

WOODRUFFE-PEACOCK, Rev. Edward Adrian (1858–1922)
b. Bottesford, Brigg, Lincs 23 July 1858 *d.* Grayingham, Lincs 3 Feb. 1922
FLS 1895. Vicar, Cadney, 1891–1920; Grayingham, 1920–22. Secretary, Lincolnshire Naturalists Union; President, 1905–6. *Checklist of Lincolnshire Plants* 1909. Contrib. to *J. Bot.*, *Naturalist*, *Trans Lincolnshire Nat. Union.*
Trans. Lincolnshire Nat. Union 1912–15, 71–80 portr.; 1922, 164–65. *Bot. Soc. Exch. Club Br. Isl. Rep.* 1922, 712–14. *J. Bot.* 1922, 161–62. *Naturalist* 1922, 137–39 portr. *Proc. Linn. Soc.* 1921–22, 49. *Quart. J. For.* v.16, 1922, 137–38. A. R. Horwood and C. W. F. Noel *Fl. Leicestershire* 1933, cxxxix. *Lincolnshire Hist. Archaeol.* 1971, 113–24 portr. E. J. Gibbons *Fl. Lincolnshire* 1975, 58–59 portr.
Herb. at Lincoln Museum. MS. ecological flora of Lincolnshire at Cambridge.

WOODS, Francis (*fl.* 1840s–1850s)
Nurseryman, Tothill Nursery, Plymouth, Devon.

WOODS, John (*c.* 1813–1897)
Nurseryman, Cumberland Street, Woodbridge, Suffolk.
J. Hort. Cottage Gdnr v.34, 1897, 231.

WOODS, Joseph (1776–1864)
b. Stoke Newington, London 24 Aug. 1776 *d.* Lewes, Sussex 9 Jan. 1864
FLS 1807. Architect, London, 1819–33. To Lewes, 1833. 'Synopsis of *Rosa*' (*Trans. Linn. Soc.* v.12, 1818, 159–234). *Tourist's Fl.* 1850. Contrib. to J. Sowerby and J. E. Smith's *English Bot.* 1301, 2823, 2886.
Trans. Linn. Soc. v.7, 1804, 101. D. Turner and L. W. Dillwyn *Botanist's Guide through England and Wales* v.1, 1805, xiv. *J. Bot.* 1864, 62–64, 96. *Proc. Linn. Soc.* 1863–64, xxxii–xli. *D.N.B.* v.62, 409–10. F. H. Davey *Fl. Cornwall* 1909, xlix. R. W. Scully *Fl. County Kerry* 1916, xiv. G. C. Druce *Fl. Buckinghamshire* 1926, xcix. A. H. Wolley-Dod *Fl. Sussex* 1937, xli–xlii. *R.S.C.* v.6, 436.
Plants at Leeds Museum, South London Botanical Institute. Letters at Kew and in N. J. Winch correspondence at Linnean Society.
Woodsia R. Br.

WOODS, Rev. Julian Edmund Tenison *see* Tenison-Woods, *Rev.* J. E.

WOODS, Richard (*fl.* 1740s–1793)
Nurseryman and landscape gardener, Chertsey, Surrey. Surveyor, N. Ockenden, Essex.

WOODVILLE, William (1752–1805)
b. Cockermouth, Cumberland 1752 *d.* London 26 March 1805
MD Edinburgh 1775. LRCP 1784. FLS 1791. Physician to Middlesex Dispensary, London, 1782. Physician to Small-pox Hospital, King's Cross, where he had a botanic garden, 1791. *Medical Botany* 1790–94 4 vols; ed. 3 1832 5 vols.

Gent. Mag. 1805 i 321, 387. W. Munk *Roll of R. College of Physicians* v.2, 1878, 345. *D.N.B.* v.62, 417. *Gdnrs Chron.* 1935 i 306–7.
Portr. at Hunt Library. Sale of Library at Sotheby 3 July 1805.

WOODWARD, George (*fl.* 1830s)
Surgeon, Bicester, Oxford. Recorded Oxfordshire plants.
G. C. Druce *Fl. Oxfordshire* 1927, cxi.

WOODWARD, John (1665–1728)
b. Derbyshire 1 May 1665 *d.* London 25 April 1728
MD Cantuar and Cantab 1695. FRCP 1702. FRS 1693. "Professor of Physick" Gresham College, 1692–1728. Geologist. 'Thoughts and Experiments on Vegetation' (*Philos. Trans. R. Soc.* v.21, 1699, 193–227) (first description of water-culture and transpiration).
J. Ward *Lives of Professors of Gresham College* 1740, 283–301. J. W. Clark and T. M. Hughes *Life and Letters of…A. Sedgwick* v.1, 1890, 165–89. *D.N.B.* v.62, 423–25. *Chronica Botanica* 1944, 84–86. *N. Western Nat.* 1934, 229–32. J. E. Dandy *Sloane Herb.* 1958, 123. *Nature* v.206, 1965, 868–70. *J. Soc. Bibl. Nat. Hist.* 1971, 399–427 portr. C. C. Gillispie *Dict. Sci. Biogr.* v.14, 1976, 500–3.
Herb at Oxford. Portr. in Dept. of Geology at Cambridge.

WOODWARD, Marcus (–1940)
d. 12 Sept. 1940
Author and journalist. *New Book of Trees* 1926. *Gerard's Herball* 1927. *Wild Flowers* 1927. *Garden Flowers* 1928. *Leaves from Gerard's Herball* 1931. *Trees of Westonbirt* 1933. *Plant Life* 1935.
Who was Who, 1929–1940 1487.

WOODWARD, Robert (–1915)
d. France 9 May 1915
Of Arley Castle, Bewdley, Worcs where he planted a fine collection of trees. *Hortus Arleyensis* 1907.
Kew Bull. 1915, 263–64.

WOODWARD, Samuel Pickworth (1821–1865)
b. Norwich, Norfolk 17 Sept. 1821 *d.* Herne Bay, Kent 11 July 1865
PhD Göttingen 1864. ALS 1842. Professor of Geology and Natural History, Royal Agricultural College, Cirencester, 1845. Assistant, Dept. of Geology, BM, 1848–65. Malacologist. 'Fl. Central Norfolk' (*Mag. Zool. Bot.* v.5, 1841, 201–6).
Geol. Mag. 1865, 383–84. *Proc. Linn. Soc.* 1865–66, lxxxvi–lxxxvii. *Quart. J. Geol. Soc. London* 1866, xxxiv–xxxvi. *Trans. Norfolk Nat. Soc.* 1881–82, 279–312 portr. *D.N.B.* v.62, 426–27. R. W. Scully *Fl. County Kerry* 1916, xiii–xiv. H. J. Riddelsdell *Fl. Gloucestershire* 1948, cxxiii.
Herb. and drawings at Cirencester. MS. flora at Norfolk and Norwich Naturalists Society. Letters at Norwich Museum.
Carduus woodwardii Wats.

WOODWARD, Thomas Jenkinson (1745–1820)
b. Huntingdon 6 March 1745 *d.* Diss, Norfolk 28 Jan. 1820
LLB Cantab 1769. FLS 1789. Of Bungay, Norfolk. Contrib. to T. Martyn's edition of P. Miller's *Gdnrs Dict.* 1807, W. Withering's *Bot. Arrangement* ed. 2 1787 and to J. Sowerby and J. E. Smith's *English Bot.* 151, 920.
Trans. Linn. Soc. v.1–3, 1791–97. *Philos. Trans. R. Soc.* v.74, 1784, 423. *Gent. Mag.* 1820 i 189, 280–81. W. M. Hind *Fl. Suffolk* 1889, 481. *D.N.B.* v.62, 427–28. *Trans. Norfolk Norwich Nat. Soc.* v.9, 1914, 659, 672–75. A. R. Horwood and C. W. F. Noel *Fl. Leicestershire* 1933, clxxxix.
Woodwardia Smith.

WOOLDRIDGE, Theodore A. (*fl.* 1890s)
Of Penang, Malaya. Collected orchids through native agents.
Gdns Bull. Straits Settlements 1927, 134.

WOOLLARD, William (*fl.* 1830s)
Nurseryman, Hanford Bridge, Ipswich, Suffolk.

WOOLLEY, Samuel (*c.* 1821–1878)
d. 5 Feb. 1878
Gardener to H. Bellenden Ker at Cheshunt, Herts; then a nurseryman.
Florist and Pomologist 1878, 64.

WOOLLGAR, Thomas (*fl.* 1760s–1820s)
Of Lewes, Sussex. W. Borrer's "earliest assistant in botany." Studied willows and sent them to J. Sowerby and J. E. Smith's *English Bot.* 1436, 1936–37, 2651. Contrib. plant localities to C. Milne and A. Gordon's *Indigenous Botany* 1793.
A. H. Wolley-Dod *Fl. Sussex* 1937, xxxix.
MS. *Fl. Lewesenses* at Brighton Museum.
Salix woollgariana Hook.

WOOLLS, Rev. William (1814–1893)
b. Winchester, Hants March 1814 *d.* Burwood near Sydney, N.S.W. 14 March 1893
PhD Göttingen. FLS 1865. To New South Wales, 1827. Master, King's School, Parramatta, Sydney, 1832. Classics teacher, Sydney College, *c.* 1836. Established private school at Parramatta. Became interested in botany through friendship with Rev. J. Walker. Corresponded with F. von Mueller. Joined Sydney Grammar School, 1857. Ordained, 1873; incumbent of Episcopalian

Church, Richmond. *Contribution to Fl. Australia* 1867. *Lectures on Vegetable Kingdom* 1879. *Plants of N.S.W.* 1885. *Plants Indigenous in Neighbourhood of Sydney* 1880; ed. 2 1891. Contrib. to *Proc. Linn. Soc. N.S.W.*

Proc. Linn. Soc. 1892–93, 27–28. *Proc. Linn. Soc. N.S.W.* 1892, 668–69. *Sydney Morning Herald* 15 March 1893. *Victorian Nat.* 1893, 185; 1932, 135–40. *J. Proc. R. Soc. N.S.W.* 1908, 130–32 portr. P. Serle *Dict. Austral. Biogr.* v.2, 1949, 509–10. *Austral. Encyclop.* v.9, 1965, 368.

Portr. at Hunt Library.

Woollsia F. Muell.

WOOLMAN, H. (*c.* 1852–1932)
d. Birmingham 8 April 1932

Nurseryman, Small Heath, Birmingham, 1881; moved to Tyseley near Acocks Green, 1893; to Shirley Nurseries, Birmingham, 1907.

Gdnrs Chron. 1932 i 301.

WOOLSHAFEN, John (*c.* 1720–1794)
b. Canterbury, Kent *c.* 1720 *d.* Canterbury 20 Sept. 1794

"An excellent herbalist" of Canterbury; son of "an eminent apothecary in that city."

Gdnrs Chron. 1919 i 147.

WOOLWARD, Florence Helen (1854–1936)
d. 3 Jan. 1936

Prepared descriptions and plates for Marquess of Lothian's *Masdevallia* 1890–96.

J. Bot. 1936, 55.

Drawings at BM(NH).

WOOSTER, David (*c.* 1824–1888)
d. Bayswater, London Sept./Oct. 1888

Assisted in revised editions of J. C. Loudon's *Hortus Britannicus* 1850 and *Encyclop. of Plants* 1855. *Synopsis of Coniferous Plants* 1850. *Alpine Plants* 1872–74 2 ser.

Gdnrs Chron. 1888 ii 393. *J. Hort. Cottage Gdnr* v.17, 1888, 336.

Letters at Kew.

WOOTTON, Charles Reginald (1897–1970)
d. 18 May 1970

Builder, Bloxwich, Staffs. Hybridised daffodils.

Daffodil and Tulip Yb. 1971, 175–76.

WORLIDGE, John (*c.* 1630–1693)

Of Petersfield, Hants. Great-nephew of J. Goodyer. Agricultural writer. *Systema Agriculturae* 1669. *Systema Horticulturae* 1677; ed. 4 1719. *Vinetum Britannicum* 1676.

G. W. Johnson *Hist. of English Gdning* 1829, 113–14. S. Felton *Portr. of English Authors on Gdning* 1830, 28–31. J. Donaldson *Agric. Biogr.* 1854, 33–34. *D.N.B.* v.63, 28. D. McDonald *Agric. Writers* 1908, 116–21.

WORMALD, Harry (*c.* 1879–1955)
b. Lofthouse, Yorks *d.* Maidstone, Kent 10 Dec. 1955

BSc London 1911. DSc 1919. Schoolteacher, Bradford and Leeds, 1900–8. Mycologist, S.E. Agricultural College, Wye, 1911–23. Head of Plant Pathology, East Malling Research Station, 1923; Assistant Director, 1936–45. President, British Mycological Society, 1940. Joint editor of *Trans. Br. Mycol. Soc.* 1931–45. *Brown Rot of Fruit Trees* 1935. *Diseases of Fruits and Hops* 1939.

Rep. E. Malling Res. Station 1955, 15–16 portr. *Nature* v.177, 1956, 649. *Trans. Br. Mycol. Soc.* 1956, 289–90 portr.

WORSDELL, Wilson Crosfield (1867–1957)
b. Altoona, U.S.A. 15 Sept. 1867 *d.* Twickenham, Middx 29 Oct. 1957

FLS 1898. To England, 1871. Studied horticulture in Dutch and English nurseries. Demonstrator in Botany, University College, London. Acting Professor of Botany, S. African College, Cape Town, 1909. Lecturer in Botany, Victoria College, Stellenbosch, 1912–13. Collected plants in S. Africa. Did research on stem structure of dicotyledons at Jodrell Laboratory, Kew. *Principles of Plant Teratology* 1915–16 2 vols. Edited 2 supplementary vols of *Index Londinensis* 1941.

Taxon 1953, 14–16 portr. *Gdnrs Chron.* 1957 ii 378. M. R. Levyns *Botanist's Mem.* 1968, 3.

S. African plants at Birmingham University. Portr. at Hunt Library.

WORSLEY, Anna *see* Russell, A.

WORSLEY, Arthington (1861–1944)
b. London 1861

Civil engineer. Travelled in Central and S. America. Lived in Isleworth, Middx from 1894 where he grew plants he had collected in America. Specialised in *Amaryllis*. *Notes of Distribution of Amaryllideae...in Grand Canary...* 1895. Contrib. to *Gdnrs Chron.*, *J. R. Hort. Soc.*

Yb. Amaryllis Soc. Amer. 1936, 9–19 portr.

WORSLEY, William (1808–)

Weaver, Middleton, Lancs. Active member of local botanical society.

J. Cash *Where there's a Will there's a Way* 1873, 132–33. *Trans. Liverpool Bot. Soc.* 1909, 108.

WORSLEY-BENISON, Henry Worsley Seymour (1845–1918)
b. Clevedon, Somerset 14 Aug. 1845 *d.* Bournemouth, Hants 14 Dec. 1918

MB London. FLS 1869. Lecturer in Botany, Westminster Hospital, 1877–89. 'Movement in Plants' (*J. Microsc. Nat. Sci. Bath* 1886, 197–209).

Proc. Linn. Soc. 1918–19, 67.

WORTHINGTON, Thomas Berkeley (–1970)
d. Kandy, Ceylon 11 Nov. 1970
Tea planter. Had herb. of Ceylon plants. *Ceylon Trees* 1959.
Herb. at Botanic Gardens, Peradeniya, Ceylon. Letters and MSS. at Kew.

WRAGGE, Clement Lindley (1852–1922)
b. Stourbridge, Worcs 18 Sept. 1852 *d.* Dec. 1922
Joined Surveyor-General's Dept., Adelaide, 1876. Established meteorological stations in Australia. Formed Capemba Botanic Gardens near Brisbane, 1889–1900. Settled in New Zealand in 1910 and founded Wragge Scientific Institute Museum and Waiata Botanic Gardens.
Who was Who, 1916–1928 1150.

WRAY, Cecil (*fl.* 1880s)
Accompanied his brother, Leonard, on expedition to Perak, 1888 and collected plants.
J. Straits Branch R. Asiatic Soc. 1890, 134. *Fl. Malesiana* v.1, 1950, 586.

WRAY, Leonard (1853–1942)
d. 14 March 1942
Superintendent, Government Hill Garden, Perak, 1881. Curator, Perak State Museum, 1883–1908. Collected plants in Taiping Hills and made expeditions to Batang Padang district, to Gunong Tahan with H. C. Robinson. Director of Museums, Federated Malay States, 1905–8. Contrib. to *Kew Bull.*
J. Bot. 1903, 307. *J. Federated Malay States Mus.* 1907, 107–61. *J. Linn. Soc.* 1908, 301–36. *Gdns Bull. Straits Settlements* 1927, 134. *Fl. Malesiana* v.1, 1950, 586–87. *Who was Who, 1941–1950* 1264.
Plants at BM(NH), Calcutta, Kew, Singapore.

WRAY, Mrs. Martha (*c.* 1775–1864)
d. Leckhampton, Glos 18 Sept. 1864
Of Oakfield, Cheltenham, Glos. Cultivated orchids, several of which were figured in *Bot. Mag.* She drew t.3875 for *Bot. Mag.*
Curtis's Bot. Mag. Dedications, 1827–1927 55–56.
Letters at Kew.

WREN, Richard Cranfield (1861–1930)
d. Westcliff-on-Sea, Essex 24 July 1930
Director of druggists, Potter and Clarke. *Potter's Cyclopaedia of Botanical Drugs and Preparations* ed. 2 1915.
Bot. Soc. Exch. Club Br. Isl. Rep. 1930, 330. *Pharm. J.* 1930, 143.

WREN, William (*fl.* 1840s)
Gardener, Horticultural Society of London. To Ascension Island in 1847 with a supply of plants and seeds to assist in land reclamation.
Gdnrs Chron. 1847, 735–36.

WRENCH, Jacob (*fl.* 1760s–1780s)
Seedsman, 126 Lower Thames Street, London.

WRENCH, Nathaniel (1682–1783)
Son of T. Wrench (*c.* 1630–1728). Nurseryman, Broomhouse Nurseries, Fulham, Middx. Nursery passed to his daughter, Elizabeth, and her husband, Daniel Fitch.
E. J. Willson *Hist. of Fulham* 1970, 245.

WRENCH, Robert (*c.* 1813–1883)
d. Surbiton, Surrey Jan. 1883
Son of Jacob Wrench. Seedsman, London Bridge.
Gdnrs Chron. 1883 i 91, 178 portr.

WRENCH, Thomas (*c.* 1630–1728)
Nurseryman, Fulham, Middx.
E. J. Willson *Hist. of Fulham* 1970, 244–45.

WRENCH, Thomas (*fl.* 1660s–1690s)
Nurseryman, Paradise Gardens, Oxford.

WRIGHT, Charles Augustus (1834–1907)
b. London 1834 *d.* Kew, Surrey 13 July 1907
FLS 1878. Lived in Malta for many years. *Times* correspondent in Mediterranean, *c.* 1865–74. Had herb. of Maltese and British plants.
Proc. Linn. Soc. 1907–8, 66–68.
Herb. at BM(NH).

WRIGHT, Charles H. (*fl.* 1840s)
Of Keswick, Cumberland. Acted as guide in the Lake District. He and his daughter collected local plants but the accuracy of both was doubtful.
Phytologist v.2, 1845, 74, 376, 428–31.

WRIGHT, Charles Henry (1864–1941)
b. Oxford 5 June 1864 *d.* Seaton, Devon 21 June 1941
ALS 1896. Assistant, Kew Herb., 1884; Assistant Keeper, 1908–29. Worked on mosses, ferns, petaloid Monocotyledons and palms. Contrib. to *Fl. Capensis*, *Fl. Tropical Africa*, *Kew Bull.*, *J. Linn. Soc.*
Gdnrs Chron. 1927 ii 42 portr.; 1941 ii 8. *Kew Bull.* 1929, 265–66; 1941, 239–40. *J. Bot.* 1929, 320. *J. Kew Guild* 1941, 93 portr. *Proc. Linn. Soc.* 1941–42, 299–300.
Plants at Exeter Museum, Kew. Portr. at Hunt Library.

WRIGHT, Edward Perceval (1834–1910)
b. Dublin 27 Dec. 1834 *d.* Dublin 2 March 1910
MA Dublin 1857. MD 1862. FLS 1859. Lecturer in Zoology, Trinity College, Dublin, 1858–68; Professor of Botany, 1869–1904;

Keeper of Herb., 1870–1910. Investigated fauna and flora of the Seychelles, 1867.

Irish Nat. 1910, 61–63 portr. *J. Hort. Cottage Gdnr* v.60, 1910, 225. *Notes from Bot. School, Trinity College, Dublin* v.2, 1910, 91–97 portr. *Nature* v.83, 1910, 73–74. *J. Bot.* 1911, 123–24 portr. *Proc. Linn. Soc.* 1909–10, 102–4. *D.N.B. Supplt 2* v.3, 709–10. *Who was Who, 1897–1916* 782. R. L. Praeger *Some Irish Nat.* 1949, 179 portr. *R.S.C.* v.6, 442; v.8, 1277; v.11, 856.

Herb. at Trinity College, Dublin. Plants and letters at Kew.

WRIGHT, Francis Bowcher (*fl.* 1800s)

FLS 1808. Of Hinton Blewett, Somerset. Discovered *Paeonia* on Steep Holme in Bristol Channel, 1803.

J. Sowerby and J. E. Smith *English Bot.* 1513, 1657.

Letters in Smith correspondence at Linnean Society.

WRIGHT, Frederick (1845–1924)

b. Sutton, Cambridgeshire 17 Sept. 1845 *d.* 2 May 1924

Bryologist. Member of South London Botanical Institute.

Rep. Br. Bryol. Soc. 1925, 182.

WRIGHT, Frederick Robert Elliston- *see* Elliston-Wright, F. R.

WRIGHT, Sir Herbert (1874–1940)

b. Lancs 10 Sept. 1874 *d.* Chalfont St. Giles, Bucks 28 Oct. 1940

Educ. Royal College of Science, London. FLS 1902. Scientific Assistant and then Acting Director, Royal Botanic Gardens, Peradeniya, Ceylon, 1900–7; also Controller of Agricultural Experiment Station. Developed rubber in Ceylon. Edited *India Rubber J.*, 1907–17. *Hevea brasiliensis or Para Rubber* 1905; ed. 3 1908. *Rubber Cultivation in British Empire* 1907. *Theobroma cacao or Cocoa* 1907. Contrib. to *Ann. R. Bot. Gdns, Peradeniya.*

Nature v.146, 1940, 677–78. *Times* 29 and 31 Oct. 1940. *Proc. Linn. Soc.* 1940–41, 302. *Who was Who, 1929–1940* 1493–94.

WRIGHT, John (1836–1916)

b. Scawby, Lincs 1836 *d.* 2 May 1916

VMH 1897. VMM 1904. Sub-editor, 1874 then Editor, 1897 of *J. Hort. Culture of Chrysanthemum* 1883. *Mushrooms for the Million* 1883; ed. 6 1889. *Garden Allotments* 1889. *Profitable Fruit-growing for Cottagers* 1889; ed. 11 1920. *Fruit Grower's Guide* 1892 3 vols. *Garden Flowers and Plants* 1895.

Gdnrs Mag. 1902, 457–58 portr.; 1916, 231–32 portr. *J. Hort. Cottage Gdnr* v.60, 1910, 371–72 portr. *Garden* 1916, 236(vi) portr. *Gdnrs Chron.* 1916 i 251–52 portr.

WRIGHT, Nathaniel (*fl.* 1610s–1630s)

Emmanuel College, Cambridge, 1623–24. MD Bourges. Physician to Oliver Cromwell in Scotland. Friend of T. Johnson.

J. Gerard *Herball* 1633, 618. C. E. Raven *English Nat.* 1947, 283.

WRIGHT, Samuel Thomas (1858–1922)

b. 21 Oct. 1858 *d.* Matlock, Derbyshire 28 April 1922

VMH 1920. First Superintendent, Royal Horticultural Society gardens at Chiswick, 1896; moved to Wisley 1903; retired 1922. *How to grow Strawberries* 1888. *Fruit-culture for Amateurs* 1897; ed. 5 1921. Contrib. to *Gdnrs Chron.*

Gdnrs Chron. 1896 i 240 portr.; 1922 i 236 portr. *Gdnrs Mag.* 1908, 350–51 portr. *Garden* 1913, 251 portr.; 1922, 232. H. R. Fletcher *Story of R. Hort. Soc., 1804–1968* 1969, 274–76, 280–81, 324–25, 398 portr.

WRIGHT, Thomas (*fl.* 1760s–1770s)

Nurseryman, Monaghan, Ireland.

Irish For. 1967, 56.

WRIGHT, Thomas (*c.* 1828–1889)

d. 25 Sept. 1889

Pteridologist. Contrib. to T. Allin's *Flowering Plants and Ferns of County Cork* 1883.

Gdnrs Chron. 1889 ii 395.

WRIGHT, Thomas (*fl.* 1830s)

Seedsman, Fennel Street, Warrington, Lancs.

WRIGHT, Walter Page (1864–1940)

b. Scawby, Lincs 3 June 1864 *d.* Folkestone, Kent 5 Feb. 1940

Prolific writer of gardening books. President, Horticultural Education Association, 1920–21. Founded and edited *Popular Gardening.* Edited *Cassell's Dictionary of Popular Gardening* 1902 2 vols. *Pictorial Practical Rose Growing* 1902. *School and Garden* 1906. *Beautiful Gardens* 1909. *Alpine Flowers and Rock Gardens* 1910. *New Gardening* 1912, etc.

Gdnrs Chron. 1940 i 145–46, 181.

WRIGHT, William (1735–1819)

b. Crieff, Perthshire March 1735 *d.* Edinburgh 19 Sept. 1819

MD St. Andrews. FRS 1778. ALS 1807. Physician-General of Jamaica, 1784. To Greenland, 1757; Jamaica, 1765–77, 1779–85; Barbados, 1796–98. Discovered *Cinchona jamaicensis* (*Philos. Trans. R. Soc.* v.67, 1777, 504–6). Collected plants in Jamaica.

D. Turner *Fuci* v.3, 1811, 32. *Mem. of Late William Wright, M.D.* 1828. *Naturalist* v.4, 1839, 399. *D.N.B.* v.63, 136–37. I. Urban *Symbolae Antillanae* v.1, 1898, 178–79; v.3, 144–45; v.7, 77. *J. Bot.* 1914, 323; 1922,

330–34. *Osiris* 1948, 116–17. *Taxon* 1970, 545–46. *R.S.C.* v.6, 446.
Herb. at BM(NH), Liverpool. Three vols of plants at Royal Botanic Garden, Edinburgh.
Fucus wrightii Turner.

WRIGLEY, Oswald Osmond (*c.* 1836–1917)
d. 11 Nov. 1917
Of Bridge Hall, Bury, Lancs. Cultivated orchids.
Orchid Rev. 1917, 250–51.

WYARD, Stanley (*c.* 1886–1946)
d. 29 Sept. 1946
MD London 1909. Physician at Cancer Hospital, London. Added new records for mosses to Surrey, Sussex, Caernarvonshire and Westmorland.
Trans. Br. Bryol. Soc. v.1, 1947, 45.

WYATT, Mrs. Mary (–*c.* 1850)
Dealer in shells, Torquay. *Algae Damnonienses* 1834–40 (exsiccatae), superintended by A. W. Griffiths, to whom she was a servant.
Mag. Nat. Hist. 1834, 95. W. H. Harvey *Manual of Br. Algae* 1841, liv. *Companion Bot. Mag.* v.2, 1837, 246. *Trans. Penzance Nat. Hist. Soc.* 1890–91, 230.
Wyattia Trevisan.

WYATT, Oliver E. P. (–1973)
d. 25 Feb. 1973
VMH 1965. Headmaster, Maidwell Hall School, Northants, 1933–63. President, Alpine Garden Society, 1967–71. Treasurer, Royal Horticultural Society, 1965–71. Grew lilies.
J. R. Hort. Soc. 1956, 294–302. *Lily Yb.* 1963, 8–10 portr. *Quart. Bull. Alpine Gdn Soc.* 1973, 185. *Times* 1 and 3 March 1973.

WYLEY, Andrew (*fl.* 1850s)
Cape Colonial Geologist, 1857. *Report on S. Namaqualand* 1857 (botanical appendix).
W. H. Harvey and O. W. Sonder *Fl. Capensis* v.1, 1859, 10*. *R.S.C.* v.6, 460.

WYN (Wynne)
Plants from Vizagapatam, Madras in Sloane Herb. at BM(NH).
J. E. Dandy *Sloane Herb.* 1958, 231.

WYNNE, Brian (–1924)
d. Feltham, Middx 2 May 1924
Gardener. On editorial staff of *Gdnrs Chron.* 1868. Editor of *Gdning World* until 1895.
Gdnrs Chron. 1924 i 280 portr.

WYNNE, John (*fl.* 1830s–1860s)
Of Hazelwood, County Sligo. Under-Secretary for Ireland. Discovered *Erica mediterranea* in Mayo, 1836. '*Adiantum capillus-veneris* in Leitrim' (*Nat. Hist. Rev.* v.4, 1857, 69).
N. Colgan and R. W. Scully *Cybele Hibernica* 1898, xxxvi. *London J. Bot.* 1845, 570.

WYTHES, George (1851–1916)
b. Worcester 1851 *d.* Folkestone, Kent 28 May 1916
VMH 1897. Gardener to Duke of Northumberland at Syon House, 1888–1906. *Book of Vegetables* 1902. *Vegetable Growing Made Easy* (with O. Thomas) 1913.
Garden v.59, 1901, 264–65 portr.; 1916, 292 portr. *Gdnrs Chron.* 1903 ii 219–20 portr.; 1916 i 316 portr. *J. Hort. Cottage Gdnr* v.53, 1906, 294–95 portr.; v.65, 1912, 416–17 portr. *Gdnrs Mag.* 1910, 259–60 portr.; 1916, 276(iii).

YALDEN, Thomas (1750–1777)
b. London 1750 *d.* Venice? 1777
BA Oxon 1772. MD Edinburgh 1774. Herb. bequeathed to J. Lightfoot and purchased with Lightfoot's herb. by Queen Charlotte; plants now at Kew.
J. Lightfoot *Fl. Scotica* v.2, 1777, 1142–48 (list of Edinburgh plants). J. Sowerby and J. E. Smith *English Bot.* 2467.
Plants and MS. 1773–74, on Scottish plants at BM(NH).

YAPP, Richard Henry (1871–1929)
b. Orleton, Herefordshire 1871 *d.* Birmingham 22 Jan. 1929
BA Cantab 1898. Curator, Cambridge University Herb., 1900–3. Botanist on Cambridge Expedition to Siamese-Malay States, 1899–1900. Professor of Botany, Aberystwyth, 1904–14; Queen's University, Belfast, 1914–19; Birmingham, 1919–29. Revised N. A. Maximov's *Plant in Relation to Water* 1929. *Botany: Junior Book for Schools* 1923. Contrib. to *Ann. Bot.*, *J. Ecol.*, *New Phytol.*
Gdns Bull. Straits Settlements 1927, 134. *Bot. Soc. Exch. Club Br. Isl. Rep.* 1928, 720–21. *J. Bot.* 1929, 85–86. *J. Ecol.* 1929, 405–8 portr. *Kew Bull.* 1929, 140. *Nature* v.123, 1929, 249–50. *Times* 25 Jan. 1929. *Who was Who, 1929–1940* 1499. *Fl. Malesiana* v.1, 1950, 588–89. *Trans. Woolhope Nat. Field Club* 1955, 259–60.
Plants at Birmingham, Cambridge, Kew.

YARNALL, Richard Keyte (1752–1826)
b. Ebrington, Glos 30 Sept. 1752 *d.* Knowsley Lancs 19 Feb. 1826
Gardener to Lord Derby at Knowsley Hall, 1796–1826.
Gdnrs Mag. 1826, 228–29.

YATES, Rev. James (1789–1871)
b. Toxteth Park, Liverpool, Lancs 30 April 1789 *d.* Highgate, London 7 May 1871
MA Glasgow 1812. FRS 1839. FLS 1822. Unitarian minister. Archaeologist. Had garden at Lauderdale House, Highgate. Had collection of cycads. Contrib. notes on *Cycadaceae* to *Proc. Linn. Soc.* 1849, 16–22; 1853, 253–55. '*Zamia gigas*' (*Proc. Yorkshire Philos Soc.* v.1, 1855, 37–42).
Gdnrs Chron. 1853, 455; 1871, 618. *Proc. Linn. Soc.* 1870–71, xci–xciii. *Proc. R. Soc.* v.20, 1872, i–iii. *Liverpool Lit. Philos. Soc. Proc.* v.26, 1872, xxxi–xxxii. S. A. T. Yates *Memorials of Family of Rev. John Yates* 1890. *D.N.B.* v.63, 295–96. *Trans. Liverpool Bot. Soc.* 1909, 109. *J. Bot.* 1921, 221–24. *R.S.C.* v.6, 465.
Plants and drawings at BM(NH). Letters at Kew. Sale at Sotheby, 24 June 1875.

YATES, Lorenzo Gordin (1837–1909)
b. East Church, Isle of Sheppey, Kent 12 Jan. 1837 *d.* Santa Barbara, California, U.S.A. 30 Jan. 1909
FLS 1888. To New York, 1851; to Michigan, Wisconsin then California, 1874. Dentist. Naturalist. With California State Geological Survey. Horticultural Commissioner, Santa Barbara, 1905–6. *Catalogue of Exotic Ferns...in L. G. Yates' Collection* 1886. *Catalogue of Ferns of N. America in L. G. Yates' Collection* 1886. '*Cheilanthes myriophylla*' (*J. Bot.* 1887, 248).
J. Soc. Bibl. Nat. Hist. 1963, 178–93 portr. *J. of the West* v.2, 1963, 377–400.
Plants at Santa Barbara Museum. MSS. at Bancroft Library. Portr. at Hunt Library.
Acrostichum yatesii Sod.

YELD, George (1845–1938)
d. Orleton, Herefordshire 2 April 1938
MA Oxon. VMH 1925. Schoolmaster, St. Peter's, York. Hybridised bearded irises and hemerocallis. First President, Iris Society, 1924. *Scrambles in Eastern Graians, 1878–1897* 1900. Contrib. to *Gdnrs Chron.*
Gdnrs Mag. 1907, 423 portr., 424. *Gdnrs Chron.* 1922 i 266 portr.; 1927 i 348 portr.; 1935 ii 92 portr.; 1938 i 246 portr., 259. *Iris Yb.* 1938, 106–8. *J. R. Hort. Soc.* 1938, 533 portr. *Quart. Bull. Alpine Gdn Soc.* 1938, 193–94 portr.
Portr. at St. Peter's School, York.

YERBURY, J. W. (*fl.* 1880s)
Major. Studied Aden fauna; also collected plants there in 1884, now at BM(NH).
J. Bot. 1884, 370. *Rec. Bot. Survey India* v.7, 1914, 17.
Albuca yerburyi Ridley.

YONGE, Charlotte Mary (1823–1901)
b. Otterbourne, Hants 11 Aug. 1823 *d.* Otterbourne 24 March 1901
Novelist. *Herb of the Field* 1853; ed. 2 1858. *Lessons from Vegetable Kingdom* 1857. *Instructive Picture Book; or, Lessons from Vegetable World* 1882.
J. Bot. 1901, 79–80, 192. C. R. Coleridge *Charlotte Mary Yonge* 1903. *D.N.B. Supplt. 2* v.3, 717–19.

YONGE, Rev. James (1748–1797)
d. 5 Dec. 1797
Of Puslinch near Plymouth, Devon. Rector, Newton Ferrers, Devon. Contrib. to plant lists in R. Polwhele's *Hist. of Devonshire* v.1, 1797, 81.
T. R. A. Briggs *Fl. Plymouth* 1880, xxix.
Herb. formerly at Puslinch House.

YONGE, Roger Harry (*fl.* 1630s)
Apothecary. Visited west of England with T. Johnson in 1634.

YOUELL, William (–1883)
d. 21 Nov. 1883
Nurseryman, Royal Nurseries, Great Yarmouth, Norfolk.
Gdnrs Chron. 1883 ii 701.

YOUNG, Miss (*fl.* 1830s)
Friend of Rev. R. T. Lowe (1802–1874). Contrib. plates to *Bot. Mag.* 1834, t.3293, 3303, 3305, 3360.

YOUNG, Alexander (1829–)
b. near Coldstream, Berwickshire 1829
Gardener at Annesley Park, Notts, 1855–98.
Gdnrs Chron. 1899 i 61 portr.

YOUNG, Alfred Prentice (1841–1920)
FLS 1884. FGS. Collected plants in Kashmir, 1883, now at BM(NH).

YOUNG, Alfred Robson (1829–1883)
b. York 14 Jan. 1829 *d.* Brooklyn, New York 12 April 1883
To U.S.A., *c.* 1843. Collected marine algae in Europe, America and Australia. Particularly knowledgeable on marine flora of New York Bay.
Bull. Torrey Bot. Club 1883, 57.

YOUNG, Rev. Andrew John (1885–1971)
b. Elgin, Morayshire 1885 *d.* 25 Nov. 1971
MA Edinburgh 1908. Vicar, Stonegate, Sussex, 1941–59. Canon, Chichester Cathedral, 1948. Poet and botanist. *A Prospect of Flowers* 1945. *A Retrospect of Flowers* 1950. *Poet and the Landscape* 1962.
L. Clark *Andrew Young: Prospect of a Poet* 1957 portr. *Times* 29 Nov. 1971; 8 Dec. 1973. *Who's Who* 1970, 3452.

YOUNG, Charles (*fl.* 1820s–1840s)
In partnership with James and Peter Young in Epsom Nursery, Epsom, Surrey.
Bot. Mag. 1830, t.3007. E. Bretschneider *Hist. European Bot. Discoveries in China* 1898, 288.

YOUNG, Donald Peter (1917–1972)
b. 20 May 1917 *d.* 18 March 1972
BSc Battersea Polytechnic 1937. PhD 1940. FRIC. Research chemist, Distillers Company. Contrib. data on Croydon to *Atlas of Br. Fl.* 1962. Studied *Epipactis* and *Oxalis*. Curator, Croydon Natural History Society, 1953–62. Edited new *Fl. Surrey* which was continued on his death by J. E. Lousley. Contrib. to *Fl. Europaea*, *Watsonia*.
Proc. Croydon Nat. Hist. Sci. Soc. 1971, 225–27 portr. *Watsonia* 1973, 293–95.

YOUNG, George (–1803)
d. Hammersmith, Middx 11 March 1803
MD. Physician to Royal Hospitals in West Indies. First Curator, St. Vincent Botanical Garden, 1765–74.
Cottage Gdnr v.9, 1853, 417. I. Urban *Symbolae Antillanae* v.3, 1902, 146.
Plants at BM(NH).

YOUNG, James (*fl.* 1820s–1836)
d. Sept. 1836
Nurseryman in partnership with Charles and Peter Young at Epsom, Surrey.
Gdnrs Mag. 1836, 612. E. Bretschneider *Hist. European Bot. Discoveries in China* 1898, 288.

YOUNG, James Forbes (1796–1860)
b. Lambeth, London April 1796 *d.* Lambeth 30 June 1860
MD Edinburgh 1817. FLS 1847. Helped D. Cooper in *Fl. Metropolitana* 1836.
Proc. Linn. Soc. 1860–61, xlv. H. Trimen and W. T. T. Dyer *Fl. Middlesex* 1869, 400.
Plants at BM(NH), Kew. Sale at Sotheby 10 April 1861.

YOUNG, James Freeland (*fl.* 1850s)
Of Hull, Yorks. Botanised in East Riding of Yorks.
Trans. Hull Sci. Field Nat. Club 1902, 14–15.
Herb. at BM(NH).

YOUNG, Rev. James Reynolds
(*c.* 1810–1884)
b. London *c.* 1810 *d.* Whitnash, Warwickshire 1884
MA Cantab 1840; Oxon 1844. Rector, Whitnash, 1846–76. Had herb. 'Catalogue of Warwickshire Plants' (with R. Baker) (*Proc. Warwickshire Nat. Hist. Soc.* 1874, 56–57).
J. E. Bagnall *Fl. Warwickshire* 1891, 506.

YOUNG, John (*c.* 1790–1862)
Nurseryman, North Town, Taunton, Somerset.

YOUNG, Maurice (*c.* 1844–1890)
d. 24 Feb. 1890
Nurseryman, Milford near Godalming, Surrey.
Gdnrs Chron. 1890 i 272.

YOUNG, Thomas (1773–1829)
b. Milverton, Somerset 13 June 1773 *d.* London 10 May 1829
MD Göttingen 1796; Cantab 1808. FRS 1794. FLS 1794. Physician, physicist and Egyptologist. Physician, St. George's Hospital, London, 1811–29. Described and drew *Opercularia paleata* in *Trans. Linn. Soc.* v.3, 1797, 30–32.
G. Peacock *Life of Thomas Young* 1855, *D.N.B.* v.63, 393–99. C. C. Gillispie *Dict. Sci. Biogr.* v.14, 1976, 562–72. *R.S.C.* v.6, 470.

YOUNG, William (*fl.* 1740s–1780s)
b. Virginia? *d.* Virginia?
Pupil of John Hill. "Botanist to their Majesties" 1764. Collected plants under guidance of A. Garden. Introduced American plants to Europe and to Kew in particular. In England, 1765–66, 1768. *Catalogue d'Abres, Arbustes et Plantes Herbacées d'Amerique* 1783 (reprint in 1916: *Botanica Neglecta*).
Bot. Mag. 1793, t.216; 1803, t.684; 1804, t.710, 748. J. E. Smith *Selection from Correspondence of Linnaeus* v.1, 1821, 512. W. Darlington *Memorials of J. Bartram and H. Marshall* 1849, 344, 510. *J. Arnold Arb.* 1930, 59–60. A. M. Coats *Quest for Plants* 1969, 279–81.
Plants and MS. *Natural History of Plants of N. and S. Carolina* at BM(NH).

YOUNG, William (*c.* 1816–1896)
d. 12 Nov. 1896
Raised florists' flowers. Assistant Secretary, Edinburgh Horticultural Society.
J. Hort. Cottage Gdnr v.33, 1896, 537.

YOUNG, William (1865–1947)
b. Kirkcaldy, Fifeshire 1865 *d.* Kirkcaldy 16 March 1947
On staff of Royal Botanic Garden, Edinburgh. President, Botanical Society of Edinburgh. Had good knowledge of alpine flora of Scottish highlands. 'List of Flowering Plants and Ferns recorded from Fife and Kinross' (*Trans. Proc. Bot. Soc. Edinburgh* v.32(1), 1936, 3–173).
Trans. Br. Bryol. Soc. 1948, 132.

YOUNG, William Henry (*c.* 1864–1938)
d. East Sheen, Surrey 29 May 1938
Kew gardener, 1886–90. Gardener to Sir

F. Wigan at Clare Lawn, East Sheen. Had nursery at Romford, Essex.
Orchid Rev. 1938, 223–24. *J. Kew Guild* 1939–40, 921–22 portr.

YOUNG, William Weston (1758–1839)
ALS 1806. Made drawings for L. W. Dillwyn's *Br. Confervae* 1802.
Proc. Linn. Soc. 1839, 36.

YOUNGE, William (–1838/9)
MD. Of Sheffield, Yorks. Fellow student and friend of J. E. Smith. Had herb. Possibly connected with Sheffield Botanic Garden.
Proc. Linn. Soc. 1839, 35. *Paxton's Mag. Bot.* v.7, 1840, 51–52. *Bot. Mag.* 1856, t.4954.
Sinningia youngeana Marnock.

YOUNGHUSBAND, Sir Francis Edward (1863–1942)
b. Murree, India 31 May 1863 *d.* Lytchett Minster, Dorset 31 July 1942
Soldier, diplomat, explorer in Asia, geographer and mystic. Indian Political Dept., 1890. Collected plants in Chitral while on relief expedition under General Gatacre. Plants went to J. F. Duthie at Dehra Dun.
Rec. Bot. Survey of India v.1, 1898, 140. *Nature* v.150, 1942, 260–61. *Who was Who, 1941–1950* 987–88.

YOUNGMAN, William (1880–1963)
b. 6 Aug. 1880 *d.* 3 June 1963
Professor of Biology, Agra College, 1910; Canning College, Lucknow, 1911. Economic Botanist, United Provinces, 1918; Central Provinces, 1922. Director of Agriculture, Ceylon. Edited *Trop. Agriculturist.*
Who was Who, 1961–1970 1241.

ZIER, John (–1796)
b. Poland *d.* London 1796
FLS 1788. Lived in London. Friend of Ehrhart. Wrote many of the descriptions in J. Dickson's *Fasciculus...Plantarum Cryptogamicarum Britanniae* 1785.
Trans. Linn. Soc. v.4, 1798, 216. *Bot. Mag.* 1811, t.1395; 1817, t.1922. *Mon. Mag.* v.30, 1810, 198. *J. Bot.* 1886, 101–4.
Plants and MSS. at BM(NH).
Zieria Smith.

ZOUCHE, Edward, 11th Baron Zouche of Harringworth (*c.* 1556–1625)
b. Molton, Lincs 1556 *d.* Hackney, London 1625
Studied botany and had botanic garden at Hackney under M. de Lobel's care. Sent seeds from Constantinople to J. Gerard. Lobel dedicated ed. 2 of his *Adversaria* 1605 to him.
M. de Lobel *Stirpium Illustrationes* 1655, 104, 112. R. Pulteney *Hist. Biogr. Sketches of Progress of Bot. in England* v.1, 1790, 98. W. Robinson *Hist. and Antiq. of...Hackney* 1842, 131–32. J. C. Loudon *Encyclop. Gdning* 1860, 275. *D.N.B.* v.63, 415–17. R. T. Gunther *Early Br. Botanists* 1922, 417. C. E. Raven *English Nat.* 1947, 235–36. *Gdnrs Chron.* 1964 ii 432 portr.

ADDENDUM

BAINES, Thomas (1820–1875)
R. F. Kennedy *Journal of Residence in Africa, 1842–1853, by Thomas Baines* 1961–64 2 vols. *Geogr. J.* 1975, 252–58. M. and J. Diemont *The Brenthurst Baines* 1975. J. R. R. Wallis *Thomas Baines* 1976 portr.

BANKS, Sir Joseph (1743–1820)
Rec. Austral. Acad. Sci. 1974, 7–24. *Notes Rec. R. Soc. London* 1974, 91–99; 1975, 205–30.

BENNET-CLARK, Thomas Archibald (1903–1975)
b. 13 Jan. 1903
BA 1923. PhD 1929. FRS 1950. Lecturer, Manchester University, 1930–36. Professor of Botany, Nottingham, 1936–44; King's College, London, 1944–62. Professor of Biology, Norwich, 1962–67. Plant physiologist. Researched on organic acid metabolism in plants, water relations in plants, and application of chromatography to the separation of plant growth hormones. Edited *J. Experimental Bot.*
Times 28 Nov. and 4 Dec. 1975.

BLACKSTONE, John (1712–1753)
S. T. Jermyn *Fl. Essex* 1974, 16. D. H. Kent *Hist. Fl. Middlesex* 1975, 16.

BOR, Norman Loftus (1893–1972)
J. Indian Bot. Soc. 1973, 342–44 portr. *J. Bombay Nat. Hist. Soc.* 1973, 532–33. *Kew Bull.* 1975, 1–10 portr.

BUCKNALL, Thomas Skip Dyot (–1804)
d. London 11 Jan. 1804
Landowner at Sittingbourne, Kent where he experimented in grafting fruit trees. MP, St. Albans. *Orchadist* 1797. 'On Fruit Trees' (A. Hunter *Georgical Essays* v.5, 1804, 531–48).
Gent. Mag. 1804 i 92.

CAVE, Norman Leslie (*c.* 1891–1974)
d. 20 Aug. 1974
Civil servant. Treasurer, British Iris Society, 1948–56. *The Iris* 1950; ed. 2 1959. *Irises for Everyone* 1960.
Iris Yb. 1974, 17–18.

CHRISTIE-MILLER, Charles Wakefield (*c.* 1878–1976)
Of Swyncombe, Oxfordshire. Chairman of family manufacturing business. Gardener; specialised in irises.
Times 27 Jan. 1976.

COMBER, Harold Frederick (1897–1969)
International Dendrological Soc. Yb. 1974, 13–18.

COOPER, James Eddowes (1864–1952)
D. H. Kent *Hist. Fl. Middlesex* 1975, 25.

DOUGLAS, Sholto Charles John Hay, 21st Earl of Morton (1907–1976)
b. 12 April 1907 *d.* 13 Feb. 1976
MA Oxon. FLS 1949. VMH 1967. Authority on horticultural literature and history. *Chelsea Physic Garden* (with D. Allen) 1965. Contrib. to *J. R. Hort. Soc.*
Times 20 Feb. 1976

DRAKE, Miss (*fl.* 1830s–1840s)
"Miss Drake, whose name appears as the artist in all of Lindley's plates almost, was present [at Lindley's home] and is, I judge, a member of his family, and perhaps a relative of Mrs. Lindley" (A. Gray *Letters* v.1, 1893, 131). Contrib. plates to J. Lindley's *Sertum Orchidaceum* 1838, J. Bateman's *Orchidaceae of Mexico and Guatemala* 1837 and over 1100 plates to *Bot. Register*.
Some orchid drawings at Kew.
Drakaea Lindl.

HEWLETT, James (1768–1836)
Of Bath, Somerset. To Isleworth, Middx *c.* 1826. Water-colour artist noted for his flower studies. Contrib. to W. Sole's *Menthae Brittanicae* 1798.
B. Henrey *Br. Bot. Hort. Literature before 1800* v.2, 1975, 152.

JESSON, Enid Mary (*afterwards* **Cotton**) (1889–1956)
b. Malvern, Victoria? Australia 1 May 1889 *d.* Farnham Common, Bucks 19 April 1956
Married A. D. Cotton 1915. Contrib. text to *Bot. Mag.* 1916, t. 8690 and to E. F. Vallentin's *Illustrations of Flowering Plants of Falkland Islands* 1921.

LEECHMAN, Alleyne (1869–*c.*1940)
b. Ceylon 16 April 1869 *d.* before 1945
MA Oxon 1914. FLS 1912. Science Lecturer, Dept. of Science and Agriculture, Georgetown, British Guiana, 1902–19. Director, Biological and Agricultural Institute, Amani, Tanganyika, 1920–23. Investigated mangrove vegetation of British Guiana; described *Rhizophora harrisonii*. 'Genus Rhizophora in British Guiana' (*Kew Bull.* 1918, 4–8).
Plants and letters at Kew.

LISTON, Henrietta, Lady (*née* **Marchant**) (–1828)
Married Sir Robert Liston, British Ambassador at Constantinople, 1812–21. Collected plants and seeds in Turkey which were given to Sir J. Banks, De Candolle and De Visiani. "Egregria coltivatrice della botanica" (Visiani).
Bot. Mag. 1822, t.2253. *Mem. Ist. Veneto Sci. Lett.* v.1, 1842, 6. *Scots Mag.* v.58, 143.
Sedum listoniae Vis.

MACINTYRE, Aeneas (fl. 1820s–1840s)
FLS 1825–43. Of Stockwell Park, Surrey, 1825; Notting Hill, 1829–31; Bouverie Street, London, 1832; West Ham, Essex, 1840. Schoolmaster. Founder Member of Botanical Society of London. Possibly author of *Compendium of English Fl.* 1829. 'Notice of plants growing spontaneously on and about Warley Common in Essex' (*Proc. Bot. Soc. London* 1839, 16–21).
Essex Nat. v.19, 1920/1, 267–69, 324. *J. Bot.* 1921, 176–78, 204–5.

MANN, Robert James (1817–1886)
b. Norwich, Norfolk 1817 *d.* Wandsworth, London 8 Aug. 1886.
MD St. Andrews 1854. Practised medicine in Norwich and Buxton, Derbyshire. In Natal, 1857–64. *Guide to Knowledge of... Vegetable...Kingdom* 1856 (botany, 42–112). 'Fl. of Central Norfolk" (*Mag. Nat. Hist.* 1840, 390–407).
DNB v.36, 43–44. *R.S.C.* v.4, 216; v.8, 319; v.10, 707; v.16, 1043.

NEWBERRY, Percy Edward (1869–1949)
b. Islington, London 23 April 1869 *d.* Godalming, Surrey 7 Aug. 1949
MA. Professor of Egyptology, Liverpool, 1906–19. Professor of Ancient Egyptian History, Cairo, 1929–33. Identified plant remains in Egyptian tombs (W. M. F. Petrie *Hawari* 1889; H. Carter *Tut-Ankh-Amon* v.2, 1927, 189–96). Contrib. articles to *Gdnrs Chron.* 1888–89 which formed basis of A. M. Amherst's *Hist. Gdning* 1895.
DNB 1941–1950 622–23. *Who was Who, 1941–1950* 841.

PEARSON, John (*fl.* 1780s–1820s)
Founded nursery Chilwell, Notts, 1782. Succeeded by his son John and grandson J. R. Pearson (1819–1876).

READE, William Winwood (1839–1875)
Visited Africa, 1862–63. Financed by Andrew Swanzy exploring voyage to W. Africa, 1868–70. Collected plants in Sierra Leone, 1868–69; Gold Coast, 1870. *African Sketchbook* 1873.
Plants at Kew.

SUBJECT INDEX

Fereday, *Rev*. J. (1813–1871)
Pollexfen, *Rev*. J. H. (1813–1899)
McCalla, W. (*c*. 1814–1849)
Cresswell, *Rev*. R. (1815–1882)
Hassall, A. H. (1817–1894)
Sanford, W. A. (1818–1902)
Ferguson, W. (1820–1887)
Gifford, I. (*c*. 1823–1891)
Clifton, G. (1823–1913)
Piquet, J. (1825–1912)
Cooke, M. C. (1825–1914)
Tellam, R. V. (1826–1908)
Watts, H. (1828–1889)
Wilson, J. B. (1828–1895)
Joshua, W. (1828–1898)
Gray, S. O. (1828–1902)
Young, A. R. (1829–1883)
Harvey, J. R. (*fl*. 1830s)
Traill, G. W. (1836–1897)
Becker, H. F. (1838–1917)
Hulme, J. R. (*fl*. 1840s)
Jeannerett, *Dr*. (*fl*. 1840s)
Mallard, *Mrs*. (*fl*. 1840s)
Sanders, G. (*fl*. 1840s–1860s)
Allom, E. A. (*fl*. 1840s–1870s)
Turner, M. (*fl*. 1840s–1880s)
Buffham, T. H. (1840–1896)
Amory, A. (1841–1921)
Holmes, E. M. (1843–1930)
Turner, W. B. (1845–1917)
Apjohn, *Mrs*. (*fl*. 1850s)
Sawers, W. (*fl*. 1850s)
Grieve, S. (1850–1932)
Griffiths, W. (1850–1936)
Vickers, A. (1852–1906)
Letts, E. A. (1852–1918)
Phillips, R. W. (1854–1926)
Brebner, G. (*c*. 1855–1904)
Crocker, E. (1858–1910)
Murray, G. R. M. (1858–1911)
Vallentin, R. (1859–1934)
Tregelles, G. F. (1859–1943)
Mauger, W. P. (*fl*. 1860s)
Goode, H. (*fl*. 1860s–1880s)
Batters, E. A. L. (1860–1907)
Harvey-Gibson, R. J. (1860–1929)
Hussey, J. L. (1862–1899)
Gepp, A. (1862–1955)
Jack, J. (1863–1955)
Turner, C. (1864–1926)
Rich, M. F. (1865–1939)
Boodle, L. A. (1865–1941)
West, G. S. (1876–1919)
Fritsch, F. E. (1879–1954)
Griffiths, B. M. (1886–1942)
Butcher, R. W. (1897–1971)
Garry, R. (–1938)
Lunam, G. (–1947)
Mathias, W. T. (1900–1954)
Meikle, C. I. (1900–1970)
Drew, K. M. (1901–1957)

ALOE
Cowell, J. (*fl*. 1700s–1730s)

AMARYLLIS
Herbert, *Rev*. W. (1778–1847)
Bearpark, E. (*fl*. 1820s)
Gowen, J. R. (–1862)
Worsley, A. (1861–1944)

AMERICA (*see also under individual countries*)
Ripley, L. A. M. (1888–1928)

AMERICA, CENTRAL
Millar, R. (*fl*. 1730s–1740s)
Downton, G. (*fl*. 1870s)
Alston, A. H. G. (1902–1958)

AMERICA, NORTH
Menzies, A. (1754–1842)
Belcher, *Sir* E. (1799–1877)
Parker, C. S. (–1869)
Doubleday, E. (1810–1849)
Melvill, J. C. (1845–1929)
Buttle, J. (*fl*. 1850s–1860s)
Cartwright, T. B. (1856–1896)

AMERICA, SOUTH
Ackerman, G.
Menzies, A. (1754–1842)
Miers, J. (1789–1879)
Jameson, W. (1796–1873)
Pearce, R. (–1868)
Chesterton, J. H. (–1883)
Lobb, W. (1809–1864)
Darwin, C. R. (1809–1882)
Beresford, *Sir* J. (*fl*. 1810s)
Spruce, R. (1817–1893)
Wallace, A. R. (1823–1913)
Barclay, G. (*fl*. 1830s–1840s)
Whymper, E. (1840–1911)
Davis, W. (1847–1930)
Sandeman, C. A. W. (1882–1951)
Sandwith, N. Y. (1901–1965)

AMSTERDAM ISLAND
Perry, W. W. (*c*. 1846–1894)

ANANAS
Decker, *Sir* M. (1679–1749)

ANDAMAN ISLANDS
Berkeley, E. S. (*c*. 1823–1898)
Kurz, W. S. (*c*. 1833–1878)
Parkinson, C. E. (1890–1945)

ANGOLA
Gladman
Kirckwood, J. (*fl*. 1690s)
Mason (*fl*. 1690s)
Welwitsch, F. M. J. (1806–1872)
Curror, A. B. (*fl*. 1830s–1840s)
Dawe, M. T. (1880–1943)

ANNONACEAE
Sinclair, J. (1913–1968)

ANTHYLLIS
Marsden-Jones, E. M. (1887–1960)

ANTIGUA *see* **WEST INDIES**

ANTARCTICA
McCormick, R. (1800–1890)
Lyall, D. (1817–1895)
Hooker, *Sir* J. D. (1817–1911)
Bruce, W. S. (1867–1921)
Brown, R. N. R. (1879–1957)

APPLES *see* **FRUIT-GROWING**

ARABIA
Wellsted, J. R. (1805–1842)
Pelly, *Sir* L. (1825–1892)
Blunt, A. I. (1837–1917)
Last, J. T. (1847/8–1933)
Bent, J. T. (1852–1897)
Lunt, W. (1871–1904)
Scott, H. (1885–1960)
Shakespear, W. H. (–1915)

ARBORICULTURE *see* **TREES**

ARCTICA
Phipps, C. J. (1744–1792)
Richardson, *Sir* J. (1787–1865)
Sabine, *Sir* E. (1788–1883)
Parry, *Sir* W. E. (1790–1855)
Scoresby, W. (1790–1857)
Back, *Sir* G. (1796–1878)
Belcher, *Sir* E. (1799–1877)
McCormick, R. (1800–1890)
King, R. (*c*. 1811–1876)
Fisher, A. (*fl*. 1820s)
Hoppner, H. P. (*fl*. 1820s)
Taylor, J. (1823–1913)
Feilden, H. W. (1838–1921)
Hart, H. C. (1847–1908)
Trevor-Battye, A. B. R. (1855–1922)
Fisher, H. (1860–1935)
Sewell, P. (1865–1928)

ARCTIUM
Evans, A. H. (1855–1943)

ARGENTINA
Hall, *Mr*. (*fl*. 1720s)
Mylam, *Mr*. (*fl*. 1720s)
Mandeville, H. J. (1773–1861)
Tweedie, J. (1775–1862)
Middleton, *Mr*. (*c*. 1780s)
Gillies, J. (1792–1834)
Baird, J. (*fl*. 1830s)
Macloskie, *Rev*. G. (1834–1920)
Cunningham, R. O. (1841–1918)
White, E. W. (1858–1884)
Kerr, J. G. (1869–1957)
Prichard, H. V. H. (1876–1922)
Comber, H. F. (1897–1969)

ARTISTS
White, J. (*fl*. 1540s–1590s)
Marshall, A. (*c*. 1639–1682)
Dunstall, J. (*fl*. 1640s–1670s)
Waller, R. (*c*. 1650–1715)
Catesby, M. (1682–1749)
Edwards, G. (1694–1773)
Price, R. (–1761)
Brass, W. (–1783)
King, W. (*fl*. 1700s)
Peachey, H. (*fl*. 1700s)
Blackwell, E. (*c*. 1700–1758)
Delany, M. (1700–1788)
Ehret, G. D. (1710–1770)
Miller, J. (1715–*c*. 1790)
Rooke, H. (*c*. 1722–1806)
Harris, M. (1730–*c*. 1788)
Hunter, J. (1737–1821)
Edwards, J. (*fl*. 1742–1790s)
Taylor, S. (1742–*c*. 1796)
Robins, T. (1743–1806)
Griffiths, M. (1743–1819)
Kennion, E. (1744–1809)
Parkinson, S. (*c*. 1745–1771)
Kilburn, W. (1745–1818)
Cleveley, J. (1747–1786)
Bolton, J. (*fl*. 1750s–1790s)
Webber, J. (*c*. 1750–1793)
Abbot, J. (1751–*c*. 1840)
Lee, A. (1753–1790)
Graves, W. (*c*. 1754–post 1827)
White, J. (*c*. 1756–1832)
Sowerby, J. (1757–1822)
Bauer, F. A. (1758–1840)
Strickland, C. (*c*. 1759–1833)
Burgis, T. (*fl*. 1760s)
Raper, G. (*c*. 1760s–1790s)
Crabtree, P. (*fl*. 1760s–1820s)
Pope, C. M. (*fl*. 1760s–1838)
Bauer, F. L. (1760–1826)
Watling, T. (1762–1806)
Strickland, J. S. (*c*. 1765–1849)
Edwards, S. T. (1768–1819)
Tobin, G. (1768–1838)
Brown, P. (*fl*. 1770s–1790s)
Miller, J. F. (*fl*. 1770s–1790s)
Nodder, F. P. (*fl*. 1770s–1800s)
Meen, M. (*fl*. 1770s–1820s)
Lewin, J. W. (*c*. 1770–1819)
Turner, M. (1774–1850)
Herbert, *Rev*. W. (1778–1847)
Hooker, W. (1779–1832)
Power, A. (*fl*. 1780s–1800s)
Sansom, F. (*fl*. 1780s–1810s)
Cooke, G. (1781–1834)
Graves, G. (1784–*c*. 1839)
Hooker, *Sir* W. J. (1785–1865)
Baker, A. E. (1786–1861)
Sowerby, J. de C. (1787–1871)
Harrison, M. (1788–1875)
MacKenzie, D. (*fl*. 1790s)
Barnard, A. (*fl*. 1790s–1800s)
Wharton, E. (*fl*. 1790s–1810s)
Andrews, H. C. (*fl*. 1790s–1830s)
Lawrance, M. (*fl*. 1790s–1830s)
Curtis, J. (1791–1862)
Withers, A. I. (*c*. 1793–1860s)
Acton, F. (*c*. 1793–1881)
Curtis, C. M. (*c*. 1795–1839)
Allport, J. (1796–*c*. 1846/7)
Guilding, *Rev*. L. (1797–1831)
Smith, F. W. (1797–1835)
Jewitt, T. O. S. (1799–1869)
Bartholomew, V. (1799–1879)
Mathews, E. (–*c*. 1811)
Telfair, A. (–1832)
Barkly, E. H. (–1857)
Swan, J. (–1872)
Miles, *Mrs*. (–1884)
De Alwis, H. (–1894)
Barnard, A. (–1899)
Gartside, *Miss* (*fl*. 1800s)
Pope, L. L. (*fl*. 1800s)
Stennett, R. (*fl*. 1800s)
Hutton, J. (*fl*. 1800s–1820s)
Cust, M. A. (1800–1882)
Andrews, J. (*c*. 1801–1876)
Stuart, J. (1802–1842)
Gould, W. B. (*c*. 1804–1853)
Twining, E. (1805–1889)
Bond, G. (*c*. 1806–1892)
Rooke, J. (1807–1872)

Becker, L. (1808–1861)
Sinclair, J. (1809–1881)
Miller, R. (*fl.* 1810s)
Brookshaw, G. (*fl.* 1810s–1820s)
Cotton, B. (*fl.* 1810s–1820s)
Strutt, J. G. (*fl.* 1810s–1850s)
McNab, J. (1810–1878)
Humphreys, H. N. (1810–1879)
Cooke, E. W. (1811–1880)
Meredith, L. A. (1812–1895)
Osborn, M. (1814–1898)
Bateman, E. L. T. (*c.* 1815–1897)
Fitch, W. H. (1817–1892)
Hooker, *Sir* J. D. (1817–1911)
Roupell, A. E. (1817–1914)
White, C. F. (1818–1896)
Abbott, C. (*fl.* 1820s)
Duncanson, T. (*fl.* 1820s)
Burgess, H. W. (*fl.* 1820s–1830s)
Clark, W. (*fl.* 1820s–1830s)
Hart, J. (*fl.* 1820s–1830s)
Hart, M. (*fl.* 1820s–1830s)
Roscoe, M. (*fl.* 1820s–1830s)
Penfold, J. W. (*fl.* 1820s–1840s)
Russell, I. (*fl.* 1820s–1840s)
Smith, E. D. (*fl.* 1820s–1840s)
Spratt, G. (*fl.* 1820s–1840s)
Bury, E. (*fl.* 1820s–1860s)
Durham, C. B. (*fl.* 1820s–1860s)
Plunkett, K. (1820–)
Sowerby, C. C. (1820–1865)
Salter, J. W. (1820–1869)
Ross, H. J. (1820–1902)
Rosenberg, M. E. (1820–1914)
Osborn, E. (1822–1877)
Angas, G. F. (1822–1886)
West, T. (1823–1891)
Sowerby, J. E. (1825–1870)
Sowerby, H. (1825–1891)
Barnard, A. M. (1825–1911)
Cooke, M. C. (1825–1914)
Linnell, M. (*c.* 1828–*c.* 1881)
Shackleton, L. (1828–1914)
Charsley, F. A. (1828–1915)
Coleman, W. S. (1829–1904)
Adams, *Miss* (*fl.* 1830s)
Alston, E. M. and C. M. (*fl.* 1830s)
Harrison, C. W. (*fl.* 1830s)
Perkins, E. E. (*fl.* 1830s)
Ronalds, E. (*fl.* 1830s)
Drake, *Miss* (*fl.* 1830s–1840s)
Holden, S. (*fl.* 1830s–1850s)
Fielding, M. M. (*fl.* 1830s–1880s)
North, M. (1830–1890)
Stopps, A. J. (1833–1931)
Calvert, C. L. W. (1834–1872)
Dresser, C. (1834–1904)
Walton, J. (1834–1914)
Smith, W. G. (1835–1917)
Macfarlane, J. L. (1836–*c.* 1913)
Lecky, S. (1837–1896)
Durham, E. (*fl.* 1840s)
Wakeling, J. (*fl.* 1840s)
Drought, I. (*fl.* 1840s–1850s)
Fitch, J. N. (1840–1927)
Allen, E. (*c.* 1842–)
De Alwis Seneviratne, W. (1843–1916)
Bell, A. J. M. (1845–1920)
Mason, M. H. (1845–1932)
Burbidge, F. W. T. (1847–1905)
Parsons, A. W. (1847–1920)
Toynbee, H. (*fl.* 1850s)
Jameson, H. G. (1852–1939)
Carlyon, H. (*c.* 1853–1924)
Smith, M. (1854–1926)
Woolward, F. H. (1854–1936)
Thiselton-Dyer, H. A. (1854–1945)
Brebner, G. (*c.* 1855–1904)
Drinkwater, H. (1855–1925)
Trower, C. G. (1855–1928)
Caparne, W. J. (1855–1940)
Highley, P. (1856–1929)
Moon, H. G. (1857–1905)
Crocker, E. (1858–1910)
Clarke, H. (1858–1920)
Lugard, C. E. (1859–1939)
Brown, M. R. (*fl.* 1860s)
Lister, G. (1860–1949)
Flockton, M. L. (1862–1953)
Morgan, R. (1863–1900)
Moxon, M. L. (1863–1920)
Flemwell, G. J. (1865–1928)
Wood, E. M. (1865–1907)
Bowles, E. A. (1865–1954)
Potter, B. (1866–1943)
Ragg, L. (1866–1945)
Bedford, E. J. (1866–1953)
Douie, F. M. E. (1866–1965)
Page, M. M. (1867–1925)
Frere, C. F. (*fl.* 1870s)
Bull, E. E. (*fl.* 1870s–1880s)
Duppa, A. F. M. (*fl.* 1870s–1880s)
Ellis, A. B. (*fl.* 1870s–1880s)
Alexander, E. (1870–)
Cole, R. V. (1870–1940)
Godfrey, H. M. (1871–1930)
Eaton, M. E. (1873–1961)
Stanley, B. (1877–1944)
Martin, *Rev.* W. K. (1877–1969)
Round, F. H. (*c.* 1878–1958)
Snelling, L. (1879–1972)
King, *Mrs.* (*fl.* 1880s–1910s)
Martin, D. (1882–1949)
Bussey, W. (1884–1969)
Hutchinson, J. (1884–1972)
Adam, R. M. (1885–1967)
Atkinson, G. (1893–1971)
Stebbing, M. A. (–1927)
Dykes, E. K. (–1933)
Boys-Smith, W. L. (–1939)
Knox, M. (–1952)
Roberts, N. (–1959)
Trevithick, W. E. (1900–1958)

ASCENSION ISLAND
Stonestreet, G. (*fl.* 1690s)
Cunningham, J. (–*c.* 1709)
Hinds, R. B. (*c.* 1812–*c.* 1847)
Wren, W. (*fl.* 1840s)
Buchanan, J. (1855–1896)

ASCLEPIADACEAE
Brown, N. E. (1849–1934)

ASTER
Ballard, E. (*c.* 1871–1952)

AURICULA
Thompson, J. (*fl.* 1750s)
Emmerton, I. (*c.* 1769–1823)
Biggs, T. (*fl.* 1780s)
Meiklejohn, A. (*c.* 1798–1885)

Lightbody, G. (–1872)
Read, J. (1809–1880)
Walker, J. (*c.* 1815–1877)
Wilson, H. (*c.* 1822–1892)
Pohlmann, E. (1825–1886)
Woodhead, T. (*c.* 1832–1882)

AUSTRALIA
Dampier, W. (1651–1715)
Nelson, D. (–1789)
Burton, D. (–1792)
Hunter, J. (1737–1821)
Parkinson, S. (*c.* 1745–1771)
Anderson, W. (1750–1778)
Menzies, A. (1754–1842)
Paterson, W. (1755–1810)
Littlejohn, R. (1756–1818) *Tasmania*
White, J. (*c.* 1756–1832)
Dawes, W. (*c.* 1758–1836)
Raper, G. (*c.* 1760s–1790s)
Bauer, F. (1760–1826)
Watling, T. (1762–1806)
Macleay, A. (1767–1848)
Macarthur, E. (1769–1850)
Lewin, J. W. (*c.* 1770–1819)
Caley, G. (1770–1829)
Brown, R. (1773–1858)
Suttor, G. (1774–1859)
Harris, G. P. R. (1775–1810) *Tasmania*
Considen, D. (*fl.* 1780s–1810s)
Shepherd, T. (1780–1835)
Meares, R. G. (1780–1862)
Drummond, J. (*c.* 1784–1863)
Bicheno, J. E. (1785–1851) *Tasmania*
Field, B. (1786–1846)
Mangles, J. (1786–1867)
Rennie, *Rev.* J. (1787–1867)
Fraser, C. (*c.* 1788–1831)
Swainson, W. (1789–1855)
Cunningham, A. (1791–1839)
King, P. P. (1791–1856)
Stirling, *Sir* J. (1791–1865)
Wilson, T. B. (1792–1843)
Mitchell, *Sir* T. L. (1792–1855)
Macleay, W. S. (1792–1865)
Collie, A. (1793–1835)
Cunningham, R. (1793–1835)
Walker, *Rev.* J. (1794–1854)
Backhouse, J. (1794–1869)
Sturt, C. (1795–1869)
Newman, F. W. (*c.* 1796–1859) *Tasmania*
Sinclair, A. (*c.* 1796–1861)
Strzelecki, *Count* P. E. de (1796–1873)
Anderson, J. (1797–1842)
Roe, J. S. (1797–1878)
Richardson, J. M. (*c.* 1797–1882)
Home, *Sir* J. E. (1798–1853)
Wickham, J. C. (1798–1864)
Clarke, *Rev.* W. B. (1798–1878)
Moore, G. F. (1798–1886)
Howitt, R. (1799–1869)
Good, P. (–1803)
Holmes, W. (–1830)
Robertson, W. N. (–1844)
Anderson, J. (–1847)
Strange, F. (–1854)
Stephens, W. (–1866)
Barker, *Mrs.* (–1876)
Stackhouse, T. (–1886)
Leycester, A. A. (–1892)
Prentice, C. (–1894)
Grey, E. L. (–1898)
Fleming, J. (*fl.* 1800s)
Gordon, J. (*fl.* 1800s)
Lhotsky, J. (*fl.* 1800s–1860s)
Bailey, J. (1800–1864)
Francis, G. W. (1800–1865)
Allport, J. (1800–1877) *Tasmania*
Macarthur, *Sir* W. (1800–1882)
Bentham, G. (1800–1884)
Story, G. F. (1800–1885) *Tasmania*
Kidd, J. (1801–1867)
Busby, J. (1801–1871)
Latrobe, C. J. (1801–1875)
Stuart, C. (1802–1877) *Tasmania*
Robertson, J. G. (1803–1862)
Bynoe, B. (*c.* 1803–1865)
Phillips, W. (1803–1871)
Arthur, J. (1804–1849)
Gould, W. B. (*c.* 1804–1853)
Denison, *Sir* W. T. (1804–1871)
Maxwell, G. (1804–1880)
Bennett, G. (1804–1893)
Molloy, G. (1805–1843)
Lawrence, R. W. (1807–1833) *Tasmania*
Milligan, J. (1807–*c.* 1883) *Tasmania*
Becker, L. (1808–1861)
McWilliam, J. O. (1808–1862)
Gunn, R. C. (1808–1881) *Tasmania*
Sinclair, J. (1809–1881)
Guilfoyle, M. (1809–1884)
Macleay, *Sir* G. (1809–1891)
Coxe, *Mr.* (*fl.* 1810s)
Stephenson, W. (*fl.* 1810s–*c.* 1863)
Gilbert, J. (*c.* 1810–1845)
Turner, *Rev.* G. E. (1810–1869)
Curdie, D. (1810–1884)
Harvey, W. H. (1811–1866) *and Tasmania*
Vernon, W. (1811–1890)
Schomburgk, *Sir* M. R. (1811–1891)
Fawcett, H. C. (1812–1890)
Meredith, L. A. (1812–1895) *Tasmania*
Grey, *Sir* G. (1812–1898)
Leichhardt, F. W. L. (1813–1848)
Sharland, F. S. (1813–1859) *Tasmania*
Fereday, *Rev.* J. (1813–1871) *Tasmania*
Bunce, D. (1813–1872)
Ewing, *Rev.* T. J. (1813–1882) *Tasmania*
Warburton, P. E. (1813–1889)
Ralph, T. S. (1813–1891)
Daintrey, E. (1814–1887)
Woolls, *Rev.* A. (1814–1893)
Bidwill, J. C. (1815–1853)
Stuart, J. M. (1815–1866)
Babbage, B. H. (1815–1878)
Bateman, E. L. T. (*c.* 1815–1897)
Waterhouse, F. G. (1815–1898)
Eyre, E. J. (1815–1901)
Dutton, F. S. (1816–1877)
Lefroy, *Sir* J. H. (1817–1890) *Tasmania*
Lyall, D. (1817–1895)
McCoy, *Sir* F. (1817–1899)
Hooker, *Sir* J. D. (1817–1911) *Tasmania*
Kennedy, E. B. C. (1818–1848)
Brewer, J. A. (1818–1886)
Knight, C. (*c.* 1818–1895)
Emmett, S. B. (1818–1898) *Tasmania*
Armstrong, *Sir* A. (1818–1899)
Sanford, W. A. (1818–1902)
Behr, H. H. (1818–1904)

Gregory, *Sir* A. C. (1819–1905)
Graham, T. (*fl.* 1820s)
Morrison, W. (*fl.* 1820s)
Baxter, W. (*fl.* 1820s–1830s)
Davidson, W. (*fl.* 1820s–1830s) *Tasmania*
McLean, J. (*fl.* 1820s–1830s)
Scott, T. (*fl.* 1820s–1830s) *Tasmania*
Dallachy, J. (*c.* 1820–1871)
Archer, W. (1820–1874) *Tasmania*
Spicer, *Rev.* W. W. (*c.* 1820–1879) *Tasmania*
Oldfield, A. F. (1820–1887)
Hill, W. (1820–1904)
Gregory, F. T. (1821–1888)
Findlay, J. (1821–1905)
Angas, G. F. (1822–1886)
Herrgott, J. F. A. D. (1823–1861)
Carron, W. (1823–1876)
Wilcox, J. F. (1823–1881)
Haviland, E. (1823–1908)
Clifton, G. (1823–1913)
Shepherd, T. W. (1824–1884)
Morton, W. L. (1824–1898)
Krichauff, F. E. H. W. (1824–1904)
Norton, J. (1824–1906)
Calvert, J. S. (1825–1884)
Landsborough, W. (1825–1886)
Mueller, *Sir* F. J. H. von (1825–1896)
Bowman, E. M. (1826–1872)
Thozet, A. (*c.* 1826–1878)
Austin, R. (1826/7–1905)
Bosisto, J. (1827–1898)
Bailey, F. M. (1827–1915)
Hannaford, S. (1828–1874) *Tasmania*
Watts, H. (1828–1889)
Allitt, W. (1828–1893)
Wilson, J. B. (1828–1895)
Rudder, A. (1828–1904)
Charsley, F. A. (1828–1915)
Slatter, J. W. (1829–1896)
Whan, *Rev.* W. T. (1829–1901)
Davies, *Rev.* R. H. (*fl.* 1830s) *Tasmania*
Mangles, G. (*fl.* 1830s)
Adamson, F. M. (*fl.* 1830s–1850s)
Smith, C. (*fl.* 1830s–1850s) *Tasmania*
Vicary, N. (*fl.* 1830s–1850s)
Kingsley, H. (1830–1876)
Allport, M. (1830–1878) *Tasmania*
Fitzgerald, R. D. (1830–1892)
Brunning, G. (1830–1893)
Howitt, A. W. (1830–1908)
Fitzalan, E. F. A. (1830–1911)
Daintree, R. (1831–1878)
Walter, C. (*c.* 1831–1907)
Bernays, L. A. (1831–1908)
Panton, J. A. (1831–1913)
Tenison-Woods, *Rev.* J. E. (1832–1889) *Tasmania*
Wilson, *Rev.* F. R. M. (1832–1903)
Smeaton, T. D. (*c.* 1832–1908)
Muir, T. (1833–1926)
Stopps, A. J. (1833–1931)
Elsey, J. R. (1834–1857)
Calvert, C. L. W. (1834–1872)
Taylor, N. (1834–1894)
O'Shanesy, J. (1834–1899)
Abbott, F. (1834–1903)
Hardy, J. (1834–1916)
De Mole, F. E. (1835–1866)
Bancroft, J. (1836–1894)
Sullivan, D. (1836–1895)
Tisdall, H. T. (*c.* 1836–1905)
Simson, A. (1836–1918) *Tasmania*
O'Shanesy, P. A. (1837–1884)
McLachlan, R. (1837–1904)
Hoffmann, G. C. (1837–1917)
Gregson, J. (1837–1919)
Veitch, J. G. (1839–1870)
Scott, J. R. (1839–1877) *Tasmania*
Collie, *Rev.* R. (1839–1892)
Bastow, R. A. (1839–1920)
Biddulph, H. S. (1839–1940)
Jeannerett, *Dr.* (*fl.* 1840s)
Mallard, *Mrs.* (*fl.* 1840s)
Whittaker, *Mr.* (*fl.* 1840s)
Waterman, W. (*fl.* 1840s–1850s)
Bennett, K. H. (*c.* 1840–1891)
Tate, R. (1840–1901)
Crawford, A. R. (1840–1912)
Guilfoyle, W. R. (1840–1912)
Holtze, M. W. (1840–1923)
French, C. (1840–1933)
Mulder, J. F. (1841–1921)
Tepper, J. G. O. (1841–1923)
Gosse, W. C. (1842–1881)
Helms, R. (1842–1914)
Harwood, G. (1842–1915)
Tate, T. (*c.* 1842–1934)
Luehmann, J. G. (1843–1904)
Brown, M. (1843–1905)
Kempe, *Rev.* H. (1844–*c.* 1907)
Johnston, R. M. (1844–1918) *Tasmania*
Tietkins, W. H. (1844–1933)
Dunn, E. J. (1844–1937)
Scortechini, *Rev.* B. (1845–1886)
Farrer, W. J. (1845–1906)
Giles, W. E. P. (1847–1897)
Coppinger, R. W. (1847–1910)
Forrest, *Sir* J. (1847–1918)
Deane, H. (1847–1924)
D'Alton, St. E. (1847–1930)
Brown, J. E. (1848–1899)
Grant, A. (1848–1906)
Lauterer, J. (1848–1911)
Keartland, G. A. (1848–1926)
McAlpine, D. (1848–1932)
Cowley, E. (1848/9–1899)
Perrin, G. S. (1849–1900)
Shirley, J. F. (1849–1922)
Brogden, J. (*fl.* 1850s?)
Clowes, G. (*fl.* 1850s)
Howitt, W. (*fl.* 1850s)
Layard, *Mr.* (*fl.* 1850s)
Mossman, S. (*fl.* 1850s)
Mylne (*fl.* 1850s)
Ferguson, W. (*fl.* 1850s–1880s)
Clunies Ross, W. J. (1850–1914)
Whitelegge, T. (1850–1927)
Veitch, P. C. M. (1850–1929)
Betche, E. (1851–1913)
Gill, W. (1851–1929)
Stirling, J. (1852–1909)
Camfield, J. H. (1852–1916)
Wragge, C. L. (1852–1922)
Smith, H. G. (1852–1924)
Turner, F. (1852–1939)
Hamilton, A. G. (1852–1941)
Lucas, A. H. S. (1853–1936)
Rodway, L. (1853–1936) *Tasmania*
Baker, R. T. (1854–1941)
Braine, A. B. (1854–1945)
Moore, A. (*c.* 1855–1884)

Hamilton, A. A. (1855–1929)
Wilson, *Rev.* J. (1855–1937)
Black, J. M. (1855–1951)
Winnecke, C. G. A. (1856–1902)
Watts, *Rev.* W. W. (1856–1920)
Musson, C. T. (1856–1928)
Tryon, H. (1856–1943)
Macmahon, P. (1857–1911)
Clements, F. M. (1857–1920)
Lea, *Rev.* T. S. (1857–1939)
Pockett, T. W. (1857–1952)
Purdie, A. (1859–1905)
Maiden, J. H. (1859–1925)
Beckler, H. (*fl.* 1860s)
Flood, J. (*fl.* 1860s)
Gulliver, B. and T. A. (*fl.* 1860s)
Murray, J. P. (*fl.* 1860s)
Sandford, *Major* (*fl.* 1860s)
Walcott, P. (*fl.* 1860s)
Spencer, *Sir* W. B. (1860–1929)
Bancroft, T. L. (1860–1933)
Hynes, S. (*c.* 1860–1938)
Ashby, E. (1861–1941)
Hussey, J. L. (1862–1899)
Flockton, M. L. (1862–1953)
Vidler, E. A. (1863–1942)
Goadby, B. T. (*c.* 1863–1944)
Forsyth, W. (1864–1910)
Boorman, J. L. (1864–1938)
Chapman, F. (1864–1943)
Bloomer, H. H. (1866–1960)
Veitch, J. H. (1868–1907)
Holtze, N. (1868–1913)
Kitson, *Sir* A. E. (1868–1937)
Darnell-Smith, G. P. (1868–1942)
Lawson, J. (*fl.* 1870s)
MacPherson, A. (*fl.* 1870s–1890s)
Wild, C. J. (*fl.* 1870s–1890s)
Brooks, S. (1870–)
Gibbs, L. S. (1870–1925)
Andrews, C. R. P. (1870–1951)
Carnegie, D. W. (1871–1900)
Gilruth, J. A. (1871–1937)
Ewart, A. J. (1872–1937)
Cheel, E. (1872–1951)
Julius, *Sir* G. A. (1873–1946)
Lawson, A. A. (1874–1927)
Blackall, W. E. (1876–1941)
Dorrien-Smith, A. A. (1876–1955)
Wilson, H. W. (1877–1955)
Davidson, A. (*fl.* 1880s)
McKibben, J. N. (*fl.* 1880s)
Evans, *Sir* G. (1883–1963)
Lane-Poole, C. E. (1885–1970)
Hyam, G. N. (1886–1958)
Dowson, W. J. (1887–1963) *Tasmania*
Osborn, T. G. B. (1887–1973)
McLuckie, J. (1890–1956)
Gardner, C. A. (1896–1970)
Comber, H. F. (1897–1969) *Tasmania*
Carlton, H. (–1917)
Coleman, E. (–1951)
Allen, C. E. F. (*fl.* 1900s–1920s)

AZORES
Landon, S. (*fl.* 1670s–1700s)
Masson, F. (1741–1805)
Hunt, T. C. (–1886)
Watson, H. C. (1804–1881)
Vidal, *Capt.* (*fl.* 1840s?)

BAHAMAS *see* **WEST INDIES**

BALKANS (*see also under individual countries*)
Turrill, W. B. (1890–1961)

BANANA *see* **MUSA**

BARBADOS *see* **WEST INDIES**

BASUTOLAND *see* **LESOTHO**

BECHUANALAND *see* **BOTSWANA**

BEGONIA
Clarke, R. T. (1813–1897)

BELGIAN CONGO *see* **CONGO**

BERMUDA
Dickinson, J. (*fl.* 1690s)
Clerk, *Rev.* W. (*fl.* 1710s–1730s)
Lefroy, *Sir* J. H. (1817–1890)
Hunter, *Rev.* R. (1823–1897)
Jones, J. M. (1828–1888)
Lane, A. W. (*fl.* 1840s)
Reade, O. A. (1848–1929)
Bishop, G. A. (*fl.* 1890s)

BETULA
Marshall, *Rev.* E. S. (1858–1919)

BHUTAN *see* **INDIAN SUB-CONTINENT**

BIGNONIACEAE
Sandwith, N. Y. (1901–1965)

BOLIVIA
Cuming, H. (1791–1865)
Pentland, J. B. (1797–1873)
Bridges, T. (1807–1865)
Castelnau, F. L. de L. de (1810–1880)

BORNEO *see* **MALAYSIA**

BOTANICAL ARTISTS *see* **ARTISTS**

BOTSWANA
Lugard, *Sir* F. D. (1858–1945)
Lugard, C. E. (1859–1939)
Lugard, E. J. (1865–1944)

BRAZIL (*see also* **FERNANDO DE NORONHA**)
Dampier, W. (1651–1715)
Alfrey, G. (*fl.* 1690s)
Burchell, W. J. (1781–1863)
Callcott, M. (1785–1842)
Swainson, W. (1789–1855)
Bowie, J. (*c.* 1789–1869)
Cunningham, A. (1791–1839)
Fox, H. S. (1791–1846)
Koster, H. (1793–1820)
Forbes, J. (1798–1823)
Don, G. (1798–1856)
Graham, T. (–1822)
Macrae, J. (–1830)
Weir, J. (–1898)
Anderson, G. (*fl.* 1800s–1817)
Bunbury, *Sir* C. J. F. (1809–1886)
Gardner, G. (1812–1849)

Weddell, H. A. (1819–1877)
Beresford, M. (*fl.* 1820s)
Chamberlayne, C. (*fl.* 1820s)
Harrison, H. (*fl.* 1820s)
Harrison, W. (*fl.* 1820s)
Hopkins, G. (*fl.* 1820s)
Warre, F. (*fl.* 1820s)
Newman, J. (*fl.* 1820s–1840s)
Paterson, J. L. (1820–1882)
Day, J. (1824–1888)
Milne-Redhead, R. (1828–1900)
March, G. (*fl.* 1830s)
Herbst, H. C. C. (*c.* 1830–1904)
Cross, R. M. (1836–1911)
Bowman, D. (1838–1868)
Gogarty, *Dr.* (*fl.* 1840s)
Cowan, J. (1842–1929)
Boxall, W. (1844–1910)
Wickham, *Sir* H. A. (1846–1928)
Wetherell, J. (*fl.* 1850s)
Trail, J. W. H. (1851–1919)
Moore, S. Le M. (1851–1931)
Dent, H. C. (1855–1909)
Lea, *Rev.* T. S. (1857–1939)
Blunt, H. (*fl.* 1860s)
Williams, C. H. (*fl.* 1860s)
Casement, R. D. (1864–1916)
Willis, J. C. (1868–1958)
Gwynne-Vaughan, D. T. (1871–1915)
Dawe, M. T. (1880–1943)
Collenette, C. L. (1888–1959)
Beadle, C. (–1917)

BRITISH GUIANA *see* GUYANA

BRITISH HONDURAS
Robertson, *Rev.* J. (*fl.* 1880s)
McNair, J. (*fl.* 1880s–1890s)
Campbell, E. J. F. (*fl.* 1880s–1920s)

BRITISH ISLES
ENGLAND

Bedfordshire
Sibley, J.
Vaux, T.
Marsh, *Rev.* T. O. (1749–1831)
Abbott, *Rev.* C. (*c.* 1761–1817)
Cooper, *Rev.* O. St. J. (*fl.* 1780s–1806)
Corder, T. (1812–1873)
McLaren, J. (1815–1888)
Crouch, *Rev.* W. (1818–1846)
Brown, H. (1824–1892)
Pollard, J. (1825–1909)
Alston, E. M. and C. M. (*fl.* 1830s)
Ransom, A. (1832–1912)
Saunders, J. (1839–1925)
Stimson, A. (1842–1915)
Hillhouse, W. (1850–1910)
Crouch, C. (1855–1944)
Higgins, D. M. (*c.* 1856–1920)
Hamson, J. (1858–1930)
Squire, P. (*fl.* 1860s)
Little, J. E. (1861–1935)
Bishop, E. B. (1864–1947)
Day, G. H. (*c.* 1884–1967)
Laflin, T. (1914–1972)

Berkshire
Watlington, J. (–1659)
Dillenius, J. J. (1684–1747)
Rudge, S. (1727–1817)
Sheffield, *Rev.* W. (*c.* 1732–1795)
Benwell, J. (*c.* 1735–1819)
Beeke, *Rev.* H. (1751–1837)
Mavor, *Rev.* W. F. (1758–1837)
Bicheno, J. E. (1784–1851)
Lamb, T. (*fl.* 1790s)
Lousley, J. (1790–1855)
Bunny, J. (1798–1885)
Gotobed, R. (–*c.* 1806)
Flower, T. B. (1817–1899)
Garnsey, *Rev.* H. E. F. (1826–1903)
Walker, F. (1829–1889)
Boswell, H. (1837–1897)
Penny, *Rev.* C. W. (1837–1898)
Hewett, *Mr.* (*fl.* 1840s)
Shadwell, C. L. (1840–1919)
Druce, G. C. (1850–1932)
Tufnail, F. (1861–1899)
Jenkinson, J. W. (1871–1915)

Buckinghamshire
Coles, W. (1626–1662)
Bentinck, M., *Duchess of Portland* (1715–1785)
Markham, M. (–1814)
Robinson, *Mrs.* (–1847)
Mill, G. G. (–1853)
Chandler, E. (1818–1884)
Britten, J. (1846–1924)
Summers, *Rev.* W. H. (1850–1906)
Druce, G. C. (1850–1932)
Burns, J. S. (*fl.* 1860s)
Ullyett, H. (*fl.* 1860s–1880s)
Martyn, *Rev.* T. W. (–1918)

Cambridgeshire
Corbyn, S. (*fl.* 1640s–1650s)
Martyn, *Rev.* T. (1735–1825)
Lyons, I. (1739–1775)
Relhan, *Rev.* R. (1754–1823)
Babington, C. C. (1808–1895)
Larbalestier, C. du B. (1838–1911)
Evans, A. H. (1855–1943)
Shrubbs, A. S. (1858–1922)
Cooke, *Rev.* P. H. (1859–1950)
West, W. (1875–1901)
Rhodes, *Rev.* P. G. M. (1885–1934)

Cheshire
Waring, *Rev.* R. H. (*c.* 1720–1794)
Crosfield, G. (1754–1820)
Bradbury, J. (1768–1823)
Buxton, R. (1786–1865)
Holland, *Sir* H. (1788–1873)
Wilson, E. S. (–1846)
Williamson, I. (–*c.* 1860)
Okell, *Mr.* (*fl.* 1800s)
Grundy, M. A. (1809–1871)
Hall, T. B. (1814–1886)
Higgins, *Rev.* H. H. (1814–1893)
Byerley, I. (*c.* 1814–1897)
Grundy, E. (1815–1894)
Harrison, W. (1821–)
Collingwood, C. (1826–1908)
Slack, R. (*fl.* 1830s)
Warren, J. B. L. (1835–1895)
Robinson, J. F. (1838–1884)
Hunt, G. E. (*c.* 1841–1873)
Webb, F. M. (1841–1880)
Harrison, J. (*fl.* 1850s)
Shillito, J. (*fl.* 1850s–1899)
Baillie, E. J. (1851–1897)

Eaves, D. (1857–1936)
Boult, W. (*fl.* 1860s–1890s)
Grundy, C. (*fl.* 1870s)
Payne, B. F. (*fl.* 1900s–1910s)

Cornwall
Carew, R. (1555–1620)
Stephens, *Rev.* L. (1654–1724/5)
Borlase, *Rev.* W. (1696–1772)
Cullum, *Sir* T. G. (1741–1831)
Thompson, *Rev.* J. T. (1752–1811)
Polwhele, *Rev.* R. (1760–1838)
Lysons, *Rev.* D. (1762–1834)
Gilbert, D. (1767–1839)
Paris, J. A. (1785–1856)
Warren, E. A. (1786–1864)
Couch, J. (1789–1870)
Jones, *Rev.* J. P. (1790–1857)
Cocks, W. P. (1791–1878)
Jacobs, *Rev.* J. (1796–1849)
Penneck, *Rev.* H. (1801–1862)
Millett, L. (1801–1871) *Isles of Scilly*
Millett, M. (1805–1855) *Isles of Scilly*
Ralfs, J. (1807–1890)
Curnow, W. (*c.* 1809–1887)
North, *Rev.* I. W. (1810–) *Isles of Scilly*
Stackhouse, E. (1811–1870)
Pascoe, F. P. (1813–1893)
Hind, *Rev.* W. M. (1815–1894)
Keys, I. W. N. (1818–1890)
Banks, G. (*fl.* 1820s–1830s)
Townsend, F. (1822–1905) *Isles of Scilly*
Couch, T. Q. (1826–1884)
Tellam, R. V. (1826–1908)
Glasson, W. A. (1828–1903)
Easton, N. (*fl.* 1830s)
White, M. (*fl.* 1830s) *Isles of Scilly*
Cunnack, J. (1831–1886)
Bennett, E. T. (1831–1908)
Bastian, H. C. (1837–1915)
Somerville, A. (1842–1907) *Isles of Scilly*
Thurston, E. (1855–1935)
Borlase, W. (1860–1948)
Davey, F. H. (1868–1915)
Bullmore, E. (*fl.* 1880s)
Boyden, *Rev.* H. (*fl.* 1880s–1900s) *Isles of Scilly*
Rilstone, F. (1881–1953)
Graham, W. J. (*fl.* 1900s)
Lousley, J. E. (1907–1976) *Isles of Scilly*

Cumberland
Lawson, *Rev.* T. (1630–1691)
Nicolson, *Rev.* W. (1655–1727)
Richardson, *Rev.* W. (–1768)
Otley, J. (1766–1856)
Winch, N. J. (1768–1838)
Hutton, T. (*fl.* 1780s–1829)
Flintoft, J. (1796–1860)
Wood, *Rev.* R. (1796–1883)
Dickinson, W. (*c.* 1799–1882)
Tweddle, D. (–1875)
Dodd, *Rev.* J. (*fl.* 1800s)
Robson, J. (1817–1884)
Malleson, *Rev.* F. A. (1819–1897)
Hodgson, W. (1824–1901)
Hindson, I. (*fl.* 1830s–1870s)
Wright, C. H. (*fl.* 1840s)
Leitch, J. (*c.* 1849–1896)
Hellon, R. (1854–1924)
Johnstone, T. S. (*fl.* 1870s–1917)
Britten, H. (1870–195

Lister, T. (*fl.* 1880s)
Coggins, G. (*fl.* 1890s)
Curwen, J. (*fl.* 1890s)
Duckworth, W. (*fl.* 1890s)

Derbyshire
Artis, E. T. (1789–1847)
Howe, W. E. (–1891)
Hagger, J. (–1895)
Harpur-Crewe, *Rev.* H. (1830–1883)
Painter, *Rev.* W. H. (1835–1910)
Linton, *Rev.* W. R. (1850–1908)

Devonshire
Pike, E. L. (–*c.* 1772)
Yonge, *Rev.* J. (1748–1797)
Shore, *Rev.* T. W. (1756–1822)
Polwhele, *Rev.* R. (1760–1838)
Neck, *Rev.* A. (1769–1852)
Wavell, W. (*fl.* 1780s–1820s)
Tozer, *Rev.* J. S. (*c.* 1790–1836)
Jones, *Rev.* J. P. (1790–1857)
Jacob, *Rev.* J. (1796–1849)
Edwards, *Rev.* Z. J. (1799–1880)
Hare, R. (–*c.* 1826)
Gudson, *Mrs.* (–1871)
Kingston, J. F. (–before 1850)
Cullen, C. S. (*fl.* 1800s)
Griffiths, A. E. (1802–1861)
Hore, *Rev.* W. S. (1807–1882)
Hutchinson, P. O. (1810–1897)
Stewart, R. (1811–1865)
Cresswell, *Rev.* R. (1815–1882)
Brent, F. (1816–1903)
Keys, I. W. N. (1818–1890)
Banks, G. (*fl.* 1820s–1830s)
Parfitt, E. (1820–1893)
Parker, C. E. (1822–1895)
Chanter, C. (1824–1882)
Hannaford, S. (1828–1874)
Ravenshaw, *Rev.* T. F. T. (1829–1882)
Cullen, W. H. (*fl.* 1830s–1840s)
Saunders, H. (1830–1914)
Briggs, T. R. A. (1836–1891)
D'Urban, W. S. M. (1837–1934)
Hiern, W. P. (1839–1925)
Jordan, R. C. R. (*fl.* 1840s)
Browne, *Rev.* W. B. (1845–1928)
Larter, C. E. (1847–1936)
Sharland, A. (1849–1917)
Ashley, W. H. (*fl.* 1850s)
Friend, *Rev.* H. (1852–1940)
Stephenson, *Rev.* T. (1855–1948)
Harris, G. T. (*c.* 1856–1938)
Tregelles, G. F. (1859–1943)
Benthall, *Rev.* C. F. (1861–1936)
Douglas, G. H. (1864–1933)
Orme, R. (1865–1934)
Brokenshire, F. A. (1866–1957)
Miller, W. D. (1868–1933)
Watts, G. A. R. (1873–1949)
Savery, G. B. (1874–1937)
Martin, *Rev.* W. K. (1877–1969)
Charge, *Rev.* J. (*fl.* 1880s)
Evans, H. A. (*fl.* 1880s)
Fraser, G. T. (1882–1942)
Falkner, H. J. (1894–1951)
Smith, R. C. (*fl.* 1900s–1920s)

Dorset
Pulteney, R. (1730–1801)
Binfield, *Rev.* E. (–before 1813)

Mansel-Pleydell, J. C. (1817–1902)
Barrett, W. B. (1833–1915)
Linton, *Rev*. E. F. (1848–1928)
Goddard, H. J. (*c*. 1864–1947)
Green, C. B. (–1918)

Durham
Winch, N. J. (1768–1838)
Backhouse, W. (1779–1844)
I'Anson, J. (1784–1821)
Thornhill, J. (*fl*. 1800s–1850s)
Hogg, J. (1800–1869)

Essex
Dale, S. (1659–1739)
Warner, R. (*c*. 1713–1775)
Forster, E. (1765–1849)
Freeman, J. (1784–1864)
Garnons, *Rev*. W. L. P. (1791–1863)
Gibson, J. M. (1794–1838)
Benson, *Rev*. T. (1802–1887)
Clarke, J. (1805–1890)
Parsons, C. (1807–1882)
Varenne, E. G. (1811–1887)
Freeman, J. (1813–1907)
Wallis, A. (1816–1856)
Coleman, *Rev*. W. H. (*c*. 1816–1863)
Gibson, G. S. (1818–1883)
MacIntyre, A. (*fl*. 1820s–1840s)
English, J. L. (1820–1888)
Lister, J. (1827–1912)
Laver, H. (1829–1917)
Bennett, A. (1843–1929)
Bentall, T. (*fl*. 1840s–1860s)
Shenstone, J. C. (1854–1935)
Thompson, P. G. (1866–1953)
Marriott, St. J. (1870–1927)
Ross, J. (1873–1962)
Prince, E. (1877–1953)
Howard, W. (1877–1954)
Moore, W. (*fl*. 1880s–1890s)
Brown, G. C. (1889–1969)
Sparrow, H. D. (1890–1959)
Peterken, J. H. G. (1893–1973)
Sewell, S. A. (*fl*. 1900s)
Jermyn, S. T. (1909–1973)

Gloucestershire
Broughton, A. (–1796/1803)
Bellers, F. (*fl*. 1720s?)
Swayne, *Rev*. G. (*c*. 1746–1827)
Baker, *Rev*. W. L. (1752–1830)
Knapp, J. L. (1767–1845)
Dyer, T. W. (*fl*. 1780s–1830s)
Rootsey, S. (1788–1855)
Smith, E. (*fl*. 1790s)
Sandys, *Rev*. G. W. (–*c*. 1840)
Prentice, C. (–1894)
Blomefield, *Rev*. L. (1800–1893)
Cundall, J. H. (1808–1884)
Thwaites, G. H. K. (1812–1882)
Broome, C. E. (1812–1886)
Witts, *Rev*. E. F. (1813–1886)
Buckman, J. (1814–1884)
Stephens, H. O. (1816–1881)
Powell, F. S. (*fl*. 1820s–1860s)
Purchas, *Rev*. W. H. (1823–1903)
Leipner, A. (1827–1894)
Barnard, R. C. (1827–1906)
Swete, E. H. (1827–1912)
St. Brody, G. A. O. (1828–1901)
Fry, D. (1834–1912)
Sessions, F. (1836–1920)
Smith, M. P. (*fl*. 1840s–1850s)
Millard, *Miss* (*fl*. 1840s–1870s)
Lucy, W. C. (*fl*. 1840s–1890s)
Cumming, L. (1843–1927)
Montgomery, A. S. (1844–1922)
Duthie, J. F. (1845–1922)
White, J. W. (1846–1932)
Harker, J. A. (1847–1894)
Bucknall, C. (1849–1921)
Atwood, M. M. (*fl*. 1850s–1860s)
Waterfall, W. B. (1850–1915)
Butt, *Rev*. W. (*c*. 1850–1917)
Reader, H. P. (1850–1929)
Boulger, G. E. S. (1853–1922)
Gibbons, H. J. J. F. (1856–1939)
Beach, H. (*fl*. 1860s)
Brady, H. (*fl*. 1860s–1870s)
Stephens, A. J. (1861–1941)
Knight, H. H. (1862–1944)
Roper, I. M. (1865–1935)
Riddelsdell, *Rev*. H. J. (1866–1941)
Hedley, G. W. (1871–1941)
Dixon, F. (1871–1951)
Burkill, H. J. (1871–1956)
Sandwith, C. I. (1871–1961)
Mellersh, W. L. (1872–1941)
Humpidge, F. C. (1874–1944)
Wintle, G. S. (*fl*. 1880s?)
Gambier-Parry, T. R. (1883–1935)
Buckell, W. R. (1856–1956)
Greenwood, W. J. H. (*fl*. 1890s–1910s)
Smith, D. D. M. (1890–1971)
Fleming, G. W. T. H. (1895–1962)
Fletcher, J. (–1916)
Harford, H. W. L. (–1921)
Coley, S. J. (–*c*. 1925)
Upton, C. (–1927)
Laurie, C. L. (–1933)
Mott, L. (–1940)
Brookes, E. M. (–1950)
Hedley, W. (–1964)
Neve, J. R. (*fl*. 1900s)
Trump, C. (*fl*. 1900s–1910s)
Sandwith, N. Y. (1901–1965)
Dudley-Smith, R. (1912–1967)

Hampshire
Bayley, W. (1529–1592)
Legg, T. (*fl*. 1760s)
Garnier, *Rev*. T. (1776–1873)
M'Gaven, D. (*fl*. 1790s)
Bell, T. (1792–1880)
Hill, R. S. (–1872)
Reeves, *Rev*. J. W. (1816–1862)
Buckell, F. (1818–1897)
Townsend, F. (1822–1905)
Palmer, C. E. (1830–1914)
Wickham, W. (1831–1897)
Bennett, E. T. (1831–1908)
Clarke, C. B. (1832–1906)
Warner, F. I. (1841–1896)
Eyre, *Rev*. W. L. W. (1841–1914)
Tatum, E. J. (1851–1929)
Rayner, J. F. (1854–1947)
Vaughan, *Rev*. J. (1855–1922)
Hilland, *Miss* (1857–1924)
Watt, H. W. B. (1879–1968)
Hall, P. M. (1894–1941)
Kelsall, *Rev*. J. E. (–1924)

Herefordshire
Billiald, R. A.
Duncumb, *Rev*. J. (*fl*. 1800s)
Crouch, *Rev*. J. F. (1809–1889)
Watkins, B. M. (1816–1892)
Purchas, *Rev*. W. H. (1823–1903)
Southall, H. (1826–1916)
Moore, H. C. (1836–1908)
Smith, W. A. (*fl*. 1840s)
Wilmott, A. T. (*fl*. 1840s)
Bickham, S. H. (1841–1933)
Ley, *Rev*. A. (1842–1911)
Towndrow, R. F. (1845–1937)
Hutchinson, *Rev*. T. (1846–1916)
Anderton Smith, W. (*fl*. 1850s)
Wait, *Rev*. W. O. (1852–1936)
Woodhouse, *Rev*. T. (*fl*. 1860s–1870s)
Binstead, *Rev*. C. H. (*c*. 1862–1941)
Armitage, E. (1865–1961)
Shawe, *Rev*. J. (*fl*. 1880s)
Marsh, F. H. B. (1881–1948)

Hertfordshire
Chambers, R. (1784–1858)
Franks, F. (1796–)
Fordham, H. (1803–1894)
Brown, I. (1803–1895)
Blake, W. J. (1805–1875)
Webb, *Rev*. R. H. (*c*. 1806–1880)
Edwards, E. (1812–1886)
Coleman, *Rev*. W. H. (*c*. 1816–1863)
Peirson, D. (1819–1899)
Marsh, M. (*fl*. 1830s–1840s)
Cottam, A. (1838–1912)
Pryor, A. R. (1839–1881)
Andrews, R. T. (1839–1928)
Hopkinson, J. (1844–1919)
Jackson, B. D. (1846–1927)
Blow, T. B. (1854–1941)
Balls, M. (*fl*. 1860s)
Little, J. E. (1861–1935)
Dymes, T. A. (1865–1944)
Cooper, C. A. (1871–1944)

Huntingdonshire
Fryer, A. (1826–1912)

Isle of Man
Forbes, E. (1815–1854)
Stowell, *Rev*. H. A. (1830–1886)
Holt, G. A. (1852–1921)
Kermode, S. A. P. (*c*. 1862–1925)
Paton, C. I. (1874–1949)
Howarth, R. (1889–1954)
Beesley, H. (*fl*. 1900s)

Isle of Wight
Snooke, W. D. (1787–1857)
Everett, E. (*fl*. 1800s)
Bromfield, W. A. (1801–1851)
Martin, G. A. (*c*. 1807–1867)
Salter, T. B. (1814–1858)
Hambrough, A. J. (*c*. 1820–1861)
More, A. G. (1830–1895)
Tucker, R. (1832–1905)
Steuart, J. H. A. (1834–1895)
Tate, G. R. (1835–1874)
Stratton, F. (1840–1916)
O'Brien, C. G. (1845–1909)
Morey, F. (1858–1925)

Kent
Jeffrey, W. R.
Mount, *Rev*. W. (1545–1602)
Johnson, T. (*c*. 1604–1644)
Bateman, *Rev*. J. (*fl*. 1660s–1720s)
Drayton, J. (1681–1749)
Jacob, E. (*c*. 1710–1788)
Forster, T. F. (1761–1825)
Hunter, R. E. (–before 1847)
Stonham, W. B. (–1896)
Jenner, E. (1803–1872)
Smith, *Rev*. G. E. (1804–1881)
Deakin, R. (1808/9–1873)
Janson, T. C. (1809–1863)
Bossey, F. (1809–1904)
Paley, F. A. (1815–1888)
Dowker, G. (1828–1899)
Cowell, M. H. (*fl*. 1830s–1840s)
Stowell, *Rev*. H. A. (1830–1886)
Pittock, G. M. (1832–1916)
Allom, E. A. (*fl*. 1840s–1870s)
Webb, F. M. (1841–1880)
Walton, G. C. (1845–1934)
Fielding, *Rev*. C. H. (1848–1918)
Beisly, S. (*fl*. 1850s–1860s)
Hanbury, F. J. (1851–1938)
Hudson, A. W. (1856–1928)
Lamb, H. (1858–1905)
Marshall, *Rev*. E. S. (1858–1919)
Granville, C. R. *Countess of* (*fl*. 1860s–1880s)
Ullyett, H. (*fl*. 1860s–1880s)
Moring, P. (1865–1930)
Stedman, F. W. (–1920)
Wallis, J. R. (1917–1944)

Lancashire
Dewhurst, J. (*fl*. 1750s–1830s)
Crosfield, G. (1754–1820)
Barton, *Rev*. W. (*c*. 1754–1829)
Atkinson, W. (1765–1821)
Crowther, J. (1768–1847)
Johns, W. (1771–1845)
Haulkyard, *Dr*. (*fl*. 1780s)
Hyde, G. (*fl*. 1780s)
Newton, J. (*fl*. 1780s)
Glazebrook, T. K. (1780–1855)
Hobson, E. (1782–1830)
Martin, J. (*c*. 1783–1855)
Helme, W. (1785–1834)
Crosfield, G. (1785–1847)
Buxton, R. (1786–1865)
Tinker, J. (1788–1871)
Kent, W. (1789–1850)
Whittle, P. A. (1789–1867)
Ward, G. (1791–1880)
Crozier, G. (1792–1847)
Horsefield, J. (1792–1854)
Tudor, R. (1798–1886)
Evans, W. (–1828)
Townley, T. (–1857)
Fisher, H. S. (–1881)
Clough, E. (–1883)
Lewis, J. H. (–before 1890)
Evans, J. (1803–1874)
Wilson, W. (1803–1876)
Dickinson, J. (*c*. 1805–1865)
Worsley, W. (1808–)
Grundy, M. A. (1809–1871)
Potts, E. (1809–1873)
Gibson, T. (*fl*. 1810s–1870s)

Jepson, W. (1812–1897)
Wood, J. B. (1813–1890)
Hodgson, E. (1814–1877)
Hall, T. B. (1814–1886)
Grundy, E. (1815–1894)
Brent, F. (1816–1903)
Sansom, T. (1816/7–1872)
Bean, W. (1817–1864)
Ashfield, C. J. (*c*. 1817–1877)
Grindon, L. H. (1818–1904)
Dobson, W. (1820–1884)
Harrison, W. (1821–)
Smith, H. E. (1823–1889)
Mason, A. (1826–1888)
Percival, J. (1828–1902)
Carr, A. (*c*. 1829–1884)
Holland, R. (1829–1893)
Hartley, W. (1829–1907)
Alcock, R. H. (1833–1885)
Whitehead, J. (1833–1896)
Peers, J. (1838–1902)
Carter, G. (1838–1903)
Bailey, C. (1838–1924)
Brown, R. (1839–1901)
Cash, J. (1839–1909)
Gerard, *Rev*. J. (1840–1912)
Webb, F. M. (1841–1880)
Kent, W. (*c*. 1842–1913)
Healey, J. C. (1843–1922)
Searle, H. (1844–1935)
Day, R. H. (1848–1928)
Harrison, J. (*fl*. 1850s)
Skellon, W. (*fl*. 1850s)
Dinsley, W. F. (1854–1941)
Ellis, J. W. (1857–1916)
Dudley, A. H. (1857–1921)
Ball, H. (1857–1925)
Eaves, D. (1857–1936)
Fisher, H. (1860–1935)
Pearsall, W. H. (1860–1936)
Wheldon, J. A. (1862–1924)
Wilson, A. (1862–1949)
Green, C. T. (1863–1940)
Clitheroe, W. (1864–1944)
Wood, E. M. (1865–1907)
Laverock, W. S. (1865–1947)
Robinson, H. (1866–1932)
Dunlop, G. A. (1868–1933)
Chapman, H. (*fl*. 1870s)
Grundy, C. (*fl*. 1870s)
Lee, W. A. (1870–1931)
Massey, J. D. (1870–1943)
Western, W. H. (1871–1948)
Travis, W. G. (1877–1958)
Green, H. E. (1886–1973)
Bunker, H. E. (1899–1969)
Beesley, H. (*fl*. 1900s)
Mort, J. (–1907)
Petty, S. L. (–1919)
Campbell, *Rev*. A. J. (–*c*. 1932)

Leicestershire

Glen, *Rev*. A. (*c*. 1666–1732)
Scampton, J. (*fl*. 1690s–1710s)
Pulteney, R. (1730–1801)
Arnold, T. (1742–1816)
Noel, H. (1743–1798)
Crabbe, *Rev*. G. (1754–1832)
Power, J. (1758–1847)
Clare, J. (1793–1864)
Bloxam, *Rev*. A. (1801–1878)
Dalby, R. (1808–1884)
Eller, *Rev*. C. I. (*fl*. 1810s–1870s)
Power, J. A. (1810–1886)
Coleman, *Rev*. W. H. (*c*. 1816–1863)
Kirby, M. (1817–1893)
Brown, E. (1818–1876)
Kirby, E. (1823–1873)
Mott, F. T. (1825–1908)
Roy, J. (1826–1893)
Cooper, E. F. (1833–1916)
Preston, *Rev*. T. A. (1838–1905)
Carter, T. (*c*. 1840–1910)
Lakin, C. (*c*. 1840–1916)
Pattison, E. (*c*. 1840–1916)
Clarke, W. A. (1841–1911)
Finch, J. E. M. (1842–1919)
Dixon, G. B. (*c*. 1850–1912)
Quilter, H. E. (*c*. 1850–1915)
Noel, C. W. F. (1850–1926)
Reader, H. P. (1850–1929)
Turner, G. C. (1858–1940)
Fisher, H. (1860–1935)
Foord-Kelcey, F. L. (*c*. 1862–1914)
Bell, W. (1862–1925)
Headly, C. B. (*c*. 1870–*c*. 1916)
Horwood, A. R. (1879–1937)
Murray, *Rev*. D. P. (1887–1967)
Mercer, G. E. (1896–1918)
Sowter, F. A. (1899–1972)

Lincolnshire

Bogg, J. (1799–1866)
Brown, J. (–1851)
Hudson, S. H. (*c*. 1827–1904)
Fowler, *Rev*. W. (1835–1912)
Bogg, T. W. (*fl*. 1850s–1870s)
Mason, *Rev*. W. W. (1853–1932)
Woodruffe-Peacock, *Rev*. E. A. (1858–1922)
Marshall, J. J. (1860–1934)
Goulding, R. W. (1868–1929)
Smith, H. B. W. (1879–)
Larder, J. (–1923)

London and Middlesex

Johnson, T. (*c*. 1604–1644)
Morley, C. L. (*c*. 1646–1702)
Blackstone, J. (1712–1753)
Cockfield, J. (*c*. 1740–1816)
La Gasca y Segura, M. (1776–1839)
Hunter, E. (*fl*. 1790s–1824)
Gibbs, J. (–*c*. 1829)
Gérard (–1840)
Baldock, R. (–*c*. 1860)
Heathfield, R. (1802–1848)
Twining, T. (1806–1895)
Pamplin, W. (1806–1899)
Davies, W. (1814–1891)
Hind, *Rev*. W. M. (1815–1894)
Cooper, D. (*c*. 1817–1842)
Ballard, E. (1820–1897)
Grugeon, A. (1826–1913)
Church, *Sir* A. H. (1834–1915)
Morris, J. (*fl*. 1840s)
Trimen, H. (1843–1896)
Melvill, J. C. (1845–1929)
Wharton, H. T. (1846–1895)
Loydell, A. (1849–1910)
Whiting, J. E. (*c*. 1850–1927)
Cornish, C. J. (1858–1906)
Champneys, M. (*fl*. 1860s–1935)

Wood, A. (*fl.* 1870s–1910s)
Britton, C. E. (1872–1944)
Green, C. B. (–1918)
Garlick, C. (–1934)

Lundy Island
Elliston-Wright, F. R. (*c.* 1879–1966)

Norfolk
Humphrey, W. (–before 1792)
Wigg, L. (1749–1829)
Crowe, J. (1750–1807)
Aram, W. (*fl.* 1770s)
Wigham, R. (1785–1855)
Munford, *Rev.* G. (*c.* 1794–1871)
Mason, S. (*fl.* 1800s)
Holmes, *Rev.* E. A. (–1886)
Trimmer, *Rev.* K. (1804–1887)
Paget, *Sir* J. (1814–1899)
Mann, R. J. (1817–1886)
Stock, D. (*fl.* 1820s–1860s)
Woodward, S. P. (1821–1865)
Burlingham, D. C. (1823–1901)
Glasspoole, H. G. (1825–1887)
Barnard, A. M. (1825–1911)
Kitton, F. (1827–1895)
Corder, O. (1828–1910)
Geldart, H. D. (1831–1902)
Long, F. (1840–1927)
Bennett, A. (1843–1929)
Preston, A. W. (1855–1931)
Hamond, C. A. (1856–1914)
Bird, *Rev.* M. C. H. (1857–1924)
Nicholson, W. A. (1858–1935)
Galpin, *Rev.* F. W. (1858–1945)
Geldart, A. M. (1862–1942)
Mayfield, A. (1868–1956)
Clarke, W. G. (1877–1925)
Bidwell, W. H. (–1909)
Robinson, F. (*fl.* 1910s–1920s)

Northamptonshire
Morton, *Rev.* J. (1670/1–1726)
Baker, A. E. (1786–1861)
Clare, J. (1793–1864)
Pitt, W. (–1823)
Berkeley, *Rev.* M. J. (1803–1889)
Lightfoot, C. A. (1807–1898)
Latham, R. G. (1812–1888)
Paley, F. A. (1815–1888)
Notcutt, W. L. (1819–1868)
Wildegose, R. (*fl.* 1820s–1830s)
Pegus, M. A. (1821–1893)
Law, W. (*c.* 1838–1916)
Druce, G. C. (1850–1932)
Dixon, H. N. (1861–1944)
Bodger, J. W. (1865–1939)
Chester, *Sir* G. (1886–1949)
Allen, H. G. (1890–1965)
Bostock, F. (–1940)

Northumberland
Wallis, *Rev.* J. (1714–1793)
Winch, N. J. (1768–1838)
Thompson, J. (*c.* 1778–1866)
Thompson, J. V. (1779–1847)
Brown, T. (1785–1862)
Johnston, G. (1797–1855)
Richardson, W. (1797–1879)
Waugh, R. (–1806)
Robertson, W. (–1846/7)
Baird, *Rev.* A. (1800–1845)
Storey, J. (*c.* 1801–1859)
Tate, G. (1805–1871)
Atthey, T. (1814–1880)
Bigge, *Rev.* J. F. (1814–1885)
Luckley, J. L. (1822–1899)
Backhouse, J. (1825–1890)
Tate, G. R. (1835–1874)

Nottinghamshire
Deering, G. C. (*c.* 1690–1749)
Kaye, *Sir* R. (1736–1809)
Cooper, T. H. (*fl.* 1750s–1840s)
Becher, *Rev.* J. T. (1770–1848)
Miller, *Rev.* J. K. (1786–1850s)
Ordoyno, T. (*fl.* 1790s–1810s)
Bohler, J. (1797–1872)
Howitt, G. (1800–1873)
Jowett, T. (*c.* 1801–1832)
Valentine, W. (*fl.* 1810s–1884)
Carr, J. W. (1862–1939)

Oxfordshire
Ashmole, E. (1617–1692)
Morison, R. (1620–1683)
Coles, W. (1626–1662)
Browne, *Rev.* W. (*c.* 1628–1678)
Plot, R. (1640–1696)
Bobart, J. (1641–1719)
Dillenius, J. J. (1684–1747)
Blackstone, J. (1712–1753)
Bellers, F. (*fl.* 1720s)
Sheffield, *Rev.* W. (*c.* 1732–1795)
Benwell, J. (*c.* 1735–1819)
Randolph, *Rev.* J. (1749–1813)
Sibthorp, J. (1758–1796)
Mavor, W. F. (1758–1837)
Baxter, W. (1787–1871)
Walker, *Rev.* R. (1791–1870)
Gulliver, G. (1804–1882)
Hyde, W. W. (1812–1880)
Beesley, T. (1818–1896)
Beck, S. C. (1821–1915)
Woodward, G. (*fl.* 1830s)
Holliday, W. H. (1834–1909)
Boswell, H. (1837–1897)
French, A. (1839–1879)
Richards, F. T. (1847–1905)
Druce, G. C. (1850–1932)
Reynolds-Moreton, H. H. (1857–1920)
Brady, H. (*fl.* 1860s–1870s)
Fox, *Rev.* E. (*fl.* 1870s–1880s)
Martyn, *Rev.* T. W. (–1918)

Rutland
Noel, E. (1731–1801)
Wingfield, A. F. (1823–1914)
Bell, *Rev.* E. (1829–1904)
Thring, L. E. (1830–1925)
Bell, T. (1836–1914)
Candler, H. (1838–1916)
Graham, *Rev.* H. L. (1844–1921)
Hall, T. K. (1848–1890)
Trevor-Jones, H. T. (1855–1901)
Lightfoot, J. P. W. (1871–1919)
Sowter, F. A. (1899–1972)

Shropshire
Brown, *Rev.* L. (1699–1749)
Babington, *Rev.* J. (1768–1826)
Aikin, A. (1773–1854)
Leighton, *Rev.* W. A. (1805–1889)

Brookes, W. P. (1809–1895)
Dickinson, F. (1816–1901)
Phillips, W. (1822–1905)
Griffiths, G. H. (*c.* 1823–1872)
Hamilton, W. P. (*c.* 1842–1910)
Blunt, T. P. (*c.* 1842–1929)
Beckwith, W. E. (1844–1892)
Benson, R. de G. (1856–1904)
Rutter, E. M. (*c.* 1890–1963)
Lloyd, L. C. (*c.* 1905–1968)

Somerset
Sole, W. (1741–1802)
Davies, J. F. (1773–1864)
Dyer, T. W. (*fl.* 1780s–1830s)
Collins, *Rev.* J. C. (1798–1867)
Jelly, *Mr.* (–before 1849)
Withers, R. (–1856)
Gibbes, *Rev.* H. (1802–1887)
Gifford, I. (*c.* 1823–1891)
Tuckwell, *Rev.* W. (1829–1919)
Southby, A. (*fl.* 1830s–1840s)
Parsons, H. F. (1846–1913)
Duck, J. N. (*fl.* 1850s)
Marshall, *Rev.* E. S. (1858–1919)
Downes, H. (1867–1937)
Moss, C. E. (1870–1930)
Boley, G. M. (*c.* 1877–1965)
Hamlin, E. J. (*c.* 1878–1966)
Dodd, A. J. (*c.* 1882–1963)
Samson, F. (–1929)
Livett, M. A. G. (*fl.* 1910s–1930s)

Staffordshire
Dickenson, *Rev.* S. (1730–1823)
Gisborne, *Rev.* T. (1758–1846)
Constable, *Sir* T. H. C. (1762–1823)
Gisborne, J. (1770–1850)
Wolseley, *Rev.* R. (1772–1815)
Wainwright, R. (*fl.* 1790s)
Forster, R. (*fl.* 1790s–1810s)
Pitt, W. (–1823)
Garner, R. (1808–1890)
Brown, E. (1818–1876)
Fraser, J. (1820–1909)
Jackson, M. A. (*fl.* 1830s–1840s)
Carrington, S. (*fl.* 1830s–1870s)
Bagnall, J. E. (1830–1918)
Blaikie, J. (1837–1912)
Hewgill, A. (*fl.* 1840s–1860s)
Douglas, *Rev.* R. C. (*fl.* 1850s)
Reader, H. P. (1850–1929)
Masefield, J. R. B. (1850–1932)
Audley, J. A. (1859–1938)
Berrisford, S. (1859–1938)
Gibbs, T. (1865–1919)
Moore, C. (1870–1944)
Deacon, *Rev.* E. (1872–1937)
Ridge, W. T. B. (1872–1943)
Curtis, *Sir* R. C. M. (1886–1954)

Suffolk
Cullum, *Rev.* J. (1733–1785)
Cullum, *Sir* T. G. (1741–1831)
Ashby, J. (1754–1828)
Crabbe, *Rev.* G. (1754–1832)
Gage, *Sir* T. (1781–1820)
Casborne, *Mrs.* (–1884)
Bunbury, *Sir* C. J. F. (1809–1886)
Hind, *Rev.* W. M. (1815–1894)
Steggall, *Rev.* W. (*fl.* 1820s–1850s)
Skepper, E. (1825–1867)
Bloomfield, *Rev.* E. N. (1827–1914)
Clarke, W. B. (*fl.* 1840s)
Batchelder, S. J. (1870–1949)

Surrey
Salmon, J. D. (*c.* 1802–1859)
Mill, J. S. (1806–1873)
Luxford, G. (1807–1854)
Cooper, D. (*c.* 1817–1842)
Brewer, J. A. (1818–1886)
Reeves, W. W. (1819–1892)
Linnell, J. (1822–1906)
Saunders, W. F. (1834–1901)
Beeby, W. H. (1849–1910)
Smith-Pearse, *Rev.* T. N. H. (1854–1943)
Monckton, H. W. (1857–1930)
Granville, C. R. *Countess* (*fl.* 1860s–1880s)
Axford, W. G. (*c.* 1861–1942)
Carruthers, S. W. (1866–1962)
Dunn, S. T. (1868–1938)
Robbins, R. W. (1871–1941)
Salmon, C. E. (1872–1930)
Capron, E. (–1907)
Lousley, J. E. (1907–1976)
Young, D. P. (1917–1972)

Sussex
Markwick, W. (1739–1813)
Cooper, T. H. (*fl.* 1750s–1840s)
Woollgar, T. (*fl.* 1760s–1820s)
Forster, T. F. (1761–1825)
Woods, J. (1776–1864)
Borrer, W. (1781–1862)
Smith, *Rev.* G. E. (1804–1881)
Unwin, W. C. (1811–1887)
Tyacke, N. (1812–1900)
Coleman, *Rev.* W. H. (*c.* 1816–1863)
Roper, F. C. S. (1819–1896)
Malleson, *Rev.* F. A. (1819–1897)
Mitten, W. (1819–1906)
Bloomfield, *Rev.* E. N. (1827–1914)
Arnold, *Rev.* F. H. (1831–1906)
Hilton, T. (1833–1912)
Smith, C. P. (1835–1892)
Bray, E. A. (*c.* 1844–1938)
Jenner, J. H. A. (*fl.* 1850s–1900s)
Guermonprez, H. L. F. (1858–1924)
Cosstick, W. (*fl.* 1860s)
Helyer, B. (*fl.* 1860s–1870s)
Richards, T. J. (*c.* 1860–1937)
Wolley-Dod, A. H. (1861–1948)
Nicholson, W. E. (1866–1945)
Bedford, E. J. (1866–1953)
Gregor, *Rev.* A. G. (1867–1954)
Cowan, D. A. (*c.* 1874–1952)
Lowne, B. T. (1878–1956)
Cooke, A. S. (1890–1957)
Peatfield, W. (–*c.* 1954)
Kaye Smith, A. D. (–1955)

Warwickshire
Holden, H. (*fl.* 1650s?)
Withering, W. (1741–1799)
Finch, L., *Countess of Aylesford* (1760–1832)
Bree, *Rev.* W. (*fl.* 1770s–1820s)
Rufford, *Rev.* W. S. (*c.* 1786–1836)
Bree, *Rev.* W. T. (1787–1863)
Perry, W. G. (1796–1863)
Cheshire, W. (–*c.* 1855)
Ick, W. (1800–1844)

Bloxam, *Rev*. A. (1801–1878)
Russell, A. (1807–1876)
Young, *Rev*. J. R. (*c*. 1810–1884)
Baker, R. (*c*. 1824–1885)
Bromwich, H. (1828–1907)
Palmer, C. E. (1830–1914)
Bagnall, J. E. (1830–1918)
Freeman, S. (*fl*. 1840s)
Thickens, *Rev*. W. (*fl*. 1840s)
Trott, H. W. (1857–)
Hardaker, W. H. (1877–1970)
Hutchinson, *Rev*. T. N. (*fl*. 1890s)
Burges, R. C. L'E. (1900–1959)
Laflin, T. (1914–1972)

Westmorland
Lawson, *Rev*. T. (1630–1691)
Dalton, J. (1766–1844)
Gough, T. (1804–1880)
Martindale, J. A. (1837–1914)
Stabler, G. (1839–1910)
Wilson, A. (1862–1949)
Britten, H. (1870–1950)
Pickard, J. F. (1876–1943)

Wiltshire
Maton, W. G. (1774–1835)
Kenrick, G. C. (1806–1869)
Smith, H. (*fl*. 1810s)
Flower, T. B. (1817–1899)
Southby, A. (*fl*. 1830s–1840s)
Preston, *Rev*. T. A. (1838–1905)
Clarke, W. A. (1841–1911)
Tatum, E. J. (1851–1929)
Goddard, *Rev*. E. H. (1854–1947)
Gwatkin, J. R. G. (1855–1939)
Todd, E. S. (1859–1949)
Heginbothom, C. D. (1874–1950)
Hurst, C. P. (–1956)
Grose, J. D. (*c*. 1901–1973)

Worcestershire
Pitt, E. (*fl*. 1650s–1670s)
Ballard, R. (*fl*. 1780s–1790s)
Jorden, G. (1783–1871)
Pitt, W. (–1823)
Lees, E. (1800–1887)
Westcombe, T. (1815–1893)
Mathews, W. (1828–1901)
Moseley, H. (*fl*. 1830s–1860s)
Amphlett, J. (1845–1918)
Towndrow, R. F. (1845–1937)
Humphreys, J. (*c*. 1850–1937)
Rea, C. (1861–1946)
Day, F. M. (1890–1962)

Yorkshire
Witham, G. (*fl*. 1610s–1680s)
Tunstal, T. (*fl*. 1620s)
Bolton, T. (–1778)
Dawson, W. (*c*. 1714–*c*. 1776)
Thornbeck, *Mr*. (*fl*. 1740s)
Teesdale, R. (*c*. 1740–1804)
Dewhurst, J. (1746–1818)
Bolton, J. (*fl*. 1750s–1790s)
Whytehead, *Rev*. W. (1757–1817)
Salt, J. (1759–1810)
Dalton, *Rev*. J. (1764–1843)
Hailstone, S. (1768–1851)
Brunton, W. (1775–1806)
Leyland, R. (1784–1847)
Atkinson, J. (1787–1828)
Bean, W. (1787–1866)
Windsor, J. (1787–1868)
Artis, E. T. (1789–1847)
Gibson, S. (1789/90–1849)
Tatham, J. (1793–1875)
Baines, H. (1793–1878)
Moore, O. A. (–1862)
Kenyon, W. (–before 1868)
Heaton, J. D. (–1880)
Nowell, J. (1802–1867)
Ward, J. (1802–1873)
Appleby, S. (1806–1870)
Shepherd, J. (1807–1859)
King, *Rev*. S. (1810–1888)
Berry, E. (1812–1869)
Ibbotson, H. (1814–1886)
Matterson, W. (1815–1890)
Inchbald, P. (1816–1896)
Howson, *Rev*. J. (1817–1866)
Thompson, S. (1818–1881)
Carr, J. (*fl*. 1820s)
Braithwaite, R. (1824–1917)
Carrington, B. (1827–1893)
Gissing, T. W. (1829–1870)
Carr, A. (*c*. 1829–1884)
Langley, L. (*fl*. 1830s)
Baker, J. G. (1834–1920)
Foggitt, W. (1835–1917)
Handey, J. (*c*. 1836–1910)
Hobkirk, C. C. P. (1837–1902)
Walker, J. (1839–1895)
Whitwell, W. (1839–1920)
Miall, L. C. (1842–1921)
Crossland, C. (1844–1916)
Parsons, H. F. (1846–1913)
Bairstow, U. (1847–1914)
Lees, F. A. (1847–1921)
Cheesman, W. N. (1847–1925)
West, W. (1848–1914)
Clarke, A. (1848–1925)
Fisher, *Rev*. R. (1848–1933)
Packer, J. J. (*fl*. 1850s)
Young, J. F. (*fl*. 1850s)
Massee, G. E. (1850–1917)
Waterfall, C. (1851–1938)
Ingham, W. (1854–1923)
Wilson, R. (1855–1927)
Thompson, R. F. (1856–1906)
Margerison, S. (1857–1917)
Robinson, J. F. (1857–1927)
Soppitt, H. T. (1858–1899)
Le Tall, B. B. (1858–1906)
Foggitt, T. J. (1858–1934)
Crowther, J. (1859–1930)
Snelgrove, E. (1859–1934)
Wilkinson, H. J. (1859–1934)
Brady, H. (*fl*. 1860s–1870s)
Cryer, J. (1860–1926)
Marshall, J. J. (1860–1934)
Bevan, D. W. (1860–1944)
Woodhead, T. W. (1863–1940)
Percival, J. (1863–1949)
Swales, J. (*c*. 1864–1908)
Burrell, W. H. (1865–1945)
Jones, R. C. F. (1865–1952)
Smith, W. G. (1866–1928)
Crump, W. B. (1868–1950)
Holmes, T. H. (1869–1944)
Britten, H. (1870–1954)
Flintoff, R. J. (1873–1941)

Bradley, A. E. (1873–1944)
Mosley, C. (1875–1933)
Brown, J. M. (1875–1951)
Cheetham, C. A. (1875–1954)
Mason, F. A. (*c*. 1878–1936)
Smith, H. B. W. (1879–)
Charge, *Rev*. J. (*fl*. 1880s)
Johnson, H. (*fl*. 1880s–1918)
Elgee, F. (1881–1944)
Walsh, H. (1881–1962)
Milsom, F. E. (1886–1945)
Taylor, J. M. (1886–1947)
Broadbent, A. (1891–1962)
Lee, P. F. (–1912)
Sturdy, H. H. (–1933)
Thwaites, W. A. (*fl*. 1900s)
Garnett, *Rev*. P. M. (1906–1967)
Robb, C. M. (1906–1975)

SCOTLAND
Preston, C. (1660–1711)
Martin, M. (*c*. 1660–1719)
Wallace, J. (*fl*. 1680s–1720s) *Orkney*
Alston, C. (1685–1760) *Edinburgh*
McCoig, M. (–1789) *Edinburgh*
Hope, J. (1725–1786)
Skene, D. (1731–1770) *Aberdeen*
Lightfoot, *Rev*. J. (1735–1788)
Dickson, J. (1738–1822)
Smellie, W. (1740–1795)
Parsons, J. (1742–1785)
Low, *Rev*. G. (1746–1795) *Orkney*
Yalden, T. (*fl*. 1750s–1770s)
Graham, *Rev*. P. (1756–1835) *Perth*
Robertson, J. (*fl*. 1760s–1770s)
Don, G. (1764–1814) *Forfar*
Lyell, C. (1767–1849) *Angus*
Maugham, R. (1769–1844) *Edinburgh*
Woodforde, J. (1771–1837) *Edinburgh*
MacKay, J. (1772–1802)
Neill, P. (1776–1851)
Landsborough, *Rev*. D. (1779–1854)
Liston, *Rev*. W. (1781–1864) *Perth*
Hopkirk, T. (1785–1841)
Fleming, *Rev*. J. (1785–1857) *West Lothian*
Brown, T. (1785–1862) *Berwick*
Hooker, *Sir* W. J. (1785–1865)
Bishop, D. (*c*. 1788–1849) *Perth*
Scott, J. R. (*c*. 1789–1821)
Gillies, J. (1792–1834) *Orkney*
Greville, R. K. (1794–1866) *Edinburgh*
Duncan, J. (1794–1881) *Aberdeen*
Johnston, G. (1797–1855) *Berwick*
Murray, A. (*c*. 1798–1838)
Duguid, A. R. (1798–1872) *Orkney*
Forrest, W. H. (1799–*c*. 1879) *Stirling*
Hall, W. (–1800) *Berwick*
Bennet, *Rev*. W. (–1805) *Edinburgh*
Robertson, J. (–1865) *Perth*
Brown, J. (–1873) *Edinburgh*
Stewart, G. A. C. (–1876)
Macfarlane, *Rev*. G. (–1884) *Berwick*
Clouston, *Rev*. C. (1800–1885) *Shetland and Orkney*
Gordon, *Rev*. G. (1801–1893) *Moray*
Duncan, *Rev*. J. (*c*. 1802–1861) *Roxburgh*
Lowe, *Rev*. R. T. (1802–1874) *Orkney*
Barty, *Rev*. J. S. (1805–1875) *Perth*
Gardiner, W. (1808–1852) *Forfar*
Hennedy, R. (1809–1877) *Lanark*
Croall, A. (1809–1885) *Braemar*
Dick, R. (1811–1866) *Caithness*
Brown, *Rev*. T. (1811–1893) *Berwick*
Howie, C. (1811–1899) *Fife*
Dickie, G. (1812–1882) *Aberdeen*
Ramsay, J. (1812–1888)
Campbell, W. H. (1814–1883)
Edward, T. (1814–1886)
Drummond-Hay, H. M. (1814–1896) *Perth*
Fraser, *Rev*. J. (1814–1902)
Douglas, F. (1815–1886) *Berwick*
Hardy, J. (1815–1898) *Berwick*
Forbes, A. (1819–1879) *Inverness*
Boswell, E. and M. (*fl*. 1820s) *Fife*
Jackson, W. (1820–1848) *Dundee*
Evans, W. W. (1820–1885)
Boswell, J. T. I. (1822–1888) *Orkney*
Sim, J. (1824–1901)
Edmondston, T. (1825–1846) *Shetland*
Landsborough, *Rev*. D. (*c*. 1826–1912) *Ayr*
Heddle, R. (*c*. 1827–1860) *Orkney*
Lyall, J. B. (*c*. 1827–1912) *Peebles*
Shearer, J. (1827–1916) *Stirling*
Buchan, A. (1829–1907)
Macnab, R. (*fl*. 1830s–1840s) *Perth*
Knox, J. (1831–1914) *Forfar*
Boyd, W. B. (1831–1918)
Craig, W. (1832–1922)
Tait, W. (1833–1904) *Aberdeen*
Brotherston, A. (1834–1891)
MacGillivray, P. H. (1834–1895) *Aberdeen*
Fergusson, *Rev*. J. (1834–1907) *Forfar*
Brown, J. W. (1836–1863)
Traill, G. W. (1836–1897)
McAndrew, J. (1836–1917) *Dumfries*
Gregorson, D. (1836/7–1916) *Arran*
Inglis, A. (1837–1875)
Sadler, J. (1837–1882)
Christie, J. (1838–1898) *Glasgow*
Bryce, J. (1838–1922) *Arran*
MacIntosh, C. (1839–1922) *Perth*
Adamson, *Mr*. (*fl*. 1840s) *Clyde Islands*
Dewar, A. (*fl*. 1840s)
Tate, R. (1840–1901) *Shetland*
Crichton, J. S. (1841–1887) *Forfar*
Fox, *Rev*. H. E. (1841–1926)
White, F. B. W. (1842–1894) *Perth*
Craig-Christie, A. (1843–1914) *Shetland*
Aitken, R. T. (1843–1915) *Berwick*
Smith, J. (1845–1930) *Ayr*
Barclay, W. (1846–1923) *Perth*
West, W. (1848–1914) *Shetland*
Beeby, W. H. (1849–1910) *Shetland*
Barrington, R. M. (1849–1915) *St. Kilda*
Watt, L. A. (1849–1939)
Anderson, A. (*c*. 1850–1932) *Berwick*
Haggart, D. A. (1850–1939) *Perth*
Trail, J. W. H. (1851–1919)
Evans, W. (1851–1922)
Fortescue, W. I. (1851–1941) *Orkney*
Spence, M. (1853–1919) *Orkney*
Bowie, W. (1853–1931) *Glasgow*
Wise, T. (1854–1932)
Cran, W. (1854–1933)
Boyd, D. A. (1855–1928)
Johnstone, R. B. (1856–1934) *Glasgow*
Johnston, H. H. (1856–1939) *Orkney and Shetland*
Aiken, *Rev*. J. J. M. L. (1857–1933) *Berwick*
Meldrum, R. H. (1858–1933) *Perth*

McGrouther, T. (1858–1941) *Falkirk*
Batters, E. A. L. (1860–1907) *Berwick*
Ewing, E. R. (1860–1951)
Scott-Elliot, G. F. (1862–1934) *Dumfries*
Ellison, G. (1862–1941) *Orkney*
Burgess, J. J. (1863–1934)
Corstorphine, M. (1863–1944) *Angus*
Jack, J. (1863–1955) *Forfar*
Young, W. (1865–1947)
Bates, G. F. (*c*. 1868–1933) *Perth*
Lee, J. R. (1868–1959)
Shanks, A. (1870–1951) *Clyde*
Ord, G. W. (1871–1899)
Hayward, I. M. (1872–1949) *Berwick*
Smith, R. (1873–1900)
Ellis, D. (1874–1937)
Corstorphine, R. H. (1874–1942) *Angus*
Balding, A. (*fl*. 1880s) *Hebrides*
Calder, M. (*fl*. 1880s–1910s)
Craib, W. G. (1882–1933) *Banff*
Patton, D. (1884–1959) *Clydesdale*
Adams, E. G. (*fl*. 1890s) *Dumfries*
Johnston, S. D. (*fl*. 1890s) *Dumfries*
Stirling, J. S. (–1900) *Stirling*
McConachie, *Rev*. G. (–1901)
Murray, A. (–1904)
Farquharson, *Rev*. J. (–1906) *Selkirk*
Sim, T. (–1912) *Aberdeen*
Renwick, J. (–1918)
Grierson, R. (–1929) *Glasgow*
Gilmour, T. (–1930) *Glasgow*
Grant, J. F. (–1930) *Caithness*
Garry, R. (–1938)
Nisbet, T. (–1946) *Argyll*
Elder, F. M. (–1964)
Grant, *Dr*. (*fl*. 1900s) *Orkney*
Milne-Redhead, H. (1906–1974)
Dennis, N. (1912–1966) *Skye*

WALES

Salusbury, *Rev*. W. (*c*. 1520–1600)
Davies, *Rev*. J. (*c*. 1567–1644)
Glyn, T. (*fl*. 1630s)
Lloyd, M. (*fl*. 1640s)
Lhuyd, E. (1660–1709)
Brewer, S. (1670–1743)
Holcombe, *Rev*. J. (1704–1770) *Pembroke*
Morris, W. (1705–1763) *Anglesey*
Green, *Rev*. W. (*fl*. 1720s–1740s)
Skinner, *Rev*. R. (*c*. 1729–1795)
Davies, *Rev*. H. (1739–1821)
Evans, *Rev*. J. (*fl*. 1760s–1810s)
Griffith, J. W. (1763–1834)
Bingley, *Rev*. W. (1774–1823)
Dillwyn, L. W. (1778–1855) *Glamorgan*
Conway, C. (*fl*. 1790s–1870s) *Monmouth*
Salwey, *Rev*. T. (1791–1877) *Merioneth*
Motley, J. (–1859) *Carmarthen*
Williams, J. (1801–1859) *Denbighshire*
Morgan, T. O. (*c*. 1801–1878) *Cardigan*
Gutch, J. W. G. (1809–1862) *Glamorgan*
Falconer, R. W. (1816–1881) *Pembroke*
Inchbald, P. (1816–1896) *Caernarvon*
Clark, J. H. (1818–) *Monmouth*
How, *Rev*. W. W. (1823–1897) *Caernarvon*
Barrett, W. B. (1833–1915) *Brecon*
Jebb, G. R. (1838–1927)
Brown, R. (1839–1901) *Flint*
Thomas, T. H. (1839–1915) *Monmouth*
Bladon, J. (*fl*. 1840s–1850s) *Monmouth*
Storrie, J. (1843–1901) *Glamorgan*
Griffith, J. E. (1843–1933) *Anglesey and Caernarvon*
Britten, J. (1846–1924) *Merioneth*
Vachell, C. T. (1848–1914) *Glamorgan*
Shoolbred, W. A. (1852–1928) *Monmouth*
Williams, J. L. (1854–1945)
Eaves, D. (1857–1936)
Jones, D. A. (1861–1936) *Merioneth*
Salter, J. H. (1862–1942) *Cardigan*
Trow, A. H. (1863–1939) *Glamorgan*
Riddelsdell, *Rev*. H. J. (1866–1941) *Glamorgan*
Ellis, D. (1874–1937)
Wallis, A. (1879–1919) *Carmarthen and Pembroke*
Vachell, E. (1879–1948) *Glamorgan*
Dallman, A. A. (*c*. 1883–1963) *Denbigh and Flint*
Hyde, H. A. (*c*. 1892–1973)
Smith, R. L. (1892–1973) *Glamorgan*
Hamilton, S. (*fl*. 1900s) *Monmouth*
Thomas, E. J. H. (–1930)
Charles, S. G. (1883–1960) *Monmouth*

IRELAND

Heaton, *Rev*. R. (*fl*. 1620s–1660s)
Rawdon, *Sir* A. (*fl*. 1660s–1695)
Nicholson, H. (*c*. 1660–1732/3) *Dublin*
Molyneux, *Sir* T. (1661–1733) *Dublin*
Mitchell, *Dr*. (*fl*. 1670s)
Bonnivert, G. (*fl*. 1670s–1700s)
Threlkeld, *Rev*. C. (1676–1728)
Keogh, *Rev*. J. (*c*. 1681–1754)
Butler, I. (1689–1755) *Down*
Vaughan, F. (*fl*. 1690s) *Wexford*
Rutty, J. (1697–1775) *Dublin*
Chemys, C. (–1733)
Smith, C. (*c*. 1715–1762)
Browne, P. (*c*. 1720–1790)
Caldwell, A. (1733–1808)
Wade, W. (1760–1825)
Brinkley, J. (1763–1835)
Stokes, W. (1763–1845)
Tighe, W. (1766–1816) *Wexford*
Templeton, J. (1766–1825)
Hincks, *Rev*. T. D. (1767–1857)
Walsh, *Rev*. R. (1772–1852) *Dublin*
MacKay, J. T. (1775–1862)
Butt, *Rev*. T. (1776–1841)
Allman, W. (1776–1846)
Mant, *Rev*. R. (1776–1848)
Underwood, J. (*fl*. 1780s–1834) *Dublin*
Gage, *Sir* T. (1781–1820)
Litton, S. (1781–1847)
Drummond, J. L. (1783–1853)
Kennedy, R. (1785–1810)
Hutchins, E. (1785–1815)
Bishop, D. (*c*. 1788–1849)
Coulter, T. (1793–1843)
Reilly, J. (*c*. 1793–1876)
Dowden, R. (1794–1861) *Cork*
Hyndman, G. C. (1796–1867)
Hincks, H. (1798–1871)
Macreight, D. C. (1799–1857)
Niven, N. (1799–1879) *Dublin*
Sealy, J. (–before 1834) *Cork*
White, J. (–1837)
Nuttall, J. (–1849/50)
Ball, A. E. (–1872)
Orr, D. (–1892)

O'Connor, G. M. (–1897)
Ball, R. (1802–1857)
Andrews, W. (1802–1880)
Moore, D. (1807–1879)
Harvey, W. H. (1811–1866)
Kane, K. S. (1811–1886)
McCosh, J. (1811–1894) *Belfast*
Oulton, *Rev*. R. (1812–1880)
Dickie, G. (1812–1882) *Ulster*
Lynam, J. (1812–1885)
Allman, G. J. (1812–1898)
McCalla, W. (*c*. 1814–1849)
Bain, J. (1815–1903)
Gage, C. (1816–1892) *Rathlin Island*
Alexander, W. T. (1818–1872)
Baily, W. H. (1819–1888)
Clinton, P. (*fl*. 1820s)
Garrett, J. R. (1820–1855)
Lefroy, H. (1820–1908)
O'Mahoney, *Rev*. T. (1823–1879)
Robinson, *Rev*. G. (*c*. 1824–1893)
Chandlee, T. (1824–1907) *Kildare*
Stewart, S. A. (1826–1910) *N.E. Ireland*
Kinahan, J. R. (1828–1863)
Carroll, I. (1828–1880) *Cork*
Wright, T. (*c*. 1828–1889) *Cork*
Kinahan, G. H. (1829–1908)
Mateer, W. (*fl*. 1830s–1840s)
Whitla, F. (*fl*. 1830s–1850s)
Wynne, J. (*fl*. 1830s–1860s)
Grainger, *Rev*. J. (1830–1891) *Belfast*
Archer, W. (1830–1897)
Phillips, W. H. (1830–1923)
Foot, F. J. (*c*. 1831–1867)
Levinge, H. C. (*c*. 1831–1896)
Kelsall, E. J. (1832–1897)
Aitchison, J. E. T. (1836–1898)
Brenan, *Rev*. S. A. (1837–1908)
Clayton-Browne, R. (1838–1906)
Davies, J. H. (1838–1909)
Lett, *Rev*. H. W. (1838–1920)
Power, T. (*fl*. 1840s)
Sanders, G. (*fl*. 1840s–1860s)
Tate, R. (1840–1901) *Belfast*
White-Spunner, L. H. D. (1840–1911)
Maffett, I. (1842–1907)
McNab, W. R. (1844–1889)
Hart, H. C. (1847–1908)
O'Brien, R. D. (1847–1917) *Limerick*
Barrington, R. M. (1849–1915)
McArdle, D. (1849–1934)
Colgan, N. (1851–1919)
Carrothers, N. (1852–1930)
Cosgrave, E. M. (1853–1925)
Shanks, J. (1854–1912) *Down*
Moore, *Sir* F. W. (1857–1950)
Waddell, *Rev*. C. H. (1858–1919)
Foster, N. H. (1858–1927)
Bullock-Webster, *Rev*. G. R. (1858–1934)
Scully, R. W. (1858–1935)
Corry, T. H. (1859–1883)
Moffat, C. B. (1859–1945)
Donaldson, G. (*fl*. 1860s)
Douglas, J. (*fl*. 1860s) *Kildare*
Tomlinson, W. J. C. (1863–1921) *Belfast*
Henry, J. (*c*. 1863–1936)
Johnson, T. (1863–1954)
McWeeney, E. J. (1864–1925) *Dublin*
Knowles, M. C. (1864–1933)
Praeger, R. L. (1865–1953)
Phillips, R. A. (1866–1945)
Delap, M. J. (1866–1953)
Bennett, S. A. (1868–1934)
Bradshaw, D. B. (1869–1944)
Dixon, H. H. (1869–1953)
Barrett-Hamilton, G. E. H. (1871–1914)
Pethybridge, G. H. (1871–1948)
Adams, J. (1872–1950)
Rea, M. W. (1875–)
Thomson, J. S. (*c*. 1876–1972)
Chase, C. D. (1878–1965)
Palmer, J. A. J. (1883–1951)
Stelfox, A. W. (1883–1972)
Megaw, *Rev*. W. R. (1885–1953)
Brunker, J. P. (1885–1970) *Wicklow*
McKay, R. (1889–1964)
O'Connor, P. (1889–1969)
Glascott, L. S. (*fl*. 1890s)
O'Kelly, P. B. (*fl*. 1890s–1930s)
Lynn, M. J. (1891–)
O'Donovan, J. E. (1898–1966) *Cork*
Redmond, D. (–1905)
Allin, *Rev*. T. (–*c*. 1909)
More, F. M. (–1909)
Leebody, M. I. (–1911)
Vowell, R. P. (–1911)
Smith, *Rev*. W. S. (–1912) *Antrim*
Sayers, W. J. (–1959)
Moon, J. M. (1901–1960)

CHANNEL ISLANDS
Lambert, J. (1619–1683) *Guernsey*
Hatton, C. (1635–*c*. 1705) *Guernsey*
Clerk, *Mr*. (*fl*. 1730s–1740s) *Jersey*
Gosselin, J. (1739–1813) *Guernsey*
Finlay, J. (*c*. 1760–1802) *Guernsey*
McGrigor, *Sir* J. (1771–1858) *Jersey*
Macculloch, J. (1773–1835)
La Gasca y Segura, M. (1776–1839) *Jersey*
Mansell, C. R. (1781–1841) *Guernsey*
Graham, R. (1786–1845) *Jersey*
Smith, H. (1786–1868) *Guernsey*
Salwey, *Rev*. T. (1791–1877) *Guernsey*
Hoskins, S. E. (1799–1888) *Guernsey*
Dickson, J. (–1874) *Jersey*
Bull, M. M. (–1879) *Sark*
Lingwood, R. M. (–1887)
Christy, W. (*c*. 1807–1839)
Clarke, L. L. (*c*. 1812–1883) *Guernsey*
Wolsey, G. (*c*. 1814–1870) *Guernsey*
Piquet, J. (1825–1912)
Larbalestier, C. du B. (1838–1911)
Marquand, E. D. (1848–1918) *Guernsey*
Hanbury, F. A. (*fl*. 1860s)
Dupuy, E. (*fl*. 1860s–1900s) *Guernsey*
Lester-Garland, L. V. (1860–1944) *Jersey*
Proudlock, R. L. (1862–1948)
Warry, W. A. (*c*. 1874–1958) *Guernsey*
Arsene, L. (1875–1959)
Guille, M. E. (–1903) *Guernsey*
Rees, K. (–1931) *Guernsey*
McCrea, R. (–1949) *Guernsey*

BRITISH SOMALILAND *see* **SOMALI REPUBLIC**

BROMELIACEAE
Baker, J. G. (1834–1920)

BRYOLOGY (*see also* **JUNGERMANNIA**)
Buddle, *Rev*. A. (*c*. 1660–1715)

Dillenius, J. J. (1684–1747)
Manningham, *Rev*. T. (1684–1750)
Scott, R. (1757–1808)
Stokes, W. (1763–1845)
Gray, S. F. (1766–1828)
Francis, *Rev*. R. B. (*c*. 1768–1850)
Eagle, F. K. (1769–1856)
Milne, J. (1776–1851)
Oglander, J. (*c*. 1778–1825)
Hobson, E. (1782–1830)
Hutchins, E. (1785–1815)
Bowman, J. E. (1785–1841)
Hooker, *Sir* W. J. (1785–1865)
Wilson, W. (1799–1871)
Stewart, J. (–1820)
Taylor, T. (–1848)
Lyle, T. (–1859)
Black, A. O. (–*c*. 1864)
Daniel, *Rev*. R. (–1864)
Edmond, J. W. (–1875)
Anderson, S. (–1878)
Orr, D. (–1892)
Weir, J. (–1898)
Howitt, G. (1800–1873)
Gray, J. E. (1800–1875)
Palgrave, T. (1804–1891)
Moore, D. (1807–1879)
Gardiner, W. (1808–1852)
Curnow, W. (*c*. 1809–1887)
Valentine, W. (*fl*. 1810s–1884)
Unwin, W. C. (1811–1887)
Howie, C. (1811–1899)
Gardner, G. (1812–1849)
Cruickshanks, J. (*c*. 1813–1847)
Hodgson, E. (1814–1877)
Barnes, J. M. (1814–1890)
Stark, R. M. (1815–1873)
Sansom, T. (1816/7–1872)
Spruce, R. (1817–1893)
Alexander, W. T. (1818–1872)
Gray, P. (1818–1899)
Mitten, W. (1819–1906)
Holl, H. B. (1820–1886)
English, J. L. (1820–1888)
Marrat, F. P. (1820–1904)
Greenwood, A. (1821–1862)
Hicks, J. B. (1823–1897)
Brown, R. (*c*. 1824–1906)
Braithwaite, R. (1824–1917)
Wood, *Rev*. H. H. (1825–1882)
Barnard, A. M. (1825–1911)
Carrington, B. (1827–1893)
Leipner, A. (1827–1894)
Fry, *Sir* E. (1827–1918)
Slater, M. B. (*c*. 1829–1918)
Bagnall, J. E. (1830–1918)
Whitehead, J. (1833–1896)
Stirton, J. (1833–1917)
Davies, G. (1834–1892)
Fergusson, *Rev*. J. (1834–1907)
Smith, C. P. (1835–1892)
Holmes, G. (1835–1910)
Stirling, *Sir* J. (1836–1916)
Sadler, J. (1837–1882)
Boswell, H. (1837–1897)
Hobkirk, C. C. P. (1837–1902)
Clayton-Browne, R. (1838–1906)
Barker, T. (1838–1907)
Davies, J. H. (1838–1909)
Lett, *Rev*. H. W. (1838–1920)
Cash, J. (1839–1909)
Stabler, G. (1839–1910)
Bastow, R. A. (1839–1920)
Plues, M. (*c*. 1840–*c*. 1903)
Renton, R. (*c*. 1840–*c*.1915)
Hunt, G. E. (*c*. 1841–1873)
Hamilton, W. P. (*c*. 1842–1910)
Holmes, E. M. (1843–1930)
Wright, F. (1845–1924)
Cleminshaw, E. (1849–1922)
Pearson, W. H. (1849–1923)
McArdle, D. (1849–1934)
Atwood, M. M. (*fl*. 1850s–1860s)
Waterfall, W. B. (1850–1915)
Whitelegge, T. (1850–1927)
Tindall, I. M. (1850–1928)
Wilson, J. C. (1850/1–1925)
Russell, T. H. (1851–1913)
Barnes, R. (1851–1918)
Holt, G. A. (1852–1921)
Bellerby, W. (1852–1936)
Jameson, H. G. (1852–1939)
Rodway, L. (1853–1936)
Ingham, W. (1854–1923)
Benson, R. de G. (1856–1904)
Macvicar, S. M. (1857–1932)
Waddell, *Rev*. C. H. (1858–1919)
Smitham, R. W. (1858–1928)
Meldrum, R. H. (1858–1933)
Sutton, A. (1859–1931)
Beach, H. (*fl*. 1860s)
Craddock, *Dr*. (*fl*. 1860s)
Marshall, J. J. (1860–1934)
Tetley, W. N. (1861–1928)
Jones, D. A. (1861–1936)
Dixon, H. N. (1861–1944)
Vaizey, J. R. (1862–1889)
Wheldon, J. A. (1862–1924)
Binstead, *Rev*. C. H. (*c*. 1862–1941)
Knight, H. H. (1862–1944)
Wilson, A. (1862–1949)
Hallowell, E. (1865–1936)
Burrell, W. H. (1865–1945)
Nicholson, W. E. (1866–1945)
Monington, H. W. (1867–1924)
Lee, J. R. (1868–1959)
Bradshaw, D. B. (1869–1944)
Duncan, J. B. (1869–1953)
Cambridge, O. P. (*fl*. 1870s)
Sowden, H. (1870–1936)
Sherrin, W. R. (1871–1955)
Murray, J. (1872–1942)
Brinkman, A. H. (1873–)
Else, J. (1874–1955)
Cheetham, C. A. (1875–1954)
Cavers, F. (1876–1936)
Thompson, A. (1876–1959)
Richards, E. A. (1880–1927)
Rilstone, F. (1881–1953)
Walsh, H. (1881–1962)
Trotter, L. B. C. (1882–1964)
Rhodes, *Rev*. P. G. M. (1885–1934)
Megaw, *Rev*. W. R. (1885–1953)
Milsom, F. E. (1886–1945)
Wyard, S. (*c*. 1886–1946)
Peterken, J. H. G. (1893–1973)
Sowter, F. A. (1899–1972)
Capron, E. (–1907)
Vowell, R. P. (–1911)
Cocks, L. J. (–1921)

Grindley, E. I. (–1948)
Beesley, H. (*fl.* 1900s)
Warburg, E. F. (1908–1966)
Saunders, E. (1912–1966)
Laflin, T. (1914–1972)

BULGARIA
Ball, C. F. (1879–1915)
Doncaster, E. D. (–*c.* 1950)

BURMA
Bulkley, E. (*c.* 1651–1714)
Hancock, W. (*fl.* 1720s)
Mason, *Rev.* F. (1799–1874)
Bulger, G. E. (–1885)
Griffith, W. (1810–1845)
Dalzell, N. A. (1817–1878)
Carey, *Rev.* F. (*fl.* 1820s)
Lobb, T. (1820–1894)
Benson, R. (1822–1894)
Parish, *Rev.* C. S. P. (1822–1897)
Blinkworth, R. (*fl.* 1830s)
Kurz, W. S. (*c.* 1833–1878)
Collett, *Sir* H. (1836–1901)
King, *Sir* G. (1840–1909)
Schlich, *Sir* W. (1840–1925)
Boxall, W. (1844–1910)
Scott, R. (*fl.* 1850s–1860s)
Hildebrand, A. H. (1852–1918)
Nisbet, J. (1853–1914)
Burke, D. (1854–1897)
Lace, J. H. (1857–1918)
Bourne, *Sir* A. G. (1859–1940)
Findlay, J. (*fl.* 1860s)
Moore, R. (1860–1899)
Proudlock, R. L. (1862–1948)
Rogers, C. G. (1864–1937)
Gage, A. T. (1871–1945)
Rodger, *Sir* A. (*c.* 1875–1950)
Cubitt, G. E. S. (*c.* 1875–1966)
Toppin, S. M. (1878–1917)
Farrer, R. J. (1880–1920)
Parker, R. N. (1884–1958)
Kingdon Ward, F. (1885–1958)
Barrington, A. H. M. (1886–1932)
Candler, E. (*fl.* 1890s)
Parkinson, C. E. (1890–1945)
Cooper, R. E. (1890–1962)
Cowan, J. M. (1892–1960)
Smales, C. B. (–1927)
Swinhoe, R. C. G. (–1927)

CACTI
Scheer, F. (*c.* 1792–1868)
Hitchin, T. (*fl.* 1810s–1830s)
Shurly, E. W. (1888–1963)
Higgins, V. (1892–1968)

CAMBODIA
Smith, E. (*c.* 1893–1930)

CAMELLIA
Curtis, S. (1779–1860)
Booth, W. B. (*c.* 1804–1874)
Chandler, A. (1804–1896)
Turner, C. H. (*fl.* 1810s–1820s)

CAMEROONS
Thomson, G. (1819–1878)
Johnston, *Sir* H. H. (1858–1927)
Kingsley, M. H. (1862–1900)
Dundas, J. (1907–1966)

CAMPANULA
Crook, H. C. (*c.* 1882–*c.* 1974)

CANADA
Hay, W. (*fl.* 1690s) *Newfoundland*
Tilden, R. (*fl.* 1700s)
Markwick, W. (1739–1813)
Masson, F. (1741–1805)
Banks, *Sir* J. (1743–1820) *Newfoundland*
Cochrane, *Sir* A. F. I. (1758–1832) *Newfoundland*
Hutchins, T. (*fl.* 1770s)
Pursh, F. T. (1774–1820)
Ramsay, C. (1786–1839)
McLeod, J. (1788–1849)
Hale, W. (*fl.* 1790s)
Drummond, T. (*c.* 1790–1835)
Goldie, J. (1793–1886)
Hincks, *Rev.* W. (1794–1871)
Cormack, W. E. (1796–1868)
Douglas, D. (1799–1834)
Banks, P. (–1838)
Wallace, R. (–1838)
Traill, C. P. (1802–1899)
Denison, *Sir* W. T. (1804–1871)
Gairdner, M. (1809–1837)
Prior, R. C. A. (1809–1902)
Edwards, J. (*fl.* 1810s–1820s)
Hepburn, J. E. (1811–1869)
Cowdry, T. (1812–)
Tolmie, W. F. (1812–1886)
Leitch, W. (1814–1864)
Lyall, D. (1817–1895)
Haviland, A. E. H. (1818–)
Maclagan, P. W. (1818–1892)
Fisher, A. (*fl.* 1820s)
Blair, T. (*fl.* 1820s–1830s)
Dawson, *Sir* J. W. (1820–1899)
Jeffrey, J. (1826–1854)
Lawson, G. (1827–1895)
Brenton, M. E. (*fl.* 1830s)
Macoun, J. (1831–1920)
Whiteaves, J. F. (1835–1909)
Saunders, W. (1836–1914)
Brown, R. (1842–1895)
MacGregor, *Sir* W. (1846–1919)
Cowdry, N. H. (1849–1925)
Waghorne, *Rev.* A. C. (1851–1900)
Fletcher, J. (1852–1908)
Christy, R. M. (1861–1928)
Macoun, J. M. (1862–1920)
Inglis, R. A. (1869–)
Adams, J. (1872–1950)
Brinkman, A. H. (1873–)
Criddle, N. (1875–1933)
Lewis, F. J. (1875–1955)
Davidson, J. (1878–1970)
Preston, I. (1881–1965)
Bisby, G. R. (1889–1958)
Groves, E. M. (1897–1956)
Huskins, C. L. (*c.* 1898–1953)
Henderson, A. H. (–1921)

CANARY ISLANDS
Masson, F. (1741–1805)
Webb, P. B. (1793–1854)
Lowe, *Rev.* R. T. (1802–1874)
Collett, *Sir* H. (1836–1901)
Murray, *Rev.* R. P. (1842–1908)
Johnston, H. H. (1856–1939)
Lister, J. J. (1857–1927)

Brooks, C. J. (1875–*c.* 1953)
Gilbert-Carter, H. (1884–1969)

CAPE VERDE ISLANDS
Lowe, *Rev.* R. T. (1802–1874)

CAREX
Goodenough, *Rev.* S. (1743–1827)
Boott, F. (1792–1863)
Carey, J. (1797–1880)
Priestley, *Sir* W. O. (1829–1900)
Peers, J. (1838–1902)
Ewing, P. (1849–1913)
Crawford, F. C. (1851–1908)
Still, A. L. (1872–1944)
Nelmes, E. (1895–1959)

CARNATION *see* **DIANTHUS**

CASTILLOA
Cross, R. M. (1836–1911)

CELEBES *see* **INDONESIA**

CENTAUREA
Marsden-Jones, E. M. (1887–1960)
Turrill, W. B. (1890–1961)

CEYLON *see* **SRI LANKA**

CHANNEL ISLANDS *see* **BRITISH ISLES**

CHAROPHYTA
Blow, T. B. (1854–1941)
Groves, H. (1855–1912)
Groves, J. (1858–1933)
Bullock-Webster, *Rev.* G. R. (1858–1934)
Allen, G. O. (1883–1963)

CHILE
Handisyd, G. (*fl.* 1690s)
Cuming, H. (1791–1865)
Gillies, J. (1792–1834)
Collie, A. (1793–1835)
Macrae, J. (–1830)
Gourlie, R. (–1832)
Mathews, A. (–1841)
Anderson, J. (–1847)
Place, F. (*fl.* 1800s–1820s)
Bridges, T. (1807–1865)
Caldcleugh, A. (*fl.* 1820s–1858)
Cruckshanks, A. (*fl.* 1820s–1850s)
King, T. (1834–1896)
Davy, G. T. (*fl.* 1840s)
Middleton, R. M. (1846–1909)
Downton, G. (*fl.* 1870s)
Gosse, P. (1879–1959)
Balfour-Gourlay, W. (*c.* 1879–1966)
Elliott, C. (1881–1969)
Comber, H. F. (1897–1969)

CHINA (*see also* **HONG KONG, MACAO**)
Keir, W. (*fl.* 1690s)
Maidstone, N. (*fl.* 1690s–1720s)
Cunningham, J. (–*c.* 1709)
Lind, J. (1736–1812)
Staunton, *Sir* G. L. (1737–1801)
Blake, J. B. (1745–1773)
Robertson, J. (*fl.* 1760s–1770s)
Bradley, H. (*fl.* 1770s)
Reeves, J. (1774–1856)
Beale, T. (*c.* 1775–1842)
Main, J. (*c.* 1775–1846)
Abel, C. (1780–1826)
Ball, S. (1780–1874)
Ker, C. H. B. (*c.* 1785–1871)
Preston, *Sir* R. (*fl.* 1790s)
Stronach, D. (*fl.* 1790s)
Haxton, J. (*fl.* 1790s–1800s)
Evans, T. (*fl.* 1790s–1810s)
Parks, J. D. (*c.* 1792–1866)
Bowring, *Sir* J. (1792–1872)
Medhurst, W. H. (1796–1857)
Home, *Sir* J. E. (1798–1853)
Vachell, *Rev.* G. H. (1799–)
Kerr, W. (–1814)
Potts, J. (–1822)
Livingstone, J. (–1829)
Hooper, J. (–1830/1)
Tarrant, W. (–1872)
Mayers, W. F. (–1878)
Henry, C. (–1894)
Drummond, J. (*fl.* 1800s)
Reeves, J. R. (1804–1877)
Cantor, T. E. (1809–1854)
Alcock, *Sir* R. (1809–1897)
Poole, J. (*fl.* 1810s–1820s)
Turner, C. H. (*fl.* 1810s–1820s)
Fortune, R. (1812–1880)
Urquhart, F. G. (1813–*c.* 1890)
Daniell, W. F. (1818–1865)
Alexander, W. T. (1818–1872)
Millett, C. (*fl.* 1820s–1830s)
Bowring, J. C. (1821–1893)
Medhurst, W. H. (1822–1885)
Sarel, H. A. (*c.* 1825–1886)
Hance, H. F. (1827–1886)
Williamson, *Rev.* A. (1829–1890)
Sampson, G. T. (1831–1897)
Christy, T. (1832–1905)
Smith, F. P. (1833–1888)
Tate, G. R. (1835–1874)
Swinhoe, R. (1836–1877)
Henderson, G. (1836–1929)
McLachlan, R. (1837–1904)
Alabaster, *Sir* C. (1838–1898)
Forbes, F. B. (1839–1908)
Braine, C. J. (*fl.* 1840s–1850s)
Bradford, E. (*fl.* 1840s–1890s)
Watters, T. (*c.* 1840–1904)
Bowra, E. C. M. (1841–1874)
Mesney, W. (1842–1919)
Gill, W. J. (1843–1882)
Ford, C. (1844–1927)
Bullock, T. L. (1845–1915)
Everard, C. W. (1846–1890s)
Perry, W. W. (*c.* 1846–1894)
Hancock, W. (1847–1914)
Carles, W. R. (1848–1929)
Cowdry, N. H. (1849–1925)
Stronach, W. G. (*fl.* 1850s–1870s)
Davenport, A. (*fl.* 1850s–1880s)
Cooper, W. M. (*fl.* 1850s–1890s)
Gregory, W. (*fl.* 1850s–1890s)
Moule, *Rev.* G. E. (*fl.* 1850s–1890s)
Maries, C. (*c.* 1851–1902)
Hosie, *Sir* A. (1853–1925)
Bourne, F. S. A. (1854–)
Dent, H. C. (1855–1909)
Appleton, H. (*c.* 1855–1930)
Henry, A. (1857–1930)
Birnie, *Mr.* (*fl.* 1860s)
Dickson, W. (*fl.* 1860s)
Fagg, F. (*fl.* 1860s)

Hay, *Mr*. (*fl*. 1860s)
Jacob, E. (*fl*. 1860s)
Margary, A. R. (*fl*. 1860s)
Parry, F. (*fl*. 1860s)
McCarthy, *Rev*. J. (*fl*. 1860s–1870s)
Shearer, G. (*fl*. 1860s–1870s)
Bushey, S. W. (*fl*. 1860s–1890s)
Kopsch, H. C. T. (*fl*. 1860s–1890s)
Parker, E. H. (*fl*. 1860s–1890s)
Scott, B. C. G. (*fl*. 1860s–1890s)
Clemens, J. (1862–1932)
Matthew, C. G. (1862–1936)
Deasy, H. H. P. (1866–1947)
Dunn, S. T. (1868–1938)
Galbraith, *Miss* (*fl*. 1870s)
Quekett, J. F. (*fl*. 1870s)
Playfair, G. M. H. (*fl*. 1870s–1880s)
Ross, *Rev*. J. (*fl*. 1870s–1890s)
Dalziel, J. M. (1872–1948)
Forrest, G. (1873–1932)
Wilson, E. H. (1876–1930)
MacGregor, D. (1877–1933)
Anderson, G. C. (*fl*. 1880s)
Beazeley, M. (*fl*. 1880s)
Webster, *Rev*. J. (*fl*. 1880s)
Wenyon, *Rev*. C. (*fl*. 1880s–1890s)
James, *Sir* H. E. M. (*fl*. 1880s–1900s)
Batchelor, *Rev*. J. (*fl*. 1880s–1910s)
Pratt, A. E. (*fl*. 1880s–1910s)
Farrer, R. J. (1880–1920)
Purdom, W. (1880–1921)
Sowerby, A. de C. (1885–1954)
Kingdon Ward, F. (1885–1958)
Smith, E. (*c*. 1893–1930)

CHRISTMAS ISLAND
Lister, J. J. (1857–1927)
Andrews, C. W. (1866–1924)
MacLear, J. F. L. P. (*fl*. 1870s)

CHRYSANTHEMUM
Salter, J. (1798–1874)
Shoesmith, H. (–1867)
James, R. (1801–1871)
Cannell, H. (1833–1914)
Newton, J. (*c*. 1837–1907)
Prickett, G. (*c*. 1844–1941)
Payne, C. H. (*c*. 1854–1925)
Jones, H. J. (*c*. 1856–1928)
Shea, C. E. (*fl*. 1860s–1900s)
Vinten, A. G. (*c*. 1871–1962)
Crane, D. B. (–1938)
Sibsbury, M. (*fl*. 1900s)

CINCHONA
Kentish, R. (1731–1792)
Lambert, A. B. (1761–1842)
McIvor, W. G. (–1876)
Howard, J. E. (1807–1883)
Ledger, C. (1818–1905)
Weddell, H. A. (1819–1877)
Markham, *Sir* C. R. (1830–1916)
Cross, R. M. (1836–1911)
King, *Sir* G. (1840–1909)

CISTUS
Warburg, *Sir* O. E. (1876–1937)

CITRUS
Carew, *Sir* F. (*fl*. 1570s–1610s)

COCOS KEELING ISLANDS
Darwin, C. R. (1809–1882)

COFFEE
Ellis, J. (*c*. 1705–1776)
Haarer, A. E. (–1970)

COLCHICUM
Bowles, E. A. (1865–1954)

COLOMBIA
Cuming, H. (1791–1865)
Purdie, W. (*c*. 1817–1857)
Bowman, D. (1838–1868)
Burke, D. (1854–1897)
Blunt, H. (*fl*. 1860s)
Sprague, T. A. (1877–1958)
Dawe, M. T. (1880–1943)
Baker, R.E.D. (1908–1954)

COMPOSITAE (*see also* **ASTER, CHRYSANTHEMUM, etc.**)
Clarke, C. B. (1832–1906)
Small, J. (1889–1955)

CONGO
Smith, C. (1785–1816)
Lockhart, D. (–1846)
Johnston, *Sir* H. H. (1858–1927)
Lynes, H. (1874–1942)
Rogers, *Rev*. F. A. (1876–1944)

CONIFERAE (*see also* **TREES**)
Lambert, A. B. (1761–1842)
Gordon, G. (1806–1879)
Douglas, R. (1813–1897)
Kent, A. H. (1828–1913)
Masters, M. T. (1833–1907)
Coltman-Rogers, C. (1854–1929)
Clinton-Baker, H. W. (1865–1935)
Cooper, N. L. (*c*. 1865–1936)
Dallimore, W. (1871–1959)
Jackson, A. B. (1876–1947)
Jay, B. A. (1911–1961)

CORSICA
Elliott, C. (1881–1969)

COSTA RICA
Lankester, C. H. (1879–1969)

COTTON *see* **GOSSYPIUM**

CRASSULA
Higgins, V. (1892–1968)

CRETE
Hawkins, J. (*c*. 1761–1841)

CRINUM
Bosanquet, L. P. (1865–1930)

CROCUS
Maw, G. (1832–1912)
Bowles, E. A. (1865–1954)

CYCADACEAE
Yates, *Rev*. J. (1789–1871)
Rattray, G. (1872–1941)

CYPERACEAE
Clarke, C. B. (1832–1906)
Bennett, A. (1843–1929)
Napper, D. M. (1930–1972)

CYPRUS
Keill, J. (1673–1719)
Hutchins, *Sir* D. E. (1850–1920)
Reid, C. (1853–1916)
Rogers, *Rev.* F. A. (1876–1944)
Merton, L. F. H. (1919–1974)

CYTOLOGY
Brebner, G. (*c.* 1855–1904)
Bateson, W. (1861–1926)
Farmer, *Sir* J. B. (1865–1944)
Saunders, E. R. (1865–1945)
Gregory, R. P. (1879–1918)
Gates, R. R. (1882–1962)
Hunter, H. (*c.* 1883–1959)
Jenkin, T. J. (1885–1965)
Caffrey, M. (*c.* 1889–1959)
Blackburn, K. B. (*c.* 1892–1968)
Newton, W. C. F. (1895–1927)
Knight, R. L. (–1972)
Barber, H. N. (1914–1971)
Haskell, G. M. L. (1920–1967)

DAHLIAS
Girdlestone, T. W. (–1899)
Buonaiuti, S. (*fl.* 1800s–1820s)
Fellowes, *Rev.* C. (*c.* 1812–1896)
Hogg, R. (1818–1897)
Bartholomew, A. C. (*c.* 1846–1940)
Barwise, J. F. (*c.* 1874–1965)
Tranter, J. R. (–1911)

DAHOMEY
Burton, *Sir* R. F. (1821–1890)

DATURA
Belling, J. (1866–1933)

DESMIDIACEAE
Ralfs, J. (1807–1890)
Roy, J. (1826–1893)
Bisset, J. P. (1839–1906)
Bisset, J. (1843–1911)
West, W. (1848–1914)
Turner, C. (1864–1926)

DIANTHUS
Hogg, T. (1771–1841)
Ely, B. (1779–1843)
Dodwell, E. S. (1819–1893)
Pigott, L. (*fl.* 1820s)
Lakin, J. (1828–1895)
Ibbett, T. (*fl.* 1830s–1840s)
Douglas, J. (1837–1911)
Brotherston, R. P. (1848–1923)
Williams, F. N. (1862–1923)
Allwood, M. C. W. (*c.* 1879–1958)
Smith, M. R. (–1908)
Rawdon, P. (–1954)

DIATOMACEAE
Johnson, C. (1782–1866)
Wigham, R. (1785–1855)
Brightwell, T. (1787–1868)
Gregory, W. (1803–1858)
Smith, *Rev.* W. (1808–1857)
Hardman, L. (1808–1896)
Carter, H. J. (1813–1895)
Atthey, T. (1814–1880)
Tulk, J. A. (1814–1896)
O'Meara, *Rev.* E. (*c.* 1815–1880)
Wallich, G. C. (1815–1899)
Harkness, R. (1816–1878)
Rylands, T. G. (1818–1900)
Roper, F. C. S. (1819–1896)
Farquharson, R. F. O. (1823–1890)
Norman, G. (1824–1882)
Sidebotham, J. (1824–1885)
Deby, J. (1826–1895)
Wonfor, T. W. (*c.* 1827–1878)
Kitton, F. (1827–1895)
Carruthers, W. (1830–1922)
Stolterfoth, H. (1836–1907)
Comber, T. R. (1837–1902)
Gill, C. H. (1841–1894)
Barratt, T. J. (1841–1914)
Baxter, W. E. (1844–1920)
Stiles, M. H. (1846–1935)
Brown, N. E. (1849–1934)
Donkins, A. S. (*fl.* 1850s–1870s)
Staunton, J. (*fl.* 1850s–1870s)
Taylor, F. B. (1851–1931)
Nelson, E. M. (1851–1938)
Philip, R. H. (1852–1912)
Payne, F. W. (1852–1927)
Wilson, J. (1853–1934)
Davidson, *Rev.* G. (*c.* 1854–1901)
Rattray, J. (1858–1900)
Sturt, G. (1860–1947)
Long, J. A. (1863–1944)
Bryan, G. H. (1864–1928)
Allen, E. J. (1866–1942)
Adams, F. (1867–1938)
Bates, G. F. (*c.* 1868–1933)
Mills, F. W. (1868–1949)
Swatman, C. C. (1884–1958)
Wise, F. C. (1884–1962)
Falkner, H. J. (1894–1951)
Upton, C. (–1927)

DIOSCOREA
Prain, *Sir* D. (1857–1944)
Burkill, I. H. (1870–1965)

DOMINICA *see* **WEST INDIES**

DROSERA
Morrison, A. (1849–1913)

EBENACEAE
Hiern, W. P. (1839–1925)

ECOLOGY
Woodhead, T. W. (1863–1940)
Oliver, F. W. (1864–1951)
Yapp, R. H. (1871–1929)
Tansley, *Sir* A. G. (1871–1955)
Evans, E. P. (1882–1959)
Pearsall, W. H. (1891–1964)
Bews, J. W. (1894–1938)
Bracher, R. (1894–1941)

ECUADOR
Jameson, W. (1796–1873)
Hall, F. (–1834)
Hartweg, C. T. (1812–1871)
Cross, R. M. (1836–1911)

EGYPT
Wilkinson, *Sir* J. G. (1797–1875)
Madden, E. (1805–1856)
Lloyd, G. (1815–1843)
Calvert, H. H. (*c.* 1816–1882)
Lord, J. K. (1818–1872)
Traill, J. (*fl.* 1820s–1853)
Hurst, H. A. (*c.* 1825–1882)
Blomfield, *Sir* R. M. (1835–1921)
Palmer, E. H. (1840–1882)
Johnston, H. H. (1856–1939)
Newberry, P. E. (1869–1949)
Meinertzhagen, R. (1878–1967)
Fish, D. S. (*c.* 1881–1912)
Balls, W. L. (1882–1960)
Bailey, M. A. (*c.* 1890–1939)
Walsingham, F. G. (*c.* 1890–1964)
Jones, G. H. (–1945)
Brown, T. W. (–*c.* 1951)
Chapman, H. L. R. (–1968)

ENGLAND *see* BRITISH ISLES

EPILOBIUM
Ash, G. M. (1900–1959)

EPIPACTIS
Young, D. P. (1917–1972)

ERICACEAE
Bauer, F. A. (1785–1840)
Sinclair, G. (1786–1834)
Andrews, H. C. (*fl.* 1790s–1830s)

ETHIOPIA
Bruce, J. (1730–1794)
Pearce, N. (1779–1820)
Salt, H. (1780–1827)
Wellby, M. S. (1866–1900)
Drake-Brockman, R. E. (1875–)
Erskine, E. N. (1885–1962)
Mooney, H. F. (1897–1964)

EUCALYPTUS
Bosisto, J. (1827–1898)
Howitt, A. W. (1830–1908)
Tait, W. C. (1844–1928)
Deane, H. (1847–1924)
Smith, H. G. (1852–1924)
Baker, R. T. (1854–1941)
Maiden, J. H. (1859–1925)

EUPHORBIACEAE
Brown, N. E. (1849–1934)

EUPHRASIA
Pugsley, H. W. (1868–1947)
Lumb, D. (1871–1951)

FALKLAND ISLANDS
Hussey, B. (*fl.* 1760s)
Jameson, R. (–1893)
Darwin, C. R. (1809–1882)
Hooker, *Sir* J. D. (1817–1911)
Edmondston, T. (1825–1846)
Vallentin, R. (1859–1934)
Vallentin, E. F. (1873–1924)
Elliott, C. (1881–1969)
Davis, W. (1899–1968)
Jesson, E. M. (1889–1956)

FAROE ISLANDS
Trevelyan, *Sir* W. C. (1797–1879)

FERNANDO DE NORONHA
Ridley, H. N. (1855–1956)
Ramage, G. A. (1864–1933)

FERNANDO PO
Boultbee, *Mr.* (*fl.* 1830s)
Hewan, D. A. (*fl.* 1860s)

FERNS
Bolton, J. (*fl.* 1750s–1790s)
Allcard, J. (*c.* 1777–1855)
Hooker, *Sir* W. J. (1785–1865)
Sim, R. (1791–1878)
Johnson, C. (1791–1880)
Norris, *Sir* W. (1793–1859)
Greville, R. K. (1794–1866)
Smith, J. (1798–1888)
Edwards, *Rev.* Z. J. (1799–1880)
Riley, J. (–1846)
Barkly, E. H. (–1857)
Duff, M. C. (–1885)
Carbonell, W. C. (–1887)
Howe, W. E. (–1891)
Francis, G. W. (1800–1865)
Bosanquet, *Rev.* E. (*c.* 1800–1872)
Newman, E. (1801–1876)
Stansfield, A. (1802–1880)
Bennett, W. (1804–1873)
Elworthy, C. (1805–)
Leith, A. H. (1807–1875)
Brackenridge, W. D. (1810–1893)
Hutchinson, P. O. (1810–1897)
Cooke, E. W. (1811–1880)
Linton, W. J. (1812–1897)
Rawson, *Sir* R. W. (1812–1899)
Ibbotson, H. (1814–1886)
Fox, E. F. (1814–1891)
Wollaston, G. B. (1814–1899)
Lyell, K. M. (1817–1915)
Ogilvie-Forbes, G. (1820–1886)
Moore, T. (1821–1887)
Wall, G. (*c.* 1821–1894)
M'Ken, M. J. (1823–1872)
Hunter, *Rev.* R. (1823–1897)
Williams, B. S. (1824–1890)
Sowerby, J. E. (1825–1870)
Sarel, H. A. (*c.* 1825–1886)
Lankester, P. (1825–1900)
Lowe, E. J. (1825–1900)
Field, H. C. (1825–1911)
Stansfield, T. (*c.* 1826–1879)
Jones, A. M. (1826–1889)
Lawson, G. (1827–1895)
Bailey, F. M. (1827–1915)
Kinahan, J. R. (1828–1863)
Allchin, *Sir* W. H. (*c.* 1828–1891)
Gissing, T. W. (1829–1870)
Morley, J. (1829–1886)
Tyerman, J. S. (*c.* 1830–1889)
Fraser, P. N. (1830–1905)
Beddome, R. H. (1830–1911)
Phillips, W. H. (1830–1923)
Levinge, H. C. (*c.* 1831–1896)
Winter, J. N. (*c.* 1831–1907)
Boyd, W. B. (1831–1918)
Hope, C. W. W. (1832–1904)
Baker, J. G. (1834–1920)
Gott, J. (*c.* 1834–1931)
Yates, L. G. (1837–1909)
D'Urban, W. S. M. (1837–1934)

Barkly, A. M. (1838–1932)
Williamson, J. (1839–1884)
Cowburn, T. B. (1839–1892)
Beckett, T. W. N. (1839–1906)
Bennett, T. E. (*fl.* 1840s)
Houlston, J. (*fl.* 1840s–1850s)
Forster, W. (*fl.* 1840s–1860s)
Courtauld, S. (1840–1899)
Plues, M. (*c.* 1840–*c.* 1903)
Bell, W. (*c.* 1840–1920)
Franklen, *Sir* T. M. (1840–1928)
Mapplebeck, J. E. (1842–*c.* 1905)
Heath, F. G. (1843–1913)
Druery, C. T. (1843–1917)
Jenman, G. S. (1845–1902)
Farquharson, M. S. (1846–1912)
Wiper, J. (*c.* 1847–1930)
Bowman, J. H. (1847–1932)
Bolton, H. (*c.* 1847–1939)
Schneider, G. (1848–1917)
Thomson, G. M. (1848–1933)
Clowes, F. (*fl.* 1850s)
Gray, R. (*fl.* 1850s)
James, J. (*fl.* 1850s)
Thompson, A. (*fl.* 1850s)
Bellairs, N. M. S. (*fl.* 1850s–1890s)
Mawson, T. W. (*c.* 1850–1876)
Lynch, R. I. (1850–1924)
Smithies, J. J. (*c.* 1850–1931)
Stansfield, W. H. (1850–1934)
Hemsley, A. (1851–1917)
Thatcher, E. J. (*c.* 1853–1934)
Moore, *Rev.* H. K. (1853–1943)
Stansfield, F. W. (1854–1937)
Blow, T. B. (1854–1941)
Bower, F. O. (1855–1948)
Palmer, W. (1856–1921)
Stansfield, H. (*c.* 1856–1928)
Sheldon, J. J. (1858–1942)
Askew, W. F. (*c.* 1858–1949)
Cranfield, W. B. (1859–1948)
Monkman, C. (*fl.* 1860s)
Clapham, A. (*fl.* 1860s–1870s)
Buchanan, *Rev.* J. (*fl.* 1860s–1880s)
Patey, G. S. (*fl.* 1860s–1880s)
Ullyett, H. (*fl.* 1860s–1880s)
Wilson, J. A. (*fl.* 1860s–1914)
Matthew, C. G. (1862–1936)
Sinclair, J. A. (*c.* 1864–1940)
Hombersley, *Rev.* A. (1866–1941)
Whiteside, R. (*c.* 1866–1960)
Wilson, W. (*c.* 1867–*c.* 1948)
Cochran, J. (1867–1961)
Bolton, R. (*c.* 1869–1949)
Hodgson, J. K. (*fl.* 1870s)
Pulham, J. R. (1873–1957)
Tucker, J. W. (*c.* 1876–1951)
Blatter, E. (1877–1934)
Davies, W. T. (*c.* 1877–1951)
Dixon, J. D. (*c.* 1877–1960)
Greenfield, P. (1880–1970)
Browne, I. M. P. (1881–1947)
Platten, E. W. (*c.* 1881–1951)
Venning, F. E. (1882–1970)
Davie, R. C. (1887–1919)
Rowlands, S. P. (1887–1957)
Buchanan, W. C. (1887–1964)
Lovelady, J. (*fl.* 1890s–1948)
Elliott *Rev.* E. A. (1890–1960)
Hyde, H. A. (*c.* 1892–1973)
Smith, E. (*c.* 1893–1930)
Payne, L. G. (1893–1949)
Williams, S. (*c.* 1898–1965)
Dadds, J. (–*c.* 1904)
Moly, J. (–1910)
Bolton, T. (–1923)
Bury, L. (–1935)
Butler, F. (–1936)
Henwood, T. E. (–1937)
Lloyd, J. (–1942)
Leighton, D. (–1943)
Alston, A. H. G. (1902–1958)
Stansfield, T. (–*c.* 1943)
Evans, G. B. (1935–1966)

FIJI *see* **PACIFIC**

FLAX *see* **LINUM**

FLOWER PAINTERS *see* **ARTISTS**

FORESTRY *see* **TREES**

FORMOSA *see* **TAIWAN**

FOSSIL BOTANY *see* **PALAEOBOTANY**

FRAGARIA
Keens, M. (*c.* 1762–1835)

FRANCE
Hungerford, J. (*fl.* 1650s–1687)
Blaikie, T. (1750–1838)
Taylor, C. (1762–1818)
Moggridge, M. (1803–1882)
Mill, J. S. (1806–1873)
Lloyd, J. (1810–1896)
Percy, J. (1817–1889)
Baker, *Rev.* T. (*fl.* 1830s)
Lyons, C. (*fl.* 1830s)
Cameron, D. (*fl.* 1830s–1850s)
Battersby, C. H. (1836–)
Moggridge, J. T. (1842–1874)
Casey, G. E. C. (1846–1912)
Raine, F. (1851–1919)

FRITILLARIA
Beck, C. (1884–1960)

FRUIT GROWING
Breedon, *Rev.* J. S. *Apples*
Knight, T. A. (1759–1838)
Braddick, J. (*c.* 1765–1828) *Apples*
Biggs, A. (1765–1848)
Cox, R. (*c.* 1776–1845) *Apples*
Hooker, W. (1779–1832)
Bland, M. (*fl.* 1800s) *Apples*
Greatorex, S. (1804–1871) *Apples*
Bull, H. G. (*c.* 1818–1885) *Apples, Pears*
Hogg, R. (1818–1897)
Ellison, *Rev.* C. C. (1835–1912) *Apples*
Bunyard, G. (1841–1919) *Apples*
Bradley, S. (*fl.* 1850s) *Apples*
Pickering, P. S. U. (1858–1920)
Bull, W. W. (*fl.* 1880s)
Hatton, *Sir* R. G. (1886–1965)
Taylor, H. V. (1887–1965) *Apples, Plums*

FUCHSIA
Bland, J. E. (*fl.* 1880s)
Lye, J. (–1906)

FUMARIA
Pugsley, H. W. (1868–1947)

FUNGI *see* **MYCOLOGY**

GALANTHUS
Bowles, E. A. (1865–1954)
Stern, *Sir* F. C. (1884–1967)
Allen, J. (–1906)

GALAPAGOS ISLANDS
Macrae, J. (–1830)
Darwin, C. R. (1809–1882)
Edmondston, T. (1825–1846)
Collenette, C. L. (1888–1959)

GAMBIA
Whithfield, T. (*fl.* 1840s)
Johnston, H. H. (1856–1939)
Dudgeon, G. C. (1867–1930)
Haydon, W. (*c.* 1871–1925)
Sampson, H. C. (1878–1953)
Brooks, A. J. (1881–)
Carter, *Sir* G. (*fl.* 1890s)
Lester, J. B. (*fl.* 1890s)

GARDENERS
Gardener, J. (*fl.* 1440s)
Chapman, J. (*fl.* 1510s–1530s)
Tradescant, J. (–1638)
Tradescant, J. (1608–1662)
Morgan, E. (*fl.* 1610s–1680s)
Meager, L. (*c.* 1620–)
Rose, J. (1621/2–1677)
Bobart, T. (*fl.* 1650s–1720s)
Atkinson, R. (*c.* 1656–1746)
Blake, S. (*fl.* 1660s)
Reid, J. (*fl.* 1680s)
Beaumont, G. (*fl.* 1680s–*c.* 1729)
Switzer, S. (1682–1745)
Ashley, R. (1682–1782)
Greening, T. (1684–1757)
Adams, J. (*fl.* 1690s)
Miller, P. (1691–1771)
Knowlton, T. (1691–1781)
Haverfield, J. (*c.* 1694–1784)
Aram, P. (–1735)
Kennedy, L. (–1743)
Greening, R. (–1758)
Greening, J. (–1770)
Kennedy, J. (–1790)
Fisher, W. (*fl.* 1700s–1743)
Miller, W. (*fl.* 1710s–1760s)
Rutter, J. (c. 1711–1772)
Carter, C. (*fl.* 1720s)
Whitmill, B. (*fl.* 1720s–1730s)
Giles, J. (*c.* 1725–1797)
Huntback, J. (*fl.* 1730s–1740s)
Dillman, J. (*fl.* 1730s–1760s)
Speechley, W. (*c.* 1733–1819)
Forsyth, W. (1737–1804)
Major, J. (*c.* 1737–1831)
Miller, J. (*fl.* 1740s)
Teesdale, R. (*c.* 1740–1804)
Jones, R. (*c.* 1740–1831)
Haverfield, J. (*c.* 1741–1820)
Williamson, C. (*c.* 1748–1835)
Maddox, J. (*c.* 1749–1828)
Hitt, T. (–*c.* 1760)
Greening, *Sir* H. T. (*fl.* 1750s–1809)
Hoy, T. (*c.* 1750–1822)
Blaikie, T. (1750–1838)
Sinclair, G. (*c.* 1750–1833)
Yarnall, R. K. (1752–1826)
Griffin, W. (*c.* 1752–1837)
Rogers, J. (1752–1842)
Knowlton, T. (1757–1837)
Beattie, W. (1758–1839)
Carter, D. (*fl.* 1760s)
Crofts, D. (*fl.* 1760s)
Taylor, A. (*fl.* 1760s)
Mawe, T. (*fl.* 1760s–1770s)
Haverfield, T. (*fl.* 1760s–1800s)
Anderson, W. (1760–1846)
Mitchell, J. (*c.* 1761–1838)
Biggs, A. (1765–1848)
Dean, W. (*c.* 1767–1881)
Dicks, J. (*fl.* 1770s)
Ellis, T. (*fl.* 1770s)
Job, W. (*fl.* 1770s)
Neale, A. (*fl.* 1770s)
Wilson, W. (*fl.* 1770s)
Meader, J. (*fl.* 1770s–1780s)
McPhail, J. (*fl.* 1770s–1800s)
Johnston, G. (1773–1835)
Niven, D. J. (*c.* 1774–1827)
Gorrie, A. (1777–1857)
Montgomery, D. (*c.* 1778–1857)
Browne, R. (*fl.* 1780s)
Kyle, T. (*fl.* 1780s)
Shaw, J. (1780s–1790s)
McNab, W. (1780–1848)
Archibald, J. (1784–1874)
Donn, W. (1787–1827)
Cameron, D. (*c.* 1787–1848)
Mills, G. (*c.* 1787–1871)
Knox, G. (*fl.* 1790s)
Menzies, R. (*fl.* 1790s)
Snow, S. (*c.* 1793–1869)
M'Intosh, C. (1794–1864)
Ingram, T. (*c.* 1796–1872)
Toward, A. (*c.* 1796–1881)
Brown, G. (1797–1874)
Williams, J. (1797–1891)
Thompson, R. (1798–1869)
Meehan, E. (*c.* 1798–1882)
Smith, J. (1798–1888)
Errington, R. (1799–1860)
Foulis, R. (1799–1877)
Graefer, J. (–1802)
Nicol, W. (–1811)
Cushing, J. (–1819/20)
Hoy, J. B. (–1843)
Edmonds, *Mr.* (–1861)
Austen, G. (–1876)
Fleming, J. (–1883)
Cliffe, P. (–1885)
Green, C. (–1886)
Vallance, G. D. (–1889)
George, E. (–1894)
Truelove, W. (–1894)
Fordyce, *Mr.* (*fl.* 1800s)
Morgan, W. (*fl.* 1800s)
Anderson-Henry, I. (1800–1884)
Drewett, J. (*c.* 1800–1885)
Nicolles, J. (*c.* 1802–1832)
Greenshields, J. (*c.* 1802–1888)
Paxton, *Sir* J. (1803–1865)
Weaver, T. (1803–1875)
Wighton, J. (1803–1878)
Smith, W. (*c.* 1804–1828)
Booth, W. B. (*c.* 1804–1874)
Turnbull, A. (1804–1886)
Frost, P. (1804–1887)
Elworthy, C. (1805–)

Halliday, J. (1806–)
Broome, S. (1806–1870)
Don, P. H. (1806–1876)
Barnes, J. (1806–1877)
Sterling, G. (*c.* 1806–1885)
Sargeson, W. (*c.* 1806–1886)
Bailey, T. (1806–1887)
Bond, G. (*c.* 1806–1892)
Bogue, G. (*c.* 1807–1893)
Fish, R. (1808–1873)
Tillery, W. (1808–1879)
Henderson, M. (1808–1892)
Buchanan, I. (1808–1893)
Medland, G. (1808–1894)
Dodds, W. (*c.* 1808–1900)
Sherare, J. (*c.* 1809–1834)
Fleming, G. (1809–1876)
Spencer, J. (1809–1881)
Forsyth, A. (*c.* 1809–1885)
Baldwin, T. (*fl.* 1810s)
MacDonald, J. (*fl.* 1810s)
Bearpark, E. (*fl.* 1810s–1820s)
Mean, J. (*fl.* 1810s–1820s)
Oldacre, I. (*fl.* 1810s–1820s)
Weeks, E. (*fl.* 1810s–1820s)
Maher, J. (*fl.* 1810s–1830s)
Wilmot, J. (*fl.* 1810s–1830s)
Forbes, A. (*fl.* 1810s–1860s)
Turner, H. (*c.* 1810–1876)
Cramb, A. (1810–1877)
Parsons, A. (1810–1880)
Macaulay, W. (1810–1900)
Caie, J. (1811–1879)
Edmonds, C. (1811–1880)
Robinson, J. (1811–1890)
Berry, E. (1812–1869)
Grieve, P. (1812–1895)
Carson, S. M. (*c.* 1814–1881)
Cox, J. (1814–1886)
Webster, J. (1814–1890)
Thomson, W. (1814–1895)
Mitchell, J. (1814–1904)
Hutchinson, W. (1815–)
Dale, J. (1815–1878)
Judd, D. (1815–1884)
Eyles, G. (1815–1887)
McLaren, J. (1815–1888)
Fowler, A. (1816–1887)
Carmichael, W. (*c.* 1816–1904)
Saul, M. (1817–1892)
Bowie, R. (1817–1901)
McElroy, J. F. (*c.* 1818–1887)
Blair, T. (1819–)
Williamson, A. (1819–1870)
Tillyard, G. B. (1819–1889)
Moult, W. (1819–1896)
Fairbairn, T. (*fl.* 1820s)
Stroud, T. B. (*fl.* 1820s)
Cooper, J. (*fl.* 1820s–1840s)
Denson, J. (*fl.* 1820s–1870s)
Anderson, A. (1820–)
Ingram, W. (1820–1894)
Batley, J. (*c.* 1820–1902)
Woolley, S. (*c.* 1821–1878)
Ingram, A. (1821–1881)
Doig, D. (1821–1886)
Smith, J. (1821–1888)
Farquhar, R. (1821–1895)
Gilbert, R. (1821–1895)
Cox, W. (1822–1883)
Miller, J. W. (*c.* 1822–1902)
Baines, T. (1823–1895)
Aughtie, R. (1823–1901)
Hill, W. (1824–1878)
Sage, G. (1824–1889)
Paterson, W. (*c.*1824–1896)
Thomson, D. (1824–1909)
Vair, J. (*c.* 1825–1887)
Smith, G. (1825–1899)
Ross, C. (*c.* 1825–1917)
Richards, J. (*c.* 1826–1877)
Ward, W. (*c.* 1826–1883)
Rust, J. (1826–1894)
Bennett, E. (*c.* 1826–1904)
George, J. (1826–1911)
Coleman, W. (1827–1908)
Laurence, J. W. (1828–)
Pressly, D. (1828–)
Niven, J. C. (1828–1881)
McIntyre, A. (1828–1887)
King, M. (1828–1901)
Miller, W. (1828–1909)
Young, A. (1829–)
Ashford, F. F. (*fl.* 1830s)
Smith, J. (*fl.* 1830s)
Macnab, R. (*fl.* 1830s–1840s)
Fowler, J. (1830–)
Carr, R. (*c.* 1830–1887)
Dean, J. G. (*c.* 1830–1900)
Pettigrew, A. (*c.* 1830–1903)
Wallis, J. (*c.* 1830–1907)
Miles, G. T. (1831–1904)
Speed, T. (1832–1883)
Cross, W. J. (1832–1885)
Woodbridge, J. (*c.* 1832–1888)
Anderson, J. (*c.* 1832–1899)
Tegg, J. (1832–1902)
Easlea, W. (*c.* 1832–1919)
Welsh, J. (*c.* 1833–1885)
Stevens, Z. (*c.* 1833–1886)
Rabone, T. H. (1833–1895)
Croucher, G. (1833–1905)
Sutherland, W. (*c.* 1833–1920)
Ballantine, H. (1833–1929)
Dixon, C. (1834–)
Knight, H. (1834–1896)
Temple, M. (1834–1902)
Penford, C. (1834–1908)
Abbey, G. (1835–)
Gater, W. A. (*c.* 1835–1900)
King, T. (1835–1902)
Barron, A. F. (1835–1903)
Latham, W. B. (1835–1914)
Moore, F. (1835–1916)
Speed, W. (*c.* 1835–1921)
Challis, T. (1835–1923)
M'Millan, A. (*c.* 1835–1924)
Baxter, J. (1836–1902)
McIndoe, J. (1836–1910)
Johnston, G. (1837–1887)
Head, W. G. (1837–1897)
Dunn, M. (1837–1899)
Smith, J. (1837–1903)
Simpson, J. (*c.* 1837–1910)
Easter, J. (*c.* 1837–1912)
Green, G. H. (1837–1918)
Wildsmith, W. (*c.* 1838–1890)
Robinson, W. (1838–1935)
Dickson, J. (*c.* 1839–)
Cairns, J. (*c.* 1839–1906)

Coomber, W. (1839–1923)
Bainbridge, R. (*fl.* 1840s)
Lawrence, G. (*fl.* 1840s)
Henderson, J. (*fl.* 1840s–1866)
Bridger, F. (*fl.* 1840s–1910s)
Cole, E. (*c.* 1840–1892)
Shrimpton, J. (1840–1899)
Ward, H. W. (1840–1916)
Bannerman, T. (1840–1920)
Stanton, G. (1840–1920)
Swan, W. (*c.* 1841–1919)
Heal, J. (*c.* 1841–1925)
Berry, E. (*c.* 1842–1903)
Shingler, W. (*c.* 1842–1925)
Moorman, J. W. (1843–)
Upjohn, W. B. (1843–)
Court, W. (1843–1888)
Inglis, D. (*c.* 1843–1921)
Thomas, O. (1843–1923)
Crump, W. (1843–1932)
Whytock, J. (1845–*c.* 1926)
Orchard, C. (*c.* 1846–)
Hudson, J. (1846–1932)
Boyd, T. (*c.* 1847–1900)
Brotherston, R. P. (1848–1923)
Ford, T. H. (1849–1932)
Dingwall, G. (*fl.* 1850s–1900s)
Taylor, W. (*fl.* 1850s–1910s)
Jeffrey, J. (*c.* 1850–1916)
Honeyman, A. (1851–1884)
Wythes, G. (1851–1916)
Hemsley, A. (1851–1917)
Molyneux, E. (1851–1921)
Whitton, J. (1851–1925)
Udale, J. (1851–1927)
Bright, F. (1853–1934)
Beckett, E. (*c.* 1853–1935)
MacKellar, A. C. (*c.* 1854–1931)
Blair, P. (*c.* 1854–1936)
Parker, R. (1856–)
Brace, J. (1856–1950)
Taylor, G. (*c.* 1857–1941)
Watson, W. (1858–1925)
Trevithick, W. E. (1858–1929)
M'Hattie, J. W. (*c.* 1859–1923)
Street, C. (1859–1946)
Perry, F. (1860–1935)
Baker, W. G. (1861–1945)
McLeod, J. F. (1863–1950)
Young, W. H. (*c.* 1864–1938)
Pitts, J. (1864–1944)
Beale, W. (*c.* 1865–1903)
Barnes, N. F. (*c.* 1865–1950)
Jordan, F. (*c.* 1865–1958)
Weeks, H. (1866–)
Herrington, A. (1866–1950)
Comber, J. (*c.* 1866–1953)
Irving, W. (1867–1934)
Pettigrew, W. W. (*c.* 1867–1947)
Weathers, P. (1869–1933)
Cook, T. H. (*c.* 1869–1947)
Curror, J. R. (*fl.* 1870s)
Smythe, W. (*fl.* 1870s)
Turton, T. (*fl.* 1870s–1910s)
Gilman, E. (*fl.* 1870s–1920s)
Pettigrew, H. A. (1871–1947)
Staward, R. (*c.* 1873–1961)
Mitchell, W. J. W. (*c.* 1874–1965)
Johnson, G. F. (*c.* 1875–1954)
Hope, J. (1875–1970)
Alexander, H. G. (*c.* 1875–1972)
Raffill, C. P. (1876–1951)
Allard, E. J. (*c.* 1877–1918)
Stewart, L. B. (1877–1934)
Puddle, F. C. (*c.* 1877–1952)
Streeter, F. (1877–1975)
Taylor, T. W. (1878–1932)
Field, E. (*c.* 1878–1970)
Bland, J. E. (*fl.* 1880s)
Mileham, G. (*fl.* 1880s–1900s)
Molyneux, N. (*fl.* 1880s–1900s)
Preston, F. G. (1882–1964)
Rose, F. J. (1885–1956)
Lees, W. H. (*fl.* 1890s)
Arnold, R. E. (*c.* 1891–1962)
Lamont, C. P. (*c.* 1892–1949)
Anderson, E. B. (*c.* 1895–1971)
Fairgrieve, P. W. (–1900)
MacKellar, R. (–1903)
Allen, J. (–1906)
Chester, W. (–1906)
Norman, G. (–1906)
Hogg, T. (–1908)
Tallack, J. C. (–1909)
Ryan, T. (–1910)
Deacon, J. (–1912)
Fyfe, W. (–1912)
Clayton, H. J. (–1914)
Chapman, A. (–1920)
Goodacre, J. H. (–1922)
Coomber, T. (–1926)
Prinsep, H. C. (–*c.* 1930)
Bedford, A. (–1934)
Markham, E. (–1937)
Shill, J. E. (–1940)
Robinson, G. W. (–1942)
Fielder, C. R. (–1946)
Perfect, B. F. (–1953)
Hawes, E. F. (*fl.* 1900s)
Willard, J. (*fl.* 1900s)
Crombie, D. (*fl.* 1900s–1920s)
Stenning, L. (1901–1965)

GENETICS *see* **CYTOLOGY**

GENTIANA
Wilkie, D. (–1961)

GHANA
Burton, *Sir* R. F. (1821–1890)
Reade, W. W. (1839–1875)
Clarke, R. (*fl.* 1840s–1860s)
Rowland, J. W. (1852–1925)
Cummins, H. A. (1864–1939)
Crowther, W. (*c.* 1867–1895)
Dudgeon, G. C. (1867–1930)
Kitson, *Sir* A. E. (1868–1937)
Willey, F. E. (1871–1898)
Atchley, S. C. (*c.* 1871–1936)
Johnson, W. H. (1875–)
Fishlock, W. C. (1875–1959)
Sampson, H. C. (1878–1953)
Smythe, C. W. (1879–1909)
Don, W. (1879–1911)
Bunting, R. H. (1879–1966)
Evans, A. E. (*c.* 1880–1951)
Chipp, T. F. (1886–1931)
Morse, E. W. (1887–1917)
Miles, A. C. (1887–1969)
Humphries, C. H. (*fl.* 1890s)

Irvine, F R. (1898–1962)
Scholes, J. (–1943)
Culham, A. B. (–1948)
Brown, T. W. (–*c.* 1951)
Hunter, T. (–1965)
Anderson, J. (*fl.* 1900s)
Howes, F. N. (1901–1973)

GIBRALTAR
Wager, *Sir* C. (1666–1743)
Brown, W. (*fl.* 1700s)
James, T. (*c.* 1720–1782)
Gage, *Sir* T. (1781–1820)
Moon, A. (–1825)
Lemann, C. M. (1806–1852)
Kelaart, E. F. (*c.* 1818–1860)
Finlay, K. (*fl.* 1820s–1880s)
Wolley-Dod, A. H. (1861–1948)
Frere, *Sir* B. H. T. (1862–1953)

GLADIOLUS
Bliss, A. J. (1859–1931)

GNETUM
Pearson, H. H. W. (1870–1916)

GOLD COAST *see* **GHANA**

GOSSYPIUM
Clarke, R. T. (1813–1897)
Irving, E. G. (1816–1855)
Flatters, A. (*c.* 1848–1929)
Watt, *Sir* G. (1851–1930)
Sampson, H. C. (1878–1953)
Balls, W. L. (1882–1960)
Evans, *Sir* G. (1883–1963)
Parnell, F. R. (*c.* 1886–1971)
Crowther, F. (1906–1946)

GRAPES *see* **VITICULTURE**

GRAMINEAE
North, R. (–*c.* 1765)
Richardson, *Rev.* W. (1740–1820)
Curtis, W. (1746–1799)
Swayne, *Rev.* G. (*c.* 1746–1827)
Keith, *Rev.* G. S. (1752–1823)
Lawrence, J. (1753–1839)
Knapp, J. L. (1767–1845)
Sinclair, G. (1786–1834)
Johnson, C. (1791–1880)
Salisbury, W. (–1823)
Murphy, E. (–1866)
Amos, W. (*fl.* 1800s)
Thornhill, J. (*fl.* 1800s–1850s)
Lloyd, G. N. (1804–1889)
Hanham, F. H. (1806–1877)
Moore, D. (1807–1879)
Parnell, R. (1810–1882)
Buckman, J. (1814–1884)
Munro, W. (1818–1880) *Bambusa*
Buchanan, J. (1819–1898)
Vasey, G. (1822–1893)
Sowerby, J. E. (1825–1870)
Lowe, E. J. (1825–1900)
Le Couteur, *Sir* J. (*fl.* 1830s–1860s)
Kurz, W. S. (*c.* 1833–1878) *Bambusa*
Mitford, A. B. F. (1837–1916) *Bambusa*
Plues, M. (*c.* 1840–*c.* 1903)
Long, F. (1840–1927)
Duthie, J. F. (1845–1922)
Gamble, J. S. (1847–1925)
Sutton, M. J. (1850–1913)
Wingate, G. (1852–1936)
Turner, F. (1852–1939)
McAlpine, A. N. (1855–1924)
MacDonald, J. (1855–1930)
Stapf, O. (1857–1933)
Maiden, J. H. (1859–1925)
Goddard, H. J. (*c.* 1864–1947)
Stapledon, *Sir* R. G. (1882–1960)
Snowden, J. D. (1886–1973)
Bor, N. L. (1893–1972)
Bews, J. W. (1894–1938)
Marquand, C. V. B. (1897–1943) *Avena*
Popenoe, D. K. (1899–1932)
Coldstream, W. (–1929)
Appleton, A. F. (*fl.* 1900s)
Napper, D. M. (1930–1972)

GRASSES *see* **GRAMINEAE**

GREECE (*see also* **CRETE**)
Covel, *Rev.* J. (1638–1722)
Wheler, *Rev. Sir* G. (1650–1724)
Keill, J. (1673–1719)
Daniel, S. (–before 1707)
Sibthorp, J. (1758–1796)
Bauer, F. (1760–1826)
Swainson, W. (1789–1855)
Talbot, W. H. F. (1800–1877)
Robb, M. A. (1829–1912)
Forsyth, C. I. (1843–1923)
Rogers, *Rev.* F. A. (1876–1944)

GREENLAND
Craycroft, *Capt.* (*fl.* 1730s)
Sabine, *Sir* E. (1788–1883)
Feilden, H. W. (1838–1921)
Whymper, E. (1840–1911)
Brown, R. (1842–1895)

GUIANA
Belcher, *Sir* E. (1799–1877)
Anderson, A. (–1811)
Broadway, W. E. (1863–1935)

GUINEA
Bartar, E. (*fl.* 1690s)
Hove, A. P. (*fl.* 1780s–1820s)
Lynes, H. (1874–1942)
Collenette, C. L. (1888–1959)
Henderson, A. H. (–1921)

GUYANA
Barkly, E. H. (–1857)
Hancock, J. (*fl.* 1800s–1840s)
Schomburgk, *Sir* R. H. (1804–1865)
Schomburgk, *Sir* M. R. (1811–1891)
Campbell, W. H. (1814–1883)
Barkly, *Sir* H. (1815–1898)
Rothery, H. C. (1817–1888)
Colley, T. (*fl.* 1820s–1830s)
Steele, M. (*fl.* 1830s)
Jenman, G. S. (1845–1902)
Sherring, R. V. (1847–1931)
Waby, J. F. (1848–1923)
Rodway, J. (1848–1926)
Thurn, *Sir* E. F. I. (*c.* 1852–1932)
Burke, D. (1854–1897)
Blow, T. B. (1854–1941)
Harrison, J. B. (1856–1926)
Ward, R. (1860–*c.* 1934)
Leechman, A. (1869–*c.* 1940)
Bartlett, A. W. (1875–1943)
Beckett, J. E. (1878–1934)

Stockdale, F. A. (1883–1949)
Bancroft, C. K. (1885–1919)
Myers, J. G. (1897–1942)
Parker, G. W. (–1904)
Sandwith, N. Y. (1901–1965)
Wolstenholme, G. E. (1907–1962)
Evans, G. B. (1935–1966)

HAITI
Mackenzie, C. (*fl.* 1820s–1830s)

HAWAII *see* **PACIFIC**

HELLEBORUS
Hope, F. J. (–1880)
Archer-Hind, T. H. (1814–1911)

HEVEA
Cross, R. M. (1836–1911)
Wickham, *Sir* H. A. (1846–1928)
Collins, J. (*fl.* 1850s–1900s)
Ridley, H. N. (1855–1956)
Wright, *Sir* H. (1874–1940)
Lock, R. H. (1879–1915)
Sharples, A. (1887–1937)

HIERACIUM
Backhouse, J. (1825–1890)
Brenan, *Rev.* S. A. (1837–1908)
Bladon, J. (*fl.* 1840s–1850s)
Linton, *Rev.* E. F. (1848–1928)
Linton, *Rev.* W. R. (1850–1908)
Hanbury, F. J. (1851–1938)
Thompson, R. F. (1856–1906)
Roffey, *Rev.* J. (1860–1927)
Pugsley, H. W. (1868–1947)

HOLLAND *see* **NETHERLANDS**

HONG KONG
Harland, W. A. (–1857)
Wilford, C. (–1893)
Bentham, G. (1800–1884)
Urquhart, F. G. (1813–*c.* 1890)
Champion, J. G. (1815–1854)
Bowring, J. C. (1821–1893)
Hance, H. F. (1827–1886)
Dickins, F. V. (1838–1915)
Dill, *Dr.* (*fl.* 1840s)
Ford, C. (1844–1927)
Lambont, *Rev.* J. (1844–1928)
O'Malley, E. W. (1847–1927)
Eyre, J. (*fl.* 1850s)
Irwin, *Rev.* J. J. (*fl.* 1860s)
Jacob, E. (*fl.* 1860s)
Dods, G. (*fl.* 1860s–1890s)
Matthew, C. G. (1862–1936)
Tutcher, W. J. (1867–1920)
Dunn, S. T. (1868–1938)
Walker, A. L. (*fl.* 1870s)
Westland, A. B. (*fl.* 1880s–1890s)
Green, H. (1887–1941)

ICELAND
Evans, E. (*fl.* 1690s)
Banks, *Sir* J. (1743–1820)
Miller, J. F. (*fl.* 1770s–1790s)
Hooker, *Sir* W. J. (1785–1865)
Carroll, I. (1828–1880)
Lindsay, W. L. (1829–1880)
Walker, *Rev.* F. A. (1841–1905)
Barrington, R. M. (1849–1915)
Proudlock, R. L. (1862–1948)

INDIAN SUB-CONTINENT
Brown, S. (–1698)
Bulkley, E. (*c.* 1651–1714)
Gifford, *Mr.* (*fl.* 1680s)
Conway, J. (*fl.* 1690s)
Fox, J. (*fl.* 1690s)
Mewse, B. (*fl.* 1690s)
Maidstone, N. (*fl.* 1690s–1720s)
Randal, *Mr.* (*fl.* 1700s)
Monson, A. (*c.* 1714–1776)
Russell, P. (1727–1805)
Koenig, J. G. (*c.* 1728–1785)
Martin, C. (1731–1800)
Kerr, J. (1738–1782)
Smith, M. R. (*c.* 1740–1819)
Kyd, R. (1746–1793)
Fleming, J. (1747–1829)
Forbes, J. (1749–1819)
Rottler, J. P. (1749–1836)
Royds, *Sir* J. (*c.* 1750–1817)
Roxburgh, W. (1751–1815)
MacKenzie, C. (1754–1821)
Hardwicke, T. (1755–1835)
Clive, H. A. (1758–1830)
Robertson, J. (*fl.* 1760s–1770s)
Carey, *Rev.* W. (1761–1834)
Hamilton, F. (1762–1829)
Amherst, *Countess* S. (1762–1838)
Moorcroft, W. (*c.* 1765–1825) *Nepal*
Colebrook, H. T. (1765–1837)
White, D. (1767–1818)
Ainslie, *Sir* W. (1767–1837)
Roxburgh, J. (*fl.* 1770s–1820s)
Heyne, B. (1770–1819)
Leycester, W. (1775–1831)
Berry, A. (*fl.* 1780s–1810s)
Hove, A. P. (*fl.* 1780s–1820s)
Swinton, G. (1780–1854)
Adam, *Sir* F. (*c.* 1781–1853)
Crawfurd, J. (1783–1868)
Gardner, E. (1784–) *Nepal*
Webb, W. S. (1784–1865)
Cullen, W. (1785–1862)
Ramsay, C. (1786–1839)
Wallich, N. (1786–1854)
Schmid, B. (1787–1857)
Sleeman, *Sir* W. H. (1788–1856)
Grant, J. W. (1788–1865)
Sykes, W. H. (1790–1872)
Bailey, *Rev.* B. (1791–1871)
Gerard, A. (1792–1839)
Masters, J. W. (*c.* 1792–1873)
Hearsey, *Sir* J. B. (1793–1865)
Jenkins, F. (1793–1866)
Gerard, J. G. (1794–1828)
Jackson, W. (1795–1822)
Gerard, P. (1795–1835)
Wight, R. (1796–1872)
Lush, C. (1797–1845)
Piddington, H. (1797–1858)
Voigt, J. O. (1798–1843)
Royle, J. F. (1798–1858)
Don, D. (1799–1841) *Nepal*
Smith, C. (–1807)
Anderson, J. (–1809)
Gwillim, *Lady* (–before 1809)
Potts, J. (–1822)
Shuter, J. (–*c.*1827)
Colquhoun, *Sir* R. (–1838) *Nepal*
Maxwell, E. (–before 1839)

Thomson, G. (–before 1855)
Brown, R. N. (–*c.* 1862)
Reid, F. A. (–1862)
McLeod, D. (–1866)
New, W. (–1873)
McIvor, W. G. (–1876)
Atkinson, W. S. (–1878/9)
Bulger, G. E. (–1885)
Davidson, A. A. (–1886)
Ward, J. R. (–1895)
Ranada, N. B. (–1897)
Gleeson, J. M. (–1899)
Richardson, D. L. (1800–1865)
Gibson, A. (1800–1867)
McClelland, J. (1800–1883)
Bentham, G. (1800–1884)
Vigne, G. T. (1801–1863)
Cathcart, J. F. (1802–1851)
Winterbottom, J. E. (1803–1854)
Elliot, *Sir* W. (1803–1887)
Denison, *Sir* W. T. (1804–1871)
Graham, J. (1805–1839)
Madden, E. (1805–1856)
Campbell, A. (1805–1874)
Munro, J. (1807–1831) *Sikkim*
Leith, A. H. (1807–1875)
Falconer, H. (1808–1865)
Ritchie, D. (1809–1866)
Parry, R. (*fl.* 1810s)
Griffith, W. (1810–1845) *Bhutan*
Law, J. S. (1810–1885)
Jerdon, T. C. (1811–1872)
Gardner, G. (1812–1849)
Edgeworth, M. P. (1812–1881)
Firminger, *Rev.* T. A. C. (1812–1884)
Balfour, E. G. (1813–1889)
Carter, H. J. (1813–1895)
Gibson, J. (1815–1875)
Jameson, W. (1815–1882)
Frere, *Sir* H. B. E. (1815–1884)
Gough, G. S. (1815–1895)
Oldham, T. (1816–1878)
Dalzell, N. A. (1817–1878)
Thomson, T. (1817–1878)
Giraud, H. J. (1817–1888)
Strachey, *Sir* R. (1817–1908)
Hooker, *Sir* J. D. (1817–1911) *Nepal*, *Sikkim*
Lyell, K. M. (1817–1915)
Munro, W. (1818–1880)
Drury, H. (1819–1872)
Waring, E. J. (1819–1891)
Fidlor, L. L. (*fl.* 1820s)
Mack, J. (*fl.* 1820s)
Govan, G. (*fl.* 1820s–1830s)
Simons, C. J. (*fl.* 1820s–1850s)
Lobb, T. (1820–1894)
Cleghorn, H. F. C. (1820–1895)
De Crespigny, E. C. (1821–1895)
Stocks, J. E. (1822–1854)
Lisboa, J. C. (*c.* 1822–1897)
Berkeley, E. S. (*c.* 1823–1898)
Brandis, *Sir* D. (1824–1907)
Hurst, H. A. (*c.* 1825–1882)
Bonavia, E. (1826–1908)
Watson, J. F. (1827–1892)
Wood, J. J. (1828–1867)
Milne-Redhead, R. (1828–1900)
Booth, T. J. (1829–)
Hume, A. O. (1829–1912)
Bruce, H. (*fl.* 1830s)
Campbell, J. (*fl.* 1830s)
Inglis, R. (*fl.* 1830s)
Pierard, F. (*fl.* 1830s)
Vicary, N. (*fl.* 1830s–1850s)
Nimmo, J. (*fl.* 1830s–1854)
Murchison, C. (1830–1879)
Loudon, W. (1830–1907)
Beddome, R. H. (1830–1911)
Bidie, G. (1830–1913)
Levinge, H. C. (*c.* 1831–1896)
Pinwell, W. S. C. (1831–1926)
Black, A. A. (1832–1865)
Anderson, T. (1832–1870)
Stewart, J. L. (*c.* 1832–1873)
Dymock, W. (1832–1892)
Hope, C. W. W. (1832–1904)
Clarke, C. B. (1832–1906)
Waterfall, W. (1832–1907)
Watson, W. (1832–1912)
Birdwood, *Sir* G. C. M. (1832–1917)
Kurz, W. S. (*c.* 1833–1878)
Anderson, J. (1833–1900)
Bell, W. (*c.* 1833–1916)
Bellew, H. W. (1834–1892)
Peal, S. E. (1834–1897)
Godwin-Austen, H. H. (1834–1924)
Mateer, *Rev.* S. (1835–1893)
Scott, J. (1836–1880)
Aitchison, J. E. T. (1836–1898)
Collett, *Sir* H. (1836–1901)
Cooke, T. (1836–1910)
Mann, G. (1836–1916)
Henderson, G. (1836–1929)
Oldham, R. (1837–1864)
Head, W. G. (1837–1897)
Birdwood, H. M. (1837–1907)
Beckett, T. W. N. (1839–1906)
Gammie, J. A. (1839–1924) *Sikkim*
Parish, W. H. (*fl.* 1840s)
Simpson, R. S. (*fl.* 1840s)
Thomas, L. D. (*fl.* 1840s)
Biddulph, J. (*fl.* 1840s–1870s)
Lawson, M. A. (1840–1896)
King, *Sir* G. (1840–1909)
Bailey, F. (1840–1912)
Schlich, *Sir* W. (1840–1925)
Henderson, F. (*c.* 1841–1895)
Treutler, W. J. (1841–1915) *Sikkim*
Young, A. P. (1841–1920)
Henry, J. M. (1841–1937)
Jamieson, A. (*c.* 1842–1895)
Ball, V. (1843–1895)
Gatacre, *Sir* W. F. (1843–1906)
Duthie, J. F. (1845–1922)
Fisher, W. R. (1846–1910)
Woodrow, G. M. (1846–1911)
Cherry, J. W. (1846–1935)
Ridley, M. (*c.* 1847–1904)
Talbot, W. A. (1847–1917)
Gamble, J. S. (1847–1925)
Bourdillon, T. F. (1849–1930)
Cotton, F. (*fl.* 1850s)
Fleming, A. (*fl.* 1850s)
Foulkes, *Rev.* T. (*fl.* 1850s)
Gomme, E. (*fl.* 1850s)
Hay, W. E. (*fl.* 1850s)
Johnson, *Rev.* E. (*fl.* 1850s)
Scott, R. (*fl.* 1850s–1860s)
Kirtikar, K. R. (1850–1917)
Hutchins, *Sir* D. E. (1850–1920)

Maries, C. (*c*. 1851–1902)
Oliver, J. W. (1851–1914)
Drummond, J. R. (1851–1921)
Watt, *Sir* G. (1851–1930)
Barclay, A. (1852–1891)
Hill, H. C. (1852–1903)
Eardley, *Sir* W. S. (1852–1929)
Wingate, G. (1852–1936)
Brace, L. J. K. (1852–1938)
Lacaita, C. C. (1853–1933) *Sikkim*
Waddell, L. A. (1854–1938) *Sikkim*
Brühl, P. (1855–)
Appleton, H. (*c*. 1855–1930)
Bamber, C. J. (1855–1941)
Conway, *Sir* W. M. (1856–1937)
Johnston, H. H. (1856–1939)
Pantling, R. (1857–1910)
Lace, J. H. (1857–1918)
Prain, *Sir* D. (1857–1944)
Hooper, D. (1858–1947)
Alcock, A. W. (1859–)
Bryant, F. B. (1859–1922)
Ryan, G. M. (*c*. 1859–1932)
Bourne, *Sir* A. G. (1859–1940)
Brown, M. R. (*fl*. 1860s)
Lushington, A. W. (*c*. 1860–1920)
Barber, C. A. (1860–1933)
Cochrane-Baillie, C. W. A. (1860–1940)
Watt, J. C. (1862–1940) *Sikkim*
Proudlock, R. L. (1862–1948)
Younghusband, *Sir* F. E. (1863–1942)
Bell, T. R. (1863–1948)
Gammie, G. A. (1864–1935) *Sikkim*
Butterworth, A. (1864–1937)
Rogers, C. G. (1864–1937)
Cummins, H. A. (1864–1939) *Sikkim*
Lane, G. T. (1867–1936)
Haines, H. H. (1867–1943)
Millard, W. S. (*c*. 1867–1952)
Veitch, J. H. (1868–1907)
Green, H. F. (1868–*c*. 1945)
Osmaston, B. B. (1868–1961)
Hayden, *Sir* H. H. (1869–1923)
Hutchinson, C. M. (1869–1941)
Dodgson, D. S. (*fl*. 1870s)
Meade, R. J. (*fl*. 1870s)
Thompson, R. H. G. (*fl*. 1870s)
Cattell, W. (*fl*. 1870s–1880s)
Duke, O. T. (*fl*. 1870s–1890s)
Stebbing, E. P. (1870–1960)
Burkill, I. H. (1870–1965)
Cave, G. H. (*c*. 1870–1965) *Nepal*, *Sikkim*
Gage, A. T. (1871–1945)
Davies, H. J. (*c*. 1871–1948)
Howard, *Sir* A. (1873–1947)
Hole, R. S. (1874–1938)
Troup, R. S. (1874–1939)
Butler, *Sir* E. J. (1874–1943)
Fischer, C. E. C. (1874–1950)
Pearson, *Sir* R. S. (1874–1958)
Burkill, E. M. (1874–1970)
Griessen, A. E. P. (1875–1935)
Smith, *Sir* W. W. (1875–1956)
Hemsley, O. T. (1876–1906)
Milne, D. (1876–1954)
Buttenshaw, W. R. (*c*. 1877–1907)
Blatter, E. (1877–1934)
Stanley, B. (1877–1944)
Leslie, J. E. (*c*. 1877–1962) *Sikkim*
Toppin, S. M. (1878–1917)
Gill, N. (*c*. 1878–1924)
Mustoe, W. R. (1878–1942)
Long, E. P. (1878–1947)
McRae, W. (1878–1952)
Sampson, H. C. (1878–1953)
Richmond, *Sir* R. D. (*c*. 1879–1948)
Chitefield (*fl*. 1880s?)
Ellis, R. (*fl*. 1880s)
Giles, G. M. J. (*fl*. 1880s)
Murray, J. A. (*fl*. 1880s)
Nairne, A. K. (*fl*. 1880s)
Pierce, E. (*fl*. 1880s)
Tanner, H. C. B. (*fl*. 1880s)
Trotter, E. W. (*fl*. 1880s)
Williams, *Rev*. J. (*fl*. 1880s)
Gleadow, F. (*fl*. 1880s–1890s)
Craib, W. G. (1882–1933)
Trevor, *Sir* C. G. (1882–1959)
Bailey, F. M. (1882–1967)
Sedgwick, L. J. (1883–1925)
Evans, *Sir* G. (1883–1963)
Woodhouse, E. J. (1884–1917)
Graham, R. J. D. (1884–1950)
Parker, R. N. (1884–1958)
Calder, C. C. (1884–1962)
Shebbeare, E. O. (1884–1964)
Burns, W. (1884–1970)
Shaw, F. J. F. (1885–1936)
Inder, R. W. (*c*. 1885–1949)
Kerr, *Rev*. F. H. W. (1885–1958)
Ludlow, F. (1885–1972) *Bhutan*, *Nepal*, *Sikkim*
Lancaster, S. P. (1886–)
Debbarman, P. M. (1887–1925)
Harper, A. G. (1889–1917)
Bisby, G. R. (1889–1958)
Cavanagh, B. (*fl*. 1890s)
Egerton, R. (*fl*. 1890s)
Elliott, C. F. (*fl*. 1890s)
Hamilton, H. (*fl*. 1890s)
Hare, E. C. (*fl*. 1890s)
Harriss, S. A. (*fl*. 1890s)
Hunter-Weston, A. G. (*fl*. 1890s)
McDonnell, J. C. (*fl*. 1890s)
Maynard, F. P. (*fl*. 1890s)
Neve, A. (*fl*. 1890s)
Picot, H. P. (*fl*. 1890s)
White, C. (*fl*. 1890s) *Sikkim*
Cooper, R. E. (1890–1962) *Bhutan*, *Sikkim*
Osborne, P. V. (1891–1943)
Barnes, E. (*c*. 1892–1941)
Cowan, J. M. (1892–1960) *Sikkim*
Bor, N. L. (1893–1972)
Mooney, H. F. (1897–1964)
Sherriff, G. (1898–1967) *Bhutan*
Lowndes, D. G. (1899–1956)
Gollan, W. (–1905)
Stephen, J. H. (–1915)
Kennedy, W. A. (–1922)
Parsons, A. (–*c*. 1923)
Coldstream, W. (–1929)
Coventry, B. O. (–1929)
Hart, G. S. (–1937)
Legge, J. (–1939)
Hartless, A. C. (–*c*. 1941)
Head, W. (–1941)
Noel, E. F. (–1950)
Mobbs, E. C. (–*c*. 1972)
Hughes-Buller, R. (*fl*. 1900s)
Keenan, R. L. (*fl*. 1900s)
Hay, A. (*fl*. 1900s–1920s)

Saunders, E. (*fl.* 1900s–1920s)
Laurie, M. V. (1901–1973)
Gee, E. P. (1904–1968)
Culbert, R. C. (*fl.* 1950s)

INDO-CHINA *see* **VIETNAM**

INDONESIA (*see also* **MALAYSIA**)
Dampier, W. (1651–1715) *Java*
Miller, C. (1739–1817) *Sumatra*
Marsden, W. (1754–1836) *Sumatra*
Campbell, C. (*c.* 1765–1808) *Sumatra*
Prince, J. (*fl.* 1770s–1820s)*Sumatra*
Horsfield, T. (1773–1859) *Java, Sumatra*
Roxburgh, W. (*fl.* 1780s–1810) *Sumatra*
Raffles, *Sir* T. S. B. (1781–1826) *Java*
Arnold, J. (1782–1818) *Sumatra*
Lumsdaine, J. (*fl.* 1790s–1810s) *Sumatra*
Cuming, H. (1791–1865) *Sumatra*
Norris, *Sir* W. (1793–1859) *Sumatra*
Jack, W. (1795–1822) *Sumatra*
Kerr, W. (–1814) *Java*
Hooper, J. (–1830/1) *Java*
Hutton, H. (–1868) *Java, Moluccas, Timor*
Curnow, R. (–1896) *Java*
Ewer, W. (*fl.* 1800s)
Griffiths, J. (*fl.* 1800s) *Sumatra*
Bennett, J. J. (1801–1876) *Java*
Newbold, T. J. (1807–1850) *Sumatra*
Lobb, T. (1820–1894) *Java*
Kurz, W. S. (*c.* 1833–1878) *Java*
Collett, *Sir* H. (1836–1901)
Hullett, R. W. (1843–1914) *Java*
Boxall, W. (1844–1910) *Java*
Hancock, W. (1847–1914) *Java, Sumatra*
Everett, A. H. (1848–1898)
Henshall, J. (*fl.* 1850s) *Java*
Curtis, C. (1853–1928) *Java, Moluccas, Sumatra*
Burke, D. (1854–1897) *Moluccas*
Guppy, H. B. (1854–1926) *Java*
Palmer, W. (1856–1921) *Java*
Hickson, S. J. (1859–1940)
Clemens, J. (1862–1932) *Java*
Hose, C. (1863–1929) *Celebes*
Burkill, E. M. (1874–1970) *Sumatra*
Brooks, C. J. (1875–*c.* 1953) *Sumatra*
Smith, E. (*c.* 1893–1930) *Celebes*
Best, G. A. (–1937) *Java*
Alston, A. H. G. (1902–1958)

IRAN
Lynch, T. K. (1818–1891)
Loftus, W. K. (*c.* 1821–1858)
Bishop, I. L. (*c.* 1832–1904)
St. John, *Sir* O. B. C. (1837–1891)
Littledale, St. G. R. (*c.* 1851–1931)
Stapf, O. (1857–1933)
Hooper, D. (1858–1947)
Cochrane-Baillie, C. W. A. (1860–1940)
Threlfall, W. (1862–1888)
Hotson, J. E. B. (1872–1944)
Bell, F. (*fl.* 1880s)
Jennings, R. H. (*fl.* 1880s)
Cowan, J. M. (1892–1960)

IRAQ
Chesney, F. R. (1789–1872)
Layard, *Sir* A. H. (1817–1894)
Loftus, W. K. (*c.* 1821–1858)
Colvill, W. H. (1838–1885)
Hyslop, J. M. (*fl.* 1850s)
Hooper, D. (1858–1947)
Rogers, *Rev.* F. A. (1876–1944)
Evans, W. E. (1882–1963)
Blakelock, R. A. (1915–1963)

IRELAND *see* **BRITISH ISLES**

IRIS
Foster, *Sir* M. (1836–1907)
Yeld, G. (1845–1938)
Lynch, R. I. (1850–1924)
Bliss, A. J. (1859–1931)
Hort, *Sir* A. F. (1864–1935)
Bowles, E. A. (1865–1954)
Chadburn, G. H. (1870–1950)
Anley, G. (*c.* 1876–1968)
Dykes, W. R. (1877–1925)
Dillistone, G. (1877–1957)
Spender, R. E. S. (*c.* 1878–1965)
Christie-Miller, C. W. (*c.* 1878–1976)
Pilkington, G. L. (*c.* 1885–1971)
Bunyard, G. N. (1886–1969)
Hadden, N. G. (*c.* 1888–1971)
Long, B. R. (1890–1962)
Cave, N. L. (*c.* 1891–1974)
Dykes, E. K. (–1933)
Miller, H. F. R. (–1962)
Howe, A. C. (–1973)
Tomalin, E. F. J. (*c.* 1916–1973)

ITALY
Willughby, F. (1635–1672)
Perin, *Dr.* (*fl.* 1640s)
Balam, A. (*fl.* 1650s–1690s)
Lawson, I. (–*c.* 1747)
Ford, J. (*fl.* 1760s–1780s)
Armitage, E. (1822–1906)
Baker, *Rev.* T. (*fl.* 1830s)
Groves, H. (1835–1891)
Bicknell, *Rev.* C. (1842–1918)
Lacaita, C. C. (1853–1933)
Hanbury, *Sir* C. (*c.* 1871–1937)
McEacharn, N. B. (1885–1964)

IVORY COAST
Ansell, J. (–1847)
Miles, A. C. (1887–1969)

JAMAICA *see* **WEST INDIES**

JAPAN
Wilford, C. (–1893)
Henry, C. (–1894)
Fortune, R. (1812–1880)
Hogg, T. (1820–1892)
Black, A. A. (1832–1865)
Blomfield, *Sir* R. M. (1835–1921)
Dickins, F. V. (1838–1915)
Hodgson, C. P. (*fl.* 1840s)
Bisset, J. (1843–1911)
Ayrton, M. (1846–1883)
Hancock, W. (1847–1914)
Maries, C. (*c.* 1851–1902)
Blow, T. B. (1854–1941)
Matthew, C. G. (1862–1936)
Veitch, J. H. (1868–1907)
Dunn, S. T. (1868–1938)
Wilson, E. H. (1876–1930)

JAVA *see* **INDONESIA**

JUNGERMANNIA
Lyell, C. (1767–1849)
Francis, *Rev*. R. B. (*c*. 1768–1850)
Hooker, *Sir* W. J. (1785–1865)
Cruikshank, J. (*c*. 1813–1847)

KENYA
Wakefield, *Rev*. T. (1836–1901)
Hutchins, *Sir* D. E. (1850–1920)
Powell, H. (1864–1920)
Gregory, J. W. (1864–1932)
Rogers, C. G. (1864–1937)
Lugard, E. J. (1865–1944)
Battiscombe, E. (*c*. 1875–1971)
James, *Sir* H. E. M. (*fl*. 1880s–1900s)
Dümmer, R. A. (*c*. 1887–1922)
Dowson, W. J. (1887–1963)
Butler, F. B. L. (–1941)
Moloney, E. R. (–1952)
Jackson, T. H. E. (1903–1968)
Dale, I. R. (*c*. 1904–1963)

KERGUELEN ISLANDS
Hooker, *Sir* J. D. (1817–1911)
Eaton, *Rev*. A. E. (1844–1929)
Kidder, J. H. (*fl*. 1870s)

KOREA
Wilford, C. (–1893)
Perry, W. W. (*c*. 1846–1894)
Carles, W. R. (1848–1929)

LABIATAE
Bentham, G. (1800–1884)

LATHYRUS
Bolton, R. (*c*. 1869–1949)
Crane, D. B. (–1938)

LESOTHO
Lugard, E. J. (1865–1944)

LIBERIA
Whyte, A. (1834–1908)
Johnston, *Sir* H. H. (1858–1927)
Collenette, C. L. (1888–1959)
Sim, D. (*fl*. 1900s)

LICHENOLOGY
Burgess, *Rev*. J. (1725–1795)
Harriman, *Rev*. J. (1760–1831)
MacGarroch, J. B. (1765–1782)
Gage, *Sir* T. (1781–1820)
Borrer, W. (1781–1862)
Leyland, R. (1784–1847)
Turner, J. (1786–1820)
Jones, T. (1790–1868)
Salwey, *Rev*. T. (1791–1877)
Mitchell, A. H. (1794–1882)
Bohler, J. (1797–1872)
Taylor, T. (–1848)
Leighton, *Rev*. W. A. (1805–1889)
Deakin, R. (1808/9–1873)
Stenhouse, J. (1809–1880)
Varenne, E. G. (1811–1887)
Knight, C. (*c*. 1818–1895)
Gray, P. (1818–1899)
Holl, H. B. (1820–1886)
Babington, *Rev*. C. (1821–1889)
Piggott, H. (1821–1913)
Jones, A. M. (1826–1889)
Carroll, I. (1828–1880)
Joshua, W. (1828–1898)
Lindsay, W. L. (1829–1880)
Mudd, W. (1830–1879)
Crombie, *Rev*. J. M. (1830–1906)
Larbalestier, C. du B. (1838–1911)
Holmes, E. M. (1843–1930)
Johnson, *Rev*. W. (1844–1919)
Hebden, T. (1849–1931)
Atwood, M. M. (*fl*. 1850s–1860s)
Smith, A. L. (1854–1937)
Knight, H. H. (1862–1944)
Knowles, M. C. (1864–1933)
Darbishire, O. V. (1870–1934)
Wattam, W. E. L. (1872–1953)
Watson, W. (1872–1960)
Calder, M. (*fl*. 1880s–1910s)
Sowter, F. A. (1899–1972)

LILIUM
Wallace, A. (1829–1899)
Baker, J. G. (1834–1920)
Elwes, H. J. (1846–1922)
Groves, A. S. (1865–1942)
Woodcock, H. B. D. (1867–1957)
Lyttel, *Rev*. E. S. (1868–1944)
Coutts, J. (1872–1952)
Bentley, W. (*c*. 1875–1953)
Wilson, E. H. (1876–1930)
Amsler, M. (*c*. 1876–1952)
Cotton, A. D. (1879–1962)
Constable, W. A. (1887–1954)
Comber, H. F. (1897–1969)
Darby, G. (–1966)
Wyatt, O. E. P. (–1973)

LIMONIUM
Salmon, C. E. (1872–1930)

LINUM
Lafferty, H. A. (1891–1954)

LORD HOWE ISLAND
Lind, *Mr*.
Milne, W. G. (–1866)
Moore, C. (1820–1905)
Fitzgerald, R. D. (1830–1892)

LOTUS
Callen, E. O. (1912–1970)

LUPINUS
Russell, G. (1857–1951)

MACAO
Lay, G. T. (*fl*. 1820s–1845)

MADAGASCAR *see* **MALAGASY**

MADEIRA
Heberden, T. (1703–1769)
Lee, S. (1791–1856)
Garnons, *Rev*. W. L. P. (1791–1863)
Lowe, *Rev*. R. T. (1802–1874)
Lemann, C. M. (1806–1852)
Christy, W. (*c*. 1807–1839)
Bunbury, *Sir* C. J. F. (1809–1886)
Hartweg, C. T. (1812–1871)
Penfold, J. W. (*fl*. 1820s–1840s)

Johnson, J. Y. (1820–1900)
Hurst, H. A. (*c*. 1825–1882)
Robley, A. J. (*fl*. 1840s)
Long, F. (1840–1927)
Mason, N. H. (*fl*. 1850s–1860s)
Johnston, H. H. (1856–1939)
Crocker, E. (1858–1910)
Gilbert-Carter, H. (1884–1969)

MAGNOLIA
Millais, J. G. (*c*. 1865–1931)

MAJORCA
Geoghegan, F. (*fl*. 1890s)

MALAGASY
Smeathman, H. (–1786)
Lyall, R. (1780s–1831)
Ellis, *Rev*. W. (1794–1872)
Bojer, W. (1795–1856)
Forbes, J. (1798–1823)
Blackburn, E. B. (–1839)
Gerrard, W. T. (–1866)
Rothery, H. C. (1817–1888)
Kitching, L. (1835–1910)
Meller, C. J. (*c*. 1836–1869)
Sibree, *Rev*. J. (1836–1929)
Forsyth-Major, C. I. (1843–1923)
Cowan, *Rev*. W. D. (1844–1924)
Baron, *Rev*. R. (1847–1907)
Last, J. T. (1847/8–1933)
Curtis, C. (1853–1928)
Blow, T. B. (1854–1941)
Fox, J. T. (*fl*. 1880s)
Parker, G. W. (–1904)

MALAWI
Kirk, *Sir* J. (1832–1922)
Whyte, A. (1834–1908)
Sharpe, *Sir* A. (1853–1935)
Johnson, *Rev*. W. J. P. (1854–1928)
Buchanan, J. (1855–1896)
Thomson, J. (1858–1895)
Purves, J. M. (1876–)
Adamson, G. (*fl*. 1890s)
McClounie, J. (*fl*. 1890s)
Shinn, A. (*c*. 1893–1938)
Cameron, K. J. (–1918)

MALAYSIA
Landon, S. (*fl*. 1670s–1700s) *Borneo*
Keir, W. (*fl*. 1690s) *Malaya*
Forrest, T. (*c*. 1729–*c*. 1802) *Borneo*
Hunter, W. (1755–1812) *Malaya*
Prince, J. (*fl*. 1770s–1820s) *Malaya*
Farquhar, W. (*c*. 1770–1839) *Malaya*
Roxburgh, W. (*fl*. 1780s–1810) *Malaya*
Ramsay, C. (1786–1839) *Malaya*
Finlayson, G. (*fl*. 1790s–1823) *Malaya*
Cuming, H. (1791–1865) *Malaya*
Jack, W. (1795–1822) *Malaya*
Smith, C. (–1807) *Malaya*
Walker, G. W. (–1844) *Malaya*
Low, J. (–1852) *Malaya*
Motley, J. (–1859) *Borneo*
Hutton, H. (–1868) *Malaya*
Niven, L. (–1876) *Malaya*
Fraser, M. (–1884/5) *Borneo*
Cantley, N. (–1888) *Malaya*
Becher, H. M. –1893) *Malaya*
Pryer, W. B. (–1899) *Borneo*
Porter, G. (*fl*. 1800s–1830s) *Malaya*
Phillips, W. E. (*fl*. 1800s–1850s) *Malaya*
Lewis, W. T. (*fl*. 1800s–1860s) *Malaya*
Brooke, *Sir* J. (1803–1868) *Sarawak*
Schomburgk, *Sir* R. H. (1804–1865) *Malaya*
Oxley, T. (1805–1886) *Malaya*
Cantor, T. E. (1809–1854) *Malaya*
Griffith, W. (1810–1845) *Malaya*
Lobb, T. (1820–1894) *Borneo*, *Malaya*
Day, J. (1824–1888) *Malaya*
Low, *Sir* H. (1824–1905) *Malaya*
Kurz, W. S. (*c*. 1833–1878) *Malaya*
Maingay, A. C. (1836–1869) *Malaya*
Hose, *Rev*. G. F. (1838–1922) *Borneo*
White, *Rev*. (*fl*. 1840s) *Malaya*
St. John, S. (*fl*. 1840s–1860s) *Borneo*
King, *Sir* G. (1840–1909) *Malaya*
Elphinstone, *Sir* G. H. D. (1841–1900) *Malaya*
Creagh, C. V. (1842–1917) *Borneo*
Hullett, R. W. (1843–1914) *Borneo*, *Malaya*
Boxall, W. (1844–1910) *Borneo*
Burbidge, F. W. T. (1847–1905) *Borneo*
Everett, A. H. (1848–1898) *Borneo*
Hervey, D. F. A. (1849–1911) *Malaya*
Barber, E. S. (*fl*. 1850s) *Borneo*
Collins, J. (*fl*. 1850s–1900s) *Malaya*
Forbes, H. O. (1851–1932) *Malaya*
Hill, H. C. (1852–1903) *Malaya*
Murton, H. J. (1853–1881) *Malaya*
Curtis, C. (1853–1928) *Borneo*, *Malaya*
Wray, L. (1853–1942) *Malaya*
Burke, D. (1854–1897) *Borneo*
Stephens, A. B. (1855–1908) *Malaya*
Ridley, H. N. (1855–1956) *Malaya*
Haviland, G. D. (1857–*c*. 1901) *Malaya*, *Sarawak*
Napier, *Sir* W. J. (1857–1945) *Malaya*
Fox, W. (1858–1934) *Malaya*
Denison, N. (*fl*. 1860s–1880s) *Malaya*
Wolferstan, L. E. P. (*fl*. 1860s–1920s) *Malaya*
Whitehead, J. (1860–1899) *Borneo*
Clemens, J. (1862–1932) *Borneo*
Matthew, C. G. (1862–1936) *Malaya*
Binstead, *Rev*. C. H. (*c*. 1862–1941) *Malaya*
Hose, C. (1863–1929) *Sarawak*
Barnes, W. D. (1865–1911) *Malaya*
Ricketts, O. F. (*c*. 1867–) *Sarawak*
Gimlette, J. D. (1867–1934) *Malaya*
Burn-Murdoch, A. M. (1868–1914) *Malaya*
Carruthers, J. B. (1869–1910) *Malaya*
Penney, F. G. (*fl*. 1870s–1910s) *Malaya*
Burkill, I. H. (1870–1965) *Malaya*
Gwynne-Vaughan, D. T. (1871–1915) *Malaya*
Yapp, R. H. (1871–1929) *Malaya*
Hose, E. S. (1871–1946) *Malaya*
Shelford, R. W. C. (1872–1912) *Sarawak*
Phillips, P. (1873–) *Malaya*
Robinson, H. C. (1874–1929) *Malaya*
Arden, S. (1874–1942) *Malaya*
Barnard, B. H. F. (1874–1953) *Malaya*
Lang, W. H. (1874–1960) *Malaya*
Burkill, E. M. (1874–1970) *Malaya*
Sands, W. N. (1875–1943) *Malaya*
Brooks, C. J. (1875–*c*. 1953) *Sarawak*
Cubitt, G. E. S. (*c*. 1875–1966) *Malaya*
Annandale, T. N. (1876–1924) *Malaya*
Kloss, C. B. (1877–) *Malaya*
Sanderson, A. R. (1877–1932) *Malaya*
Campbell, J. W. (1878–1929) *Malaya*

Main, T. W. (*c.* 1879–1944) *Malaya*
Alvins, M. V. (*fl.* 1880s) *Malaya*
Durnford, J. (*fl.* 1880s) *Malaya*
Wray, C. (*fl.* 1880s) *Malaya*
Goodenough, J. S. (*fl.* 1880s–1900s) *Malaya*
Abrams, J. (*fl.* 1880s–1910s) *Malaya*
Bryant, A. T. (*fl.* 1880s–1910s) *Malaya*
Derry, R. (*fl.* 1880s–1910s) *Malaya*
Hewitt, J. (1880–1961) *Sarawak*
Le Doux, J. A. (*c.* 1881–1961) *Malaya*
Brooks, F. T. (1882–1952) *Malaya*
Long, F. R. (1884–1961) *Malaya*
Bancroft, C. K. (1885–1919) *Malaya*
Sanger-Davies, A. E. (1885–1945) *Malaya*
Wood, W. L. (1885–1951) *Malaya*
Moulton, J. C. (1886–1926) *Borneo*
Chipp, T. F. (1886–1931) *Malaya*
Mead, J. P. (1886–1951) *Malaya*
Sharples, A. (1887–1937) *Malaya*
Collenette, C. L. (1888–1959) *Malaya*
Watson, J. G. (1889–1950) *Malaya*
Bartlett, E. (*fl.* 1890s) *Borneo*
Feilding, J. B. (*fl.* 1890s) *Malaya*
Lake, H. W. (*fl.* 1890s) *Malaya*
Wooldridge, T. A. (*fl.* 1890s) *Malaya*
Kelsall, H. J. (*fl.* 1890s–1920s) *Malaya*
Milsum, J. N. (*c.* 1890–1945) *Malaya*
Carr, C. E. (1892–1936) *Borneo, Malaya*
Smith, E. (*c.* 1893–1930) *Malaya*
Dennys, N. B. (–1900) *Borneo, Malaya*
Machado, A. D. (–1910) *Malaya*
McNair, J. F. A. (–1910) *Malaya*
Isaac, J. S. (–1918) *Malaya*
Venning, A. R. (–1927) *Malaya*
Best, G. A. (–1937) *Malaya*
Spare, G. H. (–1940/5) *Malaya*
Arnot, D. B. (–1942) *Malaya*
Kinsey, W. E. (–1943) *Malaya*
Lambourne, J. (–1965) *Malaya*
Banfield, F. S. (–1967) *Malaya*
Bland, L. S. (*fl.* 1900s) *Malaya*
Craddock, W. H. (*fl.* 1900s) *Malaya*
Hose, G. (*fl.* 1900s) *Malaya*
Robertson-Glasgow, C. P. (*fl.* 1900s) *Malaya*
Down, St. V. B. (*fl.* 1900s–1910s) *Malaya*
McGill, H. (*fl.* 1900s–1910s) *Malaya*
Nauen, J. C. (1903–1943) *Malaya*
Symington, C. F. (1905–1943) *Malaya*
Landon, F. H. (1909–1956) *Malaya*
Anderson, J. W. (*fl.* 1910s–1920s) *Malaya*
Gilliland, H. B. (1911–1965) *Malaya*
Sinclair, J. (1913–1968) *Borneo, Malaya*
Merton, L. F. H. (1919–1974) *Malaya*
Wood, G. H. S. (1927–1957) *Borneo*
Evans, G. B. (1935–1966) *Malaya*

MALDIVE ISLANDS
Darwin, C. R. (1809–1882)

MALTA
Madden, E. (1805–1856)
Wright, C. A. (1834–1907)
Reade, O. A. (1848–1929)
Dickson, E. D. (–1900)

MAMMILLARIA
Shurly, E. W. (1888–1963)

MASCARENE ISLANDS
Oliver, S. P. (1838–1907)

MAURITIUS
Hardwicke, T. (1755–1835)
McGrigor, *Sir* J. (1771–1858)
Carmichael, D. (1772–1827)
Cole, *Sir* G. L. (1772–1842)
Telfair, C. (1778–1833)
Babington, B. G. (1794–1866)
Bojer, W. (1795–1856)
Telfair, A. (–1832)
Blackburn, E. B. (–1839)
Bouton, L. (–1879)
Cantley, N. (–1888)
Duncan, J. (1802–1876)
Gardner, G. (1812–1849)
Ayres, P. B. (1813–1863)
Barkly, *Sir* H. (1815–1898)
Newman, J. (*fl.* 1820s–1840s)
Baker, J. G. (1834–1920)
Horne, J. (1835–1905)
Meller, C. J. (*c.* 1836–1869)
Barkly, A. M. (1838–1932)
Grey, J. (*fl.* 1850s?)
Curtis, C. (1853–1928)
Johnston, H. H. (1856–1939)
Scott, W. (1859–1897)
Bourne, G. C. (1861–1933)
Cattell, W. (*fl.* 1870s–1880s)
O'Connor, C. A. (1883–1963)

MEDICINAL BOTANY
Fleming, J. (1747–1829)
Woodville, W. (1752–1805)
Stokes, J. (1755–1831)
Duncan, A. (1773–1832)
Graves, G. (1784–*c.* 1839)
Coffin, A. I. (*c.* 1791–1866)
Burnett, G. T. (1800–1835)
Evans, J. (1803–1874)
Castle, T. (*c.* 1804–*c.* 1840)
Stephenson, J. (*fl.* 1820s–1850s)
Hanbury, D. (1825–1875)
Churchill, J. M. (*fl.* 1830s–1840s)
Dymock, W. (1832–1892)
Jackson, J. R. (1837–1920)
Clifton, E. S. (1841–1929)
Clements, F. M. (1857–1920)

MENTHA
Sole, W. (1741–1802)
Fraser, J. (1854–1935)
Still, A. L. (1872–1944)
Graham, R. A. H. (1915–1958)

MESEMBRYANTHEMUM
Haworth, A. H. (1768–1833)
Brown, N. E. (1849–1934)

MEXICO
Fifield, S. (*fl.* 1690s–1700s)
Houstoun, W. (*c.* 1695–1733)
Parkinson, J. (*c.* 1772–1847)
Bullock, W. (*fl.* 1790s–1840s)
Collie, A. (1793–1835)
Coulter, T. (1793–1843)
Sinclair, A. (*c.* 1796–1861)
Cowan, J. (–1823)
Graham, G. J. (1803–1878)
Hartweg, C. T. (1812–1871)
Staples, R. P. (*fl.* 1820s)
Mackenzie, C. (*fl.* 1820s–1830s)

Lay, G. T. (*fl.* 1820s–1845)
Blair, *Mr.* (*fl.* 1830s)
Dickson, G. F. (*fl.* 1830s)
Ross, J. (*fl.* 1830s–1840s)
Palmer, E. (1831–1911)
Potts, J. (*fl.* 1840s–1850s)
Hancock, W. (1847–1914)
Doyle, P. W. (*fl.* 1850s)
Hickson, S. J. (1859–1940)
Hinton, G. B. (1882–1943)

MIMOSA
Lindsay, J. (*fl.* 1780s–1803)

MINORCA
Cleghorn, G. (1716–1789)

MOROCCO
Jones, J. (–1731)
Lowe, *Rev.* R. T. (1802–1874)
Hooker, *Sir* J. D. (1817–1911)
Ball, J. (1818–1889)
Leared, A. (1822–1879)
Maw, G. (1832–1912)
Thomson, J. (1858–1895)
Lynes, H. (1874–1942)
Crump, E. (–1927/8)

MOSSES *see* **BRYOLOGY**

MOZAMBIQUE
Forbes, J. (1798–1823)
Kirk, *Sir* J. (1832–1922)
Waller, *Rev.* H. (1833–1896)
Meller, C. J. (*c.* 1836–1869)
Alexander, J. A. (1854–)
Johnson, *Rev.* W. J. P. (1854–1928)
Rockley, A. M. (1865–1941)
Moss, C. E. (1870–1930)
Honey, T. (1872–1937)
Johnson, W. H. (1875–)
Swynnerton, C. F. M. (1877–1938)
Dawe, M. T. (1880–1943)
Appleton, A. F. (*fl.* 1900s)
Allen, C. E. F. (*fl.* 1900s–1920s)

MUSA
Fawcett, W. (1851–1926)

MYCETOZOA
Cran, W. (1854–1933)
Adams, A. (1866–1919)
Sanderson, A. R. (1877–1932)

MYCOLOGY
Dandridge, J. (*fl.* 1660s–1746)
Massey, R. M. (*c.* 1678–1743)
Bolton, J. (*fl.* 1750s–1790s)
Sowerby, J. (1757–1822)
Forster, B. M. (1764–1829)
Purton, T. (1768–1833)
Flintoff, T. (*fl.* 1780s)
Crotch, *Rev.* W. R. (1799–1877)
Arden, M. E. (–1851)
Wilson, R. (*fl.* 1800s)
Bloxam, *Rev.* A. (1801–1878)
Berkeley, *Rev.* M. J. (1803–1889)
Badham, C. D. (1805–1857)
Brittain, T. (1806–1884)
Broome, C. E. (1812–1886)
Maddox, R. L. (1816–1902)
Bull, H. G. (*c.* 1818–1885)
Currey, F. (1819–1881)
English, J. L. (1820–1888)
Parfitt, E. (1820–1893)
Beck, S. C. (1821–1915)
Phillips, W. (1822–1905)
Barnard, F. (1823–1912)
Keith, *Rev.* J. (1825–1905)
Cooke, M. C. (1825–1914)
Clark, J. A. (1826–1890)
Lister, A. (1830–1908)
Carter, H. V. (1831–1897)
Vize, *Rev.* J. E. (1831–1916)
Du Port, *Rev.* J. M. (1832–1899)
Wheeler, E. (1833–1909)
King, T. (1834–1896)
Smith, W. G. (1835–1917)
Fox, W. T. (1836–1879)
Stevenson, *Rev.* J. (1836–1903)
Murray, D. (*fl.* 1840s)
Hey, T. (*c.* 1840–1919)
Eyre, *Rev.* W. L. W. (1841–1914)
Crossland, C. (1844–1916)
Scortechini, *Rev.* B. (1845–1886)
Paul, *Rev.* D. (1845–1929)
Cheesman, W. N. (1847–1925)
Clarke, A. (1848–1925)
McAlpine, D. (1848–1932)
Grove, W. B. (1848–1938)
Plowright, C. B. (1849–1910)
Needham, J. (1849–1913)
Perceval, C. H. S. (1849–1920)
Bucknall, C. (1849–1921)
Hebden, T. (1849–1931)
Hussey, A. M. (–before 1859)
Massee, G. E. (1850–1917)
Pim, G. (1851–1906)
Roebuck, W. D. (1851–1919)
Trail, J. W. H. (1851–1919)
Barber, A. (1852–1891)
Bellerby, W. (1852–1936)
Ward, H. M. (1854–1906)
Smith, A. L. (1854–1937)
Menzies, J. (1854–1945)
Rayner, J. F. (1854–1947)
Boyd, D. A. (1855–1928)
Elliott, W. T. (1855–1938)
Johnstone, R. B. (1856–1934)
Brook, G. (1857–1893)
Ellis, J. W. (1857–1916)
Paulson, R. (1857–1935)
Soppitt, H. T. (1858–1899)
Potter, *Rev.* M. C. (1858–1948)
Lister, G. (1860–1949)
Rea, C. (1861–1946)
Wager, H. W. T. (1862–1929)
Green, C. T. (1863–1940)
McWeeney, E. J. (1864–1925)
Day, E. M. (1865–1934)
Jones, R. C. F. (1865–1952)
Potter, B. (1866–1943)
Carruthers, J. B. (1869–1910)
Chapman, A. C. (1869–1932)
Bayliss-Elliott, J. (1869–*c.* 1957)
Chalmers, A. J. (1870–1920)
Darbishire, O. V. (1870–1934)
Schwartz, E. J. (1870–1939)
Petch, T. (1870–1948)
Swanton, E. W. (1870–1958)

Cooper, C. A. (1871–1944)
Pethybridge, G. H. (1871–1948)
Salmon, E. S. (1872–1959)
Howard, *Sir* A. (1873–1947)
Butler, *Sir* E. J. (1874–1943)
Buller, A. H. R. (1874–1944)
Biffen, *Sir* R. H. (1874–1949)
Ashby, S. F. (1874–1954)
Pearson, A. A. (1874–1954)
Cayley, D. M. (1874–1955)
Burkill, E. M. (1874–1970)
Allen, W. B. (1875–1922)
Bartlett, A. W. (1875–1943)
Smith, J. H. (1875–1952)
Alcock, N. L. L. (*c.* 1875–1972)
Hawley, *Sir* H. C. W. (1876–1923)
Mason, F. A. (*c.* 1878–1936)
McRae, W. (1878–1952)
Lewton-Brain, L. (1879–1922)
Wormald, H. (*c.* 1879–1955)
Cotton, A. D. (1879–1962)
Bunting, R. H. (1879–1966)
Gwynne-Vaughan, H. C. I. (1879–1967)
Pole-Evans, I. B. (1879–1968)
Sutcliffe, H. (1880–1930)
Brooks, F. T. (1882–1952)
Wilson, M. (1882–1960)
Stockdale, F. A. (1883–1949)
Duncan, J. T. (1884–1958)
Bancroft, C. K. (1885–1919)
Lechmere, A. E. (1885–1919)
Rhodes, *Rev*. P. G. M. (1885–1934)
Shaw, F. J. F. (1885–1936)
Walkden, H. (1885–1949)
Ramsbottom, J. (1885–1974)
Wakefield, E. M. (1886–1972)
Hart, J. W. (1887–1916)
Sharples, A. (1887–1937)
Murphy, P. A. (1887–1938)
Dowson, W. J. (1887–1963)
Farquharson, C. O. (1888–1918)
Barnes, B. F. (1888–1965)
Brown, W. (1888–1975)
Hoggan, I. A. (1889–1936)
Bisby, G. R. (1889–1958)
Brierley, W. B. (1889–1963)
McKay, R. (1889–1964)
Whitehead, T. (1889–1964)
Blackwell, E. M. (1889–1973)
Buddin, W. (1890–1962)
Broadbent, A. (1891–1962)
Cartwright, K. St. G. (1891–1964)
Wiltshire, S. P. (1891–1967)
Grainger, J. (*c.* 1891–1969)
Ware, W. M. (1892–1955)
Briton-Jones, H. R. (1893–1936)
Bracher, R. (1894–1941)
Smith, A. (1894–1948)
Storey, H. H. (1894–1969)
Smith, G. (1895–1967)
Groves, E. M. (1897–1956)
Butcher, R. W. (1897–1971)
Dillon-Weston, W. A. R. (1899–1953)
Green, D. E. (*c.* 1899–1968)
Hodgson, A. S. (–1934)
Jones, G. H. (–1945)
Batts, C. C. V. (–1960)
Hansford, C. G. (1900–1966)
Moore, W. C. (1900–1967)
Cook, W. R. I. (1901–1952)
Loughnane, J. B. (1905–1970)
Ashworth, D. (1908–1944)
Baker, R. E. D. (1908–1954)
Massee, I. (*fl.* 1910s)
Peck, A. E. (*fl.* 1910s)
Buxton, E. W. (1926–1964)

MYRISTICA
Sinclair, J. (1913–1968)

NAIADACEAE
Bennett, A. (1843–1929)

NARCISSUS
Leeds, E. (1802–1877)
Backhouse, W. (1807–1869)
Pickstone, W. (1823–)
Barr, P. (1826–1909)
Hartland, W. B. (1836–1912)
Bourne, *Rev*. S. E. (*c.* 1845–1907)
Haydon, *Rev*. G. P. (*c.* 1846–1913)
Engleheart, *Rev*. G. H. (*c.* 1851–1936)
Backhouse, R. O. (*c.* 1854–1940)
Jacob, *Rev*. J. (1858–1926)
Bliss, A. J. (1859–1931)
Cranfield, W. B. (1859–1948)
Barr, P. R. (*c.* 1862–1944)
Williams, P. D. (1865–1935)
Bowles, E. A. (1865–1954)
Brodie, I. (1868–1943)
Pugsley, H. W. (1868–1947)
Wilson, A. M. (1868–1953)
Leak, G. W. (1868–1963)
Arkwright, *Sir* J. S. (1872–1954)
Mitchell, W. F. (*c.* 1875–1949)
Backhouse, W. O. (1885–1962)
Wilson, G. L. (1885–1962)
Blanchard, D. (1887–1969)
Richardson, J. L. (1890–1961)
Ramsbottom, J. K. (1891–1925)
Long, A. D. (1895–1964)
Wootton, C. R. (1897–1970)
Coey, J. (–1921)
Lower, N. Y. (–1926)
Bourne, C. (*fl.* 1910s)

NATURE-PRINTING
Branson, F. (1809–1895)
Bradbury, H. (1831–1860)

NEPENTHES
Court, W. (1843–1888)

NEPAL *see* **INDIAN SUB-CONTINENT**

NETHERLANDS
Bonnivert, G. (*fl.* 1670s–1700s)

NEW BRITAIN
Goadby, B. T. (*c.* 1863–1944)

NEW CALEDONIA
Home, *Sir* J. E. (1798–1853)

NEW GUINEA
Dampier, W. (1651–1715)
Macleay, *Sir* W. J. (1820–1891)
Wilcox, J. F. (1823–1881)
Gill, *Rev*. W. W. (1828–1896)

MacFarlane, *Rev*. S. (1837–1911)
Lawes, W. G. (1839–1907)
Chalmers, *Rev*. J. B. (1841–1901)
MacGregor, *Sir* W. (1846–1919)
Veitch, P. C. M. (1850–1929)
Forbes, H. O. (1851–1932)
Le Hunte, *Sir* G. R. (1852–1929)
Burke, D. (1854–1897)
Bevan, T. F. (1860–)
Cochrane-Baillie, C. W. A. N. R. (1860–1940)
English, A. C. (1861/3–1945)
Clemens, J. (1862–1932)
Barton, F. R. (1865–1947)
Turner, *Rev*. W. Y. (*fl*. 1870s)
Goldie, A. (*fl*. 1870s–1880s)
Gibbs, L. S. (1870–1925)
Seligman, C. G. (1873–1940)
Turner, R. L. (1875–)
Wollaston, A. F. R. (1875–1930)
Bridge, C. A. G. (*fl*. 1880s)
Cheesman, L. E. (1881–1969)
Bryce, G. (1885–)
Lane-Poole, C. E. (1885–1970)
MacDonald, J. (*fl*. 1890s)
Musgrave, *Mrs*. (*fl*. 1890s)
Carr, C. E. (1892–1936)
Sinclair, J. (1913–1968)

NEW HEBRIDES
Armstrong, *Sir* A. (1818–1899)
Morrison, A. (1849–1913)
Cheesman, L. E. (1881–1969)

NEW ZEALAND
Menzies, A. (1754–1842)
Fraser, C. (*c*. 1788–1831)
Swainson, W. (1789–1855)
Cunningham, A. (1791–1839)
Sinclair, A. (*c*. 1796–1861)
Cormack, W. E. (1796–1868)
Home, *Sir* J. E. (1798–1853)
Curl, S. M. (–1890)
Traill, C. (–1898)
Bennett, G. (1804–1893)
Colenso, *Rev*. W. (1811–1899)
Grey, *Sir* G. (1812–1898)
Monro, *Sir* D. (1813–1877)
Ralph, T. S. (1813–1891)
Pascoe, F. P. (1813–1893)
Edgerley, J. (*c*. 1814–1849)
Bidwill, J. C. (1815–1853)
Lyall, D. (1817–1895)
Hooker, *Sir* J. D. (1817–1911)
Knight, C. (*c*. 1818–1895)
Mason, T. (1818–1903)
Buchanan, J. (1819–1898)
Travers, W. T. L. (1819–1903)
Dieffenbach, E. (*fl*. 1820s–1840s)
Jolliffe, J. (1822–1887)
Haast, *Sir* J. F. J. von (1824–1887)
Potts, T. H. (1824–1888)
Brown, R. (*c*. 1824–1906)
Field, H. C. (1825–1911)
Kirk, T. (1828–1898)
Lindsay, W. L. (1829–1880)
Williams, *Rev*. W. L. (1829–1916)
Davis, R. (*fl*. 1830s)
Short, T. K. (*fl*. 1830s)
Hamilton, *Mr*. (*fl*. 1830s–1840s)
Travers, H. (*fl*. 1830s–1870s)
Bell, W. (*c*. 1833–1916)
Hector, *Sir* J. (1834–1907)
Skey, W. (1835–1900)
Hutton, F. W. (1836–1905)
Enys, J. D. (1837–1912)
Adams, J. (1839–1906)
Beckett, T. W. N. (1839–1906)
Adams, T. W. (1841–1919)
Helms, R. (1842–1914)
Cheeseman, T. F. (1846–1923)
Petrie, D. (1846–1925)
Thomson, G. M. (1848–1933)
Bolton, D. (*fl*. 1850s)
Mossman, S. (*fl*. 1850s)
Townson, W. (1850–1926)
Veitch, P. C. M. (1850–1929)
Wragge, C. L. (1852–1922)
Crosby-Smith, J. (1853–1930)
Hamilton, A. (1854–1913)
Cockayne, L. (1855–1934)
Cartwright, T. B. (1856–1896)
Carse, H. (1857–1930)
Atkinson, *Rev*. H. D. (*fl*. 1860s–1890s)
Dixon, H. N. (1861–1944)
Jennings, A. V. (1864–1903)
Turner, E. P. (1865–1937)
Lawson, J. (*fl*. 1870s)
Aston, B. C. (1871–1951)
Brooks, C. J. (1875–*c*. 1953)
Dorrien-Smith, A. A. (1876–1955)
Mason, F. M. (1882–1932)
Begg, C. C. (–1928)
Mansfield, B. P. (–1949)

NICARAGUA
Tate, R. (1840–1901)

NICOBAR ISLANDS
Berkeley, E. S. (*c*. 1823–1898)

NICOTIANA
Brodigan, T. (–1849)

NIGERIA
Coombs, C. (*fl*. 1690s)
Kirckwood, J. (*fl*. 1690s)
Oudney, W. (1790–1824)
Barter, C. (–1859)
Millson, A. (–1896)
Irving, E. G. (1816–1855)
Thomson, *Rev*. W. C. (*fl*. 1820s–1870s)
Baikie, W. B. (1825–1864)
Oldfield, R. A. K. (*fl*. 1830s–1850s)
Mann, G. (1836–1916)
MacGregor, *Sir* W. (1846–1919)
Denton, *Sir* G. C. (1851–1928)
Rowland, J. W. (1852–1925)
Elliott, W. R. (1860–1908)
Woodruff, G. (1864–1891)
Casement, R. D. (1864–1916)
Lugard, E. J. (1865–1944)
Billington, H. W. L. (1867–1897)
Dudgeon, G. C. (1867–1930)
Holland, J. H. (1869–1950)
Millen, H. (1871–1908)
Talbot, D. A. (1871–1916)
Talbot, P. A. (1877–1945)
Foster, E. W. (1878–1921)
Sampson, H. C. (1878–1953)
Don, W. (1879–1911)
Goldie, *Rev*. H. (*fl*. 1880s)

McNair, J. (*fl.* 1880s–1890s)
Hislop, A. (*c.* 1880–1945)
Burton, G. (*c.* 1880–1960)
Lamb, P. H. (*c.* 1883–1937)
Miles, A. C. (1887–1969)
Farquharson, C. O. (1888–1918)
Carter, *Sir* G. (*fl.* 1890s)
Lloyd, H. B. (*fl.* 1890s–1900s)
Bell, A. R. (1892–1925)
Dodd, H. (–1912)
Dawodu, T. B. (–1920)
Evans, F. J. (–1928)
Thompson, H. N. (–1938)
Jones, G. H. (–1945)
Culham, A. B. (–1948)
Lawton, J. R. S. (–1970)
Stanfield, D. P. (–1971)
Symington, C. F. (1905–1943)
Dundas, J. (1907–1966)
Jones, A. P. D. (1918–1946)
Davey, J. T. (1923–1959)

NORFOLK ISLAND
Paterson, W. (1755–1810)
Home, *Sir* J. E. (1798–1853)
Stuart, J. (1802–1842)
Lyall, D. (1817–1895)
Comins, *Rev.* R. B. (*fl.* 1880s–1890s)

NORWAY
Wheeler, R. (*fl.* 1690s)
Christy, W. (*c.* 1807–1839)

NURSERYMEN, SEEDSMEN, MARKET GARDENERS
ENGLAND

Bedfordshire
Gibbs, T. (*fl.* 1790s–1820s) *Ampthill*
Sheppard, S. (*fl.* 1830s) *Bedford*
Laxton, T. (*c.* 1830–1893) *Bedford*
Wallace, W. E. (1851–1941) *Dunstable*
Laxton, E. A. L. (*c.* 1869–1951) *Bedford*

Berkshire
Pendar, W. (*fl.* 1760s–1770s) *Woolhampton*
Swallow, J. (*fl.* 1760s–1810s) *Reading*
Sutton, J. (1777–1863) *Reading*
Barrow, R. (*fl.* 1780s) *Windsor*
Kenyon, J. (*fl.* 1780s) *Windsor*
Sheeting, R. (*fl.* 1780s) *Windsor*
Burn, J. (*fl.* 1780s–1790s) *Windsor*
Badcocke, R. (*fl.* 1790s) *Abingdon*
Couldrey, T. (*fl.* 1790s) *Abingdon*
Poole, H. (*fl.* 1790s) *Reading*
Standish, J. (1814–1875) *Ascot*
Sutton, M. H. (1815–1901) *Reading*
Sutton, A. (*c.* 1818–1897) *Reading*
Priest, M. (*fl.* 1830s) *Reading*
Phippen, G. (1836–1885) *Reading*
Owen, R. (*c.* 1839–1897) *Maidenhead*
Tyso, C. (*fl.* 1840s–1850s) *Wallingford*
Sutton, M. J. (1850–1913) *Reading*
Sutton, A. W. (1854–1925) *Reading*
Sutton, L. G. (1863–1932) *Reading*
King, L. G. (*c.* 1868–1903) *Reading*
Sutton, M. H. F. (1875–1930) *Reading*
Fidler, J. C. (–1903) *Reading*
Ratcliffe, R. (–1965) *Didcot*
Sutton, L. N. (–1965) *Reading*
Sutton, E. P. F. (–1972) *Reading*

Buckinghamshire
Leech, W. (*fl.* 1780s) *High Wycombe*
Howland, J. (*fl.* 1790s) *High Wycombe*
Brown, C. (–1836) *Slough*
Brown, T. (1804–1886) *Slough*
Turner, C. (1818–1885) *Slough*
Austin, R. (*fl.* 1830s) *Burnham*
Nicholls, J. (*fl.* 1830s) *Newport Pagnell*
Tyler, E. (*fl.* 1830s) *Marlow*
Williams, J. (*fl.* 1830s) *Burnham*
Winch, J. (*fl.* 1830s) *Burnham*
Allen, W. (*fl.* 1830s–1840s) *Wycombe*
Boddy, J. (*fl.* 1830s–1840s) *Beaconsfield*
Mavor, A. (*fl.* 1830s–1840s) *Aylesbury*
Taylor, J. (*fl.* 1830s–1840s) *Wendover*
Austin, J. (*fl.* 1840s) *Burnham*
Cutter, W. (*fl.* 1840s) *Slough*
Dawney, J. (*fl.* 1840s) *Aylesbury*
Ferguson, D. (*fl.* 1840s) *Aylesbury*
Fraser, M. (*fl.* 1840s) *Aylesbury*
Nicholls, A. (*fl.* 1840s) *Newport Pagnell*
Turner, H. (*c.* 1848–1906) *Slough*
Weatherley, C. (*fl.* 1850s) *Aylesbury*
Black, J. M. (1870–1946) *Slough*
Allgrove, J. C. (–1930) *Slough*
Turner, A. (–1930) *Slough*
Flory, S. W. (–1954) *Slough*
Bourne, C. (*fl.* 1910s) *Bletchley*

Cambridgeshire
Clarke, R. (*c.* 1757–1836) *Cambridge*
Brewer, M. (*fl.* 1830s) *Cambridge*
Widnall, S. (*fl.* 1830s) *Grantchester*
Bester, J. (1847–) *Cambridge*
Wood, W. L. (1885–1951) *Duxford*

Cheshire
Caldwell, W. (1766–1844) *Knutsford*
Rogers, G. (*fl.* 1780s–1799) *Chester*
Caldwell, W. (1789–1852) *Knutsford*
Barnes, I. (*fl.* 1790s) *Northwich*
Bell, J. (*fl.* 1790s) *Chester*
Mullock, I. (*fl.* 1790s) *Nantwich*
Plant, A. (*fl.* 1790s) *Stockport*
Reid, M. (*fl.* 1790s–1830s) *Middlewich*
Dickson, F. (1793–1866) *Chester*
Dickson, J. H. (*c.* 1795–1867) *Chester*
Caldwell, J. (1797–1840) *Knutsford*
Carr, J. (–*c.* 1803) *Knutsford*
Nickson, J. (–1809) *Knutsford*
Mullock, P. (*fl.* 1800s–1820s) *Nantwich*
Abraham, J. (*fl.* 1800s–1830s) *Chester*
Picken, J. (*c.* 1806–1835) *Knutsford*
Plant, T. (*fl.* 1810s–1820s) *Cheadle*
Mullock, C. (*fl.* 1820s–1830s) *Nantwich*
Field, S. (*fl.* 1820s–1840s) *Boughton*
Weaver, T. (*fl.* 1820s–1840s) *Handbridge*
Caldwell, W. G. (1824–1873) *Knutsford*
Dickson, F. A. (*c.* 1826–1888) *Chester*
Thorley, S. (*fl.* 1830s) *Nantwich*
Weaver, M. (*fl.* 1830s) *Handbridge*
Dickson, T. (*c.* 1835–1877) *Upton*
Dickson, G. A. (*c.* 1835–1909) *Chester*
Dickson, W. A. (*c.* 1837–1891) *Chester*
Abraham, S. (*fl.* 1840s–1850s) *Chester*
Baillie, E. J. (1851–1897) *Chester*
Caldwell, A. (1852–1934) *Knutsford*
Caldwell, W. (1855–1918) *Knutsford*
Caldwell, A. (1865–1939) *Knutsford*
Garner, W. J. (*c.* 1873–1945) *Hale*
Caldwell, W. (1887–1953) *Knutsford*

Cornwall
Nicholls, R. (*fl.* 1790s) *Bodmin*
Boddy, J. H. (–1894) *Land's End*
Gill, R. E. (1875–1942) *Kernick*
Mitchinson, J. (–1901) *Truro*

Cumberland
Greener, J. (*fl.* 1770s–1800s) *Workington*
Clark, T. (*fl.* 1780s–1810s) *Keswick*
Atkinson, C. (*fl.* 1790s) *Keswick*
Denison, H. (*fl.* 1790s) *Carlisle*
Sander, J. (*fl.* 1790s–1810s) *Keswick*
Anson, T. (*fl.* 1810s–1830s) *Carlisle*
Carruthers, J. (*fl.* 1820s–1830s) *Carlisle*
Greener, P. (*fl.* 1820s–1830s) *Workington*
Moses, J. (*fl.* 1820s–1830s) *Penrith*
Fairbairn, G. (*c.* 1846–) *Carlisle*
Askew, W. F. (*c.* 1858–1949) *Borrowdale*
Hayes, T. R. (1864–1927) *Keswick*

Derbyshire
Hopkinson, W. (*fl.* 1790s) *Derby*
Wilson, J. (*fl.* 1790s–1820s) *Derby*
Barron, W. (1800–1891) *Borrowash*
Cooling, E. (1808–1885) *Derby*
Godwin, W. (*fl.* 1820s) *Clifton*
Mason, J. (*fl.* 1820s) *Derby*
Palmer, J. (*fl.* 1820s) *Derby*
Wilson, J. (*fl.* 1820s–1840s) *Derby*
Barron, J. (1844–1906) *Elvaston*

Devonshire
Lucombe, W. (*c.* 1696–1794) *Exeter*
Ford, J. (*c.* 1730s–1796) *Exeter*
Veitch, J. (1752–1839) *Exeter*
Ford, W. (1760–1829) *Exeter*
Bragg, T. (*fl.* 1790s) *Honiton*
Veitch, J. (1792–1863) *Exeter*
Randall, *Mr.* (–1896) *Exeter*
Pince, R. T. (*c.* 1804–1871) *Exeter*
Curtis, H. (*c.* 1819–1889) *Torquay*
Addiscott, W. (*fl.* 1820s–1850s) *Exeter*
Rendle, W. E. (1820–1881) *Plymouth*
Veitch, R. T. (1823–1885) *Exeter*
Woodman, W. R. (*c.* 1829–1891) *Exeter*
Pontey, A. (*fl.* 1830s–1840s) *Plymouth*
Frost, G. (*c.* 1836–1912) *Bampton*
Woods, F. (*fl.* 1840s–1850s) *Plymouth*
Pope, H. (*fl.* 1850s) *Exeter*
Roberts, J. (*fl.* 1850s) *Plymouth*
Veitch, P. C. M. (1850–1929) *Exeter*
Godfrey, W. J. (*fl.* 1880s–1900s) *Exmouth*
Rittershausen, P. R. C. (*c.* 1895–1965) *Newton Abbot*

Dorset
Galpine, J. K. (*fl.* 1790s) *Blandford Forum*
Lance, C. (*fl.* 1790s) *Blandford Forum*
White, R. (*c.* 1810–1877) *Parkstone*
Tutcher, W. (*fl.* 1840s) *Bridport*
Gill, E. (*fl.* 1850s) *Blandford Forum*

Durham
Joyce, W. (–1767) *Gateshead*
Dale, G. (–1781) *Gateshead*
Falla, W. (*c.* 1739–1804) *Jarrow*
Joyce, S. (*fl.* 1750s–1770s) *Gateshead*
Falla, W. (1761–1830) *Jarrow*
Clarke, J. (*fl.* 1770s) *Houghton-le-Spring*
Joyce, J. (*fl.* 1790s) *Gateshead*
Starker, T. (*fl.* 1790s) *Gateshead*
Falla, J. (*fl.* 1790s–1800s) *Jarrow*
Falla, W. (1799–1836) *Gateshead*
Forbes, W. (*fl.* 1820s) *Gateshead*
Finney, S. (*fl.* 1830s–1840s) *Gateshead*

Essex
Aldus, J. (–1767) *Colchester*
Sebborn, J. (–1780) *Colchester*
Hay, J. (–1792) *Leytonstone*
Waynman, T. (–1795/6) *Colchester*
Sorrell, T. (–1797) *Springfield*
Holt, A. (1710s–1750) *Leytonstone*
Essex, T. (*c.* 1716–1799) *Colchester*
Turner, S. (*c.* 1728–1776) *Leyton*
Auston, E. (*c.* 1738–1806) *Colchester*
Pamplin, W. (1741–1805) *Walthamstow*
Cant, W. (1742–1805) *Colchester*
Siborne, R. (*fl.* 1750s–1770s) *Leyton*
Hill, J. (*c.* 1761–1832) *Leytonstone*
Auston, E. (*c.* 1765–1820) *Colchester*
Hughes, J. (*fl.* 1770s–1780s) *Leyton*
Gonner, J. (*fl.* 1770s–1790s) *Colchester*
Essex, J. (*fl.* 1770s–1800s) *Colchester*
Agnis, J. (*fl.* 1770s–1808) *Colchester*
Perkins, W. (*fl.* 1770s–*c.* 1825) *Leyton*
Cant, W. (1779–1831) *Colchester*
Curtis, S. (1779–1860) *Coggeshall*
Mearns, J. (*fl.* 1780s–1790s) *Chelmsford*
Pamplin, J. (1785–1865) *Leyton*
Askew, J. (*fl.* 1790s) *Maldon*
Bruce, J. (*fl.* 1790s) *Dedham*
Cook, J. (*fl.* 1790s) *Halstead*
Cotton, G. (*fl.* 1790s) *Romford*
Crick, J. (*fl.* 1790s) *Rochford*
Fraser, F. (*c.* 1790–*c.* 1849) *Leyton*
Auston, E. (1796–1877) *Colchester*
Cant, G. (–1805) *Colchester*
Sorrell, J. (–1811) *Springfield*
Sorrell, T. (–1843) *Chelmsford*
Sorrell, T. (–1849) *Chelmsford*
Saltmarsh, J. (–1872) *Chelmsford*
Stacey, W. (–1888) *Dunmow*
Chater, W. (1802–1885) *Saffron Walden*
Protheroe, A. (*c.* 1803–1885) *Leytonstone*
Taber, G. (*c.* 1819–1895) *Rivenhall*
Fraser, J. (1820–1863) *Leyton*
Rawlings, G. (*c.* 1820–1903) *Romford*
Fraser, J. (1821–1900) *Leyton*
Saltmarsh, T. J. (*c.* 1827–1899) *Chelmsford*
Cant, B. R. (1827–1900) *Colchester*
Hill, C. (*fl.* 1830s) *Leytonstone*
Perkins, E. (*fl.* 1830s–1860s) *Leyton*
Ward, J. (1831–1903) *Leytonstone*
Prior, D. (*c.* 1832–1916) *Colchester*
Rawlings, J. (*c* 1855–1890) *Romford*
Rawlings, A. (*c.* 1857–1911) *Romford*
Cant, F. (*c.* 1857–1928) *Colchester*
Wallace, R. W. (1867–1955) *Colchester*
Cant, C. E. (*fl.* 1890s–1910s) *Colchester*
Auston, G. E. (–1914) *Colchester*
Vert, J. (–1929) *Saffron Walden*
King, E. W. (–1930) *Coggeshall*
Seabrook, W. (*fl.* 1900s) *Chelmsford*
Prior, W. D. (*fl.* 1910s) *Colchester*
Riding, J. B. (*fl.* 1910s) *Chingford*

Gloucestershire
Berry, J. (–1727) *Tytherington*
Clark, H. (*c.* 1702–1778) *Chipping Camden*
Wheeler, J. (*fl.* 1750s–1760s) *Gloucester*
King, T. (*fl.* 1760s–1780s) *Bristol*

Hansard, P. (*fl.* 1780s) *Bristol*
Sandford, J. (*fl.* 1780s) *Chipping Camden*
Collett, H. (*fl.* 1790s) *Tewkesbury*
Holbert, R. (*fl.* 1790s) *Gloucester*
Lauder, A. (*fl.* 1790s) *Bristol*
Sweet, J. (*fl.* 1790s) *Bristol*
Lauder, P. (*fl.* 1790s–1810s) *Bristol*
Maddock(s), J. (*fl.* 1790s–1830s) *Bristol*
Foster, W. (*c.* 1799–1877) *Stroud*
Wheeler, J. C. (–1860) *Gloucester*
Mayes, M. (*c.* 1801–1858) *Bristol*
Wheeler, E. (*fl.* 1810s–1820s) *Gloucester*
Maule, W. (*fl.* 1810s–1860s) *Bristol*
Heath, W. (*c.* 1810–1892) *Cheltenham*
Jefferies, J. (*c.* 1818–1904) *Cirencester*
Jessop, C. H. (*fl.* 1820s) *Cheltenham*
Miller, J. (*fl.* 1820s–1830s) *Bristol*
Wheeler, J. D. (*fl.* 1820s–1840s) *Gloucester*
Maule, A. J. (*c.* 1821–1884) *Bristol*
Cypher, J. (1827–1901) *Cheltenham*
Baker, C. (*fl.* 1830s) *Bristol*
Hodges, S. (*fl.* 1830s) *Cheltenham*
Gregory, W. (*fl.* 1830s–1840s) *Cirencester*
Garraway, J. (*fl.* 1830s–1850s) *Bristol*
Baylis, C. (*c.* 1838–1904) *Bourton-on-the-Water*
Arnott, R. (*fl.* 1840s) *Charlton Kings*
Jefferies, W. J. (1844–) *Cirencester*
Cypher, J. J. (*c.* 1854–1928) *Cheltenham*

Hampshire
Irwin, G. (*fl.* 1740s–1770s) *Southampton*
Parsons, T. (*fl.* 1780s) *Fareham*
Armstrong, J. (*fl.* 1780s–1790s) *North Warnborough*
Keen, I. (*fl.* 1780s–1810s) *Southampton*
Arnell, J. (*fl.* 1790s) *Winchester*
Collins, T. (*fl.* 1790s) *Havant*
Dash, T. (*fl.* 1790s) *Gosport*
Hacker, J. (*fl.* 1790s) *Basingstoke*
Marshall, J. (*fl.* 1790s) *Romsey*
Pink, W. (*fl.* 1790s) *Winchester*
Page, W. B. (1790–1871) *Southampton*
Windebank, W. (–1878) *Southampton*
Sawyer, R. (*c.* 1812–1907) *Southampton*
Rogers, W. H. (1818–1898) *Southampton*
Kingsbury, J. (*c.* 1821–1884) *Southampton*
Toogood, W. (*c.* 1827–1892) *Southampton*
Mott, W. (*fl.* 1830s) *Basingstoke*
Shilling, J. (*fl.* 1830s) *Odiham*
Taylor, W. (*fl.* 1830s) *Southampton*
Ingram, J. (*fl.* 1830s–1840s) *Southampton*
Hillier, E. (1840–1929) *Winchester*
Drover, W. (*c.* 1841–) *Fareham*
Henry, J. M. (1841–1937) *Hartley Wintney*
Hillier, E. L. (1865–1944) *Winchester*
Auton, W. J. (1875–1931) *Botley*
Mathias, H. (–1912) *Medstead*

Hertfordshire
Lucas, R. (–1733/4) *Cheshunt*
Clark, H. (–1782/3) *Barnet*
Emmerton, I. (*c.* 1736–1789) *Barnet*
Emmerton, T. (*fl.* 1760s) *Barnet*
Emmerton, I. (*c.* 1769–1823) *Barnet*
Bridgeman, C. (*fl.* 1790s) *Hertford*
Jeeves, J. (*fl.* 1790s) *Hitchin*
Cornwell, W. (*fl.* 1790s–1820s) *Barnet*
Rivers, T. (1798–1877) *Sawbridgeworth*
Murray, R. (*fl.* 1800s–1820s) *Hertford*
Lane, J. E. (*c.* 1807–1889) *Berkhamsted*
Cornwell, G. (*fl.* 1820s–1830s) *Barnet*
Kemp, G. (*fl.* 1820s–1840s) *Barnet*
Paul, W. (1822–1905) *Waltham Cross*
Cutbush, J. (1827–1885) *Barnet*
Rivers, T. F. (1831–1899) *Sawbridgeworth*
Foden, W. (*c.* 1841–1916) *Hemel Hempstead*
Paul, G. (1841–1921) *Cheshunt*
Sander, H. F. C. (1847–1920) *St. Albans*
Rochford, E. (1854–1919) *Cheshunt*
Harkness, R. (1855–1920) *Hitchin*
Rochford, J. (1856–1932) *Broxbourne*
Rivers, T. A. H. (1863–1915) *Sawbridgeworth*
Paul, A. W. (*fl.* 1870s–1910s) *Waltham Cross*
Sander, F. K. (1876–1951) *St. Albans*
Rochford, T. S. (1877–1918) *Cheshunt*
Bates, G. (*c.* 1880–1971) *Sawbridgeworth*
Elliott, C. (1881–1969) *Stevenage*
Lane, F. Q. (–1907) *Berkhamsted*
Paul, G. L. (*fl.* 1910s) *Cheshunt*

Huntingdonshire
Wood, J. (*c.* 1792–1830) *Huntingdon*
Ingram, J. (*c.* 1822–1876) *Huntingdon*
Hutchins, J. (*fl.* 1830s) *Tilbrook*

Isle of Wight
Cave, E. (*c.* 1822–1905) *Newport*

Kent
Russell, J. (*c.* 1731–1794) *Lewisham*
Stidolph, G. (1733/4–1818) *Bromley*
Pringle, W. (*c.* 1742–1813) *Sydenham*
Stidolph, G. (1760–1848) *Bromley*
Edmeades, R. (*fl.* 1770s–1780s) *Deptford*
Stidolph, W. (1772–1855) *Bromley*
Willmott, J. (1775–1834) *Lewisham*
Gregory, A. (*fl.* 1780s) *Faversham*
Hawkes, A. (*fl.* 1780s) *Sandwich*
Stidolph, H. (*fl.* 1780s–1800s) *Bromley*
Aldersey, J. (*fl.* 1790s) *Sittingbourne*
Collins, R. (*fl.* 1790s) *Maidstone*
Grist, W. (*fl.* 1790s) *Maidstone*
Bunyard, J. (*fl.* 1790s–1810s) *Maidstone*
Fraser, J. (*fl.* 1790s–1860s) *Ramsgate*
Sim, R. (1791–1878) *Foots Cray*
Parks, J. D. (*c.* 1792–1866) *Dartford*
Stidolph, D. (1793–1840s) *Bromley*
Masters, W. (1796–1874) *Canterbury*
Maller, B. (–1884) *Lee*
Masters, J. (*fl.* 1800s) *Canterbury*
Bunyard, T. (*c.* 1804–1880) *Maidstone*
Cripps, T. (*c.* 1809–1888) *Tunbridge Wells*
Cormack, W. (*fl.* 1810s–1840s) *Deptford*
Lane, G. (*c.* 1810–1885) *St. Mary's Cray*
Ayres, W. P. (1815–1875) *Blackheath*
Epps, W. J. (*c.* 1817–1885) *Maidstone*
Seale, W. (*c.* 1820–1885) *Sevenoaks*
Frost, T. (*c.* 1823–1882) *Maidstone*
Sim, R. (*c.* 1828–1882) *Foots Cray*
Cottell, J. (*fl.* 1830s) *Westerham*
Smith, J. (*fl.* 1830s–1840s) *Westerham*
Longley, C. (*c.* 1830–1915) *Rainham*
Cannell, H. (1833–1914) *Swanley*
Cattell, J. (*fl.* 1840s) *Westerham*
Stidolph, J. (*fl.* 1840s) *Bromley*
Cripps, W. T. (*c.* 1840–1871) *Tunbridge Wells*
Mount, G. (*c.* 1844–1927) *Canterbury*
Blick, C. (*c.* 1856–1919) *Hayes*
Wallace, R. W. (1867–1955) *Tunbridge Wells*
Bunyard, E. A. (1878–1939) *Maidstone*
Constable, W. A. (1887–1954) *Tunbridge Wells*
Lawrence, H. C. (–1917) *Chatham*

Reuthe, G. (–1942) *Keston*
Murrell, O. (–1957) *Orpington*
Cannell, R. (*fl.* 1910s) *Swanley*

Lancashire
Caldwell, J. (–1795) *Liverpool*
Blundell, J. (–*c.* 1798) *Ormskirk*
Middlewood, W. (*fl.* 1770s) *Manchester*
Clarke, J. (*fl.* 1770s–1780s) *Liverpool*
McNiven, C. (*fl.* 1770s–1815) *Manchester*
Cropper, J. (1773–1840) *Warrington*
Clarke, T. (*fl.* 1780s) *Liverpool*
Bridge, W. (*fl.* 1780s–1790s) *Liverpool*
Pinkerton, W. (*fl.* 1780s–1790s) *Wigan*
Whalley, T. (*fl.* 1780s–1810s) *Liverpool*
McNiven, P. (*fl.* 1780s–1818) *Manchester*
Brockbank, T. (*fl.* 1790s) *Hawkshead*
Caldwell, T. (*fl.* 1790s) *Liverpool*
Collins, W. (*fl.* 1790s) *Liverpool*
Cooke, A. (*fl.* 1790s) *Manchester*
Cope, E. (*fl.* 1790s) *Chorley*
Cope, R. (*fl.* 1790s) *Chorley*
Downham, W. (*fl.* 1790s) *Liverpool*
Ledgerwood, J. (*fl.* 1790s) *Ulverstone*
Wilkinson, T. (*fl.* 1790s–1810s) *Barton-upon-Irwell*
Hankin, J. (*fl.* 1790s–1820s) *Ormskirk*
Caldwell, W. (–*c.* 1813) *Knowsley*
Cheetham, J. (–1890) *Rochdale*
Thorley, W. (*fl.* 1800s–1810s) *Manchester*
Vaughan, G. (*fl.* 1800s–1830s) *Manchester*
Cunningham, G. (*c.* 1800–1891) *Liverpool*
Leeds, E. (1802–1877) *Manchester*
Bridgeford, J. (*fl.* 1810s) *Manchester*
Collier, A. (*fl.* 1810s) *Liverpool*
Crossley, S. (*fl.* 1810s) *Manchester*
Eaton, J. (*fl.* 1810s) *Warrington*
Fairclough, B. (*fl.* 1810s) *Liverpool*
Mercer, J. (*fl.* 1810s) *Liverpool*
Salmon, T. (*fl.* 1810s) *Warrington*
Standley, J. (*fl.* 1810s) *Manchester*
Waltham, W. (*fl.* 1810s) *Manchester*
Cooper, J. (1810s–1820s) *Wigan*
Cunningham, G. (*fl.* 1810s–1830s) *Liverpool*
Boardman, R. (*fl.* 1810s–1840s) *Wigan*
Rylance, C. (*c.* 1811–1884) *Ormskirk*
Cole, W. (*c.* 1812–1864) *Manchester*
Davies, I. (1812–1897) *Ormskirk*
Ker, R. P. (1816–1886) *Liverpool*
Cooper, W. (*fl.* 1820s) *Wigan*
Davies, P. (*fl.* 1820s) *Warrington*
Davies, T. (1829–1902) *Wavertree*
Lodge, W. (*fl.* 1830s) *Manchester*
Skirving, W. (*fl.* 1830s) *Liverpool*
Wright, T. (*fl.* 1830s) *Warrington*
Toll, G. (*c.* 1835–1884) *Manchester*
Shand, W. (*c.* 1838–1903) *Lancaster*
Ker, R. W. (1839–1910) *Liverpool*
Bruce, A. J. A. (*c.* 1843–) *Chorlton*
Stansfield, W. H. (1850–1934) *Southport*
Davies, W. (*c.* 1854–1889) *Ormskirk*
Mahood, R. W. (*c.* 1882–1970) *Ormskirk*
Grubb, C. W. (–1954) *Bolton-le-Sands*

Leicestershire
Hubbard, W. (–1787) *Church Langton*
Harrison, J. (–1788) *Leicester*
Hanbury, *Rev.* W. (1725–1778) *Church Langton*
Billings, J. (*fl.* 1790s) *Hinckley*
Cort, J. (*fl.* 1790s) *Leicester*
Cox, W. (*fl.* 1790s) *Leicester*
Davenport, R. (*fl.* 1790s) *Ashby de la Zouch*
Wilson, A. (*fl.* 1790s) *Market Harborough*
Harrison, G. (–1808) *Leicester*
Harrison, J. (–1839) *Leicester*
Harrison, T. (–1890) *Leicester*
Brown, T. (*fl.* 1820s) *Measham*
Sweet, J. (*c.* 1839–1924) *Whetstone*
Warner, T. (*fl.* 1840s) *Leicester*
Woodcock, W. K. (*c.* 1841–1897) *Humberstone*
Harrison, J. (*fl.* 1860s–1910s) *Leicester*
Hudson, J. (*fl.* 1880s–1900s) *Leicester*
Harrison, W. A. (–1916) *Leicester*

Lincolnshire
Wood, W. (*fl.* 1780s–1790s) *Grantham*
Atkin, W. (*fl.* 1790s) *Alford*
Cannon, M. (*fl.* 1790s) *Louth*
Fountain, R. (*fl.* 1790s) *Waltham*
Lowe, C. (*fl.* 1790s) *Market Deeping*
Luck, P. (*fl.* 1790s) *Louth*
M'Callan, D. (*fl.* 1790s) *Gainsborough*
Crowder, A. (1792–1873) *Cagthorpe*
Crowder, W. (–1836) *Cagthorpe*
Pennell, R. (*fl.* 1820s–1869) *Lincoln*
Pennell, C. (*c.* 1821–1891) *Lincoln*
Tuxworth, G. (*fl.* 1830s) *Louth*
Sharpe, C. (*c.* 1830–1897) *Sleaford*
Johnson, A. (*c.* 1840–1899) *Boston*
White, J. T. (*c.* 1848–1930) *Spalding*

London and Middlesex see also Essex, Hertfordshire, Kent, Surrey]
Cawsway, *Mr.* (*fl.* 1500s) *Houndsditch*
Banbury, H. (1540–1609/10) *Westminster*
Banbury, A. (1598–1664/5) *Westminster*
Pointer, V. (–1619) *Twickenham*
Millen, J. (–1635) *Cripplegate*
Lucas, W. (–1679) *Strand*
Looker, R. (–1685) *Brompton*
Crofton, R. (1603–1630s) *Twickenham*
Gurle, L. (*c.* 1621–1685) *Whitechapel*
Wrench, T. (*c.* 1630–1728) *Fulham*
Long, J. (*fl.* 1650s) *Billingsgate*
Wise, H. (1653–1738) *Brompton*
Ricketts, G. (*fl.* 1660s–1706) *Hoxton*
Cook, M. (*fl.* 1660s–1715) *Brompton*
Fairchild, T. (1667–1729) *Hoxton*
Ball, F. (*fl.* 1670s) *Brentford*
Crouch, *Mr.* (*fl.* 1670s) *Bishopsgate*
Darby, W. (*fl.* 1670s–1710s) *Hoxton*
Fuller, E. (*fl.* 1670s–1720s) *Strand*
Furber, R. (*c.* 1674–1756) *Kensington*
Blackwell, C. (*fl.* 1680s) *Holborn*
Stacy, T. (*fl.* 1680s) *Bishopsgate*
Ricketts, J. (*fl.* 1680s–1710s) *Hoxton*
Weston, F. (*fl.* 1680s–1710s) *Strand*
Mason, P. (1680–1730) *Isleworth*
Fuller, J. (*fl.* 1690s) *Strand*
Pearson, *Mr.* (*fl.* 1690s) *Hoxton*
Turner, J. (*fl.* 1690s–1730s) *Strand*
Cowell, J. (*fl.* 1690s–1730) *Hoxton*
Gray, C. (*c.* 1693/4–1764) *Fulham*
Gray, S. (1694–1766) *Pall Mall*
Woodman, H. (*c.* 1698–1758) *Strand-on-the-Green*
London, G. (–1714) *Kensington*
Mason, P. (–1719) *Isleworth*
Parkinson, J. (–1719) *Lambeth*
Reynardson, S. (–1721) *Hillingdon*
Moore, G. (–1729) *Twickenham*
Simpson, J. –1740) *Shoreditch*

Gray, W. (–1745) *Fulham*
Gray, S. (–1766) *Pall Mall*
North. R. (–1766) *Lambeth*
Scott, J. (–1770) *Chiswick*
Hewitt, H. (–1771) *Kensington*
Foster, J. (–1781) *Strand*
Luker, W. (–1784) *City Road*
Eddie, A. (–1788) *Strand*
Hewitt, H. (–1791) *Kensington*
James, J. T. (–1791) *Lambeth*
Bailey, I. (–*c*. 1793) *Bishopgate*
Gordon, J. (*c*. 1708–1780) *Mile End*
Bacon, S. (1709–1734) *Hoxton*
Kirkham, T. (*c*. 1709–1785) *Strand*
Minier, C. (*c*. 1710–1790) *Strand*
Maddock, J. (*c*. 1715–1786) *Walworth*
Lee, J. (1715–1795) *Hammersmith*
Watson, W. (*c*. 1718–1793) *Islington*
Booth, J. (*fl*. 1720s) *Fleet Street*
Bunney, J. (*fl*. 1720s) *Uxbridge*
Chapman, *Mr*. (*fl*. 1720s) *Hoxton*
Lowe, O. (*fl*. 1720s) *Battersea*
Masters, G. (*fl*. 1720s) *Chiswick*
Parker, N. (*fl*. 1720s) *Chiswick*
James, M. (*fl*. 1720s–1730s) *Lambeth*
Fuller, A. (*fl*. 1720s–1740s) *Strand*
Thompson, J. (*fl*. 1720s–*c*. 1758) *Chelsea*
Driver, S. (*fl*. 1720s–1779) *Southwark*
Kennedy, L. (*c*. 1721–1782) *Hammersmith*
Grimwood, D. (*c*. 1725–1796) *Kensington*
Burchell, W. (*c*. 1725–1800) *Fulham*
Ronalds, H. (*c*. 1726–1788) *Isleworth*
Abercrombie, J. (1726–1806) *Hackney*
Howey, W. (*c*. 1729–1792) *Putney*
Allerton, J. (*fl*. 1730s) *Knightsbridge*
Alston, J. (*fl*. 1730s) *Chelsea*
Bincks, W. (*fl*. 1730s) *Thames Street*
Cole, R. (*fl*. 1730s) *Battersea*
Hood, W. (*fl*. 1730s) *Hyde Park Corner*
Swinhoe, J. (*fl*. 1730s–1750s) *Brompton*
Rocque, B. (*fl*. 1730s–1760s) *Fulham*
Busch, J. (*fl*. 1730s–1790s) *Hackney*
Minier, C. (*c*. 1738–1808) *Strand*
Gray, S. (1739–1771) *Pall Mall*
Loddiges, C. (*c*. 1739–1826) *Hackney*
Ashe, T. (*fl*. 1740s–1750s) *Twickenham*
Swinden, N. (*fl*. 1740s–1770s) *Brentford*
Powell, N. (*fl*. 1740s–1773) *Holborn*
Teesdale, R. (*c*. 1740–1804) *Strand*
Whitlock, J. (*c*. 1740–1827) *Chelsea*
Coleman, W. (*c*. 1743–1808) *Tottenham*
Curtis, W. (1746–1799) *Brompton, Chelsea, Lambeth*
Colvill, J. (*c*. 1746–1822) *Chelsea*
Renton, J. (*c*. 1747–1810) *Hoxton*
Barritt, P. (*c*. 1748–1798) *Kensington*
Fitch, D. (*c*. 1748–1818) *Fulham*
Clarke, E. (*fl*. 1750s) *Strand*
Webb, J. (*fl*. 1750s–1760s) *Westminster*
Topping, T. (*fl*. 1750s–1770s) *London Bridge*
Williamson, J. (*fl*. 1750s–1780s) *Kensington*
Dupin, P. (*fl*. 1750s–1790s) *Strand*
Lane, J. (*fl*. 1750s–1790s) *Clerkenwell*
Malcolm, W. (*fl*. 1750s–1798) *Stockwell*
Fraser, J. (1750–1811) *Chelsea*
Burchell, (*c*. M. 1753–1828) *Fulham*
Thomson, A. (*c*. 1753–1832) *Mile End Road*
Lee, J. (1754–1824) *Hammersmith*
Whitley, R. (*c*. 1754–1835) *Fulham, Kensington*
Davey, T. (*c*. 1758–1833) *Chelsea*
Ronalds, H. (1759–1833) *Brentford*
Kennedy, J. (1759–1842) *Hammersmith*
White, W. (*fl*. 1760s) *Mile End*
Aslet, J. (*fl*. 1760s–1770s) *Isleworth*
Garraway, S. (*fl*. 1760s–1770s) *Fleet Street*
Sheilds, J. (*fl*. 1760s–1780s) *Lambeth*
Wrench, J. (*fl*. 1760s–1780s) *Lower Thames Street*
Cross, E. (*fl*. 1760s–1790s) *Fleet Street*
Gammock, A. (*fl*. 1760s–1790s) *Hoxton*
Bowstread, W. (*fl*. 1760s–1800s) *Highgate*
Field, J. (*fl*. 1760s–1800s) *Lower Thames Street*
Kirk, J. (*fl*. 1760s–1820s) *Brompton*
Gray, R. (*fl*. 1760s–1840s) *Kensington*
Cobbett, W. (1762–1835) *Kensington*
Keens, M. (*c*. 1762–1835) *Isleworth*
Maddock, J. (1763–1825) *Walworth*
Ross, W. (*c*. 1763–1825) *Stoke Newington*
Money, D. (*c*. 1763–1833) *Hampstead Road*
Rollisson, W. (*c*. 1765–1842) *Upper Tooting*
Milne, T. (*c*. 1767–1838) *Fulham*
Malcolm, W. (*c*. 1768–1835) *Stockwell*
Pamplin, W. (1768–1844) *Chelsea*
Blackwell, W. (*fl*. 1770s) *Covent Garden*
Cowie, J. (*fl*. 1770s) *Westminster*
Edwards, D. (*fl*. 1770s) *Lower Thames Street*
West, D. (*fl*. 1770s) *St. Pancras*
Smith, W. (*fl*. 1770s–1789) *Kensington*
Watson, J. (*fl*. 1770s–1790s) *Islington*
Hay, W. (*fl*. 1770s–1800s) *St. Georges Field*
Hairs, J. (*fl*. 1770s–1810s) *Haymarket*
Shailer, H. (*fl*. 1770s–1820s) *Kensington*
Forsyth, W. (1772–1835) *Fenchurch Street*
Colvill, J. (*c*. 1777–1832) *Chelsea*
Knight, J. (*c*. 1777–1855) *Chelsea*
Curtis, S. (1779–1860) *Walworth*
Ashe, W. (*fl*. 1780s) *Twickenham*
Bendel, J. (*fl*. 1780s) *Walworth*
Brinkworth, *Mr*. (*fl*. 1780s) *Whitechapel*
Cunningham, A. (*fl*. 1780s) *Paddington*
Davies, W. (*fl*. 1780s) *Snow Hill*
Fisher, J. (*fl*. 1780s) *Pall Mall*
Hanson, I. (*fl*. 1780s) *Thames Street*
Jeffreys, J. (*fl*. 1780s) *Kensington*
Lucas, E. (*fl*. 1780s) *High Holborn*
Mitchelson, G. (*fl*. 1780s) *Kennington*
Neal, G. (*fl*. 1780s) *Camberwell*
Shailer, A. (*fl*. 1780s) *Kensington*
Thoburn, P. (*fl*. 1780s) *Brompton*
Watts, D. (*fl*. 1780s) *St. James's Street*
Ashe, J. (*fl*. 1780s–1790s) *Twickenham*
Cook, J. (*fl*. 1780s–1790s) *Kensington*
Mackie, R. (*fl*. 1780s–1790s) *Islington*
Thatcher, S. (*fl*. 1780s–1790s) *Fleet Street*
Townley, E. W. (*fl*. 1780s–1790s) *Walworth*
Wilson, J. (*fl*. 1780s–1790s) *Smithfield*
Driver, A. P. (*fl*. 1780s–1800s) *Southwark*
Richards, T. (*fl*. 1780s–1800s) *Hackney*
Shields, A. (*fl*. 1780s–1805) *Brentford*
Gray, J. (*fl*. 1780s–1810s) *Kensington*
Mason, J. (*fl*. 1780s–1810s) *Fleet Street*
Shailer, J. (*fl*. 1780s–1810s) *Kensington*
Duthie, W. (*fl*. 1780s–1820s) *Bethnal Green*
Williams, R. (*fl*. 1780s–*c*. 1826) *Chiswick*
Smith, S. (*fl*. 1780s–1830s) *Hackney*
Smith, E. (*fl*. 1780s–1840s) *Hackney*
Shepherd, T. (1780–1835) *Hackney*
Crastin, C. (*c*. 1782–1849) *Islington*
Henderson, E. G. (*c*. 1782–1876) *St. John's Wood*
Sweet, R. (1783–1835) *Chelsea*

Loddiges, G. (1784–1846) *Hackney*
Charlwood, G. (*c.* 1784–1861) *Covent Garden*
Allport, J. (*fl.* 1790s) *Shoreditch*
Barr, T. (*fl.* 1790s) *Islington*
Beach, G. (*fl.* 1790s) *Southwark*
Blackwell, J. (*fl.* 1790s) *Covent Garden*
Cooper, N. (*fl.* 1790s) *Canterbury Square*
Cuthbert, J. (*fl.* 1790s) *Southgate*
Davidson, J. (*fl.* 1790s) *Southwark*
Dermer, T. (*fl.* 1790s) *Fenchurch Street*
Harris, R. (*fl.* 1790s) *Lower Thames Street*
Harrison, J. (*fl.* 1790s) *Kensington*
Hogarth, A. (*fl.* 1790s) *Islington*
Lewis, D. (*fl.* 1790s) *Cornhill*
Mawdesley, J. (*fl.* 1790s) *Enfield*
Mitchell, G. (*fl.* 1790s) *New Bond Street*
White, J. (*fl.* 1790s) *Whitechapel*
Driver, W. (*fl.* 1790s–1800s) *Southwark*
North, W. (*fl.* 1790s–1800s) *Lambeth*
Watson, T. (*fl.* 1790s–1800s) *Islington*
Bowie, A. (*fl.* 1790s–1820s) *Portman Square*
Bridge, T. (*fl.* 1790s–1820s) *Trinity Square*
Buchanan, J. (*fl.* 1790s–1820s) *Camberwell*
Gibbs, T. (*fl.* 1790s–1820s) *Old Brompton Road*
Hunt, J. (*fl.* 1790s–1820s) *Borough*
Harrison, S. (*fl.* 1790s–1830s) *Kensington*
Henderson, A. (*fl.* 1790s–1840s) *Edgware Road*
Veitch, J. (1792–1863) *Chelsea*
Low, H. (1793–1863) *Clapton*
Ansell, T. (*c.* 1793–1867) *Camden Town*
Henderson, J. A. (*c.* 1795–1872) *Edgware Road*
Appleby, T. (*c.* 1795–1875) *Edgware Road*
Mackay, J. B. (1795–1888) *Clapton*
Sangster, J. (*c.* 1796–1881) *Newington Butts*
Salter, J. (1798–1874) *Hammersmith*
Hally, J. (*c.* 1798–1879) *Blackheath*
Hurst, W. (*c.* 1799–1868) *Leadenhall Street*
Rollisson, G. (*c.* 1799–1879) *Upper Tooting*
Ronalds, R. (*c.* 1799–1880) *Brentford*
Campbell, J. (–1804) *Hampstead*
Salisbury, W. (–1823) *Chelsea*
Weare, T. (–1829) *Kensington*
Brames, P. (–1834) *Brompton, Fulham*
Brown, R. (–*c.* 1837) *Kensington*
Beck, E. (–1861) *Isleworth*
Osborn, W. (–1872) *Fulham*
Allen, D. (*fl.* 1800s) *Hammersmith*
Allport, W. (*fl.* 1800s) *Shoreditch*
Bailey, T. (*fl.* 1800s) *Covent Garden*
Barrett, R. (*fl.* 1800s) *Hornsey*
Bassington, J. (*fl.* 1800s) *Smithfield*
Batt, J. (*fl.* 1800s) *Bethnal Green*
Brooks, T. (*fl.* 1800s) *New Bond Street*
Dale, T. (*fl.* 1800s) *Islington*
Fairbairn, J. (*fl.* 1800s) *Clapham*
Harris, T. (*fl.* 1800s) *Battersea*
Malcolm, A. (*fl.* 1800s) *Stockwell*
Napier, *Mr.* (*fl.* 1800s) *Vauxhall*
Shailer, F. (*fl.* 1800s) *Kensington*
Watts, W. (*fl.* 1800s) *Lambeth*
Allport, J. (*fl.* 1800s–1810s) *Holborn Hill*
Fair, R. (*fl.* 1800s–1820s) *Borough*
Gardner, J. (*fl.* 1800s–1820s) *City Road*
Graham, R. (*fl.* 1800s–1820s) *Camden Town*
Tubb, W. (*fl.* 1800s–1820s) *Chelsea*
Turner, J. (*fl.* 1800s–1820s) *New Bond Street*
Couldry, W. (*fl.* 1800s–1830s) *Camberwell*
Middlemist, J. (*fl.* 1800s–1830s) *Hammersmith*
Minier, W. (*fl.* 1800s–1830s) *Strand*
Ross, J. (*fl.* 1800s–1830s) *Hackney*
Bassington, T. (*fl.* 1800s–1840s) *Islington*
Jenkins, T. (1800–1832) *Marylebone*
Rollisson, W. (*c.* 1802–1875) *Upper Tooting*
Chandler, A. (1804–1896) *Vauxhall*
Glendinning, R. (1805–1862) *Chiswick*
Thomson, J. W. (*c.* 1805–1895) *Hammersmith*
Lee, J. (*c.* 1805–1899) *Hammersmith*
Nash, D. (*c.* 1806–1874) *Strand*
Lee, C. (1808–1881) *Hammersmith*
Barnes, W. (1809–1869) *Camberwell*
Laing, R. (*c.* 1809–1887) *Twickenham*
Bell, T. B. (*fl.* 1810s) *Isleworth*
Butler, J. (1810s) *Covent Garden*
Coaker, J. (*fl.* 1810s) *Kensington*
Fisher, F. (*fl.* 1810s) *Paddington*
Goldsbury, J. (*fl.* 1810s) *Whitechapel*
Hall, J. (*fl.* 1810s) *St. Pancras*
Howard, *Mrs.* (*fl.* 1810s) *Chelsea*
Munro, J. (*fl.* 1810s) *Lambeth*
Allnut, *Mr.* (*fl.* 1810s–1820s) *Clapham*
Allport, G. (*fl.* 1810s–1820s) *Shoreditch*
Bassington, G. (*fl.* 1810s–1820s) *Islington*
Cleveley, J. (*fl.* 1810s–1820s) *Stoke Newington*
Grange, J. (*fl.* 1810s–1820s) *Hackney*
Hay, J. (*fl.* 1810s–1820s) *Lambeth*
Bristow, W. (*fl.* 1810s–1830s) *Kensington*
Buckingham, E. B. (*fl.* 1810s–1830s) *Lambeth*
Phillipps, L. (*fl.* 1810s–1830s) *Vauxhall*
Smith, G. (*c.* 1812–1883) *Hornsey Road*
Wrench, R. (*c.* 1813–1883) *London Bridge*
Chitty, W. (*c.* 1814–1894) *Stamford Hill*
Veitch, J. (1815–1869) *Chelsea*
Fromow, W. (1815–1886) *Chiswick*
Clay, S. (*c.* 1816–1899) *Stratford*
Osborn, T. (*c.* 1819–1872) *Fulham*
Rochford, M. (*c.* 1819–1883) *Tottenham*
Allen, J. (*fl.* 1820s) *King's Road*
Andrews, I. (*fl.* 1820s) *Lambeth*
Axton, A. (*fl.* 1820s) *Lisson Grove*
Barratt, J. (*fl.* 1820s) *Mornington Place*
Brown, T. (*fl.* 1820s) *St. Pancras*
Cochran, J. (*fl.* 1820s) *Grosvenor Square*
Coleman, C. (*fl.* 1820s) *Tottenham*
Coleman, S. (*fl.* 1820s) *Tottenham*
Coles, B. (*fl.* 1820s) *Chelsea*
Collins, J. (*fl.* 1820s) *Marylebone*
Connelly, T. (*fl.* 1820s) *Hammersmith*
Ellingham, J. (*fl.* 1820s) *Queens Gate*
Farnes, W. W. (*fl.* 1820s) *West Smithfield*
Fowler, G. (*fl.* 1820s) *St. George's*
Fraser, J. T. (*fl.* 1820s) *Chelsea*
Holmes, R. (*fl.* 1820s) *Lambeth*
Howard, M. (*fl.* 1820s) *Chelsea*
Jennings, J. (*fl.* 1820s) *Hampstead Road*
King, J. (*fl.* 1820s) *Hackney*
Lawrance, W. (*fl.* 1820s) *New Road*
Ludgater, J. (*fl.* 1820s) *Shoreditch*
Mackenzie, D. (*fl.* 1820s) *Pimlico*
MacPherson, A. (*fl.* 1820s) *Oxford Street*
Mitchell, W. (*fl.* 1820s) *Borough Road*
More, J. (*fl.* 1820s) *Chelsea*
Neale, S. (*fl.* 1820s) *Hackney*
Tansley, J. (*fl.* 1820s) *Walworth*
Vogan, J. (*fl.* 1820s) *Southwark*
Brookes, S. (*fl.* 1820s–1830s) *Islington*
Clarke, H. (*fl.* 1820s–1830s) *Oxford Street*
MacKay, J. (*fl.* 1820s–1830s) *Hackney*
Noble, W. (*fl.* 1820s–1830s) *Fleet Street*
Tate, J. C. (*fl.* 1820s–1830s) *Chelsea*
Batt, G. (*fl.* 1820s–1840s) *Strand*

Bunney, G. H. (*fl.* 1820s–1840s) *Islington*
Colley, W. (*fl.* 1820s–1840s) *Hammersmith*
Forrest, R. (*fl.* 1820s–1840s) *Kensington*
Cutbush, W. (*fl.* 1820s–1850s) *Highgate*
Groom, H. (*fl.* 1820s–1850s) *Walworth*
Smith, G. (*fl.* 1820s–1850s) *Islington*
Holmes, W. (1820–1878) *Hackney*
Laing, J. (1823–1900) *Stanstead Park*
Williams, B. S. (1824–1890) *Holloway*
Ware, T. S. (*c.* 1824–1901) *Tottenham*
Low, *Sir* H. (1824–1905) *Clapton*
Weatherill, R. (*c.* 1825–1883) *Finchley*
Low, S. H. (1826–1890) *Clapton*
Barr, P. (1826–1909) *Covent Garden*
Cutbush, J. (1827–1885) *Highgate, Finchley*
Naylor, J. (*c.* 1827–1901) *Harrow*
Nutting, W. J. (*c.* 1827–1910) *Southwark*
Bull, W. (1828–1902) *Chelsea*
Rhodes, O. (*c.* 1829–1869) *Sydenham*
Alexander, W. (*fl.* 1830s) *Islington*
Boorn, J. (*fl.* 1830s) *Bermondsey*
Bunyard, E. (*fl.* 1830s) *City Road*
Catleugh, W. (*fl.* 1830s) *Chelsea*
Clisby, G. (*fl.* 1830s) *Strand*
Cook, T. (*fl.* 1830s) *Bermondsey*
Denyer, E. (*fl.* 1830s) *Brixton*
Elliott, J. (*fl.* 1830s) *Putney*
Enkel, R. (*fl.* 1830s) *Islington*
Gellan, T. (*fl.* 1830s) *Hackney*
Hardy, J. (*fl.* 1830s) *Edmonton*
Hayhow, R. (*fl.* 1830s) *Camberwell*
Henchman, J. (*fl.* 1830s) *Edmonton*
Hopwood, E. (*fl.* 1830s) *Twickenham*
Kennon, J. (*fl.* 1830s) *Bagnigge Wells Road*
Kernan, J. (*fl.* 1830s) *Covent Garden*
Mcarthur, P. (*fl.* 1830s) *Edgware Road*
Maine, S. (*fl.* 1830s) *Strand*
Markham, W. (*fl.* 1830s) *Stockwell*
Mitchell, R. (*fl.* 1830s) *Hackney*
Owen, J. (*fl.* 1830s) *Bermondsey*
Page, T. (*fl.* 1830s) *Edmonton*
Thomson, J. (*fl.* 1830s) *Mile End Road*
Waredraper, T. (*fl.* 1830s) *Hackney*
Buchanan, W. J. (*fl.* 1830s–1840s) *Camberwell*
Catleugh, W. (*fl.* 1830s–1840s) *Chelsea*
Dulley, W. (*fl.* 1830s–1840s) *Hackney*
Gaines, N. (*fl.* 1830s–1840s) *Battersea*
Smith, J. (*fl.* 1830s–1840s) *Hackney*
Tuck, J. H. (*fl.* 1830s–1840s) *Sloane Street*
Carter, J. (*fl.* 1830s–*c.* 1856) *High Holborn*
Forsyth, A. (*c.* 1830–1898) *Stoke Newington*
Peed, J. (*c.* 1830–1902) *Wandsworth*
Hurst, W. (*c.* 1831–1882) *Houndsditch*
Dobson, J. (*c.* 1832–1878) *Isleworth*
Wills, J. (1832–1895) *Kensington*
Dickson, T. A. (*c.* 1834–1899) *Covent Garden*
Ainsworth, S. (*c.* 1834–1904) *High Holborn*
Cole, W. (1834–1904) *Feltham*
Denning, W. (1837–1910) *Hampton*
Bause, C. F. (*c.* 1839–1895) *Norwood*
Cuthbert, G. (1839–1914) *Southgate*
Adamson, W. (*fl.* 1840s) *Stoke Newington*
Boff, J. (*fl.* 1840s) *Islington*
Jerrett, W. (*fl.* 1840s) *Hackney*
Lockhart, T. and C. (*fl.* 1840s) *Strand*
Lynn, W. (*fl.* 1840s) *Hackney*
Ponsford, S. (*fl.* 1840s) *Brixton*
Veitch, *Sir* H. J. (1840–1924) *Chelsea*
Perry, A. (1841–1913) *Enfield*
Turner, T. P. (1843–1908) *Hammersmith*
Lowe, J. (1843–1929) *Uxbridge*
May, H. B. (*c.* 1845–1936) *Edmonton*
Poupart, W. (*c.* 1846–1936) *Twickenham*
Watkins, A. (*c.* 1846–1937) *Strand*
Rochford, T. (1849–1901) *Tottenham*
Cullingford, W. (*fl.* 1850s) *Islington*
Parker, R. (*fl.* 1850s–1880s) *Islington*
Rochford, J. (1851–1919) *Enfield*
Holmes, W. (*c.* 1852–1890) *Hackney*
Kay, P. E. (*c.* 1853–1909) *Finchley*
Fromow, J. J. (*c.* 1855–1903) *Chiswick*
Jenkins, E. H. (*c.* 1855–1921) *Hampton*
Jones, H. J. (*c.* 1856–1928) *Lewisham*
Peed, T. (1857–1926) *Wandsworth*
Burley, J. (*fl.* 1860s) *Bayswater*
Low, H. (*c.* 1861–1893) *Clapton*
Barr, P. R. (*c.* 1862–1944) *Covent Garden*
Low, S. H. (*c.* 1863–1952) *Clapton*
Shawyer, G. E. (*c.* 1864–1943) *Uxbridge*
Weathers, J. (1867–1928) *Isleworth*
Perry, A. (1871–1953) *Enfield*
Rochford, J. P. (*c.* 1882–1965) *Tottenham*
Williams, H. (*fl.* 1890s) *Islington*
Stevens, G. (–1902) *Putney*
Jordan, C. (–1907) *Isleworth*
Bull, W. (–1913) *Chelsea*
Neal, R. (–1915) *Wandsworth*
Page, W. H. (–1928) *Hampton*
Stevenson, T. (–1938) *Hillingdon*
Bull, E. (*fl.* 1900s) *Chelsea*

Norfolk

Lindley, G. (*c.* 1769–1835) *Catton*
Mackie, J. (*fl.* 1770s–1797) *Norwich*
Mackie, S. (*fl.* 1770s–1833) *Norwich*
Absolon, W. (*fl.* 1780s) *Great Yarmouth*
Holl, A. (*fl.* 1780s) *Norwich*
Bell, I. (*fl.* 1790s) *Downham*
Mackie, W. A. (–1817) *Norwich*
Mackie, J. (–1818) *Norwich*
Harrison, J. (–*c.* 1855) *Downham*
Youell, W. (–1883) *Great Yarmouth*
Grigor, J. (*c.* 1811–1848) *Norwich*
Harrison, G. (*fl.* 1830s) *Downham*
Roe, J. F. (*fl.* 1830s) *Norwich*
Mackie, F. (*fl.* 1830s–1840s) *Norwich*
Dover, G. (*fl.* 1830s–1850s) *Norwich*
Reynolds, W. (*fl.* 1830s–1850s) *Norwich*
Thurtle, G. (*fl.* 1840s) *Norwich*
Bell, J. (*fl.* 1840s–1850s) *Norwich*
Ewing, J. W. (*fl.* 1850s) *Norwich*
Hussey, W. (*fl.* 1850s) *Norwich*
Jannoch, T. C. W. (1850–1925) *Dersingham*

Northamptonshire

King, S. (*fl.* 1780s) *Daventry*
Adams, J. (*fl.* 1790s) *Northampton*
Haynes, T. (*fl.* 1790s–1810s) *Oundle*
Atkins, J. (*c.* 1802–1884) *Northampton*
Law, J. (*fl.* 1810s) *Drapery*
Stoddart, J. (*fl.* 1810s) *Northampton*
Turnbull, R. (*fl.* 1810s–1820s) *Daventry*
Law, B. (*fl.* 1810s–1830s) *Northampton*
Law, H. (*fl.* 1820s–1830s) *Drapery*
Perkins, J. (*fl.* 1820s–1840s) *Northampton*
Perkins, T. (*c.* 1822–1880) *Northampton*
Cole, J. W. (*c.* 1863–1925) *Peterborough*
Burch, G. (*fl.* 1890s) *Peterborough*

Northumberland
Trotter, J. (*fl.* 1780s–1790s) *Newcastle*
Callender, M. J. and W. R. (*fl.* 1790s) *Newcastle*
Hunter, R. (*fl.* 1790s) *Hexham*
Elliott, W. (*fl.* 1800s) *Newcastle*
Hogg, J. (*fl.* 1830s) *Newcastle*
I'Anson, C. (*fl.* 1840s) *Newcastle*
Fell, W. (*c.* 1847–1903) *Hexham*
Matheson, T. (*fl.* 1870s) *Morpeth*

Nottinghamshire
Leeson, W. (–1722) *Blyth*
Noble, F. (–1756) *Newark*
Ordoyno, G. (*c.* 1723–1795) *Newark*
Cowlishaw, H. (*fl.* 1740s–1777) *Blyth*
Fox, J. (*fl.* 1770s) *Blyth*
Winter, J. (*fl.* 1770s) *Blyth*
Holt, S. (*fl.* 1780s) *Newark*
Pearson, J. (*fl.* 1780s–1820s) *Chilwell*
Aram, G. (*fl.* 1790s) *Mansfield*
Clay, J. (*fl.* 1790s) *Mansfield*
Dalman, R. (*fl.* 1790s) *Newark*
Felgate, G. (*fl.* 1790s) *Mansfield*
Hooley, H. (*fl.* 1790s) *Mansfield*
Hooley, T. (*fl.* 1790s) *Mansfield*
Hunter, W. (*fl.* 1790s) *Mansfield*
Aram, J. (*fl.* 1790s–1800s) *Mansfield*
Palethorpe, T. (*fl.* 1790s–1800s) *Newark*
Rayner, J. (*fl.* 1790s–1800s) *Nottingham*
Speechley, W. (–1804) *Newark*
Clarke, G. (*fl.* 1800s) *Newark*
Barron, W. (1800–1891) *Nottingham*
Pennington, F. (*c.* 1801–1850s) *East Retford*
Frettingham, H. (*c.* 1818–1884) *Beeston*
Pearson, J. (1819–1876) *Chilwell*
Clark, J. (*fl.* 1820s) *East Retford*
Clark, T. (*fl.* 1820s–1850s) *East Retford*
Adams, C. (*fl.* 1830s) *Nottingham*
Clark, G. (*fl.* 1830s) *Newark*
Clark, W. (*fl.* 1830s) *Newark*
Daft, R. (*fl.* 1830s) *Nottingham*
Dalman, T. (*fl.* 1830s) *Newark*
Fox, J. C. (*fl.* 1830s) *Newark*
Girton, J. (*fl.* 1830s) *Newark*
Mills, C. (*fl.* 1830s) *Blyth*
Palethorpe, J. (*fl.* 1830s) *Newark*
Withers, G. (*fl.* 1830s) *Newark*
Merryweather, H. (1839–) *Southwell*
Wood, J. F. (*fl.* 1840s) *Nottingham*
Pearson, C. E. (1856–1929) *Lowdham*
Merryweather, E. A. (1872–1924) *Southwell*
Pearson, J. D. (*fl.* 1880s) *Lowdham*
Wheatcroft, A. (1895–1965) *Nottingham*
Pearson, A. H. (–1930) *Lowdham*

Oxfordshire
Wrench, T. (*fl.* 1660s–1690s) *Oxford*
Tagg, T. (1695–1750s) *Oxford*
Tagg, E. (*c.* 1696–1779) *Oxford*
Tagg, J. (*fl.* 1750s–1790s) *Oxford*
Willis, J. (*fl.* 1790s) *Burford*
Tagg, T. (*fl.* 1800s–1837) *Oxford*
Walker, J. (–1895) *Thame*
Hemmings, J. (*fl.* 1820s) *Oxford*
Chaundy, R. (*fl.* 1820s–1840s) *Oxford*
Bates, J. (*fl.* 1830s) *Oxford*
Clisby, J. (*fl.* 1830s) *Thame*
Perry, P. J. (*c.* 1836–1881) *Banbury*
Day, W. (*fl.* 1840s) *Oxford*
Dunbar, J. (*fl.* 1840s) *Oxford*
Jeffery, S. (*fl.* 1840s) *Oxford*
Lynn, W. (*fl.* 1840s) *Henley*
McLaren, M. S. (*c.* 1892–*c.* 1947) *Wheatley*
Mattock, J. R. (*fl.* 1910s) *Oxford*

Rutland
Banton, E. (*fl.* 1830s) *Oakham*

Shropshire
Rea, J. (*fl.* 1620s–1677) *Kinlet*
Dally, T. (*fl.* 1790s) *Bridgnorth*
Bigg, C. (*fl.* 1820s–1830s) *Shrewsbury*
Allen, T. (*fl.* 1820s–1840s) *Oswestry*
Eckford, H. (1823–1905) *Wem*
Adams, T. (*fl.* 1830s) *Newport*
Cox, R. (*fl.* 1830s–1840s) *Ludlow*
Instone, H. (*fl.* 1830s–1840s) *Shrewsbury*
Murrell, P. (*fl.* 1910s)

Somerset
Pullen, J. (*fl.* 1740s) *Bedminster*
Poole, J. (*c.* 1777–*c.* 1827) *Taunton*
Harris, J. (*fl.* 1780s–1810s) *Taunton*
Frances, J. (*fl.* 1790s) *Castle Carey*
Penny, T. (*fl.* 1790s) *Castle Carey*
Tucker, J. (*fl.* 1790s) *Wells*
Young, J. (*c.* 1790–1862) *Taunton*
Drummond, W. C. (1793–1868) *Bath*
Salter, J. (*fl.* 1800s) *Bath*
Kelway, J. (1815–1899) *Langport*
Eastlake, N. (*fl.* 1820s) *Bridgwater*
Giddings, R. (*fl.* 1820s) *Wells*
Holloway, R. (*fl.* 1820s) *Wells*
Lake, H. (*fl.* 1820s) *Bridgwater*
Pierce, E. (*fl.* 1830s–1840s) *Yeovil*
Stephens, W. (*fl.* 1830s–1840s) *Taunton*
Kelway, W. (1839–) *Langport*
Wheeler, J. (*fl.* 1840s) *Bath*
Griffin, J. (*fl.* 1850s) *Bath*
Tiley, E. (*fl.* 1850s) *Bath*
Langdon, C. F. (*c.* 1868–1947) *Newton St. Loe*
Blackmore, J. B. (*fl.* 1870s–1900s) *Twerton-on-Avon*
Langdon, A. G. (1893–1972) *Bath*
Ware, W. T. (–1917) *Inglescombe*
Cooling, G. (*fl.* 1900s) *Bath*
Cooling, W. F. (*fl.* 1900s) *Bath*

Staffordshire
Bates, R. (*fl.* 1680s) *Ashley*
Heap, W. (*fl.* 1780s–1790s) *Cheadle*
Hammond, H. (*fl.* 1780s–1810s) *Rugeley*
Bradley, J. (*fl.* 1790s) *Leek*
Bramall, J. (*fl.* 1790s) *Lichfield*
Haywood, J. (*fl.* 1790s) *Leek*
Johnston, J. (*fl.* 1790s) *Wolverhampton*
Haywood, D. (*fl.* 1790s–1810s) *Burslem*
Smith, W. (*fl.* 1790s–1840s) *Burton*
Plant, R. W. (–*c.* 1858) *Cheadle*
Welton, T. (*fl.* 1820s) *Cheadle*
Plant, J. (*fl.* 1820s–1840s) *Cheadle*
Thomas, W. (1823–1895) *Wolverhampton*
Lowe, R. (*c.* 1827–1908) *Wolverhampton*
Sherratt, J. (*c.* 1829–1882) *Biddulph*

Suffolk
Wood, W. (*fl.* 1740s–1780s) *Woodbridge*
Hogg, M. (*c.* 1774–1832) *Bury St. Edmunds*
Clark, P. (*fl.* 1790s) *Woodbridge*
Constable, T. (*fl.* 1790s) *Long Melford*
Jeffries, R. (*fl.* 1810s–1830s) *Ipswich*
Woods, J. (*c.* 1813–1897) *Woodbridge*

Thompson, W. (1823–1903) *Ipswich*
Girling, S. (*fl.* 1830s) *Stowmarket*
Goodwin, W. (*fl.* 1830s) *Ipswich*
Kedie, W. (*fl.* 1830s) *Brandon*
Wild, T. (*fl.* 1830s) *Ipswich*
Woollard, W. (*fl.* 1830s) *Ipswich*
Culham, W. (*fl.* 1840s) *Woodbridge*
Cundy, C. (1849–1933) *Sudbury*
Notcutt, R. C. (1869–1938) *Woodbridge*

Surrey
Hunt, F. (–1662) *Putney*
Hunt, F. (1652–1713) *Putney*
Cox, W. (1680–1722) *Kew*
Hunt, F. (1691–1763) *Putney*
Cox, W. (–1704) *Kew*
Hunt, S. (–1763) *Putney*
Hunt, F. (*c.* 1729–1775) *Putney*
Butt, R. (*fl.* 1730s–1750s) *Kew*
Cree, J. (*c.* 1738–1816) *Chertsey*
Woods, R. (*fl.* 1740s–1793) *Chertsey*
Waterer, M. (1745–1822?) *Woking*
Scott, H. (*fl.* 1750s–1760s) *Weybridge*
Clarke, J. (*fl.* 1760s) *Dorking*
Jackman, W. (1763–) *Woking*
Eddie, G. (*fl.* 1780s) *East Sheen*
Mitchelson, G. (*fl.* 1780s–1800s) *Kingston*
Cree, W. (*fl.* 1780s–1810s) *Chertsey*
Osborn, R. (*c.* 1780–1866) *Fulham*
Waterer, J. (*c.* 1783–1868) *Bagshot*
Willmer, J. T. (*c.* 1786–1879) *Sunbury*
Bradley, J. (*fl.* 1790s) *Dorking*
Giles, J. (*fl.* 1790s) *Farnham*
Wilson, H. (*fl.* 1790s) *Hersham*
Wood, J. (*fl.* 1790s) *Dorking*
Spon, *Mr.* (*fl.* 1790s–1800s) *Egham*
Taylor, J. (*fl.* 1790s–1820s) *Bagshot*
Fuller, J. (*c.* 1797–1879) *Chertsey*
Scott, W. (–1827) *Dorking*
Penny, G. (–1838) *Milford*
Cree, J. (*c.* 1800–1858) *Chertsey*
Jackman, G. (1801–1869) *Woking*
Lunt, W. (1805–1883) *Mitcham*
Mongredien, A. (*c.* 1806–1888) *Farnham*
Steel, G. (1809–1891) *Richmond*
Smith, W. (*fl.* 1810s–1840s) *Norbiton*
Donald, R. (*fl.* 1810s–1850s) *Woking*
Godfrey, R. (*c.* 1812–1874) *Woking*
Kinghorn, F. R. (1813–1887) *Richmond*
Standish, J. (1814–1875) *Bagshot*
Thomson, D. S. (*c.* 1818–1905) *Wimbledon*
Young, J. (*fl.* 1820s–1836) *Epsom*
Young, C. (*fl.* 1820s–1840s) *Epsom*
Cobbett, H. (*fl.* 1820s–1850s) *Horsell*
Waterer, F. (1822–1871) *Bagshot*
Waterer, A. (1822–1896) *Woking*
Ivery, J. (*c.* 1823–1872) *Dorking*
Henderson, A. (*c.* 1826–1879) *Croydon*
Smith, G. (*fl.* 1830s) *Farnham*
Burley, J. (*fl.* 1830s–1840s) *Limpsfield*
Herbst, H. C. C. (*c.* 1830–1904) *Richmond*
Tanton, R. (*c.* 1835–1878) *Epsom*
Jackman, G. (1837–1887) *Woking*
Douglas, J. (1837–1911) *Great Bookham*
Walker, J. (1837–1911) *Ham*
Brown, W. (*c.* 1838–1893) *Richmond*
Cooper, R. (*fl.* 1840s) *Croydon*
Henbrey, R. (*fl.* 1840s) *Croydon*
Jackson, G. (*fl.* 1840s–1860s) *Kingston*
Noble, C. (*fl.* 1840s–1880s) *Bagshot*
Young, M. (*c.* 1844–1890) *Milford*
Wells, W. (1848–1916) *Earlswood*
Dickinson, T. (*fl.* 1850s) *Guildford*
Waterer, A. (*c.* 1850–1924) *Woking*
Jackson, T. (*c.* 1851–1888) *Kingston*
Slocock, W. C. (*c.* 1853–1926) *Woking*
White, H. (*c.* 1857–1938) *Windlesham*
Russell, L. R. (1863–1942) *Windlesham*
Jackman, A. G. (1866–1926) *Woking*
Waterer, F. G. (*c.* 1868–1945) *Bagshot*
Oldham, W. R. (*c.* 1870–1949) *Windlesham*
Secrett, F. A. (1886–1964) *Milford*
Rittershausen, P. R. C. (*c.* 1895–1965) *Croydon*
Russell, J. L. (1897–1976) *Windlesham*
Leonard, H. S. (–1902) *Guildford*
Mortimer, S. (–1923) *Farnham*
Wells, B. (1900–*c.* 1942) *Merstham*
Slocock, O. C. A. (*c.* 1907–1970) *Woking*

Sussex
Newman, W. (–1789) *Chichester*
Small, M. (*fl.* 1760s–1798) *Chichester*
Bingham, J. (*fl.* 1790s) *Eastbourne*
Cox, T. (*fl.* 1790s) *Eastbourne*
Gilbert, T. (–1891) *Hastings*
Silverlock, H. (*fl.* 1800s–1840s) *Chichester*
Cheal, J. (1800–1896) *Crawley*
Thomson, J. W. (*c.* 1805–1895) *Haywards Heath*
Mitchell, J. (*c.* 1808–1873) *Maresfield*
Laker, W. (*fl.* 1820s) *Horsham*
Napper, M. (*fl.* 1820s) *Horsham*
Stronach, J. (*fl.* 1820s) *Lewes*
Tugwell, J. (*fl.* 1820s) *Horsham*
Balchin, W. (*c.* 1824–1901) *Brighton*
Cameron, J. (*fl.* 1830s) *Uckfield*
Fuller, E. H. (*fl.* 1830s) *Worthing*
Stanford, H. (*fl.* 1830s) *St. Leonards*
Wood, C. (*fl.* 1830s–1840s) *Uckfield*
Wood, W. (*fl.* 1830s–1840s) *Uckfield*
Miles, W. (*c.* 1834–1883) *Brighton*
Piper, G. W. (*c.* 1838–1912) *Uckfield*
Box, J. (*c.* 1839–1912) *Haywards Heath*
McBean, J. (1840–1910) *Cooksbridge*
Cheal, J. (*c.* 1848–1935) *Crawley*
Charlesworth, J. (*c.* 1851–1920) *Haywards Heath*
Low, E. V. (1866–1931) *Wivelsfield*
Allwood, M. C. W. (*c.* 1879–1958) *Haywards Heath*
Ingwersen, W. E. T. (1883–1960) *East Grinstead*
Wells, B. (–1904) *Crawley*
Stredwick, J. (–1917) *St. Leonards-on-Sea*
McBean, A. A. (–1942) *Cooksbridge*
Davis, N. (*fl.* 1900s) *Framfield*

Warwickshire
Brunton, J. (*c.* 1721–1803) *Birmingham*
Pope, L. (*c.* 1740–1825) *Birmingham*
Whittingham, C. (*fl.* 1780s–1790s) *Coventry*
Hunter, J. A. (*fl.* 1780s–1820s) *Birmingham*
Bettridge, W. (*fl.* 1790s) *Warwick*
Blakesley, J. (*fl.* 1790s) *Birmingham*
Forbes, A. (*fl.* 1790s) *Birmingham*
Nicholls, T. (*fl.* 1790s) *Birmingham*
Weare, J. (*fl.* 1790s–1830s) *Coventry*
Frost, J. (*c.* 1797–1835) *Lillington*
Pope, A. (–*c.* 1853) *Birmingham*
Forbes, J. (*fl.* 1800s) *Birmingham*
Brown, J. (*fl.* 1810s) *Leamington*
Pope, J. (*fl.* 1810s–1830s) *Birmingham*
Cullis, J. (*fl.* 1810s–1849) *Leamington*
Dean, W. (1825–1895) *Birmingham*

Day, E. (*fl.* 1830s) *Shipston-on-Stour*
Perkins, A. (*c.* 1833–1905) *Coventry*
Sibray, H. (*c.* 1837–1894) *Birmingham*
Mander, T. (*fl.* 1840s–1890s) *Leamington*
Pope, J. (*c.* 1847–1918) *Birmingham*
Sydenham, R. (1848–1913) *Birmingham*
Woolman, H. (*c.* 1852–1932) *Birmingham*
Treen, W. H. (*fl.* 1860s) *Rugby*
Whateley, H. (–1915) *Kenilworth*

Westmorland
Furnass, W. (*fl.* 1780s–1790s) *Kendal*
Henderson, A. (*fl.* 1780s–1790s) *Kendal*
Petrie, R. (*fl.* 1790s) *Kendal*
Grier, J. (*c.* 1806–1879) *Ambleside*
Mawson, R. R. (*c.* 1864–1910) *Windermere*

Wiltshire
Phillips, J. (–*c.* 1747) *Devizes*
Geary, A. C. (–1792) *Salisbury*
Phillips, S. (*fl.* 1740s) *Devizes*
Biggs, T. (*fl.* 1780s) *Salisbury*
Ogborne, T. (*fl.* 1780s) *Wootton Bassett*
Brewer, T. (*fl.* 1790s) *Chippenham*
Brinsden, J. (*fl.* 1790s) *Marlborough*
Brownjohn, J. (*fl.* 1790s) *Salisbury*
Cross, W. (*fl.* 1790s) *Chippenham*
Figgins, R. (*fl.* 1790s–1820s) *Devizes*
Wheeler, G. (*c.* 1791–1878) *Warminster*
Williams, W. H. (–1894) *Salisbury*
Lavington, R. (*fl.* 1800s) *Devizes*
Milne, J. (*fl.* 1800s–1810s) *Fonthill*
Keynes, J. (*c.* 1806–1878) *Salisbury*
Blake, *Mr.* (*fl.* 1830s) *Devizes*
Heale, W. (*fl.* 1830s) *Calne*
Moody, T. (*fl.* 1830s–1840s) *Salisbury*
Cross, W. J. (1832–1885) *Salisbury*
Squibb, R. W. (*fl.* 1840s) *Salisbury*

Worcestershire
Smith, R. (1748–1810) *Worcester*
Batty, J. (*fl.* 1780s) *Worcester*
Smith, R. (1780–1848) *Worcester*
Biggs, J. (*fl.* 1790s) *Worcester*
Fernhall, J. (*fl.* 1790s) *Kidderminster*
Hammond, T. (*fl.* 1790s) *Worcester*
Webb, E. (*c.* 1844–1913) *Stourbridge*
White, J. H. (*c.* 1848–1907) *Worcester*

Yorkshire
Telford, G. (1687–1711) *York*
Telford, J. (1689–1771) *York*
Telford, G. (–1704) *York*
Perfect, J. (–1763) *Pontefract*
Simpson, R. (–*c.* 1783) *Knaresborough*
Perfect, J. (1717–1762) *Pontefract*
Perfect, W. (1718–1785) *Pontefract*
Bearpark, C. (*c.* 1724–1798) *York*
Crowder, A. (*c.* 1734–1831) *Doncaster*
Abel, J. (1740–1810) *York*
Telford, J. (1744–1830) *York*
Rigg, T. (*c.* 1746–1835) *York*
Perfect, J. (1749–1800) *Pontefract*
Telford, G. (1749–1834) *York*
Thompson, C. (*fl.* 1750s–1782) *Pickhill*
Barnes, T. (*fl.* 1750s–1790s) *Leeds*
Bearpark, R. Y. (1757–1827) *York*
Thompson, W. (1759–1811) *Thirsk*
Bearpark, B. (1760–1826) *York*
Perfect, G. (1768–1822) *Pontefract*
Littlewood, J. (*fl.* 1770s–1825) *Handsworth*
Barratt, J. (*c.* 1770–*c.* 1829) *Wakefield*
Bearpark, J. (1773–1836) *York*
Rigg, J. (1777–1833) *York*
Ely, B. (1779–1843) *Rothwell Haigh*
Calvert, J. (*fl.* 1780s) *Bedale*
Cressey, W. (*fl.* 1780s) *Hull*
Dunhill, G. (*fl.* 1780s) *Pontefract*
Hawley, F. (*fl.* 1780s) *Pontefract*
Sigston, B. (*fl.* 1780s–1790s) *Beverley*
Bean, W. (*fl.* 1780s–1800s) *Scarborough*
Crowder, R. W. (*fl.* 1780s–1800s) *Doncaster*
Martin, A. (*fl.* 1780s–1810s) *Cottingham*
Pontey, W. (*fl.* 1780s–1831) *Lepton in Kirkheaton*
Crowder, W. L. (*c.* 1780–1850s) *Doncaster*
Crowder, H. (*c.* 1784–1850s) *Doncaster*
Crowder, M. (*c.* 1787–1850s) *Doncaster*
Major, J. (*c.* 1787–1866) *Leeds*
Callender, E. R. (*fl.* 1790s) *Leeds*
Dawson, W. (*fl.* 1790s) *Whitby*
Farrady, J. (*fl.* 1790s) *Settle*
Innis, J. (*fl.* 1790s) *Bridlington*
Pontey, F. (*fl.* 1790s–1810s) *Huddersfield*
Pontey, A. (*fl.* 1790s–1821) *Leeds*
Hanks, J. (*fl.* 1790s–1830s) *Pontefract*
Backhouse, W. (1792–1845) *York*
Lupton, J. W. (*c.* 1793–1877) *York*
Backhouse, J. (1794–1869) *York*
Bearpark, R. (1797–1869) *York*
Cuthbert, J. (–1830) *Rotherham*
Crowder, W. (–1831) *Doncaster*
Clark, W. (*fl.* 1800s) *Keighley*
Oxley, T. (*fl.* 1800s–1820s) *Pontefract*
Tindall, G. (*fl.* 1800s–1830s) *Beverley*
Swailes, J. (*fl.* 1800s–1840s) *Beverley*
Barratt, W. (*fl.* 1800s–1860s) *Wakefield*
Marnock, R. (1800–1889) *Sheffield*
Anderson, J. (*c.* 1802–1871) *York*
Stansfield, A. (1802–1880) *Todmorden*
Dixon, E. P. (1804–1887) *Hull*
Appleby, S. (1806–1870) *Doncaster*
Willison, W. (*c.* 1806–1875) *Whitby*
Bell, W. (*fl.* 1810s) *Bradford*
Haley, J. (*fl.* 1810s) *Leeds*
Lister, R. (*fl.* 1810s) *Bradford*
Morrel, R. (*fl.* 1810s) *Bradford*
Phillipson, H. (*fl.* 1810s) *Collingham*
Taylor, J. (*fl.* 1810s) *Leeds*
Wood, W. (*fl.* 1810s) *Bradford*
Swailes, R. (*fl.* 1810s–1820s) *Beverley*
Crowcroft, T. (*fl.* 1810s–*c.* 1860) *Doncaster*
Edward, G. (*c.* 1816–1875) *York*
Lord, R. (1818–1886) *Todmorden*
Adcock, W. (*fl.* 1820s) *Ripon*
Armitage, E. (*fl.* 1820s) *Kirkburton*
Banning, W. (*fl.* 1820s) *Pickhill*
Barnes, J. (*fl.* 1820s) *Leeds*
Boynton, J. (*fl.* 1820s) *Beverley*
Butterfield, W. (*fl.* 1820s) *Wensley*
Clark, J. (*fl.* 1820s) *Silsden*
Cowburn, R. (*fl.* 1820s) *Otley*
Grant, W. (*fl.* 1820s) *Sheffield*
Hall, A. (*fl.* 1820s) *Doncaster*
Hawley, W. (*fl.* 1820s) *Barnsley*
Hirst, R. M. (*fl.* 1820s) *Sheffield*
Horner, R. (*fl.* 1820s) *Welburn in Bulmer*
Laycock, J. (*fl.* 1820s) *Halifax*
Lea, J. (*fl.* 1820s) *Halifax*
Meek, M. (*fl.* 1820s) *Northallerton*
Nicholson, J. (*fl.* 1820s) *Sheffield*
Pontey, M. (*fl.* 1820s) *Leeds*

Tattershall, G. (*fl.* 1820s) *Silkstone*
Urwin, J. (*fl.* 1820s) *Richmond*
Willison, A. (*fl.* 1820s) *Whitby*
Abbott, G. (*fl.* 1820s–1830s) *Knaresborough*
Aram, T. (*fl.* 1820s–1830s) *Huddersfield*
Tindall, W. (*fl.* 1820s–1830s) *Beverley*
Caven, J. (*fl.* 1820s–1840s) *Bedale*
Pontey, J. (*fl.* 1820s–1840s) *Huddersfield*
Slater, J. (*fl.* 1820s–1840s) *Malton*
Throp, W. (*fl.* 1820s–1840s) *Halifax*
Clarkson, G. (*fl.* 1820s–1854) *York*
Fisher, C. (1823–1902) *Sheffield*
Backhouse, J. (1825–1890) *York*
Carter, J. (*c.* 1825–1894) *Keighley*
Dean, W. (1825–1895) *Bradford*
Stansfield, T. (*c.* 1826–1879) *Todmorden*
May, H. (*c.* 1827–1880) *Bedale*
Bearpark, H. E. (*c.* 1829–1861) *York*
Abbott, T. (*fl.* 1830s) *Knaresborough*
Aram, R. (*fl.* 1830s) *Huddersfield*
Barron, J. (*fl.* 1830s) *Sheffield*
Garner, J. (*fl.* 1830s) *Bawtry*
Hedley, R. (*fl.* 1830s) *Yarm*
Horsfield, T. (*fl.* 1830s) *Halifax*
Jones, J. (*fl.* 1830s) *Pontefract*
MacBeth, C. (*fl.* 1830s) *Goole*
Major, G. (*fl.* 1830s) *Leeds*
Major, H. (*fl.* 1830s) *Leeds*
Mitton, R. (*fl.* 1830s) *Pontefract*
Pontey, C. (*fl.* 1830s) *Leeds*
Turner, R. (*fl.* 1830s) *Sheffield*
Gaukroger, J. (*fl.* 1830s–1840s) *Halifax*
Washington, T. (*fl.* 1830s–1840s) *Halifax*
Edwards, J. (*fl.* 1830s–1850s) *York*
Pontey, H. P. (*fl.* 1830s–1850s) *Lepton in Kirkheaton*
Rider, W. L. (*fl.* 1830s–1850s) *Leeds*
Pontey, W. (*fl.* 1830s–1860s) *Lepton in Kirkheaton*
Boston, W. (*c.* 1830–) *Bedale*
Simpson, A. (*c.* 1830–1898) *York*
Eastwood, C. (1839–1895) *Luddenden*
Booker, J. (*fl.* 1840s) *Halifax*
Foster, J. (*fl.* 1840s) *Sheffield*
Hedley, E. (*fl.* 1840s) *Yarm*
Jackson, W. (*fl.* 1840s) *Bedale*
May, W. (*fl.* 1840s) *Bedale*
Menzies, J. (*fl.* 1840s–1850s) *Halifax*
Swailes, T. (*fl.* 1840s–1870s) *Beverley*
Campbell, A. (1845–1947) *Harrogate*
Holmes, E. (*fl.* 1850s) *Sheffield*
Godwin, F. (*fl.* 1850s–1860s) *Sheffield*
Scholey, J. (*fl.* 1850s–1860s) *Pontefract*
Nelson, W. (1852–1922) *Bradway*
Edwards, F. C. (*c.* 1857–1934) *Leeds*
Keeling, A. J. (*c.* 1858–1920) *Bradford*
Aram, W. (*fl.* 1860s) *Huddersfield*
Dixon, W. E. (*fl.* 1860s) *Beverley*
Kearsley, J. (*fl.* 1860s) *Leeds*
Webster, H. (*fl.* 1860s) *Leeds*
Broadhead, C. H. (*c.* 1860–1920) *Thongsbridge*
Backhouse, J. (1861–1945) *York*
Hatcher, W. H. (*c.* 1869–1951) *Leeds*
Crowther, W. (*fl.* 1870s) *Doncaster*
Mansell, W. (*c.* 1871–1948) *Leeds*
Barwise, J. F. (*c.* 1874–1965) *Barnsley*
Baker, F. (1895–1954) *Leeds*
Keeling, A. J. (–1962) *Bradford*
Hatcher, L. W. (–1964) *Leeds*
Harkness, J. (*fl.* 1900s) *Bedale*

SCOTLAND

Ferguson, H. (*fl.* 1680s) *Edinburgh*
McAslan, D. (–1741) *Glasgow*
Clark, J. (–1762) *Edinburgh*
Clephane, A. (*fl.* 1700s–1730s) *Edinburgh*
McAslan, J. (*fl.* 1710s) *Glasgow*
Boutcher, W. (*fl.* 1710s–1730s) *Edinburgh*
Miller, W. (*fl.* 1710s–1760s) *Edinburgh*
Dickson, R. (*fl.* 1720s) *Hawick*
Eagle, A. (*fl.* 1730s) *Edinburgh*
Boutcher, W. (*fl.* 1730s–1780s) *Edinburgh*
Drummond, P. (*fl.* 1740s–1760s) *Edinburgh*
Dickson, A. (*fl.* 1740s–1770s) *Hawick*
Borthwick, W. (*fl.* 1750s) *Edinburgh*
Gordon, J. (*fl.* 1750s–1770s) *Edinburgh*
McAslan, J. (*fl.* 1750s–1815) *Glasgow*
Austin, R. (*c.* 1754–1830) *Glasgow*
Richmond, J. (*fl.* 1760s–1770s) *Edinburgh*
Anderson, R. (*fl.* 1760s–1800s) *Edinburgh*
Smith, J. (*c.* 1760–1848) *Monkwood*
Hood, W. (*c.* 1763–1828) *Dumfries*
Don, G. (1764–1814) *Doo Hillock*
Fowlds, A. (*c.* 1767–1842) *Kilmarnock*
Brown, R. (*c.* 1767–1845) *Perth*
Lawson, P. (*fl.* 1770s–1821) *Edinburgh*
Austin, J. (1776–1849) *Glasgow*
Cunningham, J. (1784–1851) *Edinburgh*
Turnbull, A. (*c.* 1790–1875) *Perth*
Lymburn, R. (*c.* 1793–1843) *Kilmarnock*
Lawson, C. (1794–1873) *Edinburgh*
Drummond, P. (*c.* 1798–1877) *Stirling*
Dickson, G. (–1825) *Edinburgh*
Henderson. A. (–1827) *Edinburgh*
Dickson, W. (–1835) *Perth*
McAslan, A. (–1841) *Glasgow*
Tait, H. (–1869) *Kelso*
Laird, W. P. (–1872) *Dundee*
Galloway, G. (–1879) *Helensburgh*
Urquhart, D. (–1880) *Dundee*
Stewart, W. (–1886) *Dundee*
Downie, J. (–1892) *Edinburgh*
Alexander, J. (*c.* 1805–1881) *Edinburgh*
Cocker, J. (*c.* 1806–1880) *Aberdeen*
Grigor, J. (*c.* 1806–1881) *Forres*
Henderson, J. (*fl.* 1810s) *Brechin*
Sang, E. (*fl.* 1810s) *Kirkcaldy*
Anderson, J. A. (*c.* 1811–1891) *Perth*
Stark, R. M. (1815–1873) *Edinburgh*
Newbigging, A. T. (*c.* 1815–1885) *Dumfries*
Howie, C. (1811–1899) *St. Andrews*
Dobbie, J. (*c.* 1817–1905) *Rothesay*
Methven, T. (*c.* 1819–1879) *Edinburgh*
Fraser, D. (*fl.* 1820s) *Inverness*
Drummond, W. (*fl.* 1820s–1840s) *Stirling*
Smith, T. (1820–1904) *Stranraer*
Laird, R. B. (1823–1895) *Edinburgh*
Milne, T. (*c.* 1823–1910) *Aberdeen*
Service, J. (*c.* 1824–1901) *Maxwelltown*
Paul, W. (1825–1880) *Paisley*
Reid, G. (*c.* 1826–1881) *Aberdeen*
Handasyde, T. and W. (*fl.* 1830s–1840s) *Musselburgh*
Cocker, J. (*c.* 1832–1897) *Aberdeen*
Begg, W. R. (1833–1923) *Aberdeen*
Fraser, H. (*c.* 1834–1904) *Edinburgh*
Campbell, M. (*c.* 1835–1915) *Blantyre*
Thomson, W. (*c.* 1836–1912) *Edinburgh*
Drummond, W. P. (1838–1906) *Edinburgh*
Grieve, J. (*c.* 1840–1924) *Edinburgh*
Croll, D. (*c.* 1841–1909) *Dundee*

Williamson, W. (1843–) *Edinburgh*
Alexander, J. (*c.* 1845–1890) *Edinburgh*
Thomson, W. (1849–1893) *Galashiels*
Austin, H. (*c.* 1849–1894) *Glasgow*
Austin, W. (*fl.* 1850s–1870s) *Glasgow*
Thomson, J. (*c.* 1852–1901) *Galashiels*
Laird, D. P. (*c.* 1853–1905) *Edinburgh*
Cocker, J. (1855–1894) *Aberdeen*
Thomson, D. W. (*c.* 1855–1915) *Edinburgh*
Laing, J. (*c.* 1857–1917) *Kelso*
Cocker, W. (*c.* 1858–1913) *Aberdeen*
Phillips, J. (1859–1917) *Edinburgh*
Hunter, J. (*fl.* 1860s–1900s) *Glasgow*
Cocker, A. M. (1860–1920) *Aberdeen*
Duthie, J. A. (1868–1920) *Aberdeen*
Grigor, A. (*fl.* 1870s) *Elgin*
Fergusson, C. (–1904) *Nairn*
Forbes, J. (–1909) *Hawick*
Service, R. (–1911) *Maxwelltown*
Methven, J. (–1913) *Edinburgh*
Thomson, J. (–1925) *Kelso*
Mair, R. Y. (*fl.* 1900s) *Prestwick*
Croll, J. (*fl.* 1910s) *Dundee*

WALES
Arnot, R. (*fl.* 1790s) *Cowbridge*
Chalmers, A. (*fl.* 1790s) *Newtown*
Hill, W. (*fl.* 1790s) *Merthyr Tydfil*
Treseder, W. (*c.* 1829–1893) *Cardiff*
Jones, E. (*fl.* 1840s) *Holywell*
Matthews, S. (*c.* 1887–1954) *Port Talbot*
Drake, G. W. (*fl.* 1900s) *Cardiff*

IRELAND
Griffith, W. (–1742) *Carlow*
Sweetenham, G. (–1746) *Dublin*
Buller, W. (–1757) *Tinnahinch*
Henderson, H. (–1778) *Dublin*
Gallwey, R. (*fl.* 1720s) *Dronwickbane*
Edgar, P. (*fl.* 1720s–1790s) *Dublin*
Moody, R. (*fl.* 1730s) *Dublin*
Stevenson, R. (*fl.* 1730s–1750s) *Dublin*
Hanlon, C. (*fl.* 1740s) *Dublin*
Kemplin, H. (*fl.* 1740s) *Cork*
Bruce, G. (*fl.* 1740s–1760s) *Dublin*
Cosby, P. (*fl.* 1740s–1760s) *Stradbally*
Sullivan, D. (*fl.* 1740s–1760s) *Cork*
Phelan, J. (*fl.* 1740s–1780s) *Dublin*
Costa, L. (*fl.* 1750s) *Dublin*
Cullin, D. (*fl.* 1750s) *Dublin*
Doyle, J. (*fl.* 1750s) *Stradbally*
Robertson, *Mr.* (*fl.* 1750s) *County Kilkenny*
Galvin, W. (1756–1832) *Mount Talbot*
Robertson, W. (*fl.* 1760s) *Thomastown*
Allen, J. (*fl.* 1760s–1770s) *Limerick*
Bull, D. (*fl.* 1760s–1770s) *Dublin*
Jones, J. (*fl.* 1760s–1770s) *Dublin*
Wright, T. (*fl.* 1760s–1770s) *Monashan*
Ahern, G. (*fl.* 1760s–1780s) *Cork*
Clarke, T. (*fl.* 1760s–1780s) *Shangarry*
Hay, E. (*fl.* 1760s–1780s) *Dublin*
Adams, P. (*fl.* 1760s–1790s) *Gormanstown*
Madden, M. (*fl.* 1760s–1790s) *Ballinasloe*
Peppard, L. (*fl.* 1760s–1790s) *Dublin*
Bourke, P. (*fl.* 1770s) *Nenagh*
Courtney, D. (*fl.* 1770s) *Tralee*
Harrold, T. (*fl.* 1770s) *Irishtown*
Jones, J. (*fl.* 1770s) *Dublin*
Madden, B. (*fl.* 1770s) *Dublin*
Bourke, T. (*fl.* 1770s–1790s) *County Mayo*
Harvey, W. (*fl.* 1770s–1790s) *Dublin*
Bray, E. (*fl.* 1770s–1800s) *Dublin*
Anderson, J. (*fl.* 1780s) *County Westmeath*
Bell, W. (*fl.* 1780s) *County Down*
Byrne, P. (*fl.* 1780s) *Kilkenny*
Clarke, J. (*fl.* 1780s) *County Roscommon*
Cottingham, G. (*fl.* 1780s) *Dublin*
Hodgins, E. (*fl.* 1780s) *Dunganstown*
Carol, P. (*fl.* 1780s–1790s) *County Meath*
Clarke, R. (*fl.* 1780s–1790s) *Shangarry*
M'Evoy, J. (*fl.* 1780s–1790s) *Callon*
Madden, F. (*fl.* 1780s–1790s) *Ballinasloe*
Middlewood, W. (*fl.* 1780s–1790s) *Dublin*
Power, R. (*fl.* 1780s–1790s) *Galway*
Carroll, P. (*fl.* 1780s–1800s) *Dublin*
Reilly, T. (*fl.* 1780s–1800s) *Ballybeg*
Simpson, B. (*fl.* 1780s–1800s) *Dublin*
Smitten, G. (*fl.* 1780s–1800s) *Dublin*
Hervey, J. (*c.* 1786–1829) *Comber*
Brophy, J. (*fl.* 1790s) *County Kilkenny*
Burnett, R. (*fl.* 1790s) *Dublin*
Costello, P. (*fl.* 1790s) *County Wicklow*
Grimwood, J. (*fl.* 1790s) *Dublin*
Piers, J. (*fl.* 1790s) *Raheny*
Robertson, D. (*fl.* 1790s–1800s) *Belfast*
Tiernan, J. (*fl.* 1790s–1800s) *Dublin*
Donegan, P. (*fl.* 1790s–1810s) *Dublin*
M'Leish, A. (–1828) *Dublin*
Robertson, J. (–1839) *Kilkenny*
Saunders, M. (–1877) *Cork*
Boylan, W. (*fl.* 1800s) *County Cork*
Casey, J. (*fl.* 1800s–1830s) *Cork*
Dickson, A. (*c.* 1802–1880) *Newtownards*
Campbell, A. (*c.* 1804–1871) *Dublin*
Drummond, D. (*c.* 1813–1904) *Dublin*
Edmondson, J. (1823–1894) *Dublin*
Drummond, W. (*fl.* 1830s) *Dublin*
Dickson, G. (1832–1914) *Newtownards*
Dickson, H. (*c.* 1834–1904) *Belfast*
Hartland, W. B. (1836–1912) *Cork*
Smith, T. (*c.* 1840–1919) *Newry*
Dickson, A. (*c.* 1857–1949) *Belfast*
McGredy, S. (*c.* 1861–1926) *Portadown*
Coey, J. (–1921) *Newcastle*

CHANNEL ISLANDS
Wolsey, G. (*c.* 1814–1870) *Guernsey*
Saunders, C. B. (1824–1893) *Jersey*
Smith, C. (*c.* 1830–1910) *Guernsey*
Langelier, R. (*fl.* 1840s) *Jersey*
Saunders, B. (*fl.* 1840s) *Jersey*
Hubert, F. J. (*fl.* 1890s–1910s) *Guernsey*
Manger, W. (*fl.* 1890s–1910s) *Guernsey*
Smith, C. (–1912) *Guernsey*

NYASALAND *see* **MALAWI**

OAK *see* **QUERCUS**

OENOTHERA
Gates, R. R. (1882–1962)

OPHRYS
Moggridge, J. T. (1842–1874)

ORANGE *see* **CITRUS**

ORCHIDS
Collinson, M. (1727–1795)

Ker, J. B. (1764–1842)
Wray, M. (*c.* 1775–1864)
Barker, G. (1776–1845)
Clowes, *Rev.* J. (1777–1846)
Hooker, *Sir* W. J. (1785–1865)
Cuming, H. (1791–1865)
Lyons, J. C. (1792–1874)
Lance, J. H. (1793–1878)
Appleby, T. (*c.* 1795–1875)
Lindley, J. (1799–1865)
Hutton, H. (–1868)
Warner, C. B. (–1869)
Chesterton, J. H. (–1883)
Curnow, R. (–1896)
Wailes, G. (*c.* 1802–1882)
Skinner, G. U. (1804–1867)
Ainsworth, R. F. (1811–1890)
Bateman, J. (1811–1897)
Warner, R. (*c.* 1814–1896)
Dominy, J. (1816–1891)
Grindon, L. H. (1818–1904)
Huntley, *Rev.* J. T. (*fl.* 1820s)
Warre, F. (*fl.* 1820s)
Colley, T. (*fl.* 1820s–1830s)
Ross, H. J. (1820–1902)
Benson, R. (1822–1894)
Parish, *Rev.* C. S. P. (1822–1897)
Berkeley, E. S. (*c.* 1823–1898)
Day, J. (1824–1888)
Williams, B. S. (1824–1890)
Low, *Sir* H. (1824–1905)
Schröder, *Sir* H. (1824–1910)
Lendy, A. F. (*c.* 1825–1889)
Broome, J. (1825–1907)
Harvey, E. (*c.* 1826–1890)
Kent, A. H. (1828–1913)
Shuttleworth, E. (1829–1909)
Winn, C. (*c.* 1829–1917)
Aldridge, G. (*fl.* 1830s)
Moss, J. (*fl.* 1830s)
Drake, *Miss* (*fl.* 1830s–1840s)
Ross, J. (*fl.* 1830s–1840s)
Fitzgerald, R. D. (1830–1892)
Morris, G. F. (1831–1909)
Lawrence, *Sir* J. J. T. (1831–1913)
Hardy, G. (*c.* 1832–1894)
Thompson, W. (*c.* 1832–1916)
Measures, R. I. (*c.* 1833–1907)
Little, H. (*c.* 1833–1914)
Ballantine, H. (1833–1929)
Bolus, H. (1834–1911)
Percival, R. P. (*c.* 1835–1885)
Gower, W. H. (1835–1894)
Marshall, W. (1835–1917)
Burkinshaw, W. P. (*c.* 1835–1918)
Philbrick, F. A. (*c.* 1836–1910)
Chamberlain, *Right Hon.* J. (1836–1914)
Wrigley, O. O. (*c.* 1836–1917)
Cobb, W. (*c.* 1836–1922)
Drewett, D. O. (*c.* 1838–1910)
Courtauld, S. (1840–1899)
Hincks, T. C. (1840–1902)
Cookson, N. C. (*c.* 1840–1909)
McBean, J. (1840–1910)
Trimen, R. (1840–1916)
Ashworth, E. (*c.* 1840–1917)
Seden, J. (1840–1921)
Veitch, *Sir* H. J. (1840–1924)
Fitch, J. N. (1840–1927)
Smee, A. H. (1841–1901)
Swan, W. (*c.* 1841–1919)
Ayling, E. (1841–1931)
Horsman, F. J. S. (*c.* 1842–1894)
Owen, G. D. (*c.* 1842–1894)
Cowan, J. (1842–1929)
O'Brien, J. (1842–1930)
Leeman, J. (*c.* 1843–1918)
Boxall, W. (1844–1910)
Cowan, *Rev.* W. D. (1844–1924)
Bolton, J. J. (1844–1928)
Tautz, F. G. (*c.* 1845–1894)
Hollington, A. J. (1845–1926)
Watson, J. (1845–1932)
Burbidge, F. W. T. (1847–1905)
Sander, H. F. C. (1847–1920)
Davis, W. (1847–1930)
Ashworth, R. (*c.* 1848–1928)
Perceval, C. H. S. (1849–1920)
Cotton, F. (*fl.* 1850s)
Henshall, J. (*fl.* 1850s)
McLeod, D. (*fl.* 1850s–1932)
Tracy, H. A. (*c.* 1850–1910)
Bromilow, H. J. (1850–1930)
Holmes, W. (*c.* 1851–1913)
Charlesworth, J. (*c.* 1851–1920)
Hanbury, F. J. (1851–1938)
Hildebrand, A. H. (1852–1918)
Bolton, W. (1852–1921)
Hassall, A. (*c.* 1852–1922)
Curtis, C. (1853–1928)
Lucas, C. J. (*c.* 1853–1928)
Neale, J. J. (*c.* 1854–1919)
Braine, A. B. (1854–1945)
Brühl, P. (1855–)
Fowler, J. G. (1855–1916)
Rolfe, R. A. (1855–1921)
Moore, G. F. (*c.* 1855–1927)
Bird, G. W. (1855–1928)
Rutherford, *Sir* J. (*c.* 1855–1932)
Hooke, N. (*c.* 1856–1935)
Pantling, R. (1857–1910)
Crawshay, de B. (*c.* 1857–1924)
Gerrish, R. (*c.* 1857–1929)
Keeling, A. J. (*c.* 1858–1920)
Purdie, A. (1859–1905)
Moss, J. S. (*c.* 1859–1913)
Colman, *Sir* J. (1859–1942)
White, W. H. (*c.* 1859–1942)
Blunt, H. (*fl.* 1860s)
Harrison, C. H. (*fl.* 1860s)
Moore, R. (1860–1899)
Holford, *Sir* G. L. (1860–1926)
Ashton, E. R. (*c.* 1860–1951)
Cradwick, W. (*c.* 1862–1937)
Bristow, E. (*c.* 1863–1934)
Broadway, W. E. (1863–1935)
Low, S. H. (*c.* 1863–1952)
Lawson, H. P. (*c.* 1863–1956)
Burton, R. F. (*c.* 1865–1922)
Low, E. V. (1866–1931)
Bedford, E. J. (1866–1953)
Smith, H. H. (1867–1935)
Curtis, C. H. (1869–1951)
Hatcher, W. H. (*c.* 1869–1951)
Downton, G. (*fl.* 1870s)
Black, J. M. (1870–1946)
Hurst, C. C. (1870–1947)
Wilson, E. K. (*c.* (1870–1947)
Ellwood, A. G. (1870–1952)

Godfery H. M. (1871–1930)
Mansell, W. (*c*. 1871–1948)
Cowan, D. A. (*c*. 1874–1952)
Sander, C. F. (1874–1957)
Gentle, A. M. (*c*. 1874–1958)
Alexander, H. G. (*c*. 1875–1972)
Sander, F. K. (*c*. 1877–1951)
Sander, L. L. (1878–1936)
Stables, W. J. (*c*. 1878–1957)
Wilson, A. G. (*c*. 1878–1957)
Lankester, C. H. (1879–1969)
Woodham, H. J. (1881–1957)
Munday, A. J. (1883–1947)
Cooper, E. W. (1884–1950)
Hills, B. (*c*. 1884–1960)
Hopkins, W. O. (*c*. 1884–1960)
Lines, O. (1884–1965)
Laycock, J. (*c*. 1887–1960)
Morris, A. G. (*c*. 1887–1967)
Coningsby, A. (*c*. 1888–1966)
Williams, H. (*fl*. 1890s)
Crombleholme, *Rev*. J. (*fl*. 1890s–1920s)
Carr, C. E. (1892–1936)
Rittershausen, P. R. C. (*c*. 1895–1965)
Tufton, H. S. T. (*c*. 1897–1961)
Moreau, R. E. (*c*. 1897–1970)
Summerhayes, V. S. (1897–1974)
Pilcher, C. (–1900)
Carder, J. (–1908)
Johnson, R. (–1919)
Swinhoe, R. C. G. (–1927)
Morris, H. G. (–1928)
Clark, J. (–1929)
Cooper, *Rev*. W. H. W. (–1929)
Gratrix, S. (–1929)
Latimer, W. (–1929)
Chapman, H. J. (–1931)
Paterson, R. (–1933)
Astley-Bell, H. (–1937)
McBean, A. A. (–1942)
Tahourdin, C. B. (–1942)
Beckton, B. J. (–1943)
Thwaites, R. G. (–1943)
Armstrong, T. (–1944)
Clarke, T. H. (–1946)
Harben, G. P. (–1949)
Flory, S. W. (–1954)
Bracey, B. O. (–1956)
Branch, C. (–1958)
Roberts, N. (–1959)
Lampard, E. R. J. (–1961)
Bradbury, G. A. (–1962)
Keeling, A. J. (–1962)
Silverthorn, G. W. (–1962)
Berry, W. G. (–1964)
Hatcher, L. W. (–1964)
Ruck, A. J. (–1965)
Dixon, J. H. (–1966)
Stirling, W. (–1967)
Warren, C. N. (–1970)
Coles, J. (*fl*. 1900s)
Knights, A. I. J. (*c*. 1900–1948)
Pierce, S. (*c*. 1902–1963)
Russon, J. B. (*c*. 1903–1965)
Brooke, J. (1908–1966)
Alexander, S. G. (1910–1944)

OXALIS
Young, D. P. (1917–1972)

PACIFIC
Forster, J. R. (1729–1798)
Banks, *Sir* J. (1743–1820)
Anderson, W. (1750–1778)
Forster, J. G. A. (1754–1794)
Bligh, W. (1754–1817)
Menzies, A. (1754–1842)
Cuming, H. (1791–1865)
Douglas, D. (1799–1834) *Hawaii*
Macrae, J. (–1830) *Hawaii*
Milne, W. G. (–1866)
Bennett, G. (1804–1893) *Tahiti*
Powell, *Rev*. T. (1809–1887) *Samoa*
Nightingale, T. (1810–1865)
Harvey, W. H. (1811–1866) *Fiji, Friendly Islands*
Hinds, R. B. (*c*. 1812–*c*. 1847)
Corson, J. (1815–1841)
Adams, A. (*c*. 1820–1878)
Moore, C. (1820–1905)
Seemann, B. C. (1825–1871) *Fiji*
Gill, *Rev*. W. W. (1828–1896)
Barclay, G. (*fl*. 1830s–1840s)
Bennett, F. D. (*fl*. 1830s–1840s) *Hawaii*
Brown, G. (1835–) *Bismark Archipelago*
Horne, J. (1835–1905) *Fiji*
Thurston, *Sir* J. B. (1836–1897) *Fiji, Solomon Islands*
Whitmee, *Rev*. S. J. (1838–1925) *Samoa*
Veitch, J. G. (1839–1870)
Lawes, W. G. (1839–1907)
Guilfoyle, W. R. (1840–1912)
Cheeseman, T. F. (1846–1923) *Cook and Kermadoc Islands*
Coppinger, R. W. (1847–1910)
Veitch, P. C. M. (1850–1929) *Fiji*
Betche, E. (1851–1913) *Samoa*
Woodford, C. M. (1852–1927)
Thurn, *Sir* E. F. I. (*c*. 1852–1932) *Fiji*
Guppy, H. B. (1854–1926) *Solomon Islands*
Cartwright, T. B. (1856–1896)
Matthew, C. G. (1862–1936) *Hawaii*
Gibbs, L. S. (1870–1925) *Fiji*
Bridge, C. A. G. (*fl*. 1880s)
Comins, *Rev*. R. B. (*fl*. 1880s–1890s) *Solomon Islands*
Cheesman, L. E. (1881–1969)
Collenette, C. L. (1888–1959)

PAEONIA
Stern, *Sir* F. C. (1884–1967)

PALAEOBOTANY
Parkinson, J. (1755–1824)
Nicol, W. (1768–1851)
Williamson, J. (1774–1877)
Witham, H. T. M. (1779–1844)
Stokes, C. (1783–1853)
Buckland, W. (1784–1856)
Artis, E. T. (1789–1847)
Crow, F. (*fl*. 1790s–1810s)
Mantell, G. A. (1790–1852)
Bigsby, J. J. (1792–1881)
Hutton, W. (1797–1860)
Bowerbank, J. S. (1797–1877)
Clarke, *Rev*. W. B. (1798–1878)
Phillips, J. (1800–1874)
Peach, C. W. (1800–1886)
Clay, C. (1801–1893)
Miller, H. (1802–1856)

Tate, G. (1805–1871)
Hancock, A. (1806–1873)
Pattison, S. R. (1809–1901)
Morris, J. (1810–1886)
Hawkshaw, *Sir* J. (1811–1891)
Binney, E. W. (1812–1881)
Atthey, T. (1814–1880)
Oldham, T. (1816–1878)
Williamson, W. C. (1816–1895)
McCoy, *Sir* F. (1817/23–1899)
Baily, W. H. (1819–1888)
Haughton, *Rev*. S. (1821–1897)
Howse, R. (1821–1901)
Spencer, J. (1834–1898)
Hicks, H. (1837–1899)
Cash, W. (1843–1914)
Nicholson, H. A. (1844–1899)
Gardner, J. S. (1844–1930)
Lebour, G. A. L. (1847–1918)
Deane, H. (1847–1924)
Kidston, R. (1852–1924)
Pegler, L. W. H. (1852–1927)
Reid, C. (1853–1916)
Scott, D. H. (1854–1934)
Lomax, J. (1857–1934)
Benson, M. J. (1859–1936)
Cadell, H. M. (1860–1934)
Reid, E. M. W. (1860–1953)
Moysey, L. (1863–1918)
Seward, *Sir* A. C. (1863–1941)
Oliver, F. W. (1864–1951)
Arber, E. A. N. (1870–1918)
Davies, D. (1870–1931)
Berridge, E. M. (1872–1947)
Lang, W. H. (1874–1960)
Prankerd, T. L. (1878–1939)
Stopes, M. C. C. (1880–1958)
Gordon, W. T. (1884–1950)
Thomas, H. H. (1885–1962)
Holden, H. S. (1887–1963)
Hemingway, W. (*fl*. 1890s–1910s)
Edwards, W. N. (1890–1956)
Simpson, J. B. (1894–1960)
Walton, J. (1895–1971)
Scott, H. V. (–1929)
Maslen, A. J. (–1954)

PALESTINE
Braylsford, J. (*fl*. 1700s)
Hooker, *Sir* J. D. (1817–1911)
Tristram, *Rev*. H. B. (1822–1906)
Fox, *Rev*. H. E. (1841–1926)
Hayne, W. A. (1847–1873)
Hart, H. C. (1847–1908)
Lowne, B. T. (*fl*. 1860s)
Meinertzhagen, R. (1878–1967)

PALMS
Blatter, E. (1877–1934)

PANAMA
Cuming, H. (1791–1865)
Hayes, S. (–1863)
Barnard, H. (*fl*. 1820s)
Cross, R. M. (1836–1911)
Hart, J. H. (1847–1911)
Popenoe, D. K. (1899–1932)

PAPAVER
Wilks, *Rev*. W. (1843–1923)

PAPUA
Tryon, H. (1856–1943)
Barton, F. R. (1865–1947)

PELARGONIUM
Beck, E. (–1861)
Denny, J. (*c*. 1819–1881)
Cannell, H. (1833–1914)

PERSIA *see* **IRAN**

PERU
Cuming, H. (1791–1865)
Pentland, J. B. (1797–1873)
Cowan, J. (–1823)
Mathews, A. (–1841)
Bridges, T. (1807–1865)
Castelnau, F. L. de L. de (1810–1880)
Hartweg, C. T. (1812–1871)
Nation, W. (1826–1907)
Maclean, J. (*fl*. 1830s–1850s)
Markham, *Sir* C. R. (1830–1916)
Fox, W. (1858–1934)
Comber, H. F. (1897–1969)

PHILIPPINES
Candi, T. (*c*. 1555–1592)
Cuming, H. (1791–1865)
Kerr, W. (–1814)
Curnow, R. (–1896)
Lay, G. T. (*fl*. 1820s–1840s)
Lobb, T. (1820–1894)
Boxall, W. (1844–1910)
Burke, D. (1854–1897)
Rolfe, R. A. (1855–1921)
Whitehead, J. (1860–1899)
Clemens, J. (1862–1932)
Matthew, C. G. (1862–1936)
Balfour, A. F. (*fl*. 1870s)
Robinson, C. B. (1871–1913)
Sinclair, J. (1913–1968)

PHOTOGRAPHY
Atkins, A. (1797–1871)
Slater, J. S. (1850–1911)
Ashton, E. R. (*c*. 1860–1951)
Waltham, T. E. (*c*. 1869–1950)
Adam, R. M. (1885–1967)
Atkinson, G. (1893–1971)
Malby, R. A. (–1924)
Wallis, E. J. (–1928)
Beckton, B. J. (–1943)
Smith, F. P. (–1945)

PHYTOGEOGRAPHY
Hooker, *Sir* J. D. (1817–1911)
Turrill, W. B. (1890–1961)

PINEAPPLE *see* **ANANAS**

PLANT ANATOMY
Hooke, R. (1635–1703)
Grew, N. (1641–1712)
Adams, G. (1720–1786)
Bower, F. O. (1855–1948)
Sargant, E. (1863–1918)
Boodle, L. A. (1865–1941)
Arber, A. (1879–1960)

PLANT BREEDING *see* **CYTOLOGY**

PLANT PATHOLOGY *see* **MYCOLOGY**

PLANT PHYSIOLOGY
Digby, *Sir* K. (1603–1665)
Mayow, J. (1643–1679)
Hales, *Rev*. S. (1677–1761)
Brotherton, T. (*fl.* 1690s)
Ingen-Housz, J. (1730–1799)
Henry, T. (1734–1816)
Ibbetson, A. (1757–1823)
Knight, T. A. (1759–1838)
Keith, *Rev*. P. (1769–1840)
Marcet, J. (1769–1858)
Brown, R. (1773–1858)
Edwards, W. F. (1776–1842)
Foster, *Sir* M. (1836–1907)
Green, J. R. (1848–1914)
Darwin, F. (1848–1925)
Vines, S. H. (1849–1934)
Buxton, B. H. (1852–1934)
Chamberlain, H. S. (1855–1927)
Pertz, D. F. M. (1859–1939)
Gardiner, W. (1859–1941)
Acton, E. H. (1862–1895)
Bottomley, W. B. (1863–1922)
Blackman, F. F. (1866–1947)
Dixon, H. H. (1869–1953)
Keeble, *Sir* F. W. (1870–1952)
Escombe, F. (1872–1935)
Russell, *Sir* E. J. (1872–1965)
Blackman, V. H. (1872–1967)
Hill, T. G. (1876–1954)
Drummond, J. M. F. (1881–1965)
Atkins, W. R. G. (1884–1959)
Stiles, W. (1886–1966)
Laidlaw, C. G. P. (1887–1915)
Hunt, C. (1888–1926)
Maskell, E. J. (1895–1958)
Haines, F. M. (1898–1964)
Onslow, M. (–1932)
Richards, F. J. (1901–1965)
Barker, J. (1901–1970)
Caldwell, J. (1903–1974)
Bennet-Clark, T. A. (1903–1975)

POMOLOGY *see* **FRUIT GROWING**

POPPY *see* **PAPAVER**

PORTUGAL
Jones, J. (–1731)
Herle, T. (*fl.* 1720s)
Masson, F. (1741–1805)
Gray, E. W. (1748–1806)
Gage, *Sir* T. (1781–1820)
Webb, P. B. (1793–1854)
Sharpe, D. (1806–1856)
Tait, A. W. (1847–1917)
Atchley, S. C. (*c.* 1871–1936)

POTAMOGETON
Bennett, A. (1843–1929)
Pearsall, W. H. (1869–1936)

POTATO
Smee, A. (1818–1877)
Salaman, R. N. (1874–1955)

POTENTILLA
Smith, G. L. (1937–1963)

PRIMULA
Balfour, *Sir* I. B. (1853–1922)
Christy, R. M. (1861–1928)
Forrest, G. (1873–1932)
Smith, *Sir* W. W. (1875–1956)
Gregory, R. P. (1879–1918)

PROTEACEAE
Knight, J. (*c.* 1777–1855)
Garside, S. (1889–1961)

PYRETHRUM
Chandler, S. E. (1880–1957)

QUERCUS
Warburg, *Sir* O. E. (1876–1937)

RANUNCULUS
Tyso, *Rev*. J. (*fl.* 1830s–1840s)
Marsden-Jones, E. M. (1887–1960)

RHODESIA
Kirk, *Sir* J. (1832–1922)
Buchanan, J. (1855–1896)
Rand, R. F. (1856–1937)
Flanagan, H. G. (1861–1919)
Monro, C. F. H. (1863–1918)
Eyles, F. (1864–1937)
Gibbs, L. S. (1870–1925)
Sandwith, C. I. (1871–1961)
Rogers, *Rev*. F. A. (1876–1944)
Swynnerton, C. F. M. (1877–1938)
Hislop, A. (*c.* 1880–1945)
Miller, O. B. (1882–1966)
Adamson, G. (*fl.* 1890s)
Nutt, *Rev*. W. H. (*fl.* 1890s)
Martin, J. D. (–*c.* 1941)
Appleton, A. F. (*fl.* 1900s)
Allen, C. E. F. (*fl.* 1900s–1920s)
Macaulay, M. A. (*fl.* 1910s)
Gilliland, H. B. (1911–1965)

RHODODENDRON
Hooker, *Sir* J. D. (1817–1911)
Grindon, L. H. (1818–1904)
Balfour, *Sir* I. B. (1853–1922)
Millais, J. G. (*c.* 1865–1931)
Forrest, G. (1873–1932)
Tagg, H. F. (*c.* 1873–1933)
Stevenson, J. B. (*c.* 1881–1950)
Rothschild, L. N. de (1882–1942)
Cowan, J. M. (1892–1960)

RODRIQUEZ ISLAND
Balfour, *Sir* I. B. (1853–1922)

ROSES
Borrer, W. (1781–1862)
Andrews, H. C. (*fl.* 1790s–1830s)
Rivers, T. (1798–1877)
Lindley, J. (1799–1865)
Baker, G. (–1885)
Bennett, H. (–1890)
Francis, E. P. (*c.* 1802–1869)
Hollingworth, J. (1805–1888)
Wilde, J. P. (1816–1899)
D'Ombrain, *Rev*. H. H. (1818–1905)
Curtis, H. (*c.* 1819–1889)
Hole, *Rev*. S. R. (1819–1904)
Paul, W. (1822–1905)
Cant, B. R. (1827–1900)
Cheales, *Rev*. A. (1828–1911)
Prince, G. (*c.* 1831–1896)

Flight, F. W. (*c.* 1834–1910)
Briggs, T. R. A. (1836–1891)
Castles, R. (*fl.* 1840s)
Webb, F. M. (1841–1880)
Paul, G. (1841–1921)
Mawley, E. (*c.* 1842–1916)
Parsons, A. W. (1847–1920)
Webster, G. (1851–1924)
Cant, F. (*c.* 1857–1928)
Willmott, E. A. (1858–1934)
Boulger, G. A. (1858–1937)
Wolley-Dod, A. H. (1861–1948)
Darlington, H. R. (*c.* 1863–1946)
Bishop, E. B. (1864–1947)
Molyneux, H. E. (1868–1916)
Paul, A. W. (*fl.* 1870s–1910s)
Hurst, C. C. (1870–1947)
Bradford, S. C. (1878–1948)
Burnside, *Rev.* F. R. (*fl.* 1880s–1890s)
Cant, C. E. (*fl.* 1890s–1910s)
Foster-Melliar, *Rev.* A. (–1904)
Pemberton, *Rev.* J. H. (–1926)
Page, J. C. (–1947)
Harkness, J. (*fl.* 1900s)
Lindsell, E. B. (*fl.* 1900s)
Miles, B. A. (1937–1970)

RUBBER *see* **HEVEA**

RUBUS
Borrer, W. (1781–1862)
Jorden, G. (1783–1871)
Lees, E. (1800–1887)
Bloxam, *Rev.* A. (1801–1878)
Babington, C. C. (1808–1895)
Salter, T. B. (1814–1858)
Hort, *Rev.* F. J. A. (1828–1892)
Powell, J. T. (1833–1904)
Warren, J. B. L. (1835–1895)
Rogers, *Rev.* W. M. (1835–1920)
Briggs, T. R. A. (1836–1891)
Brenan, *Rev.* S. A. (1837–1908)
Webb, F. M. (1841–1880)
Cumming, L. (1843–1927)
Linton, *Rev.* E. F. (1848–1928)
Gilbert, E. G. (1849–1915)
Linton, *Rev.* W. R. (1850–1908)
Trower, C. G. (1855–1928)
Fisher, H. (1860–1935)
Bradley, A. E. (1873–1944)
Avery, C. (1880–1960)
Rilstone, F. (1881–1953)
Woodhead, J. E. (1883–1967)
Watson, W. C. R. (1885–1954)
Haskell, G. M. L. (1920–1967)

RUMEX
Warren, J. B. L. (1835–1895)

RUPICAPNOS
Pugsley, H. W. (1868–1947)

RUSSIA
Busch, J. (*fl.* 1730s–1790s)
Busch, J. C. (*c.* 1759–1838)
Meader, J. (*fl.* 1770s–1780s)
Hove, A. P. (*fl.* 1780s–1820s)
Atkinson, T. W. (1799–1861)
Prescott, J. D. (–1837)
Feilden, H. W. (1838–1921)
Littledale, St. G. R. (*c.* 1851–1931)
Czaplicka, *Miss* (–1921)

SABAH *see* **MALAYSIA**

ST. HELENA
Stonestreet, G. (*fl.* 1690s)
Cunningham, J. (–*c.* 1709)
Roxburgh, W. (1751–1815)
Hardwicke, T. (1755–1835)
Beatson, A. (1759–1833)
Robertson, J. (*fl.* 1760s–1770s)
Burchell, W. J. (1781–1863)
Cuming, H. (1791–1865)
Hinds, R. B. (*c.* 1812–*c.* 1847)
Haughton, J. (1836–1889)
Oliver, S. P. (1838–1907)
Wollaston, T. V. (*fl.* 1870s)

ST. KITTS *see* **WEST INDIES**

ST. LUCIA *see* **WEST INDIES**

S. TOMÉ
Rattray, J. (*fl.* 1830s–1880s)

ST. VINCENT *see* **WEST INDIES**

SALIX
Crowe, J. (1750–1807)
Forbes, W. (1773–1861)
Borrer, W. (1781–1862)
Carey, J. (1797–1880)
Leefe, *Rev.* J. E. (–1889)
Anderson, G. (*fl.* 1800s–1817)
Darwall, *Rev.* L. (1813–1897)
White, F. B. W. (1842–1894)
Linton, *Rev.* E. F. (1848–1928)
Linton, *Rev.* W. R. (1850–1908)
Fraser, J. (1854–1935)
Bradley, A. E. (1873–1944)

SANDWICH ISLANDS *see* **PACIFIC**

SARAWAK *see* **MALAYSIA**

SAXIFRAGA
Haworth, A. H. (1768–1833)
Bree, *Rev.* W. T. (1787–1863)
Brockbank, W. (1830–1896)
Marsden-Jones, E. M. (1887–1960)
Malby, R. A. (–1924)

SCOTLAND *see* **BRITISH ISLES**

SEDUM
Praeger, R. L. (1865–1953)

SEEDSMEN *see* **NURSERYMEN**

SEMPERVIVUM
Praeger, R. L. (1865–1953)
Wale, R. S. (–1952)

SEYCHELLES
Hinds, R. B. (*c.* 1812–*c.* 1847)
Kirk, *Sir* J. (1832–1922)
Wright, E. P. (1834–1910)
Horne, J. (1835–1905)
Clarke, G. (*fl.* 1840s)
Coppinger, R. W. (1847–1910)
Scott, W. (1859–1897)
Milsum, J. N. (*c.* 1890–1945)

SIAM *see* **THAILAND**

SICILY
Swainson, W. (1789–1855)
Hogg, J. (1800–1869)

SIERRA LEONE
Mathews, J. (*fl.* 1780s)
Smeathman, H. (–1786)
Don, G. (1798–1856)
Turner, *Miss* (*fl.* 1820s)
Oldfield, R. A. K. (*fl.* 1830s–1850s)
Reade, W. W. (1839–1875)
Withfield, T. (*fl.* 1840s)
Clarke, R. (*fl.* 1840s–1860s)
Burbidge, F. W. T. (1847–1905)
Johnston, H. H. (1856–1939)
Aylmer, G. P. V. (1856–before 1954)
Scott-Elliot, G. F. (1862–1934)
Willey, F. E. (1871–1898)
Haydon, W. (*c.* 1871–1925)
Smythe, C. W. (1879–1909)
Bunting, R. H. (1879–1966)
Garrett, G. H. (*fl.* 1880s)
Dawe, M. T. (1880–1943)
Hunter, J. H. (*c.* 1885–1924)
Lane-Poole, C. E. (1885–1970)

SIKKIM *see* **INDIAN SUB-CONTINENT**

SILENE
Marsden-Jones, E. M. (1887–1960)
Turrill, W. B. (1890–1961)

SOCOTRA
Wellsted, J. R. (1805–1842)
Nimmo, J. (*fl.* 1830s–1854)
Perry, W. W. (*c.* 1846–1894)
Bent, J. T. (1852–1897)
Balfour, *Sir* I. B. (1853–1922)

SOLOMON ISLANDS *see* **PACIFIC**

SOMALI REPUBLIC
Kirk, *Sir* J. (1832–1922)
James, F. L. (1851–1890)
Drake-Brockman, R. E. (1875–)
Collenette, C. L. (1888–1959)
Cole, E. (*fl.* 1890s)
Gunnis, F. G. (*fl.* 1890s)
Lort-Phillips, E. (*fl.* 1890s)
Appleton, A. F. (*fl.* 1900s)

SORBUS
Fox, W. S. (1875–1962)

SOUTH AFRICA
Adair, P. (*fl.* 1670s–1690s)
Gifford, *Mr.* (*fl.* 1680s)
Brown, A. (*fl.* 1690s)
Conway, J. (*fl.* 1690s)
Fox, J. (*fl.* 1690s)
Stonestreet, G. (*fl.* 1690s)
Lewis, *Rev.* G. (*fl.* 1690s–1700s)
Cunningham, J. (–*c.* 1709)
Brass, W. (–1783)
Brown, W. (*fl.* 1700s)
Monson, A. (*c.* 1714–1776)
Bell, G. (*fl.* 1730s)
Gordon, R. J. (1741–1795)
Masson, F. (1741–1805)
Paterson, W. (1755–1810)
Hardwicke, T. (1755–1835)
Robertson, J. (*fl.* 1760s–1770s)
Roxburgh, J. (*fl.* 1770s–1820s)
Carmichael, D. (1772–1827)
Niven, J. (*c.* 1774–1827)
Burchell, W. J. (1781–1863)
Wallich, N. (1786–1854)
Bowie, J. (*c.* 1789–1869)
Barnard, A. (*fl.* 1790s–1800s)
Bowdich, T. E. (1791–1824)
Forbes, J. (1798–1823)
Upjohn, J. (*c.* 1799–1883)
Peddie, J. (–1840)
Plant, R. W. (–*c.* 1858)
Gerrard, W. T. (–1866)
Shaw, J. (–1891)
Jameson, R. (–1893)
Cathcart, J. F. (1802–1851)
Pappe, C. W. L. (1802–1862)
Alexander, *Sir* J. E. (1803–1885)
Brown, *Rev.* J. C. (1808–1895)
Bunbury, *Sir* C. J. F. (1809–1886)
Prior, R. C. A. (1809–1902)
Castelnau, F. L. de L. de (1810–1880)
Harvey, W. H. (1811–1866)
Stanger, W. (1812–1854)
Grey, *Sir* G. (1812–1898)
Rawson, *Sir* R. W. (1812–1899)
Colenso, *Rev.* J. W. (1814–1883)
Atherstone, W. G. (1814–1898)
Barkly, *Sir* H. (1815–1898)
Cooper, T. (1815–1913)
Roupell, A. E. (1817–1914)
Barber, M. E. (1818–1899)
Synnet, W. (*fl.* 1820s)
Sanderson, J. (1820/1–1881)
Baines, T. (1822–1875)
Sutherland, P. C. (1822–1900)
M'Ken, M. J. (1823–1872)
Wood, J. M. (1827–1915)
Owen, *Miss* (*fl.* 1830s)
Burke, J. (*fl.* 1830s–1840s)
MacOwan, P. (1830–1909)
Chapman, J. (1831–1872)
Guthrie, F. (1831–1899)
Bolus, H. (1834–1911)
D'Urban, W. S. M. (1837–1934)
Becker, H. F. (1838–1917)
Bryce, J. (1838–1922)
Brownlee, *Rev.* J. (*fl.* 1840s)
McGibbon, J. (*fl.* 1840s–1860s)
Oates, F. (1840–1875)
Trimen, R. (1840–1916)
Long, F. (1840–1927)
Eaton, *Rev.* A. E.(1844–1929)
Mason, M. H. (1845–1932)
Brown, N. E. (1849–1934)
Fannin, G. (*fl.* 1850s)
Garden, R. J. (*fl.* 1850s)
Wyley, A. (*fl.* 1850s)
Hutton, H. (*fl.* 1850s–1896)
Hutchins, *Sir* D. E. (1850–1920)
Tyson, W. (1851–1920)
Nelson, W. (1852–1922)
Leighton, J. (1855–1930)
Davis, R. A. (1855–1940)
Jameson, J. S. (1856–1888)
Sim, T. R. (1856–1938)
Johnston, H. H. (1856–1939)
Thorncroft, G. (1857–1934)

Armstrong, *Miss* (*fl.* 1860s)
Buchanan, *Rev.* J. (*fl.* 1860s–1880s)
Wolley-Dod, A. H. (1861–1948)
Worsdell, W. C. (1867–1957)
Bain, T. (*fl.* 1870s)
Frere, C. F. (*fl.* 1870s)
Lawson, J. (*fl.* 1870s)
Pearson, H. H. W. (1870–1916)
Moss, C. E. (1870–1930)
Burtt Davy, J. (1870–1940)
Rattray, G. (1872–1941)
Potter, H. A. (1874–1949)
Holley, H. (*c.* 1875–1928)
Stirrat, A. H. (*c.* 1877–1910)
Potts, G. (1877–1948)
Pole-Evans, I. B. (1879–1968)
Saunders, K. (*fl.* 1880s–1890s)
Hislop, A. (*c.* 1880–1945)
Newberry, W. J. (*c.* 1880–1954)
Hewitt, J. (1880–1961)
Marriott, W. E. (1880–1965)
Kidd, H. H. (*c.* 1883–1936)
Salter, T. M. (1883–1969)
Thoday, M. G. (*c.* 1884–1943)
Long, F. R. (1884–1961)
Hutchinson, J. (1884–1972)
Voss, M. (1885–1953)
Adamson, R. S. (1885–1965)
Murray, *Rev.* D. P. (1887–1967)
Garside, S. (1889–1961)
Barber, L. M. (*fl.* 1890s)
Bowden, A. (*fl.* 1890s)
Maynard, A. W. (1890–1944)
Bews, J. W. (1894–1938)
Smith, N. J. G. (1899–)
Bowker, J. H. (–1900)
Cameron, K. J. (–1918)
Milford, H. A. (–1940)
Mathews, J. W. (–1949)
Howlett, C. J. (–1951)
Appleton, A. F. (*fl.* 1900s)
Thorns, F. W. (*c.* 1904–1971)
Bruce, E. A. (1905–1954)
Prosser, L. N. (1910–1970)
Wonfor, D. J. (1910–1971)
Gilliland, H. B. (1911–1965)

SPAIN
Landon, S. (*fl.* 1670s–1700s)
Lecaan, J. P. (*fl.* 1690s–1710s)
Jones, J. (–1731)
Bowles, W. (1705–1780)
Masson, F. (1741–1805)
La Gasca y Segura, M. (1776–1839)
Widdrington, S. E. (1787–1856)
Webb, P. B. (1793–1854)
Daubeny, C. G. B. (1795–1867)
Ellman, *Rev.* E. (1854–1929)
Lofthouse, T. A. (1868–1944)
Stocken, C. M. (*c.* 1922–1966)

SPITZBERGEN
Eaton, *Rev.* A. E. (1844–1929)
Pike, A. (*fl.* 1890s)

SRI LANKA
Knox, R. (1640/1–1720)
Koenig, J. G. (*c.* 1728–1785)
North, F. (*c.* 1766–1827)
Cuming, H. (1791–1865)
Roe, J. S. (1797–1878)
Kerr, W. (–1814)
Moon, A. (–1825)
Macrae, J. (–1830)
Watson, J. G. (–1838)
Normansell, H. T. (–1843)
Walker, G. W. (–1844)
Ondaatje, W. C. (–1888)
De Alwis, H. (–1894)
Joinville, J. (*fl.* 1800s)
Bennett, J. W. (*fl.* 1800s–1840s)
Harvey, W. H. (1811–1866)
Glenie, *Rev.* S. O. (1811–1875)
Gardner, G. (1812–1849)
Edgeworth, M. P. (1812–1881)
Thwaites, G. H. K. (1812–1882)
Champion, J. G. (1815–1854)
Walker, A. (*fl.* 1820s)
Walker, A. W. (*fl.* 1820s–1840s)
Ferguson, W. (1820–1887)
Wall, G. (*c.* 1821–1894)
Lear, J. G. (*fl.* 1830s)
Beddome, R. H. (1830–1911)
Beckett, T. W. N. (1839–1906)
Drought, I. (*fl.* 1840s–1850s)
Trimen, H. (1843–1896)
De Alwis Seneviratne, W. (1843–1916)
Jowitt, J. F. (1846–1915)
Cameron, W. (*fl.* 1850s)
Hartog, M. M. (1851–1924)
Blow, T. B. (1854–1941)
Lewis, F. (1857–1930)
Potter, *Rev.* M. C. (1858–1948)
Willis, J. C. (1868–1958)
Carruthers, J. B. (1869–1910)
Macmillan, H. F. (1869–1948)
Nock, W. (*fl.* 1870s–1900s)
Petch, T. (1870–1948)
Wright, *Sir* H. (1874–1940)
Lang, W. H. (1874–1960)
Lock, R. H. (1879–1915)
Bryce, G. (1885–)
Bisby, G. R. (1889–1958)
Worthington, T. B. (–1970)
Alston, A. H. G. (1902–1958)

STAPELIA
Masson, F. (1741–1805/6)
Barkly, *Sir* H. (1815–1898)
Barber, M. E. (1818–1899)
Brown, N. E. (1849–1934)

STRAITS OF MAGELLAN
Anderson, R. (1818–1856)

STRAWBERRY *see* **FRAGARIA**

SUDAN
Freeman, A. (–1876)
Petherick, J. (1813–1882)
Bent, J. T. (1852–1897)
Aylmer, G. P. V. (1856–before 1954)
Broun, A. F. (1858–)
Lynes, H. (1874–1942)
Sillitoe, F. S. (1877–1957)
Crossland, C. (1878–1943)
Sampson, H. C. (1878–1953)
Chipp, T. F. (1886–1931)
Buchanan, A. (1886–1954)
Bailey, M. A. (*c.* 1890–1939)

Newbold, *Sir* D. (1894–1945)
Kennedy-Cooke, B. (1894–1963)
Myers, J. G. (1897–1942)
Cartwright, T. (–1951)
Andrews, F. W. (–1961)
Knight, R. L. (–1972)
Brown, H. (*fl.* 1900s)
Robbie, J. (1904–1972)
Crowther, F. (1906–1946)

SUGAR
Barber, C. A. (1860–1933)

SUMATRA *see* **INDONESIA**

SURINAM
Lance, J. H. (1793–1878)

SWITZERLAND
Blaikie, T. (1750–1838)
Davall, E. (1763–1798)
Johnson, W. B. (*c.* 1764–1830)
Brown, P. J. (1785–1842)
Daubeny, C. G. B. (1795–1867)
Shuttleworth, R. J. (1810–1874)
Percy, J. (1817–1889)
Baker, *Rev.* T. (*fl.* 1830s)
Gale, *Rev.* J. S. (1835–1915)
Talbot, W. A. (1847–1917)
Flemwell, G. J. (1865–1928)
Rogers, *Rev.* F. A. (1876–1944)
Ball, C. F. (1879–1915)

SYRIA
Huntingdon, *Rev.* R. (*fl.* 1680s)
Russell, A. (*c.* 1715–1768)
Russell, P. (1727–1805)
Iliff, W. T. (–1876)
Hooker, *Sir* J. D. (1817–1911)
Milne-Redhead, R. (1828–1900)
Lowne, B. T. (*fl.* 1860s)
Meinertzhagen, R. (1878–1967)

TAIWAN
Wilford, C. (–1893)
Collingwood, C. (1826–1908)
Watters, T. (*c.* 1840–1904)
Ford, C. (1844–1927)
Hancock, W. (1847–1914)
Hosie, *Sir* A. (1853–1925)
Bourne, F. S. A. (1854–)
Henry, A. (1857–1930)
Dunn, S. T. (1868–1938)
Campbell, *Rev.* W. (*fl.* 1870s)
Price, W. R. (1886–1975)

TANGANYIKA *see* **TANZANIA**

TANGIER
Spottswood, R. (*fl.* 1670s)

TANZANIA (*see also* **ZANZIBAR**)
New, *Rev.* C. (1840–1875)
Cameron, V. L. (1844–1894)
Carson, A. (1850–1896)
Johnston, *Sir* H. H. (1858–1927)
Lynes, H. (1874–1942)
Swynnerton, C. F. M. (1877–1938)
Cotton, A. D. (1879–1962)
Musk, H. (–1935)
Haarer, A. E. (–1970)
Burtt, B. D. (1902–1938)

TEA
Lettsom, J. C. (1744–1815)
Insch, J. (1877–1951)

TENERIFFE
Bunbury, *Sir* C. J. F. (1809–1886)

THAILAND
Koenig, J. G. (*c.* 1728–1785)
Finlayson, G. (*fl.* 1790s–1823)
Alabaster, H. (–1884)
Smiles, F. H. (–1895)
Hunter, R. (*fl.* 1850s)
Murton, H. J. (1853–1881)
Collins, E. E. (*c.* 1858–)
Bourne, *Sir* A. G. (1859–1940)
Keith, *Sir* A. (1866–1955)
Gwynne-Vaughan, D. T. (1871–1915)
Garrett, H. B. G. (*c.* 1871–1959)
Robinson, H. C. (1874–1929)
Kerr, A. F. G. (1877–1942)
Craib, W. G. (1882–1933)
Marcan, A. (1883–1953)
Kingdon Ward, F. (1885–1958)
Candler, E. (*fl.* 1890s)
Goldham, C. (*fl.* 1890s)
Smith, E. (*c.* 1893–1930)
Barnett, E. C. (–1970)

THYMUS
Jorden, G. (1783–1871)

TIBET
Winterbottom, J. E. (1803–1854)
Henderson, G. (1836–1929)
Littledale, St. G. R. (*c.* 1851–1931)
Hosie, *Sir* A. (1853–1925)
Waddell, L. A. (1854–1938)
Wellby, M. S. (1866–1900)
Deasy, H. H. P. (1866–1947)
Cave, G. H. (*c.* 1870–1965)
Forrest, G. (1873–1932)
Leslie, J. E. (*c.* 1877–1962)
Kingdon Ward, F. (1885–1958)
Ludlow, F. (1885–1972)
Hobson, H. E. (*fl.* 1890s)
Malcolm, N. (*fl.* 1890s)
Pike, A. (*fl.* 1890s)
Thorold, W. G. (*fl.* 1890s)
Sherriff, G. (1898–1967)

TIMOR (*see also* **INDONESIA**)
Dampier, W. (1651–1715)
Cunningham, A. (1791–1839)
Richardson, J. M. (*c.* 1797–1882)
Armstrong, J. (–1847)

TOBACCO *see* **NICOTIANA**

TOGOLAND
Kitson, *Sir* A. E. (1868–1937)
Crossland, M. C. (*fl.* 1890s)
Scholes, J. (–1943)

TREES (*see also* **CONIFERAE, QUERCUS, SALIX, ULMUS**)
Mascall, L. (–1589)
Evelyn, J. (1620–1706)
Thynne, T., *Viscount* Weymouth (1640–1714)
Hamilton, T. (1680–1735)
Campbell, A., *Duke of Argyll* (1682–1761)

Marsham, R. (1707/8–1797)
Hamilton, T. (1734–1794)
Clarke, J. (*fl.* 1760s)
Watson, P. W. (1761–1830)
Forbes, J. (1773–1861)
Billington, W. (1776–1861)
Wallis, J. (*fl.* 1780s–1830s)
Selby, P. J. (1788–1867)
Bohn, H. G. (1796–1884)
Jeffrey, J. (–1886)
Strutt, J. G. (*fl.* 1800s–1850s)
Brown, *Rev.* J. C. (1808–1895)
Lyon, P. (*fl.* 1810s)
Laslett, T. (1811–1887)
Ravenscroft, C. J. (1816–1890)
Brandis, *Sir* D. (1824–1907)
Lowe, J. (1830–1902)
Schlich, *Sir* W. (1840–1925)
Adams, T. W. (1841–1919)
Heath, F. G. (1843–1913)
Elwes, H. J. (1846–1922)
Brown, J. E. (1848–1899)
Nisbet, J. (1853–1914)
Henry, A. (1857–1930)
Nelson, J. (*fl.* 1860s)
Lushington, A. W. (*c.* 1860–1920)
Holford, *Sir* G. L. (1860–1926)
Somerville, *Sir* W. (1860–1932)
Bean, W. J. (1863–1947)
Ragg, L. (1866–1945)
Burn-Murdoch, A. M. (1868–1914)
Stebbing, E. P. (1870–1960)
Borthwick, A. W. (1872–1937)
Julius, *Sir* G. A. (1873–1946)
Troup, R. S. (1874–1939)
Pearson, *Sir* R. S. (1874–1958)
Rodger, *Sir* A. (*c.* 1875–1950)
Jackson, A. B. (1876–1947)
Osborn, A. (1878–1964)
Trevor, *Sir* C. G. (1882–1959)
Lane-Poole, C. E. (1885–1970)
Chipp, T. F. (1886–1931)
Mead, J. P. (1886–1951)
Webster, A. D. (*fl.* 1890s–1920s)
Jerram, M. R. K. (*fl.* 1890s–1945)
Steven, H. M. (1893–1969)
Anderson, M. L. (1895–1961)
Mackenzie, D. F. (–1910)
Renwick, J. (–1918)
Makins, F. K. (–1956)
Lesueur, A. D. C. (–*c.* 1969)
Laurie, M. V. (1901–1973)
Dale, I. R. (*c.* 1904–1963)
Swabey, C. (1906–1972)

TRINIDAD *see* **WEST INDIES**

TRIPOLI
Dickson, J. (1779–1847)

TRISTAN DA CUNHA
Carmichael, D. (1772–1827)

TULIPA
Clark, W. (*c.* 1763–1831)
Jacob, *Rev.* J. (1858–1926)
Hall, *Sir* A. D. (1864–1942)
Dykes, W. R. (1877–1925)
Newton, W. C. F. (1895–1927)
Dykes, E. K. (–1933)

TURKEY
Covel, *Rev.* J. (1638–1722)
Sympson, A. (*fl.* 1690s)
Clerk, W. (*fl.* 1690s–1710s)
Walsh, *Rev.* R. (1772–1852)
Fellows, *Sir* C. (1799–1860)
Liston, *Lady* H. (–1828)
Forbes, E. (1815–1854)
Whittall, E. (1851–1917)
Balfour-Gourlay, W. (*c.* 1879–1966)

UGANDA
Grant, J. A. (1827–1892)
Whyte, A. (1834–1908)
Johnston, *Sir* H. H. (1858–1927)
Mahon, J. (1870–1906)
Bagshawe, *Sir* A. W. G. (1871–1950)
Wollaston, A. F. R. (1875–1930)
Wilson, *Rev.* C. T. (*fl.* 1880s)
James, *Sir* H. E. M. (*fl.* 1880s–1900s)
Dawe, M. T. (1880–1943)
Lamb, P. H. (*c.* 1883–1937)
Maitland, T. D. (1885–)
Snowden, J. D. (1886–1973)
Dümmer, R. A. (*c.* 1887–1922)
Fyffe, R. (*fl.* 1900s–1920s)
Hansford, C. G. (1900–1966)
Dale, I. R. (*c.* 1904–1963)
Wood, G. H. S. (1927–1957)

ULMUS
Christy, R. M. (1861–1928)

U.S.S.R. *see* **RUSSIA**

U.S.A.
Hariot, T. (1560–1721) *Virginia*
Tradescant, J. (1608–1662) *Virginia*
Josselyn, J. (*fl.* 1630s–1670s) *Massachusetts*
Gibbs, G. (*fl.* 1640s) *Virginia*
Hughes, W. (*fl.* 1650s–1680s) *Florida*
Banister, *Rev.* J. (1650–1692) *Virginia*
Harlow, J. (*fl.* 1660s–1680s) *Virginia*
Clayton, *Rev.* J. (*fl.* 1670s–1690s) *Virginia*
More, T. (*fl.* 1670s–1720s) *New England*
Bohun, E. (1672–1734) *Carolina*
Logan, J. (1674–1751) *Pennsylvania*
Witt, C. (1675–1765) *Pennsylvania*
Vernon, W. (*fl.* 1680s–1710s) *Maryland*
Winthrop, J. (1681–1747) *Massachusetts*
Catesby, M. (1682–1749) *Carolina, Florida, Virginia*
Colden, C. (1688–1776) *Pennsylvania*
Coombs, C. (*fl.* 1690s) *Maryland*
Road, J. (*fl.* 1690s) *Virginia*
Jones, *Rev.* H. (*fl.* 1690s–1700s) *Maryland*
Marshall, J. (*fl.* 1690s–1700s) *Virginia*
Rutherford, R. (*fl.* 1690s–1700s) *Carolina*
Collinson, P. (1694–1768)
Clayton, J. (1694–1773) *Virginia*
Oglethorpe, J. E. (1696–1785) *Georgia*
Bartram, J. (1699–1777) *Carolina, Pennsylvania*
Dale, T. (1699/1700–1750) *Carolina*
Lawson, J. (–1711) *Carolina*
Krieg, D. (–1713) *Maryland*
Gregg, J. (–1795) *Carolina*
Brown, W. (*fl.* 1700s) *Virginia*
Ellis, R. (*fl.* 1700s) *Carolina*
Franklyn, G. (*fl.* 1700s) *Carolina*
Gerard, P. (*fl.* 1700s) *Virginia*
Halsteed, W. (*fl.* 1700s) *Carolina*

Lord, J. (*fl.* 1700s) *Carolina*
Smart, J. (*fl.* 1700s) *Maryland*
Greenway, J. (*c.* 1703–1794) *Virginia*
Clerk, *Rev.* W. (*fl.* 1710s–1730s) *Carolina, Virginia*
Mitchell, J. (1711–1768) *Virginia*
Dudley, P. (*fl.* 1720s–1730s) *New England*
Tennent, J. (*fl.* 1720s–1760s) *Virginia*
Romans, B. (*c.* 1720–1783) *Florida*
Marshall, H. (1722–1801) *Pennsylvania*
Farquhar, J. (1724–1766) *New York*
Scott, A. (*fl.* 1730s) *Maryland*
Brickell, J. (*fl.* 1730s–1740s) *Carolina*
Garden, A. (1730–1791) *Carolina*
Cree, J. (*c.* 1738–1816) *Carolina*
Bartram, W. (1739–1823) *Alabama, Carolina*
Young, W. (*fl.* 1740s–1780s) *Virginia*
Walter, T. (*c.* 1740–1789) *Carolina*
Masson, F. (1741–1805) *New York*
Kuhn, A. (1741–1817) *Pennsylvania*
Brickell, J. (*c.* 1749–1809) *Georgia*
Fraser, J. (1750–1811) *Pennsylvania*
Abbot, J. (1751–*c.* 1840) *Georgia*
Landreth, D. (1752–1836) *Pennsylvania*
Muehlenberg, *Rev.* G. H. E. (1753–1815)
Marshall, M. (1758–1813) *Pennsylvania*
Clifton, W. (*fl.* 1760s) *Florida*
Johnson, W. B. (*c.* 1764–1830)
Lyon, J. (*c.* 1765–1814) *Carolina, Florida, Georgia*
Brown, R. (*c.* 1767–1845)
Bradbury, J. (1768–1823)
Hosack, D. (1769–1835) *New York*
Thorburn, G. (1773–1863) *New York*
Pursh, F. T. (1774–1820)
McMahon, B. (*c.* 1775–1816) *Pennsylvania*
Rogers, P. K. (1776–1828)
Hogg, T. (1778–1855) *New York*
Hobson, W. (*fl.* 1780s–1830s) *California*
Steinhauer, D. (1785–1852) *Pennsylvania*
Nuttall, T. (1786–1859)
Fraser, J. (*fl.* 1790s–1860s)
Drummond, T. (*c.* 1790–1835)
Boott, F. (1792–1863)
Collie, A. (1793–1835) *California*
Coulter, T. (1793–1843) *Arizona, California*
Watson, G. (1795–1858) *Pennsylvania*
Daubeny, C. G. B. (1795–1867)
Murray, A. M. (1795–1884) *Florida*
Barratt, J. (1796–1882) *New York*
Carey, J. (1797–1800) *Carolina*
Douglas, D. (1799–1834) *California*
Squibb, R. (–before 1817) *Carolina*
Coultas, H. (–1877) *Pennsylvania*
Henry, C. (–1894) *Colorado*
Shaw, H. (1800–1889) *Missouri*
Scouler, J. (1804–1871) *Oregon*
Tuomey, M. (1805–1857) *Florida*
Sherwood, J. (*c.* 1806–1883) *Pennsylvania*
Bridges, T. (1807–1865) *California*
Lobb, W. (1809–1864) *California, Oregon*
Castelnau, F. L. de L. de (1810–1880) *Florida*
Brackenridge, W. D. (1810–1893)
Harvey, W. H. (1811–1866) *Florida*
Hinds, R. B. (*c.* 1812–*c.* 1847) *California*
Gordon, A. (1813–*c.* 1873) *Colorado, Nebraska*
Douglas, R. (1813–1897)
Barfoot, J. L. (1816–1882) *Utah*
Antisell, T. (1817–1893)
Lyall, D. (1817–1895)
Hooker, *Sir* J. D. (1817–1911) *Rocky Mountains*
Vick, J. (1818–1882) *New York*
Ball, J. (1818–1889) *Rocky Mountains*
Hutchings, J. M. (1818–1902) *California*
Saul, J. (1819–1897) *Washington*
Lay, G. T. (*fl.* 1820s–1845) *California*
Hogg, T. (1820–1892) *Carolina, Virginia*
Henderson, P. (1822–1890) *New Jersey*
Vasey, G. (1822–1893)
Saunders, W. (1822–1900)
Parry, C. C. (1823–1890)
Davidson, G. (1825–1911) *Alaska, California*
Jeffrey, J. (1826–1854) *California, Oregon*
Meehan, T. (1826–1901) *Pennsylvania*
Scott, R. R. (1827–1877) *Pennsylvania*
Sanders, E. (1827–1907) *Illinois*
Smith, W. R. (1828–1912) *Washington*
Young, A. R. (1829–1883) *New York*
Knight, H. (*fl.* 1830s) *Florida*
Burke, J. (*fl.* 1830s–1840s) *Idaho, Rocky Mountains*
Palmer, E. (1831–1911)
Blackie, G. S. (1834–1881) *Tennessee*
Macloskie, *Rev.* G. (1834–1920) *New Jersey*
Yates, L. G. (1837–1909) *California*
Stretch, R. H. (1837–1926) *Nevada*
Watson, S. H. (1837–1928) *Wisconsin*
D'Urban, W. S. M. (1837–1934) *California*
Muir, J. (1838–1914)
Williamson, J. (1839–1884) *Kentucky*
Scupham, J. R. (1840–1927) *California*
Dawson, J. T. (1841–1916) *Massachusetts*
Bell, W. A. (1841–1921) *Colorado*
Matthews, W. (1843–1905)
Melvill, J. C. (1845–1929) *Florida*
Bodger, J. (1846–1924) *California*
McLaren, J. (1846–1943) *California*
Spence, J. (1848–) *California*
Barrington, R. M. (1849–1915) *Rocky Mountains*
Murray, W. (*fl.* 1850s) *California*
Beardsley, A. F. (*fl.* 1850s–1860s) *California*
Falconer, W. (1850–1928)
Macfarlane, J. M. (1855–1943) *Pennsylvania*
Palmer, W. (1856–1921)
Greata, L. A. (1857–1911) *California*
Jones, J. W. (1859–1923) *Pennsylvania*
Dunbar, J. (1859–1927)
Hickson, S. J. (1859–1940) *Arizona*
Davidson, A. (1860–1932) *California*
Roland, T. (*c.* 1861–1929) *Massachusetts*
Letts, A. (1862–1923) *California*
Clemens, J. (1862–1932)
Canning, E. J. (1863–1921) *Massachusetts*
Cocks, R. W. S. (1863–1926) *Louisiana*
Chase, J. S. (1864–1923) *California*
Bioletti, F. T. (1865–1939) *California*
Anderson, F. W. (1866–1891) *Montana*
Belling, J. (1866–1933)
Cockerell, T. D. A. (1866–1948) *Colorado*
Griffiths, D. (1867–1935)
MacElwee, A. (1869–1923) *Pennsylvania*
Henshaw, J. W. (1869–1937) *Rocky Mountains*
Coates, L. (*fl.* 1870s) *California*
Payne, T. (1872–1963) *California*
Kennedy, P. B. (1874–1930) *Nevada*
Evans, H. (1874–1960) *California*
Palmer, E. J. (1875–1962)
Cusack, M. E. (*fl.* 1880s–1890s) *Colorado*
Kerr, M. E. (1883–1950) *California*

Pring, G. H. (1885–1974) *Pennsylvania*
Judd, W. H. (1888–1946)
Hanbury, S. (*fl.* 1890s) *Rocky Mountains*
Keys, A. (–1958) *Florida*
Warren, C. N. (–1970) *California*
Jones, G. N. (1904–1970) *Illinois*

URUGUAY
Baird, J. (*fl.* 1830s)
Aplin, O. V. (*fl.* 1880s ?)

VENEZUELA
Porter, *Sir* R. K. (1777–1842)
Purdie, W. (*c.* 1817–1857)
Crueger, H. (1818–1864)
Fanning, D. (*fl.* 1820s)
Henchman, J. (*fl.* 1830s)
Birschel, F. W. (*fl.* 1850s)
Broadway, W. E. (1863–1935)
Sprague, T. A. (1877–1958)

VIETNAM
Clemens, J. (1862–1932)

VIOLA
Gregory, E. S. (1840–1932)
Drabble, E. (1877–1933)
Crane, D. B. (–1938)

VITICULTURE
Hoare, C. (*c.* 1789–1849)
Thomson, W. (1814–1895)

WALES *see* **BRITISH ISLES**

WELWITSCHIA
Pearson, H. H. W. (1870–1916)

WEST INDIES
Willisel, T. (–*c.* 1675) *Jamaica*
Horsnell, G. (*c.* 1625–1697) *Antigua*
Halley, E. (1656–1742) *Trinidad*
Harlow, J. (*fl.* 1660s–1680s) *Jamaica*
Sloane, *Sir* H. (1660–1753) *Jamaica*
Barham, H. (1670–1726) *Jamaica*
Catesby, M. (1682–1749) *Jamaica*
Brodie, J. (*fl.* 1690s)
Fenwick, R. (*fl.* 1690s) *Jamaica*
Hamilton, C. (*fl.* 1690s) *Barbados*
Handisyd, G. (*fl.* 1690s)
Reed, J. (*fl.* 1690s) *Barbados*
Houstoun, W. (*c.* 1695–1733)
Walker, T. (*fl.* 1700s) *Bahamas*
Grigg, T. (*fl.* 1700s–1710s) *Antigua*
Cressy (–*c.* 1763) *Leeward Islands*
Robinson, A. (–1768) *Jamaica*
Smeathman, H. (–1786)
Clarke, T. (–1792) *Jamaica*
Broughton, A. (–1796/1803) *Jamaica*
Hughes, *Rev.* G. (1707–) *Barbados*
Peplow, *Mrs.* (*fl.* 1710s ?) *Barbados*
Walduck, T. (*fl.* 1710s) *Barbados*
Burnet, J. (*fl.* 1710s–1730s)
Clerk, *Rev.* W. (*fl.* 1710s–1730s) *Antigua, Monserrat*
Douglas, J. (*fl.* 1710s–1730s) *Antigua*
Browne, P. (*c.* 1720–1790) *Jamaica*
Melville, R. (1723–1809) *St. Vincent*
Dale, F. (*fl.* 1730s) *Bahamas*
Millar, R. (*fl.* 1730s–1740s)
Long, E. (1734–1813) *Jamaica*
Wright, W. (1735–1819) *Jamaica*
Lindsay, *Rev.* J. (*fl.* 1750s–1788) *Jamaica*
Dancer, T. (*c.* 1750–1811) *Jamaica*
Humberston, F. M. (1754–1815) *Barbados*
Wallen, M. (*fl.* 1770s–1780s) *Jamaica*
East, H. (*fl.* 1770s–1790s) *Jamaica*
Ryan, J. (*fl.* 1770s–1790s) *Montserrat, Trinidad*
Riddell, M. (*fl.* 1770s–1800s) *Antigua*
Caley, G. (1770–1829) *St. Vincent*
Bancroft, E. N. (1772–1842) *Jamaica*
Higson, T. (1773–1836) *Jamaica*
Gardner, *Commodore* (*fl.* 1780s) *Jamaica*
Lindsay, J. (*fl.* 1780s–1803) *Jamaica*
Hamilton, W. (1783–1856)
Wiles, J. (*fl.* 1790s–1800s) *Jamaica*
Evans, T. (*fl.* 1790s–1810s)
Cuming, H. (1791–1865) *Jamaica*
Heward, R. (1791–1877) *Jamaica*
Daubeny, C. G. B. (1795–1867)
Marsh, W. T. (*c.* 1795–*c.* 1872) *Jamaica*
Guilding, *Rev.* L. (1797–1831) *St. Vincent*
Don, G. (1798–1856)
Nicholson, T. (1799–1877) *Antigua*
Young, G. (–1803) *St. Vincent*
Anderson, A. (–1811) *St. Vincent*
Tobin, J. W. (–1814)
Lochhead, W. (–1815) *St. Vincent*
Woodford, *Sir* R. (–1828) *Trinidad*
Macrae, J. (–1830) *St. Vincent*
Maycock, J. D. (–1837) *Barbados*
Lockhart, D. (–1846) *Trinidad*
Barkly, E. H. (–1857) *Jamaica*
Parker, C. S. (–1869)
Iliff, W. T. (–1876)
Hockin, J. (–before 1885) *Dominica*
Taylor, F. H. (–before 1886) *Jamaica*
Fleming, J. (*fl.* 1800s)
Anderson, G. (*fl.* 1800s–1810s) *Barbados*
Macfadyen, J. (1800–1850) *Jamaica*
Wilson, N. (1809–1874) *Jamaica*
Prior, R. C. A. (1809–1902) *Jamaica*
Lunan, J. (*fl.* 1810s) *Jamaica*
Distin, H. (*fl.* 1810s–1840s) *Jamaica*
Gosse, P. H. (1810–1888) *Jamaica*
Imray, J. (1811–1880) *Dominica*
Hartweg, C. T. (1812–1871) *Jamaica*
Grisebach, A. H. R. (1814–1879)
Ansted, D. T. (1814–1880)
Higgins, *Rev.* H. H. (1814–1893)
McNab, G. (1815–1859) *Jamaica*
Barkly, *Sir* H. (1815–1898) *Jamaica*
Purdie, W. (*c.* 1817–1857) *Jamaica, Trinidad*
Rothery, H. C. (1817–1888) *Dominica*
Crueger, H. (1818–1864) *Jamaica, Trinidad*
Daniell, W. F. (1818–1865)
Munro, W. (1818–1880) *Barbados*
Kingsley, *Rev.* C. (1819–1875) *Trinidad*
Collins, *Dr.* (*fl.* 1820s ?) *Barbados*
Elliott, *Mr.* (*fl.* 1820s) *St. Vincent*
Morrison, W. (*fl.* 1820s) *Barbados, Trinidad*
Waters, *Rev.* J. (*fl.* 1820s–1850s) *Jamaica*
Finlay, K. (*fl.* 1820s–1880s) *Antigua, Dominica, Grenada*
M'Ken, M. J. (1823–1872) *Jamaica*
Milne-Readhead, R. (1828–1900)
Aldridge, G. (*fl.* 1830s) *Trinidad*
Cowan, T. (*fl.* 1830s) *Jamaica*
Elsey, J. R. (1834–1857) *St. Kitts*
Gray, J. (*c.* 1835–1895) *St. Lucia*
Robinson, *Sir* W. (1836–1912) *Bahamas*
Feilden, H. W. (1838–1921) *Barbados*
Lane, A. W. (*fl.* 1840s)
Bradford, E. (*fl.* 1840s–1890s) *Trinidad*

Blake, *Sir* H. A. (1840–) *Jamaica*
Thomson, R. (1840–1908) *Jamaica*
Prestoe, H. (1842–1923) *Trinidad*
Syme, G. (1844–) *Jamaica*
Morris, D. (1844–1933)
Jenman, G. S. (1845–1902) *Jamaica*
Hart, J. H. (1847–1911) *Jamaica, Trinidad*
Hancock, W. (1847–1914) *Jamaica*
O'Malley, E. W. (1847–1927) *Jamaica*
Sherring, R. V. (1847–1931)
Waby, J. F. (1848–1923) *Barbados*
Fawcett, W. (1851–1926) *Jamaica*
Brace, L. J. K. (1852–1938) *Bahamas*
Guppy, H. B. (1854–1926)
Cran, W. (1854–1933) *Antigua*
Bovell, J. R. (1855–1928) *Barbados*
Cartwright, T. B. (1856–1896) *Trinidad*
Harrison, J. B. (1856–1926) *Barbados*
Watts, *Sir* F. (1859–1930)
Smith, G. W. (*fl.* 1860s–1900s)
Elliott, W. R. (1860–1908)
Harris, W. (1860–1920) *Jamaica*
Barber, C. A. (1860–1933) *Leeward Islands*
Gardiner, J. (1861–1900) *Bahamas*
Cradwick, W. (*c.* 1862–1937) *Jamaica*
Broadway, W. E. (1863–1935)
Powell, H. (1864–1920) *St. Vincent*
Gregory, J. W. (1864–1932)
Ramage, G. A. (1864–1933) *Dominica*
Rendle, A. B. (1865–1938) *Jamaica*
Hombersley, *Rev.* A. (1866–1941) *Trinidad*
Jones, J. (1867–1934) *Dominica*
Green, H. F. (1868–*c.* 1945) *Dominica*
Carruthers, J. B. (1869–1910) *Trinidad*
Blake, E. (*fl.* 1870s–1890s) *Jamaica*
Nock, W. (*fl.* 1870s–1900s) *Jamaica*
Lunt, W. (1871–1904) *Trinidad*
Millen, H. (1871–1908) *Tobago*
Norman, C. (1872–1947) *Jamaica*
Jordan, A. J. (*c.* 1873–1906) *Antigua, Montserrat, Trinidad*
Freeman, W. G. (1874–) *Trinidad*
Sands, W. N. (1875–1943) *Antigua, St. Vincent*
Fishlock, W. C. (1875–1959)
Walsh, J. J. (*fl.* 1880s)
Thompson, W. J. (*fl.* 1880s–1890s) *Jamaica*
Campbell, E. J. F. (*fl.* 1880s–1920s) *Jamaica*
Nowell, W. (1880–1968)
Brooks, A. J. (1881–) *Dominica*
Robson, W. (1882–1923) *Montserrat*
Evans, *Sir* G. (1883–1963) *Trinidad*
Wheeler, L. R. (1888–1948) *Antigua*
Harcourt, F. G. (1889–1970) *Antigua, Dominica*
Lodge, F. A. (*fl.* 1890s) *Trinidad*
Tillson, A. G. (*fl.* 1890s) *Antigua*
Mason, T. G. (1890–1959)
Williams, R. O. (1891–1967) *Trinidad, Tobago*
Downes, E. J. (1893–1957) *Jamaica*
Henderson, A. H. (–1921) *Barbados*
Evans, F. J. (–1928) *Trinidad*
Baker, R. E. D. (1908–1954) *Trinidad*

WILLOW *see* **SALIX**

YUGOSLAVIA
Doncaster, E. D. (–*c.* 1950)

ZAIRE *see* **CONGO**

ZAMBIA
Kirk, *Sir* J. (1832–1922)

ZANZIBAR
Bojer, W. (1795–1856)
Frere, *Sir* H. B. E. (1815–1884)
Kirk, *Sir* J. (1832–1922)
Last, J. T. (1847/8–1933)
Lyne, R. N. (1864–1961)
Williams, R. O. (1891–1967)